2010 Twenty-Fifth Annual IEEE Applied Power Electronics Conference and Exposition

(APEC 2010)

Palm Springs, California, USA
21 – 25 February 2010

Pages 1-767

IEEE Catalog Number: CFP10APE-PRT
ISBN: 978-1-4244-4782-4

Copyright © 2010 by the Institute of Electrical and Electronic Engineers, Inc
All Rights Reserved

Copyright and Reprint Permissions: Abstracting is permitted with credit to the source. Libraries are permitted to photocopy beyond the limit of U.S. copyright law for private use of patrons those articles in this volume that carry a code at the bottom of the first page, provided the per-copy fee indicated in the code is paid through Copyright Clearance Center, 222 Rosewood Drive, Danvers, MA 01923.

For other copying, reprint or republication permission, write to IEEE Copyrights Manager, IEEE Service Center, 445 Hoes Lane, Piscataway, NJ 08854. All rights reserved.

***This publication is a representation of what appears in the IEEE Digital Libraries. Some format issues inherent in the e-media version may also appear in this print version.**

IEEE Catalog Number: CFP10APE-PRT
ISBN 13:   978-1-4244-4782-4
Library of Congress No.: 90-643607
ISSN: 1048-2334

Additional Copies of This Publication Are Available From:

Curran Associates, Inc
57 Morehouse Lane
Red Hook, NY 12571 USA
Phone: (845) 758-0400
Fax: (845) 758-2633
E-mail: curran@proceedings.com
Web: www.proceedings.com

TABLE OF CONTENTS

Session A1L-A: DC-DC Converter I
Tuesday, February 23, 8:30 - 10:10
Session Chairs: Van Niemela, *Fairchild Semiconductor*
Haidong Yu, *Phoenix International*

Minimum Deviation Digital Controller IC for Single and Two Phase DC-DC Switch-Mode Power Supplies 1
Aleksandar Radić, *University of Toronto, Canada*
Zdravko Lukić, *University of Toronto, Canada*
Aleksandar Prodić, *University of Toronto, Canada*
Robert de Nie, *NXP Semiconductors, Netherlands*

Modeling and Design Considerations of Coupled Inductor Converters 7
Guangyong Zhu, *Auscom Engineering, Inc., United States*
Kunrong Wang, *Dell, Inc., United States*

Design Procedure for High Frequency Operation of the Modified Series Resonant APWM Converter with Improved Efficiency and Reduced Size 14
Darryl J. Tschirhart, *Queen's University, Canada*
Praveen K. Jain, *Queen's University, Canada*

Expiremantal Results and Study of a Modified Adaptive Bus Voltage Controller 19
Jaber A. Abu Qahouq, *University of Alabama, United States*
Gautam Muralidhar, *University of Alabama, United States*

Session A1L-B: AC-DC Power Factor Correction Topologies I
Tuesday, February 23, 8:30 - 10:10
Session Chairs: Gerry Moschopoulos, *University of Western Ontario*
Omer Onar, *Illinois Institute of Technology*

Bridgeless Buck PFC Rectifier 23
Yungtaek Jang, *Delta Products Corporation, United States*
Milan M. Jovanović, *Delta Products Corporation, United States*

An Active-Clamped Full-Wave Zero-Current-Switched Quasi-Resonant Boost Converter in Power Factor Correction Application 30
E. Firmansyah, *Kyushu University, Japan*
S. Abe, *Kyushu University, Japan*
M. Shoyama, *Kyushu University, Japan*
S. Tomioka, *TDK-Lambda Corporation, Japan*
T. Ninomiya, *Nagasaki University, Japan*

Novel Adaptive Master-Slave Method for Interleaved Boundary Conduction Mode (BCM) PFC Converters 36
Hangseok Choi, *Fairchild Semiconductor, United States*

A Novel Bridgeless Single-Stage Half-Bridge AC-DC Converter 42
Woo-Young Choi, *Virginia Polytechnic Institute and State University, United States*
Wen-Song Yu, *Virginia Polytechnic Institute and State University, United States*
Jih-Sheng Lai, *Virginia Polytechnic Institute and State University, United States*

Session A1L-C: Power Electronics for Utility Interface I
Tuesday, February 23, 8:30 - 10:10
Session Chairs: Zareh Soghomonian, *BMT Syntek Technologies*
                     Jin Wang, *Ohio State University*

Power Quality Improvement at Medium-Voltage Grids Using Hexagram Active Power Filter 47
Jun Wen, *University of California, Irvine, United States*
Liang Zhou, *University of California, Irvine, United States*
Keyue Smedley, *University of California, Irvine, United States*

A Generalized Capacitor Voltage Balancing Scheme for Flying Capacitor Multilevel Converters 58
Mostafa Khazraei, *Missouri University of Science and Technology, United States*
Hossein Sepahvand, *Missouri University of Science and Technology, United States*
Keith Corzine, *Missouri University of Science and Technology, United States*
Mehdi Ferdowsi, *Missouri University of Science and Technology, United States*

An Active Damping Technique for a Current Source Inverter Employing a Virtual Negative Inductance 63
Ahmed Salah Morsy, *Texas A&M University at Qatar, Qatar*
Shehab Ahmed, *Texas A&M University at Qatar, Qatar*
Prasad Enjeti, *Texas A&M University at Qatar, Qatar*
Ahmed Massoud, *Qatar University, Qatar*

Maximum Solar Power Transfer in Multi-Port Power Electronic Interface 68
Wei Jiang, *University of Texas at Arlington, United States*
Babak Fahimi, *University of Texas at Arlington, United States*

Session A1L-D: Passive Devices I
Tuesday, February 23, 8:30 - 10:10
Session Chairs: Laura Lyle, *Wright Patterson Air Force Base*
                     Mike Schutten, *General Electric*

SMD Inductors Based on Soft-Magnetic Powder Compacts 74
Etsuo Otsuki, *Toho Zinc Co., Ltd., Japan*
Kenichiro Ishii, *Toho Zinc Co., Ltd., Japan*
Shinya Nakano, *Toho Zinc Co., Ltd., Japan*

High Density Low Profile Coupled Inductor Design for Integrated Point-of-Load Converter 79
Qiang Li, *Virginia Polytechnic Institute and State University, United States*
Yan Dong, *Virginia Polytechnic Institute and State University, United States*
Fred C. Lee, *Virginia Polytechnic Institute and State University, United States*

Relationship of Quality Factor and Hollow Winding Structure of Coreless Printed Spiral Winding (CPSW) Inductor 86
Y.P. Su, *Virginia Polytechnic Institute and State University, United States*
Xun Liu, *ConvenientPower HK Ltd., China*
C.K. Lee, *Hong Kong Polytechnic University, China*
S.Y.R. Hui, *City University of Hong Kong, China*

Modeling of Adaptable-Diameter Burners Formed by Concentric Planar Windings for Domestic Induction Heating Applications 92
Jesus Acero, *University of Zaragoza, Spain*
Claudio Carretero, *University of Zaragoza, Spain*
Ignacio Millan, *University of Zaragoza, Spain*
Oscar Lucía, *University of Zaragoza, Spain*
Jose-Miguel Burdío, *University of Zaragoza, Spain*
Rafael Alonso, *University of Zaragoza, Spain*

Session A1L-E: Controls in Motor Drives I
Tuesday, February 23, 8:30 - 10:10
Session Chairs: Jonathan Kimball, *Missouri S&T*

Flux Concentration and Pole Shaping in a Single Phase Hybrid Switched Reluctance Motor Drive 98
Uffe Jakobsen, *Aalborg University, Denmark*
Kaiyuan Lu, *Aalborg University, Denmark*

Parameter Independent Maximum Torque Per Ampere (MTPA) Control of IPM Machine Based on Signal Injection 103
Sungmin Kim, *Seoul National University, Korea, South*
Young-Doo Yoon, *Seoul National University, Korea, South*
Seung-Ki Sul, *Seoul National University, Korea, South*
Kozo Ide, *Yaskawa Electric Corporation, Japan*
Koji Tomita, *Yaskawa Electric Corporation, Japan*

Performance Analysis of Three-Phase Capacitor Motor in Frequency Control System 109
Zheng-Feng Ming, *Xidian University, China*
Guang-Zheng Ni, *Zhejiang University, China*
Bing-Zhong Yang, *Chongqing University, China*

Efficiency Improvement by Changeover of Phase Windings of Multiphase Permanent Magnet Synchronous Motor with Outer-Rotor Type 112
Young-Gook Kim, *Pusan National University, Korea, South*
Chae-Bong Bae, *Pusan National University, Korea, South*
Jang-Mok Kim, *Pusan National University, Korea, South*
Hyun-Cheol Kim, *Agency for Defense Development, Korea, South*

Session A1L-F: Digital Controls in DC-DC Converters I
Tuesday, February 23, 8:30 - 10:10
Session Chairs: Dragan Maksimović, *University of Colorado at Boulder*
Jason Neely, *Purdue University*

Digital Power Controller with Non-Linear Variable Switching Frequency 120
Jaber A. Abu Qahouq, *University of Alabama, United States*

**Digital Charge Balance Controller with an Auxiliary Circuit for Superior Unloading
Transient Performance of Buck Converters** 124
Eric Meyer, *Queen's University, Canada*
Dong Wang, *Queen's University, Canada*
Liang Jia, *Queen's University, Canada*
Yan-Fei Liu, *Queen's University, Canada*

One-Step Digital Dead-Time Correction for DC-DC Converters 132
April Zhao, *University of Toronto, Canada*
Armin Akhavan Fomani, *University of Toronto, Canada*
Wai Tung Ng, *University of Toronto, Canada*

**The Practical Aspects of Utilizing Digital Power Controller for
Monitoring of Power Supply Operation** 138
Oleg Volfson, *Intersil Corporation, United States*

Session A1L-G: Wind Power
Tuesday, February 23, 8:30 - 10:10
Session Chairs: Morgan Kiani, *University of Texas at Arlington*

**A Unity Power Factor, Maximum Power Point Tracking Battery Charger for
Low Power Wind Turbines** 143
Gustavo Gamboa, *University of Central Florida, United States*
John Elmes, *University of Central Florida, United States*
Christopher Hamilton, *University of Central Florida, United States*
Jonathan Baker, *University of Central Florida, United States*
Michael Pepper, *University of Central Florida, United States*
Issa Batarseh, *University of Central Florida, United States*

**Maximum Power Point Tracking of a Wind Energy Conversion System Using
Adaptive Nonlinear Approach** 149
Majid Pahlevaninezhad, *Queen's University, Canada*
Suzan Eren, *Queen's University, Canada*
Alireza Bakhshai, *Queen's University, Canada*
Praveen Jain, *Queen's University, Canada*

A Hybrid Wind-Solar Energy System: a New Rectifier Stage Topology 155
Joanne Hui, *Queen's University, Canada*
Alireza Bakhshai, *Queen's University, Canada*
Praveen Jain, *Queen's University, Canada*

Dynamic Operation and Control of a Hybrid Wind-Diesel Stand Alone Power Systems 162
A.M.O. Haruni, *University of Tasmania, Australia*
A. Gargoom, *University of Tasmania, Australia*
M.E. Haque, *University of Tasmania, Australia*
M. Negnevitsky, *University of Tasmania, Australia*

Session A2L-A: DC-DC Converter II
Tuesday, February 23, 10:40 - 11:55
Session Chairs: Van Niemela, *Fairchild Semiconductor*
Haidong Yu, *Phoenix International*

SystemC-AMS Modeling and Simulation of Digitally Controlled DC-DC Converters 170
Matteo Agostinelli, *University of Klagenfurt, Austria*
Robert Priewasser, *University of Klagenfurt, Austria*
Mario Huemer, *University of Klagenfurt, Austria*
Stefano Marsili, *Infineon Technologies Austria AG, Austria*
Dietmar Straeussnigg, *Infineon Technologies Austria AG, Austria*

Modeling of Digitally Controlled Voltage Regulator Modules 176
Yi Sun, *Linear Technology, United States*
Fred C. Lee, *Virginia Polytechnic Institute and State University, United States*
Jian Li, *Linear Technology, United States*

Design and Comparison of Digital Control Loops Analytical Models, Laboratory
Measurements, and Simulation Results 183
Philip Cooke, *Infineon Technologies, United States*
Thomas G. Wilson, Jr., *SIMPLIS Technologies, United States*
Rohan Samsi, *Primarion, United States*

Session A2L-B: AC-DC Power Factor Correction Topologies II
Tuesday, February 23, 10:40 - 11:55
Session Chairs: Gerry Moschopoulos, *University of Western Ontario*
Omer Onar, *Illinois Institute of Technology*

Digital Control for Efficiency Improvements in Interleaved Boost PFC Rectifiers 188
Fu-Zen Chen, *University of Colorado at Boulder, United States*
Dragan Maksimović, *University of Colorado at Boulder, United States*

Reduction of the Output Capacitor in Power Factor Correctors by
Distorting the Line Input Current 196
Diego G. Lamar, *Universidad de Oviedo, Spain*
Javier Sebastián, *Universidad de Oviedo, Spain*
Manuel Arias, *Universidad de Oviedo, Spain*
Arturo Fernández, *Universidad de Oviedo, Spain*

Universal-Input Single-Stage PFC Flyback with Variable Boost Inductance for
High-Brightness LED Applications 203
Yuequan Hu, *Delta Products Corporation, United States*
Laszlo Huber, *Delta Products Corporation, United States*
Milan M. Jovanović, *Delta Products Corporation, United States*

Session A2L-C: Power Electronics for Utility Interface II
Tuesday, February 23, 10:40 - 11:55
Session Chairs: Zareh Soghomonian, *BMT Syntek Technologies*
Jin Wang, *Ohio State University*

High Frequency High Efficiency Bidirectional DC-DC Converter Module Design for 10 kVA Solid State Transformer 210
Haifeng Fan, *Florida State University, United States*
Hui Li, *Florida State University, United States*

Synchronization of Three-Phase Converters and Virtual Microgrid Implementation Utilizing the Power-Hardware-in-the-Loop Concept 216
O. Vodyakho, *Florida State University, United States*
C.S. Edrington, *Florida State University, United States*
M. Steurer, *Florida State University, United States*
S. Azongha, *Florida State University, United States*
F. Fleming, *Florida State University, United States*

A Single-Stage Grid-Connected Inverter with Wide Range Reactive Power Compensation Using Energy Storage System (Ess) 223
Liming Liu, *Florida State University, United States*
Zhichao Wu, *Florida State University, United States*
Hui Li, *Florida State University, United States*

Session A2L-D: Passive Devices II
Tuesday, February 23, 10:40 - 11:55
Session Chairs: Laura Lyle, *Wright Patterson Air Force Base*
Mike Schutten, *General Electric*

Polymer Bonded Soft Magnetics for EMI Filter Applications in Power Electronics 231
S. Egelkraut, *University of Erlangen-Nürnberg, Germany*
L. Frey, *University of Erlangen-Nürnberg, Germany*
M. Rauch, *Fraunhofer Institute for Integrated Systems and Device Technology, Germany*
A. Schletz, *Fraunhofer Institute for Integrated Systems and Device Technology, Germany*
M. März, *Fraunhofer Institute for Integrated Systems and Device Technology, Germany*

Lead-Acid Battery Modeling and State of Charge Monitoring 239
J.F. Araujo Leão, *Universidade Federal de Campina Grande, Brazil*
L.V. Hartmann, *Universidade Federal de Campina Grande, Brazil*
M.B.R. Corrêa, *Universidade Federal de Campina Grande, Brazil*
A.M.N. Lima, *Universidade Federal de Campina Grande, Brazil*

Voltage and Current Ripple Considerations for Improving Lifetime of Ultra-Capacitors Used for Energy Buffer Applications at Converter Inputs 244
Supratim Basu, *Bose Research Pvt. Ltd., India*
Tore M. Undeland, *Norwegian University of Science and Technology, Norway*

Session A2L-E: Controls in Motor Drives II
Tuesday, February 23, 10:40 - 11:55
Session Chairs: Jonathan Kimball, *Missouri S&T*

Implementation and Operational Investigations of Bipolar Gate Drivers 248
Jean-Christophe Crebier, *Grenoble Institute of Technology, France*
Manh Hung Tran, *Grenoble Institute of Technology, France*
Jean Barbaroux, *Grenoble Institute of Technology, France*
Pierre-Olivier Jeannin, *Grenoble Institute of Technology, France*

A Method for Impact Assessment of Faults on the Performance of Field-Oriented Control Drives: a First Step to Reliability Modeling 256
Ali M. Bazzi, *University of Illinois at Urbana-Champaign, United States*
Alejandro Dominguez-Garcia, *University of Illinois at Urbana-Champaign, United States*
Philip T. Krein, *University of Illinois at Urbana-Champaign, United States*

A Fault Tolerant Control System for Hexagram Inverter Motor Drive 264
Liang Zhou, *University of California, Irvine, United States*
Keyue Smedley, *University of California, Irvine, United States*

Session A2L-F: Digital Controls in DC-DC Converters II
Tuesday, February 23, 10:40 - 11:55
Session Chairs: Dragan Maksimović, *University of Colorado at Boulder*
　　　　　　　　 Jason Neely, *Purdue University*

Power Analog to Digital Converter for Voltage Scaling Applications 271
M.C. Gonzalez, *Universidad Politécnica de Madrid, Spain*
M. Vasić, *Universidad Politécnica de Madrid, Spain*
P. Alou, *Universidad Politécnica de Madrid, Spain*
O. Garcia, *Universidad Politécnica de Madrid, Spain*
J.A. Oliver, *Universidad Politécnica de Madrid, Spain*
J.A. Cobos, *Universidad Politécnica de Madrid, Spain*
H. Visairo, *Intel Corporation, Mexico*

A Digital Pulse-Width Modulator for Phase-Shift Operation of Full-Bridge Isolated DC-DC Converters 277
L. Corradini, *University of Colorado at Boulder, United States*
D. Maksimović, *University of Colorado at Boulder, United States*

Digitally Controlled Integrated Pseudo-CCM SIMO Converter with Adaptive Freewheel Current Modulation 284
Yi Zhang, *University of Arizona, United States*
Dongsheng Ma, *University of Arizona, United States*

Session A2L-G: Fuel Cells
Tuesday, February 23, 10:40 - 11:55
Session Chairs: Morgan Kiani, *University of Texas at Arlington*

Analysis of Pulse-Link DC-AC Converter for Fuel Cells Applications Operated in Zero-Current-Slope Mode 289
Kentaro Fukushima, *Kyushu University, Japan*
Isami Norigoe, *I.N. Laboratory, Japan*
Masahito Shoyama, *Kyushu University, Japan*
Tamotsu Ninomiya, *Nagasaki University, Japan*
Yosuke Harada, *Ebara Densan Ltd., Japan*
Kenta Tsukakoshi, *Ebara Densan Ltd., Japan*

A Minimum Power-Processing Stage Fuel Cell Energy System Based on a Boost-Inverter with a Bi-Directional Back-Up Battery Storage 295
Minsoo Jang, *University of Sydney, Australia*
Vassilios G. Agelidis, *University of Sydney, Australia*

Power Conditioning System for Fuel Cell with 2-Stage DC-DC Converter 303
Byung M. Han, *Myongji University, Korea, South*
Jun-Young Lee, *Myongji University, Korea, South*
Yu-Seok Jeong, *Myongji University, Korea, South*

Session B1L-A: DC-DC Converter III
Wednesday, February 24, 8:30 - 10:10
Session Chairs: Alireza Khaligh, *Illinois Institute of Technology*
Sheldon Williamson, *Concordia University*

Real-Time FPGA-Based Hardware-in-the-Loop Development Test-Bench for Multiple Output Power Converters 309
O. Lucía, *University of Zaragoza, Spain*
O. Jiménez, *University of Zaragoza, Spain*
L.A. Barragán, *University of Zaragoza, Spain*
I. Urriza, *University of Zaragoza, Spain*
J.M. Burdío, *University of Zaragoza, Spain*
D. Navarro, *University of Zaragoza, Spain*

Oversampled Digital Controller IC Based on Successive Load-Change Estimation for DC-DC Converters 315
Zdravko Lukić, *University of Toronto, Canada*
Aleksandar Radić, *University of Toronto, Canada*
Aleksandar Prodić, *University of Toronto, Canada*
Simon Effler, *University of Limerick, Ireland*

Novel Nonlinear Control of Dual Active Bridge Using Simplified Converter Model 321
Diogenes D. Molina Cardozo, *University of Arkansas, United States*
Juan Carlos Balda, *University of Arkansas, United States*
Derik Trowler, *University of Arkansas, United States*
H. Alan Mantooth, *University of Arkansas, United States*

A Novel Digital Single-Wire Quasi-Democratic Stress Share Scheme for Paralleled Switching Converters 328
Karl Rinne, *Powervation Ltd., Ireland*
Anthony Kelly, *Powervation Ltd., Ireland*
Eamon O'Malley, *Powervation Ltd., Ireland*

Session B1L-B: AC-DC Conversion Control Strategies
Wednesday, February 24, 8:30 - 10:10
Session Chairs: Alireza Khaligh, *Illinois Institute of Technology*
Omer Onar, *Illinois Institute of Technology*

Minimum-Sensing Current Control of Three-Phase PFC Converters 336
Zhonghui Bing, *Rensselaer Polytechnic Institute, United States*
Jian Sun, *Rensselaer Polytechnic Institute, United States*

Direct Power Control of a Dual Converter Operating As Synchronous Rectifier 343
José Restrepo, *Universidad Simón Bolívar, Venezuela*
José M. Aller, *Universidad Simón Bolívar, Venezuela*
Alexander Bueno, *Universidad Simón Bolívar, Venezuela*
Julio C. Viola, *Universidad Simón Bolívar, Venezuela*
Alberto Berzoy, *Universidad Simón Bolívar, Venezuela*
Thomas Habetler, *Georgia Institute of Technology, United States*

A Low-Cost Adaptive Multi-Mode Digital Control Solution Maximizing AC-DC Power Supply Efficiency 349
Yong Li, *iWatt Inc., United States*
Jerry Zheng, *iWatt Inc., United States*

Average Modeling and Control for Three-Phase Three-Level Non-Regenerate Rectifier with Unbalanced DC Loads 355
Rixin Lai, *GE Global Research Center, United States*
Fred Wang, *University of Tennessee - Knoxville and Oak Ridge National Laboratory, United States*
Rolando Burgos, *ABB Inc., United States*
Dushan Boroyevich, *Virginia Polytechnic Institute and State University, United States*

Session B1L-C: Active Power Filter
Wednesday, February 24, 8:30 - 10:10
Session Chairs: Jingjun Liu, *Xi'an Jiaotong Univ.*
Jin Wang, *Ohio State University*

A Waveform Control Technique for High Power Shunt Active Power Filter Based on Repetitive Control Algorithm 361
Zhiqiang Wang, *Zhejiang University, China*
Chuan Xie, *Zhejiang University, China*
Chao He, *Zhejiang University, China*
Guozhu Chen, *Zhejiang University, China*

A Combined Series-Parallel Active Filter System Implementation Using Generalized Non-Active Power Theory 367

Mehmet Ucar, *Kocaeli University, Turkey*
Sule Ozdemir, *Kocaeli University, Turkey*
Engin Ozdemir, *Kocaeli University, Turkey*

A Novel Control Method for Unified Power Quality Conditioner (UPQC) Under Non-Ideal Mains Voltage and Unbalanced Load Conditions 374

Metin Kesler, *Kocaeli University, Turkey*
Engin Ozdemir, *Kocaeli University, Turkey*

Resonant Current Regulation for Transformerless Hybrid Active Filter to Suppress Harmonic Resonances in Industrial Power Systems 380

Tzung-Lin Lee, *National Sun Yat-sen University, Taiwan*
Yen-Ching Wang, *National Sun Yat-sen University, Taiwan*
Josep M. Guerrero, *Technical University of Catalonia, Spain*

Session B1L-D: Semiconductor Devices
Wednesday, February 24, 8:30 - 10:10
Session Chairs: Carl Blake, *Transphorm*
Chuck Mullett, *ON Semiconductor*

Performance Evaluation of High Voltage Super Junction MOSFETs for Zero-Voltage Soft-Switching Inverter Applications 387

Sung-Yeul Park, *University of Connecticut, United States*
Pengwei Sun, *Virginia Polytechnic Institute and State University, United States*
Wensong Yu, *Virginia Polytechnic Institute and State University, United States*
Jih-Sheng Lai, *Virginia Polytechnic Institute and State University, United States*

New 1.7kV IGBT Chip with Fine Pattern and Optimized Buffer Layer 392

John F. Donlon, *Powerex, Inc., United States*
Eric R. Motto, *Powerex, Inc., United States*
K. Satoh, *Mitsubishi Electric Corp, Japan*
K. Suzuki, *Mitsubishi Electric Corp, Japan*
Y. Yoshihiura, *Mitsubishi Electric Corp, Japan*
T. Takahashi, *Mitsubishi Electric Corp, Japan*

Novel Thermally Enhanced Power Package 398

Juan A. Herbsommer, *Texas Instruments, United States*
Jonathan Noquil, *Texas Instruments, Philippines*
Chris Bull, *Texas Instruments, United States*
Osvaldo Lopez, *Texas Instruments, United States*

Recent Advances in Silicon Carbide MOSFET Power Devices 401

Ljubisa D. Stevanovic, *GE Global Research, United States*
Kevin S. Matocha, *GE Global Research, United States*
Peter A. Losee, *GE Global Research, United States*
John S. Glaser, *GE Global Research, United States*
Jeffrey J. Nasadoski, *GE Global Research, United States*
Stephen D. Arthur, *GE Global Research, United States*

Session B1L-E: Sensorless Techniques in Motor Drives
Wednesday, February 24, 8:30 - 10:10
Session Chairs: Patrick Chapman, *University of Illinois*

Start-Up Transient Improvement for Sensorless Control Approach of PM Motor 408
Dong Jiang, *Virginia Polytechnic Institute and State University, United States*
Rixin Lai, *Virginia Polytechnic Institute and State University, United States*
Fred Wang, *University of Tennessee - Knoxville, United States*
Rolando Burgos, *Virginia Polytechnic Institute and State University, United States*
Dushan Boroyevich, *Virginia Polytechnic Institute and State University, United States*

**Sensorless Position Control of Skewed Rotor Induction Machines Based on
Multi Saliency Extraction** 414
T.M. Wolbank, *Vienna University of Technology, Austria*
M.K. Metwally, *Vienna University of Technology, Austria*

**Fuzzy Gain Scheduling PI Controller for a Sensorless Four Switch Three
Phase BLDC Motor** 420
Chung-Wen Hung, *National Yunlin University of Science and Technology, Taiwan*
Jen-Ta Su, *National Taiwan University, Taiwan*
Chih-Wen Liu, *National Taiwan University, Taiwan*
Cheng-Tsung Lin, *DynaPack Co., Ltd., Taiwan*
Jhih-Han Chen, *National Yunlin University of Science and Technology, Taiwan*

Equivalent EMF Based Position Observers for Sensorless Synchronous Machines 425
Jingbo Liu, *Rockwell Automation, United States*
Thomas Nondahl, *Rockwell Automation, United States*
Peter Schmidt, *Rockwell Automation, United States*
Semyon Royak, *Rockwell Automation, United States*
Mark Harbaugh, *Rockwell Automation, United States*

Session B1L-F: Modeling, Simulation & Control I
Wednesday, February 24, 8:30 - 10:10
Session Chairs: Mahesh Krishnamurthy, *Illinois Institue of Technology*

An Improved Winding Loss Analytical Model of Flyback Transformer 433
Wei Yuan, *Zhejiang University, China*
Xiucheng Huang, *Zhejiang University, China*
Peipei Meng, *Zhejiang University, China*
Guoxing Zhang, *Zhejiang University, China*
Junming Zhang, *Zhejiang University, China*

**Identification of the Material Properties Used in Domestic Induction Heating
Appliances for System-Level Simulation and Design Purposes** 439
Jesus Acero, *University of Zaragoza, Spain*
Oscar Lucía, *University of Zaragoza, Spain*
Ignacio Millan, *University of Zaragoza, Spain*
Luis Angel Barragán, *University of Zaragoza, Spain*
Jose-Miguel Burdío, *University of Zaragoza, Spain*
Rafael Alonso, *University of Zaragoza, Spain*

A Retrofit 60 Hz Current Sensor for Non-Intrusive Power Monitoring at the Circuit Breaker 444

Zachary Clifford, *Massachusetts Institute of Technology, United States*
John J. Cooley, *Massachusetts Institute of Technology, United States*
Al-Thaddeus Avestruz, *Massachusetts Institute of Technology, United States*
Zack Remscrim, *Massachusetts Institute of Technology, United States*
Dan Vickery, *Massachusetts Institute of Technology, United States*
Steven B. Leeb, *Massachusetts Institute of Technology, United States*

Session B1L-G: Vehicle Electronics I
Wednesday, February 24, 8:30 - 10:10
Session Chairs: Ali Emadi, *Illinois Institute of Technology*

Feasibility of Capacitor Voltage Regulation and Output Voltage Harmonic Minimization in Cascaded H-Bridge Converters 452

Hossein Sepahvand, *Missouri University of Science and Technology, United States*
Mostafa Khazarei, *Missouri University of Science and Technology, United States*
Mehdi Ferdowsi, *Missouri University of Science and Technology, United States*
Keith Corzine, *Missouri University of Science and Technology, United States*

Examination of a PHEV Bidirectional Charger System for V2G Reactive Power Compensation 458

Mithat C. Kisacikoglu, *University of Tennessee, United States*
Burak Ozpineci, *Oak Ridge National Laboratory, United States*
Leon M. Tolbert, *University of Tennessee and Oak Ridge National Laboratory, United States*

Optimal Selection and Design of the Supercapacitor Module for Fuel Cell Vehicles 466

Sang-Hyun Kim, *Soongsil University, Korea, South*
Tae-Hoon Kim, *Soongsil University, Korea, South*
Wook Kim, *Soongsil University, Korea, South*
Jong-Hak Lee, *Soongsil University, Korea, South*
Woojin Choi, *Soongsil University, Korea, South*

Efficiency Evaluation of a 55kW Soft-Switching Module Based Inverter for High Temperature Hybrid Electric Vehicle Drives Application 474

Pengwei Sun, *Virginia Polytechnic Institute and State University, United States*
Jih-Sheng Lai, *Virginia Polytechnic Institute and State University, United States*
Hao Qian, *Virginia Polytechnic Institute and State University, United States*
Wensong Yu, *Virginia Polytechnic Institute and State University, United States*
Chris Smith, *Azure Dynamics Inc., United States*
John Bates, *Azure Dynamics Inc., United States*
Beat Arnet, *Azure Dynamics Inc., United States*
Alexander Litvinov, *Powerex Inc., United States*
Scott Leslie, *Powerex Inc., United States*

Session B2L-A: DC-DC Converter IV
Wednesday, February 24, 14:00 - 15:40
Session Chairs: Jin Wang, *Ohio State University*
Wayne Weaver, *Michigan Technological University*

Real-Time Hybrid Model Predictive Control of a Boost Converter with Constant Power Load 480
Jason Neely, *Purdue University, United States*
Steve Pekarek, *Purdue University, United States*
Ray DeCarlo, *Purdue University, United States*
Nir Vaks, *Purdue University, United States*

Predictive Control of Buck Converter Using Nonlinear Output Capacitor Current Programming 491
Victor Sui-pung Cheung, *City University of Hong Kong, China*
Henry Shu-hung Chung, *City University of Hong Kong, China*
Huai Wang, *City University of Hong Kong, China*

Analysis of a High Performance Voltage Regulator with Non-Linear Multi-Mode Control: Bandwidth and Large Transient Response 499
S. Pan, *Queen's University, Canada*
P.K. Jain, *Queen's University, Canada*

Multi-Output Synchronously-Rectified Forward Converter with Load Transient Considered ... 507
K.I. Hwu, *National Taipei University of Technology, Taiwan*
Y.T. Yau, *National Taipei University of Technology, Taiwan*

Session B2L-B: System Integration I
Wednesday, February 24, 14:00 - 15:40
Session Chairs: Shamala Chickamenahalli, *Intel*

Symmetric Current Balancing Circuit for Multiple DC Loads 512
Sungjin Choi, *Samsung Electronics Co., Ltd., Korea, South*
Pankaj Agarwal, *Samsung Electronics Co., Ltd., Korea, South*
Teahoon Kim, *Samsung Electronics Co., Ltd., Korea, South*
Joonhyun Yang, *Samsung Electronics Co., Ltd., Korea, South*
Baikhee Han, *Samsung Electronics Co., Ltd., Korea, South*

A Simple Method for Configuring Multi-PWM Channels for Multi-Level Converter Applications Based on PWM IP Core 519
Haibing Hu, *Nanjing University of Aeronautics and Astronautics, China*
Xiaodong Ding, *Nanjing Guojun Electric Co., Ltd., China*
Tao Xue, *Nanjing Sute Electric Co., Ltd., China*
Wenxi Yao, *Zhejiang University, China*
Zhengyu Lu, *Zhejiang University, China*

Technology Roadmapping for Power Supply in Package (PSiP) and Power Supply on Chip (PwrSoC) 525

Raymond Foley, *University College Cork, Ireland*
Finbarr Waldron, *Tyndall National Institute, Ireland*
John Slowey, *University College Cork, Ireland*
Arnold Alderman, *Anagenesis Inc., United States*
Brian Narveson, *Texas Instruments, United States*
Cian Ó'Mathúna, *Tyndall National Institute, Ireland*

Technology Road Map for High Frequency Integrated DC-DC Converter 533

Qiang Li, *Virginia Polytechnic Institute and State University, United States*
Michele Lim, *Virginia Polytechnic Institute and State University, United States*
Julu Sun, *Virginia Polytechnic Institute and State University, United States*
Arthur Ball, *Virginia Polytechnic Institute and State University, United States*
Yucheng Ying, *Virginia Polytechnic Institute and State University, United States*
Fred C. Lee, *Virginia Polytechnic Institute and State University, United States*
K.D.T. Ngo, *Virginia Polytechnic Institute and State University, United States*

Session B2L-C: Resonant DC-DC Converters I
Wednesday, February 24, 14:00 - 15:40
Session Chairs: Dustin Becker, *Emerson Network Power*
Russell Spyker, *USAF*

A New Valley-Detection Method for the Quasi-Resonance Switching 540

Gwan-Bon Koo, *Fairchild Semiconductor, Korea, South*
Sang-Cheol Moon, *Fairchild Semiconductor, Korea, South*
Jin-Tae Kim, *Fairchild Semiconductor, Korea, South*

Secondary-Side Control of a Constant Frequency Series Resonant Converter Using Dual-Edge PWM 544

Darryl J. Tschirhart, *Queen's University, Canada*
Praveen K. Jain, *Queen's University, Canada*

A Non-Insulated Resonant Boost Converter 550

Peng Shuai, *RWTH Aachen University, Germany*
Yales R. De Novaes, *ABB, Switzerland*
Francisco Canales, *ABB, Switzerland*
Ivo Barbi, *Federal University of Santa Catarina, Brazil*

Analysis and Design of a Low-Profile Resonant LCC Converter 557

A. Pawellek, *University of Erlangen-Nürnberg, Germany*
A. Bucher, *University of Erlangen-Nürnberg, Germany*
T. Duerbaum, *University of Erlangen-Nürnberg, Germany*

Session B2L-D: Miscellaneous Applications
Wednesday, February 24, 14:00 - 15:40
Session Chairs: Alejandro Dominguez-Garcia, *University of Illinois*

ZVS and ZCS DC-DC PWM Full-Bridge Fuel Cell Converters 564
Ahmad Mousavi, *University of Western Ontario, Canada*
Pritam Das, *University of Western Ontario, Canada*
Gerry Moschopoulos, *University of Western Ontario, Canada*

Effective Switching Mode Power Supplies Common Mode Noise Cancellation
Technique with Zero Equipotential Transformer Models 571
Yick Po Chan, *University of Hong Kong, China*
Man Hay Pong, *University of Hong Kong, China*
Ngai Kit Poon, *University of Hong Kong, China*
Chui Pong Liu, *University of Hong Kong, China*

50W Power Device (PD) Power in Power Over Ethernet (PoE) System with Input
Current Balance in Four-Pair Architecture with Two DC-DC Converters 575
Haimeng Wu, *Zhejiang University, China*
Zhengshi Wang, *Zhejiang University, China*
Jiande Wu, *Zhejiang University, China*
Xiangning He, *Zhejiang University, China*
Yan Deng, *Zhejiang University, China*

High-Resolution Physically-Windowed Sensors for Power Electronics Applications 580
Warit Wichakool, *Massachusetts Institute of Technology, United States*
James Paris, *Massachusetts Institute of Technology, United States*
Al-Thaddeus Avestruz, *Massachusetts Institute of Technology, United States*
Steven B. Leeb, *Massachusetts Institute of Technology, United States*

Session B2L-E: LED Lighting I
Wednesday, February 24, 14:00 - 15:40
Session Chairs: Regan Zane, *University of Colorado*

Edison Revisited: Impact of DC Distribution on the Cost of
LED Lighting and Distributed Generation 588
Brinda A. Thomas, *Carnegie Mellon University, United States*

A Novel Passive Off-Line Light-Emitting Diode (LED) Driver with Long Lifetime 594
S.Y.R. Hui, *City University of Hong Kong, China*
S.N. Li, *City University of Hong Kong, China*
X.H. Tao, *City University of Hong Kong, China*
W. Chen, *City University of Hong Kong, China*
W.M. Ng, *City University of Hong Kong, China*

Improving Current Regulation for Offline LED Driver 601
Jianwen Shao, *STMicroelectronics, United States*

LED Driver Circuit with Inherent PFC 605
D. Aguilar, *University of Minnesota, United States*
C.P. Henze, *Analog Power Design Inc., United States*

Session B2L-F: Power Electronics for Utility Interface III
Wednesday, February 24, 14:00 - 15:40
Session Chairs: Hui Li, *Florida State University*
Miaosen Shen, *United Technologies Research Center*

A New Circuit Design and Control to Reduce Input Harmonic Current for a Three-Phase AC Machine Drive System Having a Very Small DC-Link Capacitor 611
Hyunjae Yoo, *Samsung Heavy Industries Co., Ltd., Korea, South*
Seung-Ki Sul, *Seoul National University, Korea, South*

State-Space Modeling, Analysis, and Implementation of Parallel Inverters for Microgrid Applications 619
Chien Liang Chen, *Virginia Polytechnic Institute and State University, United States*
Jih-Sheng Lai, *Virginia Polytechnic Institute and State University, United States*
Daniel Martin, *Virginia Polytechnic Institute and State University, United States*
Yuang-Shung Lee, *Fu-Jen Catholic University, Taiwan*

Efficiency Improvement of Grid-Tied Inverters at Low Input Power Using Pulse Skipping Control Strategy 627
Haibing Hu, *University of Central Florida, United States*
Wisam Al-Hoor, *University of Central Florida, United States*
Nasser Kutkut, *University of Central Florida, United States*
Issa Batarseh, *University of Central Florida, United States*
John Shen, *University of Central Florida, United States*

Phase Locked Loop for Unbalanced Utility Conditions 634
Carlos D. Rodríguez-Valdez, *Rockwell Automation, United States*
Russ J. Kerkman, *Rockwell Automation, United States*

Session B2L-G: Isolated DC-DC Converters I
Wednesday, February 24, 14:00 - 15:40
Session Chairs: Alexis Kwasinski, *The University of Texas at Austin*
Sheldon Williamson, *Concordia University*

Analysis and Design Considerations for EMI and Losses of RCD Snubber in Flyback Converter 642
Peipei Meng, *Zhejiang University, China*
Xinke Wu, *Zhejiang University, China*
Jianyou Yang, *Zhejiang University, China*
Henglin Chen, *Zhejiang University, China*
Zhaoming Qian, *Zhejiang University, China*

A High Output Power Density 400/400V Isolated DC-DC Converter with Hybrid Pair of SJ-MOSFET and SiC-SBD for Power Supply of Data Center 648
Rejeki Simanjorang, *National Institute of Advanced Industrial Science and Technology, Japan*
Hiroshi Yamaguchi, *National Institute of Advanced Industrial Science and Technology, Japan*
Hiromichi Ohashi, *National Institute of Advanced Industrial Science and Technology, Japan*
Takashi Takeda, *NTT Facilities Inc., Japan*
Mikio Yamazaki, *NTT Facilities Inc., Japan*
H. Murai, *NTT Facilities Inc., Japan*

A 500 W Push-Pull DC-DC Power Converter with a 30 MHz Switching Frequency 654
John S. Glaser, *GE Global Research, United States*
Juan M. Rivas, *GE Global Research, United States*

Input-Series Connnected High Frequency DC-DC Converters with One Transformer 662
Deshang Sha, *Beijing Institute of Technology, China*
Zhiqiang Guo, *Beijing Institute of Technology, China*
XiaoZhong Liao, *Beijing Institute of Technology, China*

Session B3L-A: Renewable Energy
Wednesday, February 24, 16:10 - 17:25
Session Chairs: Chris Edrington, *Florida State University*
Alex Huang, *North Carolina State University*

**Simple Photovoltaic Solar Cell Dynamic Sliding Mode Controlled Maximum Power Point
Tracker for Battery Charging Applications** 666
Emil A. Jimenez-Brea, *University of Puerto Rico-Mayaguez, Puerto Rico*
Eduardo I. Ortiz-Rivera, *University of Puerto Rico-Mayaguez, Puerto Rico*
Andres Salazar-Llinas, *University of Puerto Rico-Mayaguez, Puerto Rico*
Jesus Gonzalez-Llorente, *University of Puerto Rico-Mayaguez, Puerto Rico*

An Enhanced Circuit-Based Model for Single-Cell Battery 672
Jiucai Zhang, *University of Nebraska-Lincoln, United States*
Song Ci, *University of Nebraska-Lincoln, United States*
Hamid Sharif, *University of Nebraska-Lincoln, United States*
Mahmoud Alahmad, *University of Nebraska-Lincoln, United States*

A High Frequency Battery Model for Current Ripple Analysis 676
Jin Wang, *Ohio State University, United States*
Ke Zou, *Ohio State University, United States*
Chingchi Chen, *Ford Motor Company, United States*
Lihua Chen, *Ford Motor Company, United States*

Session B3L-B: System Integration II
Wednesday, February 24, 16:10 - 17:25
Session Chairs: Shamala Chickamenahalli, *Intel*

**A Novel Power Line Communication Technique Based on
Power Electronics Circuit Topology** 681
Jiande Wu, *Zhejiang University, China*
Chushan Li, *Zhejiang University, China*
Xiangning He, *Zhejiang University, China*

**Compact Temperature Compensation of Inductive Fly-back Clamps for Integrated
Power Switches Using a High-Voltage Base-Current-Compensated V_{be} Multiplier** 686
Timothy P. Duryea, *Texas Instruments, United States*
Hoi Lee, *University of Texas at Dallas, United States*

Optimal Design for the Damping Resistor in RCD-R Snubber to Suppress Common-Mode Noise 691

Peipei Meng, *Zhejiang University, China*
Henglin Chen, *Zhejiang University, China*
Sheng Zheng, *Zhejiang University, China*
Xinke Wu, *Zhejiang University, China*
Zhaoming Qian, *Zhejiang University, China*

Session B3L-C: Resonant DC-DC Converters II
Wednesday, February 24, 16:10 - 17:25
Session Chairs: Dustin Becker, *Emerson Network Power*
Russell Spyker, *USAF*

A High-Efficient LLCC Series-Parallel Resonant Converter 696

Christian P. Dick, *RWTH Aachen University, Germany*
Furkan Kaan Titiz, *RWTH Aachen University, Germany*
Rik De Doncker, *RWTH Aachen University, Germany*

Accurate Switching Loss Model and Optimal Design of a Current Source Driver Considering the Current Diversion Problem 702

Jizhen Fu, *Queen's University, Canada*
Zhiliang Zhang, *Nanjing University of Aeronautics and Astronautics, China*
Andrew Dickson, *Queen's University, Canada*
Yan-Fei Liu, *Queen's University, Canada*
P.C. Sen, *Queen's University, Canada*

Bidirectional Operation of Resonant Voltage Divider 710

K.I. Hwu, *National Taipei University of Technology, Taiwan*
Y.T. Yau, *National Taipei University of Technology, Taiwan*

Session B3L-D: RF Applications
Wednesday, February 24, 16:10 - 17:25
Session Chairs: Alejandro Dominguez-Garcia, *University of Illinois*

Multiple-Input Buck Converter Optimized for Accurate Envelope Tracking in RF Power Amplifiers 715

M. Rodríguez, *University of Oviedo, Spain*
P.F. Miaja, *University of Oviedo, Spain*
A. Rodríguez, *University of Oviedo, Spain*
J. Sebastián, *University of Oviedo, Spain*

Switching Capacities Based Envelope Amplifier for High Efficiency RF Amplifiers 723

M. Vasić, *Universidad Politécnica de Madrid, Spain*
O. García, *Universidad Politécnica de Madrid, Spain*
J.A. Oliver, *Universidad Politécnica de Madrid, Spain*
P. Alou, *Universidad Politécnica de Madrid, Spain*
D. Diaz, *Universidad Politécnica de Madrid, Spain*
J.A. Cobos, *Universidad Politécnica de Madrid, Spain*

High Efficiency Power Amplifier for High Frequency Radio Transmitters 729
M. Vasić, *Universidad Politécnica de Madrid, Spain*
O. García, *Universidad Politécnica de Madrid, Spain*
J.A. Oliver, *Universidad Politécnica de Madrid, Spain*
P. Alou, *Universidad Politécnica de Madrid, Spain*
D. Diaz, *Universidad Politécnica de Madrid, Spain*
J.A. Cobos, *Universidad Politécnica de Madrid, Spain*
A. Gimeno, *Universidad Politecnica de Madrid, Spain*
J.M. Pardo, *Universidad Politécnica de Madrid, Spain*
C. Benavente, *Universidad Politécnica de Madrid, Spain*
F.J. Ortega, *Universidad Politécnica de Madrid, Spain*

Session B3L-E: LED Lighting II
Wednesday, February 24, 16:10 - 17:25
Session Chairs: Regan Zane, *University of Colorado*

**Applying One-Comparator Counter-Based Sampling to Current Sharing Control of
Multi-Channel LED Strings** 737
K.I. Hwu, *National Taipei University of Technology, Taiwan*
Y.T. Yau, *National Taipei University of Technology, Taiwan*

**High Frequency PWM Dimming Technique for High Power Factor Converters
in LED Lighting** 743
D. Gacio, *University of Oviedo, Spain*
J.M. Alonso, *University of Oviedo, Spain*
J. Garcia, *University of Oviedo, Spain*
L. Campa, *University of Oviedo, Spain*
M. Crespo, *University of Oviedo, Spain*
M. Rico-Secades, *University of Oviedo, Spain*

A RGB-Driver for LED Display Panels 750
Jaber Hasan, *University of Arkansas, United States*
Do Hung Nguyen, *University of Arkansas, United States*
Simon S. Ang, *University of Arkansas, United States*

Session B3L-F: Power Electronics for Utility Interface IIII
Wednesday, February 24, 16:10 - 17:25
Session Chairs: Hui Li, *Florida State University*
Miaosen Shen, *United Technologies Research Center*

**A Low Investment Single-Phase to Three-Phase Converter
Operating with Reduced Losses** 755
José A.A. Dias, *Instituto Federal de Educação, Ciência e Tecnologia da Paraíba, Brazil*
Euzeli C. dos Santos, *Universidade Federal de Campina Grande, Brazil*
Cursino B. Jacobina, *Universidade Federal de Campina Grande, Brazil*

Voltage and Power Balance Control for a Cascaded Multilevel Solid State Transformer 761
Tiefu Zhao, *North Carolina State University, United States*
Gangyao Wang, *North Carolina State University, United States*
Jie Zeng, *North Carolina State University, United States*
Sumit Dutta, *North Carolina State University, United States*
Subhashish Bhattacharya, *North Carolina State University, United States*
Alex Q. Huang, *North Carolina State University, United States*

Grid-Connected Voltage Source Inverter for Renewable Energy Conversion System with Sensorless Current Control 768
Suzan Eren, *Queen's University, Canada*
Majid Pahlevaninezhad, *Queen's University, Canada*
Alireza Bakhshai, *Queen's University, Canada*
Praveen Jain, *Queen's University, Canada*

Session B3L-G: Isolated DC-DC Converters II
Wednesday, February 24, 16:10 - 17:25
Session Chairs: Alexis Kwasinski, *The University of Texas at Austin*
Sheldon Williamson, *Concordia University*

Design of an 99%-Efficient, 5kW, Phase-Shift PWM DC-DC Converter for Telecom Applications 773
U. Badstuebner, *ETH Zurich, Switzerland*
J. Biela, *ETH Zurich, Switzerland*
J.W. Kolar, *ETH Zurich, Switzerland*

DC-DC Transformer Multiphase Converter with Transformer Coupling for Two-Stage Architecture 781
M.C. Gonzalez, *Universidad Politécnica de Madrid, Spain*
P. Alou, *Universidad Politécnica de Madrid, Spain*
O. Garcia, *Universidad Politécnica de Madrid, Spain*
J.A. Oliver, *Universidad Politécnica de Madrid, Spain*
J.A. Cobos, *Universidad Politécnica de Madrid, Spain*
H. Visairo, *Intel Corporation, Mexico*

A Comparison of Classical Two Phase (2L) and Transformer – Coupled (XL) Interleaved Boost Converters for Fuel Cell Applications 787
Kevin J. Hartnett, *University College Cork, Ireland*
Marek S. Rylko, *University College Cork, Ireland*
John G. Hayes, *University College Cork, Ireland*
Michael G. Egan, *University College Cork, Ireland*

Session C1L-A: Applications of DC-DC Converter I
Thursday, February 25, 8:30 - 10:10
Session Chairs: Chuck Mullett, *ON Semiconductor*
Kevin Parmenter, *Freescale*

Design Considerations for Narrow Vdc Based Power Delivery Architecture in Mobile Computing System 794
Xiaoguo Liang, *Intel Asia-Pacific Research & Development Ltd., China*
Gnanavel Jayakanthan, *Intel Asia-Pacific Research & Development Ltd., China*
Meng Wang, *Intel Asia-Pacific Research & Development Ltd., China*

Active Clamp Boost Converter with Switched Capacitor and Coupled Inductor 801
Yi Zhao, *Zhejiang University, China*
Wuhua Li, *Zhejiang University, China*
Bo Yang, *Zhejiang University, China*
Xiangning He, *Zhejiang University, China*

**Unified Modulation for Three-Phase Current-Fed Bidirectional DC-DC Converter
Under Varied Input Voltage** 807
Zhan Wang, *Florida State University, United States*
Hui Li, *Florida State University, United States*

**Integrated Switched-Capacitor Voltage Doubler with Clock Transition Periods
Boosting and Transfer Blocking Techniques** 813
Phong Ngo, *University of Arizona, United States*
Dongsheng Ma, *University of Arizona, United States*

Session C1L-B: AC-DC Conversion Misc. Topics I
Thursday, February 25, 8:30 - 10:10
Session Chairs: Frank Cirolia, *Emerson Network Power*
Alireza Khaligh, *Illinois Institute of Technology*

A Novel Class of Multipulse Converters Based on High-Frequency-Operated Transformers ... 818
Sheng Zheng, *Zhejiang University, China*
Dong Chen, *Zhejiang University, China*
Hai Lin, *Zhejiang University, China*
Yousheng Wang, *Zhejiang University, China*
Zhaoming Qian, *Zhejiang University, China*
Fang Z. Peng, *Michigan State University, United States*

A High Efficiency Flyback Converter with New Active Clamp Technique 823
Xiucheng Huang, *Zhejiang University, China*
Weijing Du, *Zhejiang University, China*
Wei Yuan, *Zhejiang University, China*
Junming Zhang, *Zhejiang University, China*
Zhaoming Qian, *Zhejiang University, China*

**Analysis and Design of a Novel Integrated Three-Phase Single-Stage
AC-DC PWM Full-Bridge Converter** 829
Dunisha Wijeratne, *University of Western Ontario, Canada*
Gerry Moschopoulos, *University of Western Ontario, Canada*

**Three-Phase Voltage Doubler Rectifier Based on Three-State Switching Cell for
Uninterruptible Power Supply Applications Using FPGA** 837
Raphael A. da Câmara, *Universidade Federal do Ceará, Brazil*
P.P. Praça, *Universidade Federal do Ceará, Brazil*
C.M.T. Cruz, *Universidade Federal do Ceará, Brazil*
R.P. Torrico-Bascopé, *Universidade Federal do Ceará, Brazil*
C.E.A. Silva, *Universidade Federal do Ceará, Brazil*
D.S. Oliveira, Jr., *Universidade Federal do Ceará, Brazil*
L.H.S.C. Barreto, *Universidade Federal do Ceará, Brazil*

Session C1L-C: Grid Interconnection I
Thursday, February 25, 8:30 - 10:10
Session Chairs: Ali Bazzi, *University of Illinois*
               Patrick Chapman, *University of Illinois*

Multi-Loop Control Algorithms for Seamless Transition of Grid-Connected Inverter 844
Qin Lei, *Michigan State University, United States*
Shuitao Yang, *Michigan State University & Zhe Jiang University, United States*
Fang Z. Peng, *Michigan state University, United States*

**Digital Controller Development for Grid-Tied Photovoltaic Inverter with
Model Based Technique** 849
Zhigang Liang, *North Carolina State University, United States*
Larry Alesi, *MegaWatt Solar Inc., United States*
Xiaohu Zhou, *North Carolina State University, United States*
Alex Q. Huang, *North Carolina State University, United States*

**High-Performance and Cost-Effective Multiple Feedback Control Strategy for
Standalone Operation of Grid-Connected Inverter** 854
Qin Lei, *Michigan State University, United States*
Shuitao Yang, *Zhejiang University, United States*
Fang Z. Peng, *Michigan State University, United States*

Current Control Optimization for Grid-Tied Inverters with Grid Impedance Estimation 861
Guoqiao Shen, *Zhejiang University, China*
Jun Zhang, *Zhejiang University, China*
Xiao Li, *Zhejiang University, China*
Chengrui Du, *Zhejiang University, China*
Dehong Xu, *Zhejiang University, China*

Session C1L-D: Inverter I
Thursday, February 25, 8:30 - 10:10
Session Chairs: Russell Spyker, *USAF*
               Haidong Yu, *Phoenix International*

A New Direct Peak DC-Link Voltage Control Strategy of Z-Source Inverters 867
Yu Tang, *Nanjing University of Aeronautics and Astronautics, China*
Jukui Wei, *Nanjing University of Aeronautics and Astronautics, China*
Shaojun Xie, *Nanjing University of Aeronautics and Astronautics, China*

High Performance Voltage Regulation of Current Source Inverters 873
S.A.S. Grogan, *Monash University, Australia*
D.G. Holmes, *Monash University, Australia*
B.P. McGrath, *Monash University, Australia*

**Development of a New Voltage Source Inverter (VSI) Average Model Including
Low Frequency Harmonics** 881
S. Ahmed, *Virginia Polytechnic Institute and State University, United States*
D. Boroyevich, *Virginia Polytechnic Institute and State University, United States*
F. Wang, *University of Tennessee - Knoxville, United States*
R. Burgos, *ABB US Corporate Research Center, United States*

Realization and Improvement of Repetitive Control in Rotating Frame for Active Power Filter System 887
Baifeng Chen, *Wuhan University, China*
Xiaoming Zha, *Wuhan University, China*
Jinwu Gong, *Wuhan University, China*
Suxuan Guo, *Wuhan University, China*
Jianjun Sun, *Wuhan University, China*

Session C1L-E: PWM in Motor Drives I
Thursday, February 25, 8:30 - 10:10
Session Chairs: Dionysios Aliprantis, *Iowa State University*

Current Constraints of PWM Rectifier Under Unbalanced Voltage Supply 895
Miroslav Chomat, *Institute of Thermomechanics, Czech Rep.*
Ludek Schreier, *Institute of Thermomechanics, Czech Rep.*
Jiri Bendl, *Institute of Thermomechanics, Czech Rep.*

Space Vector PWM for a Direct Matrix Converter Based Open-End Winding AC Drives with Enhanced Capabilities 901
Ranjan K. Gupta, *University of Minnesota, United States*
Apurva Somani, *University of Minnesota, United States*
Krushna K. Mohapatra, *University of Minnesota, United States*
Ned Mohan, *University of Minnesota, United States*

Evaluation of the Hybrid Four-Level Converter Employing Half-Bridge Modules for Two Different Modulation Schemes 909
Alessandro L. Batschauer, *Santa Catarina State University, Brazil*
Arnaldo J. Perin, *Federal University of Santa Catarina, Brazil*
Samir A. Mussa, *Federal University of Santa Catarina, Brazil*
Marcelo L. Heldwein, *Federal University of Santa Catarina, Brazil*

A Comparative Study of Space Vector PWM Strategy for Dual Three-Phase Permanent-Magnet Synchronous Motor Drives 915
Yanhui He, *Xi'an Jiaotong University, China*
Yue Wang, *Xi'an Jiaotong University, China*
Jinlong Wu, *Xi'an Jiaotong University, China*
Yupeng Feng, *Xi'an Jiaotong University, China*
Jinjun Liu, *Xi'an Jiaotong University, China*

Session C1L-F: Magnetics in DC-DC Converters
Thursday, February 25, 8:30 - 10:10
Session Chairs: Arnold Alderman, *PSMA*

A Novel Coupled Inductor for Interleaved Converters 920
Qianhong Chen, *Nanjing University of Aeronautics and Astronautics, China*
Ligang Xu, *Nanjing university of aeronautics and astronautics, China*
Xiaoyong Ren, *Nanjing University of Aeronautics and Astronautics, China*
Lingling Cao, *Nanjing University of Aeronautics and Astronautics, China*
Xinbo Ruan, *Nanjing University of Aeronautics and Astronautics, China*

Transformer's Capacitance Effect on the Operation of Triangular-Current Shaped Soft-Switched Converters 928
Ilya Zeltser, *Ben-Gurion University of the Negev, Israel*
Sam Ben-Yaakov, *Ben-Gurion University of the Negev, Israel*

An Input and Output Ripple Free Converter with a Four-Winding Coupled Inductor 935
Zhuomin Feng, *Zhejiang University, China*
Zhe Zhang, *Zhejiang university, China*
Duo Li, *Zhejiang University, China*
Min Chen, *Zhejiang University, China*
Zhaoming Qian, *Zhejiang University, China*

Investigation on Transformer Design of High Frequency High Efficiency DC-DC Converters 940
Dianbo Fu, *Virginia Polytechnic Institute and State University, United States*
Fred C. Lee, *Virginia Polytechnic Institute and State University, United States*
Shuo Wang, *Virginia Polytechnic Institute and State University, United States*

Session C1L-G: Photovoltaics I
Thursday, February 25, 8:30 - 10:10
Session Chairs: Robert Balog, *Texas A&M University*

A DSP-Based Single-Stage Maximum Power Point Tracking PV Inverter 948
Wen Long Yu, *National Taiwan University of Science and Technology, Taiwan*
Ting-Peng Lee, *National Taiwan University of Science and Technology, Taiwan*
Guan-Hong Wu, *National Taiwan University of Science and Technology, Taiwan*
Qing Su Chen, *National Taiwan University of Science and Technology, Taiwan*
Huang-Jen Chiu, *National Taiwan University of Science and Technology, Taiwan*
Yu-Kang Lo, *National Taiwan University of Science and Technology, Taiwan*
Frank Shih, *Macroblock Inc., Taiwan*

A Simple Mixed-Signal MPPT Circuit for Photovoltaic Applications 953
P. Mattavelli, *University of Padova, Italy*
S. Saggini, *University of Udine, Italy*
E. Orietti, *University of Padova, Italy*
G. Spiazzi, *University of Padova, Italy*

Low-Power Maximum Power Point Tracker with Digital Control for Thermophotovoltaic Generators 961
Robert C.N. Pilawa-Podgurski, *Massachusetts Institute of Technology, United States*
Nathan A. Pallo, *Massachusetts Institute of Technology, United States*
Walker R. Chan, *Massachusetts Institute of Technology, United States*
David J. Perreault, *Massachusetts Institute of Technology, United States*
Ivan L. Celanovic, *Massachusetts Institute of Technology, United States*

11-Level Cascaded H-Bridge Grid-Tied Inverter Interface with Solar Panels 968
Faete Filho, *University of Tennessee, United States*
Yue Cao, *University of Tennessee, United States*
Leon M. Tolbert, *University of Tennessee, United States*

Session C2L-A: Applications of DC-DC Converter II
Thursday, February 25, 10:40 - 11:30
Session Chairs: Chuck Mullett, *ON Semiconductor*
Kevin Parmenter, *Freescale*

A Life Prediction Scheme for Electrolytic Capacitors in Power Converters without Current Sensor 973
H.M. Pang, *University of Hong Kong, China*
M.H. Bryan Pong, *University of Hong Kong, China*

Load-Interactive Steered-Inductor DC-DC Converter with Minimized Output Filter Capacitance 980
S.M. Ahsanuzzaman, *University of Toronto, Canada*
Amir Parayandeh, *University of Toronto, Canada*
Aleksandar Prodić, *University of Toronto, Canada*
Dragan Maksimović, *University of Colorado at Boulder, United States*

Session C2L-B: AC-DC Conversion Misc. Topics II
Thursday, February 25, 10:40 - 11:30
Session Chairs: Frank Cirolia, *Emerson Network Power*
Alireza Khaligh, *Illinois Institute of Technology*

EMI Filter Design for High Switching Frequency Three-Phase/Level PWM Rectifier Systems 986
M. Hartmann, *ETH Zurich, Switzerland*
H. Ertl, *Vienna University of Technology, Austria*
J.W. Kolar, *ETH Zurich, Switzerland*

Self-Driven AC-DC Synchronous Rectifier for Power Applications – A Direct Energy-Efficient Replacement for Traditional Diode Rectifier 994
W.X. Zhong, *City University of Hong Kong, China*
W.C. Ho, *ConvenientPower HK Ltd., China*
X. Liu, *ConvenientPower HK Ltd., China*
S.Y.R. Hui, *City University of Hong Kong, China*

Session C2L-C: Grid Interconnection II
Thursday, February 25, 10:40 - 11:30
Session Chairs: Ali Bazzi, *University of Illinois*

A Robust Control Scheme for Grid-Connected Voltage Source Inverters 1002
Shuitao Yang, *Zhejiang University and Michigan State University, China*
Qin Lei, *Michigan State University, United States*
Fang Z. Peng, *Michigan State University, United States*
Zhaoming Qian, *Zhejiang University, China*

Application of Active NPC Converter on Generator Side for MW Direct-Driven Wind Turbine 1010
Jun Li, *North Carolina State University, United States*
Alex Q. Huang, *North Carolina State University, United States*
Subhashish Bhattacharya, *North Carolina State University, United States*
Wei Jing, *China University of Mining and Technology, United States*

Session C2L-D: Inverter II
Thursday, February 25, 10:40 - 11:30
Session Chairs: Russell Spyker, *USAF*
Haidong Yu, *Phoenix International*

Nonlinear Modeling of Switched Reluctance Motor Using Different Methods 1018
Jun Cai, *Nanjing University of Aeronautics and Astronautics, China*
Zhiquan Deng, *Nanjing University of Aeronautics and Astronautics, China*
Zeyuan Liu, *Nanjing University of Aeronautics and Astronautics, China*

Simplified Synchronous Reference Frame Control of the
Three Phase Grid Connected Inverter 1026
Abad Lorduy, *Carlos III University of Madrid, Spain*
Antonio Lázaro, *Carlos III University of Madrid, Spain*
Andrés Barrado, *Carlos III University of Madrid, Spain*
Cristina Fernández, *Carlos III University of Madrid, Spain*
Isabel Quesada, *Carlos III University of Madrid, Spain*
Carlos Lucena, *Carlos III University of Madrid, Spain*

Session C2L-E: PWM in Motor Drives II
Thursday, February 25, 10:40 - 11:30
Session Chairs: Dionysios Aliprantis, *Iowa State University*

A Novel Direct Digital SPWM Method for Multilevel Voltage Source Inverters 1034
Wanmin Fei, *Nanjing Normal University, China*
Yanli Zhang, *Nanjing Normal University, China*
Bin Wu, *Ryerson University, Canada*

Weight Oriented Optimal PWM in Low Modulation Indexes for
Multilevel Inverters with Unbalanced DC Sources 1038
Damoun Ahmadi, *Ohio State University, United States*
Ke Zou, *Ohio State University, United States*
Jin Wang, *Ohio State University, United States*

Session C2L-F: Measurement and Testing
Thursday, February 25, 10:40 - 11:30
Session Chairs: Patrick Chapman, *University of Illinois*

Oscillation-Test Technique for Buck Voltage Regulator 1043
Jing-Yi Huang, *National Cheng-Kung University, Taiwan*
Chun-Hsun Wu, *National Cheng-Kung University, Taiwan*
Le-Ren Chang-Chien, *National Cheng-Kung University, Taiwan*
Soon-Jyh Chang, *National Cheng-Kung University, Taiwan*

Core Loss Predictions for General PWM Waveforms from a
Simplified Set of Measured Data 1048
Charles R. Sullivan, *Thayer School of Engineering at Dartmouth, United States*
John H. Harris, *Thayer School of Engineering at Dartmouth, United States*
Edward Herbert, *FMTT, Inc., United States*

Session C2L-G: Photovoltaics II
Thursday, February 25, 10:40 - 11:30
Session Chairs: Robert Balog, *Texas A&M University*

High-Efficiency Inverter with H6-Type Configuration for Photovoltaic Non-Isolated AC Module Applications 1056
Wensong Yu, *Virginia Polytechnic Institute and State University, United States*
Jih-Sheng Lai, *Virginia Polytechnic Institute and State University, United States*
Hao Qian, *Virginia Polytechnic Institute and State University, United States*
Chris Hutchens, *Virginia Polytechnic Institute and State University, United States*
Jianhui Zhang, *National Semiconductor Corporation, United States*
Gianpaolo Lisi, *National Semiconductor Corporation, United States*
Ali Djabbari, *National Semiconductor Corporation, United States*
Greg Smith, *National Semiconductor Corporation, United States*
Tim Hegarty, *National Semiconductor Corporation, United States*

Analyzing the Optimal Matching of DC Motors to Photovoltaic Modules via DC-DC Converters 1062
Jesus Gonzalez-Llorente, *University of Puerto Rico-Mayaguez, Puerto Rico*
Eduardo I. Ortiz-Rivera, *University of Puerto Rico-Mayaguez, Puerto Rico*
Andres Salazar-Llinas, *University of Puerto Rico-Mayaguez, Puerto Rico*
Emil Jimenez-Brea, *University of Puerto Rico-Mayaguez, Puerto Rico*

Session C3L-A: Load Management Interface I
Thursday, February 25, 14:00 - 15:40
Session Chairs: Siamak Abedinpour, *Freescale*
Jonathan Kimball, *Missouri S&T*

Performance Analysis of an Interleaved High Step-Up Converter with Voltage Multiplier Cell 1069
Wuhua Li, *Zhejiang University, China*
Yi Zhao, *Zhejiang University, China*
Yan Deng, *Zhejiang University, China*
Xiangning He, *Zhejiang University, China*

FPGA-Based Multi-Phase Digital Pulse Width Modulator with Dual-Edge Modulation 1075
Martin Scharrer, *University of Limerick, Ireland*
Mark Halton, *University of Limerick, Ireland*
Tony Scanlan, *University of Limerick, Ireland*
Karl Rinne, *University of Limerick, Ireland*

Phase Doubler for High Power Voltage Regulators 1081
Chun Cheung, *Intersil Corporation, United States*
Weihong Qiu, *Intersil Corporation, United States*
Emil Chen, *Intersil Corporation, United States*
Greg Miller, *Intersil Corporation, United States*

Automatic Multi-Phase Digital Pulse Width Modulator 1087
Simon Effler, *University of Limerick, Ireland*
Mark Halton, *University of Limerick, Ireland*
Karl Rinne, *University of Limerick, Ireland*

Session C3L-B: Power Electronics in Motor Drives I
Thursday, February 25, 14:00 - 15:40
Session Chairs: Chris Edrington, *Florida State University*
Patrick Chapman, *University of Illinois*

A Simple Current Sharing Scheme for Dual Three-Phase Permanent-Magnet Synchronous Motor Drives 1093
Yanhui He, *Xi'an Jiaotong University, China*
Yue Wang, *Xi'an Jiaotong University, China*
Jinlong Wu, *Xi'an Jiaotong University, China*
Yupeng Feng, *Xi'an Jiaotong University, China*
Jinjun Liu, *Xi'an Jiaotong University, China*

Multilevel Current Source Inverter Topologies Based on the Duality Principle 1097
Jianyu Bao, *Ningbo Institute of Technology, Zhejiang University, China*
Weibing Bao, *Zhejiang University of Science and Technology, China*
Siran Wang, *Zhejiang University, China*
Zhongchao Zhang, *Zhejiang University, China*

3-Level Power Converter with High-Voltage SiC-PiN Diode and Hard-Gate-Driving of IEGT for Future High-Voltage Power Conversion Systems 1101
Kazuto Takao, *Toshiba Corporation, Japan*
Yasunori Tanaka, *National Institute of Advanced Industrial Science and Technology, Japan*
Kyungmin Sung, *Ibaraki National College of Technology, Japan*
Keiji Wada, *Tokyo Metoropolitan University, Japan*
Takashi Shinohe, *Toshiba Corporation, Japan*
Takeo Kanai, *Toshiba Mitsubishi-Electric Industrial Systems Corporation, Japan*
Hiromichi Ohashi, *National Institute of Advanced Industrial Science and Technology, Japan*

18 kW Three Phase Inverter System Using Hermetically Sealed SiC Phase-Leg Power Modules 1108
Hui Zhang, *Tuskegee University, United States*
Leon M. Tolbert, *University of Tennessee, United States*
Jung Hee Han, *Global Power Electronics, United States*
Madhu S. Chinthavali, *Oak Ridge National Laboratory, United States*
Fred Barlow, *University of Idaho, United States*

Session C3L-C: DC-DC Converter V
Thursday, February 25, 14:00 - 15:40
Session Chairs: Frank Ciriola, *Emerson*
Arnold Alderman, *PSMA*

Multiphase Optimal Response Mixed-Signal Current-Programmed Mode Controller 1113
Jurgen Alico, *University of Toronto, Canada*
Aleksandar Prodić, *University of Toronto, Canada*

Switching Loss Analysis of Closed-Loop Gate Drive 1119
Lihua Chen, *Michigan State University, United States*
Fang Z. Peng, *Michigan State University, United States*

Modeling and Analysis of Closed-Loop Gate Drive 1124
Lihua Chen, *Michigan State University, United States*
Baoming Ge, *Michigan State University, United States*
Fang Z. Peng, *Michigan State University, United States*

**Black-Box Modeling of DC-DC Converters Based on Transient Response
Analysis and Parametric Identification Methods** 1131
V. Valdivia, *Carlos III University of Madrid, Spain*
A. Barrado, *Carlos III University of Madrid, Spain*
A. Lázaro, *Carlos III University of Madrid, Spain*
C. Fernández, *Carlos III University of Madrid, Spain*
P. Zumel, *Carlos III University of Madrid, Spain*

Session C3L-D: Transportation
Thursday, February 25, 14:00 - 15:40
Session Chairs: Jaber Abu Qahouq, *University of Alabama*
Dionysios Aliprantis, *Iowa State University*

**Harmonic and Balance Compensation Using Instantaneous Active and
Reactive Power Control on Electric Railway Systems** 1139
A. Bueno, *Universidad Simón Bolívar, Venezuela*
J.M. Aller, *Universidad Simón Bolívar, Venezuela*
J. Restrepo, *Universidad Simón Bolívar, Venezuela*
T. Habetler, *Georgia Institute of Technology, United States*

**Review of Non-Isolated Bi-Directional DC-DC Converters for Plug-in Hybrid Electric
Vehicle Charge Station Application at Municipal Parking Decks** 1145
Yu Du, *North Carolina State University, United States*
Xiaohu Zhou, *North Carolina State University, United States*
Sanzhong Bai, *North Carolina State University, United States*
Srdjan Lukic, *North Carolina State University, United States*
Alex Huang, *North Carolina State University, United States*

**Control of Plug-in Hybrid Electric Vehicles for Mobile Power Generation and
Grid Support Applications** 1152
Gui-Jia Su, *Oak Ridge National Laboratory, United States*
Lixin Tang, *Oak Ridge National Laboratory, United States*

Interface Issues of Mining Haul Trucks Operating on Trolley Systems 1158
Joy Mazumdar, *Siemens Industry Inc., United States*
Walter Koellner, *Siemens Industry Inc., United States*
Rohit Moghe, *Georgia Institute of Technology, United States*

Session C3L-E: Power Converter Applications I
Thursday, February 25, 14:00 - 15:40
Session Chairs: Vajapeyam Sukumar, *Maxim Integrated Products*

Regenerative AC Electronic Load with One-Cycle Control 1166
In Wha Jeong, *University of California, Irvine, United States*
Mikhail Slepchenkov, *University of California, Irvine, United States*
Keyue Smedley, *University of California, Irvine, United States*
Franco Maddaleno, *University of California, Irvine, United States*

A High Efficiency Regulated Charge Pump Over Wide Input and Load Range 1172
Rong Guo, *North Carolina State University, United States*
Liyu Yang, *North Carolina State University, United States*
Alex Huang, *North Carolina State University, United States*
John Endredy, *RF Micro Devices, United States*

High Performance, High-Power Capacitor Charging:
Focus on Pulse-to-Pulse Repeatability 1177
A. Pokryvailo, *Spellman High Voltage Electronics Corporation, United States*
C. Carp, *Spellman High Voltage Electronics Corporation, United States*
C. Scapellati, *Spellman High Voltage Electronics Corporation, United States*

Generalized AC-DC Single-Phase Boost Rectifier 1183
C.B. Jacobina, *Universidade Federal de Campina Grande, Brazil*
Euzeli dos Santos, *Universidade Federal de Campina Grande, Brazil*
Nady Rocha, *Universidade Federal de Campina Grande, Brazil*

Session C3L-F: Utility Interface Applications
Thursday, February 25, 14:00 - 15:40
Session Chairs: Ali Davoudi, *University of Illinois*

Parallel Connection of Two Shunt Active Power Filters with Losses Optimization 1191
E.C. dos Santos, Jr., *Universidade Federal de Campina Grande, Brazil*
C.B. Jacobina, *Universidade Federal de Campina Grande, Brazil*
A.M. Maciel, *Universidade Federal de Campina Grande, Brazil*

Design and Implementation of an Improved Controller for Parallel-Connected
400 Hz Frequency Converters 1197
B. Tamyurek, *Eskisehir Osmangazi University, Turkey*
E. Birdane, *Kaynak Electronic Machine Industry and Trade Co. Ltd., Turkey*
Adil Ceyhan, *Kaynak Electronic Machine Industry and Trade Co. Ltd., Turkey*

Study on the Impact of the Complex Impedance on the Droop Control Method
for the Parallel Inverters 1204
Wei Yao, *Zhejiang university, China*
Mingzhi Gao, *Zhejiang University, China*
Zheng Ren, *Zhejiang university, China*
Min Chen, *Zhejiang University, China*
Zhaoming Qian, *Zhejiang University, China*

A Three-Phase Adaptive Approach to Extract Harmonic and Reactive Currents 1209
D. Yazdani, *Queen's University, Canada*
A. Bakhshai, *Queen's University, Canada*
P.K. Jain, *Queen's University, Canada*

Session C3L-G: Soft Switching Techniques I
Thursday, February 25, 14:00 - 15:40
Session Chairs: Jason Neely, *Purdue University*
Wayne Weaver, *Michigan Technological University*

Analysis, Optimized Design and Adaptive Control of a ZCS Full-Bridge Converter Without Voltage Over-Stress on the Switches 1214
Xin Zhang, *Nanjing University of Aeronautics and Astronautics, China*
Henry Shu-hung Chung, *City University of Hong Kong, China*
Xinbo Ruan, *Huazhong University of Science and Technology, China*
Adrian Ioinovici, *Holon Institute of Technology, Israel*

Analysis and Design of a Novel ZVS-PWM DC-DC Converter for Bidirectional Applications with Steep Conversion Ratio 1222
Pritam Das, *University of Western Ontario, Canada*
Ahmad Mousavi, *University of Western Ontario, Canada*
Gerry Moschopoulos, *University of Western Ontario, Canada*

Three-Level Phase-Shift ZVS-PWM DC-DC Converter with High Frequency Transformer for High Performance Arc Welding Machines 1230
Tomokazu Mishima, *Kure National College of Technology, Japan*
Hisayuki Sugimura, *Daihen Corporation, Japan*
Khairy Fathy Sayed, *Kyungnam University, Korea, South*
Soon Kurl Kwon, *Kyungnam University, Korea, South*
Mutsuo Nakaoka, *Kyungnam University and Yamaguchi University, Japan*

Fully Soft-Switched Bidirectional Resonant DC-DC Converter with a New CLLC Tank 1238
Wei Chen, *Zhejiang University, China*
Siran Wang, *Zhejiang University, China*
Xiaoyuan Hong, *Zhejiang University, China*
Zhengyu Lu, *Zhejiang University, China*
Shaoshi Ye, *Delta Electronics (Shanghai) Co., LTD., China*

Session C4L-A: Load Management Interface II
Thursday, February 25, 16:10 - 17:25
Session Chairs: Siamak Abedinpour, *Freescale*
Jonathan Kimball, *Missouri S&T*

Optimal Phase Changing Frequency Determination for Multiphase Voltage Regulator Modules 1243
Anand Ramamurthy, *North Carolina State University, United States*
Subhashish Bhattacharya, *North Carolina State University, United States*
Chris Thompson, *Intersil Corporation, United States*
Jon Day, *Intersil Corporation, United States*

A New Digital Adaptive Voltage Positioning Technique with Dynamically Varying Voltage and Current References 1248
S. Pan, *Queen's University, Canada*
P.K. Jain, *Queen's University, Canada*

A Three-Level Buck Converter and Digital Controller for Improving Load Transient Response 1256
Zhenyu Zhao, *Exar Corp, Canada*
Aleksandar Prodić, *University of Toronto, Canada*

Session C4L-B: Power Electronics in Motor Drives II
Thursday, February 25, 16:10 - 17:25
Session Chairs: Chris Edrington, *Florida State University*
Patrick Chapman, *University of Illinois*

Trends in MW-Rated VSI Technology and Reliability for Adjustable Speed Drives 1261
Hiromi Hosoda, *Toshiba Mitsubishi-Electric Industrial Systems Corporation, Japan*
Mostafa Al Mamun, *Toshiba Mitsubishi-Electric Industrial Systems Corporation, Japan*
Teruo Yoshino, *Toshiba Mitsubishi-Electric Industrial Systems Corporation, Japan*

Development of a Compact 750KVA Three-Phase NPC Three-Level Universal Inverter Module with Specifically Designed Busbar 1266
Jun Wang, *Zhejiang University, China*
Binjian Yang, *Zhejiang University, China*
Jing Zhao, *Zhejiang University, China*
Yan Deng, *Zhejiang University, China*
Xiangning He, *Zhejiang University, China*
Xu Zhixin, *Zhejiang University of Science and Technology, China*

Common Mode Voltage in DC-Fed Motor Drive System and its Impact on the EMI Filter 1272
Fang Luo, *Virginia Polytechnic Institute and State University &*
    Huazhong University of Science and Technology, United States
Shuo Wang, *GE Aviation Systems, United States*
Fred Wang, *University of Tennessee - Knoxville and Oak Ridge National Laboratory, United States*
Dushan Boroyevich, *Virginia Polytechnic Institute and State University, United States*
Nicolas Gazel, *Virginia Polytechnic Institute and State University, United States*
Yong Kang, *Huazhong University of Science and Technology, China*

Session C4L-C: DC-DC Converter VI
Thursday, February 25, 16:10 - 17:25
Session Chairs: Arnold Alderman, *PSMA*
Frank Cirolia, *Emerson Network Power*

Black-Box Modeling of Three Phase Voltage Source Inverters Based on Transient Response Analysis 1279
V. Valdivia, *Carlos III University of Madrid, Spain*
A. Lázaro, *Carlos III University of Madrid, Spain*
A. Barrado, *Carlos III University of Madrid, Spain*
P. Zumel, *Carlos III University of Madrid, Spain*
C. Fernández, *Carlos III University of Madrid, Spain*
M. Sanz, *Carlos III University of Madrid, Spain*

Digital Autotuning of DC-DC Converters Based on Model Reference Impulse Response ... 1287
A. Costabeber, *University of Padova, Italy*
P. Mattavelli, *University of Padova, Italy*
S. Saggini, *University of Udine, Italy*
A. Bianco, *STMicroelectronics, Italy*

High-Fidelity and High-Speed Modeling and Simulation for Power Conversion Systems ... 1295
Chunchun Xu, *GE Global Research, United States*
Luis Garces, *GE Global Research, United States*
Paul Szczesny, *GE Global Research, United States*

Session C4L-D: Aerospace
Thursday, February 25, 16:10 - 17:25
Session Chairs: Jaber Abu Qahouq, *University of Alabama*
Dionysios Aliprantis, *Iowa State University*

Electrical Power Distribution System (HV270DC), for Application in More Electric Aircraft .. 1300
D. Izquierdo, *EADS, Spain*
R. Azcona, *EADS, Spain*
F.J. López del Cerro, *EADS, Spain*
Carlos Fernández, *EADS, Spain*
Bernardo Delicado, *EADS, Spain*

Supercapacitor-Based Energy Management for Future Aircraft Systems 1306
R. Todd, *University of Manchester, United Kingdom*
D. Wu, *University of Manchester, United Kingdom*
J.A. dos Santos Girio, *University of Manchester, United Kingdom*
M. Poucand, *University of Manchester, United Kingdom*
A.J. Forsyth, *University of Manchester, United Kingdom*

Buck Boost Regulator (B²R) for Spacecraft Solar Array Power Conversion 1313
Olivier Mourra, *European Space Agency, Netherlands*
Arturo Fernandez, *European Space Agency, Netherlands*
Ferdinando Tonicello, *European Space Agency, Netherlands*

Session C4L-E: Power Converter Applications II
Thursday, February 25, 16:10 - 17:25
Session Chairs: Vajapeyam Sukumar, *Maxim Integrated Products*

Quadratic Power Conversion for Industrial Applications 1320
Gerry Moschopoulos, *University of Western Ontario, Canada*

Multiple-Output Resonant Inverter Topology for Multi-Inductor Loads 1328
O. Lucía, *University of Zaragoza, Spain*
J.M. Burdío, *University of Zaragoza, Spain*
I. Millán, *University of Zaragoza, Spain*
J. Acero, *University of Zaragoza, Spain*

**Variable Frequency Pulse Density Modulation[1] for Efficient High Frequency
Operation of Series Resonant Converters Operating As Voltage Regulators** 1334
Darryl J. Tschirhart, *Queen's University, Canada*
Praveen K. Jain, *Queen's University, Canada*

Session C4L-F: General Lighting
Thursday, February 25, 16:10 - 17:25
Session Chairs: Ali Davoudi, *University of Illinois*

Flexible-Controlled High Power-Density Automotive HID Electronic Ballast Using Full-Digital Control Mode 1340
Xinyi Yang, *Zhejiang University, China*
Biwen Xu, *Zhejiang University, China*
Chongguang Ma, *Zhejiang University, China*
Min Chen, *Zhejiang University, China*
Zhaoming Qian, *Zhejiang University, China*

A "Class-A2" Ultra-Low-Loss Magnetic Ballast for T5 Fluorescent Lamps 1346
S.Y.R. Hui, *City University of Hong Kong, China*
D.Y. Lin, *City University of Hong Kong, China*
W.M. Ng, *City University of Hong Kong, China*
W. Yan, *City University of Hong Kong, China*

Simple Triac Dimmable Compact Fluorescent Lamp Ballast and Light Emitting Diode Driver 1352
Andre Tjokrorahardjo, *International Rectifier, United States*

Session C4L-G: Soft Switching Techniques II
Thursday, February 25, 16:10 - 17:25
Session Chairs: Jason Neely, *Purdue University*
Wayne Weaver, *Michigan Technological University*

High Efficiency Soft-Switched Step-Up DC-DC Converter with Hybrid Mode LLC+C Resonant Tank 1358
Wei Chen, *Zhejiang University and Delta Electronics (Shanghai) Co., LTD., China*
Xiaoyuan Hong, *Zhejiang University, China*
Siran Wang, *Zhejiang University, China*
Zhengyu Lu, *Zhejiang University, China*
Shaoshi Ye, *Delta Electronics (Shanghai) Co., LTD., China*

A Family of Zero Current Switching Switched-Capacitor DC-DC Converters 1365
Dong Cao, *Michigan State University, United States*
Fang Zheng Peng, *Michigan State University, United States*

Analysis and Design of the Half Bridge Magnetizing Inductor Resonant(L_mC) DC-DC Converter 1373
B.-C. Hyeon, *Seoul National University, Korea, South*
B.-H. Cho, *Seoul National University, Korea, South*

Session B4P-H: AC-DC Conversion
Thursday, February 25, 11:30 - 13:30

A High Power Density Single Phase PWM Rectifier with Active Ripple Energy Storage 1378
Ruxi Wang, *Virginia Polytechnic Institute and State University, United States*
Fred Wang, *University of Tennessee - Knoxville and Oak Ridge National Laboratory, United States*
Dushan Boroyevich, *Virginia Polytechnic Institute and State University, United States*
Puqi Ning, *Virginia Polytechnic Institute and State University, United States*

Design Considerations for High Efficiency Buck PFC with Half-Bridge Regulation Stage .. 1384
Bernard Keogh, *Texas Instruments (Cork) Ltd., Ireland*
George Young, *Texas Instruments (Cork) Ltd., Ireland*
Hagen Wegner, *Texas Instruments (Cork) Ltd., Ireland*
Colin Gillmor, *Texas Instruments (Cork) Ltd., Ireland*

**A Novel Variable Frequency Soft Switching Method for Flyback
Converter with Synchronous Rectifier** 1392
Xiucheng Huang, *Zhejiang University, China*
Weijing Du, *Zhejiang University, China*
Wei Yuan, *Zhejiang University, China*
Junming Zhang, *Zhejiang University, China*
Zhaoming Qian, *Zhejiang University, China*

Optimal Design of a Compact 99.3% Efficient Single-Phase PFC Rectifier 1397
J. Biela, *ETH Zurich, Switzerland*
J.W. Kolar, *ETH Zurich, Switzerland*
G. Deboy, *Infineon Technologies Austria AG, Austria*

DCM Boost PFC Converter with High Input PF 1405
Kai Yao, *Nanjing university of aeronautics and astronautics, China*
Xinbo Ruan, *Nanjing University of Aeronautics and Astronautics, China*
Xiaojing Mao, *Nanjing University of Aeronautics and Astronautics, China*
Zhihong Ye, *Lite-on Technology Corp., China*

**Interleaved Forward Converter with Ripple-Free Circuit for
Humane Killer Poultry Applications** 1413
S.-Y. Tseng, *Chang Gung University, Taiwan*
T.-Y. Chiang, *Chang Gung University, Taiwan*
K.-C. Wang, *Chang Gung University, Taiwan*
S.-A. Chuang, *Chang Gung University, Taiwan*

**The Optimal Control Strategy for Rectifier Side of Low Switching
Frequency Back-to-Back Converter** 1419
Kai Tan, *Chinese Academy of Sciences, China*
Qiongxuan Ge, *Chinese Academy of Sciences, China*
Zhenggang Yin, *Chinese Academy of Sciences, China*
Congwei Liu, *Chinese Academy of Sciences, China*
Yaohua Li, *Chinese Academy of Sciences, China*

**Transformer Structure and its Effects on Common Mode EMI Noise in
Isolated Power Converters** 1424
Pengju Kong, *Virginia Polytechnic Institute and State University, United States*
Fred C. Lee, *Virginia Polytechnic Institute and State University, United States*

A Single-Stage Single-Phase Bi-Directional Grid Interface Circuit with Digital Lookup Table Based Control 1430
Evan Reutzel, *University of California, Berkeley, United States*
Seth Sanders, *University of California, Berkeley, United States*

Session B4P-J: DC-DC Converter VII
Thursday, February 25, 11:30 - 13:30

A High Performance Dual Output DC-DC Converter Combined the Phase Shift Full Bridge and LLC Resonant Half Bridge with the Shared Lagging Leg 1435
Yu Chen, *Huazhong University of Science and Technology, China*
Xuejun Pei, *Huazhong University of Science and Technology, China*
Li Peng, *Huazhong University of Science and Technology, China*
Yong Kang, *Huazhong University of Science and Technology, China*

Dual Output DC-DC Converter with Shared ZCS Lagging Leg 1441
Yu Chen, *Huazhong University of Science and Technology, China*
Li Peng, *Huazhong University of Science and Technology, China*
Xuejun Pei, *Huazhong University of Science and Technology, China*
Yong Kang, *Huazhong University of Science and Technology, China*

A Novel ZVS Full-Bridge Converter with Auxiliary Circuit 1448
Zhong Chen, *Nanjing University of Aeronautics and Astronautics, China*
Biao Ji, *Nanjing University of Aeronautics and Astronautics, China*
Feng Ji, *Nanjing University of Aeronautics and Astronautics, China*
Lei Shi, *Nanjing University of Aeronautics and Astronautics, China*

An Active Clamp ZVT Converter with Input-Parallel and Output-Series Configuration 1454
Yi Zhao, *Zhejiang University, China*
Wuhua Li, *Zhejiang University, China*
Weichen Li, *Zhejiang University, China*
Xiangning He, *Zhejiang University, China*

A Parallel Front-End LCL Resonant Push-Pull Converter with a Coupled Inductor for Automotive Applications 1460
Yuan Yisheng, *East China Jiaotong University, China*
Chen Min, *Zhejiang University, China*
Qian Zhaoming, *Zhejiang University, China*

A Novel Full Bridge Dual Output DC-DC Converter with Complementary Pulse Widths and Frequency Modulation 1464
Yu Chen, *Huazhong University of Science and Technology, China*
Xuejun Pei, *Huazhong University of Science and Technology, China*
Li Peng, *Huazhong University of Science and Technology, China*
Yong Kang, *Huazhong University of Science and Technology, China*

Analysis and Design Considerations of an Improved ZVS Full-Bridge DC-DC Converter ... 1471
Zhong Chen, *Nanjing University of Aeronautics and Astronautics, China*
Biao Ji, *Nanjing University of Aeronautics and Astronautics, China*
Feng Ji, *Nanjing University of Aeronautics and Astronautics, China*
Lei Shi, *Nanjing University of Aeronautics and Astronautics, China*

**A New Resonant Gate Driver for Switching Loss Reduction of
High Side Switch in Buck Converter** 1477
Xin Zhou, *North Carolina State University, United States*
Zhigang Liang, *North Carolina State University, United States*
Alex Huang, *North Carolina State University, United States*

**Switching Loss Analysis Considering Parasitic Loop Inductance with
Current Source Drivers for Buck Converters** 1482
Zhiliang Zhang, *Nanjing University of Aeronautics and Astronautics, China*
Jizhen Fu, *Queen's University, Canada*
Yan-Fei Liu, *Queen's University, Canada*
P.C. Sen, *Queen's University, Canada*

**Improved Asymmetric Space Vector Modulation for Voltage Source
Converters with Low Carrier Ratio** 1487
Di Zhang, *Virginia Polytechnic Institute and State University, United States*
Fred Wang, *University of Tennessee - Knoxville and Oak Ridge National Laboratory, United States*
Said El-Barbari, *GE Global Research, Germany*
Juan Sabate, *GE Global Research, United States*
Dushan Boroyevich, *Virginia Polytechnic Institute and State University, United States*

**A Hybrid Switching Scheme for LLC Series-Resonant Half-Bridge DC-DC
Converter in a Wide Load Range** 1494
Woo-Young Choi, *Virginia Polytechnic Institute and State University, United States*
Bong-Hwan Kwon, *Pohang University of Science and Technology, Korea, South*
Jih-Sheng Lai, *Virginia Polytechnic Institute and State University, United States*

Session B4P-K: Motor Drives & Inverters I
Thursday, February 25, 11:30 - 13:30

**Industrial Servo Applications of Linear Induction Motors Based on
Dynamic Maximum Force Control** 1498
Haidong Yu, *Phoenix International, United States*
Babak Fahimi, *University of Texas at Arlington, United States*

A Soft-Switching Interleaved Three-Level Inverter 1503
Yuan Yisheng, *East China Jiaotong University, China*
Chen Min, *Zhejiang University, China*
Qian Zhaoming, *Zhejiang University, China*

**Reducing Common-Mode Voltage in Three-Phase Sine-Triangle PWM with
Interleaved Carriers** 1508
Jonathan W. Kimball, *Missouri University of Science and Technology, United States*
Maciej Zawodniok, *Missouri University of Science and Technology, United States*

**Dynamic DC-Bus Voltage Control Strategies for a Three-Phase High Power
Shunt Active Power Filter** 1514
Zhiqiang Wang, *Zhejiang University, China*
Chuan Xie, *Zhejiang University, China*
Jing Zhang, *Zhejiang University, China*
Guozhu Chen, *Zhejiang University, China*

A Simplified Three Phase Three-Level Zero-Current-Transition Active Neutral-Point-Clamped Converter with Three Auxiliary Switches 1521
Jin Li, *Xi'an Jiaotong University and Virginia Polytechnic Institute and State University, China*
Jinjun Liu, *Xi'an Jiaotong University, China*
Dushan Boroyevich, *Virginia Polytechnic Institute and State University, China*

Comparison and Implementation of a 3-Level NPC Voltage Link Back-to-Back Converter with SiC and Si Diodes 1527
Mario Schweizer, *ETH Zurich, Switzerland*
Thomas Friedli, *ETH Zurich, Switzerland*
Johann W. Kolar, *ETH Zurich, Switzerland*

A Novel PWM Control Method to Eliminate the Effect of Dead Time on the Output Waveform for Hybrid Clamped Multilevel Inverters 1534
Jing Zhao, *Zhejiang University, China*
Xiangning He, *Zhejiang University, China*
Yunlong Han, *Zhejiang University, China*
Yan Chen, *Zhejiang University, China*
Rongxiang Zhao, *Zhejiang University, China*

Study on Wide Range Robust Speed Sensorless Control of Medium Voltage Induction Motor 1542
Siran Wang, *Zhejiang University, China*
Zhengyu Lu, *Zhejiang University, China*

Fault Detection and Diagnostics for Non-Intrusive Monitoring Using Motor Harmonics 1547
Uzoma A. Orji, *Massachusetts Institute of Technology, United States*
Zachary Remscrim, *Massachusetts Institute of Technology, United States*
Christopher Laughman, *Massachusetts Institute of Technology, United States*
Steven B. Leeb, *Massachusetts Institute of Technology, United States*
Warit Wichakool, *Massachusetts Institute of Technology, United States*
Christopher Schantz, *Massachusetts Institute of Technology, United States*
Robert Cox, *Massachusetts Institute of Technology, United States*
James Paris, *Massachusetts Institute of Technology, United States*
James L. Kirtley, Jr., *Massachusetts Institute of Technology, United States*
Les K. Norford, *Massachusetts Institute of Technology, United States*

Reliability Evaluation of Three-Level Inverters 1555
Yi Ding, *Nanyang Technological University, Singapore*
Poh Chiang Loh, *Nanyang Technological University, Singapore*
Kuan Khoon Tan, *Nanyang Technological University, Singapore*
Peng Wang, *Nanyang Technological University, Singapore*
Feng Gao, *Nanyang Technological University, Singapore*

Parallel Operation of PWM Inverters for High Speed Motor Drive System 1561
Un-Kwan Cho, *Seoul National University, Korea, South*
Jung-Sik Yim, *Seoul National University, Korea, South*
Seung-Ki Sul, *Seoul National University, Korea, South*

Session B4P-L: Active Components
Thursday, February 25, 11:30 - 13:30

Reverse Conduction of a 100 a SiC DMOSFET Module in High-Power Applications 1568
R.A. Wood, *US Army Research Lab, United States*
D.P. Urciuoli, *US Army Research Lab, United States*
T.E. Salem, *US Naval Academy, United States*
R. Green, *US Army Research Lab, United States*

Investigation of 1.2 kV SiC MOSFET for High Frequency High Power Applications 1572
Honggang Sheng, *Monolithic Power Systems, United States*
Zheng Chen, *Virginia Polytechnic Institute and State University, United States*
Fred Wang, *University of Tennessee - Knoxville, United States*
Alan Millner, *MKS Instruments, United States*

**Comparative Analysis of Power Stage Losses for Synchronous Buck Converter in
Diode Emulation Mode Vs. Continuous Conduction Mode at Light Load Condition** 1578
Yang Chen, *International Rectifier Corp., United States*
Peyman Asadi, *International Rectifier Corp., United States*
Parviz Parto, *International Rectifier Corp., United States*

**Controllable dv/dt Behaviour of the SiC MOSFET/JFET Cascode an Alternative
Hard Commutated Switch for Telecom Applications** 1584
Daniel Aggeler, *ETH Zurich, Switzerland*
Juergen Biela, *ETH Zurich, Switzerland*
Johann W. Kolar, *ETH Zurich, Switzerland*

Integral Micro-Channel Liquid Cooling for Power Electronics 1591
Ljubisa D. Stevanovic, *GE Global Research, United States*
Richard A. Beaupre, *GE Global Research, United States*
Arun V. Gowda, *GE Global Research, United States*
Adam G. Pautsch, *GE Global Research, United States*
Stephen A. Solovitz, *Washington State University Vancouver, United States*

3000V, 25A Pulse Power Asymmetrical Highly Interdigitated SiC Thyristors 1598
Ahmed Elasser, *GE Global Research, United States*
Peter Losee, *GE Global Research, United States*
Stephen Arthur, *GE Global Research, United States*
Zachary Stum, *GE Global Research, United States*
Jerome Garrett, *GE Global Research, United States*
Michael Schutten, *GE Global Research, United States*

Session B4P-M: System Integration III
Thursday, February 25, 11:30 - 13:30

Low Inductance Power Module with Blade Connector 1603
Ljubisa D. Stevanovic, *GE Global Research, United States*
Richard A. Beauprc, *GE Global Research, United States*
Eladio C. Delgado, *GE Global Research, United States*
Arun V. Gowda, *GE Global Research, United States*

Design of Multi-Turn LTCC Inductors for High Frequency DC-DC Converters 1610

Laili Wang, *Xi'an Jiaotong University, China*
Yunqing Pei, *Xi'an Jiaotong University, China*
Xu Yang, *Xi'an Jiaotong University, China*
Xizhi Cui, *Xi'an Jiaotong University, China*
Zhaoan Wang, *Xi'an Jiaotong University, China*
Guopeng Zhao, *Xi'an Jiaotong University, China*

Session B4P-N: Utility Interface
Thursday, February 25, 11:30 - 13:30

**Topological Research and Comparison of Low Harmonic Input Three-Phase
Rectifier with Passive Auxiliary Circuit** 1616

Zhong Chen, *Nanjing University of Aeronautics and Astronautics, China*
Yingpeng Luo, *Nanjing University of Aeronautics and Astronautics, China*
Yinyu Zhu, *Nanjing University of Aeronautics and Astronautics, China*

**The Reactive Power Compensation and Harmonic Filtering and the
Over-Voltage Analysis of the ITER Power Supply System** 1622

L. Xu, *Chinese Academy of Sciences, China*
Z. Sheng, *Chinese Academy of Sciences, China*
P. Fu, *Chinese Academy of Sciences, China*
G. Gao, *Chinese Academy of Sciences, China*
I. Benfatto, *Iter Organization, France*
A.D. Mankani, *Iter Organization, France*
J. Tao, *Iter Organization, France*

Optimal Design Method of Three-Phase Rectifier with Near-Sinusoidal Input Currents 1627

Zhong Chen, *Nanjing University of Aeronautics and Astronautics, China*
Yingpeng Luo, *Nanjing University of Aeronautics and Astronautics, China*
Yinyu Zhu, *Nanjing University of Aeronautics and Astronautics, China*
Shunqing Wang, *Nanjing University of Aeronautics and Astronautics, China*

**An Analysis on the Influence of Interface Inductor to STATCOM System with
Phase and Amplitude Control and Corresponding Design Considerations** 1633

Guopeng Zhao, *Xi'an Jiaotong University, China*
Jinjun Liu, *Xi'an Jiaotong University, China*

**Vector Oriented Control of Voltage Source PWM Inverter As a Dynamic VAR
Compensator for Wind Energy Conversion System Connected to Utility Grid** 1640

Mahmoud M.N. Amin, *Florida International University, United States*
O.A. Mohammed, *Florida International University, United States*

**Control System Design for Bi-Directional Power Transfer in Single-Phase
Back-to-Back Converter Based on the Linear Operating Region** 1651

Janeth Alcalá, *Universidad Autónoma de San Luis Potosi, Mexico*
Víctor Cárdenas, *Universidad Autónoma de San Luis Potosi, Mexico*
Emanuel Rosas, *Universidad Autónoma de San Luis Potosi, Mexico*
Ciro Núñez, *Universidad Autónoma de San Luis Potosi, Mexico*

Comparative Analysis of Low-Pass Output Filter for Single-Phase Grid-Connected Photovoltaic Inverter 1659
Hanju Cha, *Chungnam National University, Korea, South*
Trung-Kien Vu, *Chungnam National University, Korea, South*

Design and Development of Generation-I Silicon based Solid State Transformer 1666
Subhashish Bhattacharya, *North Carolina State University, United States*
Tiefu Zhao, *North Carolina State University, United States*
Gangyao Wang, *North Carolina State University, United States*
Sumit Dutta, *North Carolina State University, United Kingdom*
Seunghun Baek, *North Carolina State University, United States*
Yu Du, *North Carolina State University, United States*
Babak Parkhideh, *North Carolina State University, United States*
Xiaohu Zhou, *North Carolina State University, United States*
Alex Q. Huang, *North Carolina State University, United States*

Power Calculation Method Used in Wireless Parallel Inverters Under Nonlinear Load Conditions 1674
Zheng Ren, *Zhejiang University, China*
Mingzhi Gao, *Zhejiang University, China*
Qiong Mo, *Zhejiang University, China*
Kun Liu, *Zhejiang University, China*
Wei Yao, *Zhejiang University, China*
Min Chen, *Zhejiang University, China*
Zhaomin Qian, *Zhejiang University, China*

A Real-Time Fault Diagnosis System for UPS Based on FFT Frequency Analysis 1678
Won-Sul Shim, *Kangwon National University, Korea, South*
Gi-Taek Kim, *Kangwon National University, Korea, South*
Ha-Jin Jung, *Powertron Engineering Co. Ltd., Korea, South*
Deuk-Soo Kim, *Powertron Engineering Co. Ltd., Korea, South*

Control Strategy for a Buck-Boost Type Direct Interface Converter Using an Indirect Matrix Converter with an Active Snubber 1684
Koji Kato, *Nagaoka University of Technology, Japan*
Jun-Ichi Itoh, *Nagaoka University of Technology, Japan*

A PI Control Algorithm of Three-Level APF with Little Static Misadjustment for Tracking Harmonic Current 1692
Yingjie He, *Xi'an Jiaotong University, China*
Jinjun Liu, *Xi'an Jiaotong University, China*
Zhaoan Wang, *Xi'an Jiaotong University, China*
Yunping Zou, *Huazhong University of Science and Technology, China*

A Novel Topology of LLC Resonant Inverter with Two Resonant Tanks for Power Conditioning System 1698
Eun-Soo Kim, *Jeonju University, Korea, South*
Kwang-Ho Lee, *Jeonju University, Korea, South*
Bong-Gun Chung, *Jeonju University, Korea, South*
Joo-Hoon Kim, *Jeonju University, Korea, South*
Moon-Ho Kye, *Powerplaza, United States*

Analysis and Realization of a Fast Repetitive Controller in Active Power Filter System 1704
Jinwu Gong, *Wuhan University, China*
Xiaoming Zha, *Wuhan University, China*
Suxuan Guo, *Wuhan University, China*
Baifeng Chen, *Wuhan University, China*
Jianjun Sun, *Wuhan University, China*

Session B4P-P: Modeling, Simulation & Control II
Thursday, February 25, 11:30 - 13:30
Session Chairs: Jonathan Kimball, *Missouri S&T*
Omer Onar, *Illinois Institute of Technology*

On Extended Kalman Filters with Augmented State Vectors for the Stator Flux Estimation in SPMSMs 1711
T.J. Vyncke, *Ghent University, Belgium*
R.K. Boel, *Ghent University, Belgium*
J.A.A. Melkebeek, *Ghent University, Belgium*

State Equations Based Resonant Converters Modeling Technique 1719
Yingqi Zhang, *GE Global Research, China*
P.C. Sen, *Queen's University, Canada*

Design Considerations and Expiremantal Results of an Adaptive Frequency Controller Under Variable Line and Load Conditions 1723
Jaber A. Abu Qahouq, *University of Alabama, United States*
Wisam Al-Hoor, *University of Central Florida, United States*
Issa Batarseh, *University of Central Florida, United States*

Modeling and Mitigation of Dynamic Load Beat-Frequency Oscillation in Multiphase Voltage Regulators with High-Gain Peak Current Control Scheme 1727
Chen-Hua Chiu, *National Taiwan University, Taiwan*
Dan Chen, *National Taiwan University, Taiwan*
Ching-Jan Chen, *National Taiwan University, Taiwan*
Wei-Hsu Chang, *RichTek Technology Corporation, Taiwan*

Half-Wave Symmetry SHE-PWM Method for Multilevel Voltage Inverters 1732
Wanmin Fei, *Nanjing Normal University, China*
Xiaoli Du, *Nanjing Normal University, China*
Bin Wu, *Ryerson University, Canada*

PI Type Dynamic Decoupling Control Scheme for PMSM High Speed Operation 1736
Hao Zhu, *Tsinghua University, China*
Xi Xiao, *Tsinghua University, China*
Yongdong Li, *Tsinghua University, China*

High Performance Positive and Negative Sequence Filters in Stationary Frame Based on Complex Transfer Function 1740
Jingxin Mao, *Beijing Jiaotong University, China*
Fei Lin, *Beijing Jiaotong University, China*
Hong Li, *Beijing Jiaotong University, China*
Xiaojie You, *Beijing Jiaotong University, China*
Trillion Q. Zheng, *Beijing Jiaotong University, China*

Simulation Study of Parameter Influence on Dynamic Voltage Rise Control 1745
Ming Li, *Xi'an Jiaotong University, China*
Xiong Fang, *Xi'an Jiaotong University, China*
Yue Wang, *Xi'an Jiaotong University, China*
Leqiang Zhang, *Xi'an Jiaotong University, China*
Ke Wang, *Xi'an Jiaotong University, China*
Guopeng Zhao, *Xi'an Jiaotong University, China*

Shaping of the Noise Spectrum in Power Electronic Converters 1749
Cristian Lascu, *University of Nevada, Reno, United States*
Andrzej M. Trzynadlowski, *University of Nevada, Reno, United States*
R. Lynn Kirlin, *University of Victoria, Canada*

**Grid Interactions and Stability Analysis of Distribution Power Network with
High Penetration of Plug-in Hybrid Electric Vehicles** 1755
Omer C. Onar, *Illinois Institute of Technology, United States*
Alireza Khaligh, *Illinois Institute of Technology, United States*

**Rapid Simulation of Fourth-Order Multi-Resonant LLCC Converters with
Capacitive Output Filter** 1763
A. Bucher, *University of Erlangen-Nürnberg, Germany*
T. Duerbaum, *University of Erlangen-Nürnberg, Germany*

**FHA-Based Voltage Gain Function with Harmonic Compensation for
LLC Resonant Converter** 1770
Hong Huang, *Texas Instruments, United States*

Session B4P-Q: Aerospace & Transportation
Thursday, February 25, 11:30 - 13:30

**Analysis and Design of LCC Resonant Inverter for the
Tranportation Systems Applications** 1778
Mohamed Youssef, *Bombardier Transportation Inc., Canada*
Jaber A. Abu Qahouq, *University of Alabama, United States*
Mohamed Orabi, *South Valley University, Egypt*

**A Multi-Resolution Control Strategy for DSP Controlled 400Hz Shunt
Active Power Filter in an Aircraft Power System** 1785
Haibing Hu, *Nanjing University of Aeronautics and Astronautics, China*
Wei Shi, *Nanjing University of Aeronautics and Astronautics, China*
Jianren Xue, *Nanjing Sute Electric Co., Ltd., China*
Ying Lu, *Nanjing University of Aeronautics and Astronautics, China*
Yan Xing, *Nanjing University of Aeronautics and Astronautics, China*

Battery Discharge Regulator for Space Applications Based on the Boost Converter 1792
A. Fernandez, *European Space Agency, Netherlands*
F. Tonicello, *European Space Agency, Netherlands*
J. Aroca, *European Space Agency, Netherlands*
O. Mourra, *European Space Agency, Netherlands*

Electromagnetic Compatibility Results for an LCC Resonant Inverter for the Tranportation Systems 1800

Mohamed Youssef, *Bombardier Transportation Inc., Canada*
Jaber A. Abu Qahouq, *University of Alabama, United States*
Mohamed Orabi, *South Valley University, Egypt*

Torque Impulse for Experimental Modal Analysis in Transmitted Vibration Study of Engine-Generators 1804

Elias Ayana, *Cummins Power Generation & University of Minnesota, United States*
Steve Seidlitz, *Cummins Power Generation, United States*
Sze Kwan Cheah, *Cummins Power Generation, United States*
Ned Mohan, *University of Minnesota, United States*

Session B4P-R: Power Converters & Applications
Thursday, February 25, 11:30 - 13:30

Review and Analysis of the AC-DC Converter of ITER Coil Power Supply 1810

P. Fu, *Institute of Plasma Physics, China*
G. Gao, *Institute of Plasma Physics, China*
L.W. Xu, *Institute of Plasma Physics, China*
Z.Q. Song, *Institute of Plasma Physics, China*
Z.C. Sheng, *Institute of Plasma Physics, China*
I. Benfatto, *ITER Organization, France*
J. Tao, *ITER Organization, France*
A.D. Mankani, *ITER Organization, France*
J.S. Oh, *National Fusion Research Institute, Korea, South*
C. Neumeyer, *Princeton Plasma Physics Laboratory, United States*

Fault Tolerance on Interleaved Inverter with Magnetic Couplers 1817

K. Guépratte, *Grenoble Electrical Engineering Laboratory, France*
D. Frey, *Grenoble Electrical Engineering Laboratory, France*
P.-O. Jeannin, *Grenoble Electrical Engineering Laboratory, France*
H. Stephan, *Grenoble Electrical Engineering Laboratory, France*
J.-P. Ferrieux, *Grenoble Electrical Engineering Laboratory, France*

Latest Practical Developments of Triplex Series Load Resonant Frequency-Operated High Frequency Inverter for Induction-Heated Low Resistivity Metallic Appliances in Consumer Built-in Cooktops 1825

Hideki Sadakata, *Panasonic Corporation, Japan*
Atsushi Fujita, *Panasonic Corporation, Japan*
Shinichiro Sumiyoshi, *Panasonic Corporation, Japan*
Hideki Omori, *Panasonic Corporation, Japan*
Bishwajit Saha, *Kyungnam University / Yamaguchi University, Korea, South*
Tarek Ahmed, *Kyungnam University / Yamaguchi University, Korea, South*
Mutsuo Nakaoka, *Kyungnam University / Yamaguchi University, Korea, South*

A Study of Novel Flyback Converter with Very Low Power Consumption at the Standby Operating Mode 1833

Eun-Soo Kim, *Jeonju University, Korea, South*
Bong-Gun Chung, *Jeonju University, Korea, South*
Sang-Ho Jang, *Jeonju University, Korea, South*
Mun-Gi Choi, *LG Innotek, Korea, South*
Moon-Ho Kye, *Powerplaza, United States*

Improved Two-Stage DC-Coupled Gate Driver for Enhancement-Mode SiC JFET 1838
Robin Kelley, *SemiSouth Laboratories Inc., United States*
Andrew Ritenour, *SemiSouth Laboratories Inc., United States*
David Sheridan, *SemiSouth Laboratories Inc., United States*
Jeff Casady, *SemiSouth Laboratories Inc., United States*

**Design and Implementation of Multi-Channel Land Fowls Stunner with
Current Sharing Controller** 1842
S.-Y. Fan, *Wufeng Institute of Technology, Taiwan*
S.-Y. Tseng, *Chang Gung University, Taiwan*
Y.-H. Su, *Chang Gung University, Taiwan*
W.-C. Wu, *Wufeng Institute of Technology, Taiwan*

High Voltage Generator Using Boost/Flyback Hybrid Converter for Stun Gun Applications .. 1849
S.-Y. Tseng, *Chang Gung University, Taiwan*
C.-M. Yang, *Chang Gung University, Taiwan*
K.-C. Wang, *Chang Gung University, Taiwan*
G.-W. Hsu, *Chang-Gung University, Taiwan*

Session C5P-H: DC-DC Converter VIII
Thursday, February 25, 11:30 - 13:30

A Method to Analysis and Design for Long Life Power Converter 1857
H.M. Pang, *University of Hong Kong, China*
M.H. Bryan Pong, *University of Hong Kong, China*

DC-DC Converter for Gate Power Supplies with an Optimal Air Transformer 1865
Christoph Marxgut, *ETH Zurich, Switzerland*
Jürgen Biela, *ETH Zurich, Switzerland*
Johann W. Kolar, *ETH Zurich, Switzerland*
Reto Steiner, *ABB, Switzerland*
Peter K. Steimer, *ABB, Switzerland*

A Digitally Controlled DC-DC Buck Converter Using Frequency Domain ADCs 1871
Hani Ahmad, *Arizona State University, United States*
Bertan Bakkaloglu, *Arizona State University, United States*

**Low-Dropout (LDO) Regulator Output Impedance Analysis and
Transient Performance Enhancement Circuit** 1875
Sungkeun Lim, *North Carolina State University, United States*
Alex Q. Huang, *North Carolina State University, United States*

A Design for Small Time-Delay Control Circuit for DPWM- POL 1879
Yoichi Ishizuka, *Nagasaki University, Japan*
Yusuke Yamada, *Nagasaki University, Japan*
Fumitoshi Hirose, *Nagasaki University, Japan*
Mariko Nishi, *Nagasaki University, Japan*
Hirofumi Matsuo, *Nagasaki University, Japan*

Low Profile LLC Series Resonant Converter with Two Transformers 1885
Eun-Soo Kim, *Jeonju University, Korea, South*
Joo-Hoon Kim, *Jeonju University, Korea, South*
Sung-In Kang, *LG Innotek, Korea, South*
Jun-Ho Park, *LG Innotek, Korea, South*
Jae-Sam Lee, *LG Innotek, Korea, South*
Dong-Young Huh, *LG Innotek, Korea, South*
Yong-Chae Jung, *Namseoul University, Korea, South*

**Adaptive Frequency Control for ZVS Synchronous Boost Converters
Operated in Average Current Mode** 1890
Ben York, *Virginia Polytechnic Institute and State University, United States*
Rae-Young Kim, *Virginia Polytechnic Institute and State University, United States*
Jih-Sheng Lai, *Virginia Polytechnic Institute and State University, United States*

**Power Saving Control Strategies and Their Implementation in DC-DC
Converter for Data and Telecommunication Power Supply** 1897
Rais Miftakhutdinov, *Texas Instruments Inc., United States*

Session C5P-J: DC-DC Converter IX
Thursday, February 25, 11:30 - 13:30

**Analysis and Optimized Design of an Efficient High-Voltage Converter
with High Output Capacity** 1904
Huai Wang, *City University of Hong Kong, China*
Henry Shu-hung Chung, *City University of Hong Kong, China*
Adrian Ioinovici, *Holon Institute of Technology, Israel*

A Novel Three-Phase Three-Level ZVS PWM DC-DC Converter 1911
Eloi Agostini Junior, *Federal University of Santa Catarina, Brazil*
Ivo Barbi, *Federal University of Santa Catarina, Brazil*

Optimize the Synchronous Rectifier for LCC Converters 1919
Feng Zheng, *Xidian University, China*
Zhengfeng Ming, *Xidian University, China*

**Digital Control Scheme for Robust Clock Tuning and PWM Phase
Synchronization in Digitally Controlled Multi-POL Applications** 1922
Eamon O'Malley, *Powervation Ltd., Ireland*
Karl Rinne, *Powervation Ltd., Ireland*
Anthony Kelly, *Powervation Ltd., Ireland*
Basil Almukhtar, *Powervation Ltd., Ireland*
Paul Kelleher, *Powervation Ltd., Ireland*

Control Scheme and Transient Performance of Sigma VR 1927
Pengjie Lai, *Virginia Polytechnic Institute and State University, United States*
Julu Sun, *Virginia Polytechnic Institute and State University, United States*
Fred C. Lee, *Virginia Polytechnic Institute and State University, United States*

**A Three-Phase Current-Fed Push-Pull DC-DC Converter with
Active Clamp for Fuel Cell Applications** 1934
Sangwon Lee, *Seoul National University of Technology, Korea, South*
Sewan Choi, *Seoul National University of Technology, Korea, South*

Resonant Voltage Divider with Startup Considered 1942
K.I. Hwu, *National Taipei University of Technology, Taiwan*
Y.T. Yau, *National Taipei University of Technology, Taiwan*

LLC Resonant Converter with Two Resonant Tanks 1949
Eun-Soo Kim, *Jeonju University, Korea, South*
Joo-Hoon Kim, *Jeonju University, Korea, South*
Kwang-Ho Lee, *Jeonju University, Korea, South*
Yong-Seog Jeon, *Jeonju University, Korea, South*
Jae-Sam Lee, *LG Innotek, Korea, South*
Dong-Young Huh, *LG Innotek, Korea, South*

Session C5P-K: Motor Drives & Inverters II
Thursday, February 25, 11:30 - 13:30

A Digital Control Strategy for Brushless DC Generators 1957
Nikola Milivojevic, *Illinois Institute of Technology, United States*
Igor Stamenkovic, *Illinois Institute of Technology, United States*
Mahesh Krishnamurthy, *Illinois Institute of Technology, United States*
Ali Emadi, *Illinois Institute of Technology, United States*

**Space Vector Based PWM Scheme Without Sector Identification for a 4-Level
Dual Inverter Fed Induction Motor Drive with Asymmetrical DC Link Voltages** 1963
G. Shiny, *College of Engineering Trivandrum, India*
M.R. Baiju, *College of Engineering Trivandrum, India*

Control Method for a Novel Converter Topology for Permanent Magnet Drives 1970
Philip Brockerhoff, *Universität der Bundeswehr München, Germany*
Martin Schulz, *Universität der Bundeswehr München, Germany*

**A Voltage Controlled Adjustable Speed PMBLDCM Drive Using a Single-Stage
PFC Half-Bridge Converter** 1976
Sanjeev Singh, *Indian Institute of Technology Delhi, India*
Bhim Singh, *Indian Institute of Technology Delhi, India*

**Comparison of HF Signal Injection Methods for Sensorless Control of
PM Synchronous Motors** 1984
Eisenhawer de M. Fernandes, *Universidade Federal de Campina Grande, Brazil*
Alexandre C. Oliveira, *Universidade Federal de Campina Grande, Brazil*
Cursino B. Jacobina, *Universidade Federal de Campina Grande, Brazil*
Antonio M.N. Lima, *Universidade Federal de Campina Grande, Brazil*

A Robust Sensorless Fault Diagnosis Algorithm for Low Cost Motor Drives 1990
Seung-deog Choi, *Texas A&M University, United States*
Bilal Akin, *Texas Instruments Inc., United States*
Mina M. Rahimian, *Texas A&M University, United States*
Hamid A. Toliyat, *Texas A&M University, United States*

High Dynamic Performance Constrained Optimal Control of Induction Motors 1995
Sébastien Mariéthoz, *ETH Zurich, Switzerland*
Alexander Domahidi, *ETH Zurich, Switzerland*
Manfred Morari, *ETH Zurich, Switzerland*

PMSM Control Based on Edge Field Measurements by Hall Sensors 2002
Sungyoon Jung, *Pohang University of Science and Technology, Korea, South*
Beomseok Lee, *Pohang University of Science and Technology, Korea, South*
Kwanghee Nam, *Pohang University of Science and Technology, Korea, South*

Bridged-T Speed Controller for High Performance Switched Reluctance Motor Drives 2007
Gregory Pasquesoone, *University of Akron, United States*
Iqbal Husain, *University of Akron, United States*
Robert J. Veillette, *University of Akron, United States*

**Reducing Losses in Multilevel Coupled Inductor Inverters Using
Interleaved Discontinuous SVPWM** 2013
Behzad Vafakhah, *University of Alberta, Canada*
Andy Knight, *University of Alberta, Canada*
John Salmon, *University of Alberta, Canada*

A Novel Elevator Load Torque Identification Method Based on Friction Mode 2021
Xiaoyuan Hong, *Zhejiang University, China*
Zhe Deng, *Zhejiang University, China*
Siran Wang, *Zhejiang University, China*
Lijun Hang, *Zhejiang University, China*
Wuhua Li, *Zhejiang University, China*
Zhengyu Lu, *Zhejiang University, China*

**A Novel Digital Current Control Strategy for Torque Ripple Reduction in
Permanent Magnet Synchronous Motor Drives** 2025
Haidong Yu, *Phoenix International, United States*

Session C5P-L: Passive Components
Thursday, February 25, 11:30 - 13:30

**Evaluation of a SiC Power Module Using Low-on-Resistance IEMOSFET and
JBS for High Power Density Power Converters** 2030
Kazuto Takao, *Toshiba Corporation, Japan*
Takashi Shinohe, *Toshiba Corporation, Japan*
Shinsuke Harada, *National Institute of Advanced Industrial Science and Technology, Japan*
Kenji Fukuda, *National Institute of Advanced Industrial Science and Technology, Japan*
Hiromichi Ohashi, *National Institute of Advanced Industrial Science and Technology, Japan*

**A Novel Integrated Power Inductor in Silicon Substrate for
Ultra-Compact Power Supplies** 2036
Mingliang Wang, *University of Florida, United States*
Jiping Li, *University of Florida, United States*
Khai D.T. Ngo, *Virginia Polytechnic Institute and State University, United States*
Huikai Xie, *University of Florida, United States*

A Class of Coupled Inductors Based on LTCC Technology 2042
Laili Wang, *Xi'an Jiaotong University, China*
Yunqing Pei, *Xi'an Jiaotong University, China*
Xu Yang, *Xi'an Jiaotong University, China*
Xizhi Cui, *Xi'an Jiaotong University, China*
Zhaoan Wang, *Xi'an Jiaotong University, China*
Guopeng Zhao, *Xi'an Jiaotong University, China*

Optimising the High Frequency Bandwidth and Immuntity to Interference of Rogowski Coils in Measurement Applications with Large local dV/dt 2050
Christopher R. Hewson, *Power Electronic Measurements Ltd., United Kingdom*
William F. Ray, *Power Electronic Measurements Ltd., United Kingdom*

PFC Inductor Selection Made Easy by "PL Product" 2057
Welly Chou, *Precision Incorporated, United States*

Evaluation of LTCC Capacitors and Inductors in DC-DC Converters 2060
Laili Wang, *Xi'an Jiaotong University, China*
Yunqing Pei, *Xi'an Jiaotong University, China*
Xu Yang, *Xi'an Jiaotong University, China*
Bo Song, *Xi'an Jiaotong University, China*
Zhaoan Wang, *Xi'an Jiaotong University, China*
Guopeng Zhao, *Xi'an Jiaotong University, China*

Session C5P-M: Vehicle Electronics II
Thursday, February 25, 11:30 - 13:30

Bi-Directional Charging Topologies for Plug-in Hybrid Electric Vehicles 2066
Dylan C. Erb, *Illinois Institute of Technology, United States*
Omer C. Onar, *Illinois Institute of Technology, United States*
Alireza Khaligh, *Illinois Institute of Technology, United States*

Session C5P-N: Renewable Energy Systems
Thursday, February 25, 11:30 - 13:30
Session Chairs: Robert Balog, *Texas A&M University*

Multi-Channel Three-Port DC-DC Converters As Maximum Power Tracker, Battery Charger and Bus Regulator 2073
Zhijun Qian, *University of Central Florida, United States*
Osama Abdel-Rahman, *ApECOR, United States*
Haibing Hu, *University of Central Florida, United States*
Issa Batarseh, *University of Central Florida, United States*

A Smart and Simple PV Charger for Portable Applications 2080
Weichen Li, *Zhejiang University, China*
Yuzhen Zheng, *Zhejiang University of Science and Technology, China*
Wuhua Li, *Zhejiang University, China*
Yi Zhao, *Zhejiang University, China*
Xiangning He, *Zhejiang University, China*

RTDS-Based Real Time Simulations of Grid-Connected Wind Turbine Generator Systems .. 2085
Gyeong-Hun Kim, *Changwon National University, Korea, South*
Young-Ju Kim, *Changwon National University, Korea, South*
Minwon Park, *Changwon National University, Korea, South*
In-Keun Yu, *Changwon National University, Korea, South*
Byeong-Mun Song, *Baylor University, United States*

Investigation of Fully Digital Controlled Li-Ion Battery Power Recovery System 2091
Siran Wang, *Zhejiang University, China*
Xia Zhou, *Zhejiang University, China*
Jifeng Chen, *Zhejiang University, China*
Wenxi Yao, *Zhejiang University, China*
Zhengyu Lu, *Zhejiang University, China*

A Novel Control System for Harmonic Compensation by Using Wind Energy
Conversion Based on DFIG Technology 2096
Grazia Todeschini, *Worcester Polytechnic Institute, United States*
Alexander E. Emanuel, *Worcester Polytechnic Institute, United States*

A Transformerless Modular Permanent Magnet Wind Generator System with
Minimum Generator Coils 2104
Xibo Yuan, *Tsinghua University, China*
Yongdong Li, *Tsinghua University, China*
Jianyun Chai, *Tsinghua University, China*

Small-Signal Modeling and Analysis of the Double-Input Buckboost Converter 2111
Deepak Somayajula, *Missouri University of Science and Technology, United States*
Mehdi Ferdowsi, *Missouri University of Science and Technology, United States*

A Novel Power Distribution Strategy for Parallel Inverters in Islanded Mode Microgrid 2116
Xuan Zhang, *Xi'an Jiaotong University, China*
Jinjun Liu, *Xi'an Jiaotong University, China*
Ting Liu, *Xi'an Jiaotong University, China*
Linyuan Zhou, *Xi'an Jiaotong University, China*

Direct Power Control of Doubly-Fed Generator Based Wind Turbine Converters to
Improve Low Voltage Ride-Through During System Imbalance 2121
Murali M. Baggu, *Missouri University of Science and Technology, United States*
Luke D. Watson, *Missouri University of Science and Technology, United States*
Jonathan W. Kimball, *Missouri University of Science and Technology, United States*
Badrul H. Chowdhury, *Missouri University of Science and Technology, United States*

Active Damping for Torsional Vibrations in PMSG Based WECS 2126
Hua Geng, *Ryerson University, Canada*
Dewei Xu, *Ryerson University, Canada*
Bin Wu, *Ryerson University, Canada*
Geng Yang, *Tsinghua University, China*

Voltage and Frequency Stabilization Using PI-Like Fuzzy Controller for the
Load Side Converters of the Stand Alone Wind Energy Systems 2132
Ameen Gargoom, *University of Tasmania, Australia*
Abu Mohammad Osman Haruni, *University of Tasmania, Australia*
Md. Enamul Haque, *University of Tasmania, Australia*
Michael Negnevitsky, *University of Tasmania, Australia*

Dual-Stage Converter to Improve Transfer Efficiency and Maximum Power Point
Tracking Feasibility in Photovoltaic Energy-Conversion Systems 2138
Sairaj V. Dhople, *University of Illinois at Urbana-Champaign, United States*
Ali Davoudi, *University of Illinois at Urbana-Champaign, United States*
Patrick L. Chapman, *University of Illinois at Urbana-Champaign, United States*

A Novel Approach of Maximizing Energy Harvesting in Photovoltaic Systems Based on Bisection Search Theorem 2143

Peng Wang, *Nanyang Technological University, Singapore*
Haipeng Zhu, *Nanyang Technological University, Singapore*
Weixiang Shen, *Nanyang Technological University, Singapore*
Fook Hoong Choo, *Nanyang Technological University, Singapore*
Poh Chiang Loh, *Nanyang Technological University, Singapore*
Kuan Khoon Tan, *Nanyang Technological University, Singapore*

Simple Control Design for a Three-Port DC-DC Converter Based PV System with Energy Storage 2149

Sixifo Falcones, *Arizona State University, United States*
Raja Ayyanar, *Arizona State University, United States*

A Self-Powered Power Management Circuit for Energy Harvested by a Piezoelectric Cantilever 2154

Na Kong, *Virginia Polytechnic Institute and State University, United States*
Travis Cochran, *Virginia Polytechnic Institute and State University, United States*
Dong Sam Ha, *Virginia Polytechnic Institute and State University, United States*
Hung-Chih Lin, *National Tsing Hua University, Taiwan*
Daniel J. Inman, *Virginia Polytechnic Institute and State University, United States*

A Maximum Power Point Tracker Implementation for Photovoltaic Cells Using Dynamic Optimal Voltage Tracking 2161

Emil Jimenez-Brea, *University of Puerto Rico-Mayaguez, Puerto Rico*
Andres Salazar-Llinas, *University of Puerto Rico-Mayaguez, Puerto Rico*
Eduardo Ortiz-Rivera, *University of Puerto Rico-Mayaguez, Puerto Rico*
Jesus Gonzalez-Llorente, *University of Puerto Rico-Mayaguez, Puerto Rico*

Development of the Novel Control Algorithm for the Small Proton Exchange Membrane Fuel Cell Stack Without External Humidification 2166

Tae-Hoon Kim, *Soongsil University, Korea, South*
Sang-Hyun Kim, *Soongsil University, Korea, South*
Wook Kim, *Soongsil University, Korea, South*
Jong-Hak Lee, *Soongsil University, Korea, South*
Woojin Choi, *Soongsil University, Korea, South*

Session C5P-P: Modeling, Simulation & Control III
Thursday, February 25, 11:30 - 13:30
Session Chairs: Jonathan Kimball, *Missouri S&T*
Omer Onar, *Illinois Institute of Technology*

Stabilization of Constant-Power Loads by Passive Impedance Damping 2174

Mauricio Céspedes, *Rensselaer Polytechnic Institute, United States*
Troy Beechner, *Rensselaer Polytechnic Institute, United States*
Lei Xing, *Rensselaer Polytechnic Institute, United States*
Jian Sun, *Rensselaer Polytechnic Institute, United States*

An Adaptive External Ramp Control of the Peak Current Controlled Buck Converters for High Control Bandwidth and Wide Operation Range 2181

Liyu Yang, *North Carolina State University, United States*
Jinseok Park, *North Carolina State University, United States*
Alex Q. Huang, *North Carolina State University, United States*

Masterless Multirate Control of Parallel DC-DC Converters 2189
Anthony Kelly, *Powervation Ltd., Ireland*
Karl Rinne, *Powervation Ltd., Ireland*
Eamon O'Malley, *Powervation Ltd., Ireland*

FPGA-Based Spectral Envelope Preprocessor for Power Monitoring and Control 2194
Zachary Remscrim, *Massachusetts Institute of Technology, United States*
James Paris, *Massachusetts Institute of Technology, United States*
Steven B. Leeb, *Massachusetts Institute of Technology, United States*
Steven R. Shaw, *Montana State University, United States*
Sabrina Neuman, *Massachusetts Institute of Technology, United States*
Christopher Schantz, *Massachusetts Institute of Technology, United States*
Sean Muller, *Massachusetts Institute of Technology, United States*
Sarah Page, *Massachusetts Institute of Technology, United States*

Sigma-Delta Modulation of Multi-Phase High Frequency Converters 2202
Jonathan W. Kimball, *Missouri University of Science and Technology, United States*
Kyle Roger Eckler, *Missouri University of Science and Technology, United States*
Luke Watson, *Missouri University of Science and Technology, United States*

**Specialized Digital Signal Processor for Control of Multi-Rail/Multi-Phase
High Switching Frequency Power Converters** 2207
James Mooney, *University of Limerick, Ireland*
Mark Halton, *University of Limerick, Ireland*
Abdulhussain E. Mahdi, *University of Limerick, Ireland*

**Computer-Aided Design for Class-E Switching Circuits Taking into
Account Optimized Inductor Designs** 2212
Natsumi Sagawa, *Chiba University, Japan*
Hiroo Sekiya, *Chiba University and Wright State University, Japan*
Marian K. Kazimierczuk, *Wright State University, United States*

Characterization of IGBT Modules for System EMI Simulation 2220
Tao Qi, *Rensselaer Polytechnic Institute, United States*
Jeff Graham, *Fairchild Controls Corporation, United States*
Jian Sun, *Rensselaer Polytechnic Institute, United States*

**A Mathematical Model for Online Electrical Characterization of Thermoelectric
Generators Using the P-I Curves at Different Temperatures** 2226
Eduardo I. Ortiz-Rivera, *University of Puerto Rico-Mayaguez, Puerto Rico*
Andres Salazar-Llinas, *University of Puerto Rico-Mayaguez, Puerto Rico*
Jesus Gonzalez-Llorente, *University of Puerto Rico-Mayaguez, Puerto Rico*

**A Novel Method for Permanent Magnet Demagnetization Fault Detection and
Treatment in Permanent Magnet Synchronous Machines** 2231
Amir Khoobroo, *University of Texas at Arlington, United States*
Babak Fahimi, *University of Texas at Arlington, United States*

Session C5P-Q: Alternative Energy Applications
Thursday, February 25, 11:30 - 13:30

Series Connection of IGBT 2238
The-Van Nguyen, *Grenoble Institute of Technology, France*
Pierre-Olivier Jeannin, *Grenoble Institute of Technology, France*
Eric Vagnon, *Grenoble Institute of Technology, France*
David Frey, *Grenoble Institute of Technology, France*
Jean-Christophe Crebier, *Grenoble Institute of Technology, France*

**Three Phase Linear Permanent Magnet Energy Scavenger Based on
Foot Horizontal Motion** 2245
Igor Stamenkovic, *Illinois Institute of Technology, United States*
Nikola Milivojevic, *Illinois Institute of Technology, United States*
Cong Zheng, *Illinois Institute of Technology, United States*
Alireza Khaligh, *Illinois Institute of Technology, United States*

**Bidirectional Communication Techniques for Wireless Battery Charging
Systems and Portable Consumer Electronics** 2251
W.P. Choi, *ConvenientPower HK Ltd. and City University of Hong Kong, China*
W.C. Ho, *ConvenientPower HK Ltd., China*
X. Liu, *ConvenientPower HK Ltd., China*
S.Y.R. Hui, *City University of Hong Kong, China*

**Proposal of a DC-DC Converter with Wide Conversion Range Used in
Photovoltaic Systems and Utility Power Grid for the Universal Voltage Range** 2258
Jonas Reginaldo de Britto, *Universidade Federal de Uberlândia, Brazil*
Fábio Vincenzi Romualdo da Silva, *Universidade Federal de Uberlândia, Brazil*
Enane Antônio Alves Coelho, *Universidade Federal de Uberlândia, Brazil*
Luiz Carlos de Freitas, *Universidade Federal de Uberlândia, Brazil*
Valdeir José Farias, *Universidade Federal de Uberlândia, Brazil*
João Batista Vieira, Jr., *Universidade Federal de Uberlândia, Brazil*

Characterization of a 5 kW Solid Oxide Fuel Cell Stack Using Power Electronic Excitation .. 2264
John J. Cooley, *Massachusetts Institute of Technology, United States*
Eric Seger, *Montana State University, United States*
Steven Leeb, *Massachusetts Institute of Technology, United States*
Steven R. Shaw, *Montana State University, United States*

**Photovoltaic Parallel Resonant DC-Link Soft Switching Inverter Using
Hysteresis Current Control** 2275
Young-Ho Kim, *Sungkyunkwan University, Korea, South*
Jun-Gu Kim, *Sungkyunkwan University, Korea, South*
Young-Hyok Ji, *Sungkyunkwan University, Korea, South*
Chung-Yuen Won, *Sungkyunkwan University, Korea, South*
Yong-Chae Jung, *Namseoul University, Korea, South*

**Supercapacitor-Based Hybrid Storage Systems for Energy Harvesting in
Wireless Sensor Networks** 2281
S. Saggini, *University of Udine, Italy*
F. Ongaro, *University of Udine, Italy*
C. Galperti, *Politecnico di Milano, Italy*
P. Mattavelli, *University of Padova, Italy*

The Faulty Module Bypass for Thermoelectric Generation 2288
Wei Qian, *Michigan State University, United States*
Fang Z. Peng, *Michigan State University, United States*
Sangmin Han, *Michigan State University, United States*

Maximum Power Point Tracking Feasibility in Photovoltaic Energy-Conversion Systems . 2294
Sairaj V. Dhople, *University of Illinois at Urbana-Champaign, United States*
Ali Davoudi, *University of Illinois at Urbana-Champaign, United States*
Gerald Nilles, *University of Illinois at Urbana-Champaign, United States*
Patrick L. Chapman, *University of Illinois at Urbana-Champaign, United States*

Session C5P-R: Lighting Applications
Thursday, February 25, 11:30 - 13:30

**Realization of a General LED Lighting System Based on a Novel Power Line
Communication Technology** 2300
Chushan Li, *Zhejiang University, China*
Jiande Wu, *Zhejiang University, China*
Xiangning He, *Zhejiang University, China*

Solid-State Lamp with Integral Occupancy Sensor 2305
John J. Cooley, *Massachusetts Institute of Technology, United States*
Dan Vickery, *Massachusetts Institute of Technology, United States*
Al-Thaddeus Avestruz, *Massachusetts Institute of Technology, United States*
Amy Englehart, *Massachusetts Institute of Technology, United States*
James Paris, *Massachusetts Institute of Technology, United States*
Steven B. Leeb, *Massachusetts Institute of Technology, United States*

**A 0.9 PF LED Driver with Small LED Current Ripple Based on
Series-Input Digitally-Controlled Converter Modules** 2314
Qingcong Hu, *University of Colorado at Boulder, United States*
Regan Zane, *University of Colorado at Boulder, United States*

**A Novel Dimmable Electronic Ballast for Compact Fluorescent Lamps Using
Phase-Cut Incandescent Lamp Dimmers with Wide Dimming Range and
Low Dimming Level Lamp Ignition Capability** 2321
John Lam, *Queen's University, Canada*
Praveen K. Jain, *Queen's University, Canada*

Author Index

Foreword

25th Annual IEEE Applied Power Electronics Conference and Exposition
February 21-25, 2010
Palm Springs Convention Center, Palm Springs, CA

It is with great pleasure that I welcome you to our celebration of the twenty fifth IEEE-APEC in Palm Springs, CA. Throughout the past quarter of the century, IEEE-APEC has excelled in showcasing creativity, providing an ideal environment for technical exchange and networking, and most importantly emphasizing on the applied nature of the power electronics. Power electronics, as an enabling technology, embodies an integral part in electrification of transportation industry, modernization of the electric power grid, integration of energy storage and renewable sources, and many other technologies whose successful implementations are crucial in moving towards energy sustainability. To address the latest achievements and breakthroughs on these topics, APEC 2010 offers a rich and comprehensive menu for power electronics professionals. The plenary will cover topics related to electric vehicles, digital power, applications of nanotechnology in power electronics, efficiency and power conversion in data centers, power conversion in distributed solar energy systems, and a retrospective on power electronics and APEC over the past 25 years. Complement to our plenary talks are a series of exciting and substantive debates. This year's rap sessions are, 'Smart Grid: A dumb idea?', 'Energy Harvesting: Is it really better than a battery', and 'Social Media: The future of advertising or just a way to communicate with friends?'. The technical sessions are selected from a base of 726 digest submissions. The selected articles are strong in technical contents, delivered by industry and academic experts from 41 countries and cover a diverse set of topics. Ten special OEM presentations are yet another highlight of APEC 2010 which has attracted considerable attention over the past few years. The exposition with close to 200 booths will display the latest innovations and high tech in power electronics industry. Finally the 18 excellent educational seminars will complete a full package for personal development and technical exchange.

Surrounded by big wind farms Palm Springs, the 2010 host city, is the ideal location for APEC. With over 100 golf courses, Palm Springs is often referred to as the "golf capitol of the world". Besides golf and excellent shopping and dining options, art is the big attraction in the Palm Springs area. There are over a hundred galleries to choose from and many are widely recognized for their ability to showcase the 'up and coming' in the art world. If you are a fan of nature, the options are unlimited. Ascending 8,000 feet in the world's largest rotating tramcars you will reach an entirely different world with incredible views at forty degrees lower temperature. Great hiking, cross country skiing are among your options. I would encourage you to take advantage of these great opportunities while here.

Finally, APEC 2010 has been made possible by tremendous support from our sponsors, organizing and steering committee members, volunteers, and reviewers. I would like to take this opportunity and thank each and every one of you.

Babak Fahimi, General Chair

APEC History

Year	Site	Dates	General Chair	Program Chair
1986	Fairmont Hotel New Orleans, Louisiana	April 28 – May 1	John G. Kassakian	R. David Middlebrook
1987	Town and Country Hotel San Diego, California	March 2 – 6	John G. Kassakian	R. David Middlebrook
1988	Fairmont Hotel New Orleans, Louisiana	February 1 – 5	William W. Burns, III	William W. Burns, III
1989	Baltimore Convention Center Baltimore, Maryland	March 13 – 17	William W. Burns, III	Robert V. White
1990	Biltmore Hotel Los Angeles, California	March 11 – 16	Robert V. White	Charles Harm
1991	Hyatt Regency Reunion Hotel Dallas, Texas	March 10 – 15	Charles Harm	Thomas M. Jahns
1992	Weston Copley Plaza Hotel Boston, Massachusetts	February 23 – 27	Thomas M. Jahns	Kevin J. Fellhoelter
1993	Town and Country Hotel San Diego, California	March 7 – 11	Kevin J. Fellhoelter	Douglas McIlvoy
1994	Disney Contemporary Hotel Orlando, Florida	February 13 – 17	Douglas McIlvoy	Thomas Latos
1995	Hyatt Regency Reunion Hotel Dallas, Texas	March 5 – 9	Thomas Latos	Charles E. Mullett
1996	Fairmont Hotel San Jose, California	March 3 – 7	Charles E. Mullett	Thomas G. Wilson, Jr.
1997	Weston Peachtree Hotel Atlanta, Georgia	February 23 – 27	Thomas G. Wilson, Jr	David Torrey
1998	Disneyland Hotel Anaheim, California	February 15 – 19	David Torrey	F. Don Tan
1999	Adams' Mark Hotel Dallas, Texas	March 14 – 18	F. Don Tan	Robert V. White
2000	Fairmont Hotel New Orleans, Louisiana	February 6 – 10	Robert V. White	R. Mark Nelms
2001	Disneyland Hotel Anaheim, California	March 4 – 8	R. Mark Nelms	V. Joseph Thottuvelil
2002	Adams' Mark Hotel Dallas, Texas	March 10 – 14	V. Joseph Thottuvelil	Bruce Miller
2003	Fontainebleau Hotel Miami Beach, Florida	February 9 – 13	Bruce Miller	Jim Kokernak
2004	Disneyland Hotel Anaheim, California	February 22 – 26	Jim Kokernak	Jason Lai
2005	Hilton Austin Austin, Texas	March 6 – 10	Jason Lai	Van Niemela
2006	Hyatt Regency Hotel Dallas, Texas	March 19 – 23	Van Niemela	Russ Spyker
2007	Disneyland Hotel Anaheim, California	February 25 – March 1	Russ Spyker	Steve Pekarek
2008	Austin Convention Center Austin, Texas	February 24 – 28	Steve Pekarek	Kevin Parmenter
2009	Marriott Wardman Park Hotel Washington, District of Columbia	February 15 – 19	Kevin Parmenter	Babak Fahimi
2010	Palm Springs Convention Center Palm Springs, California	February 21 – 25	Babak Fahimi	Patrick Chapman

The APEC Conference Committee

General Chair
Babak Fahimi, University of Texas-Arlington

Program Chair
Patrick Chapman, University of Illinois at Urbana-Champaign

Assistant Program Chair
Frank Cirolia, Emerson

Finance Chair
Mark Nelms, Auburn University

Seminar Chair
Tim Haskew, University of Alabama

Assistant Seminar Chair
Bill Peterson, E&M Power

Exhibits Chair
Don Woodard, Venable Industries

Assistant Exhibits Chair
Chuck Mullet, ON Semiconductor

Publicity Chair
Greg Evans, Welcomm, Inc.

Assistant Publicity Chair
Matt McKinney, Texas Instruments

Rap Sessions Chair
Van Niemela, Fairchild Semiconductor

OEM Initiative Chair
Ada Cheng

Assistant OEM Initiative Chair
Joshua Israelsohn, International Rectifier

Publications Chair
Jonathan Kimball, Missouri University of Science and
    Technology

MicroMouse Chair
David Otten, Massachusetts Institute of Technology

Guest Program Chair
Jane Wilson

Past General Chair
Kevin Parmenter, Freescale Semiconductor

Members at Large
Jim Templeton, Maxim
Russ Spyker, US Air Force
Aung Tu, Fairchild Semiconductor
Siamak Abedinpour, Freescale Semiconductor
Alireza Khaligh, Illinois Institute of Technology

Conference Management
Pam Wagner, Courtesy Associates

Tonya Freeland, Courtesy Associates

Amy Roth, Courtesy Associates

Tom Wehner, ePapers.org

Tia Fulmer, The Printing House

The Sponsors

Power Sources Manufacturers Association

Dusty Becker
Chairman of the Board

Frank Cirolia
Vice President

Kevin Parmenter
President

Rob Hill
Secretary/Treasurer

IEEE Power Electronics Society

Deepak Divan
President

John Shen
Vice President – Products

F. Dong Tan
Vice President – Operations

Donna Florek
Administrator

Ralph Kennel
Vice President – Meetings

J.A.(Braham) Ferreira
Treasurer

IEEE Industry Applications Society

Thomas Nondahl
President

Steven A. Larson
Treasurer

Bruno Lequesne
President-Elect

S. Mark Halpin
Past President

Blake Lloyd
Vice President

APEC Steering Committee Members with Sponsoring Organizations

Babak Fahimi
University of Texas-Arlington
(IAS, General Chair)

Frank Cirolia
Emerson
(PSMA)

Bruce Miller
Dell
(PSMA)

Mark Nelms
Auburn University
(IAS)

Kevin Parmenter
Freescale Semiconductor
(PSMA, Past General Chair)

F. Dong Tan
Northrop Grumman
(PELS)

Chuck Mullet
ON Semiconductor
(PSMA)

Kevin Fellhoelter
Solara, Inc.
(PELS)

Steve Pekarek
Purdue University
(PELS)

Russ Spyker
U.S. Air Force
(IAS)

APEC 2010 Program Committee Topic Chairs

AC-DC Converters
Alireza Khaligh

Motor Drives and Inverters
Leila Parsa
Chris Edrington

Modeling, Simulation and Control
Haidong Yu
Alireza Khaligh

DC-DC Converters
Alexis Kwasinski
Jonathan Kimball
Russ Spyker

System Integration
Jonathan Kimball

Renewable Energy Systems
Rob Balog

Power Electronics Applications
Patrick Chapman
Laura Steffek

Power Electronics for Utility Interface
Jin Wang

Devices & Components
Russ Spyker

Manufacturing & Business
Laura Steffek

Vehicular Electronics
Chris Edrington

APEC 2010 Reviewers

Sam Abdel-Rahman
Asghar Abedini
Jaber Abu Qahouq
Mohammed Agamy
Anant Agarwal
Damoun Ahmadi Khatir
Mahrous Ahmed
Johan Akerblad
Jarno Alahuhtala
Eduard Alarcon
Mohammed S. Al-Numay
Adrian Zsombor Amanci
Marco Amrhein
Michael A.E. Andersen
Jaime Arau
Mohammed Arefeen
Manuel Arias
Sardis Azongha
Uwe Badstuebner
Murali Baggu
M.R. Baiju
Jonathan Baker
Saritha Balathandayuthapani
Anahita Banaei
Robert Barnet
Rick Barnett
Luiz Henrique Barreto
Maged Fouad Barsom
Ali Bazzi
Dustin Becker
Stoyan Bekiarov

Nick Benavides
Riley Bennett
Berker Bilgin
Zhonghui Bing
Benjamin Blunier
Rui Bo
Theodore Bohn
David Bouquain
John Bower
Ted Brekken
Rolando Burgos
Dong Cao
Jian Cao
Mauricio Cespedes
Arindam Chakraborty
Pat Chapman
Chien-Liang Chen
Dan Chen
Fu-Zen Chen
Lihua Chen
Min Chen
Anatoly Cherepakhin
Shamala Chickamenahalli
Frank Cirolia
Luisa Coppola
Keith Corzine
Yao Da
Jingya Dai
Dhaval Dalal
Ali Davoudi
Eric Dede

Sairaj Dhople
Alejandro Dominguez-Garcia
Zhong Du
Chris Edrington
Dylan Erb
Trishan Esram
Greg Evans
Haifeng Fan
Xiong Fang
Thomas Farkas
Guang Feng
Mehdi Ferdowsi
Fletcher Fleming
Andrew Friedl
Dianbo Fu
Hideaki Fujita
Carlos Gallo
Gustavo Gamboa
Feng Gao
Oscar Garcia
Cahit Gezgin
John Glaser
Chunying Gong
Bo Guan
Kevin Guepratte
Liping Guo
Wennan Guo
Ranjan Kumar Gupta
Yusuf Gurkaynak
Brian Hacker
Jung Hee Han

Yehui Han
Jason Hannon
Richard Hannon
Maja Harfman Todorovic
Michael Harris
Michael Hartmann
Amin Hasanzadeh
Satoshi Hashino
Sanjay Havanur
Xiangning He
Marcel Hendrix
Edward Herbert
Benoit Herve
Carl Ngai-Man Ho
Torbjörn Holmberg
Grahame Holmes
Douglas Hopkins
Tom Hopkins
Joseph Horzepa
Chung-Chuan Hou
John Houldsworth
Qingcong Hu
Weihao Hu
Yuequan Hu
Alex (qin) Huang
Chun-Shih Huang
Lilly Huang
Yi Huang
Laszlo Huber
Santa Concepcion Huerta Olivares
Ron Hui
Ger Hurley
K. I. Hwu
Adrian Ioinovici
Brian Irving
Alexander Isurin
Junichi Itoh
Praveen Jain
Lars T. Jakobsen
Pete James
Yungtaek Jang
In Wha Jeong
Wei Jiang
Takushi Jimichi
Brian Johnson
Brett Jordan
Ron Josephson
Jeehoon Jung
Yonghan Kang
Magnus Karlsson
Koji Kato

Cliff Keys
Alireza Khaligh
Pardis Khayyer
Amir Khoobroo
Jonathan Kimball
Mithat Can Kisacikoglu
Christian Klumpner
Johann W. Kolar
Gregory Koskowich
Matthew Kowalyshen
Philip Krein
Mahesh Krishnamurthy
Yingying Kuai
Rick Kuehn
Satish Kumar
Alexis Kwasinski
Jason Lai
Rixin Lai
Diego Lamar
Emanuel Landsman
Ronald Leblanc
Byoung-Kuk Lee
Tzung-Lin Lee
Young Joo Lee
Matz Lenells
Dan Lenskold
Alex Levran
Cong Li
Hui Li
Jin Li
Yiyong Li
Yuan Li
Zhihao Li
Tsorng-Juu (peter) Liang
Zhigang Liang
Haiwen Liu
Kwang H. Liu
Wenduo Liu
Yu Liu
Bing Lu
Xi Lu
Zdravko Lukic
Laura Lyle
Wilfredo Machaca
Friedrich Maile
Dragan Maksimovic
Andrey Malinin
Goran Mandic
Frank Mannarino
Hengchun Mao
Xiaolin Mao

Jim Marinos
Barry Mather
Paolo Mattavelli
Sudip Mazumder
James McFarland
Gustavo Meahhenn
Stefano Michelis
Rais Miftakhutdinov
Liviu Mihalache
Nikola Milivojevic
John Miller
John M. Miller
Toshitaka Minamisawa
Behrooz Mirafzal
Robert Mischel
Geev Mokryani
Seung-Ryul Moon
Jeffrey Morroni
Seyed Ahmad Mosavi Esterabadi
Joseph Mossoba
Eric Motto
Olivier Mourra
Priscilla Mulhall
Chuck Mullett
Shravana Musunuri
Surya Musunuri
James Nagashima
Xi Nan
Brian Narveson
Jason Neely
Mark Nelms
Vietson Nguyen
Peyman Niazi
Zhong Nie
Van Niemela
Behrooz Nikbakhtian
Jeff Nilles
Puqi Ning
Thomas Nondahl
Tim O'Connell
Terence O'Donnell
Jesus Oliver
Omer Onar
Shigeru Onoda
Mohamed Orabi
Eduardo Ortiz-Rivera
Dick Oswald
Wen Ouyang
Burak Ozpineci
Thurein Paing
Shangzhi Pan

Zhiguo Pan
Yuri Panov
Kevin Parmenter
Amir Parayandeh
Ki-Bum Park
Sang-Hyun Park
Sung-Yeul Park
Ernie Parker
Leila Parsa
L Prasad Paruchuri
Steve Pekarek
Fang Z Peng
Mor Mordechai Peretz
William Peterson
Nguyen Phung Quang
Johnson Pi
Robert Pilawa-Podgurski
Grant Pitel
Venkata Anand Kishor Prahaba
Lewei Qian
Wei Qian
Zhaoming Qian
Zhijun Qian
Liyan Qu
Weiguo Que
Brian Raczkowski
Aleksandar Radic
Farzad Rajaei Salmasi
Ray Ridley
Juan Rivas
Brian Robert
Miguel Rodriguez
Michael Ropp
Xinbo Ruan
Maurizio Salato
Marina Sanz
Ryan Schnell
Michael Schutten
Dhaval Shah
Baiming (shawn) Shao
Jianwen Shao
Greg Singh
Alex Skorcz

Salim Solomon
Deepak Somayajula
Byeong-Mun Song
Chunping Song
Zakdy Sorchini
Jim Spangler
Russ Spyker
Anurag Srivastava
Igor Stamenkovic
Victor Stefanovic
Ljubisa Stevanovic
Andreas Stiedl
David Strasser
Jian Sun
Yi Sun
Banharn Sutthiphorns
Kazuto Takao
Salman Talebi
Dong Tan
Rildo Taveira de Olivera
Timothy Thacker
Aung Thet Tu
Michael Thompson
Satish Thuta
Rohit Tirumala
Leon Tolbert
Hamid Toliyat
Ferdinando Tonicello
Reinaldo Tonkoski Jr
Dimitri Torregrossa
David Torrey
Darryl Tschirhart
Sheng-Yu Tseng
Baskar Vairamohan
Sravan Vanaparthy
Dean Venable
Fabio Vicenzi
Oleg Vodyakho
Heather Volesky
Oleg Volfson
Peter Wambsganss
Bingsen Wang
Dan Wang

Fei Wang
Hao Wang
Huai Wang
Jin Wang
Wei Wang
Xianwei Wang
Wayne Weaver
Gregory West
Matt Wilkowski
Sheldon Williamson
Sanjaka G. Wirasingh
Ernest Wittenbreder
Andrzej Wojtasik
Thomas Wolbank
Xiaofeng Wu
Ying-Jhih Wu
Guochun Xiao
Shaojun Xie
Jing Xu
Yan Xu
Xu Yang
Wenxi Yao
Y.T. Yau
Ben York
George Young
Haidong Yu
Wensong Yu
Chang Yuan
Yingqin Yuan
Regan Zane
Zheng Zao
Di Zhang
Hui Zhang
Jiucai Zhang
Yingqi Zhang
Zhendong Zhang
Zheng Zhao
Cong Zheng
Liang Zhou
Yan Zhou
Xuancai Zhu
Ke Zou

APEC 2010 Exhibitors

At the time the Proceedings went to print, the companies below were planning to exhibit at APEC 2010. The actual exhibitors at APEC 2010 may differ slightly from this list.

Adams Magnetic Products Co.
Agilent Technologies
Allied International
Alpha & Omega Semiconductor
American Furukawa Inc.
American Superconductor
Ametherm, Inc.
Amphenol Interconnect Products
Anderson Power Products
Ansoft, LLC
Avnet Electronics Marketing
Beijing New Chuang Si Fang
Electronics Col., LTD.
BI Technologies/Magnetics
Components Division
Ceramic Magnetics, Inc./Kolektor
Magma
Champs Technologies
Chang Sung Corporation
Chroma Systems Solutions, Inc.
CogniPower, LLC
Coilcraft
Coil Winding International &
Electrical Insulation Magazine
Core Technology Group, Inc.
Cornell Dubilier
CPS Technologies
Cramer Coil & Transformer Co. Inc.
Cree, Inc.
DALSA Semiconductor
Datatronics
Dearborn Electronics Inc.
Dexter Magnetic Technologies, Inc.
EBG, LLC
ECI (Electronic Coils)
EFC/WESCO
Eldre Corporation
ElectroMagneticWorks Inc.
Electronic Concepts, Inc.
Electronics Systems Packaging
Elna Magnetics
Emerson Network Power - Embedded
Power
EPCOS
EXAR Corporation
Fairchild Semiconductor
FCI
Ferroxcube USA, Inc.
Filter Concepts, Inc.
Freescale Semiconductor
GMW Associates

Gold Phoenix Technology Ltd.
Hengdian Group DMEGC Magnetics
Co., LTD
HIMAG Solutions
HV Components / CKE
HVR Advanced Power Components,
Inc.
ICE Components, Inc.
Illinois Capacitor, Inc.
IMS Research
Infineon/Primarion Technologies
Infolytica Corporation
International Rectifier
Intersil Corporation
Isotek Corporation
ITW Paktron
Kaschke Components GmbH
KDM Magnetic Powder Core
Company
KEMET
Kikusui America
Lemsys SA
Lodestone Pacific
Magnetics
Magna-Power Electronics
Magsoft Corporation
MaxQ Technology, LLC.
MESAGO PCIM GmbH
Methode - Network Bus Products
Micrel, Inc.
Micrometals, Inc.
Microsemi Corporation
MMG
Monolithic Power Systems, Inc.
Murata Power Solutions
National Semiconductor
New England Wire Technologies
Newtons4th Ltd.
NH Research, Inc.
NORWE, Inc.
NXP Semiconductors
Ohmite Manufacturing Company
ON Semiconductor
OZTEK Corporation
Payton America
Pearson Electronics, Inc.
Planar Quality Corporation
Plexim
Positronic Industries, Inc.
Power Electronic Measurements Ltd.

Power Electronics Technology
Magazine
Powerex, Inc.
Powersim, Inc.
Powervation
Precision Inc.
Qspeed Semiconductor
RAF Tabtronics Inc.
Renco Electronics Inc.
Ridley Engineering, Inc.
R.L. Components Ltd.
Rogers Corporation
R-Theta Thermal Solutions
Rubadue Wire Company
SanRex Corporation
Sapa Thermal Management
SBE Electronics, Inc
Semikron
SemiSouth Laboratories, Inc.
Sienna Technologies, Inc.
Simplis Technologies
STMicroelectronics, Inc.
TDI Power
TDK Corporation
TDK-Lambda Americas
Tektronix Inc.
Texas Instruments
Thermik
Tocos America
Toho Zinc Co., Ltd.
Toshiba America Electronic
Components, Inc.
Tower Semiconductor
TranSic
Transim Technology
Transtek Magnetics
Tyco Electronics
United Chemi-Con
uPI Semiconductor
V•I Chip, Inc.
VAC Magnetics
Venable Industries, Inc.
Vette Corp.
VISHAY SILICONIX
Voltage Multipliers
Voltech Instruments, Inc.
Wurth Electronics, Inc.
X-Fab Group
Xitron Technologies Inc.
Yokogawa Corporation of America
ZesZimmer Electronic Systems GmbH

APEC 2010 Professional Education Seminars

At the time the proceedings went to print, the Professional Education Seminars listed below were scheduled for presentation at APEC 2010. The actual seminars presented may have differed slightly from this list.

Professional Education Seminars, Session One
Sunday, February 21, 9:30 AM – 1:00 PM

S.1.   **Controlling Conducted and Radiated EMI Issues in Power Electronics Design**
        Supratim Basu, Bose Research (P) Ltd.

S.2.   **Current State and Future Improvements of GaN Based Power Conversion**
        Tim McDonald, International Rectifier

S.3.   **Virtual Prototyping of Power Supply Designs**
        Thomas Wilson, Simplis Technologies, Inc.

S.4.   **Designing Compensators for the Control of Switching Power Supplies**
        Christophe Basso, ON Semiconductor

S.5.   **New Trends in Magnetic Technology**
        Dan Jitaru, Delta Corporate Services

S.6.   **Power Electronics in Battery Powered Applications: Safety, Charging, Monitoring, and Power Conversion**
        Jinrong Qian, Texas Instruments

Professional Education Seminars, Session Two
Sunday, February 21, 2:30 PM – 6:00 PM

S.7.   **EMC for Power Supply Designers**
        Ernest Wittenbreder, Technical Witts, Inc.

S.8.   **Application Characteristics of IGBT Power Modules**
        John Donlon, Powerex

S.9.   **Simulation, Control, and Measurement of DC-DC Converters**
        Ray Ridley, Ridley Engineering

S.10.   **Thermal Management for Power Electronics Applications**
        Ahmed Zaghlol, R-Theta Thermal Solutions, Inc.

S.11.   **The Semiconductor Cycle Bites**
        Ada Cheng, Accolade

S.12.   **Power Conversion Reliability**
        Don Gerstle, Murata Power Solutions

Professional Education Seminars, Session Three
Monday, February 22, 8:30 AM – Noon

S.13.   **EMI Causes, Measurement, and Reduction Techniques for Switch-Mode Power Converters**
        Michael Schutten, GE Global Research

S.14.   **Silicon Carbide Characteristics and Application**
        Robert Callanan, Cree, Inc.

S.15.   **CAD of Power and Control Systems of Electrical Machines**
        Tanvir Rahman, Infolytica Corporation

S.16.   **Analog and Digital Feedback Loop Design**
        Dean Venable, Ablepower Corporation

S.17.   **Systematic Design for High Efficiency**
        Dhaval Dalal, ACP Technologies

S.18.   **Power Electronics in a Smart-Grid Distribution System**
        Doug Hopkins, University at Buffalo

Minimum Deviation Digital Controller IC for Single and Two Phase DC-DC Switch-Mode Power Supplies

Aleksandar Radić, Zdravko Lukić, and Aleksandar Prodić

Laboratory for Power Management and Integrated SMPS
ECE Department, University of Toronto, Toronto, CANADA
{radicale, lukic, prodic}@power.ele.utoronto.ca

Robert de Nie

NXP Semiconductors
Nijmegen, NETHERLANDS
Rob.de.Nie@nxp.com

Abstract— A digital PWM voltage mode controller integrated circuit (IC) for high-frequency dc-dc switching converters achieving virtually minimum possible, i.e. optimum, output voltage deviation to load transients is introduced. The IC is implemented with simple hardware, requiring small silicon area, and can operate as a single-phase or a two-phase controller. To minimize the area and eliminate known mode transition problems of the optimal response controllers, two novel blocks are combined. Namely, an asynchronous track-and-hold analog-to-digital converter (ADC) and a "large-small" signal compensator are implemented. The ADC utilizes a pre-amplifier and only four comparators having approximately eight times smaller silicon area and power consumption than an equivalent windowed flash architecture. The "large-small" signal compensator consists of two parts, a digital PID minimizing small variations and a zero-current detection-based compensator suppressing large load transients. The large-signal compensator requires no extra calculations and has a low sensitivity to parameter variations. It utilizes a synchronization algorithm and the PID calculation results to obtain a bumpless mode transition and stable response to successive load transients.

The IC occupying only 0.26 mm² silicon area is implemented in a CMOS 0.18μm process and its minimum deviation response is verified with a single and dual-phase 12 V-to-1.8 V, 500 kHz 60/120 W buck converter.

I. INTRODUCTION

Switch-mode power supplies (SMPS) used in consumer electronics, portable applications, and computers, are required to meet stringent voltage regulation requirements [1], [2] using a cost-effective implementation occupying a small volume. The regulation is usually achieved with on-chip integrated controllers, which, in analog implementation, occupy a silicon area as small as 0.6 mm² [3]. The integrated controllers most frequently utilize voltage mode pulse-width modulation, where a PID compensator provides an accurate output voltage regulation and robustness of the system over a wide range of operating conditions. However, due to inaccurate high-frequency dynamics of averaged small-signal models of the converters, usually used in the conventional PID design [4], the response time of the compensator is fairly limited. As a result, the reactive components of the power stage output filter are usually oversized. Furthermore, since the obtained linear

This work of Laboratory for Power Management and Integrated SMPS is sponsored by NXP Semiconductors, Eindhoven, Netherlands.

time-invariant (LTI) model is based on the small-signal variations assumption, it might not be fully valid for large excitations, frequently occurring in modern SMPS during large load changes.

Since the controller speed is closely related to the output voltage deviation, i.e. the size of the filtering components, various analog and digital linear [5]-[7] and nonlinear [8]-[17] control methods for improving the speed of the controller, have been investigated.

Arguably, the most promising results are demonstrated though various digital implementations of the optimal control [11]-[16], where the recovery from a transient is achieved through a single on-off switching action.

Even though the optimal control methods have demonstrated superior performance compared to conventional solutions, they have not been widely adopted due to one or more of the following four problems: i) Instability caused by transitions between steady state and dynamic modes, i.e. chattering; ii) High sensitivity to quantization effects and converter parameter variations, iii) Overly complex hardware required for implementation requiring a large silicon area for implementation; and iv) Inability to react to consecutive load steps occurring during the dynamic mode.

The main goal of this paper is to introduce a novel digital controller IC that achieves virtually minimal possible, i.e. the optimal, deviation during load transients and is implemented with fairly simple hardware, occupying a small silicon area,

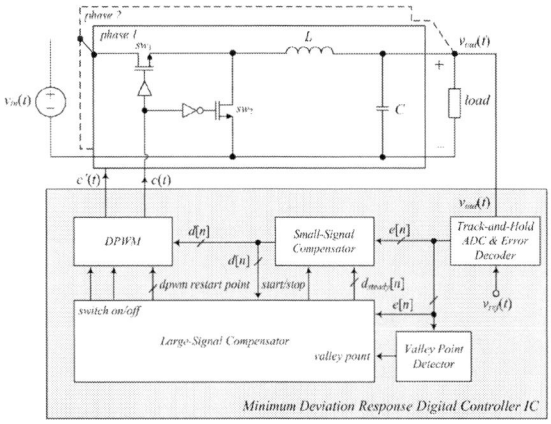

Figure 1. A dual-phase buck converter controlled by the combined "large-small" signal digital controller.

978-1-4244-4782-4/10 $26.00 © 2010 IEEE

without suffering from the above mentioned problems. Depending on the load conditions, the controller, shown in Fig.1, operates either as a small-signal or a large-signal compensator and provides smooth, i.e. bumpless, transition between the static and dynamic modes. The IC also provides the optimal deviation response to narrowly time-spaced load transients and has a low sensitivity to power stage parameter variations and quantization effects. Furthermore, the IC can either regulate operation of a single phase converter or that of a two-phase interleaved topology. To reduce the power consumption and silicon size of the controller while maintaining sufficient accuracy and speed a novel Track-and-Hold ADC architecture is also developed and implemented.

In the following section, the operation of the IC with a single-phase buck converter is explained and then extended to a dual-phase case. Special attention is given to mode transition problems and the resulting overshoot. Section III discusses problems related to overly complex hardware required for the implementation of optimal digital controllers and describes novel hardware-efficient architectures of major controller blocks. Particular attention is devoted to the application specific ADC design that, compared to conventional architectures, takes about 8 times smaller silicon area. Section IV presents experimental results verifying the operation of the controller with both single phase and two-phase buck converter.

II. PRINCIPLE OF OPERATION

The controller of Fig.1 has two control modes. Around steady state, in small-signal mode, the operation of the power stage is governed by a conventional voltage mode digital PWM regulator [21, 23]. Based on the output voltage error value $e[n]$, produced by the ADC, a control signal for digital-pulse width modulator (DPWM) $d[n]$ is created, to keep the output regulated.

The large signal compensator constantly monitors $e[n]$ and $d[n]$, and, while in steady-state ($e[n]$=0), calculates a duty ratio value $d_{steady}[n]$ by averaging $d[n]$ over several switching cycles. As it will be explained soon, this value is later used to obtain a bumpless transition between two compensators while a large load transient is suppressed.

Operation of the large signal compensator is fairly simple. During transients, when $e[n]$ exceeds a specified threshold value, it takes the active control role. It immediately turns on or off the main switch, depending on the sign of $e[n]$ (type of the transient), and disables the operation of the small-signal compensator. The controller's state remains unchanged until the inductor current $i_L(t)$ is equal to the load current $i_{load}(t)$. At that point, capacitor current changes its sign and reaches its peak or valley point. For a given converter topology, this value is also the minimum achievable voltage deviation, i.e. the smallest possible undershoot/overshoot. At this point a synchronization algorithm reactivates the DPWM, and subsequently the PID compensator, such that a smooth transition between the modes is achieved.

It can be seen that reaching the peak or valley point is a trivial task; however, due to quantization and sampling errors the detection of the peak or valley point will always be delayed and due to the utilization of small inductances (typically less than 500nH) may result in relatively large

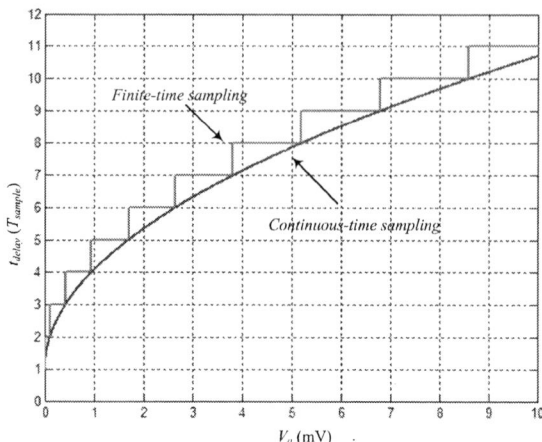

Figure. 2. The influence of ADC quantization step V_q and sampling rate on the worst-case delay time t_{delay} generating voltage overshoot. (T_{sample} is 62.5 ns)

charge injection and consequent stability problems. The following section illustrates the problem and provides quantitative motivation for the controller presented in this paper.

A. Mode Transition

In the presence of in-accurate valley or peak point detection any subsequent control actions dependent on it will only contribute to a greater error. This is particularly true for existing time-optimal controllers [11]-[16] which attempt to compensate the lost capacitor charge through a single on/off control action. This action is usually followed by second mode of operation, where a conventional PID compensator is active to provide stable steady-state operation.

These time-optimal controllers are usually burdened with detecting both the correct valley point and voltage deviation in order to calculate the proper t_{on} and t_{off} times; however, it can be shown that a worst-case delay equal to

$$t_{delay} = \sqrt{\frac{2 \cdot L \cdot C \cdot V_q}{V_{in} - V_{out}}} + t_{ps} \qquad (1)$$

will always be present, where V_q is the quantization voltage and t_{ps} is the combined power stage, sampling and computation delay. Due to the random nature of disturbances in the converter circuit, the actual delay is not constant and cannot be simply compensated.

The relationship found in (1) is illustrated in Fig. 2 where the immense hardware requirements for a proper implementation of the time-optimal control law become apparent.

A typical result of the delay is shown in Fig. 3, where a 4mV quantization step ADC is used. The error due to the valley point detection delay is compound by the t_{on}/t_{off} control action and subsequent PID takeover, denoted by the extra injected charge ΔQ_1 and ΔQ_2 respectively. Often, the resulting overshoots/undershoots cause chattering problem where the controller goes through multiple changes between its two modes of operations.

In the method presented here, rather than achieving recovery to steady state in the minimum possible time, the

978-1-4244-4782-4/10 $26.00 © 2010 IEEE

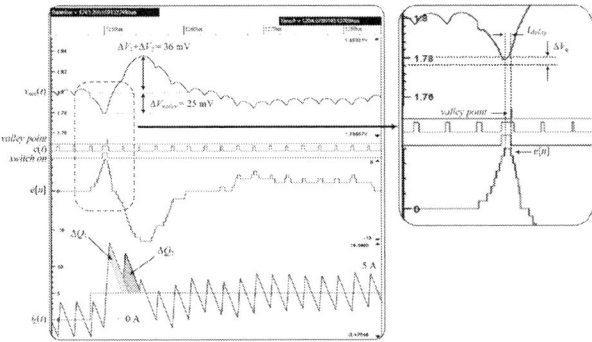

Figure 3. The voltage overshoot caused by incorrect valley point detection, power stage delay, and synchronization during 5-A load step applied to a buck converter with L of 0.47 µH and C of 400 µF.

aim is to achieve recovery with minimum possible voltage deviation and bump-less transition between two modes.

To obtain this, by minimizing the impact of inaccurate valley point detection, in the controller from Fig.1, the transition between the large and small signal compensator is always performed at the valley point in a synchronized manner.

For the light-to-heavy load transient the smooth transition between two compensators is achieved by matching the average value of $i_L(t)$ to the load current. As shown in Fig. 4, this is obtained by extending the initial on-time of the control signal $c(t)$ by t_{on} and by inserting the switch off time t_{off}. On/off times, t_{on} and t_{off}, are dynamically calculated as:

$$t_{on} = \frac{D \cdot T_{sw}}{2} \qquad (2)$$

$$t_{off} = (1-D) \cdot T_{sw} \qquad (3)$$

where D is the extracted steady duty ratio $d_{steady}[n]$ and T_{sw} is the switching period. During this short sequence, internal error registers of the small-signal compensators are cleared while the register holding the previous duty-ratio value $d[n-1]$ is updated with $d_{steady}[n]$ before it is restarted. In case of the heavy-to- light load transient, the same goal can be achieved by extending switch-off time by:

$$t_{off} = \frac{(1-D)}{2} \cdot T_{sw} \qquad (4)$$

·The use of the extracted steady state duty ratio removes the need to know the power stage parameters and also conveniently takes into account all the losses.

B. Extension to a Two-Phase Buck Converter

In some applications (e.g.. microprocessors), it may be necessary to use two interleaved converter phases to supply twice larger load current. For that case, the controller from Fig. 1 is modified to provide the same functionality as in the single-phase case. Control signal $c_1(t)$ for phase 1 is identically generated as in the single-phase case. On the other hand, for interleaved phase 2, upon exit from the valley point,

control signal $c_2(t)$ is altered based on the sequence used for the opposite transient.

III. PRACTICAL ON-CHIP IMPLEMENTATION

The introduced large-small signal digital controller, as well as other state-of-the-art fast transient response digital solutions [11]-[16], require accurate and quick A/D conversions, in order to ensure appropriate and timely control actions. To provide such characteristics, in [11], an application specific 6-bit ADC, as a part of an optimal-time controller, is presented. The presented ADC provides fast conversion time and small quantization steps, but also has a large power consumption and silicon area of about 0.5 mm^2, which is comparable in size to a complete analog controller, making it unsuitable for numerous low-power applications. In addition, the large number of comparators provide no benefit during the valley and peak point detection periods, when only one comparator is toggled. To solve this problem and take advantage of the importance of the valley point detection, a new Track-and-Hold ADC architecture is developed and presented.

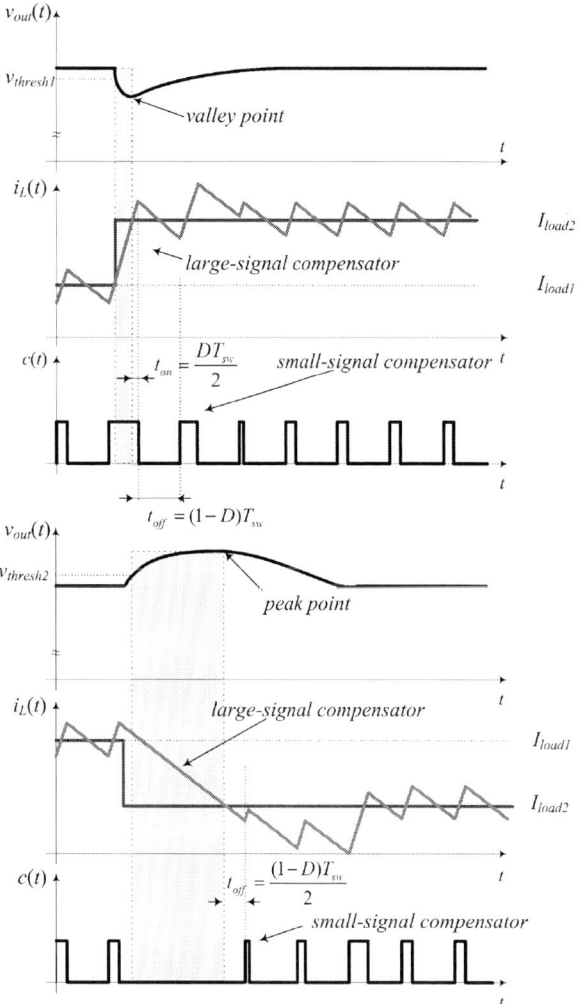

Figure 4. Principle of operation of the "large-small" signal compensator during light-to-heavy (top) and heavy-to-light (bottom)

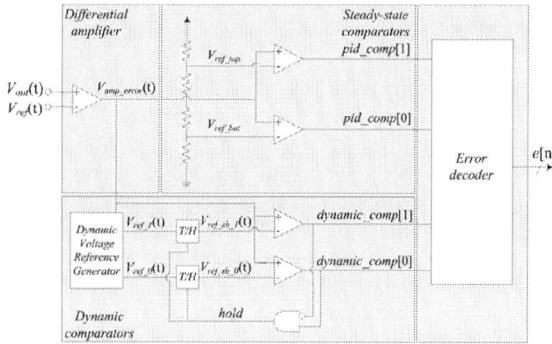

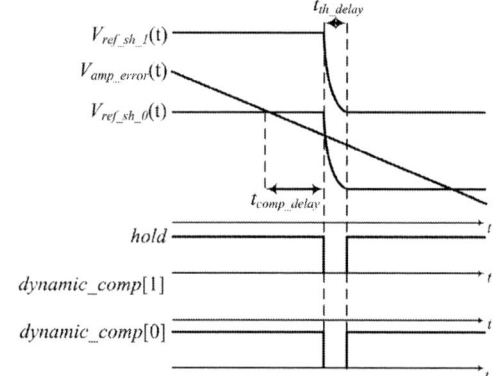

Figure 5. Block diagram (top) and a single track-and-hold transition (bottom) of the Track-and-Hold ADC.

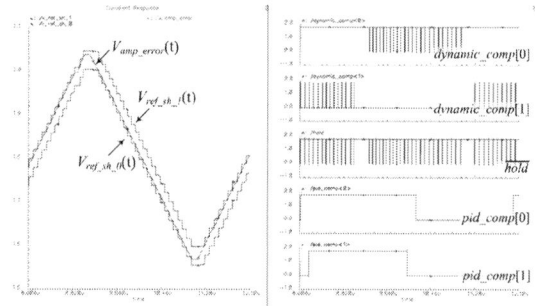

Figure 6. ADC simulation results: key analog signals (left) and digital logic outputs (right).

Conventional ADC architectures trade-off speed, accuracy, silicon area and power in order to satisfy a wide range of applications. These tradeoffs are best illustrated by the successive approximation and Flash ADCs. The successive approximation ADC (SAR) combines a single comparator, DAC and successive approximation register to generate iteratively converging input voltage approximations [23]. Speed and accuracy are sacrificed in exchange for smaller silicon area and lower power consumption. On the other hand, the Flash ADC combines $2^n - 1$ comparators in parallel, ensuring an n-bit output with a fixed conversion time, at the expense of increased power loss and silicon area [20,23,24].

The proposed Track-and-Hold ADC delivers conversion times and accuracy comparable to those of the Flash ADC architecture at a fraction of power and silicon area, combing the best of the SAR and Flash architectures. The block diagram of the novel ADC and a single track and hold transition are shown in Fig. 5. The Cadence HSpice simulation results are also shown in Fig. 6.

The detection of the peak and valley are at the core of the novel Track-and-Hold ADC. As a result, the Track-and-Hold ADC operates by continuously tracking the amplified error signal, $V_{amp\_error}(t)$, using a small window around it. The window is composed of two signals, $V_{ref\_sh\_1}(t)$ and $V_{ref\_sh\_0}(t)$, which are sampled and held $V_{amp\_error}(t)$ signal values with $+kV_q$ and $-kV_q$ offsets (The k denotes the differential amplifier amplification). The reference signals are generated using the Dynamic Voltage Reference Generator and two Sample-and-Hold circuits.

When $V_{amp\_error}(t)$ is outside the window, detected using the dynamic comparator outputs and two simple gates, the reference signals are re-sampled asynchronously. During the reference sampling process, the digital error representation is updated to reflect the $\pm V_q$ change using the dynamic_comp signal edges. Due to the comparator and sample-and-hold delays, denoted by t_{comp_delay} and t_{th_delay}, the digital error representation will accumulate an offset. To mitigate this, two steady-state comparators are used to detect the offset in the decoded digital error and re-calibrate it whenever the error signal crosses the zero point.

Furthermore, to minimize requirements on the comparators, a pre-amplifying stage is introduced. The pre-amplifier utilizes a bandwidth reduced two-stage operational transconductance amplifier [23] in order to filter high-frequency noise and generate an amplified error signal. As a result the comparators effective quantization step is larger, the required conversion time slower and the power consumption requirement reduced.

The Cadence Hspice simulation results of the Track-and-Hold flash ADC with a pre-amplification of 5 are presented in Fig. 6. It can be observed that proper tracking (left) of the fast changing input signal is achieved and that its digital outputs are generated as desired. The area and current consumption of the proposed ADC is compared with an in-house built 7-bit Flash ADC in Table I. A eight-fold reduction is obtained in both area and current consumption. A photograph of the chip, implemented using a 0.18μm CMOS process, is shown in Fig. 7. All digital blocks including a programming and debugging unit utilize less than 4500 logic gates.

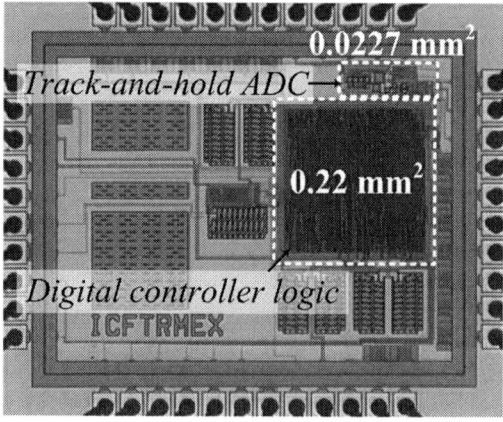

Figure 7. The large-small signal digital controller IC.

TABLE I. COMPARISON OF THE TRACK-AND-HOLD ADC AND CONVENTIONAL 7-BIT FLASH ADC

	Track-and-Hold ADC	Flash ADC
Conversion time (ns)	15	15
Area (mm^2)	0.0227	0.2
Current (mA)	0.24	2

IV. EXPERIMENTAL RESULTS

The operation of the large-small signal digital controller, from Fig. 1, is verified with a 60-W single-phase and 120-W dual-phase converter switching at 500 kHz. The parameters of a single-phase converter are given in Table II and are representative of modern point-of-load (POL) converters [18,19]. To compare the performance with a conventional PID, initially, the large-signal compensator is disabled and the operation of the small-signal compensator verified.

TABLE II. SINGLE-PHASE CONVERTER PARAMETERS

V_{in}	12 V
V_{out}	1.8 V
f_{sw}	500 kHz
L	0.47 µH
C	400 µF
ESR	0.5 mΩ
I_{load} (max)	30 A

Next, the large-signal compensator is enabled and the comparison with the previous result is shown in Fig. 8. The bandwidth of the small-signal compensator is 1/10th of the switching frequency. When the large-signal compensator is enabled the voltage deviation is reduced by a factor of three as illustrated in Fig. 9. The voltage deviation can further be reduced with a faster transient detection system as illustrated by the red waveforms. In addition, the peak transient inductor current is also reduced which prevents undesirable inductor saturation. These improvements are the result of the instantaneous control action of the large-signal compensator which increases the inductor current to the new steady-state load current. Once the valley point is detected the transition between the large-signal and small-signal compensator is performed as explained in the principle of operation. In the case of incorrect valley point detection due to capacitor ESR or quantization errors, an overshoot is detected and the t_{off} time is appropriately increased ensuring a bumpless transition.

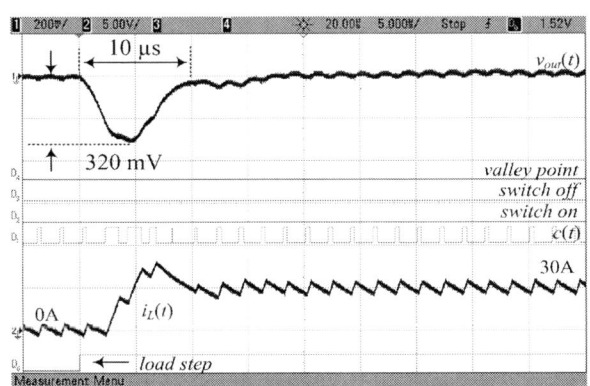

Figure 8. The response of the small-signal compensator for a load step from 0 A to 30 A.

The controller behavior in the presence of consecutive load transients, from 0 A to 12 A and 12 A to 30 A, is demonstrated in Fig. 10. The large-small signal controller effortlessly transitions between the large-signal and small-signal compensators under any load conditions, resulting in guaranteed minimum voltage deviation. In addition, the dual-phase controller output voltage and interleaved phase 2 inductor current waveforms are also shown in Fig. 11, verifying the successful minimum-deviation and bumpless transition. It should be noted that the presented experimental results show transient performance comparable to the time-

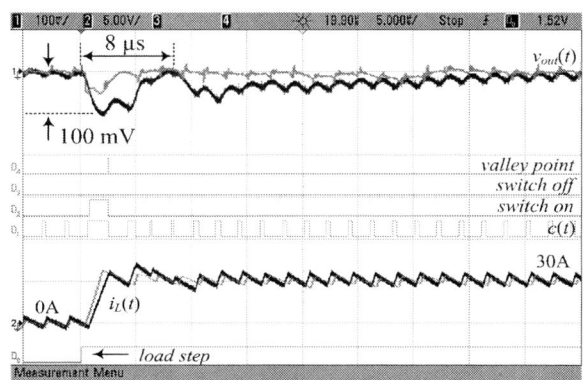

Figure 9. The response of the combined large-small signal compensator for a load step from 0 A to 30 A without (black) and with (red) fast transient detection.

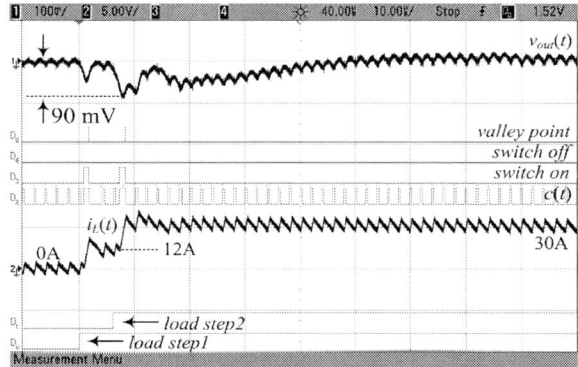

Figure 10. The response of the combined large-small signal controller under consecutive load transients.

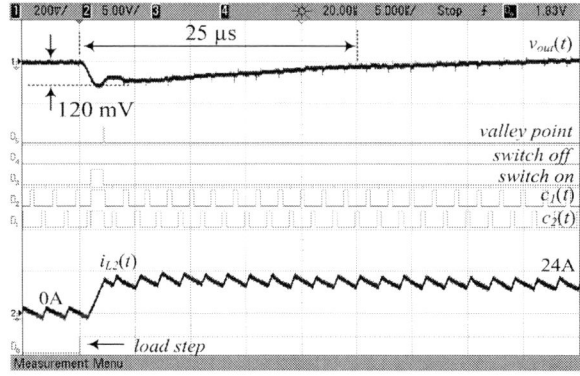

Figure 11. The response of the combined large-small signal controller for a dual-phase converter for a load step from 0A to 50A.

optimal controllers. The maximum deviation is not larger than that of the time-optimal systems and in a number of cases total recovery time is even shorter, due to the absence of overshoots/undershoots and previously mentioned chattering problems related to practical implementation problems.

V. CONCLUSIONS

The combined large-small signal controller IC that provides minimum output voltage deviation is presented. The controller implements a very simple control law and utilizes novel hardware efficient architecture of a Track-and-Hold ADC to detect and measure rapid load disturbances. The Track-and-Hold ADC has a conversion time of only 15 ns, about 8 times smaller area than the conventional ADCs, and a quantization step of only 4 mV. It's simple design allows for significant area and power reduction, enabling the use of optimal control architectures in low-power SMPS.

To overcome the bandwidth limitations of the averaged small-signal model, the controller IC implements two dedicated compensators: for the large-signal and small-signal operation. The solution for the bumpless transition at the valley point between two modes of operation taking into account the inductor current ripple is introduced. The controller was implemented in a CMOS 0.18 µm process occupying 0.26 mm^2. Due to the controllers simple architecture all digital logic blocks were implemented with less than 4500 gates. The controller operation is verified experimentally with modern POL converters under various load test conditions, including the consecutive load disturbance case. The controller was also verified with a dual-phase buck converter. All experimental results exhibit optimal voltage deviation for a given power stage and load step size, resulting in greater than three times reduction of transient output voltage deviation, equivalent to the same reduction of output capacitor value, compared to conventional PID compensators.

REFERENCES

[1] S. Saggini, M. Ghioni, and A. Geraci, "An innovative digital control architecture for low-voltage, high-current DC-DC converters with tight voltage regulation," *IEEE Trans. Power Electron.*, vol. 19, pp. 210–218, Jan. 2004.

[2] "Voltage regulator module (VRM) and enterprise voltage regulator-down (EVRD) 11.0," Intel Corp., Oregon, USA.

[3] Kuo-Hsing Cheng, Chia-Wei Su, and Hsin-Hsin Ko, "A high-accuracy and high-efficiency on-chip current sensing for current-mode control CMOS DC-DC buck converter", *IEEE 15th International Electronics, Circuits and Systems Conference*, 2008, pp. 458-461.

[4] R. W. Erickson and D. Maksimovic, *Fundamentals of Power Electronics*. New York, NY:Springer Sience+Business Media Inc., 2001.

[5] S. Saggini, P. Mattavelli, M. Ghioni, and M. Redaelli, "Mixed-signal voltage-mode control for DC-DC converters with inherent analog derivative action," *IEEE Trans. Power Electron.*,vol. 23, pp. 1485–1493, May 2008.

[6] S. Saggini, P. Mattavelli, G. Garcea, and M. Ghioni, "A mixed-signal synchronous/asynchronous control for high-frequency DC-DC boost converters," *IEEE Trans. Ind. Electron.*, vol. 55, pp. 2053–2060, May 2008.

[7] L. Corradini, P. Mattavelli, E. Tedeschi, and D. Trevisan, "High-bandwidth multisampled digitally controlled DCDC converters using ripple compensation," *IEEE Trans. Ind. Electron.*,vol. 55, pp. 1501–1508, Apr. 2008.

[8] Kelvin Ka-Sing Leung and Henry Shu-Hung Chung, "Dynamic hysteresis band control of the buck converter with fast transient response," *IEEE Trans. Circuits Syst. II*, vol. 52, pp. 398–402, July 2005.

[9] A. Soto, P. Alou, and J.A. Cobos, "Nonlinear digital control breaks bandwidth limitations," in *Proc. IEEE Applied Power Electronics Conf.*, 2006, pp. 724–730.

[10] Santa C. Huerta, P. Alou, J. A. Olivier, O. Garcia, J. A. Cobos, and A. Abou-Alfotouh, "A very fast control based on hysteresis of the cout current with a frequency loop to operate at constant frequency," in *Proc. IEEE Applied Power Electronics Conf.*, 2009, pp. 799–805.

[11] G. Feng, E. Meyer, and Y.-F. Liu, "A new digital control algorithm to achieve optimal dynamic performance in DC-to-DC converters," *IEEE Trans. Power Electron.*, vol. 22, pp. 1489–1498, July 2007.

[12] Zhenyu Zhao and A. Prodic, "Continuous-time digital controller for high-frequency DC-DC converters," *IEEE Trans. Power Electron.*, vol. 23, pp. 564–573, Mar. 2008.

[13] E. Meyer, Zhiliang Zhang, and Y.-F. Liu, "An optimal control method for buck converters using a practical capacitor charge balance technique," *IEEE Trans. Power Electron.*, vol. 23, pp. 1802–1812, July 2008.

[14] V. Yousefzadeh, A. Babazadeh, B. Ramachandran, E. Alarcon, L. Pao, and D. Maksimovic, "Proximate time-optimal digital control for synchronous buck DC-DC converters," *IEEE Trans. Power Electron.*, vol. 23, pp. 2018–2026, July 2008.

[15] A. Costabeber, L. Corradini, P. Mattavelli, and S. Saggini, "Time optimal, parameters-insensitive digital controller for DC-DC buck converters," in *Proc. IEEE Power Electronics Specialist Conf.*, 2008, pp. 1243–1249.

[16] L. Corradini, A. Costabeber, P. Mattavelli, and S. Saggini, "Time optimal, parameters-insensitive digital controller for VRM applications with adaptive voltage positioning," in *Proc. IEEE Workshop on Computers in Power Elecronics*, 2008, pp. 1–8.

[17] S. Effler, A. Kelly, M. Halton, T. Kruger, and K. Rinne, "Digital control law using a novel load current estimator principle for improved transient response," in *Proc. IEEE Power Electronics Specialist Conf.*, 2008, pp. 4585–4589.

[18] "PIP212-12M data sheet," NXP Semiconductors, Eindhoven, Netherlands.

[19] "FDMF8700 data sheet," Fairchild Semiconductors, San Jose, USA.

[20] R. J. van de Plassche, J. H. Huijsing and W. M. C Sanse, *Analog Circuit Design: High-Speed Analog-to-Digital Converters*. New York, NY: Springer, 2000.

[21] B. J. Patella, A. Prodic, A. Zirger, and D. Maksimovic, "High-frequency digital PWM controller IC for DC-DC converters," *IEEE Trans. on Power Electron.*, vol. 18, pp. 438 – 446, Jan. 2003.

[22] A. Prodic and D. Maksimovic, "Design of a digital PID regulator based on look-up tables for control of high-frequency DC-DC converters," in *Proc. IEEE Workshop on Computers in Power Electronics*, 2002, pp. 18-22.

[23] D. A. Johns and K. Martin, *Analog Integrated Circuit Design*, Toronto, ON: John Wiley & Sons, Inc., 1997.

[24] J. Yoo, D. Lee, K. Choi, and A. Tangel, "Future-ready ultrafast 8bit CMOS ADC for system-on-chip applications", *IEEE 14th International ASIC/SOC Conference*, 2001, pp. 455-459.

Modeling and Design Considerations of Coupled Inductor Converters

Guangyong Zhu
Auscom Engineering, Inc.
A subsidiary of Compal Electronics, Inc.
One Dell Way, Round Rock, TX 78682

Kunrong Wang
Enterprise Systems Group
Dell, Inc.
One Dell Way, Round Rock, TX 78682

Abstract-In this part of the sequel on the modeling and analysis of coupled inductors and coupled inductor based multi-phase switching converters, the recently developed symmetrical coupled inductor model is first extended to include the inductor winding dc resistance (DCR). The extended model is then used to analyze the influence of the coupling on the DCR based current sensing schemes popularly used in multi-phase switching regulators. It is found that the time-constant matching condition in coupled inductor converters needs to be modified to include the coupling coefficient. The proposed model is also used to derive the small-signal control-to-output transfer function of the converters incorporating coupled inductors, with which the effect of coupling on the dynamic behaviors of the converter power stage, such as resonant frequency and damping factor, can be easily evaluated.

I. INTRODUCTION

Coupled inductor as a special form of multiple winding coupled magnetic structure, or transformer, has been around since the early years of electrical engineering. In modern high-frequency switching converters, it has been used in many applications such as multi-output, cross-regulated converters [1, 4], ripple cancellation and multiple magnetic component integration [2, 6], snubbering [3], and transformer/inductor integration [5], etc. More recently, it also found applications in multi-phase buck regulators [7-18]. Moreover, the basic coupled magnetic structure was extended to practical implementations beyond three phases [10-13]. It is generally accepted that reverse coupling between the coupled windings is the preferred coupling scheme [7, 10], and fast transient response can be achieved especially when the control scheme allows multiple phases to overlap in case of fast load transient [20]. Some also made plausible argument that coupled inductor based converter also improves conversion efficiency, but this seems to be valid only under the assumption that in order to achieve the same transient response, a coupled-inductor based converter needs to switch only at a reduced switching frequency. It is noted that the phase current ripple cancelling effect at the output of a regular multi-phase switching regulator is lost in the coupled-inductor based one, leading to higher ripple voltage at the output [20].

Most of the modeling and analysis works on coupled inductors were based on the well-established transformer models, such as the T- or π-equivalent model [10, 14, 15, 19], and the cantilever model [6]. Although these models are largely valid, they are asymmetrical relative to the roughly symmetrical magnetic structure largely used in multiphase buck regulators because the magnetizing inductance appears only in one of the coupled windings. The asymmetry in the model usually makes the analysis very cumbersome. An equivalent inductance model based on the concept of self and mutual inductances in basic circuit theory was introduced in [7, 8], but it still fell short in crystallizing the general relationship involved in coupled inductors. With the understanding that coupled inductors are magnetic structures with loose coupling and that the coupling coefficient plays an important role in the resulting circuit performance, a general symmetrical and invariant model for multi-winding coupled inductors was recently derived [20]. It was based on the same self- and mutual-inductance concept as in [7, 8], but has the advantage in analyzing all the key circuit behaviors of the coupled inductor based multi-phase regulators, such as steady-state ripple currents, and transient response, etc. The model can be easily extended to multiple (more than two) winding cases, and all the circuit parameters involved in the model, such as leakage inductances and the coupling coefficient, can be easily determined based on terminal inductance measurements.

In this paper, the symmetrical model presented in [20] is first extended to include the winding dc resistance (DCR). It is then used to analyze the influence of the coupling on the DCR based current sensing schemes and to derive the small-signal control-to-output model of the converters incorporating coupled inductors.

II. REVIEW AND EXTENSION OF COUPLED INDUCTOR MODELS

Fig. 1 shows the well-known general transformer model of two inversely coupled inductors. It has four parameters: magnetizing inductor L_M, turns ratio n, leakage inductors L_{k1} and L_{k2}, with only three of them independent.

Since this model is asymmetrical in structure, it is unable to provide enough analytical insight into how the coupling coefficient affects a converter's steady state and transient performances. A new symmetrical transformer model for coupled inductors was developed in [20]. While it was discussed extensively in [20] that the symmetrical model is easy to use and suitable for analyzing converters utilizing coupled inductors, some practical design considerations in coupled inductor converters were not discussed, such as lossless current sensing based on inductor DCR [21, 22], and

978-1-4244-4782-4/10 $26.00 © 2010 IEEE

the effect of coupling coefficient on the small-signal model of the converter. In this section, a symmetrical transformer model with non-zero winding resistances will be derived first, which will serve as the basis for the discussions in the subsequent sections.

For simplicity, two inversely coupled identical windings, each with a self inductance of L and a DCR of R_L, as shown in Fig. 2(a), are assumed. The model can be analytically derived from the following expression governing two inversely coupled windings:

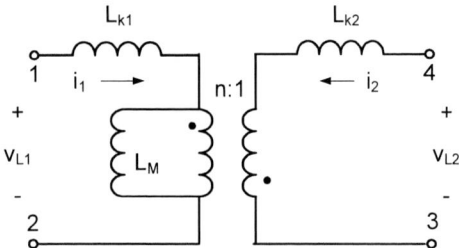

Figure 1. A general transformer model of two inversely coupled inductors.

$$v_{L1} = L\frac{di_1}{dt} - M\frac{di_2}{dt} + R_L i_1$$
$$v_{L2} = -M\frac{di_1}{dt} + L\frac{di_2}{dt} + R_L i_2 \qquad (1)$$

where M is the mutual inductance. The expression can also be re-arranged as:

$$v_{L1} = L_k\frac{di_1}{dt} - kv_{L2} + R(i_1 + ki_2)$$
$$v_{L2} = -kv_{L1} + L_k\frac{di_2}{dt} + R(ki_1 + i_2) \qquad (2)$$

where L_k is the leakage inductance defined as $L_k = (1-k^2)L$ and k is the coupling coefficient defined as $k=M/L$.

From (2), one can easily come up with a new symmetrical equivalent circuit as shown in Fig. 2(b).

It can be seen that the model shown in Fig. 2(b) takes on a quite different form from the traditional transformer models where R_L is simply a series resistor on both the primary and secondary sides of a transformer. The magnetizing inductor, which leads to asymmetry in other model, does not explicitly appear in the new model. It should also be pointed out that L_k in Fig. 2(b) is the inductance that is directly measureable and can be obtained by measuring across one winding while the other winding is shorted. It is different from the leakage inductances, L_{k1} and L_{k2}, shown in Fig. 1.

One of the distinctive features of the new equivalent circuit model for two coupled inductors given in Fig. 2(b) is that it is very convenient to derive the dynamic models of

converters utilizing coupled inductors. The condition for DCR based current sensing can also be easily determined according to (2) and Fig. 2(b).

III. APPLICATIONS OF THE DERIVED MODEL

3.1 Inductor DCR Based Current Sensing in Coupled Inductor Converters

Lossless inductor DCR based current sensing has become a popular method in multiphase, high current switching regulators such as those for CPU, memory and graphics [21, 22]. The condition to accurately extract inductor current information is well understood in converters with discrete, or uncoupled, inductors. However, such a condition, which is of practical engineering values, is not well established in converters utilizing coupled inductors.

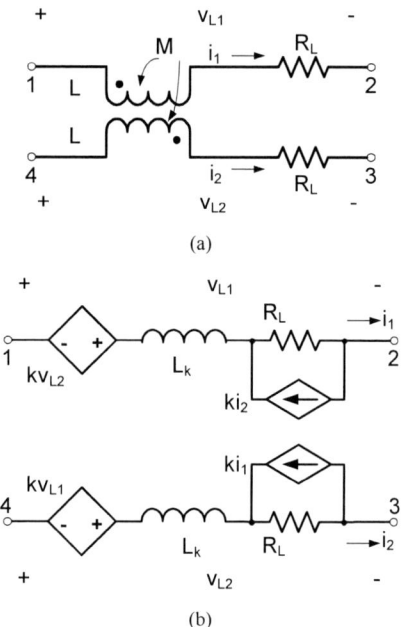

(a)

(b)

Figure 2. (a) Two inversely coupled inductor windings with series resistors, and (b) its new equivalent circuit model.

Fig. 3(a) shows the general lossless inductor DCR current sensing circuit in a two-phase interleaved buck converter with discrete inductors. Time-domain waveforms of the phase 1 inductor current, $i_1(t)$, and the sensed voltage across the capacitor, $v_{c1}(t)$, are given in Fig. 3(b). It is well-known that with proper design, i.e. when the time constant matches between the inductor and the current sensing r, C network, or $rC=L/R_L$, the inductor current waveform can be extracted from the voltage waveform across the sensing capacitor, i.e. $v_{c1}(t)= i_1(t)*R_L$.

A two-phase buck converter circuit using the lossless inductor DCR current sensing circuit with inductors coupled to each other is also shown in Fig. 4 (a), together with the time-domain inductor winding current waveforms [20], $i_1(t)$ and $i_2(t)$, as well as voltage waveforms across the sensing

978-1-4244-4782-4/10 $26.00 © 2010 IEEE

capacitors, $v_{C1}(t)$ and $v_{C2}(t)$, in Fig. 4(b). It is obvious that the sensed capacitor voltage is no longer a replica of the winding current in coupled inductors, no matter how the time constants with the inductor and the r, C network are selected, as the frequency of inductor current is twice the frequency of the voltages across the sensing capacitors.

In order to extend the lossless inductor DCR current sensing technique to applications with coupled inductors, the following analysis is carried out in s-domain. According to (2),

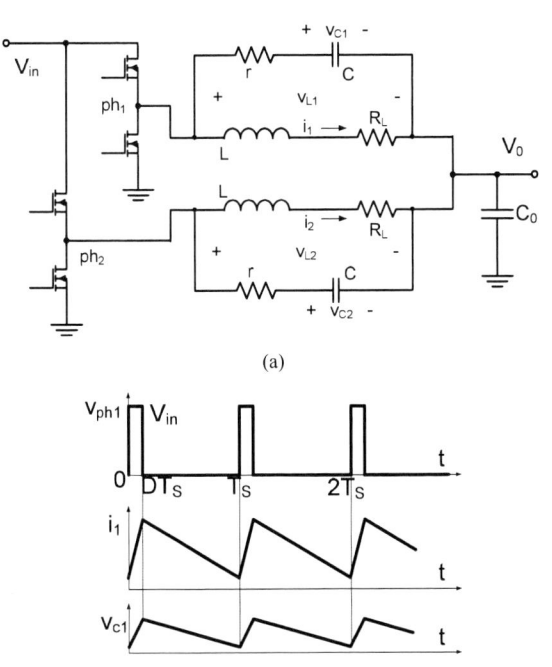

Figure 3. Lossless inductor current sensing with uncoupled inductors: (a) circuit diagram; (b) inductor current and capacitor voltage waveforms.

$$V_{L1}(s) = sL_k I_1(s) - kV_{L2}(s) + R_L\left(I_1(s) + kI_2(s)\right) \\ V_{L2}(s) = -kV_{L1}(s) + sL_k I_2(s) + R_L\left(kI_1(s) + I_2(s)\right), \quad (3)$$

and from Fig. 4(a),

$$V_{C1}(s) = \frac{V_{L1}(s)}{1+srC}, \quad V_{C2}(s) = \frac{V_{L2}(s)}{1+srC}. \quad (4)$$

Hence, the sum of the voltages across the sensing

capacitors can be obtained as

$$V_{C1}(s) + V_{C2}(s) = R_L \frac{I_1(s)+I_2(s)}{1+srC}\left(1 + s\frac{L_k}{R_L(1+k)}\right). \quad (5)$$

In the equation above, if r and C are selected such that

$$rC = \frac{L_k}{R_L(1+k)} = \frac{(1-k)L}{R_L}, \quad (6)$$

then (5) becomes

$$V_{C1}(s) + V_{C2}(s) = R_L\left(I_1(s) + I_2(s)\right), \quad (7)$$

or,

$$v_{C1}(t) + v_{C2}(t) = R_L\left(i_1(t) + i_2(t)\right). \quad (8)$$

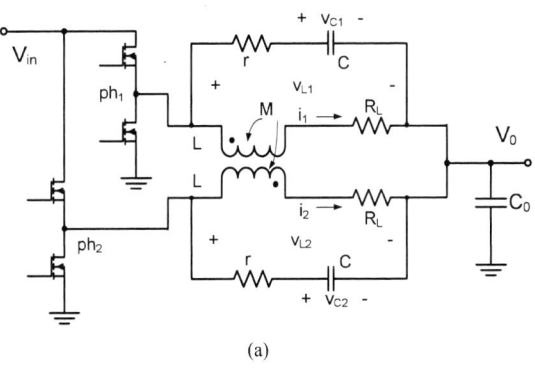

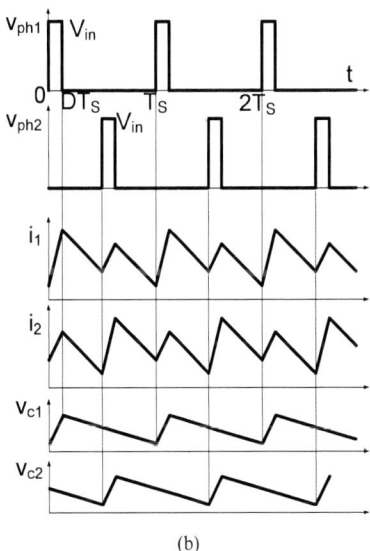

Figure 4. Lossless inductor current sensing with coupled inductors: (a) circuit diagram; (b) inductor current and capacitor voltage waveforms.

Equation (6) is the new time constant matching condition for DCR based current sensing in coupled inductor converters, which differs from that in uncoupled inductor converters. It is apparent that the coupling coefficient has a significant impact on the selection of the r and C values. (7) and (8) show that once (6) is satisfied, the sum of the voltages across the two sensing capacitors is the exact replica of the sum of the currents in both windings.

The conclusion expressed in (7) and (8) above, was also discovered in [16, 19]. The difference is that the time constant matching condition obtained in (6) includes the effect of the coupling coefficient, which is usually not close to 1 in coupled inductors, and a clear definition of the leakage inductance given in [20] and (2), and directly measurable with no need to convert or partition.

It should be pointed out that even though the sensed capacitor voltage does not represent the individual inductor winding current in coupled inductors, the result obtained from (6), (7) and (8) is still of significant practical importance. In many applications, obtaining the total current information is critical to implement output voltage positioning, or load-line, and over-current protection.

3.2 Dynamic Modeling of Coupled Inductor Converters

Figure 5 shows the equivalent circuit of a two-phase coupled inductor buck converter where the coupled inductors are replaced by the model given in Fig. 2(b). If we denote v_{ph1} and v_{ph2} as the switching node or phase node voltages (referred to ground), and v_{x1} and v_{x2} as the voltages (referred to ground) at the fictitious nodes x_1 and x_2, then from Fig. 5, we have,

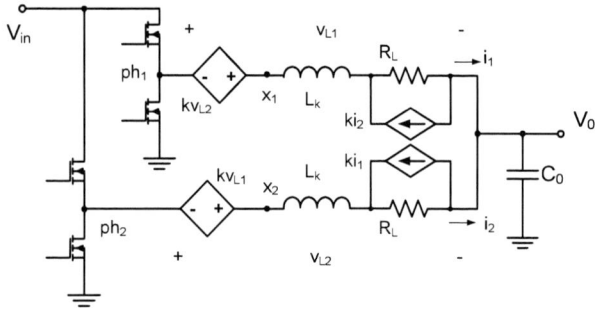

Figure 5. An equivalent circuit of a two-phase coupled inductor buck converter.

$$v_{L1} = v_{ph1} - V_0$$
$$v_{L2} = v_{ph2} - V_0 \qquad (9)$$

and,

$$v_{x1} = v_{ph1} + kv_{L2} = (v_{ph1} + kv_{ph2}) - kV_0$$
$$v_{x2} = v_{ph2} + kv_{L1} = (kv_{ph1} + v_{ph2}) - kV_0 \qquad (10)$$

In order to derive the dynamic model and the control-to-output small signal transfer function of the converter in Fig. 5, averaging modeling approach [23] is adopted in the following discussion. According to (10), the average phase node voltage, $\bar{v}_x$, can be obtained as

$$\bar{v}_x = \bar{v}_{x1} = \bar{v}_{x2} = d(1+k)\bar{v}_{in} - k\bar{v}_0 , \qquad (11)$$

where the symbol "‾" is used to denote the average of the corresponding variable. Fig. 6(a) shows the same equivalent circuit model of the converter in Fig. 5, while its equivalent averaging circuit is given in Fig. 6(b).

Assuming $d = D + \hat{d}(t)$, $\bar{v}_0 = V_0 + \hat{v}_0(t)$ and $\bar{v}_{in} = V_{in}$, where $\hat{d}(t)$ is the small-signal duty cycle perturbation and $\hat{v}_0(t)$ is the resulting perturbation on the output voltage, then the control-to-output transfer function of the power stage, $G_d(s)$, can be derived as

$$G_d(s) = \frac{V_0(s)}{\hat{d}(s)}$$
$$= \frac{R_0 V_{in}}{R_0 + R_C} \frac{(1+k)(1+sC_0 R_C)}{\frac{L_k}{2} C_0 (s^2 + 2\xi\omega_0 s + \omega_0^2)} , \qquad (12)$$

where

$$\omega_0 \triangleq \sqrt{(1+k)\frac{R_0 + R_L/2}{R_0 + R_C} \bigg/ \frac{L_k}{2} C_0} , \text{ and}$$

$$\xi \triangleq \frac{\frac{L_k}{2(R_0 + R_C)} + (1+k)(R_0 \,//\, R_C + \frac{R_L}{2})C_0}{\omega_0 L_k C_0}$$

are the resonant frequency and damping factor, respectively.

If $R_0 \gg R_C$ and $R_0 \gg R_L$, which is the case in most applications, (12) can be simplified as

$$G_d(s) = \frac{V_0(s)}{\hat{d}(s)}$$
$$= \frac{V_{in}(1+k)(1+sC_0 R_C)}{s^2 \frac{L_k}{2} C_0 + s\left(\frac{L_k}{2R_0} + (1+k)(R_C + \frac{R_L}{2})C_0\right) + (1+k)} , \qquad (13)$$

and the resonant frequency of the converter power stage transfer function can be simplified as

$$\omega_0 = \sqrt{\frac{2(1+k)}{L_k C_0}} = \sqrt{\frac{2}{(1-k)LC_0}} . \qquad (14)$$

For comparison purposes, the resonant frequency of a two-phase buck converter with the inductors uncoupled ($k=0$), ω_{dis}, is

$$\omega_{dis} = \sqrt{\frac{2}{L_{dis} C_0}} , \qquad (15)$$

where L_{dis} is the inductance of the uncoupled inductors.

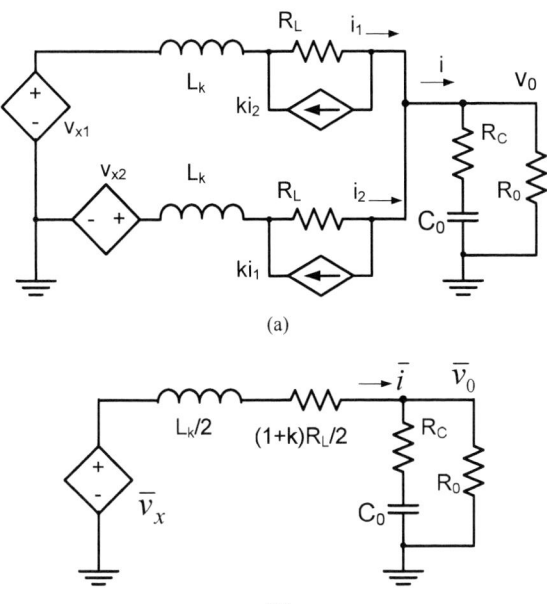

(a)

(b)

Figure 6. (a) A simplified converter circuit model with coupled inductors, and (b) its average circuit model.

In order to compare the multi-phase power converters with and without the inductors coupled, the following two different design cases can be considered: a) design based on equal magnetic component size, and b) design based on equal transient respond speed. The former implies that L_{dis} of the uncoupled inductor equals the self inductance of each winding in the coupled one, or $L_{dis}=L$, while the latter means L_{dis} equals the leakage inductance of the coupled inductor, or $L_{dis}=L_k$.

From (14) and (15), it is observed that, regardless of $L_{dis}=L_k$ or $L_{dis}=L$, there exists:

$$\omega_0 > \omega_{dis} \qquad (16)$$

for any $k>0$.

Equation (16) means that, due to coupling, the resonant frequency in converters with coupled inductors is always greater than that in converters with uncoupled inductors, regardless of $L_{dis}=L_k$ or $L_{dis}=L$. This conclusion implies that it is possible to design a control loop with higher crossover frequency, or bandwidth, in a regulator utilizing coupled inductors.

IV. SIMULATION AND EXPERIMENTAL VERIFICATION

The DCR based total current sensing scheme in coupled inductor converters as discussed in the previous section was verified through PSpice simulation. The circuit parameters used in the simulation, referred to Fig. 4(a), are listed as follows:

Input voltage: V_{in}=12 V
Output voltage: V_o=1.2 V

Output current: I = 20 A
Self inductance: L=1 µH
Coupling coefficient: k=0.6
Leakage inductance: L_k=0.64 µH
Inductor DCR: R_L=1 mΩ
Sensing network parameter: r =4 kΩ, C=0.1 µF
Per-phase switching frequency: f_s = 250 kHz

In this case, the sensing time-constant, rC=0.4 mS, was chosen according to (6) to match the time constant of the coupled inductor, $L_k/R_L(1+k)$=0.4 mS. The simulated results of the individual phase currents, i_1 and i_2, and summed phase current, i, sensing capacitor voltages, v_{C1} and v_{C2}, and their sum, v_C, are shown in Fig. 7(a) in steady-state operation, and in Fig. 7(b) during load transient, respectively. It is obvious that the sum of the voltages on the individual sensing capacitors replicates the summed phase currents exactly in both situations, and the scaling factor is R_L=1mΩ.

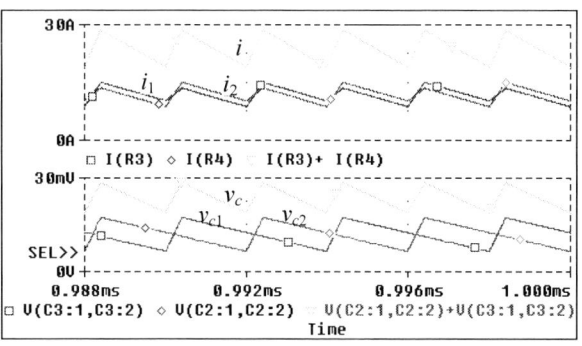

(a)

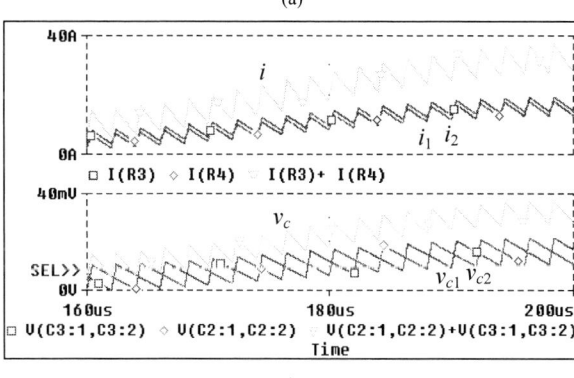

(b)

Figure 7. Simulated results of DCR based total current sensing scheme: (a) Steady-state; (b) during transients.

The same test board as used in [20] was used to measure the power converter control-to-output transfer function. The controller used on the test board shown in Fig. 8 is ISL6266, which utilizes the so-called Robust Ripple Regulator (R^3) modulator. According to the controller manufacturer [24], the small signal transfer function, or modulation gain, of the modulator, $G_m(s)$, is given by,

$$G_m(s) = \frac{\hat{d}(s)}{V_{comp}(s)} = \frac{2}{\dfrac{1.5 \times 10^5 (V_{in} - V_0)}{f_s} + \dfrac{3.2 V_{in} \times 10^5}{1.067s + 10^5}}, \quad (14)$$

where $V_{comp}(s)$ is the error compensator output and f_s is the per-phase switching frequency of the converter.

The coupled inductor is LC1740-R30R09A from NEC/Tokin, which specifies a typical self inductance of 310 nH. The measured parameters (with a short external wire connected to one terminal of each winding to fit a current probe) are L=353.5 nH, k=0.622, and L_k=216.7 nH.

Figure 8. Picture of the two-phase coupled inductor regulator test board.

Fig. 9 shows the measured Bode-plot of the transfer function from $V_{comp}(s)$ to $V_0(s)$. The magnitude and phase vs. frequency are compared with the model prediction, $G_m(s)G_d(s)$, which is also plotted in the same figure. The following measured parameters and operation conditions were used in the theoretical calculation and experimental measurement:

Input voltage: V_{in}=12 V
Output voltage: V_0=1.15 V

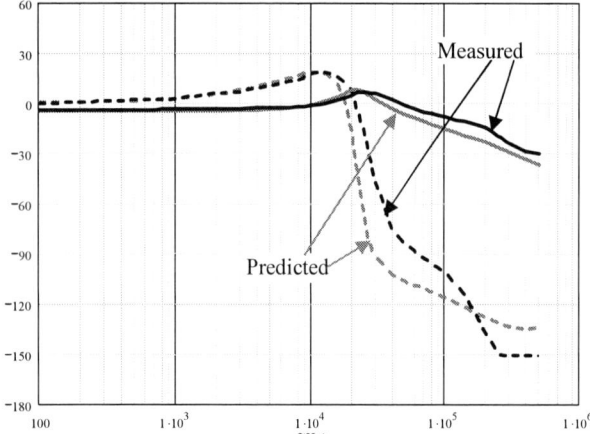

Figure 9. Measured and predicted Bode-plots of the transfer function $G_m(s)G_d(s)$

Output current: I = 20 A
Inductor DCR: R_L=0.4 mΩ
Output capacitors: C_0=36x22 μF (MLCC), R_c=0.2mΩ
Measured equivalent parasitic loop resistance: 6.04 mΩ
Per-phase switching frequency: f_s=470 kHz

Based on the information above, the load resistance R_0=57.5 mΩ. The test board uses all ceramic capacitors in parallel as the output capacitor. The equivalent ESR R_C is very low. Since $R_0 \gg R_C$ and $R_0 \gg R_L$, (13) was adopted in the calculation of the Bode-plot in Fig. 9.

It can be seen in Fig. 9 that the low-frequency gain and phase between the predicted and measured results match well, while the measured resonant frequency, ω_0, is slightly higher than predicted, which is most probably the outcome of the capacitance reduction effect under output bias voltage. The actual capacitance of the high-value X5R rated MLCC output capacitors is usually reduced by at least 25% under dc bias voltage. With the inclusion of the equivalent parasitic on-board loop resistance in the model, which was actually measured with the injection of a current source, the resonant peak also matches well. Beyond ω_0, the predicted and measured results follow the same general trend, but the measured results roll off at a higher frequency, which is partly due to the same capacitance reduction phenomenon and could be partly attributed to the inaccuracy involved in the transfer function of the complex modulator used in the controller.

Fig. 10 also shows the Bode-plots of the converter power stage control-to-output transfer functions, $G_d(s)$, with and without the inductors coupled using the same parameters given above. The gain peaks at ~21.3 kHz with coupled inductors. With discrete inductors, it peaks at ~16.1 kHz when $L_{dis}=L_k$=216.7 nH and at ~12.5 kHz when $L_{dis}=L$=353.5 nH, both of which is lower than that with the coupled one. It confirms the prediction made in (16).

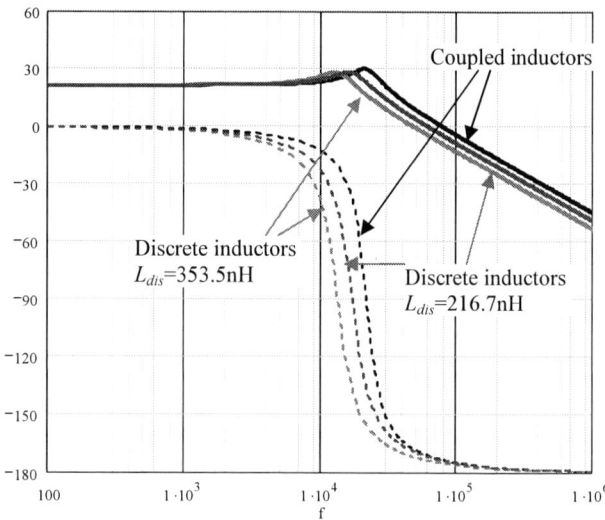

Figure 10. Bode-plots of the control-to-output transfer functions of the power stage with and without inductors coupled.

V. CONCLUSIONS

In this paper, the previously developed symmetrical coupled inductor model is first extended to include the DCR. The extended model is then used to analyze the influence of the coupling coefficient on the DCR based current sensing schemes popularly used in multi-phase switching regulators. It is found that the well-known time-constant matching condition in uncoupled inductor converters should be modified to include the effect of k when inductors are coupled, and only the summation of the sensed phase currents is valid. Finally, the proposed coupled inductor model is also used to derive the small-signal control-to-output model of the converters incorporating coupled inductors. The derivation shows that the resonant frequency of the power stage is effectively increased through coupling, indicating in practical applications a higher bandwidth loop design is achievable in regulators incorporating coupled inductors.

REFERENCES

[1] S. Cuk, and R.D. Middlebrook, "Coupled-inductor and other extensions of a new optimum topology switching dc-dc converter," *Advances in switched-Mode Power Conversion, Vols. I and II*, pp. 331-347.

[2] S. Cuk, "Switched dc-to-dc converter with zero input and output current ripple," *Proc. of IEEE IAS Annual Meeting 1978*, pp. 1131-1146.

[3] G. W. Wester, "An improved push-pull voltage fed converter using a tapped output-filter inductor," *Record of IEEE PESC 1983*, pp. 366-376.

[4] S Cuk, and Z. Zhang, "Coupled-inductor analysis and design," *Record of IEEE PESC 1986*, pp. 655-665.

[5] W. Chen, G. Hua, D. Sable, and F.C. Lee, "Design of high efficiency, low profile, low voltage converter with integrated magnetics," *Proc. of IEEE APEC 1997*, pp. 911-917.

[6] D. Maksimovic, R. Erickson, and C. Griesbach, "Modeling of cross-regulation in converters containing coupled inductors," *Proc. of IEEE APEC 1998*, pp. 350-356.

[7] P.-L. Wong, Q. Wu, P. Xu, B. Yang, and F.C. Lee, "Investigating coupling inductor in interleaving QSW VRM," *Proc. of IEEE APEC 2000*, pp. 973-978.

[8] P.-L. Wong, F.C. Lee, X. Jia, and J.D. van Wyk, "A novel modeling concept for multi-coupling core structures," *Proc. of IEEE APEC 2001*, pp. 102-108.

[9] P.-L. Wong, P. Xu, B. Yang, and F.C. Lee, "Performance improvements of interleaving VRMs with coupling inductors," *IEEE Trans. on Power Electronics*, vol. 16, no. 4, July 2001, pp. 499-507.

[10] J. Li, C.R. Sullivan, and A Schultz, "Coupled-inductor design optimization for fast-response low-voltage dc-dc converters," *Proc. of IEEE APEC 2002*, pp. 817-823.

[11] J. Li, A. Stratakos, A. Schultz, and C.R. Sullivan, "Using coupled inductors to enhance transient performance of multi-phase buck converters," *Proc. of IEEE APEC 2004*, pp. 1289-1293.

[12] A.M. Schultz, and C.R. Sullivan, "Voltage converter with coupled inductive windings and associated methods," *U.S. Patent 6,362,986*, Mar. 26, 2002, Volterra Semiconductor Corp.

[13] A.V. Ledenev, G.G. Gurov, and R.M. Porter, "Multiple power converter system using combining transformers," *U.S. Patent 6,545,450 B1*, Apr. 8, 2003, Advanced Energy Industries, Inc.

[14] W. Wu, N. Lee, and G. Schuellein, "Multi-phase buck converter design with two-phase coupled inductors," *Proc. of IEEE APEC 2006*, pp. 487-49.

[15] J. Gallagher, "Coupled inductors improve multiphase buck efficiency," *Power Electronics Technology Magazine*, Jan. 2006, pp. 36-42.

[16] Y. Dong, M. Xu, and F.C. Lee, "DCR current sensing method for achieving adaptive voltage positioning (AVP) in voltage regulators with coupled inductors," *Record of IEEE PESC 2006*, pp. 1-7.

[17] Y. Dong, F.C. Lee, and M. Xu, "Evaluation of coupled inductor voltage regulators," *Proc. of IEEE APEC 2008*, pp. 831-837.

[18] Y. Dong, Y. Yang, F.C. Lee, and M. Xu, "The short winding path coupled inductor voltage regulators," *Proc. of IEEE APEC 2008*, pp. 1446-1452.

[19] S. Xiao, W. Qiu, T. Wu, and I. Batarseh, "Investigating effects of magnetizing inductance on coupled-inductor voltage regulators," *Proc. of IEEE APEC 2008*, pp. 1569-1574.

[20] G. Zhu, B. McDonald, and K. Wang, "Modeling and analysis of coupled inductors in power converters," *Proc. of IEEE APEC 2009*, pp. 83-89.

[21] D. Goder, "System to protect switch mode dc/dc converters against overload current," *U.S. Patent 6,127,814*, Oct. 3, 2000, Switch Power, Inc.

[22] X. Zhou, P. Xu, and F.C. Lee, "A high power density, high efficiency and fast transient voltage regulator module with a novel current sensing and current sharing technique," *Proc. of IEEE APEC 1999*, pp. 289 - 294.

[23] V. Vorperian, "Simplified analysis of PWM converters using the model of the PWM switch, part I: continuous current mode," *IEEE Trans. on Aerospace and Electronic Systems*, vol. 26, no. 3, May 1990, pp. 490-496.

[24] Private communications with engineering staff of Intersil Corp., Nov. 2009, available upon request.

978-1-4244-4782-4/10 $26.00 © 2010 IEEE

Design Procedure for High Frequency Operation of the Modified Series Resonant APWM Converter with Improved Efficiency and Reduced Size

Darryl J. Tschirhart, *Grad. Student Member, IEEE*, and Praveen K. Jain, *Fellow, IEEE*
Centre for Power Electronics Research (ePOWER)
Dept. of Electrical & Computer Engineering, Queen's University
Kingston, Ontario, Canada
darryl.tschirhart@ieee.org, praveen.jain@queensu.ca

Abstract— In this paper, a generalized analysis for the auxiliary network in a modified series resonant asymmetrical pulse-width-modulated (APWM) converter is performed to produce a design procedure that ensures ZVS is achieved for any converter design. New equations that correctly predict the magnitude of auxiliary current are obtained by accounting for the trapezoidal nature of the waveforms associated with high frequency operation, and the dead time between the switches in the half bridge. A design example of a 48V/1.2V, 25A converter operating at 1MHz is chosen to highlight the validity of the proposed design and that superior results can be achieved if the resonant tank is designed in tandem with the auxiliary network. Experimental results verify that ZVS is achieved, and that the proposed design reduces the auxiliary inductor by close to a factor of 3.

I. INTRODUCTION

In order to physically shrink the size of power supplies, high frequency operation is necessary to reduce the size of reactive components. Further, for MOSFET-based converters, zero voltage switching (ZVS) is required for efficient operation to reduce or eliminate heat sink requirements. Resonant converters have been well documented to meet both of these requirements [1]-[4]. The series resonant asymmetric pulse-width-modulated converter (SR-APWM) is a half-bridge load resonant topology that achieves ZVS while operating at constant frequency [4]. Its resonant inductor can be formed entirely by the transformer leakage inductance, and its rectifiers achieve zero current switching (ZCS). Constant frequency operation simplifies EMC and magnetic component design. All these benefits make it an ideal candidate for low voltage, low power supplies. In addition to the aforementioned merits, the output filter of the SR-APWM is purely capacitive, thereby enabling it to respond quickly to load transients. Thus, using the SR-APWM as a 48V voltage regulator (VR) [5], would not only improve system efficiency by reducing the load on the intermediate bus; but also provide improved transient performance over the buck converters currently powering microprocessors, digital signal processors

(DSP), and field programmable gate arrays (FPGA). However, with the reduction of supply voltages required by high speed digital circuits, a high transformer turns ratio is required for the SR-APWM. This reduces the reflected load current, which potentially leads to a loss of ZVS, even if external capacitors are not added across the switches.

To ensure ZVS is achieved, a passive auxiliary network originally applied to the PSM full-bridge [6] was applied to the SR-APWM in order to increase its input voltage range [7]. The design procedure presented in [7] presents a trade-off between inductor size and conduction loss. However, it does so under the assumption that the dead-time between the gating signals is much less than the switching period; and that the output capacitance of the switches is negligible. As a result, there may be some circuit configurations or operating frequencies where the analysis does not hold. This occurs most notably at higher frequencies (in the megahertz range), and in converters that have a high turns ratio. At high frequency the dead-time required to achieve ZVS constitutes a greater percentage of the switching period, and the charge associated with the output capacitance of the switches cannot be neglected.

With present-day high-speed digital devices requiring low supply voltages, and the desire to switch at high frequency, the modified SR-APWM used in 48V VR applications is prone to the two issues discussed above. It is therefore imperative to ensure generalized equations are available to guarantee proper design. As will be shown, improved results are obtained by designing the auxiliary network and resonant tank concurrently.

II. STEADY-STATE OPERATION

A. Resonant Tank

The modified series resonant APWM converter is shown in Fig. 1, with the key waveforms in Fig. 2. The dead times between the gating signals of the two switches are exaggerated for clarity. Referring to the waveforms on the top axis of

---

Funding provided by Ontario Centres of Excellence, Ontario Research Funds, and Natural Sciences and Engineering Research Council of Canada

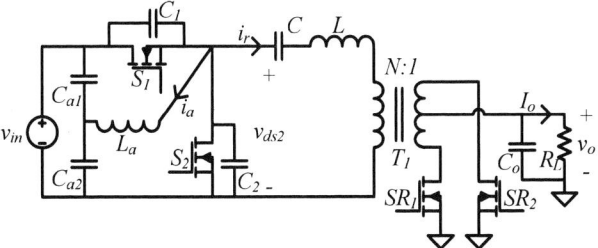

Fig. 1: Modified series resonant APWM schematic

Fig. 2, the resonant tank is excited by a unipolar trapezoidal wave created by the chopper formed by complementary switches S_1 and S_2. The capacitor C resonates with L, and also acts as a dc block to prevent transformer saturation. The sinusoidal resonant current is stepped up by the transformer, rectified by synchronous rectifiers SR_1 and SR_2, and then filtered by C_o. Line regulation is achieved by varying the duty cycle (D) of S_1. Load regulation can also be achieved through varying the primary-side duty cycle or through secondary-side control. The resonant current is defined by (1) where V_{sn} is the Fourier series expansion of the chopper voltage and ω_0 is the radian operating frequency. The terms related to the chopper voltage are given by (2) and (3), while those related to the resonant tank are given by (5) and (6).

$$i_r(t) = \frac{V_{sn}}{|Z_{eq}|}\sin(n\omega_0 t + \theta_n - \phi) \qquad (1)$$

$$V_{sn} = \sum_{n=1}^{\infty} \frac{\sqrt{2}V_{in}\sqrt{1-\cos(2n\pi D_{ch})}}{n\pi} \qquad (2)$$

$$\theta_n = \tan^{-1}\left[\frac{\sin(2n\pi D_{ch})}{1-\cos(2n\pi D_{ch})}\right] \qquad (3)$$

As switching frequency is increased, the time necessary to allow the switches to commutate and achieve ZVS becomes a greater portion of the switching cycle; thereby making the assumption that the drain-source voltage is a vertical edge inaccurate. The new definition of duty cycle accounting for finite charge time of the drain-source capacitance is given by (4) where T is the switching period, and $T_{ch1} = t_2 - t_0$, and $T_{ch2} = t_5 - t_3$ are the snubber capacitor charge times.

$$D_{ch} = \frac{t_{on}}{T} + \frac{T_{ch1} + T_{ch2}}{2T} \qquad (4)$$

$$|Z_{eq}| = R_{ac}\sqrt{1 + Q^2\left(\omega - \frac{1}{\omega}\right)^2} \qquad (5)$$

$$\phi = \angle Z_{eq} = \tan^{-1}\left[Q\left(\omega - \frac{1}{\omega}\right)\right] \qquad (6)$$

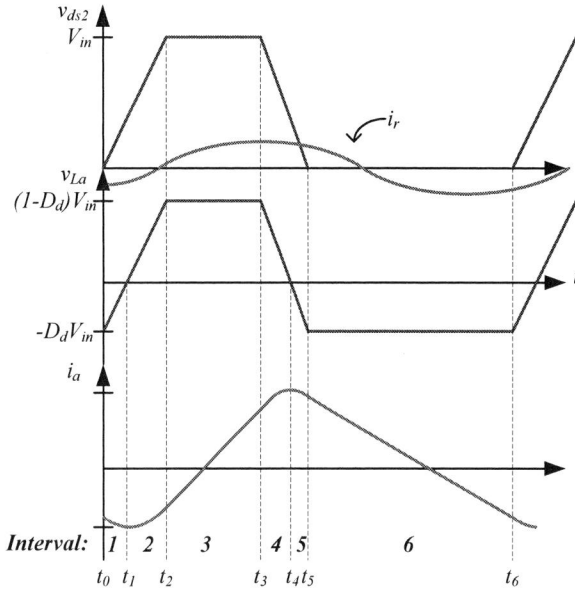

Fig. 2: Operating waveforms of the modified SR-APWM converter

In (5) and (6), R_{ac} is the equivalent ac resistance of the load and rectifiers referred to the transformer primary; $\omega = \omega_0/\omega_r$; and $Q = \omega_r L/R_{ac}$, where $\omega_r = 1/\sqrt{LC}$.

B. Auxiliary Circuit

The auxiliary circuit experiences 6 operating intervals, which are described below.

1) Interval 1 ($t_0 \leq t < t_1$):
This interval begins with switch S_2 turning off, and its voltage slowly rising to achieve zero voltage turn-off. The rate of rise is limited by the snubber capacitance across the two switches ($C_1 + C_2$), which is composed of the output capacitance plus any additional capacitors. The current through the auxiliary inductor decreases, and the interval ends when the voltage across the auxiliary inductor is zero, and $i_a = I_{pk-}$ (7).

$$I_{pk-} = \frac{-D_{ch}V_{in}}{2L_a}\left[(1-D_{ch})(T - T_{ch1})\right] \qquad (7)$$

2) Interval 2 ($t_1 \leq t < t_2$):
The voltage across S_2 continues to rise and the auxiliary inductor current rises from its negative peak value. The interval ends when the voltage across S_2 equals V_{in}.

3) Interval 3 ($t_2 \leq t < t_3$):
S_1 is turned on under zero voltage to begin this interval. The current through L_a ramps up linearly until S_1 is turned off at the end of the interval.

4) Interval 4 ($t_3 < t < t_4$):
With both switches off, the voltage across S_2 begins to fall; meaning the voltage across S_1 begins to rise. As with switch 2, the rate of rise is limited by the snubber capacitance so zero-voltage turn-off is achieved. The auxiliary current rises

and reaches its positive peak value I_{pk+} (8) at the end of the interval.

$$I_{pk+} = \frac{D_{ch}V_{in}}{2L_a}\left[(1-D_{ch})(T-T_{ch2})\right] \qquad (8)$$

5) Interval 5 ($t_4 \le t < t_5$):

In this interval, the voltage across the auxiliary inductor falls from zero, and the auxiliary current falls from its positive peak. The interval ends when the drain-source voltage of S_2 reaches zero. At this instant, the voltage across S_1 is equal to the input voltage.

6) Interval 6 ($t_5 \le t < t_6$):

Switch 2 is turned on under zero voltage to begin this interval. With S_2 on, the auxiliary inductor current ramps down linearly. This interval ends when S_2 is turned off, and the cycle is repeated.

III. DESIGN PROCEDURE

A. Resonant Tank

In low power, high frequency applications, the design of the resonant tank is typically a tradeoff between conduction loss and component stress, constrained by the leakage inductance of the transformer. High Q implies high stress, and increased ω leads to increased circulating current. In Fig. 3, the turn-off currents of the primary-side switches are shown for a given set of tank parameters. For S_2 to achieve ZVS, the turn-off current of S_1 has to be positive. For S_1 to achieve ZVS, the turn-off current of S_2 must be negative. As shown in Fig. 3, S_2 always has the potential to achieve ZVS, while S_1 loses it at $D_{ch} = 0.44$. Since D_{ch} regulates against line variations, ZVS of S_1 is lost at roughly 44V. It is therefore up to the auxiliary circuit to provide sufficient energy to discharge the output capacitance of the switches. Increasing ω has the effect of shifting the curves away from each other; which complicates the optimization of the auxiliary network.

B. Auxiliary Inductor and Snubber Capacitors

The key equation governing the achievement of ZVS is given by (9); which states that in a given charge interval, the charge held by the snubber capacitors must be less than the charge removed by the auxiliary inductor and resonant tank. For some operating points the resonant current aids in discharging the snubber capacitors, while at other operating points it works against the auxiliary network.

$$C_{sb}V_{in,max} \le \int_{T_{ch}} i_a dt + \int_{T_{ch}} i_r dt \qquad (9)$$

Plots of the results of this equation are shown in Fig. 4 and Fig. 5, for T_{ch1} and T_{ch2}, respectively at the maximum input voltage. While the charge itself is independent of V_{in}, the contribution of the resonant current is dependent on the duty cycle of the drive train; which is a function of the input voltage. The influence of the turn-off currents is demonstrated in the figures by the reduced charge time experienced during T_{ch2} for a given inductor/capacitor combination.

To achieve zero voltage turn-off, the rate of rise of the drain-source voltage should be at least twice as long as the fall time of the current in the switch. Thus, Fig. 5 provides the minimum component values for true ZVS. Since T_{ch1} is greater than T_{ch2} for a given configuration, it follows that both switches will achieve lossless transitions at both turn-on and turn-off.

As the input voltage and/or load is reduced, the turn-off current of S_1 will reduce. Thus, T_{ch2} will increase from the full-load, high-line level; thereby further reducing the turn-off loss of the switch. Conversely, increased duty cycle will cause the resonant current to work with the auxiliary current during T_{ch1}. This will reduce the charge time from the high-line value. However, the worst-case will be at low-line, full-load in which the duty cycle is maximum at ideally 50%. This leads to symmetric operation and $T_{ch1} = T_{ch2}$. From the above discussion on the second charge interval, it can be concluded that all operating conditions can achieve true ZVS if the presented design procedure is followed.

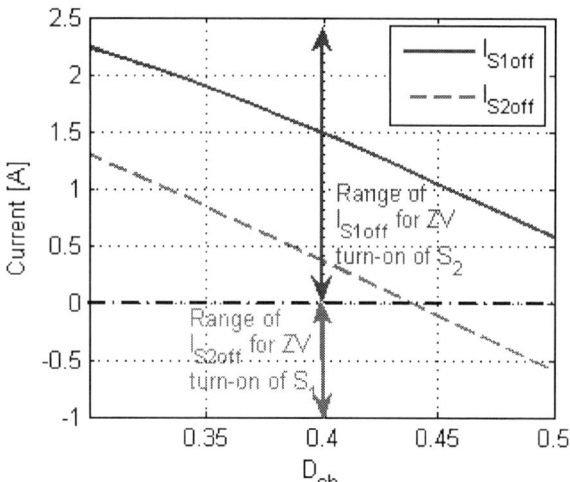

Fig. 3: Turn-off currents at full-load with input voltage variation (ω= 1.05, Q = 2, Output: 30W at 1.2V)

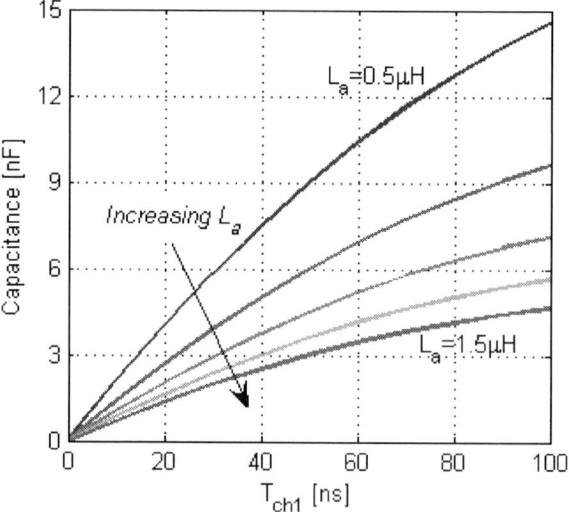

Fig. 4: Auxiliary inductor and snubber capacitance at maximum input voltage for ZVS during T_{ch1}

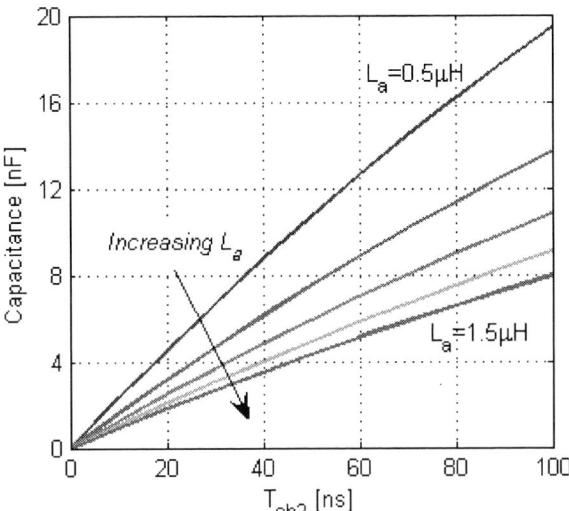

Fig. 5: Auxiliary inductor and snubber capacitance at maximum input voltage for ZVS during T_{ch2}

In addition to achieving ZVS, conduction loss, size, and cost are important factors to consider. Therefore, the auxiliary inductor and snubber capacitor selection should strike a compromise of these added issues. While lower inductance values reduce the size of the magnetic component, they increase the required snubber capacitance, and increase conduction loss. The latter requires switches with higher current ratings, and potentially large heat sinks; thereby negating any savings. Larger auxiliary inductors reduce conduction loss and snubber capacitor requirements at the expense of magnetic component size.

C. Design Example

To highlight the validity of the proposed design procedure for the auxiliary network, a modified series resonant APWM converter is to be designed to meet the specifications in Table I.

TABLE I.   CONVERTER SPECIFICATIONS FOR DESIGN EXAMPLE

Parameter	Value
Input Voltage [V]	43-53
Output Voltage [V]	1.2
Output Power [W]	30
Switching Frequency [MHz]	1

With such a large conversion ratio, a 15:1 transformer turns ratio is required. The resonant parameters $Q = 1.8$ and $\omega = 1.05$ were selected, corresponding to tank values $C = 9.9nF$, and $L = 2.65\mu H$. The tank is selected with the design of the auxiliary network in mind. Instead of increasing circulating current to maximize the range of operating points ZVS is achieved, the tank is designed to maximize power transfer. The auxiliary network is then solely responsible for ensuring soft-switching. This approach improves the working efficiency in the following two ways. First, the resonant tank current is minimized; thereby minimizing the associated conduction loss. Second, the variation of switch turn-off

current between the two switches and the extreme input voltages is reduced. This lowers the required peak auxiliary current. This not only reduces conduction loss associated with the auxiliary circuit, but also simplifies the design to ensure zero voltage turn-off is always achieved for both switches.

Selection of the auxiliary network is dependent on the capabilities of the switches. It is assumed that the current fall time of the switches is 20ns. So for ZVS, the charge time of the drain-source voltage must be greater than or equal to 40ns. The auxiliary inductor is selected to be $1\mu H$, with an extra 5nF of snubber capacitance added across the switches.

IV. RESULTS

A. Simulation Results

The converter designed in the previous section has been simulated in the spice simulator SIMetrix. The full-load waveforms at the two extreme input voltages are shown in Fig. 6 and Fig. 7. The figures have been annotated to show that both switches achieve zero voltage turn-off and turn-on. As discussed in the previous section, the two charge intervals at low-line are almost the same due to the nearly symmetric ac component of the drive voltage at this operating point. At high-line T_{ch2} is slightly greater than twice the fall time of the switch; thus ensuring zero-voltage turn-off of S_1 even at this extreme operating point.

B. Experimental Results

An experimental prototype of a modified series resonant APWM converter has been built to meet the specifications of Table I. The components used are given in Table II.

TABLE II.   LIST OF RELEVANT COMPONENTS USED

Component	Part number and description
S_1, S_2	Si7454DP (Vishay 100V, 7.8A MOSFET, current fall-time = 20ns)
$C_1 + C_2$	2x2.2nF + 1x 470pF (100V ceramic capacitors in 0805 package)
L_a	SER1360-102 (Coilcraft $1\mu H$ SMT inductor)

A comparison of the proposed design with the original is presented in Table III. The resonant inductor is formed by the transformer leakage, and therefore fixed. The previous design relied on the resonant current to aid in ZVS, and hence requires a larger auxiliary inductor to minimize conduction loss. This leads to the inclusion of turn-off loss for some operating points. With the proposed design, the auxiliary inductor is reduced by more than 60%. Moreover, zero voltage transitions are guaranteed at any operating point at both turn-on and turn-off.

TABLE III.   DESIGN COMPARISON

Parameter	Proposed Design	Design in [6]
ω, Q	1.05, 1.8	1.1, 1.6
L, C	$2.65\mu H$, 9.9nF	$2.65\mu H$, 11.6nF
L_a	$1\mu H$	$2.65\mu H$

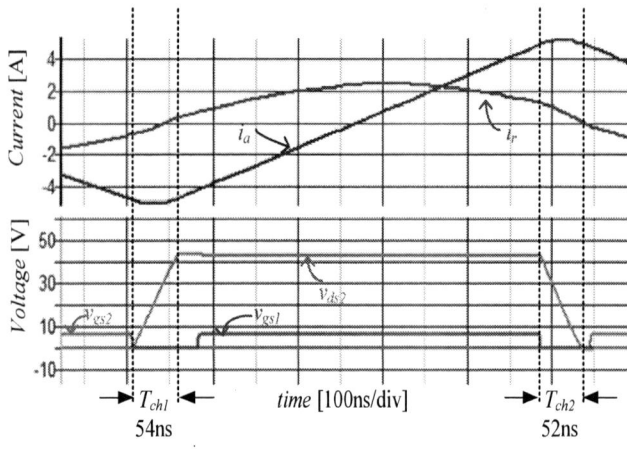

Fig. 6: Full-load simulation waveforms at 43V

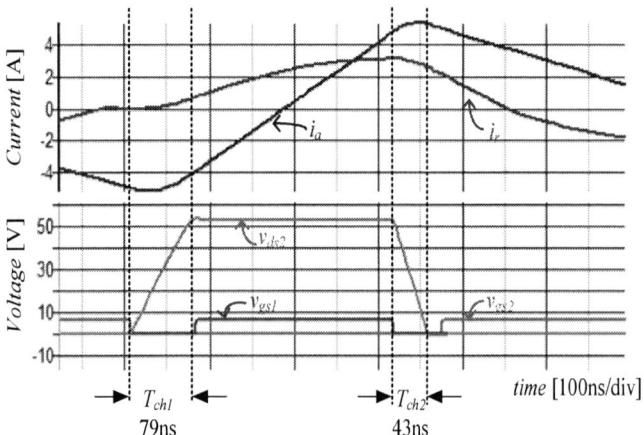

Fig. 7: Full-load simulation waveforms at 53V

Experimental waveforms are shown at no-load for the two extreme input voltages. ZVS is achieved for both cases.

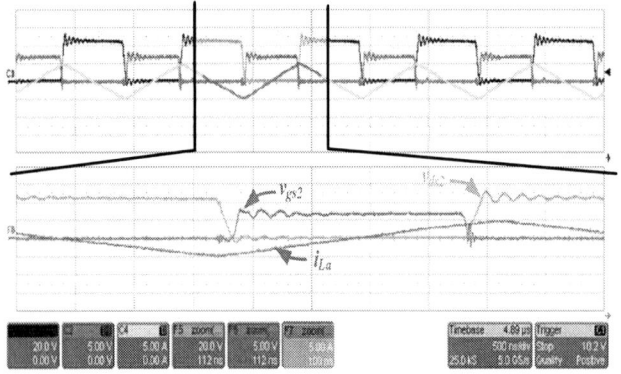

Fig. 8: Experimental waveforms at 43V (no-load)

Fig. 9: Experimental waveforms at 53V (no-load)

V. CONCLUSION

In this paper, a more thorough analysis and design procedure for the auxiliary network of a modified series resonant APWM converter has been presented. The proposed procedure achieves ZVS for all line and load conditions, and is applicable to any operating frequency, and any resonant tank configuration. It was shown that simultaneous design of the resonant tank with the auxiliary network yields optimal performance of the converter by operating closer to the resonant frequency to maximize power transfer capabilities; and allow ZVS to be achieved solely by the auxiliary network. This reduces the size of the auxiliary inductor, and ensures ZVS is achievable at every operating point. Experimental results of a 1MHz prototype were presented to prove the validity of the proposed design.

REFERENCES

[1] R.L. Steigerwald, "A Comparison of Half-Bridge Resonant Converter Topologies," IEEE Trans. on Power Electronics, vol. 3, pp. 174-182, April 1988.

[2] M.K. Kazimierczuk and C. Wu, "Frequency-Controlled Series-Resonant Converter with Synchronous Rectifier," IEEE Trans. Aerosp. Electron. Sys., vol. 33, no. 3, pp. 939-948, July 1997.

[3] M.Z. Youssef, and P.K. Jain, "Design and Performance of a Resonant LLC 48V Voltage Regulator Module with a Self-Sustained Oscillation Controller," in Proc. Applied Power Elec. Conf., Feb. 2007, pp. 141-147.

[4] P.K. Jain, A. St-Martin, and G. Edwards, "Asymmetrical Pulse-Width-Modulated Resonant DC/DC Converter Topologies," IEEE Trans. on Power Electronics, vol. 11, pp. 413-422, May 1996.

[5] M. Ye, P. Xu, B. Yang, and F.C. Lee, "Investigation of Topology Candidates for 48V VRM," in Proc. IEEE Applied Power Elec. Conf. and Expo., 2002, pp. 699-705.

[6] P.K. Jain, W. Kang, H. Soin, and Y. Xi, "Analysis and Design Considerations of a Load and Line Independent Zero Voltage Switching Full Bridge DC/DC Converter Topology," IEEE Trans. Power Electronics, vol. 17, no. 5, pp. 649-657, Sep. 2002.

[7] S. Mangat, M. Qiu, and P. Jain, "A Modified Asymmetrical Pulse-Width-Modulated Resonant DC/DC Converter Topology," IEEE Trans. Power Electronics, vol. 19, no. 1, pp. 104-111, Jan. 2004.

Expiremantal Results and Study of A Modified Adaptive Bus Voltage Controller

Jaber A. Abu Qahouq and Gautam Muralidhar

The University of Alabama
Department of Electrical and Computer Engineering
Tuscaloosa, Alabama 35487, USA

Abstract — **Two-stage power conversion is used in many applications. The determination of the bus voltage in two-stage converter is important to achieve high overall power conversion efficiency. Conventional two-stage converters utilize either a fixed DC bus voltage or a variable but predetermined DC bus voltage, which do not necessarily result in an optimum operation with optimum bus voltages under variable conditions. The paper presents a modified adaptive bus voltage controller for two-stage power converter. The controller adaptively converges to the optimum bus voltage that yields to the maximum power conversion efficiency under variable operating conditions. The modified adaptive two-stage bus voltage controller is evaluated using results obtained from a proof of concept experimental prototype.**

I. INTRODUCTION

Two-stage power conversion has been discussed and used in several applications, including the two-stage Voltage Regulator (VR) applications in communications and computing platforms [1-6]. The two-stage approach adaption may be motivated by different reasons such as the required large voltage step-down (or step-up in some applications) as a result of increasing the VR input voltage to accommodate the load current increase in order to achieve high total power efficiency and performance [1-6]. Some examples are the 8.7-19Vin to 0.8Vo in mobile or desktop CPU VR applications, 48Vin-1.5Vo in server CPU VR applications, 300~400Vin front–end dc-dc converters and 36~75Vin Telecom modules.

As shown in previous publications and research [1-6], there are some cases or applications where replacing a single-stage topology with a two-stage topology results in improved power efficiency and/or dynamic performance [1-2, 5, 6, 9, 10]. When the two-stage approach is used, the intermediate bus voltage (Vbus) between the two stages has an optimum value which needs to be selected for optimum efficiency (and performance). This Vbus value is either a predetermined constant value [1, 4] or a predetermined variable function that follows a predetermined set of Vbus values as a function of the load current for example [2]. In both of these methods, the first-stage and second-stage each has its own closed-loop voltage regulation controller, but in the first method the stable closed loop control design is simpler since the Vbus is fixed. A third two-stage method utilizes a first-stage with constant duty cycle open-loop control [3, 5], which is simpler but it may have less efficiency performance especially under large input voltage and load current variations.

Since these two-stage methods have a predetermined Vbus value(s), in many real applications they will result in improved efficiency within certain operating conditions range, but not across a wide range of operating conditions. This is especially true under wide input voltage range, under wide range of output voltage values such is those set by the VID (Voltage Identification) code in the CPU applications, under wide load range, and/or under varying power stage component characteristics and parasitics from the effect of temperature variations and aging for example. It can be shown that the optimum value of Vbus is very sensitive to operating conditions and if it is slightly off from what it should be, it may result in lower efficiency for the two-stage approach compared to the single-stage approach even if the initial design showed otherwise, just because the later operating conditions are different from the initial operating conditions assumed during the design stage. On the other hand, when there are several factors in the design (wide input voltage range, wide output voltage range, wide load range, and component characteristics variations), it is very difficult to obtain an optimum value or set of optimum values to use for Vbus for each combination of these variables.

Based on this, there is a motivation to look into an alternative control scheme that alleviates or solve the challenges discussed above in the Vbus design for two-stage converter approach. In this paper, based on the previous work [1-6, 9, 10], we assume that it is a given fact that two-stage approach is a better candidate than single-stage approach in some applications and design cases, and the focus will be on presenting and developing a new control scheme for two-stage topology.

In this paper, experimental prototype results and study is presented for a modified Adaptive Dynamic Bus Voltage (ADBV) controller of [14], which adaptively detects the optimum bus voltage values under variable operating conditions while maximizing the power conversion efficiency. The presented results of this paper are based on a modified ADBV control algorithm that is more suitable for practical experimental implementations. This paper also presents how different operating conditions of the converter affect the operation, accuracy and speed of the controller.

978-1-4244-4782-4/10 $26.00 © 2010 IEEE

Next section presents the modified adaptive bus voltage controller algorithm. Section III presents the closed loop experimental results obtained from a proof of concept experimental prototype. The conclusion is given in Section IV.

II. MODIFIED ADAPTIVE BUS VOLTAGE OPTIMIZATION CONTROLLER

Fig. 1 shows a block diagram of a two-stage topology approach with the proposed ADBV controller. The second-stage and its conventional control loop are designed to meet the stringent load requirements (ex. High switching frequency and high bandwidth). The first-stage proposed controller uses the input current information to adjust dynamically Vbus to a value that will result in maximum efficiency by minimizing the input current/power as represented in Fig. 2. Vbus is adjusted by perturbing (adjusting) the reference voltage of the first stage Vref-bus which will effectively changes the duty-cycle Dbus of the first stage or it can be adjusted by directly varying Dbus. The later result in eliminating the need for voltage regulation closed loop control and hence it is simpler.

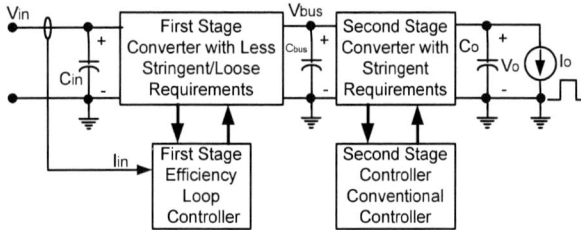

Fig. 1: Two-stage approach

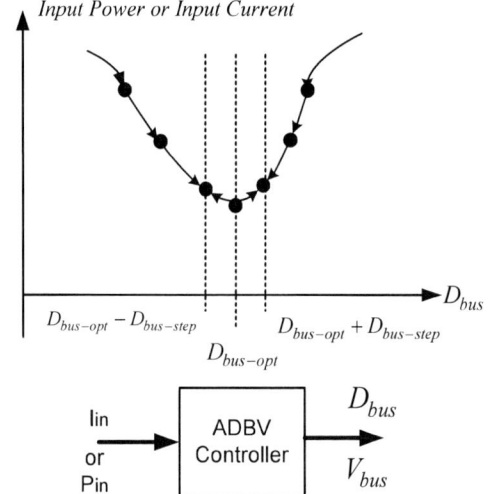

Fig. 2: ADBV operation concept representation

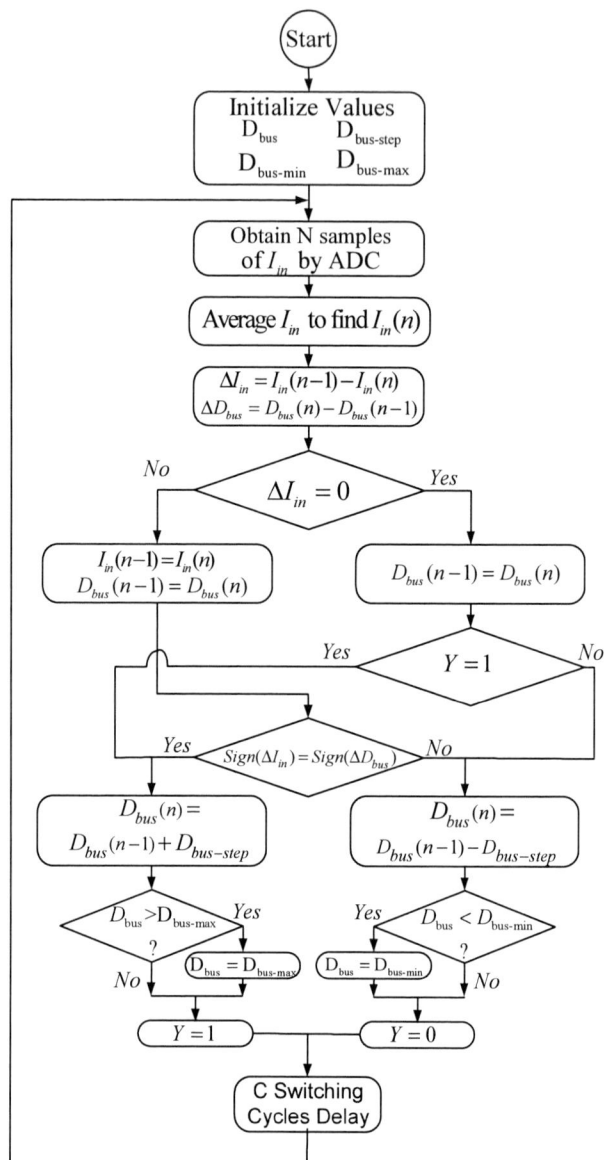

Fig. 3: The Modified Experimental ADBV main algorithm

Perturbing Vbus or Dbus until obtaining the minimum Iin (Iin-min) can be implemented by different filters or algorithms. Fig. 3 shows a modified algorithm (compared to the one introduced in [14]) that is more suitable for experimental implementation, which is a main objective of this paper. The present input current and Dbus are stored before incrementing or decrementing Dbus by a given step size. After the increment/decrement takes place, the controller waits for a given time delay (to ensure new steady-state operation) and then stores the new input current and Dbus values. Based on the input current change and Vbus (or Dbus) change signs, the next action to increment or decrement Dbus is decided such that the convergence to the input current minima is achieved.

Fig. 3 algorithm has a new feature that is related to the condition when the change in the input current is zero as a result in incrementing or decrementing Dbus. This condition may occur for different reasons that include, but not limited to: (1) The step size is not sufficient to create a change in the input current under loading conditions or input voltage conditions, (2) The input current ADC resolution is not sufficient to detect the change in the input current under certain operating conditions and (3) the noise and ripple effects resulted in not measuring a change in the input current. The way the modified algorithm takes care of this issue, as can be observed from Fig. 3, is by enforcing the following condition: If there is no change in the input current as a result of perturbing (incrementing or decrementing) the Dbus in an algorithm iteration, the controller will continue to perturb Dbus in the next iteration in the same direction as in the previous iteration until there is a change in the input current. In order to realize this, the Iin(n-1) is not replaced by new value of Iin(n) if there is no change in the input current until the change occurs. The variable "Y" is used to store the perturbation direction of the last iteration as shown in Fig. 3.

III. CLOSED LOOP EXPERIMENTAL RESULTS

A proof-of-concept (POC) experimental prototype is built in the laboratory. The experimental setup consists of the power stage and the ADBV controller. The two-stage power converter specifications are: The first-stage is with the following specifications: Two-phase buck converter topology with Vin = 8.7-19V, Vbus = 3-10V, Lo1 = 1.6μH, switching frequency of 150kHz per phase, two parallel HAT2168 upper switches for each phase, and two parallel HAT2165 lower switches for each phase. The second-stage is with the following specifications: Two-phase buck converter topology with Vo = 0.8-1.5V, Lo2 = 0.1 μH, maximum load current Io = 70A, switching frequency of 1MHz per phase, and each phase uses the iP2005 module [13].

The first stage ADBV controller is implemented in a programmable digital controller while the second stage closed loop controller is with an analog implementation. It is worth noting that the second stage is designed with high switching frequency and fast dynamic performance closed loop in order to meet fast load requirements of less than 2mΩ load line performance for a 65A load step with 1000A/μs slew rate. The first stage, which is the focus of this paper, is designed with an intention to achieve high power conversion efficiency by varying the bus voltage.

Fig. 4 shows a sample experimental waveform for the bus voltage under the proposed ADBV controller. Fig. 5 shows the second-stage steady-state and dynamic regulation characteristics under a first-stage controlled by ADBV controller. The 2% $D_{bus-step}$ used is small enough not to disturb the high performance second-stage controller. This 2% $D_{bus-step}$ will only generate less than 0.24V bus voltage change at 12V input voltage.

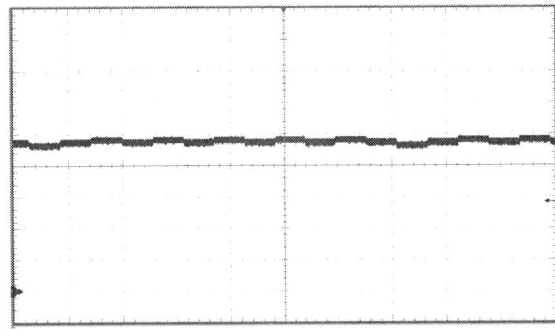

10msec/div
2V/div, Vo=1.5V, Io=65A

Fig. 4: ADBV operation experimental results

Fig. 6 shows experimental results at 60A/1.5V output that compares the controller speed (number of required iterations) for different duty cycle step sizes (perturbation steps). Based on open loop efficiency measurement from the POC prototype, the optimum bus duty cycle is about 0.81 resulting in about 81.5% efficiency. Therefore, it can be observed that smaller step size results in slower convergence speed but with the potential to converge more closely to the optimum bus duty cycle, and vice versa. The efficiency achieved in each step size case is shown on the figure which indicates that with larger step size the controller converges to lower efficiency since there is a larger convergence error in D_{bus}. This implies a design tradeoff. The obtained experimental results agree with the theoretical and simulation results presented in [14].

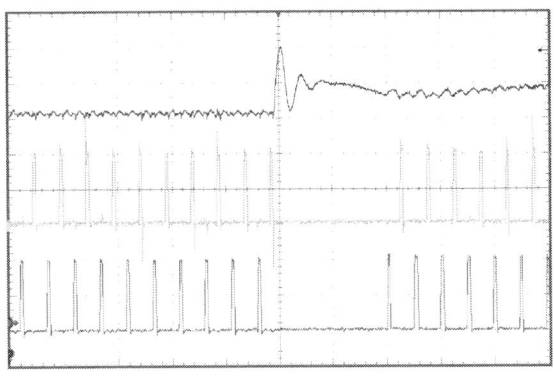

2μsec/div
Top to bottom: 70mV/div., 5V/div, 5V/div.

Fig. 5: Experimental results of second stage closed loop performance (65A step load transient response) under an ADBV-controlled first stage

978-1-4244-4782-4/10 $26.00 © 2010 IEEE   21

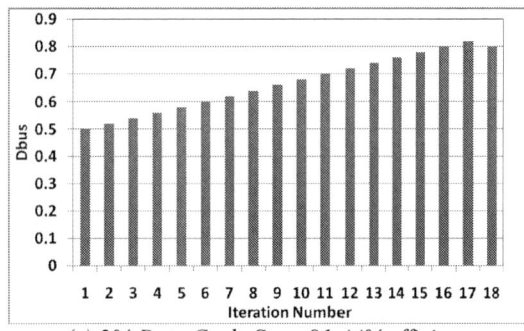

(a) 2% Duty Cycle Step, 81.44% efficiency

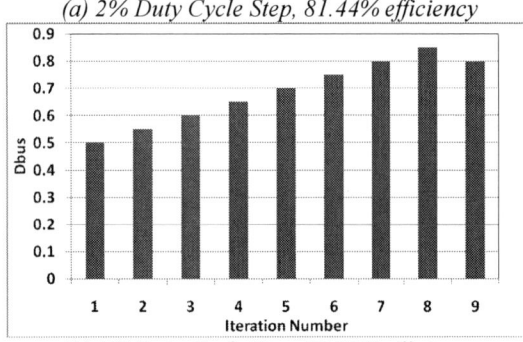

(b) 5% Duty Cycle Step, 81.37% efficiency

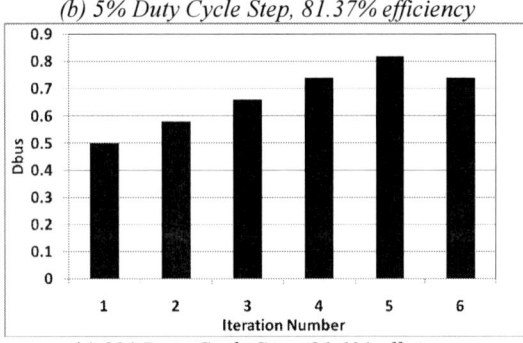

(c) 8% Duty Cycle Step, 81.1% efficiency

Fig. 6: ADBV experimental comparison under different duty cycle step sizes vs. number of iterations (speed) at 1.5V/60A

IV. CONCLUSION

The paper presents a modified adaptive bus voltage controller for two-stage power converter. The controller adaptively converges to the optimum bus voltage that yields to the maximum power conversion efficiency under variable operating conditions. The modified controller algorithm takes care of the case when there is no current change detected as a result of a single step size increment in the bus voltage. Experimental results are presented to demonstrate the controller operation. The results are in good agreement with the theoretical and simulation results presented in this paper and in [14].

References

[1] Yuancheng Ren; Ming Xu; Kaiwei Yao; Yu Meng; Lee, F.C.; "Two-stage approach for 12-V VR," IEEE Transactions on Power Electronics, Vol. 19, No. 6, Page(s):1498 - 1506 , November 2004.

[2] Kisun Lee; Jia Wei; Ming Xu; Lee, F.C.; "Adaptive Bus Voltage Positioning System for Two Stage Laptop Voltage Regulators," Power Electronics Specialists Conference, 2007. PESC 2007. IEEE, Page(s):2 – 8, 17-21 June 2007.

[3] Hong Mao, Jaber Abu-Qahouq, Shiguo Luo, and Issa Batarseh, "Zero-Voltage-Switching (ZVS) Two-Stage Approaches with Output Current Sharing for 48V Input DC-DC Converter," Nineteenth Annual IEEE Applied Power Electronics Conference and Exposition, APEC'2004, Vol. 2, Pages: 1078-1082, February 2004.

[4] Yuancheng Ren, Ming Xu, Kaiwei Yao and Fred C. Lee, "Two-Stage 48V Power Pod Exploration for 64-Bit Microprocessor," Applied Power Electronics Conference and Exposition, 2003. APEC '03. Eighteenth Annual IEEE Volume 1, Page(s):426 – 431, 9-13 Feb. 2003.

[5] P. Alou, J. A. Cobos, R. Prieto, O. García and J. Uceda, "A two stage Voltage Regulator Module with fast transient response capability," Power Electronics Specialist Conference, 2003. PESC '03. 2003 IEEE 34th Annual Volume 1, Page(s):138 – 143, 15-19 June 2003.

[6] Jia Wei and Fred C. Lee, "Two-Stage Voltage Regulator for Laptop Computer CPUs and the Corresponding Advanced Control Schemes to Improve Light-Load Performance," IEEE APEC'04.

[7] Jaber Abu-Qahouq, Hong Mao, Hussam J. Al-Atrash, and Issa Batarseh, "Maximum Efficiency Point Tracking (MEPT) Method and Dead Time Control," IEEE Transactions on Power Electronics, Vol. 21, Issue 5, Pages: 1273-1281, September 2006.

[8] Jaber Abu-Qahouq, Wisam Al-Hoor, Wasfy Michael, Lilly Huang and Issa Batarseh, "Analysis and Design of an Adaptive-Step-Size Digital Controller For Switching Frequency Auto-Tuning," IEEE Transactions on Circuits and Systems I - Regular Papers, Vol. 56, No. 12, December 2009.

[9] Yuri Panov and M. Jovanovic, " Design and Performance Evaluation of Low-Voltage/High-Current DC/DC On-Board Modules," Applied Power Electronics Conference and Exposition, 1999 (APEC '99). Fourteenth Annual, Volume 1 , 1999.

[10] Y. Panvo and M. M. Jovanovic, "Design consideration for 12-V/1.5-V, 50-A voltage regulator modules," in Proc. IEEE Applied Power Electronics Conf. Expo, Nov. 2001, pp. 776–783.

[11] W. B. Mikhael and F. H. Wu, "A unified approach for generating optimum gradient FIR adaptive algorithms with time-varying convergence factors," J. Circuits Systems and Computers, Vol. 1, no. 1, pp. 19-42. 1991.

[12] Bernard Widrow, Samuel D. Stearns, "Adaptive Signal Processing". Prentice-Hall, 1985.

[13] International Rectifier, "Reference Design – IRDCiP2005A-A" Feb. 04, 2008.

[14] Jaber Abu-Qahouq and Gautam Muralidhar, "Control Scheme for High-Efficiency High-Performance Two-Stage Power Converters," Proceedings of the IEEE Applied Power Electronics Conference and Exhibition, APEC'2009, Page(s):1226 – 1232, February 2009.

Bridgeless Buck PFC Rectifier

Yungtaek Jang and Milan M. Jovanović

Power Electronics Laboratory
Delta Products Corporation
P.O. Box 12173, 5101 Davis Drive
Research Triangle Park, NC 27709

Abstract — A new bridgeless buck PFC rectifier that substantially improves efficiency at low line of the universal line range is introduced. By eliminating input bridge diodes, the proposed rectifier's efficiency is further improved. Moreover, the rectifier doubles its output voltage, which extends useable energy of the bulk capacitor after a drop-out of the line voltage.

The operation and performance of the proposed circuit was verified on a 700-W, universal-line experimental prototype operating at 65 kHz. The measured efficiencies at 50% load from 115-V and 230-V line are both close to 96.4%. The efficiency difference between low line and high line is less than 0.5% at full load. A second-stage half-bridge converter was also included to show that the combined power stages easily meet Climate Saver Computing Initiative Gold Standard.

I. INTRODUCTION

Driven by economic reasons and environmental concerns, maintaining high efficiency across the entire load and input-voltage range of today's power supplies is in the forefront of customer's performance requirements. Specifically, meeting and exceeding U.S. Environmental Protection Agency's (EPA) Energy Star [1] and Climate Saver Computing Initiative (CSCI) [2] efficiency specifications have become a standard requirement for both multiple- and single-output off-line power supplies. Generally, the EPA and CSCI specifications define minimum efficiencies at 100%, 50%, and 20% of full load with a peak efficiency at 50% load. For example, for the highest-performance tier of single-output power supplies with a 12-V output, i.e., for the Platinum level power supplies, the required minimum efficiencies at 100%, 50%, and 20% load, measured at 230-V line, are 92%, 94%, and 91% respectively.

In universal-line (90-264-V) applications, maintaining a high efficiency across the entire line range poses a major challenge for ac/dc rectifiers that require power-factor correction (PFC). For decades, a bridge diode rectifier followed by a boost converter has been the most commonly used PFC circuit because of its simplicity and good power factor (PF) performance. However, a boost PFC front-end exhibits 1-3% lower efficiency at 100-V line compared to that at 230-V line. This drop of efficiency at low line can be attributed to an increased input current that produces higher losses in semiconductors and input EMI filter components.

Another drawback of the universal-line boost PFC front end is related to its relatively high output voltage, typically in the 380-400-V range. This high voltage not only has a detrimental effect on the switching losses of the boost

converter, but also on the switching losses of the primary switches of the downstream dc/dc output stage and the size and efficiency of its isolation transformer. Because switching losses dominate at light loads, the light-load efficiency of a power supply exhibits a steep fall-off as the load current decreases.

At lower power levels, i.e., below 850 W, the drawbacks of the universal-line boost PFC front-end may partly be overcome by implementing the PFC front-end with a buck topology. As it has been demonstrated in [3], the universal-line buck PFC front end with an output voltage in the 80-V range maintains a high-efficiency across the entire line range. In addition, a lower input voltage to the dc/dc output stage has beneficial effects on its light-load performance because lower-voltage-rated semiconductor devices can be used for the dc/dc stage and because lower input voltage reduces the loss and size of the transformer.

The buck PFC converter operation in both DCM and CCM mode was described first in [4], whereas additional analysis and circuit refinements were described in [5]-[12]. Because the buck PFC converter does not shape the line current around the zero crossings of the line voltage, i.e., during the time intervals when the line voltage is lower than the output voltage, it exhibits increased total harmonic distortion (THD) and a lower power factor (PF) compared to its boost counterpart. As a result, in applications where IEC61000-3-2 and corresponding Japanese specifications (JIS-C-61000-3-2) need to be met, the buck converter PFC employment is limited to lower power levels.

In this paper, a bridgeless buck PFC rectifier that further improves the low-line (115-V) efficiency of the buck front end by reducing the conduction loss through minimization of the number of simultaneously conducting semiconductor components is introduced. Because the proposed bridgeless buck rectifier also works as a voltage doubler, it can be designed to meet harmonic limit specifications with an output voltage that is twice that of a conventional buck PFC rectifier. As a result, the proposed rectifier also shows better hold-up time performance. Although the output voltage is doubled, the switching losses of the primary switches of the downstream dc/dc output stage still significantly lower than that of the boost PFC counter part.

To verify the operation and performance of the proposed circuit, a 700-W, universal-line experimental prototype operating at 65 kHz was built. The measured efficiencies at 50% load over the input voltage range from 115-V to 230-V

978-1-4244-4782-4/10 $26.00 © 2010 IEEE

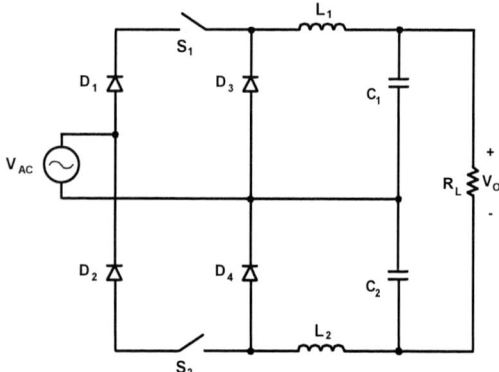

Fig. 1.   Proposed bridgeless buck PFC rectifier.

are more than 96%. In addition, the full-load efficiency difference between low line and high line is less than 0.5%. Including a half bridge dc-dc converter 12-V output stage, the measured total efficiency is well above the CSCI Gold Level efficiency targets of 115-V and 230-V line.

II. Bridgeless Buck PFC Rectifier with Voltage Doubler Output

The proposed PFC rectifier, shown in Fig. 1, employs two back-to-back connected buck converters that operate in alternative halves of the line-voltage cycle. The buck converter illustrated in Fig. 2 only operates during positive half cycles of line voltage V_{AC} and consists of a unidirectional switch implemented by diode D_1 in series with switch S_1, freewheeling diode D_3, filter inductor L_1, and output capacitor C_1. During its operation, the voltage across capacitor C_1, which must be selected lower than the peak of line voltage, is regulated by pulse-width-modulation (PWM) of switch S_1. Similarly, the buck converter consisting of the unidirectional switch implemented by diode D_2 in series with switch S_2, freewheeling diode D_4, filter inductor L_2, and output capacitor C_2 operates only during negative half cycles of line voltage V_{AC}, as shown in Fig. 3. During its operation, the voltage across capacitor C_2 is regulated by PWM of switch S_2.

As seen from Figs. 2 and 3, the input current always flow through only one diode during the conduction of a switch, i.e., either D_1 or D_2. Efficiency is further improved by eliminating input bridge diodes in which two diodes carry the input current. An additional advantage of the proposed circuit is its inrush current control capability. Since the switches are located between the input and the output capacitors, switches S_1 and S_2 can actively control the input inrush current during start up.

Output voltage V_O of the PFC rectifier, which is the sum of the voltages across output capacitors C_1 and C_2, is given by

$$V_O = 2DV_{IN} \ , \qquad (1)$$

where D is the duty cycle and V_{IN} is the instantaneous rectified ac input voltage. Because of the buck topology, the relationship shown in Eq. (1) is valid for input voltages V_{IN} greater than twice the output voltage, i.e., for $V_{IN} > 2V_O$.

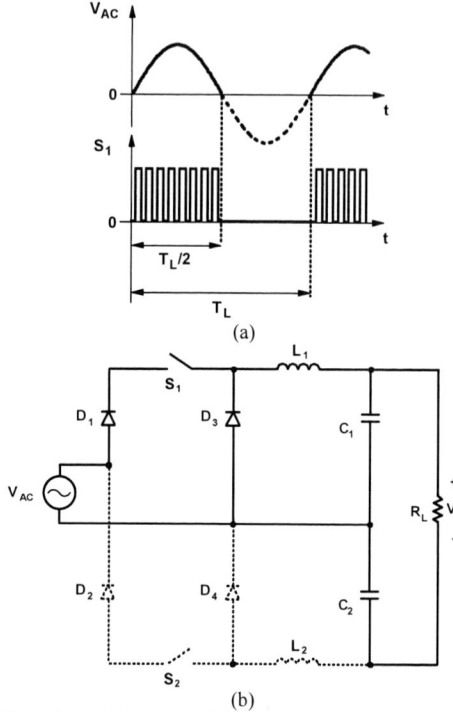

Fig. 2.   Operation of the proposed bridgeless buck PFC rectifier during the period when the line voltage is positive.

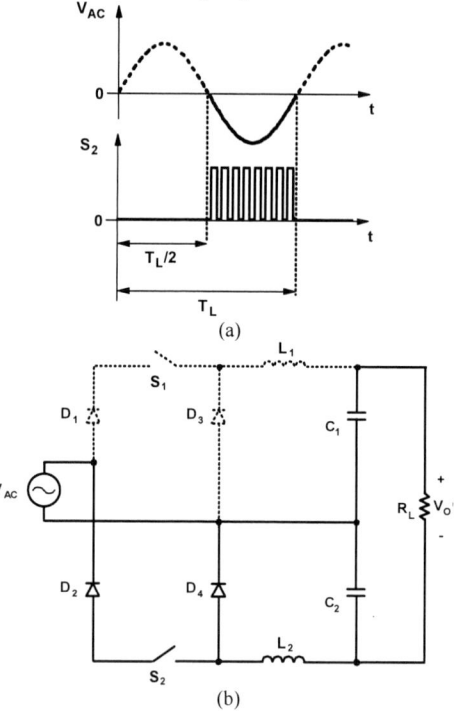

Fig. 3.   Operation of the proposed bridgeless buck PFC rectifier during the period when the line voltage is negative.

When input voltage V_{IN} falls below $2V_O$, the converters do not deliver energy from the input to the output so the load current is maintained solely by the output capacitors.

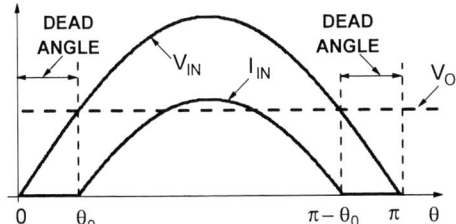

Fig. 4. Ideal input voltage and input current waveforms of a PFC buck rectifier.

Because the PFC buck rectifier does not shape the line current during the time intervals when the line voltage is lower than the output voltage, as shown in Fig. 4, there is a strong trade-off between THD and PF performance and output voltage selection. Namely, the output voltage should be maximized to minimize the size of the energy-storage capacitors for a given hold-up time. However, increasing the output voltage increases the THD and lowers the PF due to the increased dead angle as shown in Fig. 4, i.e., the time the buck converter does not operate during a half-line cycle. It was found that for power levels below 850 W, output voltage should be kept below 160 V to meet the IEC61000-3-2 harmonic requirements.

As demonstrated in [3], the clamped-current-mode control [13]-[15] is an effective, simple, and low-cost approach for controlling the buck PFC converter. The clamped-current-mode control can be easily extended to the bridgeless buck PFC front end since during each half-cycle only one buck converter in the bridgeless PFC operates at a time to regulate the voltage across its corresponding output capacitor.

As known from the general peak-current-mode theory, to ensure the stability of the current loop in the clamped-current-mode control circuit operating in CCM with a duty cycle over 50%, the slope of the compensation (external) ramp S_e should be at least 50% of the maximum down slope of the inductor current $S_{f,max}$, i.e.,

$$S_e = k_S \cdot S_{f,max}, \quad k_S \geq 0.5. \qquad (2)$$

Furthermore, as described in [3], optimum design cannot be achieved with a single value for k_S, i.e., minimize THD of input current and attain a high PF in the entire universal-input range. In universal-line applications, optimal design can only be achieved by a variable k_S that is increasing with input voltage. As found in [3], the optimal range for k_S is between 1 and 2 for nominal low line (115 V) and between 3 and 5 for nominal high line (230 V).

It also should be noted that the proposed bridgeless PFC rectifier's design criteria, as described in Eq. (2), guarantees the voltage balance of output capacitors C_1 and C_2. In fact, as long as constant k_S is higher than 0.5, the voltage balance of output capacitors C_1 and C_2 is automatically achieved.

Four topological variations of the proposed bridgeless buck PFC rectifier are shown in Fig. 5. As shown in Fig. 5(a), inductors L_1 and L_2 in the PFC rectifier in Fig. 1 can be replaced with a single inductor connected at the midpoint of

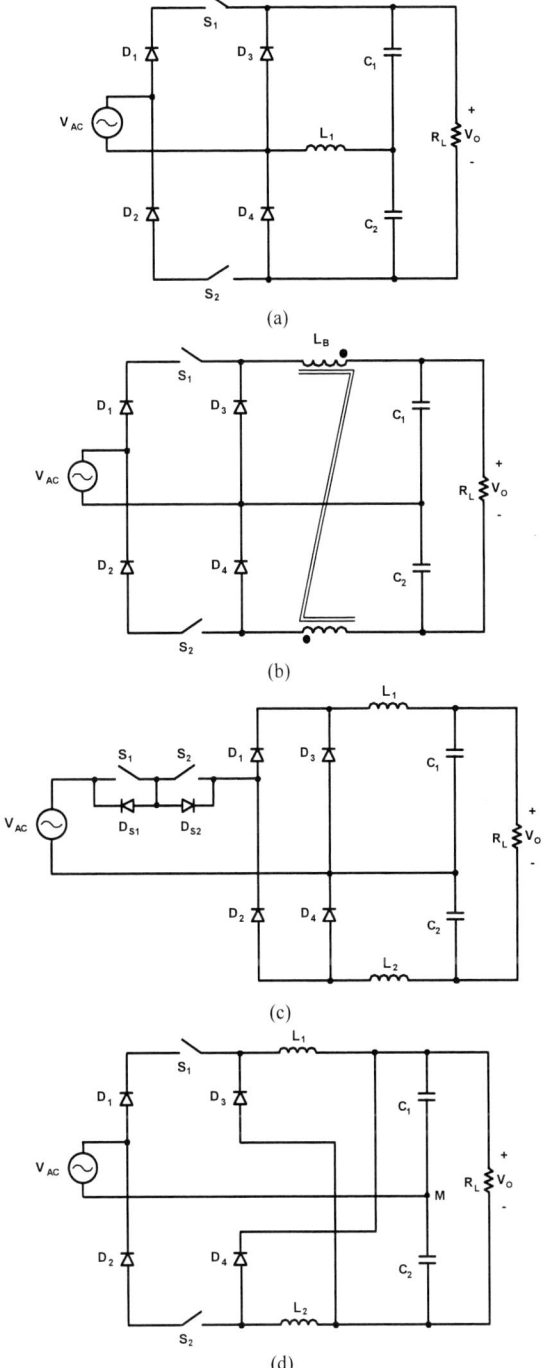

Fig. 5. Topology variations of the proposed bridgeless buck PFC rectifier. The rectifier with (a) a single inductor, (b) a coupled inductor, (c) a bi-directional switch, and (d) non-linear gain.

capacitors C_1 and capacitor C_2 and the return of the input source. Also, the number of magnetic components can be reduced to a single component by coupling inductors L_1 and L_2 in the rectifier in Fig. 1, as shown in Fig. 5(b).

Another topological variation can be obtained by moving switches S_1 and S_2 in the PFC rectifier in Fig. 1 to the ac side,

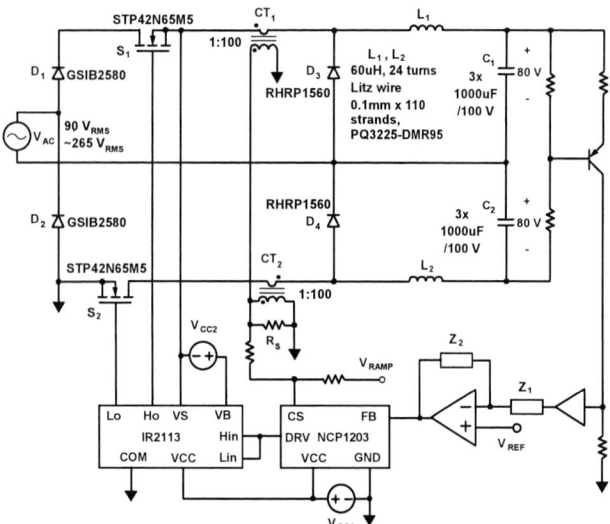

Fig. 6. Experimental prototype circuit of the proposed bridgeless buck PFC rectifier.

as shown in Fig. 5(c). In this implementation, a bi-directional switch is formed by the serial connection of switches S_1 and S_2 with their anti-parallel diodes D_{S1} and D_{S2}.

Yet another variation of the proposed bridgeless buck PFC rectifier is in Fig. 5(d). In this circuit, the anodes of freewheeling diodes D_3 and D_4 are connected directly to the negative and positive output rails, respectively, instead of to the midpoint of the output capacitors as in Fig. 1. It is interesting to note that the circuit in Fig. 5(d) exhibits a non-linear gain characteristic given by

$$V_O = \frac{2D}{1+(1-D)^2} V_{IN} \ . \qquad (3)$$

According to Eq. (3), if duty cycle D is near unity, i.e., when input voltage V_{IN} is close to half of output voltage V_O, the input-to-output gain is similar to that shown in Eq. (1). However, if duty cycle D is near zero, i.e., when input voltage V_{IN} is much greater than output voltage V_O, the input-to-output gain becomes

$$V_O = DV_{IN} \ , \qquad (4)$$

which is similar to the input-to-output gain of a conventional buck converter.

Finally, if reverse voltage blocking switches that allow unidirectional current flow are utilized for switches S_1 and S_2 in Fig. 1 and Figs. 5(a), 5(b), and 5(d), diodes D_1 and D_2 can be eliminated.

III. EXPERIMENTAL RESULTS

The performance of the proposed rectifier in Fig. 1 was evaluated on a 65-kHz, 700-W prototype circuit that was designed to operate from a universal ac-line input (85 V_{RMS}-264 V_{RMS}) with a 160-V output.

Figure 6 shows the schematic diagram and component details of the experimental prototype circuit. Since the drain voltage of switches S_1 and S_2 are clamped to the voltage difference between the input voltage and output capacitor

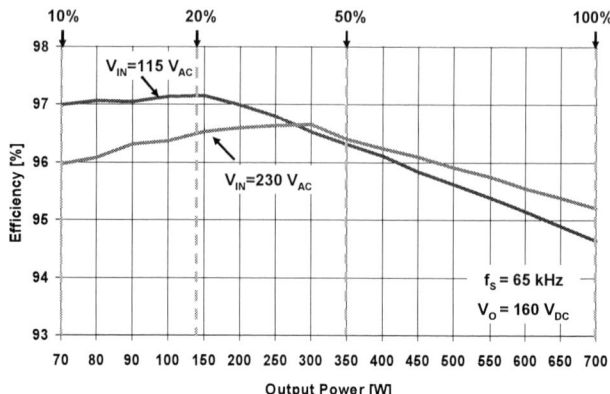

Fig. 7. Measured efficiency of the proposed bridgeless buck PFC rectifier.

voltage, the peak voltage stress on switch S_1 and S_2 can be as high as 380 V, which is the peak input voltage at the maximum line. The peak current stress on switch S, which occurs at full load and low line, is approximately 9 A. Therefore, a STP42N65M5 MOSFET (V_{DSS} = 650 V, R_{DS} = 0.079 Ω) from ST was used for each buck switch. Since output diodes D_3 and D_4 must block both the same peak voltage stress and conduct the same peak current as the switches, an RHRP1560 diode (V_{RRM} = 600 V, I_{FAVM} = 15 A) from Fairchild was used as boost diode D. It should be noted that the employed output diode is a low-cost conventional silicon diode since the reverse-recovery related loss in the proposed rectifier is much smaller than that of its boost counterpart, which frequently uses expensive silicon-carbide diodes. In fact, the voltage across the switches and diodes are much lower than those of a boost rectifier at low line operation, and the turn-on loss and the reverse-recovery-related losses are significantly lower.

To obtain the desired inductance of output inductor L_1 and L_2 of approximately 60 µH and also to achieve high efficiency at light-load, the output inductor was built using a pair of ferrite cores (PQ-3225, DMR95) and 24 turns of Litz wire (0.1mm, 110 strands). Litz wires were employed to reduce fringe effects near the gap area of the inductors.

Three aluminum capacitors (1000 µF, 100 VDC) were used for output capacitors C_1 and C_2 for their ability to meet the hold-up time requirement (20 mS at 50% load and 12 mS at full load).

As shown in Fig. 6, the bulk capacitor voltage that is the voltage across series connected capacitors C_1 and C_2 was regulated by a single controller (NCP1203 from On-Semi). Switches S_1 and S_2 were operated simultaneously by the same gate signal from the PWM controller. Although both switches were always gated, only one switch carried positive current and delivered power to the output, i.e., switch S_1 on which the positive input voltage was induced, as shown in Fig. 2. The other switch on which the negative input voltage is induced, i.e., switch S_2 in Fig. 2, did not influence the operation since diode D_2, which is connected in series with switch S_2, blocked the current. It should be noted that the voltage across each capacitor C_1 or C_2 can be independently

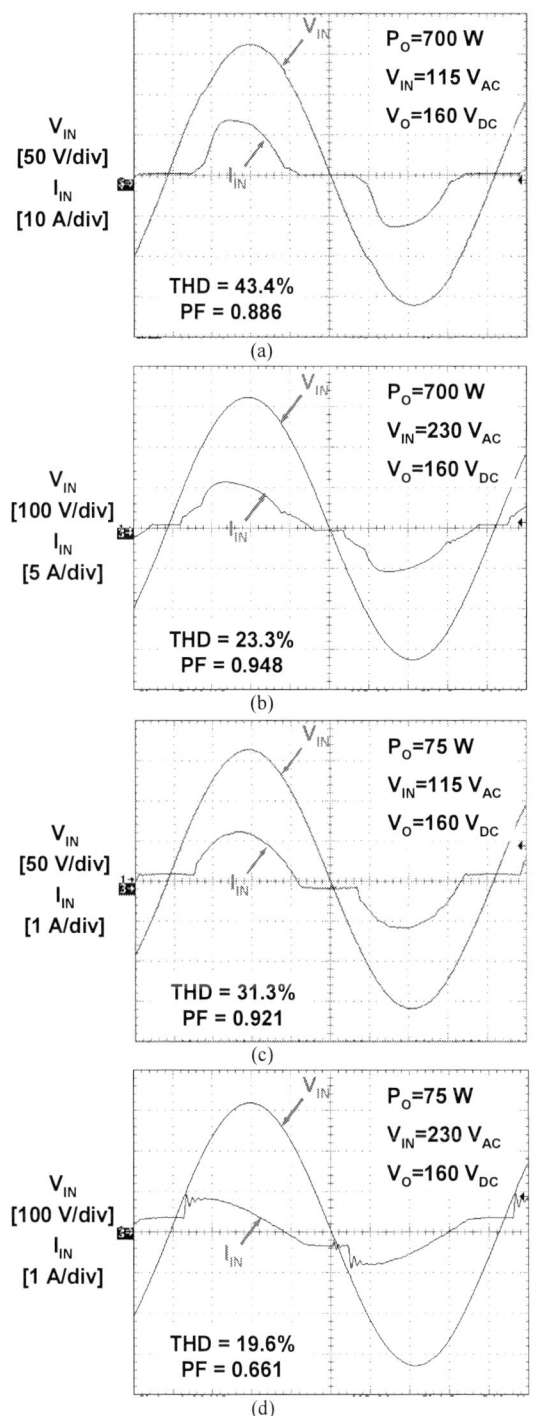

Fig. 8. Measured input voltage and current waveforms of the proposed bridgeless buck PFC rectifier when the output power is 700 W from (a) 115 V_{AC} and (b) 230 V_{AC} input voltage and 75 W from (c) 115 V_{AC} and (d) 230 V_{AC} input voltage.

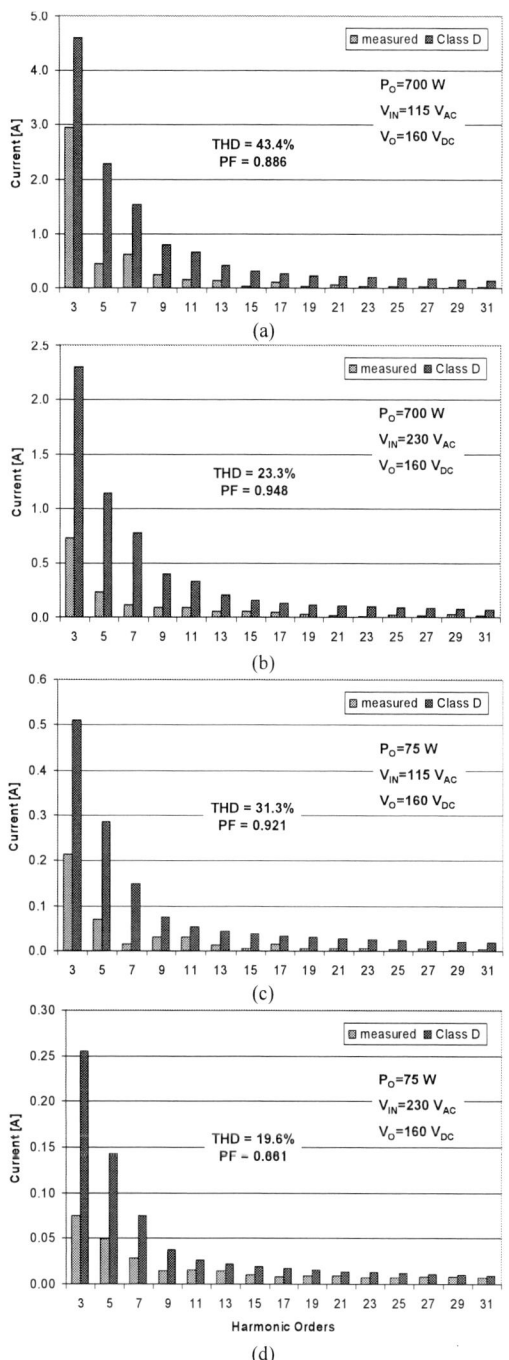

Fig. 9. Measured harmonic components of the input current at 700 W and 75 W output power. Class D requirements of IEC61000-3-2 are also plotted.

regulated by two controllers as conceptually described in Figs. 2 and 3.

Figure 7 shows the measured efficiency of the proposed bridgeless buck PFC rectifier. It should be noted that the low-line efficiency is higher than the high-line efficiency over the load range below 40%. The efficiency difference between low line and high line is less than 0.5% over the load range above 50%, which is desirable for thermal optimization.

Figure 8 shows the measured input voltage and input current waveforms of the proposed PFC rectifier when the output power is 700 W and 75 W from low and high line. The

978-1-4244-4782-4/10 $26.00 © 2010 IEEE   27

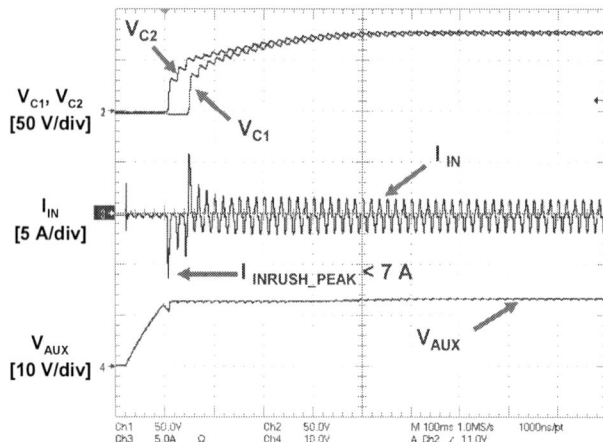

Fig. 10. Measured input current I_{IN}, output capacitor voltages V_{C1}-V_{C2} and control voltage V_{AUX} of the experimental prototype circuit during start up.

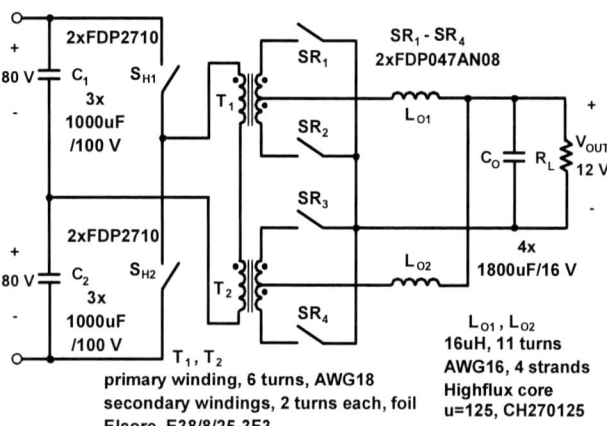

Fig. 11. Experimental half-bridge dc-dc 2nd stage converter. Input capacitors C_1 and C_2 are the same capacitors as the output capacitors of the front-end rectifier shown in Fig. 6.

measured total harmonic distortion (THD) and power factor (PF) of the rectifier are also shown in the figures. Measured harmonic components of the input current at 700 W and 75 W output power are shown in Fig. 9. Class D requirements of IEC61000-3-2 are also compared. All of the harmonic currents meet the related Class D requirements over the entire load and line ranges. Because the input current is actively controlled by switch S_1 and S_2 of the proposed buck rectifier, the inrush current during start up is well controlled as shown in Fig. 10.

To verify the performance of the entire power supply using the proposed front-end rectifier, a conventional half-bridge converter with synchronous rectifiers was implemented as the second stage converter that operates at 65-kHz switching frequency and delivers 12 V_{DC} output voltage. Although any isolated dc/dc converter topology can be used for the second stage, a half-bridge dc/dc converter is a more suitable topology as the second stage converter for the proposed

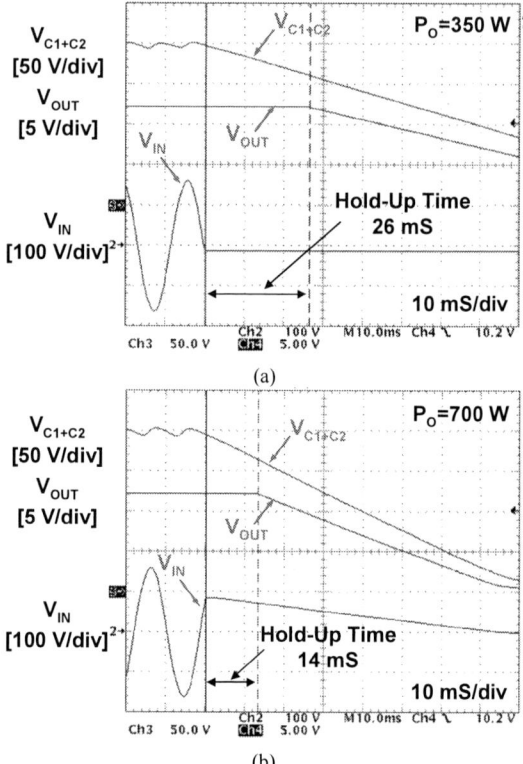

(a)

(b)

Fig. 12. Measured bulk capacitor voltage V_{C1+C2} that is the voltage across series connected capacitors C_1 and C_2, output voltage V_{OUT}, and ac input voltage V_{IN} at (a) 50% load and (b) 100% load during a hold-up time.

bridgeless buck PFC rectifier because capacitors C_1 and C_2, shown in Fig. 6, are used as two bulk capacitors of the half-bridge converter.

Figure 11 shows the experimental prototype circuit and the employed components. The second-stage half-bridge converter was implemented with two FDP2710 MOSFETs from Fairchild for each of bridge switches S_{H1} and S_{H2} and two parallel FDP047AN08AD MOSFETs from Fairchild for each of synchronous rectifier switches S_{R1-R4}. Transformer TR was built using a pair of ferrite cores (EI 38/8/25-3F3) with six turns of triple-insulated magnet wire (AWG# 18) for the primary winding and two turns of copper foil for each of the secondary windings. Output filter inductors L_{O1} and L_{O2} were built using a toroidal high flux core (CH270125) from Chang-Sung and 11 turns of magnet wire (4×AWG #16). Four low voltage aluminum capacitors (1800 μF, 16 VDC) were used for output capacitor C_O.

Figure 12 show the measured hold-up times at 50% load and full load conditions. Bulk capacitor voltage V_{C1+C2}, which is measured across the series connected capacitors C_1 and C_2 of the front-end rectifier, and output voltage V_{OUT} of the dc-dc second stage converter are shown in Fig. 12 together with input voltage V_{IN}. The measured hold-up times are approximately 26 mS and 14 mS at 50% load and full load conditions, respectively.

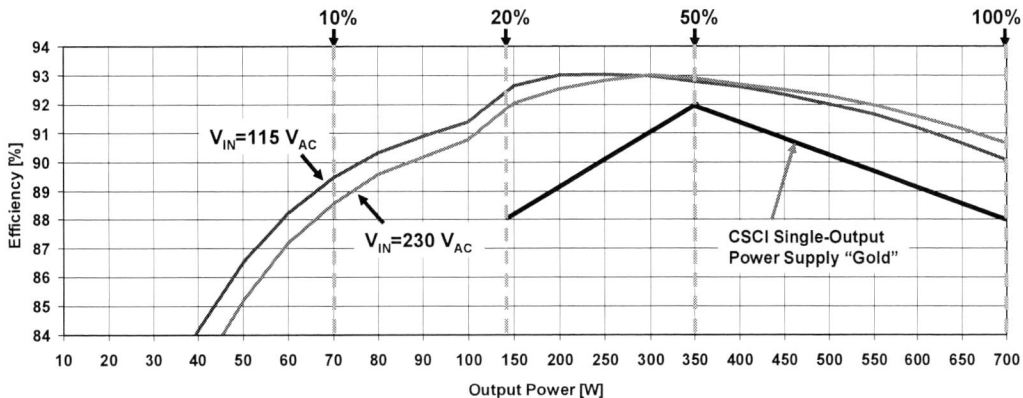

Fig. 13. Measured total efficiency of the proposed bridgeless buck PFC rectifier and half bridge 2nd stage converter. The power supply delivers 12 V dc output from 115 V and 230 V ac inputs. Efficiency requirements of Climate Saver Computing Initiative (CSCI) "gold" specification are also plotted.

The measured total efficiency of the proposed bridgeless buck PFC rectifier and half bridge 2nd stage converter is plotted in Fig. 13. The power supply that delivers 12 V dc output from 115 V and 230 V ac inputs meets the efficiency requirements of CSCI Gold specifications over the entire load and input ranges.

IV. SUMMARY

In this paper, a new bridgeless buck PFC rectifier that substantially improves the efficiency at low line has been introduced. The proposed rectifier doubles the rectifier output voltage, which extends useable energy after a drop-out of the line voltage. Moreover, by eliminating input bridge diodes, efficiency is further improved.

The operation and performance of the proposed circuit was verified on a 700-W, universal-line experimental prototype operating at 65 kHz. The measured efficiencies at 50% load from 115-V and 230-V line are close to 96.4%. The efficiency difference between low line and high line is less than 0.5% at full load. Finally, a half bridge dc-dc converter is added as a second stage converter. The measured total efficiency is well above the CSCI specifications at both 115-V and 230-V line.

ACKNOWLEDGEMENT

The authors want to thank David L. Dillman and Juan Ruiz, Support Engineers from the Power Electronics Laboratory, Delta Products Corporation, for their assistance in constructing the experimental converters and collecting data.

REFERENCES

[1] Environmental Protection Agency (EPA), "Energy Star Program requirements for single voltage external ac-dc and ac-ac power supplies," available at
http://www.energystar.gov/ia/partners/product_specs/program_reqs/EPS_Eligibility_Criteria.pdf

[2] Climate Savers Computing Initiative, White Paper, available at
http://www.climatesaverscomputing.org/docs/20655_Green_Whitepaper_0601307_rv.pdf

[3] L. Huber, L. Gang, and M.M. Jovanović, "Design-Oriented Analysis and Performance Evaluation of Buck PFC Front-End," *IEEE Applied Power Electronics Conf. (APEC) Proc.*, pp.1170-1176, Feb. 2008.

[4] H. Endo, T. Yamashita, and T. Sugiura, "A high-power-factor buck converter," *IEEE Power Electronics Specialists Conference (PESC) Rec.*, pp. 1071-1076, June 1992.

[5] R. Redl and L. Balogh, "RMS, dc, peak, and harmonic currents in high-frequency power-factor correctors with capacitive energy storage," *IEEE Applied Power Electronics Conf. (APEC) Proc.*, pp.533-540, Feb. 1992.

[6] Y.W. Lo and R.J. King, "High performance ripple feedback for the buck unity-power-factor rectifier," *IEEE Transactions on Power Electronics*, vol. 10, no.2, pp.158-163, March 1995.

[7] Y.S. Lee, S.J. Wang, and S.Y.R. Hui, "Modeling, analysis, and application of buck converters in discontinuous-input-voltage mode operation", *IEEE Transactions on Power Electronics*, vol. 12, no.2, pp.350-360, March 1997.

[8] G. Spiazzi, "Analysis of buck converters used as power factor preregulators," *IEEE Power Electronics Specialists Conference (PESC) Rec.*, pp. 564-570, June 1997.

[9] V. Grigore and J. Kyyrä, "High power factor rectifier based on buck converter operating in discontinuous capacitor voltage mode", *IEEE Transactions on Power Electronics*, vol. 15, no.6, pp.1241-1249, Nov. 2000.

[10] C. Bing, X. Yun-Xiang, H. Feng, and C. Jiang-Hui, "A novel single-phase buck pfc converter based on one-cycle control," *CES/IEEE International Power Electronics and Motion Control Conf. (IPEMC)*, pp.1401-1405, Aug. 2006.

[11] G. Young, G. Tomlins, and A. Keogh, "An acdc converter," World Intellectual Property Organization, International Publication Number WO 2006/046220 A1, May 4, 2006.

[12] W.W. Weaver and P.T. Krein, "Analysis and applications of a current-sourced buck converter," *IEEE Applied Power Electronics Conf. (APEC) Proc.*, pp.1664-1670, Feb. 2007.

[13] D. Maksimović, "Design of the clamped-current high-power-factor boost rectifier," *IEEE Transactions on Industry Applications*, vol. 31, no.5, pp.986-992, September/October 1995.

[14] R. Redl, A.S. Kislovski, and B.P. Erisman, "Input-current-clamping: an inexpensive novel control technique to achieve compliance with harmonic regulations," *IEEE Applied Power Electronics Conf. (APEC) Proc.*, pp.145-151, March 1996.

[15] L. Huber and M.M. Jovanović, "Design-oriented analysis and performance evaluation of clamped-current-boost input-current shaper for universal-input-voltage range," *IEEE Transactions on Power Electronics*, vol. 13, no.3, pp.528-537, May 1998.

An Active-Clamped Full-Wave Zero-Current-Switched Quasi-Resonant Boost Converter in Power Factor Correction Application

E. Firmansyah, S. Abe, M. Shoyama
Dept. of Electrical and Electronic Systems Engineering
Graduate School of ISEE, Kyushu University
Fukuoka, Japan

S. Tomioka
SPS R&D Division
TDK-Lambda Corporation
Fukuoka, Japan

T. Ninomiya
Energy Electronics Laboratory
Faculty of Engineering, Nagasaki University
Nagasaki, Japan

Abstract— **In this paper, a power factor correction (PFC) based on active-clamped full-wave quasi-resonant (ZCS-QR) boost converter is presented. This circuit is characterized by zero-voltage-switched (ZVS) and zero-current-switched (ZCS) that results in higher efficiency, opens possibility to incorporate higher switching frequency, and has some potency to reduce converter's conducted EMI. The working principle and steady state performance of the proposed converter are presented. A 100 V dc input, 180 W maximum output, and 440 kHz resonant frequency experimental circuit has been built. Maximum efficiency of 93% and the proof of compliance to the IEC61000-3-2 class D have been confirmed by experiment.**

I. INTRODUCTION

Boost converter is normally applied to a power factor correction (PFC) circuit for its high performance and simplicity [1]. However, its hard-switching nature rise some issues related to reverse recovery of the catch diode, main switch parasitic capacitance loss, and electromagnetic interference (EMI) problem.

Reference [2, 3, 4] presents half-wave zero-current-switch quasi-resonant (ZCS-QR) based PFC as an alternative to the abovementioned boost-based circuit. Realization of ZCS-QR to a PFC circuit makes the application of higher switching frequency more feasible. Moreover, this technique is claimed to generate less EMI due to: 1) reduce the excitation of the parasitic elements [5]; 2) inherently apply frequency modulation scheme that may further reduce the noise level [6, 7]. Instead of half-wave, [8] use the full-wave ZCS-QR topology in its PFC. The topology was selected for its consistent timing consideration. This makes the converter control effort less demanding.

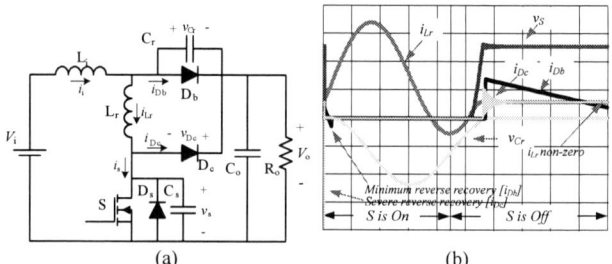

Figure 1. (a) schematic diagram and (b) waveforms of a Diode-clamped full-wave ZCS-QR boost converter [8].

A clamp diode D_c has been added to the circuit of [8] in order to alleviate the voltage ringing problem. A problem normally occurred in a full-wave ZCS-QR boost circuit during main switch S turn-off period. The modified circuit is shown on Fig. 1. (a).

However, D_c slightly changes the operating condition of the ZCS-QR switch as depicted by Fig. 1. (b). That figure reveals that: (a) high reverse recovery current occurs on D_c and (b) i_{Lr} is non-zero during switch turn-on transition. Those problems result in higher losses and more EMI emission.

Circuit in Fig. 2. (a) uses an active-clamp scheme to solve the abovementioned problems [9]. The circuit creates waveform as depicted in Fig. 2.(b). The figure shows that this solution also provides zero-voltage-switch (ZVS) capability during main switch turn-on transition. In this paper, the circuit of [9] is used as base of the proposed PFC.

978-1-4244-4782-4/10 $26.00 © 2010 IEEE

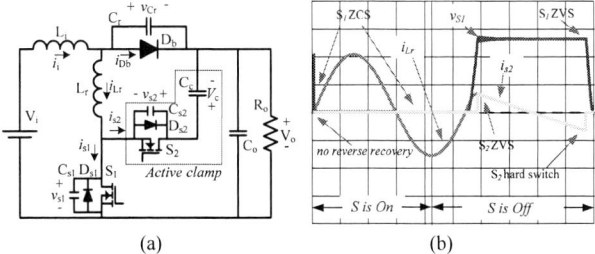

(a)   (b)

Figure 2. (a) the schematic diagram of an active-clamped full-wave ZCS-QR boost converter, and (b) its key waveforms.

II. THE ACTIVE-CLAMP CIRCUIT OPERATION STAGES

The active-clamped converter's operation stages could be analyzed from Fig. 3 and its corresponding circuit configurations shown on Fig. 4. (a) to (f). Generally, the involved process could be classified into two: A) related to the ZCS-QR switch operation and B) related to the active-clamp circuit.

A. The ZCS-QR Switch Operation Stages

Reference [10] gives detailed expression regarding full-wave ZCS-QR circuit operations. In that reference, all equations are normalized to four resonant tank parameters as follows:

$$f_0 = \frac{1}{2\pi\sqrt{L_r C_r}} = \frac{\omega_0}{2\pi}, \tag{1}$$

$$\omega_0 t = \theta, \tag{2}$$

$$R_0 = \sqrt{\frac{L_r}{C_r}}, \tag{3}$$

$$J_s = I_i \frac{R_0}{V_o}. \tag{4}$$

Basically, ZCS-QR switch operation is divided by four periods, α, β, δ, and ξ, as shown in Fig. 3.

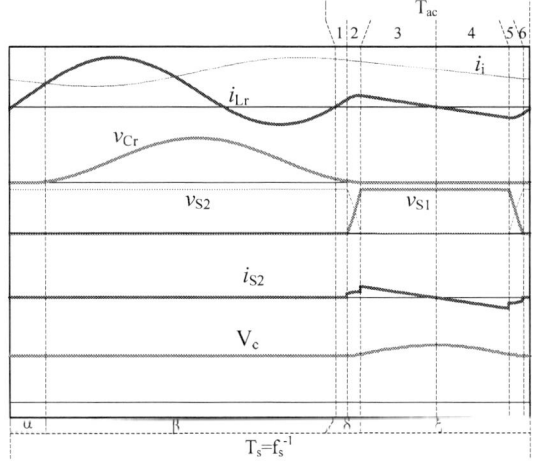

Figure 3. The active-clamped full-wave ZCS-QR boost con-verter operation stages.

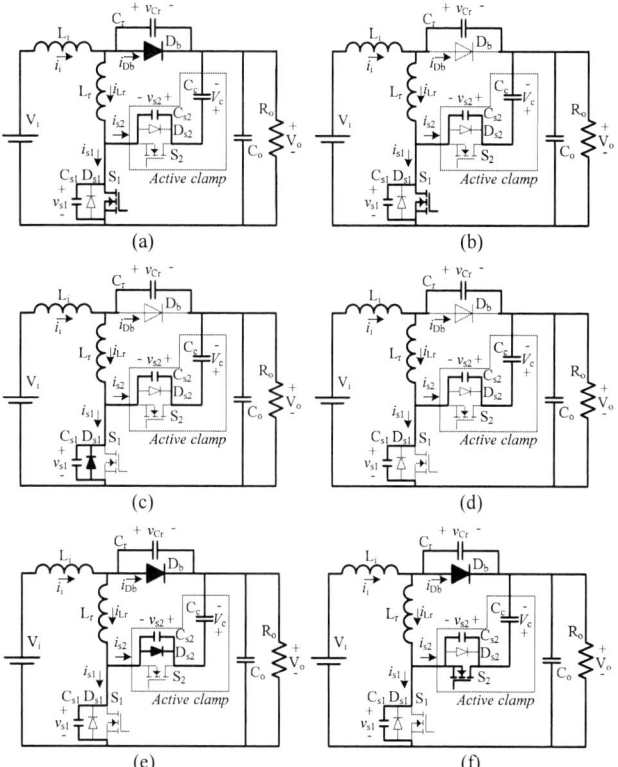

(a)   (b)

(c)   (d)

(e)   (f)

Figure 4. Circuit operation stages of the proposed converter.

Up to the end of $\omega_0 t = \alpha$, circuit configuration on Fig. 4. (a) occurs. During this period, S_1 is turned on. It makes i_{s1} increase linearly. This circuit configuration ends when i_{s1} equals to i_i. Period of α could be determined by:

$$\alpha = J_s. \tag{5}$$

The second period, β, is the resonant period. During this time, L_r and C_r are under resonant condition. Basically, this second period could be divided into positive and negative phase by considering phase of i_{Lr}. The positive phase of i_{Lr} corresponds to circuit on Fig. 4. (b). While the negative one corresponds to circuit on Fig. 4. (c). i_{Lr} equals to zero at the end each of those periods, that is when $\omega_0 t = \alpha + \beta$. For a full-wave topology, β can be found by:

$$\beta = 2\pi - \sin^{-1}(J_s). \tag{6}$$

The third period, δ, corresponds to circuit on Fig. 4. (d). During this time, v_{Cr} is discharged linearly up to zero. δ can be solved by applying equation (7) below,

$$\delta = \frac{1}{J_s}\left(1 + \sqrt{1 - J_s^2}\right). \tag{7}$$

The last period ξ is inserted in order to regulate average energy value processed by the ZCS-QR switch. The longer ξ, the smaller the processed average energy value.

From above explanations, one switching period consists of:

$$\omega_0 T_s = \alpha + \beta + \delta + \xi = 2\pi \frac{f_0}{f_s} = 2\pi \frac{1}{F_s}. \tag{8}$$

Where, f_s is the switching frequency and F_s is

$$F_s = \frac{f_s}{f_0}. \tag{9}$$

Maximum switching frequency is

$$\omega_0 T_s \geq \alpha + \beta + \delta = J_s + \left(2\pi - \sin^{-1}(J_s)\right) + \left(\frac{1}{J_s}\left(1 + \sqrt{1 - J_s^2}\right)\right). \tag{10}$$

Those timing considerations can be used to define a dimensionless variable μ as equivalent of duty cycle D that normally used in PWM converter. In order to define μ, it is important to determine the average value of i_{Lr}. Fig. 3 shows that the ZCS-QR switch process the energy only during α and β period. Therefore, the average value of i_{Lr} is defined as:

$$\left\langle i_{Lr}(t)\right\rangle_{T_s} = \frac{1}{T_s}\int_t^{t+T_s} i_{Lr}(t)dt = \frac{q_1 + q_2}{T_s}. \tag{11}$$

Where q_1 is the energy processed on period α and q_2 on period β. From this, μ ideally will be:

$$\mu = \frac{\left\langle i_{Lr}(t)\right\rangle_{T_s}}{i_i} \approx F_s. \tag{12}$$

As μ is a direct equivalent of D, boost converter steady state transfer function can be approximated by,

$$\frac{V_o}{V_i} = \frac{1}{1-\mu}. \tag{13}$$

B. The Active-Clamp Circuit Operation Stages

In an ideal ZCS-QR circuit, current on S_1 ceased completely at the end of β as shown in Fig. 3. However, the real circuit realizes finite reverse recovery current from D_{S1} and parasitic capacitance C_{s1}. Those factors make i_{Lr} rises once more shortly after the end of β. This additional current makes energy storage on L_r not zero. Without any proper reset effort, this energy generates severe voltage ringing.

In the proposed topology, to alleviate voltage ringing problem, the end of β is the beginning of the active-clamp circuit operation. The active-clamp operation stages are explained as follows;

During period 1, i_{Lr} raise once more due to reverse-recovery charge q_{rr} of D_{S1}. This event stores energy as much as q_{rr} to L_r. During this period, circuit configuration is still depicted by Fig. 4. (c).

Period 1 over after all q_{rr} of D_{S1} being eliminated. It makes the beginning of period 2 where parasitic capacitance C_{s1} and C_{s2} is charged and discharged respectively. The circuit configuration changes to Fig. 4. (d). Charge and discharge currents occur under resonant condition of:

$$f_{01} = \frac{1}{2\pi\sqrt{L_r(C_{s1} + C_{s2})}}. \tag{14}$$

$$R_{0_1} = \sqrt{\frac{L_r}{(C_{s1} + C_{s2})}}. \tag{15}$$

Those occurrences make voltage on S_1 increase until reaching V_o+V_c while voltage on S_2 decrease up to zero. It is important to note that this incident stores additional energy on L_r equal to:

$$q_{c_s} = (C_{s1} + C_{s2})(v_o + v_c). \tag{16}$$

After S_1 and S_2 reaches V_o+V_c and zero respectively, period 2 over and period 3 started. During this period, energy stored on L_r is dumped to C_c through D_{s2} under resonant condition of:

$$f_{0c} = \frac{1}{2\pi\sqrt{L_r C_c}}. \tag{17}$$

$$R_{0_c} = \sqrt{\frac{L_r}{C_c}}. \tag{18}$$

During period 3, circuit configuration is indicated by Fig. 4. (e). To take advantages of ZVS condition, S_2 should be turned-on during this period. Circuit configuration then changes to what is shown in Fig. 4. (f) after S_2 is turned-on.

After all energy on L_r dumped to C_c, the clamp current i_{s2} will be zero momentarily. This indicates the beginning of period 4. During this period, if S_2 is already turned on, i_{s2} will be flown in opposite direction towards L_r. This current returns under resonant condition like being stated in (17) and (18).

At certain period before the beginning of the next switching cycle, S_2 should be turned off. This instance is the beginning of period 5. During this momment, charge on L_r discharges C_{s1} and charges C_{s2}. After all charges on those capacitors discharged and charged respectively, the remaining energy on L_r will flow through S_1 body diode to C_o. It is evident here that S_1 experiences ZVS transition during next switching cycle as the C_{s1} is already discharged at the beginning of the next switching cycle.

C. Influence of The Active-Clamp Circuit to μ

The average value of i_{Lr} on Fig. 3 during active-clamped period (T_{ac}) can be found by:

$$\left\langle i_{Lr}(t)\right\rangle_{T_{ac}} = \frac{1}{T_{ac}}\int_0^{T_{ac}} i_{Lr}(t)dt = 0. \tag{19}$$

That was because

$$\left\langle i_{Lr}(t)\right\rangle_{T_{ac1}+T_{ac2}+T_{ac3}} = -\left\langle i_{Lr}(t)\right\rangle_{T_{ac4}+T_{ac5}+T_{ac6}}. \tag{20}$$

Therefore, it can be concluded here that the active-clamp circuit do not have any influence on μ. Therefore, equation (12) and (13) is still valid in order to calculate ideal steady state transfer function of the proposed boost converter.

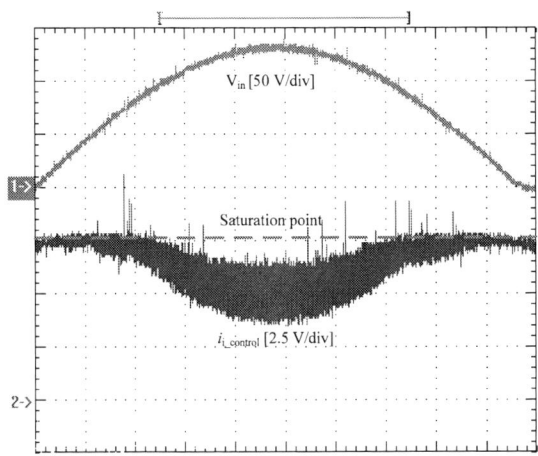

Figure 5. The current programming command waveform related to the input voltage condition.

D. Influence of The Active-Clamp Circuit to Maximum Switching Frequency

New switching cycle should not be started before all of active clamping sequence finished. However, it is not easy to calculate the minimum processing time of the active clamp circuit (T_{ac_min}). In the experimental circuit, T_{ac_min} was determined based on observation. It is concluded that.

$$T_{ac\_min} \approx 5 \cdot T_{ac_1}. \quad (21)$$

Where, T_{ac_1} is also highly non-linear and should be determined based on empirical experimentation.

The additional period of T_{ac_min} makes (10) becomes,

$$\omega_0 T_s \approx J_s + \left(2\pi - \sin^{-1}(J_s)\right) + \left(\frac{1}{J_s}\left(1 + \sqrt{1 - J_s^2}\right)\right) + 5 \cdot T_{ac_1}. \quad (22)$$

In practice, it makes

$$F_s = \frac{f_s}{f_0} \approx \mu \approx 0.8, \quad (23)$$

Or

$$\frac{V_o}{V_i} = \frac{1}{1 - \mu} \approx 5. \quad (24)$$

This condition limits the power factor performance of the proposed converter, as the converter voltage gain should be very high near the input voltage zero crossing area. Due to this constraint, input current of the PFC will contain slightly higher third harmonic spectra due to higher cusp distortion.

The phenomenon is illustrated in Fig. 5. It is shown that the current programming command reach saturation for quite significant portion of time to keep balance between output voltage regulation and good power factor input current.

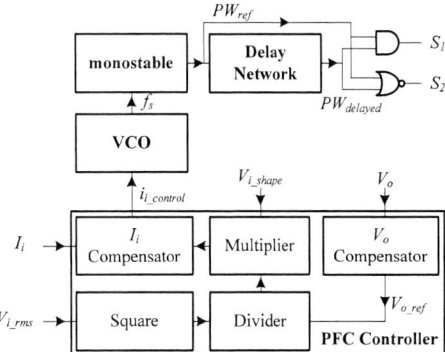

Figure 6. Block diagram of the proposed PFC control circuit.

III. THE PFC CONTROL SCHEME

A. Basic Control Technique

Reference [11] pointed out that the order of small-signal control to output characteristics of a ZCS-QR converter is similar to a conventional hard-switched converter. This makes the well-proven multiplier-based current-averaged PFC control scheme become preferred control candidate [12]. Here, assumptions similar to [8] are used in solving the control equations. It means current amplifier gain (G_{CA}) and cut-off frequency of the controller (f_c) are calculated based on designed minimum switching frequency of the converter.

Fig. 6 shows the block diagram of the proposed PFC circuit. This control scheme is based on commercially available PFC control IC. It consists of two control-loops; the inner current loop and the outer voltage-loop. For conventional hard-switched PFC circuit, all functions related to control circuit are normally contained inside a single chip IC. In this kind of IC, output of the average current controller is internally fetched to a PWM module. However, a ZCS-QR converter is controlled by frequency modulation instead of PWM. Therefore, in the proposed PFC topology, the PWM module is bypassed and an external connection to a voltage-controlled oscillator (VCO) is made.

B. Control Waveform Timing Considerations

To realize a PFC circuit based on the proposed converter, four timing parameters should be determined correctly. They are f_s, t_{on}, t_{d1}, and t_{d2}, as shown in Fig. 7.

Output of VCO, f_s is proportional to the output of PFC controller, $i_{i\_control}$. It should be assured that value f_s will not be more than maximum limit determined by (22).

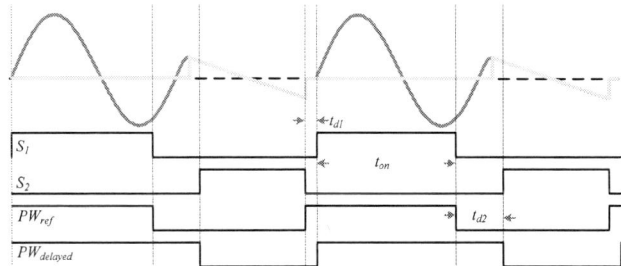

Figure 7. Timing consideration of the proposed converter.

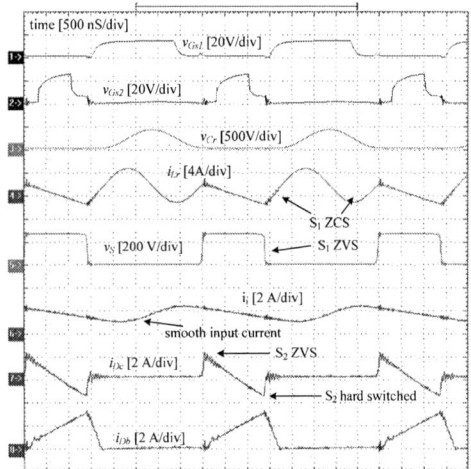

Figure 8. Two switching-cycle waveforms inside the proposed converter.

The rising edge of f_s triggers the monostable to generate a fixed wide pulse, PW_{ref}. PW_{ref} width is summation of t_{d1} and t_{on}. Where, t_{d1} is the delay time before the main switch S_1 turned-on and t_{on} is the turn-on period of S_1.

PW_{ref} is then fetched to the delay network which generates $PW_{delayed}$. The turn-on instant time of the $PW_{delayed}$ is the turn-on instant time of PW_{ref} delayed by t_{d1}. The turn-off instant time of the $PW_{delayed}$ is the turn-off instant time of PW_{ref} delayed by t_{d2}. Where, t_{d2} is the delay time before S_2 turned on.

In order to generate the timing signal for S_1 and S_2, $PW_{delayed}$ and PW_{ref} are then "and"-ed and "nor"-ed by logic circuit. Then, S_1 and S_2 can be sent to a gate drive circuit. As the proposed converter consists of a high-side switch S_2, a boot-straped gate-drive circuit should be used to drive the switch. Waveform related to the timing requirement taken from experimental set-up can be shown in Fig. 8.

IV. EXPERIMENTAL RESULT

An experimental setup has been built to confirm the aforementioned concept. Specification of the proposed converter is listed in Table 1.

Fig. 9. (a) shows the input voltage and current waveform of the proposed PFC. The current waveform is further analyzed to acquire its harmonic contents. Its harmonics histogram is then depicted in Fig. 9. (b). This lather figure shows that the input current passes the IEC61000-3-2 class D standard.

TABLE I
PARAMETERS LIST OF THE PROPOSED CONVERTER PROTOTYPE

Resonant tank inductor	L_r	21 uH
Resonant tank capacitor	Cr	6.2 nF
Resonant tank frequency	f_0	441 kHz
Resonant tank impedance	R_0	58 Ω
Input inductor	L_i	220 uH
Output capacitor	C_o	330 uF
Clamp capacitor	C_c	520 nF
Input voltage	V_i	100 V
Output voltage	V_o	270 V
Maximum Output Power	$P_{o\ max}$	300 W

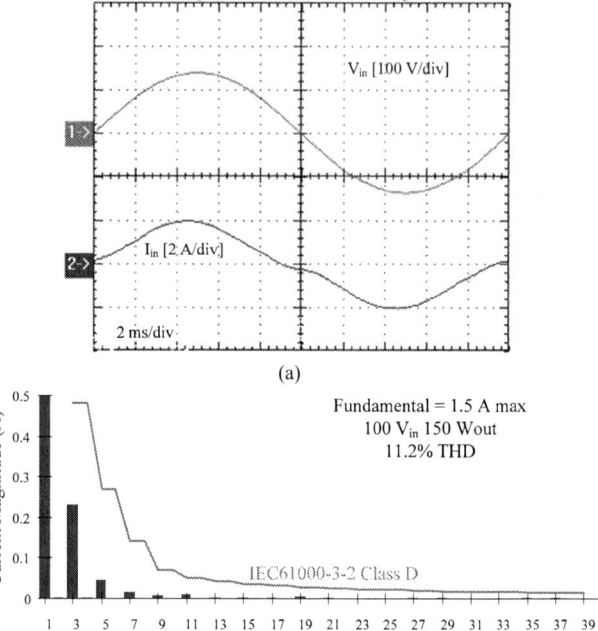

(a)

(b)

Figure 9.   (a) Input voltage and current and (b) harmonic current histogram of the proposed converter.

Fig. 10 shows the converter efficiency in relation to the percentage of its maximum power rating. The figure confirm efficiency characteristics of a quasi-resonant converter, which tends to be less efficient when it is lightly loaded. The maximum efficiency of the proposed PFC is about 93%. This point can be achieved when it works at about 80% of its output power rating.

EMI characteristic measurements also have been done. Fig. 11 shows EMI signature of the proposed converter without any input filter, 100 V_i, 270 V_o, 449 Ω load, and Rohde-Schwarz LISN ESH2-Z5. From the figure it is revealed that the proposed boost converter gives lower conducted noise floor compared to the former [9] topology. This character contributes towards lower average conducted noise energy. Other than that, at some points, the proposed boost converter also provides lower peak EMI value. Those good points may relax the effort of conduction noise filtering.

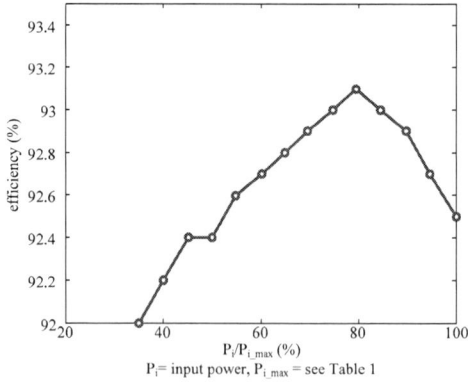

Figure 10. Efficiency of the proposssed converter.

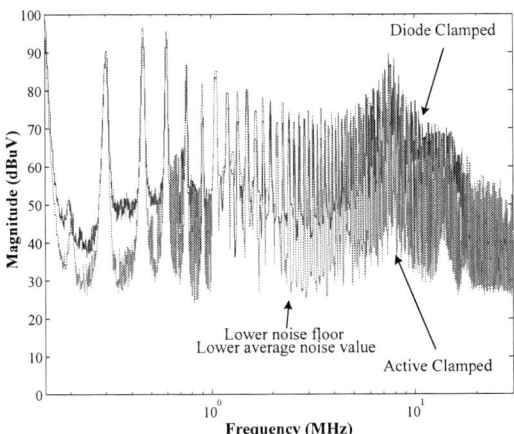

Figure 11. Conducted EMI characteristics of the proposed converter.

V. CONCLUSION

A topology called active-clamped full-wave zero-current-switched quasi-resonant boost converters has been demonstrated. Its basic operation principles, involved equations, and experimental results have been presented. This new topology has proven to be able to alleviate the voltage ringing phenomena during switch turn-off period on a full-wave ZCS-QR boost converter. Beside that, active clamp circuit also gives additional benefit by adding ZVS characteristic to current converter. It has been shown that this technique performance is better if compared to former simple diode clamp topology in terms of efficiency, voltage regulation, and conducted EMI characteristics. This topology offers good candidate in a boost power factor correction application.

REFERENCES

[1] ON Semiconductor, "Power Factor Correction (PFC) Handbook-Choosing the Right Power Factor Control-ler Solution", Rev. 2, Aug‒2004.

[2] Barbi, I., Oliveira da Silva, S.A., "Sinusoidal line current rectification at unity power factor withboost quasi-resonant converters," Applied Power Electronics Conference and Exposition, 1990 Conference Proceedings, 1990, 11-16 Mar 1990, Page(s):553 – 562.

[3] Sebastian, J., Uceda, J., Cobos, J.A., and Gil, P., "Using zero-current-switched quasiresonant converters as power factor preregulator," International Conference on Industrial Electronics, Control and Instrumentation Proceedings IECON 1991, 28 Oct.-1 Nov. 1991 Page(s):225 - 230 vol.1

[4] Sebastian, J., Martinez, J.A., Alonso, J.M., and Cobos, J.A., "Voltage-follower control in zero-current-switched quasi-resonant power factor preregulators," IEEE Transactions on Power Electronics Volume 13, Is-sue 4, July 1998 Page(s):727 – 738.

[5] B. Mammano and B. Carsten, "Understanding and Optimizing Electromagnetic Compatibility in Switchmode Power Supplies", Unitrode Power Supply Design Seminar, SEM1500, 2002.

[6] A. Santolaria, J. Balcells, D. Gonzalez, J. Gago, "Evaluation of Switching Frequency Modulation in EMI Emission Reduction applied to Power Converters", Industrial Electronics Society, 2003. IECON '03. The 29th Annual Conference of the IEEE, Vol.3, 2-6 Nov. 2003. Page(s):2306 - 2311

[7] F. Lin, D.Y.Chen, "Reduction of Power Supply EMI Emission by Switching Frequency Modulation", Power Electronics, IEEE Transactions, Volume 9, Issue 1, Jan. 1994, Page(s):132 – 137

[8] E. Firmansyah, S. Tomioka, S. Abe, M. Shoyama, T. Ninomiya, "Zero Current Switch-Quasi Resonant Boost Converter Performance in Power Factor Correction Application" ,Proc. of APEC 2009, pp. 1165-1169..

[9] Firmansyah, E., Tomioka, S., Abe, S., Shoyama, M., Ninomiya, T., "Steady state characteristics of active-clamped full-wave zero-current-switched quasi-resonant boost converters", IEEE 6th International 2009 Power Electronics and Motion Control Conference-IPEMC '09, 17-20 May 2009 Page(s):556 – 560.

[10] R. W. Erickson, D. Maksimovic, "Chapter 20 : Soft Switching", in Fundamentals of Power Electronics, second edition, Massachusetts: Kluwer Academic Publishers, 2001.

[11] A.Szabo, M. Kamsara, E.S. Ward, "A unified method for the small-signal modelling of multi-resonant and quasi-resonant converters", Proceedings of the IEEE International Symposium on Circuits and Systems, 1998. ISCAS '98. Volume 3, 31 May-3 June 1998 Page(s):522 - 525 vol.3.

[12] L. Dixon, "Average Current Mode Control of Switching Power Supplies", Unitrode Power Supply Design Seminar, SEM700, 1990.

Novel Adaptive Master-Slave Method for Interleaved Boundary Conduction Mode (BCM) PFC Converters

Hangseok Choi

Fairchild Semiconductor

8 Commerce drive Suite 3A

Bedford, NH 03110, USA

Abstract- **This paper proposes a novel adaptive master-slave interleaving method for boundary conduction mode (BCM) PFC converters. The natural period of every switching cycle of each channel is measured and compared to adaptively determine the master converter. The proposed method guarantees stable interleaving operation in any transient. The proposed method has been implemented and tested on a 400W interleaved BCM boost PFC prototype converter with a dedicated control IC.**

I. INTRODUCTION

The boundary conduction mode (BCM) boost power factor correction (PFC) converter has been the most attractive topology for low-power levels because it can achieve better efficiency with lower cost than continuous conduction mode (CCM) boost PFC converter [1]–[3]. These benefits result from the elimination of the reverse-recovery losses of the boost diode and zero-voltage switching (ZVS) or near ZVS (also called valley switching) of boost switch. However, BCM approach exhibits a relatively large peak inductor current which is twice of its average value and inevitably requires a larger differential mode electro-magnetic interference (EMI) filter in the input side than that of CCM approach. This offsets the benefits of BCM approach and the applicable practical power level has been limited below 300W. Another shortcoming of BCM operation is that its switching frequency varies with the instantaneous line voltage and output load condition. The switching frequency becomes extremely high especially for high line and light load condition, which deteriorates the efficiency severely since not the real ZVS but just valley switching of boost switch is achieved for high line.

Recently, interleaved Boundary conduction mode (BCM) power factor correction (PFC) circuits have become more popular because of their ability to reduce the input current ripple and, consequently, the size of EMI filter extending their applicable practical power level above 300W [4]-[14]. In addition, the output current ripple can be also significantly reduced by the ripple cancellation of interleaving, which allows longer life time of the output capacitor. Another benefit of interleaved approach is that the light load efficiency can be improved by shutting down one channel of the interleaved converters at light load condition, which is known as phase management. By shutting down one channel, the power that the other channel should handle becomes twice making the switching frequency half. This technique is very effective in improving the light load efficiency at high line condition.

Interleaving technique itself has been widely used for constant frequency operation where the interleaving is relatively easy to implement since the turn-on timing is predetermined by the fixed switching frequency. However, the interleaving BCM PFC is challenging since the switching frequency varies with the instantaneous line voltage and output load condition. Especially, the startup and the operation around line zero crossing are the most challenging conditions for interleaving because of the abrupt change of the switching frequency of each channel during transient.

In general, the interleaving techniques can be classified into two categories: open loop master-slave method ([4]-[9]) and PLL based closed loop method ([10]-[14]). In the open loop master-slave method, the master converter runs in standalone BCM, whereas, the slave converter is phase shifted by a time delay equal to half the switching period of the master determined from its previous switching cycle. The half-switching cycle phase shift can be applied to the turn-on instant or turn-off instant. For both cases, either voltage mode control or current mode control can be applied. Since this method is based on the assumption that the master and slave converters are identical, it has a difficulty in guaranteeing stable BCM operation for both converters against mismatch of components and switching timing. A complete analysis of the behavior of the open loop master slave interleaving BCM PFC has been presented in [8], which shows that among the open loop interleaving methods, only the current mode control with a phase shift applied on the turn-on instant can provide stable operation against disturbance.

The closed loop method measures the phase difference between the converters and adjusts the phase difference by using a phase-locked-loop (PLL) approach [10]-[14]. For BCM operation, changing the on-time of the boost switch also changes the switching period and phase difference between two converters. This method can guarantee BCM operation of each channel regardless of the mismatch of components and switching timing. The dynamic behavior of closed loop method has been well analyzed in [14]. Since a low pass filtering is required to obtain the phase difference, the closed loop method responds to the phase shift disturbance relatively slowly and it takes several tens of cycles to correct the disturbance. Another limitation of this method is that it works only when both channels are in BCM. Once BCM is lost and the converter runs with its minimum frequency, which always occurs during startup, the interleaving can be lost.

978-1-4244-4782-4/10 $26.00 © 2010 IEEE

In this paper, a novel adaptive master-slave method for interleaved boundary conduction mode (BCM) PFC converters is proposed. The natural period of every switching cycle of each converter is measured and compared, which adaptively determines which converter should be a master converter. The proposed method responds to the phase shift disturbance very fast and synchronization is achieved within one switching cycle, guaranteeing a stable interleaving operation during any transient. The proposed scheme was implemented and tested on a 400W interleaved BCM boost PFC prototype converter with a dedicated control IC.

II. OPERATION PRINCIPLE OF ADAPTIVE MASTER-SLAVE METHOD

Figure 1 shows the simplified circuit diagram of interleaved boost PFC converter. The zero current detection (ZCD) signal for each inductor is obtained using the voltage across the auxiliary winding of boost inductor. Basically, the proposed method is based on the voltage mode mater-slave method with a turn-on instant synchronization. However, the master and slave are not predetermined and adaptively changed according to the operation of each channel in each switching cycle.

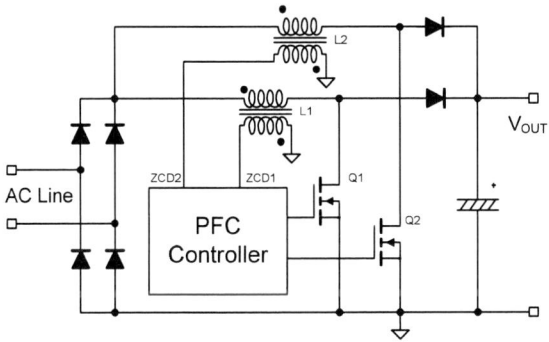

Figure. 1 Simplified circuit diagram of interleaved BCM boost PFC

Figure 2 shows the block diagram of the conventional voltage mode master-slave interleaving method with a turn-on instant synchronization. The turn-on instant of master channel is determined by its own ZCD signal, which allows master to operate in standalone BCM. Whereas, the turn-on instant of slave channel is determined by the phase shift signal (PS), which is shifted from the turn-on instant of master by a time delay equal to half the switching period of the master determined from its previous switching cycle. The turn-off instants of master and slave are determined by their own PWM comparators. The two PWM comparators share the error amplifier output voltage (V_{EA}) for their inverting inputs such that the on-times should be same for both channels. The advantage of voltage mode master slave method is that its BCM operation is not perturbed by the mismatch of boost inductors of master and slave. However, once the ON-time or turn-on instant is perturbed, the BCM operation is perturbed and the slave enters into CCM operation as shown in Figure 3

and 4. Once the slave enters into CCM, it does not come back to BCM operation even after the perturbation is removed.

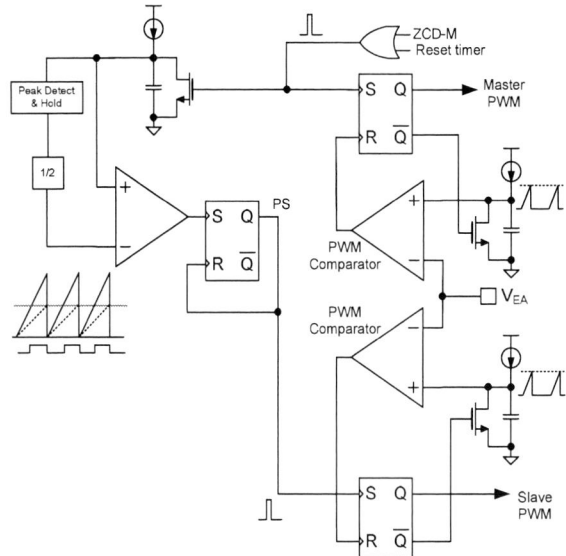

Figure. 2 Block diagram of conventional voltage mode master-slave interleaving method with a turn-on instant synchronization

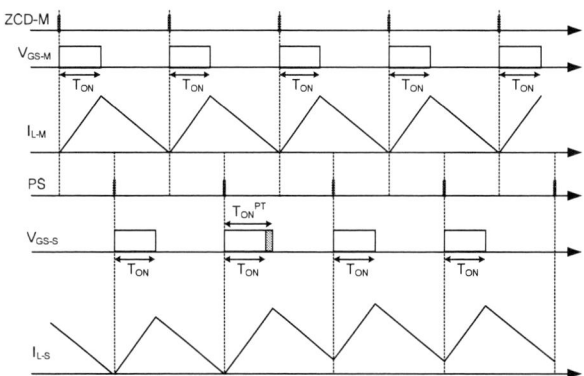

Figure. 3 Effect of ON-time perturbation on the operation of the slave with voltage mode control

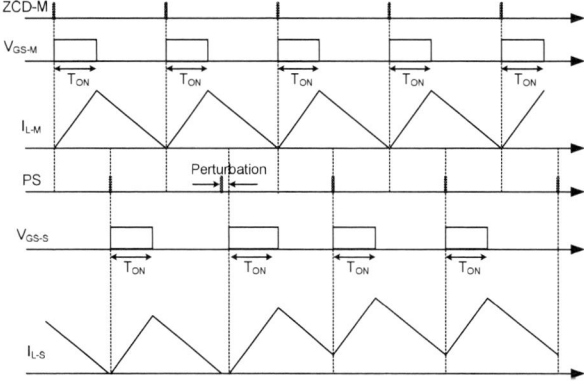

Figure. 4 Effect of turn-on instant perturbation on the operation of the slave with voltage mode control

Figure 5 shows the internal block diagram for the proposed interleaving method. For every switching cycle, the natural period of each channel is measured, which is defined as a time duration from rising edge of gate drive signal to rising edge of zero current detection (ZCD) signal. Using the natural period of the previous cycle, the phase shift signal (PS), which is delayed by half of the natural period of the previous switching cycle with respect to the turn-on instant is generated. Combining ZCD signal and phase shift signal, synchronization out (S-OUT) signal is generated. This is also the synchronization in (S-IN) signal for the other channel. In this way, two channels are cross coupled using S-IN and S-OUT signals and the master and slave channels are adaptively determined.

Among phase shift signal given from the other channel and its own ZCD signal, the signal comes later determines the turn-on instant of each channel. This makes the channel with a longer natural period become the master converter that runs as a stand-alone BCM boost converter determining the turn-on instant of the other slave channel. The turn-off instants of master and slave are determined by their own PWM comparators in the same way as the conventional voltage mode method. The two PWM comparators share the error amplifier output voltage (V_{EA}) for their inverting input such that the on-times should be same for each channel.

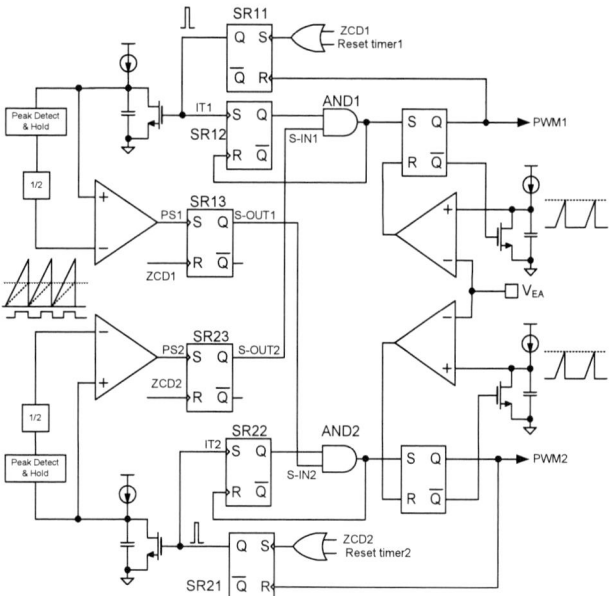

Figure. 5 Block diagram of the proposed daptive interleaving circuit

Figure 6 and 7 show how the proposed method responds to the perturbation and adjust their turn-on instants adaptively. To simplify the explanation of the circuit operation, it is assumed that the two converters are identical and there is no mismatch in the power stage. In Figure 6, the ON-times for both channels are same and both channels operate in BCM before a perturbation is applied to the ON-time of channel 2. From (n+1)-th switching cycle of channel 2, the ON-time is perturbed and increased, which results in increased natural period of channel 2. This makes ZCD2(n+2) occur later than S-OUT1(n+2) and the channel 2 is turned on by ZCD2(n+2) signal for (n+2)-th switching cycle. Whereas, the ZCD1(n+3) occurs prior to S-OUT2(n+2) and the channel 1 is turned on by S-OUT2(n+2) for (n+3)-th switching cycle. Thus, the channel 1 becomes the slave and operates in DCM. It is worthwhile to note that the turn on delay of channel 1 for the (n+3)-th is larger than the delays of switching cycles after (n+3)-th. This is the process how the proposed method auto-adjust their timing to maintain interleaved operation. From (n+5)-th switching cycle, the perturbation of ON-time of channel 2 is removed. Then, channel 2 operates in DCM for one switching cycle to adjust its turn-on timing and both channels enter into BCM with perfect interleaving.

Figure 7 shows how the proposed method finds out the proper interleaving timing when the two channels come out of minimum frequency operation. For most of the BCM PFC controllers, a reset timer should be employed to properly initiate the PWM operation during startup. Because the turn-on switching is triggered by the ZCD signal that requires a switching action of the previous cycle, the PWM operation cannot be initiated without the reset timer. This reset timer is also required when ZCD signal is missing, which usually occurs during the startup when the output voltage is not established yet and is too close to the peak of line voltage. In that condition, there is no enough voltage across the inductor to reset the inductor current to zero while the boost diode is turned off. Thus, the ZCD signal is missing and the turn-on switching of next cycle cannot be initiated without the reset timer as depicted in Figure 7. For the (n+1)-th switching cycle, the channel 1 inductor current reaches zero generating ZCD1 signal. However, the channel 2 still operates in CCM and turn-on of channel 2 is triggered by the reset timer. Therefore, channel 1 ignores its own ZCD signal and operates in DCM to maintain the interleaving operation. When the channel 2 inductor current reaches zero in the (n+3)-th switching cycle generating the ZCD2 signal, the channel 2 begins to operate in BCM. It is worthwhile to note that the channel 2 operates in BCM in (n+5)-th switching cycle to adjust the switching timing for the proper interleaving timing.

978-1-4244-4782-4/10 $26.00 © 2010 IEEE   38

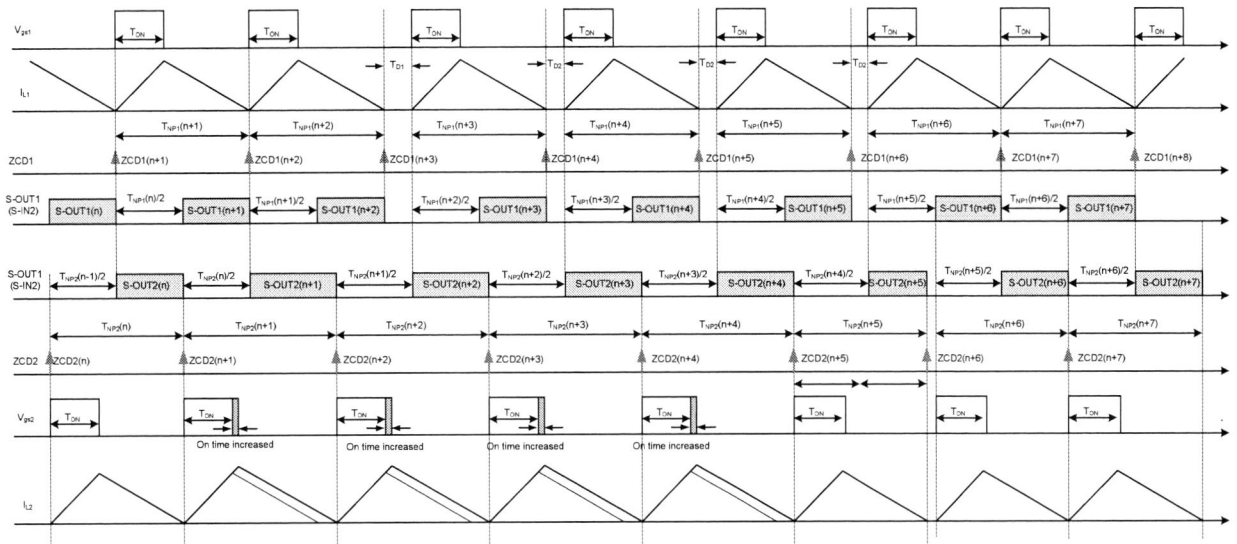

Figure. 6 The effect of ON-time perturbation on the interleaving operation

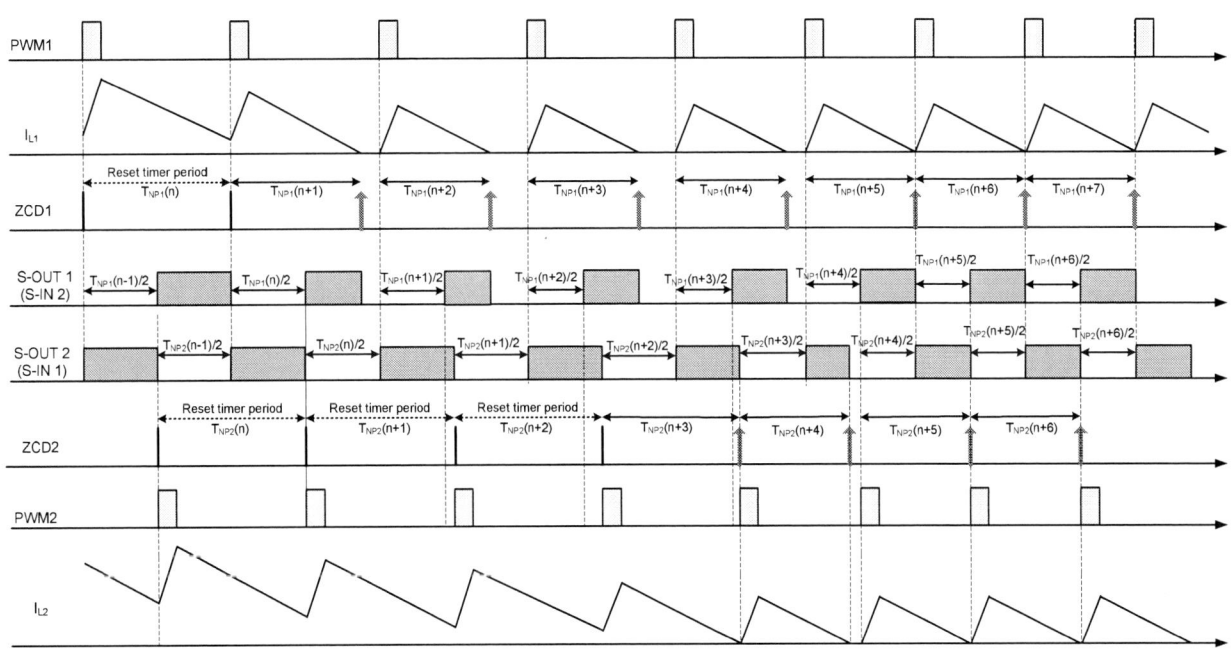

Figure. 7 The effect of turn-on instant perturbation on the interleaving operation

III. EXPERIMENTAL RESULTS

In order to show the validity of the proposed method, a 400W interleaved BCM boost PFC prototype as shown in Figure 8 has been built and tested. The proposed interleaving method has been implemented in an integrated circuit (FAN9612). The control IC (FAN9612) detects the zero current instant using the boost inductor auxiliary winding voltage when the slope (dv/dt) of the winding voltage becomes zero. Since the auxiliary winding voltage resonates

without being clamped by zero voltage switching when the instantaneous line voltage is larger than the half of the output voltage, it can be seen whether a channel is turned on by its own ZCD signal or not by simply observing the inductor auxiliary winding voltage. Figure 9 shows the interleaving waveforms of the proposed adaptive master slave interleaving method when the instantaneous line voltage is larger than half of the output voltage. Since the inductor auxiliary winding voltage reflects the drain-to-source voltage, the drain to source voltage of each channel is observed instead of auxiliary winding voltages. As discussed in the previous section, the channel with a longer natural period takes the role of master and determines the turn-on instant of the other channel (slave). Thus, the master turns on at the valley of the drain-to-source voltage by its own ZCD signal. Whereas, the slave turns later than its own ZCD signal, which causes the slave to operate in DCM. As can be seen in Figure 9, the master and slave are adaptively changed continuously.

Figure 10 shows how the proposed method maintains the interleaving of two channels around the AC line zero crossing. Around line zero crossing, the inductor current is very small and the zero current detection using the inductor auxiliary winding is lost. Thus, the each channel runs with its own minimum frequency given by its reset timer. The proposed method can maintain the interleaving around the line zero crossing regardless of sudden change of the switching frequency. Figure 11 and 12 shows how the proposed method maintains the interleaving operation when the converter enters into CCM during startup. Once the converter enters into CCM, ZCD is lost and each channel runs with its own minimum frequency given by its reset timer. The proposed method can maintain the interleaving around the line zero crossing as well.

Figure 13 and 14 show operation waveforms at full load with 115Vac and 230Vac input voltages, respectively. As can be seen, the ripple of input current is reduced by the interleaving operation.

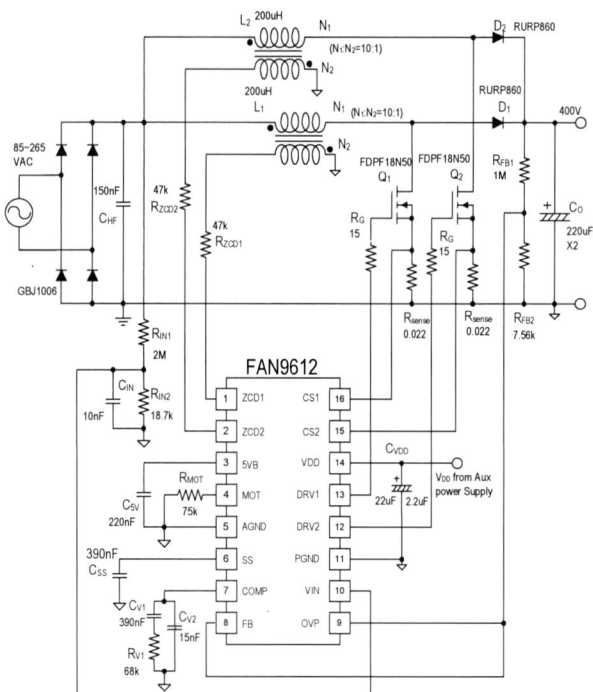

Figure. 8 Schematic of 400W interleaved BCM boost PFC converter with proposed adaptive master-slave method

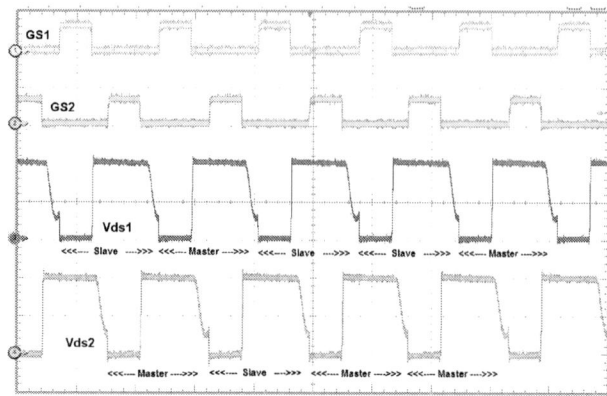

Figure. 9 Alternation of master and slave
(Master: valley switching, Slave: DCM switching)

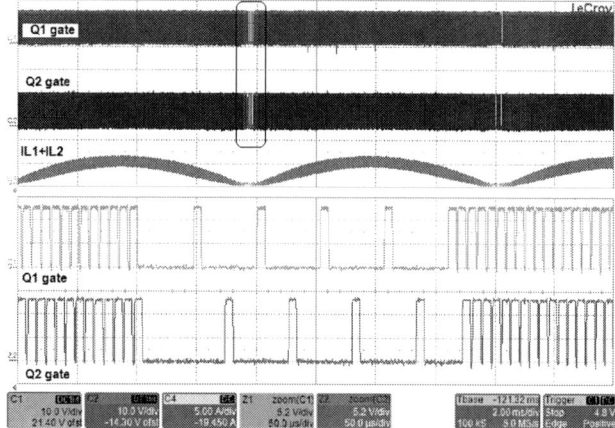

Figure. 10 Interleaving waveforms around line zero crossing

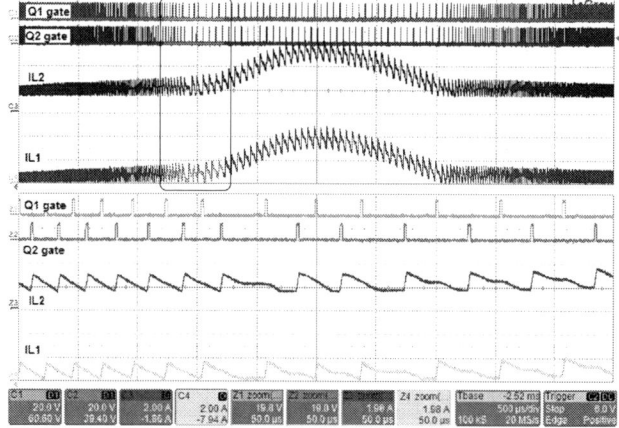

Figure. 11 Interleaving waveforms when the converters enter into CCM during startup

978-1-4244-4782-4/10 $26.00 © 2010 IEEE

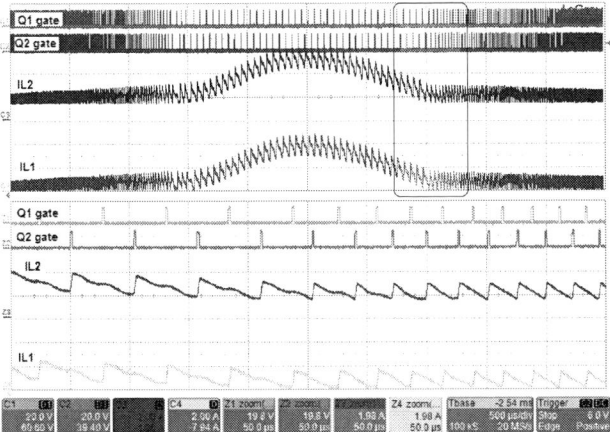

Figure. 12 Interleaving waveforms when the converters come out of CCM during startup

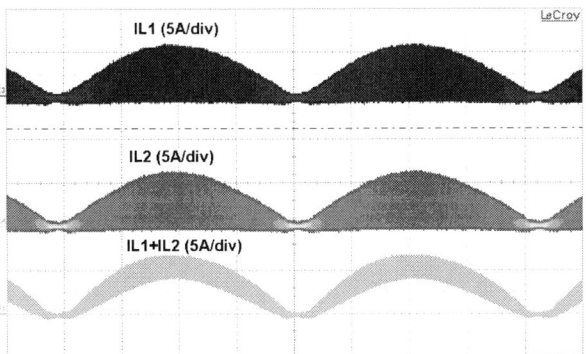

Figure.13 Ripple cancellation with interleaving operation (low line, V_{AC}=115Vrms)

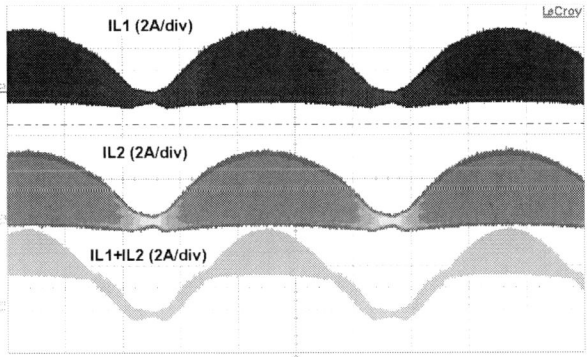

Figure. 14 Ripple cancellation with interleaving operation (high line, V_{AC}=230Vrms)

IV. CONCLUSION

This paper has proposed a novel adaptive master-slave method for interleaved boundary conduction mode (BCM) PFC converters. The natural period of every switching cycle of each converter is measured and compared, which adaptively determines which converter should be a master converter. The proposed method has a very fast response to the phase shift disturbance and the disturbance is compensated within one switching cycle, guaranteeing interleaving operation in any transient. The proposed scheme was implemented and tested on a 400W interleaved BCM boost PFC converter with a dedicated control IC.

REFERENCES

[1] Claudio Adragna, Laszlo Huber, Brian T. Irving and Milan M. Jovanovic, "Analysis and performance Evaluation of Interleaved DCM/CCM Boundary Boost PFC Converters Around Zero-Crossing of Line Voltage" APEC 2009, pp.1151-1157

[2] J.S. Lai and D. Chen, "Design consideration for power factor correction boost converter operating at the boundary of continuous conduction mode and discontinuous conduction mode," IEEE Applied Power Electronics Conf. (APEC) Proc., pp. 267-273, Mar. 1993.

[3] J.W. Kim, S.M. Choi, and K.T. Kim, "Variable on-time control of the critical conduction mode boost power factor correction converter to improve zero-crossing distortion," *IEEE Power Electronics and Drive Systems Conf. (PEDS) Proc.*, pp. 1542-1546, Nov. 2005.

[4] T. Ishii and Y. Mizutani, "Power factor correction using interleaving technique for critical mode switching converters," *IEEE Power Electronics Specialists Conf. (PESC) Proc.*, pp. 905-910, May 1998.

[5] T. Ishii and Y. Mizutani, "Variable frequency switching of synchronized interleaved switching converters," U.S. Patent 5,905,369, May 18, 1999.

[6] B.T. Irving, Y. Jang, and M.M. Jovanović, "A comparative study of soft-switched CCM boost rectifiers and interleaved variable-frequency DCM boost rectifier," *IEEE Applied Power Electronics Conf. (APEC) Proc.*, pp. 171-177, Feb. 2000.

[7] T.F. Wu, J.R. Tsai, Y.M. Chen, and Z.H. Tsai, "Integrated circuits of a PFC controller for interleaved critical-mode boost converters," *IEEE Applied Power Electronics Conf. (APEC) Proc.*, pp. 1347-1350, Feb. 2007.

[8] L. Huber, B.T. Irving, and M.M. Jovanović, "Open-loop control methods for interleaved DCM/CCM boundary boost PFC converters," *IEEE Trans. Power Electronics*, vol. 23, no. 4, pp. 1649-1657, July 2008.

[9] L. Huber, B. T. Irving, C. Adragna, and M. M. Jovanović, "Implementation of Open-Loop Control for Interleaved DCM/CCM Boundary Boost PFC Converters," *IEEE Applied Power Electronics Conf. (APEC) Proc.*, pp. 1010-1016, Feb. 2008.

[10] M.S. Elmore, "Input current ripple cancellation in synchronized, parallel connected critically continuous boost converters," *IEEE Applied Power Electronics Conf. (APEC) Proc.*, pp. 152-158, Mar. 1996.

[11] M.S. Elmore and K.A. Wallace, "Zero voltage switching supplies connected in parallel," U.S. Patent 5,793,191, Aug. 11, 1998.

[12] B. Lu, "A Novel Control Method for Interleaved Transition Mode PFC," *IEEE Applied Power Electronics Conf. (APEC) Proc.*, pp. 697-701, Feb. 2008.

[13] X. Xu, and A. Huang, "A Novel Closed Loop Interleaving Strategy of Multiphase Critical Mode Boost PFC Converters," *IEEE Applied Power Electronics Conf. (APEC) Proc.*, pp. 1033-1038, Feb. 2008.

[14] L. Huber, B.T. Irving, and M.M. Jovanović, "Closed-Loop Control Methods for Interleaved DCM/CCM Boundary Boost PFC Converters," *IEEE Applied Power Electronics Conf. (APEC) Proc.*, pp. 991-997, Feb. 2009.

[15] "Design Consideration for Interleaved Boundary Conduction Mode PFC using FAN9612," *Fairchild Semiconductor, Application Note AN6086*, June 2008, available at http://www.fairchildsemi.com/an/AN/AN-6086.pdf

A Novel Bridgeless Single-Stage Half-Bridge AC/DC Converter

Woo-Young Choi[1], Wen-Song Yu, and Jih-Sheng (Jason) Lai

Virginia Polytechnic Institute and State University
Future Energy Electronics Center
106 Plantation Road, Blacksburg, VA, 24061, USA
E-mail: wychoi@vt.edu[1]

Abstract — **This paper proposes a new bridgeless single-stage half-bridge ac-dc converter. The proposed converter integrates the operation of the bridgeless power factor correction (PFC) boost rectifier and the asymmetrical pulse-width modulation (APWM) half-bridge dc-dc converter. The proposed converter provides high power factor and direct power conversion from the line voltage to an isolated dc output voltage without using the full-bridge diode rectifier. Conduction losses are lowered with a simple circuit structure. Switching losses are also reduced by achieving zero-voltage switching (ZVS) of the power switches. The effectiveness of the proposed converter is verified on a 250 W (48 V/5.2 A) experimental prototype.**

I. INTRODUCTION

The research for single-stage ac-dc converters has been an active research topic for power factor correction (PFC) circuits in the power electronics. A number of single-stage PFC ac-dc converters have been introduced in the literature. Among them, discontinuous-conduction-mode (DCM) single-stage PFC ac-dc converters are widely used for their simple and efficient structures [1]-[8]. Generally, two power stages of the PFC circuit and dc-dc converter are simplified by sharing a common switch [1]-[4] or a pair of switches [5]-[8]. Most single-stage PFC ac-dc converters use single-switch dc-dc converter topologies such as flyback [1], [2] and forward converters [3], [4]. However, the single-stage single-switch ac-dc converters operate under hard-switching condition. The voltage stresses of switching devices and power conversion efficiency have not been optimized yet. The practical use of the single-stage single-switch ac-dc converters has been limited for low-power applications with power levels lower than 80 W.

Single-stage soft-switching ac-dc converters have been developed to improve the performance of single-stage PFC ac-dc converters [5]-[8]. Single-stage soft-switching ac-dc converters based on the half-bridge converter topology are attractive because they provide low component count and zero-voltage switching (ZVS) operation of the power switches [5], [6]. Similar efforts have been put in optimizing and improving the performance of the converter by using active-clamping techniques [7], [8]. The majority of these development efforts have been focused on reducing switching power losses on the power conversion efficiency. However, so far no single-stage ac-dc converter without using the full-bridge diode rectifier has been reported. The single-stage ac-dc converters still use the full-bridge diode rectifier, which causes high conduction losses. The full-bridge diode rectifier suffers from significant conduction losses especially at low

line voltage. Thus, a bridgeless single-stage ac-dc converter should be studied to reduce conduction losses and component counts.

This paper proposes a new bridgeless single-stage half-bridge ac-dc converter. The proposed converter integrates the operation of the bridgeless PFC boost rectifier [9]-[12] and the asymmetrical pulse-width modulation (APWM) half-bridge dc-dc converter [13]. The proposed converter provides high power factor and direct power conversion from the line voltage to an isolated dc output voltage without using the full-bridge diode rectifier. By allowing the boost inductor to operate in DCM, PFC and fast output voltage regulation are performed simultaneously by the APWM control of power switches. Conduction losses are lowered by essentially eliminating the full-bridge diode rectifier. Switching losses are also reduced by achieving zero-voltage switching (ZVS) of the power switches. Thus, the proposed approach not only reduces the number of circuit components, but also makes it possible to increase power efficiency of single-stage PFC ac-dc converter. The performance of the proposed converter is evaluated by the experimental results based on a 250 W (48 V/5.2 A) converter prototype. The proposed converter achieves a high-efficiency of 93 % with almost unity power factor at 90 V_{rms} line voltage.

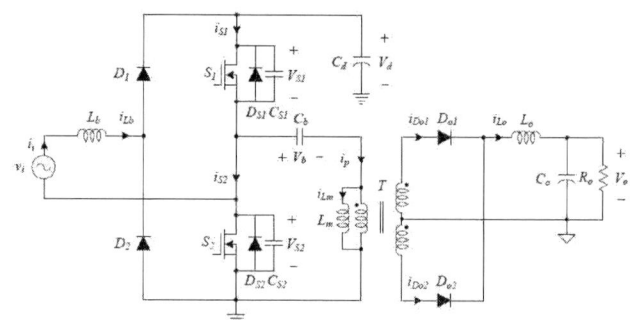

Fig. 1. Circuit diagram of the proposed converter.

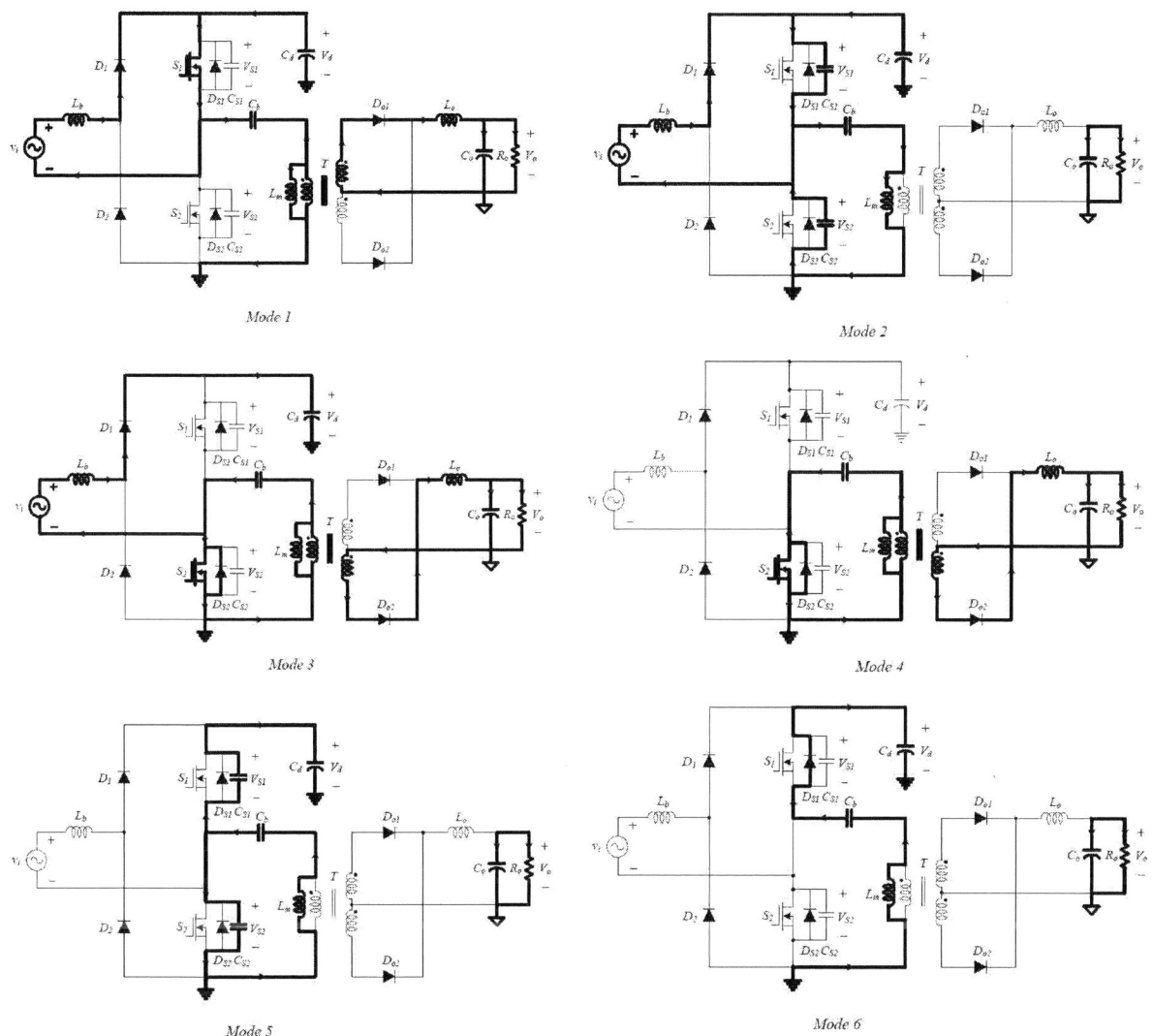

Fig. 2. Operating modes of the proposed converter in a positive half-line cycle.

II. OPERATION PRINCIPLE

Fig. 1 shows a circuit diagram of the proposed converter. The bridgeless PFC boost rectifier consists of the boost inductor L_b, dc-link capacitor C_d, and switching devices D_1, D_2, S_1, and S_2. D_1 and D_2 are slow-recovery diodes. S_1 and S_2 are MOSFETs with output capacitors C_{S1} and C_{S2} ($C_S = C_{S1} = C_{S2}$), respectively. D_{S1} and D_{S2} are body diodes of S_1 and S_2, respectively. The APWM half-bridge dc-dc converter consists of the dc-link capacitor C_d, S_1 and S_2, blocking capacitor C_b, transformer T, output diodes D_{o1} and D_{o2}, output filter inductor L_o, and output capacitor C_o. By sharing C_d, S_1 and S_2, the proposed converter integrates the operation of the bridgeless PFC boost rectifier and the APWM half-bridge dc-dc converter. S_1 and S_2 are controlled asymmetrically. By allowing the boost inductor to operate in DCM, PFC and fast

output voltage regulation are performed simultaneously by the APWM control of power switches.

For both positive and negative half-line cycle of v_i, the proposed converter has symmetric operation. In the positive half-line cycle, S_1 is controlled with duty ratio D. Then, the conduction times of the switches S_1 and S_2 are DT_s and $(1-D)T_s$, respectively. When S_1 is turned on, the input current i_i flows through L_b, D_1, and S_1. When S_1 is turned off, the input current i_i flows through L_b, D_1, C_d, S_2, and D_{S2}. In the negative half-line cycle, S_2 is controlled with duty ratio D. Then, the conduction times of the switches S_1 and S_2 are $(1-D)T_s$ and DT_s, respectively. When S_2 is turned on, the input current i_i flows through S_1, D_1, and L_b. When S_2 is turned off, the input current i_i flows through S_1, D_{S1}, C_d, D_1, and L_b. The transformer T has the magnetizing inductor L_m and leakage inductor L_{lk} with the turns ratio of $1 : n$.

Fig. 2 shows the operating modes of the proposed converter during T_s for the positive half-line cycle. Only the operation principle for the positive half-line cycle is described in this section. Due to the symmetric operation, the operation principle for the negative half-line cycle is not described here. The capacitors C_d and C_o are large enough so that the voltages V_d and V_o are assumed to be constant. $|v_i|$ is considered constant during one switching period T_s ($=f_s$).

Mode 1 $[t_0, t_1]$: At $t = t_0$, S_1 is turned on. The input current i_i flows through L_b, D_1, and S_1. The boost inductor L_b stores energy from the line voltage. The voltage across L_m is $V_d - V_b$. The primary current i_p increases as

$$i_p(t) = i_p(t_0) + \frac{V_d - V_b}{L_m}(t - t_0). \tag{1}$$

The transformer T transfers energy to the output through the output diode D_{o1}. The switch current i_{S1} is the sum of boost inductor current i_{Lb} and the primary current i_p.

Mode 2 $[t_1, t_2]$: At $t = t_1$, S_1 is turned off. As the primary current i_p charges C_{S1} and discharges C_{S2}, the voltage V_{S2} across S_2 decreases from V_d to zero. Since the time interval in this mode is negligible compared to T_s, the primary current i_p and boost inductor current i_{Lb} are considered to be constant. When the voltage V_{S2} across S_2 is zero, the primary current i_p begins to flow the body diode D_{S2} of S_2.

Mode 3 $[t_2, t_3]$: At $t = t_2$, S_2 is turned on. ZVS of S_2 is achieved because the voltage V_{S2} across S_2 is zero. The input current i_i flows through L_b, D_1, C_d, S_2, and D_{S2}. The energy stored in the boost inductor L_b is released to the dc-link capacitor C_d. The voltage across L_m is $-V_b$. The primary current i_p decreases as

$$i_p(t) = i_p(t_2) - \frac{V_b}{L_m}(t - t_2). \tag{2}$$

The transformer T transfers energy to the output through the output diode D_{o2}. The switch current i_{S2} is the sum of boost inductor current i_{Lb} and the primary current i_p as

Mode 4 $[t_3, t_4]$: At $t = t_3$, the boost inductor current i_{Lb} is zero. From the volt-second balance on the boost inductor L_b, we have the following relation as

$$|v_i|DT_s = (V_d - |v_i|)\Delta T_s. \tag{3}$$

By simplifying (3), we have the time interval ΔT_s during this mode as

$$\Delta T_s = \frac{|v_i|DT_s}{(V_d - |v_i|)}. \tag{4}$$

Mode 5 $[t_4, t_5]$: At $t = t_4$, S_2 is turned off. As the primary current i_p charges C_{S2} and discharges C_{S1}, the voltage V_{S1} across S_1 decreases from V_d to zero. Since the time interval in this mode is negligible compared to T_s, the primary current i_p and boost inductor current i_{Lb} are considered to be constant. When the voltage V_{S1} across S_1 is zero, the primary current i_p begins to flow the body diode D_{S1} of S_1.

Mode 6 $[t_5, t_6]$: At $t = t_5$, the voltage V_{S1} across S_1 is zero. The primary current i_p begins to flow the body diode D_{S1} of S_1. ZVS of S_1 can be achieved when S_1 is turned on again.

In a positive half-line cycle, from the volt-second balance on L_m during T_s, the voltage V_b across blocking capacitor C_b is expressed as

$$V_b = DV_d. \tag{5}$$

Using (5), from the volt-second balance on L_o during T_s, we have the relation between the dc-link voltage and the output voltage in a positive half-line cycle as

$$\frac{V_o}{V_d} = nD(1 - D). \tag{6}$$

III. EXPERIMENTAL RESULTS

A 250 W converter prototype was built and tested under the universal line voltage. The proposed converter has the following parameters as line voltage $v_i = 90$ V_{rms}, output voltage $V_o = 48$ V, switching frequency $f_s = 50$ kHz, boost inductor $L_b = 100$ µH, blocking capacitor $C_b = 1$ µF, dc-link capacitor $C_d = 440$ µF, output capacitor $C_o = 680$ µF, magnetizing inductor $L_m = 100$ µH, output filter inductor $L_o = 30$ µH. For switching devices, power switches $S_1 = S_2 = 20N60C3$, $D_1 = D_2 = $ SFR305PT, and $D_{o1} = D_{o2} = $ MBR20100 are used. The controller is implemented by using a single-chip microcontroller, Microchip dsPIC30F3011. Output voltage is measured by using 10-bit analog-to-digital (A/D) converter in the microcontroller.

The circuit design was simulated using PSIM 6.0; the schematic circuit is shown in Fig. 3. Fig. 4 shows the simulation results when the proposed converter supplies 250 W output power. Fig. 4(a) shows the line voltage v_i, boost inductor current i_{Lb}, and dc-link voltage V_d at $v_i = 90$ V_{rms}. Fig. 4(b) shows the voltage and current waveforms of power switches S_1 and S_2, respectively. Fig. 5 shows the experimental results when the proposed converter supplies 250 W output power. Fig. 5(a) shows the line current i_i at $v_i = 90$ V_{rms}. Fig. 5(b) shows the primary current i_p and voltage V_{S1} of the switch S_1 at $v_i = 90$ V_{rms}. The proposed bridgeless single-stage ac-dc converter provides high power factor and direct power conversion from the line voltage to an isolated dc output voltage without the suing full-bridge diode rectifier.

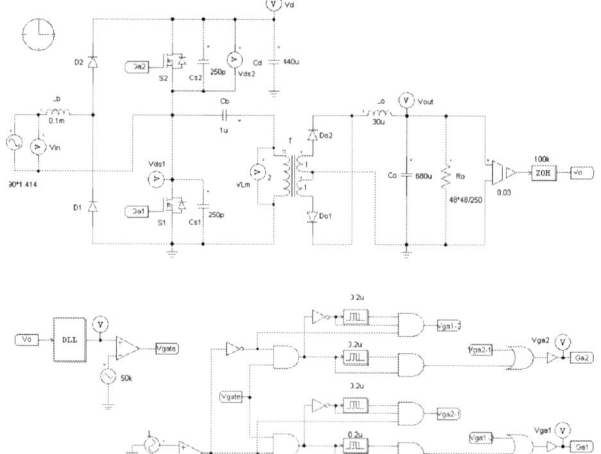

Fig. 3. Circuit diagram of the simulation circuit.

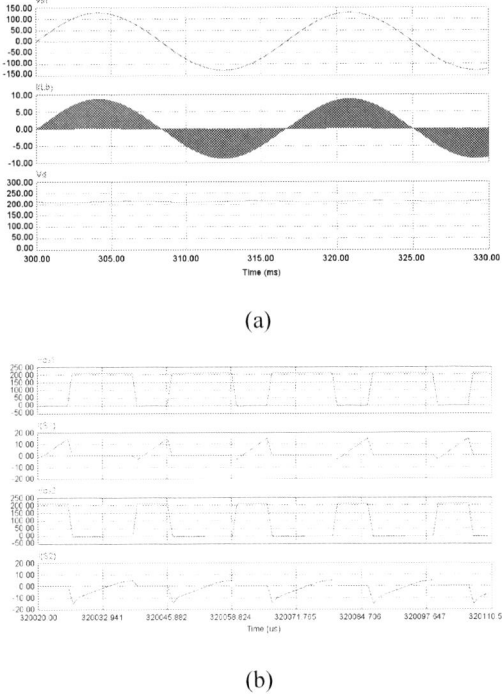

(a)

(b)

Fig. 4. Simulation results.

The proposed converter achieves a high-efficiency of 93 % with almost unity power factor at 90 V_{rms} line voltage. Compared to the previous approaches (single-stage design [7] and two-stage design [10]), the proposed approach increase the power efficiency and reduce component counts by lowering conduction losses and by eliminating the full-bridge diode rectifier in the single-stage PFC ac-dc converters. More detailed efficiency comparison, experimental waveforms and circuit design guideline will be discussed in further work.

IV. CONCLUSION

In this paper, a new bridgeless single-stage half-bridge ac-dc converter has been proposed. As a new bridgeless single-stage PFC ac-dc power conversion scheme, the proposed converter integrates the operation of the bridgeless PFC boost rectifier and the APWM half-bridge dc-dc converter. Without using any full-bridge diode rectifier, the proposed converter achieves high power factor and direct power conversion from the line voltage to an isolated dc output voltage. The proposed converter has the following features for the single-stage PFC ac-dc converter as

1. Low conduction losses by essentially eliminating the full-bridge diode rectifier;

2. Reduced component counts by integrating two power conversion stages;

3. Low switching losses by the ZVS operation of power switches;

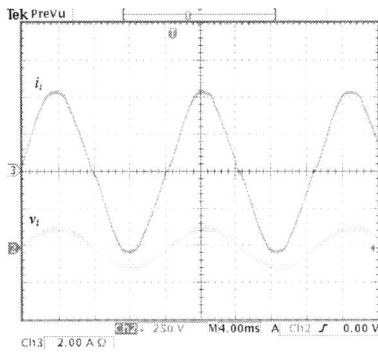

(a)

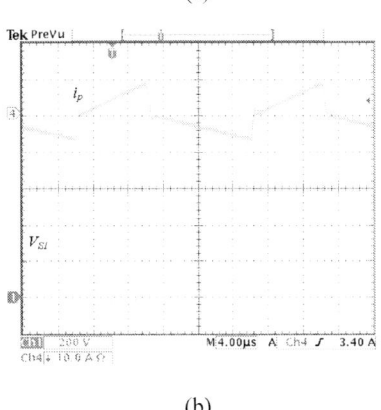

(b)

Fig. 5. Experimental results.

The performance of the proposed converter was evaluated by the experimental results based on a 250 W (48 V/5.2 A) converter prototype. The proposed converter achieves a high-efficiency of 93 % with almost unity power factor at 90 V_{rms} line voltage.

V. REFERENCES

[1] J. Y. Lee, "Single-stage AC/DC converter with input current dead-zone control for wide input voltage ranges," *IEEE Transactions on Industrial Electronics*, Vol. 54, No. 2, pp. 724-732, Apr. 2007.

[2] S. Luo, W. Qiu, W. Wu, and I. Batarseh, "Flyboost power factor correction cell and a new family of single-stage AC/DC converters," *IEEE Transactions on Power Electronics*, Vol. 20, No. 1, pp. 25-34, Jan. 2005.

[3] C. Qiao and K. M. Smedley, "A topology survey of single-stage power factor corrector with a boost type input current shaper," *IEEE Transactions on Power Electronics*, Vol. 16, No. 3, pp. 360-368, May 2001.

[4] H. E. Tacca, "Power factor correction using merged flyback-forward converters," *IEEE Transactions on Power Electronics*, Vol. 15, No. 4, pp. 585-594, Jul. 2000.

[5] R. T. Chen, Y. Y. Chen, and Y. R. Yang, "Single-stage asymmetrical half-bridge regulator with ripple reduction technique," *IEEE Transactions on Power Electronics*, Vol. 23, No. 3, pp. 1358-1369, May 2008.

[6] T. Shimizu, K. Wada, and N. Nakamura, "A novel single-stage half-bridge AC-DC converter with high power factor," *IEEE Transactions on Industrial Electronics*, Vol. 48, No. 6, pp. 1219-1225, Dec. 2001.

[7] W. Y. Choi, *et al*, "Single-stage soft-switching converter with boost type of active clamp for wide input voltage ranges," *IEEE Transactions on Power Electronics*, Vol. 24, No. 3, pp. 730-741, Mar. 2009.

[8] Y. M. Liu and L. K. Chang, "Single-stage soft-switching AC-DC converter with input current shaping for universal line applications," *IEEE Transactions on Industrial Electronics*, Vol. 56, No. 2, pp. 467-479, Feb. 2009.

[9] W. Y. Choi, *et al*, "Bridgeless boost rectifier with low conduction losses and reduced diode reverse-recovery problems," *IEEE Transactions on Industrial Electronics*, Vol. 54, No. 2, pp. 769-780, Apr. 2009.

[10] W. Y. Choi, J. M. Kwon, and B. H. Kwon, "Efficient LED back-light power supply for liquid-crystal display" *IET Proceeding on Electric Power Applications*, Vol. 1, No. 2, pp. 133-142, Mar. 2007.

[11] W. Y. Choi, J. M. Kwon, and B. H. Kwon, "Bridgeless dual-boost rectifier with reduced diode reverse-recovery problems for power-factor correction" *IET Proceeding on Power Electronics*, Vol. 1, No. 2, pp. 194-202, Jun. 2008.

[12] L. Huber, Y. T. Jang, and M. M. Jovanovic, "Performance evaluation of bridgeless PFC boost rectiifers", *IEEE Transactions on Power Electronics*, Vol. 23, No. 3, pp. 1381-1390, May 2008.

[13] C. E. Kim and G. W. Moon, "Input voltage feedforward circuit minimizing current stress of voltage doubler rectifier asymmetrical half-bridge converter," *IEEE Transactions on Industrial Electronics*, Vol. 55, No. 5, pp. 2222-2224, May 2008.

Power Quality Improvement at Medium-Voltage Grids Using Hexagram Active Power Filter

Jun Wen, Liang Zhou and Keyue Smedley
Power Electronics Lab, Univ. of California
Irvine, CA, USA

Abstract—A new Hexagram Active Power Filter (APF) is proposed for harmonics elimination at medium-voltage grids. The new converter is composed of six interconnected three-phase standard two-level voltage source converter modules with advantages including modular structure with easy construction and maintenance, no voltage unbalance problem, reduced voltage stress, symmetrical structure with automatic and equal voltage and current sharing, and easy control with well-developed two-level control techniques, etc. In this paper, One-Cycle Control – an excellent control scheme for two-level converters is extended to realize the control of the Hexagram converter for active power filter function. Both simulation and experiment have been conducted and verified the proposed converter and proposed control scheme.

I. INTRODUCTION

Traditional diode/thyristor rectifiers with capacitive and inductive current draw, as used in electronics appliances and motor drive systems, impose nonlinear loads to the ac grids. These nonlinear loads generate harmonic and reactive current, which leads to low power factor, low energy efficiency, and harmful disturbance to other appliances [1]. Power Factor Correction (PFC) [2-3] and Active Power Filter (APF) [4-5] are viable solutions to eliminate the harmonics and improve the power factor. With the PFC approach, an input current-shaping processes all the power and corrects the current to unity power factor. As a contrast, an APF is connected in parallel (shunt APF) or in series (series APF) with the nonlinear load and provides only the harmonic and reactive power to cancel the one generated by the nonlinear loads. In this case, only a small portion of the energy is processed, which may result in overall higher energy efficiency and higher power processing capability. References [6-14] provide examples of shunt APFs with various control strategies.

Limited by the voltage ratings of semiconductor switches, APFs based on two-level Voltage Source Converter (VSC) can not be directly connected to the medium-voltage grids which range from 2.3kV to 13.8kV. As a result, multilevel converters [15-21] are emerged to accommodate this situation. References [22-29] are examples of multilevel APFs, among which the one based on the cascaded H-bridge converter is attractive due to its many advantages such as modular

structure with easy construction and maintenance, and simple control.

However, determined by the single-phase structure, the cascade H-bridge converter suffers from large dc energy storage requirement and large number of components. In light of this, a new multilevel converter – Hexagram converter, as indicated in Fig. 1, has been proposed recently [30-35], which also has the modular structure, but is based on the three-phase modules, therefore, has reduced dc energy storage requirement and reduce component count in nature.

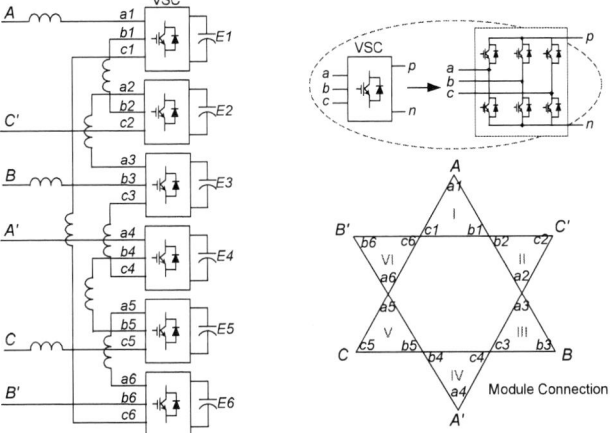

Fig. 1. A new multilevel converter – Hexagram converter.

This paper will investigate the Hexagram converter for APF application to improve the power quality at the medium-voltage grids. The paper will start from the APF system configuration in Section II. Then, One-Cycle Control (OCC) scheme for two-level APF [36] will be briefly reviewed in Section III. In Section IV, the control scheme for the Hexagram APF based on two-level OCC control technique will be proposed and analyzed in detail. And Section V and VI respectively give the simulation and experimental results. Finally, a brief conclusion will be provided.

II. HEXAGRAM CONVERTER APF SYSTEM

Hexagram converter has six ac terminals. When connected to the grid, an isolation transformer is required to provide six

978-1-4244-4782-4/10 $26.00 © 2010 IEEE

terminal three-phase source (neutral floating) or star-connected six-phase source, as indicated by Fig. 2(a) and (b). The primary of the transformer can be either Y-connected or Δ-connected.

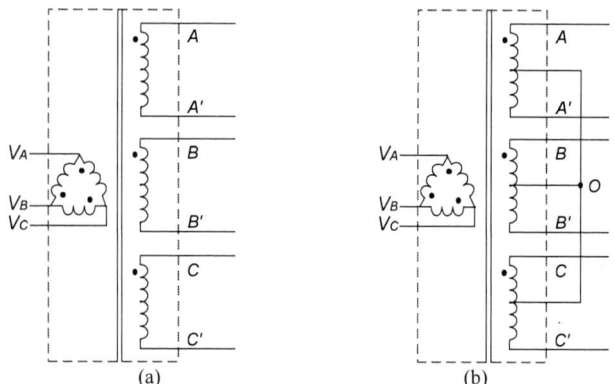

(a)   (b)

Fig. 2. Input isolation transformer configurations.
(a) Three-phase configuration (neutral floating). (b) Six-phase configuration.

This paper will highlight the Hexagram converter connected to the three-phase transformer as given in Fig. 2(a). In this case, only three phase currents need to be controlled, while at least five phase currents need to be controlled if connected to the transformer in Fig. 2(b).

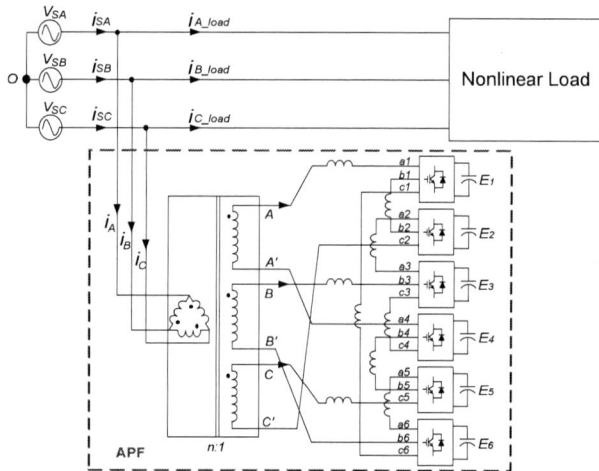

Fig. 3. A complete Hexagram converter APF system.

A complete system with the Hexagram converter as shunt APF is depicted in Fig. 3. With suitable control, the Hexagram converter will provide the harmonics and reactive currents to the grid so that the source currents are always sinusoidal and at unity power factor regardless of the load condition.

III. REVIEW OF OCC BIPOLAR CONTROL FOR TWO-LEVEL APF

This section will briefly review the OCC control technique for the two-level voltage source converter for the APF applications.

Fig. 4 shows the three-phase VSC and its switching average model [36].

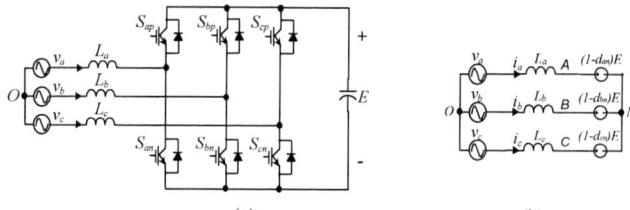

(a)   (b)

Fig. 4. Three-phase VSC and its switching average model.
(a) Three-phase VSC (b) Switching average model.

According to the converter switching average model, the input and output of the converter has the following relationship:

$$\begin{bmatrix} -\dfrac{2}{3} & \dfrac{1}{3} & \dfrac{1}{3} \\ \dfrac{1}{3} & -\dfrac{2}{3} & \dfrac{1}{3} \\ \dfrac{1}{3} & \dfrac{1}{3} & -\dfrac{2}{3} \end{bmatrix} \cdot \begin{bmatrix} d_{an} \\ d_{bn} \\ d_{cn} \end{bmatrix} = \frac{1}{E} \cdot \begin{bmatrix} v_a \\ v_b \\ v_c \end{bmatrix} \tag{1}$$

where d_{an}, d_{bn} and d_{cn} are respectively the duty ratio signals of the bottom switches S_{an}, S_{bn} and S_{cn}, v_a, v_b and v_c are respectively the input voltages of the converter, and E is the output voltage of the converter.

One possible solution for (1) is as follows,

$$\begin{bmatrix} d_{an} \\ d_{bn} \\ d_{cn} \end{bmatrix} = 0.5 - \frac{1}{E} \cdot \begin{bmatrix} v_a \\ v_b \\ v_c \end{bmatrix} \tag{2}$$

The control goal is to realize the following relationship between the input voltages and source currents,

$$\begin{bmatrix} v_a \\ v_b \\ v_c \end{bmatrix} = R_e \cdot \begin{bmatrix} i_a \\ i_b \\ i_c \end{bmatrix} \tag{3}$$

where R_e is the emulated resistance, and i_a, i_b and i_c are the input source currents.

Combining (2) and (3) results in the control equation,

$$R_s \cdot \begin{bmatrix} i_a \\ i_b \\ i_c \end{bmatrix} = V_m \cdot \begin{bmatrix} 1 - 2d_{an} \\ 1 - 2d_{bn} \\ 1 - 2d_{cn} \end{bmatrix} \tag{4}$$

$$V_m = \frac{0.5E \cdot R_s}{R_e} \tag{5}$$

where R_s is the equivalent current sensing resistance, and V_m is the output of the feedback error compensator.

Above control equation can be implemented by OCC control. Fig. 5 depicts the OCC bipolar controller of the two-level VSC for APF application [36]. The source currents are sensed and compared with carrier signals to generate proper driving signals. With OCC control, the harmonics in the three-phase input source currents will be eliminated and unity power factor can be achieved.

978-1-4244-4782-4/10 $26.00 © 2010 IEEE

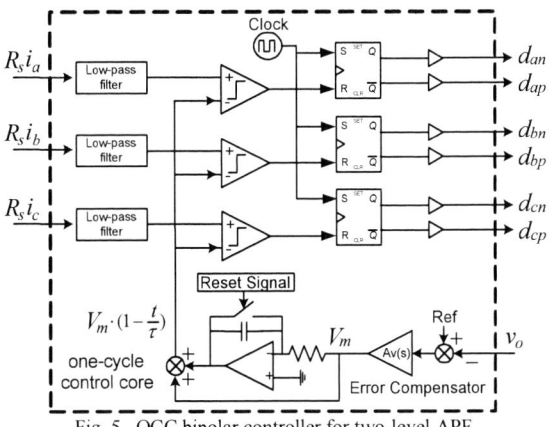

Fig. 5. OCC bipolar controller for two-level APF.

Fig. 6 gives a variation of the controller implementation, which senses the load currents and the compensation currents from the APF to duplicate the information of the input source currents. The result is the same as that can be achieved by the controller in Fig. 5.

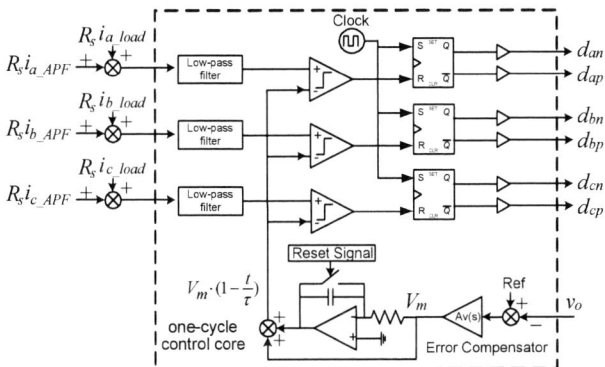

Fig. 6. A variation of OCC bipolar controller for two-level APF.

IV.   PROPOSED CONTROL OF HEXAGRAM APF

A.   Proposed Control Scheme

Similar as the control of two-level APF, the control goal of the Hexagram APF is to compensate the harmonics or reactive currents in the source currents caused by the nonlinear loads and achieve unity power factor at the source side. In addition, the control of Hexagram APF should also ensure that the six modules will operate symmetrically and equally share the input voltage and current. Fig. 7 depicts the proposed control scheme of the Hexagram APF, which is composed of three two-level OCC controllers as indicated by Fig. 6. The first controller senses the phase currents of Module I ia1, ib1, ic1 and controls Module I and IV, the second one senses the phase currents of Module III ia3, ib3, ic3 and controls Module III and VI, and the third one senses the phase currents of Module V ia5, ib5, ic5 and controls Module V and II. The three controllers share the common load currents information iA_load, iB_load and iC_load.

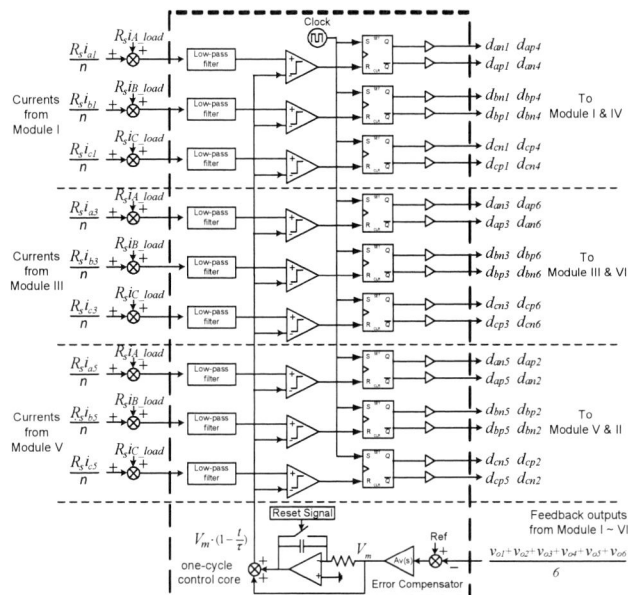

Fig. 7. Proposed control scheme of Hexagram APF.

Take the first controller for illustration, the duty ratios dan1, dbn1 and dcn1 are respectively generated by comparing the carrier signals with the summation of the reflected sensed phase currents of Module I ia1/n, ib1/n and ic1/n and the corresponding load currents iA_load, iB_load and iC_load, and used to drive the bottom switches of Module I San1, Sbn1, Scn1, and the corresponding top switches of Module IV Sap4, Sbp4, Scp4. The variable n is the turns ratio of the isolation transformer as indicated in Fig. 3. The complementary signals of the duty ratios dan1, dbn1 and dcn1 are respectively used to drive the top switches of Module I Sap1, Sbp1, Scp1 and the corresponding bottom switches of Module IV San4, Sbn4, Scn4. Similarly, the duty ratios generated from the second and third controllers are respectively used to control the switches of Module III and VI and the switches of Module V and II.

The closed-loop dc bus voltage control is realized with the averaged feedback dc bus voltages of Module I~VI.

The control equations in Fig. 7 are as follows, where n is the turns ratio of the input isolation transformer, R_s is the equivalent current sensing resistance, and V_m is the output of the feedback error compensator.

Introducing auxiliary variables to represent the summation of the reflected converter currents from Module I, III and V and the corresponding load currents,

$$
\begin{cases}
R_s \cdot
\begin{bmatrix}
\dfrac{1}{n} \cdot i_{a1} + i_{A\_load} \\[2mm]
\dfrac{1}{n} \cdot i_{b1} + i_{B\_load} \\[2mm]
\dfrac{1}{n} \cdot i_{c1} + i_{C\_load}
\end{bmatrix}
= V_m \cdot
\begin{bmatrix}
1 - 2d_{an1} \\
1 - 2d_{bn1} \\
1 - 2d_{cn1}
\end{bmatrix} \\[12mm]
R_s \cdot
\begin{bmatrix}
\dfrac{1}{n} \cdot i_{a3} + i_{A\_load} \\[2mm]
\dfrac{1}{n} \cdot i_{b3} + i_{B\_load} \\[2mm]
\dfrac{1}{n} \cdot i_{c3} + i_{C\_load}
\end{bmatrix}
= V_m \cdot
\begin{bmatrix}
1 - 2d_{an3} \\
1 - 2d_{bn3} \\
1 - 2d_{cn3}
\end{bmatrix} \\[12mm]
R_s \cdot
\begin{bmatrix}
\dfrac{1}{n} \cdot i_{a5} + i_{A\_load} \\[2mm]
\dfrac{1}{n} \cdot i_{b5} + i_{B\_load} \\[2mm]
\dfrac{1}{n} \cdot i_{c5} + i_{C\_load}
\end{bmatrix}
= V_m \cdot
\begin{bmatrix}
1 - 2d_{an5} \\
1 - 2d_{bn5} \\
1 - 2d_{cn5}
\end{bmatrix}
\end{cases}
\tag{6}
$$

$$
\begin{cases}
\begin{bmatrix}
i_{a1}' \\ i_{b1}' \\ i_{c1}'
\end{bmatrix}
=
\begin{bmatrix}
\dfrac{1}{n} \cdot i_{a1} + i_{A\_load} \\[2mm]
\dfrac{1}{n} \cdot i_{b1} + i_{B\_load} \\[2mm]
\dfrac{1}{n} \cdot i_{c1} + i_{C\_load}
\end{bmatrix} \\[12mm]
\begin{bmatrix}
i_{a3}' \\ i_{b3}' \\ i_{c3}'
\end{bmatrix}
=
\begin{bmatrix}
\dfrac{1}{n} \cdot i_{a3} + i_{A\_load} \\[2mm]
\dfrac{1}{n} \cdot i_{b3} + i_{B\_load} \\[2mm]
\dfrac{1}{n} \cdot i_{c3} + i_{C\_load}
\end{bmatrix} \\[12mm]
\begin{bmatrix}
i_{a5}' \\ i_{b5}' \\ i_{c5}'
\end{bmatrix}
=
\begin{bmatrix}
\dfrac{1}{n} \cdot i_{a5} + i_{A\_load} \\[2mm]
\dfrac{1}{n} \cdot i_{b5} + i_{B\_load} \\[2mm]
\dfrac{1}{n} \cdot i_{c5} + i_{C\_load}
\end{bmatrix}
\end{cases}
\tag{7}
$$

Then the control equation (6) can be simplified as following:

$$
\begin{cases}
R_s \cdot
\begin{bmatrix}
i_{a1}' \\ i_{b1}' \\ i_{c1}'
\end{bmatrix}
= V_m \cdot
\begin{bmatrix}
1 - 2d_{an1} \\
1 - 2d_{bn1} \\
1 - 2d_{cn1}
\end{bmatrix} \\[10mm]
R_s \cdot
\begin{bmatrix}
i_{a3}' \\ i_{b3}' \\ i_{c3}'
\end{bmatrix}
= V_m \cdot
\begin{bmatrix}
1 - 2d_{an3} \\
1 - 2d_{bn3} \\
1 - 2d_{cn3}
\end{bmatrix} \\[10mm]
R_s \cdot
\begin{bmatrix}
i_{a5}' \\ i_{b5}' \\ i_{c5}'
\end{bmatrix}
= V_m \cdot
\begin{bmatrix}
1 - 2d_{an5} \\
1 - 2d_{bn5} \\
1 - 2d_{cn5}
\end{bmatrix}
\end{cases}
\tag{8}
$$

B. Analysis of the Controlled Hexagram APF

a) Equivalent module resistor network

Take Module I first for analysis. According to the converter switching average model, by introducing the auxiliary voltage sources va1, vb1 and vc1, the input and output of converter Module I has the following relationship,

$$
\begin{bmatrix}
-\dfrac{2}{3} & \dfrac{1}{3} & \dfrac{1}{3} \\[2mm]
\dfrac{1}{3} & -\dfrac{2}{3} & \dfrac{1}{3} \\[2mm]
\dfrac{1}{3} & \dfrac{1}{3} & -\dfrac{2}{3}
\end{bmatrix}
\cdot
\begin{bmatrix}
d_{an1} \\ d_{bn1} \\ d_{cn1}
\end{bmatrix}
= \dfrac{1}{E_1} \cdot
\begin{bmatrix}
v_{a1} \\ v_{b1} \\ v_{c1}
\end{bmatrix}
\tag{9}
$$

where E1 is the dc bus voltage of Module I.

From the control equation (8),

$$
\begin{bmatrix}
d_{an1} \\ d_{bn1} \\ d_{cn1}
\end{bmatrix}
= 1 - \dfrac{0.5R_s}{V_m} \cdot
\begin{bmatrix}
i_{a1}' \\ i_{b1}' \\ i_{c1}'
\end{bmatrix}
\tag{10}
$$

Substituting (10) into (9) and using the fact that $i_{a1}+i_{b1}+i_{c1}=0$ and $i_{A\_load}+i_{B\_load}+i_{C\_load}=0$ yields,

$$
\begin{bmatrix}
v_{a1} \\ v_{b1} \\ v_{c1}
\end{bmatrix}
= \dfrac{0.5R_s \cdot E_1}{V_m} \cdot
\begin{bmatrix}
i_{a1}' \\ i_{b1}' \\ i_{c1}'
\end{bmatrix}
\tag{11}
$$

According to (11), the auxiliary voltages and the auxiliary currents of Module I have a linear relationship; therefore, Module I can be modeled as a resistor network, as indicated by Fig. 8, and the equivalent resistor Re1 satisfies the following equation.

$$
R_{e1} = \dfrac{0.5E_1 \cdot R_s}{V_m}
\tag{12}
$$

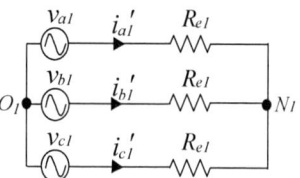

Fig. 8. Equivalent resistor network representing the relationship between the auxiliary voltage and current of Module I.

978-1-4244-4782-4/10 $26.00 © 2010 IEEE

Module IV has the same control signals as Module I. Due to the symmetrical circuit structure, Module IV can be modeled as the same resistor network as Module I.

Similarly, Module III and VI, Module V and II can be respectively modeled as resistor networks as well.

$$R_{e3} = \frac{0.5E_3 \cdot R_s}{V_m} \tag{13}$$

$$R_{e5} = \frac{0.5E_5 \cdot R_s}{V_m} \tag{14}$$

b) Converter current analysis

Assume that the three-phase input currents of the Hexagram converter are as follows,

$$\begin{bmatrix} I_{a1} \\ I_{b3} \\ I_{c5} \end{bmatrix} = \begin{bmatrix} I_a \\ I_b \\ I_c \end{bmatrix} \tag{15}$$

Due to the interconnection, the currents inside the Hexagram converter satisfy the following relationships:

$$\begin{bmatrix} I_{a4} \\ I_{b6} \\ I_{c2} \end{bmatrix} = - \begin{bmatrix} I_{a1} \\ I_{b3} \\ I_{c5} \end{bmatrix} \tag{16}$$

$$\begin{bmatrix} I_{b1} \\ I_{a2} \\ I_{c3} \\ I_{b4} \\ I_{a5} \\ I_{c6} \end{bmatrix} = - \begin{bmatrix} I_{b2} \\ I_{a3} \\ I_{c4} \\ I_{b5} \\ I_{a6} \\ I_{c1} \end{bmatrix} \tag{17}$$

The three-phase currents inside any one module satisfy the following equation:

$$\begin{bmatrix} I_{a1} + I_{b1} + I_{c1} \\ I_{a2} + I_{b2} + I_{c2} \\ I_{a3} + I_{b3} + I_{c3} \\ I_{a4} + I_{b4} + I_{c4} \\ I_{a5} + I_{b5} + I_{c5} \\ I_{a6} + I_{b6} + I_{c6} \end{bmatrix} = 0 \tag{18}$$

Limited by the inductors, the circulating current inside the Hexagram converter is low and can be neglected,

$$I_{b1} + I_{a2} + I_{c3} + I_{b4} + I_{a5} + I_{c6} = 0 \tag{19}$$

Combining the above equations, the phase currents of all the six modules can be derived.

$$\begin{cases} \begin{bmatrix} I_{a1} \\ I_{b1} \\ I_{c1} \end{bmatrix} = - \begin{bmatrix} I_{a4} \\ I_{b4} \\ I_{c4} \end{bmatrix} = \frac{1}{2} \cdot \begin{bmatrix} 2I_a \\ -I_a + I_b - I_c \\ -I_a - I_b + I_c \end{bmatrix} \\[20pt] \begin{bmatrix} I_{a3} \\ I_{b3} \\ I_{c3} \end{bmatrix} = - \begin{bmatrix} I_{a6} \\ I_{b6} \\ I_{c6} \end{bmatrix} = \frac{1}{2} \cdot \begin{bmatrix} I_a - I_b - I_c \\ 2I_b \\ -I_a - I_b + I_c \end{bmatrix} \\[20pt] \begin{bmatrix} I_{a5} \\ I_{b5} \\ I_{c5} \end{bmatrix} = - \begin{bmatrix} I_{a2} \\ I_{b2} \\ I_{c2} \end{bmatrix} = \frac{1}{2} \cdot \begin{bmatrix} I_a - I_b - I_c \\ -I_a + I_b - I_c \\ 2I_c \end{bmatrix} \end{cases} \tag{20}$$

Similar as (7), introducing auxiliary variables as given below,

$$\begin{cases} \begin{bmatrix} i_{a2}' \\ i_{b2}' \\ i_{c2}' \end{bmatrix} = \begin{bmatrix} \frac{1}{n} \cdot i_{a2} - i_{A\_load} \\ \frac{1}{n} \cdot i_{b2} - i_{B\_load} \\ \frac{1}{n} \cdot i_{c2} - i_{C\_load} \end{bmatrix} \\[24pt] \begin{bmatrix} i_{a4}' \\ i_{b4}' \\ i_{c4}' \end{bmatrix} = \begin{bmatrix} \frac{1}{n} \cdot i_{a4} - i_{A\_load} \\ \frac{1}{n} \cdot i_{b4} - i_{B\_load} \\ \frac{1}{n} \cdot i_{c4} - i_{C\_load} \end{bmatrix} \\[24pt] \begin{bmatrix} i_{a6}' \\ i_{b6}' \\ i_{c6}' \end{bmatrix} = \begin{bmatrix} \frac{1}{n} \cdot i_{a6} - i_{A\_load} \\ \frac{1}{n} \cdot i_{b6} - i_{B\_load} \\ \frac{1}{n} \cdot i_{c6} - i_{C\_load} \end{bmatrix} \\[24pt] \begin{bmatrix} i_a' \\ i_b' \\ i_c' \end{bmatrix} = \begin{bmatrix} \frac{1}{n} \cdot i_a + i_{A\_load} \\ \frac{1}{n} \cdot i_b + i_{B\_load} \\ \frac{1}{n} \cdot i_c + i_{C\_load} \end{bmatrix} \end{cases} \tag{21}$$

Although (7) and (21) describe the instantaneous currents, the corresponding current vectors also satisfy the two equations. Combining (7), (20) and (21) yields,

$$
\begin{cases}
\begin{bmatrix} I_{a1}' \\ I_{b1}' \\ I_{c1}' \end{bmatrix} = -\begin{bmatrix} I_{a4}' \\ I_{b4}' \\ I_{c4}' \end{bmatrix} = \dfrac{1}{2}\cdot\begin{bmatrix} 2I_a' \\ -I_a'+I_b'-I_c' \\ -I_a'-I_b'+I_c' \end{bmatrix} \\[4ex]
\begin{bmatrix} I_{a3}' \\ I_{b3}' \\ I_{c3}' \end{bmatrix} = -\begin{bmatrix} I_{a6}' \\ I_{b6}' \\ I_{c6}' \end{bmatrix} = \dfrac{1}{2}\cdot\begin{bmatrix} I_a'-I_b'-I_c' \\ 2I_b' \\ -I_a'-I_b'+I_c' \end{bmatrix} \\[4ex]
\begin{bmatrix} I_{a5}' \\ I_{b5}' \\ I_{c5}' \end{bmatrix} = -\begin{bmatrix} I_{a2}' \\ I_{b2}' \\ I_{c2}' \end{bmatrix} = \dfrac{1}{2}\cdot\begin{bmatrix} I_a'-I_b'-I_c' \\ -I_a'+I_b'-I_c' \\ 2I_c' \end{bmatrix}
\end{cases} \tag{22}
$$

c) Equivalent converter resistor network

Neglecting the voltages across the inductors, according to the module interconnection, the voltage relationship inside the Hexagram converter is as following:

$$
\begin{cases}
v_a = v_{A-A'} = v_{a1b1}+v_{b2a2}+v_{a3c3}+v_{c4a4} = v_{a1c1}+v_{c6a6}+v_{a5b5}+v_{b4a4} \\
v_b = v_{B-B'} = v_{b3c3}+v_{c4b4}+v_{b5a5}+v_{a6b6} = v_{b3a3}+v_{a2b2}+v_{b1c1}+v_{c6b6} \\
v_c = v_{C-C'} = v_{c5a5}+v_{a6c6}+v_{c1b1}+v_{b2c2} = v_{c5b5}+v_{b4c4}+v_{c3a3}+v_{a2c2}
\end{cases} \tag{23}
$$

Above equation can also be used to describe the corresponding voltage vectors.

Combining the equivalent resistor networks of Module I~VI as described by (12) ~ (14), the relationship of the auxiliary currents as described by (22) and the relationship of the auxiliary voltages as described by (23), the Hexagram converter can be modeled as a resistor network, as shown in Fig. 9 (a). And Fig. 9 (b) is the simplified circuit.

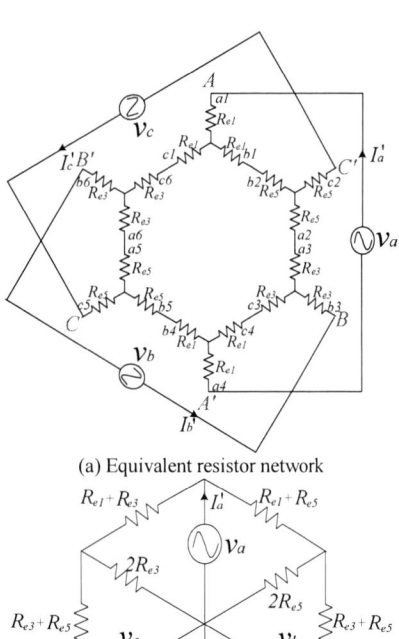

(a) Equivalent resistor network

(b) Simplified resistor network

Fig. 9. Equivalent resistor network of the Hexagram converter.

d) Analysis of the converter resistor network

According to Fig. 9 (b), the relationship between the input voltages and the emulated resistors and input currents can be derived as follows,

$$
\begin{bmatrix} V_a \\ V_b \\ V_c \end{bmatrix} = \begin{bmatrix} 3R_{e1}+R_{e3}+R_{e5} & -R_{e5} & -R_{e3} \\ -R_{e5} & R_{e1}+3R_{e3}+R_{e5} & -R_{e1} \\ -R_{e3} & -R_{e1} & R_{e1}+R_{e3}+3R_{e5} \end{bmatrix}\cdot\begin{bmatrix} I_a' \\ I_b' \\ I_c' \end{bmatrix} \tag{24}
$$

Neglecting the power loss of each converter module, the input power of each converter module is equal to its output power, which yields,

$$
\begin{bmatrix} (I_{a1}'I_{a1}^{*}{}'+I_{b1}'I_{b1}^{*}{}'+I_{c1}'I_{c1}^{*}{}')\cdot R_{e1} \\ (I_{a3}'I_{a3}^{*}{}'+I_{b3}'I_{b3}^{*}{}'+I_{c3}'I_{c3}^{*}{}')\cdot R_{e3} \\ (I_{a5}'I_{a5}^{*}{}'+I_{b5}'I_{b5}^{*}{}'+I_{c5}'I_{c5}^{*}{}')\cdot R_{e5} \end{bmatrix} = \begin{bmatrix} E_1^2/R \\ E_3^2/R \\ E_5^2/R \end{bmatrix} = \frac{V_m^2}{R_s^2\cdot R}\begin{bmatrix} R_{e1}^2 \\ R_{e3}^2 \\ R_{e5}^2 \end{bmatrix} \tag{25}
$$

Simplifying (25) yields,

$$
\begin{bmatrix} R_{e1} \\ R_{e3} \\ R_{e5} \end{bmatrix} = \frac{R_s^2\cdot R}{2V_m^2}\cdot\begin{bmatrix} 3I_a'I_a^{*}{}'+I_b'I_b^{*}{}'+I_c'I_c^{*}{}'-I_b'I_c^{*}{}'-I_c'I_b^{*}{}' \\ I_a'I_a^{*}{}'+3I_b'I_b^{*}{}'+I_c'I_c^{*}{}'-I_a'I_c^{*}{}'-I_c'I_a^{*}{}' \\ I_a'I_a^{*}{}'+I_b'I_b^{*}{}'+3I_c'I_c^{*}{}'-I_a'I_b^{*}{}'-I_b'I_a^{*}{}' \end{bmatrix} \tag{26}
$$

From (24),

$$
V_a+V_b+V_c = 3R_{e1}\cdot I_a'+3R_{e3}\cdot I_b'+3R_{e5}\cdot I_c' \tag{27}
$$

Substituting (26) into (27) yields,

$$
\begin{aligned}
&V_a+V_b+V_c \\
&= \frac{3R_s^2\cdot R}{2V_m^2}\cdot[3(I_a'I_a^{*}{}'+I_b'I_b^{*}{}'+I_c'I_c^{*}{}')(I_a'+I_b'+I_c') \\
&\quad -2(I_a'I_b'+I_b'I_c'+I_c'I_a')(I_a'+I_b'+I_c')^{*}]
\end{aligned} \tag{28}
$$

In three-phase symmetrical system which contains only positive and negative sequence components,

$$
V_a+V_b+V_c = 0 \tag{29}
$$

Combining (28) and (29) yields,

$$
\begin{aligned}
&3(I_a'I_a^{*}{}'+I_b'I_b^{*}{}'+I_c'I_c^{*}{}')\cdot(I_a'+I_b'+I_c') \\
&-2(I_a'I_b'+I_b'I_c'+I_c'I_a')\cdot(I_a'+I_b'+I_c')^{*} = 0
\end{aligned} \tag{30}
$$

Reorganize (30) gives,

$$
\begin{aligned}
&[3(I_a'I_a^{*}{}'+I_b'I_b^{*}{}'+I_c'I_c^{*}{}')-(I_a'+I_b'+I_c')\cdot(I_a'+I_b'+I_c')^{*}]\cdot(I_a'+I_b'+I_c') \\
&+(I_a'^2+I_b'^2+I_c'^2)\cdot(I_a'+I_b'+I_c')^{*} = 0
\end{aligned} \tag{31}
$$

To satisfy (31), the following relationship must be true.

$$
I_a'+I_b'+I_c' = 0 \tag{32}
$$

This result indicates that in the three-phase symmetrical system that contains only positive and negative sequence components, the sum of the input currents of the Hexagram converter is zero.

e) Hexagram APF control results

According to (32), when the input of the Hexagram converter is a three-phase symmetrical system, the three input currents of the converter add up to zero.

Substituting (32) back to (26) yields,

$$R_{e1} = R_{e3} = R_{e5} = R_e \tag{33}$$

And substituting (33) to (24) yields,

$$\begin{bmatrix} V_a \\ V_b \\ V_c \end{bmatrix} = 6R_e \cdot \begin{bmatrix} I_a{}' \\ I_b{}' \\ I_c{}' \end{bmatrix} \tag{34}$$

Based on (34), the relationship between the input source voltages and currents can be derived.

$$\begin{bmatrix} V_{SA} \\ V_{SB} \\ V_{SC} \end{bmatrix} = n \cdot \begin{bmatrix} V_a \\ V_b \\ V_c \end{bmatrix} = 6nR_e \cdot \begin{bmatrix} \frac{1}{n} \cdot I_a + I_{A\_load} \\ \frac{1}{n} \cdot I_b + I_{B\_load} \\ \frac{1}{n} \cdot I_c + I_{C\_load} \end{bmatrix} = 6nR_e \cdot \begin{bmatrix} I_{SA} \\ I_{SB} \\ I_{SC} \end{bmatrix} \tag{35}$$

This result indicates that the input source currents will be in phase with the corresponding source voltages under the proposed control scheme.

Substituting (32) to (22) yields,

$$\begin{bmatrix} I_{a1}{}' \\ I_{b1}{}' \\ I_{c1}{}' \end{bmatrix} = \begin{bmatrix} I_{a3}{}' \\ I_{b3}{}' \\ I_{c3}{}' \end{bmatrix} = \begin{bmatrix} I_{a5}{}' \\ I_{b5}{}' \\ I_{c5}{}' \end{bmatrix} = -\begin{bmatrix} I_{a2}{}' \\ I_{b2}{}' \\ I_{c2}{}' \end{bmatrix} = -\begin{bmatrix} I_{a4}{}' \\ I_{b4}{}' \\ I_{c4}{}' \end{bmatrix} = -\begin{bmatrix} I_{a6}{}' \\ I_{b6}{}' \\ I_{c6}{}' \end{bmatrix} = \begin{bmatrix} I_a{}' \\ I_b{}' \\ I_c{}' \end{bmatrix} \tag{36}$$

Combining (7), (21), (35) and (36) yields,

$$\begin{bmatrix} I_{a1} \\ I_{b1} \\ I_{c1} \end{bmatrix} = \begin{bmatrix} I_{a3} \\ I_{b3} \\ I_{c3} \end{bmatrix} = \begin{bmatrix} I_{a5} \\ I_{b5} \\ I_{c5} \end{bmatrix} = -\begin{bmatrix} I_{a2} \\ I_{b2} \\ I_{c2} \end{bmatrix} = -\begin{bmatrix} I_{a4} \\ I_{b4} \\ I_{c4} \end{bmatrix} = -\begin{bmatrix} I_{a6} \\ I_{b6} \\ I_{c6} \end{bmatrix} = \begin{bmatrix} I_a \\ I_b \\ I_c \end{bmatrix} = n \cdot \begin{bmatrix} I_{SA} - I_{A\_load} \\ I_{SB} - I_{B\_load} \\ I_{SC} - I_{C\_load} \end{bmatrix} \tag{37}$$

This result indicates that the input currents of the Hexagram converter are the compensation currents equal to the subtraction of the sinusoidal source currents and the non-linear load currents. Besides, Module I, III and V of the Hexagram converter have the same phase currents as the input currents of the converter, while Module II, IV and VI have the reversed ones.

Based on (11) ~ (14), (33) ~ (34) and (36),

$$\begin{bmatrix} V_{a1} \\ V_{b1} \\ V_{c1} \end{bmatrix} = \begin{bmatrix} V_{a3} \\ V_{b3} \\ V_{c3} \end{bmatrix} = \begin{bmatrix} V_{a5} \\ V_{b5} \\ V_{c5} \end{bmatrix} = -\begin{bmatrix} V_{a2} \\ V_{b2} \\ V_{c2} \end{bmatrix} = -\begin{bmatrix} V_{a4} \\ V_{b4} \\ V_{c4} \end{bmatrix} = -\begin{bmatrix} V_{a6} \\ V_{b6} \\ V_{c6} \end{bmatrix} = R_e \cdot \begin{bmatrix} I_a{}' \\ I_b{}' \\ I_c{}' \end{bmatrix} = \frac{1}{6} \cdot \begin{bmatrix} V_a \\ V_b \\ V_c \end{bmatrix} = \frac{1}{6n} \cdot \begin{bmatrix} V_{SA} \\ V_{SB} \\ V_{SC} \end{bmatrix} \tag{38}$$

This result indicates that the six modules inside the Hexagram converter equally share the input voltages of the converter.

From (12) ~ (14) and (33),

$$E_1 = E_2 = E_3 = E_4 = E_5 = E_6 \tag{39}$$

This indicates that the six dc bus voltages of the Hexagram converter are the same.

Equations (35) and (37) ~ (39) give the results of the Hexagram APF with the proposed control scheme in Fig. 7 when the input of the Hexagram converter is the three-phase symmetrical system containing no zero sequence components as described by (29). This condition is the most likely operation condition of the Hexagram converter, since the voltages supplied by a symmetrical three-phase transformer do not contain any zero sequence components but positive and negative sequence components no matter how the grid is unbalanced.

With the proposed control scheme, the Hexagram converter will compensate the non-linear load currents so that the input source currents will be sinusoidal and in phase with the corresponding source voltages. Besides, the six modules of the converter have the same phase currents, the same average input voltages, the same dc bus voltages and thus the same power as well. The input voltage of each module is one sixth of that of the converter so that the voltage stress of the semiconductor switches is reduced to one sixth. The instantaneous power through each module is constant.

When the transformer has some asymmetry, the equivalent source voltages of the Hexagram converter will contain some zero-sequence components. Under this condition, the converter behavior will still follow the rules described by the voltage equation in (24) and the energy equation in (25). When the zero-sequence components are small, the results are close to the derived equations (35) and (37) ~ (39) under the symmetrical systems.

V. SIMULATION VERIFICATION

A 1MW 4.16kV Hexagram APF system has been built in PSIM7.0. The non-linear load is composed of a three-phase diode rectifier with load at 40Ω. The Hexagram converter has eighteen inductors at 1mH each (three inductors per module). The dc bus capacitor for each module is 820μF each. The turns ratio of the isolation transformer for the Hexagram converter is 1.1, and the secondary windings of the transformer is configured as a six-terminal three-phase voltage source with neutral floating. The switching frequency of the Hexagram converter is 10 kHz.

Fig. 10 shows the simulation results. Fig. 10 (a) shows the input source voltages VSA, VSB, VSC, input source currents ISA, ISB, ISC, load currents IA_load, IB_load, IC_load, and the currents from the Hexagram converter IA_APF, IB_APF, IC_APF. With the compensation currents from the Hexagram converter, the source currents are sinusoidal and in phase with the corresponding source voltages. Fig. 10 (b) shows the currents of the Hexagram converter IA_APF, IB_APF, IC_APF, currents of Module I Ia1, Ib1, Ic1 and currents of Module IV Ia4, Ib4, Ic4. The currents of Module I are the same as the currents of the Hexagram converter, while the currents of Module IV are reversed. The other four modules have the similar results. Fig. 10 (c) shows the six dc bus voltages, which are identical and are at 1300V. Fig. 10 (d) shows the input line voltages VA-VB and the line voltages of

Module I Va1-Vb1. The input line voltage is six times the line voltages of Module I, and other modules have similar result. Since the voltage of one module is only one sixth of the input voltage, the voltage stress of the converter is reduced to one sixth compared to a single two-level VSC at the same input voltage. The simulation results have well verified the proposed Hexagram APF and proposed control scheme.

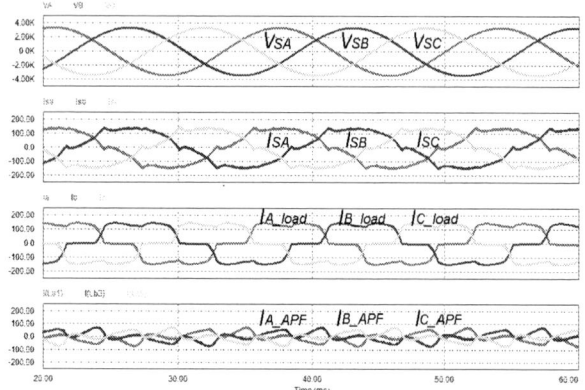

(a) Source voltages, source currents, load currents and currents from Hexagram converter

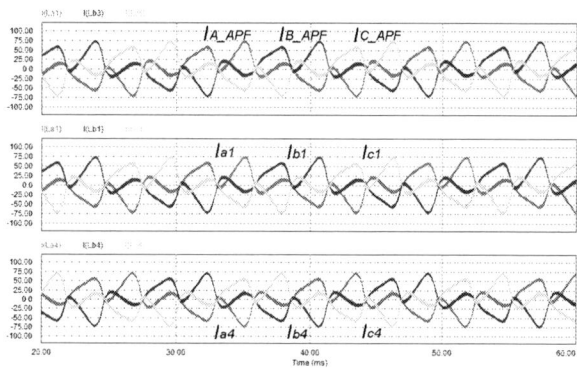

(b) Currents of Hexagram converter, Module I and Module IV

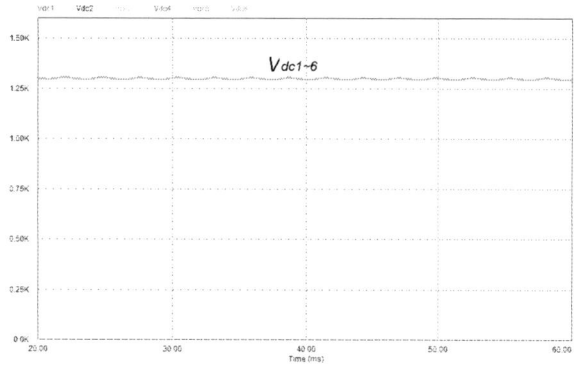

(c) Six dc bus voltages

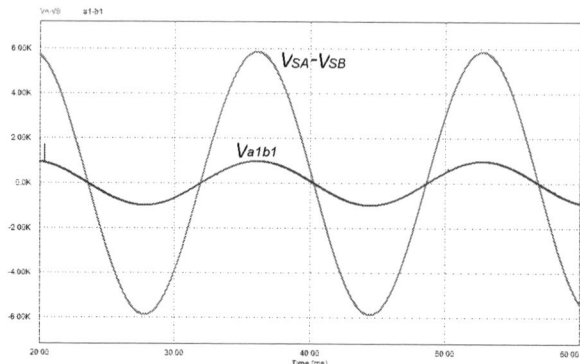

(d) Line voltages of Hexagram converter and line voltages of Module I

Fig. 10. Simulation results of 1MW Hexagram APF system.

VI. EXPERIMENTAL VERIFICATION

A 1kW Hexagram converter prototype has been built for verification. The prototype has eighteen inductors at 1.5mH each and six dc bus capacitors at 470μH each. The nonlinear load is composed of a three-phase diode rectifier and a 40Ω resistor. The isolation transformer of the Hexagram converter is 3:1. The switching frequency of the converter is 16 kHz.

Fig. 11 shows the experimental results. Fig. 11 (a) shows the input source voltage VSA and three-phase input source currents ISA, ISB, ISC. The input source voltage is 80V line-to-neutral RMS, and the input source currents are 4A RMS. The input source currents are in phase with the corresponding source voltages. Fig. 11 (b) shows the input source voltage VSA, input source currents ISA, load current IA_load and the current from the Hexagram converter IA_APF. With the compensation currents from the Hexagram converter, the input source currents are sinusoidal and in phase with the corresponding source voltages. Fig. 11 (c) shows the input phase voltage VAA' and the three-phase input currents Ia, Ib and Ic of the Hexagram converter. Since the turns ratio of the isolation transformer is 3:1, the input phase voltage of the Hexagram converter is 240V RMS. Fig. 11 (d) and (e) respectively shows the input phase voltage VAA' of the Hexagram converter with the currents of Module I Ia1, Ib1, Ic1 and the currents of Module IV Ia4, Ib4, Ic4. The currents of Module I are the same as the input currents of the Hexagram converter, while the currents of Module IV are in the reversed direction. The currents of the other four modules have similar results. Fig. 11 (f) shows the input line voltage VAA'-VBB' of the converter and the corresponding line voltage of Module I Va1b1. The input line voltage is around 420V RMS, and the line voltage of Module I is around 70V. Correspondingly, the input phase voltage is 240V and the phase voltage of Module I is 40V. The voltage of one module is one sixth of the input voltage, so that the voltage stress of the semiconductor switches is reduced to one sixth. Fig. 11 (g) shows the six dc bus voltages, which are identical and are controlled at 170V each.

The experimental results have fully verified the Hexagram APF and the proposed control scheme. With the proposed control scheme, the currents from the Hexagram converter compensate the non-linear load currents, and the input source currents are in phase with the corresponding input source voltages. The voltage stress of the converter is reduced to one sixth and the six modules equally share the output power.

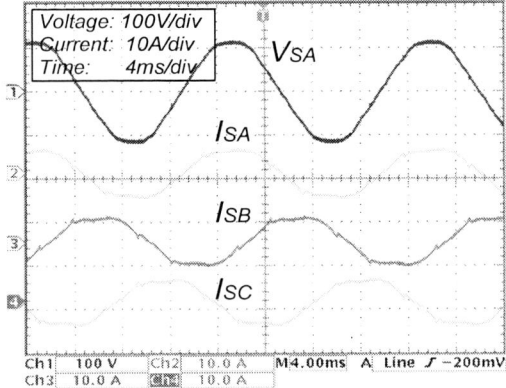

(a) Source voltage and currents

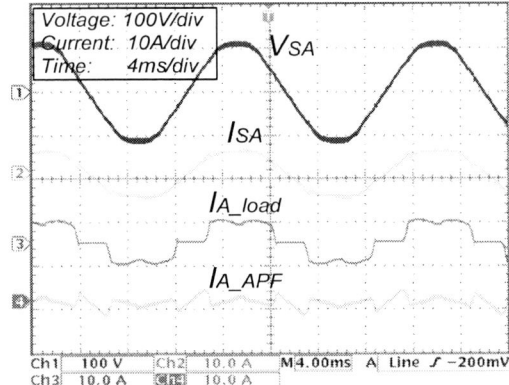

(b) Source voltage, source current, load current and current from Hexagram converter

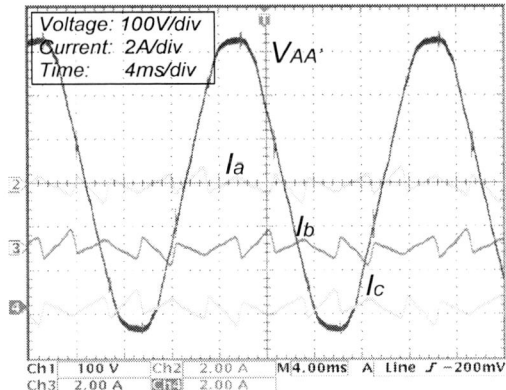

(c) Input voltage and currents of Hexagram converter

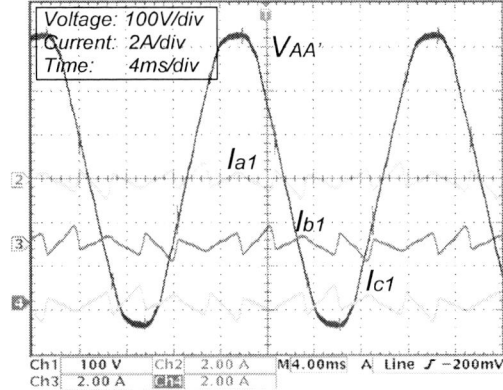

(d) Input voltage of Hexagram converter and currents of Module I

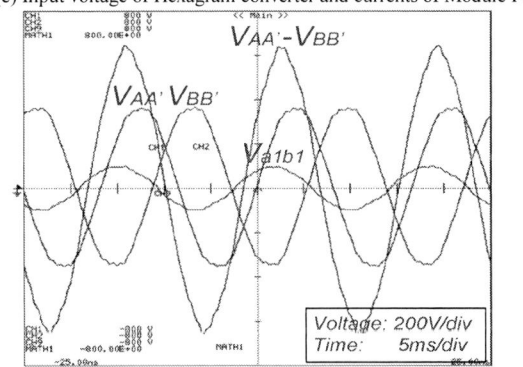

(e) Input voltage of Hexagram converter and currents of Module IV

(f) Input line voltage of Hexagram converter and line voltage of Module I

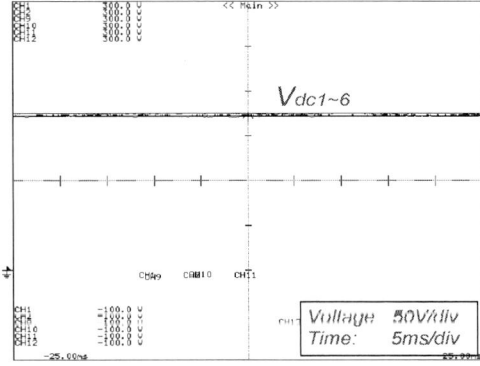

(g) Six dc bus voltages

Fig. 11. Experimental results of 1kW Hexagram APF system.

VII. CONCLUSION

This paper proposed a Hexagram APF for power quality improvement at the medium-voltage grids. One-cycle control for the two-level APF is extended to control the Hexagram APF. The proposed control scheme adopts three two-level OCC bipolar controllers that effectively control all the six modules to achieve harmonics current cancellation in the line, while equally share the voltage and current. The control of the six dc bus voltages is achieved by the closed-loop control of the sensed and averaged feedback dc voltage signals from the six modules and dc voltage balancing is automatically satisfied due to the symmetrical structure.

With the proposed control scheme, the currents of the Hexagram converter compensate the non-linear load currents, and the input source currents are respectively in phase with the corresponding input source voltages. The six modules of the converter are symmetrical with the same voltages and currents when the input voltages of the converter are symmetrical with no zero sequence components. The voltage of one module is one sixth of the overall voltage of the converter so that the voltage stress of the converter is reduced to one sixth compared to a two-level VSC at the same input voltage. The six dc buses of the modules are equal, and the six modules equally share the output power.

A 1MW Hexagram APF system and a 1kW experimental prototype have been respectively built in simulation and experiment and both have well verified the proposed converter and control method for the APF applications.

REFERENCES

[1] IEEE Working Group on Power System Harmonics, "Power system harmonics: an overview," *IEEE Trans. Power App. Syst.*, vol. PAS-102, pp. 2445-2460, Aug 1983.

[2] H. Mao, C.Y. Lee, D. Boroyevich, and S. Hiti, "Review of high-performance three-phase power-factor correction circuits," *IEEE Trans. on Industrial Electronics*, vol. 44, issue. 4, pp. 437-446, Aug 1997.

[3] B. Singh, B.N. Singh, and A. Chandra etc, "A review of three-phase improved power quality AC-DC converters," IEEE Trans. on Industrial Electronics, vol. 51, no. 3, pp. 641-660, Jun 2004.

[4] M.M. Jovanovic and Y. Jang, "State-of-the-art, single-phase, active power-factor-correction techniques for high-power applications – an overview," IEEE Trans. on Industrial Electronics, vol. 52, issue. 3, pp. 701-708, June 2005.

[5] B. Singh, K. Al-Haddad, and A. Chandra, "A review of active filters for power quality improvement," IEEE Trans. on Industrial Electronics, vol. 46, no. 5, pp. 960-971, Oct 1999.

[6] D.A. Torrey and A.M. Al-Zamel, "A single-phase active power filter for multiple nonlinear loads," IEEE APEC'94, vol. 2, pp. 901-908, 1994.

[7] F. Pottker and I. Barbi, "Power factor correction of nonlinear loads employing a single-phase active power filter: control strategy, design methodology and experimentation," IEEE PESC'97, vol. 1, pp. 412-417, 1997.

[8] S.-J. Huang and J.-C. Wu, "Control algorithm for three-phase three-wired active power filters under nonideal mains voltages," IEEE Trans. on Power Electronics, vol. 14, pp. 735-760, July 1999.

[9] J.W. Dixon, J.M. Contardo and L.A. Moran, "A fuzzy-controlled active front-end rectifier with current harmonic filtering characteristics and minimum sensing variables," IEEE Trans. on Power Electronics, vol. 14, pp. 724-729, July 1999.

[10] R.S. Herrera, P. Salmerýýn, and H. Kim, "Instantaneous reactive power theory applied to active power filter compensation: different

approaches, assessment, and experimental results," IEEE Trans. on Industrial Electronics, vol. 55, issue. 1, pp. 184-196, Jan 2008.

[11] S.A. Gonzalez, R. Garcia-Retegui, and M. Benedetti, "Harmonic computation technique suitable for active power filters," IEEE Trans. on Industrial Electronics, vol. 54, issue. 5, pp. 2791-2796, Oct 2007.

[12] K.M. Cho, W.S. Oh, Y.T. Kim, and H.J. Kim, "A new switching strategy for pulse width modulation (PWM) power converters," IEEE Trans. on Industrial Electronics, vol. 54, issue. 1, pp. 330-337, Feb 2007.

[13] L. Asiminoaei, P. Rodriguez, F. Blaabjerg, and M. Malinowski, "Reduction of switching losses in active power filters with a new generalized discontinuous-PWM strategy," IEEE Trans. on Industrial Electronics, vol. 55, issue. 1, pp. 467-471, Jan 2008.

[14] R. Grino, R. Cardoner, R. Costa-Castello, and E. Fossas, "Digital repetitive control of a three-phase four-wire shunt active filter," IEEE Trans. on Industrial Electronics, vol. 54, issue. 3, pp. 1495-1503, June 2007.

[15] B. Suh, G. Sinha, and M.D. Manjrekar, T.A. Lipo, "Multilevel power conversion – an overview of topologies and modulation strategies," Optimization of Electrical and Electronic Equipments, 1998. OPTIM'98. vol. 2, pp. AD-11 – AD-24, May 14-15 1998.

[16] J. Rodríguez, J.S. Lai, and F.Z. Peng, "Multilevel inverters: a survey of topologies, controls, and applications," IEEE Trans. on Industrial Electronics, vol. 49, pp. 724-738, Aug 2002.

[17] D. Krug, M. Malinowski, and S. Bernet, "Design and comparison of medium voltage multi-level converters for industry applications," Conference Record of IAS Annual Meeting, 2004, vol. 2, pp. 781-790, 2004.

[18] C. Hochgraf, R. Lasseter, D. Divan, and T.A. Lipo, "Comparison of multilevel inverters for static Var compensation," Conference Record of IAS Annual Meeting, 1994, pp. 921-928.

[19] J. S. Lai, and F. Z. Peng, "Multilevel converters – a new breed of power converters," IEEE Trans. on Industry Applications, vol. 32, no. 3, May/June 1996.

[20] J. Rodriguez, S. Bernet, B. Wu, J.O. Pontt, and S. Kouro, "Multilevel voltage-source-converter topologies for industrial medium-voltage drives," IEEE Trans. on Industrial Electronics, vol. 54, issue. 6, pp. 2930-2945, Dec 2007.

[21] J.M. Carrasco, L.G. Franquelo, J.T. Bialasiewicz etc, "Power-Electronic system for the grid integration of renewable energy sources: a survey," IEEE Trans. on Industrial Electronics, vol. 53, issue. 4, pp. 1002-1016, June 2006.

[22] F.Z. Peng and J.S. Lai, "A static VAR generator using a staircase waveform multilevel voltage-source converter," in Proc. Power Quality Conf., 1994, pp. 58-66.

[23] S. Rahmani and K. Al-Haddad, "A single phase multilevel hybrid power filter for electrified railway applications," IEEE International Symposium on Industrial Electronics, 2006, vol. 2, pp. 925-930, Jul 2006.

[24] H. Miranda, V. Cardenas, J. Perez, and G. Nunez, "A hybrid multilevel inverter for shunt active filter using space-vector control," IEEE 35th Annual Power Electronics Specialists Conference, 2004. PESC 04. vol. 5, pp. 3541-3546, Jun 2004.

[25] Y-H. Kim, S-H. Kim, K-H. Lee, "A new hybrid power filter using multilevel inverters," IEEE 4th International Power Electronics and Motion Control Conference, 2004. vol.1, pp. 210-214, 2004.

[26] L. Wang and W. Wu, "Shunt active power filter with sample time staggered space vector modulation based cascaded multilevel converters," IEEE 5th International Power Electronics and Motion Control Conference, 2004. vol.2, pp. 1-5, Aug 2006.

[27] J. Li, C. Hu, L. Wang, and Z. Zhang, "APF based on multilevel voltage source cascade converter with carrier phase shifted SPWM," TENCON 2003. Conference on Convergent Technologies for Asia-Pacific Region, vol. 1, pp. 264-267, Oct 2003.

[28] F.Z. Peng, J.W. McKeever, and D.J. Adams, "Cascaded multilevel inverters for utility applications," 23rd International Conference on Industrial Electronics, Control and Instrumentation, 1997. IECON 97. vol. 2, pp. 437-442, Nov 1997.

978-1-4244-4782-4/10 $26.00 © 2010 IEEE

[29] M.E. Ortuzar, R.E. Carmi, J.W. Dixon, and L. Moran, "Voltage-source active power filter based on multilevel converter and ultra capacitor DC link," IEEE Trans. on Industrial Electronics, vol. 53, issue. 2, pp. 477-485, Apr 2006.

[30] J. Wen and K. Smedley, "New converters for high power applications," U.S. Patent, University of California, Irvine, 2006.

[31] J. Wen and K. Smedley, "Synthesis of multilevel converters based on single- and/or three-phase converter building blocks," IEEE Trans. on Power Electronics, submitted for publication.

[32] J. Wen and K. Smedley, "Hexagram rectifier – active front end for medium voltage adjustable speed drive systems," IEEE Transmission and Distribution Conference and Exposition, Apr 21-24. 2008, Chicago USA.

[33] J. Wen and K. Smedley, "Hexagram converter for static VAR compensation," IEEE Transmission and Distribution Conference and Exposition, Apr 21-24. 2008, Chicago USA.

[34] J. Wen and K. Smedley, "A new multilevel inverter – Hexagram inverter for medium-voltage adjustable speed drive systems. II. Three-phase motor drive," IEEE Power Electronics Specialists Conference, PESC'07, June. 2007.

[35] J. Wen and K. Smedley, "A new multilevel inverter – Hexagram inverter for medium-voltage adjustable speed drive system. I. Six-phase motor drive," International Conference on Power Engineering, Energy and Electric Drives, POWERENG'07, Apr. 2007.

[36] C. Qiao and K. Smedley, "Three-phase bipolar mode active power filter," IEEE Trans. on Industry Applications, vol. 38, issue. 1, Jan-Feb 2002, pp. 149-158.

A Generalized Capacitor Voltage Balancing Scheme for Flying Capacitor Multilevel Converters

Mostafa Khazraei, Hossein Sepahvand, Keith Corzine, and Mehdi Ferdowsi
Missouri University of Science and Technology
Rolla, Missouri USA
http://power.mst.edu

Abstract— **Multilevel power electronic converters are the converter of choice in medium-voltage applications due to their reduced switch voltage stress, better harmonic performance, and lower switching losses. Although it has received little attention, the flying-capacitor multilevel converter has a distinct advantage in terms of its ease of capacitor voltage balancing. A number of techniques have been presented in the literature for capacitor voltage balancing, some relying on "self-balancing" properties. However, self balancing cannot guarantee balancing of capacitor voltages in practical applications. Other researchers present closed-loop control schemes which force voltage balancing of capacitors. In this paper, a new closed loop control scheme is proposed which regulates the capacitor voltages for a multilevel flying capacitor converter. The proposed scheme is based on the converter equations and involves implementing simple rules. In particular, multiple duty cycles are defined and modulated in direct response to the capacitor voltages. Through simulation, the method is shown to work on four, eight and nine-level flying capacitor inverters.**

I. INTRODUCTION

Multilevel converters have become a popular option in medium-voltage applications. While common two-level converters utilize direct series connection of switches to meet medium-voltage requirements, multilevel converters allow higher voltage handling capability with reduced harmonic distortion and lower switching power losses [1]. The flying capacitor converter (FCC) was introduced as a viable multilevel converter topology in 1992 [2, 3]. Although the FCC topology is not as common as other structures, it has some distinct advantages over the diode-clamped topology including the absence of clamping diodes and the ability to regulate the flying-capacitor voltages through redundant state selection even if the number of voltage levels is greater than three [4]. Different methods have been introduced in the literature to maintain capacitor voltage balancing. The simplest approach is to rely on the "self-balancing" property of the FCCs. It has been demonstrated that by satisfying certain conditions, a simple open loop control guarantees natural balancing of the flying capacitors [5]. An RLC filter tuned at the switching frequency and connected in parallel with the load can be used to achieve natural balancing under all conditions [6]. However, the extra RLC filter increases the cost and power losses while it decreases the dynamic response.

The phase shifted sinusoidal PWM (PS-SPWM) method is considered an effective control method for the multilevel FCC since it benefits from self-balancing property when applied to an ideal and symmetrical circuit. Compared to other PWM methods, it is easier to balance the capacitor voltage in a relatively short time [7-9]. However, to obtain voltage balancing, switching cells have to operate at the same duty cycle, the power devices must have the same characteristics, and the load current has to be symmetrical. Satisfaction of these conditions may not be guaranteed. Recently, several voltage balancing strategies have been presented which are based on the combination of the PS–SPWM method with external control loops. These include methods which modify the duty cycles, choose different switching states, add auxiliary voltages, etc. [10].

In [10], a control scheme for the multilevel FCC based on manipulating the modulating waveform in the PS-SPWM method is presented. Based on the polarity of the phase current, a relatively small square wave is added to the original modulation signals (typically sinusoidal) controlling the converter switches. This way, the flying capacitor voltages can be directly regulated. However, this method has a negative effect on the output waveform [11]. In [12], a closed loop control system to maintain voltage balancing for a three-level FCC has been proposed. In this method, the voltage regulation is maintained by modifying the duty cycles in proportion to the unbalanced portion. However, describing the nonlinear relationship between the output current and the capacitor voltages is not straightforward for converters with a high number of levels. In [11], a modified PS–SPWM control method for a five-level FCC has been presented. This closed-loop method consists of PI controllers and a voltage balancing algorithm. In order to compensate for the deviation of the capacitor voltages, this algorithm utilizes the redundancy of the switching states of the FCC. It also adjusts the time duration of selected switching states and modifies the switching instants of PS–PWM.

In this paper, a new modifying PS-SPWM control method is proposed for capacitor voltage balancing. The concept of this new method is based on the manipulation of the sinusoidal modulator waveforms in order to compensate for capacitor voltages. The proposed method is most similar to

978-1-4244-4782-4/10 $26.00 © 2010 IEEE

the method in [11], but with a relatively straightforward implementation. In this paper, in Section II, flying capacitor converter modeling will be discussed. In Section III, natural voltage balancing based on the PS-SPWM method will be described. In Section IV, the new control method will be introduced and finally in Section V the simulation results will be presented.

II. FLYING CAPACITOR CONVERTER MODELING

Fig. 1 shows two four-level flying capacitor poles utilized in a single-phase inverter application. One unique property of the FCC is that the capacitor voltages can be regulated by the redundant state selection within one pole, i.e., voltages v_{ac1} and v_{ac2} can be controlled using the adjacent six transistors. Therefore, the FCC can be considered on a single pole basis and generalized to n levels as shown in Fig. 2. In this general notion, the capacitors are labeled from the inside out ranging from 1 to n-2. The top transistors are labeled starting at the inside and ranging from 1 to n-1. In this paper, the dead-time states of the transistors are ignored so that the lower transistors are commanded to be the complement of the upper transistors. The inputs of the general one power pole model, as in Fig. 2, are dc voltage v_{dc}, current i_{dc}, and the transistor signals. The outputs are ac voltage v_{ag}, ac current i_{as}, and the flying capacitor voltages. The first step is to compute the voltage across and current through each transistor as:

$$v_{Tai} = \left(1 - T_{ai}\right)\left(v_{aci} - v_{ac(i-1)}\right) \qquad (1)$$

$$i_{Tai} = T_{ai}\, i_{as} \qquad (2)$$

where i=1, 2, …$(n$-1). These expressions neglect device voltage drops. However, this can be easily added to the expression, but it is not included here for clarity. Once the transistor voltages and currents are established, the capacitor currents can be calculated from

$$i_{aci} = i_{Ta(j+1)} - i_{Taj} \qquad (3)$$

where j=1, 2, …$(n$-2). Dividing by the respective capacitance and integrating yields the capacitor voltages which are needed to evaluate (1). Finally, the ac side voltage is computed using (1) as:

$$v_{ag} = v_{dc} - \sum_{i=1}^{n-1} v_{Tai} \qquad (4)$$

and the contribution to the power pole from dc side current i_{adc} is calculated using:

$$i_{adc} = i_{Ta(n-1)} \qquad (5)$$

These straightforward expressions can be used to code a flying capacitor model in a way that is efficient and expandable to any number of voltage levels. Some simulation packages allow the integration of vectors of variables which further simplify the capacitor voltage calculation.

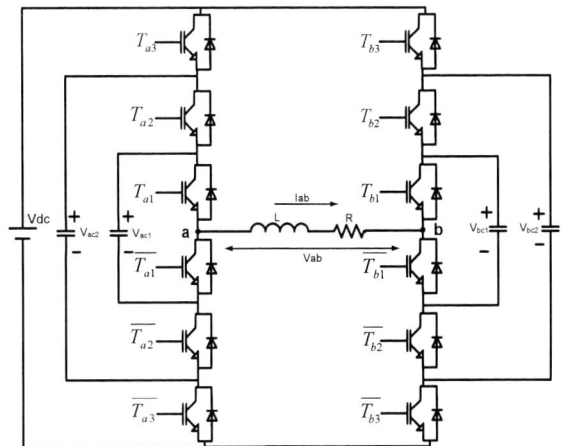

Fig. 1. The four-level flying capacitor inverter topology.

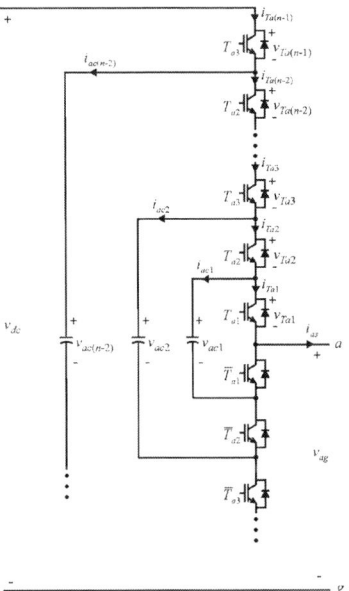

Fig. 2. The n-level flying-capacitor converter.

III. NATURAL VOLTAGE BALANCING BASED ON PS-SPWM METHOD

The most common PWM method used for FCCs is the PS-SPWM which can maintain balanced capacitor voltages by applying equal-in-duration charging and discharging switch states [4]. For an n-level FCC, the sinusoidal modulation signal is compared with n-1 triangular carrier signals that are phase shifted by $360/(n$-1) degrees. The generated PWM signals control the corresponding switches. In Fig. 3, the three triangular carrier waveforms each with $360°/3$ phase shift accompanied with sinusoidal modulation waveform are shown. By comparing these carriers and the modulation waveform, the PWM signals for a four-level FCC are produced. Under natural voltage balancing circumstances, all of the duty adjacent cycles for the switches are equal and the output voltage would be in a regulated shape as shown in Fig. 3. However, under practical circumstances, this open-loop control cannot guarantee voltage balancing. Fig. 4, shows the capacitor voltages of a four-level flying capacitor inverter when the load suddenly changes from L=40 mH, R=20 Ω to L=10 mH, R=4 Ω at t=1 sec. As seen, the capacitor voltages start to deviate drastically after the step change in load and consequently balancing is not maintained. The first trace in Fig. 5 shows the line-to-line voltage immediately after the load change and the second trace shows it some time after the load change. Clearly, because in the instant when load change happens the deviation of the capacitor have not been started yet the first trace shows the line-to-line voltage in a better shape compared to the second trace.

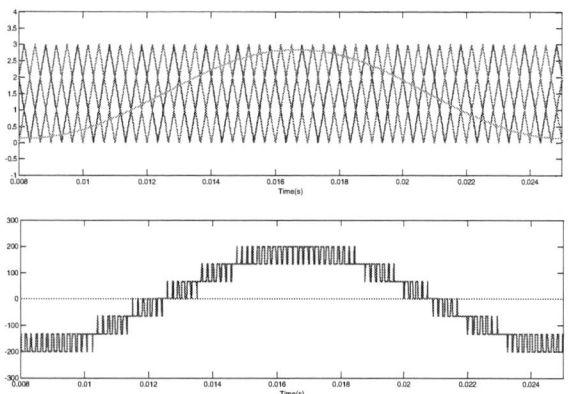

Fig. 3. PS-SPWM for the four-level FCC with the resulting line to line voltage.

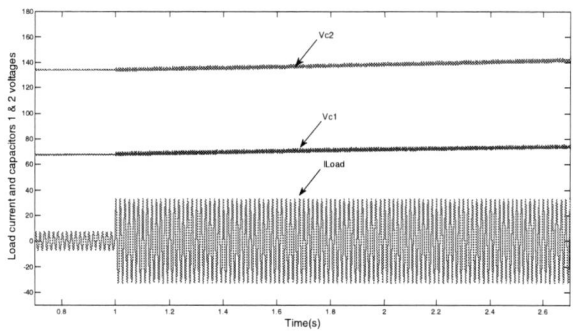

Fig. 4. Voltage of capacitors for the four-level inverter based on natural voltage balancing when a step load change occurs at t=1 sec.

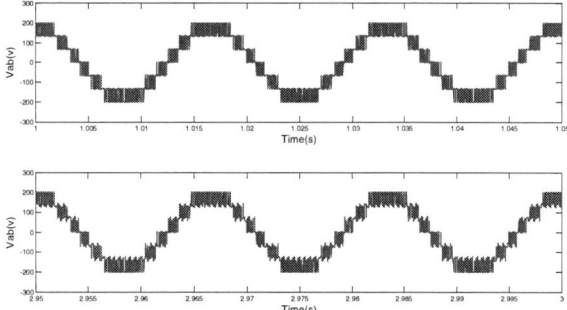

Fig. 5. Line-to-line voltage; Top) just after a sudden change in load and Bottom) some time after the sudden change in load.

IV. FLYING CAPACITOR VOLTAGE CONTROL

To describe the proposed control scheme, the presented equations in section II are developed for the four-level FCC in Fig. 1. Hence, one can write:

$$i_{ac1} = i_{Ta2} - i_{Ta1} \, , \, i_{ac2} = i_{Ta3} - i_{Ta2}$$

$$i_{ac1} = (T_{a2} - T_{a1})i_{ab} \, , \, i_{ac2} = (T_{a3} - T_{a2})i_{ab} \quad (6)$$

$$i_{ac1} = (d_{T_{a2}} - d_{T_{a1}})i_{ab} \, , \, i_{ac2} = (d_{T_{a3}} - d_{T_{a2}})i_{ab}$$

Considering (6), it is clear that the value of the change of duty cycles for switches 2 and 1 have direct and reverse effect on the value of the current of capacitor 1. That is, by increasing the duty cycle of switch 2 and decreasing the duty cycle of switch 1, capacitor 1 current will increase. This fact can help control the voltage of the capacitor by modifying the duty cycle. Changing the duty cycle can be done by modifying the sinusoidal modulation waveform or the triangular carrier waveforms.

Using a simple analysis it is easily shown how the control of voltage of the capacitor can be done by modifying the duty cycles. Considering Fig. 2 as one leg of an n-level flying-capacitor converter one can write:

$$\hat{i}_{aci} = (\hat{i}_{T_{a(i+1)}} - \hat{i}_{T_{a(i)}})i_{asr} = (d_{T_{a(i+1)}} - d_{T_{a(i)}})i_{asr} \quad (7)$$

where duty cycles are in sinusoidal modulation waveform format and can be written as:

$$d_{T_{ai}} = \frac{n-1}{2}m_i\cos(\omega t) + \frac{n-1}{2} \quad (8)$$

and the i_{asr} current is considered as:

$$i_{asr} = \sqrt{2}\, I_s \cos(\omega t + \phi) \quad (9)$$

By substituting the equations (8-9) into (7) and using the moving average technique one can write:

$$\hat{i}_{aci} = (d_{T_{a(i+1)}} - d_{T_{a(i)}})i_{asr} =$$

$$\frac{m_{i+1} - m_i}{2}\sqrt{2}\, I_s \cos(\omega t + \phi)\cos(\omega t) = \quad (10)$$

$$\frac{m_{i+1} - m_i}{4}\sqrt{2}\, I_s \left[\cos(\phi) + \cos(2\omega t + \phi)\right]$$

The steady-state long-term average of (10) yields:

$$\bar{i}_{aci} = \frac{m_{i+1} - m_i}{4}\sqrt{2}\, I_s \cos(\phi) \quad (11)$$

The change of the voltage of capacitors can be related to (11) as follows:

$$\Delta V_{aci} = \frac{\bar{i}_{aci}}{C} = \frac{m_{i+1} - m_i}{4C}\sqrt{2}\, I_s \cos(\phi) \quad (12)$$

From (12) it is clear that the voltage of capacitors can be regulated by modifying the amplitude of sinusoidal modulation waveforms. Moreover, the change of the voltage of capacitors is dependent on the power factor and the amplitude of the i_{asr} current and not the direction.

In order to regulate the voltage of capacitors based on (12) a control scheme is proposed in (13). Equation (13) can be described using the control diagram in Fig. 6.

$$\begin{cases} m_j = M + m_{j,j} & j=1 \\ m_j = M + m_{j,j} + m_{j,j-1} & 2 \le j \le n-2 \\ m_j = M + m_{j,j-1} & j=n-1 \end{cases} \quad (13)$$

Fig. 6 shows the proposed control system based on (13) for an n-level converter. As it is clear in this figure, $m_{j,j-1}$ is the change in the modulation index which is produced by the voltage loop of capacitor j-1 (for $j>1$) and affects the modulation index of switch j. Inverter modulation index M is a constant value less than 1 and m_j is the modulation index for switch j which is updated continuously. Each voltage loop consists of a PI controller which helps regulate j^{th} capacitor voltage V_{Cj} to be equal to V_{refj}. Now, there are n-1 sinusoidal modulation waveforms as follows:

$$d_{T_{a1}} = \frac{n-1}{2} m_1 \cos(\omega t) + \frac{n-1}{2},$$

$$d_{T_{a2}} = \frac{n-1}{2} m_2 \cos(\omega t) + \frac{n-1}{2}, \quad (14)$$

$$....$$

$$d_{T_{a(n-1)}} = \frac{n-1}{2} m_{n-1} \cos(\omega t) + \frac{n-1}{2}$$

Based on the control demand, the modulation indexes (m_1, m_2, ..., m_{n-1}) are modified then the duty cycles are changed and consequently the capacitor voltages are regulated. These waveforms produce the switching pulses.

Note that in this paper the FCC has been considered as an inverter with RL load. For the inverter with the capacitive load the same equation as in (12) is valid. However, for an FCC as a rectifier the direction of i_{asr} current is reversed. Therefore, all the control rules must be multiplied with a negative sign.

V. SIMULATION RESULTS

The proposed control method has been applied to control the four-level FCC in Fig. 1. Fig. 7 shows line-to-line voltage and load current with a power factor of 0.81 lagging. Fig. 8 shows the voltage of both capacitors in leg a when a step change in the load similar to that of in Section III is applied. As this figure shows, the capacitor voltages are regulated before and even after the step change. Comparing the capacitor voltages in Figs. 4 and 8 clearly shows the effectiveness of the proposed control method. The advantage of the proposed control method is that it can be applied to FCCs with any number of levels. However, two points must be noticed when higher level FCCs are considered. One is the start up strategy for capacitor voltages. As Fig. 6 suggests, in the proposed method the control loops associated with each two adjacent capacitor greatly affect each other with a reverse sign. This creates some difficulties during the start up process. Hence, a start up strategy needs to be implemented to solve this problem. The technique applied herein is to start applying reference voltage for each two

adjacent capacitor with a delay between them. As an example of FCC with higher number of levels an eight-level FCC has been simulated. Fig. 9 shows the voltage of the capacitors for a simulated eight-level FCC using mentioned start up strategy. Fig. 10 shows the line-to-line voltage for the same converter.

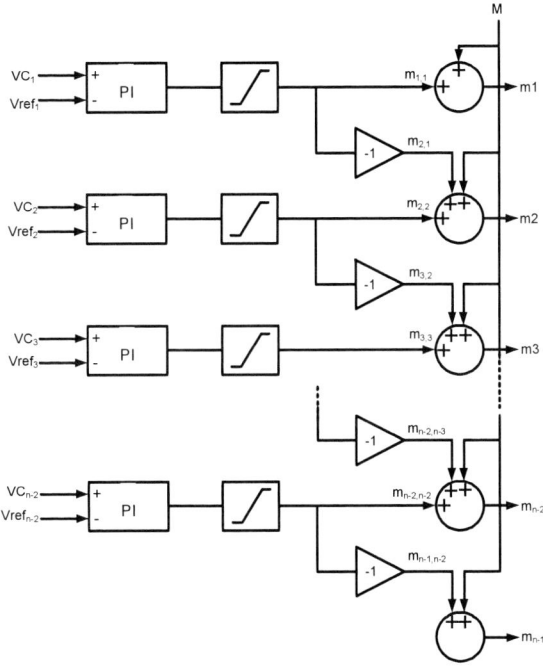

Fig. 6. Block diagram of the proposed control scheme.

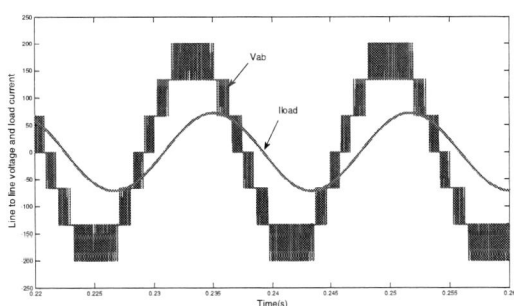

Fig. 7. Line-to-line voltage and load current with a power factor of 0.81 lagging.

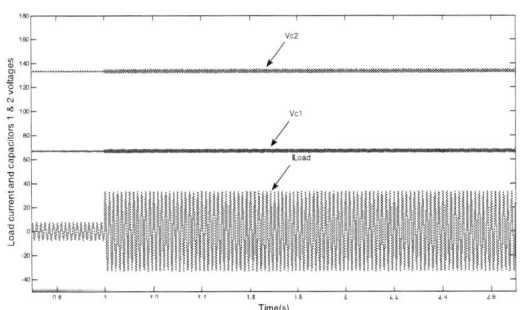

Fig. 8. Capacitors voltage of the four-level FCC under proposed control method when a step load change occurs at t=1 sec.

The second note is related to the number of levels of the converter when it is an odd number. In this case, the triangular waveform must be modified. This modification is not related to the proposed control methods. In fact, it relates to an inherent problem associated with PS-SPWM. Fig. 11 shows the modified triangular waveform that is used for the PS-SPWM control method when n is odd. This triangular waveform accompanied with seven similar ones (each with a $2\pi/8$ phase shift) are used to control a nine-level FCC with the proposed control method. Fig. 12 shows line-to-line voltage of FCC. As this figure shows, the proposed control method works for FCCs with any desired number of voltage levels.

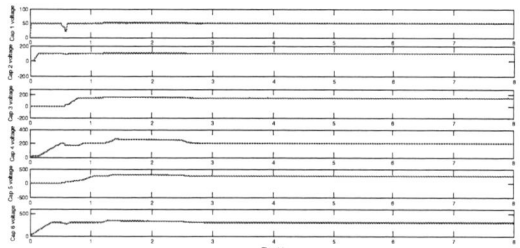

Fig. 9. Voltage of all the capacitors for an eight-level FCC under the proposed control method.

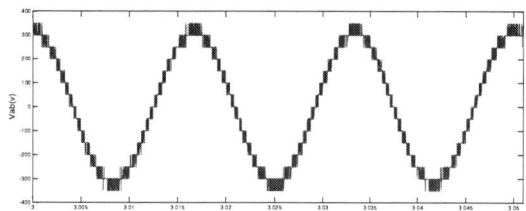

Fig. 10. Line to line volotage for an eight-level FCC under the proposed control method.

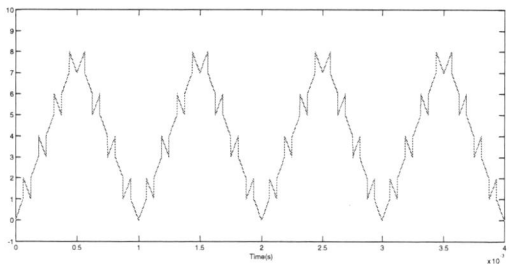

Fig. 11. Modified triangular carrier waveform

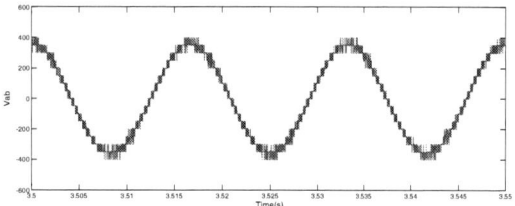

Fig. 12. Line-to-line voltage of a nine-level FCC under the proposed control method.

VI. CONCLUSION

In this paper, a new control scheme is proposed which can regulate the capacitor voltages in a flying-capacitor multi-level converter with any desired number of levels. This control scheme is defined based on the general equations of the converter. Compared with other balancing methods, the implementation of this control method is very straightforward. The control scheme is simulated for four, eight, and nine-level flying capacitor inverters. The simulation results clearly show the effectiveness of this method in maintaining balanced capacitor voltages.

REFERENCES

[1] T. A. Meynard, "Modeling of multilevel converters," *IEEE Transactions on Industrial Electronics*, volume 44, pages 356–364, June 1997.

[2] T. A. Meynard and H. Foch, "Multilevel choppers for high voltage applications," *EPE J.* volume 2, number 1, pages 45-50, March 1992.

[3] T. A. Meynard, M. Fadel, and N. Aouda, "Modeling of multilevel converters," *IEEE Transactions on. Industrial Electronics* volume 44, pages 356-364, June 1997.

[4] X. Yuan, H. Stemmler, and I. Barbi, "Self-balancing of the clamping capacitor-voltages in the multilevel capacitor-clamping-inverter under sub-harmonic PWM modulation," *IEEE Transactions on Power Electronics*, volume 16, number 2, pages 256–263, Mar. 2001.

[5] T.A. Meynard, H. Foch, P. Thomas, J. Courault, R. Jakob and M. Nahrstaedt, "Multicell converters: basic concepts and industry applications," *IEEE Transactions on Industrial Electronics*, volume 49, number 5, pages 955-964, October 2002.

[6] A. Shukla, A. Ghosh, A. Joshi, "Capacitor voltage balancing schemes in flying capacitor multilevel inverters," *IEEE Power Electronics Specialists Conference*, volume 6, number 1, pages 2367-2372, June 2007.

[7] C. Feng, J. Liang, V. G. Agelidis, "A novel voltage balancing control method for flying capacitor multilevel converters," *Industrial Electronics Society Conference*, volume 2, pages 1179-1184., November 2003.

[8] R. Wilkinson, H. du Mouton, and T. A. Meynard, "Natural balance of multicell converters: The two-cell case," *IEEE Transactions on Power Electronics*, volume 21, number 6, pages 1649–1657, November 2006.

[9] R. Wilkinson, H. du Mouton, and T. A. Meynard, "Natural balance of multicell converters: The general case," *IEEE Transactions on Power Electronics*, volume 21, number 6, pages 1658–1666, November 2006.

[10] L. Xu and V. G. Agelidis, "Active capacitor voltage control of flying capacitor multilevel converters," *IEE Proc. Electr. Power Appl.* volume 151, number 3, pages 313- 320, May 2004.

[11] C. Feng, J. Liang, V. G. Agelidis, "Modified Phase-Shifted PWM control for flying capacitor multilevel converters," *IEEE Transactions on Power Electronics* volume 22, number 1, pages 178–185, January 2007.

[12] B. M. Song, J. S. Lai, C. Y. Jeong, and D. W. Yoo, "A soft-switching high-voltage active power filter with flying capacitors for urban maglev system applications," *IEEE/IAS Annual. Meeting.* volume 3, pages 1461–1468, 2001.

An Active Damping Technique for a Current Source Inverter Employing a Virtual Negative Inductance

Ahmed Salah Morsy, Shehab Ahmed and Prasad Enjeti
Electrical and Computer Engineering Department
Texas A&M University at Qatar
Doha, Qatar
{ahmed.salah, shehab.ahmed, prasad.enjeti}@qatar.tamu.edu

Ahmed Massoud
Electrical and Computer Engineering Department
Qatar University
Doha, Qatar
ahmed.massoud@qatar.tamu.edu

Abstract —A grid connected CSI requires a CL filter stage. In this work, the isolation transformer leakage inductance constitutes this filter inductance. This CL filter is lightly damped for efficiency constraints. Hence, it is subject to resonance if excited by inverter current harmonics, pre-existing grid voltage harmonics, or due to step changes in power. Previous work on active damping employed a virtual resistance at high frequencies, including the resonance frequency. However, this method shows limited gain and phase stability margins. In this paper an active damping technique employing a virtual negative inductance for the CSI is proposed. The introduction of a virtual negative inductance around the resonance frequency causes active frequency shifting. The advantages of the proposed system are better time response and increased stability margins. Simulation results for a grid connected CSI have been presented to substantiate the proposed technique.

I. INTRODUCTION

The CSI is widely used in high power applications and adjustable speed drives [1-3] due to its motor friendly voltage and current waveforms, lower torque pulsations, and reduced EMI and stress on the motor insulation. In addition, the CSI is used for grid connected DC-AC power converters for fuel cells [4, 5] owing to its inherently boosting capability, and with wind farm collection grids due to its proven short circuit protection and parallel operation capability. Different types of resonance caused by passive lightly damped CL filters used in a CSI were investigated in [6]; furthermore, passive [7] and active [6] damping techniques were discussed. Compared to passive damping techniques, active damping techniques introduce lower power loss and reduced size and cost, due to the absence of actual resistors.

Active damping is associated with problems in stability margins of the feedback loop especially when operating at high resonance frequencies that require high loop gains and higher switching frequency for the same damping ratio (CSI has to operate in closed loop [9]). In this paper an active damping technique employing a virtual negative inductance for a CSI is introduced. This virtual inductance is formed by positive feedback from the output current. The introduction of a virtual negative inductance at the resonance frequency causes active

frequency shifting; resulting in faster and smoother damping along with increased stability margins.

II. PASSIVE FILTER DESIGN

Since the CSI is a dual of the VSI, its passive filter is characterized by a shunt capacitor in parallel with the CSI to assist in the commutation of switching devices and to eliminate high frequency components of the modulated current waveform. The isolation transformer leakage inductance represents the filter's inductive component. The low inductance value typical of high power applications constitutes a challenge to control design, especially when considering the filter cut off frequency constraint and compliance with IEEE requirements for harmonic control in power systems [11].

In [8], design of a passive filter to provide unity power factor at full load is studied, but power factor decreases significantly at light loading. Design of a unity power factor filter requires a high value of inductance especially at medium voltage levels, since capacitive reactive power is proportional to grid voltage squared, while inductive reactive power is proportional to output current squared. Therefore, power factor control is implemented actively by inverter.

III. RESONANCE MODES

Fig. 1 shows a typical CSI connected to the grid. The lightly damped CL filter resonates at its cut off frequency. This resonance can be excited by CSI harmonic current (parallel resonance) or by pre-existing grid voltage harmonics (series resonance) [1, 6, 7].

IV. PASSIVE DAMPING TECHNIQUES

Several ways of damping were investigated in [6]. The most common techniques are summarized in Fig. 2. They were applied in LC filters used with a VSI. However, they can be directly applied to the CSI case by duality. Generally, using passive damping increases size, cost, and losses of a filter. Therefore active damping is preferred. *Cases (a, b)* in Fig. 2 are characterized by reduced attenuation above cut off frequency. The drawback of *Case (c)*, are the large losses. *Case (d)* includes many filter components, hence, increased size and cost.

978-1-4244-4782-4/10 $26.00 © 2010 IEEE

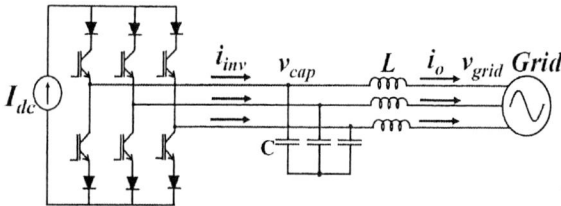

Figure 1. Grid connected CSI with CL passive filter

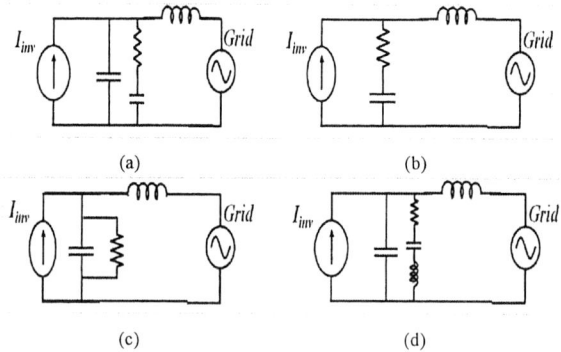

(a) (b)

(c) (d)

Figure 2. Passive damping techniques

V. ACTIVE DAMPING

Active damping performs the required attenuation of resonance with lower cost, losses and size, since it emulates the presence of a damper resistance [6] by injection of current component in phase with capacitor voltage. This virtual resistance is used in a selected band of frequencies (resonance) to prevent interference with fundamental current control. Spectral separation between the fundamental and resonance frequencies makes this possible. By decreasing the resistance (r_d), the damping ratio increases. However, this increase is limited by phase and gain margins [5]. It is worthy to point out the effect of modulation and delay on these stability margins. The proposed control law in (1) depends on both capacitor voltage, with negative feedback, and inductor current, with positive feedback. The feedback law is on the overall frequency bandwidth just for simplicity of discussion.

$$ i_{inv\,(abc)} = i^{ref}_{inv\,(abc)} - \frac{1}{r_d} v_{cap\,(abc)} + K\, i_{o\,(abc)} \qquad (1) $$

In Fig. 3, the colored semicircles represent root-loci (damper resistance variation) at K = 0, 0.45, 0.9. The vertical doted lines represent increasing gain, K, at rd = 15, 6, 3.5, 2.4. Hence, combining voltage and current feedback achieves higher damping ratios due to the reduction (shift) in natural frequency. Positive feedback of inductor current emulates the presence of a negative inductance since a component of the inverter current is out-of-phase with filter inductor current as shown in Fig. 4.

VI. ACTIVE DAMPING LOOP (VIRTUAL RESISTANCE)

After transforming voltage and current variables from a stationary reference frame to a synchronous one, the resonance

frequency will be mapped to two values (its corresponding positive and negative sequence). In previous work, a high–pass filter (virtual resistance) was used for active damping, Fig. 5. This implies trying to damp high frequency harmonics which are uncontrollable according to Shannon's theorem. Therefore, a band-pass filter (virtual resistance), Fig. 6, should be used to focus on specific resonance frequencies making damping gain increase applicable (3 times for the given parameters) without violating stability margins (Gain margin = ∞). The band-pass filter bandwidth covers the resonance region, i.e. zero at origin and two poles around ($\omega_{resonance} \pm 2*\omega_{fundamental}$). The filter gain ($1/r_d$) is adjusted to provide maximum damping factor, Fig. 6.

VII. ACTIVE FREQUENCY SHIFT LOOP (VIRTUAL NEGATIVE INDUCTANCE)

Positive feedback is usually related to instability in control problems. However, in this work, it provides higher stability margins, Fig.8, than negative feedback, Fig. 7. The reason is that, applying negative feedback on a high frequency current loop forces the filter to respond faster than its bandwidth (corner frequency) [10], which is beyond the capability of the discrete controller (delay) and the PWM inverter (sharp switching). Therefore, negative feedback in the high frequency current loop further stimulates the resonance frequency.

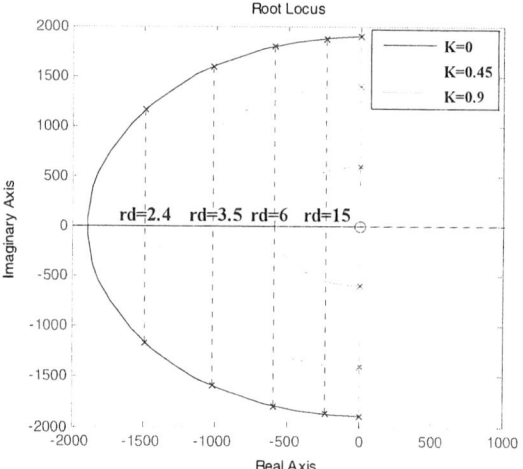

Figure 3. Root-locus at different values of (rd & K)

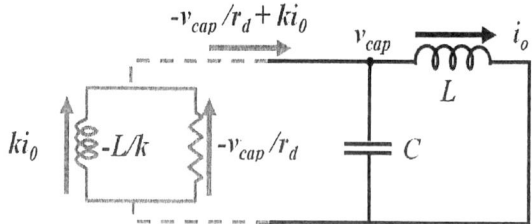

Figure 4. Virtual resistance and negative inductance

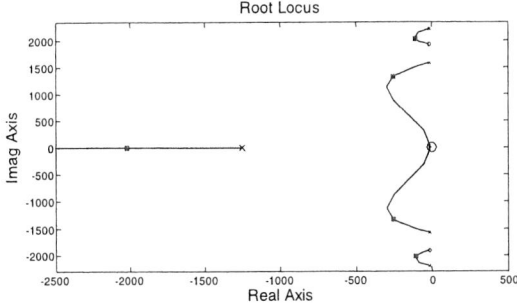

Figure 5. Root locus of active damping (High–pass resistive)

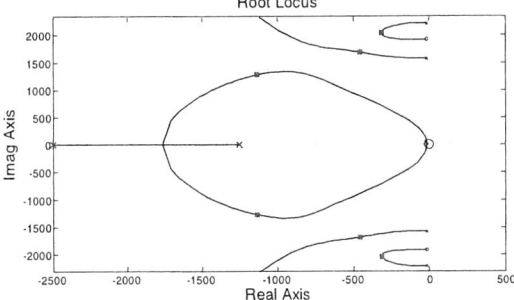

Figure 6. Root locus of active damping (Band–pass resistive)

Applying both feedback loops, the root locus is shown in Fig. 9. The two circled groups of poles & zeros are system dominant and can't be moved significantly. Fortunately closed loop zeros are near those poles so they will minimize their effect. Here, the same band-pass filter, of the active damping loop, is used. The gain (K) of the positive feedback on high frequency inductor current is tuned to reach maximum damping factor.

VIII. REFERENCE CURRENT ESTIMATION AND DECOUPLING

Equations (2, 3) define active and reactive output power in terms of grid voltage and current transformed to a (DQ) synchronous reference frame. If $v_{grid\,(d)}$ is aligned with the stationary voltage $v_{grid\,(a)}$, through a sensitive phased locked loop, PLL, simply $v_{grid\,(q)} = 0$, and the output reference current is computed by (4) and (5). However, due to susceptibility of this system to resonance, it's preferred to use a robust or even virtually generated PLL and compute the reference current from (6), and (7). In this way, phase and frequency errors from the PLL won't affect the control process. For better estimation of *inverter reference current,* the capacitor's steady state current can be included by (8), and (9).

$$P = \frac{3}{2} \left(v_{grid\,(d)}\, i_{o\,(d)} + v_{grid\,(q)}\, i_{o\,(q)} \right) \tag{2}$$

$$Q = \frac{3}{2} \left(v_{grid\,(q)}\, i_{o\,(d)} - v_{grid\,(d)}\, i_{o\,(q)} \right) \tag{3}$$

$$i_{o\,(d)}^{ref} = \frac{2}{3}\, P^{ref} / v_{grid\,(d)} \tag{4}$$

$$i_{o\,(q)}^{ref} = \frac{-2}{3}\, Q^{ref} / v_{grid\,(d)} \tag{5}$$

$$i_{o\,(d)}^{ref} = \frac{2}{3\left(v_{grid\,(d)}^2 + v_{grid\,(q)}^2\right)} \left(v_{grid\,(d)} P^{ref} + v_{grid\,(q)}\, Q^{ref}\right) \tag{6}$$

$$i_{o\,(q)}^{ref} = \frac{2}{3\left(v_{grid\,(d)}^2 + v_{grid\,(q)}^2\right)} \left(v_{grid\,(q)} P^{ref} - v_{grid\,(d)}\, Q^{ref}\right) \tag{7}$$

$$i_{inv\,(d)}^{ref} = i_{o\,(d)}^{ref} - v_{grid\,(q)}\ (\omega C) \tag{8}$$

$$i_{inv\,(q)}^{ref} = i_{o\,(q)}^{ref} + v_{grid\,(d)}\ (\omega C) \tag{9}$$

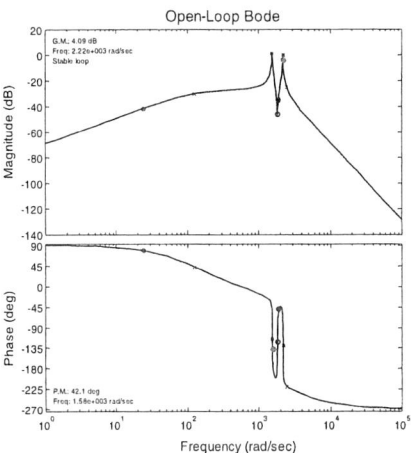

Figure 7. Bode plot of band–pass positive inductance
(Gain margin = 4 db) (Phase margin = 42°)

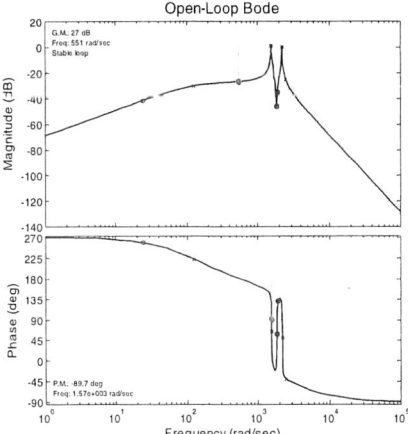

Figure 8. Bode plot of band–pass negative inductance
(Gain margin = 27 db) (Phase margin = 90°)

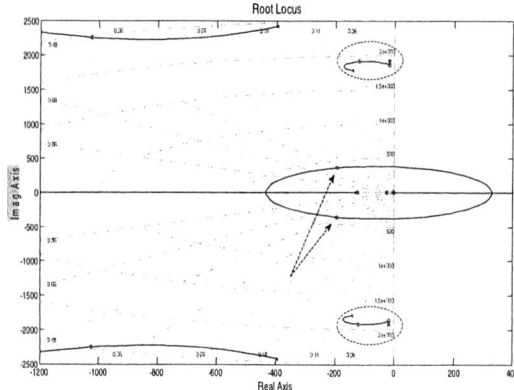

Figure 9. Root locus of both feedback loops

IX. OUTPUT CURRENT REGULATOR DESIGN

The regulator consists of a feed-forward path that is accurately estimated and an integral controller for steady state error compensation. It has to be noted that this regulator's bandwidth is smaller, i.e. not overlapping with that of the active damping loops. The use of the feed-forward path (static decoupling) accelerates the response without involvement of the bandwidth overlap problem. A detailed block diagram, in Fig. 10, shows control scheme including a band-pass resistance loop, a band-pass virtual negative inductance loop and integral controller (compensator) with a static decoupling loop.

X. SIMULATION RESULTS

The designed controllers were discretized with a zero-order-hold at a sampling frequency twice the inverter switching

frequency. Effectiveness of the active frequency shifting is more significant when the passive filter corner frequency increases. The following figures (Fig. 11, 12) demonstrate the time response for a CL filter tuned at 380 Hz, which shows how the introduction of negative virtual inductance has increased the system damping by active frequency shifting, thus raising filter inertia. Table 1 presents the CSI parameters implemented for simulation.

TABLE I.   PARAMETERS OF PROPOSED MODEL:

Line voltage	15 KV
Switching frequency of converter is	2KHz
Leakage inductance of isolation transformer	0.8 mH
DC link current (peak line current)	1200 A
Passive filter cut-off frequency	300 Hz
Capacitor value	220 µF

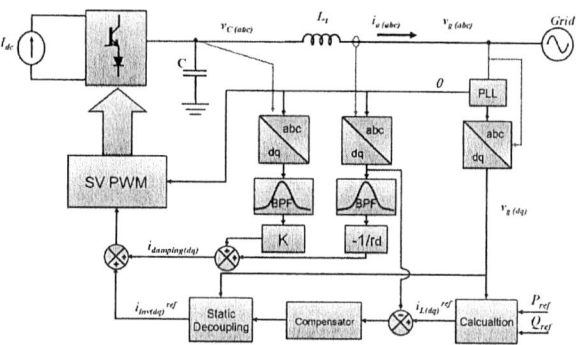

Figure 10. Proposed Control scheme

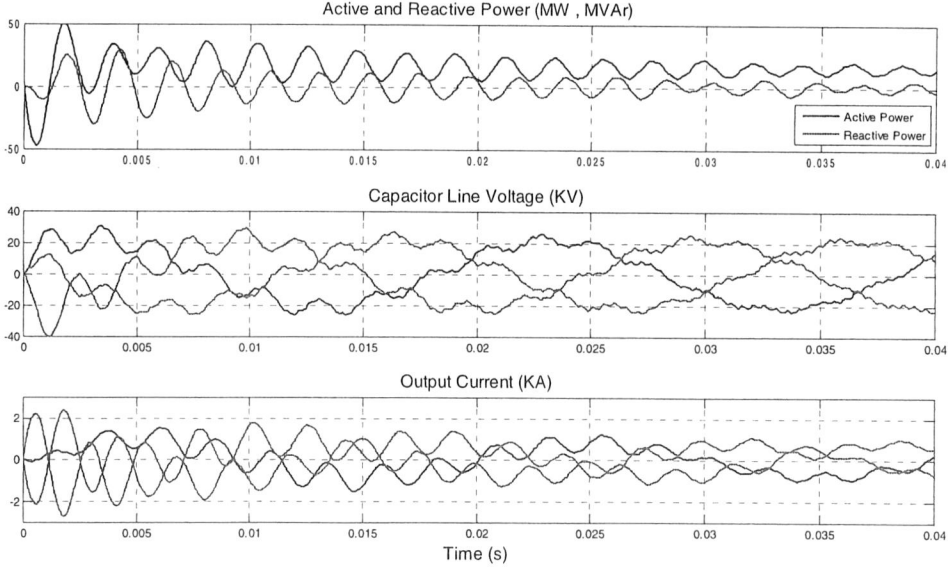

Figure 11. Active damping (virtual resistance only)

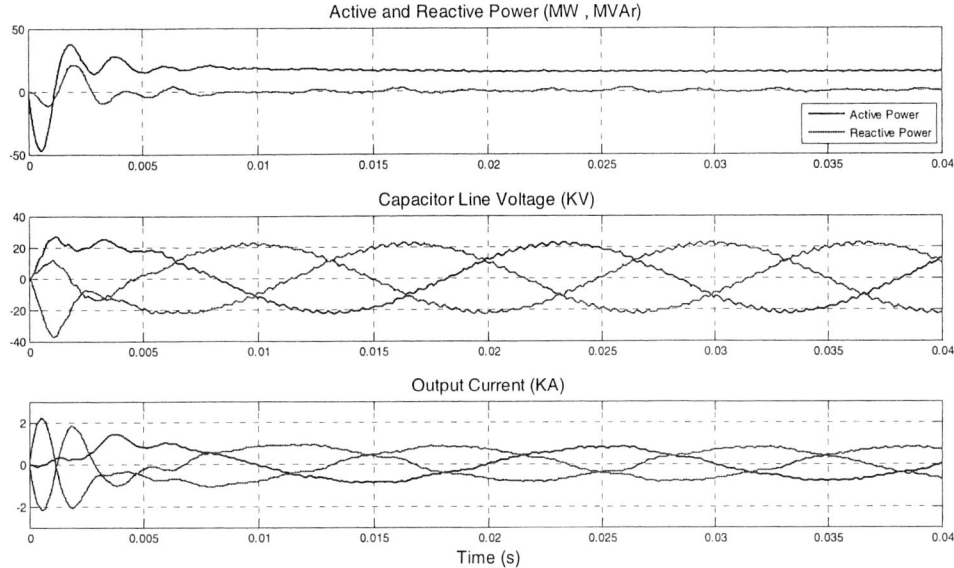

Figure 12. Active damping (virtual resistance) + Active frequency shift (virtual negative inductance)

XII. CONCLUSION

Using positive feedback is usually associated with poor dynamic response and instability. However, in this work it results in better time response with increased stability margins, because it shifts the filter natural frequency to a smaller value and smoothens the inverter current. This control technique employs a virtual negative inductance at the resonance frequency in parallel with the filter inductance, resulting in a larger inductance, impeding harmonics. Root locus and Bode plots, along with simulation results prove the superiority of the proposed technique.

REFERENCES

[1] Bin Wu, "High-Power Converters and AC Drives", Wiley-IEEE Press, 2006

[2] Salo, M.; Tuusa, H., "Experimental results of the current-source PWM inverter fed induction motor drive with an open-loop stator current control," Applied Power Electronics Conference and Exposition, 2003. APEC '03. Eighteenth Annual IEEE , vol.2, no., pp. 839-845 vol.2, 9-13 Feb. 2003

[3] Salo, M.; Tuusa, H., "A high performance PWM current source inverter fed induction motor drive with a novel motor current control method," Power Electronics Specialists Conference, 1999. PESC 99. 30th Annual IEEE , vol.1, no., pp.506-511 vol.1, Aug 1999

[4] Mohr, M.; Fuchs, F.W., "Comparison of three phase current source inverters and voltage source inverters linked with DC to DC boost converters for fuel cell generation systems," Power Electronics and Applications, 2005 European Conference on , vol., no., pp.10 pp.-P.10, 0-0 0

[5] M. Mohr, B. Bierchoft and F. W. Fuchs, "Dimensioning of a current Source inverter for the feed-in of electrical energy from fuel cells to the mains". Paper 41, presented at the Nordic Workshop on Power and Industrial Electronics, NORPIE 2004, Trondheim, Norway, June 14-16. 2004.

[6] J. Wiseman and B. Wu, "Active Damping Control of a High Power PWM Current Source Rectifier for Line Current THD Reduction", IEEE Power Electronics Specialist Conference, pp. 552–557, 2004.

[7] Ahmed, K.H.; Finney, S.J.; Williams, B.W., "Passive Filter Design for Three-Phase Inverter Interfacing in Distributed Generation," Compatibility in Power Electronics, 2007. CPE '07 , vol., no., pp.1-9, May 29 2007-June 1 2007

[8] Abdelsalam, A.K.; Masoud, M.I.; Finney, S.J.; Williams, B.W., "Comparative study of AC side passive and active filters for medium voltage PWM current source rectifiers," Power Electronics, Machines and Drives, 2008. PEMD 2008. 4th IET Conference on , vol., no., pp.578-582, 2-4 April 2008

[9] Poh Chiang Loh; Holmes, D.G., "Analysis of multi-loop control strategies for power conversion applications," Industry Applications Conference, 2003. 38th IAS Annual Meeting. Conference Record of the , vol.3, no., pp. 1778-1785 vol.3, 12-16 Oct. 2003

[10] Kopasakis, George, "Feedback Control Systems Loop Shaping Design with Practical Considerations", NASA, Glenn Research Center, Cleveland, Ohio, Sep 2007.

[11] IEEE Standards 519-1992, Recommended Practices and Requirements for Harmonic Control in Electric Power Systems, 1992.

Maximum Solar Power Transfer in Multi-port Power Electronic Interface

Wei Jiang, *Student Member IEEE,* Babak Fahimi, *Senior Member IEEE*

Renewable Energy and Vehicular Technology Laboratory
Department of Electrical Engineering
University of Texas at Arlington
416 S. Yates Street Nedderman Hall Room 130
Arlington, TX, 76019
[wei.jiang], [fahimi] @uta.edu

Abstract— **Ambient energy is a good supplement to existing hybrid power systems. Among the available options, solar is one of the most readily obtainable and technologically compatible sources. However, uncertainty of sun radiation intensity and changing atmospheric conditions impose challenges in efficient usage of sun power. Significant progress has been made during the last decades in optimally harvesting solar energy for single input power system. However, little literature focuses on the control and coordination of solar energy conversion with other actively controlled sources and storages. This paper proposes the operation and control of solar power conversion in Multi-port Power Electronic Interface (MPEI) for different modes of operation: load sharing mode and battery recovery mode. The control structure for solar maximum power point tracking (MPPT) in load sharing mode is proposed for optimal harvesting from solar panel and to provide immunity to load dynamics. Current-Mode Maximum Power Transfer (CMMPT) method is proposed to transfer maximum power to battery during charge recovery. The control system is implemented in a TMS320F2812 DSP and experimental results are presented to prove the effectiveness of proposed control system.**

I. INTRODUCTION

Solar power is considered as one of the best supplement candidates to existing power networks for several reasons:

- Easy manufacturing and compatible to existing technologies,

- Least impact to ambient ecological system,

- Quiet and safe.

Since the existing lines of solar cell products for civilian use are of low efficiency (mostly under 18%), the economical usage of solar power becomes very important for solar based power system. Solar power conversion usually takes place in two forms: solar-to-chemical and solar-to-electricity, the typical applications of which are solar based battery charger and solar source inverter.

Single-source two-stage structures, in Figure 1(a) are commonly used in the solar power processing, where the front-end stage interfacing with solar panel performs input current or voltage regulation to ensure the tracking of maximum power point of solar power. Due to direct regulation of input current/voltage, the output voltage of the solar interface stage is uncontrollable. Therefore, in order to generate usable ac voltage from solar panel, the dc-link voltage stabilization has to be incorporated into the design of second stage inverter.

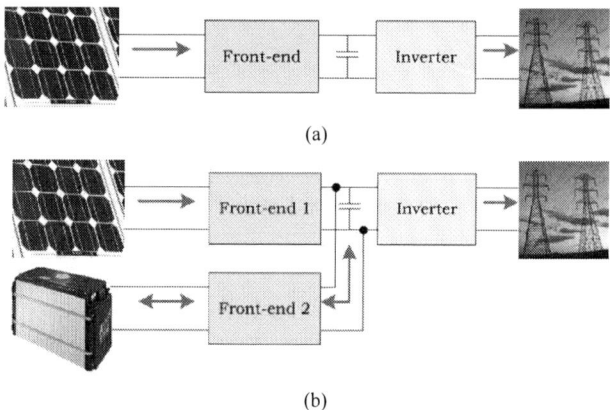

Figure 1. Solar power processing: (a) single source two-stage, (b) two sources two-stage

In multi-converter front-end system, as Figure 1(b) the dc-link voltage fluctuation introduced by solar power tracking can be compensated by other paralleled converters under voltage control; solar converter and inverter can be dedicated to their own tasks to ensure maximum power delivery. The mono-tasking manner of local controller design can achieve high performance in parallel operation; however, if no load demand comes from ac side and only Maximum Power Transfer (MPT) is needed between solar panel and energy storage, the discrete organization of local controller will not be

978-1-4244-4782-4/10 $26.00 © 2010 IEEE

the best choice, since the coordination between two local controller might not be able to guarantee power balance between harvesting and consumption hence will render capacitor voltage fluctuation.

The concept of Multi-port Power Electronic Interface (MPEI) created in past work [1] offers interfaces for multiple source and storage as well as integrated control system structure for optimal power management. Although MPEI modes of operation were proposed in previous publication, the control aspect of solar power processing was not fully investigated. This paper investigates the two important operation modes of solar power processing:

- Load sharing mode,

- Battery recovery mode.

The control system design for each of operation mode are investigated and proposed to achieve following objectives:

- To ensure maximum power harvesting from solar panel in both steady state and load transient,

- Transfer maximum power from solar panel to battery storage when there is no ac power demand

The hardware will be constructed with TMS320F2812 DSP based digital control, the corresponding experimental waveforms will be provided to prove the effectiveness of control design.

II. BACKGROUND OF MPEI

The concept of MPEI is created in [1], where the definition is given as: *A Multi-port Power Electronic Interface (MPEI) is a self-sustainable multiple input/output static power electronic converter which is capable of interfacing with different sources, storages and loads, the integrated control system of MPEI enables both excellent system dynamic and steady state performance which renders optimal renewable energy harvesting, optimal energy management and optimal and economical utility grid interactions in a deregulated power market.*

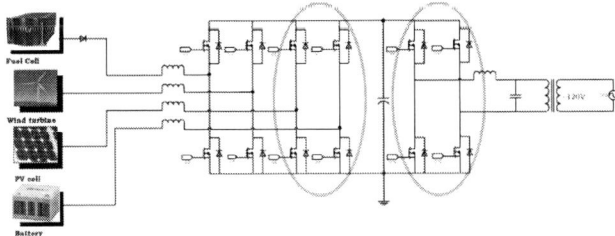

Figure 2. Multi-port Power Electronic Interface: five-port Implementation

A six-phase-leg structure was used for MPEI which interface three renewable sources: fuel cell, solar panel, wind turbine (BLDC), one energy storage: battery and single phase ac load. As indicated in Figure (2) the upper switch for unidirectional converters are disabled, only paralleled fast diodes are used. The ports interfacing with battery storage and ac load are both bidirectional. The objects under investigation are circled in the graph, including the phase-legs for solar, battery and inverter stage.

III. SOLAR-BATTERY LOAD SHARING MODE

The generic control structure has been proposed for optimal power sharing in literature [1], however, the detailed control design dealing with solar power conversion and energy storage is not addressed in detail, hence will be introduced in the following two chapters..

The maximum power point tracking technique is extensively studied during the past decades. Different micro-controller based methods [2] [3] have been proposed, for example: incremental inductance method [4], analytical method [5], perturbation & observation method [6], hill-climbing method [7], open circuit voltage/ short-circuit current method [8] and loss-free resistor method [9] etc. P&O method is chosen as the power tracking method in this paper due to its compatibility to existing control structure. The P&O routine is presented in Figure 3, where current reference is perturbed at timely bases; change of power as well as voltage output from solar panel is observed to decide the location of current power point; the reference current is further updated to approach the maximum power point.

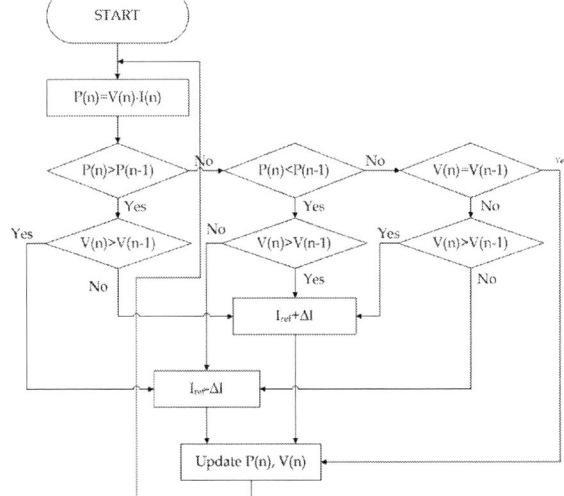

Figure 3. Maximum power point tracking routine

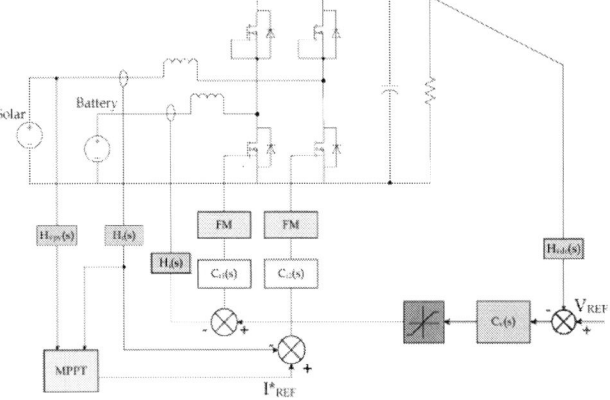

Figure 4. Maximum power transfer by means of load share

978-1-4244-4782-4/10 $26.00 © 2010 IEEE   69

The solar panel output current is directly controlled for maximum power tracking purpose; therefore, the solar switching cell output voltage is not controllable. In a load sharing scenario, the output voltage can be controlled by battery switching cell. The control structure is proposed in Figure (4). As indicated in the figure, for solar switching cell, the current reference is directly fed from Maximum Power Point Tracking (MPPT) routine; the solar switching cell is under current control. The battery switching cell is under Average Current Mode (ACM) control: the dc-link voltage is regulated by battery voltage controller, whose output is used as current reference for inner battery current loop. This control system offers both controlled power source as well as controlled voltage source; given limited power harvested from solar panel, the voltage source (battery switching cell) will provide the rest of power-in-demand dynamically.

Control loop for solar current and dc-link voltage can be designed separately for two reasons: first, one switching cell contributes to solar power harvesting while the other regulates the dc-link voltage; second, the disturbance injected from current controlled switching cell only modifies the system response at high frequency range [10] comparing to voltage loop cross-over frequency.

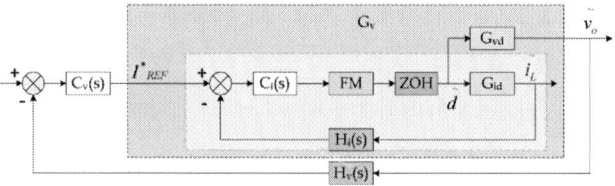

Figure 5. Average current mode control diagram

A generic ACM control block diagram is presented in Figure (5) which involves an outside voltage loop and an inner current loop. In case of direct solar current control, the design target only resides inside the blue block. The control-to-current transfer function can be obtained from small signal average model of boost converter which is presented in Equation (1).

$$G_{id}(s) = \frac{\tilde{i}_L(s)}{\tilde{d}} = \frac{V_C RC_1 s + V_C + RI_L(1-D)}{RLCs^2 + Ls + R(1-D)^2} \tag{1}$$

$$G_{vd}(s) = \frac{\tilde{v}_D(s)}{\tilde{d}} = \frac{V_C R(1-D) - RLI_L s}{RLCs^2 + Ls + R(1-D)^2} \tag{2}$$

Since the dc-link voltage is controlled by battery switching cell, which is to be addressed below, the current loop cross-over frequency can be set high. A PI controller is used in solar current loop design. The controller zero is placed at system resonant frequency and the integrator gain k_i is selected to be 200rad/sec. The compensated solar current loop has a cross-over frequency of 1.07kHz with gain margin of 12.2dB and phase margin of 55.1° as shown in the bode plot in Figure (6).

Battery switching cell design follows the ACM control structure given in Figure (5). The loop gain for compensated current loop is expressed in Equation (3), where $C_i(s)$ is a PI controller, G_{id} is control-to-battery current transfer function, G_{vd} is the control-to-voltage transfer function, as shown in

Equation (2). The outside voltage loop is designed with a well compensated inner current loop; the voltage loop gain is given by Equation (5), where $G_v(s)$ given in Equation (5)

$$G_{iBatt} = C_i(s) \cdot FM \cdot ZOH \cdot G_{id}(s) \cdot H_i(s) \tag{3}$$

$$G_{vBatt} = C_v(s) \cdot G_v(s) \cdot H_v(s) \tag{4}$$

$$G_v = \frac{C_i(s) \cdot FM \cdot ZOH \cdot G_{vd}(s)}{1 + C_i(s) \cdot FM \cdot ZOH \cdot G_{id}(s) \cdot H_i(s)} \tag{5}$$

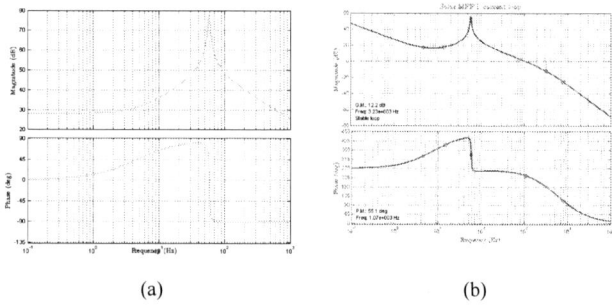

(a)   (b)

Figure 6. Solar current loop design in load sharing mode: (a) uncompensated current loop, (b) compensated current loop

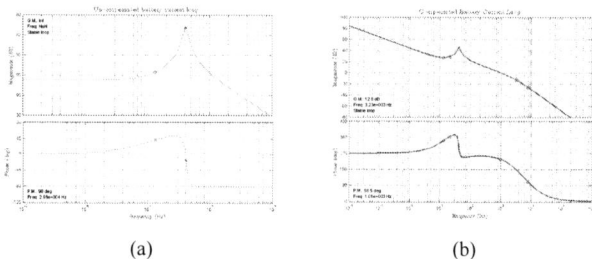

(a)   (b)

Figure 7. Battery current loop design in load sharing mode: (a) uncompensated current loop, (b) compensated current loop

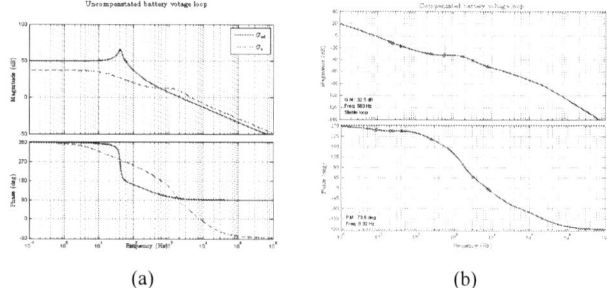

(a)   (b)

Figure 8. Battery voltage loop design in load sharing mode: (a) uncompensated voltage loop, (b) compensated voltage loop

Since in load sharing mode the single phase inverter is the load of dc-dc switching cells, the current loop and voltage loop cross-over frequency for battery switching cell has to be located on each side of 120Hz frequency with at least half a decade separation. PI controllers are applied in both inner current loop and outside voltage loop. By choosing proper zeros and gain, the cross-over frequency of 1.01kHz, 12.8dB of gain margin and 56.5° of phase margin can be achieved for battery current loop, as indicated in Figure (7. The voltage loop cross-over frequency is chosen to be 8.32Hz, with

32.5dB of gain margin and 73.5° of phase margin, as indicated in Figure (8).

IV. MAXIMUM POWER TRANSFER IN BATTERY RECOVERY MODE

In battery recovery mode operation, there is no demand from ac side; the harvested maximum power from solar panel is directly transferred to battery storage. Since solar panel voltage varies with sun irradiation intensity, buck or boost operation will needed at different irradiation levels. Independent control of the solar power and battery charging current might cause power imbalance and lead to dc-link voltage fluctuation. A Current-Mode Maximum Power Transfer (CMMPT) control structure is proposed for this particular operation mode as in Figure (9), charging current reference is generated by MPPT routine; the charging current controller generates the inner current reference for solar output current; the resulted PWM signal will gate two active switches synchronously. The proposed controller has follow advantages:

- Flexible input/output voltage level, the steady-state voltage gain is given in Equation (6),

- Dynamic link between input current and charging current hence the input power and output power,

- Solve instability problem of boost switching cell and reduce the number of control loops,

- Indirectly stabilize dc-link capacitor voltage for over-voltage hazard.

$$\frac{V_O}{V_I} = \frac{I_{IN}}{I_O} = \frac{D}{1-D} \tag{6}$$

In order to find the transfer function of power stage, small signal model for the double-leg structure is derived in Figure (9).

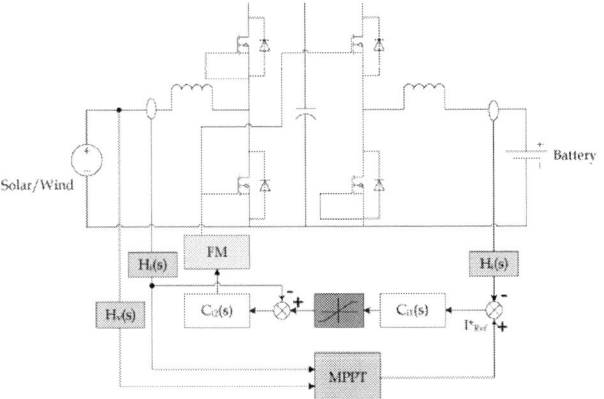

Figure 9. Control system for current-mode maximum power transfer (CMMPT)

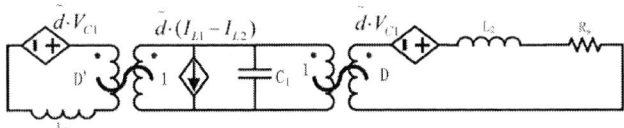

Figure 10. Small signal mode for circuit configuration in MPT

Three equations based on KCL and KVL can be formed to solve for transfer functions,

$$\begin{cases} \tilde{i}_{L1} L_1 s + \tilde{v}_{C1} D' - \tilde{d} V_{C1} = 0 \\ \tilde{i}_{L1} D' - \tilde{i}_{L2} D - \tilde{v}_{C1} C_1 s - \tilde{d}(I_{L1} - I_{L2}) = 0 \\ \tilde{i}_{L2}(L_2 s + R_s) - \tilde{v}_{C1} D - \tilde{d} V_{C1} = 0 \end{cases} \tag{7}$$

The control-to-solar current and control-to-charging current can be derived as,

$$\frac{\tilde{i}_{L1}(s)}{\tilde{d}} = \frac{kV_{C1}C_1 s + kD'(I_{L1} - I_{L2})}{kL_1 C_1 s^2 + L_1 D^2 s + kD'^2} \tag{8}$$

$$\frac{\tilde{i}_{L2}(s)}{\tilde{d}} = \frac{V_{C1} L_1 C_1 s^2 - DL_1(I_{L1} - I_{L2})s - D'V_{C1}}{kL_1 C_1 s^2 + L_1 D^2 s + kD'^2} \tag{9}$$

Where $k = L_2 s + R_s$

The control system structure for CMMPT is presented in form of transfer function blocks, which is described in Figure (11). G_{iL1d} is the control-to-solar current transfer function in Equation (8), and G_{iL2d} is the control-to-charging current transfer function in Equation (9).

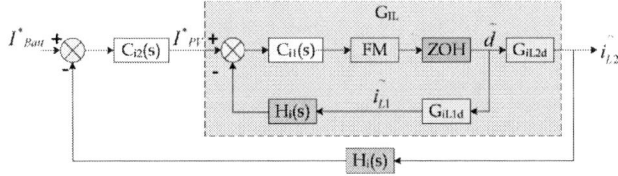

Figure 11. Control system for current-mode maximum power transfer (CMMPT)

The open loop transfer function G_{iL1} for inner solar current loop under PI compensation can be expressed by,

$$G_{iL1} = C_{i1}(s) \cdot FM \cdot ZOH \cdot G_{iL1d}(s) \cdot H_i(s) \tag{10}$$

under PI controller compensation with the zero placed at resonant frequency of $G_{iL1d}(s)$, a stable inner current loop can be achieved. The cross-over frequency is set at 40Hz, 14dB of gain margin and 24° of phase margin can be achieved. As can be observed, the cross-over frequency of solar current loop in MPT mode is significantly lower than in load sharing mode; there are two reasons for this: first, in MPT mode, control-to-current transfer functions for the power stage is of higher order, the frequency domain response differs significantly from that in load sharing mode, conservative design are applied; second, in MPT mode the load is battery, which will contribute to negligible load dynamics; therefore, low bandwidth controller can well serve the purpose.

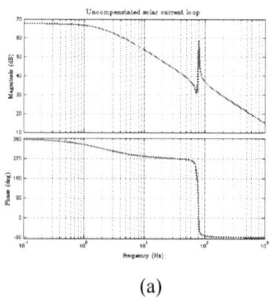 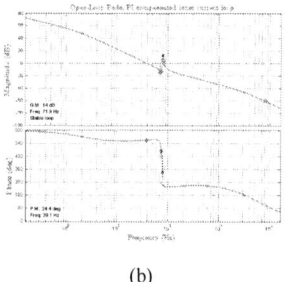

| (a) | (b) |

Figure 12. Solar current loop design in CMMPT: (a) uncompensated inner current loop, (b) compensated inner current loop

With a compensated inner solar current loop, the open-loop transfer function for battery charging current loop can be formulated as in Equation (11).

$$G_{iL2} = C_{i2}(s) \cdot G_{IL}(s) \cdot H_i(s) \tag{11}$$

where G_{IL} is the equivalent control-to-charging current plant with inner current loop under PI controller compensation. The detailed expression for G_{IL} is given as,

$$G_{IL}(s) = \frac{C_{i1}(s) \cdot FM \cdot ZOH}{1 + C_{i1}(s) \cdot FM \cdot ZOH \cdot G_{iL1d}(s) \cdot H_i(s)} \tag{12}$$

A classical PI controller is used to shape the loop response of the plant G_{IL}. With integral gain $k_i=5$ and zero at $w_z=600$rad/s, the cross-over frequency of compensated outside current loop is set at 1.5Hz, with 30dB gain margin and 90° of phase margin.

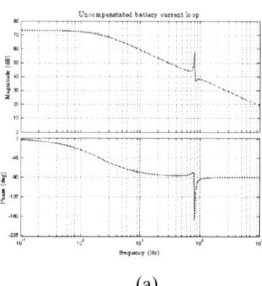 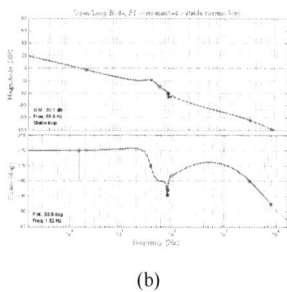

| (a) | (b) |

Figure 13. Battery charging current loop design in CMMPT: (a) uncompensated inner current loop, (b) compensated inner current loop

V. EXPERIMENTAL RESULTS

The proposed control systems are implemented in TMS320F2812 DSP and tested on the system as in Figure (14). Figure (15) shows the start-up phase of MPPT in load sharing mode, which takes approximately 10sec to reach maximum power point. Figure 15(b) demonstrates the effort of voltage control in load sharing operation: solar switching cell is under direct current control for maximum power tracking; at the mean time, battery switching cell is under ACM control to stabilize the dc-link voltage. During a pulse load of 300W, the tracking of solar power is barely affected by the load dynamics; battery switching cell handles the load dynamics by providing a pulse current.

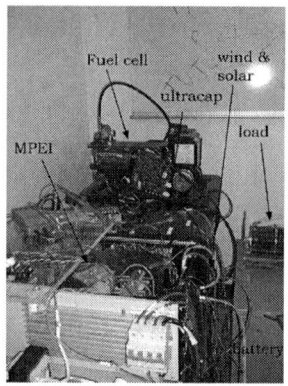

Figure 14. MPEI testbed

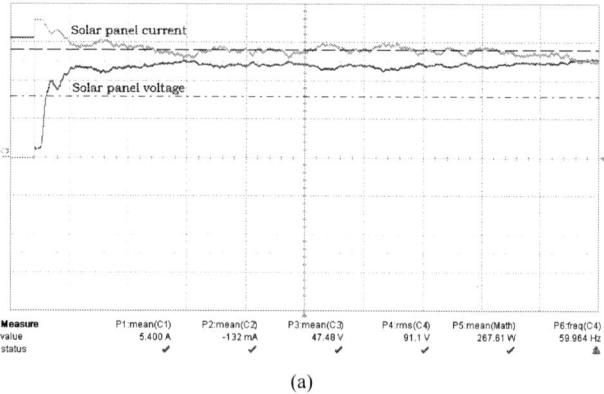

(a)

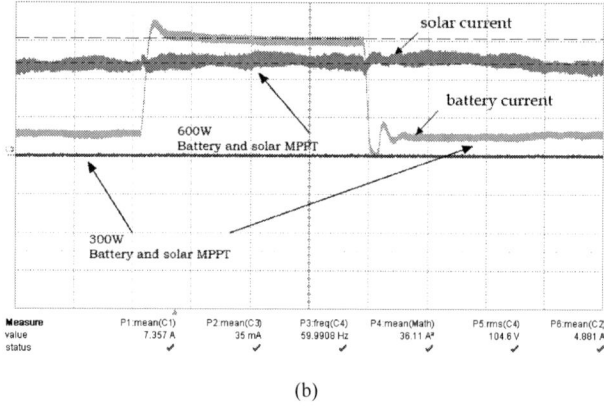

(b)

Figure 15. Solar power processing in transient: (a) Solar MPPT startup phase (t=5sec/div, 20V/div, 2A/div), (b) Pulse load tests during load sharing with solar MPPT

Figure (16) presents the MPPT start-up phase in CMMPT mode. As can be observed, with a dynamic link between charging current and solar current, the start-up process is faster than in the load sharing mode.

Figure (17) shows the steady state test for load sharing and maximum power transfer operation. Both operations are tested on the same time of the day and gives the very close power output from the solar panel (<1%).

978-1-4244-4782-4/10 $26.00 © 2010 IEEE

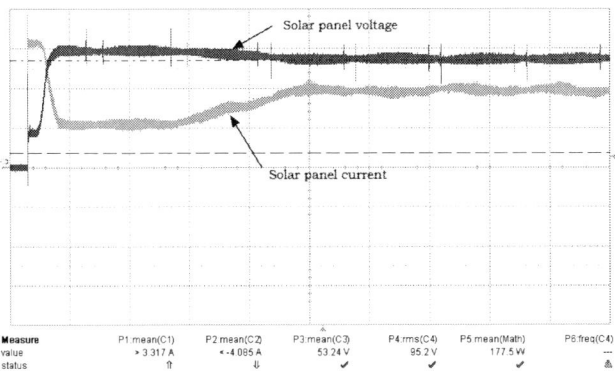

Measure	P1:mean(C1)	P2:mean(C2)	P3:mean(C3)	P4:rms(C4)	P5:mean(Math)	P6:freq(C4)
value	> 3.317 A	< -4.085 A	53.24 V	95.2 V	177.5 W	—
status	⇑	⇓	✓			△

Figure 16. Solar power tracking start-up in CMMPT mode (t=2sec/div, 20V/div, 2A/div)

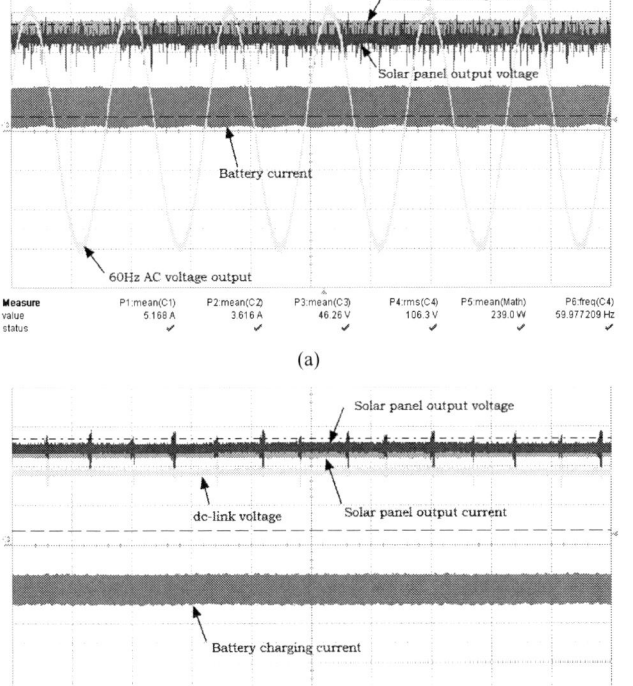

Measure	P1:mean(C1)	P2:mean(C2)	P3:mean(C3)	P4:rms(C4)	P5:mean(Math)	P6:freq(C4)
value	5.168 A	3.616 A	46.26 V	106.3 V	239.0 W	59.977209 Hz
status	✓	✓	✓	✓	✓	✓

(a)

Measure	P1:mean(C1)	P2:mean(C2)	P3:mean(C3)	P4:rms(C4)	P5:mean(Math)	P6:freq(C4)
value	4.791 A	-5.354 A	49.18 V	91.3 V	235.4 W	6.854309 Hz
status	✓	✓	✓	✓	✓	⚡

(b)

Figure 17. Solar power processing in steady state: (a) MPT by load share, (b) MPT by CMMPT control

VI. CONCLUSION

This paper proposes the system design for solar power processing in Multi-port Power Electronic Interface. System behavior is identified under both load sharing mode and battery recovery modes of operation. The control structure is proposed for load sharing mode, providing dynamic power sharing during both steady state and load dynamics. Current-Mode Maximum Power Transfer method is proposed for battery recovery mode to ensure efficient power storage. The control loop designs are presented in details for each control structure. The control systems are implemented in a DSP and tested on laboratory prototype; experimental results indicate the effectiveness of system design.

REFERENCES

[1] W. Jiang and B. Fahimi, "Multi-port power electronic interface for renewable energy sources," *24th Annual IEEE Applied Power Electronics Conference and Exposition*, pp. 347--352, Feb. 2009.)

[2] B. K. Bose, P. M. Szczesny, and R. L. Steigerwald, "Microcomputer control of a residential photovoltaic power conditioning system," *IEEE Trans. Ind. Appl.*, vol. 21, no. 5, pp. 1182--1191, Sep. 1985.

[3] C. Hua, J. Lin, and C. Shen, "Implementation of a dsp-controlled photovoltaic system with peak power tracking," *IEEE Trans. Ind. Electron.*, vol. 45, no. 1, pp. 99--107, Feb. 1998.

[4] K. H. Hussein, I. Muta, T. Hoshino, and M. Osakada, "Maximum photovoltaic power tracking: an algorithm for rapidly changing atmospheric conditions," *IEE Proc. Generation, Transmission and Distribution*, vol. 142, no. 1, pp. 1350--2360, Jan. 1995.

[5] E. I. Ortiz-Rivera and F. Z. Peng, "Analytical model for a photovoltaic module using the electrical characteristics provided by the manufacturer data sheet," *36th IEEE Power Electronics Specialists Conference*,pp.2087--2091,Jun.2005

[6] N. Femia, G. Petrone, G. Spagnuolo, and M. Vitelli, "Optimization of perturb and observe maximum power point tracking method," *IEEE Trans. Power Electron.*, vol. 20, no. 4, pp. 963 -- 973, Jul. 2005.

[7] T. Esram and P. L. Chapman, "Comparison of photovoltaic array maximum power point tracking techniques," *IEEE Trans. Energy Convers.*, vol. 22, no. 2, pp. 439--449, Jun. 2007.

[8] M. A. S. Masoum, H. Dehbonei, and E. F. Fuchs, "Theoretical and experimental analyses of photovoltaic systems with voltage and current-based maximum power-point tracking," *IEEE Trans. Energy Convers.*, vol. 17, no. 4, pp. 514--522, Dec. 2002.

[9] N. Femia, G. Petrone, G. Spagnuolo, and M. Vitelli, "A pure realization of loss-free resistor," *IEEE Trans. Circuits Syst.*, vol 51, no. 8, pp 1639--1647, Aug. 2004.

[10] W. Jiang, "Multi-port power electronic interface for renewable energy Sources," Ph.D. dissertation, University of Texas at Arlington, Arlington, TX, Sep. 2009.

SMD Inductors Based on Soft-Magnetic Powder Compacts

Etsuo Otsuki
otsuki-etsuo@toho-zinc.co.jp

Kenichiro Ishii
ishii-kenichiro@toho-zinc.co.jp

Shinya Nakano
nakano-shinya@toho-zinc.co.jp

TOHO ZINC CO., LTD. TOHO ZINC TECHNICAL CENTER
Naka 387, Fujioka, Gunma, JAPAN
Phone +81-274-22-1416 Fax +81-274-22-7852

Abstract— **This paper presents SMD inductors composed of soft-magnetic metal powder compacts based on newly developed materials and processes. High saturation magnetization of the core materials results in high DC-Bias characteristics and down-sizing inductors, which can never be achieved by conventional ferrite core inductors. These SMD inductors can realize high efficiency in a circuit due to low loss characteristics in comparison with metal composite type inductors as well as ferrite inductors.**

I. INTRODUCTION

Multi-functioning and down-sizing with keeping high efficiency in the electronic devices have forced unsolvable problems of increase in electric current, high efficiency etc. to power supplies. The devices in the power supplies have been demanded same task as well. Especially, recent trend of enhancing current in power supplies is big headache for the inductors. SMD inductors with ferrite cores have been faced the limitations to respond to customer demands, increase in electrical current capacity etc., due to their low saturation magnetization.

On the other hand, a metal powder compacted core (dust core) is well known to have high saturation magnetization (as shown in Fig. 1) and the Curie point. Thus, this material is attractive as SMD inductor cores.

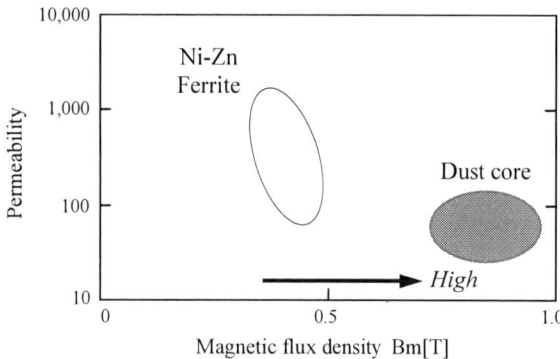

Fig. 1 Comparison of magnetic flux density

However several limitations had hindered them to be applied to a practical use.

Such as;

(1) Low electrical resistivity of dust cores necessitates the insulation material, which result in enlarging a device size.

(2) Low mechanical strength of the dust cores results in the fracture of products in the assembly lines as well as use in field, consequently the reasonable yield of products could not be obtained. (Fig. 2)

(3) It was impossible to obtain a reasonable yield of small size dust cores by conventional powder pressing process.

(4) Low productivity of pressing process due to high pressing pressure

(5) Low tolerance of shaping the dust cores by conventional process

II. MATERIAL AND PROCESS DEVELOPMENT

Recently, we have developed the new dust core material with solving all of the above-mentioned problems. That is, the dust cores with the high resistivity and high fracture strength can be produced by new production processes with a reasonable yield. As shown in Table 1, the new dust core has a very high resistivity of Mega-ohm meters which can eliminate the use of insulating plastics. And Fig. 2 shows that the new dust core can keep its shape. By applying this new material to SMD inductors, we have developed new type of SMD inductor, which can achieve down-sizing as well as several high performances of DC-Bias capacity, temperature characteristics etc. In this paper, the performance of the new SMD inductors and their effect in the power supply circuits will be presented.

Core	Resistivity[$\Omega \cdot$m]
Conventional material	4.0
Developed material	5.9×10^6

Table 1 Resistivity of core

Conventional material Developed material

Fig. 2 Comparison of strength of the dust core

III. FEATURE OF PRODUCTS

There are many varieties in electrical specifications of the inductor (switching power choke coil) depending on the design of DC-DC converters as power supplies, such as home electronics, vehicular equipment, and portable devices. We have developed two series of SMD power inductors (Dust Drum® Coil) by classifying the electric current capacity of DC-DC conveters, that is, TCM series is middle current (10A or less) in high inductance, and TCH series is large current (10A or more) in low inductance, as followings for details.

(1) TCM series

< Feature> L : 1.8~45µH, I : 1.6~6.8A Size : 8×8×4mm

<Process> The pressed rod shape dust core with Sendust powders is machined to make a trench on this peripheral surface for coil, and after heat-treated, Cu wire is wound on it. Outer sleeve core is prepared separately, of which material is ferrites or Sendust depending on the usage. After assembling the wire wound inner core and outer sleeve core, process is finished with terminal forming. Their outer features are shown in Fig. 3.

(2) TCH series

<Feature> L : 0.2~2.2µH, I : 8~26A Size : 8×9×4mm

<Process>Rectangular shape pressed core is usually prepared. The trenches for coil window and wire to terminal are machined on the bottom face of the pressed core, and then residual stress caused by mechanical process is relieved with heat-treatment. The edge-wise wound coil is prepared separately. The inductor is assembled with coil and 2 pieces of cores, and terminal forming. Their outer features are shown in Fig. 3.

TCM series   TCH series

Fig. 3 Dust Drum® Coil

Outer Shape of TCM series and TCH series can be change on the demand of a circuit. Furthermore, a low profile type is also under development newly.

IV. PERFORMANCES OF PRODUCTS

A. DC-Bias Characteristics

Fig. 4 shows the DC-Bias characteristics of Dust Drum® Coil TCM series (TCM-0840-1R8) in comparison with similar size ferrite inductor (8mm square and 4mm height) at the 20°C and 100°C. At 20°C, Dust Drum® Coil can sustain a high level of the inductance at even high DC current range, while ferrite inductor shows the sharp drop of the inductance due to magnetic saturation. At 100°C, the difference of their characteristics becomes remarkable. This means that Dust Drum® Coil shows high stability to over current and has fewer risks of losing control by temperature increase as compared with ferrite inductors. That is, Dust Drum® Coil can offer high reliability and minimize the margin of performances to the design of a circuit.

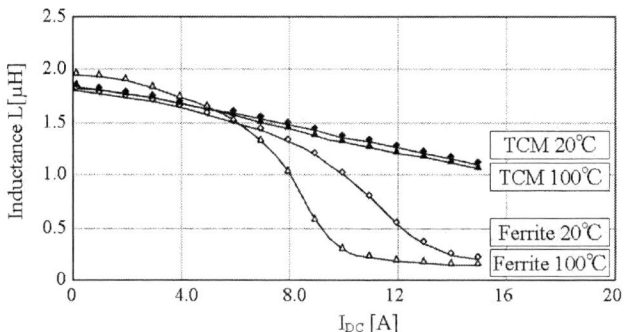

Fig. 4 DC-Bias characteristics

Sample	Dimensions[mm]	L[µH]	DCR[mΩ]
TCM -0840-1R8	7.9×8.4×4.0	1.8	9.3
Ferrite inductor	8.3×8.3×4.0	1.8	13

Table 2 Specification of inductor samples

B. Waveforms in buck-chopper circuit

The conventional buck-chopper circuit is shown in Fig. 5. This circuit is well known as low price as well as high efficiency due to simple structure, it is vastly used as a Step-Down type DC-DC converter. The Dust Drum® Coil TCM series (TCM-0840-4R2) and ferrite inductor with almost same size and rated current are mounted in this buck-chopper circuit and evaluated by waveform and efficiency.

978-1-4244-4782-4/10 $26.00 © 2010 IEEE

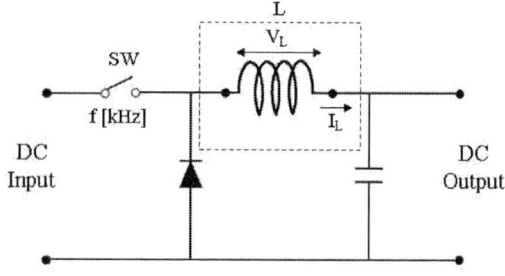

[Operating conditions]
f = 300kHz   Duty 50%
Input voltage 9.0[V]
Output voltage 4.5[V]
Output current 4.3[A]

Fig. 5 Buck-chopper circuit

Fig. 6 shows the voltage and the current waveforms of a ferrite inductor in buck-chopper circuit by driven by rectangular voltage waveform V_L. The current waveform is seen to be distorted. Fig. 7 shows the magnetization curve of the ferrite inductor computed from inductor voltage V_L and current I_L, which has non-linear loop with a big hysteresis. These non-linear characteristics will bring about the decrease in efficiency and overheat in DC-DC converters.

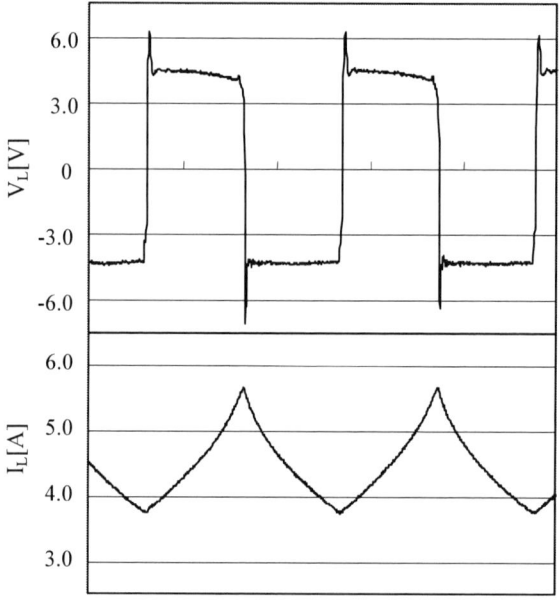

Fig. 6 Waveforms of Ferrite inductor

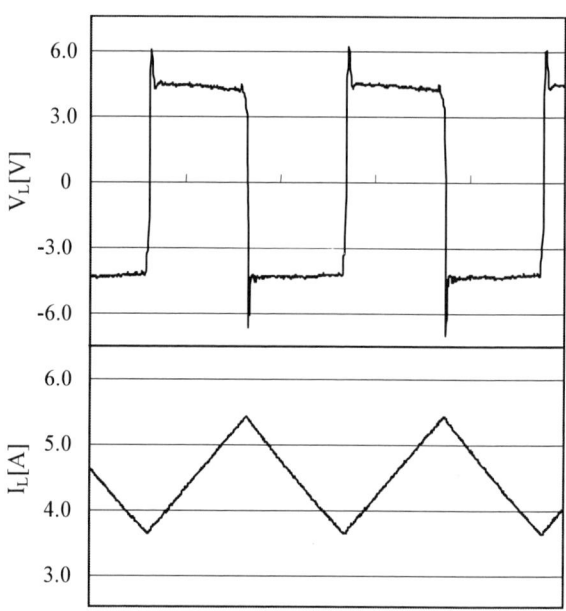

Fig. 8 Waveforms of Dust Drum® Coil

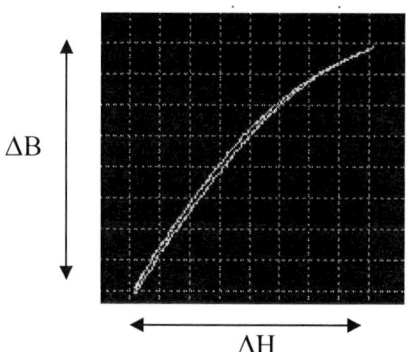

Fig. 7 Minor loop shape of Ferrite inductor

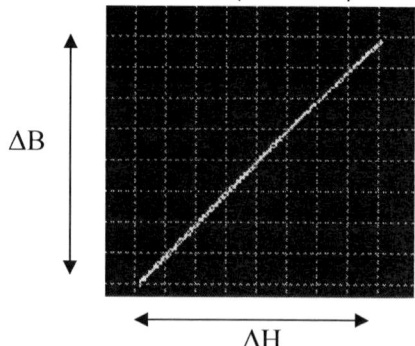

Fig. 9 Minor loop shape of Dust Drum® Coil

978-1-4244-4782-4/10 $26.00 © 2010 IEEE

Fig. 8 and 9 show the voltage and the current wave form of Dust Drum® Coil in buck-chopper circuit and magnetization curve, respectively. As seen in figures, there is no distortion in waveform and linear magnetization with very small hysteresis. Therefore Dust Drum® Coil can make good operation and high efficiency in the DC-DC converters.

Fig. 10 shows the DC-Bias characteristics for both inductor sample with similar size and rated current as shown in Table 3. As seen in this figure, less-distorted waveform as well as linear magnetization curve by Dust Drum® coil can be attributed the excellent characteristics.

Sample	Dimensions[mm]	$L[\mu H]$ at I_T	$I_T[A]$ $\Delta T=40°C$
TCM -0840-4R2	7.9×8.4×4.0	3.6	4.2
Ferrite inductor	8.3×8.3×4.0	5.0	4.0

Table 3 Specification of inductor samples

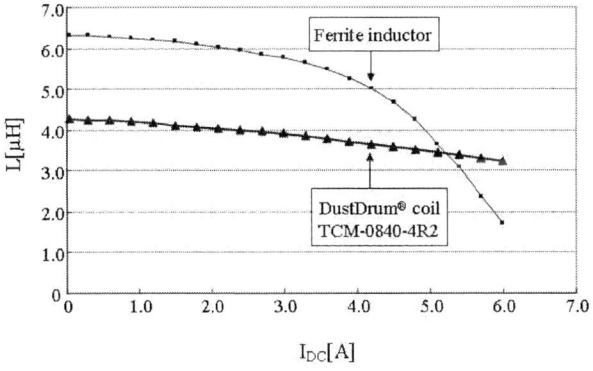

Fig. 10 DC-Bias characteristics

C. Efficiency in DC-DC Converter (TCM series)

Fig. 11 shows the efficiency in DC-DC converter assembled with the inductor samples same as Table 2. Both samples have the similar size, inductance and DC resistance. The DC-DC converter with Dust Drum® Coil TCM series (TCM-0840-1R8) is found to show higher efficiency than that of ferrite inductor for the full range of output current.

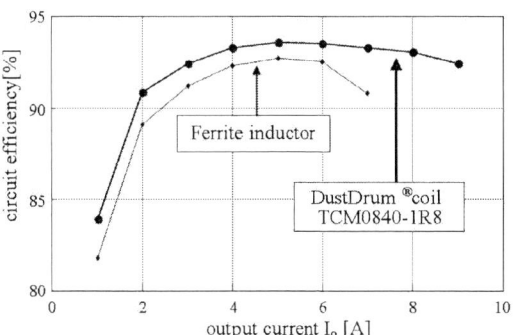

Fig. 11 Efficiency in DC-DC converter 1

D. Efficiency in DC-DC Converter (TCH series)

Fig. 12 shows the efficiency of DC-DC converter with Dust Drum® Coil TCH series (TCH-0840-R47) in comparison with ferrite inductor and metal composite type inductor. Difference in the efficiency in the circuit between Dust Drum® Coil and metal composite type inductor can be explained with the power loss, as shown in Table 4.

Although the power loss of ferrite material has lower power loss than Dust Drum® material, efficiency in DC-DC converter is reversed. This can be attributed to the non-gapped structure in Dust Drum® Coil, while the gap in the ferrite inductors causes an additional power loss.

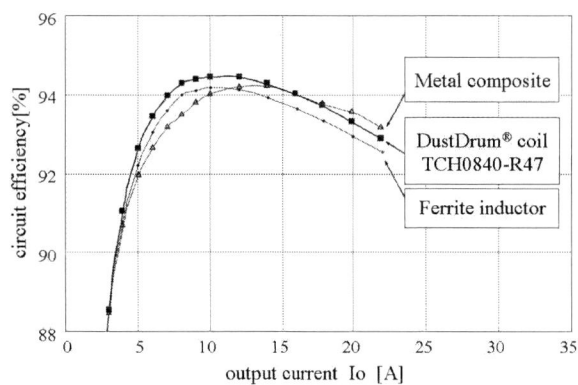

Fig. 12   Efficiency in DC-DC converter 2

Material	DustDrum®	Metal composite type	Ni-Zn ferrite
Power loss [kW/m³]	800	3,200	600

at 100kHz 100mT   R.T.

Table 4 Power loss in inductor materials

Dust Drum® Coil can reduce the volume of 40% with keeping similar efficiency to others, as shown in Table 5.

Sample	Dimensions[mm]	L[μH]	DCR[mΩ]
TCH -0840-R47	8.0×9.0×4.0	0.47	1.4
Ferrite inductor	10.6×10.5×4.4	0.45	2.2
Metal composite inductor	8.3×8.3×4.0	0.45	1.1

Table 5 Specification of inductor samples

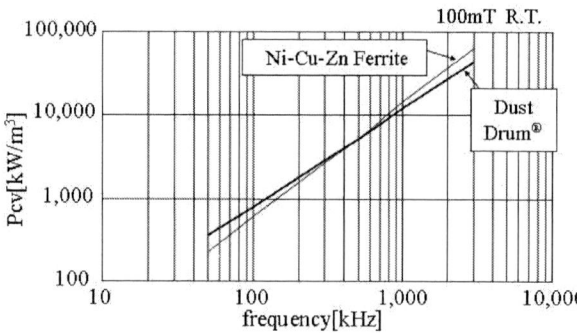

Fig. 13 Characteristics of power loss

V. CONCLUSIONS

The development of new dust core material based on magnetic metal powder can afford to achieve the new type of SMD inductor, Dust Drum® Coil.

These results, such as good temperature characteristics, large current correspondence, high efficiency, etc. in Dust Drum® Coil, are sure of leading to a big advantage in the design of DC-DC converters.

Moreover, the new material used in Dust Drum® Coil exhibits better power loss characteristics than Ni-Cu-Zn ferrites at over 500kHz. Thus, Dust Drum® Coil is thought suitable to high frequency-driven DC-DC converters rather than the ferrite inductors.

References

[1]   S. Nakano, "The technical trend of power inductor," *Dempa shinbun High Technology,* DEMPA PABLICATIONS, INC., vol. 1191, 4 June 2009 (in japanese)

High Density Low Profile Coupled Inductor Design for Integrated Point-of-Load Converter

Qiang Li, Yan Dong, Fred C. Lee
Center for Power Electronics Systems
Virginia Polytechnic Institute and State University
Blacksburg, VA 24061 USA

Abstract—Low profile integrated Point-of-Load (POL) converter is today's industry trend for portable electronic applications. Magnetics is the major challenge and bottleneck for achieving low profile high power density integrated POL. So, how to design a low profile magnetic becomes one of the key technologies for integrated POL. Inverse coupling is one of the possible methods to reduce inductor size due to the dc flux cancelling effect. Several integrated low profile coupled inductor structures with different flux patterns (vertical flux and lateral flux) are proposed in this paper based on low temperature co-fired ceramics (LTCC) technology. This paper also reveals that the lateral flux coupled inductor structure can have higher inductance density than vertical flux structure. Two LTCC coupled inductor prototypes are designed and fabricated to verify the theoretical analysis. A 1.5MHz, 5V to 1.2V, 3D integrated buck converter with LTCC coupled inductor substrate is also fabricated. The power density of this integrated converter is as high as 700W/in^3.

I. INTRODUCTION

With the increased popularity of portable electronics, improved integrated solutions are desired to improve the power density of point-of-load (POL) converter. Therefore, research in integrated POL has triggered widespread interest around the world both in industry and the academia. Fig.1 shows some low profile high power density POL with different current level. Most of these low profile high power density POLs are integrated. From Fig.1, we can see the power density of POL will decrease when current level increase. One major bottleneck for increasing the power density for high current application is the magnetic components. Another one is the thermal. This paper focuses on the first issue and tries to increase converter power density by reducing magnetic size. Due to the high aspect ratio, it is not easy to design a low profile integrated inductor that works as good as a discrete inductor. In order to improve the performance of low profile inductor, our previous work focuses on studying different low profile inductor structures. In terms of the flux path pattern, these different structures can be classified into two types. One type has vertical flux pattern; another one has lateral flux pattern [1]. The vertical flux pattern means the magnetic flux path plane is perpendicular with the substrate, like a structure with spiral or meander coil. The lateral flux pattern means the magnetic flux path plane is parallel with the substrate, like a structure with toroidal coil.

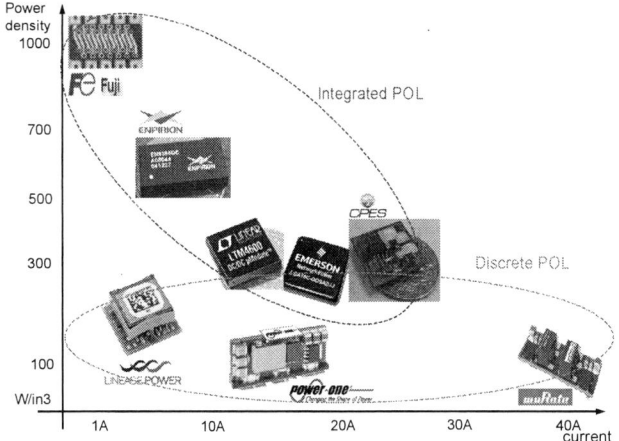

Figure 1. Low profile high power density POL.

Fig. 2 shows two low profile inductor examples with different flux patterns. Our previous study results reveal that the inductance density of vertical flux inductor will be impacted by core thickness; the maximal inductance density will decrease when core thickness decreases. But, the inductance density of lateral flux inductor will not be impacted by core thickness. So, it is better to use lateral flux structure when designing a low profile inductor because of its higher maximal inductance density. By designing and fabricating low profile LTCC inductor substrates with different flux patterns for a 1.5MHz, 5V to 1.2V, 3D integrated buck converter, our previous study shows that the lateral flux inductor prototype can help to save around 30% of the footprint to achieve around 300W/in^3 power density [1]. This POL prototype is also shown in Fig. 1 under the CPES logo. At that time, with 20A output current, CPES integrated POL prototype has almost the highest power density. But, it is still not as high as some low current POL, which can achieve more than 500W/in^3 power density.

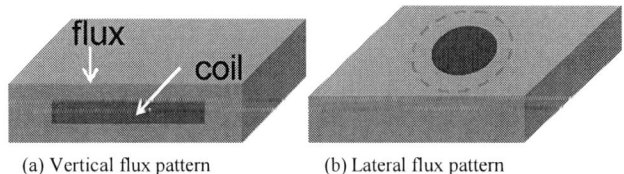

(a) Vertical flux pattern   (b) Lateral flux pattern

Figure 2. Different low profile inductors.

978-1-4244-4782-4/10 $26.00 © 2010 IEEE

As mentioned before, inductor size is the major bottleneck for further increasing converter power density for high current application. Because the magnetic core should store large dc energy caused by dc current, it is very difficult to design a very small inductor for high current application. Multi-phase converter with inverse coupled inductor is one possible method to solve this problem. Inverse coupled inductor has dc flux cancelling effect [2], so the magnetic core is no longer need to store large dc energy. Fig. 3 shows the core structure and equivalent magnetic circuit of a discrete inverse coupled inductor. It can be seen that the magnetic fluxes generated by two phase currents cancel each other in the outer legs of the core. This flux cancellation effect will help to reduce coupled inductor size. The coupling coefficient α and dc flux ϕ_{dc} are impacted by reluctance R_{side} and R_{mid}, which are controlled by air gap g_{side} and g_{mid} [2]. α and ϕ_{dc} equations are written in the following.

$$\alpha = -\frac{R_{mid}}{R_{side} + R_{mid}} \quad (1)$$

$$\phi_{dc} = \frac{N \cdot I_{dc}}{R_{side} + 2R_{mid}} \quad (2)$$

where, N is the turn number of coil, I_{dc} is the dc current.

This paper proposes some low profile integrated coupled inductor structures to help increasing power density of integrated POL. Both vertical flux coupled inductor and lateral flux coupled inductor are studied. By using vertical flux LTCC coupled inductor, the power density of a 1.5MHz, 5V to 1.2V integrated POL can be increased to 500W/in^3 even with 40A output current. Our study also shows that the lateral flux coupled inductor can reduce core thickness to the half value of vertical flux coupled inductor thickness. As a result, the integrated POL with LTCC lateral flux coupled inductor can achieve power density as high as 700W/in^3.

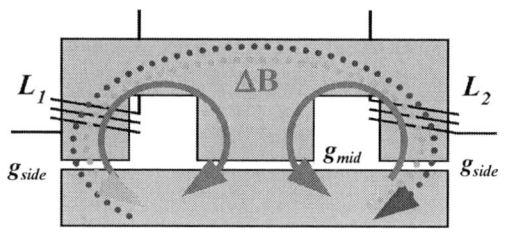

(a) Core structure example

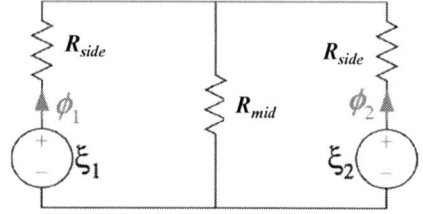

(b) Equivalent magnetic circuit

Figure 3. Discrete inverse coupled inductor.

II. LOW PROFILE COUPLED INDUCTOR STRUCTURES WITH VERTICAL FLUX PATTERN

Fig. 4 shows two possible LTCC vertical flux inverse coupled inductor structures: the top-and-bottom-winding coupled inductor structure (structure 1) and the side-by-side-winding coupled inductor structure (structure 2). From Fig. 4, it can be seen that these LTCC coupled inductor structures don't have any air gap to control coupling. The coupling coefficient can be controlled by controlling dimension d. It can be seen that, the winding of structure 1 has larger aspect ratio than that of structure 2 due to the core thickness limitation. As a result, the magnetic mean path length of structure 1 normally is longer than that of structure 2, which means structure 2 is much more possible to have higher inductance density than structure 1. So, for vertical flux coupled inductor structure, this paper focuses on structure 2.

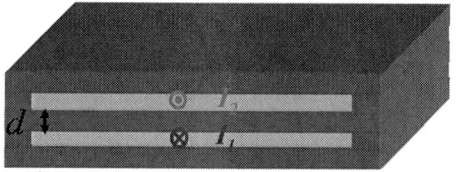

(a) Structure 1

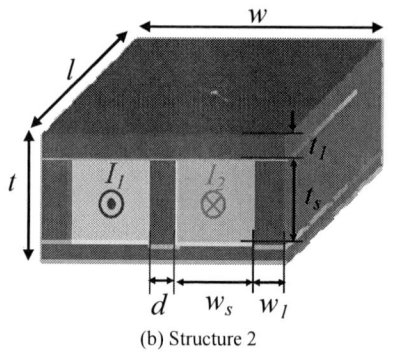

(b) Structure 2

Figure 4. Concept drawing of vertical flux inverse coupled inductor.

Fig. 5 shows the Maxwell 2D FEA simulation results of dc flux density for LTCC integrated coupled inductor structure 2 with different d value. From Fig. 5, it can be seen that coupling coefficient is increased when reduces d; and the dc flux density is also reduced when reduces d to increase inverse coupling coefficient. This is because the stronger inverse coupling has more dc flux cancellation.

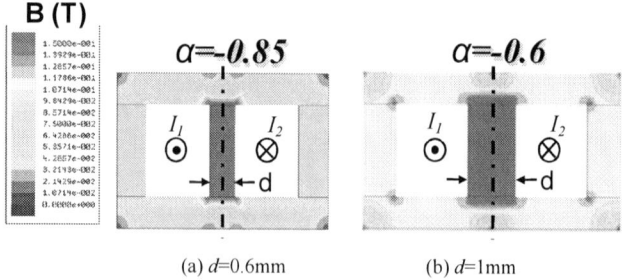

(a) d=0.6mm   (b) d=1mm

Figure 5. FEA simulation results of dc flux distribution with structure 2.

From Fig.5, it also can be seen that the dc flux distribution of LTCC integrated coupled inductor structure 2 is uniform in two areas: middle leg area and the other out part area. So, coupled inductor structure 2 can be roughly divided into these two parts. Each area roughly operates at one B-H curve operating point and has one permeability value. μ_{mid} is the permeability of middle leg area; μ_{out} is the permeability of other out part area. Fig.6 shows the B-H curve of the LTCC material that is used for coupled inductor fabrication and also an example with two permeability segments simplification. Assume the two permeability segments assumption is valid. The reluctance model of the LTCC coupled inductor structure 2 can be built as that of discrete coupled inductor shown in Fig.3 (b). The dimension definitions of the coupled inductor structure 2 are shown in Fig. 4 (b). Then, R_{mid} and R_{side} equations are written in the following.

$$R_{mid} = \frac{t_s}{\mu_0 \cdot \mu_{mid} \cdot d \cdot l} \tag{3}$$

$$R_{side} = \frac{2w_s}{\mu_0 \cdot \mu_{out} \cdot t_1 \cdot l} + \frac{t_s}{\mu_0 \cdot \mu_{out} \cdot w_1 \cdot l} \tag{4}$$

Based on the reluctance model, the self inductance L_{self}, Leakage inductance L_k and coupling coefficient α can be calculated as following.

$$L_{self} = \frac{R_{side} + R_{mid}}{R_{side}(R_{side} + 2R_{mid})} = K_1 \cdot l \tag{5}$$

where $K_1 = \dfrac{\mu_{out} \cdot \mu_0}{(\dfrac{2w_s}{t_1} + \dfrac{t_s}{w_1})[1 + \dfrac{1}{1 + (\dfrac{2w_s}{t_1} + \dfrac{t_s}{w_1})\dfrac{\mu_{mid}}{\mu_{out}}\dfrac{d}{t_s}}]}$

$$L_k = \frac{1}{R_{side} + 2R_{mid}} \tag{6}$$

$$\alpha = -\frac{R_{mid}}{R_{side} + R_{mid}} = \frac{1}{1 + K_2 \cdot d} \tag{7}$$

where $K_2 = (\dfrac{2w_s}{t_1} + \dfrac{t_s}{w_1}) \cdot \dfrac{1}{t_s} \cdot \dfrac{\mu_{mid}}{\mu_{out}}$

From (7), it also can be seen that d can be used to control the coupling between the phase currents. In order to verify this reluctance model, calculation results are compared with simulation results by using Maxwell FEA software. The dimension of structure 2 is: d=0.6mm, w_s=1.5mm, t_s=2mm, w_1=0.7mm, t_1=0.7mm and l=19.5mm. The current defined as I_1=I_2=20A. Fig. 7 shows both the calculation and simulation results. From Fig.7, it can be seen that the reluctance model's precision is acceptable. So, this model will be used to design the LTCC integrated coupled inductor structure 2.

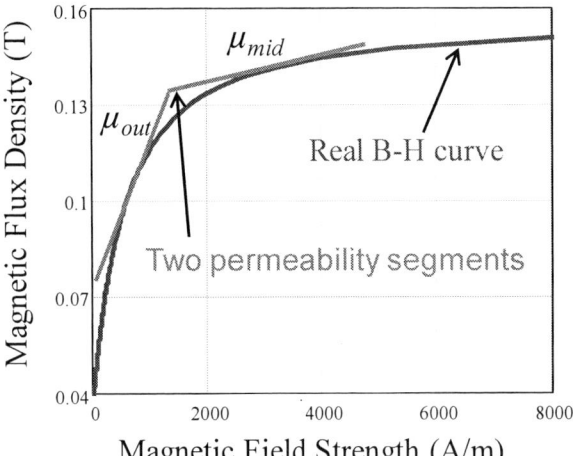

Figure 6. B-H curve of LTCC material.

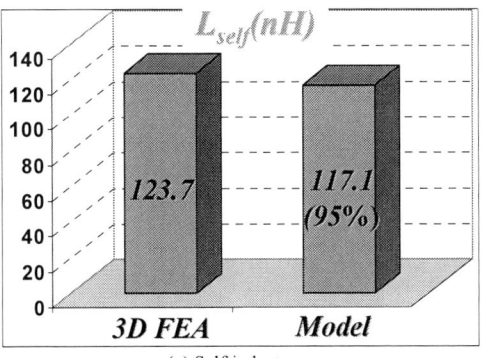

(a) Self inductance

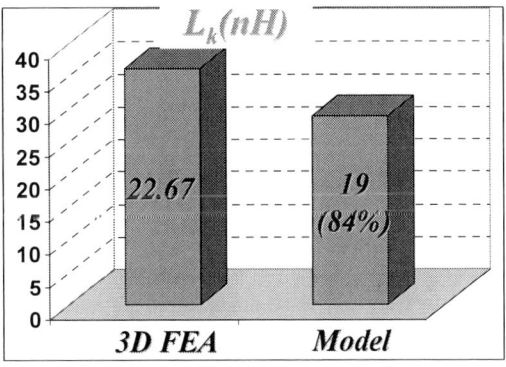

(b) Leakage inductance

Figure 7. Inductance calculation and simulation result for structure 2. (d=0.6mm, w_s=1.5mm, t_s=2mm, w_1=t_1=0.7mm, l=19.5mm, I_1=I_2=20A.)

III. LOW PROFILE COUPLED INDUCTOR STRUCTURES WITH LATERAL FLUX PATTERN

Fig. 8 shows top view of 4 lateral flux inverse coupled inductors. The blue dots represent the embedded coils; the solid lines represent the top surface windings; the dash lines represent the bottom surface windings. Structure 3 is the single turn coil lateral flux structure for each inductor. Structure 4, structure 5 and structure 6 all are 2-turn coil lateral flux structure for each inductor. Coupling coefficient of these lateral flux structures also all can be controlled by controlling dimension d. Structure 4 (2-turn coil) can be classified as the extension of structure 3 (1-turn coil). Actually, the 2-turn coil structure can be further extended to any number turn coil structures, such as 3-turn coil and 4-turn coil structures. With this structure, all the flux of one inductor has some coupling effect with the flux of another inductor. This is the key difference between structure 4 and structure 5, which only has coupled flux in the middle part. The flux around coil vias on the two sides of structure 5 is not coupled. So, there is no dc flux cancellation effect on the two sides of structure 5, which is not good for high inductance density design. For structure 6, because it has interleaved coils between two inductors, besides coupling effect between two inductors, the dimension d also will impact the coupling effect between the coils of each inductor. Reducing d will decrease coupling between two inductors, but it also will decrease coupling effect inside each inductor, which is also not good for high inductance density design. So, the structure 6 is not very suitable for coupled inductor with week coupling, but it is a good candidate for coupled inductor with strong coupling or transformer design. So, for lateral flux coupled inductor, this paper only focuses on structure 4.

Fig. 9 shows the Maxwell 2D FEA simulation results of dc flux density for LTCC integrated coupled inductor structure 4 with different d value. From Fig. 9, it also can be seen that the flux distribution of LTCC integrated coupled inductor structure 4 is not as uniform as structure 2. So, it is very difficult to build a reluctance model to help designing structure 4. As a result, Maxwell FEA 2D simulation is used to help designing this LTCC lateral flux coupled inductor (structure 4). Fig. 10 shows the simulation results of coupling coefficient with different dimension of structure 4. The dimension definition is in Fig.8 (b). From Fig. 10, it can be seen that, just like structure 2 (vertical flux), when d increase, coupling coefficient of structure 4 (lateral flux) decreases. The dimension g also will impact coupling coefficient. Increasing g will increase coupling coefficient. However, in real design, g normally is much larger than d, so dimension g can be used to control the footprint of structure 4 first, then, with given g value, dimension d can be used to control the coupling coefficient. The relationship between g and inductor footprint of structure 4 is shown in the following.

$$Footprint = (4 \cdot g + 4 \cdot r_v) \cdot (2 \cdot g + 8 \cdot r_v + d)$$
$$\approx (4 \cdot g + 4 \cdot r_v) \cdot (2 \cdot g + 8 \cdot r_v) \tag{8}$$

where, r_v is the radius of the coil via, which is determined by inductor current.

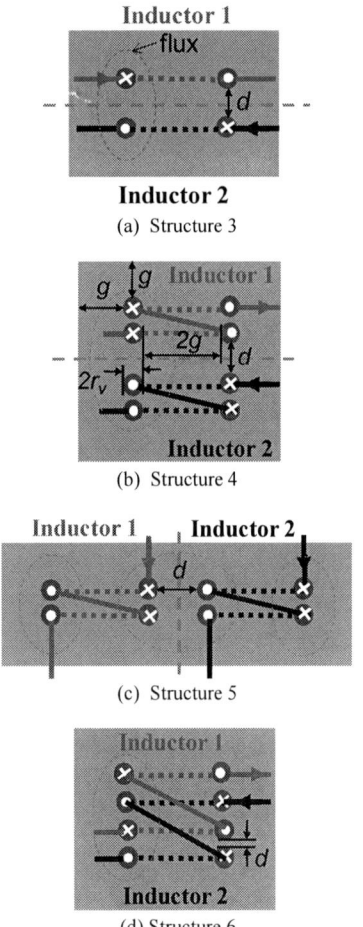

(a) Structure 3

(b) Structure 4

(c) Structure 5

(d) Structure 6

Figure 8. Concept drawing of lateral flux coupled inductor. (Top view)

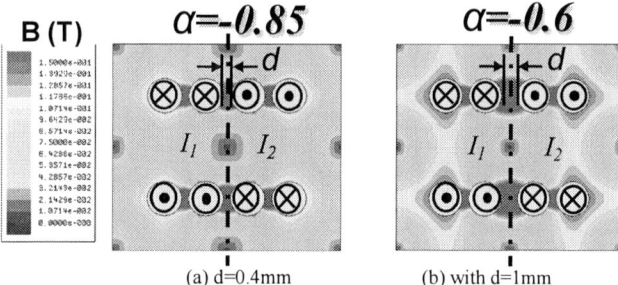

(a) d=0.4mm   (b) with d=1mm

Figure 9. FEA simulation results of dc flux distribution with structure 4.

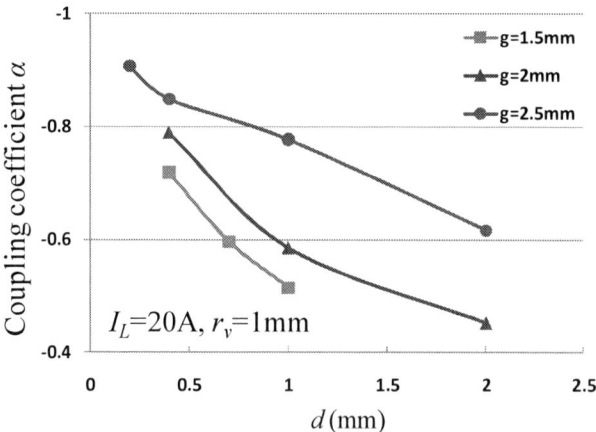

Figure 10. FEA simulation results of coupling coefficient for structure 4 with 2-turn coil.

Fig. 11 shows the FEA simulation results of steady state inductance Lss density for this lateral flux coupled inductor structure. With given g value, there is an optimal coupling coefficient α to give the maximal Lss density. The basic reason for these inverse-U shape curves is that increasing α can increase dc flux cancelling effect to help reducing inductor size, but if α is increased to be too large, the coupled inductor will work more like a transformer, which is not very efficient for storing energy. As a result, there is an optimal α value to give the maximal Lss inductance density.

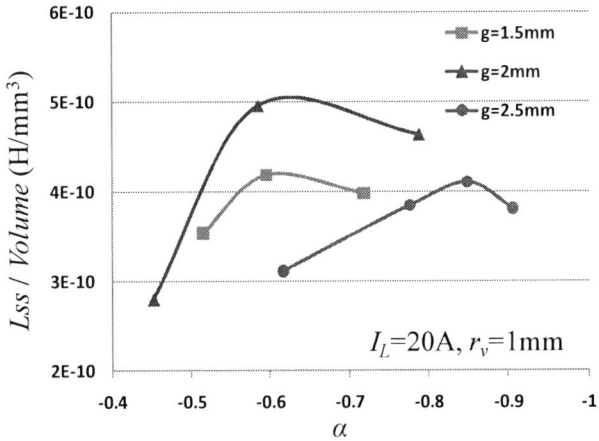

Figure 11. FEA simulation results of steady state inductance density for structure 4 with 2-turn coil.

IV. LTCC COUPLED INDUCTOR DESIGN PROCEDURE AND FABRICATION PROCESS

In order to verify previous analysis and simulation results, two low profile coupled inductors are designed and fabricated based on LTCC technology. These two low profile inductors work as the substrate for an integrated 1.5MHz, 5V to 1.2V, 40A two-phase buck converter. The converter integration concept and fabrication process are the same as [3]. One inductor substrate is using vertical flux structure 2; another one is using lateral flux structure 4 with 2-turn coil. Both of these two inductor substrates are designed based on given steady state inductance (Lss=68nH with I_L=20A) and footprint (FP=160mm^2), which are determined by converter active layer design.

There are 5 steps for design vertical flux structure 2:
1) Determine winding cross-section area according to inductor current;
2) Preselect w and l according to given footprint;
3) Preselect α, then determine w_s, t_s, d, w_l and t_l by using reluctance model. In order to avoid core cracking, w_l should be larger than 2mm.
4) Select different α, then repeat the step 3;
5) Select different w and l, then repeat the step 3 and 4.
After these 5 steps, several design results with different coupling coefficient and core thickness t can be compared to

make a final decision. Fig. 12 shows 10 design results for this structure. From Fig. 12, it can be seen that with given w value, there is an optimal coupling coefficient to give the thinnest core. Increase w from 6mm to 8mm the minimal core thickness will be reduced. However, the minimal core thickness is almost the same when further increase w from 8mm to 10mm, but the optimal coupling coefficient for achieving this core thickness will change from -0.6 to -0.4, which means the coupling effect become weaker. So, design case 5 is chosen for final fabrication.

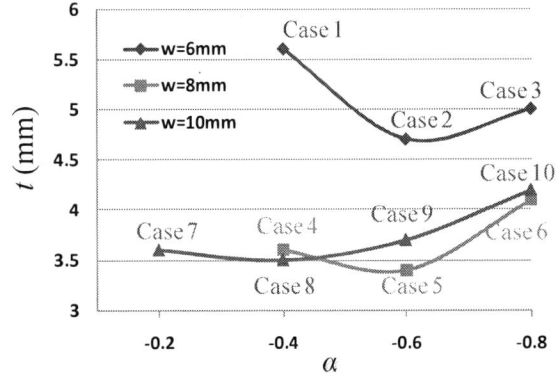

Figure 12. Design results of vertical flux structure 2 with different coupling coefficient and w.

For lateral flux structure 4, there are also 5 design steps:
1) Determine winding cross-section area according to inductor current;
2) Calculate g value according to given footprint;
3) Select α value according to the simulation results of relationship between steady state inductance density and α (Fig. 11 is an example);
4) Determine d value according to the simulation results of relationship between α and d (Fig. 10 is an example);
5) Calculate core thickness according to given steady state inductance.

For this particular example: Lss=68nH, inductor footprint FP=160mm2, inductor current =20A, the radius of coil via is chosen as r_v=1mm; then, according to (8), g can be calculated as g=2mm. According to the curve "g=2mm" in Fig. 11, if α=-0.6 is chosen, the maximal steady state inductance density can be achieved. However, α=-0.8 is also another good choice, because with this case the steady state inductance density is just a little bit lower, but the coupling coefficient is higher. If α=-0.8 case is chosen, according to the curve "g=2mm" in Fig.10 and Fig. 11, it can be known that d equals 0.4mm, and the steady state inductance density $Lss_{density}$ equals 0.47nH/mm^3. Because Lss=68nH, the core thickness t can be calculated as following (Because there are two inductors, the Lss should timed by factor 2).

$$t = \frac{2 \cdot Lss}{Lss_{density} \cdot FP} = \frac{2 \cdot 68\text{nH}}{0.47\text{nH/mm}^3 \cdot 160\text{mm}^2} = 1.8\text{mm}$$

978-1-4244-4782-4/10 $26.00 © 2010 IEEE

The detailed fabrication process for LTCC low profile inductor was introduced in [4]. Fig.13 shows some major fabrication steps for low profile LTCC coupled inductor. For vertical flux structure, the procedure is: 1) laminate LTCC bottom layer; 2) laminate LTCC middle layer with slots; 3) screen print silver paste inside slots to build coils; 4) laminate top LTCC layer; 5) sinter inductor around 900°C. The vertical flux coupled inductor prototype is shown in Fig. 14(a). For lateral flux structure, the procedure is: 1) laminate whole LTCC core; 2) build vias on LTCC core; 3) screen print silver paste inside vias to build coils; 4) sinter this structure around 900°C (The real prototype after this step is shown in Fig.14(b).); 5) connect via coils by using copper trace. Comparing with vertical flux structure, the lateral flux structure has easier fabrication process. First, the lamination process is reduced to only once; second, silver paste coils have more exposure area, which is good for air dissipating during sintering. For vertical flux structure, many small holes should be drilled on the surface to help dissipating air during sintering (see Fig. 14(a)); otherwise, vertical structure is very easy to be cracked during sintering.

Fig.15 shows the test results of inductor current and device drain to source voltage for a 2-phase buck converter with this lateral flux coupled inductor prototype. From Fig.15, it can be seen that the inductor current ripple is smaller at light load condition (I_L=0A) comparing with that at heavy load condition (I_L=20A), which means the inductance of this LTCC coupled inductor becomes larger when dc current becomes smaller. This is because the B-H curve of this LTCC material is very non-linear (shown in Fig.6). So, different dc current will cause different permeability due to different saturation condition. According to the inductor current waveform, the Lss and Ltr of this coupled inductor can be calculated [2]. Fig. 16 shows the steady state inductance with different phase current. From Fig. 16 it can be seen that the vertical flux and lateral flux coupled inductors have almost the same steady state inductance curve. Both of them have around 70nH-80nH steady state inductance with 20A phase current. And both of them have nonlinear inductance due to the nonlinear B-H curve. At light load condition, these LTCC coupled inductors can achieve steady state inductance as high as 400nH, which is very good for increasing light load efficiency.

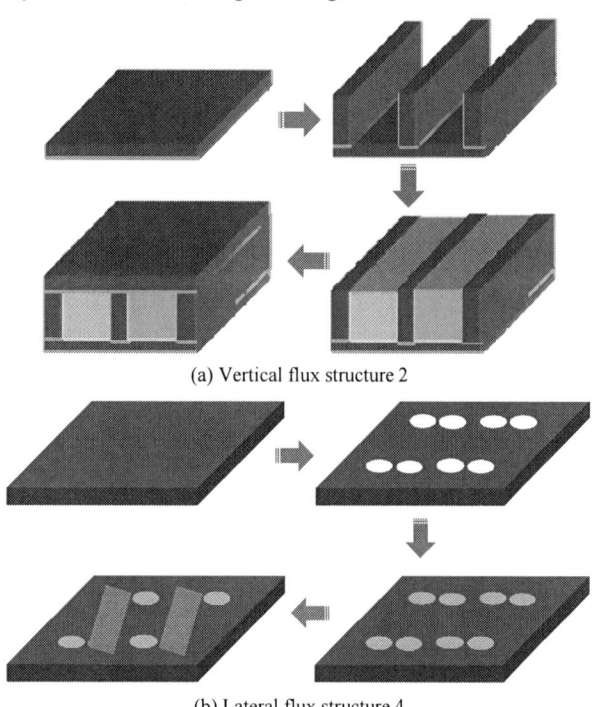

(a) Vertical flux structure 2

(b) Lateral flux structure 4

Figure 13. Fabrication steps for LTCC coupled inductor.

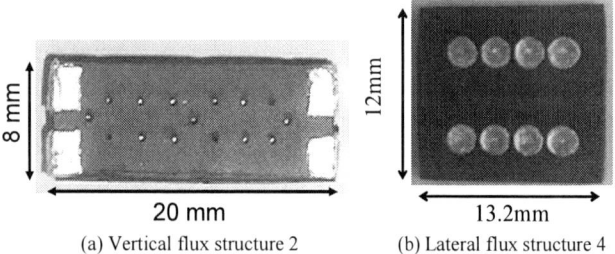

(a) Vertical flux structure 2   (b) Lateral flux structure 4

Figure 14. Prototypes of LTCC coupled inductor.

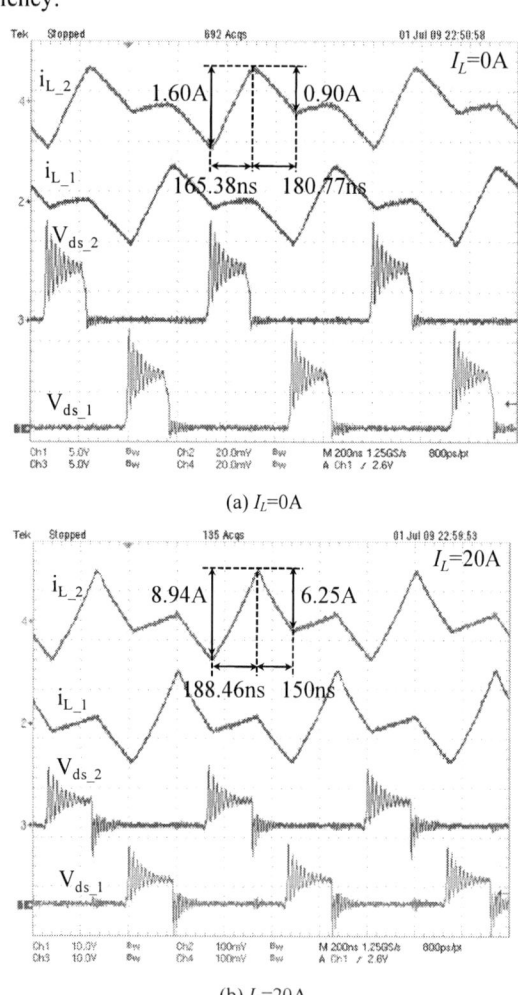

(a) I_L=0A

(b) I_L=20A

Figure 15. Test results of inductor current and device drain to source voltage.

978-1-4244-4782-4/10 $26.00 © 2010 IEEE

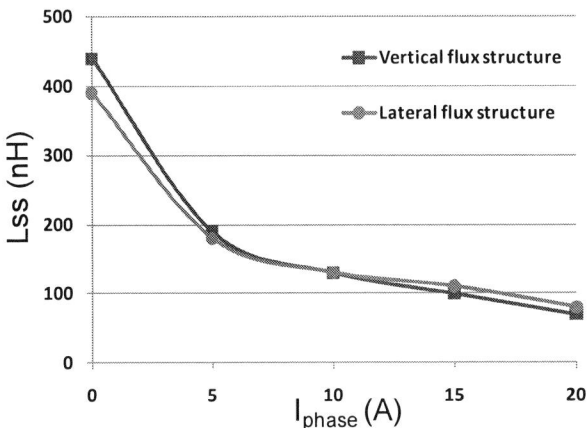

Figure 16. Steady state inductance of LTCC coupled inductor prototypes.

TABLE I. is the comparison results between lateral flux coupled inductor and vertical flux coupled inductor prototypes. It can be seen that the lateral flux and vertical flux coupled inductor has almost the same steady state inductance, transient inductance and footprint. The core thickness of lateral flux structure is only half of that of vertical flux structure. However, the coil DCR of lateral flux structure is larger than that of vertical flux structure due to the more complex coil structure. Fig. 17 shows the integrated buck converter with LTCC coupled inductor substrate. With vertical flux coupled inductor this converter can have power density as high as 500W/in^3; with lateral flux coupled inductor the power density of this converter can be further increased to 700W/in^3.

TABLE I. TEST RESULTS OF LOW PROFILE COUPLED INDUCTOR

Coupled inductor	Footprint	Core thickness	Lss (I_L=20A)	Ltr (I_L=20A)	Coil DCR
Lateral flux	158mm^2	1.8mm	80nH	29nH	0.6m
Vertical flux	160mm^2	3.4mm	70nH	28nH	0.4m

Figure 17. Integrated buck converter with LTCC coupled inductor substrate.

V.CONCLUSIONS

In order to improve the power density of integrated POL converter, this paper proposes some low profile LTCC coupled inductor structures with different flux patterns (vertical flux and lateral flux) to help further reduce inductor size. Our study reveals that the lateral flux coupled inductor structure can have higher inductance density than vertical flux structure. Two LTCC coupled inductor prototypes are designed and fabricated to verify the theoretical analysis. The lateral flux coupled inductor prototype has almost the same inductance and footprint as vertical flux coupled inductor prototype, but only half the core thickness. A 1.5MHz, 5V to 1.2V, 3D integrated buck converter with LTCC coupled inductor substrate is also fabricated. By using lateral flux coupled inductor, the power density of this integrated converter is as high as 700W/in^3.

ACKNOWLEDGMENT

This work was conducted with use of Maxwell 2D software, donated in kind by Ansoft Corporation of the CPES Industry Partnership Program.

REFERENCES

[1] Qiang Li, Fred C. Lee, "High Inductance Density Low-Profile Inductor Structure for Integrated Point-of-Load Converter" APEC 2009, 15-19 Feb. 2009, pp. 1011 – 1017.

[2] P. Wong, "Performance improvement of multi-channel interleaving voltage regulator modules with integrated coupling inductors", VPI&SU dissertation, Mar., 2001.

[3] Arthur Ball, Michele Lim, David Gilham, Fred C. Lee, "System design of a 3D integrated non-isolated Point Of Load converter", APEC 2008, pp. 181 - 186 .

[4] Michele H. Lim, Jacobus, D. van Wyk, F. C. Lee, and Khai D. T. Ngo, "A Class of Ceramic-Based Chip Inductors for Hybrid Integration in Power Supplies", IEEE Transactions on Power Electronics, vol. 23, no. 3, pp.1556-1564, may 2008.

[5] P. Wong, Q. Wu, P. Xu, B. Yang, and F.C. Lee, "Investigating Coupling inductor in interleaving QSW VRM," Proc. IEEE APEC conf., 2000, pp.973-978.

[6] Jieli Li, Charles R. Sullivan, Aaron Schultz, "Coupled inductor design optimization for fast-response low-voltage DC-DC converters", in Proceedings of APEC 2002 - Applied Power Electronics Conf., pp. 817– 823 vol.2.

[7] Ming Xu, Yucheng Ying, Qiang Li, "Novel coupled-inductor multiphase VRs," in Proc. IEEE APEC, 2007.

[8] Eberhard Waffenschmidt, Bernd Ackermann, J. A. Ferreira, "Design Method and Material Technologies for Passives in Printed Circuit Board Embedded Circuits", IEEE TRANSACTIONS ON POWER ELECTRONICS, VOL. 20, NO. 3, MAY 2005, pp.576-584.

[9] Robert Hahn, Steffen Krumbholz, Herbert Reichl, "Low Profile Power Inductors Based on Ferromagnetic LTCC Technology", Electronic Components and Technology Conference 2006, pp.528-533.

[10] Tsutomu Mikura, Koichi Nakahara, Kota Ikeda, Ken Furukuwa, K. Onitsuka, "New Substrate for Micro DC-DC Converter", Electronic Components and Technology Conference 2006, pp.1326-1330.

[11] M. J. Prieto, A. M. Pernia, J. M. Lopera, J. A. Martin, F. Nuno, "Design and analysis of thick-film integrated inductors for power converters", IEEE Trans. Industry Applications, Volume 38, Issue 2, March-April 2002, pp. 543 - 552.

978-1-4244-4782-4/10 $26.00 © 2010 IEEE

Relationship of Quality Factor and Hollow Winding Structure of Coreless Printed Spiral Winding (CPSW) Inductor

Y. P. Su[1], *Student Member, IEEE*, Xun Liu[2], *Member, IEEE*,
C. K. Lee[3] *Member, IEEE*, and S. Y. R. Hui[4], *Fellow, IEEE*
1. Virginia Polytechnic Institute and State University, Blacksburg, VA, United States
2. ConvenientPower HK Limited, Hong Kong
3. The Hong Kong Polytechnic University, Hong Kong
4. Center for Power Electronics, City University of Hong Kong, Hong Kong

(Email: eeronhui@cityu.edu.hk)

Abstract—The principle of using hollow spiral winding is not novel, but the study on this topic is far from complete. In this paper, how hollow the central region of the Coreless Printed Spiral Winding (CPSW) inductor should be in order to achieve the maximal quality factor value Q_{max} is explored. A new parameter, namely the ratio of the inner hollow radius and the outer winding radius $\tau = R_{in} / R_{out}$, is proposed as an indicator for optimization and used to quantify how hollow a spiral winding is. With the aid of Finite Element Analysis (FEA), the relationship between τ and Q_{max}, which depends on the operating frequency and the dimensional parameters of CPSW inductor, is established. For a specific operating frequency, it is discovered that if the conductor width is comparable with the skin depth, or the conductors are placed relatively far away from each others, the hollow design of the CPSW inductor has little improvement on Q but reduces the inductance. On the contrary, if the conductor width is much larger than the skin depth and the conductors are placed relatively close, the hollow spiral design is recommended. The optimal range of τ with which the Q_{max} can be achieved is found to be around 0.45 to 0.55.

I. INTRODUCTION

The increasing demand for slim portable electronic appliances, such as notebook and palmtop computers, highlights the significance of the low-profile low-power power converters. Lots of efforts have been put into planar integrated power passive modules design [1][2], which aims at reducing the volume and vertical dimension of the power electronic circuits. Increasing the switching frequency leads to a reduction in the required energy storage and permits the use of smaller passive components. The magnetic core of the inductor can be eliminated if the operating frequency is sufficiently high. Coreless inductors pave the way to fully integrated power converters [3]. Understanding of the coreless PCB transformer theory [4][5], optimal operation [6] and applications [7] helps eliminate previous misunderstandings that coreless PCB transformer might have unacceptable low magnetic coupling, low voltage gain and high EMI radiation problems. Specifically, a resonant technique has been incorporated into the use of the coreless PCB transformers so as to achieve a high voltage

gain (to overcome the apparent low magnetic coupling) and take advantage of the leakage inductance (to turn the apparent disadvantage into an advantage) [8][9]. The EMI radiation problems also can be solved through including various EM shielding structures into the coreless PCB transformers [10][11]. So the coreless printed spiral winding (CPSW) becomes a desirable alternative for power inductors and transformers integration. Furthermore, due to its low profile and excellent "contactless" properties, coreless PCB transformers has been extensively used in wireless power transmission applications, in which the load can be movable with respect to the energy transmitter. [12-14].

For power conversion and power transmission, the efficiency of the magnetic component is an important factor. Several studies have sought to reduce the power consumption in CPSW and improve its quality factor (Q). Design approach for winding's layout with geometric radii was addressed in [15] in order to achieve low DC resistance. Eddy current loss due to the high frequency effects can be suppressed by the subdivided conductors [16], planar litz structure [17] and layout optimization [18]. Especially when CPSW inductors are used as the transmitter or receiver of the wireless power transmission system, the eddy current loss suppression becomes much more significant, because all the current components in the windings are high frequency AC. These techniques in

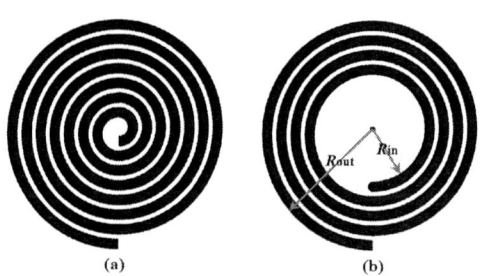

Fig. 1. Spiral winding types
(a) full spiral and (b) hollow spiral

Fig. 2. Magnetic field intensity (H) of spiral winding
scanned by EMC scanner [21]

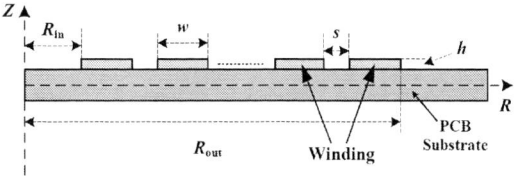

Fig. 3. Cross-sectional view of the single layer CPSW inductor
in half R-Z plane

[16-18], however, do not offer simple design and manufacture process. In [19], a proposal has been made to simply remove some inner turns, resulting in a centrally hollow spiral winding as shown in Fig. 1(b). However, the design approach of the hollow spiral winding is incomplete. The optimal ratio of the radius of the inner hollow region and the radius of the outer winding for maximizing the quality factor Q has not been analyzed, namely how hollow the spiral winding should be is unrevealed. This project fills this gap by using finite-element analysis to evaluate the quality factors of a series of CPSW inductors, with the objective of determining the relationship of the optimal quality factor and the hollowness of CPSW.

II. HOLLOW SPIRAL WINDING DESIGN PRINCIPLE

The equivalent AC resistance of the CPSW can be increased substantially by skin effect and proximity effect at high frequency. The detailed mechanisms of skin effect and proximity have been analyzed in [20]. With a critical assumption, one-dimensional (1D) analytical model for isolated single foil conductor in [20] demonstrates that the eddy current power loss for each metal conductor is highly related to the frequency and external magnetic field penetrating the conductor perpendicularly. The trend is that the higher the frequency and the magnitude of the external magnetic field are, the more significant eddy current loss becomes. For some CPSWs which are filled with turns going to the center of the coil (i.e. full winding in Fig. 1(a)), the magnetic field distribution is non-linear in a "convex" manner that its highest magnitude occurs in the central region of the spiral winding as shown as in Fig. 2. Because the total magnetic field intensity (H) of central area is the superposition of magnetic field intensity of each turn. So that the inner turns have relatively high contribution to the total power loss of the winding due to their high AC resistance, but low contribution to the inductance due to the small area they enclose. In other words, their presence causes a dramatic deterioration of the overall quality factor. In order to suppress the current crowding effect in metal traces, a simple but effective way is to remove the metal traces out of the central region where H is large. It means that some inner turns should be eliminated from the winding structure. This well-known design philosophy results in a hollow spiral winding.

III. CPSW WITH OPTIMAL HOLLOWNESS

The one-dimensional (1D) analytical model for isolated single foil conductor interprets the dependence of eddy current loss on frequency (f) and external magnetic field intensity (H), and it can accurately predict high frequency loss when the winding structure is relatively simple. Nevertheless, for multi-turn CPSW inductor, the interaction among turns becomes prominent and the magnetic flux distribution is much more complex than that of isolated single foil conductor, especially at high frequency. Therefore, its high frequency loss cannot be evaluated accurately by 1D model. Finite Element Analysis (FEA) has been proven over many years as a useful tool to precisely predict the high-frequency metal AC resistance. In this study, the Ansoft[TM] 10 is employed to evaluate AC loss as well as the quality factor of the CPSW inductor. In order to seek the optimal hollowness for spiral windings with different dimensional parameters, a series of single-layer CPSW inductors are built up as practical prototypes for investigation. Their half cross-sectional view and dimensions are shown in Fig. 3. A new parameter, defined as the ratio between inner radius and outer radius ($\tau = R_{in} / R_{out}$), is introduced here to represent the hollow scale of a winding. The higher value of τ is, the more hollow the central area of a winding becomes.

A. Study 1: spirals with different conductor width (w)

In this first study, four single-layer CPSW inductor models are developed with the FEA software. Their dimensional parameters are listed in Table 1. The outer radius of the windings, R_{out}, and the thickness of the trace, h, are kept as constant for all of them, while the trace width, w, is decreased from 1.0 mm to 0.4 mm with the step of 0.2 mm. We also keep the ratio of trace width (w) to trace separation (s) at a constant value (i.e. 2). The number of turns of the windings is initially designed to be the maximum value according to the other parameters. So

TABLE I
DIMENSIONAL PARAMETERS OF THE TESTED CPSW INDUCTORS
IN STUDY 1

	$R_{out}/(mm)$	$h/(\mu m)$	$w/(mm)$	$s/(mm)$	No. of turns
1#	15	105	1.0	0.5	10
2#	15	105	0.8	0.4	12
3#	15	105	0.6	0.3	16
4#	15	105	0.4	0.2	24

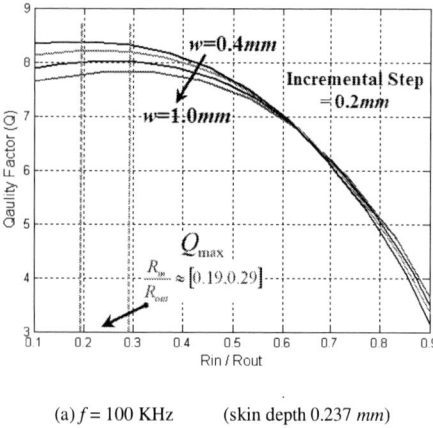

(a) f = 100 KHz   (skin depth 0.237 mm)

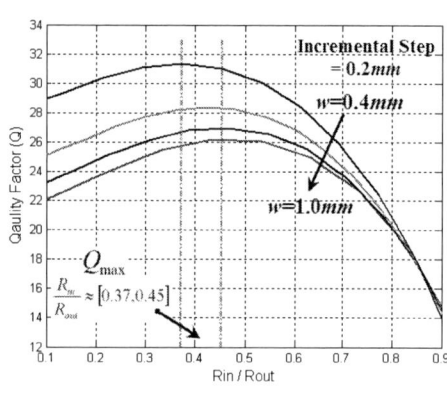

(b) f = 500 KHz   (skin depth 0.106 mm)

Fig. 4. Quality factor value as a function of τ with different conductor width

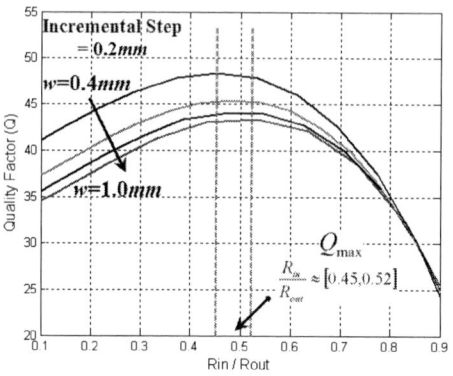

(c) f = 1 MHz   (skin depth 0.075 mm)

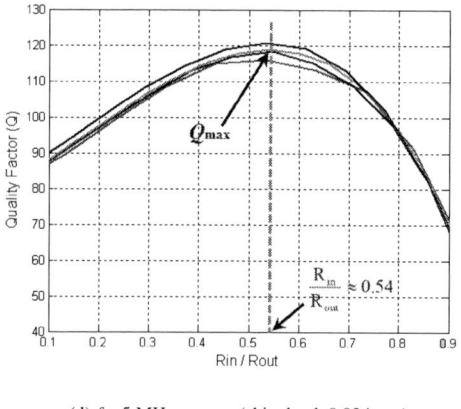

(d) f = 5 MHz   (skin depth 0.034 mm)

Fig. 4. Quality factor value as a function of τ with different conductor width

these four windings can be regarded as full CPSW inductor.

Firstly, FEA is carried out in order to obtain the Q value of each full winding. Then we remove the innermost turns out of the coils one by one. The analysis is repeated as long as every innermost turn is removed. The variation of quality factor can be plotted as the winding is gradually transformed from "full" to "hollow". The Q value of the winding as a function of τ can be plotted in Fig. 4, and the operating frequency of 100 kHz, 500 kHz, 1 MHz and 5 MHz are considered in Fig. 4(a) to Fig. 4(d), respectively. It can be found that at low frequency (100 kHz), there is little improvement on the quality factor when inner turns are removed. However, with the frequency increasing into several megahertz, the hollow layout can enhance the winding quality factor prominently.

Actually, the skin depth in the copper conductor for the frequency of 100 kHz, 500 kHz, 1 MHz and 5 MHz can be calculated by (1) as 0.237 mm, 0.106 mm, 0.075 mm and 0.034 mm, respectively.

$$\delta = \sqrt{\frac{\rho}{\pi \cdot \mu \cdot f}} \qquad (1)$$

where ρ and μ are the resistivity and the permeability of the copper material respectively, and f is the operating frequency. After the comparison of the simulation results for different frequency, it can be demonstrated that, if the conductor width is comparable to the skin depth at a given frequency (i.e. w is 2-5 times of δ), the hollow winding layout does not increase the Q value, but will decrease the winding inductance, as shown in Fig. 4(a). Therefore, the full winding layout is preferred at this relatively low frequency. On the contrary, if the traces width is at least 10 times larger than skin depth, the hollow winding layout is better due to the significant improvement on the quality factor, as illustrated in Fig. 4(d). The maximum Q is always achieved when τ is equal to 0.54, because all the curves almost follow the same path under this condition. Furthermore, the quality factor of the winding with narrow traces is always better than that with wide traces. In

978-1-4244-4782-4/10 $26.00 © 2010 IEEE

TABLE II
DIMENSIONAL PARAMETERS OF THE TESTED CPSW INDUCTORS
IN STUDY 2

	R_{out}/(mm)	h/(μm)	w/(mm)	s/(mm)	No. of turns
5#	15	105	0.6	0.3	16
6#	15	105	0.6	0.6	12
7#	15	105	0.6	0.9	10

summary, a CPSW with narrow trace width and optimal hollowness is always preferred. Said narrowness is normally determined by manufacturing capability, application current level and thermal limitation. Said hollowness should be optimized by considering skin depth at the operating frequency.

B. Study 2: spirals with different conductor separation (s)

In the second study, the other three single-layer CPSW inductors, whose dimensional parameters are listed in Table 2, are analyzed. What we keep the same as study 1

are the outer radii of the windings, R_{out}, and the thickness of the trace, h. A major difference from study 1 is the trace width, w, which is kept as 0.6 mm for all of inductor samples. The trace separation, s, is increased from 0.3 mm to 0.9 mm with step of 0.3 mm (i.e. from 0.5*w to w and then to 1.5*w). Following the same process in study 1, these three CPSW inductors are also initially designed to be the full windings, then the variation of quality factor can be obtained as the inner turns are removed one by one. The Q values of these windings as a function of τ at 100 kHz, 500 kHz, 1 MHz and 5 MHz operating frequency are plotted in Fig. 5(a) to Fig. 5(d). The maximum Q value has been marked by a circular dot for each curve.

The series of curves show that at relatively low operating frequency, such as 100 kHz in Fig. 5(a), the trace separation (s) has little effect on the τ value with which Q_{max} can be achieved. The optimal τ values are roughly equal to 0.22 for all the analyzed winding structures. The hollow winding structure is undesired under this circumstance, because removing some inner turns does not improve the quality factor of the winding,

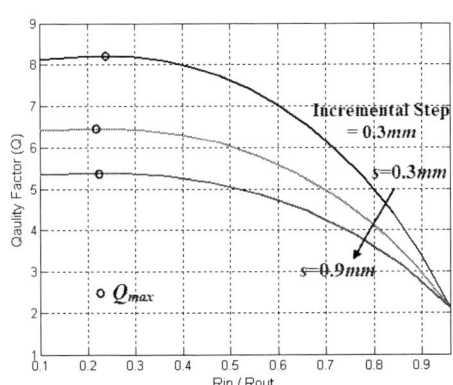

(a) f = 100 KHz   (skin depth 0.237 mm)

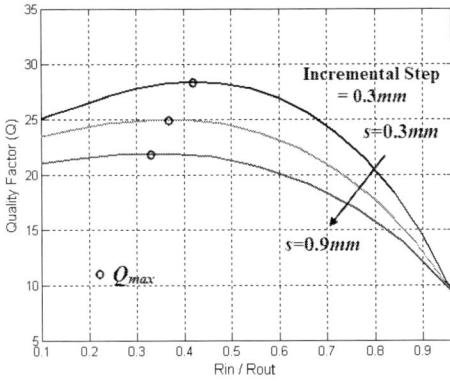

(b) f = 500 KHz   (skin depth 0.106 mm)

Fig. 5. Quality factor value as a function of τ with different conductor separation

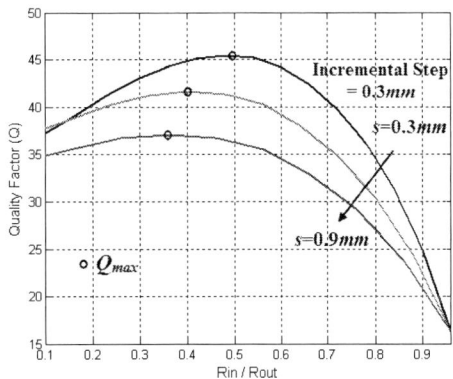

(c) f = 1 MHz   (skin depth 0.075 mm)

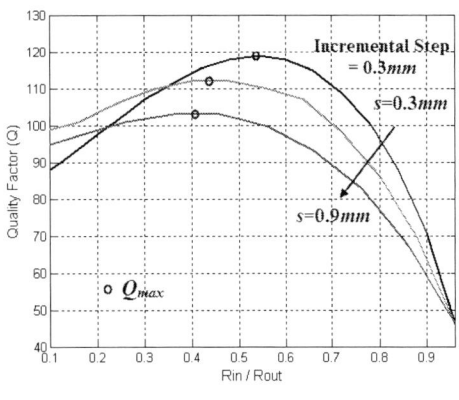

(d) f = 5 MHz   (skin depth 0.034 mm)

Fig. 5. Quality factor value as a function of τ with different conductor separation

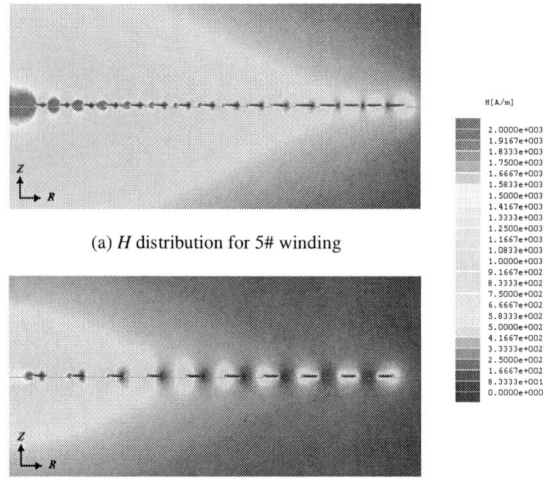

(a) *H* distribution for 5# winding

(a) *H* distribution for 7# winding

Fig. 6. *H* distribution from simulation for compact and loose winding structure

but rather decreases its inductance. However, as the operating frequency increasing up to megahertz, the design principle is highly dependent on the trace separation (s). Taking frequency of 5 MHz in Fig. 5(d) as an example, if the winding has a compact structure, namely every turn of the winding is placed relatively close to each other compared with the trace width (w=0.6 *mm* and s=0.3 *mm*), hollow structure CPSW is preferred due to the significant improvement of quality factor. In contrast, if the trace separation of the winding is equal to or larger than the trace width (w=0.6 *mm*, s=0.6 *mm* and w=0.6 *mm*, s=0.9 *mm*), namely the winding has a "loose" (or non-compact) structure, a hollow CPSW design cannot improve the quality factor value, thus should not be used. This can be explained by the reexamination of the magnetic flux intensity (*H*) distribution of CPSW. Fig. 6 shows the simulated *H* distribution for 5# and 7# windings when excitation peak currents are both 1 A, from which it can be seen that the *H* value distribution of loose winding is slightly more even than that of compact winding, and also the central *H* value of loose winding is smaller than that of compact winding due to the less number of turns. Therefore, the hollow structure can be utilized for compact winding, but not appropriate for loose winding at very high frequency.

Moreover, for a given trace width, the maximum quality factor value that can be achieved in a compact winding is always larger than that in loose winding. Therefore, the conclusion here is similar to that obtained in study 1: Within the manufacturing capability, the application current level and the thermal limitation, one should design a CPSW as compact as possible. Whether the hollow structure should be employed or not is again determined by the relationship between conductor width and skin depth of the conductor material at that given frequency.

As a summary of Study 1 and 2, we come to the design philosophy of a CPSW inductor, with which the quality factor of the CPSW inductor can be optimized. For a particular application (i.e. if the footprint, operating frequency, power and current level and thermal limitation of the CPSW inductor are all given), the winding layout should be designed with the following steps:

1. Firstly, the conductor width and the conductor separation both should be as small as possible with the consideration of the particular application requirements.

2. Secondly, if the conductor width value obtained from step one is much larger than the skin depth of the conductor at given frequency, the hollow winding structure with $\tau \approx 0.54$ is recommended. On the contrary, if the conductor width value is comparable with the skin depth, the full winding structure is preferred.

IV. CONCLUSIONS

With the objective to complete the hollow design approach for power CPSW inductor, a new parameter based on the innermost and outmost radii of the winding is introduced as measure of hollowness of the winding in this paper. The choice of hollow or full winding structural designs in order to maximize its quality factor is discussed. Two typical FEA studies on different series of windings indicate that the trace width and trace separation should be as small as possible within the thermal limitation and manufacturing capability. Optimal hollowness of the winding depends on the ratio of the conductor width to the skin depth at the given frequency. This design principle results in great quality factor improvement and power loss reduction of the planar magnetic component. The outcome of this paper can also be used to study and design the low profile planar integrated converter with high efficiency.

ACKNOWLEDGMENT

The authors are grateful to the support of the City University of Hong Kong under the Strategic Research Grant 7002218.

REFERENCES

[1] I. W. Hofsajer, J. A. Ferreira, D. van Wyk, "Design and analysis of planar integrated L-C-T components for converters," *IEEE Transactions on Power Electronics*, vol. 15, no. 6, pp. 1221-1227, Nov. 2000.

[2] J. T. Strydom, D. van Wyk, "Electromagnetic design optimization of planar integrated power passive modules," in *Proc. Power Electronics Specialists Conference*, vol. 2, Jun. 2002, pp. 573-578.

[3] R. C. N. Pilawa-Podgurski, A. D. Sagneri, J. M. Rivas, D. I. Anderson, D. J. Perreault, "Very high frequency resonant boost converters," in *Proc. Power Electronics Specialists Conference*, Jun. 2007, pp. 2718-2724.

[4] C. Fernandez, O. Garcia, R. Prieto, J. A. Cobos, S. Gabriels, G.Van Der Borght, "Design issues of a core-less transformer for a

contact-less application," in *Proc. Applied Power Electronics Conference and Exposition*, vol. 1, Mar. 2002, pp. 339-345.

[5] S. C. Tang, S. Y. R. Hui, H. Chung, "Characterization of coreless printed circuit board (PCB) transformers," *IEEE Transactions on Power Electronics*, vol. 15, no. 6, pp. 1275-1282, Nov. 2000.

[6] E. Dallago, M. Passoni, G. Venchi, "Design and optimization of a high insulation voltage DC/DC power supply with coreless PCB transformer," in *Proc. International Conference on Industrial Technology*, vol. 2, Dec. 2004, pp. 596-601.

[7] P. Luniewski and U. Jansen, "Unsymmetrical gate voltage drive for high power 1200V IGBT[4] modules based on coreless transformer technology driver," in *Proc. Power Electronics and Motion Control Conference*, Sep. 2008, pp. 88-96.

[8] H. Abe, H. Sakamoto and K. Harada, "A noncontact charger using resonant converter with parallel capacitor of the secondary coil," *IEEE Transactions on Industrial Applications*, vol. 36, no. 2, pp. 444-451, Mar. 2000.

[9] Chwei-Sen Wang, G. A. Covic, O. H. Stielau, "Power transfer capability and bifurcation phenomena of loosely coupled inductive power transfer systems," *IEEE Transactions on Industrial Electronics*, vol. 51, no. 1, pp. 148-157, Feb. 2004.

[10] S. Judek and K. Karwowski, "Supply of electric vehicles via magnetically coupled air coils," in *Proc. Power Electronics and Motion Control Conference*, Sep. 2008, pp. 1497-1504.

[11] Y. P. Su, X. Liu, S. Y. R. Hui, "Extended theory on the inductance calculation of planar spiral windings including the effect of double-layer electromagnetic shield," *IEEE Transactions on Power Electronics*, vol. 23, no. 4, pp. 2052-2061, Jul. 2008.

[12] N. Hemche and A. Jaafari, "Wireless transmission of power using a PCB transformer with mobile secondary," in *Proc. Mediterranean Electrotechnical Conference*, May. 2008, pp. 629-634.

[13] Y. P. Su, X. Liu, S. Y. R. Hui, "Mutual inductance calculation of movable planar coils on parallel surfaces," *IEEE Transactions on Power Electronics*, vol. 24, no. 4, pp. 1115-1124, Apr. 2009.

[14] M. Takahashi, K. Watanabe, F. Sato and H. Matsuki, "Signal transmission system for high frequency magnetic telemetry for an artificial heart," *IEEE Transactions on Magnetics*, vol. 37, no. 4, pp. 2921-2924, Jul. 2001.

[15] X. Huang and K. D. T. Ngo, "Design technique for a spiral planar winding with geometric radii," *IEEE Transactions on Aerospace and Electronic Systems*, vol. 32, no. 2, pp. 825-830, Apr. 1996.

[16] M. Peter, H. Hein, F. Oehler, P. Baureis, "Planar inductors with subdivided conductors for reducing eddy current effects," *2003 Topical Meeting on Silicon Monolithic Integrated Circuits in RF Systems*, Apr. 2003, pp.104-106.

[17] Shen Wang, M. A. de Rooij, W. G. Odendaal, J. D. van Wyk, D. Boroyevich, "Reduction of high-frequency conduction losses using a planar litz structure," *IEEE Transactions on Power Electronics*, vol. 20, no. 2, pp. 261-267, Mar. 2005.

[18] J. M. Lopez-Villegas, J. Samitier, C. Cane, P. Losantos, J. Bausells, "Improvement of the quality factor of RF integrated inductors by layout optimization," *IEEE Transactions on Microwave Theory and Techniques*, vol. 48, no. 1, pp. 76-83, Jan. 2001.

[19] J. Craninckx and M. S. J. Steyaert, "A 1.8-GHz low-phase-noise CMOS VCO using optimized hollow spiral inductors," *IEEE Journal of Solid-State Circuits*, vol. 32, no. 5, pp. 736-744, 1997.

[20] Shen Wang, "Modeling and design of planar integrated magnetic components," Master Thesis, Virginia Polytechnic Institute and State University, USA, Jul. 2003.

[21] X. Liu and S. Y. R. Hui, "Optimal design of a hybrid winding structure for planar contactless battery charging platform," *IEEE Transactions on Power Electronics*, vol. 23, no. 1, pp. 455-463, Jan. 2008

Modeling of adaptable-diameter burners formed by concentric planar windings for domestic induction heating applications

Jesus Acero, Claudio Carretero, Ignacio Millan,
Oscar Lucia, Jose-Miguel Burdio
Dept. of Electronic Engineering and Communications
University of Zaragoza
Maria de Luna 1, Zaragoza, Spain
jacero@unizar.es

Rafael Alonso
Dept. of Applied Physics
University of Zaragoza
Pedro Cerbuna 9, Zaragoza, Spain

Abstract—**Adaptable-diameter inductors are being implemented in domestic induction hobs in order to increase the range of suitable pot's diameters and to achieve a better use of the installed power electronics. Such inductors are arranged by means of several concentric planar windings, usually up to two or three units, each one of them comprising several Litz-wire turns. Normally, one resonant inverter is dedicated to supply each winding. In this paper, a model of these inductors in terms of their impedance matrix is derived. The self-impedance and the coupling between the windings are analyzed on the basis on the transformer analogy. The analysis also includes the losses in the Litz wires generated by the currents in each winding as well as the losses produced by the windings over its neighbors.**

I. INTRODUCTION

In the last years, the induction hobs are attracting an increasing number of users; mainly due to its specific features as high power rates, quick response, and automatic pot detection. Moreover, induction cookers represent an efficient alternative to the traditional cookers. The usual induction hob consists of several burners, each one of them having a single flat-type winding which is supplied by a resonant inverter. IGBT half-bridge series resonant inverter, including ZVS operation, is one of the most popular [1]. Normally, inverters are operated from 20 kHz to 100 kHz and, in current implementations they can reach up to 3.5 kW [2].

Nowadays, some improvements are being developed and applied to the induction cookers. Adaptable-diameter burners are being developed in response to the demand of large size burners (up to 32 cm) suitable for large pots but which should also work for small pots. Two or more concentric windings, each one of them fed by an inverter (see Fig. 1), forming a large-size burner, are being implemented. The activation of the windings is planned according to the diameter of the pot.

The use of concentric multiple-winding inductors has clear benefits: a given burner can operate with a wider range of

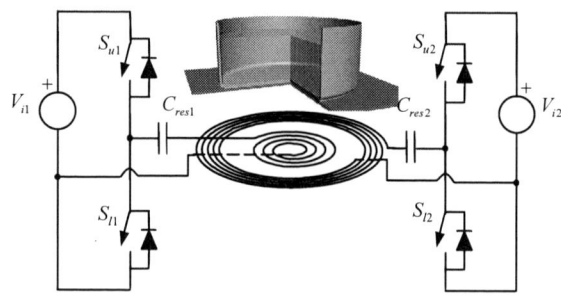

Fig. 1. Schematic representation of a burner consisting of two concentric windings each one of them supplied by a resonant half-bridge inverter.

pots, higher utilization ratio of electronics, higher maximum power when only one of the windings is active, and finally, it is possible to share some components of the power converters. However, the use of concentric windings presents some drawbacks, mainly derived from the tight coupling between them. Designing each winding and preserving the ZVS operation of the inverters are two issues whose study requires models which must include the coupling. Besides, proximity losses are favored by the concentric and planar arrangement of the windings.

The purpose of this paper is to develop a model of litz-wire concentric multiple-winding inductors in terms of matrix impedances whose terms represent the physical phenomena in both the vessel and the windings (equivalent inductance, dissipation in the pot, and losses in the windings). The model is based on the transformer analogy. In addition, the modeling also considers the losses in the litz-wire windings which in this case will have a special significance due to the proximity losses between concentric windings. Despite the model is mainly intended for induction cooking, it could be extended to other applications.

This work was supported in part by the Spanish Ministry of Science and Innovation (MICINN) under Project TEC2007-64188, in part by Diputación General de Aragón (DGA) under Project PI008/08, and in part by the Bosch and Siemens Home Appliances Group. The authors are with the Aragon Institute for Engineering Research (I3A).

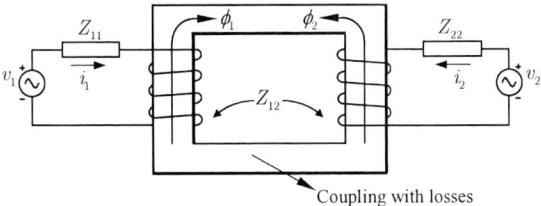

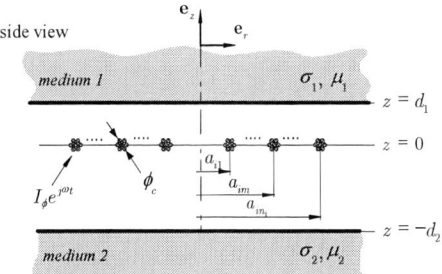

Fig. 2. Basic structure of a transformer used for modeling an induction system consisting of two concentric planar windings sharing the same pot.

Fig. 3. Basic induction system consisting of several Litz-wire turns placed between two half spaces.

II. MODELING OF LITZ-WIRE CONCENTRIC MULTIPLE-WINDING INDUCTORS

A. Transformer analogy and diagonal terms

The modeling of the coupled inductor-pan system has often been dealt by means of the transformer analogy [3], [4]. This approach is also suitable to characterize the system in terms of the *matrix impedance*. In the classical case shown in Fig. 2, the voltages of the sources can be expressed as

$$\begin{bmatrix} v_1 \\ v_2 \end{bmatrix} = \begin{bmatrix} Z_{11} & Z_{12} \\ Z_{21} & Z_{22} \end{bmatrix} \begin{bmatrix} i_1 \\ i_2 \end{bmatrix} \quad (1)$$

The term Z_{ii}, corresponds to the impedance of a single i^{th} winding placed between two media, one of them representing the pot and the other the ferrite which usually is placed to improve the coupling. This system is shown in Fig. 3. In arrangements with discrete ferrite bars instead of a homogeneous slab, an equivalent relative magnetic permeability can be found [6]. Such system was analytically solved in the past [3], [7] and its impedance corresponds to the series connection of equivalents resistor and inductor

$$Z_{ii} = R_{eq,ii} + j\omega L_{eq,ii} \quad (2)$$

Moreover, both $R_{eq,ii}$, $L_{eq,ii}$ can be expressed as the sum of the contribution of the winding and the media:

$$R_{eq,ii} = R_{o,ii} + \Delta R_{ii} , \quad L_{eq,ii} = L_{o,ii} + \Delta L_{ii} \quad (3)$$

where $R_{o,ii}$ represents the losses in the cables of the ith winding, ΔR_{ii} is a resistance associated to the inductive heating of the media, $L_{o,ii}$ is the self-inductance of the ith winding and ΔL_{ii} represents the contribution of the media. Usually, ΔL_{ii} is negative in presence of conductive media due to the induced eddy currents. In addition, $R_{o,ii}$ can be also divided into conduction resistance $R_{cond,ii}$, and induction (proximity) resistance. In the case of litz wires, the last one can be further divided into internal and external induction losses $R_{ind\_int,ii}$, $R_{ind\_ext,ii}$, respectively, thus

Integral expressions for ΔR_{ii}, $L_{o,ii}$ ΔL_{ii} and formulas to calculate $R_{cond,ii}$, $R_{ind\_int,ii}$, $R_{ind\_ext,ii}$ can be directly collected from [7], [8], and [9], respectively.

B. Characterization of the mutual impedance terms

Unlike the most part of the transformers, the flux path in domestic induction systems, which includes the pot, presents high losses. Indeed, maximizing these losses with respect to others is a priority goal of designers. The coupling, classically characterized by the *mutual inductance* M defined as $M = d\phi_{ij} / di_i$ [5] where ϕ_{ij} is the flux linked by the winding j generated by the winding i, must be replaced by the *mutual impedance* Z_{ij}, as it is shown in Fig. 2.

Fig. 4 shows two concentric windings placed between two linear homogeneous half spaces. The media are characterized by their electromagnetic properties. Windings i^{th} and j^{th} consist of a set of n_i and n_j concentric turns, respectively. The turns of the ith winding carry a sinusoidal filamentary current of amplitude $I_{\phi,i}$, at angular frequency ω. The terms Z_{ij} are obtained by evaluating the derivative of the flux of the magnetic induction **B** over the jth winding.

From [7], the z component of the magnetic field generated by a circular filamentary current of radius a_{im} (belonging to the ith winding) at any point (r,z) such as $0 < z < d_1$ is

$$H_{z,a_{im}}(r,z) = a_{im} \frac{I_{\phi,i}}{2} \int_0^\infty \beta e^{-\beta z} \left(1 + \phi_1 e^{2\beta(z-d_1)} \right)$$
$$\times \frac{\left(1 + \phi_2 e^{-2\beta d_2} \right)}{\left(1 - \phi_1 \phi_2 e^{-2\beta(d_1+d_2)} \right)} J_1 (\beta a_{im}) J_0 (\beta r) d\beta \quad (4)$$

with β being the integration variable of the Fourier-Bessel transform [10]. Moreover J_0, J_1 are the Bessel function of first kind and order 0 and 1 respectively; and ϕ_1, ϕ_2 are parameters dependent of the media properties [7]. Other geometrical parameters are also shown in Fig. 4.

Let $\Phi_{a_{jk},a_{im}}$ be the flux of $B_z(r,z)$ generated by the current at a_{im} through the surface limited by the turn placed at $(r = a_{jk}, z = 0)$, belonging to the jth winding. This flux is calculated by means of

978-1-4244-4782-4/10 $26.00 © 2010 IEEE

$$\Phi_{a_{jk},a_{im}} = \iint_S B_z ds = \mu_0 \int_0^{2\pi} \int_0^{a_{jk}} H_z(r,z=0) r \, dr \, d\phi \quad (5)$$

Applying the superposition principle, the total flux through the turn placed at $r = a_{jk}$., $\Phi_{a_{jk}}$, is obtained by adding the flux created by the rest of the turns of the ith winding. Moreover, the global flux through the jth winding, Φ_{ij}, is obtained by summing the flux of every k^{th} turn, i.e.

$$\Phi_{ij} = \sum_{k=1}^{n_j} \Phi_{a_{jk}} = \sum_{k=1}^{n_j} \sum_{m=1}^{n_i} \Phi_{a_{jk},a_{im}} \quad (6)$$

Using (4) and (6) the global flux results

$$\Phi_{ij} = I_{\phi,i} \pi \mu_0 \int_0^\infty \frac{\left(1 + \phi_1 e^{-2\beta d_1}\right)\left(1 + \phi_2 e^{-2\beta d_2}\right)}{\left(1 - \phi_1 \phi_2 e^{-2\beta(d_1+d_2)}\right)} G\left(\beta, a_{im}, a_{jk}\right) d\beta \quad (7)$$

where $G\left(\beta, a_{im}, a_{jk}\right)$ is a function depending on the geometry and the number of turns of the windings and defined as

$$G\left(\beta, a_{im}, a_{jk}\right) = \sum_{k=1}^{n_j} \sum_{m=1}^{n_i} a_{im} J_1\left(\beta a_{im}\right) a_{jk} J_1\left(\beta a_{jk}\right) \quad (8)$$

Taking into account that $Z_{ij} = V_{ij}/I_{\phi,i}$ and according to the Faraday-Lenz's law and also considering harmonic time-variation of the currents of the ith winding, the non-diagonal terms of the characteristic impedance matrix result

$$Z_{ij} = \underbrace{j\omega\pi\mu_0 \int_0^\infty G\left(\beta, a_{im}, a_{jk}\right) d\beta}_{Z_{o,ij}} +$$
$$\underbrace{j\omega\pi\mu_0 \int_0^\infty \frac{\phi_1 e^{-2\beta d_1} + \phi_2 e^{-2\beta d_2} + 2\phi_1 \phi_2 e^{-2\beta(d_1+d_2)}}{\left(1 - \phi_1 \phi_2 e^{-2\beta(d_1+d_2)}\right)} G\left(\beta, a_{im}, a_{jk}\right) d\beta}_{\Delta Z_{ij}} \quad (9)$$

It can be demonstrated that the first term, $Z_{o,ij}$ is a purely imaginary impedance, thus it is identified as the mutual inductance M_{ij} between the ith and the jth windings in air. The second term ΔZ_{ij}, represents the contribution of the media. Since parameters ϕ_1, ϕ_2, are complex, ΔZ_{ij} will have a resistive and inductive components

$$\Delta R_{ij} = \text{Re}\left[\Delta Z_{ij}\right]; \Delta L_{ij} = \text{Im}\left[\Delta Z_{ij}/\omega\right] \quad (10)$$

Therefore, when there are a dissipative medium it could be interpreted that the coupling is not only inductive but also *resistive*. On the opposite, if both media are non dissipative $\left(\sigma_1 = \sigma_2 = 0\right)$, ΔZ_{ij} becomes purely inductive.

III. ANALYSIS OF THE LOSSES IN THE WINDINGS

The losses in a litz-wire winding have three components [9]. First, losses due to Joule effect by the carrying current in the wire, second, losses due to the eddy currents induced in each strand due to its vicinity to other strands, (the so called *internal proximity losses*), third, losses due to the eddy

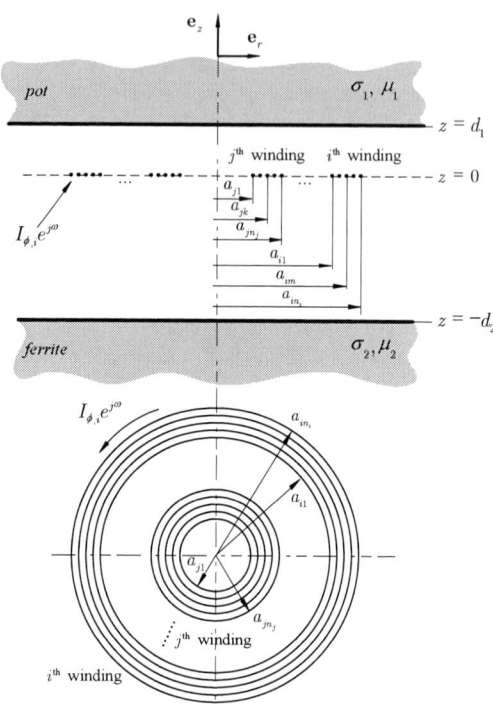

Fig. 4. Representation of a system comprising two concentric filamentary windings between two half spaces.

currents induced in the whole winding by the magnetic flux created by itself (the *external proximity losses*). The first two ones have been represented by the conduction resistance $R_{cond,ii}$ and the internal induction resistance, $R_{ind\_int,ii}$, which are associated to the diagonal terms of (1) due to they only depend of the current driven by each winding. Formulas of these terms can be collected in [9].

However, the external proximity losses shows some peculiarities which are worth to analyze in detail. Fig. 5 shows the ith winding carrying a sinusoidal current of amplitude $I_{\phi,i}$ in the presence of other windings. According to this figure, a magnetic field $\mathbf{H}_{o,i,a_{ik}}$ is generated over the kth turn of the self winding and a magnetic field $\mathbf{H}_{o,i,a_{jk}}$ is also applied over the kth of the neighbor winding. Thus, the ith winding induces proximity losses both in its turns an also in the turns of the rest of the windings. Because the set of these proximity losses are generated by the current driven by the ith winding, they can be represented by a resistive term $R_{ind\_ext,ii}$ associated to this winding. So, the external proximity resistance associated to the ith winding can be calculated as the sum of the losses generated in its turns $R_{ind\_ext,ii,i}$ and the losses generated over the turns of the rest of the windings $R_{ind\_ext,ii,j}$. According to the formula of the proximity losses of a planar winding [9], this resistance results

$$R_{ind\_ext,ii} = \sum_{j=1}^{n_{bob}} R_{ind\_ext,ii,j} = n_o \frac{2\pi^2 \xi r_o \Phi_{ind}}{\sigma_{Cu}} \sum_{j=1}^{n_{bob}} \left[\sum_{k=1}^{n_j} \left[a_{jk} \left\langle H_{o,i,a_{jk}}^2 \right\rangle \right] \right] \quad (11)$$

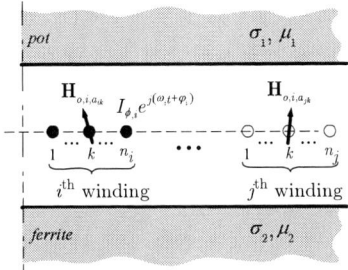

Fig. 5. Magnetic field generated by the ith winding over its turns and over the turns of the neighbor windings.

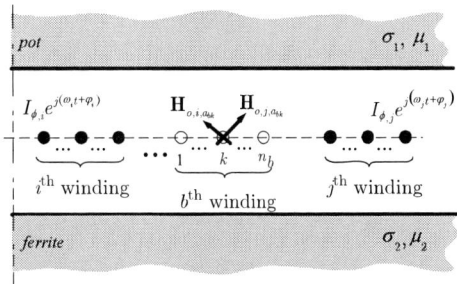

Fig. 6. Magnetic fields generated over a turn of the bth winding by the carried currents through the ith and j^{th} windings.

where n_{bob} is the number of the windings of the inductor, n_o is the number of strands of the Litz wire, r_o is the radius of the strand, $\xi = \sqrt{2}/\delta$ with δ the skin depth, $\left\langle H_{o,i,a_{jk}}^2 \right\rangle$ is the average of the squared magnetic field generated by the ith winding over the kth turn belonging to the jth winding and placed at a_{jk}. Φ_{ind} is a parameter characteristic of the proximity losses in round conductors [9] and σ_{Cu} is the conductivity of the copper. In the last equation, the case $i = j$, represents the losses of the ith winding induced by itself. $R_{ind\_ext,ii}$ is a resistance assigned to the ith diagonal element of the matrix impedance.

The non-diagonal terms of the matrix impedance can also include resistive terms associated to the effect of two fields simultaneously coupled over a turn. Non-diagonal terms are responsible of the different proximity resistances achieved at different current phases in the windings. For instance, in concentric windings proximity losses are penalized for the in-phase connection of windings with respect to the opposing-phase connection. An appropriate modeling of such terms is needed to complete adequately the model of the induction system.

In order to understand the non-diagonal terms of the proximity resistance, it is useful to turn to the physical origin of the proximity losses. The induced losses over a turn depend on the squared field over the turn, as it can be seen in (11). Indeed, the calculation of the proximity resistance reduces to a proper calculation of the squared field over the turns.

The squared field H_o^2 over a turn results of multiplying the field in the turn by its conjugated $H_o^2 = \mathbf{H}_o \cdot \mathbf{H}_o^*$. Physically, the origin of this product is the microscopic Ohm's Law $\mathbf{E} \cdot \mathbf{J}_{ind}^*$, in other words, the origin of H_o^2 is the product of the voltage in the turn by the conjugated of the induced current in its conductors. In the case of a single winding, the voltage and the induced current have the same origin; however, in the case of multiple windings the induced current and the voltage in a winding could have different sources, which could result in cross terms of the proximity losses to be assigned to the non-diagonal terms of the matrix impedance. Moreover, in the case of the phase of both $\mathbf{E}$ and $\mathbf{J}_{ind}$ was not the same, real part of the product $\mathbf{E} \cdot \mathbf{J}_{ind}^*$ must be considered.

Let $\mathbf{H}_{o,i,a_{bk}}$ the magnetic field generated by the ith winding over the kth turn belonging to the bth winding and placed at a_{bk}. Similarly $\mathbf{H}_{o,j,a_{bk}}$ is the magnetic field generated by the jth winding over the same turn as it shown in Fig. 6. The total field over this turn will be:

$$\mathbf{H}_{o,a_{bk}} = \mathbf{H}_{o,i,a_{bk}} + \mathbf{H}_{o,j,a_{bk}} \qquad (12)$$

The product $\mathbf{H}_{o,a_{bk}} \cdot \mathbf{H}_{o,a_{bk}}^*$ produces three kinds of terms. First, the term $\mathbf{H}_{o,i,a_{bk}} \cdot \mathbf{H}_{o,i,a_{bk}}^*$ is equal to $H_{o,i,a_{bk}}^2$, resulting in a proximity term associated to diagonal term of the ith winding, as it has been seen above. Second, over the term $\mathbf{H}_{o,j,a_{bk}} \cdot \mathbf{H}_{o,j,a_{bk}}^*$ similar reasoning can be applied. Third, cross products $\mathbf{H}_{o,i,a_{bk}} \cdot \mathbf{H}_{o,j,a_{bk}}^*$, $\mathbf{H}_{o,i,a_{bk}}^* \cdot \mathbf{H}_{o,j,a_{bk}}$ will be associated to proximity resistances in the non-diagonal terms of the impedance matrix. Moreover, considering the following equality

$$\text{Re}\left(\mathbf{H}_{o,i,a_{bk}} \cdot \mathbf{H}_{o,j,a_{bk}}^*\right) = \text{Re}\left(\mathbf{H}_{o,i,a_{bk}}^* \cdot \mathbf{H}_{o,j,a_{bk}}\right) \qquad (13)$$

it can be proved that the impedance matrix remains symmetrical.

Finally, cross proximity resistances results to be as follows:

$$R_{ind\_ext,ij} = n_o \frac{2\pi^2 \xi r_o \Phi_{ind}}{\sigma_{Cu}} \sum_{b=1}^{n_{bob}} \left[\sum_{k=1}^{n_b} \left[a_{bk} \left\langle \text{Re}\left(\mathbf{H}_{o,i,a_{bk}} \cdot \mathbf{H}_{o,j,a_{bk}}^*\right) \right\rangle \right] \right] \qquad (14)$$

The magnetic fields appearing in this section can be calculated by means of (4) considering an amplitude of the current of 1 A. As it can be seen, this calculation also includes the influence of the material properties and the ferrite characteristics.

IV. EXPERIMENTAL VERIFICATION

An experimental setup was built to verify the models. Frequency-dependent resistance and inductance are measured by means of a precision LCR meter (Agilent E4980A). The explored frequency range is comprised between 1 kHz and 1 MHz. The inductor consists of two concentric windings of $n_1 = 17$ and $n_2 = 9$ turns, respectively (see Fig. 7 (a)). Internal and external radii for each winding are $a_{1,1} = 25$ mm

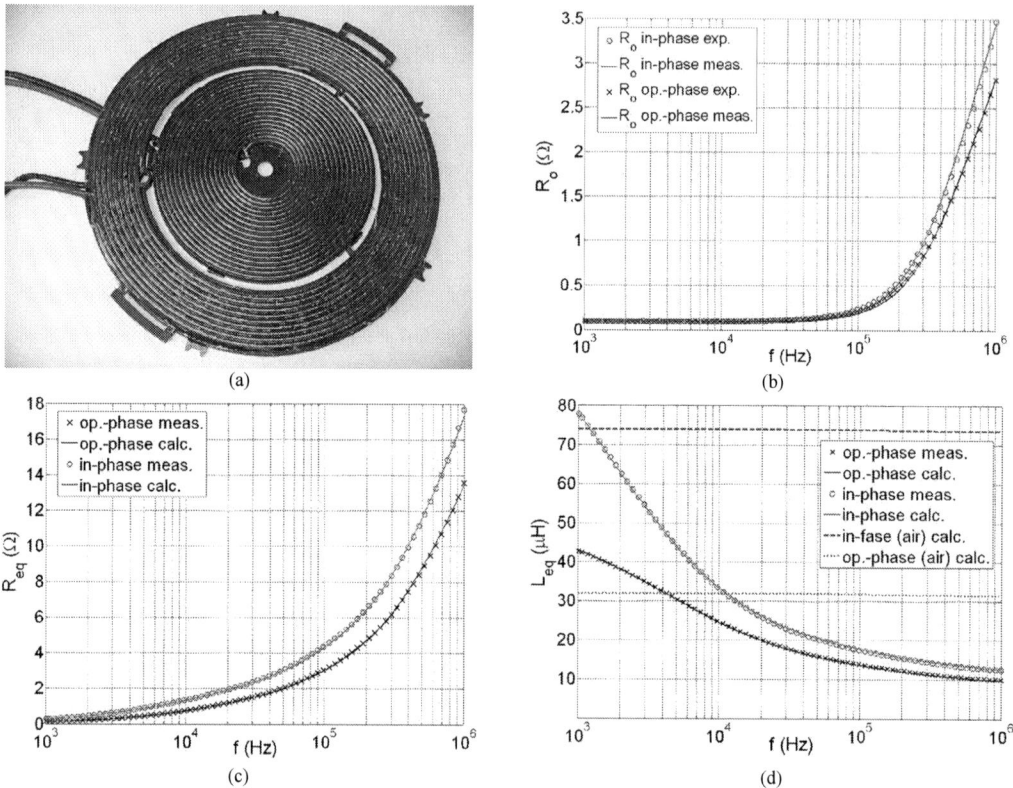

Fig. 7. Comparison of calculated and experimental results: (a) inductor arrangement; (b) resistance of the non-loaded windings for the in-phase and opposing-phase connections; (c) resistance of the loaded windings with ferrite slabs for both connections; (d) inductance of the loaded windings with ferrite slabs for both connections.

$a_{2,1} = 100$ mm, $a_{1,17} = 90$ mm y $a_{2,9} = 137$ mm, respectively. Both windings are wound with a Litz-wire of $n_o = 38$ strands each one of diameter $\phi_o = 0.3$ mm. Tests were performed connecting the windings in-phase and opposing phase, due to the interest in feeding the windings according to currents in-phase or in opposition-phase in the real operation. Besides, the non-diagonal terms of the impedance matrix can be obtained subtracting the results of both experiments. Resistance R_o in Fig. 7 (b) represents the frequency-dependent resistance of windings in the air according to the two mentioned connections. As it can be seen, at DC conditions, the resistance is similar for both connections; however, at higher frequencies the proximity resistance changes due to the different magnetic field generated in each connection.

The windings were also measured with load and ferrite slabs of 3C90 material. The load consists of a disk of a ferromagnetic steel of $\sigma = 8 \cdot 10^6$ $(\Omega m)^{-1}$, $\mu_r = 150$. Equivalent resistance for the loaded inductor is presented in Fig. 7 (c). As it can be seen, this resistance depends on the phase of the currents, which can be exploited as power control parameter. Comparing Fig. 7 (b) and Fig. 7 (c) it can be appreciated the resistive contribution of the load. Similarly, the equivalent inductance is presented in Fig. 7 (d), where the

inductance of the windings in the air for the considered connections is also shown. These results highlight the different magnetic field generated between both connections, which confirm the tight coupling between the windings.

In general, a good agreement between theory and experiment has been observed.

V. CONCLUSION

A model of the impedance of concentric planar windings for domestic induction heating appliances is derived. The model is posed in impedance matrix form, based on the transformer analogy. The non-diagonal terms of the impedance matrix have been generalized to take into account the particularities of the induction system. A generalization of the proximity losses in the case of several windings have also been proposed, which has permitted to explain the different losses achieved at different phases of the currents in the windings. Furthermore, an exhaustive experimental verification has been performed, on the basis on the in-phase and opposing phase connection of coupled windings.

The developed model permits both designing the windings (number of turns and wire's yarn) and extracting the impedance of the system to perform time-domain simulations including the electronics.

REFERENCES

[1] S. Llorente, F. Monterde, J.M. Burdio, and J. Acero, "A comparative study of resonant inverter topologies used in induction cookers," in *IEEE Applied Power Electronics Conf. (APEC) Rec.* 2002, pp.1168-1174.

[2] J. Acero, et al., "The domestic induction heating appliance: An overview of recent research," in *IEEE Applied Power Electronics Conf. (APEC) Rec.* 2008, pp. 651-657.

[3] W.G. Hurley and J.G. Kassakian, "Induction heating of circular ferromagnetic plates," *IEEE Trans. Magn.*, vol. mag-15, no 3, pp 1174-1181, July 1979.

[4] H.W.E. Koertzen, J.D. Van Wyk, J.A. Ferreira, "Investigating the influence of material properties on the efficiency of an induction heating load transformer using FEM simulation," *IEEE Industry Applications Society Annual Meeting (IAS) Rec.* 1995, pp. 868-873.

[5] J.A. Edminister, *Electric Circuits*. New York: Schaum Publishing Co., 1965.

[6] J. Acero, R. Alonso, J.M. Burdío, L.A. Barragán, "Enhancement of induction heating performance by sandwiched planar windings," *Electronics Letters*, vol. 42, pp. 241-242, Feb. 2006.

[7] J. Acero, R. Alonso, J. M. Burdío, L. A. Barragán, S. Llorente, "Electromagnetic induction of planar windings with cylindrical symmetry between two half-spaces," *Journal of Applied Physics*, vol. 103, pp. 104905(8), May 2008.

[8] J. Acero, R. Alonso, L.A. Barragán, and J.M. Burdío, "Modeling of planar spiral inductors between two multilayer media for induction heating applications," *IEEE Trans. Magn.*, vol. 42, pp. 3719-3729, Nov. 2006.

[9] J. Acero, R. Alonso, J.M. Burdio, L.A. Barragán, and D. Puyal, "Frequency-dependent resistance in litz-wire planar windings for domestic induction heating appliances," *IEEE Trans. Power Electronics*, vol. 21, pp. 856-866, July 2006.

[10] J.A. Stratton, *Electromagnetic Theory*. New York: McGraw-Hill, 1941.

[11] M. Abramowitz and I.A. Stegun, *Handbook of Mathematical Functions*. New York: Dover Publications, 1970.

[12] J.C. Maxwell, *A Treatise on Electricity and Magnetism*. Oxford: Oxford University Press, 1998.

Flux Concentration and Pole Shaping in a Single Phase Hybrid Switched Reluctance Motor Drive

Uffe Jakobsen and Kaiyuan Lu
Institute of Energy Technology
Aalborg University
Aalborg, Denmark
Email: uja@iet.aau.dk

Abstract— The Single phase hybrid switched reluctance motor (HSRM) may be a good candidate for low-cost drives used for pump applications. This paper presents a new design of the HSRM with improved starting torque achieved by stator pole shaping, and a better arrangement of the embedded stator permanent magnets with flux concentration effects. Analysis and simulation results of the proposed HSRM drive are validated using experimental results on a prototype HSRM.

I. INTRODUCTION

Cost minimization of drive systems used for pump applications has always been the main focus in the industries. During the last decade, there is an increasing demand on the system efficiency improvement. Previous work reported in [1] [4] has demonstrated a promising performance for a single phase hybrid switched reluctance motor (HSRM) drive system. This paper further investigates the HSRM technology, with an attempt to improve its starting torque and drive system efficiency. A stator pole-shaping method is used in this paper for effectively increasing the motor starting torque. Detailed investigation on the influence of different arrangements of embedded stator permanent magnets on torque production and torque ripple is also covered.

II. TORQUE PRODUCTION IN HYBRID SWITCHED RELUCTANCE MOTORS

In the hybrid switched reluctance motor (HSRM), the main torque component is generated due to the variation in the reluctance.

To better describe the design parameters that affect torque production in the HSRM, the principle for torque production is briefly described first. The energy transferred to the magnetic system (W_e) may be described as:

$$
\begin{aligned}
W_e &= W_{supply} - W_{cu} \\
&= \int \left(u \cdot i - R \cdot i^2 \right) dt \\
&= \int i\, d\Psi(\theta, i)
\end{aligned}
\tag{1}
$$

Where W_{supply} is the energy from the supply and W_{cu} is the copper loss.

Ignoring iron losses, the field energy ($W_{field}(\Psi, \theta)$) is given by:

$$
W_{field}(\Psi, \theta) = W_e(\theta, i) - W_{em}(\theta) - W_{cog}(\theta)
\tag{2}
$$

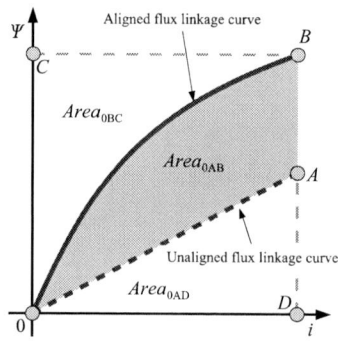

Fig. 1. Determination of the electromagnetic torque for a switched reluctance motor.

W_{em} is the energy delivered to the mechanical subsystem of the machine from the magnetic subsystem.

W_{cog} is the energy stored in the magnets. If the load of the magnet is changed, ie. by varying the air-gap, the stored energy of the magnet would change, which gives rise to cogging torque (permanent magnet reluctance torque).

Based on the flux linkage vs. current curves shown in Fig. 1, at the minimum and maximum inductance positions (known as aligned an unaligned positions, respectively), the calculation of the average reluctance torque may be carried out as:

$$
\begin{aligned}
\Delta W_{em} &= \Delta W_e - \Delta W_{field} \\
&= Area_{OAB} \\
&= \tau_{em} \cdot \Delta \theta
\end{aligned}
\tag{3}
$$

The electromagnetic torque calculation for a hybrid switched reluctance motor is given based on Fig. 2. For a hybrid SRM, with permanent magnet field present at the unaligned position, the flux linkage vs. current curve at the unaligned position will then have an offset flux linkage value when the current is zero. This is shown in Fig. 2. Similarly, the average torque may be calculated by:

$$
\begin{aligned}
\Delta W_{em} &= \Delta W_e - \Delta W_{field} \\
&= Area_{DABC} \\
&= \tau_{em} \cdot \Delta \theta
\end{aligned}
\tag{4}
$$

978-1-4244-4782-4/10 $26.00 © 2010 IEEE

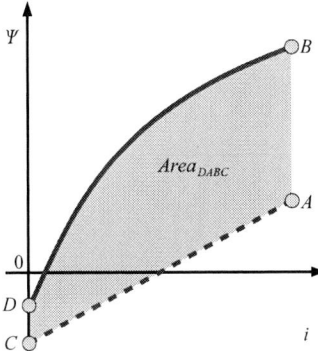

Fig. 2. Determination of the electromagnetic torque for a hybrid switched reluctance motor. Notice that both the unaligned and aligned flux linkage may not start at 0 due to flux from the permanent magnets.

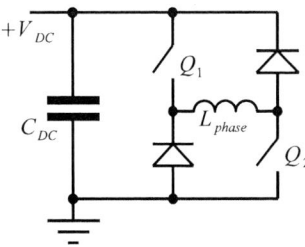

Fig. 3. Inverter for single phase HSRM. L_{phase} represents the phase inductance. The DC-link capacitor is used to store magnetic energy temporarily, when the machine is defluxed.

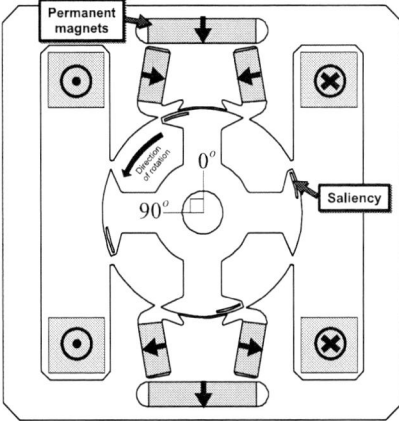

Fig. 4. Proposed motor, here with rotor shown at an angle of 0 degrees. Circles enclosing either dots or crosses indicate the direction the coil current. The two coils are connected in series. Due to the saliency in the rotor the normal parking position of the rotor is normally at 1-5 degrees. Manufacturing sets a lower limit on how thin the cutouts in the laminations can be. This limitation means that the saliency is not thin enough to generate a flat top torque when the coil is energized. Thick black arrows indicate magnetization direction of the permanent magnets. Ferrite magnets were used in the motor.

Where the energy associated with the cogging torque, W_{cog}, is not involved because the average cogging torque value is always zero.

As it may be observed from Fig.2, with permanent magnet flux present, the unaligned flux-linkage-current curve has been moved toward the negative direction of the flux linkage axis. The aligned curve is also shifted downwards, but not as much as for the unaligned curve. The total enclosed area by these two flux-linkage-current curves, denoted as $Area_{DABC}$, is equivalent to the energy used for torque production, which has gained an increase due to the additional permanent magnet field.

III. PROPOSED MOTOR DRIVE

The proposed motor drive has a single phase asymmetrical inverter as it is presented in Fig 3. The motor structure is shown in Fig. 4. The machine stator has two permanent magnet poles (PM-poles) and four reluctance poles. The rotor has four poles with a saturable saliency. This ensures that the rotor is parked by the permanent magnets at a position that is able to self start. When the two series connected coils are energized, a flux is generated that opposes the flux produced by the permanent magnets and thus attracts the rotor poles to the reluctance poles. When the coil is defluxed, positive cogging torque produced by the permanent magnets will be present and drag the rotor to rotate forward, until the parking position is reached again. The motor has a square frame that allows square bobbins to be inserted before the rotor is inserted, simplifying assembly. To ensure that the machine only operates in the desired direction, the rotor has to be properly parked as discussed below.

A. Poleshaping to Improve Starting Torque

In a pump drive, the rotor may be parked in the same position for a long time, therefore due to the environment, the stationary friction may be high. It is thus important that the machine has sufficient starting torque to overcome the stationary friction. The saliency in the rotor poles makes the rotor park at an angle of 1-5 degrees depending on the static

friction. To improve the starting torque an pole extension for two of the reluctance poles is used as it is shown in Fig. 5.

This helps shift the torque curve along the negative rotational direction and the torque production at near zero position is significantly improved. This may be observed from Fig. 6 and Fig. 7. Results are obtained from FEA. For the prototype machine, by pole shaping, the torque at 0 degrees is increased from less than 0.1Nm to 0.25Nm. In total this improves the starting torque in the region from - 0.5 to 3.5 degrees. The pole elongation has little impact on cogging torque and negligable effect on the output power. The elongation is evaluated using finite element analysis, and the elongated pole is implemented in the prototype.

B. Impact of Flux Concentration Permanent Magnet Arrengement on the Torque Production

The permanent magnet flux plays an important role in increasing the positive part of the cogging torque and in minimization of the torque ripple. Three different permanent

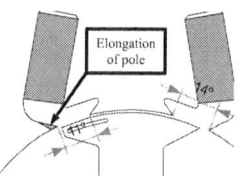

Fig. 5. Two of the four reluctance poles are elongated, so the stator poles are no longer axially symmetric.

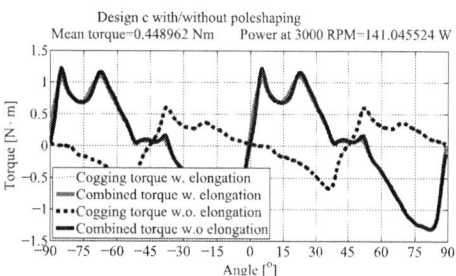

Fig. 6. The torque production from the cogging (phase current=0A) and the combined torque (phase current=4A), with and without the pole elongation.

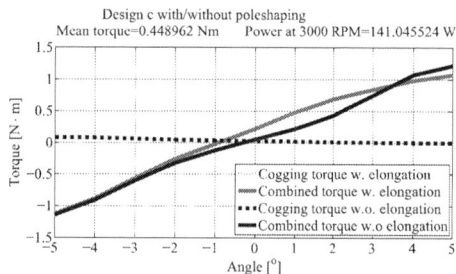

Fig. 7. Zoom of Fig. 6 in the region near the parking region of 1-5 degrees. The starting torque is improved in the region from -0.5 to 3.5 degrees. The cogging torque is so similar that the curves for the cogging torque overlaps eachother.

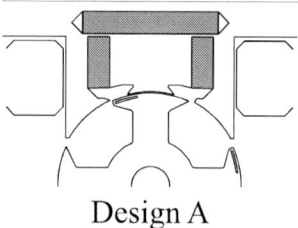

Design A

Fig. 8. Design A for a flux concentration arrangement using permanent magnets. Pole elongation is applied this design.

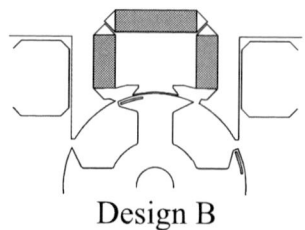

Design B

Fig. 9. Design B for a flux concentration arrangement using permanent magnets. Pole elongation is applied this design.

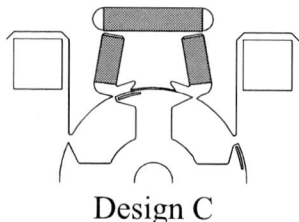

Design C

Fig. 10. Design C for a flux concentration arrangement using permanent magnets. Pole elongation is applied this design.

magnet arrangements for achieving a flux concentration design were compared. Designs for the three presented permanent magnet arrangements are shown in Fig. 8, Fig. 9 and Fig. 10.

The two flux-linkage-current curves at the minimum and maximum inductance positions, and instantaneous torque obtained from FEA, are shown in Fig. 9, Fig. 10, and Fig. 11 respectively. Using the instantaneous torque obtained, the averaged torque was calculated. The most interesting performance of these designs is the output power for a given input current. To have a clear comparison, the average torque obtained was multiplied by the rated speed of the motor to calculate is output power. The average torque and output power obtained were indicated in the titles of the Fig. 11-13-(b). Since the flux from the coil interacts with the flux from the permanent magnets, the saturation knee-point is also influenced by the flux concentration arrangement, and this is demonstrated in Fig. 11-13-(a).

For the design A, referring to the FEA simulation results shown in Fig. 11, it has lower output power than the other two designs (100 W vs. 140W.). Compared to the other designs, this design uses the most PM material, and in a non-efficient way.

Design B has almost the same output power as the Design C. But, Design B saturates at a lower current value compared to Design C. Design B has a greater PM flux linkage offset located on the flux linkage axis than that for Design C, as may be observed from Fig. 9 and 10. But due to the earlier saturation in Design B, the output power is not improved compared to Design C. Since the motor is to be used as a variable speed drive, it needs to operate at different current levels. Lower injected current means that the negative torque will be significant during turn-off of the phase current. Negative

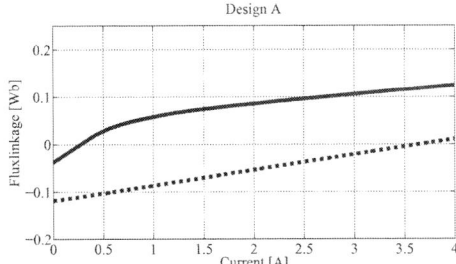
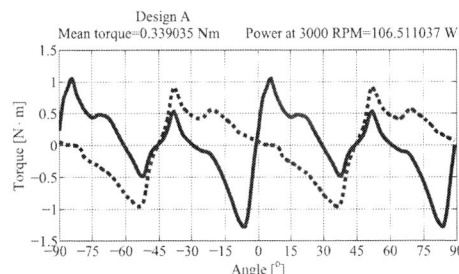

(a) flux linkage for unaligned and aligend rotor position. The full line indicates flux linkage when the rotor is aligned with coil reluctance poles, and the dotted line indicate flux linkage when the rotor is unaligned with the coil reluctance poles.

(b) The torque production from the cogging (phase current=0A, dotted line) and the combined torque (phase current=4A, full line), for design A

Fig. 11.   Flux linkage and torque profile for the design A from FEA

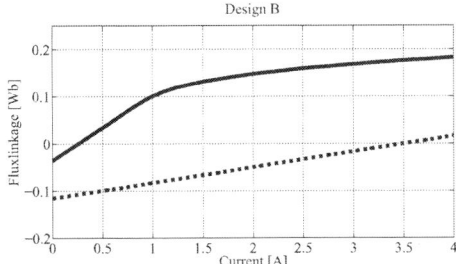
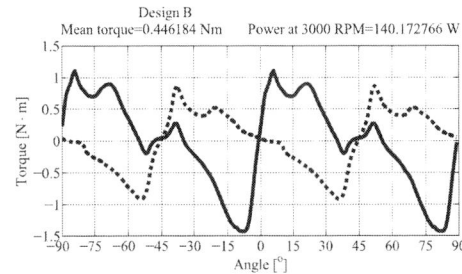

(a) flux linkage for unaligned and aligend rotor position. The full line indicates flux linkage when the rotor is aligned with coil reluctance poles, and the dotted line indicate flux linkage when the rotor is unaligned with the coil reluctance poles.

(b) The torque production from the cogging (phase current=0A, dotted line) and the combined torque (phase current=4A, full line), for design B

Fig. 12.   Flux linkage and torque profile for the design B from FEA

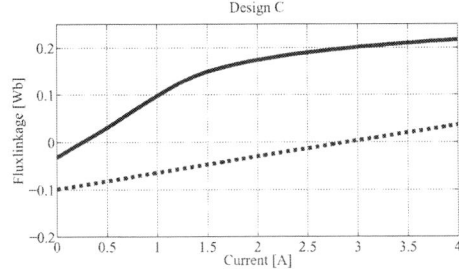
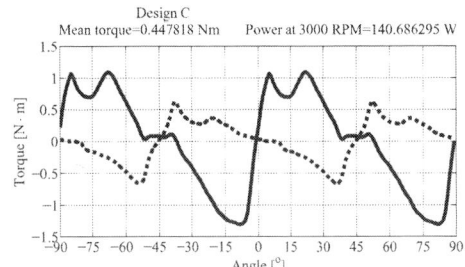

(a) flux linkage for unaligned and aligend rotor position. The full line indicates flux linkage when the rotor is aligned with coil reluctance poles, and the dotted line indicate flux linkage when the rotor is unaligned with the coil reluctance poles.

(b) The torque production from the cogging (phase current=0A, dotted line) and the combined torque (phase current=4A, full line), for design C

Fig. 13.   Flux linkage and torque profile for the design C from FEA

978-1-4244-4782-4/10 $26.00 © 2010 IEEE

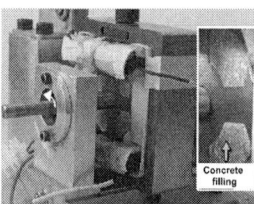

Fig. 14. The prototype motor in its test frame. The rotor is here parked at an angle of approximately 4 degrees.

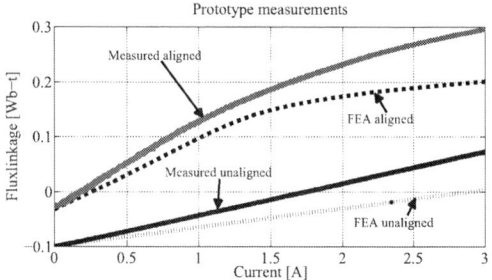

Fig. 15. Measured flux linkage up to 3A for the prototype motor. The current was limited by AC-source. End effects not in the FEA model is the main reason for the differences.

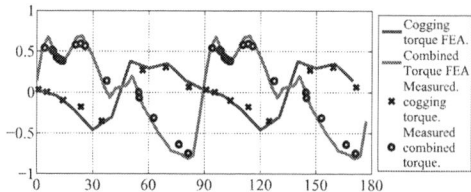

Fig. 16. The figure shows the measured static torque and from FEA predicted static torque for a phase current of 0 A and 3 A. The maximum current is limited by the DC-supply.

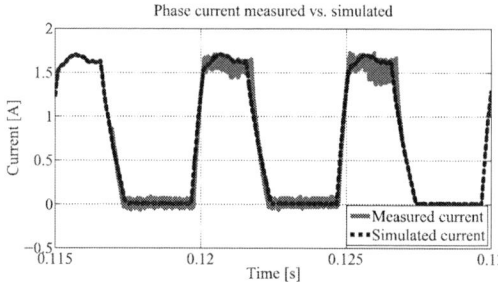

Fig. 17. Measured phase current compared with simulated current from a dynamical non-linear model of the drive. The measured motor uses a non linear time variant speed control that handles the torque ripple and still achieve accurate speed control. The speed control method is described in [5].

torque is generated if the current is turned of too quick, or if the current level is not sufficient for a given angle to generate negative torque. For design B, there is negative torque present close to the turn off point even at the nominal current.

Design C can theoretically produce positive torque for an entire stroke and has a higher saturation knee point. Part of the reason why design C has a higher saturation point is because of increased area for the flux from the coils. The increased area decreases the flux density and thus moves the saturation knee point slightly upwards. Design C is implemented in the prototype.

IV. MEASUREMENT RESULTS ON THE PROTOTYPE MOTOR

Design C with pole elongation was manufactured and tested. The prototype motor can be seen in Fig. 14.

The flux linkage measurement result for the prototype is shown in Fig. 15. As may be observed from Fig. 15, measured flux linkage is higher than the flux linkage given by FEA. This is due to the leakage reluctance (e.g. due to the end-effects), that cannot be modelled by 2D-FEA, but is present in the measurements. An approximately constant leakage inductance is used to model this difference in the motor model used for drive system performance simulation.

The torque measurement in Fig. 16 show an acceptable agreement with the predicted torque, though it seems to be slightly lower than predicted. The torque production may be affected by the flux from the permanent magnets and the six pieces of magnets may have different levels of magnetization, causing the difference between FEA and measurement.

Measured efficiency for the motor and its drive electronics is 74%, where the motor is 79.5% efficient at an output

power of 32 W. The test was performed with digital control on a DSP, with a current loop at 50 kHz. The constant switching frequency was constant at 50 kHz. The measured and simulated phase current is compared in Fig. 17.

V. CONCLUSION

A single phase hybrid switched reluctance motor was presented with a novel stator pole shaping method to increase its starting torque and a new arrangement of permanent magnets for flux concentration. It was shown that stator pole shaping may effectively improve the starting torque for a single phase hybrid switched reluctance motor. It was demonstrated that the configuration of permanent magnets in different flux concentration arrangements influences the shape and magnitude of the torque. Saturation plays a crucial part in determining the output torque profile, and needs to be considered in the design of hybrid switched reluctance motors.

REFERENCES

[1] Roland Sudler and Jean-Francois Schwab, *German patent de2707684: ein phasen schrittmotor*, 1978.

[2] Vilmos Trk, *Swedish patent se467852: Elektrisk motor*, 1992.

[3] Walter Wissmach, *Single-Phase Switched Reluctance Motors - Design and Application*, PhD thesis, cole PolyTechnique Fdrale de Lausanne, 2003.

[4] K.Y. Lu, P.O. Rasmussen, S.J. Watkins, and F. Blaabjerg, *A New Low-Cost Hybrid Switched Reluctance Motor for Adjustable-Speed Pump Applications*, 2006 IEEE Industry Applications Conference, vol. 2, pp. 849-854, 2006.

[5] Uffe Jakobsen and Jin Woo Ahn, *Non Linear, Time Variant Speed Control of a Single Phase Hybrid Switched Reluctance Motor*, 31st IEEE Intelect Conference, 2009.

Parameter Independent Maximum Torque per Ampere (MTPA) Control of IPM Machine Based on Signal Injection

Sungmin Kim, Young-Doo Yoon and Seung-Ki Sul
School of Electric Engineering & Computer Sciences
Seoul National University
Seoul, Korea
ksmin@eepel.snu.ac.kr, sulsk@plaza.snu.ac.kr

Kozo Ide, and Koji Tomita
Motion Control R&D Group
Yaskawa Electric Corp.
Kitakyushu, Japan
kozo@yaskawa.co.jp, tomita@yaskawa.co.jp

Abstract— **This paper presents a new maximum torque per ampere (MTPA) control method for Interior Permanent Magnet Synchronous Machine (IPMSM) drives. The proposed method uses the conventional speed control scheme, where speed control loop produces the magnitude of the stator current reference. According to the current angle in the rotor reference frame, θ, the current reference is decomposed to the d-/q-axis current references. To operate IPMSM in the MTPA mode, this paper presents the new MTPA tracking method using signal injection. This method works based on the inherent definition of MTPA, which is that the torque variation due to the current angle variation should be zero at the specific torque. The proposed method detects the accurate current angle where the magnitude of the stator current is the minimum at the specific torque without any pre-made look-up tables and machine parameters.**

I.   INTRODUCTION

Interior Permanent Magnet Synchronous Machines (IPMSM) have gained an increasing popularity in recent years for a variety of industrial applications, because of their high power density, high efficiency and possibility of flux weakening operation [1]. The efficiency of IPMSM is one of the performance characteristic which is quite important for machine drives especially powered by the battery [2]. Therefore, there have been many researches to improve the efficiency of IPMSM operation. In those researches, the Maximum Torque Per Ampere (MTPA) operation has been considered as a basic method to control the IPMSM in highly efficient operating mode [2].

Because IPMSM's permanent magnets are buried inside the rotor core, the q-axis inductance in the rotor reference frame is conspicuously larger than the d-axis inductance. This difference between the q-axis inductance and d-axis inductance contribute to the additional torque production so called as the reluctance torque. Even if the magnitude of the stator current is fixed, the produced torque varies according to the input current vector (exactly, the current angle in the rotor reference frame current plane). Under the specific torque, if the less current flows, the copper loss is getting less and the operating efficiency is getting higher. Therefore, to operate

IPMSM in high efficiency, the controller should find the MTPA operating point.

As yet, many research have been reported to decrease the copper loss to improve the efficiency. Most of them have found the MTPA operating point using the machine parameter [2], [3]. However, these methods were vulnerable to parameter variations regardless of the huge efforts to compensate parameter variation by using on-line estimation techniques [4]-[6] or pre-made look-up tables [7]-[9]. Meanwhile, another methods based on seeking of the minimum of the current magnitude at the specific torque without machine parameter have been reported [10]-[13]. However, the performance of these methods was restricted according to the speed variation, load torque variation and other operating conditions.

In this paper, a totally new method is proposed for the MTPA operation. This method works based on the inherent characteristics of MTPA, that the torque variation according to the current angle variation should be zero on the MTPA operating point. This method is robust to parameter variation and does not use any pre-made look-up tables. Hence it is easy to apply to any off-the-shelf IPMSM.

II.   PROPOSED MTPA TRACKING METHOD USING SIGNAL INJECTION

A.   Basic Principle

In the rotor reference frame current plane of IPMSM, there are many current reference pairs (i_{ds}^{r*}, i_{qs}^{r*}) to generate the specific torque. To operate the IPMSM in the MTPA, the current reference whose magnitude is the minimum at the specific torque should be found. Fig. 1 shows the constant torque locus in the rotor reference frame current plane. Torque of IPMSM can be derived as

$$T_e = \frac{3}{2}\frac{P}{2}\left\{\lambda_f i_{qs}^r + \left(L_{ds} - L_{qs}\right)i_{ds}^r i_{qs}^r\right\} \qquad (1)$$

where P is the number of pole of IPMSM, λ_f is the permanent magnet flux linkage, L_{ds} and L_{qs} are d-axis and q-axis inductance, respectively. The MTPA operating point is the nearest point to the origin in the current plane. Therefore, the differentiation of the torque with respect to the angle of the current in the rotor reference frame, $\partial T_e / \partial \theta$, should be zero at the MTPA point as

$$\frac{\partial T_e}{\partial \theta} = \frac{3P}{4} I_S \left\{ \left(L_{ds} - L_{qs} \right) I_S \left(2\cos^2 \theta - 1 \right) + \lambda_f \cos \theta \right\} = 0 \quad (2)$$

where I_S is the magnitude of the input current and θ is the angle of the current in the rotor reference frame. This is the MTPA criterion equation of IPMSM. If the value of $\partial T_e / \partial \theta$ is evaluated, the control system can recognize whether the present operating point is on the MTPA operating point or not even though the torque of IPMSM is not exactly measured.

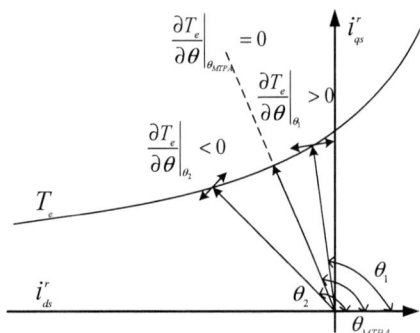

Figure 1.Locus of the constant torque in the current vector plane

B. *Evaluation of $\partial T_e / \partial \theta$*

Because it is very difficult to calculate or measure the torque in the industrial field, the proposed MTPA tracking method evaluates $\partial T_e / \partial \theta$ using the concept of a signal injection. The proposed method injects a high frequency small signal to the current reference angle, θ, as

$$\theta = \theta_{avg} + \theta_h = \theta_{avg} + A_{mag} \sin\left(f_h \times 2\pi t \right) \quad (3)$$

where A_{mag} and f_h are the magnitude and the frequency of the injected signal, respectively. The frequency of the injected signal, f_h should be large enough compared to the bandwidth of the speed control loop or torque control loop, and the speed control loop or the torque control loop does not respond to the injected signal. On the other hand, the frequency should be small enough compared to the inverter switching frequency to modulate the injected signal. The magnitude of the signal should be small enough not to result in the speed variation.

By injecting the signal to the current angle, θ, the control system could experience the variation of the mechanical power related to the torque variation due to the angle variation. The calculated instantaneous input power by the terminal voltage and current can be used as an index of the torque variation. The instantaneous input power includes not only the mechanical power but also the losses due to the stator resistance and the reactive power related with the inductance of the IPMSM. Therefore, to determine the MTPA operating point, the mechanical power in the measured input power should be picked up. The power due to the injected signal can be analyzed as follows.

From the injected signal, d-/q-axis current can be described as (4) and (5) under the assumption that A_{mag} is much smaller than θ_{avg}.

$$\begin{aligned}
i_{ds}^r &= I_S \cos(\theta_{avg} + A_{mag} \sin \omega_h t) \\
&\approx I_S \cos \theta_{avg} - I_S A_{mag} \sin \theta_{avg} \sin \omega_h t = i_{dsf}^r + i_{dsh}^r.
\end{aligned} \quad (4)$$

$$\begin{aligned}
i_{qs}^r &= I_S \sin(\theta_{avg} + A_{mag} \sin \omega_h t) \\
&\approx I_S \sin \theta_{avg} + I_S A_{mag} \cos \theta_{avg} \sin \omega_h t = i_{qsf}^r + i_{qsh}^r.
\end{aligned} \quad (5)$$

The electric input power can be calculated as (6) which consists of copper loss, reactive power, and mechanical power. Considering the high frequency current signal, the copper loss, P_{copper}, reactive power, $P_{reactive}$, and mechanical power, P_{mech} can be expressed as (6)-(9). The copper loss due to the stator resistance is not affected by the injected signal as in (7), because the magnitude of the input current is constant regardless of the injected signal. However, the reactive power, (8) and the mechanical power, (9) contain the power variation whose frequency is identical to that of the injected signal. However, the phase of the injected signal frequency component in the reactive power is out of the phase to that of the injected signal itself. While, in the mechanical power in (9), the phase of the injected signal frequency component is in phase to that of the injected signal, and further more the component in the mechanical power is proportional to the MTPA criterion equation in (2).

$$\begin{aligned}
P_e &= P_{copper} + P_{reactive} + P_{mech} \\
&= \frac{3}{2} \left\{ \begin{array}{l} R_S \left(i_{ds}^{r2} + i_{qs}^{r2} \right) + L_{ds} \dfrac{di_{ds}^r}{dt} i_{ds}^r + L_{qs} \dfrac{di_{qs}^r}{dt} i_{qs}^r \\ + \omega_r \lambda_f i_{qs}^r + \omega_r \left(L_{ds} - L_{qs} \right) i_{ds}^r i_{qs}^r \end{array} \right\}.
\end{aligned} \quad (6)$$

$$P_{copper} = R_S I_S^2. \quad (7)$$

$$\begin{aligned}
P_{reactive} &\approx -\frac{1}{2} \left(L_{ds} - L_{qs} \right) I_S^2 A_{mag} \omega_h \sin 2\theta_{avg} \cos \omega_h t \\
&\quad + \frac{1}{2} \left(L_{ds} \sin^2 \theta_{avg} + L_{qs} \cos^2 \theta_{avg} \right) I_S^2 A_{mag}^2 \omega_h \sin 2\omega_h t.
\end{aligned} \quad (8)$$

$$P_{mech} \approx \omega_r \left\{ \frac{1}{2} \left(L_{ds} - L_{qs} \right) I_S^2 \sin 2\theta_{avg} + \lambda_f I_S \sin \theta_{avg} \right\}$$
$$+ \left\{ \left(L_{ds} - L_{qs} \right) I_S^2 \cos 2\theta_{avg} + \lambda_f I_S \cos \theta_{avg} \right\} \omega_r A_{mag} \sin \omega_h t \quad (9)$$
$$+ \frac{1}{4} \omega_r \left(L_{ds} - L_{qs} \right) I_S^2 A_{mag}^2 \sin 2\theta_{avg} \cos 2\omega_h t.$$

To evaluate the MTPA criterion, that is the differentiation of the torque with respect to the angle of the current in the rotor reference frame, $\partial T_e / \partial \theta$, a simple signal processing block can be used as in Fig. 2.

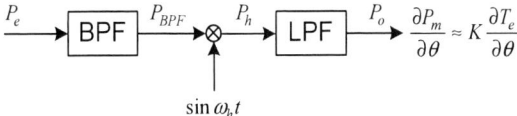

Figure 2. Signal process to extract the MTPA criterion

To extract the component of the injected frequency, f_h, from the measured power, the power is processed with the band-pass-filter whose center frequency, f_c, is the injected frequency. Using the band-pass-filter, the component of the injected frequency, P_{BPF}, in the electric input power can be extract. The result includes not only the component from the mechanical power but also the component from the reactive power. The component from the reactive power is orthogonal to the injected signal as seen from (8) and (9). Hence, to extract the component only from the mechanical power, the output of the hand-pass-filter can be multiplied with the injected sinusoidal signal. After then, the results should be filtered out by a low-pass-filter whose cut-off frequency is far below from the injected frequency, f_h, to cut off the orthogonal term from the results. Through this multiplication and low-pass-filtering, the component proportional to $\partial T_e / \partial \theta$ can be obtained as P_o described in (10)

$$P_0 = \frac{3}{4} \omega_r A_{mag} I_S \left\{ \begin{array}{l} \left(L_{ds} - L_{qs} \right) I_S \cos 2\theta_{avg} \\ + \lambda_f \cos \theta_{avg} \end{array} \right\} \propto \frac{\partial T_e}{\partial \theta}. \quad (10)$$

The criterion in (2), the differentiation of the torque with respect to the current reference angle is identical to P_o in (10) except proportional constant. Therefore, if the output of the signal processing is controlled as null, the MTPA operation is always guaranteed regardless of the variations of parameters of IPMSM.

From Fig. 3 to Fig. 5, the relationship between the current reference angles, the signal during the signal processing procedure, and the actual torque are demonstrated. The current reference angle, θ, varies from 1.6rad to 2.3rad in Fig. 3 under the condition of the constant magnitude of the stator current and the constant speed. The output signal from the band-pass-filter, P_{BPF}, the signal multiplied by the injected

sinusoidal signal, P_h, and the output signal, P_o, of the low-pass-filter are shown in Fig. 4. And in Fig. 5, it is shown that the torque varies according to the current reference angle as shown in Fig. 3. In spite of the constant magnitude of the stator current, the torque varies according to the current reference angle. The maximum torque is 66Nm at $\theta = 2.0$rad.

As mentioned previously, the output of the signal processing procedure, P_o, is near zero around the MTPA operating point, 2.0rad. The result, P_o, of the signal processing is positive whenever the current reference angle is below 2.0rad. In the other hand, P_o is negative whenever the current reference angle is above the angle of the MTPA operating point, 2.0rad.

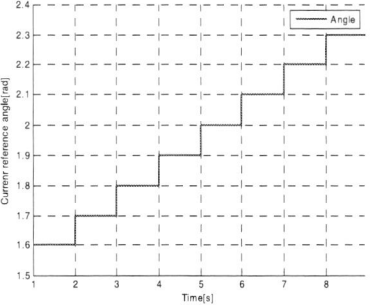

Figure 3. Variation of current reference angle, θ

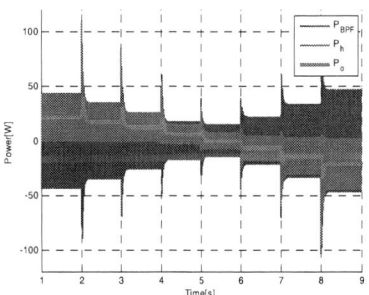

Figure 4. Variation of the signals in the signal processing according to the current reference angle variation in Fig. 3.

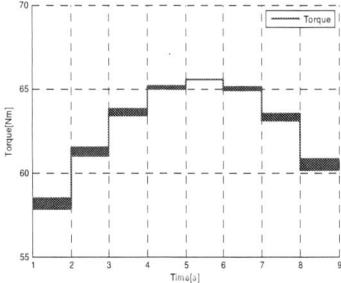

Figure 5. Torque variation according to the current angle variation in Fig. 3

C. Understanding of Signal Injection Method

The argument in the section B and C can be understood by Taylor series expansion. The MTPA tracking method using signal injection has its ground in the MTPA definition; the torque variation according to the current angle variation is zero at the MTPA operating point of the IPMSM. To use this definition, the torque variation has to be detected in the system. Measuring the torque, however, is very difficult and is almost impossible in the real industrial field. Therefore, a new variable is necessary to detect the torque variation, and the electric power variation can be substituted to the torque variation in this method.

$$P_m = T_e \omega_{rm}, \quad T_e(\theta) = \frac{1}{\omega_{rm}} P_m(\theta). \tag{11}$$

Under constant speed, if the torque varies, the mechanical power varies proportionally to the torque as given by (11). To get the information about the torque variation according to the current angle variation, the high frequency sinusoidal signal is injected into the current reference in the rotor reference frame. As for the injected signal, the torque can be derived as (12)-(13) by Taylor series expansion.

$$T_e(\theta + \Delta\theta) = T_e(\theta) + \frac{\partial T_e}{\partial \theta}\Delta\theta + \frac{\partial}{\partial \theta}\left(\frac{\partial T_e}{\partial \theta}\right)\Delta\theta^2 + \cdots. \tag{12}$$

$$
\begin{aligned}
T_e(\theta + A\sin\omega_h t) = {}& T_e(\theta) + \frac{\partial T_e}{\partial \theta} A\sin\omega_h t \\
& + \frac{\partial}{\partial \theta}\left(\frac{\partial T_e}{\partial \theta}\right) A^2 \sin^2\omega_h t + \cdots.
\end{aligned} \tag{13}
$$

Because the magnitude of the injected signal is very small and the frequency of the injected signal is high enough compared to the bandwidth of the speed regulation loop, the speed variation due to the injected signal can be neglected. Therefore, the torque would be proportional to the mechanical power as (11). And the higher order terms including the second order term in (13) can be neglected. Through the signal processing, the electric power can be translated to the torque variation according to the current angle variation as (14).

$$
\begin{aligned}
& \text{Signal Processing}\{P_e\} \\
& \propto \text{Signal Processing}\left\{T_e(\theta) + \frac{\partial T_e}{\partial \theta} A\sin\omega_h t\right\} \\
& = \frac{\partial T_e}{\partial \theta} A \propto \frac{\partial T_e}{\partial \theta}.
\end{aligned} \tag{14}
$$

D. High Frequency Current Injection

Because the proposed method relies on the power variation value at the injected signal frequency, the power variation should be happen only from the injected signal. However, the input power can vary by many reasons such as the speed reference change, torque disturbance, load variation, and so on. Therefore, if the signal frequency of the injected signal is high enough, the power variation due to the injected signal can be differentiated from that due to the other reasons. In this paper, 300Hz is used as the frequency of the injected signal. To inject

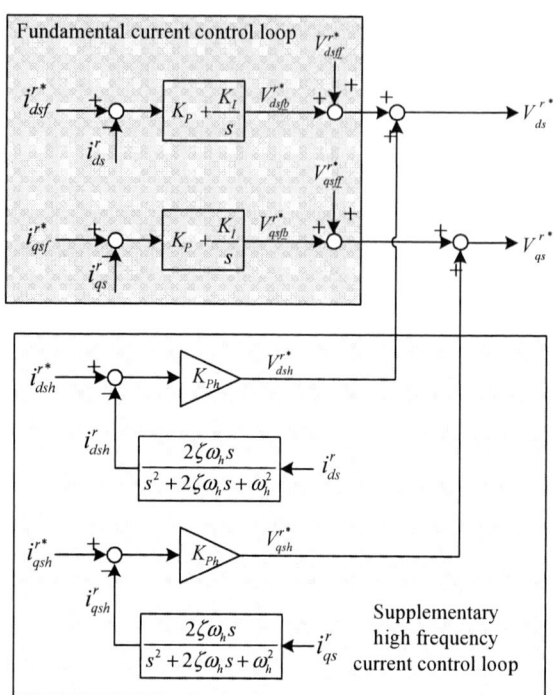

Figure 6. Current control loop

such a high frequency current signal, the supplementary high frequency current control loop is augmented to the basic current control loop as in Fig. 6. The d-/q-axis current references are decomposed to the fundamental component, $i_{dsf}^{r*}, i_{qsf}^{r*}$ and the high frequency component, $i_{dsh}^{r*}, i_{qsh}^{r*}$ as (4) and (5).

As a supplementary high frequency current control loop, the proportional regulator was incorporated as Fig. 7. The Bode plot of this control loop is shown in Fig. 7(b). At injected signal frequency, 300Hz, the amplitude gain was 0dB and the phase delay, 0deg.

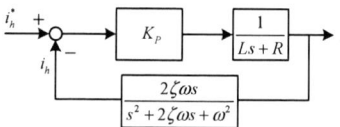

(a) High frequency signal control loop block

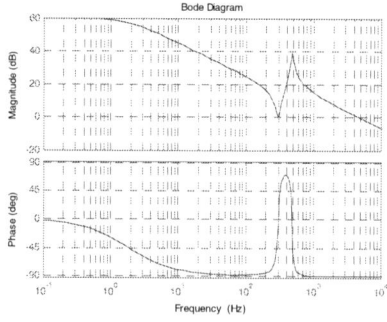

(b) Bode plot of high frequency signal control loop

Figure 7. High frequency signal control loop

978-1-4244-4782-4/10 $26.00 © 2010 IEEE

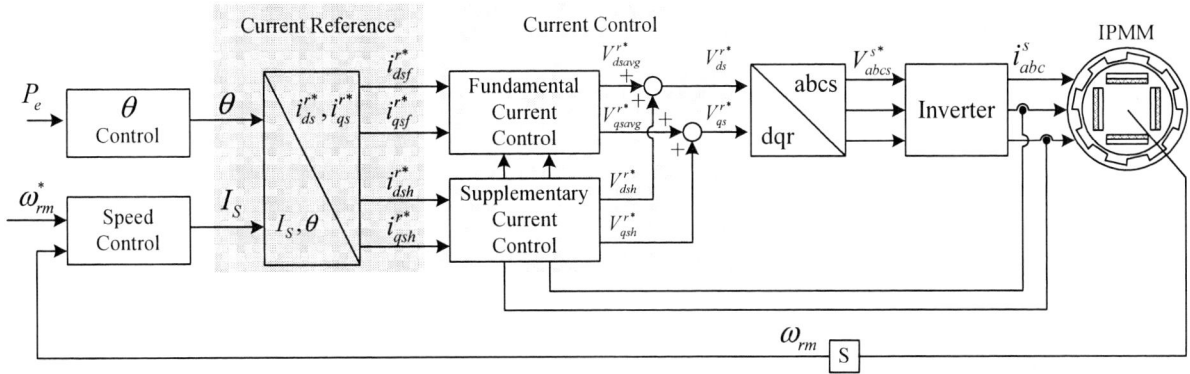

Figure 8. Control system configuration

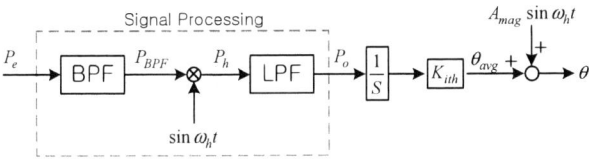

Figure 9. Current reference angle, θ control block

E. Control System Configuration

The speed control system including the proposed tracking method is represented in Fig. 8. The current reference angle, θ, comes from an integral regulator whose block diagram including the signal processing block is shown in Fig. 9. The integral regulator works to nullify the MTPA criterion value, P_o, as zero. The current references including the high frequency injected signal are derived from the current reference block in Fig. 8 according to (4) and (5). After that, the current control block which consists of the fundamental control loop and the supplementary control loop generates the output voltage references as shown in Fig. 6.

III. EXPERIMENTAL RESULTS

To verify the feasibility of the proposed MTPA tracking method, 11kW IPMSM was pre-tested at 900r/min, 1750r/min and at different load torque conditions, from 5Nm to 45Nm by 5Nm. At each torque and speed, the magnitude of the stator current was recorded in the various current reference angle, θ, in order to find MTPA operating point manually. After these pre-tests, the proposed MTPA tracking method was applied to the drive system at each operating point to compare the MTPA operating point by the proposed signal injection method to the point by pre-tests. The results are shown in Fig. 10 and Fig. 11. These MTPA operating points by the proposed method matched quite well to those by the pre-tests. To demonstrate autonomous tracking capability of the proposed method clearly, the current angle is manually adjusted from 0.5πrad to 0.6πrad and back to 0.5πrad at 1700r/min. After that the proposed method was engaged at the same speed at two different load torques, 20Nm and 35Nm, respectively. The results are shown in Fig. 12 and Fig. 13, respectively. In Fig. 12, the high frequency current signal was injected from 10s

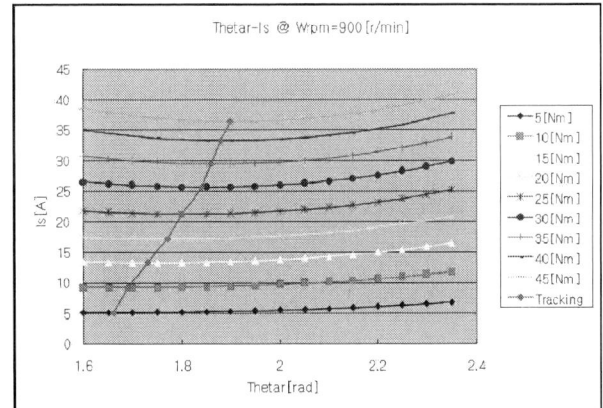

Figure 10. Current angle and magnitude at 900r/min

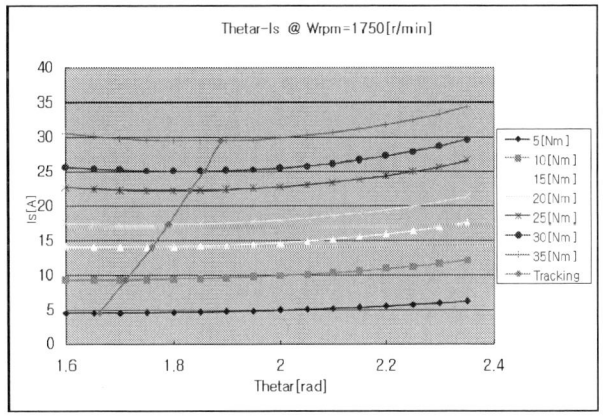

Figure 11. Current angle and magnitude at 1750r/min

and the MTPA operating point was tracked. When the speed was 1700r/min and load torque was 20Nm, the current angle of the manual MTPA operating point was 1.78rad and the tracked MTPA current angle was 1.82 rad. The difference of the current magnitudes is less than 0.05A, which was very small to be compared the magnitude of the stator current at the operating point, 18.8A. In Fig. 13, which is the case of 35Nm load torque, the MTPA current angle of the manual operation and the proposed method was 1.85rad and 1.89rad, respectively. And current magnitude difference between the

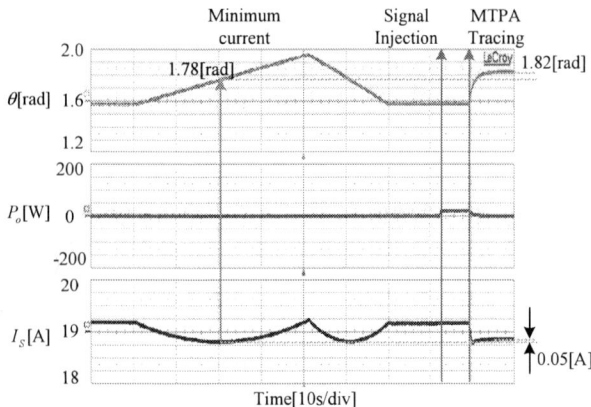

Figure 12. Current angle, MTPA criterion value and current magnitude at 1700r/min and load torque 20Nm

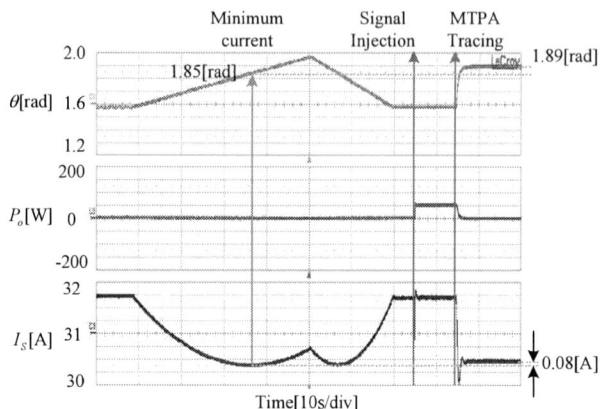

Figure 13. Current angle, MTPA criterion value and current magnitude at 1700r/min and load torque 35N·m

manual operation and the proposed method was less than 0.1A, which is very small again to be compared the magnitude of the stator current, 30.3A.

IV. CONCLUSIONS

The proposed MTPA tracking method exploits the inherent characteristics of the MTPA operating point, which is that the torque variation due to the current angle variation is zero at the MTPA point. Injecting a small high frequency signal into the current reference angle, the torque variation would be detected through the instantaneous input power of IPMSM. The differentiation of torque in terms of the angle of the stator current vector in rotor reference frame can be calculated through the signal processing of the input power. As shown in the experimental results, this proposed method could track the MTPA operating point of IPMSM without any pre-made look-up tables, machine parameters and parameter estimation method. The error bound of tracking is less than 0.5% in terms of the magnitude of the current.

REFERENCES

[1] N. Bianchi, and T. M. Jahns, *Design, Analysis, and Control of Interior PM Synchronous Machines*, Tutorial course notes of IAS2004, 2004.

[2] T. M. Jahns, G. B. Kliman, and T. W. Neumann, "Interior permanent-magnet synchronous motors for adjustable-speed drives," *IEEE Transactions on Industry Applications*, vol. IA-22, no. 4, pp. 738-747, Jul./Aug. 1986.

[3] J.M. Kim, and S.K. Sul, "Speed control of interior permanent magnet synchronous motor drive for the flux weakening operation," *IEEE Transactions on Industry Applications*, vol. 33, no. 1, pp. 43-48, Jan./Feb. 1997.

[4] H.B. Kim, J. Hartwig, and R.D. Lorenz, "Using on-line parameter estimation to improve efficiency of IPM machine drives," " in *IEEE 2002 Power Electronics Specialists Conference*, 2002, pp. 815-820.

[5] Y.I. Mohamed, and T.K. Lee, "Adaptive self-tuning MTPA vector controller for IPMSM drive system," *IEEE Transactions on Energy Conversion*, vol. 21, no. 3, pp. 636-644, Sep. 2006.

[6] P. Niazi, H.A. Toliyat, and A. Goodarzi "Robust maximum torque per ampere (MTPA) control of PM-assisted SynRM for traction application," *IEEE Transactions on Vehicular Technology*, vol. 56, no. 4, pp. 1538-1545, Jul. 2007.

[7] S. Morimoto, K. Hatanaka, Y. Tong, Y. Takeda, and T. Hirasa "High performance servo drive system of salient pole permanent magnet synchronous motor," in *IEEE 1991 Industry Applications Society Annual Meeting*, 1991, pp. 463-468.

[8] S. Morimoto, M. Sanada, and Y. Takeda "Effects and compensation of magnetic saturation in flux-weakening controlled permanent magnet synchronous motor drives," *IEEE Transactions on Industry Applications*, vol. 30, no. 6, pp. 1632-1637, Dec. 1994.

[9] G. Kang, J. Lim, K. Nam, H.B Ihm, and H.G. Kim "A MTPA control scheme for an IPM synchronous motor considering magnet flux variation caused by temperature," in *IEEE 2004 Applied Power Electronics Conference and Exposition*, 2004, pp. 1617-1621.

[10] Z.Q. Zhu, Y.S. Chen, and D. Howe "Online optimal flux-weakening control of permanent-magnet brushless AC drives," *IEEE Transactions on Industry Applications*, vol. 36, no. 6, pp. 1661-1668, Nov./Dec. 2000.

[11] S. Bolognani, L. Sgarbossa, and M. Zordan "Self-tuning of MTPA current vector generation scheme in IPM synchronous motor drives," in *European Conference on Power Electronics and Applications*, 2007, pp. 1-10.

[12] S. Bolognani, R. Petrella, A. Prearo, and L. Sgarbossa "Automatic tracking of MTPA trajectory in IPM motor drives based on AC current injection," in *IEEE 2009 Energy Conversion Congress and Exposition*, 2009, pp. 2340-2346.

[13] D. Anton, Y.K Kim, S.J Lee, and S.T Lee "Robust self-tuning MTPA algorithm for IPMSM drives," in *IEEE 2008 Industrial Elctronics*, 2008, pp. 1355-1360.

Performance Analysis of Three-phase Capacitor Motor in Frequency Control System

Zheng-Feng Ming
Department of Electrical Engineering
Xidian University
Xi'an, China
mingzf@xidian.edu.cn

Guang-Zheng Ni
Department of Electrical Engineering
Zhejiang University
Zhejiang, China

Bing-Zhong Yang
Department of Electrical Engineering
Chongqing University
Chongqing, China

Abstract—Three-phase capacitor motor with main winding and auxiliary winding vertically arranged in space can effectively weaken the harmonic MMF (magnetic motive force), and raise efficiency and power factor. In this paper, performance of new three-phase capacitor motor in frequency control system is analyzed. Through the contrast experiment between the new three-phase capacitor motor and three-phase induction motor, it has an in-depth research on torque characteristic, current characteristic and power factor. And then it shows the superiority of the new three-phase capacitor motor according to experimental data.

I. INTODUCTION

In the early 1980s, when the Wanlass three-phase induction motor was proposed, it claimed that it can effectively weaken the harmonic magnetic potential, and raise the motor efficiency and power factor. So it aroused the academic attention. But, theoretical analysis and experimental verification show that, the Wanlass three-phase induction motor have a little effect on weakening harmonic MMF. Only because of the role of capacitance in motor, it got the power factor improvement [1]. Until the end of the century, improved model compared to the Wanlass three-phase induction motor was proposed [2-5]. The model is just the new three-phase capacitor motor in this paper. The main characteristics of this motor are:

1）By selecting the appropriate phase winding to improve the efficiency of motor.

2）Optimal series capacitance value enable power factor be equal to 1 at rated load operation.

3）Due to different series capacitance values, it can change the motor starting torque value.

According to theoretical analysis and experimental data, the three-phase capacitor motor improves motor performance compared with conventional motor. One of the advantages is the improvement of the efficiency and power factor, the other is the elimination of self-excitation overpressure phenomena and the reduction of motor vibration noise. Therefore, the application of this new motor in the low-speed fan and pump motors will yield social and economic benefits. Base on this,

it is very important to research the operating characteristic of this motor in frequency control system. This paper gives an in-depth research and experimental verification to three-phase capacitor motor in frequency control system.

II. PRINCIPLE OF HARMONIC ELIMINATION IN 3-PHASE CAPACITOR MOTOR

In three-phase capacitor motor, the stator has two sets of symmetrical three-phase windings. Phase winding with series capacitance is called auxiliary winding, expressed by c. The main winding is expressed by m. Two sets of winding have a phase difference of ninety in space and parallel excitation. Winding connection is shown in Fig.1. A_m, B_m, C_m is the three-phase main winding, A_c, B_c, C_c is the three-phase auxiliary winding, air-gap rotating magnetic field is the anti-clockwise direction.

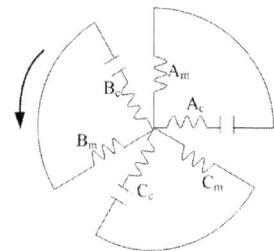

Fig.1 Stator winding connection diagram
of three-phase capacitor motor

A. Analysis of Air-gap MMF Fundamental Wave in Three-phase Capacitor Motor

In Fig.1 phase difference between the main winding and auxiliary winding is 90°. A_m、A_c、B_m、B_c and C_m、C_c each set winding produces fundamental wave expressed as follow（the main winding lags behind the auxiliary winding and the lag angle is ϕ）:

$$f_{A1}(t,\alpha) = F_{\varphi 1}\cos\alpha\sin\omega t + F_{\varphi 1}'\cos(\alpha + \pi/2)\sin(\omega t + \phi) \quad (1)$$

$$f_{B1}(t,\alpha) = F_{\varphi 1}\cos(\alpha - 2\pi/3)\sin(\omega t - 2\pi/3) \\ + F_{\varphi 1}'\cos(\alpha - \pi/6)\sin(\omega t - 2\pi/3 + \phi) \tag{2}$$

$$f_{C1}(t,\alpha) = F_{\varphi 1}\cos(\alpha - 4\pi/3)\sin(\omega t - 4\pi/3) \\ + F_{\varphi 1}'\cos(\alpha - 5\pi/6)\sin(\omega t - 4\pi/3 + \phi) \tag{3}$$

$$f_1(t,\alpha) = f_{A1}(t,\alpha) + f_{B1}(t,\alpha) + f_{C1}(t,\alpha) \\ = 1.5F_{\varphi 1}\sin(\omega t - \alpha) + 1.5F_{\varphi 1}'\sin(\omega t - \alpha + \phi - 2\pi/2) \tag{4}$$

If the turns ratio k and the series capacitance C are rational, the relationship between main winding and auxiliary winding is expressed as follow: (where W_m and W_c represent the number of turns for main winding and the auxiliary winding , K_m and K_c are the main and the auxiliary winding coefficients):

$$\dot{I}_{Ac} = \frac{j\,\dot{I}_{Am}}{k} \quad ; \quad k = \frac{W_c K_c}{W_m K_m} \tag{5}$$

This time the running state of motor is symmetrical and the phase difference ϕ is equal to 90°. Define F_{Am}, F_{Bm}, F_{Cm} as amplitudes of the main winding fundamental wave magnetic motive force; define F_{Ac}, F_{Bc}, F_{Cc} as the amplitudes of the auxiliary winding fundamental wave magnetic motive force.

$$F_{Am} = F_{Bm} = F_{Cm} = 0.9\frac{W_m I_{Am} k_m}{p} \quad ;$$
$$F_{Ac} = F_{Bc} = F_{Cc} = 0.9\frac{W_c I_{Ac} k_c}{p} \tag{6}$$

$$F_{Am} = F_{Bm} = F_{Cm} = F_{\varphi 1} \quad ; \\ F_{Ac} = F_{Bc} = F_{Cc} = F_{\varphi 1}' \quad ; \\ F_{\varphi 1} = F_{\varphi 1}' \tag{7}$$

$$f_1(t,\alpha) = f_{A1}(t,\alpha) + f_{B1}(t,\alpha) + f_{C1}(t,\alpha) \\ = 1.5F_{\varphi 1}\sin(\omega t - \alpha) + 1.5F_{\varphi 1}'\sin(\omega t - \alpha + \phi - 2\pi/2) \\ = 3F_{\varphi 1}\sin(\omega t - \alpha) \tag{8}$$

This shows that at the symmetrical running state, the fundamental wave which is produced by magnetic motive force has a fixed amplitude and forward rotation direction.

B. Analysis of Harmonic Air-gap MMF in Three-phase Capacitor Motor

The expressions of third harmonic air-gap MMF:

$$f_{A3}(t,\alpha) = F_{\varphi 3}\cos 3\alpha \sin \omega t + \\ F_{\varphi 3}'\cos 3(\alpha + \pi/2)\sin(\omega t + \phi) \tag{9}$$

$$f_{B3}(t,\alpha) = F_{\varphi 3}\cos 3(\alpha - 2\pi/3)\sin(\omega t - 2\pi/3) \\ + F_{\varphi 3}'\cos 3(\alpha - \pi/6)\sin(\omega t - 2\pi/3 + \phi) \tag{10}$$

$$f_{C3}(t,\alpha) = F_{\varphi 3}\cos 3(\alpha - 4\pi/3)\sin(\omega t - 4\pi/3) + \\ F_{\varphi 3}'\cos 3(\alpha - 5\pi/6)\sin(\omega t - 4\pi/3 + \phi) \tag{11}$$

$$f_3(t,\alpha) = f_{A3}(t,\alpha) + f_{B3}(t,\alpha) + f_{C3}(t,\alpha) = 0 \tag{12}$$

So the three-phase capacitor motor similar to common induction motor can completely eliminate the third harmonic. the fifth and seventh harmonic can be expressed:

$$f_5(t,\alpha) = f_{A5}(t,\alpha) + f_{B5}(t,\alpha) + f_{C5}(t,\alpha) = \\ 1.5F_{\varphi 5}\sin(\omega t + 5\alpha) + 1.5F_{\varphi 5}'\sin(\omega t + 5\alpha + \phi + \pi/2) \tag{13}$$

$$f_7(t,\alpha) = f_{A7}(t,\alpha) + f_{B7}(t,\alpha) + f_{C7}(t,\alpha) = \\ 1.5F_{\varphi 7}\sin(\omega t - 7\alpha) + 1.5F_{\varphi 7}'\sin(\omega t - 7\alpha + \phi + \pi/2) \tag{14}$$

Because of the equation (7), the expressions show as follow:

$$f_5(t,\alpha) = f_{A5}(t,\alpha) + f_{B5}(t,\alpha) + f_{C5}(t,\alpha) = 0 \\ f_7(t,\alpha) = f_{A7}(t,\alpha) + f_{B7}(t,\alpha) + f_{C7}(t,\alpha) = 0 \tag{15}$$

Thus three-phase capacitor motor can eliminate fifth 、 seventh harmonic and effectively improve torque and raise the efficiency of motor.

III. PERFORMANCE ANALYSIS OF 3-PHASE CAPACITOR MOTOR IN FREQUENCY CONTROL SYSTEM

Three-phase capacitor motor power and torque equation can be described as follows:

$$P_{mec} = \frac{m I_2^2 r_2 (1-s)}{s} = I_f^2 R_f (1-s) \tag{16}$$

$$P_2 = P_{mec} - p_{fw} - p_s \tag{17}$$

$$P_1 = P_{mec} + p_{Cu1} + p_{Cu2} + p_{Fe1} \tag{18}$$

$$\eta = \frac{P_2}{P_1} * 100\% \tag{19}$$

$$\cos\varphi = \frac{P_1}{3U_N I_L} \tag{20}$$

P_{mec} is mechanical power, P_2 is output power, P_1 is input power, η is efficiency, φ is power factor. UN is rated phase voltage, I_L is line current. By the analysis above, friction torque produces mechanical loss power (P_{fw}). In addition, for the effect of stator and rotor slotting and harmonic MMF, there is an additional power loss (Ps). Thus mechanical power minus mechanical loss power (P_{fw}) and additional power loss (P_s) equals the output power (P_2). The expression above divided by mechanical angular velocity Ω on both sides, it obtains the equation $T_2=T_{em}-T_{fw}-T_s$. T_{em} is the electromagnetic torque. The turns ratio k and the series capacitance C should meet the conditions:

$$k = \left(3 x_1 + 6 X_f\right)\Big/\left(r_1 + 6 R_f\right)$$
$$x_c = \left(1 + k^2\right)\left(3 x_1 + 6 X_f\right) \tag{21}$$

Thus frequency is the important factor for turns ratio (k) and series capacitance (C).

A. Experiment and Results Analysis Common Three-phase Asynchronous Motor and Three-phase Capacitor Motor with Electric Power Supply(50HZ)

Fig.2 is experimental results of power factor, stator current and efficiency with electric power supply. Two motors have the same rated voltage (380V) and rated power (5.5KW) when frequency is 50HZ. Experimental results show that choosing suitable turns ratio (k) and capacitance value, the three-phase capacitor motor gets a significant improvement compared to common asynchronous motor : it raises power factor and when the motor runs with rated load the power factor is close to 1; the stator current and efficiency also get improvement.

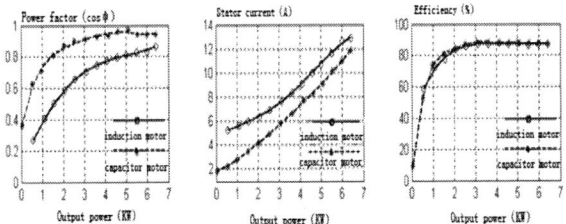

Fig.2 Experiment results of common three-phase asynchronous motor and new three-phase capacitor motor with electric power supply (50HZ)

B. Experiment and Results Analysis of Common Three-phase Asynchronous Motor and New Three-phase Capacitor Motor with Inverter Supply

The inverter is produced by ROCKWELL, the rated voltage is 380V,rated power is 7.5KW. Power factor, stator current and efficiency are showed in Fig.3 when frequency is 50HZ and Fig.4 when frequency is 40HZ.

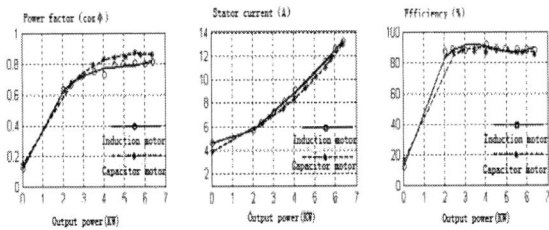

Fig.3 Experiment results of common three-phase asynchronous motor and new three-phase capacitor motor with inverter supply (50HZ)

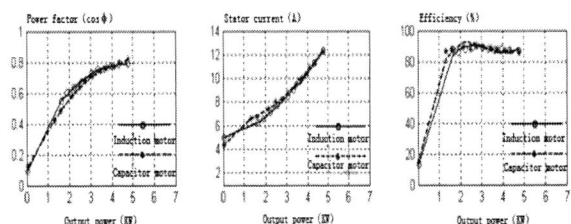

Fig.4 Experiment results of common three-phase asynchronous motor and new three-phase capacitor motor with inverter supply (40HZ)

CONCLUSION

In this paper, air-gap magnetic motive force is analyzed in three-phase capacitor motor and it draws a conclusion that the fundamental wave produced by magnetic motive force has a fixed amplitude and a forward rotation direction at the symmetrical running state ; the paper analyzes principle of fifth and seventh harmonic elimination in three-phase capacitor motor. So the new motor reduces the additional power loss and improves the efficiency. Finally, it gives experimental research to verify the theoretical analysis in frequency control system.

REFERENCES

[1] Stephem D. Umans Herbert L. Hess， Modeling and Analysis of The Wanlass Three-Phase Induction Motor Configuration . IEEE Trans on Power Apparatus and Systems, vol.-PAS-102, No. 9, 1983

[2] Xiong Suming ， Yang Bingzhong and Ni Guangzheng. Starting Performance Analysis of Three-phase Capacitor Motor for Beam Pumping Unit. Trans on ICEMS 2008. pp.3489-3492.

[3] Jimoh Adisa A. Nicolae Dan V. A study of improving the power factor of a three-phase induction motor using a static switched capacitor. 12th International Power Electronics and Motion Control Conference, pp.1088-1093, 2006

[4] Bor-Ren Lin, Chun-Hao Huang and Zheng-Zhang Yang. THREE-PHASE POWER FACTOR CORRECTOR BASED ON CAPACITOR-CLAMPED TOPOLOGY. Proceedings - IEEE International Symposium on Circuits and Systems. pp. 3643 -3646 .2005

[5] N. Bianchi, S. Bolognani, F.Tonel Thermal Analysis of a Run-Capacitor Single-Phase Induction Motor. IEEE Trans Ind Appl. pp. 457-465.2003

Efficiency improvement by Changeover of Phase Windings of Multiphase Permanent Magnet Synchronous Motor with Outer-Rotor type

Young-Gook Kim
Chae-Bong Bae, Jang-Mok Kim
Dept. of Electrical Engineering, Pusan National University
Busan, 609-735, Korea
rain@pusan.ac.kr

Hyun-Cheol Kim
Agency for Defense Development
Chin hae, South Korea
hckim@add.re.kr

Abstract – In this paper, a new control algorithm is proposed to improve the efficiency of multiphase permanent magnet synchronous motor with outer-rotor type (MPMSM). The improved efficiency is obtained by using the change of the phase number from 6-phase motor to 3-phase motor and reducing the switching loss by using the bi-direction switch between two H-bridge inverter of six phase to change the number of the phase. The simulation and experimental results are presented to demonstrate the feasibility and advantages of the proposed algorithm.

I. INTRODUCTION

High phase order or Multi-phase (phase order more than three) machine drives are gaining growing attention in recent years, due to multiphase motor drives posse many other advantages over the traditional three-phase motor drives such as reducing the amplitude and increasing the frequency of torque pulsation, reducing the rotor harmonic currents and the stator current per phase without increasing the voltage per phase, lowering the dc-link current harmonics, higher reliability and an improvement of the fault tolerance. By increasing the number of phases it is also possible to increase the torque per rms ampere for the same volume machine. The additional degrees of freedom in multi phase system also enable us to inject harmonic current of supply multi motors from a single inverter. The high phase order drive is likely to remain limited to specialized applications where high reliability is demanded such as electric/hybrid vehicles, aerospace applications, ship propulsion, and high power application where a combination of several solid state devices form one leg of the drive. Therefore, the requirement of n separate drive units in a multi-phase system is not oppressive for large drives since many of the necessary components are presented in the contemporary designs [1-3].

In this paper, the efficiency improvement of multi-phase permanent magnet synchronous motor(MPMSM) can be obtained by using the changing of the phase number according to the rotor speed and reducing the switching losses of the inverters at three phase operation or the low speed range. At the low speed, the number of the phase winding is three, and at the high speed, the number is six because the application of this motor is the electric ship propulsion. The load of the electric ship is propositional to the cube of the rotor speed. In the low speed range(or 3-phase operation), the number of the switching devices is 12, but three switching elements are added to change the number of the phase from six phases to three phases. So the total switching elements of the independent three phases are 15. And the stator current of three phase is the same to the six phases windings, but the amplitude of three phase back-emf are twice than the six phases because two three phases are directly connected by cascade. Therefore total power loss can be reduced by changing the number of the phase and reducing the number of the switching elements especially in the low speed range. To improve the additional efficiency, the new switching method by using bi-direction switch is proposed. The experimental results verify the usefulness of the proposed control algorithm.

II. MPMSM WITH OUTER-ROTOR TYPE

A. Configuration of MPMSM

MPMSM is composed of the separated six phase windings as shown in Fig. 1. And each phase is connected to separate six H-bridge inverters [7]. In general, H-bridge inverter system can be used for the single phase of ac machine or the chopper of a DC machine. In this paper, H-brides are utilized for the inverter of MPMSM as shown in Fig.1. Total H-bridge is six because MPMSM is six phase machine.

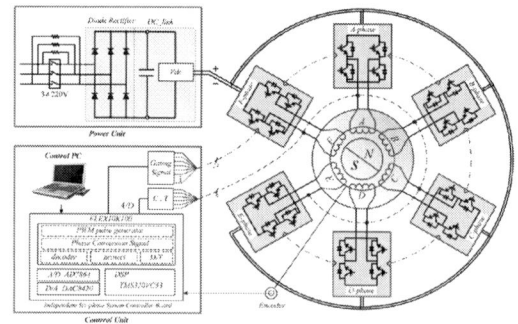

Fig. 1.   The configuration of MPMSM

B. Mathematical modeling of MPMSM

Fig. 2 shows the equivalent circuit of the stator winding of MPMSM. The end of each phase is not connected to neutral point unlike the conventional three phase ac machine. The winding of MPMSM can be divided into two groups of three phase motors to easily implement the vector control of this motor. One group is V_a, V_c, and V_e. The other is V_b, V_d, and V_f. So the control system of MPMSM can be considered as a parallel operation of two- three phase motors.

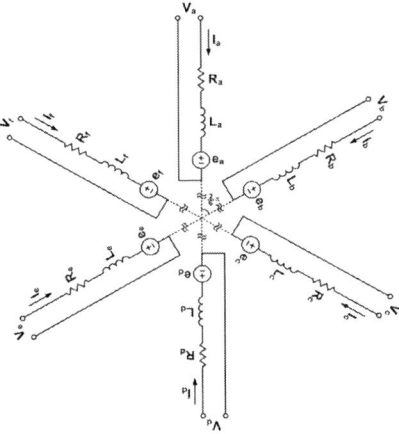

Fig. 2. Equivalent winding circuit of MPMSM

Fig.3 shows six phase back-emf of MPMSM which has the sinusoidal waveform. And the difference angle of each phase is 60 degree.

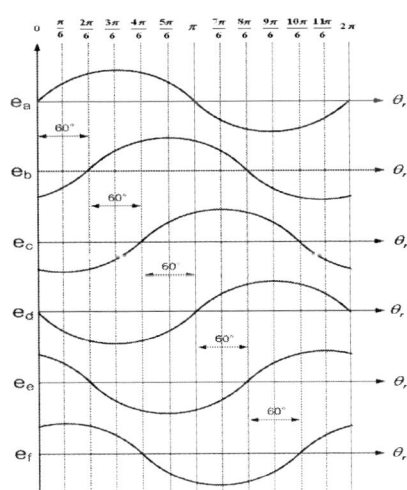

Fig. 3. Waveform of back-emf

Voltage equation of MPMSM is given at (1).

$$v_x = A i_x + B \frac{d}{dx} i_x + e_x \qquad (1)$$

where x means the specific phase, i is current, e is back-emf of the electric parameters and A, B are matrix defined as

$$A = \begin{bmatrix} R & 0 & 0 & 0 & 0 & 0 \\ 0 & R & 0 & 0 & 0 & 0 \\ 0 & 0 & R & 0 & 0 & 0 \\ 0 & 0 & 0 & R & 0 & 0 \\ 0 & 0 & 0 & 0 & R & 0 \\ 0 & 0 & 0 & 0 & 0 & R \end{bmatrix} \quad B = \begin{bmatrix} L & 0 & 0 & 0 & 0 & 0 \\ 0 & L & 0 & 0 & 0 & 0 \\ 0 & 0 & L & 0 & 0 & 0 \\ 0 & 0 & 0 & L & 0 & 0 \\ 0 & 0 & 0 & 0 & L & 0 \\ 0 & 0 & 0 & 0 & 0 & L \end{bmatrix}$$

where L is the synchronous inductance, and R is the stator resistance. If θ_e is the rotor position, back-emf can be expressed as in (2).

$$e_x = k_e(\theta_e)\omega_m \qquad (2)$$

The torque equation can be derived from the input power as shown in (3).

$$T_e = \frac{P_e}{\omega_m} = \frac{e_a i_a + e_b i_b + e_c i_c + e_d i_d + e_e i_e + e_f i_f}{\omega_m} \qquad (3)$$

where T_e is produced torque, P_e is output power, respectively. And ω_m is rotational angle speed of the rotor. Mechanical motion equation of the motor can be expressed as in (4). This is the same equation form of the conventional three phases motor.

$$\frac{d}{dt}\omega_m = -\frac{B_m}{J_m}\omega_m - \frac{T_L}{J_m} + \frac{T_e}{J_m} \qquad (4)$$

The block diagram of 6-phase of MPMSM is shown in Fig. 4.

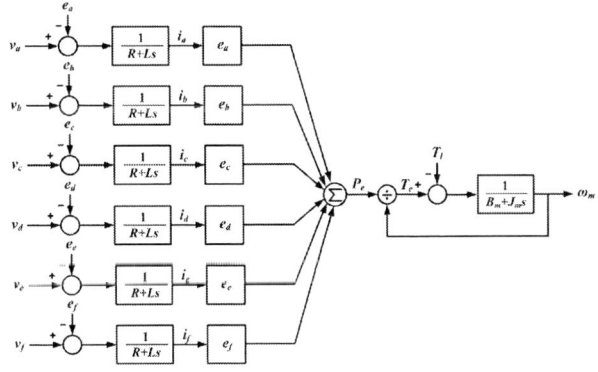

Fig. 4. Block diagram of MPMSM

III. OPERATION CHARACTERISTIC

A. Analysis of power losses in inverter [6]

The losses of the inverter can be classified as conduction loss, off state loss, and switching loss. Since the leakage current of the off-state of the device is negligibly small, the power loss during the off-state can be neglected in real inverter system. So, in the power loss of the switching device, the dominant term is conduction loss and switching loss.

Conduction losses occur at the state of the switching device at the on-state and the current follows through the switching device. Therefore, power dissipation during conduction is computed by multiplying the on state saturation voltage by on-state current.

$$p_{on} = |i_c| \cdot v_{on} \qquad (5)$$

where, v_{on} is the on-state saturation voltage and i_c is the load current.

Simple expression of the average conduction loss of each device can be expressed as follows:

$$P_{ave.conduction\_loss} = \frac{1}{2\pi} \int_{\alpha}^{\beta} P_{on} \cdot k \, d\theta \qquad (6)$$

where, α, β are defined as the start and the end of the conduction state interval of each device over one period, and k is on-state ratio of conducting device, respectively.

Switching loss is the power dissipation during turn-on and turn–off switching transitions. In high frequency, PWM switching losses can be substantial and must be considered to improve the efficiency of the inverter system.

Simple expression of the average switching loss for diode and controllable switch can be expressed follows:

$$P_{D\_ave.switching\_loss} = \frac{1}{2\pi} \int_{\alpha}^{\beta} E_{rec} \cdot |i_c| \cdot f_c \, d\theta \qquad (12)$$

$$P_{S\_ave.switching\_loss} = \frac{1}{2\pi} \int_{\alpha}^{\beta} (E_{on} + E_{off}) \cdot |i_c| \cdot f_c \, d\theta \qquad (13)$$

Where f_c is the frequency of carrier wave, E_{rec} [J/A], E_{on} [J/A] and E_{off} [J/A] are reverse-recovery energy coefficient, turn-on switching energy and turn-off switching energy coefficient, respectively.

The efficiency of the inverter system is nearly affected by load current, carrier wave frequency, the saturation voltage, and energy coefficients. If the same device is used, the saturation voltage and energy coefficients are also same. And the same of the carrier wave frequency will be used for the measurement of the efficiency. So, these factors can be neglected for the test.

B. Operation concept of MPSM with outer-rotor type

Fig. 5(a) shows 6-phase inverter system. This is composed of 6 H-bridge inverters and the total number of the switching elements is 24. If there is the variation of the phase number from 6 to 3-phase inverter systems, Fig. 5(b) shows the transformed 3-phase inverter systems from 6-phase. As shown from Fig. 5(b), the bi-directional switching elements are inserted between a pair of 6 phase H-bridge inverters so that load1 and load 2 are connected in series. This series connection makes 3-phase inverter circuit from 6-phase inverter circuit. Therefore, the total switching elements of three phase inverter is 15, and two diodes are added for the condition path of the bi directional switch.

3-phase system has less power losses than 6-phase system because the power losses of both the conduction and switching are increased in proportion to the number of switching devices.

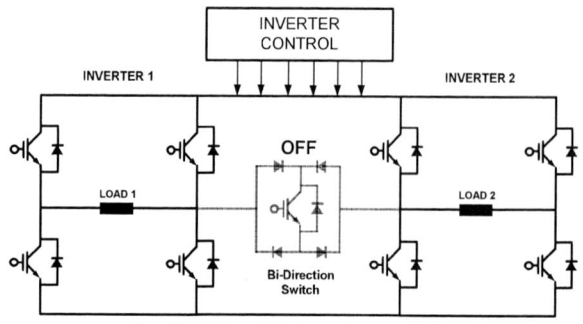

(a) Configuration of 6-phase of MPMSM

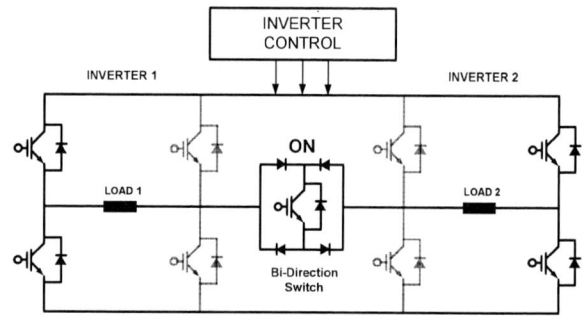

(b) Configuration of 3-phase of MPMSM

Fig. 5. Configuration of changeover phase windings

Fig. 6 shows the concept of the variation of the phase number of the machine from 3 to 6-phases according to the variation of the ship speed and the required power of the specific speed. As known from this figure, in the low speed range, three phase operation is effective, but in the high speed range, six phase operation is more useful because this motor will be used for the eclectic propulsion of the ship. In the ship application, the load or the required power is propisitional to the cubic of the ship speed. The maximum power and the terminal voltage of 3-phase system are obtained at the changeover point. After this changeover point, the changeover operation is needed.

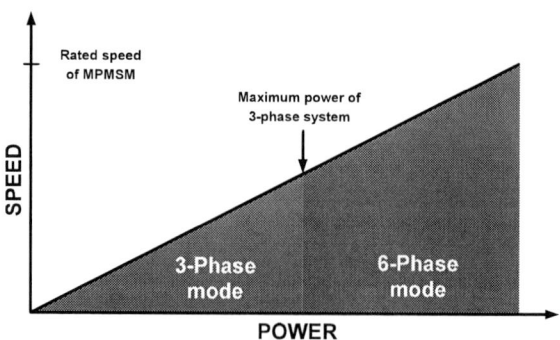

Fig. 6. Concept of changing the number of the phase system

Fig 7(a) shows the simple equivalent circuit of 6 phase windings system including the phase current, the stator resistance, the inductance and back-emf. The transformed equivalent circuit of 3-phase system can be expressed as shown in Fig 7. (b). And the load current of 3-phase system is the same to the six –phase inverter systems, but the amplitude of three phase back-emf is twice than six phase because two phases are directly connected by cascade.

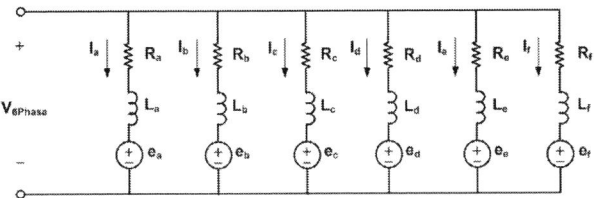

(a)   The equivalent circuit of 6-phase of MPMSM

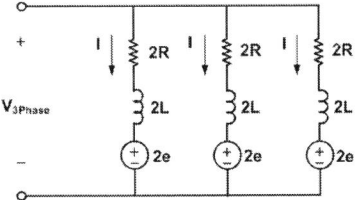

(b)   Transformed equivalent circuit of 3-phase of MPMSM

Fig. 7.   The equivalent circuit of changeover of the phase windings system from six phases to the three phase.

And the phase voltage of each winding can be expressed as follows:

$$V_{6-phase} = R_X I_X + L_X \frac{dI_X}{dt} + e_X \qquad (14)$$

$$V_{3-phase} = 2(R_X I_X + L_X \frac{dI_X}{dt} + e_X) \approx 2V_{6-phase} \quad (15)$$

Where $I_x = I$, $R_x = R$, $L_x = L$, $e_x = e$, and x means specific phase.

From (14) and (15), the phase voltage of 3-phase system is about twice than 6-phase system because two phases are directly connected by cascade.

The limitation of phase voltage is equal to dc-link voltage, V_{DC} because each phase is connected to separate six H-bridge inverters. Phase voltage of both 3-phase system and 6-phase system are limited to dc-link voltage V_{DC} as shown in Fig. 7(a) and (b), respectively.

Fig. 8 shows the marginal voltage of the dc-link for the operation of 3-phase mode and 6-phase mode on the basis of the relationship of the speed and the required power at the specific speed.

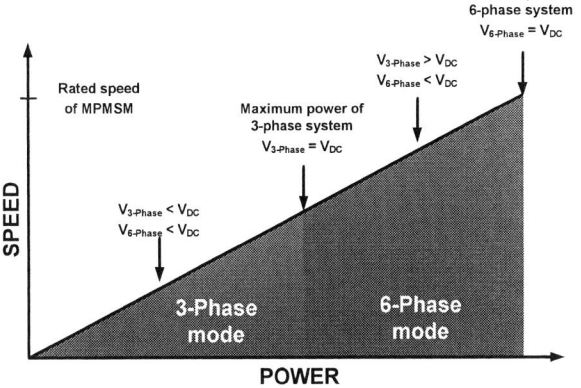

Fig. 8.   The condition of operating mode in the changeover of phase windings system.

From Fig. 9, in the low speed range, three phase operation is useful, but in the high speed range, six phase operation is more effective due to the margin of the phase voltage. Therefore, the changeover of the phase windings is very effective to improve the efficiency of the total inverter system.

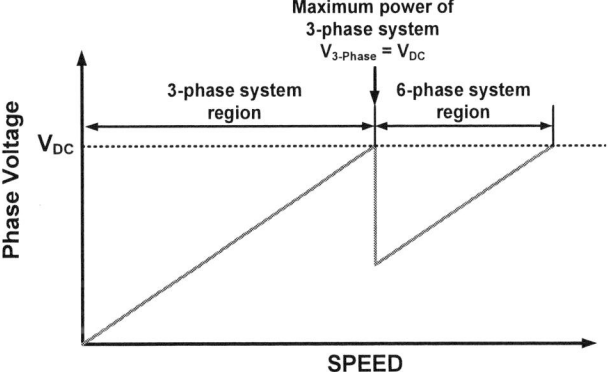

Fig. 9.   The phase voltage according to motor speed

C.   New switching method for reducing switching loss

The connection switch is needed for the changeover of the phase number. The mechanical relay can be used for the connection path between two H-bridge inverters. But this relay has many problems of the low response and the mechanical contact parts of the copper.

Therefore, in this paper, bi-directional switch is proposed for this connection between two H-bridge inverters. This bi-directional switch is consisted of 4 diodes and one IGBT as shown in Fig. 10. For the additional improvement of the efficiency of the total inverter system, this switch provides not only the conduction path of the changeover but also is the main switching element in three phase inverter system.

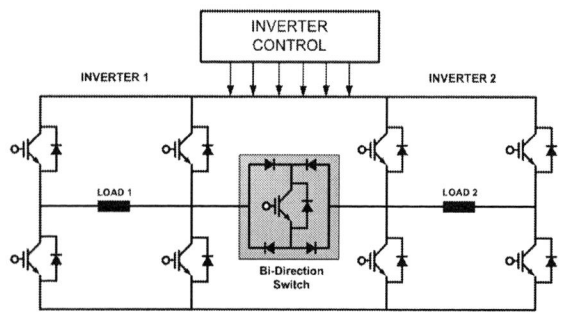

Fig. 10. A configuration of system for changeover of phase windings

1) Operation modes of 6-phase inverter system

Fig. 11 shows the operation modes of 6-phase inverters system of MPMSM. One is positive on and off modes as shown in Fig. 11 (a) and (b). The other is negative on and off modes as shown in Fig. 11 (c) and (d), respectively. In 6-phase mode, a bi-direction switch S_c is always off as shown in Fig. 11.

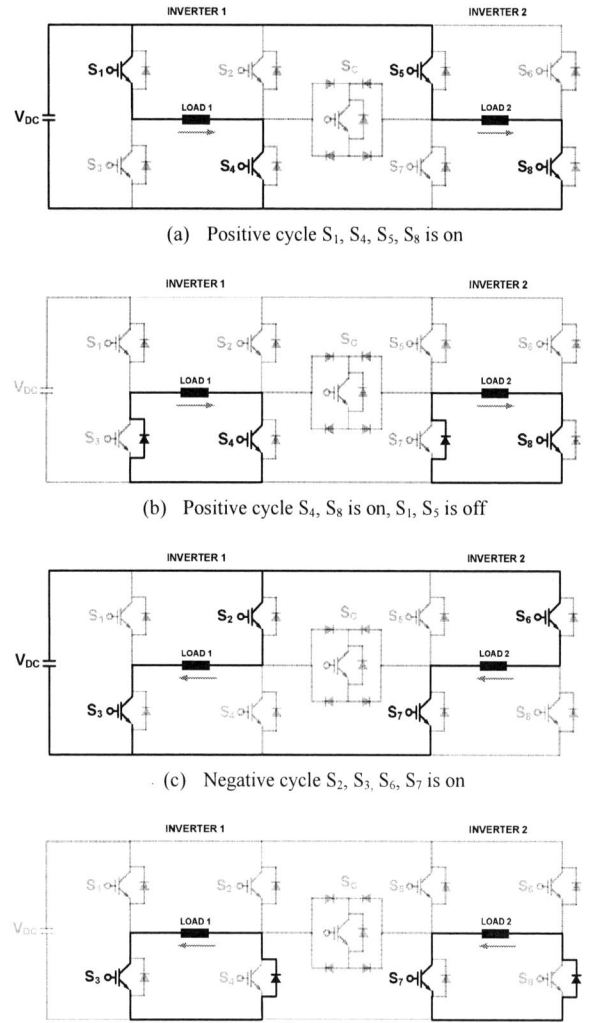

(a)   Positive cycle S_1, S_4, S_5, S_8 is on

(b)   Positive cycle S_4, S_8 is on, S_1, S_5 is off

(c)   Negative cycle S_2, S_3, S_6, S_7 is on

(d)   Negative cycle S_3, S_7 is on, S_2 S_6 is off

Fig. 11. Operation modes of 6-phase system

2) Operation modes of 3-phase inverter system

Fig. 12 shows the operation modes of 3-phase inverter system. In 3-phase mode, a bi-direction switch is used for not only main PWM switching element, but also switch for changeover of phase windings. The other switching elements, S_1 and S_8, provide not the switching elements but the conduction path as shown in Fig. 12(a) and (c).

Fig. 12(a) shows the proposed switching method by using bi-directional switch. In this paper, S_1 and S_8 provide only commutative path, the main switching element is S_c. By using the proposed switching method, total number of the switching elements is not total switches number of 3-phase inverter system, 15 but only total bi-direction switches of 3-phase inverter system, 3. Therefore, the switching loss of three inverters system can be reduced considerably because most of switching loss is happened only at S_c.

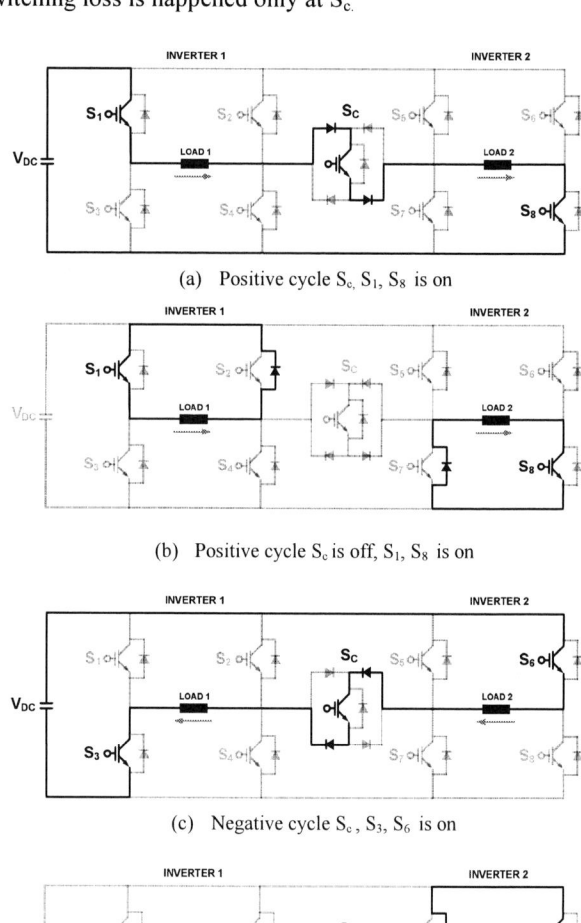

(a)   Positive cycle S_c, S_1, S_8 is on

(b)   Positive cycle S_c is off, S_1, S_8 is on

(c)   Negative cycle S_c, S_3, S_6 is on

(d)   Negative cycle S_c is off, S_3, S_6 is on

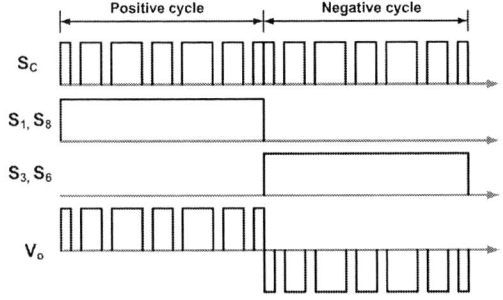

(e) Detailed waveforms of the proposed PWM method

Fig. 12. Operation modes of 3-phase system

IV. SIMULATION RESULT

MATLAB/Simulink is used for the verification of the proposed algorithm. The limitation voltage of dc-link, V_{DC}, is 100[v].

Fig. 13 shows the simulation waveforms of rotor speed, the produced torque, power, phase current, back-emf and phase voltage when the motor operates at 600[rpm]. As known from this waveform, the changeover of the phase windings happens at 330[rpm] when the phase voltage of 3-phase system is equal to dc-link voltage V_{DC}. The maximum power of 3-phase inverter system is produced at this speed.

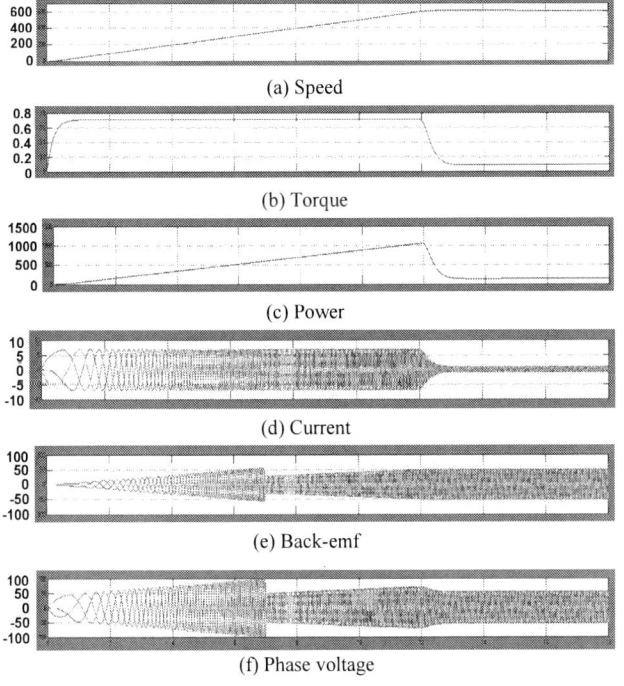

(a) Speed

(b) Torque

(c) Power

(d) Current

(e) Back-emf

(f) Phase voltage

Fig. 13. Simulation result of changeover of phase windings system

Fig. 14 (a) is the sinusoidal reference voltage for the operation of the three phases. Fig. 14 (b) shows the switching waveform of the bi-directional switch. Fig.14(c) shows the positive switching waveform of S1 and S8 to provide the conduction path for the operation of the three phases as

shown in Fig. 12(a) and (b). Fig. 14(d) shows the negative switching waveform of S3 and S6 to provide the conduction path for the operation of the three phases as shown in Fig. 12(c) and (d). And Fig. 14 (e) shows one of output voltages of three phase inverter systems. As know from these resultant waveforms, the three phase inverter system using the proposed switching method operates well

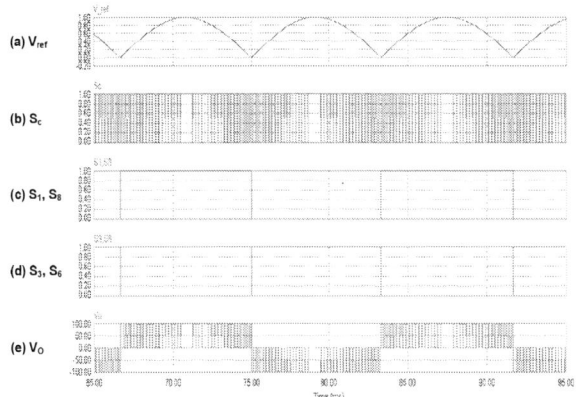

Fig. 14. The simulation result of proposed PWM method

V. EXPERIMENT RESULT

The MPMSM is used for the experimentation as shown in Fig. 15. The type of inverter is H-bridge. And the number of this inverter is six as shown in the front of Fig. 15.

Fig. 15. Configuration of the experimental setup

Fig. 16. shows the experimental waveform of a-phase current, voltage and rotor speed from 0 to 600[rpm]. The changeover of phase windings happens at 300[rpm]. This speed is about 50% of the rated speed. The phase voltage is nearly V_{DC} at the changeover point as shown in Fig. 16 (b).

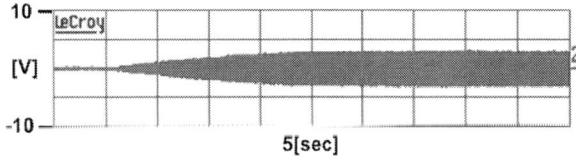

(a) The phase current in 3-phase system

978-1-4244-4782-4/10 $26.00 © 2010 IEEE

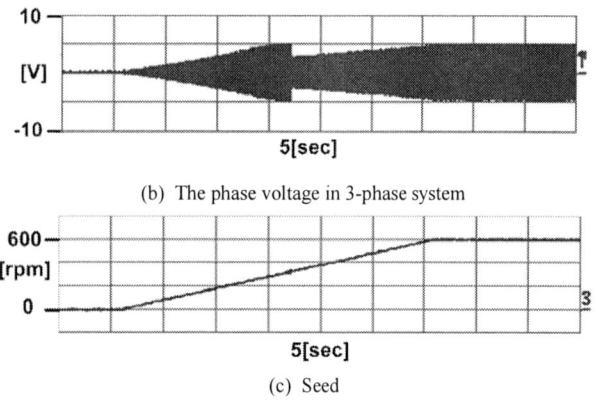

(b) The phase voltage in 3-phase system

(c) Seed

Fig. 16. Phase current and voltage according to the rotor speed.

The control block diagram is used for the measurement of the efficiency of MPMSM as shown Fig. 17.

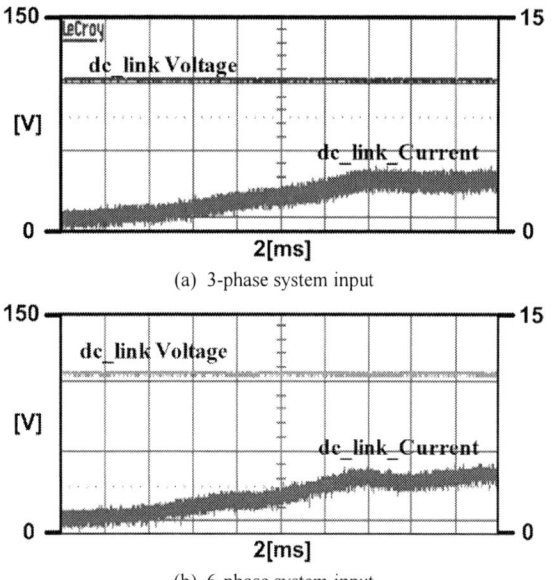

Fig. 17. Determine the efficiency of MPMSM

Fig. 18 and 19 are measured the voltage and current at the dc-link and load of each system. These are used to decide the input and output power for the measurement of the efficiency of MPMSM.

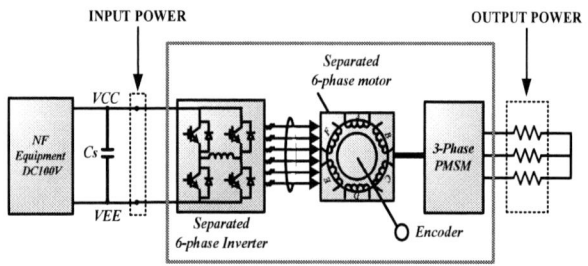

(a) 3-phase system input

(b) 6-phase system input

Fig. 18. The dc-link voltage and current in MPMSM

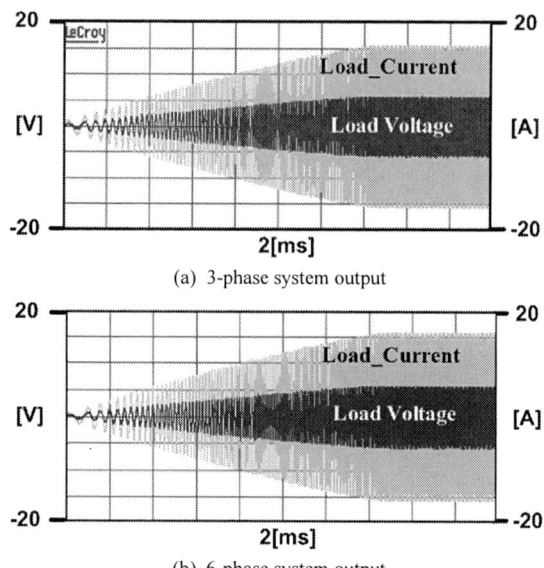

(a) 3-phase system output

(b) 6-phase system output

Fig. 19. The voltage and current of a phase at the load

Fig. 20 shows the resulting efficiency curve of MPMSM according to the rotor speed. From this figure, the 3-phase system has high efficiency than 6-phase system in the low speed range. And the 6-phase system of MPMSM is used to produce the full torque in the high speed range. So the total efficiency of MPMSM can be improved by changing the number of phase in the low speed range.

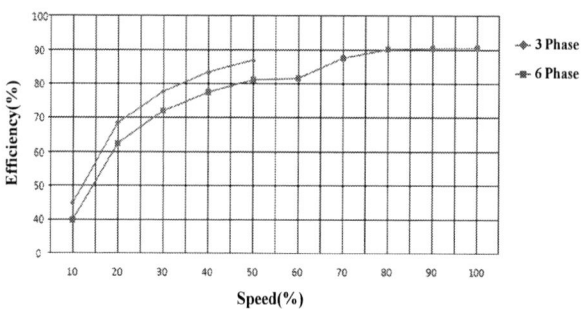

Fig. 20. The resulting efficiency curve of MPMSM according to the rotor speed

VI. CONCLUSION

In this paper, a new control algorithm was proposed to improve the efficiency of multi-phase inverter system. The improvement of the efficiency was obtained by using the change of the phase number from six to three phase windings. And bi-directional switching element was introduced to provide not only conduction path but also the main switching function of three phase inverter system. The total efficiency of three phase inverter is much high than six phase inverter system. And in the three phase inverter, the improvement of the efficiency could be obtained because the switching elements are not H-bridge switching element but bi-directional switching elements.

And the usefulness of the proposed algorithm verified through computer simulation and the experimental results.

REFERENCE

[1] G.K Singh, "Multi-Phase induction machine drive research-a survey," Electric Power Systems Research, Volume 61, Number 2, 28 March 2002 , pp. 139-147(9).

[2] Casadei, D.; Dujic, D.; Levi, E.; Serra, G.; Tani, A.; Zarri, L, "General Modulation Strategy for Seven-Phase Inverters With Independent Control of Multiple Voltage Space Vectors," Industrial Electronics, IEEE Transactions on, Volume 55, Issue 5, May 2008 Page(s):1921 – 1932.

[3] Kumar Singh, R, "Multiphase Inverter Topology and its Modulation Technique for Optimal Harmonic Output," Power Electronics, Drives and Energy Systems, 2006. PEDES '06. International Conference on, 12-15 Dec. 2006 Page(s):1 - 10

[4] Maswood, A.I, "A switching loss study in SPWM IGBT inverter," Power and Energy Conference, 2008. PECon 2008. IEEE 2nd International, 1-3 Dec. 2008 Page(s):609 – 613.

[5] Ji-Sheng Lai, "Power conditioning circuit topologies," Industrial Electronic Magazine, IEEE. Volume 3, Issue 2, June 2009 Page(S) : 24-34.

[6] Tae-Jin Kim, Dae-Wook Kang, Yo-Han Lee, Dong-Seok Hyun, "The Analysis of Conduction and Switching Losses in Multi-Level Inverter System," Power Electronics Specialists Conference, 2001. PESC. 2001 IEEE 32nd Annual, Volume 3, 17-21 June 2001 Page(s):1363 - 1368 vol. 3.

[7] Sanmin Wei, Bin Wu, Richard Cheung, "A Novel SVM Algorithm for Reducing Oscillationis in Cascaded H-Bridge Multilevel Inverters," Industrial Electronics Society, 2003. IECON '03. The 29th Annual Conference of the IEEE.

[8] T. Gopalarathnam, H. A. Toliyat, and J. C. Moreira, "Multi-Phase Fault-Tolerant Brushless DC Motor Drives," IEEE Industrial Application Society Annual Meeting, pp.1683-1688, October 2000.

[9] H.A. Toliyat, L.Y. Xue, T.A. Lipo, "A Five Phase Reluctance Motor with High Specific Torque," IEEE Trans. on Industry Applications, Vol. 28, No. 3, pp. 659-667, 1992.

[10] H.A. Toliyat, S. Waikar, T.A. Lipo, "Analysis and Simulation of Five Phase Synchronous Reluctance Machines Including Third Harmonic of Air-Gap MMF,"IEEE Transactions on Industry Applications, Vol. 34, No.2, pp. 332-339, 1998.

Digital Power Controller with Non-Linear Variable Switching Frequency

Jaber A. Abu Qahouq

The University of Alabama
Department of Electrical and Computer Engineering
Tuscaloosa, Alabama 35487, USA

Abstract — **Digital controller for a power converter with non-linear variable switching frequency is presented in this paper. Regression analysis is used to obtain approximated function of the non-linear variable switching needed to yield output voltage ripple in DCM that is close to the CCM output voltage ripple. The need to sense and sample the peak inductor current and the input voltage is avoided in the presented digital controller. The approximated function has the load current as an input and the non-linear switching frequency as its output and it is used by the digital controller to set the switching frequency during DCM operation of the power converter.**

I. INTRODUCTION

Low output voltage ripple of the power converter is desired in many applications to satisfy the load requirements [1-9] such as Digital Signal Processing (DSP) Integrated Circuits (ICs) and single and multi-core microprocessors. On the other hand, in order to improve the power conversion efficiency at light load current values, variable switching frequency combined with Discontinuous Conduction Mode (DCM) operation can be used [1-9].

Variable switching frequency operation can be realized by different methods such as hysteretic control and conventional Pulse Width Modulation (PWM) linear variable frequency control [1-3, 8, 9]. The later method is common, however, it results in increased output voltage ripple. In [1], a method and analog controller implementation that utilize non-linear variable switching frequency are presented to maintain fixed output voltage ripple in DCM that is equal to the ripple value in Continuous Conduction Mode (CCM). This method requires detecting the peak inductor current value and it requires the knowledge of the input voltage (if it is variable).

Digital controllers for power converters have been increasingly used and researched because of the potential advantages they bring [10-16]. This paper presents a digital controller with non-linear variable switching frequency without the need to detect and sample the peak inductor current value and the input voltage. This results in less hardware requirements and less required ADCs (Analog-to-Digital Converters). The Non-linear function is approximated using an equation based on regression analysis applied to selected data points [17-19]. The selected data points are chosen from the originally desired non-linear switching frequency curve for fixed output voltage ripple in DCM that is equal to the CCM ripple. As discussed later in this paper, the points are alternatively selected from the two curves for the lowest and highest input voltages such that the approximated non-linear equation is able to cover all the input voltage range.

Next section presents the regression analysis used to obtain the equation that approximates the required non-linear variable switching frequency function. Section III discusses other options to obtain the regression analysis-based non-linear switching frequency function. Section IV presents experimental results and the conclusion is given in Section V.

II. REGRESSION ANALYSIS-BASED FUNCTION DEVELOPMENT FOR THE NON-LINEAR VARIABLE SWITCHING FREQUENCY

Fig. 1 shows a DC-DC buck converter. The output voltage ripple in CCM and DCM is given by:

$$V_{o-pp} = \begin{cases} \dfrac{(1-D) \cdot V_o}{8 \cdot L_o \cdot C_o \cdot f_{CCM}^{2}} & , CCM \\[4mm] \dfrac{D_1 \cdot (I_{L\max.DCM} - I_o)^2}{2 \cdot C_o \cdot I_{L\max.DCM} \cdot f_{DCM}} & , DCM \end{cases} \tag{1}$$

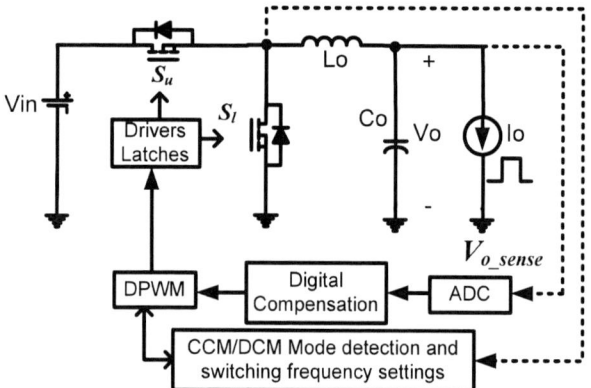

Fig. 1: DC-DC Buck Converter Block diagram with digital controller

978-1-4244-4782-4/10 $26.00 © 2010 IEEE

Where V_{o-pp} is the output voltage peak to peak ripple, V_o is the output voltage, L_o is the output inductance, C_o is the output capacitance, D is the high-side FET duty cycle (the ratio between the FET On-time duration and the total switching cycle time period), D_1 is the duty cycle defined as the ratio between the time period starting from the beginning of the switching cycle (when the high-side FET is turned-ON) and ending when the low-side FET is turned-OFF and the total switching cycle time period, f_{CCM} is the CCM switching frequency, f_{DCM} is the DCM switching frequency, and $I_{L\max.DCM} = V_o \cdot (D1-D)/L_o \cdot f_{DCM}$ is the peak inductor current value in DCM.

Consider a buck converter with the following specifications: $V_{in} = 8V - 12V$, $V_o = 3.3V$, $I_{o-\max} = 8A$, $L_o = 0.6\mu H$, $f_{CCM} = 500kHz$, and $C_o = 4 \times 22\mu F$. Fig. 2 shows switching frequency and ripple comparisons between conventional (linear) and non-linear variable frequency operation. As discussed in [1], compared to the conventional case the non-linear case results in lower output voltage ripple in DCM that is equal to the output voltage ripple in CCM.

Implementing the non-linear variable switching frequency described in [1] and shown in Fig. 2 requires added sensing and sampling of the input voltage and the accurate sensing, sampling and detection of the inductor's peak current. While this is visible, it adds cost, size and power loss to the digital controller.

In this paper, instead, a regression-based function using regression analysis [17-19] is developed and used to approximately realize the non-linear variable switching frequency. "Regression analysis includes any techniques for modeling and analyzing several variables, when the focus is on the relationship between a dependent variable and one or more independent variables" [18]. The regression analysis can be of different types such as linear regression and exponential regression [17, 19].

The converter design example has a minimum input voltage of 8V and a maximum input voltage of 12V. The non-linear switching frequency curves are shown in Fig. 2(a). It is obvious that any input voltage between 8V and 12V will result in a frequency curve that is in between the 8V and the 12V curves. There are different possible ways to perform regression analysis. The method used in this paper is as follows: As the current varies from almost 0A to about 4A (DCM range for 12V input voltage), several current-frequency points are selected alternatively, one from the 12V frequency curve and the other from the 8V frequency curve shown in Fig. 2(a). This is shown in Table 1 and plotted in Fig. 3. Then regression analysis to find the approximated function based on second order polynomial is performed. This mixed data selection to obtain the regression analysis based function is used to obtain a function that can be used for all the input voltage range for the design example given here.

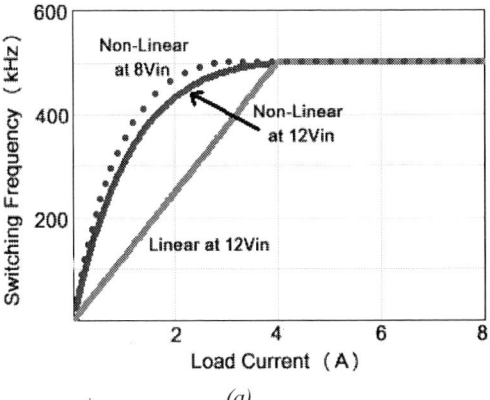

(a)

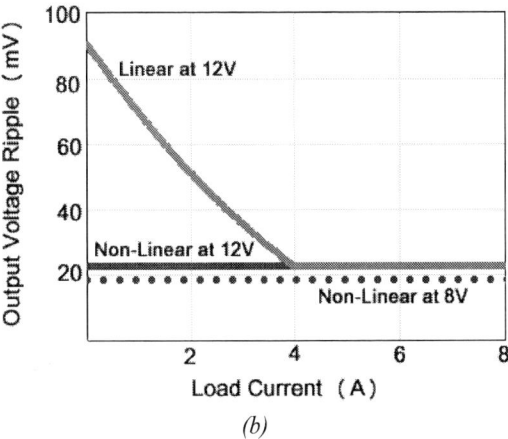

(b)

Fig. 2: *Frequency and ripple comparison simulation results of an example buck converter design under conventional and non-linear variable frequency operation: (a) switching frequency and (b) output voltage ripple*

Table 1: *Selected Points for Regression Analysis*

Io (A)	8Vin	12Vin	Selected
0.05	28000	**23500**	**23500**
0.1	**55000**	45000	**55000**
0.5	223000	**187000**	**187000**
1	**352000**	306000	**352000**
2	470000	**432000**	**432000**
3	**500000**	482000	**500000**
4	500000	**500000**	**500000**

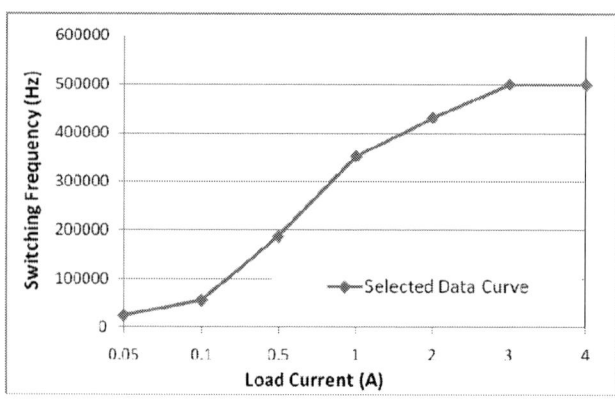

Fig. 3: Selected Data Curve Plot

The obtained second order regression analysis equations are:

$$f_{DCM1}(I_o) = -50189 \cdot I_o^2 + 314660 \cdot I_o + 31948 \quad Hz \qquad (2)$$

$$f_{DCM2}(I_o) = -50 \cdot I_o^2 + 315 \cdot I_o + 32 \quad kHz \qquad (3)$$

Eq. (3) is just an approximation of Eq. (2) by changing the answer from Hz to kHz and omitting the some extra digits. This is to simplify the calculations. To obtain the regression function formula here, Microsoft Excel® [17, 19] is used.

Eq. (3) can be used in the digital controller to realize the non-linear variable switching frequency controller in DCM. Fig. 4 shows the plot of Eq. (3) and the resulted output voltage ripple. The ripple resulted from Eq. (3) is fluctuating closely around the desired value which is equal to the CCM ripple. It is lower than the ripple in the case of conventional linear variable switching frequency control.

III. OTHER REGRESSION-BASED EQUATIONS OPTIONS

The previous section has demonstrated the regression analysis-based non-linear variable switching frequency control using second order polynomial and using only one single function in the DCM operating range. Higher order polynomial equation and/or other function types could also be used to try to obtain more accurate approximation of the original non-linear variable switching frequency curve. However, this may come at the expense of more complicated digital controller design, larger size, higher cost and higher power consumption.

Another possible option is to use more than one equation, an equation for each part of the DCM range. For example, obtain and use an equation for the DCM load range of lower than 2A and another equation for higher than or equal to 2A. The controller will then use one of the two equations based on the load current value.

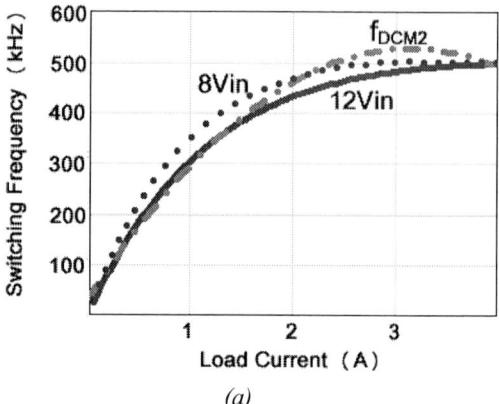

(a)

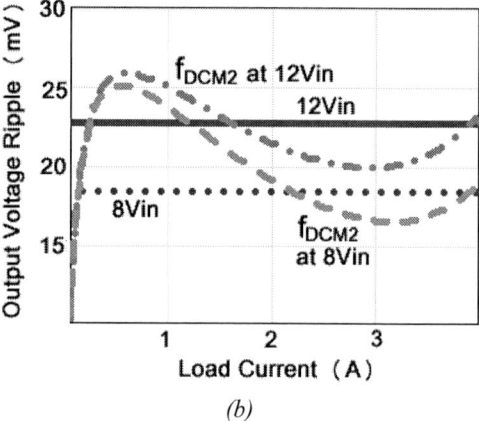

(b)

Fig. 4: Frequency and ripple simulation results comparisons for an example buck converter under non-linear variable switching frequency with and without using the regression analysis equation

IV. EXPERIMANTAL WORK

Fig. 5 shows a sample plot of the output voltage ripple resulted from the non-linear variable frequency implemented using Eq. (3) at 12V input voltage as compared to the linear switching frequency case. It can be observed that while the output voltage ripple during DCM operation is not exactly fixed to the CCM value as in [1], it fluctuates closely around the desired CCM output voltage ripple value and it is lower than the case of linear variable switching frequency.

Fig. 6 shows the resulted efficiency comparison. As expected, there some efficiency tradeoff for the non-linear switching frequency method compared to the linear switching frequency method. However, the efficiency is still much better when compared to the fixed switching frequency case.

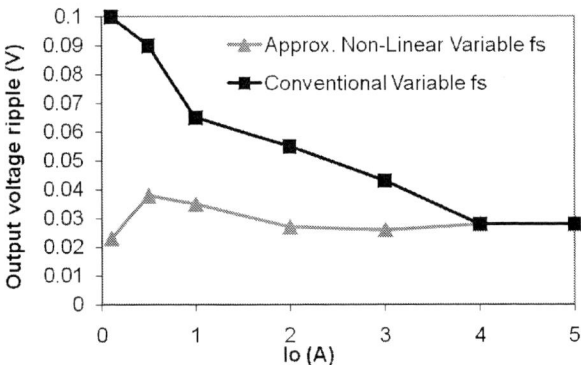

Fig. 5: Efficiency Results and Comparison

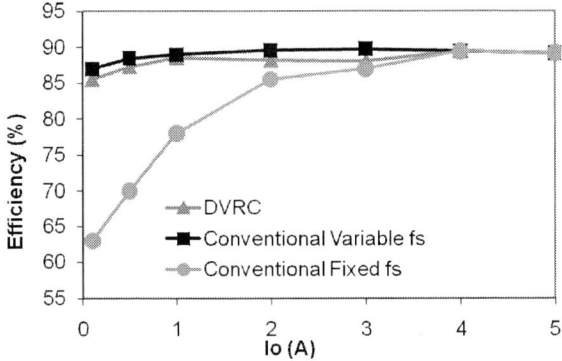

Fig. 6: Output Voltage Ripple Results and Comparison

ACKNOWLEDGEMENT

This work was supported in part by the National Science Foundation (NSF) of USA under grant number 0927104 and in part by The University of Alabama.

V. CONCLUSION

Digital power converter controller with non-linear variable switching frequency is presented in this paper. The Non-linear function is approximated using a function based on regression analysis applied to selected data points. The selected data points are chosen from the originally desired non-linear switching frequency curve for fixed output voltage ripple in DCM that is equal to the CCM ripple. The points are alternatively selected from the two curves for the lowest and highest input voltages such that the approximated non-linear equation is able to cover wide input voltage range without the need to sensing the input voltage. The peak inductor current does not need to be sensed too. This results in less hardware requirements and less required ADCs. While the frequency function is developed for a digital controller implementation, it could also be used as a simpler approximated analog implementation of the non-linear variable switching frequency function compared to the one used in [1].

REFERENCES

[1] J.A. Abu-Qahouq, O. Abdel-Rahman, L. Huang, I. Batarseh, "On Load Adaptive Control of Voltage Regulators for Power Managed Loads: Control Schemes to Improve Converter Efficiency and Performance," IEEE Transactions on Power Electronics, vol. 22, September 2007.

[2] H. Arbetter, R. Erickson, and D. Maksimovid, "DC-DC Converter Design for Battery-Operated Systems," IEEE Power Electronics Specialists Conference, 1995, pp.103-109.

[3] T.G. Wang, B. Tomescu, and F.C. Lee, "Achieving High Efficiency for a Low voltage DC/DC Converter with 1% to 100% Load Range" VPEC Seminar'96.

[4] B. Arbetter, and D. Maksimovic, "Control Method for Low-Voltage DC Power Supply In Battary-Powered Systems with Power Management," IEEE Power Electronics Specialists Conference, 1997, pp. 1198-1204.

[5] T.G. Wang, Z. Xunwei, and F.C. Lee, "A low voltage high efficiency and high power density DC/DC converter ," IEEE Power Electronics Specialists Conference, 1997, pp.240–245.

[6] Z. Xunwei,T.G Wang, and F.C Lee, "Optimizing design for low voltage DC-DC converters, " IEEE Applied Power Electronics Conference and Exposition Proceedings, 1997, pp.612-616.

[7] X. Xhou, M. Donati, L. Amoroso, and F. Lee, "Improved Light-Load Efficiency for Synchromous Rectifier Voltage Regulator Module," IEEE Transactions on Power Electronics, Volume 15, pp. 826 – 834, September 2000.

[8] M. Gildersleeve, H. Forghani-zadeh, and G. Rincon-Mora, "A Comprehensive Power Analysis and a highly efficient Mode-Hopping DC-DC Converter," IEEE Proceedings of Asia-Pacific Conference on ASIC, 2002, pp. 153–156.

[9] Jaber Abu-Qahouq, Hong Mao, and Issa Batarseh, "Multiphase Voltage-Mode Hysteretic Controlled Dc-Dc Converter with Novel Current Sharing," IEEE Transactions on Power Electronics, vol. 19, no. 6, pages: 1397-1407,November 2004.

[10] Jaber Abu-Qahouq, Wisam Al-Hoor, Wasfy Michael, Lilly Huang and Issa Batarseh, "Analysis and Design of an Adaptive-Step-Size Digital Controller For Switching Frequency Auto-Tuning," IEEE Transactions on Circuits and Systems I - Regular Papers, Vol. 56, No. 12, December 2009.

[11] B.J. Patella, A. Prodic, A. Zirger and D Maksimovic "High-Frequency Digital PWM Controller IC for DC-DC Converters," IEEE Transactions on Power Electronics, VOL, 18, January 2003.

[12] A. Syed, E. Ahmed, D. Maksimovic, E. Alarcon, "Digital pulse width modulator architectures" IEEE Power Electronics Specialists Conference, PESC'04, Vol.6, pages: 4689-4695, 2004.

[13] J. Abu-Qahouq, H. Mao, H. Al-Atrash, and I. Batarseh, "Maximum Efficiency Point Tracking (MEPT) Method and Dead Time Control," IEEE Transactions on Power Electronics, Vol. 21, Issue 5, pages: 1273-1281, Sept. 2006.

[14] V. Yousefzadeh and D. Maksimovic, "Sensorless optimization of dead times in DC-DC converters with synchronous rectifiers," IEEE Transactions on Power Electronics, Vol. 21, Issue 4, pages: 994-1002, July 2006.

[15] Jaber Abu-Qahouq, Lilly Huang, and Douglas Huard, "Sensor-less Current Sharing Analysis and Scheme for Multiphase Converters," IEEE Transactions on Power Electronics, Vol. 23, No. 5, pp. 2237-2247, September 2008.

[16] B. Miao, R. Zane, and D. Maksimovic, "Automated Digital Controller Design for Switching Converters," IEEE Power Electronics Specialists Conference, Pages: 2729-2735, June 2005.

[17] Microsoft Excel Help, "Search keyword: Regression," Microsoft Office 2007.

[18] Wikipedia, "Regression analysis," Online at http://www.wikipedia.org/, as of November 2009.

[19] Microsoft Office Online, "Perform a regression analysis," Online at http://office.microsoft.com/en-us/default.aspx, as of Novemver 2009.

978-1-4244-4782-4/10 $26.00 © 2010 IEEE

Digital Charge Balance Controller with an Auxiliary Circuit for Superior Unloading Transient Performance of Buck Converters

Eric Meyer, Dong Wang, Liang Jia, Yan-Fei Liu
Department of Electrical & Computer Engineering
Queen's University
Kingston, Ontario, Canada
eric.meyer@amd.com, dong.wang@queensu.ca, liang.jia@queensu.ca, yanfei.liu@queensu.ca

Abstract— **In this paper, a digital charge balance controller is presented which is capable of controlling a Buck converter and an auxiliary circuit to achieve an excellent unloading transient response. The auxiliary circuit significantly reduces the voltage overshoot caused by an unloading transient while the digital charge balance controller reduces the settling time of the converter. The controller is capable of implementing load-line regulation and yields a smooth transition from one loading condition to another. Simulation and experimental verification is performed and demonstrates significant transient improvement over previously-proposed solutions.**

I. INTRODUCTION

As the capabilities of high-performance digital devices continue to exponentially expand, the demand on the power electronics industry to supply such devices becomes increasingly complex. Load transients of digital devices are becoming larger while physical real-estate constraints are becoming tighter preventing the tried-and-true method of adding capacitors to improve the transient performance of Buck converters. Thus, extensive research has been conducted developing controllers which improve the transient performance of Buck converters to their physical limits.

In [1]-[11], controllers have been presented which utilize second-order sliding surfaces, pre-calculated switching time intervals or capacitor charge balance methodologies to reduce the voltage deviation and settling time of a Buck converter, undergoing a load transient, to its virtually optimal level.

However ,it is demonstrated in [1]-[2] that for low duty cycle conversion applications (e.g. 12VDC→1.5VDC), the optimal voltage overshoot caused by a step-down load current transient may be more than 5 times as large as the corresponding voltage undershoot caused by a positive current step of equal magnitude, as illustrated in Figure 1. Therefore, to adhere to voltage specifications, capacitor selection must be based on the larger voltage overshoot condition.

Thus in [12]-[22], various auxiliary circuits for the Buck converter have been proposed to improve the transient performance of a converter undergoing high-to-low load

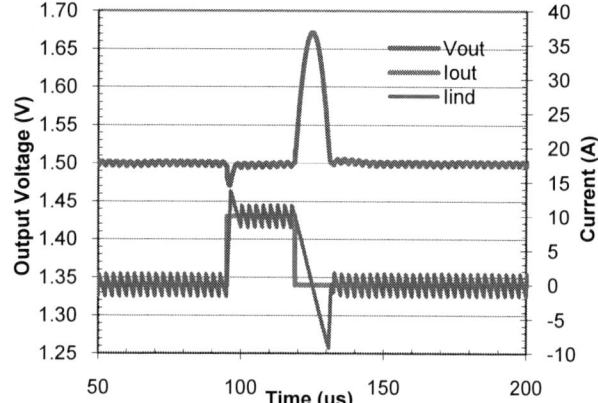

Figure 1   Assymetrical transient response to positive/negative load current step change (charge balance controller response)

current changes. Methods include temporarily inversing the converter's input voltage, temporarily disconnecting the inductor from the load, or diverting a portion of the inductor current to the input of the Buck converter through a separate switching circuit.

For example, the unloading transient response is improved in [12] by utilizing the high-frequency switching auxiliary circuit (illustrated in Figure 2) which rapidly transfers current from the output of the Buck converter to its input.

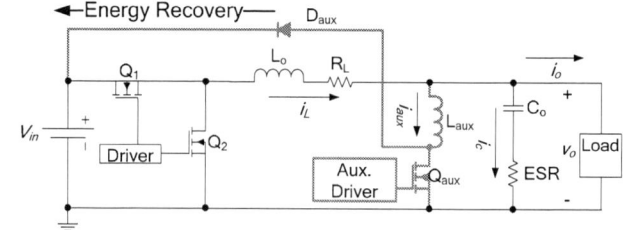

Figure 2   Implementaiton of high frequency auxiliary circuit

While such methods do improve the unloading transient performance of a Buck converter, there has been no attempt to simultaneously reduce the voltage overshoot *and* minimize the settling time through control methods such as those presented in [1]-[11]. As an example, Figure 3 shows a simulated comparison of the unloading transient response of the non-linear charge balance controller (presented in [1]-[2]), versus the auxiliary circuit and control method, presented in [12].

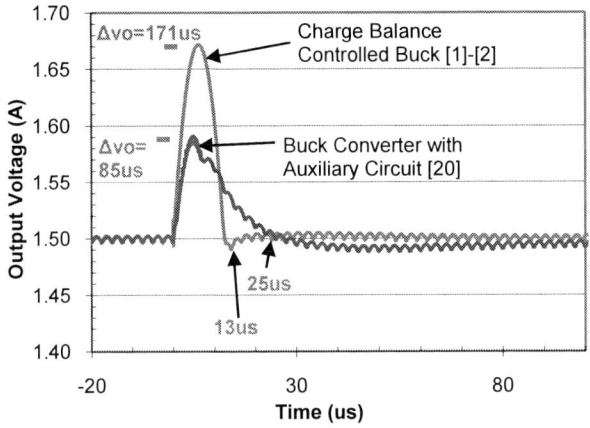

Figure 3   Charge Balance Control Respone [1]-[2] vs. Auxiliary Circuit Response [20]

It is shown in Figure 3 that while the addition of the aforementioned auxiliary circuit significantly reduces the voltage overshoot caused by an unloading transient, the settling time of the charge balance controller is far superior.

Furthermore, previously-proposed methods do not address applications in which load-line regulation (a.k.a. adaptive voltage positioning AVP) is required.

In this paper, a digital charge balance controller is proposed which combines the auxiliary circuit, presented in [12] and illustrated in Figure 2, with the control methodology presented in [1]-[2] to yield a converter with superior unloading transient performance. The proposed method actively reduces the output voltage overshoot caused by an unloading transient while minimizing the settling time to virtually-optimal levels (an achievement not demonstrated in previous literature).

Since the detailed implementation of the auxiliary circuit is covered extensively in [12], this paper will focus primarily on the proposed modified charge balance control method.

II.   Concept of Operation

The operation of the proposed method will be described without and with the use of load-line regulation.

A.   Operation without Load-Line Regulation

Figure 4 illustrates the proposed controller's reaction to a rapid unloading transient without load-line regulation.

The high-level operation can be described in 5 steps:
1. The converter is controlled by a linear voltage-mode control scheme during steady-state conditions.
2. Immediately following an unloading transient, the controller will set the Buck converter's PWM signal low and set the auxiliary circuit's PWM signal high.

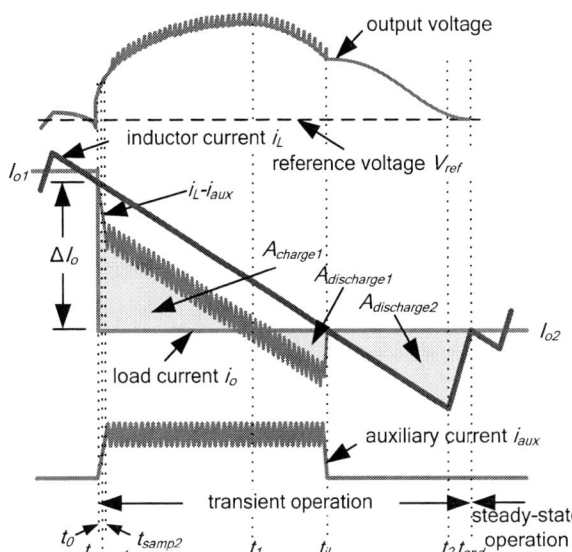

Figure 4   Proposed controller reaction to an unloading transient (w/o load-line regulation)

3. The controller will estimate the magnitude of the unloading transient $|\Delta I_o|$ and set the peak auxiliary current I_{aux_peak} to an appropriate level based on $|\Delta I_o|$. At this point, the auxiliary circuit will begin switching operation, transferring current from the Buck converter's output to its input.
4. At the moment that the inductor current i_L first equals the new load current I_{o2} (at t_{iL}), the auxiliary circuit will be deactivated. However, the Buck converter's PWM signal will continue to remain low.
5. The Buck converter's PWM signal will be set high at t_2 causing the inductor current to increase toward I_{o2}. t_2 should be such that the net capacitor charge is equal to zero at the exact moment that the inductor current equals the new load current for a second time. In other words, referring to Figure 4, $A_{charge1} = A_{discharge1} + A_{discharge2}$ when $i_L = I_{o2}$ at t_{end}. This will ensure that the output voltage and the inductor current equal their respective steady-state values simultaneously at t_{end}.

B.   Operation with Load-Line Regulation

Two cases must be addressed when charge balance control, the auxiliary circuit and load-line regulation are employed.

1)   Case #1 ($v_o > V_{o2}$ when $i_L = I_{o2}$)

Case #1 is illustrated in Figure 5.

As shown, after the moment that i_L first equals I_{o2} ($t \geq t_{iL}$), additional charge must be removed from the capacitor such that the output voltage can decrease to its new steady-voltage V_{o2}. Therefore, the PWM control signal will remain low until t_2. t_2 is such that the charge balance equation (1) is true.

$$
\begin{aligned}
&A_{charge1} - A_{discharge1} - A_{discharge2} \\
&= \int_{t_0}^{t_1} (i_L - i_{aux} - i_o)\, dt - \int_{t_1}^{t_{iL}} (i_o - i_{aux} - i_L)\, dt \\
&- \int_{t_{iL}}^{t_2} (i_o - i_L)\, dt = (I_{o1} - I_{o2}) \cdot R_{droop} \cdot C_o
\end{aligned}
\tag{1}
$$

978-1-4244-4782-4/10 $26.00 © 2010 IEEE   125

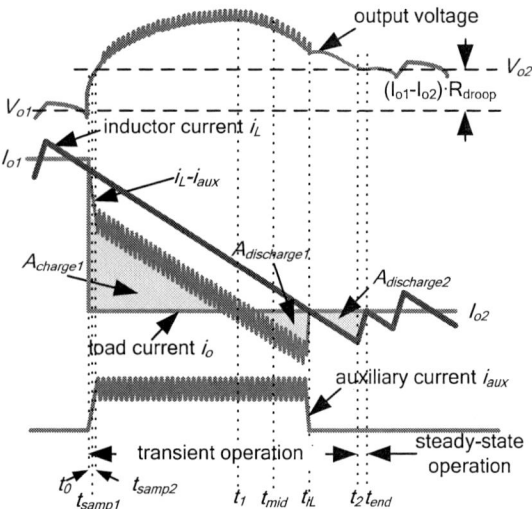

Figure 5   Controller reaction to an unloading transient with load-line regulation (Case #1)

Where $A_{charge1}$, $A_{discharge1}$, $A_{discharge2}$ are the capacitor current integral areas shown in Figure 5.

2)   *Case #2 ($v_o = V_{o2}$ before $i_L = I_{o2}$)*

Case #2 is shown in Figure 6.

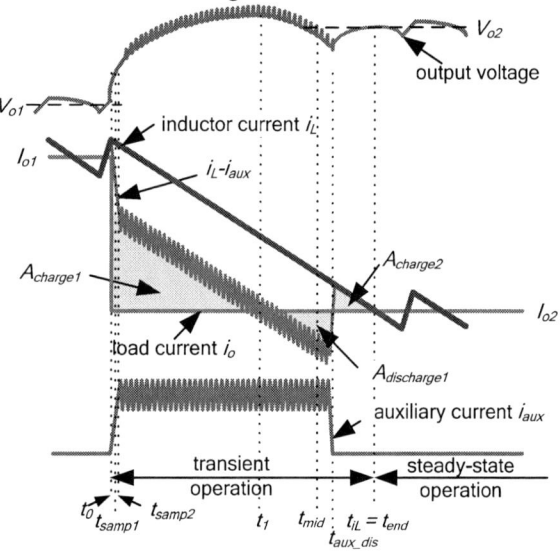

Figure 6   Controller reaction to an unloading transient with load-line regulation (Case #2)

As shown, during Case #2, the output voltage v_o equals its new steady-state voltage V_{o2} before i_L equals I_{o2}. In this case, the auxiliary circuit will be de-activated *prior* to i_L equalling I_{o2} (at t_{aux_dis}). t_{aux_dis} will be such that equation (2) is satisfied.

$$A_{charge1} - A_{discharge1} + A_{charge2}$$
$$= \int_{t_0}^{t_1} (i_L - i_{aux} - i_o)\, dt - \int_{t_1}^{t_{aux\_dis}} (i_o - i_{aux} - i_L)\, dt$$
$$+ \int_{t_{aux\_dis}}^{t_{end}} (i_L - i_o)\, dt = (I_{o1} - I_{o2}) \cdot R_{droop} \cdot C_o \tag{2}$$

The Buck converter's PWM signal will be held low until the inductor current equals the new load current (at t_{iL}). At this point, the linear controller will retain control.

III.   MATHEMATICAL ANALYSIS OF CHARGE BALANCE CONTROLLER WITH AUXILIARY CIRCUIT

This section will derive the charge balance equations necessary to implement digital charge balance control such that a Buck converter, with the proposed auxiliary circuit, will recover from an unloading transient with decreased settling time. The charge balance equations are presented without and with load-line regulation employed.

A.   Without Load-Line Regulation

Referring to Figure 4, it is noted that there is one positive integral area of capacitor current $A_{charge1}$ and two negative integral areas of capacitor current $A_{discharge1}$ and $A_{discharge2}$. The capacitor charge area $A_{charge1}$ is derived in (3).

$$A_{charge1} = \iint_{t_0}^{t_1} m_2\, dtdt \tag{3}$$

m_2 is the falling slew rate of the inductor current i_L ($m_2 \approx V_o/L_o$). t_1 represents the first moment that the capacitor current i_c equals zero (i.e. when $i_L - I_{aux\_avg} = I_{o2}$).

Referring to Figure 4, it is shown that the auxiliary circuit is de-activated when the inductor current equals the new load current (at t_{iL}). It is assumed that when the auxiliary circuit is de-activated, that the auxiliary current i_{aux} decreases to zero in negligible time. This is a fair assumption since the falling i_{aux} slew rate is much faster than the falling i_L slew rate.

For the charge balance equations, the ripple of the auxiliary current is neglected since the high frequency auxiliary switching causes the ripple's effect to be neutralized.

With the above assumptions, $A_{discharge1}$ is calculated in (4).

$$A_{discharge1} = \iint_{t_1}^{t_{iL}} m_2\, dtdt \tag{4}$$

Through   geometric   observation   and   simplification, $A_{discharge2}$ is expressed in (5).

$$A_{discharge2} = \iint_{t_{iL}}^{t_2} \frac{m_1 \cdot m_2 + m_2^2}{m_1}\, dtdt \tag{5}$$

m_1 is the rising slew rate of the inductor current i_L ($m_1 \approx (V_{in}-V_o)/L_o$). In order to ensure that v_o and i_L equal their respective steady-state values simultaneously, the net capacitor charge over the transient period must equal zero, as expressed in (6).

$$A_{charge1} - A_{discharge1} - A_{discharge2} = 0$$
$$\iint_{t_0}^{t_1} m_2\, dtdt - \iint_{t_1}^{t_{iL}} m_2\, dtdt - \iint_{t_{iL}}^{t_2} \frac{m_1 \cdot m_2 + m_2^2}{m_1}\, dtdt = 0 \tag{6}$$

By dividing both sides of (6) by the constant m_2 and substituting the known values for m_1 and m_2, multiplying both sides by $(V_{in}-V_o)$ and simplifying, equation (7) is created.

$$(V_{in} - V_o) \cdot \iint_{t_0}^{t_1} dt\, dt - (V_{in} - V_o) \cdot \iint_{t_1}^{t_{iL}} dt\, dt - V_{in}$$
$$\cdot \iint_{t_{iL}}^{t_2} dt\, dt = 0 \tag{7}$$

978-1-4244-4782-4/10 $26.00 © 2010 IEEE

Therefore, by using (7), it is possible to determine the moment that the Buck converter's PWM control signal should be set high (at t_2), by the use of two accumulators, connected in series. The waveforms of the double accumulator are shown in Figure 6. This methodology is similar to the double accumulator method presented in [2].

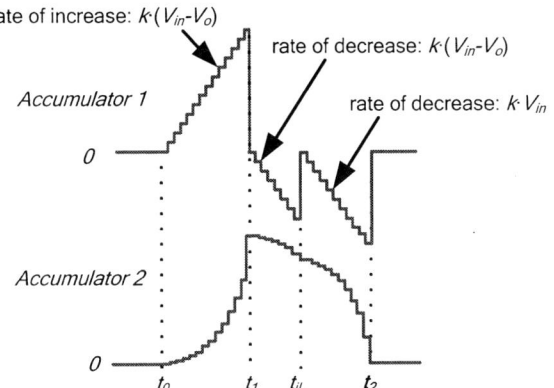

Figure 7   Digital double accumulator to determine t_2 without load-line regulation

As shown, when the output of *accumulator 2* returns to zero, t_2 is determined and the Buck converter's PWM signal is set high. This control strategy is suitable when load-line regulation is not employed; however, it must be modified when load-line regulation is employed.

B. With Load-Line Regulation

This subsection will be divided into Case #1 and Case #2, as previously described in Section II.

1)   Case #1 ($v_o > V_{o2}$ when $i_L = I_{o2}$)

The capacitor charge regions for Case #1 are illustrated in Figure 5. As shown, a time instant t_{mid} has been identified in Figure 5. t_{mid} will be used in the charge balance calculation and occurs when (8) is true.

$$i_L - I_{o2} = \frac{1}{2} \cdot I_{aux_{avg}} \quad when\ t = t_{mid} \tag{8}$$

As shown, t_{mid} bisects time instants t_1 and t_{iL}. The use of t_{mid} will be discussed in this section. The method of detecting t_{mid} will be discussed in Section IV.

The charge balance equation can be obtained by modifying (7), as expressed in (8).

$$(V_{in} - V_o) \cdot \iint_{t_0}^{t_1} dt\,dt - (V_{in} - V_o) \cdot \iint_{t_1}^{t_{iL}} dt\,dt - V_{in}$$
$$\cdot \iint_{t_{iL}}^{t_2} dt\,dt = (V_{in} - V_o) \cdot R_{droop} \cdot C_o \cdot \int_{t_0}^{t_{iL}} dt \tag{9}$$

The right side of equation (9) represents the required capacitor charge offset to implement load-line regulation. As expressed, the output of an additional accumulator (the *load-line accumulator*) is compared with the output of the double accumulator to determine t_2.

Since t_{mid} bisects time instances t_1 and t_{iL} and the input of the *load-line accumulator* is a constant, the final value of the *load-line accumulator* is obtained at time instant t_{mid}, as expressed in (10).

$$\int_{t_0}^{t_{iL}} dt = \int_{t_0}^{t_1} dt + \int_{t_1}^{t_{mid}} dt + \int_{t_{mid}}^{t_{iL}} dt = \int_{t_0}^{t_1} dt + 2 \cdot \int_{t_1}^{t_{mid}} dt \tag{10}$$

By substituting (10) into (9), the charge balance equation for load-line regulation (Case #1) is derived in (11).

$$(V_{in} - V_o) \cdot \iint_{t_0}^{t_1} dt\,dt - (V_{in} - V_o) \cdot \iint_{t_1}^{t_{iL}} dt\,dt - V_{in}$$
$$\cdot \iint_{t_{iL}}^{t_2} dt\,dt = (V_{in} - V_o) \cdot R_{droop} \cdot C_o \cdot \left(\int_{t_0}^{t_1} dt + 2 \cdot \int_{t_1}^{t_{mid}} dt \right) \tag{11}$$

Thus, t_2 can be determined by using the aforementioned double accumulator (left side of (11)) and comparing its output with the output of a single accumulator (right side of (11)).

2)   Case #2 ($v_o = V_{o2}$ before $i_L = I_{o2}$)

The capacitor charge regions for Case #2 are illustrated in Figure 6. It is shown that, for Case #2, the moment the auxiliary circuit is de-activated t_{aux_dis} occurs before t_{iL} to allow charge area $A_{charge2}$ to charge the output capacitor.

If the auxiliary circuit were de-activated at t_{mid}, the capacitor current ($i_L - I_{o2}$) would equal $I_{o2} + I_{aux\ avg}$, which can also be expressed in terms of the falling inductor slew rate, as shown in (12).

$$i_L(t_{mid}) - I_{o2} = \int_{t_1}^{t_{mid}} m_2 dt \tag{12}$$

Following t_{mid}, the charge area $A_{charge2}$ begins to decrease as the inductor current approaches the new load current at a slew rate of m_2. Thus, $A_{charge2}$ is expressed in (13).

$$A_{charge2} = A_{discharge1}$$
$$- \int_{t_{mid}}^{t_{aux\_dis}} \left(i_L(t_{mid}) - I_{o2} - m_2 \cdot \int_{t_{mid}}^{t_{aux\_dis}} dt \right) d\tau \tag{13}$$

By substituting (3),(4),(10),(13), into (2), the charge balance equation for Case #2 load line regulation is expressed in (14).

$$\iint_{t_0}^{t_1} m_2\, dt\,dt - \iint_{t_1}^{t_{aux\_dis}} m_2\, dt\,dt + \iint_{t_1}^{t_{mid}} m_2 dt\, dt$$
$$- \int_{t_{mid}}^{t_{aux\_dis}} \left(\int_{t_1}^{t_{mid}} m_2 dt - \int_{t_{mid}}^{t_{aux\_dis}} m_2 dt \right) d\tau$$
$$= R_{droop} \cdot C_o \cdot \left(\int_{t_0}^{t_1} m_2 dt + 2 \cdot \int_{t_1}^{t_{mid}} m_2 dt \right) \tag{14}$$

By collecting like terms, (14) is simplified, as shown in (15).

$$\iint_{t_0}^{t_1} dt\,dt - \int_{t_{mid}}^{t_{aux\_dis}} \left(2 \cdot \int_{t_1}^{t_{mid}} dt \right) dt$$
$$= R_{droop} \cdot C_o \cdot \left(\int_{t_0}^{t_1} dt + 2 \cdot \int_{t_1}^{t_{mid}} dt \right) \tag{15}$$

By using (15), the output of a double accumulator (left side) can be compared to the output of a single accumulator (right side) to determine the moment to deactivate the auxiliary circuit t_{aux_dis}.

In order to minimize the digital gate count of the controller, it is beneficial to design one double accumulator that can be used for Case #1, Case #2 and cases without load-line regulation. In order to achieve this, charge balance equation (15) can be modified by multiplying both sides by $(V_{in} - V_o)$, as expressed in (16).

978-1-4244-4782-4/10 $26.00 © 2010 IEEE

$$(V_{in} - V_o) \cdot \iint_{t_0}^{t_1} dt\, dt - (V_{in} - V_o) \cdot \int_{t_{mid}}^{t_{aux\_dis}} \left(2 \cdot \int_{t_1}^{t_{mid}} dt \right) d\tau$$
$$= R_{droop} \cdot C_o \cdot (V_{in} - V_o) \cdot \left(\int_{t_0}^{t_1} dt + 2 \cdot \int_{t_1}^{t_{mid}} dt \right) \qquad (16)$$

By utilizing the combination of (11) and (16), one double accumulator can be used for all cases, as shown in Figure 8.

Figure 8 illustrates the load-line accumulator output for Case #2, Case #1 and no load-line (from top to bottom).

As shown, if the output of *accumulator 2* decreases below that of the *load-line accumulator* before t_{iL}, Case #2 is detected and the auxiliary circuit is de-activated, at time t_{aux_dis}.

If t_{iL} occurs before the output of *accumulator 2* decreases below the output of the *load-line accumulator*, the auxiliary circuit is de-activated at t_{iL}, as shown in Figure 4 and Figure 5. In these cases, the Buck converter's PWM signal will remain low until output of *accumulator 2* decreases below the *load-line accumulator*'s output at time t_2.

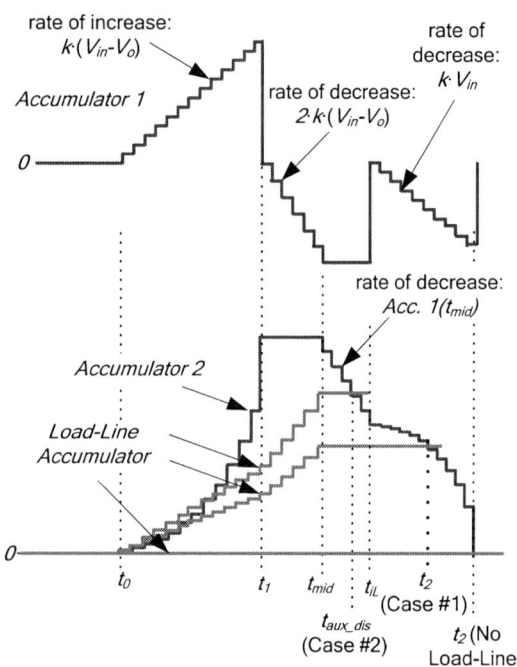

Figure 8 Digital double for all possible cases

IV. DETAILED OPERATION AND IMPLEMENTATION OF CHARGE BALANCE CONTROLLER WITH AUXILIARY CIRCIUT

The high-level system diagram of the digital charge balance controller with auxiliary circuit for a synchronous Buck converter is illustrated in Figure 9.

This section summarizes the operation of the proposed charge balance controller with the auxiliary circuit. The operation of the controller can be summarized in eight steps.

Step 1 ($t=t_0$)

The Buck converter is controlled by a digital linear controller during steady-state operation. Following an unloading transient, the output of the analog load transient detector (see Figure 9) will rapidly exceed the transient threshold.

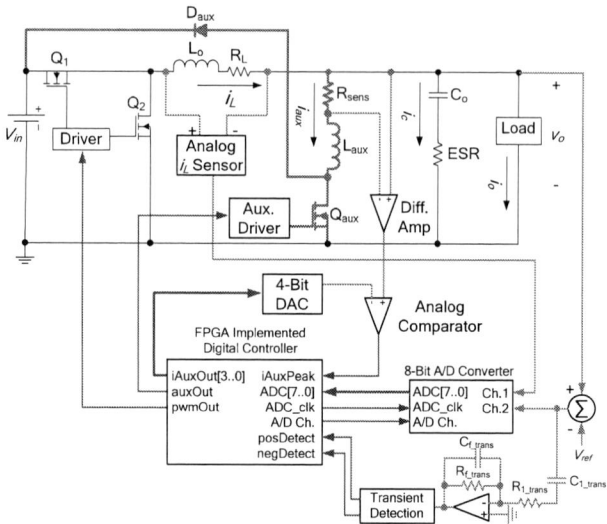

Figure 9   System-level block diagram of digitally-implemented charge balance controller with auxiliary circuit

This will cause the controller to immediately enter transient mode. The linear controller will be frozen and the charge balance controller will take control of the converter.

The PWM signal of the Buck converter will be initially set low and the PWM signal of the auxiliary circuit will be initially set high.

As shown in Figure 8, the output of *accumulator 1* will begin to increase linearly with a slope of $k \cdot (V_{in}-V_o)$ and the output of *accumulator 2* will begin to increase exponentially. If load-line regulation is required, the *load-line accumulator* will begin to increase linearly with a slope of $k^2 \cdot R_{droop} \cdot C_o \cdot (V_{in}-V_o)$.

Step 2 ($t_{samp1} \leq t \leq t_{samp2}$)

As shown in Figure 4-Figure 6, two output voltage samples are acquired at t_{samp1} and t_{samp2} to estimate the load transient magnitude. Using this information and equation (17), an appropriate value of I_{aux_peak} is selected from an LUT. The digital value I_{aux_peak} is passed to the system's DAC and the i_c cross-over predictor. At t_{samp2}, the auxiliary circuit will begin high-frequency switching operation using a peak-current, constant off-time controller. The appropriate peak current is user defined based on the desired output voltage response.

$$\Delta I_o = C_o \cdot \frac{\Delta v_{o\_samp}}{T_{samp}}$$
$$+ \left[V_o \cdot \left(\frac{1}{L_{aux}} + \frac{1}{L_o} \right) \cdot \left(t_{samp2} - {}^{1}/_{2} \cdot T_{samp} - t_0 \right) + ESR \right] \qquad (17)$$

It is shown in Figure 9, that an analog comparator is used to detect the peak auxiliary current I_{aux_peak}. The detection signal is used by the digital controller. The mixed-signal implementation is necessary due to the high slew rate of the auxiliary current.

Step 3 ($t=t_1$)

The capacitor current crosses zero for the first time at t_1. In order to predict the time instances t_1, t_{mid} and t_{iL}, a capacitor current zero cross-over predictor is utilized. This method is introduced in [2]; however in the proposed method, the predictor is capable of compensating for the known average auxiliary current I_{aux_avg}. The predictor acquires output voltage

978-1-4244-4782-4/10 $26.00 © 2010 IEEE   128

samples (at a frequency equivalent to auxiliary circuit switching frequency f_{aux}) for a short period after t_0. From these samples, it is possible to calculate the derivative, estimate the capacitor current slope and magnitude allowing for the fine resolution prediction of t_1, t_{mid} and t_{iL}, as shown in Figure 10.

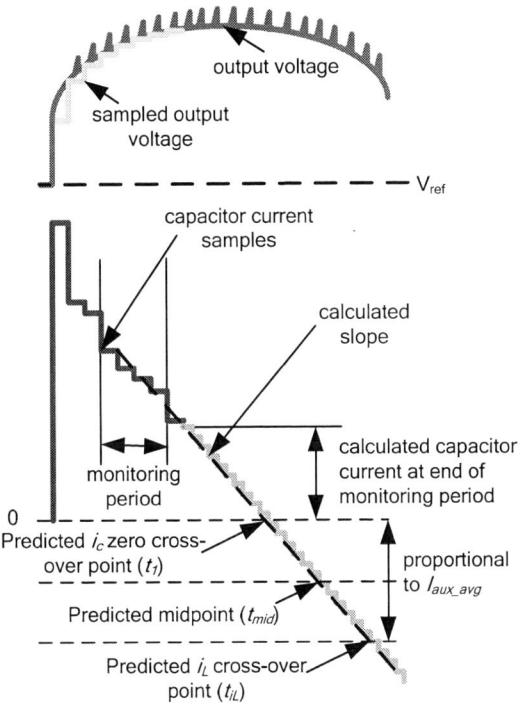

Figure 10 Prediction of t_1, t_{mid} and t_{iL} by capacitor current zero cross-over predictor

Referring to Figure 8, at t_1, *accumulator 1* is reset to zero. The input of *accumulator 1* is set to $-2 \cdot (V_{in} - V_o)$ and the enable input of *accumulator 2* is set low. Thus, the output of *accumulator 1* will begin to decrease at a rate of $2 \cdot k \cdot (V_{in} - V_o)$ and the output of *accumulator 2* will remain constant.

The channel select of the ADC is set to the inductor current sensor at this point, and a sample of the inductor current is taken and passed to the linear controller. This will be used by the linear controller for load-line regulation.

Referring to Figure 8, the input of the *load-line accumulator* is switched to $2 \cdot k \cdot R_{droop} \cdot C_o \cdot (V_{in} - V_o)$ at t_1.

Step 4 ($t=t_{mid}$)

When the i_c zero cross-over predictor indicates $t=t_{mid}$, it is known that $i_L - I_{o2} = \frac{1}{2} \cdot I_{aux\_avg}$. Referring to Figure 8, at $t=t_{mid}$, the *enable* input of *accumulator 1* is set low and the enable input of *accumulator 2* is set high. *Accumulator 2* is set to decrement mode causing its output to decrease at a linear rate equal to $k \cdot Acc1(t_{mid})$ (where $Acc1(t_{mid})$ equals the output of *accumulator 1* at $t=t_{mid}$).

As shown in Figure 8, the enable input of the *load-line accumulator* is set low at t_{mid}, causing its input to remain constant following t_{mid}.

Step 5 ($t=t_{aux\_dis}$) (Case #2 only)

As shown in Figure 8, if the output of *accumulator 2* decreases below that of the *load-line accumulator* before the inductor current equals the new load current (at t_{iL}) then Case

#2 is detected. If this occurs, the auxiliary circuit is de-activated at this moment ($t=t_{aux\_dis}$), as shown in Figure 6. The PWM signal of the Buck converter continues to be held low.

Step 6 ($t=t_{iL}$)

When the inductor current equals the new load current for the first time, $t=t_{iL}$.

If Case #2 was previously detected, this moment signifies the end of the load transient and the linear controller re-takes control of the Buck converter.

If Case #2 was not previously detected, *accumulator 1* is cleared and its input is switched to $-V_{in}$. This causes *accumulator 1*'s output to decrease at a rate of $k \cdot V_{in}$ and the output of *accumulator 2* to decrease at an exponential rate. As shown in Figure 4 and Figure 5, the auxiliary circuit is de-activated and the PWM control signal is held low following t_{iL}.

Step 7 ($t=t_2$) (No Load-Line Regulation or Case #1)

When the output of *accumulator 2* decreases below that of the *load-line accumulator* (at t_2), the PWM signal of the Buck converter is switched high and the inductor current begins to increase toward the new load current, as shown in Figure 4 and Figure 5.

Step 8 ($t=t_{end}$)

When the inductor current equals the new load current for the first time (for Case #2) or the inductor current equals the new load current for a second time, it is determined that the transient is over. The second current cross-over is detected using a digital accumulator (*accumulator 3*), as introduced in [2]. The linear controller is unfrozen and retakes control of the Buck converter.

V. SIMULATION RESULTS

The following simulation was conducted under the ideal case, without considering timing delays, digital quantization effects, etc. The purpose of the simulation was to demonstrate the effectiveness of the charge balance controller with the auxiliary circuit over (1) a linear analog controller and (2) the charge balance controller alone.

The parameters of the simulated Buck converter were as follows: V_{in}=12V, V_o=1.5V, f_{sw}=400kHz, L_o=1uH, C_o=180uF, ESR=0.5mΩ, ESL=100pH. The auxiliary circuit parameters were: L_{aux}=100nH, $f_{aux} \approx$2MHz.

The transient response was simulated with load-line regulation. The output impedance R_{droop} was set to 5mΩ. Figure 11 illustrates a simulated comparison between a voltage-mode controlled converter, a digital charge balance controlled Buck converter (without auxiliary circuit) [2] and the proposed digital charge balance controller with auxiliary circuit. Each converter undergoes a 10A→0A load step transient.

It is illustrated that the voltage deviation magnitude and the settling time is improved significantly over previously-proposed solutions. The Buck converter with the proposed controller and auxiliary circuit has an output voltage overshoot of 18 mV (68mV-50mV) over the new steady-state voltage (68mV over the original steady-state voltage) and a settling time of 7us.

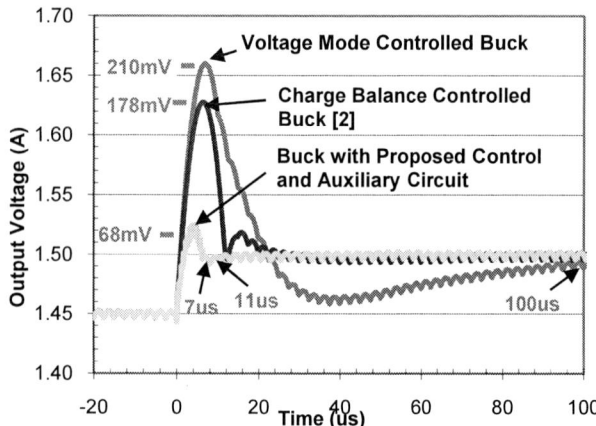

Figure 11 Simulated response to a 10A→0A load current step change with load line regulation (I_{aux_avg} = 3.8A)

VI. EXPERIMENTAL RESULTS

The proposed controller was implemented on an Altera Cyclone II FPGA. The Buck converter and auxiliary circuit parameters were identical to those of the simulation.

Figure 12 illustrates the controller's reaction to an 11.5A→0A load step (without load line regulation). For this unloading magnitude, the auxiliary current was set to approximately I_{aux_avg}=3.5A. For reference, the time instants t_0-t_{end} were super-imposed on the scope display to better illustrate the controller's behavior.

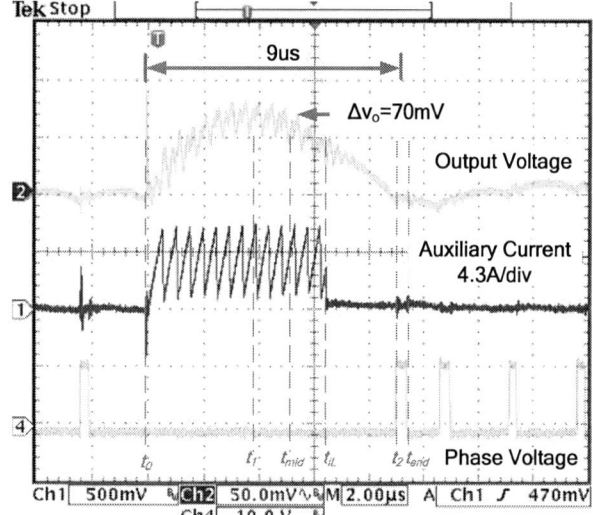

Figure 12 Digital charge balance controller's response to a 11.5A→0A load step without load line regulation

As shown in Figure 13, the auxiliary circuit is de-activated when the inductor current first equals the new load current (at t_{iL}). However, the Buck converter's PWM signal is kept low until t_2 is determined. This is to allow additional charge to be removed from the output capacitor such that the output voltage equals the reference voltage at the exact moment that the inductor current equals its new steady-state value.

The converter is capable of recovering from the load transient within 9us with a voltage overshoot of 70mV.

Figure 13 illustrates the unloading transient response of the digital charge balance controller (introduced in [2]) without the use of the auxiliary circuit.

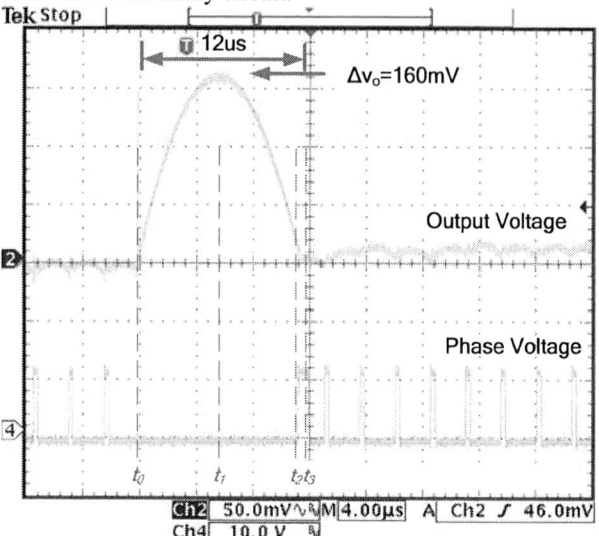

Figure 13 Digital charge balance controller's response (without auxiliary circuit) to a 11.5A→0A load step without load line regulation

As observed, the use of the auxiliary circuit improves the settling time by 25% (12us→9us) and improves the voltage overshoot by 56% (160mV→70mV) over that of the digital charge balance controller alone.

Figure 14 illustrates the controller's reaction to an 11.5A→0A load step (with load line regulation). Figure 15 illustrates the inductor current (measured from the analog inductor current sensor).

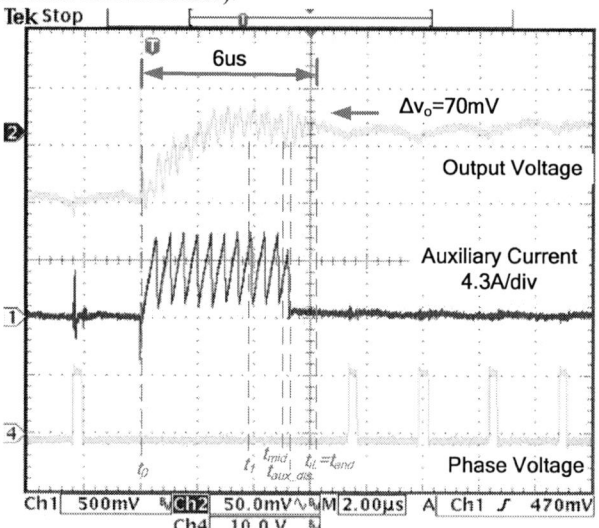

Figure 14 Digital charge balance controller's response to an 11.5A→0A load step with load line regulation

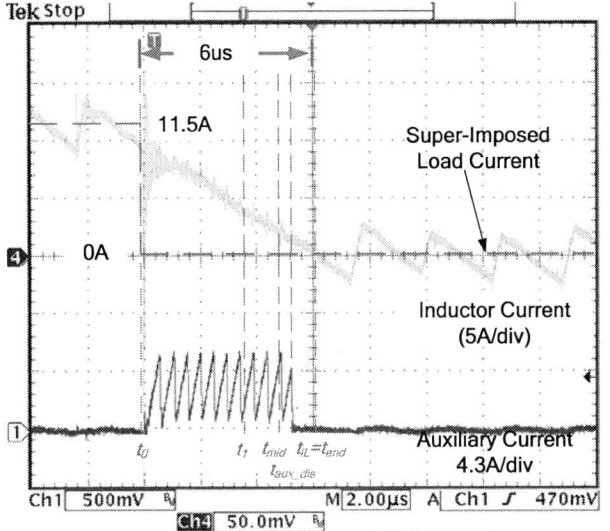

Figure 15 Digital charge balance controller's response to an 11.5A→0A load step with load line regulation (inductor current and auxiliary current)

As illustrated, Case #2 occurs and the auxiliary circuit is deactivated before t_{iL}. The auxiliary current is deactivated at t_{aux_dis} to allow the capacitor charge areas to appropriately balance by time t_{iL}. This results in a smooth transition as the output voltage equals its new steady-state value at the exact moment that the inductor current equals the new load current. The converter is able to recover from the unloading step within 6us and with only a 10mV (70mV-60mV) overshoot beyond the final steady-state voltage.

VII. CONCLUSIONS

In this chapter a novel digital charge balance control method is described capable of reducing the voltage overshoot of a Buck converter (through the use of an auxiliary circuit) *and* implementing load line regulation. The proposed digital controller does not require multipliers, dividers or two-dimensional LUTs, significantly reducing the IC real-estate required.

The use of the auxiliary circuit significantly reduces the voltage overshoot (due to an unloading transient) beyond the physical capabilities of the Buck converter alone. It is shown that by implementing the charge balance principle with the auxiliary circuit, that the settling time can be also significantly improved over previously proposed solutions. In addition, it is demonstrated that the controller can be extended to applications which require load-line regulation.

REFERENCES

[1] E. Meyer, Z. Zhang, Y-F. Liu, "An Optimal Control Method for Buck Converters Using a Practical Capacitor Charge Balance Technique", *IEEE Transactions on Power Electronics*, vol. 23, no. 4, July 2008, pp. 1802-1812

[2] E. Meyer, Z. Zhang, Y.F. Liu, " Digital Charge Balance Controller with Low Gate Count to Improve the Transient Response of Buck Converters", *IEEE Energy Conversion Congress and Exposition (ECCE)*, San Joes, California, September 2009, pp. 3320-3327.

[3] K.K.S Leung, H.S.H. Chung, "A Comparative Study of Boundary Control With First- and Second-Order Switching Surfaces for Buck

Converters Operating in DCM", *IEEE Transactions on Power Electronics*, vol. 22, no. 4, July 2007, pp 1196-1209

[4] T. Geyer, G. Papafotiou, R. Frasca, M. Morari, "Constrained Optimal Control of Step-Down DC-DC Converter", *IEEE Transactions on Power Electronics*, Volume 23, Issue 5, September 2008, pp. 2454-2464

[5] M. Oronez, M.T. Iqbal, J.E. Quaicoe, "Selection of a Curved Switching Surface for Buck Converters", *IEEE Transactions on Power Electronics*, Volume 21, Issue 4, July 2006, pp 1148-1153

[6] A. Soto, A. de Castro, P. Alou, J.A. Cobos, J. Uceda, A. Lotfi, "Analysis of the Buck Converter for Scaling the Supply Voltage of Digital Circuits", *IEEE Transactions on Power Electronics*, vol. 22, no. 4, July 2007, pp. 1196-1209

[7] G. Feng, E. Meyer, Y.F. Liu, "A New Digital Control Algorithm to Achieve Optimal Dynamic Response Performance in DC-to-DC Converters", *IEEE Transactions on Power Electronics*, vol. 22, no. 4, July 2007, pp. 1489-1498

[8] Z. Zhao, A. Prodic, "Continuous-Time Digital Controller for High-Frequency DC-DC Converters", *IEEE Transactions on Power Electronics*, vol. 23, no. 2 March 2008, pp. 564-573

[9] S. Effler, A. Kelly, M. Halton, T. Kruger, K. Rinne, "Digital Control Law using a Novel Load Current Estimator Principle for Improved Transient Response", *IEEE Power Electronics Specialists Conference, 2008*, pp. 4585-4589

[10] A. Costabeber, L. Corradini, P. Mattavelli, S. Saggini, "Time Optimal, Parameters-Insensitive Digital Controller for DC-DC Buck Converters", *IEEE Power Electronics Specialists Conference*, 2008, pp. 1243-1249

[11] V. Yousefzedah, A. Babazadeh, B. Ramachandran, E. Alarcon, L. Pao, D. Maksimovic, "Proximate Time-Optimal Control for Synchronous Buck DC-DC Converters", *IEEE Transactions on Power Electronics*, vol. 23, no 4, July '08, pp. 2018-2026

[12] E. Meyer, Z. Zhang, Y.F. Liu, "Controlled Auxiliary Circuit with Measured Response for Reduction of Output Voltage Overshoot in Buck Converters", *IEEE Applied Power Electronics Conference (APEC)*, 2009, pp. 1367-1373

[13] R. Singh, A. Khambadkone, "A Buck Derived Topology with Improved Step-Down Transient Performance", *IEEE Transactions on Power Electronics*, Volume 23, Issue 6, November 2008, pp. 2855-2866

[14] M. Rico, J. Uceda, J. Sebastian, F. Aldana, "Static and Dynamic Modeling of Tapped-Inductor DC-to-DC Converters", *IEEE Power Electronics Specialists Conference (PESC)*, 1987, pp. 281-288

[15] D.D-C Lu, J.C.P. Liu, F.N.K. Pong, B.M.H. Pong, "A Single Phase Voltage Regulator Module (VRM) With Stepping Inductance for Fast Transient Response", *IEEE Transactions on Power Electronics*, Volume 22, Issue 2, March 2007, pp. 417-424

[16] A. Stupar, Z. Lukic, A. Prodic, "Digitally-Controlled Steered-Inductor Buck Converter for Improving Heavy-to-Light Load Transient Response", *IEEE Power Electronics Specialists Conference (PESC)*, 2008, pp. 3950-3954

[17] X. Wang, I. Batarseh, S.A. Chickamennahalli, E. Standford, "VR Transient Improvement at High Slew Rate Load—Active Transient Voltage Compensator", *IEEE Transactions on Power Electronics*, Volume 22, Issue 4, July 2007, pp. 1472-1479

[18] H. Zhou, X. Wang, T. Wu, I. Batarseh, "Magnetics Design for Active Transient Voltage Compensator", *IEEE Applied Power Electronics Conference (APEC)*, 2006

[19] O. Abdel-Rahman, I. Batarseh, "Transient Response Improvement in DC-DC Converters Using Output Capacitor Current for Faster Transient Detection", *IEEE Power Electronics Specialists Conference (PESC)*, 2007, pp. 157-160

[20] X. Wang, L. Qingshui, I. Batarseh, "Transient Response Improvement in Isolated DC-DC Converter with Current Injection Circuit", *IEEE Applied Power Electronics Conference (APEC)*, 2005, pp. 706-710

[21] A. Barrado, A. Lazaro, R. Vazquez, V. Salas, E. Olias, "The Fast Response Double Buck DC-DC Converter (FRDB): Operation and Output Filter Influence", *IEEE Transactions on Power Electronics*, Volume 20, Issue 6, November 2005, pp. 1261-1270

[22] A.M. Wu, S.R. Sanders, "An Active Clamp Circuit for Voltage Regulation Module (VRM) Applications", *IEEE Transactions on Power Electronics*, vol. 16, no. 5, September 2001, pp. 623-634

One-Step Digital Dead-time Correction For DC-DC Converters

April (Yang) Zhao, Armin Akhavan Fomani, Wai Tung Ng

The Edward S. Rogers Sr. Electrical and Computer Engineering Department, University of Toronto
10 King's College Road, Toronto, Ontario, Canada M5S3G4
E-mail: ngwt@vrg.utoronto.ca

Abstract — **This paper introduces a novel one-step digital control technique that can dynamically optimize the dead-times for the turn-on and turn-off of the power MOSFETs in DC-DC converters. A NOR gate and a delay-line circuit are used to detect and measure the duration of the unwanted low-side MOSFET body-diode conduction. Based on this measurement, the optimum dead-time is calculated on-the-fly and the DPWM controller will respond immediately to maximize the conversion efficiency in the next switching cycle. This approach is well suited for digital IC implementation. Experimental results from a digitally controlled 6V to 1V, 10A synchronous buck converter verified the efficiency improvement and the practical implementation of the proposed one-step dead-time correction algorithm. This one-step dead-time correction can improve the converter's efficiency by 2 to 4%, depending on output current, output voltage and switching frequency.**

I. INTRODUCTION

Due to the finite turn-on and turn-off delays of power MOSFETs, dead-time is required to eliminate the conduction loss arise from the simultaneous conduction of the high-side (HS) and low-side (LS) switches. The length of the dead-time affects the power conversion efficiency in a significant way. Insufficient dead-time will result in shoot-through current via the HS and LS switches, while excessively long dead-time will result in unwanted body-diode conduction loss and reverse recovery loss in the LS power MOSFET.

The optimum dead-time varies with circuit parameters, operation conditions and temperature. Traditional DC-DC converters that use fixed dead-time control schemes do not take varying operating conditions into account. In order to address the switching delay that exists between the DPWM signal and the turning on and off of the power MOSFETs, an adaptive delay technique was proposed to provide on-the-fly dead-time adjustment for different MOSFETs and accommodate temperature variations [1], [2]. However, this technique relies on fast comparators to match the critical instants and lacks the compensation for the MOSFET gate charging time. Predictive gate drive scheme [3] uses a very slow linear incremental method to search for the optimum

dead-time. Analogue predictive delay-locked loop schemes [4], [5] can mitigate the dependence on fast comparators to some degree and offer faster response time, at the expense of complex analogue circuits. Recently, a dead-time searching algorithm was proposed [6]. Unfortunately, it requires many switching cycles to obtain the optimum dead-time, rendering this scheme unsuitable for applications with frequently changing loads.

This paper proposes a one-step digital correction algorithm to provide quick and accurate adjustment of the dead-time. The proposed approach is fully synthesizable, requires minimum hardware resources and is well suited for FPGA and ASIC implementation. The fundamental building blocks of the one-step correction controller will be presented in following sections.

II. SYSTEM OPERATION

The proposed dead-time control system as shown in Fig. 1 is a fully digital implementation. The body-diode conduction detection circuit (NOR gates) is realized with discrete components in the output stage. The control system comprises of a pulse width measurement module, an optimum dead-time generator module and a hybrid DPWM module.

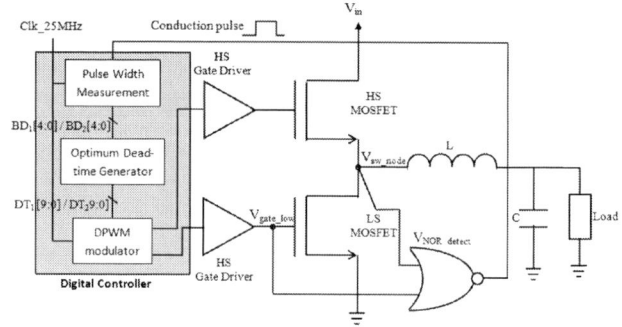

Figure 1. The dead-time correction algorithm is implemented digitally, consisting of a pulse width measurement module, an optimum dead-time generator module and a hybrid DPWM module.

978-1-4244-4782-4/10 $26.00 © 2010 IEEE

The controller operates as follows: the body-diode conduction detection circuit monitors the gate-source voltage, $V_{gate-low}$ and the drain-source voltage, $V_{sw-node}$ of the LS power MOSFET. Whenever body-diode conduction is detected, a pulse will be generated and pass onto the pulse width measurement module. The conduction pulse width is proportional to the duration of the body-diode conduction. It can be measured with a resolution of 1.25ns. This produces a 5 bits digital signal for the optimum dead-time generator. Based on the body-diode conduction time measurement and the current dead-time values, the optimum dead-time is then calculated, and the hybrid DPWM module will generate the appropriate HS and LS DPWM signals in next switching cycle.

III. FUNCTIONALITY DESCRIPTION

A. Turn-on and Turn-off Delays

The HS level shifter, bootstrap circuit, HS and LS gate driver circuits in DC-DC converters can introduce different propagation delays that alter the actual dead-time seen by the gate electrodes of the HS and LS power MOSFETs. In some cases, the dead-time deviation (offset) between the actual gate signals and the DPWM signals can be quite substantial as shown in Fig. 2. With the presence of the dead-time offset, the gate drive signals may require positive or negative dead-times. In addition, the dead-times required for the rising and falling edge of the switching signals are not the same. Fig. 3 illustrates the four possible cases of dead-time with respect to the HS signals that need to be accounted.

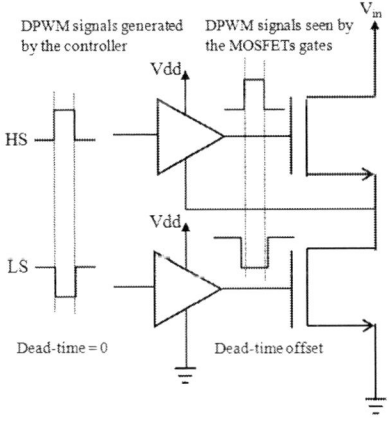

Figure 2. Dead-time offset caused by the difference in propagation delays between the HS and LS gate drivers.

B. One-Step Dead-time Correction Algorithm

A flow chart for the proposed one-step dead-time correction algorithm is as shown in Fig. 4. The algorithm starts with an initial safe dead-time value of 5 ns. It begins by measuring the actual body diode conduction time and then changes the dead-time to the optimum value in the next switching cycle. During the transitions (0-1 and 1-0) of the HS DPWM signal, a NOR gate senses the gate-source

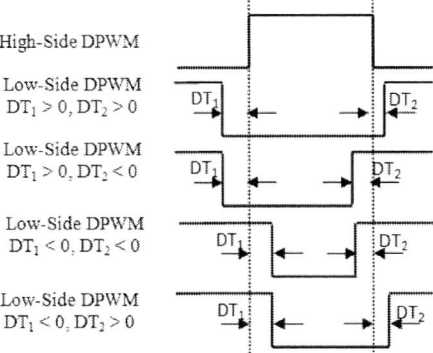

Figure 3. Timing diagram for positive and negative dead-times seen by the MOSFET gates.

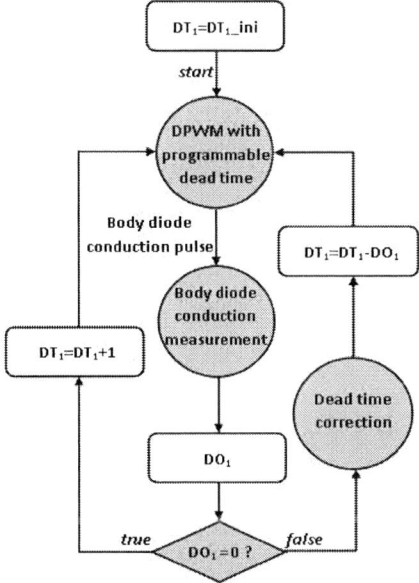

Figure 4. A flow chart representation of the one-step dead-time optimization algorithm.

voltage, $V_{gate-low}$ and the drain-source voltage, $V_{sw-node}$ from the LS power MOSFET. Whenever there is a delay between the transitions of these two signals, body-diode conduction is assumed [3]. This delay time is the excess dead-time that needs to be measured and subtracted from the current dead-time values (this is done digitally via a subtraction circuit in the FPGA implementation). The new dead-time is then used to update the DPWM signals for the next switching cycle. Zero output from the NOR gate that monitors $V_{gate-low}$ and $V_{sw-node}$ may indicate (1) insufficient dead-time condition or (2) optimum dead-time condition. Under this situation, the controller will increase the dead-time by 1-bit in the next cycle. In case (1), dead-time will continuously increase until the optimum value is obtained. Case (2) implies that the system is operating at the cross-conduction boundary, dead-time dithers within a 1-bit window. Finer adjustment of the dead-time ensures the maximization of conversion efficiency. This optimization method is independently applied to both the turn-on and turn-off edge of the gate drive signals.

978-1-4244-4782-4/10 $26.00 © 2010 IEEE   133

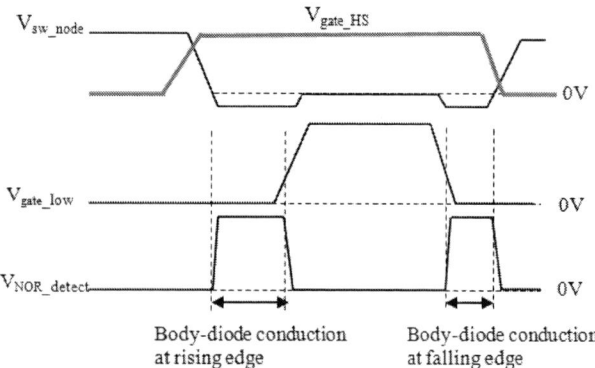

Figure 5. Detection of body-diode conduction using a simple NOR gate.

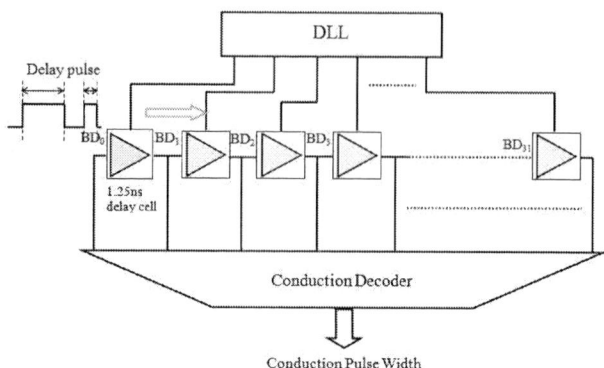

Figure 6. Body-diode conduction pulse width measurement using a delay line and delay lock loop.

C. Body-diode Conduction Detection

The presence of body-diode conduction in the LS power MOSFET is an indication of excessively long dead-time. A simple way to detect body-diode conduction is to monitor the voltages $V_{sw-node}$ and $V_{gate-low}$. As shown in Fig. 5, under ideal condition, the LS power MOSFET should be turned on as soon as the HS power MOSFET is turned off (causing $V_{sw-node}$ to go low). If it is not the case, the inductor current will flow through the LS MOSFET body-diode, resulting in unwanted conduction loss. Moreover, when the LS power MOSFET is turned on, and if the voltage drop across it is less than V_F (synchronous rectifier body-diode forward voltage drop), it will force the body-diode to turn off. The turn-off transient therefore unavoidably causes additional reverse recovery loss. A two-input NOR gate with high voltage capability can be used to monitor these signals. Compared with the traditional comparator detection scheme, NOR gate detection is much faster and less susceptible to noise at the switching node. The switching signals in Fig. 5 illustrates the NOR gate output as a result of non-optimized dead-time.

D. Body-diode Conduction Pulse Width Measurment

Accurate knowledge of the body-diode conduction time is required to deduce the optimal dead-time. A delay-line circuit which measures the body-diode conduction time with 1.25 ns resolution is as shown in Fig. 6. The delay of each individual delay cell is adjusted by a digital delay lock loop (DLL) [7] to ensure linearity of the delay signals, as well as the constant delays in individual delay cell regardless of process and temperature variation.

Since each delay cell of the 32 stages generates a 1.25ns delay, the digital DLL will require a 40ns calibration pulse to control the same amount of overall delay in the delay line. In addition, the body-diode conduction pulse is measured using the same delay line. As a result, the calibration pulse and the body-diode conduction pulse should be combined before passing through the delay line. Fig. 7 demonstrates the generation of the delay pulse.

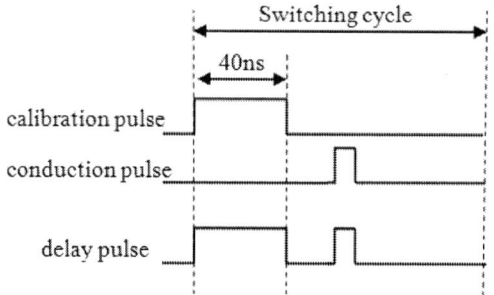

Figure 7. Timing waveforms of the 40ns delay line calibration pulse and the body-diode conduction pulse. They are combined to generate the delay pulse in each clock cycle.

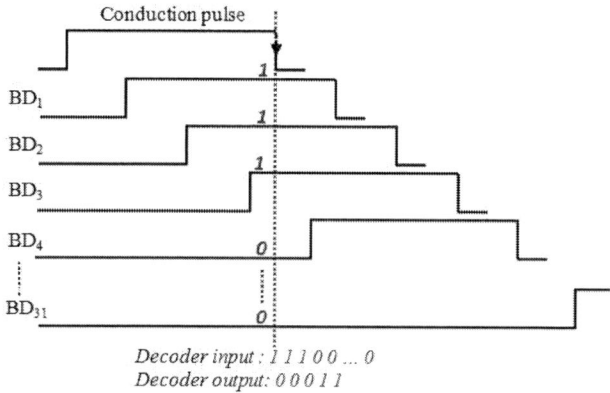

Figure 8. Timing diagram for the body-diode conduction pulse width measurement decoder.

As the delay pulse travels along the delay-line, the falling edge of the body-diode conduction pulse, $V\_NOR(detect)$ will trigger the conduction decoder. This decoder converts the thermometer code into a binary representation of the pulse width. The working principle is as illustrated in Fig. 8. Since the same delay-line also is used to generate the DPMW signal, this binary number can be subtracted from the intended duty cycle directly. A hybrid DPWM scheme is used to provide a practical trade-off between the high clock frequency requirements of counter-based DPWMs and the large hardware requirements of delay-line based DPWMs [8]. The 10-bit DPWM with a switching frequency, f_{sw} = 781 kHz, is implemented with a 5-bit counter and a 5-bit delay-line topology.

IV. SYSTEM VERIFICATION AND EXPERIMENTAL RESULTS

The proposed dead-time correction algorithm is verified experimentally in two steps using an FPGA implementation for the controller and discrete output power MOSFETs with gate driver ICs. First, the conversion efficiency with varying dead-times is measured experimentally using a LabView testing module, the measurement set up is as shown in Fig. 9.

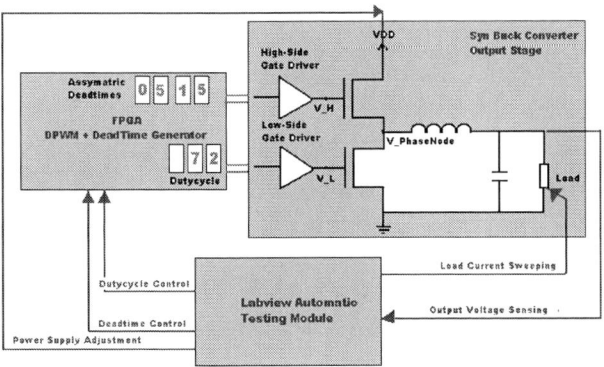

Figure 9. Experimental setup for the optimum dead-time identification.

The steady state efficiency of the converter is measured with different dead-times for varying load current. Output current is allowed to sweep from 2A to 10A at 0.5A intervals. Optimum dead-times are identified as the dead-times that are associated with the maximum efficiency for each load conditions. Fig. 10 shows the measured power conversion efficiency as a function of output current for symmetric dead time -10ns, 0ns, 10ns and 25ns, as well as the efficiency for

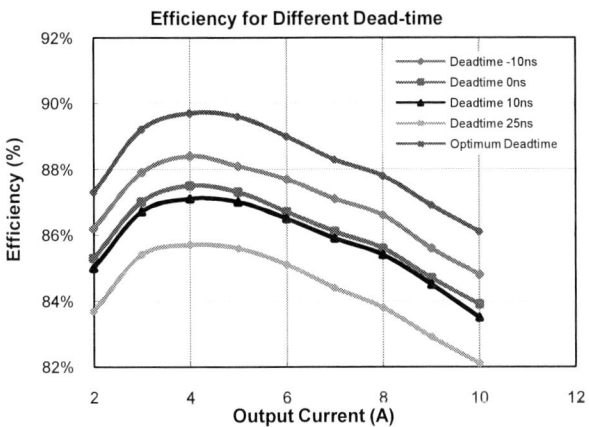

Figure 10. Power conversion efficiency comparison for various dead-times.

deduced optimum dead-time conditions. The comparison graph indicates the advantages of using optimum dead-time over a particular fixed dead-time.

Second, the efficiency is tested with different load current using the one-step dead-time correction algorithm. Optimum dead-times are adjusted at the rising and falling edge for the HS MOSFET using the same correction algorithm. For simplicity, the following experiment will be tested on the rising edge only.

A. Conversion Efficiency Test Durying Steady State

Dead-time correction algorithm is first tested under constant current conditions. Fig. 11 shows the waveform for non-optimum dead-time case when converter is delivering a load current of 2A. *V_NOR*(*detect*) is the output of NOR gate, signaling the presence of body-diode conductions. *V_NOR*(*test*) is the re-shaped *V_NOR*(*detect*) pulse after passing through two inverters to restore logic levels.

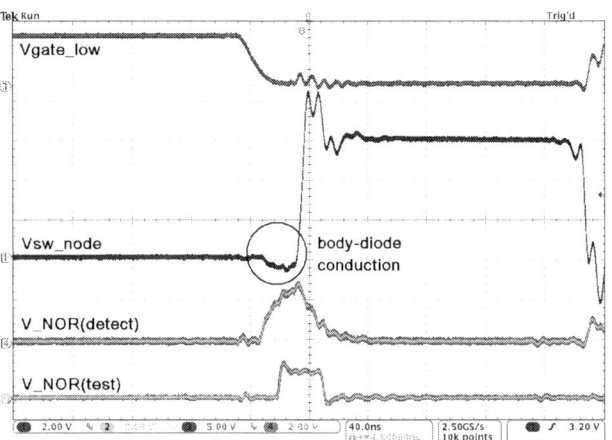

Figure 11. The NOR gate body-diode conduction detection waveform (without load change) shows a non-optimum dead-time condition.

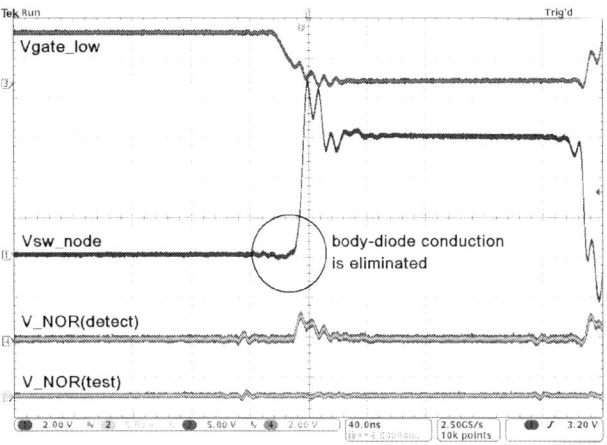

Figure 12. Body-diode conduction detection waveform (without load change) shows body-diode condition is eliminated with dead-time correction algorithm.

V_NOR(test) is monitored by the Optimum Dead-time Generator in the digital controller.

Fig. 12 shows the body-diode conduction is eliminated when the dead-time correction algorithm is applied. Steady-state efficiency is then measured under this condition. Similar tests are conducted for load current varying from 0.5A to 10A at 0.5A interval. Experimental results (Fig. 13) show that with the body-diode conduction eliminated, the power conversion efficiency is improved by 2% to 4%.

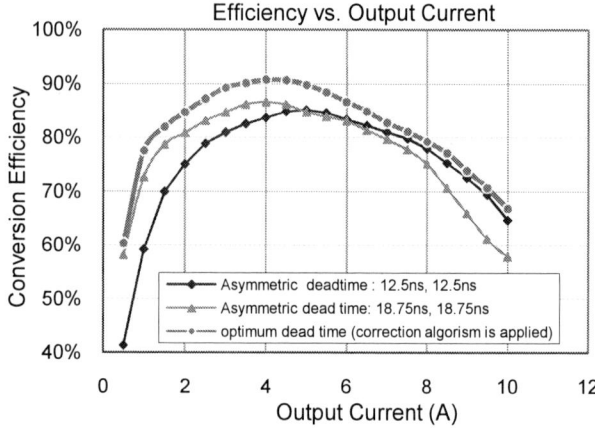

Figure 13. Power conversion efficiency is increased by 2% to 4% compared with non-optimum dead-times case.

B. Conversion efficiency test durying load transient

Different converter close loop control schemes have different dynamic characteristic, therefore transient response time largely depends on the topologies of the controller, as well as design specifications. In modern controller designs, such as PID, PI, hysteretic controllers [9], the converter usually requires a few switching cycles to reach stead-state, especially after a large load transient. In order to eliminate the influence of the various controllers, the conversion efficiency during load transient is tested under open loop conditions.

Fig. 14 shows the existence of body-diode conduction with the change of load. When the *load_change* signal is activated, the load current is changed from 0.1A to 4A. As the inductor current starts to rise, body-diode conduction begins to appear. The gradual change of the inductor currents requires different optimum dead-time for each current level. In this case, as the output current increases, shorter dead-times are required. However, in this part of the experiment, the dead-time is fixed. As a result, the LS switch's body-diode starts to conduct. The body-diode conduction time also gradually increases as the inductor current rises slowly to the new load current. In order to show the gradual change in body-diode conduction upon reaching steady state, the oscilloscope capture time scale is set as 2µs/div. Fig. 15 is a closed-up view at the 14th switching cycle from beginning of the *load_change* activated.

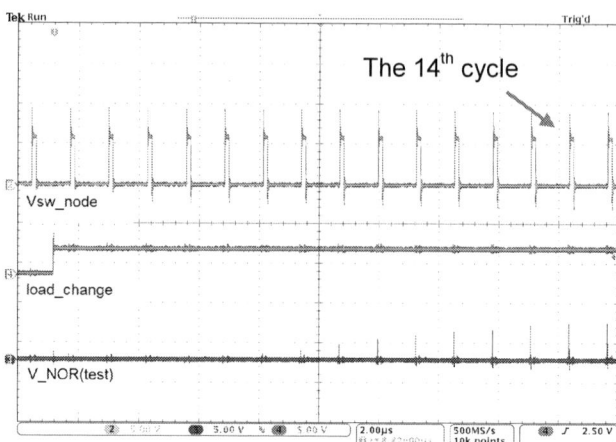

Figure 14. Switching waveforms show that body-diode conduction gradually increases when current experiences a sudden increase from 0.1A to 4A.

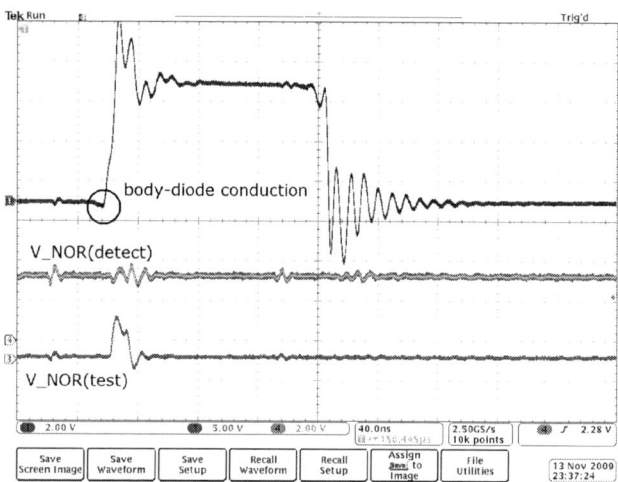

Figure 15. a close-up view of the body-diode conduction at the 14th switching cycle after the load current starts to change.

The gradual change in the inductor current for the same 0.1A to 4A load change is as shown in Fig. 16. The amount of time required to reach the new steady state is approximately 20 switching cycles. This verifies the observation of the progressive appearance of the body-diode conduction after the load change in Fig. 14.

With the Optimum Dead-time Generator activated, the switching waveforms in Fig. 17 shows that body-diode conduction can always be eliminated during and after the load change.

In comparison, other previously reported slow dead-time searching algorithms [3, 6] require multiple switching cycles to response to the changes in load condition, making them unsuitable for frequent load change applications. Moreover, when the system is operated in close loop, the transition time will be reduced depending on particular control scheme. However current variation still exists for certain period of time, and the optimum dead-times are also needed to ensure the conversion efficiency during load change.

978-1-4244-4782-4/10 $26.00 © 2010 IEEE

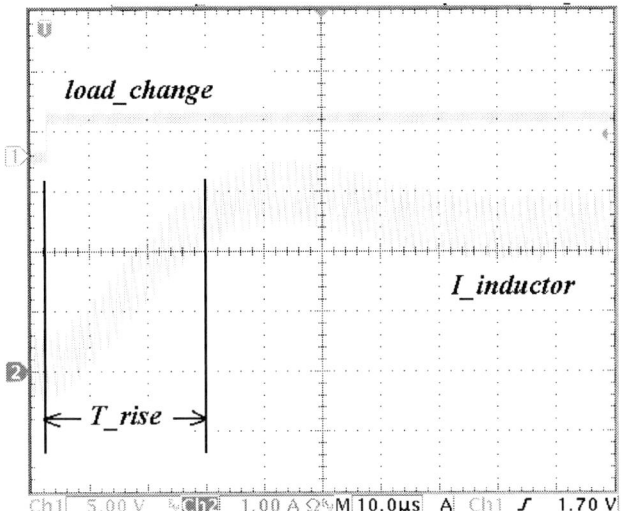

Figure 16. The inductor current waveform increases gradually during a load change from 0.1A to 4A.

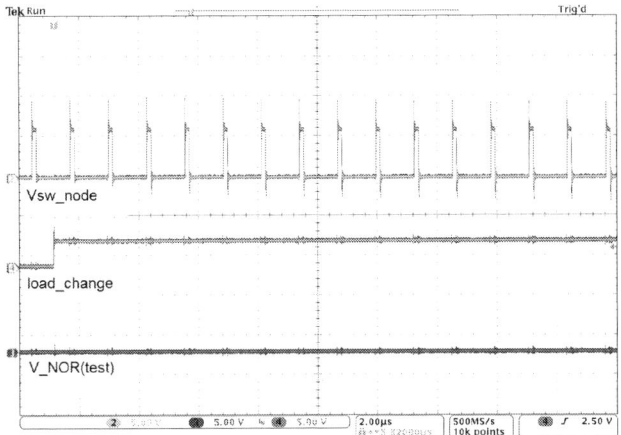

Figure 17. Switching waveforms show that body-diode conduction is eliminated with dead-time correction algorithm during 0.1A to 4A load change.

V. CONCLUSIONS

A one step dead-time correction algorithm is introduced in this paper. The proposed method can detect and measure body-diode conduction and optimize the dead-time on the fly. In addition to the elimination of body-diode conduction loss, the reverse recovery loss is completely avoided since the body-diode never becomes fully saturated. This method successfully avoids the drawbacks of many existing dead-time control schemes. It constantly optimizes the dead-time during any load condition such that the maximum possible conversion efficiency is always ensured. A higher switching frequency can therefore be used due to the significant reduction of power dissipation. As a result, the further minimization of output stage becomes possible.

VI. ACKNOWLEDGMENT

This work was initiated in the summer of 2008 with the gracious technical and financial support of Ciclon Semiconductors (now Texas Instrument Ltd.). The authors would also like to gratefully acknowledge the financial support from Auto 21, a Network of Centers of Excellence in Canada and Fuji Electric Holdings, Japan.

REFERENCES

[1] "Designing fast response synchronous buck regulators using the TPS5210" APPL. Rep., SLVA044, Texas Instruments, 1999.J. Clerk Maxwell, A Treatise on Electricity and Magnetism, 3rd ed., vol. 2. Oxford: Clarendon, 1892, pp.68–73.

[2] A. Stratakos, S. Sanders, and R. Brodersen, "A low-voltage CMOS DC-DC converter for a portable battery-operated system," in *IEEE Power Electronics Specialists Conference, 1993* Rec., vol. 1, pp. 619-626, June 1994.K. Elissa, "Title of paper if known," unpublished.

[3] S. Mapus, "Predictive gate drive boosts synchronous DC/DC power converter efficiency," APPL. Rep. SLUA281, Texas Instruments, Apr. 2003.

[4] O. Trescases, W.T. Ng, and S. Chen, "Precision gate drive timing in a zero-voltage switching DC-DC converter," in *Proc. IEEE Int. Symp. Power Semicond. Devices ICs*, 2004, pp. 55-58.

[5] B. Acker, C.R. Sullivan, and S.R. Sanders, "Synchronous rectification with adaptive timing control," in *IEEE Power Electronics Specialists Conference, 1995*, pp. 88-95.

[6] V. Yousefzadeh, and D. Maksimovic, "Sensorless optimization of dead-times in DC-DC converters with synchronous rectifiers," *IEEE Transactions on Power Electronics*, vol. 21, no.4, July 2006.

[7] E.O. Malley, K. Rinne, "A programmable digital pulse width modulator providing versatile pulse patterns and supporting switching frequencies beyond 15 MHz," in *IEEE Applied Power Electronics Conference*, 2004, Vol. 1, pp. 53-59, Feb. 2004.

[8] V. Yousefzadeh, T. Takayama, and D. Maksimovic, "Hybrid DPWM with Digital Delay-Locked Loop," in *IEEE COMPEL Workshop*, July 2006.

[9] R. Miftakhutdinov "Synchronous buck regulator design using the TI TPS5311 high-frequency hysterectic controller," Analog Applications Journal, Texas Instruments, Nov. 1999.

"The Practical aspects of utilizing Digital Power Controller for monitoring of Power Supply operation"

Oleg Volfson/Intersil Corporation
Intersil Corporation
Waltham, Ma USA

ABSTRACT

Utilization of digital control in the design of power controllers provides designers with the capability to monitor multiple operational parameters of a power converter. This functionality is often harder to achieve with converter designs based on analog controllers. The reason for this difference is that the information used for internal control purposes in the case of a digital controller is digitized, and thus is readily available for reading by an external device. In the case of an analog controller, this is not the case, so achieving monitoring and measurement functionality with an analog controller requires significant increases in system complexity. Monitor and measurement capabilities are in many cases very much welcomed by both power and system designers.
To be able to fully utilize this functionality, however, and to be able to realize reliable results from measurements made internally by the controller poses a set of complex challenges that most of engineers who design power converters have little familiarity with.
The purpose of this paper is to review some of the measurement techniques, the design considerations involved in the utilization of these techniques, and also to assess what are reasonable expectations for the accuracy of such measurements. To achieve this goal we are going to review examples of how Input Voltage, Output Voltage, Output Current and Temperature sensing are implemented, as well as the influence of external components, board layout and operating range on the results of such measurements. We are also going to review what kind of information from the controller IC manufacturer is required to properly realize the maximum achievable accuracy of readings for a specific design.

I. INTRODUCTION

When looking what parameters are most important to power supply designers and system engineers, the most typical ones that we can think of are the following:

1. Output Voltage

2. Output Current

3. Operating efficiency

4. Operating Temperature

5. Input Voltage

6. Switching Frequency

7. Output ripple Voltage

8. Transient Voltage deviations

Other measurements can be added to this list, but those eight items are the most typical and sufficient for the purpose of a discussion in this paper. That list is also representative of a set of challenges posed by the introduction of measurement capabilities to a Power Converter design. Upon reviewing this list we can further categorize these parameters according to the following list:

1. Direct vs. Derivative measurements. Output voltage is an example of direct measurement, while efficiency is an example of a derivative one, since it requires calculations based on directly measured parameters

2. High Bandwidth and Low Bandwidth measurements. Depending on how a data obtained from a measurement of specific parameters is utilized, high bandwidth might or might not be required. Sometimes same measurement information

can result in two different sets of data. Example of such a situation is Output Current measurement. Relatively high bandwidth of this measurement is required for converter's cycle to cycle Over-Current protection, while much lower bandwidth required for current monitoring purposes

3. Degree of dependence on External factors such as temperature and tolerance of external components

Those are large topics and while the nature of this paper allows only limited review of these calculations, we will review several typical examples. These examples will cover the considerations that need to be taken into account to achieve the best utilization of measurement capabilities of controllers based on a digital implementation, as well as cover the cost vs. performance compromises associated with selection and application of different measurement techniques.

II. VOLTAGE MEASUREMENTS

Block Diagram of a Power Controller optimized for operation in Non-Isolated Step-Down (Buck) Converter is presented on Figure 1 below. An ability to accurately measure input and output voltages of a converter is among key features that are usually required for converter operation, and is a good example of how direct measurements can be treated. Typically, a voltage signal would be processed through a multiplexor, A/D converter with or without Programmable Gain Amplifier (PGA) and likely a digital filter. Consequently, the following factors would come into play when determining the best achievable accuracy for a specific design:

1. Resolution of A/D converter

2. Range of a signal being measured

3. Accuracy of reference voltage used

4. Affects of additional signal filtering/averaging

Let's use typical Point-of-Load (POL) DC/DC converter with following input and output parameters as an example:

Input voltage range- 3.3V to 12V
Output Voltage range- 0.6V to 5V (0.54V to 5.5V including 10 % margins)

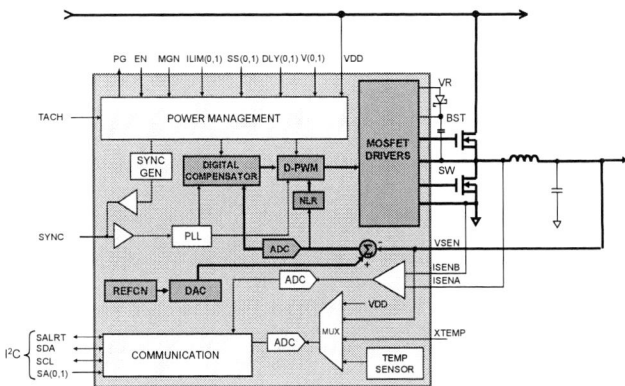

Fig. 1 Block Diagram of a Digital Power Controller

According to the criteria that were defined previously, we can see that both input and output voltage signals are measured directly. We can also see that the minimum to maximum span in relative terms is much larger for output voltage measurements than for input voltage signal. There is also significant difference in the measurement requirements applicable to input and output voltage signals due to the fact that the output voltage is likely needed to be sampled at a much faster rate to support sufficient bandwidth for feedback loop. Let's also take a look at the resolution requirement needed while measuring those voltages. Assuming that a 12 bit A/D converter is used with a voltage range of 20V, the best resolution available can be calculated per the formula below:

$$\text{Vstep} = \text{Vrange} / 2^{12} = 20/4096 = 0.0488 \text{ V}$$

While 50 mV resolution is probably good enough number for Input Voltage measurements, it would not satisfy the output voltage resolution requirements for most POL applications. If the voltage range for the Vout reading is reduced to 6V, the resulting resolution would be about 15mV, which is probably satisfactory for many applications (0.25% of full range, and 0.3% of actual output Voltage range).
Further improvement in reading resolution can be achieved by using PGA (Programmable gain Amplifier) and averaging of multiple readings.

III. CURRENT MEASUREMENTS

In DC/DC converters current is usually measured as a voltage drop across either a dedicated resistance or parasitic impedance such as the Rds-on of a switching element or winding resistance (DCR) of an inductor. These types of readings are clearly an example of derivative measurements. Several factors come into play to affect the accuracy of these readings including the temperature of current sense element, converter's circuitry ability to filter out switching noise and techniques used to do so, tolerances of current sense element and the device's ability to support calibrating procedures that can nullify those factors. All those variables are added on top of similar ones that were discussed in the previous section on Voltage measurements, including bit resolution and usable range. When utilizing current measurement techniques, it is also important to verify that sufficient number of samples is available during each dedicated time slot for measurement. Using the same example of a POL DC/DC converter let's look what are some reasonable expectation for current measurement accuracy and how different factors affect it. Since the same controller design needs to be capable of supporting different design it would be reasonable to assume that the range of rated output current that it is supported can be anywhere from as low as 2-3A to as high as 30A. To effectively protect the power supply and its load from over-current faults, the current sense circuitry has to be able to process multiple current samples during one switching cycle. Continuously running 10 or 12 bit A/D conversion with sampling rate of several MHz requires amounts of power dissipation that can hardly be tolerated in such an application due to reduced total efficiency and thermal considerations. One way to deal with this problem is to reduce the resolution of such current sense A/D converters, while still maintaining high sample rate. Assuming a 6 bit implementation with 200 mV (+/- 100mV) input voltage range, a designer will need to select a FET or Output Inductor to achieve the best compromise between having high voltage across the sense element (Rds(on) or DCR) for better current sense resolution and noise immunity, while at the same time minimizing voltage drop across the sense element to achieve better efficiency. Care should also be taken not to produce a signal in excess of maximum input range allowed on the input of current sense amplifier. If a FET with 5mOhm Rds-on impedance is used to sense current for a converter with rated output current of 20A, just nominal DC current would produce a signal of 100 mV (Vsense= I*Rsense). This is without taking in account any ripple current with magnitude usually between 20 and 50% of rated Iout, any overload conditions and/or the effect of increasing impedance of the current sense element with heating. On the other hand, if the same scheme is used for a converter with a rated current of 3A, the total voltage drop associated with DC current would be only 15mV which provides a lot of margin for input range. However, each resolution step would be $200/2^6 = 3.125$ mV. That means that controllers would not see any current below 0.625A and there would be only four full resolution steps available for entire output current range.

Those considerations can be helpful in determining a best compromise between accuracy of a current sense circuit and minimizing losses associated with voltage drop across current sense element. It should be also made clear that while those calculations are definitely true for a fast cycle-by-cycle measurements that is important for Over-Current protection, if a data from current sense is used to monitor an output current of a Converter by a remote device, several consecutive current readings can be accumulated and averaged. This technique allows not only for filtering out a noise, but since measurements are performed using a regular saw-tooth shaped signal, it can also result in increase in resolution of such measurements. On the graph on Fig. 2 below we can see actual results of implementing averaging of multiple current reading measured by reading a voltage across DCR of output inductor with DC resistance of approximately 1mOhm. Improvement in read-out resolution due to implementation of averaging techniques is quite clearly visible on this plot.

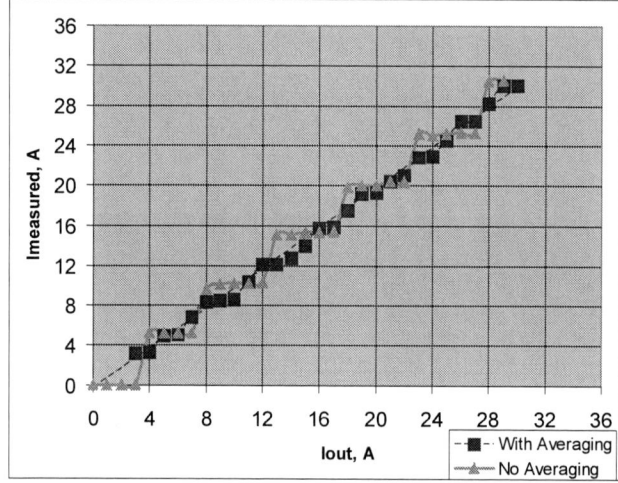

Figure 2. Output Current Readings obtained from Digitally Controlled DC/DC Buck converter

Since usually current is measured indirectly by measuring a voltage across resistive element such as Switching Device (FET), Parasitic Resistance of an Inductor, etc, an impedance of such current sense element is changing with its temperature, according to a following equation:

$$R = R_0[\alpha(T - T_0) + 1]$$

Were:

R - an actual resistance of a resistive element at given temperature

Ro- Resistance at reference temperature (usually 25 DegC)

α- Thermal Resistance coefficient for a resistive element

T- Actual temperature of a resistive element

To- Reference temperature

Since thermal coefficient is different for different sense elements (it is about 39ppm for copper for example) and it is rarely possible to measure a precise temperature of such of an element due to the presence of isolation between a temperature sensor and a current sense device, to achieve consistent current reading over load and ambient temperature range, some scheme of temperature compensation of a current sense element impedance drift has to be utilized as part of the controller's current sense algorithm. An important point here being that proper coefficient has to be selected depending on where temperature is sensed, to achieve a good correlation with current sense impedance changes over temperature.

978-1-4244-4782-4/10 $26.00 © 2010 IEEE

IV. CALIBRATION AND CONTROLLER ARCHITECTURE

Two other inter-related factors that influence measurement capabilities of a Power Converter are an internal architecture of a controller IC and its ability to calibrate out tolerances of external to controller components and environmental factors.

When talking about digital controllers architecture includes a combination of functions implemented on Hardware level and those that can be achieved using Software (Firmware). One of results of how that architecture is implemented is that different signals might be routed to a same measurement circuit (Analog-to-Digital converter, Op-Amp, etc) through a Multiplexor (MUX) as shown on Figure 1. As a consequence, a specific signal might not be monitored all the time but only at curtain intervals of time. Knowing the length of such intervals might help to determine bandwidth of measurements and better understand a behavior of converter.

Also related to this is an ability to calibrate measurements to reduce an influence caused by variations of values of external components utilized to measure various operating parameters as well as by changes in environmental conditions (temperature being most typical example).

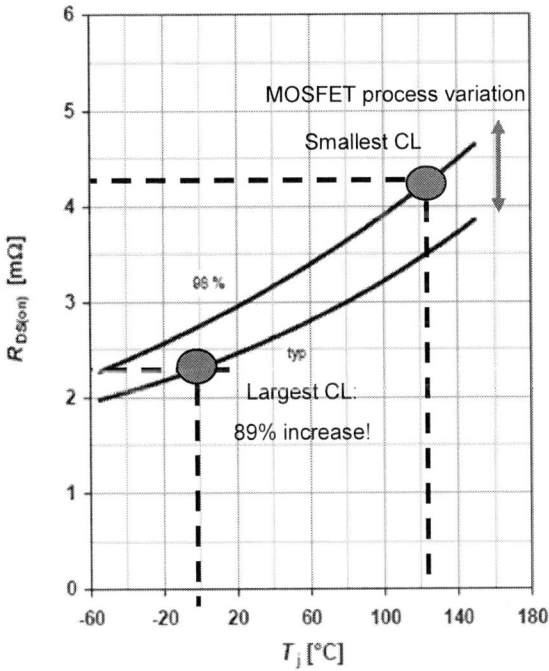

Figure 3. Over-Current Protection threshold vs. effects of switching FET process variation and its operating temperature

Red dots on a Figure 3 above show how Over-Current Protection threshold that is triggered by a voltage measured across switching Power FET can be affected by variations in FET's On-State impedance due to its manufacturing process variations and by variations in its operating temperature. We can observe that combined effect of FETs die temperature

difference and tolerance in it ON-State resistance can result in Minimum and Maximum value of Output Over-Current Threshold that would trigger a protection response of 89%. Depending on a specific design, this large variable can be completely or partially calibrated out by updating calibration data in controller's Non-Volatile memory and utilizing proper Temperature Correction coefficient as described in CURRENT MEASUREMENT chapter.

V. CONCLUSIONS

The following factors have to be reviewed and considered in order to be able to assess what the predicted accuracy of measurements obtained from a Digital Power Controller will be:

1. Data on Measurement Accuracy that is available from controller is not always readily available on the part's data sheet

2. Depending on the type of measurement being taken, finding what that accuracy actually is can be a multistep process requiring knowledge of at least some of the following factors:

   a. Resolution of A/D convertors used to sample specific parameter

   b. Sampling rate of A/D

   c. Whether multiplexing of multiple measured signals is used and if yes, what are the measurement intervals

   d. If Programmable Gain Amplifier (PGA) is utilized, what are signal input ranges on its input

   e. What are external factors that might cause a drift in reading and whether compensation and/or calibration procedures are available to reduce an impact of such a drift

3. Often compromise has to be made between accuracy of measurements and other factors such as cost and converter efficiency

REFERENCES

[1] D. Maksimovic, R.Zane, "Digital Control of SMPS," seminar presented at IEEE APEC 2006

[2] D. Maksimović, R. Zane, R. Erickson, "Advances in practical high-performance digital control," Digital Power Forum, September 2005

[3] N. Mohan, T. Underland, W. Robbins, Power Electronics, Second Edition, John Wiley & Sons, New York, , 1995.

[4] O. Volfson, K. Dehnel, "Digital Power, Beyond Acceptance," seminar presented at IEEE APEC 2008

A Unity Power Factor, Maximum Power Point Tracking Battery Charger for Low Power Wind Turbines

Gustavo Gamboa, John Elmes, Christopher Hamilton, Jonathan Baker, Michael Pepper, and Issa Batarseh

School of Electrical and
Computer Engineering
University of Central Florida
Orlando, Florida 32816
Email: john.elmes@knights.ucf.edu

Abstract—This paper proposes a unique implementation of power factor correction (PFC) and maximum power point tracking (MPPT) for low power wind turbines. For a given wind condition, there is a unique electrical load which will harvest the maximum power from a wind turbine, the proposed control algorithm actively tracks this electrical loading condition for maximum power. An active 3-phase rectifier (VIENNA) converter is used to rectify the 3-phase AC voltage with near unity power factor which is critical in this application where the series resistance of the turbine is very high. An experimental 300W prototype was designed and tested to verify the design. Experimental results showed a significant increase in power extracted from the low power wind turbine when PFC and MPPT were implemented.

I. INTRODUCTION

Low power wind turbines (100 W - 1000 W) are typically fixed blade 3-phase permanent magnet machines. Unlike larger turbine systems, the pitch angle of the blades are fixed, so the only way to maximize the harvested energy is to vary the electrical loading on the turbine. By varying the electrical loading, the power delivered by the turbine can be maximized without any knowledge of the wind conditions or the turbine characteristics. The proposed maximum power point tracking (MPPT) battery charger for low power wind turbines also features unity power factor control so as to reduce the ohmic losses in the relatively high series resistance of the turbine. In [1], [2], and [3], different PFC techniques have been proposed. Also, [4] [5] explains methods for maximum power transfer. However, these papers only describe implementation for high power wind turbines. MPPT has been implemented for low power wind turbines but with the requirement of an anemometer as described in [6]. However, the introduction of an anemometer can significantly increase the overall cost of the low power turbine. An alternative approach was proposed in [7] where a six diode rectifier at the input stage and a dummy load was used.

As shown in Fig. 1, there exists a specific load resistance which will result in the maximum amount of harvested power. The implemented MPPT algorithm implements the perturb-and-observe algorithm, which is a popular choice for the

comparable photovoltaic system. All algorithms will be implemented without the need of an anemometer or a dummy load. Also, an active rectifier is used which achieves unity power factor while also reducing the voltage drop of the rectification stage in comparison to the typical 6-diode rectifier.

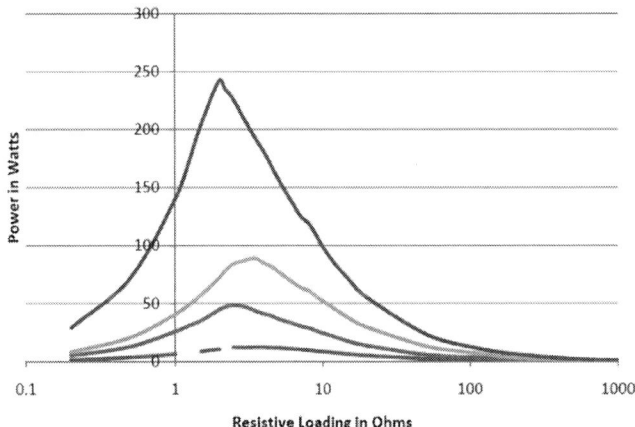

Fig. 1: Power curve of experimental wind turbine under different simulated wind conditions

Depending on the load and wind speed, the wind turbine may have trouble keeping a clean sinusoidal waveform. The proposed control algorithm also features a three phase digital phase locked loop (DPLL) control to further improve the implementation of PFC. This paper presents one approach for frequency and phase detection and produces a clean sinusoidal wave without the need of a filter.

II. REALIZING PFC FOR LOW POWER WIND TURBINE APPLICATIONS

Due to the high impedance of a wind turbine, to not utilize any PFC techniques would introduce high I^2R losses. Therefore, if the power converter is designed to behave like a purely resistive load, reducing the high peak currents, minimizing the I^2R losses which will maximize the power extracted from the wind turbine. The battery charger implements a unique approach of controlling the conductance (G) of the unity power factor 3-phase AC-DC stage using the VIENNA topology. With PFC, the VIENNA converter allowed more power extraction, by increasing the efficiency of the converter as well as the wind turbine.

The sensed battery voltage and current determines the controlled conductance value. Over-voltage battery protection is implemented using an output voltage and output current regulation (OVR and OCR, respectively). When the battery voltage reaches the desired reference value, it limits the reference current going into the OCR controller. The OCR then tries to reduce the maximum current allowed to the battery by increasing the conductance value. On the other hand, when the battery voltage is below the reference value, the MPPT algorithm tracks the optimum conductance value. Therefore, the rotor speed is optimized in order to allow maximum power transfer from a given wind speed. The maximum of these two conductance values is selected so that when the battery requires current limiting, the converter will command a high conductance value, G, which corresponds to a slower rotor speed. This approach is illustrated in Fig. 2.

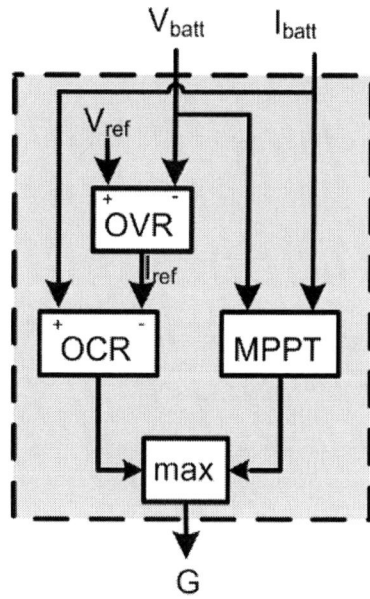

Fig. 2: Conductance (G) controller

After the proper conductance is calculated, it is received by the individual controllers for each phase as shown in Fig. 3. The duty cycle of the individual phase is controlled and shapes the input current as a scalar of the input sinusoidal waveform. This scalar is the conductance value. Therefore, as the battery gets charged, the converter limits the amount of power that is pushed to the bus. When the maximum power is no longer needed, the increased conductance value causes the turbine to slow the rotor speed.

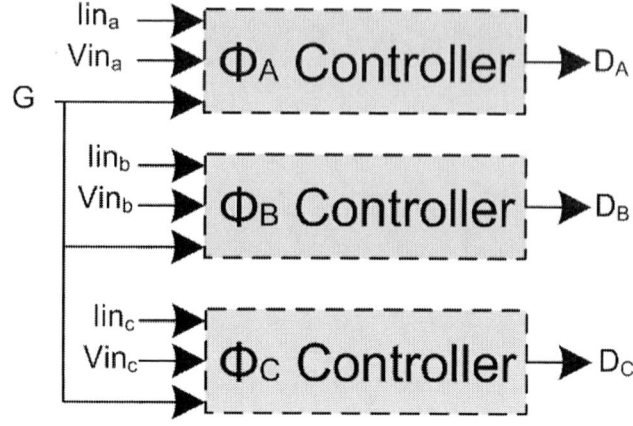

Fig. 3: PFC control block diagram

III. PROPOSED CONTROL STRUCTURE

With the implementation of PFC in a low power wind turbine (300W), it can be challenging to sense a clean sinusoidal waveform. Sometimes, noise can affect the stability of any control strategy that is implemented in the converter. Due to the high ESR of the turbine, the input voltage waveform is highly affected by the amount of current extracted from the turbine. The sinusoidal current reference is a factor of the input voltage and it is used to shape the input current. Noise in the sensed signal can cause the controller to become unstable. Another important issue is stability at low RMS voltage due to either low wind speeds or heavy loading. It becomes difficult to scale a small RMS value to create the input current controller reference to shape the current when PFC is desired under these conditions. The system needs to have more reliable operation during the heavy load condition because this operating point is used to keep the wind turbine in the current limit mode thus slowing the wind turbine down.

A new control structure is proposed for low power wind turbine battery chargers where it introduces the implementation of a digital phase-locked-loop (PLL) generates three separate sinusoidal waveforms, each with a 120 degrees phase shift. These waveforms are used as the current references for the input current controllers on the active rectifier stage. A digital approach is preferred in this research instead of analog because noise can be cancelled without any extra hardware. Therefore, heavy loading and low wind speeds will not affect the stability of the controller thus making it more reliable while maintaining a close to unity power factor.

A three phase DSP based PLL implementation is discussed in [8]. However, based on analog PLL structures, a simulation was formulated in Simulink that confirms the successful operation of the PLL and the general structure is shown in Fig. 4.

A lookup table approach for the digital oscillator was selected because of the high quality sine wave at particularly low frequencies as well as it's ease of increasing accuracy along the sin wave by adding more sample points. A disadvantage

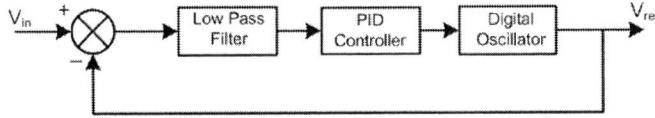

Fig. 4: Phase Locked Loop control diagram

to this method is the non-linearity that exists in a lookup table based oscillator. However, the oscillator can be approximated to be a linear device along the lower frequency band.

The implementation of the PLL system is a hybrid approach, using a timer based zero crossing detection algorithm for coarse frequency tracking, and a phase-error detecting controller, to minimize the phase shift between the sensed input voltage and the digital oscillator reference. The coarse frequency detection is excecuted using a hysteresis windowed zero crossing detection algorithm to detect the frequency of both the input current as well as the digital oscillator. From this, we can directly compute the proper frequency for the digital oscillator to operate at such that the two frequencies are matched. This frequency is changed slightly using a phase error detection controller to minimize the measured phase difference. The phase difference measurement is done using a zero crossing detection algorithm as well. By introducing the control algorithm shown in Fig. 4, the controller can use a fine tune adjustment method to make small changes to the frequency so that the two waveforms are perfectly locked on without completely relying on zero crossing detection.

The new PLL control structure implemented into the existing PFC control structure as shown in Fig. 5. Note how the PLL controller is fed into the PFC controller just as the original algorithm does allowing us to use the same PFC algorithm without any modifications.

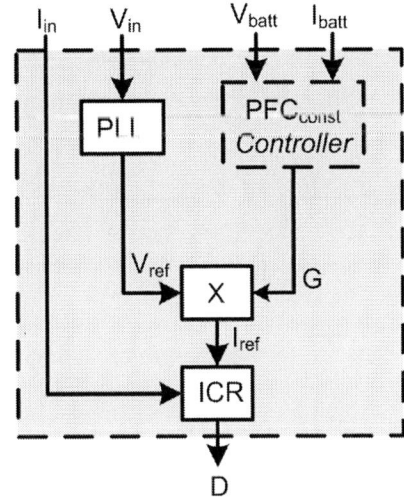

Fig. 5: PLL system interaction with PFC system

IV. TOPOLOGY

To reduce the voltage drop of the rectifying stage and to actively control the power factor, a controllable rectifier (VI-

ENNA rectifier) is used for the first stage of the system. The VIENNA operates in boost mode and assures a bus voltage that is always higher than the battery voltage. The second stage is a synchronous buck DC/DC converter controlled by the IVR loop. The topology realized for this research along with a block diagram describing the controls for one of the phase (Phase C), is shown in Fig. 6.

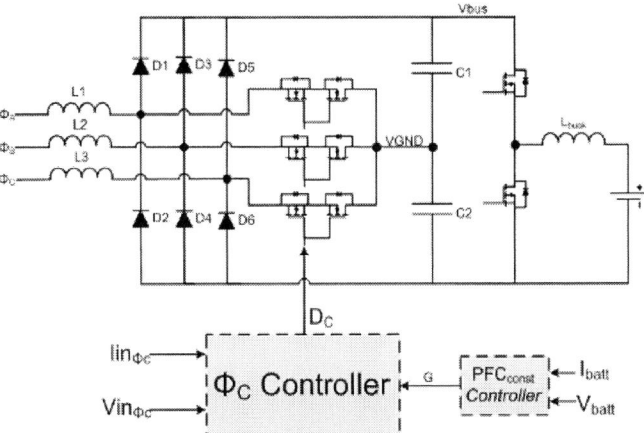

Fig. 6: VIENNA rectifier with buck topology showing a controller for one of the phases

As explained previously, each phase has its individual controller that shapes the phase current to be a scalar of the digital PLL referenced locked in phase to the input voltage. The proper conductance value is determined by the sensed information from the load. As the battery current demand decreases, higher values in the conductance value will force the wind turbine to rotate at a slower speed while still supplying the maximum power permitted by the load battery. This topology also allows the end user to increase the desired battery voltage. Because it uses an intermediate bus, the load voltage could be as high as 72V as long as the bus has a voltage value greater than the load. This gives flexibility in the application for the topology and the proposed algorithm.

V. EXPERIMENTAL RESULTS

A. Power Factor Correction in low power wind turbines

In a system which utilizes the typical 6-diode rectifier to generate the DC voltage which charges the battery, there are high current spikes resulting in low overall power factor. This is due to the clamping of the input voltage by the battery. Even in cases where there is a second DC-DC stage to charge the battery from the rectified bus voltage, the current will spike as the large bus capacitance is quickly charged with the rising AC voltage. With high current spikes in the input current, the high ESR of a wind turbine can significantly affect the overall efficiency of the AC generator. In Fig. 7, high current spikes were observed when the converter was operated as a conventional diode bridge rectifier with a 470uF bus capacitor. These high current spikes cause significant wire loss.

978-1-4244-4782-4/10 $26.00 © 2010 IEEE   145

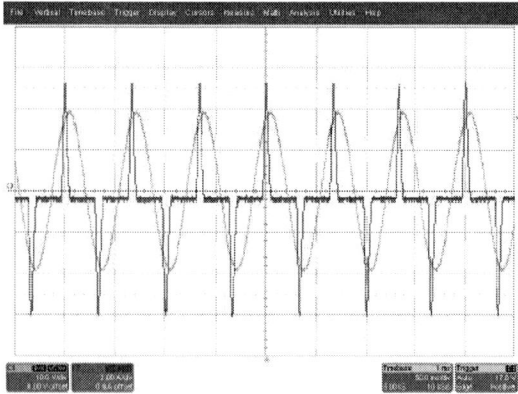

Fig. 7: Conventional 6-diode rectifier

With the implementation of PFC, the input current is in phase with the voltage thus achieving near unity power factor as seen in Fig. 8. The high power factor eliminates the current spikes and allows the turbine to operate more efficiently.

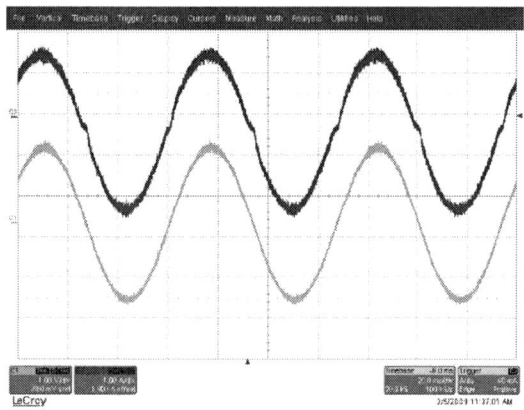

Fig. 8: Active PFC rectification

The effect of PFC in a low power wind turbine is shown in Table I. With a power factor of 0.98 with PFC versus 0.94 with constant duty cycle control, a significant increase in input and output power was observed. Also, due to the elimination of the current spikes, the efficiency of the converter increased as well.

TABLE I: Overall system efficiency w/active PFC vs. constant duty cycle for the AC/DC stage

	Constant Duty Cycle Control	Active PFC	% Increase
PF	0.94	0.98	4.08 %
P_{in}	187.90 W	191.59 W	1.93 %
η	91.10%	92.89%	1.93 %
P_{out}	171.18 W	177.97 W	1.93 %

B. Phase Locked Loop

The PLL implemented in this research is able to lock onto a frequency as low as 8Hz and can go as high as 35Hz. Fig. 9 shows how it successfully locks to an 8Hz signal. On the

other end of the spectrum, Fig. 10 shows the phase locking at 35Hz. These results were taken from the PLL system that uses zero crossing detection strictly as it's frequency and phase controller.

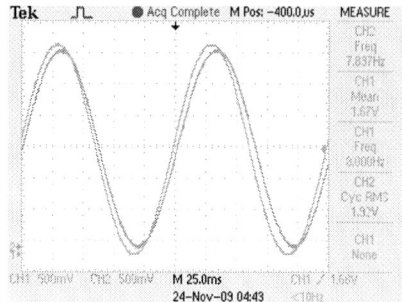

Fig. 9: PLL results at 8 Hz: (blue) digital oscillator, (orange) input voltage

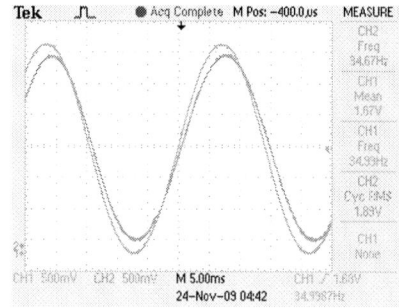

Fig. 10: PLL results at 35 Hz: (blue) digital oscillator, (orange) input voltage

The simulation results for the simulation diagram shown in Fig. 4 yield the results shown below in Fig. 11. Note how the initial sine wave starts out with high distortion but quickly locks on with a low THD sinusoidal wave.

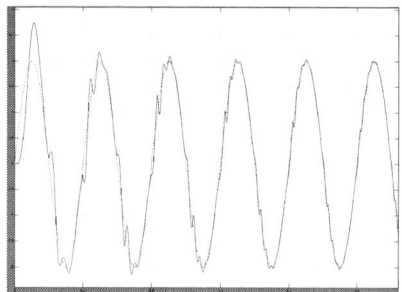

Fig. 11: PLL simulation results: (red) input voltage, (black) digital oscillator

C. Topology Prototype

The top and bottom view of a 300W experimental prototype, including the VIENNA rectifier with the buck converter is shown in Fig. 12 and 13, respectively.

Fig. 12: Experimental 300W prototype (top view)

Fig. 13: Experimental 300W prototype (bottom view)

This design utilizes three isolated MOSFET drivers, each powered by a bootstrapping diode and capacitor, which takes the energy from the rectified turbine input voltage. This design removes the need for floating isolate power supplies to drive the bidirectional blocking MOSFETs of the VIENNA converter. The enclosed prototype is shown in Fig. 14

Fig. 14: Experimental 300W prototype (enclosed)

D. MPPT

A conventional perturb-and-observe MPPT algorithm was experimentally verified, and is shown in Fig. 15 . When the load voltage is below the desired reference, the digital controller continues to make changes to the PFC constant. For a conventional hill climbing algorithm, the old power is compared with the new power. If the new power is greater than the old power, the controller will keep making changes in the same direction. However, if the new power is less, it switches the direction of change in the PFC constant.

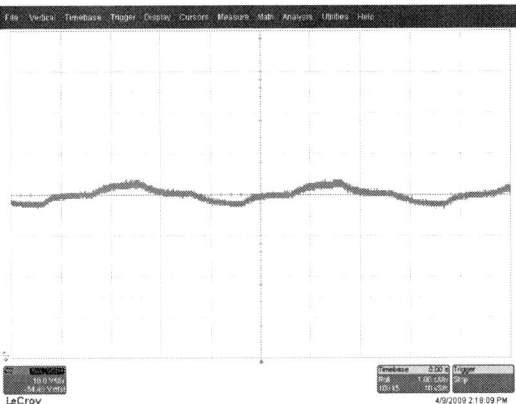

Fig. 15: Hill climbing MPPT steps

At a given wind speed of 11m/s, the MPPT was capable of extracting 325W of power from a 300W rated wind turbine. For every step in the PCF constant there is a change in the output battery current. The rate of change in this PFC constant must be slow enough for the wind turbine to react to the new value. If these changes occur too quickly, the controller might become unstable.

E. Efficiency and reliability

It was important to test the efficiency of the topology where the proposed algorithm was tested on. Results show that all the control loops were stable and the energy transfer was done at high efficiency. The test conditions are as follows:

- Constant Bus Voltage of 40 V
- Open loop switching on the rectifier so that test conditions can be kept constant at varying wind speeds
- 12 V battery output

Table II shows the efficiency results for this test case. Note that these results do not reflect the effectiveness of the MPPT algorithm or the PFC algorithm as these two algorithms work to increase the input power to the converter, not the overall efficiency of the power electronics.

TABLE II: Overall system power efficiency testing

Power In (W)	Power Out (W)	Efficiency (%)
103.38	94.06	91
114.09	104.09	91
134.39	123.1	92
172.17	158.76	92
210.18	195.16	93
243.69	223.36	92
247.82	228.78	92

Reliability was also tested on this prototype when it was tested with a two hour test of sourcing 200W to a 12V battery. The prototype performed at efficiencies greater than 90%. The proposed algorithm is also able to increase the allowable battery voltage.

F. Test bench

These test results were implemented in a test bench where a 300W turbine was coupled to a DC motor. The rotational speed properly controlled the rotational speed of a 90V DC motor in order to simulate real characteristics of a wind turbine. The test bench is shown in Fig. 16.

Fig. 16: 300W wind turbine test bench

VI. SUMMARY AND CONCLUSION

Through theoretical and experimental verification, it was confirmed that an actively controlled AC-DC PFC rectifier can significantly improve the overall power transfer of a low power wind turbine converter in contrast to the traditional 6-diode rectifier. Furthermore, an optimal converter topology and control structure has been presented which demonstrate high efficiency, near unity power factor, and the ability to actively track the maximum power point of the wind turbine. These benefits are in addition the elimination of the need for a dummy load or an anemometer with the implementation of the proposed algorithm, as well as it allows flexibility with batteries of various voltages and chemistries.

ACKNOWLEDGMENT

This work is partially funded by the National Science Foundation - IRES Award # 0652048

REFERENCES

[1] C. Silva, R. Bascope, and D. Oliveira, "Three-phase power factor correction rectifier applied to wind energy conversion systems," in *Applied Power Electronics Conference and Exposition, 2008. APEC 2008. Twenty-Third Annual IEEE*, Feb. 2008, pp. 768–773.

[2] N. Kimura, T. Hamada, M. Sonoda, T. Morizane, K. Taniguchi, and Y. Nishida, "Suppression of current peak of pfc converter connected to induction generator for wind power generation excited by voltage source converter," in *Power Electronics and Motion Control Conference, 2009. IPEMC '09. IEEE 6th International*, May 2009, pp. 2269–2274.

[3] F. dos Reis, K. Tan, and S. Islam, "Using pfc for harmonic mitigation in wind turbine energy conversion systems," in *Industrial Electronics Society, 2004. IECON 2004. 30th Annual Conference of IEEE*, vol. 3, Nov. 2004, pp. 3100–3105 Vol. 3.

[4] J. Yaoqin, Y. Zhongqing, and C. Binggang, "A new maximum power point tracking control scheme for wind generation," in *Power System Technology, 2002. Proceedings. PowerCon 2002. International Conference on*, vol. 1, Oct 2002, pp. 144–148 vol.1.

[5] J. Thongam, P. Bouchard, H. Ezzaidi, and M. Ouhrouche, "Wind speed sensorless maximum power point tracking control of variable speed wind energy conversion systems," in *Electric Machines and Drives Conference, 2009. IEMDC '09. IEEE International*, May 2009, pp. 1832–1837.

[6] N. Mutoh and A. Nagasawa, "A maximum power point tracking control method suitable for compact wind power generators," in *Power Electronics Specialists Conference, 2006. PESC '06. 37th IEEE*, June 2006, pp. 1–7.

[7] E. Koutroulis and K. Kalaitzakis, "Design of a maximum power tracking system for wind-energy-conversion applications," *Industrial Electronics, IEEE Transactions on*, vol. 53, no. 2, pp. 486–494, April 2006.

[8] S. Mussa and H. Mohr, "Three-phase digital pll for synchronizing on three-phase/switch/level boost rectifier by dsp," in *Power Electronics Specialists Conference, 2004. PESC 04. 2004 IEEE 35th Annual*, vol. 5, June 2004, pp. 3659–3664 Vol.5.

Maximum Power Point Tracking of a Wind Energy Conversion System Using Adaptive Nonlinear Approach

Majid Pahlevaninezhad, Suzan Eren, Alireza Bakhshai, Praveen Jain

Queen's University

ECE Dept.

Kingston, Ontario, Canada

7mp@queensu.ca, 2se1@queensu.ca, alireza.bakhshai@queensu.ca, praveen.jain@queensu.ca

Abstract— **This paper introduces a new control method to track the maximum power point for a Wind Energy Conversion System (WECS). This WECS is based on a permanent magnet synchronous generator (PMSG) fed by a matrix converter. Since the mechanical power generated by the wind turbine is a function of its shaft speed at a given wind velocity, the proposed controller provides the desired voltage at the output of the matrix converter so as to control the generator speed. This controller is based on the nonlinear adaptive backstepping approach which is well suited for this system. This method is able to effectively accommodate the effects of system uncertainties. Theoretical discussions and performance analysis verify the feasibility and performance of the proposed approach.**

I. INTRODUCTION

Nowadays, renewable energies are becoming increasingly important as alternative energy sources. Many factors such as diminishing fossil-fuel resources, energy security concerns, and increased global warming increase the need for renewable energy sources. Wind is one of the most abundant renewable energy sources and it can be harnessed by wind turbines. This energy is transformed into mechanical energy by wind turbines and then converted to electrical energy by generators. Thus, WECSs consist of a wind turbine, an electric generator and a power converter [1-2].

Different types of components can be used to realize a WECS. Many different configurations of WECSs have been reported in the literature [3]. The common candidates for the generator are Doubly-Fed Induction Generators (DFIG), Squirrel Cage Induction Generators (SCIG), and Permanent Magnet Synchronous Generators (PMSG). The conventional back-to-back voltage source converter is usually used to connect the generator to the grid. In this converter, the DC-link capacitor provides the decoupling between the generator and grid; however, it is bulky and has a pretty short life time. The matrix converter realizes a direct AC/AC conversion, and

it is a good candidate for this application. This converter can control the magnitude, frequency and phase angle of the output voltage as well as the input power factor. Despite the attractive features of the matrix converter, it suffers from some problems such as low voltage gain, complex control, bi-directional switches and lack of ride-through capability [4-7]. However, there are also many papers that are reporting solutions to mitigate these difficulties. Thus, the matrix converter seems to be a promising approach for AC/AC conversion [8].

WECSs have so many uncertainties due to the erratic nature of wind-based systems. Therefore, the controller should accommodate the effects of uncertainties and keep the system stable against a large variation of system parameters. The conventional PI-based controllers cannot fully satisfy stability and performance requirements. On the other hand, the system is highly nonlinear and has a large range of operating points. Thus, linearization around one operating point cannot be employed to design the controller. Nonlinear control methods can be used to effectively solve this problem. The integrator backstepping approach is a nonlinear method which is perfectly suited for PMSG-based systems. In this paper, an adaptive integrator backstepping approach is used to suppress the effects of uncertainties and keep the system stable for a large variation of parameters.

II. PROPOSED WIND ENERGY CONVERSION SYSTEM

A. General Block Diagram

Fig. 1 illustrates the proposed scheme for the WECS. The wind blades rotate the shaft which is connected to the generator through a gear box. The mechanical energy is converted to electrical energy by the generator and the matrix converter is the interface between the generator and grid. The nonlinear controller provides proper switching signals for the matrix converter so as to extract maximum power from the wind turbine. It also accommodates the system uncertainties through an adaptive algorithm.

To design the nonlinear controller the WECS model should be derived. Fig.2 illustrates the WECS block diagram. First of all, the wind turbine model needs to be derived. The wind speed is the main input signal of the wind turbine system. The speed, obviously, depends on the given site and atmospheric conditions. Therefore, its variation is absolutely erratic and difficult to model. The wind turbine model given in [9] is

$$\Gamma_w = \frac{P_w}{\omega_T} = \frac{1}{2}.\pi.\rho.v^2.R^3.C_\Gamma(\lambda) \ ,$$

$$C_p(\lambda,\theta) = c_1(\frac{c_2}{\lambda_i} - c_3\beta - c_4)e^{\frac{-c_5}{\lambda_i}} + c_6\lambda \ , \qquad (1)$$

$$\frac{1}{\lambda_i} = \frac{1}{\lambda + 0.08\beta} - \frac{0.035}{\beta^3 + 1}$$

where β is the rotor blade pitch angle, v is the wind velocity, ω_T is the shaft angular speed, $C_\Gamma = C_p/\lambda$ is the torque coefficient, R is the blade length of the wind turbine, ρ is the air density and $\lambda = R\omega_T/v$ is the tip speed ratio of the wind turbine, and C_p is a function of the tip speed ratio and blade pith angle. The drive train model is given by [10]

$$\omega_g = n_g\omega_T \ ,$$

$$J_h\frac{d\omega_g}{dt} = \frac{\eta}{n_g}\cdot T_W - T_g \ , \qquad (2)$$

$$J_h = (J_1 + J_W)\cdot\frac{\eta}{n_g} + J_2 + J_g$$

where J_1 and J_2 are the inertias of the multiplier gearings. Fig. 3 illustrates the wind turbine model.

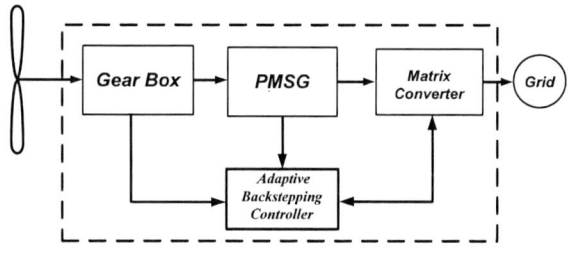

Figure 1. WECS block diagram

The matrix converter does not have any storage element, and thus has no dynamics. Thus the WECS is modeled as a third order system defined by:

$$\frac{di_d}{dt} = -a_1i_d + p\omega i_q + a_2u_d$$

$$\frac{di_q}{dt} = -a_1i_q + p\omega i_d - a_3\omega_g + a_2u_q \qquad (3)$$

$$\frac{d\omega_g}{dt} = a_6 - a_5\omega + a_4i_q$$

where $a_1 = \frac{R}{L}$, $a_2 = \frac{1}{L}$, $a_3 = \frac{P\phi}{L}$, $a_4 = \frac{3P\phi}{2J_h}$, $a_5 = \frac{B}{J_h}$ and, $a_6 = \frac{n\Gamma}{J_hn_g}$.

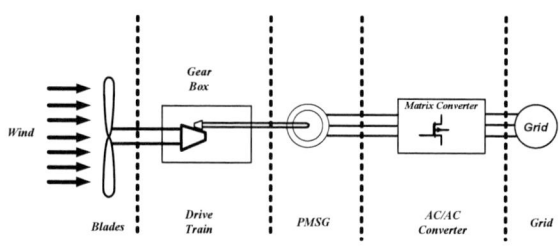

Figure 2. Different parts of WECS

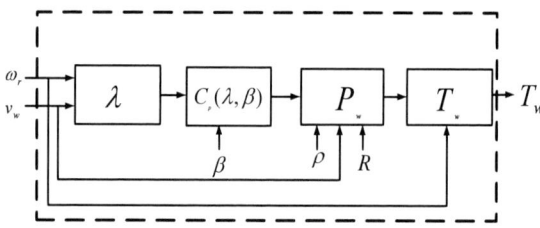

Figure 3. Wind turbine model

B. Maximum Power Point Tracking (MPPT)

Optimal operation of the WECS is to extract the maximum power from wind. At a fixed wind velocity, the maximum achievable power occurs at a specific shaft speed. Thus, optimal control of the WECS means that it has to track the optimal value of the shaft rotational speed, namely,

$$\omega_T(opt) = \frac{v.\lambda_{opt}}{R} \ .$$

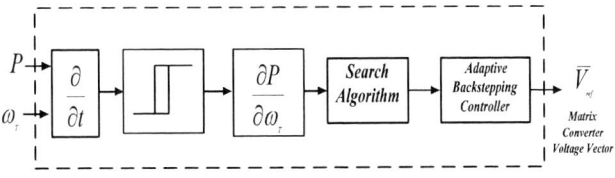

Figure 4. MPPT Block Diagram

Due to the erratic nature of wind, nonlinear variant behavior of the WECS, poor reliability of the measured information and unknown parameters, the control method used to track the optimal speed should be robust enough against these issues. Fig. 4 shows the proposed maximum power point tracker for the WECS.

The approach is basically based on the calculation of the power gradient. Since the power curve is a convex curve, there is no local maximum. Thus, the algorithm can easily find the maximum point based on the power gradient. Eventually, the search algorithm provides the reference value for the shaft rotational speed. This reference value is the command signal for the adaptive integrator backstepping controller. This controller produces the desired voltage vector for the matrix converter so as to track the speed reference and hence track the maximum power point for the wind turbine. This controller is robust against system uncertainties.

C. Adaptive Nonlinear Controller Design for WECS

The proposed controller in this paper is based on a nonlinear approach called integrator backstepping. The concept of the integrator backstepping approach is fully explained in [10]. The control objective is that the generator rotational speed tracks the reference value derived by the MPPT algorithm. Therefore, the error is defined using the rotational speed $e = \omega_{ref} - \omega$ and the first Lyapanov function is defined as

$$V_1 = \frac{1}{2}e^2 \qquad (4)$$

The time derivative of V_1 is given by

$$\dot{V}_1 = e\dot{e} = e\left(\dot{\omega}_{ref} - a_6 + a_5\omega + a_4 i_q\right) \qquad (5)$$

i_q is considered to be the virtual control input. Thus, its reference is obtained as

$$i_q^* = \alpha(t) = \frac{1}{a_4}\left[-k_1 - \omega_{ref} + a_6 - a_5\omega\right] \qquad (6)$$

Because the parameters are unknown the virtual control is given by

$$i_q^* = \alpha(t) = \frac{1}{\hat{a}_4}\left[-k_1 - \omega_{ref} + \hat{a}_6 - \hat{a}_5\omega\right] \qquad (7)$$

where $\hat{a}_i$ are the estimated parameters.

On the other hand, the reference value for i_d is zero, because the maximum power can be extracted from the PMSG if the d-component of the current is zero . Thus, e_d and e_q are defined by

$$e_d = i_d^* - i_d = -i_d \,,$$

$$e_q = i_q^* - i_q = \alpha(t) = \frac{1}{\hat{a}_4}\left[-k_1 - \omega_{ref} + \hat{a}_6 - \hat{a}_5\omega\right] - i_q \quad (8)$$

Finally the Lyapanov function is defined as

$$V_2 = \frac{1}{2}\left[e^2 + e_d^{\,2} + e_q^{\,2} + \sum_{i=1}^{6}\frac{1}{\mu_i}\tilde{a}_i^{\,2}\right] \qquad (9)$$

where $\tilde{a}_i$ is the estimated error $\tilde{a}_i = a_i - \hat{a}_i$.

Control laws and adaptive laws are derived by differentiating the Lyapanov function with respect to time.

$$u_d = \frac{1}{\hat{a}_2}\left(\hat{a}_1 i_d - p\omega i_q + k_2 e_d\right),$$

$$u_q = \frac{1}{\hat{a}_2}\begin{pmatrix} k_3 e_q - \ddot{a}_4 e + \ddot{a}_1 i_q + p\omega i_d + \hat{a}_3\omega + \hat{a}_5 i_q \\ -\dfrac{\hat{a}_5^{\,2}}{\hat{a}_4}\omega + \dfrac{\hat{a}_5^{\,2}}{\hat{a}_4}\hat{a}_6 - k_1 i_q + \dfrac{k_1}{\hat{a}_4}\hat{a}_5\omega - \dfrac{k_1}{\hat{a}_4}\hat{a}_6 \end{pmatrix} \quad (10)$$

$$\dot{\hat{a}}_1 = \mu_1\left(e_d i_d + e_q i_q\right),$$

$$\dot{\hat{a}}_2 = \mu_2\left(-e_d u_d - e_q u_q\right),$$

$$\dot{\hat{a}}_3 = \mu_3\left(\omega e_q\right),$$

$$\dot{\hat{a}}_4 = \mu_4\left(-e i_q + \frac{\hat{a}_5}{\hat{a}_4}e_q i_q - \frac{k_1}{\hat{a}_4}e_q i_q\right)$$

$$\dot{\hat{a}}_5 = \mu_5\left(e\omega - \frac{\hat{a}_5}{\hat{a}_4}\omega e_q + \frac{k_1}{\hat{a}_4}\omega e_q\right),$$

$$\dot{\hat{a}}_6 = \mu_6\left(e - \frac{\hat{a}_5}{\hat{a}_4}e_q + \frac{k_1}{\hat{a}_4}e_q\right) \qquad (11)$$

The Lyapanov function derivative is given by

$$\dot{V}_2 = -k_1 e^2 - k_2 e_q^2 - k_3 e_d^2 \qquad (12)$$

The controller is implemented by Eq. (10) and (11) which is written as a C-script to implement in DSP.

D. Stability Analysis

According to (12), the Lyapunov function derivative is not negative definite and is just negative semidefinite. Thus the adaptive backstepping lemma only guarantees global boundedness of a_i and regulation of e, e_d, e_q but not the regulation of $\tilde{a}_i$. Therefore, the estimated parameters do not necessarily converge to the actual values. This can be proven by Barbalet's lemma. By integrating (12) over the interval [0,t] we have

$$\int_0^t \dot{V}_2 \, dt = V(t) - V(0) - \int_0^t \underbrace{\left(k_1 e^2 + k_2 e_q^2 + k_3 e_d^2\right)}_{E(T)} dt = V(t) - V(0) \quad,$$

$$\int_0^t E(t) dt = V(0) - V(t) \qquad (13)$$

Therefore, $\displaystyle\lim_{t\to\infty}\int_0^\infty E(\tau)d\tau$ exists and is finite. It is just required to show that $E(t)$ is uniformly continuous. The time derivative of $\dot{V}(t)$ is $\ddot{V} = -2k_1 e \dot{e} - 2k_2 e_q \dot{e}_q - 2k_3 e_d \dot{e}_d$ which is bounded, hence $\dot{V}$ is uniformly continuous. Invoking Barbaret's lemma, the errors converge to zero $\displaystyle\lim_{t\to\infty} E(t) = 0$.

Thus, the objective of tracking is achieved under parameter uncertainties and disturbances.

III. MATRIX CONVERTER

The matrix converter is an array of semiconductors which interfaces between two multi-phase systems with different frequencies. Fig. 5 shows a three-phase-to-three-phase matrix converter with 9 bi-directional switches. The operation principles of the matrix converter and its modulation schemes are comprehensively explained in [25]. In this paper, a space vector modulation technique is employed as the modulation scheme for the matrix converter. According to Fig. 5, the output voltage reference vector and input current reference vector are applied by the MPPT to the matrix converter modulator.

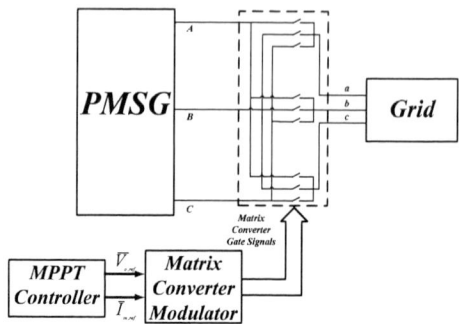

Figure 5. Matrix Converter Block Diagram

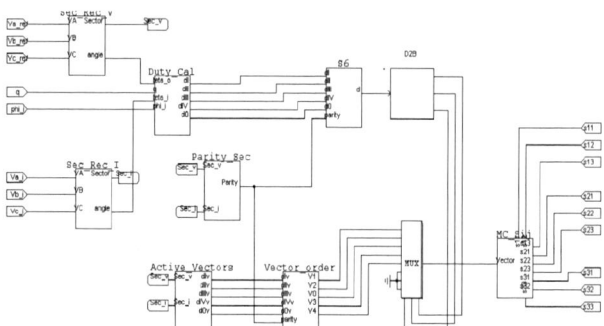

Figure 6. SVM implementation for matrix converter

The matrix converter produces a set of voltages to control the PMSG angular speed so that the maximum power is extracted from the wind turbine.

IV. PERFORMANCE ANALYSIS

Fig. 7 shows the simulated WECS. The performance analysis is carried out for a low power prototype. However, they can easily be applied to the high power systems. The PMSG parameters are given in Table I. Fig. 8 illustrates the output phase voltage and Fig. 9 shows the output line-to-line voltage. Simulations are carried out with the proposed controller and with the conventional PI controller to compare the performances and to show the improvements of the proposed controller. Fig. 10 and Fig. 11 illustrate the shaft speed tracking for a step change in the wind speed with the proposed controller and the conventional PI controller respectively. According to this figure, the proposed controller is much faster than the conventional PI one. The MPPT is shown in Fig. 12 and Fig. 13 for step and sinusoidal changes in the wind speed profile respectively.

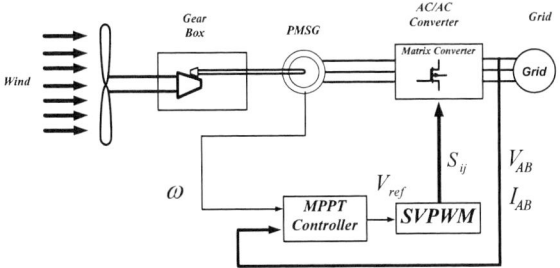

Figure 7.   WECS Closed-Loop Block Diagram

TABLE I.   SYSTEM PARAMETERS

WECS	Converter Parameters		
	Generator Parameters	P_o	f_{sw}
Value	R_s=1.93; L_d=L_q=0.8mH; P=2; J=2m; B=0.0002	1KW	23KHz

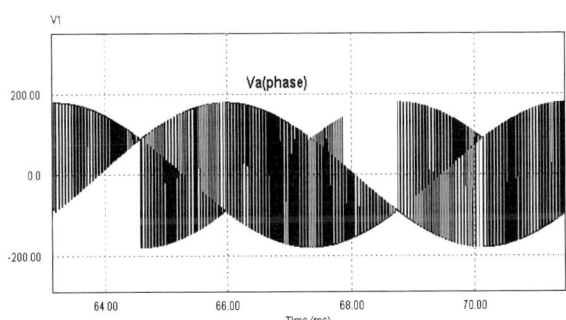

Figure 8.   Output Phase Voltage (v)

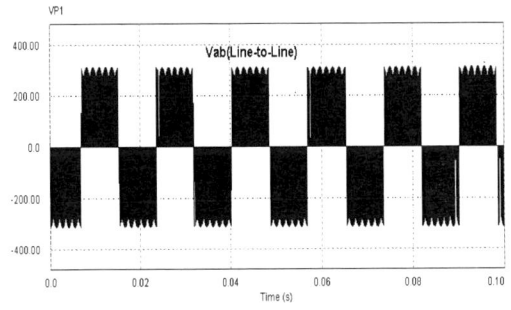

Figure 9.   Output Line Voltage (v)

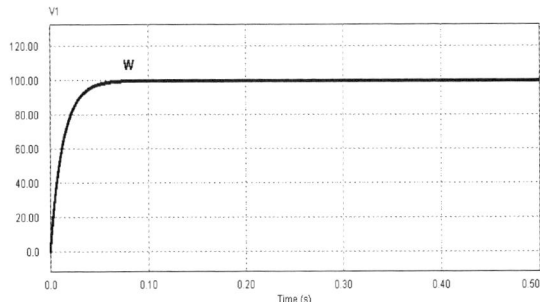

Figure 10. Shaft Speed Response with the Proposed Controller (rad/s)

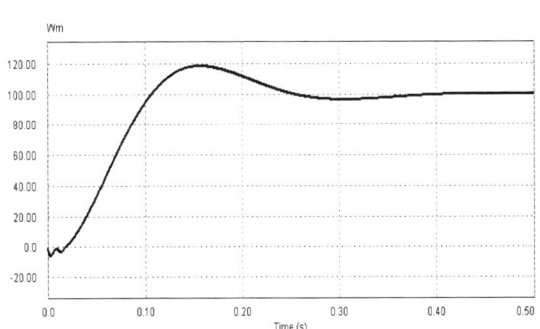

Figure 11. Shaft Speed Response with the PI Controller (rad/s)

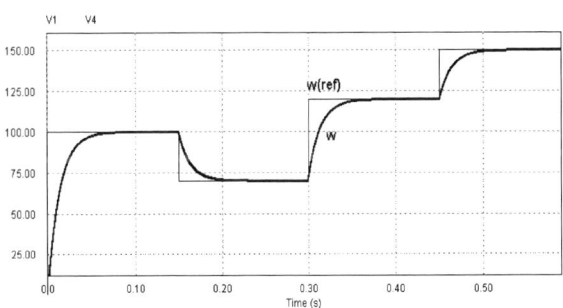

Figure 12. Shaft Speed Response to the step changes in the wind speed (rad/s)

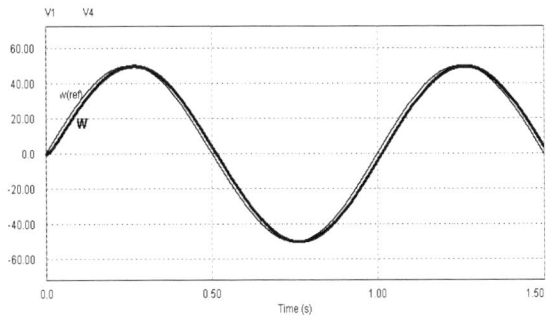

Figure 13. Shaft Speed Response to the low frequency sinusoidal change in the wind speed (rad/s)

978-1-4244-4782-4/10 $26.00 © 2010 IEEE

V. CONCLUSION

This paper presented an adaptive nonlinear approach for maximum power point tracking of a WECS. The proposed approach is also robust against system parameter uncertainties. In addition, the closed loop stability has been analyzed for the proposed controller through a nonlinear analysis. A matrix converter is utilized as an interface between the wind turbine and the load. Direct space vector modulation is used to control the matrix converter switches. Theoretical analysis and simulation results are given to verify the feasibility and performance of the proposed approach.

REFERENCES

[1] Blaabjerg, F. Zhe Chen Kjaer, S.B., "Power electronics as efficient interface in dispersed power generation systems" , IEEE Transactions on Power Electronics Sept. 2004 Vol. 19, page(s): 1184- 1194.

[2] J. M. Carrasco, L. G. Franquelo, J. T. Bialasiewicz, E. Galvan, R. C. P. Guisado, M. A. Martin Prats, J. I. Leon, N. M. Alfonso, "Power electronic systems for grid integration of renewable energy sources: a survey" IEEE Trans. Industrial Electronics, vol. 53, no. 4, pp. 1002-1016, August. 2006.

[3] F. Blaabjerg Z. Chen R. Teodorescu F. Iov , "Power Electronics in Wind Turbine Systems" Power Electronics and Motion Control Conference, 2006. IPEMC '06. CES/IEEE 5th International, page(s): 1-11 Aug. 2006.

[4] Wheeler, P.W. Rodriguez, J. Clare, J.C. Empringham, L. Weinstein, A. , "Matrix converters: a technology review ", IEEE Transactions on Industrial Electronics, Vol. 49 page. 276-288, April 2002.

[5] Domenico Casadei, , Giovanni Serra, , Angelo Tani, and Luca Zarri, "Matrix Converter Modulation Strategies: A New General Approach Based on Space-Vector Representation of the Switch State" IEEE Transactions on Industrial Electronics, Vol. 49, No. 2, April 2002.

[6] Christian Klumpner, Ion Boldea and Frede Blaabjerg, "Limited Ride-Through Capabilities for Direct Frequency Converters" IEEE Transactions on Power Electronics, Vol. 16, No. 6, November 2001.

[7] Kwak, S. Toliyat, H.A., "An Approach to Fault-Tolerant Three-Phase Matrix Converter Drives" IEEE Transaction on Energy Conversion,Vol. 22, Issue 4, Page(s):855 – 863, Dec. 2007.

[8] [8] Klumpner, C. Blaabjerg, F. Thogersen, P. ," Evaluation of the converter topologies suited for integrated motor drives" , Industry 38th Applications Conference, IAS, page(s): 890- 897, vol.2 2003.

[9] Munteau, I., Bratcu, A. I. Cutululis, N. A., Ceanga, E., "Optimal Control of Wind Energy Systems" Book, Springer-Verlag London 2008.

[10] Krstic, Kanellakopoulos, Kokotovic, " Nonlinear & Adaptive Control Design" Book, John Wiley & Sons, 1995.

[11] Parks, P. ," Liapunov redesign of model reference adaptive control systems" IEEE Transactions on Automatic Control, July 1966, 11, 362-367 .

[12] Kuo-Kai Shyu; Ming-Ji Yang; Yen-Mo Chen; Yi-Fei Lin;, "Model Reference Adaptive Control Design for a Shunt Active-Power-Filter System" IEEE Transactions on Industrial Electronics, Volume 55, Issue 1, Jan. 2008 Page(s):97 – 106.

[13] Kim, K.-H.; "Model reference adaptive control-based adaptive current control scheme of a PM synchronous motor with an improved servo performance Power" Electric Applications, IET Volume 3, Issue 1, January 2009 Page(s):8 – 18.

[14] Liu Hsu; Costa, R.R.; Lizarralde, F.; "Lyapunov/Passivity-Based Adaptive Control of Relative Degree Two MIMO Systems With an Application to Visual Servoing" Transactions on Automatic Control, IEEE Volume 52, Issue 2, Feb. 2007 Page(s):364 – 371.

A Hybrid Wind-Solar Energy System: A New Rectifier Stage Topology

Joanne Hui*, *IEEE Student Member*, Alireza Bakhshai, *IEEE Senior Member*, and Praveen K. Jain, *IEEE Fellow*

Department of Electrical and Computer Engineering
Queen's Center for Energy and Power Electronics Research (ePOWER), Queen's University
Kingston, Ontario, Canada
*email: 2cyjh@queensu.ca

Abstract—Environmentally friendly solutions are becoming more prominent than ever as a result of concern regarding the state of our deteriorating planet. This paper presents a new system configuration of the front-end rectifier stage for a hybrid wind/photovoltaic energy system. This configuration allows the two sources to supply the load separately or simultaneously depending on the availability of the energy sources. The inherent nature of this Cuk-SEPIC fused converter, additional input filters are not necessary to filter out high frequency harmonics. Harmonic content is detrimental for the generator lifespan, heating issues, and efficiency. The fused multi-input rectifier stage also allows Maximum Power Point Tracking (MPPT) to be used to extract maximum power from the wind and sun when it is available. An adaptive MPPT algorithm will be used for the wind system and a standard perturb and observe method will be used for the PV system. Operational analysis of the proposed system will be discussed in this paper. Simulation results are given to highlight the merits of the proposed circuit.

I. INTRODUCTION

With increasing concern of global warming and the depletion of fossil fuel reserves, many are looking at sustainable energy solutions to preserve the earth for the future generations. Other than hydro power, wind and photovoltaic energy holds the most potential to meet our energy demands. Alone, wind energy is capable of supplying large amounts of power but its presence is highly unpredictable as it can be here one moment and gone in another. Similarly, solar energy is present throughout the day but the solar irradiation levels vary due to sun intensity and unpredictable shadows cast by clouds, birds, trees, etc. The common inherent drawback of wind and photovoltaic systems are their intermittent natures that make them unreliable. However, by combining these two intermittent sources and by incorporating maximum power point tracking (MPPT) algorithms, the system's power transfer efficiency and reliability can be improved significantly.

When a source is unavailable or insufficient in meeting the load demands, the other energy source can compensate for the difference. Several hybrid wind/PV power systems with MPPT control have been proposed and discussed in works [1]-[5]. Most of the systems in literature use a separate DC/DC boost converter connected in parallel in the rectifier stage as shown in Figure 1 to perform the MPPT control for each of the renewable energy power sources [1]-[4]. A simpler multi-input structure has been suggested by [5] that combine the sources from the DC-end while still achieving MPPT for each renewable source. The structure proposed by [5] is a fusion of the buck and buck-boost converter. The systems in literature require passive input filters to remove the high frequency current harmonics injected into wind turbine generators [6]. The harmonic content in the generator current decreases its lifespan and increases the power loss due to heating [6].

In this paper, an alternative multi-input rectifier structure is proposed for hybrid wind/solar energy systems. The proposed design is a fusion of the Cuk and SEPIC converters. The features of the proposed topology are: 1) the inherent nature of these two converters eliminates the need for separate input filters for PFC [7]-[8]; 2) it can support step up/down operations for each renewable source (can support wide ranges of PV and wind input); 3) MPPT can be realized for each source; 4) individual and simultaneous operation is supported. The circuit operating principles will be discussed in this paper. Simulation results are provided to verify with the feasibility of the proposed system.

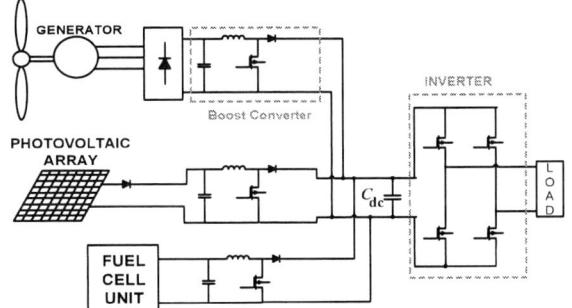

Figure 1: Hybrid system with multi-connected boost converter

II. PROPOSED MULTI-INPUT RECTIFIER STAGE

A system diagram of the proposed rectifier stage of a hybrid energy system is shown in Figure 2, where one of the inputs is connected to the output of the PV array and the other input connected to the output of a generator. The fusion of the two converters is achieved by reconfiguring the two existing diodes from each converter and the shared utilization of the Cuk output inductor by the SEPIC converter. This configuration allows each converter to operate normally individually in the event that one source is unavailable. Figure 3 illustrates the case when only the wind source is available. In this case, D_1 turns off and D_2 turns on; the proposed circuit becomes a SEPIC converter and the input to output voltage relationship is given by (1). On the other hand, if only the PV source is available, then D_2 turns off and D_1 will always be on and the circuit becomes a Cuk converter as shown in Figure 4. The input to output voltage relationship is given by (2). In both cases, both converters have step-up/down capability, which provide more design flexibility in the system if duty ratio control is utilized to perform MPPT control.

$$\frac{V_{dc}}{V_W} = \frac{d_2}{1-d_2} \tag{1}$$

$$\frac{V_{dc}}{V_{PV}} = \frac{d_1}{1-d_1} \tag{2}$$

Figure 5 illustrates the various switching states of the proposed converter. If the turn on duration of M_1 is longer than M_2, then the switching states will be state I, II, IV. Similarly, the switching states will be state I, III, IV if the switch conduction periods are vice versa. To provide a better explanation, the inductor current waveforms of each switching state are given as follows assuming that $d_2 > d_1$; hence only states I, III, IV are discussed in this example. In the following, $I_{i,PV}$ is the average input current from the PV source; $I_{i,W}$ is the RMS input current after the rectifier (wind case); and I_{dc} is the average system output current. The key waveforms that illustrate the switching states in this example are shown in Figure 6. The mathematical expression that relates the total output voltage and the two input sources will be illustrated in the next section.

State I (M_1 on, M_2 on):

$$i_{L1} = I_{i,PV} + \frac{V_{PV}}{L_1}t \qquad\qquad 0 < t < d_1 T_s$$

$$i_{L2} = I_{dc} + \left(\frac{v_{c1}+v_{c2}}{L_2}\right)t \qquad\qquad 0 < t < d_1 T_s$$

$$i_{L3} = I_{i,W} + \frac{V_W}{L_3}t \qquad\qquad 0 < t < d_1 T_s$$

State III (M_1 off, M_2 on):

$$i_{L1} = I_{i,PV} + \left(\frac{V_{PV}-v_{c1}}{L_1}\right)t \qquad\qquad d_1 T_s < t < d_2 T_s$$

$$i_{L2} = I_{dc} + \frac{v_{c2}}{L_2}t \qquad\qquad d_1 T_s < t < d_2 T_s$$

$$i_{L3} = I_{i,W} + \frac{V_W}{L_3}t \qquad\qquad d_1 T_s < t < d_2 T_s$$

State IV (M_1 off, M_2 off):

$$i_{L1} = I_{i,PV} + \left(\frac{V_{PV}-v_{c1}}{L_1}\right)t \qquad\qquad d_2 T_s < t < T_s$$

$$i_{L2} = I_{dc} - \frac{V_{dc}}{L_2}t \qquad\qquad d_2 T_s < t < T_s$$

$$i_{L3} = I_{i,W} + \left(\frac{V_W-v_{c2}-V_{dc}}{L_3}\right)t \qquad\qquad d_2 T_s < t < T_s$$

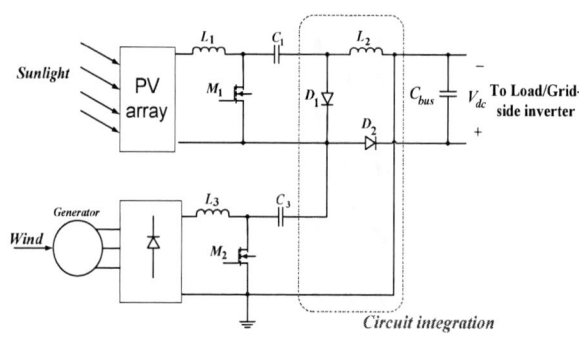

Figure 2: Proposed rectifier stage for a Hybrid wind/PV system

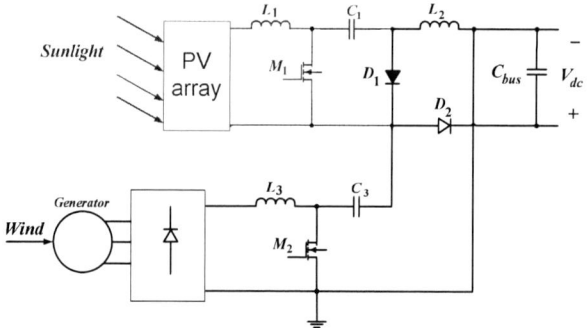

Figure 3: Only wind source is operational (SEPIC)

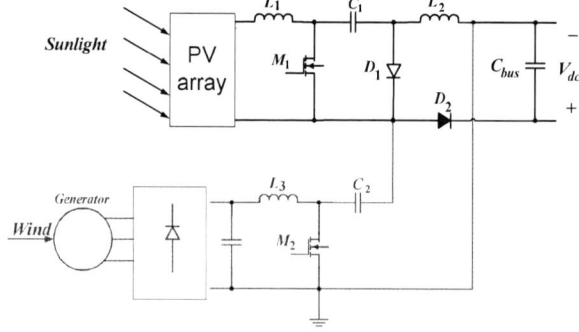

Figure 4: Only PV source is operation (Cuk)

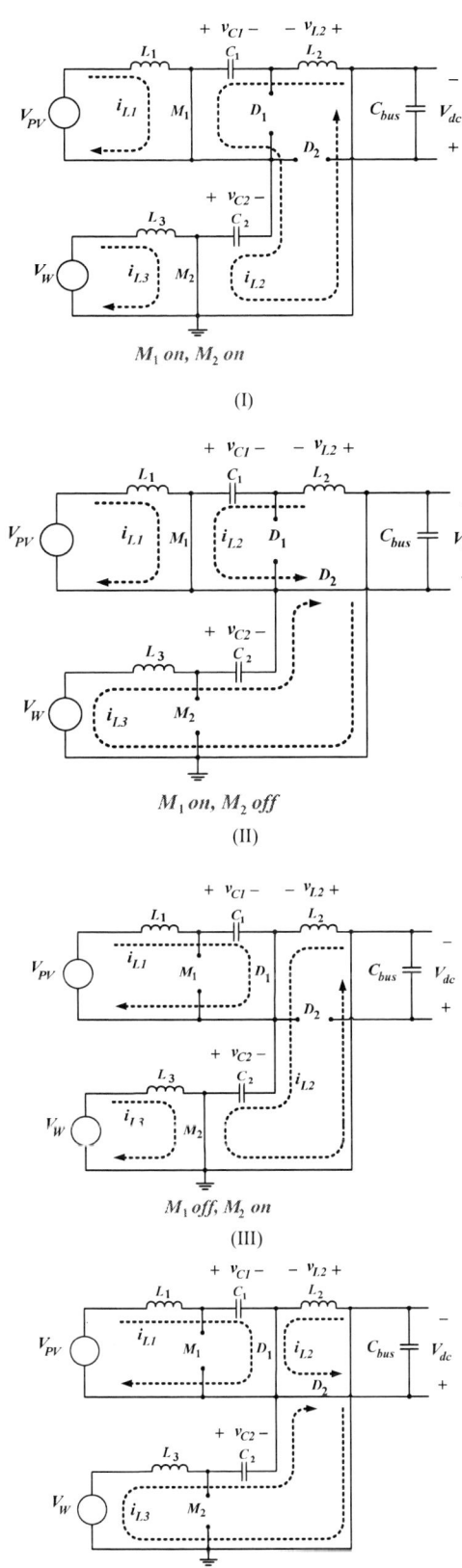

(I)

M_1 on, M_2 on

(II)

M_1 on, M_2 off

(III)

M_1 off, M_2 on

(IV)

M_1 off, M_2 off

Figure 5 (I-IV): switching states within a switching cycle

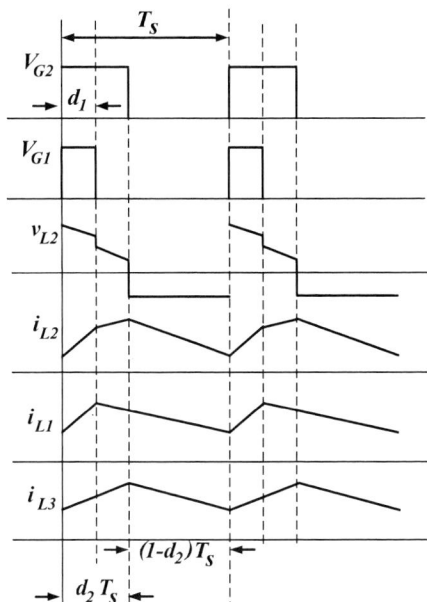

Figure 6: Proposed circuit inductor waveforms

III. ANALYSIS OF PROPOSED CIRCUIT

To find an expression for the output DC bus voltage, V_{dc}, the volt-balance of the output inductor, L_2, is examined according to Figure 6 with $d_2 > d_1$. Since the net change in the voltage of L_2 is zero, applying volt-balance to L_2 results in (3). The expression that relates the average output DC voltage (V_{dc}) to the capacitor voltages (v_{c1} and v_{c2}) is then obtained as shown in (4), where v_{c1} and v_{c2} can then be obtained by applying volt-balance to L_1 and L_3 [9]. The final expression that relates the average output voltage and the two input sources (V_W and V_{PV}) is then given by (5). It is observed that V_{dc} is simply the sum of the two output voltages of the Cuk and SEPIC converter. This further implies that V_{dc} can be controlled by d_1 and d_2 individually or simultaneously.

$$(v_{c1} + v_{c2})d_1 T_s + (v_{c2})(d_2 - d_1)T_s + (1 - d_2)(-V_{dc})T_s = 0 \quad (3)$$

$$V_{dc} = \left(\frac{d_1}{1 - d_2}\right)v_{c1} + \left(\frac{d_2}{1 - d_2}\right)v_{c2} \quad (4)$$

$$V_{dc} = \left(\frac{d_1}{1 - d_1}\right)V_{PV} + \left(\frac{d_2}{1 - d_2}\right)V_w \quad (5)$$

The switches voltage and current characteristics are also provided in this section. The voltage stress is given by (6) and (7) respectively. As for the current stress, it is observed from Figure 6 that the peak current always occurs at the end of the on-time of the MOSFET. Both the Cuk and SEPIC MOSFET current consists of both the input current and the capacitors (C_1 or C_2) current. The peak current stress of M_1 and M_2 are given by (8) and (10) respectively. L_{eq1} and L_{eq2}, given by (9) and (11), represent the equivalent inductance of Cuk and SEPIC converter respectively.

978-1-4244-4782-4/10 $26.00 © 2010 IEEE

The PV output current, which is also equal to the average input current of the Cuk converter is given in (12). It can be observed that the average inductor current is a function of its respective duty cycle (d_1). Therefore by adjusting the respective duty cycles for each energy source, maximum power point tracking can be achieved.

$$v_{ds1} = V_{pv}\left(1 + \frac{d_1}{1 - d_1}\right) \tag{6}$$

$$v_{ds2} = V_W\left(1 + \frac{d_2}{1 - d_2}\right) \tag{7}$$

$$i_{ds1,pk} = I_{i,PV} + I_{dc,avg} + \frac{V_{PV}d_1T_s}{2L_{eq1}} \tag{8}$$

$$L_{eq1} = \frac{L_1L_2}{L_1 + L_2} \tag{9}$$

$$i_{ds2,pk} = I_{i,W} + I_{dc,avg} + \frac{V_W d_2 T_s}{2L_{eq2}} \tag{10}$$

$$L_{eq2} = \frac{L_3 L_2}{L_3 + L_2} \tag{11}$$

$$I_{i,PV} = \frac{P_o}{V_{dc}}\frac{d_1}{1 - d_1} \tag{12}$$

IV. MPPT CONTROL OF PROPOSED CIRCUIT

A common inherent drawback of wind and PV systems is the intermittent nature of their energy sources. Wind energy is capable of supplying large amounts of power but its presence is highly unpredictable as it can be here one moment and gone in another. Solar energy is present throughout the day, but the solar irradiation levels vary due to sun intensity and unpredictable shadows cast by clouds, birds, trees, etc. These drawbacks tend to make these renewable systems inefficient. However, by incorporating maximum power point tracking (MPPT) algorithms, the systems' power transfer efficiency can be improved significantly.

To describe a wind turbine's power characteristic, equation (13) describes the mechanical power that is generated by the wind [6].

$$p_m = 0.5\rho A C_p(\lambda, \beta)v_w^3 \tag{13}$$

Where
ρ = *air density,*
A = *rotor swept area,*
$C_p(\lambda, \beta)$ = *power coefficient function*
λ = *tip speed ratio,*
β = *pitch angle,*
v_w = *wind speed*

The power coefficient (C_p) is a nonlinear function that represents the efficiency of the wind turbine to convert wind energy into mechanical energy. It is dependent on two

variables, the tip speed ratio (TSR) and the pitch angle. The TSR, λ, refers to a ratio of the turbine angular speed over the wind speed. The mathematical representation of the TSR is given by (14) [10]. The pitch angle, β, refers to the angle in which the turbine blades are aligned with respect to its longitudinal axis.

$$\lambda = \frac{R\,\omega_b}{v_w} \tag{14}$$

Where
R = *turbine radius,*
ω_b = *angular rotational speed*

Figure 7 and 8 are illustrations of a power coefficient curve and power curve for a typical fixed pitch (β =0) horizontal axis wind turbine. It can be seen from figure 7 and 8 that the power curves for each wind speed has a shape similar to that of the power coefficient curve. Because the TSR is a ratio between the turbine rotational speed and the wind speed, it follows that each wind speed would have a different corresponding optimal rotational speed that gives the optimal TSR. For each turbine there is an optimal TSR value that corresponds to a maximum value of the power coefficient ($C_{p,max}$) and therefore the maximum power. Therefore by controlling rotational speed, (by means of adjusting the electrical loading of the turbine generator) maximum power can be obtained for different wind speeds.

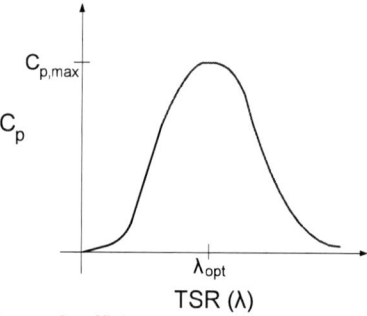

Figure 7: Power Coefficient Curve for a typical wind turbine

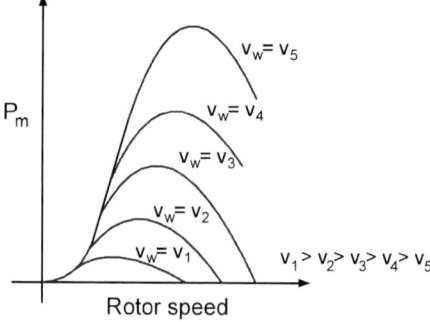

Figure 8: Power Curves for a typical wind turbine

A solar cell is comprised of a P-N junction semiconductor that produces currents via the photovoltaic effect. PV arrays are constructed by placing numerous solar cells connected in series and in parallel [5]. A PV cell is a diode of a large-area forward bias with a photovoltage and the equivalent circuit is shown by Figure 9 [11]. The current-voltage characteristic of a solar cell is derived in [12] and [13] as follows:

$$I = I_{ph} - I_D \qquad (15)$$

$$I = I_{ph} - I_0 \left[\exp\left(\frac{q(V + R_s I)}{A k_B T} \right) - 1 \right] - \frac{V + R_s I}{R_{sh}} \qquad (16)$$

Where
I_{ph} = *photocurrent,*
I_D = *diode current,*
I_0 = *saturation current,*
A = *ideality factor,*
q = *electronic charge 1.6x10-9,*
k_B = *Boltzmann's gas constant (1.38x10^{-23}),*
T = *cell temperature,*
R_s = *series resistance,*
R_{sh} = *shunt resistance,*
I = *cell current,*
V = *cell voltage*

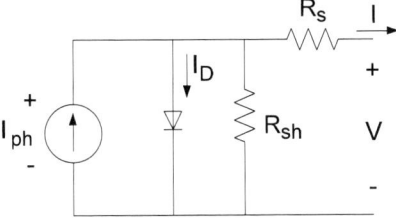

Figure 9: PV cell equivalent circuit

Typically, the shunt resistance (R_{sh}) is very large and the series resistance (R_s) is very small [5]. Therefore, it is common to neglect these resistances in order to simplify the solar cell model. The resultant ideal voltage-current characteristic of a photovoltaic cell is given by (17) and illustrated by Figure 10. [5]

$$I = I_{ph} - I_0 \left(\exp\left(\frac{qV}{kT} \right) - 1 \right) \qquad (17)$$

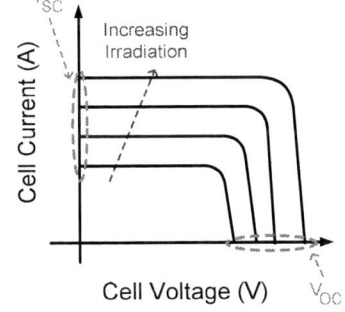

Figure 10: PV cell voltage-current characteristic

The typical output power characteristics of a PV array under various degrees of irradiation is illustrated by Figure 11. It can be observed in Figure 11 that there is a particular optimal voltage for each irradiation level that corresponds to maximum output power. Therefore by adjusting the output current (or voltage) of the PV array, maximum power from the array can be drawn.

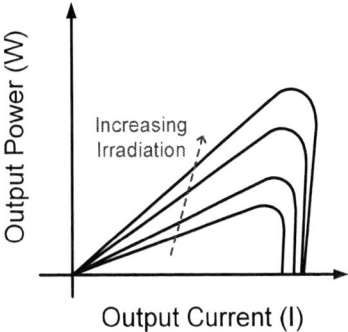

Figure 11: PV cell power characteristics

Due to the similarities of the shape of the wind and PV array power curves, a similar maximum power point tracking scheme known as the hill climb search (HCS) strategy is often applied to these energy sources to extract maximum power. The HCS strategy perturbs the operating point of the system and observes the output. If the direction of the perturbation (e.g an increase or decrease in the output voltage of a PV array) results in a positive change in the output power, then the control algorithm will continue in the direction of the previous perturbation. Conversely, if a negative change in the output power is observed, then the control algorithm will reverse the direction of the pervious perturbation step. In the case that the change in power is close to zero (within a specified range) then the algorithm will invoke no changes to the system operating point since it corresponds to the maximum power point (the peak of the power curves).

The MPPT scheme employed in this paper is a version of the HCS strategy. Figure 12 is the flow chart that illustrates the implemented MPPT scheme.

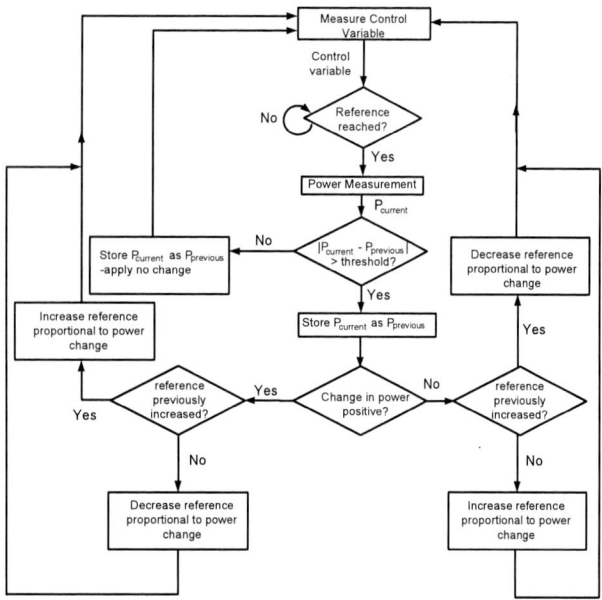

Figure 12: General MPPT Flow Chart for wind and PV

V. SIMULATION RESULTS

In this section, simulation results from PSIM 8.0.7 is given to verify that the proposed multi-input rectifier stage can support individual as well as simultaneous operation. The specifications for the design example are given in TABLE I. Figure 13 illustrates the system under the condition where the wind source has failed and only the PV source (Cuk converter mode) is supplying power to the load. Figure 14 illustrates the system where only the wind turbine generates power to the load (SEPIC converter mode). Finally, Figure 15 illustrates the simultaneous operation (Cuk-SEPIC fusion mode) of the two sources where M_2 has a longer conduction cycle (converter states I, IV and III—see Figure 5).

TABLE I. Design Specifications

Output power (W)	3kW
Output voltage	500V
Switching frequency	20kHz

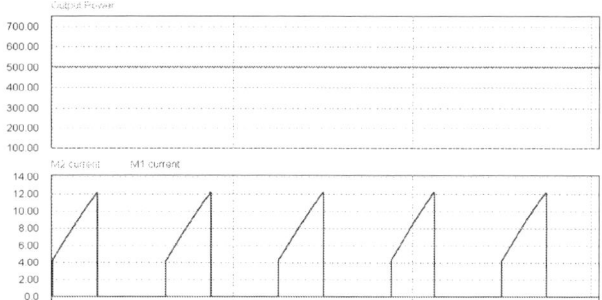

Figure 13 : Individual operation with only PV source (Cuk operation)
Top: Output power, Bottom: Switch currents (M_1 and M_2)

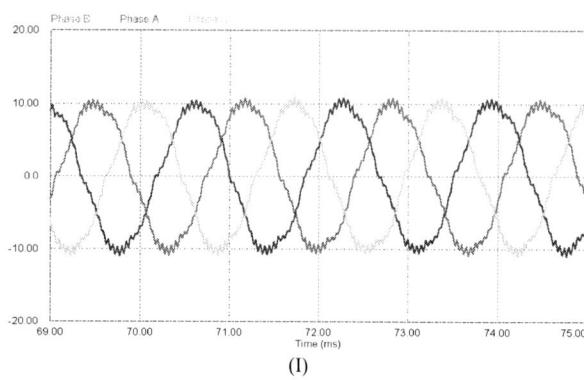

(I)

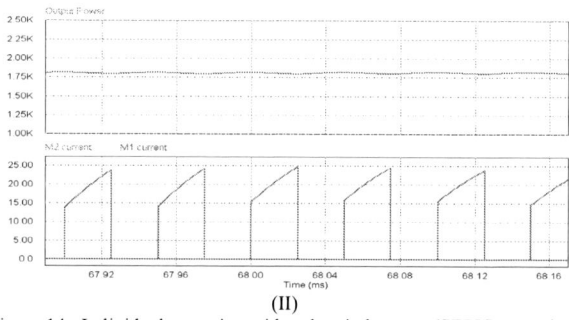

(II)

Figure 14 : Individual operation with only wind source (SEPIC operation)
(I) The injected three phase generator current; (II) Top: Output power, Bottom: Switch currents (M_1 and M_2)

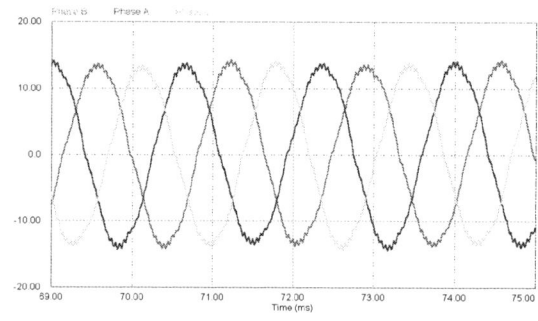

(I)

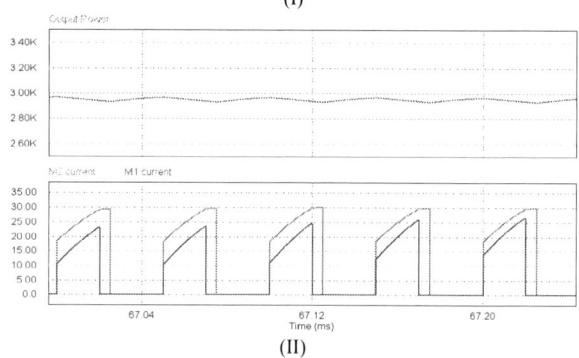

(II)

Figure 15 : Simultaneous operation with both wind and PV source (Fusion mode with Cuk and SEPIC)
(I) The injected three phase generator current; (II) Top: Output power, Bottom: Switch currents (M_1 and M_2)

Figure 16 and 17 illustrates the MPPT operation of the PV component of the system (Cuk operation) and the Wind component of the system (SEPIC operation) respectively.

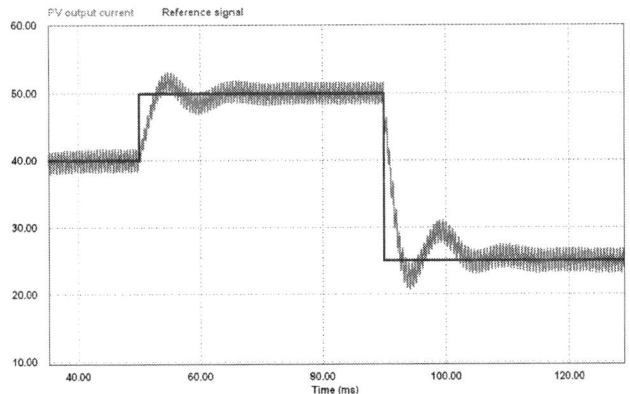

Figure 16 : Solar MPPT – PV output current and reference current signal (Cuk operation)

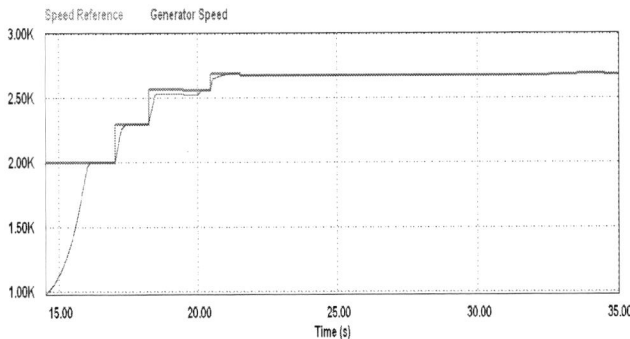

Figure 17 : Wind MPPT – Generator speed and reference speed signal (SEPIC operation)

VI. CONCLUSION

In this paper a new multi-input Cuk-SEPIC rectifier stage for hybrid wind/solar energy systems has been presented. The features of this circuit are: 1) additional input filters are not necessary to filter out high frequency harmonics; 2) both renewable sources can be stepped up/down (supports wide ranges of PV and wind input); 3) MPPT can be realized for each source; 4) individual and simultaneous operation is supported. Simulation results have been presented to verify the features of the proposed topology.

REFERENCES

[1] S.K. Kim, J.H Jeon, C.H. Cho, J.B. Ahn, and S.H. Kwon, "Dynamic Modeling and Control of a Grid-Connected Hybrid Generation System with Versatile Power Transfer," *IEEE Transactions on Industrial Electronics*, vol. 55, pp. 1677-1688, April 2008.

[2] D. Das, R. Esmaili, L. Xu, D. Nichols, "An Optimal Design of a Grid Connected Hybrid Wind/Photovoltaic/Fuel Cell System for Distributed Energy Production," *in Proc. IEEE Industrial Electronics Conference*, pp. 2499-2504, Nov. 2005.

[3] N. A. Ahmed, M. Miyatake, and A. K. Al-Othman, "Power fluctuations suppression of stand-alone hybrid generation combining solar photovoltaic/wind turbine and fuel cell systems," *in Proc. Of Energy Conversion and Management, Vol 49, pp. 2711-2719, October 2008.*

[4] S. Jain, and V. Agarwal, "An Integrated Hybrid Power Supply for Distributed Generation Applications Fed by Nonconventional Energy Sources," *IEEE Transactions on Energy Conversion, vol. 23, June 2008.*

[5] Y.M. Chen, Y.C. Liu, S.C. Hung, and C.S. Cheng, "Multi-Input Inverter for Grid-Connected Hybrid PV/Wind Power System," *IEEE Transactions on Power Electronics*, vol. 22, May 2007.

[6] dos Reis, F.S., Tan, K. and Islam, S., "Using PFC for harmonic mitigation in wind turbine energy conversion systems" *in Proc. of the IECON 2004 Conference*, pp. 3100- 3105, Nov. 2004

[7] R. W. Erickson, "Some Topologies of High Quality Rectifiers" *in the Proc. of the First International Conference on Energy, Power, and Motion Control*, May 1997.

[8] D. S. L. Simonetti, J. Sebasti'an, and J. Uceda, "The Discontinuous Conduction Mode Sepic and ' Cuk Power Factor Preregulators: Analysis and Design" *IEEE Trans. On Industrial Electronics*, vol. 44, no. 5, 1997

[9] N. Mohan, T. Undeland, and W Robbins, "Power Electronics: Converters, Applications, and Design," John Wiley & Sons, Inc., 2003.

[10] J. Marques, H. Pinheiro, H. Grundling, J. Pinheiro, and H. Hey, "A Survey on Variable-Speed Wind Turbine System," Proceedings of Brazilian Conference of Electronics of Power, vol. 1, pp. 732-738, 2003.

[11] F. Lassier and T. G. Ang, "Photovoltaic Engineering Handbook" 1990

[12] Global Wind Energy Council (GWEC), "Global wind 2008 report," June 2009.

[13] L. Pang, H. Wang, Y. Li, J. Wang, and Z. Wang, "Analysis of Photovoltaic Charging System Based on MPPT," Proceedings of Pacific-Asia Workshop on Computational Intelligence and Industrial Application 2008 (PACIIA '08), Dec 2008, pp. 498-501.

Dynamic Operation and Control of a Hybrid Wind-Diesel Stand Alone Power Systems

A. M. O. Haruni, A. Gargoom, M. E. Haque, and M. Negnevitsky

Centre of Renewable Energy and Power Systems (CREPS)
University of Tasmania,
Hobart, Tasmania, Australia
e-mail: amharuni@utas.edu.au

Abstract— This paper presents the dynamic operation and control strategies of a hybrid wind-diesel-battery energy storage based power supply system for isolated communities are investigated. Control strategies for voltage and frequency stabilization and efficient power flow among the hybrid system components are developed. The voltage and frequency of the hybrid wind-diesel system is controlled either by a load side inverter or by diesel generation depending on the wind conditions. During high penetration of wind, the wind turbine supplies the required power to the load. A battery energy storage system is connected to the dc-link to balance the power generated from the wind turbine and the power demand by load. Under low wind conditions, a diesel generator is used with wind energy conversion system to generate the required power to the load. A power sharing technique is developed to allocate power generation for diesel generator in low wind conditions. Results show that the control strategies work very well under dynamic and steady state condition to supply power to the load.

Index Terms— diesel generator; dump load; droop mode; isochonous mode; permanent magnet synchronous generator; power electronic converters; wind–diesel hybrid power system; wind energy conversion system.

I. INTRODUCTION

Diesel generators are normally used to supply power to remote areas where grid connection is not available. However, power generation from diesel generators is not environmental friendly as they produce pollutant gases and the price of the diesel is on the rise. Considering the environmental and economic aspect of diesel generator, it is very essential to generate power from cost-effective environmental friendly renewable energy sources such as wind, solar, hydro.

The renewable energy source such as wind is available in remote areas. However, power from the wind is fluctuating and weather dependent. Therefore, they can be used as a standalone power supply source for an isolated community. A variable speed wind turbine together with a suitable power electronic converter with voltage and frequency controller can be used to make the wind power suitable for the load. A diesel generator can also be used with the wind turbine to

supply adequate power to the load under low wind conditions. However, the dynamics of wind turbine and diesel generator are quite different. An energy storage system can be used to supply power during the transients before the diesel fully take up the shared load. An efficient control and coordination strategy is mandatory to make the hybrid (wind-diesel-battery storage-dump load) system work properly.

In a typical wind-diesel hybrid system, due to the random fluctuation of wind speed, the power generation from wind fluctuates and often mismatches with the load demand. In such situation, the diesel generator has to supply the power for a short period of time. However, the frequent start/stop of diesel generator is not recommended. In order to avoid the frequent start/stop of the diesel generator, an energy storage system can be used in the dc link voltage of the wind energy system. The start/stop time of diesel generator can be reduced from 30 time per hour to 2 times per hour if an energy storage system capable of supplying 2 minutes of load demand is used [1]. Moreover, during wind guest, the energy storage system can improve the transient performance of the system by smoothing the dc link voltage [2].

In wind energy conversion system, the use of variable speed wind turbines is preferable over fixed speed wind turbines as they offer various advantages such as increased energy capture, operation at maximum power point, better efficiency and power quality [3-6]. However, the presence of gear box in the variable speed wind turbine makes then unreliable as often suffers failure and it needs regular maintenance. The use of permanent magnet synchronous generators can increase the reliability of variable speed wind turbine as it does not require the gear box. Moreover, the use of permanent magnet in the rotor of synchronous generators makes them self excited. As a result, permanent magnet synchronous generators do not need to draw the reactive power from the system. Consequently, the system operates more efficiently with a higher power factor compared with induction generator based wind turbine [7]. In this work a permanent magnet synchronous generator based variable speed wind turbine is used.

Previous works on hybrid wind-diesel system show that

The project is sponsored by Australian Research Council (ARC) and Hydro Tasmania.

the voltage and frequency of the system are controlled by the continuous operation of a diesel generator, while a wind turbine is used to balance the power by controlling the pitch angle [8]. However, continuous operation of a diesel generator is not preferable due to higher emissions and low efficiency of the diesel at low load. The diesel generator has to run at least 30% of its rated load for efficient operation.

In this paper, an efficient control and co-ordination strategy is proposed among the different component of hybrid wind-diesel system under different wind and load conditions. The main objective of the proposed control system is to ensure the stable voltage and frequency of the system as well as to ensure an efficient power management among the energy sources under changing wind and load conditions. In the proposed control system, the voltage and frequency of the system is stabilized by the load side inverter in different wind and load conditions. The power management among the wind energy conversion system, the diesel generator and the energy storage system is controlled in two stages. The first stage

involves the dc link voltage control in excess wind conditions. This control strategy ensures the utilization of excess power during high wind condition by charging battery storage. This control strategy also allows discharging of the battery to supply the power to the load for a short time period during transients. The second stage involves the efficient control of diesel generator power flow in low wind conditions which ensures the minimum fuel consumption when diesel generator is working with wind energy conversion system.

II. SYSTEM OVERVIEW AND MODELING

The proposed wind-diesel hybrid system shown in Fig. 1 consists of the following:

- wind energy conversion system ,
- diesel generator system,
- battery energy storage system, and
- dump load.

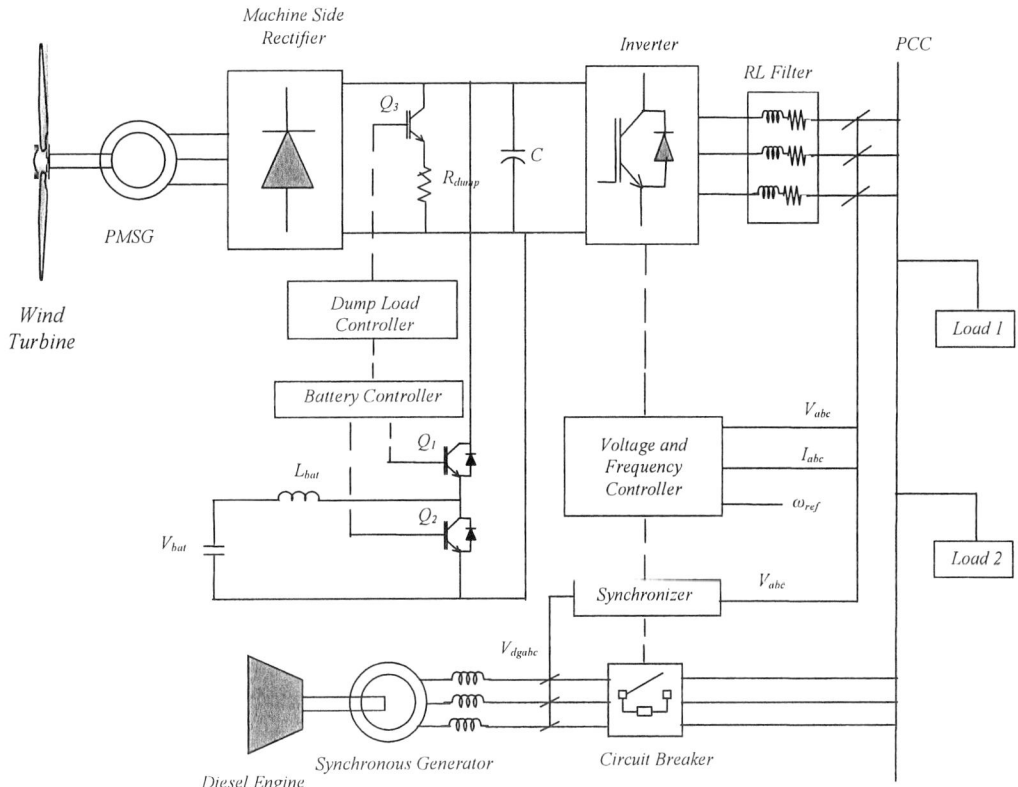

Figure 1. Proposed system Overview of a wind-diesel hybrid system.

A. Wind Turbine:

The power from wind captured in wind turbine can be expressed as follows [9, 10]:

$$P_{wind} = \frac{1}{2}\rho A v^3 \tag{1}$$

where, ρ is the air density, A is the rotor swept area, and v is the wind speed.

A variable speed wind turbines have three main regions of operation. In region 1, wind energy is not sufficient to start the turbine. Region 2 is an operation mode with objective to maximize wind energy capture. In region 3, because of high wind speed, turbine limits the capture of wind power to maintain a safe electrical and mechanical operation. The expression of the operation is given as follows:

$$P_{wind} = \begin{cases} 0; & 0 > v > v_o \\ \frac{1}{2}\rho A v^3; & v_o > v > v_i \\ P_{rated} \end{cases} \tag{2}$$

The wind power co-efficient (C_p) is defined as the ratio of aerodynamic rotor power (P) to power available from wind (P_{wind}) as shown below:

$$C_p = \frac{P}{P_{wind}} \tag{3}$$

The aerodynamic rotor power is given by [12]:

$$P_{aero} = \tau_{aero}\omega_r \tag{4}$$

where, τ_{aero} is the aerodynamic torque applied to the rotor by wind, ω_r the rotor speed.

The angular acceleration of rotor $(\dot{\omega}_r)$ is given by

$$\dot{\omega}_r = \frac{1}{J}(\tau_{aero} - \tau) \tag{5}$$

where, J is the combined rotational inertia of the rotor, gearbox, generator, and shafts and τ is the mechanical torque.

B. Parmanent Magnet Synchronous Generator

The dynamic voltage equations of PMSG are expressed in d-q reference frame is shown as follows [11]:

$$v_d = -i_d R_s - \omega_r \lambda_q + p\lambda_d \tag{6}$$

$$v_q = -i_q R_s + \omega_r \lambda_d + p\lambda_q \tag{7}$$

where, v_d and v_q are the d and q axis component of stator voltage; λ_d and λ_q are the d and q axis stator flux linkages; R_s is the stator resistant, i_d and i_q are the d and q axis components of stator current; ω_r is the rotor speed in $rad/\sec$; and p is the operator d/dt.

Stator flux linkage $(\lambda_d$ and $\lambda_q)$ can be expressed as follows:

$$\lambda_d = -L_d i_d + \lambda_f \tag{8}$$

$$\lambda_q = -L_q i_{q_f} \tag{9}$$

The torque equation of PMSG can be expressed as follows:

$$\begin{aligned} T_g &= -\frac{3}{2}P(\lambda_d i_q - \lambda_q i_d) \\ &= -\frac{3}{2}P\{\lambda_f i_q + (L_d - L_q)i_d i_q\} \end{aligned} \tag{10}$$

C. Diesel Engine and Governor System

A diesel engine consists of a governor, and an engine as shown in Fig. 2. The governor is a combination of a speed regulator and an actuator.

The differential equations describing the diesel engine and speed regulation are shown as follows [12, 13]:

$$\frac{dP_C}{dt} = -\frac{K_1}{\omega_{ref}}\Delta\omega \tag{11}$$

$$\frac{dm_B}{dt} = \frac{1}{\tau_2}\left(K_2 P_C - \frac{K_2}{\omega_{ref}R}\Delta\omega - m_B\right) \tag{12}$$

where, m_B is the diesel engine fuel consumption rate (kg/sec); K_1 is the governor summing-loop amplification factor, R is the diesel engine permanent speed droop;

The engine is a combustion system which is represented by a gain K_2 and a dead time τ_1. The dead time can be expressed as

$$\tau_2 = \frac{60s_t}{2Nn} + \frac{60}{4N} \tag{13}$$

where, $s_t = 4$ for four stroke engine; N is the speed in rpm; and n is the number of cylinders. After combustion, the mechanical power (p_m) of the engine is developed which is defined in the following:

$$p_m = C_1 m_B \varepsilon \tag{14}$$

where, C_1 is proportionality constant. ε is the efficiency.

In a diesel generator, per unit mechanical torque (T_{Dm}) generated by the engine is shown as follows:

$$T_{Dm} = \frac{P_m}{\omega_m T_b} = C_2 P_k \tag{15}$$

where, T_b is the base torque; and C_2 is the proportional constant.

The electrical rotor angle (δ) is related to the electrical angular velocity shown as follows:

$$\frac{d\delta}{dt} = \omega - \omega_O = \Delta\omega \tag{16}$$

The mechanical motion equation is shown as:

$$\frac{d\omega}{dt} = \frac{\omega_O}{2H}\left(T_{Dm} - T_{De} - \frac{D}{\omega_O}\Delta\omega\right) \tag{17}$$

where, T_{De} is the electrical torque; D is the load damping coefficient and $D = \frac{\partial P_L}{\partial f}$ and H is the inertia constant of the generator.

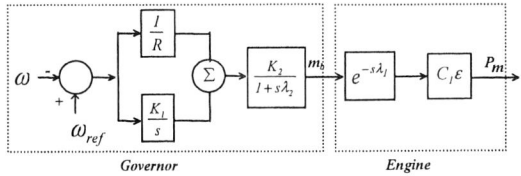

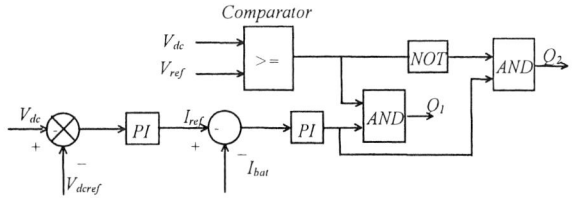

Figure 2. Block diagram of diesel generator.

Figure 4. Battery Controller.

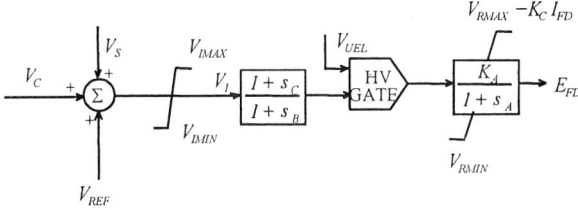

Figure 3. Block diagram of excitation system

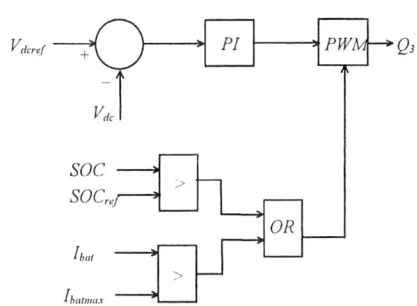

Figure 5. Dump Load Controller.

D. Excitation System:

The excitation system used in diesel generator is Type AC4A excitation model taken from IEEE standard 421.5 shown in Fig. 3 [14].

III. DC LINK VOLTAGE CONTROL DURING HIGH WIND CONDITIONS

In high wind conditions, excess power produced from the wind turbine can be either stored by the emergency storage system or dissipated in a dump load. An efficient energy management system is necessary between an energy storage system and a dump load to ensure the maximum utilization of excess wind. The energy management monitors the state of charge (SOC) of battery and battery charging current at any instant of time. The dump load controller works if the SOC of battery is higher than its reference value or the instantaneous battery current exceeds its upper limit. The description of battery charger and dump load controller is given in the following sub-sections.

A. Battery Controller:

A battery controller is basically a buck-boost bi-directional dc to dc converter. During charging the controller act as a buck converter while it act as a boost converter during discharge period. The main objective of battery controller is to store excess energy in the battery during high wind condition and discharge the energy in temporary wind fall. The controller compares the dc link voltage with a reference value, and error is passed through a PI controller to obtain a reference battery current. The reference battery current is compared with the measured battery current. The error is passed through another PI controller and generates the necessary PWM signal for Q1 and Q2 as shown in fig. 4.

B. Dump Load Controller:

The objective of dump load controller is to dissipate excess power during high wind condition if the battery energy storage is full. The dump load controller compared the dc-link voltage with a reference value and generates error signal. Based on error signal, an appropriate PWM signal is generated which controls the chopper switch (Q_3).

The dump load resistance (R_{dump}) has a linear relationship with the duty cycle as a function of over-voltage. The value of dump load resistance at any instant is given by:

$$R_{dump} = \frac{V_{dc}^2}{P_{g\max} Q_3} \qquad (18)$$

where, $P_{g\max}$ is the maximum power the wind energy conversion system can produce, V_{dc} is the dc link voltage.

At no load condition, all the generated power can be dissipated in the dump load where $Q_3 = 1$ leading to dump load resistor as given below:

$$R_{dump} = \frac{V_{dc}^2}{P_{g\max}} \qquad (19)$$

IV. LOAD SIDE VOLTAGE AND FREQUENCY CONTROL

The output voltage and frequency of the WECS is regulated by the load side converter. A vector control scheme is developed based on the rotating reference frame [15]. The frequency of the system is controlled by generating a reference angular velocity which defines the electrical system

978-1-4244-4782-4/10 $26.00 © 2010 IEEE   165

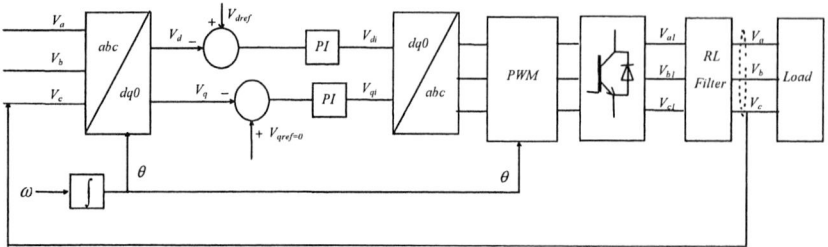

Figure 6. Output Voltage and Frequency Controller.

frequency of the system.

In Fig. 6, the voltage balance across the LR filter:

$$\begin{bmatrix} V_a \\ V_b \\ V_c \end{bmatrix} = R_f \begin{bmatrix} i_a \\ i_b \\ i_c \end{bmatrix} + L_f \frac{d}{dt} \begin{bmatrix} i_a \\ i_b \\ i_c \end{bmatrix} + \begin{bmatrix} V_{a1} \\ V_{b1} \\ V_{c1} \end{bmatrix} \qquad (20)$$

where, R_f and L_f are the resistance and inductance of LR filter, and i_a, i_b, i_c are three phase load current.

The d and q axis components of load voltage are shown as below:

$$v_d = v_{di} - i_d R_f - L_f \frac{di_d}{dt} + \omega L_f i_q \qquad (21)$$

$$v_q = v_{qi} - i_q R_f - L_f \frac{di_q}{dt} - \omega L_f i_d \qquad (22)$$

The voltage v_{di} and v_{qi} are fed to PWM signal generator in order to control the output voltage.

V. POWER FLOW CONTROL OF DIESEL GENERATOR

The power controller of diesel generator ensures the required power generation from diesel generator during no wind or low wind conditions in order to maintain voltage and frequency. The controller allows the diesel generator to run in two different modes; one is isochronous mode and the other is droop mode [16].

During no wind conditions, the diesel generator runs on isochronous mode shown as in Fig. 7(a). In isochronous mode of operation, the governor produces required power in order to keep a constant system frequency. The control strategy of isochronous mode is similar as described in section II(C).

In low wind conditions, the diesel generator runs on droop mode. In this mood of operation, the frequency of the system is controlled by the load side inverter of the wind energy conversion system while diesel generator is used to supply the offset power. The power control strategy of droop mode shown in Fig. 7(b) compares the reference power with the output power. The error is multiplied by a droop constant and it is added to the frequency error of the governor. The error

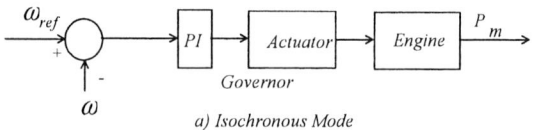

a) Isochronous Mode

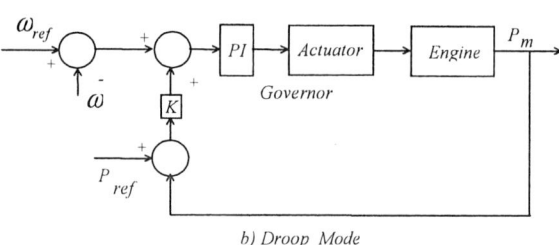

b) Droop Mode

Figure 7. Diesel Generator Operation (a) in isochronous mode and (b) in droop mode

signal is fed to another PI controller. Based on the signal from PI controller, the actuator controls the necessary fuel flow to produce the required power for the system.

VI. SYSTEM OPERATION AND PERFORMANCE

The hybrid system shown in Fig. 1 is implemented in Matlab/Simpower environment. The performance of the system is simulated on variable wind and load conditions. The parameters of the wind energy conversion system, the diesel generator, the energy storage system and the dump load used in the simulation studies are shown in Table 1. Three case studies are presented to justify the performance of the model in different wind and load conditions.

A. Case 1 — Sufficient Wind Condition

This case demonstrates time domain simulation of hybrid wind-diesel system under different load conditions during sufficient wind conditions. The hypothetical wind speed is shown in Fig. 8(a) which changes from 10 m/sec to 11 m/sec at t = 4 sec; then 11 m/sec to 9.5 m/sec at t = 7 sec; then 9.5

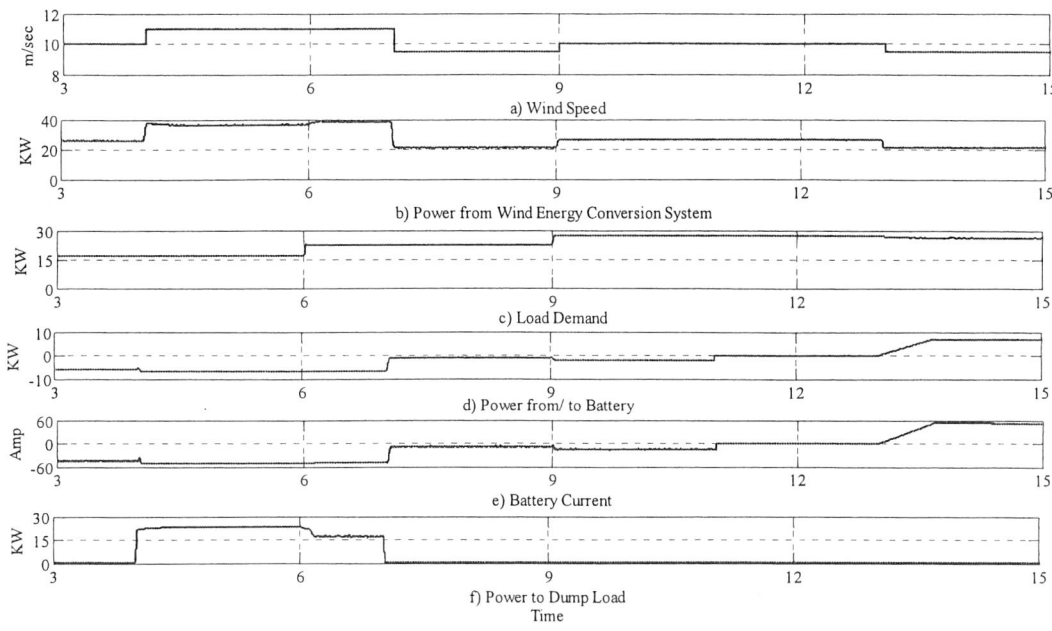

Figure 8. Power management response in sufficient wind conditions.

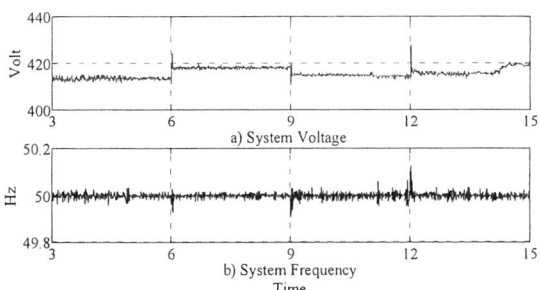

Figure 9. Voltage and frequency response during insufficient wind conditions.

m/sec to 10 m/sec at t = 9 sec; then 10 m/sec to 9.5 m/sec at t = 13 sec. The respective power from wind energy conversion system is shown in Fig. 8(b). The load demand changes from18 KW to 22 KW at t – 6sec, then 28 KW at t = 9 sec as shown in Fig. 8(c). Fig 8(d) and 8 (e) shows the charging and discharging power and current of energy storage system, respectively. From Fig. 8(a) – 8(c), it is found that from t = 3 to t = 13, the wind energy conversion system produces more power than load requirement. As a result, the battery storage system store excess power from wind from t = 3 sec to t = 13 sec. It is also found that, the battery current exceeds its upper limit from t = 4 sec to t = 7 sec. During this time, the dump load controller works and consumes excess power from wind as shown in Fig. 8(f). From Fig. 8(a) – 8(c) it is found that from t = 13 sec to t = 15 sec, the power from wind energy conversion system is less than the load demand as a consequence of temporary wind fall. As a result, the battery

storage system provides required power temporarily. Fig. 9 shows that the voltage and frequencies of the system are almost constant during the wind and load changes.

B. Case 2 — Insufficient Wind Condition:

This case study demonstrates the performance of a wind-diesel hybrid system in insufficient wind conditions. In insufficient wind conditions, the diesel generator works a with wind turbine to meet the power demand. In this case study, the wind speed is varied from 8 m/sec to 7.5 m/sec on t = 6 sec; then 7.5 m/sec to 7 m/sec on t =12 sec as shown in Fig. 10(a). The power from wind energy conversion system changes as the wind speed changes shown in Fig. 10(b). The load demand changes from 25 KW to 22 KW at t=5 sec; then 22 KW to 30 KW at t = 8 sec; then 30 KW to 26 KW at 12 sec as shown in Fig. 10 (c). From Fig. 10(a) -10(c), it can be found that the power generation from the wind energy conversion system is always lower than the load demand. As a result, initially, the emergency storage system supplies the offset power from t = 3 sec to t = 6 sec. As the low wind condition persist for longer time, the diesel generator acts. The diesel generator is connected to the system through the synchronizer when the voltage and frequency of the both system is similar. From Fig. 10(d), it is revealed that, the diesel generator supplies required the offset power from t= 6 sec. From Fig. 11, it is also found that the voltage and frequency of the system are almost constant despite wind and load changes.

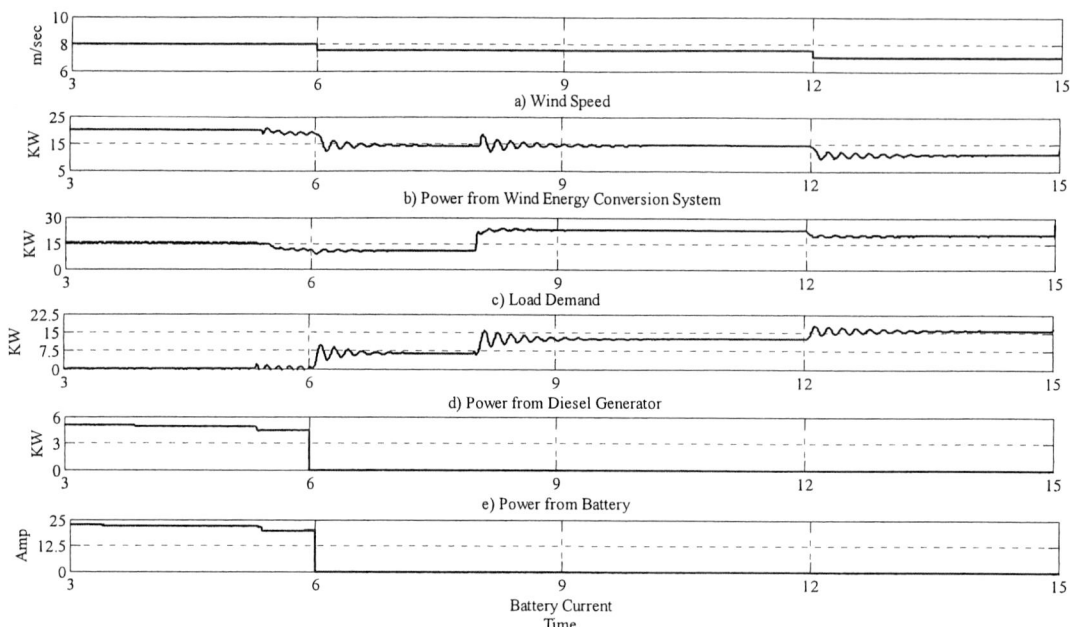

Figure 10. Power management response in insufficient wind conditions.

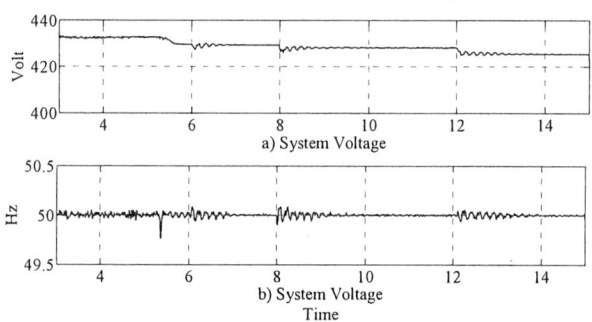

Figure 11. Voltage and frequency response during insufficient wind conditions.

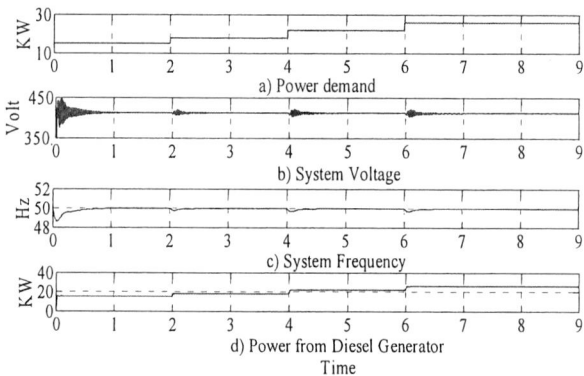

Figure 12. Diesel generator response in no wind conditions.

C. Case 3 — No Wind Condition

This case study demonstrates the performance of a wind-diesel hybrid system in no wind conditions. During no wind conditions, the diesel generator supplies the required power.

The load varies in this case from 15 KW to 18 KW at t = 2 sec; then 18 KW to 22 KW at t = 4 sec; then 22 KW to 27 KW at t = 6 sec. as shown in Fig. 12 (a). The voltage, frequency and power response of the diesel generator as a result of different loading conditions are also shown in Fig. 12(b) and Fig. 12(c), respectively. From Fig. 12(c) and Fig. 12(d), it can be revealed that the diesel generator can maintain almost constant voltage and frequency despite of load fluctuations.

Moreover, the power generation of diesel generator can match the load demand as shown in Fig. 12(d).

VII.   CONCLUSION

A control strategy for voltage and frequency stabilisation of a hybrid wind-diesel power system for different wind and loading conditions is proposed. An efficient power management among a wind energy conversion system, a diesel generator and a battery energy storage system is also proposed for different wind and load conditions. Three case studies are demonstrated to justify the application of proposed model. From the case studies, it is revealed that voltage and frequency can be controlled in the wind-diesel hybrid system despite different wind and load conditions. Moreover, an efficient power sharing among the energy sources are successfully demonstrated for different wind and load conditions.

TABLE I. SIMULATION PARAMETERS

Permament Magnet Synchronous Generator	
Number of Pole Pairs	4
Rated Speed (rpm)	3000
Rated Power (kw)	35
Stator Resistance (ohm)	0.05
Direct Inductance (mh)	0.0635
Quadrature Inductance (mh)	0.0635
Torque Constant (N.m)	1.152
Inertia	0.011
Wind Turbine	
Rated Power (kW)	35
Base Wind Speed (m/s)	9
RL Filter	
Series Inductance (mh)	13
Shunt Resistance (mohm)	20
Synchronous Generator	
Rated Power (kW)	31.3
Rated Speed (m/s)	1500
Rated Voltage (volt)	400
No of Pole pair	2
Inertia	0.08671
Emergency Storage System	
Rated Voltage (volt)	180
Rated Capacity (amp-hour)	50
Dump Load	
Dump Resistance (Ohm)	25

REFERENCES

[1] R. Cardenas, R. Pena, M. Perez, J. Clare, G Asher, and F. Vargas, "Vector Control of Front-End Converters for Variable-Speed Wind-Diesel Systems" IEEE Transaction on Industrial Electronics, Vol. 53, No. 4, pp. 1127-1136, August, 2006.

[2] D. J. Lee, and L. Wang, "Small-Signal Stability Analysis of an Autonomous Hybrid Renewable Energy Power Generation/ Energy Storage System Part 1: Time-Domain Simulation", IEEE Transaction of Energy Conversion, Vol. 23, No. 1, pp. 311-320, March 2008.

[3] M. De Broe, S. Drouilhet, and V. Gevorgian, "A peak power tracker for small wind turbines in battery charging applications," IEEE Transaction on Energy Conversion., Vol. 14, No. 4, pp. 1630–1635, Dec. 1999.

[4] R. Datta and V. T. Ranganathan, "A method of tracking the peak power points for a variable speed wind energy conversion system", IEEE Transaction on Energy Conversion, Vol. 18, No 1, pp. 163–168, March. 2003.

[5] K. Tan and S. Islam, "Optimal control strategies in energy conversion off PMSG wind turbine system without mechanical sensors," IEEE Transaction on Energy Conversion., Vol. 19, No. 2, pp. 392–399, June.   2004.

[6] S. Morimoto, H. Nakayama, M. Sanada, and Y. Takeda, "Sensorless Output Maximization Control for Variable- Speed Wind Generation System Using IPMSG", IEEE Transaction on Industrial Application., Vol. 41, No. 1, pp. 60-67, January. 2005.

[7] Chinchilla, S. Arnaltes, and J. C. Burgos, "Control of Permanent-Magnet Generators Applied to Variable- Speed Wind-Energy Systems Connected to the Grid" IEEE Transaction on Energy Conversion., vol 21, No. 1, pp. 130-135, March 2006.

[8] Z. Chen and Y. Hu, "A hybrid generation system using variable speed wind turbines and diesel units," in Proceeding of IEEE IECON, Nov. 2–6, 2003, vol. 3, pp. 2729–2734.

[9] K. E. Johnson, L. Y. Pao, M. J. Balas, and L. J. Fingersh "Control of variable-speed wind turbines: standard and adaptive techniques for maximizing energy capture", IEEE Control System Magazine, Vol. 26, No. 3, pages. 70-81, June 2006,

[10] I. J. Iglesias, L. Garcia, A. Agudo, I. Cruz, and L. Arribas, "Design and simulation of a stand-alone wind diesel generator with a flywheel energy storage system to supply the required active and reactive power," in Proceedings of IEEE PESC, 2000, pp. 1381–1386.

[11] P. C. Klause, O. Wesynczuk, and S. D. Sudhoff, "Analysis of Electric Machinary and Drive System" 2nd Edition, John Willey and Sons, Inc. Publication, ISBN: 0-471-14326-X.

[12] G. S. Stavrakakis, and G. N. Kariniotakis, "A general simulation algorithm of the accurate assessement of isolated diesel-wind turbine systems interaction, Part I: A generatl multi-machine power system model" EEE Transactions on Energy Conversion, Vol. 10, No. 3, September 1995, pp 577 – 583.

[13] G. S. Stavrakakis, and G. N. Kariniotakis, "A general simulation algorithm of the accurate assessement of isolated diesel-wind turbine systems interaction, Part II: Implementation of the Algorithm and Case-Study with Induction Generators" EEE Transactions on Energy Conversion, Vol. 10, No. 3, pp 584 – 590, September 1995.

[14] IEEE Std 421.5™-2005, "IEEE Recommended Practice for Excitation System Models for Power System Stability Studies", IEEE Power Engineering Society, 21 April 2006.

[15] M. E. Haque, K. M. Muttaqi, and M. Negnivitsky, "Control of a stand alone variable speed wind turbine with a permanent magnet synchronous generator" IEEE Power and Energy Society General Meeting, pp. 1 – 9, 20-24 July 2008

[16] R. J. Best, D. J. Morrow, D. J. McGowan, and P. A. Crossley, "Synchronous Islanded Operation of a Diesel Generator" IEEE Transactions on Power Systems, Vol. 22, No. 4, Novermber 2007, pp. 2170-2176.

SystemC-AMS modeling and simulation of digitally controlled DC-DC converters

Matteo Agostinelli, Robert Priewasser, Mario Huemer
Networked and Embedded Systems – University of Klagenfurt
9020 Klagenfurt, Austria
email: Matteo.Agostinelli@uni-klu.ac.at

Stefano Marsili, Dietmar Straeussnigg
Infineon Technologies Austria AG
9500 Villach, Austria
email: Stefano.Marsili@infineon.com

Abstract—In this paper, an innovative method to model and simulate DC-DC converters with a digital or mixed-signal control loop is proposed using the SystemC-AMS hardware-description language. The proposed method was employed to model a specific test case, consisting of a Buck converter with a digital PID regulator. The reliability of the model was checked by comparing the results with MATLAB/Simulink simulations. The SystemC-AMS approach was found to be well suited to model the proposed system and very efficient from a computational point of view, since the simulation time can be strongly reduced with respect to other solutions (e.g. MATLAB/Simulink).

Index Terms—Modeling, SystemC, Analog and Mixed-Signal simulation and design, Digital control.

I. INTRODUCTION

Digital control loops are increasingly adopted to achieve regulation of the output voltage of DC-DC converters for several reasons. The main advantages of digital solutions with respect to the analog counterparts are programmability, versatility and reduced power consumption. Moreover, reduced sensitivity to noise and analog variations and easiness of integration with other digital systems are further advantages of digital loops [1]–[3].

From the prototyping to the verification phase of such systems, modeling capabilities and computational efficiency are key aspects of the simulation environment. From the examination of several previous works, it can be inferred that one of the most popular modeling and simulation tools used in the recent past is the MATLAB/Simulink environment (see [2], [4], [5]). While this is a feasible approach, in this paper we present an alternative solution, the SystemC language [6] along with the analog and mixed-signal (AMS) extension [7]–[10], which was found to be better suited for the modeling of such systems and can also yield a substantial reduction of simulation time.

In order to identify advantages and properties of the proposed approach, a specific test case has been modeled and simulated in this work. A schematic representation of the system is reported in Fig. 1, in which its mixed-signal structure is noticeable. It consists of a Buck converter with a digital proportional-integral-derivative (PID) controller, a simple yet popular way to regulate DC-DC converters [11]. The converter's parameters have been chosen in order to reproduce a mobile application and have been taken from a real-world commercial product [12]. An additional control

loop, consisting of a mixed-signal PID controller, has also been simulated in order to highlight the mixed-signal modeling capabilities offered by SystemC-AMS.

II. SYSTEMC-AMS

SystemC is a hardware-description language built on top of standard C++ and, as such, has some unique characteristics when compared to other hardware-modeling approaches, such as MATLAB/Simulink. A major difference between SystemC and MATLAB/Simulink lies in the fact that the SystemC code is compiled into an executable file, whereas in MATLAB/Simulink the source code is interpreted. Moreover, being an extension of C++, SystemC allows to re-use existing C/C++ code or to employ existing external C/C++ libraries. For instance, several libraries that support advanced numerical methods can be exploited. Furthermore, all the object-oriented features of C++, such as inheritance and polymorphism, can be effectively used to model hardware [13], due to the fact that all the building blocks (or *modules*) of a system are implemented as classes.

While SystemC was initially employed to model digital systems only, the SystemC-AMS extension offers the possibility to introduce system-level design and modeling of analog and mixed-signal systems by enabling the use of dedicated simulation kernels synchronized with the standard SystemC kernel [8], [10]. Due to this fact, the AMS extension of the SystemC language permits to model part of the system as a linear electrical network. Thus, the DC-DC converter model can be built by simply specifying the electrical netlist of the converter. On the contrary, a set of differential equations must be manually introduced in MATLAB in order to model the converter, assuming that no additional toolboxes (e.g. SimPowerSystem) are used. Hence, the refinement of the DC-DC converter model (e.g. the inclusion of the equivalent-series inductance of the capacitor) is easier in the SystemC-AMS implementation (by altering the netlist) with respect to the MATLAB one, where a new set of differential equations has to be solved.

As a further remark, the SystemC-AMS language can be effectively employed at different levels of design abstraction [8], and the most suitable description method can be adopted for a given module. On the other hand, system-level tools such as MATLAB/Simulink, which are commonly used to model

978-1-4244-4782-4/10 $26.00 © 2010 IEEE

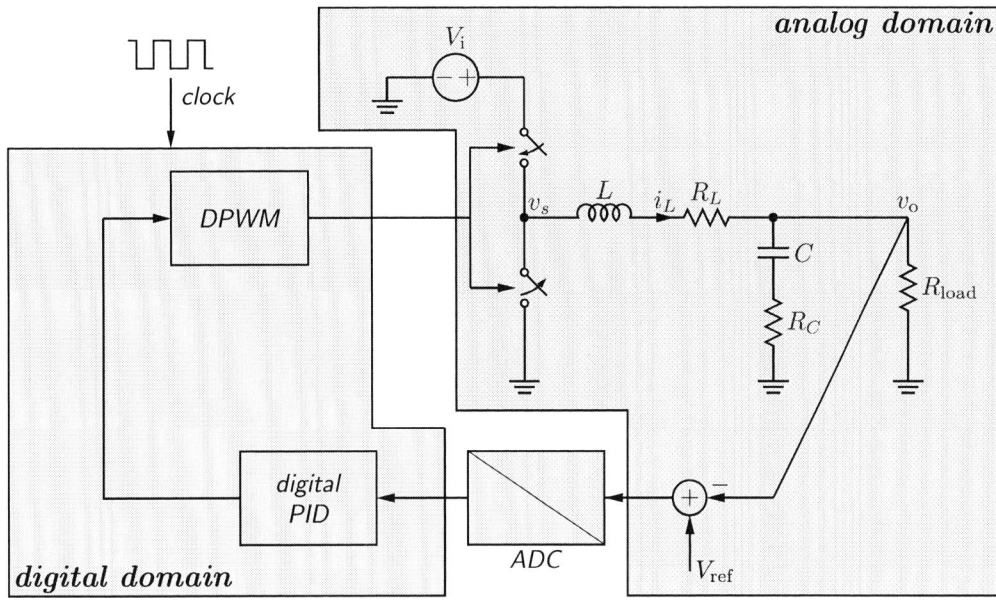

Fig. 1. Graphical representation of the system.

power converters, are capable of capturing continuous-time behavior but they do not target the design of AMS systems at an architecture-level.

III. TEST CASE ANALYSIS: A BUCK CONVERTER WITH A DIGITAL PID REGULATOR

The test case presented in this paper is an AMS system, consisting of a Buck converter (represented in the analog domain) with a digital-domain PID voltage-mode controller (see Fig. 1). The system parameters have been taken from a real-world product [12] for mobile applications. Realistic values of the parasitics of the components of the converter are also included in the model. In this system, the analog-to-digital converter (ADC) and the digital pulse-width modulator (DPWM) represent the interfaces between the two domains. The ADC is a Double-Sampling Averaging ADC, thus two equidistant points per switching period are sampled and then the average is computed and passed to the PID regulator. The output of the PID block, i.e. the duty cycle, is then fed to the DPWM, which generates the signal used to drive the switches of the power-stage.

A. SystemC-AMS implementation

Some code excerpts are reported to illustrate briefly how to model the building blocks of the system in the SystemC-AMS language. In Lst. 1 an excerpt from the `AveragingADC` class declaration is shown. The `AveragingADC` model is defined such that the `sample_and_quantize()` function is executed on the positive edge of the `trigger` signal (see lines 11-12), which occurs twice every switching period. The implementation of this function is reported in lines 19-49 and it is briefly illustrated in the following. The average of the last two samples, taken from the continuous-time signal `in` which represents the error on the output voltage, is computed and

written on the `out` port, producing one value per switching period. The output port `out`, which is declared in line 5, is a standard SystemC port. The port `data_ready`, declared in line 6, is used for the synchronization with the PID block.

The code that implements the PID module is reported in Lst. 2. This module is implemented as a "conventional" SystemC class and its `update()` function is executed on every positive edge of the `data_ready` signal, as it can be seen on lines 15-16. It can be easily seen that this block implements the following discrete-time transfer function:

$$C(z) = g\frac{(z - z_1)(z - z_2)}{(z - p_1)(z - p_2)} \qquad (1)$$

which is the discrete-time expression of a PID controller with an additional high-frequency pole p_2, in order to ensure the properness of the transfer function $C(z)$ [11], [14].

On the other hand, the `Buck` model, which is reported in Lst. 3, is built by subclassing from the `DcDcConverter` base class, which in turn inherits from `sc_module`. The netlist of the circuit is provided by re-implementing the `architecture()` function. It is worth noting that it is possible to automatically generate the netlist from a schematic of the circuit. The definition of the high-side switch of the Buck converter is reported in Lst. 3. The connections to the electrical nodes (which are declared in line 6) are defined in line 18, while the control signal is assigned in line 19. In this case, the control signal is a standard SystemC boolean signal, as it can be seen in line 10, and it is generated by the DPWM block. Other parameters, such as the on- and off-resistances, are set in lines 20-21. A voltage source, that models the battery voltage, is then defined in lines 23-25. The voltage source is connected to the corresponding electrical nodes in line 24, while the magnitude of the voltage generated by this element

```
1   class AveragingADC: public sc_module {
2       public:
3       sca_sdf_in<double>  in;
4       sc_in<bool>         trigger;
5       sc_out<double>      out;
6       sc_out<bool>        data_ready;
7
8       void sample_and_quantize();
9
10      SC_CTOR(AveragingADC) {
11          SC_METHOD(sample_and_quantize);
12              sensitive << trigger.pos();
13      }
14
15      double vq;       // Quantization step
16      double n_bits;   // Number of bits
17  };
18
19  void AveragingADC::sample_and_quantize()
20  {
21      // read the sample
22      double sample = in.read();
23      // set first_sample flag
24      first_sample = !first_sample;
25
26      // quantization
27      double sample_quantized = floor(sample/vq);
28
29      // saturation
30      if (sample_quantized >= up_lim)
31          sample_quantized = up_lim;
32      if (sample_quantized < lo_lim)
33          sample_quantized = lo_lim;
34
35      // if it's the second sample,
36      // write the average on the output port
37      if (!first_sample) {
38          double average = 0.5*
39              (0.5+sample_quantized + previous_sample)*vq;
40
41          out.write(average);
42          data_ready.write(true);
43      } else {       // otherwise do nothing
44          data_ready.write(false);
45      }
46
47      // save sample
48      previous_sample = sample_quantized;
49  }
```

Listing 1.   Code excerpt from `AveragingADC` class

is set by a signal `vin` in line 25. This signal can be generated in an appropriate testbench and could be used to simulate line jumps. The definition of the other elements of the converter is similar to the voltage source case reported in lines 23-25, thus it has been omitted.

Using a `DcDcConverter` base class is beneficial because portions of code can be shared among different DC-DC topologies (e.g. Buck, Boost, Buck-Boost) and can be included in the common base class, thus allowing code reuse. The base class can also be used to define a common interface for different DC-DC topologies, thus simplifying the substitution of a topology with another one.

B. Comparison with MATLAB/Simulink models

The same system has been modeled in the MAT-LAB/Simulink environment in order to evaluate the accuracy of the SystemC-AMS implementation. The Buck converter has been modeled in MATLAB through its differential equations, discretized in time using the Euler method. An open-loop

```
1   class PID: public sc_module
2   {
3       public:
4           sc_in<double>   in;
5           sc_in<bool>     data_ready;
6           sc_out<double>  out;
7
8           double pid_p1,pid_p2,pid_z1,pid_z2,pid_gain;
9
10          void update();
11
12          SC_CTOR(PID)
13          {
14              // read parameters from file (omitted)
15              SC_METHOD(update);
16              sensitive << data_ready.pos();
17          }
18
19      private:
20          double x1,x2;
21  };
22
23  void PID::update()
24  {
25      double error = in.read();
26      double v = error+x1; // intermediate variable
27      double w = v+x2;     // intermediate variable
28      double output = pid_gain*w;
29
30      // compute next state
31      x1 = pid_p1*(x1 + error) - pid_z1*error;
32      x2 = pid_p2*(x2 + v) - pid_z2*v;
33
34      out.write(output);
35  }
```

Listing 2.   Code excerpt from `PID` class

```
1   class Buck: public DcDcConverter {
2       public:
3           sca_sc_rswitch *ls, *hs; // Switches
4           sca_sdf2v *vbat; // Battery
5           sca_l *l; sca_c *c; // Coil and cap
6           sca_elec_node n1, n2; // Nodes
7           ...
8           sca_sdf_in<double> vin; /* Battery voltage
9                                      (inherited) */
10          sc_in<bool> pwm; /* PWM signal
11                              (inherited) */
12          ...
13  };
14
15  void Buck::architecture()
16  {
17      hs = new sca_sc_rswitch("high-side_sw");
18      hs->p(n1); hs->n(n2);
19      hs->ctrl(pwm);
20      hs->off_val = false;
21      hs->ron = p.Ronp;    hs->roff = 1e12;
22
23      vbat = new sca_sdf2v("v_bat");
24      vbat->p(n1); vbat->n(gnd);
25      vbat->ctrl(vin);
26
27      ...
28  }
```

Listing 3.   Code excerpt from `Buck` class

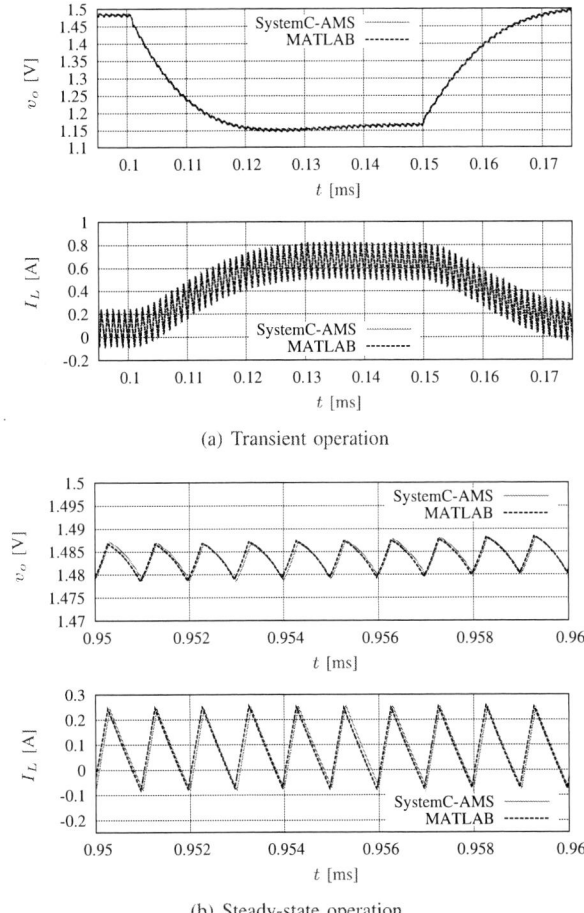

(a) Transient operation

(b) Steady-state operation

Fig. 2. Comparison of the output voltage and coil current waveforms obtained from the SystemC-AMS and MATLAB models of the open-loop system. A load jump (drop) is occurring at $t = 0.1$ ms ($t = 0.15$ ms).

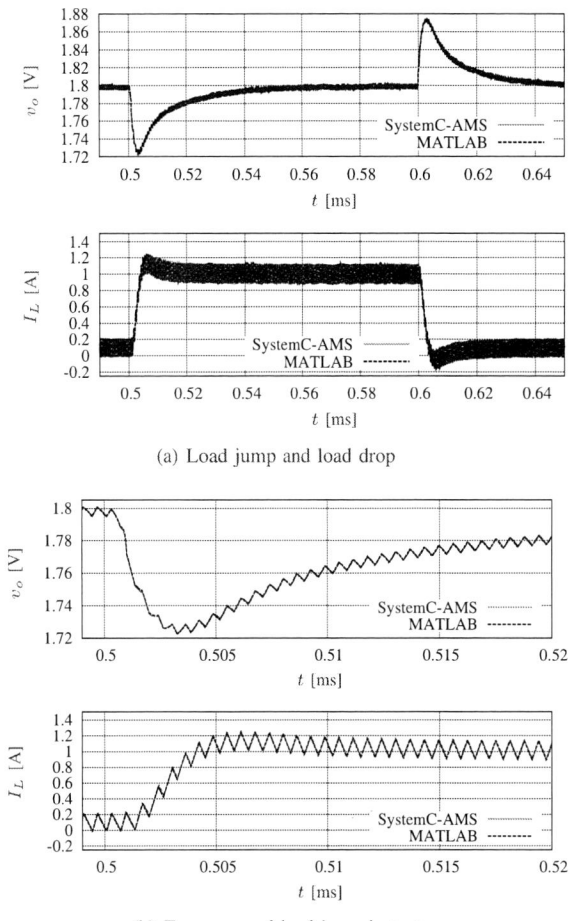

(a) Load jump and load drop

(b) Zoom around load jump instant

Fig. 3. Comparison of the output voltage and coil current waveforms obtained from the SystemC-AMS and MATLAB/Simulink models of the closed-loop system. A load jump (drop) with an amplitude of 1 A is occurring at $t = 0.5$ ms ($t = 0.6$ ms).

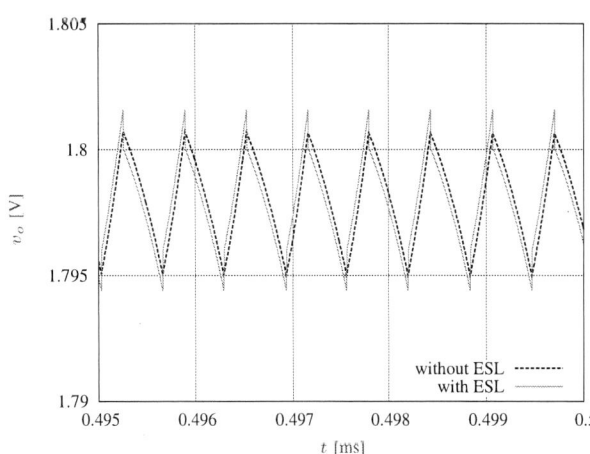

Fig. 4. Influence of the equivalent series inductance of the output capacitor on the output voltage waveform.

simulation of the DC-DC converter has been run to check the agreement between the two different approaches. The results are plotted in Fig. 2, from which it can be concluded that the SystemC-AMS model is fully reliable. Then the closed-loop system, including the digital PID regulator, has been simulated in both environments and the corresponding results are reported in Fig. 3. Again, a perfect matching between the models can be observed.

As previously anticipated in Sec. II, one of the advantages of the proposed SystemC-AMS approach lies in the easiness of refining the Buck converter model. As an example, the Equivalent Series Inductance (ESL) of the output capacitor can be easily added to the SystemC-AMS model, by including an additional sca_l element and changing the netlist accordingly. To achieve the same result in the MATLAB/Simulink environment, the effort would be greater because a new set of differential equations must be solved. Fig. 4 shows how the introduction of an ESL of 1 nH in the Buck model affects the output voltage waveform.

The execution times of the simulations are compared in Tab. I in order to assess the speed performance of the different

978-1-4244-4782-4/10 $26.00 © 2010 IEEE

TABLE I
SIMULATION EXECUTION TIMES (IN SECONDS).

	open-loop	closed-loop
SystemC-AMS	0.17	0.75
MATLAB/Simulink	1.04	7.04
MATLAB/Simulink (Rapid Accelerator)	2.11	2.37
MATLAB/Simulink (SimPowerSystems)	1.64	7.05
MATLAB/Simulink (PLECS)	1.08	7.01

implementations. The SystemC-AMS implementation yields the best simulation times, due to the fact that the code is compiled into an executable file (however, the compilation time has not been taken into account). The Rapid Accelerator mode of MATLAB/Simulink has also been introduced in order to provide a fair comparison with the SystemC case. In fact, when the Rapid Accelerator mode is enabled, a binary file is generated and executed by MATLAB, just as with the SystemC code. This is the reason why the gap between the SystemC-AMS and MATLAB simulation times is reduced. Instead, the Rapid Accelerator mode is not advantageous for the open-loop simulation because of the added overhead, i.e. the connection between the compiled binary file and MATLAB.

For the sake of completeness, two additional MATLAB/Simulink models were compared with the other solutions. The first one was built by using the SimPowerSystems toolbox, which allows the description of an electrical network by providing a schematic. In order to make a fair comparison, the option to use a discrete-time solver for the electrical network (with the same time-step) has been activated. The second one is instead based on the PLECS toolbox [15], a popular tool that can be used to simulate power electronics systems in the MATLAB environment. In this case, the discrete-time Buck model has been replaced by a circuit representation, which is allowed by the toolbox. The resulting waveforms were found to be indistinguishable to the ones obtained by using the discretized differential equations in MATLAB for both models, thus they are not shown here. The execution times of the simulations are instead reported in Tab. I, from which it can be concluded that the SimPowerSystem and PLECS toolboxes exhibit a similar performance if compared to the MATLAB discrete-time model.

IV. MIXED-SIGNAL CONTROL LOOP

The proposed SystemC-AMS approach is particularly suited to model mixed-signal control loops, i.e. part of the controller is represented in the digital and part in the analog domain. The digital PID architecture that has been presented in Sec. III, which is schematically reproduced in Fig. 5(a), can be modified by moving some parts to the analog domain. The controller architecture that is evaluated in this section is a mixed-signal voltage-mode PID controller with analog derivative action [16], as reported in Fig. 5(b), where the DC-DC converter has not been drawn for simplicity.

As it can be seen from Fig. 5, the derivative action is transferred to the analog domain in order to allow a faster

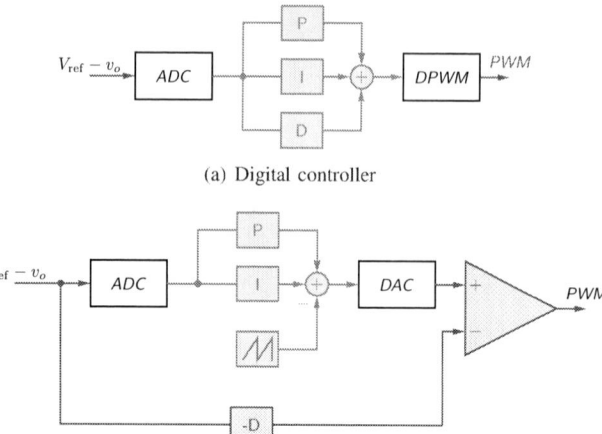

(a) Digital controller

(b) Mixed-signal controller

Fig. 5. Graphical representation of the (a) digital and (b) mixed-signal PID controller. Digital domain blocks (PI-controller, ramp generator) are painted in red, analog blocks (derivative action, comparator) in blue.

reaction to a line or load jump. In fact, the quantization and especially the delay introduced by the Analog-to-Digital converter do not affect the derivative part of the controller. The proportional and integral terms are instead mantained in the digital domain, by means of an ADC and a DAC. The ADC architecture has been mantained identical to the one used in the purely digital controller (as in Lst. 1). The digital PWM modulator has now been replaced by a (digital) ramp generator and an analog comparator. The system's parameters have been mantained identical to the digital PID case, while the controller coefficients have been slightly optimized to better exploit the new controller architecture.

Simulation results are reported in Fig. 6, where the output voltage and inductor current waveforms are shown. It can be observed that, having shifted the derivative action to the analog domain, the mixed-signal controller is capable of a faster reaction to a load jump, thus reducing the under- and overshoots. This effect can be clearly seen in Fig. 6(b) where the inductor current waveform immediately starts rising after the load jump, which is occurring at $t = 0.5$ ms. On the other hand, the digital implementation shows a delay of an additional switching period.

It is worth noting that the execution time of the simulations for the mixed-signal PID controller are comparable to the purely digital controller. In fact, the average execution time was found to be equal to 0.87 seconds.

V. CONCLUSIONS

In this paper, an innovative method to model and simulate DC-DC converters with a digital or mixed-signal control loop is proposed, using the SystemC-AMS language. The accuracy of the SystemC-AMS model has been evaluated by comparing the results with the ones obtained within the MATLAB/Simulink environment, and a satisfactory agreement between the different implementations has been proven. For completeness' sake, additional MATLAB/Simulink simulation

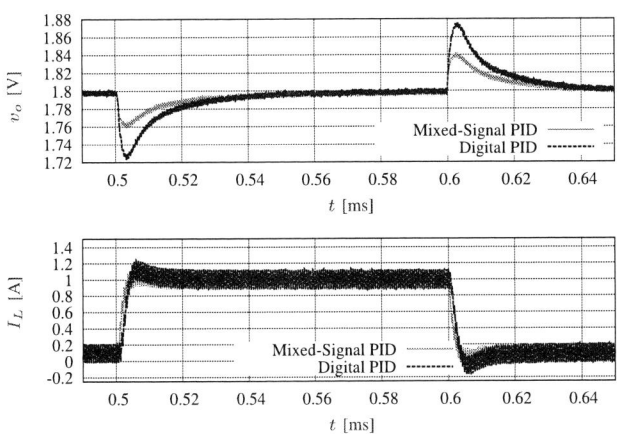

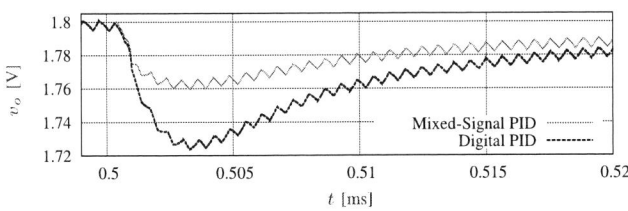

(a) Load jump and load drop

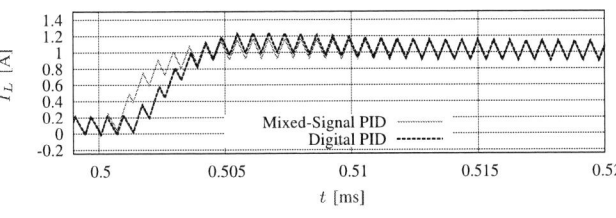

(b) Zoom around load jump instant

Fig. 6. Comparison of the output voltage and coil current waveforms obtained from the mixed-signal and the digital control loops. A load jump (drop) with an amplitude of 1 A is occurring at $t = 0.5$ ms ($t = 0.6$ ms).

tools, such as the SimPowerSystem and PLECS toolboxes, have been included in the comparison.

From the presented results, it can be concluded that the SystemC-AMS environment is capable of providing better performance in terms of simulation time and it proved to be very well-suited to model the chosen test case. An additional mixed-signal control loop, which is capable of improving the dynamic performance of the system, has also been successfully modelled and simulated with the proposed SystemC-AMS approach.

ACKNOWLEDGEMENT

This work was supported by Lakeside Labs GmbH, Klagenfurt, Austria and was funded by the European Regional Development Fund and the Carinthian Economic Promotion Fund (KWF) under grant 20214/16470/23854.

REFERENCES

[1] "Special issue on digital control in power electronics," *IEEE Trans. Power Electron.*, vol. 18, no. 1, pp. 293–503, Jan. 2003.

[2] S. Saggini, M. Ghioni, and A. Geraci, "An innovative digital control architecture for low-voltage, high-current DC-DC converters with tight voltage regulation," *IEEE Trans. Power Electron.*, vol. 19, no. 1, pp. 210–218, Jan. 2004.

[3] B. Patella, A. Prodic, A. Zirger, and D. Maksimović, "High-frequency digital PWM controller IC for DC-DC converters," *IEEE Trans. Power Electron.*, vol. 18, no. 1, pp. 438–446, Jan. 2003.

[4] L. Corradini, S. Saggini, and P. Mattavelli, "Analysis of a high-bandwidth event-based digital controller for DC-DC converters," in *Proc. IEEE Power Electron. Specialists Conf.*, Jun. 2008, pp. 4578–4584.

[5] J. Morroni, R. Zane, and D. Maksimović, "Design and implementation of an adaptive tuning system based on desired phase margin for digitally controlled DC-DC converters," *IEEE Trans. Power Electron.*, vol. 24, no. 2, pp. 559–564, Feb. 2009.

[6] "Open SystemC initiative." [Online]. Available: http://www.systemc.org

[7] K. Einwich, A. Vachoux, C. Grimm, and M. Bernasconi, "SystemC AMS extensions draft 1." [Online]. Available: http://www.systemc-ams.org/

[8] C. Grimm, M. Bernasconi, A. Vachoux, and K. Einwich, "An introduction to modeling embedded Analog/Mixed-Signal systems using SystemC AMS extensions," Jun. 2008. [Online]. Available: http://www.systemc-ams.org/

[9] A. Vachoux, C. Grimm, and K. Einwich, "Towards analog and mixed-signal SOC design with SystemC-AMS," in *Proc. IEEE Intl. Workshop on Electron. Design, Test and Applications*, Jan. 2004, pp. 97–102.

[10] A. Vachoux, C. Grimm, and K. Einwich, "Extending SystemC to support mixed discrete-continuous system modeling and simulation," in *Proc. IEEE Intl. Symp. on Circuits and Systems*, vol. 5, May 2005, pp. 5166–5169.

[11] R. Erickson and D. Maksimović, *Fundamentals of Power Electronics*, 2nd ed. Springer, 2001.

[12] Linear Technologies, *LTC3404 Step-Down Regulator Datasheet*. [Online]. Available: http://cds.linear.com/docs/Datasheet/3404fb.pdf

[13] L. Pomante, "Exploiting polymorphism in HW design: a case study in the ATM domain," in *Proc. of IEEE/ACM/IFIP Intl. Conf. on Hardware/software codesign and system synthesis*, 2004, pp. 81–85.

[14] R. Priewasser, M. Agostinelli, S. Marsili, D. Straeussnigg, and M. Huemer, "Comparative study of linear and non-linear integrated control schemes applied to a buck converter for mobile applications," in *Proc. of 17th Austrian Workshop on Microelectronics (Austrochip)*, Oct. 2009, pp. 51 – 56.

[15] Plexim GmbH, *PLECS toolbox*. [Online]. Available: http://www.plexim.com/

[16] S. Saggini, P. Mattavelli, M. Ghioni, and M. Redaelli, "Mixed-signal voltage-mode control for DC-DC converters with inherent analog derivative action," *IEEE Trans. Power Electron.*, vol. 23, no. 3, pp. 1485–1493, May 2008.

978-1-4244-4782-4/10 $26.00 © 2010 IEEE

Modeling of Digitally Controlled Voltage Regulator Modules

Yi Sun
Linear Technology Corporate
Milpitas, CA, USA
ysun@linear.com

Fred C. Lee
Center for Power Electronics Systems
Virginia Tech
Blacksburg, VA, USA
fclee@vt.edu

Jian Li
Linear Technology Corporate
Milpitas, CA, USA
jian.li@linear.com

Abstract—**This paper proposed small signal models of digital VRMs. At first, the ADC's conversion delay and digital compensator's calculation delay are neglected. The focus is placed on the small signal model of the current sampling and the DPWM unit. It is shown that even with a "fast" controller, the current sampling and DPWM will still introduce some delay to the loop. Then the conversion and calculation delays are considered. Two time periods, T_{1ff} and T_{1rr}, are employed to describe the delay effects in the control loop. It is observed that the total delay in the loop is an integral number of sampling periods, which is never reported by any other literatures. The proposed model consolidates these to one delay term and the value can be found through a pre-determined lookup table. Design guidelines of digital VRMs are provided. Simulation and experimental results verify the validity of this model.**

I. INTRODUCTION

It can be expected that digital controllers will be increasingly used in low voltage, high-current and high frequency voltage regulator modules (VRMs) where conventional analog controllers are currently preferred because of the cost and performance reasons [1]. However, delay effect is one major concern for the digital controlled VRMs, since it may hurt the converters' dynamic performances and bring the stability issues. There exist several types of delays in the digital feedback loop, including the voltage/current ADC conversion delay, digital compensator calculation delay, Digital Pulse-Width-Modulator (DPWM) delay as well as some propagation delays. Usually these delays are inside the digital controller and it is hard to know the exact values. There are several papers talking about the small signal models of the digital voltage mode control [2], [3], [4], [5], [6] and [7]. In [2] and [3], a frequency-domain approach based on Laplace-domain modulator modeling and the modified -transform has been described proposed an exact discrete-time model that correctly takes into account sampling, modulator effects and delays in the digital control loop. However, according to [4], the approach is straightforwardly applicable only to buck-type converters. In [5], [6] and [7], the model for multi-sampled DPWM unit in a digital DC/DC converter is proposed. These models are valid only if all the

delay terms are known exactly since each delay is considered separately. In reality, this is not easy.

This work proposed a new small signal model of digital VRMs. Section II proposed the small signal model without considering the T_{con} and T_{cal}. The focus is placed on the modeling of the current sampling and the DPWM delay effects. In Section III, the conversion and calculation delay are considered into the model. The proposed model does not require accurate values of all the delays but only requires a range. Design guidelines for a digital VRM, simulation and experimental verifications are provided in Section IV. Finally, a summary is given in Section V.

II. SMALL SIGNAL MODEL OF DIGITALLY CONTROLLED VRM WITHOUT CONVERSION AND CALCULATION DELAY

A typical digitally controlled VRM is shown in Fig. 1. The power stage is usually a multi-phase buck converter. The blocks inside the dashed box describe the digital controller. For the voltage feedback loop, the sampling frequency for the voltage ADC is as: $F_{sv} = 2 \cdot F_{sw} \cdot N$, where F_{sw} is the switching frequency, N is phase number. This means that the error voltage will be sampled twice per phase's switching period. The calculation and DPWM's updating frequency is equal to F_{sv}. The DPWM used here is a digital double-edge modulator. The current ADC operates in this manner: for each phase, the current will be sampled once per switching period and the sampling instant is at the middle of OFF time. The sampling frequency for the phase current is equal to F_{sw}. Then the phase current will be added to obtain the sampled load current. Since all phases are interleaved, the total load current's equivalent sampling frequency is $F_{si} = N \cdot F_{sw}$. For the current loop, the calculation and DPWM frequency will be aligned with the load current sampling, $N \cdot F_{sw}$.

Since all phases are assumed to be evenly interleaved, and identical for both the power stage and control blocks. Therefore, all the phases are equivalent. Then borrowing the concept from analog control, the small signal model of a multi-phase buck can be simplified to be a single phase buck. Fig. 2 shows the example of the current loop.

978-1-4244-4782-4/10 $26.00 © 2010 IEEE

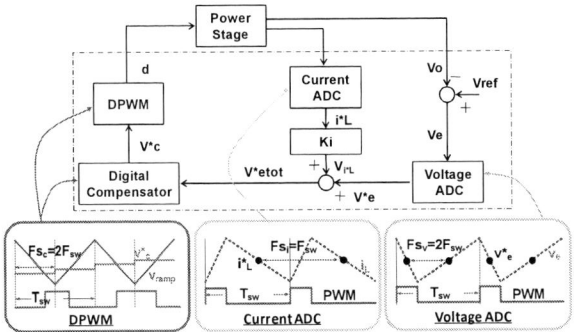

Figure 1. Structure of a digital VRM

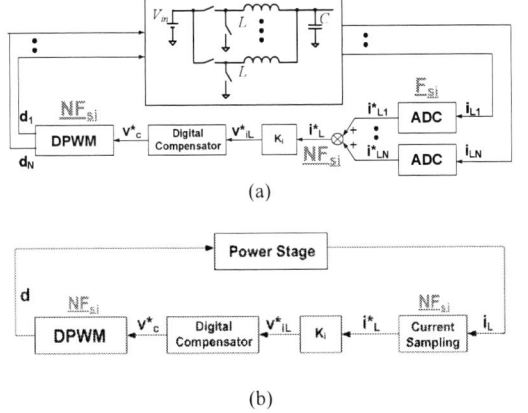

(a)

(b)

Figure 2. Simplification of digitally controlled multi-phase buck converters. (a) Block diagram of a multi-phase buck; (b) simplified block diagram of a multi-sampled single phase buck.

Now the objective of modeling is to find the transfer functions of current sampling (from i_L to i^*_L) and DPWM (from v^*_c to d). The assumptions for the current loop modeling are: voltage loop is open; sampling instant for each phase's current is at the middle of OFF time; T_{con} and T_{cal} are neglected; the quantization effect is neglected.

A. Small Signal Model of Current Sampling

The digital VRM samples the phases' currents and adding them together to obtain the load current. A 2-phase example is given in Fig. 3.

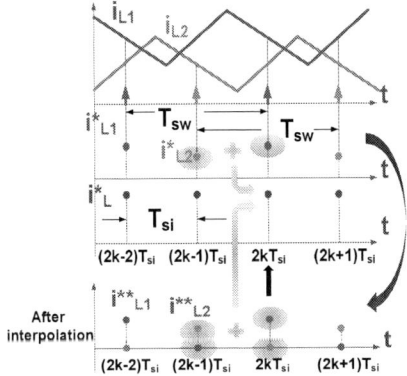

Figure 3. Current sampling and the interpolation method.

i_{L1} and i_{L2} are the phases' currents; i^*_{L1} and i^*_{L2} are the sampled phases' currents; and i^*_L is the sampled load current. Since sampling frequency for i^*_{L1} and i^*_{L2} (F_{sw}) is different from that of i^*_L ($2 \cdot F_{sw}$), it is not straightforward to apply Laplace analysis or Z-transform to this sampling process. The interpolation method is a common tool in digital signal processing to increase the sampling frequency of a digital signal [8]. In our application, we employ the interpolation method to sampled phases' current (i^*_{L1} and i^*_{L2}) to increase its sampling frequency equal to that of i^*_L. Since i^*_L's sampling frequency is twice of that of i^*_{L1} and i^*_{L2}, we only need to add one zero point between two adjacent sampling instant of i^*_{L1} and i^*_{L2}. In Fig. 3, i^{**}_{L1} and i^{**}_{L2} are the new signals after the interpolation. It is clearly shown that the sampling frequency for i^{**}_{L1} and i^{**}_{L2} is equal to that of i^*_L. Now, we can apply the classical z-transform analysis to the current sampling process. The time domain expressions of the interpolated signals are calculated as:

$$i^{**}_{L1}(nT_{si}) = \begin{cases} i_{L1}(t)\delta(t-nT_{si}) & n=2k \\ 0 & n=2k-1 \end{cases} \quad (1)$$

$$i^{**}_{L2}(nT_{si}) = \begin{cases} 0 & n=2k \\ i_{L2}(t)\delta(t-nT_{si}) & n=2k-1 \end{cases} \quad (2)$$

Therefore, the total load current can be calculated by (1) and (2):

$$i^*_L(n) = i^{**}_{L1}(n) + i^{**}_{L1}(n-1) + i^{**}_{L2}(n) + i^{**}_{L2}(n-1) \quad (3)$$

It can be seen from (3), $i^*_L(n)$ can be described directly by the interpolated signals in the time domain with one single equation. Similarly, we can write the expression of sampled total load current at $(n-1)T_{si}$ instant, as:

$$i^*_L(n-1) = i^{**}_{L1}(n-1) + i^{**}_{L1}(n-2) + i^{**}_{L2}(n-1) + i^{**}_{L2}(n-2) \quad (4)$$

The differential equation of i^*_L can be obtained by combining (3) and (4):

$$i^*_L(n) - i^*_L(n-1) = i^{**}_{L1}(n) - i^{**}_{L1}(n-2) + i^{**}_{L2}(n) - i^{**}_{L2}(n-2) \quad (5)$$

The z-transformation of (5) is calculated as:

$$i^*_L(z) = \left(1 + z^{-1}\right)\left(i^{**}_{L1}(z) + i^{**}_{L2}(z)\right) \quad (6)$$

Eq. (6) reveals the z-domain relationship between the interpolated signals and the total sampled signals. z^{-1} denotes a unit delay of a sampling period. It is observed here that when adding phase currents to get i^*_L, there is a delay effect in it. It is helpful to translate (6) into the Laplace-domain. The definition of z is: $z = e^{sT_{si}}$, where T_{si} denotes the sampling

978-1-4244-4782-4/10 $26.00 © 2010 IEEE   177

period. With this definition, (6) can be directly transferred into the S-domain equation as:

$$i_L^*(s) = \left(1 + e^{-sT_{si}}\right)\left(i_{L1}^{**}(s) + i_{L2}^{**}(s)\right) \quad (7)$$

With (7), the S-domain relationship between i^*L and $i^{**}L$ is presented. However, (7) cannot be directly put in the loop model since the relationship between the interpolated signals and the original analog signal (i_L) are not clear yet. The next step is to build a bridge between these two. Following the definition of the Laplace-transformation, $i^{**}_{L1}(s)$ can be directly derived with the time domain expression (1), as:

$$i_{L1}^{**}(s) = \int_{t=-\infty}^{\infty} i_{L1}^{**}(t)e^{-st}dt = \int_{t=-\infty}^{\infty} i_{L1}(t)\sum_{k=-\infty}^{\infty}\delta(t-2kT_{si})e^{-st}dt \quad (8)$$

According to the theory of a sample-data system [8], the impulse series can be organized with Fourier series, as:

$$i_{L1}^{**}(s) = \int_{t=-\infty}^{\infty} i_{L1}(t)\sum_{k=-\infty}^{\infty}\delta(t-2kT_{si})e^{-st}dt = \int_{t=-\infty}^{\infty} i_{L1}(t)\left(\frac{1}{T_{sw}}\sum_{n=-\infty}^{\infty}e^{jn\frac{2\pi}{T_{sw}}t}\right)e^{-st}dt \quad (9)$$

$$= \frac{1}{T_{sw}}\sum_{n=-\infty}^{\infty}\int_{t=-\infty}^{\infty} i_{L1}(t)e^{-\left(s-jn\frac{2\pi}{T_{sw}}\right)t}dt = \frac{1}{T_{sw}}\sum_{n=-\infty}^{\infty}i_{L1}\left(s-jn\frac{2\pi}{T_{sw}}\right)$$

Similar derivation can be applied for phase 2's interpolated current, i^{**}_{L2}:

$$i_{L2}^{**}(s) = \frac{1}{T_{sw}}\sum_{n=-\infty}^{\infty}i_{L2}\left(s-jn\frac{2\pi}{T_{sw}}\right) \quad (10)$$

Eq. (9) and (10) reveal the relationship of the interpolated phase's currents and real phase current in the Laplace-domain. Note here that these equations are valid only for up to half of the sampling frequency. According to the sample-data system's theory, aliasing effects will take effects for frequency components higher than the Nyquist frequency. Moreover, for digitally controlled VRMs, crossover frequency of the voltage loop is much lower than this frequency, normally from 1/10 to 1/3 of the switching frequency. Hence, it is safe to limit (9) and (10) to Nyquist frequency, which will give:

$$i_{L1}^{**}(s) \approx \frac{1}{T_{sw}}i_{L1}(s) \quad (11)$$

$$i_{L2}^{**}(s) \approx \frac{1}{T_{sw}}i_{L2}(s) \quad (12)$$

Substituting (11) and (12) back into (7) and the relationship between the analog phase currents and sampled load current in Laplace-domain is achieved:

$$i_L^*(s) = \frac{1}{T_{sw}}\left(1 + e^{-sT_{si}}\right)\left(i_{L1}(s) + i_{L2}(s)\right) \quad (13)$$

For the multi-phase buck converters, the load current is obtained by adding all phase's current, as:

$$i_L(s) = i_{L1}(s) + i_{L2}(s) \quad (14)$$

Therefore, substitute (14) into (13) and move the $i_L(s)$ to the left side of the equation, the transfer function from load current $i_L(s)$ to the sampled load current $i^*_L(s)$ in Laplace-domain is finally achieved:

$$G_{sample}(s) = \frac{i_L^*(s)}{i_L(s)} = \frac{1}{T_{sw}}\left(1 + e^{-sT_{si}}\right) = \frac{1}{2T_{si}}\left(1 + e^{-sT_{si}}\right) \quad (15)$$

$G_{sample}(s)$ describes the transfer function of the "current sampling" blocks in Fig. 2(b). It has a very straightforward and simple physical meaning: at each i^*L sampling instant, the total sampled load current is the sum of one phase's current from this instant and the other phase's from previous sampling instant, which will include one sampling period delay. Therefore, the transfer function from i_L to i^*L will contain a delay effect.

All the previous derivations above are based on 2-phase case. It is easy to extend this modeling method for multi-phase case. For an N-phase buck converter, the current sampling's transfer function is calculated as:

$$G_{sample}(s) = \frac{i_L^*(s)}{i_L(s)} = \frac{1}{T_{sw}}\left(1 + e^{-s \cdot T_{si}} + e^{-s \cdot 2T_{si}} + \cdots + e^{-s \cdot (N-1)T_{si}}\right) \quad (16)$$

Eq. (17) can be simplified with Euler Equation, as:

$$G_{sample}(s) = \frac{i_L^*(s)}{i_L(s)} \approx \frac{1}{T_{sw}}e^{-s \cdot 0.5 \cdot (N-1) \cdot Tsi} \quad (17)$$

B. Small Signal Model of DPWM in Current Loop

Another delay effect in the current feedback loop is DPWM delay. There are several literatures talking about the modeling of multiple sampling DPWM delay in [5-7]. Here, the conclusions from these literatures are borrowed, as:

$$G_{DPWM}(s) = \frac{\hat{d}(s)}{\hat{v}_c^*(s)} \approx T_{si}e^{-s \cdot 0.5 \cdot Tsi} \quad (18)$$

C. Small Signal Model of Voltage Loop

The small signal model of the digital voltage loop can be derived similarly to that of the current loop. The block diagram of the digital voltage loop is shown in Fig. 3.

978-1-4244-4782-4/10 $26.00 © 2010 IEEE

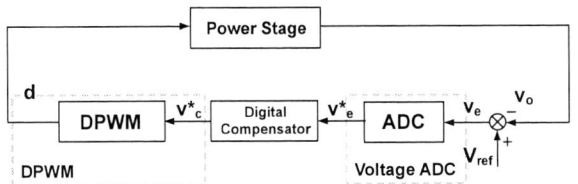

Figure 4. Block diagram of the digital voltage loop without T_{con} and T_{cal}

Compared with current sampling method, the total voltage is directly sampled. Therefore there is no delay effect due to sampling. Based on the sampling theory and ignoring the non-linear quantization effects, the voltage ADC can be modeled as a pure gain, as:

$$\frac{v_e^*(s)}{v_e(s)} = \frac{1}{T_{sv}} \quad (19)$$

where $T_{sv} = 1/(2 \cdot N \cdot T_{sw})$. The transfer function of DPWM can be derived similarly to the current loop case, which is shown in Fig. 3. The transfer function of DPWM can be found in [2], as:

$$G_{DPWM}(s) = \frac{d(s)}{v_c^*(s)} \approx T_{sv} e^{-s\frac{T_{sv}}{2}} \quad (20)$$

Considering the transfer functions of the voltage ADC and the DPWM, the small signal model of the digital voltage loop is obtained as shown in Fig. 5.

D. Complete Small Signal Model Without Conversion and Calculation Delay

In previous sections, the small signal models of current loop and voltage loop are derived respectively. In a complete digital VRMs circuit, the two loops co-exist in the feedback control. The complete small signal model of the digital VRMs without conversion and calculation delay is obtained as shown in Fig. 6.

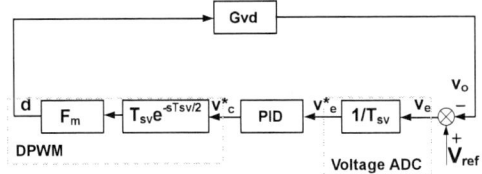

Figure 5. Small signal model of digital voltage loop without T_{con} and T_{cal}

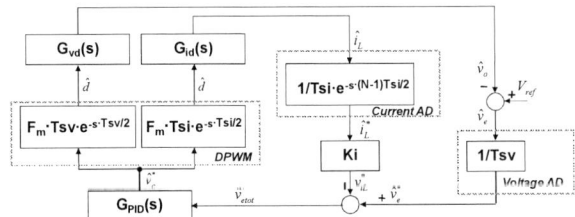

Figure 6. Small signal model of digital VRM without T_{con} and T_{cal}

III. SMALL SIGNAL MODEL OF DIGITALLY CONTROLLED VRM WITH CONVERSION AND CALCULATION DELAY

The model derived in Section II is valid for the applications where F_{sw} is relatively low while the digital controller's speed is fast enough, and in these applications, conversion and calculation time can be neglected. However, this is not always true for today's most today's products targeted in high frequency DC-DC converters. In most VRM applications, T_{con} and T_{cal} cannot be neglected.

Fig. 7 gives the small signal model of current loop in digital VRMs considering T_{con} and T_{cal} reported in [2]. Fig. 8 shows the key waveforms of this case. To use this model, accurate values of T_{con}, T_{cal}, T_{1f} and T_{1r} (defined in [2-3]) should be known. Also, T_{1f} and T_{1r} will change due to different phase number and duty cycle. Therefore, it is not easy to use this model for the design. This work proposed a new method to model the digital loop which has a straightforward and clear meaning and is easy to use.

Two new time periods are defined in Fig. 7: T_{1ff}, which is from the falling edge of a duty cycle to its effective sampling instant and T_{1rr}, which is from the rising edge of a duty cycle to its effective sampling instant. The expression of T_{1ff} and T_{1rr} are as:

$$T_{1ff} = T_{1f} + T_{con} + T_{cal} \qquad T_{1rr} = T_{1r} + T_{con} + T_{cal} \quad (21)$$

Substitute (21) into Fig.7 and collect all the delay term, then the total delay in the current loop is:

$$e^{-s(0.5T_{si}+T_{con})} \cdot e^{-sT_{cal}} \cdot e^{-s0.5(T_{1ff}+T_{1rr}-2T_{con}-2T_{cal})}$$
$$= e^{-s \cdot 0.5T_{si}} \cdot e^{-s0.5(T_{1ff}+T_{1rr})} \quad (22)$$

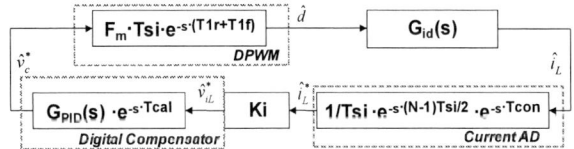

Figure 7. Small signal model of digital VRM without T_{con} and T_{cal}

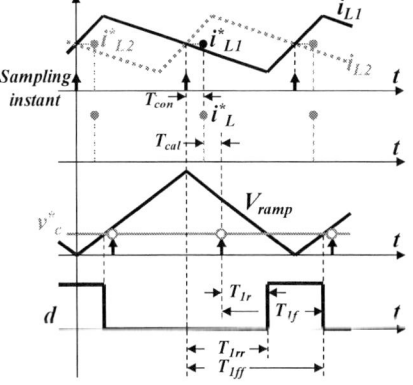

Figure 8. Key waveforms of digital VRM with T_{con} and T_{cal}

978-1-4244-4782-4/10 $26.00 © 2010 IEEE

It is found that the total delay in the current loop contains two parts: one is $0.5T_{si}$, which comes from adding the phase currents to get the load current; the other is average of T_{lff} and T_{lrr}. Therefore, the problem becomes simpler. To know the exact time delay in the current loop, it only needs to find the time period from falling edge, rising edge to their corresponding sampling instants, not constrained to exact value of T_{con} and T_{cal}. Actually, there are some fixed relationships between the sum of T_{lff} and T_{lrr} depending on different T_{cal} and T_{con}. Here we use a 2-phase and $D<0.5$ case as an example shown in Fig. 9 to explore this relationship.

In Fig. 9, $T_c = T_{cal} + T_{con}$. If T_c is in the range of $[0, 0.5DT_{sw}]$ as shown in (a), the falling edge and rising edge are determined by their closest sampling instants; If T_c is in the range of $[0.5DT_{sw}, 0.5(1-D)T_{sw}]$ as shown in (b), the falling edge will be determined by previous sampling instant due to longer T_c. If T_c is in the range of $[0.5(1-D)T_{sw}, 0.5(1+D)T_{sw}]$ as shown in (c), the rising edge should be updated by previous sampling instant accordingly. The results are summarized in Table 1. It is found that when T_c is within one range, T_{lff} and T_{lrr} are constant values and the sum is an integral number of Tsi. Hence, the total delay should be considered in the loop is $0.5kT_{si}$ plus the sampling delay, where k is an integer, $k \geq 1$. This means that you only need to find a range of $T_{con}+T_{cal}$, but not accurate values, and the total delay is known. This derivation is for the 2-phase case, and the similar conclusions can be applied for any phase cases.

The complete small signal model of the digital VRMs with conversion and calculation delay is obtained as shown in Fig. 10. The total delay considered in the loop can be found in TABLE I.

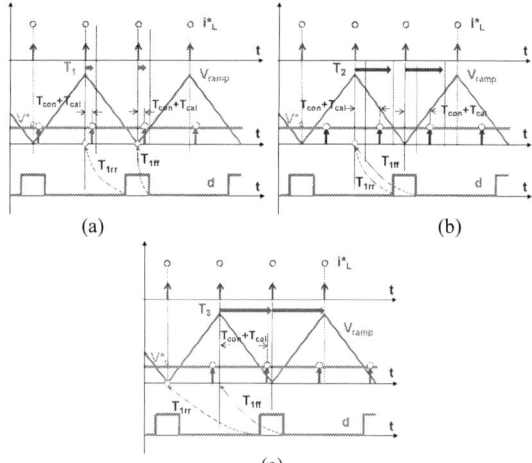

Figure 9. Relationship of T_{lff} and T_{2ff}. (a) $T_c<0.5DT_{sw}$, (b) $0.5DT_{sw}<T_c<0.5(1-D)T_{sw}$, (c) $0.5(1-D)T_{sw}<T_c<0.5(1+D)T_{sw}$

TABLE I. T_{lff} and T_{lrr} for 2-phase and D<0.5

	T_{lff}	T_{lrr}	$T_{lff}+T_{lrr}$
$0 \leq T_c \leq T_1 (T_1=0.5DT_{sw})$	$0.5DT_{sw}$	$T_{si}-0.5DT_{sw}$	T_{si}
$T_1 \leq T_c \leq T_2 (T_2=0.5(1-D)T_{sw})$	$T_{si}+0.5DT_{sw}$	$T_{si}-0.5DT_{sw}$	$2T_{si}$
$T_2 \leq T_c \leq T_3 (T_3=0.5(1+D)T_{sw})$	$T_{si}+0.5DT_{sw}$	$2T_{si}-0.5DT_{sw}$	T_{si}
•••	•••	•••	$k \cdot T_{si}$

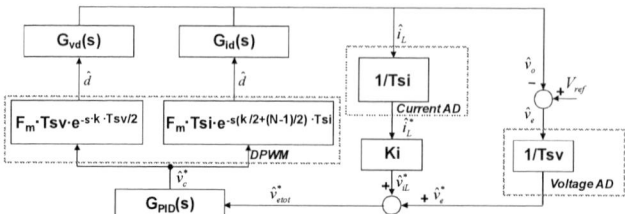

Figure 10. Complete small signal model of digital VRMs with T_{con} and T_{cal}.

IV. DEISNG GUIDELINES OF DIGITALLY CONTROLLED VRM

As a special power supply for the microprocessor, the VRM must maintain a low output voltage within a tight tolerance range during operation with a large current step change and high slew rate. To meet such transient requirements, the VRM must use many output capacitors, which increase its size and cost. To reduce the demand of capacitors, Adaptive voltage position (AVP) is a necessary function for VRM control design. The basic idea to achieve AVP is to design the output impedance of the VRM to be a constant value. The digital active droop control is employed in the designed digital VRMs. Still borrowing the concept of analog control, the digital active-droop control is intended to achieve constant output impedance. To achieve this goal, a high bandwidth current loop design is required [9].

The design guidelines for the digital active-droop control are as:

(1) Based on the phase number and steady state duty cycle as well as the worst case of $(T_{cal}+T_{con})$, select the corresponding delay for the current loop model. The delay can be found either from TABLE I. or the curves shown in Fig. 11, which is an equivalent graphic interpretation.

(2) Select $K_i = R_{droop}$. This is to guarantee the correct DC value to meet the load line requirement.

(3) Use a proper compensator to achieve the high bandwidth current loop design. Guarantee that the phase margin is larger than 60deg and gain margin is larger than 6dB. Normally, if the delay is larger than one switching period, a 3-pole 2-zero compensator is enough, as:

$$A_V(s) = K \cdot \frac{(1+s/\omega_{z1}) \cdot (1+s/\omega_{z2})}{s \cdot (1+s/\omega_{p1}) \cdot (1+s/\omega_{p2})} \quad (23)$$

(4) Use bilinear transformation method to transfer the Laplace-domain $A_v(s)$ to the discrete form. Then, the parameters of the discrete compensator are obtained.

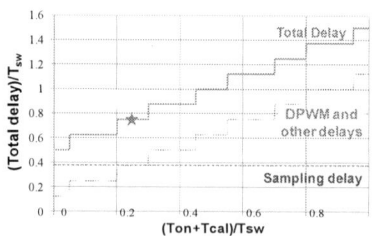

Figure 11. $(T_{con}+T_{cal})$ and related total delay in the current loop.

A. Design Example 1

The first example is a digital 4-phase VRM of with no conversion and calculation delay. Some key parameters are summarized in Table II. For the current loop, the total delay is $2T_{si}$, consisting of $T_{si}/2$ as the DPWM delay and $3/2T_{si}$ as the current sampling delay. For the voltage loop, only DPWM delay ($T_{sv}/2$) is considered. Substituting these values into model provided in Fig. 6, and following the design steps (2) – (4), we can get a good digital AVP control design. The current loop design result is shown in Fig. 10. The bandwidth is 120kHz, and the phase margin is 70deg. This design is then verified by SIMPLIS simulation. Fig. 11 (a) shows the T_2 comparison between the model and the simulation results. The solid line is the model the dashed line is the simulation result. It can be found that the model is pretty accurate. Fig.21 (b) shows the time domain load transient waveforms.

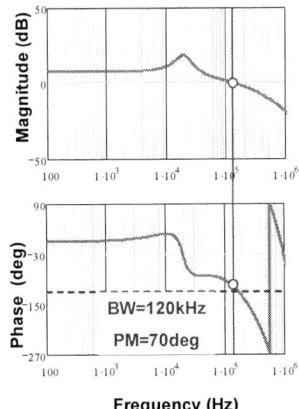

Figure 12. Bode plot of current loop gain (T_i)

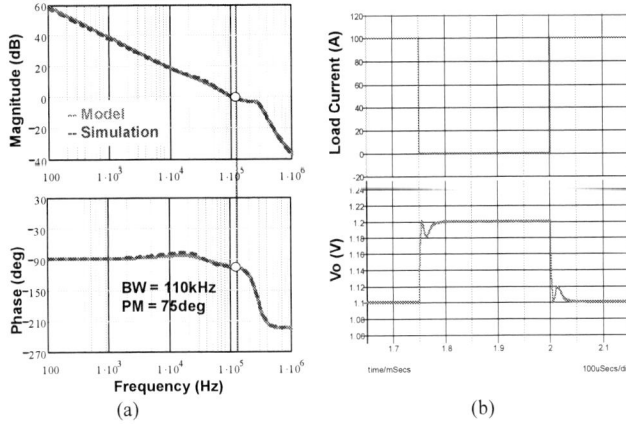

Figure 13. Simulation results of design for 4-phase digital VRM. (a), comparison of bode plot of T2; (b) time domain load transient simulation.

B. Design Example 2

The second design example is a digitally controlled 12V self-driven VRM. The 12V self-driven VRM is proposed in [10] with the merit of high efficiency. The operation principle of this circuit is similar to a 4-phase interleaved buck converter with steady state duty cycle equal to 0.4. For simplicity, we can directly use a 4-phase buck converter's small signal model for the power stage. A commercial product of digital VR 11 controller was employed. The circuit is shown in Fig. 12 and the parameters are summarized in Table III.

In this application, we need to use model provided by Fig. 11. With around 300ns conversion and calculation delay, the total delay considered in the loop model can be found in Fig. 23. The conversion and calculation time is located at the star point. Based on this curve, the total delay for the current loop is around $0.7T_{sw}$. This value will be used for the controller design. The design is verified by the experiment results. Fig.15 (a) shows the loop gain comparisons between the model and the experiment results. The model matches the experiment measurements well. Fig. 15 (b) shows the time domain load dynamics waveforms. Io step is 100A and di/dt is 2A/ns. The output voltage overshoot/undershoot during the load transient can meet the VR 11.0 dynamic specifications.

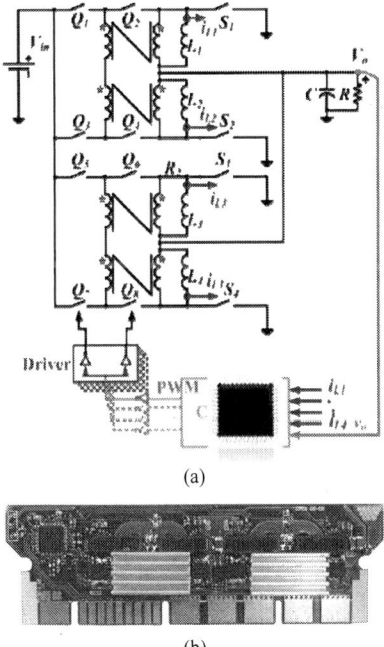

Figure 14. (a) Circuit diagram of digitally controlled 12V self-driven VRM (b) Picture of digitally controlled 12V self-driven VRM

TABLE II. Parameters of the digital 4-phase VRM

Input voltage	12V	Switching frequency	700kHz
Output voltage	1.2V	Voltage ADC sampling frequency	5.6MHz
R_{droop}	1mΩ	Current ADC sampling frequency	2.8MHz
Inductor	100nH	Capacitor	2mF
$T_{con}+T_{cal}$	0	Steady state Duty Cycle	0.1
Desired bandwidth, $f_c=1/(2\pi \cdot R_{droop} \cdot C)$		110kHz	

TABLE III. Parameters of the digital 12V self-driven VRM

Input voltage	12V	Switching frequency	700kHz
Output voltage	1.2V	Voltage ADC sampling frequency	5.6MHz
R_{droop}	1.25mΩ	Current ADC sampling frequency	2.8MHz
Inductor	60nH	Capacitor	2mF
$T_{con}+T_{cal}$	≈300ns	Steady state Duty Cycle	0.4
Desired bandwidth, $f_c=1/(2\pi \cdot R_{droop} \cdot C)$		140kHz	

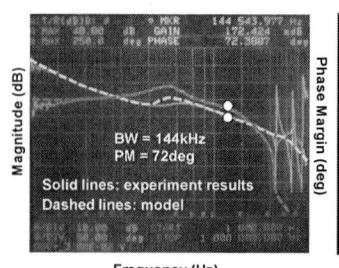 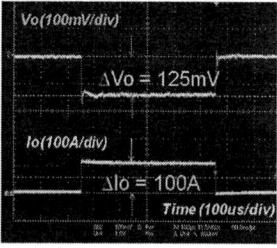

(a) Loop gain measurement   (b) Load transient waveforms

Figure 15. Experiment results of digitally controlled 12V self-driven VRM

V. SUMMARY

This paper presents a small signal model for digitally controlled VRMs. This work analyzes the sampling, modulation and delay effects inside the control loop, and uses a single delay term to describe all the delay effects. The analysis is derived based on the double-edge modulation and can be easily extended to trailing-edge and leading-edge modulation methods. The design guideline for the digital AVP control is provided. The simulation and experimental results show the validity of the proposed model.

REFERENCES

[1] Maksimovic, D., R. Zane, and R. Erickson, "Impact of digital control in power electronics". in ISPSD 2004.

[2] Van de Sype, D.M., et al, "Small-signal Laplace-domain analysis of uniformly-sampled pulse-width modulators", in PESC 2004.

[3] Van de Sype, D.M., et al, "Small-signal z-domain analysis of digitally controlled converters", in PESC 2004.

[4] Maksimovic, D. and R. Zane, "Small-Signal Discrete-Time Modeling of Digitally Controlled PWM Converters", IEEE Trans.on Power Electronics, 2007. 22(6): p. 2552-2556.

[5] Corradini, L. and P. Mattavelli, "Analysis of Multiple Sampling Technique for Digitally Controlled dc-dc Converters", in PESC 2006.

[6] Corradini, L. and P. Mattavelli, "Modeling of Multisampled Pulse Width Modulators for Digitally Controlled DC-DC Converters", IEEE Trans. on Power Electronics, 2008. 23(4): p. 1839-1847.

[7] Corradini, L., et al., "High-Bandwidth Multisampled Digitally Controlled DC-DC Converters Using Ripple Compensation", IEEE. Transactions on Industrial Electronics, 2008. 55(4): p. 1501-1508.

[8] Franklin, G.F., J.D. Powell, and M. Workman, "Digital Control of Dynamics Systems", Third ed. 1997: Addison Wesley Longman, Inc.

[9] Kaiwei, Y., et al., "Design considerations for VRM transient response based on the output impedance", IEEE Trans. on Power Electronics, 2003. 18(6): p. 1270-1277.

[10] Jinghai, Z., et al., "A self-driven soft-switching voltage regulator for future microprocessors", IEEE Trans. on Power Electronics, 2005. 20(4): p. 806-814.

Design and Comparison of Digital Control Loops Analytical Models, Laboratory Measurements, and Simulation Results

Philip Cooke, Infineon Technologies, Thomas G. Wilson, Jr., SIMPLIS Technologies, and Rohan Samsi, Primarion, (an Infineon Technologies company)

Abstract - A buck converter with a digital controller is modeled analytically and compared to laboratory measurements and simulation results. These models will help the practicing engineer in understanding the new digital controllers. This presentation provides engineers less familiar with digital control some tools and insight into how to characterize the behavior of these systems. The models are validated by comparing their results with experimental measurements. An analytical synthesis approach will be given to estimate the control loop PID values to design and assure the stability of these systems. This approach reduces design iteration and the resulting design can be further optimized in the laboratory and by way of the simulator. Additional simulation techniques and methods are presented which will help both young and more experienced engineers evaluate the robustness of these control systems. Finally, real life practical tips are provided to avoid common simulation and modeling traps.

Index Terms - Digital control, control loop modeling, power converter simulation

I. INTRODUCTION

Mixed-signal controllers are becoming more popular at lower power levels due to the reduction of cost and the improved functionality and feature sets which have a positive effect on overall system performance. This paper discusses one such digital controller and provides analytical, simulation, and experimental results for the design and analysis of the control loop. The paper is organized as follows; the first section presents the IC block diagram and application circuit; this is followed by the control loop model and a proposed synthesis design approach for a representative mixed-signal controller; and finally, simulated and experimental results are given to show the performance of the controller.

II. IC BLOCK DIAGRAM AND APPLICATION CIRCUIT

The IC block digram is shown in Fig. 1 with the corresponding application circuit given in Fig. 2. As with most digital controllers the analog output voltage is sensed and filtered, it then is converted into a digital word by an analog-to-digital converter (ADC).

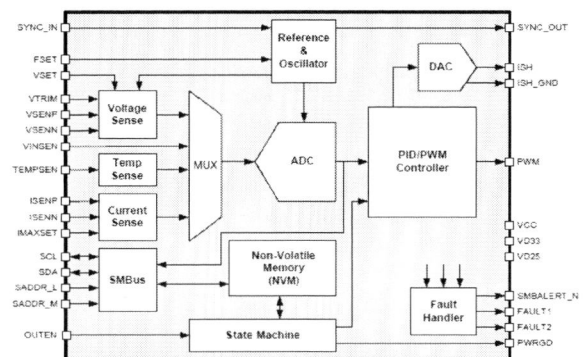

Fig. 1: IC Block Diagram of the PX7510

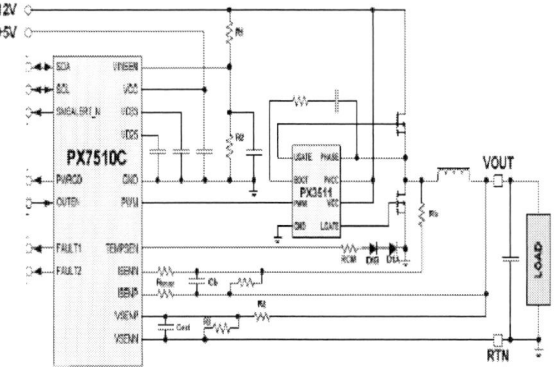

Fig. 2: Application Circuit for the PX7510

In this controller the current and temperature of the inductor are also converted into the digital domain. These signals are channeled to the ADC via a multiplexer as shown in Fig. 1. Also provided is a communications bus to both program the IC and provide real-time telemetry of the measured values. Use of this bus is optional, for example, applications exist that don't rely on telemetry data. This IC has a state machine based controller, which has dedicated digital logic to control the pulse-width modulation (PWM) of the power converter. A digital-to-analog converter (DAC) is used to drive a current share bus which allows up to four of these controllers to be interleaved for multiphase applications. Finally a fault handler detects system faults and reacts quickly based on a pre-programmed response action. An example of a single phase synchronous buck converter was used in this paper with a switching frequency of 518 kHz, an inductance of 300 nH, and a total output capacitance of C (bulk) = 1220 μF and C (ceramic) = 320 μF. In the next section we discuss the control loop model.

III. CONTROL LOOP MODEL

The discrete buck converter power stage control-to-output transfer function including the feedback gain H is given by [1]:

$$G_{OC-H}(z) = \frac{n_1 z + n_0}{z^2 + d_1 z + d_0} H \qquad (1)$$

where n_1, n_0, d_1, and d_0 are known coefficients calculated from the converter component values and the sampling rate. This transfer function models the dynamics of the power converter in the discrete domain. It can also be approximated by converting the analog transfer function

$$G_{OC-H}(s) = K_{OC} \frac{1 + s/\omega_{ESR}}{1 + s/Q\omega_O + (s/\omega_O)^2} H \qquad (2)$$

into the discrete domain using the MatLab or Octave c2d command. In using the c2d command or doing it by hand one of the several approximation techniques, Euler's approximation also called backward difference {replace s with $(1-z^{-1})/T_s$}, matched pole-zero mapping { $z = e^{-s \cdot T_s}$ }, or Tustin's method also called the bilinear transform {replace s with $2(1-z^{-1})/[T_s(1+z^{-1})]$} may be used. The sampling period is T_s. Alternately, a more accurate method, which is used in this paper, is the matrix exponential best described in [1], it then accounts for the total delay, t_D, in the digital control loop.

In (2), K_{OC} is equal to the input voltage (12 V dc), ω_{ESR} = $1/(R_{ESR}C)$, ω_O^2 = $(1+R_{DCR}/R)/\{LC(1+R_{ESR}/R)\}$, Q = $Q_{LOAD} \cdot Q_{LOSS}/(Q_{LOAD}+Q_{LOSS})$, Q_{LOAD} = $R(C/L)^{0.5}$ and Q_{LOSS} = $(L/C)^{0.5}/(R_{ESR}+R_{DCR})$. In these equations, R is the load resistance, L the inductance, R_{DCR} is the dc resistance of the inductor, C the total output capacitance, and R_{ESR} is the equivalent series resistance of this capacitor. Also note that we have included the feedback gain H in (1) & (2), which is unity for a 1 V output in this control IC. The PX7510 has 4 times oversampling in the voltage loop, so one uses a sampling period of T_s = 1/(4·f_s), where f_s = 518 kHz is the power converter switching frequency.

The controller transfer function may be written as

$$G_C(z) = A \frac{az^2 + bz + c}{z^2 - (1+K_{FD})z + K_{FD}}$$

$$= A \left[K_I \cdot \frac{1}{1-z^{-1}} + (K_P + K_D \cdot (1-z^{-1})) \cdot \frac{1}{1-K_{FD}z^{-1}} \right] \qquad (3)$$

which has two equivalent parametric forms. The first expression in (3) uses combined terms, the second expression is a factored form using the common K_P, K_D, and K_I coefficients which are determined by the design procedure. K_{FD} is a pole location also selected

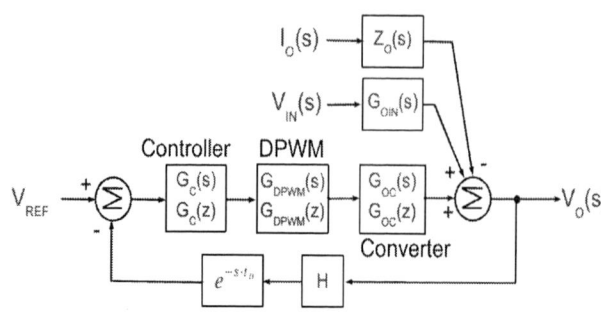

Fig. 3: Control Loop Model, $e^{-s \cdot t_D}$ is the Total Delay in the Control Loop, we Assume $G_{DPWM} \approx 1$ in this Paper, G_C is the Controller Transfer Function and G_{OC} is the Control-to-Output Transfer Function

in the design procedure. The term A is a fixed gain in this controller IC and is equal to 40. A system diagram is shown in Fig. 3 which includes a lumped $e^{-s t_d}$ term for the total delay in the loop resulting from the ADC and any delay from the Digital PWM (DPWM). The goal is for the designer to calculate a, b, c, and K_{FD} given the desired crossover frequency, f_C, and phase margin for the overall loop.

One way to do this, as will be shown next, is to solve the loop gain equation at the crossover frequency for the unknown parameters [5]. In order to simplify the analytical procedure and to have the same procedure cover both the real and complex-zero conjugate pair controller case we combine the two controller zeroes into a second order expression determined from its resonance, $\omega_x = 2 \cdot \pi \cdot f_x$ and its quality factor Q_x (these terms will be seen again in the equation for the controller transfer function below). These are easy to select once the resonance and quality factor of the power converter are known (ω_O and Q in equation 2). The resulting phase margin of the final design may be calculated once the controller parameters are found.

The last equation needed is the loop gain, T(s) = T(2·π·f_C), given by

$$|G_C(z_C)G_{OC-H}(z_C)| = 1 \qquad (4)$$

in the discrete domain. At the continuous time domain ("analog") crossover frequency f_C, this analog crossover frequency is converted to the digital domain crossover as z_C, using $z_C = e^{-j\omega_c \cdot T_s}$ with $\omega_C = 2 \cdot \pi \cdot f_C$.

Given the equations above, the design procedure will now be discussed. For the controller, you end up with either two real zeroes or a complex conjugate zero pair. Both cases are easy to implement with digital control. The advantage of this design approach is that both real and complex zero pair cases are dealt with by the same equation set and the problem of too many unknowns is addressed by providing guidance for the choice of the resonance (ω_x) and quality factor (Q_x) of the controller zeroes which are more easily selected from "gut feel" and experience.

1) Select the desired analog crossover frequency f_C, which is the overall control loop bandwidth, and then calculate the system resonance f_O

$$f_O = \frac{1}{2\pi}\sqrt{\frac{1}{LC}\frac{(1+R_{DCR}/R)}{(1+R_{ESR}/R)}} \qquad (5)$$

2) Set the analog post filter pole, f_{PA2}, to approximately $3 \cdot f_C$, and find K_{FD}. A reasonable starting range for f_{PA2} is anywhere between $f_C/2$ and $3 \cdot f_C$. For K_{FD}, select one of the following values {0.125, 0.25, 0.375, 0.50, 0.625, 0.75, 0.875, 1.00}. An initial choice of $K_{FD} = 0.75$ may be used.

3) Start by setting the resonant frequency of the controller to $f_X = 0.85 \cdot f_O$ and its quality factor to $Q_X = 0.7$. These determine the controller zero locations. Next find the required loop-gain to have $T(z)$ crossover at f_C. Note that the value chosen for f_X should be equal to or less than f_O, for design margin, to support component variation in production (but not too low)

$$f_X = 0.85 f_O \qquad (6)$$

4) Find α (α, β, and ε are intermediate variables to ease the calculations), from (1)

$$\alpha = \frac{|n_1 z_C + n_0|}{|z_C^2 + d_1 z_C + d_0|}H \qquad (7)$$

where $z_C = e^{-j\omega_c \cdot T_S}$, T_s is the sampling period, and $\omega_C = 2 \cdot \pi \cdot f_C$

5) Now calculate

$$\beta = |z_C^2 - (1+K_{FD})z_C + K_{FD}| \qquad (8)$$

6) Given that the system characteristic equation poles in continuous time maps into discrete time as

$$1 + s/Q_X\omega_X + (s/\omega_X)^2 \;\rightarrow\; (z - z_{ZN1})(z - z_{ZN2}),$$
$$\omega_X = 2 \cdot \pi \cdot f_X \qquad (9)$$

z_{ZN1} and z_{ZN2} are the zeroes from the analog system mapped into the unit circle. Now the last unknown, ε, is calculated from, with the values of z_{ZN1} and z_{ZN2},

$$\varepsilon = |z_C - z_{ZN1}| \cdot |z_C - z_{ZN2}| \qquad (10)$$

7) Finally the a, b, and c values are calculated from [5]

$$a = \frac{\beta}{\varepsilon \alpha A},$$
$$b = -2 \cdot a \cdot r \cdot \cos\left[2 \cdot \pi \cdot f_X \cdot T_S \sqrt{1 - 1/(2 \cdot Q_X)^2}\right],$$
$$c = a \cdot r^2 \qquad (11)$$

where $r = e^{-\pi \cdot f_X \cdot T_S/Q_X}$ is used to simplify the expressions in (11) as was done in [5]

8) For the alternatively controller expression in (3), find K_D, K_I, and K_P in that order from the a, b, and c values found above

$$K_D = c, \quad K_I = \frac{a+b+K_D}{1-K_{FD}}, \quad K_P = a - K_I - K_D \quad (12)$$

These equations can be put into a spreadsheet, Mathcad, MatLab, or Octave to ease the calculations. Note that it is a closed form solution to help reduce the usual trial and error approaches. After the IC is programmed with the coefficients the converter time domain and frequency domain characteristics can be simulated and measured in the laboratory. These results are provided in the next section.

IV. SIMULATION AND EXPERIMENTAL RESULTS

A simulation model was created in SIMPLIS™ and time domain results are shown in black traces of figures 4 through 7 (in black & white these will be the solid curve without noise). In each figure the converter is subjected to either a 10 A step load increase or a 10 A step load decrease. Also a 0 A and 10 A static load current is used to show no-load vs. loaded performance. These simulation results are compared with an experimental demonstration board that was fabricated for a 12 V input to 1.0 V output application at a maximum output current of 25 A. Multiple transients were taken in the laboratory for each case to show that there is variability in the measurements. In order to validate a model, a designer needs to take several measurements to get a sense of the range of this variability. In each of these figures the solid "noise free" black traces are the simulation results. The other traces are three separate transients captured by the oscilloscope.

As for the simulations, the component values in the actual hardware rarely match the nominal component values assumed in the simulation model, including simple things like the switching frequency, which can vary from one circuit to the next. Once these values are measured, the model can be updated and the new simulations waveforms may be better compared to the measured experimental results. The goal here is to validate and calibrate the simulation model. Once this is achieved, then parameter sweeps such as Monte-Carlo analysis can be done in the simulation tools to build confidence on the robustness of the design in a production environment. This is one of the many powerful benefits of developing these analytical and simulation models using a step-by-step calibrate and verify procedure.

It should be mentioned here that the analytical model from which the design procedure is used to calculate the starting values for the controller, is an approximate model. Its important that the designer has an design-oriented analysis [6] view into the control system and these expressions aid in that understanding. The power of the computer simulation tools, once they have been validated and calibrated, is to provide additional understanding and computer simulations to highlight the robustness or weakness of the design.

978-1-4244-4782-4/10 $26.00 © 2010 IEEE

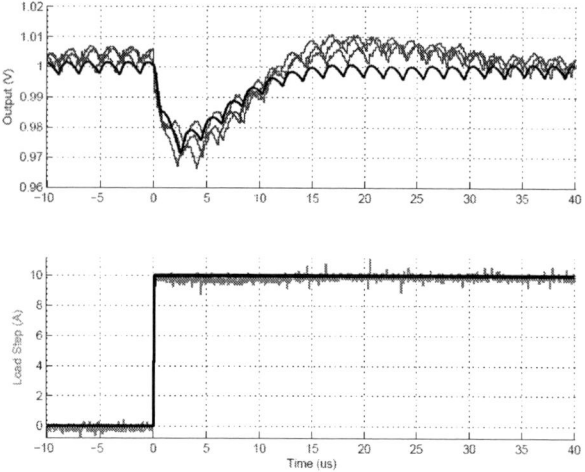

Fig. 4: Top, output voltage; Bottom, load current. Comparison of simulated and experiment results for a 0 A to 10 A step load. Smooth black curves are simulation results.

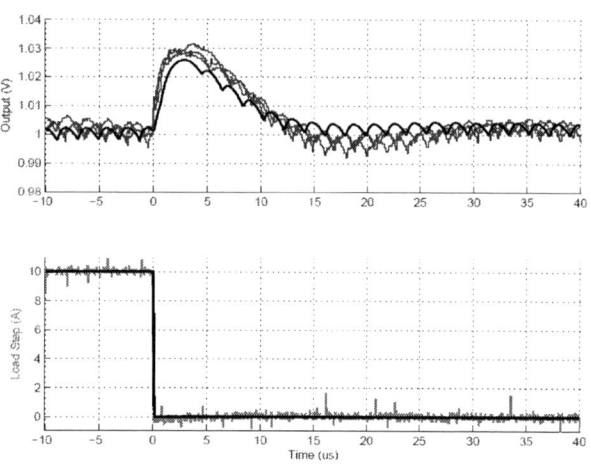

Fig. 5: Top, output voltage; Bottom, load current. Comparison of simulated and experiment results for a 10 A to 0 A step load. Smooth black curves are simulation results.

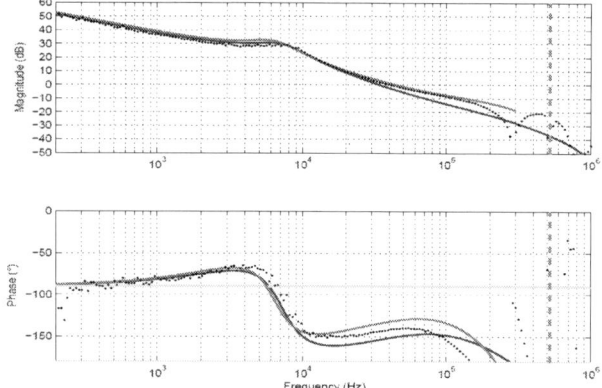

Fig. 8: Simulated Loop Gain and Phase (red), Analytical Model (blue), and Experimental Results (dotted line); Loop Gain $T = G_C(z) G_{OC-H}(z)$

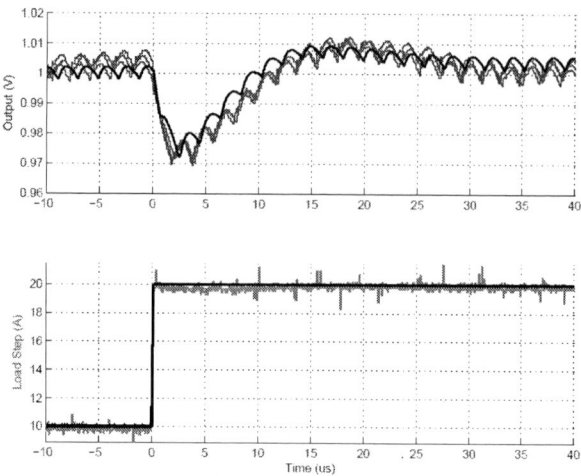

Fig. 6: Top, output voltage; Bottom, load current. Comparison of simulated and experiment results for a 10 A to 20 A step load. Smooth black curves are simulation results.

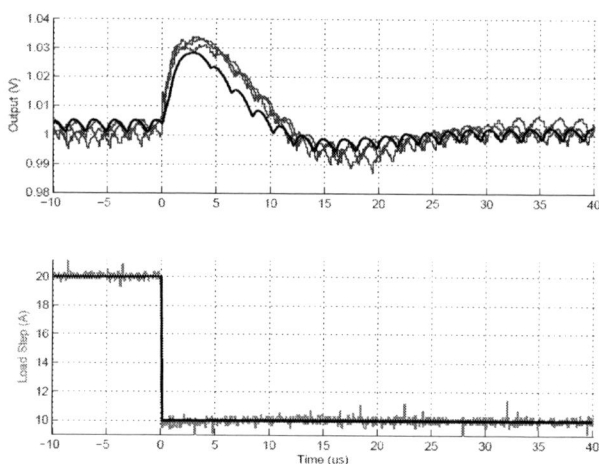

Fig. 7: Top, output voltage; Bottom, load current. Comparison of simulated and experiment results for a 20 A to 10 A step load. Smooth black curves are simulation results.

The frequency domain loop gain and phase were also measured as shown in Fig. 8. The dotted line is the measured data, the solid lines closest to the measured results are from SIMPLIS™, and the last curve is a simplified analytical model plotted using MatLab (the same analytical equations given above). There are minor differences between the experimental results and simulations, but slightly more differences to the simplified analytical model. This actually occurs quite often. In fact, it usually requires a bit of investigation into power converter component values to better calibrate their simulation models to be more representative of the actual circuit and measured data. The biggest factors are the lumped values of inductance, capacitance and their associated DCR and ESR parasitic values, not to mention the switching frequency and the MOSFET dynamic and static characteristics. The ESR, DCR, and the MOSFET's $R_{ds,on}$ are accounted for in the analytical model but the dynamic characteristics are not. This is about the right level of modeling which keeps the equations simple enough that design insight is gained.

V. CONCLUSION

In this paper an analytical design procedure and simulation models were presented for a digital controller. The important transfer functions for this voltage mode controlled application were given and the theory was shown along with the experimental results - with suggestions to improve the accuracy of the comparisons. Computer analysis tools were used in addition to the analytical results to further validate the understanding of the dynamics of these systems. The models presented here enable a fast design-oriented analysis approach for synthesis and all the non-linear behavior can then be checked in the laboratory when the converter is running or in the calibrated SIMPLIS™ simulator model. The digital controllers on the market today are easy to use in applications once these models and techniques are understood. These simulation models also extend a young designers' insight into these digital systems and can also be used to evaluate the robustness of the design as well.

VI. REFERENCES

[1] D. Maksimovic, R. Zane, R. Erickson, "Digital Control in Switched Mode Power Supplies", summer 2005 short course.

[2] H. Peng, A. Prodic, E. Alarcon, D. Maksimovic, "Modeling of Quantization Effects in Digitally Controlled Dc-dc Converters", IEEE Trans. Power Electronics, vol. 22, no. 1, Jan. 2007.

[3] L. Corradini, P. Mattavelli, "Modeling of Multisampled Pulse Width Modulators for Digitally Controlled Dc-dc Converters", IEEE Trans. Power Electronics, vol. 23, no. 4, Jul. 2008.

[4] D. Van de Sype, K. De Gusseme, F. De Belie, A. Van den Bossche, J. Melkebeek, "Small-Signal z-Domain Analysis of Digital Controlled Converters", IEEE Trans. Power Electronics, vol. 21, no. 2, Mar. 2006.

[5] V. Yousefzadeh, N. Wang, Z. Popovic, D. Maksimovic, "A Digitally Controlled DC/DC Converter for an RF Power Amplifier", IEEE Trans. Power Electronics, vol. 21, no. 1, Jan. 2006.

[6] Design-Oriented Analysis Rules and Tools, http://www.RDMiddlebrook.com.

Digital Control for Efficiency Improvements in Interleaved Boost PFC Rectifiers

Fu-Zen Chen and Dragan Maksimović
Colorado Power Electronics Center
ECEE Department, University of Colorado,
Boulder, CO 80309-0425
{fchen, maksimov}@colorado.edu

Abstract— This paper presents a simple passive power sharing approach to paralleling power-factor correction (PFC) modules. A digital controller senses only the total inductor current, and drives the modules with matched phase-shifted control signals. Advantages of the approach include simplified current sensing and control, support for arbitrarily large number of paralleled modules, and minimization of conduction losses at heavy loads. Increased current stresses resulting from unequal power sharing are evaluated in terms of parameter mismatches. To maintain high efficiency over very wide range of loads, the digital controller further includes adaptive near-zero-voltage switching and adaptive frequency operation in discontinuous conduction mode (DCM), as well as phase shedding at light loads. Experimental results are shown for a 600W two-phase boost PFC rectifier.

I. INTRODUCTION

To improve power system modularity, and to reduce input current ripple, paralleling of phase interleaved power factor correction (PFC) boost rectifier modules has been adopted in single-phase AC-DC rectifiers [1-8]. A two-phase example is shown in Fig. 1. In the approaches reported so far, power sharing among the paralleled modules operating in continuous conduction mode (CCM) has been based on active current sharing control that requires individual current sensing in each module [6-8]. Since the modules share a common ground

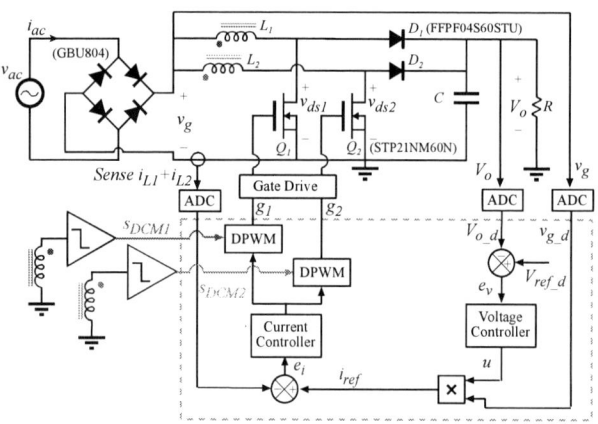

Fig. 1 Experimental digitally controlled 600 W interleaved PFC rectifier with passive power sharing (a two-phase example).

This work has been sponsored through the Colorado Power Electronics Center (CoPEC).

terminal, typically at the negative terminal of the output filter capacitor, and are supplied from the same AC line input voltage, the most common approach to current sensing based on an input-side sensing resistance (shown in Fig.1) gives a sum of the inductor currents. Additional per-module switch or diode current sensing circuits are therefore required to achieve active equal power sharing.

This paper examines an alternative, simpler approach based on passive power sharing using only one current sensing circuit (as shown in Fig.1) and a digital controller driving the power MOSFETs with phase shifted control signals having identical duty ratios. Such passive power sharing approach has earlier been proposed for multi-phase DC-DC microprocessor power supplies [9], where it was shown to yield minimum overall conduction losses and improved efficiency (at heavy loads) at the expense of unequal distribution of currents among the modules. An objective in this paper is to examine this trade-off in the context of multi-phase PFCs such as the two-phase example shown in Fig.1.

A further objective is to apply digital control techniques to optimize efficiency not just at heavy loads, but over a very wide range of loads. Extending high efficiency operation to light load is gaining increasing importance due to increasingly stringent energy efficiency standards. The approach includes adaptive switching and adaptive frequency operation [10], as well as phase shedding [7-8].

Section II discusses passive power sharing and efficiency optimization at heavy loads when the PFC modules operate in continuous conduction mode (CCM). The light load efficiency improvement approaches, adaptive switching and adaptive frequency, under discontinuous conduction mode (DCM), or mixed mode (DCM/CCM operation), are addressed in Section III. Section IV demonstrates experimental results for a 600W digitally controlled two-phase boost PFC rectifier. Conclusions are given in Section V.

II. PASSIVE POWER SHARING

A digital controller can produce perfectly matched phase shifted control signals for each module. This property has enabled effective passive current sharing in multi-phase DC-DC converters [9]. This section examines the outcomes in terms of conduction losses and current mismatch of the passive power sharing approach when the paralleled PFC modules are operated with identical (but phase shifted) switch control signals.

A. Conduction Losses

An ideal single-phase PFC presents a resistive load to the AC line [11]. As shown in Fig. 2, a two-phase PFC rectifier is modeled as two emulated resistances (R_{e1}, R_{e2}) on the input side with corresponding controlled power sources P_{ac1} and P_{ac2}, respectively. To simplify the analysis of passive power sharing, conduction losses are approximately modeled as equivalent series resistances R_{L1} and R_{L2}. Furthermore, the simplified analysis assumes that the inductors are well matched, $L_1 = L_2$.

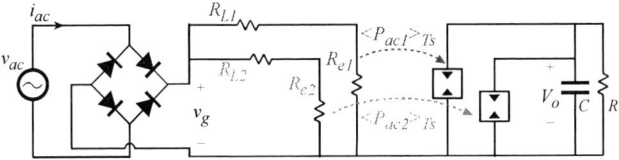

Fig. 2 Model of a two-phase boost PFC rectifier, including conduction losses

In CCM operation, under small ripple assumption, the conduction loss (P_{cond}) can be found as

$$P_{cond} \approx (\frac{1}{R_{e1}^2}R_{L1} + \frac{1}{R_{e2}^2}R_{L2})V_{ac,rms}^2 \qquad (1)$$

where $V_{ac,rms}$ is the root-mean-square (RMS) value of the input ac line voltage. It follows that the conduction loss (1) is minimized if the PFC module emulated resistances satisfy the following condition:

$$\frac{R_{e1}}{R_{e2}} = \frac{R_{L1}}{R_{L2}} \qquad (2)$$

In Fig. 3, the normalized conduction loss, which is the conduction loss with respect to the minimum conduction loss over different R_e and R_L ratios, illustrates how the minimum conduction loss occurs when (2) is satisfied.

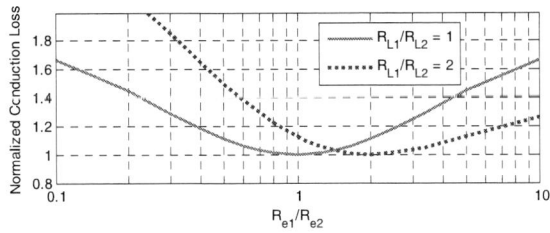

Fig. 3 Normalized conduction loss with R_e and R_L mismatch.

In multi-phase PFC rectifiers with active equal power sharing, current controllers for each module are implemented to make the current split evenly among the modules. With evenly shared power, the emulated resistances are the same for all modules, $R_{e1} = R_{e2}$. As a result, in the presence of conduction loss mismatches ($R_{L1} \neq R_{L2}$), equal current sharing does not achieve minimum conduction loss.

If the controller controls the power switches with identical duty ratios, while controlling the total input current, it can be shown that the resulting emulated resistances meet (2), which

means that the passive power sharing approach minimizes the total conduction losses.

A more detailed analysis can be performed based on the models averaged over a switching period T_s, as shown in Fig. 4.

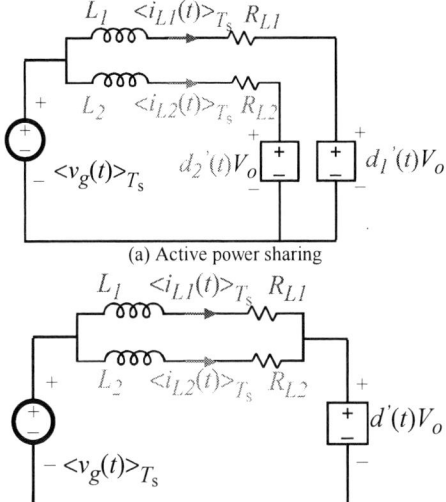

(a) Active power sharing

(b) Passive power sharing

Fig. 4 Two-phase PFC models averaged over a switching period.

In the PFC rectifier with active power sharing, each phase shares equal current (Fig. 4(a)). Assuming the inductor values are the same in all phases, and taking into account only conduction losses, overall efficiency of the PFC rectifier with active power sharing can be found by integrating the power loss over the line period,

$$\eta_{active} = 1 - \frac{R_L}{2R_e} \qquad (3)$$

where $R_L = (R_{L1}+R_{L2})/2$ is the nominal equivalent series resistance, and R_e is the total emulated resistance.

In the passive power sharing PFC rectifier, each phase operates at the same duty ratio (Fig. 4(b)). Assuming the inductor values are the same in all phases and that the time constant of the inductor ($\tau_L = L/R_L$) is much shorter than one half of the line period (T_L), overall efficiency of the passive power sharing PFC rectifier becomes a function of the equivalent series resistance mismatch ($\Delta R_L = |R_{L1}-R_{L2}|$),

$$\eta_{passive} \approx 1 - \frac{R_L}{2R_e} + \frac{\Delta R_L^2}{8R_e R_L} \qquad (4)$$

From (3) and (4), the reduction in conduction loss by passive power sharing is shown in Fig. 5 as a function of the relative R_L mismatch $\Delta R_L/R_L$. One may note that, although the reduction in conduction losses due to passive power sharing is relatively small, the approach can result in efficiency improvements at heavy loads where conduction losses dominate.

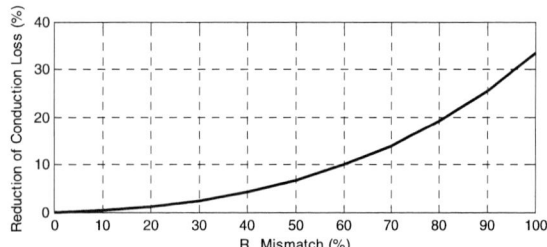

Fig.5 Reduction of conduction loss using passive power sharing in the two-phase boost PFC rectifier in CCM.

B. Current Mismatch due to Passive Power Sharing

In the presence of component mismatches, passive power sharing helps to reduce conduction losses. A disadvantage is that each phase processes different amount of power, and the resulting current mismatch increases the current stresses on the components. With passive power sharing, the current mismatch can be related to a mismatch in the equivalent series resistances, a mismatch in the inductance values, and the phase shift between the PFC modules. This section presents approximate analyses of each of these three effects separately.

From the model in Fig. 4(b), considering the R_L mismatch only, the maximum relative current difference can be found as

$$\max[\Delta i_L] \approx \frac{V_M}{2R_e}\frac{\Delta R_L}{R_L} \tag{5}$$

where V_M is the peak ac line voltage. The current-mismatch penalty is illustrated in Fig. 6, which shows the maximum current mismatch as a function of the R_L mismatch.

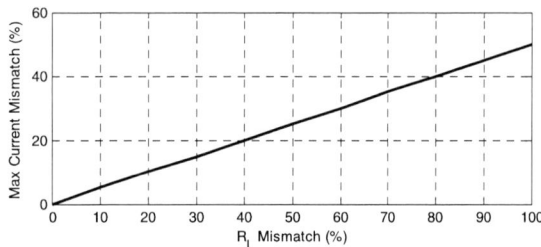

Fig.6 Maximum current mismatch using passive power sharing in the two-phase boost PFC rectifier in CCM.

Considering the inductance mismatch only, the maximum current difference can be found as

$$\max[\Delta i_L] \approx \frac{V_M}{2R_e}(\tau_L\omega_L)\frac{\Delta L}{L} \tag{6}$$

where L is the nominal inductance (mean value of all the inductances) and τ_L is the nominal inductor time constant.

Under an assumption that the inductor time constant τ_L is much shorter than one half of the line period T_L, the maximum current mismatch due to inductor mismatch (6) is much smaller than the current mismatch due to R_L mismatch (5). This conclusion is in contrast to the case when PFC rectifiers operate in discontinuous conduction mode (DCM) or in transition mode (CCM/DCM boundary) [3, 4], when the current mismatch is caused mainly by the inductor value mismatch.

Even in the case when the modules are perfectly matched, a current mismatch occurs in PFC rectifiers with passive power sharing due to phase-shifted control signals in combination with time-varying input voltage. By noting that the large-signal models averaged over a switching period (Fig. 4) are linear, an s-domain approach based on the closed-loop model shown in Fig. 7 can be applied to examine this effect. The phase shift between the phases is modeled by a transfer function $G_d(s)$. The total inductor current i_L is well regulated by the current controller G_c to track the reference current ($i_{ref} = v_g/R_e$).

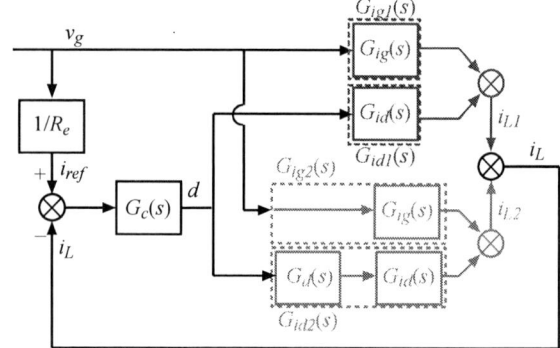

Fig.7 Two-phase interleaved PFC rectifier model, including phase shift modeled by $G_d(s)$.

From the model in Fig. 7, the current mismatch can be found as

$$\Delta i_L(s) = \frac{v_g(s)}{R_e}\left[\frac{(G_d(s)-1)(2G_{ig}(s)R_e-1)}{\frac{1}{G_{id}(s)G_c(s)}+(1+G_d(s))}\right] \tag{7}$$

From (7), it can be observed that the current mismatch is due to the phase shift ($G_d(s) \neq 1$), and the time-varying input voltage $v_g(s)$. Assuming a well-regulated input current, a first-order approximation for the delay transfer function $G_d(s)$, and a sinusoidal ac input voltage at 60 Hz, the maximum current mismatch is shown in Fig. 8 as a function of R_L/R_e for several values of the input voltage $V_{ac,rms}$, and inductance L. For heavy loads with dominant conduction loss, the current mismatch due to phase shift is relatively small.

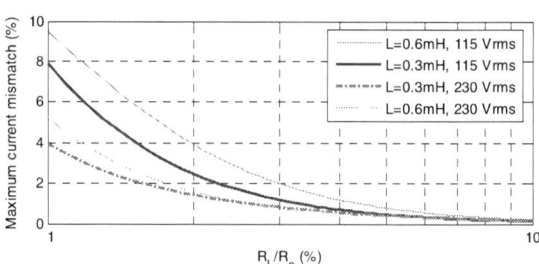

Fig.8 Maximum current mismatch using passive power sharing in the two-phase interleaved boost PFC rectifier in CCM ($600W$, $f_s=100kHz$, $f_L=60Hz$)

C. Over Current Protection

Although passive power sharing minimizes the conduction losses, as described in Section II.B, the resulting current mismatch discussed in Section II.C may result in additional current stresses. It is therefore of interest to consider ways to provide per-module over-current protection without compromising the current-sensing simplicity and modularity of the passive power sharing approach.

An over-current detection circuit proposed in this paper is based on monitoring the charge-up time for the switching node capacitance (C_x) as an indication of the peak inductor current. Upon transistor turn off, it takes time to charge the switching node voltage v_{ds} up to the output voltage after which the diode turns on. For a given v_g and C_x, the total required charge Q_c supplied by the inductor to charge C_x from 0 to v_g is fixed. The charge up time t_c is therefore inversely proportional to the inductor peak current $i_{L,peak}$.

$$Q_c = \int_0^{v_g} C_x(v_{ds})dv_{ds} = i_{L,peak} \cdot t_c \quad (8)$$

As shown in Fig. 1, the DCM detection comparators are included to detect polarity of the voltage across the inductor [10]. The same comparators can be used to determine the time when v_{ds} reaches v_g after the MOSFET is turned off. Digital controller simply counts the time interval t_d between the gate signal g high-to-low transition and the corresponding transition in the DCM comparator signal S_{DCM}. It should be noted that t_d is a sum of the switching node charge-up time (t_c), gate drive delay ($t_{d,GD}$), and the comparator delay ($t_{d,COMP}$).

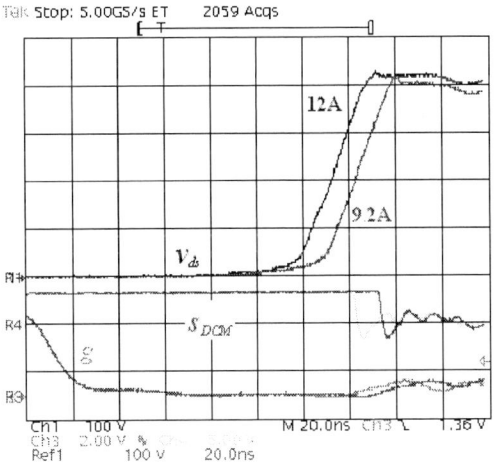

Fig.9 Transistor turn-off interval (v_g =100V, $i_{L,peak}$ = 12A, 9.2A)

The last two terms are highly dependent on the component selections and their variations, so a calibration is required in a practical implementation of the proposed over-current protection.

Fig. 9 shows examples of v_{ds} waveforms for different inductor peak currents. Most of t_d is when v_{ds} is low, which is due to the larger transistor drain to source capacitance at low v_{ds}. When v_{ds} is high, the charging-up slope is much steeper. The peak current occurs around the peak line voltage $v_g = V_M$, which corresponds to a lower capacitance as v_{ds} reaches v_g. Therefore, t_d increases only slightly with increasing input voltage, which means that the calibration can be performed at just one voltage.

Time resolution of the digital controller system clock (10 ns in the experimental prototype) is too low for precise current detection. In order to measure t_d more precisely, a delay line based timer has been constructed as shown in Fig. 10, improving the timing resolution to about 2 ns. Experimental results showing t_d as a function of $i_{L,peak}$ at $v_g = 100V$ are shown in Fig. 11. This result can be used to to calibrate the proposed over-current protection function.

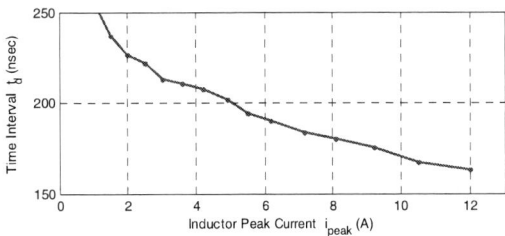

Fig.11 Time interval (t_d) as a function of the inductor peak current ($i_{L,peak}$) (v_g=100V, Q(STP21NM60N), D(FFPF04S60STU))

III. LIGHT LOAD EFFICIENCY IMPROVEMENTS

In order to improve overall efficiency, phase shedding approaches have been developed for multi-phase interleaved PFC rectifiers [7, 8]. The main idea of phase shedding is to reduce the switching loss when PFC rectifier is processing less power. Most phase shedding PFC rectifiers reduce the number of active phases based on the power command (u), which makes the number of active phases constant over the line period. Programmability of a digital controller makes the required scaling of the power command and the current loop gain easy.

In addition to phase shedding, since the power processed by the PFC rectifier changes within a line period, approaches based on varying the switching frequency within a line period have been proposed [8, 10]. The approach applied in this paper is similar to the adaptive frequency approach reported in [10],

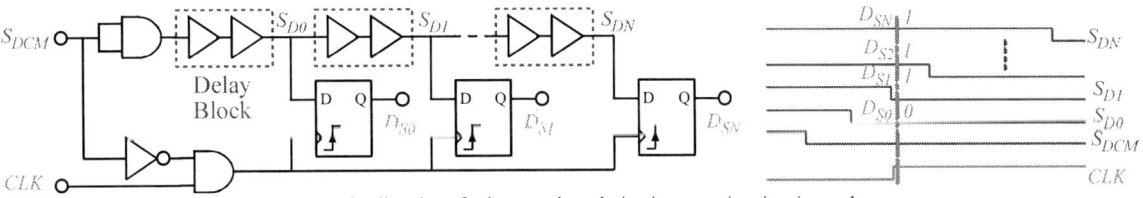

Fig.10 Delay line timer for improved resolution in measuring time interval t_d

which maintains constant-frequency CCM operation at heavy loads (as needed for passive power sharing), and adaptively reduces the switching frequency in DCM based on the power command and line voltage. In all cases, the adaptive switching approach [10] of turning on the power MOSFET at the minimum of v_{ds} ringing in DCM is also applied.

This section discusses additional challenges related to phase shifting and current sensing in applying the adaptive frequency approach in the multi-phase PFC configuration.

A. Phase Shifting with Adaptive Frequency Operation

The adaptive frequency approach changes the operating frequency to reduce switching loss at light loads [10]. A constant time shift between the phases would result in additional current ripple. The approach implemented in the experimental prototype is based on a digital pulse-width modulator (DPWM) in a master phase operating as described in [10], while DPWM's in the slave phases replicate the master phase turn-on/turn-off intervals with an adaptive phase shift. The slave phases adjust turn-off time to achieve the required phase shift. The operation is illustrated by the waveforms in CCM and in DCM shown in Fig. 12(a) and (b), respectively.

B. Current Sensing

As shown in Fig. 1, the current analog-to-digital converter (ADC) samples the total current in the middle of the transistor conduction interval. In CCM, with an even phase shift, the sensed current represents the total average current, as the two-phase example in Fig. 13(a) shows. With an uneven phase shift, there would be an offset between the average inductor current and the sensed current, which would increase the current harmonic distortion.

In DCM, the current sensing correction presented in [10] cannot be applied directly in the multi-phase configuration. In deep DCM, the sensed current may only represent part of the inductor current as shown in the example of Fig. 13(b). With an uneven phase shift in DCM, even larger current sensing errors can be expected. However, it should be noted that the effects of the current sensing correction and the current sensing error in DCM are in the same direction, which means that the overall error between the real average current and the sensed current can be relatively small. Furthermore, most current distortion happens in deep DCM around zero crossing of the line voltage, so that the overall effect on the input current distortion is small. Finally, as noted above, phase shedding reduces the number of active phases at light loads. Once the system operates with a single active phase, the DCM current correction described in [10] applies, and the light-load current harmonic distortion is reduced.

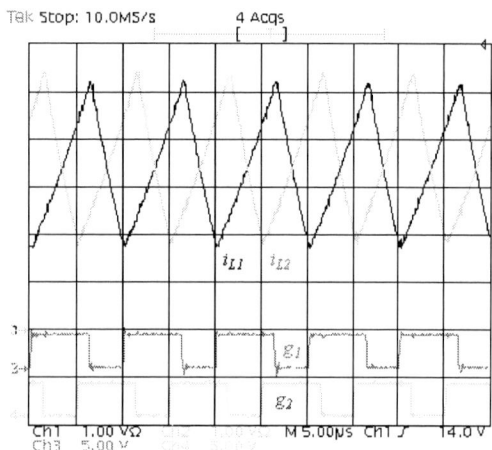

(a)

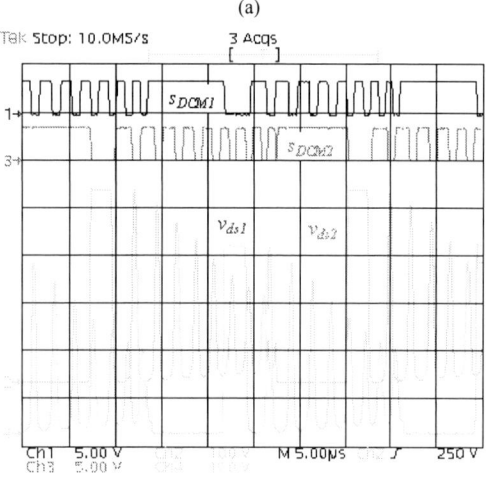

(b)

Fig.12 DPWM operation in both CCM and DCM (1-master, 2-slave)
(a) Inductor current (i_L) and gate drive signal (g) in CCM
(b) Drain node voltage (v_{ds}) and comparator signal (g) in DCM

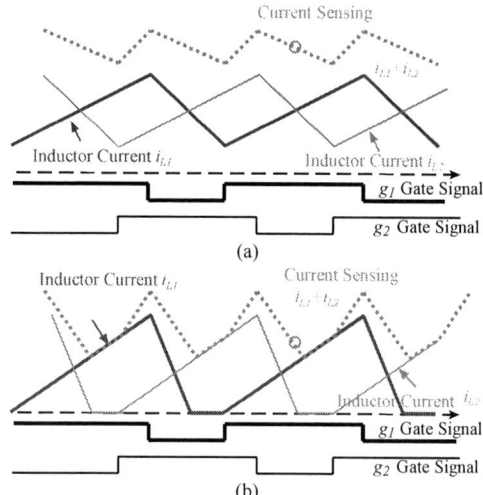

Fig.13 Current sense in passive power sharing approach
(two-phase interleaved example)
(a) At CCM
(b) At DCM

A final comment relates to application of the adaptive switching, which adjusts the switching period slightly to achieve switching at the lowest v_{ds} in DCM [10]. In the multi-phase configuration, if all DPWM's are performing adaptive switching, the phase shift between the phases can be slightly off. Similar phenomena happen in transition mode interleaved PFC rectifiers, which have been addressed and discussed in [3, 4]. A PLL approach may not apply because of the discrete nature of the timing related to an integer number of DCM oscillation periods, resulting in sudden jumps in the switching periods. Fig. 14 shows how current sensing can be highly dependent on the discontinuous conduction period (T_{dcm}) in the previous switching period. This effect may cause some oscillations in the current loop, and a light increase in input current distortion.

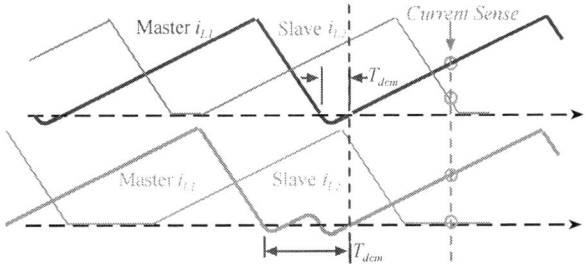

Fig. 14 Current sensing in the passive power sharing approach
(a two-phase interleaved example)
(Master - adaptive switching DPWM; Slave - follower DPWM)

IV. EXPERIMENTAL RESULTS

A 600W two-phase boost PFC rectifier (f_s = 100kHz, $L_1 = L_2 = 320\ \mu$H, $C = 440\ \mu$F) has been built with an FPGA platform implementing the digital controller. The experimental prototype is shown in Fig. 15.

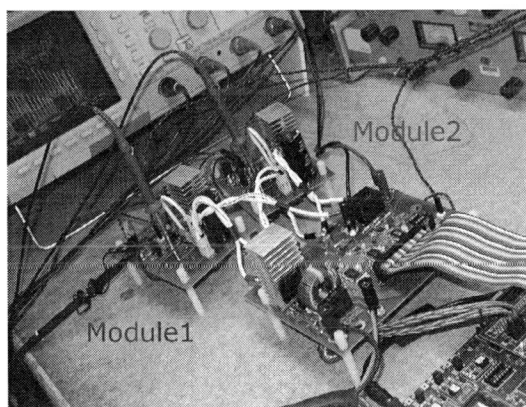

Fig. 15 Experimental setup for 600W digitally controlled
two phase boost PFC rectifier

For heavy load operation, the controller operates at constant frequency with two active phases interleaved and evenly phase-shifted by 180°. The operating waveforms are shown in Fig. 16.

For moderate loads, two active phases operate in both CCM and DCM over a line period. The controller starts to reduce the switching frequency (Fig. 17(a)) following the adaptive frequency approach. Because the DCM current sensing error discussed in Section III.B cannot be fully corrected, some

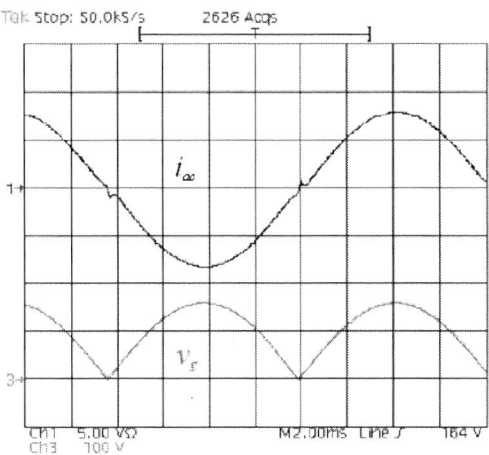

Fig. 16 Rectifier line voltage (v_g) and line current (i_{ac})
(100kHz, 600W, two active phases interleaved)

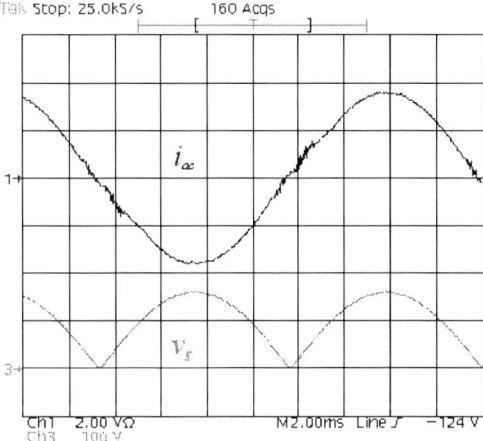

(a)

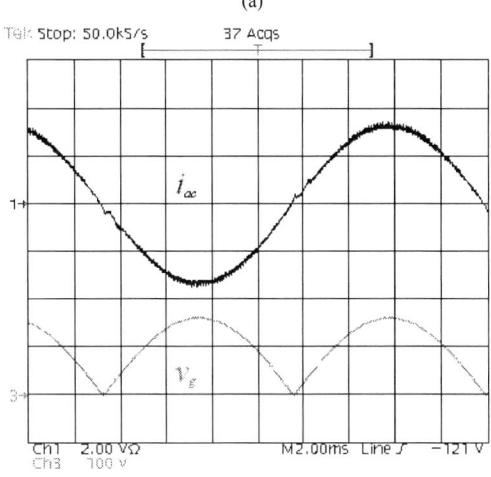

(b)

Fig. 17 Rectifier line voltage (v_g) and line current (i_{ac}) (100kHz, 270W)
(a) Two active phases interleaved with adaptive frequency
(b) Single active phase due to phase shedding.

978-1-4244-4782-4/10 $26.00 © 2010 IEEE

additional current distortion can be observed. When the controller drops one phase, the remaining active phase operates in CCM as shown in Fig. 17(b).

Once phase shedding results in a single active phase, both adaptive switching and adaptive frequency are activated with DCM current correction as described in [10] and illustrated by the waveforms in Fig. 18(a). Compared to the two-phase constant frequency interleaved case (Fig. 18(b)), the current distortion is reduced.

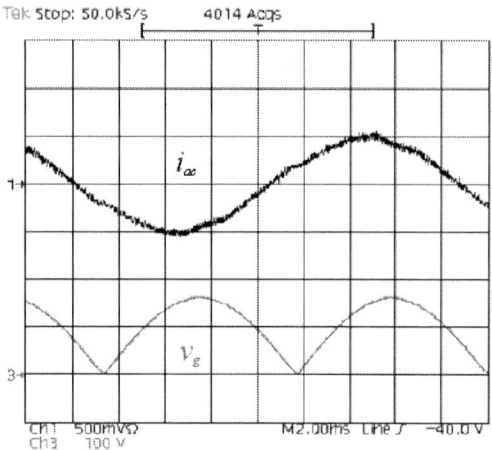

(a)

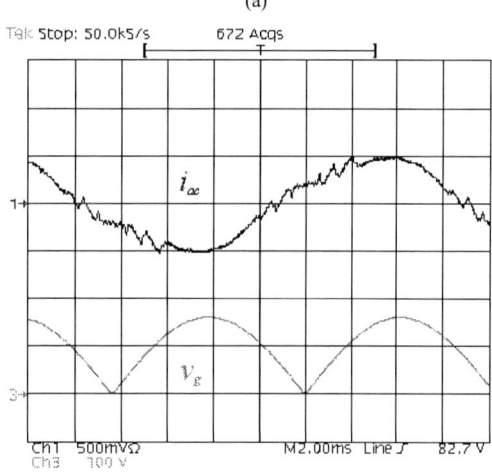

(b)

Fig. 18 Rectifier line voltage (v_g) and line current (i_{ac}) (100kHz, 30W)
(a) Single active phase with adaptive switching and adaptive frequency
(b) Two active phases interleaved with constant frequency operation

Efficiency improvements are illustrated in Fig. 19. The adaptive switching and the adaptive frequency approaches reduce switching losses at light loads, while phase shedding further improves the light load efficiency. The experimental efficiency results indicate that phase shedding should be activated when the power drops below 270 W, leaving one phase active. At very light load (30 W), efficiency is improved

by about 3% compared to the conventional two-phase system operating at constant switching frequency.

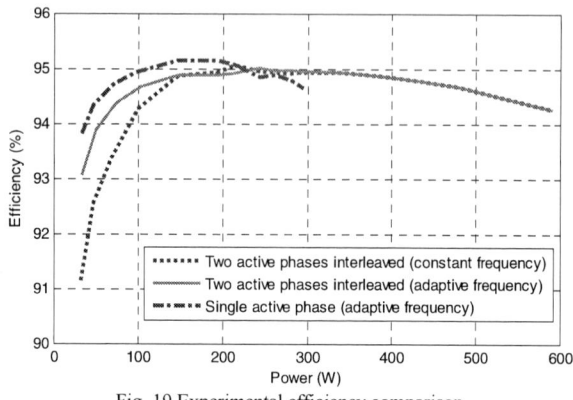

Fig. 19 Experimental efficiency comparison
($V_{g\text{-}rms} = 115$ V, $f_{s,max} = 100$ kHz, $f_{s,min} = 40$ kHz)

V. CONCLUSIONS

A simple passive power sharing approach to paralleling PFC modules is discussed in this paper. A digital controller operates PFC modules with matched phase-shifted control signals and regulates the total input current by sensing only the total inductor current. This approach simplifies current sensing and control, and can support arbitrarily large number of paralleled modules. The paper further discusses tradeoffs between conduction loss minimization and current mismatches. It is shown that in the presence of component mismatches the passive power sharing approach results in minimum conduction losses at the expense of unequal current sharing and increased current stresses. An approach is introduced to achieve over-current protection in each phase based on detecting the timing of the switch voltage after turn off as an indication of the peak inductor current.

To improve efficiency over wide range of loads, the controller implements phase shedding, and adaptive frequency operation for multiphase operation. In addition, adaptive near-zero-voltage switching is enabled when the system operates with a single active phase in DCM at very light loads. A light-load efficiency improvement of 3% is demonstrated compared to conventional constant-frequency multi-phase operation on an experimental 600W two-phase boost PFC rectifier.

REFERENCES

[1] B. A. Miwa, D. M. Otten and M. E. Schlecht, "High Efficiency Power Factor Correction Using Interleaving Techniques," in *Proc. IEEE APEC* 1992, pp. 557-668.

[2] C. H. Chan and M. H. Pong, "Input Current Analysis of Interleaved Boost Converters Operating in Discontinuous-Inductor-Current Mode," in *Proc. IEEE PESC* 1997, pp. 392-398.

[3] L. Huber, B. T. Irving and M. M. Jovanovic, "Open-Loop Control Methods for Interleaved DCM/CCM Boundary Boost PFC Converters," in *IEEE Trans. on Power Electron.*, vol. 23, no. 4, pp. 1649–1657, July 2008.

[4] L. Huber, B. T. Irving, and M. M. Jovanovic, "Closed-Loop Control Methods for Interleaved DCM/CCM Boundary Boost PFC Converters," in *Proc. IEEE APEC* 2009, pp. 991-997.

[5] C. Wang, M. Xu, F. C. Lee and Z. Luo, "Light Load Efficiency Improvement for Multi-Channel PFC," in *Proc. IEEE PESC* 2008, pp. 4080-4085.

[6] S. Choudhury and J. P. Noon, "A DSP Based Digitally Controlled Interleaved PFC Converter," in *Proc. IEEE APEC* 2005, pp. 648-654.

[7] J. Zhang, Y. Zou, Y.Zhang and J. Tang, "DSP Implementation of Digitally Controlled SMPS," in *Proc. IEEE IECON* 2007, pp. 1484-1488.

[8] T. Grote, H. Figge, N. Frohleke, W. Beulen, F. Schafmeister, P. Ide and J. Bocker, "Semi-Digital Interleaved PFC Control with Optimized Light Load Efficiency," in *Proc. IEEE APEC* 2009, pp. 1722-1727.

[9] A. V. Peterchev, J. Xiao and S. R. Senders, "Architecture and IC Implementation of a Digital VRM Controller," in *IEEE Trans. on Power Electron.*, vol. 18, no. 1, pp. 356–364, Jan. 2003.

[10] F. Chen and D. Maksimovic, "Digital Control for Improved Efficiency and Reduced Harmonic Distortion over Wide Load Range in Boost PFC Rectifiers," in *Proc. IEEE APEC* 2009, pp. 760-766.

[11] R. W. Erickson and D. Maksimovic, *Fundamental of Power Electronics*, 2nd Edition, Springer 2000.

978-1-4244-4782-4/10 $26.00 © 2010 IEEE

Reduction of the Output Capacitor in Power Factor Correctors by Distorting the Line Input Current

Diego G. Lamar, Javier Sebastián, Manuel Arias and Arturo Fernández
Universidad de Oviedo. Grupo de Sistemas Electrónicos de Alimentación (SEA)
Edificio Departamental nº 3. Campus Universitario de Viesques. 33204 Gijón. SPAIN
gonzalezdiego@uniovi.es

Abstract- Active Power Factor Correctors (PFCs) are needed to design ac-dc power supplies with universal input voltage range and sinusoidal input current. The classical method to control PFCs consists in two feedback loops and an analog multiplier. Hence, the input current is sinusoidal and it is in phase with the input voltage. However, a bulk capacitor is needed to balance the input power and the output power. Due to its high capacitance, an electrolytic capacitor is traditionally used as a bulk capacitor in PFCs. As a consequence, the lifetime of the ac-dc power supply is limited by the electrolytic capacitor's, which becomes insufficient to some applications (e.g. High-Brightness Light Emitting Diodes, HB-LEDs). This paper proposes a reduction of the output voltage ripple (which allows reduction of the output capacitance) by distorting the input current, but maintaining the harmonic continent compatible with EN 61000-3-2 regulations. Also, a control strategy with a low-cost microcontroller is developed to put the proposed study into practice. Finally, the theoretical results are validated in a 500 W prototype.

I. INTRODUCTION

In order to limit the harmonic content on the line current of mains-connected equipment, the use of an active Power Factor Corrector (PFC) as a first stage of the two-stage solution [1-3] is almost mandatory. Figure 1a shows a general scheme of an active PFC controlled by two feedback loops, which is the most popular circuitry to control this type of power converters. In this figure, the inner feedback loop is an input-current feedback loop and the outer one is an output-voltage feedback loop. The current loop makes the line current follow a reference signal, in phase with the input voltage, which is obtained by multiplying a rectified sinusoidal waveform (obtained from the line voltage) by $v_A(t)$. The output voltage of the voltage loop ($v_A(t)$) is a dc voltage due to the low-pass filter placed in the voltage loop in order to obtain a sinusoidal line input current. Therefore, the pulsating input power is a square cosine function. In this case a storage capacitor with large capacitance is required to balance the instantaneous power difference between the pulsating input power and the constant output power. A large capacitance is needed, and, therefore, an electrolytic capacitor is often used as the storage capacitor. However, it is known that due to its liquid electrolyte, the lifetime of electrolytic capacitor is very limited. At this point, the output capacitor can be an obstacle to design an ac-dc power supply for long-lifetime loads, for example, ac-dc drivers for High-Brightness Light-Emitting Diodes (HB-LEDs) [4-5] or long-lifetime power supplies, for example, non-accessible or remote equipment. Therefore, the power supply manufacturers are looking for reducing this capacitance in

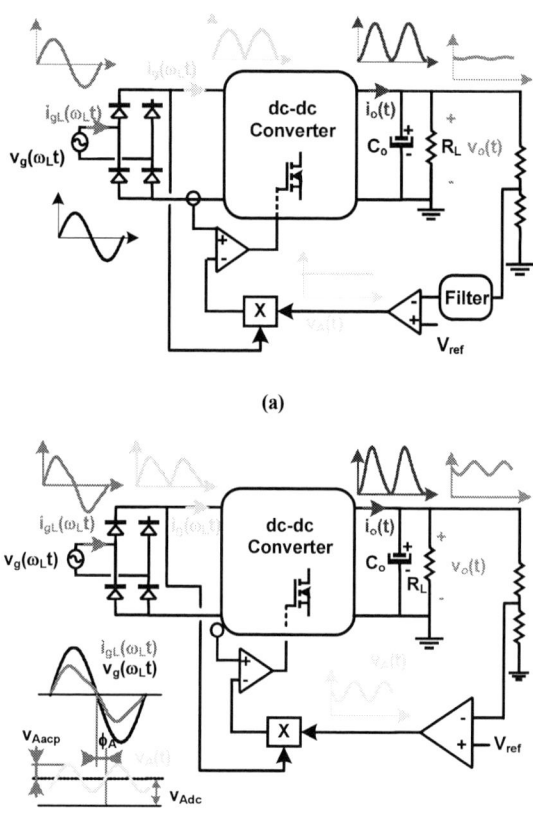

(a)

(b)

Fig. 1: a) PFC with slow output-voltage feedback loop. b) PFC with fast output-voltage feedback loop.

order to avoid the use of an electronic capacitor. According to this idea, nowadays many of them offer lifetime warranty in the range of 5 to 10 years of lifetime.

The objective of this paper is to propose a method to reduce the storage capacitance, thus other capacitor technologies could be adopted instead of electrolytic capacitor to achieve long lifetime [6]. Taking into account the energy transfer process of the PFC defined by its main waveforms (Fig. 1b), if the pulsating output current waveform (which is equal to the pulsating input power) was adequately modified, then the amplitude of the output voltage ripple can be modified. The question is how the pulsating input power can be reduced. A possible solution consists on distorting the line input current. Following this idea, the second section of this paper presents a study that analyzes the tradeoff between the line input current distortion and the output voltage ripple reduction in PFC. However, if the input current is distorted, then the compliance with EN 61000-3-2 regulations is not assured. To solve this problem, the limits of the output ripple reduction compatible with international regulations are presented in the third section of this paper. A simple control strategy based on a low-cost microcontroller that distorts the input current is presented in the fourth section of this paper. This control strategy allows the output voltage ripple reduction distorting the line input current. Finally, all the theoretical results presented in this paper are validated in a 500-W boost PFC prototype.

II. REVIEWING THE STATIC MODELLING OF PFCS WITH FAST OUTPUT-VOLTAGE FEEDBACK LOOP.

The objective of this section is to determinate the evolution of the output voltage ripple when the line input current of the PFC is distorted. The static model of PFC with fast output-voltage feedback loop presented in [9] can be useful for this study because it considers distortion in the line input current. This model describes the static behavior of the PFCs when the bandwidth of the output voltage feedback loop is increased and, therefore, considerable voltage ripple appears in the control signal $v_A(t)$, distorting the input current (see Fig. 1b). In this case, the voltage and the current at the input of the power stage shown in Fig. 1b can be written as:

$$v_g(\omega_L t) = v_{gp}\left|\sin(\omega_L t)\right|, \qquad (1)$$

$$i_g(\omega_L t) = \frac{v_{gp}\left|\sin(\omega_L t)\right| v_A(t)}{K_M}, \qquad (2)$$

where v_{gp} is the peak value of $v_g(\omega_L t)$, ω_L is the line angular frequency, $v_A(t)$ is the output voltage of the error amplifier and K_M is a constant.
The voltage $v_A(t)$ can be re-written as (see Fig.1b):

$$v_A(t) = v_{Adc} + v_{Aac}(t), \qquad (3)$$

$$v_{Aac}(t) = v_{Aacp}\sin(2\omega_L t - \phi_L), \qquad (4)$$

where v_{Adc} is the dc component of $v_A(t)$, $v_{Aac}(t)$ is its ac component, v_{Aacp} is the peak value of $v_{Aac}(t)$ and ϕ_L is its phase lag angle. It should be noted that we have assumed that $v_A(t)$ only has an ac component of twice the line frequency because the rest of possible harmonics are strongly attenuated by the bulk capacitor C_B.

The pulsating input power $p_g(\omega_L t)$ can be obtained by multiplying the values of $v_g(\omega_L t)$ and $i_g(\omega_L t)$ obtained from (1) and (2):

$$p_g(\omega_L t) = v_g(\omega_L t)i_g(\omega_L t), \qquad (5)$$

$$p_g(\omega_L t) = \frac{v_{gp}^2 v_{Adc}}{K_M}\sin^2(\omega_L t)\cdot(1 + k\sin(2\omega_L t - \phi_L)), \qquad (6)$$

where $k = v_{Aacp}/v_{Adc}$ is the relative ripple of v_A. The pulsating output power (the power delivered by the power stage in Fig. 1b) can be obtained by multiplying the output voltage v_o by the

current $i_o(t)$ injected by the power stage into the output cell made up of the bulk capacitor C_B and the load R_L:

$$p_{oi}(\omega_L t) = v_o i_o(\omega_L t). \qquad (7)$$

After establishing the balance between $p_g(\omega_L t)$ and $p_{oi}(\omega_L t)$ we obtain (i.e. : $p_g(\omega_L t)=p_{oi}(\omega_L t)$).

$$i_o(\omega_L t) = \frac{v_{gp}^2 v_{Adc}}{v_o K_M}\sin^2(\omega_L t)\cdot(1 + k\sin(2\omega_L t - \varphi_L)). \qquad (8)$$

The average value of $p_g(\omega_L t)$ in half a line cycle will be:

$$p_{gav} = \frac{\omega_L}{\pi}\int_0^{\frac{\pi}{\omega_L}} p_g(\omega_L t)dt = \frac{v_{gp}^2 v_{Adc}}{4K_M}(2 + k\sin\phi_L). \qquad (9)$$

From (2-4) and (9), the value of the input line current $i_g(\omega_L t)$ will be:

$$i_{gL}(\omega_L t) = \frac{4v_o^2}{v_{gp}R_L(2 + k\sin\phi_L)}\cdot$$
$$\cdot\left[\sin(\omega_L t) + \frac{k}{2}\cos(\omega_L t - \phi_L) + \frac{k}{2}\cos(3\omega_L t - \phi_L)\right] \qquad (10)$$

Taking into account (2-5) and (9) the pulsating input power $p_g(t)$ will be

$$p_g(\omega_L t) = \frac{v_o^2}{R_L} + \frac{2v_o^2}{R_L(2 + k\sin\phi_L)}\cdot$$
$$\cdot\left[k\sin(2\omega_L t - \phi_L) - \cos(2\omega_L t) - \frac{k}{2}\text{sen}(4\omega_L t - \phi_L)\right] \qquad (11)$$

The harmonic content of $i_o(\omega_L t)$ is easily obtained from (7), (11) and the balance between input and output power, by applying basic trigonometric relationships:

$$i_{odc} = \frac{v_o}{R_L}, \qquad (12)$$

$$i_{o2}(\omega_L t) = \frac{2v_o[k\sin(2\omega_L t - \phi_L) - \cos(2\omega_L t)]}{R_L(2 + k\sin\phi_L)}, \qquad (13)$$

$$i_{o4}(\omega_L t) = \frac{-v_o k}{R_L(2 + k\sin\phi_L)}\sin(4\omega_L t - \phi_L). \qquad (14)$$

where i_{odc}, $i_{o2}(\omega_L t)$ and $i_{o4}(\omega_L t)$ are the dc component, the second and the fourth harmonic of $i_o(\omega_L t)$, respectively.
The value of the output voltage ripple can be calculated by multiplying the value of $i_{o2}(\omega_L t)$ and $i_{o4}(\omega_L t)$ by the impedance constituted by C_B and R_L connected in parallel. However, the impedance of C_B at twice and at fourth the line frequency must be much lower than R_L in order to maintain the output voltage ripple in a reasonable value and, hence, the parallel impedance of C_B and R_L can be approximated by the impedance of C_B. Thus, we obtain from (13-14):

$$v_o(t) = v_o + v_{oac}(t) = v_o + \frac{2v_o}{2\omega_L \cdot C_B \cdot R_L(2 + k\sin\phi_L)}\cdot$$
$$\cdot\left[-k\cos(2\omega_L t - \phi_L) - \text{sen}(2\omega_L t) + \frac{k}{2}\cos(4\omega_L t - \phi_L)\right] \qquad (15)$$

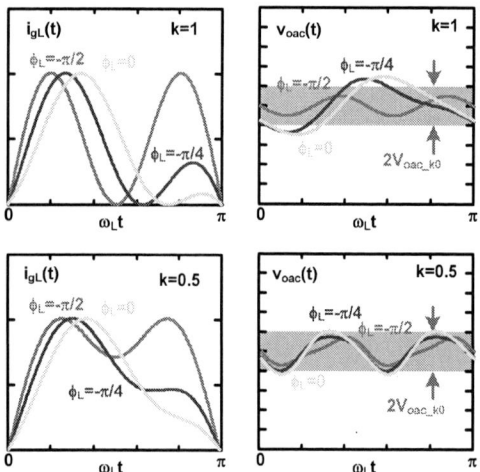

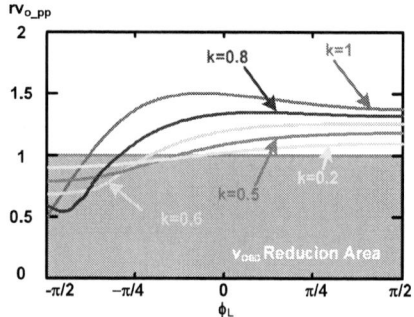

Fig. 2: Line current (left) and output voltage ripple(right) for the same output power and different values of k and ϕ_L.

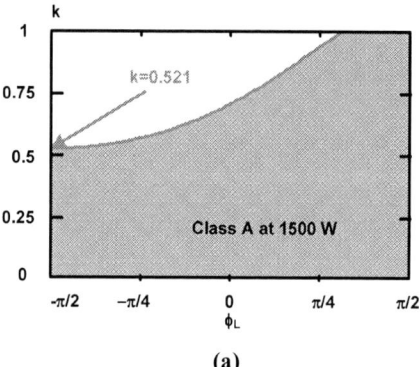

Fig. 3: Normalized peak to peak amplitude of the output voltage ripple in PFCs for different values of k and ϕ_L.

The expression of the output voltage ripple without distorting the input current (i.e. k=0) can be easily calculated:

$$v_{oac\_k0} = \frac{v_o}{2\omega_L \cdot C_B \cdot R_L}. \qquad (16)$$

Figure 2 shows the line current waveforms normalized at its peak value and the output voltage ripple compared to v_{oac_k0} (grey area), in both cases for the same output power and different values of k and ϕ_L. As can be seen, the output voltage ripple can be reduced distorting adequately the input current.

III. LIMITS OF THE OUTPUT VOLTAGE RIPPLE REDUCTION AND COMPLIANCE WITH EN 61000-3-2 REGULATIONS

From (15), the peak to peak amplitude of the output voltage ripple can be calculated as a function of ϕ_L and k. Also, this amplitude can be normalized to the amplitude of the output voltage ripple of a PFC without distortion in the line current (14):

$$rv_{o\_pp} = \frac{\max\{v_{oac}(t)\} - \min\{v_{oac}(t)\}}{2v_{oac\_k0}} \qquad (17)$$

Figure 3 shows rv_{o_pp} for different designs of the PFC with input current distortion. As this figure shows, reduction of the output voltage ripple (grey area) is obtained for negative ϕ_L values. When ϕ_L is close to -90° the highest reduction in the output voltage ripple is obtained for a given k value. However, the proposed study is still not finished. This is because the line-current harmonic content corresponding to the values of k and ϕ_L obtained in Fig. 3 have not been checked yet. In fact, this harmonic content should be low enough to guarantee the compliance with EN 61000-3-2 regulations. As it is very well known, any piece of equipment can be classified into four classes according to EN 61000-3-2 regulations [7-8]:

Class A and Class B.

In these classes the limits of $i_{gL}(t)$ are absolute values. Hence, the PFC will comply with the regulations up to a maximum input power for each class and for each set of values of ϕ_L and k. This maximum power for Class A and Class B can be calculated by using (9):

Class A:
$$p_{g\,max} = \frac{529(2 + k\sin\phi_L)}{k}, \qquad (18)$$

Class B:
$$p_{g\,max} = \frac{793.5(2 + k\sin\phi_L)}{k}. \qquad (19)$$

As an example, Fig. 4a shows the area (in grey color) of compliance with regulations in Class A for a 1500 W PFC. Figure

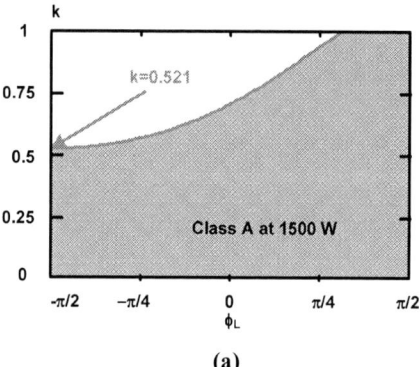

(a)

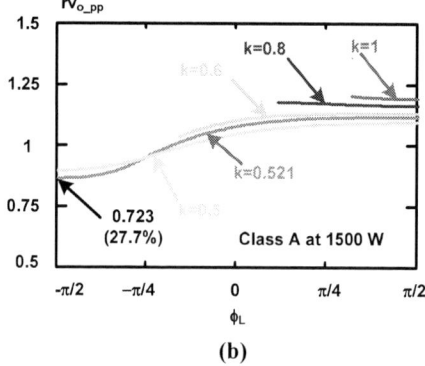

(b)

Fig. 4: a) Combinations of k and ϕ_L for compliance in Class A at 1500 W. b) rv_{o_pp} compatible with regulations in Class A at 1500 W for different k and ϕ_L values.

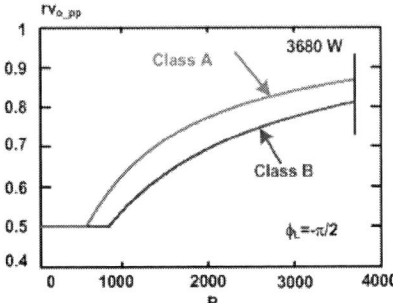

Fig. 5: rv_{o_pp} for $\phi_L=-\pi/2$ compatible with regulations in Class A and Class B as a function of the output power.

4b shows rv_{o_pp} values that comply with regulations for different designs at the same input power. As can be seen, the maximum output voltage ripple reduction compatible with Class A regulations (i.e. 27.7%) is obtained for k=0.525 and $\phi_L=-\pi/2$. In fact, the maximum output voltage ripple reduction at any output power always occurs at $\phi_L=-\pi/2$. Therefore, the maximum output voltage ripple reduction versus the output power can be easily calculated in Class A and Class B (Fig. 5). As this figure shows, a 50% output voltage ripple reduction can be obtained for output power levels above 600 W in Class A and above 800 W in Class B.

Class C

In this class, the limit imposed on the third harmonic depends on the PF and on the rms value of the first harmonic of the line current. The expression of the inequality that defines the compliance with regulations in this class can be calculated by using (10):

$$0.212 \geq \frac{k\sqrt{2+k^2+2k\sin\phi_L}}{(2+k\sin\phi_L)\sqrt{4+k^2+4k\sin\phi_L}}. \quad (20)$$

This inequality defines the area of compliance shown in Fig. 6a. As given in this figure, the PFC always complies with the regulations in for Class C equipment if k is lower than 0.448, whereas it never complies with them if k is higher than 0.82. For values of k between 0.45 and 0.82, the compliance depends on the value of ϕ_L. Figure 6b shows rv_{o_pp} values versus k and ϕ_L for PFC designs that comply with Class C regulations. The values of rv_{o_pp} given in Fig. 6b shows that reductions of 23.8 % can be obtained for PFCs classified in Class C.

Class D

In this class of equipment, the limit imposed on each harmonic by the regulations is proportional to the power handled by the PFC. In other words, the quotient between the rms value of any harmonic divided by the input power must be below the limit specified by the regulations. For the third harmonic, this limit is 3.4 mA/W (rms value). Thus, applying this condition, we obtain the expression of compliance with regulations in Class D:

$$\frac{\sqrt{2}}{v_{gp}} \cdot \frac{k}{(2+k\sin\phi_L)} \leq 3.4 \cdot 10^{-3}. \quad (21)$$

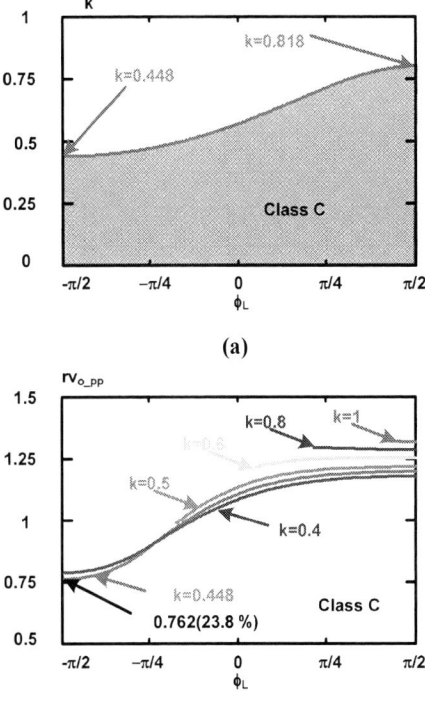

Fig. 6. a) Area of compliance in Class C. b) rv_{o_pp} compatible with regulations in Class C for different values of k and ϕ_L

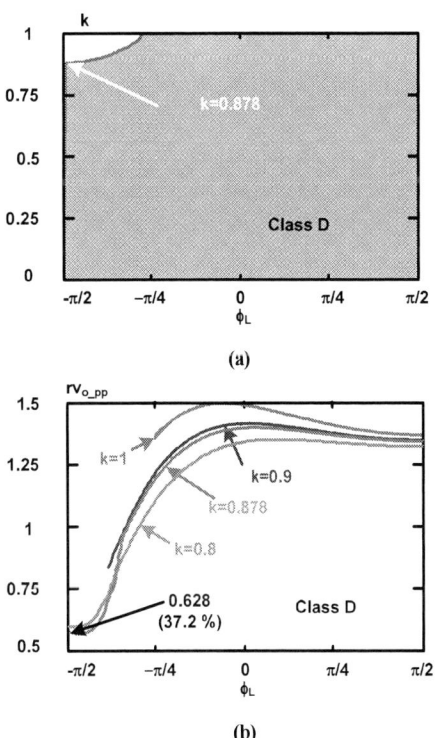

Fig. 7. a) Area of compliance in Class D. b) rv_{o_pp} compatible with regulations in Class D for different values of k and ϕ_L.

This inequality defines the area of compliance plotted in Fig. 7a. As this figure shows, the relative value of the third harmonic is below the limit imposed by EN 61000-3-2 for almost any design condition. In fact, the converter fails to comply with the regulations only if ϕ_L is between -90° and -45° and, at the same time, k is higher than 0.878. Also, Fig. 7b shows the rv_{o_pp} values complying with international regulations in Class D versus k and ϕ_L PFC designs. Maximum reductions of 37.8 % can be obtained for PFCs classified in this class.

IV. CONTROL STRATEGY BASED ON A DISTORTED SINUSOIDAL FIXED REFERENCE GENERATED BY A LOW-COST MICROCONTROLLER.

In order to obtain the highest output voltage ripple reduction the value of ϕ_L must be -90° and the k value varies with the classification of the piece of equipment (between 0.448 and 1, Fig. 4b, Fig. 6b and Fig. 7b). The ϕ_L value (which is the phase

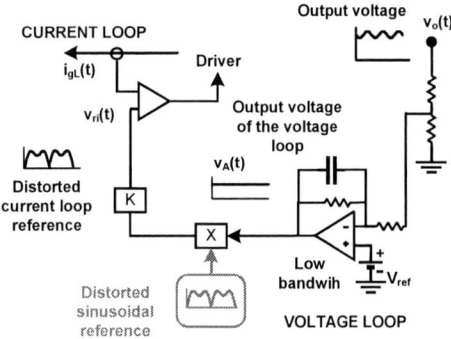

Fig. 8: A new control strategy based on a distorted sinusoidal reference in PFC.

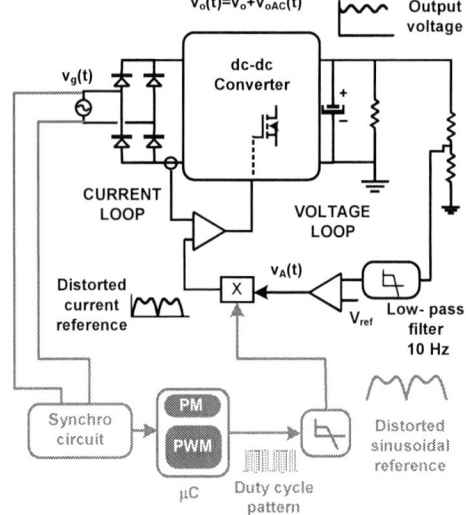

Fig. 9: Control strategy based on a distorted sinusoidal fixed reference generated by a low-cost microcontroller.

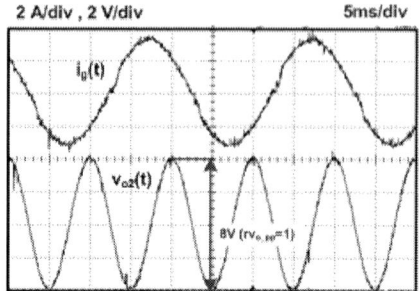

Fig. 10: Input current and the output voltage ripple for a traditional PFC design.

lag of $v_A(t)$) depends on the bandwidth of the output voltage regulator [9]. However, the standard regulator design of a PFC with fast output voltage feedback loop does not allow a ϕ_L value of -90° in $v_A(t)$ (i.e. lead angle of +90°). Therefore, a new control strategy must be adopted to distort adequately the line current in order to obtain the maximum output voltage ripple reduction.

If a distorted sinusoidal reference is introduced as a fixed pattern in the PFC multiplier (instead of the traditional sinusoidal reference sensed from the rectifier input voltage, Fig. 8), then the input current will follow the distorted sinusoidal reference. This distorted reference must be calculated in order to obtain the desired distortion in the input current. Therefore, the maximum possible output voltage reduction can be obtained.

This pattern can be generated by a Pulse Width Modulation (PWM) module of a low cost microcontroller (µC), as can be seen in Fig.9. The microcontroller reference is smaller than the traditional reference took from the rectifier input voltage. Nevertheless, these multiplier control chips (i.e. UC3854B) sense the $|v_g(t)|$ reference as a current reference. Therefore, if the value of current limit resistance [10] used in the traditional control is substituted by the adequate value, then the chip levels are maintained.

On the other hand, the fixed and $2f_L$ repetitive pattern has to be synchronized with the line. The distorted sinusoidal reference and the rectifier input voltage have to be "in phase". Then a synchronization circuit has to be implemented. Normally, this kind of circuits detects significant points of the input voltage. In this case, zero voltage points of $|v_g(t)|$ have been detected. When these points are detected, the microcontroller launches the fixed reference. Therefore, the reference signal is synchronized every 10 ms (European line).

Regarding the algorithm, it is based on a look-up table with normalized values of the duty cycle. With this system, the program goes through the table from top to bottom when the synchronized point is detected. Apart from the PWM generation, the microcontroller can also perform another systems: protection system, supervision system, soft start of the converter, etc..

VI. EXPERIMENTAL RESULTS.

A 500-W PFC prototype, based on a boost topology, was developed in order to verify the proposed study. The main specifications of the PFC boost converter are the following: 85-265 V input voltage, 400 V output voltage; output capacitor is of 500 μF (to obtain a 1% of output voltage ripple in traditional design, k=0) and 10- kHz switching frequency. In this case a low cost microcontroller (PIC 16F627) working at 20 MHz has been used. Its cost is about 1 €.

First, the PFC experimental prototype was built with a traditional design. As can be seen in Fig. 10, the input current is sinusoidal and the output voltage ripple is 1% in amplitude (8 V peak to peak voltage). Therefore, the experimental output voltage ripple complies with aforementioned specifications.

Then, two PFC designs were implemented to maximize the output voltage ripple reduction in Class C and Class A. Figure 11 and Fig. 12 shows the experimental results of the input current, the output voltage ripple and the distorted sinusoidal reference synchronized with the input voltage. As can be deduced from Fig 11a, a 21.2 % of output voltage ripple reduction is obtain in Class C (from 8 V to 6.3 V). Also, in Fig.12a a 47.5 % of output voltage ripple reduction is obtained in Class A (from 8V to 4.2 V). These reduction results match with the study presented in this paper in Class C (Fig. 6a) and Class D (Fig. 5). The experimental waveforms of the line current and the output voltage match with theoretical ones too. Table 1 shows the experimental harmonic content of these designs versus the theoretical one.

Finally, Fig. 13 shows the harmonic content of both designs versus the limits of Class C (Fig. 13a) and Class A (Fig. 13b). As can be seen, the input current is distorted into the limits imposed by EN 61000-3-2 regulations. Therefore, the distorted sinusoidal reference used in both designs generates the desired input current in the edge of the regulation limits.

VI. CONCLUSIONS.

The study of the output voltage ripple reduction distorting the input current in PFCs has been carried out in this paper. Also the limits of maximum reduction of the output voltage ripple compatible with EN 61000-3-2 regulation have been defined. In order to apply this study, a new control strategy for PFCs has been presented: if a properly distorted sinusoidal reference (based on the static analysis presented) is employed as a fixed pattern instead of the traditional reference, then the input current will be distorted allowing the maximum output voltage ripple reduction. The fixed pattern can be generated in a low cost microcontroller. A 500W prototype has been built and tested, and the experimental results are presented to verify the validity of the proposed method.

TABLE I

EXPERIMENTAL HARMONIC CONTENT VS. THEORETICAL HARMONIC CONTENT

	Class C		Class A	
	Experimental results	Theoretical results	Experimental results	Theoretical results
1st. harmonic (A_{rms})	2.23	2.261	2.31	2.261
3rd. harmonic (A_{rms})	0.637	0.651	2.295	2.261

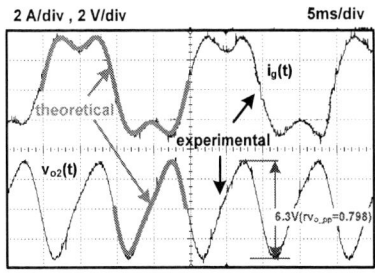

(a)

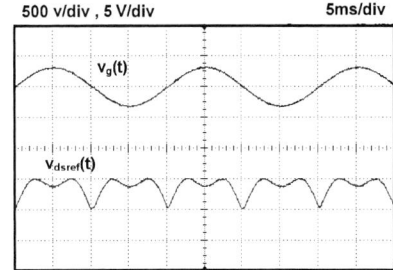

(b)

Fig. 11 a) Input current and output voltage ripple for a Class C design. b) Distorted sinusoidal reference synchronized with the input voltage.

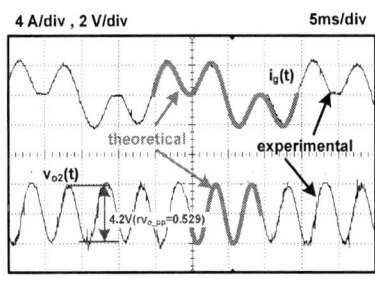

(a)

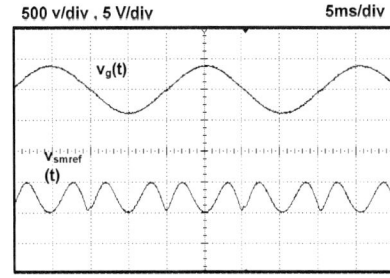

(b)

Fig. 12 a) Input current and the output voltage ripple for a Class A design. b) Distorted sinusoidal reference synchronized with the input voltage.

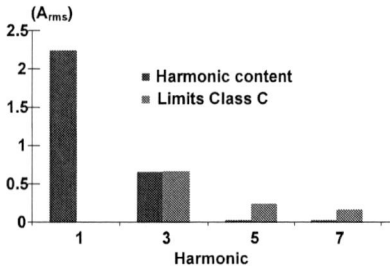

(a)

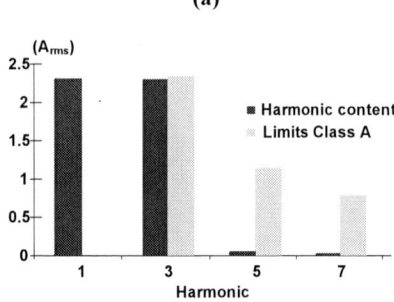

(b)

Fig. 13 Experimental harmonic content of the PFC vs EN 61000-3-2 regulations in a) Class C. b) Class A..

In summary, this paper proposes a method to reduce the storage capacitance in PFCs. The results show that reductions of 50% in Class A and B, 23% in Class C and 38 % in Class D can be achieved.

ACKNOWLEDGMENT

This work has been supported by the Spanish Ministry of Education and Science under Project TEC2007-66917/MIC and the Grant AP2008-03380.

REFERENCES

[1] M. J. Kocher and R. L. Steigerwald, "An ac-to-dc converter with high quality input waveforms", IEEE Trans. Ind. Appl., vol. 19, no. 4 1983, pp. 586-599.

[2] L. H. Dixon, "High power factor preregulators for off-line power supplies", Unitrode Power Supply Design seminar, 1990, pp I2-1 to I2-16.A.

[3] Garcia, O.; Cobos, J.A.; Prieto, R.; Alou, P.; Uceda, J., "Single phase power factor correction: a survey," Power Electronics, IEEE Transactions on , vol.18, no.3, pp. 749-755, May 2003.

[4] Chen, C.-C.; Wu, C.-Y.; Chen, Y.-M. and Wu, T.-F "Sequential Color LED Backlight Driving System for LCD Panels", IEEE Trans. on Power Electron., Vol. 22, no. 3, May 2007, pp. 919-925

[5] Kening Zhou; Jian Guo Zhang; Yuvarajan, S. and Da Feng Weng, T.-F "Quasi-Active Power Factor Correction Circuit for HB LED Driver", IEEE Trans. on Power Electron., Vol. 23, no. 3, May 2008, pp. 1410-1415.

[6] Linlin Gu; Xinbo Ruan; Ming Xu; Kai Yao, "Means of Eliminating Electrolytic Capacitor in AC/DC Power Supplies for LED Lightings," IEEE Trans. Power Electron., Vol.24, no.5, pp.1399-1408, May 2009.

[7] Electromagnetic compatibility (EMC)-part 3: Limits-section 2: Limits for harmonic current emissions (equipment input current<16A per phase), IEC1000-3-2 Document, 1995.

[8] Draft of the proposed CLC Common Modification to IEC 61000-3-2 Ed. 2.0:

[9] J. Sebastián, D.G. Lamar, M.M. Hernando, A. Rodríguez, A. Fernández "Steady-state analysis of power factor correctors with a fast output-voltage feedback loop" IEEE APEC, Washington (USA), February 2009, pp 970-976. 3

[10] P. C. Todd, "UC3854 controlled power factor correction circuit design," Unitrode Product & Applications Handbook 1995-1996, Unitrode Corporation, Merrimack, NH, pp. 10;303–10;322

Universal-Input Single-Stage PFC Flyback with Variable Boost Inductance for High-Brightness LED Applications

Yuequan Hu, Laszlo Huber, and Milan M. Jovanović

Delta Power Electronics Laboratory
Delta Products Corporation
P.O. Box 12173, 5101 Davis Drive
Research Triangle Park, NC 27709, USA
Email: yhu@deltartp.com

Abstract — **This paper presents a single-stage flyback power-factor-correction circuit with a variable boost inductance for high-brightness light-emitting-diode applications for the universal input voltage (90-270 Vrms). The proposed circuit overcomes the limitations of the conventional single-stage PFC flyback with a constant boost inductance, which cannot be designed to achieve a practical bulk-capacitor voltage level (i.e., less than 450 V) at high line while meeting the IEC 61000-3-2 Class C line current harmonic limits at low line. According to the proposed variable boost inductance method, the boost inductance is constant in the high-voltage range and it is reduced in the low-voltage range, resulting in discontinuous-conduction-mode operation and a low total harmonic distortion (THD) in both the high-voltage and low-voltage ranges. Measurements obtained on a 24-V/91-W experimental prototype are as follows: PF = 0.9873, THD = 12%, and efficiency = 88% at nominal low line (120 V$_{rms}$); and PF = 0.9474, THD = 10.39%, and efficiency = 91% at nominal high line (230 V$_{rms}$). The line current harmonics satisfy the IEC 61000-3-2 Class C limits with enough margin.**

I. INTRODUCTION

The technology and performance of high-brightness light-emitting diodes (HB LEDs) has undergone significant improvements driven by new applications in liquid-crystal-display (LCD) backlighting, automobiles, traffic lights, and general-purpose lighting [1]-[3]. As a solid state light source which does not contain mercury, HB LEDs have been widely accepted because of their superior longevity, low-maintenance requirements, and continuously-improving luminance with a great potential to replace existing lighting sources such as incandescent and fluorescent lamps in the future.

For LED drivers with an output power over 25 W in general lighting applications, the line current harmonics have to satisfy the limits set by IEC 61000-3-2 Class C regulations [4]. With passive power-factor-correction (PFC), which uses only inductors and capacitors, it is difficult to meet such requirements, and the size of the components is large.

An LED driver with active PFC, which is implemented with two stages, is shown in Fig. 1. The first stage can

achieve a near unity power factor and a low THD at the universal input voltage, while the second stage is used for the dc/dc conversion. However, the circuit in Fig. 1 requires two independently controlled power switches and two control circuits, leading to a high component count, increased cost, and larger size. In low-power lighting applications, where cost is the dominant issue, such an approach loses appeal.

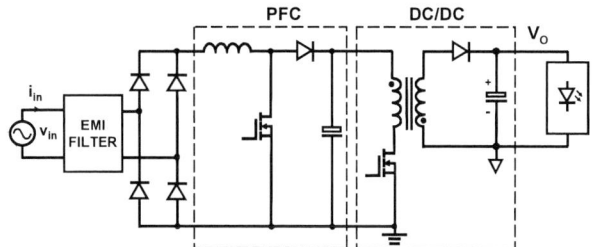

Fig. 1. Conventional two-stage LED driver.

Another active PFC implementation employs a single-stage ac/dc converter [5]-[13], where the PFC stage is integrated with the dc/dc stage, resulting in a reduced complexity and cost. The single-stage PFC ac/dc converter can be implemented without and with a bulk capacitor at the primary side, as illustrated in Figs. 2 and 3, respectively. Although the single-stage PFC circuit in Fig. 2 [5] has the advantage of a low component count, its output voltage has a high ripple at twice the line frequency unless very large output capacitors are used. For an LED load, a small variation in the driving voltage can lead to a large variation in the LED current. A large ripple of the LED current would seriously affect the longevity of the LEDs [12]. Therefore, the approach in Fig. 2 often requires a post-regulator, which adds cost and lowers the efficiency.

The single-stage PFC flyback topology shown in Fig. 3 [11] presents one of the most cost-effective single-stage solutions. In this converter, the PFC stage operates in discontinuous conduction mode (DCM), while the dc/dc stage operates at the DCM/CCM (continuous-conduction-mode) boundary. A low input-current harmonic distortion

978-1-4244-4782-4/10 $26.00 © 2010 IEEE

can be achieved due to the inherent property of the DCM boost converter to draw a near sinusoidal current if its duty cycle is held relatively constant during a half line cycle. However, voltage V_B across bulk capacitor C_B is not regulated and at high line it can increase to impractical levels. To reduce the bulk capacitor voltage, one terminal of the boost inductor winding is connected to a tapping point of the primary winding of the flyback transformer, which provides a negative magnetic feedback [7]. However, the tapping of the flyback primary winding also results in a zero-crossing distortion of the line current. In fact, as long as the instantaneous line voltage is lower than the voltage at the tapping point, no current is drawn from the input, which reduces the power factor and increases the line-current harmonics.

The single-stage PFC flyback topology shown in Fig. 3 has been successfully applied in adapter/charger applications for the universal line voltage, where the line current harmonics need to meet the IEC 61000-3-2 Class D limits, which are less stringent than the IEC 61000-3-2 Class C limits.

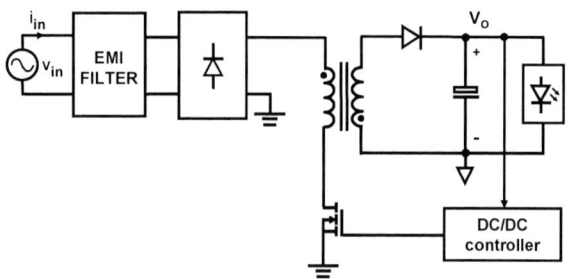

Fig. 2. Single-stage flyback LED driver without energy-storage capacitor at primary side.

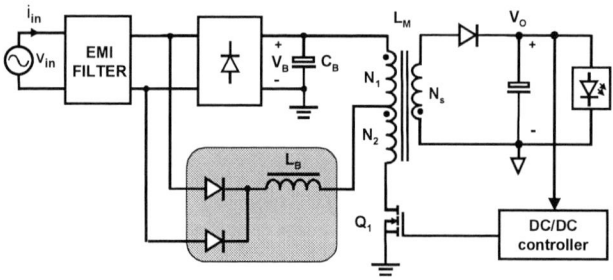

Fig. 3. Single-stage flyback LED driver with energy-storage capacitor at primary side.

It was shown in [14] that the single-stage PFC flyback in Fig. 3 with a constant boost inductance cannot be designed to achieve a practical bulk-capacitor voltage level (i.e., less than 450 V) at high line while meeting the IEC 61000-3-2 Class C line-current harmonic limits at low line. To overcome these limitations, a variable boost inductance is required, i.e., a high boost inductance at high line to limit the bulk-capacitor voltage and a lower boost inductance at low line to ensure DCM operation and a low THD. In fact, at low line, when a constant boost inductance is used, the inductor will enter CCM operation around the peak of the line voltage, and the

line current waveform will have a bulge around its peak value [14], resulting in an increased THD. Furthermore, if the bulk-capacitor voltage is slightly lower than the peak of the rectified line voltage, the peak charging of the bulk capacitor through the bridge rectifier will also result in a bulge in the line current waveform [7] with an increased THD.

In this paper, it is shown that by optimizing the tapping point of the primary-winding of the flyback transformer in Fig. 3 and by employing a novel technique to reduce the boost inductance at low line, a high power factor and a low THD with relatively high efficiency can be achieved such that the line current harmonics satisfy the IEC 61000-3-2 Class C limits.

II. VARIABLE BOOST INDUCTANCE

A. Concept

As voltage V_B across bulk-capacitor C_B in Fig. 3 is not regulated and varies with both the input voltage and output power, the design of the magnetic components significantly affects the bulk-capacitor voltage level. Generally, a higher boost inductance L_B leads to a lower voltage V_B. If the boost inductance increases during steady-state operation, the input power initially decreases because of a lower input current. The difference between the output power and input power has to be supplied from the bulk capacitor, causing a drop of the bulk-capacitor voltage. Meanwhile, as the bulk-capacitor voltage decreases, the duty cycle of main switch Q_1 increases to keep the output voltage regulated, resulting in an increase of the input power until a new balance between the input and output power is reached. A higher boost inductance can limit voltage V_B to an acceptable level and ensure DCM operation at high line. However, at low line, if the boost inductance is larger than the maximum value for DCM operation, the boost inductor will operate in CCM around the peak of the rectified line voltage, resulting in a severe distortion of the line current [14]. Therefore, a variable boost inductance is required, i.e., a high boost inductance at high line and a lower boost inductance at low line.

Various methods for achieving a variable inductance were reported in [15]-[23], but their applications are limited either because of a high power loss [15]-[20], or complexity of the implementation [21]-[22], or because they can only meet IEC 61000-3-2 Class D current harmonic limits [23].

The concept of the variable boost inductance proposed in this paper is shown in Fig. 4. The basic PFC boost inductor is implemented with an EE core and winding N_{LB}. A half core (E) with winding N_{BIAS} is closely attached to the bottom part of the EE core. The boost inductance L_B is controlled by a bias current I_{BIAS}. The control circuit includes switch Q_{BIAS}, a dc bias control circuit, and an input voltage sensing circuit. At low line, switch Q_{BIAS} is open by sensing the input voltage, and the bias current flows through bias winding N_{BIAS}, inducing a magnetic flux Φ_{BIAS} that is added to the main magnetic flux Φ_{LB} at the bottom part of the boost-inductor EE core. As a result, at the bottom part of the EE core, the effective permeability is reduced, and consequently, the boost inductance is decreased [17]. The reduction of the boost inductance is proportional to the applied bias current. At high line, switch Q_{BIAS} is closed, shorting bias winding N_{BIAS}. As a result, there is no bias current through winding N_{BIAS}, and the boost inductance does not change.

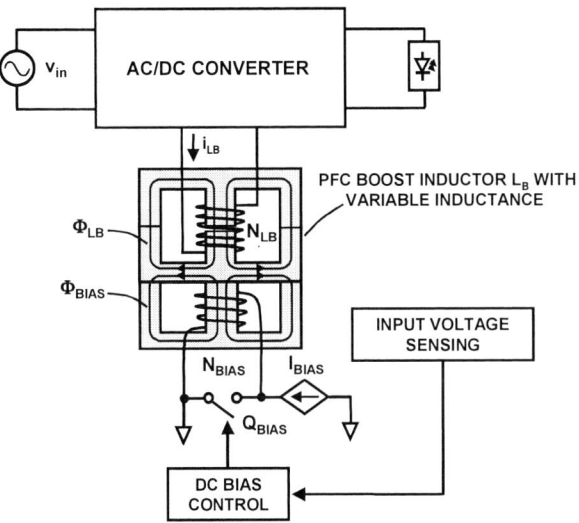

Fig. 4. Concept of proposed single-stage PFC with a variable boost inductance.

B. Control circuit

Figure 5 shows the detailed schematic of the proposed single-stage PFC flyback HB LED driver. The input voltage is sensed by a circuit comprising winding N_3 wound around the boost-inductor EE core, diode D_8, and capacitor C_1. The load current is used as the dc bias current to reduce complexity and loss of efficiency. Switch Q_2 connected in series with the LED load is the bias switch Q_{BIAS} in Fig. 4.

When main switch Q_1 is turned on, diode D_8 is forward biased, peak charging capacitor C_1 with a maximum voltage

$$V_{CIMAX} = (\sqrt{2}V_{IN} - \frac{N_1}{N_P}V_B)\frac{N_3}{N_{LB}}, \qquad (1)$$

where N_1, N_P and N_{LB} are the number of turns of the feedback winding, primary winding of the flyback transformer, and the boost-inductor winding, respectively. A proper turns number N_3 is chosen so that the voltage across capacitor C_1 turns on Zener diode ZD_1 only at high line (180-270 Vrms). When ZD_1 is turned on, switch Q_4 is turned on and switch Q_3 is turned off. As a result, the gate-to-source voltage of MOSFET Q_2 is high and Q_2 is turned on. The load current flows through switch Q_2 and the bias current of bias winding N_{BIAS} is approximately zero. Therefore, the boost inductance remains unchanged. It should be noted that the turn-on resistance of switch Q_2 needs to be negligible compared to the resistance of bias winding N_{BIAS} to prevent a substantial current flowing through the bias winding at high line. Otherwise, the effective boost inductance would become lower and voltage V_B would increase to an undesirable level. At low line, the voltage across capacitor C_1 is lower than the turn-on voltage of ZD_1, Q_4 is turned off and Q_3 is turned on. As a result, the gate-to-source voltage of MOSFET Q_2 is low and Q_2 is turned off. The entire load current flows through the bias winding. Therefore, the boost inductance is reduced.

C. Design considerations

The design of the flyback circuit in Fig. 5 without the PFC part is the same as the design of the conventional flyback circuit. Key design parameters of the PFC part of the flyback circuit in Fig. 5 are number of turns N_1 and boost inductance L_B. The design goal is to achieve a proper PFC operation, i.e., the line current to meet the IEC 61000-3-2 Class C limits, and to limit bulk-capacitor voltage V_B below 400 V.

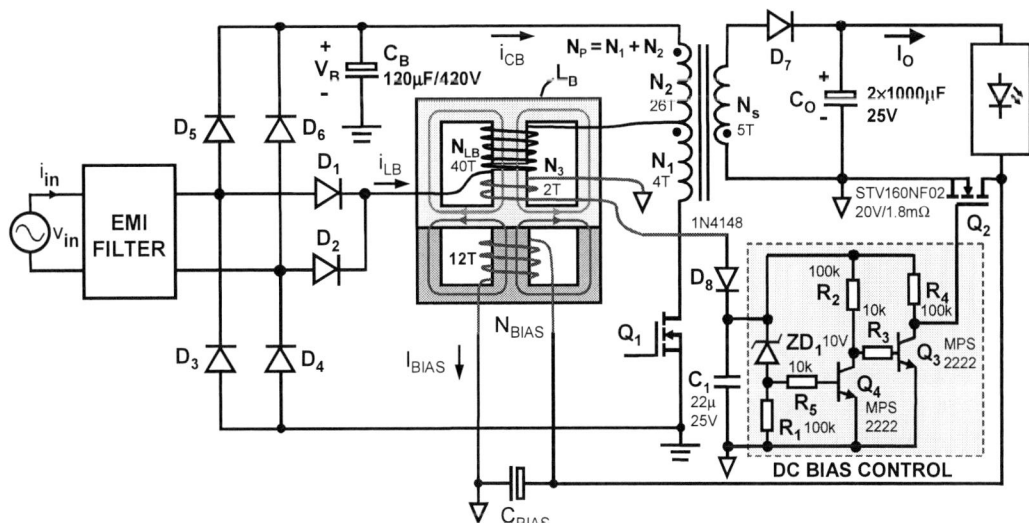

Fig. 5. Schematic of proposed single-stage PFC flyback with a variable boost inductance for HB LED applications.

It was shown in [14] that bulk-capacitor voltage V_B is a function of input voltage v_{IN}, ratio of inductances L_B/L_M, ratio of number of turns N_1/N_P, and output voltage V_o, i.e.,

$$V_B = f(v_{IN}, L_B/L_M, N_1/N_P, V_O). \qquad (2)$$

The bulk-capacitor voltage increases with increasing rms value of the line voltage, and it decreases with increasing turns ratio N_1/N_P and increasing ratio of inductances L_B/L_M, as illustrated in Figs. 6 and 7, respectively.

For proper PFC operation, in order to meet the IEC 61000-3-2 Class C limits, the zero-crossing distortion of the line current due to the tapping of the primary winding should be optimized, and the boost inductor should be prevented from operating in CCM.

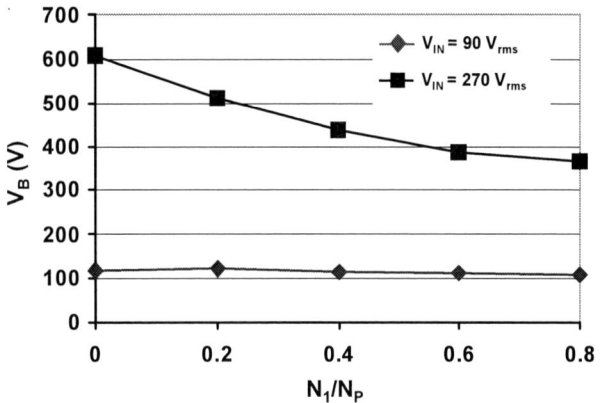

Fig. 6. Calculated voltage V_B vs. ratio N_1/N_P (L_B/L_M=0.25).

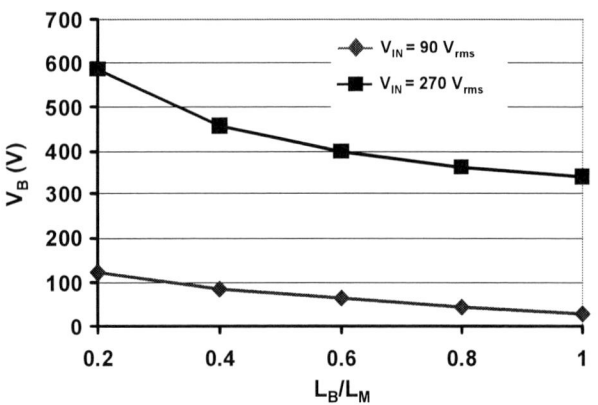

Fig. 7. Calculated voltage V_B vs. ratio L_B/L_M (N_1/N_P=0.133).

Following the same procedure for the analysis of the bulk-capacitor voltage as in [14] and assuming that the boost inductor operates in DCM, the line current waveform and the corresponding THD can be easily calculated. For example, calculated line current waveforms at three different turns ratios N_1/N_P, at nominal low line (V_{IN} = 120 Vrms), are shown in Fig. 8. It follows from Fig. 8 that with increasing turns ratio N_1/N_P the dead-angle of the line current around zero crossing increases, resulting in an increased THD. As another example, Fig. 9 shows calculated THD vs. N_1/N_P for three different values of L_B, at nominal high line (V_{IN} = 230

Vrms). It follows from Fig. 9 that THD significantly increases with increasing N_1/N_P, whereas the presented variation of L_B does not have significant effect on THD. It can be seen in Fig. 9 that for a THD lower than 20%, which is a typical requirement for lighting applications, turns ratio N_1/N_P should be smaller than 0.15.

In order to ensure DCM operation of boost inductor L_B, time T_{RES_LB}, i.e., the time to completely reset the boost-inductor core, should be shorter than the turn-off time, T_{OFF_Q1}, of switch Q_1 around the peak of the line voltage (worst case), i.e.,

$$T_{RES\_LB} \leq T_{OFF\_Q1}. \qquad (3)$$

In this way, the current flowing through the boost inductor decreases to zero before switch Q_1 is turned on, ensuring a high power factor and a low THD.

The turn-off time, T_{OFF_Q1}, of switch Q_1 operating at CCM/DCM boundary can be expressed as

$$T_{OFF\_Q1} = (1 - \frac{V_O N_P/N_S}{V_B + V_O N_P/N_S})T_S, \qquad (4)$$

where, T_S is the switching period of switch Q_1.

The time to completely reset the boost-inductor core can be expressed as

$$T_{RES\_LB} = \frac{(v_{IN}^{rec} - N_1 V_B/N_P)N_P V_O/N_S}{[V_B + (1 - N_1/N_P)N_P V_O/N_S - v_{IN}^{rec}](V_B + N_P V_O/N_S)}T_S,$$

$$(5)$$

where, v_{IN}^{rec} is the instantaneous rectified line voltage.

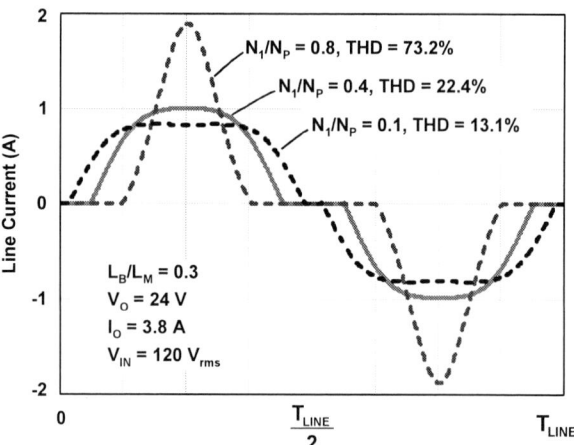

Fig. 8. Calculated line current waveforms for different ratios of N_1/N_P.

Using (2)-(5), the maximum ratio L_B/L_M vs. N_1/N_P that will ensure operation of L_B in DCM can be calculated. In Fig. 10, the calculated maximum ratio L_B/L_M vs. N_1/N_P that will ensure operation of L_B in DCM is presented at nominal low line (120 Vrms). Figure 10 also includes the calculated minimum L_B/L_M vs. N_1/N_P that will ensure limiting the bulk-capacitor voltage below 400 V at the upper end of the low line-voltage range (140 Vrms).

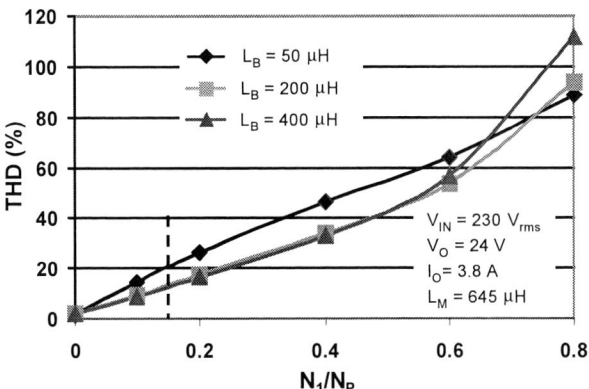

Fig. 9. Calculated THD vs. ratio N_1/N_P at V_{IN} = 230 Vrms.

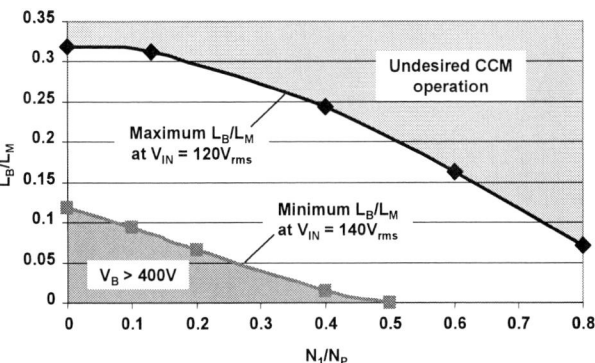

Fig. 10. Desired range for ratio L_B/L_M vs. ratio N_1/N_P in the low line-voltage range.

The calculated maximum and minimum ratios L_B/L_M vs. N_1/N_P in the high line-voltage range are presented in Fig. 11. As follows from Fig. 11, in the high line-voltage range, the possible range of ratio L_B/L_M is much narrower than at the low line-voltage range. It also follows from Fig. 11 that the minimum possible turns ratio N_1/N_P is 0.1.

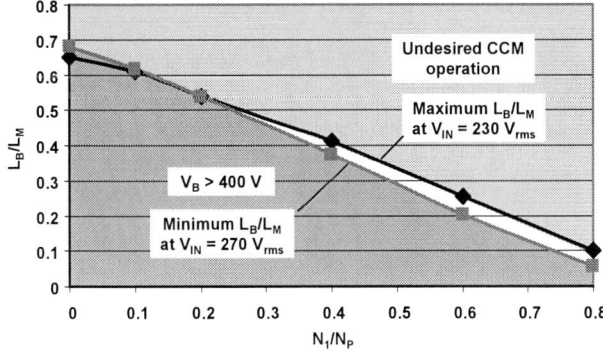

Fig. 11. Desired range for ratio L_B/L_M vs. ratio N_1/N_P in the high line-voltage range.

It can be clearly seen from Figs. 10 and 11 that different ratios of inductances L_B/L_M are required in the low line-voltage range and in the high line-voltage range, i.e., that a variable boost inductance is required in the universal line-voltage range.

Based on the calculated results presented in Figs. 6-9, turns ratio N_1/N_P = 0.13 (N_1 = 4, N_P = 30) is selected for the final design. It follows from Fig. 11 that for N_1/N_P = 0.13 in the high line-voltage range, ratio L_B/L_M should be around 0.6, i.e., the desired boost inductance is 390 uH for the selected L_M = 645 uH.

Finally, it follows from Fig. 10 that for N_1/N_P = 0.13 in the low line-voltage range, ratio L_B/L_M should be smaller than 0.32, i.e., the desired boost inductance should be smaller than 206 uH for the selected L_M = 645 uH.

III. EXPERIMENTAL RESULTS

To verify the proposed variable boost-inductance technique, a 24-V/91-W single-stage PFC flyback prototype for HB LED applications was built. Figure 12 shows a photograph of the variable boost inductor with a dc bias winding, while Fig. 13 shows the measured boost inductance vs. dc bias current. As shown in Fig. 13, the effective boost inductance drops faster with increasing dc bias current when the turns number N_{BIAS} of the bias winding is higher, and therefore, it requires a lower dc bias current. However, a higher turns number N_{BIAS} leads to a higher resistance, hence, higher winding loss for the same load current. Moreover, a lower boost inductance generally results in a lower overall efficiency of the LED driver. Therefore, turns number N_{BIAS} should be minimized to ensure low-enough boost inductance and DCM operation at low line while maintaining relatively high efficiency. A turns number N_{BIAS} of 12 was selected for the final design. The dc resistance of winding N_{BIAS} is 20 mΩ while the turn-on resistance of bias control switch Q_2 is 1.8 mΩ (<<20 mΩ). At low line, switch Q_2 is turned off, and the entire load current I_O = 3.8 A flows through the bias winding resulting in a power loss of 0.29 W, i.e., a 0.3% decrease of efficiency. At high line, bias switch Q_2 is turned on and bias winding N_{BIAS} is essentially shorted since its dc resistance is much higher than the turn-on resistance of switch Q_2, resulting in a power loss of 22 mW at I_O = 3.8 A.

The measured line voltage and line current waveforms at full load are shown in Fig. 14. At nominal high line (230 Vrms), THD = 10.39%, PF = 0.9474, V_B = 327 V, and efficiency = 91%; while at nominal low line (120 Vrms), THD = 12.24%, PF = 0.9873, V_B = 193 V, and efficiency = 88% were obtained. Figure 15 shows that the measured line-current harmonics are below the IEC 61000-3-2 Class C limits with enough margin. The measured efficiency vs. output power is shown in Fig. 16.

Measurements with an actual LED load were also performed. Four LED strings each with 7 series-connected white LEDs (Philips Lumileds, LXHL-LW3C) were paralleled and directly driven by the proposed PFC flyback prototype with an output voltage of 24 V. The measured LED current and output voltage ripple are shown in Fig. 17. The peak-to-peak output voltage ripple is 30 mV, resulting in a low LED current ripple (1.6% of the average current). Therefore, the proposed PFC flyback circuit is suitable for directly driving LED strings, and no post-regulators are necessary, which is a significant advantage over the conventional PFC flyback circuit without an energy-storage capacitor at the primary side.

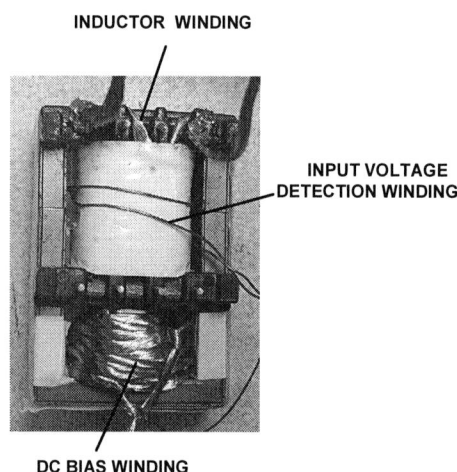

Fig. 12. Photograph of the variable boost inductor.

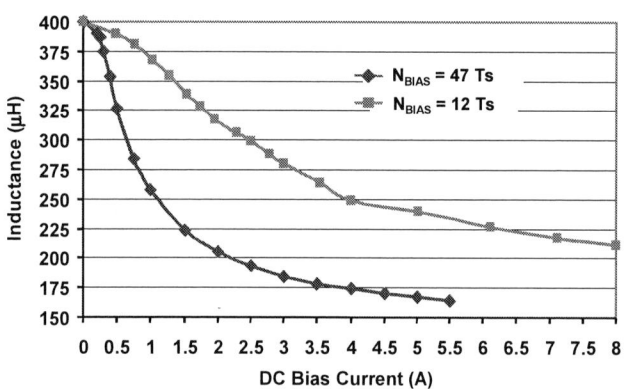

Fig. 13. Measured boost inductance vs. dc bias current.

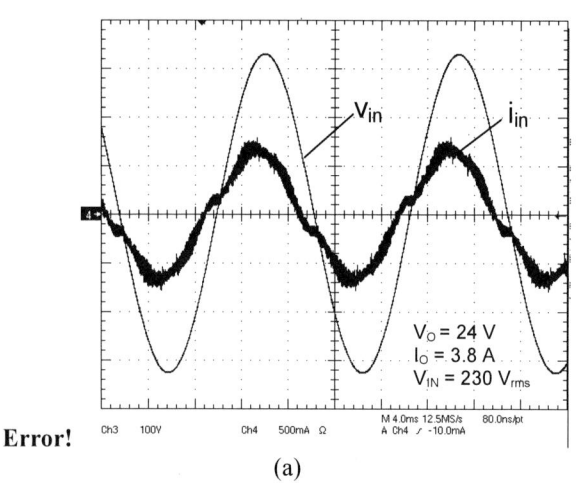

Error!

(a)

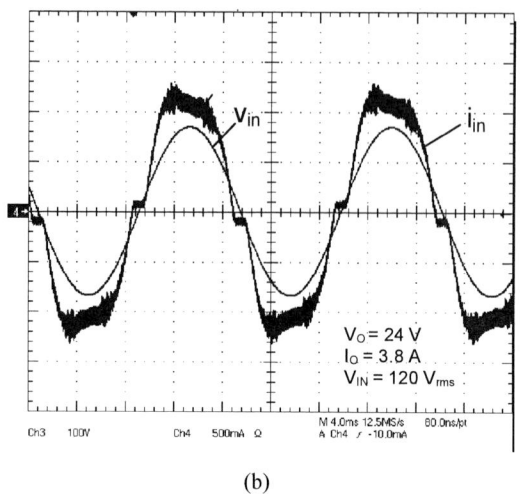

(b)

Fig. 14. Measured line current (500 mA/div.) and line voltage (100 V/div.) waveforms, (a) V_{IN} = 230 Vrms; (b) V_{IN} = 120 Vrms.

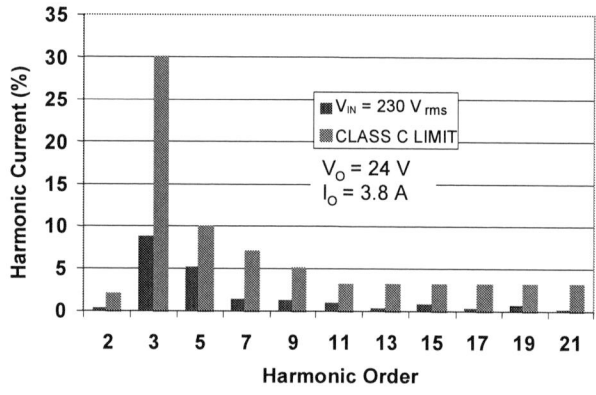

(a)

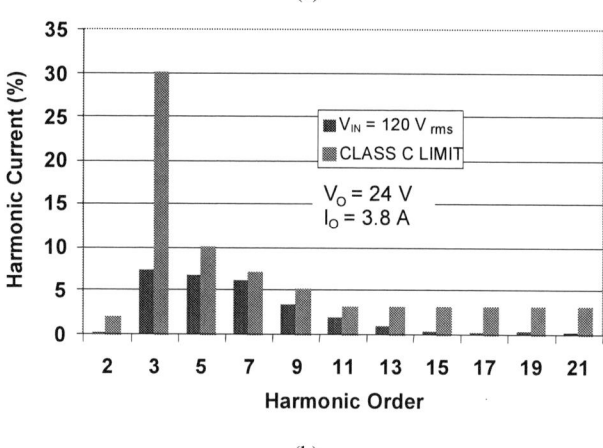

(b)

Fig. 15. Measured line current harmonics, (a) V_{IN} = 230 Vrms; (b) V_{IN} = 120 Vrms.

978-1-4244-4782-4/10 $26.00 © 2010 IEEE   208

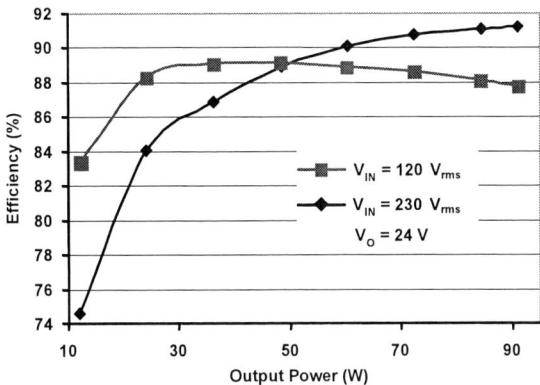

Fig. 16. Measured efficiency vs. output power.

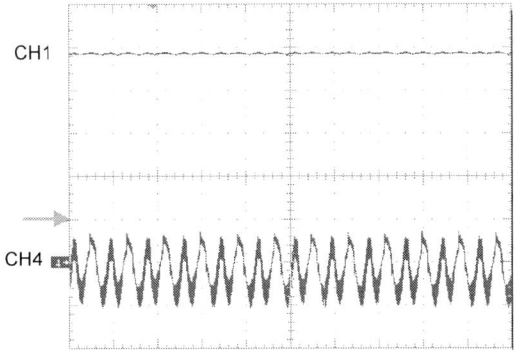

Fig. 17. Measured LED current and output voltage ripple at V_{IN} = 230 Vrms. CH1: LED current (1 A/div.); CH4: Output voltage ripple (20 mV/div.); Time scale: 10 ms/div.

IV. SUMMARY

A single-stage PFC flyback with a variable boost inductance for HB LED applications for the universal input voltage is presented in this paper. Experimental results obtained on a 24-V/91-W prototype show that the proposed PFC converter achieves an efficiency of 88%, a power factor of 0.9873 and a THD of 12.24% at nominal low line (120 Vrms), and an efficiency of 91%, a power factor of 0.9474 and a THD of 10.39% at nominal high line (230 Vrms). Line-current harmonics satisfy the IEC 61000-3-2 Class C limits with enough margin.

REFERENCES

[1] J. Y. Tsao, "Solid-state lighting: lamps, chips, and materials for tomorrow," IEEE Circuits and Devices Magazine, vol. 20, no. 3, pp. 28 - 37, May-June 2004.

[2] N. Narendran and Y. Gu, "Life of LED-based white light sources," Journal of Display Technology, vol. 1, no. 1, pp. 167 - 171, Sept. 2005.

[3] T. Komine and M. Nakagawa, "Fundamental analysis for visible-light communication system using LED lights," IEEE Trans. on Consumer Electronics, vol. 50, no. 1, pp. 100 - 107, Feb. 2004.

[4] Electromagnetic Compatibility (EMC), Part 3-2: Limits–Limits for harmonic current emissions (equipment input current ≤ 16 A per phase), International Standard IEC 61000-3-2, 2001.

[5] ON Semiconductor, "90 W, universal input, single stage, PFC converter," www.onsemi.com/pub_link/ Collateral/ AND8124-D.PDF, Dec. 2003.

[6] R. Redl, L. Balogh, and N. O. Sokal, "A new family of single-stage isolated power-factor correctors with fast regulation of the output voltage," Proc. IEEE Power Electronics Specialists Conf., 1994, pp.1137-1144.

[7] L. Huber and M. M. Jovanovic, "Single-stage single-switch input-current-shaping technique with reduced switching loss," IEEE Trans. Power Electron., vol. 15, no. 4, pp. 681-687, July 2000.

[8] C. Qiao and K. M. Smedley, "A topology survey of single-stage power factor corrector with a boost type input-current-shaper," IEEE Trans. Power Electron., vol. 16, no. 3, pp. 360–368, May 2001.

[9] Q. Zhao, F. C. Lee, and F. Tsai, "Voltage and current stress reduction in single-stage power factor correction ac/dc converters with bulk capacitor voltage feedback," IEEE Trans. Power Electron., vol. 17, no. 4, pp. 477 - 483, July 2002.

[10] G. Spiazzi, S. Buso and G. Meneghesso, "Analysis of a high-power-factor electronic ballast for high brightness light emitting diodes," IEEE Power Electronics Specialists Conference (PESC) Proc., pp. 1494 - 1499, 11 - 14 Sept. 2005.

[11] L. Huber and M. M. Jovanovic, "AC/DC flyback converter," U. S. Patent No. 6950319, Sept. 2005.

[12] T. F. Pan, H. J. Chiu, S. J. Cheng, and S. Y. Chyng, "An improved single-stage PFC flyback converter for high-luminance lighting LED lamps," The 8th International Conference on Electronic Measurement and Instruments, vol. 4, pp. 212 - 215. Aug. 2007.

[13] K. Zhou, J. G. Zhang, and S. Yuvarajan, "Quasi-active power factor correction circuit for HB LED driver," IEEE Trans. on Power Electron., vol. 23, no. 3, pp. 1410 - 1415, May 2008.

[14] Y. Hu, L. Huber and M. M. Jovanovic, "Single-stage flyback power-factor-correction front-end for HB LED application," Proc. of IAS 2009, Oct. 2009.

[15] C. A. Willis, "Ballast control device," U.S. Patent No.3,873,910, Mar. 25, 1975.

[16] R. T. Elms, "Variable inductance ballast apparatus for HID lamp," U.S. Patent No. 4,162,428, July 24, 1979.

[17] S. F. Lim and A. M. Khambadkone, "Non linear inductor design for improving light load efficiency of boost PFC," IEEE ECCE 2009 Proceedings, pp. 1339 – 1346, 2009.

[18] S. B. Yaakov and M. M. Peretz, "A self-adjusting sinusoidal power source suitable for driving capacitive loads," IEEE Trans. on Power Electron., vol. 21, no.4, pp. 890 – 898, July 2006.

[19] D. Medini and S. B. Yaakov, "A current-controlled variable-inductor for high frequency resonant power circuits," Conference Proceedings of Applied Power Electronics Conference and Exposition, vol. 1, pp. 219-225, 1994.

[20] C. Q. Lee, K. Siri, A. K. Upadhyay, "Parallel resonant converter with zero voltage switching," U. S. Patent No. 4,992,919, Dec. 1991.

[21] R. E. Hammond, E. F. Rynne, and L. J. Johnson, "Voltage controlled variable inductor," U. S. Patent 5,999,077, Dec. 7, 1999.

[22] G. Roberge and A. Doyon, "Variable inductor," U.S. Patent No. 4,393,157, July 12, 1983.

[23] W. H. Wölfle and W. G. Hurley, "Quasi-active power factor correction with a variable inductive filter: theory, design and practice," IEEE Trans. on Power Electron., vol. 18, no.1, PP. 248-255, Jan. 2003.

High Frequency High Efficiency Bidirectional DC-DC Converter Module Design for 10 kVA Solid State Transformer

Haifeng Fan, Hui Li
Center for Advanced Power Systems
Florida State University
Tallahassee, FL 32310, USA
fan@caps.fsu.edu, hli@caps.fsu.edu

Abstract—This paper presents the development of modular dual-half-bridge (MDHB) bidirectional dc-dc converter as the dc-dc stage of 10 kVA single phase solid state transformer (SST) for future renewable electric energy distribution and intelligent power management systems. The dc-dc converter, connected to 12 kV DC bus generated by an ac-dc rectifier interfacing with 7.2 kV electric utility grid, is to provide galvanic isolation function as well as 400 V DC bus for DC loads. The dc-dc converter consists of multiple low-voltage modules connected in input-series and output-parallel mode so that low-voltage commercial silicon MOSFETs, which usually have low conduction losses and high switching speed, can be adopted. Besides bidirectional power flow capability, the phase-shift dual-half-bridge (DHB) can realize zero-voltage-switching for all the switching devices without auxiliary switch devices, which enables the high switching frequency operation with low switching losses. As a result, high efficiency and high power density can be achieved. Other advantages of DHB topology have also been investigated for this application. A planar transformer adopting printed-circuit-board (PCB) winding is designed to realize high voltage solid isolation and identical parameters in multiple modules. The power loss of each main component has been analyzed for DHB converter under high frequency operation. Finally, the experimental results of two modules operating at 50 kHz switching frequency are presented with 97% efficiency.

I.   INTRODUCTION

The future intelligent electric energy distribution and management systems are expected to integrate highly distributed and scalable alternative generating sources, wherein solid state transformer (SST) is one of the key elements and intended to replace the conventional line-frequency (50/60 Hz) transformer based on iron/steel cores and copper/aluminum coil [1-6]. SST can achieve high power density, low weight, and low volume with good power quality. Various configurations for SST were reported in [1-5], of which the ac-dc-dc-ac configuration has the advantages to provide power factor correction, reactive power, and an additional dc bus.

The dc-dc conversion is a key stage in the ac-dc-dc-ac SST configuration. High-voltage silicon/post-silicon devices based multilevel dc-dc converters have been reported for SST application [4-5]. However, high-voltage insulated gate bipolar transistors (IGBTs) usually have high switching losses at switching frequency higher than 10 kHz, and post-silicon devices are still at laboratory level. Moreover, the reported dc-dc stages have diode rectifier on the output end, and therefore fail to provide bidirectional power flow path.

This paper proposes a modular dual-half-bridge (MDHB) bidirectional dc-dc converter with multiple dc-dc converter modules cascaded together as the dc-dc stage of a 10 kVA single phase SST. The cascaded MDHB converters apply low-voltage MOSFETs to achieve soft switching operating which enables high switching frequency operation with low switching losses. As a result, high efficiency and high power density can be achieved in either power flow direction. Other advantages of MDHB topology have also been investigated for this application. A planar transformer adopting interleaved printed-circuit-board (PCB) winding is designed to realize reduced ac resistance, high voltage solid isolation, and identical parameters in multiple modules. The power loss of each main component for MDHB converter has been analyzed. Finally, the experimental results of two modules operating at 50 kHz switching frequency are presented with 97% efficiency.

II.   SYSTEMS DESCRIPTION

Fig.1 shows the block diagram of 3-stage 10 kVA single phase SST consisting of an ac-dc rectifier, an isolated dc-dc converter, and a dc-ac inverter. The ac-dc rectifier interfacing with the 7.2 kV electric utility grid is to provide power factor correction function while converting 7.2 kV ac to 12 kV dc. The dc-dc converter, the key stage of SST, provides high

---

This work was supported by ERC Program of the National Science Foundation under Award Number EEC-08212121.

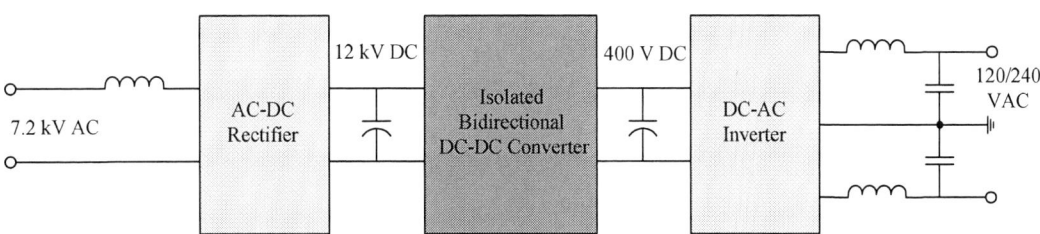

Figure 1. The block digram of 3-stage 10 kVA single phase SST.

frequency galvanic isolation and converts 12 kV high voltage dc to 400 V low voltage dc as well. The 400 V dc is then converted to 120/240 V low-voltage ac for end-use application through a dc-ac inverter.

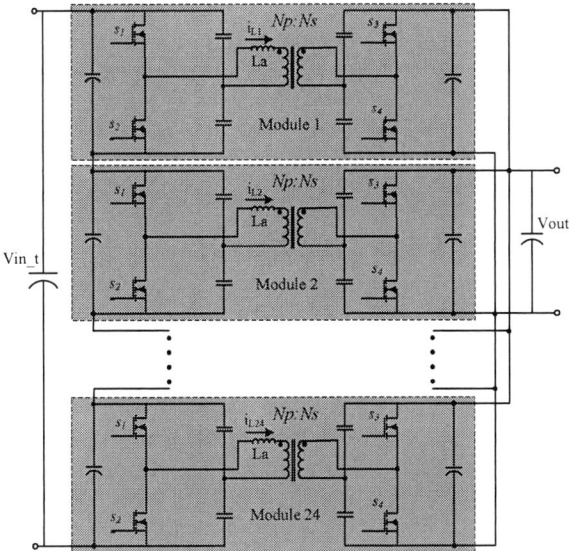

Figure 2. Modular dual-half-bridge bidirectional dc-dc converter.

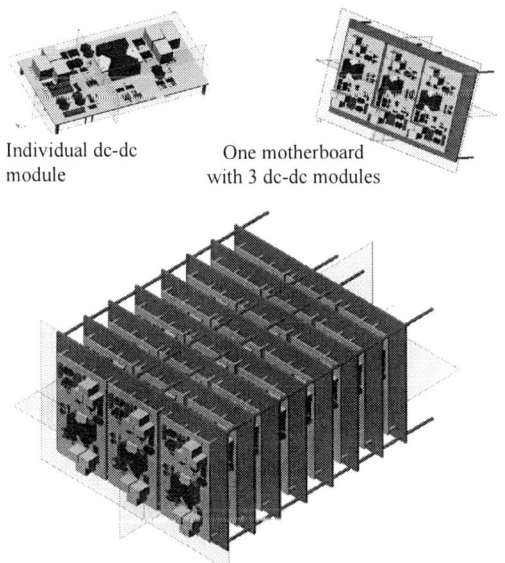

Figure 3. 3-D structure of the proposed MDHB converter.

As shown in Fig.2, the proposed MDHB converter for the dc-dc conversion stage consists of multiple low-voltage bidirectional dc-dc converter modules connected in input-series and output-parallel mode. The input and output voltage of each module are chosen as 500 V and 400 V respectively. Thus, the low-voltage commercial silicon MOSFETs with low conduction losses and high switching speed can be selected as the switching device. In order to interface with the 12 kV dc voltage from rectification stage, an 8-layer structure with 3 modules on each layer was shown in Fig.3. Each module is a bidirectional dc-dc converter, which adopts phase-shift technique to realize zero-voltage-switching (ZVS) operation mode for all switching devices without auxiliary switch devices in either direction of power flow [7-9], and therefore enables the high switching frequency operation with low switching losses. Although a total of 24 modules will be used to interface high voltage, the utilization of low-voltage device along with ZVS operation mode results in high efficiency, high frequency, good thermal performance and eventually high power density of the dc-dc conversion stage. As a result, the SST can achieve much smaller size than conventional line frequency (50/60 Hz) transformer by adopting the proposed high frequency high efficiency MDHB design.

III. DC-DC CONVERTER MODULE DESIGN

A. Topology Selection

Dual active full bridge (DAB) and dual half bridge (DHB) are two popular topologies among phase-shift ZVS bidirectional dc-dc converters. Fig.4 and Table I compares the operational conditions of DAB and DHB converters. Transformer flux swing of DHB topology is only half of DAB's when same effective cross sectional area of the transformer are adopted at same switching frequency. The DHB converter achieves smaller transformer core loss. This will be described in detail in the following transformer design section. Moreover, A DHB's use of half the number of switching devices as DAB, results in a more economical implementation especially in this multiple modules structure. The phase-shift DHB is therefore selected for the dc-dc converter module in this paper.

The output power of phase-shift DHB can be expressed as:

$$P_{out} = n(V_{in}/2)(V_{out}/2)\varphi(\pi - \varphi)/2\pi^2 L_a \qquad (1)$$

where L_a is the sum of leakage inductance of the transformer and the external auxiliary inductance, and φ is phase shift angle. Then, the output current can be given by:

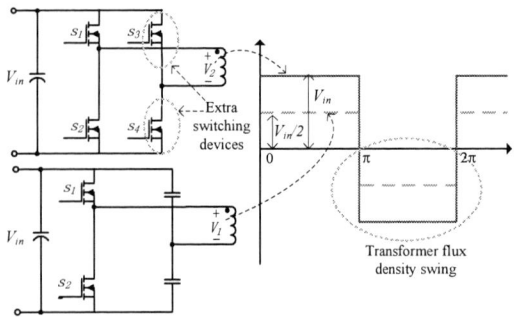

Figure 4.   Comparison between DAB and DHB.

TABLE I.   OPERATIONAL CONDITIONS COMPARISON

Item	DHB	DAB
Turns ratio (n)	V_{in}/V_{out}	V_{in}/V_{out}
Duty cycle (D)	0.5	0.5
Transformer flux swing (ΔB)	$(V_{in}/2)D/nA_ef$	$V_{in}D/nA_ef$
Quantity of switching device	4	8

V_{in} : input voltage, V_{out} : output voltage, A_e : effective cross sectional area of transformer, f : switching frequency, n : transformer turn ratio

$$I_{out} = P_{out}/V_{out} = nV_{in}\varphi(\pi - \varphi)/8\pi^2 L_a \qquad (2)$$

From (2), it can be seen that the output current I_{out} is independent from the output voltage V_{out}. This characteristic is similar to that of a current-source converter, which is an important advantage for this modular structure with parallel connection on output side. This feature will help the control design of the cascaded MDHB converter modules.

B. High Frequency Transformer Design

Planar transformer with coils encapsulated within multi-layer PCB can achieve lower profile and higher power density than conventional wire-wound transformer especially for the multiple modules system MDHB. In addition, the windings of transformer are etched within the PCB and thus are completely repeatable; this can make the windings of the transformer identical in multiple modules and contribute to the balance among these modules. Furthermore, the planar transformer utilizes solid insulation excluding air from the construction to minimize corona and partial discharge and therefore enhance reliability of SST. However, it is difficult to find a planar core suitable for this high voltage application requiring large cross sectional area. In this paper, a pair of PC40 PQ107/87/70 ferrite cores is modified to much lower profile while keeping the desired cross sectional area. After modification, the total window height of the transformer is reduced from 56 mm to 4.55 mm. The final transformer prototype is shown in Fig. 5, the primary to secondary turn ratio is 15:12, and the core loss can be calculated by the following empirical formula:

$$P_{cl} = V_e C_m f^x B_{ac}{}^y \qquad (3)$$

where V_e is effective core volume of transformer, C_m, x, and y are coefficient related to core material, B_{ac} is maximum flux density and can be expressed as:

$$B_{ac} = V_T D/2N_p A_e f \qquad (4)$$

where N_p is the primary number of turns, and V_T is the applied voltage on the primary side of transformer. As shown in Fig. 4, V_T equals to V_{in} for DAB and $V_{in}/2$ for DHB. For 80°C, C_m = 2.0, x=1.46, y = 2.57. Fig. 6 shows the transformer core loss with respect to V_T and f. The higher the frequency f and the lower V_T, the lower the core loss is. For 50 kHz operation, core loss of DAB with V_T = 500 V is 10.56 W, while the core loss of DHB with V_T = 250 is only 1.778 W, which verifies the analysis that DHB has much lower core loss than DAB.

Both skin effect and proximity effect will increase high frequency copper losses in transformer winding [10-11], and therefore these effects must be taken into account when designing the transformer winding. PCB winding offers the flexibility to achieve the winding structure as desired. In this paper, 10-layer PCB with 2 oz copper is adopted for the transformer winding. As shown in Fig.7, the winding are triple

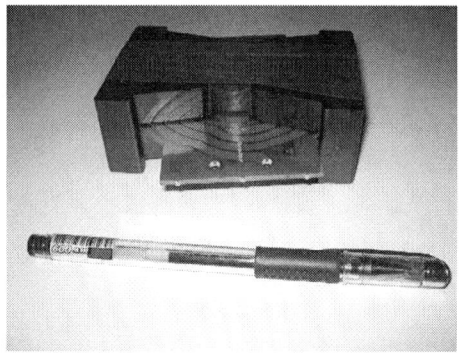

Figure 5.   The planar transformer.

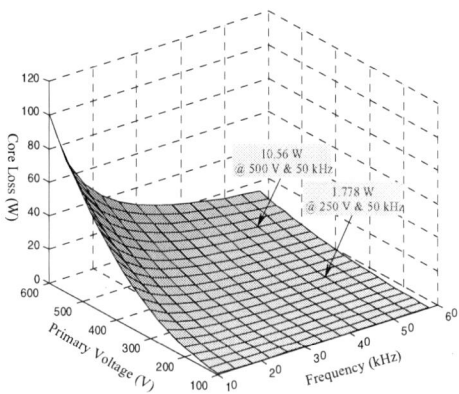

Figure 6.   Transformer core loss with respect to frequency and primary voltage.

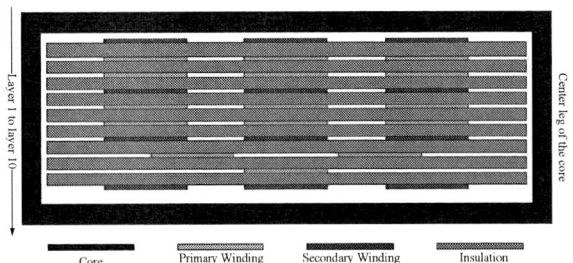

Figure 7. Cross section of interleaved transformer winding.

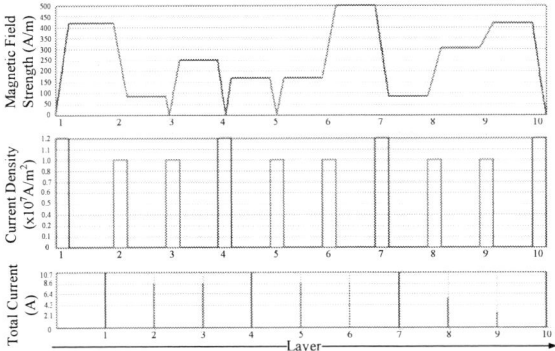

Figure 8. Plot of J,H and I distribution of transformer.

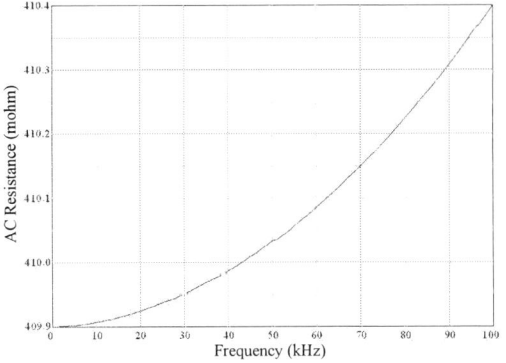

Figure 9. AC resistance with respect to frequency.

interleaved to reduce ac resistance. Fig. 8 shows the simulation results of current and magnetic field strength distribution in each layer, and Fig. 9 shows the total ac resistance of the transformer winding with respect to switching frequency. The current in each layer is evenly distributed in the copper, and low ac resistance at 50 kHz, which is very close to the dc resistance, is achieved.

C. Loss Analysis

Since the Phase-shift DHB converter realizes zero-voltage turn on for all switching devices, the turn-off losses and conduction losses of the switching devices are considered as the only power loss for the devices. Fig. 10 shows steady state

waveforms of one switching cycle and zoomed switching waveforms of S_1. The current of S_1 of one switching cycle can be expressed as follows:

$$
\begin{cases}
i_{S1} = -\varphi V_{in}/4\pi f L_a + (V_{in}/L_a)t & 0 \leq t < T_s(\varphi/2\pi) \\
i_{S1} = \varphi V_{in}/4\pi f L_a & T_s(\varphi/2\pi) \leq t \leq T_s/2 \\
i_{S1} = 0 & T_s/2 < t < T_s
\end{cases}
\tag{5}
$$

where T_s is switching period, and the rms value of current through S_1 can be given as:

$$
I_{rms} = \sqrt{\int_0^{T_s}(i_{S1})^2\,dt\Big/T_s} = (\varphi V_{in}/4\pi f L_a)\sqrt{(3\pi - 2\varphi)/6\pi}
\tag{6}
$$

The conduction loss of S_1 can be calculated as:

$$
P_{conduction} = R_{on}(I_{rms})^2 = R_{on}(\varphi V_{in}/4\pi f L_a)^2(3\pi - 2\varphi)/6\pi
\tag{7}
$$

where R_{on} is on state resistance of S_1.

As shown in Fig.10, at the moment S_1 is turned off, i_{L1}, the current of L_a, can be approximately considered as constant during the turn-off interval and will charge C_{o1}, the output capacitor of S_1, and discharge C_{o2}, the output capacitor of S_2. The turn off loss of S_1 therefore can be obtained as:

$$
\begin{aligned}
P_{off} &= f\int_0^{t_f} v_{ds}i_{s1} \\
&= f\left[i_{s1}(T_s/2)\right]^2 t_f^2\Big/48c_{o1} = \left(\varphi V_{in}t_f\right)^2\Big/768\pi f L_a c_{o1}
\end{aligned}
\tag{8}
$$

where t_f is the fall time of S_1.

The loss calculation method of S_1 can apply to S_2, S_3, and S_4 since DHB is symmetrical. Core loss and copper loss of inductor L_a can be calculated using the method introduced in transformer design section. The loss breakdown for the 500 W dc-dc converter module is shown in Fig. 11.

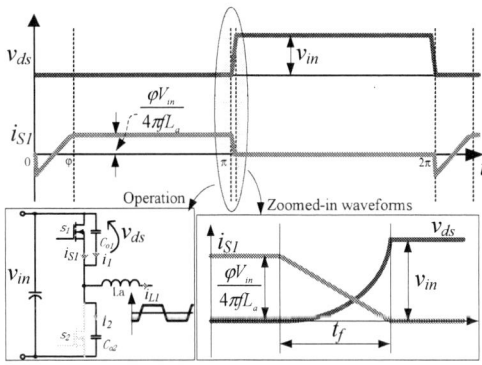

Figure 10. Operation and waveforms of S_1.

978-1-4244-4782-4/10 $26.00 © 2010 IEEE

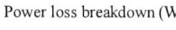

Power loss breakdown (W)

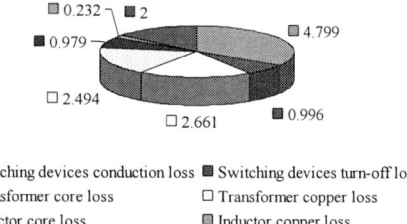

- ▨ Switching devices conduction loss ■ Switching devices turn-off loss
- ☐ Transformer core loss ☐ Transformer copper loss
- ■ Inductor core loss ▨ Inductor copper loss
- ▨ Others

Figure 11. Power loss breakdown.

IV. EXPERIMENTAL RESULTS

Two 500 W dc-dc converter modules shown in Fig. 12 have been built and tested to verify the high frequency and high efficiency operation. The specifications and circuit parameters of each dc-dc converter module are shown in Table II.

TABLE II.   KEY SPECIFICATIONS AND CIRCUIT PARAMETERS

V_{in} (V)	V_{out} (V)	Rated Output Power (W)	Primary MOSFET	Secondary MOSFET	Commutation Inductance(μH)
500	400	500	IXFT 24N80P	IXFT 36N60P	90

Fig. 13 shows the experimental waveforms of drain-source voltage (V_{ds}) and gate driver signal (V_{gs}) of the low side MOSFET S_4 on the secondary of one dc-dc converter module, and their zoom-in waveforms with 500 V input and 400V output. As shown in Fig.13, the ZVS operation mode can be achieved. Fig. 14 shows the measured efficiency with respect to the output power of one dc-dc converter module, 97% efficiency can be achieved at rated full load 500 W. Fig.15 shows the inductor current waveforms of two cascaded dc-dc

shows the inductor current waveforms of two cascaded dc-dc converter modules sharing 500 V input voltage and 500 W output power. Fig. 16 and Fig. 17 show output current sharing results and input voltage sharing results of two cascaded dc-dc converter modules, respectively.

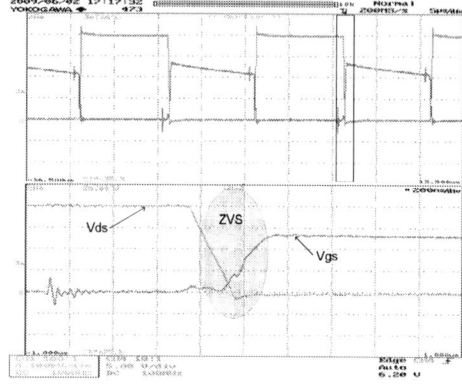

Figure 13. V_{ds} and V_{gs} of S_4.

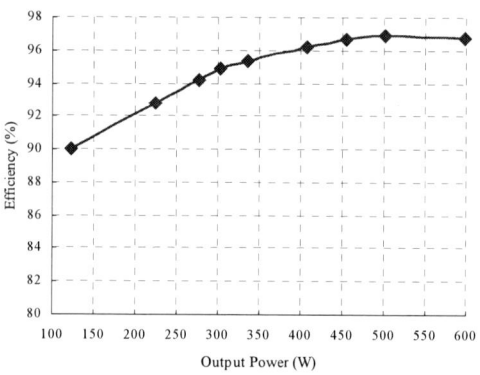

Figure 14. Efficiency versus output power.

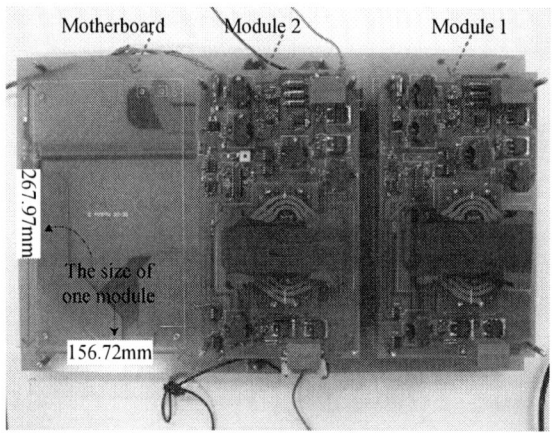

Figure 12. Photo of the prototype.

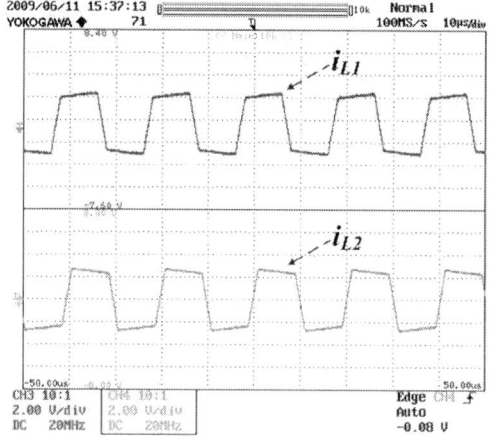

Figure 15. Inductor current waveforms.

978-1-4244-4782-4/10 $26.00 © 2010 IEEE

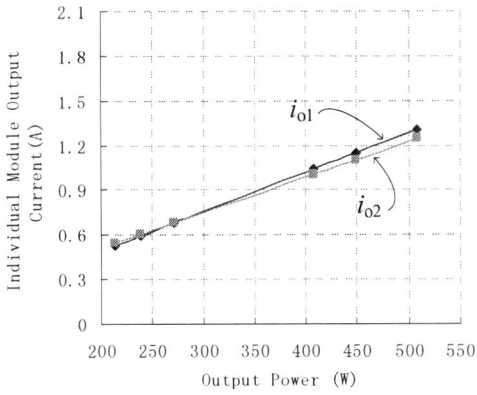

Figure 16. Output current sharing.

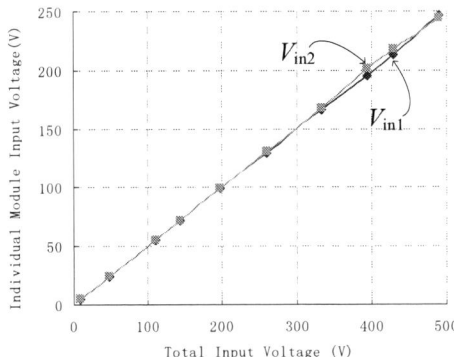

Figure 17. Input voltage sharing.

V. CONCLUSION

High frequency bidirectional dc-dc converter module based on low voltage switching device has been presented for 10 kVA single phase solid state transformer. Phase-shift ZVS technique along with low voltage switching device enables the high frequency operation of the dc-dc converter module while keeping low switching loss and conduction loss. In addition,

planar transformer with interleaved windings has been designed and implemented to obtain low core loss, optimized high frequency copper loss, low profile, and solid insulation which results in enhanced reliability. As a result, the high frequency, high efficiency, and high power density can be achieved. Two 50 kHz bidirectional dc-dc converter modules have been built successfully, and the measured efficiency is 97% at rated power; the two cascaded dc-dc converter modules can share input voltage and output current very well.

REFERENCES

[1] Kang, M., Enjeti, P., and I. Pitel, "Analysis and Design of Electronic Transformers for Electric Power Distribution System," in *IEEE Transactions on Power Electronics,* Nov. 1999.

[2] Manjrekar MD, Kieferndorf R, Venkataramanan G. Power electronics transformers for utility applications. Conference Record of the 2000 *IEEE-IAS Annual Meeting*; 2000, vol. 4, p. 2496–502.

[3] Ronan, Jr., E., Sudhoff, S., Glover, S., and D. Galloway, "A Power Electronic-Based Distribution Transformer," *IEEE Transactions on Power Delivery*, April 2002, pp. 537-543.

[4] Jih-Sheng Lai, Maitra, A., Mansoor, A., and Goodman, F, "Multilevel Intelligent Universal Transformer for Medium Voltage Applications," in *Proc. IEEE IAC conf.*, October 2005, Vol.3, pp. 1893 - 1899.

[5] L. Y. Yang, T. F. Zhao, J. Wang, and A. Q. Huang, "Design and Analysis of a 270 kW Five-Level DC/DC Converter for Solid State Transformer Using 10 kV SiC Power Devices," in *Proc. 38th IEEE Annual. PESC*, Jun. 2007, vol. 1, pp. 245–251.

[6] E. C. Aeloiza, P. N. Enjeti, L.A. Moran, I. Pitel,"Next Generation Distribution Transformer: To Address Power Quality for Critical Loads," in *Proceedings of Power Electronics Specialists Conference*, June 2003, Acapulco, Mexico, pp. 1266 – 1271.

[7] R. W. De Doncker, D. M. Divan, and M. H. Kheraluwala, "A three phase soft-switched high-power density dc/dc converter for high-power applications," *IEEE Trans. Ind. Appl.*, vol. 27, no. 1, pp. 63–73, Jan./Feb. 1991.

[8] H. Li, and F. Z. Peng, "Modeling of a new ZVS bi-directional dc-dc converter," *IEEE Trans. Aerospace and Electro. Systems*, vol. 40, no. 1, pp. 272-283, 2004.

[9] S. Inoue and H. Akagi, "A Bidirectional Isolated DC-DC Converter as a Core Circuit of the Next-Generation Medium-Voltage Power Conversion System," *IEEE Trans. Power Electronics*, vol. 22, no. 2, pp. 535–542, Mar. 2007.

[10] W. Hurley, E. Gath, and J. Breslin, "Optimizing the ac resistance of multilayer transformer windings with arbitrary current waveforms," *IEEE Trans. Power Electron.*, vol. 15, no. 2, pp. 369–376, Mar. 2000.

[11] P. L. Dowell, "Effect of eddy currents in transformer windings," *Proc.Inst. Electr. Eng.*, vol. 113, no. 8, pp. 1387–1394, Aug. 1966.

Synchronization of Three-Phase Converters and Virtual Microgrid Implementation Utilizing the Power-Hardware-in-the-Loop Concept

O. Vodyakho, *Member, IEEE*, C. S. Edrington, *Senior Member, IEEE* M. Steurer, *Senior Member, IEEE*,
S. Azongha, F. Fleming, *Student Member IEEE*
Florida State University, Center for Advanced Power Systems
Tallahassee, FL 32310
email: {vodyakho, edrington, steurer, azongha, fleming}@caps.fsu.edu

Abstract—This paper addresses the timely issues of synchronization and application of three-phase power converters connected in parallel utilizing the Power-Hardware-in-the-Loop concept. Without proper synchronization, distinguishing the currents circulating between the converters are unclear. The paper centers on control methodology for achieving precise phase synchronization for equal load sharing, with minimum current circulation between the paralleled power converter modules, and robust dynamic system control under different transient conditions. One of the possible applications for the configuration presented in this paper is the conceptual virtual microgrid, which utilizes the reactive power compensation ability of the Static Synchronous Compensator (STATCOM). The microgrid behavior and load dynamics are simulated with a real-time digital simulator which generates appropriate control commands to a power electronics based voltage amplifier interfaced via a cascaded LC-LC type filter to a variable speed drive (VSD). This is necessary as reactive power control is a critical consideration in improving the power quality of power systems. To compensate for reactive power, the STATCOM controller will be developed and integrated into the proposed virtual microgrid system. This concept provides a solution for de-risking these costs as it utilizes the PHIL concept in conjunction with high-fidelity microgrid model and detailed load dynamics. Selected experimental results on two, 25-kVA and 15-kVA, converters in parallel are presented.

I. INTRODUCTION

In many cases, it is favorable to connect power converters in parallel. For instance, this could be in systems with high reliability requirements or in a system where a low demand factor is expected, such that the full power rating of the installed converter rating is minimized. However, special precautions must be prepared in order to make the converters share the common load equally, due to the nonphysical relation between output power and switching frequency in a solid-state converter [1]-[2].

The main control target is to insure that the units in parallel share the common load. Active, reactive, and harmonic powers must be shared equally. The conventional approach to parallel solid-state power converters requires interconnections between the converters to achieve balanced load sharing, for example, by having a voltage-controlled "master" unit and several current-controlled "slave" units. However, to achieve true redundancy, all units must be able to operate independently. This matter has been discussed in [3]. Linear

balanced loads can be shared equally by using droop coefficients that make the frequency and the voltage amplitude proportional to the active and reactive power, respectively. The principle of sharing loads by droop coefficients is well established in the utility sector. The proposed parallel-connected converters system utilizes both of the approaches described above in order to achieve and evaluate the synchronization of the three-phase converters.

When voltage-controlled variable source inverters (VSI) are parallel connected, inherent differences in ac voltage amplitude tend to generate reactive power flow; phase difference causes active power flow (circulating currents) between converter modules. Control schemes and algorithms are required to continually correct these variances and achieve stable operation of the system.

Utilizing the power-hardware-in-the-loop (PHIL) concept in conjunction with high-fidelity models and detailed load dynamics, as well as the different modulation strategies, switching frequencies, dc-link voltage values, etc. for three-phase power converters will be investigated in order to ensure the stable and proper synchronization of the parallel power modules. In addition, this paper will describe one of the possible applications of the proposed configuration with three-phase power converters connected in parallel utilizing PHIL - a Virtual Microgrid (VMG) utilizing the reactive power compensation ability of the Static Synchronous Compensator (STATCOM).

The idea of a VMG is to reproduce a relatively weak power system with a nonlinear load and STATCOM in order to investigate the power quality issues and power electronics control.

The primary motivation in this work is to develop a test bed which can be utilized to conduct research experiments using/involving: advanced control system strategies, power quality improvement, novel/new drive system topologies, prototype electric machines, or a combination of the aforementioned. Furthermore, this eliminates the need for physical implementation of the hardware, and the system can be entirely modeled within the virtual environment, ultimately lowering cost and risk [4]-[5].

In the following section, the fundamental theory of connecting ac power units in parallel will be described by referencing the principle of sharing linear balanced loads

978-1-4244-4782-4/10 $26.00 © 2010 IEEE

adapted from the utility control theory. Also, the results of experimental tests will show the value of the proposed concept as applied to the two 25-kVA and 15-kVA converters connected in parallel.

II. POWER CONVERTERS CONFIGURATION AND CONTROL STRATEGIES

Consider two solid-state three-phase power converters connected through a delta/wye step-up transformer as shown in Fig. 1. The setup consists of two Power Electronics Building Blocks (PEBB), PEBB_I and PEBB_II, connected to the three-phase 480V-system and 208V-system, respectively. The PEBBs consist of an Active Front End Unit (AFU) and VSI. The complex powers delivered from PEBB_I and PEBB_II to the load (in this case a delta/wye step-up transformer) are then given by:

$$\overline{S}_x = P_x + j \cdot Q_x = \overline{V}_{PCC} \cdot \overline{I}^*_x \qquad (1)$$

Where, S_x is the complex power vector of converter x [1]-[2], P_x is the active power of converter x, Q_x the reactive power of converter x, I_x the output current vector of converter x, and V_{PCC} is the voltage vector of the point of common coupling.

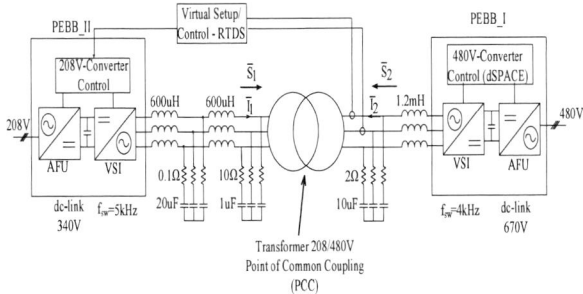

Figure 1. Configuration of three-phase converters.

The currents delivered from the two converters are given by [1]:

$$\overline{I}_1 = \left[\frac{\overline{V}_1 - \overline{V}}{jX} \right] = \left[\frac{V_1 (\cos \phi_1 + j \sin \phi_1) - V_{PCC}}{jX} \right] \qquad (2)$$

$$\overline{I}_2 = \left[\frac{\overline{V}_2 - \overline{V}}{jX} \right] = \left[\frac{V_2 (\cos \phi_2 + j \sin \phi_2) - V_{PCC}}{jX} \right] \qquad (3)$$

where X is the interconnection reactance between the converters and the common point, Φ_x is the power angle (angle between converter output and the common point) of converter x, V_x the output voltage vector of converter x; V is the amplitude of the output voltage vector, and V_{PCC} the amplitude of the voltage at the point of common coupling .

From (2) and (3) and Fig. 1, it can be seen that the active power flow is dominated by the power angles, while the voltage amplitudes primarily influence the reactive power flow. Since the two converter systems are based on solid-state inverters, the frequency is not itself load dependent, and the voltage is relatively stiff due to the voltage control of the units. If the two units are not properly synchronized at the

time of connection due to component tolerances (switching frequencies, filter parameters, interconnection impedance variation) and are paralleled without additional synchronization control, a large circulating current flow will result.

Through experimentation, it was shown that, due to the harmonic components produced by the PWM switching in the inverter, high frequency noise appeared on the VSI voltage output and causes additional problems in parallel operated VSIs. This switching transient causes a ripple in the voltage seen by the transformer at PCC. Typically, only series inductors are used as a filter, which interface the PWM-inverter and the power grid. In this case, the size of the inductor for optimal switching frequency should be chosen appropriately to reduce the ripple (harmonics) around the switching frequency. To accommodate for this ripple phenomena, an LC-ripple filter (PEBB_I) and an LC-LC-ripple filter (PEBB_II), were placed between the output of the both inverters and the input of the transformer, as shown in Fig. 1. The resonant frequency calculation using these filter parameters is given in (4).

$$f_{res} = \frac{1}{2\pi \sqrt{LC}} \qquad (4)$$

The LC-filters provide advantages in costs and dynamic performance since smaller inductors can be used, compared to L-filters, in order to achieve the necessary damping of the switching harmonics. However, LC-filter design is complex and needs to consider many constraints, such as the current ripple through inductors, total impedance of the filter, resonance phenomenon, reactive power absorbed by filter capacitors, etc. In addition, the LC-filters may cause steady-state and transient distortion in the output voltage due to resonances. This distortion can be reduced, but not solved, if in PWM converters the main resonance frequency is selected in a range where no harmonics of the output voltage exist. With this filter, the switching frequency of the converter has to be high enough to obtain sufficient harmonic attenuation. In conventional PWM controls, the resonance frequency can be chosen easily as a frequency lower than the definable lowest pulse frequency, which is roughly the switching frequency in the case of a two-level inverter.

The systems incorporating LC-filters are second order, thus a peak amplitude response exists at the resonant frequency of the LC-filter. This requires more care when designing the LC-filter parameters and current control strategy in order to maintain system stability since the filters tend to oscillate with the filter resonance frequency.

The most popular method is to insert a damping resistor in the capacitor shunt branch of the LC-filter as shown in Fig. 1. The frequency response of the LC-filter with a damping resistor is shown in Fig. 2.

As shown in Fig. 2, the damped filter has more attenuation at the resonant frequency, but has less attenuation in the high frequency region as opposed to that of the non-damped filter. In addition, this approach results in considerable power loss.

978-1-4244-4782-4/10 $26.00 © 2010 IEEE   217

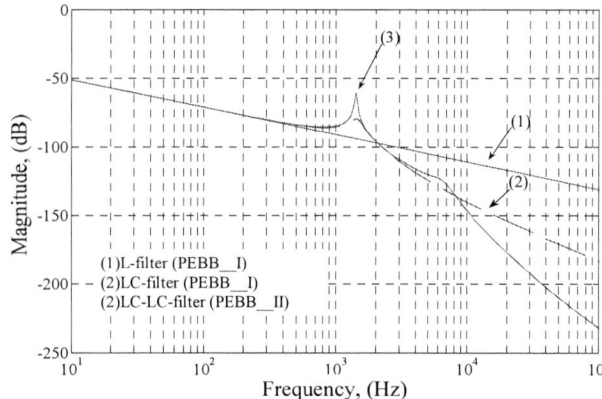

Figure 2. Transfer functions of an L-filter and a LC-filter in PEBB_I and PEBB_II.

A. AFU control strategy

The AFU control circuit consists of six fully controlled IGBT switches, three line inductors on the ac side, and a capacitor on the dc side. For a constant dc-link voltage, the six IGBTs are switched to produce a PWM voltage waveform at the input of the AFU. This allows for control of the voltage across the line inductor, which in turn allows for regulation of the line current. Basically, the AFU can be represented by a controlled voltage source, since the dc-link voltage imposed across the electrolytic capacitor is reflected at the rectifier input terminals, modulated by the switching function. To obtain a unity power factor, the line current needs to be in phase with the line voltage, allowing the voltage across the line inductance to lead the line voltage by 90° electrically [3].

A block diagram of the AFU control is shown in Fig. 3. This controller contains two loops: the outer loop, which regulates the error between the reference dc-link voltage and the actual value, and the inner loop, which controls the current in the synchronous rotating coordinate system (i.e., dq-frame). The reference value for the q-axis current is set to constant zero unless a certain amount of reactive power is requested. Due to the interaction between the d- and q-axis currents, a feed forward crosslink (ωL) is added in order to compensate for the coupling. The PI parameters are tuned according to the symmetric optimum method [6]. Due to the symmetry of the d- and q-axis currents, their PI parameters are the same. The conventional sine-triangle PWM is used as a modulation strategy for the AFU.

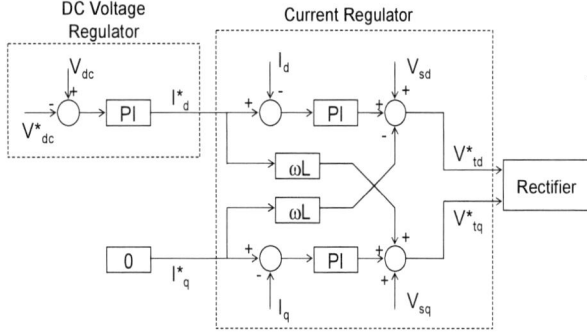

Figure 3. Block diagram of the AFU control.

PEBB_I and PEBB_II contain the same AFU so the control strategy described above is utilized in both PEBBs. Dc-link voltage for PEBB_I is 670 V and dc-link voltage for PEBB_II is 340 V as shown in Fig. 1. In contrast, the VSIs in the PEBBs are controlled by different control strategies (for better evaluation of the synchronization issues) which will be described in the next section.

B. VSI control strategies

In order to investigate the synchronization issues of three-phase converters in a parallel connection, three different VSI control strategies are implemented.

The inverter within the PEBB_I is controlled via two modulation strategies: space-vector PWM (SVM) and conventional sine-triangle PWM (SPWM). Fig. 4 shows a flow chart describing the space vector modulation SVM technique. The abc-voltage is the reference signal for the SVM modulator. The corresponding space-vector coordinates qα(t) and qß(t), referring to the α- and ß-axes, are obtained from the stationary coordinates by coordinate transformation abc to αß. The αß equivalents are then used to determine which of the αß-plane's six sectors, each 60° apart, the voltage vector lies within. This output then feeds a switching time calculator, which in turn calculates the timing at which the voltage vector is to be applied to the inverter and the block input of the voltage vector. A ramp generator serves as a clock for the switching time calculator. Finally, a gate logic block compares the voltage vector timing, the block input, and the ramp generator's clock signal to control the IGBT switches of the inverter at the correct time.

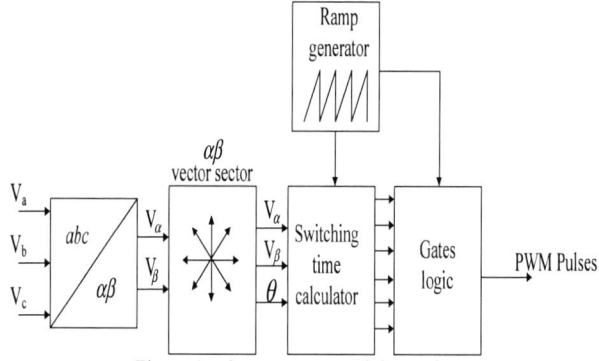

Figure 4. Space vector modulator of a PEBB_I.

The second modulation technique used to issue PWM duty cycle commands to the inverter legs is a conventional sine-triangle PWM. The basic principle of SPWM is shown in Fig. 5. Briefly described, each of the phase reference voltages v* are compared to a triangle carrier function Vm. Based on the relationship between the reference values and the triangular carrier function, the switches are controlled to produce a PWM pulse pattern. For uniform testing purposes, a modulation index of 0.1 was used as the symmetrical triangle carrier depth. The frequency of the carrier function is the IGBT switching frequency, and the amplitude is related to the dc-link voltage. In [7], the direct relation between space vector and carrier PWM is explained.

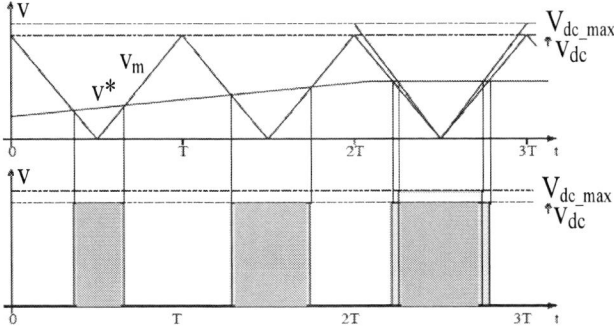

Figure 5. Basic principle of SPWM modulator of PEBB_I.

Both VSI's control strategies are implemented using a control system rapid prototyping environment such as dSPACE. For both control strategies for VSI in PEBB_I, the switching frequency is set at 4 kHz.

In contrast to PEBB_I, the VSI of PEBB_II is controlled by a Reference-Carrier-Modulator (RCM), which uses a Reference-Carrier method for calculation of three-phase patterns [9]. There are two frequency integrators calculating the reference and carrier angles. The carrier angle calculation is only required if the reference and carrier are asynchronous. Support for the carrier integrator is configurable by a CPU-interface. The reference angle (one for each phase) is used to interpolate a reference value. The reference function value is scaled by an index and added offset value as illustrated in Fig. 6.

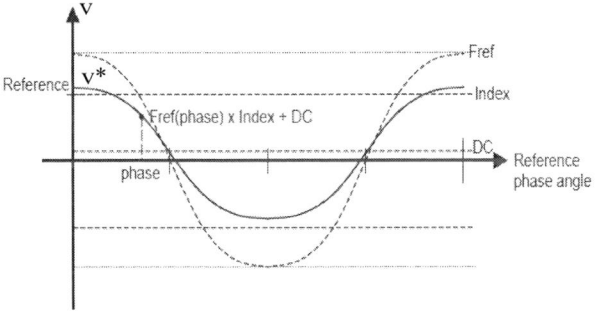

Figure 6. Basic principle of SPWM modulator of PEBB_I.

The RCM supports symmetrical switching patterns only. The function interpolation takes advantage of pattern symmetry to store only ¼ of the function period for the reference functions. The switching frequency of the VSI in PEBB_II is set to 5 kHz.

Utilizing the PHIL-concept, the PEBB_I and PEBB_II will be synchronized as shown in Section III. The PHIL-concept implementation is described in the next Section.

III. PHIL IMPLEMENTATION

In order to achieve proper synchronization, additional control is implemented in RTDS. The correct line phase-angle is very important information in the proposed system and in other grid-connected equipment such as controlled rectifiers, active filters, dynamic voltage restorers, and also in emerging distributed generation systems such as photovoltaic power

plants. In parallel VSI arrangements, a very precise synchronization is also required prior to each VSI connection to the grid in order to avoid catastrophic transients. To estimate the phase-angle, open-loop and closed-loop methods are available. The closed-loop methods are commonly known as phase-locked loops (PLLs). The closed loop method is implemented in RTDS. The figures of merit of a PLL are the steady state phase-angle error, speed of response to phase, frequency and voltage amplitude disturbances, harmonic rejection, and line imbalance rejection in case of three-phase systems.

The used PLL algorithm is based on a fictitious electrical power (power-based PLL (pPLL)), which is presented in [8]. The selected structure has a simple digital implementation and therefore low computational burden. The block diagram of the synchronization control is illustrated in Fig. 7.

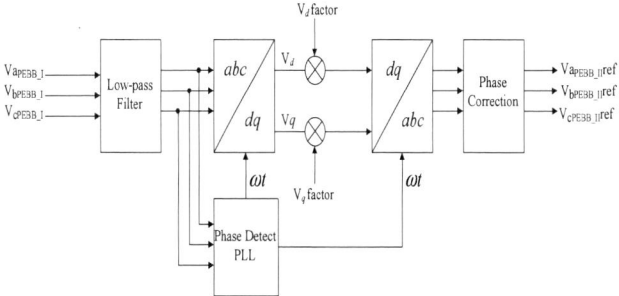

Figure 7. Basic principle of SPWM modulator of PEBB_I.

The measured PEBB_I output three-phase voltages are filtered with a low-pass filter and the output-voltage space-vector coordinates, v_α (t) and v_β (t), are calculated from the measured abc-voltages, $v_{aPEBB\_I}(t)$, $v_{bPEBB\_I}(t)$, and $v_{cPEBB\_I}(t)$ first and then transformed to dq-coordinates. The measured PEBB_I output three-phase voltages are also passed through a PLL for synchronization of the voltages. The PLL block provides the phase information for dq-abc and abc-dq coordinate transformations. In order to evaluate the synchronization issues more precisely, the PEBB_I output-voltage coordinates v_d(t) and v_q(t) are multiplied by a factor for the reference voltage amplitude settings. The scaled voltages v_d(t) and v_q(t), are inputs of the coordinate transformation block dq-abc. The scaled and synchronized PEBB_I abc-voltages are then passed through a phase correction block and are the reference voltages for PEBB_II.

The proposed system is evaluated on two, 25-kVA and 15-kVA, converters in parallel and selected simulation and experimental results are presented in the Section IV.

IV. SIMULATION AND EXPERIMENTAL RESULTS

Simulation is performed to evaluate the proposed system as shown in Fig. 1. For the simulation results only SVM is utilized as a control strategy for both power electronics buildings blocks PEBB_I and PEBB_II. Fig. 8 shows the simulation results of the proposed converter configuration.

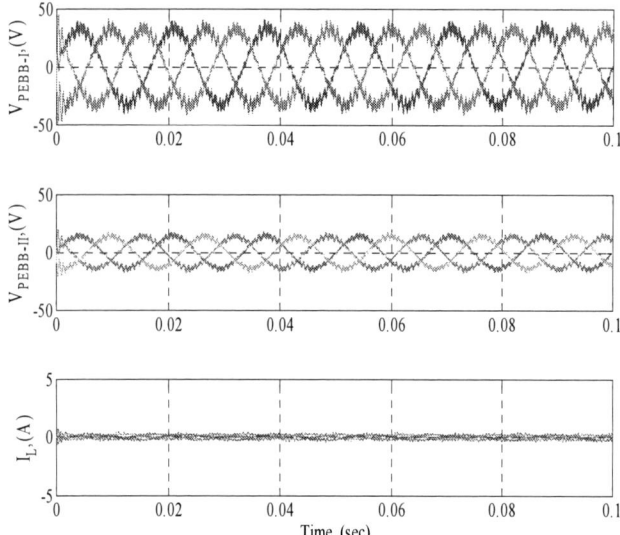

Figure 8. Simulation results of the proposed system configuration with SPWM.

As shown in Fig. 8, synchronization is achieved with implementation of the PLL, and a very small current circulates between the PEBBs.. The output three-phase voltages of PEBB_I are rather small (ca. 45 V) in order to achieve the synchronization even with a small resolution of measured voltages.

In Fig. 9, the top wave is the transformer primary at 480V, the middle is its secondary or low side at 208V, and the bottom is the PWM signal out of the PEBB_I. Without the ripple filter installed, high frequency noise is apparent in both transformer voltages, and will compromise the synchronization of the PEBBs.

The next set of figures show the selected experimental results for the parallel-operated PEBBs with different modulation and control strategies as described above. Fig. 10 and Fig. 11 show the reference voltage for the PEBB_I (480V-side converter) and for the PEBB_II (208V-side converter). Fig. 10 shows a phase shift in references and Fig. 11 shows the phase compensation implemented in the RTDS synchronization algorithm (Fig. 7). Current draw into the voltage amplifier (PEBB_II) was accomplished via a series of source voltage increments which are shown in Fig. 12. Initially, the PEBB_I inverter was set to source 0V, and the voltage amplifier's output also set at zero. The voltage amplitude was slowly increased, creating a voltage difference between sources resulting in current flow.

The voltage amplifier's output was then slowly ramped to reference 100% of the PEBB_I inverter output from the previous 0%, eliminating the current flow. The phase angle of each source has been aligned via previous work through a PLL and phase compensation algorithm.

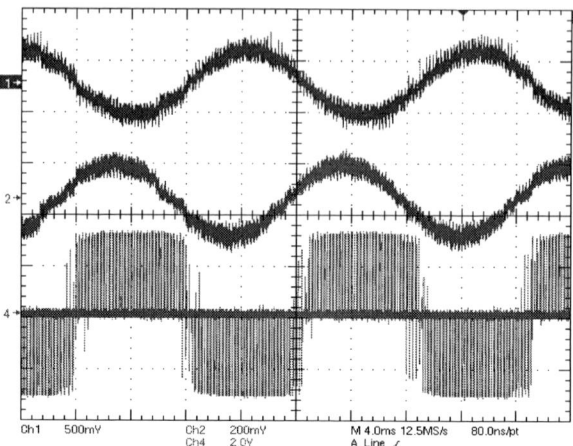

Figure 9. PEBB_I to delta/wye step-up transformer interface voltages with LC-ripple filter (SVM; CH1:50V/div; CH2: 20V/div; CH4:200V/div)).

Fig. 12 shows the current drawn as a function of the difference in voltage source magnitude. At 0.5 seconds, a trip condition triggers as the current draw magnitude exceeds the allowed limit (set at 20A). Thus, with the PEBB_I output at 50V and a 60% reference limiter for the amplifier output, a current flow of sufficient magnitude to trip the system is created. Fig. 13 (obtained via a Tektronix oscilloscope) shows each of the source voltages in phase with one another while in operation. The amplifier reference ramp is again set at 60% and no current is flowing. Fig. 13 shows the accomplished synchronization for the converters with the different modulation strategies (for PEBB_I the conventional PWM and SVM are used), switching frequencies, dc-link voltages and system voltages.

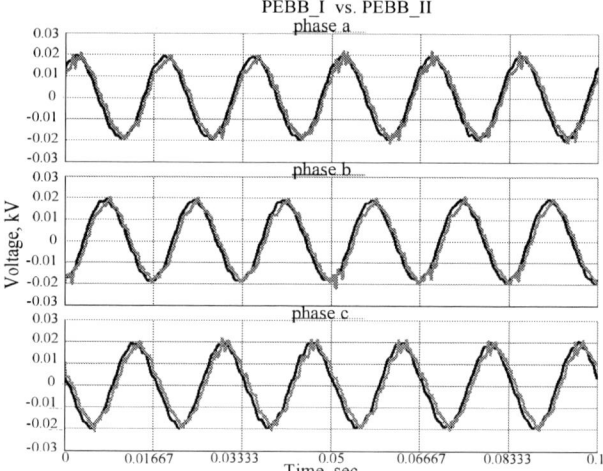

Figure 10. Phase shift in references (SVM).

978-1-4244-4782-4/10 $26.00 © 2010 IEEE

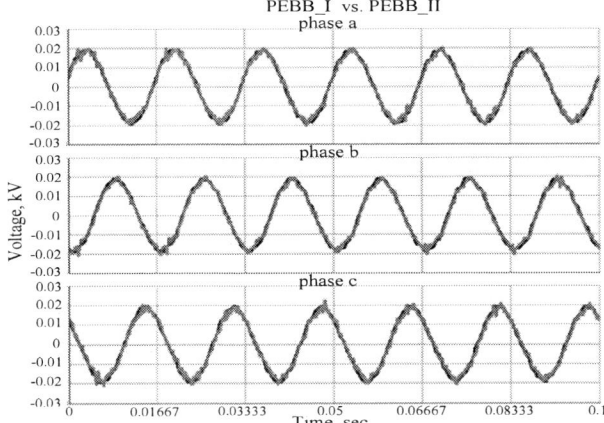

Figure 11. Phase compensation in references (SVM) using the Phase Correction block.

The results provided in this section illustrate the parallel operation of the PEBBs with the different parameters only if an additional synchronization control is implemented. In the main, it should be noted that if PEBBs are properly synchronized, their parallel operation does not depend on PEBBs characteristics and system parameters.

The next section describes the application of the proposed system configuration with the proper synchronization of PEBBs.

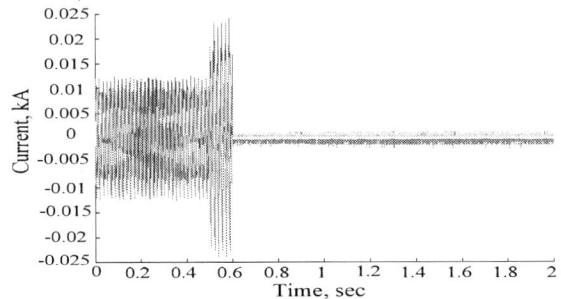

Figure 12. Current drawn due non-synchronization (SVM)

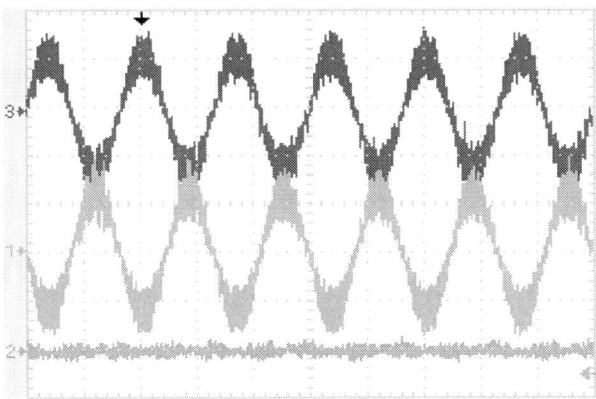

Figure 13. Converters voltages and drawn current (conventional PWM; CH1,CH3: 20V/div; CH2:5A/div).

V. APPLICATION OF THE PROPOSED SYSTEM CONFIGURATION

The main focus of this paper is to investigate the synchronization issues of the two three-phase converters under different control strategies, dc-link voltage, and filter parameters. The synchronization is successfully achieved as shown in previous section. However, other applications such as STACOM need to be analyzed and described in this paper in order to provide the outline for future work.

An application of the system configuration shown in Fig. 14 exhibits how this type of simulation can be very effective in evaluating the performance of a microgrid with the reactive power compensation ability of the STATCOM. The original circuit simulated in RTDS consisted of a 100 kVA gas turbine generator used to mimic a relatively weak power system, a delta/wye step-up transformer, a 40 kW / 0.1 s pulse load, and a motor load. The latter was removed from the simulated environment and replaced by a real 20 hp induction motor (MT1) driven by a variable speed drive (VSD1). A second motor drive set (TF2, VSD2, MT2), identical to the first one, operates as a dynamometer and provides the counter torque to adjust the motor load. A power electronic building block (PEBB_II converter) based PWM type converter amplifies the simulated microgrid voltage and provides power to the motor. For the simulated system to "see" the motor load, a current source controlled by the actual measured load current is modeled in the RTDS to represent the motor. If such an interface is ideal, this PHIL system should perform exactly the same as the original circuit [5].

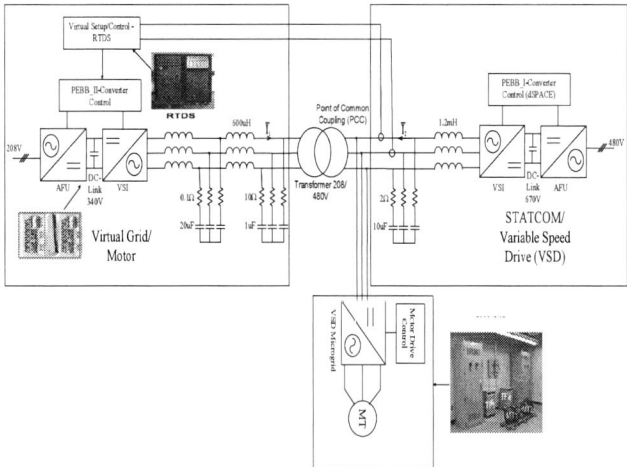

Figure 14. VMG implementation utilizing PHIL.

For the STATCOM controller, a two-loop PI control algorithm is utilized and a set of PI controllers are implemented and optimized based on [1]. With the conventional PI controller design, the new control block diagram is shown in Fig. 15. Compared with conventional PI controllers, the inner current controller is much simpler. Since this is a multivariable control system, the coupling effect is automatically accounted for and the feed forward term is no longer required.

978-1-4244-4782-4/10 $26.00 © 2010 IEEE

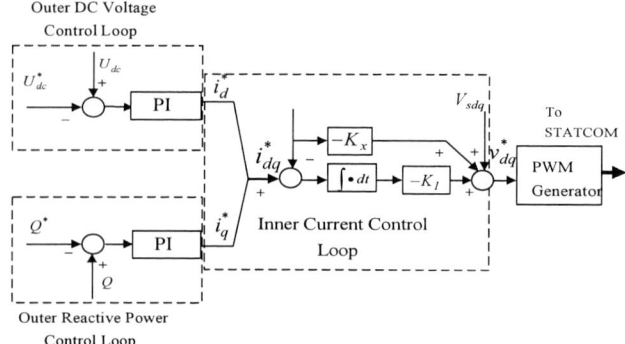

Figure 15. Detailed control block of the control for STATCOM.

Fig. 16 shows the step response of reactive power Q. To test the anti-disturbance ability of both conventional PI and two-loop PI controllers, a 0.2 pu grid voltage sag with a duration of 0.2s was initiated at 0.8 s. Although the overshoot is higher, it can be observed that the PI / LQR controller has a faster dynamic response and less oscillation than the PI controller. Both controllers display a good anti-disturbance ability.

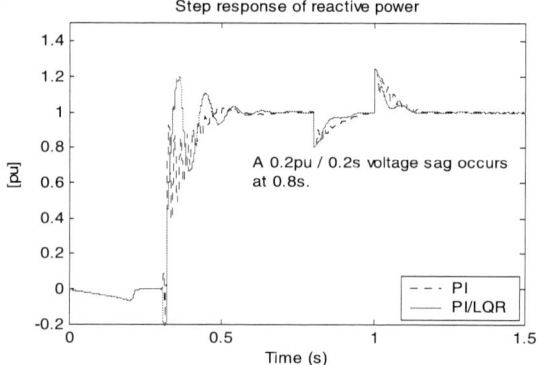

Figure 16. Reactive power responses of the two systems with disturbance.

Fig. 17 shows, 20V grid voltage sag with duration of 1.5s utilizing the PHIL concept in the proposed conceptual virtual microgrid with the reactive power compensation ability of the STATCOM.

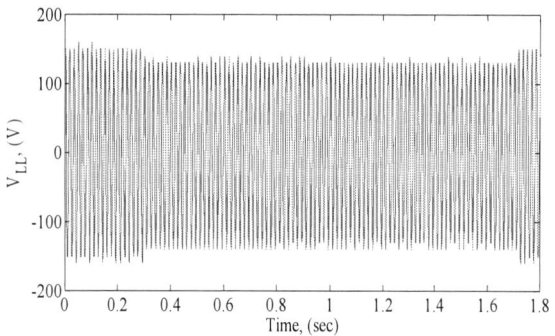

Figure 17. Reactive power responses of the two systems with disturbance.

Our first-stage work regarding STATCOM implementation still lacks the detailed experiments. Further effort is needed to validate the STATCOM control in the proposed test bed utilizing the Power Hardware-in-the-Loop concept. In addition, future work will focus on improvement of the very simple interfacing method utilized herein in order to obtain better current tracking and accommodate a wider range of test scenarios.

VI. CONCLUSIONS

In this paper, synchronization issues of three-phase power converters connected in parallel and utilizing the PHIL concept are discussed. Precise phase synchronization for equal load sharing with minimum current circulation between the paralleled power converter modules with different power rating, modulation strategies, etc., are achieved and experimentally verified. In addition, this paper has presented the outline of work for the proposed system configuration such as: a conceptual virtual microgrid, which utilizes the reactive power compensation ability of the STATCOM based on real-time digital simulation and PHIL concepts. An overview of the system was given, the basic premise of the research effort explained, justification of the research was provided, the current stage of development and data presented, and the most salient and challenging issues were addressed.

REFERENCES

[1] U. Borup, F. Blaabjerg, P.N. Enjeti, "Sharing of nonlinear load in parallel connected three-phase converters," IEEE Transactions on Industry Applications, vol. 37, Issue 6, pp.1817-1823, Nov.-Dec. 2001.

[2] J.-F.Chen and C.-L.Chu, "Combination voltage-controlled and current controlled PWM inverters for parallel operation of UPS," in *Proc. IEEE IECON'93*, vol. 2, pp. 1111–1116, 1993.

[3] V. Blasko, V. Kaura, "A New Mathematical Model and Control of a Three-Phase AC-DC Voltage Source Converter," IEEE Transactions on Power Electronics, vol. 12, Issue 1, pp.116-123, Jan 1997.

[4] W. Ren, M. Steurer, T. L. Baldwin, "Improve the Stability of Power Hardware-in-the-Loop Simulation by Selecting Appropriate Interface Algorithm", IEEE Transaction of Industry Applications, vol. 44, Issue 4, pp. 1286-1290, July-Aug. 2008.

[5] H. J. Slater, D. J. Atkinson, A. G. Jack, "Real-time emulation for power equipment development, II. The virtual machine," IEE Proceedings of Electric Power Applications, vol. 145, Issue 3, pp. 153-158, May 1998.

[6] X. Wu, S. Lentijo, A. Deshmuk, A. Monti, F. Ponci, "Design and implementation of a power-hardware-in-the-loop interface: a nonlinear load case study," in Proc. Twentieth Annual IEEE Applied Power Electronics Conference and Exposition, APEC 2005. vol. 2, pp. 1332-1338, March 2005.

[7] K. Zhou, D. Wan, "Relationship between space-vector modulation and three-phase carrier-based PWM: A comprehensive analysis", IEEE Trans. on Ind. Electronics, Vol. 49, Issue I, February 2002, pp. 181-196.

[8] L. G. B. Rolim et al., "Analysis and software implementation of a robust synchronizing PLL circuit based on the pq theory," IEEE Trans. Ind. Electron., vol. 53, Issue 6, pp. 1919–1926, Dec. 2006.

[9] IP Reference-Carrier-Modulator Rev 3, ABB, 2004.

[10] W. Ren, L. Qian, D.Cartes, M. Steurer, "Multivariable control method in STATCOM application for performance improvement", in Proc. Fortieth IAS Annual Meeting, IAS'05, vol.3, pp.2246-2250, 2005

A Single-stage Grid-connected Inverter with Wide Range Reactive Power Compensation using Energy Storage System (ESS)

Liming Liu, Zhichao Wu, Hui Li

Center for Advanced Power Systems
Florida State University
Tallahassee, FL, USA
liming@caps.fsu.edu, zcwu@caps.fsu.edu, hli@caps.fsu.edu

Abstract --This paper presents a single-stage grid-connected inverter with energy storage system (ESS) for small distributed power generation (DG) system application. A reactive power allocation (RPA) strategy is developed so the proposed inverter can provide real and reactive power in wide range. In particular, the ESS can provide not only harmonic compensation but also reactive power to enhance the system stability. An appropriate reactive power allocation coefficient is designed to avoid duty cycle saturation and over modulation in order to maintain good power quality. Furthermore, a control system, including proportional resonant control, capacitor voltage balance control and RPA control, is developed in grid-connected mode. Finally, a 3.5 kW single-stage grid-connected inverter with proposed control strategy is implemented in the laboratory. Both simulation and experimental results are presented to verify the validity of the proposed technology

I. INTRODUCTION

Grid-connected inverters with energy storage system (ESS) for small distributed power generation (DG) system have been gaining popularity due to the following advantages: (1) power quality can be improved; (2) voltage regulation capability can be improved; and (3) grid stability can be enhanced. However, the reported grid-connected inverters with ESS [1-6] require multiple conversion stages as shown in Fig.1. The dc side shunted topology has been used in [1-3], which can achieve flexible real power management but suffer the limited reactive power compensation due to the limited AC output voltage. The ac side shunted topology has been used in [4-6], which can implement the wide real and reactive power management but increase the complex and cost of the whole system. Cascaded multilevel inverters with ESS [7-10] have been used to achieve single-stage energy conversion, but suffer inadequate dc voltage usage and limited analysis on real and reactive power allocation.

This paper proposes a single stage grid-interactive inverter with ESS as shown in Fig.2. The advantages of the inverter are: (1) improved efficiency and lower cost by single stage conversion; (2) transformerless design; (3) higher dc boost ratio; (4) reduced switching loss by using hybrid phase-shift PWM. In addition, the topology can provide: (1)

a wide range of real and reactive power compensation; (2) fast dynamics for real and reactive power requirement; and (3) rejection capability against grid voltage disturbance and grid current distortion.

The main difficulties to implement wide range reactive power compensation and maintain power quality are: (1) how to separate the real and reactive component of inverter output voltage; (2) how to properly distribute the reactive power between distributed energy source (DES) and ESS; (3) how to achieve the voltage balance within ESS.

This paper firstly addresses the system topology and power flow distribution. A reactive power allocation strategy is then developed to implement wide range reactive power compensation and maintain high power quality. A vector diagram is derived to illustrate the reactive power

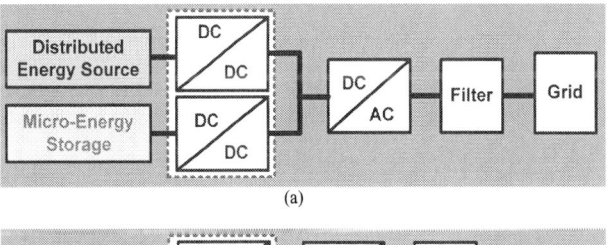

(a)

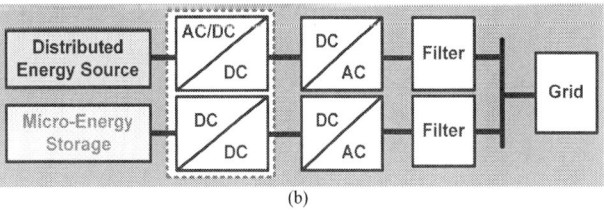

(b)

Fig.1 Reported grid-connected inverters with ESS: (a) DC side shunted topology; and (b) AC side shunted topology

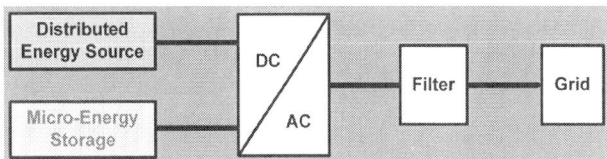

Fig.2 Proposed single-stage grid-connected inverter with ESS

978-1-4244-4782-4/10 $26.00 © 2010 IEEE

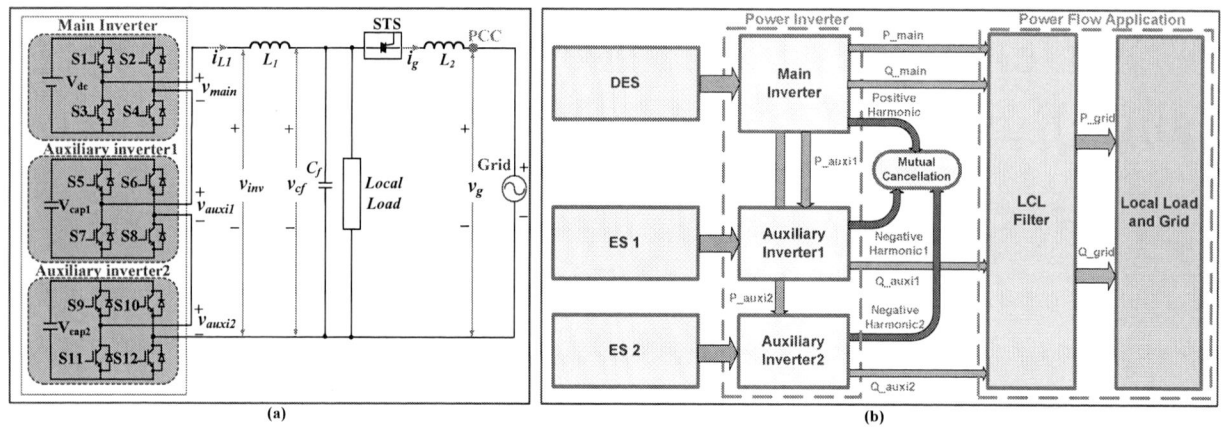

Fig.3 Proposed DG System with ESS: (a) Topology; and (b) Real and reactive power allocation between DES and ESS

distribution principle between DES and ESS. An appropriate reactive power allocation coefficient is chosen by 3D plot analysis in different conditions. The control system is then proposed to achieve wide reactive power compensation, voltage balance control of ESS, fast dynamics and zero steady state error. The proposed system is simulated using Matlab/Simulink and PSIM simulation software. A 3.5 kW grid-connected inverter has been built in the laboratory. Finally, simulation and preliminary experimental results confirm the validity of the proposed control.

II. SYSTEM DESCRIPTION

The single phase single-stage DG system is shown in Fig.3 (a). The "main" inverter is connected to DES and the "auxiliary" inverter cells are interfaced with energy storage elements, which are capacitors in this paper. The voltage ratio between V_{dc} and V_{cap} is 2:1. The "main" and "auxiliary" inverters switch at fundamental and PWM frequency, respectively. The research has revealed that the cascaded auxiliary cell number of 2 is the optimized number considering the trade off among the cost, power quality and reactive power compensation capability.

The real and reactive power allocation between DES and ESS is shown in Fig.3 (b). P_grid and Q_grid are P and Q delivered to the grid. P_main and Q_main are P and Q

generated from the main inverter cell. Q_auxi1 and Q_auxi2 are Q from auxiliary inverter cells. The main inverter provides all the real power and part of reactive power Q to the grid. Auxiliary inverters provide the rest of reactive power. P_auxi1 and P_auxi2 are delivered from main inverter to auxiliary inverter cells to charge ESS during start-up, compensate the power loss and maintain the ESS voltage during grid-connected mode. In addition, the low-order harmonic voltages generated by the main inverter resulting from fundamental switching frequency get cancelled by the equivalent negative harmonic voltage generated from auxiliary inverters. The DG system is able to operate in both stand-alone mode and grid-connected mode through a static transfer switch (STS). However this paper is focused on grid-connected mode.

III. REACTIVE POWER ALLOCATION STRATEGY

One advantage of cascaded structure is that the dc voltage of each inverter cell can be reduced. However, the reduced dc voltage will affect reactive power generation capability. In order to achieve wide range reactive power compensation, it is desired that the reactive power is provided by all the inverter cells instead of a single inverter cell. How to distribute the reactive power among inverter cells is designed based on the proposed RPA strategy.

A. Vector Diagram

A vector diagram is shown in Fig.4 to illustrate the calculation of reactive power distribution between DES and ESS. The rotation frequency of the *pq* frame is the system frequency. $\dot{I}_{L1}$, $\dot{V}_{inv}$ and $\dot{V}_g$ are the vectors of i_{L1}, v_{inv}, and v_g respectively where the latter can be referred in Fig.3 (a). $\dot{V}_{main\_F}$ is the fundamental component of main inverter output voltage. $\dot{V}_{invq}$ and $\dot{V}_{invp}$ is the q-axis and p-axis component of $\dot{V}_{inv}$ respectively in the vector diagram. Since the real power is entirely provided by the main inverter, the p-axis component of $\dot{V}_{main\_F}$ is the same as that of $\dot{V}_{inv}$. k is defined as reactive power allocation coefficient (RPAC). α is the phase shift angle between $\dot{V}_{inv}$ and $\dot{I}_{L1}$. β is the angle

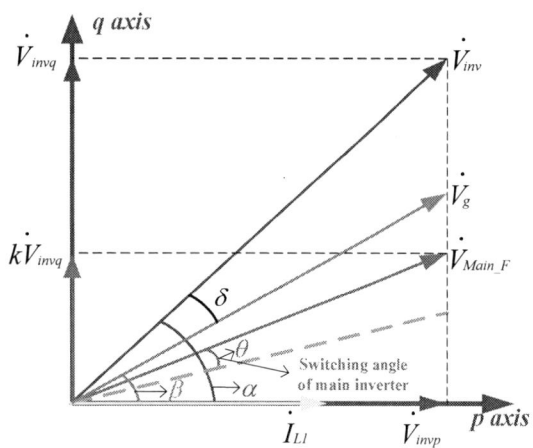

Fig.4 RPA vector diagram

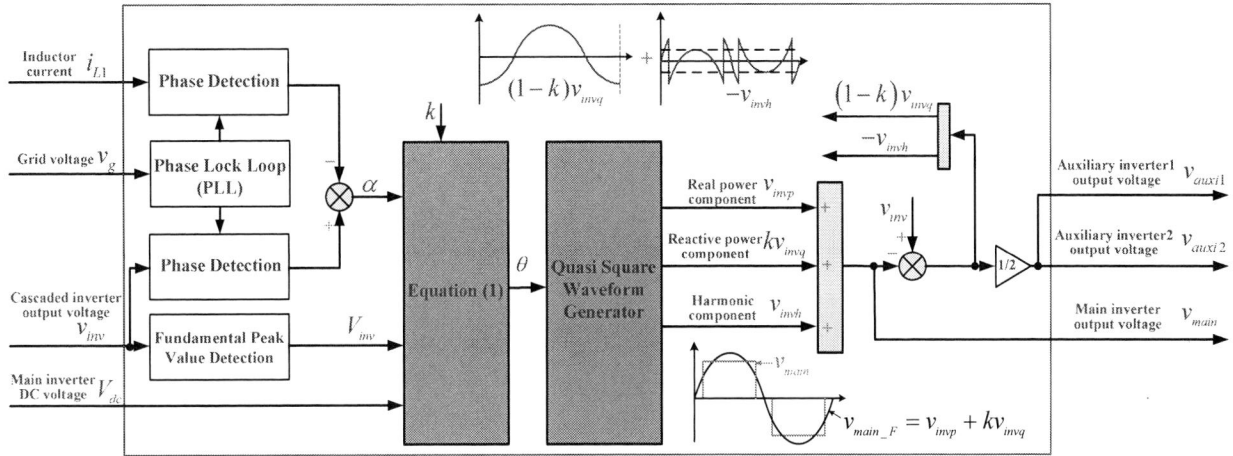

Fig.5 Proposed RPA strategy

between $\dot{V}_g$ and $\dot{I}_{L1}$. δ is the phase shift between $\dot{V}_{inv}$ and $\dot{V}_g$. θ is the switching angle of main inverter.

In grid-connected mode, the real and reactive power delivered to grid, as well as grid voltage, are known. The average real and reactive power delivered to the grid can be given by:

$$\begin{cases} P\_grid = \dfrac{V_g V_{inv}}{2\omega L}\sin\delta \\ Q\_grid = \dfrac{V_g}{2\omega L}\left(V_{inv}\cos\delta - V_g\right) \end{cases} \quad (1)$$

where V_g is the amplitude of grid voltage; V_{inv} is the amplitude of cascaded inverter output voltage; ω is the fundamental frequency; $L=L_1+L_2$ is the total filter inductor. The filter capacitor is neglected.

According to (1), the phase shift between grid voltage and cascaded inverter output voltage can be calculated as:

$$\delta = \tan^{-1}\left(\frac{P\_grid \times 2\omega L}{Q\_grid \times 2\omega L + V_g^2}\right) \quad (2)$$

The average real and reactive power delivered to grid can also be represented in (3):

$$\begin{cases} P\_grid = \dfrac{V_g I_{L1}}{2}\cos\beta \\ Q\_grid = \dfrac{V_g I_{L1}}{2}\sin\beta \end{cases} \quad (3)$$

where I_{L1} is the amplitude of inductor current.

So the angle between grid voltage and inductor current can be obtained in (4):

$$\beta = \tan^{-1}\left(\frac{Q\_grid}{P\_grid}\right) \quad (4)$$

The phase shift between cascaded inverter output voltage and inductor current can be obtained by adding (2) to (4) as:

$$\alpha = \beta + \delta \quad (5)$$

Based on (1), the V_{inv} can be derived in (6):

$$V_{inv} = \sqrt{\left(\frac{Q\_grid \times 2\omega L}{V_g} + V_g\right)^2 + \left(\frac{P\_grid \times 2\omega L}{V_g}\right)^2} \quad (6)$$

Fig.4 shows that the selection of θ with respect to the reactive power allocation and θ is derived in (7):

$$\theta = \cos^{-1}\left[\frac{V_{inv}}{V_{dc}} \times \frac{\pi}{4} \times \sqrt{\left(\cos\alpha\right)^2 + \left(k\sin\alpha\right)^2}\right] \quad (7)$$

where V_{dc} is the main inverter dc input voltage.

B. RPA Strategy

The reactive power allocation strategy with hybrid fundamental and PWM control is proposed in Fig.5. The "PLL" module generates the frequency and phase angle of v_g so i_{L1} and v_{inv} can be synchronized. The two "phase detection" modules are used to detect the phase of i_{L1} and v_{inv} by subtracting the v_g phase information produced by PLL. Therefore α can be derived. V_{inv} can be extracted from v_{inv} through the "fundamental peak value detection" module. After obtaining θ, the main inverter outputs the quasi square wave voltage v_{main}, including real power component v_{invp}, reactive power component kv_{invq} and harmonic component v_{invh}. As shown in Fig.5, the harmonic component v_{invh} can be

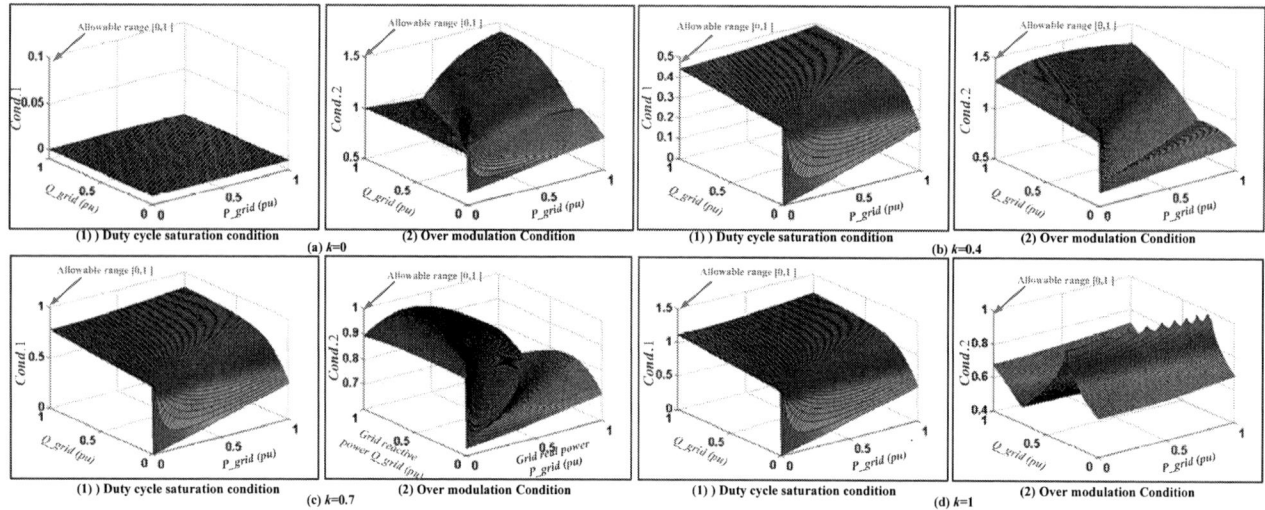

Fig.6 Operation conditions of single-stage unit with different k when P and Q changes from 0 to 1.0 pu.

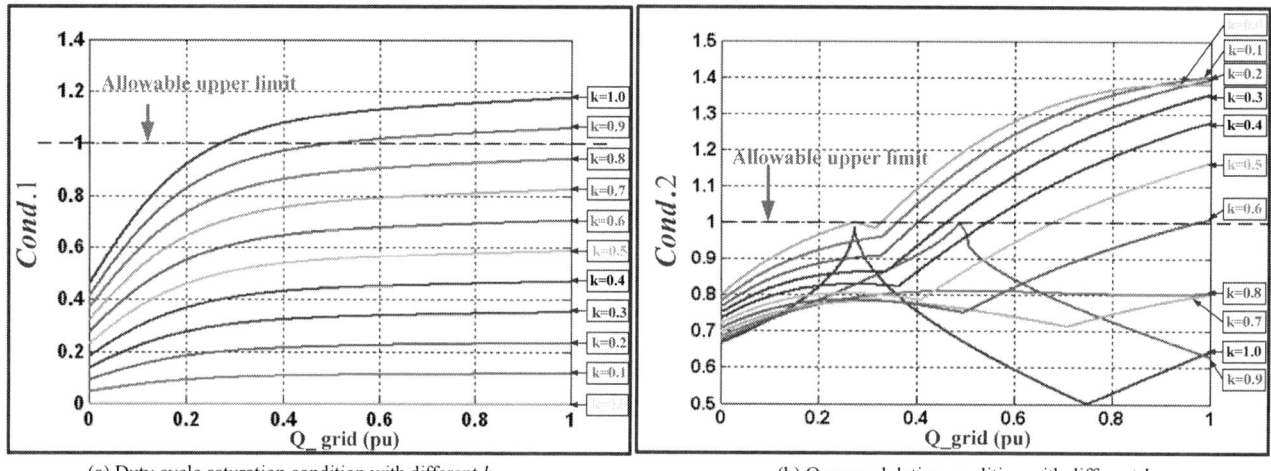

(a) Duty cycle saturation condition with different k

(b) Over modulation condition with different k

Fig.7 Operation conditions of single-stage unit with different k when P=1.0 pu. and Q changes from 0 to 1.0 pu.

canceled by negative v_{invh} from auxiliary inverter with PWM control to improve cascaded inverter output voltage quality. The auxiliary inverter also supplies the rest of reactive power component $(1-k)v_{invq}$ of v_{inv}. In this way, the reactive power can be distributed between main and auxiliary inverter cells.

C. Selection of k

k decides weighted q-axis voltage component in main inverter cell and auxiliary inverter cells, thus determines the reactive power in each cell. k can be varied from 0~1 to generate a wide range Q_grid under a required P_grid. However, inappropriate k will cause duty cycle saturation or over modulation resulting in degraded power quality. In addition, the appropriate range of k varies with P_grid. The design guidelines of selecting k is described as follows when P_grid and Q_grid vary from 0~1 pu, respectively.

Firstly, in order to avoid the duty cycle saturation, the assigned q component of v_{main}, i.e., kv_{invq}, should be no bigger than the available maximum magnitude of the q component

of v_{main}, which is calculated as $\sqrt{\left(4V_{dc}/\pi\right)^2-\left(V_{invp}\right)^2}$. The inequality condition in (8) is therefore derived and needs to be satisfied as the first limitation condition defined as *Cond.1*:

$$0 \le \frac{kV_{invq}}{\sqrt{\left(4V_{dc}/\pi\right)^2-\left(V_{invp}\right)^2}} \le 1 \qquad (8)$$

Secondly, in order to avoid over modulation, the peak value of $v_{auxi1} + v_{auxi2}$ should be no bigger than $V_{cap1} + V_{cap2}$, as the second limitation condition defined as *Cond.2*, which is shown in (9).

$$\frac{\max\left|(1-k)V_{invq}-V_{invh}\right|}{\left(V_{cap1}+V_{cap2}\right)} \le 1 \qquad (9)$$

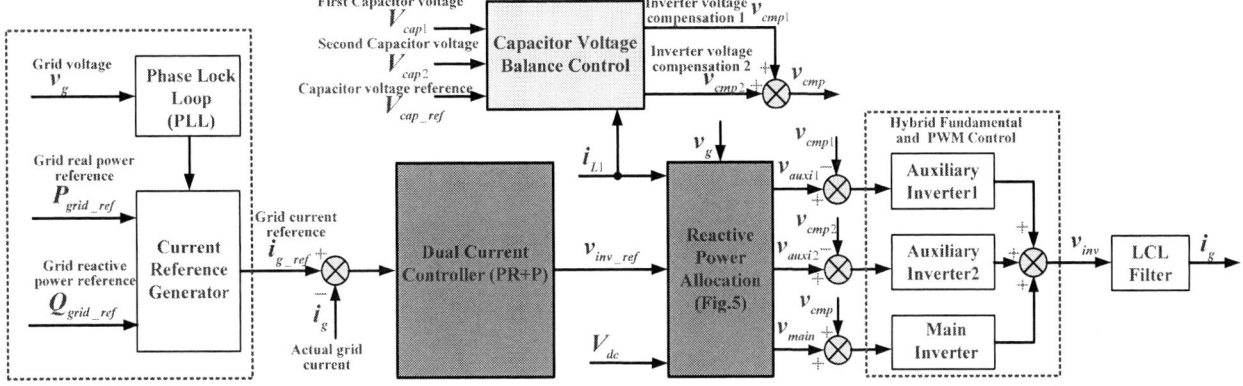

Fig.8 Control system block of single-stage DG unit in grid-connected mode

An appropriate k can be obtained based on (8-9). Four cases with different k shown in Fig.6 (a-1, b-1, c-1, and d-1) illustrate *Cond.1* variation as *P_grid* and *Q_grid* vary from 0 to 1.0 pu. Fig.6 (a-2, b-2, c-2, and d-2) illustrates *Cond.2* variation with different k under wide *P_grid* and *Q_grid* variation range. As shown in the 3D plot, when 'k' = 0, 0.4 and 1, (8) and (9) can not be satisfied simultaneously. The 'k' = 0.7 is a proper reactive power allocation coefficient under wide real and reactive power ranges.

In order to verify the above analysis further, the grid real power *P_grid* is fixed to 1.0pu with varied grid reactive power *Q_grid* and 'k' as depicted in Fig.7. The 'k' changes from 0 to 1 by 0.1. As shown in Fig.7 (a), the allowable range of *Cond.1* is in the interval [0, 1]. In this case, the reactive power allocation with '$k=0.9$' and '$k=1.0$' can not meet duty cycle saturation limitation condition in the whole grid reactive power. Fig.7 (b) shows the allowable range of *Cond.2* is in the interval [0, 1]. If the '$k{\le}0.6$, the over modulation limitation condition can not be satisfied in the whole grid reactive power.

IV. CONTROL SYSTEM DESIGN

Fig.8 shows the control system design of proposed DG unit operating in grid-connected mode. The grid current reference i_{g_ref} is generated by the 'current reference generator' module. An inner proportional (P) controller is cascaded with an outer loop proportional plus resonant (PR)

TABLE I: SYSTEM PARAMETERS

Parameters		Symbol	Value
DG system	DC link voltage	V_{dc}	140V (1.0 pu)
	Capacitor Voltage	V_{cap1}, V_{cap2}	70V (0.5 pu)
	Capacitor size	C_{cap1}, C_{cap2}	40 mF
	Filter Inductor	L_1, L_2	0.8mH (0.103 pu)
	Filter Capacitor	C_f	12uF (0.013 pu)
Grid	Rated real power	$P_{\_grid}$	3.5 kW (1.0 pu)
	Rated reactive power	$Q_{\_grid}$	3.5kVAR (1.0 pu)
	Rated RMS phase voltage	V_g	120V (1.0 pu)

controller to control the grid current i_g to track i_{g_ref} with zero steady-state error. The "reactive power allocation" module receives v_{inv_ref} from current controller and generates the references of v_{main}, v_{auxi1} and v_{auxi2}. The details of this module has been explained in section III and shown in Fig.5. The 'capacitor voltage balance control' module is developed to achieve two capacitors' voltage, V_{cap1} and V_{cap2}, to track the reference V_{cap_ref}. It receives the inductor current i_{L1} and generates the inverter voltage compensation components v_{cmp1} and v_{cmp2}. The sum $v_{cmp} = v_{cmp1} + v_{cmp2}$ is added to v_{main}, and sent to the hybrid fundamental and PWM control module to generate the main inverter voltage. Similarly, the v_{auxi1} - v_{cmp1} and v_{auxi2} - v_{cmp2} are sent to the control module respectively to generate the corresponding auxiliary inverter voltages.

V. SIMULATION AND EXPERIMENTAL RESULTS

The performance of proposed RPA strategy is firstly tested in simulation. The system parameters are shown in Table 1. In order to investigate the effect of 'k' on power quality, the grid current and its FFT results with different 'k' are compared as real and reactive power delivered to grid are fixed to 0.5pu and 1.0pu, respectively in Fig.9. As shown in Fig.9 (a), the grid current with 'k'=0.4 has a total harmonic distortion (THD) of 16.37%. However, THD can be decreased to 0.47% as 'k'=0.7 in Fig.9 (b). It is clear that 'k'=0.7 is a proper reactive power allocation coefficient.

Fig.10-12 shows the simulation results with 'k'=0.7. Fig.10 shows the two capacitor voltages, V_{cap1} and V_{cap2}, in self-startup and grid-connected mode. Before the normal operation of the DG system, the two capacitors are charged to the 0.5pu by self-startup. After 3s, the DG system operates at grid-connected mode. It can be seen from Fig.10 that the capacitors are kept approximately constant at 0.5pu while real and reactive power delivered to grid vary.

Fig.11 illustrates the dynamic response to step changes in the real power to grid reference *P*_grid* from 0 to 1.0pu at 4s and back to 0.5pu at 6s, reactive power to grid reference *Q*_grid* from 0 to 1.0pu at 5s and back to 0.5pu at 7s, and then to -0.5pu at 8s. It is obvious that the real and reactive power, *P_grid* and *Q_grid*, can track smoothly and fast their references.

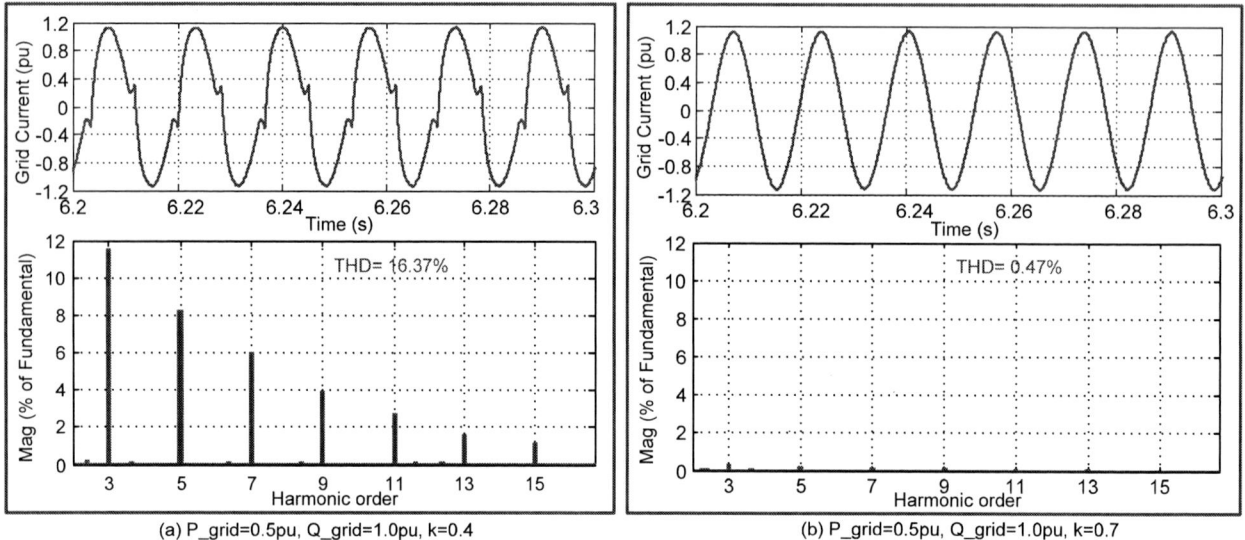

(a) P_grid=0.5pu, Q_grid=1.0pu, k=0.4

(b) P_grid=0.5pu, Q_grid=1.0pu, k=0.7

Fig.9 Grid current and total harmonic distortion as *P_grid*=0.5pu, *Q_grid*=1.0pu with different '*k*'

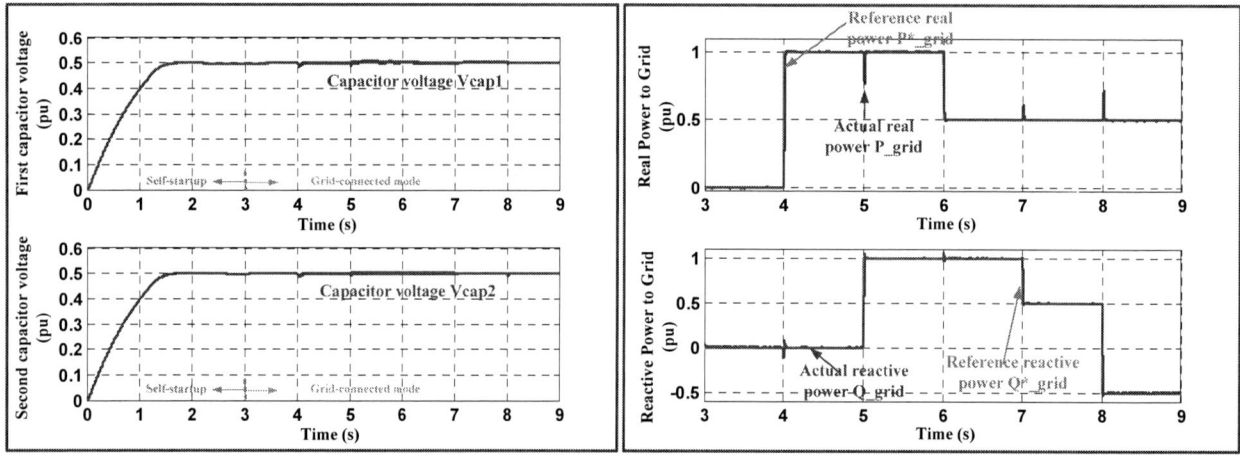

Fig.10 Two capacitor voltages in self-startup and grid-connected mode

Fig.11 The real and reactive power delivered to grid in grid-connected mode

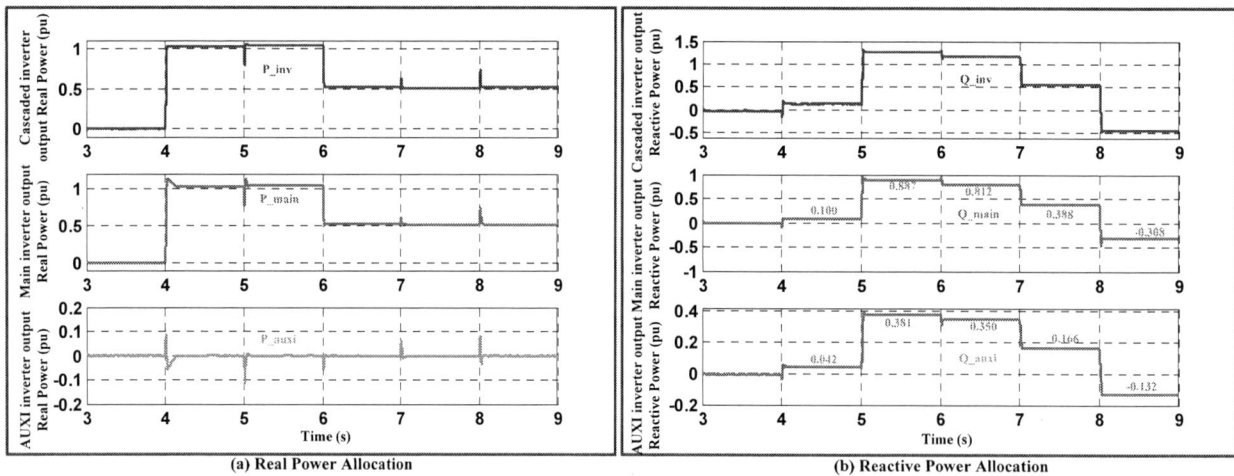

(a) Real Power Allocation

(b) Reactive Power Allocation

Fig.12 The real and reactive power from cascaded inverter, main inverter and auxiliary inverter with '*k*'=0.7

978-1-4244-4782-4/10 $26.00 © 2010 IEEE

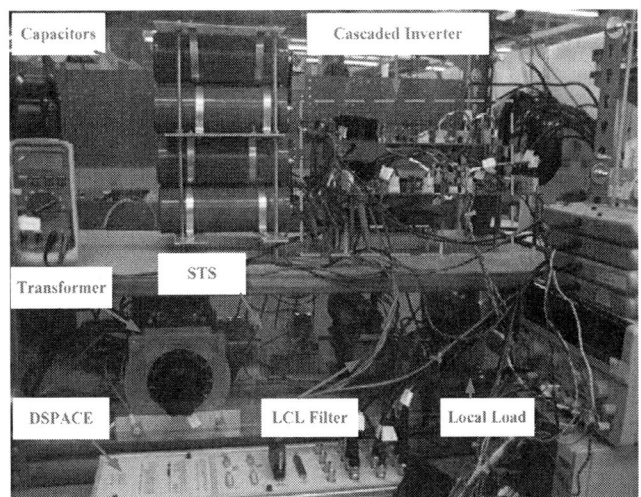

Fig.13 Experimental setup

Fig.14 Experimental results of real and reactive power delivered to grid P_grid, Q_grid, and their reference P*_grid, Q*_grid

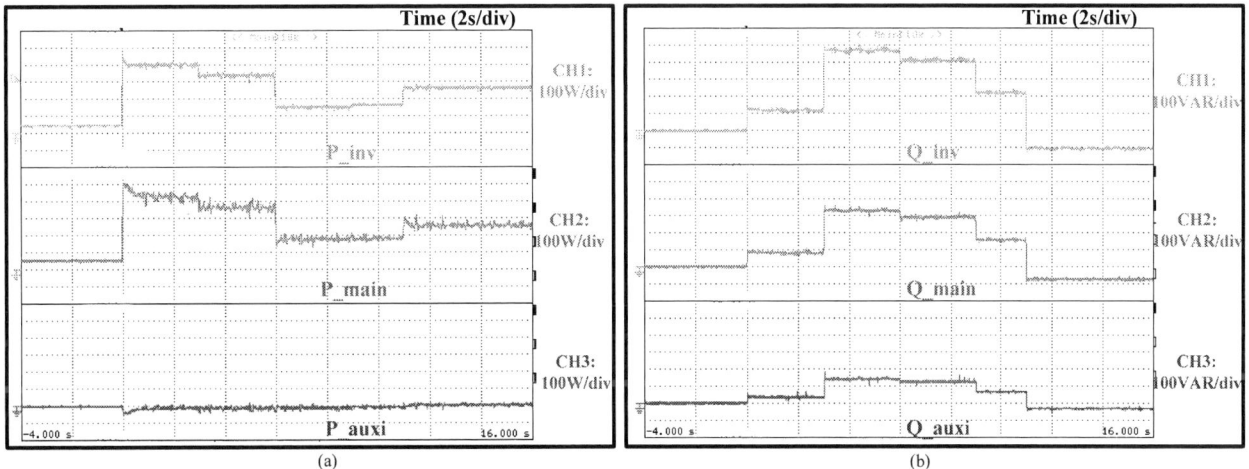

(a)   (b)

Fig.15 Experimental results of power allocation: (a) Real power allocation of cascaded inverter, main inverter and auxiliary inverter, P_inv, P_main, P_auxi, (b) Reactive power allocation of cascaded inverter, main inverter and auxiliary inverter, Q_inv, Q_main, Q_auxi

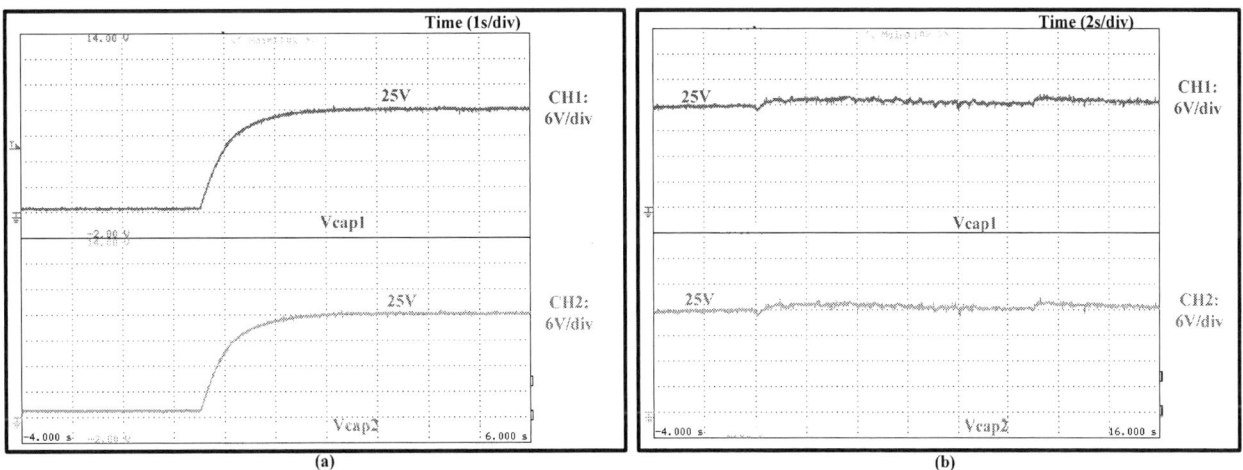

(a)   (b)

Fig.16 Experimental results of two capacitor voltages, V_{cap1} and V_{cap2}: (a) Self-startup; (b) Grid-connected mode

978-1-4244-4782-4/10 $26.00 © 2010 IEEE   229

Fig.12 depicts the real and reactive power from cascaded inverter, main inverter and auxiliary inverter while the *P_grid* and *Q_grid* vary. At steady-state, the real power provided by main inverter *P_main* is equal to the real power from cascaded inverter *P_inv* as shown in (a). The real power from auxiliary inverter *P_auxi* changes only during power transition to keep the capacitor voltage constant. The *P_inv* is a bit more than *P_grid* due to the system loss. As shown in (b), the reactive power from cascaded inverter *Q_inv* is also more than *Q_grid* due to the system loss. The ration between reactive power from main inverter *Q_main* and auxiliary inverter *Q_auxi* is always 0.7 as *P_grid* and *Q_grid* vary.

A 3.5kW hardware prototype has been built in the laboratory and is shown in Fig.13. The IGBT FMG2G100US60 has been chosen as main inverter switch operating at fundamental frequency. The MOSFET SUP85N15-21 has been used in the auxiliary inverters operating at 2.5 kHz. The control algorithms are implemented in dSPACE DS1104 controller. The experimental results at 350W with '*k*'=0.7 are presented from Fig.14 to Fig.16.

Fig.14 shows the response of real and reactive power to grid at step changes in grid-connected mode. *P*_grid* increases from 50W to 350W and back to 175W. *Q*_grid* increases from 0 to 350VAR and then drops to 175VAR, finally reaches to -175VAR. The real and reactive power distribution between main inverter and auxiliary inverters are shown in Fig.15. It can be seen from (a) that the *P_inv* is very close to *P_main* at steady state. *P_auxi* changes a small amount only to keep capacitor voltage constant. As anticipated, the ratio between *Q_main* and *Q_auxi* is maintained to be 0.7 in (b). *Q_inv* is more than *Q_grid* due to the system reactive power loss. Fig.16 (a) shows that the two capacitor voltages charges to 25V fast during startup process. Capacitor voltages are stabilized to 25V in grid-connected mode when power varies in Fig.16 (b).

VI. CONCLUSION

The grid-connected inverter proposed in this paper can provide real and reactive power in wide range with the assist of ESS. In particular, the auxiliary inverters can provide not only harmonic compensation but also reactive power to enhance the system stability. However, inappropriate reactive power allocation coefficient '*k*' can cause duty cycle saturation or over modulation, and thus lead to output voltage distortion. A RPA strategy was developed and a proper '*k*' was derived from RPA analysis in this paper to meet a wide variation of real and reactive power without degrading power quality. In addition, capacitor voltages balance control was developed and integrated with RPA control. The simulation and experimental results confirmed that the proposed RPA strategy worked well under grid-connected mode.

REFERENCES

[1] Y. Li, D.M. Vilathgamuwa, P. C. Loh, "Design, Analysis, and Real-Time Testing of a Controller for Multibus Microgrid System", *IEEE Trans. Power Electron.*, vol. 19, no.5, pp.1195-1204, Sep. 2004.

[2] H. Tao, J. L. Duarte, M.A.M. Hendrix, "Control of Grid-Interactive Inverters as Used in Small Distributed Generators" , in *Conf. Rec. IEEE IAS Annu. Meeting*, Sep. 2007, vol. 3, pp. 1574-1581.

[3] X. Zhu, D. Xu, P. Wu, G. Shen, P. Chen, " Energy management design for a 5kW fuel cell distributed power system," " in *Proc. IEEE APEC*, Feb. 2008, vol. 1, pp. 291–297.

[4] D. Georgakis, S. Paptjanassiou, N. Hatziargyriou, A. Engler, C. Hardt, "Operation of a Prototype Micro-grid System Based on Micro-sources Equipped with Fast-acting Power Electronics Interfaces", in *Proc. IEEE PESC*, Jun. 2004, vol. 4, pp. 2521–2526.

[5] K. Rajashekara, "Hybrid Fuel-Cell Strategies for Clean Power Generation", *IEEE Trans. Ind. Appl.*, vol. 41, no.3, pp.682 -689, May/Jun. 2005.

[6] M. Xu, C. Wang, Y. Qiu, B. Lu, F. C.Lee, G. Kopasakis, " Control and Simulation for Hybrid Solid Oxide Fuel Cell Power Systems", in *Proc. IEEE APEC*, Mar. 2006, vol. 2, pp. 1269–1274.

[7] M.D. Mabhrekar, P.k. Steimer, and T.A. Lipo, "Hybrid Multilevel Power Conversion System: A Competitive Solution for High-Power Applications," *IEEE Trans. Ind. Appl.*, vol. 36, no.3, pp. 834–841, May/June 2000.

[8] J. Rodriguez, J.S. Lai, and F.Z. Peng, "Multilevel inverters: A Survey of Topologies, Controls and Applications," *IEEE Trans. Ind. Electron.*, vol. 49, no.4, pp. 724–738, Aug. 2002.

[9] Z. Du, L. M. Tolbert, J. N. Chiasson, and B. Özpineci, "A Cascade Multilevel Inverter Using a Single DC Source," in *Proc. IEEE APEC*, Mar. 2006, vol. 1, pp. 426–430.

[10] L. Maharjan, S. Inoue, H. Akagi, "A Transformerless Energy Storage System Based on a Cascade Multilevel PWM Converter with Star Configuration," *IEEE Trans. Ind. Appl.*, vol. 44, no.5, pp. 1621–1630, Sep./Oct. 2008.

Polymer Bonded Soft Magnetics for EMI Filter Applications in Power Electronics

S. Egelkraut, L. Frey
Chair of Electron Devices
University of Erlangen-Nuremberg
Erlangen, Germany
Email: sven.egelkraut@leb.eei.uni-erlangen.de
http://www.leb.eei.uni-erlangen.de

M. Rauch, A. Schletz, M. März
Fraunhofer Institute for Integrated
Systems and Device Technology (IISB)
Erlangen, Germany
http://www.iisb.fraunhofer.de/

Abstract—In this study, polymer bonded soft magnetic materials (PBSMM) were investigated for the application as a magnetic core and electromagnetic shielding material in inductive devices for EMI filter applications. The nature of switch mode power converters makes them a potential source of EMI noise emission. EMI filters are generally necessary to ensure electromagnetic compatibility of converters to the other electronic equipment. Conventional discrete EMI filters usually comprise passive components with different volume and form factors. The manufacturing of conventional inductive components requires different processing and packaging technologies, of which many include cost intensive processing steps. Due to the parasitics of the discrete components and their interconnections the effective filter frequency range is limited. As a result discrete EMI filters are usually not integrable into an arbitrary formed volume and show relative high production costs. This study aims on solving this issue by the integration of inductive EMI filter components using polymer bonded soft magnetics. PBSMMs were produced using thermoplastic polyamide 6 matrix materials. The filler materials were chosen from the wide range of different soft magnetics. The magnetic properties were characterized using injection molded ring core test specimens and a computer controlled hysteresis recorder as well as an impedance analyzer. Inductive devices with PBSMM as magnetic core have great potentials in automotive applications that have to meet a high geometric flexibility and highest power densities.

I. INTRODUCTION

Power electronics has been continuously improved by new semiconductor devices and materials, an continuously increasing switching frequency and advanced integration technologies. The need for high integration level, high performance, and reduced production costs of power electronics is the driving force for polymer bonded soft magnetics inductor technologies. Especially in automotive applications the space requirements and the possibility to form the devices without any restrictions in the outer form is one outstanding argument for the use of polymer bonded inductive components. Recent studies on polymer bonded soft magnetic materials showed the perspective to produce soft magnetics with a high saturation flux density, useful permeability values, and low coercitivity for low frequency applications [1]. Metallic materials like Fe-Si3 [2], [3] or nano crystalline FeSiBCuNb [4], [5] show good magnetic properties with the capability of processing these materials with conventional polymer processing technologies. Polymer bonded soft magnetics were investigated for planar

transformer production [6] and verify the possibility to build up flat and integrated power converter for example for gate driver applications.

Any switching electronic device is a potential EM noise source. High level electromagnetic disturbances may cause electronic systems to malfunction in a common electromagnetic environment [7].

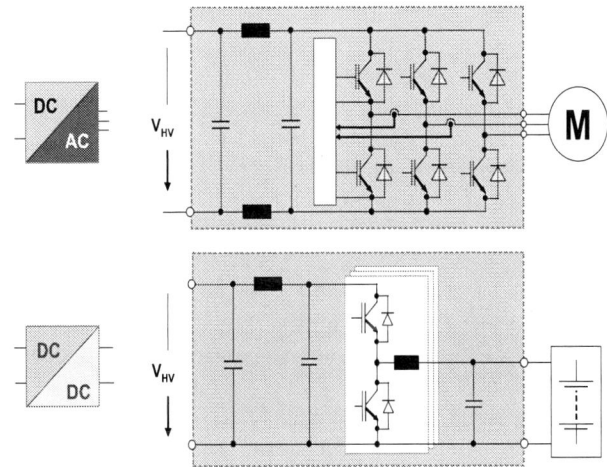

Fig. 1. EMI filter applications in power electronics

Conventionally, EMI filters are normally implemented by using discrete conventional components. The total volume of such an assembly is dominated by the passive components and the death volume between the chunky devices. Schematics of two typical EMI filter structures are shown in Fig. 1. Film capacitors enable the integration of capacitive devices into a volume but with strictly restrictions on the outer form of this. So the effective degrees of freedom in the form of these devices are limited by the processing technology but they are much larger than for example of electrolytic capacitors. The integrated hybrid drive presented in [8] e.g. uses a ringshaped dc link capacitor in order to achieve an optimum filling of the available package volume. This capacitor has been developed in cooperation with the Epcos AG and provides a capacitance of $500\mu F$ (450V).

Recent studies on integrated inductive devices focus on

978-1-4244-4782-4/10 $26.00 © 2010 IEEE

power converters and therefore mostly on planar structures and low power applications. [9], [10]. The aim of this study was to produce and characterize polymer bonded soft magnetics with application adapted magnetic properties and to realize inductive devices for high frequency EMI filter applications using these materials. The following demands and targets have been specified for the exemplary EMI filter made of polymer bonded soft magnetics:

- An optimal use of the available space
- The potential to produce the device at lower production costs with highly efficient production processes
- An ampacity for high DC currents up to 125 A
- An attenuation sufficient to push the noise below the limits for conducted emissions as defined in several standards

II. EMI CHARACTERIZATION AND FILTER TOPOLOGY

The integrated hybrid drive described in [8] was characterized regarding its EMC behavior in order to estimate the necessary attenuation characteristic of the EMI filter. Only conducted differential mode emissions were considered, because any other emissions are highly depending on the overall system configuration. The drive unit was mounted on an emotor test bench for the characterization under various torque and speed load conditions. To avoid any interference with heavyduty electronic power supplies, the drive was fed from a NiCd battery pack, configured for a nominal voltage of 288 V, and capable to provide energy of up to 8 kWh. The DC link voltage noise was measured close to the DC link capacitor using both a spectrum analyzer (R & S FSP3 with FSP-B29) and an oszilloscope (Tektronix TDS5034B). An AC coupling capacitor and an attenuator (20dB) were inserted to protect the spectrum analyzer, the additional attenuation is considered in Fig. 2. The DC link current was measured using a Tektronix current probe TCP303 with TCPA300. Fig. 2 shows the DC link voltage ripple under maximum load, with a speed of 1500 rpm corresponding to a fundamental frequency (f_0) of the inverter output current of 200 Hz; the switching frequency (f_s) of the inverter is 8 kHz.

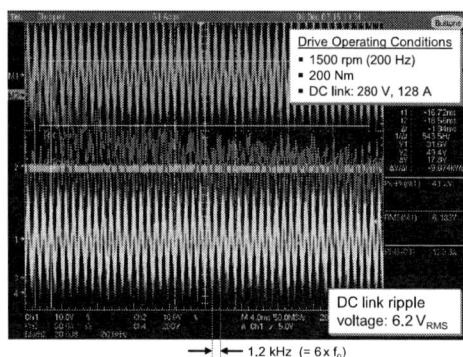

Fig. 2. DC link ripple voltage

Fig. 3 shows the spectrum up to 10 MHz. According to the limiting lines given in Fig. 3, an attenuation of about 22 dB

at 2.2 MHz is necessary to push the noise below the limits defined, e.g., in the Daimler standard DC 10614.

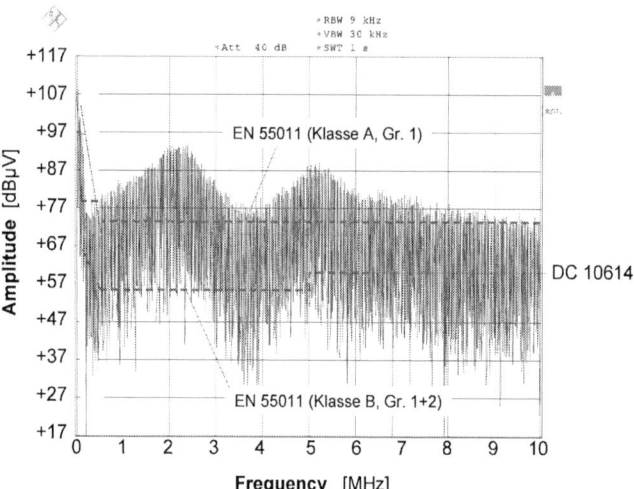

Fig. 3. Spectrum of the conducted DC link voltage noise with some limiting lines

Fig. 4 depicts the AC equivalent circuit of the drive inverter including the chosen filter topology, the high-voltage cable L_{La} and L_{Lb}, and the electrical parameters of the battery system v_{bat} and R_i. L_{1a} and L_{1b} are the filter inductors presented in this paper.

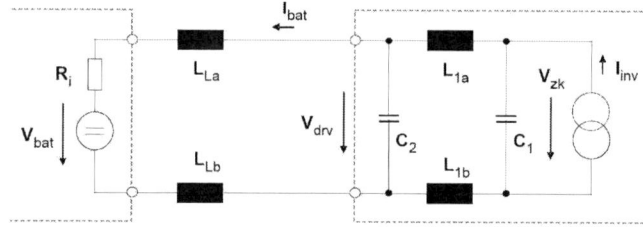

Fig. 4. Inverter with chosen EMI filter topology and HV supply equivalent circuit. C_1 is the main DC link capacitor. The available space allows a maximum capacitance of 60μF for the filter capacitor C_2

III. MATERIALS AND PROCESSES

This chapter presents the used materials, processes and the measurement equipment.

A. Filler Materials

Magnetic materials can be divided into different groups depending on their magnetic properties. Iron, cobalt, and nickel based metallic alloys like FeSi, NiFe, and CoFe show a high saturation flux density (2.35 T for CoFe) and a high permeability at low frequencies. These properties are well suited for e.g. EMI filters and low frequency chokes. The most effective manner to produce metallic powders is the water or gas atomization. The final particles have a spherical geometry shown in Fig. 5 resulting in a low viscosity of the polymer - filler compound. For our PBSMM investigations, an iron powder (FeSi6.8 from Höganäs) was used. The saturation inductance of bulk material test specimens made of this material is 1.6 T.

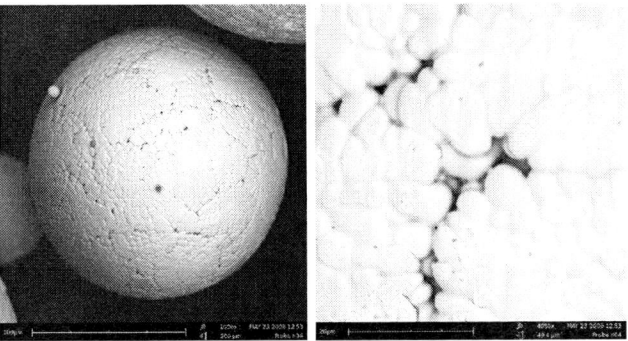

Fig. 5. SEM pictures of the water atomized FeSi6.8 powder right: surface structure

B. Polymers and Processing

The polymer matrix materials was an Ultramid B27 from BASF, an semicrystalline thermoplastic material. It has a low viscosity and good thermal and mechanical properties regarding the present application. A PA6 containing 20vol.%, 40vol.%, 50vol.%, 60vol.% and 65vol.% filler powder was prepared in a twin screw extruder. Polymer injection molding is a production process for manufacturing components from thermoplastic and thermosetting plastic materials and was uesd for the plastic molding. An injection molding tool was used for ring core test specimens' production. A picture of the molding tool is shown in Fig. 6 left.

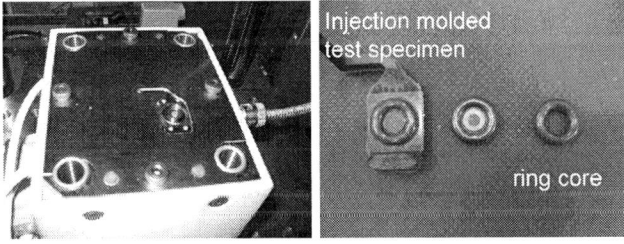

Fig. 6. Injection molding tool and test specimens

Polymer pressure molding was used for the inductive device manufacturing in order to reduce the costs for a complex injection molding tool. This process consists of filling the tool with the polymer compound and pressing the final device in a vacuum oven.

C. Magnetic Characterization

Toroidal cores were chosen due to the homogenous field distribution resulting in an accurate measurement. The magnetic permeability was measured with a precision impedance analyzer for frequencies up to 100 MHz (Agilent 4294A). The magnetic losses were measured with a computer controlled hysteresis recorder similar to that described in [11]. A schematic of this recorder is shown in Fig. 7.

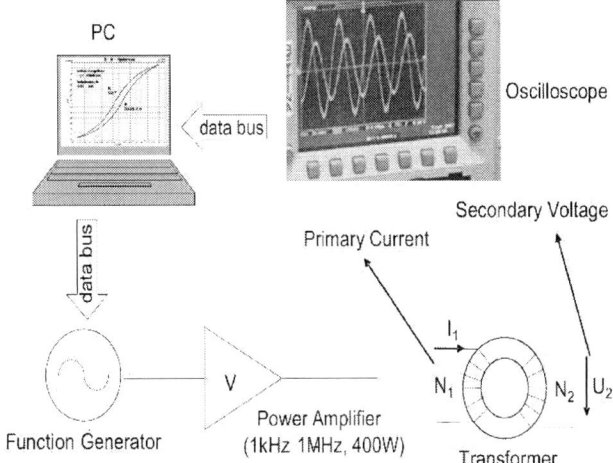

Fig. 7. Hysteresis recorder

The specific magnetic hysteresis power losses per volume are related to the area of the hysteresis curve and the frequency, and can be expressed by the following equation:

$$\frac{P}{V} = f \cdot \oint H \partial B \qquad (1)$$

where f is the frequency of the applied field, H the magnetic field strength, and B the magnetic flux density. The saturation flux density was measured using a vibrating sample magnetometer (VSM).

IV. MATERIAL CHARACTERIZATION RESULTS

The measurements of the magnetic properties were performed using toroidal test specimens with an effective magnetic length of 40 mm and a core cross-section area of 11.3 mm^2 with a winding comprising 10 turns. The test specimens are shown in Fig. 6 right.

A. Frequency Characterization

The permeability of the polymer bonded soft magnetics was calculated from the measured inductance of the ring cores. The investigations showed only a small deviation in the value of the inductance over a number of 30 test specimens. Therefore constant filler contents in all test specimens can be assumed. The permeability of the polymer bonded soft magnetics depends on the filler content and increases with increasing filler fraction up to 30 at a measurement frequency of 10 kHz. The polymer compounds with the spherical FeSi6.8 filler particles show a low viscosity of the polymer melt. This results in high processable filler fractions up to 65 vol.%. The resulting permeability stays quite constant up to a frequency of around 1 MHz for highest filler fractions. The permeability measurements of the test specimens are given in Fig. 8.

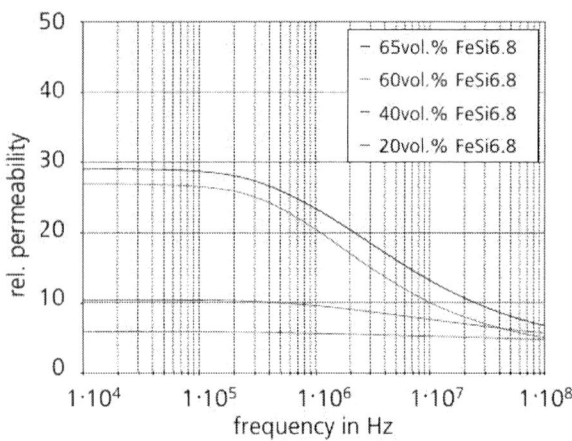

Fig. 8. Rel. permeabilities of the polymer compounds

B. Saturation Flux Density

The measurement was done using a vibrating sample magnetometer (VSM). For this purpose the test specimens were physically vibrated sinusoidally in an outer magnetic field. The induced voltage in the pickup coil was measured and is proportional to the sample's magnetic moment.

The volume of the ring cores was assumed to be totally magnetic even if the compound shows a defined filler fraction of the magnetic material. This was done in order to measure the saturation inductance of the total compound volume and not of the filler material itself. The measured values are given in Tab. I. In comparison to other magnetic materials for high

TABLE I

B_{Sat} OF THE POLYMER COMPOUND AT A GIVEN FILLER FRACTION

Filler fraction	20	40	50	60	65
B_{Sat} FeSi measured	0.29	0.59	0.72	0.89	0.97

frequency applications this saturation values are between the ferrites with nearly 500 mT and the pressed iron powder cores with up to 1.6 T.

C. Power Losses

In the case of magnetic materials, losses are attributed to three physical mechanisms [14]. The different energy losses per cycle are given by:

$$W = W_h + W_{cl} + W_{exc} \qquad (2)$$

W_h represents the hysteresis losses for quasi static frequencies which are assumed to be constant with frequency. W_{cl} represents the eddy current losses which are directly connected with the electrical conductivity of the magnetic material. W_{exc} represents the losses due to the high frequency dynamic movements of the magnetic domains. The measured losses of the FeSi filled PBSMMs are shown in Fig. 9, all losses

are referred to the total core volume. The magnetic area was defined as the geometric crosssectional area of the toroidal core. At a given macroscopic flux density, PBSMMs generally reveal higher power losses per volume than the corresponding soft magnetic raw materials. Since the magnetic particles are dispersed in a polymer matrix, the effective magnetic cross section is lower depending on the filler content. In addition,

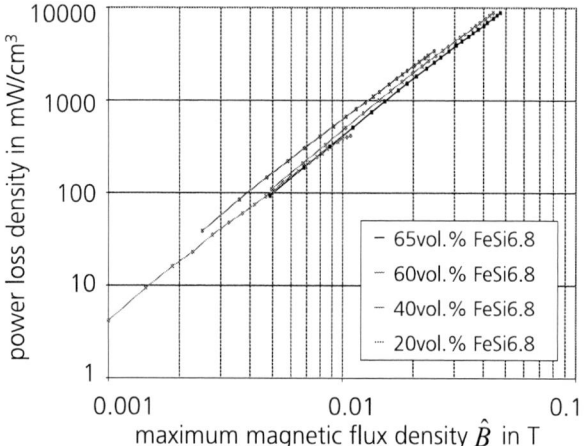

Fig. 9. Power losses of FeSi6.8 filled PBSMMs

compared to solid soft magnetics there is no homogeneous flux density across the total cross section of a PBSMM toroidal core. This leads to an increased magnetic flux density in the magnetic particles, and thus to increased power losses. This fact gains influence especially for particles which are in touch with each other. The magnetic flux takes lines of least magnetic resistance and hence the flux density exaggerates in the points of contact between the particles. This flux density concentration is expected to be a dominating power loss source for the spherical filler particles at a filler fraction around 40 vol.% and results in an increasing of all loss mechanismens. Especially eddy current losses increase with increasing electrical conductivity. At lower filler fractions the particles are not in touch with each other and each particle is covered by an insulating polymer layer. With increasing filler content more and more particles get in contact.

At filler fractions around 40 vol.% magnetic and electric paths through the magnetic material arise from the particles touching each other. These magnetic and electric percolation networks show only a small effective magnetic and electric aera. At higher filler contents the effective magnetic area increases and therefore the power losses due to the magnetic flux density concentration decrease. Pictures of the particle distribution at different filler fractions are given Fig 10.

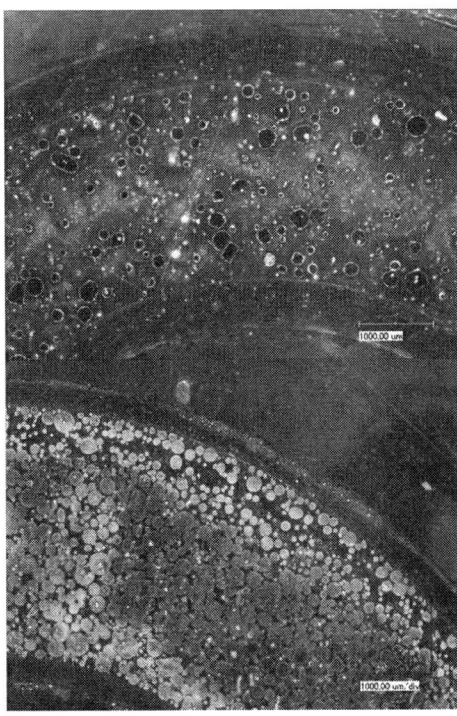

Fig. 10. Cross sections of the FeSi filled polymer ring cores at 20 vol.% (top) and 60 vol.% (bottom)

V. ELECTROMAGNETIC DESIGN OF THE EMI FILTER DEVICE

Inductive devices made of PBSMM don't feature predefined and catalogued geometries. For different winding arrangements in the given construction space the effective magnetic parameters like the effective magnetic area or the magnetic length, the inductance and the maximum flux density were calculated using an electromagnetic FI (finite integration) simulation software (CST-EM-Studio).

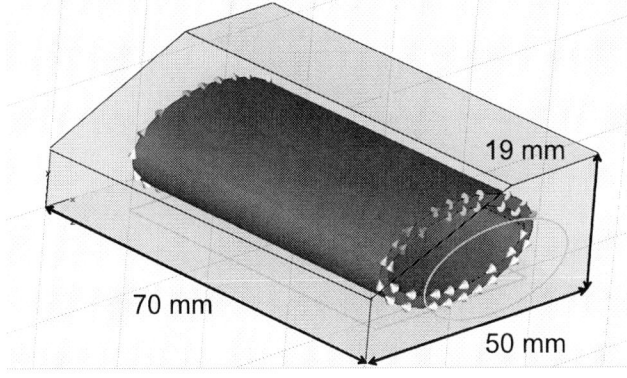

Fig. 11. Simulation geometry A

The complexity of the manufacturable design was simplified

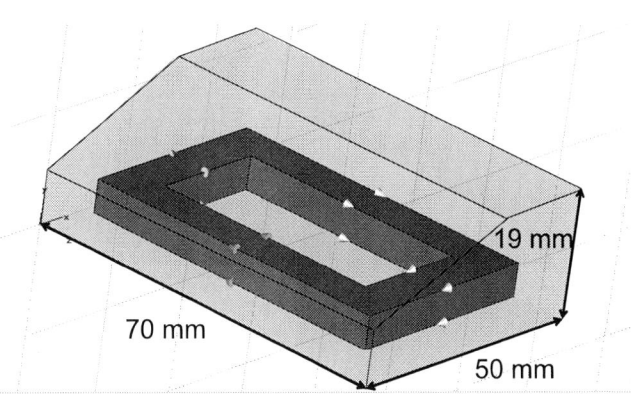

Fig. 12. Simulation geometry B

in order to reduce the simulation time. The windings of the inductor were simulated using a predefined function reducing the physical windings to a current carrying volume with a chosen number of turns. The goals of these simulations were:

- Calculation of the inductance and the maximum flux density of the magnetic material in the predefined volume.
- Choosing that winding geometry with an optimal B-field distribution

Winding geometries analyzed in the following are:

- Spiral winding normal to the base plate with a magnetic flux orientation in the longitudinal axle of the inductor as shown in Fig. 11.
- Rectangular winding parallel to the base plate with a magnetic flux orientation normal to the base plate as shown in Fig. 12.

The simulation results show that the simulation geometry A realizes a more homogenous field distribution. The effective magnetic area is nearly constant over the total magnetic length. Therefore the magnetic flux density shows no local concentration. The simulation geometry B shows higher maximum flux densities due to local flux concentration resulting in higher power losses. In addition these simu-lations disqualify the Vitroperm fillers for the given application due to the reduced saturation inductance in comparison to the FeSi filled polymer. The simulation results are shown in Fig. 13.

In addition to the flux density the inductance was calculated for the two winding geometries in order to estimate the required filler fraction for the soft magnetic polymer. The simulation results are shown in Fig. 14.

The simulation illustrates that permeability values higher than 22 are sufficient to produce an inductor with an inductance higher than our target value of 5 μH. The usage of a material with a permeability of 28 like it was measured for the 60 vol.% FeSi filled polymer results in an inductance of 6.5 μH for the simulation geometry A. The simulation results for this material predict a maximum flux density lower than 0.6 T which is much lower than the measured saturation

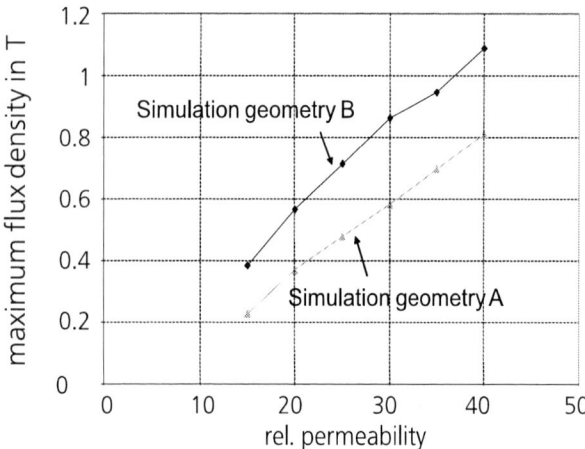

Fig. 13. Max. flux density B_{max} in T vs. permeability

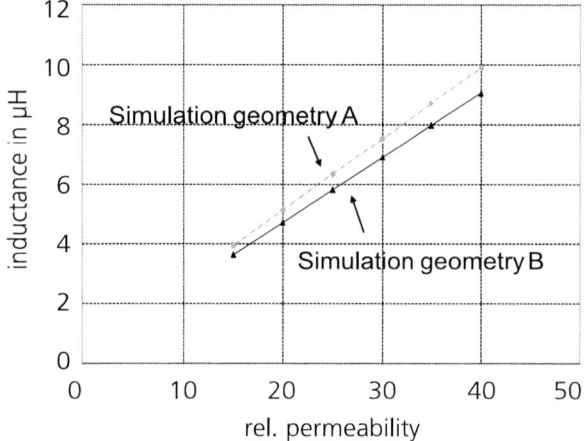

Fig. 14. Simulated inductance L vs. permeability

inductance of 0.89 T. Therefore the simulation results verify the possibility to produce powerful inductive devices for EMI filters using polymer bonded soft magnetics.

VI. MECHANICAL DESIGN

An elliptic crosssection of the coil was chosen in the electromagnetic simulations as it utilizes the available volume best. The radii of the coil shown in Fig. 11 are a=16.5 mm and b=7.5 mm (measured from the core centre to the centre of the wire). The wire has a cross section of 6x2 mm^2 as it is required for carrying a mean current of 125 A. To satisfy the process and reliability requirements a wall thickness of 2 mm from the coil surface to the outer device wall was chosen. The area of the final manufactured ellipse was reduced insignificantly in order to ensure this wall thickness. To ensure high voltage insulation the copper wire was insulated by a polymer finish. A plastic cap was constructed in order to realize the electric contact between the device and the electronic environment using two screws. The total mechanical design is given in Fig. 18. For the realization of a prototype choke, a polymer pressure molding

tool was constructed based on the mechanical design of the inductive device.

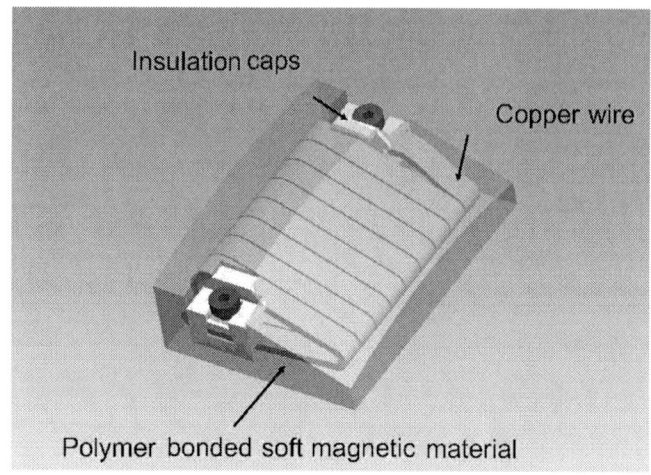

Fig. 15. Mechanical design

VII. THERMAL DESIGN

The aim of the thermal simulation was the calculation of the maximum allowable power losses to ensure a maximum operating temperature of 125°C for the polymer matrix. The thermal conductivity of the polymer compound strongly depends on the filler, the polymer matrix and the temperature. Values from 0.5 W/mK up to 10 W/mK are known for highly filled polymers in the literature [15]. Due to the high filler fraction and the spherical filler geometry a linear increasing of the thermal conductivity from 1 W/mK up to 4 W/mK with an increasing filler fraction 20 vol.% up to 65vol.% was expected [16] and taken as a input parameter for the simulations. For the simulation the temperature of the water cooled heat sink was set to 90°C in order to simulate the later application. The ohmic losses at an rms current of 125 A were calculated to 20 W. The contact layer between the inductive device and the cooling channel was modeled with a thermal conductance value of k=10 W/mm^2 K (assuming a layer of conductive adhesive with a thickness of d=0.05 mm and a thermal conductivity of λ=5 W/m K). The contact layer between the encapsulation and the coil has a thermal conductance value of k=0.4 W/mm^2 K (assuming a layer of isolation resin with a thickness of d=0.5 mm and a thermal conductivity of λ=0.2 W/m K). For the simulation the magnetic power losses were varied from 0 W up to 80 W depending on the load state of the synchronous drive.

The simulation results shown in Fig. 16 illustrate that a thermal conductivity of 3 W/mK is sufficient to dissipate a total power loss of 80 W and a magnetic power loss of 60 W. The possibility to produce inductive devices closely thermally coupled to heat sink structures as a result of the great degree in geometric freedom is a great advantage of polymer bonded soft magnetics.

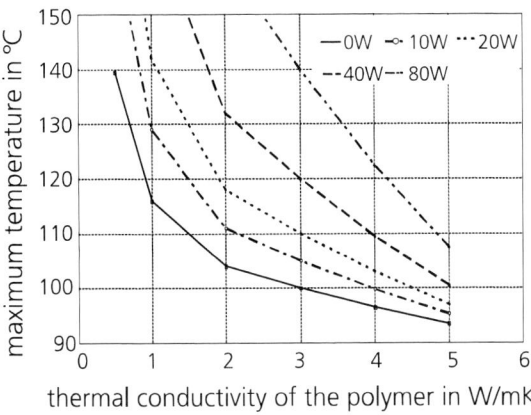

Fig. 16. Thermal simulation results

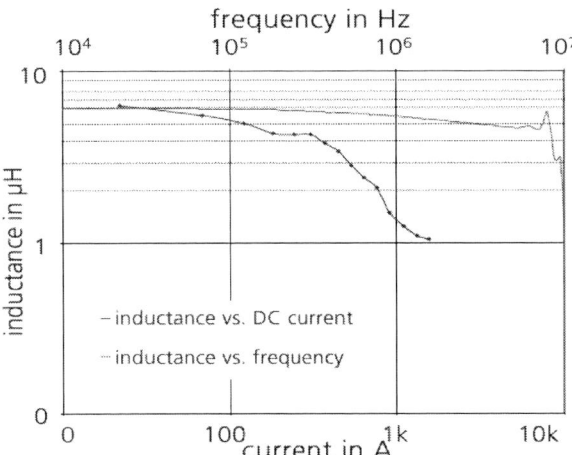

Fig. 18. Electrical characterization results

VIII. MEASURMENT RESULTS

A photograph of the produced device is given in Fig. 17.

Fig. 17. Photograph of the inductive filter device made of PBSMM

The inductance characterization results like they are shown in Fig. 18 clearly demonstrate the high frequency applicability for the recommended inductive device. The inductance reaches a maximum value of 6.1 μH. According to the simulation this corresponds to a permeability of 26 and therefore a filler content of 58vol.% of the iron silicon filler material. The inductance decrease slightly to 4.2 μH at 7.2 MHz and shows a resonant frequency around 8MHz.

The presented inductive device shows a soft saturation of the magnetic material, due to the fact that each soft magnetic iron particle shows an individual flux density distribution depending on its geometric parameters, its location in the device volume and its distance to other particles. At a DC current of 125 A, as it is required by the application, an inductance of 5 μH is attached. The saturation of the total soft magnetic volume requires a current higher than 1000 A.

The saturation behaviour of the presented device is show in Fig. 18.

IX. CONCLUSION

This paper presents the characterization of polymer bonded soft magnetics and a design flow for an inductive device for EMI filter applications made of these materials. The EMI filter is used to attenuate the conducted electromagnetic noise on the DC link of an inverter of a hybrid drive. The concept of inductive devices using polymer bonded soft magnetics could be presented. These materials were investigated in order to manufacture filter inductors using production processes of polymer technology and therefore to realize complex formed devices nearly without restrictions in the outer form for automotive applications. It could be shown that polymer compounds filled with high filler fractions of soft magnetic particles fulfill soft magnetic requirements for power electronics filter applications regarding the parameters permeability, power losses and saturation flux density. A mechanical, electromagnetic and thermal design flow was done for a concrete demonstrator with a very complex construction space. All these simulations show the possibility to manufacture inductive devices using soft magnetic polymer compounds. The advantages of polymer bonded soft magnetics for the presented application are:

- The air gap is distributed in the total core volume. This results in a reduced stray field and a very soft saturation.
- The polymer EMI filter fits into the predefined space and fills in the blank between the water cooled heat sink and the outer clutch box. So no additional mechanical fixation or filling polymer is needed for a reliable assembly.
- The EMI filter is directly attached on the water cooled heat sink and shows planar surfaces. Therefore an optimal distribution of thermal energy is ensured.
- The power losses of the soft magnetic polymer are higher than for ferrite or metallic bulk materials producing a higher attenuation at high frequencies for the EMI filter.

978-1-4244-4782-4/10 $26.00 © 2010 IEEE

ACKNOWLEDGMENT

The author would like to thank the Deutsche Forschungs Gemeinschaft (DFG) and the Sonderforschungsbereich 694 (SFB) as well as the European Center for Power Electronics (ECPE) for the financial support of this work.

REFERENCES

[1] M. Anhalt, B. Weidenfeller, "Dynamic losses in FeSi filled polymer bonded soft magnetic composites", J. Magn. Magn. Mater., vol.304, pp. 549-551, 2006

[2] M. Wulfa, L. Anestiev, L. Dup, L. Froyen, J. Melkebeek, "Magnetic properties and loss separation in iron powder soft magnetic composite materials", Journal of Applied Physics, vol. 91, no. 7845, (2002)

[3] B. Weidenfeller, M. Anhalt, W. Riehemann, "Variation of magnetic properties of composites filled with soft magnetic FeCoV particles by particle alignment in a magnetic field", J. Magn. Magn. Mater, vol. 320, pp. 362-365, (2008)

[4] R. Lebourgeois, S. Brenguer, C. Ramiarinjaonab, T. Waeckerl, "Analysis of the initial complex permeability versus frequency of soft nanocrystalline ribbons and derived composites", J. Magn. Magn. Mater., vol. 254-255, pp. 191-194, (2003)

[5] F. Alves, C. Ramiarinjaona, S. Brenguer, R. Lebourgeois, and T. Waeckerl, "High-frequency behavior of magnetic composites based on FeSiBCuNb particles for power electronics", IEEE Trans.Magn. vol.38, issue. 5, pp. 3135 - 3137, (2002)

[6] S. Egelkraut, M.März, H. Ryssel, "Polymer bonded soft magnetic particles for planar inductive devices", Proc. of the 5th International Conference on Integrated Power Systems (CIPS), Nuremberg, Germany, pp. 167-174, (2008)

[7] R. Chen, J.D. van Wyk, S. Wang, W.G. Odendaal, "Application of Structural Winding Capacitance Cancellation for Integrated EMI Filters by Embedding Conductive Layers", IEEE Industry Applications Conference, vol. 4, pp. 2679- 2686, (2004)

[8] M. März, M. H. Poech, E. Schimanek, A. Schletz, "Mechatronic Integration into the Hy-brid Powertrain - The Thermal Challenge", Proc. 1th International Conference on Automotive Power Electronics (APE) 2006

[9] E.J. Brandon, E.E. Wesseling, V. Chang, W.B. Kuhn, "Printed microinductors on flexible substrates for power applications", Trans. Compon. Packag. Technol., IEEE, Volume 26, Issue 3, Sept. 2003

[10] E. Waffenschmidt, "Printed circuit board integrated multi output transformer", Proc. 5th International Conference on Integrated Power Systems, Nuremberg, 2008.

[11] DIN EN 60404-6 European Standard

[12] M. März, E. Schimanek, M. Billmann, "Towards an Integrated Drive for Hybrid Traction", CEPS Workshop: From Success to Significance (2005)

[13] A.W. Kelly, F.P. Symonds, "Plastic-iron-powder distributed air gap magnetic material",IEEE Power Electronics Specialists Conference (PESC), (1990)

[14] A. Boglietti, A. Cavagnino, M. Lazzari, and M. Pastorelli, "Predicting iron losses in soft magetic materials with arbitrary voltage supply - An Engineering Approach", IEEE Trans. Magn. vol. 39, no. 2, pp. 981-989 (2003)

[15] S. N. Maiti and K. Chosh, "Thermal Characteristics of Silver Powder Filled Polypropylene Composites", Journal of Applied Polymer Science, vol. 52, pp. 1091 - 1103, (1994).

[16] H. Serkan Tekce, D. Kumlutas, I. H. Tavman, "Effect of Particle Shape on Thermal Conductivity of Copper Reinforced Polymer", Composites, Journal of Reinforced Plastics and Composites, vol. 26, no. 1, (2007)

Lead-Acid Battery Modeling and State of Charge Monitoring

J.F. Araujo Leão, L.V. Hartmann, M.B.R. Corrêa, A.M.N. Lima

Departamento de Engenharia Elétrica

Universidade Federal de Campina Grande

Caixa Postal 10.105

58109-970 Campina Grande - PB - Brazil

E-mails: [joableao; lucas.hartmann]@gmail.com;[mbrcorrea;amnlima]@dee.ufcg.edu.br

Abstract - In this paper batteries' models and state of charge monitoring procedures are evaluated. Modeling of batteries are intended to assist designing of power electronics devices, while monitoring procedures are intended for a more effective State Of Charge (SOC) computing. It is worth noting that this still a hard task when working with lead acid batteries. Two models and three procedures were evaluated by using experimental data, and classified accordingly with their effectiveness.

I. Introduction

Lead-Acid battery is one of the most ancient "portable" and rechargeable energy storage device. After more than 140 years, it is employed in several applications. Even with a lot of new rechargeable batteries technologies, it is still preferred in many circumstances. Photovoltaic and UPS systems are examples where lead-acid batteries are adopted. It has the lowest cost when compared with other battery types with equal power density. Even if low energy density is a drawback for such storage device, it looks like that the use of this rechargeable storage device will remain for a long time. In spite of lead-acid battery had been discovered a long time ago, there still existing room for improvements in modeling [1–9] and monitoring [10–19].

Besides modeling, another great challenge to whom works with lead acid batteries is to compute the amount of available energy, in another words, State of Charge (SOC). If this information is properly monitored, it could be possible to achieve a long battery lifetime. Accordingly with [13], "battery is aged by charging and discharging cycles; this process degrades the chemical composition of the battery. An undercharged battery has sulphation and stratification effects that shorten the lifetime of the battery. Overcharging causes gassing and water loss." Another variable that has a great influence over battery electrical characteristics is temperature. In summary, a battery can be seen as a very complex non-linear system that needs to be effectively monitored along its whole lifetime. Battery monitoring can be based on current, voltage and temperature measurement, or with additional support from a mathematical model. In this paper, monitoring based model will not be addressed.

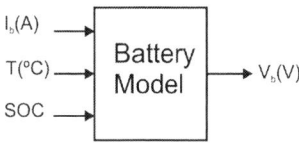

Fig. 1.   Block diagram for a battery.

Nevertheless, it is worth noting that battery model can be very useful to help in designing interface circuits [6]. Independently of choice and application it is possible to verify that several monitoring options are available. In this paper, battery models and monitoring methods are evaluated and classified with respect to their usability and effectiveness.

II. Battery Modeling

A. Long term and Short term Models

Modeling of lead-acid batteries is very useful for a better knowledge about batteries behavior, and for the design of power electronics devices. In general, battery-models allow to compute terminal voltage based on its current, SOC and ambient temperature. Therefore, effect of voltage dump due to start up current can be compensated. A functional block diagram for a battery is presented in Fig. 1. Two models based on variable electrical parameters are presented in Figs. 2 and 3. These models are adequate for long and short term discharge/charge modeling, respectively. Indeed they are better to describe discharge behavior. Models' parameters can be computed based on experimental data.

From [7] and [8] it is possible to verify how to compute circuit parameters, for model presented in Fig. 2, given data such as the one presented in Fig. 4. Knowledge of points V_0, V_1, V_2, V_3 and V_4 can be adopted to compute a set of constants. All of them related with circuit electrical parameters. These constants are presented in Table (I). Assuming a discharge process, we have $R_p = \infty$ and $R_2 = 0$. Final values of circuit

978-1-4244-4782-4/10 $26.00 © 2010 IEEE

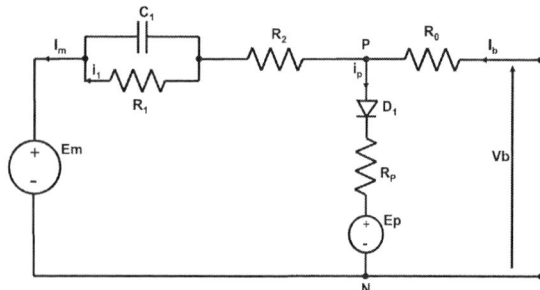

Fig. 2. Second order equivalent circuit.

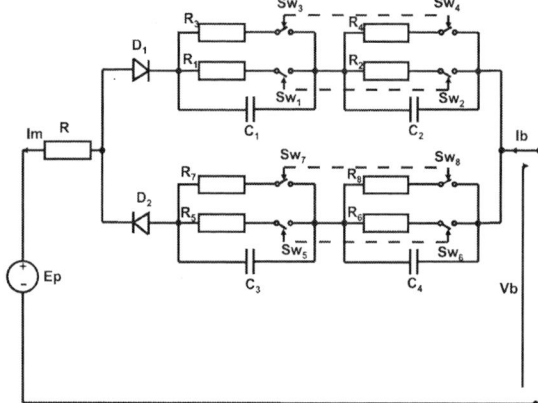

Fig. 3. Forth order equivalent circuit.

parameters, as a function of SOC, can be computed as folow

$$R_1 = -R_{10} \ln (SOC) \tag{1}$$

$$R_0 = R_{00} [1 + A (1 - SOC)] \tag{2}$$

$$E_m = E_{m0} - K_c (273 + \theta) (1 - SOC) \tag{3}$$

In [5], this battery model has been considered as very simple and effective.

Details about how to compute circuit parameters for the model presented in 3 can be found in [9]. Accordingly with the authors the model is valid for lead acid batteries and, takes into account if battery is in charging/discharging state, or in the rest period. The parameters of the model depend on the discharging or the charging current magnitude. Details about computing circuit parameters are included in the paper. For the previously described battery (150Ah), parameters values for discharging period are given in Table (II). For the resting period, after discharging, capacitances do not change. Nevertheless, new resistence values need to be calculated – see Table (III).

In Fig. 5, on bottom, it is shown experimental data from

TABLE I

PARAMETER OF LONG TERM BATTERY MODEL.

$E_{mo}(V)$	$K_c(V/K)$	A	$R_{00}\ \Omega$	$R_{10}\ \Omega$	$T\ (s)$
12.6324	0.0036	-0.0948	0.0485	0.0125	1834

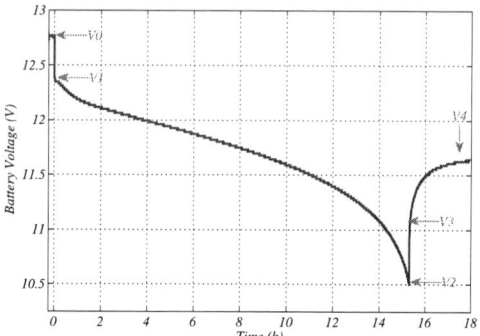

Fig. 4. Voltage profile for a discharge under constant current.

TABLE II

PARAMETER OF SHORT TERM BATTERY MODEL (DISCHARGING PERIOD).

$R\ (\Omega)$	$R_2\ (\Omega)$	$R_1\ (\Omega)$	$C_2\ (F)$	$C_1\ (F)$
0.0337	0.0055	0.0055	5.123e5	1.2e5

a complete discharge operation and simulation data generated from both models. Note that long term model provides completely wrong result. In Fig. 5, on top, it is shown experimental data from a short discharge and, again, simulation data generated from both models. Now, results from short term model is wrong. As it can be seen, depending on the scenario, it is possible to used the most adequate model. Nevertheless, if we try to change the role of these two models their dynamics will be strongly depreciated. This depreciation can be verified throughout results shown in Fig. (5).

TABLE III

PARAMETER OF SHORT TERM BATTERY MODEL (RESTING PERIOD)

$R\ (\Omega)$	$R_4\ (\Omega)$	$R_3\ (\Omega)$
0.0337	0.0055	0.0155

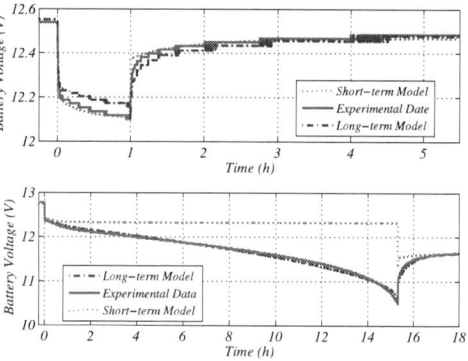

Fig. 5. Comparison about long and short term models.

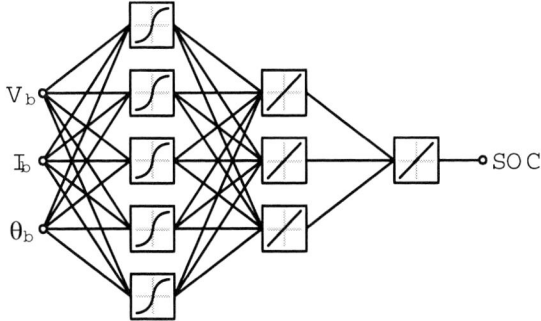

Fig. 6. Layout of the artificial neural network.

III. STATE OF CHARGE MONITORING

Battery SOC is a very relevant information to make user aware about available energy. Battery monitoring procedures to compute battery SOC is not a novelty, but until now, we are far away from the final solution. In this paper, three different SOC monitoring methods are evaluated and compared among themselves. They are based on Ah counter, open circuit voltage and artificial neural network. The first one is very effective for constant current discharge but, it is very sensible to error when several charge/discharge are executed in a row. The second one can be effectively adopted when open circuit voltage, after battery resting period, is not a problem. The third method is based on using an artificial neural network. It is well known that artificial neural network can be very effective when data for its training are reliable.

Assuming that Ah counter and open circuit voltage are easily formulated, more details will be provided only about the third method.

A. Artificial Neural Network Design

In this paper, neural network design was based on the fact that battery current, voltage and temperature are usually available variables. Then, after a proper evaluation of these variables it should be possible to compute battery SOC. Two models had been proposed and evaluated. The first one is the Artificial Neural Network (ANN) shown in Fig. 6. This ANN is based on a typical feed-forward topology comprised of three layers of neurons: an input layer, a hidden layer and an output layer. The input layer contains five neurons fed from the three measured variables, and uses a tangent-sigmoid output activation function. The hidden layer contains three neurons fed from the input layer's outputs, and uses a pure-linear output activation function. Finally the output layer is a single neuron fed from the hidden layer's outputs, also with a pure linear output activation function.

In order to have data for training, several discharge test at constant temperature and current had been run to generate these data. In Figs. 7(a) and 7(b) it is shown a surface generated from experimental data. These data where measured from a 150Ah battery partially submersed in a temperature regulated

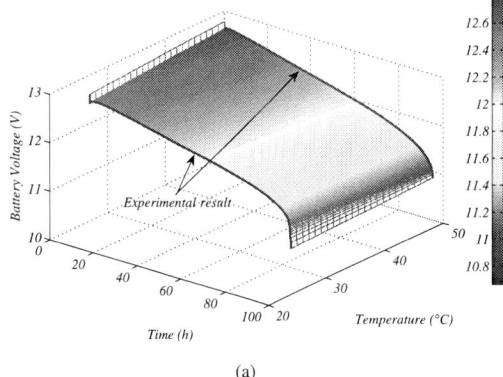

(a)

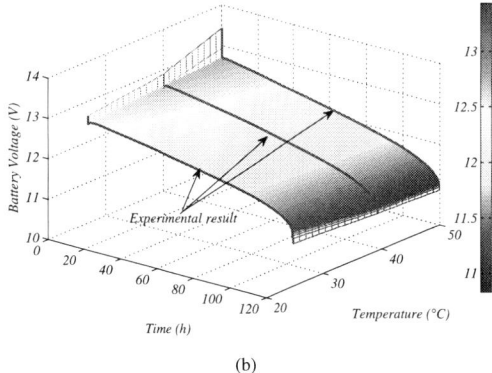

(b)

Fig. 7. Battery terminal voltage under constant current discharge. (a) I=-1.8A; (b)I=-1.47A.

water chamber. The complementary data are generated by means of interpolation.

Once neural network training had been finished it was possible to confirm that it was very precise in situations where battery discharge had a profile that looks like the one adopted for training. Therefore, a new pulse based discharge was performed to evaluate the SOC monitoring methods. Battery voltage, current and temperature are shown in Fig. 8. From Fig. 9 all SOC methods can be compared. As it can be seen the pulse discharge drove the neural network to wrong conclusion when dynamic is included in discharge process. This behavior is easily explained regarding the ANN's inputs and topology. Basically, there are no paths or inputs which store past values. This imposes a limitation to the ANN, once the current drops back to zero the ANN is unable to figure out if the current has been zero for a few seconds or several hours, leading to significant error values even when the ANN is trained directly for the pulsed discharge.

In order to overcome such restriction some modification was implemented in the neural network. When the battery was previously modeled, as an electrical circuit, its dynamic response was represented by two RC networks, each one with a different time constant. In order to bring this dynamics

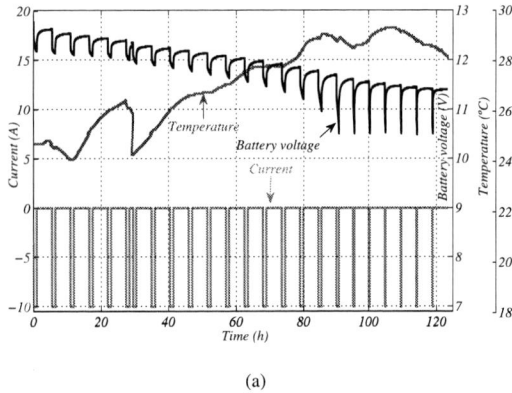

(a)

Fig. 8. Battery data along a pulsed discharge.

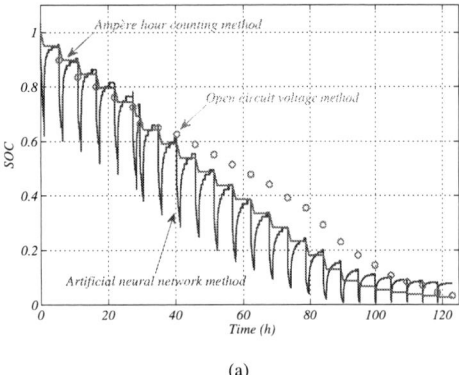

(a)

Fig. 9. Battery SOC computed from evaluated methods.

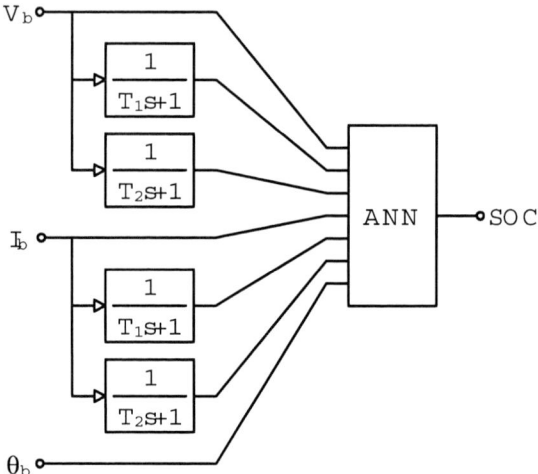

Fig. 10. Layout of the modified artificial neural network.

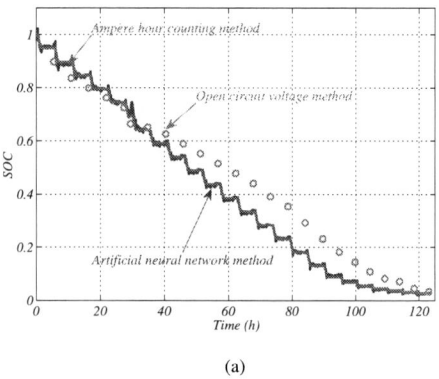

(a)

Fig. 11. Battery SOC computed from evaluated methods, including a modified ANN.

information into the ANN four additional inputs were defined: the battery voltage filtered through each of the time constants, and the battery current filtered through each of the time constants. The resulting ANN model is shown in Fig. 10, and is basically the same as in Fig. 6 but, now with 7 inputs. The time constants T_1 and T_2 are obtained by curve fitting procedure with data from voltage recovery period, at the end of a discharge – see (4). Once the values of T_1 and T_2 are complete, the ANN is trained for the pulsed discharge, so that there are sufficient dynamics for it to learn from.

$$V_b(t) = B - K_1 \exp\left(\frac{-t}{T_1}\right) - K_2 \exp\left(\frac{-t}{T_2}\right) \qquad (4)$$

Improvement on the response quality from Fig. 9 to Fig. 11 is clear, which leads to the conclusion that now the ANN has sufficient information for a proper estimation of the SOC. In order to validate such improvement, the ANN must succeed in a different condition from the one verified at training data set. A different set of experimental data was then obtained with two partial charge-discharge cycles, applying the ANN for estimating SOC during both the discharging sections of

the data set. The results obtained are shown in Fig. 12.

From Fig. 12 one can realize that the ANN output follows the Ah-estimated SOC to some extent, but disagrees during zero-current sections. At these sections the ANN output is near to the open-circuit-voltage-estimated SOC, which is known to be a good estimative during extended zero-current periods. It should be noted, however, that at no point the ANN was ever trained using open circuit voltage SOC estimation. During the second discharge the ANN output again seems reasonable and in agreement with open circuit voltage SOC estimation, while the ampere-hour SOC estimation is far from the expected values due to accumulated integration error, and to the fact that not all of the charging current effectively changed the SOC.

IV. CONCLUSION

In this paper, battery models for long and short charge/discharge was presented. From this models it is possi-

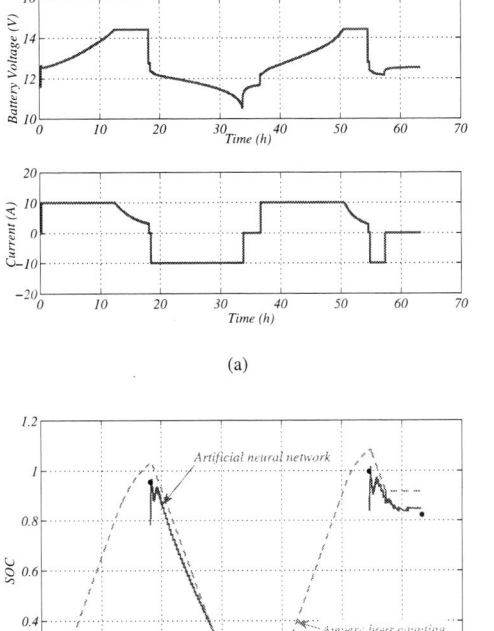

(a)

(b)

Fig. 12. Battery *SOC* computed from evaluated methods along two charge and discharge cycles.

ble to conclude that the one for short charge/discharge is better when we need a model to aid in designing power electronics devices. Besides that, a great concern it was about *SOC* prediction when a battery is submitted to charge/discharge cycles in a row that includes partial charge. In this case, it is easy to verify a cumulative error when we use *Ah* method. Also, if open circuit voltage measurement cannot be performed none of these methods are reliable. Therefore, the proposed neural network can be a good option to inform about battery *SOC* during discharging and resting periods.

REFERENCES

[1] C. Zhan, X.C. Wu, S. Kromlidis, V. Ramachandaramurthy, M. Barnes, N. Jenkins, and A.J. Ruddell. Two electrical models of the lead-acid battery used in a dynamic voltage restorer. In *Generation, Transmission and Distribution, IEE Proceedings-*, volume 150, pages 175–182, March 2003.

[2] Z.M. Salameh, M.A. Casacca, and W.A. Lynch. A mathematical model for lead-acid batteries. *IEEE Trans. Eneg. Conv.*, 7(1):93–98, March 1992.

[3] H.L Chan. A new battery model for use with battery energy storage systems and electric vehicles power systems. In *Proc. of PES Meeting*, pages 470 – 475, 2000.

[4] K. Kutluay, Y. Cadirci, Y.S. Ozkazanc, and I. Cadirci. A new online state-of-charge estimation and monitoring system for sealed lead-acid batteries in telecommunication power supplies. *IEEE Trans. Ind. Electron.*, 52(5):1315–1327, October 2005.

[5] Robyn Jackey The MathWorks Inc. A simple, effective lead-acid battery modeling process for electrical system component selection. In *http://www.mathworks.de/industries/auto/technicalliterature.html*, September 2007.

[6] Yoon-Ho Kim and Hoi-Doo Ha. Design of interface circuits with electrical battery models. *IEEE Trans. Ind. Electron.*, 44(1):81–86, February 1997.

[7] M. Ceraolo. New dynamical models of lead-acid batteries. *IEEE Trans. Power Systems*, 15(4):1184–1190, November 2000.

[8] S. Barsali and M. Ceraolo. Dynamical models of lead-acid batteries: implementation issues. *IEEE Trans. Energ. Conv.*, 17(1):16–23, March 2002.

[9] S. Mischie and D. Stoiciu. A new and improved model of a lead acid battery. *FACTA UNIVERSITATIS (NIS)*, 20(2):187–202, Aug. 2007.

[10] Islam S. Duryea, S. and W. Lawrance. A battery management system for stand-alone photovoltaic energy systems. *IEEE Industry Applications Magazine*, 2001.

[11] M. Coleman, C.B. Zhu, C.K. Lee, and W.G. Hurley. A combined SOC estimation method under varied ambient temperature for a lead-acid battery. In *Proc. IEEE APEC*, volume 2, pages 991–997, March 2008.

[12] M. Ragsdale, J. Brunet, and B. Fahimi. A novel battery identification method based on pattern recognition. In *Proc. of VPPC.*, pages 1–6, September 2008.

[13] M. Coleman, W.G. Hurley, and C.K. Lee. An improved battery characterization method using a two-pulse load test. *IEEE Trans. Energ. Conv.*, 23(2):708–713, March 2008.

[14] B.S. Bhangu, P. Bentley, D.A. Stone, and C.M. Bingham. Nonlinear observers for predicting state-of-charge and state-of-health of lead-acid batteries for hybrid-electric vehicles. *IEEE Trans. Vehic. Tech.*, 54(3):783–794, May 2005.

[15] M. Coleman, C.K.Lee, C.Zhu, and W.G. Hurley. State-of-charge determination from EMF voltage estimation: Using impedance, terminal voltage, and current for lead-acid and lithium-ion batteries. *IEEE Trans. Ind. Electron.*, 54(5):2550–2557, October 2007.

[16] D.J. Deepti and V. Ramanarayanan. State of charge of lead acid battery. In *IEEE Proc. of IICPE*, pages 89–93, Chennai,, December 2006.

[17] C.B. Zhu, M. Coleman, and W.G. Hurley. State of charge determination in a lead-acid battery: combined EMF estimation and ah-balance approach. In *IEEE Proc. of PESC*, volume 3, pages 1908–1914, June 2004.

[18] T. Tsujikawa and T. Matsushima. Remote monitoring of VRLA batteries for telecommunications systems. *Journal of Power Sources*, 168:99–104, Available online at www.sciencedirect.com 2007.

[19] Y. Morita, S. Yamamoto, S.H. Lee, and N. Mizuno. On-line detection of state-of-charge in lead acid battery using both neural network and on-line identi cation. In *IEEE Proc. of Industrial IECON*, pages 3379–3384, Paris,, November 2006.

Voltage and Current Ripple Considerations for Improving Lifetime of Ultra-Capacitors Used for Energy Buffer Applications at Converter Inputs

Supratim Basu
BOSE RESEARCH PVT. LTD.
#1/1 & 44, 2nd Main, 2A Cross,
R.M.V. Stage-II, 3rd Block, New BEL Road,
Bangalore 560 094, India
Tel.: +91 80 2341 9658 / 23419278
E-Mail: boseresearch@vsnl.net

Tore. M. Undeland
NORWEGIAN UNIVERSITY OF SCIENCE AND TECHNOLOGY
Department of Electrical Power Engineering,
Trondheim, Norway
Tel.:+47 73 59 42 44,
E-Mail: Tore.Undeland@elkraft.ntnu.no

Abstract— While Ultra Capacitors offer high power density, high cycling capability and mechanical robustness. Voltage and current ripple from downstream PWM converters result in their heating and lifetime reduction. This paper discusses how a simple novel design scheme of using a very low value series inductance significantly increases capacitor lifetime.

I. INTRODUCTION

Ultra-capacitors or electric double layer capacitors (ELDC), are electrical energy storage devices, which offer high power density, extremely high charge and discharge cycling capability and mechanical robustness [1]. Due to these features, Ultra-Capacitors have a high potential of being used in industrial applications. To improve their performance, reliability and lifetime, efficient charge balancing circuits [2], power circuits that do not overcharge or overheat these capacitors due to high ripple voltage or current, are very important. When using Ultra-Capacitors as an energy storage buffer for downstream PWM converters, the resulting voltage and current ripple in them can cause temperature rise and lifetime reduction of the Ultra-Capacitor. Based on extensive practical measurements made on an Ultra-Capacitor bank, a novel design scheme that significantly reduces their operating voltage ripple and operating ripple current resulting in significant improvement of their lifetime performance, is presented in this paper.

II. THE PROBLEM IN SPECIFIC

Both overcharging and temperature rise reduces the lifetime of Ultra-Capacitors. Though the standard temperature rating for Ultra-Capacitors is –25°C to 70°C, ambient temperature rise in combination with high charging voltage, can reduce their lifetime significantly. In general, raising the ambient temperature by just 10°C will decrease their lifetime

by at least a factor of two [3]. Thus the maximum operating voltage of an Ultra-Capacitor should be reduced with increasing ambient temperature. Overheating of Ultra-Capacitors occur due to either the high charging/discharging ripple current from the downstream converter or charging over voltage, leading to increased gas generation, decreased lifetime, leakage, venting or rupture. Though for highest energy storage the Ultra-Capacitor must be charged to its maximum rated nominal working voltage, care needs to be taken that the charging voltage ripple does not overcharge the capacitor. Moreover, as Ultra-Capacitors have a higher ESR compared to aluminum electrolytic capacitors, they are more susceptible to internal heat generation when exposed to higher ripple current. In order to ensure a long lifetime, it is [3] thus recommended that the maximum ripple current should not increase the surface temperature of the Ultra-Capacitor by more than 3°C above ambient and thus minimize the operating ripple current and ripple voltage on these capacitors.

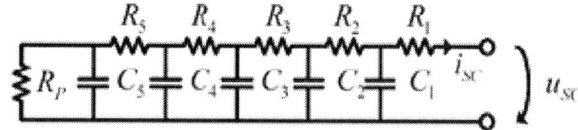

Figure 1a. Model of Ultra-capacitor

In addition to the above issues, the nominal capacitance of an Ultra-Capacitor is applicable only at dc with its capacitance dropping rapidly to near zero at higher frequency [4]. Fig. 1a shows a representative Ultra-Capacitor model where the leakage resistance is represented as R_P, the highest and lowest capacitances are respectively represented by C5 and C1 respectively and R1 to R5 are the Ultra-Capacitor's equivalent

---

Bose Research Pvt. Ltd www.boseresearch.com

978-1-4244-4782-4/10 $26.00 © 2010 IEEE   244

series resistance (ESR). As apparent from the frequency (F_{SC}) response graph given in Fig. 1b, an Ultra-Capacitor cannot significantly attenuate the high frequency voltage ripple generated by a switching PWM converter connected

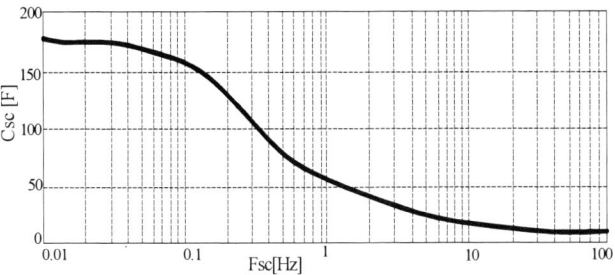

Figure 1b. Capacitance vs. frequencyplot of Ultra-capacitor

downstream since the large capacitance (C_{SC}) measured at dc reduces to a very low value beyond just 100 Hz.

Ultra-capacitors can be simultaneously charged and also used as an energy storage buffer by paralleling them to an energy source, i.e. battery, fuel cell, DC-DC converter [5], etc. The voltage and current ripple caused by the charging converter or the converter that these capacitors intend to provide power back up, can cause over charging or temperature rise of the capacitor.

III.   STRATEGIES FOR REDUCING RIPPLE VOLTAGE AND RIPPLE CURRENT

As discussed so far, reduction of both operating voltage ripple and charging/discharging ripple current is necessary for increasing the lifetime of Ultra-Capacitors. Though one solution would be to increase the filter inductance (L1 in Fig. 3) and or reduce the switching frequency of the buck derived DC-DC converter that is usually used for charging or for power conversion downstream, this will significantly increase converter size and cost. Moreover increasing inductance requires higher turns and this increases both the radiated fields from the inductor and the inter-winding capacitance of the inductor. A representation of a practical inductor's inter-winding capacitance series resistance is shown in Fig. 2.

Ideal Inductor

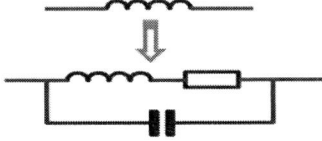

Practical Inductor

Figure 2.   Interwinding Inductor Capacitance

These radiated fields and the feed through noise through the inter-winding capacitance [6] from the inductor mainly

couple to surrounding circuits and increase EMI. Thus a better solution would be to use additional filter circuits that attenuate both the voltage ripple and ripple current and thus allowing higher frequency operation of the converter.

Ultra-capacitors are usually always used for energy storage or energy buffer applications. Their poor high frequency response makes them completely unsuitable for high frequency applications and is therefore more suitable for dc circuits. It is thus proposed that Ultra-capacitors should be connected to any high frequency charging converter with a small inductance of about 20 µH in series to it. The circuit model of the proposed scheme is shown in Fig. 3. L1 is the main converter filter inductance that sets the ripple current magnitude while L2 is the proposed small inductance of about 20 µH.

It is intended that L2 will attenuate the operating ripple current through the Ultra-capacitors and mitigate temperature rise issues in them due to heating caused by the charging/discharging ripple current through the Ultra-capacitors. Further since L2 has few turns due to its small inductance value, its copper losses and cost/size will be low.

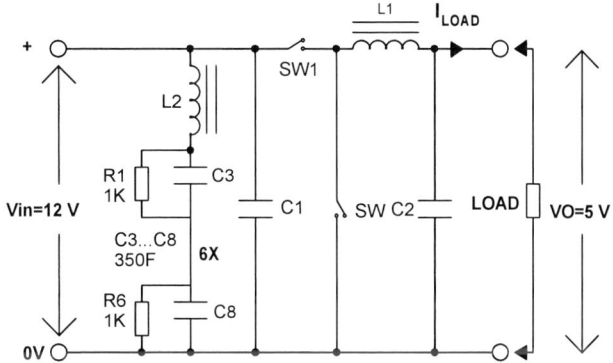

Figure 3.   Proposed Charging Scheme

IV.   EXPERIMENTAL RESULTS

To develop a better understanding about the influence of high ripple current through Ultra-Capacitors, measurements were made for a synchronous buck converter with a 58 F Ultra-Capacitor and a current limited 12 V dc source connected at its input. The converter circuit model is shown in Fig. 4. The 58 F Ultra-Capacitor comprised of six 350 F (C1-C6) Ultra-Capacitors connected in series with charge balancing resistors (R1-R6). All measurements were made when the input current reached its minimum steady state value indicating that the Ultra-Capacitors were almost fully charged. R7 kept the converter in CCM and also provided a discharge path for the Ultra-Capacitors. The downstream converter's switching frequency was set to about 100 kHz while the control circuit regulated the output voltage to about 5 V by adjusting the operating duty cycle of the converter to about 40%.

By controlling switch SW1, the effect on the operating voltage/current ripple of an ultra-capacitor with and without the low value inductance in series to the capacitor, was investigated.

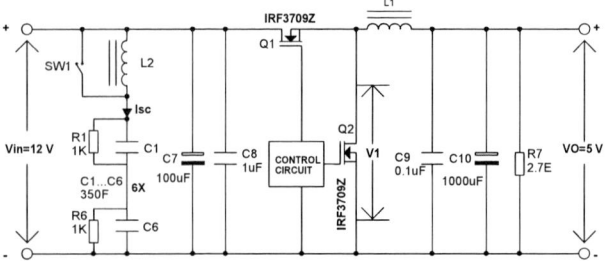

Figure 4.   Buck converter circuit model with Ultra-Capacitor at the input.

These are presented in the oscillograms given in Fig. 5 and Fig. 6. In all these oscillograms Channel 1 shows V_1. Fig. 5a shows the charging voltage ripple in Channel 2 and the charging ripple current I_{SC} through the capacitor is shown in Fig. 6a. As seen from these oscillograms, the large voltage ripple or the ripple current generated by the downstream converter, demonstrates the poor frequency response of Ultra-Capacitors. However on connecting a single 20 μH inductance (L2) in series to the Ultra-Capacitor, the voltage ripple and ripple current reduced significantly. With the 20 μH connected, Fig. 5b shows the charging voltage ripple in Channel 2 and the charging ripple current I_{SC} through the capacitor is shown in Fig. 6b.

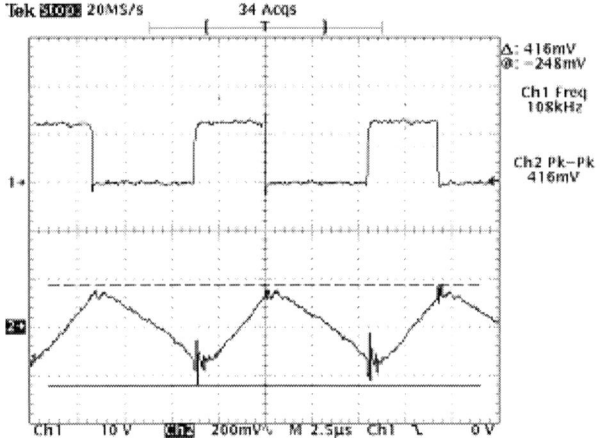

Figure 5a. Operating Ripple Voltage without L2

Thus with only the Ultra-capacitor connected, the high operating voltage ripple and the ripple current through the Ultra-Capacitor would cause overcharging and heating of the capacitor. On connecting the 20 μH inductance (L2) in series with the Ultra-Capacitor, not only could the voltage ripple be reduced significantly but the ripple current could also be reduced.

While the reduction in the operating ripple voltage was about 25%, the ripple current responsible for capacitor heating reduced even more significantly. It can be seen in Fig. 6 that though the peak to peak ripple current reduced by about 50%,

the reduction in the average value of the ripple current was even more significant

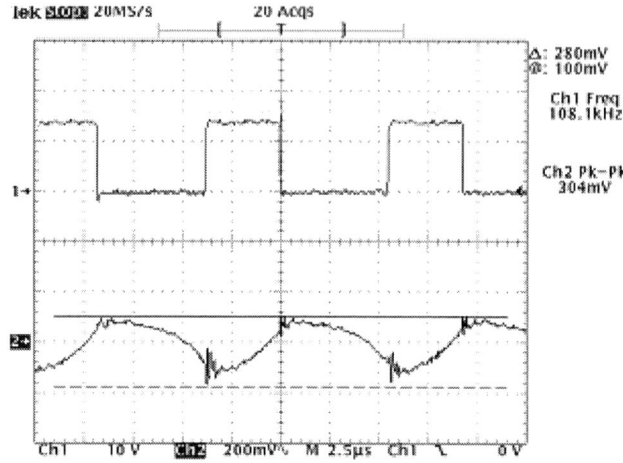

Figure 5b. Operating Ripple Voltage with L2

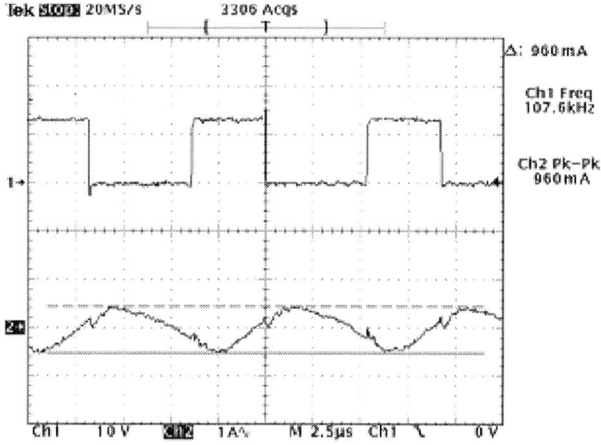

Figure 6a. Operating Ripple Current without L2

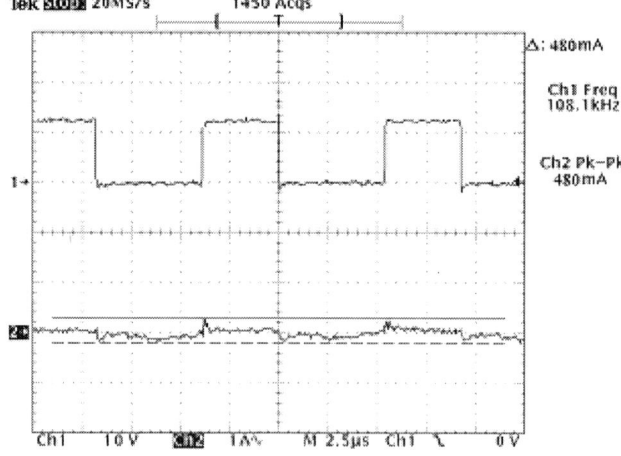

Figure 6b. Operating Ripple Current with L2

V. CONCLUSION

This paper discusses how downstream PWM converters can generate high voltage and current ripple in Ultra-Capacitors used as a energy storage buffer and thus result in their heating and lifetime reduction. Based on extensive measurements, the proposed simple novel design scheme of using a very low value inductance in series to an Ultra-Capacitor helped significantly reduce both the voltage ripple and ripple current of an Ultra-Capacitor by more than 40%, resulting in increase in capacitor lifetime. The change in capacitance and ESR of an Ultra-Capacitor with frequency is also highlighted.

REFERENCES

[1] B.E. Conway, Electrochemical Supercapacitors – Scientific Fundamentals and Technological Applications. London: Kluwer Academic/Plenum Publishers, 1999.

[2] Linzen, D.; Buller, S.; Karden, E.; De Doncker, R.W.; Analysis and evaluation of charge-balancing circuits on performance, reliability, and lifetime of supercapacitor systems Industry Applications, IEEE Transactions on Industry Applications Volume 41, Issue 5, Sept.-Oct. 2005 Page(s):1135 – 1141.

[3] Cooper-Bussmann Application Guidelines, Available at: http://www.cooperet.com/library/products/PS-5507%20 Guidelines.pdf

[4] O. Garcia, "DC/DC-Wandler für die Leistungsverteilung in einem Elektrofahrzeug mit Brennstoffzellen und Superkondensatoren," D.Tech. Dissertation, Prof. Leistungselektronik & Messtechnik, ETH, Zürich, 2002.

[5] Carl Klaes Maximum Charging of an Ultra-Capacitor Using Switch Mode Rectifiers in a Regeneration Cycle Vehicle Power and Propulsion, 2005 IEEE Conference 7-9 Sept. 2005 Page(s):5 pp

[6] Basu. S; Undeland. T. M., A Novel EMI Reduction Design Scheme for Continuous Mode PFC Converters, (NORPIE 2006), Lund, Sweden, 12-14 June 2006.

Implementation and operational investigations of bipolar gate drivers

Jean-Christophe CREBIER*, Manh Hung TRAN*, Jean BARBAROUX*, Pierre-Olivier JEANNIN*

*Grenoble Institute of Technology, Grenoble Electrical Engineering Lab (G2ELab),
ENSE3 - BP 46 - 38402 Saint-Martin-d'Hères Cedex, FRANCE
Email: crebier@g2elab.grenoble-inp.fr, tranm@g2elab.grenoble-inp.fr

Abstract—This paper deals with the investigation of simple implementation and design of bipolar gate driver for high side power transistor control. Bipolar gate signals are usually preferred for high switching dynamic control and power device shielding. Nevertheless, bipolar supplies are harder to implement and rely on numerous components that may result in significant reduction of overall converter robustness. An alternative solution is to use unipolar supplies which are more complex but integrable with specific gate driver circuits. The addition of resonant structure gives the possibilities for involving the efficiency of the gate driver and reducing further gate driver supply requirements. The paper presents theses issues based on several gate drivers and their supplies.

I. INTRODUCTION

Gate drivers for power devices such as IGBTs or MOSFETs are increasingly asked to offer high performances while being simple to implement and highly reliable. In order to provide operational flexibility, security and effective dynamic control, gate drivers are usually implemented with bipolar supplies. By using bipolar gate driver signals, charge and discharge sequences can be well controlled while ensuring optimal ON state polarization and effective natural shielding during the OFF state. Besides, a number of studies are engaged toward the operational optimization and the integration of the gate drivers in order to simplify their implementations and to maximize their performances [1][2]. In particular, it can be noticed that many research results have been published with attempt to reduce the power consumption of the drivers for very high switching frequency applications [3][4]. More specifically, resonant converters are designed to drive each power device in a converter trying to exchange energy between the sources and the gate rather than dissipating periodically the energy fed to the gate. However, the practical implementation of floating bipolar sources to power supply the gate drivers of high side transistors remain difficult to minimize and to integrate. Researches have been carried out in order to identify simple and cost effective solutions but they are not always suitable for effective operation of modern gate drivers [5][6][7]. Other researches have been carried out trying to integrate

gate driver supplies in order to simplify their implementation and to minimize the impact of common mode currents [8][9][10]. Today, it remains difficult to implement high efficiency resonant or bipolar gate drivers with simple and integrable gate driver supplies. As a result, we decided to investigate the implementation and the operation of bipolar gate drivers with respect to their power supplies in order to identify what could be the best available solutions.

The following sections of the paper are dedicated to study and to analyze the operational combination of the gate driver and its isolated power supply especially for driving high side transistors where electrical isolation is required. The first part introduces the state of the art concerning the gate driver and the gate driver supplies. This is studied considering power converter topologies based on the implantation of regular high side n-type transistors for medium and high voltage (100V up to 600V) application and low to medium power(10 to 10kW). A similar study will also consider the case of CMOS and derived converter topologies [11][12] for low to medium voltages (below 15V and around 100V) and low to medium power levels (1 to 100W). The following section focuses on the combination of the bootstrap driver supply and specific gate driver topologies in order to perform high efficiency bipolar control signals. The last section offers a comparable investigation based on the self supplied gate driver powering solution presented already in [13]. The investigation results will be compared in the last section of the paper based on effectiveness, integrability and ease of implementation.

II. SUPPLY AND DRIVING TOPOLOGIES

A. Gate driver circuits

Gate driver topologies can be classified as a function of the supplies they are associated with. Basically, the supply is either unipolar or bipolar. Bipolar gate driver signals are usually preferred because they offer better turn OFF dynamics as well as a natural and effective EMI shielding inside the power devices. Therefore, even if needed performances and operational behavior can be reached with

978-1-4244-4782-4/10 $26.00 © 2010 IEEE

unipolar gate drivers, we focus our investigations on the solutions that can generate and apply bipolar control signals to the gate capacitance of the power devices. Considering the association of the gate driver with a bipolar supply, a simple push-pull circuit is enough to drive the power switch with bipolar signals. The topology of the gate drivers is given on Figure 1. With such a topology, the gate capacitance can be driven from the voltage sources but accurate current feeding can also be implemented. This allows the driver to control the gate charge and to shape the switching transition.

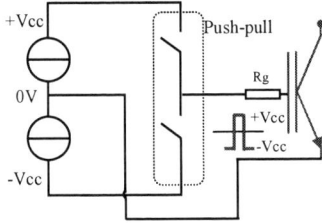

Figure 1. Bipolar gate driver consist of bipolar supply and push-pull topology

In this case, the gate capacitance is fed from the positive supply and then discharged applying the negative source between gate and source (or emitter). Hence, the energy stored in the gate capacitance is dissipated in the negative source each switching cycle. The power consumption of the gate driver is then a function of the gate charge, the switching frequency and the gate driver's supply voltage. At reasonable switching frequencies (between 10 kHz and 50 kHz), the power consumption of the gate drivers remains small and the gate driver supplies may be implemented without trouble since their efficiency level is not critical. As an example, let us consider a practical case with a well known power device, the IRF740 having a gate capacitance of 63nC driven at 35kHz, under +15/-15V and supplied by sources having 30% efficiencies. The total gate driver's power consumption is in the range of 0.5W[3] knowing that this kind of device is usually used to convert electric power in the range of 500 up to 1kW. The gate signal can be shaped as needed if the push pull structure is operated in current mode. A simple implementation is depicted on Figure 2. If possible, the negative supply can be lowered to -5V which is usually enough to enable current control and natural shielding. This improves slightly the gate driver's power consumption.

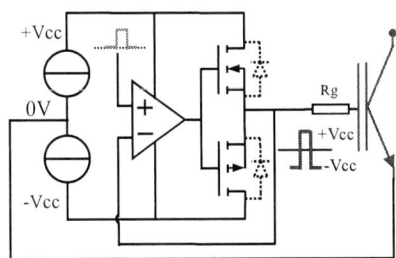

Figure 2. Bipolar gate driver with bipolar supply and current amplifier topology

At higher switching frequencies and for larger power devices, the gate driver's power consumption may become excessive and it requires either the implementation of effective gate driver supplies or a drastic reduction in the consumption levels of the gate drivers. Efficient gate driver supplies will be costly and will require specific designs. They are much harder to integrate and lead to increased failure risk for the whole converter. On the other hand, more efficient driving solutions for high frequency operation have been studied based on regular bipolar supplies and push pull structure [14]. Figure 3 presents one of these solutions where the push pull structure is associated to a resonant topology that "recycles" part of the energy withdrawn from the gate capacitance each switching transition.

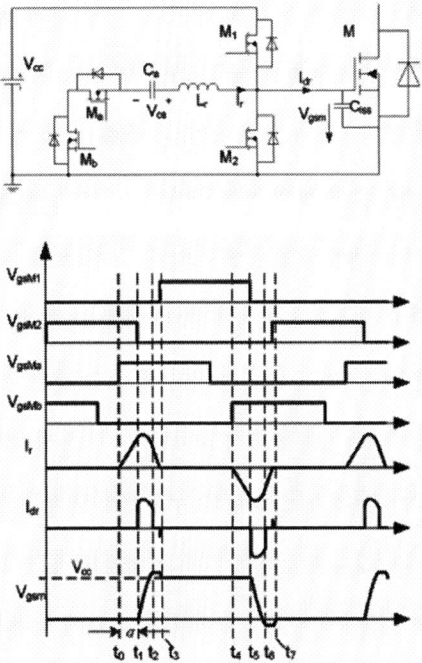

Figure 3. Resonance MOSFET driver circuit [14]

This structure allows the gate driver to save a large part of the energy needed in order to charge or to discharge the gate capacitance of the power device. Thus, it lowers the power needed and decreases the requirements on the power supply while improving the overall efficiency of the power converter.

If a unipolar supply is considered, the topology of the gate driver must be modified. At first, a second voltage supply can be generated from the first one with the help of a buck-boost chopper. This structure is described in Figure 4.

Another solution consists in using a resonant LC structure to discharge the gate capacitance and to invert the voltage across the gate to source terminals. This gate driver topology is shown on Figure 5. Theoretically, half of the energy consumption is saved in this case and a negative potential is applied to the gate terminals during the OFF state. However, the dynamics are limited by the value of the LC circuit or the pulse discharge current making this solution either difficult to integrated or possibly susceptible to external parasitic.

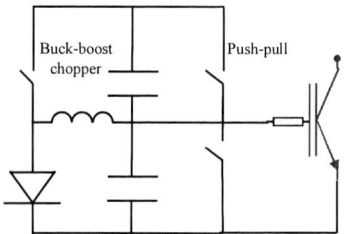

Figure 4.   Unipolar supply associative with buck-boost chopper and push pull topology

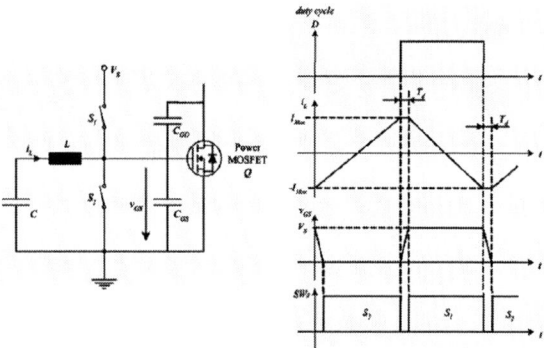

Figure 5.   Resonant gate driver[15]

The most flexible and effective solution relies on the implementation of a full bridge converter to drive the gate capacitance. In this configuration, one inverter harm middle point is connected to the gate terminal whereas the other inverter harm middle point is connected to the source terminal of the power device. The DC+ and DC- terminals are connected to the plus and minus terminals of the unipolar gate driver source. Hence, it is possible to implement resonant converter structure in order to optimize the gate driver power consumption. The structure can also be used to drive the current to the gate, suitable for precise and specific gate charge evolution. Figure 6 below presents the topology in case of a regular implementation. It can be noticed that the floating source voltage reference is not anymore tied to the power device voltage reference i.e. the source terminal in case of a power MOSFET or the emitter terminal in case of an IGBT. This topology is of interest thanks to its monolithic integration possibility and reduced number of passive components (only one storage capacitor).

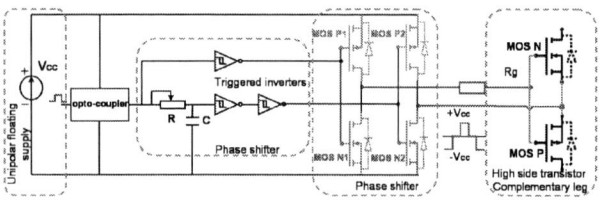

Figure 6.   Unipolar supply associative with full bridge converter that drives a complementary inverter leg

This ends our brief review of the main gate driver topologies that can be used to apply bipolar gate signals under high frequency operation and depending on the gate supply types. We focus now on the gate driver supplies with a short overview of the main topologies.

B.   Isolated gate driver power supplies

High side power devices or complementary P-N structures [12] require an isolated floating voltage source that can be referenced to the bottom terminal of the power device, which needs to be driven. Basically, we identify four approaches that are regularly used or presented in the literature. The most basic one is the linear regulator which is simple to implement and exhibits extremely low efficiency levels. The energy is usually taken from the power side through a resistor and regulated thanks to an avalanche diode. The second one is the regular DC to DC isolated converter using a HF or piezoelectric transformer [16]. Numerous studies have been carried out on the design and the optimization of such small power supplies [17]. Recently, several research works are related to the integration of these low voltage and low power converters that still offer high voltage isolation levels. We can cite a commercially available device [18] and research works [19][20][21][22]. These supplies can be duplicated to implement bipolar supplies, more suitable for bipolar gate signal generation. However, these integrated converters have limited efficiency levels of 33% for the industrial ones and around 50 to 60% for the lab demonstrators. Figure 7 depicts the topology of such converter and a possible 3D integrated assembly. Isolated supplies based on discrete components can be more efficient but require specific designs [23]. Besides, their robustness may affect the global reliability of the converter if it is not integrated.

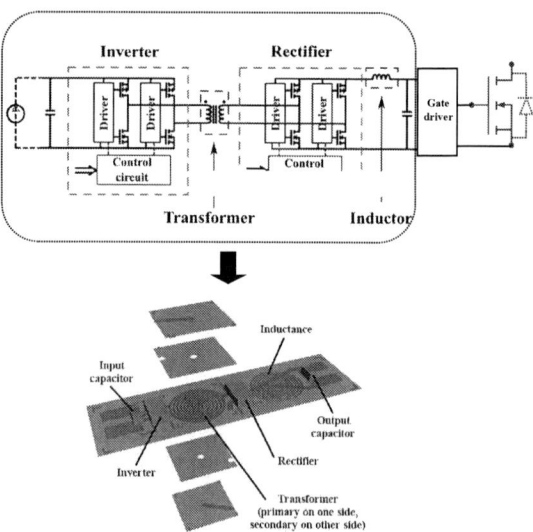

Figure 7.   Integrated isolated DC/DC micro-converter[19].

978-1-4244-4782-4/10 $26.00 © 2010 IEEE   250

Another popular driver supply is the bootstrap topology that takes advantage of the converter operation to charge up periodically a capacitor tank, which is used to supply the floating gate driver of the high side power devices. The topology of the bootstrap technique is given on Figure 8. Each time the bottom power switch is conducting, a circuit formed by the bootstrap diode, the storage capacitor, the low side gate driver supply and the bottom device allowing the recharge of the storage device. When the bottom switch is OFF, the bootstrap diode is blocked and the high side gate driver is supplied thanks to the energy stored in the storage capacitor Cs. This requires periodical recharge sequences and therefore permanent converter operation. The classical bootstrap technique corresponds to a unipolar supply.

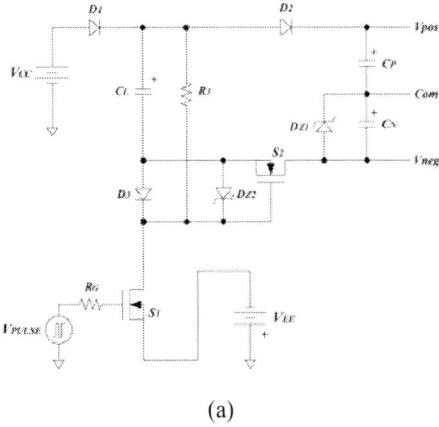

(a)

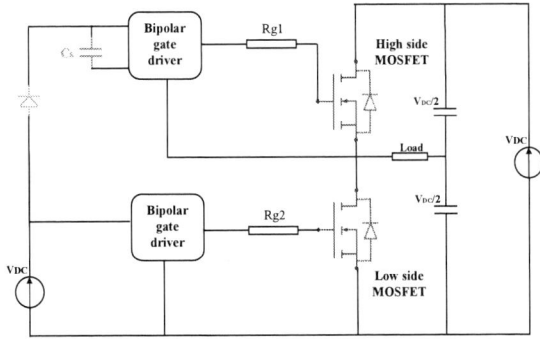

Figure 8. Conventional bootstrap technique topology

This bootstrap technique can be improved as it is presented in [7]. The first improvement is a self controlled bootstrap that operates independently to the power side, simplifying the design requirement and removing the possible operational limits. An extension is proposed in [7] in order to "emulate" a bipolar driver supply based on the unipolar topology. Although this approach requires a large extend in component number, it seems to offer a possible solution. Its topology is presented on Figure 9(a)

Another solution is patented on an extension of the unipolar structure in order to derive a bipolar structure. Figure 9(b) below shows its topology. The approach is similar to the classical topology exception made to the additional switch which is required to open the negative supply loop while the bottom switch is turned ON.

The last topology considered here is the self-supply technique that is implemented across the power device to be driven. The classical topology of this technique is depicted on Figure 10. This supply is also based on a periodical recharge of a storage capacitor and a pulse linear regulation approach [24]. Thanks to this particular operation, the efficiency of the supply can be optimized while being simple to integrate. This gate driver supply is unipolar type and can be extended to bipolar type thanks to a symmetric structure depicted on Figure 11.

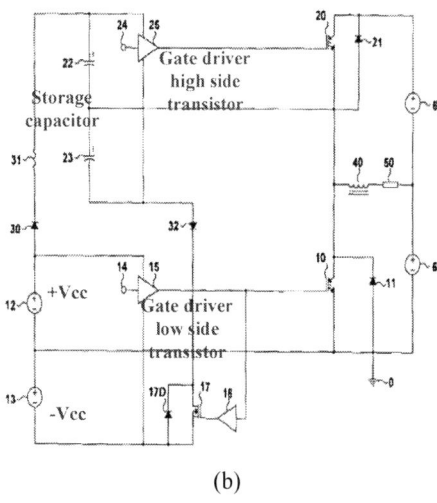

(b)

Figure 9. (a) Self-boost charge pump topology with positive- and negative-bias voltage supply capability [7]. (b) Boostrap technique with bipolar supply[6]

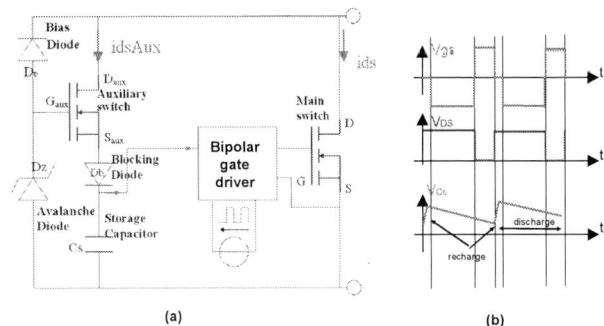

(a)   (b)

Figure 10. Self-supply technique topology [24]

978-1-4244-4782-4/10 $26.00 © 2010 IEEE

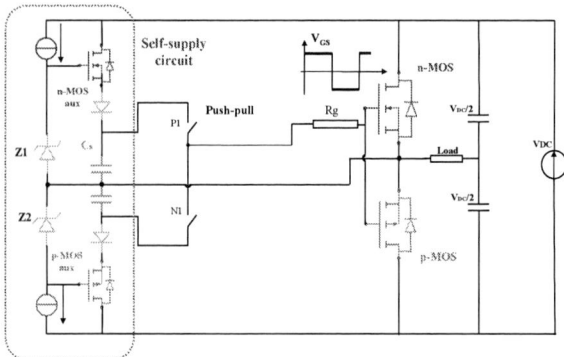

Figure 11. Symetric topology of bipolar self-supply technique

III. DRIVERS PLUS BOOTSTRAP POWER SUPPLIES

The voltage regular bootstrap technique provides a floating unipolar source to the high side switch gate driver. Based on the overview presented in the previous section, this type of supply must be added to specific gate driver topologies in order to control the power switch with bipolar gate signals. The other approach consists in using the bipolar bootstrap supply plus a push pull gate driver structure. In all cases, a major limitation of this approach for high frequency operations may remain the fact that a minimum time is required to recharge the storage capacitor each switching cycle, leading to a minimum low side switch duty cycle and leading to a limited maximum switching frequency.

A. Bipolar bootstrap plus push-pull topology

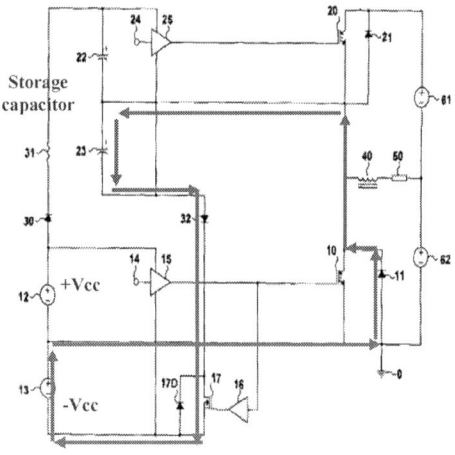

Figure 12. Bipolar Boostrap technique with negative recharge loop[6]

The topology of the bipolar bootstrap technique is presented in Figure 9. The bipolar supply is composed of two unipolar bootstrap supplies exception made the addition of a high voltage switch inserted in series with the negative bootstrap loop. It is required in order to prevent current flowing from the power side down to the negative gate driver supplies while the high side switch is in its ON state. It is controlled with the same control signal as the one fed to the low side power switch gate driver keeping the

global control of the structure simple. However, the added high voltage switch needs to maintain opend the negative bootstrap loop when needed that remains not simple to drive, because it is also at a floating potential making this solution not simple to implement from a hardware point of view (Figure 12). Nevertheless, this gate driver supply can be associated with a push pull topology in order to perform bipolar control signals to the high side switch.

B. Unipolar bootstrap plus full bridges

Considering unipolar bootstrap technique to supply a gate driver that must perform bipolar gate signals, resonant gate drivers may be a solution. A combination of this type of driver supply plus the resonant circuit presented in Figure 5 is straightforward. It may minimize the implementation cost and complexity exception made for the required passive devices. A structure that is more complex could be considered. As an example, the combination of the unipolar bootstrap technique plus a full bridge gate drive inverter could be interesting. Figure 13 below presents the full topology of this approach.

However, a precise analysis of the global behavior shows that this is not as straightforward as it seems. In a similar manner as in the previous section, the combination of these two topologies ends up offering a leakage current path between the power circuit and the gate driver. This is detailed on Figure 13 with the visualization of the fault current path when the top switch is turned OFF.

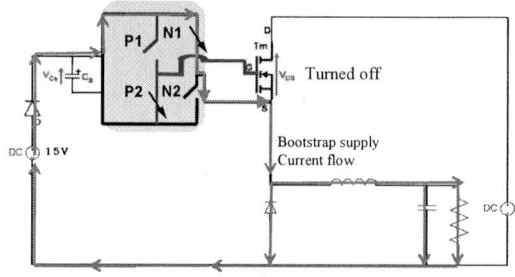

Figure 13. Bootstrap supply technique associative with full bridge gate driver

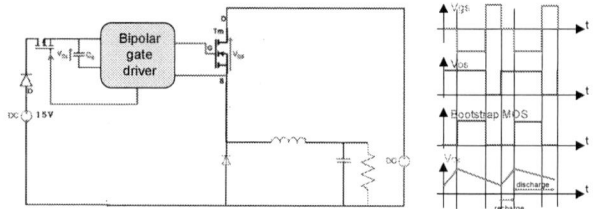

Figure 14. Boostrap bipolar supply with additional current limiting transistor

In a similar manner, as in part A, it is therefore required to insert a transistor in the fault current loop in order to open the loop when needed (Figure 14). Preferably, this switch can be a P type transistor that must be turned OFF when the high side transistor is turned ON. It does not increase too much the complexity of the structure may be monolithically integrated within the gate driver circuit die. Indeed, it is a low voltage device rating.

C. Limits of these approaches

The unipolar bootstrap supply technique is already known for its ease of implementation and operation. It is also well known for its operational limitations due to its interactions with the power parts. If solutions are provided to emulate autonomous operation and bipolar supply [7][6], these gate driver supply solutions which rely on numerous components are not always simple to integrate. This leads to an increase in cost, design and a possible reduction of reliability making these approaches not always competitive with integrated DC to DC supplies such as [18].

IV. DRIVERS PLUS SELF SUPPLY TECHNIQUE

Another high side gate driver supply technique that can be considered is the self powering technique presented in Figure 10. This type of unipolar power supply can be integrated without extra cost. Its evolution and combination with a haddock gate driver to perform bipolar gate signals is now going to be investigated. If the association with resonant gate drivers such as those presented in figure 5 and 6 is more or less straightforward, the association with the full bridge gate driver is more complex.

A. Unipolar self supply plus full bridge gate driver

The unipolar self-powering technique is based on the pulsed linear regulation technique that periodically, picks up some energy from the power side at each high side transistor turn OFF [25]. The self-supply is set to be sensitive only to positive dv/dt and recycles a part of commutation energy to supply the gate driver. Basically, a path is temporary offered for the load current in parallel with the switch which is turning OFF. This path includes the gate driver storage capacitor. Once it is recharged, the path is closed and the switching transition is quickly terminated [25]. Based on this operation, the addition of a full bridge gate driver is not as straightforward as it seems. Indeed, while the high side power device is turned OFF by the bipolar commutation of the full bridge gate driver, an attention must be paid in order to maintain available the path for the load current to flow through the storage capacitor. The figure 15 below presents the full topology including the gate driver and the driver supply. It is shown that during the turn OFF sequence, the bridge harm connected to the source terminal of the power transistor must be at first maintained into a specific configuration before switching in order to apply the correct voltage level to the gate to source terminals of the power switch to drive. Therefore, the full bridge converter must operate under phase shift control in order to be correctly combined with the self power supply technique. Figure 16 depicts simulation results showing this particular switching transition procedure.

This approach has been implemented on a PCB in order to test and to validate this functionality. Figure 17 below gives a set of practical results demonstrating the operation of the whole gate driver. It appears that the power flow through the self supply technique is in the range of 1 W, including its own efficiency. This quite large power consumption could be minimized if resonant gate drivers were implemented.

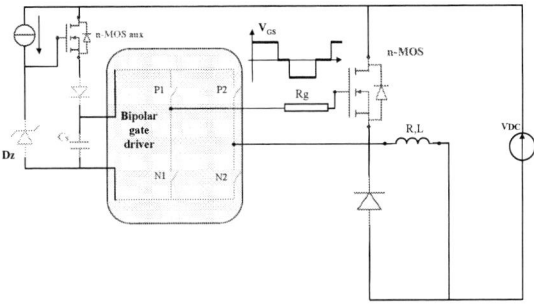

Figure 15. Self-supply in combination with full bridge gate driver

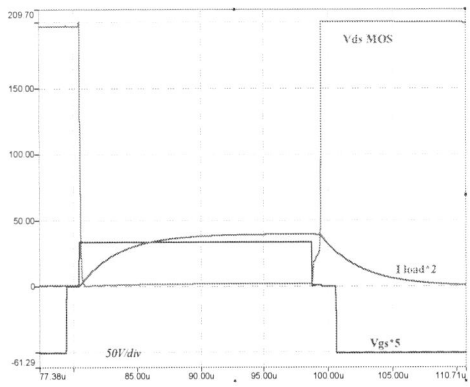

Figure 16. Simulation results of a Buck converter with full gate driver topology

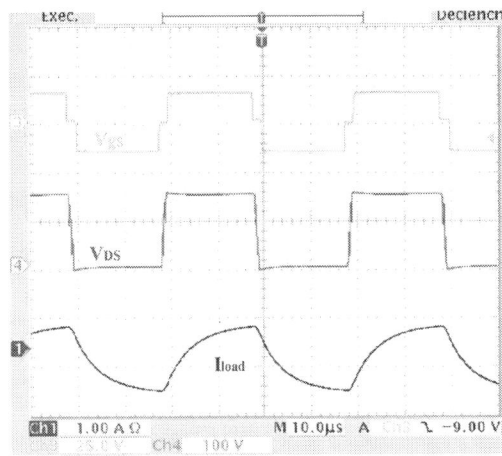

Figure 17. Pratical results of a half-bridge inverter with full gate driver topology

Besides, the phase shift control of the gate bridge may appear less flexible to manage the turn ON turn OFF of the power device since the three levels operation does not allow to pull back the gate current as desired. One way to optimize its effect is to monitor the recharge sequence of the storage capacitor in order to minimize the zero level duration and to keep switching dynamics at their best levels. This can be implemented based on the schematic presented on Figure 15. In this case, the switching transition of the second harm of the driver bridge is engaged only but also shortly after the capacitor has been recharged.

Nevertheless, this approach remains highly integrable and may offer satisfactory operation from both efficiency and dynamics point of views.

B. Bipolar self supply plus push pull gate driver

The evolution of the self supplied technique can also be investigated. It is based on a symmetrical supply technique trying to take advantage of the two switching transitions that occur each switching period of a commutation cell. The resulting topology was depicted on Figure 11. This topology remains sensitive only to positive dv/dt but this time coming from both active power devices. The driver supply structure is more complex but the gate driver implementation can be simplified to a push pull structure including or not a resonant structure in order to save on power consumption. Figure 18 shows experimental results coming from this topology.

The only bad points of this structure are that it is harder to integrate since two storage capacitors are required. In addition, the most critical point must consider that the load current does not switch from positive to negative within each switching cycle in all applications. In this way, one of the driver supply (the positive one if the load current enter the middle point and the negative supply if the load current exits the middle point, as shown on Figure 18) will not be fed periodically by the load current but by the power converter supply. Therefore, the efficiency of such gate driver supply may be largely reduced limiting the operation of the structure to low power, low to medium voltage applications.

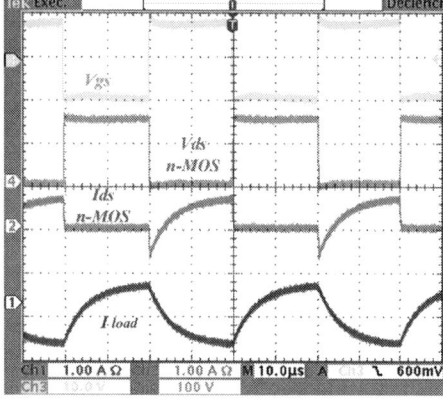

Figure 18. Experimental results of bipolar self supply associative with push pull gate driver

C. Symetrical unipolar self supply plus full bridge.

In order to overcome these limitations, a last configuration can be considered. It relies on the topology Figure 19 where the bipolar gate driver self supply has been merged to a combined structure where only one capacitor remains. This capacitor can be recharged periodically by both driver supplies or only by one of them as a function of converter operation. In this case, the three level driver bridge operation is required but the gate driver supply operation can be optimized. Its integration remains possible if complementary devices are used. Below Figure 20 is given a set of simulation results.

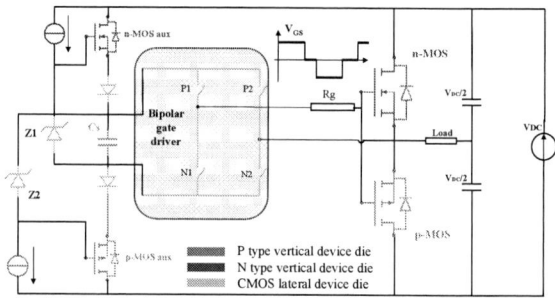

Figure 19. Symetrical unipolar self supply with full bridge gate driver topology

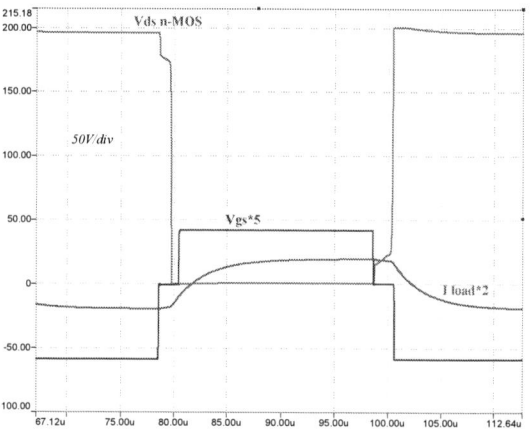

Figure 20. Simulation results of symetrical self supply topology with gate driver

V. CONCLUSION

The paper presents an overview of possible combination of gate drivers and driver supplies for the implementation of bipolar control gate signals. It appears that simple and flexible unipolar gate driver powering techniques are not always simple to emulate toward bipolar supplies. This leads to more complex schematics and numerous additional components not always simple to integrate. Nevertheless, it appears that:

-   The bootstrap technique can be modified to implement bipolar gate driver supply and self-operation. Associated with resonant gate driver circuits, it may be a good solution for medium to high power application and high switching frequencies. The solutions remain hard to integrate.

- The self-powering supply technique can be modified to implement bipolar gate driver supply. The solutions are quite simple to integrate but their efficiency and dynamic behavior may become a limitation in some cases.

The following steps must now consider deeply the efficiency and dynamic characterization of the solutions, introducing the resonant topology in the gate drivers to optimize the operation of these circuits. On the other hand, the integration remains also an issue in order to maintain the robustness of the global converter.

REFERENCES

[1] Zhihua Yang, Sheng Ye, et Yan-Fei Liu, "A New Dual-Channel Resonant Gate Drive Circuit for Low Gate Drive Loss and Low Switching Loss," *Power Electronics, IEEE Transactions on*, vol. 23, 2008, pp. 1574-1583.

[2] Zhiliang Zhang, W. Eberle, Ping Lin, Yan-Fei Liu, et P. Sen, "A new hybrid gate drive scheme for high frequency buck voltage regulators," *Power Electronics Specialists Conference, 2008. PESC 2008. IEEE*, 2008, pp. 2498-2504.

[3] J. Strydom, M. de Rooij, et J. van Wyk, "A comparison of fundamental gate-driver topologies for high frequency applications," *Applied Power Electronics Conference and Exposition, 2004. APEC '04. Nineteenth Annual IEEE*, 2004, pp. 1045-1052 vol.2.

[4] R. Pilawa-Podgurski, A. Sagneri, J. Rivas, D. Anderson, et D. Perreault, "Very-High-Frequency Resonant Boost Converters," *Power Electronics, IEEE Transactions on*, vol. 24, 2009, pp. 1654-1665.

[5] C. Klumpner et N. Shattock, "A Cost-Effective Solution to Power the Gate Drivers of Multilevel Inverters using the Bootstrap Power Supply Technique," *Applied Power Electronics Conference and Exposition, 2009. APEC 2009. Twenty-Fourth Annual IEEE*, 2009, pp. 1773-1779.

[6] Rick West, "Bipolar boostrap top switch gate driver for half-bridge semiconductor power topologies," U.S. Patent US7248093.

[7] Shihong Park et T. Jahns, "A self-boost charge pump topology for a gate drive high-side power supply," *Power Electronics, IEEE Transactions on*, vol. 20, 2005, pp. 300-307.

[8] N. Rouger et J. Crebier, "Toward Generic Fully IntegratedGate Driver Power Supplies," *Power Electronics, IEEE Transactions on*, vol. 23, 2008, pp. 2106-2114.

[9] B. Nguyen, J. Crebier, R. Mitova, L. Aubard, et C. Schaeffer, "AC switches with integrated gate driver supplies," *Power Electronics and Applications, 2005 European Conference on*, 2005, pp. 9 pp.- P.9.

[10] Baoxing Chen, "Isolated half-bridge gate driver with integrated high-side supply," *Power Electronics Specialists Conference, 2008. PESC 2008. IEEE*, 2008, pp. 3615-3618.

[11] Manh Hung Tran et J. Crebier, "Complementary MOS structures for common mode EMI reduction," *Power Electronics and Applications, 2009. EPE '09. 13th European Conference on*, 2009, pp. 1-10.

[12] M.H. Tran, J. Crebier, et C. Schaeffer, "Quantification of benefits and drawbacks in power conversion based on complementary MOS structures," *Energy Conversion Congress and Exposition, 2009. ECCE. IEEE*, 2009, pp. 3423-3430.

[13] N. Rouger, J. Crebier, R. Mitova, L. Aubard, et C. Schaeffer, "Fully integrated driver power supply for insulated gate transistors," *Power Semiconductor Devices and IC's, 2006 IEEE International Symposium on*, 2006, pp. 1-4.

[14] S. Pan et P. Jain, "A New Pulse Resonant MOSFET Gate Driver with Efficient Energy Recovery," *Power Electronics Specialists Conference, 2006. PESC '06. 37th IEEE*, 2006, pp. 1-5.

[15] T. Lopez, G. Sauerlaender, T. Duerbaum, et T. Tolle, "A detailed analysis of a resonant gate driver for PWM applications," *Applied Power Electronics Conference and Exposition, 2003. APEC '03. Eighteenth Annual IEEE*, 2003, pp. 873-878 vol.2.

[16] D. Vasic, F. Costa, et E. Sarraute, "Piezoelectric transformer for integrated MOSFET and IGBT gate driver," *Power Electronics, IEEE Transactions on*, vol. 21, 2006, pp. 56-65.

[17] S.P. Vlahu, "High frequency pulse transformer for an IGBT gate drive ," U.S. Patent USRE38082, Avril 22, 2003.

[18] B. Chen, "Fully integrated isolated DC-to-DC converter and half bridge gate driver with integral power supply," Cork, Ireland, 2008: .

[19] O. Deleage, J. Crebier, M. Brunet, Y. Lembeye, et H. Tran Manh, "Design and realization of highly integrated isolated DC/DC micro-converter," *Energy Conversion Congress and Exposition, 2009. ECCE. IEEE*, 2009, pp. 3690-3697.

[20] J. Popovic et J. Ferreira, "Design and evaluation of highly integrated dc-dc converters for automotive applications," *Industry Applications Conference, 2005. Fourtieth IAS Annual Meeting. Conference Record of the 2005*, 2005, pp. 1152-1159 Vol. 2.

[21] Z. Hayashi, Y. Katayama, M. Edo, et H. Nishio, "High efficiency DC-DC converter chip size module with integrated soft ferrite," *Magnetics Conference, 2003. INTERMAG 2003. IEEE International*, 2003, pp. FC-04.

[22] V. Kursun, S. Narendra, V. De, et E. Friedman, "High input voltage step-down DC-DC converters for integration in a low voltage CMOS process," *Quality Electronic Design, 2004. Proceedings. 5th International Symposium on*, 2004, pp. 517-521.

[23] J. Popovic et J. Ferreira, "Converter concepts to increase the integration level," *Power Electronics, IEEE Transactions on*, vol. 20, 2005, pp. 558-565.

[24] N. Rouger, J. Crebier, et S. Catellani, "High-Efficiency and Fully Integrated Self-Powering Technique for Intelligent Switch-Based Flyback Converters," *Industry Applications, IEEE Transactions on*, vol. 44, 2008, pp. 826-835.

[25] J. Crebier et N. Rouger, "Loss Free Gate Driver Unipolar Power Supply for High Side Power Transistors," *Power Electronics, IEEE Transactions on*, vol. 23, 2008, pp. 1565-1573.

A Method for Impact Assessment of Faults on the Performance of Field-Oriented Control Drives: A First Step to Reliability Modeling

Ali M. Bazzi, Alejandro Dominguez-Garcia, and Philip T. Krein
Grainger Center for Electric Machinery and Electromechanics
Department of Electrical and Computer Engineering
University of Illinois at Urbana-Champaign
Urbana, IL, 61801, USA

Abstract— In this paper, the effects of certain component faults on the performance of three-phase inverter-fed induction motors are analyzed under indirect field-oriented control. Simulations of faults in the current sensors, speed encoder, three-phase inverter, and motor are presented. Sample hardware faults verifying the simulation results are also presented. Performance requirements are set based on typical performance measures of electric vehicles. These requirements are then used to determine whether or not the system survives a fault. A simple fault detection and isolation scheme using multiple speed encoders is also tested. The construction of a Markov reliability model of the overall system is a direct application of the results, and allows quantifying global reliability measures.

I. INTRODUCTION

Induction motors have been a main part of several traction systems and are currently being utilized in electric vehicles (EVs), e.g. Tesla Roadster™ [1]. They have been established as highly reliable for hybrid electric vehicles (HEVs) [2]. The main initiatives behind EVs and HEVs have been environmental friendliness and fuel savings. Safety and reliability also must be considered carefully by manufacturers. In this paper, the focus is on motor drive reliability in EVs and EHVs. This might be a limiting factor for the widespread acceptance of these vehicles, as both manufacturers and users demand reliability without compromise on performance.

Generally speaking, reliability of a motor drive can be explored via the emerging field of thermo- electrical analysis [3]. To achieve high reliability levels and fault-tolerant operation, which is important for safety considerations, two main elements should be engineered into the motor drive: i) component redundancy, and ii) fault detection and isolation (FDI) mechanisms. In this regard, a typical induction motor drive for a general application is shown in Fig. 1, with no redundancy or details about FDI. There are well-developed methods to evaluate the performance of such a drive under all operational conditions in the absence of faults, however, literature lacks systematic methods that quantify the system ability to tolerate faults and model its overall reliability. This poses a fundamental problem for designers.

In this paper, steps are taken to address this problem, i.e., *a systematic method to analyze reliability and fault-tolerance in motor drives for EV and HEV applications*. In particular, models are provided for faults that can occur in key components of an indirect field-oriented control (IFOC) induction motor drive, e.g., current sensors and speed encoders. A simulation-based framework to analyze the effect of these faults on system performance is developed. The results of this work can be directly applied to build a Markov reliability model of the drive, thus quantifying global reliability measures.

Literature shows significant work in all aspects of design for reliability and fault-tolerance in induction motor drives. Definition of fault models, which is key for fault-tolerant design, is extensively discussed, including inverter faults [4], control faults [5], supply faults [6, 7], and motor faults [8]. Aspects of fault detection and isolation have been extensively analyzed (see, e.g., [4, 9-16]). For example, pattern recognition is used in [13], while a short-time Fourier transform is used in [16] to identify faults. Redundancy has also been investigated with extensive work on multiphase motors [17] and split-wound motors [18]. At the control level, fault-tolerant control algorithms for permanent magnet synchronous machines (PMSMs) and induction machines were presented in [19-22]. Development of reliable communication and control hardware used in electric drives has also been investigated [23]. Other strategies for improving the reliability of a system include preventive maintenance [24, 25], component derating, and component count reduction.

978-1-4244-4782-4/10 $26.00 © 2010 IEEE

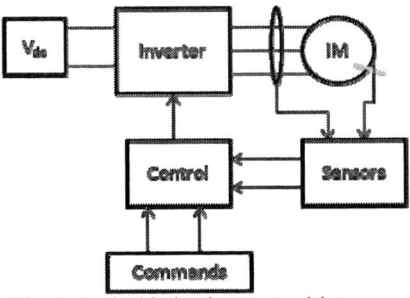

Fig. 1. Typical induction motor drive.

Even though extensive work has been conducted on fault-tolerant drives and motor control design, literature on systematic methods for modeling and analysis of fault effects on system operation and reliability is limited. Such methods are key to assess whether or not a design meets reliability and fault tolerance requirements for all possible operational conditions, and to compare different design choices. Available work on reliability modeling usually targets the failure rate of the overall drive system. For example, the reliability of the motor and supply is presented in [26]. Component reliability analyses are presented in [27-29]. In [28], faults in control, power electronics, and the motor are addressed, while in [27] control faults are ignored and faults in the transformer and line filters are considered. The work presented in [29] also ignores control, but considers cooling faults. In terms of system reliability modeling, the application of Markov models, which are among the most powerful reliability modeling tools, to motor drives is rare. In [30], only power electronics failures are considered with open circuits, short circuits, and other component failure modes. The induction motor is considered as part of a larger power system for which a Markov model is developed in [31]. An excellent attempt to develop a Markov model of an induction motor drive is available in [32]. Even though sensors can suffer from significant faults, these faults and performance bounds are ignored in [32]. Multiple sensor faults in automotive applications and a multilayer control scheme that restructures faulted sensor signals are presented in [33].

In summary, a complete framework for fault modeling and reliability analysis of a motor drive, including the motor, control, sensors, and power electronics, is not available. This framework would have significant value in assisting designers of fault-tolerant motor drives. This paper sets the foundations for establishing such a framework, focusing on component fault modeling and the subsequent analysis of the impact of faults on overall drive performance. In this regard, there is significant work in the literature on modeling physical motor faults, e.g. broken rotor bars and stator winding short-circuits; however, fault modeling of non-physical effects in control and communication elements, e.g., sensors and controllers, and their impacts on system performance are usually not addressed.

Component fault models as well as system performance measures, the key elements of the proposed methodology, are discussed in Section II. Section III discusses the implementation of fault models and performance measures in

a simulation environment. In Section IV, both simulation and experimental fault injection results are presented. Section V includes an overview of the expected Markov reliability model, and Section VI concludes with remarks and future work.

II. METHODOLOGY

The proposed methodology uses a model of motor-drive dynamics plus additional features to model component failures. These features include component failure modes (and associated failure rates) and how these failure modes affect the dynamic behavior of the component. Depending on the type of component fault, time of fault, and sequence in the event of multiple faults, the system will evolve from the nominal unfaulted configuration to other configurations with different dynamics. Each possible system configuration will be analyzed to check whether its dynamic properties meet system performance requirements. Once all configurations have been evaluated, the performance metrics for each configuration and the probabilities of going from the nominal configuration to any other configuration are merged into a Markov model.

A. Fault Modeling

A system under no faults is considered to be in its nominal configuration. The transitions between configurations occur stochastically, triggered by random faults. Faults identified in literature include those within the motor, power electronics, control, and supply. Faults in sensors are also essential but have not been elaborated upon in FOC drives. While faults in other major components such as digital signal processors and interface electronics also have severe consequences, these are relatively unlikely except as consequences of other faults. Here, they are lumped into controller and sensor faults rather than treated separately.

The paper focuses on common faults that affect the performance of an IFOC drive from [34], and analyzes the status of the drive under such faults. Sensor faults include omission, incorrect gain, bias, constant value, and noise in current sensors and speed encoders. Inverter faults are also addressed and include short circuit (SC) to ground, SC to the dc voltage bus, and open circuit (OC) [4, 17, 30]. Motor faults studied here are phase-to-phase faults and broken rotor bars, which are the most common faults in induction machines. Motor OC and SC to ground faults can be modeled on the inverter side. Another squirrel-cage induction motor fault is a broken end ring in a motor, which follows a similar analysis as a broken bar [35]. The faults to be analyzed here are tabulated in Table I. The discussion addresses only single faults, although as the methodology develops, it will be important to evaluate the effects of multiple sequential faults to quantify system fault tolerance. The modeling process is simplified by symmetry. For example, a fault in inverter phase-leg a or its current sensor has the same impact as in phases b or c. This inherent symmetry could help with fault detection and isolation in motor drives by utilizing multiphase motors, reconstructing faulted sensor signals from healthy signals, etc.

B. Performance Evaluation

Performance metrics are measurable properties of the system that quantify how well it performs its desired functions. Performance metrics can be related to system functionality, e.g., in a tracking-system, relevant performance metrics are the tracking error, the overshoot, or the settling time. However, non-functional metrics can be considered as well, e.g., in the tracking system, the power consumption may be important..

General performance metrics and associated requirements are presented for driving conveyors in [36]. Some of these can be extended to EVs. These requirements are typical for acceptable performance, including safety requirements, e.g., ability of the system to return to desired operation within 500 ms. A vehicle travelling at 65 mph (105 km/h) moves about 15 m during this interval, which is faster than a driver's reaction time.. Sample performance bounds are tabulated in Table II. These bounds are then scaled for a 1.5 hp induction machine used for experiments. The current bound is 10 A which is about twice the peak rated current. Note that the torque bounds are set and taken into consideration only in simulations, as they only provide results for open-loop torque control, while experiments, as explained later, provide results for closed-loop control.

For each system configuration arising from different faults, performance metrics are analyzed. Performance metrics are useful in determining whether each possible system configuration is declared as failed or non-failed, but it is necessary to define an aggregated measure of reliability, which can be later accomplished by constructing a Markov reliability model.

TABLE I
FAULTS IN THE FOC DRIVE

	Speed Encoder	Current Sensor	Phase Leg	Motor
Fault Types	• Omission • Gain • Bias • Constant • Noise	• Omission • Gain • Bias • Constant • Noise	• OC • SC to ground • SC to dc bus	• Phase to phase fault • Broken rotor bar

TABLE II
PERFORMANCE METRICS AND ASSOCIATED REQUIREMENTS

	Performance Requirements
Speed	Command speed ± 50 rpm
Current	Current peak not exceeding 12A
Settling time	Less than 500ms
Torque (Simulations)	Nominal load torque ±0.5 N·m

TABLE III
VALUES USED WITH FAULTS IN SENSORS

Fault Type	Speed Encoder	Current Sensor
Gain	1.5	1.5
Bias	+10 rpm	+1 A
Constant	900 rpm	3 A
Noise	±10 rpm	±0.5 A

III. IMPLEMENTATION OF THE PROPOSED METHODOLOGY IN SIMULATION AND EXPERIMENT

Modeling faults in simulations is an essential step to emulate real-life operation of a faulted IFOC drive. It can be risky to inject certain faults in experiments.

A. Simulation Environment

Simulations provide a safe environment to evaluate even the most extreme faults, provided a simulation has been validated in hardware. Some commercial drives have fault detection and isolation, which would not be helpful if the target is to observe drive performance under faults, such as a phase-to-phase fault. In simulations, faults are mainly of two kinds: sensor faults where the sensor feedback signal can be faulted, and circuit faults, where the power electronics and motor electrical connections are modified in the faulted configurations.

Sensor faults can be modeled as shown in Fig. 2, which shows a simulation model for a sensor fault in MATLAB/Simulink. This modeling approach, proposed in [37], provides the flexibility of changing the gain, bias, constant, and noise values, in addition to the time of the fault injection. Values used for these faults are shown in Table III. The gain change fault results in a 50% increase in encoder and current sensor nominal gains. The bias faults imply the sensor output gives a biased value of the true quantity. Constant faults imply that sensor output gets stuck at a fixed value, while noise faults imply the sensor output is corrupted by additive Gaussian noise. These values could be swept over a wide range, but constants are used for this analysis based on logical expectations. Circuit faults can be modeled with switches that generate the faulted configurations. The controlled power electronics switches available in the SimPowerSystems toolbox provide a possible approach. A drawback of the SimPowerSystems toolbox is that any transition from a short circuit to an open circuit results in rapid changes in currents and voltages that could lead to numerical errors. This problem can be avoided with suitable parallel or series resistances, although in the real system, stray inductances and capacitances limit the rates of change during faults. Broken rotor bars require a different fault model in which rotor currents, resistances, and inductances are modified when the fault happens. A simple model used here ignores the change of rotor inductance and focuses on the change of rotor resistance [38]. Therefore, the induction motor in Fig. 3 employs a wound rotor model in which additional resistance is connected to represent a broken rotor bar fault.

Another approach to model circuit faults is the modification of the switching pattern. For example, an open circuit in the upper switch in an inverter phase-leg can be modeled simply by switch turn-off. These models give an appropriate idea of system behavior under faults and avoid severe failures in the experimental setup.

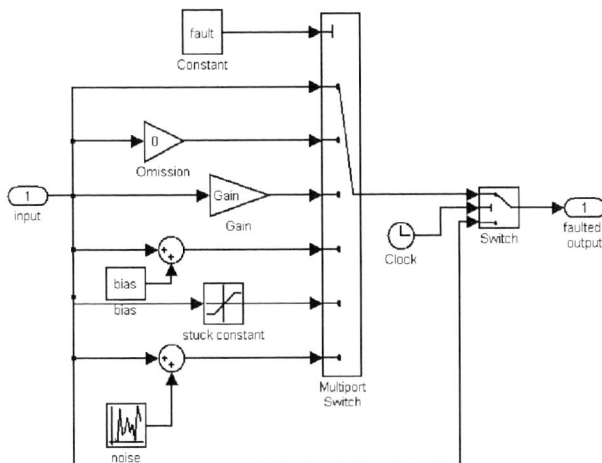

Fig. 2. Sensor fault model.

B. Experimental Environment

The experimental setup includes all control, power electronics, motor, load, and measurements. The control and some measurements are available on an eZdspF2812 platform [39] which is based on a Texas Instruments TMS320F2812 DSP. This control platform is integrated into the Grainger Center Modular Inverter [40] which also includes an inverter power stage rated at 400V and 100A. A dynamometer is used to set the load torque. Measurements include speed, torque, currents, voltages, and input and output power. All control and measurement devices, including the dynamometer, are controlled using MATLAB/Simulink through the Simulink Real-Time Workshop. Communication between the software and the DSP occurs through real-time data exchange (RTDX) via a parallel port. The experimental setup is shown in Fig. 4.

IV. SIMULATION AND EXPERIMENTAL RESULTS

An IFOC motor drive with a 1.5 hp motor was simulated in MATLAB/Simulink using the SimPowerSystems toolbox, as shown in Fig. 3. A load quadratic in speed emulates single quadrant operation of an EV. The speed command is 1000 rpm, which results in steady-state torque of about 2 N·m. The red blocks in Fig. 3 are the locations of the faults. All 15 faults shown in Table I were simulated, with emphasis on torque, speed, and current responses. The faults were injected at t=2 s.

Several simulation and experimental results are shown in Figures 5 through 10. The experimental setup is shown in Fig. 11. Waveforms in the experimental results are as follows: The top trace is the speed, scaled at 500 rpm/div, the center trace is the torque at 2 N·m/div, and the bottom trace is current at 10 A/div. Faults leading to failures with low risk of system damage, e.g. encoder omission, were experimentally validated. Table IV includes the system status after each fault. Current sensor faults and electrical faults were imposed on phase "a". The broken rotor bar was modeled as a 0.1 Ω

increase in rotor resistance. It is clear from Table IV that more than one third of the faults cause the system to fail in the sense of not meeting performance requirements. This is expected because IFOC, without any redundancy, uses four feedback signals: three for currents and one for speed; thus, with one of these four signals failing, the motor drive could fail. Figures 5 through 10 show that the simulation model accurately reflects the experimental setup, which simplifies the simulation of severe faults and failures. The general dynamic performance shown in simulations and the steady-state shapes and values of the speed and current waveforms match those in experiments.

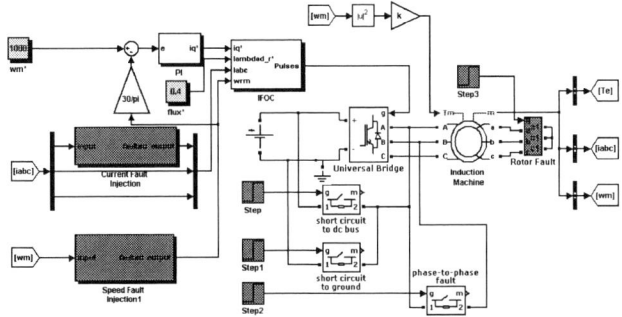

Fig. 3. Simulink model of the IFOC motor drive.

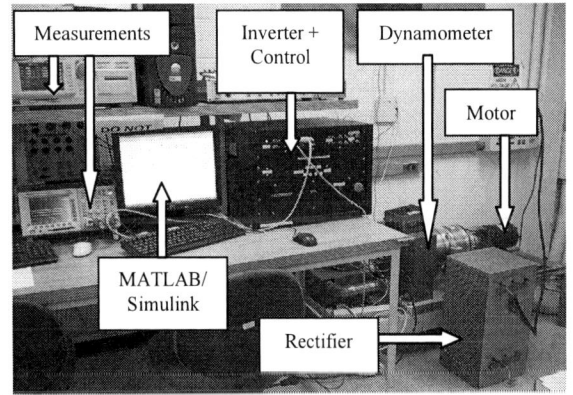

Fig. 4. Experimental setup.

These results suggest that it should be possible to enhance a conventional IFOC algorithm with redundancy and fault detection to cover faults. For example, a 2-of-3 voting mechanism with three speed encoders will eliminate speed encoder problems, as shown in Fig. 11. Even though adding two more speed encoders might not be economically feasible, the functions could be provided by speed estimators. The reference speed could also be used as one of the three signals compared in a voting mechanism given an appropriate tolerance band among signals. Similar voting mechanisms can be implemented for current sensors. Alternatively, a faulty current can be reconstructed from healthy phase currents or from the dc link current and switching states [33].

978-1-4244-4782-4/10 $26.00 © 2010 IEEE   259

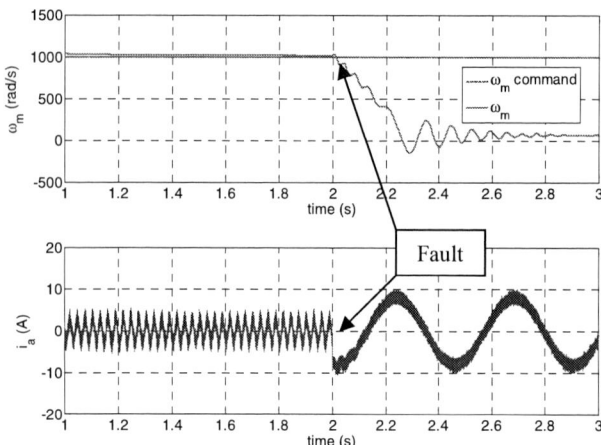

Fig. 5. System failure after speed encoder omission (simulation results).

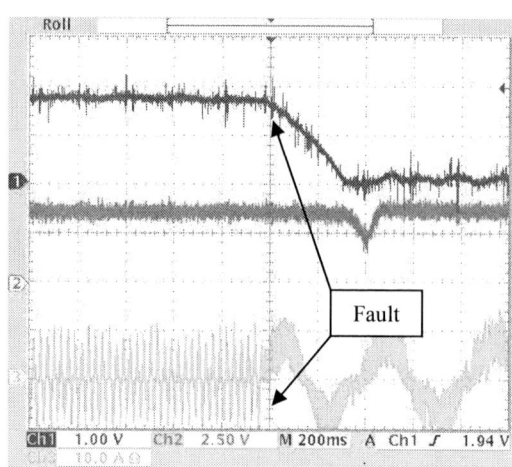

Fig. 8. System failure after speed encoder omission (experimental results).

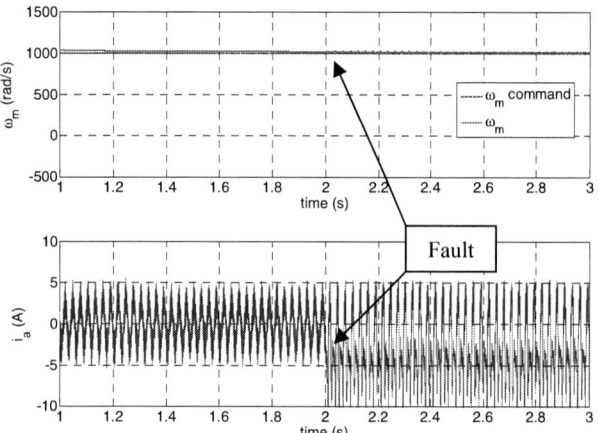

Fig. 6. System survival within performance bounds after current sensor constant (simulation results).

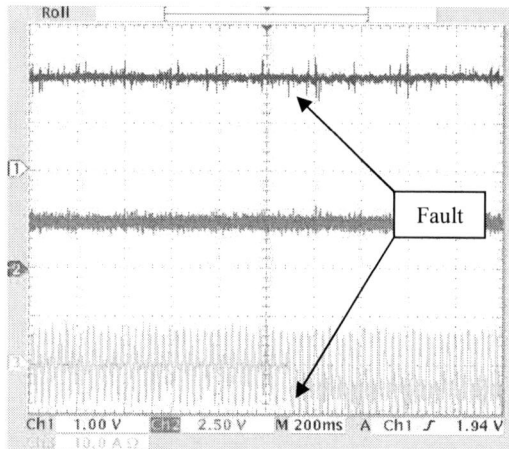

Fig. 9. System survival within performance bounds after current sensor constant (experimental results).

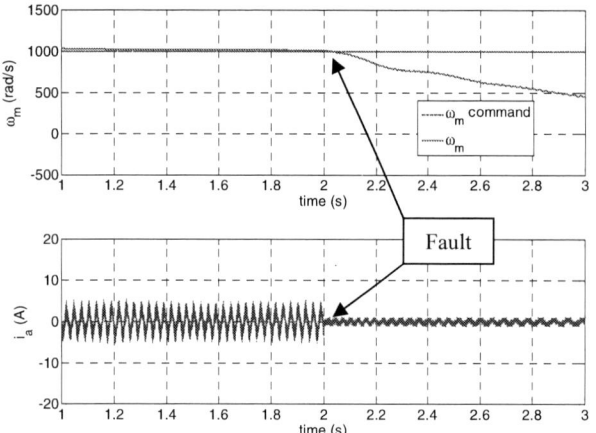

Fig. 7. System failure after an open circuit on phase "a" (simulation results).

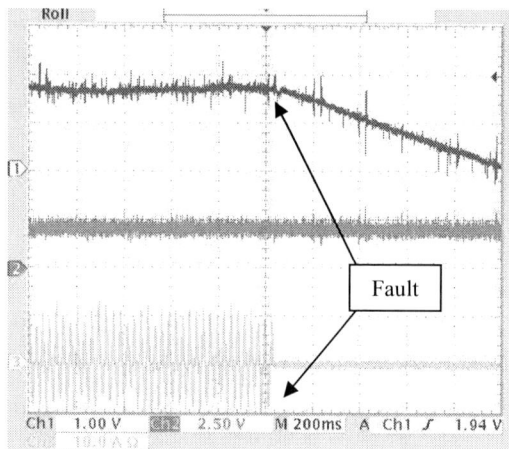

Fig. 10. System failure after open circuit on phase "a" (experimental results).

978-1-4244-4782-4/10 $26.00 © 2010 IEEE

TABLE IV
SYSTEM RESPONSE TO FAULTS BASED BOUNDS IN TABLE III

Fault	Status
Speed Encoder: Omission, Gain, Noise	Failed
Speed Encoder: Bias, Constant	OK
Current Sensor: Constant	Failed
Current Sensor: Gain, Bias, Omission, Noise	OK
OC, SC phase-to-phase	Failed
SC to dc bus, SC to ground , Broken rotor bar	OK

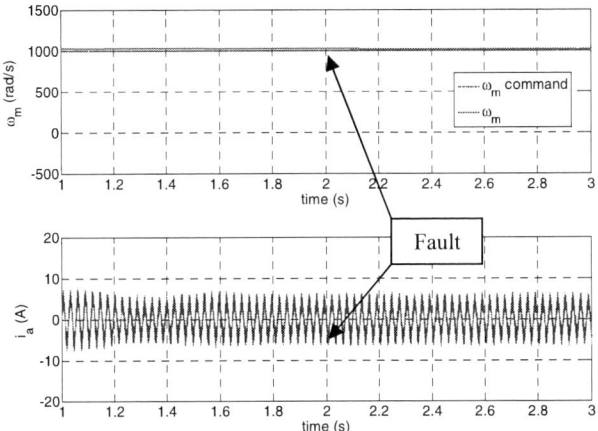

Fig. 11. System survival after speed encoder omission with a 2-of-3 voting system.

V. MARKOV RELIABILITY MODEL

The overall drive reliability can be obtained by formulating a Markov reliability model [41]. Unlike combinatorial reliability models such as fault trees or reliability block diagrams, state-dependent component failure rates can be naturally included in a Markov reliability model. Such a model can be represented graphically in terms of a state-transition diagram of the motor-drive system status (failed or operational) for each configuration reached after a unique sequence of component faults. The motor-drive system can evolve from the fault-free configuration to other configurations depending on the status of the components within the motor-drive system. The nodes of the state-transition diagram represent the status (failed or operational) of each system configuration, and the edges represent transitions between configurations triggered by component faults and described by the failure rate of the faulted component causing the transition. There are two types of nodes: absorbing nodes and non-absorbing nodes. The system will fulfill its function whenever it is in a non-absorbing node. System reliability is quantified as the probability the system is in any non-absorbing state, which can be obtained by solving the Chapman-Kolmogorov equations associated with the state-transition diagram [41].

The analysis conducted in Section IV focused on single fault occurrences. Table IV shows the system status after

each fault considered. Figure 12 shows a portion of the Markov reliability model corresponding to the speed encoder faults in Table IV and the failure rates shown in Table V [42]. The green states represent post-fault operational states, whereas the red states correspond to failed system states. It is important to note that different faults yield very different results. Although not represented here, the sequences of faults involving more than one component can be incorporated in the Markov model.

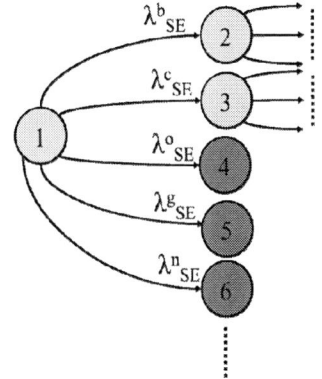

Fig. 12. Snapshot of the motor-drive Markov reliability model.

TABLE V
FAILURE RATES FOR SPEED ENCODER FAULTS

Fault	Failure Rate (failures/h)
Omission	$\lambda^{O}_{SE} = 0.74 \times 10^{-6}$
Gain	$\lambda^{G}_{SE} = 0.19 \times 10^{-6}$
Bias	$\lambda^{B}_{SE} = 0.42 \times 10^{-6}$
Constant	$\lambda^{C}_{SE} = 0.19 \times 10^{-6}$
Noise	$\lambda^{N}_{SE} = 0.09 \times 10^{-6}$

VI. CONCLUDING REMARKS

Engineering methods for analysis of faults, failures, and general reliability of an IFOC induction motor drive can be used for system performance improvements. EVs and HEVs require special safety considerations, thus posing an essential challenge. All single faults and the resulting failures or survivals were analyzed based on preset performance bounds. Simulations of faults in the system were presented with experimental validation of several failures. Simple fault detection and isolation was presented based on a redundant speed sensor example.

ACKNOWLEDGMENT

This work is supported by the Grainger Center for Electric Machinery and Electromechanics at the University of Illinois.

APPENDIX

Motor Parameter	Value
Rated power	1.5 hp
Rated speed	1750 rpm
Number of poles (P)	4
Referred rotor resistance ($R_r{}'$)	0.7309 Ω
Stator resistance (R_s)	1.5293 Ω
Referred rotor leakage inductance ($L_{lr}{}'$)	0.005343 H
Stator leakage inductance (L_{ls})	0.00356 H
Magnetizing inductance (L_m)	0.19778 H
Core Loss (R_c)	505 Ω
Inertia (J)	0.01 Kg.m^2

REFERENCES

[1] "http://www.teslamotors.com/performance/perf_specs.php."

[2] L. Chang, "Comparison of AC drives for electric vehicles-a report on experts' opinion survey," *IEEE Aerospace and Electronic Systems Magazine*, vol. 9, pp. 7-11, 1994.

[3] R. D. Lorenz, "The future of electric drives: where are we headed?," in *Proc. Int'l. Conf. Power Electronics and Variable Speed Drives*, 2000, pp. 1-6.

[4] D. Kastha and B. K. Bose, "Investigation of fault modes of voltage-fed inverter system for induction motor drive," *IEEE Trans. Industry Applications*, vol. 30, pp. 1028-1038, 1994.

[5] J. Pontt, J. Rodriguez, J. Rebolledo, L. S. Martin, E. Cid, and G. Figueroa, "High-power LCI grinding mill drive under faulty conditions," in *Rec. IEEE Industry Applications Soc. Annual Meet.e*, 2005, pp. 670-673.

[6] R. M. Tallam, D. W. Schlegel, and F. L. Hoadley, "Failure mode for AC drives on high resistance grounded systems," in *Proc. IEEE Applied Power Electronics Conference and Exposition*, 2006, pp. 1587-1591.

[7] Y. Yuexin and A. Y. Wu, "Transient response of electric drives under utility upset conditions," in *Pulp and Paper Industry Technical Conference*, 1996, pp. 77-85.

[8] V. P. Shevchuk, "Investigations of the operation reliability increase of the alternating current electric machines in diamond extractive industries," in *International Scientific and Practical Conference of Students, Post-graduates and Young Scientists: Modern Technique and Technologies*, 2002, pp. 103-104.

[9] A. Fekih and F. N. Chowdhury, "A Fault Tolerant Control Design for Induction Motors," in *IEEE International Conference on Systems, Man and Cybernetics*, 2005, pp. 1320-1325.

[10] S. Green, D. J. Atkinson, A. G. Jack, B. C. Mecrow, and A. King, "Sensorless operation of a fault tolerant PM drive," *IEE Proc. Electric Power Applications*, vol. 150, pp. 117-125, 2003.

[11] O. Jasim, C. Gerada, M. Sumner, and J. Arellano-Padilla, "Investigation of induction machine phase open circuit faults using a simplified equivalent circuit model," in *Proc. International Conference on Electrical Machines*, 2008, pp. 1-6.

[12] K. S. Lee and J. S. Ryu, "Instrument fault detection and compensation scheme for direct torque controlled induction motor drives," *IEE Proceedings on Control Theory and Applications*, vol. 150, pp. 376-382, 2003.

[13] O. Ondel, G. Clerc, E. Boutleux, and E. Blanco, "Fault Detection and Diagnosis in a Set “Inverter–Induction Machine” Through Multidimensional Membership Function and Pattern Recognition," *IEEE Trans. Energy Conversion*, vol. 24, pp. 431-441, 2009.

[14] R. L. A. Ribeiro, C. B. Jacobina, E. R. C. Da Silva, and A. M. N. Lima, "Compensation strategies in the PWM-VSI topology for a fault tolerant induction motor drive system," in *IEEE International Symposium on Diagnostics for Electric Machines, Power Electronics and Drives*, 2003, pp. 211-216.

[15] R. B. Sepe, Jr., B. Fahimi, C. Morrison, and J. M. Miller, "Fault tolerant operation of induction motor drives with automatic controller reconfiguration," in *IEEE International Electric Machines and Drives Conference*, 2001, pp. 156-162.

[16] W. G. Zanardelli, E. G. Strangas, and S. Aviyente, "Identification of Intermittent Electrical and Mechanical Faults in Permanent-Magnet AC Drives Based on Time and Frequency Analysis," *IEEE Transactions on Industry Applications*, vol. 43, pp. 971-980, 2007.

[17] M. T. Abolhassani and H. A. Toliyat, "Fault tolerant permanent magnet motor drives for electric vehicles," in *IEEE International Electric Machines and Drives Conference*, 2009, pp. 1146-1152.

[18] J. C. Salmon and B. W. Williams, "A split-wound induction motor design to improve the reliability of PWM inverter drives," *IEEE Transactions on Industry Applications*, vol. 26, pp. 143-150, 1990.

[19] J. W. Bennett, A. G. Jack, B. C. Mecrow, D. J. Atkinson, C. Sewell, and G. Mason, "Fault-tolerant control architecture for an electrical actuator," in *IEEE Annual Power Electronics Specialists Conference*, 2004, pp. 4371-4377.

[20] D. Kastha and B. K. Bose, "Fault mode single-phase operation of a variable frequency induction motor drive and improvement of pulsating torque characteristics," *IEEE Trans Industrial Electronics*, vol. 41, pp. 426-433, 1994.

[21] O. Wallmark, L. Harnefors, and O. Carlson, "Control Algorithms for a Fault-Tolerant PMSM Drive," *IEEE Transactions on Industrial Electronics*, vol. 54, pp. 1973-1980, 2007.

[22] B. W. Williams, "High reliability 3-phase variable-frequency inverter," *IEE Proc. Electric Power Applications*, vol. 129, pp. 353-354, 1982.

[23] S. Bolognani, L. Peretti, L. Ssgarbossa, and M. Zigliotto, "Improvements in Power Line Communication Reliability for Electric Drives by Random PWM Techniques," in *IEEE Industrial Electronics, IECON 2006 - 32nd Annual Conference on*, 2006, pp. 2307-2312.

[24] G. F. D'Addio, S. Savio, and P. Firpo, "Optimized reliability centered maintenance of vehicles electrical drives for high speed railway applications," in *IEEE International Symposium on Industrial Electronics*, 1997, pp. 555-560 vol.2.

[25] F. A. DeWinter, R. Paes, R. Vermaas, and C. Gilks, "Maximizing large drive availability," *IEEE Industry Applications Magazine*, vol. 8, pp. 66-75, 2002.

[26] J. A. Oliver and D. Poteet, "High-speed, high-horsepower electric motors for pipeline compressors: available ASD technology, reliability, harmonic control," *IEEE Trans. Energy Conversion*, vol. 10, pp. 470-476, 1995.

[27] R. Bozzo, V. Fazio, and S. Savio, "Power electronics reliability and stochastic performances of innovative ac traction drives: a comparative analysis," in *IEEE Power Tech Conference Proceedings*, 2003, p. 7.

[28] R. D. Klug and M. Griggs, "Reliability and availability of megawatt drive concepts," in *International Conference on Power System Technology*, 2004, pp. 665-671.

[29] P. Wikstrom, L. A. Terens, and H. Kobi, "Reliability, availability, and maintainability of high-power variable-speed drive systems," *IEEE Transactions on Industry Applications*, vol. 36, pp. 231-241, 2000.

[30] R. Letchmanan, J. T. Economou, A. Tsourdos, I. A. Ashokaraj, and B. A. White, "Fault Evaluation of Relative-Coupled BLDC Drives for Multi-Facet Mobile Robot with Distributed Speed Factors," in *IEEE Vehicle Power and Propulsion Conference*, 2006, pp. 1-6.

[31] M. H. J. Bollen and P. M. E. Dirix, "Simple model for post-fault motor behaviour for reliability/power quality assessment of industrial power systems," *IEE Proceedings Generation, Transmission and Distribution*, vol. 143, pp. 56-60, 1996.

[32] M. Molaei, H. Oraee, and M. Fotuhi-Firuzabad, "Markov Model of Drive-Motor Systems for Reliability Calculation," in *IEEE International Symposium on Industrial Electronics*, 2006, pp. 2286-2291.

[33] W. Hainan, S. Pekarek, and B. Fahimi, "Multilayer control of an induction motor drive:A strategic step for automotive applications," *IEEE Trans. Power Electronics*, vol. 21, pp. 676-686, 2006.

[34] P. C. Krause, O. Wasynczuk, and S. D. Sudhoff, *Analysis of Electric Machinery and Drive Systems*, 2nd ed. New York: Wiley - IEEE Press, 2002.

[35] L. Xiaogang, L. Yuefeng, H. A. Toliyat, A. El-Antably, and T. A. Lipo, "Multiple coupled circuit modeling of induction machines," *IEEE Trans. Industry Applications*, vol. 31, pp. 311-318, 1995.

[36] K. D. Jackson, M. D. McCulloch, and C. F. Landy, "A study of the suitability of electric drives to the task of driving conveyors," in *IEEE Industry Applications Society Annual Meeting*, 1993, pp. 488-495.

[37] A. D. Domínguez-García, J. G. Kassakian, J. E. Schindall, and J. J. Zinchuk, "An Integrated Methodology for the Dynamic Performance and Reliability Evaluation of Fault-Tolerant Systems," *Journal of Reliability Engineering and System Safety,* vol. 93, pp. 1628-1649, Nov. 2008.

[38] A. F. Alshandoli, "Model-Predicted Induction Motor Behaviour under Different Operating Conditions," in *International Conference on Electrical Engineering,* 2007, pp. 1-7.

[39] "eZdsp(TM) F2812 Technical Reference 506265-0001 Rev. F," Spectrum Digital, Stafford, TX 2003.

[40] J. Kimball, M. Amerhein, A. Kwasinski, J. Mossoba, B. Nee, Z. Sorchini, W. Weaver, J. Weels, and G. Zhang, "Modular Inverter for Advanced Control Applications," Technical Report CEME-TR-200-01, University of Illinois May 2006, .

[41] M. Rausand and A. Høyland, *System Reliability Theory: Models, Statistical Methods, and Applications,* 2 ed. Hoboken, NJ: Wiley, 2005.

[42] "IEEE standard reliability data for pumps and drivers, valve actuators, and valves," *ANSI/IEEE Std 500-1984,* 1984.

A Fault Tolerant Control System for Hexagram Inverter Motor Drive

Liang Zhou and Keyue Smedley
Power Electronics Laboratory
University of California-Irvine, CA, USA

Abstract—In this paper, a fault tolerant control method for hexagram inverter motor drive is proposed. Due to its unique topology, the hexagram inverter is able to tolerate certain degree of switch failure with a proper control method. The proposed method consists of fault detection, fault isolation and post fault control. A simple fault isolation method is to use fuses in the DC links to disable the whole inverter module with switch failure. When a fault is detected in one inverter module, it is isolated by turning on all switches temporarily to blow out the fuse in the DC link, the gate drive signals for this faulty inverter module and its interconnecting legs are then disabled. After one inverter module is disabled, the hexagram inverter works in the post fault two-phase mode. The post fault control algorithm is initiated to control the two remaining output currents in order to maintain a smooth torque operation for the motor drive. Simulations and a small-scale PMSM motor drive experiment verified the proposed fault tolerant control system design.

I. INTRODUCTION

Variable Speed Drives (VSDs) have brought significant advantages in process control, efficiencies, soft-start, and energy savings to the industry. In recent years, the demand for higher power equipment has reached the megawatt level. Adjustable ac drives in the megawatt range are usually connected to the Medium Voltage (MV) network and employs multi-level converters. Due to the high power nature and the fact that multilevel inverters use many components, these motor drives are prone to various types of failures, thus the reliability of motor drive is a great concern. In general, when one failure occurs, the drive system has to stop for an unscheduled maintenance. In recent years, many researchers have focused their effort on the development of fault-tolerant ac drive systems in order to maintain smooth torque operation after a fault happens. Short-torque transient and even permanently reduced performance after fault are accepted. Typically, a fault tolerant method involves fault detection, fault isolation and post fault reconfiguration and control. Several fault detection techniques have been provided in [1]-[6]. For 2-level three-phase motor drive, the faulty switch can be isolated by blowing out fuses, which usually required an additional SCR [7]. Multi-level inverters offer redundancy degrees which can be utilized to enable the operation with

faulty elements. For the cascaded H-bridge inverter, the faulty inverter unit is isolated by using a contactor to bypass it [8]. In this case, additional contactors have to be used for fault tolerant control purpose. In the commercial area, according to [9], most MV VSD systems are developed to have N-1 redundancy, i.e., they can tolerate one switch device fault or one inverter leg failure without fully stopping the motor. In the event, an alarm is tripped, and a repair can be made during a scheduled maintenance.

In recent years, the UCI Power Electronics Laboratory proposed a new topology, the hexagram converter for medium voltage applications [10]-[14]. This converter has many advantages over the conventional multilevel converters such as module structure, reduced component count, constant power flow to each module, reduced energy storage requirement, and improved fault tolerance. In this paper, a fault tolerant control method is proposed that is simple and does not require extra power components like SCR or contactor. Since fuses in the DC links are usually used for protection purpose, it could also be used to disconnect the DC power when a switch failed in this inverter module soon it is detected by monitoring the three-phase voltage and current in this inverter. When a fault is confirmed, all the switches are turned on temporarily in the faulty inverter in order to blow out the fuse in the DC link. After the DC power supply is disconnected, the gate driving signals to this faulty inverter as well as several inverter legs that are connected to this faulty inverter are disabled. Then the three-phase motor drive enters post fault two-phase operation. With proper control strategy, the hexagram inverter motor driver is able to maintain a smooth torque operation.

II. HEXAGRAM INVERTER MOTOR DRIVE SYSTEM

Fig.1 gives the schematic of a hexagram inverter. Six three-phase inverters, named inverter module 1-6, are interconnected via six inductors, where the inductors can be coupled on one core [12]. Fig.2 illustrates a complete system of the Hexagram Inverter for a three-phase motor drive, which is composed of an input isolation transformer with six secondary windings, six diode rectifiers, six dc capacitors, and one hexagram inverter. The transformer is wounded such that

978-1-4244-4782-4/10 $26.00 © 2010 IEEE

the secondary windings are phase-shifted to form 18-pulse or even 36-pulse rectification to reduce the current harmonics in the transformer primary winding. There are six output terminals in the hexagram inverter, which is suitable for open-end winding AC machines. Open-end winding machine provides more degree of freedom in control, it only requires open the three winding connection of a traditional three-phase motor, so no special design of electric machine is required. The hexagram inverter is perfect for this type of motor structure.

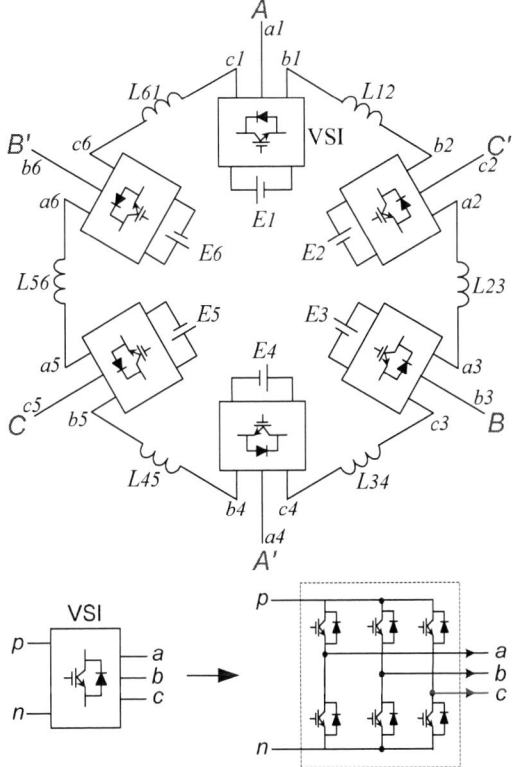

Figure 1: Hexagram Inverter

A. Normal operation of hexagram inverter

In normal operation, the output voltage are controlled according to voltage phasor diagram in Fig.3(a).

$$
\begin{bmatrix} v_{a1o1} \\ v_{b1o1} \\ v_{c1o1} \end{bmatrix} = \begin{bmatrix} v_{a3o3} \\ v_{b3o3} \\ v_{c3o3} \end{bmatrix} = \begin{bmatrix} v_{a5o5} \\ v_{b5o5} \\ v_{c5o5} \end{bmatrix} = -\begin{bmatrix} v_{a2o2} \\ v_{b2o2} \\ v_{c2o2} \end{bmatrix} = -\begin{bmatrix} v_{a4o4} \\ v_{b4o4} \\ v_{c4o4} \end{bmatrix} = -\begin{bmatrix} v_{a6o6} \\ v_{b6o6} \\ v_{c6o6} \end{bmatrix}
$$

$$
= \begin{bmatrix} \sqrt{2}V\cos(\omega t) \\ \sqrt{2}V\cos(\omega t - 2\pi/3) \\ \sqrt{2}V\cos(\omega t + 2\pi/3) \end{bmatrix} \tag{1}
$$

Neglecting the voltage drop on six inductors, the output voltage of hexagram inverter is derived as:

$$
\begin{cases} v_{AA'} = v_{c4a4} + v_{a3c3} + v_{b2a2} + v_{a1b1} = v_{b4a4} + v_{a5b5} + v_{c6a6} + v_{a1c1} \\ v_{BB'} = v_{a6b6} + v_{b5a5} + v_{c4b4} + v_{b3c3} = v_{c6b6} + v_{b1c1} + v_{a2b2} + v_{b3a3} \\ v_{CC'} = v_{b2c2} + v_{c1b1} + v_{a6c6} + v_{c5a5} = v_{a2c2} + v_{c3a3} + v_{b4c4} + v_{c5b5} \end{cases} \tag{2}
$$

The output voltage increase by six times as shown in Fig.3(b). It is worth noting that for each phase there are two expressions to derive the output voltage of the hexagram inverter, because a loop is formed by the inverter's inner legs. The purpose of the interconnecting legs is to create a symmetrical condition so that the current flow in each module and each phase is balanced. This interconnecting loop provides a built-in redundancy for hexagram inverter.

The three-phase output current is determined by the load, which could be written as:

$$
\begin{bmatrix} i_{a1} \\ i_{b3} \\ i_{c5} \end{bmatrix} = -\begin{bmatrix} i_{a4} \\ i_{b6} \\ i_{c2} \end{bmatrix} = \begin{bmatrix} \sqrt{2}I\cos(\omega t - \theta_L) \\ \sqrt{2}I\cos(\omega t - \theta_L - 2\pi/3) \\ \sqrt{2}I\cos(\omega t - \theta_L + 2\pi/3) \end{bmatrix} \tag{3}
$$

where I is the RMS value of the output currents and θ_L is the phase angle of the current lagging behind the corresponding voltage.

The current in each inverter is determined by the following equation for each inverter

$$
i_{aj} + i_{bj} + i_{cj} = 0, \qquad j = 1, ..., 6 \tag{4}
$$

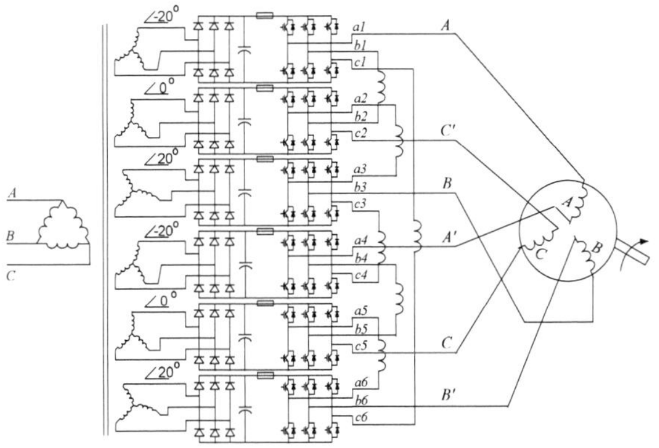

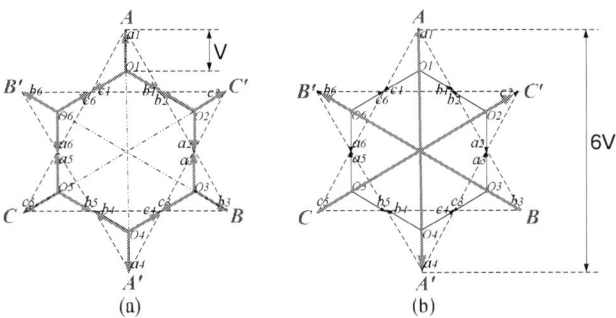

Figure 2: Hexagram Inverter

And the connecting phase current are just opposite:

(a) Voltages of VSI modules   (b) Voltages of the hexagram inverter.

Fig. 3. Voltage phasor diagram of the hexagram inverter.

$$\begin{bmatrix} i_{b1} \\ i_{a2} \\ i_{c3} \\ i_{b4} \\ i_{a5} \\ i_{c6} \end{bmatrix} = - \begin{bmatrix} i_{b2} \\ i_{a3} \\ i_{c4} \\ i_{b5} \\ i_{a6} \\ i_{c1} \end{bmatrix} \tag{5}$$

The loop current is suppressed by six inductors and is assumed zero here:

$$i_{b1} + i_{a2} + i_{c3} + i_{b4} + i_{a5} + i_{c6} = 0 \tag{6}$$

From (3)-(6), the currents in each inverter are derived as:

$$\begin{bmatrix} i_{a1} \\ i_{b1} \\ i_{c1} \end{bmatrix} = \begin{bmatrix} i_{a3} \\ i_{b3} \\ i_{c3} \end{bmatrix} = \begin{bmatrix} i_{a5} \\ i_{b5} \\ i_{c5} \end{bmatrix} = - \begin{bmatrix} i_{a2} \\ i_{b2} \\ i_{c2} \end{bmatrix} = - \begin{bmatrix} i_{a4} \\ i_{b4} \\ i_{c4} \end{bmatrix} = - \begin{bmatrix} i_{a6} \\ i_{b6} \\ i_{c6} \end{bmatrix}$$
$$= \begin{bmatrix} \sqrt{2}\mathrm{I}\cos(\omega t - \theta_L) \\ \sqrt{2}\mathrm{I}\cos(\omega t - \theta_L - 2\pi/3) \\ \sqrt{2}\mathrm{I}\cos(\omega t - \theta_L + 2\pi/3) \end{bmatrix} \tag{7}$$

In practical system design, to get all the current information for current control and protection purpose, four current sensors are needed, that is three output current sensor and one phase current in the inner leg. Other currents can all be determined by repetitively using equation (4) & (5). The three-phase output current is used for control purpose, the fourth current is used for monitoring loop current and could also be used for fault detection purpose.

B. Model for open-end winding machine

The mathematical model of three-phase open-end winding PMSM in the ABC frame is written as in (8), here the stator current is actually independent and the sum of three-phase of current is not necessarily zero, which provides more flexibility to handle fault.

$$u_a = R_{sa}i_a + \frac{d\psi_{sa}}{dt}$$
$$u_b = R_{sb}i_b + \frac{d\psi_{sb}}{dt} \tag{8}$$
$$u_c = R_{sc}i_c + \frac{d\psi_{sc}}{dt}$$
$$\begin{bmatrix} \psi_{sa} \\ \psi_{sb} \\ \psi_{sc} \end{bmatrix} = \begin{bmatrix} L_{aa} & L_{ab} & L_{ac} \\ L_{ba} & L_{bb} & L_{bc} \\ L_{ca} & L_{cb} & L_{cc} \end{bmatrix} \begin{bmatrix} i_a \\ i_b \\ i_c \end{bmatrix} + \psi_f \begin{bmatrix} \cos\theta \\ \cos(\theta - 120) \\ \cos(\theta + 120) \end{bmatrix}$$

The mathematical equation in dq0 can also be used, as describe in (9):

$$u_d = R_s i_d + L_d \frac{di_d}{dt} - \omega L_q i_q$$
$$u_q = R_s i_q + L_q \frac{di_q}{dt} + \omega L_d i_d + \omega \psi_f \tag{9}$$
$$u_0 = R_s i_0 + L_0 \frac{di_0}{dt}$$
$$T_e = 1.5 n_p \left(\psi_d i_q - \psi_q i_d \right) = 1.5 n_p \left[\left(L_d - L_q \right) i_d i_q + \psi_f i_q \right]$$

In normal operation, voltages and currents are all symmetrical and are controlled under dq reference frame like conventional vector control of motor drive. In order to simulate the system operation under various faults, since the electric circuit and input voltage current is not symmetrical, the original model under ABC frame is preferred.

A simulation model has been built in Matlab using simulink and SimPowerSystem, the three-phase open-end winding machine is built in simulink since it is not available in the matlab library. A hexagram inverter is built using components from SimPowerSystem library. The hexagram inverter and numerical motor model is interfaced by controlled voltage sources and current measurements. Three-phase currents is measured from physical domain and input to the simulink numerical motor model, while in the simulink numerical motor model back-EMF value is calculated and determine the back-EMF, which is a controlled voltage source in the physical domain. Interfacing the simulink model with the hexagram inverter model in this way will make sure that the back-EMF is taken into consideration when simulating inverter under fault conditions. Table 1 gives the simulation parameters used through this paper, which are also the experiment parameters for a small-scale experiment PMSM fault tolerant control system setup.

TABLE I.   MOTOR DRIVE PARAMETERS

DC bus	30V
Stator Resistance	2.5Ω
Switching frequency	10KHz
Magnetizing inductance Lm	7.1mH
Leakage inductance	0.1mH
Interconnecting inductace	50uH
Pole pairs	4
Rotor flux	0.3T

III.   FAULT DETECTION AND ISOLATION

In a practical motor drive system, many types of faults could happen, for instance fault in motor winding, input diode bridge, the three-phase inverter, or control and gate drive board. In this paper, switching device fault in hexagram inverter including a switch short-circuit fault and a switch open-circuit fault will be discussed in this paper.

A. Short-circuit fault and its detection

Short-circuit fault of one IGBT will short-circuit the DC bus, the result is disastrous if not protected. When a short circuit happens, the other switch in the same inverter leg must be turned off immediately. Short-circuit protection techniques like desaturation detection or sensing IGBT current is already widely used in the industry since short-circuit protection of DC link is a common problem for many power converters. After switch short-circuit fault, the output of one phase is always connected to one DC bus terminal and connected to the other DC bus via a diode. The output phase voltage is fixed

978-1-4244-4782-4/10 $26.00 © 2010 IEEE

and uncontrollable, further steps have to be taken to isolate the fault for a continuous fault tolerant operation.

B. Open-circuit fault and its detection

Open-circuit fault detection for 2-level inverter has been discussed in several papers [1] [2]. The open-circuit fault detection could be categorized into two groups: voltage-based fault detection and current based fault detection. Currents in hexagram inverter are already sensed in order for control purpose, which could also be used for fault detection purposed. Take inverter module 1 for example. Assuming that the lower switch San is open-circuited, the post fault circuit is showed in Fig.4.

When current i_{a1} is positive, the open-circuit of switch San will not affect the PWM operation since the conducting path is the diode Dan. When current i_{a1} is negative and during the ON interval of San, the output of phase a is connected to negative DC bus via a reverse diode instead of directly connected to negative DC bus. From the above analysis, the conducting of San is essential when the output current is negative. This reverse diode connection will affect the operation of this inverter module.

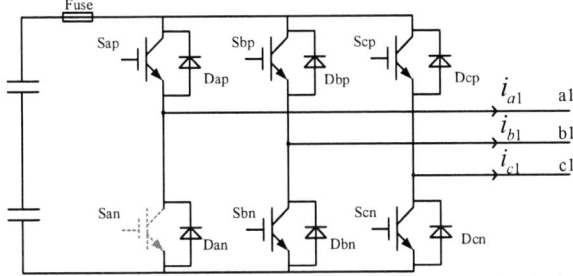

Fig.4 post fault circuit after lower switch San is open-circuited

When the output terminal is connected to the negative DC bus via a reverse diode, the conduction of diode depends on the voltage across the diode. As shown in Fig.5, in this situation, other inverter legs are all working properly, except leg a1 in inverter module1. The output pole voltage with reference to terminal A' is the back-EMF in phase A assuming that diode Dan is not conducting. In normal operation, the back-EMF is slightly smaller than the output voltage $v_{AA'}$ considering the voltage drop on winding resistors. Hence, the voltage across the diode cathode a1 and DC bus mid-point o1 is approximately the output phase voltage v_{o1a1} in healthy normal operation. The voltage of the diode anode with reference to the dc bus mid-point is a constant $-Vdc/2$. From the above analysis, the diode is still possible to conduct especially when a lower DC bus voltage is used. Moreover, when PMSM motor drive is working in high-speed flux weakening mode, where the back-EMF is significantly higher than the normal operation, the diode will conduct and work as diode rectifier to feedback the energy from the motor back to the DC link until the back-EMF voltage drops. The uncontrolled conducting of freewheeling diode for 2-level inverter is also reported in [15] [16]. Under rated speed, if the DC bus is higher as SPWM modulation is used, the diode is

not able to conduct during to follow a negative current control reference, at the same time causing a fluctuating torque. In order to make sure that the diode is not conducting uncontrolled, the open-circuit fault has to be isolated.

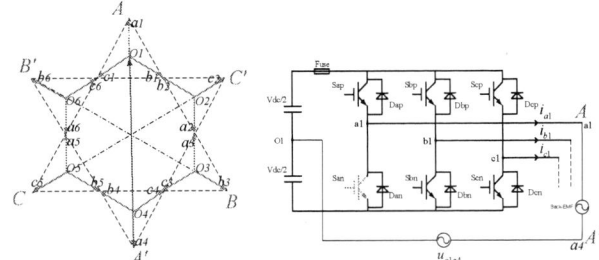

Fig.5 analysis of the diode conduction when San is open-circuited.

When output current is positive, the output pole voltage is our control reference. However, when the current is zero in the negative half-cycle, the pole voltage is different from our control reference since it is not able to connect to the negative DC bus during PWM switching. An error will be observed and this could be used for detection purpose. When the diode is not conducting, the output is still connected to the motor winding terminals. In such situation, the output voltage is determined by the back EMF instead of the control reference.

C. Simulation of open circuit fault

Simulation was conducted to show the error signal when switch San is open-circuited at time 1.0s, the hexagram inverter is driving the motor at a speed of 480RPM.

Fig.6 shows the output current when San is open-circuited, the output current A has only half cycle of a sinusoidal waveform because of this fault. The sum of three phase output current is not zero since this is an open-end winding machine. Fig.7 shows the output current of inverter module1. Fig.8 shows the voltage control reference and the actual output pole voltage. From the simulation result, it is obvious that by monitoring the three-phase output currents and voltages, open-circuit fault can be detected. Voltage sensors are used to sensor the voltage information. Current information is also used in order to help detect and confirm the fault since current information is already available. Due to inherent equation as in (5), the current waveform is the same in both c1, c6 and there is no way to determine which inverter module contains faulty switches. As a result, current detection technique alone is not sufficient to locate the position of fault switch.

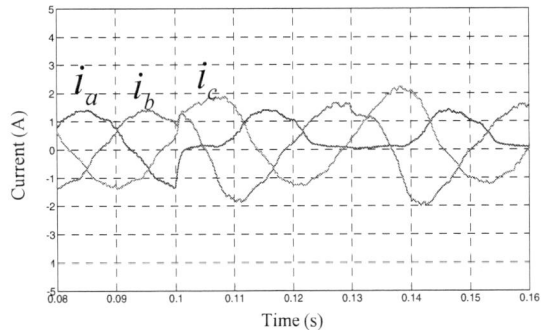

Fig.6 simulated three-phase output current with San open-circuit fault

978-1-4244-4782-4/10 $26.00 © 2010 IEEE

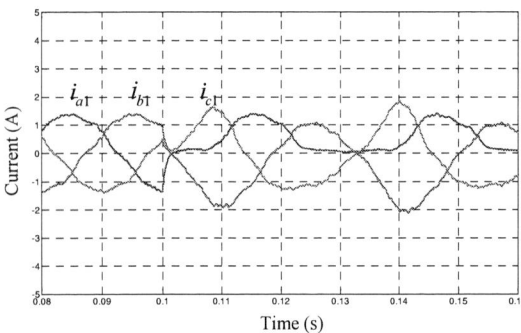

Fig.7 simulated three-phase output current in inverter module1 with San open-circuit fault

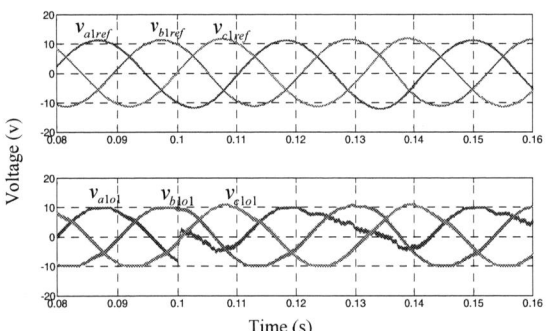

Fig.8 simulated three-phase output voltage and control reference voltage in inverter module1 with San open-circuit fault

D. Fault isolation

After a switch fault is detected, the faulty cell is isolated by blowing out the fuse in the DC bus and then disabling all necessary gate drive signals. As discussed in last section, after a switch open-circuit fault, the anti-parallel diode may conduct in high speed operation when the back-EMF is high or the DC bus voltage is low. After a short-circuit fault, the output is connected to the positive DC bus or negative DC bus permanently. In either case, the DC bus power source must be disconnected. After the DC link fuse is blown, the gate drive signal for the faulty inverter module and the gate drive signal for inverter legs that are connecting to this inverter will all be disabled. As for the previous example, when the switch San open-circuit fault is detected, the DC bus for inverter module1 is disconnected by blowing out the fuse in the dc link. After the DC power is disconnected, gate driving signal for inverter module 1 is disabled, gate drive signal for inverter leg c6, b2, and a4 are all disabled. Fig.9 shows the post fault phasor diagram when faulty cell is disconnected from the hexagram inverter in this situation, the dashed line shows that the inverter leg is disabled and not generating voltage.

In this case, there are two remaining output phase voltage, which is able to control two remaining independent current so as to maintain a smooth torque operation. The remaining two-phase output voltage is still generated as:

$$\begin{cases} v_{BB'} = v_{a6b6} + v_{b5a5} + v_{c4b4} + v_{b3c3} \\ v_{CC'} = v_{a2c2} + v_{c3a3} + v_{b4c4} + v_{c5b5} \end{cases} \tag{10}$$

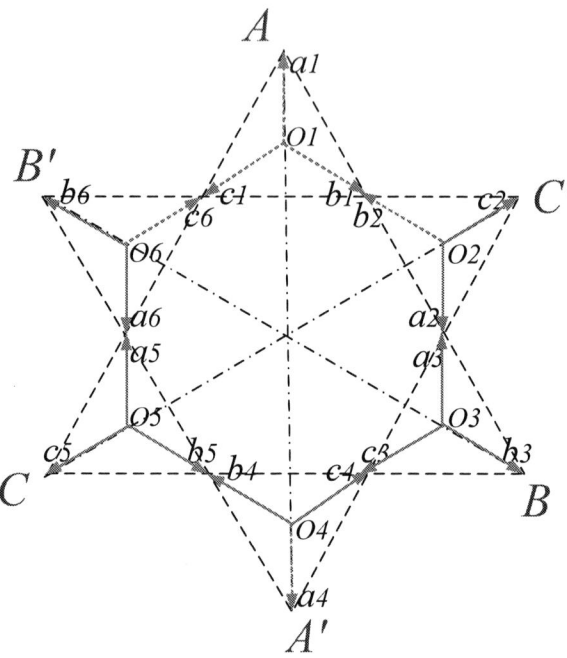

Fig.9 Post fault voltage phasor diagram after inverter modue 1 is disabled

From the point of view of controlling torque, the objective is to maintain the same dq current in this "two-phase" operation, since they represent the torque and flux demanded by the speed loop. The currents in the remaining two healthy phases have to produce the same dq current components, as well as the same αβ currents, in the stationary reference frame.

For vector control of PMSM, i_d is controlled to be 0, and i_q is controlled to be I_q, which is determined by the outer speed loop.

$$\begin{bmatrix} i_a \\ i_b \\ i_c \end{bmatrix} = \begin{bmatrix} 1 & 0 & 1 \\ -1/2 & \sqrt{3}/2 & 1 \\ -1/2 & -\sqrt{3}/2 & 1 \end{bmatrix} \begin{bmatrix} i_\alpha \\ i_\beta \\ i_0 \end{bmatrix} \tag{11}$$

$$\begin{bmatrix} i_\alpha \\ i_\beta \end{bmatrix} = \begin{bmatrix} \cos\theta & -\sin\theta \\ \sin\theta & \cos\theta \end{bmatrix} \begin{bmatrix} i_d \\ i_q \end{bmatrix} = \begin{bmatrix} -I_q\sin\theta \\ I_q\cos\theta \end{bmatrix} \tag{12}$$

In the post fault situation discussed here, phase A is disconnected from the converter, which means $i_a = 0$, since the current in dq reference frame i_d, i_q, are controlled to be unchanged by a proper fault-tolerant control, the zero-sequence component will be:

$$i_0 = I_q\sin\theta \tag{13}$$

The remaining two-phase currents will be:

$$i_b = -\frac{1}{2}i_\alpha + \frac{\sqrt{3}}{2}i_\beta + i_0 = \sqrt{3}I_q\cos(\theta - \frac{\pi}{3}) \tag{14}$$

$$i_c = -\frac{1}{2}i_\alpha - \frac{\sqrt{3}}{2}i_\beta + i_0 = \sqrt{3}I_q\cos(\theta - \frac{2\pi}{3}) \qquad (15)$$

The amplitude of the remaining two-phase current must increase by $\sqrt{3}$ times to generate the same torque, and its phase shift is 60 degrees, instead of the 120 degrees normal in three-phase operation.

In practice, it is important to maintain the same current stress and voltage stress in order to keep the remaining phases in the safe operation condition, therefore the output ability of hexagram inverter must be reduced to $1/\sqrt{3}$ of its rated output capability, which is acceptable during the urgent situation when some non-critical loads can be removed.

In two-phase operation, zero-sequence current arises in the system. This zero-sequence current will induce a zero-sequence voltage. In dq reference frame, the post fault controller uses a PI plus resonant controller to reject harmonic disturbance caused by the presence of zero-sequence current and voltage [14]. In the experimental setup, a single DSP was sufficient to control the hexagram inverter [11]. Although the hexagram inverter has six modules, the symmetrical three-phase output can be viewed as the output summation of two three-phase drives, so that the control scheme of the three-phase motor drive could be directly used to control the three-phase inverter. The implementation scheme, illustrated in Figs. 10 and 11, shows a PI plus resonant controller being used in both d and q current controllers.

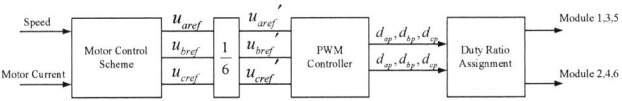

Fig. 10. Integral control scheme of the hexagram inverter.

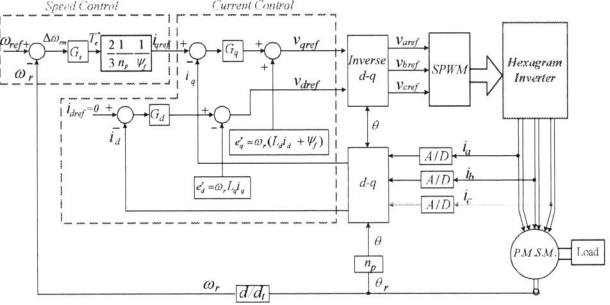

Fig. 11. Post fault control diagram of PMSM motor for the hexagram inverter

In the software design of fault tolerant control system, six back-up post fault programs are there since the fault could happen in one of these six inverter modules. When a fault is detected, the normal control algorithm will switch to one of the six back-up fault tolerant program.

IV. SIMULATION AND EXPERIMENT RESULTS

Simulation was conducted to see the transient when the hexagram inverter changed from three-phase operation to two-phase operation with one inverter disabled. Fig.12 shows the three-phase output current transient. After a short dynamic, the remaining two phase currents become 60 degrees apart and the

amplitude increases by $\sqrt{3}$ times. Fig.13 shows the speed dynamics, which proves that the hexagram inverter is able to maintain a smooth operation after one inverter is disabled after a fault.

A small-scale experiment was also conducted to verify the proposed fault tolerant control. The system parameter is given in table 1. The speed is set to be 360RPM. In the experimental prototype, the lower switch of inverter module 1 is intentionally turn-off to create an open-switch fault. Fig.14 shows the transient when the San is open-circuited. The fault is detected by monitoring voltage and current signal and then the inverter module 1 is disabled by blowing out the DC link. The current is unable to following the reference in half of the cycle. The DC link is disconnected and after some transient, the output current in phase A become zero. Two remaining current are properly controlled and the steady state is shown in Fig.16. The current amplitude do increases by $\sqrt{3}$ times as expected. Fig.15 shows the speed waveform. The speed was read from the coder in the DSP program, and the speed waveform was also recorded in the TI Code-Composer-Studio software the speed drops down in the transient but is smooth after post fault control is initiated.

V. CONCLUSION

In this paper, a fault tolerant control method for hexagram inverter motor drive is proposed. Since the three phase output currents of hexagram inverter are independent, it is possible to control the two remaining output currents to maintain a smooth torque operation, in the post fault situation when one phase is isolated due to the fault. The unique interconnecting nature of the hexagram inverter makes it capable of generating two-phase output voltages even though one inverter module is failed. Basis on this, a fault isolation and post fault control method is presented in this paper. In order to isolate the faulty inverter module, a fuse in the DC link is blown out by simultaneously turning on all the IGBTs in the module. After the faulty module is isolated, the hexagram will generate two phase output voltages. By proper post fault control, the output currents are regulated to maintain a smooth torque operation. From the simulation and experiment result it is clear that the fault tolerant control system is effective. For practically system, in order to maintain the same current stress for the switches, output capacity will be reduced to allow a "limp-home" mode with smooth operation, which is acceptable until arrangement is made for a schedule repair.

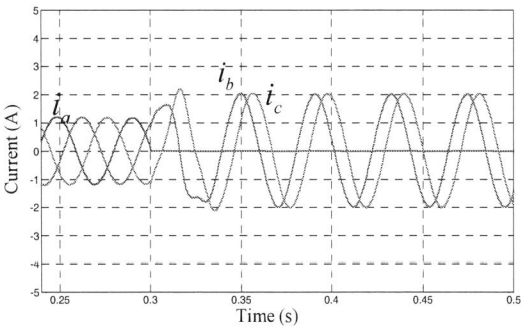

Fig.12 simulated three-phase current during a fault transient

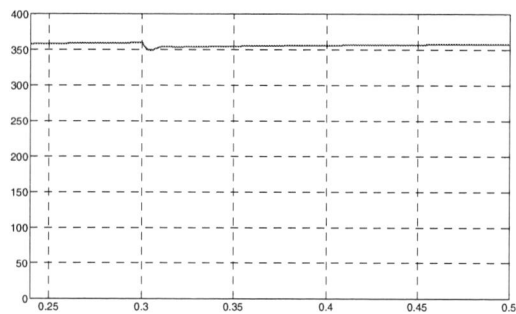

Fig.13 simulated speed waveform during a fault transient

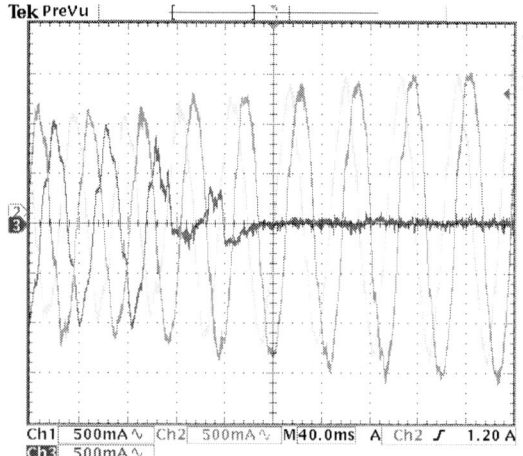

Fig.14 Experimental output current waveform during inverter module1 San open-circuit fault transient

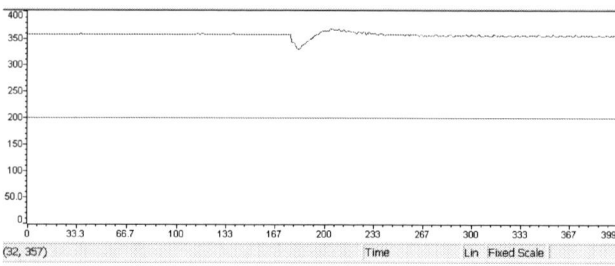

Fig.15 speed waveform during inverter module 1 San open-circuit fault transient

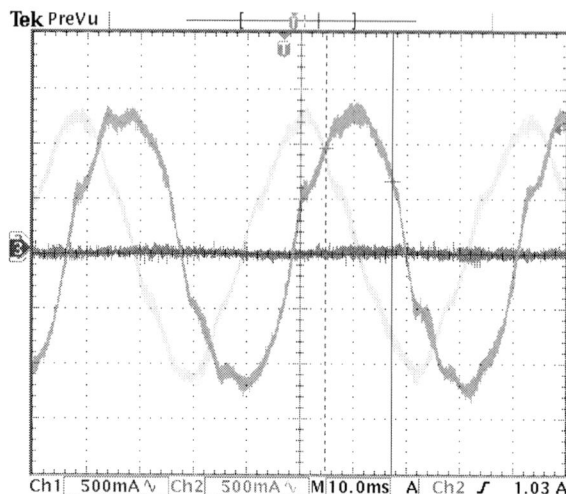

Fig.16 steady-states two-phase current after inverter module 1 disabled.

REFERENCES

[1]   R.L.A. ribeiro, C.B. Jacobina, E.R.C. da Silva, and A.M.N. Lima, "Fault detection of open-switch damage in voltage-fed PWM motor drive systems", IEEE Trans. Power Electron., vol.18, no.2, pp.587-593. Mar. 2003.

[2]   R. Peuget, S. Courtine, and J. Rognon, "Fault detection and isolation on a PWM inverter by knowledge-based model", IEEE Trans, Ind. Applicat. Vol.34, Nov/Dec 1998, pp.1318-1326.

[3]   S. Khomfoi and L.M. Tolbert, "Fault diagnostic system for a multilevel inverter using a neural network," IEEE Trans. Power Electron., vol. 22, pp. 1062-1069, May 2007.

[4]   S. Khomfoi and L.M. Tolbert, "Fault diagnosis and reconfiguration for multilevel inverter drive using AI-based techniques," IEEE Trans. Ind. Electron., vol. 54, pp. 2954-2968, Dec. 2007.

[5]   Khomfoi, S.; Tolbert, L.M., "Fault Detection and Reconfiguration Technique for Cascaded H-bridge 11-level Inverter Drives Operating under Faulty Condition" PEDS '07. Nov. 2007 pp. 1035 – 1042

[6]   Brando, G.; Dannier, A.; Del Pizzo, A.; Rizzo, R.; "Quick identification technique of fault conditions in cascaded H-Bridge multilevel converters," International Aegean Conference on Electrical Machines and Power Electronics, 10-12 Sept. 2007 pp. 491 - 497

[7]   R.L. de Araujo Ribeiro, C.B. Jacobina, E.R.C. da Silva and A.M.N. Lima, "Fault-Tolerant Voltage-Fed PWM Inverter AC Motor Drive Systems," IEEE Trans. Ind. Applicat., vol.51, pp.439-446, April 2004

[8]   J. Rodriguez, P.W. Hammond, J.Pontt, R.Musalem, P. Lezana and M.J. Escobar, "Operation of a Medium-Voltage Drive Under Faulty Conditions", IEEE Trans. Ind. Electron., vol.52, pp.1080-1085, Aug.2005

[9]   R.A. Hanna and S. Prabhu, "Medium voltage drives-users' and manufacturers' experiences," IEEE Trans. on Ind. Appl., vol. 33, pp. 1407-1415, Nov./Dec. 1997.

[10] J. Wen and K. Smedley, "Hexagram rectifier—active front end of hexagram inverter for medium-voltage variable-speed drives," IEEE Trans. Power Electron., vol. 23, pp. 3014-3024, Nov. 2008.

[11] J. Wen and K. Smedley, "Hexagram inverter for medium-voltage six-phase variable-speed drives," IEEE Trans. Ind. Electron., vol. 55, pp. 2437-2481 June 2008.

[12] J. Wen, L. Zhou, and K. Smedley, "Improved hexagram inverter for medium-voltage variable speed drive," in Proc. 34th Annu. Conf. IEEE Industrial Electronics, pp.3266-3271, Nov. 2008.

[13] J. Wen and K. Smedley, "A new multilevel inverter - hexagram inverter for medium voltage adjustable speed drive systems part II. three-phase motor drive," in 2007 Proc. IEEE Power Electronics Specialists Conf., pp.1571-1577

[14] L. Zhou and K. Smedley, "Post Fault Control Strategy for the Hexagram Inverter Motor Drive", accepted for publication in IEEE Trans. Ind. Electron.

[15] T.M. Jahns, V. Caliskan, "Uncontrolled Generator Operation of Interior PM Synchronous Machines Following High-Speed Inverter Shutdown," IEEE Trans. Ind. Applicat., vol.35, no.6, pp. 1347-1357, Nov./Dec. 1999

[16] N.Biachi, S. Bolognani, and M. Zigliotto, "Analysis of PM Synchronous Motor Drive Failures during Flux Weakening Operation," IEEE PESC, 1996

Power Analog to Digital Converter for Voltage Scaling Applications

M.C.Gonzalez, M.Vasić, P.Alou, O.Garcia,
J.A. Oliver and J.A.Cobos
Centro de Electrónica Industrial
Universidad Politécnica de Madrid
Madrid, España
carmen.gsanchez@upm.es

H.Visairo
Systems Research Center, Mexico
Intel Corporation
Guadalajara, México
horacio.visairo-cruz@intel.com

Abstract—**In order to optimize energy efficiency, some applications require adapting supply voltage according to the work load requirements. For example, in high performance digital systems and in RF systems, voltage scaling and modulation techniques have been adopted in order to achieve a more efficient processing of the energy. These techniques are based in rapidly adjusting the system supply voltage level. In order to achieve this, a topology which is capable of achieving very fast changes of the output voltage is needed. In this paper a PWM multiphase topology whose phases are coupled by using transformers is proposed to be used in an envelope elimination and restoration (EER) technique. The proposed topology can achieve very fast changes between discrete voltage steps so it can be considered as a power analog to digital converter.**

I. INTRODUCTION

More efficient processing of the energy has been a matter of interest in many fields of power electronics since energy savings reflect directly in cost savings and higher autonomy of mobile devices. For instance, communication and digital systems are two applications where a more efficient processing of the energy means higher autonomy of mobile devices. Techniques such as DVS (Dynamic Voltage Scaling) are employed in order to reduce the power consumption of high performance digital systems; while in RF systems, EER (Envelope Elimination and Restoration) and envelope tracking techniques are employed. In these techniques, the power supply plays an important role, since the energy savings rely partly on the power supply ability to rapidly adjust the output voltage. Apart from adequately adjusting the output voltage, power supplies must accomplish other requirements in order to enable voltage modulation. Some of these requirements are the following ([1], [4]):

- Small size

- Fast dynamic response

- High efficiency over a wide load range

- Fast voltage transitions

- While in DVS it is also important for the power supply to exhibit very good static regulation, power supplies for RF systems should not interfere with the output spectrum of the transmitter.

A simplified block diagram of a digital system with DVS architecture and an RF system with envelope elimination and restoration (EER) technique are shown in Figure 1. In this paper, a PWM multiphase converter is proposed as a part of a power supply in a RF system. As stated, the reason to implement voltage modulation in RF systems is to improve the efficiency of the processing of the energy. In order to do this, the solutions proposed in state of the art can be PWM topologies as in [1,5], multilevel converters, as in [4] or hybrid solutions that employ linear regulators [6].

The main drawbacks of using a PWM converter for this kind of applications rely on the complexity of the design of the output filter and the control [2]. The output filter should have an adequate size in order to accomplish certain requirements of current and voltage ripples but should also allow very fast changes in the output voltage level. These requirements can be considered to be contradictory. A converter operating with very high operating frequency (MHz range) is necessary, along with a very fast control loop. These increase the complexity of the PWM power supply.

The PWM multiphase topology proposed in this paper, could avoid some of these drawbacks. The proposed topology is a multiphase converter with magnetic coupling among the phases. Its operating principle (presented in [3]) will be reviewed in next section. It is important to point out that in this converter the phases are coupled with transformers instead of coupled inductors ([7]). This implies that the storage of energy in the converter is minimized and that the energy transfer between the input and the output is very fast, so the output voltage can be changed with a very high slew rate. When this concept is applied, the converter can accomplish an adequate filtering of the output along with a high bandwidth while working in open loop.

978-1-4244-4782-4/10 $26.00 © 2010 IEEE

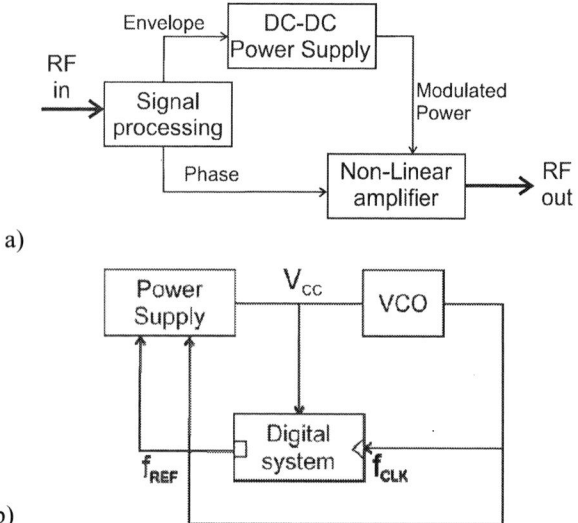

a)

b)

Figure 1. a) Simplified scheme of an RF system: Power Supply for the linear amplifier must supply a modulated output voltage in order to maximize the overall efficiency of the system. **b)** System architecture for DVS implementation [1]. In both systems, adjusting the power supply output can lead to a more efficient processing of the energy.

Under these conditions of minimum energy storage, the dynamic response is decoupled from the operating frequency and some of the limitations of a PWM converter when used for saving power in RF systems can be avoided.

The paper is organized as follows. The application of the proposed concept as a fast output voltage adjusting power supply is presented in Section II, along with the operating principle of the topology. Section III presents an example of an implementation of the proposed topology; measurements of the output voltage, following a 500 kHz sinusoidal waveform reference are also presented. In Section IV the conclusion of the paper is presented.

II. POWER ANALOG TO DIGITAL CONVERTER: OPERATING PRINCIPLE

As mentioned above, when using PWM converters for voltage modulation techniques, such as EER, one of the main drawbacks is the complex design of the feedback and the output filter [2,8]. This design is a direct tradeoff between the achievable bandwidth of the converter and the filtering of the output ripple [1].

It is well known that tight filtering needs can be easily met with a high value of inductor that will allow the reduction of current ripple. However, the energy stored in this inductor will limit the energy transfer when a load change or voltage step is demanded to the converter. It is also well known that the frequency rising allows the use of small inductor and capacitor values, while accomplishing tight filtering and enough bandwidth. The main drawback when rising operating frequency of power converters is that switching losses also increase, degrading the efficiency of the converter and hence the overall efficiency of the system where the power supply is placed. In order to find a better trade-off among dynamic response, filtering and efficiency, multiphase converters were proposed in state of the art [9]. Thanks to current ripple cancellation in multiphase converters, smaller inductance values can be used, while still accomplishing filtering needs.

The proposed topology can be described as a multi-phase converter with magnetic coupling among its phases. In it, the phases of the converter are coupled using discrete transformers rather than coupled inductors. A schematic of the topology is shown in Figure 2.b. In this figure, the ideal representation of the transformers that couple the phases is presented; the set of these transformers can be treated as a unique magnetic structure. An adequate control of the energy flowing from the input should ensure that the sum of the input voltages to the magnetic structure (v_1, v_2, v_3 and v_4 in Figure 2b) should be held constant for every instant of time.

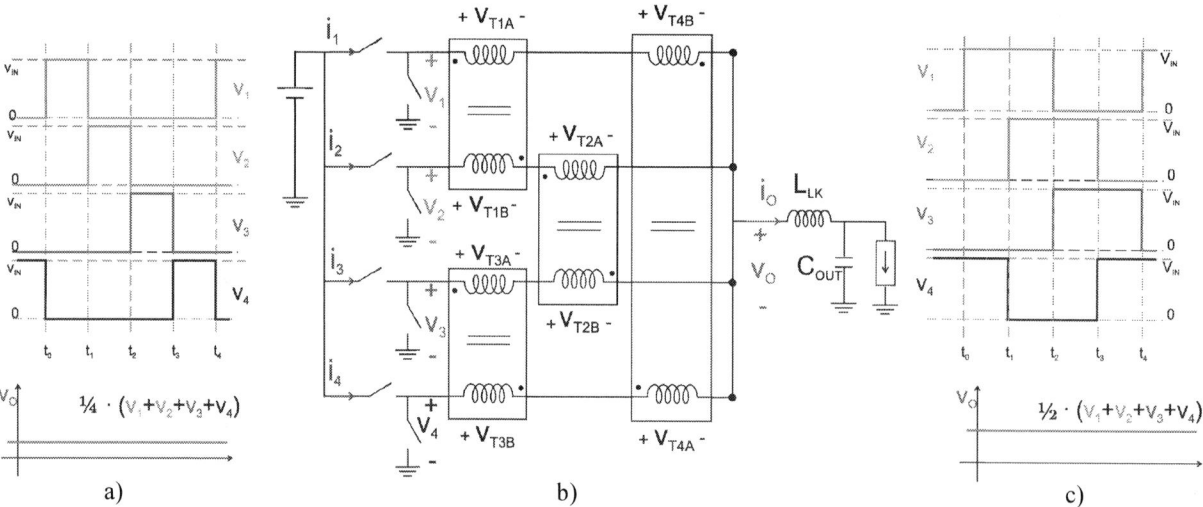

a)   b)   c)

Figure 2. a) Operating waveforms for proposed topology when d=25% the sum of this waveforms results in a constant voltage for every instant of time. b) Schematic diagram of proposed topology with four phases. c) operating voltage waveforms for d=50%

978-1-4244-4782-4/10 $26.00 © 2010 IEEE

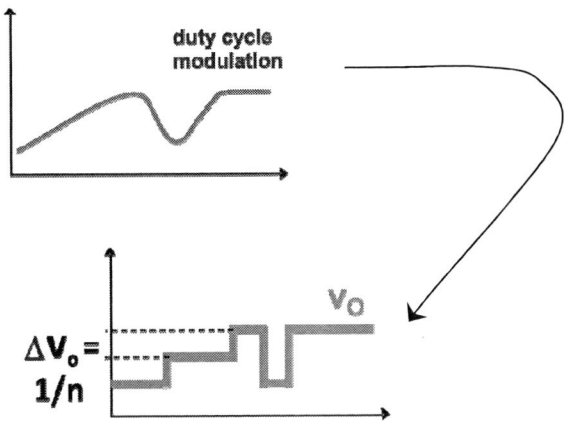

Figure 3. The output voltage of the proposed topology can be changed only within discrete values, this enables the use of the proposed topology as a power ADC where the resolution is given by 1/n (where n is the number of phases.)

The sum of the voltages at the input will result in a constant voltage at the output of the magnetic structure (v_C).

The general principle of operation can be described by the following equations:

$$v_C = \frac{1}{n} \cdot \sum v_i(t) \qquad (1)$$

$$i_i = \frac{i_o}{n} \qquad (2)$$

where *n* is the total number of phases. The validation of this concept is presented in [3] where it is applied to a two-phase converter. In order to use transformers to couple the phases of the converter, the energy flow to the magnetic structure should be kept constant for every instant of time. This means that the proposed topology can operate only at the duty cycles where the sum of the input voltages to the magnetic structure is constant. These operating points can be referred as operation nodes. The number of available operation nodes is related to the number of the phases in the converter; the available operation nodes are given by:

$$d=k/n \qquad (1)$$

where *k* is an integer number whose values range between *1* and *n-1*; *k* represents the number of phases which are simultaneously transferring energy from the input.

This topology operates in open loop; hence, the changes in the output voltage are achieved by changing the value of *k* between *1* and *n-1*. The minimum change between consecutive operation nodes will be set by the number of phases. This value is given by *1/n*. Figure 3 illustrates this concept. If duty cycle of the converter is modulated with an analog signal, the output of the power supply will follow it within discrete steps. Because of this feature, it can be said that this topology can be considered as a Power Analog to Digital Converter where the resolution depends on the number of phases.

The application of this control strategy to a four-phase topology is illustrated in Figure 2. Since it is a four-phase topology, there are three operating nodes with duty cycles: 25%, 50% and 75%. Figure 2a shows the case where duty cycle is 25%. The operation of the topology over one switching cycle has been divided into four instants of time:

• During t_0-t_1, phase 1 is connected to V_{IN} and transfers energy from the input to the magnetic structure, while the other phases are connected to ground. At t_1, phase 1 changes from V_{IN} to ground, and at the same time, phase 2 changes from ground to V_{IN}, since the energy flow must remain constant for every instant of time.

• During t_1-t_2, phase 2 transfers energy from V_{IN} to the magnetic structure, making V_2 equal to V_{IN}.

The same process is repeated for phases 2 and 3 during t_2-t_3 and t_3-t_4. Table I summarizes the values of the input voltages to the magnetic structure for every instant of time, and Table II shows the value of the voltages of each transformer for the same instants of time.

In this operation point and for every time interval, there is only one phase transferring energy to the magnetic structure at every instant of time, hence in equation (1), k=1 and the resulting duty cycle is ¼ (25%). The sum of the input voltages ($v_1+v_2+v_3+v_4$) results in a constant input voltage to the magnetic structure for every instant of time. If the input voltage to the magnetic structure is constant, v_O will be also constant and equal to:

$$(v1+v2+v3+v4)/n \qquad (2)$$

In this case and for a duty cycle of 25%, $v_O=1/4 \cdot V_{IN}$.

The next operation node corresponds to d=50%. The operating principle is the same than that for d=25%. Except that, instead of one phase transferring energy, there are two phases simultaneously connected to V_{IN} and transferring energy to the magnetic structure in order to keep the sum of the input voltages constant for every instant of time. If two cells are connected to V_{IN} the input voltage to the magnetic structure is the double than that of the former case:

$$2 \cdot (v1+v2+v3+v4)/n \qquad (3)$$

and since duty cycle equals 50%, $v_O=1/2 \cdot V_{IN}$.

When duty cycle is equal to 75% there would be three phases transferring energy to the magnetic structure for every instant of time, and $v_O=3/4 \cdot V_{IN}$. For these three duty cycles, v_O remains constant for every instant of time before the output filter of the converter and it is ideally not necessary to design the size of its components in order to filter the output ripple. In fact, the output inductor is comprised only by the equivalent leakage inductance of the set of transformers (represented in Figure 2 by L_{LK}). The value of this inductance can be minimized (to tenths of nH) if the adequate interleaving technique is chosen when placing the windings in the transformer.

978-1-4244-4782-4/10 $26.00 © 2010 IEEE

TABLE I. PHASE VOLTAGES FOR EVERY INSTANT OF TIME

Time Instant	Input Voltages to the magnetic structure			
	v1	v2	v3	v4
t0-t1	V_{IN}	0	0	0
t1-t2	0	V_{IN}	0	0
t2-t3	0	0	V_{IN}	0
t3-t4	0	0	0	V_{IN}

TABLE II. TRANSFORMER VOLTAGES FOR EVERY INSTANT OF TIME, CORRESPONDING TO D=25%

Time Instant	Transformer Voltages			
	VT1	VT2	VT3	VT4
t0-t1	$3/8 \cdot V_{IN}$	$1/8 \cdot V_{IN}$	$-1/8 \cdot V_{IN}$	$-3/8 V_{IN}$
t1-t2	$-3/8 V_{IN}$	$3/8 \cdot V_{IN}$	$1/8 \cdot V_{IN}$	$-1/8 \cdot V_{IN}$
t2-t3	$-1/8 \cdot V_{IN}$	$-3/8 V_{IN}$	$3/8 \cdot V_{IN}$	$1/8 \cdot V_{IN}$
t3-t4	$1/8 \cdot V_{IN}$	$-1/8 \cdot V_{IN}$	$-3/8 \cdot V_{IN}$	$3/8 \cdot V_{IN}$

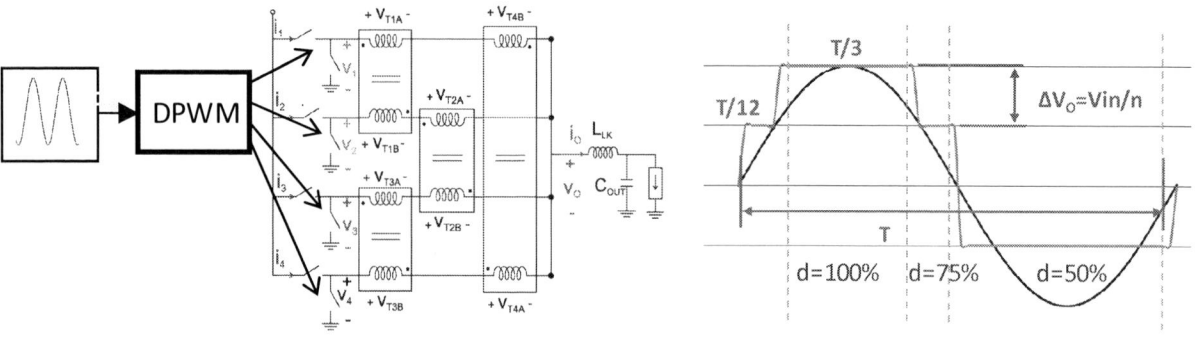

Figure 4. A sinusoidal reference is used to modulate the duty cycle applied to the converter, the resulting output voltage follows the sinusoidal reference within discrete steps.

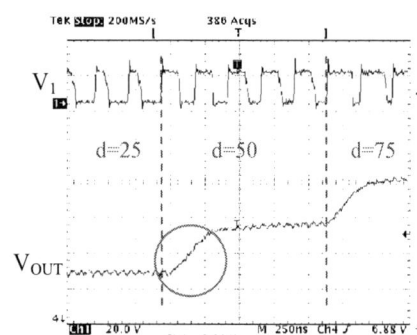

Figure 5. dv/dt of approx. 8 V/µs is achieved under no load condition, f_{SW}=4MHz and an output voltage change from 3 V to 6 V and an output capacitor of 1µF. V_1 represents the voltage of phase 1 as seen by the input of the magnetic structure. V_{OUT} represents the output voltage of the converter

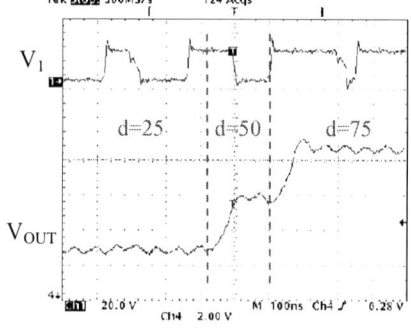

Figure 6. Output voltage waveform of the converter for f_{SW}=2MHz and a capacitor of 47nF. The frequency of the sinusoidal waveform that modulates the duty cycle is 170KHz V_1 represents the voltage of phase 1 as seen by the input of the magnetic structure. V_{OUT} represents the output voltage of the converter

III. APPLICATION EXAMPLE: ENVELOPE POWER SUPPLY FOR RF AMPLIFIER

In order to validate this concept in the area of modulated power supplies for RF applications, the RF system presented in [4], where the EER technique is implemented, was selected in order to evaluate the application of the proposed topology as a part of the aforementioned RF system.

The configuration of the envelope amplifier of the system presented in [4] is comprised of three stages: a multiple output dc-dc converter that provides stable dc voltages; a multilevel converter, which is used to adjust different voltage levels and a linear regulator in series with the multilevel converter. In the same way, a system for an envelope amplifier could be built with the proposed topology plus the linear regulator.

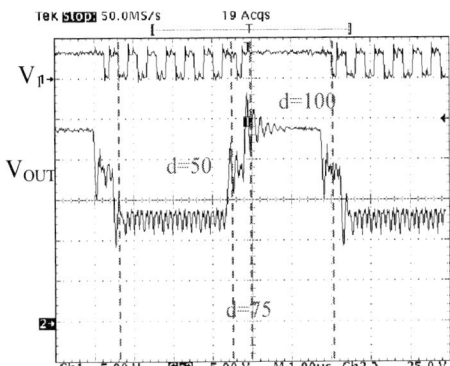

Figure 7. Output voltage waveform of the converter for f_{SW}=4MHz, sinusoidal waveform is 170kHz. V_1 represents the voltage of phase 1 as seen by the input of the magnetic structure. V_{OUT} represents the output voltage of the converter

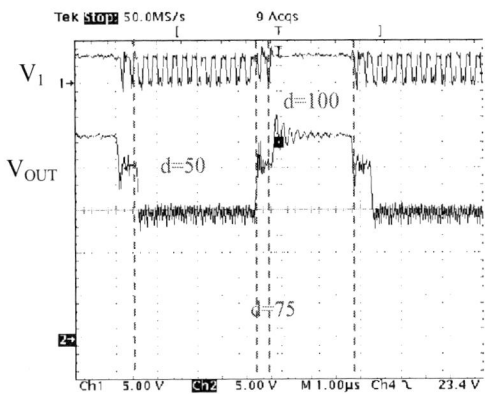

Figure 8. dv/dt of approx. 16 V/μs is achieved under no load condition, fSW=4MHz and an output voltage change from 3 V to 6 V an output capacitor of 33uF. V_1 represents the voltage of phase 1 as seen by the input of the magnetic structure. V_{OUT} represents the output voltage of the converter

Although the filter of the converter can be designed independently from the filtering needs, and the dynamic response of the system is decoupled from switching frequency, the maximum frequency of the sinusoidal reference that modulates the duty cycle is still limited by the switching frequency of the converter. So, at the operating frequency of 2 MHz, the maximum frequency of the sine wave that can be tracked is around 200 kHz.

A block diagram of the configuration of the power supply as a part of an RF system power supply is shown in Figure 4. It can be seen that the system operates in open loop and that a sinusoidal reference is used to modulate the duty cycle. The resulting output voltage modulation is also shown in this figure.

Since this topology operates in open loop, the slew rate of the output voltage can be approximated as the response of a second order system where its main components are the output inductor and capacitor. The value of the output inductor is fixed by the equivalent leakage inductance of the transformers set, and then the value of the output capacitor will define if the system response to a duty cycle change is fast and under-damped or slower and over-damped. Figure 5 and Figure 6 show two slew rates measured at f_{SW}=4 MHz, these slew rates correspond to different output capacitor values. It can be seen that the fastest slew rate is achieved with a 47nF MLC Capacitor. It is important to point out that in this application, current load steps are not a concern.

The configuration of the system is as follows: V_{IN}=24 V, f_{SW}=2 MHz, P_{OUTMAX}=30W. The duty cycle is modulated by a sinusoidal waveform of 170 kHz; the resulting sequence of duty cycles is as follows: 50%-75%-100%-75% and the measured efficiency at 2 MHz is around 87%. The configuration of the magnetic structure corresponds to that shown in Figure 3b. The results of modulating the duty cycle with a sinusoidal waveform are shown in Figure 7 and Figure 8, the resulting output waveform is shown at f_{SW}=2 MHz and f_{SW}=4 MHz, respectively. The configuration of the final system includes a linear regulator. This linear regulator filters

the noise that comes from the converter and provides fine adjustment of the output voltage of the envelope amplifier.

IV. CONCLUSIONS

In this paper, a PWM multiphase topology based on transformer-coupling was presented along with an application for voltage modulation. In this multiphase topology, the coupling between phases is done by using transformers instead of coupling inductors. Since transformers do not store energy, the transfer of the energy between the input and the output is done very fast, which allows a very fast change of the output voltage.

This topology can be useful when integrated in RF systems where power modulation techniques, such as envelope elimination and restoration (EER) are employed in order to increase the system efficiency. A demonstrator, where the duty cycle is modulated with a 170 kHz sinusoidal waveform has been presented, showing an efficiency of 87% and a correct tracking of the modulating waveform. The output power for this application is 30 W.

ACKNOWLEDGMENT

The authors would like to thank Narciso Ferreros Sánchez whose participation in the development and measurement of the aforementioned prototype were very valuable to this work.

REFERENCES

[1] A.Soto, P.Alou, J.A.Cobos and J.Uceda, "The future DC-DC Converter as an Enabler of Low Energy Consumption Systems with Dynamic Voltage Scaling", in IECON 02. Industrial Electronics Society, IEEE 2002 28th Annual Conference.

[2] L.Marco, E.Alarcón and D.Maksimovic, "Effects of switching power converter non-idealities in Envelope Elimination and Restoration

technique", in Circuits and Systems, 2006. ISCAS 2006. Proceedings. 2006 IEEE International Symposium on

[3] M.C.Gonzalez, L.Laguna, P.Alou, O.Garcia, J.A.Cobos and H.Visairo, "New control strategy for energy conversion based on coupled magnetic structures", Power Electronics Specialists Conference, 2008. PESC 2008

[4] M.Vasic, O.Garcia, J.A.Oliver, P.Alou, D.Diaz and J.A.Cobos, "Multilevel Power Supply for High Efficiency RF Amplifiers", Applied Power Electronics Conference and Exposition, 2009. APEC 2009. Twenty-Fourth Annual IEEE pp 1233 – 1238

[5] Mikkel C. W. Høyerby and Michael A. E. Andersen; "Ultrafast Tracking Power Supply With Fourth-Order Output Filter and Fixed-Frequency Hysteretic Control", IEEE Transactions On Power Electronics, Vol. 23, No. 5, September 2008

[6] V. Yousefzadeh, E. Alarcon, D. Maksimovic,; "Efficiency optimization in linear assisted switching power converters for envelope tracking in RF power amplifiers", IEEE International Symposium on Circuits and Systems, ISCAS 2005, pages:1302-1305 Vol. 2

[7] Pit-Leong Wong, Peng Xu, P.Yang, and F.C.Lee, "Performance improvements of interleaving VRMs with coupling inductors". IEEE Transactions on Power Electronics, Jul 2001

[8] Mikkel C. W. Høyerby, Michael A. E. Andersen, "Optimized Envelope Tracking Power Supply for Tetra2 Base Station RF Power Amplifier". Applied Power Electronics Conference and Exposition, 2008. APEC 2008

[9] Oliver J.A., Zumel P., Garcia O., Cobos J.A. and Uceda J. "Passive component analysis in interleaved buck converters", Applied PowerElectronics Conference and Exposition, 2004. APEC '04. Nineteenth Annual IEEE

A Digital Pulse-Width Modulator for Phase-Shift Operation of Full-Bridge Isolated DC-DC Converters

L. Corradini, D. Maksimović

Colorado Power Electronics Center
Department of Electrical, Computer and Energy Engineering
University of Colorado at Boulder
Boulder, CO – United States
{luca.corradini, maksimov}@colorado.edu

Abstract— This paper presents the concept, implementation and testing of a digital pulse-width modulator for phase-shift driving of full-bridge isolated DC-DC converters. The proposed phase-shift digital modulator employs a hybrid counter/delay line structure in order to achieve a 1.25 ns phase-shift resolution with a switching frequency of about 400 kHz on the secondary side, using a clock frequency of 25 MHz. Provisions for programmable dead times for switching of the bridge legs are included. The modulator is tested as a part of a digital controller for a full-bridge DC-DC isolated converter supplied from a 320 V DC bus with a 20 V-180 V, 2 A programmable output.

I. INTRODUCTION

Phase-shift operation of isolated DC-DC converters based on full-bridge topologies is nowadays common in high-voltage, high-power applications as it allows a significant reduction of the converter switching losses through zero-voltage switching (ZVS) of all four electronic devices [1]-[2]. While many analog IC controllers for phase-shift control are available on the market [3], there are relatively few commercial examples of digital ICs with phase-shift modulation capabilities [4]. In the literature, applications employing digital phase shift modulation implemented via commercial DSP or microcontrollers are reported in [5]-[6].

The paper discusses digital realization of a pulse-width modulator to be employed for phase-shift operation of digitally controlled isolated converters based on full-bridge topologies. The proposed structure and principles of operation for an 11 bit, 400 kHz switching frequency digital phase-shift modulator are presented in Section II; specific provisions for programmable dead time generation are also discussed. Open-loop experimental tests are first shown to demonstrate the correct digital phase-shifting and dead-time generation. Section III is then devoted to the experimental verification of the proposed modulator in closed-loop configuration as a part of the digitally-controlled 180 V, 2 A isolated DC-DC converter shown in Fig. 1. Details addressing the digital compensator design and implementation are also outlined.

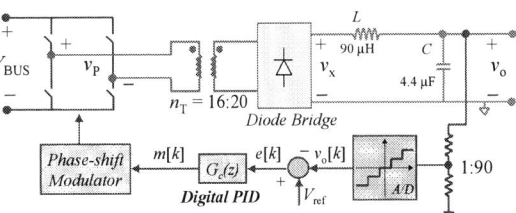

Fig. 1 – Digitally controlled 180 V, 2 A isolated DC-DC converter employing the proposed digital phase-shift modulator

II. PHASE-SHIFT DIGITAL PULSE-WIDTH MODULATOR

A. Simplified structure for digital phase-shifting of the driving signals

A simplified block diagram for digital generation of phase-shifted driving signals is shown in Fig. 2, along with the main associated waveforms. A core digital pulse-width modulator processes an input modulating signal m and produces a conventional pulse-width modulated output S_{PWM} with switching frequency equal to f_s; signal S_{PWM} is then employed to generate the switching signals S_A, S_B, S_C and S_D for the two legs A-B and C-D of the bridge.

Following a scheme commonly employed in analog phase-shift control ICs, signals S_A and S_C are derived from the positive and negative edges of S_{DPWM}, respectively, by employing two toggle flip-flops (T-FF); as a result, S_A and S_C are two phase-shifted, 50% duty cycle square waveforms switching at $f_s/2$.

In the simplified architecture shown in Fig. 2(a), S_B and S_D are merely inverted versions of S_A and S_C, respectively. In a practical implementation, however, a dead time T_D must be generated between the turn-off and turn-on command of the switches belonging to the same leg in order to avoid shoot-through currents and to optimize zero-voltage switching operation. Implementation of dead time generation for the switching signals therefore requires the two T-FF shown in Fig. 2(a) to be replaced with slightly more complex Finite State Machines (FSM). The development of the dead-time generation logic will be

978-1-4244-4782-4/10 $26.00 © 2010 IEEE

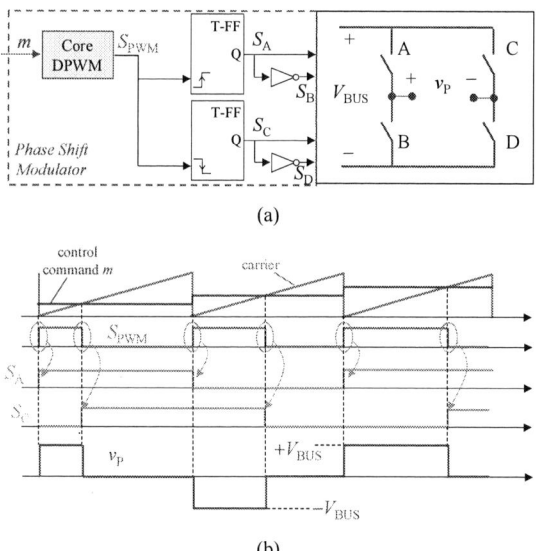

(a)

(b)

Fig. 2 – (a) Simplified block diagram for digital phase-shifting of the drive signals and (b) associated ideal waveforms

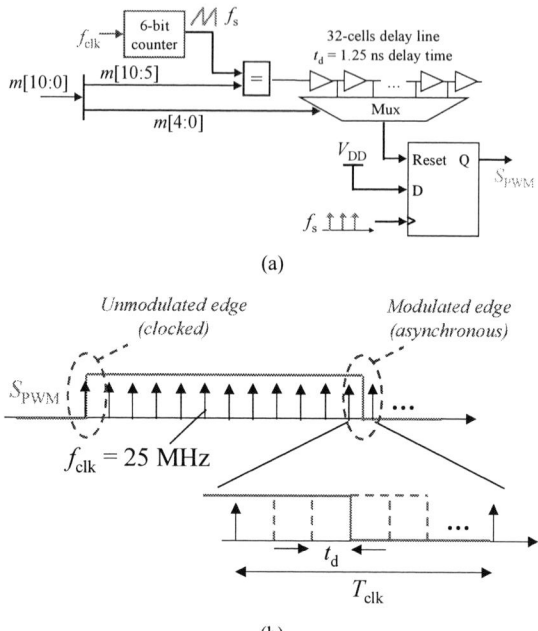

(a)

(b)

Fig. 3 – (a) Architecture of the core hybrid counter/delay line modulator and (b) modulator output S_{PWM}

addressed in Section II.C, after a brief discussion of the architecture of the core modulator.

B. Core Modulator

In this work, a core trailing edge modulator is implemented as a hybrid counter/delay line structure similar to what was presented in [7], as shown in the simplified block diagram of Fig. 3(a). The major advantage of the hybrid structure is the achievement of a fine time resolution without the need for a correspondingly high clock frequency. The core modulator employed in this work is designed for a $f_s = 390.625$ kHz switching rate and achieves a 1.25 ns timing resolution by combining a 6-bit counter clocked at $f_{clk} = 25$ MHz with a 5-bit delay line consisting of 32 cascaded delay cells, each having a delay $t_d = 1.25$ ns.

Principle of operation of the hybrid modulator can be summarized as follows. The 11-bit control command m is split into a 6-bit msb portion $m[11:5]$ and a 5-bit lsb portion $m[4:0]$; a 6-bit counter clocked at f_{clk} divides each switching cycle into 64 time slots, and is used to generate a logic pulse located at the specific time slot indicated by the msb command. The generated pulse propagates through a delay line consisting of 32 cascaded delay cells, each controlled by a digital delay locked loop (not shown in Fig. 3) to have an effective delay of $t_d = 1.25$ ns, as described in [5]. The cell outputs are tapped out and input to a 32-to-1 multiplexer whose control word is set by the lsb portion of the command. The appropriate tap is therefore selected and routed to the MUX output according to the lsb portion of the control command. The PWM signal S_{PWM} is asserted at the beginning of each switching cycle and reset by the MUX output.

As shown in Fig. 3(b), the modulated – i.e. falling – edge of S_{PWM} is determined with a time resolution given by the cell delay t_d; as can be inferred from Fig. 2(b), t_d also represents the phase-shift resolution of the overall modulator. Please refer to [7] for further details concerning the hybrid DPWM implementation.

C. Dead Time Generation: leg A-B

In what follows, the simplified architecture shown in Fig. 2(a) is further developed in order to include provisions for dead time generation.

The desired switching behavior for signals S_A and S_B is exemplified in Fig. 4. In the A→B transition example shown, switch A is initially in the on state; at the rising edge of the PWM signal S_{PWM} generated by the core modulator, S_A is immediately toggled to the off state and the leg enters an intermediate state with both switches being off. This state lasts for N_D clock periods, corresponding to the desired dead time $T_D = N_D \cdot T_{clk}$. It should be noted that this choice allows the dead time T_D to be programmable with a time resolution equal to the system clock period $T_{clk} = 1/f_{clk} = 40$ ns. At the end of the dead-time, switching signal S_B is toggled on and switch B is turned on, completing the A→B transition.

The digital logic required to generate the foregoing behavior is shown in Fig. 5. The circuitry shown replaces the upper T-FF originally shown in Fig. 2(a). With reference to the A→B transition described above, a Finite-State Machine resets signal S_A at the rising edge of S_{PWM}, enabling counter B at the same time. When the counter reaches the externally programmed threshold N_D, signal S_B

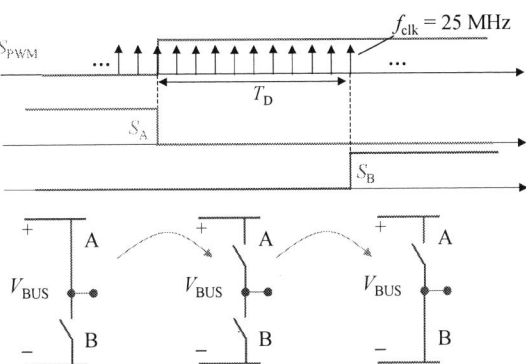

Fig. 4 – Desired switching behavior of signals S_A and S_B during a A→B transition

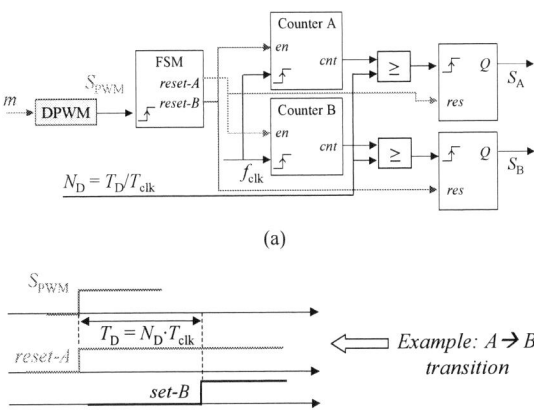

Fig. 5 – (a) Digital logic for dead-time generation in leg A-B and (b) associated waveforms

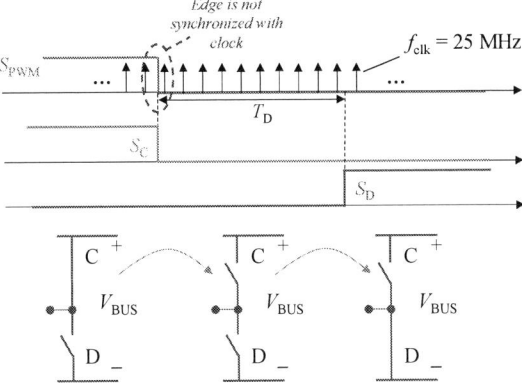

Fig. 6 – Desired switching behavior of signal S_C and S_D during a C→D transition

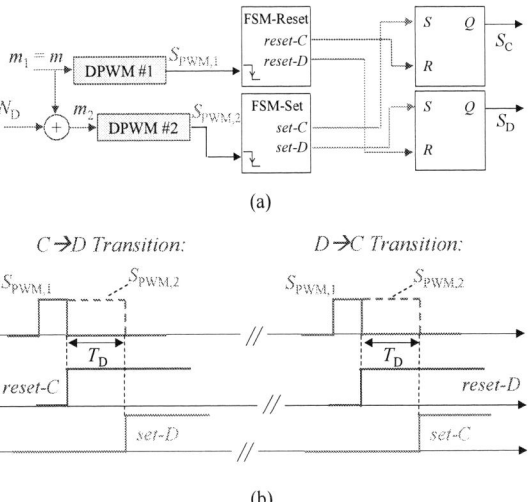

Fig. 7 – (a) Digital logic for dead-time generation in leg C-D and (b) associated waveforms

is switched on. Transition B→A is handled in a completely symmetric manner.

D. Dead-time generation: leg C-D

The implementation of the dead time generation described in Section II.C is only suitable for driving the A-B leg, whereas dead time generation in leg C-D brings in a slightly more subtle design issue.

As far as leg A-B is concerned, transitions of S_A – and therefore of S_B – are determined by the *unmodulated* (i.e. rising) edge of S_{PWM}. Since this edge is always clocked, the required dead time can be simply measured by means of counters as illustrated in Fig. 5. On the other hand, since S_C and S_D are toggled by the *modulated* (i.e. falling) edge of S_{PWM}, and since this edge is asynchronous with respect to the clock signal – as exemplified in Fig. 6 – the above method would generate a dead-time defined only to within T_{clk}, with an error that would become duty-cycle dependent.

In applications in which this error can be tolerated, the same circuitry shown in Fig. 5(a) could be replicated for the C-D leg. In this work, we outline an alternative solution for dead-time generation in leg C-D, illustrated in Fig. 7(a).

Two distinct instances of the core modulator are employed; the rising edges of $S_{PWM,1}$ and $S_{PWM,2}$ are synchronized, but the two modulators are driven by two different commands $m_1 = m$ and $m_2 = m_1 + m_D$. The programmable quantity m_D delays the falling edge of $S_{PWM,2}$ with respect to $S_{PWM,1}$ by a quantity equal to the desired dead time T_D – that is, if m_D is expressed in number of system clock cycles, then $m_D = T_D/T_{clk} = N_D$. With two available falling edges exactly delayed by T_D, dead time generation for leg C-D can be accomplished by two FSMs clocked by the negative edges of $S_{PWM,1}$ and $S_{PWM,2}$; Fig. 7(a) and Fig. 7(b) illustrate the overall hardware arrangement and associated waveforms for driving the C-D leg.

978-1-4244-4782-4/10 $26.00 © 2010 IEEE

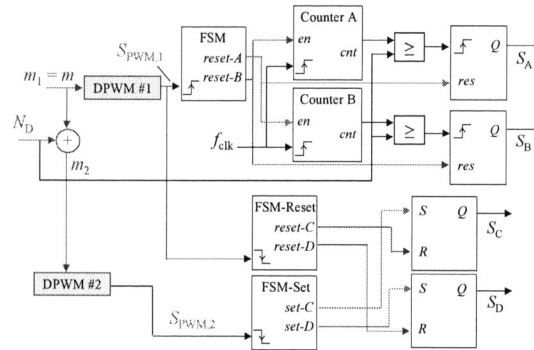

Fig. 8 – Overall structure of the proposed digital phase-shift modulator with programmable dead-time generation provisions

E. Overall structure of digital phase-shift modulator

Fig. 8 shows a complete diagram of the proposed digital phase-shift modulator, combining the provisions outlined in the previous sections. The state diagrams of the four FSM are not explicitly reported, since they can be inferred from the waveforms sketched in Fig. 5(b) and Fig. 7(b). The modulator clock rate $f_{clk} = 25$ MHz is employed as the system clock for the two instances of the core modulators DPWM#1 and DPWM#2 and to clock counters A and B which generate the dead-time for leg A-B. The dead time T_D is digitally programmable through the 6-bit word N_D, and can be set with a resolution equal to the system clock period $T_{clk} = 40$ ns.

F. Open-loop tests

Fig. 9(a) illustrates operation of the modulator with a $M = 49\%$ phase shift input command and the dead time programmed to $T_D = 5 \cdot T_{clk} = 200$ ns. Fig. 9(b) and 9(c) provide close-up views of signals $S_{PWM,1}$, $S_{PWM,2}$, S_A, S_B, S_C and S_D at $M = 3$ %; in particular, Fig. 9(c) demonstrates the correct dead time generation in leg C-D achieved by the provision illustrated in Fig. 7(a) during a C→D transition: turn-off of the active device C is commanded by the negative edge of $S_{PWM,1}$, while the turn-on of switch D is synchronized with the negative edge of $S_{PWM,2}$, delayed by $T_D = 200$ ns. Without the dedicated provision for driving the C-D leg described in Section II.D, the dead-time generation error in leg C-D could be as high as 40 ns, i.e. 20 % of the nominal programmed value.

III. Simulation and Experimental Results

The digital pulse-width modulator described in Section II has been experimentally tested as a part of the digitally controlled high-voltage DC-DC converter shown in Fig. 1. Details concerning the controller design are provided in this section, along with closed-loop simulation and experimental tests aimed at verifying the design and the operation of the controller.

The converter derives the input DC voltage $V_{BUS} = 317$ V ± 12 % by rectification of the AC line voltage – not explicitly shown in Fig. 1. Both 110 V ± 10 % and 230 V ± 10 % inputs are accepted. The DC bus voltage is then inverted by a phase-shifted full-bridge, stepped down by a $n_T = 16{:}20$ turns-ratio

transformer and rectified back by a diode bridge to produce a pulse-width modulated waveform $v_x(t)$. Voltage v_x is then filtered to produce the output voltage v_o; the output LC filter parameters are $L = 90$ μH and $C = 2 \times 2.2$ μF.

The converter is digitally controlled in order produce a 20 V-180 V programmable output voltage in the 0-2 A current range. The output voltage v_o is first scaled down by a 1:90 resistive divider, then sampled and A/D quantized over a 2 V full-scale range by a 12-bit pipeline A/D converter; the minimum equivalent output voltage quantization bin is therefore $\Delta q_{AD} = 44$ mV. The A/D converter is clocked at $f_{A/D} = 12.5$ MHz and exhibits a 5-cycles latency delay.

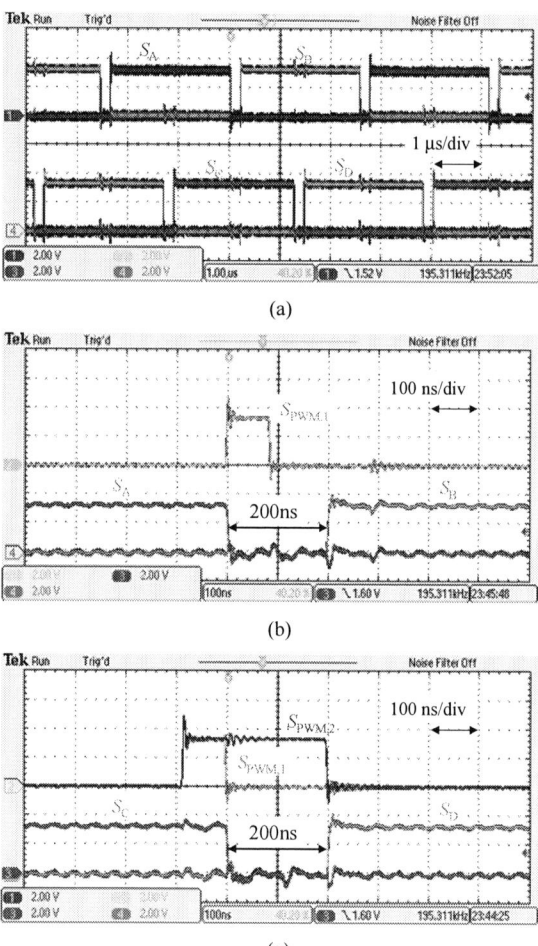

Fig. 9 – Experimental results for the proposed digital phase-shifted modulator: (a) leg switching commands for $M = 49$ %, (b) close-up view of dead time generation in leg A-B (A→B transition) and (c) close-up view of dead time generation in leg C-D (C→D transition)

A digital PID compensator is employed to regulate the output voltage. The digitized output voltage is first subtracted from the 12-bit digital setpoint V_{ref}; the resulting digital error signal is then downsampled to the converter switching rate $f_s \approx 390$ kHz by the digital compensator, which evaluates the control command $m[k]$ on a switching-cycle basis. In the experimental prototype the digital PID controller is VHDL-coded and implemented on a Xilinx Virtex-4 FPGA device hosted on a commercial development board interfaced with the power converter.

A. Compensator design and implementation details

For the control of the high-voltage converter described above, a digital PID compensator was designed for a 20 kHz control bandwidth with 50° phase margin.

The design involves a preliminary z-domain modeling of the control-to-output transfer function of the converter.

The continuous-time transfer function $G_{vd}(s)$ between the small-signal duty cycle $\hat{d}_x$ of the pulse-width modulated voltage v_x and v_o is:

$$G_{vd}(s) \equiv \frac{\hat{v}_o}{\hat{d}_x} = \frac{n_T \cdot V_{BUS}}{1 + \dfrac{s}{\omega_0 \cdot Q} + \dfrac{s^2}{\omega_0^2}} \tag{1}$$

where $\omega_0 = (LC)^{-1/2} \approx 50$ krad/s is the resonant angular frequency of the LC filter, and $Q \approx 1.5$ is the estimated Q-factor of the power stage.

Since the phase-shift operation of the full-bridge translates into an effective PWM modulation of voltage v_x, the small-signal behavior of the digital phase-shift modulator itself will be modeled as that of a conventional digital pulse-width modulator. A uniformly-sampled trailing edge modulator introduces a small-signal delay Δt_{PWM} which depends on the steady-state duty cycle D and the choice of the sampling instant [8]. Since the converter is expected to operate anywhere between 20 V and 180 V, i.e. with duty cycles ranging between $D_{min} \approx 8\ \%$ and $D_{max} \approx 70\ \%$, a worst-case design has been developed based on the latter operating point.

As far as the sampling instant is concerned, it has been located five system clock cycles before the beginning of each switching period. This has been estimated to allocate enough time for the digital PID compensator to fully calculate the control command based on the sampled error signal.

Summarizing the foregoing design choices, the DPWM-related delay is found to be:

$$\Delta t_{PWM} = D_{max} \cdot T_s + 5 \cdot T_{clk} \approx 2\ \mu s \tag{2}$$

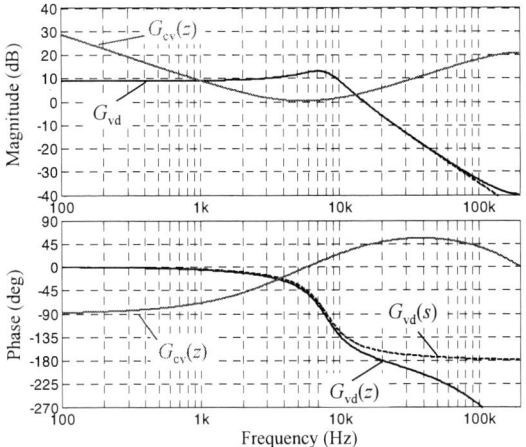

Fig. 10 – Bode diagrams of the plant transfer function and of the PID transfer function

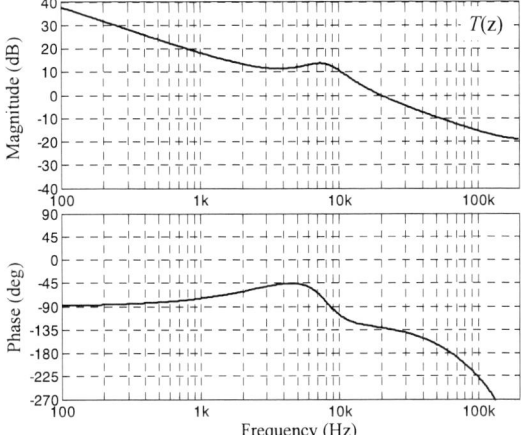

Fig. 11 – Bode diagrams of the designed voltage loop gain

The last item to consider before deriving the z-domain model of the plant is the latency delay of the A/D converter, which corresponds to five A/D clock cycles as anticipated:

$$\Delta t_{A/D} = 5 \cdot T_{A/D} = 400\ ns \tag{3}$$

At this point the z-domain equivalent of the plant can be found. Taking into account the scaling factor $H = 1/90$ of the resistive sensing network, one has [8]:

$$G_{vd}(z) \equiv \frac{\hat{v}_o}{\hat{m}} = Z_{T_s}\left[H \cdot G_{vd}(s)e^{-s(\Delta t_{PWM} + \Delta t_{A/D})}\right] \tag{4}$$

where $Z_1[\cdot]$ denotes the z-transform operator with sampling period T. Bode diagrams of the plant (4) are shown in Fig. 10, together with the frequency response $G_{cv}(z)$ of the digital PID designed for a 20 kHz bandwidth

978-1-4244-4782-4/10 $26.00 © 2010 IEEE

and 50° phase margin. In Fig. 10, the frequency response of the continuous-time plant (1) has been additionally included in order to better appreciate the importance of modeling the delays (2) and (3). In particular, the phase contribution of the foregoing delays at the crossover frequency is about 17°, with the main contribution coming from the DPWM delay effect. The resulting z-domain voltage loop gain $T(z)$ is shown in Fig. 11.

Following a common practice in the design of digital control loops for DC-DC converter, additional criteria were considered to ensure a steady-state converter operation free of limit cycling [9]. A first well-known no-limit cycling criterion requires that a DPWM bin always falls within the A/D zero error bin. In other words, the DPWM resolution must be sufficiently finer compared with the A/D resolution. In the proposed design, the equivalent DPWM bin equals $\Delta q_{\mathrm{DPWM}} = n_{\mathrm{T}} \cdot V_{\mathrm{BUS}}/2^{11} \approx 120$ mV, larger than the 44 mV A/D bin. Therefore, inside the digital controller three A/D bits were dropped from the digitized voltage and from the digital setpoint, bringing the equivalent A/D resolution to ~350 mV, enough to satisfy the mentioned no-limit cycling criterion, and yet small enough to guarantee accurate output voltage regulation.

A second no-limit cycling criterion requires the PID integral gain K_{I} to be sufficiently small to guarantee fine positioning of the output voltage in steady-state. Expressing this criterion in the context of the proposed design, one has:

$$K_I \cdot V_{BUS} \cdot n_T \cdot H < 1 \qquad (5)$$

In the designed controller, the left-hand member in (5) evaluates to ~0.12; since the design verifies the foregoing criterion with large margin, no further corrections were made to the integral gain K_{I}.

The digital PID compensator has been implemented in the additive form shown in Fig. 12, where the proportional, integral and derivative terms m_{P}, m_{I} and m_{D} are computed separately and then added to obtain the overall control action m. Additional provisions included in the implementation of the digital compensator are the limitation of the integrator state variable and of the PID control command, as well as an anti-windup circuitry which freezes the integral term as soon as the PID control command saturates. As shown in Fig. 12, this feature is implemented by deriving a saturation flag sat from the limiting block of the PID command, and employing the flag to control the input of the PID integrator. Under normal operation, the voltage error is passed to the integrator, whereas a zero signal is passed under saturation condition, therefore stopping the integration. An anti-windup provision is particularly needed in this application due to the programmable nature of the output voltage: large, sudden variations in the digital setpoint V_{ref} in fact induce large error variations and subsequent saturation of the control command. The anti-windup provision prevents excessive swings in the integrator state during these transients, eliminating the consequent voltage overshoot.

B. Closed-loop simulation and experimental results

The digitally-controlled converter shown in Fig. 1 has been first modeled in the Matlab/Simulink environment in order to verify the controller design. Fig. 13(a) illustrates the simulated closed-loop response of the system shown in Fig. 1 to a 70 V reference step-up with an about 63 Ω resistive load, corresponding to an approximately 20 W to 178 W output power variation.

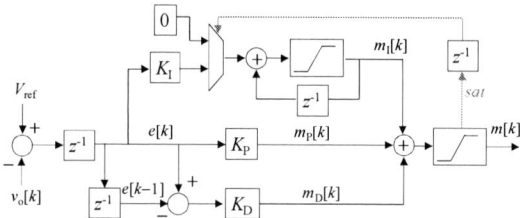

Fig. 12 – Structure of digital PID compensator

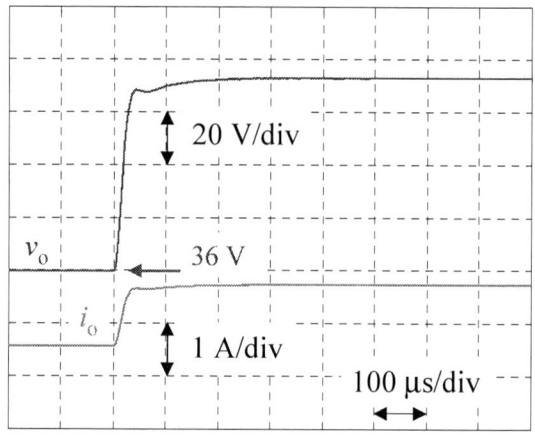

(a)

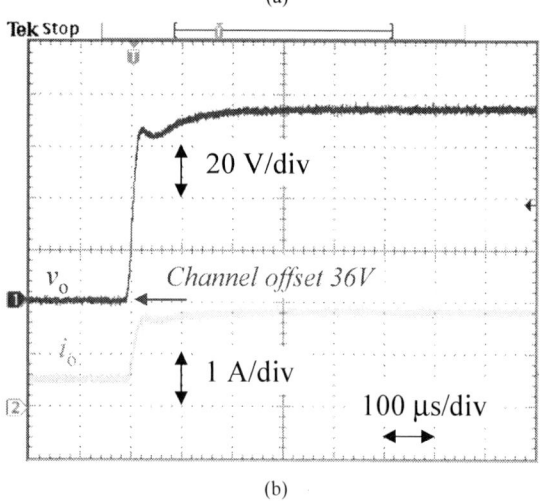

(b)

Fig. 13 – (a) Simulated and (b) experimental closed-loop transient response to a 70 V reference step-up; v_o: 20 V/div, i_o: 1 A/div, time scale 100 µs/div

The corresponding experimental response, measured under the same conditions, is illustrated in Fig. 13(b); the comparison validates the design of the control loop and the correct operation of the proposed digital phase-shift modulator under transient conditions. It can be noted that the anti-windup provision prevents the output voltage from overshooting, as anticipated.

IV. CONCLUSIONS

A digital modulator for phase shift driving of full-bridge isolated DC-DC converter is presented. The modulator achieves a 1.25 ns phase shift resolution at 400 kHz switching rate by employing a core modulator which combines a hybrid counter/delay line PWM modulator. The system clock frequency is 25 MHz. Provisions for phase-shift generation and programmable dead times are presented. The modulator has been experimentally tested and successfully employed in a digitally controlled full-bridge transformer-isolated converter having 317 V $\pm$ 12 % input DC voltage and programmable 20 V-180 V, 2 A output.

REFERENCES

[1] J. A. Sabate, V. Vlatkovic, R. B. Ridley, F. C. Lee, B. H. Cho, "Design Considerations for high-voltage high-power full-bridge zero-voltage-switched PWM converter," in *Proc. 5th IEEE Applied Power Electronics Conference and Exposition* (APEC'90), pp. 275-284, March 11-16 1990

[2] V. Vlatkovic, J. A. Sabate, R. B. Ridley, F. C. Lee, B. H. Cho, "Small-Signal Analysis of the Phase-Shifted PWM Converter," *IEEE Trans. Power Electron.*, vol. 7, no. 1, pp. 128-135, Jan. 1992

[3] B. Andreycak, "Phase Shifted, Zero Voltage Transition Design Considerations and the UC3875 PWM Controller," *UNITRODE Application Note U-136A* (document SLUA107) [Online]. Available: http://www.ti.com/litv/pdf/slua107

[4] "Digital Controller for Isolated Power Supply Applications," *Analog Devices ADP1043 Datasheet* [Online]. Available: www.analog.com/static/imported-files/data_sheets/ADP1043.pdf

[5] H. Tao, J. L. Duarte, M. A. M. Hendrix, "High-Resolution Phase-Shift and Digital Implementation of a Fuel Cell Powered UPS System," in *Proc. 2007 European Conference on Power Electronics and Applications*, pp. 1-10 2-5 Sep. 2007

[6] M. Cacciato, A. Consoli, N. Aiello, R. Attanasio, F. Gennaro, G. Macina, "A Digitally Controlled Double Stage Soft-switching Converter for Grid-connected Photovoltaic Applications," in *Proc. 23rd IEEE Applied Power Electronics Conference and Exposition*, pp. 141-147, 24-28 Feb. 2008

[7] V. Yousefzadeh, T. Takayama, D. Maksimovic, "Hybrid DPWM With Digital Delay-Locked Loop," in *Proc. IEEE Workshop on Computers in Power Electronics* (COMPEL 2006), pp. 142-148, July 16-19 2006, Troy, NY

[8] D. M. Van de Sype, F. De Gusseme, K. De Belie, A.R. Van den Bossche, J.A. Melkebeek, "Small-signal z-domain analysis of digitally controlled converters," *IEEE Trans. Power Electron.*, vol. 21, no. 2, pp. 470-478, Mar. 2006

[9] A.V. Peterchev, S.R. Sanders, "Quantization resolution and limit cycling in digitally controlled PWM converters," *IEEE Trans. Power Electron.*, vol. 18, no. 1, pp. 301-308, Jan. 2003

Digitally Controlled Integrated Pseudo-CCM SIMO Converter with Adaptive Freewheel Current Modulation

Yi Zhang and Dongsheng Ma

Integrated System Design Laboratory
Department of Electrical and Computer Engineering
The University of Arizona
Tucson, AZ 85721, USA
Email: dma@email.arizona.edu

Abstract—**In this paper, a digitally controlled integrated single-inductor multiple-output (SIMO) converter operating in pseudo-CCM (PCCM) mode is presented. With an adaptive freewheel current modulation (AFCM) technique, conduction loss of the freewheel switch is significantly reduced in unbalanced load conditions. Moreover, cross-regulation and switching noise are both effectively restricted. A prototype of a single-inductor dual-output (SIDO) boost converter was designed with a 130-nm CMOS process. The two outputs are regulated at 2.5 V and 1.8 V respectively, at a switching frequency of 500 KHz. The simulation results indicate that the expectant specifications are well implemented.**

I. INTRODUCTION

Dynamic voltage scaling (**DVS**) techniques with multiple supply voltages have been proven as the most effective ways to reduce power consumption in a large variety of electronic systems, which require different supply voltages to power various functional modules [1-3]. DC-DC converters with multiple outputs play a key role in such a technique. In traditional implementations, multiple power outputs are regulated by several independent single-output switching DC-DC converters [4-6], leading to excessive power and volume due to the employment of multiple inductors and power devices. Several single-inductor multiple-output (**SIMO**) switching converters [7-12] were reported in recent years as much cost-effective solution.

In a SIMO switching converter, system volume can be significantly reduced by removing redundant inductors and by time-sharing power switches. Moreover, the EMI and cross-couplings are effectively reduced because only one inductor is employed. However, there exist several drawbacks in prior arts. In the designs of [7-9], the converters operate in continuous conduction mode (**CCM**). As illustrated in Fig. 1(a), because the discharge period of the inductor current in every phase ends until the corresponding sub-converter operation phase expires, the inductor current at the end of each

phase is uncertain. As a result, the initial value of the inductor current to the second sub-converter is dependent to the previous one. If a sudden load change occurs in one phase, it will inevitably affect the subsequent phases, causing severe cross-regulation problems [10].

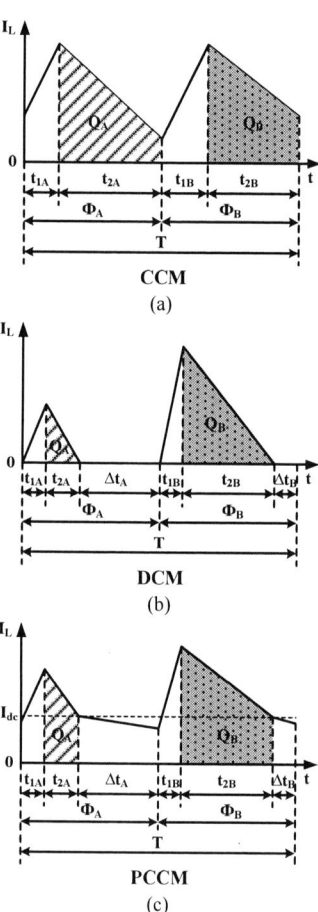

Figure 1. Inductor current waveforms of a SIMO DC-DC converter operating in (a) CCM, (b) DCM, and (c) PCCM, respectively.

This work is jointly sponsored by the *Semiconductor Research Corporation* under the contracts SRC GRC 1836.015 and 1836.017, and by the *US National Science Foundation* (NSF) under the contact NSF CCF-0844557.

978-1-4244-4782-4/10 $26.00 © 2010 IEEE

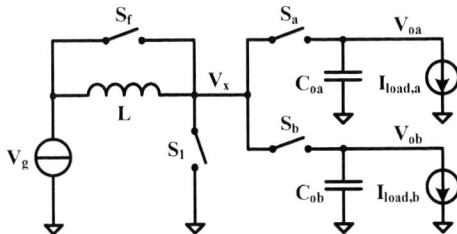

Figure 2. Architecture of SIMO converter presented in [12].

In [10-12], due to the adoption of discontinuous conduction mode (**DCM**), the cross-regulation between different sub-converters is suppressed by using time-multiplexing (TM) technique, as shown in Fig. 1(b). Because the inductor current returns back to zero before the end of each phase in DCM, the load regulation of one sub-converter does not disturb the others. However, at heavy load conditions, large peak inductor current value must be reached to deliver sufficient power to the loads, leading to undesirable switching noise and current stress to the system.

The work in [13] successfully introduced a pseudo-CCM (**PCCM**) operation to overcome the drawbacks in [10-11]. Instead of going back to zero, the inductor current drops to a predefined DC level (I_{dc}), as shown in Fig. 1(c). Therefore, the large peak inductor current can be avoided in heavy load conditions. Both the switching noise and current stress are significantly reduced while keeping low cross-regulation. However, due to the non-zero turn-on resistance of freewheel switch S_f in Fig. 2, the inductor current in Fig. 1(c) decreases gradually during freewheel switching periods, causing extra conduction power loss. In particular, at unbalanced load conditions, a freewheel switching period could last much longer at light-load output (Δt_A in Fig. 1(c)) than the heavy-load one (Δt_B in Fig. 1(c)), leading to the increase on conduction loss and switching noise.

In this paper, an adaptive freewheel current modulation (**AFCM**) technique is proposed to reduce conduction power loss of SIMO converters operating in PCCM. It utilizes a high frequency clock signal to measure the duration of each freewheel switching period. Accordingly, the proportion of switching phases for all sub-converters is redistributed and the value of I_{dc} is adjusted. In Section II, the potential issues in unbalanced load conditions are discussed. The AFCM technique is introduced in Section III. Section IV describes the system architecture and schematics of AFCM controller. The performance verification results are shown in Section V. Finally, we conclude this research in Section VI.

II. ISSUES OF PCCM SIMO CONVERTERS AT UNBALANCED LOAD CONDITIONS

For a SIMO switching converter at balanced load conditions, as illustrated in Fig. 3(a), the freewheel switching time for each output is moderate. The converter then operates well in PCCM. However, at unbalanced load conditions, the freewheel switch conduction time for light-load outputs become much longer, as depicted in Fig. 3(b). The value of $\Delta D_a T_S$ is much larger than $\Delta D_b T_S$ due to the unbalanced load conditions. Assuming the turn-on resistance of freewheel

switch S_f is $R_{on,f}$ and the initial value of freewheel switching currents is I_{dc}, the expression of inductor current I_L during freewheel switching period $\Delta D_a T_S$ is given by

$$I_L(t) = I_{dc} \cdot e^{-\left(\frac{R_{on,f}}{L}\right)t}. \tag{1}$$

And the conduction loss of S_f is expressed as

$$E_{a,f} = \int_0^{\Delta D_a T_S} I_L^2(t) \cdot R_{on,f}\, dt = \frac{I_{dc}^2 \cdot L}{2}\left(1 - e^{-\frac{2R_{on,f}}{L}\Delta D_a T_S}\right). \tag{2}$$

Eqn. (2) reveals that the conduction loss is highly related to the I_{dc} value and freewheel switch conduction time. As a result, when freewheel switch conduction time becomes much longer and I_{dc} stays high, the conduction loss of S_f gets much higher. These undesirable results also cause unnecessarily large voltage and current ripples, and cannot be overcome by reducing the switching period. Otherwise, the freewheel switching period in Φ_b ($\Delta D_b T_S$) would become so short that the converter returns to CCM and severe cross regulation will repeatedly occur. As I_{dc} stays high to power the heavy-load outputs, power switches are turned on/off with large dynamic currents. This can generate severe switching noise, coupling throughout the entire chip and jeopardizing the system stability.

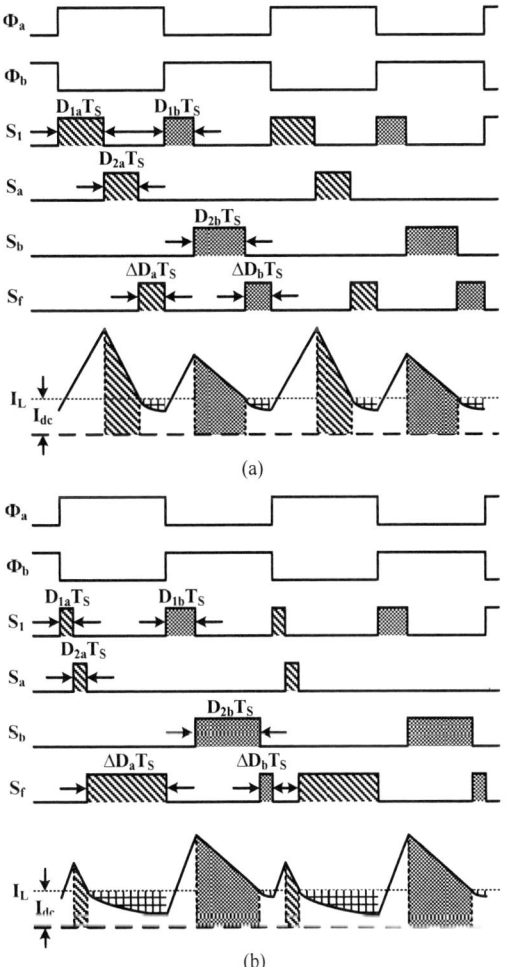

Figure 3. Timing diagrams of a PCCM SIMO converter at (a) balanced load conditions, and (b) unbalanced load conditions.

978-1-4244-4782-4/10 $26.00 © 2010 IEEE

III. THE PROPOSED AFCM TECHNIQUE

The proposed adaptive modulation on freewheel switching current includes two tasks: redistribution of each subinterval and I_{dc} adjustment. For example, in a PCCM SIDO boost converter, due to the employment of time-multiplexing (TM) method, the total time of two subintervals is constant. In order to balance the length of $\Delta D_a T_S$ and $\Delta D_b T_S$, the proportion of Φ_a should be less than Φ_b. In the proposed AFCM technique, a high frequency clock signal is used to count the duration of each freewheel switching period. Assuming the switching frequency is f_s and the clock frequency is $n \times f_s$, then the resolution equals $1/n$. If the length of $\Delta D_a T_S$ is shorter than $\Delta D_b T_S$, the proportion of Φ_a will be increased by $1/n$ and accordingly, the proportion of Φ_b will be decreased by $1/n$, and vice versa. The final values of ΔD_a and ΔD_b should satisfy

$$\left| \Delta D_a - \Delta D_b \right| \leq \frac{1}{n}. \tag{3}$$

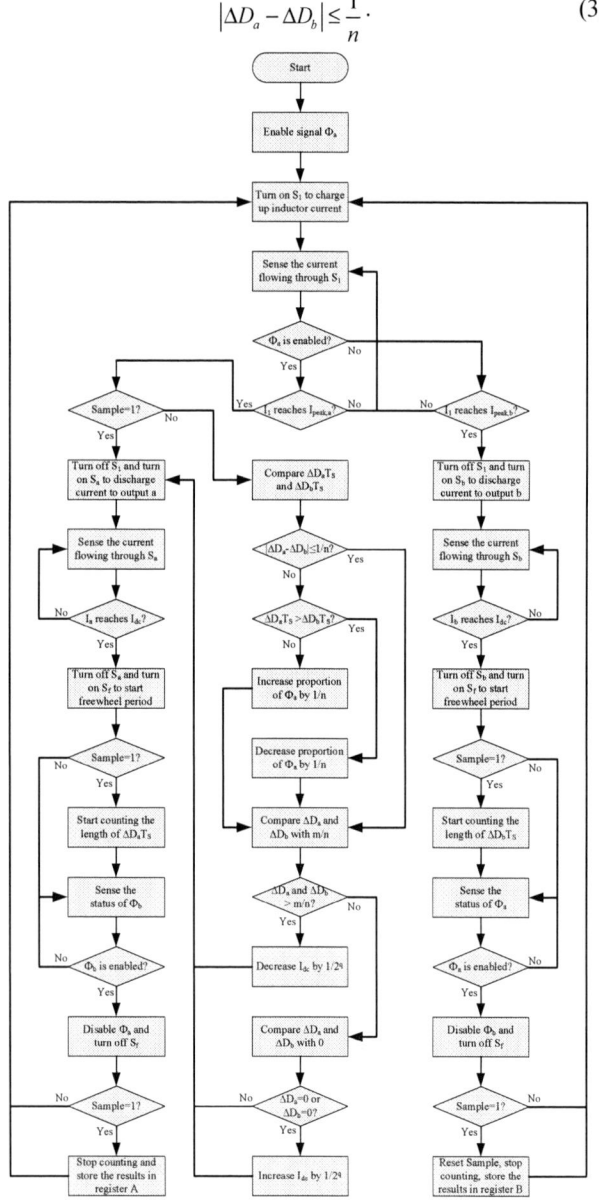

Figure 4. Control flow chart of AFCM technique.

Besides the subinterval redistribution, I_{dc} should also be adjusted to reduce the conduction loss of freewheel switch and switching noise. The shorter the freewheel switching period is, the less the conduction loss is. However, the freewheel switching time in each subinterval should not be too short to keep low cross-regulation. A reasonable value of variable m should be selected to realize

$$\frac{1}{n} \leq \Delta D_a T_S \leq \frac{m}{n}, \quad \frac{1}{n} \leq \Delta D_b T_S \leq \frac{m}{n}. \tag{4}$$

Note that the freewheel switching time in each subinterval is reduced when I_{dc} decreases, and vice versa. After the subinterval redistribution, if the proportion of freewheel switching periods in both outputs are larger than m/n, I_{dc} is decreased to reduce the conduction loss and switching noise. Similarly, if either of the two freewheel switching periods equals to zero, I_{dc} is increased to retain low cross-regulation. In the AFCM technique, reference voltage (V_{ref_dc}) for I_{dc} is generated by an integrated q-bit DAC. Therefore, the resolution of I_{dc} value is $1/2^q$, meaning that I_{dc} is increased or decrease by $1/2^q$ at each bit. After I_{dc} adjustment, the process goes back to subinterval redistribution again. This circulation repeats until Eqns. (3) and (4) are satisfied in steady-state. The control flow chart of AFCM technique is given in Fig. 4.

IV. SYSTEM ARCHITECTURE & CIRCUIT IMPLEMENTATION

The block diagram of PCCM SIMO boost converter with AFCM technique [14] is shown in Fig. 5. The inductor L and the NMOS power switch M_n are shared by the two sub-converters, which operate in Φ_a and Φ_b alternately. At the beginning of each switching cycle, the NMOS power switch M_n is turned on and the inductor current I_L ramps up with the slope V_g/L to $I_{peak,a}$, which is defined by the output of error amplifier EA_a. Then M_n is turned off and the PMOS power switch $M_{p,a}$ is turned on by control signal $S_{p,a}$. The inductor current flows through $M_{p,a}$ into the corresponding output V_{oa} and ramps down at a rate of $(V_g-V_{oa})/L$. When I_L reaches I_{dc}, which is decided by V_{ref_dc} signal, a freewheel switching period is started by turning off $M_{p,a}$ and turning on the PMOS freewheel switch $M_{p,fw}$ with control signal $S_{p,fw}$. The inductor current circulates through L and $M_{p,fw}$ and drops gradually owing to the non-zero resistance of the freewheel switch. This freewheel switching current continues until Φ_a expires and Φ_b starts. The above described switching actions repeat in Φ_b until the whole switching cycle ends.

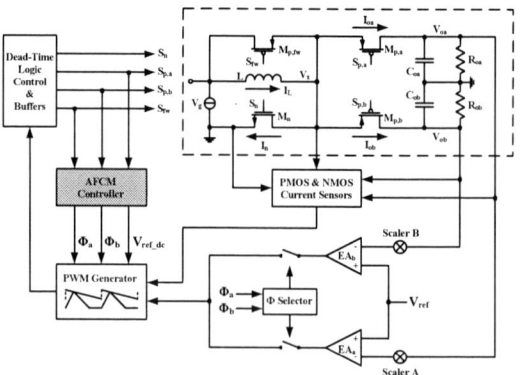

Figure 5. Block diagram of the proposed system.

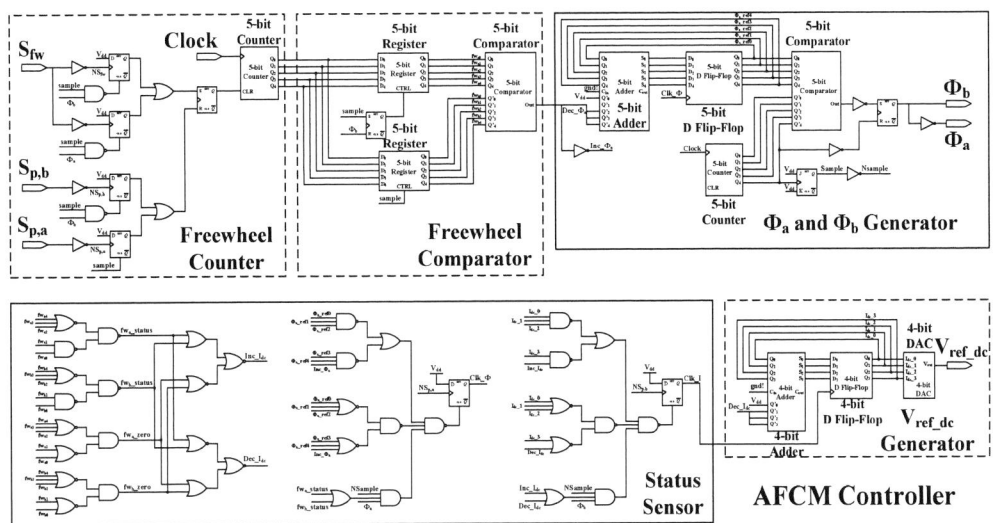

Figure 6. Logic gate level circuit schematic of the AFCM controller.

Based on the status of $S_{p,a}$, $S_{p,b}$ and S_{fw}, the Φ_a, Φ_b and V_{ref_dc} signals are adaptively modulated by the AFCM controller, which implements AFCM technique in the proposed boost DC-DC converter, to reduce the conduction loss of freewheel switch and switching noise in unbalanced load conditions while keeping low cross-regulation. The logic gate based schematic of AFCM controller is shown in Fig. 6. Here, the freewheel counter is used to measure the duration of each freewheel switching period. The counting results are then compared by the freewheel comparator to decide how to adjust the phase durations and I_{dc} value, as illustrated in the control flow chart in Fig. 4. In addition, the function of status sensor is providing protection to the system in order to avoid malfunctions. Finally, based on the control signals produced by freewheel comparator and status sensor, the new Φ_a, Φ_b and V_{ref_dc} signals are created by the corresponding generators for the next switching cycle.

V. PERFORMANCE VERIFICATION

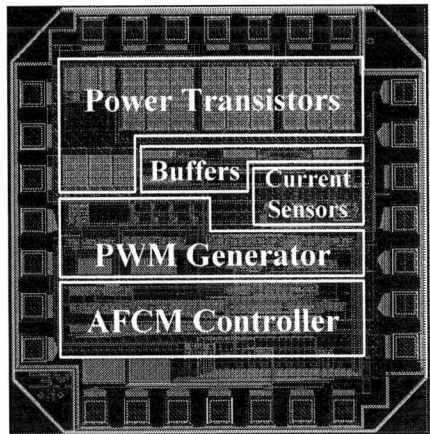

Figure 7. Chip layout.

The proposed converter was designed and simulated with a 130-nm CMOS process. Fig. 7 shows the chip layout. It

occupies 1 mm^2 silicon area. All the simulation results shown in the following parts are fully transistor-based through Synopsis HSPICE. In this design, we set the switching frequency of the converter at 500 KHz. The values of n, m and q are chosen as 32, 2 and 4, respectively. Therefore, the Φ resolution equals 1/32 and the I_{dc} resolution is 1/16. The two outputs are regulated at 1.8 V and 2.5 V respectively, while the input voltage is 1.5 V.

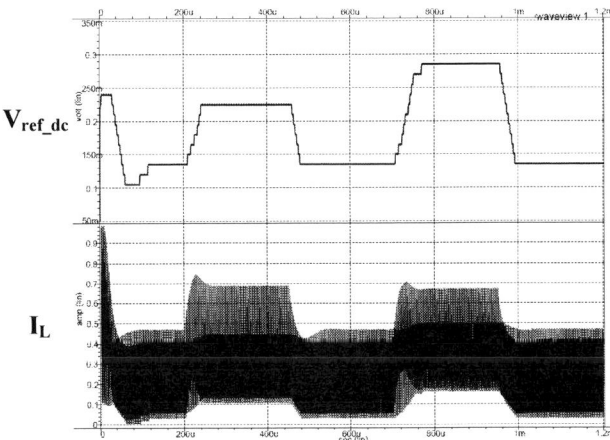

Figure 8. Transient waveforms of V_{ref_dc} and I_L signals.

The load transient waveforms of inductor current I_L and V_{ref_dc} are shown in Fig. 8. When $I_{load,a}$ changes between 50 mA and 100 mA at the instants of 200 μs and 450 μs, V_{ref_dc} changes between 135 mV and 225 mV accordingly. Similarly, $I_{load,b}$ varies between 100 mA and 200 mA at the instants of 700 μs and 950 μs, V_{ref_dc} changes between 135 mV and 285 mV correspondingly. Hence the value of I_{dc} is decided by V_{ref_dc}, the simulation results clearly indicate that the freewheel switching current is adaptively controlled according to different load conditions.

The simulation results on cross-regulation are shown in Fig. 9. When $I_{load,a}$ changes between 50 mA and 100 mA, the voltage ripple of V_{ob}, which is caused by cross-regulation, is well limited within 12 mV. Similar results are observed in V_{oa} when $I_{load,b}$ changes between 100 mA and 200 mA.

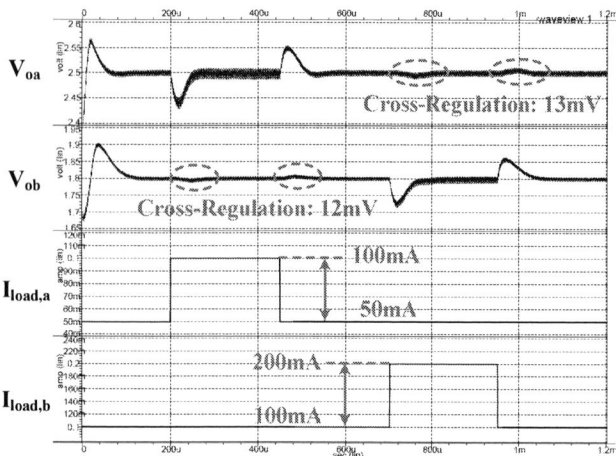

Figure 9. Cross-regulation simulation during load transients.

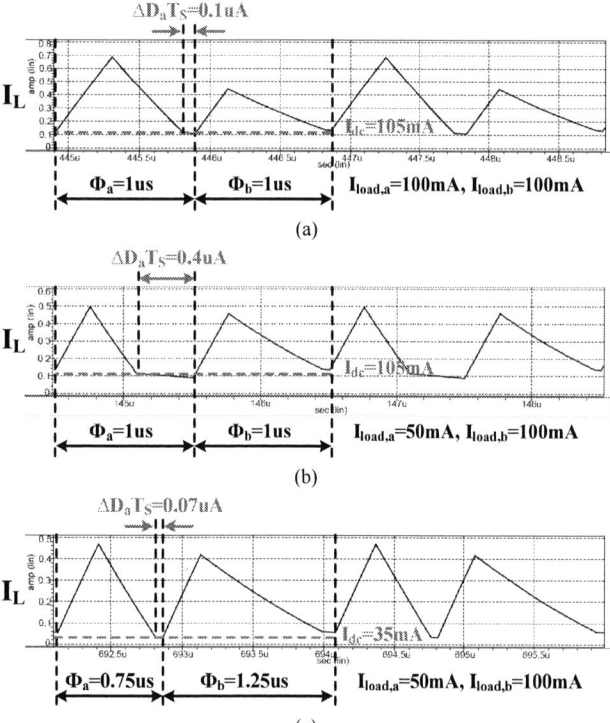

Figure 10. Waveforms of the inductor current at different operation and load conditions: (a) with AFCM, $I_{load,a}$=100mA, $I_{load,b}$=100mA, (b) without AFCM, $I_{load,a}$=50mA, $I_{load,b}$=100mA; (c) with AFCM, $I_{load,a}$=50mA, $I_{load,b}$=100mA.

Another important function is the automatic minimization of freewheel switch conduction loss. Based on Eqn. (2), when the inductor value and turn-on resistance of freewheel switch are set, the conduction loss of freewheel switch is decided by the duration of freewheel switching period and I_{dc} value. In this design, we set the inductor value to 1 μH and turn-on resistance of freewheel switch to 100 mΩ. Fig. 10(a) shows the steady-state waveform of inductor current when both $I_{load,a}$

and $I_{load,b}$ are equal to 100 mA. When $I_{load,a}$ changes from 100 mA to 50 mA, if the phase duration and I_{dc} value are not adaptively modulated, a long freewheel switching period with high I_{dc} value will be observed in Φ_a, as shown in Fig. 10(b). By employing the AFCM technique, both the freewheel switching duration and I_{dc} value can be adaptively controlled according to various load conditions. When $I_{load,a}$ changes from 100 mA to 50 mA, the waveform of inductor current applying AFCM technique is depicted in Fig. 10(c). Based on the simulation data and Eqn. (2), the conduction loss of freewheel switch is reduced by 98% compared with the case without the AFCM technique.

VI. CONCLUSIONS

In this paper, an AFCM technique focusing on adaptive modulation of freewheel switching current value and time for SIMO converters was introduced. The conduction loss of freewheel switch and switching noise in unbalanced load conditions are effectively restricted. Low cross-regulation is also achieved simultaneously.

REFERENCES

[1] J.-M. Chang and M. Pedram, "Energy minimization using multiple supply voltages," *IEEE Trans. VLSI Syst.*, vol. 5, pp. 436-443, Dec. 1997.

[2] T. D. Burd, T. A. Pering, A. J. Stratakos, R. W. Brodersen, "A dynamic voltage scaled microprocessor system," *IEEE J. Solid-State Circuits*, vol 35, no. 11, pp. 1007-1014, Nov. 2000.

[3] T. Ishihara and K. Asada, "A system level memory power optimization technique using multiple supply and threshold voltages," in *Proc. Asian and South Pacific Design Automation Conf.*, Jan. 2001, pp. 456-461.

[4] M. Brown, *Practical Switching Power Supply Design*, Academic, San Diego, 1990.

[5] R. W. Erickson, *Fundamentals of Power Electronics*, Kluwer, Boston, 1999.

[6] A. P. Dancy, R. Amirtharajah, and A. P. Chandrakasan, "High-efficiency multiple-output dc-dc conversion for low-voltage systems," *IEEE Trans. VLSI Syst.*, vol. 8, pp. 252-263, Jun. 2000.

[7] M. W. May, M. R. May, and J. E. Willis, "A synchronous dual-output switching dc-dc converter using multibit noise-shaped switch control," *IEEE Int. Solid-State Circuits Conf. Dig. Tech. Papers*, Feb. 2001, pp. 358-359.

[8] D. Goder and H. Santo, "*Multiple output regulator with time sequencing*," U.S. Patent 5 617 015, Apr. 1, 1997.

[9] T. Li, "*Single inductor multiple output boost regulator*," U.S. Patent 6 075 295, Jun. 13, 2000.

[10] D. Ma, W.-H. Ki, C. Y. Tsui, and P. K. T. Mok, "A 1.8-V single-inductor dual-output switching converter for power reduction techniques," *IEEE Symp. VLSI Circuits*, pp. 137-140, June 2001.

[11] W.-H. Ki, D. Ma, "Single-inductor multiple-output switching converters," *IEEE Power Electronics Specialists Conf.*, pp.226-231, June 2001.

[12] D. Ma, W-H Ki, C-Y Tsui and P. K. T. Mok, "Single-inductor dual-output CMOS switching converters in discontinuous-conduction mode with time-multiplexing control", *IEEE J. of Solid-State Circuits*, Vol. 38, No. 1, pp. 89-100, Jan. 2003.

[13] D. Ma, W-H Ki, C-Y Tsui, "A Pseudo-CCM/DCM SIMO switching converter with freewheel switching," *IEEE J. Solid-State Circuits*, Vol. 38, No. 6, pp. 1007-1014, June 2003.

[14] D. Ma, Y. Zhang, "Adaptive freewheel switching control mechanisms and circuits for DC-DC converters operating in PCCM," *U. S. Provisional Patent Application*, in pending.

Analysis of Pulse-link DC-AC Converter for Fuel Cells Applications Operated in Zero-Current-Slope Mode

Kentaro Fukushima[1)5)], Isami Norigoe[2)], Masahito Shoyama[1)], Tamotsu Ninomiya[3)],
Yosuke Harada[4)] and Kenta Tsukakoshi[4)]

1) Kyushu University, 744 Moto-oka, Nishi-ku, Fukuoka, Japan
2) I. N. Lab., 3-13-9, Midori-machi, Koganei City, Tokyo, Japan
3) Nagasaki University, 1-14 Bunkyo-machi, Nagasaki, Japan
4) EBARA DENSAN .LTD, 11-1 Asahi-machi, Ota-ku, Tokyo, Japan
5) JSPS Research Fellow

Abstract— **In fuel cells applications, current-ripple reduction is essential for conversion efficiency and life span. This paper analyzes the pulse-link DC-AC converter for fuel cells applications operated in zero-current-slope mode. As the result, in zero-current-slope operation mode, input-current-ripple is reduced. Furthermore, in this operation mode, the parameters of series LC circuit which is worked as ripple canceling are less values.**

I. INTRODUCTION

Recently, energy consumption problem such as global heating and fossil fuel shortage has become a serious problem all over the world. In Japan, about 30% of amount of CO_2 emission is wasted by energy conversion department [1]. Therefore, renewable energy system is desired strongly. Fuel cells are one of the renewable energy systems. When fuel cells generate electricity, a large amount of thermal energy arises in the same time. The cogeneration system using both electricity and thermal energy is now researched and developed actively around the world. In particular, a home-use cogeneration system with fuel cells is developed from the stream of what is called distributed power system or micro-grid power systems. When cogeneration system is set near home, the distribution loss can be reduced. Now, home power source is AC, on the other hand, the voltage provided by fuel cells is DC. Therefore, a DC-AC converter is essential for the home-use cogeneration system [2-3]. In fuel cells applications, small input current-ripple is essential because the fuel cell chemical reaction time is much slower than commercial frequency. Current-ripple may damage to fuel efficiency and life cycle in fuel cells [4-6].

The construction of the conventional DC-AC converter is two components - first stage is boost converter and second stage is PWM inverter [7-9]. Conventional DC-AC converter

is interpolated large capacitor between boost converter stage and PWM inverter stage. That capacitor disturbs the size reduction of this unit. To overcome this problem, authors have proposed a novel topology called as Pulse-link DC-AC converter shown in Fig. 1 [10]. The proposed topology provides boosted-voltage pulse directly to PWM inverter. This topology does not require large capacitor between two stages. Instead, small values of inductor and capacitor are connected series and inserted between two stages in parallel [11]. This paper examines the relationship between the inductance, capacitance values and input current-ripple. As the result, when the inductor current has zero slope state in one switching period, input current-ripple is reduced. This paper analyzes the operation mode called as zero-current-slope mode. In this operation mode, input current-ripple is reduced.

II. PULSE-LINK DC-AC CONVERTER

Fig. 1 shows the pulse-link DC-AC converter topology. This topology has two stages, and this converter provides boosted pulsed voltage directly to PWM inverter. And between two stages, series LC circuit is connected in parallel in order to reduce current-ripple. The value of the capacitor using this LC circuit is less than the conventional one. This converter has 5 switches. Switch Q_1 controls the boost pulse voltage from input voltage. And, from S_1 to S_4 are PWM inverter switches. S_1 and S_4 are regulated to make output voltage sinusoidal waveform, on the other hand, S_2 and S_3 are controlled the positive/negative of output voltage. And control combination of S_1, S_3 and S_2, S_4 is operated symmetry. Q_1 and S_1/S_4 are synchronous at rising time.

978-1-4244-4782-4/10 $26.00 © 2010 IEEE

A. INPUT CURRENT-RIPPLE CONSIDERATION

Firstly, it examines the characteristics of input current-ripple changed by inductance and capacitance of L_2 and C_3. Fig. 2 shows the relationship between L_2, C_3 values and input current-ripple. From this figure it is considered that there are two domains by the difference of the characteristics for the current-ripple. One is larger than 1A, the other is less than 1A. Furthermore, this difference is dependent on inductance value of L_2. Furthermore, these two domains are dependent on inductance values from Fig. 2. Here, the domain which is less than 400uH is defined as Domain I, and the domain which is more than 400uH is defined as Domain II. The operation waveforms in each domain are considered.

As the result, the difference of waveform is observed at the current of L_2. Fig. 3 shows the waveforms of the input voltage to PWM inverter (v_{inv_in}), and the current of L_2 (i_{L2}) in Domain I (L_2 is 200uH, C_3 is 40uF). And Fig. 4 shows those waveforms in Domain II (L_2 is 700uH, C_3 is 40uF). In Domain I, it is observed that in one switching cycle, i_{L2} has the period that the slope is kept flat. On the other hand, in Domain II, this period is not observed but constant conduction mode. Therefore, it is considered that the characteristics domain of the current-ripple is operation mode difference. Here, it is called as zero-current-slope mode that the operation mode which has zero current slope state. At next chapter, zero-current-slope mode is analyzed.

III. ZERO-CURRENT-SLOPE MODE

Fig. 4 shows the state transition diagram of Zero-Current-Slope mode and the current of L_1 and L_2. Fig. 5 shows the equivalent circuit of each state. In Zero-Current-Slope mode, inductor current L1 which is converted to the secondary side is equal to inductor current of L2 at state 4. Therefore, inductor current cannot be regarded as independent variable. Then, average values of inductor currents are described by unified state-space averaging method. And here, duty ratio of switch Q_1 is defined as D_{Q1}, and duty ratio of S_1 is defined as d_s. d_s is changed to make output voltage commercial sinusoidal waveforms as shown below:

$$d_s(t) = d_{s\_max} \cdot \left| \sin(2\pi \cdot 50t) \right| \qquad (1)$$

Moreover, the relationship of D_{Q1} and d_{s_max} has limited condition, because PWM inverter is provided voltage only when Q_1 is ON.

$$D_{Q_1} \geq d_{s\_max} \qquad (2)$$

In the above condition, pulse output voltage is regarded as constant voltage viewing from PWM inverter.

A. State 1 (Q_1:ON, S_1:ON, S_3: ON)

Fig. 5(a) shows the equivalent circuit at state 1. Here, the initial DC component value of inductor current is defined as I_{DC}. The current flown to L_1 is defined as i_{L1}, and the value which is converted to secondary side is defined as i'_{L1}. State equation at state 1 is written as below:

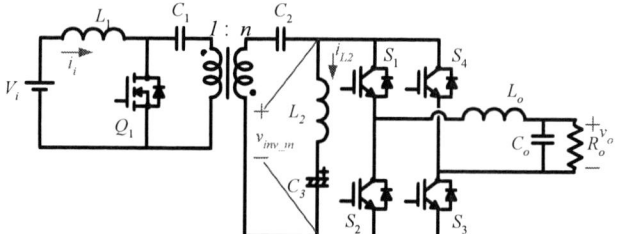

Fig. 1. Pulse-link DC-AC Converter

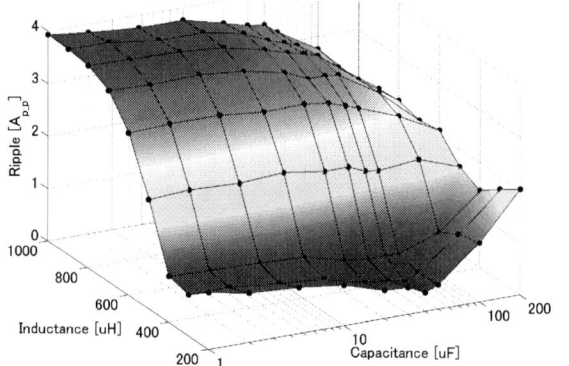

Figure 1. Relationship between $L_2 C_3$ values and input current-ripple.

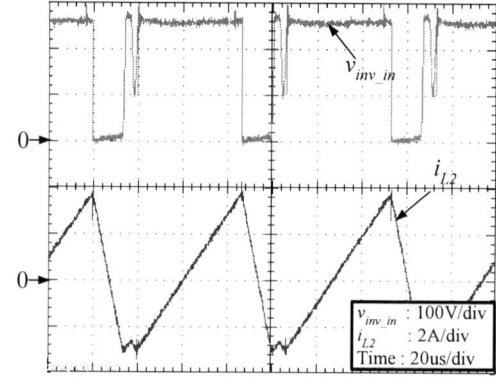

Figure 2. Relationship between $L_2 C_3$ values and input current-ripple.

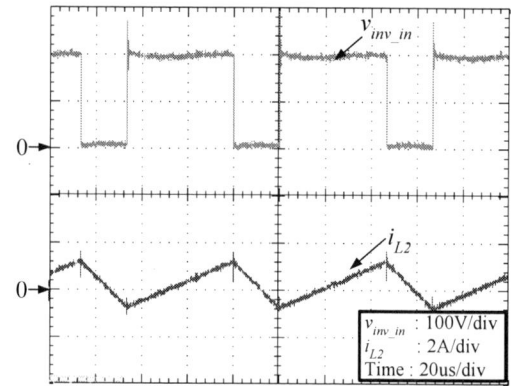

Figure 3. Relationship between $L_2 C_3$ values and input current-ripple.

$$i'_{L1} = \frac{nV_i}{L_1}t + I_{DC}$$

$$i_{L2} = \frac{\hat{v}_{C'} - \hat{v}_{C3}}{L_2}t - I_{DC}$$

$$v_{Lo} = \hat{v}_{C'} - \hat{v}_o \qquad (3)$$

$$i_{C'} = -i_{L2} - \hat{i}_{Lo}$$

$$i_{C3} = i_{L2}$$

$$i_{Co} = \hat{i}_{Lo} - \frac{\hat{v}_o}{R}$$

And here, $C' = \dfrac{C_1 C_2}{C_1 + n^2 C_2}$.

B. State 2 (Q_1:ON, S_1:OFF, S_3: ON)

Fig. 5(b) shows the equivalent circuit at state 2. From this figure, state equation at state 2 is written as below:

$$i'_{L1} = \frac{nV_i}{L_1}t + I_{DC}$$

$$i_{L2} = \frac{\hat{v}_{C'} - \hat{v}_{C3}}{L_2}t - I_{DC}$$

$$v_{Lo} = -\hat{v}_o \qquad (4)$$

$$i_{C'} = -i_{L2}$$

$$i_{C3} = i_{L2}$$

$$i_{Co} = \hat{i}_{Lo} - \frac{\hat{v}_o}{R}$$

C. State 3 (Q_1:OFF, S_1:OFF, S_3: ON)

Fig. 5(c) shows the equivalent circuit at state 3. From this figure, state equation at state 3 is written as below:

$$i_{L1} = \frac{nV_i - v_{C'}}{L_1}(t - DT_s) + \frac{nV_i}{L_1}DT_s + I_{DC}$$

$$i_{L2} = -\frac{\hat{v}_{C3}}{L_2}(t - DTs) + \frac{\hat{v}_{C'} - \hat{v}_{C3}}{L_2}DT_s - I_{DC}$$

$$v_{Lo} = -\hat{v}_o \qquad (5)$$

$$i_{C'} = i_{L2}$$

$$i_{C3} = i_{L2}$$

$$i_{Co} = \hat{i}_{Lo} - \frac{\hat{v}_o}{R}$$

When t is $(D_{Q1}+d)T_s$, $i'_{L1}=i_{L2}$. So, d is written as below:

$$d = \frac{nV_i}{\hat{v}_{C'} - nV_i}D_{Q1} \qquad (6)$$

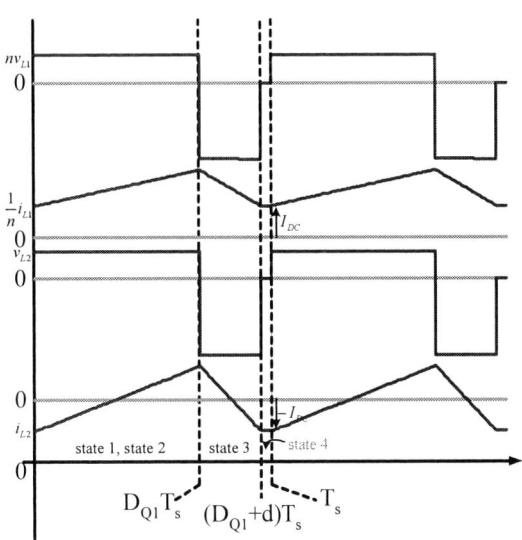

Figure 4. State transition diagram of the operation mode.

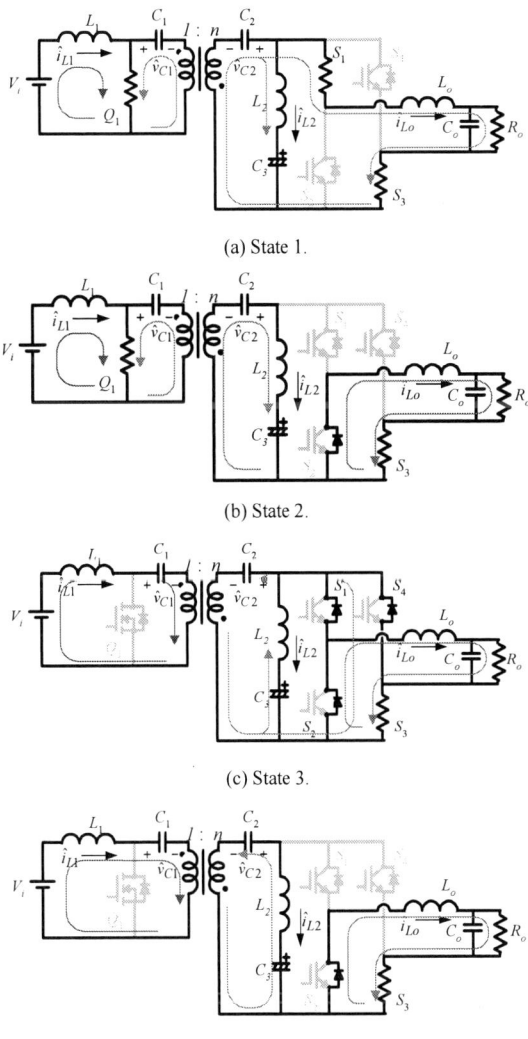

(a) State 1.

(b) State 2.

(c) State 3.

(d) State 4.

Figure 5. Equivalent circuit of each state.

D. State 4 (Q_1:OFF, S_1:OFF, D_1:OFF, S_3: ON)

Fig. 5(d) shows the equivalent circuit at state 4. From this figure, state equation at state 4 is written as below:

$$\left.\begin{array}{l} i_{L1} = I_{DC} \\ i_{L2} = -I_{DC} \\ v_{Lo} = -\hat{v}_o \\ i_{C'} = i_{L2} = I_{DC} \\ i_{C3} = i_{L2} = -I_{DC} \\ i_{Co} = \hat{i}_{Lo} - \dfrac{\hat{v}_o}{R} \end{array}\right\} \tag{7}$$

Furthermore, the relationship with the voltage of each capacitor is written as below from this state:

$$nV_i = \hat{v}_{C'} - \hat{v}_{C3} \tag{8}$$

E. Steady State

From those 4 states, steady-state equations are described. Averaging current of i_{L2} is written as below:

$$\bar{i}_{L2} = \frac{nV_i}{2L_2}(D+d)^2 T_s^2 - \frac{\hat{v}_{C'}}{2L_2} d^2 T_s^2 - I_{DC} \tag{9}$$

In this operation mode, $\bar{i}_{L2}$ is zero in one switching period, so I_{DC} is written as below:

$$I_{DC} = \frac{nV_i}{2L_2}(D+d)^2 T_s^2 - \frac{\hat{v}_{C'}}{2L_2} d^2 T_s^2 \tag{10}$$

Furthermore, substituting equation (6) into this equation, I_{DC} is described as follow:

$$I_{DC} = \frac{\hat{v}_{C'} nV_i D_{Q1}^2 T_s}{2L_2(\hat{v}_{C'} - nV_i)} \tag{11}$$

In the same way, averaging current of $i_{C'}$ is written as below:

$$\bar{i}_{C'} = \frac{1}{2}\left(\frac{1}{n^2 L_1} + \frac{1}{L_2}\right)\frac{V_i^2}{\hat{v}_{C'} - V_i} D^2 T_s - \frac{ds^2}{R}\hat{v}_{C'} \tag{12}$$

In steady state condition, $\bar{i}_{C'}$ is zero. From this condition, the voltage of C_3 in steady state is described as below:

$$V_{C'} = \frac{V_i}{2} + \frac{V_i}{2}\sqrt{1 + \frac{2RD^2 T_s}{ds^2}\left(\frac{1}{n^2 L_1} + \frac{1}{L_2}\right)} \tag{13}$$

From this equation, output voltage is written as below:

$$V_o = d_s V_{C'} = \frac{V_i}{2}\left(d_s + \sqrt{d_s^2 + 2RD^2 T_s\left(\frac{1}{n^2 L_1} + \frac{1}{L_2}\right)}\right) \tag{14}$$

Furthermore, averaging current of i_{L1} is written as below:

$$\bar{i}_{L1} = \frac{n^2 d_s^2}{2R_o}V_i + \frac{n^2 D_{Q1}^2 T_s V_i}{2}\left(\frac{1}{n^2 L_1} + \frac{1}{L_2}\right) + $$
$$+ \frac{n^2 V_i}{2R_o}\sqrt{ds^4 + 2R_o d_s^2 D_{Q1}^2 T_s\left(\frac{1}{n^2 L_1} + \frac{1}{L_2}\right)} \tag{15}$$

In this equation, it is considered that second term of this equation is independent from d_s and load resistor. It means that I_{L1} has DC component, and the values is decided by the constant values. As the result, in this operation mode it is considered that input current-ripple is reduced.

Moreover, the boundary condition between constant conduction mode and this zero-current-slope mode is determined by the equation of $V_{C'}$. In constant conduction mode, $V_{C'}$ is described as below:

$$V_{C'} = \frac{n}{1 - D_{Q1}} V_i \tag{16}$$

The boundary condition is described below:

$$\frac{(1 + D_{Q1})^2}{(1 - D_{Q1})^2} = 1 + \frac{2RD_{Q1}^2 T_s}{d_s^2}\left(\frac{1}{n^2 L_1} + \frac{1}{L_2}\right) \tag{17}$$

From this equation, it is considered that the boundary condition is dependent on inductance.

IV. EXPERIMENTAL CONSIDERATION

A. Experimental parameters settings

In order to evaluate the performance of the circuit, the experimental circuit is implemented with the specifications and parameters in Table I. From table I, C_1 is 3mF, and it is aluminum electrolytic capacitor. C_1 is decided from the allowable current. Primary-side is flown large current, so capacitance of C_1 becomes large value. However, primary-side is low voltage, so the size of aluminum electrolytic capacitor is not so large even if the value is large because withstand-voltage is low. Therefore, large value of aluminum electrolytic capacitor is used at C_1 in this experiment.

Table I. Experimental values

Symbol	Description	value
Vi	Input voltage	20[V]
L1	Input inductance	470[uH]
LM	Magnetizing inductance	2.6[mH]
C1	primary-side capacitance	3[mF]
C2	Secondary-side capacitance	330[uF]
n	Turn ratio	3
Lo	Output inductance	3[mH]
Co	Outout capacitance	9.4[uF]
fs	Switching frequency	30[kHz]
Ro	Output resistance	100[ohm]

B. Relationship between L_2 and current-ripple

Fig. 6 shows the experimental measurement of input current-ripple and output voltage changed by inductance L_2. C_3 is 40uF.

From this result, it is considered that input current-ripple is rising at more than 400uH; on the other hand, input current-ripple is small at less than 400uH. This result means that at less than 400uH, operation mode is Zero-Current-Slope mode. On the other hand, operation mode is Constant-Conduction mode at more than 400uH. Furthermore, output voltage is increasing less than 400uH, and is almost constant from more than 400uH. The boundary condition is estimated by equation (17). Analytical result when d_s is regarded as root mean square (461uH) is agreed well. Moreover, inductance value at the experimental bottom of the current-ripple value (300uH) is similar to the analytical result of the boundary condition when d_s is regarded as d_{s_max} (223uH). Moreover, analytical result of output voltage is shown orange line in Fig. 6. From this result, voltage characteristic is also agreed well with experimental result.

C. Relationship between L_2 and current-ripple

Fig. 7 shows the experimental measurement of input current-ripple by capacitance C_3. Here, inductance L_2 is decided as 300uH. From the result, it is considered that there is a bottom value when C_3 is 40uF. Furthermore, as capacitance value is more than 40uF, input current-ripple becomes large. On the other hand, when C_3 is less than 40uF, input current-ripple is kept small. Moreover, the scale of the values is logarithm scale. In less than 40uF, the capacitor can be used film-type capacitor. As the result, it can be improved the reliability.

Finally, Fig. 8 shows the experimental waveforms of output voltage and input current at the least ripple value when L_2 is 300uH and C_3 is 40uF. From the waveforms, it is observed that input current is almost flat and ripple is $0.3A_{p\_p}$. If the parameters are chosen to be Zero-Current-Slope mode, input current-ripple is reduced. Capacitance C_3 is also small values, and C_3 can be applied film-capacitor.

V. CONCLUSION

On the pulse-link DC-AC converter for fuel cells applications, this paper examined the relationship between series L_2C_3 parameter and input current-ripple. As the result, Zero-current-slope mode which means current-slope has zero state in one switching cycle is reduced the input current-ripple. The static characteristics operated in zero-current-slope mode were analyzed. Input current equation has the DC term which is independent from the inverter side condition. As the result, input current-ripple is less. Furthermore, the boundary condition between this operation mode and continuous conduction mode is described. From the experimental result, analytical result was agreed well. This operation mode is achieved current-ripple reduction by small inductance and capacitance values.

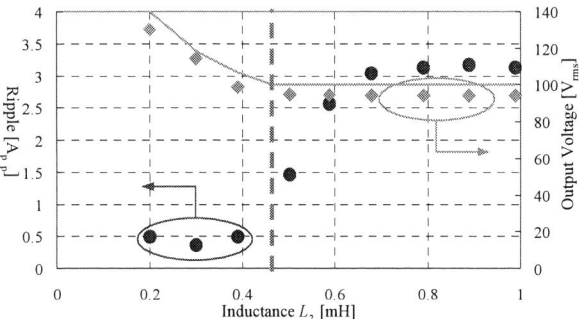

Figure 6.   Inductance L_2 vs. Input current-ripple and Output voltage.

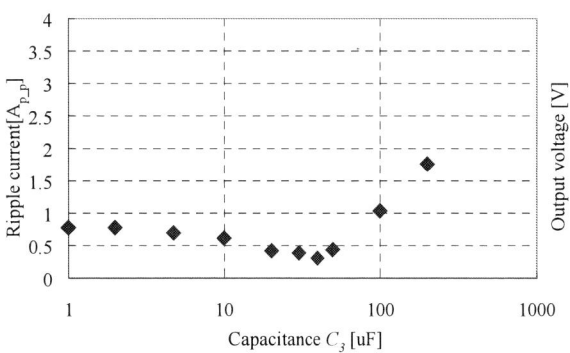

Figure 7.   Capacitance C_3 vs. Input current-ripple.

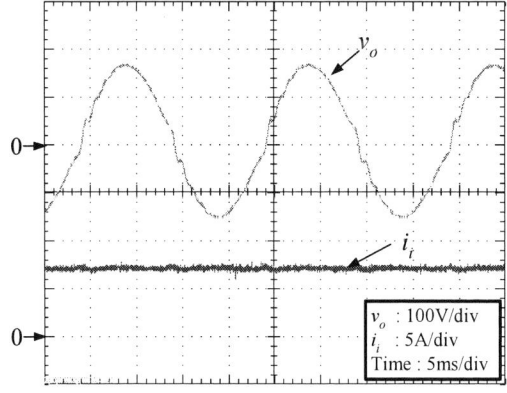

Figure 8.   Experimental waveforms of Output voltage and Input current.

The effect of Capacitance C_3 is that less than 40uF current-ripple is less. Therefore, C_3 can be used film-capacitor. From those result, the advanced reduction method can be reduced the size of this unit and improved reliability of this unit.

REFERENCES

[1] "The GHGs Emissions Data of Japan", Greenhouse Gas Inventory Office of Japan, by way of Japan Center for Climate Change Actions, http://www.jccca.org/

[2] J. Mazumdar, I. Batarseh, N. Kutkut and O. Demirci, "High Frequency Low Cost DC-AC Inverter Design with Fuel Cell Source for Home Applications," Proc. of IAS 2002, pp. 789-794, 2002.

[3] S.K. Mazumder, R.K. Burra, R. Huang, and Vince Arguelles, "A Low-cost Single-stage Isolated Differential Ĉuk Inverter for Fuel-cell Application," Proc. of IEEE PESC 2008, pp. 4426-4431, Jun. 2008.

[4] W. Choi, P.N. Enjeti and J.W. Howze, "Development of an Equivalent Circuit Model of a Fuel Cell to Evaluate the Effects of Inverter Ripple Current," Proc. of IEEE APEC 2004, pp. 255-361, Feb. 2004.

[5] R. S. Gemmen, "Analysis for the Effect of Inverter Ripple Current on Fuel Cell Operating Condition," Journal of Fluids Engineering, Vol. 125, Issue 3, pp. 576-585, May 2003.

[6] W. Shireen, R. A. Kulkarni, M. Arefeen, "Analysis and minimization of input ripple current in PWM inverters for designing reliable fuel cell power systems," Journal of Power Sources, Vol. 156, pp. 448-454, 2006.

[7] M. H. Todorovic, L. Palma, and P. N. Enjeti, "Design of a Wide Input Range DC-DC Converter With a Robust Power Control Scheme Suitable for Fuel Cell Power Conversion," IEEE transaction on IE, Vol. 55, No. 3, pp. 1247-1255, Mar. 2008.

[8] S. Sumiyoshi, H. Omori, and Y. Nishida, "Power Conditioner Consisting of Utility Interactive Inverter and Soft-Switching DC-DC Converter for Fuel-Cell Cogeneration System," Proc. of PCC Nagoya 2007, pp. 455-462, Apr. 2007.

[9] J. Wang, F. Z. Peng, J. Anderson, A. Joseph, R. Buffenbarger, "Low Cost Fuel Cell Converter System for Residential Power Generation," IEEE transaction on PE, Vol. 19, No. 5, pp.1315-1322, Sep. 2004.

[10] K. Fukushima, T. Ninomiya, S. Abe, I. Norigoe, Y. Harada, K. Tsukakoshi, and Z. Dai, "Steady-State Characteristics of a novel DC-AC Converter for Fuel Cells," Proc. of IEEE INTELEC 2007, pp.904-908, 2007.

[11] K. Fukushima, T. Ninomiya, M. Shoyama, I. Norigoe, Y. Harada, and K. Tsukakoshi, "Consideration for Input Current-Ripple Reduction on a Novel Pulse-link DC-AC Converter for Fuel Cells," Proc. of PESC 2008, pp. 1284-1289, Jun. 2008.

[12] B. Singh, G. D. Chaturvedi, "Modeling, Simulation and Development of Isolated Cuk AC-DC Converter in DCM and CCM Operation," IETE journal of research, Vol. 54, Issue 6, pp. 414-420, Nov-Dec 2008.

A Minimum Power-Processing Stage Fuel Cell Energy System Based on A Boost-Inverter with A Bi-Directional Back-Up Battery Storage

Minsoo Jang   Vassilios G. Agelidis

School of Electrical and Information Engineering
The University of Sydney
NSW, 2006, Australia
Email: minsoo@ee.usyd.edu.au and v.agelidis@ee.usyd.edu.au

Abstract—When low-voltage unregulated fuel cell (FC) output is conditioned to generate AC power, two stages are required: a boost stage and an inversion one. In this paper, the boost-inverter topology that achieves both boosting and inversion functions in a single-stage is used to develop an FC-based energy system which offers high conversion efficiency, low-cost and compactness. The proposed system incorporates additional battery-based energy storage and a DC-DC bi-directional converter to support instantaneous load changes. The output voltage of the boost-inverter is voltage-mode controlled and the DC-DC bi-directional converter is current-mode controlled. The load low-frequency current ripple is supplied by the battery which minimizes the effects of such ripple being drawn directly from the FC itself. Analysis, simulation and experimental results are presented to confirm the operational performance of the proposed system.

I. Introduction

In general, renewable energy systems based on photovoltaic (PV) and fuel cell (FC) generation sources need to be regulated and in many applications must be supported through additional energy storage unit to achieve high quality supply of power [1]-[3]. When such systems are used to power AC loads or be connected with the electricity grid, an inversion stage is also required.

The typical output of any FC is low and variable DC voltage with respect to the load current. For instance, based on the current-voltage characteristics of a Proton Exchange Membrane FC (PEMFC) power module, the voltage varies between 26 to 43 V_{DC} depending upon the level of the output current [5], [6]. Moreover, the slow response due to the natural electrochemical reactions required for the balance of enthalpy must be taken into account when designing the FC converter system [6], [7]. This is especially crucial, when the power drawn from the FC exceeds the maximum permissible power, as in this case, the FC module may not only fail to supply the demanded power to the load but also shut down or be damaged [6], [8]-[11]. For instance, when a load is added specific problems were founded such as fuel starvation

phenomenon and slow increased oxygen flow for about 0.8s [10]. Therefore, the power converter needs to ensure that the demanded power remains within the limit of the maximum availability [10], [11].

A two-stage fuel cell power conditioning system to deliver AC power has been commonly considered and studied in many technical papers [2], [3], [8]-[14]. The system usually includes transformer type DC-DC boost converter stage and DC to AC inverter stage with auxiliary energy unit in Fig. 1 [8], [13], [14]. This type of power conditioning system has inevitable drawbacks such as being bulky, costly and inefficient because it actually has three power conversion stages (DC to high-frequency AC, then DC and low frequency AC) [3].

In order to minimize the problems with a two-stage fuel cell power conditioning system, a topology with reduced power processing and conversion stages is required. A topology that is suitable for AC loads and is powered from DC sources able to boost and invert the voltage at the same time has been proposed in [15]-[17]. The double loop control scheme of this topology has also been proposed for better performance even during transient conditions [18]-[20].

However, if such topology was to be used for an AC load to be powered by an FC, the FC would be exposed to a number of problems such as load variation, slow respond and

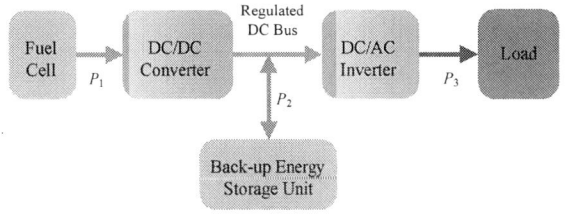

Figure 1. Block diagram for a conventional fuel cell energy system including dc-to-dc boost converter and dc-to-ac inverter (two main power stages) with energy storage as a back-up unit.

This work is supported by an Australian Postgraduate Award from the Australian Government and a Norman I Price Scholarship from the University of Sydney.

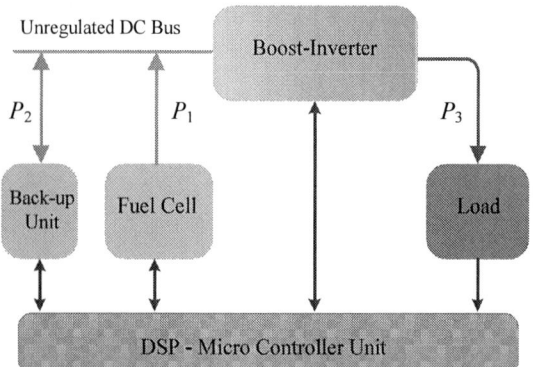

Figure 2. Block diagram for the proposed fuel cell energy system with one main power stage. The back-up unit and the FC power module are connected in the unregulated DC bus.

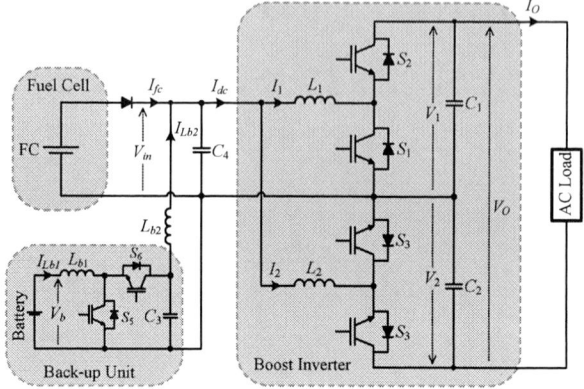

Figure 3. Proposed fuel cell energy system consisting of the boost-inverter and a back-up unit.

current ripple. In this case, an energy storage back-up unit would be required to address the previously mentioned problems [13].

The objective of this paper is to propose an FC energy system with the lowest possible energy conversion stages. In particular, the proposed system, based on the boost-inverter with a back-up energy storage unit, solves the previously mentioned problems, i.e., the low and variable output voltage of the FC and its slow response. The boost-inverter utilizes two identical bi-directional boost converters and delivers in a single-stage boosting and inversion functions. This results in a high power conversion efficiency, reduced converter size and low cost. Additionally, the back-up unit supplies the low frequency current harmonics hence minimizing the stresses on the FC should it was to supply such currents. The control of the boost-inverter is moderately complex to handle and sliding-mode control [15]-[17] or double-loop voltage and current control schemes [18]-[20] may be adopted in this system.

The paper is organized in the following way. In Section II, the proposed FC energy system is introduced including the converter topology and the control algorithm of the boost inverter as well as the back-up unit and its power converter design. In Section III, analysis and simulation results are presented to validate the performance of the system. We present experimental results from a fully digitally controlled laboratory prototype with a Nexa™ 1.2kW PEMFC power module in Section IV during load variations. Finally, we present our conclusions in Section V.

II. PROPOSED FUEL CELL ENERGY SYSTEM

A. Description of the FC energy system

Fig. 1 shows a popular FC energy system which includes two main power conversion stages between the FC and the load: a DC/DC boost converter and a DC/AC inverter. Due to stability considerations, high power applications and low input voltage have considered the two power stage approach. The proposed single main power stage should provide several advantages such as reduced number of components, high power conversion efficiency, and low cost if it is used in low power applications.

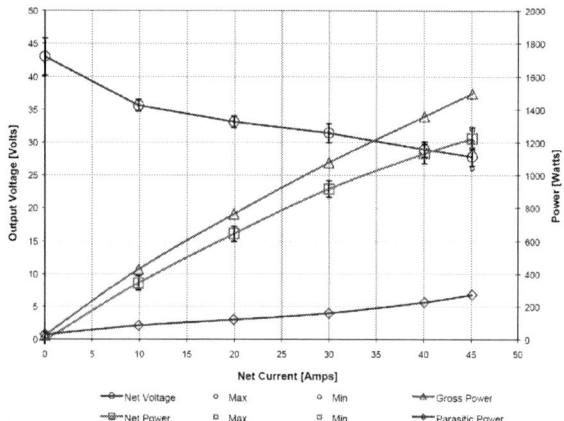

Figure 4. Illustration of the beginning of life (BOL) polarization characteristics of the Nexa™ 1.2kW PEMFC power module: voltage-current and power-current characteristics with parasitic power graph. The net output power ranges from zero being idle to 1200 W at rated power. The net output current ranges from zero to 46 A across the operating range. The output voltage varies with the operating load according to the polarization characteristics of the fuel cell stack. The normal idle voltages of the Nexa™ system are approximately 43 V_{DC}. At rated power, the Nexa™ system output voltage ranges from 26 V_{DC} to 29 V_{DC} at beginning-of-life.

The proposed FC energy system consists of two power converters: the boost-inverter and the back-up bi-directional unit as shown in Figs. 2 and 3 The output of the boost-inverter is connected to the load while the input side is supplied by the FC and the back-up unit, and both connected to the same unregulated dc bus. The back-up unit incorporates a current-mode controlled bi-directional boost converter with battery-based energy storage to support the FC power generation and two voltage-controlled boost converters making up the boost-inverter stage.

Fig. 4 is provided by the manufacturer of the PEMFC and illustrates the system parasitic load with net current and net output power for the Nexa™ power module. At rated power approximately 250 W of auxiliary load is required to power other components such as the air pump, the cooling fan, the onboard sensors, the actuators and the controllers [5].

outputs are described by

$$V_1 = V_{dc} + \frac{1}{2} \cdot A_1 \cdot \sin\theta \qquad (1)$$

$$V_2 = V_{dc} + \frac{1}{2} \cdot A_2 \cdot \sin(\theta - 180) \qquad (2)$$

$$V_O = V_1 - V_2 = A \cdot \sin\theta, \ when \ A = A_1 = A_2 \qquad (3)$$

where V_1 and V_2 are the output voltages of each boost converters and A is the peak amplitude of the boost-inverter output voltage, V_o. To increase the efficiency of the converter with variable input voltage the minimum dc-bias (V_{dc}) for the converters can be determined by

$$V_{dc} > \frac{A}{2} + V_{in} \qquad (4)$$

where V_{in} is the input voltage as variable of the FC output voltage.

From (3), it is observed that the required output is as desired, i.e. AC only. This concept and several control methods have been discussed in numerous technical papers [15]-[23].

Fig. 5 illustrates the relationship between the individual duty cycle and each output voltage reference signals. Based on the averaging concept for the boost converter, the voltage relationship for the continuous conduction mode (CCM) is given by

$$V_1 = \frac{V_{in}}{1 - d_1} \quad , \quad V_2 = \frac{V_{in}}{1 - d_2} \qquad (5)$$

$$\frac{V_1}{V_{in}} = \frac{1}{1 - d_1} \quad , \quad \frac{V_2}{V_{in}} = \frac{1}{1 - d_2} \qquad (6)$$

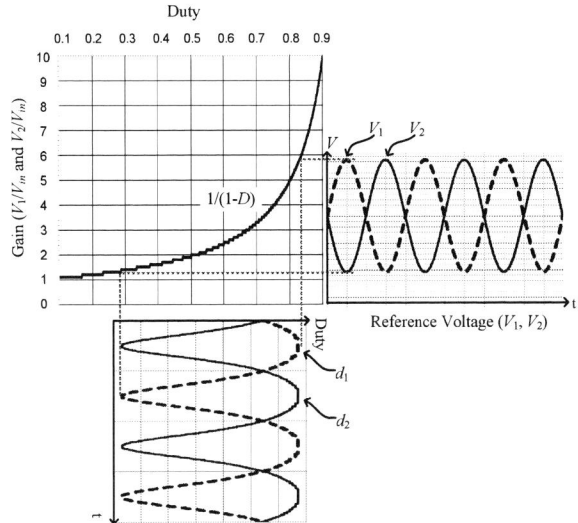

Figure 5. DC gain graph of the boost converter with reference voltages and duty cycles. It also illustrates the relationship between reference voltage and duty cycles to operate each boost converter.

The FC energy system must dynamically adjust to varying input voltage while maintaining constant power operation. Voltage and current limits need to be imposed at the input of the converter to protect the FC from damage due to excessive loading and transients. The power has to be ramped up and down so that the FC can react appropriately, avoiding transients and extending its life. The converter also has to meet the maximum ripple current requirements of the FC [6].

B. Boost-inverter

The boost-inverter consists of two bi-directional boost converters and their outputs are connected in series as shown in Fig. 3 Each boost converter generates a dc bias with deliberate ac output voltage (a dc-biased sinusoidal waveform as an output), so that each converter generates a unipolar voltage greater than the FC dc voltage with a variable duty cycle condition. Each converter output and the combined

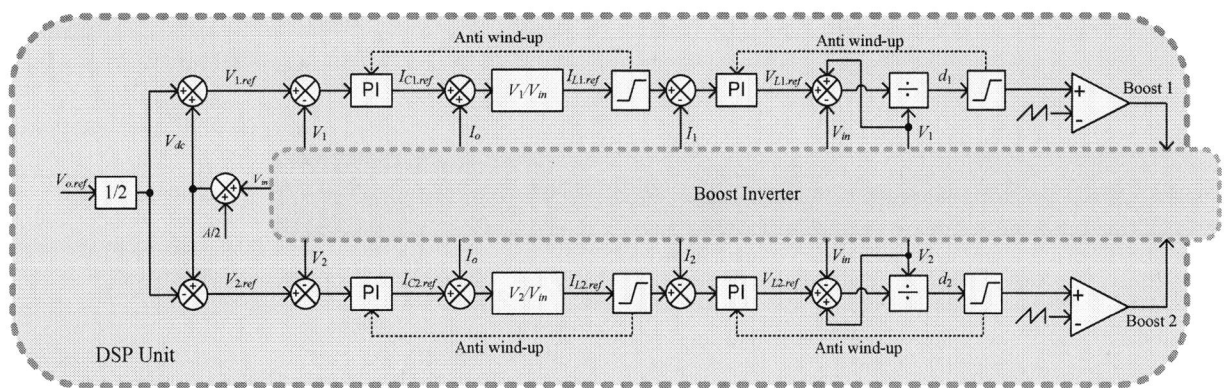

Figure 6. Block diagram for the boost inverter controller: two identical controllers for each of the two boost converters have different voltage reference including ac component and dc-bias. The control algorithm consists of inner current loops for inductor current control and outer voltage loops to regulate the output voltage.

978-1-4244-4782-4/10 $26.00 © 2010 IEEE

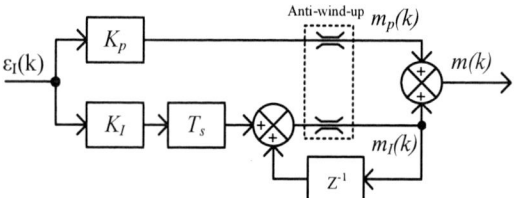

Figure 7. Block diagram representation of the digital PI controller with anti-wind-up technique

TABLE I. BACK-UP UNIT OPERATIONS

P_3 Increase ($P_1+P_2 \rightarrow P_3$)	P_3 Decrease ($P_1 \rightarrow P_2+P_3$)	Normal ($P_1=P_3$)
Discharge ↓ Charge ↓ Normal	Charge ↓ Normal	Normal

where d_1 and d_2 are the duty cycles of the boost converters respectively. Fig. 5 also illustrates that the zero output voltage of the boost-inverter is achieved at approximately $d_1 = d_2 = 0.72$. The reference signals are generated by the DSP unit itself as (1) and (2). After compensation, the duty cycles are calculated in every sampling time, T_s. The control algorithm for the two boost converters is presented in detail in Fig. 6.

In this paper, a double-loop control scheme is chosen for the boost-inverter control being the most appropriate to control the individual boost converters covering the wide range of operating points. This control method is based on the averaged continuous-time model of the boost topology and has several advantages with special conditions that may not be provided by the sliding mode control, such as nonlinear loads, abrupt load variations and transient short circuit situations. Using the control method the inverter maintains a stable operating condition by means of limiting the inductor current. Because of this ability to keep the system under control even in these situations, the inverter achieves a very reliable operation [18]-[20].

The boost-inverter (Fig. 3) is based on a voltage-mode control. The voltages of the C_1 and C_2 are controlled by two PI regulators and the currents of the L_1 and L_2 are controlled to achieve a stable operation in special conditions such as sudden load change and unexpected short circuit.

The control block diagram for the boost-inverter is shown in Fig. 6. The output voltage reference is divided to generate the two individual output voltage references of the two boost converters with the dc-bias (V_{dc}). The dc-bias can be obtained by adding input voltage (V_{in}) to the half of the peak output amplitude. V_{dc} is also used to minimize the output voltages of the converters and the switching losses in the variable input voltage condition. The PI controller on the right side of the diagram is for inner current control loop that should be designed to allow at least 50°-phase margin and a high bandwidth. The left side PI controller is for controlling the outer output voltage control loop that should be designed with the same phase margin and lower bandwidth compared with the inner loop. The anti-wind-up is used for saturation and to limit the inductor current. The control block diagram in Fig. 6 including the digital PI controller with the anti-wind-up (Fig. 7) has been implemented in the DSP unit.

The output voltage reference is determined by

$$V_{O.ref} = A \cdot \sin(2\pi f t), \quad f = 50 Hz, \quad A = 100\sqrt{2}V \quad (7)$$

where $V_{O.ref}$ is the reference voltage and f is the fundamental frequency for the boost-inverter.

Then $V_{1.ref}$ and $V_{2.ref}$ are calculated by (1) and (2). The digital PI controller in Fig. 7 has been implemented in the DSP unit and the equations are as follows:

$$m_I(k) = K_I \cdot T_S \cdot \varepsilon_I(k) + m_I(k-1) \quad (8)$$

$$m(k) = m_p(k) + m_I(k) = K_p \cdot \varepsilon_I(k) + m_I(k) \quad (9)$$

where $\varepsilon_I(k)$ represents the current error at instant kT_s and $m(k)$ is the output. K_P and K_I are the proportional and integral gains respectively. The PI controller is adapted by the anti-wind-up technique removing significant reduction of performance because of the well-known phenomenon of the integrator windup with saturation of the actuators [24], [25].

C. Back-up unit

The back-up unit is designed to support the slow response of the FC and is shown in Fig. 3. The back-up unit comprises of a current-mode controlled bi-directional boost converter and a battery as the energy storage medium. For instance, when a 1kW load is added from a no-load condition, the back-up unit immediately provides the 1kW power from the battery to the load as shown in Table I. On the other hand, when the load is disconnected suddenly, the surplus power from the FC could be recovered and stored into the battery to increase the overall efficiency of the energy system. Two generic 12 V lead-acid batteries are introduced in this unit for energy storage to deal with the need to provide fast response and a relatively low cost solution. The proposed back-up unit performs properly not only the support function for the FC module during transients but also is used as storage when any surplus power delivered by the FC is recovered.

In order to control the output current of the back-up unit, the inner current control loop of the boost-inverter is used. The reference of I_{Lb1} is taken from I_{dc} through a high-pass filtering and the demanded current I_{demand} relating the load change in Fig. 8. Detecting only the ac component from the dc input current I_{dc} for the current reference is used to eliminate ac ripple current into the FC power module while the dc component is used to determine the amount of demanded current. The elliptic 3rd order digital low pass filter (LPF) has been chosen and designed because of its narrow transition band and for being the most efficient of the IIR filters. The implementation has been integrated into the DSP.

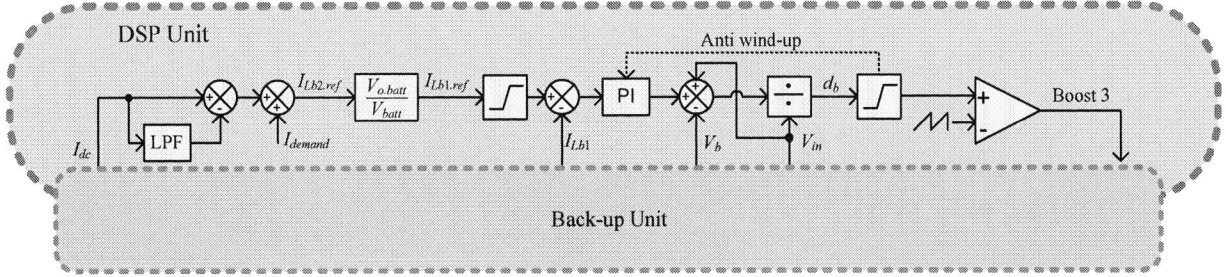

Figure 8. Block diagram for the controller of the back-up unit: the current control loop of the bi-directional boost converter is tracking the reference current from dc input current of the boost inverter, I_{dc}.

TABLE II.   DESIGN SPECIFICATIONS

FC output voltage	26 – 43 V_{DC}
AC output voltage	100 V_{AC} RMS, Single phase, 50 Hz
Switching frequency	20 kHz
Output power	1.2kW
V_{in}	26V (min)
R_a (resistance of L_1 and L_2)	$\approx 10m\Omega$
V_1(t)	175.4V (max)
V_2(t)	36V (min)
Δt_1 (maximum on time)	42.5µs (max at 20kHz)
Δi_{Lmax}	5% of $i_{L(max)}$
ΔV_c	5% of V_{1max}
R_1 (load)	8.3Ω (1.2kW)

TABLE III.   SPECIFICATIONS FOR THE SIMULATED SYSTEM

Output voltage	Step down from 41 to 30 V_{DC}
AC Output voltage	100 V_{AC} RMS, Single phase, 50 Hz
Switching frequency	20 kHz
Output power	Step change from 40W to 500W

D. Power component design

The power components of the proposed system were designed with the parameters given in Table II.

To calculate the inductance of L_1 and L_2 the following equations are used [15]

$$i_L(\max) = \frac{V_{in} - \sqrt{V_{in}^2 - 4R_a\left(-V_1(t)\right)\cdot\left(\dfrac{V_2(t) - V_1(t)}{R_1}\right)}}{2R_a} \quad (10)$$

$$\Delta i_L(t) = \frac{\left(V_{in} - R_a i_L(t)\right)\cdot\Delta t_1}{L} \quad (11)$$

where $i_L(\max)$ is maximum current and Δi_L is high-frequency ripple current of the inductors for the boost-inverter. Equations (10) and (11) illustrate the alternating with operation frequency and a high-frequency ripple caused by switching respectively. The maximum inductor current ripple Δi_{Lmax} is chosen to be equal to 5% of the maximum inductor

current, as calculated from (10) when the V_1 is maximum and V_2 is minimum. From (10) and (11) the minimum inductance is calculated as 177µH and 200µH which are the chosen values for L_1 and L_2. The ripple voltage of the C_1 and C_2 is given by [15]

$$\Delta V_c = \left(\frac{V_1(t) - V_2(t)}{C\cdot R_1}\right)\cdot\Delta t_1 \quad (12)$$

30µF is obtained from (12) and two 30µF 800V rated metalized polypropylene film capacitors are used for C_1 and C_2.

During transient conditions, the back-up unit should provide all the power required by the load. In this case, the maximum inductor current of the boost-inverter should appear in the inductor L_{b2}. Therefore, the maximum inductance of L_{b2} can be calculated by (11) and $\Delta i_L(t)max$ need to be larger than the maximum inductor current of the boost-inverter in order to track the maximum slope of the current. The maximum inductance is obtained from (11) as 51µH and the values of L_{b1} and L_{b2} are chosen to be 20µH. The values of the capacitors C_3 and C_4 of the back-up unit are chosen to be 30µF.

III.   ANALYSIS AND SIMULATION RESULTS

The proposed FC energy system (Fig. 2), has been designed, simulated, experimentally build as a laboratory prototype and tested to validate its overall performance.

The simulations have been performed with the PSIM software to validate the analytical results [26]. The simulation results show the operation of the boost-inverter and the back-

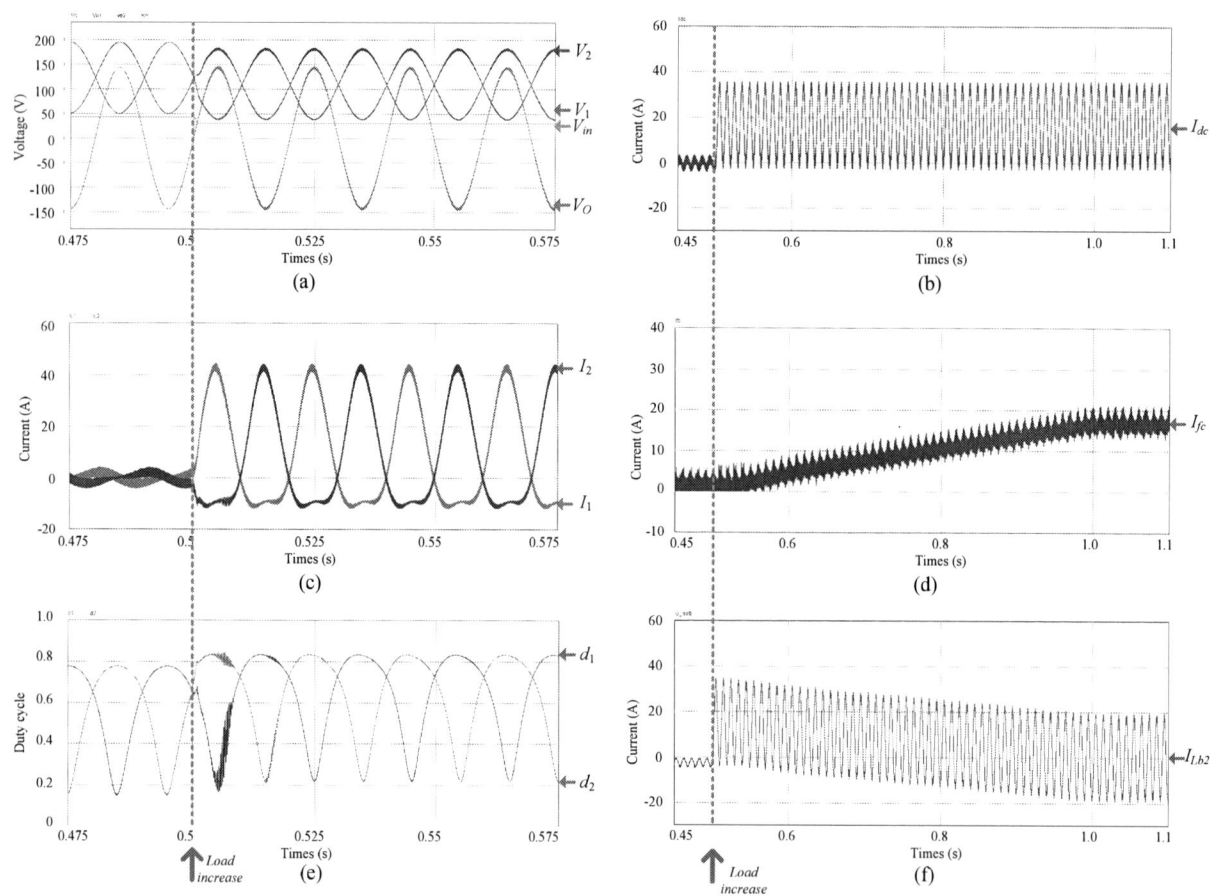

Figure 9. Simulation results for the boost-inverter and the back-up unit. (a) Output voltages of the boost-inverter. (b) Boost inverter input current, I_{dc}. (c) Current waveforms of L_1 and L_2. (d) Fuel cell output current, I_{fc}. (e) Duty cycles of the boost-inverter. (f) Output current of the back-up unit, I_{Lb2}.

up unit. In particular, Fig. 9(a) illustrates the output voltages of each of the two boost converters (V_1 and V_2), the input voltage V_{in} same as the FC output voltage and the final output voltage of the boost-inverter V_O. The input currents of each boost converter flowing through the inductors L_1 and L_2 are shown in Fig. 9(c). Fig. 9(e) shows the duty cycles (d_1 and d_2) of each boost converter that are varying between approximately 0.15 and 0.85. Figs. 9(b), (d) and (f) illustrate the waveforms of the dc total output current I_{dc} (which is equal to the inverter input current), the FC output current I_{fc}, and the output current I_{Lb2} of the backup unit respectively. Fig. 9(d) and (f) also illustrate how the back-up unit supports the FC power in transients when the load is increased. When full-load is required from the no-load operating point, the entire power is provided by the back-up unit to the load as shown in Fig. 9(f). Then, the power drawn from the battery starts decreasing moderately allowing gentle step-up to deliver power which should increase up to meet the demanded load power. Moreover, the back-up unit protects the FC from potential damage by eliminating the ripple current due to the boost operation. If there was no back-up unit, the FC output current waveform should be the same as Fig. 9(b). The high frequency output ripple current of the FC

can be canceled by a passive filter placed between the FC and the boost-inverter.

TABLE IV. SPECIFICATIONS FOR THE PROTOTYPE

Switching frequency	20 kHz
Rated power	1.2kW
Power stack	General purpose SEMISTACK-IGBT
Controller	eZdsp TMS320F28335
Voltage transducers	LEM LV25-P
Current transducers	LEM HAL50s
Energy storage	Two 12V lead acid batteries
$L_1=L_2$	200µH
$L_{b1}=L_{b2}$	20µH
$C_1=C_2=C_b$	30µF

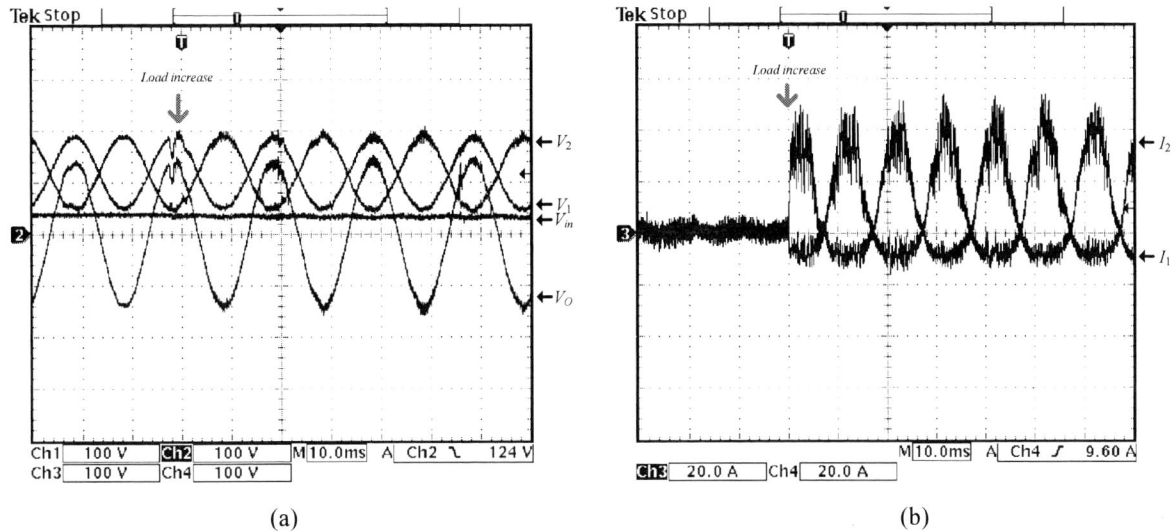

(a)   (b)

Figure 10. Experimental results ($P=0$ to 0.5kW) (a) Each output voltage of boost converters (V_1 and V_2), output voltage of the boost inverter (V_O) and dc input voltage, V_{in} (b) Current waveforms of L_1 and L_2.

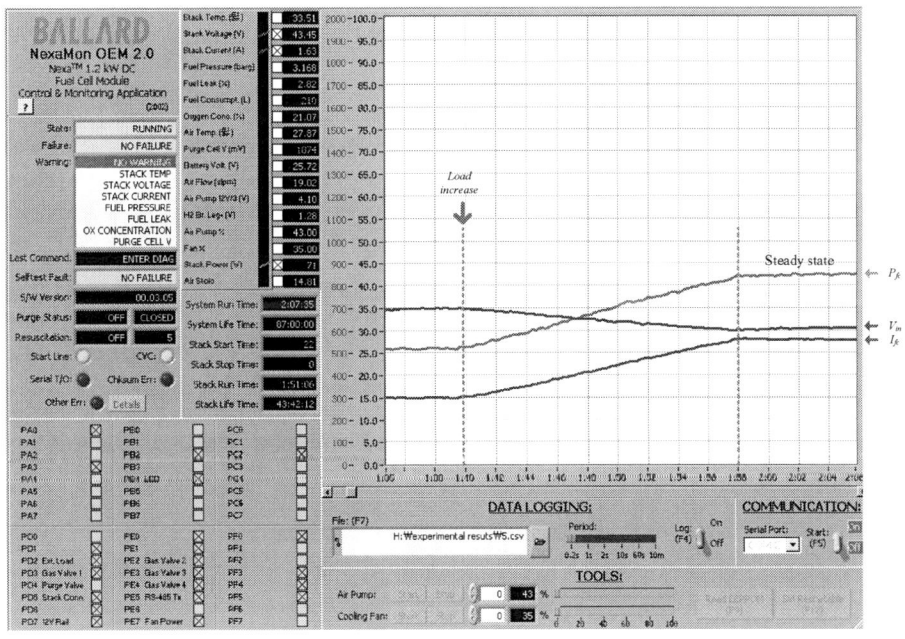

Figure 11. Nexa^{TM} power module monitoring for the fuel cell operation

IV. EXPERIMENTAL RESULTS

The parameters of the prototype for the boost-inverter and the back-up unit are summarized in Table IV. The power stack consists of three IGBT modules that are used to build the boost-inverter for two modules and back-up unit for one module. The controller unit has been selected for a number of reasons such as low cost, embedded floating point unit, high speed, on chip analog to digital converter and high performance PWM unit.

Experimental results presented in Fig. 10 show good performance of the boost-inverter operation with the load changing between no-load and 0.5kW-resistive. Specifically, Fig. 10(a) illustrates the input and output voltages of the boost-inverter. The input voltage decreases after a load increase in some period and the dc component of each boost converter is decreased as predicted by (4). The current waveforms of the two different inductors are shown in Fig. 10(b).

978-1-4244-4782-4/10 $26.00 © 2010 IEEE   301

Fig. 11 is a snapshot generated by the Nexa™ power module monitoring software during the load changing. As shown in the screen, the FC power changed from approximately 500W to 850W in 18 seconds which is a moderate amount of time. This figure also illustrates how the back-up unit supports the FC, since the FC is afforded a slow power up operation.

V.   CONCLUSION

A minimum stage power-processing FC energy system based on the boost-inverter topology with a back-up battery-based energy storage unit is proposed in this paper. The simulated and the laboratory test results presented in the paper have verified the operation characteristics of the proposed energy system. In summary, the proposed FC energy system has a number of attractive features, such as single main power stage with high efficiency, simplified topology, low cost and stand alone operation.

REFERENCES

[1] S.B. Kjaer, J.K. Pedersen, F. Blaabjerg, "A review of single-phase grid-connected inverters for photovoltaic modules," *IEEE Trans. on Industry Applications*, vol. 41 no. 5, pp. 1292 – 1306, Sept.-Oct. 2005.

[2] J.-S. Lai, "Power conditioning circuit topologies" *IEEE Industrial Electronics Magazine*, Vol. 3, 2, June 2009, pp. 24-34.

[3] M.E. Schenck, J.-S. Lai and K. Stanton, "Fuel cell and power conditioning system interactions," *Proc. of the APEC 2005*. vol. 1, pp. 114 – 120, 6-10 March 2005.

[4] S. K. Mazumder, R. K. Burra, and K. Acharya, "A ripple-mitigating and energy-efficient fuel cell power-conditioning system," *IEEE Trans. Power Electron.*, vol. 22, no. 4, pp.1437-1452, July 2007.

[5] Nexa™ Power module user guide, MAN5100078, 2003, Ballard Power System, Inc.

[6] J. Anzicek and M. Thompson, "DC-DC boost converter design for Kettering University's GEM fuel cell vehicle," in *Electrical Insulation Conference and Electrical Manufacturing Expo, 2005. Proceedings*, Oct. 26-26 2005, pp. 307 – 316

[7] J.M. Correa, F.A. Farret, J.R. Gomes, and M. G. Simoes, "Simulation of fuel cell stacks using a computer-controlled power rectifier with the purposes of actual high power injection applications," *IEEE Trans. on Industry Applications*, vol. 39 no. 4, pp. 1136-1142, July/Aug. 2003.

[8] J. Lee, J. Jo, S. Choi, and S. Han, "A 10-kW SOFC low-voltage battery hybrid power conditioning system for residential Use," *IEEE Trans. Power Electron.*, vol. 21, no. 2, pp. 575-585, June 2006.

[9] P.Thounthong, S. Rael, and B. Davat, "Utilizing fuel cell and supercapacitors for automotive hybrid electrical system," *Proc. of the 2005 IEEE Applied Power Electronics Conference and Exposition (APEC05)*, Texas, 6-10 Mar. 2005, pp. 90-96.

[10] P. Sethakul, S. Rael, B. Davat, and P. Thounthong, "Fuel cell high-power applications," *IEEE Industrial Electronics Magazine*, vol. 3, no. 1, pp. 32-46, March 2009.

[11] M. W. Ellis, M. R. Von Spakovsky, and D. J. Nelson, "Fuel cell systems: efficient, flexible energy conversion for the 21st century," *Proc. of the IEEE*, vol. 39, no. 12, pp. 1808-1818, Dec. 2001.

[12] C. Pan, C. Lai, "A high efficiency high step-up Converter with low switch voltage stress for fuel cell system applications," *IEEE Trans. Industrial Electron.*, Accepted for future publication, 2009.

[13] X. Yu, M. R. Starke, L. M. Tolbert, and B. Ozpineci; "Fuel cell power conditioning for electric power applications: a summary," *IET-Electric Power Appl.*, vol. 1, no. 5, pp. 643-656, Sept. 2007.

[14] A. Vazquez-Blanco, C. Aguilar-Castillo, F. Canales-Abarca and J. Arau-Roffiel, "Two-stage and integrated fuel cell power conditioner: Performance comparison," *Proc. of IEEE APEC 2009*. vol 15-19, pp. 452–458, Feb. 2009

[15] R. O. Cáceres and I. Barbi, "A boost dc-ac converter: Analysis, design, and experimentation," *IEEE Trans. Power Electron.*, vol. 14, no. 1, pp.134-141, Jan. 1999.

[16] R. O. Cáceres and I. Barbi, "A boost dc-ac converter: analysis, design and experimentation," in *Proc. IEEE IECON'95 Conf.*, Orlando, FL, Nov. 5-11, 1995, pp. 546-551.

[17] R. O. Cáceres and I. Barbi, "Sliding mode controller for the boost inverter," in *Proc. IEEE CIEP'96*, Cuernavaca, Mexico, Oct. 14-17, 1996, pp. 247-252.

[18] P. Sanchis, A. Ursea, E. Gubía, and L. Marroyo, "Boost dc-ac inverter: A new control strategy," *IEEE Trans. Power Electron.*, vol. 20, no. 2, pp.343-353, Mar. 2005.

[19] P. Sanchis, O. Alonso, and L. Marroyo, "Variable operating point robust control strategy for boost converters," in *Proc. 9th Eur. Conf. Power Electronics Applications (EPE'01)*, Graz, Austria, Aug. 27-29, 2001.

[20] P. Sanchis, O. Alonso, L. Marroyo, T. Meynard, and E. Lefeuvre, "A new control strategy for the boost dc-ac inverter," in *Proc. IEEE PESC'01 Conf.*, Vancouver, Canada, Jun. 17-21, 2001, pp. 974-979.

[21] T. Liang, J. Shyu, and J. Chen, "A novel DC/AC boost inverter," in *37th International Conference on Electrical and Computer Engineering 2002 (IECEC'02)*, July 29-31, 2002, pp. 629-634.

[22] A. A. S. Jhan and K. M. Rahman, "Voltage mode control of single phase boost inverter," in *5th ICECE'08*, Dec. 20-22, 2008, pp. 665-670.

[23] B. Kalaivani, V. K. Chinnaiyan, and J. Jerome, "A novel control strategy for the boost dc-ac inverter," in *Proc. India International Conference on Power Electronics 2006 (IICPE'06)*, India, Dec. 19-21, 2006, pp. 341-344.

[24] A. Scottedward Hodel, C.E. Hall, "Variable-structure PID control to prevent integrator windup," *IEEE Trans. Industrial Electron.*, vol. 48, no. 2, pp. 442–451, April 2001.

[25] A. Visioli, "Modified anti-windup scheme for PID controllers," *Proc. of the 2003 IEE Control Theory and Applications*, vol. 150, no. 1, pp. 49-54, Jan. 2003.

[26] PSIM Software, POWERSIM INC.

Power conditioning System for Fuel Cell with 2-Stage DC-DC Converter

Byung M. Han, Jun-Young Lee, and Yu-Seok Jeong
Dept. of Electrical Engineering
Myongji University
Gyeonggi-do, South Korea
erichan@mju.ac.kr, pdpljy@mju.ac.kr, jeong@mju.ac.kr

Abstract— **This paper proposes a grid-tied power conditioning system for the fuel cell, which consists of an LLC resonant DC-DC converter and a 3-phase inverter. The LLC resonant converter boosts the fuel cell voltage of 26-48V up to 400V, using the hard-switching boost converter and the high-frequency unregulated LLC resonant converter. The operation of proposed power conditioning system was verified through simulations with PSCAD/EMTDC software. The feasibility of hardware implementation was verified through experimental works with a laboratory prototype, which was built with 1.2kW PEM fuel-cell stack, 1kW high gain step-up converter, and 2kW PWM inverter. The proposed system can be utilized to commercialize a real interconnection system for the fuel-cell power generation.**

I. INTRODUCTION

Fuel cell is a clean energy source to generate the electricity like a solar cell. Many kinds of fuel cell were developed for supplying the electricity to the car or the home. PEM(proton exchange membrane) fuel cell, which has simple structure and high power density, is considered as a DC power source for the distributed power generation and for the passenger car.[1]

Fuel cell has non-linear characteristic in electrical operation due to the polarization phenomena of electro-chemical reaction. The terminal voltage at the rated load drops to the half value of the terminal voltage at no load. So, a DC-DC converter with high efficiency and high amplification is definitely required to boost the low terminal voltage up to the high DC link voltage.[2],[3]

Full-bridge converter, push-pull converter, or boost converter has been widely used as a DC-DC converter for the fuel cell. Full-bridge converter has a disadvantage of high switching loss due to large number of switching units. Push-pull converter has lower switching loss due to lower number of switching units, but it has 88% efficiency because it requires double winding structure in the primary side. Boost converter has lower switching loss due to small number of switching units, but it has lower voltage boosting ratio of 3 to 4 times. Recently, multi-stage boost converter without

transformer was developed to obtain high voltage boosting ratio. But its efficiency is located between 86~90%. Isolated boost converter was also developed to obtain high voltage boosting ratio, of which the efficiency is located between 86~90%. But it requires large number of switching units and transformers. In order to increase the efficiency and to reduce the number of components, various type converters were proposed and are being proposed by many researchers [4],[5]

This paper proposes a new structure of DC-DC converter for fuel cell application, which is composed of a hard-switching boost converter cascaded with an unregulated LLC resonant converter. The proposed DC-DC converter was coupled with the general inverter to configure the power conditioning system for fuel cell, which can be interconnected with the power grid. The operational feasibility of proposed power conditioning system for fuel cell was confirmed by computer simulations with PSCAD/EMTDC and experimental results with a prototype.

II. PROPOSED SYSTEM

Fig. 1 shows the configuration of proposed power conditioning system including the whole system controller. Power conditioning system for fuel cell requires a high amplification, high efficiency DC-DC converter because it has a severe voltage variation between at no-load and full load condition. In order to satisfy this condition this paper proposes a new DC-DC converter composed of boost converter and 2-stage LLC half-bridge with high-frequency transformer. The proposed converter has high efficiency because it operates in soft switching mode with resonance. It has a simple control structure to regulate the output voltage of 400V by controlling the duty ratio of boost converter.

Whole system controller is divided into the control part for DC-DC converter and the control part for inverter. The control part for the DC-DC converter is to maintain the DC output voltage constant, while the control part for the inverter is to control the active power P and the reactive power Q.

This work was financially supported by the advanced human resource development program of MKE through the Research Center in Myongji University.

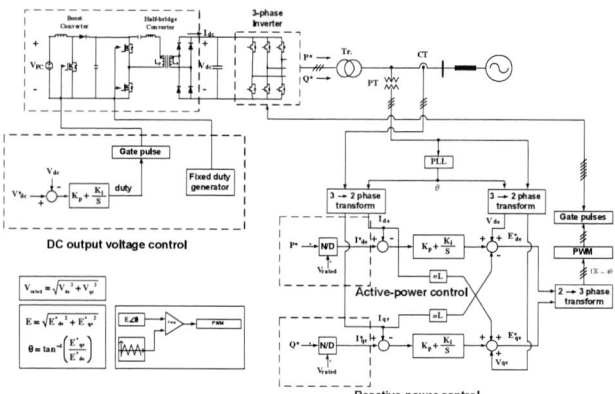

Figure 1. Configuration of Proposed System

The control part for DC-DC converter compares the measured DC capacitor voltage with the reference value and generates an error signal. The error signal is sent to the PI control to determine the duty ratio. The boost converter maintains the output voltage to 80V, and the high-frequency converter increases the 80V up to the 400V. The control part for the inverter measures the 3-phase voltage and current and performs the d-q transform with the phase-locked angle. It generates the reference values of d-axis current and q-axis current dividing the reference values of active power P and the reactive power Q by the rated voltage. These d and q reference currents are sent to the current control for generating the d and q reference voltage of inverter. These d and q reference voltages are used for generating the PWM pulses by inverse d-q transform.

III. FUEL-CELL MODELING

In PEM fuel cell the hydrogen gas is supplied to the anode through a platinum catalyst to be ionized into a hydrogen proton and an electron. The hydrogen proton moves to the cathode through the solid polymer membrane and is combined with the oxygen supplied to the cathode. Through this electron-chemical reaction, fuel cell generates electricity and heat, and water as a by-product.

The ideal fuel cell voltage is same as the equilibrium voltage represented by Nernst model which is based on Gip's free energy. The actual fuel cell voltage is represented by the reduction characteristic of equilibrium voltage due to polarization phenomenon.

The unit cell voltage of fuel cell is represented by subtracting three polarization losses from the equilibrium voltage with respect to the exchange current density. The unit cell voltage shows non-linear characteristic and is expressed by equation (1).

$$E_{cell} = E_{rev} - E_{act} - E_{con} - E_{ohm} \qquad (1)$$

Where, E_{rev} is the equilibrium voltage, E_{act} is the activated polarization loss, E_{con} is the concentration polarization loss, and E_{ohm} is the ohmic polarization loss.

The activated polarization loss, which is due to the difference of reaction speed on the electrode, is represented by equation (2).

Where, i_o is the exchange current density related to the normal and reverse reactions between the electrolyte and the electrode.

The exchange current is dependent on the pressure, the catalyst, the activated energy, and the temperature. If this value is reduced, the activated polarization E_{act} is reduced and the output voltage of fuel cell E_{cell} is increased. So, the exchange current is very important parameter in fuel cell.

The concentration polarization loss E_{con} is due to the gradient difference of reaction material concentration, which is represented by equation (4). The ohmic polarization loss E_{ohm}, which is composed of electrolyte resistance, electrode resistance, and the lead wire resistance, is represented by equation (5).

$$E_{act} = \frac{RT}{\alpha F} \ln\left(\frac{i + i_{loss}}{i_o}\right) \qquad (2)$$

$$i_o = i_o^{ref} a_c L_c \left(\frac{P_r}{P_r^{ref}}\right) \exp\left[-\frac{E_c}{RT}\left(1 - \frac{T}{T_{ref}}\right)\right] \qquad (3)$$

$$E_{con} = \frac{RT}{nF} \ln\left(\frac{i_L}{i_L - i}\right) \qquad (4)$$

$$E_{ohm} = iR_i \qquad (5)$$

So, the output voltage characteristic of fuel cell is represented by the above five equations. If all the parameters described in these equations are known for a specific fuel cell, its output voltage with respect to output current is easily analyzed.

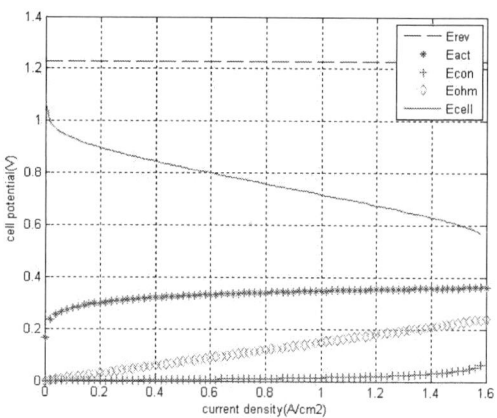

Figure 2. Output voltage characteristic of Fuel-cell

Fig. 2 shows output characteristic curves for unit cell of a typical fuel cell. The actual cell output E_{cell} is reduced with non-linear manner by three polarization components from the equilibrium voltage.

978-1-4244-4782-4/10 $26.00 © 2010 IEEE

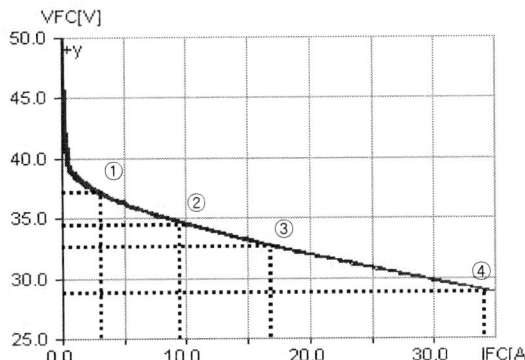

Figure 3. Output characteristics modeling of fuel cell stack

In the actual fuel cell, many unit cells are connected in series as a stack structure to build up the terminal voltage. So, the voltage and current characteristics of stack is determined multiplying the unit cell voltage by the number of cells and multiplying the unit cell current density by the area of cell.

Fig. 3 shows the output voltage characteristic of fuel cell stack that is composed of 47 cells. The y-axis shows the actual output voltage, while the x-axis shows the fuel cell current. This curve is required to design the DC-DC converter for fuel cell power conditioning system.

IV. 2-STAGE DC-DC CONVERTER

Fig. 4 shows the configuration of 2-stage DC-DC converter, which is composed of a hard switching boost converter and half-bridge LLC resonant converter. The controller is composed of an output voltage sensor, boost converter control, and fixed-duty gate pulse generator. The proposed 2-stage DC-DC converter offers several benefits to reduce the complexity of controller design and to decrease the resonant converter transformer by high frequency operation over 100kHz. Also By separating functions, each converter can be designed at an optimal operating point, which helps to increase the overall efficiency.

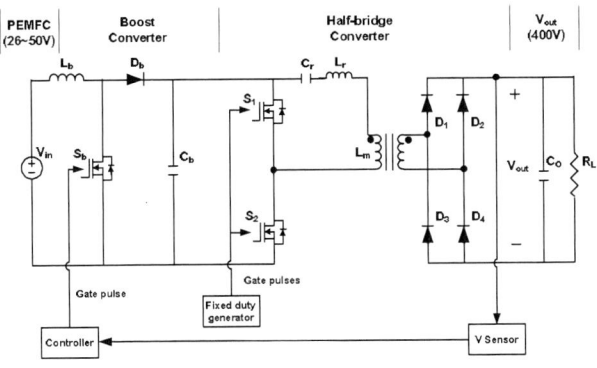

Figure 4. Circuit Diagram of 2-stage DC-DC Converter

The input voltage is determined to be 24~48V by considering the output voltage characteristic of fuel cell. This voltage is boosted up to 80V by controlling the duty ratio of boost converter. The half-bridge LLC resonant converter boosts the input voltage of 80V up to 400V by operating a fixed duty ratio.

The control part of boost converter compares the reference voltage 400V with the measured converter output voltage, and the error signal is passed through the PI control to generate the gate pulse for the boost converter.

When the switch S_b of boost converter is ON, current flows through the reactor L_b and energy is charged in the reactor. When the switch S_b of boost converter is OFF, this energy is discharged into the capacitor C_b through the diode D_b.

Fig. 5 shows operation modes of the LLC resonant converter, which is composed of two MOSFET switches, resonant capacitor C_r, leakage inductance of transformer L_r and magnetizing inductance L_m. The secondary side of transformer is connected to the full-bridge diode rectifier.

Fig. 6 shows the voltage and current waveforms at each component of the LLC resonant converter. The converter operation can be divided into the four modes according to the time interval.

Operation Mode 1 (t_0-t_1)

Fig. 5(a) shows the operation mode in the powering section period, which starts at the instant when the MOSFET S_2 turns on. The resonant current flows through MOSFET S_2 and the energy is transferred to the secondary side of transformer. The resonant capacitor C_r is charged and the resonant frequency f_r is determined by equation (6), because the magnetizing inductance L_m does not involve in resonant.

$$f_r = \frac{1}{2\pi\sqrt{LC}} \tag{6}$$

In the secondary side, diode D_1 and D_4 are on conduction and the current through the magnetizing inductance L_m is linearly increased.

Operation Mode 2 (t_1-t_2)

Fig. 5(b) shows the operation mode in the dead time section period, which starts at the instant when the switch S_2 turns off. The current that was flown through the switch S_2 flows through the diode inside the switch S_1. This allows the zero-voltage switching condition at switch S_1. At this section the magnetizing current does not increase any more, and the energy transfer to the secondary side of transformer is cut off.

Operation Mode 3 (t_2-t_3)

Fig. 5(c) shows the operation mode in the powering section period, which starts at the instant when the switch S_1 turns on. The energy that is charged in C_r transferred to the secondary side of transformer, in which the resonant frequency is determined by equation (6). In the secondary side, diode D_2 and D_3 are on conduction, and the current through the magnetizing inductance L_m is linearly decreased.

Operation Mode 4 (t_3-t_4)

Fig. 5(d) shows the operation mode in the dead time section period, which starts at the instant when the switch S_1

turns off. The current that was flown through the switch S_1 flows through the diode inside the switch S_2. This allows the zero-voltage switching condition at switch S_2. At this section the magnetizing current does not increase any more, and the energy transfer to the secondary side of transformer is cut off.

If the switching frequency of LLC converter f_{sL} is selected as the same value of the resonant frequency f_r, calculated as $1/(2\pi\sqrt{L_r C_r})$, to reduce the switching loss, it can be simplified by

$$\frac{V_o}{V_{in}} = \frac{V_b}{V_{in}} \cdot \frac{V_o}{V_b} = \frac{1}{(1-D)} \cdot \frac{n}{2} \qquad (7)$$

where D means the duty-ratio of the boost converter.

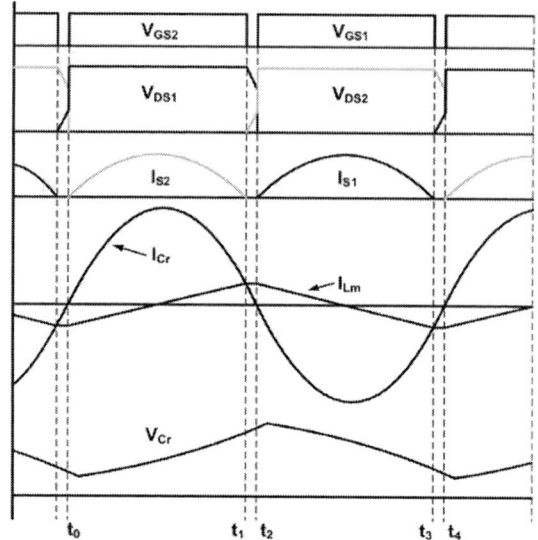

Figure 6.   Operation analysis of LLC Resonant Converter

V.   COMPUTER SIMULATION

In order to analyze the operation of proposed fuel-cell power conditioning system, Fuel-cell modeling was first carried out. The unit-cell electrical characteristic of fuel-cell is represented by mathematical model described in [1]. The fuel cell current is determined multiplying the cell current density by the cell area. The fuel-cell stack voltage is determined multiplying the cell voltage by the number of stack.

Computer simulations were carried out with PSCAD software to verify the operation of proposed power conditioning system. The simulation model consists of a fuel-cell model, 2-stage DC-DC converter, grid-tied inverter, and digital controller. The fuel-cell model and the digital controller were represented by user-defined models programmed with C-codes, while the 2-stage DC-DC converter, grid-tied inverter, and the 3-phase voltage source were represented by built-in models in PSCAD/EMTDC software.

In order to confirm the voltage and current variations of fuel cell according to the variation of active power, a simulation scenario was selected as shown in Table 1. The active power varies with step manner from 100W up to 1kW, and then down to 100W, while the reactive power varies from 0 up to300Var, and then down to 0Var during 0~8 sec.

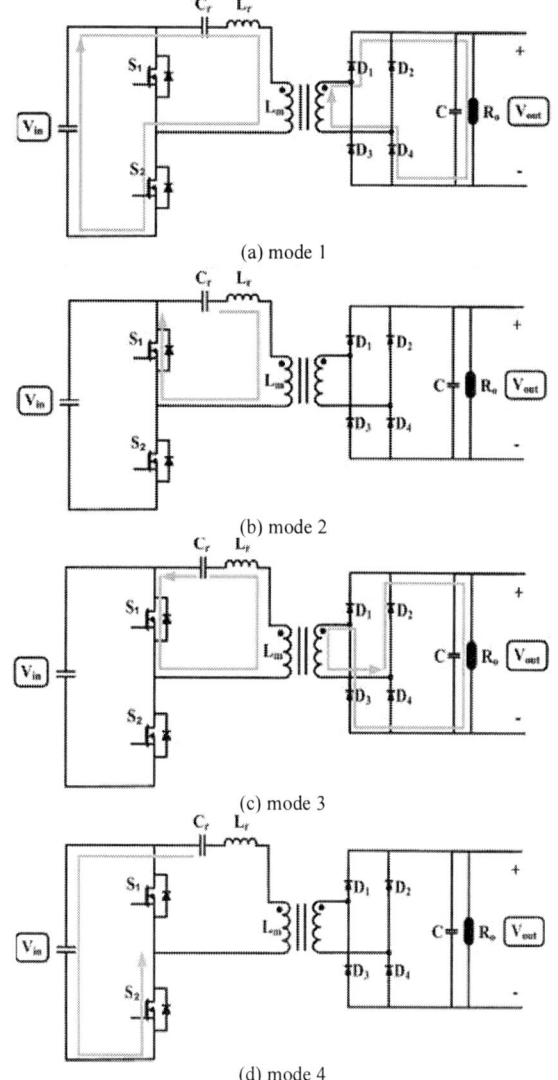

(a) mode 1

(b) mode 2

(c) mode 3

(d) mode 4

Figure 5.   Operation of LLC Resonant Converter

These four operation modes are sequentially repeated according to the switching frequency of 100kHz. Although the switching frequency is rather high, the switching loss is quite small because of the zero-voltage switching scheme.

TABLE I.   OPERATION SCENARIO FOR COMPUTER SIMULATION

Time [s]	1 ①	2 ②	3 ③	4 ④	5	6	7	8
P [W]	100	300	500	1000	1000	500	300	100
Q [Var]	0	300	100	0	0	300	100	0

Fig. 7 shows the simulation results to verify the operation of proposed system. The operation point at each output power was shown in Fig. 3. Fig. 7(a) shows the operation voltage of fuel cell, which changes according to the variation of active power delivered to the grid.

Fig. 7(b) shows the operation current of fuel cell, which changes according to the variation of active power delivered to the grid similar to the case of voltage variation.

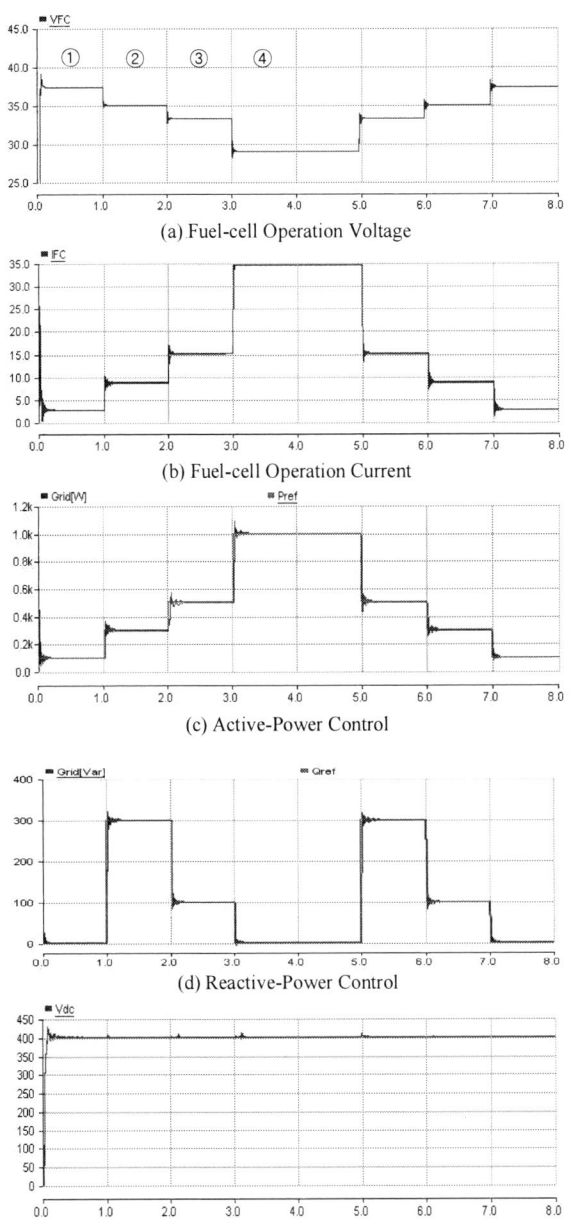

(a) Fuel-cell Operation Voltage

(b) Fuel-cell Operation Current

(c) Active-Power Control

(d) Reactive-Power Control

(e) DC Output Voltage Control

Figure 7. Simulation results of proposed system

Fig. 7(c) shows the tracking performance of active power. It is clear that the transient phenomena can be stabilized within 1sec and the steady-state tracking performance seems to be accurate. Fig. 7(d) shows the tracking performance of reactive power. It is clear that the measured value of reactive power tracks the reference value accurately and the transient phenomena is not so severe. Fig. 7(e) shows the tracking performance of DC output voltage. The DC output voltage is maintained at 400V without severe transients.

VI. EXPERIMENTAL WORK

Based on the simulation results, a prototype of proposed system was built and tested to confirm the feasibility of hardware implementation. The fuel-cell power unit used in the experiment is the 1.2kW Ballard Nexa PEM Module. The controller for the 2-stage DC-DC converter which was designed and built with OP amps, adjusts the duty ratio of boost converter to maintain the output voltage. The additional voltage amplification is carried out by the fixed duty ratio of LLC resonant converter and the winding ratio of high-frequency transformer. The boost converter stage was designed with 30kHz of switching frequency and the LLC stage was done with 100kHz of switching frequency.

The controller for the grid-tied inverter was designed and built with a floating-point DSP(Digital Signal Processor), TMS320vc33-150 by TI and EPLD(Erasable Programmable Logic Device), EP1K100QC208 by Altera. The control board has 24ch of ADC, 4ch of DAC, 4ch of Digital Input, 4ch of Digital Output, 1module of Encoder pulse input, 1port of RS232, and 2port of RS485.

The actual fuel-cell stack can not track the fast variation of active power in the grid because the chemical reaction in the fuel cell stack is much slower relatively. In order to confirm a safe and reliable operation, the duration of active power variation was determined by 100sec in the experimental work.

TABLE II.   OPERATION SCENARIO FOR HARDWARE EXPERIMENT

Time [s]	100	200	300	400	500	600	700	800
P [W]	100	300	500	1000	1000	500	300	100
Q [Var]	0	300	100	0	0	300	100	0

Fig. 8 shows the experimental results to verify the operation of proposed system. Fig. 8(a) shows the operation voltage and current of fuel cell module with 10V(A)/div. The time div was selected by 100s same as the duration period of active or reactive power. As the active power increases, the fuel cell voltage decreases while the fuel cell current increase. Fig. 8(b) shows that the measured active power tracks the reference value accurately without severe transient. Fig. 8(c) shows the output voltage and current variation of fuel cell. The output current has same variations as the active power while the capacitor voltage is maintained at 400V. This confirms that the DC-DC converter can accurately control the output voltage without the variation of active power. Fig. 8(d) shows that the measured reactive power tracks the reference value accurately without severe transient.

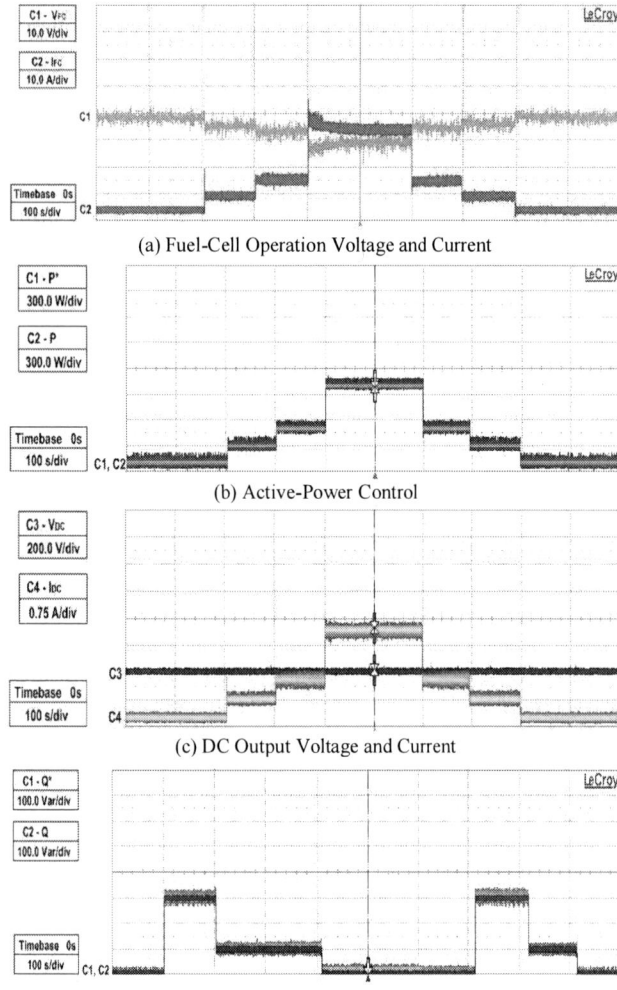

(a) Fuel-Cell Operation Voltage and Current

(b) Active-Power Control

(c) DC Output Voltage and Current

(d) Reactive-Power Control

Figure 8. Experimental Results of Hardware Prototype

Fig. 9 shows the voltage-current characteristic curve of the fuel cell module used in the experiment. The no-load voltage of fuel cell is about 50V and the terminal voltage goes down while the output current increases, as the active power increases. At the active power of 1kW, the fuel cell voltage is 27.3V, and the fuel cell current is 34.41A.

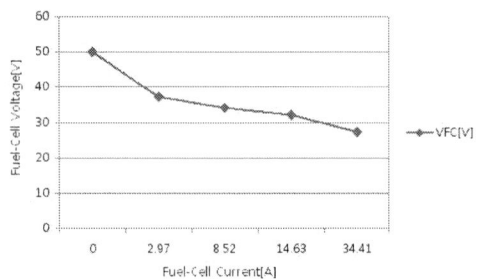

Figure 9. Measured Fuel Cell Voltage and Current

Fig. 10 shows the efficiency of 2-stage DC-DC converter which was measured through experimental works. The

efficiency was measured with respect to the lowest input voltage 24V and the highest input voltage 48V. The efficiency of proposed converter is 93.5% when the input voltage is 24V and the output power is 1kW. Since the actual fuel cell voltage is 27.4V at 1kW output, it can be confirmed that the efficiency of DC-DC converter is approximately 94%. Assuming the efficiency of grid-tied inverter is 97%, the total efficiency of proposed system could be about 91%.

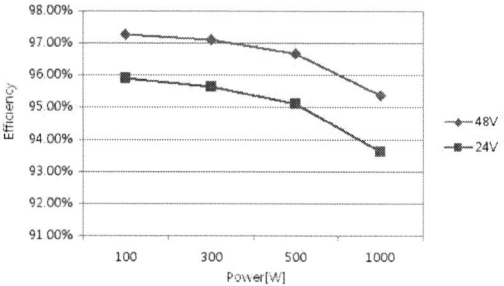

Figure 10. Measured Efficiency of 2-Stage DC-DC Converter

VII. CONCLUSION

In this paper a new power conditioning system to supply the generated power from the fuel cell to the power grid. The proposed power conditioning system consists of a LLC resonant DC-DC converter and 3-phase inverter. The LLC resonant converter boosts the fuel cell voltage of 26-48V up to 400V, using the hard-switching boost converter and the high-frequency LLC resonant converter. The operation of proposed power conditioning system was verified through simulations with PSCAD/EMTDC software by checking the active and reactive power control capability. The feasibility of hardware implementation was verified through experimental works with a laboratory prototype, which was built with 1.2kW PEM fuel-cell stack, 1kW LLC resonant converter, and 2kW PWM inverter. The proposed system can be utilized to commercialize a real interconnection system for the fuel-cell power generation. Also, it can be applied for implementing the micro-grid.

REFERENCES

[1] Gregor Hoogers, "FUEL CELL TECHNOLOGY HANDBOOK", CRC Press, 2003.

[2] Mousavi, A.; Das, P.; Moschopoulos, G, "A ZCS-PWM Full-Bridge Boost Converter for Fuel-Cell Applications", Applied Power Electronics Conference and Exposition, 2009. APEC 2009. Twenty-Fourth Annual IEEE 15-19 Feb. 2009 Page(s):459 – 464.

[3] Wingelaar, P.J.H. Duarte, J.L. Hendrix, M.A.M "Dynamic Characteristics of PEM Fuel Cells", Power Electronics Specialists Conference, 2005. PESC '05. IEEE 36th, 16-16 June 2005 Page(s):1635 - 1641.

[4] Bo Yang and Fred C. Lee. Alpha J. Zhang, Guisong huang, "LLC Resonant Converter for Front End DC/DC Conversion", Applied Power Electronics Conference and Exposition, 2002. APEC 2002. Seventeenth Annual IEEE Volume 2, 10-14 March 2002 Page(s):1108 - 1112.

[5] Rathore, A.K, Bhat, A.K.S. Oruganti, R, "A Comparison of Soft-Switched DC-DC Converters for Fuel Cell to Utility Interface Application", Power Conversion Conference - Nagoya, 2007. PCC '07 2-5 April 2007 Page(s):588 – 594.

Real-time FPGA-based Hardware-in-the-Loop Development Test-Bench for Multiple Output Power Converters

O. Lucía, O. Jiménez, L.A. Barragán, I. Urriza, J.M. Burdío, and D. Navarro

Department of Electronic Engineering and Communications
University of Zaragoza
50018 Zaragoza. SPAIN.
E-mail: {olucia, ojimenez, barragan, urriza, burdio, denis}@unizar.es

Abstract—The implementation of multiple-inductor power converters requires often the development of specific purpose control architectures to obtain the most of the converter. These are usually based on a processor, which provides software flexibility, and specific purpose hardware, which provides customized functionalities. In addition, recent trends suggest the integration of both functionalities in a single chip using an embedded processor and customized hardware, providing a System-on-Programmable-Chip (SoPC) solution.

The aim of this paper is to implement a SoPC system which provides flexibility and customized hardware, and develop a real-time FPGA-based development test-bench based on the Hardware-In-the-Loop (HIL) simulation technique, which accelerates simulation, reduces design cycle times, and allows software and power converter modulation schemes development. The complete system is integrated by a multiple-inductor power converter and an FPGA, which contains a MicroBlaze embedded processor and a specific purpose Digital Pulse Width Modulator (DPWM). HIL at power level is applied by implementing the power converter model into the FPGA. It has been designed with the VHDL-2008 float_pkg package, which allows a straight-forward implementation. As a result, the digital and the analog power converter signals can be traced by means of ChipScope tool, and the processor software can be traced by means of a software debugger tool.

I. INTRODUCTION

The implementation of multiple-inductor power converters requires often the development of specific purpose control architectures to obtain the most of the converter. These are usually based on a processor, which provides software flexibility, and specific purpose hardware, which provides customized functionalities. In addition, recent trends suggest the integration of both functionalities in a single chip using an embedded processor and customized hardware, providing a System-on-Programmable-Chip (SoPC) solution [1].

The implementation and simulation of those systems can be a challenging and time-consuming task. Hardware-In-the-Loop (HIL) simulation has been extensively used for controller evaluation with reduced simulation times [2], [3]. It is based on adding the power electronics system, or at least one actual sub-system, into the loop. Some works have been developed using an FPGA in the loop to implement the controller and/or the plant model [4-9]. In [4]-[6], the power converter model is implemented in fixed-point on an FPGA; and the digital controller runs on an external DSP [4], [5] or CPU [6]. In [7] the FPGA is used as a high-speed peripheral I/O device. In [8] the controller is implemented in the FPGA, but the power converter model runs on a PC. Only in [9], the plant model and the power converter are implemented in fixed-point into the same FPGA, although without the use of any FPGA embedded processor.

The aim of this paper is to provide a real-time FPGA-based development test-bench based on the Hardware-In-the-Loop (HIL) simulation technique for the proposed SoPC system. It accelerates simulation of the whole system, reduces design cycle times and allows software and power converter modulation schemes development avoiding the risk of damaging real prototypes. Besides, the use of the embedded processor accelerates the development process since a change of firmware does not require synthesizing again the whole system.

This paper is organized as follows. Section II gives a brief introduction to the multi-load power converter used in this work and Section III details the proposed FPGA-based control architecture. Section IV explains the HIL simulation technique application, including the power converter modeling and FPGA implementation. Finally, the HIL simulation results are shown in Section V and the conclusions of this work are outlined in Section VI.

This work was supported in part by the Spanish Ministry of Science and Innovation (MICINN) under Project TEC2007-64188 and FPU Grant AP2007 03276, in part by Diputación General de Aragón (DGA) under Project PI008/08, and in part by the Bosch and Siemens Home Appliances Group.

The authors are with the Department of Electronic Engineering and Communications, University of Zaragoza, and the Aragon Institute for Engineering Research (I3A).

II. POWER CONVERTER ARCHITECTURE

The power converter considered for this work is a multiple-output series resonant inverter (MOSRI) which allows supplying the required output power to several loads simultaneously [10]. For this work, we will consider a 4-load converter. This provides a cost-effective solution for multiple inductor systems.

The control patterns required to optimize the converter operation does not fit usually the classical PWM modulators present in most processors or DSPs. For this reason a control architecture based on specific purpose digital hardware has been considered. The complete system (Fig. 1) is made up by the power converter and a versatile control architecture based on a SoPC implemented into an FPGA. It allows testing modulation patterns directly from a PC. The control architecture is based on the MicroBlaze softcore processor [11], [12] and several ad-hoc peripherals [13].

III. CONTROL ARCHITECTURE

The control architecture implementation has been generated using Xilinx Embedded Development Kit (EDK), which provides the required tools to create the embedded system. It also allows the use of Intellectual Property (IP) cores from the available libraries, provides the required templates for creating new peripherals and allows hardware project simulation and software debugging. The next lines describe the soft-core and DPWM peripheral implementation.

Xilinx MicroBlaze soft-core has been used for the proposed architecture. It is a 32-bit Reduced Instruction Set Computer (RISC) which allows peripheral attaching by means of the bus standard Processor Local Bus (PLB). MicroBlaze has been configured with the following settings: 50 MHz processor clock frequency, 50 MHz bus clock frequency and 16 kB of local data and instruction memory. The selected peripherals are an IP UART module and a customized DPWM. The software running on MicroBlaze has been written in C language.

The generation and dynamic adaptation of the proposed pulse width modulations are not feasible by means of conventional PWM peripherals. For this reason, specific purpose DPWMs are often implemented in programmable logic devices to drive inverters [13]. This provides an efficient way to obtain advanced switching patterns with improved performance [14].

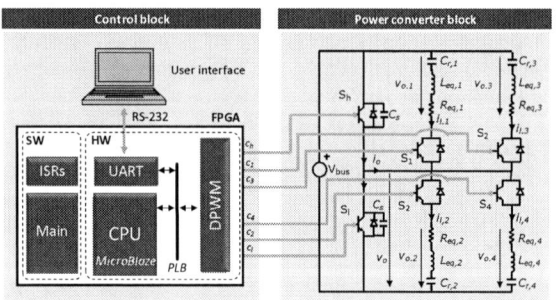

Fig. 1. Block diagram of the developed system: control and power converter blocks.

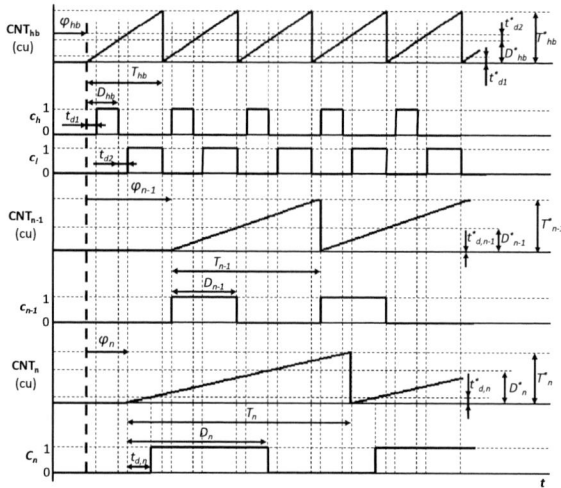

Fig. 2. Main control parameters and waveforms of the multiphase digital pulse width modulator.

TABLE I
CONTROL PARAMETERS FOR MULTIPLE-OUPUT POWER CONVERTER

Converter block	Param.	Range	Range (clock cycles)	# Bits
Inverter	T_{hb}	4-250 µs	200-12500	14
	D_{hb}	0-250 µs	0-12500	14
	t_{d1}	0.5-250 µs	25-12500	14
	t_{d2}	0.5-250 µs	25-12500	14
Multiplexed load	T_i	4-250 µs	200-12500	14
	D_i	0-250 µs	0-12500	14
	φ_i	0-250 µs	0-12500	14

To achieve the aims of this paper, a multiphase DPWM [15], [16] has been designed in VHDL hardware description language. The control parameters given in Fig. 2 and their ranges are summarized in Table I, where a 50 MHz clock frequency has been assumed. The implementation has been carried out through several phase-shifted counters and comparators to define the gating signals. In addition, synchronization operation is ensured during critical operations such as start-up, turn-off, modulation changes, and reset, to ensure converter reliability. The peripheral modulation parameters are mapped into the MB memory, providing a straight-forward power converter modulation configuration.

During the initial design stages, the simulation of the complete system is required in order to optimize and test the proper operation of both the power converter components and the modulation scheme. However, this task becomes time-consuming due to the presence of both a soft-core processor (programmed in C) and the power converter analog circuitry. For this reason, a complete simulation test-bench based on the HIL simulation technique is detailed in the next Section.

IV. HARDWARE IN THE LOOP APPLICATION

To achieve the aim of this paper, HIL at power converter level is applied, including the power converter model in the FPGA implementation. In addition, a supervisor block has been added to help modulation development by warning when certain conditions such as non-ZVS switching are reached. Finally, hardware (ChipScope) and software debugging tools are applied to complete the development test-bench. Fig. 3 summarizes the proposed development scheme.

To analyze and develop modulation schemes it is necessary to obtain a simple power-converter equivalent circuit. Provided that the inverter operates in continuous mode, the loads are parallel-connected to a voltage source inverter. Therefore, voltage applied and current through the loads are independent, and each load can be analyzed individually. Fig. 4 (a) shows the equivalent circuits for the loads. Those whose S_i switches are switched-on are applied $k_i V_{bus}$ voltage. The k_i parameter value depends on the inverter switches S_h and S_l status and the load position. Its values are shown in Fig. 4 (b). By the opposite, the loads whose S_i switches are switched-off stays disconnected once the remaining $C_{r,i}$ charge circulates through S_i diodes D_i. As $C_s \ll C_r$, their influence will be neglected for this analysis.

The aforementioned equivalent circuits allow obtaining the behavioral model of the multiple-output converter. The FPGA implementation is carried out by means of a Finite States Machine (FSM). Four states based on the modeling developed are considered (Fig. 4 (a)). State CF1 represent the current flowing through the load with the T_i transistor switched-on, and state CF2 applies when the current flows through D_i and T_i transistor is switched-off. Besides, state CF0 models switched-off loads and state CF3 has been added to consider the situation in which a transistor has been switched-off with positive current.

Applying Kirchhoff's laws to each network configuration, a continuous time state-space equation is obtained for each configuration. The transition conditions are included in the model, which requires a prior knowledge of the converter operation. First of all, the state-space description for each load is given by

$$
\begin{aligned}
\dot{\mathbf{x}}_i(t) &= \mathbf{A}_{s,i} \cdot \mathbf{x}_i(t) + \mathbf{B}_{s,i} \cdot \mathbf{u}_i(t) \\
\mathbf{y}_i(t) &= \mathbf{C}_{s,i} \cdot \mathbf{x}_i(t) + \mathbf{D}_{s,i} \cdot \mathbf{u}_i(t)
\end{aligned}
\tag{1}
$$

where subscript i denotes the load number, subscript s denotes the circuit configuration, $\mathbf{x}_i(t)$ is the energy state vector, and $\mathbf{u}_i(t) = k_i \cdot V_{bus}$ is the input vector. The selected energy state variables are the current through the inductor i_i and the voltage in the resonant capacitor $v_{c,i}$. Then $\mathbf{x}_i(t) = \begin{pmatrix} i_i & v_{c,i} \end{pmatrix}^{\mathrm{T}}$. Matrices $\mathbf{A}_{s,i}$ and $\mathbf{B}_{s,i}$ depends on the equivalent converter configuration and their values are given in the following expressions, where $s \in [1,4]$ value depends on the network configuration.

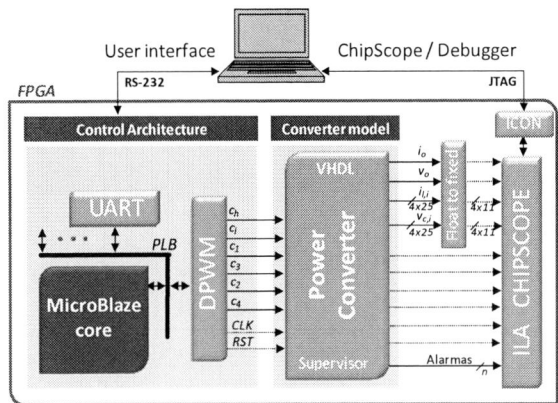

Fig. 3.   Hardware-In-the-Loop simulation test-bench blocks diagram.

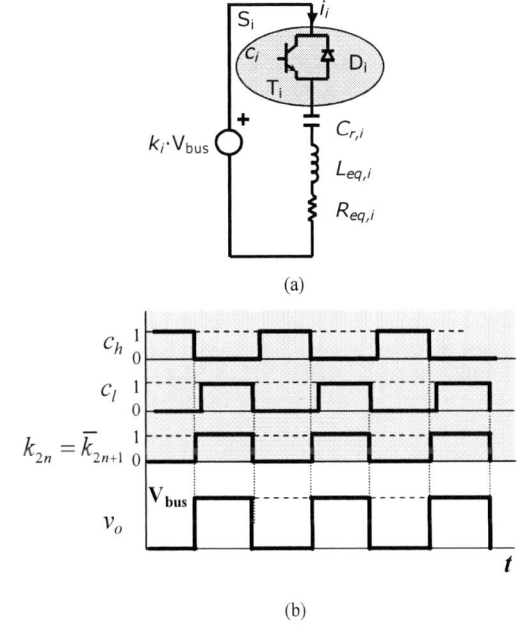

Fig. 4.   Power converter model: (a) equivalent circuits and (b) k parameter values.

$$
\mathbf{A}_{1,i} = \mathbf{A}_{2,i} = \begin{pmatrix} -\dfrac{R_{eq,i}}{L_{eq,i}} & -\dfrac{1}{L_{eq,i}} \\[2mm] \dfrac{1}{C_{r,i}} & 0 \end{pmatrix}
\tag{2}
$$

$$
\mathbf{A}_{0,i} = \mathbf{A}_{3,i} = \begin{pmatrix} 0 & 0 \\ 0 & 0 \end{pmatrix}
$$

$$
\mathbf{B}_{1,i} = \mathbf{B}_{2,i} = \begin{pmatrix} \dfrac{1}{L_{eq,i}} \\[2mm] 0 \end{pmatrix}
\tag{3}
$$

$$
\mathbf{B}_{0,i} = \mathbf{B}_{3,i} = \begin{pmatrix} 0 \\ 0 \end{pmatrix}
$$

Besides, for hardware implementation, the power converter model must be discretized following (4).

$$\mathbf{x}(k) = \mathbf{F}_{s,i} \cdot \mathbf{x}_i(k-1) + \mathbf{G}_{s,i} \cdot \mathbf{u}_i(k-1) \qquad (4)$$

Provided that the matrix A is non-singular, the space-state variables can be calculated using:

$$\mathbf{x}_i(k) = e^{\mathbf{A}_{s,i}T_s}\mathbf{x}_i(k-1) + \mathbf{A}_{s,i}^{-1}(e^{\mathbf{A}_{s,i}T_s}-\mathbf{I})\mathbf{B}_{s,i}\cdot\mathbf{u}_i(k-1). \,(5)$$

The sampling period T_s is selected by simulation in order to obtain accurate converter waveforms while allowing real-time calculation. Peak and rms current values have been considered to evaluate the waveforms accuracy due to the correlation with the switching losses and power delivered respectively, which are important design factors.

The evaluation of the optimum T_s has been carried out by means of mixed signal simulation through AdvanceMS co-simulation tool [17], [18]. Both the VHDL converter model and the ELDO-spice model have been simulated in parallel and compared to reach optimum T_s. Fig. 5 shows the mixed signal simulation structure. MicroBlaze is not included at this point in order to decrease the simulation time. As a conclusion, it has been selected $T_s = 100$ ns to obtain simulation error below 1% for both parameters.

To make synthesizable the model, the real type constant and signals are transformed into floating point objects using VHDL-2008 float_pkg package [19]. Floating point numbers are represented using an exponent and a mantissa following the format shown in Fig. 6, where s is the sign bit, e is the exponent, and f is the unsigned fractional part of the mantissa. The floating-point is "normalized", since it uses a hidden bit, so that the mantissa value is $1+f$. The exponent is biased with $bias = 2^{ne-1}-1$. Then, the value of the normalized number is:

$$(-1)^s \cdot (1.f) \cdot 2^{e-bias} \qquad (6)$$

We will assume a uniform word length format ne and nf for each constant and signal in the synthesizable VHDL model of the power converter. Since the precision is associated with parameter nf, and to speed-up the process, we restrict ourselves to consider $ne = 8$ and $nf = 16$. These parameters are selected to take the most of the FPGA embedded multipliers. The results have been validated applying the previously used simulation scheme (Fig. 5).

Matrices $\mathbf{F}_j$, $\mathbf{G}_j$, are pre-computed for each configuration and stored as constants, this way computation time is reduced at the expense of higher memory requirements. Finally, the conditions evaluated by the supervisor block are the ZVS inverter operation and the proper commutation of the inductive loads. These conditions depend on the control signals, the current i_o and the FSM state, and they are computed using (7) and (8).

$$hb\_notzvs = ((\overline{c_h(k)}\, \& \, c_h(k-1))\, \& \,(i_o < 0)) \,||$$
$$|| \,(((\overline{c_l(k)}\, \& \, c_l(k-1))\, \& \,(i_o > 0)) \qquad (7)$$

$$sw\_xli\_broken = (STATE_i == CF3) \qquad (8)$$

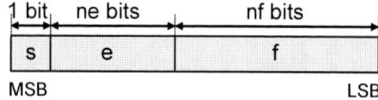

Fig. 6.   Floating point representation.

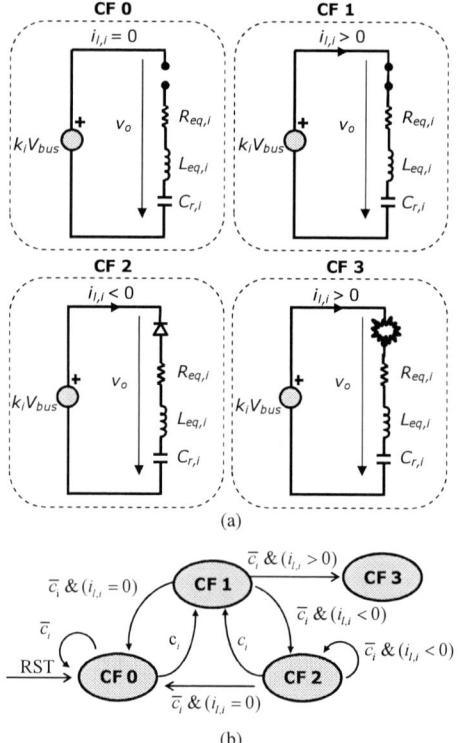

Fig. 7.   (a) Equivalent network configurations for each load and (b) finite state machine and transition conditions for each load.

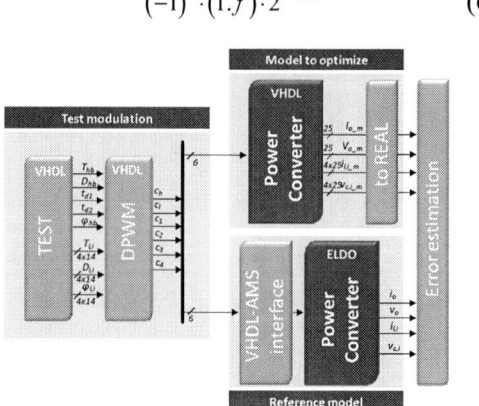

Fig. 5.   Mixed signal simulation blocks diagram used to select T_s.

978-1-4244-4782-4/10 $26.00 © 2010 IEEE   312

V. HIL SIMULATION RESULTS

In this work, the *DN8000K10PCI* board from *Dini Group* has been used. It is an ASIC prototyping engine based on Xilinx Virtex4 FPGA. The MB system has been synthesized into one 4VLX200 using XST 9.2i from Xilinx, and the power converter model using Precision RTL Plus 2007a.18 from Mentor Graphics. The selected clock frequency is 50 MHz. The final FPGA implementation uses 4,249 out of 178,176 registers, and 29,566 out of 178,176 LUTs. As a result, 18,298 out of 89,088 Slices are occupied. The DSP48s has been used as two's complement 18x18-bit multipliers, employing 27 out of 96 available.

Communications with the PC allows changing modulation parameters on-line to provide a convenient power converter test-bench. A user-friendly graphical user interface in MATLAB has been developed. It enables changing modulation parameters and, in addition, it shows a modulation preview, and pre-process and check the parameters. The power converter model waveforms have been traced using ChipScope hardware debugging tool. The power converter parameters are summarized in Table II, and Fig. 8 shows the power converter model waveforms and alarms and the laboratory converter prototype. These results are consistent with the previous mixed signal simulation and non-synthesizable power converter model implementation. The power converter currents can be traced, and the alarms generated by the supervisor block allow the detection of undesired converter operation. Besides, the software running on MB can be safely debugged.

VI. CONCLUSIONS

The development of control architectures for multiple-inductor power converters is a complex task which requires exhaustive simulation of the complete system, including analog and digital sub-systems. In this paper, a customized System-on-a-Programmable-Chip and a complete real-time FPGA-based development test-bench, based on the Hardware-In-the-Loop (HIL) simulation technique, has been proposed.

The complete system contains a 4-loads multiple-inductor resonant inverter, and an FPGA with MicroBlaze embedded processor and a customized Digital Pulse Width Modulator.

TABLE II
POWER CONVERTER PARAMETERS

Converter block	Parameter	Value
High-side load 1	$R_{eq,1}$	5 Ω
	$L_{eq,1}$	35 µH
	$C_{r,1}$	220 nF
High-side load 3	$R_{eq,3}$	5 Ω
	$L_{eq,3}$	35 µH
	$C_{r,3}$	220 nF
Low-side load 2	$R_{eq,2}$	4 Ω
	$L_{eq,2}$	25 µH
	$C_{r,2}$	220 nF
Low-side load 4	$R_{eq,4}$	4 Ω
	$L_{eq,4}$	25 µH
	$C_{r,4}$	220 nF

The power converter continuous model has been obtained and discretized. Then, the HIL technique has been applied by modeling the power converter in VHDL. The sampling period has been determined by means of mixed-signal simulation, and the VHDL-2008 floating point package has been used for FPGA implementation.

The HIL simulation results show that the proposed test-bench allows tracing both MicroBlaze and peripheral digital signals, and power converter model signals by means of ChipScope hardware debugging tool. In addition, the embedded processor software execution can be traced through software debugging tools. As a conclusion, the proposed system provides a versatile and fast method to develop ad-hoc control architectures, avoiding the need of time-consuming mixed-signal simulations and the risk of damaging the actual power converter implementation.

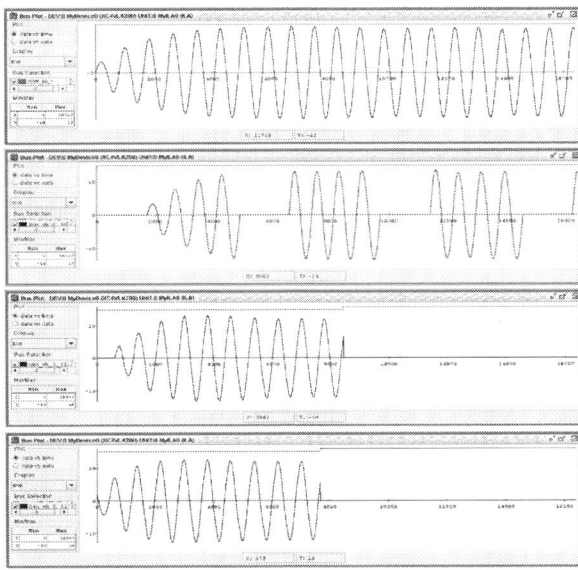

(a)

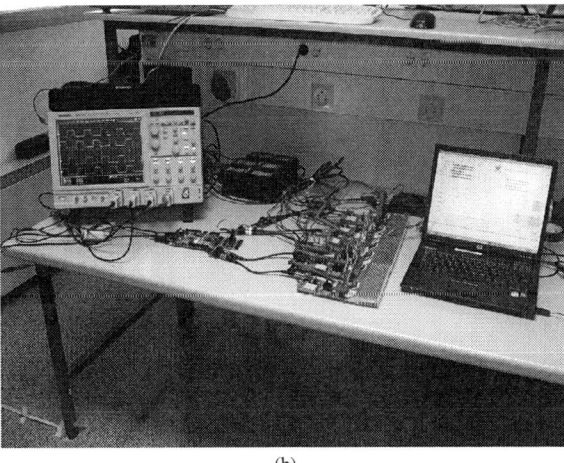

(b)

Fig. 8. Simulation results: (a) Power converter model waveforms and alarms obtained with ChipScope. From top to bottom i_{1_m}, i_{3_m}, i_{2_m}, i_{4_m}, (b) power converter prototype.

REFERENCES

[1] S. Sanchez-Solano, A. J. Cabrera, I. Baturone, F. J. Moreno-Velo, and M. Brox, "FPGA implementation of embedded fuzzy controllers for robotic applications," *IEEE Transactions on Industrial Electronics,* vol. 54, no. 4, pp. 1937-1945, August 2007.

[2] A. Bouscayrol, "Different types of Hardware-In-the-Loop simulation for electric drives," in *IEEE International Symposium on Industrial Electronics,* 2008, pp. 2146-2151.

[3] Z. Weidong, S. Pekarek, J. Jatskevich, O. Wasynczuk, and D. Delisle, "A model-in-the-loop interface to emulate source dynamics in a zonal DC distribution system," *IEEE Transactions on Power Electronics,* vol. 20, no. 2, pp. 438-445, March 2005.

[4] R. Ruelland, G. Gateau, T. A. Meynard, and J. C. Hapiot, "Design of FPGA-based emulator for series multicell converters using co-simulation tools," *IEEE Transactions on Power Electronics,* vol. 18, no. 1, pp. 455-463, January 2003.

[5] A. M. Lienhardt, G. Gateau, and T. A. Meynard, "Digital sliding-mode observer implementation using FPGA," *IEEE Transactions on Industrial Electronics,* vol. 54, no. 4, pp. 1865-1875, August 2007.

[6] C. Dufour, V. Lapointe, J. Belanger, and S. Abourida, "Hardware-in-the-loop closed-loop experiments with an FPGA-based permanent magnet synchronous motor drive system and a rapidly prototyped controller," in *IEEE International Symposium on Industrial Electronics,* 2008, pp. 2152-2158.

[7] P. Lok-Fu, M. O. Faruque, N. Xin, and V. Dinavahi, "A versatile cluster-based real-time digital simulator for power engineering research," *IEEE Transactions on Power Systems,* vol. 21, no. 2, pp. 455-465, May 2006.

[8] S. Karimi, A. Gaillard, P. Poure, and S. Saadate, "FPGA-based real-time power converter failure diagnosis for wind energy conversion systems," *IEEE Transactions on Industrial Electronics,* vol. 55, no. 12, pp. 4299-4308, December 2008.

[9] G. G. Parma and V. Dinavahi, "Real-time digital hardware simulation of power electronics and drives," *IEEE Transactions on Power Delivery,* vol. 22, no. 2, pp. 1235-1246, April 2007.

[10] O. Lucía, J. M. Burdio, I. Millán, J. Acero, and D. Puyal, "Multiple-output resonant inverter topology for multi-inductor loads," in *Applied Power Electronics Conference and Exposition APEC10,* 2010.

[11] J. G. Tong, I. D. L. Anderson, and M. A. S. Khalid, "Soft-Core Processors for Embedded Systems," in *Microelectronics, 2006. ICM '06. International Conference on,* 2006, pp. 170-173.

[12] K. Ying-Shieh and T. Ming-Hung, "FPGA-based speed control IC for PMSM drive with adaptive fuzzy control," *IEEE Transactions on Power Electronics,* vol. 22, no. 6, pp. 2476-2486, November 2007.

[13] P. C. Loh, F. Gao, and F. Blaabjerg, "Topological and modulation design of three-level Z-source inverters," *IEEE Transactions on Power Electronics,* vol. 23, no. 5, pp. 2268-2277, September 2008.

[14] S. C. Huerta, A. de Castro, O. Garcia, and J. A. Cobos, "FPGA-based digital pulsewidth modulator with time resolution under 2 ns," *IEEE Transactions on Power Electronics,* vol. 23, no. 6, pp. 3135-3141, November 2008.

[15] R. Naderi and A. Rahmati, "Phase-shifted carrier PWM technique for general cascaded inverters," *IEEE Transactions on Power Electronics,* vol. 23, no. 3, pp. 1257-1269, May 2008.

[16] R. Foley, R. Kavanagh, W. Marnane, and M. Egan, "Multiphase digital pulsewidth modulator," *IEEE Transactions on Power Electronics,* vol. 21, no. 3, pp. 842-846, May 2006.

[17] J. R. Pimentel and A. Rojas-Moreno, "VHDL-AMS modeling and simulation support for SoC design and implementation of AC motor drives," in *IEEE International Symposium on Power Electronics, Electrical Drives, Automation and Motion,* 2008, pp. 638-643.

[18] I. Urriza, L. A. Barragan, J. I. Artigas, J. Acero, D. Navarro, and J. M. Burdio, "Using Mixed-Signal Simulation to Design a Digital Power Measurement System for Induction Heating Home Appliances," in *Industrial Electronics, 2007. ISIE 2007. IEEE International Symposium on,* 2007, pp. 1447-1451.

[19] "IEEE Standard VHDL Language Reference Manual," in *IEEE Standard P1076,* 2008.

Oversampled Digital Controller IC Based on Successive Load-Change Estimation for DC-DC Converters

Zdravko Lukić, Aleksandar Radić, and Aleksandar Prodić
Laboratory for Power Management and Integrated SMPS,
ECE Department
University of Toronto, 10 King's College Road
Toronto, ON, M5S 3G4, CANADA
E-mail: prodic@ele.utoronto.ca

Simon Effler
Department of Electronic and Computer Engineering
University of Limerick
Limerick, Ireland
E-mail: simon.effler@ul.ie

Abstract – **This paper presents an oversampled digital controller IC for low-power switch-mode power supplies (SMPS) that achieves fast response with a minimized addition of switching losses due to its control actions. To reduce the voltage deviation and improve converter response time, the controller samples the output voltage four times per switching cycle. A single voltage sample is processed by the conventional digital PID compensator to provide tight voltage regulation. Three additional samples are used to calculate the change in duty ratio value that results in the fastest possible response. If the load disturbance is significant, potentially causing large deviation, additional switching pulses are injected based on the estimated load change. To prevent operation at frequencies larger than the switching frequency, while still taking into account the results of oversampling, a novel oversampled digital pulse-width modulator (ODPWM) is introduced. The ODPWM adds or subtracts additional pulses such that the extra pulses are most effective in achieving fast recovery. The controller is implemented on-chip in CMOS 0.18μm technology. The complete controller occupies 0.53 mm^2 of silicon area and takes only 5500 logic gates for the implementation. Its functionality and effectiveness are demonstrated on a 12-V-to-1.8-V, 60-W buck converter switching at 500 KHz. Compared to a fast PID compensator, having crossover frequency equal to 1/10th of the switching rate, the new IC reduces the voltage deviation by two times.**

I. INTRODUCTION

Digital controllers [1]-[14] for dc-dc converters usually regulate the output voltage by taking the samples produced by an analog-to-digital converter (ADC). To improve the converter efficiency, minimize the power consumption of the controller circuit and reduce the hardware complexity, the ADC and accompanying digital pulse-width modulator usually update their values once per switching cycle. Previously, several such low-power, high-frequency digital controller IC implementations [1]-[5] were presented. However, compared to analog IC controllers [15], the dynamic response of those is usually inferior, negatively affecting the size of the power stage components, especially the output capacitor.

Recent linear [7] and non-linear digital controllers [8]-[14] have demonstrated that oversampling of the output voltage results in significant response improvements and reduction of power stage components. The linear controllers usually benefit from modest oversampling rates only, due to the converter

model limitations, which for linearized systems is valid for frequencies significantly smaller than the switching frequency. The dynamic response of linear controllers can potentially be improved by significantly increasing the switching frequency during transients [16,17]. However, such a solution would introduce significant switching losses.

On the other side, non-linear controllers do not have model-related limitations [8]-[14] but their implementation still usually require operation at relatively high oversampling rates and, consequently, fairly complex hardware. Often, the complexity and the silicon area occupied by such an ADC [14] exceed that of a complete analog controller [18], making digital implementations unpractical in low-power cost-sensitive applications. Furthermore, during load transients, these methods also often require the switching action of the power transistors to be performed at the rate significantly larger than that used in steady state causing increased switching losses. As a result, for frequent load changes, the converter efficiency is lower compared to conventional implementations [1]-[5].

The main goal of this paper is to introduce a non-linear oversampling controller, shown in Fig.1, which achieves

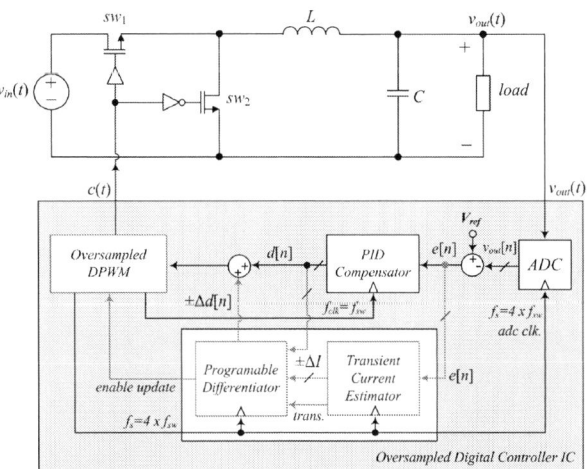

Fig. 1. The oversampled digital controller IC regulating the operation of the buck converter.

This work of Laboratory for Power Management and Integrated SMPS is sponsored by Exar Corporation, Fremont, CA, USA.

significant output voltage transient improvements compared to conventional PID based controllers [1]-[5] using a very modest increase in oversampling rate and the minimum number of additional switching actions of the power transistors. During transients, the controller estimates the change in the load current [10,13] and accordingly modifies the control action such that the calculated changes of the duty ratio are averaged and "glued" to the DPWM. As a result, the switching frequency is not increased, but yet the effect of oversampling is fully utilized.

In the following section the controller operation based on successive load-change estimation is explained. In Section III we address the practical implementation of the key controller blocks, in particular oversampled DPWM with glued logic, and present the controller IC. In the final section, experimental results obtained with a buck converter prototype verifying the controller's operation are demonstrated.

II. PRINCIPLE OF OPERATION

Figures 1 and 2 can be used for explaining how the control pulses in this controller are generated. To minimize the delays existing in once-per-cycle sampled systems [1]-[5], the oversampling controller of Fig.1 takes four samples of the output voltage errors signal $e[n]$ during each switching cycle. Those four samples are processed by two functional blocks. The first block consists of *Programmable Differentiator* and a *Transient Current Estimator* (Fig.1). This block is active during transients only, to improve dynamic response. It takes all four error samples and, as shown in Fig.2.a), during output voltage deviations, produces $\Delta d[n]$ values, corresponding to increase of the duty ratio control variable $d[n]$. The value $\Delta d[n]$ is calculated with the programmable differentiator, which parameters are dynamically changed by the estimator. The estimator uses a modification of capacitor charge balance

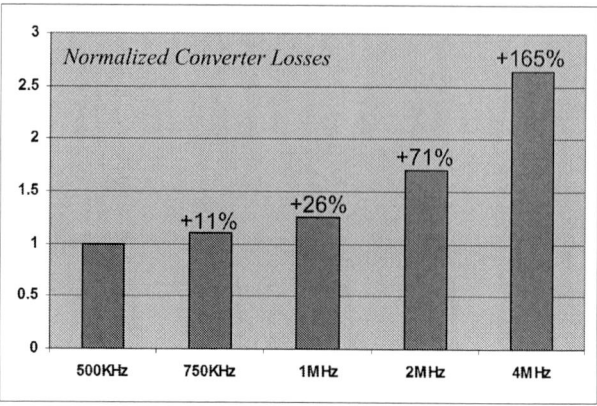

Switching Frequency

Fig. 3. Normalized converter losses versus the frequency of the switching actions for a typical industrial converter designed to operate at 500 kHz [19].

algorithm [8]-[12] to estimate proper differentiator gain, ideally resulting in proximity time-optimal response.

The second block is a PID compensator that takes every 4th sample and by producing duty ratio control signal $d[n]$ keeps system stable in steady-state conditions.

Ideally, during transients, this system could operate at the 4 times higher rate than the switching frequency to obtain a fast response. However, this implementation is not completely practical. As it can be seen from Fig.3, such an operation would incur additional switching losses and, for highly dynamic loads, significantly reduce the converter efficiency. Another practical problem is related to random quantization effects significantly affecting accuracy of $\Delta d[n]$ calculations.

To solve for both of the previously mentioned problems sequence of Fig.2.a) is modified as shown in Fig.2.b). Instead of producing $\Delta d[n]$ pulses at the 4 times the switching rate they

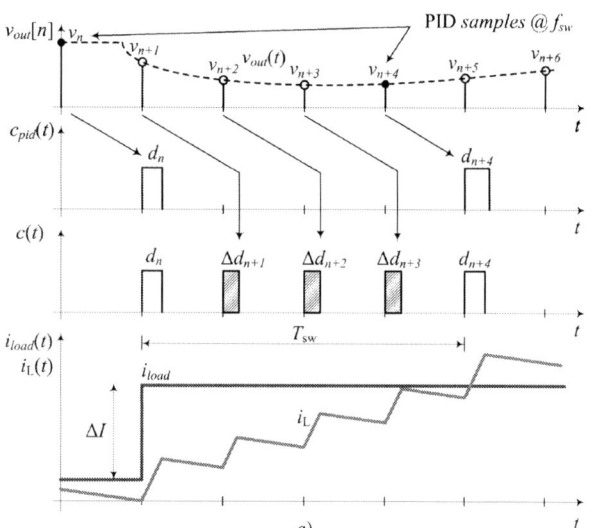

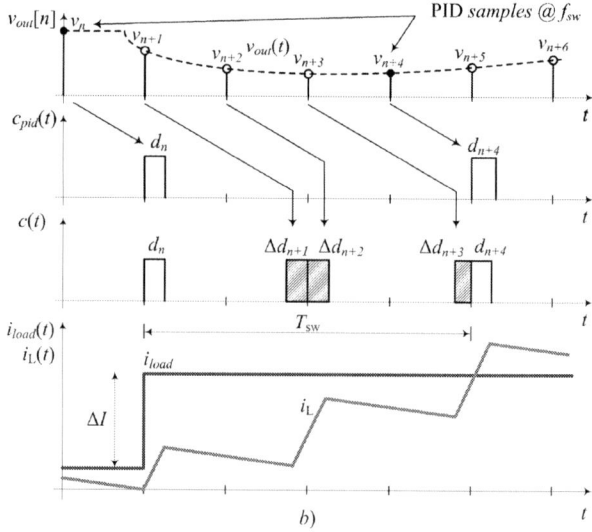

Fig. 2. The operation of the oversampled digital controller during transient: a) multiple control actions based on successive load-change estimation b) control actions are "glued" to limit switching activity during transients to $2f_{sw}$.

978-1-4244-4782-4/10 $26.00 © 2010 IEEE

are created every other switching cycle, while still the information about the calculated values is maintained. To minimize the random error effect, the average value of several pulses is calculated and their average value produced every other sampling cycle. At the sampling instants where the $\Delta d[n]$ pulses coincide with those produced by the PID compensator, the increments are "glued" to the created pulse-width modulated signals. In this way, the switching rate of the system during transient is reduced to $2f_{sw}$, where f_{sw} is the nominal frequency of the converter, reserved for the steady state operation.

The created pulses are sent to an oversampled digital pulse-width modulator (ODPWM). The ODPWM provides updates of the calculated values at the twice switching frequency. The ODPWM actively monitors the switching pulses of Fig.2.a), and, accordingly, readjusts their relative position to reduce the effective switching frequency, as shown in Fig.2.b). Its operation is described in the following section.

III. PRACTICAL IMPLEMENTATION

A. Oversampled DPWM

To resolve the problem of frequent switching actions, shown in Fig. 2.a), and reduce power stage switching losses during transients, the ODPWM attaches the oversampled pulses, such that the effect of the oversampling calculations is maximized

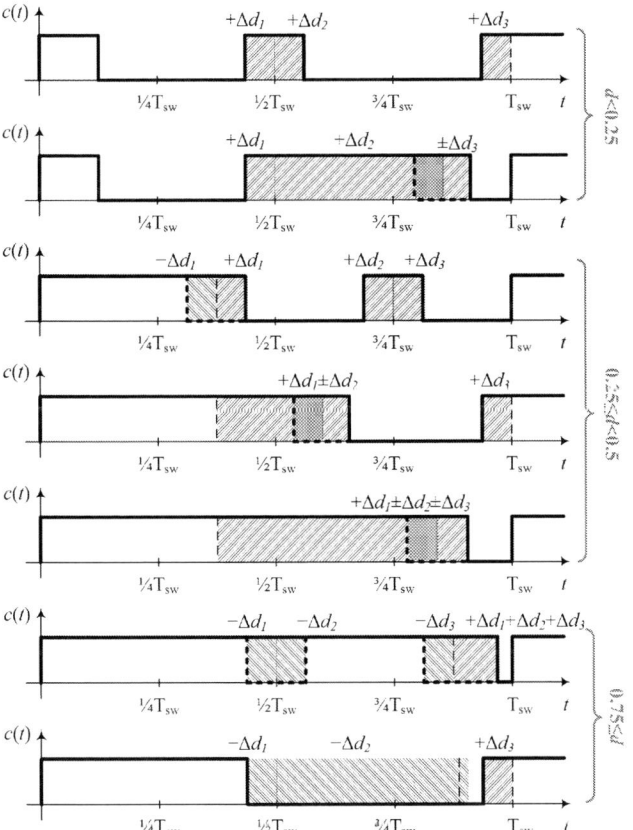

Fig. 4. Switching waveforms generated by the ODPWM for different duty ratio values.

without a significant increase of switching frequency.

The principle of operation is illustrated in Fig. 4 and can be explained through the following example. For PID-calculated duty ratio values less than 0.25, additional pulses Δd_1 and Δd_2 are merged at the middle of the switching period while Δd_3 is appended to the rising edge of the next generated PID pulse. If Δd_1 and Δd_2 are significantly large such that they extend beyond ¾T_{sw}, Δd_3 is merged with the falling edge of Δd_2 in order to reduce the delay of the control action and improve the response. Therefore the effective switching frequency is limited to $2f_{sw}$. A similar approach is used for duty ratios above 0.25. The only difference is that negative oversampled $-\Delta d$ pulses can be now generated by subtracting them from the original pulse. This is beneficial for minimizing the voltage deviation during heavy-to-light load steps.

From Fig. 4, it can be observed that the ODPWM, while reducing the effective switching frequency, also minimizes the control action delay. This reduction in control delay contributes significantly to the reduction in output voltage deviation.

B. Conditions for Transient Operation

Modern power converters [20,21] usually have a substantial output voltage ripple due to the output capacitor ESR and the small inductance necessary for fast transient response. To appropriately react to load disturbances, the required ADC quantization step is small, typically in the range of a few millivolts [11-12,14]. This is several orders of magnitude smaller than the steady-state output voltage ripple [7]. To make the controller insensitive to the sampled voltage ripple, a trivial solution would be to increase the ADC quantization step above the ripple magnitude. The penalty of the increased quantization step is a slower transient response. Alternatively, ripple-compensation techniques [7] may be utilized at the price of more complex controller hardware. To resolve this problem in a simple manner, threshold conditions for transient operation with a small ADC quantization step are introduced and derived here.

To prevent the triggering of the system by small high-frequency variations of the load current and output voltage ripple, minimum injected pulse-width thresholds of the oversampling controller are calculated based on the worst-case steady-state ripple. In order to make these thresholds robust, both the capacitor ESR and the inductor current ripple are taken into account during this analysis. The change in the output voltage in steady-state is given by:

$$\frac{dv_{out}}{dt} = R_{esr} \cdot \frac{di_L}{dt} + \frac{1}{C} \cdot i_c, \qquad (1)$$

where R_{esr} is the capacitor ESR value. From (1) it is clear that the maximum/minimum output voltage rates of change, causing maximum Δe, occur when the capacitor current is maximum/minimum and is given by:

$$\Delta e_{max} = \frac{-\dfrac{dv_{out}}{dt} \cdot T_{sample}}{V_q} + Q_{error}, \qquad (2)$$

978-1-4244-4782-4/10 $26.00 © 2010 IEEE

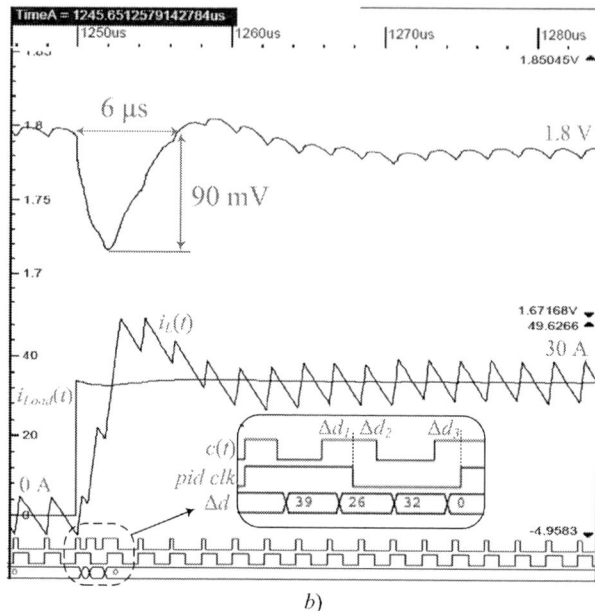

a)   *b)*

Fig. 5. The oversampled controller IC: *a*) chip die photo *b*) mixed-signal simulation for a load step of 30 A with a buck converter switching at 500 KHz and having inductor and capacitor values of *L*=325 nH and *C*=600 μF.

where Q_{error} is the maximum quantization error due to the ADC sampling (Q_{error} = 1). Therefore the minimum injected duty value thresholds then become:

$$\Delta d_{min} = \pm \Delta e_{max} \cdot c_{1/2}, \qquad (3)$$

where coefficients c_1 and c_2 are calculated based on the charge-balance principle [8]-[12] and the rising/falling inductor current slope respectively [22]. On the other hand, in order to reject the influence of large noise, the above analysis can be repeated to determine the maximum $\pm \Delta d_{max}$ threshold values. The value of $\pm \Delta d_{max}$ is obtained from the sum of the maximum load step and the inductor current ripple value, which is simply substituted into (1) in place of i_c. If the rate of change of the load step is known it can also be included in the analysis; however, it may be beneficial to assume it is zero such that the initial output voltage drop due to R_{esr} is also filtered out. The values of the $\pm \Delta d_{min}$ and $\pm \Delta d_{max}$ thresholds calculated by (3) are set within the *programmable differentiator*.

C. IC Implementation

The controller architecture from Fig. 1 is fabricated on-chip in CMOS 0.18μm technology and the chip die is shown in Fig. 5.*a*). A summary of the key IC parameters is provided in Table I. The controller occupies 0.53 mm² of active silicon area. The digital portion of the controller is implemented in Verilog HDL and after synthesis it consists of 5500 logic gates. The operation of the controller is verified with a mixed-signal simulation. For a 30-A light-to-heavy load step, the results are shown in Fig. 5.*b*) where a 90 mV output voltage deviation and 6 μs settling time are observed. As it can be seen in Fig. 5.*b*), to minimize switching activity and improve efficiency, the

additional pulses are effectively "glued" while they are produced only until the voltage deviation is suppressed.

TABLE I.   OVERSAMPLING CONTROLLER CHIP SUMMARY

Technology	TSMC 0.18μm CMOS
Supply voltage	1.8 V / 3.3 V
ODPWM resolution	8 bits
ODPWM nominal frequency	500 KHz
ADC resolution	4 mV
ADC sampling frequency	2 MHz
ADC conversion time	300 ns
Controller complexity	5500 gates
Controller area (digital part)	0.31 mm²
ADC area	0.22 mm²
Total active chip area	0.53 mm²

IV. EXPERIMENTAL SYSTEM AND RESULTS

An experimental system verifying the operation of the oversampled controller IC was built based on the diagram shown in Fig. 1. The power stage is a 60-W, 12-V-to-1.8-V buck converter switching at 500 KHz. The inductor value *L* is 325 nH and the output capacitor value *C* is 600 μF. The on-chip ADC has a 4 mV resolution and a 300 ns conversion time. The PID compensator coefficients are externally programmed onto the chip to obtain a bandwidth higher than 1/10th of the switching frequency. Initially, controller blocks responsible for the non-linear operation are disabled and controller response is verified.

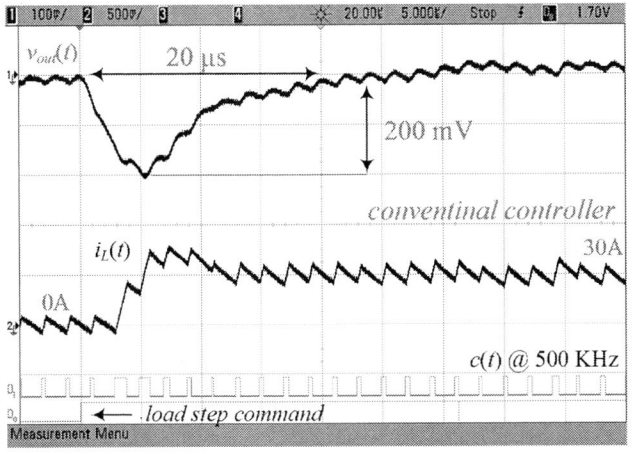 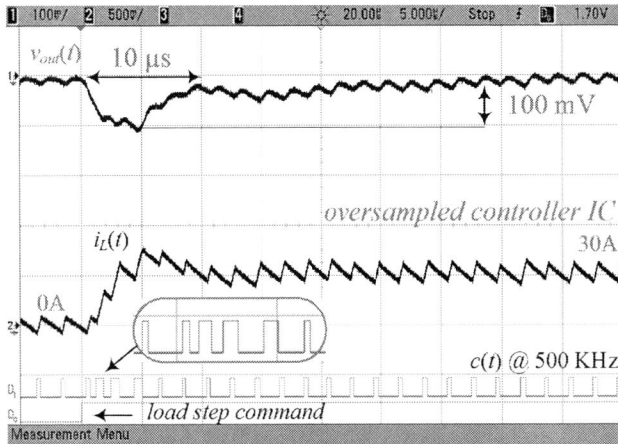

Fig.6. The controller response to a 30-A load step with *a*) Conventional controller (left) *b*) Oversampled controller IC (right) – Ch1: Output converter voltage (100mV/div); Ch2: actual inductor current $i_L(t)$; D1- switching control signal, D0- load step command. Time scale is 5μs/div.

Fig. 6.*a*) shows the obtained response with a load step of 30 A with the PID compensator only. Even though the PID compensator reacts aggressively, increasing the inductor current to a value near the load step in one control action, due to the one cycle delay the voltage deviation is large and equal to 200 mV. The settling time is around 20 μs.

In the next step, the non-linear parts of the controller are enabled and the obtained controller response with the identical load step is tested as shown in Fig. 6.*b*). In this case, as soon as the load step is detected, by taking three additional samples, Δd pulse values are calculated by the programmable differentiator block. To minimize the number of switching actions and improve the converter efficiency the ODPWM attaches the pulses as described previously. As a result, only one additional switching sequence is added as shown in Fig. 6.*b*). Fig. 6.*b*) demonstrates also that the additional pulses are injected only until the initial voltage deviation is stopped, i.e. the inductor current has reached approximately the output load current. Therefore, during most of the settling period and in steady-state the converter switches at the nominal 500 KHz. The obtained voltage deviation is reduced by a factor of two (50%) compared to the PID compensator only and is 100 mV. The settling time is reduced to approximately 10 μs.

V. CONCLUSION

A practical oversampling digital controller IC for dc-dc converter is presented. The controller operates at a modest 4x oversampling rate. The controller utilizes parallel processing of error signal, where a PID compensator and a programmable differentiator are combined to provide stable operation and fast transient response. To minimize the switching losses transients and quantization effect while maintaining all advantages of the oversampling, "glue logic" and application specific oversampling digital pulse-width modulator are introduced.

Also, conditions for eliminating the unintentional triggering of the controller, due to capacitor voltage ripple are derived. The effectiveness of the IC is verified on an experimental system, demonstrating fast transient response and stable operation in all operating conditions.

REFERENCES

[1] B. Patella, A. Prodić, A. Zirger and D. Maksimović, "High-frequency digital PWM controller IC for DC/DC converters," *IEEE Transactions on Power Electronics*, Special Issue on Digital Control, Jan. 2003, Vol.18, Issue1, pp. 438-446.

[2] J.Xiao, A.V. Peterchev, J. Zhang, and S.R. Sanders, "A 4-µA quiescent-current dual-mode digitally controlled buck converter IC for cellular phone applications," *IEEE Journal of Solid-State Circuits*, vol. 39, pp. 2342–2348, Dec. 2004.

[3] A.V. Peterchev, J. Xiao and S.R. Sanders, "Architecture and implementation of a digital VRM controller," *IEEE Transactions on Power Electronics*, vol.18, issue 1, pp. 356 – 364, Jan. 2003.

[4] Z. Lukić, N. Rahman, and A. Prodić, "Multi-bit Σ-Δ PWM digital controller IC for DC–DC converters operating at switching frequencies beyond 10 MHz," *IEEE Trans. Power Electron.*, vol. 22, pp. 1693-1707, Sept. 2007.

[5] A. Parayandeh, and A. Prodić, "Programmable analog-to-digital converter for low-power DC-DC SMPS," *IEEE Trans. Power Electron.*, vol. 23, pp. 1719-1730, Jan. 2008.

[6] H. Hu, V. Yousefzadeh and D. D. Maksimović, "Nonuniform A/D quantization for improved dynamic responses of digitally controlled DC–DC converters," *IEEE Transactions on Power Electronics*, vol.23, issue 4, pp. 1998 – 2005, Jul. 2008.

[7] L. Corradini, P. Mattavelli, E. Tedeschi, D. Trevisan, "High-bandwidth multisampled digitally controlled DC–DC converters using ripple compensation," *IEEE Trans. Industrial Electron.*, Volume 55, Issue 4, April 2008, pp. 1501 - 1508.

[8] G. Feng, E. Meyer, and Y.-F. Liu, "A new digital control algorithm to achieve optimal dynamic performance in DC-to-DC converters," *IEEE Trans. Power Electron.*, vol. 22, pp. 1489–1498, July 2007.

[9] Z. Zhao, and A. Prodić , "Continuous-time digital controller for high-frequency DC-DC converters," *IEEE Trans. Power Electron.*, vol. 23, pp. 564-573, May 2008.

[10] V. Yousefzadeh, A. Babazadeh, B. Ramachandran, E. Alarcon, L. Pao, and D. Maksimović, "Proximate time-optimal digital control for synchronous buck DC-to-DC converters," *IEEE Trans. Power Electron.*, Volume 22, Issue 4, Jul. 2007, pp. 1489 – 1498.

[11] A. Costabeber, L. Corradini, P. Mattavelli, and S. Saggini, "Time optimal, parameters-insensitive digital controller for DC-DC buck converters," in *Proc. IEEE Power Electronics Specialist Conf.*, 2008, pp. 1243–1249.

[12] L. Corradini, A. Costabeber, P. Mattavelli, and S. Saggini, "Time optimal, parameters-insensitive digital controller for VRM applications with adaptive voltage positioning," in *Proc. IEEE Workshop on Computers in Power Elecronics*, 2008, pp. 1–8.

978-1-4244-4782-4/10 $26.00 © 2010 IEEE

[13] S. Effler, A. Kelly, M. Halton, T. Kruger, and K. Rinne, "Digital control law using a novel load current estimator principle for improved transient response," in *Proc. IEEE Power Electronics Specialist Conf.*, 2008, pp. 4585–4589.

[14] L. Corradini, E. Orietti, P. Mattavelli, and S. Saggini, "Digital hysteretic voltage-mode control for DC-DC converters based on asynchronous sampling," *IEEE Trans. Power Electron.*, vol. 24, pp. 201–211, Jan. 2009.

[15] "A voltage regulator module (VRM) using the HIP6004 PWM controller application note," Intersil Corp, Milpitas, USA.

[16] Xiaoming Duan, A. Q. Huang, "Current-mode variable-frequency control architecture for high-current low-voltage DC-DC converters," *IEEE Transactions on Power Electronics*, July 2006, Vol.21, Issue1 4, pp. 1133-1137.

[17] Lilly Kumar, M. Pavan , "Variable switching frequency voltage regulator to optimize power loss", US Patent 6639391.

[18] Kuo-Hsing Cheng, Chia-Wei Su, and Hsin-Hsin Ko, "A high-accuracy and high-efficiency on-chip current sensing for current-mode control CMOS DC-DC buck converter", *IEEE 15th International Electronics, Circuits and Systems Conference*, 2008, pp. 458-461.

[19] "AN9672 - PIP212-12M Design Guide application note," NXP Semiconductors, Eindhoven, Netherlands.

[20] "PIP212-12M data sheet," NXP Semiconductors, Eindhoven, Netherlands.

[21] "FDMF8700 data sheet," Fairchild Semiconductors, San Jose, USA.

[22] R. W. Erickson and D. Maksimovic, *Fundamentals of Power Electronics.* New York, NY:Springer Sience+Business Media Inc., 2001.

Novel Nonlinear Control of Dual Active Bridge Using Simplified Converter Model

Diogenes D. Molina Cardozo, Juan Carlos Balda, Derik Trowler, H. Alan Mantooth

National Center for Reliable Electric Power Transmission

University of Arkansas, Fayetteville AR

Abstract— This paper addresses the control of a Dual Active Bridge (DAB) dc/dc converter. Closed-loop control of the output voltage is achieved by a novel nonlinear technique that separates the control algorithm into a linearization stage and a linear control law. The resulting DAB is capable of bidirectional power flow, and the transformer isolation makes it suitable for a wide variety of applications. The size of the isolation transformer is greatly reduced due to high-frequency (HF) operation resulting in high power densities. Also, switching losses are reduced since zero-voltage switching (ZVS) takes place over a range of operating conditions. Simulation and experimental results are presented for a 400V-400V, 10-kHz, and 10kW prototype.

I. INTRODUCTION

The DAB topology introduced by [1], [2] offers many advantages in dc/dc conversion. The main ones are: bidirectional power flow, galvanic isolation, step up/down capabilities and ZVS depending on the operating condition [1], [2], [3], [4], [5]. The bidirectional power flow capabilities make it an attractive choice in many applications; e.g., energy storage systems [4]. Performing the isolation using a HF transformer increases the system power density by replacing the larger line-frequency transformers at the cost of increased complexity. Also, the step up/down capability allows for large differences between the input and output voltages since the voltage conversion ratio depends not only on the control of the power switches, but also on the transformer turns ratio, making this topology suitable for applications requiring large boost/buck capabilities. Recently, the potential for high power densities of this converter was demonstrated in [6], a feature that increases its applicability to systems where size and weight are critical, such as in electric vehicles.

In spite of the many advantages offered by this converter, the closed-loop control of the output (capacitor) voltage is still in the process of refinement by the research community. In general, linear PI controllers have been shown to provide acceptable results [7]. However, the DAB is nonlinear in nature, and therefore nonlinear control techniques are likely to result in improvements to the overall DAB system. In [8], an accurate small-signal DAB model was developed, and digital control techniques were implemented to regulate the output voltage. In [9], a DAB average model was presented whose validity for frequencies an order of magnitude lower than the switching frequency was demonstrated using frequency- and time-domain studies. Furthermore, [9] linearized this converter model around an operating point in order to use linear controller design techniques to achieve output voltage regulation. The work presented in this paper takes a novel approach to the control of the DAB. Satisfactory tracking of the output voltage is achieved by dividing the control task into a real-time linearization stage and a linear controller stage. The algorithm requires measurements of only the input and output voltages; however, improvements to the converter response are possible if a load current measurement is taken and used for feed-forward in the controller.

The remainder of this paper is organized as follows: Section II presents DAB basic concepts; Section III provides the DAB average model; Section IV develops the proposed control approach; Section V depicts simulation results; Section VI presents practical results to demonstrate the feasibility of the proposed ideas; Section VII presents the DAB design equations, and Section VIII provides the main conclusions.

II. THE DUAL ACTIVE BRIDGE

Fig. 1 presents the DAB circuit configuration consisting of two full dc-ac bridges connected by a HF transformer on their ac sides. Several switching schemes have been developed to improve the DAB operation [10], [11]; however, this paper implements the switching scheme on which the only control input to the system is the phase shift between the two converters. In this switching scheme the power switches are treated as diagonal pairs (pairs S1x-S4x, and S2x-S3x) with constant 50% duty cycle for all pairs, and the switching signals for the pairs within each bridge are complimentary (with the exception of the blanking time); that is, the signal for pair S1x-S4x is complimentary of the signal for pair S2x-S3x. Essentially, it is possible to control the magnitude and direction of power flow in the system by changing the phase shift between the bridges. Fig. 2 shows a simplified circuit that could be used to analyze the behavior of the DAB.

978-1-4244-4782-4/10 $26.00 © 2010 IEEE

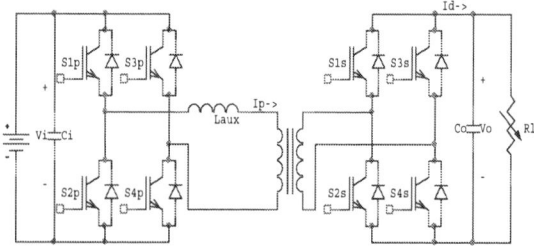

Figure 1: Dual Active Bridge with Variable Resistive Load

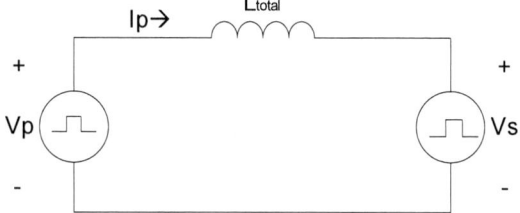

Figure 2: Simplified DAB Circuit

Assume that the amplitude of the square wave voltage pulse applied by each "equivalent" source is identical. Using the above figure, it is clear that there is no voltage applied to the inductor and the current Ip remains unchanged when the phase shift between the two square wave pulses is zero. When a phase shift is introduced, then pulses of width defined by the amount of phase shift are applied across the inductor causing changes (increase or decrease) in the current Ip. Depending on the direction of the phase shift the power flow direction can be controlled.

Figs. 3 and 4 show experimental switching waveforms for the DAB to substantiate the viability of the model in Fig. 2. In Fig. 3, Ch.1 and Ch. 2 are the voltages on the ac sides of the primary and secondary full bridges of the DAB, respectively. The pulses have a steady-state amplitude of 350 V. Ch. 3 is the primary side current Ip.

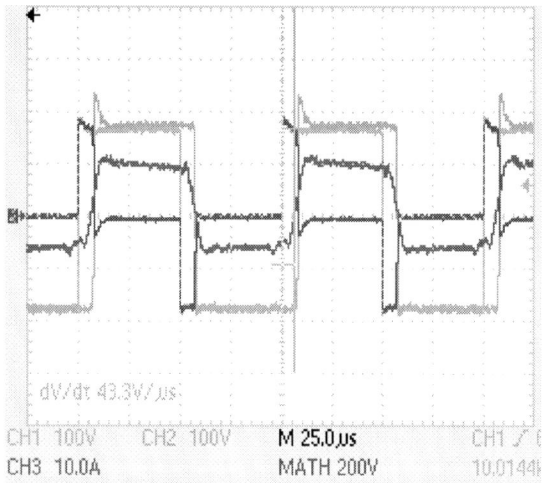

Figure 3: Switching Waveforms for Primary to Secondary Power Transfer

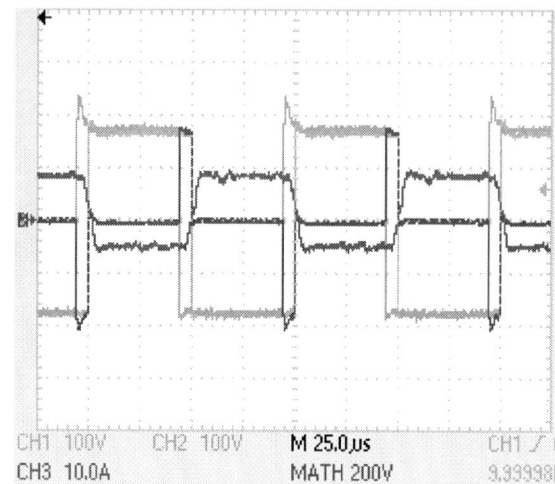

Figure 4: Switching Waveforms for Secondary to Primary Power Transfer

For ease of understanding the subtraction of Ch. 1 and Ch. 2 (the voltage across the total inductance in the ac link) is shown in the MATH channel. The selected phase shift is causing power to flow from the primary to the secondary side; primary voltage leads the secondary voltage. Fig. 4 presents the waveforms when the sign of the phase shift is reversed. As expected the power flow direction is now from the secondary to the primary side. In depth analysis of the different modes of operation for this converter has been addressed in [1]- [5].

III.   DUAL ACTIVE BRIDGE AVERAGE MODEL

The power transfer between the two bridges as a function of the phase shift α and other variables in the system has been shown to be given by [1], [9]:

$$P_{out} = \frac{V_{in}V'_{out}}{2L_{total}f_{sw}}\alpha(1 - \alpha); \quad (1)$$

where V_{in} is the input voltage, V'_{out} is the output voltage referred to the input side, L_{total} is the total series inductance and f_{sw} is the switching frequency. Using (1) it can be shown that the average current flowing out of the output bridge (see Id in Fig. 1) during half of a switching cycle can be given by:

$$I_{d\_avg} = \frac{V_{in}}{2L_{total}f_{sw}}\alpha(1 - \alpha). \quad (2)$$

Using (2) it is possible to derive the simplified average model shown in Fig. 5. Frequency- and time-domain studies were performed by the authors of [9] to validate the model. In particular, the authors of this paper validated the model for frequencies an order of magnitude lower than the switching frequency using time-domain simulations. One of the simulations used for the validation of the model is shown in Fig. 6. A full switched model was simulated in parallel to the averaged model. The effects of blanking time were not accounted for. The main disturbances applied during the time-domain simulation were: input voltage, load resistance, and phase shift variations. Note that the waveforms of the full model and the simplified model overlap almost perfectly.

978-1-4244-4782-4/10 $26.00 © 2010 IEEE   322

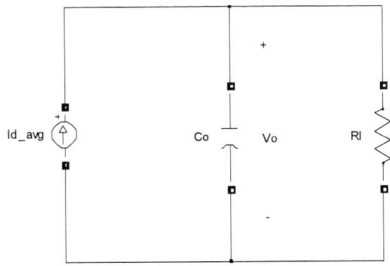

Figure 5: Simplified DAB Model

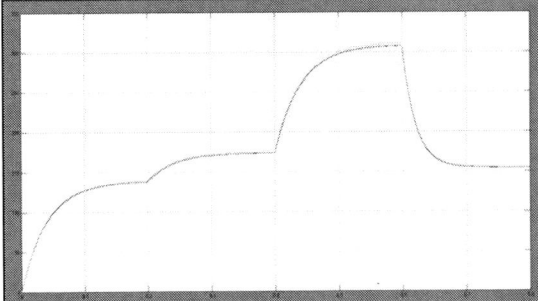

Figure 6: Time Domain Simulations for DAB Model Validation

Using the simplified model it is possible to write the differential equation shown below to describe the dynamic behavior of the converter (from KCL):

$$\frac{dV_{out}}{dt} = -\frac{1}{R_{load}C_o}V_{out} + \frac{1}{C_o}\left(\frac{V_{in}}{2L_{total}f_{sw}}\alpha(1-\alpha)\right). \quad (3)$$

IV. CONTROL OF THE DAB

A. Feedback Linearization

With reference to (3), the novel linearization is done by defining the auxiliary input $V_{aux} = V_{in}\alpha(1-\alpha)$ as the input to the system resulting in:

$$\frac{dV_{out}}{dt} = -\frac{1}{R_{load}C_o}V_{out} + \frac{1}{2\,C_oL_{total}f_{sw}}V_{aux}. \quad (4)$$

Equation (4) can be used to design a linear controller and its output "arithmetically manipulated" in real time to obtain the required phase shift α that generates the desired V_{aux}. The required phase shift is obtained using the equality defined above, which defines V_{aux} as:

$$V_{aux} = V_{in}\alpha(1-\alpha). \quad (5)$$

Solving for the phase shift α (there are two solutions since (5) is a quadratic equation):

$$\alpha_1 = \frac{1 + \sqrt{1 - 4\frac{V_{aux}}{V_{in}}}}{2} \,, \alpha_2 = \frac{1 - \sqrt{1 - 4\frac{V_{aux}}{V_{in}}}}{2} .$$

At this point it is important to note that the phase shift α can vary from 0 to 360°; however, for simplicity this quantity is converted to the per unit system, resulting in a range for α of 0 to 1. Furthermore, this phase spread can be "redistributed" so that it swings between -0.5 to 0.5 (-180° to 180°) to ease the controller implementation task. In this form it

is possible to say that $\alpha > 0$ results in power flow from the primary to the secondary side of the HF transformer (refer to Fig. 3). As one might expect, $\alpha < 0$ would result in power flow from the secondary to the primary side of the HF transformer (refer to Fig. 4). After evaluating the two solutions for the linearizing equations one can observe that the $\sqrt{1 - 4\frac{V_{aux}}{V_{in}}}$ term will never become negative, and so α_1 is not capable of producing solutions lower than 0.5 and so all solutions would be outside the predefined range of $-0.5 < \alpha < 0.5$, making α_2 the solution of choice.

B. Linear Controller

The control objective is to regulate the output (capacitor) voltage V_{out} with the gains being selected to satisfy certain damping and bandwidth specifications. The controller structure selected to achieve zero steady-state error for a step reference command is defined in the equations below:

$$V_{aux} = -K_1V_{out} - K_2 \int (V_{out}^{ref} - V_{out})dt. \quad (6)$$

This controller structure results in a system of the form:

$$\frac{dV_{out}}{dt} = -\frac{1}{R_{load}C_o}V_{out} + \frac{1}{2\,C_oL_{total}f_{sw}} \times$$

$$\times \left(-K_1V_{out} - K_2 \int (V_{out}^{ref} - V_{out})dt\right). \quad (7)$$

When transformed to the Laplace domain the closed-loop transfer function for the system above can be expressed as:

$$\frac{V_{out}}{V_{out}^{ref}} = \frac{\dfrac{-K_2}{2\,C_oL_{total}f_{sw}}}{s^2 + \left(\dfrac{K_1}{2\,C_oL_{total}f_{sw}} + \dfrac{1}{C_oR_{load}}\right)s + \dfrac{-K_2}{2\,C_oL_{total}f_{sw}}} \quad (8)$$

Using the nominal second-order system to determine the controller gains by comparing it to the system of (8) results in:

$$K_1 = \left(2\xi\omega_n - \frac{1}{C_oR_{load}}\right)2\,C_oL_{total}f_{sw} \quad (9)$$

$$K_2 - -2\,C_oL_{total}f_{sw}\omega_n^2 \quad (10)$$

Once the formulas for the controller gains were developed, the next step of the control design task consisted of choosing the desired design specifications for the converter. In lieu of more stringent requirements the specifications were set to:

$$\omega_b = 2\pi10 \text{ [rad/s]},$$

$$\text{Percent Overshoot} = \text{P. O.} = 2\%.$$

The two performance specifications defined above are related to the damping ratio ξ and the natural frequency ω_n in equations (9) and (10) by equations (11) (only valid for damping ratios lower than 0.8) and (12) [12]. Once the parameters of the system (inductances, capacitances, etc.) are known, Eq. (9)-(12) are used to calculate the gains K_1 and K_2 for the linear controller.

$$\omega_n = \frac{\omega_b}{-1.19 \times \xi + 1.85} \quad (11)$$

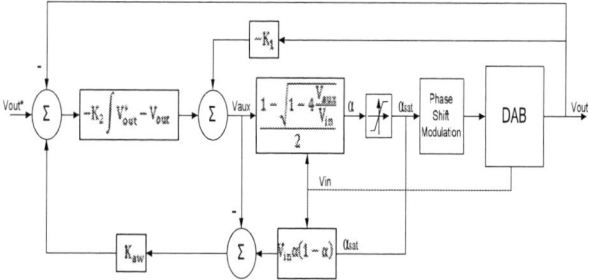

Figure 7: Closed Loop DAB Block Diagram

$$\xi = \sqrt{\frac{\ln{(P.O./100)^2}}{\pi^2 + \ln{(P.O./100)^2}}} \quad (11)$$

The block diagram for the implemented control algorithm is presented in Fig. 7. Anti-windup was added to the integral term to avoid problems due to control input saturation during startup and other transient conditions. The gain for the anti-windup term was selected heuristically through simulations resulting in $K_{aw}=10$. The feed-forward signal of the load current will be added in future runs of the controller design task. Note that only the input and output capacitor voltages need measuring.

V. SIMULATION RESULTS

Table 1 presents the system parameters for the simulations and the laboratory experimental setup developed to test the DAB control approach.

Table 1: DAB System Parameters

Transformer Leakage Inductance	64 µH
Auxiliary Inductance	125 µH
Rated DC Input Voltage	400 Vdc
Rated DC Output Voltage	400 Vdc
DC Input Capacitance	1350 µF
DC Output Capacitance	1350 µF
Load Resistance	40 Ω
Switching Frequency	10 kHz
Rated Power	10 kW

The simulation results are shown in Fig 8. The phase shift angle between the two bridges is at the top, the capacitor output voltage is in the middle, which is the variable being regulated, and the load current is at the bottom. The capacitor output voltage remains around its regulated value during all mentioned dynamic situations.

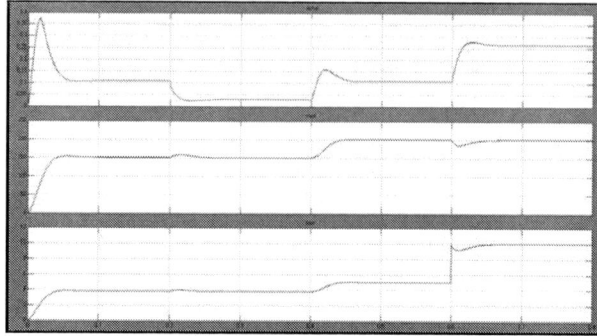

Figure 8: Simulation Results for Closed Loop Controlled DAB

Simulations were run in MATLAB Simulink™ using the SimPowerSystems toolbox for 0.8 s. The main dynamic situations happen in the following sequence: at t = 0 s the output capacitor initial condition is set to 0 V, and $V_{in} = 150$ V, $V_{out}^{ref} = 150$ V, and $R_{load} = 40$ Ω. At t = 0.2 s the input voltage supply steps up to $V_{in} = 200$ V. At t = 0.4 s the reference command is stepped up to $V_{out}^{ref} = 200$ V. At t = 0.6s the load resistance is stepped down to $R_{load} = 20$ Ω, causing a 100% rise in the load current.

VI. EXPERIMENTAL RESULTS

An experimental setup developed to test the validity of the DAB control approach is displayed in Fig. 9. The circuit parameters for the setup are presented in Table 1 and the control software was implemented in a TI F2808 DSP. The potential for high power density was not exploited in this setup since that was not the focus of the project.

A. Tests with a Resistive Load

Fig. 11 shows the system response to a step increase in the dc load current at a regulated voltage of 400 V (load steps from 3.5 kW to 7 kW). The voltage controller maintains regulation of the output dc voltage during the load transient.

Figure 9: Experimental Setup

978-1-4244-4782-4/10 $26.00 © 2010 IEEE

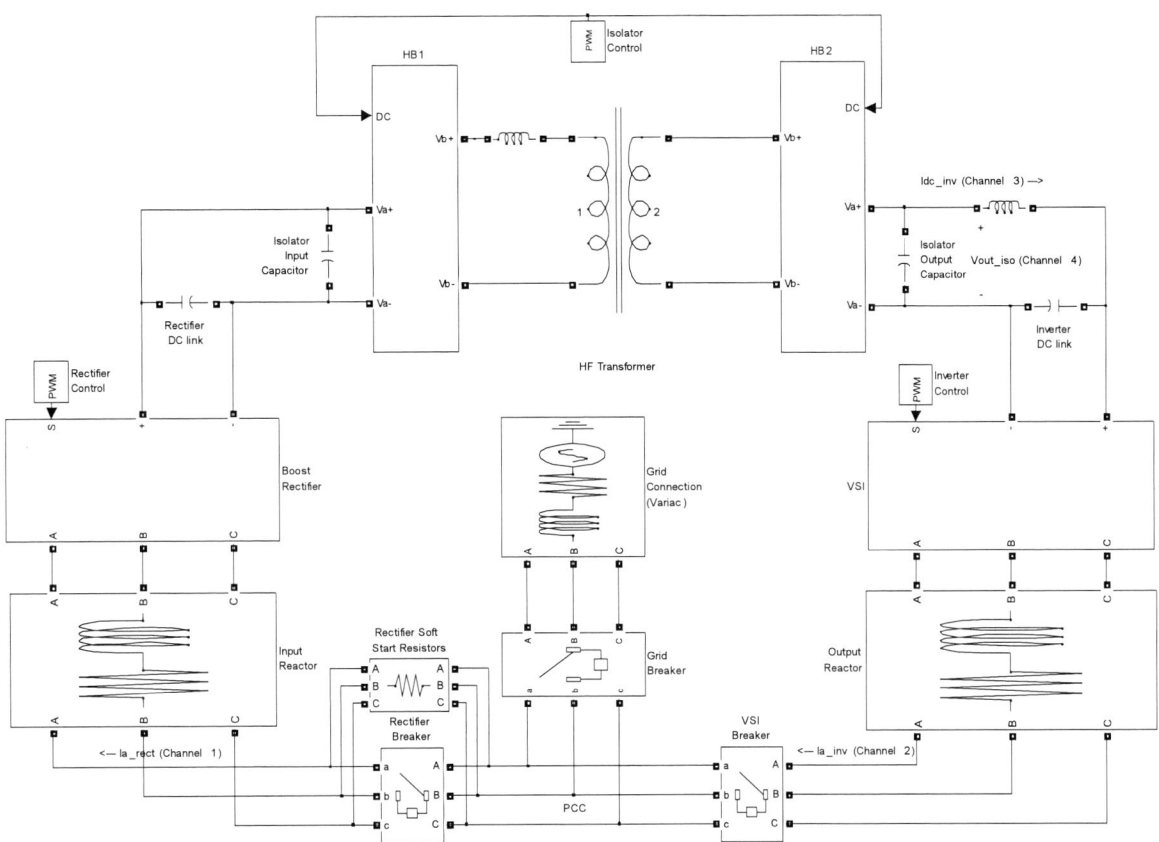

Figure 10: Diagram of Experimental Setup for DAB Tests with Active Bidirectional Power Electronic Load

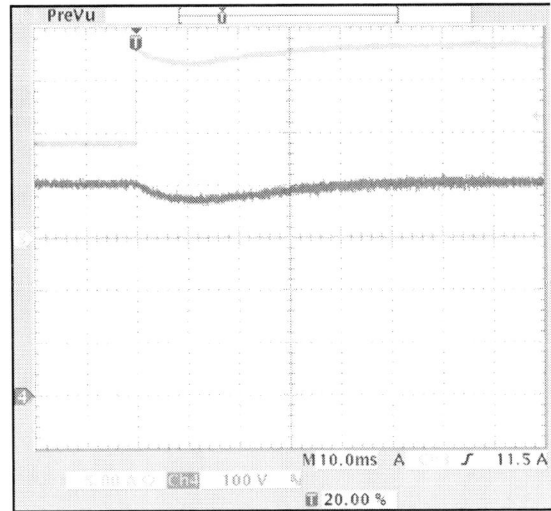

Figure 11: Load Step Change Transient

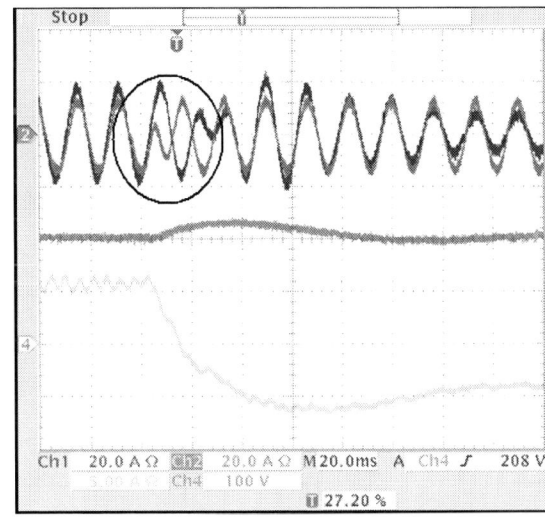

Figure 12: Power Flow Reversal – Load to Source

B. Tests with Active Bidirectional Power Electronic Load

Since the DAB is expected to perform under conditions of bidirectional power flow and to drive nonlinear loads it was important to evaluate its behavior under those conditions. After all, during the controller design it was assumed that the load was constant and resistive, so there is no guarantee that the converter will behave satisfactorily when more realistic loads are connected to it.

Fig. 10 shows the diagram of the laboratory environment implemented for testing the DAB. A bidirectional three-phase boost rectifier was used as the voltage source feeding the DAB. The rectifier ensures that the voltage supplied to the DAB is a constant dc voltage. For the load realization a bidirectional voltage source inverter (VSI) was connected at the dc output of the DAB. The VSI has a current-control closed loop. Both of these converters were operated at unity power factor during the tests. Note that Fig. 10 not only presents the system configuration, but also provides the oscilloscope channel setup for the waveforms shown next. Ch. 1 and Ch. 2 are rectifier and inverter phase A currents, respectively. Ch. 3 is the dc current drawn from the DAB and Ch. 4 is the DAB output voltage.

Fig. 12 presents the behavior of the system when the VSI switches from acting as a load to acting as a source (see the encircled 180° change in the inverter current). With reference to Fig. 10 the power flow changes from flowing clockwise to flowing counter-clockwise. The DAB is capable of holding its output regulated at 200 V even during this transient. This is a testimony of the regenerative capabilities of the DAB.

Fig. 13 shows the behavior of the system when the power flow reversal is in the opposite direction. In this case the VSI changes from acting as a source to acting as a load. Again, the DAB is capable of holding the output capacitor voltage regulated during this transient, illustrating the suitability of the developed control algorithm.

VII. DAB DESIGN EQUATIONS

Some of the design equations developed in [5] are repeated in this section for completeness. The three equations that were found to be most helpful during this design were the maximum attainable power, the maximum current, and the ripple in the output voltage. The maximum attainable power is given by:

$$P_{out\_max} = \frac{V_{in}V'_{out}}{2L_{total}f_{sw}} 0.25 \, .$$

If the connected load draws more power than the maximum value the converter will no longer be capable of regulating the output voltage.

The maximum current equation guides the selection of active devices for the converter:

$$I_{max} = \frac{1}{4L_{total}f_{sw}} [V_{out} - (1 - 2\alpha)V_{in}] \, .$$

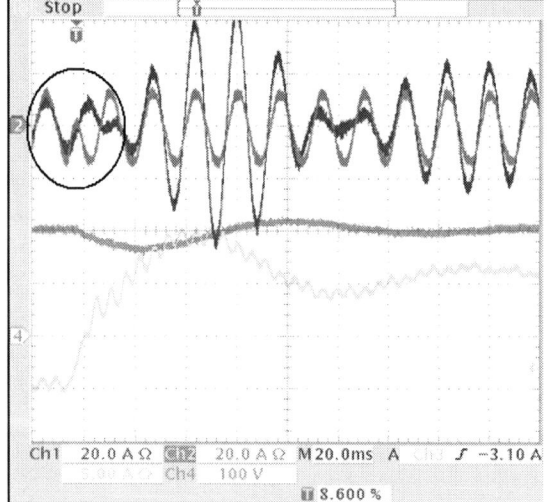

Figure 13: Power Flow Reversal – Source to Load

In [5] the authors address the issues of SOA design. The capacitor bank size selection can be guided by the equation given below which defines the ripple in the output voltage:

$$\Delta V_{out} = \frac{1}{32f_{sw}^2 L_{total}C_o} \times$$
$$\frac{[\alpha(1 - \alpha)R_{load} - 2(1 - 2\alpha^2)f_{sw}L_{total}]^2}{[\alpha^2(1 - \alpha)^2 R_{load}{}^2 - 2f_{sw}L_{total}\alpha(1 - \alpha)R_{load}]} \, .$$

The component selection for the DAB is a complex tradeoff where achieving higher power will be driven by reductions on the leakage inductance or in the switching frequency; however, such reductions are likely to result in exceeding the rating of the semiconductor devices so much care must be given to this selection process.

VIII. CONCLUSIONS

This paper was dedicated to the design, control, and implementation of a closed-loop voltage controlled bidirectional isolated dc-dc converter known as the Dual Active Bridge (DAB). A novel nonlinear control strategy was introduced and its validity was proven both in simulations and through experimental work. The control strategy was based on a simplified converter model that provides acceptable accuracy for frequencies an order of magnitude lower than the switching frequency. By feedback linearizing the system it was possible to apply linear control strategies to design the DAB controller. The bandwidth of the closed-loop system is likely to be significantly increased if a feed-forward signal of the load current is provided to the controller. This idea will be implemented in future iterations of the controller design task. The resulting closed-loop controlled DAB was tested not only with a resistive load, as assumed during the controller design, but also with a nonlinear bidirectional active power electronic load. Examining Fig. 10 one can see that there are many possibilities for combining the DAB with other converter topologies to create a variety of useful complex topologies, but those will be addressed in future work.

ACKNOWLEDGEMENTS

Mr. Diogenes D. Molina Cardozo is grateful for the funding received from NCREPT member companies for his research work.

REFERENCES

[1] R. W. De Doncker, D. M. Divan, M. H. Kheraluwala. "A Three-Phase Soft-Switched High Power Density DC/DC Converter for High Power Applications," *Conference Record of the 1988 IEEE Transactions on Industry Applications Society Annual Meeting*, Vol. 1, 2 – 7 October, Pages: 796 – 805.

[2] M. H. Kheraluwala, R. W. Gascoigne, D. M. Divan and Baumann, E. D. "Performance Characterization of a High-Power Dual Active Bridge dc-to-dc Converter," *IEEE Transactions on Industry Applications,* Volume 28, Issue 6, November-December 1992, Page(s):1294 – 1301.

[3] Shigenori Inoue, Hirofumi Akagi. "A Bi-Directional Isolated DC/DC Converter as a Core Circuit of the Next-Generation Medium-Voltage Power Conversion System," *37th IEEE Power Electronics Specialists Conference,* 18 – 22 June 2006, Pages: 1 – 7.

[4] Shigeroni Inoue and Hirofumi Akagi. "A Bi-Directional DC/DC Converter for an Energy Storage System," *22nd Annual IEEE Applied Power Electronics Conference,* February 25 – March 1 2007, Pages: 761 – 767.

[5] C. Mi, H. Bai, C. Wang and Gargies, S. "Operation, design and control of dual H-bridge based isolated bidirectional DC-DC converter," *IET Power Electronics,* Vol. 1, Issue 4, December 2008, Pages: 507 – 517.

[6] Pavlovsky, M., de Haan, S. and Ferreira, J.A. "Reaching High Power Density in Multikilowatt DC–DC Converters With Galvanic Isolation," *IEEE Transactions on Power Electronics,* Volume 24, Issue 3, March 2009, Page(s):603 – 612.

[7] Segaran, D., Holmes, D.G. and McGrath, B.P. "Comparative analysis of single and three-phase dual active bridge bidirectional DC-DC converters," *Australasian Universities Power Engineering Conference (2008. AUPEC '08),* 14-17 December 2008, Page(s):1 – 6.

[8] Krismer, F. and Kolar, J.W. "Accurate small-signal model for an automotive bidirectional Dual Active Bridge converter," *11th Workshop on Control and Modeling for Power Electronics (COMPEL 2008),* 17-20 August 2008, Page(s):1 – 10.

[9] Harish K. Krishnamurthy, Raja Ayyanar. "Building Block Converter Module for Universal (AC-DC, DC-AC, DC-DC) Fully Modular Power Conversion Architecture," *Power Electronics Specialists Conference,* 17 – 21 June 2007, Pages: 483 – 489.

[10] Jain, A.K. and Ayyanar, R. "PWM control of dual active bridge: comprehensive analysis and experimental verification," *34th Annual Conference of IEEE Industrial Electronics (IECON 2008),* 10-13 November 2008, Page(s):909 – 915.

[11] Haihua, Zhou and Khambadkone, A.M. "Hybrid Modulation for Dual Active Bridge Bi-Directional Converter With Extended Power Range For Ultracapacitor Application," *IEEE Industry Applications Society Annual Meeting (2008 IAS '08),* 5-9 October 2008, Page(s):1 – 8.

[12] Dorf, Richard C. and Bishop, Robert H. "Modern Control Systems," Upper Saddle River, NJ: Person Prentice Hall, 2005. 0-13-127765-0

A Novel Digital Single-Wire Quasi-Democratic Stress Share Scheme For Paralleled Switching Converters

Karl Rinne, Anthony Kelly, Eamon O'Malley
Research and Development
Powervation Ltd.
Limerick, Ireland
{karl.rinne, anthony.kelly, eamon.omalley}@powervation.com

Abstract— A novel Digital Stress Share (DSS) scheme, useful for paralleled switch-mode power converters (SMPCs), is presented. Due to its unique feature set DSS lends itself particularly well to modern power architectures where load currents (or - more generically - converter stresses) need to be actively balanced between an arbitrary number of digitally controlled DC-DC switching converters. The DSS scheme is suitable for single-wire implementation offering a low-cost and robust platform. DSS is master-less and quasi-democratic, eliminating all known drawbacks of analog current share lines, and offering significant improvements over competing digital current share methods. It features inter-device stress share communication with fully predictable timing, regardless of the number of SMPCs working in parallel. Data throughput as well as data storage requirements are minimized. Measurement results confirm excellent stress share performance under various operating conditions. The DSS scheme is versatile, robust, fault-tolerant and scalable. DSS defines an electrical bus interface, as well as an isochronous protocol.

I. Introduction

Paralleled switching converters have been used in power delivery systems for decades. More recently, modern power architectures, including Distributed Power Architectures (DPAs), Intermediate Bus Architectures (IBAs) and Centralized Control Architectures (CCAs) deploy paralleled DC-DC converters at all power levels. Generally, paralleled units are non-identical, due to differences in voltage references, component variations, unmatched propagation delays, and system imbalances (impedances to load, environmental conditions, etc.). Hence, with the desire to operate power converters in parallel comes the requirement to ensure that the units share stresses (current stress, and/or thermal stress) equally and stably. A large variety of passive and active current share schemes exists, differing in complexity, scalability and performance [3]. Advantages of paralleling switching converters include improved power conversion efficiency, higher level of redundancy, simplified thermal management, improved system reliability, ease of power system maintenance, and so forth. The availability of high-performance stress share schemes creates a path forward for standardized power blocks and scalable multi-POL (point-

of-load) modular delivery systems, supporting future trends in power system architectures. A summary overview of existing current sharing methods is offered in Section II., while Section III. outlines the novel DSS scheme. Section IV. explains how the basic DSS scheme can be extended in order to further improve system reliability and contribute to smart system configuration. Measurement results are provided in Section V.

II. Brief Overview of Current Share Methods

Current share schemes can be broadly divided into two categories: passive and active current share methods. Most passive current share schemes employ one of a number of possible droop methods. There is no need to distribute stress share signals between the participating SMPCs, hence no requirement for a current share bus. Individual SMPCs program their output impedances in order to achieve current sharing. Droop methods do not serve the needs of typical power delivery systems as load regulation is seriously impacted.

Instead, modern multi-SMPC systems require active current share schemes. Active current share schemes employ one of a number of possible control structures (outer loop regulation, inner loop regulation, and external regulation). The common core principle of outer loop regulation schemes is to adjust the output voltage set-point reference of the inner voltage control loop of individual SMPCs in order to remove current imbalances detected by the outer loop. In order to do this, individual SMPCs must share reference current information through a current share bus. Information on the current share bus may either be provided by a single master (master-slave method), or by all devices connected to the bus (democratic master-less).

The master/slave (M/S) method may either employ a predetermined, automatically determined, or rotating master. The master provides current share reference information to all slaves, and the slaves adjust their inner loops in an attempt to match their own output current to this reference. The drawbacks of the M/S method are well known. The current share master does not actively participate in the system response to a current imbalance. Rather, the master's output current is indirectly modified by the slaves. In a way, the master is used as a current sensor only, rather than an actuator. Because of this, the dynamic response to a current imbalance is not optimum. Furthermore, system response to excitations in the current share loop depends strongly on the number of slave modules connected in parallel, leading to conservatively compensated current share loops, degrading dynamic performance of the system. Predetermined masters introduce single points of failure into power systems. System maintenance is hampered in case the master module needs to be hot-swapped.

The shortcomings of M/S schemes can be addressed by master-less democratic schemes. Here, all participating SMPCs submit their output current information on to the common current share bus, which in turn reflects the average slave output current in the system. This allows the slaves to match their local output current to the average output current. Analog democratic current share schemes are well covered in literature ([1], [6], [7], [8] and [9]), and popular in application. Unfortunately, in larger scale power delivery systems, the current share bus tends to be susceptible to noise and ground-potential differences, severely limiting scalability and/or complicating system design and interconnects through the use of fully differential analog current share signaling.

Advantages of digital SMPC controllers are well known: availability of sophisticated adaptive control algorithms, in-system programmability, flexibility, robustness, communication, improved fault management, and so forth. Utilizing established serial interface standards (SMBUS, SPI, etc.) directly supported by most digital SMPC controllers for the purpose of digital current share is undesirable as bandwidth limitation apply, and the necessity for separate clock lines arises (buses are synchronous). Utilizing standardized protocols (PMBUS) is ineffective as peer-to-peer communication is not catered for. Running current share traffic and power management traffic concurrently over the same bus/protocol is prohibitive as data contention would be unacceptable.

Digital SMPC controllers do not lend themselves well for the support of analog democratic load share buses as extra data converters would be required (DAC to generate injection signal, ADC to quantize share signal), leading to increased cost, complexity and power dissipation. Despite the obvious system advantages of democratic load share schemes, digital SMPC controllers remained stuck in the M/S load share domain [2], [5], [13], [18], [19].

Value-discrete PWM-encoded M/S load share schemes (with fixed, rotating or automatic masters) have been a popular choice for a long time in commercially available telecommunication rectifiers ([2], [13]), and more recently also reconsidered for low-power converters [18]. The scheme is straight-forward to implement, eliminating most problems of analog share schemes (ground potential differences, noise susceptibility). It lends itself well to integration into digital SMPC controllers, but is sensitive to data distortion due to capacitive loading effects along the current share bus (which may significantly affect rise time, duty cycle, and therefore distort data).

Another M/S load share scheme was proposed in [19], offering a method for embedding clock information in the data stream (through return-to-zero data RZ encoding) in order to achieve a single-wire digital stress-share bus. The scheme is attractive because of its simple bus, ease of clock recovery and collision detection, but remains somewhat constrained in its usefulness due to its single-purpose protocol and narrow encoding word-length (8b). The recommended data transmission rate is 100kb/s leading to a packet repetition rate of 10k·packets/s (with potential audio susceptibility impacts).

A notable exception is [4], where a master-less chain control method suitable for digital SMPC controllers is presented. Moving-average current information is shared via a digital communication bus, with time slots allocated to each of the participating slaves allowing each of them to receive the average current information from the bus, to update it based on local output current, and provide the updated average current back to the bus. The scheme reduces the computation and storage overhead. The fundamental limitation of this method is the latency due to the moving average filter and its impacts on dynamic load share performance. This latency is proportional the number of SMPCs participating in the load share scheme. Implementation details of the share bus are not provided, but it it can be assumed that satisfactory system performance will either require a parallel bus, or a daisy-chained serial bus.

Is a low-latency master-less democratic single-wire digital current share scheme feasible? The following section shows that it is.

III. THE NOVEL DIGITAL STRESS SHARE SCHEME

The principle of operation of the novel DSS scheme (in its purest implementation) is presented in this section. The DSS described in this section focuses on master-less quasi-democratic single-wire current share. Section IV. outlines how the DSS scheme can be extended beyond current share.

The primary challenge to be addressed in a single-wire democratic digital scheme is the variable timing latency. In a traditional time-multiplexed digital democratic scheme the latency of transmission and moving average filter depends on the number of slaves connected to the bus. A subsequent challenge is that the total number of connected slaves may not be known to the individual slaves (potentially requiring system-specific configuration). In fact, the number of slaves may change dynamically during operation.

For the purpose of DSS these challenges are reduced based on the following assumption: Instead of allowing all slaves in the power system to submit their individual output current contribution, a set of relatively homogeneous slaves is assumed, and only the marginal slaves (i.e. those slaves which contribute the highest and the lowest local output current, Io_{max} and Io_{min}) are asked to share their information with others.

978-1-4244-4782-4/10 $26.00 © 2010 IEEE

Now, assume that the average system current per slave can be approximated as

$$Io_{avg} \approx \frac{Io_{max} + Io_{min}}{2} \qquad (1)$$

For the vast majority of systems this is a valid assumption, and the side-effects of this simplification are minor. This is particularly true if erroneous slaves can be detected and isolated from the system.

The problem is now reduced to a challenge whereby only extreme local currents Io_{max} and Io_{min} in a system with an arbitrary number of slaves need to be shared via the bus. Preferably, this information is made available to all slaves concurrently (no daisy chain) and with lowest possible latency (by providing alternating information packets for Io_{max} and Io_{min} over the bus). The latency of transmission and averaging is now fixed, regardless of the number of slaves on the bus. Also, knowledge about the number of slaves connected to the bus is not required. The DSS scheme uses a dedicated single-wire physical interface tapped by an arbitrary number of DSS slaves. A power delivery system employing the DSS bus interconnecting n paralleled SMPCs is shown in Fig. 1.

As a separate DSS clock line would be highly undesirable, the DSS interface needs to be based on asynchronous data

transmission, with clock recovery circuitry employed in each of the DSS receivers. The absence of a clock line means that the serial data-rate on the DSS line is pre-agreed with variations limited to the catch range of the clock recovery circuit. Physically, each slave is connected to the DSS bus through a simple bus transceiver, consisting of an open-drain transmitter with internal pull-up resistor, and a receiver (Schmitt-trigger for improved noise margins, and acceptance of a wide V_{DSS} range 1.6-3.6V).

Serial data transmitted via the DSS line is arranged in DSS packets. The structure of a DSS packet is illustrated in Fig. 3. Transmission speed is a trade-off between bandwidth requirements for current share, transceiver power dissipation, interference considerations, and cost of implementation. The recommended nominal data-rate of DSS is 500kb/s (based on T_{bit}=2μs), and a data packet rate of 20k·packets/s. Start bit **dss_start** and stop bits **dss_stop[5:0]** allow slaves to synchronize themselves to the bit-stream on the bus. Synchronization bits **dss_sync[3:0]** enforce bus state transitions and ensure that long static sequences are eliminated. Checksum bit **dss_checksum** allows single-bit transmission errors to be detected. Synchronization bits **dss_sync[3:0]** each take on the 1's complement value of the immediately preceding bit (e.g. **dss_sync[3]**=~**dss_info[0]**). 10-bit (10b) data quantities are transmitted in field **dss_data[9:0]**. Various DSS packet types exist. They can be distinguished by identifier **dss_info[2:0]**. Currently used packet types are summarized in Table I.

TABLE I.   DSS Packets Identified by DSS Info Bitfield

dss_info[2:0]	DSS Packet Information	
	DSS Packet ID	DSS Packet Description
011b	DSS_SLAVE	Slave device ID (address) and status
100b	DSS_TMIN	Slave local temperature (min)
101b	DSS_TMAX	Slave local temperature (max)
110b	DSS_IOMIN	Slave local output current (min)
111b	DSS_IOMAX	Slave local output current (max)
others	n/a	reserved for future extensions

An arbitration scheme is required so that bus contention can be recognized and resolved by the DSS slaves. Signals on the bus are either dominant ("0" created by a open-drain NMOS output to ground) or weak (also "recessive", pull-up resistor to VDD). For any given time during transmission a DSS slave transmitting the strong state will win arbitration over another DSS slave attempting to transmit the weak state. After a slave is hot-plugged into a system, it first of all synchronizes itself to the ongoing traffic (if any) on the DSS bus. Slaves synchronize themselves to both raw transfer rate, as well as packet transmission. Due to the presence of synchronization bits **dss_sync[3:0]** sufficient clock information is embedded in the data stream.

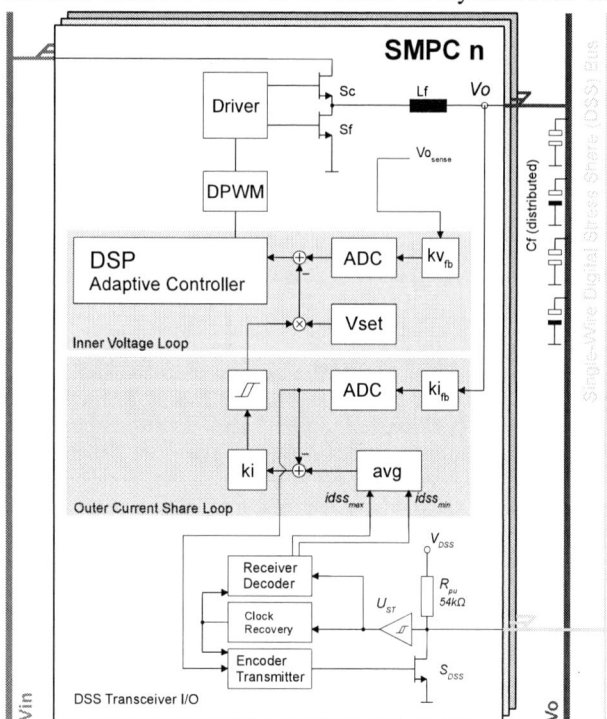

Figure 1.   Paralleled SMPC Controlled by Adaptive Digital Controller Employing the Single-Wire Quasi-Democratic DSS Bus

Uniquely, in order to function as a quasi-democratic bus allowing all slaves to determine the average slave output current Io_{avg} as per (1), it must be ensured that the two marginal DSS slaves in the system will win arbitration in an alternating fashion. DSS slaves receiving the information must be able to detect if Io_{max} or Io_{min} information is currently being transmitted. The novelty of the DSS scheme lies in its simple dynamic data encoding scheme supporting this specific arbitration requirement. Each packet type is identified by **dss_info[2:0]**, with DSS_IOMIN and DSS_IOMAX declared as two distinct packet types.

For each packet to be transmitted all slaves attempt to submit their own encoded local output current Io_{local} in the form of a DSS_IOMIN or DSS_IOMAX packet. DSS slaves encode their Io_{local} as 10b quantities. The encoding is dynamic, depending on whether a DSS_IOMIN packet or a DSS_IOMAX packet is to be transmitted. For DSS_IOMIN packets, data encoding is direct, with lowest current resulting in lowest (and thus most dominant) transmission code, ensuring that the slave carrying the lowest local output current in the system (i.e. the output current closest to $-\infty$) will win arbitration. For DSS_IOMAX packets, data encoding is bit-inverted (1's complement), ensuring that 0's and 1's within the code change their transmission dominance. Thus, for DSS_IOMAX packets, the highest output current Io_{max} (i.e. the output current closest to $+\infty$) will now result in the most dominant transmission code. Transmission codes are illustrated in Fig. 2.

Consider the following example: Two slaves 1 and 2 (prior to current balancing) carrying local currents Io_1 and Io_2, respectively ($Io_1 < Io_2$). For DSS_IOMIN packets slave 1 will win arbitration as its code $dss\_data_{n1min}[9:0]$ will be more dominant than $dss\_data_{n2min}[9:0]$. Conversely, for DSS_IOMAX packets, slave 2 will always win arbitration as its code $dss\_data_{n2max}[9:0]$ will be more dominant than $dss\_data_{n1max}[9:0]$. All DSS slaves connected to the common bus receive DSS information concurrently, allowing them to determine Io_{avg} simultaneously. Under any static or transient conditions the timing latency for determining a new (quasi-)average system current is independent of the number of slaves connected allowing for scalable systems without impacting static or dynamic current share performance.

During normal operation, packets DSS_IOMAX and DSS_IOMIN are submitted to the bus in an alternating fashion, allowing all slaves to update their Io_{avg} at the packet transmission rate. It should be noted that zero current is not explicitly shown in Fig. 2 (Io_1 and/or Io_2 may be positive or negative). A signed-to-unsigned conversion is taking place during the encoding process, ensuring that the DSS scheme works equally well in power delivery systems where local currents may be bidirectional (e.g. bus terminator applications). It should also be noted that all local currents (encoded/decoded by the DSS slaves) are normalized to the rated slave output currents. Thus, slaves with different output current ratings can readily participate in the DSS scheme, and will share currents in the expected relative (rather than absolute) fashion.

A special (but entirely valid) case arises if only one slave is connected to the DSS bus. This slave will continuously succeed in transmitting its own local output current as Io_{max} as well as Io_{min}. Its output current will inherently match the determined Io_{avg}.

All slaves (except those which are in the OFF state, or those with a detected fault) connected to the DSS bus will attempt packet transmission at every opportunity they get. Each slave will start packet transmission whenever it regards the bus as idle (in the high state for a duration of more than $6 \cdot T_{bit} = 12\mu s$), or whenever it concludes that another (faster) slave has commenced packet transmission. In effect, the fastest slave in the system (i.e. the slave with the highest clock frequency) will determine the beginning of the start bit of each packet. This event is detected by all other (slower) slaves, and they all join in to transmit **dss_start**. As all slaves are synchronized to packet transmission, all slaves continue to also concurrently transmit (without conflicts) **dss_info[2:0]** as well as **dss_sync[3]**. As soon as actual data **dss_data[9:0]** is transmitted, arbitration events may occur, leading to those slaves transmitting the weak state to drop out for the remainder of this DSS packet. It is entirely possible (and valid without side effects) that two slaves concurrently submit the same packet, containing the same data value.

The requirements for the electrical DSS interface are easily satisfied. Bits of the incoming DSS data stream are sampled approximately half-way through the bit duration $T_{bit}/2 = 1\mu s$ following the expected/received transition. Ignoring track capacitance, the recommended integrated DSS pull-up resistors $R_{pu} = 54k\Omega$ combined with typical pad/package capacitances of 3-5pF provide sufficient detection margin. Assuming a DSS bus track implemented as an unbalanced PCB microstrip using a 250μm wide outer-layer track separated through a 4mil PCB prepreg from an internal ground plane leads to a track capacitance (per unit length) of about $C_{track} = 20pF/m$.

If the DSS bus significantly exceeds 10cm in length, an external pull-up resistor should be considered. If used in this scenario, its value should be approximately

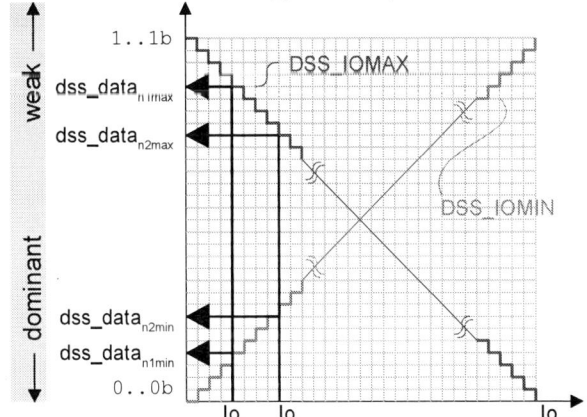

Figure 2.   DSS Dynamic Current Encoding Scheme. Mapping of Signed Output Current Io to Unsigned Transmission Code dss_data[9:0]

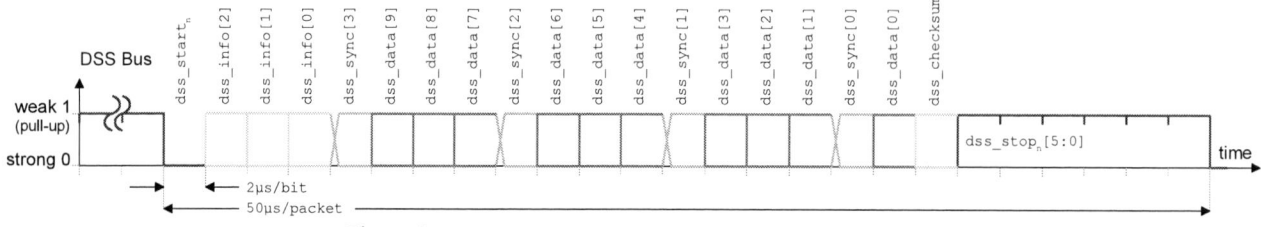

Figure 3.   DSS Packet Structure and Transmission Timing

$$R_{ext} \approx \frac{T_{bit}}{4 \cdot C_{track} \cdot l_{track}} \qquad (2)$$

Propagation delay (per unit length) will be in the region of 5ns/m and usually negligible. Power dissipation due to DSS is not typically a problem, but can be estimated assuming random DSS traffic. Approximately 13 line state transitions per packet are expected (compared to 22 in method [19], and 2 in PWM-based schemes [2] and [18]).

It should be noted that through simple slave configuration, the presented DSS scheme can easily be modified to cater for alternative current share requirements. By restricting each slave to transmission of DSS_IOMAX (or DSS_IOMIN) packages only, M/S current share with automatic master can be accomplished. By restricting all slaves (except one) to DSS reception only, M/S current share with predetermined master can be chosen.

IV. BEYOND CURRENT SHARE

This section describes and illustrates how the DSS scheme can be extended beyond the quasi-democratic current share method outlined in Section III. This extension is possible due to the flexibility of the DSS protocol. It will be shown that the DSS extensions can contribute to further improvements in system reliability, and can be deployed in a power delivery system with smart master-less configuration.

Reliability is recognized as an essential need in electronic systems. This is especially true for power conversion systems supplying critical equipment in telecommunication, computing, medical and industrial applications, where continuous system availability is expected by end-users and operators. Reliability is considered as a means for reducing costs from system operators, where the logistics and cost of repair or replacement is prohibitive. The achievement of reliability is an integral part of power system engineering.

It is important to assess and compare system reliability in a quantitative (rather than anecdotal) fashion. Established methods for quantifying inherent reliability of electronic systems are well known (e.g. [21]), allowing system designers to estimate random failure rates and resulting system reliability parameters MTBF and MTTF ("mean time between/to failures") based on component stress levels.

The general procedure for determining system failure rate λ_{system} is to sum calculated component failure rates λ_p

$$\lambda_{system} = \sum_{components} \lambda_p \qquad (3)$$

with

$$\lambda_p = \lambda_b \prod \pi = \lambda_b \cdot \pi_T \cdot \pi_A \cdot \pi_S \cdot \pi_C \cdot \pi_Q \cdot \pi_E \qquad (4)$$

where λ_b is the base parts failure rate extracted from a comprehensive empirically-derived database. The database is established over extended periods of time, covering a large variety of electronic components, and taking into account various stress factors π (temperature π_T, application π_A, stress π_S, construction π_C, quality grouping π_Q, and environmental π_E).

In SMPCs, the dominant components contributing to failure rate are power switches and magnetic/capacitive energy transfer/storage components (inductors, capacitors). As per [21], power mosfets in typical POL applications (rated output power Po<50W) are assigned an application factor of π_A=4.0, with temperature factor π_T rising from 2.3 (at 70°C) to 3.2 (at 90°C), an increase of approximately 40% for a ΔT=20°C temperature rise. Other than as a contributor to switch junction temperature, no direct penalty is assigned to increases in on-state current as long as the switch is operated within its ratings. Power inductors rated for a 125°C maximum operating temperature see their base failure rate λ_p increase by more than 230% for the same temperature rise. Again, no direct failure rate impact is assigned to inductor current, other than the temperature rise caused by it. Similar observations can be made for filter capacitors. As far as these critical SMPC components are concerned, it is more important to actively manage their thermal stresses, than their current stress.

SMPCs employed in power conversion systems typically see different thermal paths to system temperature reference points, and therefore experience significantly different thermal stress, even if output currents are actively balanced. Depending on particular implementation trade-offs, SMPC stress S can be expressed as a combination of SMPC thermal stress and output current stress, normalized to rated stress S_{rated}, as

$$S = f(T, Io)/S_{rated} \qquad (5)$$

In practice, it makes sense to actively balance combined stress levels S, rather than output currents between participating SMPCs. As a consequence of sharing stress levels S, SMPC failure rates (rather than output currents) are more balanced, reducing the overall system failure rate.

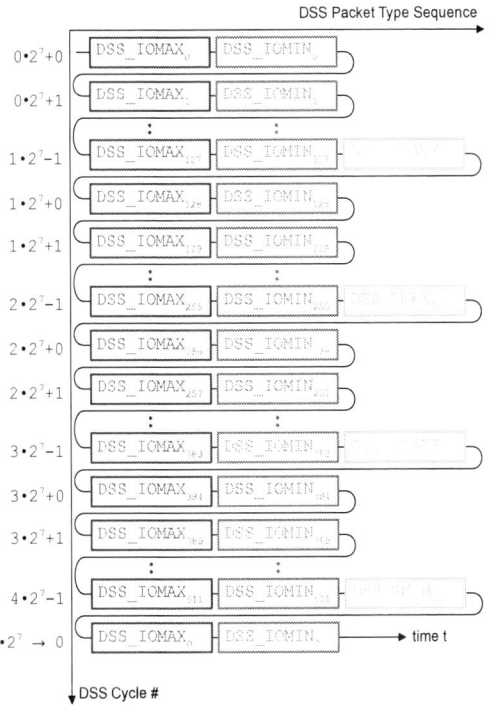

Figure 4.   DSS Extended Packet Sequence With DSS_SLAVE, DSS_TMAX, DSS_TMIN Packets Inserted

The DSS protocol supports this by setting aside additional DSS packets and bandwidth. In addition to the transmission of DSS_IOMAX and DSS_IOMIN packets, DSS slaves may also share information about their thermal stress using extended DSS packets, following the same dynamic encoding scheme presented in Section III. In contrast to output currents, thermal stress is a slow-moving quantity. Therefore, it makes sense to transmit thermal packets (DSS_TMAX, DSS_TMIN, as per Table I) at a much reduced rate, thereby leaving maximum bandwidth to the transmission of DSS_IOMAX and DSS_IOMIN packets. The extended DSS protocol allows the insertion of an extra DSS packet following the transmission of 2^7 DSS_IOMAX and DSS_IOMIN packets. Assuming that DSS slaves also utilize sharing DSS_SLAVE packets (further details below), DSS slaves receive a set of DSS_TMAX and DSS_TMIN packet once in every complete DSS cycle period

consisting of $4 \cdot 2^7 \cdot 2 + 4 = 1028$ DSS packets. This allows the slaves to determine the average thermal slave stress in the system once every $1028 \cdot 50us = 51.4ms$. The extended DSS packets are shown in Fig. 4.

Smart master-less SMPC configuration is also possible through the use of extended DSS packet DSS_SLAVE (see Table I, and Fig. 4). DSS_SLAVE packets allow each and every DSS slave in the system to identify itself to all other slaves, and share with them its unique slave ID (typically the slave address, as for PMBus purposes), 7b encoded, and basic operational status, 3b encoded. All slaves connected can gather information as to how many slaves are participating in the DSS stress share scheme. This information can be used by the slaves to adjust some of their operational parameters (e.g. phase alignment, in conjunction with a synchronization feature, or advanced inter-slave phase dropping) without intervention from a host or master.

Compared to all other DSS packets, slaves must apply a modified arbitration scheme for DSS_SLAVE packets in order to ensure that all slaves (not just the ones with highest or lowest stress) get a chance to periodically submit their DSS_SLAVE packet. During the reserved time slow for DSS_SLAVE packets in the DSS cycle, the slave with the lowest device address will win arbitration, and will submit its DSS_SLAVE packet. Once successfully submitted, this slave will no longer compete for arbitration during the following time slots for DSS_SLAVE packets, giving the lower-priority slaves a chance to submit their DSS_SLAVE packet. This will continue until all slaves have successfully submitted a DSS_SLAVE packet, and no slave will compete during the reserved time slot (i.e. the DSS_SLAVE time slot will not be taken by any slave). All slaves recognize this condition, and a new DSS_SLAVE cycle can start.

Consider the following example: A system consists of 3 slaves with slave IDs 10, 99, and 125. During one full DSS cycle period (1028 packets), two DSS slaves get a chance to submit their DSS slave packet (cycle numbers #127 and #383). During the 1st DSS_SLAVE packet time slot, DSS slave 10 will win arbitration. Once done, slave 10 will not attempt to transmit another DSS_SLAVE packet until it recognizes that the DSS_SLAVE time slot remains vacant. The 2nd DSS_SLAVE packet time slot will be taken by slave 99. The 3rd DSS_SLAVE packet time slot will be taken by slave 125. The 4th DSS_SLAVE packet time slot remains vacant, and will be skipped. A new round of DSS_SLAVE packet submission begins subsequently. A complete DSS_SLAVE packet period depends on the number of slaves n present in the system. It will take $(n+1) \cdot 2 \cdot 2^7 \cdot 2 + 2 = (n+1) \cdot 514$ packets, or $(n+1) \cdot 514 \cdot 50us = (n+1) \cdot 25.7ms$. For the given example, it will take 102.8ms for all slaves to become aware of each other, and share their basic operational status.

It should be noted that the extended DSS capabilities described in this section are optional, and not required for quasi-democratic current share. Slaves with and without handling capabilities for extended DSS packets can co-exist in a given system. In order for this to work, DSS slaves must ignore those DSS packets they cannot handle. Slaves which are only designed to accept packets relating to output current (DSS_IOMAX, DSS_IOMIN) will ignore extended packets, and (due to the priority of extended DSS packets as per Table I) will automatically loose arbitration against fully-featured slaves whenever they attempt to submit extended DSS packets (DSS_TMAX, DSS_TMIN, DSS_SLAVE).

V. MEASUREMENT RESULTS

Measurement results were obtained from two distinct distributed power delivery systems employing identical paralleled DC-DC converters each controlled by a CMOS digital adaptive SMPC controller (Powervation PV3002) with integrated support of single-wire DSS bus and protocol. Power stage parameters are listed in Table II:

TABLE II.   DC-DC CONVERTER PARAMETERS FOR MEASUREMENTS

Parameter	Value
Input voltage Vin	12V
Output voltage Vo	1.5V
Output filter inductor L_f	470nH
Output filter capacitor C_f	4·560µF (Oscon) *in parallel with* 4·100µF (MLC 1206)
Rated output current per converter Io_{max}	30A

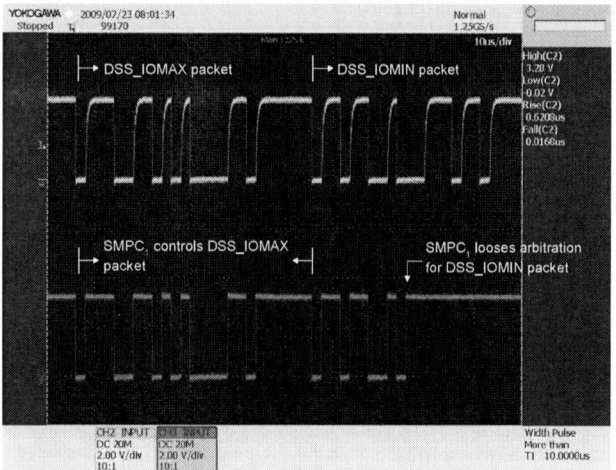

Figure 5.   System$_1$: Arbitration on Single-Wire DSS Bus Between SMPC$_1$ (Io$_1$=20A) and SMPC$_2$ (Io$_2$=10A). Ch2 (green) DSS bus. Ch3 (purple) SMPC$_1$ transmitter output

Both systems utilized a single-wire digital DSS bus implemented as a plain PCB outer-layer micro-strip track, (length l=20cm, characteristic impedance Z=75Ω) with internal DSS pull-up resistors assisted by a single external pull-up resistors of 47kΩ.

System$_1$ is intended to demonstrate the DSS transmission and arbitration scheme in operation, and employed two DC-DC converters (SMPC$_1$ and SMPC$_2$). System$_1$ uses separate loads at distinct load current levels in order to force the system into a known DSS arbitration scenario. Local load currents for SMPC$_1$ and SMPC$_2$ respectively were Io$_1$=20A, and Io$_2$=10A. As seen in Fig. 5, SMPC$_1$ periodically wins arbitration and consistently succeeds in transmitting its local output current framed as DSS_IOMAX packet (as Io$_1$>Io$_2$), while consistently loosing arbitration to SMPC$_2$ during DSS_IOMIN packets.

It can be seen that DSS transmitter of SMPC$_1$ detects an arbitration-lost event approximately half-way through the attempted transmission of a DSS_IOMIN packet and immediately withdraws from the DSS line, leaving SMPC$_2$ submit its local current as DSS_IOMIN packet. Both converters periodically receive system Io$_{max}$ and Io$_{min}$ information and can determine the average POL output current Io$_{avg}$=15A in the system. However, in this test setup, their respective controller cannot adjust local output current to match the average system current as load currents are fixed (electronic constant-current loads).

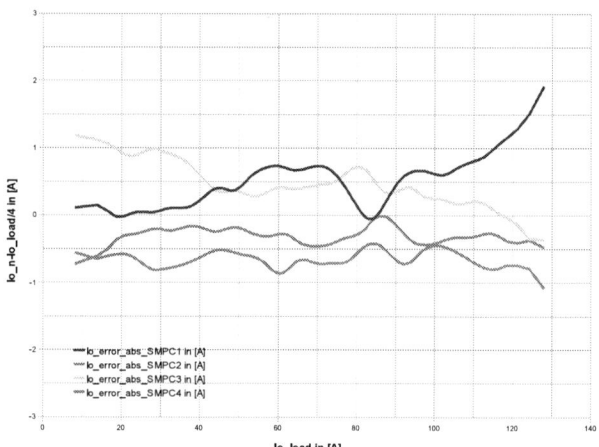

Figure 6.   System$_2$: DSS Static Current Share Performance. Total Io$_{load}$=0-120A delivered by 4 DUTs (SMPC$_1$-SMPC$_4$)

System$_2$ employed four identical paralleled DC-DC converters (SMPC$_1$-SMPC$_4$) mounted on a system board, with a total rated system maximum output current of Io$_{load}$=120A into a common load distributed across the system board. Impedances to the distributed loads were relatively mismatched (ranging from 10-20mΩ) in order to verify static and dynamic performance of the stress share algorithm. Fig. 6 shows static absolute current mismatch (defined as difference of POL output current from theoretically expected Io$_{local}$=Io$_{load}$/4) over the full system load current 0-120A. Mismatches are predominantly by limitations in current sense accuracy in SMPC$_1$-SMPC$_4$.

More details regarding controller design and trade-offs, as well as dynamic current share measurements, are provided in [20].

VI. CONCLUSIONS

This paper introduces a quasi-democratic stress share scheme called DSS. The DSS scheme supports a master-less quasi-democratic stress share scheme for paralleled switching converters featuring a single-wire digital interconnect between the participating slaves.

The DSS scheme defines a simple electrical bus interface, as well as an isochronous protocol. The DSS bus is a single-wire interface, interconnecting an arbitrary number of slaves, allowing them to exchange stress-related information. Like other bus systems, it utilizes distinct driver strengths (weak, dominant) for the two distinct bus states (high, low), in order to achieve wired-AND logic and serve the arbitration needs of the DSS protocol. Uniquely, DSS achieves quasi-democratic behavior by allowing marginal slaves to win packet arbitration and submit their stress information packets in an alternating fashion. Each slave connected to DSS receives these information packets concurrently, allowing the slaves to determine the quasi-average stress level in the system. This, in turn, allows the slaves to adjust their output, in order to match the average system stress. The paper also presented more advanced features of DSS supporting advanced master-less system configuration.

REFERENCES

[1] K.T. Small, "Single wire current share paralleling of power supplies", U.S. Patent 4,717,833, 1988

[2] G. Retz, K. Rinne, "Load balancing of several independently working modules of a power supply installation", European Patent EP 0 958 645 B1, 1998

[3] S. Luo, Z. Ye, R.-L. Lin, F. C. Lee "A classification and evaluation of paralleling methods for power supply modules", Proc. 30th Annual IEEE Power Electronics Specialists Conference PESC 99, 1999, 2, 901-908

[4] Y. Zhang, R. Zane, D. Maksimovic, "Current sharing in digitally controlled masterless multi-phase DC-DC converters", Proc. IEEE 36th Power Electronics Specialists Conference PESC '05, 2005, 2722-2728

[5] K. Kutluay, I. Cadirci, A. Yafavi, Y. Cadirci, "Digital control of universal telecommunication power supplies using dual 8-bit microcontrollers", 37th IAS Annual Meeting Industry Applications Conference Conference Record of the, 2002, 2, 1197-1204

[6] M. M. Jovanovic, D. E. Crow, L. Fang-Yi, "A novel, low-cost implementation of ``democratic'' load-current sharing of paralleled

converter modules", IEEE Transactions on Power Electronics, 1996, 11, 604-611

[7] Y. Panov, M. M. Jovanovic, "Stability and dynamic performance of current-sharing control for paralleled voltage regulator modules", IEEE Transactions on Power Electronics, 2002, 17, 172-179

[8] F. Musavi, K. Al-Haddad, H. Y. Kanaan, "A novel large signal modelling and dynamic analysis of paralleled DC/DC converters with automatic load sharing control", Proc. IEEE International Conference on Industrial Technology IEEE ICIT '04, 2004, 1, 536-541

[9] Y. Panov, M. M. Jovanovic, "Stability and dynamic performance of current-sharing control for paralleled voltage regulator modules", Proc. Sixteenth Annual IEEE Applied Power Electronics Conference and Exposition APEC 2001, 2001, 2, 765-771

[10] Y. Panov, M. M. Jovanovic, "Loop gain measurement of paralleled dc-dc converters with average-current-sharing control", Proc. Twenty-Third Annual IEEE Applied Power Electronics Conference and Exposition APEC 2008, 2008, 1048-105

[11] J. Rajagopalan, K. Xing, Y. Guo, F. C. Lee, B. Manners, "Modeling and dynamic analysis of paralleled DC/DC converters with master-slave current sharing control" Proc. 1996. Eleventh Annual Applied Power Electronics Conference and Exposition APEC '96, 1996, 2, 678-684

[12] H. H. C. Iu, C. K. Tse, "Instability and bifurcation in parallel-connected buck converters under a master-slave current sharing scheme", Proc. IEEE 31st Annual Power Electronics Specialists Conference PESC 00, 2000, 2, 708-713

[13] K. Rinne, K. Theml, J. Duigan, O. McCarthy, "A digitally controlled zero-voltage-switched fullbridge converter", PCIM1995, 317-324

[14] H. Shan, Y. Kang, S. Duan, Y. Zhang, M. Yu, Y. Liu, G. Chen, F. Luo, "Research on a novel digital parallel current sharing control technique of modularized UPS", Proc. ICEMS Electrical Machines and Systems International Conference on, 2007, 106-109

[15] A. Kelly, "Current share in multiphase DCDC converters using digital filtering techniques", IEEE Transactions on Power Electronics, 2009, 24, 212-220

[16] J. A. A. Qahouq, L. Huang, D. Huard, A. Hallberg, "Novel current sharing schemes for multiphase converters with digital controller implementation", Proc. APEC 2007 - Twenty Second Annual IEEE Applied Power Electronics Conference, 2007, 148-156

[17] B. Sahu, "Analysis and design of a fully-integrated current sharing scheme for multi-phase adaptive on-time modulated switching regulators", Proc. IEEE Power Electronics Specialists Conference PESC 2008, 2008, 3829-3835

[18] A. Chapuis, "A digital current share scheme for POL regulators", Proc. Digital Power Forum, 2006

[19] A. Bakker, L. McGarry, "Current share using digital techniques", Proc. Digital Power Forum, 2006

[20] A. Kelly, E. O'Malley, K. Rinne, "Masterless multi-rate control of parallel DC-DC converters", Proc. APEC 2010

[21] Handbook MIL-HDBK-217F, "Reliability Prediction of Electronic Equipment", 1991

Minimum-Sensing Current Control of Three-Phase PFC Converters

Zhonghui Bing   and   Jian Sun

Department of Electrical, Computer and Systems Engineering
Rensselaer Polytechnic Institute, Troy, NY 12180-3590, USA

Abstract– **A nonlinear average current control method using only dc-rail current for three-phase power-factor-corrected (PFC) converters was introduced recently. This paper discusses the effects of dead-time, sector synchronization, input voltage harmonic distortion and sensing delay on its control performance. A dead-time compensation scheme without additional sensing requirement that is suitable to the nonlinear average current control is presented. Based on the analysis of the current control performance considering effects of different factors, design guidelines are provided for practical implementation of the control method. Measurement and simulation results are included to support the conclusions.**

I. INTRODUCTION

Power-factor-correction (PFC) is an effective way to solve harmonic distortion problems. Various current control methods have been reported for three-phase PFC converters [1-3]. Implementing these control methods requires sensing of the phase currents demanding at least two current sensors with wide bandwidth. Techniques to reconstruct phase currents from sensed dc-rail current have been reported [4~7]. This is advantageous for low-cost and low-power applications, and many IGBT modules already have built-in dc current sensors such that no additional current sensors would even be required.

One limitation of these methods is phase current detection dead-zone when the voltage vector moves from one sector to the next or when the modulation index is very low. In either case, the active state of one phase is too short to be measurable from the dc-rail current. The PWM algorithm can be modified to guarantee the minimum length of conduction, but adds more complexity to the control [4, 7].

Another problem with phase current reconstructing is the one switching cycle delay introduced by the reconstructing algorithm. With a modified PWM scheme [6], such delay can be reduced to half a switching cycle. The delay can also be compensated for by state observer [5]. However, performance of such methods can be greatly degraded by parameter variations.

Reference [8] presented a new nonlinear average current control method for three-phase PFC converters (see Fig. 1) by using the dc-rail current directly as feedback. By dividing a line cycle into six sectors and operating two phases in each sector, a simple control circuit consisting of a single integrator and two comparators with additional logic circuits can be used to obtain the PWM control signals that drive the phase currents to have unity power factor.

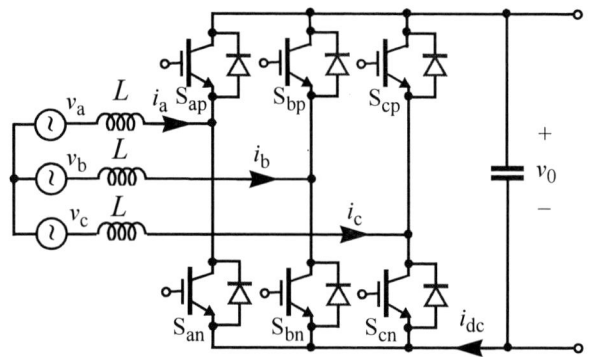

Fig. 1. A three-phase boost-type PWM rectifier.

This paper discusses issues related to the nonlinear average control method that were not addressed in [8]. One of these issues is the switching dead-time inserted in order to avoid direct shoot-through in each bridge leg. Such dead-time introduces additional nonlinear effect causing voltage loss and current distortion [10]. Reported methods to mitigate its effect include compensation by PWM [11] and avoiding complementary operation of the high-side and low-side switches [12]. However, these methods require sensing the phase currents or additional circuits to detect the current polarity, which are not suitable for the new control method. A new compensation scheme that requires no additional circuit is presented in this paper. Additionally, this work also investigates the effects of sector synchronization and sensing delay on current control method in the new control method to provide design guidelines for practical applications.

The rest of the paper is organized as follows: In Section II, the developed nonlinear average current control method for three-phase PFC converters is briefly reviewed. It is followed by the discussion of the dead-time effect and the proposed compensation method. Sector synchronization and other factors affecting current control performance are analyzed in Section IV. Section V summarizes the paper.

II. NONLINEAR AVERAGE CONTROL

The nonlinear control method for three-phase PFC converters was developed based on the control principle described in [9] for single-phase PFC converters and dual-phase current control. In this method, the operation of a PFC converter over a line fundamental cycle is divided into six sectors, each spanning over a 60° interval (shown in Fig. 2). For easy reference, the six sectors are

978-1-4244-4782-4/10 $26.00 © 2010 IEEE

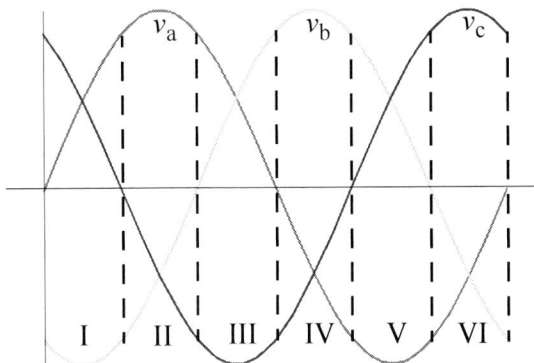

Fig. 2. Definition of six sectors in a line cycle

TABLE I DC-RAIL CURRENT RELATIONSHIP TO THE INPUT CURRENTS

S_{an}	S_{cn}	i_{dc}
0	0	$i_a + i_c$
1	0	i_c
0	1	i_a
1	1	0

TABLE II DESIGNATION OF PHASE VOLTAGES IN EACH SECTOR

Sector	I	II	III	IV	V	VI
v_x	v_a	$-v_b$	v_b	$-v_c$	v_c	$-v_a$
v_y	v_c	$-v_c$	v_a	$-v_a$	v_b	$-v_b$

referred to as Sector I ~ VI in Fig. 2. In each sector, the two phases that have the smallest voltages (therefore the smallest currents under unity power factor operation assumption) are controlled, while the third phase current is uncontrolled and flows through the high or the low side diode depending on the sector of operation.

The current control method is formed by noting that a three-phase PFC converter is equivalent to two parallel single-phase boost converters in each sector. Hence, the control principle presented in [9] for single-phase PFC converters can be generalized to the three-phase converters. Further, each of the controlled phase currents is part of the dc rail current, i_{dc}, whenever the corresponding actively-controlled switch is off (see Table I for Sector I). The dc-rail current contains information of the phase currents that are to be controlled, hence could be used in place of the actual phase currents as feedback.

Although the control works the same way in each sector, the phases to be controlled change periodically. In order to unify the representation of the control algorithm, two notional voltages v_x and v_y as defined in Table II are used. The off-time duty ratios of the switches in the two corresponding controlled phases are

denoted as d_x and d_y. Under ideal conditions, d_x and d_y can be calculated as follows where v_0 is the dc output voltage and g_e is a constant generated by dc output voltage controller):

$$d_{xff} = \frac{1}{v_0}\left[2v_x + v_y - g_e L \frac{d(2v_x + v_y)}{dt}\right] \tag{1}$$

$$d_{yff} = \frac{1}{v_0}\left[v_x + 2v_y - g_e L \frac{d(v_x + 2v_y)}{dt}\right]. \tag{2}$$

Define further $d_M = \max\{d_x, d_y\}$, $d_m = \min\{d_x, d_y\}$, $d_{Mff} = \max\{d_{xff}, d_{yff}\}$, $d_{mff} = \min\{d_{xff}, d_{yff}\}$, and

$$i_m = \begin{cases} g_e v_x & \text{if } (d_{xff} \le d_{yff}) \\ g_e v_y & \text{if } (d_{xff} > d_{yff}) \end{cases}.$$

Based on these, the control method is described by:

$$\frac{1}{T_s}\int_0^{d_m T_s}(i_0 - i_{dc})dt = d_{mff}(i_0 - g_e v_x - g_e v_y), \tag{3}$$

$$\frac{1}{T_s}\int_0^{d_M T_s}(i_0 - i_{dc})dt = d_{Mff}(i_0 - g_e v_x - g_e v_y) + |d_{xff} - d_{yff}|i_m \tag{4}$$

where $i_0 = \alpha(v_x + v_y)$. Detailed operation of the control and its implementation are presented in [8].

III. SWITCHING DEAD-TIME AND COMPENSATION

Dead-time is commonly inserted before the gate turning-on instant to avoid 'shoot-through' in a bridge leg. The effect of this dead-time is more pronounced at high switching frequency [10]. This dead-time also affects the performance of the nonlinear current control method discussed above. The control method is implemented on a three-phase PFC prototype. The parameters of the PFC converter power stage are as follows: rated power 2 kW, input RMS voltage $V_{in} = 110$ V, output voltage $V_0 = 350$ V, and boost inductor per phase $L = 1$ mH. An IRAMY20UP60B three-phase IGBT module with integrated gate drive circuits is used. The dc-rail current is sensed by using the 17 mΩ built-in current sensing resistor of the IGBT module. An Altera DB2 Evaluation board with Altera Cyclone II series FPGA is used to implement all the logic and control functions required by the control. The digital control functions are implemented using the Nios II soft processor [14].

Fig. 3a) shows measured waveforms from the prototype under the condition of 60 Hz line, 36 kHz switching frequency and 600 W output power, which is about 20% of the rated power. The dead-time is set at 1 μs for the safety of the IGBT module, which represents 3.6% of a switching cycle. As can been seen, the current waveform exhibits significant distortion, especially around sector transition regions (phase zero-crossings) where the length of a duty ratio is comparable to the dead-time. Fig. 3b) shows the spectrum of the input current and the computed THD, which

978-1-4244-4782-4/10 $26.00 © 2010 IEEE

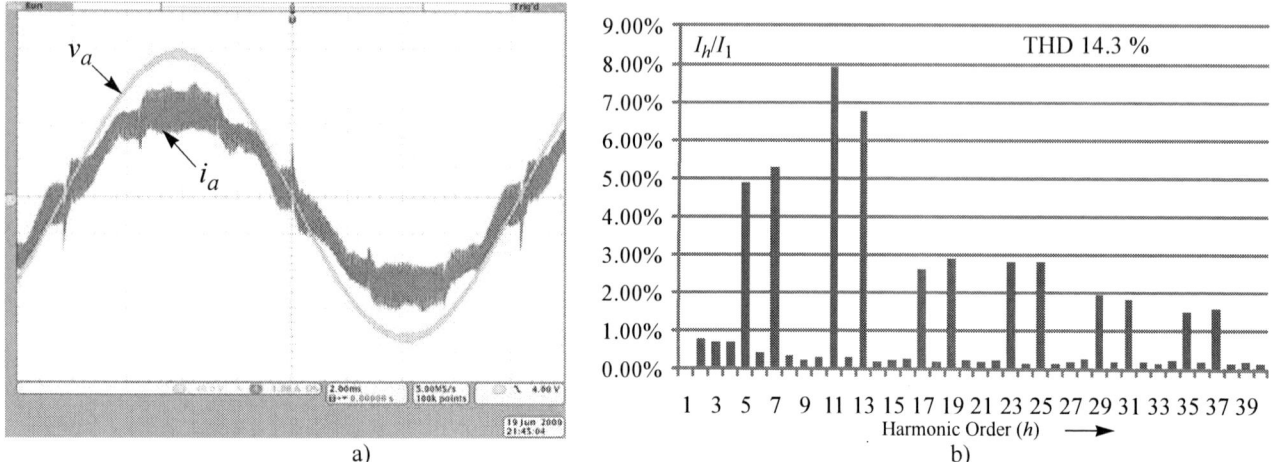

Fig. 3. Measurement of the prototype converter with 60 Hz line, 36 kHz switching frequency and 600 W output power without dead-time compensation: a) phase voltage and current waveform; b) input current spectrum.

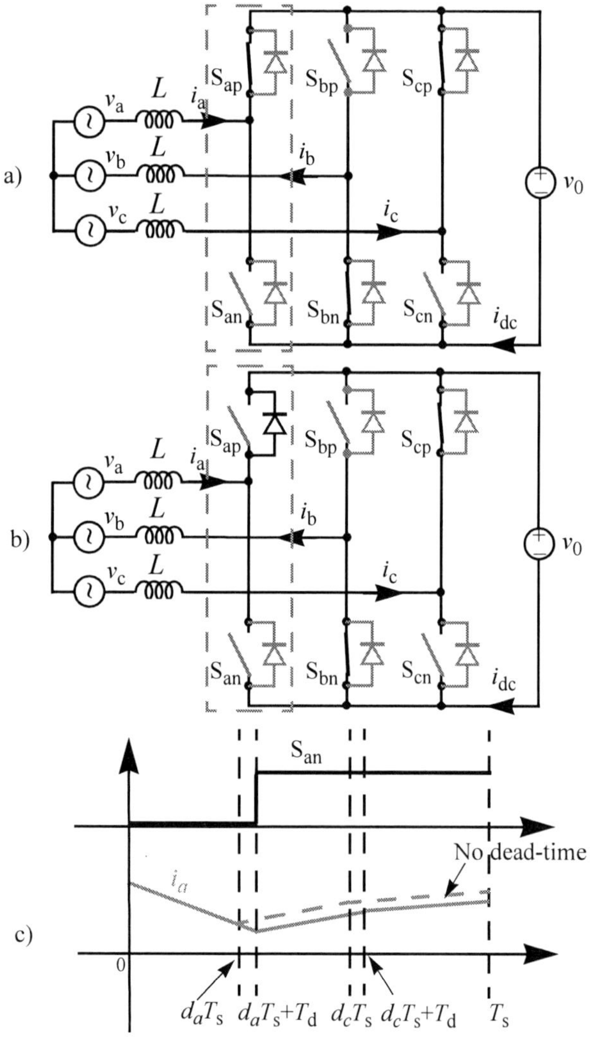

Fig. 4. Equivalent circuit and current commutation during switch dead-time when the phase current doesn't change direction in a switching cycle: a) prior to dead-time interval, b) in dead-time interval, and c) phase current waveform.

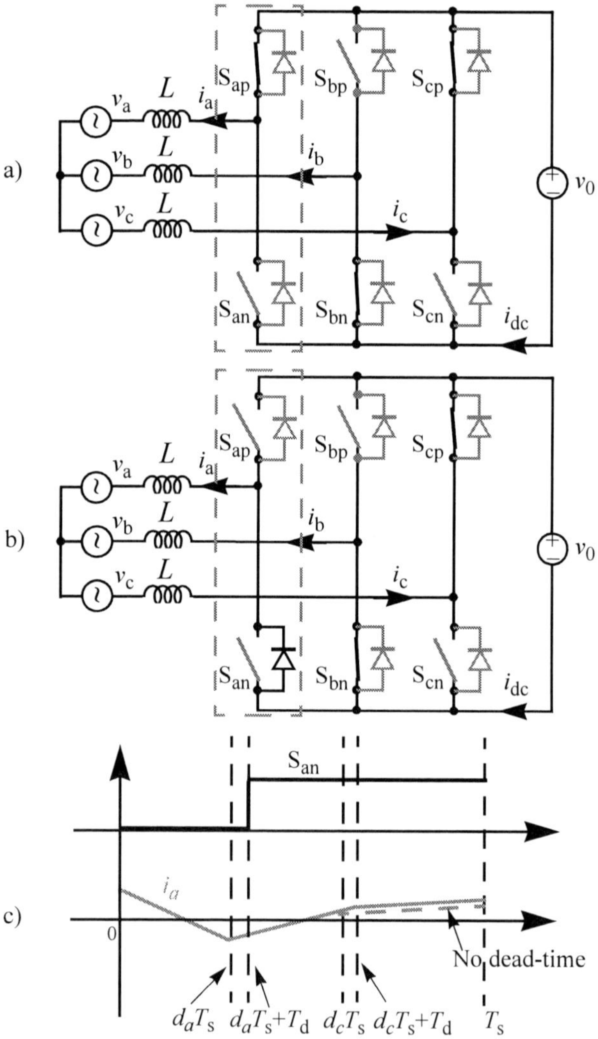

Fig. 5. Equivalent circuit and current commutation during switch dead-time when the phase current changes direction in a switching cycle: a) prior to dead-time interval, b) in dead-time interval, and c) phase current waveform

indicate significant harmonic distortion in the phase current.

Existing dead-time compensation methods discussed in Section I cannot be applied here because 1) current polarity, which is required in existing methods, cannot be determined from dc-rail current sensed; 2) there is no explicit PWM function in the nonlinear current control that can be modified to compensate for the dead-time.

To develop a compensation scheme for the new current control method, we will first analyze the effect of dead-time on current control performance. The effects are found to depend on the current direction. In the interval when the phase current (e.g. i_a) is positive (see Fig. 4), it flows through the upper switch (S_{ap}) when it's turned on, and through its parallel diode when it's turned off and before the lower switch is turned on. Therefore, the inductor current in that phase is subject to additional volt-second of $v_0 T_d$ compared to the case when there is no dead-time. On the other hand, when the phase current is negative (see Fig. 5), (which is only the case near the sector transition points), the current will flow through the parallel diode of the lower side switch as soon as he upper switch is turned off. In that case, the dead-time has no effect on the phase current response.

Based on this analysis, the dead-time effect can be compensated for in Sector I by reducing the duty ratio of the upper switch by T_d. This can be implemented by subtracting T_d / T_s from the feedforward duty ratios defined by (1) and (2):

$$d_{\text{ff}}' = d_{\text{ff}} - T_d / T_s \tag{5}$$

That is,

$$d_{x\text{ff}} = \frac{1}{v_0}\left[2v_x + v_y - g_e L \frac{d(2v_x + v_y)}{dt}\right] - \frac{T_d}{T_s} \tag{6}$$

$$d_{y\text{ff}} = \frac{1}{v_0}\left[v_x + 2v_y - g_e L \frac{d(v_x + 2v_y)}{dt}\right] - \frac{T_d}{T_s}. \tag{7}$$

For the reasons discussed above, compensation scheme (5) should be disabled near the phase current the zero-crossing points. This would require complexity associated determination of the current polarity. In order to overcome difficulty of current polarity detection, the maximum current ripple Δi_M in one switching cycle can be estimated:

$$\Delta i_M = \frac{v_0 T_s [2 d_{m\text{ff}}^2 - 2 d_{m\text{ff}} d_{M\text{ff}} + d_{M\text{ff}}^2 - d_{M\text{ff}}]}{6L} \tag{8}$$

and compared to the reference to determine if the phase current would change direction in a switching cycle.

The overall compensation scheme doesn't require additional sensing signals or circuit modifications. The computation overhead only includes the current ripple estimation (8) and several simple logical operations. The detailed compensation scheme operates as follows:

1) Calculate the feedforward duty ratios from (1) and (2);

2) Estimate the current ripple value based on (8) and compare it with the current reference $g_e v_x$ and $g_e v_y$;

3) If the ripple is larger than one current reference, then update the corresponding duty ratio by introducing the dead-time compensation as (5);

4) Calculate the two comparator reference signals based on the right-hand side of (3) and (4), and update their outputs.

The proposed compensation scheme has been implemented on the three-phase PFC prototype to verify its performance. The operation condition is the same as that for the measurement shown in Fig. 3. After introducing the dead-time compensation, the resulting current waveform is shown in Fig. 6a). Compared to the

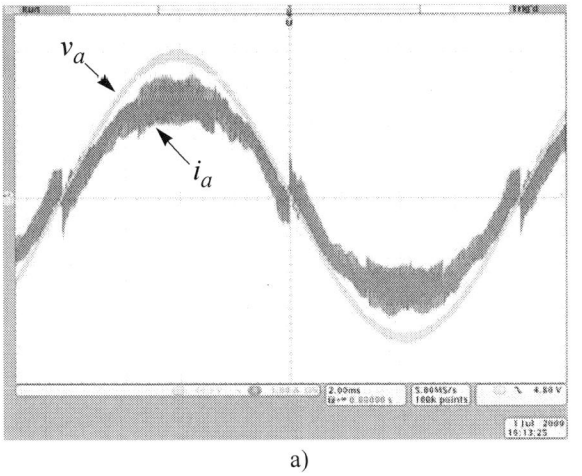

a)

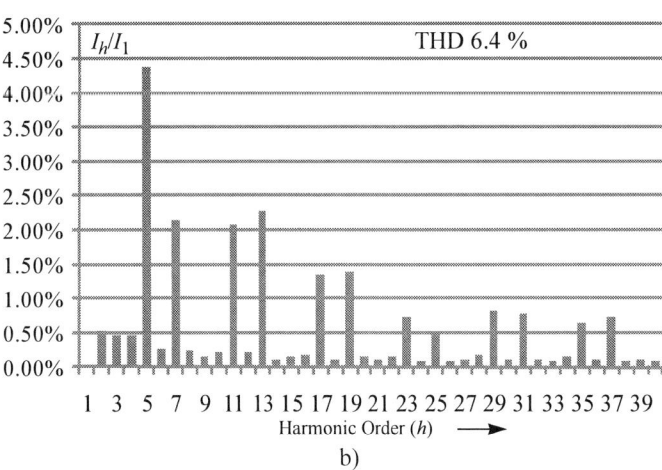

b)

Fig. 6. Measurement of the prototype converter with 60 Hz line, 36 kHz switching frequency and 600 W output power with dead-time compensation: a) phase voltage and current waveform; b) input current spectrum..

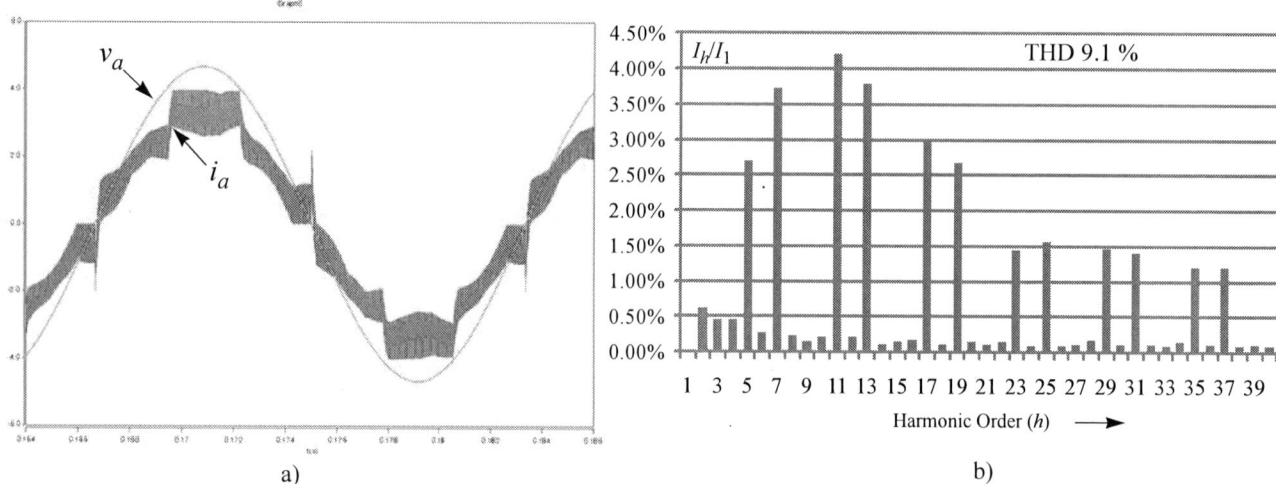

a)

b)

Fig. 7. Measurement of the prototype converter with 60 Hz line, 36 kHz switching frequency and 600 W output power with dead-time compensation but without disable function near zero-crossing points: a) phase voltage and current waveform; b) input current spectrum..

current waveform without dead-time compensation, the distortion caused by dead-time effect, especially around the control sector transitions, is compensated and reduced after inclusion of the scheme proposed into the control implementation. The spectra of the current waveforms are also analyzed and depicted in Fig. 6b). It can be seen that adding compensation helps to bring down 7th, 11th, 13th, 23th and 25th harmonics dramatically so that the THD is greatly reduced.

The simulated current waveform when compensation scheme (5) is not disabled near zero-crossing points is also shown in Fig. 7a). As can be seen, the distortion near zero-crossing deteriorates as the compensation in effect introduces additional undesired volt-second changes to the control response in that region. The spectrum of the phase current shown in Fig. 7b) indicates significant harmonic distortion in the phase current.

IV. OTHER FACTORS AFFECTING CONTROL PERFORMANCE

The current control performance is also affected by several other factors as discussed below. Operation of the PWM signal distribution circuit would be disrupted if the logic signals representing the sector the circuit is operating in change within a switching cycle. This would result in gate control signals intended for certain switches being distributed to other switches which should conduct in the following sector, leading to control errors in phase currents. The effect is more pronounced when the switching frequency is low. To avoid degradation of current control performance due to this effect, the sector detection circuit can be synchronized to the switching clock such that its output doesn't change until the end of a switching cycle even if the actual sector transition takes place inside the switching cycle. Alternatively, the sector detection circuit outputs can be masked by a logic circuit such that their state changes don't affect the operation of the gate signal distribution circuit until the end of a switching cycle.

Harmonic distortion in the input voltages also affects perfor-

mance of the nonlinear current control method. Fig. 8a) shows one phase voltage waveform and three-phase current waveforms when the line contains line 5% 5th harmonic of the 400 Hz fundamental. It can be observed that the current waveforms are controlled in proportion to the distorted phase voltage instead of being sinusoidal. Fourier analysis of the input current (shown in Fig. 9) indicate that the 5th harmonic is close to 5% while other harmonics are suppressed. This is because the current reference signals and the control signals are calculated to be proportional to the sensed input voltages regardless of the presence of harmonic distortion.

The control algorithm can be modified if sinusoidal input current are desired when the line is distorted. This can be accomplished by extracting the fundamental component of the phase voltages using PLL or other techniques, and use the extracted fundamental voltage in place of the actual phase voltage in the control equations, as given below, where v_{x1} and v_{y1} represent the transformed fundamental components of the input voltages:

$$d_{x\text{ff}} = \frac{1}{v_0}\left[2v_x + v_y - g_e L \frac{d(2v_{x1} + v_{y1})}{dt}\right] \qquad (9)$$

$$d_{y\text{ff}} = \frac{1}{v_0}\left[v_x + 2v_y - g_e L \frac{d(v_{x1} + 2v_{y1})}{dt}\right] \qquad (10)$$

$$\frac{1}{r}\int_0^{d_m T_s}(i_0 - i_{\text{dc}})dt = d_{m\text{ff}}(i_0 - g_e v_{x1} - g_e v_{y1}, \qquad (11)$$

$$\frac{1}{T_s}\int_0^{d_M T_s}(i_0 - i_{\text{dc}})dt = d_{M\text{ff}}(i_0 - g_e v_{x1} - g_e v_{y1})$$
$$+ |d_{x\text{ff}} - d_{y\text{ff}}|i_m \qquad (12)$$

Fig. 8b) shows one set of simulation results with this technique. The fundamental components are obtained by PLL, which are

978-1-4244-4782-4/10 $26.00 © 2010 IEEE

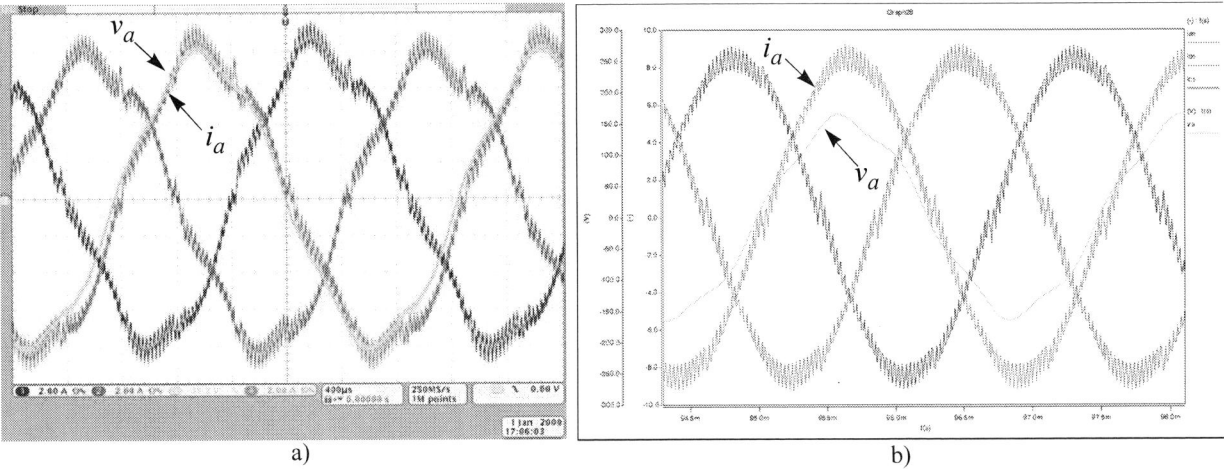

a)   b)

Fig. 8. One phase input voltage and current waveforms of the prototype converter with 400 Hz line frequency, 36 kHz switching frequency and 1850 W output power: There is 5% 5th harmonic distortion in the input line voltages. a) Measured waveforms with original control algorithm; b) simulated waveforms with modified control algorithm.

further transformed using the definition in Table II for each sector. The control are carried out based on (9-12). As can be seen, the 5th voltage harmonic is suppressed and current waveforms maintain sinusoidal as expected.

AD converters that have multiple channels sharing one sample-and-holder may be used to sense the three phase input voltages and dc output voltage to reduce the cost. The sequential sensing of the phase voltages causes delays among phase voltage samples. Such delay is equivalent to having unbalanced phase voltages, and is found to affect the balance of phase currents, especially when the line frequency is high. In order to limit such effect, the sampling delay must be limited. Fig. 10a) shows phase current measurement when the ADC has 2.5 µs delay. It can be seen that although phase a current is controlled to be almost sinusoidal, the other two phase current are much distorted. The spectrum of phase

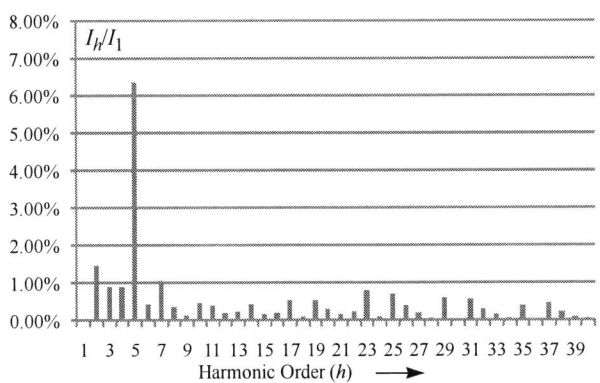

Fig. 9. Current spectra of input current in Fig. 8a)

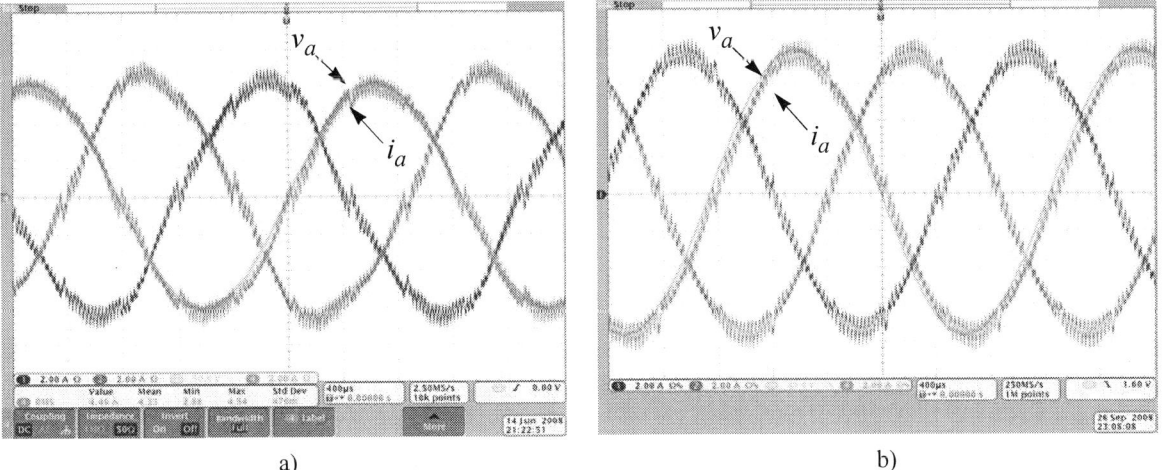

a)   b)

Fig. 10. Measured input voltages and current waveforms of the prototype converter with 400 Hz line, 36 kHz switching frequency and 1850 W output power. a) 2.5 µs ADC delay; b) 1.1 µs ADC delay.

978-1-4244-4782-4/10 $26.00 © 2010 IEEE   341

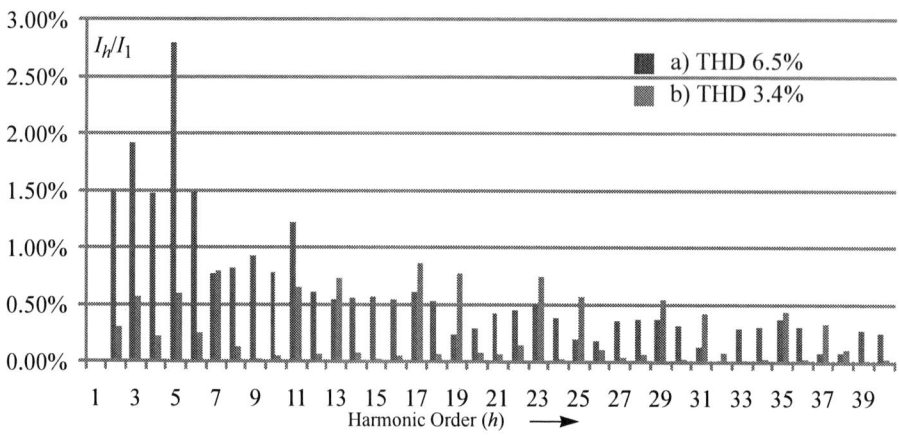

Fig. 11. Spectra comparison of phase c current waveforms in Fig. 10

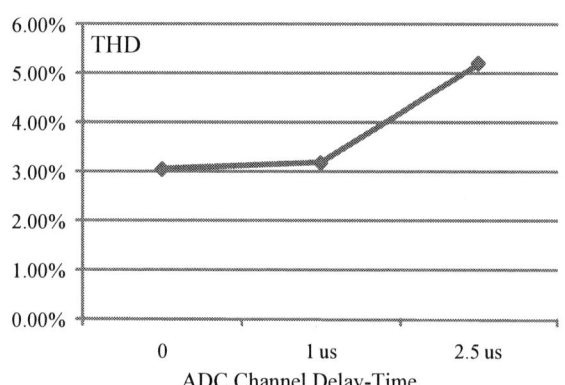

Fig. 12. THD comparisons under different ADC channel delay-times

c current is given in Fig. 11. Fig. 10b) shows the measurement when the delay is reduced to 1 μs. As the spectral comparison in Fig. 11 shows, significant reduction in current harmonics is obtained.

Operation of the control with no ADC delay was simulated in Saber and the resulting input current THD is computed and compared to the other two cases in Fig. 12. The comparison indicates that 1 μs delay is acceptable for the selected control parameters.

V. SUMMARY

The new nonlinear current control method reported lately for three-phase PFC converters by using the dc-rail current directly as feedback yields a simple control circuits and excellent current control performance. This paper continues on discussions on the issues that should be taken care of for the performance of the control method. Among them, the switching dead-time effect is first addressed, and a compensation scheme that works with the nonlinear control method without additional sensing requirement or circuit changes is introduced and verified by experimental measurements. Other implementation issues such as sector

detection and ADC sensing delay are also discussed, and design guidelines are provided based on the associated conclusions. Also, in order to obtain sinusoidal current under harmonic distorted line condition, the control should be changed by including extracted line fundamental components and the algorithm are modified accordingly.

REFERENCES

[1] H. Mao, D. Boroyevich, A. Ravindra, and F. C. Lee, "Analysis and design of high frequency three-phase boost rectifiers," in *Records of IEEE APEC '96*, vol.2, pp. 538-544,1996.

[2] V. Blasko, and V. Kaura, "A new mathematical model and control of a three-phase AC-DC voltage source converter," *IEEE Transactions on Power Electronics*, vol. 12, no.1, pp. 116-123, 1997.

[3] C. Qiao, and K. M. Smedley, "A general three-phase PFC controller for rectifiers with a parallel-connected dual boost topology," *IEEE Transactions on Power Electronics*, vol. 17, no. 6, pp. 925-934, 2002.

[4] W. Lee, D. Hyun, and T. Lee, "A novel control method for three-phase PWM rectifiers using a single current sensor," *IEEE Transactions on Power Electronics*, vol. 15, no. 5, pp. 861-870, 2000.

[5] W. Lee, T. Lee, and D. Hyun, "Comparison of single-sensor current control in the DC link for three-phase voltage-source PWM converters," *IEEE Transactions on Industrial Electronics*, vol. 48, no. 3, pp. 491-505, 2001.

[6] H. Kim and T. M. Jahns, "Phase current reconstruction for AC motor drives using a DC link single current sensor and measurement voltage vectors," *IEEE Transactions on Power Electronics*, vol. 21, no. 5, pp. 1413-1419, 2006.

[7] F. Blaabjerg, J. K. Pedersen, U. Jaeger, and P. Thoegersen, "Single current sensor technique in the DC link of three-phase PWM-VS inverters: A review and a novel solution," *IEEE Transactions on Industrial Applications*, vol. 33, no. 5, pp. 1241-1253, 1997.

[8] Z. Bing, X. Du, and J. Sun, "Three-phase PFC current control using DC-rail current as feedback", in *Records of IEEE ECCE'09*, pp. 1212-1219, 2009.

[9] M. Chen, A. Mathew, and J. Sun, "Nonlinear current control of single-phase PFC converters," *IEEE Transactions on Power Electronics*, vol. 22, no. 6, pp. 2187-2194, 2007.

[10] S. Jeong, and M. Park, "The analysis and compensation of dead-time effects in PWM inverters," *IEEE Transactions on industrial Electronics*, vol. 38, no. 2, pp. 108-114, 1991.

[11] A. R. Munoz and T. A. Lipo, "On-line dead-time compensation technique for open-loop PWM-VSI drives," *IEEE Transactions on Power Electronics*, vol. 14, no. 4, pp. 683-689, 1999.

[12] L. Chen and F. Z. Peng, "Dead-time elimination for voltage source inverters," *IEEE Transactions on Power Electronics*, vol. 23, no. 2, pp. 574-580, 2008.

[13] D. Lee and D. Lim, "AC voltage and current sensorless control of three-phase PWM rectifiers," *IEEE Transactions on Power Electronics*, vol. 17, no. 6, pp. 966-972, 2002.

[14] Altera Corp., *Nios II Software Developer's Handbook*, 2007

Direct Power Control of a Dual Converter Operating as Synchronous Rectifier

José Restrepo[*], José M. Aller[*], Alexander Bueno[*], Julio C. Viola[*], Alberto Berzoy[*] and Thomas Habetler[**]
[*]Universidad Simón Bolívar, Caracas, Venezuela
Email: {restrepo, jaller, bueno, jcviola, aberzoy}@usb.ve
[**]School of Electrical and Computer Engineering
Georgia Institute of Technology, Atlanta, Georgia 30332–0250
Email: thabetler@ece.gatech.edu

Abstract—This work presents a dual converter employed as a rectifier with power factor regulation and bidirectional power flow. The active and reactive power flowing into the converter is controlled using an optimized direct power control algorithm. The multilevel structure of the converter is exploited to control the voltage level in each sub converter by selecting the modulation method from one commonly found in the literature, with the option of clamping one of the sub converters. These modulation methods are used to control the power taken by each sub converter, providing limited DC link voltage regulation. The system is first simulated in SIMULINK and the results were experimentally validated using a digital signal processor (DSP) based test rig.

I. INTRODUCTION

Nowadays, multilevel inverters continue to be a topic of intense research [1]–[4], and several modulation techniques have been reported with numerous advantages over conventional two-level inverters [5]–[8]. One important advantage is the possibility to improve harmonic content for the synthesized voltage, with a reduced number of commutations [5]. Another advantage of multilevel converters is the possibility to reach higher voltage levels and higher power rating with power devices having lower breakdown voltages [5], [9]. The increase in components in multilevel converters results in a corresponding increase in the number of valid commutation states, and this in smoother changes in the state variables of the system and its ensuing reduction in the output voltage dv/dt.

Among many existing multilevel topologies, the dual converter structure has the advantage that multilevel operation can be obtained by using two standard two-level converters. Since it was proposed [10], this topology has been used mainly for control of induction motors with open-end stator windings [3], [4], [11]. Others applications for this topology are found in photovoltaic generation systems [12], [13]. Multilevel topologies have been used also as VAR compensators [14], but there are no reported applications of the dual converter working as an active front end (AFE) rectifier with controlled power factor. The possibility of its use as an active filter was suggested in [12], however.

In this work, a synchronous rectifier is obtained by connecting the center tap of each sub converter's leg to the open end of a three-phase transformer secondary windings. The

converter's block diagram is shown in Fig. 1(a) where the direct power control (DPC) algorithm proposed in [15] is used to control the dual converter. In this version of the DPC algorithm, the optimum voltage space vector to be synthesized by the dual converter, in each control cycle, is obtained in a straight forward manner by using a closed form formula, derived directly from the expressions of the instantaneous active and reactive power at the converter's input. Additionally, since the dual converter's structure has two independent DC links, it was shown that different modulation methods can be used to control the power flow into each DC link.

The principle of alternate-sub-hexagonal center PWM switching strategy proposed in [3] for standard SVM-PWM is extended to the generalized space vector PWM method in this paper.

The paper is organized as follows: section II shows a description of the dual converter's space vector computation derived from the active and reactive power demands. Section III presents the general PWM strategy used to obtain the rectifier's voltage vector. Section IV shows a generalization of the principle of alternate sub-hexagonal center initially proposed in [16]. Finally in section V, the proposed scheme is analyzed by simulations and by an experimental test.

II. SPACE VECTOR COMPUTATION

The control algorithm used in this work relies in the computation of the dual converter space vector $\vec{v}_r$ required by the active and reactive power demands. From Fig. 1(b) the system can be modeled with the following first order differential equation.

$$\vec{v}_s(t) = R_{AF}\vec{i}_s(t) + L_{AF}\frac{d\vec{i}_s(t)}{dt} + \vec{v}_r(t) \qquad (1)$$

A discrete version of the system model can be obtained by using a first order approximation of (1).

$$\vec{v}_s(k) = R_{AF}\vec{i}s(k) + \frac{L_{AF}}{T_s}\left[\vec{i}_s(k+1) - \vec{i}_s(k)\right] + \vec{v}_r(k) \quad (2)$$

where T_s is the sampling time.

The instantaneous active and reactive power in the system can be computed using the system's state variables described

978-1-4244-4782-4/10 $26.00 © 2010 IEEE

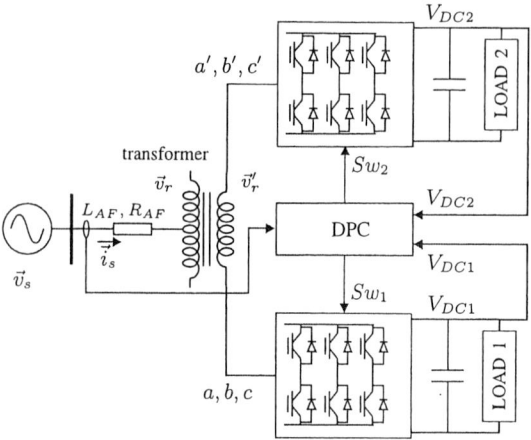

(a) Converter diagram

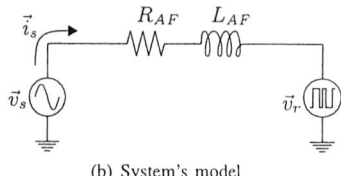

(b) System's model

Fig. 1. (a) Single phase diagram of the dual converter. (b) system's model.

in (x, y) coordinates as follows,

$$
\begin{aligned}
p &= v_{sx} i_{sx} + v_{sy} i_{sy} \\
q &= v_{sy} i_{sx} - v_{sx} i_{sy}
\end{aligned}
\tag{3}
$$

For a required instantaneous active and reactive power, the system current can be obtained as

$$
\begin{aligned}
i_{sx} &= \frac{p \cdot v_{sx} + q \cdot v_{sy}}{v_{sx}^2 + v_{sy}^2} \\
i_{sy} &= \frac{p \cdot v_{sy} - q \cdot v_{sx}}{v_{sx}^2 + v_{sy}^2}
\end{aligned}
\tag{4}
$$

For the next control step $(k + 1)$ the dual converter voltage needed for impressing system current defining the required active and reactive power flow is given by the following expressions [15].

$$
\vec{v}_r(k+1) = \vec{v}_s(k+1) - \frac{L_{AF}}{T_s} \left[\vec{i}_s(k+1) - \vec{i}_s(k) \right] - R_{AF} \vec{i}_s(k)
\tag{5}
$$

An approximate value of $\vec{v}_s(k + 1)$ can be obtained by rotating $\vec{v}_s(k)$ by $e^{j\omega T_s}$.

III. GENERALIZED SPACE VECTOR PWM METHOD

The dual converter's space vector, $\vec{v}_r$, is obtained using a generalized space vector modulation algorithm on each sub converter. This algorithm uses the following intermediate variables,

$$
f_x = v_{rx}; \qquad f_y = \frac{v_{ry}}{\sqrt{3}}
\tag{6}
$$

Using these variables, a sector selector $N(f_x, f_y)$ can be defined as

$$
N(f_x, f_y) = \left\lfloor \frac{3\theta}{\pi} \right\rfloor =
$$
$$
= 2.5 - \operatorname{sgn}(f_y) \left[(f_x > f_y) + (f_x > -f_y) + 0.5 \right]
\tag{7}
$$

In two level three phase converters the null vector can be obtained using states $(0,0,0)$ or $(1,1,1)$, and the amount of time employed to synthesize vector zero using either $(0,0,0)$ or $(1,1,1)$ is defined by δ; if $\delta = 0$ the null vector is synthesized using only state $(1,1,1)$ and if $\delta = 1$ the null vector is obtained using only $(0,0,0)$. For the modulation, a sector selector $N(f_x, f_y)$ locates the space vector to be synthesized in one of the sectors in the hexagonal space, and defines the expressions needed for computing the required duty cycles D_a, D_b and D_c, according to Table I. In Table I, δ is used to select the modulation method.

TABLE I
EXPRESSIONS FOR DUTY CYCLES REQUIRED FOR THE IMPLEMENTATION
OF THE MODULATION ALGORITHM.

N	D_a	D_b	D_c
0	$\delta (f_x + f_y - 1) + 1$	$D_a - f_x + f_y$	$D_a - f_x - f_y$
1	$D_b + f_x - f_y$	$\delta (2f_y - 1) + 1$	$D_b - 2f_y$
2	$D_b + f_x - f_y$	$\delta (-f_x + f_y - 1) + 1$	$D_b - 2f_y$
3	$D_c + f_x + f_y$	$D_c + 2f_y$	$\delta (-f_x - f_y - 1) + 1$
4	$D_c + f_x + f_y$	$D_c + 2f_y$	$\delta (-2f_y - 1) + 1$
5	$\delta (f_x - f_y - 1) + 1$	$D_a - f_x + f_y$	$D_a - f_x - f_y$

Table I can be used directly as the algorithm for implementing the generalized SVPWM as a function of the null vector ratio δ. Table II shows the values of δ required to implement several commonly used modulation methods.

TABLE II
MODULATION METHODS FOR DIFFERENT VALUES OF δ.

Modulation	δ
$DPWM_{min}$	1
$DPWM_{max}$	0
SVPWM	$\frac{1}{2}$
$DPWM_0$	$\frac{1}{2} \left[1 + (-1)^{n_1} \right]$
$DPWM_1$	$\frac{1}{2} \left[1 + (-1)^{n_2} \right]$
$DPWM_2$	$\frac{1}{2} \left[1 + (-1)^{(n_1+1)} \right]$
$DPWM_3$	$\frac{1}{2} \left[1 + (-1)^{(n_2+1)} \right]$
For (x, y) coordinates:	
$n_1 = N = 2.5 - \operatorname{sign}(f_y) \left[(f_x > f_y) + (f_x > -f_y) + 0.5 \right]$	
$n_2 = 3.5 - \operatorname{sign}(f_x + 3f_y) \left[(f_x > 0) + (f_x > 3f_y) + 0.5 \right]$	

IV. GENERALIZED PRINCIPLE OF ALTERNATE SUB-HEXAGONAL CENTER

In a dual converter, space vectors can be synthesized using an infinite number of switching strategies for both sub converters. This plethora of combinations makes possible the imposition of different constrains on the dual converter's operation. For example, in the selection of a modulation strategy with reduced number of commutations, power flow regulation is done by controlling each sub converter contribution in the synthesis of the final space vector [3], [17], allowing for DC link voltage regulation. Additionally, different modulation methods can be used in each sub converter and a constraint on thermal stress on the power devices or common mode voltage reduction can be implemented.

In this work, a generalization of the alternate sub-hexagonal center PWM presented in [3], [16] is proposed. The duty cycle demands in each sub converter are computed by first selecting the modulation method from Table II, with the proper selection of δ. With the modulation method selected, general 2-level duty cycles are computed and a "sub converter clamping selector" ($SCCS$) value is computed using the following expression,

$$SCCS = (D_a > 0.5) + (D_b > 0.5) + (D_c > 0.5) \qquad (8)$$

For reducing the common mode voltage sub converter 1 is clamped if $SCCS < 2$ otherwise sub converter 2 is the one clamped [16]. If sub converter 2 is clamped, a further reduction in the common mode voltage can be achieved by employing $DPWM_{min}$ in sub converter 1. On the other hand, if sub converter 1 is clamped the reduction in common mode voltage can be increased by using $DPWM_{max}$ in sub converter 2. Figure 2(a) shows an example of space vector synthesis for $\delta = 1$, with sub converter 2 clamped and sub converter 1 using PWM to synthesize the remaining part of the space vector. The base vector for sub converter 1 are marked as $\{a, b, c\}$, and for sub converter 2 as $\{a', b', c'\}$ and they are the clamping values allowed for each sub converter. Figure 2(b) shows the clamping zones for both sub converters when using the modulation methods presented in [18]. It is important to notice that a change in the definition of $SCCS$ could produce additional clamping zones patterns. A useful constraint that can be imposed in the modulation process is how the power flows into each sub converter. The contribution to the space vector provided by each sub converter can have a different magnitude, while the current flowing into each sub converter is the same. Therefore, some control of the active power going into each sub converter can be obtained [4].

V. SIMULATION AND EXPERIMENTAL RESULTS

The dual converter system was simulated using MATLAB-SIMULINK to verify operation of the dual converter operating as a controlled rectifier under direct power control. Additionally, the simulations allow to verify the effect of different modulation methods over the DC link voltage imbalance on both sub converters.

For the experimental test, the proposed algorithm was implemented on a custom build floating-point DSP-(ADSP-21369) based test rig shown in Fig. 3(a). The power stage in each sub converter uses six 50 A, 1200V insulated gate bipolar transistors (IGBTs) with a 2200 μF, 450 V capacitor in the dc link and the sampling and switching frequency was 10 kHz.

Fig. 3(b) shows that only two out of six converter branches are under PWM while the remaining are not switching at any

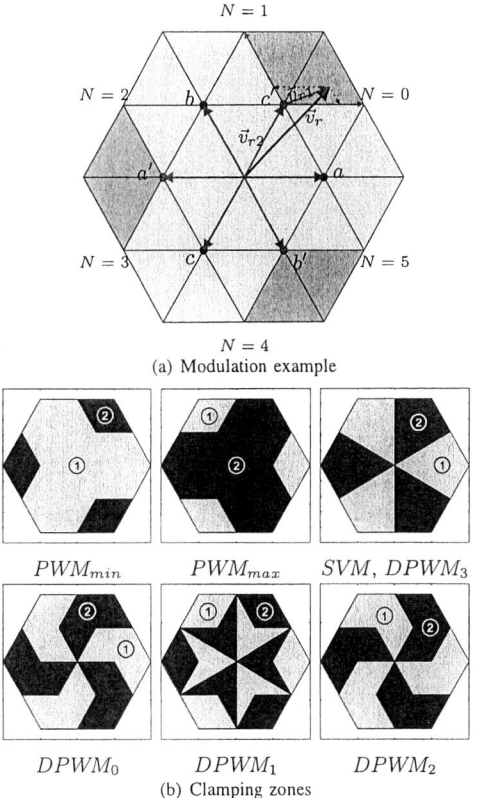

(a) Modulation example

(b) Clamping zones

Fig. 2. Modulation example and clamping zones for sub converters 1 and 2.

given control cycle. This happens only for the discontinuous modulation methods. Also, the clamped sub converter is operating with only one of its branches in high state.

The effect of the modulation method on the sub converters DC link voltage was first simulated, and the results for the modulation methods presented in Table II are shown in Fig. 4. These simulations show that for balanced loads, for continuous SVM only $\delta = 0.5$ produces balanced outputs. For discontinuous SVM, only $DPWM_{min}$ and $DPWM_{max}$ produce imbalances with opposite effect on the DC link voltages. This feature can be exploited to attain some degree of power flow control while clamping one of the sub converters using an algorithm equivalent to the one presented in [4].

Figure 5 shows the system line voltage ($\vec{v}_s$), the voltage at the dual converter's input ($\vec{v}_r'$) and the system current ($\vec{i}_s$) for unity power factor operation. Figure 6 shows the harmonic content of the supply current $\vec{i}_s$ when the dual converter is operating using $DPWM_1$ and the power demands are for unity power factor operation, the resulting THD obtained in this experiment is 3.4%. Figure 7 shows the experimental DC link voltage for two modulation methods, in Fig. 7(a) shows the DC-link voltage evolution in each sub converter for the $DPWM_{max}$ method, and Fig. 7(b) shows the corresponding results for SVM ($\delta = 0.5$). These results are in agreement with the simulations presented in Fig. 4. Small mismatches in the voltage levels are due to tolerance errors in parameter values for both sub converters.

978-1-4244-4782-4/10 $26.00 © 2010 IEEE   345

(a) test rig

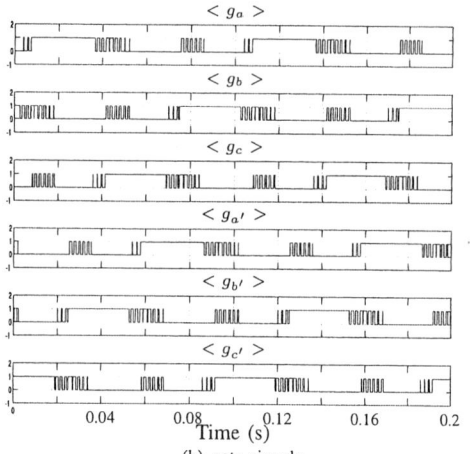

(b) gate signals

Fig. 3. Experimental test-rig and example of gate signals for $DPWM_2$.

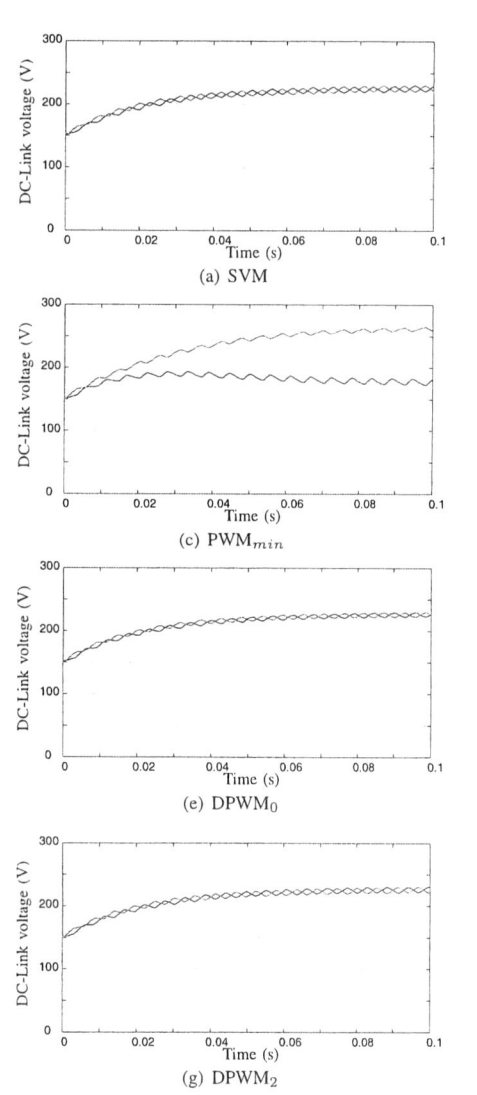

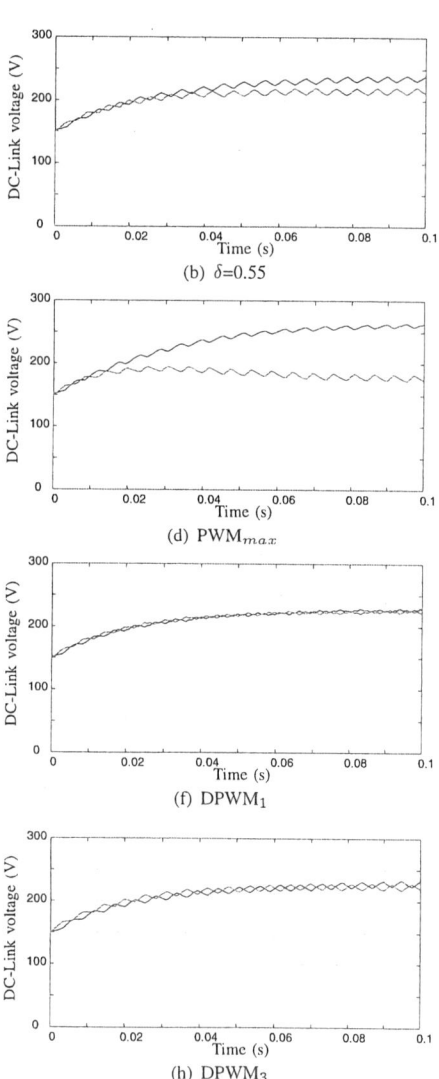

Fig. 4. Results for effect of modulation method on the sub converters DC-link voltage (simulation).

978-1-4244-4782-4/10 $26.00 © 2010 IEEE   346

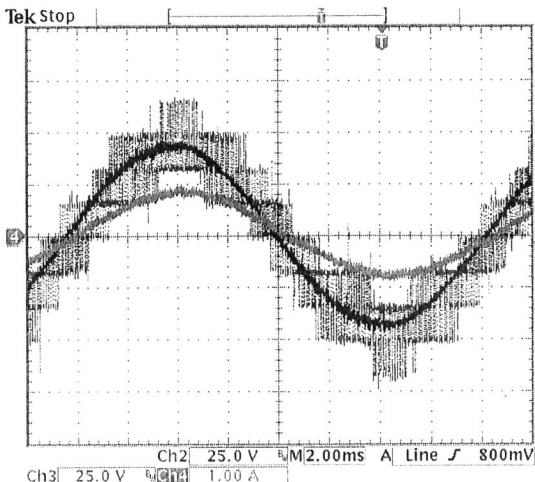

Fig. 5. Experimental waveforms.

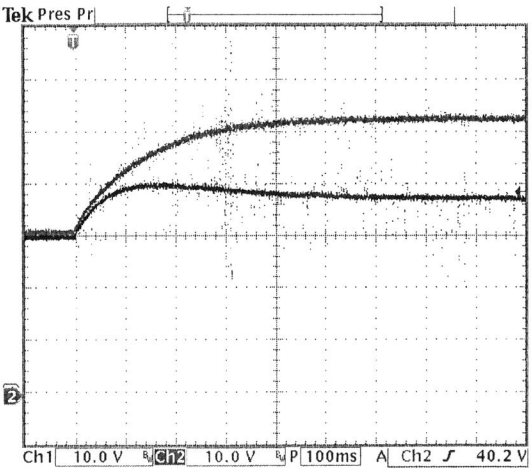

(a) PWM$_{max}$

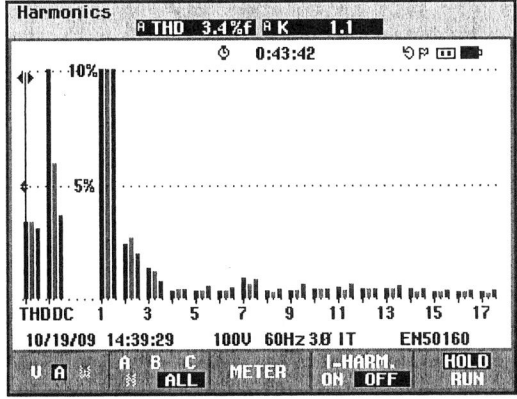

Fig. 6. Experimental results for unity power factor using $DPWM_1$.

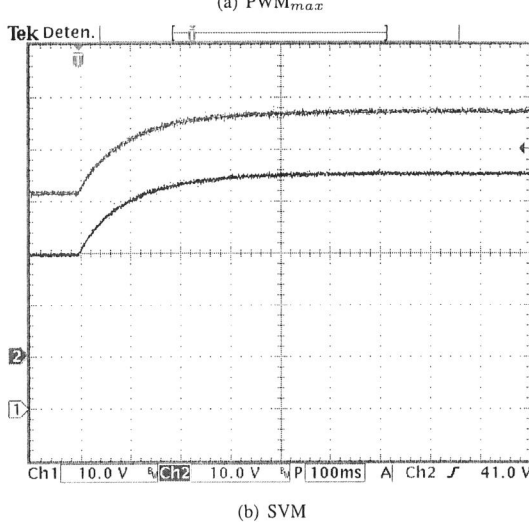

(b) SVM

Fig. 7. Results for effect of modulation method on the sub converters DC-link voltage (experimental).

VI. CONCLUSION

The dual converter's topology has been tested as a controlled rectifier, for increased power conversion employing lower voltage switching devices. The algorithm used in the control of the dual converter was an optimized version of the direct power control (DPC). Also, a generalization of the principle of alternate sub-hexagonal center have been presented, with the generation of different clamping zone patterns for commonly used modulation methods. Simulations and experimental results show that different modulation strategies can be employed in the regulation of DC links in both sub converters. Some modulation strategies allow the converter to operate in a balanced or imbalanced fashion. Also, the flexibility of the modulation strategies result in a reduction on the number of simultaneous switchings in the sub converters. The $\frac{dv}{dt}$ is reduced by the increased number of levels generated by the dual converter topology.

ACKNOWLEDGMENT

The authors want to express their gratitude to the Dean of Research and Development Bureau (DID) of the Simón

Bolívar University, for the annual financial support provided to the GSIEP (registered as GID-04 in the DID) to perform this work.

REFERENCES

[1] Y. Cheng, C. Qian, M. L. Crow, S. Pekarek, and S. Atcitty, "A comparison of diode-clamped and cascaded multilevel converters for a STATCOM with energy storage," *IEEE Transactions on Industrial Electronics*, vol. 53, no. 5, pp. 1512–1521, Oct. 2006.
[2] L. G. Franquelo, J. Rodriguez, J. I. Leon, S. Kouro, R. Portillo, and M. A. M. Prats, "The age of multilevel converters arrives," *IEEE Industrial Electronics Magazine*, vol. 2, no. 2, pp. 28–39, June 2008.
[3] V. T. Somasekhar, S. Srinivas, and K. K. Kumar, "Effect of zero-vector placement in a dual-inverter fed open-end winding induction motor drive with alternate sub-hexagonal center PWM switching scheme," *IEEE Trans. Power Electron.*, vol. 23, no. 3, pp. 1584–1591, May 2008.
[4] D. Casadei, G. Grandi, A. Lega, and C. Rossi, "Multilevel operation and input power balancing for a dual two-level inverter with insulated DC sources," *IEEE Trans. Ind. Applicat.*, vol. 44, no. 6, pp. 1815–1824, Nov./Dec. 2008.
[5] J. Rodriguez, J.-S. Lai, and F. Z. Peng, "Multilevel inverters: a survey of topologies, controls, and applications," *IEEE Transactions on Industrial Electronics*, vol. 49, no. 4, pp. 724–738, Aug. 2002.

978-1-4244-4782-4/10 $26.00 © 2010 IEEE   347

[6] V. T. Somasekhar, S. Srinivas, B. Prakash Reddy, C. Nagarjuna Reddy, and K. Sivakumar, "Pulse width-modulated switching strategy for the dynamic balancing of zero-sequence current for a dual-inverter fed open-end winding induction motor drive," *IET Electric Power Applications*, vol. 1, pp. 591–600, July 2007.

[7] B. P. McGrath, D. G. Holmes, and T. Lipo, "Optimized space vector switching sequences for multilevel inverters," *IEEE Transactions on Power Electronics*, vol. 18, no. 6, pp. 1293–1301, Nov. 2003.

[8] O. Lopez, J. Alvarez, J. Doval-Gandoy, and F. D. Freijedo, "Multilevel multiphase space vector PWM algorithm," *IEEE Transactions on Industrial Electronics*, vol. 55, no. 5, pp. 1933–1942, May 2008.

[9] J.-S. Lai and F. Z. Peng, "Multilevel converters-a new breed of power converters," *IEEE Trans. Ind. Applicat.*, vol. 32, no. 3, pp. 509–517, May/June 1996.

[10] H. Stemmler and P. Guggenbach, "Configurations of high-power voltage source inverter drives," in *Power Electronics and Applications, 1993., Fifth European Conference on*, Brighton, Sept. 1993, pp. 7–14.

[11] V. T. Somasekhar, S. Srinivas, and K. K. Kumar, "Effect of zero-vector placement in a dual-inverter fed open-end winding induction-motor drive with a decoupled space-vector PWM strategy," *IEEE Trans. Ind. Electron.*, vol. 55, no. 6, pp. 2497–2505, June 2008.

[12] G. Grandi, D. Ostojic, C. Rossi, and A. Lega, "Control strategy for a multilevel inverter in grid-connected photovoltaic applications," in *Electrical Machines and Power Electronics, 2007. ACEMP '07. International*

[13] V. Oleschuk, J. Tlusty, and V. Valouch, "Photovoltaic power conversion system based on cascaded inverters with synchronized space-vector modulation," in *Proc. International Conference on Renewable Energies and Power Quality (ICREPQ09) 2009)*, Valencia, Spain, Apr. 2009.

[14] J. Dixon, L. Moran, E. Rodriguez, and R. Domke, "Reactive power compensation technologies: State-of-the-art review," *Proceedings of the IEEE*, vol. 93, no. 12, pp. 2144–2164, Dec. 2005.

[15] J. A. Restrepo, J. M. Aller, J. C. Viola, A. Bueno, and T. G. Habetler, "Optimum space vector computation technique for direct power control," *IEEE Trans. Power Electron.*, vol. 24, no. 6, pp. 1637–1645, June 2009.

[16] V. T. Somasekhar, S. Srinivas, and K. Gopakumar, "A space vector based PWM switching scheme for the reduction of common-mode voltages for a dual inverter fed open-end winding induction motor drive," in *Power Electronics Specialists Conference, 2005. PESC '05. IEEE 36th*, Recife, June 2005, pp. 816–821.

[17] G. Grandi, C. Rossi, A. Lega, and D. Casadei, "Power balancing of a multilevel converter with two insulated supplies for three-phase six-wire loads," in *Power Electronics and Applications, 2005 European Conference on*, Dresden.

[18] A. M. Hava, R. J. Kerkman, and T. A. Lipo, "Simple analytical and graphical methods for carrier-based PWM-VSI drives," *IEEE Trans. Power Electron.*, vol. 14, no. 1, pp. 49–61, Jan. 1999.

A Low-Cost Adaptive Multi-Mode Digital Control Solution Maximizing AC/DC Power Supply Efficiency

Yong Li and Jerry Zheng

iWatt
Los Gatos, CA 95032

Abstract--- **This paper presents a new low-cost digital control solution that maximizes the AC/DC flyback power supply efficiency. This intelligent digital approach achieves the combined benefits of high performance, low cost and high reliability in a single controller. It introduces unique multiple PWM and PFM operational modes adaptively based on the power supply load changes. While the multi-mode PWM/PFM control significantly improves the light-load efficiency and thus the overall average efficiency, it does not bring compromise to other system performance, such as audible noise, voltage ripples or regulations. It also seamlessly integrated an improved quasi-resonant switching scheme that enables valley-mode turn on in every switching cycle without causing modification to the main PWM/PFM control schemes. A digital integrated circuit (IC) that implements this solution, namely iW1696, has been fabricated and introduced to the industry recently. In addition to outlining the approach, this paper provides experimental results obtained on a 3-W (5V/550mA) cell phone charger that is built with the iW1696.**

I. PROBLEMS TO SOLVE

World wide energy standards (US-EPA 2.0, EU-CoC, etc.) are being tightened, among which one critical specification is the power supply active-mode Average Efficiency [1]. The Average Efficiency is specified by testing the power supply at 25%, 50%, 75% and 100% of the rated current output and then computing the average value of the efficiencies obtained at these four testing conditions, and in most times power supplies typically operate only in the light to medium load range. Therefore improving the light-load efficiency is very important to meet the energy standards as well as eventually to reduce the global warming effect.

In most switching power converters, switching losses play a greater role as the load is decreased. Conversely, conduction losses play a greater role as the load is increased. Old-time PWM converters use constant PWM frequencies throughout the entire load range, suffering from poor efficiency in light load conditions [2]. Various conventional power converters use PWM at heavy load and reduce the switching frequency F_{SW} at light load by means of pulse-frequency-modulation (PFM) control [3]-[7]. But most PWM to PFM transitions occur at a load level far below 25% of the rated current. This kind of PWM/PFM control does not help to improve the efficiency at 25% of the rated current; therefore, it does not improve the overall average efficiency. Some designs attempted to increase the PWM to PFM transition at a higher load level, but suffered

from serious performance problems such as excessive voltage ripples due to unsmooth transition between the PWM and PFM modes.

Other conventional power converters operate in a PFM mode throughout the entire range of load conditions [8]-[9]. The switching frequency normally decreases from the maximum at the full load down to a few kilo-Hz at very light load. But operating only in the PFM mode suffer from audible noises when the switching frequency drops to around 16 kHz (the human ear audible range) while there is still significant energy remaining; this is unacceptable for consumer products such as low-power AC/DC adaptor/chargers for cell phones, PDAs and digital still cameras. In order to avoid the audible noise issue, some conventional PWM/PFM power converters forcibly clamp the switching frequency at a level above 20kHz when the PFM frequency is decreasing to close to about 20kHz, or use cycle skipping schemes at light load conditions. They normally cause undesirable voltage regulation and ripple issues at light load conditions, and most importantly, they can not meet the low no-load standby power consumption requirement, which is another critical specification of the energy standards.

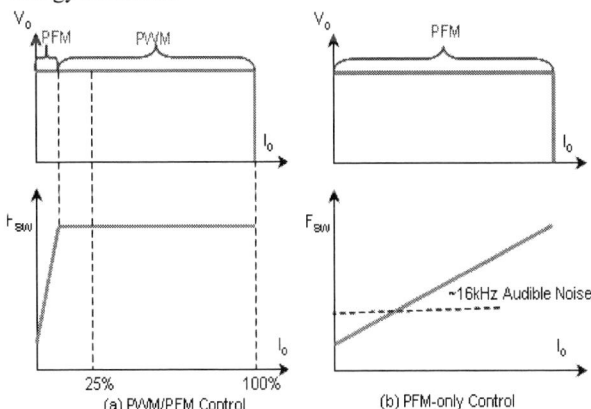

Figure 1. Conventional PWM converter frequency control, represented by the output voltage V_O and output current I_O as a VI curve.

In addition to the frequency control, quasi-resonant or valley-mode switching (VMS) has been used in flyback converters to reduce the switching loss and electro-magnetic-interference (EMI) [10]-[12]. But the conventional VMS schemes typically do not work with constant-frequency PWM controls. Instead, they require variable frequency control only, and normally the

978-1-4244-4782-4/10 $26.00 © 2010 IEEE   349

switching frequency can vary as low as 40 kHz and as high as 130 kHz within the load change range for the sole purpose of achieving VMS; or else the VMS may be lost. This is not suitable for slow switching devices such as BJT's and can bring difficulties to EMI filter design.

This paper addresses these problems in the conventional low-power AC/DC flyback power supply designs by providing a low-cost high-efficiency high-performance digital control solution [13]-[15]. It introduces unique multiple PWM and PFM operational modes adaptively based on the power supply load changes. While the multi-mode PWM/PFM control significantly improves the light-load efficiency and thus the overall average efficiency, it does not bring compromise to other system performance, such as audible noise, voltage ripples or regulations. It also seamlessly integrates an improved quasi-resonant switching scheme that enables valley-mode turn on in every switching cycle without causing modification to the main PWM/PFM control schemes. A digital integrated circuit (IC) that implements this innovative solution, namely iW1696, has been fabricated and introduced to the industry recently. In addition to outlining the approach, this paper provides experimental results obtained on a 3-W (5V/550mA) cell phone charger that is built with the iW1696 digital IC.

II. THE PROPOSED ADAPTIVE MULTI-MODE DIGITAL CONTROL SOLUTION

A) The Power Supply System Configuration

The power supply topology chosen for the target applications is an AC/DC flyback converter operating in discontinuous conduction mode (DCM) by means of peak current mode control. The AC input is universal line 85Vac-264Vac, and the secondary DC output is typically in the range of 5V to 10V. The proposed digital controller uses a primary-side feedback technique to eliminate opto-isolator feedback and secondary regulation circuits required in traditional designs. Figure 2 shows a typical complete power supply circuit with the proposed digital control solution. The power stage is comprised of power transformer T_1, power BJT Q_1 and control IC. The transformer includes a primary winding, a secondary winding, and an auxiliary winding. The output voltage at the secondary side, V_{OUT}, is reflected across the auxiliary winding. Referring to the functional diagram shown in Figure 3, V_{SENSE} is an analog input configured to receive the voltage across the auxiliary winding of the flyback converter via a resistive voltage divider comprised of R_3 and R_4, and through the Signal Conditioning block, V_{SENSE} is converted to a digital signal V_{FB} that precisely represents the secondary output voltage V_{OUT}, and is used for the primary-side control and regulation. The I_{SENSE} is an analog input pin configured to sense the primary-side current in the form of an analog voltage, which is the product of the instantaneous BJT emitter current times the sensing resistor R_S. This I_{SENSE} pin and the input logic provide cycle-by-cycle peak current control and limit. Based on the voltage

and current feedback information, the Digital Control Logic block generates the switching commands to turn on and off the BJT. The V_{CC} provides power supply to the internal digital/analog circuits, and it is powered by the line voltage through R_1 and R_2 at start-up, and by the voltage across the auxiliary winding in normal operation.

One unique feature of this solution is that the control IC directly drives a power BJT Q_1. Fundamentally BJT device fabrication involves simpler process than power MOSFET's, so BJT's are with much lower cost than power MOSFET's, in particular for high voltage (>700V) and low power (<5W) applications. Also, BJT's inherently have slower switching turn on and off speed, thus less di/dt and dv/dt, which can allow easier EMI design. For sure the BJT driver design is more challenging and complicated. The iW1696 IC not only controls the on and off of BJT, but it also has a built-in digitally-controlled BJT driver that dynamically adjusts the BJT base current amplitude cycle-by-cycle in real time based on the load change. This ensures that the BJT is always working in an optimal switching condition.

The schematic shown in Figure 2 is not a simplified version. Because of the slow di/dt of BJT switching, snubber circuits can be eliminated for low-power designs with the iW1696. Typically a complete low-power flyback AC/DC power supply built with iW1696 needs only around 22-25 parts. This results in a very low-cost solution for high-volume applications such as cell phone chargers.

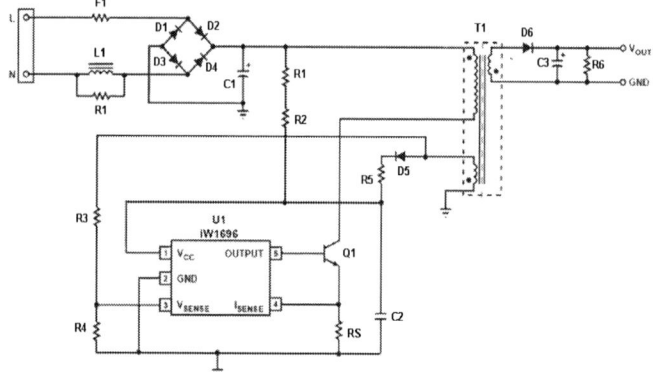

Figure 2. Typical power supply circuit built with the proposed digital solution.

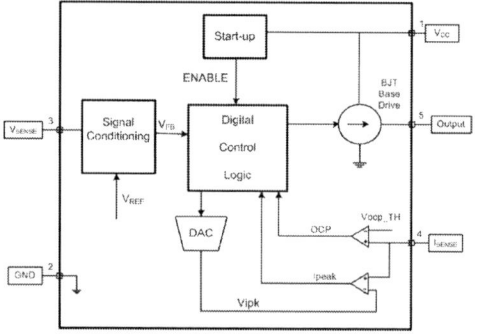

Figure 3. iW1696 functional diagram.

B) Principal of the Adaptive Multi-Mode PWM/PFM Control

In order to improve the efficiency, an adaptive multi-mode PWM/PFM control is proposed and implemented to dynamically change the power supply switching frequency based on the load change, without bringing compromise to other system performance, such as audible noises, voltage ripples or regulations.

As illustrated in Figure 4, during the constant voltage (CV) operation, the controller operates in a first PWM mode in heavy load conditions when the output current I_O is greater than a certain level, such as approximately 50% of the specified maximum load current. In the PWM mode, the switching frequency keeps around constant, with certain dithering to reduce EMI, and the voltage regulation is accomplished by adjusting the switching ON time. For the low-power AC/DC power supply designs, the typical PWM switching frequency can be around 40kHz-65kHz. As the output load current I_O is reduced, the ON time is decreased. The controller detects the load information in real time. When the load is detected to be less than the given level, such as below about 50% of the maximum load, the controller transitions to a first PFM mode. During the PFM mode, the power switch is turned on for a set duration under a given instantaneous rectified AC input voltage, but its off time is modulated by the load current. With the load current decreases, the off time increases and thus the switching frequency decreases. This first PFM mode mostly helps to improve the efficiency at the light to medium load range. As described in Section I, world wide energy standards specify the average efficiency of the power converter based on the averaging of the efficiencies at four loading points (25% load, 50% load, 75% load, and 100% load). In order to satisfy such standards, it is advantageous to control the power converter operating in the PFM mode around the 25% load level.

With the load current decreasing, when the switching frequency F_{SW} approaches to human ear audio band, the controller transitions from the first PFM mode to a second PWM mode (namely "deep" PWM or DPWM). During the DPWM mode, F_{SW} keeps constant around 20kHz in order to avoid audible noise, which otherwise would be generated mainly by the electro-mechanical vibration in the transformer of the power converter. As in any PWM control, the switching ON time becomes smaller when the load decreases in the DPWM mode.

As the load becomes even lighter, however, maintaining DPWM mode is undesirable for at least the following reasons: (i) the fixed switching frequency results in higher switching loss; (ii) a low power consumption at the no-load standby conditions cannot be achieved; and (iii) the minimum switching ON time limit forces the power converter to generate an output voltage higher than desired. Therefore, as the load current is further reduced to a certain level, the controller transitions to a second PFM mode (namely "deep" PFM or DPFM) where the switching

frequency is controlled as in any PFM mode, and it can be smoothly reduced down to only a very hundreds of Hertz at the no-load condition. This DPFM mode helps to improve the efficiency at very light load, and more importantly, to achieve low no-load standby power consumption, which is another critical specification in the energy standards. The load level for DPWM to DPFM transition can be controlled to at a substantial low level, such as less than around 5% of the maximum load. Although F_{SW} is dropped across the audible frequency range during the DPFM mode, the current has been already reduced to a negligible level in the DPWM mode before transitioning to the DPFM mode. Therefore, the power converter produces little or practically no audible noise over the entire operation range, while achieving high efficiency across varying load conditions.

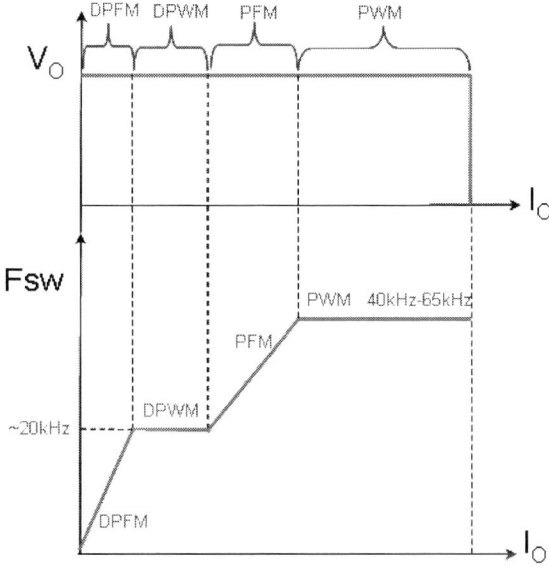

Figure 4. Illustration of the proposed adaptive multi-mode PWM/PFM control.

Adding the unique intervening DPWM mode between the PFM mode and the DPFM mode is advantageous for the additional reason that the DPWM mode provides smooth transition between the PFM and DPFM modes. By operating in the DPWM mode, abrupt jumps in the switching ON times and primary peak currents in the power converter are reduced or removed. Therefore, excessive voltage ripples and unstable regulation of output power is avoided. Further, by adding the intervening DPWM mode, the switching frequency can be kept above audible frequencies until the energy transfer per switching cycle is low enough to not cause the audible noise.

C) Implementation of the Multi-Mode PWM/PFM Control

Clearly the proposed adaptive multi-mode PWM/PFM control is quite unusual and complicated compared to the conventional PWM/PFM controls. This would be very difficult, if not impossible, to be implemented by analog circuits. Thanks to the innovative digital control architecture and algorithms, together with advanced ASIC technology, however, this adaptive multi-mode PWM/PFM control scheme has been successfully implemented and fabricated in a single 5-pin digital control IC with tiny SOT-23 package, iW1696. The experimental results in Section III will demonstrate its superior performance.

D) Improved Valley-Mode Switching (VMS)

VMS is useful to reduce the switching turn-on loss and dv/dt, thus reducing EMI (at frequency range of switching noises). As illustrated in Figure 5, in a flyback converter, there are multiple ringings in the power switch V_{CE} after this switch is turned off, due to the resonance between the transformer magnetizing inductance and parasitic capacitance seen from BJT collector to emitter in parallel with the equivalent parallel capacitance of the transformer windings. Conventional VMS schemes usually stick to turn on the switch at the First valley after the switch is turned off. This results on operating at near the boundary of CCM and DCM. Since the transformer reset time varies with the load change, these conventional VMS schemes requires variable frequency modulation for the only purpose of achieving VMS, and normally the switching frequency can go to as high as 130kHz (and then is typically clamped at around 130kHz for EMI reasons) and as low as 40kHz. This is not suitable for slow switching devices such as BJT's and can bring difficulties to EMI filter design.

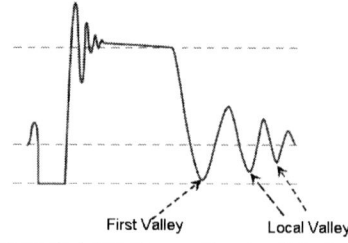

Figure 5. Typical BJT Vce waveform in a flyback converter.

While operating in the adaptive multi-mode PWM/PFM control, the proposed iW1696 digital solution also seamlessly integrates an improved VMS scheme to the flyback converter control [15].

During the switch OFF time, a desired switch turn-on instant is determined for the next cycle based on the main PWM/PFM control scheme. In the meanwhile, the digital logic detects the instants corresponding to every local minimums (valleys) of the V_{CE} across the BJT in real time. The BJT is then turned on at the first local valley following the desired switch turn-on instant (but not stick to the first valley after the switch is turned off). As a result, this improved VMS enables quasi-zero-voltage turn-on in every switching cycle in both CV and CC operations, without changing the fundamental main switching frequencies or main control loop design or forcing the switching frequency to go too high. Thus, it can work with constant PWM control; moreover, it can work under PWM, PFM, or any other appropriate switching schemes and still achieve the benefits of reducing switching loss and EMI. This feature is particularly suitable for BJT's. Furthermore, due to the cycle-by-cycle variation of switching instants, the

improved VMS can provide natural frequency dithering which helps to reduce EMI. Because the digital detection logic works in real time, this improved VMS is insensitive to transformer magnetizing/leakage inductance and parasitic capacitance variations.

III. EXPERIMENTAL RESULTS

Following the proposed solution, a digital control IC iW1696 has been designed, fabricated and tested. Figure 6 shows the photos of a 3-W (5V/550mA) cell phone charger designed with iW1696, built on a very-cost single-layer PCB. As shown in Figure 7, the measured VI curve represents very tight and smooth constant voltage (CV) and constant current (CC) regulation with the primary-side-only feedback, and the switching waveforms associated with the VI curve indicate that the switching frequency does change in the entire load range, following the desired adaptive multi-mode PWM/PFM control. There is no excessive voltage ripples during the mode transitions. From the detailed switching waveforms shown in Figure 8, it can be seen clearly that under different load conditions the BJT turns on under the local lowest V_{CE} in each switching cycle, which indicates the improved VMS is achieved at every operational modes, including both CV and CC. No audible noise could be heard during the test.

Figure 9 compares the measured power supply efficiency with the proposed digital control against a conventional control. Both use the same power stage circuit and components. It can be seen the proposed control significantly improves the light-load efficiency as well as the overall average efficiency. Since the switching losses become higher and dominant under higher voltage, this measured efficiency improvement is more effective at 230Vac where the 25%-load efficiency is improved by up to 10%, and the average efficiency is improved by over 3%, both in absolute numbers.

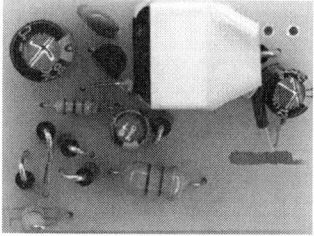

(a) Top side

Digital Control IC iW1696

(b) Bottom side

Figure 6. Photo of a 3-W (5V/550mA) cell phone charger PCB built with the schematic shown in Figure 2.

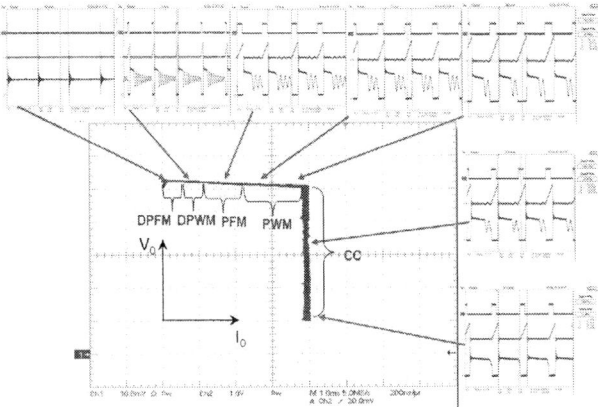

Figure 7. Measured VI curve and switching waveforms at different load conditions.

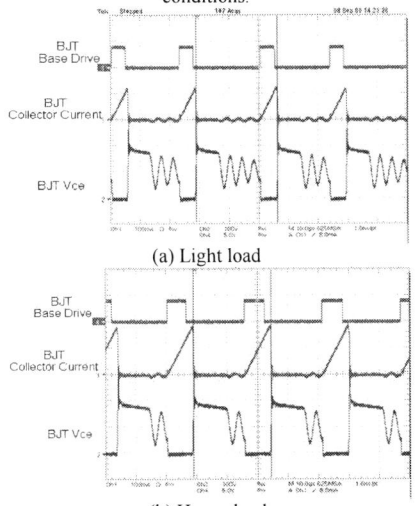

(a) Light load

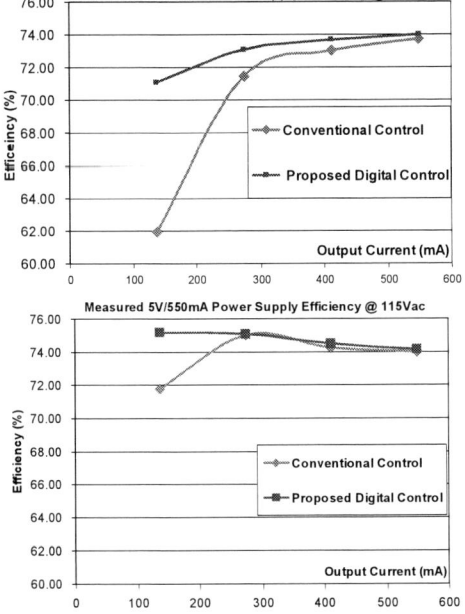

(b) Heavy load

Figure 8. Detailed BJT switching waveforms at different load conditions.

Figure 9. Measured power supply efficiency

Figure 10 shows without adding Y-Cap or common-mode choke to the PCB, the measured EMI of the iW1696 demo board, for both conducted and radiated, easily comply with the regulations. This demonstrates the superior EMI performance of the BJT drive solution.

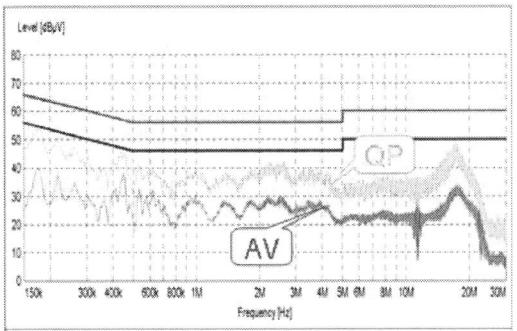

(a) Conducted EMI (230V/50Hz, live; QP: quasi-peak, AV: Average)

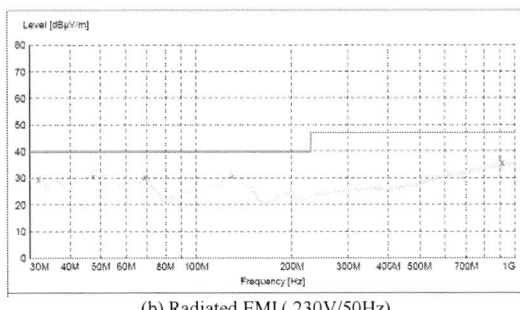

(b) Radiated EMI (230V/50Hz)

Figure 10. Measured EMI performance of iW1696 demo board. No Y-Cap or common-mode choke were used.

IV. CONCLUSION

A new low-cost digital control solution is proposed, implemented, experimentally verified, and introduced to the industry. Thanks to the innovative adaptive multi-mode PWM/PFM control seamlessly integrated with the improved valley-mode switching, without causing compromise to other system performance, including audible noises, voltage ripples and regulations, the iW1696-based BJT drive solution offers the lowest total system bill-of-material cost, and achieves the highest efficiency and lowest EMI for low-power AC/DC power supply designs available in the industry.

It should be noted that although the iW1696 and its application circuit are based on the primary-side feedback control, the same principle of adaptive multi-mode PWM/PFM operation and improved VMS is also applicable to alternative designs based on the conventional secondary-side feedback control. Further, a MOSFET switch may be used in place of BJT switch for different applications.

This work is protected by U.S. Provisional Patent Applications No. 61/140,605, 61/141,059, and 61/141,600.

978-1-4244-4782-4/10 $26.00 © 2010 IEEE   353

REFERENCES

[1] ENERGY STAR Program Requirements for Single Voltage AC-DC and AC-AC Power Supplies, Eligibility Criteria (Version 2.0), FINAL, pp. 3-4.

[2] UC3842 current mode PWM controller datasheet", *Texas Instruments Datasheet*, June 2007.

[3] M. Day, et. al., "A practical guide to low power efficiency measurement," *NETWORK SYSTEMS DESIGN LINE*, April 25, 2007.

[4] E. Kok, "Latest AC/DC integrated power IC for low standby medium power applications," *ENDAsia*, Feature, 08 May 2009.

[5] M. Ishitobi, et. al., "Pulse width and pulse frequency modulation pattern controlled ZVS inverter type AC-DC power converter with lowered utility AC grid sideharmonic current components for magnetron drive," in *IEEE 2002 Power Electronics Specialists Conference*, Vol. 4, pp. 2062 – 2067.

[6] J. Xiao, et. al., "A 4-μA quiescent-current dual-mode digitally controlled buck converter IC for cellular phone applications," in *IEEE JOURNAL OF SOLID-STATE CIRCUITS*, Vol. 39, No. 12, Dec. 2004.

[7] M. Nakaoka, et., al., " Pulse width and pulse frequency modulation pattern controlled active clamp ZVS inverter link AC-DC power converter utility AC side active power filtering function for consumer magnetron driver," in *IEEE IECON 2007 Conference*, pp. 1968-1971.

[8] Y. Zhu, et. al., "Method and system for pulse frequency modulated switching mode power supplies," U.S. Provisional Patent Application No. 60/943,498, filed Jun. 12, 2007.

[9] H. Huang, "Maximizing AC/DC efficiency from full-load to no-load," *Power Management Design Line*, Nov. 30, 2007.

[10] Y. Panov, et. al., "Adaptive off-time control for variable-frequency, soft-switched flyback converter at light loads," in *IEEE 1999 Power Electronics Specialists Conference*, Vol. 1, pp. 457-462.

[11] W. Langeslag, et. al., "VLSI design and application of a high-voltage-compatible SoC–ASIC in bipolar CMOS/DMOS technology for AC–DC rectifiers," *IEEE Trans. on Industrial Electronics*, vol. 54, no. 5, pp. 2626-2641, Oct 2007.

[12] J. Harper, "Using quasi-resonant and resonant converters," *Power Management Design Line*, Oct. 16, 2008.

[13] Y. Li, et. al., "Controller for switching power converter driving BJT based on primary side adaptive digital control," U.S. Provisional Patent Application No. 61/140,605, filed Dec. 23, 2008.

[14] Y. Li, et. al., "Adaptive multi-mode digital control improving light load efficiency in switching power converters," U.S. Provisional Patent Application No. 61/141,059, filed Dec. 29, 2008.

[15] Y. Li., et. al., "Improved valley-mode switching schemes for switching power converters," U.S. Provisional Patent Application No. 61/141,600, filed Dec. 30, 2008.

978-1-4244-4782-4/10 $26.00 © 2010 IEEE

Average Modeling and Control for Three-Phase Three-Level Non-Regenerate Rectifier with Unbalanced DC Loads

Rixin Lai[1], Fred Wang[2,5], Rolando Burgos[3], Dushan Boroyevich[4]

1) GE Global Research Center, Niskayuna, NY 12309, USA
2) Department of EECS, the University of Tennessee, Knoxville, TN 37996, USA
3) ABB Inc., U.S. Corporate Research Center, Raleigh, NC 27606, USA
4) Center for Power Electronics Systems, Virginia Tech, Blacksburg, VA 24061, USA
5) Oak Ridge National Lab, Oak Ridge, TN 37830, USA

Abstract— **This paper proposes a new average model and a control approach for the three-phase three-level non-regenerate rectifier (Vienna-type rectifier) with unbalanced dc load. State space analysis is first carried out to achieve the relationship between the voltage unbalance, load conditions and the control duty cycle. With the implementation of an optimum zero-sequence component, a simplified average model for the dc output stage with unbalanced load is obtained. Based on the developed model, a new control approach and the criteria of control parameter selection are presented. The simulation and experiment results validate the proposed control scheme.**

I. INTRODUCTION

The non-regenerative three-level boost rectifier, known as a Vienna-type rectifier [1-2], as shown in Fig. 1, is characterized by a reduced number of active switching devices, a high input power factor, and low device voltage stress, which make it a suitable topology for medium- and high-power applications with high power density [3-4]. In addition, due to the three-level neutral point clamped (NPC) structure, the Vienna-type rectifier can feed two individual loads with partial and/or full dc-link voltage. Therefore the Vienna-rectifier potentially can reduce the cost and provide more design flexibility for the system architecture in the applications, such as telecommunication equipments, where the dc-bus structure with multiple voltage levels is considered.

However, an essential requirement of three-level NPC structure is to maintain the balance of the two dc-link capacitor voltages. Any unbalance can cause low frequency harmonics voltage on the ac side, and lead to higher voltage stress on the bridge devices and capacitors. For realization of feeding two individual load systems at the same time, the uneven power distribution to the two capacitor voltages could

occur and therefore the voltage balance control is a challenge. In the past several years lots of research work has been put on the Vienna-type rectifier, including the average model and the control scheme development. The previous modeling efforts focused on the d-q representation for the rectifier system. Due to the intrinsic time variant nature of the dc-link neutral point current [2], low frequency state model [5-7] was built by averaging the dc-link neutral point operation in a complete ac line cycle. And the control loop design for the voltage balance is absent in the developed control approaches [8-10] due to the limited frequency range of the available models. A non-linear control approach with neutral point regulation is proposed in [11], but both dc-link voltages and currents need to be measured and the control itself is complex. In [12], an average model with extended frequency range is proposed and a simple control approach is developed. However, the model is developed based on balanced dc load condition.

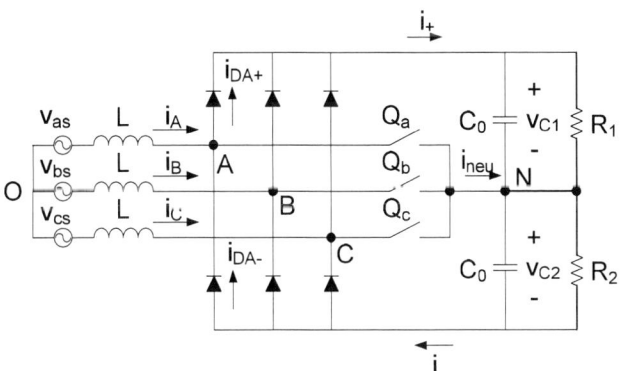

Fig. 1. Rectifier topology.

In this paper, an average model for the Vienna-type rectifier with unbalanced dc load is proposed. The optimal zero component concept [13] is applied in the model manipulation, with which the dc-link unbalance behavior can be approximated by a first order system in the d-q coordinates.

978-1-4244-4782-4/10 $26.00 © 2010 IEEE

And the impact of the unbalance dc loads is modeled by a current source. Since the averaging is implemented over one switching cycle, the proposed model is valid up to half of the switching frequency. A carrier-based control approach is also developed based on the model. And the criteria for control parameter selection are provided to achieve a given bandwidth requirement. Simulations and experiments are presented in Section IV. The results prove the feasibility of the proposed control scheme.

II. AVERAGE MODELING

As shown in Fig. 1, the Vienna-type rectifier consists of six diodes and three bi-directional switching units, Q_a, Q_b and Q_c. The three active switching units are controlled to ensure sinusoidal ac current and steady dc-link voltage. The Vienna-type rectifier is current forced commutated. For example, if Q_a is on, phase A will be clamped to the dc-link neutral point. If Q_a is off, phase A will be clamped to the positive bus or the negative bus, depending on the current polarity. The same operation principle applies to the other phases. Therefore the switch voltages V_{AN}, V_{BN} and V_{CN} are determined by both the switching pattern and the current polarity, which, after some algebraic transformation, can be obtained by

$$
\begin{cases}
V_{AN} = \dfrac{V_{dc}}{2} \cdot [\mathrm{sgn}(i_A) + \dfrac{\Delta v}{V_{dc}}] \cdot (1 - S_a) \\[2mm]
V_{BN} = \dfrac{V_{dc}}{2} \cdot [\mathrm{sgn}(i_B) + \dfrac{\Delta v}{V_{dc}}] \cdot (1 - S_b) \\[2mm]
V_{CN} = \dfrac{V_{dc}}{2} \cdot [\mathrm{sgn}(i_C) + \dfrac{\Delta v}{V_{dc}}] \cdot (1 - S_c)
\end{cases}
\tag{1}
$$

where *sgn* is the signum function, and

$$
S_{a,b,c} = \begin{cases} 0 & \text{if } Q_{a,b,c} \text{ is turned off} \\ 1 & \text{if } Q_{a,b,c} \text{ is turned on} \end{cases}
\tag{2}
$$

$$
\begin{cases} V_{dc} = v_{C1} + v_{C2} \\ \Delta v = v_{C1} - v_{C2} \end{cases}
\tag{3}
$$

Normally Δv is much smaller than V_{dc} and therefore the $\Delta v / V_{dc}$ component in (1) can be ignored, which leads to

$$
\begin{cases}
V_{AN} = \dfrac{V_{dc}}{2} \cdot \mathrm{sgn}(i_A) \cdot (1 - S_a) \\[2mm]
V_{BN} = \dfrac{V_{dc}}{2} \cdot \mathrm{sgn}(i_B) \cdot (1 - S_b) \\[2mm]
V_{CN} = \dfrac{V_{dc}}{2} \cdot \mathrm{sgn}(i_C) \cdot (1 - S_c)
\end{cases}
\tag{4}
$$

The phase leg duty cycles can then be defined as

$$
d_{a,b,c} = (1 - K_{a,b,c}) \cdot \mathrm{sgn}(i_{A,B,C}) = d'_{a,b,c} + d_0
\tag{5}
$$

where $K_{a,b,c}$ represents the average on-time within one switching cycle for Q_a, Q_b and Q_c respectively, $d'_{a,b,c}$ are the sinusoidal components and d_0 is the zero sequence duty cycle. Ignoring the impact of the voltage unbalance, the state space average model for the ac input stage is given by [12]

$$
\begin{cases}
V_{sd} = L \dfrac{di_d}{dt} - \omega_0 \cdot L \cdot i_q + \dfrac{V_{dc}}{2} \cdot d'_d \\[2mm]
V_{sq} = L \dfrac{di_q}{dt} + \omega_0 \cdot L \cdot i_d + \dfrac{V_{dc}}{2} \cdot d'_q
\end{cases}
\tag{6}
$$

For the dc output stage, the state equations are given by

$$
\begin{cases}
C \dfrac{dv_{C1}}{dt} = i_+ - \dfrac{v_{C1}}{R_1} \\[2mm]
C \dfrac{dv_{C2}}{dt} = i_- - \dfrac{v_{C2}}{R_2}
\end{cases}
\tag{7}
$$

Per Kirchhoff's law and the switching patterns, the relationship between the dc bus current and the input current is given by [12]

$$
\begin{cases}
i_+ + i_- = d_a \cdot i_A + d_b \cdot i_B + d_c \cdot i_C \\
i_{neu} = i_- - i_+ = K_a \cdot i_A + K_b \cdot i_B + K_c \cdot i_C
\end{cases}
\tag{8}
$$

For three-wire system it follows that

$$
i_A + i_B + i_C = 0
\tag{9}
$$

With (5) and (9), the dc bus current in (8) can be represented by

$$
\begin{cases}
i_+ + i_- = d'_a \cdot i_a + d'_b \cdot i_b + d'_c \cdot i_c = d'_d \cdot i_d + d'_d \cdot i_d \\
i_{neu} = -|i_A| \cdot d'_a - |i_B| \cdot d'_b - |i_C| \cdot d'_c - d_0(|i_A| + |i_B| + |i_C|)
\end{cases}
\tag{10}
$$

Assume

$$
\begin{cases}
R_0 = \dfrac{1}{2}(R_1 + R_2) \\[2mm]
\Delta R = \dfrac{1}{2}(R_1 - R_2)
\end{cases}
\tag{11}
$$

Since R_1 and R_2 are positive, (11) indicates that $\Delta R \leq R_0$. After substituting (3), (11) into (7) and some algebraic manipulation and approximation, state space equation for dc output stage can be obtained as

$$C_0 \frac{dV_{dc}}{dt} = i_+ + i_- - \frac{V_{dc} \cdot R_0 - \Delta v \cdot \Delta R}{R_0^2 - \Delta R^2} \qquad (12)$$

$$C_0 \frac{d\Delta v}{dt} = i_+ - i_- - \frac{\Delta v}{R_0} + \frac{V_{dc} \cdot \Delta R}{R_0^2} \qquad (13)$$

Since Δv is much smaller than V_{dc}, and ΔR is smaller than R_0, (12) can be simplified as

$$C_0 \frac{dV_{dc}}{dt} \approx i_+ + i_- - \frac{V_{dc}}{R_0 - \frac{\Delta R^2}{R_0}} \qquad (14)$$

Define the zero sequence component as

$$d_0 = d'_0 + \Delta d_0 = -\frac{|i_A| \cdot d'_a + |i_B| \cdot d'_b + |i_C| \cdot d'_c}{|i_A| + |i_B| + |i_C|} + \Delta d_0 \quad (15)$$

where d'_0 is the optimal zero component [12], and Δd_0 is the controlled zero component, which is used to balance the dc-link voltage.

Substitution of (10) and (15) into (13) and (14) leads to the average model of the dc output stage as

$$C \frac{dV_{dc}}{dt} = d'_d \cdot i_d + d'_d \cdot i_d - \frac{V_{dc}}{R_0 - \frac{\Delta R^2}{R_0}} \qquad (16)$$

$$C \frac{d\Delta v}{dt} = 2\Delta d_0 \cdot \frac{|i_A| + |i_B| + |i_C|}{2} - \frac{\Delta v}{R_0} + \frac{V_{dc} \cdot \Delta R}{R_0^2} \qquad (17)$$

Ignoring the harmonic components of $\dfrac{|i_A| + |i_B| + |i_C|}{2}$, (17) can be further simplified as

$$C \frac{d\Delta v}{dt} \approx \frac{2\sqrt{6}}{\pi} i_d \cdot \Delta d_0 - \frac{\Delta v}{R_0} + \frac{V_{dc} \cdot \Delta R}{R_0^2} \qquad (18)$$

With (6), (16) and (18) the average model of the Vienna-type rectifier can be achieved, as shown in Fig. 2. The red dash frame highlights the model of the dc-link unbalance voltage. As can be seen, a simple first order relationship is established between the controlled zero-sequence component Δd_0 and the dc-link voltage unbalance Δv. And the impact of the unbalance dc load resistors is now explicitly represented by a current source. The other parts of the model are similar to that of the conventional 2-level voltage source converter,

except that the dc-link resistor has an adjustment term ($\frac{\Delta R^2}{R_0}$).

Based on this model, the neutral point balance controller can be easily designed under a given operating point. Fig. 3 shows the diagram for the voltage balance control loop. The blocks in the dash frame represent the proposed model for the dc-link voltage unbalance. In order to achieve a control bandwidth of ω_n, the controller parameters for the PI controller are given by (10). I_d is the steady state d-channel current.

$$\begin{cases} K_p = \dfrac{\pi \cdot \omega_n \cdot C_0}{2\sqrt{6} \cdot I_d} \\ K_i = \dfrac{\pi \cdot \omega_n}{2\sqrt{6} \cdot I_d \cdot R_0} \end{cases} \qquad (19)$$

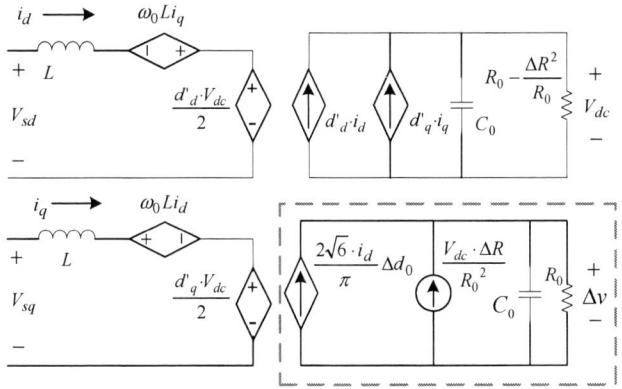

Fig. 2. State space average model.

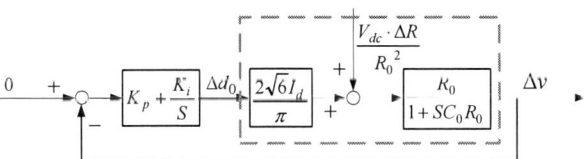

Fig. 3. Neutral point voltage control loop.

III. CONTROL APPROACH DEVELOPMENT

As found during the model derivation, the average model of the input stage for the Vienna-type rectifier is similar to the 2-level voltage source rectifier. Therefore the conventional multi-loop controller in the synchronous reference frame (d-q frame) [14] commonly used in the three-phase boost rectifiers can be applied for the Vienna-type rectifier to control the input current and the total dc-link voltage. This multi-loop controller will generate the d and q components for the voltage command. The zero component is then achieved by (15), which consists of the feed forward part d'_0 (optimal zero

978-1-4244-4782-4/10 $26.00 © 2010 IEEE   357

component) and the feed back part Δd_0. In the control implementation, the optimal zero component can be approximated by [12]

$$d'_0 \approx -\frac{M}{4}\cos(3\omega_0 t) \qquad (20)$$

Fig. 4 shows the full diagram of the proposed controller. The d-q controller represents the standard multi-loop control in d-q coordinates. But instead of using the complicated three-level space vector modulator, d_d and d_q are directly converted into *abc*-coordinates. And then combined with the zero component from the dc-link voltage balance control loop, the phase leg duty cycles d_a, d_b and d_c can be achieved.

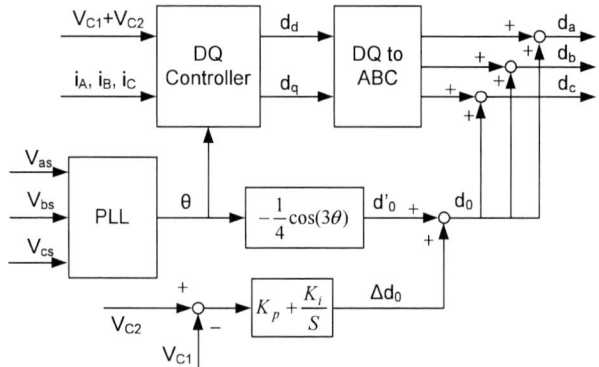

Fig. 4. Control scheme diagram.

IV. SIMULATION AND EXPERIMENT

A detailed Saber model has been built to verify the control scheme. The inductors are 160 µH each and the two dc link capacitors connected in series are 40 µF each. The switching frequency is 40 kHz. The input voltage is 60 Vrms / 400 Hz and the dc-link voltage is 180 V. A 15 Ω resistor is placed across capacitor C_1 and a 20 Ω resistor is across capacitor C_2. The steady state simulation results are shown in Fig. 5. The duty cycles are shown in Fig. 6. As can be seen in the figure, there is a dc offset in the zero component d_0. It is due to the effect of the unbalance loads, which is indicated by the dc current source $\dfrac{V_{dc} \cdot \Delta R}{R_0^{\,2}}$ in the average model.

An experimental prototype, as shown in Fig. 7, was constructed with SiC Schottky diodes and Si MOSFETs, using a DSP-FPGA based digital controller to implement the proposed control strategy. Due to superior switching performance, SiC Schottky diodes are very suitable for high density applications [15-16]. The testing conditions are the same as the ones used in simulation. The rectifier starts from diode rectification mode (passive mode). The experimental waveforms for the start up transient are shown in Fig. 8. As can be seen, at diode rectification mode, V_{C1} is lower than V_{C2} due to the unbalance load condition. Once the rectifier runs in active mode, the two voltages are balanced and the input currents become sinusoidal. The steady state waveforms are

shown in Fig. 9. Fig. 10 shows the experimental result for the load step change under balance condition, with the load increasing from 500 W to 1 kW within 100 µs. The figure shows that the system is stable.

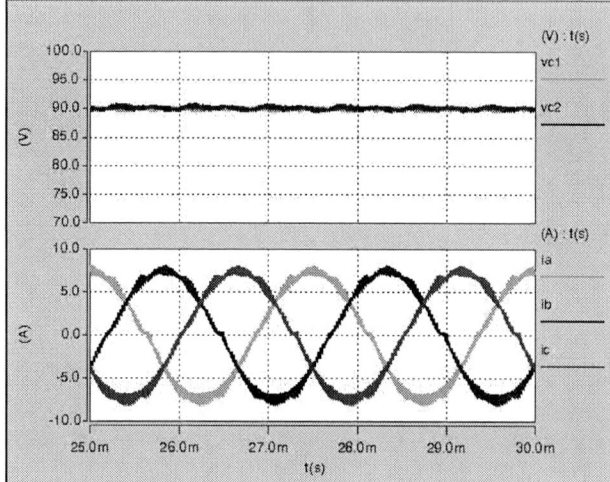

Fig. 5. Simulation results for dc voltages and ac input currents.

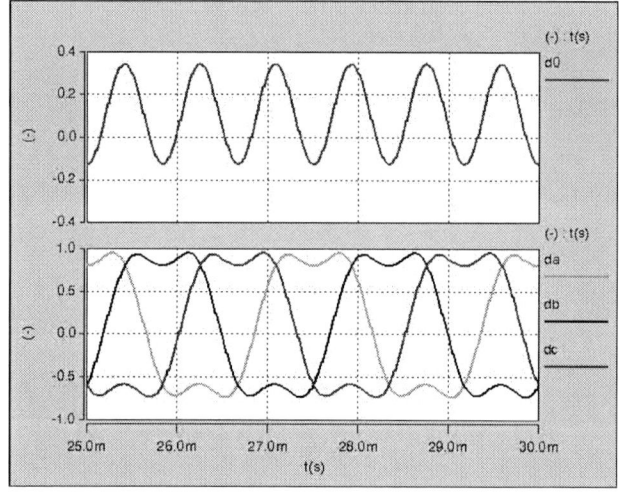

Fig. 6. Simulation results for duty cycles.

Fig. 7. Experimental system.

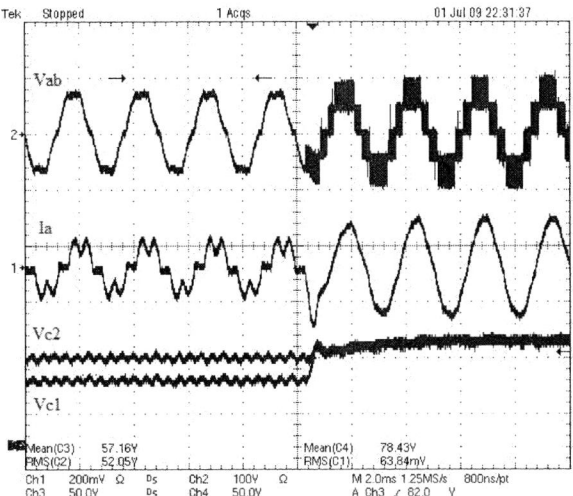

Fig. 8. Start up transient.

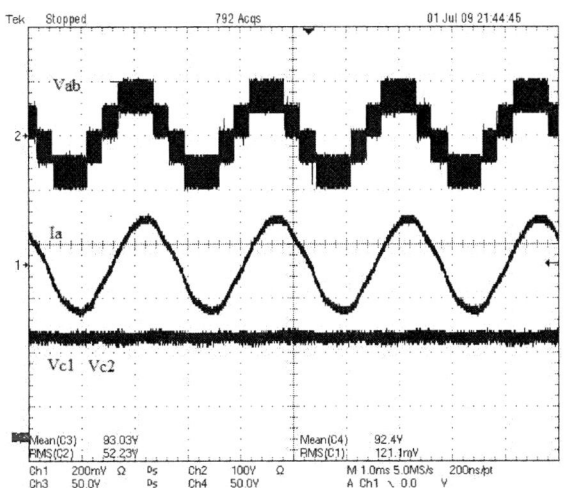

Fig. 9. Steady state results.

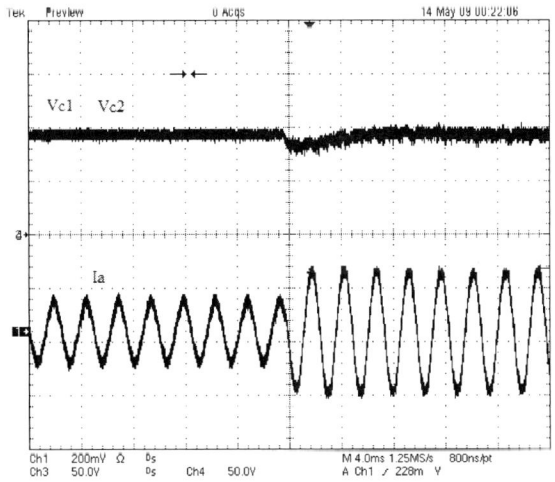

Fig. 10. Load step test under balance condition.

V. CONCLUSION

This paper presented a new average model for the non-regenerative three-level boost (Vienna-type) rectifier operating under unbalanced dc load condition. The effect of the unbalance load to the neutral point voltage can be represented by a current source in this model. The proposed model features extended frequency range using the conventional d-q representation and therefore can be applied to the high performance controller design. Based on this model a carrier-based control approach is presented and the criteria for the control parameter selection are also provided. The feasibility proposed model and control scheme are verified by both simulation and experimental results.

REFERENCES

[1] J. W. Kolar, and F. C. Zach, "A novel three-phase utility interface minimizing line current harmonics of high-power telecommunications rectifier modules", in *Proc. INTELEC'94*, Vancouver, BC, Canada, pp. 367-374, Nov. 1994.

[2] J. W. Kolar, and U. Drofenik, "A new switching loss reduced discontinuous PWM scheme for a unidirectional three-phase/switch/level boost-type PWM (Vienna) rectifier," *Proc. 21st INTELEC*, Jun. 1999.

[3] R. Lai, F. Wang, R. Burgos, Y. Pei, D. Boroyevich, B. Wang, T. A. Lipo, V. D. Immanuel, and K. J. Karimi, "A systematic topology evaluation methodology for high-density three-phase PWM ac-ac converters," *IEEE Trans. Power Electron.*, vol. 23, no. 6, pp. 2665 – 2680, Nov. 2008.

[4] G. Gong, M. L. Heldwein, U. Drofenik, J. Minibock, K. Mino, and J. W. Kolar, "Comparative evaluation of three-phase high-power-factor AC-DC converter concepts for application in future more electric aircraft," *IEEE Trans. Ind. Electron.*, vol. 52, no. 3, pp. 727–737, Jun. 2005.

[5] H. Y. Kanaan, K. Al-Haddad, R. Chaffai, L. Duguay, and F. Fnaiech, "A new low-frequency state model of a three-phase three-switch three-level fixed-frequency PWM rectifier", *INTELEC 2001*, pp. 384-391, Oct. 2001.

[6] N. B. H. Youssef, F. Fnaiech, and K. Al-Haddad, "Small signal modeling and control design of a three-phase AC/DC Vienna converter", *Proc. IECON 2003*, pp. 656-661, Nov. 2003.

[7] H. Y. Kanaan, K. Al-Haddad, and F. Fnaiech, "Modelling and control of three-phase/switch/level fixed-frequency PWM rectifier: state-space averaged model", *IEE Proc. Electric Power Applications 2005*, pp. 551-557, May 2005.

[8] C. Qiao, K. M. Smedley, "Three-phase unity-power-factor star-connected switch (VIENNA) rectifier with unified constant-frequency integration control", *IEEE Trans. on Power Electronics*, vol. 18, .no. 4, pp. 952-957, Jul. 2003.

[9] B. Wang, G. Venkataramanan, A. Bendre, "Unity power factor control for three phase three level rectifiers without current sensors", *IEEE Trans. on Industry Applications*, vol. 43, no. 5, pp. 1341-1348, Sep. 2007.

[10] T. Viitanen and H. Tuusa, "Three-level space vector modulation – an application to a space vector controlled unidirectional three-phase/level/switch Vienna I rectifier," *European Conference on Power Electronics and Applications (EPE)*, 2003.

[11] H. Y. Kanaan, K. Al-Haddad, and F. Fnaiech; "DC load unbalance and mains disturbances effects on a three-phase three-switch three-level boost rectifier", *Proc. ISIE'03*, pp. 1043-1048, Jun. 2003.

[12] R. Lai, F. Wang, R. Burgos, and D. Boroyevich, "Modeling and control for non-regenerative three-level boost rectifier considering dc-link voltage balance", *Proc. IECON'08*, pp. 827-832, Oct. 2008.

[13] R. Lai, F. Wang, R. Burgos, and D. Boroyevich, "Voltage balance control of non-regenerative three-level boost rectifier using carrier based pulse width modulation", *Proc. PESC'08*, pp. 3137-3142, Jun. 2008.

[14] V. Blasko, and V. Kaura, "A new mathematical model and control of a three-phase AC-DC voltage source converter", *IEEE Trans. on Power Electronics*, vol. 12, .no. 1, pp. 116-123, Jan. 1997.

[15] D. Fu, Y. Qiu, Y. Sun, F.C. Lee, "A 700kHz High-Efficiency High-Power-Density Three-Level Parallel Resonant DC-DC Converter for High-Voltage Charging Applications," in Proc. APEC'07, pp. 962-968, Mar. 2007.

[16] Y. Wang, P.A. Losee, and T.P. Chow, "4H-SiC Vertical RESURF Schottky Rectifiers and MOSFETs", 2007 International Journal of High Speed Electronics and Systems, vol. 17, no. 1, pp. 55-59, 2007.

A Waveform Control Technique for High Power Shunt Active Power Filter Based on Repetitive Control Algorithm

Zhiqiang Wang, Chuan Xie, Chao He and Guozhu Chen
College of Electrical Engineering Zhejiang University
Hangzhou, Zhejiang 310027, China
Email: gzchen@zju.edu.cn

Abstract—**Shunt Active Power Filters (SAPF) could restrain harmonic current in distribution grid effectively and their compensation precision relies heavily on the design of current loop controllers. Since the reference and feedback signal of the current controller are composed by different harmonic components and the bandwidth of conventional Proportion-Integral (PI) controller is limited, this controller cannot control the output current waveform without any steady error. In this paper, the limitation of conventional PI controller in the application of high power SAPF is analyzed firstly. Then the paper puts forward a double-loop composite controller composed by a PI inner loop and repetitive controller external loop to improve the output current waveform with the control object of LCL filters and the design method is given in detail. The controller could not only acquire high compensation precision when used to suppress odd harmonic, even harmonic current as well as negative sequence parts created by unbalanced load, but also guarantee excellent dynamic response performance. Finally, simulation and experimental results verify the composite controller gains extraordinary effects in high power SAPF.**

I. INTRODUCTION

With the rapid development of power electronics, there is an increasing harmonic level in distribution grids due to voltage and current waveform distortion. Shunt Active Power Filters(SAPF), controlled as harmonic current sources, are suitable to compensate current type harmonic sources [1].

PI controllers are widely used in industrial application SAPFs. Ordinary PI controllers can track DC signals accurately, while they cannot control the output current waveform composed by different harmonics without any steady error in AC application, e.g. SAPFs. In addition, in high power SAPF the bandwidth of PI controller cannot be very large since the switching frequency of devices is limited. Consequently, conventional PI controller is greatly restricted in high power active power filters.

In order to realize high compensation at different harmonic frequencies, paper [2] and [3] proposed proportional-sinusoidal signal integrators and multiple rotating integrators respectively. However, both methods based on selective

harmonic compensation approach needs to insert a controller at every harmonic frequency. Thus, they are very complex in implementation.

To deal with the problem, repetitive controller was used to current loop control. Repetitive control based on internal model can eliminate all periodic errors in a stable closed loop and is easy to be realized; however it has poor dynamic characteristics. Paper [4] proposed modified negative internal model to compensate odd harmonics. Although these methods play an important role in enhancing dynamic response speed and saving memory space, they cannot inhibit even harmonics. Paper [5]-[8] have studied the application of repetitive control method in single phase and three phase lower power APF based on single L output filters and design methods are given. However, these methods cannot be applied directly to three phase high power APF based on LCL filters which are difficult to be stabilized. Paper [9] has applied simplified positive internal model as well as active damping in APF, which not only save memory spaces but also reduce power loss to some extent, whereas it cannot effectively mitigate even harmonics and negative sequence harmonics created by unbalanced load on one hand. On the other hand, overall cost of the system is promoted significantly as a result of over many additional sensors.

In this paper, a double loop composite controller, based on the LCL filter, composed by a PI inner loop and repetitive controller external loop is put forward in order to combine fast dynamic response of PI controllers and high steady state precision of repetitive controllers. It could guarantee excellent dynamic response performance and acquire high compensation precision when used to inhibit odd harmonic, even harmonic current as well as negative sequence harmonic current. Moreover, the complexity and cost of the SAPF are reduced significantly. Feasibility and effect are demonstrated by experimental results in conclusion.

II. CONVENTIONAL CONTROL STRATEGY

A. SAPF Configuration

The main circuit of three phases SAPF is shown in Fig.1. Where, u_g is the grid phase voltage, L_g is the grid side

---

The authors would like to thank the sponsorship of NCET Program of Ministry of Education, China (#060512)

978-1-4244-4782-4/10 $26.00 © 2010 IEEE

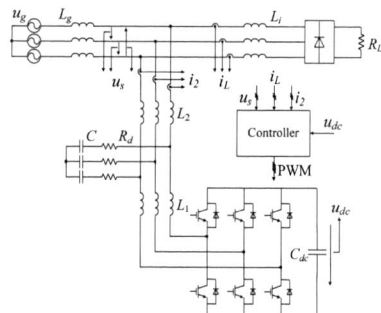

Fig. 1. Configurations of the shunt active power filter with LCL filter.

inductance, L_i is the load side inductance, C_{dc} the DC capacitor, R_L is the load resistor, L_1, L_2, C and R_d represents the inverter side inductor, grid side inductor, filter capacitors and damping resistor of the LCL filter respectively. i_L, i_2, u_s, u_{dc} stands for the load current, compensation current, system voltage and DC voltage respectively.

B. Conventional PI controller

Fig. 2 shows the block diagram of APF current loop in which conventional PI controller is adopted with the control object of LCL filters.

Where, i_{2r} is the reference compensation current, u_s is the feed-forward voltage, z^{-1} is one beat delay of digital control. The transfer function from output current i_2 to u_i is

$$G(S) = \frac{i_2}{u_i} = \frac{R_d CS + 1}{L_1(L_2 + L_g)CS^3 + (L_1 + L_2 + L_g)R_d CS^2 + + (L_1 + L_2 + L_g)S} \cdot (1)$$

The resonance frequency is

$$f_{osc} = \sqrt{L_1 + (L_2 + L_g)} / 2\pi \sqrt{L_1(L_2 + L_g)C} . \quad (2)$$

The typical bode diagram is shown in curve A of Fig. 3.

Comparing to single L filters, LCL filters have such advantages as restraining switching frequency ripple into power grid, mitigating negative effects of grid inductance and so on. Nevertheless, it also involves some drawbacks in high power APF application as follow: 1) LCL filters are difficult to be stabilized on account of large resonance peak and phase lag. 2) Stable though it could be, the bandwidth of system would be restricted to decrease switching frequency ripple injection under condition of relatively low device switching frequency. Therefore, steady state compensation cannot be guaranteed. 3) For the sake of high compensation precision, it is necessary to set such special component as a notch filter to counteract the resonance peak.

Evidently, to address these difficulties the key point is suppressing resonance peak. The common methods include passive damping with a damping resistor in series with a filter capacitor and active damping with feedback of current flowing

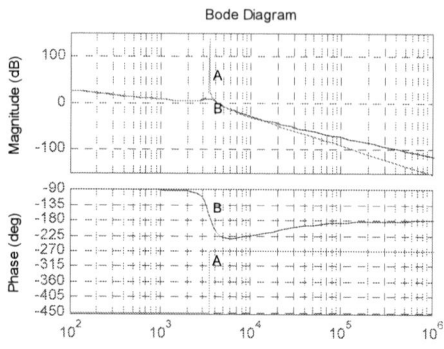

Fig. 3. LCL filter Bode diagram.

through a filter capacitor. Considering too many sensors are needed in the latter method which greatly increases overall cost and complexity of the system, passive damping is adopted in the system. The resistor should be as small as possible to suppress switching frequency ripple and reduce power loss greatly, which can be guaranteed by choosing a low bandwidth PI controller properly. The bode diagram of the LCL filter with passive damping is shown in curve B of Fig. 3. The discrete transfer function of PI controller can be given as

$$PI(z) = K_p + K_I / (z-1). \quad (3)$$

The open loop transfer function of system is

$$F(z) = PI(z)G(z)z^{-1} \quad (4)$$

where, G (z) is the discrete transfer function of control object. The closed loop transfer function of system can be written as

$$P(z) = F(z)/(1 + F(z)). \quad (5)$$

Fig. 4 shows the closed bode diagram of the system. Among the range of compensation current frequency (0-2.5kHz), there are large phase lags between the reference and output current which deteriorate compensation effects sharply.

III. COMPOSITE CONTROL STRATEGY

A. Principles of Plug-in Repetitive Control

Repetitive control is introduced into current loop for the sake of offsetting low compensation precision resulted from low bandwidth of the system. The typical repetitive control system diagram is shown in Fig. 5 in which the dashed block is a repetitive controller.

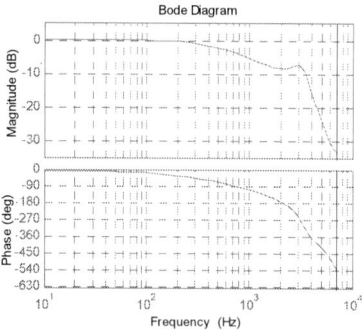

Fig. 4. Closed-loop Bode diagram of conventional PI control strategy.

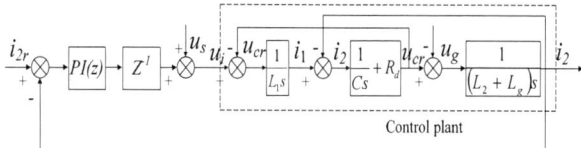

Fig. 2. Current loop control block diagram with conventional PI controller.

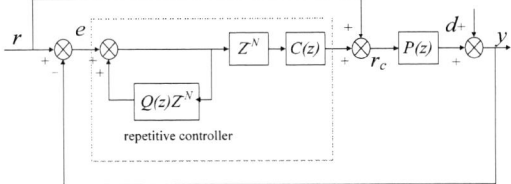

Fig. 5. Repetitive control system diagram.

Where, z^{-N} is the time delay unit (N denotes the number of sampling in one fundamental period), $Q(z)$ can be a low pass filter or close-to-unity constant, $P(z)$ is a control object, $C(z)$ is a compensator. Supposing the system is stable, the error transfer function can be derived

$$e(z)=\frac{(1-Q(z)z^{-N})(1-P(z))}{1-z^{-N}(Q(z)-C(z)P(z))}r(z)+\frac{(1-Q(z)z^{-N})}{1-z^{-N}(Q(z)-C(z)P(z))}d(z). \quad (6)$$

A sufficient condition for system stability can be derived with small gain theorem [10]

$$\left|H(e^{j\omega T})\right|<1 \quad (7)$$

$$H(e^{j\omega T})=Q(e^{j\omega T})-C(e^{j\omega T})P(e^{j\omega T}),\omega\in[0,\pi/T]$$

where, T is the sampling time in the system. If error frequency is a multiple of fundamental frequency, there is $z^{-N}=1$. Moreover, if $Q(z)$ is cancelled, that is $Q(z)=1$, then there is $e(z)=0$, which indicates the basic principle of error elimination.

B. Design of compensator C(z)

Usually, repetitive controller is plugged into conventional current loop with PI controller to attain excellent steady state precision and dynamic response. The double loop composite controller composed by a PI inner loop and repetitive control outer loop is depicted as Fig. 6.

According to the analysis above, the system can attain the best performance when there is $Q(z)=1$ and $H(z)=0$, that is $C(z)P(z)=1$. Obviously, this can be satisfied by $C(z)=P(z)^{-1}$ at any frequency. Nevertheless, it is not feasible in practical application since a myriad of factors such as dead-time, parameter variation and inaccurate system model may play a deleterious role in system compensation precision and stability.

Considering the compensation current frequency ranges from 0 to 2.5 kHz, it seems a wise choice to maintain good effects at medial and low frequencies effect only. In addition, at high frequencies the controller gain falls off sharply to avoid any chance of breaking the system stability condition. Consequently, the desired bode diagram of $C(z)P(z)$ should be zero gain and zero phase shift at medial and low frequencies, gain decreasing quickly and phase shift trying to keep zero at high frequencies.

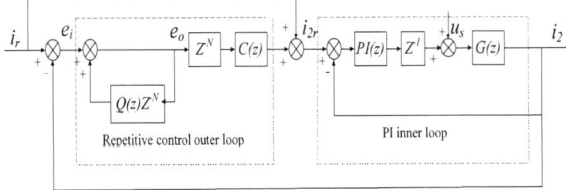

Fig. 6. Double-loop composite controller diagram.

With passive damping in the branch of filter capacitor, resonance peak in the current loop is suppressed effectively, which is critical for the stability of overall system. Therefore, it is not necessary to setup special link, like a notch filter to keep zero gain and zero phase shift at medial and low frequencies. In this paper, only a conventional second-order filter S(z) is required for high frequencies attenuation, which greatly simplifies the control design. The transfer function of typical second-order filter is given as

$$S(s)=\frac{\omega_n^2}{s^2+2\xi\omega_n s+\omega_n^2}. \quad (8)$$

The design of $S(z)$ is a tradeoff between enough attenuation at high frequencies with a fairly low cutoff point of the second-order filter and unnecessary attenuation at lower frequencies with a moderately high one. In this paper, the cutoff frequency and damping ratio are $\omega_n=3000\pi,\xi=0.4$ respectively. The magnitude frequency characteristic of $S(z)P(z)$ is shown in Fig. 7 which indicates zero gain and zero phase shift at medial and low frequencies, gain descending abruptly at high frequencies as is depicted above.

In addition, time advance unit z^k is adopted so as to compensate the phase lag of compensator $S(z)$ as well as $P(z)$. Even though there are some phase errors canceling out at high frequencies, the characteristics of overall system would not be deteriorated owing to severe attenuation at these frequencies. The phase frequency characteristics of $S(z)P(z)$ and z^{-6} is depicted in Fig. 8. In the figure, the curve of z^6 and $S(z)P(z)$ is very similar which guarantees perfect phase compensation.

The compensator $C(z)$ is ultimately defined as $C(z)=K_r*z^k*S(z)$. Where, K_r is the proportional, ranging from 0 to 1. A smaller K_r brings a larger stability margin while a higher one brings smaller steady-state error and faster error convergence. Therefore, supposing $K_r=1$ under condition of sufficient stability margin. In the system the discrete closed loop transfer function is given as:

$$P(z)=\frac{0.08433z^3+0.09644z^2-0.155z-0.02435}{z^5-2.026z^4+1.798z^3-1.253z^2+0.5065z-0.02435}. \quad (9)$$

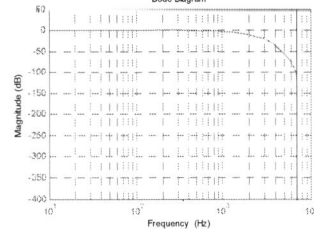

Fig. 7. The amplitude-frequency characteristic of $S(z)P(z)$.

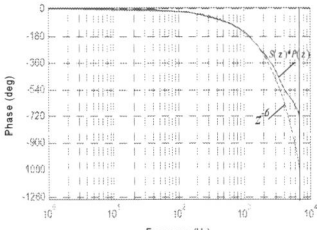

Fig. 8. The phase-frequency characteristics of $S(z)P(z)$ and z^{-6}.

According to the analysis above, $S(z)$ and $C(z)$ can be designed as:

$$S(z) = \frac{0.08195z^2 + 0.1639z + 0.08195}{z^2 - 1.283z + 0.6105}. \quad (10)$$

$$C(z) = \frac{0.08195z^2 + 0.1639z + 0.08195}{z^2 - 1.283z + 0.6105} z^6. \quad (11)$$

C. Design of Q(z)

Usually, $Q(z)$ can be a low pass filter or close-to-unity constant. Comparing to a low pass filter, a constant is easier to check stability of a system and realize in digital control system. Thus, a constant $Q(z)$ is adopted in the control system. Moreover, the value of $Q(z)$ is a tradeoff between high compensation precision with a larger one and strong robustness of the system with a smaller one. In this paper, $Q(z)$ is selected as 0.9.

D. Analysis of Stability

As is analyzed above, a sufficient condition for system stability can be given as

$$H(e^{j\omega T}) = Q(e^{j\omega T}) - C(e^{j\omega T})P(e^{j\omega T}), \omega \in [0, \pi/T]. \quad (12)$$

Using the design results above, $H(z)$ is plotted with MATLAB in Fig. 9. In the figure, the locus of $H(z)$ is always in the unity circle, which satisfies the stability condition. Hence, the system is stable. In addition, there is a fairly large distance between the locus of $H(z)$ and the unity circle, which denotes a enough stability margin.

IV. COMPOSITE CONTROL STRATEGY

A. Experimental circuit and parameters

To verify the design, simulation and experiments are conducted on the three phase SAPF, as is depicted as Fig. 1. Parameters of main circuit are: u_{dc} =700 V, C_{dc} =17.55mF, L_1=0.056mH, L_2=0.020 mH, C=120uF, R_d=0.1 Ohm, u_g =380 V, f_g=50 Hz, rated capacity S_c=260 kVA.. The control scheme is implemented with a TMS320F2812DSP.

B. Steady-state waveforms

Experiments are conducted on the SAPF with three phase balanced load in steady state. Fig. 10 shows experimental waveforms and grid current spectrogram of single-PI control system, in which i_L, i_2 and i_s represent the load, compensation, power grid current respectively. The THD of power grid current descends from 27.5% to 18.2%.

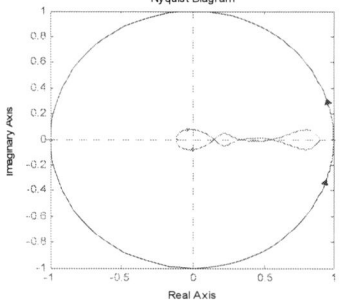

Fig. 9. The nyquist diagram of $H(z)$.

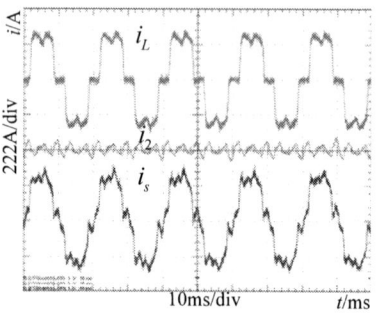

(a)Current waveforms

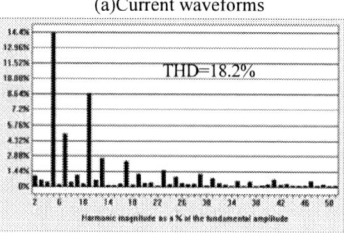

(b) Grid current spectrogram

Fig. 10. The experimental waveforms and grid current spectrogram of single-PI control system.

Although THD decreases moderately, the grid current waveform is still terrible due to a large proportion of low frequency harmonics, especially during the commutation of load current.

Fig. 11 shows experimental waveforms and power grid current spectrogram with the composite controller. The THD of power grid current descends sharply from 27.5% to 2.8%. Moreover, comparing the two power grid current spectrograms, both odd and even harmonic current are inhibited effectively.

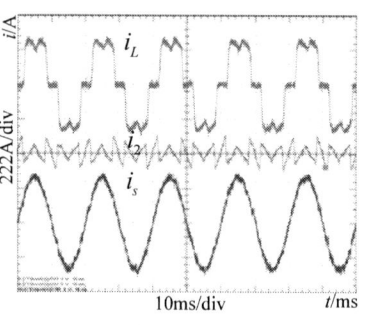

(a)Current waveforms

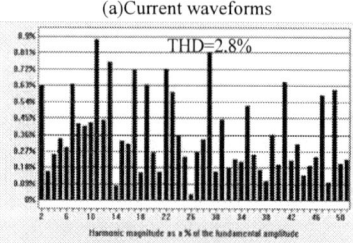

(b) Grid current spectrogram

Fig. 11. The experimental waveforms and grid current spectrogram of composite control system.

In order to verify the ability of reactive compensation, reactive current waveforms with the two control strategies are shown in Fig. 12 and Fig. 13. The THD of reactive current reduces sharply from 10.8% to 2.1%. In addition, as experimental results shown above, both odd and even harmonic current are suppressed effectively.

In order to validate the compensation ability under three phase unbalanced load, an additional 33 ohm resistor is connected between phase B and C in the load side. Fig. 14 and Fig. 15 show the three phase load current and grid current waveforms with the composite controller under unbalance loads. The grid current waveforms denote that the system acquires high compensation precision when used to inhibit negative sequence harmonic current.

C. Dynamic waveforms

Fig. 16 and Fig. 17 show the dynamic waveforms of DC link voltage (AC-coupling, 700VDC) and grid side current as the load current increases from 48A to 96A (rms) and decreases from 96A to 48A(rms) respectively. The fluctuation of DC voltage and grid side current is very small. From Fig. 16 to Fig. 17, it can be seen that the system has acquired fast dynamic responses since the inner PI loop could sense and follow the load change immediately.

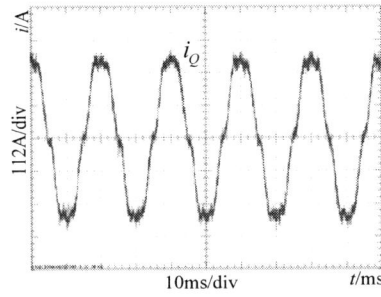

(a) Reactive current waveform

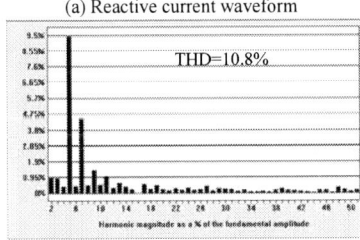

(b) Reactive current spectrogram

Fig. 12. The reactive waveform and spectrogram of single-PI control system.

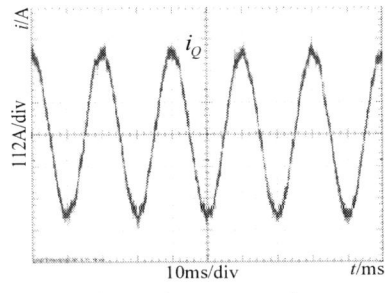

(a) Reactive current waveform

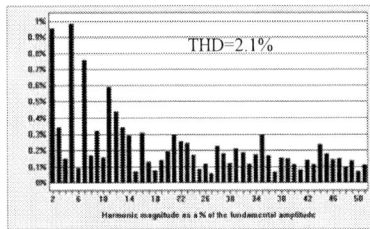

(b) Reactive current spectrogram

Fig. 13. The reactive waveform and spectrogram of composite control system.

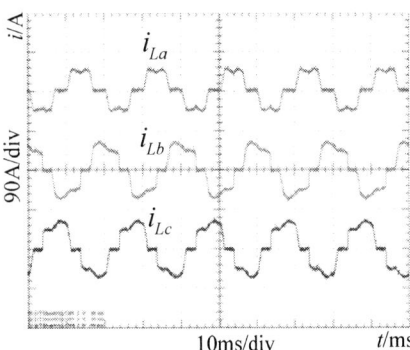

Fig. 14. Three phase load current waveforms with composite controller and unbalanced loads.

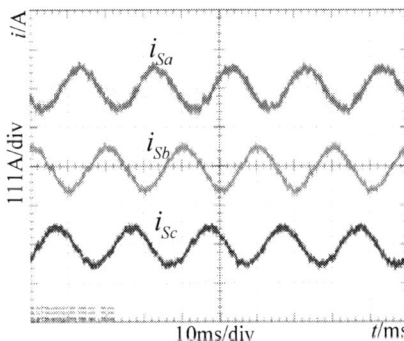

Fig. 15. Three phase grid current waveforms with composite controller and unbalanced loads.

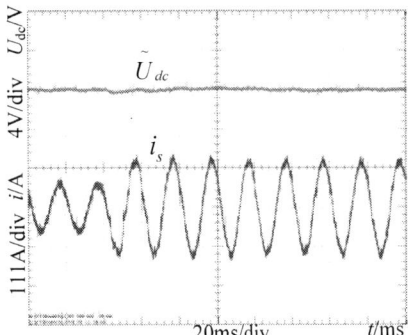

Fig. 16. Dynamic waveforms with load increased.

978-1-4244-4782-4/10 $26.00 © 2010 IEEE

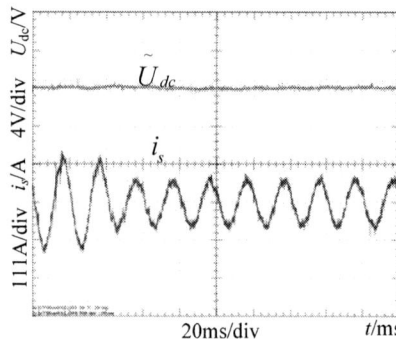

Fig. 17. Dynamic waveforms with load reduced

V. CONCLUSIONS

This paper proposed a double-loop composite controller composed by a PI inner loop and repetitive controller external loop with the control plant of LCL filters, which has reduced the THD of power grid current significantly. The inner loop plays a critical role in the fast response of the system, while the external loop aims at enhancing the waveform quality.

Experimental results verify that the controller could inhibit odd harmonic, even harmonic current and negative sequence harmonic current effectively and excellent dynamic response performance is realized simultaneously.

The proposed current loop composite controller is suitable for the high power SAPF application, which can also be applied to other power electronic equipments, like static var generator (SVG), static synchronous compensator (STATCOM).

REFERENCES

[1] Akagi.H, "New trends in active filters for power conditioning, " *IEEE Transactions on Industry Application*, vol. 32, no. 6, pp. 1312-1322, 1996.

[2] R.I.Bojoi, G.Griva, V.Bostan, M.Guerriero, F.Farina, and F.Profumo. "Current control strategy for power conditioners using sinusoidal signal integrators in synchronous reference frame, " *IEEE Transactions on Power Electronics*, vol. 20, no. 6, pp. 1402−1412, Nov. 2005.

[3] M. Bojrup, P. Karlsson, M. Alakula, and L. Gertmar, "A multiple rotating integrator controller for active filters, "in *Proc. EPE Conf.*, 1999.CD-ROM.

[4] G. Escobar, P. R. Martínez, J. Leyva-Ramos, and P. Mattavelli, "A negative feedback repetitive control scheme for harmonic compensation, " *IEEE Transactions on Industrial Electronics*, vol. 53, no. 4, pp. 1383-1386, 2006.

[5] Wei Xueliang, Dai Ke, Fang Xin, Kang Yong, "Performance analysis and improvement of output for three phase shunt active power filter," *Proceedings of the CSEE*, vol. 27, no. 28, pp. 113-119, 2007.

[6] Aurelio García Cerrada, Omar Pinzón Ardila, Vicente Feliu Batll, Pedro Roncero Sánchez, and Pablo García-González, "Application of a repetitive controller for a three-phase active power filter," *IEEE Transactions on Power Electronics*, vol. 22, no. 1, pp. 237-246, 2007.

[7] R. Griñó, R. Cardoner, R. Costa-Castelló, and E. Fossa, "Digital repetitive control of a three-phase four-wire shunt active filter," *IEEE Transactions on Industrial Electronics*, vol. 54, no. 3, pp. 1495-1503, 2007.

[8] R. C. Castelló, R. Griñó, and E. Fossas, "Odd-harmonic digital repetitive control of a single-phase current active filter," *IEEE Transactions on Power Electronics*, vol. 19, no. 4, pp. 1060-1068, 2004.

[9] Wu Jian, He Na, Xu Dian-guo, "Application of repetitive control technique in shunt active power filter," *Proceedings of the CSEE*, vol. 28, no. 18, pp. 66-72, 2008.

[10] Chuan Xie, Zhiqiang Wang, Guozhu Chen, "A simple method of realization of low power loss and high compensation precision active power filter," in *IEEE 2009 Sustainable Power Generation and Supply*, 2009, pp. 178-183.

A Combined Series-Parallel Active Filter System Implementation Using Generalized Non-Active Power Theory

Mehmet Ucar, Sule Ozdemir and Engin Ozdemir

Kocaeli University, Faculty of Technology, 41380, Umuttepe, Kocaeli, Turkey

e-mails: {mucar, sozaslan, eozdemir}@kocaeli.edu.tr

Abstract—**In this paper, a generalized non-active power theory based control strategy is implemented in a 3-phase 4-wire combined series-parallel active filter (CSPAF) system for periodic and non-periodic waveforms compensation. The CSPAF system consists of a series active filter (SAF) and a parallel active filter (PAF) combination connected a common dc-link. The generalized non-active power theory is valid for single-phase and multi-phase systems, as well as periodic and non-periodic waveforms. The theory was applied in previous studies for current control in the PAF. In this study the theory is used for current and voltage control in the CSPAF system. The CSPAF system is simulated in Matlab/Simulink and an experimental setup is also built, so that different cases can be studied in simulations or experiments. The simulation and experimental results verify that the generalized non-active power theory is suitable for periodic and non-periodic current and voltage waveforms compensation in the CSPAF system.**

I. INTRODUCTION

The widespread use of non-linear loads and power electronic converters has increased the generation of non-sinusoidal and non-periodic currents and voltages in electric power systems. Generally, power electronic converters generate harmonic components which frequencies that are integer multiplies of the line frequency. However, in some cases, such as controlled 3-phase rectifiers, arc furnaces and welding machines are typical loads, the line currents may contain both frequency lower than the line frequency and frequency higher than the line frequency but not the integer multiple of line frequency [1]-[4]. These currents interact with the impedance of the power distribution system and disturb voltage waveforms at point of common coupling (PCC) that can affect other loads. These waveforms are considered as non-periodic for the period of the currents is not equal to the period of the line voltage [1], [2].

The effects of non-periodic components of voltages and currents are similar to that caused by harmonics. They may contribute power loss, disturbances, measurement errors and control malfunctions, thus degradation of the supply quality in distribution systems [2]. Additionally, voltage sags are one of most important power quality problems in the distribution

system and usually caused by fault conditions or by the starting of large electric motors [5].

Various non-active power theories in the time domain have been discussed [6]. The generalized non-active power theory was applied compensation of the non-sinusoidal and non-periodic load current for parallel active filter (PAF) [7], [8] and static synchronous compensator (STATCOM) [9]. This paper presents the application of the generalized non-active power theory for the compensation of periodic (but non-sinusoidal) and non-periodic currents and voltages with the combined series-parallel active filter (CSPAF) system. The simulation and experimental results showed that the theory proposed in this paper is applicable to the non-active power compensation of periodic load currents and source voltages with harmonics and non-periodic load currents and source voltages in 3-phase 4-wire systems.

The CSPAF system consists of back-to-back connection of the series active filter (SAF) and the PAF with a common dc-link. The CSPAF system function is to compensate for all current related problems such as reactive power compensation, power factor improvement, current harmonic compensation, and load unbalance compensation. It regulates the dc-link voltage using the PAF. Besides, it can compensate all voltage related problems, such as voltage harmonics, voltage sag, flicker and regulate the load voltage using the SAF [10], [11]. Fig. 1 shows the general power circuit configuration of the CSPAF system.

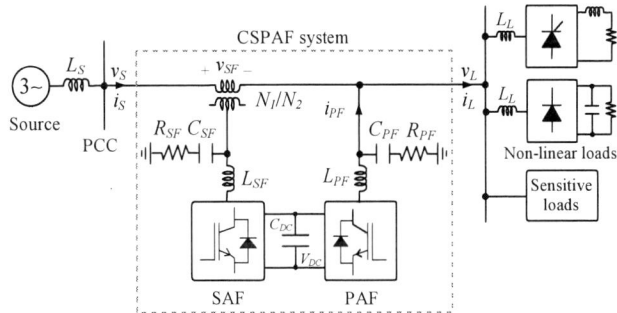

Fig. 1. General power circuit configuration of the CSPAF system.

This work is supported by TUBITAK Research Fund., (No. 108E083)

978-1-4244-4782-4/10 $26.00 © 2010 IEEE

II. GENERALIZED NON-ACTIVE POWER THEORY

The generalized non-active power theory [7] is based on Fryze's definition of non-active power [12] and is an extension of the theory proposed in [13]. Voltage vector $v(t)$ and current vector $i(t)$ in a 3-phase system,

$$v(t) = [v_1(t), v_2(t), v_3(t)]^T, \qquad (1)$$

$$i(t) = [i_1(t), i_2(t), i_3(t)]^T. \qquad (2)$$

The instantaneous power $p(t)$ and the average power $P(t)$ is defined as the average value of the instantaneous power $p(t)$ over the averaging interval $[t-T_c, t]$, that is

$$p(t) = v^T(t)i(t) = \sum_{p=1}^{3} v_p(t)i_p(t), \qquad (3)$$

$$P(t) = \frac{1}{T_c} \int_{t-T_c}^{t} p(\tau)\,d\tau. \qquad (4)$$

The instantaneous active current $i_a(t)$ and instantaneous non-active current $i_n(t)$ are given in (5) and (6).

$$i_a(t) = \frac{P(t)}{V_p^2(t)} v_p(t) \qquad (5)$$

$$i_n(t) = i(t) - i_a(t) \qquad (6)$$

In (5), voltage $v_p(t)$ is the reference voltage, which is chosen on the basis of the characteristics of the system and the desired compensation results. $V_p(t)$ is the corresponding rms value of the reference voltage $v_p(t)$, that is

$$V_p(t) = \sqrt{\frac{1}{T_c} \int_{t-T_c}^{t} v_p^T(\tau)v_p(\tau)\,d\tau}. \qquad (7)$$

The instantaneous non-active power $p_n(t)$ and average non-active power $P_n(t)$ are defined by averaging the instantaneous powers over time interval $[t-T_c, t]$,

$$p_n(t) = v^T(t)i_n(t) = \sum_{p=1}^{m} v_p(t)i_{np}(t), \qquad (8)$$

$$P_n(t) = \frac{1}{T_c} \int_{t-T_c}^{t} p_n(\tau)\,d\tau. \qquad (9)$$

In the generalized non-active power theory, the standard definitions for an ideal 3-phase, sinusoidal power system use the fundamental period T to define the rms values and average active power and non-active power. If there are only harmonics in the load current, T_c does not change the compensation results as long as it is an integral multiple of $T/2$, where T is the fundamental period of the system.

However, in other cases, such as a 3-phase load with sub-harmonics, or a non-periodic load, T_c has significant influence on the compensation results, and the power and energy storage rating of the compensator's components [7].

III. CONTROL OF THE CSPAF SYSTEM

The 3-phase 4-wire CSPAF system is realized two 3-leg voltage source inverter (VSI) with split dc-link capacitor and used the generalized non-active power theory based current and voltage control techniques.

A. SAF Control Technique

Control block diagram of the SAF is shown in Fig. 2. In the method the positive sequence detector generates auxiliary control signals $(i_{a1+}, i_{b1+}, i_{c1+})$ used as a reference current $i_p(t)$ for the generalized no-active power theory. The source voltages are input of the positive-sequence detector that includes a phase locked loop (PLL) function [14]. The output signals of the positive-sequence detector are i_{a1+}, i_{b1+} and i_{c1+}, which have unity amplitude and are in phase with the fundamental positive-sequence component of the source voltages $(v_{Sa1+}, v_{Sb1+}, v_{Sc1+})$. Effective value of the reference current $I_p(t)$ is given in (10).

$$I_p(t) = \sqrt{\frac{1}{T_c} \int_{t-T_c}^{t} i_p^T(\tau)i_p(\tau)\,d\tau} \qquad (10)$$

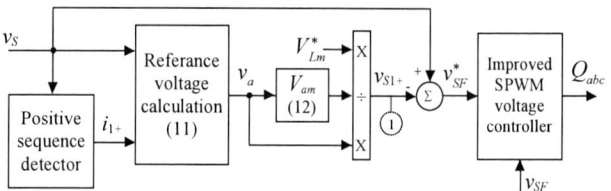

Fig. 2. Control block diagram of the SAF.

The average power calculated given (4) by using the reference currents and the source voltages. The sinusoidal load voltage $(v_a(t))$ is derived by using (11) [15]. As clearly shown in Fig. 2, the $v_a(t)$ is divided by their amplitude (V_{am}) calculated by (12) and multiplied the desired load voltage magnitude (V_{Lm}) for converting the $v_a(t)$ to the desired load voltage (v_{S1+}). Then, the compensation reference voltages of the SAF are derived by (13) and compared SAF voltages. Thus SAF switching signals are obtained by using the improved sinusoidal pulse width modulation (SPWM) [11].

$$v_a(t) = \frac{P(t)}{I_p^2(t)} i_p(t) \qquad (11)$$

$$V_{am} = \frac{2}{3} \sqrt{v_{aa}^2 + v_{ab}^2 + v_{ac}^2} \qquad (12)$$

$$v_{SF}^*(t) = v_S(t) - v_{S1+}(t) \qquad (13)$$

B. PAF Control Technique

The average power calculated given (4) by using load currents and fundamental positive sequence source voltages (v_{Sa1+}, v_{Sb1+}, v_{Sc1+}) over the averaging interval [$t-T_c$, t]. Desired sinusoidal load currents (i_{La1+}, i_{Lb1+}, i_{Lc1+}) is derived by using (5) and instantaneous non-active current $i_n(t)$ is calculated as in (6). Also, the additional active current $i_{ca}(t)$ required to meet the losses in (14) is drawn from the source by regulating the dc-link voltage v_{DC} to the reference V_{DC}. A PI controller is used to regulate the dc-link voltage v_{DC}. The error between the actual dc voltage and its reference value is treated in the PI controller and the output is multiplied by a sinusoidal fundamental template of unity amplitude for each phase of the three phases. In addition, as shown in Fig. 3, the difference between V_{dc1} and V_{dc2} is applied to the PI controller. Thus, equal voltage sharing between the capacitors is accomplished. The compensation reference currents of the PAF are obtained by (15). The reference currents are compared the PAF currents and applied to hysteresis current controller. Thus, the PAF switching signals are obtained. Control block diagram of the PAF is shown in Fig. 3.

$$i_{ca}(t) = (v_{S1+}[K_{P1}(V_{DC} - v_{DC}) + K_{I1}\int_0^t (V_{DC} - v_{DC})dt])$$
$$+ (K_{P2}(v_{DC1} - v_{DC2}) + K_{I2}\int_0^t (v_{DC1} - v_{DC2})dt) \tag{14}$$

$$i_{PF}^*(t) = i_n(t) - i_{ca}(t) \tag{15}$$

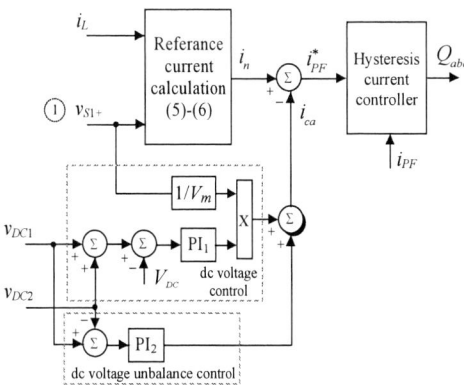

Fig. 3. Control block diagram of the PAF.

IV. SIMULATION AND EXPERIMENTAL RESULTS

The CSPAF system prototype is designed and developed in laboratory to validate the generalized non-active power theory proposed in the paper. The power circuit and control block diagram of the CSPAF system implementation is given in Fig. 4. The non-linear load-1 (which contains a 3-phase half controlled thyristor rectifier with firing angle 30° and a single-phase diode rectifier are used as nonlinear loads) is the load that requires ideal source voltages. The non-linear load-2 (which contains a 3-phase diode rectifier) is connected to the PCC to create source voltage distortion and imitates the effect of other loads on a radial network. The 3-phase source voltages with distortion are synthesized by increasing system impedance from 59 µH to 2.2 mH and connecting the non-linear load-2 to PCC as shown in Fig. 4.

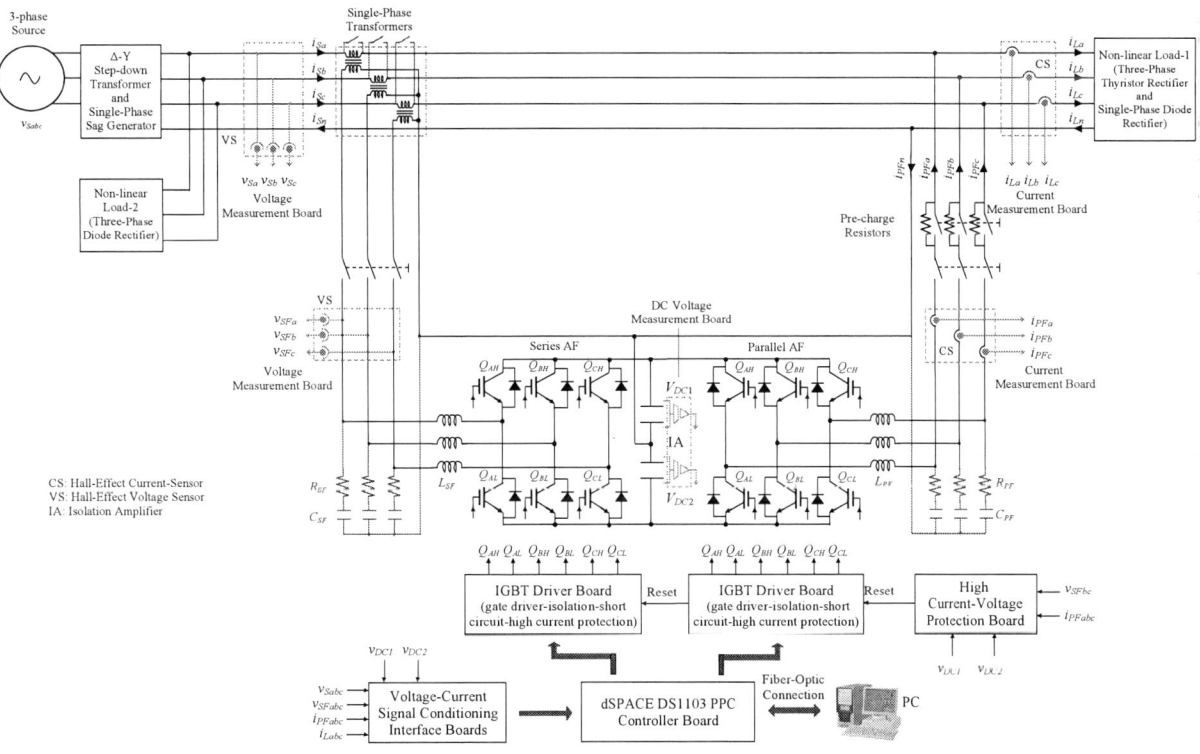

Fig. 4. Power circuit and control block diagram of the CSPAF system implementation.

978-1-4244-4782-4/10 $26.00 © 2010 IEEE

Additionally, the voltage-sag generator was employed to simulate the single-phase source voltage sag for phase-a in the laboratory. The 3-phase step-down transformer is used for supply voltage to the CSPAF system and testing the experimental voltage sag problem. The power circuit configuration of the CSPAF system combines 3-phase 4-wire SAF and PAF. Two voltage source 3-leg IGBT converters sharing a common dc-link are used. The dc-link includes two capacitor with the midpoint connected to the neutral wire of the supply system. The dc-link voltage is adjusted at 400 V. The ac side of the SAF is connected through single-phase injection transformers in series with the input supply lines. The PAF is connected in parallel with the output of the system through an inductor. The CSPAF system parameters are given in Table I.

Both AF are digitally controlled using a dSPACE DS1103 controller board, includes a real-time processor and the necessary I/O interfaces that allow carry-out the control operation. Owing to the switching of the parallel and the series VSI's, the compensating currents and voltages have unwanted high-order harmonics that can be removed by small high-pass passive filters represented by R_{PF}, C_{PF} and R_{SF}, C_{SF}.

The generalized non-active power theory based compensation system is simulated and an experimental setup is also built, so that different cases can be studied in simulations or experiments. The first three cases for periodic current and voltage compensation (subsections A–C) are tested in the experimental setup and the last two cases for (subsections D and E) are simulated in Matlab/Simulink software since they are difficult to be carried out in an experimental setup. The compensation of periodic currents and voltages with fundamental period T, using a compensation period T_c that is a multiple of $T/2$ is enough for complete compensation [7].

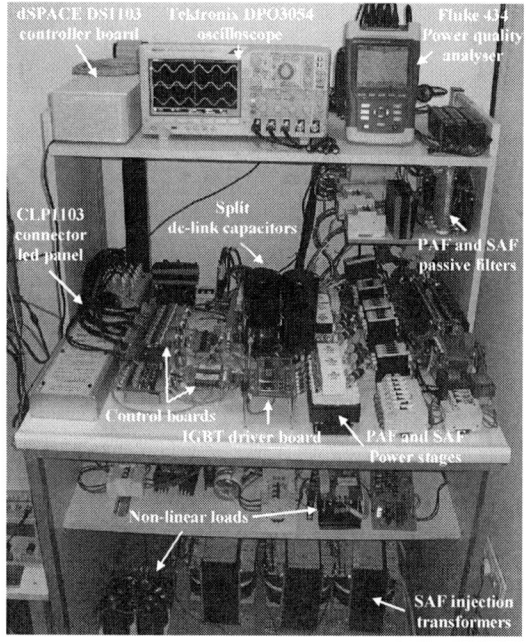

Fig. 5. The experimental test setup photograph.

TABLE I
THE CSPAF SYSTEM PARAMETERS

Components		Symbol	Parameters
Power source	Voltage, frequency	V_{Sabc}, f_s,	110V, 50Hz,
	Impedance	L_s	59μH
DC-link	Capacitors	C_1, C_2	4700μF, 4700μF
	Reference voltage	V_{DC}	400V
PAF	Filter	L_{PF}, R_{PF}, C_{PF}	3mH, 5Ω, 30μF
	Switchching frequency	f_{SWp}	8kHz
SAF	Filter	L_{SF}, R_{SF}, C_{SF}	2.5mH, 2Ω, 150μF
	Switchching frequency	f_{SWs}	10kHz
	Injection transformer	N_1/N_2, S	2, 5.4kVA
Non-linear loads (rectifiers)	3-phase thyristor	L_L, L_{DC}, R_{DC}	3mH, 5.7mH, 12Ω
	1-phase diode	L_{L2}, C_{DC}, R_{DC}	2mH, 330μF, 45Ω
	3-phase diode	C_{DC}, R_{DC}	8800μF, 15Ω

A. *Unbalanced Non-linear Load Current Compensation*

The experimental results of unbalanced non-linear load current compensation under ideal source voltages are shown in Fig. 6.

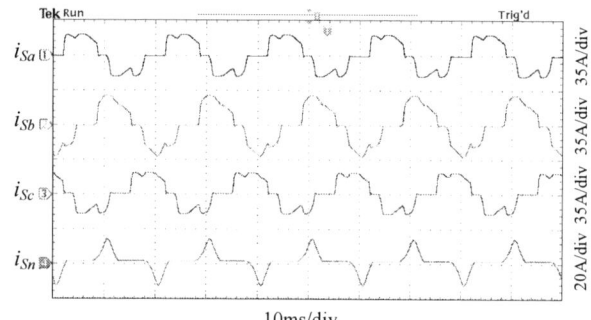

(a) Source currents before compensation.

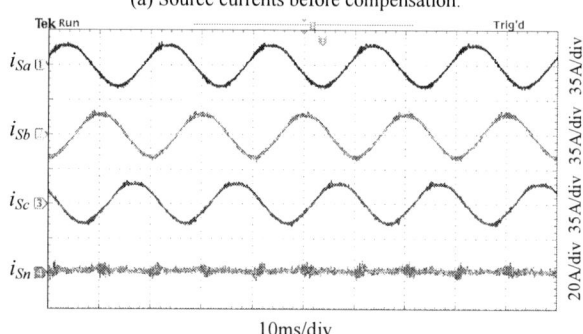

(b) Source currents after compensation.

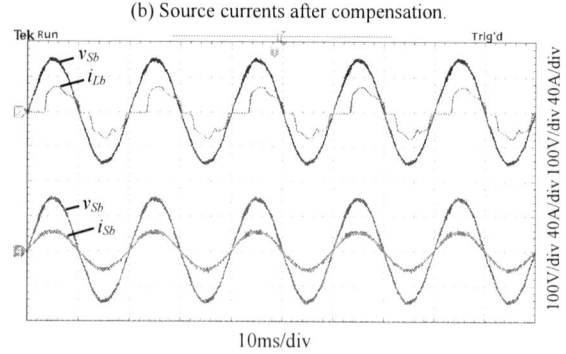

(c) Reactive power compensation.

Fig. 6. Experimental results: Unbalanced non-linear load current compensation under ideal source voltages.

Fig. 6(a) shows the unbalanced non-linear source currents before compensation. After compensation choosing the period as $T_c=T/2$ source currents are almost sinusoidal, balanced and have very low total harmonic distortion (THD) as shown in Fig. 6(b). Moreover, the neutral line current is obviously diminished. Fig. 6(c) shows the experimental waveforms of the phase difference between source voltages and source currents for the reactive power compensation; source voltage and load current (upper waveform) and source voltage and current (lower waveform). The PAF compensates the load reactive power, thus source currents are in phase with its phase voltage and making the unity power factor source current. The compensation results are summarized in Table II.

TABLE II
SUMMARY OF EXPERIMENTAL RESULTS FOR
THE LOAD CURRENT COMPENSATION

Source currents (i_S)		Before	After
RMS (A)	phase-a	14.2	14.2
	phase-b	17.6	14.2
	phase-c	14.3	14.1
	neutral	4.9	1.2
THD (%)	phase-a	33.8	4.4
	phase-b	29.8	4.1
	phase-c	33.6	4.5
PF		0.89	0.99

B. Source Voltage Harmonic Compensation

Fig. 7 shows the experimental results of the distorted source voltages compensation, while the load currents are non-linear and unbalanced.

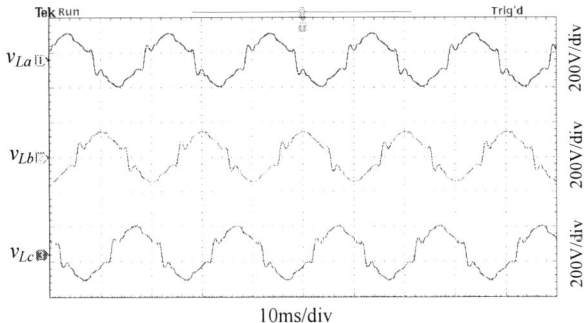

(a) Load voltages before compensation.

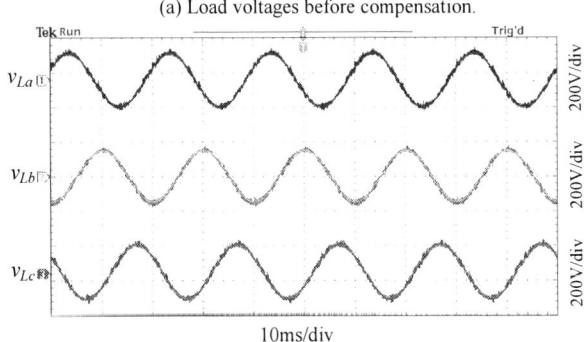

(b) Load voltages after compensation.

Fig. 7. Experimental results: Distorted source voltage compensation.

Non-linear loads draw highly distorted currents from the utility as well as causing distortion of the voltages. The 3-phase distorted load voltages before compensation are demonstrated in Fig. 7(a). After compensation choosing the period as $T_c=T/2$, the source voltages with distortion is compensated to the sinusoidal waveforms are shown in Fig. 7(b). The THD of the load voltages, which was approximately 9.3% before compensation, is approximately 4.4% after compensation. The compensation results are summarized in Table III.

TABLE III
SUMMARY OF EXPERIMENTAL RESULTS FOR THE
DISTORTED SOURCE VOLTAGE COMPENSATION

Load voltages (v_L)		Before	After
RMS (V)	phase-a	104.1	110.5
	phase-b	103.4	110.1
	phase-c	104.2	109.2
THD (%)	phase-a	9.2	4.3
	phase-b	9.1	4.6
	phase-c	9.6	4.4

C. Source Voltage Sag Compensation

Voltage sags are one of the most important power quality problems because of its impact on malfunctioning electrical equipment. Voltage sags are typically caused by remote faults such as a single line to ground fault on the power system or due to starting of large induction motors. Fig. 8 shows the experimental waveforms under single-phase voltage sag with a depth of 50% choosing the period as $T_c=T/2$. In the Fig. 8, from top to bottom, phase-a source voltage, compensated load voltage, compensated source current and load current are showed. The load terminal voltage is regulated and almost constant nominal value during voltage sag of phase-a using the CSPAF system. The required power for the compensation of the load voltage is supplied from the source. Thus, the source currents increased. The compensation results are summarized in Table IV.

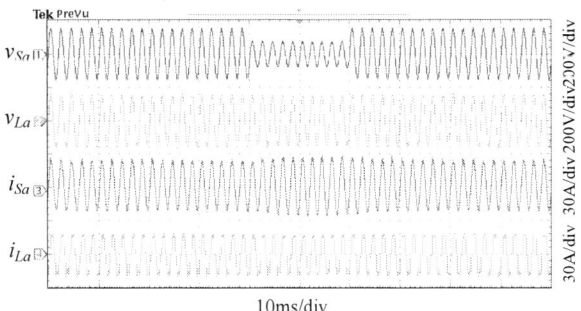

Fig. 8. Experimental results: Source voltage sag compensation: source voltage, load voltage, source current and load current waveforms.

TABLE IV
SUMMARY OF EXPERIMENTAL RESULTS FOR THE
SINGLE-PHASE VOLTAGE SAG COMPENSATION

Load voltages (v_L)		Before	After
RMS (V)	phase-a	52.1	108.3
	phase-b	105.8	108.8
	phase-c	106.8	109.6

D. Sub-Harmonic Current and Voltage Compensation

The sub-harmonic currents (frequency lower than fundamental frequency) are typically generated by power electronic converters [7]. The main feature of these non-periodic currents is that the currents may have a repetitive period. When the fundamental frequency of the source voltage is an odd multiple of the sub-harmonic frequency, the minimum T_c for complete compensation is 1/2 of the common period of both f_s and f_{sub}. When f_s are an even multiple of f_{sub}, the minimum T_c for complete compensation is the common period of both f_s and f_{sub} [8]. In this study, source voltage and load current contains sub-harmonics of 10 Hz frequency and 20% amplitude are given in Table V. The sub-harmonic current and voltage compensation simulation results are shown in Fig. 9 and Fig. 10, respectively.

TABLE V
THREE-PHASE SOURCE VOLTAGE AND
LOAD CURRENT VALUES

Parameters	Fundamental	Sub-harmonic
Freq. (Hz)	50	10
Currents	15 A	% 20
Voltages	110 V	% 20

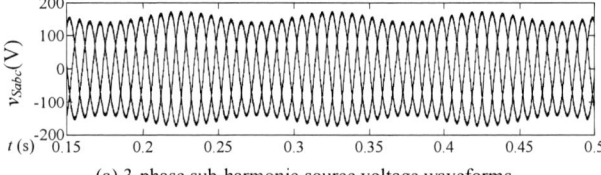

(a) 3-phase sub-harmonic source voltage waveforms.

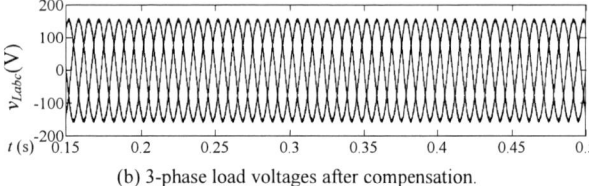

(b) 3-phase load voltages after compensation.

Fig. 9. Simulation results: Sub-harmonic voltage compensation.

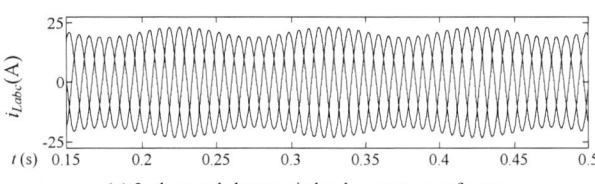

(a) 3-phase sub-harmonic load current waveforms.

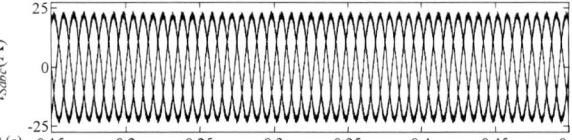

(b) 3-phase source currents after compensation.

Fig. 10. Simulation results: Sub-harmonic current compensation.

The sub-harmonic component can be completely compensated by choosing T_c=2.5T, and the source currents

and load voltages are balanced and sinusoidal. The CSPAF system is able to suppress all the sub-harmonic component of the load current and the voltage at the load terminals is constant amplitude after compensation.

E. Stochastic Non-Periodic Current and Voltage Compensation

The arc furnace load currents may contain stochastic non-periodic currents (frequency higher than fundamental frequency but not an integer multiple of it). Theoretically, the period T of a non-periodic load is infinite [7]. In a non-periodic system, the instantaneous current varies with different averaging interval T_c, which is different from the periodic cases. The source current could be a pure sine wave if T_c goes to infinity. However, this is not practical in a power system, and T_c is chosen to have a finite value. The non-active components in these loads cannot be completely compensated by choosing T_c as $T/2$ or T, or even several multiples of T. Choosing that period as may result in an acceptable both source current and load voltage which are quite close to a sine wave. If T_c is large enough, increasing T_c further will not typically improve the compensation results significantly [8].

In this work, 3-phase source voltage and load current components are given in Table VI [16]. Fig. 11 and Fig. 12 shows simulation results of the stochastic non-periodic voltage and current compensation choosing the period as T_c=5T. After compensation, load voltages and source currents are balanced and almost sinusoidal with low THD as shown in Fig 11(b) and Fig 12(b). In addition, source neutral current have been reduced considerably.

TABLE VI
THREE-PHASE SOURCE VOLTAGE AND
LOAD CURRENT COMPONENTS

Parameters	Fund.	Components (%)				
Freq. (Hz)	50	104	117	134	147	250
Currents	15 A	30	40	20	20	50
Voltages	110 V	7.5	10	5	5	12.5

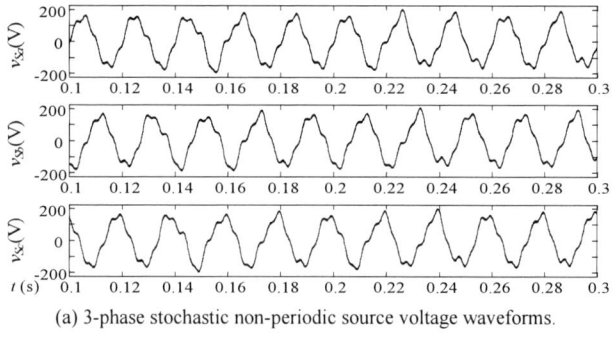

(a) 3-phase stochastic non-periodic source voltage waveforms.

(b) 3-phase load voltages after compensation.

Fig. 11. Simulation results: Stochastic non-periodic voltage compensation.

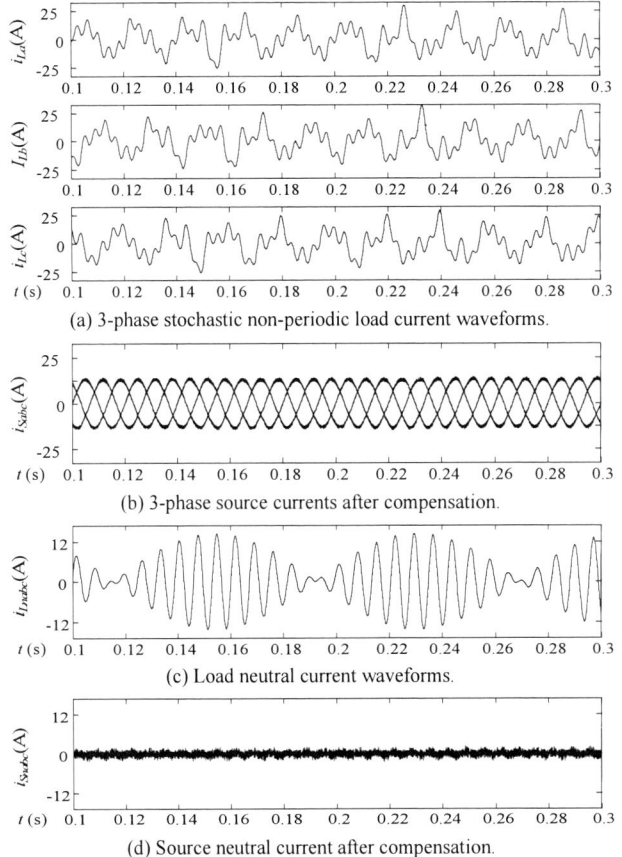

(a) 3-phase stochastic non-periodic load current waveforms.

(b) 3-phase source currents after compensation.

(c) Load neutral current waveforms.

(d) Source neutral current after compensation.

Fig. 12. Simulation results: Stochastic non-periodic current compensation.

V. CONCLUSION

The presence of non-linear, time-variant, disturbing loads connected to the electric power system is responsible for the presence of periodic and non-periodic disturbances on the line currents and voltages. In this paper, the generalized non-active power theory, which is applicable to sinusoidal or non-sinusoidal, periodic or non-periodic, balanced or unbalanced electrical systems, is presented. It has been applied to the 3-phase 4-wire CSPAF system. This theory is adapted to different compensation objectives by changing the averaging interval T_c. The CSPAF experimental setup system was built and tested in the laboratory. Three cases, unbalanced nonlinear load currents, distorted source voltages and source voltage sag with unbalanced non-linear load currents compensation are tested in the experiments. The sub-harmonic and the stochastic non-periodic current and voltage

compensation are simulated in Matlab/Simulink. The simulation and experimental results showed that the theory proposed in the CSPAF system was applicable to non-active power compensation of periodic and non-periodic waveforms in 3-phase 4-wire systems.

VI. REFERENCES

[1] E. H. Watanabe and M. Aredes, "Compensation of nonperiodic currents using the instantaneous power theory," *IEEE Power Engineering Soc. Summer Meeting*, pp. 994–999, 2000.

[2] L. S. Czarnecki, "Non-periodic currents: their properties, identification and compensation fundamentals," *IEEE Power Engineering Soc. Summer Meeting*, pp. 971-976, 2000.

[3] H. Akagi, "Active filters and energy storage systems operated under nonperiodic conditions," *IEEE Power Engineering Soc. Summer Meeting*, Seattle, pp. 965-970, 2000.

[4] S. A. Farghal, M. S. Kandil and Elmitwally, "Evaluation of a shunt active power conditioner with a modified control scheme under nonperiodic conditions," *IEE Proc. Generation, Transmission and Distribution*, vol. 149, no. 6, pp. 726-732, Nov. 2002.

[5] M. F. McGranaghan, D. R. Mueller and M. J. Samotyj, "Voltage sags in industrial systems," *IEEE Trans. Ind. Appl.*, vol. 29, no. 2, pp. 397-403, 1993.

[6] L. M. Tolbert and T. G. Habetler, "Comparison of time-based non-active power definitions for active filtering," *IEEE Int. Power Electron. Congress*, pp. 73–79, Oct. 15-19, 2000.

[7] Y. Xu, L. M. Tolbert, F. Z. Peng, J. N. Chiasson and J. Chen, "Compensation-based non-active power definition," *IEEE Power Electr. Letter*, vol. 1, no. 2, pp. 45-50, 2003.

[8] Y. Xu, L. M. Tolbert, J. N. Chiasson, J. B. Campbell and F. Z. Peng, "Active filter implementation using a generalized nonactive power theory", *IEEE Industry Applications Conference*, pp. 153-160, 2006.

[9] Y. Xu, L. M. Tolbert, J. N. Chiasson, J. B. Campbell and F. Z. Peng, "A generalised instantaneous non-active power theory for STATCOM," *IET Electric Power Applications*, pp. 853-861, 2007.

[10] H. Fujita and H. Akagi, "The unified power quality conditioner: the integration of series and shunt active filters," *IEEE Trans. on Power Electr.*, vol. 13, no. 2, 1998.

[11] M. Aredes, K. Heumann, and E. H. Walandble, "An universal active power line conditioner," *IEEE Trans. Power Del.*, vol. 13, no. 2, pp. 545-551, Apr. 1998.

[12] S. Fryze, "Active, reactive, and apparent power in non-sinusoidal systems," *Przeglad Elektrot.*, vol. 7, pp. 193-203 (in Polish), 1931.

[13] F. Z. Peng and L. M. Tolbert, "Compensation of non-active current in power systems-definitions from compensation standpoint," *IEEE Power Eng. Soc. Summer Meeting*, pp. 983-987, 2000.

[14] G. W. Chang and W. C. Chen "A new reference compensation voltage strategy for series active power filter control," *IEEE Trans. on Power Delivery*, vol. 21, no. 3, pp. 1754-1756, July 2006.

[15] M. Ucar, S. Ozdemir and E. Ozdemir, "A control strategy for combined series-parallel active filter system under non-periodic conditions," *International Conference on Renewable Energies and Power Quality, ICREPQ'09*, Valencia (Spain), 15-17 Apr. 2009.

[16] IEEE Interharmonic Task Force, "Interharmonic in power systems," Cigre 36.05/CIRED 2 CC02, Voltage Quality Working Group, 1997.

A Novel Control Method for Unified Power Quality Conditioner (UPQC) Under Non-Ideal Mains Voltage and Unbalanced Load Conditions

Metin Kesler
Kocaeli University Technical Education Faculty, 41380
Umuttepe Kocaeli Turkey
metinkesler@kocaeli.edu.tr

Engin Ozdemir
Kocaeli University Technical Education Faculty, 41380
Umuttepe Kocaeli Turkey
eozdemir@kocaeli.edu.tr

Abstract--**This paper presents a new control method to compensate the power quality problems through a three-phase unified power quality conditioner (UPQC) under non-ideal mains voltage and unbalanced load conditions. The performance of proposed control system was analyzed using simulations with Matlab/Simulink program, and experimental results with the hardware prototype. The proposed UPQC system can improve the power quality at the point of common coupling (PCC) on power distribution system under non-ideal mains voltage and unbalanced load conditions.**

I. INTRODUCTION

Unified power quality control was widely studied by many researchers as an eventual method to improve power quality of electrical distribution system [1-3]. The function of unified power quality conditioner is to compensate supply voltage flicker/imbalance, reactive power, negative-sequence current, and harmonics. In other words, the UPQC has the capability of improving power quality at the point of installation on power distribution systems or industrial power systems. Therefore, the UPQC is expected to be one of the most powerful solutions to large capacity loads sensitive to supply voltage flicker/imbalance [2]. The UPQC consisting of the combination of a series active power filter (APF) and shunt APF can also compensate the voltage interruption if it has some energy storage or battery in the dc link [3].

The shunt APF is usually connected across the loads to compensate for all current-related problems such as the reactive power compensation, power factor improvement, current harmonic compensation, and load unbalance compensation [1-2], whereas the series APF is connected in a series with the line through series transformers. It acts as controlled voltage source and can compensate all voltage-related problems, such as voltage harmonics, voltage sag, voltage swell, flicker, etc.

In this paper a new control algorithm for the UPQC system is optimized without measuring transformer voltage, load and filter current, so that system performance is improved. The proposed control technique has been evaluated and tested under non-ideal mains voltage and unbalanced load conditions using Matlab/Simulink software. The proposed method is also validated through experimental study.

II. UPQC CONTROL ALGORITHM

The UPQC consists of two voltage source inverters connected back to back with each other sharing a common dc link. One inverter is controlled as a variable voltage source in the series APF, and the other as a variable current source in the shunt APF. Fig. 1 shows a basic system configuration of a general UPQC consisting of the combination of a series APF and shunt APF. The main aim of the series APF is harmonic isolation between load and supply; it has the capability of voltage flicker/ imbalance compensation as well as voltage regulation and harmonic compensation at the utility-consumer PCC. The shunt APF is used to absorb current harmonics, compensate for reactive power and negative-sequence current, and regulate the dc-link voltage between both APFs. The proposed UPQC control algorithm block diagram in Matlab/Simulink simulation software is shown in Fig. 2.

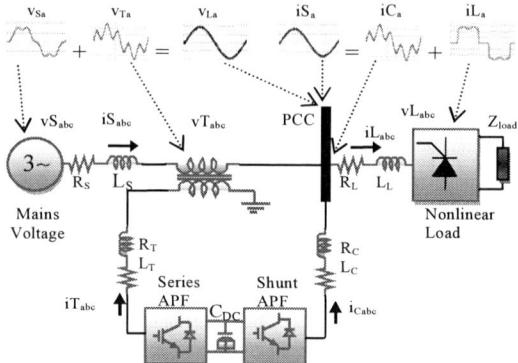

Fig. 1. Unified power quality conditioner configuration.

A. *Reference Voltage Signal Generation for Series APF*

The function of the series APF is to compensate the voltage disturbance in the source side, which is due to the fault in the distribution line at the PCC. The series APF control algorithm calculates the reference value to be injected by the series APF transformers, comparing the positive-sequence component with the load side line voltages. The proposed series APF reference voltage signal generation algorithm is shown in Fig. 3. In equation (1), supply voltages v_{Sabc} are transformed to d-q-0 coordinates.

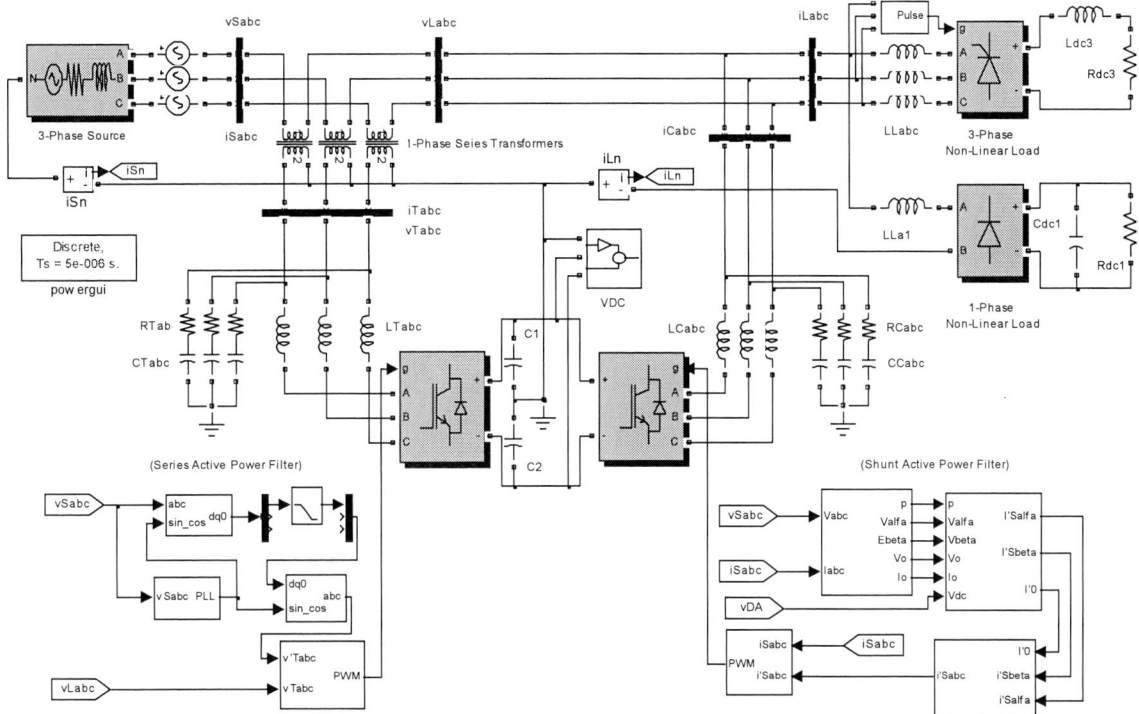

Fig. 2. The proposed UPQC control algorithm block diagram in MATLAB Simulink.

$$\begin{bmatrix} v_{S0} \\ v_{Sd} \\ v_{Sq} \end{bmatrix} = \frac{2}{3} \begin{bmatrix} \frac{1}{2} & \frac{1}{2} & \frac{1}{2} \\ \sin(wt) & \sin(wt - 2\frac{\pi}{3}) & \sin(wt + 2\frac{\pi}{3}) \\ \cos(wt) & \cos(wt - 2\frac{\pi}{3}) & \cos(wt + 2\frac{\pi}{3}) \end{bmatrix} \begin{bmatrix} v_{Sa} \\ v_{Sb} \\ v_{Sc} \end{bmatrix} \quad (1)$$

The voltage in d axes (v_{Sd}) given in (2) consists of average and oscillating components of source voltages ($\overline{v}_{Sd}$ and $\widetilde{v}_{Sd}$). The average voltage $\overline{v}_{Sd}$ is calculated by using second order LPF (low pass filter).

$$v_{Sd} = \overline{v}_{Sd} + \widetilde{v}_{Sd} \quad (2)$$

The load side reference voltages v_{Labc}^{*} are calculated as given in equation (3). The switching signals are assessed by comparing reference voltages (v_{Labc}^{*}) and the load voltages (v_{Labc}) and via sinusoidal PWM controller.

$$\begin{bmatrix} v_{La}^{*} \\ v_{Lb}^{*} \\ v_{Lc}^{*} \end{bmatrix} = \frac{2}{3} \begin{bmatrix} \sin(wt) & \cos(wt) & 1 \\ \sin(wt - 2\frac{\pi}{3}) & \cos(wt + 2\frac{\pi}{3}) & 1 \\ \sin(wt + 2\frac{\pi}{3}) & \cos(wt + 2\frac{\pi}{3}) & 1 \end{bmatrix} \begin{bmatrix} \overline{v}_{Sd} \\ 0 \\ 0 \end{bmatrix} \quad (3)$$

These produced three-phase load reference voltages are compared with load line voltages and errors are then processed by sinusoidal PWM controller to generate the required switching signals for series APF IGBT switches.

B. Reference Current Signal Generation for Shunt APF

The shunt APF described in this paper used to compensate the current harmonics and reactive power generated by the nonlinear load. The shunt APF reference current signal generation block diagram is shown in Fig. 3. The instantaneous reactive power (p-q) theory is used to control of shunt APF in real time. In this theory, the instantaneous three-phase currents and voltages are transformed to α-β-0 coordinates as shown in equation (4) and (5).

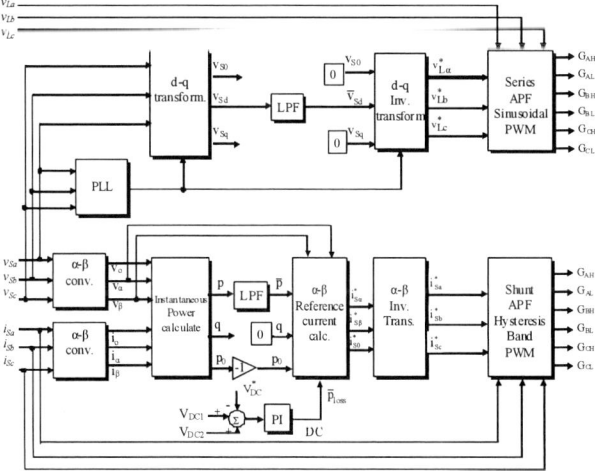

Fig. 3. Series APF reference voltage and shunt APF reference current signal generation block diagram.

978-1-4244-4782-4/10 $26.00 © 2010 IEEE   375

$$\begin{bmatrix} v_0 \\ v_\alpha \\ v_\beta \end{bmatrix} = \sqrt{\frac{2}{3}} \begin{bmatrix} 1/\sqrt{2} & 1/\sqrt{2} & 1/\sqrt{2} \\ 1 & -1/2 & -1/2 \\ 0 & \sqrt{3}/2 & -\sqrt{3}/2 \end{bmatrix} \begin{bmatrix} v_{Sa} \\ v_{Sb} \\ v_{Sc} \end{bmatrix} \quad (4)$$

$$\begin{bmatrix} i_0 \\ i_\alpha \\ i_\beta \end{bmatrix} = \sqrt{\frac{2}{3}} \begin{bmatrix} 1/\sqrt{2} & 1/\sqrt{2} & 1/\sqrt{2} \\ 1 & -1/2 & -1/2 \\ 0 & \sqrt{3}/2 & -\sqrt{3}/2 \end{bmatrix} \begin{bmatrix} i_{Sa} \\ i_{Sb} \\ i_{Sc} \end{bmatrix} \quad (5)$$

The source side instantaneous real and imaginary power components are calculated by using source currents and phase-neutral voltages as given in (6). The instantaneous real and imaginary powers include both oscillating and average components as shown in (7). Average components of p and q consist of positive sequence components ($\bar{p}$ and $\bar{q}$) of source current. The oscillating components ($\tilde{p}$ and $\tilde{q}$) of p and q include harmonic and negative sequence components of source currents [4]. In order to reduce neutral current, p_0 is calculated by using average and oscillating components of imaginary power and oscillating component of the real power; as given in (8) if both harmonic and reactive power compensation is required. $i_{s\alpha}{}^*$, $i_{s\beta}{}^*$ and $i_{s0}{}^*$ are the reference currents of shunt APF in α-β-0 coordinates. These currents are transformed to three-phase system as shown in (9).

$$\begin{bmatrix} p \\ q \end{bmatrix} = \begin{bmatrix} v_\alpha & v_\beta \\ -v_\beta & v_\alpha \end{bmatrix} \begin{bmatrix} i_\alpha \\ i_\beta \end{bmatrix} \quad (6)$$

$$p_0 = v_0 * i_0 \quad ; \quad p = \bar{p} + \tilde{p} \quad (7)$$

$$\begin{bmatrix} i_{S\alpha}^* \\ i_{S\beta}^* \end{bmatrix} = \frac{1}{v_\alpha^2 + v_\beta^2} \begin{bmatrix} v_\alpha & -v_\beta \\ v_\beta & v_\alpha \end{bmatrix} \begin{bmatrix} \bar{p} + p_0 + \bar{p}_{loss} \\ 0 \end{bmatrix} \quad (8)$$

$$\begin{bmatrix} i^*_{Sa} \\ i^*_{Sb} \\ i^*_{Sc} \end{bmatrix} = \sqrt{\frac{2}{3}} \begin{bmatrix} 1/\sqrt{2} & 1 & 0 \\ 1/\sqrt{2} & -1/2 & \sqrt{3}/2 \\ 1/\sqrt{2} & -1/2 & -\sqrt{3}/2 \end{bmatrix} \begin{bmatrix} i^*_{S0} \\ i^*_{S\alpha} \\ i^*_{S\beta} \end{bmatrix} \quad (9)$$

The reference currents are calculated in order to compensate neutral, harmonic and reactive currents in the load. These reference source current signals are then compared with sensed three-phase source currents, and the errors are processed by hysteresis band PWM controller to generate the required switching signals for the shunt APF switches [6].

III. SIMULATOIN RESULTS

In this study, a new control algorithm for the UPQC is evaluated by using simulation results given in Matlab/Simulink software under non-ideal mains voltage and unbalanced load current conditions. The simulated UPQC system parameters are given in Table I. In simulation studies, the results are specified before and after UPQC system are operated. In addition, when the UPQC system is operated, the load has changed and dynamic response of the system is tested. The proposed control method has been examined under non-ideal mains voltage and unbalanced load current conditions. Before harmonic compensation, the THD of the supply current is 26.23%. The obtained results show that the proposed control technique allows the 3.4% mitigation of all harmonic components.

TABLE I.   UPQC SYSTEM PARAMETERS

	Parameters		Value
Source	Voltage	v_{Sabc}	380 V_{rms}
	Frequency	f	50 Hz
Load	3-Phase ac Line Inductance	L_{Labc}	2 mH
	1-Phase ac Line Inductance	L_{La1}	1 mHΩ
	3-Phase dc Inductance	L_{dc3}	10 mH
	3-Phase dc Resistor	R_{dc3}	30 Ω
	1-Phase dc Resistor	R_{dc1}	87,5 Ω
	1-Phase dc Capacitor	C_{dc1}	240 μF
dc-link	Voltage	V_{DC}	700 V
	Capacitor 1/2	$C_1 C_2$	2200 μF
Shunt APF	ac Line Inductance	L_{Cabc}	3.5 mH
	Filter Resistor	R_{Cabc}	5 Ω
	Filter Capacitor	C_{Cabc}	10 μF
	Switching Frequency	f_{pwm}	~15 kHz
Series APF	ac Line Inductance	L_{Tabc}	1.5 mH
	Filter Resistor	R_{Tabc}	5 Ω
	Filter Capacitor	C_{Tabc}	20 μF
	Switching Frequency	f_{pwm}	12 kHz

Simulation results for the load and source voltages under unbalanced-distorted mains voltage conditions are shown in Fig. 4. Load current compensation simulation results under non-ideal (unbalanced-distorted) mains voltage conditions are given in Fig. 5.

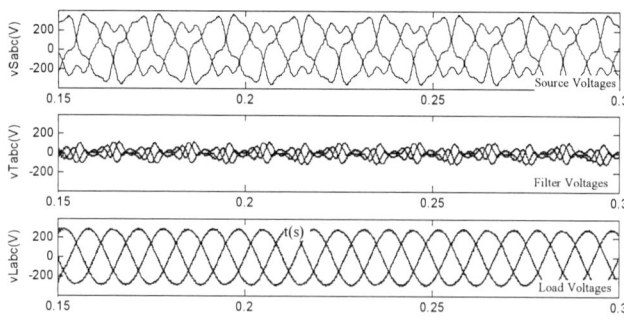

Fig. 4. Simulation results for unbalanced and distorted mains voltage condition.

The neutral current compensation results are given in Fig. 6. The proposed UPQC control algorithm has ability to compensate both harmonics and reactive power of the load and neutral current is also eliminated. The proposed control technique has been evaluated and tested under dynamical and steady-state load conditions. Simulation results for under load changing are shown in Fig. 7.

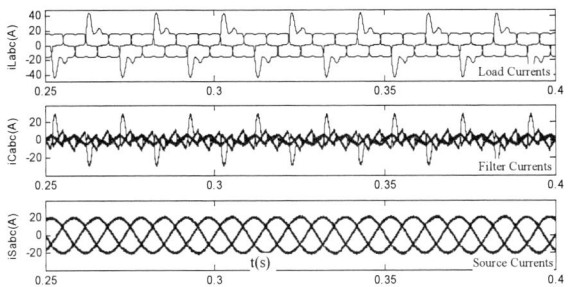

Fig. 5. Simulation results for unbalanced and non-linear load current condition.

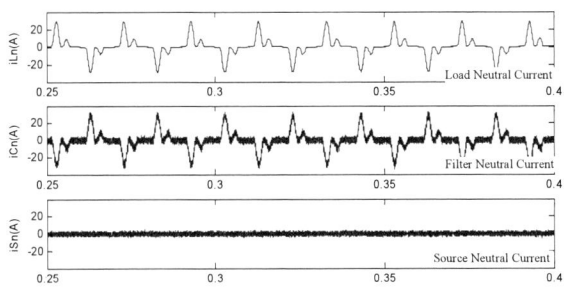

Fig. 6. Simulation results for neutral current compensation.

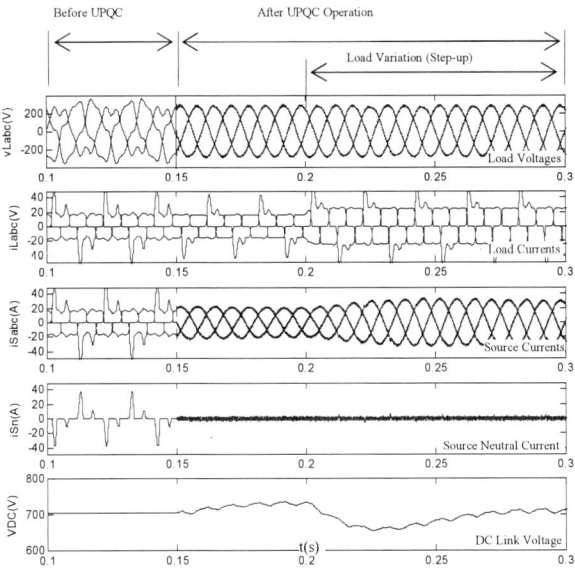

Fig. 7. Simulation results for operational performance of the UPQC system.

IV. EXPERIMENTAL TEST RESULTS

Fig. 8 shows an experimental system configuration photograph of the proposed UPQC system. The aim of the UPQC system is not only to compensate for the current

harmonics produced by a diode-bridge rectifier of 10 kVA, but also to eliminate the voltage harmonics contained in the receiving terminal voltage of the load. The UPQC consists of two back to back connected voltage source inverters and three DSP processors for controlling shunt and series APF's and computer communication for all system control functions. The dc link of both APFs is connected to a common dc capacitor of 1100 microfarad and 700 V dc. All of the circuit parameters and experimental conditions are set up exactly the same as the simulation conditions. Although the proposed control scheme cannot be studied experimentally for unbalanced mains voltage conditions, an optimal control can be designed to eliminate this problem, which will have been discussed as a future work.

Fig. 8. The photograph of the proposed UPQC system.

The source and load voltages are sensed using LEM LV 25P voltage sensors, whereas, all the currents are sensed using LEM LA-55P Hall-Effect current sensors. The series and shunt inverters are built using SEMIX 101GD128Ds six-pack IGBT switches from Semikron. CONCEPT 6SD106EI and Semikron SKHI 61 IGBT drivers are used for series and shunt APF respectively. The IGBT driver modules have short circuit and over current protection functions for every IGBT and provides electrical isolation for all PWM signals applied to the digital signal processor (DSP). The proposed experimental control system consists of three DSP cards from TI (TMS320F28335). Three DSP cards are designed to control shunt and series APF and one of them is responsible for all system operation and power quality analysis. Both inverters use the variable frequency hysteresis band controller.

Fig. 9 shows source voltage and current waveforms before filtering. After compensation, source current becomes sinusoidal and in phase with the source voltage; hence, both harmonics and reactive power are compensated simultaneously. Before harmonic compensation, the THD of the supply current is 29.13% and after the harmonic compensation, it is reduced to 5.3% which complies with the IEEE 519 harmonic standards. Fig. 10 and Fig. 11 show experimental results for source voltage (v_{Sa}), filter current (i_{Ca}) and source current (i_{Sa}) after filter operation respectively.

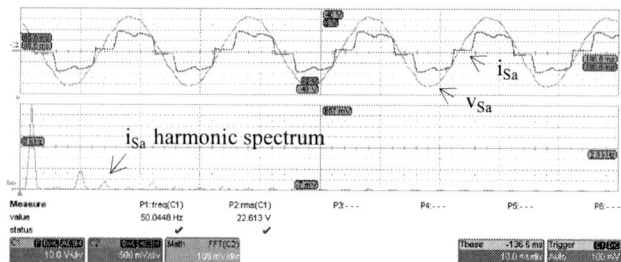

Fig. 9. Experimental results for source voltage (v_{Sa}) and source current (i_{Sa}) before filter operation.

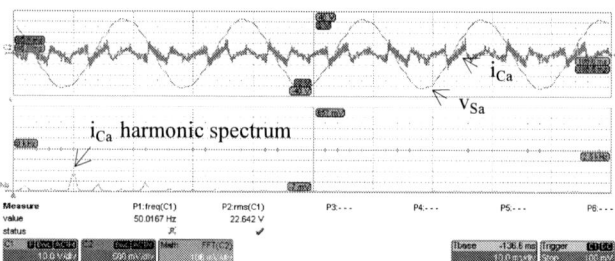

Fig. 10. Experimental results for source voltage (v_{Sa}) and filter current (i_{Ca}) after filter operation.

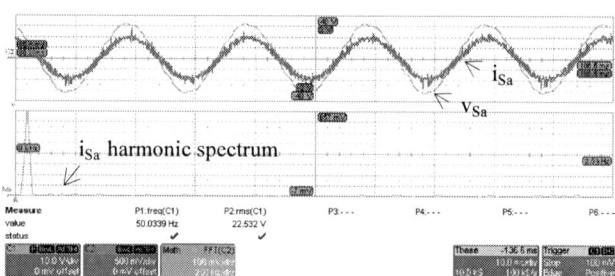

Fig. 11. Experimental results for source voltage (v_{Sa}) and source current (i_{Sa}) after filter operation.

Fig. 12 shows experimental results for three-phase source currents (i_{Sabc}) before and after filter operation. Fig. 13 shows experimental results for the dc link voltage and the source current (i_{Sa}) before and after load variation (load step-up), the shunt APF tested under dynamical and steady-state load conditions under load changing. Fig. 14 shows the experimental results for source currents (i_{Sabc}) and neutral current (i_{Sn}) before and after filter operation. Fig. 15 shows results for load neutral (i_{Ln}), filter neutral (i_{Cn}) and source neutral current (i_{Sn}) before and after filter operation.

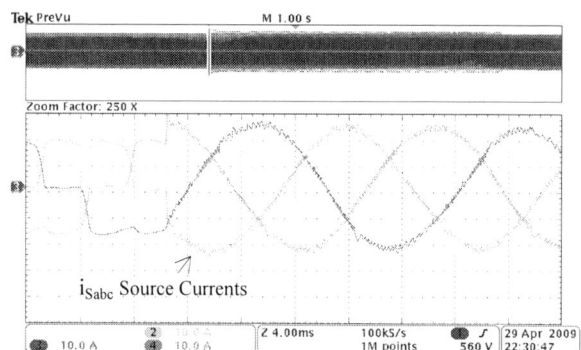

Fig. 12. Experimental results for source current (i_{Sabc}) before and after filter operation.

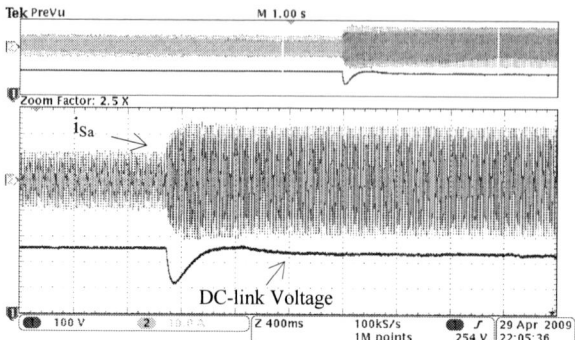

Fig.13. Experimental results for dc link voltage and source current (i_{Sa}) before and after load variation (load step-up).

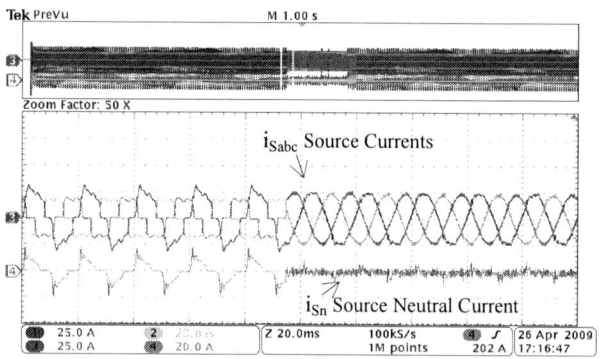

Fig.14. Experimental results for source current (i_{Sabc}) and neutral current i_{Sn} before and after filter operation

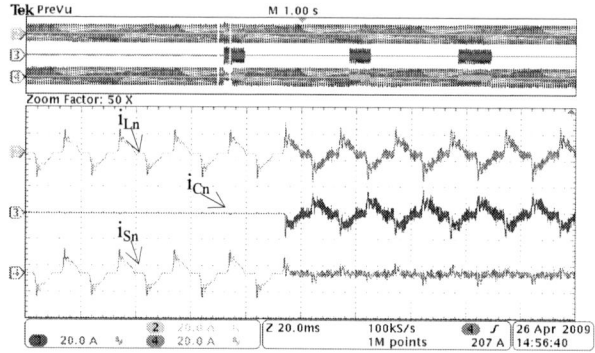

Fig.15. Experimental results for load neutral (i_{Ln}), filter neutral (i_{Cn}) and source neutral current (i_{Sn}).

These experimental results given above shows that the harmonic compensation features of the UPQC, by appropriate control of the shunt APF can be done effectively. The shunt APF with reduced current measurement based control method can be compensating neutral, harmonic and reactive currents effectively, in the unbalanced and distorted load conditions. The series APF experimental results for mains and load voltages before filter operation is shown in Fig. 16. Fig. 17 shows the experimental results for the load voltages in three-phase form before and after filter operation.

978-1-4244-4782-4/10 $26.00 © 2010 IEEE   378

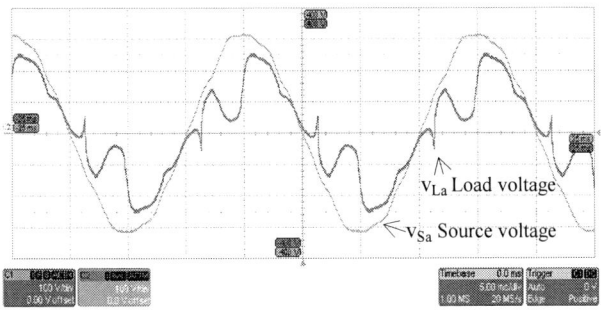

Fig. 16. Experimental results for mains and load voltages before filter operation.

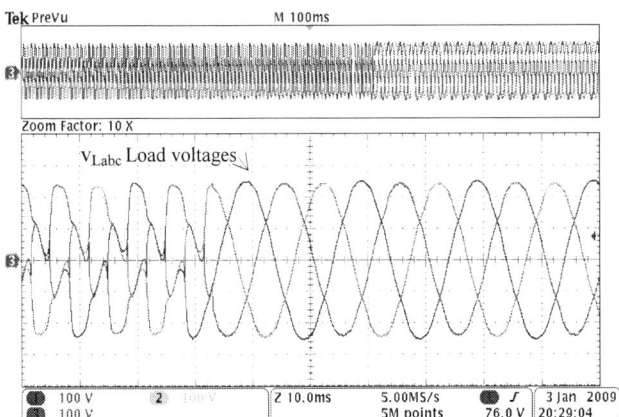

Fig. 17. Experimental results for load voltages in three-phase form before and after filter operation.

V. CONCLUSION

This paper describes a new control strategy used in the UPQC system, which mainly compensate reactive power and voltage and current harmonics in the load under non-ideal mains voltage and unbalanced load current conditions. The proposed control strategy use only loads and mains voltage measurements for series APF based on the synchronous reference frame theory. The instantaneous reactive power theory is used for shunt APF control algorithm by measuring mains voltage and currents. The conventional methods require measurements of the load, source and filter voltages and currents.

The simulation results show that, when unbalanced and nonlinear load current or unbalanced and distorted mains voltage conditions, the above control algorithms eliminate the impact of distortion and unbalance of load current on the power line, making the power factor unity. Meanwhile, the series APF isolates the loads voltages and source voltage, the shunt APF provides three-phase balanced and rated currents for the loads.

The experimental results obtained from a laboratory model of 10 kVA, along with a theoretical analysis, are shown to verify the viability and effectiveness of the proposed UPQC control method.

ACKNOWLEDGEMENT

This study is supported financially by TUBITAK research fund number 108E083 and Kocaeli University Scientific Research Fund.

This work is also supported by Concept Inc. (Concept IGBT driver), Semikron Inc. (IGBT and IGBT driver), LEM Inc. (voltage and current sensor) and TI Inc. (F28335 eZdsp), which is gratefully acknowledged. The authors gratefully acknowledge the contributions of Halim Ozmen (from Semikron Turkey) and Robert Owen (from TI).

REFERENCES

[1] H. Akagi, and H. Fujita, "A new power line conditional for harmonic compensation in power systems," *IEEE Trans. Power Del.*, vol. 10, no. 3, pp. 1570–1575, Jul. 1995.

[2] H. Fujita, and H. Akagi, "The unified power quality conditioner: The integration of series and shunt-active filters," *IEEE Trans. Power Electron.*, vol. 13, no. 2, pp. 315–322, Mar. 1998.

[3] H. Akagi, E. H. Watanabe and M. Aredes, Instantaneous Power Theory and Applications to Power Conditioning. Wiley-IEEE Press. April 2007.

[4] D. Graovac, A. *Katic*, and A. Rufer, "Power Quality Problems Compensation with Universal Power Quality Conditioning System," *IEEE Transaction on Power Delivery*, vol. 22, no. 2, 2007.

[5] B. Han, B. Bae, H. Kim, and S. Baek, "Combined Operation of Unified Power-Quality Conditioner with Distributed Generation," *IEEE Transaction on Power Delivery*, vol. 21, no. 1, pp. 330-338, 2006.

[6] M. Aredes, "A combined series and shunt active power filter," in Proc. IEEE/KTH Stockholm Power Tech Conf., Stockholm, Sweden, pp. 18–22, June 1995.

[7] Y. Chen, X. Zha, and J. Wang, "Unified power quality conditioner (UPQC): The theory, modeling and application," *Proc. Power System Technology Power Con Int. Conf.*, vol. 3, pp. 1329–1333, 2000.

[8] F. Z. Peng, J.W. McKeever, and D. J. Adams, "A power line conditioner using cascade multilevel inverters for distribution systems," *IEEE Trans.Ind. Appl.*, vol. 34, no. 6, pp. 1293–1298, Nov./Dec. 1998.

[9] G. M. Lee, D.C. Lee and J. K. Seok, "Control of series active power filter compensating for source voltage unbalance and current harmonics," *IEEE Transaction on Industrial Electronics*, vol. 51, no. 1, pp. 132- 139, Feb. 2004.

[10] V. Khadkikar, A. Chandra, "A New Control Philosophy for a Unified Power Quality Conditioner (UPQC) to Coordinate Load-Reactive Power Demand Between Shunt and Series Inverters," *IEEE Trans. on Power Delivery*, vol.23, no.4, pp. 2522-2534, 2008.

[11] M. Aredes, H. Akagi, E.H. Watanabe, E. V. Salgado, L. F. Encarnacao, "Comparisons Between the p-q and p-q-r Theories in Three-Phase Four-Wire Systems," *IEEE Transactions on Power Electronics*, vol. 24, no. 4, pp. 924-933, April, 2009.

[12] A. Esfandiari, M. Parniani, A. Emadi, H. Mokhtari, "Application of the Unified Power Quality Conditioner for Mitigating Electric Arc Furnace Disturbances," *International Journal of Power and Energy Systems*, vol. 28, no. 4, pp. 363-371, 2008.

Resonant Current Regulation for Transformerless Hybrid Active Filter to Suppress Harmonic Resonances in Industrial Power Systems

Tzung-Lin Lee*   Yen-Ching Wang*   Josep M. Guerrero**

* Dept. of Electrical Engineering, National Sun Yat-sen University, Kaohsiung, TAIWAN
** Dept. of Automatic Control Systems and Computer Engineering, Technical University of Catalonia, Barcelona, SPAIN

Abstract—Severe harmonic distortion, due to unintentional series or parallel resonance of the passive filters or power factor correction capacitors, is a significant issue in the industrial power system. The transformerless hybrid active filter, which operates at reduced kVA rating and switching ripples, is a promising filtering solution in high-power applications. This paper presents resonant current regulation for the transformerless hybrid active filter to enhance its suppressing capability of harmonic resonances. The proposed current regulator is composed of various band-pass filters in parallel connection to resonate at harmonic frequencies for accurately controlling the active filter as variable harmonic conductance. The current tracking capability of the active filter and the associated damping performance can be improved without increasing overall gain of controller. Therefore, the harmonic resonance can be definitely avoided and the damping conductance can be dynamically adjusted to conform with the harmonics limitation. In addition, the required damping conductance to maintain the harmonic voltage at an allowable level is reduced, compared with proportional control only. Operational principles are detailed and computer simulations are provided to validate the effectiveness of the proposed approach.

Keywords

Hybrid active filter, active power filter, harmonic resonance

I. Introduction

In contrast with active front-end converters, diode or thyristor rectifiers strongly dominates in high-power applications of power electronics, such as adjustable speed drives, due to either lower component cost or less control complexity. These equipment results in a large amount of harmonic current injecting into the power system, which may cause excessive harmonic voltage distortion and even give rise to malfunction of sensitive equipment in the vicinity of the harmonic source. Multi- or single-tuned passive filter is usually installed at the secondary side of the distribution transformer in the industrial power system to draw dominant harmonic current and also provide power factor correction for inductive loads [1], [2]. However, unintentional series and/or parallel resonance, due to the passive filters and nonlinear loads and/or the utility as shown in Fig. 1, may result in excessive harmonic voltage amplification and even nullify functionality of the passive filters [3], [4]. Extra engineering work, therefore, must be consumed to maintain required filtering performances and avoid possible harmonic resonance as well.

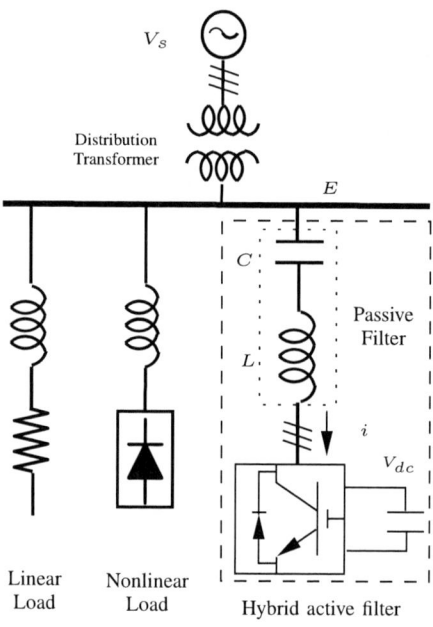

Fig. 1.   One-line circuit diagram of the proposed hybrid active filter in the industrial power system.

Various active filtering approaches have been presented previously to address the harmonic problem in the power system [5]. The most popular active power filters are intended for compensating the harmonic current of nonlinear loads, but they may not effectively approach the harmonic resonance issues resulting from the passive filters or the power factor correction capacitors [6]. Bhattacharya and his coworkers proposed a hybrid series active filter to isolate harmonics between power system and harmonic source [7]. Fujita and his coworkers proposed a hybrid shunt active filter to suppress the fifth harmonic resonance between the utility and a capacitor bank [8]. Detjen and his coworkers proposed a hybrid filter in series with a capacitor bank by a coupling transformer to suppress harmonic resonance and compensate harmonic current [9]. These methods provide effective harmonics suppression functionality; however, extra added passive components, such as matching transformers or tuned passive filters, may become a critical issue in terms of installation space and cost. A transformerless hybrid active filter for damping the harmonic

978-1-4244-4782-4/10 $26.00 © 2010 IEEE   380

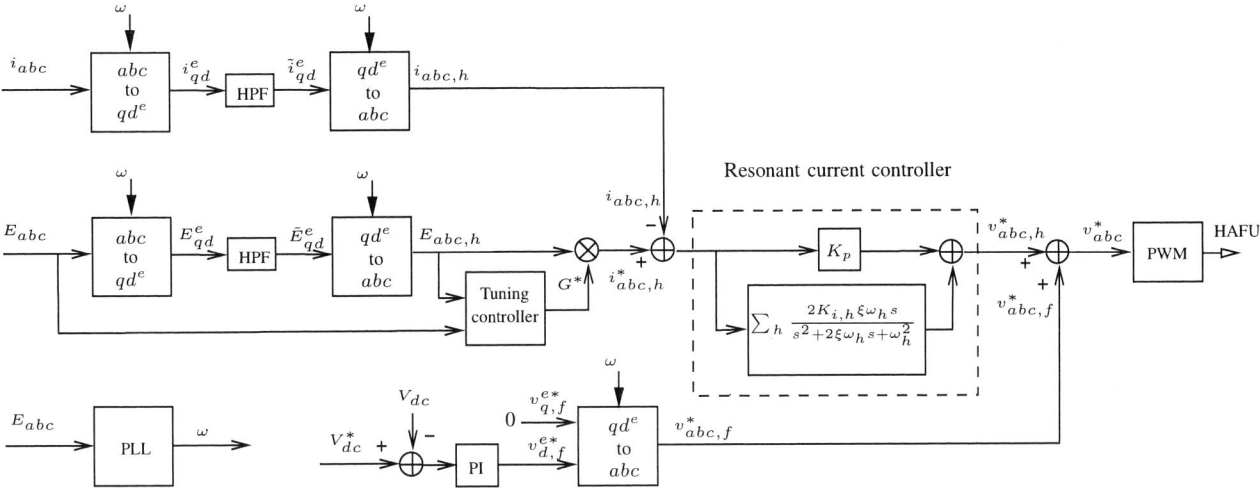

Fig. 2. Control block diagram of the proposed HAFU.

resonance in distribution power systems was presented [10]. Since the series capacitor sustains the fundamental component of the grid voltage, the active filter can be operated with a very low dc bus voltage compared with the pure shunt active filter, which is the significant advantage of reducing both the rated kVA capacity and the switching ripples. However, the damping performance is impeded due to limited bandwidth of the proportional current control. Distributed active filters with voltage detection feature were proposed to cope with the harmonic resonance on the capacitor bank, but a droop-controlled algorithm is required to coordinate the operation of multiple active filters [11].

This paper proposes resonant current regulation to enhance the damping performances of the transformerless hybrid active filter to suppress the harmonic resonance in industrial power system [12], [13]. The harmonic regulator, which is composed of various band-pass filters in parallel connection, is tuned to resonate at harmonic frequencies for accurately controlling the active filter as variable harmonic conductance. Based on this algorithm, the resonances resulting from the passive filters or the power factor correction capacitors can be definitely avoided, and also the damping conductance can be dynamically adjusted to maintain harmonic voltage distortion conforming with the harmonics limitation, such as IEEE std. 519-1992 [14]. Compared with the proportional current control, the proposed resonant regulator introduces additional gain at specific harmonic frequencies to enhance the current tracking capability of the active filter and the associated damping performance without increasing overall gain of controller. The required damping conductance to maintain the harmonic voltage at an allowable level is also reduced.

II. OPERATION PRINCIPLES

Fig. 1 shows a simplified one-line diagram of the proposed hybrid active filter. The hybrid active filter unit (HAFU) is composed of a conventional three-phase voltage source

inverter and a power factor correction capacitor C in series connection at the secondary side of the distribution transformer in industrial power system. The HAFU operates as variable conductance at harmonic frequencies as given,

$$i^*_{abc,h} = G^* \cdot E_{abc,h} \qquad (1)$$

where $i^*_{abc,h}$ represents the harmonic current command and $E_{abc,h}$ is the harmonic voltage component at the installation point of the HAFU. The conductance command G^* is defined as a variable gain to determine the harmonic current to be drawn from the grid for suppressing voltage harmonics. The control algorithm of Fig. 2, which includes harmonic detection, resonant current regulation, and conductance tuning control, is detailed as follows.

A. Harmonic detection

Both the harmonic voltage component $E_{abc,h}$ at the installation location of the HAFU and the harmonic current component $i_{abc,h}$ of the HAFU can be determined by using the synchronous reference frame (SRF) transformation [7] as shown in Fig. 2, where phase-locked loop (PLL) is required for grid synchronization. In the SRF, the fundamental component becomes a dc value , whereas the harmonic component is a ac value. Therefore, the harmonic voltage component $\tilde{E}^e_{qd}$ and the harmonic current component $\tilde{i}^e_{qd}$ in the SRF can be simply extracted by using high-pass filters (HPFs). After applying the inverse SRF transformation, $E_{abc,h}$ and $i_{abc,h}$ in the three-phase system are derived. Subsequently, the harmonic current command $i^*_{abc,h}$ is generated by multiplying the voltage harmonics $E_{abc,h}$ and the conductance command G^*, which is dynamically adjusted by the tuning controller.

B. Resonant current regulation

The resonant harmonic control (RHC) in Fig. 2 is proposed to enhance current tracking capability of the hybrid active filter. The RHC is defined as follows [13]:

$$RHC(s) = k_p + \sum_h \frac{2K_{i,h}\xi\omega_h s}{s^2 + 2\xi\omega_h s + \omega_h^2} \quad (2)$$

where h represents the order of the harmonic frequency, k_p is the proportional gain, and $k_{i,h}$ is the integral gain for each harmonic frequency. The RHC is tuned to resonate at harmonic frequencies ω_h with damping ratio ξ. Therefore, various narrow gain peaks centered at harmonic frequencies are introduced to improve current tracking performance, which will be further discussed in the frequency domain analysis of the next section.

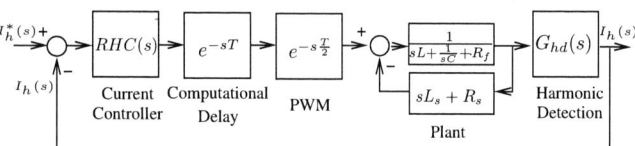

Fig. 3.   Current control block diagram of the proposed HAFU.

Fig. 3 shows current control block diagram of the proposed HAFU, where both line resistance R_s and filter resistance R_f are included. Computational delay of digital signal processing is equal to one sampling delay T and PWM delay approximates to half sampling delay $\frac{T}{2}$. Since the high-pass filter of the harmonic detection in the SRF is equivalent to the band-reject filter in the stationary frame, its transfer function can be expressed as:

$$G_{hd}(s) = \frac{(s - j\omega_{fun})T_h}{1 + (s - j\omega_{fun})T_h} \quad (3)$$

where ω_{fun} is the fundamental frequency and T_h is time constant of the high-pass filter. Hence, system stability and current tracking capability are simply evaluated from both open-loop and closed-loop gains.

Based on the harmonic current command $i^*_{abc,h}$, the measured harmonic current $i_{abc,h}$, the harmonic voltage command $v^*_{abc,h}$ can be derived. Since the series capacitor draws the fundamental reactive current from the grid, the dc voltage of the HAFU can be regulated by using a proportional-integral (PI) controller to adjust the fundamental reactive voltage $v^{e*}_{abc,f}$ of the HAFU. According to the voltage command v^*_{abc}, the space vector PWM is employed to synthesize the required output voltage of the inverter.

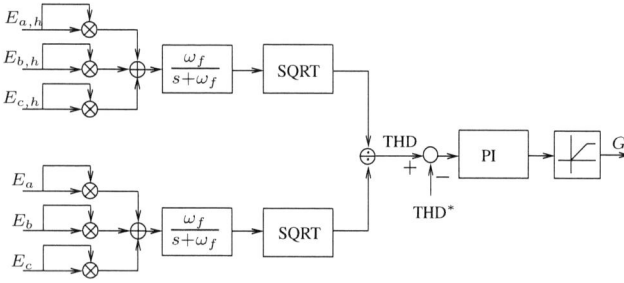

Fig. 4.   Conductance tuning control.

C. Conductance tuning control

Fig. 4 shows the conductance tuning control of the HAFU. The conductance command G^* is determined according to the voltage THD at the HAFU installation point E_{abc}. The derivation of THD can be approximately evaluated by using two low-pass filters (LPFs) with cutoff frequency ω_f, which are to filter out ripple components in the calculation. The error between the allowable THD and the measured THD is then fed into the PI regulator to adjust G^*. Based on this control, the damping capability of the HAFU can be dynamically tuned to provide effective damping capability and maintain harmonic voltage distortion at an allowable level based on the harmonic voltage limit in IEEE std. 519-1992.

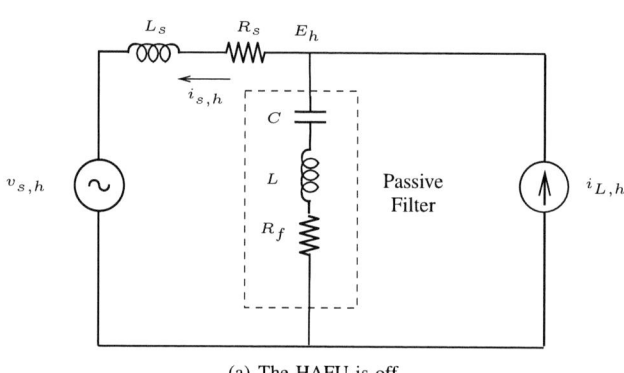

(a) The HAFU is off.

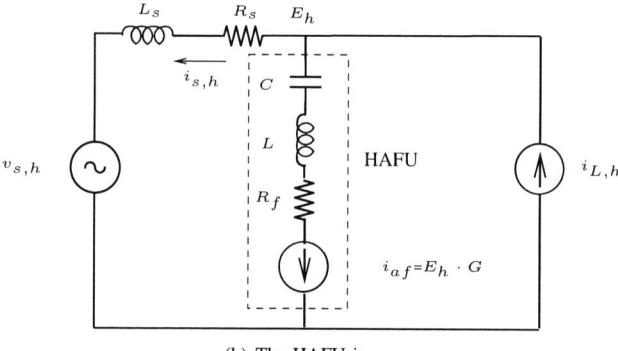

(b) The HAFU is on.

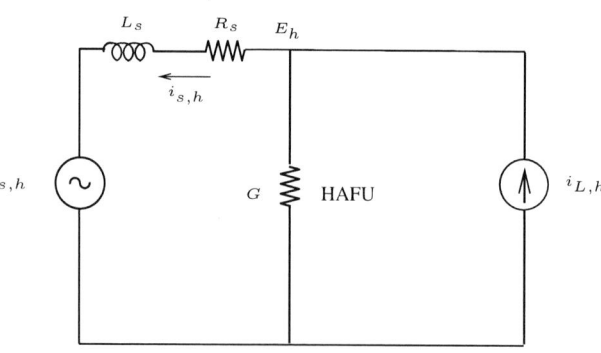

(c) The simplified circuit of Fig. 5(b) assuming the HAFU as ideal conductance G.

Fig. 5.   Simplified single-phase equivalent circuit of the HAFU at harmonic frequencies in the industrial power system.

$$E_h = \frac{(1 - \omega^2 LC + j\omega R_f C)v_{s,h}}{1 - \omega^2(L_s + L)C + j\omega(R_f + R_s)C} + \frac{(R_s - \omega^2(L_s R_f + LR_s)) + j(-\omega^3 L_s LC + \omega L_s + \omega CR_s R_f))i_{L,h}}{1 - \omega^2(L_s + L)C + j\omega(R_f + R_s)C}$$

$$i_{s,h} = \frac{v_{s,h}}{1 - \omega^2(L_s + L)C + j\omega R_f C} + \frac{(1 - \omega^2 LC + jR_f C)i_{L,h}}{1 - \omega^2(L_s + L)C + j\omega R_f C} \tag{4}$$

D. Damping performance analysis

A simplified single-phase equivalent circuit of the HAFU at harmonic frequencies is shown in Fig. 5. $v_{s,h}$ and $i_{L,h}$ represent the background harmonic voltage of the power system and the harmonic current producing by nonlinear loads, respectively. Line resistance R_s and filter resistance R_f are also included in this model. Note that linear loadings are omitted due to worst-case consideration of harmonic resonances. When the HAFU is off, i.e. v^*_{abc}=0, the passive filter is directly connected to the load bus as in Fig. 5(a). The harmonic voltage E_h and the harmonic current $i_{s,h}$ can be expressed as (4), where the resonant frequency is

$$f_{res} = \frac{1}{2\pi\sqrt{(L_s + L)C}}. \tag{5}$$

Assuming the HAFU is controlled as ideal harmonic conductance as given in (1), the equivalent circuit in Fig. 5(b) can be simplified as Fig. 5(c) after the HAFU is in operation. The harmonic voltage E_h and the harmonic current $i_{s,h}$, therefore, can be expressed as follows,

$$E_h = \frac{j\omega L_s \cdot i_{L,h} + v_{s,h}}{1 + j\omega L_s G + R_s G}$$

$$i_{s,h} = \frac{i_{L,h} - G v_{s,h}}{1 + j\omega L_s G + R_s G}. \tag{6}$$

Obviously, the resonances of both E_h and $i_{s,h}$ in (4) no longer occur, and their magnitudes can be reduced by increasing conductance G.

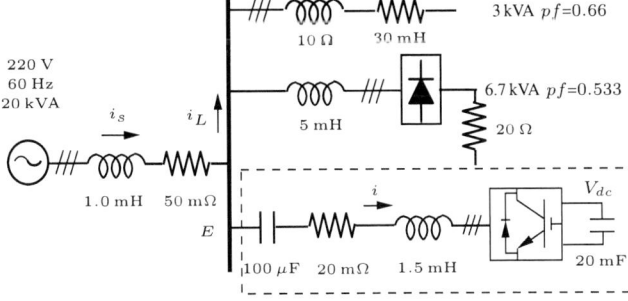

Fig. 6. Simulation circuit.

III. SIMULATION RESULTS

Fig. 6 shows the simulation circuit and the associated circuit parameters are given as follows.

- Power system: 220 V(line-to-line), 60 Hz, 20 kVA, L_s=1.0 mH(16 %), R_s=0.05 Ω(2 %).

- The passive filter, which is tuned at seventh harmonic frequency with quality factor 20, provides power factor improvement for both linear and nonlinear loads. L_f=1.5 mH(23 %), C_f=100 μF(9 %).
- Linear and nonlinear loads are rated at 3 kVA pf=0.66, 6.7 kVA pf=0.533, respectively.
- The AFU is implemented by conventional three-phase voltage source inverter with the switching frequency 10 kHz and the reference dc bus voltage 50 V. The OFF state of the HAFU corresponds to turning on three upper switches, but turning off three lower switches.
- The reference voltage THD is set as 3% based on the individual harmonic voltage limit of IEEE std. 519-1992.
- The RHC is implemented with four resonant frequencies (5th, 7th, 11th, 13th). Parameters are k_p=5, $k_{i,5}$=10, $k_{i,7}$=2, $k_{i,11}$=5, $k_{i,13}$=5, ξ=0.01.

A. Time domain analysis

Before the HAFU is started, i.e. in the OFF state, the resonance frequency (f_{res}=318 Hz) between the passive filter and the utility causes large fifth harmonic current circulating between the source current i_s and the filter current i, as shown in Fig. 7(a). The voltage THD of E exceeds 8%. This result shows the passive filter loses its filtering functionality and even causes excessive harmonic amplification. Fig. 7(b) shows the operation of the proposed HAFU, in which the voltage THD of E is significantly improved and maintained at 3.0% with the conductance command G^*=0.28Ω^{-1}. The harmonic resonance and the circulating harmonic current no longer occur. Fig. 8 illustrates the harmonic components of i_s, i, i_L, E, respectively. Fifth harmonic current of i_s is reduced from 5.3A to 1.6A, and fifth harmonic voltage of E is reduced from 10V to 3V. Since the inverter is simply operated at V_{dc}=50V, the HAFU consumes about 250 VA, which is approximately 1.25% of the system rating. This is a significant advantage of the HAFU, in terms of reduction of both the active filter kVA capacity and the associated switching ripples.

Fig. 9 shows the conductance command and the voltage THD in case of increasing nonlinear loading from 6.7 kVA to 9.7 kVA at 3s. Since voltage distortion is enlarged, G^* is raised to 0.53Ω^{-1} for maintaining the voltage THD at 3%. The steady-state waveforms of E, i_s, i_L, and i in Fig. 10 also verify the effectiveness of the proposed HAFU as the nonlinear load increases .

B. Frequency domain analysis

Fig. 11(a) shows the open-loop gain of the current control. The $RHC(s)$ produces four resonant peaks at 5th, 7th, 11th, and 13th harmonic frequencies without affecting both

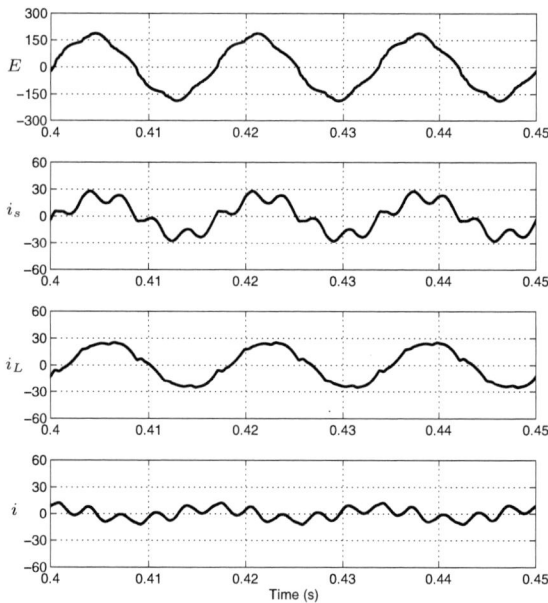

(a) The HAFU is off.

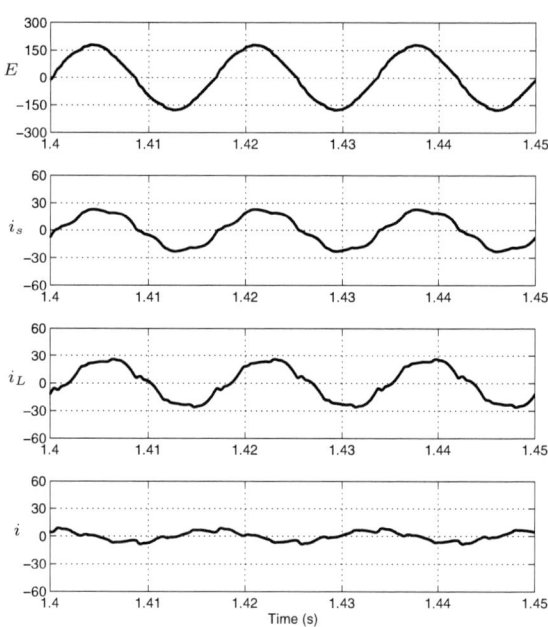

(b) The HAFU is on.

Fig. 7. Grid voltage E, source current i_s, load current i_L, and filter current i.

bandwidth and phase margin. The HAFU also functions like pure harmonic conductance due to the $RHC(s)$ providing additional phase modification at harmonic frequencies. Therefore, current tracking capability is significantly improved as illustrated in closed-loop gain of Fig. 11(b).

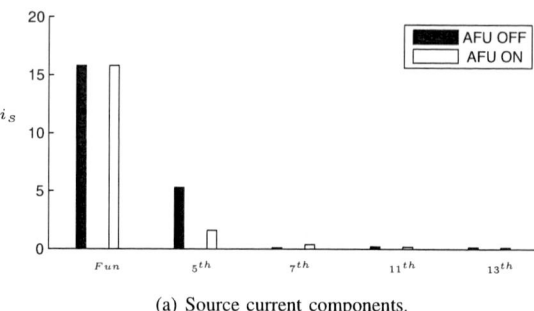

(a) Source current components.

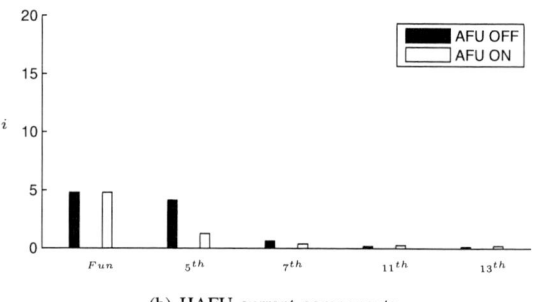

(b) HAFU current components.

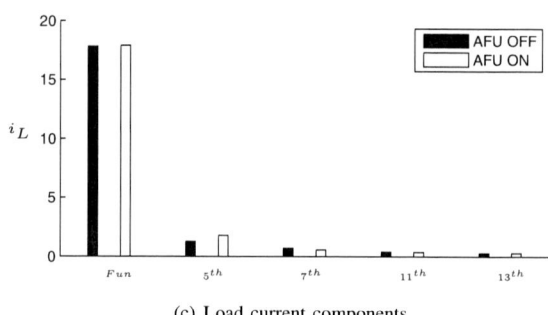

(c) Load current components.

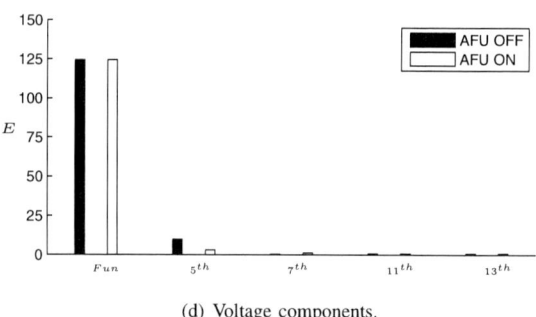

(d) Voltage components.

Fig. 8. Harmonic components before and after the HAFU is in operation.

Damping performances based on (4) and (6) are shown in Fig. 12. The resonant peak (318 Hz), due to the line impedance and the passive filter, is located near the fifth harmonic frequency. After the HAFU is in operation, the resonant phenomenon would fully disappear and the damping performance is strongly dependent on the conductance provided by the active filter. As demonstrated in Fig. 12(a) and Fig. 12(b), the magnitude of the harmonic impedance $\frac{E_h}{i_{L,h}}$ and the magnitude of the harmonic current amplification $\frac{i_{s,h}}{i_{L,h}}$ are effectively suppressed with increasing conductance

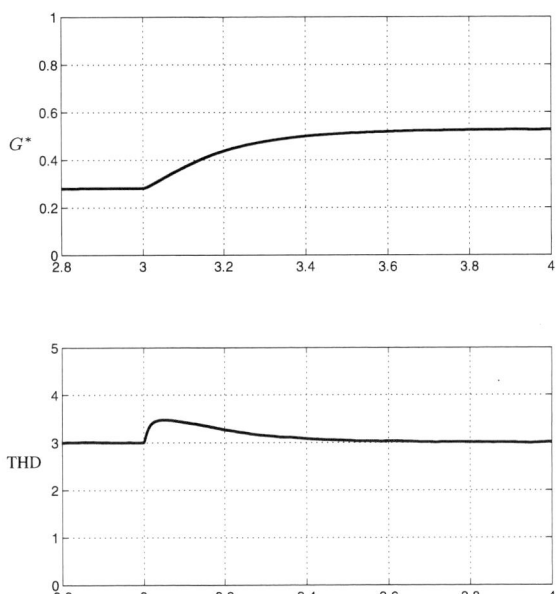

Fig. 9. Step response of conductance command and voltage THD as the nonlinear load is increased at 3s.

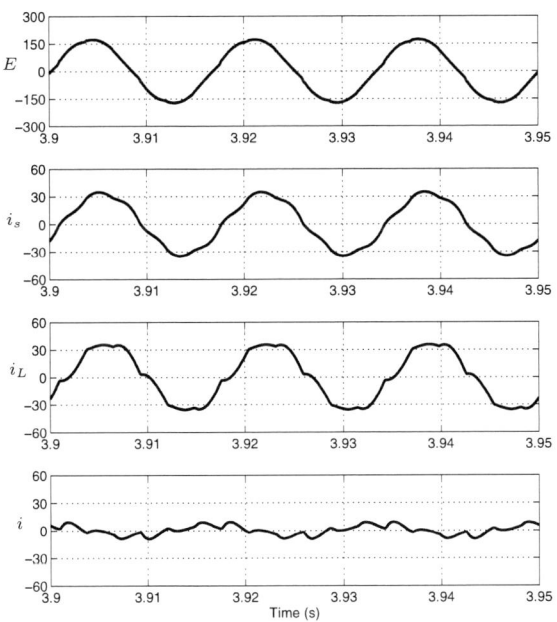

Fig. 10. Grid voltage E, source current i_s, load current i_L, and filter current i as the nonlinear loading is at 9.7 kVA.

command. Note that the HAFU exhibits high impedance for harmonic frequencies and the passive filter simply provides reactive power compensation at G^*=0.

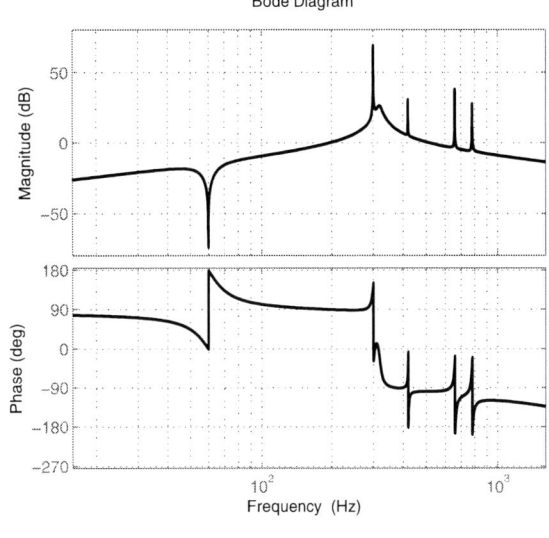

(a) Open-loop gain.

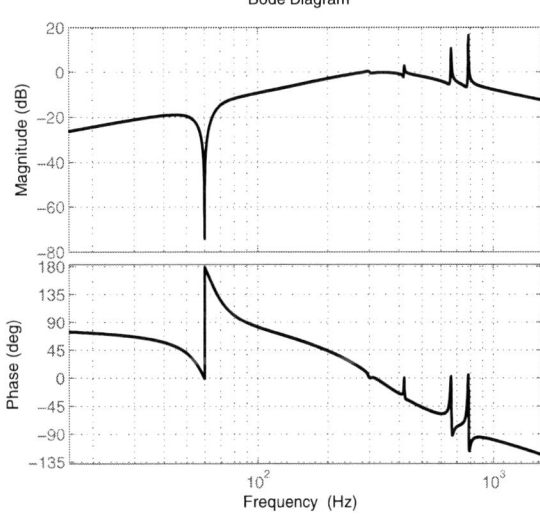

(b) Closed-loop gain.

Fig. 11. Frequency domain analysis of current control.

IV. SUMMARY

This paper presents harmonic current regulation for the transformerless hybrid active filter to enhance the harmonic suppression capability in industrial power system. The proposed current regulator is composed of parallel-connected band-pass filters tuned at various harmonic frequencies, so the active filter can be accurately operated as variable harmonic conductance based on voltage distortion of the active filter installation location. Therefore, harmonic resonances can be definitely avoided and the damping conductance can be dynamically adjusted to conform voltage quality with harmonics regulation. Design considerations based on frequency-domain analysis are detailed and time-domain simulations validate the effectiveness of the proposed approach.

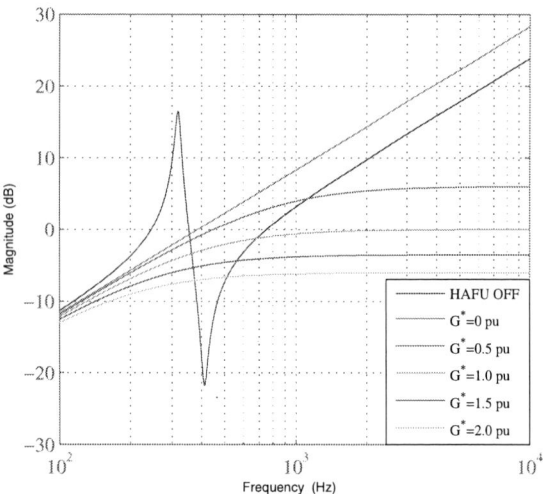

(a) Magnitude plot of harmonic impedance $\frac{E_h}{i_{L,h}}$.

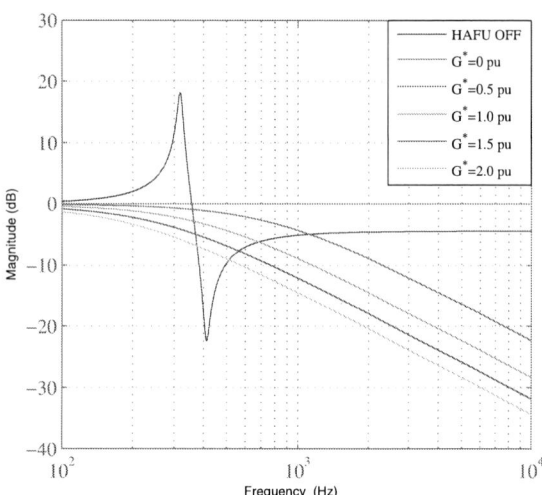

(b) Magnitude plot of harmonic current amplification $\frac{i_{s,h}}{i_{L,h}}$.

Fig. 12. Damping performance analysis of the HAFU.

In power electronics applications, low-pass filters or EMI filters are usually deployed at the grid side of the inverter for alleviating switching ripples into the power system. These filters may cause unintentional harmonic resonances with the leakage inductance of the power system as shown in Fig. 13, in which an equivalent capacitor C_{emi} is installed at the loading bus. At this situation, the harmonic voltage E_h and the harmonic current $i_{s,h}$ can be expressed as:

$$E_h = \frac{(j\omega L_s + R_s)i_{L,h} + v_{s,h}}{1 + R_s G - \omega^2 L_s C_{emi} + j\omega(L_s G + C_{emi} R_s)}$$

$$i_{s,h} = \frac{i_{L,h} - (G + j\omega C_{emi})v_{s,h}}{1 + R_s G - \omega^2 L_s C_{emi} + j\omega(L_s G + C_{emi} R_s)}. \tag{7}$$

Obviously, both E_h and $i_{s,h}$ can also be suppressed by

controlling the harmonic conductance G when the harmonic resonances occur between C_{emi} and L_s.

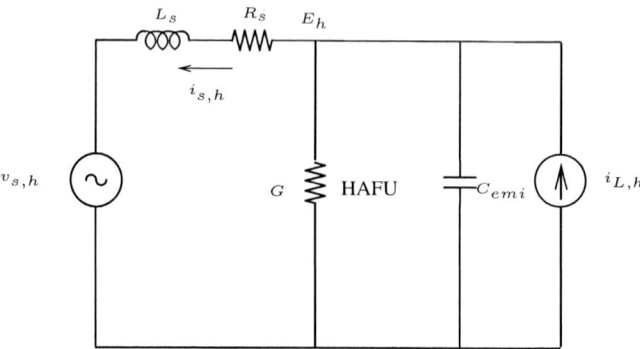

Fig. 13. The simplified circuit of the HAFU considering the capacitive filter C_{emi}.

ACKNOWLEDGMENT

This research is funded by the National Science Council of TAIWAN under grant NSC 98-2221-E-110-078.

REFERENCES

[1] R. L. Almonte and A. W. Ashley, "Harmonics at utility industrial interface: a real world example," *IEEE Trans. Ind. Appl.*, vol. 31, no. 6, pp. 1419–1426, Nov./Dec. 1995.

[2] R. H. Simpson, "Misapplication of power capacitors in distribution systems with nonlinear loads–three case histories," *IEEE Trans. Ind. Appl.*, vol. 41, no. 1, pp. 134–143, Jan. 2005.

[3] G. Lemieux, "Power system harmonic resonance-a documented case," *IEEE Trans. Ind. Appl.*, vol. 26, no. 3, pp. 483–488, May/Jun. 1990.

[4] E. J. Currence, J. E. Plizga, and H. N. Nelson, "Harmonic resonance at a medium-sized industrial plant," *IEEE Trans. Ind. Appl.*, vol. 31, no. 3, pp. 682–690, May/Jun. 1995.

[5] H. Akagi, "Active harmonic filters," *Proc. IEEE*, vol. 93, no. 12, pp. 2128–2141, Dec. 2005.

[6] F. Z. Peng, "Application issues of active power filters," *IEEE Ind. Appl. Mag.*, pp. 21–30, Sep./Oct. 2001.

[7] S. Bhattacharya and D. Divan, "Design and implementation of a hybrid series active filter system," in *IEEE 26th Annual Power Electronics Specialists Conference*, 1995, pp. 189–195.

[8] H. Fujita, T. Yamasaki, and H. Akagi, "A hybrid active filter for damping of harmonic resonance in industrial power systems," *IEEE Trans. Power Electron.*, vol. 15, no. 2, pp. 215–222, Mar. 2000.

[9] D. Detjen, J. Jacobs, R. W. De Doncker, and H.-G. Mall, "A new hybrid filter to dampen resonances and compensation harmonic currents in industrial power systems with power factor correction equipment," *IEEE Trans. Power Electron.*, vol. 16, no. 6, pp. 821–827, Nov. 2001.

[10] R. Inzunza and H. Akagi, "A 6.6-kV transformerless shunt hybrid active filter for installation on a power distribution system," *IEEE Trans. Power Electron.*, vol. 20, no. 4, pp. 893–900, July 2005.

[11] S.-Y. Kuo, T.-L. Lee, C.-A. Chen, P.-T. Cheng, and C.-T. Pan, "Distributed active filters for harmonic resonance suppression in industrial facilities," in *Power Conversion Conference - Nagoya*, 2007, pp. 391–397.

[12] D. N. Zmood, D. G. Holmes, and G. H. Bode, "Frequency-domain analysis of three-phase linear current regulators," *IEEE Trans. Ind. Appl.*, vol. 37, no. 2, pp. 601–610, Mar./Apr. 2001.

[13] M. Castilla, J. Miret, J. Matas, L. G. De Vicuna, and J. M. Guerrero, "Linear current control scheme with series resonant harmonic compensator for single-phase grid-connected photovoltaic inverters," *IEEE Trans. Ind. Electron.*, vol. 55, no. 7, pp. 2724–2733, July 2008.

[14] *IEEE Recommended practices and requirements for harmonic control in electrical power systems*, IEEE Std. 519-1992, 1993.

Performance Evaluation of High Voltage Super Junction MOSFETs for Zero-Voltage Soft-Switching Inverter Applications

Sung-Yeul Park
University of Connecticut
Center for Clean Energy Engineering
Storrs, Connecticut 06269-2157, USA

Pengwei Sun, Wensong Yu, and Jih-Sheng Lai
Virginia Polytechnic Institute and State University
Future Energy Electronics Center
Blacksburg, VA 24061-0111, USA

Abstract—**This paper evaluates three different 600V-level super junction (SJ) MOSFETs employed in zero-voltage soft-switching inverter applications. Inverter efficiency was measured and compared with the same inverter test setup by only changing different MOSFETs. Besides high efficiency requirement, better switching performance is also highly appreciated in various inverter applications. Comparison test was done on each super junction MOSFET to investigate its body diode reverse recovery and associated problems. Based on performance of both efficiency and reverse recovery related issues, a high voltage super junction MOSFET selection for soft-switching inverter was suggested.**

I. INTRODUCTION

Conventional power MOSFETs have a large conduction loss due to the high on-state resistance, which is caused by a high voltage blocking capability, which is the main barrier for power MOSFETs to play a role in high voltage (400V or more) applications [1]. The super junction MOSFETs has been introduced as a breakthrough technology [2-6], and Siemens introduced the first generation of CoolMOS[TM] in 1999 [3] followed up by ST Microelectronics launch of its MDmesh[TM] in 2001 [6]. Both super junction MOSFET technologies claimed about 80% reduction over conventional MOSFETs in terms of on resistance. Since then, almost a decade passed, high voltage power MOSFET technology improves further. For example, newly deployed ST's MDmesh[TM] V already reduced its R$_{DS}$(on) by 80% compared to its very first predecessor [7]. Though papers [8] reported new semi-super junction technology improves both R$_{DS}$(on) and the softness of body diode over super junction

counterpart, those devices are still at the stage of research and development.

Zero-voltage soft-switching (ZVS) technology [9-11] enables the use of high voltage power MOSFETs in inverter applications, which avoid the intrinsic body diode reverse recovery, induced large turn-on loss and possible device damage problems existing in hard-switching inverter operations. The coupled-magnetic ZVS topology [9] with variable timing control [12] adopted for test and comparison of different MOSFETs in this paper eliminates the turn-on loss completely. In addition, the turn-off loss, though not significant for MOSFETs, are further reduced by paralleled resonant capacitor [9]. Therefore, the inverter system efficiency mainly relies on conduction loss of MOSFETs, which translates into lower $R_{DS(on)}$, higher efficiency.

Although high efficiency is a critical criterion to judge an inverter performance, switching characteristics of different MOSFETs and associated problems are also an important concern when it comes to device selection. Using MOSFET body diode as free-wheeling diode results in the problems generated by its reverse recovery [13-16]. Problems include parasitic ringing of device voltage and increased resonant current. Severe ringing of device voltage could lead to very high spike and thus damage the device and other circuit components. In addition, it could introduce EMI problems, which are unexpected in soft-switching inverters. Increased resonant current will lead to over-design of coupled magnetic, more auxiliary circuit loss and reduced effective duty cycle and thus harmonic distortion for output current. All of these problems could jeopardize the system reliability,

978-1-4244-4782-4/10 $26.00 © 2010 IEEE

harm system efficiency and bring on EMI troubles to inverter system and other electronic equipment.

Based on the above analysis, in order to select the best candidate from different power MOSFETs for soft-switching inverter system, we need to evaluate both efficiency performance as well as switching performance of those devices. In this paper, the evaluation will be thus carried on two parts. First, three different 600-V MOSFETs were put into the same ZVS inverter for efficiency measurement and comparison, including CoolMOS^{TM} IPW60R045CP [17] from Infineon Technologies, CoolMOS^{TM} IXKK85N60C [18] from IXYS Corporation, and MDmesh^{TM} STW77N65M5 [19] from ST Microelectronics. Then, a double pulse generator will be set up for body diode reverse recovery test of each device under soft-switching condition. Device voltage and diode reverse recovery process will be documented and analyzed. Finally suggestion on trade-off between two evaluation criteria and better power MOSFET selection will be made for inverter applications.

II. POWER MOSFETs SPECIFICATION

Table 1 shows the specifications of three MOSFETs under evaluation [17-19]. Fig. 1 shows a conduction test of the three MOSFETs. The voltage drop is measured between drain and source with a fixed 15V applied across gate and source and a constant power source applied through drain and source.

Table. 1 Specifications of MOSFETs under evaluation

Device Spec	IPW60R045CP	IXKK85N60C	STW77N65M5
V_{DS}	650V	600V	650V
I_D	60A	85A	66A
$R_{DS(on)}$	45mΩ	36mΩ	33mΩ
Package	TO-247	TO-264	TO-247

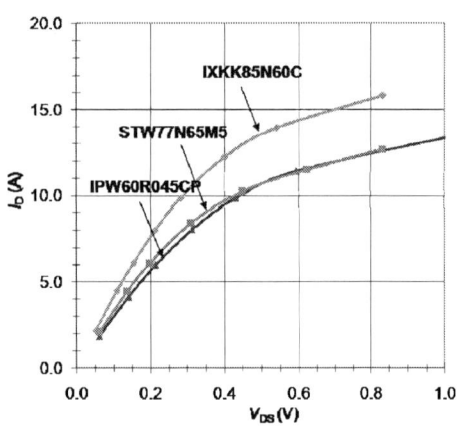

Fig. 1 Conduction test of three MOSFETs

III. INVERTER EFFICIENCY EVALUATION

Fig. 2 shows the full bridge single-phase soft-switching inverter configuration with main MOSFETs (S_1 to S_4), connected auxiliary IGBTs (S_{x1} to S_{x4}), auxiliary diodes (D_{x1} to D_{x4}) through couple magnetic auxiliary resonant circuit. Auxiliary IGBTs and diodes provide resonant current in order to achieve zero voltage switching of main MOSFETs.

Fig. 3 is the photograph for the inverter test setup including dc input capacitor bank, digital signal processor board, and output LC filter board. In Fig. 3, all test setups are the same except power stage board.

Fig. 4 is the one switching-on transition of MOSFETs to show ZVS is achieved. Because ZVS depends only on topology and variable timing control [12], there's no turn-on loss for all three MOSFETs under the whole load conditions. In addition, due to the lossless snubber capacitors (C1 to C4), turn-off loss for MOSFETs are negligible. As a result, the inverter efficiency is mostly determined by conduction loss.

Fig. 5 shows the measured inverter efficiency for three MOSFETs with 20kHz switching frequency and bipolar PWM pattern. Fig. 6 shows the measured inverter efficiency for three MOSFETs with 20kHz switching frequency and unipolar PWM pattern. Fig. 7 shows the measured inverter efficiency for three MOSFETs with 10kHz switching frequency and unipolar PWM pattern.

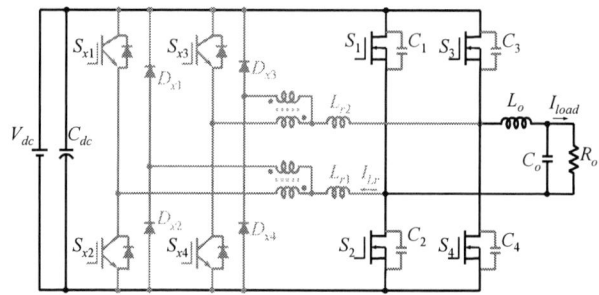

Fig. 2 Soft-switching inverter configurations

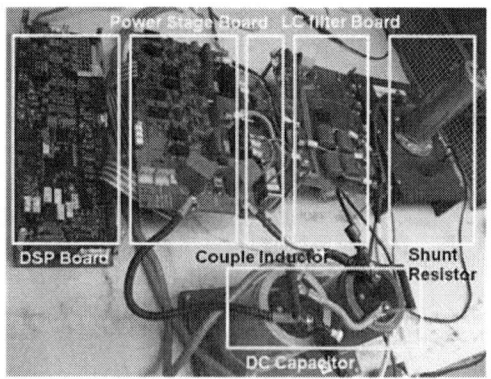

Fig. 3 Soft-switching inverter test setup

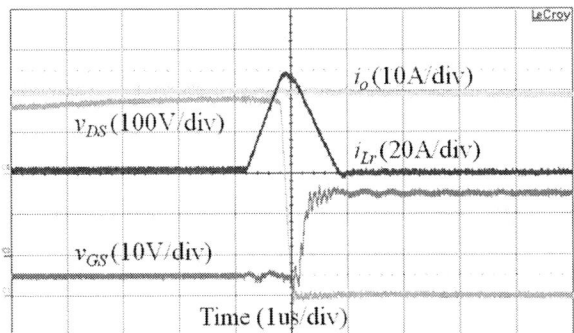

Fig. 4 ZVS transition of MOSFETs

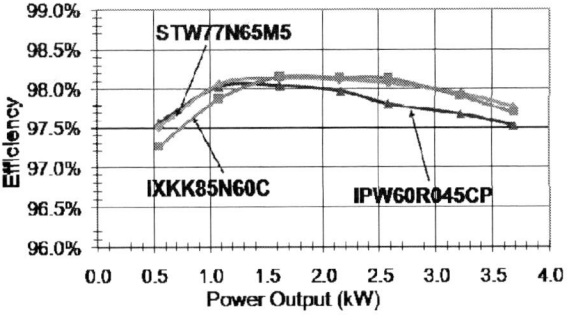

Fig. 5 Efficiency results with f_{sw}=20kHz and bipolar PWM pattern

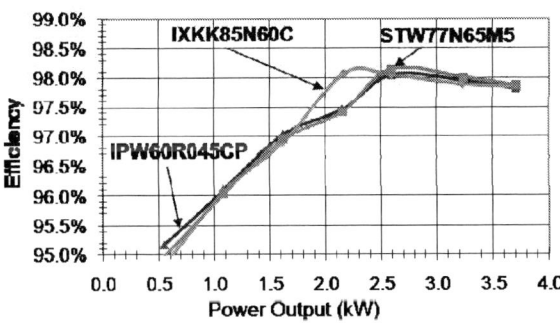

Fig. 6 Efficiency results with f_{sw}=20kHz and unipolar PWM pattern

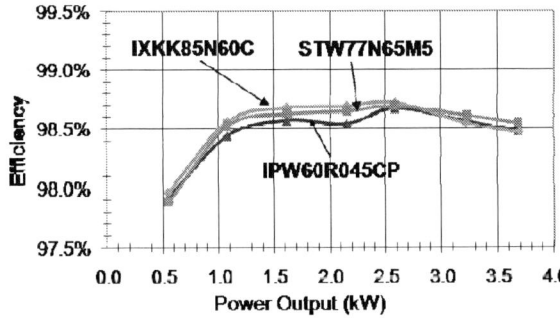

Fig. 7 Efficiency results with f_{sw}=10kHz and unipolar PWM pattern

The efficiency from IXKK85N60C and STW77N65M5 are almost the same, which leads the one from IPW60R045CP. This proves that smaller R_{DS}(on) results in higher efficiency in this topology. LC filter has two times of switching loss in the 20kHz switching frequency and unipolar PWM pattern condition.

IV. BODY DIODE REVERSE RECOVERY EVALUATION

Fig. 8 is the double pulse generator setup for MOSFET body diode reverse recovery evaluations. The device under test (DUT) is MOSFET body diode by shorting the gate and source. Body diode reverse recovery current i_{bd}, body diode voltage, V_{bd} and load current, i_L were measured for each evaluated MOSFET with two resonant inductor L_r values. Smaller values allow higher resonant current build-up for high power inverter applications. Fig.9 demonstrates the tested results with marked peak reverse recovery current, I_{rrm} and reverse recovery time t_{rr}, t_a, t_b for each measurement. The reverse current was measured by Rogowski current probe. In order to minimize the common mode noise, we used non-isolated probe for the measurement of body diode voltage.

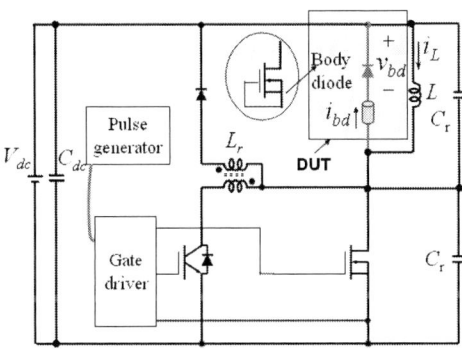

Fig. 8 Reverse recovery current test setup

As can be seen from Fig. 9, the reverse recovery process has an impact on device voltage ringing and thus generated EMI problems for inverter system and other electronic equipment. The larger the reverse current and the longer recovery time translate into more reverse recovery losses. For the coupled-magnetic assisted ZVS inverter operation, the peak reverse recovery current, I_{rrm}, will superimpose on the auxiliary resonant current branches, which means more current going through auxiliary switches and resonant inductors.

It leads directly to auxiliary circuit over-design and additional losses. At high power inverter applications, it could result in non-reset of coupled-magnetic. In addition, this prolonged current period, t_a will reduce the effective duty cycles, which causes some problems in control design and associated output harmonics problems.

978-1-4244-4782-4/10 $26.00 © 2010 IEEE

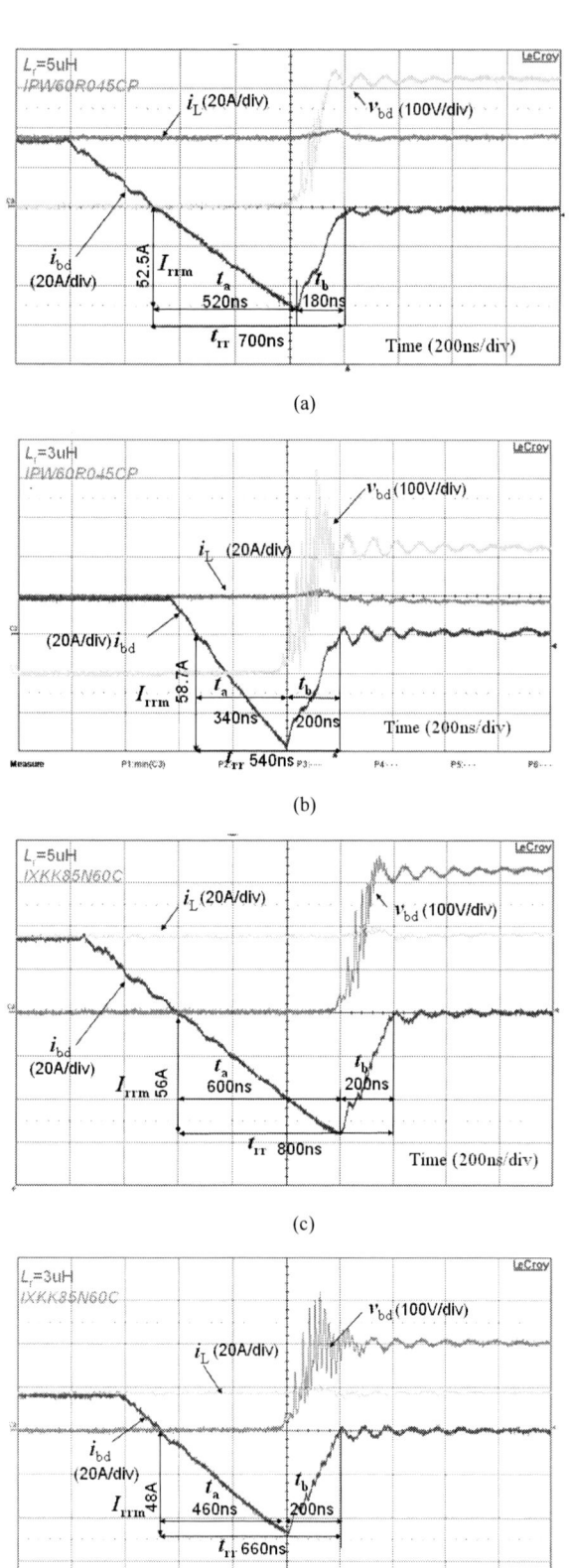

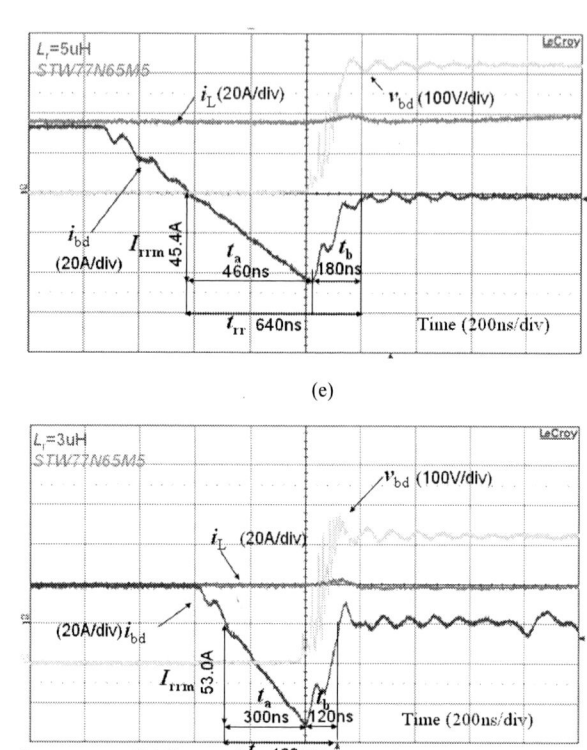

Fig. 9 Body diode reverse recovery test of different MOSFETs: (a) 5μH IPW60R045CP, (b) 3μH IPW60R045CP, (c) 5μH IXKK85N60C, (d) 3μH IXKK85N60C, (e) 5μH STW77N65M5, (f) 3μH STW77N65M5.

STW77N65M5 has the fastest recovery time and lowest peak reverse current value in both small and large resonant inductor tests. It has less voltage ringing, especially obvious at 3μH, making it more appropriate for high power inverter designs. On the other hand, the worst reverse recovery performance of IXKK85N60C leads to severe device voltage ringing and damage of the device at 3μH at bus voltage 325V. The switching performance of IPW60R045CP is in between with STW77N65M5 and IXKK85N60C. Table 2 makes the summary of the test and calculates the reverse recovery charge for each case.

From Table 2, we can clearly see that the reverse recovery time of STW77N65M5 is 30% less than that of IXKK85N60C. The peak reverse current is cut down 20%, and the reverse recovery charge is reduced 40% for STW77N65M5 compared to IXKK85N60C. STW77N65M5 also improves over IPW60R045CP by 30% on reverse charge.

Table. 2 Reverse recovery performance of MOSFETs

$T=25^{\circ}C$ $V_{dc}=325V$ $I_F=20A(3uH)$ $I_F=32A(5uH)$	IPW60R045CP		IXKK85N60C		STW77N65M5	
	3uH	5uH	3uH ($^*V_{dc}=200V$)	5uH	3uH	5uH
$I_{rrm}(A)$	58.7	52.5	48.0	56.0	53.0	45.4
t_a (ns)	340	520	460	600	300	460
t_b (ns)	200	180	200	200	120	180
t_{rr} (ns)	540	700	660	800	420	640
Q_{rr} (nC)	15849	18375	15840	22400	11130	14528

V. CONCLUSION

Three different high voltage power MOSFETs were evaluated under soft-switching inverter operations. Two comparison tests were conducted on different expectations of soft-switching inverter design and application. Based on the inverter efficiency performance and body diode reverse recovery performance, STW77N65M5 showed superiority in both evaluations over other high voltage power super junction MOSFETs. Therefore, in soft-switching inverter or other converter applications where stress on higher efficiency and overall system performance, STW77N65M5 is recommended.

REFERENCES

[1] R.W. Erickson and D. Maksinmovic, *Fundamentals of Power Electronics*, Kluwer Academic Publishers, Norwell, 2001.

[2] T. Fujihira, Y. Miyasaka, "Simulated superior performances of semiconductor superjunction devices," in *Proc. of IEEE International Symposium on Power Semiconductor Devices and ICs*, Jun. 1998, Kyoto Japan, pp. 423-426.

[3] G. Deboy, et al., "A new generation of high voltage MOSFETs breaks the limit line of silicon," in *Proc. of IEEE International Electron Devices Meeting*, Dec. 1998, San Francisco, CA, pp. 683-685.

[4] P.M. Shenoy, et al., "Analysis of the effect of charge imbalance on the static and dynamic characteristics of the super junction MOSFET," in *Proc. of IEEE International Symposium on Power Semiconductor Devices and ICs*, May. 1999, Toronto, Ont pp. 99-102.

[5] T. Minato, et al., "Which is cooler, trench or multi-epitaxy? Cutting edge approach for the silicon limit by the super trench power MOS-FET (STM)," in *Proc. of IEEE International Symposium on Power Semiconductor Devices and ICs*, May. 2000, Toulouse, France, pp. 73-76.

[6] M. Saggio, D. Fagone and S. Musumeci, "MDmesh[TM]: innovative technology for high voltage Power MOSFETs," in *Proc. of IEEE International Symposium on Power Semiconductor Devices and ICs*, May. 2000, Toulouse, France, pp. 65-68.

[7] ST's high-voltage power MOSFET roadmap, 2009, ST Micro-electronics. http://www.st.com/stonline/press/news/backgrounders/mdmesh_09.pdf

[8] W. Saito, et al., "Semisuperjunction MOSFETs: new design concept for lower on-resistance and softer reverse-recovery body diode," *IEEE Trans. on Electron Devices*, vol. 50, no. 8, pp. 1801-1806, Aug. 2003.

[9] J.-S. Lai, et al., "Source and load adaptive design for a high-power soft-switching inverter," *IEEE Tran. on Power Electronics*, vol. 21, no. 6, pp. 1667-1675, Nov. 2006.

[10] J. Zhang and J.-S. Lai, "A synchronous rectification featured soft-switching inverter using CoolMOS," in *IEEE Applied Power Electronics Conference and Exposition*, Mar. 2006, Dallas, Texas, pp. 19-23.

[11] J.-S. Lai and J. Zhang, "Efficiency design considerations for a wide-range operated high-power soft-switching inverter," in *Proc. of IEEE Industrial Electronics Society Annual Conference*, Nov. 2005, Raleigh, NC, pp. 1-6.

[12] J.-S. Lai, W. Yu, and S. Park, "Variable timing control for wide current range zero-voltage soft-switching inverters," in *Proc. of IEEE Applied Power Electronics Conference and Exposition*, Feb. 2009, Washington DC, pp. 407-412.

[13] L. Saro, K. Dierberger and R. Redl, "High-voltage MOSFET behavior in soft-switching converters: analysis and reliability improvements," in *Prof. of IEEE International Telecommunications Energy Conference*, Oct. 1998, San Francisco, CA, pp. 30-40.

[14] P. Singh, "Power MOSFET failure mechanisms," in *Proc. of IEEE International Telecommunications Energy Conference*, Sep. 2004, Chicago, IL, pp. 499-502.

[15] K. Hongrae, et al., "Minimization of reverse recovery effects in hard-switched inverters using CoolMOS power switches," in *Rec. of IEEE Industry Applications Conference*, Oct. 2001, Chicago, IL, pp. 641-647.

[16] X. Huang, et al., "Characterization of paralleled super junction MOSFET devices under hardand soft-switching conditions," in *Proc. of IEEE Power Electronics Specialists Conference*, Jun. 2001, Vancouver, BC, pp. 2145-2150.

[17] CoolMOS[TM], "IPW60R045CP," datasheet rev 2.2, 2008, Infineon Technologies.

[18] CoolMOS[TM], "IXKK85N60C," datasheet, 2008, IXYS Corporation.

[19] MDmesh[TM], "STW77N65M5," datasheet rev.1, 2009, ST Microelectronics.

New 1.7kV IGBT Chip with Fine Pattern and Optimized Buffer Layer

John F. Donlon, Eric R. Motto
Powerex, Inc.
173 Pavilion Lane
Youngwood, PA 15697 USA

K. Satoh, K. Suzuki, Y. Yoshihiura, T. Takahashi
Mitsubishi Electric Corporation
1-1-1Imajukuhigashi Nishi-ku
Fukuoka JAPAN

Abstract— **Since the introduction of the IGBT, improvements in power loss and efficiency have been achieved by applying new technologies. In this paper, refinements in fine pattern processing technology and optimization of the low impurity profile of the buffer layer using thin wafer technology are proposed to further reduce the power loss and improve efficiency in 1.7kV IGBT chips.**

I. INTRODUCTION

Responding to widespread needs for energy conservation, IGBT (Insulated Gate Bipolar Transistor) devices are widely used in high power applications, such as inverter motor drives, uninterruptable power supplies, electric vehicles, and alternative energy. Improvement in IGBT characteristics is essential for decreasing V_{CE}(sat) (on-state conduction loss) and E_{SW}(off) (turn-off switching energy loss) to achieve improved efficiency. The major ways to improve IGBT characteristics are optimization of vertical carrier concentration profile by using a trench-gate structure and the reduction of the substrate thickness of LPT (Light Punch Through) devices. After many years of iterative improvements including the replacement of planar with trench gate structures, the current high performance IGBT is the CSTBT^{TM} (Carrier Stored Trench-gate Bipolar Transistor) [1-3]. This paper proposes further improvement in the CSTBT^{TM} by use of a carrier enhancement emitter structure and advanced thin wafer technology to achieve a more efficient 1.7kV IGBT.

The CSTBT^{TM} IGBT consists of a triple layered structure in the region of the trench gate, i.e., N-Emitter/P-Base/CS-layer. The CS-layer under the P-Base has the effect of reducing V_{CE}(sat). Recently, the second generation of CSTBT^{TM} which adopts thin wafer process technology achieved a further improvement in device performance, especially by optimizing the backside collector profile on mid-high voltage region [2]. For the development of the next generation 1200V IGBT, the concept of CSTBT^{TM}(III) was proposed and developed in terms of a finer pattern for the trench gate structure and a retrograde doping profile in the CS-layer [6]. The CSTBT^{TM}(III) shows good V_{CE}(sat)-E_{SW}(off) trade-off relationship and large Short Circuit Safe Operating Area (SCSOA). The success of CSTBT^{TM}(III) can be attributed to the optimization of the doping distribution in the triple-layer structured emitter region of the CSTBT^{TM} and the adoption of advanced ULSI wafer process technology. In this paper, a new CSTBT^{TM} (III) at 1.7kV for which performances are further improved is proposed.

II. DEVICE STRUCTURE

Figures 1(a) and (b) show the conventional CSTBT^{TM}(II) and the new CSTBT^{TM}(III) for 1.7kV class, respectively. Table 1 compares their differences in geometry and manufacturing process. The new 1.7kV CSTBT^{TM}(III) has finer pattern process and the structure of trench gate and dummy trench partly separated. The retrograde doping profile which prevents the CS-layer from adversely affecting the MOS channel region in the middle triple layer is described in detail in [6]. Design optimization of both the arrangement of the trench gates and dummy trenches and the ratio of the trench gates to the dummy trenches has an injection enhancement effect, enhances the electron injection from N-Emitter, and accumulates the collector side carriers in the n-drift layer more than with the conventional structure. Figure 2 shows a simulation result of the hole concentration distribution of the new and conventional structure from the emitter side at on-state. In the new CSTBT^{TM}(III), the design ratio of the arrangement of the trench gates and dummy trenches is different in the type A and Type B devices. A number of comparative iterations of this nature were made to determine the optimum ratio. The resulting new structure has higher hole concentration than the conventional structure, which results in the reduction of conduction losses. As shown in Table I the wafer thickness is reduced about 24% resulting in improvement in the trade-off between Vce(sat) and Esw(off) while maintaining the wide SCSOA specification. References [6] and [7] detail how the SCSOA performance is enhanced by the distributed capacitance effect of the dummy trenches.

Several variations of MOS structure were evaluated, i.e., trench gate ratio versus dummy trenches in a fixed trench pitch. Particular attention was paid to the effects on gate capacitance (Cies, Coes, and Cres), Vce(sat), and Esw(off) as shown in Table II. Structure B was selected for the next generation chip after careful consideration of these factors.

III. ELECTRICAL CHARACTERISTICS

Figure 3 shows the experimental result of V_{CE}-Ic characteristics of the new structure in comparison with the conventional structure at 125°C. The new structure is lower by 0.4V at Jc=120A/cm^2 than the conventional CSTBT^{TM}(III) is at Jc=100A/cm^2. In addition to the wafer thickness, the bulk resistivity has been optimized in the new structure to improve the trade-off relationship between V_{CE}(sat) and E_{SW}(off) while maintaining SCSOA (short circuit safe operating area). Figure 4 indicates the trade-off characteristics between V_{CE}(sat) and E_{SW}(off) of the new CSTBT^{TM}(III) at Jc=120A/cm^2 and CSTBT^{TM}(II) at Jc=100A/cm^2. In contrast to CSTBT^{TM}(II), CSTBT^{TM}(III) displays a much better trade-off characteristic. Figure 5 gives the Vce vs. Jc characteristic for the new chip at 25 and 175°C. Turn-on and turn-off waveforms are shown in Figures 6 and 7. It is particularly noted that there is no oscillation in the turn-off waveform due to optimization of the bulk resistivity and N-drift doping concentration. Figure 8 shows the waveforms at SCSOA test in the new CSTBT^{TM}(III) which has been rated at 185A. The new CSTBT^{TM}(III) exhibits wide SCSOA with more than 14µs short circuit withstand time under the test conditions of V_{GE}=15V/-15V and V_{CE}=1100V which provides margin over the standard 10µs specification. Key characteristics of the 1700V 6th generation CSTBT^{TM}(III) are listed in Table III.

A new diode chip was also developed to coordinate with the new CSTBT^{TM}(III) IGBT chip. Figure 9 is a photograph of the new IGBT and FWDi chips showing their relative size and physical dimensions. The forward voltage characteristic of the new diode is given in Figure 10. Reverse recovery waveforms for the new diode are shown in figure 11.

IV. PERFORMANCE ENHANCEMENT

IGBT application requirements have evolved to require a higher maximum operating junction temperature of 175°C to accommodate increased design margin, overload requirements, and elevated cooling temperatures. The new CSTBT^{TM}(III) and FWDi chips have been rated for 175°C. This increased temperature rating requires that steps be taken to offset the adverse effect it has on power cycling, particularly for the wire bonds that contact the chip itself. The new CSTBT^{TM}(III) chip utilizes a cell plug structure and

an aluminum metallization process that produces a more planar surface for wire bond attachment. This gives an increased bond contact area, a stronger bond with increased peel strength, and higher reliability at the increased temperature excursion due to the 175°C junction temperature rating. Figure 12 shows the improvement in flatness of the emitter electrode surface of the new 6th generation chip. Figure 13 shows the improved peel strength of the 6th generation over the 5th generation as a function of relative ultrasonic bonding power.

V. CONCLUSION

The performance of the proposed 1.7kV CSTBT^{TM}(III) was examined by simulation and experiment. Improvement in the trade-off relationship between V_{CE}(sat) and E_{SW}(off) over conventional 5th generation was demonstrated while maintaining short circuit withstand capability. Applying the new carrier-enhancement emitter structure and advanced thin wafer technology, the proposed CSTBT^{TM}(III) achieves reduction in losses while meeting the 1.7kV blocking voltage class. The new CSTBT^{TM}(III) chips show great potential for improved efficiency in industrial applications of power semiconductor electronics and will be applied to 6th generation 1.7kV power modules in the near future.

REFERENCES

[1] H. Takahashi, H. Haruguhi, H. Hagino, and T. Yamada, "Carrier Stored Trench-gate Bipolar Transistor (CSTBT) – A Novel Power Device for High Voltage Application," ISPSD'96, pp.349-352 (1996)

[2] H. Nakamura, K. Nakamura, S. Kusunoki, H. Takahashi, Y. Tomomatsu, et al., "Wide Cell Pitch 1200V NPT CSTBTs with Short Circuit Ruggedness," ISPSD'01, pp.299-302 (2001)

[3] K. Nakamura, S. Kusunoki, H. Nakamura, Y. Ishimura, Y. Tomomatsu, and T. Minato, "Advanced Wide Cell Pitch CSTBTs having Light Punch-Through (LPT) Structures," ISPSD'02, pp.277-280 (2002)

[4] S. Iura, E. Suekawa, K. Morishita, M. Koga, E. Thal, "New 1700V IGBT Modules with CSTBT", PCIM Europe 2004

[5] Nicholas Clark, John Donlon, and Shinichi Iura, "New 1700V A-Series IGBT Modules with CSTBT and Improved FWDi," Power Electronics Technology Conference Record (2006)

[6] T. Takahashi, Y. Tomomatsu, and K. Satoh, "CSTBT^{TM}(III) as the Next Generation IGBT," ISPSD'08, pp.72-75 (2008)

[7] John F. Donlon and Katsumi Satoh , "IGBT Chip Improvements for Industrial Motor Drives," Electric Motor and Drive Technology Conference (2009)

[8] Katsumi Satoh, Tetsuo Takahashi, Hidenori Fujii, Manabu Yoshino, and John F. Donlon, "New Chip Design Technology for Next Generation Power Module," PCIM Europe Conference Record (2008)

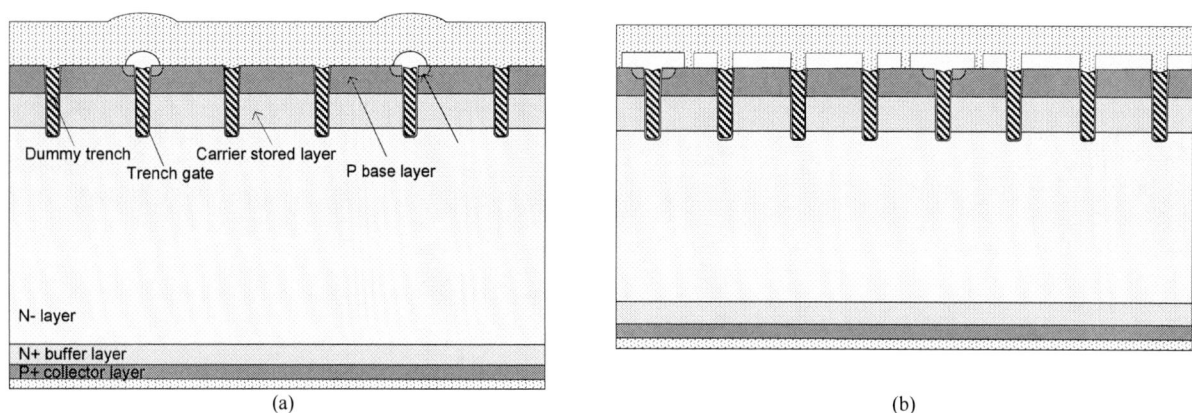

(a)   (b)

Figure 1: Cross section of (a) conventional CSTBT^{TM} and (b) CSTBT$^{TM(}$III)

Table I: Comparison of conventional CSTBT^{TM} and proposed CSTBT$^{TM(}$III)

Structure Item	CSTBT^{TM}	CSTBT^{TM}(III)
trench pitch	1.00	0.60
doping profile	standard	retro grade
wafer thickness	1.00	0.76
cell structure	conventional	optimize p-base and dummy trench contact
emitter electrode	conventional	planarization

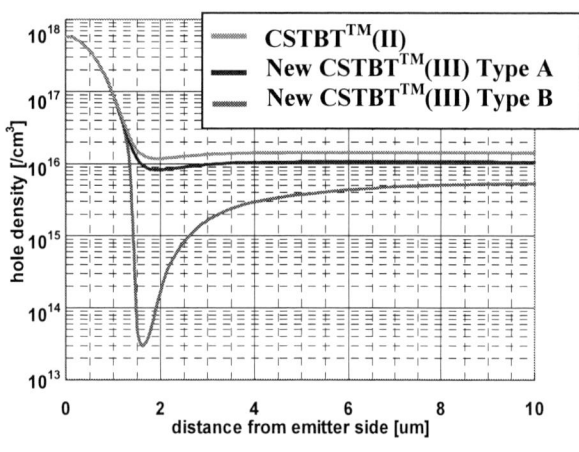

Figure 2: Simulation result of hole distribution in the on state

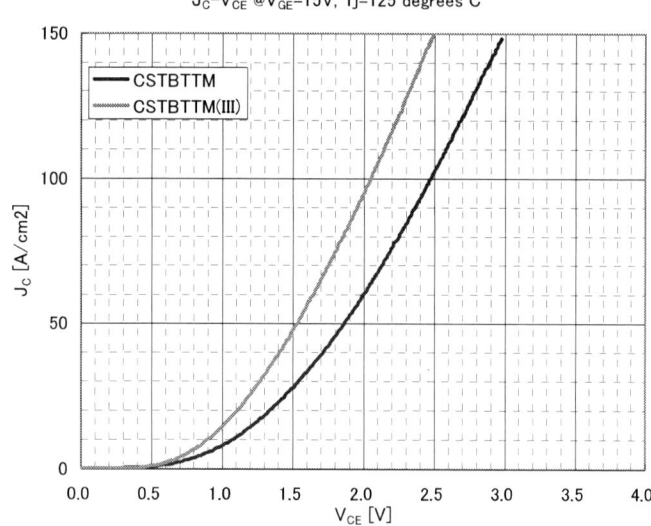

Figure 3: Comparison of V_{CE}-I_C characteristics

978-1-4244-4782-4/10 $26.00 © 2010 IEEE   394

Table II: Variation of MOS structures evaluated

MOS structure	A	B	C	D
Cies [nF]	7.1	5.4	4.4	3.8
Coes [nF]	0.25	0.24	0.25	0.24
Cres [nF]	0.07	0.05	0.04	0.03
V_{CE}(sat) [V] @Jc=120A/cm^2, Tj=125°C	2.49	2.44	2.36	2.32
E_{SW}(off) [mJ/A] @Jc=120A/cm^2, V_{CE}=850V, Tj=125°C	0.191	0.199	0.225	0.231

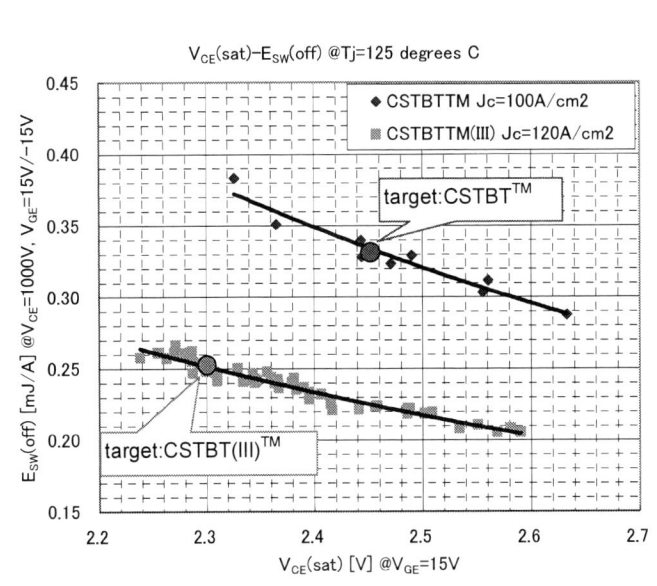

Figure 4: V_{CE}(sat)-E_{SW} trade-off comparison

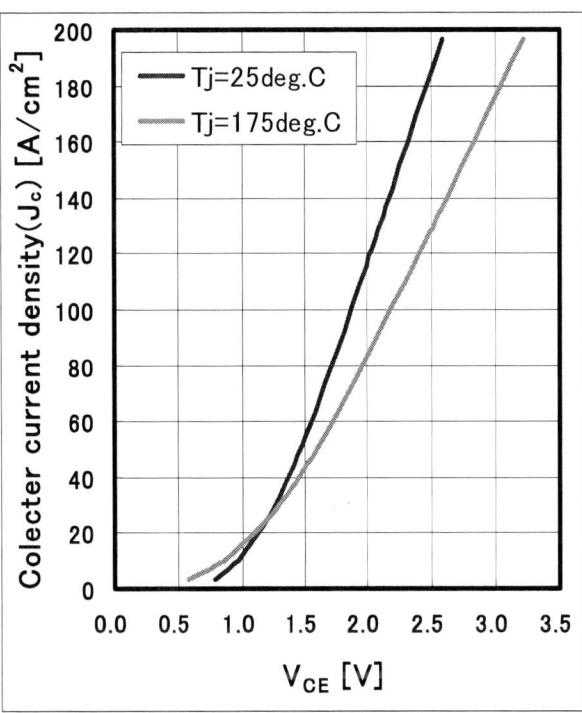

Figure 5: VCE-JC characteristics

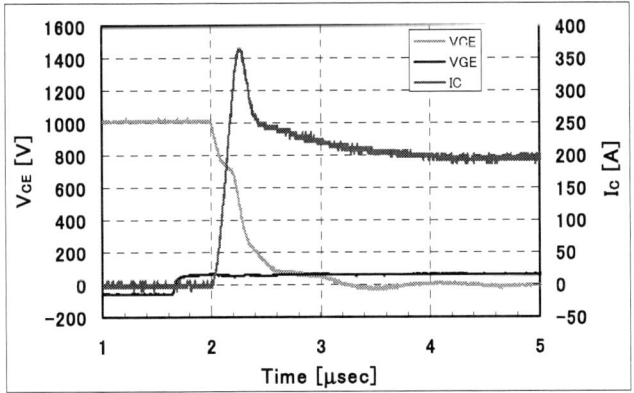

Figure 6: Turn-on waveforms

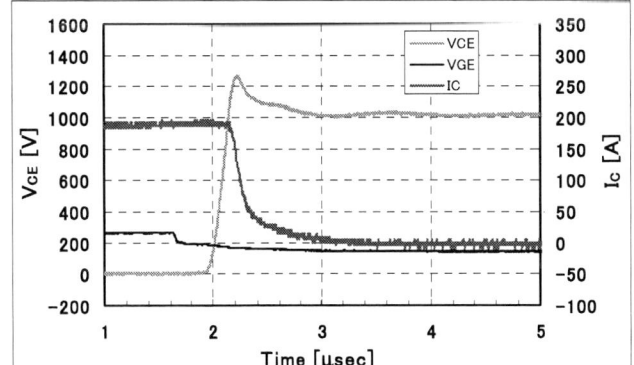

Figure 7: Turn-off waveforms

V_{CC}=1000V, V_{GE}=+15V/-15V, $R_{G(ON)}$/$R_{G(OFF)}$=5Ω, Tj=175°C

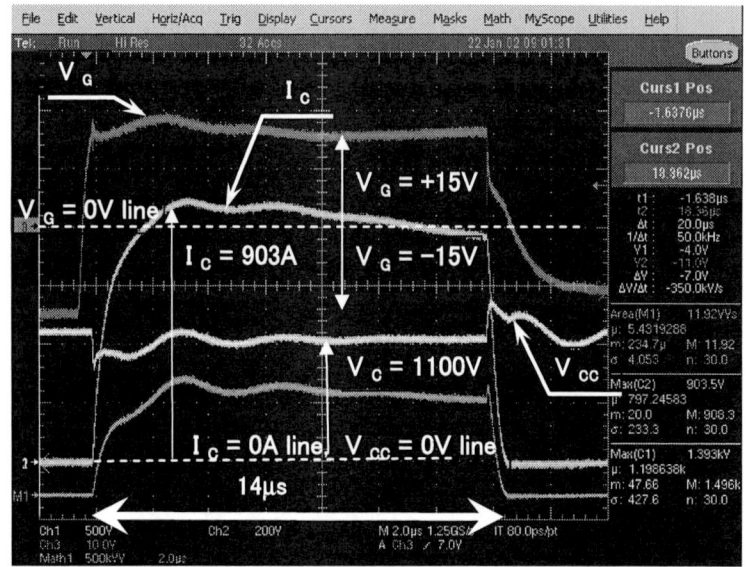

Figure 8: Short circuit test waveform

Table III: Characteristics of the new 1700V CSTBT^{TM}(III) chip

Items	Condition	Typ.	Unit
J_C(sat)		120	A/cm^2
V_{CE}(sat)	Tj=125°C, Jc(sat)	2.35	V
V_{GE}(th)		6.0	V
Eoff	Vcc=1000V,V_{GE}=15V/-15V Tj=125°C,Inductive load Jc(sat)	0.25	mJ/p/A

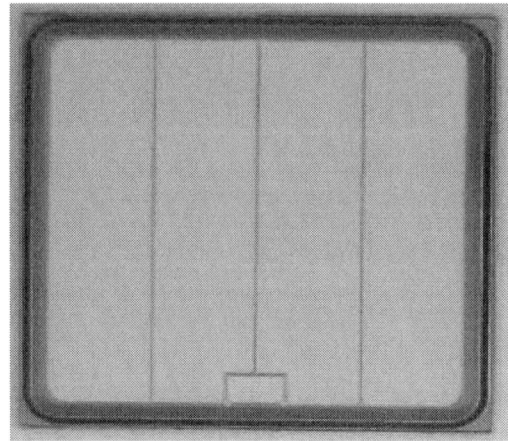

New 1700V, 185A IGBT chip   New 1700V, 185A FWDi chip

Figure 9: Photographs of the new IGBT and FWDi chips

978-1-4244-4782-4/10 $26.00 © 2010 IEEE

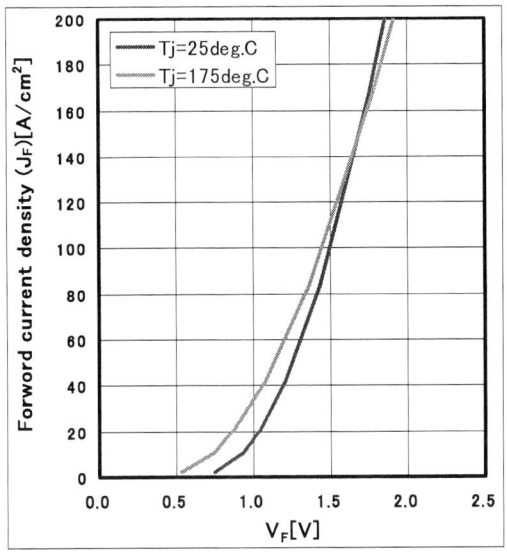

Figure 10: V_F-J_C characteristic at Tj=25 and 175°C

Figure 11: Reverse recovery waveform at VCC=1000V, Tj=175°C

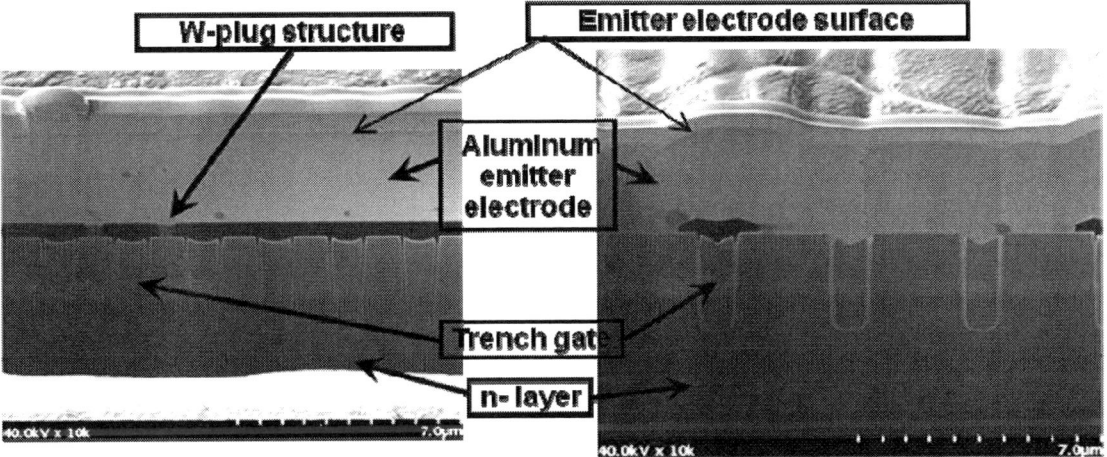

Figure 12: Comparison of emitter electrode surface profile

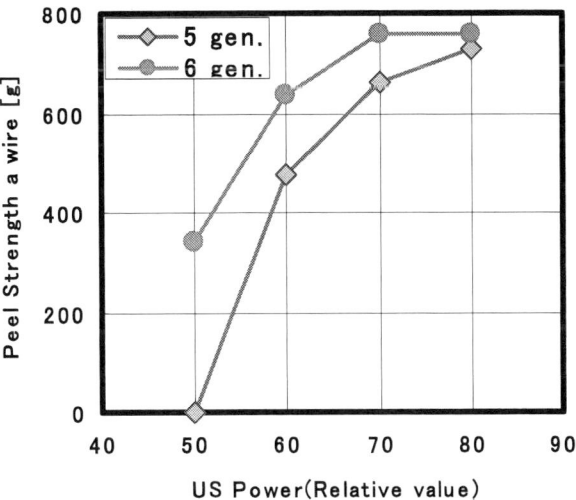

Figure 13: Wire peel strength comparison vs. bond power

Novel Thermally Enhanced Power Package

Juan A. Herbsommer, Jonathan Noquil, Chris Bull and Osvaldo Lopez

Texas Instruments

Bethlehem, PA, USA

jherbsommer@ti.com

Abstract— Heat generated in microelectronic devices as a result of dissipated power is a major issue in power electronics applications resulting in elevated application PC board temperatures. In order to minimize the down ward heat transfer to the application board an efficient method enabling the upward flow of heat from the silicon die to the top of the microelectronic package and subsequently transferred to the environment via forced convection needs to be employed [1]. The problem is that most of the current packaging technologies have a very poor junction-to-top thermal resistance so it is very difficult to have a substantial portion of the heat flowing to the top of the device [2]. In this paper we present a novel power package design that enables heat conduction to the top surface of the microelectronic package through the use of a high thermal conductivity path which reduces by more than a factor of ten the junction-to-top thermal resistance compared to standard solutions. The thermal resistance junction-to-top is found to be as low as 1 C/W, which is comparable with thermal resistance junction to board. This allows for a significant portion of the dissipated energy in the die to be conducted to the topside of the package where natural or forced convection can transfer the heat to the air. We discuss the design, manufacturability, performance and reliability of the package as well as thermal measurements which demonstrates the ability of the package to dissipate the heat. We also compare this solution with existing solution sin the marketplace.

I. INTRODUCTION

Continuous integration of microelectronic devices and shrinkage in electronic board dimensions are driving power densities to levels that endanger the performance and reliability of the electronic systems. This increase in power dissipation density is particularly worse in power application boards. Engineers and System designers have to design solutions that maintain temperatures of key components under certain limits to ensure proper electrical performance and reliability. For example many computing applications have a strict rule of 100C for the maximum temperature the board can reach in operation. One solution to this problem is to conduct the generated heat to the external environment via natural or forced convection, instead of conducting it to the board. Since the heat is generated internally in the

microelectronic devices it is critical to have a package solution that allows the heat to be easily transferred to the ambient. A package technology, referred to as DUAL COOL[TM], was developed that allows standard QFN type of devices to achieve very low junction-to-top thermal resistance (Θj-t). In the first section the technology is described along with manufacturability aspects. In the second section the results of the thermal measurements of the package are shown and comparing the results with standard package solutions available in the marketplace.

II. TECHNOLOGY DESCRIPTION

Figure 1 shows a construction diagram of a device using a DUAL COOL[TM] technology. The backside electrode of MOSFET die is attached to the leadframe of the package with solder and a copper clip is soldered to the topside electrode. In a standard QFN package the device is over molded with low thermal conductivity mold compound making the thermal resistance to the top of the device (Θj-t) high.

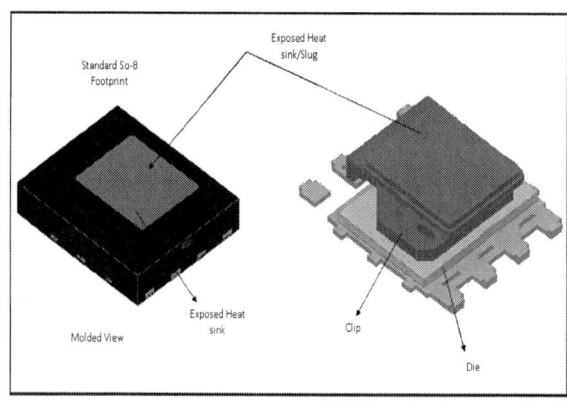

Figure 1: Construction diagram of a dual cool device

In this case the high thermal conductivity of the cooper leadframe will force most of the heat to be conducted downwards to the application board. The DUAL COOL[TM] technology consists of an exposed heat sink or heat slug connected to the clip providing for a low thermal resistance path for the heat to be conducted to the topside of the device. Using this technology thermal resistances to the top of the device (Θj-t) are comparable to thermal resistance junction to case (Θj-c), so similar amounts of heat can be dissipated to the environment provided an appropriate heat sink is attached to the top of the device as we show in Figure 3. The DUAL COOL^{TM} devices are manufactured with same set of tools used for the standard QFN device (SO-8 footprint compatible) so no major changes are needed in the QFN production line to manufacture devices with DUAL COOL^{TM} technology. Design consideration were taken into account for small non-coplanarities between the heat slug and the mold cavity producing mold flashes/ mold bleeds in the event the heat slug is not in full contact with the mold cavity. This was solved by tightening the height tolerances of the internal components of the device and controlling the bond line thickness of the die attach. In addition, a mechanical buffing step was added at the end of the process flow to eliminate residual mold compound flashes over the exposed heat sink. The product is fully qualified and passed Moisture Sensitivity Level 1 classification.

III. THERMAL PERFORMANCE

Figure 2 shows the thermal resistance junction to top for three 5x 6 mm QFN devices with the same silicon die but different constructions: a) the wire bond package has seven 8 mil Al wires connecting the drain to the pins b) the clip package used a 10 mil thick clip covering 70% of the die area to connect the drain with the external pins and c) the same clip package but in this case we used a dual cool technology.

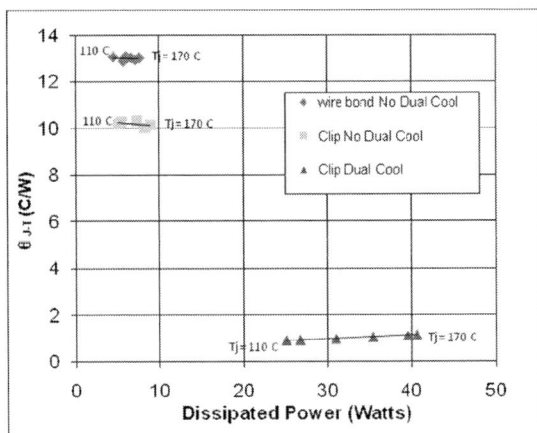

Figure 2: Thermal resistance junction-top vs power

The results (Fig 2.) show a dramatic improvement when the DUAL COOL^{TM} technology is used by reducing this thermal resistance by a factor of ten. The wire bond solution has the highest Θj-t (13 C/W) reflecting the poor capability the wires and the mold compound to conduct the heat to the top side. The clip package has a lower Θj-t (10.2 C/W) because the thick Cu clip on top of the die helps to conduct the heat to the topside. As expected the device that uses the DUAL COOL^{TM} technology has the lowest thermal resistance

junction-to-top of package (Θj-t = 1 C/W) similar to the junction to case thermal resistance (Θj-c) (exposed pad of the leadframe on the backside of the package).

To evaluate the system level impact of the DUAL COOL^{TM} technology, simulations were performed using a model that considers the device mounted on a typical board and with a heat sink mounted on top of the device. To make the calculations, datasheet values for all the thermal resistances associated with the devices (Θj-t and Θj-c) and standard values for thermal resistances board to air Θ_{B-A} = 7°C/W and thermal resistance device top to air Θ_{T-A} = 2.5°C/W were used.

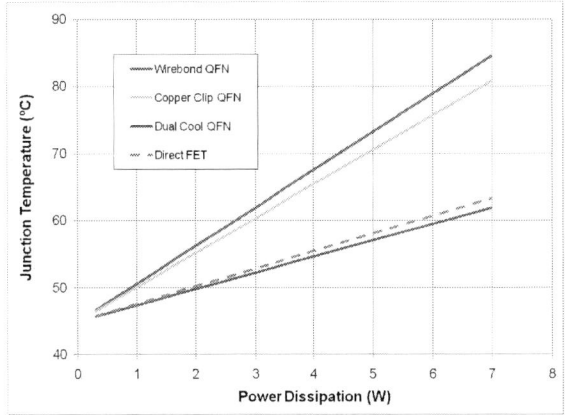

Fig 3: Thermal performance comparison for different packaging technologies.

Fig 3 shows the results of the simulation for different packaging technologies [3]. The plot shows the junction temperature as a function of the dissipated power. As you can see the improvement in thermal performance using the DUAL COOL^{TM} technology is significant. For a typical power dissipation of 5 Watts the reduction in the junction temperature is 16 C or almost 25%. The DUAL COOL^{TM} solution has excellent thermal characteristics because the power dissipated in the active device channel on the top side of the die is in close proximity to the Cu clip, so heat moving to the top side only needs to conduct through the Cu clip and heat slug.

To further demonstrate the effect of the DUAL COOL^{TM} technology on the thermal performance of a device we performed an experiment with the objective of comparing a standard device with a device that uses the DUAL COOL^{TM} technology. Two MOSFET transistors with the same die were packaged in a clip QFN 5x6 mm. Device "A" utilized the DUAL COOL^{TM} technology while device "B" did not. This is the only difference between both devices. The devices were mounted to identical standard application boards and attached to heat sinks commonly used by in systems. The temperature of the junction was measured using an *Analysis-Tech* thermal measurement system that allows the junction temperature to be measured by using the forward diode voltage of the transistor as a thermometer (after a proper calibration). A thermal IR camera was also used to measure the external temperature of the system. In the first part of the experiment we forced a enough continuous current through the transistors to dissipate a power of Pdiss=2.1W (this is a typical dissipation of these transistors in the real application) while a fan of 300 lfm is forcing convection parallel to the fins of the heat sink. Figure 4 shows the results. For the same conditions the device using the DUAL COOL™ technology has a junction temperature of Tj=58°C in comparison with Tj= 71°C of the device that does not. This is a reduction of 13°C or a reduction of 20% in junction temperature for the same conditions in a real application board. The IR image also shows the difference in temperature of the heat sink of both systems: device "A" (DUAL COOL™) shows a hotter heat sink than device B (standard solution) indicating that more heat is being conducted to the top in the system using the DUAL COOL™ concept.

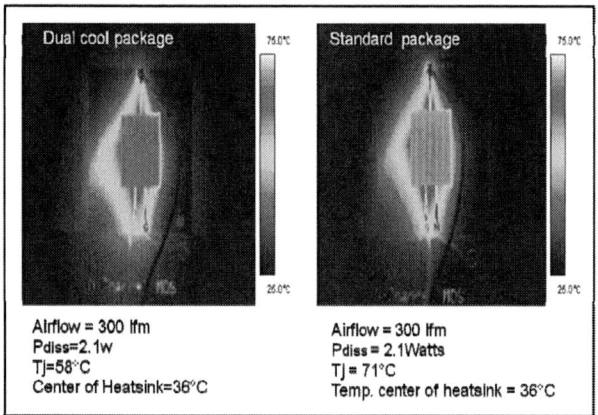

Airflow = 300 lfm
Pdiss=2.1w
TJ=58°C
Center of Heatsink=36°C

Airflow = 300 lfm
Pdiss = 2.1Watts
TJ = 71°C
Temp. center of heatsink = 36°C

Fig 4: Results of the experiment to compare the dual cool solution with the standard technology at system level.

Figure 5 shows how the effectiveness of the DUAL COOL™ technology changes if the fan off is turned off. As you can see the improvement of 13°C or 20% with the fan on is reduced to 8°C or 7% with the forced convection off. Clearly the maximum advantage of the DUAL COOL™ technology is obtained in a forced convection situation.

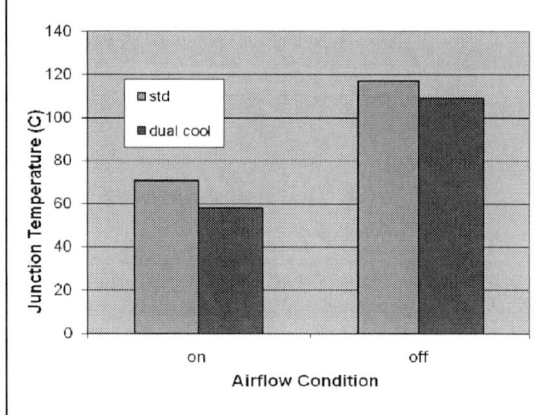

Fig 5: Advantages of using dual cool technology with or without forced convection.

Another way to take advantage of the DUAL COOL™ technology is to increase the current capability of the device keeping the same junction temperature. In Fig. 6 the results of such a condition are shown. The dissipated power Pdiss is adjusted for both systems until the same junction temperature was measured in both devices "A" and "B". It can be seen that with the fan on, the current in device "A" can be increased up to a 30% and still have the same junction temperature as device "B".

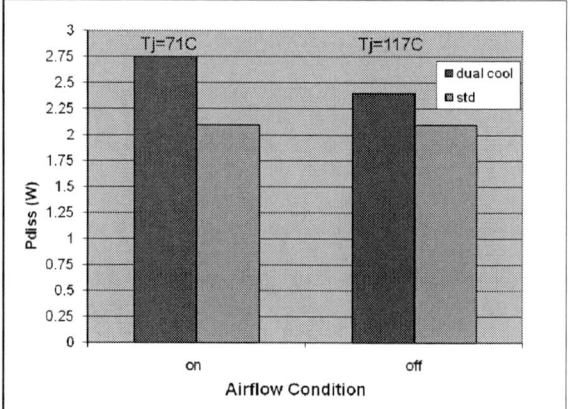

Fig 6: Current in the dual cool device can be increased keeping the same junction temperature

IV. CONCLUSIONS

An innovative and cost effective high power packaging technology that reduces the junction-to top thermal resistance to a level comparable to the thermal resistance junction to case has been presented.

Thermal measurements confirm the dramatic improvement in performance in comparison with standard wire bond and clip package technologies were shown. Systems level simulations that show the improvement from a real application point of view were also demonstrated. At the system design level the DUAL COOL™ technology can enable the following benefits:

1) Higher current densities for a given junction temperature.

2) Lower junction temperatures and lower PC board temperatures resulting in increased long term reliability.

REFERENCES

[1] Zhang, H.Y. Pinjala, D. Poi-Siong Teo , "Thermal management of high power dissipation electronic packages: from air cooling to liquid cooling," Electronics Packaging Technology, 2003 5th Conference (EPTC 2003) 10-12 Dec. 2003 pages 620- 625

[2] A. Chan and J. Wei, "Study on alternative cooling methods beyond next generation microprocessors", Proc. IPACK 2003, pp.1-6,2003, Maui, Hawaii.

[3] Andrew Sawle, Carl Blake and Dragan Mariæ, "Novel Power MOSFET Packaging Technology Doubles Power Density in Synchronous Buck Converters for Next Generation Microprocessors," APEC 2002.

Recent Advances in Silicon Carbide MOSFET Power Devices

Ljubisa D. Stevanovic, Kevin S. Matocha, Peter A. Losee, John S. Glaser, Jeffrey J. Nasadoski, and Stephen D. Arthur

General Electric Global Research Center
Niskayuna, NY 12309
stevanov@ge.com

Abstract — Emerging silicon carbide (SiC) MOSFET power devices promise to displace silicon IGBTs from the majority of challenging power electronics applications by enabling superior efficiency and power density, as well as capability to operate at higher temperatures. This paper reports on the recent progress in development of 1200V SiC power MOSFETs. Two different chip sizes were fabricated and tested: 15A (0.225cm x 0.45cm) and 30A (0.45cm x 0.45cm) devices. First, the 30A MOSFETs were packaged as discrete components and static and switching measurements were performed. The device blocking voltage was 1200V and typical on-resistance was less than 50 mΩ with gate-source voltages of 0V and 20V, respectively. The total switching losses were 0.6 mJ, over five times lower than the competing devices. Next, a buck converter was built for evaluating long-term stability of the MOSFETs and typical switching waveforms are presented. Finally, the 15A MOSFETs were used for fabrication of 150A all-SiC modules. The module on-resistance values were in the range of 10 mΩ, resulting in the best-in-class on-state voltage values of 1.5V at nominal current. The module switching losses were 2.3 mJ during turn-on and 1 mJ during turn-off, also significantly better than competing designs. The results validate performance advantages of the SiC MOSFETs, moving them a step closer to power electronics applications.

I. INTRODUCTION

Silicon IGBTs are used in majority of today's power electronics applications with device voltage ratings between 1.2kV and 6.5kV. Developed almost thirty years ago [1], [2] and continually refined ever since, the IGBTs have reached performance entitlement of both material and device structure. A quantum leap in device performance requires either a better material or a better device structure. One such possibility are wide-bandgap (WBG) semiconductor materials, such as silicon carbide (SiC), gallium nitride (GaN) and diamond. Due to intrinsic material properties, WBG semiconductor devices have entitlement superior to equivalent silicon devices. Specifically:

- wider energy bandgap results in much lower leakage currents and significantly higher operating temperatures of WBG devices; radiation hardness is also improved

- higher critical electric field means that blocking layers of WBG devices can be thinner and with higher doping concentrations, resulting in orders-of-magnitude lower on-resistance values compared to equivalent silicon devices

- higher electron saturation velocity leads to higher operating frequencies

- higher thermal conductivity (e.g., SiC and diamond) improves heat spreading and allows operation at higher power densities.

Table 1 summarizes the key material properties of silicon and the three WBG semiconductor materials.

TABLE I.   SUMMARY OF KEY SEMICONDUCTOR MATERIAL PROPERTIES

Parameter	Si	4H-SiC	GaN	Diamond
Energy bandgap, E_g (eV)	1.1	3.3	3.4	5.5
Critical electric field, E_c (MV/cm)	0.25	2.2	3	10
Electron drift velocity, v_{sat} (cm/s)	$1x10^7$	$2x10^7$	$2.2x10^7$	$2.7x10^7$
Thermal conductivity, λ (W/cm-K)	1.5	4.9	1.3	22

For these reasons, the WBG devices have been identified as the future of power semiconductor industry. The SiC power devices have received the most attention for the past several decades, including large investments under government R&D programs [3]. They reached the tipping point with the commercial launch of Schottky diode by Infineon [4] and Cree [5] in 2001. The diode product introduction validated the material's commercial feasibility and started to fulfill its considerable application promise. It also intensified the SiC power switch R&D efforts. Of all known SiC switch structures, the power MOSFET is the best candidate to replace the silicon IGBT in demanding power electronics applications. When compared to the IGBT, the SiC MOSFET offers much lower losses and operation at higher temperatures [6-8]. The key roadblock to SiC MOSFET productization has been reliability of gate oxide. As in silicon, availability of the native

978-1-4244-4782-4/10 $26.00 © 2010 IEEE   401

SiO$_2$ oxide makes SiC uniquely suitable for implementation of MOS-gated devices. However, the MOSFET gate oxide growth and subsequent device fabrication steps are significantly different between silicon and SiC MOSFETs. One obvious difference is the presence of carbon atoms in the SiC crystal lattice and their detrimental impact at the SiC-SiO$_2$ interface. The other differences are higher processing temperatures during the gate oxide growth and subsequent fabrication steps. It was, therefore, not surprising that direct application of previously developed silicon gate oxidation processes gave very poor SiC MOSFET results. Because the early results were so discouraging, the gate oxide was perceived as a showstopper to commercialization of SiC MOSFETs. However, more recent research activities [9] demonstrated reliable gate oxides, clearing the key remaining challenge to SiC MOSFET productization. The focus can now be shifted to optimizing device structure to reach its performance entitlement.

This paper reports on the recent development of 1200V SiC power MOSFETs at General Electric Company. Section II describes static and switching characteristics of the 1200V MOSFET discrete devices. In addition, design of a test setup for switching characterization is described. Section III shows a prototype buck converter for long-term evaluation of discrete devices. Section IV of the paper explores fabrication of multi-chip power modules using 15A MOSFETs. Implementation of switching test setup is described and the module static and switching characteristics are reported. Finally, Section V concludes the paper.

II. SiC MOSFET DISCRETE DEVICES

A. Fabrication of Discrete MOSFETs

Vertical n-channel power MOSFETs with chip sizes of 0.45cm x 0.45cm (0.15cm^2 active area) and 0.225cm x 0.45cm (0.068cm^2 active area) were fabricated on 4H-SiC wafers of 7.5cm diameter. On-wafer device characterization was performed, wafers were diced and a subset of functional devices was selected for packaging and further evaluation. Two types of packages were used: discrete and multi-chip power modules. For the discrete components, surface-mount TO268 and through-hole TO247 packages were used. Fig. 1 is a photograph of TO268 parts with one 30A, 1200V chip per package.

Figure 1. Prototype 30A, 1200V GE SiC MOSFET parts with one 0.45cm x 0.45cm chip per TO268 package.

Static characteristics of the packaged devices were measured using a curve tracer. Fig. 2 shows room temperature on-state and off-state characteristics of the MOSFETs.

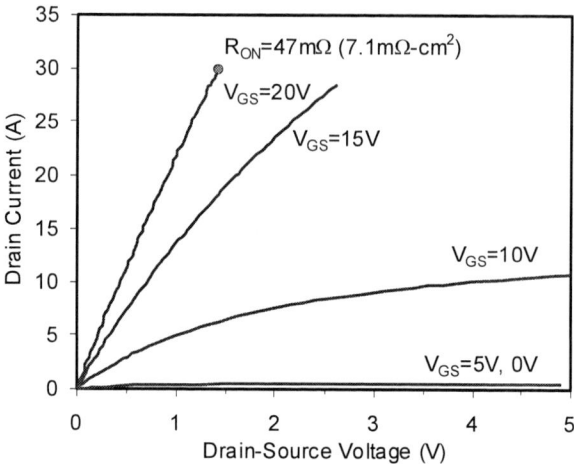

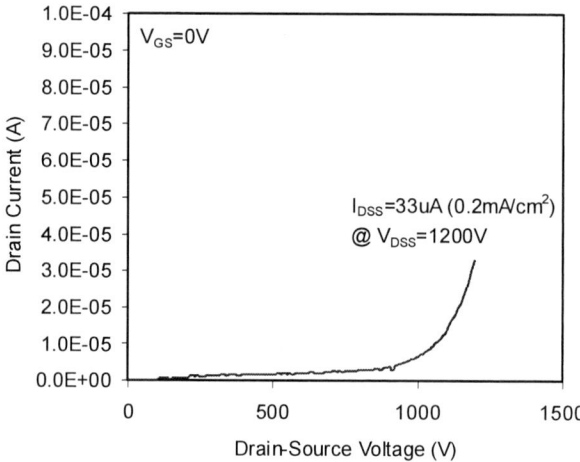

Figure 2. On-state (top) and blocking (bottom) characteristics of the packaged 30A, 1200V GE SiC MOSFET at T$_j$=25°C.

According to the upper graph, the device on-resistance is R$_{ON}$ = 47 mΩ (specific on-resistance: R$_{ON\_SP}$ = 7.1 mΩ-cm^2) at nominal gate-source voltage V$_{GS}$ = 20V and drain current I$_{DS}$ = 30A. The MOSFET on-state voltage at nominal current is V$_{DS\_ON}$ < 1.5V, the best-in-class value for a 1200V switch. The lower portion of Fig. 2 shows room-temperature off-state (blocking) characteristics at the rated voltage of 1200V. The leakage current is I$_{DSS}$ = 33μA (or 0.2mA/cm^2) at gate-source voltage V$_{GS}$ = 0V.

B. Switching Tester

Next, a switching tester, shown in Fig. 3, was developed for characterization of discrete devices. A low inductance layout was implemented using a multilayer circuit board with a number of high frequency multilayer ceramic capacitors, seen in foreground. The device under test (DUT) and the upper freewheeling diode are located in the middle of the photo. An adjustable gate driver was used for testing the MOSFETs. Located to the right of the DUT (oriented edge-wise and orthogonally to the main circuit board), the driver is capable of producing up to +20V and -10V output waveforms. A 0.1 Ω coaxial low inductance shunt resistor from T&M Research, model SBNC-5-1, was used to measure source current of the DUT. The shunt is located directly behind the DUT. Drain-to-source voltage measurement was made at a testpoint located to the left of the DUT, shown with a voltage probe inserted. An air-core inductor, whose leads can be seen at the left edge of the photo, conducted the load current. Also not shown in the image are power supplies, multi-meters and oscilloscope.

Figure 3.   Photograph of the testbed for switching loss measurement of discrete GE SiC MOSFET devices reported in this paper.

C.   Switching Test Results

A double-pulse inductive switching of the TO268-packaged MOSFET with SiC Schottky freewheeling diode was performed using the test setup in Fig. 3. Fig. 4 shows the room temperature switching waveforms at I_D = 30A and V_{DS} = 600V with a 0V to 20V gate drive. Timescales are 20 ns/div and the switching times are approximately 20 ns. The upper screenshot shows relevant waveforms during the turn-on transient, including switching energy E_{ON} (60 μJ/div). The lower screenshot shows the turn-off waveforms, including the energy E_{OFF} (50 μJ/div). The measured losses are: E_{ON} = 0.35 mJ and E_{OFF} = 0.24 mJ.

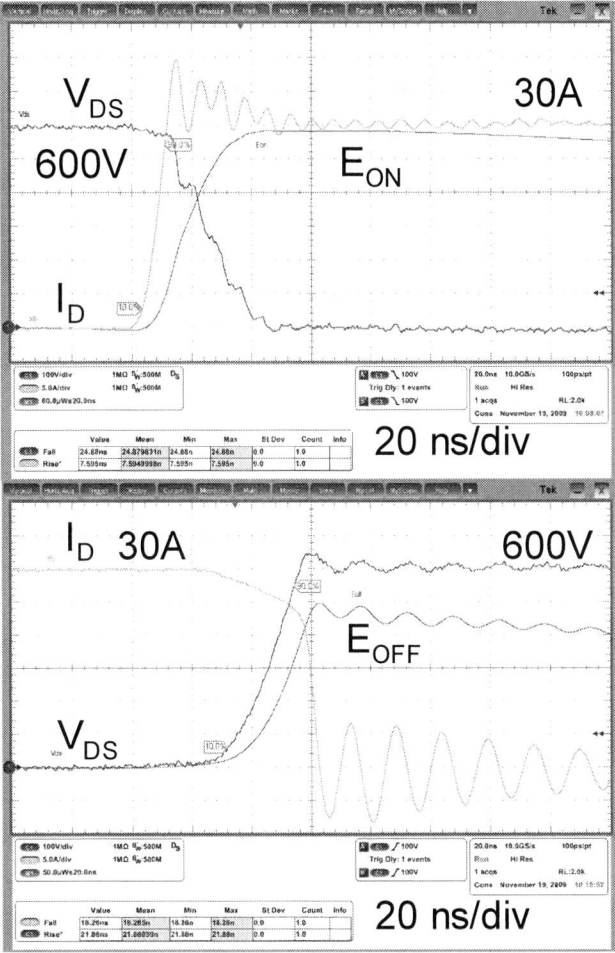

Figure 4.   Turn-on (upper) and turn-off (lower) inductive switching waveforms of the TO268-packaged SiC MOSFET DUT with SiC Schottky freewheeling diode. Test conditions: V_{DS} = 600V, I_D = 30A, T_j = 25°C., gate drive voltage 0V to 20V. Time scale: 20 ns per division. Measured switching energies are: E_{ON} = 0.35 mJ and E_{OFF} = 0.24 mJ per pulse.

Such extremely low switching loss values can be put in perspective by comparison with competing devices, such as fast silicon IGBTs and SiC MOSFETs developed by other R&D groups. For example, published switching loss values for a high speed discrete IGBT from Infineon [10] (with 25A, 1200V rating) are: E_{ON} = 2.6 mJ and E_{OFF} = 1.7 mJ. The IGBT losses were measured with a silicon freewheeling diode and under the following test condition: I_C = 25A, V_{CE} = 600V and junction temperature T_j = 175°C. Also, switching performance of a 60A SiC MOSFET with SiC Schottky freewheeling diode was recently reported in [11]. The switching losses, E_{ON} = 4.6 mJ and E_{OFF} = 4.4 mJ, were measured under the following test conditions: I_D = 65A, V_{DS} = 750V and junction temperature T_j = 150°C. Fig. 5 shows the switching loss comparison between the 30A GE MOSFET and the two competing designs mentioned above. The losses were linearly scaled to the following test conditions: 30A, 600V, T_j = 175°C. Remarkably, the GE SiC MOSFET offers 5.7 to 8.8 times lower switching losses than the two competing designs.

978-1-4244-4782-4/10 $26.00 © 2010 IEEE

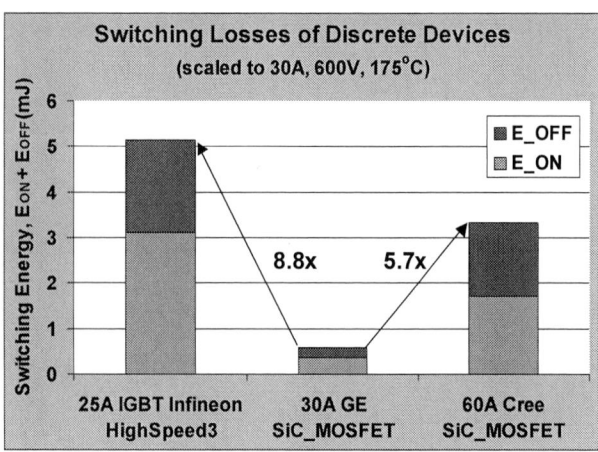

Figure 5. Switching loss comparison between the 30A GE MOSFET and two competing discrete devices: a 25A high speed IGBT from Infineon, and a 60A SiC MOSFET from Cree. Losses were linearly scaled to the following test conditions: 30A, 600V, $T_j = 175°C$.

III. CONTINOUS SWITCHING TESTBED

In order to ascertain long-term device performance under realistic operating conditions, and to provide a baseline performance standard to compare different power semiconductors, an adaptable switching converter testbed was developed. The testbed contains all parts to implement hard-switched buck or boost converters, and includes integrated sensing of key waveforms. Special care was taken to reduce parasitic components due to layout and operation in order to produce near-ideal behavior and accommodate switching frequencies up to 500 kHz. The intent of this effort was to ensure that differences in performance between DUTs could be readily traced to the devices with minimal confounding of the data due to converter parasitics. Fig. 6 shows the testbed implemented as a buck converter with a TO247-packaged 15A MOSFET and a Cree C2D10120 SiC Schottky diode.

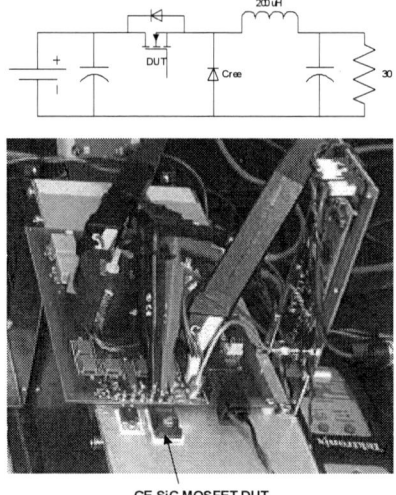

GE SiC MOSFET DUT

Figure 6. Switching converter testbed for long-term evaluation of SiC MOSFETs.

Fig. 7 shows waveforms of the all-SiC buck converter switching at 200 kHz into a 3 kW resistive load bank. The top waveform is a 0 – 20V gate drive and the bottom waveform is a drain-source voltage at 480V DC input. The time-scale is 1 μs per division. The waveforms are almost completely free of high frequency oscillations.

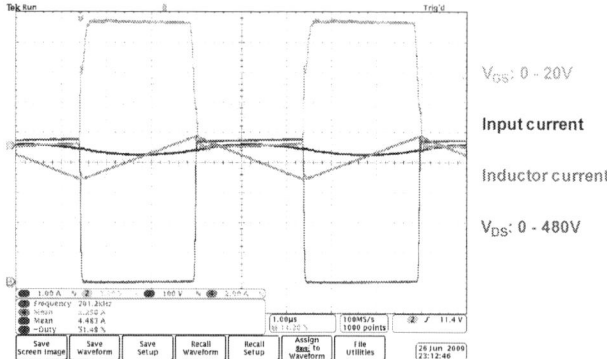

Figure 7. Typical switching waveforms of the all-SiC buck converter switching at 200 kHz. Test parameters: input voltage 480V, 50% duty cycle and 2kW output power.

In addition, a controller has been developed to allow two testbeds to operate together as buck and boost converters operating in pump-back mode, i.e. both converters are tied to the same bus and connected by one inductor with the inductor current under closed-loop control. This configuration allows long-term testing with low overall power consumption.

IV. SiC MOSFET MULTI-CHIP MODULES

A subset of the smaller, 0.225cm x 0.45cm sized, devices was used for fabrication of multi-chip power modules. The 15A devices with matching gate threshold and on-resistance characteristics were binned in sets of ten. A low inductance GE module with low inductance blade power connector [12] was selected in order to take full advantage of the SiC MOSFETs fast switching speed. The 150A-rated blade modules were built by wirebonding ten MOSFETs and fifteen anti-parallel SiC Schottky diodes. Fig. 8 shows a photo of the module. A set of ten modules was fabricated for this project. The lower part of Fig. 8 shows the module's room temperature on-state characteristics. At the rated current of 150A and the nominal gate-source voltage of 20V, the module on-resistance is $R_{ON} = 10.1$ mΩ and the specific on-resistance is $R_{ON\_SP} = 6.8$ mΩ-cm^2. The module on-state voltage at 150A is $V_{DS\_ON} = 1.52V$. Compared to fast IGBT modules with on-state voltages in the range of 3.2 to 5 Volts (for example, [13] and [14]), the all-SiC modules have two to three times lower conduction losses.

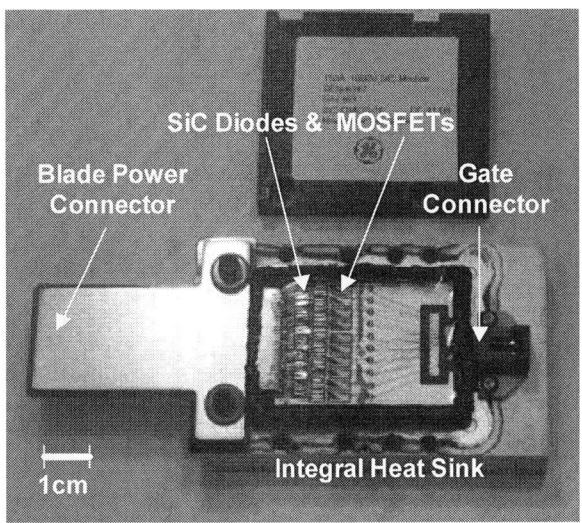

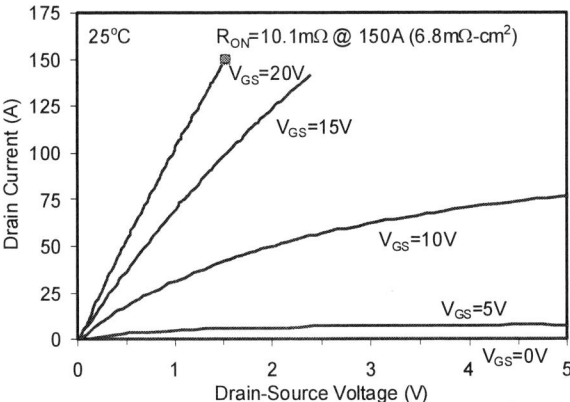

Figure 8. Photograph of all-SiC power module with low inductance blade power connector. The 150A, 1000V module has ten GE SiC MOSFETs and fifteen SiC diodes. Shown below are on-state characteristics at $T_j = 25^{\circ}$C. At the rated current of 150A, the module on-resistance is 10 mΩ.

A. Module Switching Tester

Next, the switching loss measurements of the all-SiC blade modules were performed using the switching test setup described in [10]. Fig. 9 shows a circuit schematic of the switching tester. A flexible gate drive capable of applying up to +20V and -10V output waveforms was used to drive the lower SiC switch Q2. Gate drive resistances were optimized to achieve the best switching performance. The upper SiC switch Q1 was biased off, so that the inductive load current I_load could freewheel through an upper Schottky diode D1. The I_load amplitude was adjusted by the on-time of the lower switch Q2. The lower module DUT current was measured using a 5 mΩ low inductance precisions shunt resistor.

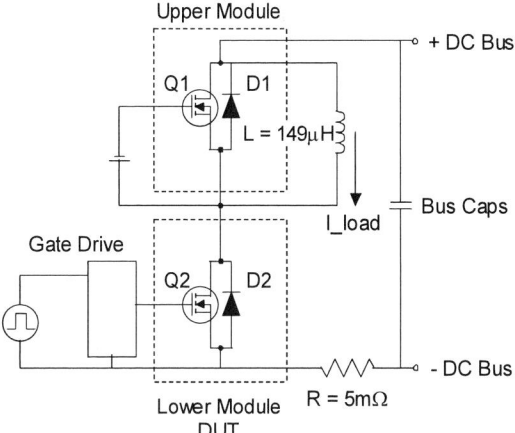

Figure 9. Circuit schematic of the switching test setup used for measuring module turn-on and turn-off losses. The SiC MOSFET Q2 of the lower module DUT was controlled by an adjustable gate drive, while the upper SiC switch Q1 was biased off, allowing the inductive load current I_load to freewheel through an upper Schottky diode D1.

B. Switching Test Results

Fig. 10 shows the room temperature switching waveforms at I_D = 150A and V_{DS} = 600V. The switching times are approximately 50 ns. The turn-off (upper) and turn-on (lower) screenshots show relevant waveforms, including the module DUT voltage V_{DS} and current I_D, gate drive voltage V_{GS} and switching energy E_{SW}. The switching loss values were: E_{OFF} = 1 mJ and E_{ON} = 2.3 mJ.

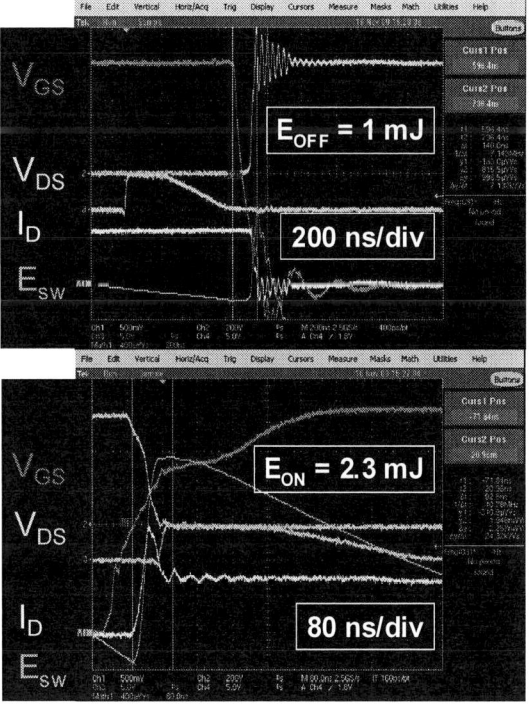

Figure 10. Inductive turn-off (upper) and turn-on (lower) waveforms for the 150A, 1000V all-SiC blade module shown in Fig. 8. The timescales are 80 ns/div and 200 ns/div, respectively. Testing was performed under the following conditions: V_{DS} = 600V, I_D = 150A, T_j=25°C.

978-1-4244-4782-4/10 $26.00 © 2010 IEEE

The on-resistance and the switching losses of the 150A module were also characterized at junction temperatures of 100°C and 175°C. Fig. 11 shows temperature variation of both characteristics. The switching losses varied less than 10% over the entire temperature range, while the on-resistance increase was less than 30%. The behavior is in contrast with characteristics of silicon power devices, such as MOSFET on-resistance and IGBT switching losses, which exhibit much stronger temperature dependence.

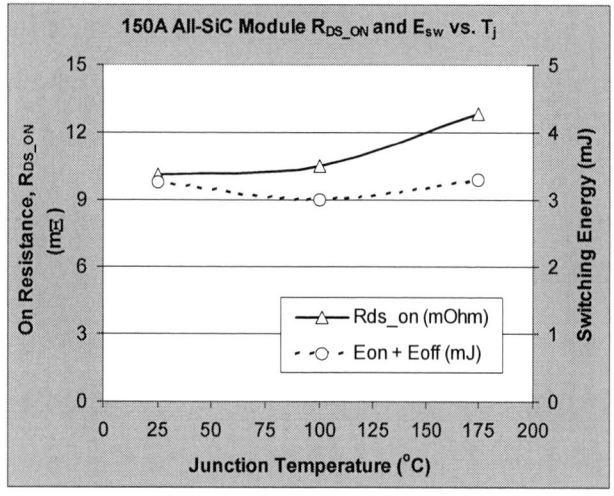

Figure 11. The on-resistance and switching losses of the 150A all-SiC module were characterized at junction temperatures up to 175°C.

Next, the switching losses of the 150A GE SiC MOSFET module were compared with two competing SiC-based modules. The first module was a 300A, 1200V hybrid module from Infineon with fast KS4 IGBTs and SiC Schottky anti-parallel diodes. The Infineon module switching losses: E_{ON} = 7.2 mJ and E_{OFF} = 15 mJ were measured under the following conditions: I_C = 300A, V_{CE} = 600V and junction temperature T_j = 125°C [12]. The second competing design was a 100A, 1200V all-SiC dual module from Cree [15], with reported switching loss values E_{ON} = 2.5 mJ and E_{OFF} = 9.7 mJ. The 100A module losses were measured under the following conditions: I_D = 100A, V_{DS} = 800V, T_j = 150°C. Fig. 12 shows comparison between the GE 150A module and the two competing designs. The reported Infineon and Cree losses were linearly scaled to the following test conditions: 150A, 600V, T_j = 125°C. The GE SiC module offers three to five times lower switching losses than the two SiC-based modules.

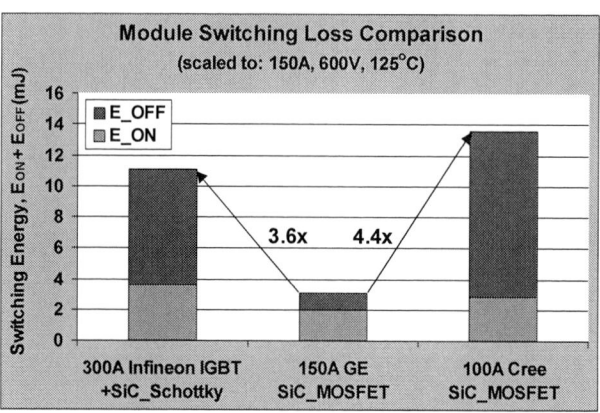

Figure 12. Switching loss comparison between the 150A GE SiC MOSFET blade module and two other competing designs: a 300A hybrid (fast Si IGBT + SiC Schottky diode) module from Infineon and a 100A SiC MOSFET module from Cree. The reported Infineon and Cree losses were linearly scaled to the following test conditions: 150A, 600V, T_j = 125°C.

V. CONCLUSIONS

This paper described the recent progress in development of the 1200V SiC power MOSFETs at General Electric Company. Two different device sizes were fabricated and tested: 15A and 30A. First, the 30A MOSFETs were packaged as discrete components and static and switching measurements were performed. The device on-state resistance was 47 mΩ and the blocking voltage 1200V. The switching tests showed turn-on and turn-off switching times in the 20 ns range and total switching losses of 0.6 mJ. Compared to switching losses of the two competing devices, the SiC MOSFET from Cree and the fast IGBT from Infineon, the GE MOSFET losses were 5.7 to 8.8 times lower. Next, a buck converter switching test setup was developed for long-term evaluation of SiC devices and typical switching waveforms were presented. Finally, the 15A MOSFETs were used for fabrication of multi-chip modules. The 150A all-SiC module on-resistance values were in the range of 10 mΩ, resulting in the best-in-class on-state voltage values of 1.5V at nominal current. The module switching losses were 2.3 mJ during turn-on and 1 mJ during turn-off. Compared to switching losses of the SiC-based modules from Infineon and Cree, the GE SiC module losses were 3.6 and 4.4 times lower. The results validate performance advantages of the SiC MOSFETs, moving them a step closer to power electronics applications.

REFERENCES

[1] B. J. Baliga, M. S. Adler, P. V. Gray, R. P. Love, and N. Zommer, "The insulated gate rectifier (IGR): A new power switching device," IEDM Tech. Dig., pp. 264-267, 1982.

[2] J. P. Russell, A. M. Goodman, L. A. Goodman, and J. M. Neilson, "The COMFET - A new high conductance MOS-gates device," IEEE Electron Device Lett., vol. EDL-4, pp. 63-65, 1983.

[3] J.C. Zolper, "Emerging silicon carbide power electronics components," Proceedings of the 20th IEEE Applied Power Electronics Conference, APEC 2005, vol. 1, pp. 11-17, Austin, TX, March 6-10, 2005.

[4] D. Stephani, "Status, prospects and commercialization of SiC power devices," IEEE Device Research Conference, p. 14, Notre Dame, IN, June 25-27, 2001.

[5] http://www.cree.com/ftp/pub/CPWR-AN02.pdf

[6] S. Harada, Y. Hayashi, K. Takao, A. Kinoshita, M. Kato, M. Okamoto, T. Kato, S. Nishazawa, T. Yatsuo, K. Fukuda, H. Ohashi and K. Arai, "Demonstration of motor drive with SiC normally-off IBMOSFET/SBD power converter," Proceedings of the 19th International Symposium on Power Semiconductor Devices and IC's, ISPSD 2007, pp. 289-292, Jeju Island, S. Korea, May 27-31, 2007.

[7] G. Majumdar, T. Oomori, "Some key researches on SiC device technologies and their predicted advantages," Proceedings of Power Conversion Intelligent Motion Conference, PCIM Europe 2009, pp. 328-333, Nuremberg, May 12-14, 2009.

[8] P. Losee, K. Matocha, S. Arthur, E. Delgado, R. Beaupre, A. Pautsch, R. Rao, J. Nasadoski, J. Garrett, Z. Stum, L. Stevanovic, R. Conte and K. Monaghan, "100 amp, 1000 volt class 4H- silicon carbide MOSFET modules," Materials Science Forum, vols. 615-617 (2009), pp. 899-902.

[9] K. Matocha, P. Losee, A. Gowda, E. Delgado, G. Dunne, R. Beaupre and L. Stevanovic, "Performance and reliability of SiC MOSFETs for high-current power modules," Proceedings of the 13th International Conference on Silicon Carbide and Related Materials, ICSCRM 2009, Nuremberg, Germany, Oct 11-16, 2009.

[10] For example, Infineon IGW25N120H3 HighSpeed3 IGBT: http://www.infineon.com/dgdl/DS_IG25N120H3_1_1_final.pdf?folderI

d=db3a30431c69a49d011c6f86019b00a1&fileId=db3a304325305e6d0 1258d0f50a8369e

[11] B. Hull, C. Jonas, S.-H. Ryu, M. Das, M. O'Loughlin, F. Husna, R. Callanan, J. Richmond, A. Agarwal, J. Palmour and S. Scozzie, "Performance of 60 A, 1200 V 4H-SiC DMOSFETs," Materials Science Forum, vols. 615-617 (2009), pp. 749-752.

[12] L.D. Stevanovic, R.A. Beaupre, E.C. Delgado and A.V. Gowda, "Low inductance power module with blade connector," Proceedings of the 25th IEEE Applied Power Electronics Conference, APEC 2010, Palm Springs, CA, Feb 21-25, 2010.

[13] For example, Infineon KS4-series IGBT module FF225R12MS4: http://www.infineon.com/dgdl/DS_FF225R12MS4_2_0.PDF?folderId= db3a3043139a1bac0113b4bd91b70542&fileId=db3a304313b8b5a6011 3ba985a0e00b0

[14] For example, Mitsubishi NFH-series IGBT module CM200DU-24NFH: http://www.pwrx.com/pwrx/docs/cm200du-24nfh.pdf

[15] J. Richmond, S. Leslie, B. Hull, M. Das, A. Agarwal and J. Palmour, "Roadmap for megawatt class power switch modules utilizing large area silicon carbide MOSFETs and JBS diodes," Proceedings of the 1st IEEE Energy Conversion Congress and Exposition, ECCE 2009, San Jose, CA, Sep 20-24, 2009.

Start-up Transient Improvement for Sensorless Control Approach of PM Motor

Dong Jiang*, Rixin Lai*, Fred Wang**, Rolando Burgos*, Dushan Boroyevich*

*Department of Electrical and Computer Engineering

Virginia Polytechnic Institute and State University

Blacksburg, VA, USA, 24060

jiang@vt.edu

** Department of Electrical Engineering and Computer Science

The University of Tennessee

Knoxville, TN

fred.wang@utk.edu

Abstract—**PM motor's speed sensorless control does not work well in the low speed. To avoid this issue, open-loop control is usually used to start the motor, and then it is switched to sensorless close-loop control. The transition between the two control modes can cause a transient during the starting process. This transient can be undesirable especially for motor drives with front-end rectifier and small energy storage components, such as small dc capacitors in the case of voltage source inverter (VSI) drive. This paper studies the principle of dc bus transient in sensorless control start-up process and proposes a method to reduce the oscillation: before closing the sensorless loop, the reference current is adjusted to continuously track the real motor rotor position to the estimated rotor position, and the speed regulator is pre-calculated to generate the q-axis reference current before closing the speed loop. An experiment is conducted using a Vienna-type rectifier and a VSI based motor drive with back-EMF observer based sensorless control on a PM motor. Experimental results show that with the proposed method, the DC bus voltage transient is obviously reduced.**

I. INTRODUCTION

Position/speed sensorless control has been applied widely in PM motor control. There are several kinds of sensorless control methods [1], which can be classified into two main groups: methods based on a fundamental wave observer and methods based on harmonic injection. The harmonic injection methods can perform well at low speeds and stand-still[2] but are difficult to implement. The fundamental wave based sensorless control methods usually do not perform well at low speeds and stand-still. Thus, during the motor's start-up process, non-sensorless methods are usually used and then the motor is switched to sensorless mode[3],[4]. At the switch time, because the control reference will changes suddenly, transients appear. Several papers, including [4] mentioned this problem, but very few papers discuss the ways to constrain this problem. If the motor drive contains a large DC capacitor, the transient will not be a serious problem.

In the past few years, high power density motor drive with active front-end rectifier has been developed to satisfy requirements for applications like airplanes. In these motor drives, the DC-link capacitor is designed to be small to limit the total volume[5],[6],[7]. The starting transient in sensorless control is more serious since the DC-link capacitor cannot fully compensate the power unbalance between the rectifier

and the inverter during the transient, especially for non-regenerative rectifier, like the Vienna-type rectifier[5],[7].

Section II of this paper introduces the basic principle of back-EMF observer-based sensorless control and its conventional start-up process. The principle of the sensorless plug-in transient is studied in section III, and an improved start-up process to weaken the transient is proposed in section IV.

Beside the proposed back-EMF observer-based sensorless method, the starting process can be applied to other observer based sensorless control' start up.

In this paper, the analysis is supported by experimental results based on a Vienna-type rectifier and a VSI based motor drive with a PM motor experimental platform. The circuit structure is shown in Fig.1, where we can see that the AC source side is controlled with unit power factor. In order to increase the switching frequency, SiC JFET is used as the active switches in the converter. The controller uses ADSP21160 together with FPGA, and the experimental results are memorized in the controller and analyzed in MATLAB. To verify the function before approaching full voltage and full power, in the test the DC bus voltage is controlled to be 90V and DC capacitor is only 17.5uF. The actual platform is shown in Fig.2. The parameter of the motor and motor drive are shown in Table.1.

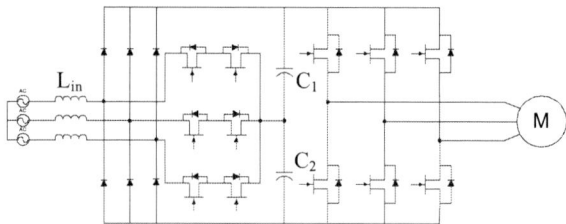

Figure.1 Vienna rectifier-VSI motor drive with PM motor

Figure.2 Experiment platform: Vienna-type rectifier & VSI

978-1-4244-4782-4/10 $26.00 © 2010 IEEE

The PM motor parameters are shown in Table.1. In the experiment, the motor is connected to a generator with a Yaskawa 626VM3 back-to-back converter as the load. The diagram of the experimental connection is shown in Fig.3 and the actual motor testbed is shown in Fig.4.

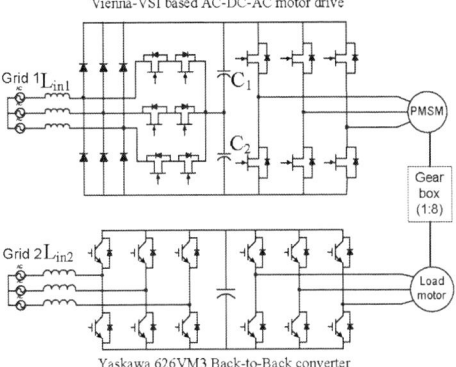

Figure.3 Diagram of the experimental platform

Figure.4 Experiment platform: motor with generator load

TABLE. 1 MOTOR AND MOTOR DRIVE PARAMETERS IN THE EXPERIMENT

Rs	0.53ohm
Ls	280uH
PM flux	0.0347Wb
DC cap	17.5uF
AC voltage in test	30V, 400Hz
DC voltage in test	90V

II. BACK-EMF OBSERVER BASED SENSORLESS CONTROL METHOD AND STARTING PROCESS

For the PM motor, i_d, i_q v_d, v_q e_d, e_q are the current, voltages and back EMF in the d-q axis. The PM motor is controlled with the back-EMF observer based sensorless method, which is shown in Fig.4.[4] The back-EMF observer is shown in Fig.5(a), $u=[v_d, v_q]^T$, $x=[i_d, i_q, e_d, e_q,]^T$, $y=[i_d, i_q]^T$. The space state equation of the PM motor is expressed in (1), it can be written in the form of (2), the output equation is shown in (3).

The observer can estimate the back-EMF $\hat{e}_{dq}$. Then the position tracker in Fig.5(b) tracks the error angle to zero, getting the estimated rotor speed $\hat{\omega}$ and rotor position $\hat{\theta}$.

$$\begin{bmatrix} \dot{i}_d \\ \dot{i}_q \\ \dot{e}_d \\ \dot{e}_q \end{bmatrix} = \begin{bmatrix} -\dfrac{R_s}{L_s}, & \hat{\omega}, & \dfrac{1}{L_s}, & 0 \\ -\hat{\omega}, & -\dfrac{R_s}{L_s}, & 0, & \dfrac{1}{L_s} \\ 0, & 0, & 0, & 0 \\ 0, & 0, & 0, & 0 \end{bmatrix} \begin{bmatrix} i_d \\ i_q \\ e_d \\ e_q \end{bmatrix} + \begin{bmatrix} \dfrac{1}{L_s}, & 0 \\ 0, & \dfrac{1}{L_s} \\ 0, & 0 \\ 0, & 0 \end{bmatrix} \begin{bmatrix} v_d \\ v_q \end{bmatrix} \quad (1)$$

$$\dot{x} = Ax + Bu \quad (2)$$

$$y = Cx = \begin{bmatrix} 1,0,0,0 \\ 0,1,0,0 \end{bmatrix} x \quad (3)$$

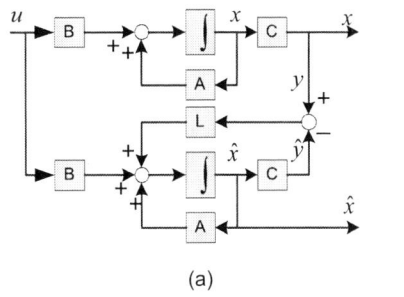

(a)

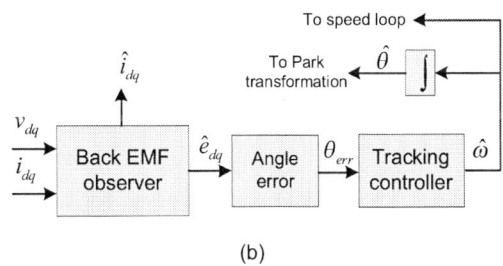

(b)

Figure. 5 Back-EMF observer based sensorless control method: (a) Back EMF observer, (b) Speed/position tracker

The motor starts with open-loop speed control. The starting process is shown in Fig.6. The starting process is divided into 5 regions:

(1) In Region 1, by injecting DC current, the motor is aligned to the 0 angle position.

(2) In Region 2, the speed increases continuously, but the observer cannot work well, so the reference currents i_{d_ref} and i_{q_ref} are given directly and rotated with the open loop speed.

(3) In Region3, the observer begins to work, generating $\hat{i}_{dq}, \hat{\omega}$ and $\hat{\theta}$, but the motor is still driven without closing speed loop.

978-1-4244-4782-4/10 $26.00 © 2010 IEEE   409

(4) When the reference speed is approaching ω_{plug}, the observer estimated $\hat{\omega}$ and $\hat{\theta}$ will be steady, at t_3, estimated position $\hat{\theta}$ replaces the open loop position and estimated speed $\hat{\omega}$ is plugged into the speed loop, the motor goes into sensorless mode in Region 4.

(5) In Region 5, the reference speed reaches the given value ω_{ref}. At time t_3, because of the sudden plug in, a transient appears and the DC bus voltage oscillates. The experimental result is shown in Fig.7, in which the transient peak in the DC voltage is $\pm 20V$.

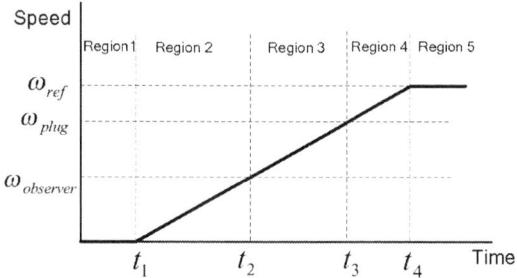

Figure.6 Conventional starting process of sensorless controlled PM motor

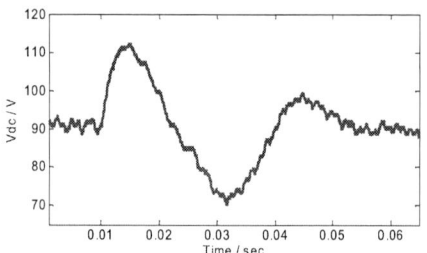

Figure.7 DC bus transience in sensorless plug in time

III. PRINCIPLE OF DC BUS VOLTAGE TRANSIENT

There are two main reasons for the transient at the sensorless plug-in time.

First of all, as discussed in section II, in Region 1 and Region 2, the motor is driven with given reference current vector $\vec{i}_{ref} = i_{d\_ref} + j \cdot i_{q\_ref}$, and the rotating speed of the vector is directly given by reference speed ω_{ref}, which means the reference angle is $\theta_{ref} = \int \omega_{ref} dt$. However, as the reference current is directly given and not based on the load torque requirement, the real rotor will not be in the reference d-q coordinate. This means that if the observer works well, the estimated rotor angle is $\theta_{est} \neq \theta_{ref}$. This principle is shown in Fig.8. With the same current vector $\vec{i}$ in Region 3, the feedback current and the observer current will be in different coordinates. This phenomenon is proved by

experimental result in Fig.9. In time t_3, θ_{est} will replace θ_{ref}, which will cause significant transient in the motor. In addition, the power balance between the rectifier and the inverter will be disturbed, causing the DC bus voltage to oscillate.

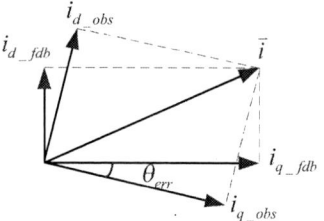

Figure.8 Current vector relationship in open loop and observer coordinate

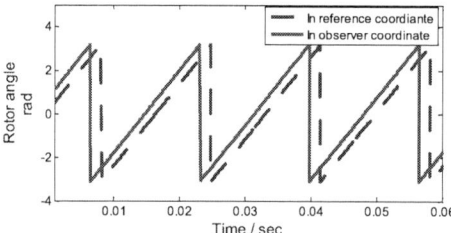

Figure.9 Rotor angle in open loop and observer

The other reason for transient is shown in Fig.10. Before t_3, i_{q_ref} is directly given to drive the motor (i_{q_ref0}), at t_3, i_{q_ref} is suddenly switched from i_{q_ref0} to the output of the speed PI regulator. The experimental result in Fig.11 proves this phenomenon: at the switch point, i_{q_ref} jumps directly from 4A to less than 0, and feedback current i_{q_fdb} tracks i_{q_ref} and breaks the power balance, causing the DC bus voltage oscillation.

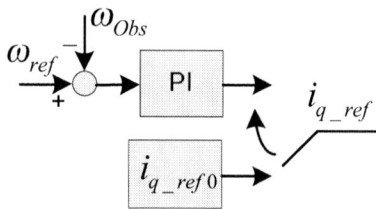

Figure.10 i_{q_ref} sudden change in speed loop plug in time

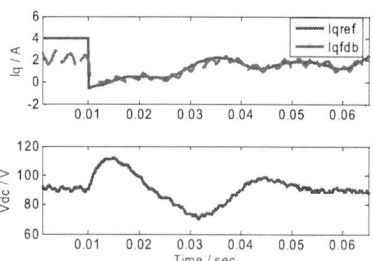

Figure.11 Experiment result of iqref sudden change in the plug in time (t3):
Top: q-axis current, Bottom: DC bus voltage

IV. PROPOSED START-UP METHOD AND COMPARISON

Based on the results of the analysis in section III, an improved start-up process is proposed, shown in Fig.12 and Table.2. This process contains seven regions. Building on with the original process in Fig.6, two new regions are added in the start-up process.

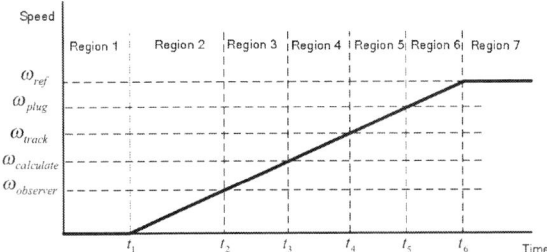

Figure. 12 Proposed start-up process of sensorless controlled PM motor

TABLE.2 FUNCTION OF EACH REGION IN THE PROPOSED
START-UP PROCESS

Region 1	Initial alignment
Region 2	Open-loop start-up, observer does not work
Region 3	Observer begins to work
Region 4	Calculate the updated i_{q_ref}
Region 5	Adjusts i_{q_ref}
Region 6	Switch to sensorless control
Region 7	Arrive at reference speed

In order to reduce the angle error in the open-loop accelerating time as much as possible, an updated i_{q_ref} is calculated in Region 4. The calculated method is based on a sample of 1000 data points of i_{q_Obs} in the beginning of Region 4. An average is calculated, and multiplied by a coefficient λ ($\lambda > 1$), shown in Eq.(4). If I_{qref_update} is applied to the q-axis, it will be near the actual torque current, so θ_{err} will be much smaller. The reason to multiply λ is to make sure the current is big enough to accelerate the motor. In this experiment, $\lambda = 1.2$. In Region 5, the reference current I_{q_ref} is gradually adjusted to I_{qref_update} with a slope. Experimental results are shown in Fig 13, where we see that in 0.1 second the feedback q-axis current gradually tracks the reference current to the updated value, and the DC bus voltage stays nearly constant in this period. Then the i_q in two different coordinates will be nearly the same, and the rotor position will be tracked to near the reference coordinate, experimental results are shown in Fig.14.

$$I_{qref\_update} = (\frac{1}{1000}\sum_{1}^{1000} I_{q\_Obs}) \times \lambda \qquad (4)$$

In addition, the calculated I_{qref_update} is also sent to the speed PI controller in Region 5 as the initial value, so that when the sensorless control is plugged in at t_5, the reference current will not suddenly drop. The experimental result is shown in Fig.15. If the initial value is not settled, i_{q_ref} will jump more than 3A. If the initial value is settled, i_{q_ref} jump

in t_5 is less than 0.5A.

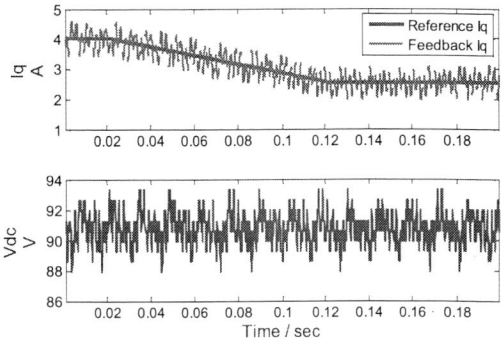

Figure.13 Experimental results: reference and feedback q-axis current and

DC bus voltage in the tracking period (t4)

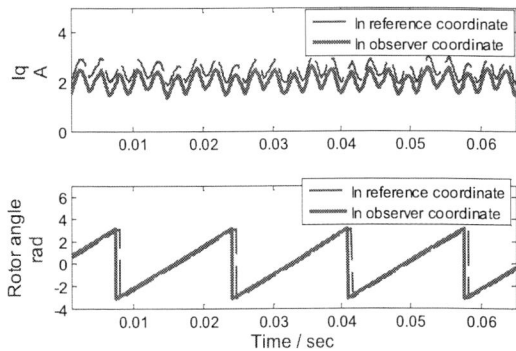

Figure.14 Iq and rotor angle in open loop and observer coordinate: In

Region 5, after Iqref tracking

A final comparison of experimental results is made in Fig.16. With the normal method (method 1), the peak transient DC bus voltage will be $\pm 20V$. With i_{q_ref} adjustment in Region 5 (method 2), the peak transient DC bus voltage will be weakened to $\pm 15V$. With i_{q_ref} adjusted and speed PI initialization (method 3), the peak transient DC bus voltage will be less than $\pm 5V$. The proposed start-up process can obviously improve the transient.

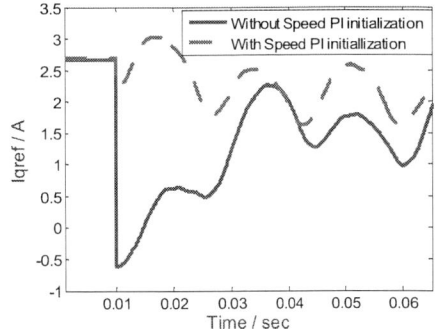

Figure.15 Iqref in the sensorless plug in time (t5):

With and without speed PI initialization

978-1-4244-4782-4/10 $26.00 © 2010 IEEE

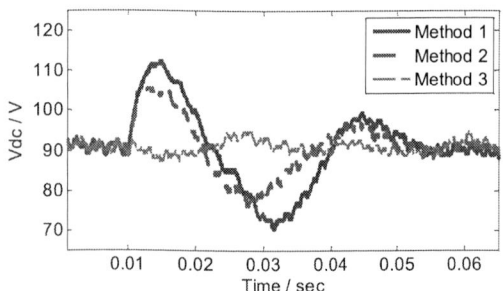

Fig.16 DC bus voltage in transient with different methods

Although the start-up principle is studied only with a low voltage 90V experiments, with the improved start-up method the DC bus voltage transient can be reduced to the accepted level when the DC link is built with rated voltage. Fig.17 and Fig.18 are the experimental results using 330V DC-link voltage with the motor speed pushed to 5500rpm with sensorless control. Fig.17 is the experimental results of DC bus voltages of the rectifier, the rectifier input voltage and input current. Fig.18 is the DC bus voltage of the inverter, inverter output voltage and current and rectifier current. The steady-state value in DSP is shown in Fig.19. The rotor speed is controlled to stay close to the reference value.

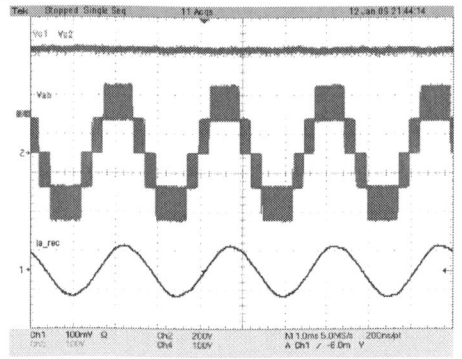

Figure.17. Final experimental results (330V DC voltage Vc1+Vc2, rectifier input voltage Vab, rectifier input current Ia_rec)

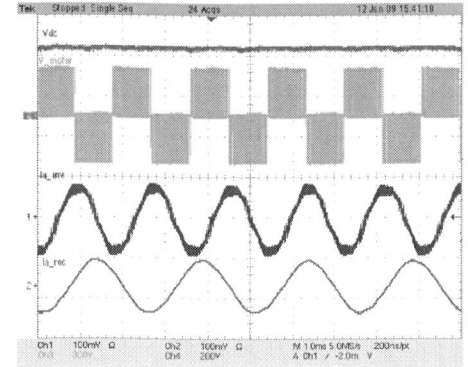

Figure.18. Final experimental results (330V DC voltage Vdc, inverter output voltage V_motor, inverter output current Ia_rec, rectifier input current Ia_rec)

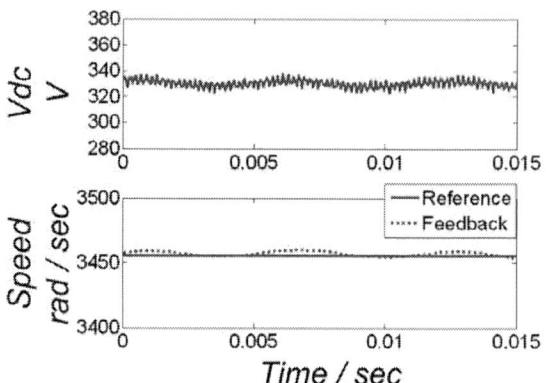

Figure.19 Steady-state DC bus voltage, reference and feedback (estimated) speed

V. CONCLUSION

In the normal start-up process of a PM motor with sensorless control, a significant transient can appear at the sensorless plug-in time, especially in a motor drive with front-end rectifier and small DC-link capacitor. Based on the Vienna-type rectifier and the VSI based motor drive and PM motor experimental platform, this paper studies the principle of the start-up transient. Two factors cause the start-up transient when the motor is switched from open loop control to sensorless control:

(1)The difference between the reference coordinate and the observer coordinate

(2) The reference current jump at the sensorless plug-in time

This paper proposes an improved start-up process: by pre-calculating the q-axis current in the observer, the open-loop reference current is gradually adjusted to the updated value, further reducing the angle error, and initializing the speed regulate output. Experimental results prove that the proposed start-up process can obviously improve the transient with sensorless control.

ACKNOWLEDGMENT

The authors would like to thank Boeing Company for supporting the research. The authors also would like to thank Darren Tremelling and T.A.Lipo from University of Wisconsin-Madison for providing the PM motor for the experiment.

REFERENCE:

[1] Holtz.J. "Sensorless control of induction machines—with or without signal injection?" *IEEE Transactions on Industrial. Electronics* , vol.53 no.1,pp:7-30, Dec. 2005

[2] Yu-seok Jeong, Robert D.Lorenz, Thomas M.Jahns,Seung-Ki Sul Initial Rotor Position Estimation of an Interior Permanent-Magnet Synchronous Machine Using Carrier-Frequency Injection Method *IEEE Transaction on Industry Application* NO.1 2005: 38~44

[3] Burgos, R.P.; Kshirsagar, P.; Lidozzi, A.; Wang, F.; Boroyevich, D. Mathematical Model and Control Design for Sensorless Vector Control of Permanent Magnet Synchronous Machines, *Computers in Power Electronics*, 2006. COMPEL '06. IEEE Workshops on

[4] Parasiliti, F.; Petrella, R.; Tursini, M. Sensorless speed control of a PM synchronous motor by sliding mode observer Industrial Electronics, 1997. ISIE '97., *Proceedings of the IEEE International Symposium on*

[5] Rixin Lai; Wang, F.; Burgos, R.; Boroyevich, D. Voltage balance control of non-regenerative three-level boost rectifier using carrier-based pulse width modulation *Power Electronics Specialists Conference*, 2008. PESC 2008. IEEE

[6] Gu, B.G.; Nam, K. A Theoretical minimum DC-link capacitance in PWM converter-inverter systems *IEE proceedings of Electric Power Applications*, Volume 152, Issue 1, pp.81-88, 2005

[7] R. Lai, Y. Pei, F. Wang, R. Burgos, D. Boroyevich, T. A. Lipo, V. Immanuel, and K. Karimi, "A systematic evaluation of AC-fed converter topologies for light weight motor drive applications using SiC semiconductor devices," Proc. IEEE Electric Machines & Drives Conference, pp. 1300-1305, 2007.

Sensorless Position Control of Skewed Rotor Induction Machines Based on Multi Saliency Extraction

T.M. Wolbank

Department of Electrical Drives and Machines
Vienna University of Technology
Vienna, AUSTRIA
e-mail: thomas.wolbank@*tuwien.ac.at*

M.K. Metwally

Department of Electrical Drives and Machines
Vienna University of Technology
Vienna, AUSTRIA

Abstract-This paper addresses sensorless position and torque control of induction machines around zero frequency using inherent saliencies of a skewed rotor machine. The estimation method employs transient excitation of voltage pulses to detect machine saliencies. The resulting signals from injection based sensorless control methods are composed of components caused by saturation, slotting, as well as inter-modulation effects. As multiple saliencies cause problems in tracking only a particular saliency, decoupling of the different signal components is applied via neural network. This leads to the estimation of both the rotor angle as well as the flux angle, which are used in combination with the stator and the rotor equations. Identifying the disturbances of the different signal components it is possible to establish an adaptive control signal merging to reduce disturbance based torque pulsations. The performance of the control is investigated at loaded operation and zero speed as well as zero fundamental frequency.

I. INTRODUCTION

Most of the scientific and technical efforts performed over the past two decades on sensorless control of AC drives have moved from being purely a research topic to being widely used in different industrial applications. It is accepted that no single sensorless method is capable of controlling all types of machines, under all operating conditions. Back-emf based techniques have been shown to be capable of providing high performance, field-oriented control in the medium to high speed range. As the speed decreases, the parameter sensitivity of these methods becomes bigger, and in the low frequency range they fail due to uncertainty of the resistive machine parameters for dc excitation of the stator [1]. Methods based on the tracking of saliencies using a high frequency carrier signal excitation (voltage or current) [2]-[7], or a transient pulse voltage excitation [8]-[10] have been developed over the past decade. These methods provide high bandwidth rotor position/speed and/or flux angle estimates, even at zero speed and fundamental frequency. In combination with the back-emf methods, sensorless control is now possible over the whole speed range of an AC drive. The method applied in this paper estimates the flux angle by applying test voltage pulses and calculating the leakage inductance which varies due to main path and leakage saturation [8], as well as slotting air-gap width modulation. This method is applied to a standard induction machine with skewed rotor, exhibiting a high spatial saturation effect with respect to the slotting effect. The machine saliencies will thus create both rotor and flux position-dependent modulations of similar magnitude in the resulting saliency control signal, which can be signal processed to yield the required position. In addition, there are also inter-modulation induced signal components present, which may act as disturbance and have to be approached using for example side band filter [11], or the so called structured neural network [12]. In this paper the compensation of saturation saliencies and inter-modulation effects were performed using artificial neural network (ANN) together with the necessary extraction of the rotor position signal [13] and the separation of the flux position signal component [14]. The reduction of disturbing harmonics is based on a compensation of non-ideal inverter/sensor properties and can be realized by optimization of the excitation pulses and/or placing of the sample instants [15]. There are however, specific operating states where the signal to noise ratio (SNR) of one of the extracted signal components may be deteriorated due to disturbances. When using both components for control purpose it is thus advantageous to weight each component according to its current reliability. Thus it is possible to reduce torque pulsation due to signal disturbances in these specific operating states. This weighting is especially important when using standard machines with multiple saliency and inter-modulation components. The weighting is based on specific quality indicators for each signal component.

II. TRANSIENT VOLTAGE EXCITATION METHOD

If an ac machine is operated on a voltage source inverter, it is persistently excited with short voltage pulses to enable control of the reference current. This leads to a transient change of the armature current, which is determined by the dc link voltage, the three transient phase reactances, the machines back emf, and the stator resistance.

The value of the dc link voltage can be assumed constant during the short time of the transient excitation due to the dc link capacitors. The share of the back emf and stator resistance on the current change can be eliminated by signal processing if at least two different switching states are observed. The

978-1-4244-4782-4/10 $26.00 © 2010 IEEE

transient current change is measured and the resulting signal after signal processing is thus only determined by the three transient phase reactances, which in turn are modulated during operation of the machine by different inherent saliencies [8]. This signal is denoted saliency control signal in the following.

III. HARMONIC SEPARATION AND SIGNAL MERGING SCHEME

The saliency control signal is a superposition of different asymmetry components and can not directly be used for sensorless control. An accurate and reliable separation of its components is essential. The separation of the different harmonic components of the saliency control signal, depicted in Fig. 1 upper diagram, is done using ANN. This separation can be used for obtaining reliable signal components for both flux- and rotor-position estimation. Full details of the technique can be found in [13] and [14]. In recent years, the ANN has found a place in real time control and system identification applications with the increase in microcontroller speed and capabilities. The ability of ANN to model nonlinear characteristics and to compensate disturbances and uncertainties makes it an ideal candidate for estimation. The structure of the ANN used in this paper is multi-layer perceptron [16]. Fig. 1 lower diagram shows the slotting component of the saliency control signal after elimination of saturation, cross-saturation effects and inter-modulation disturbance components.

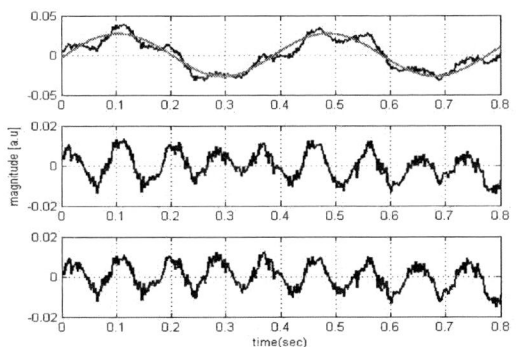

Fig. 1. Real component of saliency control signal at 90% rated load. Upper: no compensation; Middle: compensation of saturation effects; Lower: compensation of saturation effects and inter-modulation.

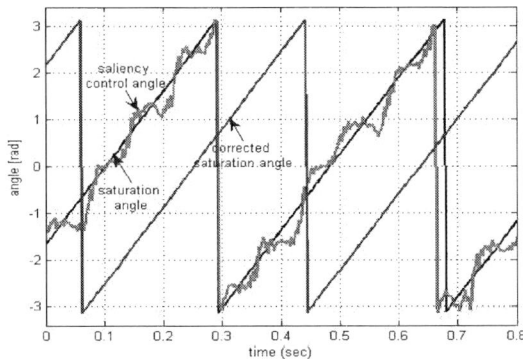

Fig. 2: Angle of saliency control signal at 90% rated load (red); estimated saturation angle (black); and corrected saturation angle (blue).

The estimated rotor position using this component is combined with the current model (rotor eq.) to get the estimated flux position [13]. Regarding the flux position estimation using the saturation saliency, Fig. 2 shows the angle of the saliency control signal before (red) and after elimination of the disturbing slotting and inter-modulation components (black). This angle is further processed using the output of an ANN (flux angle deviation) to get the corrected saturation angle (blue) that corresponds the estimated flux angle, which is the difference between saturation saliency angle and twice the flux angle obtained under sensored mode.

This angle is further processed using the fundamental wave information of the stator equation (voltage model) [14].

In order to establish an adaptive signal merging of the two estimated flux angles "$\gamma_{\psi A}$" (based on saturation signal in combination with voltage model (stator-eq.)) and "$\gamma_{\psi B}$" (based on slotting signal in combination with current model (rotor-eq.)), it is necessary to realize some sort of quality indicator for each of the signal components as in Fig. 3. This is done by comparing the instantaneous values of signal magnitude and angle of slotting and saturation signals as shown in Fig. 4. The slotting signal magnitude and angle are shown in the two upper most traces and saturation magnitude and angle in the two lower traces in Fig. 4 at 90% rated load for a standard induction machine with skewed rotor. It is shown that at time instant t=3.6 sec there is one missed slot when considering only the slotting signal. In this case the signal merging will increase the influence of the saturation signal. The quality indicator of the slotting and saturation signal in combination with the signal merging are used to determine the reliabilities of each of the flux angles "γ_{ψ}" calculated by stator and rotor equation. The most reliable one is finally used for field oriented control (FOC) as depicted in Fig. 3.

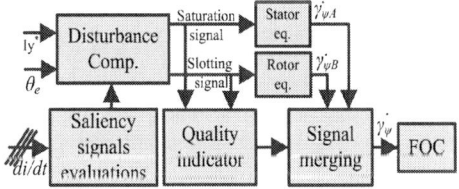

Fig. 3: Disturbances compensation and signals merging scheme.

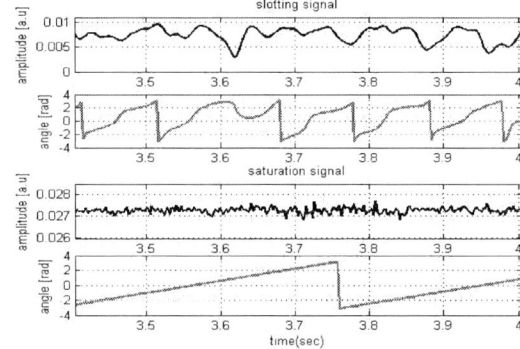

Fig. 4: From top to bottom: slotting magnitude and angle; saturation magnitude and angle at 90% rated load. (operation with 15rpm speed)

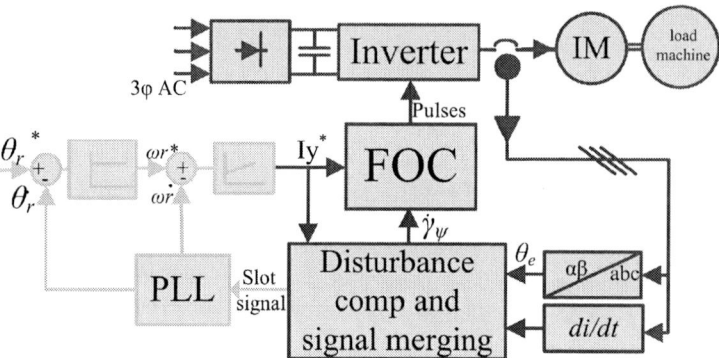

Fig. 5: Sensorless control scheme

IV. SENSORLESS CONTROLLED DRIVE

Figure 5 shows the sensorless controlled drive using both saliencies and the proposed signals merging scheme. The individual signals disturbances are identified using ANN. The slotting signal in combination with the phase locked loop (PLL) is used to estimate the rotor angle "θ_r". Depending on the quality of the slotting and the saturation signal as well as the point of operation the adaptive signal merging of the two estimated flux angles is used to combine the reliabilities of the flux angles estimated by the stator and rotor equations and eventually to reconstruct a missed slot.

This merged flux angle "γ_ψ" is finally used for sensorless (FOC) as depicted in Fig. 5. Considering the fact that the position signal is delivering only incremental information it is possible to apply even position control (indicated with gray color) to a loaded induction machine.

V. EXPERIMENTAL RESULTS

The experimental results shown are from an induction machine with skewed rotor coupled to a load machine as shown in Fig. 5. The machine under test was operated under sensorless torque controlled conditions using the proposed excitation and separation algorithm. The drive speed is imposed by the speed controlled load dynamometer.

In another arrangement the machine under test is loaded at zero mechanical speed with a weighted bar to prove position control under loaded operation at standstill.

The parameters of the machine are given in appendix I. The control is done on a digital signal processor board plugged into a computer programmable in Matlab/Simulink. It performs the vector control and separation algorithm using ANN. The pulse sequences and instances of injections of the pulses are being calculated on a field programmable gate array system (FPGA) plugged into another computer using the programming language Labview. There is a communication board between the two systems for transferring and receiving data. The induction motor was fed by a voltage source inverter with PWM operating at 5kHz. Three industrial current sensors were used for the current measurements.

An optional position signal is available from an encoder with 1024 pulses resolution as a reference signal.

A. Flux Angle Estimation using Slotting Signal

The following results are taken under sensorless field oriented torque control (based on the slotting signal in combination with the current model (rotor-eq.).

Figure 6 demonstrates the accuracy of the harmonic separation during changing saturation levels due to load change. The induction machine is operated with zero mechanical speed when a load change from 50% rated load at time instant approximately t= 6 sec to 90% rated load and the load changes from 90% rated load at time instant t= 35 sec back to 50% rated load in a ramp function to avoid problems with the Shannon frequency due to the control dynamics of the coupled load machine (sampling frequency of the rotor position signal is about 500 Hz).

The upper diagram shows the reference rotor position angle from the position encoder (black) and the estimated (mechanical) rotor position using slotting signal (red). The difference between these two angles is depicted in the lower diagram (black) in degree. The phase difference between the estimated and reference position, which is about 102.5° in the lower diagram in Fig. 6, results from a zero position offset as only incremental information is contained in the slotting signal. The quality of estimated position signal and the constant phase difference is clearly seen. The control is able to accurately determine the mechanical position within a tolerance of less than ±1° mechanical.

Figure 7 shows the same operating state but for the rotor flux angles. The upper diagram in the figure shows the flux angle obtained from the current model (red) and the flux angle from the sensor- based (reference) current model (black). The signal tracks the angle of the current model, with only limited error as shown in Fig. 7 (middle diagram).

For the same measurement setup, the error between the reference flux angle and the estimated flux angle (based on the saturation signal in combination with stabilized voltage model (stator-eq.) is depicted in Fig. 7 (lower diagram) as shown the angle error here stays within a few degrees.

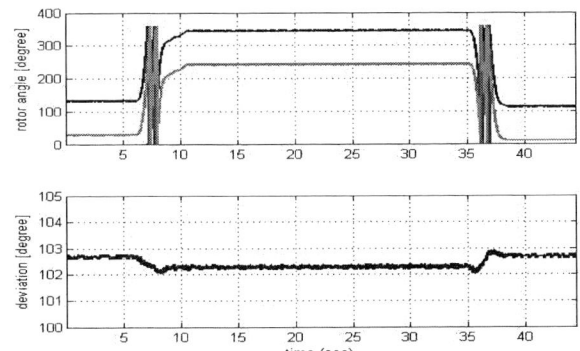

Fig. 6: Sensorless operation using current model and slotting saliency at load change from 50% to 90% rated load and from 90% to 50% rated load at standstill. Upper: Reference rotor angle (black), estimated rotor angle (red); Lower: error between reference and estimated rotor angles (°).

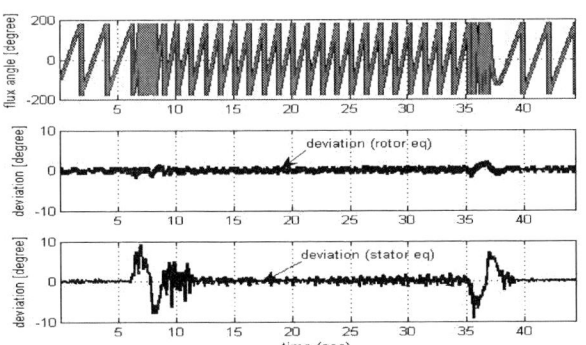

Fig. 7: Sensorless operation using current model and slotting saliency at load change from 50% to 90% rated load and from 90% to 50% rated load at standstill. Upper: Reference flux angle (sensor-based) (black), and flux angle from current model (red). Middle: error between reference and estimated angles using current model (°). Lower: error between reference and estimated angles using voltage model (°).

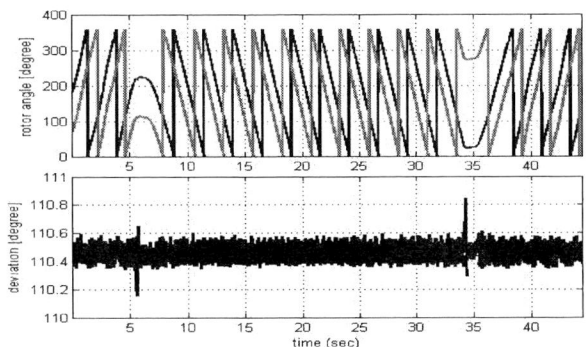

Fig. 8: Sensorless operation using current model and slotting saliency at 90% load and speed reversal ±33.5 rpm.
Upper: Reference (black) and estimated (red) rotor angle; Lower: error between reference and estimated rotor angle (°).

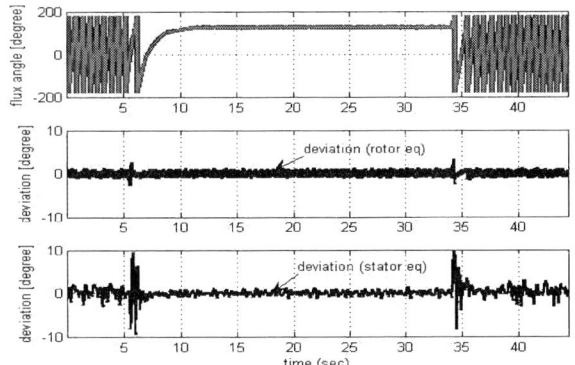

Fig. 9: Sensorless operation using current model and slotting saliency at 90% load and speed reversal ±33.5 rpm.
Upper: Reference flux angle (sensor-based) (black), and flux angle from current model (red). Middle: error between reference and estimated angles using current model (°). Lower: error between reference and estimated angles using voltage model (°).

In Figs.8-9 the drive performs a speed reversal ±33.5 rpm under 90% rated load condition. Fig. 8 (upper diagram) shows the rotor position angle from the position encoder (black) and estimated (mechanical) rotor position using the slotting signal (red). The difference between these angles is depicted in Fig. 8 (lower diagram) in degree. At -33.5 rpm, this load condition corresponds to the drive operating around zero excitation frequency.

Since this speed equals the slip frequency at that load level, the result is zero flux frequency with a constant rotor flux position as shown in Fig. 9 (upper diagram). The upper diagram in the figure shows the flux angle obtained from the current model (red) and the flux angle from the sensor-based (reference) current model (black). The signal tracks the angle of the current model, with only limited error as shown in Fig. 9 (middle diagram).

For the same measurement setup, the error between the reference flux angle and the flux angle (based on the saturation signal in combination with stabilized voltage model (stator-eq.) is depicted in Fig. 9 (lower diagram). As was shown for the control using the rotor equation, the angle error here also stays within a few degrees.

In Fig. 8 (lower diagram) the offset value of ~110.5° again results from an initial position error as the slotting saliency component delivers only incremental information. Nevertheless the control is able to accurately determine the mechanical position within a tolerance of less than ±1.5° mechanical. The same comparison is also shown in Fig. 9 where the deviation of the estimated flux angle is approximately ±1.5° electrical. From Fig. 8 and Fig. 9 it can be seen that the tracking of the rotor speed is very accurate using the compensation technique during transient and steady state operation.

B. Flux Angle Estimation using Saturation Signal

Next results are taken under sensorless field oriented torque control (based on the saturation signal in combination with stabilized voltage model (rotor-eq.).

The induction machine is operated with zero mechanical speed when a load change from 50% rated load at time instant approximately t= 7 sec to 90% rated load and the load changes from 90% rated load at time instant t= 33 sec to 50% rated load in a ramp function to avoid problems with the Shannon frequency due to the control dynamics of the coupled load machine. Figure 10 shows the rotor flux angles. The upper diagram in the figure shows the flux angle obtained from the voltage model (red) and the flux angle from the sensor-based (reference) current model (black). The signal tracks the angle of the current model, with only limited error as shown in Fig. 10 (middle diagram). For the same measurement setup, the error between the reference flux angle and the estimated flux angle (based on the slotting signal in combination with the current model (rotor-eq.) is depicted in Fig. 10 (lower diagram) as shown the angle error here stays within a few degrees.

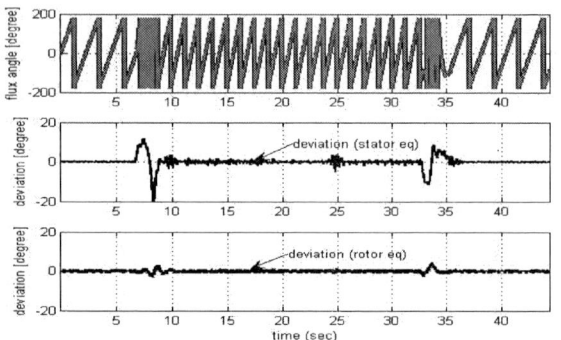

Fig. 10: Sensorless operation using voltage model and saturation saliency at load change from 50% to 90% rated load and from 90% to 50% rated load at standstill. Upper: Reference flux angle (sensor-based) (black), and flux angle from voltage model (red). Middle: error between reference and estimated angles using voltage model (°). Lower: error between reference and estimated angles using current model (°).

In Fig. 11 the drive performs a speed reversal ±33.5 rpm under 90% rated load condition. The upper diagram in the figure shows the flux angle obtained from the voltage model (red) and the flux angle from the sensor-based (reference) current model (black). The signal tracks the angle of the current model, with only limited error as shown in Fig. 11 (middle diagram). For the same measurement setup, the error between the reference flux angle and the flux angle (based on the slotting signal in combination with current model (rotor-eq.) is depicted in Fig. 11 (lower diagram). As was shown for control using the voltage equation, the angle error here also stays within a few degrees.

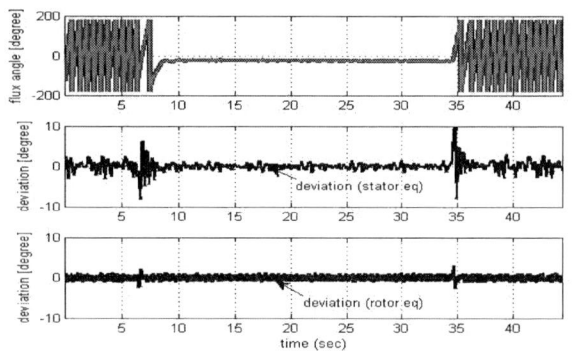

Fig. 11: Sensorless operation using voltage model and saturation saliency at 90% load and speed reversal ±33.5 rpm.
Upper: Reference flux angle (sensor-based) (black), and flux angle from voltage model (red). Middle: error between reference and estimated angles using voltage model (°). Lower: error between reference and estimated angles using current model (°).

C. Sensorless Position Control

Figure 12 shows the dynamic behavior of sensorless position control at no load with a change in the reference position of ±150°. The deviation between actual and estimated rotor position is about 1° mechanical during a transient position change and zero in steady state as shown in the lower diagram.

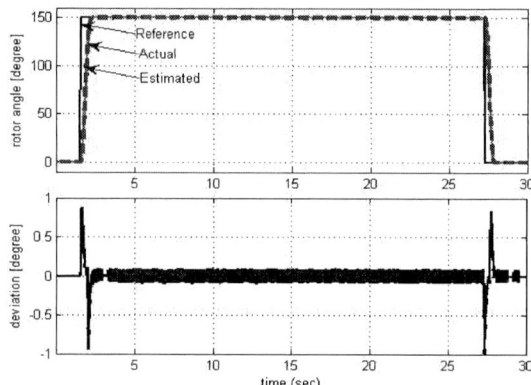

Fig. 12: Sensorless position control using current model and slotting saliency at no load and change in reference position of ±150°.
Upper: Reference rotor angle (black), estimated rotor angle (blue), and actual rotor angle (red); Lower: error between actual and estimated rotor angles in (°).

Figure 13 shows the dynamic behavior of the sensorless position control at 60% static load using a bar with load weight fixed to the rotor.

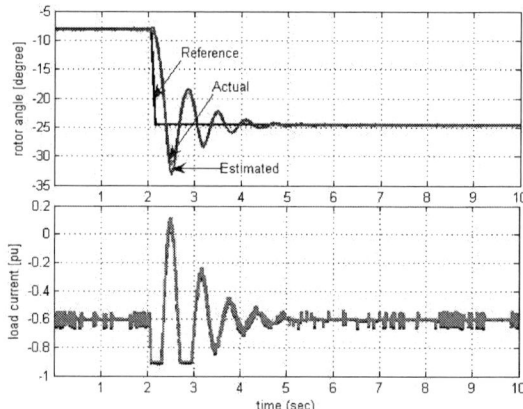

Fig. 13: Sensorless position control using current model and slotting saliency at 60% rated load and change in reference position from -8 to -24°.
Upper: Reference rotor angle (black), estimated rotor angle (blue), and actual rotor angle (red) at 60% static load; Lower: load current (p.u.).

The weight is balanced by the position control with a transient change of the reference position from -8 to -24 mechanical degrees (lifting the bar in negative direction). The deviation between actual and estimated rotor position is below ±1.5°. The response time is limited by the bandwidth of the speed controller. The position controller is proportional only. Good performance of the system is achieved noting the current controller limitation to about 90% load current imposed by the inverter as shown in the lower diagram of Fig. 13.

It should be noted that the results presented indicate the potential of the excitation method in combination with the signal processing.

VI. CONCLUSION

This paper has shown sensorless position and torque control of an full pitched stator winding induction machine with skewed rotor using transient pulse excitation and a separation algorithm. It is based on the signals obtained from an excitation of the machine with voltage pulses. The resulting signals contain a superposition of signal components caused by all saliencies present in the machine including saturation, slotting, inter-modulation and offset signal components. By identifying actual disturbance levels of the different saliency components it is possible to always use that component with the highest reliability thus reducing torque oscillations caused by signal disturbances. The paper shows how tracking of all saliencies lead to good torque and position control of machine in the zero and low frequency range at high torque levels.

APPENDIX I

Machine parameters of the applied induction machine
Nominal Current: 30A
Nominal Voltage: 280V
Nominal Frequency: 75Hz
Rated Power: 11kW
4-poles, 36 stator teeth, skewed rotor, 44 rotor bars.

ACKNOWLEDGMENT

The authors gratefully acknowledge the financial support of the Austrian Science Foundation - "Fonds zur Förderung der wissenschaftlichen Forschung" (FWF) - under grant no. P19967-N14.

REFERENCES

[1] Holtz. J., "Sensorless Control of Induction Motor Drives," *Proceedings of the IEEE*, vol. 90, No. 8, Aug. 2002.

[2] Degner, M. W., Lorenz, R.D., "Position Estimation in Induction Machines Utilizing Rotor Bar Slot Harmonics and Carrier Frequency Signal Injection", *IEEE Trans. on Ind. Appl.*, May 2000, Vol.36, no.3, pp. 736-742.

[3] Teske, N., Asher, G.M., Sumner, M., Bradley, K.J., "Analysis and Suppression of High Frequency Inverter Modulation on Sensorless Position Controlled Induction Machine Drives", *IEEE Trans. on Ind. Appl.*, Vol. 39, no. 1, Jan. 2003, pp. 10-18.

[4] Consoli, A., Scarcella, G., Testa, A., "A New Zero-Frequency Flux-Position Detection Approach for Direct-Field-Oriented-Control Drives", *IEEE Trans. on Ind. Appl.*, May 2000, Vol. 36, no. 3, pp.797-1004.

[5] Jansen, P.L., Premerlani, W.J., Garces, L.J., "System and Method for Sensorless Rotor Tracking of Induction Machines", US Patent no. 6,388,420 B1, May 2002.

[6] Briz, F., Degner, M.W., Garcia, P., Lorenz, R.D.,"Comparison of Saliency Based Sensorless Control Techniques for AC Machines", *Proc. IEEE-IAS Annual Metting*, Salt Lake City, Oct. 2003, CD-ROM.

[7] Ha, J., and Sul, S.K., "Sensorless Field Orientation Control of an Induction Machine by High Frequency Signal Injection", *Proc. IEEE-IAS Annual Metting*, New Orleans, Oct. 1997, pp 426-432.

[8] M. Schroedl, "Sensorless Control of Ac Machines at Low Speed and Standstill Based on the INFORM Method" (1996) *Proceedings of 31st IAS Annual Meeting* San Diego, Vol1, pp.270-277.

[9] J.Holtz, H. Pan, "Elimination of Saturation Effects in Sensorless Position Controlled Induction Motors", *Proceedings of IEEE Industry Applications Annual Meeting*, Vol3, pp.1695-1702, (2002).

[10] C. Spiteri. Staines, G.M. Asher, and M. Sumner, "Sensorless Control of Induction Machines at Zero and Low Frequency using Zero Sequence Currents*", IEEE IAS Annual Meeting*, 2004.

[11] Q. Gao, G. M. Asher, and M. Sumner, "Sensorless Position and Speed Control of Induction Motors Using High Frequency Injection and Without Offline Precommissioning", *IEEE Trans. on Industrial Electronics*, Oct 2007, Vol. 54 pp. 2474-2481.

[12] Pablo Garcia, Fernando Briz, Dejan Racn and Robert D. Lorenz, "Saliency Tracking-based Sensorless Control of AC Machines Using Structured Neural Networks", *the 40th Annual Meeting of IEEE, IAS* 2005.

[13] Th.M. Wolbank, and M.K. Metwally, "Zero speed sensorless control of induction machines using rotor saliencies", the 43rd *Annual Meeting of IEEE, IAS* 2008.

[14] Th.M. Wolbank, and M.K. Metwally, "Speed Sensorless Flux and Position Control of Induction Machines Based on Pulse Injection and Multiple Saliency Extraction", *Proceedings of IEEE Industrial Electronics Society* (IECON 2008), pp,1403-1408.

[15] Th.M. Wolbank, and J.L. Machl, "Influence of Inverter-nonlinearity and Measurement Setup on Zero Speed Sensorless Control of AC Machines Based on Voltage Pulse Injection", *Proceedings of IEEE Industrial Electronics Society* (IECON 2005), pp,1568-1573.

[16] Bimal K. Bose, "Neural Network Applications in Power Electronics and Motor Drives—An Introduction and Perspective," *IEEE Trans. on Industrial Electronics*, vol. 54, no. 1, pp. 14-33, Jan 2007.

978-1-4244-4782-4/10 $26.00 © 2010 IEEE

Fuzzy Gain Scheduling PI Controller for a Sensorless Four Switch Three Phase BLDC Motor

Chung-Wen Hung[1]   Jen-Ta Su[2]   Chih-Wen Liu[2]   Cheng-Tsung Lin[3]   Jhih-Han Chen[1]

Department of Electrical Engineering,[1]
National Yunlin University of
Science and Technology Yunlin,
Taiwan, R.O.C.

Department of Electrical Engineering[2]
National Taiwan University Taipei,
Taiwan, R.O.C

DynaPack Co., Ltd.[3]
Taoyuan, Taiwan, R.O.C.

Abstract— A sensorless method for six–space-vector four Switch three phase BLDC motor driver is described in this paper. Due to the nature of low resolution of position sensing, the speed feedback is variable sampling. A fuzzy gain scheduling PI controller is proposed in this paper. It is based on three selected PI controllers in fixed sampling time intervals, high-, median- and low-speed, and combined by simplified fuzzy logic. The experimental results are provided to show this controller is efficient and workable.

I. INTRODUCTION

Due to high efficiency, high power factor, high torque, simple control, and lower maintenance, the brushless dc (BLDC) motor is becoming popular. Conventionally, BLDC motors are excited by a six-switch inverter. However, for cost-effective design, some researchers [1]–[6] proposed attractive power inverters with only four switches, as shown in Fig. 1, to reduce switches and freewheeling diode count and conduction losses. The four-space-vector scheme was used in [4], and in [5] the six commutation modes based on current control. Normally, position sensors are used to achieve commutation control of BLDC motors. For reduction of cost, sensorless control for six-switch three-phase BLDC motors has had many successful applications [7]–[11], and most are based on the zero-crossing point of voltage waveforms from unexcited windings.

Due to no floating winding for four-switch three-phase BLDC motors by using the conventional four-space-vector strategy [4], it is impossible to achieve sensorless control schemes. [12] proposed a sensorless control scheme for the FSTP BLDC motors based on six-space-vector and voltage source strategy. No matter Hall sensor base or sensorless scheme, position sensing originally is used for commutation, then the speed measurement is based on time interval between two commutation signals. The speed sampling updating is dependent on speed and not uniform, so the normal digital controller based on uniform system clock could not handle this variable sampling phenomenon and some response may not be as expecting [13]. In sensorless FSTP BLDC motor driver, due to 3 times-longer times interval of speed feedback, this effect is more serious. This paper proposes a fuzzy gain scheduling control to solve this problem; three different speed PI controllers are combined based on fuzzy logic.

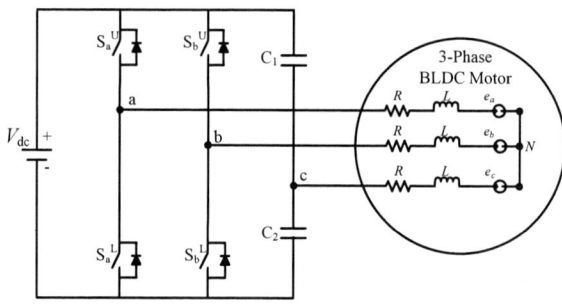

Fig 1. Configuration of four-switch three-phase inverter.

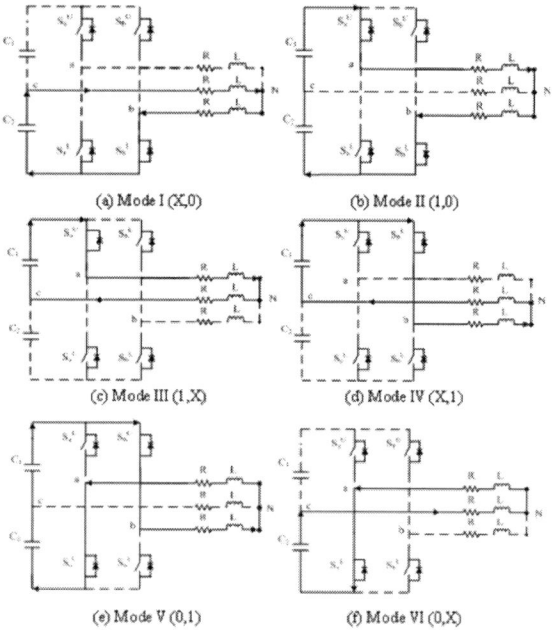

(a) Mode I (X,0)   (b) Mode II (1,0)

(c) Mode III (1,X)   (d) Mode IV (X,1)

(e) Mode V (0,1)   (f) Mode VI (0,X)

Fig 2. Six commutating modes of voltage PWM scheme for FSTP inverter.

978-1-4244-4782-4/10 $26.00 © 2010 IEEE   420

II. NOVEL PWM SCHEME FOR FSTP BLDC MOTOR DRIVES

Rectangular stator current is required to produce a constant electric torque, in BLDC with a trapezoidal back EMF. In [12], the proposed voltage pulse width modulation (PWM)scheme for FSTP inverter requires six commutation modes which are (X,0), (1,0), (1,X), (X,1), (0,1) and, (0,X), as shown in Fig. 2. The symbols in parenthesis denote the switch states of S_a^U, S_a^L, S_b^U and S_b^L (phases A and B). "X" denotes the OFF state for both the high- and low-side switching devices in the same leg, "1" and "0" denote the ON state for the high- and Low- side switching device. Two modes need to be noted: In Mode II, the conventional voltage PWM scheme what is construct of stages (1,0) and (X,0) is not suitable for the FSTP BLDC motor, because a discharging loop between the capacitor and the low-side switch is formed. And it causes non-rectangular stator current waveforms which are harmful for constant torque, as shown in Fig. 3(c).

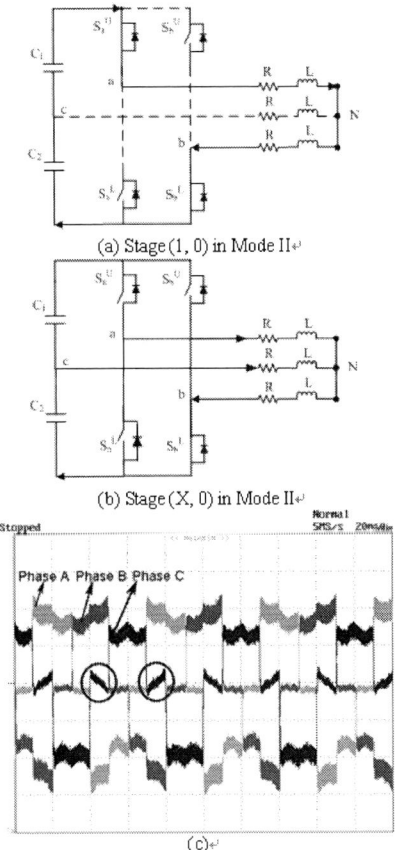

(a) Stage (1, 0) in Mode II

(b) Stage (X, 0) in Mode II

(c)

Fig 3. Operation stages of FSTP inverter using conventional PWM scheme in Mode II (a) stage (1,0), (b) stage (X,0) and (c) the experimental results of stator current waveforms in Mode II.

Similar situations also occur in Mode V. As shown in Fig. 4(a), novel asymmetric voltage PWM is proposed to overcome this drawback by introducing a new stage (X, X) in Modes II and V to turn off all power devices to prevent the capacitor discharging from the low-side switch. The extra stage in Mode

II is shown in Fig. 4(b), and the experimental stator rectangular current waveforms are shown in Fig. 4(c).

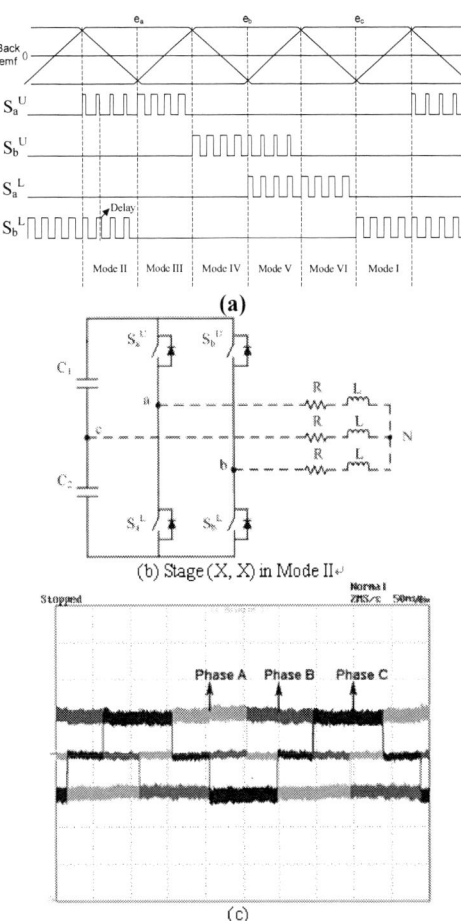

(a)

(b) Stage (X, X) in Mode II

(c)

Fig 4. Operation stages of FSTP using novel PWM scheme in Mode II: (a)PWM wavefrom, (b) Extra stage (X,X) and (c) The experimental results of stator current waveforms.

Similar situations apply to Mode V. Furthermore, the supply voltages in Modes II and V are double of those in the other four Modes while the PWM duty cycles in Modes I, III, IV and VI are double of those in the Mode II and V. The FSTP BLDC motor drives using the novel voltage PWM scheme have two phases to detect the back EMF, but the split capacitors cause the voltage waveform of back EMF to be triangular like as shown in Fig. 5. So, the conventional sensorless methods using in six-switch three-phase inverter could not work in the FSTP BLDC motors. However, from Fig. 5, one could find that two waveform crossings matched the two Hall signals (101 and 010) at the same time, respectively. Therefore, the two crossings are used to estimate rotor position estimation for sensorless commutation purposes, and other 4 commutation points are based on extrapolations.

978-1-4244-4782-4/10 $26.00 © 2010 IEEE   421

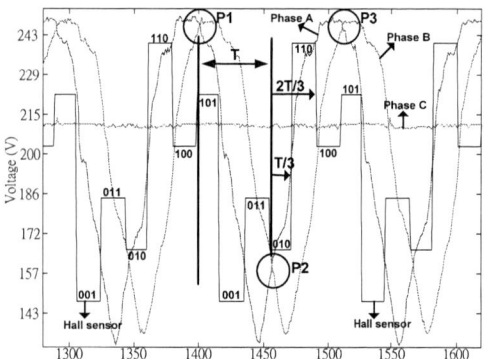

Fig 5. The voltage waveforms for BLDC motor using FSTP inverter and the relationships between waveform crossings and Hall signals.

III. FUZZY GAIN SCHEDULING PI CONTROL

In Section I, the speed variable sampling phenomenon of BLDC motor is introduced. The speed sensing in most BLDC motors is based on position sensing which is used to decide commutation timing, so the resolution is very low and the speed sampling time interval is dependent on rotor speed. Normally, for 2-pole of BLDC motor, there are only 6 commutation states in every revolution. That is, a motor runs from 300 to 3000 rpm, the speed sampling intervals are 1/30 to 1/300 Sec. When sampling time interval is variable, the parameters of discrete time plant will be not uniform and the conventional digital PI control could not handle. A robust controller is necessary, a fuzzy gain scheduling PI control as shown in Fig. 6 is proposed. Three PI controllers, PI_{HS}, PI_{MS} and PI_{LS}, are selected to handle high-, median- and low-speed situation, what are 500, 1500 and 2500 rpm. And the fuzzy logic based on rotor speed (ω) is used to combine the output of PI controllers.

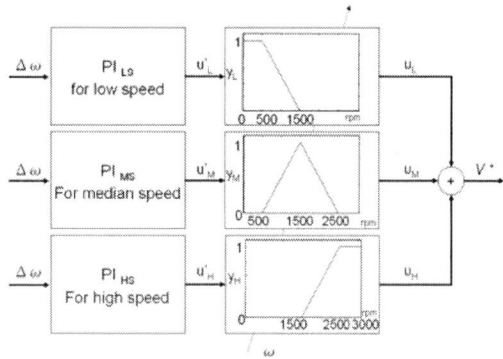

Fig 6. A fuzzy gain scheduling PI control.

A. Fuzzication

For reduction of MCU cost in implementation, a simply fussy rule is proposed in this paper. Due to focus on variable sampling, so the fuzzication is based on speed. The input membership functions of the fuzzy sets are trapezoidal and triangular exception as shown in Fig. 7, and label "LS", "MS", and "HS" are represent "low-", "median-", and "high-speed".

The membership function of "LS", "MS", and "HS" is showed in (1.1), (1.2) and (1.3).

$$y_L(\omega_r) = \begin{cases} 1, & \text{for} \quad \omega_r < 500 \\ \dfrac{1500 - \omega_r}{1000}, & \text{for} \quad 500 \le \omega_r \le 1500 \\ 0, & \text{for} \quad \omega_r > 1500 \end{cases} \tag{1.1}$$

$$y_M(\omega_r) = \begin{cases} 0, & \text{for} \quad \omega < 500 \\ \dfrac{\omega_r - 500}{1000} & \text{for} \quad 500 \le \omega_r \le 1500 \\ \dfrac{2500 - \omega_r}{1000} & \text{for} \quad 1500 \le \omega_r \le 2500 \\ 0, & \text{for} \quad \omega_r > 2500 \end{cases} \tag{1.2}$$

$$y_H(\omega_r) = \begin{cases} 0, & \text{for} \quad \omega_r < 1500 \\ \dfrac{\omega_r - 1500}{1000}, & \text{for} \quad 1500 \le \omega_r \le 2500 \\ 1, & \text{for} \quad \omega_r > 2500 \end{cases} \tag{1.3}$$

$$y_L + y_M + y_H = 1 \tag{1.4}$$

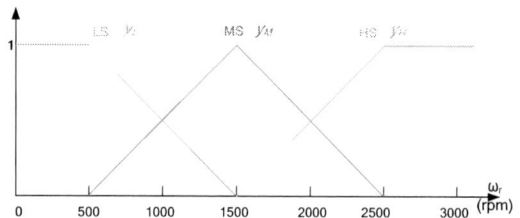

Fig 7. Membership of fuzzy logic for different speed.

B. Fuzzy control rule

To simplify implementation, the complex control rules are skipped as shown in Fig. 7. The errors of speed directly feed to three fixed PI controllers to get controller out.

C. Defuzzication

Weighted average is used in defuzzication as (1) to effortlessly scale the control output coming from three different PI controller which are best for Low-, Median- and High-Speed sampling time intervals. The scaling is also based on dynamic speed of motor and create the most suitable control variable V^*.

$$V^* = \frac{y_L \cdot u_L + y_M \cdot u_M + y_H \cdot u_H}{y_L + y_M + y_H} \tag{2}$$

IV. EXPERIMENTAL RESULTS

The three sets of PI parameters are selected to get best responses in different speed command and list in Table 1. The block diagram of experimental system is Fig 8, a low pass filter (LPF) is used to filter noise before phase voltage send to DSP. The ratings of the motor are described in Table 2 and the controller is implemented on a TI DSP TMS320F243 digital signal processor.

978-1-4244-4782-4/10 $26.00 © 2010 IEEE

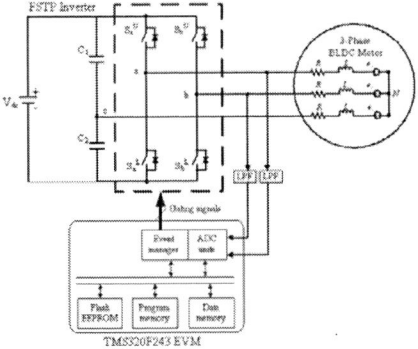

Fig 8. Block diagram of experimental.

Table 1. The parameters of Kp & Ki in PI_{HS}, PI_{MS} & PI_{LS}

Low	K_P	0.6875	Median	K_P	0.9375	High	K_P	0.2539
Speed	K_I	1	Speed	K_I	1.1718	Speed	K_I	0.8828

Table 2. Ratings spec. of the experimental motor

Specifications	Units
Number of poles	4
Moment of inertia, J	0.000195982 Kg-m^2
Viscous damping constant, D	0.000001 N-m-s/rad
Torque constant, K_t	0.3484 V-s/rad
Back-EMF constant, K_e	0.3484 V-s/rad
Armature inductance, L	24 mH
Terminal resistance, R	5.17 Ω

The peripherals of DSP are used to implement the system: 1.ADC unit is used to get the voltages of A & B phases and decide commutation & speed. 2. PWM unit is applied to create the corresponding PWM waveforms according to the control command specified by the fuzzy gain scheduling PI controller.

Due to sensorless scheme, the motor needs a startup mode before into sensorless mode of 750 rpm: longer commutations and higher PWM duty are used to deriver motor before controller can get clear voltage crossing of phase A and B.

The experimental results are shows in Fig.9: In sensorless mode, controller gets two step speed commands, 1000 and 2000 rpm. Based on fast response and lowest overshoot, the fuzzy gain scheduling PI controller shows the better performances than other three fixed PI controller.

The experimental results are enlarged to show in Fig 10. From (a), in the low- and mid-speed state, the fuzzy gain scheduling PI controller has lowest overshoot and response speed is similar to other control method. (b) shows the fuzzy gain scheduling PI controller is a better method, Due to the lowest overshoot and fast response in the high-speed state. So,

the fuzzy gain scheduling PI controller could run in the whole speed situation

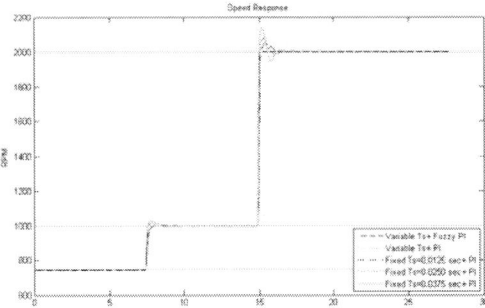

Fig 9. Speed Responses in 1000 & 2000 rpm step commands.

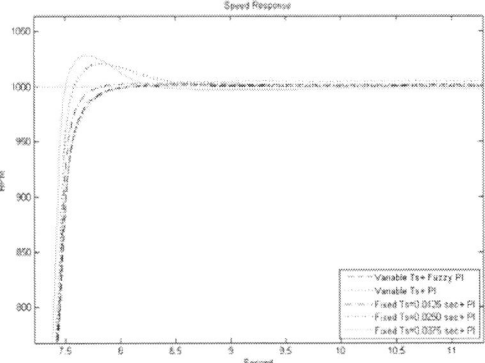

(a)Speed Responses in 1000 rpm step command

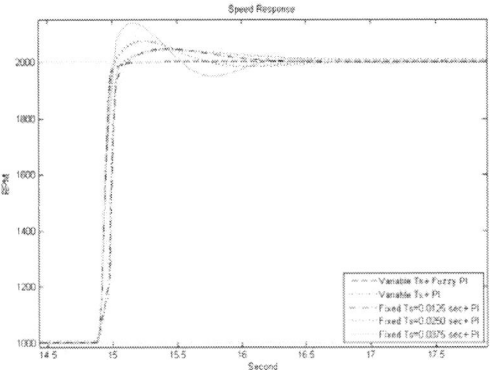

(b) Speed Responses in 2000 rpm step command
Fig 10. Speed Responses in 1000 & 2000 rpm step command

V. CONCLUSIONS

A sensorless FSTP BLDC motors driver based on six-space-vector is described in this paper, what is a strategy of cost down. Due to low resolution of position sensing scheme, the speed variable sampling effect is rising. A fuzzy gain scheduling PI controller is proposed, three PI controllers separately for different sampling time intervals in low-, median- and high-speed, are combined simply to create a workable control output based on fuzzy logic. The experimental results show that the fuzzy-gain scheduling controller proposed in this paper is workable and robust in variable-sampling situation.

Acknowledgment

This work is partly supported by the National Science Council, Taiwan, under contract NSC 98-2218-E-224-007.

REFERENCES

[1] C. B. Jacobina, E. R. C. da Silva, A. M. N. Lima, and R. L. A. Ribeiro, Vector and scalar control of a four switch three phase inverter," in *Proc. IEEE Ind. Appl. Conf., 1995*, vol. 3, pp. 2422–2429.

[2] M. Azab and A. L. Orille, "Novel flux and torque control of induction motor drive using four switch three phase inverter," in *Proc. IEEE Annu. Conf. Ind. Electron. Soc., 2001*, vol. 2, pp. 1268–1273.

[3] Z. Jiang, D. Xu, and Z. Xiangjuan, "A study of the four-switch low cost inverter that uses the magnetic flux control method," in *Proc. IEEE Power Electron. Motion Control Conf., 2004*, vol. 3, pp. 1368–1371.

[4] J.-H. Lee, S.-C: Ahn, and D.-S. Hyun, "A BLDCM drive with trapezoidal back EMF using four-switch three phase inverter," in *Proc. IEEE Ind. Appl., 2000*, vol. 3, pp. 1705–1709.

[5] B.-K. Lee, T.-H. Kim, and M. Ehsani, "On the feasibility of four-switch three-phase BLDC motor drives for low cost commercial applications: Topology and control," *IEEE Trans. Power Electron.*, vol. 8, no. 1, pt. 1, pp. 164–172, Jan. 2003.

[6] M. B. de Rossiter Corrêa, C. B. Jacobina, E. R. C. da Silva, and A. M. N. Lim, "A general PWM strategy for four-switch three-phase inverters," *IEEE Trans. Power Electron.*, vol. 21, no. 6, pp. 1618–1627, Nov. 2006.

[7] R.-L. Lin, M.-T. Hu, S.-C. Chen, and C.-Y. Lee, "Using phase-current sensing circuit as the position sensor for brushless dc motors without shaft position sensor," in *Proc. IEEE Annu. Conf. Ind. Electron. Soc., 1989*, vol. 1, pp. 215–218.

[8] J. P. Johnson, M. Ehsani, and Y. Guzelgunler, "Review of sensorless methods for brushless DC," in *Proc. IEEE Ind. Appl., 1999*, vol. 1, pp. 143–150.

[9] J. P. Johnson and M. Ehsani, "Sensorless brushless dc control using a current waveform anomaly," in *Proc. IEEE Ind. Appl., 1999*, vol. 1, pp. 151–158.

[10] J. Shao and D. Nolan, "Further improvement of direct back EMF detection for sensorless brushless dc (BLDC) motor drives," in *Proc. IEEE Appl. Power Electron. Conf. Expo, 2005*, vol. 2, pp. 933–937.

[11] S. Ogasawara and H. Akagi, "An approach to position sensorless drive for brushless dc motors," *IEEE Trans. Ind. Appl.*, vol. 27, no. 5, pp. 928–933, Sep. 1991.

[12] Cheng-Tsung Lin, Chung-Wen Hung, and Chih-Wen Liu, "Position Sensorless Control for Four-Switch Three-Phase Brushless DC Motor Drives," *IEEE Tran. on Power Electronic.*, vol. 23, no. 1, pp. 438-444, Jan. 2008

[13] Chung-Wen Hung, Cheng-Tsung Lin, and Chih-Wen Liu, "An Efficient Simulation Technique for the Variable Sampling Effect of BLDG Motor Applications," in *Proc. IEEE Annu. Conf. Ind. Electron. Soc.,2007*, pp. P.1175 – 1179, Taipei, Taiwan

Equivalent EMF Based Position Observers for Sensorless Synchronous Machines

Jingbo Liu, Thomas Nondahl, and Peter Schmidt
Rockwell Automation
1201 South Second St
Milwaukee, WI 53204, USA
jliu2@ra.rockwell.com, tanondahl@ra.rockwell.com,
pbschmidt@ra.rockwell.com

Semyon Royak and Mark Harbaugh
Rockwell Automation
1 Allen-Bradley Drive
Mayfield Heights, OH 44124, USA
sroyak@ra.rockwell.com, mmharbaugh@ra.rockwell.com

Abstract—**This paper provides a simple, robust, and universal position observer for position sensorless control of synchronous machines. The observer is designed using an equivalent EMF model of a synchronous machine or, alternately, using a sliding mode controller based on the equivalent EMF model. The sliding mode observer provides fast convergence and low sensitivity to parameter variations. Experimental results with a 5-hp Interior Permanent Magnet (IPM) machine have validated the effectiveness of the proposed equivalent EMF theory.**

I. INTRODUCTION

Research on sensorless control strategies for synchronous machines has gained much attention because of several disadvantages of using a position sensor in a drive system, such as added cost, reliability problems, increased maintenance requirements and need of shaft extension and mounting arrangements, etc [1]-[6]. The most commonly used sensorless control methods can be divided into two strategies: (1) back electromotive force (EMF) method; and (2) high frequency injection (HFI). The HFI method takes advantage of the anisotropic properties of the machines, which makes it a viable sensorless scheme for synchronous machines at low speeds including zero speed. Unfortunately, the HFI method should only be used in low speed ranges because of extra losses, transient disturbances, and limited control bandwidth, etc. On the other hand, back EMF based sensorless methods fail at low and zero speeds because the rotor position estimation fundamentally relies on back-EMF or speed dependent voltages. Therefore the back-EMF based position observers are generally used for machines operating within the medium and high speed ranges.

These EMF observers may not be suitable for all types of synchronous motors. For example, some observers may require that the d axis and q axis inductance values be equal, making them suitable only for non-salient pole synchronous machines [2]. Other models may make assumptions that are not valid for all operating conditions. For example, a commonly mentioned EMF based position estimation method for salient synchronous machines is called extended EMF [4]-[5] The extended EMF model assumes that the estimated speed error and the rate of change of the extended EMF are both zero. While such assumptions may be valid under constant operating conditions, the assumptions are not always valid during transient conditions.

This paper provides a simple and universal equivalent EMF position observer for sensorless synchronous machines. The equivalent EMF observer does not need machine velocity as an observer input and it can be used with any type of synchronous machine.

Furthermore, researchers have shown increasing interest in sliding mode observers (SMO) for sensorless control of electric machines due to their attractive features such as robustness, order reduction, fast convergence, low sensitivity to disturbance and parameter variations etc. [6]-[8]. Most SMOs in the literature are for non salient PMSMs. Very few SMOs are designed for salient synchronous machines and those have very complicated mathematics [6],[8]. Based on the dynamic models utilizing equivalent EMF, a novel SMO is proposed in this paper that can be utilized for any type of synchronous machines, salient or non-salient.

II. REVIEW OF EMF-BASED SENSORLESS METHODS FOR SYNCHRONOUSE MACHINES

Fig. 1 shows the basic structure of a salient 2-pole PM machine. The α-axis is aligned to the stator phase a axis. The d-q frame corresponds to the synchronously rotating reference frame. The d-axis is aligned with the N-pole of the rotor and the q-axis is 90 degrees apart from the d-axis.

A salient machine model in synchronous rotating reference (d-q) frame can be expressed by

$$\begin{bmatrix} v_d \\ v_q \end{bmatrix} = \begin{bmatrix} R + \dfrac{d}{dt}L_d & -\omega L_q \\ \omega L_d & R + \dfrac{d}{dt}L_q \end{bmatrix} \cdot \begin{bmatrix} i_d \\ i_q \end{bmatrix} + \omega \lambda_{pm} \begin{bmatrix} 0 \\ 1 \end{bmatrix} \qquad (1)$$

978-1-4244-4782-4/10 $26.00 © 2010 IEEE

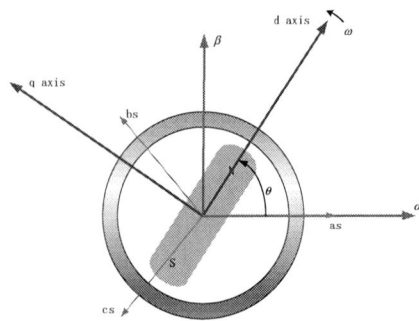

Figure 1. Salient PM synchronous machine

where

v_d and v_q are the d- and q-axis stator voltages;

i_d and i_q are the d- and q-axis stator currents;

R is the stator resistance;

L_d and L_q are d- and q-axis stator self inductances;

λ_{pm} is per-phase permanent magnet flux linkage.

However, the d-q model cannot be utilized in a position sensorless control system because the rotor position is not detected. Transforming the d-q model to the stationary reference frame yields equations for a salient-pole synchronous machine in the α-β coordinates.

$$\begin{bmatrix} v_\alpha \\ v_\beta \end{bmatrix} = \begin{bmatrix} R + \frac{d}{dt}(L_0 + L_1\cos 2\theta) & \frac{d}{dt}L_1\sin 2\theta \\ \frac{d}{dt}L_1\sin 2\theta & R + \frac{d}{dt}(L_0 - L_1\cos 2\theta) \end{bmatrix} \cdot \begin{bmatrix} i_\alpha \\ i_\beta \end{bmatrix} + \omega\lambda_{pm}\begin{bmatrix} -\sin\theta \\ \cos\theta \end{bmatrix} \quad (2)$$

where

$$L_0 = \frac{L_q + L_d}{2} \quad (3)$$

and

$$L_1 = \frac{L_d - L_q}{2} \quad (4)$$

The first term on the right side of (2) contains unknown variables 2θ caused by machine saliency. That makes the mathematical model of a salient PM machine much more complicated than its non-salient counterpart.

A. EMF observer for Non Salient SPMSM

Equations for a non-salient SPMSM in stationary (α-β) reference frame are [2]:

$$\begin{bmatrix} v_\alpha \\ v_\beta \end{bmatrix} = \begin{bmatrix} R + L\frac{d}{dt} & 0 \\ 0 & R + L\frac{d}{dt} \end{bmatrix} \begin{bmatrix} i_\alpha \\ i_\beta \end{bmatrix} + \omega\lambda_{pm}\begin{bmatrix} -\sin\theta \\ \cos\theta \end{bmatrix} \quad (5)$$

where $L_d = L_q = L$.

The first term on the right side of (5) contains known variables. The only unknown variables are in the EMF term (the second term) that contains the rotor position. If we define

$$\lambda_\alpha = -i_\alpha \cdot L + \int(v_\alpha - i_\alpha \cdot R) \quad (6)$$

$$\lambda_\beta = -i_\beta \cdot L + \int(v_\beta - i_\beta \cdot R) \quad (7)$$

then

$$\hat{\theta} = \tan^{-1}\left(\frac{\lambda_\beta}{\lambda_\alpha}\right) \quad (8)$$

Apparently, this model can not be used with salient synchronous machines.

B. Extended EMF Observer

The most commonly mentioned EMF based position estimation method for a salient PM machine is called extended EMF [4]-[5]. The d-q voltage (1) is re-arranged to:

$$\begin{bmatrix} v_d \\ v_q \end{bmatrix} = \begin{bmatrix} R + \frac{d}{dt}L_d & -\omega L_q \\ \omega L_q & R + \frac{d}{dt}L_d \end{bmatrix} \cdot \begin{bmatrix} i_d \\ i_q \end{bmatrix} + \begin{bmatrix} 0 \\ E_{ext} \end{bmatrix} \quad (9)$$

Similarly, the first term on the right side of (9) contains known variables. The second term is defined as the extended EMF term in [4]-[5]:

$$E_{ext} = \omega \cdot [\lambda_{pm} + (L_d - L_q) \cdot i_d] - (L_d - L_q)\frac{d}{dt}i_q \quad (10)$$

Transforming (10) into γ-δ reference frame (γ-δ frame is defined as an estimated reference frame in [4]) gives

$$\begin{bmatrix} \hat{e}_\gamma \\ \hat{e}_\delta \end{bmatrix} = \begin{bmatrix} R + L_d \cdot \frac{d}{dt} & -\omega \cdot L_q \\ \omega \cdot L_q & R + L_d \cdot \frac{d}{dt} \end{bmatrix} \begin{bmatrix} i_\gamma \\ i_\delta \end{bmatrix} + E_{ex}\begin{bmatrix} -\sin\theta_e \\ \cos\theta_e \end{bmatrix} + (\hat{\omega} - \omega)L_d\begin{bmatrix} -i_\delta \\ i_\gamma \end{bmatrix} \quad (11)$$

We can see that the first term on the right side of (11) contains known variables while the second term contains an extended EMF that can be used to extract rotor position. Note that the third term, the error between estimated speed and actual speed, is assumed to be zero.

There are several inherent limitations of the extended EMF method. The assumptions of the extended EMF are not always good for all operating conditions. The assumption that speed error is zero is almost always correct. However, the assumption that the rate of change of the extended EMF is zero is not always right. In fact those derivative terms could be very large. Additionally, this method requires machine velocity as an input which is also undesirable because the velocity is unknown and has to be calculated via the estimated position.

III. EQUIVALENT EMF THEORY

Rearrange (2) to

$$v_\alpha = R \cdot i_\alpha + \frac{d}{dt}[(L_0 - L_1 + L_1\cos 2\theta) \cdot i_\alpha] + \frac{d}{dt}[L_1 \cdot i_\alpha] + \frac{d}{dt}[L_1\sin 2\theta \cdot i_\beta] - \omega\lambda_{pm}\sin\theta \quad (12)$$

$$v_\beta = R \cdot i_\beta + \frac{d}{dt}[(L_0 - L_1 - L_1\cos 2\theta) \cdot i_\beta] + \frac{d}{dt}[L_1 \cdot i_\beta] + \frac{d}{dt}[L_1\sin 2\theta \cdot i_\alpha] + \omega\lambda_{pm}\cos\theta \quad (13)$$

Defining position dependent variables gives

$$v_\alpha = R \cdot i_\alpha + \frac{d}{dt} L_q i_\alpha + \frac{d}{dt} \lambda_\alpha{}' \tag{14}$$

and

$$v_\beta = R \cdot i_\beta + \frac{d}{dt} L_q i_\beta + \frac{d}{dt} \lambda_\beta{}' \tag{15}$$

where

$$\frac{d}{dt} \lambda_\alpha{}' = \frac{d}{dt}(L_1 \cos 2\theta \cdot i_\alpha) + \frac{d}{dt}(L_1 \cdot i_\alpha) + \frac{d}{dt}(L_1 \sin 2\theta \cdot i_\beta) - \omega \lambda_{pm} \sin \theta \tag{16}$$

$$\frac{d}{dt} \lambda_\beta{}' = \frac{d}{dt}(-L_1 \cdot \cos 2\theta \cdot i_\beta) + \frac{d}{dt}(L_1 \cdot i_\beta) + \frac{d}{dt}(L_1 \sin 2\theta \cdot i_\alpha) + \omega \lambda_{pm} \cos \theta \tag{17}$$

Note that

$$i_\alpha = i_d \cos \theta - i_q \sin \theta \tag{18}$$

$$i_\beta = i_d \sin \theta + i_q \cos \theta \tag{19}$$

We define the equivalent EMF terms as

$$e_\alpha{}' = \frac{d}{dt} \lambda_\alpha{}' = -(2L_1 \cdot i_d \omega + \omega \lambda_{pm}) \sin \theta \tag{20}$$

$$e_\beta{}' = \frac{d}{dt} \lambda_\beta{}' = (2L_1 \cdot i_d \omega + \omega \lambda_{pm}) \cos \theta \tag{21}$$

Integrating both sides of (20) and (21) with respect to time and assuming that machine velocity changes slowly over a sampling period ($\dot\omega \approx 0$), we obtain

$$\lambda_\alpha{}' = (2L_1 \cdot i_d + \lambda_{pm}) \cdot \cos \theta \tag{22}$$

$$\lambda_\beta{}' = (2L_1 \cdot i_d + \lambda_{pm}) \cdot \sin \theta \tag{23}$$

Taking the arctangent of $\lambda_\beta{}' / \lambda_\alpha{}'$ will give θ. Note that the assumption that the machine velocity changes slowly over a sampling period can be easily met for all operating conditions.

The term $2L_1 i_d + \lambda_{pm}$ should be nonzero for salient synchronous machines such that equivalent EMF exists, i.e.

$$i_d \neq \frac{\lambda_{pm}}{I_q - I_d} \tag{24}$$

In fact for IPMSM ($L_d < L_q$), (24) will never be a physical operation point because i_d is always commanded to be zero or negative.

In short, equations for a synchronous machine in the stationary (α-β) reference frame can be expressed using the equivalent EMF, i.e

$$\begin{bmatrix} v_\alpha \\ v_\beta \end{bmatrix} = \begin{bmatrix} R + \dfrac{d}{dt} L_q & 0 \\ 0 & R + \dfrac{d}{dt} L_q \end{bmatrix} \begin{bmatrix} i_\alpha \\ i_\beta \end{bmatrix} + \begin{bmatrix} \dfrac{d}{dt} \lambda_\alpha{}' \\ \dfrac{d}{dt} \lambda_\beta{}' \end{bmatrix} \tag{25}$$

The estimated position is

$$\bar\theta = \tan^{-1} \left(\frac{\lambda_\beta{}'}{\lambda_\alpha{}'} \right) \tag{26}$$

The relationship of the equivalent EMF and the actual EMF can be clearly explained by a phasor diagram (see Fig.

2). The equivalent EMF vector $\bar{E}'$ is always aligned with the actual EMF vector $\bar{E}$. Consequently one may utilize the equivalent EMF to extract the rotor position. The difference between the two EMF vectors is caused by machine saliency ($j\omega(L_d - L_q)\bar{I}_d$). The equivalent EMF vector will be the same as the actual EMF vector under two circumstances: (1) non-salient PMSM; and (2) i_d is zero.

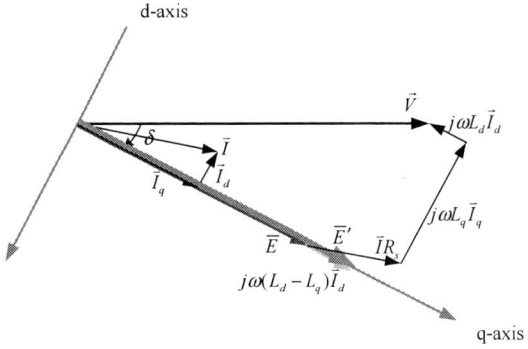

Figure 2. Phasor diagram of a synchronous machine

The actual EMF vector can be expressed as

$$\bar{E} = \bar{E}' - j\bar{I}_d(X_d - X_q) = \omega \lambda_{pm} \tag{27}$$

Equation (27) can also be utilized to estimate the permanent flux linkage λ_{pm} for a PM synchronous machine based on the equivalent EMF model.

TABLE I. COMPARISON

Features	EMF observer for non-salient machine	Extended EMF observer	Equivalent EMF observer
Voltage equations	Eq (5)	Eq (9)	Eq (25)
Format	-Simple -Integrator term	-Generic format for all types of synchronous machines -Derivatite terms -Velocity as input	-Generic format for all types of synchronous machines -Integrator term
Assumptions	Can be easily met for all operation modes	May not be valid for all operation modes	Can be easily met for all operation modes
Input parameters	L,R	L_d, L_q, ω, R	R, L_q
Limitations	-Sensitive to machine parameter variations -Only for non-salient PMSM	-Might not work well for transients -Sensitive to machine parameter variations	-Sensitive to machine parameter variations - Eq (24)

Table I shows the comparison among three EMF methods. As a summary, the advantages of the proposed equivalent EMF observer are: (1) simple and generic equations for all types of synchronous machines; (2) only two machine parameters (L_q and R) are needed; (3) velocity is not needed as

978-1-4244-4782-4/10 $26.00 © 2010 IEEE

an observer input; (4) the assumption ($\dot{\omega} \approx 0$) can be easily met; (5) works well for both steady state and transients.

The equivalent EMF model can be used by any type of synchronous motor: (1) Surface PMSM ($L_d = L_q$); (2) Wound rotor synchronous machine ($L_d > L_q$); (3) Interior PMSM ($L_d < L_q$); (4) Synchronous reluctance machine ($\lambda_{PM} = 0$).

IV. SLIDING MODE OBSERVER BASED ON EQUIVALENT EMF

The sliding mode control theory is an effective nonlinear robust control method providing system dynamics with an invariance property to uncertainties if sliding mode occurs [7]-[8]. The first step of a design of a sliding mode controller is to select a sliding surface that models the desired closed-loop performance in state variable space. Then, design the control such that the system state trajectories are forced toward the sliding surface. Once the system trajectory reaches the sliding surface, it stays on it and slides along towards the origin. The system trajectory sliding along the sliding surface to the origin is the sliding mode. A sliding mode observer is generally designed with inputs as discontinuous functions of the error between the estimated and measured outputs [7]-[8]. By enforcing the sliding mode, the system states can be reconstructed.

Let us design a sliding mode observer by

$$\begin{bmatrix} \dot{\hat{i}}_\alpha \\ \dot{\hat{i}}_\beta \end{bmatrix} = \frac{1}{L_q}\begin{bmatrix} v_\alpha \\ v_\beta \end{bmatrix} - \frac{R}{L_q}\begin{bmatrix} \hat{i}_\alpha \\ \hat{i}_\beta \end{bmatrix} - \frac{K_{SM}}{L_q}\begin{bmatrix} sign(\hat{i}_\alpha - i_\alpha) \\ sign(\hat{i}_\beta - i_\beta) \end{bmatrix} \quad (28)$$

where:

K_{SM} is a constant observer gain;

$\hat{i}_\alpha$ is the observed α-axis current;

$\hat{i}_\beta$ is the observed β-axis current.

Subtracting the model equation (25) from (28) yields the error dynamics along the sliding surface

$$\begin{bmatrix} \dot{\bar{i}}_\alpha \\ \dot{\bar{i}}_\beta \end{bmatrix} = -\frac{R}{L_q}\begin{bmatrix} \bar{i}_\alpha \\ \bar{i}_\beta \end{bmatrix} + \frac{1}{L_q}\begin{bmatrix} e'_\alpha \\ e'_\beta \end{bmatrix} - \frac{K_{SM}}{L_q}\begin{bmatrix} sign(\bar{i}_\alpha) \\ sign(\bar{i}_\beta) \end{bmatrix} \quad (29)$$

where observation errors are

$$\bar{i}_\alpha = \hat{i}_\alpha - i_\alpha \quad (30)$$

$$\bar{i}_\beta = \hat{i}_\beta - i_\beta \quad (31)$$

Let us select the Lyapunov function as.

$$V = \frac{1}{2}(\bar{i}_\alpha^{\,2} + \bar{i}_\beta^{\,2}) \quad (32)$$

then

$$\dot{V} = \bar{i}_\alpha \cdot \dot{\bar{i}}_\alpha + \bar{i}_\beta \cdot \dot{\bar{i}}_\beta = -\frac{R}{L_q}(\bar{i}_\alpha^{\,2} + \bar{i}_\beta^{\,2}) + \frac{1}{L_q}(e'_\alpha \cdot \bar{i}_\alpha + e'_\beta \cdot \bar{i}_\beta) - \frac{K_{SM}}{L_q}(|\bar{i}_\alpha| + |\bar{i}_\beta|) \quad (33)$$

Equation (33) shows that if K_{SM} is large enough, i.e

$$K_{SM} > \max\{|e'_\alpha|, |e'_\beta|\} \quad (34)$$

Then $\dot{V} < 0$ is always guaranteed until $\bar{i}_\alpha = 0$ and $\bar{i}_\beta = 0$. Note that equivalent EMF components are bounded. Once the system trajectory reaches the sliding surface, it stays on it and slides along it to the origin. The estimated current will converge to the actual current when sliding mode occurs with the state trajectory confined to the sliding manifold after a finite time interval.

The sliding mode forces convergence of the observed current values to the measured current values. In order to force convergence of the observed current values with the measured current values, the desired error values between the observed and actual current values in the stationary reference frame, $\bar{i}_\alpha$ and $\bar{i}_\beta$, are set to zero and an equivalent control method [7], as shown in Fig. 3, is applied. By setting the error values, $\bar{i}_\alpha$ and $\bar{i}_\beta$, to zero, an expression for the equivalent EMF is obtained, according to (35) and (36).

$$[K_{SM} sign(\bar{i}_\alpha)]_{eq} = e'_\alpha \quad (35)$$

$$[K_{SM} sign(\bar{i}_\beta)]_{eq} = e'_\beta \quad (36)$$

Low pass filters are utilized to extract e'_α and e'_β for the position estimation. Finally, the rotor position is obtained by

$$\hat{\theta} = -\tan^{-1}\left(\frac{e'_\alpha}{e'_\beta}\right) = -\tan^{-1}\left(\frac{[K_{SM} sign(\bar{i}_\alpha)]_{eq}}{[K_{SM} sign(\bar{i}_\beta)]_{eq}}\right) \quad (37)$$

Fig.3 shows a block diagram of the implementation of the proposed SMO. Preferably, the observed motor current from the sliding mode model is in a stationary reference frame, having an alpha and a beta component. The difference between the observed and actual motor currents for each of the alpha and beta components is passed through a sign function. A low pass filter is then applied to the output of the sign function to obtain the equivalent EMF values, e'α and e'β, in the stationary reference frame. The rotor position is finally estimated by determining the negative of the arctangent of the alpha component of the equivalent EMF value over the beta component as shown in the figure.

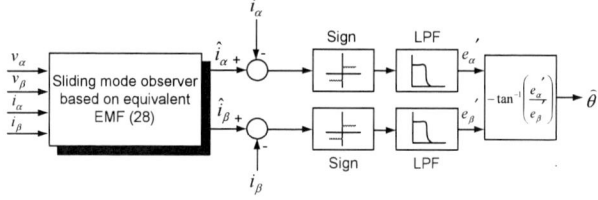

Figure 3. Implementation of a SMO based on equivalent EMF

V. EXPERIMENTAL RESULTS

The test fixture of a 5-hp IPM motor coupled to a DC load machine is shown in Fig. 4. The control algorithm is built within the MATLAB/Simulink environment combined with the real-time interface provided by dSPACE. A 7.5 hp AC drive is utilized as the power structure. Three phase currents and DC bus voltage are sensed and sent to the DS2004 A/D board for the feedback control. Gate signals are produced via the DS5101 digital waveform board and sent to the interface board of the power structure.

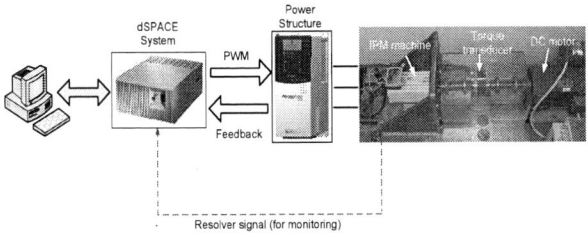

Figure 4. Test fixture

It should be noted the resolver signals (position & velocity) are for display only and were not used by any closed loop control. In all of the test results shown in this section, the estimated position and velocity are utilized to close the current and velocity loops. The parameters of the 5-hp IPM machine are shown in Table II.

TABLE II. 5-HP IPM MACHINE PARAMETERS

Parameter	Value (unit)
Base speed	1800 (rpm)
Rated torque	20 (N.m)
Rated voltage	460 (V,line-line)
Rated current	5.8 (A, rms)
Stator resistance	1.75 (ohm)
L_d (no-load)	0.033 (H)
L_q (no-load)	0.127 (H)
λ_{pm}	0.66 (wb)
Poles	4

As is well known, EMF based methods fail at low and zero speeds. During tests the 5-hp IPM motor was started using high frequency injection (HFI) and then switched to an equivalent EMF based position observer at about 15% of the machine base speed. The principle of the HFI at low speed is independent of the EMF-based observers. It is therefore omitted from this paper. A 1st order HPF with 3 Hz cut off frequency was utilized to remove the DC offsets caused by the integrator.

Fig. 5 shows the sensorless control diagram utilizing the proposed equivalent EMF observers. The stationary reference frame current and voltage signals are inputs to the position observer. The current regulator outputs a voltage reference in the synchronous reference frame which is transformed back into the stationary reference frame. A reference frame transformation generates the d-axis and q-axis current feedback signals (i_d and i_q) for use by the current regulators.

A. Experimental Results with Equivalent EMF Observer

Fig. 6 shows the experimental results using the equivalent EMF observer to close the current loops (torque mode) with $i_q{}^*$ =8A and $i_d{}^*$= 0A. The IPM machine was started firstly and

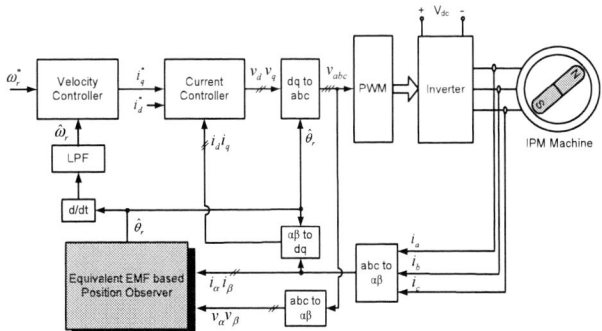

Figure 5. Control block diagram using equivalent EMF observers

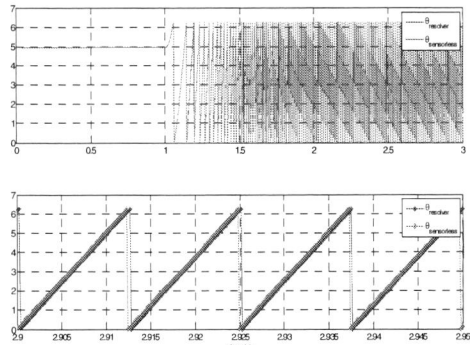

Figure 6. Experimental results in torque mode (no load) using the equivalent EMF observer with $i_d{}^*$ =0A & $i_q{}^*$ =8A (blue trace: resolver position 1 rad/div; red trace: estimated position 1 rad/div)

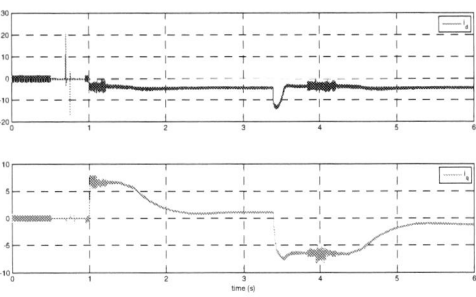

Figure 7. Experimental results in torque mode (no load) with $i_d{}^*$ = -4.3A & $i_q{}^*$ changed from +7A to -7A (top trace: i_d feedback; bottom trace: i_q feedback

then switched to the estimated position calculated by the equivalent EMF observer. The results in Fig.6 show that the estimated position matches the actual position well.

Figs. 7 & 8 show the experimental results using the equivalent EMF observer to close the current loops (torque mode) when the i_q command was changed from +7A to -7A with the i_d command -4.3A (corresponding maximum torque per ampere current commands at the rated phase current of 8.2A). Note that the IPM machine velocity is over 300 rpm when the proposed equivalent EMF observer was used to

978-1-4244-4782-4/10 $26.00 © 2010 IEEE

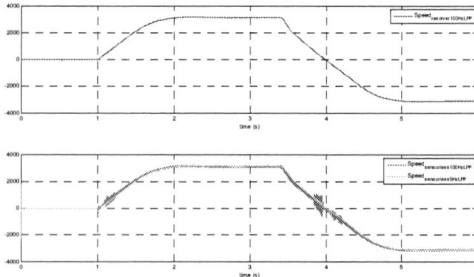

Figure 8. Experimental results in torque mode (no load) with $i_d{}^* = -4.3A$ & $i_q{}^*$ from +7A to -7A (top trace in blue: resolver velocity after a 100Hz LPF; bottom red trace: estimated velocity after a 100Hz LPF; bottom green trace: estimated velocity after a 5Hz LPF)

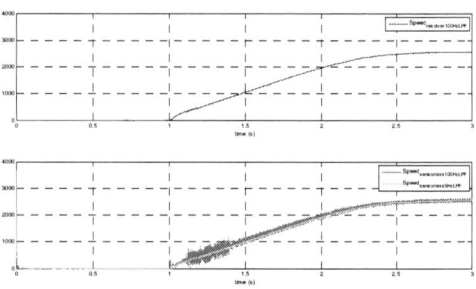

Figure 9. Experimental results in torque mode with 10 N.m load with $i_d{}^* = -4.3A$ & $i_q{}^*$ +7A (top trace in blue: resolver velocity after a 100Hz LPF; bottom red trace: estimated velocity after a 100Hz LPF; bottom green trace: estimated velocity after a 5Hz LPF)

close the control loops. The estimated velocity after a first order 100 Hz LPF (bottom trace in red in Fig. 7) and a first order 5 Hz LPF (bottom trace in green in Fig. 7) are plotted for display purposes. Note that the injected high frequency voltages for the low speed sensorless algorithm were completely turned off when the machine velocity is over 800 rpm.

Fig. 9 shows the experimental results using the equivalent EMF observer to close the current loops (torque mode) with $i_q{}^* = 7A$ and $i_d{}^* = -4.3A$ with 10 N.m load applied on the shaft.

Fig. 10 shows the experimental results when the IPM machine was controlled in velocity mode (using estimated position to close the current loops and the estimated velocity to close the velocity loop) with no load. The velocity command is 3000 rpm.

Fig. 11 shows the experimental results when the IPM machine was controlled in velocity mode (using estimated position to close the current loops and estimated velocity to close the velocity loop) at 1000 rpm with 16 N.m load. The estimated flux linkages by two different models are also shown in Fig. 11. One is with equivalent EMF model by (27) and the other one with the d-q model by (1). The estimated flux linkages also agree well with the machine nameplate data.

The results in Figs. 5 through 11 have verified that the proposed equivalent estimated position works very well.

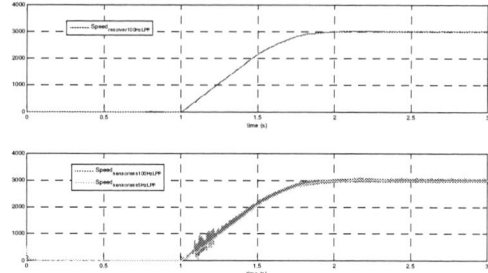

Figure 10. Experimental results in velocity mode with a velocity command at 3000 rpm, no load (top trace in blue: resolver velocity after a 100Hz LPF; bottom red trace: estimated velocity after a 100Hz LPF; bottom green trace: estimated velocity after a 5Hz LPF)

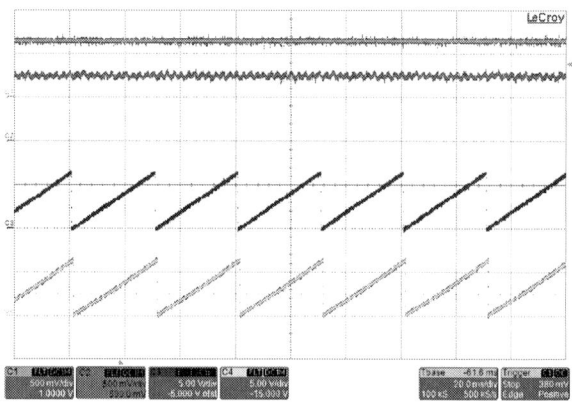

Figure 11. Experimental results in velocity mode at 1000 rpm (steady state) with 16 N.m load (Ch1 estimated flux linkage λ_{EMF} by Eq (27), 0.5 wb/div; Ch2: estimated flux linkage λ_{d-q} by d-q model Eq (1) 0.5 wb/div; Ch3: estimated position by equivalent EMF observer θ_{EMF} 5 rad/div; Ch4: resolver position $\theta_{resolver}$ 5 rad/div

B. Experimental Results with SMO

We tested with both position observers: (1) the equivalent EMF observer and (2) the sliding mode observer based on equivalent EMF. As an attempt to compare the performance of the two position observers, we tested with both observers at very low speeds when the EMF observer generally does not work well. During tests, the IPM motor was turned by a DC motor at various speeds. The IPM motor was controlled in torque mode while both $i_d{}^*$ and $i_q{}^*$ set to zero. The sliding mode gain K_{SM} was set at 3000. Figs. 12 & 13 show the test results in torque mode when the IPM machine was running at 20 rpm with the equivalent EMF observer and SMO, respectively. Figs. 14 & 15 show the test results when the IPM machine was running at 50 rpm.

It is clear that the SMO shows improved performance as compared to the equivalent EMF observer when the machine was operating at low speeds. In fact, the equivalent EMF observer takes much longer time to converge to the actual position due the integrator and the use of a HPF to remove the DC offset caused by the pure integrator. On the other hand, the sliding mode position observer converges much faster,

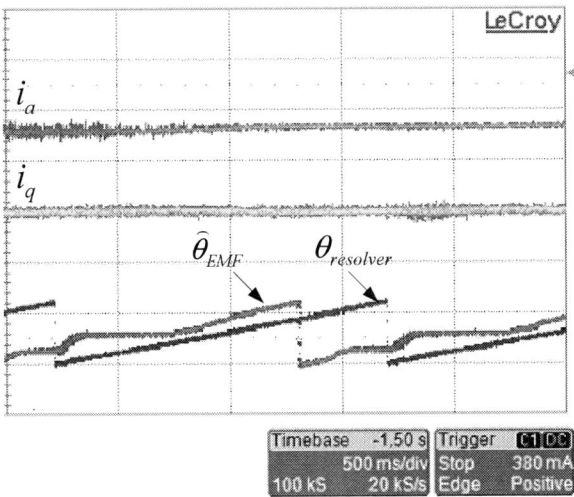

Figure 12. Sensorless operation in torque mode @ 20 rpm using the equivalent EMF observer to close the current loop. Traces from top to bottom are: phase a current i_u(1A/div); i_q feedback (1A/div); estimated position by equivalent EMF observer θ_{EMF} (5 rad/div); and $\theta_{resolver}$ (5 rad/div).

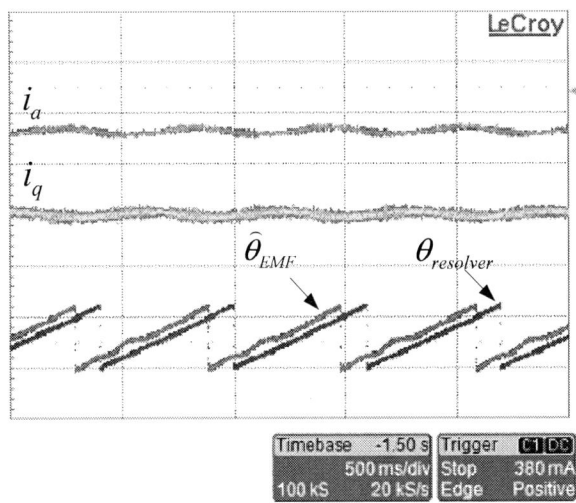

Figure 14. Sensorless operation in torque mode @ 50 rpm using the equivalent EMF observer to close the current loop. Traces from top to bottom are: phase a current i_u(1A/div); i_q feedback (1A/div); estimated position by equivalent EMF observer θ_{EMF} (5 rad/div); and $\theta_{resolver}$ (5 rad/div).

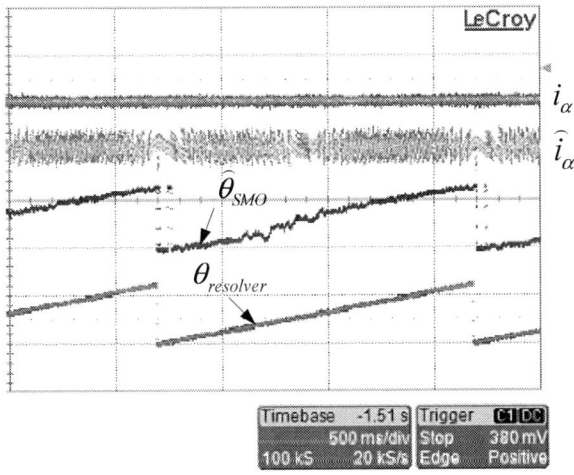

Figure 13. Sensorless operation in torque mode @ 20 rpm using SMO to close the current loop. Traces from top to bottom are: α axis current i_α (1 A/div); observed i_α by SMO (1 A/div); estimated position by SMO θ_{SMO} (5 rad/div); and $\theta_{resolver}$ (5 rad/div).

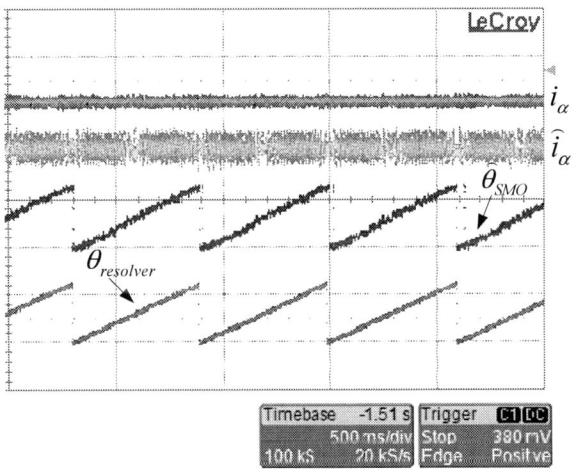

Figure 15. Sensorless operation in torque mode @ 50 rpm using SMO to close the current loop. Traces from top to bottom are: α axis current i_α (1 A/div); observed i_α by SMO (1 A/div); estimated position by SMO θ_{SMO} (5 rad/div); and $\theta_{resolver}$ (5 rad/div).

almost instantaneously, as compared to the equivalent EMF observer.

VI. CONCLUSIONS

This paper provides a simple, robust, and universal position observer for position sensorless control of synchronous machines. The observer may be implemented using an equivalent EMF model of a synchronous machine or, alternately, using a sliding mode controller based on the equivalent EMF model of the synchronous machine. The observer may be used on any type of synchronous machine, including salient or non-salient pole machines such as a permanent magnet, interior permanent magnet, wound rotor, or reluctance synchronous machine. No knowledge of velocity is required as an input to the observer and an estimated position may be calculated using a subset of the machine parameters.

Sliding mode observers for sensorless control of electric machines have gained much research interest due to many of their attractive features such robustness, order reduction and fast convergence. In this paper, a novel sliding mode observer is designed based on the equivalent EMF theory. Unlike other sliding observers for salient synchronous machines which are generally very complicated, the proposed SMO is very simply for implementation.

Experimental results with a 5hp IPM machine have validated the effectiveness of the proposed equivalent EMF method as well as the proposed SMO based on equivalent EMF.

References

[1] M. J. Corey and R. D. Lorenz, "Rotor position and velocity estimation for a salient-pole permanent magnet synchronous machine at standstill and high speeds", IEEE *Trans. on Ind. Appl.*, vol. 34, no. 4, pp. 784-789, Jul./Aug. 1998.

[2] C. Silva, G. M. Asher, and M. Sumner, "Hybrid Rotor Position Observer for Wide Speed-Range Sensorless PM Motor Drives Including Zero Speed', IEEE *Trans. on Ind. Electron.*, vol. 53, pp. 373-378, Apr. 2006.

[3] P.B. Schmidt, M.L. Gasperi, and T. A. Nondahl, "Method and apparatus for rotor angle detection", U.S. Patent 6172498, Jan. 9, 2001

[4] S. Morimoto, K. Kawamoto, M. Sanada, Y. Takeda., "Sensorless Control Strategy for Salient-Pole PMSM Based on Extended EMF in Rotating Reference Frame", IEEE *Trans. on Ind. Appl.*, vol. 38, no. 3, pp. 1054-1061, Jul./Aug. 2002

[5] Z. Chen, M. Tomita, S. Doki, and S. Okuma, "An extended electromotive force model for sensorless control of interior permanent-magnet synchronous motors", ", IEEE *Trans. on Ind. Electron.*, vol. 50, no. 3, pp. 288-295, Apr. 2003.

[6] M. Cernat, V. Comnac, M. Cotorogea, P. Korondi, S. Ryvkin, and R.-M. Cernat, "Sliding mode control of interior permanent magnet synchronous motors", in Proc. Power Electronics Congress CIEP, pp. 48 – 53, Oct. 2000.

[7] Vadim Utkin et al, "Sliding mode control in electromechanical systems", 1st Edition, Taylor & Francis, 1999.

[8] Y. Zhang and V.I. Utkin, "Sliding Mode Observers for Electric Machines-An Overview", IEEE IECON'02, vol. 3, pp. 1842-1847, Nov. 2002.

An Improved Winding Loss Analytical Model of Flyback Transformer

Wei Yuan, Xiucheng Huang, Peipei Meng, Guoxing Zhang, Junming Zhang

College of Electrical Engineering

Zhejiang University, Hangzhou, 310027, China

Email: zhangjm@zju.edu.cn

Abstract—The winding loss of Flyback transformer is difficult to calculate because the current flowing in both sides is not simultaneous, and the current phase shift at each harmonic between primary side and secondary side is not fixed at 180°. This paper proposes a novel winding loss analysis model by considering the phase shift angle of the two current and building a new magnetic field in the region between each layer. The detailed theoretical analysis for a Flyback transformer with the proposed model is presented in this paper, the transition period is analyzed in detail to explain its effect on interleaving structure. Finally the analysis results based on the proposed model is verified with both FEA simulation and experimental results.

I. INTRODUCTION

The AC conduction losses of the switching mode power transformer windings increases significantly with a higher frequency due to the eddy current effects, which include the losses induced by the skin and proximity effects. Eddy current losses have been previously treated in the literature. The basis for these publications is the paper by Dowell [1]. Dowell's formula has been founded to reliably predicted the increased resistance in cylindrical windings where the foil or layer thickness is less than 10% of the radius curvature. This formula has been utilized and developed by many applications such as the method in [2],[3],[4]. Paper [5] drives power loss expressions for each layer based on general field solutions for the distribution of current density in the layers of an infinitely long, cylindrical current sheet. And paper [6] simplify the methods above by the series expansions and apply it to any periodic waveform.

However, the contributions mentioned above could not be directly used in the Flyback type transformer, because the current flowing in both sides is not simultaneous, when decomposed into Fourier components, the harmonic field strength amplitudes on either side of each turn could not be obtained because the phase shift of two current is not fixed at 180° out of phase, thus the ac loss calculation of the Flyback transformer is always considered as trial and error, especially when the interleaving structure is applied. In [7], the copper loss of Flyback transformer in CCM operation was analyzed by plotting the time variation of the magnetic field, consequently the ratio of ac fields could be got by

subtracting the dc component and neglecting the current ripple, which is inappropriate to use when there is large current ripple, as the harmonic filed of each layer after Fourier decomposition do not has fixed phase shift, especially in the DCM mode. Paper [8] proposed a better winding loss analytical model by decomposing the total winding current into two components: transformer current, and inductor current. This is verified to be more accurate in the CCM mode. However, it is not suitable for DCM mode either, as the primary and the secondary current is not exactly 180° out-of-phase. The equivalent frequency method was also introduced by the authors to converter the DCM frequency to a boundary DCM frequency to use the same analytical method, but it is not explicit and introduces errors.

Another way to analyze the Flyback transformer has been introduced in [9], mentioned as common analytical model (CAM) here. CAM claimed that substantial circulating currents flow even when a foil winding is in off-state. This conclusion was based on the use of a 2-D finite element model in which the off-state foil was modeled as being open circuit while the on-state winding was modeled as having a continuous, steady state sinusoidal current. It is absolutely correct in recognizing the importance of 2-D effects in allowing significant loss to be induced in the off-state winding. However, the main drawback of this method is the neglecting of the transition period when the primary current falls to zero and the secondary current rises, as in this period, current flows through both sides simultaneously. It has also been claimed in [10],[11],[12] that the loss of this period takes a large part of the total winding loss by the time domain FEA models. Meanwhile, the assumption that a Flyback transformer could be approximated by forcing a sinusoidal current seems not suitable, as the winding loss is proved to be overestimated [10].

This paper proposed a novel winging loss analytical model (PAM) for the Flyback transformer by considering the phase shift angle of the two current, and building a new magnetic field in the region between each layer, and the corresponding loss for each harmonic could be determined based on the new magnetic field. The ac losses in DCM mode has been calculated by PAM, and also verified by the FEA simulation as well as the experimental results. Besides, PAM is also used to prove

978-1-4244-4782-4/10 $26.00 © 2010 IEEE

the reason that the interleaving techniques have better performance than the non-interleaving method in reducing the ac conduction losses.

II. THE PROPOSED WINDING LOSS ALAYSIS METHOD FOR FLYBACK IN DCM MODE

A. Nomenclature

dp : Thickness of the primary layer or foil
ds : Thickness of the secondary layer or foil
b_w : Winding width
H : Magnetic field intensity
a : Number of primary layer, counting from left to right
b : Number of secondary layer, counting from left to right
$H(a,b)$: Magnetic field in the outer side of layer (a,b)
$l(a,b)$: Length of the turn at Layer(a,b)
i : Harmonic number
Ipp : Peak current value of the primary side
Isp : Peak current value of the secondary side
$Ip(i)$: Rms current value of the ith harmonic of one primary layer
$Is(i)$: Rms current value of the ith harmonic of one secondary layer
j : $\sqrt{-1}$
J : Current density (A/m^2)
$\varphi(i)$: Phase angle between the primary and secondary current at ith harmonic
$\omega(i)$: Angular frequency of the ith harmonic
η : Conductor spacing factor
u_0 : Permeability of free space
T : Period of the current waveform
σ : Conductivity
σ_w : Equivalent conductivity of the round conductors
δ_o : $\sqrt{2/2\pi f \mu_o \sigma}$, skin depth at fundamental frequency
Q : The power loss per square meter

B. The Flyback Transformer

In the Flyback topology, as shown in Fig. 1(a), for a Discontinuous mode, we can identify four periods. The first period when the primary switch is on: D_1*T. And the transition period $Dr*T$ when the energy transfers from the primary to secondary, I_p falls and I_s rises, it should be noted that both sides conduct current in this period. The third period D_2*T when the secondary conducts, the last period $(1-D_1-Dr-D_2)*T$ when neither of the two side conducts current.

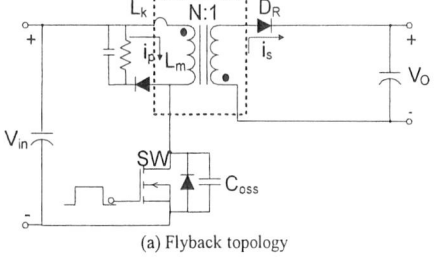

(a) Flyback topology

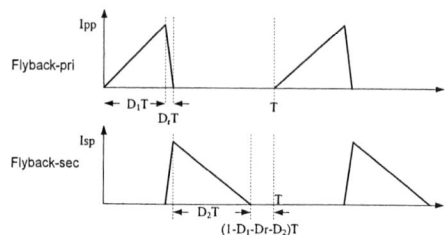

(b) Current waveform under DCM mode
Fig 1 Flyback topology and current waveforms under DCM

A traditional way to calculate the ac losses caused with non-sinusoidal current waveforms is given as follows [6]:

*1).*Decompose the current of the two sides into Fourier components.

*2).*Calculate the losses at each harmonic frequency.

*3).*Calculate the sum of each harmonic ac loss as the harmonic components are orthogonal.

However, in the Flyback type transformer, the loss calculation at each harmonic frequency would be more complex because the current phase shift in two windings are not always 180° out of phase.

Fig.2 shows the phase shift of the primary and secondary current, the traditional transformer loss calculation method all based on a in or 180° out of phase, as the relationship between $I_p(i)$ and $I_s'(i)$ shown in fig.2(a), unfortunately, in the Flyback type transformer, the phase shift angle $\varphi(i)$ between $I_p(i)$ and $I_s(i)$ is not always 0° or 180°. Table I presents a typical phase shift angle of a Fourier decomposition in DCM Flyback with f_s=40kHz, from which we can see that none of the ith harmonic current has a in or 180° out of phase, hence the magnetic field of the ith harmonic at the region of each layer could not be deduced using Ampere's law as shown in fig.2(b), thus the ith harmonic magnetic field $H(i)$ in the region between each layer could not be directly presented by the MMF diagram either [1],[6],[7].

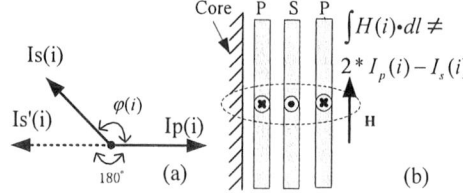

Fig.2 (a) Phase shift angle of the primary and secondary current and (b) the transformer geometry and Ampere's Law

TABLE I   Phase shift of Ip and Is of the Fourier components

Frequency(kHz)	Ip (A)	Is (A)	$\varphi(i)$
40	0.4112	0.4045	247.5
80	0.1742	0.1815	-179.3
120	0.1172	0.1129	204.5
160	0.0863	0.0898	-179.5
200	0.0685	0.066	194.2

C. The Proposed Analytical Model-PAM

In the PAM, the secondary current is supposed to be delayed for a certain phase, which makes the winding multi-dimensional and causes more loss. Based on the Fourier decomposition, the delayed phase at each harmonic could be derived, as long as the phase shift angle of each harmonic is compensated to get a complex type magnetic field in the region between each layer, the conduction loss of the Flyback transformer could be obtained.

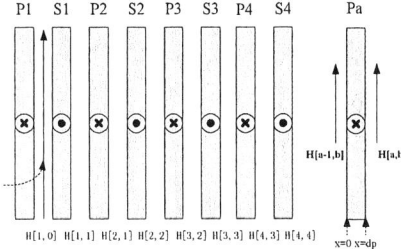

Fig.3 Interleaving transformer structure-analytical model

An interleaving structure of a Flyback transformer shown in fig.3 is adopted as an analytical example, each layer is considered to be a foil layer. The curvature of a layer is assumed to be small enough, thus the problem can be solved in Cartesian coordinates rather than cylindrical coordinates. Windings which consist of round conductors, or foils which do not extend the full winding window, may be treated as foils with equivalent thickness d and effective conductivity[6].As is shown in fig.4.

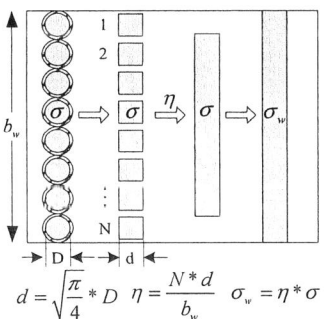

$$d = \sqrt{\frac{\pi}{4}} * D \quad \eta = \frac{N*d}{b_w} \quad \sigma_w = \eta * \sigma$$

Fig. 4 Spacing factor of round conductor

Taking the phase angle of the two currents into consideration, secondary current is decomposed into two parts: the real part in or 180° out of phase, and the imaginary part, as shown in fig.5, under this decomposition, the complex type ith magnetic field of each layer in the primary side could be deduced as follow:

$$H_p(a,b,i) = \frac{a * Ip(i) + b * Is(i) * \cos(\varphi(i)) + j * b * Is(i) * \sin(\varphi(i))}{b_w}$$

(1)

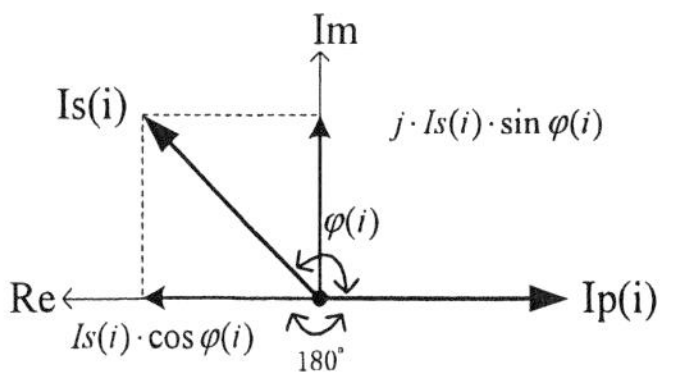

Fig. 5 Current factor decomposition of the secondary current

Where a, b are the coordinate of the primary and secondary layers as shown in fig.3,which is used to define the layer as well as it's winding structure. For example, **H[2,1]** indicates that there are two primary and one secondary layer included to form the magnetic field at the region between P2 and S2 in fig.3.

With the Cartesian coordinates, the H field with the function of x could be deduced from the differential equation below

$$\frac{d^2 H_p}{dx^2} - j\omega\mu_0\sigma_w H = 0$$

(2)

And then we get the current density distribution from (3)

$$J_p(x) = -\frac{dH_p(x)}{dx}$$

(3)

With $H_{(x=0)}$=H[a-1,b] and $H_{(x=dp)}$=H[a,b] deduced from (1),we get the current density with the function of x

$$J_p(a,b,i,x) = \frac{-k_p(i)}{\sinh(k_p(i)*d_p)}[H(a,b,i)*\cosh[k_p(i)*x]$$
$$-H(a-1,b,i)*\cosh[k_p(i)*(d_p-x)]]$$

(4)

Where $k_p = \sqrt{j\omega(i)\eta_p u_o \sigma}$

(5)

And the power dissipated per square meter of this layer is of the form:

$$Q_p(a,b,i) = \frac{1}{\eta_p * \sigma} * \int_0^{d_p} (|J(a,b,i,x)|)^2 dx$$

(6)

And the power dissipation of the layer (a,b) would be:

$$P_p(a,b) - \sum_{i=1}^{n} b_w * l(a,b) * Q_p(a,b,i)$$

(7)

The winding loss of the secondary side could be deduced in the same way by only adjusting the coordinates of the primary and secondary layers. Finally, the total winding loss could be obtained by summing up the losses at each layer.

III. ANALYSIS AND FEA SIMULATION RESULTS

Maxwell 2D simulation tools is used to verify the PAM, a PQ2620 core is adopted as an example, the distribute air gap length is 0.4 mm, solid round wire is adopted to make the windings of the transformer. Since the air gap length is quite small comparing to the winding width b_w, the fringing loss and the by-pass loss is not considered into the calculation, one could turn to [13] if needed.

Two kinds of winding structures are calculated and

simulated to verify the effectiveness of PAM with various parameters as mentioned in table II. Meanwhile, Fig.6 shows the details of the two transformers' structure: interleaving and non-interleaving. Two kinds of solid wires are adopted to build the layers with almost the same conductor spacing factor η to find the influences of the wire diameter. Note that the diameters of the two wires adopted in this comparison are both much thinner than $2\delta_o$ to minimize the affection of the skin effect, so that the current is uniformly distributed through all conductors at low frequency.

TABLE II   Parameters of the four transformers

	T1: I_0.21	T2: I_0.31	T3: N_0.21	T4: N_0.31
Turn Radio	48:8	48:8	48:8	48:8
Wire Diameter (mm)	0.21	0.31	0.21	0.31
Parallel Number	P(3) S(16)	P(2) S(12)	P(3) S(16)	P(2) S(12)
Wire Numbers per layer	P(36) S(32)	P(24) S(24)	P(36) S(32)	P(24) S(24)
Total Layers	P(4) S(4)	P(4) S(4)	P(4) S(4)	P(4) S(4)
Interleaving(Y/N)	Y	Y	N	N

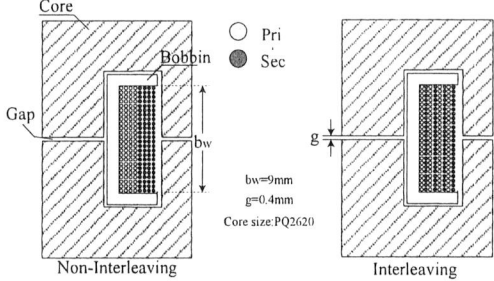

Fig.6 2-D winding structure of the simulated transformers

Two cases of current waveform parameters used in the simulation and theoretical analysis is listed below in table III according to fig. 1(b).

TABLE III   Parameters of the Two Cases

	Case1	Case2
Primary peak current(I_{pp})	3.03A	2.23A
D1	0.488	0.302
Secondary peak current(I_{sp})	18.18A	13.38A
D2	0.468	0.625
Dr(Interleaving)	0.005	0.009
Dr(Non-Interleaving)	0.012	0.02
Switching frequency(F_{sw})	39kHz	67kHz

The two comparison cases are set according to a real experimental result of a 64W Flyback working in DCM mode. Case1 takes a higher peak current but lower switching frequency, while Case 2 takes a lower peak current but higher switching frequency.

And it should be noted that the leakage inductance has something to do with the transition period. Interleaving could shorten the transition period by reducing the leakage inductance. This has been taken into consideration with different Dr. The two cases with the four transformers are designed to: 1).Comparing the accuracy of PAM with CAM, and validate it with FEA time domain method. 2).Find whether a thinner wire could have less conduction loss, and also find its relationship with the switching frequency as well as the winding structure. 3).Find whether interleaving could effectively reduce ac conduction losses in the Flyback type transformer comparing to non-interleaving structure.

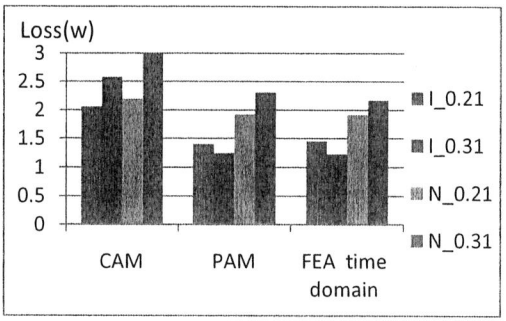

Fig.7 (a) Comparison of the two methods with FEA @Case1.

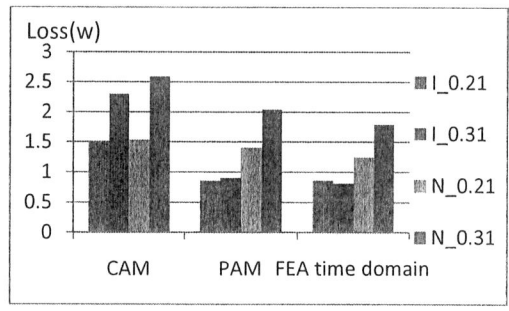

Fig.7 (b) Comparison of the two methods with FEA@Case2.

The calculated results of the PAM at n=40 are given in fig.7 (a) and (b), which are also compared to the CAM and the FEA time domain results, the time step is set at 100 ns, which would be enough to get a satisfying accuracy.

As has been mentioned in [10], the assumption to approximate the ac loss of a Flyback transformer by forcing a equivalent sinusoidal current to conduct is exaggerated, hence the actual waveform as fig. 1(b) is adopted in the comparison by FEA tools with the primary on and secondary off, and vice versa. The difference between CAM and the time domain FEA is mainly the transition period, as the time domain FEA method let both sides conduct in that period.

From fig.7 we can see that the calculated losses of the PAM almost match with the FEA time domain results both in two cases, which has a better accuracy than the

978-1-4244-4782-4/10 $26.00 © 2010 IEEE   436

CAM, as CAM overestimates the conduction loss in both cases even the actual waveform is used, especially where thicker wires are applied. Interleaving seems to have little effect on reducing the winding losses.

However, when the transition period is taken into consideration in the FEA time domain, the situation where interleaving is applied changes. Fig.8 shows the FEA time domain results of power loss versus time with T2(I_0.31) and T4(N_0.31) both works in Case1, it could be seen that the main difference between the two structure is the loss in the transition period, which means the conduction loss during the Dr period in fig.1(b) could be reduced dramatically with interleaving structure though this period is very shot. The reason is mainly because the two sides' currents in this period change quickly and simultaneously, which makes considerable eddy current loss. As the slew rate di/dt of the two winding current in the Flyback transformer is opposite and simultaneous, there would be reduce effect on the magnetic field of two sides mutually. Obviously, the reduce effect of the H field in the interleaving would be much better than the non-interleaving, which is not considered in CAM

Meanwhile, the reduce effect of the H field could be reflected in the current components after Fourier decomposition, which means the PAM could calculate the ac loss in view of the transition period, and the results in fig.7 have validated this.

Based on the calculated results with PAM, we can find some interesting phenomenon which may helps to optimize the transformer design:

1). Thinner wire with 0.21*3 causes more conduction loss than 0.31*2 due to small cross conduction area in the interleaving structure in Case1, but the difference decreases when the frequency gets higher as the ac loss increases, the ac loss of the two wires in Case 2 are almost equivalent.

2). For the transformer with a non-interleaving structure, the transformers with a thinner wire get an overall less conduction loss at both cases, and the difference increases when the switching frequency gets higher, this can be the result of the ac loss which increases with the rise of the frequency.

3). Interleaving does have less conduction loss than Non-interleaving. More than 1 watt could be saved in both cases with the wire Φ 0.31. Fig.9 shows the FEA results of *current density J* with T2 and T4 both working at Case1, which is very impressive to tell the difference between the two structures.

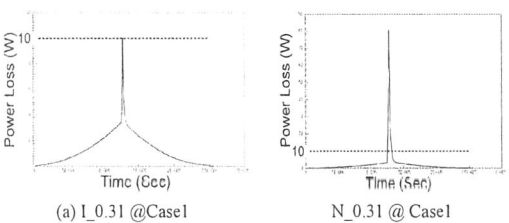

(a) I_0.31 @Case1 N_0.31 @ Case1

Fig.8 Power loss vs time in time domain FEA

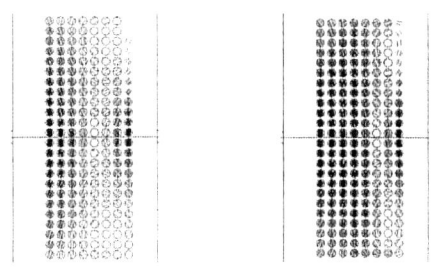

(a) Interleaving_T2:0.31   (b) Non-interleaving_T4:0.31

Fig.9 Current density of interleaving and non-interleaving in Case1

IV. EXPERIMENTAL RESULTS

To verify the analysis results with the PAM, four transformers (shown in Fig. 10) with structures given in Fig. 6 and the parameters in table III and built, and the tested parameters are given in Table IV. To simplify the test difficulties of a power transformer, these transformers are all tested at 40kHz (case 1) and 67kHz (case 2) on a 16V/4A Quasi-Resonant Flyback converter to see the difference of the power losses by keeping the remain circuit the same, it should be noted that parameters in Case1 and Case2 mentioned above are designed according to this experiment.

Fig.10 Picture of the four tested transformers

Table IV   Tested parameters of the four transformers

Winding Structure	T1	T2	T3	T4
L_m(uH)	412.5	412.7	412.5	412.8
L_k(uH)@40kHz	2.0	1.97	10.67	10.68
L_k(uH)@66.6kHz	1.8	1.81	9.97	9.95

L_{ka}=7.8uH   Interleaving   Non-interleaving

Fig.11 Added L_{ka} to balance the leakage between two structures

As the leakage inductance has much to do with the loss, an extra inductance L_{ka}=7.8uH is connected in series in the interleaving transformers to make sure that the four transformers have the same leakage inductance in fig.11.

The experimental total loss of the converter with the four transformers is presented in fig.12. Compared to the analysis given in Fig.7, the PAM has an excellent fit with the experimental results considering the influence of wire

978-1-4244-4782-4/10 $26.00 © 2010 IEEE   437

diameter, frequency as well as the winding structure, as the profile of the fig.12 is similar to fig.7. For example, a wire with $\Phi\,0.31$ could save about 0.2w AC conduction loss compared with $\Phi\,0.21$ in the interleaving structure. And an interleaving structure could almost have a 0.7w conduction loss saving compared with Non-interleaving structure both using a $\Phi\,0.31$ wire. Clearly both the two cases show that interleaving structure can lead to less conduction losses in the Flyback type transformer.

It should be noted that there would be some errors induced as the PAM is based on the one-dimensional field with certain assumptions, summarized in [14], especially when round conductors is replaced with equivalent foils in the calculation though the spacing factor is considered. And the 2-D FEA tool also has been proved with errors in [15]. Thus the results among the PAM, the 2-D FEA time domain simulation and the experimental would be a little bit inconsistent. Nevertheless, the identified errors are judged to be tolerable. And the PAM could be further used to calculate the winding loss in the situation of the coupled inductors which also have unfixed current angles between the two sides of transformer [16]-[17].

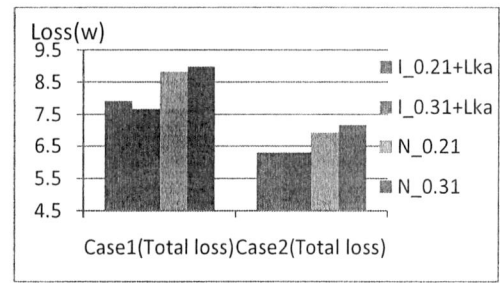

Fig.12 Measured total loss of the converter with 4 transformer -adding an equivalent inductance L_{ka} in the interleaving structure

V. CONCLUSION

This paper proposes a new winding loss analytical model for the Flyback transformer. As the current is not flowing in both windings simultaneously, the current in the two sides at each harmonic after Fourier decomposition is not 180° out of phase, thus the magnetic field of the harmonic component in the region between each layer could not be directly got by the Ampere's law as well as the MMF diagram. The PAM considers the phase shift angle of the two current and generates a complex type magnetic field to calculate the ac winding losses, meanwhile, the transition period is also analyzed and included into the Fourier decomposition so as to better predict the ac losses. The PAM is validated to be much more accurate than the CAM by The FEA simulation and the experimental results. Meanwhile, it also proves that interleaving structure does obviously have less ac winding loss in the Flyback transformer, inconsistent with the results from CAM.

ACKNOWLEDGEMENT

This work is supported by China National Science Fund, No. 50907061.

REFERENCES

[1] P.L. Dowell, "Effects of eddy currents in transformer windings", *Proceedings of the IEEE*, vol. 113, , pp. 1387–1394, Aug. 1966.

[2] P. S. Venkatraman, "Winding Eddy Current Losses in Switch Mode Power Transformers Due to Rectangular Wave Currents," *Proceedings of* Powercon11, 1984, Section A, pp. 1-11.

[3] Bruce Carsten, "High Frequency Conductor Losses in Switch mode Magnetics." High-frequency Power Converter Conference Record, pp. 155-176, May 1986.

[4] Xi Nan and Charles R. Sullivan," An Improved Calculation of Proximity-Effect Loss in High-Frequency Windings of Round Conductors" *IEEE PESC*,vol.2, pp. 853-860,June 2003.

[5] M. P. Perry, "Multiple Layer Series Connected Winding Design for Minimum Losses," *IEEE Trans. Power Electron.* , vol. PAS-98, No.1, pp. 116-123, Jan. 1979.

[6] William Gerard Hurley, Eugene Gath, and John G. Breslin, "Optimizing the AC Resistance of Multilayer Transformer Windings with Arbitrary Current Waveforms", *IEEE Trans. Power Electron.*,vol.15, no. 2, March 2000.

[7] J. Vandelac and P. D. Ziogas, "A novel approach for minimizing high frequency transformer copper losses," *IEEE Trans. Power Electron.*, vol.3, no. 3, pp. 166–176, July 1988.

[8] Zengyi Lu and Wei Chen "Novel Winding Loss Analytical Model of Flyback Transformer", *IEEE PESC*, pp.1213-1218, June 2006.

[9] R. Prieto, J.A. Cobos, O.Garcia, P. Alou and J. Uceda, "High Frequency Resistance in Flyback Type Transformers", *IEEE APEC*, vol. 2, pp. 714–19, Feb 2000.

[10] Doug Lavers, Eric Lavers," Waveform Dependent Switching Losses in Flyback Transformer Foil Windings", *IEEE APEC*, vol. 3, pp. 116–112, June 2002.

[11] J. M. Lopera, M. J. Prieto, F. Nuno, A. M. Pernia, and J. Sebastian, "A quick way to determine the optimum layer size and their disposition in magnetic structures", *IEEE PESC*, pp. 1150–1156, June 1997.

[12] Charles R. Sullivan, Tarek Abdallah," Optimization of a flyback transformer winding considering two-dimensional field effects, cost and loss" *IEEE APEC*, pp. 1-7, March 2001.

[13] Wei Chen, Jiannong He, Henglian Luo, et.al, "Winding Loss Analysis and New Air-Gap Arrangement for High-frequency Inductor" *IEEE PESC*, pp. 2084-2089, June 2001.

[14] Urling, A.M. Niemela, V.A. Skutt, G.R. Wilson, T.G. "Characterizing high-frequency effects in transformer windings-a guide to several significant articles", *IEEE. APEC*, pp. 373-385, March 1989.

[15] I.D. Lavers, ED. Lavm, "An Accuracy Assessment of 2-D vs. 3-D Finite Element Models For Ferrite Core, Sheet Wound Transformers", *IEEE. APEC*, pp. 158-164, March 10-14, 2002

[16] X. C. Wang, I. Batarseh, "Active Transient Voltage Compensator for VR Transient Improvement at High Slew Rate Load", *IEEE Trans. Power Electron*, 2007,Vol. 2, pp: 1472-1479

[17] K. Yao, Y. Qiu, M. Xu, and F. C. Lee, "A Novel Winding-Coupled Buck Converter for High-Frequency, High-Step-Down DC-DC Conversion", *IEEE Trans. Power Electron,* 2005, Vol. 20, pp: 1017-1024

Identification of the material properties used in domestic induction heating appliances for system-level simulation and design purposes

Jesus Acero, Oscar Lucia, Ignacio Millan, Luis Angel
Barragan, Jose-Miguel Burdio
Dept. of Electronic Engineering and Communications
University of Zaragoza
Maria de Luna 1, Zaragoza, Spain
jacero@unizar.es

Rafael Alonso
Dept. of Applied Physics
University of Zaragoza
Pedro Cerbuna 9, Zaragoza, Spain

Abstract— Designing the converter for an induction cooker, the main difficult lies in the variability of the load, due to the wide range of vessels available in the market. Such issue is equivalent to the problem of identifying the vessel's properties. In this paper an identifying method of such properties is proposed with the objective of system-level simulation and design. The method is based on a combination of analytical electromagnetic calculations, and Fourier-series-based waveform synthesis.

I. INTRODUCTION

Induction cookers constitute the main domestic application of the induction heating phenomenon. Basically, a domestic induction arrangement consists of a planar inductor situated below a metallic vessel and supplied by a resonant inverter operating between 20 kHz and 100 kHz. The electrical equivalent of the coupled inductor-vessel system, used in simulations, consists of the series connection of a resistance and inductance (see Fig. 1) R_{eq}, L_{eq} [1]-[4]. Normally, the simulation of a domestic induction system becomes uncertain due to the variability of the pots, whose electromagnetic parameters often are not known.

Extracting the equivalent circuit of the inductive load, Finite Element Analysis (FEA) tools have often been used [2], [5]. The use of the electromagnetic analyses for the generation of a model valid to a circuit simulator is not evident especially if the properties of the involved materials are not known. In one usual approach, the frequency-dependent resistance and inductance obtained from electromagnetic harmonic analyses leads to linear equivalent circuits composed of series or/and parallel connected resistances and inductances [6]-[7] (Foster's networks). However, the problem of extracting Foster's networks can be cumbersome if the frequency-dependent curves are not well conformed [8]. Other approaches use time-domain FEA solvers or even non-linear time-domain simulations; however, the amount of data necessary to post-processing normally is huge [7].

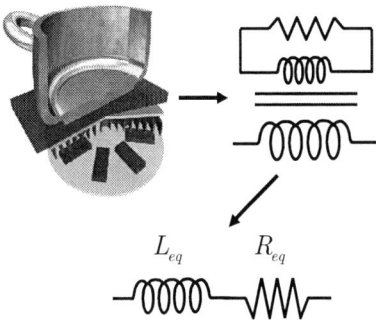

Fig. 1. The induction load and its equivalent circuit.

In addition, a great deal has been written on the subject of system identification and time-domain waveform fitting, with the objective of generating black-box models [9]-[11]. Nevertheless, conclusions from these approaches only apply to the prototype under test, and more general information beyond the specific experiment hardly can be extracted.

In the case of the inductive load, it is useful extracting the vessel's properties in order to perform electromagnetic simulations in which the effect of several parameters could be investigated. Electromagnetic properties of the materials (electrical conductivity σ and relative magnetic permeability μ_r) can be extracted from physical measurements. In the case of μ_r measuring, probes must be conformed as small rings (Rowland's ring technique), [12]. Having in mind that the mechanization of the bottom's pot leads to a change of its magnetic properties [13], in this paper an identification method preserving the pot's properties is proposed. The method is based on usual techniques in the power electronics field, such as electromagnetic calculations, waveform synthesis by means of Fourier series and oscilloscope measurements.

This work was supported in part by the Spanish Ministry of Science and Innovation (MICINN) under Project TEC2007-64188, in part by Diputación General de Aragón (DGA) under Project PI008/08, and in part by the Bosch and Siemens Home Appliances Group. The authors are with the Aragon Institute for Engineering Research (I3A).

978-1-4244-4782-4/10 $26.00 © 2010 IEEE

II. EQUIVALENT ELECTRICAL PARAMETERS OF THE INDUCTION SYSTEM

Fig. 2 shows the arrangement of a domestic induction system. Analytical models of these systems, including the dependence of several variables such as frequency, pot properties σ, μ_r, geometry, number of turns, ferrite properties, and winding yarn were proposed on the basis on its cylindrical symmetry [3]-[4]. Moreover, analytical models of losses for the most used kind of cables, including multistranded isolated strands (Litz wires) were also proposed [14]-[15].

Analytical models permit to investigate the influence of some parameters over R_{eq}, L_{eq}. From the point of view of the time-domain waveform synthesis, the most interesting results are R_{eq} and L_{eq} as a function of the frequency. From [3]-[4] integral expressions of such parameters are the following

$$R_{eq} = \text{Re}\left\{ j\omega\mu_0\pi \int_0^\infty \frac{\phi_1 e^{-2\alpha d} + \phi_4 e^{-2\alpha h} + 2\phi_1\phi_4 e^{-2\alpha(d+h)}}{1 - \phi_1\phi_4 e^{-2\alpha(d+h)}} \text{T}(\alpha)d\alpha \right\} \quad (1)$$

$$L_{eq} = \text{Im}\left\{ \mu_0\pi \int_0^\infty \frac{\phi_1 e^{-2\alpha d} + \phi_4 e^{-2\alpha h} + 2\phi_1\phi_4 e^{-2\alpha(d+h)}}{1 - \phi_1\phi_4 e^{-2\alpha(d+h)}} \text{T}(\alpha)d\alpha \right\} \quad (2)$$

where parameters ϕ_1, ϕ_4, $\text{T}(\alpha)$ and α are related with the properties of the induction system. The definition of such parameters is provided in [3]-[4].

Expressions (1) and (2) were deduced on the basis on the cylindrical symmetry of the system and also considering an ideal model of the windings in which the real conductors are replaced by filamentary currents. Nevertheless, the losses in the conductors can be taken into account considering the DC losses, the skin effect and the proximity losses of wires [14]-[15]. Assuming that these losses can be represented by a resistive term, they can be directly added to (1). In any case, the induction systems are properly designed to minimize the losses in windings, so, efficiencies better that the 95 % are actually achieved [16].

In Fig. 3 a plot of $R_{eq}(\omega)$, $L_{eq}(\omega)$ is presented, being

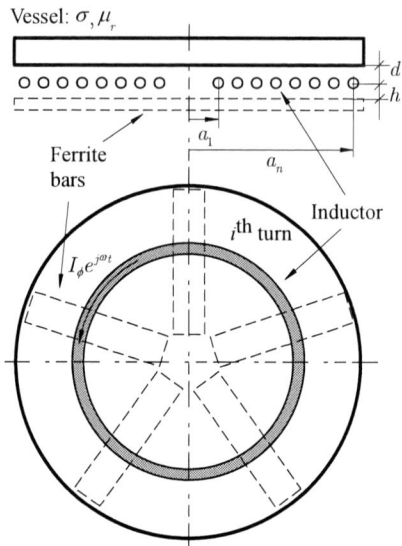

Fig. 2. Schematic representation of a domestic induction system.

σ, μ_r the swept parameters. These results correspond to an inductor of $n = 16$ turns having an external diameter of $\phi_{ext} = 210$ mm. The cable was a litz-wire of $n_o = 54$ strands of diameter $\phi_o = 0.3$ mm. Sweeping σ, μ_r parameters permits to estimate the electrical equivalent of a wide range of materials. As it can be seen, the impedance of the induction system is quite different depending on the properties of the materials. Well-conductor and non-magnetic materials, as the aluminum, exhibit both lower equivalent resistance and inductance. Using these curves, different kinds of materials, ranging from diamagnetic (i.e. copper) to ferromagnetic ones could be considered.

The impedances at the different frequencies shown in Fig. 3 will be used in the next Section in order to synthesize the waveforms presented in the system by means of the Fourier series technique.

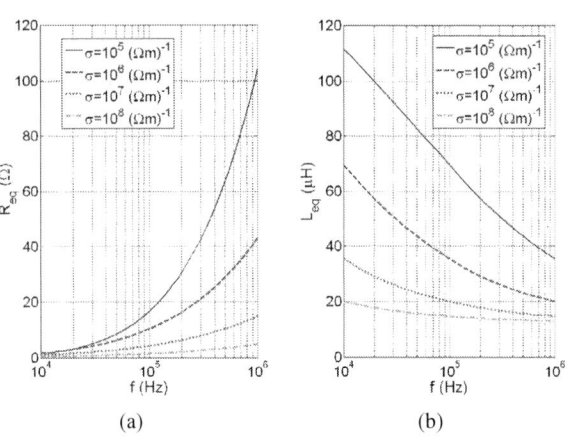

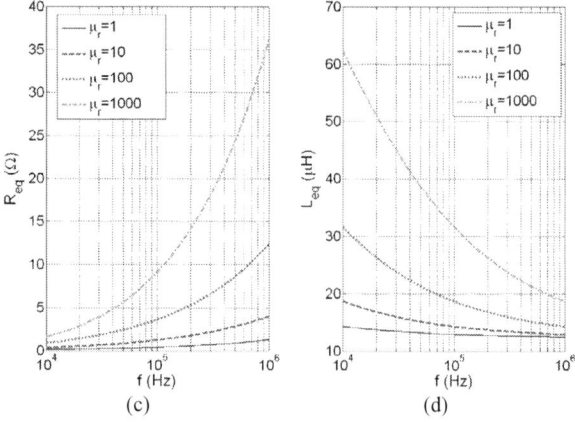

Fig. 3. Influence of σ, μ_r over $R_{eq}(\omega)$, $L_{eq}(\omega)$, calculated results from the analytical models for an induction system: (a), (b) dependence according to σ; (c), (d) dependence according to μ_r.

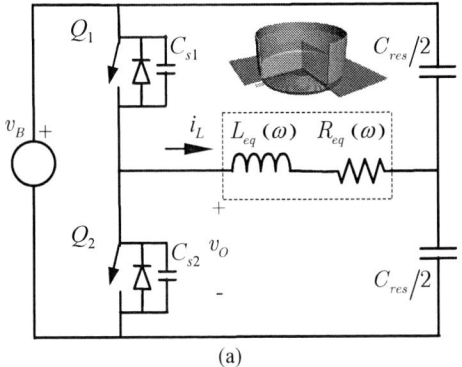
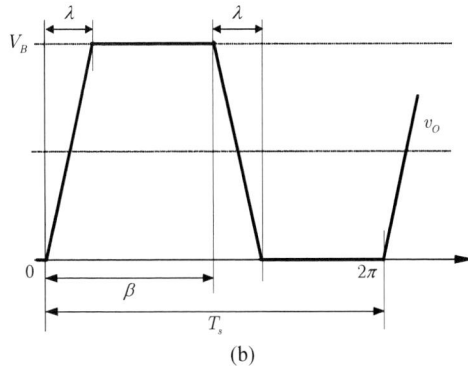

Fig. 4. (a) Half-bridge series resonant inverter. (b) Output voltage trapezoidal waveform.

III. APPLICATION TO THE TIME-DOMAIN WAVEFORM SYNTHESIS

Resonant inverter topologies are commonly used for induction hobs. At present, the half-bridge series resonant inverter (shown in Fig. 4 (a)) is the most popular one due to its robustness and cost savings [17]. The power is controlled by both varying the switching frequency and the duty cycle under zero-voltage-switching (ZVS) conditions. Snubber capacitors are added in parallel with the switches to achieve soft switching when turning off the devices.

$R_{eq}(\omega)$ and $L_{eq}(\omega)$ represent de equivalent impedance of the coupled inductor-pan system. The resonant frequency is fixed at those in which the nominal power is delivered. Usually, the value of R_{eq} and L_{eq} at this frequency is used for design purposes. The approach followed in this paper includes the frequency dependence of R_{eq} and L_{eq} in order to design the inverter with a more realistic model of the load.

Fig. 4 (b) shows a trapezoidal waveform characterized by the period T_s, the control angle β, and ramp angle λ. This waveform is similar to the voltage applied to the resonant tank, v_O in Fig. 4 (a), including the effect of the snubber capacitors. So, λ represents the charge and discharge intervals of the snubber, assuming a ZVS operation, and β is a control angle coincident with the duty cycle, D, expressed in radians, of the switches Q_1 and Q_2. Assuming that the input voltage v_B is a dc value V_B, the time variation in steady state of the voltage v_O may be represented by the following Fourier series, where the amplitude and phase of the hth harmonic of v_O are denoted as $\widehat{V}_{Oh}$ and ϕ_{vh}, respectively

$$\widehat{V}_{Oh} = \frac{V_B}{h\pi}\sqrt{a_h^2 + b_h^2}, \quad \phi_{vh} = \tan^{-1}\frac{a_h}{b_h} \tag{3}$$

$$a_h = \frac{\cos(h\lambda) + \cos(h\beta) - \cos[h(\lambda+\beta)] - 1}{h\lambda} \tag{4}$$

$$b_h = \frac{\sin(h\lambda) + \sin(h\beta) - \cos[h(\lambda+\beta)]}{h\lambda} \tag{5}$$

The amplitude and phase of the load current for the hth harmonic, $\widehat{I}_{Lh}$, ϕ_{ih} are found to be respectively

$$\widehat{I}_{Lh} = \frac{\widehat{V}_{Oh}}{R_{eq\_h}\sqrt{1 + Q_h^2\left[\psi h\omega_n - \frac{1}{\psi h\omega_n}\right]^2}} \tag{6}$$

$$\phi_{ih} = \phi_{vh} - \tan^{-1}\left[Q_h\left(\psi h\omega_n - \frac{1}{\psi h\omega_n}\right)\right] \tag{7}$$

where

$$\psi = \sqrt{\frac{L_{eq\_h}}{L_{eq\_res}}}; \quad Q_h = \frac{\sqrt{L_{eq\_h}/C_{res}}}{R_{eq\_h}} \tag{8}$$

$$\omega_n = \frac{\omega_s}{\omega_{res}}; \quad \omega_{res} = \left[\sqrt{L_{eq\_res}C_{res}}\right]^{-1} \tag{9}$$

being R_{eq_h}, L_{eq_h} the equivalent resistance and inductance at the frequency correspondent to the hth harmonic, L_{eq_res} the inductance at the resonant frequency (extracted from the electromagnetic model above described), ω_s the angular switching frequency, ω_{res} the angular resonant frequency, and C_{res} the resonant capacitor (Fig. 4 (a)). Waveforms are generated summing up to fifteen harmonics.

Considering an *rms* mains voltage of 230 V, results (3)-(9) allow calculating the current requirements to deliver the nominal power for the two evaluated loads for the same C_{res}. Fig. 5 (a) shows the results for the ferromagnetic one and Fig. 5 (b) for the aluminum one. Figures are plotted at the same scale to highlight the current requirements of aluminum load compared with the ferromagnetic one.

Moreover taking into account that in an induction cooker the delivered power is controlled by both the switching frequency and the duty cycle, curves of the normalized power as a function of the normalized frequency (with respect to the resonant ones) are used for design purposes. Fig. 5 (c) shows MATLAB calculated curves for two different duty cycles.

978-1-4244-4782-4/10 $26.00 © 2010 IEEE

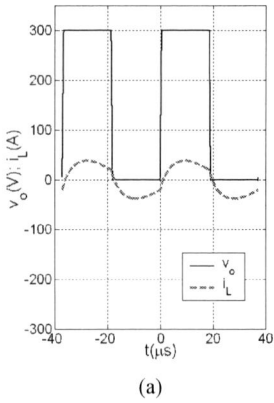 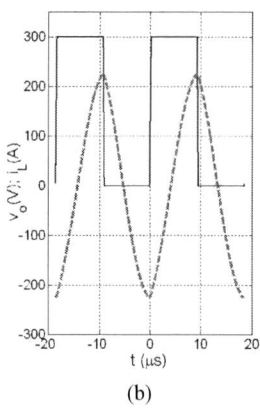 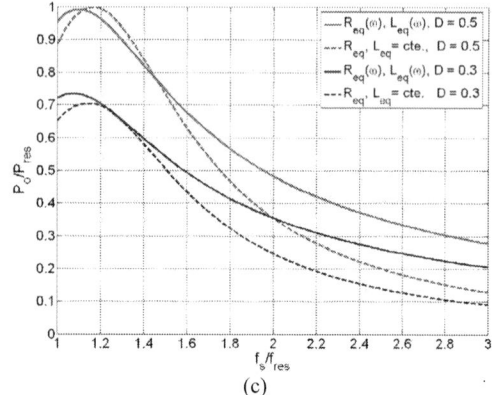

(a) (b) (c)

Fig. 5. (a) Calculated voltage and current waveforms for the ferromagnetic load, (b) calculated voltage and current waveforms for the aluminum load, (c) Frequency-dependent normalized output power for two duty cycles.

These curves, in fact, represent the reachable power at the different frequencies for the ferromagnetic material.

A comparison between the classical load approach (consisting of constant R_{eq} and L_{eq}) and the proposed model is also presented. As it can be observed, the predicted power using the constant R_{eq} and L_{eq} parameters is lesser than the calculated with the proposed model. A more complete model of the load is achieved in this case due to the contribution of the harmonics which are not been considered in the classical approach.

IV. IDENTIFICATION CRITERIA

Selecting the properties of the material under test, the followed criteria are rather focused on the power electronics requirements, for design purposes, instead of a criterion based on a perfect curve fitting. Calculated and measures waveforms are compared on the basis of the peak current and the *rms* current in the load.

A simple algorithm for finding the electrical conductivity and magnetic permeability has been performed. The algorithm sweeps such parameters and generates time-domain current waveforms by means of (3)-(9). The peak and the *rms* values of the calculated currents are compared with experiments. Parameters σ, μ_r are selected according to the better adjustment of both *rms* and peak values. Different scenarios having different switching frequencies or duty cycles are used to verify the selected properties.

A representative set of materials have been chosen for extracting its properties. These materials range form the more ferromagnetic (as the material used by the brand *Sillit*) and the lesser ferromagnetic (as the brand's pots *Hackman*). An intermediate material (manufactured under Zenith brand) was also tested. Electromagnetic properties of these materials are presented in Table I. The knowledge of these parameters allows further simulations, for instance to design induction systems with other geometry or different number of turns for heating these materials. These parameters can also be used for Finite Element simulations.

TABLE I. PROPERTIES OF THREE REPRESENTATIVE MATERIALS

	Hackman	*Zenith*	*Sillit*
σ $(\Omega \cdot m)^{-1}$	$8 \cdot 10^6$	$5 \cdot 10^6$	$2 \cdot 10^6$
μ_r	105	250	580

V. COMPARISON BETWEEN CALCULATED AND MEASURED RESULTS

The experimental verification consists of comparing calculated and measured results in the time domain for different switching frequencies and several duty cycles. A half-bridge inverter with FAIRCHILD 20N60 IGBT, was used to feed the inductor loaded with the materials previously studied.

The used inductor has identical properties and yarn parameters of those presented in Section II for simulation purposes. The pot under test is heated in a commercial induction hob which is delivering the nominal power, i.e. close to the resonant conditions. Current waveforms are captured once the boiling water state is reached, and therefore the system is thermally stable.

Actual waveforms and the calculated ones are presented in Fig. 6 for several switching frequencies and duty cycles. The calculated results use the parameters presented in Table I for every case. In general, an acceptable agreement is obtained for the different excitation conditions at is shown in Fig. 6.

VI. CONCLUSION

A method for extracting the electromagnetic parameters of the materials used in induction heating cookers is presented. Instead of physical measurements, this method exploits the analytical calculation of the frequency-dependent electrical parameters and the waveform synthesis by means of Fourier series. The criterion for selecting the parameters is the comparison between the calculated and measured peak and *rms* values of the load current. The same model is able to describe the load in a variety of different situations without any need of readjusting the parameters as it has been shown

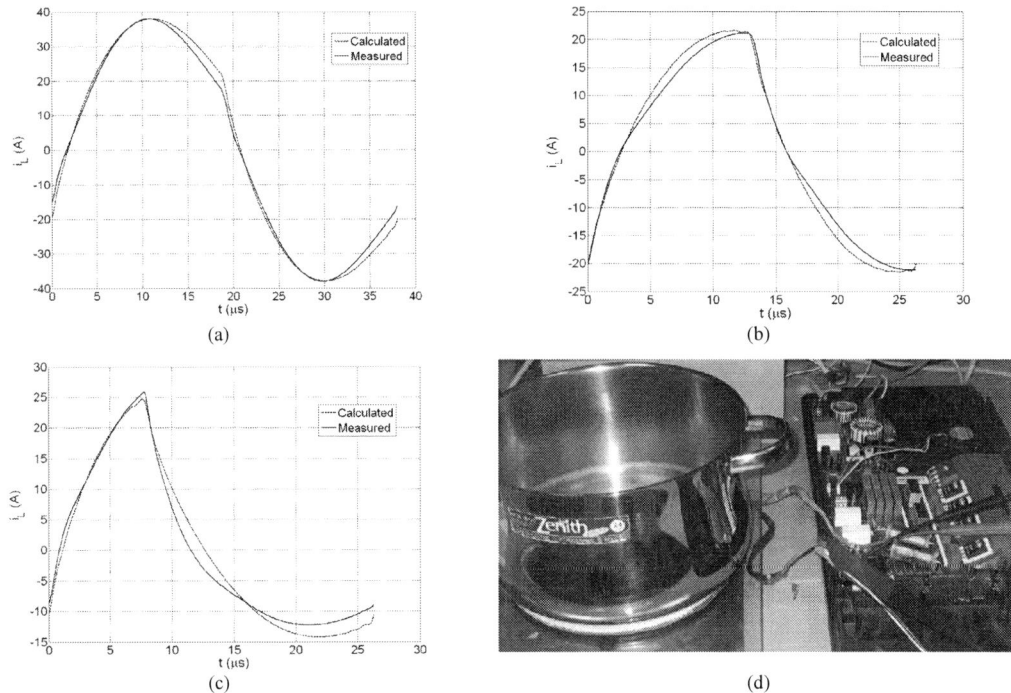

Fig. 6. Some comparisons of resultant current waveforms between actual measurement and simulation varying the switching frequency f_s and the duty cycle D. (a) $f_s = 26$ kHz , $D = 0.5$; (b) $f_s = 38$ kHz , $D = 0.5$; (c) $f_s = 38$ kHz , $D = 0.3$; (d) an image of the experimental setup.

The whole set of calculations have been implemented in MATLAB in order to simplify the computational costs.

REFERENCES

[1] W.G. Hurley and J.G. Kassakian, "Induction heating of circular ferromagnetic plates," *IEEE Trans. Magn.*, vol. mag-15, no 3, pp 1174-1181, July 1979.

[2] H.W.E. Koertzen, J.D. Van Wyk, J.A. Ferreira, "Investigating the influence of material properties on the efficiency of an induction heating load transformer using FEM simulation," *IEEE Industry Applications Society Annual Meeting (IAS) Rec.* 1995, pp. 868-873.

[3] J. Acero, R. Alonso, J.M. Burdío, L.A. Barragán, and D. Puyal, "Analytical equivalent impedance for a planar circular induction heating system," *IEEE Trans. Magn.*, vol. 42, pp. 84-86, Jan. 2006.

[4] J. Acero, R. Alonso, J.M. Burdío, L.A. Barragán, S. Llorente, "Electromagnetic induction of planar windings with cylindrical symmetry between two half-spaces," *Journal of Applied Physics*, vol. 103, 104905(8), May 2008.

[5] J-K. Byun, K. Choi, H-S. Roh, S-Y. Hahn, "Optimal design procedure for a practical induction heating cooker," *IEEE Transactions on Magnetics*, vol. 36, no. 4, pp. 1390-1393, July 2000.

[6] L. Heinemann, R. Schulze, P. Wallmeier, and H. Grotstollen, "Modeling of high frequency inductors," in *Power Electronics Specialists Conference, PESC '94* vol. 2, pp. 876-883, 1994.

[7] R. Prieto, R. Asensi, C. Fernández, J.A. Oliver, J.A. Cobos, "Bridging the gap between FEA field solution and the magnetic component model," *IEEE Trans. Power Electronics*, vol. 22, pp. 943-951, May 2007.

[8] S.H. Min, M. Swaminathan, "Construction of Broadband Passive Macromodels form Frequency Data for Simulation of Distributed Interconnect Networks," *IEEE Trans. Electromagnetic Compatibility*, vol. 46, no. 4, Nov. 2004, pp. 544-558.

[9] D. Puyal, C. Bernal, J.M. Burdío, I. Millán, J. Acero, "A new dynamic electrical model of domestic induction heating loads," in *IEEE Applied Power Electronics Conf. (APEC) Rec.* 2008, pp.409-414.

[10] V. Valdivia, A. Barrado, A. Lazaro, P.Zumel, C. Raga, "Easy modeling and identification procedure for "black box" behavioral models of power electronics converters with reduced order based on transient response analysis," *IEEE Applied Power Electronics Conference (APEC) Rec.* 2009, pp. 318-324.

[11] L. Ljung, *System Identification: Theory for the User*, 2nd ed. Englewood Cliffs, NJ: Prentice-Hall, 1999.

[12] R. Huang; D. Zhang, "Using a single toroidal sample to determine the intrinsic complex permeability and permittivity of Mn–Zn ferrites," *IEEE Trans. Magn.*, vol. 43, no 10, pp 3807-3815, Oct. 2007.

[13] B.D. Cullity, *Introduction to Magnetic Materials*, Addison Wesley, 1972.

[14] J. Acero, R. Alonso, J.M. Burdío, L.A. Barragán, and D. Puyal, "Frequency-dependent resistance in litz-wire planar windings for domestic induction heating appliances," *IEEE Trans. Power Electronics*, vol. 21, pp. 856-866, July 2006.

[15] J. Acero, R. Alonso, J.M. Burdío, L.A. Barragán, and C. Carretero, "A model of losses in twisted-multistranded wires for planar windings used in domestic induction heating appliances," in *IEEE Applied Power Electronics Conf. (APEC) Rec.*, 2007, pp. 1247-1253.

[16] J. Acero, et al., "The domestic induction heating appliance: an overview of recent research," ,in *IEEE Applied Power Electronics Conf. (APEC) Rec.*, 2008, pp. 651-657.

[17] S. Llorente, F. Monterde, J.M. Burdio, and J. Acero, "A comparative study of resonant inverter topologies used in induction cookers," in *IEEE Applied Power Electronics Conf. (APEC) Rec.* 2002, pp.1168-1174.

A Retrofit 60 Hz Current Sensor for Non-Intrusive Power Monitoring at the Circuit Breaker

Zachary Clifford *Member, IEEE*, John J. Cooley *Student Member, IEEE*, Al-Thaddeus Avestruz *Member, IEEE*,
Zack Remscrim, Dan Vickery, and Steven B. Leeb *Fellow, IEEE*

Abstract—**We present a new sensor for power monitoring that measures current flow in a circuit breaker without permanent modification of the breaker panel or the circuit breaker itself. The sensor consists of three parts: an inductive pickup for sensing current from the breaker face, an inductive link designed to transmit power through the steel breaker panel door, and a passive, balanced JFET modulator circuit for transmitting information through the inductive link. The demodulated breaker current signal is available outside of the breaker panel door. This sensor provides a solution for low-cost, non-intrusive retrofit of any circuit breaker panel for centralized power monitoring.**

Index Terms—**Home power monitoring, non-intrusive load monitoring, load diagnostics, circuit breaker**

I. INTRODUCTION AND MOTIVATION

Non-Intrusive Load Monitoring (NILM) is an approach to electrical system diagnostics and power monitoring that has a potentially much lower sensor count than other load-specific monitoring systems. A NILM identifies and monitors individual loads on a power distribution system by measuring the frequency content of transient events in the current signals from a centralized location. Work with this technology was demonstrated in [1]–[12].

The current signal for the NILM is typically measured using a magnetic field sensor wrapped around the utility feed for the circuit to be monitored. Such a sensor may be impractical for some retrofit applications, especially in the home where skilled labor would be required to separate line and neutral in order to deploy a wrap-around magnetic field sensor, or in industrial environments where electrical service cannot be interrupted. The sensor presented in this paper is an alternative to the wrap-around magnetic field sensor. It measures the current in the utility feed by sensing the magnetic field at the face of any circuit breaker in a standard breaker panel, where the line and neutral are already separated. A major challenge, overcome by the system presented here, is communicating through the steel breaker panel door, which typically must be closed to comply with safety regulations.

II. SYSTEM OVERVIEW

The sensor shown in Figure 1(a) consists of three parts: an inductive pickup for sensing current from the breaker face (Breaker Pickup), an inductive link designed to transmit power through the steel breaker panel door (Through-door Inductive Link), and a balanced JFET modulator circuit for transmitting information through the inductive link (JFET Mixer).

The outer coil in Figure 1(a) is driven with a high-frequency sinusoidal carrier voltage. That voltage couples to the inner

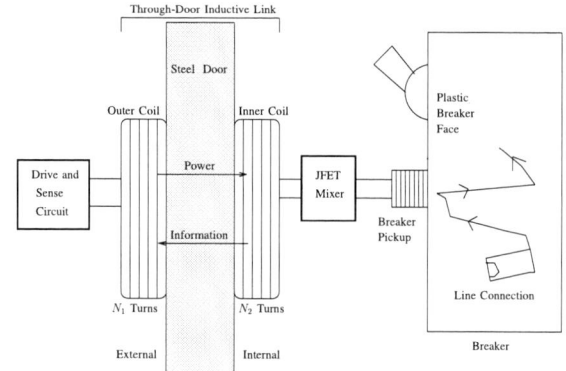

(a) Sensor block diagram

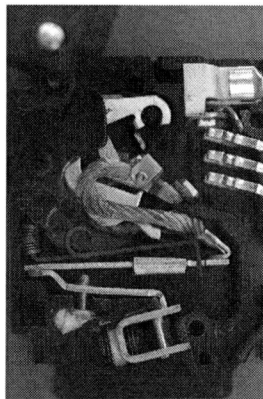

(b) Circuit Breaker Cross-section

Fig. 1. The current sensor measures magnetic fields at the face of the circuit breaker and modulates a high frequency carrier signal to transmit that information through the panel door.

coil through the inductive link and drives the JFET mixer. The JFET mixer controls the amount of current drawn from the inner coil according to the low-frequency (60 Hz) current signal measured by the breaker pickup. The result is a modulation between the high frequency carrier signal and the low-frequency (60 Hz) signal measured at the breaker face. The external sense circuit in Figure 1(a) monitors the current drawn through the inductive link to extract the resulting modulated signal. The JFET modulator is fully powered by the applied carrier, and the entire system works without modification to the breaker panel or the circuit breaker itself. With the modulated signal available to the sense circuit external to the door, the current through the main breaker may be analyzed with a

NILM or other power monitoring system for load identification and diagnostics.

III. BREAKER PICKUP

This section describes the inductive sensor referred to as the "breaker pickup" in Figure 1(a). The current path inside a typical circuit breaker passes by the lower face as suggested by Figures 1(a) and 1(b). The breaker pickup, positioned outside the breaker on its lower face, was designed to focus and measure the magnetic field resulting from current flowing inside the breaker.

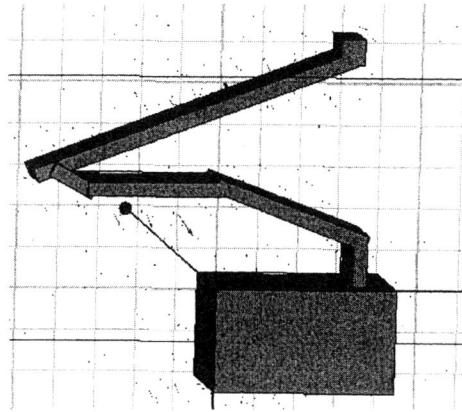

Fig. 2.   Maxwell 3D model of the circuit breaker

Ansoft's Maxwell 3D was used to model the magnetic fields generated by the current-carrying member inside a typical circuit breaker. The geometric model of the current-carrying member shown in Figure 2 was developed to match the geometry common to several brands of circuit breakers. The simulated magnetic fields were used to identify the appropriate location of the breaker pickup. Various breaker pickup shapes were considered to yield sufficient concentration of magnetic flux in the pickup core. Simulations using the finite element modeling software, FEMM, verified that a half toroid of high permeability material placed on the breaker face was suitable. A typical FEMM simulation is depicted in Figure 3.

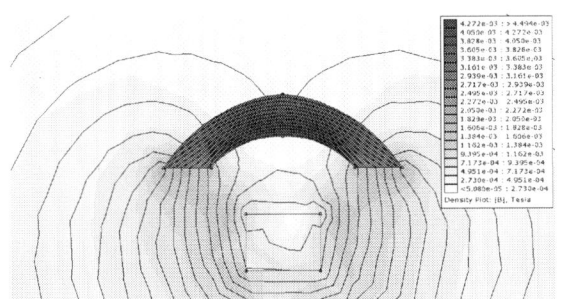

Fig. 3.   Finite Element Magnetic Model (FEMM) of magnetic flux through the breaker. The plastic breaker is ignored because it is neither conductive nor affected by the magnetic field.

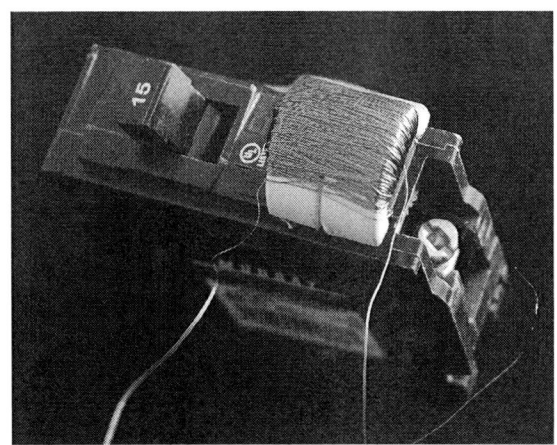

Fig. 4.   Breaker pickup photograph.

To form the magnetic yoke, a core[1] with relative permeability of 10,000 was first cut in half. The two halves were then joined for increased cross-sectional area. A total of 1,200 turns of 34 AWG magnet wire were wrapped around the yoke. A photograph of the breaker pickup affixed to the breaker face is shown in Figure 4. Maximizing the number of turns, cross-sectional area and relative permeability was important for increasing the inductance of the breaker pickup. The resulting inductive impedance at 60 Hz was sufficient to provide suitable voltage signals corresponding to the breaker current.

IV. JFET MIXER

The four-quadrant balanced JFET modulator (mixer), shown in Figure 5, was designed to transmit information from the breaker pickup through the inductive link and out of the breaker panel. This circuit consists of two JFET devices for modulation control and two resistors to improve linearity, but it does not require a DC power supply.

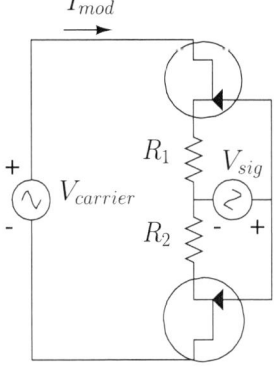

Fig. 5.   Adaptive Referencing Balanced two-JFET Modulator circuit enables simultaneous powering and modulation with no DC bus.

The JFET mixer can be modeled as a time-varying load on the carrier voltage source, $V_{carrier}$ in Figure 5, that

[1]Ferroxcube TX25/15/10-3E6 [13]

corresponds to the voltage on the inner coil in Figure 1(a). The load presented to $V_{carrier}$ in Figure 5 varies with the control signal, V_{sig}, applied to the JFET gates and leads to a corresponding modulation of the current I_{mod}. The two-JFET mixer circuit is particularly advantageous for this application because it requires a minimal amount of circuitry inside the breaker door and lends itself to a low-cost solution.

The JFET is a normally-on device that requires a negative gate-to-source voltage, V_{gs}, to turn it off. It may be modeled as a symmetric device, so that the drain and source are interchangeable and determined by which leg of the JFET has the lower potential. On positive half cycles ($V_{carrier} > 0$), the source of each device is the lower leg and the drain, the upper leg. The gate-source voltage of the lower device is the positive-valued drain-source voltage of the lower device added to the positive-valued drop across R_2 and V_{sig}. If the 60 Hz signal, V_{sig}, is sufficiently small, the lower device maintains a strictly positive gate-source voltage for most of the positive half cycle of the carrier signal. Thus the lower device can be taken to be fully on during that time, well-modeled by a small resistance. Meanwhile, the upper device has a gate-to-source voltage that is V_{sig} minus the positive drop across R_1. Again, if the 60Hz signal, V_{sig}, is sufficiently small, the upper device maintains a gate-source voltage that may not be strictly positive so that, instead of behaving as a small resistor, it is controlled so that its current varies according to the 60 Hz signal measured at the breaker face. In general, the gate-source voltage for the lower device contains a positive dc offset compared to that of the upper device even if they are not strictly positive and strictly negative respectively.

The roles of the two devices reverse when the polarity of the carrier signal reverses. The result is a modulation of the carrier signal current by the 60 Hz signal. The JFET mixer is adaptively-referencing because during both positive and negative half-cycles of $V_{carrier}$, one device is fully-on, referencing the source of the other device to low-potential end of the mixer circuit.

The behavior described above can be validated by simulations (LTSPICE) of the circuit in Figure 5. Figure 6(a) shows simulations of the gate-to-source voltages for the upper and lower devices with for a fixed positive carrier voltage. The gate-to-source voltage of the lower device is more positive than that of the upper device. Figure 6(b) shows the resulting 60Hz-modulated current when the circuit is excited with an ac-carrier voltage of 5kHz. Note that the quantities in the simulations of Figure 6 were chosen to provide illustrative visualizations of the behavior but do not necessarily correspond to design values or signal levels in our experimental setup. The plots of Figure 6 are also identical upon reversing the JFET devices with respect to the drain and source terminals, consistent with the drain-source symmetry of the JFETs described above.

The experimental setup is built with PN4117A JFET devices from Fairchild semiconductor, and 1.2 kΩ resistors to improve linearity. Modulation behavior was confirmed experimentally.

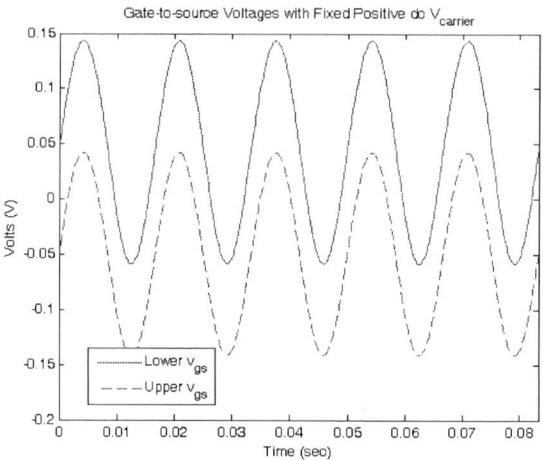

(a) Gate-to-source voltages with a fixed dc $V_{carrier} = 10$V.

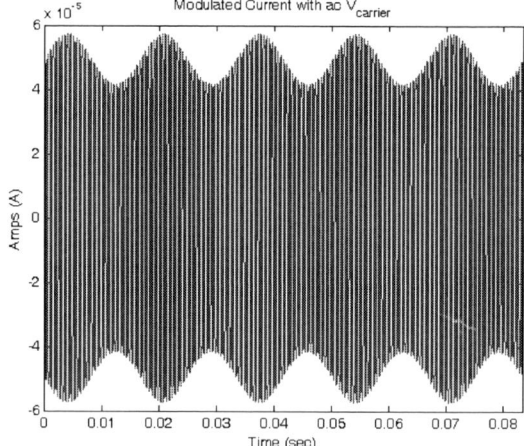

(b) Modulated current, I_{mod}, with a 2kHz 10V amplitude $V_{carrier}$.

Fig. 6. Balanced JFET Mixer, adaptive referencing and modulation behavior. PN4117 devices, $R_1 = R_2 = 1k\Omega$, $|V_{sig}| = 0.1$V, $f_{sig} = 60$Hz.

V. THROUGH-DOOR INDUCTIVE LINK

The inductive link across the steel door shown in Figure 1(a) consists of two resonant coils wrapped around samarium cobalt magnets with a N_1:N_2 turns ratio. These two coils and the steel door form a magnetic circuit linking the two coils. We take the magnetic reluctance model of Figure 7 to model the behavior of the inductive link. Reluctance is given in Equation (1) where l is the magnetic path length and A is the magnetic area.

$$R = \frac{l}{\mu A} \qquad (1)$$

The MMF generators $N_1 I_p$ and $N_2 I_2$ model the effect of the two coils and the reluctances capture the magnetic paths in the pieces of the physical system. Most notably, the reluctance of the path through the steel door, R_{steel}, is generally small because of the high permeability of steel. This presents a design challenge for the inductive link because R_{steel} tends to shunt magnetic flux away from the desired magnetic path. That is, the steel door acts as a shield to the magnetic fields. The inductive link was designed to overcome the challenge

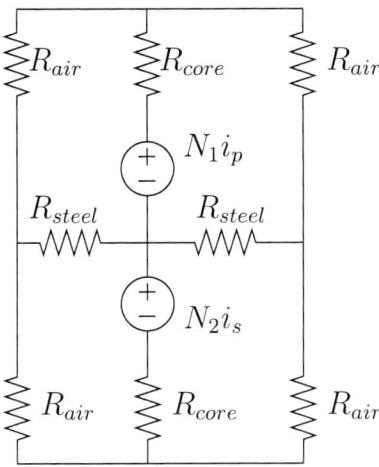

Fig. 7.   Reluctance model of through door transmission

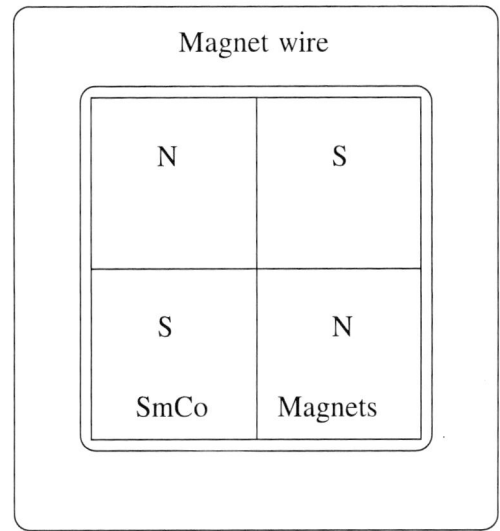

Fig. 8.   Top view of transmission coil configuration

described above by selecting (empirically) an optimum carrier frequency. At high frequency, the permeability of the steel diminishes thereby decreasing the shielding effect due to the ferromagnetism in the steel [14]. However, at high frequency, the shielding effect due to eddy currents induced in the steel intensifies. Therefore, we expect an optimum frequency where the shielding due to high-frequency eddy currents and low-frequency magnetic permeability is minimized. We attempt, in our experimental setup, to further reduce the permeability of the steel using strong magnets. The arrangement of the magnets (Figure 8) is intended to saturate the magnetic domains in the steel thereby increasing the reluctance R_{steel} in the model of Figure 7. Moreover, the magnets provide a convenient means of securing the device to the door. Finally, the coil geometry itself is broad and flat yielding a large A and a small l for the reluctance described in equation (1).

The signal of interest to the sense circuitry of Figure 1(a) is the current drawn by the JFET mixer on the inside of the door. That current is the high-frequency carrier modulated with the low-frequency (60 Hz) signal sensed from the breaker face. A large turns ratio, $\frac{N_2}{N_1}$, yields a large voltage gain to the inner coil to develop the necessary drain-source voltages in the JFET mixer. Also, a large turns ratio amplifies the current drawn by the JFET mixer to the outer coil. The number of turns on the inner coil may consist of as many turns of wire as allowed by the physical space, while the number of turns on the outer coil is lower-bounded by the current drive capability of the power source driving it.

A useful carrier frequency for the through-door inductive link was found empirically as follows. The demodulated 60Hz signal output was viewed on an oscilloscope while the carrier frequency was adjusted. For each adjustment of the carrier frequency, the capacitance comprising the sense impedance Z_{sense} was adjusted to maximize the demodulated signal at

that frequency. A suitable carrier frequency of approximately 220kHz was identified for our experimental setup. This method of carrier frequency selection can be automated if the carrier frequency and primary side current sense impedance are electronically adjustable.

VI. SENSE AND DEMODULATION CIRCUIT

The sense circuit is a DSP-based I/Q demodulation circuit. It consists of a power front-end responsible for driving the outer coil, an analog signal chain for demodulating the signal, and a DSP for performing post processing.

A. Power front-end

The power front-end, shown in Figure 9, is a push-pull driver composed of two BJT devices. The bases of these BJT devices are driven using a high voltage decompensated operational amplifier in a high gain configuration. A square wave at the carrier frequency is ac-coupled to the noninverting input of the amplifier. The operational amplifier then increases the voltage to a level suitable for driving the push-pull driver.

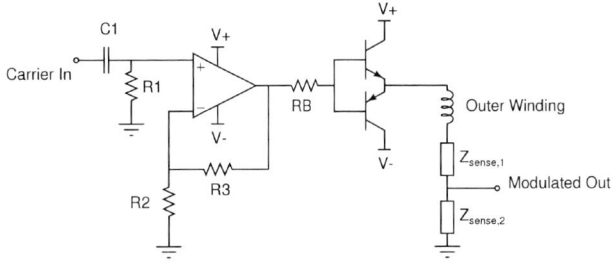

Fig. 9.   A simplified schematic of the coil drive power front-end.

The push-pull driver is connected to the series combination of the outer coil and two sense impedances. The voltage between the two sense impedances is taken as the input to the analog filter chain. The total sense impedance is matched to the coil impedance at the carrier frequency, but the ratio of the two impedances is chosen to deliver acceptable voltage levels to the analog signal chain.

B. Signal Processing and I/Q Demodulation

A block diagram of the analog signal chain is shown in Figure 10. The analog front end measures the voltage across the coil sense impedance, V_{sense}. In-phase and quadrature reference signals multiply the measured signal resulting in the I and Q channels shown. Multiplication is achieved with a full-bridge consisting of analog switches. The demodulated signal is followed by a low-pass filter that attenuates the remaining high frequency content. While the signal of interest is demodulated to 60 Hz, the unsuppressed carrier in the input signal is demodulated to dc. Removing the resulting dc offset with a highpass filter allows for a large gain before sampling. A final low-pass filter reduces aliasing in the sampled signal, while an offset of 1.65 V centers the signal in the input voltage range of the 12-bit ADC[2]. The I and Q channels are combined using a dsPIC33 to form the final output signal. Here, we discuss some tradeoffs in various implementations of I/Q demodulation while highlighting the method used in our experimental setup.

An amplitude modulated signal with unsuppressed-carrier content can be represented by

$$R(t) = A \cos (\omega_c t + \phi)$$
$$+ \frac{M}{2} (\cos ((\omega_c + \omega_m)t + \phi) + \cos ((\omega_c - \omega_m)t + \phi)),$$
$$\text{(2)}$$

where ω_c is the carrier frequency, A is the unsuppressed carrier amplitude, $M/2$ is the amplitude of the modulated signal, ω_m is the modulation frequency, and ϕ is a phase offset. The system presented here is inherently an unsuppressed carrier system because the parallel current path through the magnetizing inductance of the through door link will always yield a carrier signal in addition to the modulated signal. The phase term, ϕ, is unknown or time-varying because it depends on the properties of the breaker panel door in addition to the properties of the through door link and the modulator circuit. I/Q demodulation is used to detect the signal and cancel out the phase term.

To produce the I and the Q-channel signals, the received signal, represented by $R(t)$ above, is multiplied by in-phase and quadrature reference signals, represented by $\cos (\omega_c t)$ and $\sin (\omega_c t)$ in Figure 10. Assuming high frequency terms have been perfectly eliminated by the ensuing low-pass filter, the final signals can be shown to be

$$I(t) = \cos (-\phi)(\frac{M}{2} \cos \omega_m t + \frac{A}{2}) \quad \text{(3)}$$

$$Q(t) = \sin (-\phi)(\frac{M}{2} \cos \omega_m t + \frac{A}{2}). \quad \text{(4)}$$

[2]Internal to the dsPIC33FJ256MC710

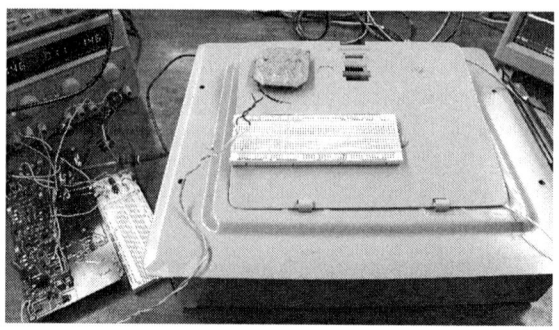

Fig. 11. Entire experimental setup with circuit breaker door closed. The outer through door coil and demodulation circuitry are visible.

The final step in I/Q demodulation would be to take the square root of the sum of the squares of $I(t)$ and $Q(t)$ in equations (3) and (4) above. The result of that operation would be the desired signal

$$S(t) = \frac{M}{2} \cos \omega_m t + \frac{A}{2}, \quad \text{(5)}$$

in which the unknown phase terms have been eliminated by combining the I and Q signals.

In our experimental setup, the signals in (3) and (4) are first high-pass filtered to remove the $A/2$ offset terms before sampling. This allows for a significant gain (x100 in Figure 10) and an improvement in the SNR of the sampled signal. However, the filtered signal is bipolar and the ensuing sum of squares operation described above would normally be stripped of the original sign information. While, in principle, this sign information could be recovered from knowledge of ϕ, in practice, the value of ϕ is unknown and may drift in time. One approach is to add an offset to the signal after the HPF and gain. This removes sign ambiguity, but will also reduce the accuracy of the I/Q demodulation result because the added offset creates an unwanted nonlinearity in the sum of squares calculation. An alternate approach would be to retain the offset (remove the HPF in Figure 10). However, in that approach, the remaining dc offset will limit the ensuing gain. A compromise between these two extremes would be to only remove a fraction of the $\frac{A}{2}$ offset. If this fraction is properly selected, the remaining offset, $\frac{A'}{2}$, may remove the sign ambiguity while still allowing for a sufficient gain without saturation. This approach would effectively control the modulation depth of the sampled signal. In the current experimental setup, the sum of squares operation is performed after adding an artificial offset which yields the tradeoff between sign ambiguity and accuracy described above.

VII. EXPERIMENTAL SETUP AND RESULTS

A photograph of the experimental setup is shown in Figure 11. An HP 6834B ac power source was used to generate test current signals. The current pickup was secured to the breaker face with leads running to a circuit attached to the inside surface of the steel door. This was connected to the

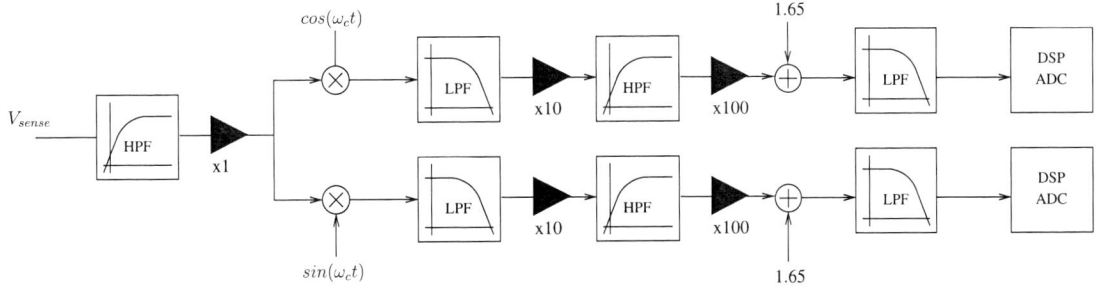

Fig. 10. Block diagram of analog filter chain

Fig. 12. Experimental setup with circuit breaker door open. The JFET Mixer, test circuit breaker, and inner coil are visible.

Fig. 13. Demodulation board attached to power supply.

14.4:1 step-up transformer, the Tamura MET-01, before being connected to the JFET modulator. This portion of the setup is shown in Figure 12. The JFET modulator was built on the same breadboard and connected to the inner coil.

The inner coil consisted of 1000 turns of 34 AWG magnet wire wrapped around four samarium cobalt magnets of dimension 1/2"x1/2"x1/4" and grade 26 MGOe arranged as shown in Figure 8. The outer coil was similarly constructed with a set of permanent magnets arranged so that the N and S poles of magnets on opposite sides of the door were facing each other. The outer coil was wound with 24 windings and then connected to the signal processing electronics described above.

Experimental results from the prototype monitoring hardware are shown in Figures 14 and 15. The HP 6834B ac source was used to provide different frequencies and amplitudes of current through the test circuit breaker monitored by the prototype sensor. The programmable ac source made it possible to control the current frequency, permitting not only 60 Hz sinusoidal currents, but also higher frequencies that might appear in distorted current waveforms with higher harmonic content, e.g., as might arise in the current drawn by an unfiltered or lightly filtered full-bridge rectifier at the front-end of a power supply.

Figure 14 shows the performance of the current sensor for two different frequencies of current, 60 Hz and 180 Hz, flowing through the breaker. The 180 Hz current represents

third harmonic content, typical in current waveforms distorted by rectification. The two current waveforms, shown in Figure 14 in graphs (a) and (b), are shown during experiments with individual frequency currents for clarity, although the sensor works well for combined currents with fundamental 60 Hz content and harmonic distortion. The current levels in both tests correspond to 5 A peak.

The middle graphs in Figure 14, graphs (c) and (d), show spectrum analyzer plots of the modulated carrier on the primary side or "outside" the door of the circuit breaker panel. As expected, the spectrum shows a strong peak at the carrier frequency. The side-lobes correspond to the modulating current signal, at 60 Hz and 180 Hz differences from the carrier frequency.

The lower graphs in Figure 14, graphs (e) and (f), show the reconstructed or demodulated current signals on the primary side as computed by the DSP microcontroller. The amplitudes of the signals are scaled by the transfer characteristic pickup core at the face of the breaker. That is, the inductance of the pickup core varies with the frequency of the sensed magnetic field and associated current, filtering the lower frequencies in favor of the higher frequencies, as shown in Figure 14. This is a predictable variation with frequency, and can be easily "inverted" with a digital filter (not employed to make Figure 14) on the primary side in the DSP microcontroller "outside" of the breaker panel. That is, the final output of a full sensor

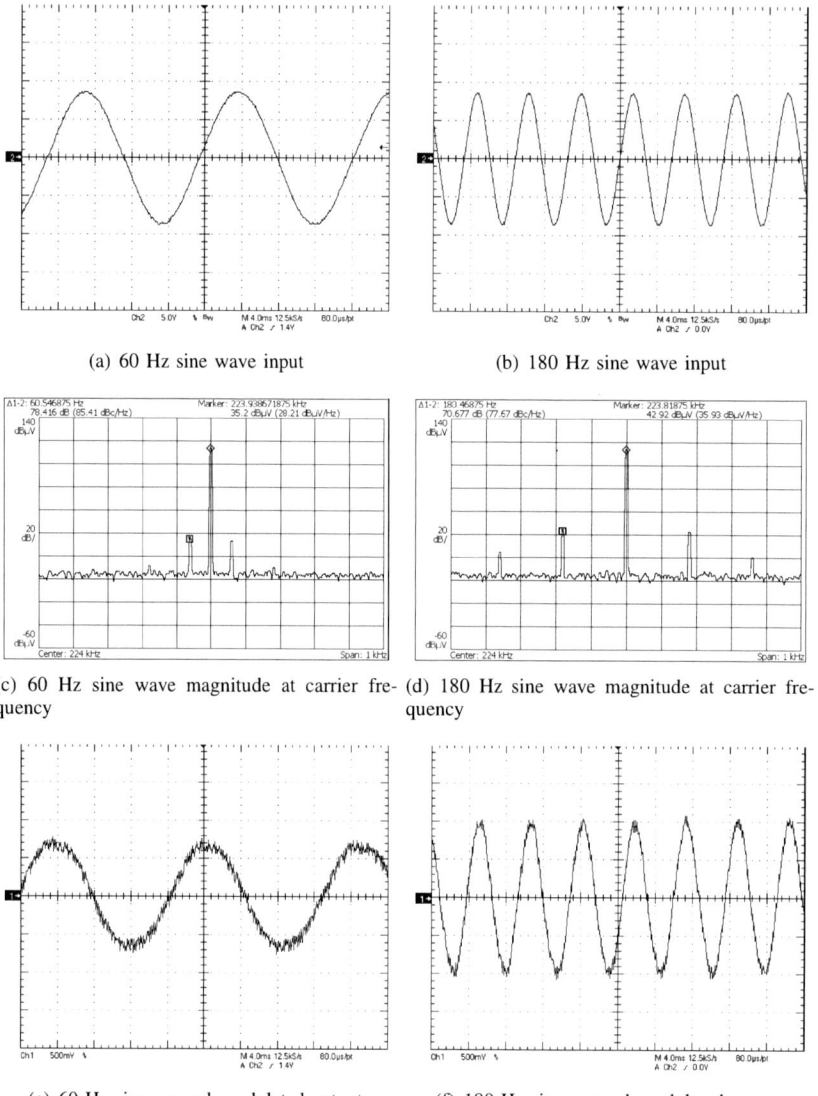

(a) 60 Hz sine wave input

(b) 180 Hz sine wave input

(c) 60 Hz sine wave magnitude at carrier frequency

(d) 180 Hz sine wave magnitude at carrier frequency

(e) 60 Hz sine wave demodulated output

(f) 180 Hz sine wave demodulated output

Fig. 14.   (a) and (b): Breaker currents generated for testing, (c) and (d): frequency spectrum of the signal received outside the door prior to demodulation, (e) and (f): demodulated signals corresponding to the breaker currents in (a) and (b).

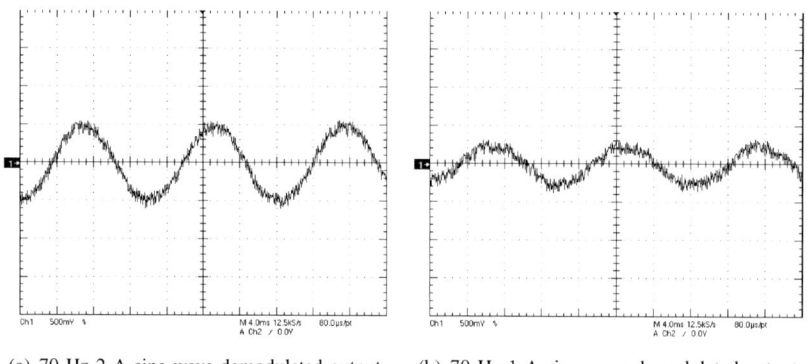

(a) 70 Hz 2 A sine wave demodulated output

(b) 70 Hz 1 A sine wave demodulated output

Fig. 15.   70 Hz test signals were generated to verify that measured signals were not simply 60 Hz ambient pickup.

978-1-4244-4782-4/10 $26.00 © 2010 IEEE

system would be scaled in software with a digital filter to recover the correct current amplitudes.

Figure 15 shows demodulation and processing of a 70 Hz signal. The first graph (a) shows the output of the sensor when a 2A, 70 Hz current is passing through the breaker. The second graph (b) shows the output of the sensor for a 1A, 70 Hz current. This test demonstrates that the output of the sensor system is robust with respect to "pick-up". That is, this test indicates that the sensor is not corrupted by other 60 Hz signals in the general environment of the monitored breaker and circuit panel. Since 70 Hz is not among the harmonics of 60 Hz, this shows that the signal being measured is actually from the test current source and not unwanted pickup from the air.

The prototype sensor has successfully detected currents as small as 100 mA in amplitude, and should be able to detect substantially smaller currents with signal processing hardware optimized for low noise and higher digital resolution.

VIII. CONCLUSION

The U.S. Department of Energy has identified "sensing and measurement" as one of the "five fundamental technologies" essential for driving the creation of a "Smart Grid" [15]. Consumers will need "simple, accessible. . . , rich, useful information" to help manage their electrical consumption without interference in their lives [15]. Both vendors and consumers will likely find innumerable ways to mine information if it is made available in a useful form. However, appropriate sensing and information delivery systems remain a chief bottleneck for many applications, and metering hardware and access to metered information will likely limit the implementation of new electric energy conservation strategies in the near future.

Closed or "clamp" core sensors wrapped around the utility feed are often used to provide current sense signals to energy monitoring systems. These sensors prohibit widespread monitoring because skilled labor must separate line and neutral in order to deploy a wrap-around magnetic field sensor, even in the case of separable core sensors. This paper proposes an alternative to traditional clamp or Hall-effect sensors. This alternative requires no skilled installation. This sensor measures the current in the utility feed by sensing the resulting magnetic field at the face of the main (or other) circuit breaker in a standard breaker panel. The sensor can be interrogated through the steel panel door with no direct electrical contact, permitting the door to remain closed to comply with safety regulations.

The proposed sensor could be a "silver bullet" for many power monitoring and control problems. The sensor is as easy to install in a retrofit situation as in new work. This approach could make it easy to provide essential, comprehensible information about opportunities and the success of efforts for energy conservation.

Acknowledgments

The authors would like to thank The Grainger Foundation and BP-MIT research alliance for their generous and necessary support and funding. This work was partially supported by the Center for Materials Science and Engineering at MIT as part of the MRSEC Program of the National Science Foundation under grant number DMR-08-19762.

REFERENCES

[1] J. S. Ramsey, S. B. Leeb, T. DeNucci, J. Paris, M. Obar, R. Cox, C. Laughman, and T. J. McCoy, "Shipboard applications of non-intrusive load monitoring," in *American Society of Naval Engineers Reconfigurability and Survivability Symposium*, Atlantic Beach, Florida, February 2005.

[2] T. DeNucci, R. Cox, S. B. Leeb, J. Paris, T. J. McCoy, C. Laughman, and W. Greene, "Diagnostic indicators for shipboard systems using non-intrusive load monitoring," in *IEEE Electric Ship Technologies Symposium*, Philadelphia, Pennsylvania, July 2005.

[3] W. Greene, J. S. Ramsey, S. B. Leeb, T. DeNucci, J. Paris, M. Obar, R. Cox, C. Laughman, and T. J. McCoy, "Non-intrusive monitoring for condition-based maintenance," in *American Society of Naval Engineers Reconfigurability and Survivability Symposium*, Atlantic Beach, Florida, February 2005.

[4] S. B. Leeb, S. R. Shaw, and J. J. L. Kirtley, "Transient event detection in spectral envelope estimates for nonintrusive load monitoring," *IEEE Transactions on Power Delivery*, vol. 10, no. 3, pp. 1200–1210, July 1995.

[5] L. K. Norford and S. B. Leeb, "Non-intrusive electrical load monitoring in commercial buildings based on steady state and transient load-detection algorithms," *Energy and Buildings*, vol. 24, pp. 51–64, 1996.

[6] U. A. Khan, S. B. Leeb, and M. C. Lee, "A multiprocessor for transient event detection," *IEEE Transactions on Power Delivery*, vol. 12, no. 1, pp. 51–60, 1997.

[7] S. R. Shaw, S. B. Leeb, L. K. Norford, and R. W. Cox, "Nonintrusive load monitoring and diagnostics in power systems," *IEEE Transactions on Instrumentation and Measurement*, vol. 57, no. 7, pp. 1445–1454, July 2008.

[8] G. R. Mitchell, R. W. Cox, J. Paris, and S. B. Leeb, "Shipboard fluid system diagnostic indicators using non-intrusive load," *Naval Engineers Journal*, vol. 119, no. 1, November 2007.

[9] J. P. Mosman, R. W. Cox, D. McKay, S. B. Leeb, and T. McCoy, "Diagnostic indicators for shipboard cycling systems using non-intrusive load monitoring," in *American Society for Naval Engineers Day 2006*, Arlington, VA, June 2006.

[10] R. W. Cox, P. Bennett, D. McKay, J. Paris, and S. B. Leeb, "Using the non-intrusive load monitor for shipboard supervisory control," in *IEEE Electric Ship Technologies Symposium*, Arlington, VA, May 2007.

[11] G. Mitchell, R. W. Cox, M. Piber, P. Bennett, J. Paris, W. Wichakool, and S. B. Leeb, "Shipboard fluid system diagnostic indicators using nonintrusive load monitoring," in *American Society for Naval Engineers Day 2007*, Arlington, VA, June 2007.

[12] E. Proper, R. W. Cox, S. B. Leeb, K. Douglas, J. Paris, W. Wichakool, L. Foulks, R. Jones, P. Branch, A. Fuller, J. Leghorn, and G. Elkins, "Field demonstration of a real-time non-intrusive monitoring system for condition-based maintenance," in *Electric Ship Design Symposium*, National Harbor, Maryland, February 2009.

[13] I. Ferroxcube, "Tx25/15/10-3e6 datasheet," Available http://www.ferroxcube.com/prod/assets/tx251510.pdf.

[14] N. Bowler, "Frequency-dependence of relative permeability in steel," *Review of Quantitative Nondestructive Evaluation*, vol. 25, pp. 1269–1276, 2006.

[15] U. S. D. of Energy, "The smart grid: An introduction," August 2009.

Feasibility of Capacitor Voltage Regulation and Output Voltage Harmonic Minimization in Cascaded H-Bridge Converters

Hossein Sepahvand, Mostafa Khazarei, Mehdi Ferdowsi, and Keith Corzine
Missouri University of Science and Technology
Rolla, MO 65409 USA
http://power.mst.edu

Abstract— **Multilevel converters have found popularity in electric-drive vehicles. In cascaded H-bridge multilevel converters, it is advantageous to use only one DC voltage source for the main H-bridge cell and supply the rest of the cells with capacitors. However, regulating the capacitors' voltage level becomes a challenging issue. This paper studies the existing constraints on regulating the capacitor voltage levels especially with respect to the elimination of the fifth and seventh harmonics. Detailed as well as simplified analyses are presented and verified. In addition, the constrained nonlinear optimization method is used to find the switching angles that regulate the capacitor voltage, eliminate the fifth harmonic, and minimize the combination of seventh and eleventh harmonics.**

I. INTRODUCTION

Multilevel converters are formed by combining two or more separate voltage sources to synthesize single or three-phase (commonly sinusoidal) voltage waveforms [1]. In general, by increasing the number of levels, the obtained voltage waveform becomes closer to its sinusoidal reference. This leads to reduced harmonic distortion when compared with conventional inverters. In addition, multilevel converters benefit from other advantage such as low electromagnetic emissions, high efficiency, lower voltage stress on switches, and modularity [1-5]. Furthermore, due to the high VA (volt-amp) ratings of multilevel converters, they are one of the best choices for electric motor drives in hybrid power-trains [2, 3].

The cascaded H-bridge multilevel converter consists of two or more H-bridge converters connected in series. This type of converter has been applied to high-power and high-quality applications including static VAR generation (SVG) [6], active filters, reactive power compensators [7], photovoltaic power conversion [8], and UPSs. In early implementations, each H-bridge converter cell had its own voltage source [9-12]. Since using separate voltage sources for

each H-bridge converter cell is costly and requires numerous practical considerations, it is proposed to use only one voltage for the main converter cell and utilize capacitors as voltage sources for the remaining converter cells [10, 13-16]. Although using capacitors has significant advantages, regulating the voltage of the capacitors becomes an issue in this case. In previous studies, it is shown that by choosing the right voltage level for the capacitors, it is possible to create redundant states which charge or discharge the capacitor. These redundant states can then be utilized to regulate the capacitor voltage at the desired level [10, 13-16]. In [13], it is assumed that the existence of these redundancies is adequate to regulate the capacitor voltage in all conditions. However, the time duration of redundant switching states has a great impact on the possibility of regulating the capacitor voltages. In other words, the voltage regulation of capacitors is only feasible in a limited range of operation.

This paper examines the constraints on the capacitor voltage regulation in cascaded H-bridge converters. Then, utilizing the obtained regulation constraints, a method is proposed to minimize the harmonics of the output voltage waveform. In this paper, a cascaded H-bridge multilevel converter with two cells is studied in which a capacitor is used as a voltage source for one of the H-bridge converter cells (see Figure 1). This converter generates seven different voltage levels. In Section II, switching angles are chosen in a way to eliminate the fifth and seventh harmonics of the output voltage waveform. In Section III, a method is proposed to find a condition to check the possibility of the regulation of the capacitor voltage for the entire range of the modulation index. Based on this method, the constrained nonlinear optimization method is used to find the switching angles that can regulate the capacitor voltage, eliminate the fifth harmonic, and at the same time minimize the combination of the seventh and eleventh harmonics.

978-1-4244-4782-4/10 $26.00 © 2010 IEEE

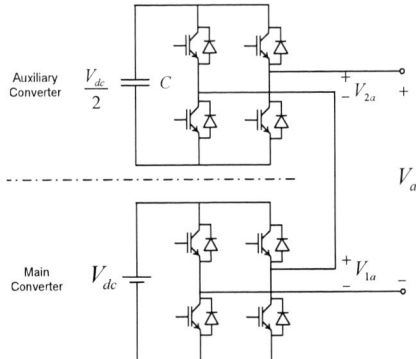

Figure 1. Circuit diagram of the cascaded H-bridge multilevel converter

II. SEVEN-LEVEL CASCADED H-BRIDGE CONVERTER

In this study, a seven-level cascaded H-bridge multilevel converter consisting of two H-bridge converter cells is considered (see Figure 1). The main H-bridge cell has an independent dc voltage source and the auxiliary converter cell is fed by a capacitor. The voltage across this capacitor is meant to be regulated at half of the voltage level of the main converter. According to Table I, there are two redundant states for the $V_{dc}/2$ and $-V_{dc}/2$ levels. In one of the redundant states the capacitor is charged while in the other one the capacitor is discharged.

A. Elimination of the Fifth and the Seventh Harmonics

The Fourier series expansion of the staircase output voltage waveform which is shown in Figure 2 is [13]:

$$V_a(\theta_1, \theta_2, \theta_3) = \frac{V_{dc}}{2} \frac{4}{\pi} \sum_{n=1,3,5,\dots}^{\infty} \frac{1}{n} \big(\cos(n\theta_1) + \cos(n\theta_2) + cose(n\theta_3) \big) * \sin(n\omega t) \quad (1)$$

where θ_1, θ_2, and θ_3 are switching angles of the output voltage waveform. Equation (1) consists of three variables (θ_1, θ_2, and θ_3) which can be chosen to satisfy three conditions. Since it is required to generate an output voltage waveform which has variable amplitude in its fundamental harmonic, one of the variables is consumed. Therefore, it is possible to satisfy two other conditions with the remaining variables.

Since in a three-phase system, the third harmonics is naturally canceled; therefore, switching angles can be chosen to eliminate the fifth and the seventh harmonics. In other words, the solution of the following system of equations yields the desired modulation index and eliminates the fifth and seventh harmonics in the output voltage waveform of the cascaded H-bridge converter. Here in, m denotes modulation index and is defined as

$$m = \frac{\pi}{2} \frac{v_m}{V_{dc}}. \quad (2)$$

TABLE I. OUTPUT VOLTAGE OF THE SEVEN-LEVEL H-BRIDGE CONVERTER

V_{1a}	V_{2a}	V_a	V_{1a}	V_{2a}	V_a
0	0	0	0	$-V_{dc}/2$	$-V_{dc}/2$
0	$V_{dc}/2$	$V_{dc}/2$	$-V_{dc}$	$V_{dc}/2$	$-V_{dc}/2$
V_{dc}	$-V_{dc}/2$	$V_{dc}/2$	$-V_{dc}$	0	$-V_{dc}$
V_{dc}	0	V_{dc}	$-V_{dc}$	$-V_{dc}/2$	$-3V_{dc}/2$
V_{dc}	$V_{dc}/2$	$3V_{dc}/2$			

where v_m is the peak voltage of the desired fundamental frequency of the converter.

$$\cos(\theta_1) + \cos(\theta_2) + \cos(\theta_3) = m \quad (3)$$
$$\cos(5\theta_1) + \cos(5\theta_2) + \cos(5\theta_3) = 0 \quad (4)$$
$$\cos(7\theta_1) + \cos(7\theta_2) + \cos(7\theta_3) = 0 \quad (5)$$

In order to solve the system of equations in (3)-(5), several approaches, including resultant theory, have been proposed [13, 17]. In addition to the discussed methods, one can use MATLAB nonlinear solvers to solve these nonlinear systems of equations. This method can be used without dealing with the complexity of the previously proposed methods [13, 17]. The results are depicted in Figure 3. As it is shown, when m is between 1.488 and 1.852, two sets of solutions exist.

III. CONSTRAINTS ON CAPACITOR VOLTAGE REGULATION

In the following calculations, without the loss of generality, the load is assumed to be resistive. In this converter, the maximum capacitor discharge occurs when the output voltage is at its maximum level. In resistive loads, the current is at its maximum level in this subinterval. Therefore, the capacitor is discharged faster than other types of loads (inductive or capacitive). In other words, if the constraint

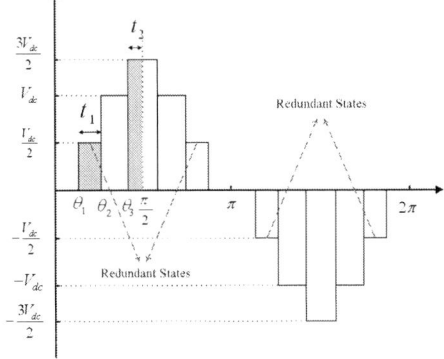

Figure 2. Staircase output voltage waveform of the seven-level cascaded H-bridge converter

978-1-4244-4782-4/10 $26.00 © 2010 IEEE

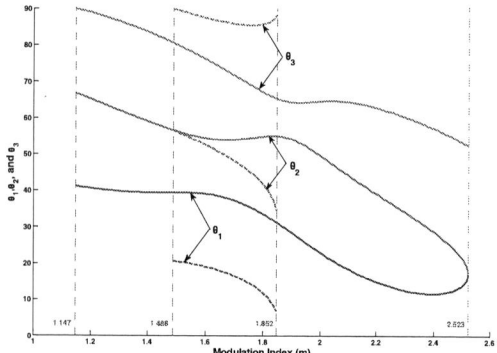

Figure 3. Solutions for the switching angles when the elimination of the fifth and seventh harmonics is intended

satisfies resistive loads it also satisfies inductive or capacitive loads.

Since the charging and discharging patterns are repeated every quarter of the period, examining only one quarter is adequate. The first quarter of the period is divided into four subintervals and the voltage level of the capacitor is investigated in each subinterval (see Figure 2).

The following two methods are presented for investigation of regulation condition of capacitor.

A. Detailed method:

The four subintervals in one quarter are as follow:

$0 \to \theta_1$ No charge or discharge subinterval: No capacitor charge or discharge occurs in this subinterval.

$\theta_1 \to \theta_2$ Charge or discharge subinterval: Although it is possible to charge or discharge the capacitor in this subinterval, here only the charging option is considered since the focus is on the capacitor voltage regulation. The left-hand side of Figure 4 shows the simplified combination of the two H-bridge converter cells during this subinterval. The capacitor voltage is given by

$$V_c = V_{dc} + \frac{V_{dc}}{2}\left(1 - e^{-\frac{t}{\tau}}\right). \tag{6}$$

The capacitor voltage increment in this subinterval can be described as

$$\Delta V_{c1} = \frac{V_{dc}}{2}\left(1 - e^{-\frac{t_1}{\tau}}\right). \tag{7}$$

By expanding the exponential term and considering that the capacitor used for the auxiliary H-bridge cell is a large capacitor; the changes in the capacitor voltage can be simplified to

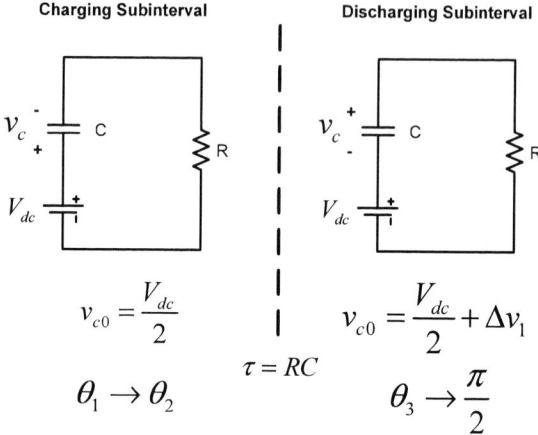

Figure 4. Simplified H-bridge converter and load in charging and discharging subintervals

$$\Delta V_{c1} = \frac{V_{dc}}{2} * \frac{t_1}{\tau} = \frac{V_{dc}}{2} * \frac{\theta_2 - \theta_1}{\omega\tau}. \tag{8}$$

$\theta_2 \to \theta_3$ No charge or discharge subinterval: In this subinterval the capacitor is not involved; therefore, no charge or discharge occurs.

$\theta_3 \to \frac{\pi}{2}$ Discharge subinterval: The capacitor is discharged by the load current in this subinterval. The right-hand side of Figure 4 shows the simplified combination of the two H-bridge converter cells during this subinterval.

$$\Delta V_{c2} = -(V_{dc} + V_{c0})\left(1 - e^{-\frac{t_2}{\tau}}\right) \to \Delta V_{c2}$$
$$= -\frac{V_{dc}}{2}\left(3 + \frac{t_1}{\tau}\right) * \frac{t_2}{\tau} \tag{9}$$

Since the capacitor used for the auxiliary H-bridge cell is a large capacitor, (9) can be simplified to

$$\Delta V_{c2} = -3\frac{V_{dc}}{2} * \frac{t_2}{\tau} = -\frac{3V_{dc}}{2} * \frac{\frac{\pi}{2} - \theta_3}{\omega\tau}. \tag{10}$$

The capacitor voltage increment during the charge subinterval should be more than the voltage decrement during the discharge subinterval to insure capacitor voltage regulation. Therefore, the capacitor voltage can be regulated if $|\Delta V_{c1}| > |\Delta V_{c2}|$. Consequently, the regulation condition can be described as

$$-\theta_1 + \theta_2 + 3\theta_3 > \frac{3\pi}{2}. \tag{11}$$

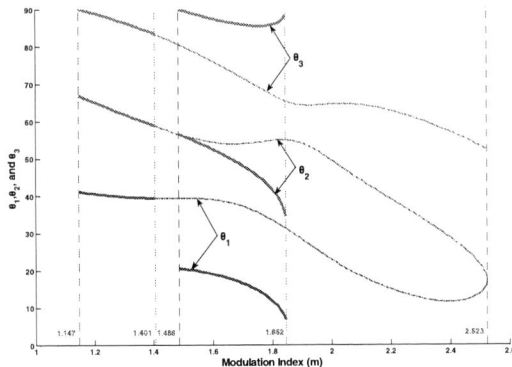

Figure 5. Switching angles for elimination of 5th and 7th harmonics
Solid: regulation is feasible **Dashed:** regulation is not feasible.

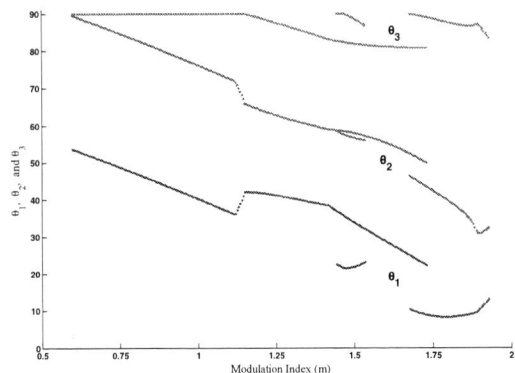

Figure 6. Switching angle solutions for elimination of the 5th harmonic and minimization of 7th and 11th harmonics

B. Simplified method:

In this method it is assumed that the output voltage (and consequently output current) is constant during each subinterval. This assumption is valid for big capacitor. Similar to the detailed method, capacitor regulation is investigated in one quarter of a period. In one quarter there are four subintervals but only $(\theta_1 \to \theta_2)$ and $(\theta_3 \to \frac{\pi}{2})$ intervals should be investigated for capacitor regulation conditions (only in these two subintervals capacitor is involved).

$\theta_1 \to \theta_2$ *Charge or discharge subinterval:*

Since it is assumed that the output voltage is constant in this subinterval, the output current (and capacitor current) is

$$I_{c1} = I_{o1} = \frac{V_{dc}}{2R}. \tag{12}$$

where R is output resistance. Change in capacitor charge in this subinterval can be described as

$$\Delta Q_1 = I_{c1} * t_1 = \frac{V_{dc}}{2R} * \frac{\theta_2 - \theta_1}{\omega}. \tag{13}$$

$\theta_3 \to \frac{\pi}{2}$ *Discharge subinterval.*

Since capacitor is discharging in this subinterval, capacitor current is negative of output current. Similarly output current (and negative of capacitor current) in this subinterval is

$$I_{c2} = -I_{o2} = -\frac{3V_{dc}}{2R}. \tag{14}$$

Changes in capacitor charge in this subinterval can be described as

$$\Delta Q_2 = I_{c2} * t_2 = -\frac{3V_{dc}}{2R} * \frac{\frac{\pi}{2} - \theta_3}{\omega}. \tag{15}$$

Capacitor voltage can be regulated if capacitor charge can be positive in one quarter of waveform. In the other word regulation of capacitor is possible if

$$\Delta Q = \Delta Q_1 + \Delta Q_2 > 0 \tag{16}$$

or

$$\frac{V_{dc}}{2R} * \frac{\theta_2 - \theta_1}{\omega} - \frac{3V_{dc}}{2R} * \frac{\frac{\pi}{2} - \theta_3}{\omega} > 0. \tag{17}$$

Simplification of (17) leads to a regulation condition which is the same as the condition found in the detailed method (11).

By applying (11) constraint on Figure 3, one finds the modulation indices that insure capacitor voltage regulation. In Figure 5, the range in which capacitor voltage regulation is feasible is shown in solid lines. The dashed lines represent the range in which the capacitor voltage cannot be regulated.

Since the regulation of the capacitor voltage is possible only in a limited range, it is not practical to use this staircase modulation method and eliminate the fifth and seventh harmonic at the same time for all of the range. However, there are several alternatives which can be used to widen the range and regulate the capacitor voltage at the same time. One alternative approach is to eliminate only the fifth harmonic in the entire range of the modulation index. The other method is to fix θ_3 to an angle close or equal to 90° in the areas that the simultaneous elimination of the fifth and seventh harmonics is not doable and then eliminate the fifth harmonic in those areas. One other approach is to consider the entire range of modulation index (m) and then obtain the set of angles which eliminate the fifth harmonic and at the same time minimize the combination of the seventh and eleventh harmonics. The latter approach is investigated here. In order to guarantee capacitor voltage regulation, in this approach, the angles should also comply with (10). The object function for this optimization problem is defined as

$$\min_{\theta_1, \theta_2, and\ \theta_3} \frac{\left(cos(7\theta_1) + cos(7\theta_2) + cos(7\theta_3)\right)^2}{7^2} + \frac{\left(cos(11\theta_1) + cos(11\theta_2) + cos(11\theta_3)\right)^2}{11^2} \tag{18}$$

Figure 6 depicts the results of nonlinear optimization using this method. As it is shown, the capacitor regulation region in

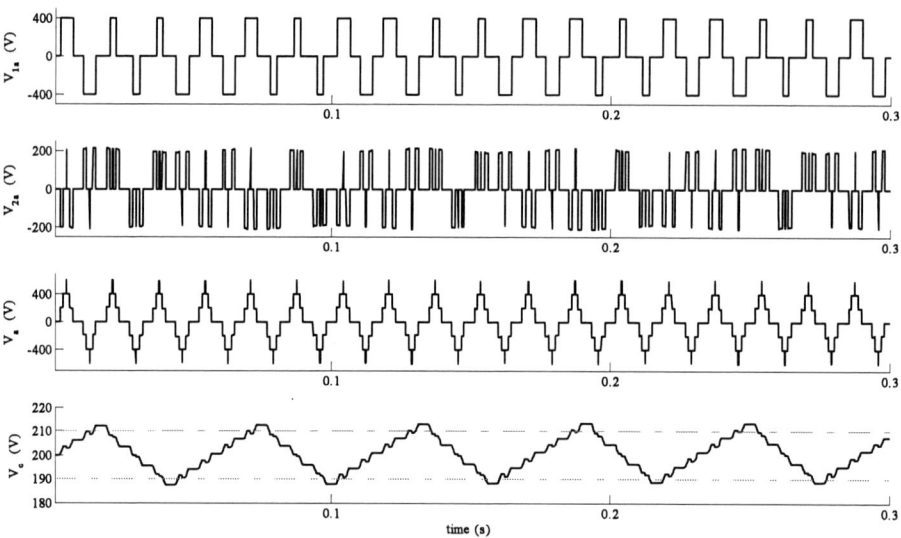

Figure 7. Voltage of converters and capacitor when capacitor voltage regulation is possible

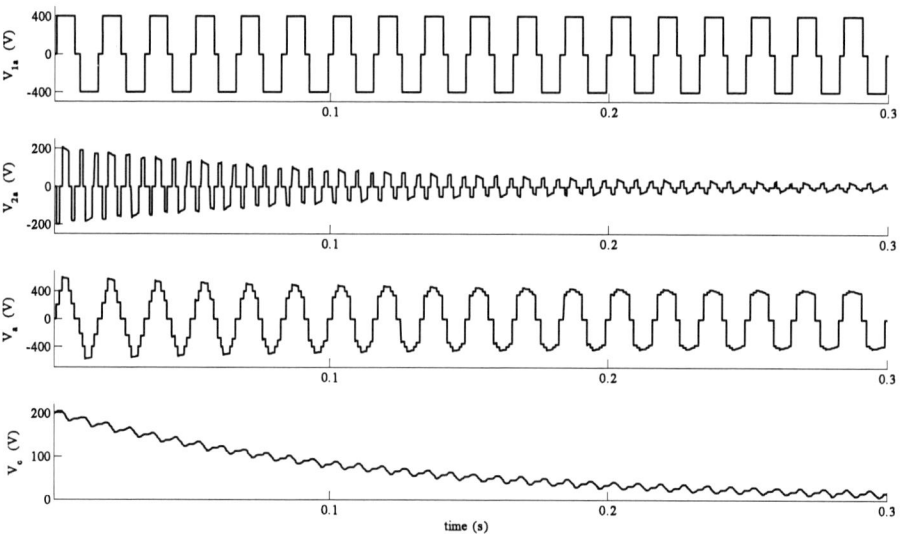

Figure 8. Voltage of converters and capacitor when capacitor voltage regulation is not possible

this approach is wider than the previous one (shown in Figure 5).

IV. SIMULATION RESULTS

The H-bridge converter is simulated in MATLAB/Simulink to verify the operation of the converter. In this simulation two cascaded H-bridge converters feed a resistive load (30Ω). The main converter is fed by a 400 V DC voltage source. A 2.1 mF capacitor is used in the auxiliary converter. The voltage of the capacitor is expected to be regulated at 200 V (half of the voltage source of the main converter). A hysteresis control is used to regulate the charge of capacitor and keep its voltage within ±5% range. In order to avoid extra switching in one period, decision for switching from charging to discharging mode (switching between two redundant statuses) is made every half of a period. In this section two sets of the angles are used to investigate the capacitor voltage regulation.

978-1-4244-4782-4/10 $26.00 © 2010 IEEE

In the first simulation, m is 1.2 then θ_1, θ_2, and θ_3 are selected to be 40.5°, 65.1°, and 88.9°, respectively (see Figure 3). In this case, according to (11), capacitor voltage regulation is expected to be possible (see also Figure 5). In this case control system is able to regulate the capacitor charge by switching from charging to discharging. In Figure 7, voltage of the main converter (V_{1a}), output voltage of auxiliary converter (V_{2a}), total output of cascade H-bridge converter (V_a), and voltage of the capacitor (V_c) are shown respectively. As it is shown, control system has been successful to regulate the charge of the capacitor.

In the second simulation, m is 2.1 then θ_1, θ_2, and θ_3 are selected to be 18.3°, 44.1°, and 64.4°, respectively (see Figure 3). Based on (11), capacitor voltage regulation is not possible for this value of m (see also Figure 5). Figure 8 depicts the operation of the converter using this set of switching angles. Even though the controller selects charging states for the capacitor, capacitor voltage eventually falls to 0 V. In Figure 8 voltage of the main converter (V_{1a}), output voltage of auxiliary converter (V_{2a}), total output of cascade H-bridge converter (V_a), and voltage of the capacitor (V_c) are shown respectively. As it is shown, the output voltage waveform gradually deteriorates as the capacitor gets discharged.

V. CONCLUSION

In this paper, a study is carried out to find the conditions in which it is possible to regulate the capacitor voltage of a cascaded seven-level H-bridge converter. Next, the obtained constraint is used to identify the range of the modulation index in which the regulation of capacitor voltage is possible when the elimination of the fifth and seventh harmonics of the output voltage is intended. In addition, utilizing nonlinear programming and based on the proposed method, a set of switching angles are obtained which not only minimize the seventh and eleventh harmonics and eliminate the fifth harmonic but also they make it possible to regulate the capacitor voltage at the same time. The proposed method leads to a wider range of modulation index in which the capacitor voltage regulation is possible.

REFERENCES

[1] J. S. Lai and F. Z. Peng, "Multilevel converters - A new breed of power converters," *IEEE Trans. Industry Applications*, vol. 32, pp. 509-517, May/June 1996.

[2] C. Cecati, A. Dell'Aquila, and M. Liserre, "Design of H-bridge multilevel active rectifier for traction systems," *IEEE Trans. Industry Applications*, vol. 39, no. 5, pp. 1541-1550, Sept./Oct. 2003.

[3] L. M. Tolbert and F. Z. Peng, "Multilevel converters for large electric drives," *IEEE Trans. Industry Applications*, vol. 34, pp. 36-44, 1999.

[4] J. Chiasson, L. M. Tolbert, and K. McKenzie, "A new approach to solving the harmonic elimination equations for a multilevel converter," in *Proc. IEEE Industry Applications Conference*, vol. 1, Oct. 2003, pp. 640-647.

[5] L. G. Franquelo, J. Rodriguez and J. I. Leon, "The age of multilevel converters arrives," *Industrial Electronics Magazine, IEEE*, vol. 2, no. 2, pp. 28-39, 2008

[6] G. Joos, X. Huang, and B. Ooi, "Direct-coupled multilevel cascaded series VAR compensators," *IEEE Trans. Industry Applications*, vol. 34, pp. 1156-1163, Sept./Oct. 1998.

[7] J. Dixon, L. Moran, E. Rodriguez, and R. Domke, "Reactive power compensation technologies: State-of-the-art review," *Proc. IEEE*, vol. 93, no. 12, Dec. 2005, pp. 2144–2164.

[8] A. R. Being, U. Kumar, and V. Ranganathan, "A novel fifteen level inverter for photovoltaic power supply system," *Proc. IEEE Industry Applications Society Ann. Meeting*, Seattle, WA, 2004, pp. 1165–1171.

[9] M. D. Manjrekar and T. A. Lipo, "A hybrid multilevel inverter topology for drive applications," in *Proc. the IEEE Applied Power Electronics Conference*, vol. 2, Feb. 1998, pp. 523-529.

[10] K. A. Corzine, F. A. Hardrick, and Y. L. Familiant, "A cascaded multi-level H-bridge inverter utilizing capacitor voltages sources," in *Proc. IASTAD Power Electronics Technology and Applications Conference*, Feb. 2003.

[11] K. A. Corzine, M. W. Wielebski, and F. Z. Peng, "Control of cascaded multilevel inverters," *IEEE Trans. Power Electronics*, vol. 19, pp. 732-738, May 2004.

[12] M. Veenstra and A. Rufer, "Control of a hybrid asymmetric multilevel inverter for competitive medium-voltage industrial drives," *IEEE Trans. Industry Applications*, vol.41, pp.665-664, March/April 2005.

[13] Z. Du, L. Tolbert, and J. Chiasson, "A cascade multilevel inverter using a single DC source," *Proc. Applied Power Electronics Conference and Exposition*, Mar. 2006, pp. 19-23.

[14] J. Liao, K. Wan, and M. Ferdowsi, "Cascaded H-bridge multilevel inverters - a reexamination", *Proc. IEEE VPPC*, Sept. 2007, pp. 203-207.

[15] Z. Du, L. M. Tolbert, B. Ozpineci, J. N. Chiasson, "Fundamental frequencey switching strategies of a seven-level hybrid cascaded H-bridge multilevel inverter,' *IEEE Trans. Power Electronics*, vol. 24, no. 1, pp. 25-33, Jan. 2009.

[16] J. A. Ulrich, A.R. Bendre, "Floating capacitor voltage regulation in diode clamped hybrid multilevel converters," *IEEE ESTS Symp.*, April 2009, pp. 197-202.

[17] L. M. Tolbert, K. McKenzie, and J. Chiasson, "Elimination of harmonics in a multilevel converter with non equal DC sources," *IEEE Trans. Industry Applications*, vol. 41, pp. 75-82, 2005.

978-1-4244-4782-4/10 $26.00 © 2010 IEEE

Examination of a PHEV Bidirectional Charger System for V2G Reactive Power Compensation

Mithat C. Kisacikoglu[1], Burak Ozpineci[2], and Leon M. Tolbert[1,2]

[1]Dept. of Electrical Engineering and Computer Science
The University of Tennessee
Knoxville, TN 37996-2100

[2]Power and Energy Systems Group
Oak Ridge National Laboratory
Oak Ridge, TN 37831

Abstract—**Plug-in hybrid electric vehicles (PHEVs) potentially have the capability to fulfill the energy storage needs of the electric grid by supplying ancillary services such as reactive power compensation, voltage regulation, and peak shaving. However, in order to allow bidirectional power transfer, the PHEV battery charger should be designed to manage such capability. While many different battery chargers have been available since the inception of the first electric vehicles (EVs), on-board, conductive chargers with bidirectional power transfer capability have recently drawn attention due to their inherent advantages in charging accessibility, ease of use, and efficiency. In this paper, a reactive power compensation case study using just the inverter dc-link capacitor is evaluated when a PHEV battery is under charging operation. Finally, the impact of providing these services on the batteries is also explained.**

Keywords - **PHEV; charger; V2G; reactive power; battery**

I. INTRODUCTION

Today, hybrid electric vehicles (HEVs) offer customers a way to increase gasoline mileage by having batteries and electric drive systems assist the internal combustion engine. However, HEVs lack the availability to go for more than just short distances at low speeds with only electric power because the battery is not capable of storing enough energy to power the vehicle for a daily commute. PHEVs provide electricity-only drive option up to a specified distance, and they can help reduce carbon emissions as well as other pollutants [1].

While PHEVs will provide economic and environmental benefits, they can also offer a potential source of energy storage which is valuable to the electric power grid. The possibility of using battery-powered vehicles to support the electric grid has been studied for more than a decade [2]. Recent papers including [3-5] have discussed several topologies and control methods that can perform bidirectional power transfer using a PHEV as a distributed energy resource. However, there has not been much technical analysis about reactive power compensation using bidirectional PHEV chargers as well as the effects of such a power support on the PHEV's battery and charger system components.

The purpose of this study is to examine a PHEV charger system to utilize it for reactive power support to the grid. The authors investigate different scenarios to deliver the stored energy from vehicle to grid (V2G) and explain the effects of this usage on the vehicle traction battery and the charger dc link capacitor. In the following section, authors discuss battery charger types briefly. Later, an analysis introduces the dynamics that govern bidirectional power flow in the system and shows how to control the on-board vehicle charger to provide reactive power to the electric grid. The battery of a PHEV can be used for ancillary services such as peak power shaving and reactive power support. However, in the simulation section of this paper, it is observed that compared to peak power shaving, reactive power regulation causes no degradation at all on battery life, since the dc link capacitor is enough for supplying full reactive power for level 1 charging and therefore the PHEV battery is not engaged in reactive power transfer.

II. ELECTRIC VEHICLE CHARGERS

Battery chargers play an important role by maintaining the condition and health of the battery while utilizing it for the best performance. A battery charger is a device that is composed of one or more power electronics circuits used to convert ac electrical energy into dc with an appropriate voltage level so as to charge a battery. It has the potential to increase charging availability of the PHEV since it can operate as a universal converter accepting different voltage and power levels. In addition, a battery charger should prevent overcharging from happening. Especially for lithium-ion batteries, the charger warrants a sophisticated charging control algorithm to avoid overcharging [6]. Also, balancing the battery cells requires special circuitry. Consequently, the charger should protect the battery from over-current, over-voltage, under-voltage, and over-temperature [7].

A PHEV battery can be charged either by a separate charging circuit or via using the traction drive that serves to power the electric motor. The first EVs used the former method. Since this option requires an extra charging circuit, it increases the total cost of the vehicle (if the charger is on the vehicle) or it requires a dedicated charging station (if the charger is off the vehicle). If the charger is on-board, it can be optimized to accept different charging levels as well as to match different vehicle battery requirements. With an on-board charger, a vehicle can be charged at any outlet that is available at home garages or workplaces with ground protection [8]. Availability of such charging places will increase the acceptance of PHEV technology.

978-1-4244-4782-4/10 $26.00 © 2010 IEEE

On the other hand, off-board chargers make use of fast charging and can charge a vehicle in a considerably shorter amount of time. It is possible to charge a battery in 10 minutes to increase its state of charge (SOC) by 50% with an off-board charger rated at 240 kW [9]. Also, according to Nissan, its Leaf electric car, which will be on the road in 2010 and mass produced in 2012, can be charged up to 80% SOC of its 24 kWh Li-ion battery pack in around 30 min at a quick charge station [10].

Since on-board chargers' power rating is limited due to space and weight restrictions on the vehicle, it takes much more time to fully charge a vehicle battery compared to off-board chargers. However, an integrated on-board charger utilizing the traction inverter can charge the battery at high power levels that reduce the charging time [11]. Using these types of chargers which are classified as Level 2+ chargers, it takes about one hour to put 80% SOC to a battery rated at 30 kWh [12]. Not only do integrated chargers connect the vehicle's battery to most available standard 120V and 240V outlets, with special configuration it also couples a PHEV to an off-board charger if faster charging is needed [11]. However, since integrated chargers use motor inductance as inverter input inductance by connecting the neutral point of the motor to the grid, the inductance of the motor may not be the optimal value for the inverter operation. Also, this design causes the majority of the losses to be the copper losses of the motor windings [5].

An additional point a charger can offer is the capability of transferring power not only from grid to vehicle but also from vehicle to grid so that each car would operate as a distributed power source.

In summary, on-board, conductive chargers with bidirectional power transferring capability have recently drawn attention due to their inherent advantages in cost, charging accessibility, ease of use and efficiency. The following section will present the theoretical analysis of an on-board, conductive charger to utilize it in bidirectional power transfer.

III. THEORETICAL SYSTEM ANALYSIS OF BIDIRECTIONAL POWER TRANSFER BETWEEN A PHEV AND THE GRID

A. Grid-Inverter

The PHEV charger that is analyzed in this study is composed of a full-bridge inverter/rectifier and a dc-dc converter. The analysis will start by investigating the interaction between the grid and the inverter. In order to understand all the dynamics, the basic ideal case is introduced with several assumptions that will make the computation much easier. During the analysis, the positive current direction will be assumed to be from grid to the inverter as shown in Fig. 1. Therefore, positive power sign (P = active power and Q = reactive power) corresponds to the power flow from grid to the inverter. The system parameters are given as follows:

$v_c(t)$ instantaneous charger voltage [V],

$v_s(t)$ instantaneous grid voltage [V],

$i_c(t)$ instantaneous charger current [A],

L_c coupling inductor [H],

δ phase difference between $v_c(t)$ and $v_s(t)$,

θ phase difference between $i_c(t)$ and $v_s(t)$.

Root mean square (rms) values of the instantaneous variables are given in capital cases throughout this study.

The grid voltage is assumed to be purely sinusoidal, and high frequency components of inverter output voltage, $v_c(t)$, is neglected for analysis purposes as shown by the following equations:

$$v_s(t) = \sqrt{2} V_s \sin(wt), \tag{1}$$

$$v_c(t) = \sqrt{2} V_c \sin(wt - \delta). \tag{2}$$

In order to ensure power transfer from charger to the utility, a coupling inductor is used and the two voltage sources are decoupled. From Fig. 1 and applying necessary mathematical transformations, the line current can be written as,

$$i_c(t) = \sqrt{2} I_c \sin(wt - \theta). \tag{3}$$

Since the default direction for active and reactive power transfer is from grid to charger, $i_c(t)$ and $v_c(t)$ are lagging the grid voltage. Also, note that the reactance is equal to

$$X_c = 2\pi f L_c, \tag{4}$$

where the system frequency, f, is 60 Hz.

Table I and the P-Q plane shown in Fig. 2 show all the different operation modes in which the system can be working. In order to conserve the amount of energy that is drawn from the battery and to keep the battery undisturbed as much as possible, operation in quadrants I and IV is preferred over working in quadrants II and III. In other words, PHEV battery will not provide active power to the grid in this study. Although the utility may prefer to be able to use the PHEV as a peak shaving power source, it may not be accepted by the vehicle manufacturers and the customers due to safety concerns, decrease in battery lifetime, and reduced available battery energy. The topology that is studied here can run in all four quadrants, but for the analysis the system dynamics will be when the charger is operating in quadrants I and IV.

From Fig. 1 it can also be written that

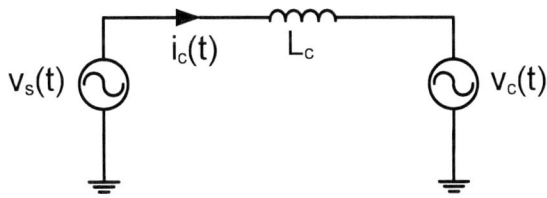

Fig. 1. Representation of grid and charger.

TABLE I. CHARGER OPERATION MODES

#	P	Q	Operation Mode of the Charger
1	Zero	Positive	Inductive
2	Zero	Negative	Capacitive
3	Positive	Zero	Charging
4	Negative	Zero	Discharging
5	Positive	Positive	Charging and inductive
6	Positive	Negative	Charging and capacitive
7	Negative	Positive	Discharging and inductive
8	Negative	Negative	Discharging and capacitive

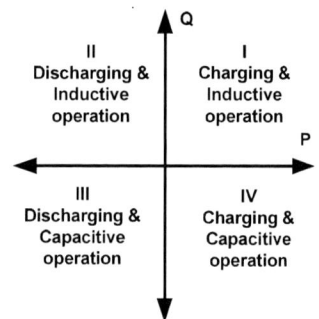

Fig. 2. P-Q plane showing charger operation modes.

$$V_s = V_c + jX_cI_c . \qquad (5)$$

Using (5), the system variables are shown in the phasor diagrams in Fig. 3 to illustrate the differences between the operation modes. Only the operation modes under discussion are explained in the phasor analysis. Some conclusions drawn from the sketches will help to understand the control algorithm. First, as illustrated in Fig. 3a and Fig. 3b, active power is provided by the grid as long as $v_c(t)$ lags $v_s(t)$, and it is sent to grid when $v_s(t)$ lags $v_c(t)$. Since $v_c(t)$ and $v_s(t)$ are sinusoidal, $i_c(t)$ is also sinusoidal as shown before. Its phase angle, θ, determines the direction of the reactive power flow. If θ is positive, reactive power is sent to the grid, and if θ is negative, reactive power is provided by the grid to the charger.

Based on the available charging infrastructure, the system will either be charged by level 1 or level 2 charging. Level 3 charging is not examined here. This analysis will be included in a future study. Therefore, the inverter current, i_c, is limited by the charging equipment to 12 A or 32 A. For all operations the control algorithm should maintain that the current stays below either of these levels. In Table II, different charging methods in North America are given for further reference.

There are two control methods to influence the magnitude and the direction of P and Q. The first option is to control the charger voltage, $v_c(t)$, and its phase angle, δ. The second option is to control the charger current, $i_c(t)$ and its phase angle, θ.

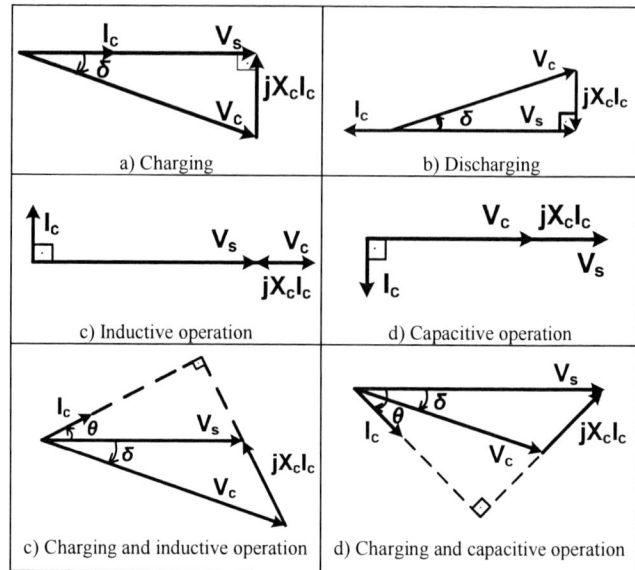

Fig. 3. Vector diagram for different operation modes.

The fundamental equations derived using these variables that govern average active and reactive power flow from grid to inverter is listed in Table III.

In summary, the variables that govern the interaction between the grid and the charger have been introduced in this section. The following section will describe the inverter operation.

B. Inverter

For this study, a full-bridge PWM inverter/rectifier is used as the first stage of the PHEV charger as shown in Fig. 4. Since the PHEV charger is operating like a current source, it is important that it complies with IEEE 1547 to present the minimum current harmonics possible. Therefore, a hysteresis-band current control PWM is used to effectively regulate the current waveform. As a result, current and its phase angle are selected to be the variables of the control algorithm.

The reason for using this topology is to have a system that is able to operate in all four quadrants of the P-Q plane. Although a half bridge inverter could also satisfy this

TABLE II. ELECTRICAL RATINGS OF DIFFERENT CHARGING METHODS IN NORTH AMERICA [13]

Charging method	Nominal supply voltage	Maximum current	Branch circuit breaker rating	Continuous input power
AC Level 1	120 V, 1-phase	12 A	15 A	1.44 kW
AC Level 2	208 to 240 V, 1-phase	32 A	40 A	6.66 to 7.68 kW
AC Level 3	208 to 600 V, 3-phase	400 A	As required	> 7.68 kW
DC charging	600 V maximum	400 A	As required	<240 kW

978-1-4244-4782-4/10 $26.00 © 2010 IEEE

TABLE III. FUNDAMENTAL EQUATIONS FOR ACTIVE AND REACTIVE POWER

Control Variable	P	Q
$v_c(t)$ and δ	$\dfrac{V_s \times V_c}{X_c}\sin(\delta)$	$\dfrac{V_s^2}{X_c}\left[1-\dfrac{V_c}{V_s}\cos(\delta)\right]$
$i_c(t)$ and θ	$V_s \times I_c \times \cos(\theta)$	$V_s \times I_c \times \sin(\theta)$

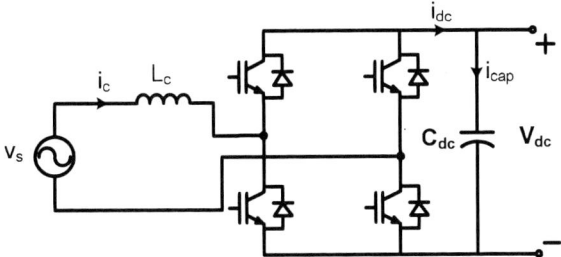

Fig. 4. Full bridge inverter charger.

operation, it requires two large capacitors to effectively regulate their junction voltage. Also, using full bridge active rectifier, the dc link voltage is doubled reducing the output current rating for the same power level.

According to the hysteresis-band current control PWM, the reference current generated by the controller is compared to the actual line current, and the switch pairs change their position accordingly.

The inverter control system shown in Fig. 5 operates with two feedback loops, one is for reactive power regulation and the other is for dc voltage regulation which indirectly facilitates active power transfer to the dc-dc converter. Based on these two feedbacks, the controller calculates the exact I_c and θ values to generate $i_c(t)^*$ as a reference waveform to be compared with actual line current.

The maximum switching frequency that is shown using hysteresis-band current control PWM is calculated as [14]

$$f_{max} = \frac{V_{dc}}{2L_c H}. \tag{6}$$

where H is the difference between upper and lower hysteresis bands and equal to 1 A. L_c is chosen to be 5 mH and V_{dc} to be 500 V. Therefore, the maximum switching frequency is calculated as

$$f_{max} = \frac{500}{2 \cdot 5 \cdot 10^{-3} \cdot 1} = 50kHz. \tag{7}$$

C. Dc Bus Components

In this section, the relationship between reactive power transfer and dc bus variables will be given. The dc parameters that will be analyzed are as follows:

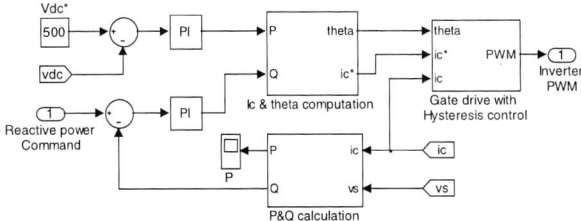

Fig. 5. Control system structure of the inverter.

V_{dc} nominal dc link voltage [V],

ΔV_{dc} rms dc ripple voltage [V],

ΔI_{cap} rms dc capacitor ripple current [A],

C_{dc} dc link capacitor [F].

The dc link capacitor's major purpose is to regulate dc voltage during battery charging. However, it can also be used for reactive power regulation. First analysis shows how the reactive power transfer affects ΔV_{dc}. As it is given in [15], in a full bridge inverter, the dc link voltage and current exhibit a ripple at double the frequency of the line voltage with the same phase of the line current. For the start of the analysis, the PWM ripples are neglected and only *2f* ripple is considered. Therefore, the instantaneous capacitor voltage and current can be defined as

$$v_{dc}(t) = V_{dc} + \sqrt{2}\Delta V_{dc}\sin(2wt+\theta), \tag{8}$$

$$i_{cap}(t) = 2\sqrt{2}Cw\Delta V_{dc}\cos(2wt+\theta). \tag{9}$$

Dc capacitor minimum and maximum voltages occur at the following time instants:

$$2wt+\theta = \frac{-\pi}{2} \Rightarrow wt_{min} = \frac{-\pi}{4}-\frac{\theta}{2} \quad \text{and,} \tag{10}$$

$$2wt+\theta = \frac{\pi}{2} \Rightarrow wt_{max} = \frac{\pi}{4}-\frac{\theta}{2} \tag{11}$$

In [16], the energy conservation principle is used assuming there is no energy loss. Similarly, it can be written that

$$\int_{wt_{min}}^{wt_{max}} v_c(t)i_c(t)\,dwt = \int_{wt_{min}}^{wt_{max}} v_{dc}(t)i_{cap}(t)\,dwt, \tag{12}$$

$$V_c I_c\left(\cos(\delta-\theta)\frac{\pi}{2}-\cos(\delta+2\theta)\right) = 2\sqrt{2}\,wCV_{dc}\,\Delta V_{dc}. \tag{13}$$

In addition to this, the net reactive power that is sent to the charger is written as

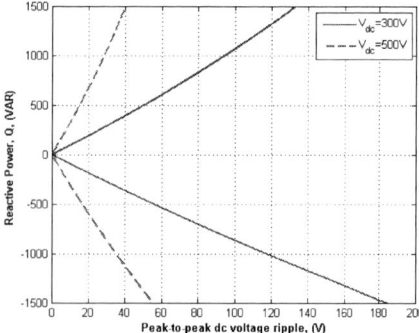

Fig. 6. Peak-to peak voltage ripple for different reactive power levels for a 500μF dc capacitor.

$$Q = V_c I_c \sin(\theta - \delta) + w L_c I_c^{\ 2} . \tag{14}$$

Using (13) and (14), the relation between reactive power and the peak-to-peak dc voltage ripple can be found as shown in Fig. 6. The reactive power produced is not directly related to the V_{dc}. However, the higher the dc voltage, the lesser the capacitor ripple current. Therefore, the system will be able to supply/sink more reactive power with higher V_{dc} for the same ΔI_{cap} levels. Similarly, the dc capacitor value does not affect the reactive power transfer. Rather, ΔV_{dc} reduces with increasing capacitor rating because the right hand side of (13) should stay constant. Moreover, in the simulation analysis section, the relation observed between reactive power and dc capacitor rms ripple current will be given.

D. Dc-dc Converter and Battery

When charging from the grid, a bidirectional dc-dc converter shown in Fig. 7 steps down the high dc-link voltage and charges the battery using constant current-constant voltage (CC-CV) charging algorithm.

A Li-ion battery model is implemented in Simulink using the model and parameters given in [17-19] to account for the charging profile of a PHEV. The equivalent circuit of a Li-ion battery cell is given in Fig. 8. The nonlinear relationship between open circuit voltage, V_{oc}, and SOC is captured using a controlled voltage source. Two RC time constants are used to mimic response to transient power. In Fig. 8, the series resistor, R_{Series}, accounts for the instantaneous voltage drop during a step change in the battery current. Also, $R_{Transient\_S}$, $C_{Transient\_S}$, $R_{Transient\_L}$, and $C_{Transient\_L}$ stand for short and long time constants that mimic the step response of the battery voltage [17].

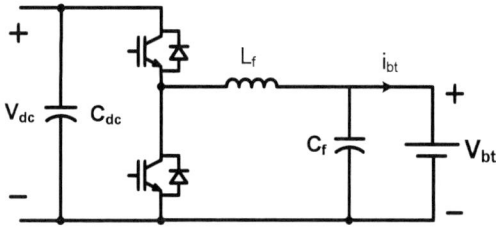

Fig. 7. Dc-dc converter and PHEV battery.

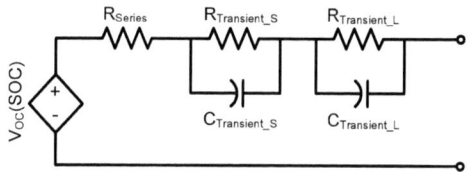

Fig. 8. Electric equivalent circuit of a Li-ion battery cell [17].

In PHEV applications, the required amount of terminal voltage and capacity of the energy storage system is obtained arranging multiple battery cells in series and parallel. The cells that are in series determine the terminal voltage of the battery system, and the number of parallel cells decides the current carrying capability of the system. The total capacity of the battery is given as

$$C_t = C_i\, n_s\, n_p , \tag{15}$$

where C_t is the total capacity (Ah); C_i is the cell capacity (Ah); n_s is the number of cells in series; and n_p is the number of cells in parallel. As given in [17], the capacity of individual cells, C_i, modeled is 0.85 Ah. The Li-ion battery cell model is scaled up to 5 kWh to account for the battery size as it is used in Toyota Prius Hymotion PHEV [20]. If each cell is assumed to be operating at 3.8 V, then 53 cells in series and 29 cells in parallel constitutes this capacity as shown below:

$$E = C_i\, n_s\, n_p\, V_t = 0.85 \cdot 53 \cdot 29 \cdot 3.8 \approx 5 \text{ kWh}. \tag{16}$$

where V_t is the nominal terminal voltage of each cell (V). The implemented battery model output signal is generated in a Simulink model and then transferred to a PLECS software block which is embedded in Simulink for power processing stage.

E. Reactive Power Support During PHEV Charging

In this section, the potential for reactive power regulation during battery charging is explored using the experimental measurements of the charging power drawn by the 2008 Toyota Prius PHEV [20]. For this purpose, the battery pack of the Toyota Prius has been depleted and recharged several times, and the resulting waveforms are given in Fig. 9.

In Fig. 9, P1, P2, and P3 stand for three different level 1 charging profiles observed when charging the PHEV. When it is first plugged in, the battery voltage level is at its minimum, and there is an excess current margin that can be utilized for reactive power generation for about 45 minutes. During this time, the battery is charged with constant current, and its terminal voltage increases gradually; the line current increases gradually too. The amount of reactive power that the system can supply until the charger reaches its maximum power is calculated by the following formula:

$$|Q| = \sqrt{\left| V_s \times I_c \right|^2 - |P|^2} . \tag{17}$$

978-1-4244-4782-4/10 $26.00 © 2010 IEEE

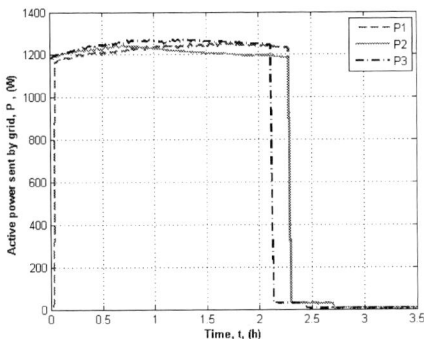

Fig. 9. Experimental data for the charging power of Toyota Prius PHEV converted by Hymotion.

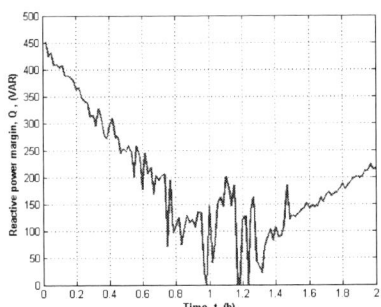

Fig. 10. Reactive power availability during constant current charging of the Toyota Prius PHEV.

Using (17), the available reactive power is given in Fig. 10 for charge profile P3. For this data, the maximum power drawn from the grid is 1.27 kW with almost 1.0 power factor. For the reactive power calculation, the apparent power is limited to be 1.27 kVA to limit the peak current.

As illustrated in Fig. 10, even during constant charging, there is an opportunity to supply 0.45 kVAR power to grid which is still 35% of the full reactive power amount that can be supplied without charging the battery.

IV. SIMULATION ANALYSIS

The purpose of the simulation study is to verify that the proposed system is able to work in the aforementioned operation modes. Also, the effect of different operation modes on the dc variables will be given. The system will be commanded to work in two quadrants although the topology is able to work in all four quadrants. Moreover, the system is compatible to work with both level 1 and level 2 charging equipment. However, since everything except the ratings will stay the same for the analysis, only level 1 will be evaluated. All the results have been achieved using a 500μF dc capacitor and 500 V dc voltage level.

Because of the long simulation time, the system is only simulated for a few seconds. A safety limit is imposed on the line current not to exceed the system limitations at all operation modes.

The simulation realizes the operation modes #1, #2, #3, #5 and #6 as given in Table I respectfully in 11 s. In Fig. 11, the

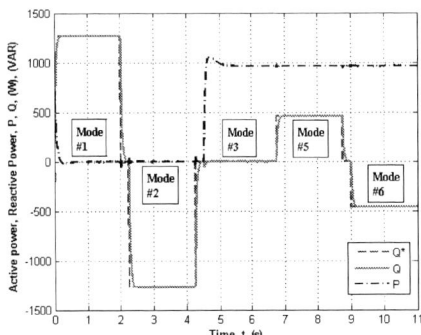

Fig. 11. Reactive power demanded by the controller and supplied by the charger.

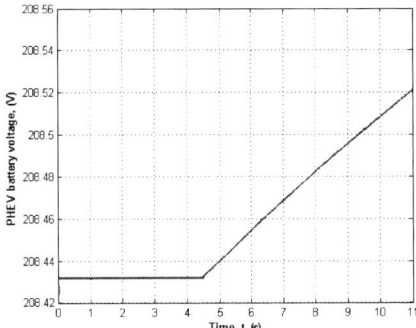

Fig. 12. PHEV battery terminal voltage.

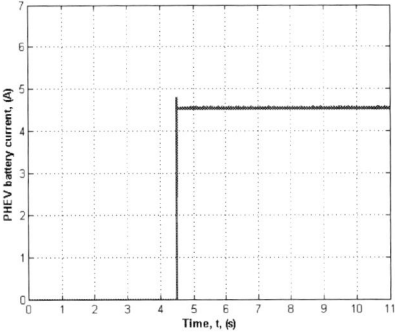

Fig. 13. PHEV battery terminal current.

reactive power command to the charger is given along with the charger's response as Q* and Q respectively. Also, in the same graph, the active power sent by the grid to charge the PHEV battery is included. Since it takes considerable amount of time to cover all the charging process of the battery, only the initial part of the constant current charging is shown. Finally, towards the end of the simulation, the bidirectional charger provides reactive power in response to the controller command when the PHEV is plugged in for charging operation. Note that, minus sign stands for capacitive operation and positive sign means inductive operation.

Following Fig. 11, in order to show that the PHEV battery is not used to supply reactive power regulation during the simulation, battery terminal voltage and current are given in Figs. 12 and 13, respectively. During the simulation, battery voltage and current have not shown any deviance from their usual profile satisfying safe operation regulations. In other words, the battery is always operated such that the current and voltage ripple presented are the same as it is during a normal charging operation.

After confirming that the designed system is able to operate at the planned operation modes without putting adverse effects on the battery, the effects of the different operation modes on the dc bus variables should be presented to investigate if the dc dynamics of the system pose a danger on the dc link capacitor.

First, the dc link capacitor voltage has shown a profile with a frequency that is double the line frequency as expected. Fig. 14 shows the regulation of dc voltage during the simulation. When the PHEV is plugged in to be charged at 4.5 s, the dc link voltage suddenly drops and then regulates itself.

Fig. 15 shows the changes in the dc capacitor peak to peak ripple voltage when the system operates at different modes. For reactive power only operation, the dc link capacitor is exposed to ~13 V peak-to-peak voltage ripple when absorbing 1.27 kVAR from grid (mode #1) and ~18 V when supplying 1.27 kVAR to grid (mode #2). Because of the coupling inductor, it requires less voltage ripple to absorb reactive power. These results confirm well with the initial analysis equations, (8) - (13) and Fig. 6. If the PHEV is used to sink reactive power from the grid during charging (mode #5), it only requires ~1 V more peak-to-peak ripple voltage and around ~2V more for capacitive operation (mode #6).

Fig. 16 also illustrates how the capacitor ripple current changes with different operation modes. The net change in the rms ripple current is small when the charger switched between different operation modes keeping the dc link capacitor in its safe operating limits.

The results confirm that, with level 1 charging, supplying ancillary services such as reactive power compensation can be achieved with an on-board, conductive, and bidirectional charger without using the PHEV battery and keeping the dc link capacitor in its operating limits.

V. CONCLUSION

The basis of this paper is to introduce the technical understanding of the V2G reactive power compensation. Therefore, V2G operation is shown by simulating different modes of operation out of which reactive power supply/sink with/without PHEV battery charging being the most important ones.

The simulation study showed that with level 1 charging, it is possible to fulfill reactive power compensation without any power demand from the battery. Moreover, the dc link capacitor of the bidirectional charger is used to supply reactive power. The results show that reactive power compensation can be accomplished with/without battery charging and it does not put stress on the dc link capacitor. The peak to peak ripple of the dc voltage and dc capacitor rms ripple current are observed for safety of the dc link capacitor.

Future study will be on engaging the PHEV battery for the reactive power support at higher power levels and showing if there are adverse effects of this operation on the battery.

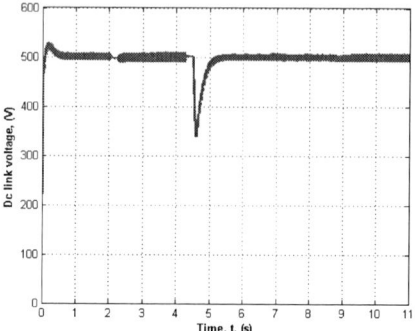

Fig. 14. Dc link voltage when charging PHEV battery.

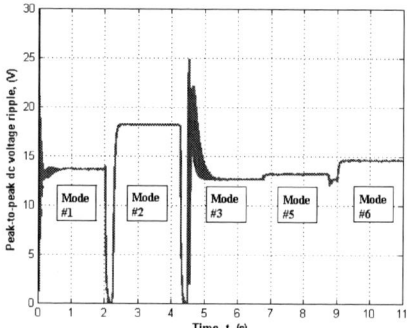

Fig. 15. Peak-to-peak ripple voltage seen on the dc link capacitor for different operation modes (#1,2,3,5, and 6 in Table I).

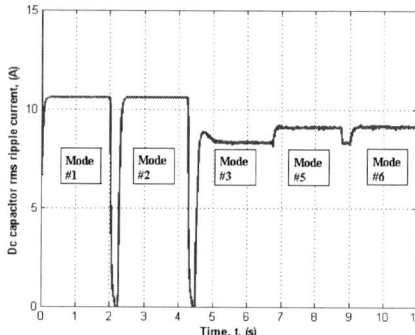

Fig. 16. Rms dc link capacitor current for different operation modes (#1,2,3,5, and 6 in Table I).

REFERENCES

[1] Electricity advisory committee, "Bottling electricity: storage as a strategic tool for managing variability and capacity concerns in the modern grid," December 2008.

[2] W. Kempton, and A. E. Letendre, "Electric vehicles as a new source for electric utilities," *Transport. Res. Part D Transport. Envir.*, vol. 2, no 3, pp. 157-175, September 1997.

[3] X. Zhou, et. al., "Design and control of grid-connected converter in bi-directional battery charger for plug-in hybrid electric vehicle application," *Vehicle Power and Propulsion Conference (VPPC'09)*, Dearborn, MI, USA, 7-10 September 2009.

[4] I. Cvetkovic, et. al. "Future home uninterruptable renewable energy system with vehicle-to-grid technology," *Energy Conversion Congress & Exposition (ECCE'09)*, San Jose, CA, USA, September 20-24, 2009.

[5] L. Tang and G.-J. Su, "A low-cost, digitally-controlled charger for plug-in hybrid electric vehicles," *Energy Conversion Congress & Exposition (ECCE'09)*, San Jose, CA, USA, September 20-24, 2009.

[6] J. Voelcker, "Lithium batteries take to the road," *IEEE Spectrum*, pp. 27–31, September 2007.

[7] M. F. M. Elias, et. al. "Lithium-ion battery charger for high energy application," *National Power and Energy Conference*, 15-16 December 2003.

[8] P. V. D. Bossche, " The electric vehicle: raising the standards," Ph.D. dissertation, Vrije Universteit Brussel, 2003.

[9] California Air Resources Board, "Staff Report: Initial Statements of Reasons – Proposed Amendments to the California Zero Emission Vehicle Regulations: Treatment of Majority Owned Small or Intermediate Volume Manufacturers and Standardization of Battery Electric Vehicle Charging Systems for the Zero Emission Vehicle Program," May 2001.

[10] Nissan Leaf Electric Car, http://www.nissanusa.com/leaf-electric-car.

[11] Reductive charging, Available online: http://www.acpropulsion.com.

[12] W. Korthof, "Level 2+: Economical Fast Charging for EVs," *17th Electric Vehicle Symposium*, Montreal, Canada, October 2000.

[13] SAE J1772 Electric vehicle conductive charge coupler and SAE J1773 Electric vehicle inductively coupled charging, Society for Automotive Engineers, Inc.

[14] L. J. Borle, "Zero average current error control methods for bidirectional ac-dc converters," Ph.D. dissertation, Curtin University of Technology, October 1999.

[15] N. Mohan, T. M. Undeland, and W. P. Robbins, "Power Electronics converters, applications and design," 3rd ed. John Wiley & Sons, Inc., 2003, pp. 214-215.

[16] Y. Xu, "A generalized instantaneous nonactive power theory for parallel nonactive power compensation," Ph.D. dissertation, Elec. and Comp. Eng. Dep., University of Tennessee, 2006.

[17] M. Chen and G. A. Rincon-Mora, "Accurate electrical battery model capable of predicting runtime and I-V performance," *IEEE Trans. Energy Convers*,. vol. 21, no. 2, pp. 504-511, June 2006.

[18] M. Knauff, et. al, "Simulink model of a lithium-ion battery for the hybrid power system testbed," *Proceedings of the ASNE Itelligent Ships Symposium*, Philadelphia, PA, USA, May 2007.

[19] Erdinc, O., Vural, B., and Uzunoglu, M. "A dynamic lithium-ion battery model considering the effects of temperature and capacity fading," *2009 International Conference on Clean Electric Power*, Capri, Italy, 9-11 June 2009.

[20] A123 Systems Hymotion Products http://www.a123systems.com/hymotion/products/N5_range_extender.

Optimal Selection and Design of the Supercapacitor Module for Fuel Cell Vehicles

Sang-Hyun Kim, Tae-Hoon Kim, Wook Kim, Jong-Hak Lee and Woojin Choi
Department of Electrical Engineering
Soongsil University
Seoul, Republic of Korea
Email: cwj777@ssu.ac.kr

Abstract—**Supercapacitor is anticipated to play an important role in increasing system efficiency by reducing the peak load for the batteries and recycling regenerated energy. As supercapacitors are produced in various voltages and capacitances by the manufacturers, accurate criteria for performance evaluation is essentially needed due to high unit price and it is also very important to raise system economic advantage by organizing optimal module on such basis. In this paper criteria for better selection of the supercapacitor through EIS (Electrochemical Impedance Spectroscopy) experiment are presented and based on the experimental results optimal method of designing supercapacitor module applied to fuel cell vehicle is proposed. The validity of the proposed criteria is proved through the computer simulation using FTP-72 urban dynamometer driving schedule.**

I. INTRODUCTION

A supercapacitor can improve the performance of a system that normally uses fuel cells or batteries because of its high power density. Its reliability, stability and environmentally friendly nature make it a good candidate for the next generation of energy storage devices [1-7]. Unlike conventional batteries, which produce energy through an electrochemical reaction, supercapacitors produce energy through a physical phenomenon, whereby energy is stored and released through the adsorption and desorption of ions, occurring at the interface of an electrode and electrolyte. This simplicity can provide several desirable characteristics for a device with excellent energy-storage properties such as a long cycle life and high specific power density. Due to the porous electrodes, supercapacitor can store far more energy than the conventional capacitor such as electrolytic capacitor, and charging and discharging can take place at extremely high rates, making it highly efficient. It can be widely used in applications where there is energy to be captured through repetitive motions and where the load leveling is required for low power density devices such as fuel cells and batteries, thereby enhancing the efficiency and reliability of the system. Also, it can compensate for the slow response of them by shaving off the overload or peak power, thereby protecting the main power source of the system.

Large-capacity supercapacitors mostly comprise many cells connected to each other. There is a wide choice of supercapacitors on the market with different technical characteristics and prices. It is therefore important to select the most efficient product and to have a good understanding of its characteristics.

There have been several attempts to analyze and model their internal properties experimentally [8-12]. However, most of these analyses involve DC tests, making it impossible to study characteristics that depend on the state of charge (SOC). This paper discusses a method that avoids this problem by measuring the impedance spectrum as a function of frequency at several different SOCs. This is done using electrochemical impedance spectroscopy (EIS), an AC test method of dividing the spectrum into real and imaginary components and calculating the capacitance from the latter component. In addition, this paper proposes criteria for evaluating the performance of supercapacitor and selecting the best product on the basis of EIS alone, a non-destructive test, by comparing the dynamic characteristics through equivalent circuit modeling.

Furthermore, the most efficient and economical product for a certain application is chosen on the basis of the experimental results and applied to the supercapacitor module design. A supercapacitor with outstanding characteristics is unfortunately very costly. Since supercapacitors are used in a module with numerous cells connected in series and parallel, economic module design is very important. EIS measurements were performed on products from different manufacturers and the most suitable was selected.

This paper performs EIS experiment on products taken from different manufacturers, selects product most suitable for fuel cell vehicle application, and proposes method of economic sizing of the module by using the characteristics of supercapacitor obtained through the analysis of the EIS results. In order to verify the usefulness of the proposed selection method, FTP-72 UDDS was used in the simulation.

II. SUPERCAPACITOR PERFORMANCE CHARACTERISTICS

A. Electrichemical Impedance Spectroscopy

For charging/discharging and current perturbation, a bipolar power supply unit (model BP4610, NF Company) was used [13]. The measurement equipment and software used for EIS were developed in our laboratory [14]. In the developed system, sensing circuit, LabVIEW 8.6 and PCI-6154 were used to measure and record voltage and current of the supercapacitor on a real-time basis. PCI-6154 is a simultaneous sampling multifunction I/O device for PCI bus computers from National Instruments [15]. It is an isolated PCI device featuring four isolated differential 16-bit analog inputs, four isolated 16-bit analog outputs, six DI lines, four DO lines, and two general-purpose 32-bit counter/timers. All A/D converters and D/A converters are capable of 250 kS/s for each channel. Digitized data were acquired by the software and the signal was extracted using a digital lock-in amplifier. The impedance spectrum was then plotted.

EIS, a method that perturbs the device and analyzes its response, was performed by limiting the magnitude of the applied perturbation current to below 2% of the charge in order to ensure linear behavior:

$$I_a < 0.02 \times \pi f C_{rated} V_{rated} \tag{1}$$

Experiments were performed at room temperature and six different SOC ranging from 0% to 100% at intervals of 20%, at frequencies ranging from 0.01 Hz to 1 kHz. Products made by four different companies were selected. Fig. 1 shows the Nyquist impedance plot and a Bode plot for the product from Company A (2.7V, 2600 F). Results indicate that with increasing charge, the real component of impedance increases and the magnitude of the imaginary component decreases.

B. Equivalent Impedance Modeling of the Supercapacitor

The impedance of a supercapacitor can be expressed in terms of an ionic resistance placed in series with the impedance of a porous electrode, Z_{pore}. An impedance model for a porous electrode has already been proposed by De Levie with a ladder circuit in which the electrode resistance component R_e and the electric double-layer capacitance C_d are connected in parallel [16]. Here, by adding the resistance component and the inductance due to external elements, the equivalent impedance model of the supercapacitor becomes a series circuit with an equivalent series inductance (ESL), an equivalent series resistance (ESR), and a porous electrode impedance, expressed as follows (2) [17]:

$$Z_{SC} = j\omega L_s + R_s + \sqrt{\left(\frac{R_e}{j\omega C_d}\right)} \coth \sqrt{j\omega R_e C_d} \tag{2}$$

Fig. 2 shows the Nyquist impedance plot of the product from Company A, measured at 80% SOC and its fitted curve. EIS indicates that in the high-frequency range (0.3–50 Hz), the higher was the frequency, the faster was the convergence of the impedance to R_s, the high-frequency equivalent series resistance (HF ESR). This was due to the porous electrode of the supercapacitor, and the real component of this range coincided with R_e, the electrode resistance. The impedance of an ideal supercapacitor would converge asymptotically to a line parallel to the imaginary axis, since only the imaginary component increases with frequency, as shown in (3) at low frequencies.

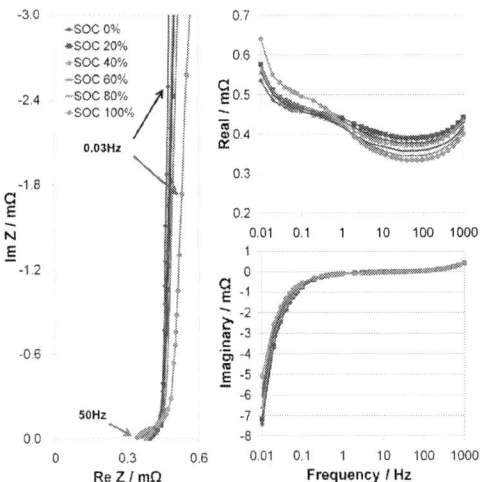

Figure 1. Nyquist impedance plot of the supercapacitor by manufacturer A [0%–100% SOC, 20 °C]

$$Z_{SC(\omega \to 0)} = R_s + \frac{R_e}{3} + \frac{1}{j\omega C_d} \tag{3}$$

However, the experimental result shows that the impedance curve is slightly inclined. This is because the electric double-layer capacitor of the porous electrode is closer to a constant phase element (CPE) than a purely capacitive component. Therefore, the equivalent circuit of supercapacitor can be expressed as in (4) [18]. At low frequencies, because the real component of the impedance increases with decreasing frequency, the impedance curve bends slightly toward the real axis. Coefficient d in (5) represents this tendency. Fig. 2 shows a good agreement between the measured impedance and the model.

$$Z_{SC} = j\omega L_s + R_s + \sqrt{\left(\frac{R_e}{(j\omega)^d Q_d}\right)} \coth \sqrt{(j\omega)^d R_e Q_d} \tag{4}$$

$$Z_{SC(\omega \to 0)} = R_s + \frac{R_e}{3} + \frac{1}{(j\omega)^d Q_d} \tag{5}$$

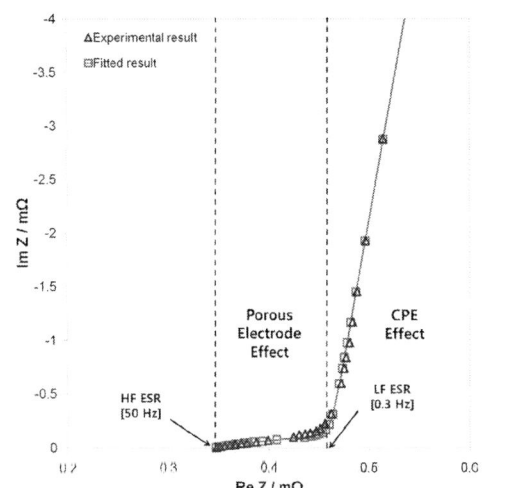

Figure 2. Nyquist impedance plot of the supercapacitor by manufacturer A and its fitted result [80% SOC, 20 °C]

978-1-4244-4782-4/10 $26.00 © 2010 IEEE

TABLE I.   FITTED PARAMETER OF THE SUPERCAPACITOR FOR EACH
MANUFACTURER [80% SOC, 20 ℃].

	L_s (nH)	R_s (mOhm)	R_e (mOhm)	Q_d	d
A	65.8	0.329	0.393	2705	0.9879
B	54.4	0.492	0.163	2258	0.9564
C	56.3	0.26	0.164	2965	0.9659
D	54.8	0.944	0.414	2279	0.9484

Table 1 lists values of parameter obtained by curve fitting
the EIS results performed at 80% SOC on each
manufacturer's product.

III.   PERFORMANCE EVALUATION

It is possible to evaluate the performance of the
supercapacitor by the equivalent circuit parameters obtained
through the electrochemical impedance spectroscopy. The
criteria for performance evaluation include; the magnitude of
the equivalent series resistance that causes power loss, the
variation of the capacitance depending on the state of charge
that causes variation of the available energy and the
coefficient d of CPE that affects self-discharge and thus
charging/discharging efficiency.

Fig. 3 shows the Nyquist impedance plots corresponding
to 80% SOC on high-capacitance supercapacitors from four
different manufacturers (Company A: 2.7 V, 2600 F;
Company B: 2.8 V, 3000 F; Company C: 2.7 V, 3500 F;
Company D: 2.5 V, 2700 F) that have similar capacitance
values. Results show that the high-frequency equivalent series
resistance is the lowest for Company C, and that the
impedance plot in the low-frequency range is most nearly
vertical for Company A. The verticality of the curve indicates
that coefficient d for the supercapacitor equivalent circuit is
close to 1, and that the porous impedance of the product from
Company A is most purely capacitive.

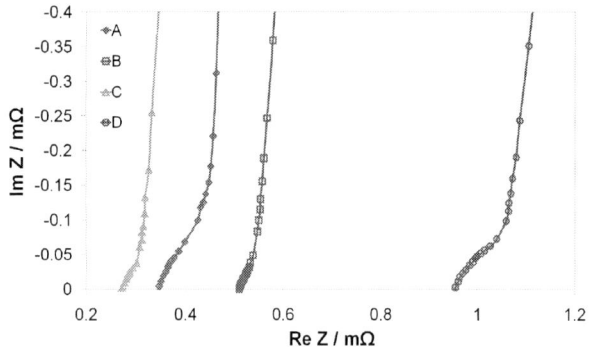

Figure 3.   Nyquist impedance plots of the supercapacitors for different
manufacturers [80% SOC, 20 ℃]

A.   Power Loss and ESR

The equivalent internal series resistance of a
supercapacitor depends on its SOC and make. Therefore, if a
supercapacitor is used dynamically at 50%–100% SOC, the
dissipated power may differ by SOC and make.

Both the series resistance and the electrode resistances in
Fig. 4 indicate that the values differ by make and increase
proportionally with SOC. The resistive component (LF ESR)

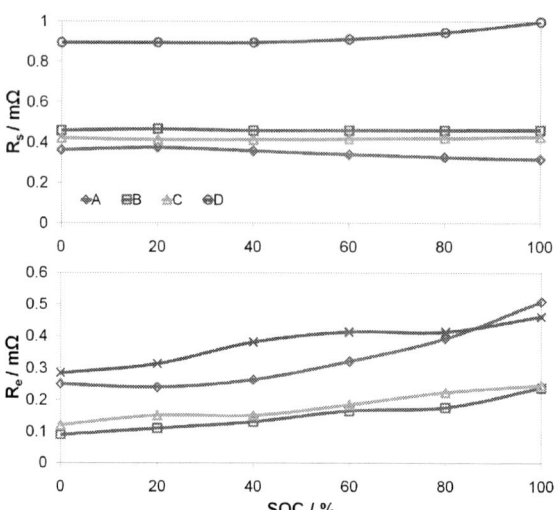

Figure 4.   Variation of the electrode resistance (R_e) and the series resistance
(R_s) at each SOC

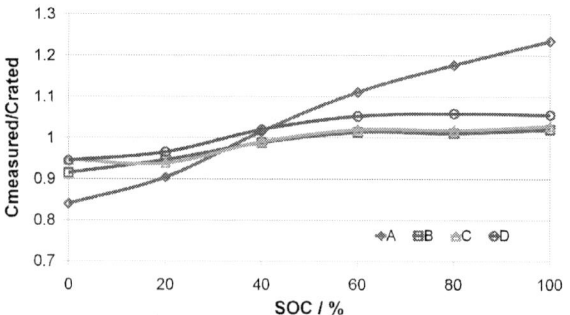

Figure 5.   Variation of the capacitance at each SOC

in the low-frequency regime and the power dissipated are
expressed as,

$$R_{SC}(\omega \to 0) = R_s + \frac{R_e}{3} \qquad (6)$$

$$P_{loss} = I^2 R_s + I_e^2 \frac{R_e}{3} \qquad (7)$$

where I is the current through the supercapacitor and I_e is
the current flowing only through the electrode. Therefore, the
greater are the HF ESR (R_s) and electrode resistance (R_e), the
greater is the power loss [19].

B.   Available Energy and Capacitance Variation

The capacitance of a supercapacitor can be calculated
from the imaginary component of the impedance determined
from EIS [20]. Since the time affecting porous electrode
differs depending on the frequency of perturbation current in
AC test, the magnitude of capacitance changes as well and
converges to constant value in the low-frequency area (<0.01
Hz). It is therefore crucial to use frequencies below 0.01 Hz
in EIS experiments.

978-1-4244-4782-4/10 $26.00 © 2010 IEEE

$$Im\, Z \cong 2\pi f L_s - \frac{1}{2\pi f C} \qquad (8)$$

Fig. 5 shows the capacitance plotted as a function of the SOC at 0.01Hz for the different manufacturers. Companies B, C and D do not show any increase in the capacitance with SOC, whereas Company A shows a 30% increase between 0% and 100% SOC. This feature is very important for calculating the available energy.

In most applications, a supercapacitor releases its charge from 100% to 50% SOC. Thus, if the capacitance is constant, the available energy is 75% of the stored energy, as demonstrated below. However, since the capacitance does change with the SOC and also differs by make, the available energy E_A should be calculated using (10).

$$E_A = \frac{1}{2} C_{rated} (V_{rated}^2 - (0.5 V_{rated})^2) \qquad (9)$$

$$E_A = \frac{1}{2} C_{SOC100\%} V_{rated}^2 \left(1 - 0.25 \times \frac{C_{SOC50\%}}{C_{SOC100\%}}\right) \qquad (10)$$

Thus, the greater is the change of the capacitance, the greater is the available energy. It is useful also to define the normalized quantity E_V:

$$E_V = 1 - 0.25 \times \frac{C_{SOC50\%}}{C_{SOC100\%}} \qquad (11)$$

C. Influence of 'd' parameter on the self-discharge and the charge/discharge efficiency

A CPE is the component of the equivalent circuit that most strongly influences the characteristics of a supercapacitor. A CPE represents the natural discharging phenomenon caused by diffusion and appears on a Nyquist plot at an angle to the right side of the vertical axis in the low-frequency region [21]. Accurate simulations of charging/discharging can then be performed. A simulation was made using Matlab/Simulink to analyze the voltage response of a CPE when coefficient d in (12) was changed. The voltage response was calculated using (13) [22]. The simulation displays the natural discharging from the fully charged state, after the CPE was charged with a current (I_c) of 10 A to a voltage of 2.7 V. Current was blocked at 2.7 V to observe changes in voltage. The wave of charging current was expressed as a step function, and the resulting expression for the response voltage is given by (14). Here, Q_d is 2600. The simulation was repeated for different values of d: 1, 0.95 and 0.9.

As seen in Fig. 6, charging took 702 s for d = 1, but 970 s for d = 0.95, and 1392 s for d = 0.9. Thus, decreasing d implies a longer charging time. It is also observed that the smaller is the value of d, the faster is the discharge.

$$Z_{CPE} = \frac{1}{(j\omega)^d Q_d} \qquad (12)$$

$$v_{CPE}(t) = I_c \times \left(\frac{t^d}{Q_d \Gamma(1+d)}\right) \qquad (13)$$

$$v(t) = V_i + I_c \times \left(\frac{t^d - (t - t_{cut-off})^d \cdot H(t - t_{cut-off})}{Q_d \cdot \Gamma(1+d)}\right) \qquad (14)$$

In order to observe the self-discharge characteristics of the supercapacitor, experiments were performed as described in Fig. 7. After charging from 0% to 100% SOC using a constant current of 10 A, the charging cable was removed and the voltage change was measured over 15,000 s. The product of Company A, with d = 0.9837 showed the smallest self-discharge with a final SOC of 96.0212%. The product of Company D, with d = 0.9418, showed the largest self-discharge to a final SOC of 83.6517%.

As shown in Fig. 8, the charging / discharging simulation was performed using the model of (5) while changing coefficient d. Here, Q_d = 2600 and LF ESR was 0.47 mΩ. The charging time increases with decreasing d, implying greater power loss, especially in the region with high SOC. The electric double-layer capacitance behaves more like a pure capacitor and shows linear charging/discharging characteristics with little loss as d approaches 1. There is less loss for lower d.

Fig. 9 compares the charging / discharging response to a square wave current to the simulation result for the product of Company A. For charging/discharging experiments, the supercapacitor voltage was maintained at 0 V using bidirectional power supply for 10 minutes before charging. As suggested by the IEC 62391-1 standard, charging was done with a current of 0.0004CratedVrated/s. The rated voltage was then maintained by floating charge for 30 minutes. Then, efficiency of charging/discharging was calculated while discharging with the same current.

As shown in the figure, charging required a time of 2,645s and discharging 2,602s, which implies power loss. As mentioned in the previous section, since the supercapacitor operates between 50% and 100% SOC, the charging/discharging efficiency must also be considered in this range. Accordingly, the efficiency was calculated as the ratio of the charging energy from 50% to 100% SOC to the discharging energy from 100% to 50% SOC. The charging and discharging energies and the charge/discharge efficiency were calculated as 8,627 J, 8,009 J, and 92.83%, respectively.

In the simulation, in order to reflect the changes in the equivalent impedance model with the SOC, each parameter extracted from the Nyquist plot, obtained at the six different SOCs in Fig. 1, were fitted onto a curve. Then, varied values of these parameters with SOCs were used to calculate the results at every iterative step of simulation as shown in Fig. 9.

IV. OPTIMAL PRODUCT SELECTION

Table 2 compares the specific energy, energy density, charge/discharge efficiency and ratio of remaining charge after self-discharge for 15000 s and capacitance change of supercapacitor by the manufacturers. The discharging energy in going from 100% to 50% SOC was again used as the reference in calculating the specific energy and the energy density.

Table 2 shows a small difference between the calculated (E_V) and the measured (E_{V_EX}) ratios of available energy that results from the higher charged condition, and hence from the higher resistive component and strong self-discharge effect. Also, product of Company A, which has the highest capacitance change, shows the highest value of the measured ratio of available energy (E_{V_EX}).

Fig. 10 compares the performances by the products by using a spider map. The product of Company A was considered to be optimal because it had small ESR, high remaining charge after self-discharge, high-efficiency charge/discharge, and high specific energy and energy density.

From an overall view of the above results, the two major elements that determine loss in a supercapacitor are coefficient d of the CPE, and the equivalent series resistance. However, the equivalent series resistance of a supercapacitor has an extremely small value of several hundred μΩ, and loss from this resistance is also small. As Fig. 8 shows, diffusion loss expressed by d is very large. Therefore, it is most important to choose a product with d closest to 1 when selecting an optimal supercapacitor. At the same time, the composition of economically feasible modules becomes possible only if a relatively large available energy can be used with a product that shows large change in capacitance by SOC.

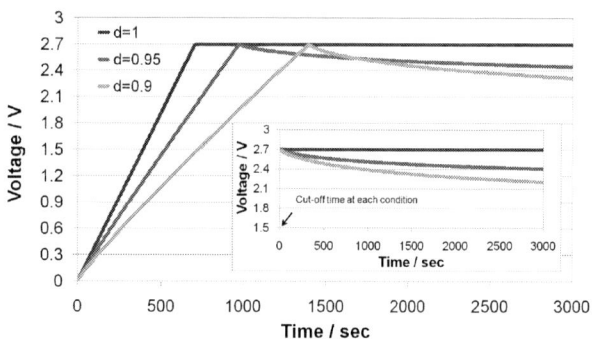

Figure 6. Simulation of charge and self-discharge voltage by the changes in coefficient d ($I_c = 10$ A, $Q_d = 2600$)

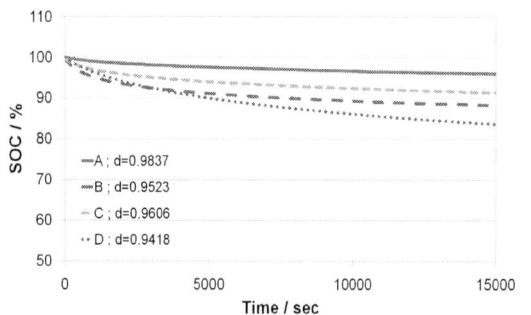

Figure 7. Self-discharge response at 100% SOC for different manufacturers

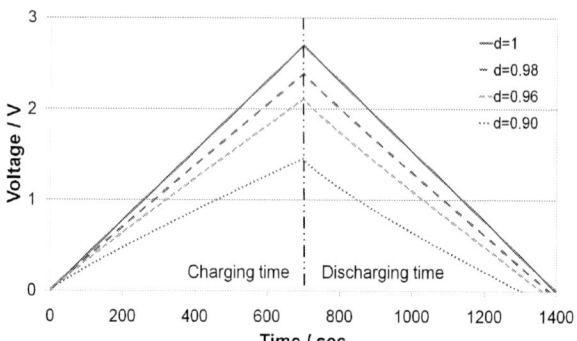

Figure 8. Simulation of charge/discharge voltage by the changes in coefficient d ($I_c = 10$ A, $Q_d = 2600$, ESR = 0.47 mΩ)

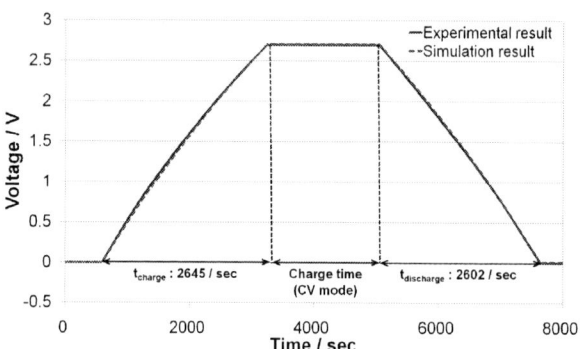

Figure 9. Result of charge/discharge experiment and simulation by manufacturer A

TABLE II. COMPARISON OF SUPERCAPACITOR CHARACTERISTICS.

Manufacturer	A	B	C	D
V_{rated} (V)	2.7	2.8	2.7	2.5
C_{rated} (F)	2600	3000	3500	2700
C_V	0.8178	0.9813	0.9775	0.9817
E_V	0.7955	0.7547	0.7556	0.7546
E_{V_EX}	0.7743	0.7429	0.7378	0.7374
Charge/discharge Efficiency (%)	92.86	86.50	79.95	74.2
Specific energy (Wh/kg)	4.6866	3.858	3.8536	3.9219
Energy density (Wh/L)	6.8282	5.1824	5.1270	4.2361
Remaining charge after self-discharge (%)	96.0212	88.2778	91.4250	83.6517

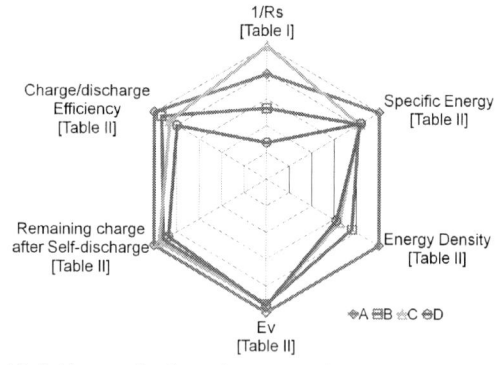

Figure 10. Spider map for the performance evaluation of the supercapacitors

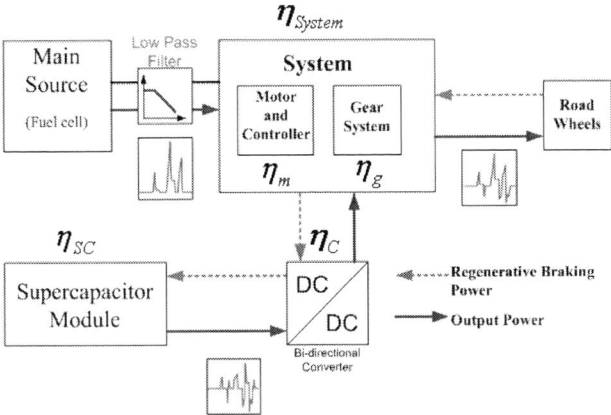

Figure 11. Block diagram of the fuel cell vehicle

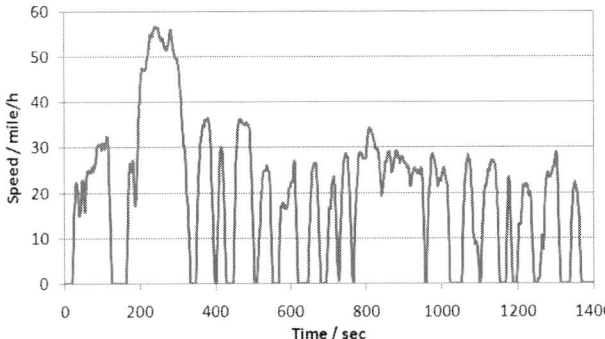

Figure 12. FPT-72 Urban Dynamometer Driving Schedule (UDDS)

V. METHOD OF ECONOMIC SIZING FOR THE SUPERCAPACITOR

To design supercapacitor module, fuel cell vehicle system was composed by using LabVIEW in Fig. 11, and FPT-72 Urban Dynamometer Driving Schedule (UDDS) was used for the simulation. In the simulation, as for the supercapacitor module to be used in fuel cell vehicle, The product of Company A which was selected as optimal product was used.

A. Fuel Cell Vehicle Simulation using FTP-72 UDDS

The specifications of fuel cell vehicle used in the simulation are set to weight 980kg, front area 'A' 1.5m^2, gear ratio 'G' 13, tire radius 'r' 0.275m, rolling resistance coefficient μ_{rr} 0.015, and air density ρ 1.25kg/m3. To find out the number of cells for composition of supercapacitor, the amount of energy needed to drive FTP-72 UDDS was calculated. The input/output power characteristics of motor needed to drive UDDS can be expressed as the product of total traction force and angular velocity as in (15) [23]. Fig.12 shows the speed profile of FTP-72 UDDS, and the total power calculated by using (15) in Fig. 13.

Supercapacitor, used with main energy source like fuel cell as auxiliary energy storage device, plays the role of handling the peak power of system. Therefore, main energy source handles only low frequency component of the needed output power by filtering it using low pass filter of 8 seconds time constant, as shown in Fig. 14. And supercapacitor takes care of high frequency component and regenerated energy from the vehicle when decelerating and braking. Therefore, the charging/discharging energy of supercapacitor is calculated as in (16).

$$P_{total} = 144.06V + 0.35V^3 + 1029\frac{dv}{dt}V \tag{15}$$

$$P_{SC} = \frac{1}{\eta_{System}}P_{peak} + \eta_{System}\eta_{SC}P_{regeneration} \tag{16}$$

For accurate calculation, in case energy is charged/discharged at supercapacitor, supercapacitor efficiency(η_{sc}) and system efficiency(η_{system}) both were considered. System efficiency all considered motor and controller efficiency(η_m), gear system efficiency(η_g) and power converter efficiency(η_c), and the value was calculated in total about 0.7. Therefore, the input/output power of supercapacitor is the value in which efficiency is applied to regenerated power, the high frequency component and low frequency component of total power and is expressed in Fig. 15. The supercapacitor module was fully charged before starting the simulation and it was charged only by the regenerative energy during the course of simulation.

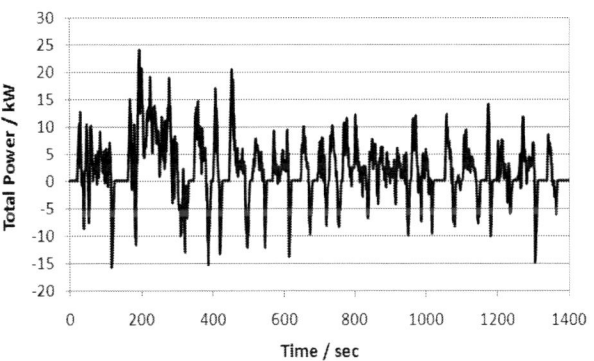

Figure 13. Total power profile for FTP-72 UDDS

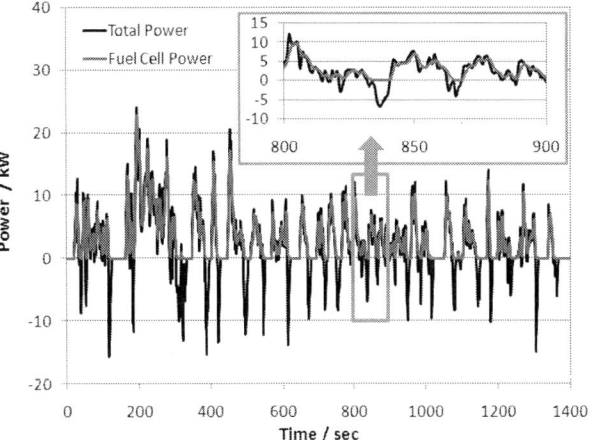

Figure 14. Total power and fuel cell power profile for FTP-72 UDDS

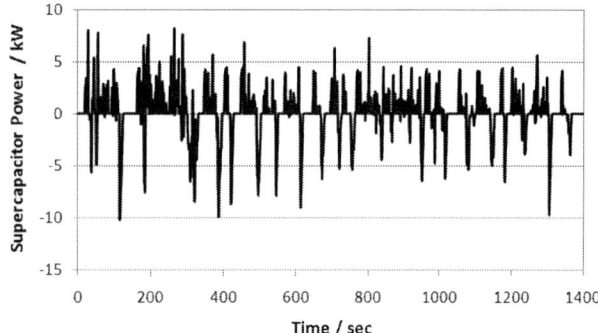

Figure 15. Power profile of the supercapacitor for FTP-72 UDDS

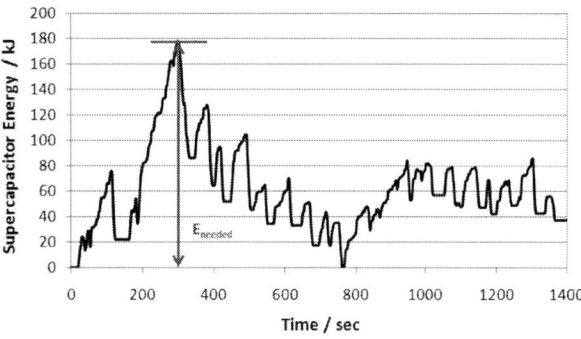

Figure 16. Energy profile of the supercapacitor for FTP-72 UDDS

B. Design and Comparison of the Supercapacitor Modules

The energy of supercapacitor can be expressed by integrating the power of Fig.15 as in Fig.16, and the difference between the minimum value and maximum value becomes the amount of energy (E_{needed}) needed for the sizing of supercapacitor module. Therefore, if the supercapacitor module is designed for the fuel cell vehicle to drive FTP-72 UDDS, 177.5kJ is needed. From this value, the number of supercapacitor cells needed is determined, and actual supercapacitor is used to get discharged only up to 50% SOC, so the number is calculated as in (17) by available energy(E_A) measured by experiment. Therefore, when using the product of Company A, driving of FTP-72 needs 23 cells.

$$N_s = \frac{E_{needed}}{E_A} \qquad (17)$$

Since the maximum output current of the supercapacitor is limited typically by 12% of short circuit current [24], it is first required to check if the maximum output current exceeds this level when implementing a module. If this is the case, parallel connection of the supercapacitors is essential to meet the maximum output current. In this case, the maximum output current of the supercapacitor for the FTP-72 UDDS is smaller than the 12% of short circuit current and thus all the supercapacitor cells can be connected in series. Thus, the terminal voltage and capacitance of the module are calculated as in (18) and (19).

$$V_{T\_Module} = N_s \times V_{Cell} \qquad (18)$$

$$C_{Module} = \frac{C_{Cell}}{N_s} \qquad (19)$$

The number of cells in a module varies by manufacturer due to the fact that the available energy values calculated as in (10) are different by the manufacturer. Thus the number of cells to compose a module also varies by the manufacturer. The number of cells required to compose a module with different supercapacitor are shown in Table III. It is noticed that the product of Company D needs 33 cells in a module while the product of Company C only needs 21 cells, which is significantly lower.

TABLE III. COMPARISON OF NUMBER OF CELLS IN A MODULE.

Manufacturer	A	B	C	D
E_{needed} [J]	177483	185009	192761	199566
E_A [J]	8005	8575	9279	6140
N_s	23	22	21	33

Fig. 17 shows the voltage and current profiles of the supercapacitor of Company A for FTP-72 UDDS. As shown in the figure, the maximum output current of the supercapacitor does not exceed the 12% of the short circuit current, 4600A. Also, the voltage of supercapacitor module does not drop below the half of the full charge voltage, verifying the suitability of the design using (10), (17) and Table III.

Fig. 18 shows the energy profiles of the supercapacitors by manufacturer during the course of one FTP-72 UDDS. The obtained results are different by the manufacturer. However, it is easily noticed that the charge of the supercapacitor after one cycle is reduced, meaning that the extra energy other than the regenerative energy is needed to restore the charge of the supercapacitor for the next cycle. Also the amounts of energy required to restore the charge of each supercapacitor are different. The product of Company D is worst and this is due to the lowest charge/discharge efficiency as previously shown in Table II. This also implies that higher rating of the fuel cell is needed for the fuel cell vehicle with the product of Company D and thus resulting in a lower fuel efficiency and higher system cost. Even though the number of cells in a module is lowest with product of Company C, product of Company A is preferred due to its high charge/discharge efficiency.

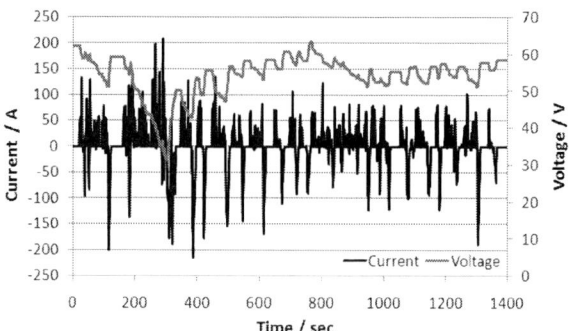

Figure 17. Voltage and current profiles of the supercapacitor of the Company A for FTP-72 UDDS

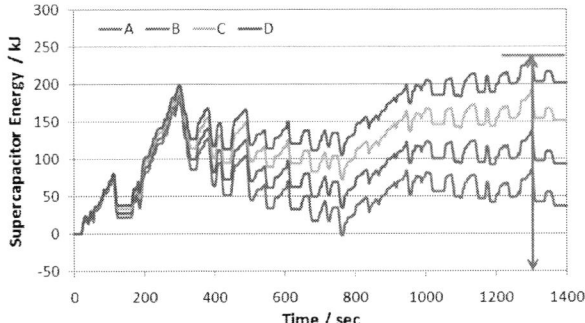

Figure 18. Energy profiles of the supercapcitors by manufacturer

VI. CONCLUSION

This paper evaluated the performance of supercapacitors by various manufacturers through EIS experiment, and compared and analyzed the characteristics by the manufacturers. In addition, by considering the difference of capacitance changes and charge/discharge efficiency by SOCs, needed energy to drive the FTP-72 UDDS driving cycle of fuel cell vehicle that uses supercapacitor module, and the numbers of cells needed for supercapacitor module composition were calculated, and through normalization and comparison, optimal product was selected. It is anticipated that if the proposed method is used, system price can be reduced and economic profit raised through optimal selection of supercapacitor in composing large capacity system that uses supercapacitor module.

ACKNOWLEDGMENT

This work has been supported by KESRI(R-2008-30), which is funded by MKE(Ministry of Knowledge Economy).

REFERENCES

[1] B.E. Conway, Electrochemical Supercapacitor: Scientific Principles and Technological Applications, Plenum, New York, NY, 1999.

[2] Maged N.F. Nashed, "Transient Performance of a Hybrid Electric Vehcle with Multiple Input DC-DC Converter" J. Power Electronics, vol. 3, pp. 230-238, October 2003.

[3] B-H Lee, D-H Shin, H-S Song, H Heo, H-J Kim, "Development of an Advanced Hybrid Energy Storage System for Hybrid Electric Vehicles" J. Power Electronics, vol. 9, pp. 51-60, 2009.

[4] E. Faggioli, P. Rena, V. Danel, X. Andrieu, R. Mallant, H. Kahlen, " Supercapacitors for the energy management of electric vehicles" J. Power Sources, vol. 84, pp. 261-269, 1999.

[5] A. D. Pasquir, I. Plitz, S. Menical, G. Amatucci, "A comparative study of Li-ion battery, supercapacitor and nonaqueous asymmetric hybrid devices for automotive applications", J. Power Sources, vol. 115, pp.171-178, 2003 .

[6] J.N. Marie Francoise, H. Gualous, R. Outbib, A. Berthon, "42V Power Net with supercapacitor and battery for automotive applications", J. Power Sources, vol. 143, pp. 275-283, 2005.

[7] P. Thounthong, S. Rael, B. Davat, "Control Strategy of Fuel Cell and Supercapacitors Association for a Distributed Generation System", IEEE Trans. Industrial Electronics, vol. 54, pp. 255-3233, December 2007.

[8] L. Zubieta, R. Bonert, "Characterization of Double-Layer Capacitors for Power Electronics Applications", IEEE Trans. Industrial Electronics, vol. 36, pp. 199-205, 2000.

[9] H. Gualous, D. Bouquain, A. Berthon, J.M. Kauffmann, "Experimental study of supercapacitor serial resistance and capacitance variations with temperature", J. Power Sources, vol. 123, pp. 86-93, 2003.

[10] M. Itagaki, S. Suzuki, I. Shitanda, K. Watanabe, H. Nakazawa, "Impedance analysis on electric double layer capacitor with transmission line model", J. Power Sources, vol. 164, pp. 415-424, 2007.

[11] P.J. Mahon, G.S. Paul, S.M. Keshishian, A.M. Vassallo, "Measurement and modelling of the high-power performance of carbon-based supercapacitors", J. Power Sources, vol. 91, pp. 68-76, 2000.

[12] W. Lajnef, J.-M. Vinassa, O. Briat, S. Azzopardi, E. Woirgard, "Characterization methods and modelling of ultracapacitors for use as peak power sources", J. Power Sources, vol. 168, pp. 553-560, 2007.

[13] NF BP4610 Instruction Manual

[14] J. Lee, W. Choi, "Development of the Low-Cost Impedance Spectroscopy System for Modeling the Electrochemical Power Sources", The 7th International Conference on Power Electronic, pp. 113-118, October 2007.

[15] NI 6124/6154 User Manual, National Instruments Corporation, 2008

[16] R. De Levie, Electrochemical Response of Porous and Rough Electrodes, Advances in Electrochemistry and Electrochemical Engineering, Vol.6, Wiley Interscience, New York, 1967.

[17] S. Buller, E. Karden, D. Kok, De Doncker, "Modeling the Dynamic Behavior of Supercapacitors Using Impedance Spectroscopy" IEEE Trans. Industry Application, vol. 38, pp. 1622-1626, November/December 2002.

[18] O. Bohlen, J. Kowal, D.U. Sauer, "Ageing behaviour of electrochemical double layer capacitors Part I. Experimental study and ageing model", J. Power Sources, vol. 172, pp. 468-475, 2007.

[19] O. Bohlen, J. Kowal, D.U. Sauer, "Ageing behaviour of electrochemical double layer capacitors Part II. Lifetime simulation model for dynamic applications", J. Power Sources, vol. 173, pp. 626-632, 2007.

[20] F .Rafik, H. Gualous, R. Gallay, A. Crausaz, A. Berthon, "Frequency, thermal and voltage supercapacitor characterization and modeling" J. Power Sources, vol. 165, pp. 928-934, 2007.

[21] H. E. Brouji, J-M. Vinassa, O. Briat, N. Bertrand, F. Woirgard, "Ultracapacitors self discharge modelling using a physical description of porous electrode impedance", IEEE VPPC, September, 2008.

[22] A. Salkind, T. Atwater, P. Singh, S. Nelatury, S. Damodar, C. Fennie Jr, D. Reisner, "Dynamic characterization of small lead0acid cells", J. Power Sources, vol. 96 , pp. 151-159, 2001.

[23] Y. Cheng, J. V. Mierlo, P. Lataire, G. Maggetto, "Test Bench of Hybrid Electric Vehicle with the Super Capacitor Based Energy Storage", IEEE, 2007.

[24] Y. Cheng, J. V. Mierlo, P. Lataire, M. Lieb, E. Verhaeven, R. Knorr, " Configuration and verification of the supercapacitor based energy storage as peak power unit in hybrid electric vehicles", In Proc. Euro. Conf. Power Electronics and Applications 2007, 2007.

Efficiency Evaluation of A 55kW Soft-Switching Module Based Inverter for High Temperature Hybrid Electric Vehicle Drives Application

Pengwei Sun, Jih-Sheng Lai,
Hao Qian, and Wensong Yu
Virginia Tech,
Blacksburg, VA 24060, USA

Chris Smith, John Bates, and
Beat Arnet
Azure Dynamics Inc.,
Woburn, MA 01801, USA

Alexander Litvinov, and
Scott Leslie
Powerex Inc.,
Youngwood, PA 15697, USA

Abstract— **This paper presents a 55kW three-phase soft-switching inverter for hybrid electric vehicle drives at high temperature conditions. Highly integrated soft-switching modules have been employed to achieve switching loss as well as conduction loss reduction. Detailed experimental evaluations of inverter efficiency have been conducted through both inductive load and motor-dynamometer load at coolant temperatures ranging from 25°C to 90°C. Efficiency measurement using power meter showed that the peak efficiency is around 99%, and it drops slightly at lower speed and higher temperature conditions. To ensure measurement fidelity, a double chamber differential calorimeter system was designed and calibrated for the inverter testing. Through long-hour testing, the measured efficiencies consistently showed 99% and higher. The soft-switching inverter has been operated reliably and demonstrated high efficiency at different temperature and test conditions.**

I. INTRODUCTION

Optimized device and system structure for high efficiency power conversion can prove to be beneficial to many industries, especially for high-temperature operation requirement. One of the driving forces of high-temperature operation is the elimination of bulky and expensive cooling systems, which are necessary to protect power electronics system from extreme working conditions [1-4]. The state-of-the-art hybrid electric vehicles have a separate cooling loop for electronics at a maximum coolant temperature of 70°C. It would be desirable to eliminate such an extra cooling loop and use the engine coolant system, which would have a maximum temperature of 105°C. Nevertheless, the performance of the semiconductor devices degrades rapidly with the increase of temperature. Despite challenges, a coolant temperature requirement of 105°C has been established for 2015 with an

intermediate step of 90°C for 2010 by FreedomCAR and Fuel Partnership Program under U.S. Department of Energy [5].

To meet the design challenges, there are several alternative ways [6], including using more silicon, by means of wide-band-gap semiconductors [7-9], improvement of thermal management techniques. The main barrier is the rising cost imposed by adopting those measures. In [6], it points out that with the help of soft-switching technique, the conventional silicon power devices have the chance of meeting high-temperature operation requirements with much more reasonable cost than other approaches. However, it only explored the use of IGBT under soft-switching for reduction of switching loss. For IGBT devices, there is a fixed voltage drop even at lower current region. In order to reduce the conduction loss, a hybrid switch in the form of MOSFET and IGBT parallel operation is proposed. The advantage of this switch combination is to have MOSFET conducting the current at low current and IGBT conducting the high current. The hybrid switch voltage drop at low currents is proportional to current, and at high currents is dominated by the IGBT and is increased as the current increases.

Using the hybrid switch in the soft-switching inverter circuit described in [10], a liquid-cooled soft-switch module has been developed. The module integrates main IGBTs, MOSFETs, auxiliary IGBTs, and diodes with capability of 400-A continuous current operation. The integration of these chips allows significant parasitic inductance and thermal resistance reduction. Using such a highly integrated liquid-cooled soft-switch module, a 55-kW three-phase soft-switching inverter was designed and assembled. The inverter efficiency was evaluated under both inductive load and motor-dynamometer load tests with coolant temperatures ranging from 25°C to 90°C. A double chamber differential calorimeter was introduced for precision inverter efficiency measurement. The soft-switching inverter was successfully operated at various temperature and test conditions. The power meter measurement from 20% to 100% output consistently shows

---

This material is based upon work supported by the U.S. Department of Energy (DOE) under Award Number DE-FC26-07NT43214.

efficiency higher than 98% under different temperature conditions, and the peak efficiency with calorimeter measurement exceeds 99%.

II. INVERTR SYSTEM ASSEMBLY AND SETUP

The three-phase soft-switching inverter consisted of three identical soft-switching modules, which are shown in shaded area in Fig.1. S_1 to S_6 are main switches composed of paralleled IGBT and MOSFET devices. S_{x1} to S_{x6} are auxiliary IGBT switches. In each module, there are also four auxiliary diodes. All the devices have the voltage rating of 600V. L_{r1} to L_{r6} are coupled magnetics with turns-ratio 1:1.35. C_1 to C_6 are resonant capacitors with value of 100nF. Fig. 2 shows the photograph of the complete integrated liquid-cooled soft-switching module based inverter. The coolant can be pumped to each individual switch through manifold, and the inlet temperature is regulated by a chiller/heater. The manifold is designed to make sure each module has the same water-cooling flow rate and same length cooling path.

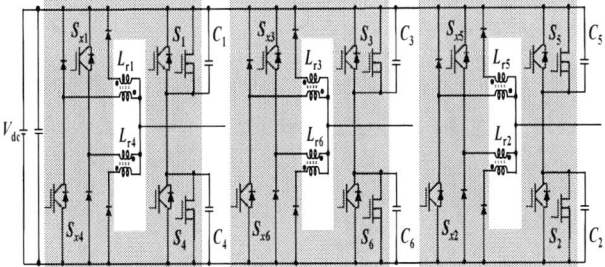

Figure 1. Circuit diagram of three-phase soft-switching module based inverter.

Figure 2. Liquid-cooled soft-switching modules.

A complete three-phase inverter has been designed and assembled using the integrated liquid-cooled soft-switching modules. Fig. 3 shows the assembled inverter. The gate drivers sit on top of each soft-switching module. They incorporate variable timing control circuit to ensure the entire load range zero voltage switching of main devices [11]. The inverter is controlled with a 10-kHz discontinuous space vector modulation using TI TMS320F2407A DSP. The DC power supply provides the DC bus voltage (325V) and power to the soft-switching inverter. An AC55^{TM} induction motor with 2500rpm nominal speed, 30kW continuous shaft-power, and 55kW peak power was connected to an eddy-current braked dynamometer through torsional coupling. Fig. 4 shows the picture of the complete motor-dynamometer system setup in the lab.

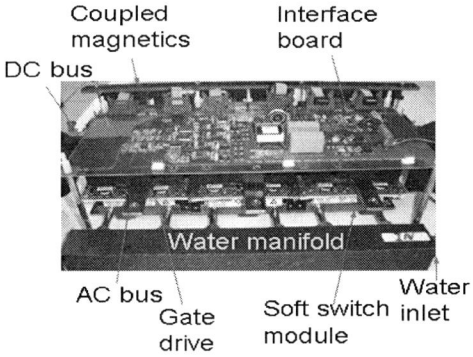

Figure 3. Module-based soft-switching inverter.

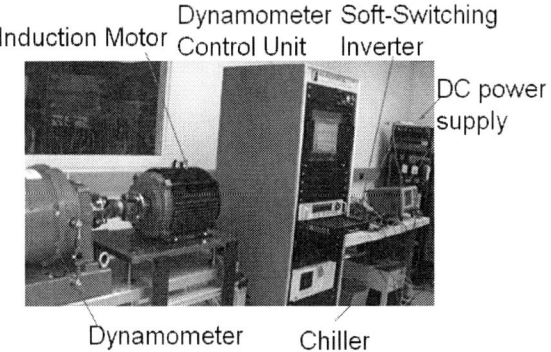

Figure 4. Soft-switching inverter motor test system.

III. EFFICIENCY TEST UNDER INDUCTIVE LOAD

The inductive load test was performed in order to push to higher output voltage and current conditions. The reactive power kVA and the line frequency represent the output power and speed of the motor load, and their losses are in the similar scale. Therefore, it is reasonable to project the efficiency with reactive power test. The load is a Δ-connected three-phase inductor. Equivalent inductance is about 4.5mH per phase. By controlling the modulation index, the output voltage, and thus the output reactive power can be controlled. The dc bus voltage was fixed at 325V, and the output line frequencies varied at 45Hz, 60Hz and 83.3Hz. Temperature was regulated at four different conditions: 25°C, 50°C, 75°C and 90°C.

Fig. 5 shows the projected efficiency based on the inductive load measured loss results at different output line frequency and different temperatures. The power factor in these cases is assumed 0.83, which is the same as what has been tested on the motor drive cases. It is noted that at the light load condition, the efficiency difference is more obvious than that at the heavy load condition. The reason is that at light loads, MOSFET shares more current, and with positive temperature coefficient, the efficiency suffers. However, at heavy loads, the LPT IGBT shares more current, and with the negative temperature coefficient [6], its efficiency hit by temperature is not as severe. The peak efficiency approaches close to 99%. Efficiency drops slightly with higher temperatures, typically 0.1% per 50°C.

978-1-4244-4782-4/10 $26.00 © 2010 IEEE

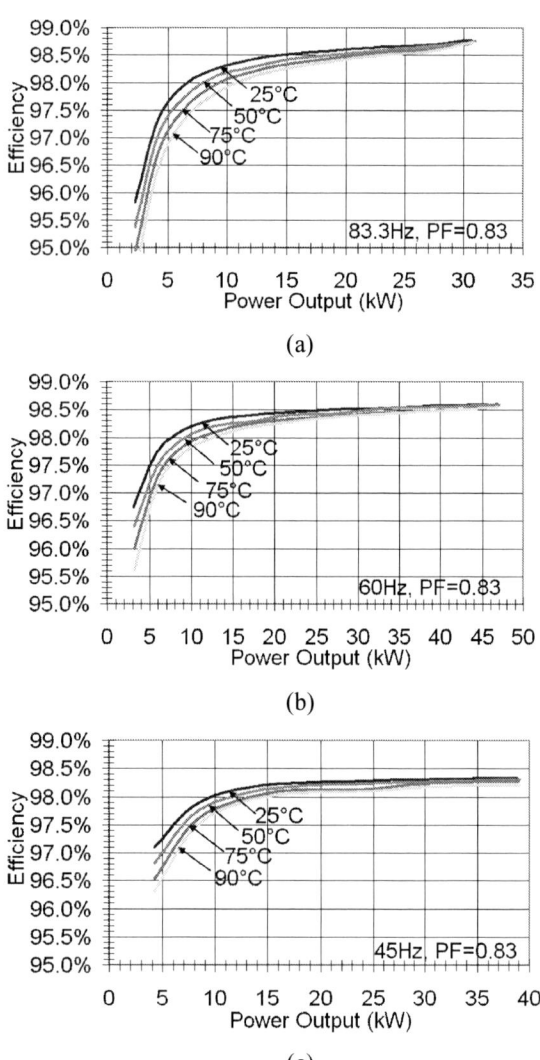

(a)

(b)

(c)

Figure 5. Efficiency comparison under different temperatures and line frequencies: (a) 83.3Hz, (b) 60Hz, (c) 45Hz.

To compare the efficiency under different frequency or motor speed conditions, the above results are rearranged to have the same temperature condition but under different frequencies. Fig. 6 shows the projected efficiencies between different output line frequencies at 25°C, 50°C, 75°C and 90°C, respectively. As can be seen, at the same output power point, the efficiency is higher at a higher output line frequency. That can be translated into higher speed with higher efficiency, which is proven by later motor tests.

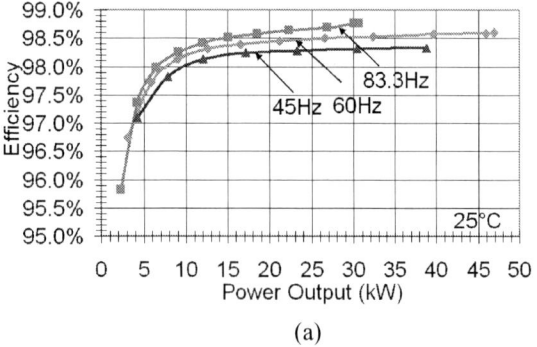

(a)

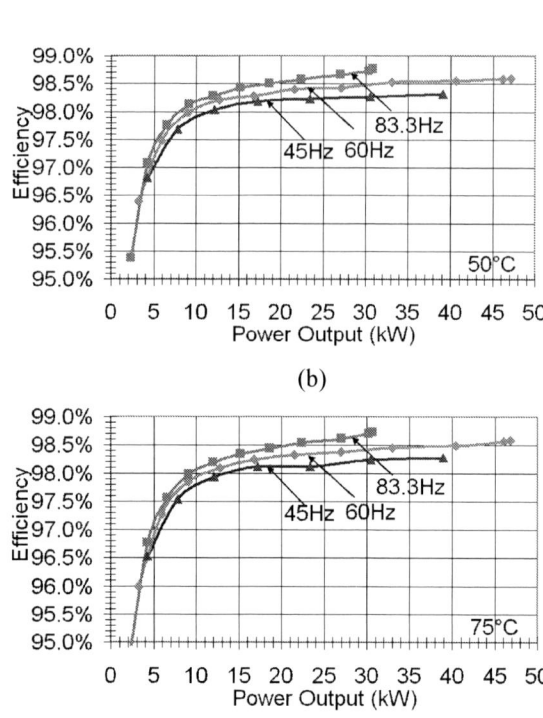

(b)

(c)

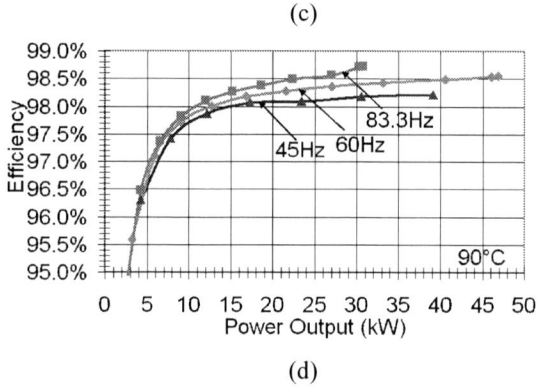

(d)

Figure 6. Efficiency comparison under different line frequencies and temperatures: (a) 25°C, (b) 50°C, (c) 75°C, and (d) 90°C.

IV. EFFICIENCY TEST UNDER MOTOR LOAD

The motor test setup is shown in Fig. 4. We tested motor at different speed conditions, 1000rpm, 1500rpm and 2000rpm with different output current values, 30A, 40A and 50A at different temperatures, 25°C, 50°C and 75°C. Due to limited DC power supply capacity in the lab, the test was conducted for up to 30% load. The high power test with calorimeter measurement was then conducted with regeneration type dynamometer, and the results will be discussed in the next section.

Table 1 shows the tested inverter efficiency at different speeds and different output currents at different temperatures with power factor 0.83. At lower output power, the low-temperature efficiency is slightly higher than the high-temperature one. At higher output power, the high temperature efficiency catches up and may surpass low temperature one. At higher motor speed and thus higher output frequency, the inverter efficiency is higher than lower speed conditions.

978-1-4244-4782-4/10 $26.00 © 2010 IEEE   476

Table 1. Efficiency measurement with motor test at different temperatures.

1000rpm (33.3Hz)	75°C		50°C		25°C	
30A	98.1%	9.8kVA	98.2%	9.8kVA	98.3%	9.7kVA
40A	98.2%	14.0kVA	98.3%	14.0kVA	98.4%	14.0kVA
50A	98.3%	18.3kVA	98.3%	18.3kVA	98.4%	18.3kVA

1500rpm (50Hz)	75°C		50°C		25°C	
30A	98.3%	12.1kVA	98.5%	12.1kVA	98.6%	12.2kVA
40A	98.6%	17.4kVA	98.6%	17.4kVA	98.7%	17.4kVA

2000rpm (66.6Hz)	75°C		50°C		25°C	
30A	98.8%	13.7kVA	98.9%	13.7kVA	98.9%	13.7kVA

Table 2 shows the efficiency comparison between inductive load test and motor test when the efficiency is reflected to 0.83 power factor. At the same output power, the motor test efficiency is higher than the pure inductive load test efficiency. The reason is during motor dynamometer test, the current is mainly conducting through MOSFET and IGBT channels, while in inductive load test, the duty cycle of the anti-paralleled diodes increases, and the efficiency is suffered slightly. Previous inductive load test shows that at 83.3Hz (30.6kVA), the efficiency is 98.8%; and at 60Hz (46.8kVA), the efficiency is 98.6%. Therefore, the peak efficiency at higher motor load can be expected to exceed 99%.

Table 2. Efficiency comparison between motor test and inductive load test at different temperatures.

	75°C		50°C		25°C	
50Hz/motor	98.3%	12.1kVA	98.5%	12.1kVA	98.6%	12.2kVA
45Hz/inductor	97.9%	12.1kVA	98.0%	12.1kVA	98.1%	12.1kVA

	75°C		50°C		25°C	
66.6Hz/motor	98.8%	13.7kVA	98.9%	13.7kVA	98.9%	13.7kVA
60Hz/inductor	98.1%	13.7kVA	98.2%	13.7kVA	98.3%	13.7kVA

V. EFFICIENCY TEST UNDER CALIROMETER

All the above efficiency measurements were done by using digital power meters with accuracy of ±0.1%. For a high efficiency inverter, the calorimeter method to measure the total power loss of the high-frequency switched inverter is considered the more accurate way to determine the efficiency [12-14]. Therefore, the calorimeter measurement was conducted for precision efficiency determination. Fig. 7 shows the diagram of the calorimeter measurement setup. We used a double chamber differential calorimeter method [15], which removes the need for measuring fluid properties and associated measurement errors. Fig. 8 shows photograph of the calorimeter used to test the inverter efficiency. Fig. 9 shows the high temperature heat exchanger and pump hooked into the back of the calorimeter.

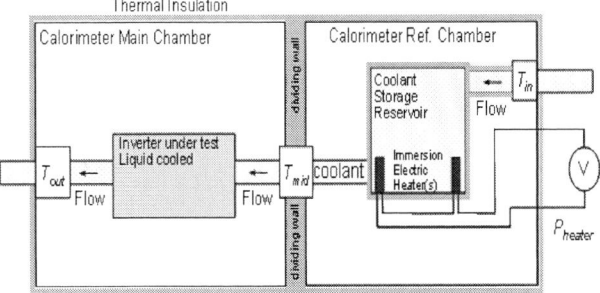

Figure 7. Diagram of the calorimeter measurement setup.

Figure 8. Calorimeter with reference chamber in foreground and inverter chamber in back.

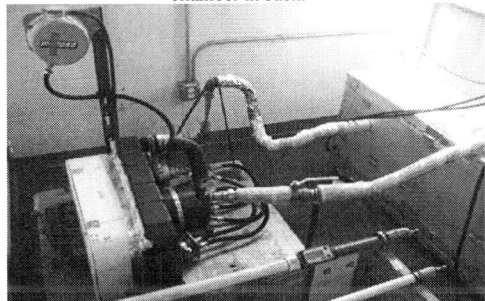

Figure 9. The high temperature heat exchanger and pump hooked into the back of the calorimeter.

Assuming that the properties of the cooling fluid in the setup remain constant across the system, a simple energy balance condition can be used to find the power losses in the inverter using the temperature rise across the first and second chambers, ΔT_1 and ΔT_2 respectively, and the power input by an adjustable heater. This balance is

$$P_{inv-loss} = P_{heater} \cdot \frac{\Delta T_1}{\Delta T_2} = P_{heater} \cdot \frac{T_{out} - T_{mid}}{T_{mid} - T_{in}} \quad (1)$$

Precision temperature sensors were placed in the system to measure points T_{out}, T_{mid}, and T_{in}. The power of the electric heater was obtained by measuring its current and voltage. The calorimeter test was performed for more than five hours to wait until the thermal condition reached its steady state. Fig. 10 to Fig. 12 shows the coolant temperatures and the measured inverter efficiency at 12kW, 18kW and 27kW, respectively. Different speeds and power factors were tested, which were indicated in the figures. As can be seen, at the initial stage, the efficiency fluctuates; after thermal balanced is well established, the efficiency flattens. From the test results, it is clear that at higher speed and higher power factor, the efficiency is higher, which proves the motor test in previous section. The calorimeter-tested inverter efficiency is 98.8% at 12kW, 99.1% at 18kW, 98.8% at 27kW.

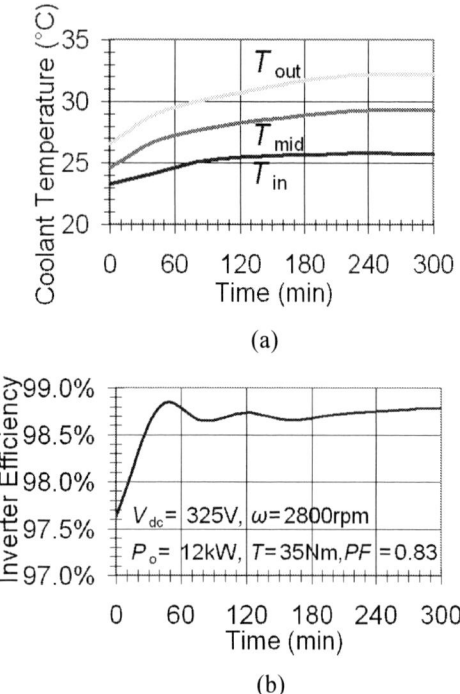

Figure 10. Calorimeter measurement at 12kW: (a) coolant temperatures, (b) inverter efficiency.

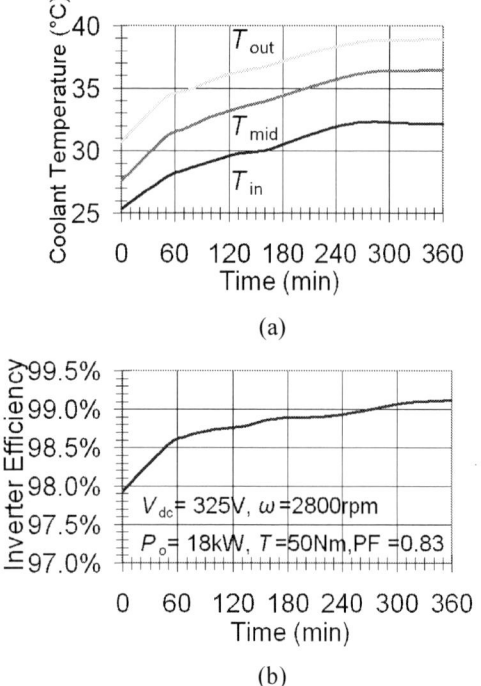

Figure 11. Calorimeter measurement at 18kW: (a) coolant temperatures, (b) inverter efficiency.

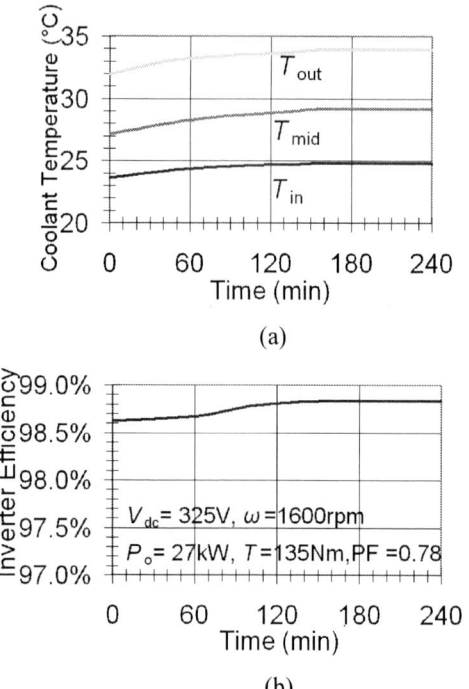

Figure 12. Calorimeter measurement at 27kW: (a) coolant temperatures, (b) inverter efficiency.

VI.   CONCULSION

A high efficiency three-phase soft-switching inverter has been developed and evaluated for high temperature hybrid electric vehicle application. The inverter is designed and assembled by three liquid-cooled soft-switching modules, which reduce both switching and conduction losses.

Complete efficiency tests have been performed at different coolant temperatures ranging from 25°C to 90°C under inductive and motor dynamometer loads. Test results indicate that peak efficiency at the rated speed is around 99%. For the same torque, efficiency drops as speed lowers, typically 0.1% per 500 rpm. For the same output, efficiency drops slightly as temperature increases, typically 0.1% per 50°C. The calorimeter test has been conducted to verify the test results with power meters. Peak efficiency of 99.1% at 30% load was observed. Experimental results prove the feasibility of soft-switching module based inverter operating at high temperature for hybrid electric vehicle application.

ACKNOWLEDGMENT

The authors would like to thank their DOE FreedomCAR Program partners for their valuable contributions. The soft-switching modules were designed and fabricated by colleagues at Powerex. The calorimeter test setup was provided by colleagues at Azure Dynamics. Special thanks go to the lab mechanical engineer, Gary Kerr, for his effort on mechanical design and assembly of the system.

REFERENCES

[1] S. Lande, "Supply and demand for high temperature electronics," in *IEEE 1999 High Temperature Electronics European Coference*, 1999, pp. 133-135.

[2] C.C. Chan and K.T. Chau, "An overview of power electronics in electric vehicles," *IEEE Transactions on Industrial Electronics*, vol. 44, no. 1, pp. 3-13, Feb. 1997.

[3] S.G. Wirasingha, N. Schofield and A. Emadi, "Plug-in hybrid electric vehicle developments in the US: trends, barriers, and economic feasibility," in *IEEE 2008 Vehicle Power and Propulsion Coference*, 2008, pp. 1-8.

[4] F. Renken and R. Knorr, "High temperatue electronic for future hybrid powertrain application," in *IEEE 2005 icle European Coference on Power Electronics and Applications*, 2005, pp. 1-7.

[5] *FreedomCAR and fuel partnership electrical and electronics technical team roadmap*, U.S. Department of Energy, Nov. 2006.

[6] J. Lai, W. Yu, H. Qian, P. Sun, P. Ralston and K. Meehan, "High temperature device characterization for hybrid electric vehicle traction inverters," in *IEEE 2009 Applied Power Electronics Conference and Exposition*, 2009, pp. 665-670.

[7] J.M. Hornberger, E. Cilio, R.M. Schupbach, A.B. Lostetter, and H.A. Mantooth, "A high- temperature multichip power module inverter utilizing silicon carbide and silicon on insulator electronics," in *IEEE 2006 Power Electronics Specialists Conference*, 2006, pp. 1-7.

[8] P. Friedrichs, "Silicon carbide power devices-status and upcoming challenges," in *IEEE 2007 Power Electronics and Applications European Conference*, 2007, pp. 1-11.

[9] B. Ozpineci, M. Chintavali and L.M. Tolbert, "A 55 kW three-phase automotive traction inverter with SiC schottky diodes," in *IEEE 2005 Vehicle Propulsion and Power Conference*, 2005, pp. 541-546.

[10] W. Yu, J.S. Lai and S.Y. Park, "An improved zero-voltage-switching inverter using two coupled magnetics in one resonant pole," in *IEEE 2009 Applied Power Electronics Conference and Exposition*, 2009, pp. 401-406.

[11] J.S. Lai, W. Yu and S.Y. Park, "Variable timing control for wide current range zero-voltage soft-switching inverters," in *IEEE 2009 Applied Power Electronics Conference and Exposition*, 2009, pp. 407-412.

[12] F. Blaabjerg, J.K. Pedersen and E. Ritchie, "Calorimetric measuring systems for characterizing high frequency power losses in power electronic components and systems," in *IEEE 2002 Industry Applications Conference*, 2002, pp. 1368-1376.

[13] P.D. Malliband, N.P. van der Duijn Schouten and R.A. McMahon, "Precision calorimetry for the accurate measurement of inverter losses," in *IEEE 2003 International Conference on Power Electronics and Drive Systems*, 2003, pp. 321-326.

[14] W.P. Cao, K.J. Bradley and A. Ferrah, "Development of a High-Precision Calorimeter for Measuring Power Loss in Electrical Machines," *IEEE Transactions on Instrumentation and Measurement*, vol. 58, no. 3, pp. 570-577, Mar. 2009.

[15] A. Jalilian, V.J. Gosbell, B.S.P. Perera, and P. Cooper, "Double chamber calorimeter (DCC): a new approach to measure induction motor harmonic losses," *IEEE Transactions on Energy Conversion*, vol. 14, no. 3, pp. 680-685, Sep. 1999.

Real-Time Hybrid Model Predictive Control of a Boost Converter with Constant Power Load

Jason Neely, *Member, IEEE*, Steve Pekarek, *Member, IEEE*, Raymond DeCarlo, *Fellow, IEEE*, Nir Vaks

School of Electrical and Computer Engineering Purdue University
West Lafayette, IN 47907

Abstract — **In this paper, a model predictive control method is implemented for improved performance of a boost converter sourcing a constant power load (CPL). The method applies results of recently developed hybrid optimal model predictive control to determine switching through real-time minimization of a user-defined performance index tailored to CPLs. In contrast to many controllers designed for boost converters, the proposed method does not utilize small-signal average value models in its development. Therefore, the controller design is not equilibrium-point specific. In addition, since the optimization is performed online, the control is more flexible than many recently proposed controllers that utilize the hybrid model to determine optimal switching strategies offline and subsequently apply utilizing a look-up table. The result is a robust, high bandwidth control that has been validated in hardware.**

I. INTRODUCTION

DC power distribution systems have found widespread use in ship [1]-[2], space [3], and hybrid electric vehicle power applications [4]-[5]. DC power distribution systems are comprised of multiple power-electronic-based converter modules, each of which can appear as a constant power load (CPL) to the converter module supplying it. Regulation of these systems has been cited as more difficult to control due to negative incremental impedance instabilities associated with CPLs [3]-[6]. Therefore, much research has been devoted to the development of stability criteria [7] and controls for both the stabilization of converters with constant power loads [4],[7]-[9] as well as active control of converter input impedance, known as *power buffering* [10] or *active damping* [11]. In particular, recent improvements have been made in the control of converters that source constant power loads by viewing the converter dynamics in terms of stored energy and instantaneous power. In [12], the authors describe a *power shaping* control strategy to regulate the desired stored energy in the output capacitor by commanding input power. In [13], the authors develop a sliding mode control strategy wherein optimal trajectories are determined in the *impedance-energy plane* through offline minimization of an objective function.

In parallel with stability work, there has been recent success in the application of Model Predictive Control (MPC) to control switching in the boost converter [14]-[16]. MPC is a discrete-time control strategy wherein the control is obtained at each sampling through finite time horizon optimization using the current (measured) state of the system as the initial state [17]-[18]. MPC requires an accurate dynamical model and a user-defined performance index (PI) to be optimized. At each sampling instant, the converter state is measured, which initiates the computation of piecewise constant switching control values over a finite horizon window so that a PI is minimized over that window. The first control in this sequence is then applied to the converter, and the process is repeated. MPC is valuable when controlling systems with constraints on the control signal and/or state [17]-[18]. For the control of switching converters, the computational burden of MPC has typically prohibited real-time solution. The control instead is often expedited with offline calculation and applied in hardware using lookup tables [14]-[16].

In this research, an MPC strategy based on recently developed hybrid optimal control theory [19] is applied in real-time to a dc-dc boost converter with constant power load. The new control differs from classical control methods that rely on average value models (AVMs) and pre-computed control laws. Instead, an embedded form of the hybrid state model of the converter is generated and used to predict converter dynamics within a nonlinear model predictive control strategy termed Hybrid Model Predictive Control (HMPC). Specifically, each switch period, the output voltage and inductor current are measured, a user-defined performance function is minimized *on-line* to generate an optimal sequence of *embedded* switch states, and the first embedded switch state in the sequence is used to generate the real-time on/off sequence of the converter switch.

The HMPC method has four advantages. First, since switching is optimized over individual switching cycles, the controller achieves a higher control bandwidth than controllers based on AVMs. Second, since HMPC does not rely on linearized small-signal approximations, the control is not limited to a narrow range of anticipated operating points. Third, HMPC allows added flexibility in defining converter performance through manipulation of the performance function. Fourth, since HMPC is a predictive control scheme, the control can better manage constraints, such as maintaining limits on inductor current or maintaining continuous conduction.

The approach described here extends the research presented in [20]-[21]. In [20]-[21], HMPC was applied in real-time to a boost converter with resistive load, wherein the underlying model was a linear switched state model. In this paper, the hybrid state model and the performance index of the HMPC are adapted for regulating a boost converter with

978-1-4244-4782-4/10 $26.00 © 2010 IEEE

constant power load, wherein the underlying model is nonlinear. In addition, a new *power flow* formulation is adopted for the performance index. Since the control must regulate the output voltage in lieu of variation in load power, a load-power estimator is derived to update the embedded (nonlinear hybrid) model used by the control. The HMPC control is demonstrated in simulation and in hardware for two multi-converter systems: (1) a boost converter driving a buck converter and (2) a boost converter supplying power to a permanent magnet synchronous machine (PMSM) drive.

The paper is organized as follows. In Section II, the state model of the boost converter with CPL is presented, and the embedded optimal control problem is formulated in continuous time. In Section III, the real-time model predictive control scheme is explained. Hardware results are given in Section IV. Conclusions are provided in Section V.

II. HYBRID OPTIMAL CONTROL OF BOOST CONVERTER WITH CONSTANT POWER LOAD (CPL)

A. Hybrid Model

The control described herein is applied to a boost converter (Fig. 1) sourcing a constant output power P_{load}. In contrast to [20]-[21], we represent the boost converter in continuous conduction mode as a nonlinear hybrid state model

$$
\begin{aligned}
\dot{\mathbf{x}} = (1-s)\cdot\big[&\mathbf{A}_0\left(\mathbf{x},P_{\text{load}}\right)\mathbf{x}+\mathbf{b}_0\left(V_s\right)\big] \\
+ s\cdot\big[&\mathbf{A}_1\left(\mathbf{x},P_{\text{load}}\right)\mathbf{x}+\mathbf{b}_1\left(V_s\right)\big]
\end{aligned}
\tag{1}
$$

where the discrete valued switch state is given $s\in\{0,1\}$ (converter *on* (s=1) or *off* (s=0)) and where

$$
\mathbf{A}_0(\mathbf{x},P_{\text{load}})=\begin{bmatrix} -\dfrac{r_L}{L} & -\dfrac{1}{L} \\ \dfrac{1}{C} & -\dfrac{P_{\text{load}}}{Cv_C^2} \end{bmatrix},\quad \mathbf{b}_0(V_s)=\begin{bmatrix} (V_s-V_d)/L \\ 0 \end{bmatrix}
\tag{2}
$$

$$
\mathbf{A}_1(\mathbf{x},P_{\text{load}})=\begin{bmatrix} -\dfrac{r_L}{L} & 0 \\ 0 & -\dfrac{P_{\text{load}}}{Cv_C^2} \end{bmatrix},\quad \mathbf{b}_1(V_s)=\begin{bmatrix} (V_s-V_{sw})/L \\ 0 \end{bmatrix}
\tag{3}
$$

where $\mathbf{x}=\begin{bmatrix} i_L & v_C \end{bmatrix}^T$ is the state, V_d and V_{sw} are the diode and switch forward voltage drops respectively, and all other variables are shown in Fig 1. For the system described herein, the switch and diode forward voltage drops are assumed to be equal and given by V_f, where $V_f = V_d = V_{sw}$. This model is a nonlinear generalization of the linear state models considered in [20]-[21] which is due to the presence of the constant power load.

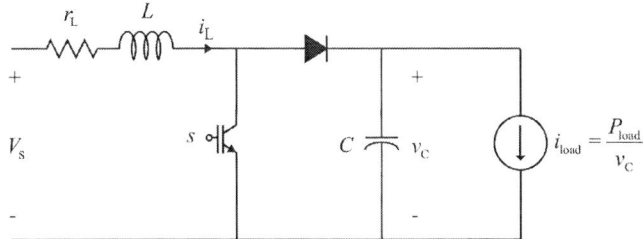

Fig. 1: Schematic of Boost Converter with CPL

B. Formulating the EOCP

The control objective is to optimally determine switching using the hybrid optimal control methodology of [19]-[21]. The procedure set forth in these references requires that we first construct the so-called embedded form of the hybrid state model of (1)-(3) by replacing the discrete valued switch state $s\in\{0,1\}$ with the continuous-valued embedded switch state $\tilde{s}\in\begin{bmatrix} 0 & 1 \end{bmatrix}$.

The second phase in the control of the boost converter involves solving for an $\tilde{s}$ that minimizes an embedded performance index of the form

$$
P_s = \int_{t_0}^{t_f}\Big[\tilde{s}\cdot F_1(\mathbf{x},\mathbf{x}^*,V_s,P_{\text{load}})+(1-\tilde{s})\cdot F_0(\mathbf{x},\mathbf{x}^*,V_s,P_{\text{load}})\Big]dt
\tag{4}
$$

subject to the embedded model. In (4), $\begin{bmatrix} t_0 & t_f \end{bmatrix}$ is a finite time window, $\mathbf{x}^* = \begin{bmatrix} I_L^* & V_C^* \end{bmatrix}^T$ is the desired state, P_{load} is load power, and the user-defined integrands $F_0, F_1 \in C^1$ represent the desired performance in modes 0 and 1 respectively. Once the optimal $\tilde{s}\in\begin{bmatrix} 0 & 1 \end{bmatrix}$ is computed, the solution is projected onto the allowable switching set $s\in\{0,1\}$ using a duty-cycle interpretation of $\tilde{s}\in\begin{bmatrix} 0 & 1 \end{bmatrix}$.

The integrands are defined in terms of power and energy balance as follows:

$$
\begin{aligned}
F_0 = F_1 &= F\left(\mathbf{x},\mathbf{x}^*,V_s,P_{\text{load}}\right) \\
&= C_P\Big[p_{\text{in}}(V_s,\mathbf{x})-p_{\text{loss}}(\mathbf{x})-P_{\text{load}}-K\left(E_{LC}^*-e_{LC}(\mathbf{x})\right)\Big]^2 \\
&\quad +C_E\left(e_{LC}(\mathbf{x})-E_{LC}^*\right)^2
\end{aligned}
\tag{5}
$$

where the term $C_E\left(e_{LC}(\mathbf{x})-E_{LC}^*\right)^2$ weights through $C_E \in \mathbb{R}^+$ the difference between the stored energy in the converter,

$$
e_{LC}(\mathbf{x})=\frac{1}{2}Li_L^2+\frac{1}{2}Cv_C^2
\tag{6}
$$

and the desired steady-state stored energy $E_{LC}^* = e_{LC}(\mathbf{x}^*)$. Thus, in steady-state, one aspect of the desired performance is to have

$$\left(e_{LC}(\mathbf{x}) - E_{LC}^*\right) = 0 \qquad (7a)$$

Assuming V_C^*, I_L^* are constant, the quantity

$$\frac{d}{dt}\left(e_{LC}(\mathbf{x}) - E_{LC}^*\right) = p_{in}(V_s, \mathbf{x}) - p_{loss}(\mathbf{x}) - P_{load} \qquad (7b)$$

represents the rate at which energy is being stored or dissipated in relation to the desired equilibrium energy, i.e. the instantaneous stored/dissipated power in regards to the equilibrium energy where $p_{in} = V_s i_L$ is instantaneous input power, $p_{loss} = r_L i_L^2 + i_L V_f$ is conduction loss, and P_{load} the instantaneous power consumed by the load. In addition to achieving a steady-state equilibrium energy as quickly as possible (equation 7a), we also desire to drive the power to the storage devices to zero, i.e. input power is going to the output and losses, but not to energy storage, and thereby arrive in steady-state operation. Hence, the equilibrium power term in the PI is $C_P\ p_{in}(V_s, \mathbf{x}) - p_{loss}(\mathbf{x}) - P_{load} - K\left(E_{LC}^* - e_{LC}(\mathbf{x})\right)^2$ which weights through $C_P \in \mathbb{R}^+$ the difference between stored/dissipated power ($p_{in}(V_s, \mathbf{x}) - p_{loss}(\mathbf{x}) - P_{load}$) and a commanded compensation power, $K\left(E_{LC}^* - e_{LC}(\mathbf{x})\right)$. Not only does the PI explicitly force $e_{LC}(\mathbf{x}) \to E_{LC}^*$, the PI through switching forces

$$p_{in}(V_s, \mathbf{x}) \to p_{loss}(\mathbf{x}) + P_{load} + K\left(E_{LC}^* - e_{LC}(\mathbf{x})\right) \qquad (8)$$

in which case

$$\frac{d}{dt}\left(E_{LC}^* - e_{LC}(\mathbf{x})\right) \cong -K\left(E_{LC}^* - e_{LC}(\mathbf{x})\right) \qquad (9)$$

through substitution of equation (8) into equation (7b). Equation (9) indicates that in steady state equilibrium is maintained robustly: when $\left(E_{LC}^* - e_{LC}(\mathbf{x})\right) > 0$ then $\frac{d}{dt}\left(E_{LC}^* - e_{LC}(\mathbf{x})\right) < 0$ and when $\left(E_{LC}^* - e_{LC}(\mathbf{x})\right) < 0$ then $\frac{d}{dt}\left(E_{LC}^* - e_{LC}(\mathbf{x})\right) > 0$, and the constant $K \in \mathbb{R}^+$ adjusts the rate of convergence. Specifically, equation (9) yields an approximate solution for the error $\left(E_{LC}^* - e_{LC}(\mathbf{x}(t))\right)$ given as:

$$\left(E_{LC}^* - e_{LC}(\mathbf{x}(t))\right) \cong \left(E_{LC}^* - e_{LC}(\mathbf{x}(t_0))\right) \cdot e^{-K(t - t_0)} \qquad (10)$$

which converges exponentially to zero.

This work (herein) shows that the dynamic performance of the converter may be improved by defining the control-to-output behavior to include energy stored in the converter. Indeed, this work supports that in [13],[22]-[24] where a consideration of converter energy allows for robust geometric and time-domain control that achieves fast response despite the non-minimum phase characteristic of the converter.

In [20]-[21], HMPC was applied to a boost converter with resistive load; therein, the integrands of the PI were defined in terms of voltage and current and took the form

$$F_0 = F_1 = C_I\left[i_L - (I_L^* + K(V_C^* - v_C))\right]^2 + C_V\left(V_C^* - v_C\right)^2 \qquad (11)$$

wherein the optimization emulated a current-mode control. The control therein demonstrated fast dynamic response and good steady state behavior at multiple operating points.

To illustrate the difference between the PI's given in (11) and in (5), it is convenient to consider the embedded state trajectories of the boost converter with constant power load in the phase plane. This is shown in Fig. 2. wherein solid lines denote the performance surfaces in (11) and the dashed curve represents the so-called constant energy curve associated with the desired equilibrium. For the control dictated by (5), the embedded state trajectory approaches the desired energy curve in the limit. In contrast, the PI given by (11) results in a control that drives $i_L \to I_L^* + K(V_C^* - v_C)$, which may result in excess energy being added to the converter when $(V_C^* - v_C)$ is large.

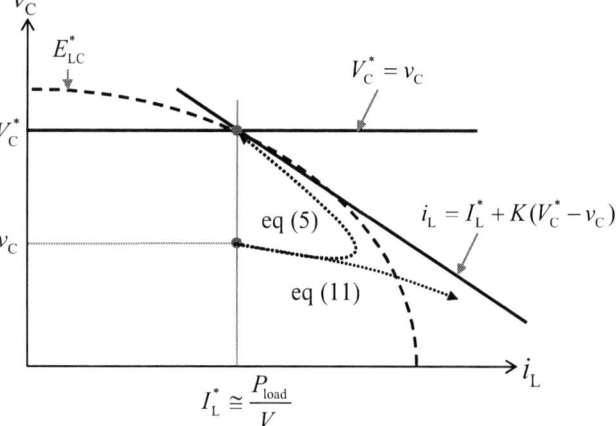

Fig. 2: Illustration of performance surfaces in state phase plane

Two final points. Since the load power is not directly measured, equation (5) uses an estimate of load power denoted $\hat{P}_{load}$; $F\left(\mathbf{x}, \mathbf{x}^*, V_s, \hat{P}_{load}\right)$ is minimized when $e_{LC} = E_{LC}^*$ and $p_{in} = \hat{P}_{load} + p_{loss}(\mathbf{x}^*)$. Herein, the estimated load power $\hat{P}_{load}$ is generated using an *output power estimator*. Assuming power tracking as in equation (8), with P_{load} replaced by $\hat{P}_{load}$, the time derivative of circuit energy is

$$\frac{d}{dt}\left(E_{LC}^* - e_{LC}(\mathbf{x})\right) \cong -K\left(E_{LC}^* - e_{LC}(\mathbf{x})\right) + \left(P_{load} - \hat{P}_{load}\right) \quad (12)$$

Thus, as $\left(\hat{P}_{load} - P_{load}\right) \to 0$, $e_{LC}(\mathbf{x}) \to E_{LC}^*$.

Finally, although the commanded output voltage V_C^* is specified, the commanded inductor current I_L^* must be calculated in order to compute E_{LC}^* by solving $p_{in}(V_s, \mathbf{x}^*) - p_{loss}(\mathbf{x}^*) - \hat{P}_{load} = 0$. However, for many applications, $p_{loss}(\mathbf{x}) \ll \hat{P}_{load}$, and the commanded inductor current may be approximated as: $I_L^* \cong \hat{P}_{load} / V_s$.

III. REAL-TIME HYBRID MODEL PREDICTIVE CONTROL

Since a practical system has some computational delay and load variation, delay compensation and a method for load power estimation are necessary. A step-by-step guide for designing the complete control is provided here, and an overview of the control scheme is illustrated in Figure 4. We assume that the source voltage V_s and the circuit parameters r_L, L, C, V_f are known and constant over the operating region. Thus using the steps set forth below, the EOCP is cast as a discrete-time non-linear MPC problem that may be solved in real-time.

A. Controller Formulation

Step 1. The discrete-time embedded model

$$\begin{aligned}
\mathbf{x}_{k+1} = &(1 - \tilde{s}_k) \cdot \left(\left[\mathbf{I} + T_s \mathbf{A_0}(\mathbf{x}_k, P_{load}(k))\right]\mathbf{x}_k + T_s \mathbf{b_0}(V_s)\right) \\
&+ \tilde{s}_k \cdot \left(\left[\mathbf{I} + T_s \mathbf{A_1}(\mathbf{x}_k, P_{load}(k))\right]\mathbf{x}_k + T_s \mathbf{b_1}(V_s)\right)
\end{aligned} \quad (13)$$

is constructed using the Forward Euler (FE) method with time step of $T_s = 1/F_s$, F_s being the switching frequency. The discrete time points are $t_k = k \cdot T_s$, $\mathbf{x}_k = \mathbf{x}(t_k)$; the discrete-time embedded switch state $\tilde{s}_k = \tilde{s}(t_k) \in [0,1]$ is considered constant over $t_k \le t < t_{k+1}$.

Although alternate discrete-time formulations may be used, the FE method is selected to minimize the numerical burden on the convex programming algorithm to be described.

If a measurement of P_{load} is unavailable, an estimate is computed as given in Step 2.

Step 2. Measurements taken at t_k and t_{k-1} are used to update the load power estimate $\hat{P}_{load}(k)$ using a simple two-step estimator set forth in equations (14)-(15) below.

By discretizing (7b) and assuming that P_{load} is constant over $t_k \le t < t_{k+1}$, an initial estimate is simultaneously made of the circuit energy and load power at t_k as follows:

$$\begin{aligned}
\begin{bmatrix} \bar{e}_{LC}(k) \\ \hat{P}_{load}(k) \end{bmatrix} &= \begin{bmatrix} 1 & -T_s \\ 0 & 1 \end{bmatrix} \begin{bmatrix} \hat{e}_{LC}(k-1) \\ \hat{P}_{load}(k-1) \end{bmatrix} \\
&+ T_s \begin{bmatrix} p_{in}\left(V_s, \dfrac{\mathbf{x}_{m,k} + \mathbf{x}_{m,k-1}}{2}\right) - p_{loss}\left(\dfrac{\mathbf{x}_{m,k} + \mathbf{x}_{m,k-1}}{2}\right) \\ 0 \end{bmatrix}
\end{aligned}$$

(14)

The estimates are then improved using a state measurement at t_k as follows:

$$\begin{bmatrix} \hat{e}_{LC}(k) \\ \hat{P}_{load}(k) \end{bmatrix} = \begin{bmatrix} \bar{e}_{LC}(k) \\ \hat{P}_{load}(k) \end{bmatrix} + \begin{bmatrix} -1 \\ T_s \kappa_p \end{bmatrix}\left(\bar{e}_{LC}(k) - e_{LC}(\mathbf{x}_{m,k})\right) \quad (15)$$

where κ_p is the estimator gain. It is noted that the input power and circuit loss calculated in (14) are evaluated using the midpoint $(\mathbf{x}_{m,k} + \mathbf{x}_{m,k-1})/2$ for reduced discretization error. The errors between estimated and real quantities are given:

$$\bar{e}_{LC}(k) = \hat{e}_{LC}(k) - e_{LC}(\mathbf{x}_k) \quad (16)$$

$$\bar{P}_{load}(k) = \hat{P}_{load}(k) - P_{load}(k) \quad (17)$$

Combining (14)-(17), the error system is given by

$$\begin{bmatrix} \bar{e}_{LC}(k) \\ \bar{P}_{load}(k) \end{bmatrix} = \begin{bmatrix} 0 & 0 \\ T_s \kappa_p & \left(1 - T_s^2 \kappa_p\right) \end{bmatrix} \begin{bmatrix} \bar{e}_{LC}(k-1) \\ \bar{P}_{load}(k-1) \end{bmatrix} \quad (18)$$

The error system (18) is stable as long as $\left|1 - T_s^2 \kappa_p\right| < 1$ which is enforced numerically.

Step 3. To compensate for computation delay, the state at t_{k+1} is estimated one switch period earlier and used as the initial state in the finite horizon time window. A portion of the time interval $[t_k, t_{k+1}]$ is then used to compute the optimal embedded switch state $\tilde{s}_{k+1}$ so that it may be applied precisely at t_{k+1}. As such, the state initializing the MPC optimization is estimated according to:

$$\begin{aligned}
\hat{\mathbf{x}}_{k+1} = &(1 - \tilde{s}_k) \cdot \left(\left[\mathbf{I} + T_s \mathbf{A_0}(\mathbf{x}_{m,k}, \hat{P}_{load}(k))\right]\mathbf{x}_{m,k} + T_s \mathbf{b_0}(V_s)\right) \\
&+ \tilde{s}_k \cdot \left(\left[\mathbf{I} + T_s \mathbf{A_1}(\mathbf{x}_{m,k}, \hat{P}_{load}(k))\right]\mathbf{x}_{m,k} + T_s \mathbf{b_1}(V_s)\right)
\end{aligned} \quad (19)$$

where $\mathbf{x}_{m,k}$ is a state measurement taken at t_k, $\tilde{s}_k$ was calculated in the previous interval, and $\hat{\mathbf{x}}_{k+1}$ is the predicted state at t_{k+1}.

Step 4. The user selects a finite horizon time window $[t_0, t_f] = [t_{k+1}, t_{k+N+1}]$ and divides it into N partitions:

$[t_{k+1}, t_{k+2}] \ldots [t_{k+N}, t_{k+N+1}]$. For the converter studied here, $N=2$; thus, the window includes partitions $[t_{k+1}, t_{k+2}]$ and $[t_{k+2}, t_{k+3}]$.

Step 5. The discrete-time embedded PI is formed by discretizing (4) using the trapezoidal rule. Since $F_0 = F_1$, the discrete-time performance index for $N = 2$ may be expressed

$$P_s = \frac{T_s}{2}\left(\hat{\mathbf{y}}_{k+1}^T \mathbf{Q}\hat{\mathbf{y}}_{k+1} + 2\hat{\mathbf{y}}_{k+2}^T \mathbf{Q}\hat{\mathbf{y}}_{k+2} + \hat{\mathbf{y}}_{k+3}^T \mathbf{Q}\hat{\mathbf{y}}_{k+3}\right) \quad (20)$$

with

$$\hat{\mathbf{y}}_k = \begin{bmatrix} p_{in}\left(V_s, \hat{\mathbf{x}}_k\right) - p_{loss}\left(\hat{\mathbf{x}}_k\right) - \hat{P}_{load} \\ e_{LC}\left(\hat{\mathbf{x}}_k\right) - E_{LC}^* \end{bmatrix} \quad (21)$$

where hatted variables denote predicted values and where

$$\mathbf{Q} = \begin{bmatrix} C_P & C_P K \\ C_P K & C_P K^2 + C_E \end{bmatrix} \quad (22)$$

with $\mathbf{Q} \succ 0$ for $C_E > 0$, which guarantees that equation (20) is strictly convex in $\hat{\mathbf{y}}$ to avoid singular solutions.

Step 6. Minimization of the PI (20) is subject to equality constraints defined by the discrete-time model (13), and inequality constraints defined by bounds on the control $\tilde{s}_{k+1}, \tilde{s}_{k+2}$ and on the state. Specifically, to maintain continuous conduction, the discrete-time embedded inductor currents are subject to a lower bound: $\hat{i}_L(k+2) \geq i_{min}$, $\hat{i}_L(k+3) \geq i_{min}$. Using (13), this lower bound may be represented as additional constraints on $\tilde{s}_{k+1}, \tilde{s}_{k+2}$. The inequality constraints are summarized as follows:

-Admissible controls

$$0 \leq \tilde{s}_{k+1}, \tilde{s}_{k+2} \leq 1 \quad (23)$$

-Controls that maintain continuous conduction

$$\tilde{s}_{k+1} \geq \frac{L}{T_s} \frac{\left(i_{min} - \hat{i}_L(k+1)\right)}{\hat{v}_C(k+1)} + \frac{\left(\hat{v}_C(k+1) + r_L \hat{i}_L(k+1) - V_s\right)}{\hat{v}_C(k+1)} \quad (24)$$

$$\tilde{s}_{k+1} + \tilde{s}_{k+2} \geq \frac{L}{T_s} \frac{\left(i_{min} - \hat{i}_L(k+1)\right)}{\hat{v}_C(k+1)} + 2\frac{\left(\hat{v}_C(k+1) + r_L \hat{i}_L(k+1) - V_s\right)}{\hat{v}_C(k+1)} \quad (25)$$

Inequality (24) is given by direct algebraic manipulation of the model. Inequality (25) requires the assumption that $v_C(k) / v_C(k+1) \approx 1$. The lower bound i_{min} requires knowledge of the switching function that will be used. As will be described in Step 8, a center-aligned PWM will be implemented. Thus, it can be shown that this lower bound is given by

$$i_{min} = \frac{V_s T_s}{2L}\frac{\left(\hat{v}_C(k+1) + r_L \hat{i}_L(k+1) - V_s\right)}{\hat{v}_C(k+1)} \quad (26)$$

The set of feasible controls is thus given by the intersection of constraint sets (23)-(25) and denoted S_{CCM}. Since the constraints (23)-(25) are each convex sets, the intersection of these, S_{CCM}, is also a convex set.

Step 7. The local control problem is thus stated as

$$\underset{\tilde{s}_{k+1}, \tilde{s}_{k+2} \in S_{CCM}}{\text{minimize}} \quad P_s \quad (27)$$

subject to the model (13) and to the inequality constraints (23)-(25). Proper selection of control parameters C_P, K and C_E can ensure that P_s is convex in $\tilde{s}_{k+1}, \tilde{s}_{k+2}$ over a given operating region. Since S_{CCM} is a convex set and P_s is determined to be a convex function of $\tilde{s}_{k+1}, \tilde{s}_{k+2}$ over the operating region, the discrete-time hybrid optimal control problem is reduced to a convex optimization problem that may be solved numerically. Specifically, the minimization over $\tilde{s}_{k+1}, \tilde{s}_{k+2}$ is accomplished using an Active Set algorithm that relies on Newton steps. The algorithm is explained in detail later in this section.

Step 8. The numerical solution of the embedded hybrid optimal control problem solves for $\tilde{s} \in [0,1]$ over the time windows $[t_{k+1}, t_{k+2}]$ and $[t_{k+2}, t_{k+3}]$. The physical switch requires a signal with $s(t) \in \{0,1\}$. Therefore, as a final step, the embedded solution for the first partition $[t_{k+1}, t_{k+2}]$ is projected onto an $s(t) \in \{0,1\}$ using a duty cycle interpretation which is provided to the gate drive of the converter. The embedded solution for the second partition $[t_{k+2}, t_{k+3}]$ is used as an initial estimate for the next window optimization. The projection algorithm used in the system presented here is illustrated in Fig. 3 [25].

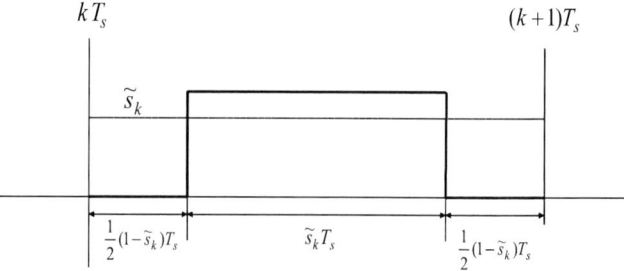

Fig. 3. Projection algorithm

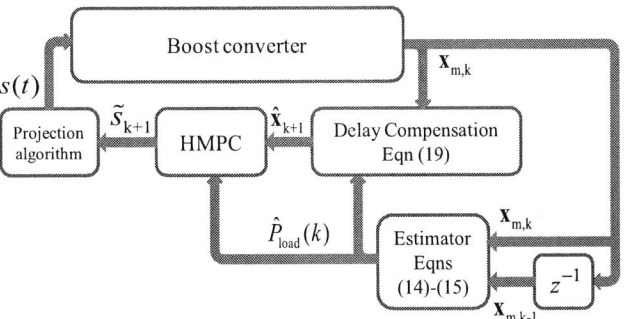

Fig. 4. Controller Block Diagram

B. Numerical Optimization Algorithm

Convex optimization problems with inequality constraints may be solved using any one of a number of methods, including Gradient Projection, Interior Point, or Active Set methods [26]. While Active Set methods have a combinatorial element that makes them undesirable for systems with many constraints, they are typically well suited for low-dimensional problems with few constraints [26]-[27]. For the optimization problem described herein, the inequality constraints are few in number and linear (with the majority being bound constraints); therefore, an Active set strategy is employed that is straight forward to implement and fast enough to solve in real-time. The Active set method employed here is similar to those given in [26],[28].

The inequality constraints given by (23)-(25) may be written in the form $g_i\left(\tilde{s}_{k+1},\tilde{s}_{k+2}\right) \le 0$ where $i \in \{1,2,...,6\}$. The constraint is considered to be *active* when $g_i\left(\tilde{s}_{k+1},\tilde{s}_{k+2}\right) = 0$ and *inactive* when $g_i\left(\tilde{s}_{k+1},\tilde{s}_{k+2}\right) < 0$. The indices of *active* constraints are members of the *working set* ϑ, ie. $g_i\left(\tilde{s}_{k+1},\tilde{s}_{k+2}\right) = 0$ for $i \in \vartheta$.

The algorithm begins from a feasible starting point with all inequality constraints *inactive* (ie. $\vartheta = \varnothing$). In each iteration of the algorithm, the system is subject only to equality constraints given by the model and by the *active* inequality constraints; the equality constrained system is approximately solved using a Newton step. If an inequality constraint is violated by a full Newton step, the step is scaled such that the iterate lies on the constraint boundary, the constraint becomes *active*, and the algorithm repeats.

The set of active constraints must be linearly independent [28]. Therefore, since the optimization is performed in $\mathbb{R}^2$, at most two constraints can be active.

If the initial value is infeasible, the algorithm begins with an orthogonal projection step, projecting the starting value back onto the interior of S_{CCM}. Beginning with a feasible initial value $\left[\left(\tilde{s}_{k+1}\right)^0 \quad \left(\tilde{s}_{k+2}\right)^0\right]^T$ and $\vartheta = \varnothing$, the algorithm starts with n=1 and is given as follows:

Iterate

1. For the nth iteration, perform one Newton step to approximately solve the equality constrained problem:
$$\underset{(\tilde{s}_{k+1})^n,(\tilde{s}_{k+2})^n}{\text{minimize}} \; P_s$$
subject to the model (13) and $g_i\left((\tilde{s}_{k+1})^n,(\tilde{s}_{k+2})^n\right) = 0$ for $i \in \vartheta$

2. Assuming $\sum_{i \in \vartheta}\lambda_i \nabla g_i - \nabla P_s \approx \mathbf{0}$, where ∇P_s is the gradient of P_s with respect to the control, calculate the Lagrange multipliers λ_i for $i \in \vartheta$, and remove the constraint with the most negative Lagrange multiplier from the working set.

$$\text{If } \exists \lambda_i < 0 \quad \text{then} \quad \vartheta = \vartheta / i \text{ where } i = \arg\min\left(\lambda_i\right)$$

3. Define:
$$\Delta\tilde{s}_{k+1} = (\tilde{s}_{k+1})^n - (\tilde{s}_{k+1})^{n-1} \tag{28}$$
$$\Delta\tilde{s}_{k+2} = (\tilde{s}_{k+2})^n - (\tilde{s}_{k+2})^{n-1} \tag{29}$$

Check feasibility of constraints; scale $\Delta\tilde{s}_{k+1}, \Delta\tilde{s}_{k+2}$ if solution is "blocked" by a constraint.

If $g_i\left((\tilde{s}_{k+1})^n,(\tilde{s}_{k+2})^n\right) < 0$ for $\forall i \notin \vartheta$
 then
 $\alpha' = 1$
 else
 $\alpha' = \max\left\{\alpha \mid g_i\left(\tilde{s}_{k+1} + \alpha\Delta\tilde{s}_{k+1}, \tilde{s}_{k+2} + \alpha\Delta\tilde{s}_{k+2}\right) \le 0 \text{ for } i \notin \vartheta\right\}$

4. Add new active constraints to working set.

$$\vartheta = \vartheta \cup i \quad \text{where} \quad g_i\left(\tilde{s}_{k+1} + \alpha'\Delta\tilde{s}_{k+1}, \tilde{s}_{k+2} + \alpha'\Delta\tilde{s}_{k+2}\right) = 0$$

5. Update the solution according to (30), and go to step 1.

$$\begin{bmatrix} \tilde{s}_{k+1} \\ \tilde{s}_{k+2} \end{bmatrix}^n = \begin{bmatrix} \tilde{s}_{k+1} \\ \tilde{s}_{k+2} \end{bmatrix}^{n-1} + \alpha' \begin{bmatrix} \Delta\tilde{s}_{k+1} \\ \Delta\tilde{s}_{k+2} \end{bmatrix} \tag{30}$$

The above steps outline the Active set algorithm used to solve the constrained HMPC optimization. The most challenging step is step 1; however, the order of the optimization may easily be reduced using elimination of variables (ie. substitution). If $\vartheta = \varnothing$, step 1 is implemented by:

$$\begin{bmatrix} \tilde{s}_{k+1} \\ \tilde{s}_{k+2} \end{bmatrix}^n = \begin{bmatrix} \tilde{s}_{k+1} \\ \tilde{s}_{k+2} \end{bmatrix}^{n-1} - \begin{bmatrix} \dfrac{\partial^2 P_s}{\partial \tilde{s}_{k+1}^2} & \dfrac{\partial^2 P_s}{\partial \tilde{s}_{k+1}\partial \tilde{s}_{k+2}} \\ \dfrac{\partial^2 P_s}{\partial \tilde{s}_{k+1}\partial \tilde{s}_{k+2}} & \dfrac{\partial^2 P_s}{\partial \tilde{s}_{k+2}^2} \end{bmatrix}^{-1} \begin{bmatrix} \dfrac{\partial P_s}{\partial \tilde{s}_{k+1}} \\ \dfrac{\partial P_s}{\partial \tilde{s}_{k+2}} \end{bmatrix} \tag{31}$$

where the model equations have been substituted directly into the performance function. If ϑ contains one constraint, the optimization reduces to a line search. If ϑ contains two constraints, $\Delta\tilde{s}_{k+1} = \Delta\tilde{s}_{k+2} = 0$. The algorithm iterates exactly three times. Thus, if the working set does not change from one iteration to the next, the solution is further refined by subsequent Newton steps. Under most circumstances, the inequality constraints will not be active, (ie. steady-state conditions). If the inequality constraints are inactive for all iterations, the algorithm reduces to Newton-Raphson. It is assumed the starting value is already "close" to the optimal value, and Newton Raphson has second order convergence; therefore, the error between the solution and the optimal value is considered insignificant after three Newton steps.

IV. EXPERIMENTAL RESULTS

The HMPC control is evaluated using two multi-converter systems: (1) a boost converter driving a buck converter and (2) a boost converter driving a permanent magnet synchronous machine (PMSM) drive. In each system, the boost converter is regulated using the above described control wherein the buck converter and PMSM drive loads are modeled as CPLs and their power demands are estimated in real-time.

The source voltage V_s, for both systems, was provided to the boost converter by a Sorensen 300-33T DC power supply. All computations for implementing the HMPC and estimator are performed by a TMS320C6711 DSP onboard the Toro PCI card by Innovative Integration. The voltage is measured by a Tektronix P5200 differential voltage probe, and the current is measured using a Tektronix A6303 current sensor with AM503B amplifier. The sensor outputs are connected directly to Analog-to-Digital channels on the Toro PCI card.

The DSP samples voltage and current measurements at time t_k, updates $\hat{P}_{load}$, predicts the state $\hat{\mathbf{x}}_{k+1}$, solves for $\tilde{s}_{k+1}$ using the Active Set algorithm, and produces an analog signal between 0 and 5V that corresponds linearly to values for the embedded switch state $\tilde{s}(t) \in [0,1]$. The embedded solution $\tilde{s}(t)$ is provided to a center-aligned PWM peripheral that implements the projection algorithm.

The physical boost converter parameters and control parameters used for all experiments are provided in Table I.

A. System 1: Boost-Driven Buck Converter

In the first system, HMPC is applied in hardware for a boost converter driving the input of a buck converter. The buck converter maintains a constant voltage across a resistive load using a proportional + integral (PI) current-mode control with a hysteresis modulator. A schematic of the system is provided in Fig. 5, and the buck converter parameters are given in Table II. Photos of the physical boost and buck converters are shown in Fig. 6.

TABLE I. BOOST CONVERTER PARAMETERS

Parameter	Value	Description
V_s	System 1: 85 V System 2: 201.6 V	Source voltage
F_s	15.68 kHz	Switching frequency
r_L	0.1 Ω	Input inductor series resistance
V_f	1.0 V	Diode/switch voltage drop
L	1.0 mH	Boost converter input inductor
C	500 μF	Boost converter output capacitor
C_P	1.0 W^{-2}	Power gain
K	150 W/J	Energy-to-power Control gain
C_E	1000 J^{-2}	Energy gain
κ_p	1.5e6 W/J	Estimator gain

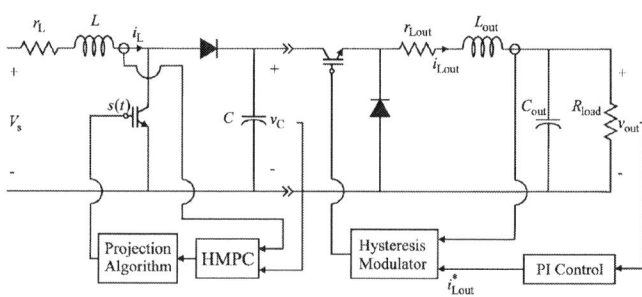

Fig. 5. Schematic of boost converter driving buck converter

TABLE II. BUCK CONVERTER PARAMETERS

Parameter	Value	Description
r_{Lout}	26 mΩ	Output inductor series resistance
L_{out}	1.99 mH	Buck converter output inductor
C_{out}	1800 μF	Buck converter output capacitance
R_{load}	Exp 1: 5 Ω Exp 2: 10 Ω/5 Ω	Buck converter load resistance
V_{out}	50 V	Buck converter output voltage
h	0.8 A	Hysteresis Current

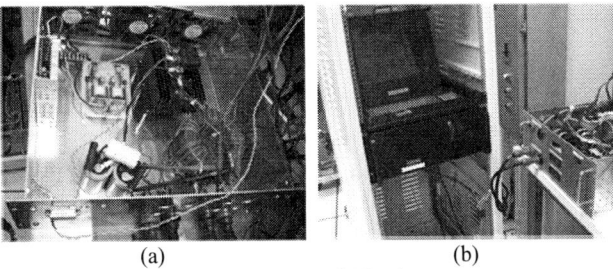

(a)   (b)

Fig. 6. Hardware (a) boost converter (b) buck converter

Two experiments were performed using System 1. In the first experiment, the boost converter was supplied with 85 V and the buck converter supplied a 5 Ω load with 50V. The boost converter was allowed to reach steady state with a commanded output voltage of $V_C^* = 100\ V$. Two step responses in commanded boost converter output voltage were done in 200 msec: $V_C^* = 100\ V$ to $V_C^* = 200\ V$ and $V_C^* = 200\ V$ to $V_C^* = 150\ V$. The voltage and current waveforms are shown in Fig. 7. In addition to voltage and current data, a Digital-to-Analog channel on the DSP was used to output the load power estimate $\hat{P}_{\text{load}}$ which is also provided in Fig. 7.

The converter is seen to double the output voltage from $V_C^* = 100$ V to $V_C^* = 200\ V$ with a rise time of 14 msec without overshoot and then steps the voltage down to $V_C^* = 150\ V$ with similar rate of convergence. This result demonstrates the fast response of HMPC as well as its ability to regulate over a wide range of operating points. In addition, it is noted that the buck converter efficiency is affected by the voltage supplied by the boost converter; Fig. 7(c) indicates that $\hat{P}_{\text{load}}$ is adjusted to compensate as the buck converter power demands change at each voltage level. A close-up of the embedded switch state is shown in Fig. 8 following the first commanded step change in output voltage. It is noted that $\tilde{s}$ varies over the whole range [0 1] during the transient.

A close-up of the inductor current is shown in Fig. 9 after the change in commanded output voltage $V_C^* = 200\ V$ to $V_C^* = 150\ V$, illustrating that the HMPC controller precisely maintains the converter in continuous conduction operation during the second transient.

A second experiment is conducted for System 1 to demonstrate the load-power estimator. The system is operated first in steady-state with $V_C^* = 150\ V$ and the load of the PS1 set at $R_{\text{load}} = 10\ \Omega$. A second 10 Ω load is switched in parallel with the output resistor, making $R_{\text{load}} = 5\ \Omega$. The measured transient response is shown in Fig. 10 for the 500 msec experiment.

A step change in the PS1 resistive load results in an output voltage disturbance on the boost converter (Fig. 10). After the change in PS1 resistive load, the estimated load power increases from a mean value of 302 W to 574 W. The voltage

falls briefly from 150V to 143V (approximately 4.7%) and the output voltage recovers in approximately 50 msec.

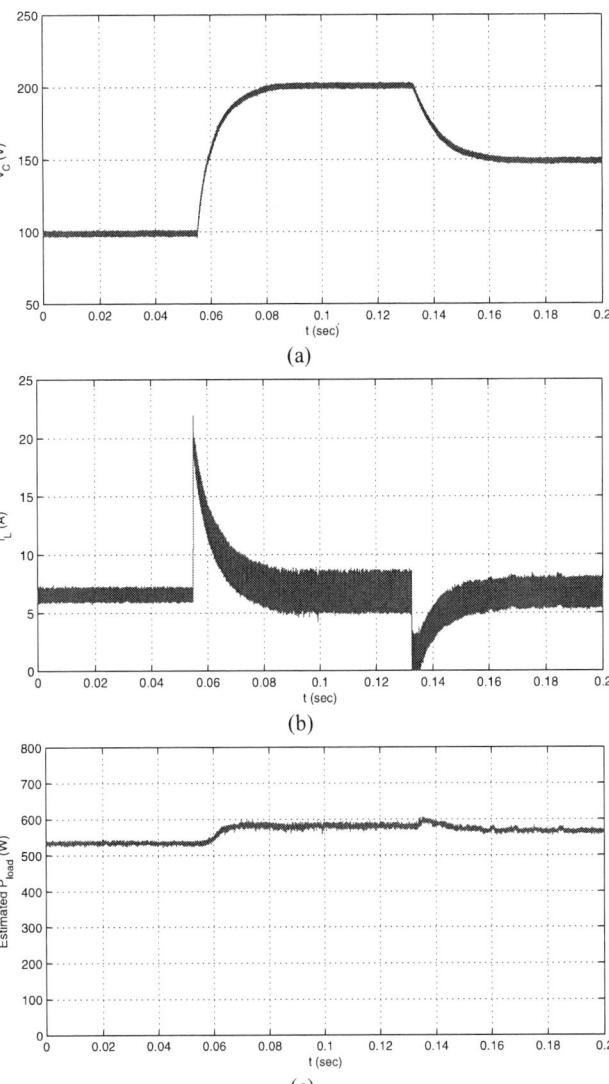

Fig. 7: Experiment 1 hardware result including (a) voltage (b) current and (c) estimated load power for step changes in commanded output voltage: $V_C^* = 100$ V to $V_C^* = 200\ V$ to $V_C^* = 150\ V$

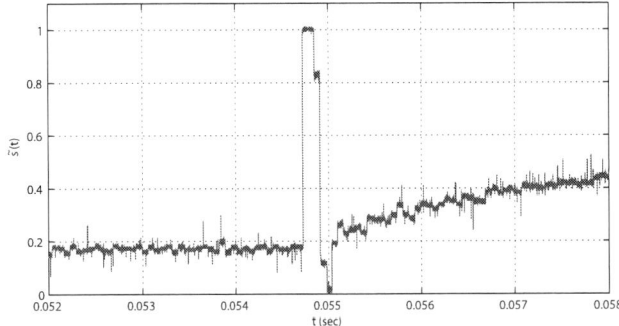

Fig. 8: Close-up of embedded switch state for step change in commanded output voltage: $V_C^* = 100$ V to $V_C^* = 200\ V$

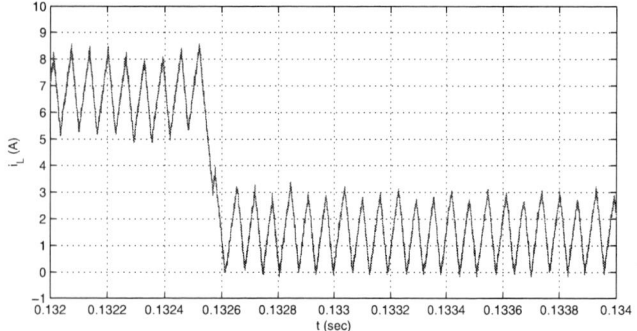

Fig. 9. Close-up of inductor current after step change in commanded output voltage: $V_C^* = 200\ V$ to $V_C^* = 150\ V$

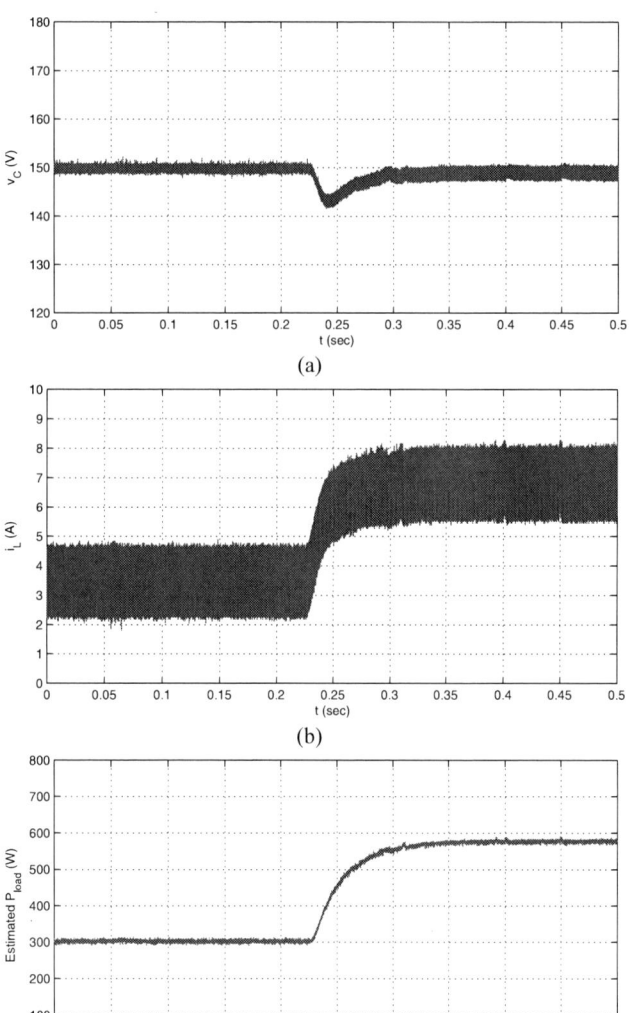

(a)

(b)

(c)

Fig. 10. Hardware result for experiment 2 including (a) boost converter output voltage (b) inductor current and (c) estimated load power for step change: $R_{\text{load}} = 10\ \Omega$ to $R_{\text{load}} = 5\ \Omega$

B. System 2: Boost-Driven PMSM Drive

In the second system, HMPC is applied in hardware to a boost converter supplying a permanent magnet synchronous machine (PMSM) drive. The PMSM is driven by a 6-pulse inverter that maintains a commanded electromagnetic torque T_e^* using a hysteresis current control described in [2]. The system schematic is shown in Fig. 11, system parameters are provided in Table III, and a photograph of the PMSM and drive is shown in Fig. 12. Certain operating conditions were selected to emulate those of a hybrid electric vehicle. To this end, the boost converter source voltage was selected to be $V_s = 201.6$ V in order to emulate 28 series connected 7.2 V Nickel-metal hydride battery modules, the PMSM mechanical rotor speed was set to 1200 RPM, and the commanded output voltage of the boost converter was $V_C^* = 300$ V [29].

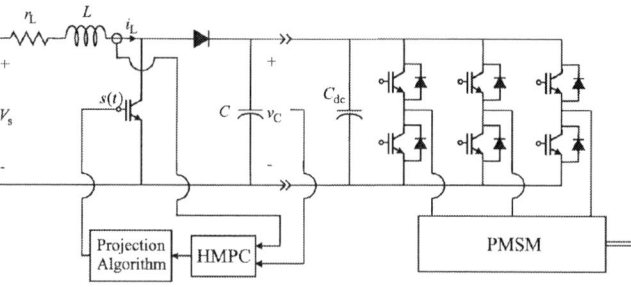

Fig. 11. Schematic of boost converter driving PMSM drive

Fig. 12. Dynamometer, PMSM, and 6-pulse Inverter

The final experiment was performed with System 2. The PMSM was controlled to provide a constant electromagnetic torque of $T_e = 10$ Nm, and the PMSM rotor was connected to a dynamometer which was controlled to maintain the commanded speed of 1200 RPM, thus maintaining a constant mechanical output power from the PMSM. An experiment was done wherein the commanded electro-magnetic torque was stepped from $T_e^* = 10$ Nm to $T_e^* = 15$ Nm. The measured boost converter state and load power estimate for 500 msec are shown in Fig. 13.

The output voltage was regulated to within 1.5 volts of the commanded 300 V in steady state at both torque levels. As with the last experiment, the change in torque resulted in a

978-1-4244-4782-4/10 $26.00 © 2010 IEEE   488

disturbance of the output voltage of the boost converter. During the disturbance, the capacitor voltage fell 14 V (approximately 4.7%), and the disturbance lasted approximately 50 msec.

After the change in commanded torque, the estimated load power increased from a mean value of 1.40 kW to a mean value of 2.15 kW.

TABLE III.   PMSM PARAMETERS

Parameter	Value	Description
r_s	0.45 Ω	Stator winding resistance
L_q	2.6 mH	Q-axis inductance
L_d	2.6 mH	D-axis inductance
λ'_m	0.1723 Vsec	Magnetic flux linkage
n_p	8	Number of poles
r_{Cdc}	40 mΩ	Output capacitor series resistance
C_{dc}	1.0 mF	DC link capacitance
ω_{rm}	1200 RPM	Mechanical rotor speed
T_e^*	10 / 15 Nm	Commanded PMSM torque

V.   CONCLUSIONS

An HMPC control strategy has been applied to a dc-dc boost converter with constant power load. Switching is determined through online solution of a model predictive control optimization. An advantage of the new control method is that the converter may be regulated about multiple operating points through online optimization of a performance metric. This is unlike other optimal control or gain scheduling methods that require extensive offline computation. Design of the MPC controller is described step-by-step and an Active Set algorithm is provided that allows real-time implementation of the control. Finally, the control is demonstrated in hardware for two multi-converter systems, wherein a single controller design shows strong performance at multiple operating points.

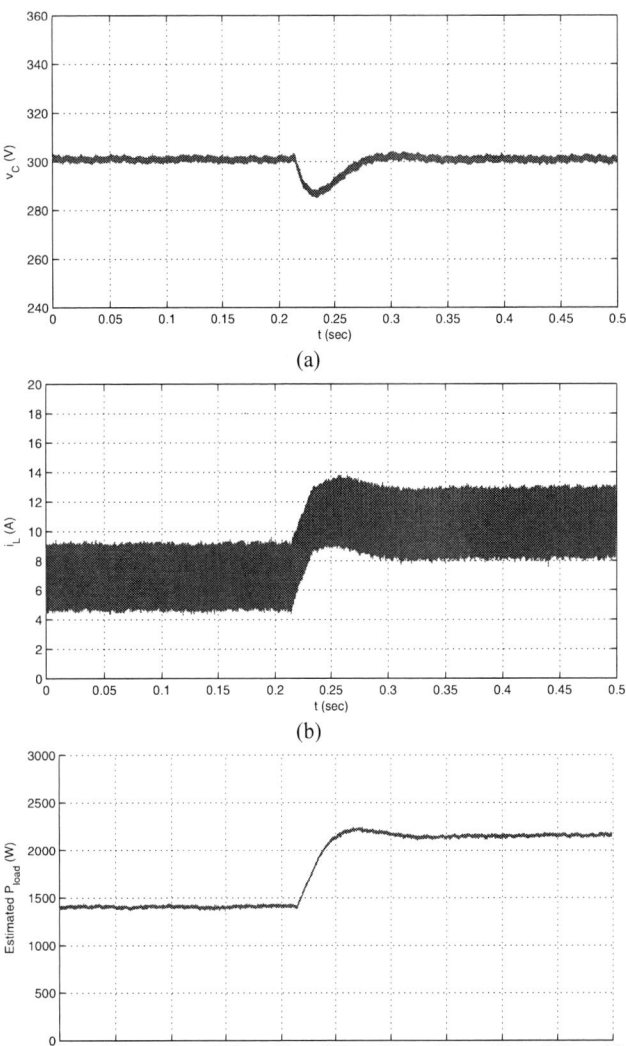

Fig. 13. for experiment 2 including (a) boost converter output voltage and (b) inductor current and (c) estimated load power for step change in commanded PMSM torque: $T_e^* = 10$ Nm to $T_e^* = 15$ Nm

REFERENCES

[1] Monti, A.; Boroyevich, D.; Cartes, D.; Dougal, R.; Ginn, H.; Monnat, G.;Pekarek, S.; Ponci, F.; Santi, E.; Sudhoff, S.; Schulz, N.; Shutt, W.; Wang, F., "Ship power system control: a technology assessment,"*Electric Ship Technologies Symposium, 2005 IEEE*, pp. 292-297, 25-27 July 2005

[2] Bash, M.; Chan, R.R.; Crider, J.; Harianto, C.; Lian, J.; Neely, J.; Pekarek, S.D.; Sudhoff, S.D.; Vaks, N., "A Medium Voltage DC Testbed for ship power system research," *Electric Ship Technologies Symposium, 2009. ESTS 2009. IEEE*, pp. 560-567, 20-22 April 2009

[3] Emadi, A.; Johnson, J.P.; Ehsani, M., "Stability analysis of large DC solid-state power systems for space, "*Aerospace and Electronic Systems Magazine, IEEE* , vol. 15, no. 2, pp. 25-30, February 2000

[4] Emadi, A.; Khaligh, A.; Rivetta, C.H.; Williamson, G.A., "Constant power loads and negative impedance instability in automotive systems: definition, modeling, stability, and control of power electronic converters and motor drives," *Vehicular Technology, IEEE Transactions on* , vol. 55, no. 4, pp. 1112-1125, July 2006

[5] Rahimi, A. M.; Emadi, A., "An Analytical Investigation of DC/DC Power Electronic Converters With Constant Power Loads in Vehicular Power Systems," *Vehicular Technology, IEEE Transactions on* , vol. 58, no. 6, pp. 2689-2702, July 2009

[6] Sudhoff, S.D.; Glover, S.F.; Lamm, P.T.; Schmucker, D.H.; Delisle, D.E., "Admittance space stability analysis of power electronic systems," *Aerospace and Electronic Systems, IEEE Transactions on* , vol. 36, no. 3, pp. 965-973, July 2000

[7] Emadi, A.; Ehsani, A., "Dynamics and control of multi-converter DC power electronic systems ," *Power Electronics Specialists Conference, 2001. PESC. 2001 IEEE 32nd Annual* , vol. 1, pp. 248-253, vol. 1, 2001

[8] Emadi, A.; Ehsani, M., "Negative impedance stabilizing controls for PWM DC-DC converters using feedback linearization techniques," *Energy Conversion Engineering Conference and Exhibit, 2000. (IECEC) 35th Intersociety* , vol. 1, pp. 613-620, 2000

[9] Glover, SF, Sudhoff SD; "An Experimentally Validated Nonlinear Stabilizing Control for Power Electronics Based Power Systems"; Proceedings of the 1998 SAE Aerospace Power Systems conference, no. P 981255, pp. 71-80, 1998

[10] Logue, D.L.; Krein, P.T., "Preventing instability in DC distribution systems by using power buffering," *Power Electronics Specialists Conference, 2001. PESC. 2001 IEEE 32nd Annual* , vol. 1, pp. 33-37, 2001

[11] Rahimi, A.M.; Emadi, A., "Active Damping in DC/DC Power Electronic Converters: A Novel Method to Overcome the Problems of Constant Power Loads," *Industrial Electronics, IEEE Transactions on* , vol. 56, no. 5, pp. 1428-1439, May 2009

[12] Jiabin Wang; Howe, D., "A Power Shaping Stabilizing Control Strategy for DC Power Systems With Constant Power Loads," *Power Electronics, IEEE Transactions on* , vol. 23, no. 6, pp. 2982-2989, November 2008

[13] Weaver, W.W.; Krein, P.T., "Optimal Geometric Control of Power Buffers," *Power Electronics, IEEE Transactions on* , vol. 24, no. 5, pp.1248-1258, May 2009

[14] Papafotiou, G.; Geyer, T.; Morari, M., "Hybrid modeling and optimal control of switch-mode dc-dc converters," Computers in Power Electronics, 2004. Proceedings. 2004 IEEE Workshop on, pp. 148-155, 15-18 August 2004

[15] Beccuti, A.G.; Papafotiou, G.; Frasca, R.; Morari, M., "Explicit Hybrid Model Predictive Control of the dc-dc Boost Converter," *Power Electronics Specialists Conference, 2007. PESC 2007. IEEE*, pp. 2503-2509, 17-21 June 2007

[16] Beccuti, A.G.; Mariethoz, S.; Cliquennois, S.; Shu Wang; Morari, M., "Explicit Model Predictive Control of DC–DC Switched-Mode Power Supplies With Extended Kalman Filtering," Industrial Electronics, IEEE Transactions on , vol. 56, no. 6, pp. 1864-1874, June 2009

[17] Mayne; Rawlings, Rao, Scokaert, "Constrained model predictive control: stability and optimality". *Automatica,* Volume 36, Issue 6, June 2000, pp. 789-814

[18] Camacho, E. F., Bordons, C., *Model Predictive Control*, 2nd ed., London, Springer-Verlag, 2004

[19] Bengea, Sorin C.; DeCarlo, Raymond A., "Optimal Control of Switching Systems", *Automatica,* Volume 41, Issue 1, January 2005, Pages 11-27

[20] Oettmeier, F.M.; Neely, J.; Pekarek, S.; DeCarlo, R.; Uthaichana, K., "MPC of Switching in a Boost Converter Using a Hybrid State Model With a Sliding Mode Observer," *Industrial Electronics, IEEE Transactions on* , vol. 56, no. 9, pp. 3453-3466, September 2009

[21] Neely, Jason; Pekarek, Steve; DeCarlo, Raymond, "Hybrid Optimal-Based Control of a Boost Converter," *Applied Power Electronics Conference and Exposition, 2009. APEC 2009. Twenty-Fourth Annual IEEE*, pp. 1129-1137, 15-19 February 2009

[22] Ting-Ting Song; Chung, H.S.-h., "Boundary Control of Boost Converters Using State-Energy Plane," *Power Electronics, IEEE Transactions on* , vol. 23, no. 2, pp. 551-563, March 2008

[23] Gupta, P.; Patra, A., "A stable energy-based control strategy for DC-DC boost converter circuits," *Power Electronics Specialists Conference, 2004. PESC 04. 2004 IEEE 35th Annual* , vol. 5, pp. 3642-3646, 20-25 June 2004

[24] Berkovich, Y.; Ioinovici, A., "Large-signal stability-oriented design of boost regulators based on a Lyapunov criterion with nonlinear integral," *Circuits and Systems I: Fundamental Theory and Applications, IEEE Transactions on* , vol. 49, no. 11, pp. 1610-1619, November 2002

[25] Valentine, R.; *Motor Control Electronics Handbook*; New York, McGraw Hill, 1998, page 26

[26] Nocedal, J., Wright, J.W., *Numerical Optimization*, 2nd ed., New York, Springer, 2006

[27] Bartlett, R.A.; Wachter, A.; Biegler, L.T., "Active set vs. interior point strategies for model predictive control," *American Control Conference, 2000. Proceedings of the 2000* , vol. 6, pp. 4229-4233, 2000

[28] Allgöwer, F., Zheng, A., *Nonlinear Model Predictive Control*, Berlin, Birkhäuser-Verlag, 2000, pp. 335-346

[29] R. H. Staunton, C. W. Ayers, J. Chiasson, T. A. Burress, and L. D. Marlino, *Evaluation of 2004 Toyota Prius Hybrid Electric Drive System,* ORNL/TM-2006-423, UT-Battelle, LLC, Oak Ridge National Laboratory, Oak Ridge, Tennessee, May 16, 2006

Predictive Control of Buck Converter Using Nonlinear Output Capacitor Current Programming

Victor Sui-pung Cheung, *Member, IEEE*, Henry Shu-hung Chung, *Member, IEEE*, Huai Wang, *Student Member, IEEE*

Centre for Power Electronics
City University of Hong Kong
Tat Chee Avenue, Kowloon Tong
Kowloon, Hong Kong

Abstract - **A predictive control of buck converter to achieve fast transient response is presented. The methodology is based on using the output capacitor current and instantaneous output voltage to predict the output voltage at the end and mid of a switching cycle to determine the state of the main switch. Two switching criteria, one for switching on and one for switching off the main switch, will be formulated. The operating principles of the control method and all possible cases in the operation will be discussed. The proposed control method has been verified with experimental results of a 50W, 12V/5V buck converter prototype. The controller is realized by analogue devices. The steady-state and transient response to large-signal disturbances will be discussed.**

I. Introduction

There is a large body of literature related to control methods for enhancing static and dynamic responses of switching converters [1]-[2]. Most of them, such as voltage-mode [3]-[7] and current-mode regulators [8]-[15], are based on firstly linearizing the power conversion stage around the operating point to formulate time-invariant transfer functions for describing the input-to-output and control-to-output characteristics. Then, the controller is designed by applying classical control theory. However, as switching converters are time-variant systems, the linearized models are only applicable for low-frequency characterizations. Thus, the dynamic response of the entire converter system is limited.

To achieve fast dynamic response in switching converters, many time-domain dynamic control techniques, such as one-cycle control (OCC) [16]-[19], boundary control [20]-[30], state trajectory control [36]-[38] and digital control [39]-[43], have been proposed. The instantaneous values of the circuit variables are sensed and used to dictate the switching instants of the switches. The OCC scheme is suitable for systems, like buck-derived converters [16], that the output has a direct response to the input fluctuation. Thus, it has good input-perturbation rejection, but its load-perturbation rejection is comparatively weak.

The boundary control, which is a geometric control method, addresses complete operation of a converter over the startup, transient, and steady-state periods [20]. Typical boundary control methods are hysteretic control or sliding-mode control (SMC). The hysteretic controller tightly regulates the inductor current at the current reference [21]. With further extension on regulating the output voltage, the SMC is the popular choice in boundary control. However, the optimal sliding surface and the stability for fast dynamic response are dependent on the supply and load characteristics [32]. It is thus difficult to design a set of well-defined control parameters for the sliding surface for different operating points. The control parameters such as the slope of the switching function are sometimes designed by using trial-and-error approach and optimizing the startup profile and switching frequency, etc. The SMC is typically applied to the fast control loop and the output is regulated by a slow proportional-plus-integral (PI) controller. The PI controller is designed by classical control theory or sophisticated design method, like the fuzzy controller in [24]. The overall system dynamics are thus limited by the slow control loop.

Instead of a linear switching surface, optimal curved switching surfaces are proposed in [27]-[31] for buck converters. The boundary control with curved switching surface has been shown to give better static and dynamic responses than the one with linear switching surface. The main concept of these methods is that they determine the appropriate time of changing the state of the switches, so that the converter reaches the steady state within a few switching actions after the converter is subject to external disturbances.

A well-designed switching surface can provide good robustness and good dynamic response for the converter. However, the advantages of these methods are offset by two drawbacks. First, the switching frequency is not fixed. Second, a hysteresis band, which is noise sensitive, is needed. Although some fixed-frequency boundary control methods have been proposed, they require an additional feedback loop for regulating the switching frequency [33]-[35].

The work described in this paper was fully supported by a grant from the CityU (Project No.: 7002460). The authors are with the Centre for Power Electronics, City University of Hong Kong, Tat Chee Avenue, Kowloon Tong, Kowloon, Hong Kong.

The state trajectory control in [36]-[38] makes the converter reach the steady state for a step change in the input or output in one on/off control. The operating principle is to calculate the circuit behaviors in each topology. Apart from dealing with many system equations, such method also requires many analogue devices in the circuit implementation.

With the advancement of digital signal processor and semiconductor, many digital control methods [39]-[43] have been proposed. The control algorithms are usually extended from the developed control methods together with the flexibility of adjusting the control parameters, in order to obtain optimal static and dynamic behaviors.

In this paper, a control method that tackles the above drawbacks and features the advantages of the time-domain dynamic control technique is proposed. The turn-on and -off criteria are based on sensing the output capacitor current and instantaneous output voltage to predict the output voltage at the end or mid of the switching cycle. A nonlinear current reference for programming the output capacitor current is formulated. Comparing with the prior-art control methods, the proposed technique has the following advantages:

1) Instead of using separate control loops to regulate the fast and slow circuit variables, the proposed method predicts the state trajectory and then determines the appropriate switching instants. The entire converter system has fast dynamic response.

2) As the control law has a set of well-defined parameters, it is unnecessary to choose the control parameters with trial-and-error approach.

3) No hysteresis band adding onto the switching surfaces is needed.

4) The switching frequency is constant. No control loop for regulating the switching frequency is needed.

A 50W, 12V/5V prototype has been built and tested. The theoretical prediction and experimental results are in close agreement. Section II will give the operating principles of the control method. Section III will give the three possible operating cases in the steady state. Finally, the experimental results will be given in Sec. IV. The steady-state characteristics and transient responses of the converter to large-signal input and output disturbances will be given.

II. Operating Principles of Control Method

Fig. 1 shows the buck converter with the block diagram of the proposed controller. The switching period of the main switch S is T. Fig. 2 shows the key waveforms. The controller turns S off at the beginning of each switching cycle and turns S on in the middle of the switching cycle (i.e., $T/2$). As depicted in Fig. 2(a), the controller determines the time instant t_1 to turn S on. In this operation, the duty cycle d of S is larger than or equal to 0.5. Similarly, as shown in Fig. 2(b),

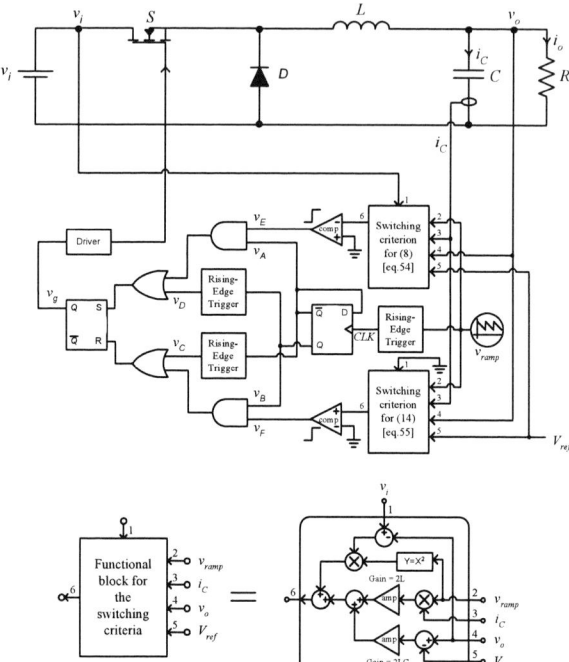

Fig. 1 Circuit schematic of the buck converter with the proposed controller.

the controller determines the time instant t_2 to turn S off. In this operation, $d \le 0.5$. Thus, the controller has two switching criteria for determining t_1 and t_2, respectively. Derivations of the switching criteria are given as follows.

A. Switching criterion for turning S on at t_1

The controller will turn S on if the predicted value of the output voltage v_o at T is less than or equal to v_{ref}. Mathematically,

$$v_o(T) \le v_{ref} \qquad (1)$$

The criterion is derived by considering the trajectory of v_o after S is on. When S is on,

$$\frac{d\, i_L}{d\, t} = \frac{v_i - v_o}{L} \qquad (2)$$

where i_L is the inductor current, v_i is the input voltage, and L is the output inductor.

As the load variation is small within the switching cycle, v_o is almost constant. Thus,

$$i_L = i_C + i_o \Rightarrow \frac{d\, i_L}{d\, t} \cong \frac{d\, i_C}{d\, t} \qquad (3)$$

where i_o is the output current.

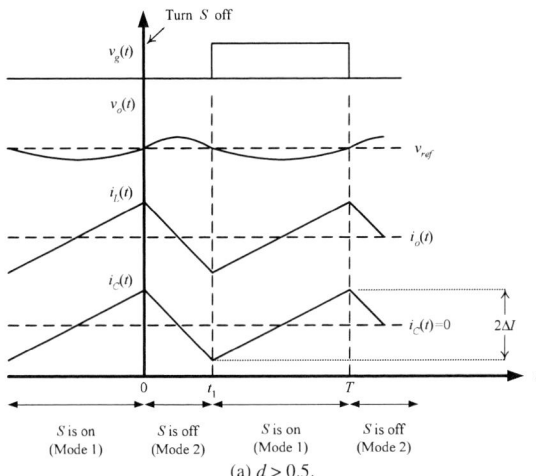

(a) $d > 0.5$.

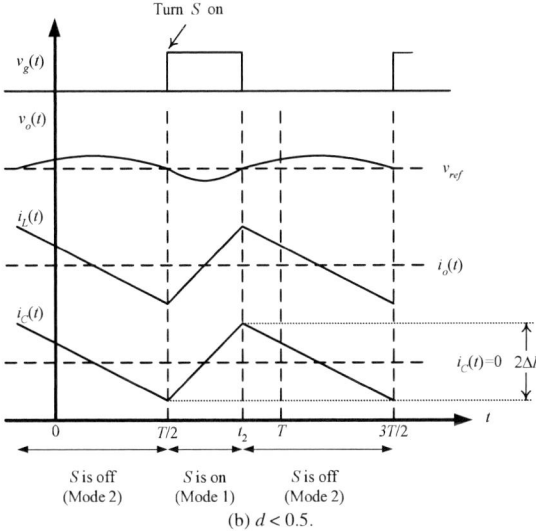

(b) $d < 0.5$.

Fig. 2 Key waveforms of the converter.

The capacitor current ripple is thus the same as the inductor current ripple. By using (2) and (3),

$$\frac{d\,i_C}{d\,t} = \frac{v_i - v_o}{L} \qquad (4)$$

Since v_o is almost constant over the switching period, $d\,i_C/d\,t$ is considered to be constant in the considered switching cycle.

As $i_C = C \dfrac{d\,v_o}{d\,t}$, it can be derived from (4) that

$$
\begin{aligned}
v_o(T) - v_o(t_1) &= K_1[i_C^{\,2}(T) - i_C^{\,2}(t_1)] \\
&= K_1\{[i_L(T) - i_o(T)]^2 - [i_L(t_1) - i_o(t_1)]^2\}
\end{aligned} \qquad (5)
$$

where $K_1 = \dfrac{L}{2\,C\,[v_i(t_1) - v_o(t_1)]}$.

By using (2),

$$i_L(T) = i_L(t_1) + \frac{1}{2\,K_1\,C}(T - t_1) \qquad (6)$$

By substituting (3) and (6) into (5),

$$v_o(T) - v_o(t_1) = \frac{i_C(t_1)}{C}(T - t_1) + \frac{1}{4\,K_1\,C^2}(T - t_1)^2 \quad (7)$$

Thus, by putting (1) into (7),

$$i_C(t_1) \le [\,V_{ref} - v_o(t_1)\,]\frac{C}{(T - t_1)} - \frac{1}{4\,K_1\,C}(T - t_1)\,(8)$$

Equation (8) is the criterion for turning S on at t_1 for $d \ge 0.5$.

B.   *Switching criterion for turning S off at t_2*

The controller will turn S off if the predicted value of the output voltage v_o at $1.5\,T$ is larger than or equal to v_{ref}. Mathematically,

$$v_o(1.5\,T) \ge v_{ref} \qquad (9)$$

The criterion is derived by considering the trajectory of v_o after S is off. When S is off,

$$\frac{d\,i_C}{d\,t} = \frac{d\,i_L}{d\,t} = -\frac{v_o}{L} \qquad (10)$$

Based on (10),

$$
\begin{aligned}
v_o(1.5\,T) - v_o(t_2) &= K_2[i_C^{\,2}(1.5\,T) - i_C^{\,2}(t_2)] \\
&= K_2\{[i_L(1.5\,T) - i_o(1.5\,T)]^2 - \\
&\quad [i_L(t_2) - i_o(t_2)]^2\}
\end{aligned} \qquad (11)
$$

where $K_2 = -\dfrac{L}{2\,C\,v_o(t_2)}$.

By using (10),

$$i_L(1.5\,T) = i_L(t_2) + \frac{1}{2\,K_2\,C}(1.5\,T - t_2) \qquad (12)$$

By substituting (10) and (12) into (11),

$$
\begin{aligned}
v_o(1.5\,T) - v_o(t_2) &= \\
\frac{i_C(t_2)}{C}(1.5\,T - t_2) &+ \frac{1}{4\,K_2\,C^2}(1.5\,T - t_2)^2
\end{aligned} \qquad (13)
$$

Thus, by putting (9) into (13),

$$i_C(t_2) \ge [V_{ref} - v_o(t_2)] \frac{C}{(1.5\,T - t_2)} - \frac{1}{4\,K_2\,C}(1.5\,T - t_2)$$

(14)

Equation (14) is the criterion for turning S off at t_2 when $d \le 0.5$.

Equations (8) and (14) are output capacitor current programming functions. The main switch will be commanded to take appropriate action when i_C satisfies the equations. The right-hand-side (RHS) expressions of (8) and (14) are in nonlinear functions with time.

Fig. 3 gives the timing diagram and key signals in Fig. 1. Selection of the switching criteria, i.e., either (8) or (14), is decided by signals v_A and v_B, which are derived from v_{ramp}. When v_A is high, (8) will be used. When v_B is high, (14) will be used. The durations of v_A and v_B are both $T/2$. v_C is used to turn S off while v_D is used to turn S on. v_E and v_F are the signal generated by the switching criteria in (8) and (14), respectively. In the actual implementation, their operations are slightly modified and are described in Sec. VI. Finally, the gate signal v_g for $d \ge 0.5$ and $d \le 0.5$ is illustrated in Fig. 3.

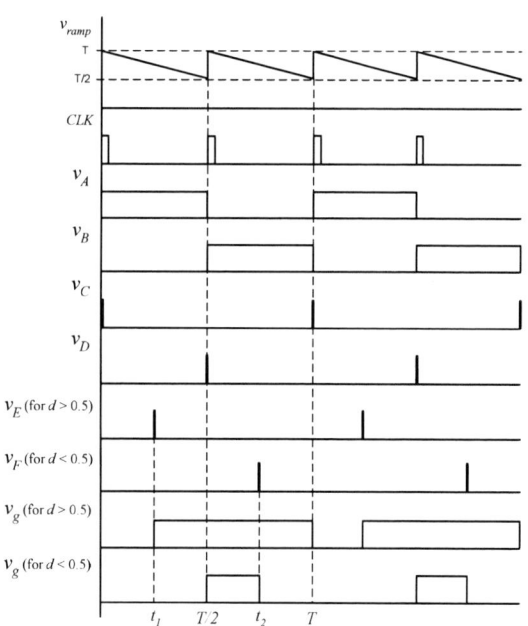

Fig. 3 Timing diagram and key signals in Fig. 1.

III. Boundary conditions for three possible cases

The switching criteria derived in Sec. II are based on the assumption that the equivalent series resistance (ESR) of the output capacitor is zero. In reality, due to the presence of ESR, there are three possible operating cases, as depicted in Fig. 4, namely Case I, Case II and Case III. In Case I, the

main switch is turned on at t_1 and turned off at T. In Case II, it is turned on at t_1 and turned off at t_2. In Case III, it is turned on at $T/2$ and turned off at t_2. Cases I and III are the similar to the ones shown in Fig. 2, i.e., either eq. (8) or eq. (14) is applied in one switching cycle at steady state. Case II is the one that both eqs. (8) and (14) are applied. The boundary conditions of the three cases are listed in Table I.

Table I – Boundary conditions of the three cases

Case	t_1	t_2	Criterion(s) used
I	$\le T/2$	$> T$	Eq. (8)
II	$\le T/2$	$\le T$	Eqs. (8) and (14)
III	$> T/2$	$\le T$	Eq. (14)

Thus, the three cases are determined by solving the values of t_1 and t_2 using eqs. (8) and (14), respectively, to obtain the boundary condition shown in Table I.

At t_1, $i_C(t_1)$ [45] is equal to

$$i_C(t_1) = -\Delta I - \frac{\Delta v_o(t_1)}{R}$$

(15)

where

$$\Delta I = \frac{(1-D)\,V_o\,T}{2\,L}$$

(16)

and

$$\Delta v_o(t_1) = \frac{2\,R\,\Delta I}{1 - e^{-\frac{T_s}{(R+r_C)C}}} \left\{ \begin{array}{l} -\dfrac{1}{2} + \dfrac{RC}{(1-D)T_s} - \\[2mm] \dfrac{RC}{D(1-D)T_s}\,e^{-\frac{(1-D)T_s}{(R+r_C)C}} + \\[2mm] \left(\dfrac{RC}{DT_s} + \dfrac{1}{2}\right)e^{-\frac{T_s}{(R+r_C)C}} \end{array} \right\}$$

(17)

By putting (15)-(17) into (8), t_1 is solved.

At t_2, $i_C(t_2)$ [45] is equal to

$$i_C(t_1) = \Delta I - \frac{\Delta v_o(t_2)}{R}$$

(18)

where

$$\Delta v_o(t_2) = \frac{2\,R\,\Delta I_L}{1 - e^{-\frac{T_s}{(R+r_C)C}}} \left\{ \begin{array}{l} \dfrac{1}{2} - \dfrac{RC}{DT_s} + \\[2mm] \dfrac{RC}{D(1-D)T_s}\,e^{-\frac{DT_s}{(R+r_C)C}} - \\[2mm] \left(\dfrac{RC}{(1-D)T_s} + \dfrac{1}{2}\right)e^{-\frac{T_s}{(R+r_C)C}} \end{array} \right\}$$

(19)

By putting (18) and (19) into (14), t_2 is solved. Fig. 4 shows the boundary of the three cases at different duty cycles and ESR of the capacitor. The component values are based on the ones given in Sec. IV.

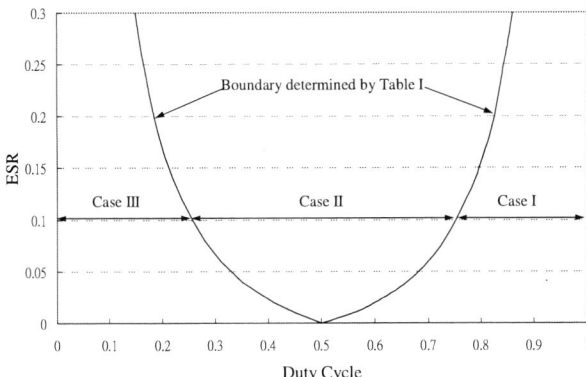

Fig. 4 Boundary of the three possible cases.

IV. Experimental Verification

Equations (8) and (14) are rearranged in the actual implementation so that the RHS of these two functions will never go to large values when t_1 and t_2 are close to T. They are modified into

$$v_o(t_1) \leq V_{ref} - \frac{1}{C} i_C(t_1)(T - t_1) - \frac{1}{4 K_1 C^2}(T - t_1)^2$$

(20)

for (8), and

$$v_o(t_2) \geq V_{ref} - \frac{1}{C} i_C(t_2)(1.5 T - t_2) - \frac{1}{4 K_2 C^2}(1.5 T - t_2)^2$$

(21)

for (14).

The functional block for the switching criteria of (8) and (14) shown in Fig. 1 are based on the operations in (20) and (21), respectively.

A 50W buck converter with $v_i = 12V$, $L = 30\mu H$, $C = 33\mu F$, and $v_{ref} = 5V$ has been built. The dc resistance of L is 0.02Ω. The equivalent series resistance of C is 0.2Ω. The switching frequency is 100kHz. The output is connected to a resistor bank. The large-signal response is tested by suddenly changing the value of the output resistor. Fig. 5 shows the transient responses when the output current is changed from 11A to 2A, and vice versa. When the output current is changed from 11A to 2A, it can be observed from the gate signal v_g that the output reaches the steady after two switching actions. There are spike generated due to the equivalent series inductance (ESL) of the output capacitor, namely ESL spike, and step due to the equivalent series resistance (ESR) of the output capacitor, namely ESR step [44]. The magnitude of the ESR step is 2V. When the output current is changed from 2A to 11A, there are a series of gate

pulses with very short turn-off signal generated. Such "turn-off" pulses are derived from the reset signal v_C in Fig. 1. As the duration of each "turn-off" pulse is very short, the resulting effect is that the switch is almost on. The inductor current i_L is monotonically increasing over the period. Thus, if the short "turn-off" pulses are ignored, the converter can reach the steady-state after two "effective" switching actions as illustrated in Fig. 5. The ESR step in this case is 1.1V. The settling time of the output voltage is less than 80μs in both cases. The behaviors are similar to the ones in [27]-[29].

Fig. 6 shows the transient responses when v_{ref} is changed from 8V to 4V, and vice versa. Thus, the steady-state duty cycle is changed between $D > 0.5$ (i.e., 8 / 12) and $D < 0.5$ (i.e., 4 / 12). Again, the output voltage can reach the steady-state in two "effective" switching actions if the pulses due to the reset signals are ignored. The settling time is less than 60μs in both cases. Of particular importance, the transition changing between $D > 0.5$ and $D < 0.5$ is smooth. The switching frequency is fixed at 100kHz before and after the load and output reference changes.

Fig. 7 shows the transient response when the load is changed from no load to full load conditions. At the no load condition, the converter is in "burst" mode that the controller periodically turns the main switch on for a short time to maintain the output voltage. Again, the converter can reach the steady-state in two "effective" switching actions.

Fig. 8 shows the steady-state waveforms when the input voltage is unregulated. The input voltage varies between 8V and 12.5V at the frequency of 100Hz. The output voltage can be tightly maintained at 5V.

Fig. 9 shows the transient waveforms when the actual values of L and C have substantial deviations from the ones used in the controller. The output current is changed from 2A to 11A. The values of L and C have ±20% differences from the values used in the controller. Results reveal that the profiles of the steady-state behaviors and transient responses are almost the same as the one with the nominal value, i.e., Fig. 5(b). In general, the settling time increases as the value of L is increased. The output voltage undershoot increases as the value of C is reduced. In all cases, the converter can revert to the steady state in two "effective" switching actions.

Fig. 10 shows the transient responses when the ESR of the capacitor is doubled, i.e., 0.4Ω. It is simulated by adding a 0.2Ω resistor in series with the capacitor. Compared Fig. 10(a) with Fig. 5(a), the ESR step is increased from 2V to 3.9V because the ESR is increased. Compared Fig. 10(b) with Fig. 5(b), the ESR step is increased from 1.1V to 2.2V. Moreover, the steady-state voltage ripple is increased due to an increase in the ESR. Nevertheless, the profiles of the transient responses in the two cases are almost the same. Hence, the ESR does not significantly affect the performance of the controller.

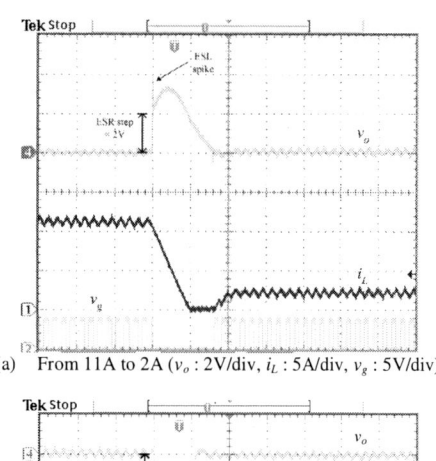

(a) From 11A to 2A (v_o : 2V/div, i_L : 5A/div, v_g : 5V/div).

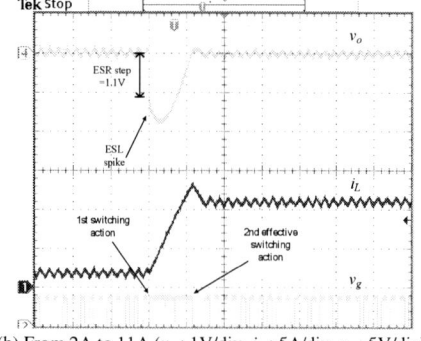

(b) From 2A to 11A (v_o : 1V/div, i_L : 5A/div, v_g : 5V/div).
Fig. 5 Transient responses under step load changes [Timebase: 40μs/div].

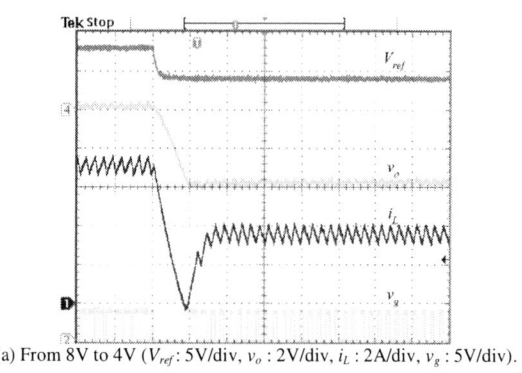

(a) From 8V to 4V (V_{ref} : 5V/div, v_o : 2V/div, i_L : 2A/div, v_g : 5V/div).

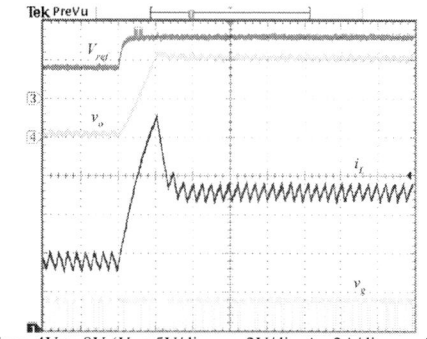

(b) From 4V to 8V (V_{ref} : 5V/div, v_o : 2V/div, i_L : 2A/div, v_g : 5V/div)
Fig. 6 Transient responses when v_{ref} is suddenly changed [Timebase: 40μs/div].

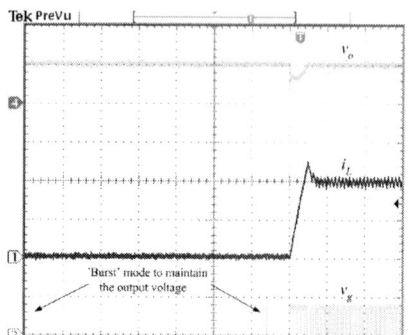

Fig. 7 Load transient when the load is changed from no load to full load. (v_o : 1V/div, i_L : 5A/div, v_g : 5V/div) [Timebase: 40μs/div]

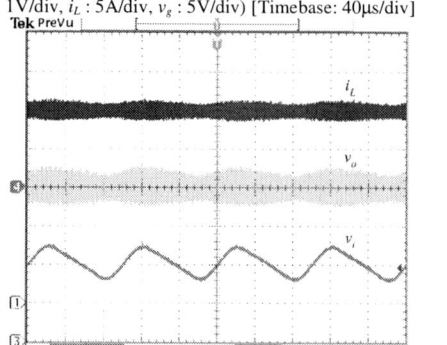

Fig. 8 Steady-state waveforms with unregulated input voltage (i_L : 2A/div, v_o : 200mV/div, v_i : 5V/div) [Timebase: 4ms/div].

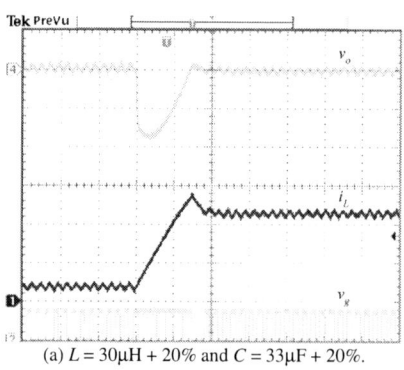

(a) $L = 30\mu H + 20\%$ and $C = 33\mu F + 20\%$.

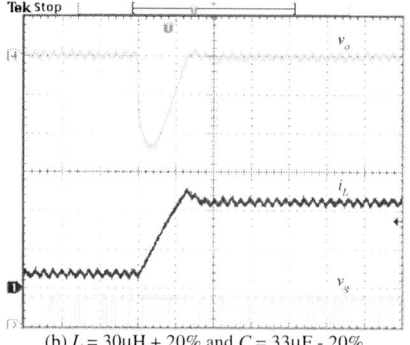

(b) $L = 30\mu H + 20\%$ and $C = 33\mu F - 20\%$.

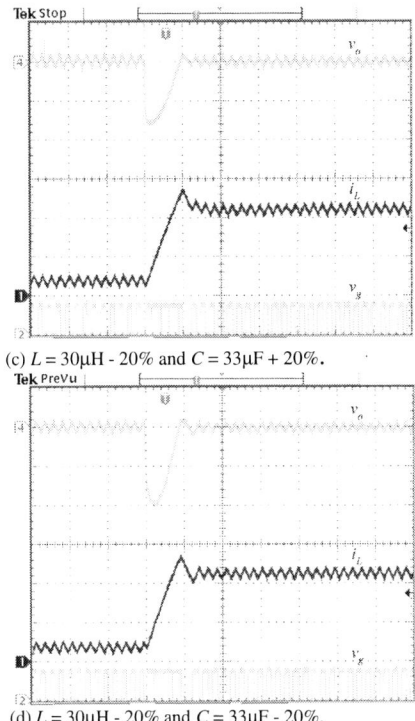

(c) $L = 30\mu H - 20\%$ and $C = 33\mu F + 20\%$.

(d) $L = 30\mu H - 20\%$ and $C = 33\mu F - 20\%$.

Fig. 9   Transient waveforms with substantial changes in the actual values of L and C (v_o : 1V/div, i_L : 5A/div, v_g : 5V/div) [Timebase: 4ms/div].

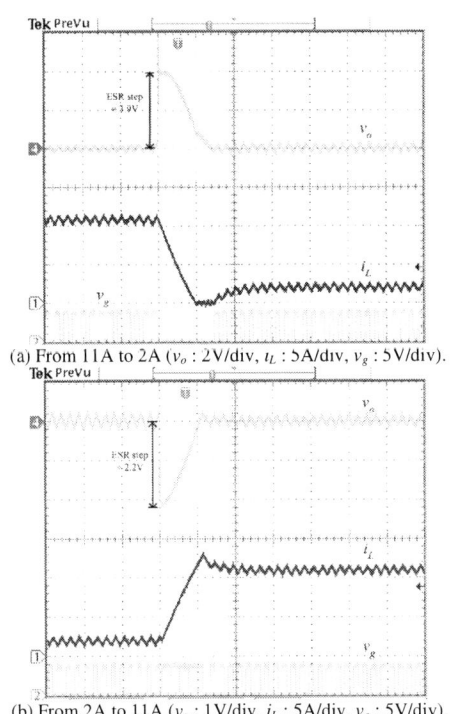

(a) From 11A to 2A (v_o : 2V/div, i_L : 5A/div, v_g : 5V/div).

(b) From 2A to 11A (v_o : 1V/div, i_L : 5A/div, v_g : 5V/div).

Fig. 10   Transient response with a step load change and the ESR of C = 0.4Ω [Timebase: 4ms/div]

V. Conclusion

A nonlinear output capacitor current programming scheme for buck converter to achieve fast transient response has been presented. The concept is based on continuously predicting the output voltage at the end or mid of a switching cycle to dictate the state of the main switch. Two switching criteria for turning on and off of the main switch are derived. The steady-state characteristics, system stability and sensitivity of the output voltage to the parametric variation have been investigated. The proposed method has been confirmed by the experimental results of a 50W prototype. Results reveal that the transient responses are similar to the boundary control with optimal switching surface in [27]-[29]. However, the proposed method does not require using hysteresis band and is in fixed switching frequency operation.

References

[1] R. W. Erickson and D. Maksimovic, *Fundamentals of Power Electronics*, Norwell, M.A.: Kluwer Academic, c2001.

[2] K. C. Wu, *Switch-Mode Power Converters: Design and Analysis*, Academic Press, 2005.

[3] R. D. Middlebrook and S. Ćuk, "A general unified approach to modeling switching converter power stages," *IEEE Power Electron. Spec. Conf.*, vol. 4, 1976, pp. 18–34.

[4] M. Veerachary, "Two-loop voltage-mode control of coupled inductor step-down buck converter," *IEE Proc. Electric Power App.*, vol. 152, no. 6, pp. 1516-1524, Nov. 2005.

[5] M. Siu, P. Mok, K. N. Leung, Y.-H, Lam, W.H. Ki, "A voltage-mode PWM buck regulator with end-point prediction," *IEEE Trans. Circuits Systs. II: Express Brief*, vol. 53, no. 4, pp. 294 – 298, Apr. 2006.

[6] E. Figueres, G. Garcera, J.M. Benavent, M. Pascual, and J. A. Martinez, "Adaptive two-loop voltage-mode control of DC-DC switching converters," *IEEE Trans. Ind. Electron.*, vol. 53, no. 1, pp. 239-253, Feb. 2006.

[7] M. Peretz and S. Ben-Yaakov, "Revisiting the closed loop response of PWM converters controlled by voltage feedback," in *Proc. Applied Power Electron. Conf. and Expo.*, 2008, Feb 2008, 58-64.

[8] C. W. Deisch, "Simple switching control method changes power converter into a current source," *in Proc. IEEE Power Electron. Spec. Conf.*, pp. 300-306, 1978.

[9] A. S. Kislovski, R. Redl, and N. O. Sokal, *Dynamic Analysis of Switching-Mode DC/DC Converters*, Van Nostrand Reinhold, New York, 1991.

[10] R. B. Ridley, "A New Continuous-Time Model for Current-Mode Control," *IEEE Trans. Power Electron.*, April, 1991, Apr. 1991.

[11] F. D. Tan and R. D. Middlebrook, "A Unified Model for Current-Programmed Converters," *IEEE Trans. Power Electron.*, vol. 10, no. 4, pp. 397-408, July 1995.

[12] K. Yao, M. Xu, Y. Meng, and F. C. Lee," Design considerations for VRM transient response based on the output impedance," *IEEE Trans. Power Electron.*, vol. 18, no. 6, pp. 1270 – 1277, Nov. 2003.

[13] V. Voperian, "Synthesis of Medium Voltage dc-to-dc Converters From Low-Voltage, High-Frequency PWM Switching Converters," *IEEE Trans. Power Electron.*, vol. 22, no. 5, pp. 1619-1635, Sept. 2007.

[14] M. Karppanen, J. Arminen, T. Suntio, K. Savela, and J. Simola, "Dynamical Modeling and Characterization of Peak-Current-Controlled Superbuck Converter," *IEEE Trans. Power Electron.*, vol. 23, no. 3, pp. 1370-1380, May 2008.

[15] A. Lachichi, S Pierfederici, J. P. Martin, and B. Davat, "Study of a Hybrid Fixed Frequency Current Controller Suitable for DC–DC Applications," *IEEE Trans. Power Electron.*, vol. 23, no. 3, pp. 1437-1448, May 2008.

[16] K. M. Smedley and S. Ćuk, "One-cycle control of switching converters," *IEEE Trans. Power Electron.*, vol. 10, no. 6, pp. 625-633, Nov. 1995.

[17] K. Smith, Z. Lai, and K. Smedley, "A new PWM controller with one-cycle response," *IEEE Trans. Power Electron.*, vol. 14, no. 1, pp. 142-

150, Jan 1999.

[18] W. Shi, H. Xu, X. Wen, and W. Wen, "One-cycle controlled DC-DC converters operating with nonlinear power load," in *Proc. Int. Conf. Electrical Machines and Systems 2005*, vol. 2, Sept. 2005, pp. 1361 – 1365.

[19] G. Chen and K. Smedley, "Steady-State and dynamic study of one-cycle-controlled three-phase power-factor correction," *IEEE Trans. Ind. Electronics*, vol. 52, no. 2, pp. 355-362, Apr. 2005.

[20] P. T. Krein, Nonlinear Phenomena in Power Electronics: Attractors, Bifurcation, Chaos, and Nonlinear Control. New York: IEEE Press, 2001, Chap. 8.

[21] Y. Liu and P. C. Sen, "Large-signal modeling of hysteretic current-programmed converters," *IEEE Trans. Power Electron.*, vol. 11, no. 3, pp. 423-430, May 1996.

[22] L. Rossetto, G. Spiazzi, P. Tenti, B. Fabiano, and C. Licitra, "Fast-response high-quality rectifier with sliding mode control," *IEEE Trans. Power Electron.*, vol. 9, no. 2, pp. 146-152, Mar. 1994.

[23] H. Sira-Ramirez, G. Escobar, and R. Ortega, "On passivity-based sliding mode control of switched DC-to-DC power converters," in *Proc. IEEE Conf. Decision and Control 1996*, Kobe, Japan, vol. 3, pp. 2525-2526, Dec. 1996.

[24] E. Vidal-Idiarte, L. Martinez-Salamero, F. Guinjoan, J. Calvente, and S. Gomariz, "Sliding and fuzzy control of a boost converter using an 8-bit microcontroller," *IEE Proceedings – Electric Power Applications*, vol. 151, no. 1, pp. 5-11, Jan. 2004.

[25] M. Lopez, L. de Vicuna, L.G., M. Castilla, P. Gaya, and O. Lopez, "Current distribution control design for paralleled DC/DC converters using sliding-mode control," *IEEE Trans. Ind. Electronics*, vol. 51, no. 2, pp. 419-428, Apr. 2004.

[26] S. Tan, Y. Lai, and C. Tse, "Implementation of pulse-width-modulation based sliding mode controller for boost converters," *IEEE Power Electronics Letter*, vol. 3, no. 4, pp. 130-135, Dec. 2005.

[27] K. Leung and H. Chung, "Derivation of a Second-Order Switching Surface in the Boundary Control of Buck Converters," *IEEE Power Electronics Letter*, vol. 2, no. 2, pp. 63-67, June 2004.

[28] K. Leung and H. S. H. Chung, "A Comparative Study of the Boundary Control of Buck Converters Using First- and Second-Order Switching Surfaces -Part I: Continuous Conduction Mode," *in Proc. IEEE PESC '05*, pp. 2133 – 2139, 2005.

[29] K. Leung and H. Chung, "A Comparative Study of Boundary Control with First- and Second-Order Switching Surfaces for Buck Converters Operating in DCM," *IEEE Trans. Power Electron.*, vol. 22, no. 4, pp. 1196-1209, Jul. 2007.

[30] M. Ordonez, M. T. Iqbal, and J. E. Quaicoe, "Selection of a curved switching surface for buck converters," *IEEE Trans. Power Electron.*, vol. 21, no. 4, pp. 1148-1153, Jul 2006.

[31] M. Ordonez, J. E. Quaicoe, and M. T. Iqbal, "Advanced boundary control of inverters using the natural switching surface: normalized geometrical derivation," *IEEE Trans. Power Electron.*, vol. 23, no. 6, pp. 2915-2930, Nov. 2008.

[32] F. Dong and R. Ramshaw, "Instabilities of a boost converter system under large parameter variations," *IEEE Trans. Power Electron.*, vol. 4, no. 4, pp. 442-449, Oct. 1989.

[33] D. Grant, "Frequency control of hysteretic power converter by adjusting hysteresis levels," *US patent 6,348,780*, Feb 2002.

[34] Alex Mihalka, "Fixed frequency hysteretic regulator," *US Patent 6,885,175*, Apr. 2005.

[35] W. T. Yan, H. Chung, Keith T. K. Au, and Carl N.M. Ho, "Fixed-frequency boundary control of buck converters with second-order switching surface," in *Proc. IEEE Power Electron. Spec. Conf.*, Rhodes, Greece, June 2008, pp. 629-635.

[36] W. W. Burns and T. G. Wilson, "State trajectories used to observe and control DC-to-DC converter," *IEEE Trans. Aerospace and Electronic Systems*, vol. 12, no. 6, pp. 706-717, Nov. 1976.

[37] W. Burns and T. G. Wilson, "Analytical derivation and evaluation of a state trajectory control law for dc-to-dc converters," in *Proc. Power Electron. Spec. Conf.*, pp. 70-85, 1977.

[38] S. D. Huffman, *et al*, "Fast-response free-running dc-to-dc converter employing a state-trajectory law," in *Proc. Power Electron. Spec. Conf.*, pp. 180-189, 1977.

[39] C. C. Fang, "Sampled-data modeling and analysis of one-cycle control and charge control," *IEEE Trans. Power Electron.*, vol. 16, no. 3, pp. 345-350, May 2001.

[40] Q. Feng, J. Hung, R. Nelms, "Digital control of a boost converter using Posicast," in *Proc. Applied Power Electron. Conf. and Exp.* 2003, vol. 2, Feb. 2003, pp. 990 – 995.

[41] C. Kranz, "Complete digital control method for PWM DCDC boost converter," in *Proc. IEEE Power Electron. Spec. Conf.*, vol. 2, June 2003, pp. 951-956.

[42] L. Guo, J. Hung, and R. Nelms, "Digital controller design for buck and boost converters using root locus techniques," in *Proc. Annual Conf in Industrial Electronics*, vol. 2, pp. 1864-1869, Nov 2003.

[43] P. Mattavelli, "Digital control of DC-DC boost converters with inductor current estimation," in *Proc. Applied Power Electron. Conf and Exp.*, 2004, vol. 1, pp. 74-80.

[44] C. Simpson, "Load Transient Testing Simplified," *National Semiconductor Application Note 1733*, Nov 7, 2007.

[45] H. Hyakutake and K. Harada, "Analysis of output voltage ripple caused by ESR of a moothing capacitor for a low-voltage high-current buck converter," Electrical Engineering in Japan, vol. 143, no. 2, pp. 59-66, 2003.

978-1-4244-4782-4/10 $26.00 © 2010 IEEE

Analysis of a High Performance Voltage Regulator with Non-Linear Multi-Mode Control: Bandwidth and Large Transient Response

S. Pan, *Member, IEEE*   P. K. Jain, *Fellow, IEEE*
Department of Electrical and Computer Engineering
Queen's University
Kingston, Ontario, Canada
Email: Shangzhi.Pan@queensu.ca

Abstract - A novel digital adaptive voltage positioning (digital AVP) technique with dual-voltage-loop was proposed in [21]. Good transient performance had been achieved without using complicated control. In this paper, a small signal model is proposed for this mixed-signal digital controller. It is revealed by this small signal model that the inner current loop is in analog and the voltage loop is in digital, thus the controller can benefit from both: having valuable features of digital control but without limitations such as limit cycle. On the other hand, dynamic behavior of the voltage regulator with this digital AVP controller under large load transient is analyzed. Decoupling between bandwidth and large transient response with this novel digital AVP controller is verified.

I. INTRODUCTION

Power consumption of today's processor's is dramatically increasing when they are delivering better and better performance. As a result, voltage regulator (VR) design has become a serious challenge. The output voltage has dropped to 1V or less while the output current is dramatically increasing up to 150A. Moreover, high dynamic load characteristics (oscillations from 100Hz to 1MHz) of the processors impose large-step load oscillations to the voltage regulators, making it more difficult to maintain the regulated voltage within the tolerance.

Most of today's VRs are running at 300 kHz with a typical bandwidth of 50 kHz. To meet the future specification, eight 560μF OSCON capacitors are needed. Both the cost and the footprint are unacceptable. Therefore, much research [1-16] has delved into VR transient design: improving VR transient performance to reduce the number of bulk capacitors and make the VR meet the even tighter output voltage regulation necessary in recent and future processors. Most of the research cited above attempts to optimize the control loop or the output impedance to improve the transient performance and reduce the number of bulk capacitors. However, all are limited by the control bandwidth. One way to solve this problem is to increase the switching frequency to achieve higher bandwidth. Unfortunately, high switching frequency operation results in large switching frequency related losses including switching loss, gate driver loss, and synchronous rectifier body diode conduction loss.

The non-linear control such as V^2 [17] and hysteretic controls [18] can provide the fast transient response to the converter without relying on a high system bandwidth under

large variations in the load current or in the voltage reference. However, they suffer from noise problems when trying to increase the bandwidth, since the output voltage ripple sensing is required. A linear-non-linear control [19-20] was proposed, but the transition between linear and nonlinear is not smooth, which makes it difficult to settle down after large load transient. It would be getting worse while it delivers power to a high dynamic processor.

In [21], a novel digital adaptive voltage positioning (digital AVP) technique with dual-voltage-loop was proposed, which operated under fixed-frequency peak current mode and achieved the AVP control by generating dynamic voltage reference and dynamic current reference. Two digital-to-analog converters (DACs) were used instead of analog-to-digital converters to reduce system complexity and core area. A straightforward control law was implemented that does not require compensator in the control loop. Good transient performance had been demonstrated. However, lots of things are not clear: switching stability, large signal behavior, controller design, and high dynamic load performance.

In this paper, a small signal model is proposed for this mixed-signal digital controller. It is revealed by this small signal model that the inner current loop is in analog and the voltage loop is in digital, thus the controller can benefit from both: having valuable features of digital control but without limitations such as limit cycle. The switching stability constraint is also derived. On the other hand, dynamic behavior of the voltage regulator with this digital AVP controller under large load transient is analyzed based on large signal equations. Then, from these equations controller design is given to achieve good dynamic performance. It is found that independent of the system bandwidth designed for system stability, the proposed non-linear digital control algorithm can help achieve very fast transient response, even in a low-frequency low-bandwidth VR. Thus, decoupling between bandwidth and large transient response can be achieved with this novel digital AVP controller.

II. DIGITAL CONTROL ARCHITECTUR

Fig.1 shows a two-phase buck converter using the dual-voltage-loop digital AVP controller. Two digital-to-analog converters are used to generate the voltage and current references. Two comparators are required for the current loop of two phases and another two comparators are required for two voltage loops: the slow loop and the fast loop. The inductor current of each phase is sensed separately and compared to the peak current reference I_{ref} produced by the current DAC. The peak current reference I_{ref} is set as the phase peak current reference. The logic signal of the comparator of each phase is used to turn off its own high-side MOSFET and turn on its own low-side MOSFET, whereas the high-side MOSFET and the low-side MOSFET are turned on and off respectively scheduled by the ADPWM. The ADPWM is implemented using the counter-comparator method, combined with asynchronous comparison signal CI. The digital control algorithm block chooses voltage comparison result CV or CVq to close the voltage loop according to different controller operation modes. The purpose of the two voltage loops is to smooth the load transient transition and improve the system stability. A dedicated load transient detection circuit is implemented to detect the transient state.

The fixed-frequency peak current mode control is applied. With both voltage and current loops closed, the output current variation ΔI_o will follow the peak current reference variation ΔI_{ref} and the voltage reference variation ΔV_{ref} will follow the output voltage variation ΔV_o [21]:

$$\Delta I_o = \Delta I_{ref} \qquad (1)$$

$$\Delta V_{ref} = \Delta V_o \qquad (2)$$

Therefore, the output impedance R_o within the control bandwidth is approximated as

$$R_o = -\frac{\Delta V_o}{\Delta I_o} = -\frac{\Delta V_{ref}}{\Delta I_{ref}} \qquad (3)$$

The desired AVP control can be achieved by separately adjusting the voltage reference and the current reference according to the desired output impedance R_o: the voltage reference V_{ref} should decrease while the current reference I_{ref} should increase, and vice versa. Their relationship $f(z)$ can be expressed as

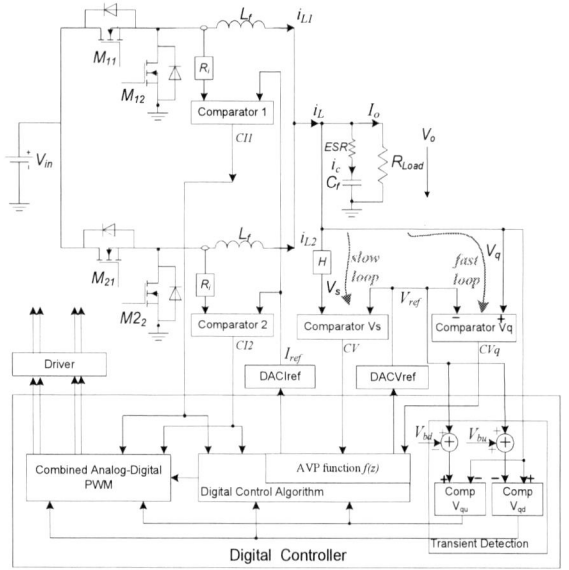

Fig. 1 a two-phase buck converter using the dual-voltage-loop digital AVP controller

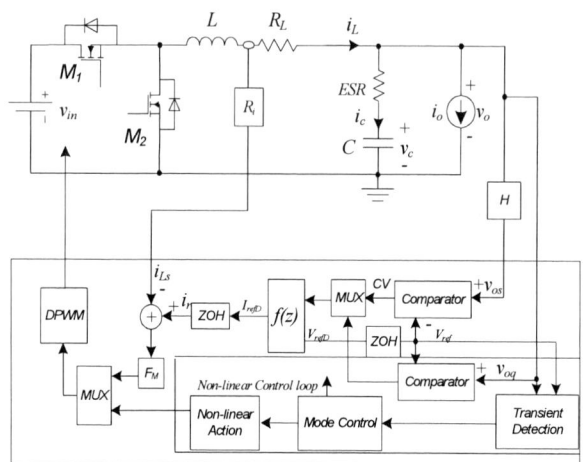

Fig. 2 simplified equivalent buck converter with the digital AVP controller

$$\Delta I_{ref} = -\frac{1}{R_o}\Delta V_{ref} \qquad (4)$$

Both the voltage reference and the current reference are always varying. The voltage reference is passively adjusted to follow the output voltage. The driving factor for the reference adjustment is the change in load currents: the difference between the load current and the average inductor current results in an output voltage variation that the voltage reference will follow/mimic.

III.   SMALL SIGNAL MODEL

Fig. 2 shows the simplified equivalent buck converter with the digital AVP controller. To derive the small signal model, the non-linear control loop can be temperately ignored since it is only for large signal transient. The equivalent series resistor R_L models the series combination of the equivalent resistance of the inductor, the losses of the MOSFETs and the losses of the power paths. The equivalent series resistor of the output capacitors is represented by ESR, while the current source i_o models the load.

Fig. 3 shows the proposed mixed-signal small signal model blocks. The transfer functions of the buck converter can be given as follows:

$$Z_o(s) = \frac{\hat{v}_o}{-\hat{i}_o} = \frac{ESR \cdot (s + \omega_{ESR})(s + \omega_L)}{s^2 + Q \cdot \omega_r \cdot s + \omega_r^2} \qquad (5)$$

$$G_{vv}(s) = \frac{\hat{v}_o}{\hat{v}_{in}} = \frac{\dfrac{D \cdot ESR}{L} \cdot (s + \omega_{ESR})}{s^2 + Q \cdot \omega_r \cdot s + \omega_r^2} \qquad (6)$$

$$G_{vd}(s) = \frac{\hat{v}_o}{\hat{d}} = \frac{\dfrac{V_{in} \cdot ESR}{L} \cdot (s + \omega_{ESR})}{s^2 + Q \cdot \omega_r \cdot s + \omega_r^2} \qquad (7)$$

$$G_{ii}(s) = \frac{\hat{i}_L}{\hat{i}_o} = \frac{\dfrac{ESR}{L} \cdot (s + \omega_{ESR})}{s^2 + Q \cdot \omega_r \cdot s + \omega_r^2} \qquad (8)$$

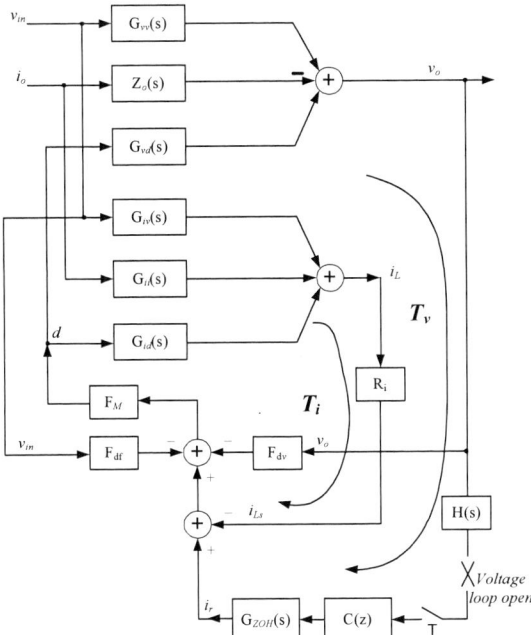

Fig. 3 the mixed-signal small signal model

$$G_{id}(s) = \frac{\hat{i}_L}{\hat{d}} = \frac{V_{in}}{L} \frac{s}{s^2 + Q \cdot \omega_r \cdot s + \omega_r^2} , \qquad (10)$$

$$\omega_{ESR} = \frac{1}{ESR \cdot C}, \quad \omega_L = \frac{R_L}{L}, \quad \omega_r = \sqrt{\frac{1}{L \cdot C}}, \quad Q = \frac{R_L + ESR}{\sqrt{L/C}} .$$

Where, Z_o represents the open-loop output impedance; G_{vv} is the open-loop line-to-output transfer function; G_{vd} is the open-loop control-to-output transfer function; G_{ii} is the open-loop load-to-inductor-current transfer function; G_{iv} is the open-loop line-to-inductor-current; G_{id} is the open-loop control-to-inductor-current transfer function.

In Fig.3, the ZOH (zero-order hold) models the DAC function, whose transfer function is represented as

$$G_{ZOH}(s) = \frac{1 - e^{-sT}}{s} , \qquad (11)$$

where T is the DAC clock period, which is much smaller than switching period T_s.

$H(s)$ is the transfer function of the output voltage sensing network (slow loop). An R_f-C_f low-pass filter is used for this sensing network with a pole at ω_f, whose transfer function is represented as

$$H(s) = \frac{1}{1 + s \cdot R_f \cdot C_f} = \frac{1}{1 + s/\omega_f} . \qquad (12)$$

The inner current loop is in analog. Therefore, it has the same perforamnce and siganl resoltion as analog peak current mode control. The average inductor current $<i_L>$ can be represented as:

$$<i_L> = \frac{1}{R_i} \cdot <i_r> - m_a \cdot d \cdot T_s - \frac{m_1 \cdot d^2 \cdot T_s}{2} - \frac{m_2 \cdot (1 - d)^2 \cdot T_s}{2} \qquad (13)$$

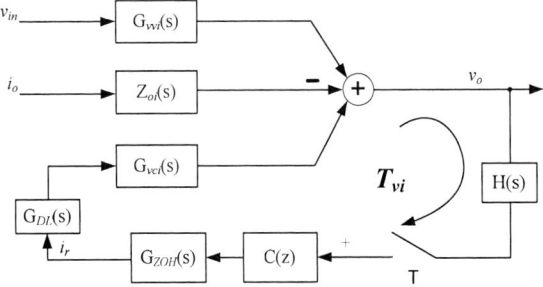

Fig. 4 The small signal model blocks with the inner current loop closed and the voltage loop open

$$m_1 = \frac{v_{in} - v_o}{L} \qquad m_2 = \frac{v_o}{L} \qquad m_1 \cdot D = m_2 \cdot (1 - D)$$

where m_a is the artificial ramp slope for the purpose of stability, which will be equal to 0 if no artificial ramp is added. The small signal control variable is expressed by:

$$\hat{d} = \frac{1}{m_a \cdot T_s \cdot R_i} (\hat{i}_r - R_i \cdot \hat{i}_L - R_i \cdot \frac{D^2 \cdot T_s}{2L} \cdot \hat{v}_{in} - R_i \cdot \frac{(1 - 2D) \cdot T_s}{2L} \cdot \hat{v}_o) \qquad (14)$$

$$= F_M \cdot \hat{i}_r - F_{df} \cdot \hat{v}_{in} - F_{dv} \cdot \hat{v}_o$$

When the inner current loop is closed and the voltage loop is open, the transfer functions of the power stage and the control loop are modified to be

$$G_{vvi}(s) = \frac{G_{vv}(s) \cdot (1 + F_M \cdot R_i \cdot G_{id}) - F_M \cdot (R_i \cdot G_{iv} + F_{df}) \cdot G_{vd}}{1 + F_M \cdot (F_{dv} \cdot G_{vd} + R_i \cdot G_{id})} , \qquad (15)$$

$$Z_{oi}(s) = \frac{Z_o(s) \cdot (1 + F_M \cdot R_i \cdot G_{id}) + F_M \cdot R_i \cdot G_{ii} \cdot G_{vd}}{1 + F_M \cdot (F_{dv} \cdot G_{vd} + R_i \cdot G_{id})} , \qquad (16)$$

$$G_{vci}(s) = \frac{F_M \cdot G_{vd}}{1 + F_M \cdot (F_{dv} \cdot G_{vd} + R_i \cdot G_{id})} , \qquad (17)$$

where G_{vvi}, Z_{oi} and G_{vci} represent the line-to-output transfer function, the output impedance and the control-to-output transfer function respectively when the inner current loop is closed.

Fig. 4 shows the transfer function blocks of the synchronous buck converter using the proposed controller with the inner current loop closed and the outer voltage loop open. $G_{DL}(s)$ is added to model the effective delay of the modulator, the gate drivers, and the power switches, and can be represented by

$$G_{DL}(s) = e^{-s \cdot T_d} \qquad (18)$$

To simplify the analysis, it is assumed that no artificial ramp Ma is added and the effects of the inductor current ripple are ignored. Therefore, $F_M \to \infty$, $F_{df} \to 0$, and $F_{dv} \to 0$. The control-to-output transfer function Gvci(s) with inner current loop closed will be reduced to a first-order system, given as

$$G_{vcis}(s) = \lim_{\substack{F_M \to \infty \\ F_{df} \to 0 \\ F_{dv} \to 0}} G_{vci}(s) = \frac{G_{vd}}{R_i \cdot G_{id}} = \frac{s/\omega_{ESR} + 1}{s \cdot C \cdot R_i} \qquad (19)$$

978-1-4244-4782-4/10 $26.00 © 2010 IEEE

On the other hand, the peak current reference i_r is generated by the digital-to-analog converter. The outer voltage loop is formed in the discrete-time domain. Thus it can have all valuable features of digital control: design flexibility, allowing the use of complicated control algorithms, and intelligent power management, etc. The control law ($f(z) = -1/R_o$) to describe the behavior of the digital AVP controller is summarized as follows [21]:

1) At each DAC clock, if logic signal CV is high ($CV = 1$), that is, the output voltage v_o exceeds the voltage reference V_{ref} set by the voltage DAC, the voltage reference V_{ref} will increase by a small amount ΔV_{ref} and the peak current reference I_{ref} will decrease by a small amount ΔI_{ref}.

2) At each DAC clock, if logic signal CV is low ($CV=0$), that is, the output voltage v_o is smaller than the voltage reference V_{ref}, the voltage reference V_{ref} will decrease by a small amount ΔV_{ref} and the peak current reference I_{ref} will increase by a small amount ΔI_{ref}.

In the digital form, the behavior of the digital AVP controller can be described as

$$\begin{cases} V_{ref}(n+1) = V_{ref}(n) + e(n) \\ \\ I_{ref}(n+1) = I_{ref}(n) + R_i \cdot f(z) \cdot e(n) \end{cases} \quad (20)$$

$$e(n) = \begin{cases} \left| \Delta V_{ref} \right| & V_{ref}(n) \le V_o \\ -\left| \Delta V_{ref} \right| & V_{ref}(n) > V_o \end{cases}$$

The discrete-time transfer function $C(z)$ of the digital controller is derived approximately as:

$$C(z) = R_i \cdot f(z) \cdot z^{-1} = -\frac{R_i}{R_o} \cdot z^{-1} \cdot \quad (21)$$

To facilitate analysis, this discrete-time transfer function $C(z)$ of the digital controller is converted into the continuous-time domain. If the sample frequency T is much higher than the switching frequency T_s, such a conversion will be quite accurate and the sampler and zero-order hold (G_{ZOH}) may be ignored with very little resulting error [22]. The continuous-time transfer function is

$$C(s) = S\{C(z)\} = -\frac{R_i}{R_o} \cdot e^{-sT} \quad (22)$$

The system open loop transfer function T_{vis} in the continuous-time domain can be simplified and found to be

$$\begin{aligned} T_{vis}(s) &= -C(s) \cdot G_{vcis}(s) \cdot H(s) \cdot G_{DL}(s) \\ &= \frac{1}{s \cdot C \cdot R_o} \cdot \frac{s/\omega_{ESR}+1}{s/\omega_f+1} e^{-s \cdot (T+T_d)} \\ &\approx \frac{1}{s \cdot C \cdot R_o} \cdot \frac{s/\omega_{ESR}+1}{s/\omega_f+1} \end{aligned} \quad (23)$$

Where $T_d + T << T_s$.

Fig.5 shows bode diagram of continuous-time system open loop transfer function at different ω_f.

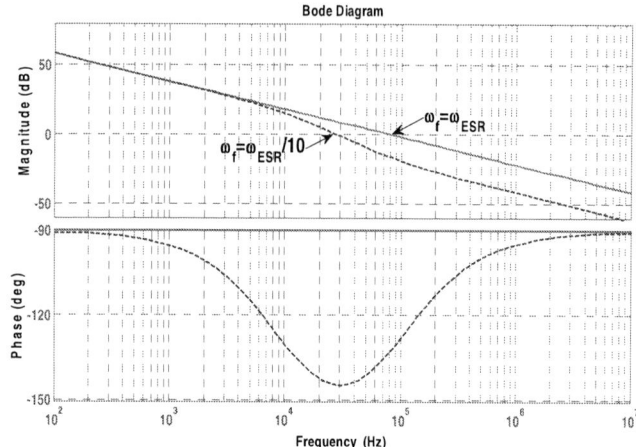

Fig. 5 Bode diagram of continuous-time system open loop transfer function at different ω_f

So, if choosing the pole ω_f of the sensing network to cancel the capacitor ESR zero ω_{ESR}, the open loop transfer function can be further simplified:

$$T_{vis}(s) \approx \frac{1}{s \cdot C \cdot R_o} \quad (24)$$

Obviously, the system crossover frequency f_c can be determined to be:

$$f_c = \frac{1}{2\pi \cdot R_o \cdot C} \quad (25)$$

This crossover frequency only depends on the designed output impedance R_o and the output filter capacitor C. And it is limited by the switching frequency and should be designed within 1/6~1/3 the equivalent switching frequency. Therefore, the output capacitor C should be selected to be sufficiently large for switching stability.

IV. LOAD TRANSIENT RESPONSE UNDER NON LINEAR CONTROL

The small signal analysis gives the switching stability constraint. But, when large load transient occurs, the controller will switch to transient mode and non-linear control takes over the control. Fig. 6 shows the control strategy of the dual voltage loop control, involving three modes of operation: normal operation mode, transient mode, and link mode. Peak current mode control is utilized in the normal operation mode and link mode. The slow voltage loop is closed in normal operation mode (steady-state), but the fast voltage loop is closed in link mode. Non-linear control is used in transient mode, while peak current mode control is disabled. The link mode makes the load transient transition smoothly without big voltage ring-back. Transitions among three types of operation modes are controlled by the mode control unit.

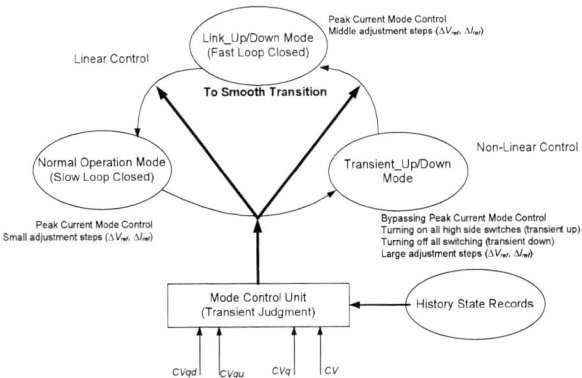

Fig. 6 Control strategy of dual voltage loop control

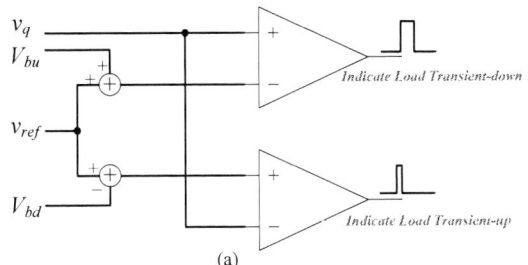

(a)

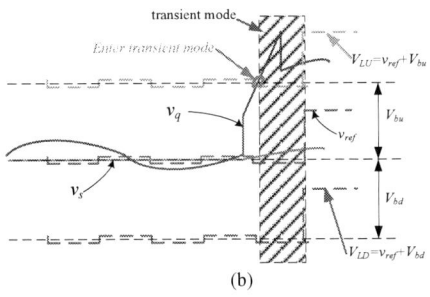

(b)

Fig. 7 Dedicated transient detection circuit and its voltage limitation gap

Fig. 7 shows the dedicated transient detection circuit and its voltage limitation gap, which always follows the voltage reference. As shown in Fig.7, a voltage limitation gap for load transient detection is formed around the voltage reference v_{ref}, and always follows the voltage reference. In the steady state, the voltage reference tightly follows the sensed output voltage v_s, which is the fed back by the slow voltage loop. once the voltage v_q from the fast loop exceeds the gap, the controller will enter transient mode. A large current/voltage reference step is used to track the new balance point. The voltage limitation gap also automatically varies in a large step. When the output voltage v_q goes back into the voltage limitation gap, the controller enters link mode. When the voltage reference crosses over the output voltage v_s again, the controller enters normal operation mode. From the above description, in transient mode and link mode, only the fast loop was closed to regulate the converter. The slow loop, which determines the system bandwidth, doesn't affect the converter behavior in transient mode and link mode. The slow loop only affect when the link mode should end. So, the system bandwidth only has very little effect on the converter large transient response speed.

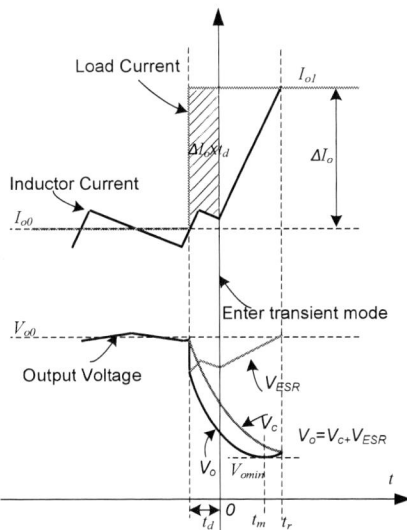

Fig. 8 the output voltage deviation during load transient

However, the variation of the voltage/current reference steps during transient mode and link mode has big impact on the controller dynamic behavior. Therefore, dynamic voltage deviation during transient is closely investigated to derive equations to select the steps for smooth transition: such as small voltage spikes and reduced voltage oscillation.

Fig.8 shows the voltage deviation during load transient. It is assumed that the load current slew rate is fast enough compared to the inductor current slew rate, the initial output voltage is V_{o0}, the initial inductor current is I_{o0}, the final inductor current is I_{o1}, and the step-up load current is $\Delta I_o = I_{o1} - I_{o0}$. The capacitor is discharging during the periods t_d and t_r. However, the minimum output voltage occurs sometime before the end of t_r due to ESR effects.

The output voltage function over the time after $t=0$ can be derived:

$$v_o(t) = v_{ESR} + v_c = -i_c(t) \cdot ESR - \int_{t_D}^{t} i_c(t)dt + V_{o0} \quad (26)$$

$$= V_{o0} - \frac{\Delta I_o \cdot t_d}{C} - ESR \cdot (\Delta I_o - \frac{V_L}{L_f}t) - \frac{\Delta I_o}{C}t + \frac{V_L}{2L_f \cdot C}t^2$$

Where $V_L = V_{in} - V_o$ for transient up and $V_L = -V_o$ for transient down, t_d approximates the transient-to-action delay and $\Delta I_o \times t_d$ approximately represents the charge variation of the capacitors.

During load transient-up mode, the voltage reference should keep up with the output voltage as quickly as possible. The current reference should also increase to track the new load current. Referring to (26), the constraint on the reference step can be expressed as

$$M_t \geq \frac{\frac{\Delta I_{o\max}}{C} \cdot (t_d + \frac{2}{f_{clk}}) + ESR \cdot \Delta I_{o\max} - V_{bd} - (ESR + \frac{1}{C \cdot f_{clk}}) \cdot \frac{2 \cdot (V_{in} - V_o)}{L_f \cdot f_{clk}} - \text{int}[t_d \cdot f_{clk}] \cdot \Delta V_{ref}}{2 \cdot \Delta V_{ref}}$$

.

During the load transient-down stage, the inductor slew rate is quite low, so the controller should stay in the transient mode before the output voltage reaches its maximum, and make the inductor current continue decreasing. The constraint on the reference step can be expressed as

$$M_t \leq \begin{cases} \dfrac{\dfrac{\Delta I_{o\max} \cdot t_D}{C} + ESR \cdot \Delta I_{o\max} - \Delta V_{bu} - \text{int}[t_d \cdot f_{clk}] \cdot \Delta V_{ref}}{\Delta V_{ref} \cdot \text{int}[T_m \cdot f_{clk}]} & L_f < L_{crit} \\[4mm] \dfrac{\dfrac{\Delta I_{o\max} \cdot t_d}{C} + \dfrac{ESR^2 \cdot C^2 \cdot V_o^{\,2} + \Delta I_{o\max}^{\,2} \cdot L_f^{\,2}}{2 L_f \cdot C \cdot V_o} - V_{bu} - \text{int}[t_d \cdot f_{clk}] \cdot \Delta V_{ref}}{\Delta V_{ref} \cdot \text{int}[T_m \cdot f_{clk}]} & L_f \geq L_{crit} \end{cases}$$

$$T_m = \begin{cases} 0 & L_f < \dfrac{ESR \cdot C \cdot V_L}{\Delta I_o} = L_{crit} \\[4mm] \dfrac{\Delta I_o \cdot L_f}{V_L} - ESR \cdot C & L_f \geq \dfrac{ESR \cdot C \cdot V_L}{\Delta I_o} = L_{crit} \end{cases}$$

During link-up mode, the voltage reference should catch up with the output voltage before the output voltage reaches its minimum and the voltage reference deviation rate should be larger than the output voltage deviation rate to avoid entering transient mode again. Therefore, the constraint on the reference step can be expressed as

$$M_L \geq \begin{cases} \dfrac{ESR \cdot C \cdot (V_{in} - V_o) - \Delta I_{o\max} \cdot L_f}{\Delta V_{ref} \cdot f_{clk} \cdot L_f \cdot C} & L_f < L_{crit} \\[4mm] \dfrac{\dfrac{\Delta I_{o\max} \cdot t_d}{C} + \dfrac{ESR^2 \cdot C^2 \cdot (V_{in} - V_o)^2 + \Delta I_{o\max}^{\,2} \cdot L_f^{\,2}}{2 L_f \cdot C \cdot (V_{in} - V_o)} - 2 \cdot \Delta V_{ref} \cdot M_t - \text{int}(t_d \cdot f_{clk}] \cdot \Delta V_{ref}}{\Delta V_{ref} \cdot \text{int}[T_m \cdot f_{clk}]} & L_f \geq L_{crit} \end{cases}$$

.

During the link-down mode, the current reference deviation rate should be larger than the inductor current slew rate to keep the inductor current decreasing, and catch up with load current quickly to avoid entering the transient mode again. The constraint on the reference step can be expressed as

$$\begin{cases} M_L \geq \max\left(\dfrac{V_o}{\Delta I_{ref} \cdot L_f \cdot f_{clk}}, \dfrac{\Delta I_{o\max}}{\Delta I_{ref}} - t \cdot f_{clk} \cdot M_t \right) \\[4mm] t \cdot f_{clk} \cdot M_t \cdot \Delta V_{ref} + V_{bu} > \dfrac{\Delta I_o \cdot t_d}{C} + ESR \cdot (\Delta I_o - \dfrac{V_L}{L_f} t) + \dfrac{\Delta I_o}{C} t - \dfrac{V_L}{2 L_f \cdot C} t^2 \end{cases}$$

Obviously, these equations are independent of the system bandwidth. It can be expected that the large signal transient response of the system designed based on these equations will not be limited by the system bandwidth.

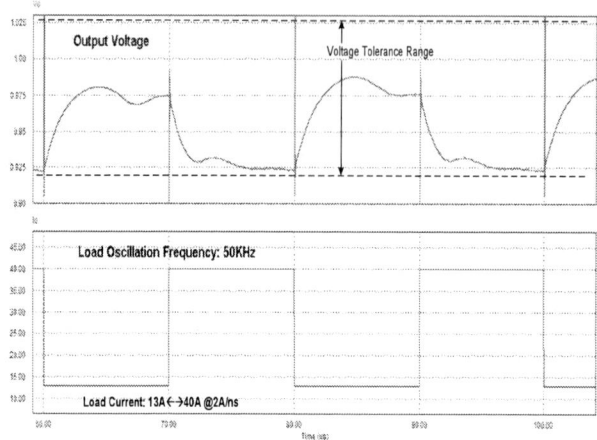

Fig. 9 The load transient response with 50kHz load oscillation (Design #1, f_c=80 kHz) (voltage scale: 25mV/div, current scale: 5A/div, time scale: 10μs)

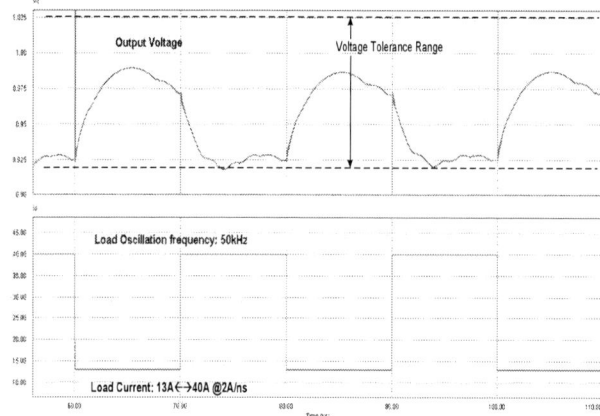

Fig. 10 The load transient response with 50kHz load oscillation (Design #2, f_c=23 kHz) (voltage scale: 25mV/div, current scale: 5A/div, time scale: 10μs)

V. SIMULATION VERIFICATIONS AND EXPERIMENTAL RESULTS

A two-phase interleaved synchronous buck converter for 12V-1V/40A VRs was designed and simulations were conducted in the co-simulation environment consisting of PSIM, ModelSim and Simulink. The phase switching frequency is 250 KHz, the phase inductor is 400nH and the total output capacitors are 1mF (ESR: 1.6mΩ, ESL: 0.6nH). The desired output impedance is set at 2mΩ. Two voltage sensing network (for the slow loop) are designed to have different system bandwidth: R_f=2kΩ, C_f=1nF (design #1), 80kHz crossover frequency; R_f=20kΩ, C_f=1nF (design #2), 23kHz crossover frequency.

Fig. 9 shows the converter (design #1, f_c=80 kHz) transient response under large step load current oscillation of 50 kHz. Fig. 10 shows the converter (design #2, f_c=23 kHz) transient response under large step load current oscillation of 50 kHz. The load transient step is 27A at the slew rate of 2A/ns. It is shown that both the peak overshoot voltage and the settle down time in both designs are very similar. Only difference is that the link mode ends earlier in design #1 than in design #2. So the transition in design #1 is smoother.

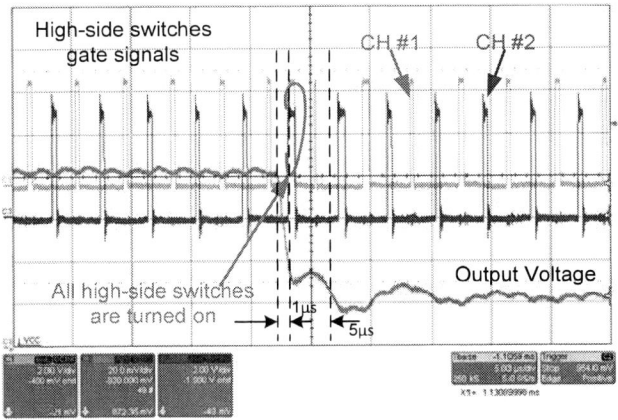

Fig. 11 Extended transient response with dual-voltage-loop control (voltage scale: 20mV/div, gate signal scale: 2V/div, time scale: 5μs/div)

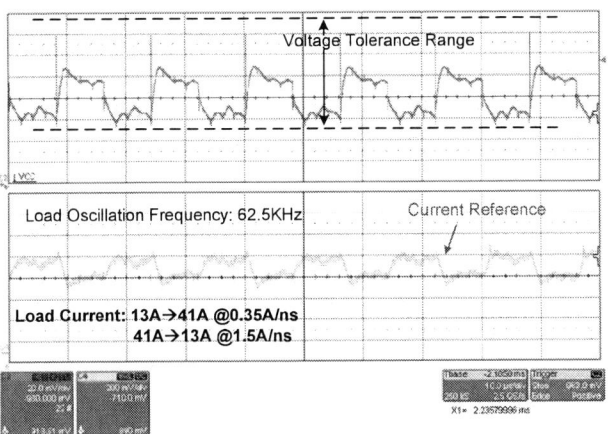

Fig. 12 The tested output voltage and current reference waveforms with 62.5 kHz load oscillation (voltage scale: 20mV/div, current scale: 20A/div, time scale: 10μs/div)

A prototype of a two-phase interleaving synchronous buck converter for 12V-1V/40A VR was tested. The phase switching frequency is 250 KHz, the phase inductor is 400nH and the output capacitors are one 330uF bulk capacitor + 32x22uF MLCC. The slow-loop voltage sensing network is R_f=2kΩ, C_f=1nF.

The digital controller was implemented in FPGA (8MHz, Stratix II EP2S60 from Altera) by VHDL. The minimum voltage reference step ΔV_{ref} is 0.84mV and the minimum current reference step ΔI_{ref} is 0.21A. Steps in transient/link-up stage: (8.4mV, 2.1A)/(3.36mV, 0.84A); Steps in transient/link-down stage: (1.68mV, 0.42A) /(10.08mV, 2.52A).

The output impedance was measured about 2.04mΩ. The measured crossover frequency was about 71.2 kHz. Fig.11 shows the load transient response (from 13A to 41A @ 350A/μs). The transient-to-action is about 1μs and the transient can be settled down less than 5μs, which cannot be achievable for a traditional analog controller with the same bandwidth. As an example to demonstrate the smooth transient response at high dynamic load oscillation (have been verified from 100Hz to 1MHz), Fig. 12 shows the output voltage and current reference waveforms with 62.5 kHz load oscillation.

VI. CONCLUSION

In this paper, the mixed-signal small model was proposed for a dual-voltage-loop digital AVP controller proposed in [21], to demonstrate how it benefits from both analog and digital loops. And dynamic large signal behavior was studied to optimize the controller, showing that the large step load transient response is independent of the system bandwidth. Despite of low bandwidth designed for switching stability, the VR with the digital AVP controller exhibits very excellent dynamic performance because of non-linear multi-mode control, making it a valuable candidate as a high-volume low-cost controller for voltage regulators.

REFERENCES

[1]. R. Redl, B.P. Erisman and Z. Zansky, "Optimizing the load transient response of the buck converter," IEEE Applied Power Electronics Conference, 1998, pp. 170-176.

[2]. S.A. Chickamenahalli, S. Mahadevan, E. Stanford and K. Merley, "Effect of target impedance and control loop design on VRM stability," in Proc. IEEE Applied Power Electronics Conference, 2002, pp. 196-202.

[3]. Kaiwei Yao, Y. Ren, and F. C. Lee, "Critical bandwidth for the load transient response of voltage regulator modules", IEEE Transaction on Power Electronics, 19(6):1454–1461, November 2004.

[4]. Kaiwei Yao, M. Xu, Y. Meng, and F. C. Lee, "Design considerations for VRM transient response based on the output impedance", IEEE Transaction on Power Electronics, pp: 1270-1277, November 2003.

[5]. K. Yao, K. Lee, M. Xu and F. C. Lee, "Optimal Design of the Active Droop Control Method for the Transient Response", IEEE Applied Power Electronics Conference and Exposition, 2003, Vol.2, Pp.718-723.

[6]. P.L. Wong, F.C. Lee, P. Xu and K. Yao, "Critical inductance in voltage regulator modules," in IEEE Transaction on Power Electronics, Vol. 17, Issue 4, July 2002, pp. 485-492.

[7]. F.N.K. Poon, C.K. Tse and J.C.P. Liu, "Very fast transient voltage regulators based on load correction," IEEE Power Electronics Specialists Conference 1999, pp. 66-71.

[8]. C.J. Mehas, K.D. Coonley and C.R. Sullivan, "Converter and inductor design for fast-response microprocessor power delivery," IEEE Power Electronics Specialists Conference 2000, pp. 1621-1626.

[9]. BA. Barrado, R. Vazquez, A. Lazaro, J. Pleite and E. Olias, "Fast transient response DC/DC converter for low output voltage," in Electronics Letters, Vol. 38, Sept. 2002, pp. 1127-1128.

[10]. L. Amoroso, M. Donati, X. Zhou and F.C. Lee, "Single shot transient suppressor (SSTS) for high current high slew rate microprocessor," IEEE Applied Power Electronics Conference, 1999, pp. 284-288.

[11]. R. Miftakhutdinov, "Analysis and optimization of synchronous buck converter at high slew-rate load current transients," IEEE Power Electronics Specialists Conference 2000, pp. 714-720.

[12]. J. Xu, X. Cao and Q. Luo, "The effects of control techniques on the transient response of switching DC-DC converters," in Proc. IEEE PEDS, 1999, pp. 794-796.

[13]. P. Wong, F.C. Lee, X. Zhou and J. Chen, "VRM transient study and output filter design for future processors," in Proc. IEEE IECON, 1998, pp. 410-415.

[14]. Angel V. Peterchev, and Seth R. Sanders "Design of Ceramic–Capacitor VRM's with Estimated Load Current Feedforward", IEEE Power Electronics Specialists Conference, 2004, Vol. 6, June 2004, pp:4325- 4332.

[15]. Xin Zhang, Gary Yao, and Alex Q. Huang, "A novel VRM control with direct load current feedback," in Proc. IEEE Applied Power Electronics Conference, 2004.

[16]. R. Miftakhutdinov, "Optimal design of interleaved synchronous buck converter at high slew-rate load current transients," IEEE Power Electronics Specialist Conference, 2001.

[17]. D. Goder and W. R. Pelletier, "V^2 architecture provides ultra-fast transient response in switch mode power supplies," in Proc. HFPC, 1996, pp. 16-23.

[18]. Wei Gu, Weihong Qiu, Wenkai Wu, and Issa Batarseh, "A Multiphase DC/DC Converter with Hysteretic Voltage Control and Current Sharing", IEEE Applied Power Electronics Conference and Exposition, Vol. 2, March 2002 pp: 670 – 674.

[19]. BA. Barrado, J. Quintero, A. Lazaro, C. Fernandez, P. Zumel, E. Olias, "Linear-Non-Linear Control Applied in Multiphase VRM", IEEE Power Electronics Specialists Conference (PESC), 2005, pp.904-909.

[20]. J. Quintero, A. Barrado, M. Sanz, A. Lazaro, E. Olias, "Experimental Validation of the Advantages provided by Linear - Non - Linear Control in Multi-phase VRM", IEEE Applied Power Electronics Conference and Exposition (APEC), February 2007.

[21]. S. Pan, P. Jain, "Novel Digital Control Architecture with Non-Linear Control Algorithms Exhibiting Very Fast Transient Response", IEEE Applied Power Electronics Conference and Exposition (APEC), February 2009

[22]. S. Banerjee and G. C. Verghese (Editors), "Nonlinear Phenomena in Power Electronics: Attractors, Bifurcations, Chaos, and Nonlinear Control", New York: IEEE Press, 2001.

Multi-Output Synchronously-Rectified Forward Converter with Load Transient Considered

K. I. Hwu, *Member, IEEE*, and Y. T. Yau, *Student Member, IEEE*

Department of Electrical Engineering, National Taipei University of Technology, Taiwan

Abstract—In this paper, an FPGA-counter-based scheme is presented herein and applied to a forward converter with single isolation stage and multiple outputs having synchronous rectification (SR). With only the required comparators and without any analog-to-digital converter (ADC), the information on feedback output voltage is entirely obtained according to a counter. Therefore, the proposed control topology for an SR forward converter can improve the load transient response and the cross regulation. Besides, to further upgrade the load transient response, the proposed nonlinear control technique is applied. In this paper, the proposed control scheme is described and some experimental results are provided to verify its effectiveness.

I. INTRODUCTION

Since the operating speed of the integrated chip is getting faster and faster, its working voltage is reduced to 2.5V or less. However, this focuses on its kernel only. As for its peripherals, the required voltage still uses 3.3V or 5V. Consequently, multiple outputs are required to achieve reduction of cost and size. And there are many types available today [1]. For a forward converter with multiple isolation stages and multiple outputs to be considered, the magnetic amplifier [2][3], belonging to the secondary-side post regulator (SSPR), is commonly used for the traditional control of multiple outputs, but inherently possesses high non-linearity so as to be difficult in control and to tend to create a huge core loss under high-frequency operation. In order to overcome this problem, the transformer control [4-6] has been proposed, which two additional windings at the secondary side are needed if one additional output is needed. Therefore, the corresponding cost and size increase. Also, there is another way by extracting the energy stored in the snubber [7] via one additional transformer, and this has the problem of how to reduce cost and size and how to increase output power.

On the other hand, for a forward converter with the single isolation stage and multiple outputs to be considered, this type is suitable for point-of-load conditions without isolation between outputs, so as to reduce cost and size. As for control of such multiple outputs, the SSPR has been proposed [8], which is capable of operating as a stand-alone buck regulator. Consequently, the improvement in the load transient response and the cross regulation is not easy to achieve. To conquer this problem, a field-programmable gate array (FPGA)-based counter-based control strategy [9] is presented herein along with the proposed PWM type and the proposed primary-side post regulator (PSPR) integrated with the main controller into one FPGA chip, and is applied to a forward converter with two outputs having the single isolation stage. Each output possesses its own controller along with synchronous rectification (SR).

Moreover, with two comparators acting like sampling and without any analog-to-digital (ADC) converter required, information on the feedback output voltage is entirely obtained according to counters, thereby rendering the load transient response fast and the cross regulation good. Furthermore, a nonlinear control technique is proposed to further enhance the load transient response. Such a control topology is very useful in industrial applications and in integrating functions into one chip.

II. OVERALL SYSTEM CONFIGURATION

As shown in Fig. 1, FPGA is used as a control kernel for two SR outputs of a forward converter with active clamping. Since two output voltages needs accurately regulating, two feedback control loops are dispensable, both of which are based on one-comparator control [9]. The actual output voltages after the high-speed comparators are transferred to the digital signal that is passed to FPGA through the high-speed photo-couplers. In FPGA, there are two proportional-integral-derivative (PID) controllers used to calculate the required duty cycle for the upcoming pulse width modulation (PWM) cycle. Besides, since the synchronous rectification method is used herein, the PWM duty cycle of the main output is not changed too much even under light load. Therefore, the control of the auxiliary output is easy and stable, and there are six PWM control signals created to drive six switches.

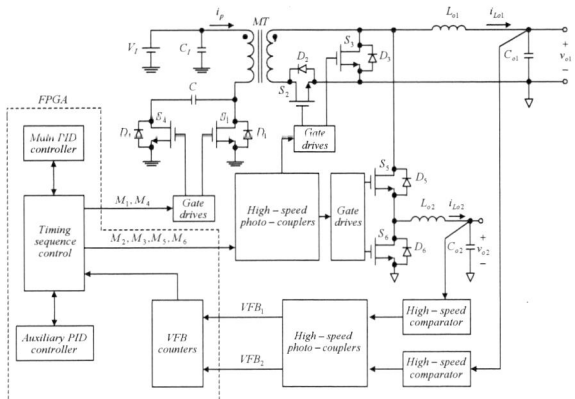

Fig. 1. Proposed overall system block diagram.

III. PROPOSED CONTROL TECHNIQUE

As shown in Figs. 2 and 3, the PWM control signals M_1 and M_5 are for the switch S_1 of the main output and the switch S_5 of

978-1-4244-4782-4/10 $26.00 © 2010 IEEE

the auxiliary output, respectively. Besides, the currents flowing through the main output inductor L_{o1}, the auxiliary output inductor L_{o2} and the resulting current at the primary side are also illustrated. Fig. 2 shows that the turn-off instants for the switches S_1 and S_5 are identical but the turn-on instants are different, whereas Fig. 3 shows that the turn-on instants for the switches S_1 and S_5 are the same but the turn-off instants are different. The former is commonly used in the SSPR under peak current mode control, so as to avoid the problem of controlling two peak currents; however the latter is different from the former due to the proposed PSPR under voltage mode control. Therefore, the latter has an advantage that the time sequence for all the switches is easily generated because M_5 is synchronized with M_1 at the rising edge of M_1. Besides, the maximum current flowing through the switch S_1 in the case of Fig. 3 is lower than that in the case of Fig. 2, such that the peak current for S_1 is reduced and hence the current stress for S_1 is decreased.

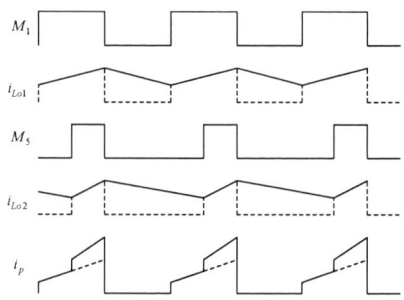

Fig. 2. Key waveforms for the conventional PWM type.

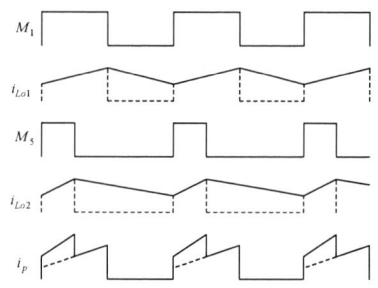

Fig. 3. Key waveforms for the proposed PWM type.

As shown in Fig. 4, one-comparator counter-based control [9] is employed herein based on FPGA with the system clock set to 100MHz, i.e., the period of system clock set to 10ns. If the output voltage is well-regulated in the steady state, the output of the comparator after the photo-coupler, in theory, has the interval of 256CLK counted, i.e., 2.56μs for the entire PWM period of 512CLK counted, i.e., 5.12μs, and hence the resulting output error $v_{o\text{-}error}$ is zero. That's to say, if the resulting value of the counter is smaller than 256CLK, $v_{o\text{-}error}$ is positive, implying the output voltage is below the output voltage reference; if the resulting value of the counter is larger than 256CLK, then $v_{o\text{-}error}$ is negative, implying the output voltage is above the output voltage reference. Two outputs have individual comparators and PID controllers. But, the timing

sequence for all PWM control signals to drive MOSFET switches is controlled by one flow chart, and this makes the control of the multi-output SR forward converter easier. By the way, the slope of v_{o1} is different from that of v_{o2} due to difference in inductor current slope between the two.

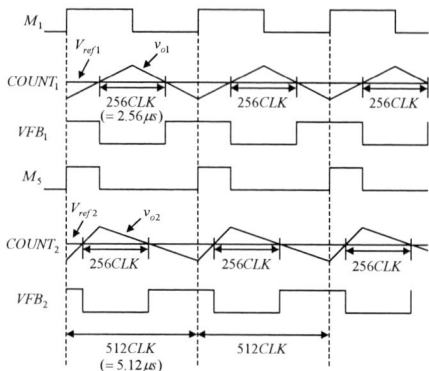

Fig. 4. One-comparator counter-based sampling.

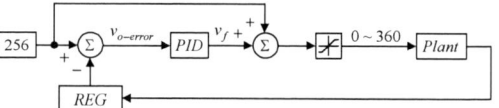

Fig. 5. Proposed control loop.

As shown in Fig. 5, how to design the control loop for each output is described. The value of REG is obtained from the sensed output voltage through one comparator and $COUNT$, and such a value is subtracted from 256 to get an output voltage error $v_{o\text{-}error}$, which is sent to the PID controller to obtain control force v_f, and added with 256 and then restricted to 360, which corresponds to the maximum duty cycle of 70%. After this, the duty cycle for the 9-bit PWM control signal is created.

As shown in Fig. 6, there are three modes for the converter to operate. One is the normal mode, another is the upload mode and the other is the download mode. How to accelerate the load transient response due to load change from no/rated load to rated/no load of the main output under the no/rated load of the auxiliary output. From Fig. 6, if the main output voltage v_{o1} is lower than $V_{o1} - 0.5\Delta v_{o1}$, where V_{o1} is the DC value of v_{o1} and Δv_{o1} is the peak-to-peak value of the ripple of v_{o1}, then the converter operates in the upload mode. In this case, the PID controller is disabled and the duty cycle for the main output, d_m, is held constant at some value, say, 0.7, thereby causing v_{o1} to be increasing until v_{o1} locates between $V_{o1} - 0.5\Delta v_{o1}$ and $V_{o1} + 0.5\Delta v_{o1}$. After this, the converter goes back to the normal mode, and hence the corresponding PID controller for the main output is enabled so as to reduce the main output voltage error in the steady state. On the other hand, if v_{o1} is higher than $V_{o1} + 0.5\Delta v_{o1}$, then the converter works in the download mode. In this case, the PID controller is disabled and the duty cycle of the main output is theoretically held zero so as to reduce the overshoot on the main output voltage. However, by doing so, the auxiliary output voltage will be zero, and hence, in order to maintain the auxiliary output voltage at the desired value as constant as possible, the duty cycle for the main output, d_m, is kept constant at the duty cycle for the auxiliary output, d_a.

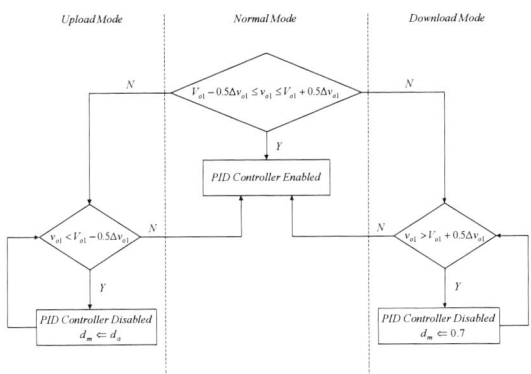

Fig. 6. Proposed response acceleration during the transient.

IV. EXPERIMENTAL RESULTS

Before the effectiveness of the proposed control scheme for two SR outputs of the forward converter is verified, there are some specifications required, to be described below: (i) turns ratio of the main transformer $n = N_p/N_s = 6/5$; (ii) rated input voltage of 12V; (iii) main rated output voltage/current of 5V/7.5A; (iv) auxiliary rated output voltage/current of 2.5V/5A; (v) switching frequency of 195kHz; and (vi) controller parameters for individual outputs with k_p, k_i and k_d set to 0.25, 0.03125 and 0.25 respectively.

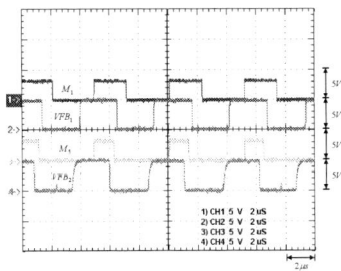

Fig. 7. Timing sequence for M_1, VFB_1, M_5 and VFB_2.

Fig. 7 shows the waveforms relevant to the PWM control signal for the main switch S_1 of the main output, M_1, the main output feedback signal after the corresponding photo-coupler, VFB_1, the PWM control signal for the main switch S5 of the auxiliary output, M_5, and the auxiliary output feedback signal after the accompanying photo-coupler, VFB_2, all of which almost meet the corresponding waveforms shown in Fig. 4. The results depict that in the steady state the duty cycles of VFB_1 and VFB_2 both are about 50%, implying each output voltage follows the prescribed output voltage reference.

Figs. 8(a) and 8(b) show the load transient load responses of the auxiliary output due to load change from 0A to 5A and from 5A to 0A respectively, under the condition that the main output is kept constant at 0A, whereas Figs. 9(a) and 9(b) show the load transient responses of the auxiliary output due to load change from 0A to 5A and from 5A to 0A respectively, under the condition that the main output is kept constant at 7.5A.

On the other hand, Figs. 10(a) and 10(b) show the load transient responses of the main output due to load change from 0A to 7.5A and from 7.5A to 0A respectively, under the condition that the auxiliary output is kept constant at 0A, whereas Figs. 11(a) and 11(b) show the load transient responses of the main output due to load change from 0A to 7.5A and from 7.5A to 0A respectively, under the condition that the auxiliary output is kept constant at 5A.

It is noted that Figs. 8 and 9 have better responses than Figs. 10 and 11. Consequently, the proposed response acceleration during the transient is presented to further upgrade the load transient responses shown in Figs. 10 and 11. With the response acceleration applied, Figs. 12(a) and 12(b) show the load transient responses of the main output due to load change from 0A to 7.5A and from 7.5A to 0A respectively, under the condition that the auxiliary output is kept constant at 0A, whereas Figs. 13(a) and 13(b) show the load transient responses of the main output due to load change from 0A to 7.5A and from 7.5A to 0A respectively, under the condition that the auxiliary output is kept constant at 5A. From the results mentioned above, it can be seen that the load transient responses can be improved further, especially for the download responses. Based on the results mentioned above, the proposed control scheme possesses fast load transient response and good cross regulation for two SR outputs of the forward converter.

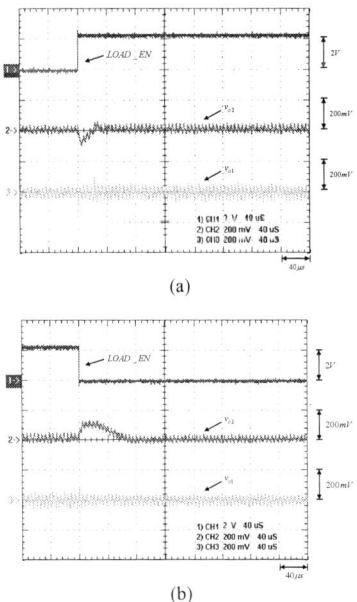

Fig. 8. Under the condition that the main output is kept constant at 0A, the load transient responses of the auxiliary output due to load change: (a) from 0A to 5A; (b) from 5A to 0A.

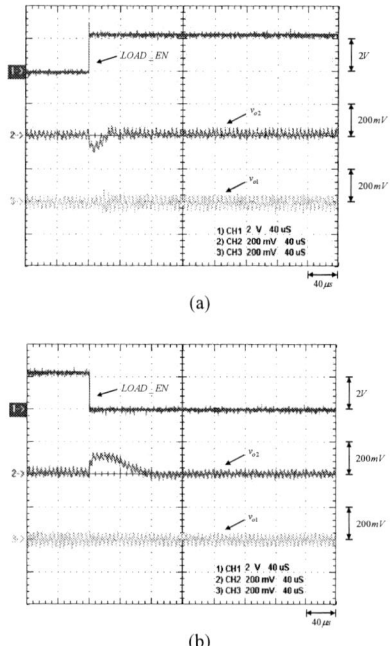

(a)

(b)

Fig. 9. Under the condition that the main output is kept constant at 7.5A, the load transient responses of the auxiliary output due to load change: (a) from 0A to 5A; (b) from 5A to 0A.

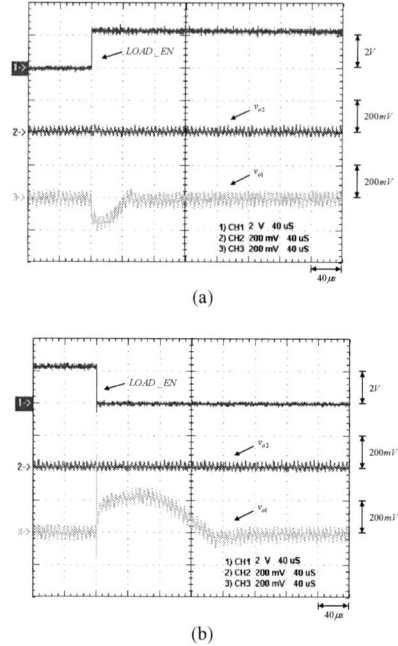

(a)

(b)

Fig. 11. Under the condition that the auxiliary output is kept constant at 5A, the load transient responses of the main output due to load change without the proposed response acceleration: (a) from 0A to 7.5A; (b) from 7.5A to 0A.

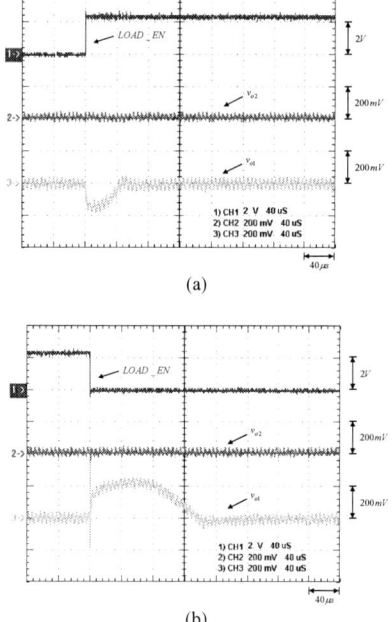

(a)

(b)

Fig. 10. Under the condition that the auxiliary output is kept constant at 0A, the load transient responses of the main output due to load change without the proposed response acceleration: (a) from 0A to 7.5A; (b) from 7.5A to 0A.

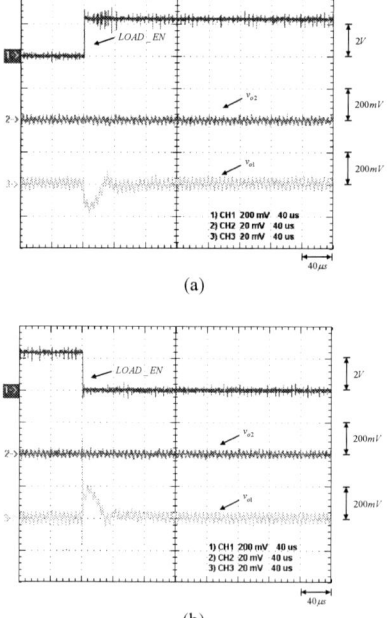

(a)

(b)

Fig. 12. Under the condition that the auxiliary output is kept constant at 0A, the load transient responses of the main output due to load change with the proposed response acceleration: (a) from 0A to 7.5A; (b) from 7.5A to 0A.

978-1-4244-4782-4/10 $26.00 © 2010 IEEE

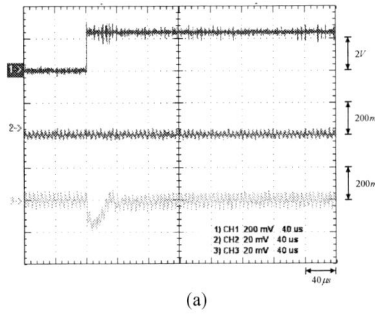

(a)

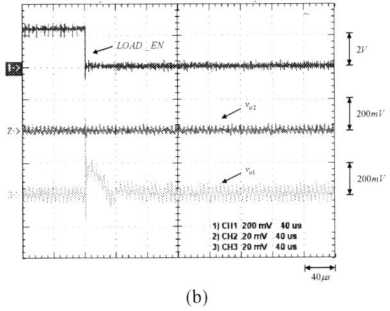

(b)

Fig. 13. Under the condition that the auxiliary output is kept constant at 5A, the load transient responses of the main output due to load change with the proposed response acceleration: (a) from 0A to 7.5A; (b) from 7.5A to 0A.

VI. CONCLUSION

An FPGA-based counter-based control strategy, together with the proposed primary-side post regulator and the proposed PWM type, is applied to two SR outputs of the forward converter, to get the fast load transient response and the good cross regulation as well as to make the main and auxiliary PID controllers integrated into one FPGA chip and hence to render the time sequence for all the switches created easily. By the way, two outputs can be extended to three or more outputs, if needed.

ACKNOWLEDGMENT

The authors would like to thank the National Science Council for supporting this work under Grant NSC-97-2221-E-027-107.

REFERENCES

[1] Dr. Ray Ridley, "Ways to generate multiple outputs," *Switching Power Magazine*, vol. 4, no. 2, 2003.

[2] Xi Youhao and P. K. Jain, "A forward converter topology with independently and precisely regulated multiple outputs," *IEEE Trans. Power Electronics*, vol. 18, no. 2, pp. 648-658, 2003.

[3] M. Gekinozu, S. Igarashi, N. Miura and K. Karube, "Multi-output type current resonant converter with magnetic amplifier control," *INTELEC'03*, pp. 410-415, 2003.

[4] A. Ferreres, J. A. Carrasco, E. Maset and J. B. Ejea, "Small-signal modeling of a controlled transformer parallel regulator as a multiple output converter high efficient post-regulator," *IEEE Trans. Power Electronics*, vol. 19, no. 1, pp. 183-191, 2004.

[5] A. Ferreres, J. A. Carrasco, E. Sanchis, J. M. Espi and E. Maset, "Application of a novel parallel regulation technique in a two output forward converter," *IEEE PESC'99*, vol. 2, pp. 914-919, 1999.

[6] J. A. Carrasco, J. B. Ejea, A. Ferreres, and E. J. Eded, "Modeling multiple-output converters with parallel post-regulators," *IEEE PESC'97*, vol. 2, pp. 916-921, 1997.

[7] M. Jinno, Chen Po-Yuan and Lin Kun-Chih, "An efficient active LC snubber for multi-output converters with fly-back synchronous rectifier," *IEEE PESC'03*, vol. 2, pp. 622-627, 2003.

[8] LM5115 datasheet, *National Semiconductor*, 2005.

[9] K. I. Hwu and Y. T. Yau, "Applying a counter-based PWM control scheme to an FPGA-based SR forward converter," *IEEE APEC'06*, vol. 3, pp.1396-1400, 2006.

Symmetric Current Balancing Circuit for Multiple DC loads

Sungjin Choi, Pankaj Agarwal, Teahoon Kim, Joonhyun Yang, and Baikhee Han

Advanced R&D Group 1, Visual Display Division, Samsung Electronics Co., Ltd.
Suwon, Gyeonggi-do, 443-742, South Korea
sjc.choi@samsung.com

Abstract — One of the key challenges in driving multiple parallel loads using a single circuit is to ensure that each load share the same current. Current imbalance occurs because these loads have varying individual characteristics, which surface due to inevitable deviations in mass production. Lighting is one representative application where multiple lamps such as LED (light-emitting diode) arrays need some sort of current balancing mechanism. While resistive balancing technique can be highly inefficient, current balancing using dedicated control IC for each channel significantly increases the cost and circuit complexity. In this paper, a novel current balancing circuit for multiple DC loads is proposed. A smart combination of an inherent symmetry of circuit and capacitive balancing mechanism enables an efficient and cost-effective current balancing. The operating principle of the proposed balancing method is analyzed in detail and an appealing generalization is made. The feasibility of the proposed balancing scheme is verified by developing a hardware prototype to drive a 100W system, having six LED arrays.

I. INTRODUCTION

With significant progress in lighting technology, efficient and cost-effective light driver circuits have gained the much needed attention in power electronics industry. Light sources are categorized into two groups. One group is ac light source such as cold cathode fluorescent lamp (CCFL), external electrode fluorescent lamp (EEFL), and hot cathode fluorescent lamp (HCFL). The other group is dc light source such as light-emitting diode (LED), organic light-emitting diode (OLED).

In most applications, driving the same current through multiple parallel lamps to ensure brightness uniformity is a key issue. While the cost effective reactive balancing techniques such as capacitive balancing method and Jin-balance method [1] are well established for ac lamps, there has been no counterpart for dc lamps. Instead, active balancing methods such as linear balancing method [2] or dedicated dc/dc converter [3] are commonly used in industry. Over the last few years, IC vendors have integrated these dc current

balancing functions into a single chip to reduce the cost but the overall implementation is still expensive.

In this paper, a new current balancing circuit for multiple dc loads is outlined. The proposed scheme is a simple, efficient, and cost-effective solution because it utilizes an inherent symmetric balancing feature only using reactive components and diode rectifier.

II. CONVENTIOAL BALANCING METHOD

In case of most lamp loads, lamp current should be stabilized to regulate the brightness. More particularly, for series-connected white LED arrays, which are commonly used dc lamps in backlights for display, the diode forward voltage and current are exponentially related – i.e. when the forward voltage of LED lamp changes slightly, its current varies dramatically.

In reality, even LEDs from the same production lot have poorly matched I-V characteristics. When constructing a system where multiple LED arrays are driven by a single drive circuit, this mismatch causes the lamps not to share the current equally. Without a balancing mechanism, LED current can be severely unbalanced even if energized by the same voltage. If the current values are different, their luminous intensity is not

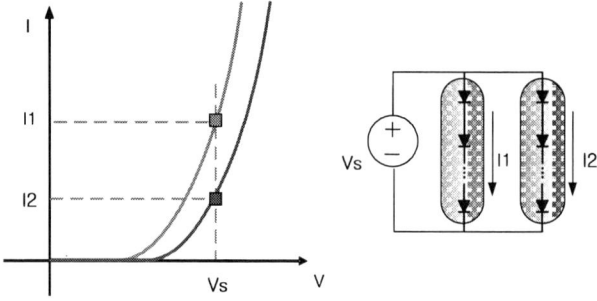

Figure 1. I-V curves for two different LED arrays and

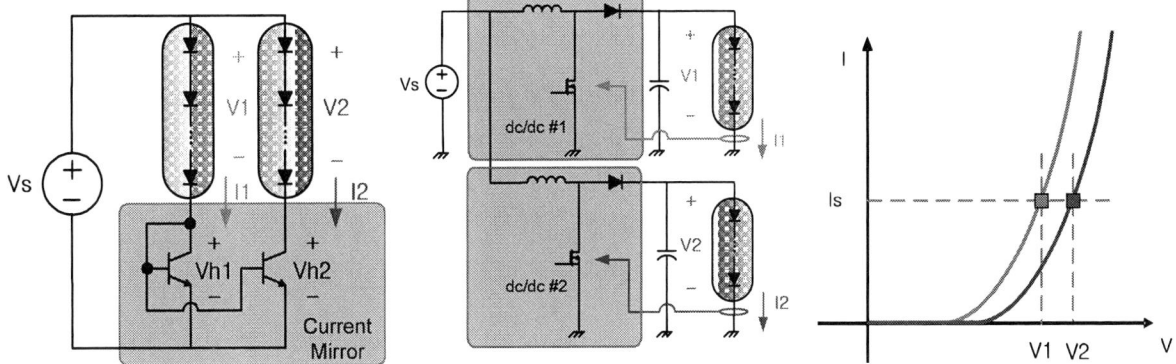

Figure 2. Conventional method (a) Linear current-mirror, (b) Dedicated dc/dc converter, (c) I-V Curves in both methods

uniform and the color spectrum can be shifted from the desired value. An LED used in this manner may also exceed the absolute maximum current in the datasheet, which will damage the device.

To solve this problem, a resistor can be placed in series with each LED to minimize current differences as shown in [4]. However, this method cannot balance the circuit without dissipating significant loss in the balancing resistors. More reasonable approaches can be categorized by the following two methods: a linear regulator method and a switching regulator method.

A. Linear current-mirror regulator method

To equalize the currents in each multiple LED array, a current mirror-based linear regulator method shown in Fig. 2(a) is very widely used. In this driving method, the load current is balanced by a β-gain of transistor switches and each transistor makes a nearly equal current in each LED arrays. With same voltage source, the transistor forces a different operating voltage on each LED arrays. The voltage difference between the supply voltage and each lamp, V1 or V2, is applied to

individual transistor. This extra voltage called "headroom voltage" causes a serious power loss, which is given by load current multiplied by the voltage. The power dissipation can take a large portion of the circuit loss either by higher load current or large deviation of the operating voltage between the lamp loads. Because of this kind of driver loss, this approach is only suitable for low current or low voltage lamp arrays. When the number of LED connections is significantly higher, or a large power is required, another method should be adopted.

B. Dedicated dc/dc converter method

To reduce the power dissipation of the balancing circuit, separate voltage source can be used to condition each lamp bias point. In this method, the number of dc-dc converters equals to the number of LED strings. The current in each LED array is sensed and its individual feedback controller maintains the output voltage to equalize the current as in Fig. 2(b). On the I-V curve the load condition is maintained by different voltage, V1 and V2, to each LED arrays. Each converter is operating optimally and its efficiency is much

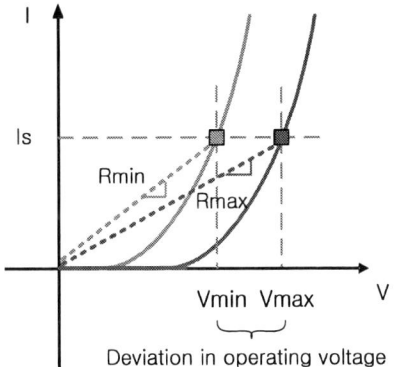

Figure 3. Definition of dc load impedance

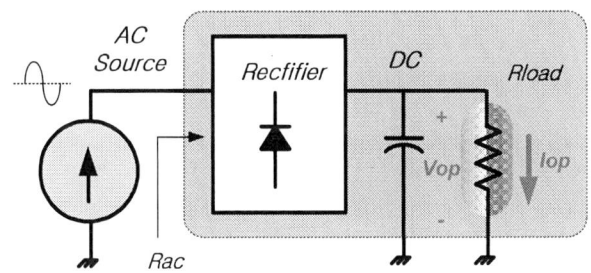

Figure 4. AC equivalent resistance including rectifier and dc load

978-1-4244-4782-4/10 $26.00 © 2010 IEEE   513

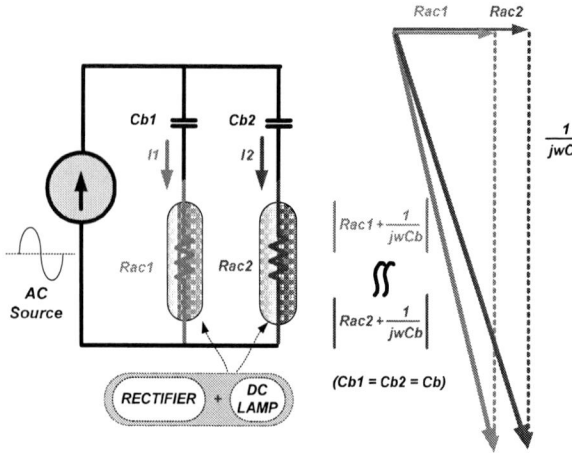

Figure 5. Reactive balancing for the rectifier-driven load

datasheet, two upper and lower load resistance values, R_{max} and R_{min}, are calculated.

B. Capacitive Balancing for paralleled dc loads

Two slightly different dc loads, R1 and R2 for example, can be balanced by placing a lossless reactive component instead of a lossy series balancing resistance. However, to utilize reactive impedance balancing, the dc load should be driven by an ac source coupled with a diode rectifier. The diode rectifier including the load can be modeled by equivalent resistance, R_{ac}, by using energy balance [5]. The basic idea is to use efficient resonant current source and then provide some balancing mechanism using ac impedance balancing. While the reactance can be made by an inductor, which is rather bulky system, this paper uses a capacitive component, which can utilize symmetric feature of the proposed circuit which will be explained later.

As shown in Fig. 5 the current balancing can be achieved by balancing capacitors and the current in each branch can be obtained as in (1) below. The impedance vector diagram shows that the ac current in each equivalent resistance is almost the same if the reactance of C_b is sufficiently larger than the load resistance itself.

better than the linear regulator method. However, the circuit is complex and far more expensive. To reduce the cost and the loss of the circuit simultaneously, a new method of driving for multiple dc loads is proposed in the next section.

$$\frac{I_2}{I_1} = \frac{\frac{1}{j\omega C_{b1}} + R_{ac1}}{\frac{1}{j\omega C_{b2}} + R_{ac2}} \tag{1}$$

III. THEORY OF REACTIVE BALANCING FOR DC LOADS

A. Modeling of load impedance

As mentioned above, LEDs from even the same production lot have poorly matched I-V characteristics due to statistical distribution. The lamp curve is also strongly dependent on the operating temperature. Lamp vendors usually specify the operating voltage tolerance in a constant test current in a specified operating temperature. Equivalent impedance for the lamp can be defined as the ratio of the operating voltage to the test current as in Fig. 3. From the

However, in real situation, the tolerance in the capacitance can also influence the balancing performance. Taking that factor into account, the errors in the current between the two branches can be estimated. If α is the tolerance rate of the balancing capacitors which is normally 5%, and β is the tolerance of the load impedance which is dependent on the load, the current deviation ratio can be calculated as in (3).

$$C_{b2} = C_{b1}(1 \pm \alpha), \quad R_{ac2} = R_{ac1}(1 \pm \beta) \tag{2}$$

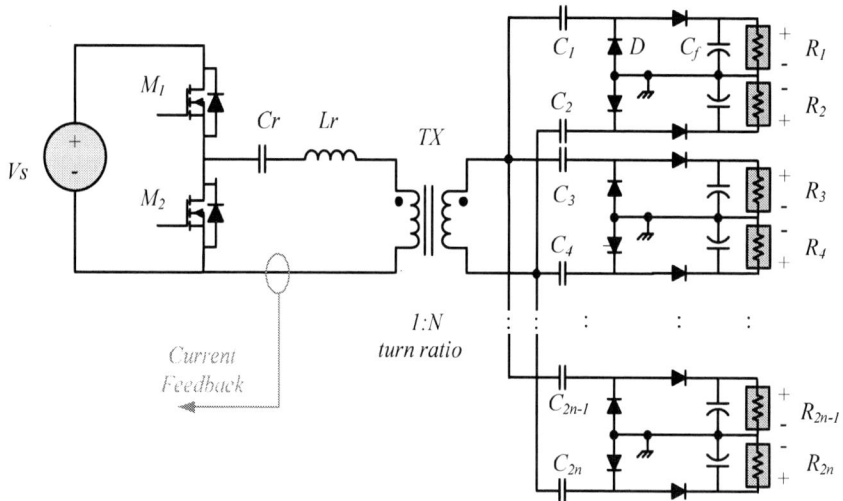

Figure 6. Proposed symmetric current balancing driver for multiple dc loads

$$D \equiv \frac{|I_1 - I_2|}{I_1} = \frac{\sqrt{\alpha^2 q^2 + \beta^2}}{\sqrt{q^2(1\pm\alpha)^2 + (1\pm\beta)^2}} \quad (3)$$

where, $q \equiv \dfrac{1}{\omega C_{b1} R_{ac1}}$ (called "reactance factor")

IV. ANALYSIS OF PRINCIPLE OF OPERATION

As a generalized implementation of the proposed balancing mechanism, Fig. 6 shows the symmetric current balancing driver for 2n number of dc loads. The source side consists of a series resonant branch and half-bridge MOSFET switch pair and operates above the resonance frequency to supply a pre-determined current to the secondary side. Secondary side consists of balancing capacitors, current-driven symmetric half-wave rectifiers, filtering capacitors, and dc lamp loads.

The fundamental principle of individual lamp current balancing is guided by the following three rules: i) total current control, ii) symmetry of even and odd branch, iii) capacitive impedance balancing mechanism.

A. Total current control

Instead of sensing the current in each branch, current control loop regulates total primary current such that the sum of the secondary side current should be constant. Let the magnitude of secondary sinusoidal current, I_T, be always constant as (4) for the following analysis.

$$i_T(t) = I_T \sin(\omega t) \quad (4)$$

Applying fundamental harmonic approximation and energy balance equation [5] to the simplified rectifier circuit in Fig.7 the ac equivalent resistance $R_{ac,k}$ is derived as (5) when individual lamp resistance is R_k.

$$R_{ac,k} = \frac{4}{\pi^2} R_k, \quad k = 1,2,\ldots,2n \quad (5)$$

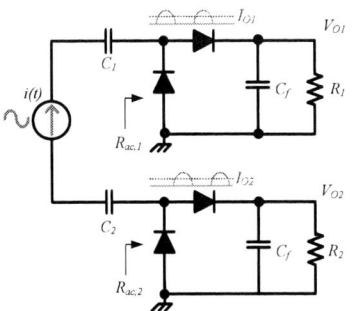

Figure 7. Derivation of ac equivalent resistance

B. Symmetry of even and odd branch

From the above results, overall equivalent circuit for the proposed circuit is depicted in Fig. 8. The circuit has even/odd symmetry in the sense that the odd branches are activated in positive cycle of the source and the even parts are energized only in negative cycle.

For odd branch, the following current relation holds.

$$i_T(t) = \begin{cases} I_k' \sin(\omega t) & (0 \le \omega t < \pi) \\ 0 & (\pi \le \omega t < 2\pi) \end{cases} \quad (6)$$

where $k = 1,3,5,..,(2n-1)$

Similarly for even branch the current is given by (7)

$$i_T(t) = \begin{cases} 0 & (0 \le \omega t < \pi) \\ I_k' \sin(\omega t) & (\pi \le \omega t < 2\pi) \end{cases} \quad (7)$$

where $k = 2,4,6,\ldots,2n$

Applying again the energy balance equation in (7), the the current on the ac side is related to the individual dc load current as shown in (8), where I_k' is the magnitude of the current that flows in $R_{ac,k}$ and I_k is the magnitude of the current in R_k.

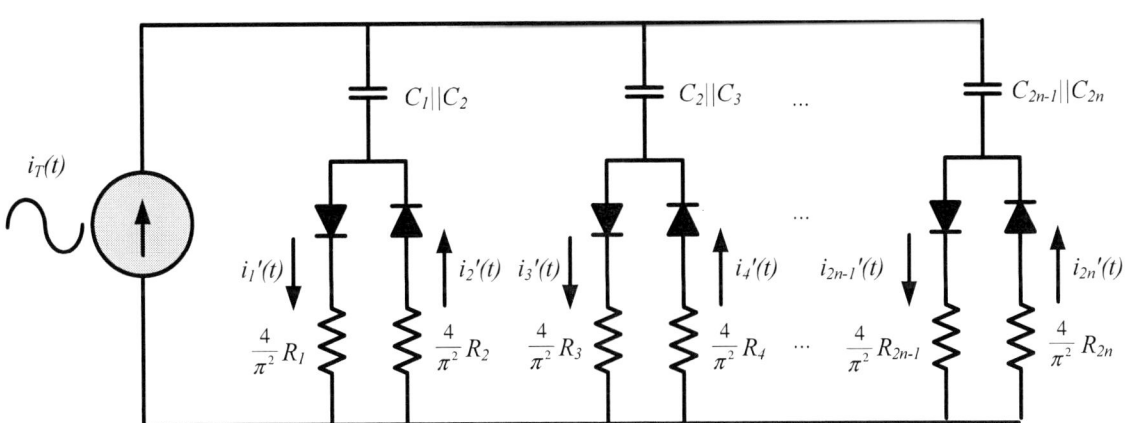

Figure 8. AC equivalent model for the proposed circuit

$$I_k' = \pi I_k, \quad k = 1,2,\ldots,2n \tag{8}$$

Due to the inherent symmetry of the half-wave rectifier, partial sum of currents in the even branch and that of currents in the odd branch have the same value as in (9).

$$\sum_{k,odd} I_k' = \sum_{k,even} I_k' = I_T \tag{9}$$

C. Capacitive impedance balancing mechanism

As can be seen from the equivalent circuit, the slightly different dc loads, $R_{ac,k}$, are balanced by series capacitors, $C_{b,eff}$, and the ac current in each equivalent resistance is almost the same if the impedance of $C_{b,eff}$ is made sufficiently larger than the load resistance, thus meaning that the following holds.

$$\left|I_i' - I_j'\right| \to 0 \text{ for even } i,j \text{ with } i \neq j \tag{10}$$

$$\left|I_i' - I_j'\right| \to 0 \text{ for odd } i,j \text{ with } i \neq j \tag{11}$$

From (9) ~ (11), all the dc load currents are balanced to have nearly the same value as in (12).

$$\left|I_i' - I_j'\right| \to 0 \text{ for } i,j = 1,2,\ldots,2n \text{ with } i \neq j \tag{12}$$

Combining (12) with (8) verifies the balancing operation (13) of the proposed circuit.

$$I_i \approx I_j \approx \frac{I_T}{n\pi} \text{ for } i,j = 1,2,\ldots,2n \text{ with } i \neq j \tag{13}$$

V. DESIGN PROCEDURE

To design the whole system, dc load impedance, R_{load}, should be found from datasheet. Most LED vendor specify the upper and lower limit for the operating voltage, V_{op}, in the constant operating current, I_{op} at the specified temperature condition. Table I shows the excerpted datasheet of the sample used in this paper. The maximum and minimum equivalent dc impedance, $R_{load,max}$ and $R_{load,min}$, can be calculated as (14). With the same load module, all of the load impedance can be regarded as sharing the same tolerance range.

$$R_{load,max} = \frac{V_{op,max}}{I_{op}} \qquad R_{load,min} = \frac{V_{op,min}}{I_{op}} \tag{14}$$

To design the balancing network, the ac equivalent resistance, R_{ac}, including rectifier is then calculated assuming sinusoidal driving condition.

$$R_{ac} = \frac{4}{\pi^2} R_{load} \tag{15}$$

For simplicity and without any loss of generality, it is assumed that the balancing components in Fig. 6 is designed with the same value as (16)

$$C_1 = C_2 = \cdots = C_{2n} = 2C_b \tag{16}$$

TABLE I. EXEMPLARY DATASHEET FOR A LED ARRAY SAMPLE

Symbol	Min	Typ	Max	Unit	Test condition (I_{op})
Vop	164.3	-	177.7	V	@ 95mA, 25°C

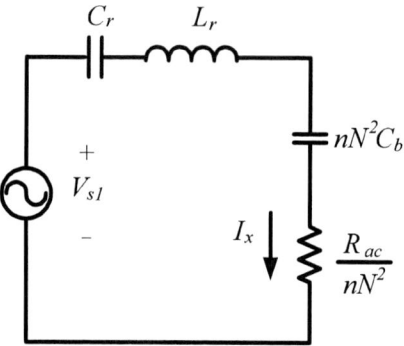

Figure 9. Simplified circuit for the overall circuit

From the required tolerance range specification in the load current which determines the overall brightness uniformity of the backlight unit, the required reactance factor, q, is calculated by (3). By choosing a relevant switching frequency, ω, effective balancing capacitor value, C_b, is calculated from the definition of the reactance factor.

To investigate the gain characteristics of the proposed converter, the simplified circuit in Fig. 9 has been derived from Fig. 6. In the mean time, all the components in the transformer secondary have been reflected to the primary side and all the 2n numbers of the identical balancing networks including rectifiers and loads are merged into equivalent components. This converter can be regarded as a series resonant network when applying fundamental frequency and high Q approximation of the resonant tank as in (17) and (18); the resonant frequency and the loaded quality factor can be defined as in (19) and (20).

$$v_{s1}(t) = \frac{2}{\pi} V_s \sin(\omega t) \tag{17}$$

$$I_x = n\pi N I_{op} \tag{18}$$

$$\omega_o = \frac{1}{\sqrt{L_r(C_r \| nN^2 C_{b,eff})}} \tag{19}$$

$$Q = \frac{nN^2}{R_{ac}} \sqrt{\frac{L_r}{(C_r \| nN^2 C_{b,eff})}} \tag{20}$$

The trans-conductance gain from input dc voltage, V_s, to output dc current, I_{op} is derived as (21).

$$\frac{I_{op}}{V_s} = \frac{N}{2 R_{load} \sqrt{1 + Q^2 (\frac{\omega}{\omega_o} - \frac{\omega_o}{\omega})^2}} \tag{21}$$

To achieve the constant load current I_{op}, even with input voltage change from $V_{s,max}$ to $V_{s,min}$ and load tolerance change condition from $R_{load,max}$ to $R_{load,min}$, the operating frequency should be changed within operating curve which is overlapped on the gain curve shown in Fig.10. Finally, the resonant tank value L_r, C_r and the transformer turn ratio, N, can be calculated from the equation (21) and the gain curve.

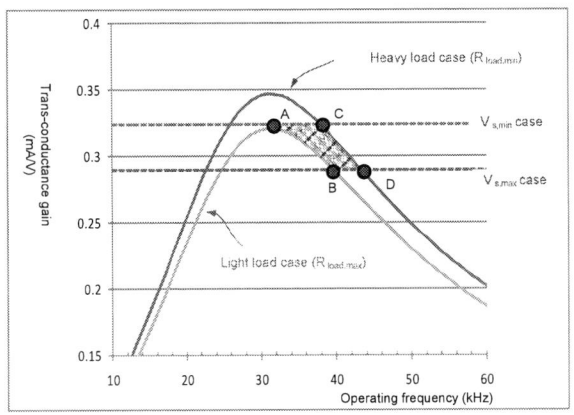

Figure 10. Gain curve for design procedure

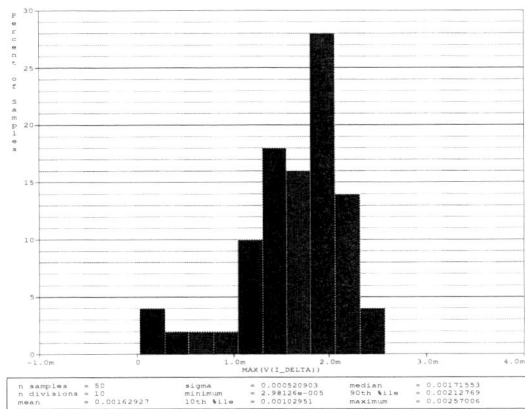

Figure 12. Performance analysis using PSpice

To obtain zero-voltage-switching in the half-bridge switches, the operating frequency should be placed above the resonant frequency. Optimally, operating point A ($V_{s,min}$, $R_{load,max}$) should be placed on the right side of the peak resonant point of the light load curve as shown in the figure. Worst case efficiency degradation occurs in the operating point D ($V_{s,max}$, $R_{load,min}$) where the operating frequency is far from the resonant frequency. In this case, different Q design can reshape the sharpness of the curve and change the circulating current level in the resonant tank to improve the efficiency.

To achieve the current regulation, total current of primary side is sensed and fed back to the frequency control loop as in Fig. 11. Detailed compensator design for the frequency control loop is well established in other literature [6] and is out of scope of this paper.

A prototype balancing converter for six-channel (n=3) LED arrays is designed by using above procedure and major component values are listed in Table II. The input voltage range is 340~380V and the operating frequency is around 40 kHz. The target current in each lamp is set to 95 mA. In the following section the performance of the proposed circuit is verified by simulation and hardware using this prototype.

VI. PERFORMANCE ESTIMATION

As mentioned before, current deviation between any two branches can be estimated by (3). It can be guessed that the worst case occurs when only one load has its minimum value and all the other loads have their maximum value. However, this estimation is valid only for the comparison between any two branches and shows approximated value for the worst case deviation.

To estimate overall balancing performance for multiple loads, simulation tools providing statistical worst-case analysis can be used. In this paper, a Monte Carlo simulation has been performed using PSpice. In the simulation, the component values in Table II are used and balancing capacitors are assumed to be constant. The upper and lower limit value $R_{load,max}$ (1.73 kΩ) and $R_{load,min}$ (1.87 kΩ) are obtained using (14) and Table I thus the device tolerance parameter (DEV) is set to 3.9% for each load resistor. The number of trial is 50 and the distribution is assumed to be Gaussian.

The results are listed in the Fig. 12 and the maximum deviation between the currents in each lamp load is nearly 2.57mA (2.7%). In real situation, because of the tolerance of other components, the current balancing can be worse than the

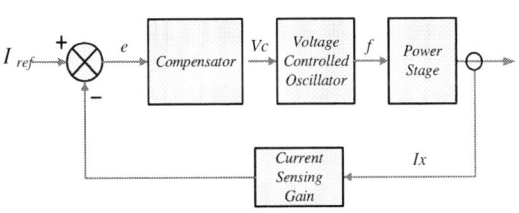

Figure 11. Feedback control loop

TABLE II. MAJOR COMPONENTS LISTS

Location	Description	Value	H/W Implementation
C_1~C_6	balancing cap.	4.7nF	600V/film
Cr	dc block cap.	1000nF	400V/film
Lr	series inductor	1.3mH	EFD4549 (gapped)
N	turn rato	1.2	
C_f	output cap.	22uF	250V/electrolytic
M_1,M_2	MOSFET	-	FDPF12N50
D	rectifier	-	1N4007

978-1-4244-4782-4/10 $26.00 © 2010 IEEE

Figure 13. Prototype hardware implementation

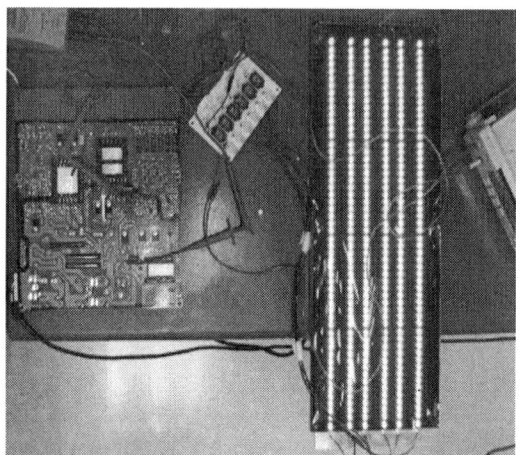

Figure 14. Six-LED array driver system

estimation. Even in this case, the slightly decreased balancing capacitance value from the original calculation, which means increased reactance factor, can be adopted to enhance the current balancing performance.

VII. EXPERIMENTAL RESULTS

To verify the operation of the proposed circuit, a 100W hardware prototype to drive six channels (n=3) of LED load (typically equivalent to 1.8kΩ @ 95mA) is constructed. A key portion of the hardware prototype is shown in Fig. 13 and Fig. 14 and the measured load currents in each load are listed in Table III. To simplify the circuit, the series inductance and the transformer can be integrated to be a leakage inductance can be utilized using LLC transformer using technique as discussed in [7].

Even with a load resistance tolerance of 7.8%, the measured current deviation can be forced to be within 2.6% with our proposed approach. The experimental results not only verify the underlying current balancing principle but also show the performance of the proposed topology.

VIII. CONCULSIONS

In this paper, a new current balancing driver circuit for multiple DC loads is proposed. A smart combination of an inherent symmetry of circuit and capacitive balancing mechanism enables an efficient and cost-effective current balancing. A detailed analysis of the operating principle is performed to generalize the proposed scheme thus showing that the balancing performance can be finely adjusted by the reactance factor. Experimental results of a 100W six-channel LED lighting system verify the feasibility of the proposed balancing scheme.

REFERENCES

[1] X. Jin, *Balancing Transformers for Ring Balancer*, U. S. Patent 7,294,971, Nov., 2007

[2] H. van der Broeck, G. Sauerlander, and M. Wendt, "Power driver topologies and control schemes for LEDs," Proc. IEEE APEC, 2007, pp. 1319-1325

[3] Y. Hu, and M. M. Jovanovic, "A novel LED driver with adaptive drive voltage," Proc. IEEE APEC, 2008, pp. 565-571

[4] M. Khatib, "Ballast Resistor Calculation – Current Matching in Parallel LEDs," Texas Instrument Application Report SLVA325, 2009

[5] M. K. Kazimierczuk, and D. Czarkowski, *Resonant Power Converters*, John Wiley & Sons, Inc., New York, 1995

[6] E. X. Yang, F. C. Lee and M. Jovanovic, "Small-Signal Modeling of Series and Parallel Resonant Converters," Proc. IEEE APEC, 1992, pp. 785-792

[7] B. Yang, *Topology Investigation for Front End DC/DC Power Conversion for Distributed Power System*, Ph.D. Dissertation, Virginia Tech, Blacksburg, VA, Sep. 2003

TABLE III. SUMMARY OF EXPERIMENTAL RESULTS

	R_{load} (kΩ)	Measured Current(mA)	Error (%) *
1	min (1.73)	97.5	+2.6%
2	max (1.87)	94.8	-0.2%
3	max (1.87)	94.9	-0.1%
4	max (1.87)	96.4	+1.5%
5	max (1.87)	95.0	0%
6	max (1.87)	95.9	+0.9%

* Deviation from the rated value of 95mA

A Simple Method for Configuring Multi-PWM Channels for Multi-level Converter Applications Based on PWM IP core

Haibing Hu
Aero-Power-Sci-Center
Nanjing University of Aeronautics and
Astronautics, Nanjing, 210016, China
huhaibing@163.com

Xiaodong Ding, Tao Xue
Nanjing SUTE ELECTRIC Co., LTD
Nanjing, 210016, China

Wenxi Yao, Zhengyu Lu
National Key Lab. of Power Electronics
Zhejiang University
Hangzhou, 310027, China

Abstract- **In this paper, based on the PWM Intellectual Property(IP) core, a simple method for configuring multi-PWM channels for multi-level converter applications is proposed. Each PWM channel has the same timing base and its own control registers. Employing building block method, multi-PWM channels can be easily assembled by connecting PWM IP cores with a simple logic circuit. Having these features, the multi-PWM channels are ready to realize different modulation strategies for multi-level converters. The configured PWM generator has a friendly interface to MCUs or DSPs. A design example, which is applied to a three-phase-three-level cascaded inverter, is given to demonstrate the simplicity of the proposed method. The design method can facilitate the control circuit design for multi-level converters, enhance system reliability and guarantee the synchronization of PWM signals.**

I. INTRODUCTION

Extensive research has conducted on pulse width modulation(PWM) techniques during the last few decades[1-2]. With advent of new topologies, some new PWM techniques are still coming up. Currently multi-level topologies have caught a great attention in the practical applications, for they improve output waveform quality of the voltage source converter by increasing waveform steps and reduce voltage stress across switches. As a result, the low dv/dt and less EMI problems can be easily solved in multi-level converters[3-4].

However, as levels increase, the number of PWM signals required for switch gates increases as well. For example, 24 PWM signals with a dead time are required in a three phase two-level cascaded inverter[5], while the three-phase-three-level inverter needs 48 PWM signals[6]. In[7], in order to increase the equivalent switch frequency and improve output waveform quality, four three-level H bridges need 32 PWM channels with certain phase shift when interleaved. At present, no single PWM chip commercially available can generate so many PWM channels. In some motor-control based DSPs, such as TMS320F2407 and TMS320F2812, there are at most 16 PWM channels available. Therefore in above applications,

these chips are required to expand PWM channels, which will complicate the control circuit design and reduce the system reliability. What is more, it is difficult to synchronize all these PWM signals with different oscillators.

Thanks to the advancement of FPGA technology, multi-PWM generator configured by FPGA is practicable and affordable. In this paper, a simple method for configuring multi-PWM channels using FPGA is proposed based on the PWM IP(Intellectual Property) core. We can use building block approach to construct arbitrary number of PWM channels when needed. Each PWM has its own control registers for setting switching period, dead time and PWM modes. Having these features, the configured multi-PWM channels can easily realize the phase shift PWM signals. A 24-channel PWM generator is designed as an example and applied to the three-phase-three-level cascaded inverter.

II. DESIGN of a UNIVERSAL DIGITAL PWM IP CORE

A. Overall Architecture

Fig.1 illustrates the general structure of the universal PWM IP core, whose function is similar to the built-in PWM generator in DSPs from Texas Instrument. The IP core consists of five registers - period register, compare register, dead time register, counter register and control register. Functional blocks in Fig.1 are explained briefly as follows.

Decoder and interface: This block is designed to facilitate the interface with DSPs or MCUs. Commands to the registers are routed through the decoder and interface circuit. The control parameters can be fed by externally interfaced DSPs or MCUs. The interface composes of 16-bit data bus, 3-bit address bus for selecting one of five registers, 1-bit chip select signal to enable access to the chip, and 2 write/read signals. The timing sequences for controlling these five registers are similar to most of commercial digital IC chips, which make it flexible to interface to off-the-shelf MCUs or DSPs.

Research was sponsored by the Doctoral Program of Higher Education of China (200802871040).

978-1-4244-4782-4/10 $26.00 © 2010 IEEE

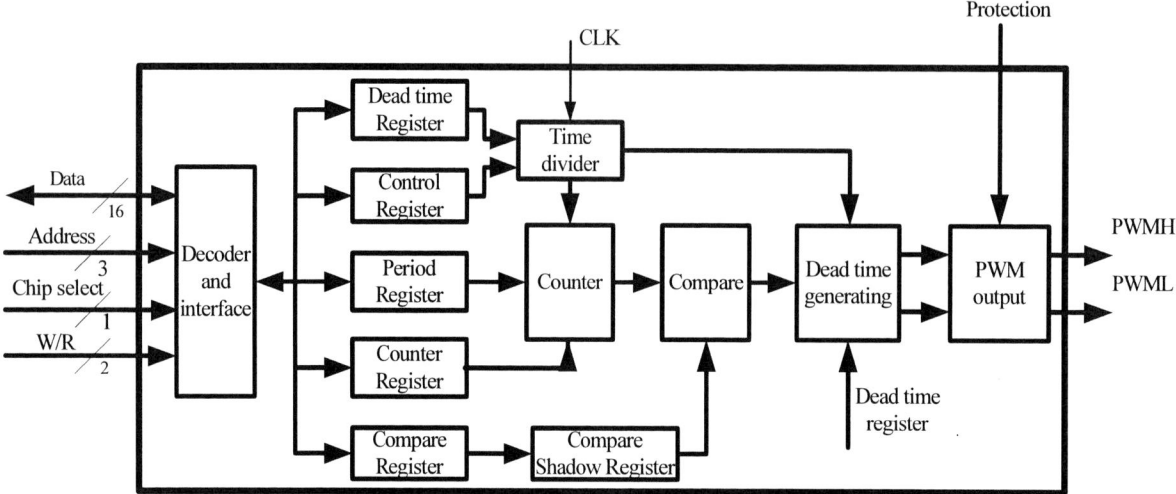

Fig.1 Overall architecture of universal PWM IP core

Counter unit: In this disign, a 16-bit counter is employed, whose counting modes can be set as continuous up-counting mode and continuous up-/down-counting mode through the control register. The timing base for the counter is decided by the scaled CLK. The scale factor, which varies from 128 down to 1, can be set by the control register as well. The period of the counter is determined by the value set in the period register.

Compare shadow register: The purpose for designing this register is to ensure that the value in the compare register can be loaded into the shadow register only when the underflow or overflow of the counter occurs, In this manner, the phenomenon of occurrence of unexpected PWM signal, which is caused by immediately loading the value of compare register into the compare unit as shown in Fig.2, can be avoided.

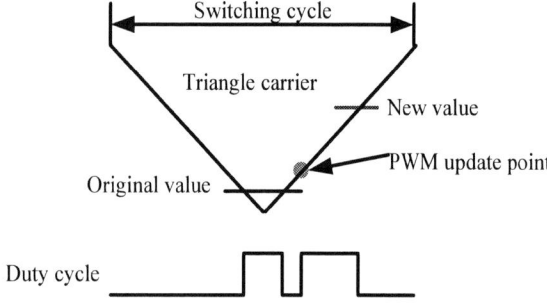

Figure 2: The case of occurrence of unexpected PWM signal

In digital control system, the updated PWM value(new PWM value) may be calculated out at a fixed point as marked as a point in Fig.2. Suppose the original PWM value is lower than the value in updating point, at which point the new value, which is greater than that at updating point, is calculated out and updated immediately, resulting in unexpected PWM signal.

Compare unit: The unit is used to generate the PWM signal by comparing the values in the counter unit and the compare shadow register.

Dead time generating circuit: This unit is relatively complicated. Next part will demonstrate the dead-time generating circuit.

PWM output unit: The unit has two functions. One is for setting PWM output modes, which are complementary PWM modes with and without dead time, and the PWM polarity selection mode, while the other is for disabling PWM signals when the protection signal or the reset bit in the control register is set.

B. Dead Time Generating Circuit

In this paper, a method is employed to create the dead time as seen in Fig.3, where CMPR, a 16-bit value, compares with PWM counter to generate PWM signal. The PWM_H and PWM_L signals are a pair of complementary PWM signals with a dead time. From Fig. 3, it is easy to draw the following two rules to generate the dead time.

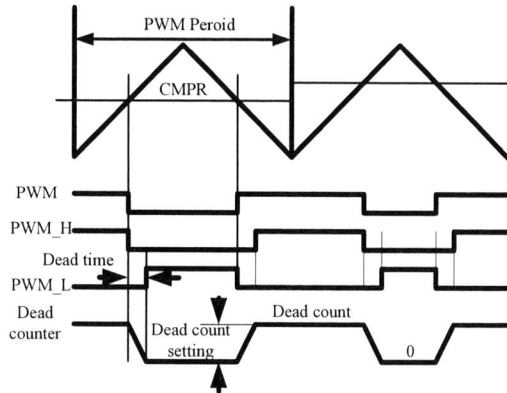

Figure 3: Timing diagram for generating dead time

(1)PWM=1. If the value in the dead time counter equals to the setting value in the dead time register, the dead time counter keeps unchanged, otherwise the dead time counter keeps on counting until it equals to the setting value.

(2)PWM=0. If the dead time counter is 0, the dead time counter stops counting, otherwise it counts down to 0.

According to above rules, the circuit to generate the dead time is not difficult to design using hardware description language (HDL). Fig. 4 depicts the simulation result of the dead time generation circuit.

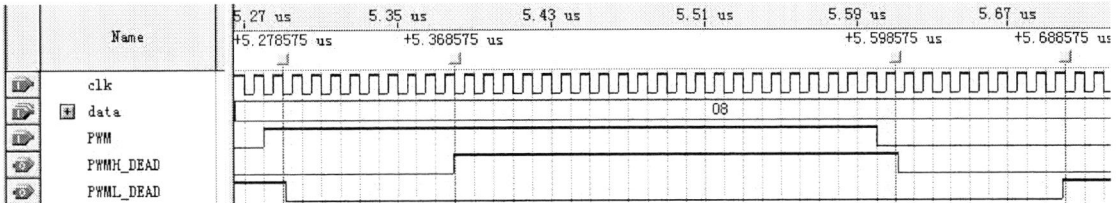

Figure 4: Simulation results of the dead time generation circuit

III. CONFIGURATION of MULTI-PWM CHANNELS USING PWM IP CORE

Fig.5 shows the basic structure of the multi-PWM channels. In this structure, the PWM IP cores are connected by a 16-bit data bus and a 5-bit control bus(address and W/R signals), while the CLK signal feeds all the IP cores, making sure that all the PWM IP cores have same timing base. Since each PWM IP core has its own control registers, a chip selection is desirable to feed or fetch commands to or from the control registers in different PWM IP cores. Therefore, the only circuit needed to be designed is a simple decoder circuit to generate chip selection signals for each PWM IP core. After finishing bus connection and a simple decoder circuit design, the multi-PWM channels are already configured and are ready to be compiled, synthesized and downloaded to the corresponding FPGAs. As seen from Fig.5, each PWM IP core is independent of others, except for having the same timing base. With these features, it is very convenient to realize different PWM modulation strategies using the configured multi-PWM channels, such as carrier phase shift PWM and etc. and is as well very easy to keep all the PWM signals in synchronization.

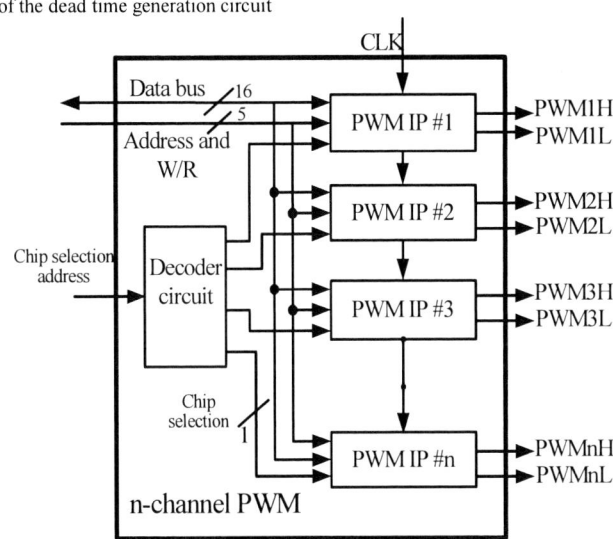

Fig.5 Multi-PWM channels constructed based on PWM IP cores

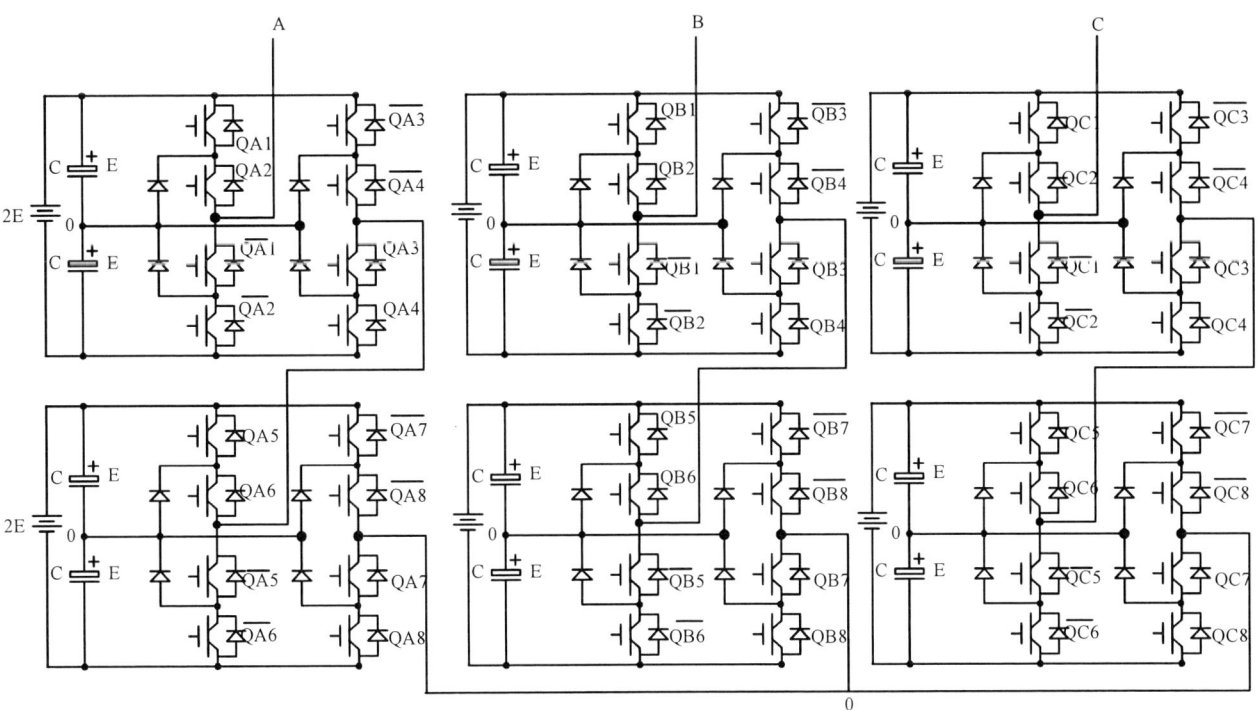

Fig.6 The topology of the three-phase-three-level inverter

978-1-4244-4782-4/10 $26.00 © 2010 IEEE   521

IV. 24 CHANNEL-PWM GENERATOR CONFIGURED FOR THREE-LEVEL CASCADED TOPOLOGY

Based on the method mentioned in Section III, we construct a 24-channel PWM generator for three-phase-three-level converter. The topology of the converter is shown in Fig. 6. The carrier phase shift PWM modulation strategy is employed. A Cyclone FPGA (EP1C6) from Altera,Inc. was chosen to implement the multi-PWM generator. The hardware resources dedicated to the PWM generator are summarized in Tab.1. The digital platform equipped with DSP and FPGA functions as the application platform for the PWM generator.

The external clock 15MHz from DSP output clock feeds to the FPGA, in which the 60MHz internal clock is generated by its internal phase-locked loop (PLL) circuit. The internal clock of DSP is set to 60MHz as well. The modulation frequency is set to 50Hz, while the carrier frequency is set to 5kHz. All these settings can be achieved by feeding setting values to corresponding registers in the multi-PWM generator

Table 1. Hardware Resource Used in the PWM Generator

EP1C6-8	Resources on chip	Used resources	Usage(%)
Les	5890	3279	55.7%
Pin	98	50	51%
Memory bits	921600	0	0
PLLs	2	1	50%

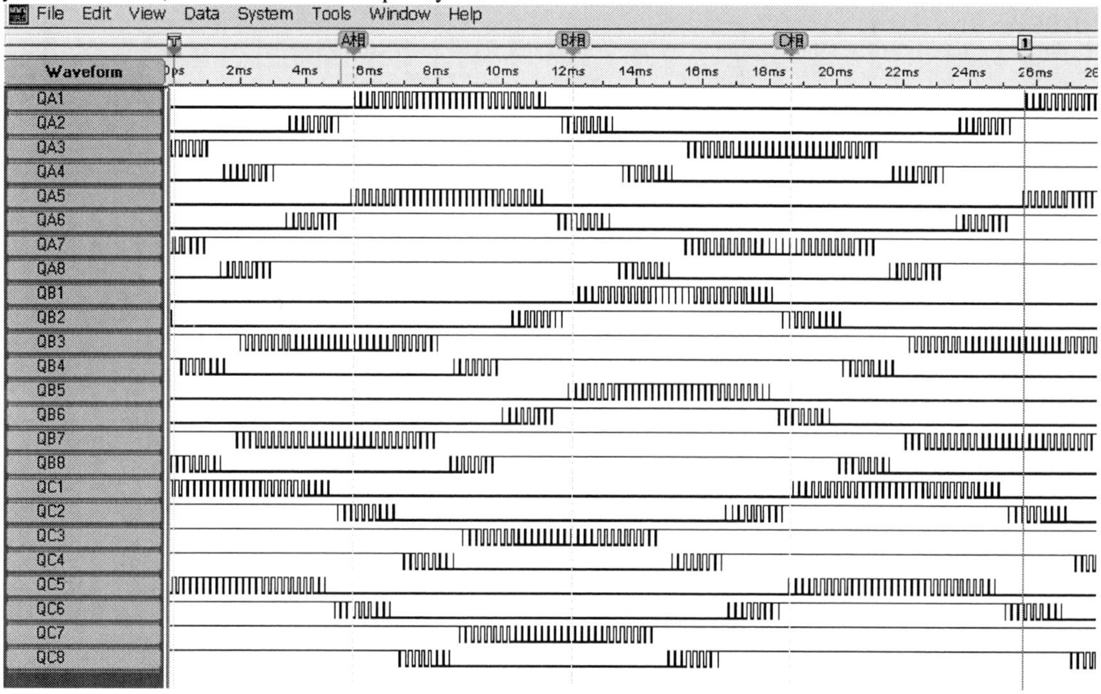

through the DSP. In the DSP algorithm, the modulation waveform can be generated using a look-up table, in which a waveform period with the length of 100 points is stored. Each period interruption will trig the DSP to increment the look-up table pointer by one and then feed the modulation index to the 24-channel PWM generator. Fig. 7 shows the 24-channel PWM waveforms measured by the logic analyzer(TLA5202). The measured results verify that the 24-PWM generator can meet the requirements of the complicated circuit topology.

Fig.7 Measured 24-PWM signals

V. ENCODING PWM SIGNALS FOR SERIES TRANSMISSION

In high power application, optical transmission of PWM signals is normally employed due to electrical isolation and good immunity to EMI. In this design, 24 PWM signals are needed, which means at least 24 optical transmission lines are required. It will complicate the circuit design, reduce the system reliability and increase cost. To address these issues, PWM signals are encoded to be transmitted in series.

The series transmission architecture is illustrated in Figure 8.As seen from Fig.8, we encode four PWM signals for each three-level H-bridge. Six signals, composing of 4 PWM signals and two control bits, are sent to the parallel-to-series chip at control side, and then are transmitted to the power stage side through optical fiber. At the receive side, received data are converted to a parallel data, and two control bits are used to register the four PWM signals. For simplification, the CLK signal is transmitted to the power stage side to keep the data synchronization when the four PWM signals have changes(rising edge or falling edge comes). The Enable signal is used to control enabling or disabling the transmission of CLK signal. Fig.9 demonstrates the relationship between Enable signal and PWM signal changes.

As seen from Fig.9(b), whenever the any of four PWM signals changes, it will trig the Enable signal to be high and the Enable signal will maintain high for six CLK cycles and go low again, during which period, the four PWM signals are transmitted to power stage side and locked to a register.

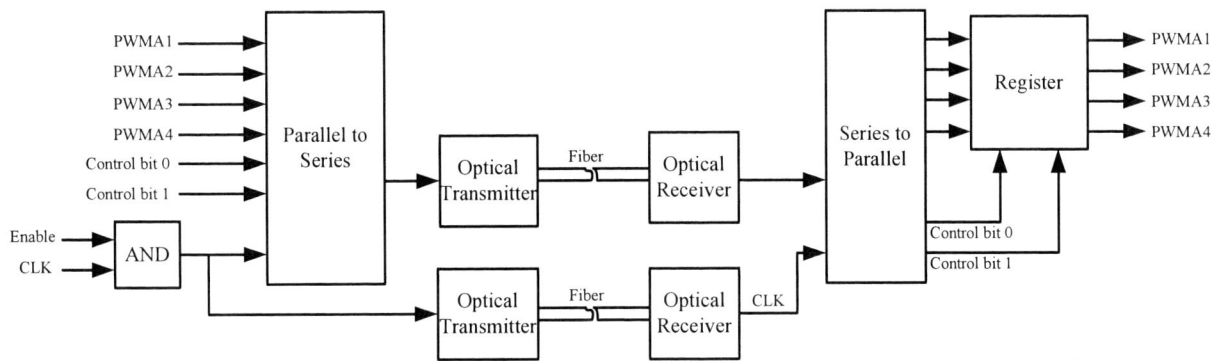

Fig.8 The series transmission architecture

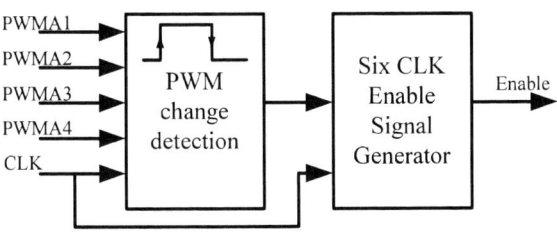

(a) Diagram of Enable generator circuit

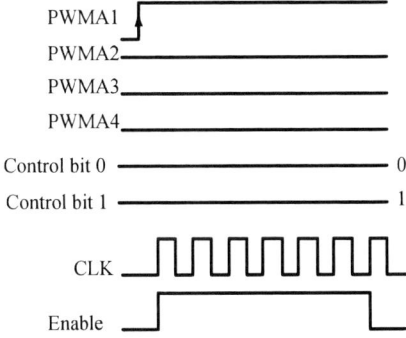

(b) Timing diagram of Enable signal and the input signals.

Figure 9. Enable signal generation

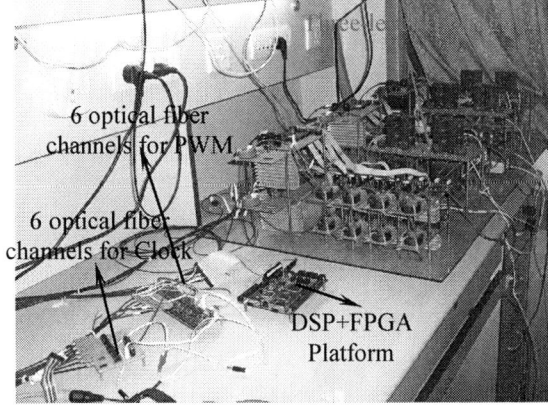

Fig.10 Experiment setup

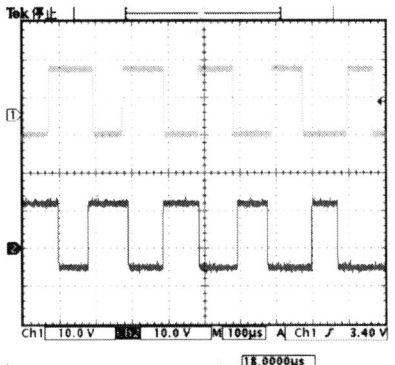

Fig.11 Drive waveforms of up- and down-bridges with 180^0 shift.

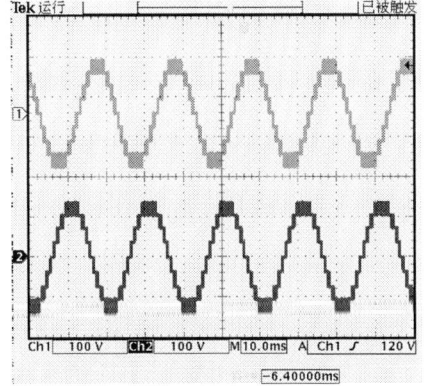

Fig.12 Voltage of phase A and B

VI. EXPERIMENTAL RESULTS

In this paper, the prototype of the three-phase-three-level converter is shown in Fig. 10. As seen from Fig. 11, the drive signals of the bridges of phase A in high and low sides have 180^0 phase shift with their frequency of 5kHz. Fig.12 illustrates the output waveforms of phase A, B under modulation index 0.916. Fig. 13 depicts the line voltage and its spectra. According to theoretical analysis, when carrier waveforms shift 180^0 in phase, the equivalent switch frequency would be doubled. As a result, the harmonics will be concentrated around the double switch frequency. As demonstrated in Fig. 13, the spectra of the line voltage are spread around the frequency of 10kHz. The experimental results are coincided with theoretical analysis perfectly. The experimental result verifies the simplicity of the design of the multi-PWM generator applied to the multi-level converter applications.

978-1-4244-4782-4/10 $26.00 © 2010 IEEE

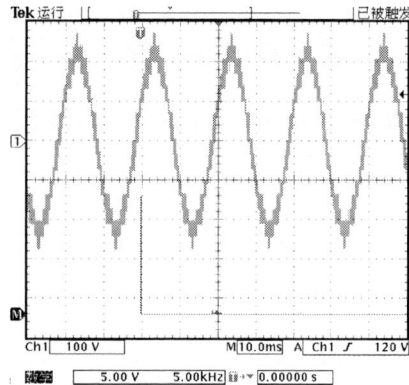

Fig.13 Line voltage and its spectra

VII. CONCLUSION

This paper presents a simple method to configure multi-PWM channels offering a large number of PWM drive signals required in the multi-level topologies. Each PWM IP core has the same timing base and its own control registers. With these features, it is very easy to realize the synchronization of PWM signals and different PWM modulation strategies. The configured multi-PWM generator is flexible and versatile in the multi-level converter applications. A design example is presented to verify simplicity and conciseness of the proposed method.

[1] Holtz, J. Pulsewidth modulation-a survey. IEEE Trans.on Industrial Electronics,Vol.39(5),Oct.10 1992,pp:410-420.

[2] Sidney R.Bowes, Derrick Holliday. Optimal Regular-Sampled PWM Inverter Control Techniques, IEEE Trans. On Industrial Electronics, Vol.54(2),Jun. 2007,pp:1547-1559.

[3] Jin-Sheng Lai, Fang Zheng Peng. Multilevel converters-a new breed of power converters. IEEE Trans. On Industry Applications, Vol.32(3),May.1996.pp:509-517.

[4] Rodriguez,J., Jin-Sheng Lai, Fang Zheng Peng. Multilevel inverters: a survey of topologies, control, and applications. IEEE Trans. On Industrial Electronics. Vol.49(4),Aug.2002,pp:724-738.

[5] Jianli Li Liqiao Wang Zhongchao Zhang, Multi-PWM Pulse Generator Based FPGA, Proceedings of the CSEE,May,2005, Vol.25(10),pp:55-59.

[6] Ding Kai. Research on Topology and Modulation Method for Hybrid Multilevel Inverter, Huazhong University of Science & Technology, Oct.2004.

[7] Cheng Shijie. Research on Power Amplifier based on DSP Control, Master Dissertation, Zhejiang University, Mar. 2005.

978-1-4244-4782-4/10 $26.00 © 2010 IEEE

Technology Roadmapping for Power Supply in Package (PSiP) and Power Supply on Chip (PwrSoC)

Raymond Foley[1], Finbarr Waldron[2], John Slowey[1], Arnold Alderman[3], Brian Narveson[4] and Sean Cian Ó'Mathúna[2]

[1]Dept of Electrical Engineering	[2]Tyndall National Institute	[3]Anagenesis Inc.	[4]Texas Instruments
University College Cork	University College Cork	El Segundo	Dallas
Cork, Ireland	Cork, Ireland	California, USA	Texas, USA

Email: raymond.foley@ucc.ie

Abstract— **This paper presents a review and summary of the PSMA "PSiP2PwrSoC" Special Project that investigated the technology and performance underpinning recent commercial developments in Power Supply in Package and Power Supply on Chip. The results of the study are based on the identification of more than 28 commercial products, 6 of which were analyzed in detail, both physically and electrically. The methodology of the project is described and some of the salient results of this benchmarking effort are presented. In this work, a representative subset of the available commercial products was selected and a comprehensive physical, electrical and thermal performance analysis was carried out. The main aims of this analysis were to identify the components, materials and assembly technologies used, and to determine if the drive towards greater integration and higher power density affected the performance of newer devices. The results of the analysis were then used to determine the current state of the technology in this application space, to show how it has developed to date and to predict how it might progress in the future. These results are presented in a generic format that does not identify individual products. This project was co-sponsored by the PSMA and member companies. The final report of the project, which includes more detailed information on the reviews described here as well as considerable trending analysis, is now available.**

I. INTRODUCTION

Over many years, consumers have become accustomed to miniaturization and increased levels of functional integration in electronic devices. For example, in the last decade, cell phones have transformed from having a simple monochrome display and basic call capability into feature-rich computers with GPS, email, internet connectivity, PDA functionality, cameras, music players and high resolution touch screens with vivid colors. Of course, this has not happened by chance; a great deal of time, effort and money has been expended in adding functionality whilst shrinking form-factors to fractions of their former size. The power delivery path to these functions has not been immune to this trend, and there is

concerted effort in the power supply industry to reduce the area and volume occupied by power sources whilst increasing their power density [1][2]. The motivation for this, in space-constrained and miniaturized applications, is clear, but reduced costs, potentially higher reliability (due to lower component counts) and simplified design-in for customers also provide compelling reasons to move towards more integrated solutions whenever possible. From a manufacturing end-user perspective, this can allow lower assembly and inventory costs, and, from a consumer end-user perspective, it amounts to smaller, more reliable and cheaper devices.

Currently the power supply industry is at an inflection point with a clear move away from the traditional power conversion modules towards products based more directly on semiconductor and microelectronics technologies. Power conversion and supply solutions are now beginning to be delivered in functionally-integrated hardware based on system-in-package (SiP) and system-on-chip (SoC) platforms. One of the factors driving this trend is the ability to achieve increased levels of integration through research advances in semiconductors, magnetics, capacitors and packaging technologies [3-10]. These trends were identified in the PSMA market and technology report "Power Supply in a Package, Power Supply on a Chip" [11] (PSiP2PwrSoC Phase I) published in February 2008 and further verified by the participants at the 1st International Workshop on Power Supply on Chip [12] held in September 2008. These provide a context for this work, in which we have undertaken the first published benchmark study of the technologies and integration trends currently in evidence in commercial PSiP and PwrSoC products [13]. Also, we have undertaken electrical and thermal performance evaluation to correlate advances in technology with metrics such as regulation and efficiency. From this work, it is clear that the trends evident to date point towards the ultimate development of efficient, highly miniaturized, point-of-use power converters in both power-supply-in-package (PSiP) and power-supply-on-chip (PwrSoC) form factors.

This work was sponsored by the Power Sources Manufacturers Association, Fairchild Semiconductor Corporation, Crane Aerospace and Electronics, Murata Power Solutions, ON Semiconductor Corporation and Leader Electronics

978-1-4244-4782-4/10 $26.00 © 2010 IEEE

The following are the technology integration definitions are used in this paper:

- **PSiP**: Separate chips (switches drivers, controller) within the same package with external inductor/passives

- **PwrSoC**: A single chip (containing switches, drivers and controller) with integrated inductor/passives

- **Power IC**: A single chip (containing switches, drivers and controller) with external inductor/passives

II. METHODOLOGY

This project had two main thrusts that were investigated simultaneously:

- Technology evaluation and benchmarking involving the assessment of the physical, electrical and thermal aspects of a selected subset of commercially available devices

- Technology trending and road-mapping based on a wide array of commercially available products

A. Technology evaluation and benchmarking

The aim of this part of the project was to establish the level of integration and miniaturization which has been achieved in state-of-the-art PSiP and PwrSoC products as well as benchmarking the electrical and thermal performance under consistent operating conditions. The evaluation and benchmarking effort consisted of the following tasks:

1) Selection of power supplies for investigation: In PSiP2PwrSoC Phase I, a number of commercially-available products were identified as being either PSiP or PwrSoC devices. During Phase II, this database was further updated with new product releases up to December 2008, yielding a total of 28 candidates that were considered for inclusion in the study. Six of these devices were then chosen for further investigation based on criteria such as introduction date, input and output voltage ranges, output current range and assumed level of integration. On this last criterion, it was decided that the product selection should include at least two examples of products which were considered to likely be "PwrSoC" devices. All of the chosen products are in the 0.6 V to 5.0 V output voltage range, however there is a considerable range of sizes, output current levels and power densities amongst them. The selection also included some established products as well as units which are relatively new to the market. The products are referred to only as products A to F. Various parameters relating to these power supplies are summarised in Table I.

2) Evaluation of Physical, Electrical and Thermal Performance: The physical, electrical and thermal performances of the selected devices were evaluated using both destructive and non-destructive testing under identical conditions where appropriate (including closely matched PCB layouts to enable a fair and accurate comparison of electrical and thermal performance).

TABLE I. SUMMARY OF SELECTED PRODUCTS

Product ID	Year of Intro.	V_{in} Range (V)	V_{out} Range (V)	I_{out} (A)	Package Size (mm)		
					L	W	H
A	2008	2.375 – 5.5	0.75-V_{in}	9.0	12.0	10..0	1.85
B	2005	4.5-28.0	0.6-5.0	12.0	15.0	15.0	2.80
C	2007	3.0-5.5	1.0-5.0	2.2	6.0	4.0	0.85
D	2003	4.5-14.0	0.8-3.63	8.0	17.8	15.2	8.13
E	2007	2.4-5.5	0.6-3.63	6.0	12.2	12.2	7.25
F	2008	7.0-13.2	0.6-2.5	10.0	15.0	15.0	3.50

The evaluation was carried out under the following headings:

a) Physical evaluation:
- substrate technology
- assembly technology
- components (ICs, capacitors, resistors, discrete semiconductors)
- magnetic/inductor technology
- encapsulation method
- termination finish
- dimensions and weight
- moisture sensitivity

b) Electrical performance evaluation:
- efficiency vs. load current
- input voltage, output voltage
- output voltage ripple
- transient analysis
- line and load regulation
- over-voltage/over-current functionality

c) Thermal performance evaluation:
- analysis of heat-transfer paths
- calculation of junction-to-case thermal resistance (Rθjc)
- infra-red imaging under load

B. Technology trending and road-mapping

Technology trending and road mapping based on a larger selection of 19 products (i.e. devices in volume production and for which datasheets are available) was undertaken. Additional trending analysis was also undertaken on the six devices identified above taking advantage of testing done at a common operating point. The main trends studied were:

- size, footprint and current density
- input/output voltage trends
- efficiency
- switching frequency
- thermal trends

- cost
- functionality
- integration level
- packaging trends
- interconnection trends

Additionally, roadmaps for advancement in switching frequency and integration of passives were forecast based on recent research findings in the various influencing technologies.

III. BENCHMARKING

Descriptions of a selection of benchmarking steps performed on the six products under test are described below.

A. Physical Evaluation

Devices were subjected to high-magnification external and internal inspection, radiographic imaging, scanning acoustic microscopy (SEM), decapsulation, internal disassembly, cross sectioning and materials analysis. Figs. 1(a) and 1(b) show cross-sections of the lead-frame substrate of Product C and the multi-layer PCB substrate of Product D, while Figs 2(a) and 2(b) show SEM images of Product A and Product F following removal of encapsulant. A summary of the physical evaluation results is provided in Appendix A where products A-F are compared side-by-side.

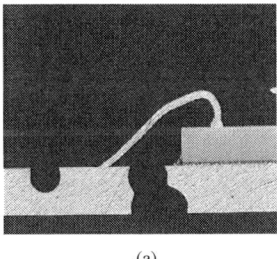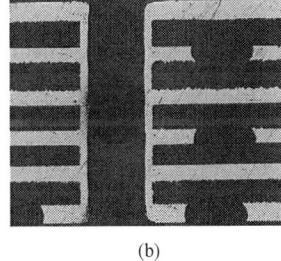

(a)	(b)

Figure 1. Photographs showing cross-sections of power supply substrates at 50X magnification. (a) Product C lead-frame showing 300 μm copper substrate and one of the gold wire-bond connections. (b) Product D 6-layer FR4 PCB substrate having 60 μm thick inner and 85 μm thick outer layers of copper

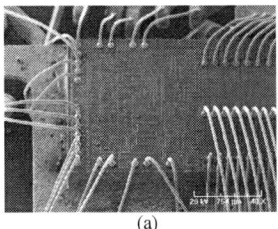

(a)	(b)

Figure 2. Scanning acoustic microscope images of decapsulated devices. (a) Product A wire-bond interconnections from power IC at 40X magnification. (b) Product F power IC and interconnections at 70X magnification

(a)

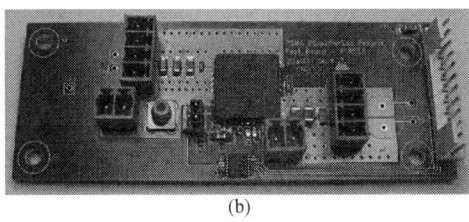

(b)

Figure 3. Sample PCB layouts for characterization of (a) Product E and (b) Product F

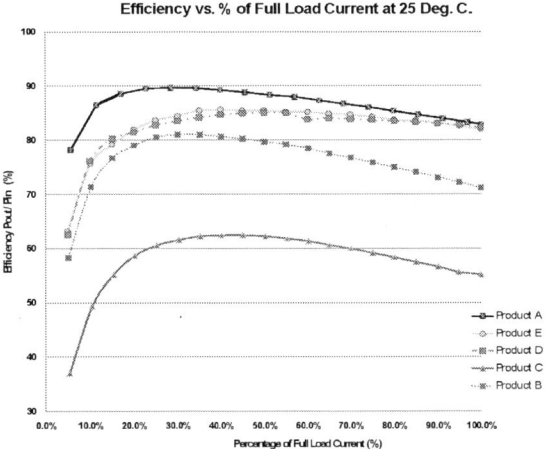

Figure 4. Relative measured efficiencies of Products A-E plotted against percentage of full rated load current

B. Electrical Evaluation

A converter test PCB was designed for each of the 6 products being evaluated. These were designed to be as similar as practicable so that any performance difference arising from board layout could be minimized. In order that the transient test results would be as consistent, manufacturers' recommendations on power density vs. PCB area metrics were followed. Electrical testing was performed in a temperature controlled chamber and the tests were carried out in an automated manner using LabVIEW software with GPIB interfacing to the test equipment. Figs. 3(a) and 3(b) show the PCB layouts for Products E and F. Fig 4 is a chart plotting the variation of measured efficiency against percentage of full load current for five of the tested devices. This method allows a normalized comparison of the products across their respective load ranges. A detailed summary of the electrical tests that were performed on the products is outlined in Appendix B.

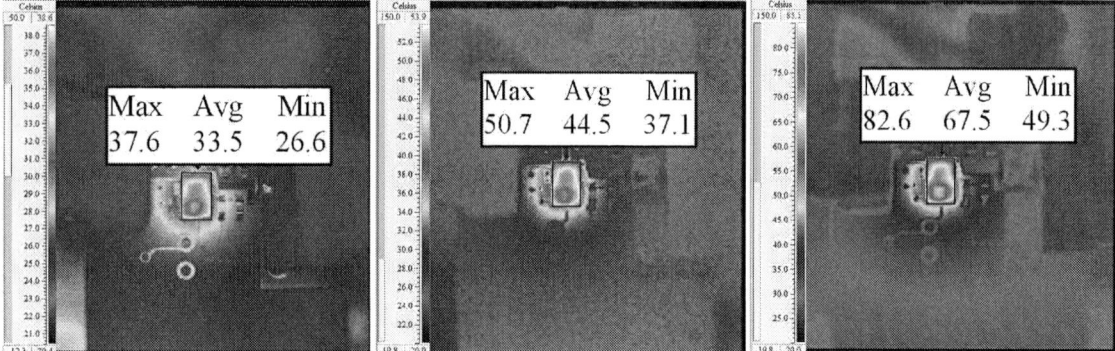

Figure 5. (l-r) IR images of product C at 33%, 66% and 99% of rated load current (Values shown are temperatures in degrees Celsius)

C. Thermal Evaluation

For each of the six selected products, a detailed analysis of the principal and secondary heat dissipation paths was undertaken. Three principal methods of heat transfer were identified, depending on the product type i.e. lead-frame-based (A, C & F), PCB-based with chip-on-board (B) and PCB-based with packaged ICs (D & E). This was followed by calculation of an Rθjc value for each product based on the information on the materials composition and dimensions gained from the physical evaluation. Finally, IR imaging of each of the products was undertaken to study the case temperature profile under a range of load conditions, as shown in Fig 5.

IV. TRENDS

In the Power Supply in a Package (PSiP) and Power Supply on a Chip (PwrSoC) Report (Phase 1), the authors derived the trend analysis from information gleaned from datasheets of products available at the time. In this report, considered Phase II of that report, we have further updated the trends based on new product releases that include 19 devices.

At present there are two major categories of suppliers. Firstly, there are those manufacturers that are primarily involved in providing power supplies. Secondly, there are those that are primarily providing semiconductor devices. Over the past 10 years, power supply manufacturers have predominantly provided power supplies. However, some adjacent industry participants have started to provide some power supplies including contract manufacturers, original equipment manufacturers (OEMs) or power supply users, and semiconductor manufacturers. Since both the PSiP and PwrSoC devices have up to 80% semiconductor content, it is natural for those manufacturers to start supplying the simpler non-isolated dc-dc type power supplies such as the PSiP and PwrSoC. From our analysis, it is clear that semiconductor manufacturers are becoming increasingly prevalent in this field.

In order to give a brief summary of the main trending results, some details are provided below under the headings outlined earlier.

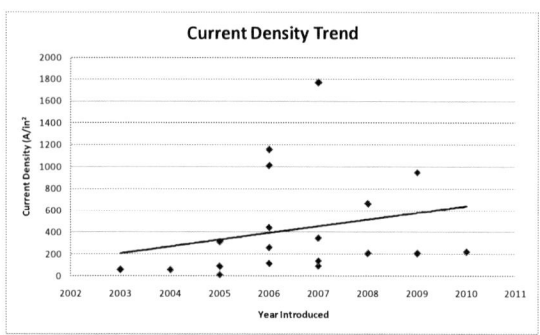

Figure 6. Current density trend for analyzed devices by year of introduction

A. Size, Footprint and Current Density

Much smaller devices are likely to be introduced and these will be constrained by inductor size and operating frequency. The significant improvement will be in current density. These improvements will depend on a) reducing the losses and b) improving the heat transfer design, though an increase in efficiency will have the greatest impact. Fig. 6 shows the current density for the 19 devices introduced into the market since 2003.

B. Input/Output Voltage

PSiP voltage ratings will continue to trend downward as the voltage requirements of microprocessors, ASICs, and field programmable gate-array (FPGA) devices continue the downward trend. We expect lower output voltage for future PSiP devices, since microprocessors, ASICs, and FPGA chips demand lower voltage. According to the International Technology Roadmap for Semiconductors (ITRS), output voltages of approximately 0.5 V will be required by 2016 [14].

The input voltage will decrease as the output voltage decreases due to the converter duty cycle limitations. In a standard synchronous buck converter, if the duty cycle becomes very small the converter cannot operate in a stable fashion, and the peak currents in the power components increase inversely with the duty cycle reduction. Recently introduced advanced topologies provide higher efficiency that the traditional buck converter topology thus offering designers with an opportunity to delay reducing the input voltage for a couple of design generations.

C. Efficiency

Two major application drivers will force the device efficiencies to continue to increase over the next few years. These drivers are power density and external pressure by users for less overall system power loss. Achieving improved overall system efficiency forces each element within the system or power architecture to be a candidate for improved efficiency. Since the final dc-dc power supply in that chain operates at the lower voltage, usually it has the least efficiency. Therefore, the DC-DC PSiP or PwrSoC will continue to be a primary candidate for enhanced efficiency. From our analysis, a key expectation is that efficiency at maximum load could increase from today's approximately 83% to up to 91% by 2013. While power supply manufacturers have had some of the highest efficiency converters in the marketplace until now, semiconductors manufacturers will likely close that gap in the future.

D. Switching Frequency

There is a correlation between increased switching frequencies and current densities as the frequency has a direct effect on the size of the power filter components. It is likely that the average switching frequency of PSiP and PwrSoC devices will trend upwards towards 10 MHz over the next 5 years. Indeed, the highest frequency device assessed had a switching frequency of 8 MHz. Nevertheless, if switching losses dominate the design then higher switching frequencies will not be reached.

E. Thermal Performance

The general trend is towards lower thermal resistance. This is related to the packaging technology as many suppliers are moving towards encapsulated high-current chip-scale packaging with many small (or a few large) pads including center package pads for better thermal transfer.

F. Cost

Though real costs are dependant on volume, product category and board real-estate, the general trend in per-unit cost for commercially available devices is downward, as might be expected.

G. Functionality

It is doubtful that the PSiP product will increase in functionality much beyond the functionality seen in today's products. There will be a lot of emphasis on digitally-controlled products which inherently provide additional functionality. The power-passive components can take up so much real estate within the PSiP device that designers have great difficulty finding enough room to include several converters, or multiphase devices. This means that the much higher functionality of power management units (PMUs) that include MOSFET power transistors will certainly retain the lead in advanced devices, in terms of functionality, for the foreseeable future.

H. Integration

We have established the level of integration of currently available PSiP devices on a scale of 1 to 10 where 1 is the lowest level of integration and 10 is the highest. Products from traditional power supply manufacturers are lower on the scale than those made by semiconductor manufacturers; however, of the devices analyzed, the highest level of integration was level 5 (power IC with a single regulator and a discrete inductor). There is clearly much scope for further integration towards a true power supply on a chip with power management and multiple phases (level 10)

I. Packaging Technology

The packaging used for PSiP products is transitioning from early open construction to dipped body to over-moulded straight-leaded and gull-wing lead packages. Most recent packages are taking advantage of chip-scale packaging technology such as QFN packages, leadless land grid array (LGA), etc. As cost is coming down for high current packages due to their utilization in mainstream ICs, the trend will be a move towards total use of chip scale packaging

J. Internal Interconnection

The trend here is a move away from SMD interconnection towards chip and wire interconnect which is less costly. With higher levels of semiconductor integration, the chip and wire construction will be a necessity for lowest cost solution. Some PSiP products already use flip-chip components. If vendors can overcome the testing challenges, flip-chip internal assembly construction may become the preferred method.

K. Influencing Technologies

A large selection of influencing technologies have been analyzed by the authors. In particular, roadmaps for switching frequency, integrated inductor and integrated capacitor technologies have been identified. Fig. 7 shows an overall parametric assessment of the influencing technologies with the various axes showing our assessment of the relative progress in key influencing areas. On this scale, 0 is the lowest relative level of progress and 10 is the highest.

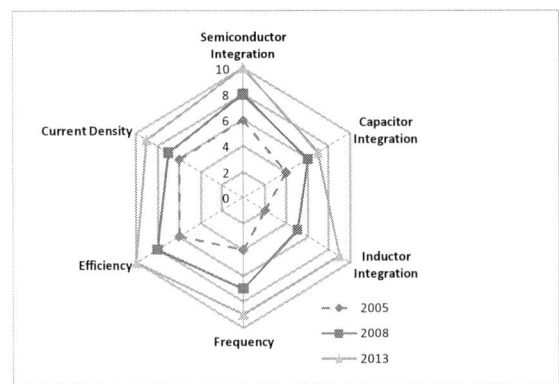

Figure 7. Overall parametric assessment of key influencing technologies.

V. CONCLUSION

This study, the first reported analysis of its kind, has provided a large amount of information on commercially available PSiP and PwrSoC devices and the technology trends that will drive this market segment into the future. Some clear conclusions have been drawn from the results; in particular the absence to-date from the market of any "true" PwrSoC device (with integrated inductor on semiconductor substrate). Also notable is the difference in integration levels from suppliers that come from a traditional power supply background and those that are from a semiconductor manufacturing background. It was found that the inexorable trend towards higher switching frequencies is justified, given the large relative size of the inductors in most of the products analyzed. From the results of the analysis carried out on the selected samples (and with reference to the findings of Phase I of this work), the authors consider future developments in this product technology over the medium term will be:

- A gradual reduction in footprint & size

- Significant improvements in current density

- A gradual downward trend in voltage ratings driven by ASICs, microprocessors, and FPGAs voltage requirements

- Pressure for enhanced efficiency or to maintain efficiency with greater density

- A steady increase in switching frequency up to ~ 10MHz

- No significant change in functionality

- A gradual decrease in cost /Amp

- A gradual increase in the use of higher-density assembly technologies (flip-chip)

- Greater use of over-molded lead frame based packaging

In the longer-term, higher levels of integration on chip will lead to the ready commercial availability of true Power-Supply-on-Chip products with very high levels of integration. It is likely that this will require greater cooperation between semiconductor manufacturers and power supply manufacturers.

Finally, it is clear that the PSiP and PwrSoC are now joining the less integrated embedded power and point of load (POL) power modules as major new contributors to the 1 watt to 150 ampere power supply landscape that power system ICs. Their market growth will be derived from two application areas a) providing quicker and simpler design solution replacing the motherboard regulator "down" or embedded

REFERENCES

[1] Musunuri, S., Chapman, P.L., Jun Zou, Chang Liu, "Design issues for monolithic DC-DC converters," Power Electronics, IEEE Transactions on, vol.20, no.3, pp. 639-649, May 2005

[2] Karnik, T., Hazucha, P., Schrom, G., Paillet, F., Gardner, D., "High-frequency DC-DC conversion : fact or fiction," Circuits and Systems, 2006. ISCAS 2006. Proceedings. 2006 IEEE International Symposium on, Greece, May 2006.

[3] O'Mathuna, S.C., O'Donnell, T., Ningning Wang, Rinne, K., "Magnetics on silicon: an enabling technology for power supply on chip," Power Electronics, IEEE Transactions on , vol.20, no.3 (Special Issue on Integrated Power Electronics), pp. 585-592, May 2005

[4] Musunuri, S.; Chapman, P.L., "Design of Low Power Monolithic DC-DC Buck Converter with Integrated Inductor," Power Electronics Specialists Conference, 2005. PESC '05. IEEE 36th, pp.1773-1779, Recife, Brasil, 16-16 June 2005.

[5] Abedinpour, S.; Bakkaloglu, B.; Kiaei, S., "A 65MHZ switching rate, two-stage interleaved synchronous buck converter with fully integrated output filter," Circuits and Systems, 2006. ISCAS 2006. Proceedings. 2006 IEEE International Symposium on, Greece, May 2006.

[6] Roozeboom, F., Kemmeren, A.L.A.M., Verhoeven, J.F.C., van den Heuvel, F.C., Klootwijk, J. , Kretschman, H., Fric, T., van Grunsven, E.C.E., Bardy, S., Bunel, C., Chevrie, D. LeCornec, F., Ledain, S., Murray, F., Philippe, P., "Passive and heterogeneous integration towards a Si-based System-in-Package concept", Thin Solid Films, Volume 504, May 2006, Pages 391-396.

[7] Hannon, J., O'Sullivan, D., Foley, R., Griffiths, J., McCarthy, K.G., Egan, M.G., "Design and optimisation of a high current, high frequency monolithic buck converter," Applied Power Electronics Conference and Exposition, 2008. APEC 2008. Twenty-Third Annual IEEE , pp.1472-1476, Austin, TX, 24-28 Feb. 2008.

[8] Meade, T., O'Sullivan, D., Foley, R., Achimescu, C., Egan, M., McCloskey, P., "Parasitic inductance effect on switching losses for a high frequency Dc-Dc converter," Applied Power Electronics Conference and Exposition, 2008. APEC 2008. Twenty-Third Annual IEEE, pp.3-9, Austin, TX, 24-28 Feb. 2008.

[9] Hannon, J., Foley, R., Griffiths, J., O'Sullivan, D., McCarthy, K.G., Egan, M.G., "A 20 MHz 200-500 mA Monolithic Buck Converter for RF Applications," Applied Power Electronics Conference and Exposition, 2009. APEC 2009. Twenty-Fourth Annual IEEE , pp.503-508, Washington, D.C., 15-19 Feb. 2009.

[10] Ó'Mathúna, S.C., "Power Supply on Chip: Has the Ship Come In?," plenary talk presented at the Applied Power Electronics Conference and Exposition, 2009. APEC 2009. Twenty-Fourth Annual IEEE, Washington, D.C., 15-19 Feb. 2009.

[11] Alderman, A.N., Panossian, V., "PSMA Marketing and Technology Report: Power Supply in a Package, Power Supply on a Chip," PSMA Packaging and PSMA Semiconductor Committees, Feb. 2008.

[12] 1st International Workshop on Power Supply on Chip, www.powersoc.org, Cork, Ireland, 22-24 Sep. 2008.

[13] Waldron, F., Foley, R., Slowey, J., Alderman, A., "PSMA Technology Report: A Review of Commercial Developments in Power Supply in a Package and Power Supply on a Chip," PSMA Packaging and PSMA Semiconductor Committees, Feb 2009

[14] International Technology Roadmap for Semiconductors 2007. [Online]. Available: http://www.itrs.net/

APPENDIX A

Technology Heading	Product A	Product B	Product C	Product D	Product E	Product F
Substrate	Cu lead frame	2-layer PCB	Cu lead frame	6-layer PCB	4-layer PCB	Cu lead frame + 4 layer PCB
Assembly Technology	Direct mounting on lead frame using Pb-free solder.	Single-sided SMT assembly using Pb-free solder. CoB mounting of power & control ICs.	Direct mounting on lead frame with silver epoxy die attach adhesive.	Double-sided SMT solder assembly.	Single-sided SMT assembly using Pb-free solder.	Direct mounting of power ICs & inductor on lead frame using Pb-free solder. Single sided SMT control PCB assembly.
Components	Total = 7 IC x 1 L x 1 C x5	Total = 33 IC x 4 Q x 2 L x 1 C x 9 R x 17	Total = 2 IC x 1 L x 1	Total = 35 IC x 3 Q x 4 L x 1 C x 9 R x 18	Total = 25 IC x 3 Q x 2 L x 1 C x 9 R x 10	Total = 13 IC x 3 L x 1 C x 5 R x 4
Inductor	Proprietary design. Central core with external winding	Surface mount Chip Inductor with centre winding.	Surface mount Chip Inductor with centre winding.	Surface mount Inductor with 2-piece core.	External unit mounted in frame over PCB with double "E" shaped core.	Surface mount Chip Inductor with centre winding.
Encapsulation	Epoxy loaded with silica filler particles.	Epoxy loaded with silica filler particles.	Epoxy loaded with silica filler particles.	Epoxy loaded with silica filler particles.	None.	Epoxy loaded with silica filler particles.
Termination Finish	Sn (5.0μm)	Ni-Pd (8.0μm) Au (0.5μm)	Ni-Pd (3.0μm) Au (0.4μm)	Sn (15.0μm)	Sn (180.0μm) bumped pads	Sn (14.0μm)
Weight & Dimensions	L= 10.0 mm W = 12.0 mm H = 1.85 mm Weight = 0.65g	L= 15.2 mm W = 15.0 mm H = 2.82 mm Weight= 1.74g	L= 4.0 mm W = 6.0 mm H = 0.85 mm Weight= 0.07g	L= 17.0 mm W = 14.6 mm H = 8.0 mm Weight = 5.91g	L= 12.0 mm W = 12.0 mm H = 7.00 mm Weight= 1.94g	L= 15.0 mm W = 15.0 mm H = 3.50 mm Weight = 2.61g
Moisture Sensitivity Level	Tested to MSL3 (260ºC) J-STD-020D-1	Tested to MSL3 (260ºC) J-STD-020D-1	Tested to MSL3 (260ºC) J-STD-020D-1	Tested to MSL3 (260ºC) J-STD-020D-1	Tested to MSL3 (260ºC) J-STD-020D-1	Tested to MSL3 (260ºC) J-STD-020D-1

978-1-4244-4782-4/10 $26.00 © 2010 IEEE

APPENDIX B

Test Details	Constants	Variables	Outcome
Efficiency vs. Load over temperature.	- Input Voltage = Nominal - Output Voltage = Nominal	- Output Current = 0 to max, 20 steps. - Temp = 0°C, 25°C, 50°C & 70°C.	Plots of efficiency vs. output current at different temperature points.
Efficiency vs. Input Voltage over temperature.	- Input Voltage = Nominal - Output Current = Maximum	- Input Voltage = Min to Max, 20 steps. - Temp = 0°C, 25°C, 50°C & 70°C.	Plots of efficiency vs. input voltage at different temperature points.
Efficiency vs. Output Voltage over temperature.	- Input Voltage = Nominal - Output Current = Maximum	- Output Voltage = Min to Max, 20 steps. - Temp = 0°C, 25°C, 50°C & 70°C.	Plots of efficiency vs. output voltage at different temperatures.
Ripple vs. Output Current over temperature.	- Input Voltage = Nominal - Output Voltage = Nominal	- Output Current = 0 to max, 20 steps. - Temp = 0°C, 25°C & 50°C.	Scope plots at different load points P-P ripple/mV at different temperature points.
Load Transient over temperature.	- Input Voltage = Nominal - Output Voltage = Nominal	- Output Current = 50% & 75%. - Temp = 0°C, 25°C & 50°C.	Plot of transient response & slew rate in A/µs at different temperatures.
Line Regulation	- Input Voltage = Nominal - Output Current = Maximum	- Input Voltage = Min to Max, 20 steps. - Temp = 25°C.	Plot of output voltage vs. input voltage.
Load Regulation.	- Input Voltage = Nominal - Output Voltage = Nominal	- Output Current = 0 to max, 20 steps. - Temp = 25°C.	Plot of output voltage vs. output current.
UVLO Functional Test.	- Input Voltage = Nominal - Output Current = Maximum - Temperature = 25°C.	- Input Voltage = 0 to max, 100mV steps. - Input voltage = max to 0, 100mV steps.	- Min turn-on voltage. - Min turn-off voltage.
Current Limit Functional Test.	- Input Voltage = Nominal - Output Voltage = Nominal	- Output current = max to 0, 100mA steps. - Temp = 25°C.	- Current limit value. - Current limit type.

Technology Road Map for High Frequency Integrated DC-DC Converter

Qiang Li, Michele Lim, Julu Sun, Arthur Ball, Yucheng Ying, Fred C. Lee, K. D. T. Ngo

Center for Power Electronics Systems
Virginia Polytechnic Institute and State University
Blacksburg, VA 24061 USA

Abstract—This preliminary road map is provided for state-of-the-art technologies and trends toward integration of point-of-load converters. This paper encompasses an extended survey of literature ranging from device technologies and magnetic materials to integration technologies and approaches. The paper is organized into three main sections. 1) Device technologies, including the trench MOSFET, lateral MOSFET and lateral trench MOSFET, are discussed along with their intended applications. The critical role of device packaging to high-frequency integration is assessed. 2) Magnetic materials: In recent years, a number of new magnetic materials have been explored in various research labs to facilitate magnetic integration for high-frequency POL applications. These data are collected and organized to help selecting magnetic material for various current levels and frequency ranges. 3) Levels of integration, which are defined with the focus on magnetic integration techniques and approaches, namely board-level, package-level and wafer-level, each with suitable current scale and frequency range.

I. INTRODUCTION

Industry currently has a large research effort underway in integrated DC/DC converter applications. As a result, we are in the forefront of the paradigm shift underway in the power supply industry, as integration requires we work more and more closely with semiconductor manufacturers. In the survey portion of our work, we are targeting highly-integrated designs for assessment; those with high switching frequencies and high power density. This analysis involves taking a close look at the limitations of current designs - in terms of the semiconductor, material, processing aspects, or a combination of these. This information can be used to determine where we are today from a technology standpoint, what the trends are, and what will be necessary to achieve ultra-high-frequency and high-current POL converters in the future.

The main limitation for achieving both high frequency and high current in many of today's DC/DC converters is the magnetic components design. The magnetic is typically the largest component in converters, and at high frequencies it becomes increasingly difficult to have high current output and remain competitive with the size of silicon. Since the magnetic component imposes such constraints, the method of analysis for making the road map is therefore largely based on magnetic component design and fabrication. As a result, the analysis is broken down into three main sections: First, it

starts with device technology. The trench MOSFET, lateral MOSFET and lateral trench MOSFET are discussed along with their intended applications. A comprehensive assessment of package parasitic parameters and their impact on circuit performance has been performed. Device selection is also discussed based on device characteristics. This information is useful to determine which devices should be chosen for a given application and how they can be integrated with the rest of the system. The second part will be magnetic materials. In recent years, there have been a large number of new materials developed to tackle the problems outlined above to facilitate magnetic integration for high-frequency POL applications. Categorizing this information is a key step to seeing which materials cover which range of frequencies, and what will be the trend in the future with the knowledge we have today. The third part of the analysis will be magnetic integration processes. There is once again a wide range of techniques for getting the most out the materials. This also plays a large role in defining what the finished inductor or transformer will look like, how it can be integrated with the rest of the system, and what its operating characteristics are. Three integration levels, namely board-level, package-level and wafer-level are defined with the focus on magnetic integration techniques and approaches. Each integration level is introduced with suitable current scale and frequency range.

II. DEVICE TECHNOLOGY

For low-voltage application, there are three basic types of semiconductor technology widely used in industry: the vertical diffusion MOS (VDMOS), the trench MOS, and the lateral diffusion MOS (LDMOS). Among these structures, the trench MOSFET has the lowest on-resistance, while the lateral MOSFET has the fastest switching speed. For sub-30V application, the trench MOS and lateral MOS are widely used. Recently, the lateral-trench MOSFET structure has been developed to achieve the benefits from both the lateral MOSFET and trench MOSFET. Fig. 1 shows the possible frequency and current range for these different MOSFETs. It can be seen that the VDMOS and trench MOS are more suitable for low–frequency, high-current applications. The LDMOS is more suitable for high frequency low-current applications. There is an application gap between the trench MOS and the LDMOS. The new coming lateral-trench MOSFET has some potential to fill this application gap.

This work was supported primarily by International Rectifier

978-1-4244-4782-4/10 $26.00 © 2010 IEEE

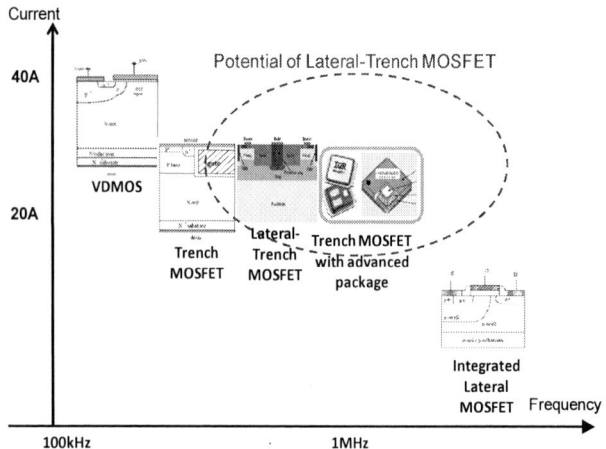

Figure 1. Possible frequency and current range for different MOSFET

Traditionally, Qgd*Rdson is used to evaluate the device performance, which is called the Figure-Of-Merit (FOM). Fig.2(a) shows a summary of FOM vs. breakdown voltage for different types of commercial MOSFETs. All devices can be used for VRM applications, and the devices for the same type of MOSFET are obtained from the same company to make a fair comparison. It can be seen that most of them are trench MOSFETs. However, The above FOM (Qgd*Rdson) is actually derived from high-voltage applications, and is not suitable for low-voltage, high-current application since it neglects the Qgs2 effect, which is related to the current rising time (for turn-on) or current falling time (for turn-off). It also does not include the impact of the gate-driving voltage. A better FOM was proposed recently by [42] that is more suitable for low-voltage high-current applications. A new device Figure-Of-Merit is proposed as:

$$FOM = (Q_{gd} + K_{gs2} \cdot Q_{gs2}) \cdot R_{dson} \qquad (1)$$

where, K_{gs2} is a variable related with driving voltage, input voltage and output current [42].

Fig. 2(b) shows the new FOM vs. breakdown voltage. It can be observed that Fig. 2(b) has some differences from Fig.2(a). For example, the 25V trench MOSFET from Vishay has a higher new FOM than Vishay's 30V devices due to its larger Qgs2. This indicates that 25V may not be an optimal design for the VRM application. Another example is the comparison between the 7V Greatwall lateral MOSFET and the 12V Ciclon lateral-trench MOSFET. In Fig. 2(a), the 7V lateral MOSFET is better than the 12V lateral-trench MOSFET. However, in Fig. 2(b), the result is reversed. The reason for this is that although the Qgd of the 7V lateral MOSFET is lower than the 12V lateral-trench MOSFET, the Qgs2 is higher. There are three trends clearly shown in Fig.2(b): 1) For each structure, the FOM decreases when the breakdown voltage decreases; 2) The general trend from the high FOM to low FOM is: Trench MOS, Lateral-Trench MOS and LDMOS; 3) With lower breakdown voltage, the LDMOS shows significant benefit in terms of FOM.

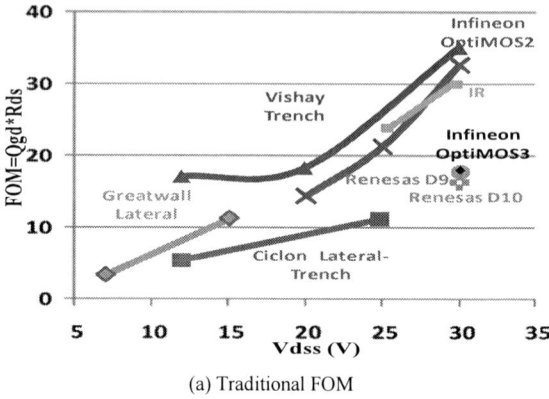

(a) Traditional FOM

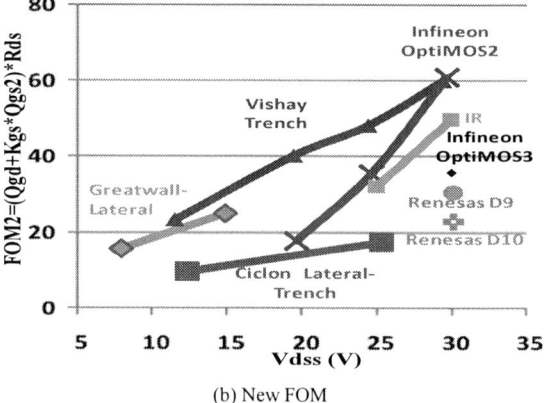

(b) New FOM

Figure 2. Figure-Of-Merit for different devices

Another important factor is the device packaging parasitic. To operate, the MOSFET must be connected to the external circuit using a certain packaging method. Different packaging methods use different ways to connect the MOSFET die to the pins; for example, wire bonding or a solid copper strap. These packaging methods introduce packaging parasitic inductors: the common-source inductor (Ls) and the drain-side parasitic inductor (Ld). Both turn-on and turn-off loss will increase by increasing the common-source parasitic inductance (Ls). The total switching loss is tremendously impacted by the common-source parasitic inductance (Ls). Increasing the drain-side parasitic inductor (Ld) would decrease the turn-on loss, but it would also increase the turn-off loss and would not significantly influence the total switching loss in some ranges because the gain from the turn-on loss and the loss from the turn-off loss cancel each other out.

As mentioned above, different packaging methods use different ways to connect the MOSFET die to the pins, resulting in different packaging parasitic inductances. For example, Renesas has three major packaging methods, SO-8, LFPAK and Dr.MOS, for its 30V MOSFET product line. In order to make a fair comparison of the loss performance between different packaging methods, it is reasonable to use the same device die during the analysis. Thus the following analysis is all based on a Renesas RJK0305DPB die with

978-1-4244-4782-4/10 $26.00 © 2010 IEEE

different packaging methods. Fig. 3 shows the loss breakdown with the different packaging methods. The circuit conditions are: Vin=12V, Vo=1.2V, Io=20A, fs=600kHz. The Ls and Ld-related loss is defined as the packaging-related loss; and the Qgs2 and Qgd-related loss, conduction loss, and gate driving loss are all part of the silicon-die-related loss. From the loss breakdown, it can be seen that by changing to the LFPAK from the SO-8 package, the packaging-related loss can be greatly reduced due to the smaller parasitic inductance, Ls. However, the Ls-related loss is still a dominant part of the total loss. The silicon-die-related loss and the packaging-related loss are almost equal. The DrMOS packaging method can further improve the performance by integrating the MOSFET and the driver. By integrating these components, it is possible to place the driver very close to the MOSFET and minimize the gate driving loop. Therefore the common-source parasitic inductance (Ls) can be greatly reduced to 0.1nH. From Fig. 3, it can be seen that the packaging-related loss of the DrMOS is no longer a dominant part of the total loss. In this case the silicon-die-related loss is more important than the packaging-related loss. Fig. 4 summarizes the survey of different applications for different devices. It can be seen that the trench MOSFET is more suitable for low-frequency applications, while a lateral MOSFET can be used for high-frequency applications. Fig. 4 also shows the impact of the device packaging. The IR DirectFET and Renesas DrMOS have basically the same trench MOSFET technology, but the packaging inductance is greatly reduced compared to a discrete MOSFET. As a result, they can run at higher frequencies (1MHz range). At the higher frequencies, two trends can be observed: 1) devices need to be monolithic–integrated with the driver to reduce parasitic inductance, and 2) almost all the devices used for higher frequencies are lateral MOSFETs because of their convenience during integration. As for the lateral-trench MOSFET, since the commercial devices are discrete with large parasitic inductances, the frequency is relatively low (600 kHz range), although its die performance is very good. However, we think the lateral-trench MOSFET is a very promising candidate for high–current, high-frequency applications.

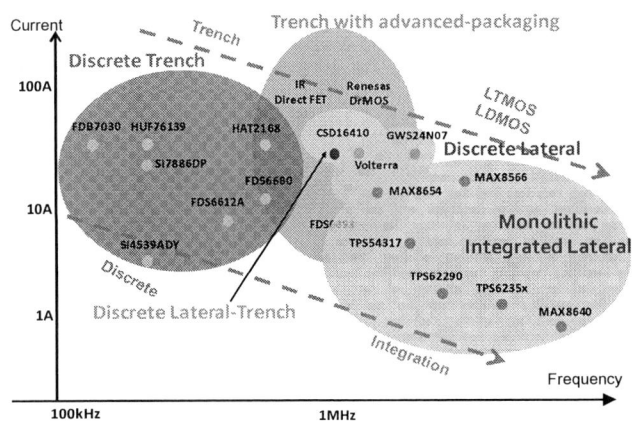

Figure 4. Survey results for the applications of different devices

III. MAGNETIC MATERIAL

Semiconductor devices have been prominent in the efforts towards improvements in power electronics systems. With miniaturization and improvements in device performance, the development of power electronics systems has progressed to a state where the active devices' impact on the system's size and cost has been taken over by the passive components. Magnetic components have been one of the bulkiest components, taking up significant circuit surface area. With further demands for miniaturization in power transformers and inductors, magnetic-core materials with low energy losses and high flux densities, permeability, and operating frequencies are in high demand. Using detailed survey results of magnetic material, this section identifies some magnetic materials that have a promising future for high-frequency integrated converter applications. Fig. 5 is the possible working frequency range for different magnetic materials. Among the commercial products, the working frequency of amorphous magnetic material and NiZn ferrite can be as high as 10MHz. Besides these commercial magnetic materials, there are many ongoing researches trying to develop some new magnetic materials for high-frequency and integration applications, such as granular film material CoZrO, polymer-bonded materials and thin film alloy CoNiFe. In granular films, magnetic particles are embedded in oxide insulating matrix phases, which results in high dc resistivity. CoZrO is one of the granular film materials. This material can have constant resistivity (590 µohm·cm) up to 20MHz. The Polymer-bonded magnetic material is composed of polymer matrices and magnetic powders which can be produced by traditional polymer processing methods. One of the important advantages is the ease of molding, such as injection molding which can save on manufacturing costs. Ferrite polymer is one of the promising polymer bonder materials for high frequency application. Since it has higher resistivity than other magnetic materials, eddy current losses are reduced. The electroplated alloy CoNiFe is first introduced for magnetic recording applications. Later on it has been used for power conversion applications.

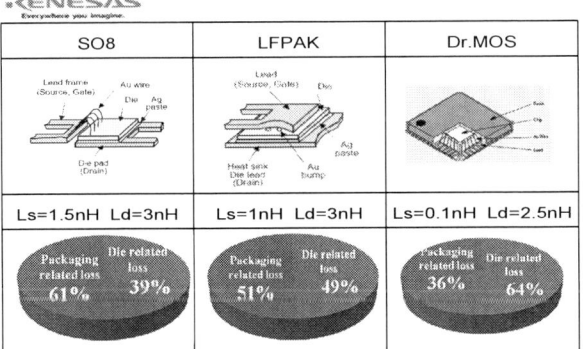

Figure 3. Device loss breakdown with different packaging

978-1-4244-4782-4/10 $26.00 © 2010 IEEE

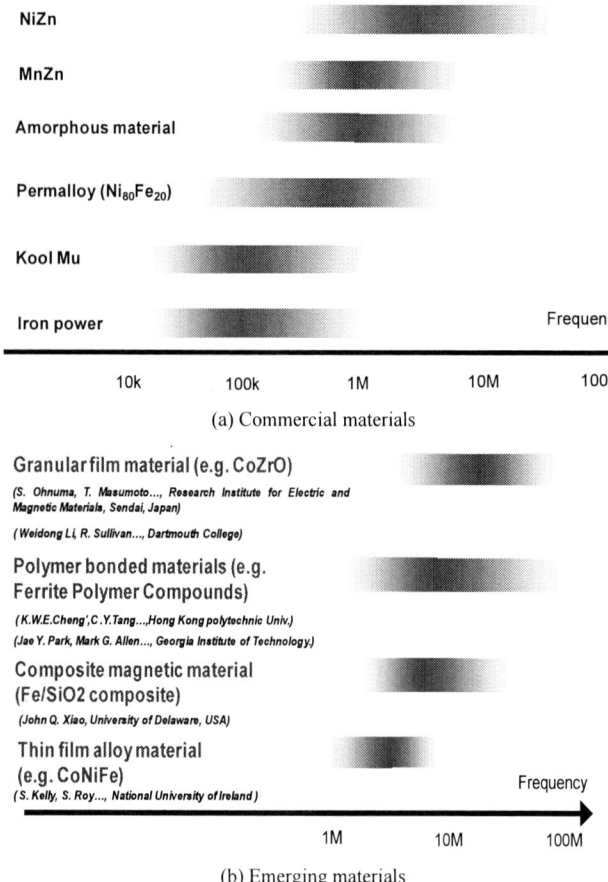

(a) Commercial materials

(b) Emerging materials

Figure 5. Frequency range for different magnetic materials

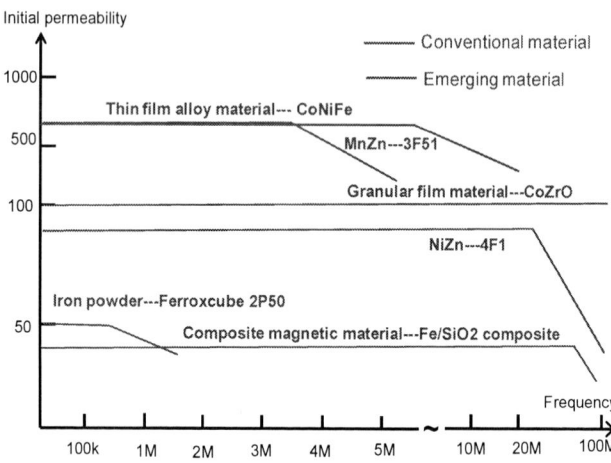

Figure 6. Frequency dependence of initial permeability

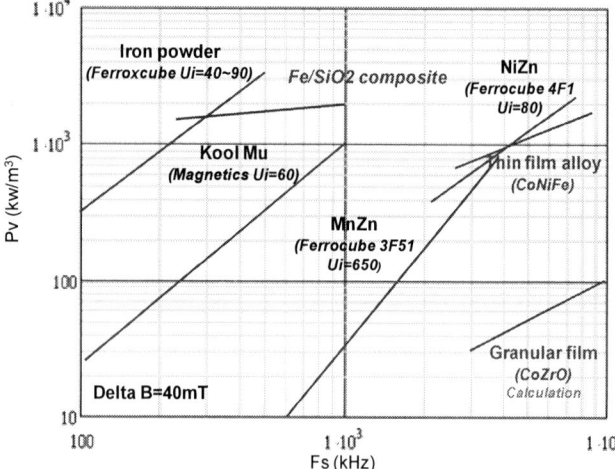

Figure 7. Core loss density of different magnetic material

Fig. 6 shows the frequency dependence of initial permeability for different magnetic materials. It is clear that the emerging material developed by ongoing research can have constant permeability at higher frequencies than conventional materials. There are several magnetic materials even can have constant permeability up to more than 20MHz, such as granular film CoZrO, NiZn ferrite 4F1 from Ferrocube and Fe/SiO2 composite material. Fig.7 shows the core loss density of different magnetic materials. It can be seen that at high frequencies, the granular film CoZrO has the lowest core loss density. NiZn ferrite 4F1 and thin film alloy CoNiFe are another two materials that have lower core loss densities. Actually, because CoNiFe is a metal alloy magnetic material, it should be designed very carefully to reduce eddy current loss. According to [39], when reduce core thickness to $2.4\mu m$, the CoNiFe can has core loss density as low as NiZn ferrite 4F1. Fig. 8 shows the survey results of the integrated converter with different magnetic materials. From Fig. 8, it can be seen that as applications go from high power to low power, the integrated converter moves from low frequencies to high frequencies. Although there have been many attempts at using new materials to further raise the converter frequency, right now it is still difficult to develop a high–power, high-frequency integrated converter.

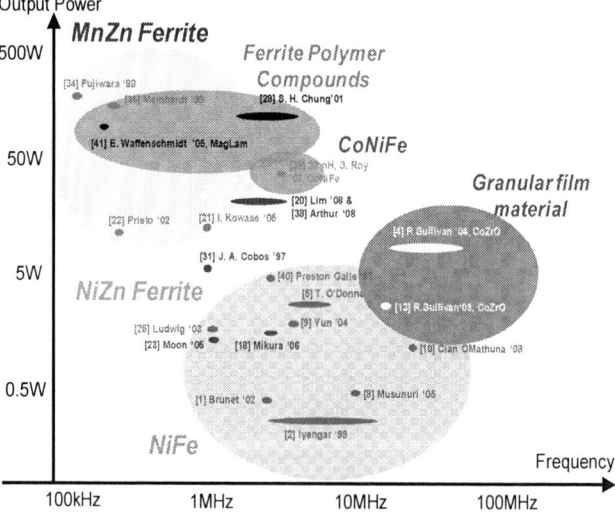

Figure 8. Integrated converter with different magnetic material

IV. INTEGRATION LEVEL

In view of the numerous techniques available for inductor integration, the integration techniques can be classified into the following categories:

1. Board-level integration

2. Package-level integration

3. Wafer-level integration

For board-level integration, the magnetic component is built on a substrate on which the active device can be mounted. Board-level integration can be classified into two categories. One is integrating the inductor / transformer in an organic-based substrate, where the processing temperature is low (<200 °C). Another package-level integration method is integrating the component in a ceramic-based substrate where the processing temperature for the component is high (> 200 °C). The mechanical properties of the organic substrate are usually not compatible with the silicon die; for example, FR4 has a coefficient of thermal expansion (CTE) of 17 ppm / °C. For ceramic-based technology, the materials involved, such as LTCC, can support higher operating temperatures and have a CTE value around 5 ppm/°C, which is close to that of silicon.

Fig. 9 shows the survey results of an integrated converter with board-level integration. It can be seen that there is currently a large overlap between organic-based and ceramic-based technology. However, the ceramic-based technology looks like it has more potential for higher frequency applications, because it can be compatible with a silicon die. For package-level integration, the magnetic component is co-packaged side-by-side with the silicon die or used as a platform for the silicon device and connected internally within the package. There are currently many industry products with integrated magnetics that come from different companies, such as Enprion, Linear Tech, National Semiconductor and Fuji. Almost all of these industry products belong to this kind of package-level integration.

For wafer-level integration, the inductor / transformer is built in or on the silicon die. The processes are compatible with silicon processing. For magnetics built on the silicon die, the magnetics can be fabricated on top of the active circuitry, which reduces the footprint. For magnetics fabricated in the silicon, the active circuitry is built beside the magnetics on the same die. Fig. 10 shows the survey results of the integrated converter with wafer-level integration. From Fig. 10, it can be seen that the current level of magnetics on silicon is generally lower than magnetic in silicon. The current range using on-silicon magnetics is more suitable for operation at less than 1 A. The current capability of the magnetic component can be improved by forming conductors of larger cross-sectional areas in the silicon.

For in-silicon magnetics, the current level can be increased to several amperes. Fig. 11 shows the total survey results of the integrated converter with different magnetic integration methods. From Fig. 11, it can be seen that from low integration levels (board-level) to high integration levels (wafer-level), the output current is reduced and the frequency is increased.

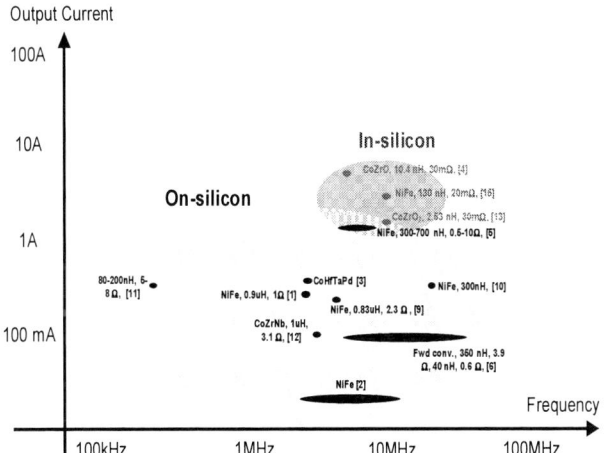

Figure 10. Integrated converter with wafer-level integration

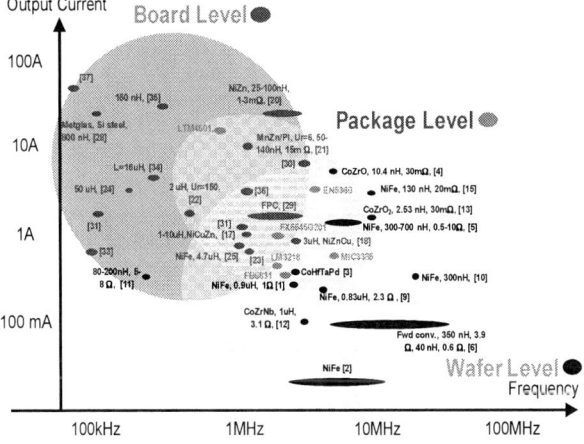

Figure 11. Integrated converter with different magnetic integration methods

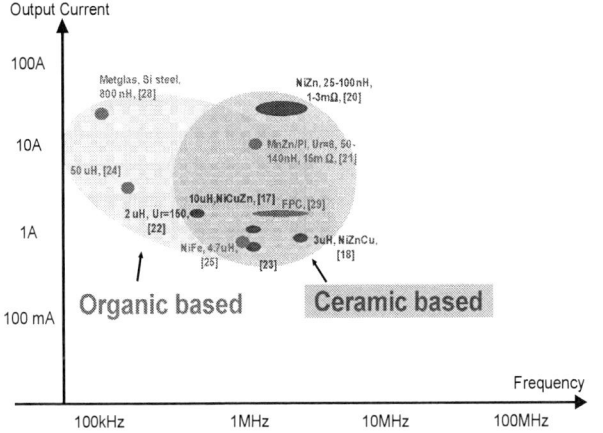

Figure 9. Integrated converter with board-level integration

V. CONCLUSIONS

The material presented thus far represents our latest survey results. The requirements that will have to be met in order to design the next generations of integrated POL converters are broken down into individual parts based on device technology, magnetic material and integration method. The device FOM evaluation figures clearly show that the lateral MOSFET or lateral-trench MOSFET have much better performance than the trench MOSFET if the device voltage rating is under 20V, especially when the device rating is 12V. This indicates that the input voltage of the converter should be around 5V instead of 12V to take advantage of the advanced device technology and to achieve higher switching frequency and better efficiency. Based on our magnetic material survey, it can be known that recently many researchers have been focused on developing new magnetic material for high-frequency integration application.

Granular film CoZrO, NiZn ferrite 4F1 and thin film alloys CoNiFe have a smaller core loss density than other materials. The permeability of these magnetic materials also is constant up to more than 20MHz. Because ferrite polymer material can further reduce eddy current loss, it should have less core density than ferrite. As a result, we think Granular film CoZrO, NiZn ferrite 4F1, ferrite polymer material and thin film alloy CoNiFe are four promising magnetic materials for high-frequency applications. Based on our integration method survey, the integration techniques can be classified into three levels, namely board-level, package-level and wafer–level integration. The board-level integration is suitable for high–current, low-frequency applications, and the wafer-level integration is suitable for low–current, high-frequency applications. The current and frequency ranges for package-level integration are between the board-level and wafer–level integration. Right now, most integrated converter products use package-level integration.

REFERENCES

[1] Magali Brunet, Terence O'Donnell, Laurent Baud, Ningning Wang, Joe O'Brien, Paul McCloskey, and Sean C. O'Mathuna, "Electrical performance of microtransformers for DC-DC converter applications." IEEE Transactions on Magnetics, vol. 38, no. 5, pp. 3174-3176, Sept. 2002.

[2] Srinivasan Iyengar, Trifon M. Liakopoulos and Chong H. Ahn, "A DC/DC boost converter toward fully on-chip integration using new micromachined planar inductors", in proc. Power Electronics Specialists Conference, 1999, pp. 72-76.

[3] Yasushi Katayama, Satoshi Sugahara, Haruo Nakazawa, Masaharu Edo, "High-Power-Density MHz-Switcing Monolithic DC-DC Converter with Thin-Film Inductor." , in proc. Power Electronics Specialists Conference, 2000, pp. 1485-1490.

[4] Parul Dhagat, Student, Satish Prabhakaran, and Charles R. Sullivan, "Comparison of magnetic materials for V-groove inductors in optimized high-frequency DC-DC converters." IEEE Transactions on Magnetics, vol. 40, no. 4, pp. 2008-2010, July 2004.

[5] T. O'Donnell, N. Wang, M. Brunet, S. Roy, A. Connell, J. Power, C. O'Mathuna, P. McCloskey, "Thin film micro-transformers for future power conversion", Applied Power Electronics Conference and Exposition, 2004, pp. 939-944.

[6] Masato Mino, Toshiaki Yachi, Akio Tago, Keiichi Yanagisawa, and Kazuhiko Sakakibara, "A new planar microtransformer for use in micro-switching converters." IEEE Transactions on Magnetics, vol. 28, no. 4, pp. 1969-1973, July 1992.

[7] Sullivan, C. R. and S. R. Sanders, "Design of microfabricated transformers and inductors for high-frequency power conversion." IEEE Transactions on Power Electronics, vol. 11, no. 2, pp. 228-238, Mar. 1996.

[8] Surya Musunuri, Patrick L. Chapman, Jun Zou, and Chang Liu, "Design issues for monolithic DC-DC converters." IEEE Transactions on Power Electronics, vol. 20, no. 3, pp. 639-649, May 2005.

[9] Eui-Jung Yun, Myunghee Jung, Chae Il Cheon, and Hyoung Gin Nam, "Microfabrication and characteristics of low-power high-performance magnetic thin-film transformers." IEEE Transactions on Magnetics, vol. 40, no. 1, pp. 65-70, Jan. 2004.

[10] T. O'Donnell, N. Wang, R. Meere, F. Rhen, S. Roy, D. O'Sullivan, C. O'Mathuna, "Microfabricated inductors for 20 MHz Dc-Dc converters", Applied Power Electronics Conference and Exposition, 2008, pp. 689-693.

[11] Ahn, C. H. and M. G. Allen (1996). "A comparison of two micromachined inductors (bar- and meander-type) for fully integrated boost DC/DC power converters." Power Electronics, IEEE Transactions on 11(2): 239-245.

[12] Toshiro Sato, Hiroshi Tomita, Atsuhito Sawabe, Tetsuo Inoue, Tetsuhiko Mizoguchi, and Masashi Sahashi, "A Magnetic Thin Film Inductor and its Application to a MHz Switching dc-dc Converter", IEEE Transactions on Magnetics, vol. 30, no. 2, pp. 217-223,Mar. 1994.

[13] Satish Prabhakaran, Charles R. Sullivan, and Kapil Venkatachalam, "Measured electrical performance of V-groove inductors for microprocessor power delivery." IEEE Transactions on Magnetics, vol. 39, no. 5, pp. 3190-3192, Sept. 2003.

[14] Jian Lu, Hongwei Jia, Andres Arias, Xun Gong and Z. John Shen, "On-Chip Bondwire Transformers for Power SOC Applications", in proc. Applied Power Electronics Conference and Exposition, 2008, pp. 199-204.

[15] Mingliang Wang; Batarseh, I.; Ngo, K.D.T.; Huikai Xie, "Design and Fabrication of Integrated Power Inductor Based on Silicon Molding Technology", Power Electronics Specialists Conf., 2007, pp. 1612-1618.

[16] J. M. Lopera, M. J. Prieto, A. M. Pemia M. de Graaf, W. Waanders L. Alvarez, "Design of integrated magnetic elements using thick-film technology", in proc. Applied Power Electronics Conference and Exposition, 1998, pp. 407-413.

[17] Robert Hahn, Steffen Krumbholz, Herbert Reichl, "Low profile power inductors based on ferromagnetic LTCC technology", in proc. Electronic Components and Technology Conference, 2006, pp. 528-533.

[18] Tsutomu Mikura, Koichi Nakahara, Kota Ikeda, Ken Furukuwa, and Katsuhiko Onitsuka, "New substrate for micro DC-DC converter", Electronic Components and Technology Conference, 2006, pp. 1326-1330.

[19] Miguel J. Prieto, Alberto M. Pernia, Juan M. Lopera, Juan Martin, Fernando Nuiio, "Thick-film integrated inductors for power converters", in proc. Industry Applications Conference, 2000, pp. 3111-3118.

[20] Michele H. Lim, Jacobus. D. van Wyk, F. C. Lee, and Khai D. T. Ngo, "A Class of Ceramic-Based Chip Inductors for Hybrid Integration in Power Supplies", IEEE Transactions on Power Electronics, vol. 23, no. 3, pp. 1556-1564, May 2008.

[21] I. Kowase, T. Sato, K. Yamasawa, Y. Miura, "A planar inductor using Mn-Zn ferrite/polyimide composite thick film for low-Voltage and large-current DC-DC converter", IEEE Transactions on Magnetics, vol 41, no. 10, pp. 3991-3993, Oct. 2005.

[22] A. M. Pernia, M. J. Prieto, J. M. Lopera, J. Reilly, S. S. Linton, C. Quinones, "Thick-film hybrid technology for low output voltage DC/DC converter", in proc. Industry Applications Conf., 2002, pp. 1315-1322.

[23] Moon, K.W.; Hong, S.H.; Kim, H.J.; Kim, J.;, "A fabrication of DC-DC converter using LTCC NiZnCu ferrite thick films", in proc. Int. Magnetics Conf., 2005, pp. 1109-1110.

[24] Eberhard Waffenschmidt, Bernd Ackermann, and J. A. Ferreira, " Design method and material technologies for passives in printed circuit Board Embedded circuits", IEEE Transactions on Power Electronics, vol. 20, no. 3, pp. 576-584, May 2005.

[25] Matthias Ludwig, Maeve Duffy, Terence O'Donnell, Paul McCloskey, and Seán Cian Ó Mathùna, "PCB integrated inductors for low power DC/DC converter", IEEE Transactions on Power Electronics, vol. 18, no. 4, pp. 937-945, July 2003.

[26] Isao Kowase, Toshiro Sato, Kiyohito Yamasawa, Member, IEEE, and Yoshimasa Miura, "A planar inductor using Mn-Zn ferrite/polyimide composite thick film for low-Voltage and large-current DC-DC converter." IEEE Transactions on Magnetics, vol 41, no. 10, pp. 3991-3993, Oct. 2005.

[27] Jae Y. Park, Mark G. Allen "Low Temperature Fabrication and Characterization of Integrated Packaging-Compatible, Ferrite-Core Magnetic Devices" APEC 1997, pp. 361-367.

[28] Zhang, Y. E. and S. R. Sanders, "In-board magnetics processes", in proc. Power Electronics Specialists Conference, 1999, pp. 561-567.

[29] Tang, S. C., S. Y. R. Hui, and Henry Shu-Hung Chung, "A low-profile power converter using printed-circuit board (PCB) power transformer with ferrite polymer composite." IEEE Transactions on Power Electronics, vol. 16, no. 4, pp. 493-498, July 2001.

[30] Brett A. Miwa, Leo F. Casey, and Martin F. Schlecht, "Copper-based hybrid fabrication of a 50 W, 5 MHz 40 V-5 V DC/DC converter." IEEE Transactions on Power Electronics, vol. 6, no. 1, pp. 2-10, Jan. 1991.

[31] J. A. Cobos, M. Rascon, L. Alvarez, S. Ollero, M. de Graf and W. Waanders, "Low profile and low output voltage DC/DC converters for on-board power distribution using planar magnetics", in proc. Industry Applications Conference, 1997, pp. 1153-1158.

[32] Rengang, C., J. D. van Wyk, Shuo Wang, and Willem Gerhardus Odendaal, "Improving the Characteristics of integrated EMI filters by embedded conductive Layers." IEEE Transactions on Power Electronics, vol. 20, no. 3, pp. 611-619, May 2005.

[33] A.B. Lostetter, F. Barlow, A. Elshabini, K. Olejniczak, and S. Ang, "Polymer thick film (PTF) and flex technologies for low cost power electronics packaging", International Workshop on Integrated Power Packaging, IWIPP 2000, pp. 33-40.

[34] T. Fujiwara, "Planar integrated magnetic component with transformer and inductor using multilayer printed wiring board", IEEE Transactions on Magnetics, vol. 34, no. 4, pp. 2051-2053, July 1998.

[35] Meinhardt, M., T. O'Donnell, Henning Schneider, John Flannery, Cian O'Mathuna, Peter Zacharias, and Thomas Krieger, "Miniaturised "Low Profile" module integrated converter for photovoltaic applications with integrated magnetic components", in proc. Applied Power Electronics Conference, 1999, pp. 305-311.

[36] Lopera J. M. Lopera, Miguel J. Prieto, Alberto M. Pern´ıa, Fernando Nu˜no, Martinus J. M. de Graaf, Jan Willem Waanders, and Lourdes Alvarez Barcia, "Design of integrated magnetic elements using thick-film technology." IEEE Transactions on Power Electronics, vol. 14, no. 3, pp. 408-414, May 1999.

[37] D. C. Hopkins, "Thick-film power hybridization of switchmode power circuits", in proc. Applied Power Electronics Conference, 1989, pp. 249-255.

[38] Ball, Arthur; Lim, Michele; Gilham, David; Lee, Fred C., "System design of a 3D integrated non-isolated Point Of Load converter," Applied Power Electronics Conference and Exposition, 2008. APEC 2008. Twenty-Third Annual IEEE, vol., no., pp.181-186, 24-28 Feb. 2008

[39] S. Kelly, C. Collins, M. Duffy, F. M. F. Rhen, S. Roy, "Core Materials for High Frequency VRM Inductors", PESC,2007, pp. 1767-1772.

[40] Preston Galle, Xiaosong Wu, Luke Milner, Seong-Hyok Kim, Peter Johnson, Peter Smeys, Peter Hopper, Kyuwoon Hwang, and Mark G. Allen, "Ultra-Compact Power Conversion Based on a CMOS-Compatible Microfabricated Power Inductor With Minimized Core Losses" ECTC, 2007.

[41] Eberhard Waffenschmidt, Bernd Ackermann, J. A. Ferreira, "Design Method and Material Technologies for Passives in Printed Circuit Board Embedded Circuits, IEEE Transactions, vol. 20, no. 3, May 2005, pp. 576-584.

[42] Yucheng Ying, "Device selection criteria----based on loss modeling and Figure of Merit", Thesis of Master of Science in Electrical Engineering of Virginia Polytechnic Institute and State University, 2008.

978-1-4244-4782-4/10 $26.00 © 2010 IEEE

A New Valley-Detection Method for the Quasi-Resonance Switching

Gwan-Bon Koo, Sang-Cheol Moon, and Jin-Tae Kim
Korea Power Conversion, Fairchild Semiconductor
82-3, Dodang-dong, Wonmi-gu, Bucheon-si, Gyeonggi-do, KOREA

Abstract-To reduce switching losses in flyback topologies with a single switch, many kinds of quasi-resonant techniques are used. One of them is a valley-switching method which finds valleys of the drain-to-source voltage of the switch and makes the next switching begin at one of the valleys. It results in high efficiency and low EMI but an additional pin is required to detect where valleys are in view point of a control IC. This paper suggests a new valley-detection method without increasing the number of pin. Operational principles and experimental results will be shown to verify the validity of the proposed method.

I. INTRODUCTION

Flyback is one of the most popular topologies in medium and low power applications due to its simplicity. It can implement isolated switching mode power supplies (SMPS) with one switching component and one transformer resulting in low manufacturing cost. However, the switching losses are unavoidable in hard-switching mode as can be seen in Fig. 1.

When it operates in discontinuous conduction mode (DCM) with a fixed frequency, the drain-to-source voltage of the MOSFET is arbitrary. So the capacitive loss of the MOSFET during switch turn-on transition cannot be predicted and minimized. Therefore many kinds of quasi-resonant techniques are used to increase total efficiency by reducing switching losses in flyback topologies [1~2]. A valley-switching technique is one of them [3~4]. When the flyback topologies operate in DCM, the drain-to-source voltage of the switch resonates after the current of the secondary rectifying diode runs dry. In a valley-switching technique, the switching component always turns on at the instant of the first minimum value of the drain-to-source voltage of the switch even though the input voltage and load current are changed [3]. It allows good EMI (Electromagnetic Interference) and thermal performance, as well as high efficiency. Its operating waveforms are shown in Fig. 2.

However, it requires an additional pin and a couple of external components such as resistors and capacitors to detect where the valleys are. Fig. 3 shows an external circuit to indirectly detect the second valley via the bias winding voltage and its key waveforms. The bias winding voltage reflects the drain-to-source voltage without a dc offset corresponding to the dc link voltage, as shown in Fig. 3(b). The shape of the drain-to-source voltage of several hundreds

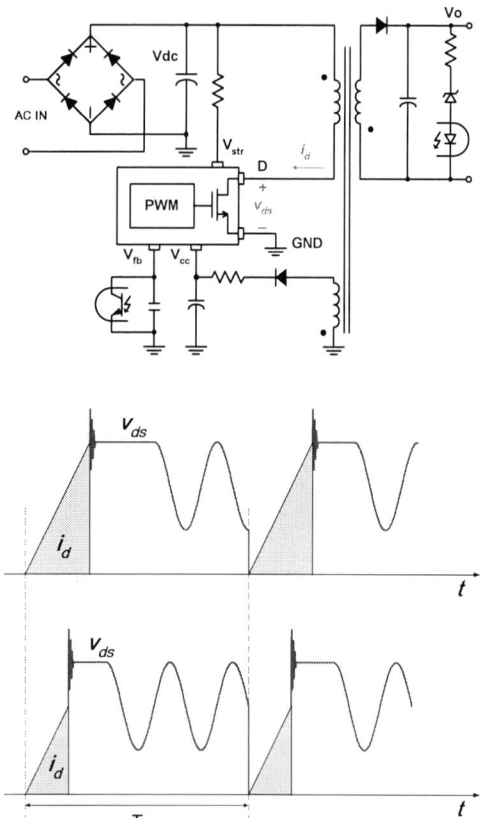

Fig. 1. A conventional flyaback converter with its key waveforms

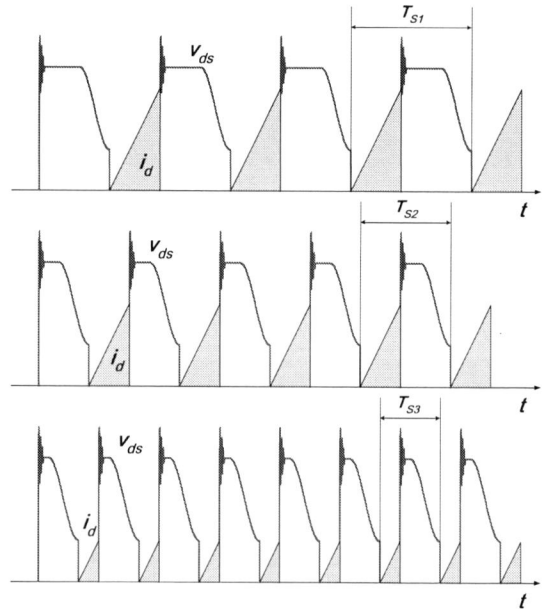

Fig. 2. Drain current and drain-to-source voltage waveforms with valley-switching operation

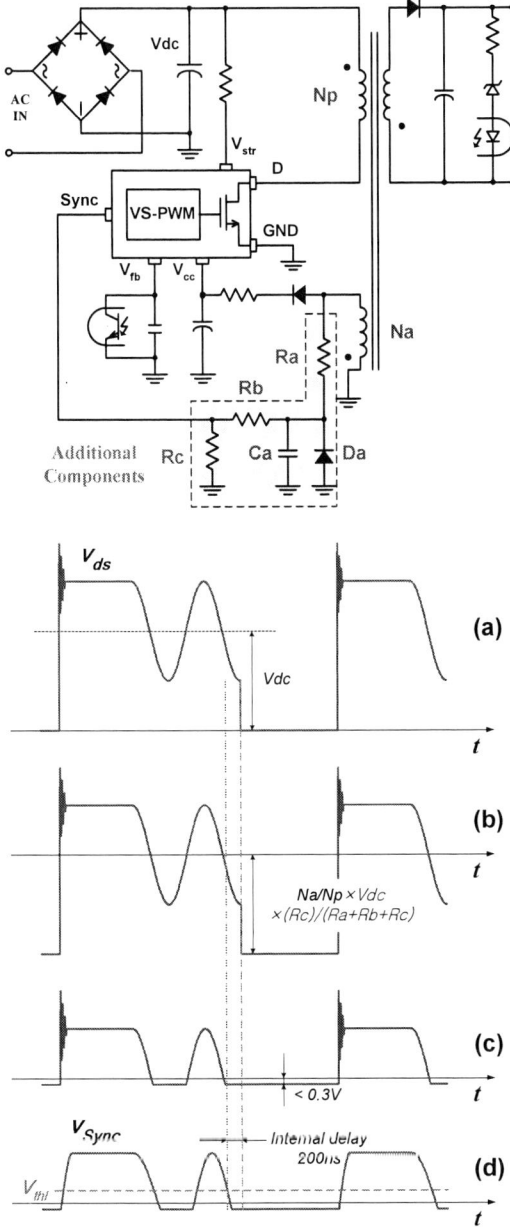

Fig. 3. External circuit for detecting valleys and key waveforms

volt is observed on the "Sync" pin with scaling down using the turns ratio Na/Np and the voltage divider Ra, Rb, and Rc, where Na and Np are the turns numbers of the bias and primary windings, respectively. Since control IC usually cannot accept a negative voltage exceeding -0.3 V as an input, a diode Da is needed to remove negative parts from Fig. 3(b). That is why a diode Da is added and the voltage divider is composed of three resistors rather than two resistors. However, the most important information, i.e. where valleys are, disappears unfortunately. Now capacitor Ca has a role. The observed voltage on the "Sync" pin is delayed by time constant made by Ca and the voltage divider,

as shown in Fig. 3(d). Considering an internal delay time around 200 ns, adjusting the instant where v_{Sync} reaches an internal low threshold voltage V_{thl} to the valley of the drain-to-source voltage of the MOSFET is possible [5].

In areas where the keen price competition exists, adding one more pin and several components to detect valleys only would be critical for manufacturers to select a control IC. Therefore it is highly required to change the method to detect valleys and to combine the "Sync" pin with one of existing protection pins such as the pin for line-sensing.

Section II introduces the concept of a new valley-detection method and one embodiment for commercial ICs. Experimental results to prove the validity of the proposed method and the conclusion are discussed in Sections III and IV, respectively.

II. THE PROPOSED METHOD TO DETECT VALLEYS

A. Basic Concept

The drain current of the MOSFET increases with a slope of Vdc/Lm during turn-on period T_1 where Vdc is the dc link voltage and Lm is the magnetizing inductance of the main transformer. After the MOSFET turns off, the current of the secondary rectifying diode will decrease with a slope of -Vo/Ls and finally reach to zero during T_2 due to a DCM operation where Vo is the output voltage and Ls is the secondary side inductance of the main transformer. The resonance between Lm and Coss begins after the run-dry of the secondary rectifying diode where Coss is the output capacitance of the MOSFET. Therefore the first valley comes after a time of a half of resonance period between Lm and Coss (T_3) goes. In flyback topologies, the magnetizing current of the main transformer is composed of the drain current of the MOSFET and the reflected current of the secondary rectifying diode. Fig. 4 shows the drain-to-source voltage of the MOSFET and the magnetizing current of the

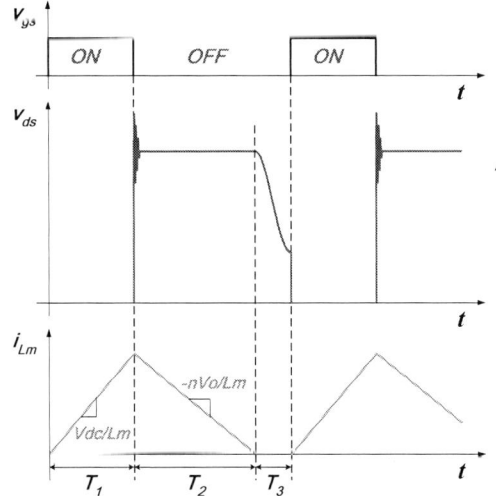

Fig. 4. Drain-to-source voltage and magnetizing current of the main transformer along with the MOSFET gate signal

978-1-4244-4782-4/10 $26.00 © 2010 IEEE

main transformer i_{Lm} along with the MOSFET gate signal v_{gs}. The magnetizing inductance is magnetized with the dc link voltage and demagnetized with the reflected output voltage. If the dc link voltage, reflected output voltage, turn-on time of the MOSFET, magnetizing inductance, and output capacitance are known, the time duration T_1, T_2, and T_3 are calculated so that the turn-on instant of the next switching is defined easily.

B. One Embodiment and Its Operational Principles

Fig. 5 shows one embodiment for the basic concept. There is an internal capacitor, C_{int} which will be charged and discharged by a current source and a current sink. The current sink is an independent one representing the reflected output voltage nVo. The current source is dependent on the dc link voltage Vdc. These two current source and sink will charge and discharge the internal capacitor by turns according to the gate signal. After charged by the current source during a turn-on time of the switch, the internal capacitor voltage decreases by the current sink during the rest cycle. When the internal capacitor voltage drops under V_{TH} which is almost zero, the comparator outputs high so that the next gate turn-on signal is produced to the switch with an internal delay time. The

internal delay time is a constant which is a slightly smaller than a half of resonance period between Lm and Coss generally used. Then the next gate turn-on signal will be generated near the first valley by tuning the external resistors to sense the dc link voltage taking the independent current sink and the internal delay time into account, as shown in Fig. 6. To meet the next turn-on signal with the first valley, the slope of the drain current should be determined as the thick solid line instead of the dotted lines. Once the external sensing resistors are tuned well at certain dc link voltage, the controller will find the instant of the next gate turn-on around the first valley over whole dc link voltage ranges.

The information about the dc link voltage through the "Line Sense" pin could be used for other protections such as a brownout function, variable current limit function, and so on. The proposed method allows the pin for sensing the dc link voltage to include the function of finding valleys. It results in saving the manufacturing cost for the control IC, and reducing external components.

III. THE EXPERIMENTAL RESULTS

To verify the performance of the embodiment described in Section II-B, a prototype is constructed and experimented with discrete components. The specifications of the prototype are shown in Table I.

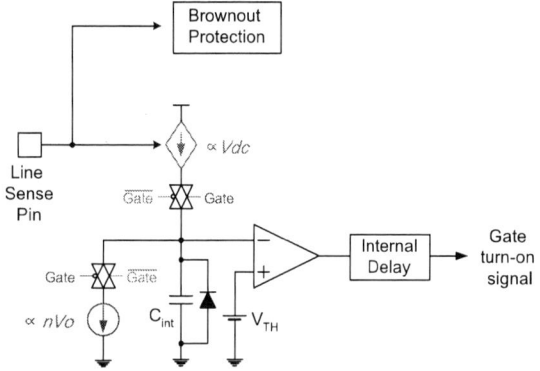

Fig. 5. One embodiment for the basic concept

TABLE I
SPECIFICATIONS OF THE PROTOTYPE

Dc Link Voltage	150 Vdc ~ 310 Vdc
Output Voltage	5 V
Output Load Current	0.5 A ~ 2 A
C_{int}	2.2 nF
Current Sink	1.94 mA
Internal Delay Time	1.44 us
Resonance Period	2.14 us
V_{TH}	1.6 V

The internal delay time has to be shorter than a half of the resonance period to tune the external sensing resistors up easily. But in the prototype, a longer internal delay time is required because V_{TH} is not zero. A possible internal capacitor to be implemented on a chip is several decades of pico-farad, in general. To make the experiments easy, 2.2 nF as the internal capacitor is selected in the prototype. Therefore the internal current sink increases as well. For the case of manufacturing a control IC, several decades of pico-farad and several decades of micro-ampere will be used as the internal capacitor and current sink, respectively.

Figs. 7 and 8 show the valley switching well-matched at different dc link voltages. Figs. 7 and 8 are measured under full load condition with 150 Vdc and 310 Vdc of link voltage, respectively. They are well agreed with the theoretical analysis.

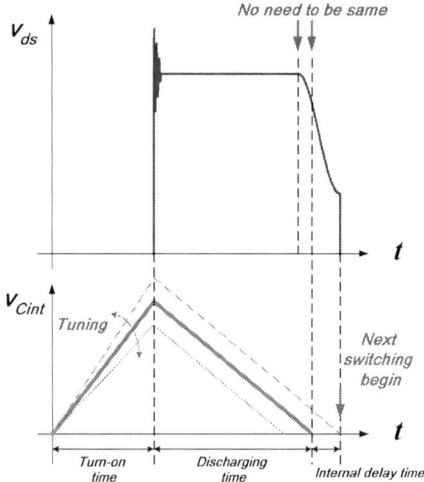

Fig. 6. How to tune the external sensing resistors up

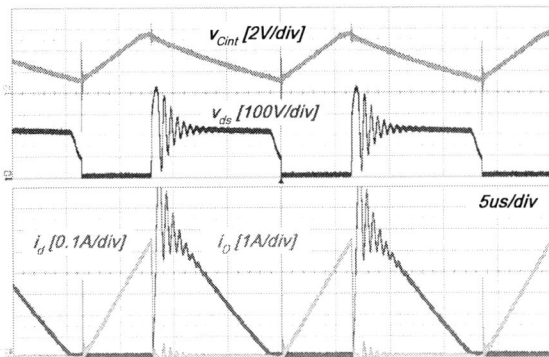

Fig. 7. Key waveforms at full load with 150 Vdc

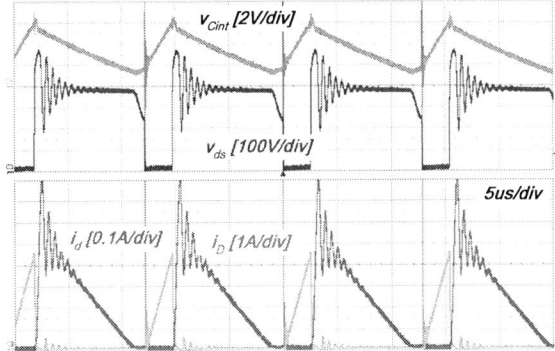

Fig. 8. Key waveforms at full load with 310 Vdc

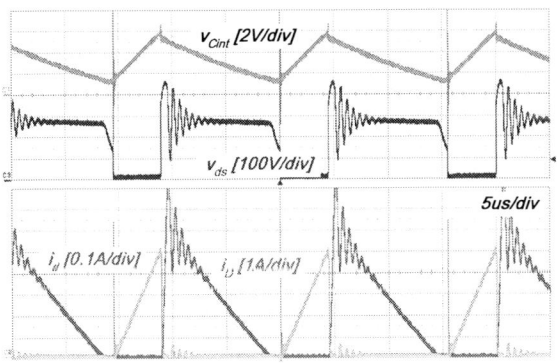

Fig. 9. Key waveforms at full load with 200 Vdc

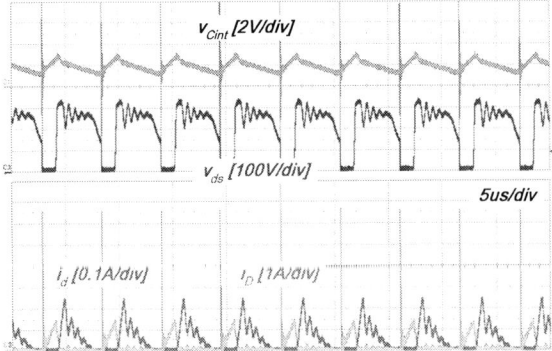

Fig. 10. Key waveforms at 25% load current of the rated condition
with 200 Vdc

Figs. 9 and 10 show the case of load current change. Fig. 9 is measured at full load condition with 200 Vdc of link voltage. The case of 25% load of the rated condition with 200 Vdc is shown in Fig. 10. It is verified that the proposed method to detect valleys works well even though the dc link voltage and load current change.

IV. CONCLUSIONS

A new valley-detection method was introduced in this paper. Instead of the existing method to indirectly detect valleys through a bias winding with a couple of external components, a method to detect the dc link voltage was suggested. It results in combining the pin which is connected to the bias winding for detecting valleys with a line sense pin so that one pin could be saved.

A prototype of 10 W was constructed and experimented to prove the validity of the proposed method. The test results were well agreed with the theoretical analysis. Finding the first valley and turning the next switching on worked well even if the dc link voltage and load current change.

A new control IC for a quasi-resonance switching could be made using the proposed method without increasing the number of pins. It will allow for the IC users to reduce the total bill of materials (BOM.)

REFERENCES

[1] K. H. Liu and F. C. Lee, "Resonant switches – a unified approach to improved performances of switching converters," in *Proceedings of the International Telecommunications Energy Conference*, 1984, pp. 344-351.

[2] D. Balocco and C. Zardini, "The half-wave quasi-resonant ZCS flyback converter as an automatic power factor preregulator: An evaluation," in *IEEE Applied Power Electronics Conference*, 1996, pp. 138-144.

[3] FSCQ-series, Green Mode Fairchild Power Switch (FPS™) datasheet, Fairchild Semiconductor.

[4] "Design Guidelines for Flyback Converters using FSQ-series Fairchild Power Switch (FPS™)," Literature No. AN-4150, Fairchild Semiconductor.

[5] FSQ510, Green Mode Fairchild Power Switch (FPS™) datasheet, Fairchild Semiconductor.

Secondary-Side Control of a Constant Frequency Series Resonant Converter using Dual-Edge PWM

Darryl J. Tschirhart, *Grad. Student Member, IEEE,* and Praveen K. Jain, *Fellow, IEEE*

Centre for Power Electronics Research (ePOWER)
Dept. of Electrical & Computer Engineering, Queen's University
Kingston, Ontario, Canada
darryl.tschirhart@ieee.org, praveen.jain@queensu.ca

Abstract—48V Voltage regulators (VR) can be used to improve the system efficiency of telecom and data centres by eliminating a converter stage in intermediate bus architectures. Small size and improved transient performance can be achieved through high switching frequency if resonant converters are adopted for these applications. This work proposes the use of a dual-edge PWM controller on the secondary-side of a series resonant converter to remove the opto-coupler from the feedback loop, and allow the achievement of fast dynamic performance. Simulation results of a 4MHz 48V/1.2V, 25A example highlight the fast load transient results achievable with standard linear control.

I. INTRODUCTION

In distributed power architectures like those in telecom and data centres, buck converters fed from a 5-12V intermediate bus provide the low voltage required by microprocessors, FPGAs, and DSPs. To improve the system efficiency 48V voltage regulators (VR) can be used [1] to reduce the loading on the intermediate bus converter, and reducing the number of conversion stages in the high-current path. To address transient requirements, and physically reduce the size of the 48V VR, high switching frequency must be achieved. Resonant converters are able to achieve soft-switching for all operating points [2]; thereby making them natural candidates for this application. With highly dynamic loads, current-type resonant converters, like the series resonant converter (SRC), are advantageous due to their capacitive output filter. Miniaturization is achieved by heat-sink reduction through lossless switching of both the primary- and secondary-side switches, and the ability to use the transformer leakage inductance solely to create the resonant inductor.

However, in practical implementation, the opto-coupler required in the feedback path of isolated topologies hinders the dynamic performance. Therefore, to achieve the full benefits of the SRC at high switching frequency, it is necessary to achieve regulation without the need to cross the isolation barrier. Two similar approaches proposed for the SRC achieve regulation by only sending a portion of the resonant current to the output; but are limited to full-bridge rectifiers and are therefore unsuitable for low voltage applications

[3],[4]. A digital approach that uses the conduction difference between the on-resistance and diode voltage drop of a MOSFET was presented in [5], but is limited by clock frequency and resolution capabilities. The same principle was applied to a series-parallel resonant converter [6] but requires a current sensor.

In this work, a fast analog control method using dual-edge PWM is proposed to achieve load regulation without the aforementioned drawbacks. Dual-edge PWM uses a triangular carrier to eliminate of the turn-on delay associated with standard trailing-edge PWM, and turn-off delay associated with leading-edge PWM [7]. By applying it to secondary-side control of a series resonant converter, fast dynamic response can be achieved, and high constant switching frequency can be practically realized.

The circuit is described in the following section, and analyzed in Section III, including the effects this control method has on efficiency. A representative 48V VR operating at 4MHz is designed in Section IV, with simulated verification provided in Section V.

II. CIRCUIT DESCRIPTION

The schematic of the series resonant converter under secondary-side control is given In Fig. 1. The primary circuit consists of an asymmetrical pulse width modulated (APWM [8]) drive train, made up of switches S_1 and S_2, and a series resonant tank made up of capacitor C and inductor L. The chopper circuit operates at constant frequency and varies the duty cycle (D) of S_1 to regulate against line variations and excite the resonant tank by a unipolar square wave with peak magnitude equal to the input voltage. The resonant components convert the driving voltage to a sinusoidal current. The resonant current i, is stepped up by the high frequency transformer and then rectified by the synchronous rectifiers SR_1 and SR_2. The gate signal generation of the SRs is determined by the sign of the resonant current, and the output voltage error signal to ensure regulation against load variations. The rectified current is filtered by C_o, and fed to the load R_L.

Funding provided by Ontario Centres of Excellence, Ontario Research Funds, and Natural Sciences and Engineering Research Council of Canada

978-1-4244-4782-4/10 $26.00 © 2010 IEEE

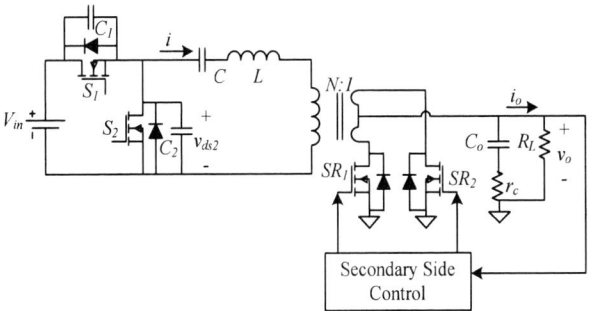

Fig. 1: Constant frequency series resonant converter with secondary-side control

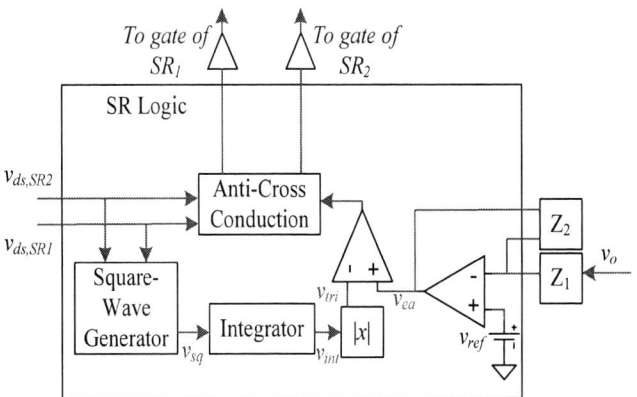

Fig. 2: Block diagram of proposed SR controller

The block diagram of the proposed controller is shown in Fig. 2, with the waveforms shown in Fig. 3. In Fig. 3, green waveforms represent signals that occur in the power circuit, and blue waveforms represent signals internal to the controller. There are two main components to the controller circuit: a PWM carrier wave generator, and an error amplifier for a compensation network. The challenge of synchronizing the SR duty cycle to the centre of the conduction period can be solved by using a triangular carrier signal with peaks that occur at the zero crossing of the resonant. This implies dual-edge PWM synchronized to the resonant current.

Referring to the wave forms in Fig. 3, the generation of the triangular carrier begins by sensing the conduction of the first SR. From that information, a square wave (v_{sq}) is produced that is in phase with the resonant current. Integrating the square wave results in a triangular waveform (v_{int}) that crosses zero at the midpoint of the conduction period, and reaches its peaks at the zero-crossing of the resonant current. By rectifying the triangular wave, the PWM carrier is produced (v_{tri}). It contains only positive peaks which occur at the zero-crossing of the resonant current, and hits zero at the middle of the conduction interval.

As with standard PWM controllers, the user has access to the inverting and output terminals of an internal op-amp in order to create a compensation network for the output voltage error. Thus, Z_1 and Z_2 are defined by the user to suit their particular application. A comparator produces gate pulses when the compensated error signal (red trace on the same axis as v_{tri}) is greater than the PWM signal. The anti-cross conduction block prevents gating only after the opposite SR stops conducting.

III. ANALYSIS

A. Steady-State

Traditionally, steady state analysis of resonant converters uses a fundamental approximation where the load is referred to the output of the resonant tank as an equivalent resistance R_{ac} under the assumption that the filter and rectifiers are lossless [2], as defined by (1).

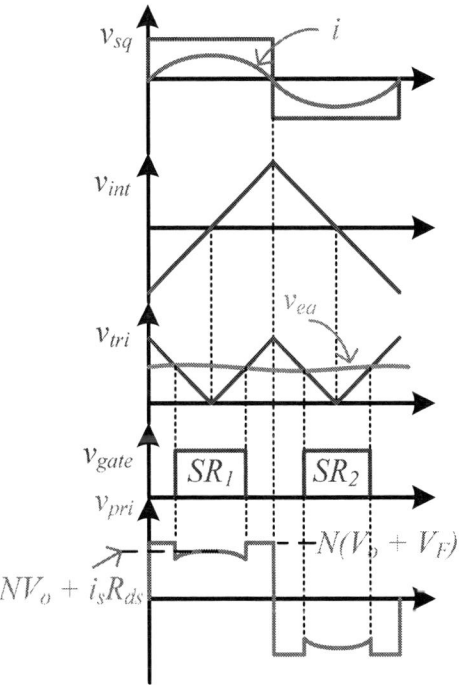

Fig. 3: Waveforms of the proposed SR controller

$$R_{ac,0} = \frac{V_{ac,rms}}{I_{ac,rms}} = \frac{8N^2}{\pi^2} R_L \qquad (1)$$

However, under secondary-side control, R_{ac} has to be redefined to account for the conduction difference between the SR channel and the body diode. The resulting ideal equivalent circuit is shown in Fig. 4.

Under this control method, the resonant current maintains a sinusoidal waveform, so its fundamental component is consistent with traditional results. As illustrated in Fig. 3, practical transformer-primary waveforms are dependent on SR parameters, thus resulting in a redefined fundamental ac voltage. The new ac equivalent resistance is defined by (2), with the conversion factor β defined by (3). $D_{SR}=t_{SR}/T_s$; t_{SR} is the on-time of the SRs; T_s is the switching period; V_F is the diode forward voltage; R_{ds} is the SR on-resistance; $\gamma = \dfrac{V_F}{V_o}$;

$\lambda = \dfrac{R_{ds}}{R_L}$; and N is the transformer turns ratio N_p/N_s. In Fig. 5, the full-load conversion factor is plotted for three different MOSFETs using datasheet values at a junction temperature of $T_j = 100°C$ with a gate drive voltage of 5V. To simplify the plot, γ is kept constant at the peak value of forward voltage for a 25A output. The conversion ratio is at its maximum when the diodes are fully conducting; and is higher with higher diode forward voltages. As the duty cycle increases and the SRs conduct a greater percentage of the switching cycle, the conversion factor reduces. Naturally, the lowest conversion factor is obtained by the switches with the lowest on-resistance. As the on-resistance approaches zero, the conversion factor β approaches unity at unity duty cycle. At this point, the new definition of R_{ac} is identical to the traditional definition.

Using the fundamental circuit of Fig. 4, and accounting for the dc values of the circuit under APWM control [8], the transfer function of the tank can be found with (5), with the variables defined by (6)-(8). The transfer function assumes the resonant tank is excited by a constant magnitude fundamental component to maximize efficiency and allow the secondary-side controller to provide regulation independent of the actions of the primary side.

$$R_{ac} = R_{ac0}\,\beta . \tag{2}$$

$$\beta = 1 + \gamma - \gamma \sin\!\left(\frac{\pi D_{SR}}{2}\right) + \frac{\lambda\pi}{2}\sin\!\left(\frac{\pi D_{SR}}{2}\right) \tag{3}$$

$$R_{ac0} = \frac{8N^2 R_L}{\pi^2} \tag{4}$$

$$\frac{V_o}{V_{in}\sqrt{1-\cos(2\pi D)}} = \frac{1}{2\sqrt{2}N\beta}\,\frac{1}{1+jQ\left(\omega-\dfrac{1}{\omega}\right)} \tag{5}$$

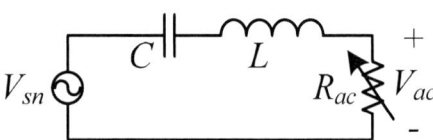

Fig. 4: Fundamental SRC under secondary-side control

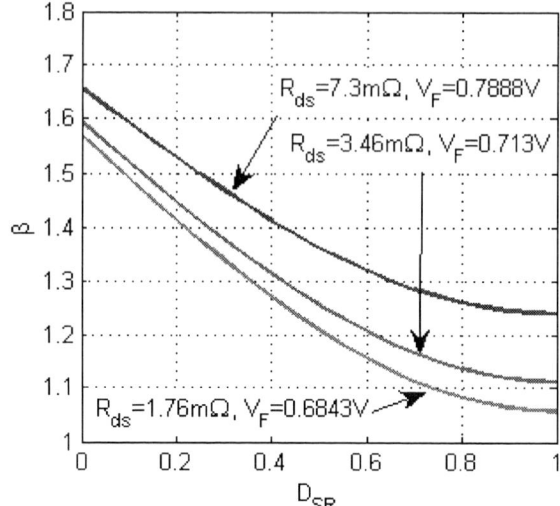

Fig. 5: Conversion factor β for different MOSFET part numbers

$$Q = \frac{\omega_r L}{R_{ac}} \tag{6}$$

$$\omega_r = \frac{1}{\sqrt{LC}} \tag{7}$$

$$\omega = \frac{\omega_0}{\omega_r} \tag{8}$$

The voltage stresses of the resonant components are given by (9) and (10); where V_s is the fundamental component of the drive voltage.

$$\frac{V_C}{V_s} = \frac{-j\dfrac{Q}{\omega}}{1+jQ\left(\omega-\dfrac{1}{\omega}\right)} \tag{9}$$

$$\frac{V_L}{V_s} = \frac{j\omega Q}{1+jQ\left(\omega-\dfrac{1}{\omega}\right)} \tag{10}$$

B. Loss Analysis

The sinusoidal nature of the load current of the SRC, combined with the current-dependent voltage drop of diodes, permits diode conduction with less severe efficiency penalties compared to voltage-type resonant, or PWM topologies. To evaluate the performance of three different MOSFETs ranging in structure from low gate charge/higher on-resistance to low on-resistance/higher gate charge, V_F curves (at $T_j = 100°C$) were taken from the datasheets to obtain quadratic equations modeling $V_F(i(t))$. These equations were then used to find the diode conduction loss at different loads and duty cycles. SR conduction loss is simply the product of the rms value of the current not conducted by the diodes and the channel resistance. In Fig. 6, the power loss as a function of diode conduction is shown at full-load and 25% load for a 25A output. To annotate the curves, the peak forward voltage is labeled on the graphs with the on-resistance of the switch. Note that at full-load, the forward voltage drop is 15-18% higher than the light-load case. Furthermore, the forward voltage of the diode is different for equal duty cycles at different loads. Using these results, efficiency curves for a 30W 48V/1.2V converter operating at 4MHz were produced in Fig. 7. Of substantial importance is the fact that rectification efficiency greater than 87% can be achieved across the load range with low R_{ds} MOSFETs. Further, the variation of efficiency is less than 1%. Therefore, secondary-side control can be employed to achieve transient benefits without compromising efficiency.

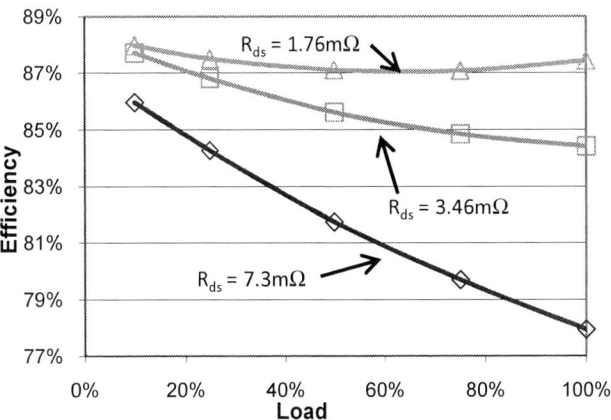

Fig. 6: Rectification conduction loss as a function of diode conduction

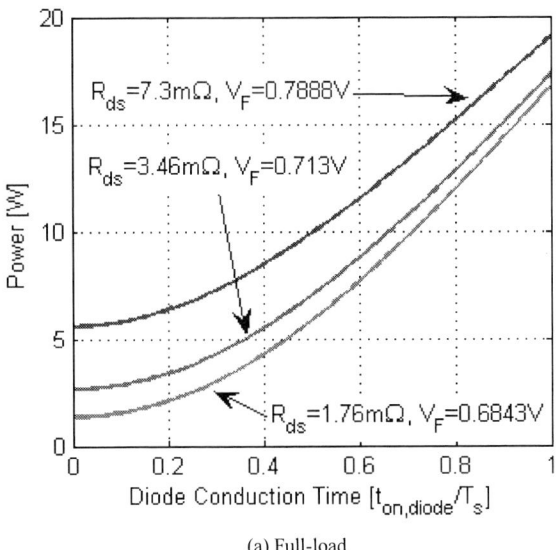

(a) Full-load

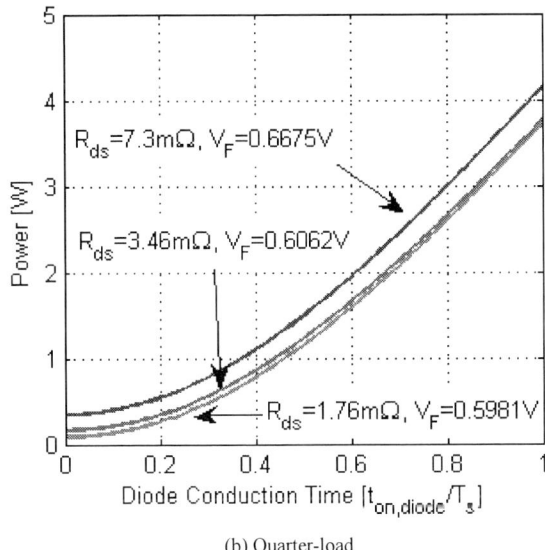

(b) Quarter-load

Fig. 7: Conduction efficiency under secondary-side regulation

IV. DESIGN

In this section, the resonant tank will be designed to meet the converter specifications given in Table I.

TABLE I: 48V CONVERTER SPECIFICATIONS

Parameter	Value
Input Voltage	43-53V
Output Voltage	1.2V
Output Power	30W
Operating Frequency	4MHz

A graphical approach to selecting the resonant tank parameters is the most insightful. Hence, it will be used here. In Fig. 8, the gain of the resonant converter is plotted against D_{SR} for different ω and Q values. It is assumed that V_F = 0.7V, and R_{ds} = 2.5mΩ. As ω increases, the influence of Q is more pronounced. As Q increases, the peak gain is reduced. Therefore, it would appear that for maximum controllability, low Q and moderate ω are desirable. Voltage stress curves of the resonant components are presented in Fig. 9 and Fig. 10. Reduced stress occurs with low Q and increased ω. Taking resonant component stress into account, along with the transfer characteristics, resonant parameters ω = 1.15 and Q = 1.25 are selected; which translates to component values of L = 518nH, and C = 3.7nF.

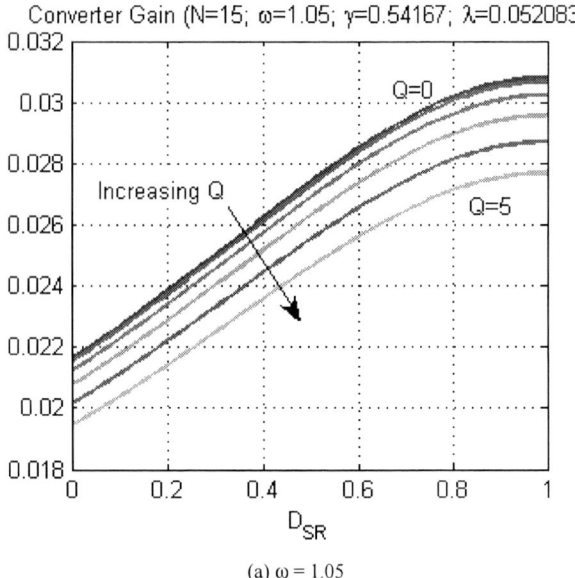

(a) ω = 1.05

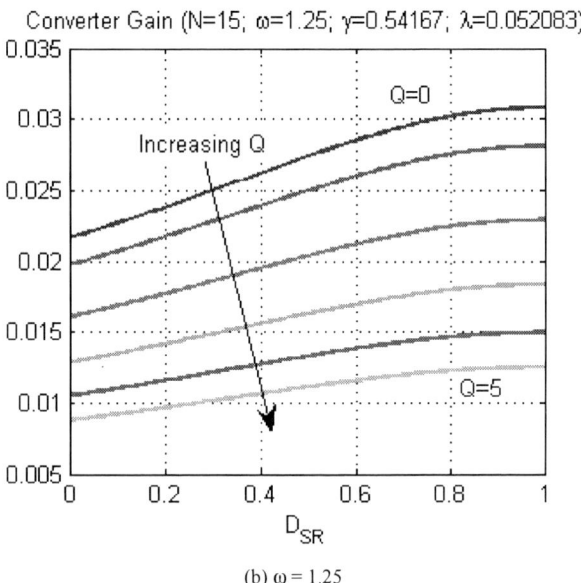

(b) ω = 1.25

Fig. 8: Converter gain for different tank parameters

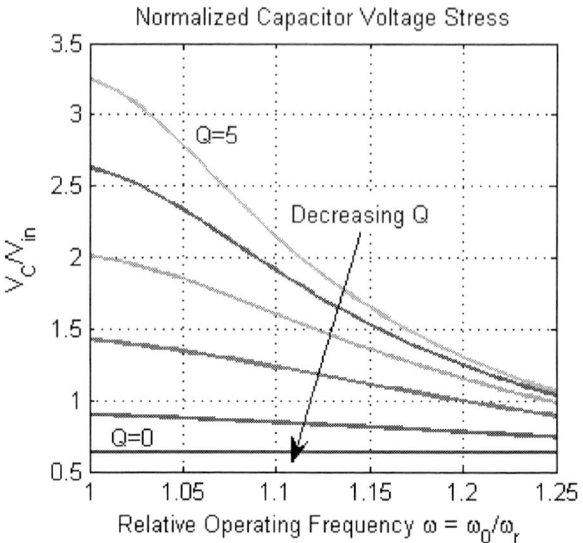

Fig. 9: Voltage stress of resonant capacitor

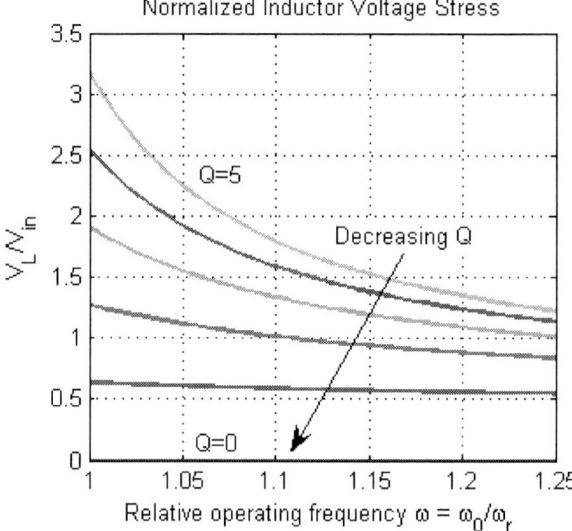

Fig. 10: Voltage stress of resonant inductor

V. RESULTS

The designed converter was simulated in spice simulator SIMetrix. In the following figure the drain-source voltage across the two SRs is on the top grid, the resonant current on the middle grid, and the triangular PWM carrier on the bottom. Notice the peak of the triangular waveform occurs at the zero crossing of the resonant current, and valley at the midpoint of the conduction period. Thus, the controller behaves as expected.

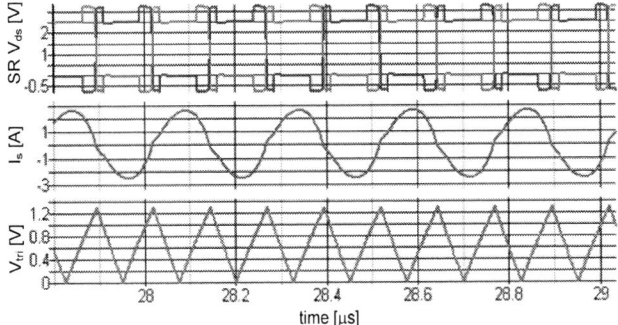

Fig. 11: Controller/converter waveforms at full-load

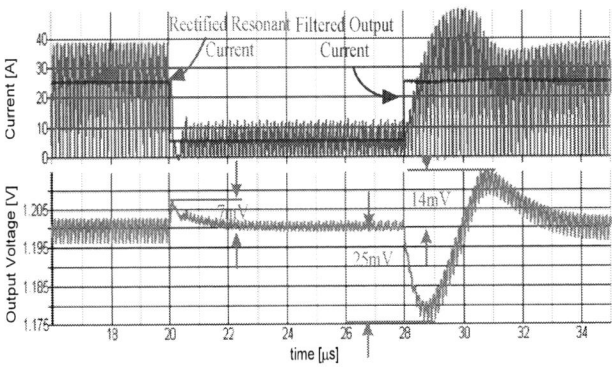

Fig. 12: Simulated load transient results

To test the transient capabilities, a 20A load step was applied with a slew rate of 100A/ns. The results are shown in Fig. 12. As annotated on the figure, a negative load step results in a 7mV (or 0.5%) overshoot, while a positive load step results in a 25mV (2.083%) undershoot, followed by a 14mV (1.16%) overshoot. The filter capacitance used in simulation was 300μF, with ESR of 150μΩ

VI. CONCLUSION

In this paper, a dual-edge PWM controller was proposed for secondary-side control of current-type resonant converters operating at constant frequency. Analysis and a design procedure were presented. A 48V/1.2V, 30W series resonant converter operating at a constant 4MHz was simulated in the spice simulator SIMetrix. The results validate the design, and show good transient performance under an 80% load step. It was also shown that high efficiency can be maintained across the load range due to the sinusoidal nature of the resonant current, and current-dependent forward voltage of diodes. Hence, fast transient performance can be achieved without a conduction loss penalty.

VII. REFERENCES

[1] M. Ye, P. Xu, B. Yang, and F.C. Lee, "Investigation of Topology Candidates for 48V VRM," in *Proc. IEEE Applied Power Elec. Conf. and Expo.*, 2002, pp. 699-705.

[2] R.L. Steigerwald, "A Comparison of Half-Bridge Resonant Converter Topologies," *IEEE Trans. Power Elec.*, vol. 3, pp. 174-182, April 1988.

[3] L. Rossetto, and G. Spiazzi, "Series resonant converter with wide load range," in *Proc. Industry App. Conf.*, 1998, pp. 1326-1331.

[4] A. Conesa, R. Pique, E. Fossas, "The serial resonant converter with controlled rectifier stage," in *Proc. Euro. Conf. on Power Elec. and App.*, 2005, pp. 1-10.

[5] S. Pan, and P.K. Jain, "Secondary-side adaptive digital controlled series resonant dc-dc converters for low voltage high current applications," in *Proc. Power Elec. Specialists Conf.*, 2008, pp. 711-717.

[6] M.Z. Youssef, and P.K. Jain, "An advanced design solution for the 48V isolated voltage regulator modules," in *Proc. IEEE Int. Symp. Industrial Elec.*, 2006, pp. 1036-1041.

[7] W. Qiu, G. Miller, and Z. Liang, "Dual-edge pulse width modulation scheme for fast transient response of multiple-phase voltage regulators," in *Proc. Power Elec. Specialists Conf.*, 2007, pp. 1563-1569.

[8] P.K. Jain, A. St-Martin, and G. Edwards, "Asymmetrical pulse-width-modulated resonant DC/DC converter topologies," *IEEE Trans. Power Elec.*, vol. 11, pp. 413-422, May 1996.

A Non-Insulated Resonant Boost Converter

Peng Shuai[*], Yales R. De Novaes[†], Francisco Canales[†] and Ivo Barbi[‡]

[*]ISEA-Institute for Power Electronics and Electrical Drives, RWTH-Aachen University, Aachen, Germany
Email: Peng.Shuai@isea.rwth-aachen.de
[†]ABB Corporate Research, Dättwil, Switzerland
Email: {yales.de-novaes, francisco.canales}@ch.abb.com
[‡]INEP-Power Electronics Institute, UFSC-Federal University of Santa Catarina, Florianópolis, SC, Brazil

Abstract— In this paper, a resonant boost converter is analyzed and verified through experimentation. Switching losses are reduced since the converter operates at ZCS (Zero Current Switching). The analysis presented here covers its operation with variable switching frequency but only below the resonant frequency. The range of its voltage gain goes from 1 to 2 and the energy conversion efficiency (including filtering at input) is around 97% for the lowest input voltage. Its envisioned that it could be applied as a front-end converter when a back-end inverter is needed, for instance with batteries, photovoltaic converters or fuel cells, perform partial output voltage regulation only.

I. INTRODUCTION

Nowadays, resonant techniques have widely been utilized in power electronics converters. Compared with the conventional PWM converters, the switching losses of the resonant converters are significantly reduced due to the soft-switching properties. This allows for increasing the switching frequency to levels as high as 1 MHz and drastically increasing the power density compared to hard switched converters. Electromagnetic interference is usually less critical for resonant converters since there are no spikes during commutations (or they are reduced). Although resonant commutations have been utilized long time ago with thyristor semiconductors, the resonant conversion evolved from resonant converters to quasi-resonant converters and multi-resonant converters [1] [2] [3]. Resonant converters contain resonant L-C networks and the voltage and current of this networks vary sinusoidally in one or more commutation intervals [4]. The commutation of the switches is usually with zero-voltage switching (ZVS) or zero current switching (ZCS).

In this paper, a two-switch non-insulated DC-DC boost converter using resonant technique is proposed, investigated and validated at 1 kW and maximum switching frequency of 100 kHz. The only switching losses of this converter occur because of the energy stored in the parasitic capacitance of the active switch. The original document where this topology is proposed among other resonant and PWM converter is [6]. But by coincidence, the commutation cell of this converter can be extracted from the PWM multilevel converter presented by [16], however the capacitor has a different functionality since in the resonant converter it is completely charged and discharged during operation. This topology can also be seen in [9] where a snubber for the NPC inverter is presented. It

becomes more clear if one looks at half of the modulation cycle of the inverter operating with no load.

In this work, the analysis is carried in detail by describing every topological stage and its equivalent equations. The equation of the static characteristic is obtained describing the voltage gain as function of the load current and switching frequency. The voltage gain of this converter can be controlled by controlling the switching frequency, the minimum value is 1 and the maximum is 2. Since the maximum gain is limited, it is envisioned that this converter could be a good option if the primary energy source would be a battery or photovoltaic string and the output of this converter would be connected to an inverter. By doing so, assuming a variating input voltage, the resonant converter could regulate the output voltage partially, limit the lowest value seen by the inverter. This is depicted in Fig. 1, where the resonant converter regulates the voltage from 1 to 2 pu, and the inverter accepts the variation from 2 to 2.5 pu without compromising its output voltage and current quality. In an application, for instance, ideally the input voltage could variate from 200 V to 500V, and the resonant converter boosts and regulates its output voltage to 400V while the input voltage is lower than 400V. When the input voltage is higher than 400V, then the converter could be by-passed.

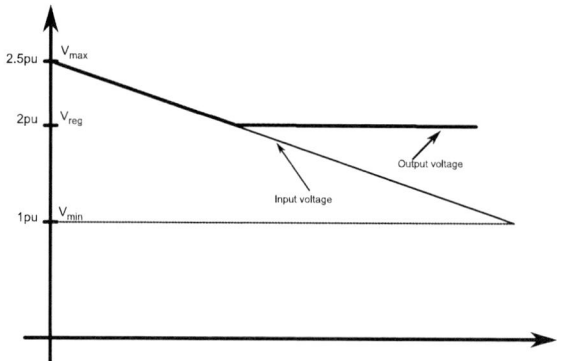

Fig. 1: Input voltage variation and regulation range.

II. CONVERTER OPERATION PRINCIPLE

The converter topology investigated in this paper is shown in Fig. 2. The resonant tank of this non-insulated converter

978-1-4244-4782-4/10 $26.00 © 2010 IEEE

is composed of the resonant inductor L_r and the resonant capacitor C_r. Two switches and two diodes are represented as ideal devices. In addition, it is assumed that the output capacitor C_o is large enough, so that the output voltage V_o is kept constant during one switching cycle. The control signals

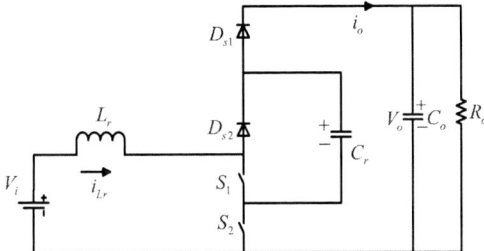

Fig. 2: Converter Topology.

of the two switches are two complementary signals with 50% duty cycle. Each switch turns on for a half of the switching cycle and in each half cycle there are three switching stages according to the operation of the resonant tank. The operation stages of the converter in a switching cycle are illustrated in Fig. 3.

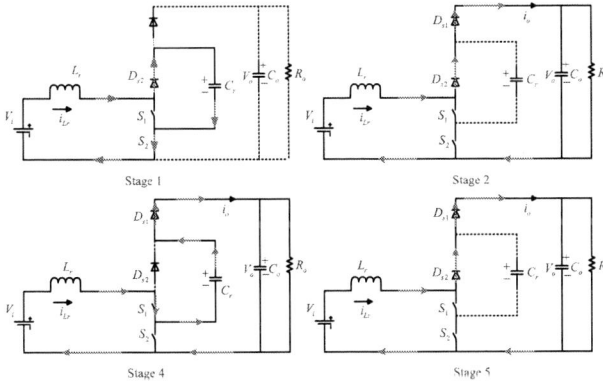

Fig. 3: Representation of the main topological stages.

A. Switching Stage 1 [t_0,t_1]

At t_0, S_1 is turned off and S_2 is switched on, diode D_{s2} is conducting. During this stage, the resonant capacitor C_r is charged by the source to the voltage level of the output voltage V_o. Due to the resonance, the current through the resonant inductor L_r increases sinusoidally from 0 to a certain value, which is supposed to be I_1. This stage can be described with the following two equations:

$$V_i = L_r \frac{di_{Lr}}{dt} + v_{Cr}(t) \tag{1}$$

$$i_{Lr}(t) = C_r \frac{dv_{Cr}(t)}{dt} \tag{2}$$

By assuming that the initial condition for this stage is:

$$\begin{cases} i_{Lr}(t_0) = 0 \\ v_{Cr}(t_0) = 0 \end{cases}$$

After mathematical transformation and calculation, one can obtain the following equations:

$$i_{Lr}(t) = \frac{V_i}{L_r} \frac{\sin(\omega_0 t)}{\omega_0} \tag{3}$$

$$v_{Cr}(t) = V_i[1 - \cos(\omega_0 t)] \tag{4}$$

Where $\omega_0 = 1/\sqrt{L_r C_r}$ is the resonant angular frequency. The inductor current can be parameterized as a function of the input voltage and the resonant circuit impedance ($Z_r = \sqrt{L_r/C_r}$), and the resonant capacitor voltage can be parameterized as a function of the input voltage, as follows:

$$\overline{i_{Lr}(t)} = \frac{i_{Lr}(t)}{V_i/Z_r} \tag{5}$$

$$\overline{v_{Cr}(t)} = \frac{v_{Cr}(t)}{V_i} \tag{6}$$

So, for this topological stage, by applying (5) and (6) to (3) and (4), the parameterized inductor current and capacitor voltage of the resonant tank are:

$$\overline{i_{Lr}(t)} = \sin(\omega_0 t) \tag{7}$$

$$\overline{v_{Cr}(t)} = 1 - \cos(\omega_0 t) \tag{8}$$

At the end of this stage, at the time instant t_1, the following equations are valid:

$$\overline{i_{Lr}(t_1)} = \overline{I_1} \tag{9}$$

$$\overline{v_{Cr}(t_1)} = 1 - \cos(\omega_0 t_1) \tag{10}$$

where I_1 is defined as the final condition of the inductor current at t_1 ($\overline{I_1}$ is parameterized). At this time instant, the resonant capacitor voltage is equal to the output voltage V_o, therefore:

$$\overline{v_{Cr}(t_1)} = G \tag{11}$$

Where $G = V_o/V_i$ is also the voltage gain of the converter. From (10) and (11) the duration of this topological stage can be calculated as:

$$\omega_0 \Delta t_{10} = \pi - \arccos(G - 1) \tag{12}$$

Where $\Delta t_{10} = t_1 - t_0$.

As usually done in resonant converter analysis [10], a vector z can be defined as per (13).

$$z = \overline{v_{Cr}(t)} + j\overline{i_{Lr}(t)} \tag{13}$$

The real part of the vector z stands for the voltage on the resonant capacitor while the imaginary part represents the current through the resonant inductor. So the first stage can be described by the following vector:

$$z_1 = 1 - \cos(\omega_0 t) + j\sin(\omega_0 t) - 1 - e^{-j\omega_0 t} \tag{14}$$

This vector shall be utilized in a next section to build a state-plane.

B. Switching Stage 2 [t_1, t_2]

At t_1, the resonant capacitor voltage v_{Cr} is equal to the output voltage V_o, the diode D_{s1} turns on. So in this stage v_{Cr} is clamped as V_o, while the current through the inductor i_{Lr} drops linelly to zero, since the output voltage is higher than the input voltage, a negative voltage is applied across L_r. By similar mathematical calculation, the vector in a state-plane can be derived as:

$$z_2 = G + j[\overline{I_1} - (G-1)\omega_0(t-t_1)] \qquad (15)$$

The duration of this stage can be calculated as following:

$$\omega_0 \Delta t_{21} = \frac{\overline{I_1}}{G-1} \qquad (16)$$

C. Stage 3 [t_2, t_3]

As the current becomes 0 at the end of the second stage, D_{s2} blocks, so there is no current through L_r, and the voltage across C_r remains at V_o as in stage 2. In this stage, no current is circulating in the circuit. The vector to describe this stage is then quite simple:

$$z_3 = G \qquad (17)$$

The end of this switching stage is half of the whole switching cycle, which means $\omega_0 t_3 = \pi$. So:

$$\omega_0 \Delta t_{32} = \pi - \omega_0 \Delta t_{10} - \omega_0 \Delta t_{21} \qquad (18)$$

D. Stage 4 [t_3, t_4]

At the beginning of this stage, S_1 is turned-on and D_{s1} starts to conduct the resonant current. D_{s2} remains blocked. The resonant capacitor is discharged, so the voltage across it drops from V_o to 0. At the same time, the current through the inductor increases from 0 to I_1. The operation of the converter is similar as in the first stage. One can obtain similar vector as for the first stage:

$$z_4 = G - 1 - \cos(\omega_0 t) - j\sin(\omega_0 t) = G - 1 - e^{-j\omega_0 t} \quad (19)$$

This stage has the same duration of the first stage:

$$\omega_0 \Delta t_{43} = \pi - \arccos(G-1) \qquad (20)$$

E. Stage 5 [t_4, t_5]

The operation of the converter in this stage is quite similar as in stage 2, difference is that the resonant capacitor voltage keeps at zero. The vector related to this stage is:

$$z_5 = j[\overline{I_1} - (G-1)\omega_0(t-t_4)] \qquad (21)$$

The duration of stage 5 is also the same as stage 2, so:

$$\omega_0 \Delta t_{54} = \frac{\overline{I_1}}{G-1} \qquad (22)$$

F. Stage 6 [t_5, t_6]

At the end of stage 5, the current drops to 0 and there is no voltage across the resonant capacitor. Thus, in this last stage there is no current through L_r and no voltage across C_r. Therefore, the vector of this stage is equal to zero:

$$z_6 = 0 \qquad (23)$$

And the duration is:

$$\omega_0 \Delta t_{65} = \pi - \omega_0 \Delta t_{43} - \omega_0 \Delta t_{54} \qquad (24)$$

G. Summary of the switching behavior

The main waveforms of voltages and currents of the components are shown in Fig. 4. The waveforms of the two switches

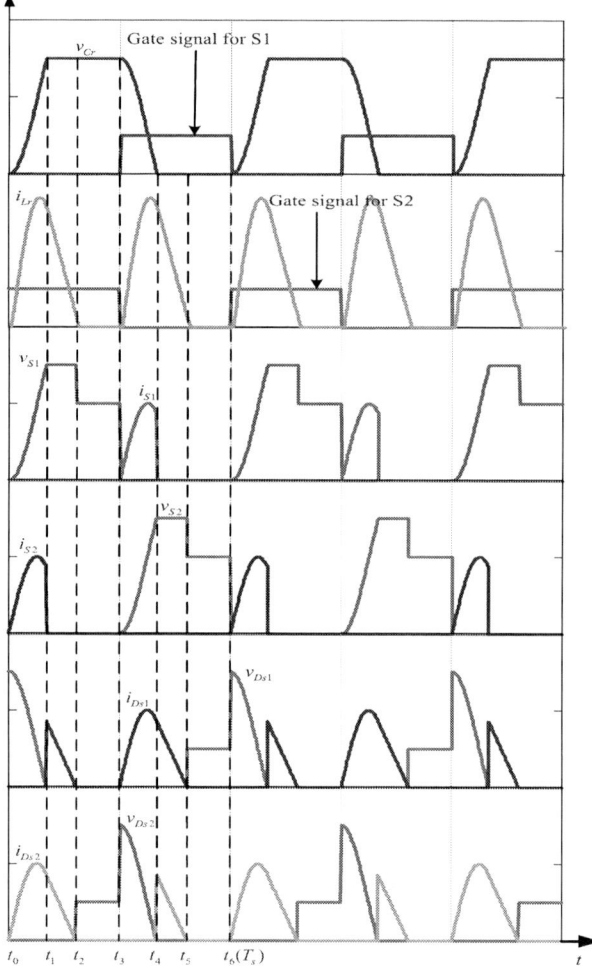

Fig. 4: Converter waveforms based on analysis with ideal components.

and two diodes are complementary with each other in each switching cycle. The ZCS (Zero Current Switching) can be clearly seen when looking to the instantaneous values of the voltages and currents of the active switches. In regarding the resonant circuit, the voltage across the resonant capacitor v_{Cr} is charged to V_o and then clamped at this value during the first half switching cycle. Then in the second half switching cycle the resonant capacitor is discharged and then v_{Cr} becomes zero. The frequency of the current through the resonant inductor i_{Lr} is twice of the switching frequency. The average value of the current through Ds_1 diode is dependent on the switching frequency, then the output voltage can be regulated by the ratio between the switching frequency and the resonant frequency.

Based on the analysis above, the complete state-plane graph for the vector z in a switching cycle can be depicted. The real axis is the parameterized resonant capacitor voltage $\overline{v_{Cr}}$, while the imaginary axis is the resonant inductor current $\overline{i_{Lr}}$, see Fig. 5. Following the direction of the arrows the variations of the current and voltage during the switching cycle can be seen. The maximum value of the inductor current can be found

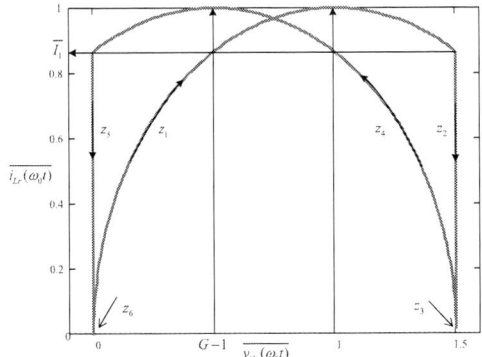

Fig. 5: State-Plane Graph.

in the graph as: $\overline{i_{Lr}(\omega_0 t)}_{max} = 1$, when $\overline{v_{Cr}(\omega_0 t)} = 1$ or $\overline{v_{Cr}(\omega_0 t)} = G - 1$.

III. CONVERTER EXTERNAL CHARACTERISTIC

Based on the switching behavior of the converter, the parameterized average output current can be calculated by the following equations:

$$i_{oAVG} = f_s \left(\left(\int_{t_1}^{t_2} i_{Lr}(t)dt + \int_{t_3}^{t_4} i_{Lr}(t)dt + \right. \right. \quad (25)$$

$$\left. \left. + \int_{t_4}^{t_5} i_{Lr}(t)dt \right) \right) \quad (26)$$

where f_s is the switching frequency of the converter.

$$f_s = \frac{1}{T_s} \quad (27)$$

$$\overline{i_{oAVG}} = \mu_0 \frac{G}{2\pi(G-1)} \quad (28)$$

In this equation $\mu_0 = 2\pi f_s/\omega_0$, which is the ratio between the switching frequency f_s and the resonant frequency f_0. The time instants can be obtained by the switching stage duration calculated previously. This equation describes the external

characteristic of the converter and is depicted in Fig. 6 for several values of μ_0. The ideal gain is limited between 1 and 2 for the full range of frequency variation. This means that the output voltage cannot be lower than the input voltage (there would be forward conduction of both diodes), and cannot be higher the twice the input voltage.

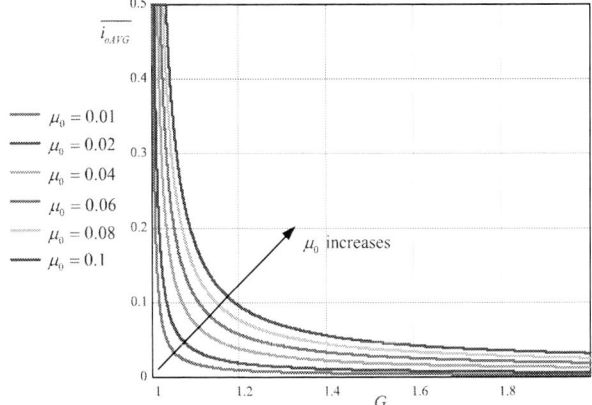

Fig. 6: Converter gain as a function of the average output current, having the frequency ratio as parameter.

It is important to highlight that in this study it is assumed that the switching frequency is always lower than the resonant frequency, i.e. $f_s \leq f_0$ or $0 < \mu_0 \leq 1$. By using the resonant impedance Z_r to parameterize the load resistance, the following ratio can be introduced:

$$r_o = R_o/Z_r \quad (29)$$

With this ratio, the following equation can be derived:

$$G = r_o \overline{i_{oAVG}} \quad (30)$$

By substituting (28) and (29) into (30), (31) can be obtained.

$$G = \frac{\mu_0}{2\pi} r_o + 1 \quad (31)$$

The relationship described by (31) is shown in Fig. 7. It can be

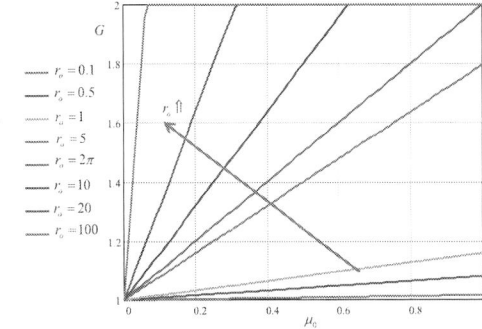

Fig. 7: Dependence of Gain G on frequency ratio and gain.

seen that, for a constant load and constant input voltage, the output voltage changes linearly with μ_0. This means the output voltage can be easily to control. Now, if it is necessary to operate the converter over the full range of gain variation, the relation (32) has to be respected. If a wide range of frequency variation is required, then r_o should be equal to 2π.

$$r_o \geq 2\pi \qquad (32)$$

IV. EXPERIMENTAL VALIDATION

In order to verify the theoretical analysis and to verify the concepts, a prototype rated at the following specifications has been built and tested:

Output power: $P_o = 1kW$

Regulated output voltage: $V_o = 400V$

Input voltage: $200V \leq V_i \leq 400V$

$r_o = 2\pi \rightarrow Z_r = R_o/r_o = 25.46\Omega$

Resonant inductor and capacitor: $L_r = 39.69\mu H$, $C_r = 61.2nF$

Resonant frequency: $f_0 = 102.1kHz$

Output capacitor: $C_o = 110\mu F$

Another constrain added to the specification is that the converter should be able to withstand an input voltage of 500 V. In this case, above 400V there will not be regulation of the output voltage and this converter could be bypassed by an additional diode or a mechanical switch. A supposed application where this makes sense would be a two stages converter where the second stage could be an inverter able to regulate its AC variables while having its input voltage variating from 400 to 500V maximum. The utilized silicon semiconductor devices were rated at 600 V as breakdown voltage.

Fig. 8 shows the resonant inductor current i_{Lr}, resonant capacitor voltage v_{Cr} and the input and output voltage for operation at maximum power and minimum input voltage. As expected, since this frequency is almost the resonant frequency, the waveforms of the resonant tank are sinusoidal. The input voltage is 200V while the output voltage is 400V, the maximum gain G=2 is reached. The peak of v_{Cr} is equal to the output voltage while the peak of i_{Lr} is approximately 7.8A as calculated from V_i/Zr.

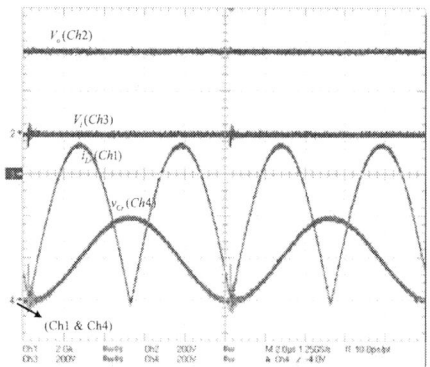

Fig. 8: i_{Lr}, v_{Cr}, V_i and V_o @ $f_s = 100kHz$, $P_o = 1kW$.

Fig 9 is showing the active switch commutations. The turn-on is depicted in Fig. 9 (a). From this figure it can be seen that there are reduced switching losses since the current increases sinusoidally from zero. The only switching losses occur due to the intrinsic capacitance of the active switch, in this case MOSFET. The turn-off behavior is shown in Fig. 9 (b). As the instantaneous values of current and voltage are not overlapping the turn-off losses can be neglected.

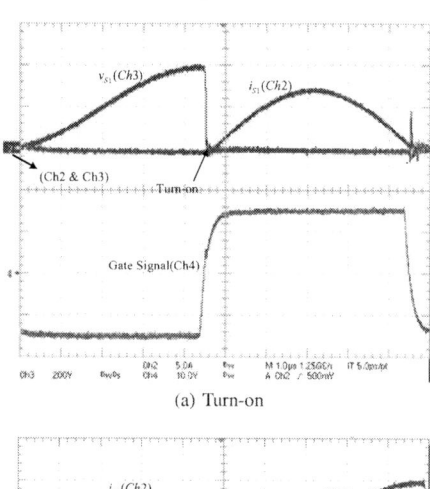

(a) Turn-on

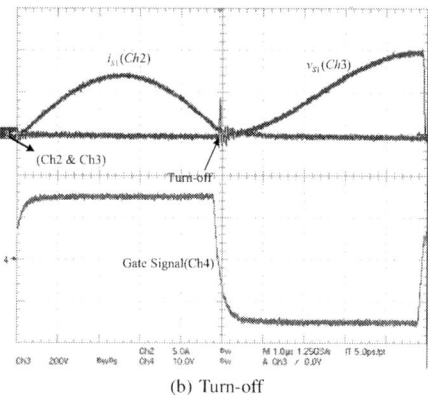

(b) Turn-off

Fig. 9: Commutation of the switching devices at $f_s = 100kHz$.

The waveforms of the diode current and reverse voltage are shown in Fig. 10. The reverse recovery of the diode can hardly be found from the waveform, so almost no loss is produced by the reverse recovery current.

Fig. 11 illustrates the behaviors of the current and voltage of MOSFET at $f_s = 50kHz$. The oscillations in the MOSFET voltage and current waveforms are due to the parasitic capacitance.

In order to verify the operation at high input voltage, the converter has been tested at $f_s = 5kHz$ and full power, the waveforms can be found in Fig. 12. At this switching frequency, the input voltage is quite close to the output voltage. As the designed specification, the input voltage can vary from 200V to 400V, while keeping the output voltage at 400V and the output power at 1kW.

The efficiency curve of this converter, considering the

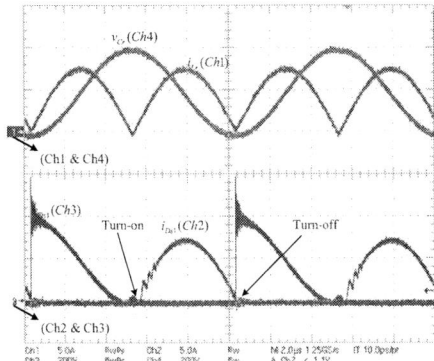

Fig. 10: Diode reverse voltage and current @ $f_s = 100kHz, P_o = 1kW$.

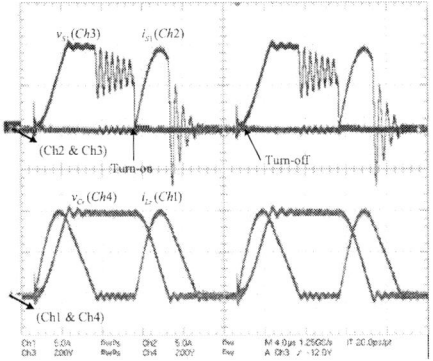

Fig. 11: MOSFET current and voltage behaviors @ $f_s = 50kHz, P_o = 1kW$.

variation of the output power while keeping the input and output voltage constants and gain equal to 2 is depicted in Fig. 13. At nominal power the efficiency is about 97 %. The highest efficiency over the load range is at around 60% of the full load, which is about 97.8%. The efficiency curve has also been obtained considering input voltage variation, while keeping the output voltage constant at 400V and output power at 1 kW. The results are presented in Fig. 14. As expected for this converter, for higher input voltage the efficiency is higher.

V. CONCLUSION

A two-switches boost resonant converter capable of stepping-up the input voltage by 2 times has been analyzed and proposed in this paper. The utilized mechanism to control the power flow is implemented by variating the switching frequency, while keeping it below the resonant frequency. Both active switch commutations are soft and the only switching losses occur due to the energy stored in the parasitic capacitance of the active switch.

Detailed analysis and experimentation shown that this converter has potential for application where high efficiency and simplicity are needed while its limited gain would not be a drawback. With an appropriate dimensioning of the resonant

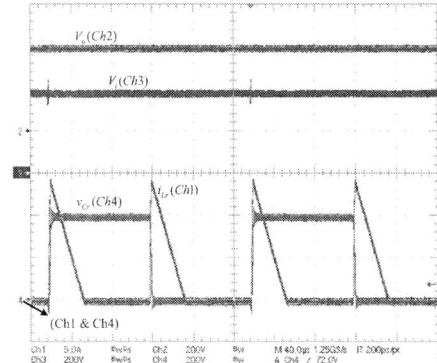

Fig. 12: i_{Lr}, v_{Cr}, V_i and V_o @ $f_s = 5kHz, P_o = 1kW$.

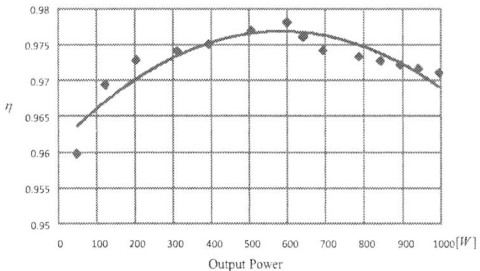

Fig. 13: Efficiency curve of 1kW prototype by power variation.

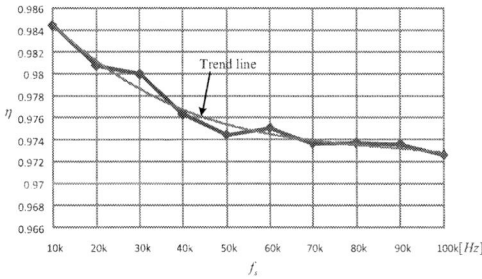

Fig. 14: Efficiency as a function of switching frequency (or input voltage) with constant output power and voltage.

components, the converter reactive energy circulation could be kept at low values, not compromising the efficiency. An efficiency around 97 % was obtained by utilizing 600 V MOSFETs (80 $m\Omega$).

Due to its reduced switching losses, this converter has potential for applications where high power density is required.

REFERENCES

[1] F.C. Lee, *"High-Frequency Quasi-Resonant and Multi-Resonant Converter Technologies"*, Proceedings of the International Conference on Industrial Electronics, Singapore, October 24-28,1988, pp.509-521.

[2] W.A. Tabisz and M.M. Jovanović and F.C. Lee, *"High-Frequency Multi-Resonant Converter Technology and Its Applications"*, Proceedings of the International Conference on Power Electronics and Variable Speed Drives, London, England, July 17-19,1990, pp.1-8.

[3] G. Hua and F.C. Lee, *"An Overview of Sof-Switching Techniques for PWM Converters"*, Proceedings of the International Conference on Power

978-1-4244-4782-4/10 $26.00 © 2010 IEEE

Electronics and Motion Control, Beijing, China, June 27-30,1994, pp.801-808.

[4] Robert W. Erickson and Dragan Maksimović, *"Fundamentals of Power Electronics"*, 2nd ed. New York, Boston, Dordrecht, London, Moscow: Kluwer Academic Publishers, 2004.

[5] Muhammad H. Rashid, *"Power Electronics Handbook"*, San Diego, San Francisco, New York, Boston, London, Sydney, Tokyo: Academic Press, 2001.

[6] Barbi, I. and Tomaselli, L. C. and Guedes, J. A. M., *"Buck, Boost and Buck-Boost resonant converters with soft switching and clamped capacitor voltage"*, Internal Report, INEP/UFSC, 1999.

[7] T.M. Undeland, *"Snubbers for Pulse Width Modulated Bridge Converters With Power Transistors or GTOs"*, Proceedings of IPEC, vol.1, Tokyo, 1983, pp.313-323.

[8] Péres, Adriano and Barbi, Ivo, *"Experimental results of the new ZVS PWM voltage source inverter with active voltage clamping and comparison with classical structures"*, Proceedings of Telecommunications Energy Conference, Phoenix, AZ, 2000, pp.173-179.

[9] De Novaes, Y. R. and Barbi, Ivo, *"Analysis, Design and Experimentation of a Snubber for the Three-level Neutral Clamped Inverter"*, Proceedings of Congresso brasileiro de eletrônica de potência, Florianópolia, 2001.

[10] M.M. Jovanović and K.H. Liu and R. Qruganti and F.C. Lee, *"State-Plane Analysis of Quasi-Resonant Converters"*, IEEE Transactions on Power Electronics, vol. 2, pp.56-73, January, 1987.

[11] K.D.T. Ngo, *"Generalization of Resonant Switches and Quasi-Resonant DC-DC Converters"*, Proceedings of IEEE Power Electronics Specialists Conference, Blaksburg, Va, 1987, pp.395-403.

[12] Jai P. Agrawal, *"Power Electronic Systems: Theory and Design"*, Upper Saddle River, NJ: Prentice Hall, 2000.

[13] Seri Lee, *"Optimum Design and Selection of Heat Sinks"*, IEEE Transactions on Components, Packaging, and Manufacturing Technology, Part A, vol. 18, pp.812-817, December, 1995.

[14] Ron Lenk, *"Practical Design of Power Supplies"*, Mountain View, California: Wiley-IEEE Press, 2005.

[15] Mark J.Nave, *"Multi-level conversion: high voltage choppers and voltage-source inverters"*, New York: Van Nostrand Reinhold, 1991.

[16] Meynard, T.A. and Foch, H., *"Power Line Filter Design for Switched-Mode Power Supplies"*, Power Electronics Specialists Conference, PESC92, 1992, vol.1, pg 397-403.

Analysis and Design of a Low-Profile Resonant LCC Converter

A. Pawellek, A. Bucher, T. Duerbaum
Friedrich-Alexander-Universität Erlangen-Nuremberg
Cauerstraße 7
91058 Erlangen

Abstract—In addition to high efficiency of switch mode power supplies, miniaturization of converters is an increasingly important aspect. In order to meet safety regulations as well as the design goal of flatness, complicated integrated magnetics are proposed in literature. Unfortunately, their design is difficult and the costs of these components are rather high. This paper focuses on the resonant LCC converter with capacitive output filter, which is a promising topology with respect to low-profile designs. The transformer necessary for mains isolation is realized on a ring core, thus regulation requirements are combined with a low-cost magnetic component. The exact analysis of the LCC converter in the time domain is presented and the solutions of the four important modes of the converter are derived. A prototype was designed in order to demonstrate the feasibility of the proposed approach with an application scenario typical for notebook adapters.

I. INTRODUCTION

Nowadays, switch mode power supplies can be found in applications for virtually all power classes. The omnipresent trend within the field of power electronics towards miniaturization is pushing switching frequencies to higher levels in order to reduce the size of the passive components [1]. Furthermore, high frequency operation is desirable with respect to a fast converter transient response, which is also an increasingly important aspect. Conventional converters based on pulse-width modulated topologies are limited in terms of high switching frequencies, as hard-switching conditions occur, causing significant switching losses [2]. Therefore, countermeasures such as soft-switching snubbers or advanced PWM topologies have to be taken against this effect for the purpose of increasing the switching frequency with these topologies. Furthermore, the almost square and triangle voltage/current waveforms of PWM converters have high frequency harmonic components, which result in a high effort on filtering to fulfill the requirements of EMI regulations. In addition to that, parasitic components gain influence at high frequencies and have to be taken into account during the design process.

A more suitable family of converters for high frequency operation are resonant converters. Their operation allows for the semiconductor devices to switch under zero-voltage (ZVS) or zero-current-conditions (ZCS) without any additional efforts. Thus, switching losses are nearly completely avoided and high switching frequencies can be achieved without sacrificing converter efficiency. The quasi-sinusoidal current waveforms are EMI friendly and additionally, parasitic elements are included as a part of the resonant tank.

Resonant converters can be categorized on the one hand by the number, kind and arrangement of the components in the resonant tank and on the other hand by the kind of the output filter. The most simple member of this family is the LC converter, with the tank containing one inductance and one capacitance. Depending on the arrangement of both reactances, two basic types can be distinguished known as series-resonant and parallel-resonant converter [3-5]. The series-resonant converter has good part load efficiency, but cannot regulate the output voltage in the case of no-load. On the other hand, the parallel-resonant converter can handle no-load conditions, with decreased efficiency at low load. In order to combine the advantages of both converters, resonant converters with a higher number of resonant elements can be used. A summary of the arrangement of three and four resonant elements is given in [6] and [7].

The trend towards miniaturization and low-profile results in high requirements for the magnetic components. A suitable resonant converter for low-profile applications is the LCC converter. By means of a ring core as transformer geometry, very low converter heights can be realized. The effort to design the magnetic component is reduced and with a classical wire wound winding, no integrated magnetics are necessary. Furthermore, ring cores are inexpensive and without an air gap, fringing field effects regarding additional proximity losses are avoided.

This paper focuses on the resonant LCC converter with capacitive output filter. The principle of operation is discussed in Section II, the analysis of the converter in the time domain for the four important modes follows. Section IV contains several evaluations of the derived equations, with special attention being paid to the boundaries between different modes as well as the switching behavior. Based on the results obtained by this analysis, a converter is designed and built in section V.

II. OPERATING PRINCIPLE

Fig. 1 illustrates the basic schematic of the LCC converter with capacitive output filter, driven by a full-bridge configuration. The resonant tank consists of the three reactive components C_s, L_s and $\acute{C}_p$. The actual transformer is represented by a

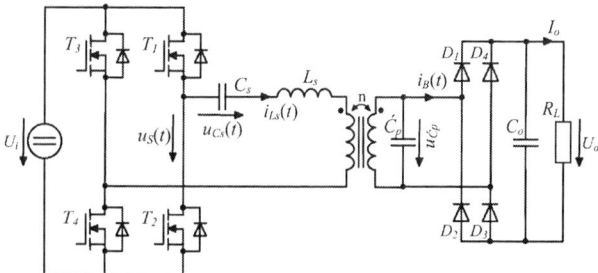

Figure 1. LCC converter with capacitive output filter

cantilever equivalent circuit, whereas the magnetizing inductance is assumed to be large enough to be neglected. The parallel capacitance is connected to the secondary of the transformer. As a result, the parasitic capacitances of the bridge rectifier and the leakage inductance of the transformer are a part of the resonant tank. With a proper design of the magnetic component, L_s can be integrated within the transformer without the need for an additional series inductor.

Thus, the parasitic stray inductance, which often causes problems in hard-switched PWM topologies, is incorporated into the operation of the converter. The full-bridge acts as inverter stage for the dc input voltage, generating a square-wave voltage with a duty cycle of 50 % which is applied to the input of the resonant tank. The diode-bridge rectifies the tank's output current $i_B(t)$. The remaining harmonic frequencies are filtered out by the output capacitance C_o. Its value is assumed to be large enough in order to neglect the remaining voltage ripple of U_o.

The occurring subintervals can be divided into two major categories. With the rectifier conducting, the voltage across the parallel capacitance is clamped to the output voltage with $u_{\acute{C}p}(t) = \pm U_o$ and energy is transferred to the output. The transition of the diode bridge is determined by the zero-crossing of the inductor current $i_L(t)$ setting off a resonance with all four diodes reverse biased. During this resonant subinterval, the complete inductor current charges the parallel capacitance $\acute{C}_p$ until $u_{\acute{C}p}(t)$ reaches the output voltage $\pm U_o$. Due to this discontinuous energy transfer, the bulk output capacitor is stressed with high rms-currents. However, this is a typical characteristic for resonant converters with capacitive output filter.

Although three resonant elements are present within the tank of the LCC converter, the behavior of the converter during subintervals with the tank resonating can be described by a second order differential equation. During subintervals with clamped parallel voltage, the parallel capacitance is not involved in the resonance. Hence, two different resonant frequencies can be identified, with this converter topology often referred to as multiresonant converter in literature.

III. ANALYSIS

Several methods exist for the purpose of analysing resonant converters [8,9]. An approximation in the frequency domain is the First-Harmonic-Approximation (FHA) [10,11] and the Extended-First-Harmonic-Approximation (eFHA) [12-14]. An exact mathematical description is the analysis in the time domain [15]. The steady-state operation of resonant converters is characterized by several switching events during

one switching cycle, caused by the input bridge and the output rectifiers. The transitions of the input bridge are controlled by a control-IC, whereas the transitions of the bridge rectifier are dependent on the states of the converter. Therefore, a resonant converter is a nonlinear, time variant system. Nevertheless, the intervals between the transitions can be represented by a linear network. The state variables are the inductor current and the voltages across the capacitances which have to be periodic under steady-state conditions. In addition, the analysis can be reduced to one half of the switching period as the waveforms of resonant converters are antisymmetric. For this paper analytical steady-state solutions are derived, thus it is necessary to determine the occurring modes of the converter in advance. For a wide frequency range, the LCC converter with capacitive output has 4 modes of operation which are of interest. Each of these modes numbered 1 to 4 consists of a sequence of three subintervals, each with a different cycle of resonant and clamped intervals.

For the analysis, the secondary side parallel capacitance $\acute{C}_p$ is referred to the transformer primary side, resulting in the equivalent capacitance $C_p = \acute{C}_p/n^2$ with the transformer turns ratio n. The equivalent circuits of the converter during the different subintervals are shown in Fig. 2 to Fig. 4. In order to perform the analysis of the complete switching cycle, the waveforms of the state variables for the three subintervals are required. During interval A, the inductance L_s and the two capacitances C_s and C_p are resonating with the upfront unknown starting values

$$i_{Ls}(t=0) = i_{Ls0} \tag{1}$$

$$u_{Cs}(t=0) = u_{Cs0} \tag{2}$$

$$u_{Cp}(t=0) = u_{Cp0}. \tag{3}$$

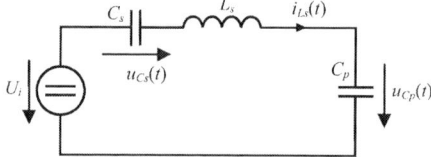

Figure 2. Equivalent curcuit, interval A

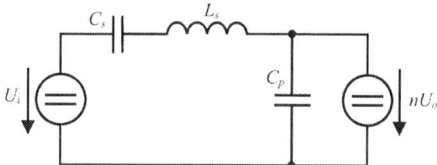

Figure 3. Equivalent curcuit, interval B

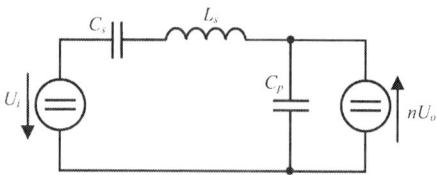

Figure 4. Equivalent curcuit, interval C

The resulting waveforms can be obtained by solving the differential equation e.g. by using the Laplace transformation:

$$i_{Ls}(t) = \frac{(U_i - u_{Cs0} - u_{Cp0})}{\sqrt{L_s/C^*}} \sin(\omega_0^* t) + i_{Ls0} \cos(\omega_0^* t) \quad (4)$$

$$u_{Cs}(t) = (U_i - u_{Cs0} - u_{Cp0})C^*/C_s [1 - \cos(\omega_0^* t)] \\ + i_{Ls0}\sqrt{L_s C^*}/C_s \sin(\omega_0^* t) + u_{Cs0} \quad (5)$$

$$u_{Cp}(t) = (U_i - u_{Cs0} - u_{Cp0})C^*/C_p [1 - \cos(\omega_0^* t)] \\ + i_{Ls0}\sqrt{L_s C^*}/C_p \sin(\omega_0^* t) + u_{Cp0} \quad (6)$$

With both capacitors C_p and C_s in series

$$C^* = C_p C_s / (C_p + C_s), \quad (7)$$

the resonance frequency for this subinterval is given by

$$\omega_0^* = 1/\sqrt{L_s C^*}. \quad (8)$$

For a better overview and an easier calculation, the following base quantities for normalization and abbreviation are introduced:

$$U_{\mathrm{norm}} = U_i \quad (9)$$

$$R_{\mathrm{norm}} = Z_0 = \sqrt{L_s/C_s} \quad (10)$$

$$I_{\mathrm{norm}} = U_{\mathrm{norm}}/R_{\mathrm{norm}} \quad (11)$$

$$\omega_0 = 2\pi f_0 = 1/\sqrt{L_s C_s} \quad (12)$$

$$t_{\mathrm{norm}} = 1/\omega_0 = \sqrt{L_s C_s} \quad (13)$$

$$F = \omega/\omega_0 = f/f_0 \quad (14)$$

$$Q = n^2 R_L / Z_0 \quad (15)$$

$$\Gamma = C_s / C_p \quad (16)$$

$$M = U_0 / U_i \quad (17)$$

$$\zeta = \sqrt{1 + C_s/C_p} = \sqrt{1 + \Gamma} \quad (18)$$

Based on this normalization, the normalized solution for interval A with (4)-(6) is given by

$$j_{Cs}(t_n) = (1 - M_{Cs0} - M_{Cp0})/\sqrt{1+\Gamma} \sin(\zeta t_n) \\ + J_{Ls0} \cos(\zeta t_n) \quad (19)$$

$$m_{Cs}(t_n) = (1 - M_{Cs0} - M_{Cp0})/(1+\Gamma)[1 - \cos(\zeta t_n)] \\ + J_{Ls0}/\sqrt{1+\Gamma} \sin(\zeta t_n) + M_{Cs0} \quad (20)$$

$$m_{Cp}(t_n) = (1 - M_{Cs0} - M_{Cp0})\Gamma/(1+\Gamma)[1 - \cos(\zeta t_n)] \\ + J_{Ls0}\Gamma/\sqrt{1+\Gamma} \sin(\zeta t_n) + M_{Cp0}. \quad (21)$$

Following the same steps for subinterval B (Fig. 3) and subinterval C (Fig. 4) one obtains

$$j_{Ls}(t_n) = (1 - M_{Cs0} \mp nM)\sin(t_n) + J_{Ls0}\cos(t_n) \quad (22)$$

$$m_{Cs}(t_n) = (1 - M_{Cs0} \mp nM)[1 - \cos(t_n)] \\ + J_{Ls0}\sin(t_n) + M_{Cs0} \quad (23)$$

$$m_{Cp}(t_n) = \pm nM. \quad (24)$$

The upper sign in (22)-(24) is valid for subinterval B, the lower sign for subinterval C.

Table I contains the interval sequence of the four modes for the first half of the switching period. At the beginning of each mode, the input bridge switches to the positive supply voltage U_i. After interval B and C, interval A must follow to charge the parallel capacitance C_p. For the analysis, relative time variables t_{n1} to t_{n3} are introduced, each starting at zero with the corresponding interval change. As a result, the normalized time variables t_{n1} and t_{n2} represent the durations of the first and the second interval. The duration of the third interval is the normalized half period less t_{n1} and t_{n2}.

A. Mode 1

Fig. 5 shows the waveforms of a complete switching cycle of the state variables for mode 1. In order to obtain a steady state solution for mode 1, a periodic solution for (19)-(24) representing the three subintervals in the time domain has to be found. Based on the waveforms' symmetry, the values of the state variables at the end of the third interval are equal with opposite sign to those at the beginning of the switching cycle. Thus, a second order system of transcendental equations is derived with the unknown durations t_{n1} and t_{n2}.

TABLE I. INTERVAL SEQUENCE

mode 1	mode 2	mode 3	mode 4	interval duration
A	C	A	B	t_{n1}
C	A	B	A	t_{n2}
A	B	A	C	$\pi/F - t_{n1} - t_{n2}$

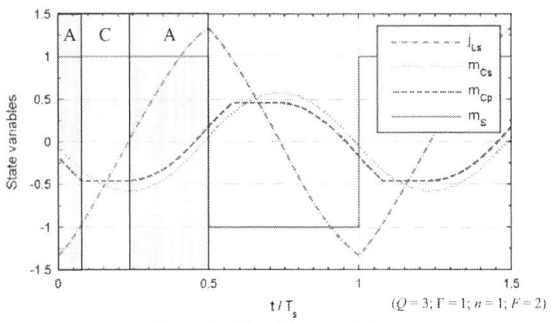

Figure 5. Waveforms mode 1

$$0 = 1 - \cos(\theta_1)\{[1+K] + \cos(\theta_2)[1-K]\}$$
$$+ 1/\sqrt{1+\Gamma}\,\sin(\theta_2)\{[1+K]\sin(\theta_3) + [1-K]\sin(\theta_1)\}$$
$$+ 0.5\left[1 - 1/\sqrt{1+\Gamma}\right]\cos(\theta_1 - \theta_2 + \theta_3) \qquad (25)$$
$$+ 0.5\left[1 + 1/\sqrt{1+\Gamma}\right]\cos(\theta_1 + \theta_2 + \theta_3)$$

$$0 = 2\sin(\theta_3) - \left[1 + \sqrt{1+\Gamma}\right]\sin(\theta_1 + \theta_2)$$
$$- \left[1 - \sqrt{1+\Gamma}\right]\sin(\theta_1 - \theta_2) \qquad (26)$$

with

$$\theta_1 = \zeta\, t_{n1} \qquad (27)$$

$$\theta_2 = t_{n2} \qquad (28)$$

$$\theta_3 = \zeta\left(\pi/F - t_{n1} - t_{n2}\right) \qquad (29)$$

and the abbreviation

$$K = 2FQ(1+\Gamma)/\pi\Gamma. \qquad (30)$$

For a given switching frequency and a given converter configuration, these durations can be determined by numeric means. The resulting output voltage for this operation point can be derived by taking the output current $i_B(t_n)$ into account with

$$M = Q/n^2\,\overline{|j_B(t_n)|} = FQ/\pi n^2 \int_0^{t_{n2}} |j_{Ls}(t_n)|\,dt_n. \qquad (31)$$

With known values for θ_1 to θ_2 the voltage transfer ratio $M = U_o/U_i$ is given by

$$M = \Gamma/[nX(1+\Gamma)]\{1 - \cos(\theta_1)[1 + \cos(\theta_2)]$$
$$+ \cos(\theta_2)\cos(\theta_1 + \theta_3) + 1/\sqrt{1+\Gamma}\,\sin(\theta_2) \qquad (32)$$
$$\cdot [\sin(\theta_1) + \sin(\theta_3) - \sin(\theta_1 + \theta_3)]\}$$

with

$$X = 1/\sqrt{1+\Gamma}\,\sin(\theta_2)\sin(\theta_3 + \theta_1)$$
$$- \cos(\theta_2)\cos(\theta_3 + \theta_1) - 1. \qquad (33)$$

B. Mode 2

Fig. 6 illustrates the waveforms of the state variables with the interval sequence according to Table I over a switching period for mode 2. Following the same argumentation as in mode 1, two equations for the unknown interval durations are obtained. The first equation

$$\cos(\theta_1) = \left[1 + \cos(\theta_2)\cos(\pi/F - \theta_2/\zeta)\right.$$
$$\left. - 1/\sqrt{1+\Gamma}\,\sin(\theta_2)\sin(\pi/F - \theta_2/\zeta)\right] \qquad (34)$$
$$\cdot \left\{[1 - 1/K] + [1 + 1/K]\cos(\theta_2)\right\}^{-1}$$

delivers θ_1 as a function of θ_2 with

$$\theta_1 = t_{n1} \qquad (35)$$

$$\theta_2 = \zeta\, t_{n2} \qquad (36)$$

$$\theta_3 = \pi/F - t_{n1} - t_{n2}. \qquad (37)$$

Substitution of (34) in

$$0 = 1/\sqrt{1+\Gamma}\,\sin(\theta_2)\cos(\theta_3)$$
$$+ \cos(\theta_2)\sin(\theta_3) - \sin(\theta_1), \qquad (38)$$

yields a single transcendental equation for θ_2. With a valid solution for θ_2, the dependent variable θ_1 can be calculated with (34). The voltage transfer ratio M for mode 2 is then given by

$$M = \Gamma\cos(\theta_1)[\cos(\theta_1 - 1)]/[nX(1+\Gamma)] \qquad (39)$$

$$X = 1/\sqrt{1+\Gamma}\,\sin(\theta_2)\sin(\theta_3 + \theta_1)$$
$$- \cos(\theta_2)\cos(\theta_3 + \theta_1) - 1. \qquad (40)$$

C. Mode 3 and Mode 4

On the first look, the interval sequence for all four modes appears to be different. Nevertheless, the modes can be divided into two groups with a characteristic sequence of subintervals, with interval A as the resonant interval and interval B and C as the clamp interval:

Mode 1 & Mode 3   resonance ⇨ clamp ⇨ resonance

Mode 2 & Mode 4   clamp ⇨ resonance ⇨ clamp

Thus, the steady state of mode 3 and mode 4 can also be described by means of the derived equations of mode 1 and mode 2 respectively. With the output diode bridge conducting in the opposite direction, a negative sign for the voltage transfer ratio M in (32) and (39) is obtained due to the reverse flow direction of $i_B(t)$.

IV. EVALUATION

Fig. 7 shows the voltage transfer ratio M given by (32) and (39). A turns ratio of $n = 8$ and a capacitor ratio of $\Gamma = 10$ is assumed. Additionally, the area with ZVS and ZCS is highlighted. In order to provide the specified output voltage under full load conditions with $Q = Q_{min}$ at the minimum input voltage ($M = M_{max}$), a suitable value for the turns ratio n and parameter Γ must be chosen. Attention must be paid to the peak value of the Q_{min} curve which must be higher than the specified maximum voltage conversion ratio in order to fulfill the specification.

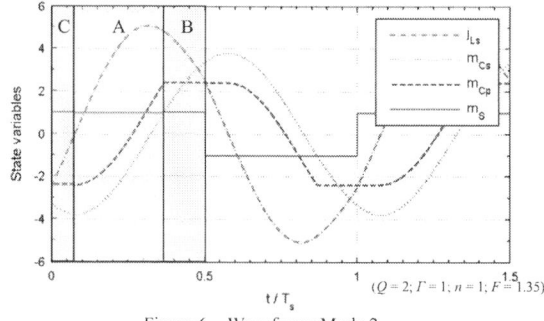

Figure 6. Waveforms Mode 2

Fig. 8 illustrates the switching behavior as a function of the normalized switching frequency F and the normalized load resistor Q. Additionally, the area of operation of an exemplary converter design with $0.095 < M < 0.152$ and a minimum load resistor of $Q_{min} = 2$ at full load is highlighted. According to the figure, the converter is operating under ZVS conditions for the complete load range.

The mode boundaries are shown in Fig. 9. The operation area in the picture is the same as in Fig. 8. For no-load conditions with $Q \to \infty$ mode 2 vanishes, with the boundaries to mode 1 and mode 3 approaching one another. The vertical asymptotic is situated at the normalized switching frequency $F = \zeta$ with $\zeta = 3.3$ for the discussed example. Together with the ZVS boundary of Fig. 8 it can be stated that ZVS is obtained for $F > \zeta$ under all load conditions. In order to obtain a regulated output voltage, the switching frequency has to be controlled for different load situations. Therefore the operating area has to be limited to mode 2 and mode 1 if ZVS is to be guaranteed. Table II summarizes the switching behavior of the four modes.

V. DESIGN

An application scenario typical for notebook adapters was chosen with an input voltage range of 250 V - 400 V and an output voltage of 19 V. The dc-bus voltage is generated by a preregulated ac-dc stage with power factor correction and minimized conducted emissions, in order to fulfill the requirements of EMI regulations. Some concepts are shown in [16-18]. The LCC converter provides a maximum continuous output power of 100 W. An important design-goal was a height constraint of 10 mm. Based on the results obtained by

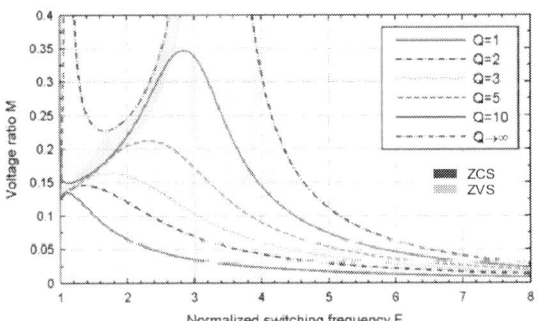

Figure 7. Voltage ratio $M = U_o/U_i$ versus normalized switching frequency F for $n = 8$, $\Gamma = 10$

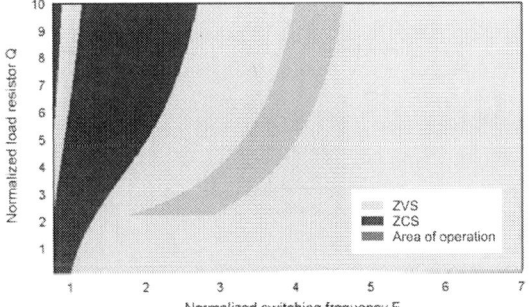

Figure 8. Switching behavior as function of normalized load resistor Q and normalized switching frequency F

TABLE II. SWITCHING BEHAVIOUR OF THE FOUR MODES

mode 1	mode 2	mode 3	mode 4
ZVS	ZVS	ZVS, ZCS	ZCS

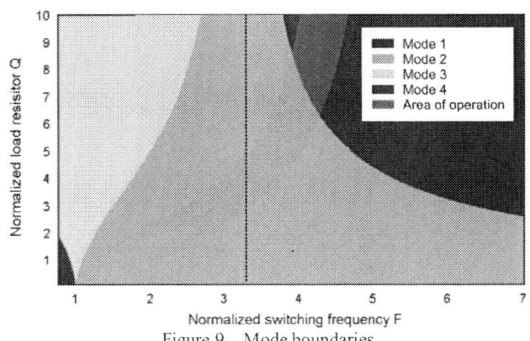

Figure 9. Mode boundaries

the methods of chapter III, a converter with the specifications according to Table III was designed and built.

For the input a half-bridge configuration was chosen, reducing the amplitude of the driving square-wave voltage to one half of the nominal DC input voltage. Fig. 10 shows the series inductor current at full load as a function of the series inductance L_s with $\Gamma = 10$. A turns ratio of $n = 7$ is chosen as design goal for the transformer together with an inductance value of L_s above 150 µH. For the transformer, a ring core TN32/19/13 [19] is used. To meet the requirements of the 10 mm height constraint, the height of the core is reduced down to 7 mm. Fig. 11 shows the arrangement of the primary side and the secondary side on the core. A rather large distance between primary and secondary is on the one hand necessary in order to fulfill safety requirements, on the other hand high values for L_s can be expected.

The transformer data is listed in Table IV. The values of the leakage inductance L_s, the magnetizing inductance L_h and transformer turns ratio n are based on the cantilever-model of a two winding transformer. An additional degree of freedom for optimization is the capacitance ratio Γ. Fig. 12 shows the series inductor rms-current at full load. Fig. 13 illustrates the

TABLE III. CONVERTER SPECIFICATION

minimal input voltage	$U_{i,min} = 250$ V
maximum input voltage	$U_{i,max} = 400$ V
output voltage	$U_o = 19$ V
output power	$P_o = 100$ W
maximum switching frequency	$f_{max} = 500$ kHz
maximum converter height	$h_{max} = 10$ mm

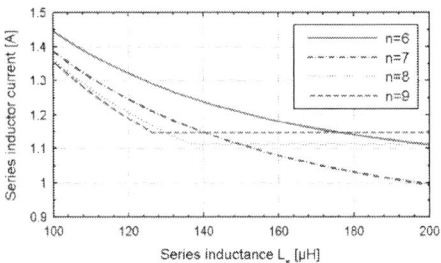

Figure 10. Series inductor current at full load versus Series inductance L_s

978-1-4244-4782-4/10 $26.00 © 2010 IEEE

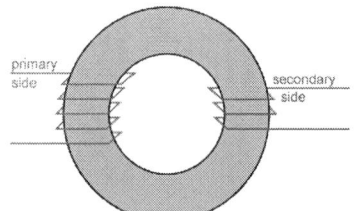

Figure 11. Winding structure

TABLE IV. TRANSFORMER DATA

Core:	TN32/19/13, reduced to 7 mm		
Material:	3F3		

Primary side	$N_p = 63$	35 x 0,1 mm	$R_P = 113.8\ m\Omega$
Secondary side	$N_s = 9$	90 x 0,1 mm	$R_S = 9.35\ m\Omega$

$L_s = 192\ \mu H$		$L_h = 5830\ \mu H$	$n = 7$

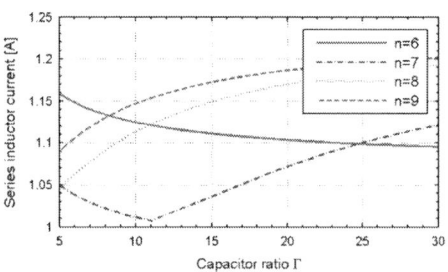

Figure 12. Series inductor rms-current at full load versus capacitor ratio Γ

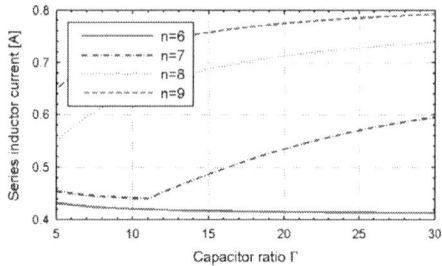

Figure 13. Series inductor rms-current at noload versus capacitor ratio Γ

series inductor rms-current at no load as a function of Γ. According to the two diagrams, a capacitor ratio of $\Gamma = 11$ is used. With denormalization, the values of the capacitors are

$$C_s = 13.6\ nF = 6.8\ nF + 6.8\ nF \qquad (41)$$

$$C_p = 66\ nF = 33\ nF + 33\ nF. \qquad (42)$$

In each case, two capacitors are parallelled, in order not to exceed the maximal current through the capacitors.

Fig. 14 shows the prototype of the designed LCC converter. Normal litz wire is used for the transformer windings to eliminate hf losses. To meet safety regulations, triple isolated wire would be necessary. This modification only has a small effect on the transformer data and thus, no significant influence on the design process can be expected. The denormalized voltage transfer function is shown below in Fig. 15.

Figure 14. Prototype

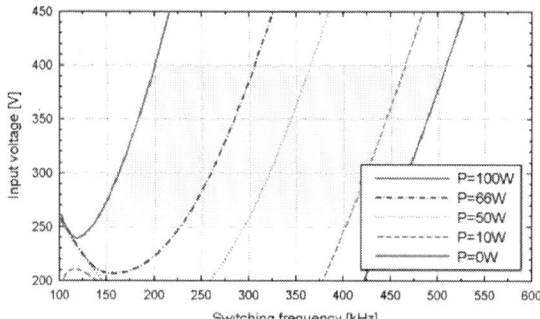

Figure 15. Voltage transfer function

Additionally, the area of operation is highlighted. The maximum switching frequency of 500 kHz is reached in the case of no load with the maximum input voltage. The minimum frequency of 133 kHz is obtained at full load conditions together with the minimum input voltage of 250 V. A wide frequency span is necessary to control the output voltage through the whole load and input voltage range. In spite of the full bridge rectification, the converter reaches an efficiency up to 92 % at the optimum operating point. The corresponding waveforms for the state variables at the operating point given in Table V are shown in Fig. 16. Since losses are not yet taken into account, a more accurate analysis would be desirable. A great share of the converter's total losses can be allocated to the rectifier diodes. This loss mechanism can be integrated into the exact analysis of resonant converters as discussed in [20]. A possible method to calculate core losses under arbitrary wave-shapes of the magnetizing current, as occurring within the LCC converter, is presented in [21]. As demonstrated in [22], hf winding losses can be predicted by means of a two-dimensional field calculation. For further improvements of the prototype these approaches should be considered for a more accurate design routine.

VI. CONCLUSION

One of the main trends driving the development of power electronics is miniaturization. The family of resonant converters are attractive for high switching frequency operation as the size of the passive components can be reduced without sacrificing converter efficiency. Offering virtually lossless switching without any additional effort and including parasitic

TABLE V. OPERATING POINT

input voltage:	$U_i = 300\ V$
output load:	$P_o = 83\ W$
switching frequency:	$f = 190\ kHz$

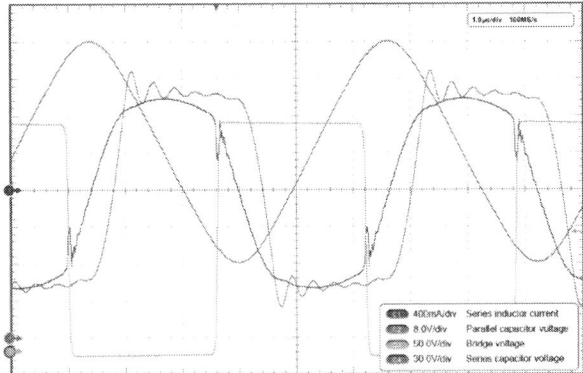
Figure 16. Waveforms at the operating point

components into the design, they are promising candidates compared to traditional hard-switching PWM converters. This paper presented the three-element multiresonant LCC converter with capacitive output filter, which is suitable for low-profile applications. The converter was analysed in the time domain and four modes of operation were identified which occur for a wide frequency range. The steady waveforms of converter can be calculated by solving a second order transcendental equation system (mode 1 and mode 3) or by solving a single nonlinear equation (mode 2 and mode 4). Based on the voltage conversion ratio, the mode boundaries and the switching behavior as well as design parameters such as rms-values can be calculated. It has been shown that ZVS is guaranteed throughout the complete load range if the resonant tank is designed properly. In the last section of the paper, a prototype was designed based on a specification typical for notebook adapters. A ring core was chosen for the magnetic component with a classical wire wound structure, thus expensive integrated magnetic components are avoided. The prototype reaches a converter height of 10 mm, but also flatter designs are possible.

VII. REFERENCES

[1] H. de Groot, E. Janssen, R. Pagano, and K. Schetters, "Design of a 1-MHz LLC Resonant Converter Based on a DSP-Driven SOI Half-Bridge Power MOS Module," IEEE Transactions on Power Electronics 22, vol. 6, 2007, pp. 2307-2320.

[2] D. Kübrich, T. Dürbaum, and A. Bucher, "Investigation of Turn-Off Behaviour under the Assumption of Linear Capacitances," PCIM Conference, May/June 2006, Nuremberg, Germany.

[3] A. Bucher, T. Dürbaum, and D. Kübrich, "Comparison of First Harmonic Approximation with exact Solution in case of a Series Resonant Converter," PCIM Conference, May/June 2006, Nuremberg, Germany.

[4] A. Bucher, T. Dürbaum, and D. Kübrich, "Comparison of Methods for the Analysis of the Parallel Resonant Converter with Capacitive Output Filter," 12th European Conference on Power Electronics and Applications EPE, September 2007, Aalborg, Denmark.

[5] A. Bucher, T. Dürbaum, D. Kübrich, and A. Stadler, "Comparison of Different Design Methods for the Parallel Resonant Converter," 12th International Power Electronics and Motion Control Conference EPE-PEMC, September 2006, Portoroz, Slovenia.

[6] R. P. Severns, "Topologies for Three-Element Resonant Converters," IEEE Transactions on Power Electronics 7, vol. 1, 1992, pp. 89-98.

[7] I. Batarseh, "Resonant Converter Topologies with Three and Four Energy Storage Elements," IEEE Transactions on Power Electronics 9, vol. 1, 1994, pp. 64-73.

[8] M. P. Foster, H. I. Sewell, D. A. Stone, and C. M. Bingham, "Review of modeling methodologies to facilitate rapid simulation of high order resonant converters," 8th European Conference on Power Electronics and Applications EPE 2001, Gratz – Austria.

[9] A. Bucher, T. Duerbaum, D. Kuebrich, and S. Hoehne, "Multiresonant LCC converter - comparison of different methods for the steady-state analysis," Power Electronics Specialists Conference, 2008, pp. 1891-1897.

[10] R. L. Steigerwald, "A Comparison of Half-Bridge Resonant Converter Topologies," IEEE Transactions on Power Electronics 3, vol. 2, 1988, pp. 174-182.

[11] M. C. Tsai, "Analysis and implementation of a full-bridge constant-frequency LCC-type parallel resonant converter," IEE Proceedings on Electric Power Applications 141, vol. 3, 1994, pp. 121-128.

[12] A. J. Forsyth, G. A. Ward, and S. V. Mollov, "Extended fundamental frequency analysis of the LCC resonant converter," IEEE Transactions on Power Electronics 18, vol. 6, 2003, pp. 1286-1292.

[13] A. K. S. Bhat, and S. B. Dewan, "A generalized approach for the steady-state analysis of resonant inverters," IEEE Transactions on Industry Applications 25, vol. 2, 1989, pp. 326-338.

[14] G. Ivensky, A. Kats, and S. Ben-Yaakov, "An RC Load Model of parallel and Series-Parallel Resonant DC-DC Converters with Capacitive Output Filter," IEEE Transactions on Power Electronics 14, vol. 3, 1999, pp. 515-521.

[15] S. D. Johnson, and R. W Erickson, "Steady-state analysis and design of the parallel resonant converter," IEEE Transactions on Power Electronics 3, vol. 1, 1988, pp. 93-104.

[16] M. Albach, "An ac-dc converter with low mains current distortion and minimized conducted emissions," 3rd European Power Electronics Conference EPE, vol. 1, 1989, Aachen, Germany, pp. 457-460

[17] M. Albach, "Conducted interference voltage of ac-dc converters," IEEE Power Electronics Specialists Conference PESC, Vancouver, Canada, 1986, pp. 203-212

[18] M. Albach, "Power supply for 2kW halogen lamp with continous power control and sinusoidal input current," Fifth International Symposium on the Science & Technology of Light Sources, York, UK, 1989, p. 151-152

[19] Ferroxcube datasheet TN32/19/13. URL: www.ferroxcube.com/prod/assets/tn321913.pdf.

[20] A. Bucher, T. Duerbaum, D. Kuebrich, and M. Schmid, "Consideration of conduction losses for the series resonant converter by means of a simple extension to the spa approach," Proc. 13th Power Electronics and Motion Control Conference EPE-PEMC 2008, 2008, pp. 244-249

[21] T. Duerbaum, and M. Albach, "Core losses in transformers with an arbitrary shaape of the magnetizing current," Proc. 6th European Power Electronics Conference EPE, vol. 3, 1995, pp. 1639-1644

[22] M. Albach, "Two-dimensional calculation of winding losses in transformers," Proc. IEEE 31st Annual Power Electronics Specialists Conference PESC 00, vol 3, 2000, pp. 1639-1644

978-1-4244-4782-4/10 $26.00 © 2010 IEEE

ZVS and ZCS DC-DC PWM Full-Bridge Fuel Cell Converters

Ahmad Mousavi
University of Western Ontario
Department of Electrical Engineering
Thompson Engineering Building
London, ON, Canada, N6A5B9
mseyeda@uwo.ca

Pritam Das
University of Western Ontario
Department of Electrical Engineering
Thompson Engineering Building
London, ON, Canada, N6A5B9
pdas2@uwo.ca

Gerry Moschopoulos
University of Western Ontario
Department of Electrical Engineering
Thompson Engineering Building
London, ON, Canada, N6A5B9
gmoschopoulos@uwo.ca

Abstract—PWM current-fed full-bridge dc-dc boost converters are typically in applications where the output voltage is considerably higher than the input voltage, such as fuel cell converters. The zero-voltage switching (ZVS) active-clamp converter is the most commonly used converter of this type. Zero-current switching (ZCS) techniques have also been used in current-fed full bridge converters, but not in fuel cell converters. In the paper, a comparison is made between the ZVS active-clamp full-bridge converter and a ZCS full-bridge converter. The operation of the two converters is discussed in detail and experimental results obtained from each converter are presented and compared.

I. INTRODUCTION

Current-fed PWM full-bridge boost converters like the one shown in Fig. 1 are very attractive as fuel cell converters. This is because they can produce high dc voltages from the low dc output voltages of fuel cells. During a typical converter switching cycle, the input current falls and the converter is an energy transfer mode when only a pair of diagonally opposed switches is on. The input current rises and no energy is transferred to the output when all switches are on and the converter is in a boosting mode.

A PWM full-bridge boost converter can be implemented with either zero-voltage switching (ZVS) [1]-[8] or zero-current switching (ZCS) [9]-[15], depending on the application. ZCS techniques, however, are rarely used in fuel cell converters. These converters are implemented with MOSFETs due to the low dc voltages that the converter switches are exposed to and the unavailability of low voltage IGBTs (<400V). Since ZCS techniques are associated with IGBTs and ZVS techniques are associated with MOSFETs, it has been naturally assumed that ZCS techniques are therefore unsuitable for fuel cell converters. The objective of this paper is to show that this assumption is not necessarily true.

In this paper, a comparative study of two PWM dc-dc full-bridge boost converters - a ZVS converter that uses the standard active-clamp technique and a new ZCS converter - is presented. The operation of the ZVS active-clamp converter is explained and its strengths and weaknesses are stated. A new and ZCS-PWM boost converter is then proposed and its operation is explained in detail. Experimental results obtained from a 50 kHz, 600 W prototype of each converter are presented and compared, and conclusions about the suitability of ZVS and ZCS in fuel cell converters are stated.

II. ZVS-PWM FULL-BRIDGE BOOST CONVERTER WITH ACTIVE CLAMP

The most widely used current-fed full-bridge converter for fuel cell converter applications is the zero-voltage-switching (ZVS) active clamp converter shown in Fig. 2 [1]-[8]. Any converter switch can be made to turn on with ZVS by turning the active clamp switch Q_{aux} just before it is to be turned on. This action discharges energy from the active clamp capacitor into the transformer leakage inductance that is then used to discharge the capacitance across the switch, thus bringing its voltage down to zero before it is turned on. The capacitor also acts as a clamp that prevents uncontrolled voltage spikes and ringing from appearing across the switches due to the interaction of the bridge switches and the transformer primary leakage inductance when the switches are turned off. The operation of the converter is as follows:

The clamp switch Q_{aux} remains on along with any two diagonally opposite switches of the full-bridge. For simplicity, assume Q_1 and Q_4 are on and energy is being transferred to the output. This discharges the energy stored in the clamp capacitor into the leakage inductance of the transformer so the current in the leakage inductor of the transformer becomes greater than the input current. The active clamp switch is turned off before the other pair of diagonally opposite switches Q_2 and Q_3 are to be turned on. Once Q_{aux} is turned off, then the difference of the current in the leakage inductor and the input current, discharges the output capacitors of Q_2 and Q_3, turns on their body diodes and reduces the voltage V_{pn} in the active clamp branch to zero. The corresponding switches (Q_2 and Q_3) can now be turned on with ZVS.

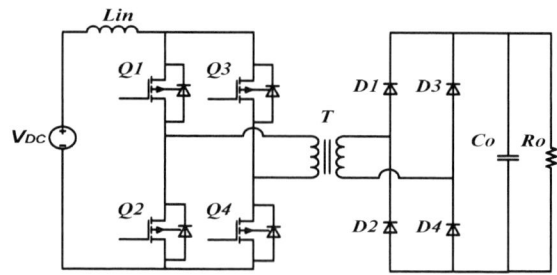

Fig.1. Current-fed PWM full-bridge boost converter.

When the current through the leakage inductor reduces to zero, the input current starts flowing through the four bridge switches and the input inductor gets energized. The input inductor charging mode ends when the other two diagonally opposite switches Q_1 and Q_4 are turned off.

After these two switches are turned off, the input current flows through the active clamp branch and the voltage across the active clamp capacitor starts increasing. The presence of the clamp capacitor reduces voltage rise and spikes across the switches Q_1 and Q_4. The clamp capacitor C_r starts being charged by the input current flowing through the body diode of Q_{aux} and clamps the primary voltage. Sometimes during this charging period of the clamp capacitor the clamp switch Q_{aux} is turned on.

Once the clamp capacitor voltage exceeds the reflected output voltage at the transformer primary, the current in the transformer primary begins to grow almost linearly until it becomes greater than the input current and transfers energy to the output. When the current in the leakage inductor exceeds the input current it changes direction and starts flowing through Q_{aux}. Energy continues to be transferred to the output and the input inductor gets discharged.

Sometime while the clamp current is flowing through switch Q_{aux} and the current in the leakage inductor is greater than the input current, switch S_a is turned off. This marks the beginning of next half of the switching cycle.

The advantages of the active-clamp ZVS full bridge boost converter include the following:

- The converter is a fixed frequency ZVS current-fed converter.

- The converter uses a very simple auxiliary circuit consisting of one auxiliary switch and one resonant capacitor to create ZVS.

- The gate drive signal for active clamp is easy to generate, being almost complimentary to the main switches of the bridge with some dead time.

- Soft-switching over extended range of load can be easily ensured.

- The clamp capacitor prevents the sudden rise of voltage across bridge switches during turn-off.

The disadvantages of the active clamp ZVS full bridge boost converter include the following:

- The active clamp branch has a considerable amount

of conduction losses as current continues to flow in the branch when any two diagonally opposite switches of the full bridge remain off.

- Energy from the active clamp contributes to circulating energy in the full-bridge, which contributes to additional conduction losses in the circuit.

- Although the clamp capacitor prevents the sudden rise of voltage across the bridge switches during turn-off, it does not prevent the bridge switches from having considerable turn-off losses. This is especially true when the converter is operating under heavy load conditions.

- The active clamp creates additional switch current stresses in the bridge switches that are higher than those found in a conventional current-fed full bridge converter.

III. ZCS-PWM FULL-BRIDGE BOOST CONVERTERS

Another approach to soft-switching in current-fed boost type dc-dc PWM full-bridge converters is using zero-current switching [9]-[15]. ZCS methods allow the full-bridge switches to turn off softly by diverting current away from them before they turn off. This soft turn-off removes the need for additional snubber capacitances to be connected across each bridge switch. There is, therefore, no need to be concerned about losses due to snubber capacitor energy being dissipated in the switches. The transformer leakage inductance and small inductances placed in the converter are used to slow down the rise in current through the switches when they are turned on.

Previously proposed converters ZCS current-fed full-bridge converters, however, have at least one of the following disadvantages:

- The converter is a resonant converter that generates a considerable amount of circulating current in the full bridge so that the switches can turn off with ZCS. This circulating current is not transferred to the load and does little but add to the conduction losses of the converter.

- The converter uses an active auxiliary circuit that is connected parallel to the full-bridge and used to divert current away from the switches in the bridge before they are turned off to achieve ZCS turn off of devices. This circuit is activated just before any switches are to be turned off and is deactivated shortly afterwards. Since the circuit is active for only a short length of time, there is less circulating current compared to the fixed frequency resonant converters, but this current is still significant. All the energy from this current is trapped in the primary side of the converter and it contributes to losses.

- Additional circulating current generated by the converter also contributes to increased peak current stresses in the full-bridge switches. Devices that can

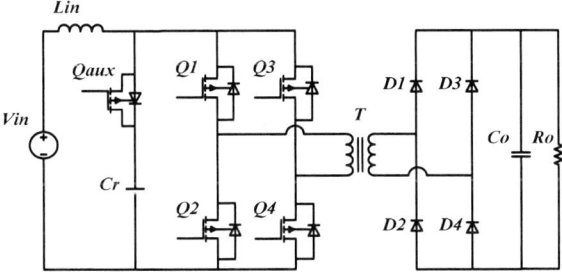

Fig. 2. Active Clamp ZVS full-bridge boost converter.

withstand higher current stresses than those found in conventional boost full-bridge converters are needed.

- Diodes are placed in series with the switches in some converters so that current does not flow through the body-diodes of the switches and circulating current is reduced. Conduction losses, however, are increased due to the voltage drops of the added diodes. Some converters avoid using series diodes by using reverse blocking IGBTs, but they are more expensive than regular IGBTs and not appropriate for low input voltage applications.

- A voltage spike and/or significant voltage ringing can appear across the main converter switches because the output switch capacitances resonates with the leakage inductance of the main transformer during turn off. This spike and ringing can appear across the secondary diodes as well. This creates a need for higher voltage rated devises, which increases the cost and the losses in the converter.

IV. PROPOSED ZCS-PWM FULL-BRIDGE CONVERTER

A new ZCS-PWM dc-dc full-bridge boost converter that has none of the above disadvantages is proposed in this paper and is shown in Fig. 3. The proposed converter has an active auxiliary circuit that is connected across its transformer primary side dc bus that helps the full-bridge switches turn off with ZCS.

The proposed converter is a standard PWM full-bridge converter with an auxiliary circuit that consists of an auxiliary switch Q_{aux}, a resonant capacitor C_r, a resonant inductor L_{r2}, a transformer with a center-tapped secondary, and two secondary diodes D_{s1} and D_{s2} connected to the output. Another small inductor L_{r1} is added at the input of the bridge to help in the ZCS turn-on and off of the full-bridge switches. The sequence of gating signals that the converter operates with in a typical switching cycle is: Q_1 and Q_4 on, all bridge switches on, then Q_2 and Q_3 on and again all switches on - in other words, an energy transfer mode when only a pair of diagonally opposed switches is on is always followed by a "boosting" mode where all the switches are on and no energy is transferred.

The basic principle behind the converter is that the auxiliary circuit is activated during the time when all full-bridge switches are on so that current can be diverted away from these switches and the appropriate pair of switch can turn off with ZCS. Energy in the auxiliary circuit can be transferred to the output through the transformer in the auxiliary circuit. Switches turn on with ZCS due to the leakage inductance of the transformer preventing current from rushing into a switch in a sudden manner.

V. MODES OF OPERATION

The modes of operation that the proposed converter goes through during half of a steady-state switching cycle are explained in this section. Typical converter waveforms are shown in Fig. 4 and the equivalent circuit modes are shown in Fig. 5.

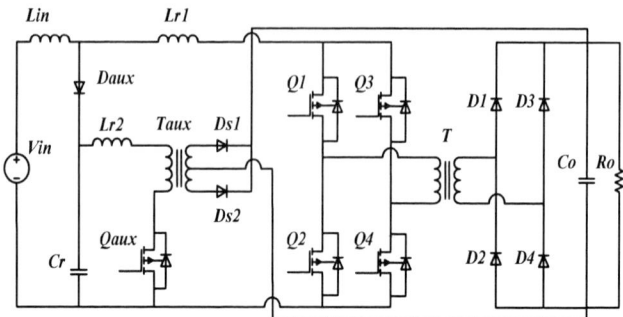

Fig. 3. Proposed ZCS full-bridge boost converter.

Mode 0 ($t_0 \le t < t_1$) (Fig. 5(a)): Switches Q_1 and Q_4 are turned at the beginning of this mode at $t=t_0$ while the other two switches Q_2 and Q_3 were already on carrying the full input current I_{in}. Due to the primary transformer leakage inductance, the transfer of current to these switches is gradual so that they turn on with ZCS. This is a commutation mode, during which a negative voltage of $-V_o/N$ appears across the leakage inductor so that current through it starts decreasing. This will result the current through Q_2 and Q_3 to decrease while current through Q_1 and Q_4 increases. At the end of this mode there is no current in the transformer primary and the bridge is shorted.

Mode 1 ($t_1 < t < t_2$) (Fig. 5(b)): At time $t = t_1$, all four switches Q_1, Q_2, Q_3 and Q_4 are on and the input current flows through all four of these devices so that Q_1-Q_2 and Q_3-Q_4 carry half of the input current ideally. The transformer primary remains shorted and the output capacitor gets discharged into the load. The input inductor has the full input voltage across it and gets energized.

Mode 2 ($t_2 < t < t_3$) (Fig. 5(c)): At $t = t_2$, the auxiliary switch Q_{aux} is turned on and C_r begins to resonate with L_{r2} and discharge. Energy in the auxiliary circuit is transferred to the load through T_{aux} and D_{s1}. This mode ends when the voltage of C_r, V_{Cr}, reaches zero and diode D_{aux} turns on. The transformer primary is clamped to $V_x = V_o N_1/N_2$ from time t_2 to t_4 and the secondary diode D_{s1} is forward biased. Circulating energy from the auxiliary circuit is transferred to the output during this time.

Mode 3 ($t_3 < t < t_4$) (Fig. 5(d)): At $t = t_3$, the voltage across C_r reaches zero and D_{aux} starts to conduct as C_r continues to resonate with L_{r2} and the voltage across it becomes negative. This negative voltage appears across the resonant inductor L_{r1} and thus current begins to be diverted away from the full-bridge switches.

Mode 4 ($t_4 < t < t_5$) (Fig. 5(e)): At $t = t_4$, the current through the main switches becomes zero and begins reversing direction by flowing through the body diodes of the switches. Switches Q_2 and Q_3 can be turned off softly at any time while current is flowing in their body diodes. Current in the auxiliary circuit is positive but decreasing.

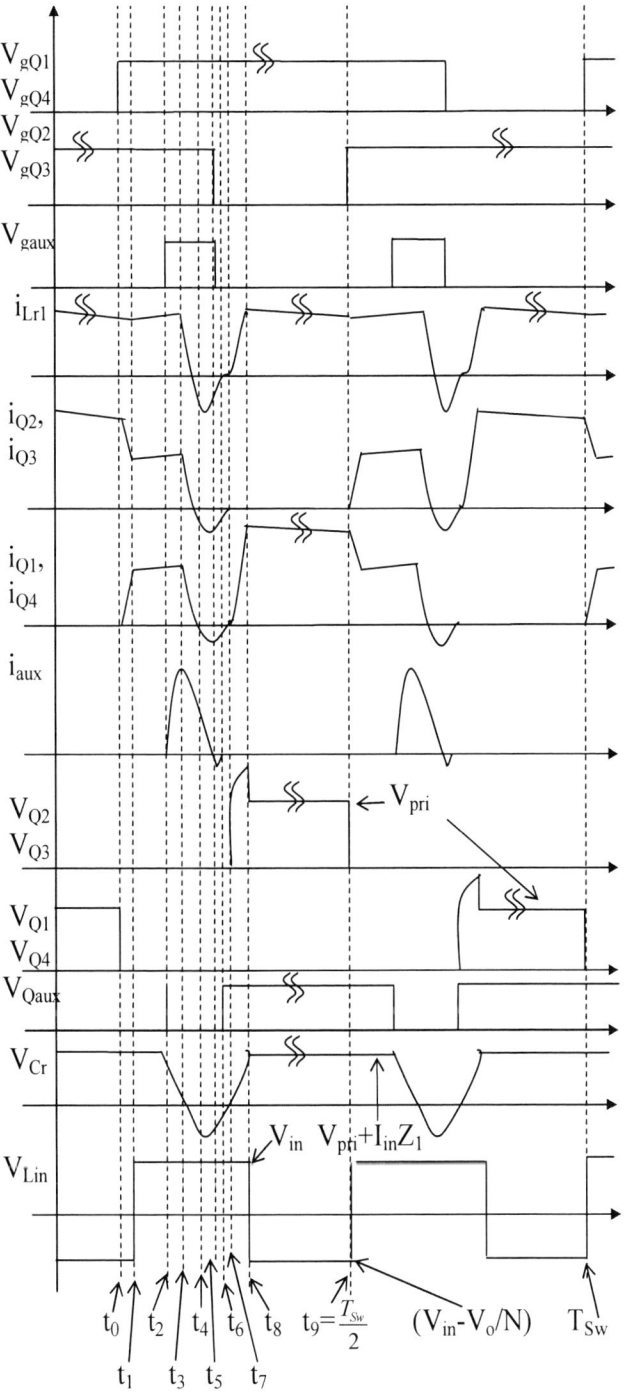

Fig. 4. Converter waveforms

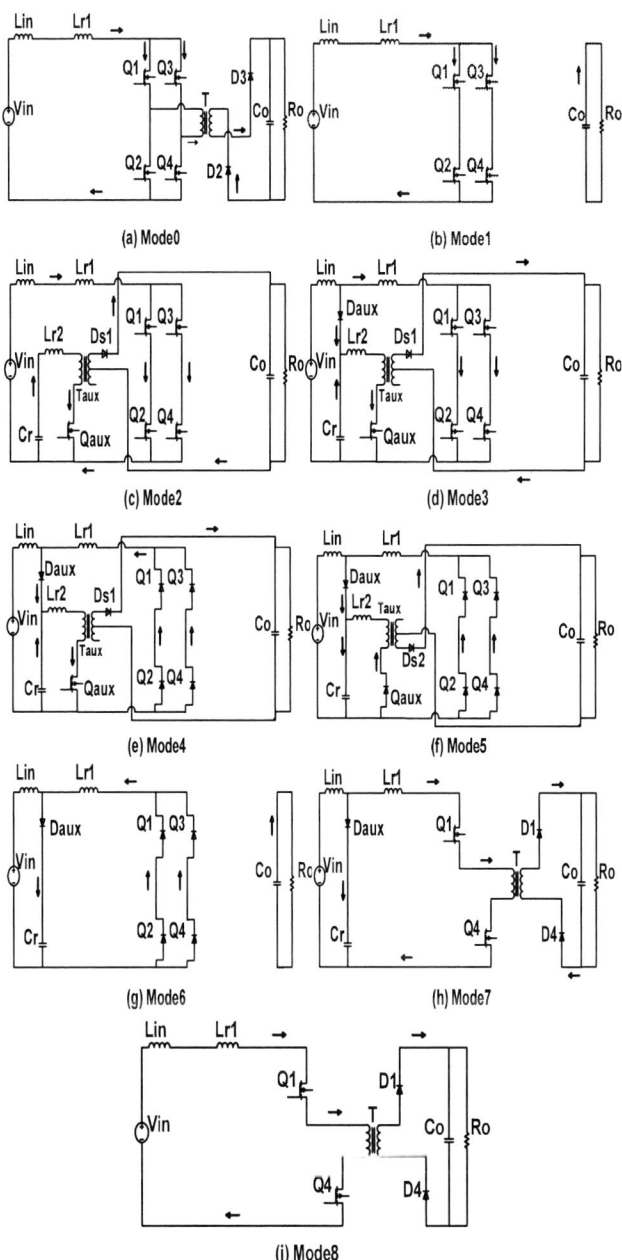

Fig. 5. Converter modes of operation

Mode 5 ($t_5 < t < t_6$) (Fig. 5(f)): At $t = t_5$, the body diode of Q_{aux} starts conducting and the switch can be turned off softly after this instant. During this mode, all the body diodes of all converter switches conduct current and the current coming out of the bridge flows through L_{r1} and D_{aux} charging up C_r. Energy is transferred from the auxiliary circuit to the load through T_{aux} and D_{s2}.

Mode 6 ($t_6 < t < t_7$) (Fig. 5(g)): At $t = t_6$, the current in the body diode of Q_{aux} goes to zero. During this mode, the voltage across C_r increases in resonance with L_{r1} while the current flows through the body diodes of the full-bridge switches. At the end of this mode the voltage across C_r reaches $V_{pri} = V_o/N$.

Mode 7 ($t_7 < t < t_8$) (Fig. 5(h)): At $t = t_7$, the current in the body diodes of the full-bridge switch becomes zero, and some input current starts to flow through L_{r1}, Q_1 and Q_4. The remaining input current continues to charge C_r and the voltage across it rises. Energy begins to be transferred to the load through D_1 and D_4. At the end of this mode the voltage across C_r charges up to V_{Cr0}.

978-1-4244-4782-4/10 $26.00 © 2010 IEEE

Mode 8 ($t_8 < t < t_9$) (Fig. 5(i)): At t = t_8, all the input current flows through the bridge switches Q_1 and Q_4, and none through C_r. The converter is in an energy transfer mode. A negative voltage of $V_{in} = -V_{pri}/N$ is incident across the input inductor. At the end of this mode at $t_9 = T_{sw}/2$, Q_2 and Q_3 are turned on and this begins the beginning of a similar switching half cycle.

VI. CONVERTER FEATURES

The proposed ZCS-PWM full-bridge boost converter has the following features:

- The presence of a transformer in the auxiliary circuit provides a path for energy that would otherwise be trapped in the auxiliary circuit. Energy can be transferred to the output instead of contributing to conduction losses.

- The auxiliary circuit is adaptive as greater the energy that would otherwise be trapped in the circuit, greater the energy that is transferred to the output. Since trapped auxiliary circuit energy is a cause of circulating current, the converter can be made to operate with little additional circulating current regardless of whether it is operating with a light load or a heavy load. This property does not exist in most other ZCS-PWM full-bridge boost converters as they have a considerable amount of circulating current when operating under light load conditions.

- Since the proposed converter has little additional circulating current regardless of load, it does not need additional diodes connected in series with the full bridge switches to prevent current flowing through their body diodes. There are therefore no conduction losses due to series blocking diodes unlike several other ZCS-PWM full-bridge boost converters.

- The peak current stress of the switches is the same as that of a switch in a conventional PWM boost full-bridge converter, as D_{aux} blocks any auxiliary circuit current from flowing into the full-bridge.

- One of the drawbacks of a conventional current-fed full bridge converter is that it lacks a dc bus capacitor across the full bridge section. This creates uncontrolled voltage spikes during the turn-off of the bridge devices due to resonance between their output capacitance and the transformer leakage inductance. In the proposed converter, the rise of this voltage hump is restricted by the presence of C_r. This allows lower rated devices to be used as the full bridge switches.

VII. EXPERIMENTAL RESULTS

Experimental prototypes of the active clamp and the ZCS converters were built to confirm the feasibility of the new converter and also to compare the efficiencies of the tow converters. The basic current fed converter prototypes in each case had the following specifications:

- Input voltage V_{in} = 12V and 24 V,

- Output voltage V_o = 300V,

- Output power P_o = 600W,

- Transformer turns ratio N=1:12

- Input inductor L_{in} =500µH,

- Switching frequency: f_{sw} = 50 kHz.

The prototype of the new ZCS converter was implemented using the following components values:

- Auxiliary circuit inductors L_{r1} =300nH, L_{r2} =900 nH,

- Auxiliary circuit capacitor C_r =360 nF.

It should be noted that the inductor L_{r2} is realized as the leakage inductance at the primary of the auxiliary transformer. The following devices were used for the semiconductors in the prototype:

- IRF540 as bridge switches

- IRF520IR as auxiliary switch S_{aux}

- Output diodes: HFA16PA60C

- Auxiliary diodes D_{aux}: FR802,

- Diodes D_{S1} and D_{s2}: GUR5H60,

- Saturable reactors implemented on Toshiba saturable cores (SA14x8x4.5) were in series with the switches used to reduce the effect of parasitic resonances during switching transitions.

The active clamp converter prototype had the following components:

- Active clamp capacitor C_{cl}=10µF

- IRF540 as bridge switches

- IRF520IR as auxiliary switch S_{aux}

- Output diodes: HFA16PA60C

It can be seen in Fig. 6(a) that the main bridge switches have a ZCS turn-on and turn-off. During turn-off, the switches have negative current as current is flowing through their body diodes. The same can be concluded about the auxiliary switch from its current and voltage waveforms in Fig. 6(b). The main switch voltage has a well behaved hump when the switch is turning off and not a spike, which is common in almost all current fed full bridge converters. It can be seen from Fig. 7 that the primary voltage waveform is similar to that of a conventional current fed full-bridge converter.

Fig. 8 shows the current and voltage waveforms of a bridge switch in the active clamp prototype. It can be seen that the bridge switches have a ZVS turn-on, but have an overlap of voltage and current during turn-off, which can create significant turn-off losses. Also, the bridge switches show increased current stress during the energy transfer modes due to the charging current of the active capacitor. Moreover it can be seen in Fig. 9 that the active clamp remains on for much longer time than that of the auxiliary

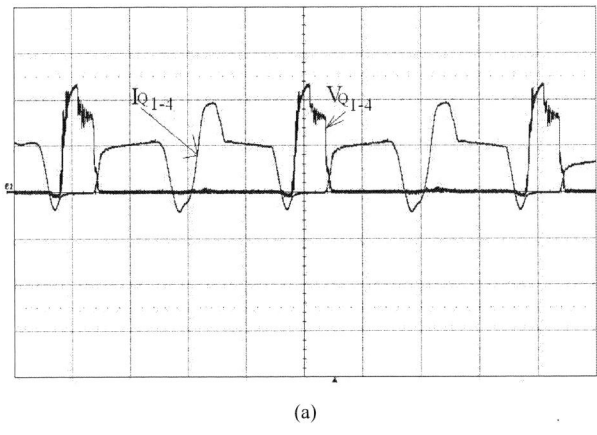

(a)

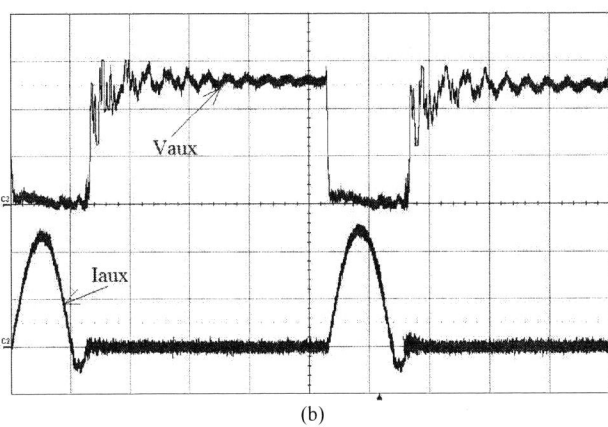

(b)

Fig. 6 .(a) Current and voltage in Switches Q_1 and Q_4 over one full switching period. (V: 30V/div, I: 15Amps/div, t:5µs/div) (b) Current and voltage in auxiliary switch over one switching period (V: 30V/div, I: 15A/div, t:2 µs/div).

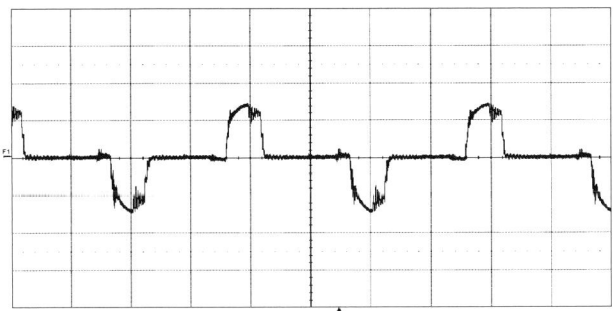

Fig. 7. Transformer Primary Voltage Waveform (V: 50V.div, t:5 µs/div).

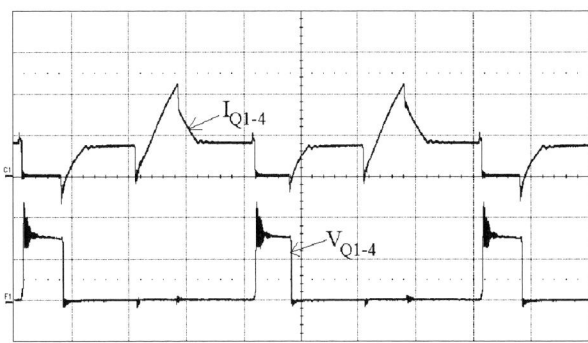

Fig. 8. Current and voltage in Switches Q_1 and Q_4 over one full switching period in the Active Clamp Converter (V: 30V/div, I: 15Amps/div, t:5µs/div).

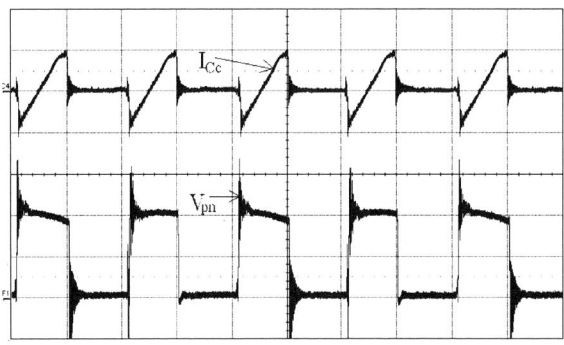

Fig. 9. Current and voltage in auxiliary switch branch in active clamp Converter (V: 30V/div, I: 15A/div, t:5 µs/div).

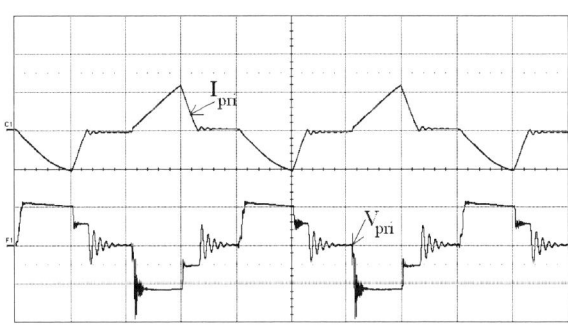

Fig.10. Primary current and voltage waveform across the transformer (V:80V/div, I: 25Amps/div, t:4 µs/div).

switch of the ZCS converter. This also creates additional conduction losses in the converter. The primary current and the voltage waveform across the transformer are shown in Fig. 10.

Efficiency measurements taken from the ZCS, the active clamp converter and an equivalent hard switching converter are shown in Fig. 11 for a 12V input. It can be seen that the proposed ZCS converter has a greater efficiency than the active clamp converter for loads over 325W. The increased efficiency is due to the fact that the converter switches turn off with ZCS and have fewer conduction losses than the active clamp converter. Moreover, there are much fewer

conduction losses in the auxiliary circuit, which remains active for a much smaller time as compared to the active-clamp switch.

In the active clamp converter, the switches have turn-off losses due to the high currents being handled by the switches during turn-off transitions. These losses become significant with increased loads. In addition to these losses is the circulating energy that remains trapped in this converter and creates additional conduction losses.

978-1-4244-4782-4/10 $26.00 © 2010 IEEE

VIII. CONCLUSION

The ZVS active clamp current-fed full-bridge converter is the standard converter used in many fuel cell applications. ZCS converters have not been considered for these applications as it is widely believed that ZVS is preferable to ZCS in all applications where MOSFETs are used. MOSFETs are the devices of choice in fuel cell converters as they are low voltage devices.

A comparative study of the ZVS active clamp full-bridge converter and a new ZCS-PWM converter was presented in this paper. The ZCS-PWM is an improvement over other previously proposed converter of the same type. In the paper, The operation of the ZVS active-clamp converter was explained and its strengths and weaknesses were stated. The new ZCS-PWM converter was then proposed and its operation was explained in detail. Experimental results obtained from a 50 kHz, 600 W prototype of each converter were presented and compared. It was seen that the ZCS converter was more efficient that the ZVS converter when the converters were operating under heavy load conditions.

ACKNOWLEDGMENT

The authors would like to acknowledge funding from the National Sciences and Research Council of Canada (NSERC) for financial support during the course of this work.

REFERENCES

[1] R. Watson and F. C. Lee, "A soft-switched, full-bridge boost converter employing an active- clamp circuit," in *Proc. IEEE PESC*, 1996, pp. 1948-1954.

[2] V. Yakushev, and S. Fraidlin, "Full-bridge isolated current fed converter with active clamp," in *Proc. IEEE APEC*, 1999, pp. 560 – 566.

[3] C. Qiao, and K. M. Smedley "An isolated full bridge boost converter with active soft switching," in *Proc. IEEE PESC*, 2001, pp. 896-903.

[4] L. Zhu, K. Wang, F.C. Lee, and J.-S. Lai "New start-up schemes for isolated full-bridge boost converters," *IEEE Trans. Power Elec.*, vol. 18, no.4, pp. 946 – 951, July 2003.

[5] C. Hanju, C. Jungwan and P.N. Enjeti, "A three-phase current-fed DC/DC Converter with active clamp for low-DC renewable energy sources," *IEEE Trans. on Power Elec.*, vol.23 no.6 pp.2784-2793, Nov. 2008.

[6] W. Li, J. Liu, J. Wu, and X. He," Design and analysis of isolated ZVT boost converters for high-efficiency and high-step-up applications," *IEEE Trans. on Power Elec.*, vol. 22, no. 6, pp. 2363-2374, Nov. 2007.

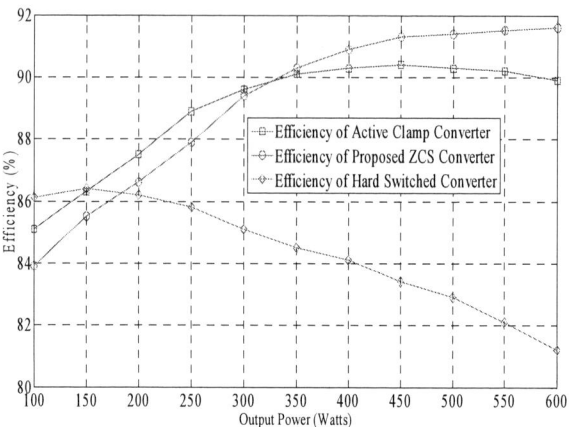

Fig.11 Efficiency measurements for the proposed converter, the active clamp converter and the conventional hard switching converter with 12V input and varying output load.

[7] S.-K. Han, H.-K.Yoon, and G.-W.Moon et al., "A new active clamping zero-voltage switching PWM current-fed half-bridge converter," *IEEE Trans. on Power Elec.*, vol. 20, no. 6, pp. 1271–1279, Nov. 2005.

[8] H. Xiao and S. Xie, "A ZVS Bidirectional DC–DC converter With Phase-Shift Plus PWM Control Scheme," *IEEE Trans. on Power Elec..*, vol. 23, no. 2, pp. 813–823, Mar. 2008.

[9] C. Iannello, S. Luo, and I. Batarseh, "Full bridge ZCS PWM converter for high-voltage high-power applications," *IEEE Trans. Aero. and Elec. Syst.*, vol. 38, no.2, pp. 515 – 526, Apr. 2002.

[10] J.-F. Chen, R.-Y. Chen, and T.-J. Liang, "Study and implementation of a single-stage current-fed boost PFC converter with ZCS for high voltage applications," *IEEE Trans. on Power Elec.*, vol. 23, no.1, pp. 379-386, Jan. 2008.

[11] H. Benqassmi, J. P. Ferrieux, and J. Barbaroux, "Current-source resonant converter in power factor correction," in *IEEE PESC'97 Conf. Proc.*, 1997, pp. 378–384.

[12] X. Zhu, D. Xu, H. Umida, K. Mino, and Z. Qian, "Current-fed phase shift controlled full bridge ZCS DC-DC converter with reverse block IGBT", in *Proc. IEEE APEC*, 2006, pp. 1605 – 1610.

[13] S. W. Leung, S. H. Chung, and T. Chan, "A ZCS isolated full-bridge boost converter with multiple inputs," in *Proc. IEEE PESC*, 2007, pp. 2542 – 2548

[14] Q. Sun, H. Wang, R. T. H. Li, H. S. H. Chung, S. Tapuchi, N. Huang, and A. Ioinovici "Modeling and analysis of a current-fed ZCS full-bridge DC/DC converter with adaptive soft-switching energy," in *Proc. IEEE APEC*, 2009, pp. 1410-1416.

[15] A. Averberg, K.R.Meyer, and A. Mertens, "Current-fed full bridge converter for fuel cell systems," in *Proc. IEEE PESC*, 2008, pp. 866-872.

Effective Switching Mode Power Supplies Common Mode Noise Cancellation Technique with Zero Equipotential Transformer Models

Yick Po Chan, Man Hay Pong, Ngai Kit Poon and Chui Pong Liu

Department of Electrical Engineering, The University of Hong Kong

Email: achan@eee.hku.hk

Abstract—In this paper a transformer construction technique is proposed that effectively cut off the Common Mode (CM) noise voltage passing across the isolated primary and secondary windings. This technique employs the Zero Equipotential Line theory to construct an anti-phase winding. It effectively cuts down CM noise by eliminating the noise voltage across the isolated primary and secondary windings. The concept of maintaining an equipotential line along the bobbin and quiet node connections are justified by analysis. A well considered transformer design with the proposed CM noise cancellation technique can achieve high conversion efficiency as well as good CM noise insulation.

I. INTRODUCTION

Electromagnetic Interference (EMI) is always a barrier in designing a high efficiency Switching-Mode Power Supplies (SMPS) due to the presence of the common-mode (CM) noise. In many power supply designs, different noise suppression schemes are always required in order to meet the EMI requirements for electronic equipments, most of which create unwanted power loss that has lead to size, efficiency and thermal issues. Nowadays there are several commonly known methods to minimize the CM noise. Here is a brief summary of schemes on minimizing the CM noise flowing through the LISN and problems about these schemes.

1) Use of CM Noise Filters – This involves time-consuming designs as suggested by Shih [1], which commonly used in many SMPS designs. Usually, in order to gain satisfactory results, a bulky CM noise suppression filter is required. This is becoming more undesirable as the product size is shrinking and the filter actually lies on the power path. Damnjanovic et al [2][3] acknowledged that the importance of the size of the CM choke filters and proposed of SMD CM choke designs [4]. Although the size of the CM choke is small, the result is only effective from 1MHz or above, therefore the low frequency cannot be suppressed. The problem remains with use of large size CM choke to tackle the low frequency end up to 1MHz. Roc'h et al [5][6] also emphasized on the importance of the CM choke filter design because it is often difficult to design a low power loss, minimal size filter. An active CM filter is proposed by Mortensen [7] to try to further reduce the CM noise. Although this way the designer has greater flexibility to fine tune the CM filter than just use the passive component alone, the effect of such active filter is not easily modeled and the gain bandwidth is severely limited by the active component.

2) Minimize the parasitic coupling capacitors from the primary winding to the secondary winding – It leads to a high leakage inductance and produces efficiency problem.

3) Bypass Capacitor connected across the primary and the secondary side - Chen et al [8] have discussed on the effects of this Y-Capacitor on common mode noise performance, but the applicable capacitance is always limited by safety standards and this method alone usually cannot provide a low enough impedance to shunt all the CM noise current flowing along this path.

4) Faraday Shielding – The method requires careful integration of a piece of conducting sheet into the transformer to shunt away noise current. This is not always effective because there are many paths from which the CM noise current can go through. The shield must be properly installed in order to meet the safety requirement.

The CM noise source in SMPS is caused by the high frequency high voltage switching on the primary MOSFET. In the example of an isolated flyback converter, the CM noise current can be imagined to mainly follow two paths as shown in figure 1, via the parasitic capacitor from the drain node of the MOSFET to the ground, or via the isolation transformer coupling path to the secondary then through the parasitic capacitor to the ground.

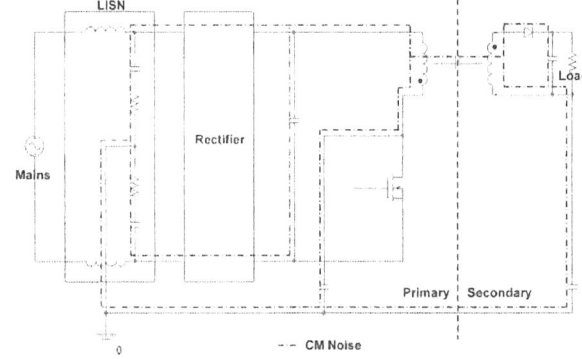

Fig. 1. Flyback Converter showing CM noise paths

There are other techniques [9-15] that have been proposed to reduce the conducted CM noise in EMI. In the first noise path described, Cochrane et al [9] employed a compensation capacitor with an anti-phase winding to passively cancel the noise current flowing through the MOSFET parasitic capacitor. However, this simple addition of the capacitor cannot stop the significant part of noise current flowing through the secondary side and returned via the ground path. Herbert [10] proposed to use two or more transformers in series to reduce the overall parasitic capacitance between the primary and secondary windings so that it minimizes the coupling between them. This requires extra magnetic components and tedious designs. The reduction of the cross-coupling between the primary and secondary side is undesirable because this would increase the leakage inductance and lead to poor conversion efficiency in many cases. Wang [11] proposed another way of cancelling the CM noise by creating negative capacitances that balance the parasitic capacitances on different points in the power converter, yet it is not easy to generate repeatable results with another prototype.

In this paper, a new model is presented to construct the transformer which can effectively cut down the flow of CM noise. This method is based on the production of a balanced anti-phase noise voltage source [12][13] with a special transformer construction arrangements. An analytical model with P-Spice equivalent circuit is also presented to explain the theory of the method. This method produces no loss and requires no extra component, which is most favorable in terms of converter energy efficiency and small physical size.

II. EQUIPOTENTIAL LINE CONCEPT – ANTI-PHASE WINDING

The Equipotential Line concept for common mode noise reduction is introduced to cancel the noise current flowing from the primary winding to the secondary winding via the coupling capacitance C_{PS}. The idea is to produce an electric field opposite to that produced by the primary winding, it is possible to reduce the potential of the secondary winding to zero. In this case no common mode current can flow though the capacitance C_{PS}. The opposite electric field is produced by an additional anti-phase winding.

The flyback converter example in figure 1 is taken and an anti-phase winding with the same number of turns to the primary winding is added, as shown in figure 2. Since only the electric field of the anti-phase winding is required, one end does not connect to anything in order to avoid power current flow.

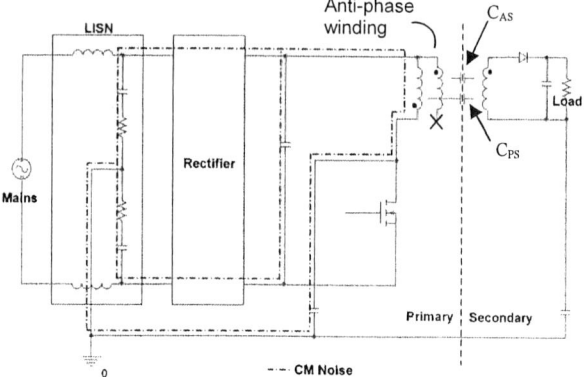

Fig.2. Flyback Converter with an anti-phase winding

Usually, the turns ratio N_{PS} between the primary and secondary winding is comparable, so the secondary winding will in fact be one of the noise voltage source as well, acting across the bobbin, similar to the anti-phase winding in the same phase because the switching action will also induce a switching voltage across the secondary winding, model is shown in figure 3a and 3b.

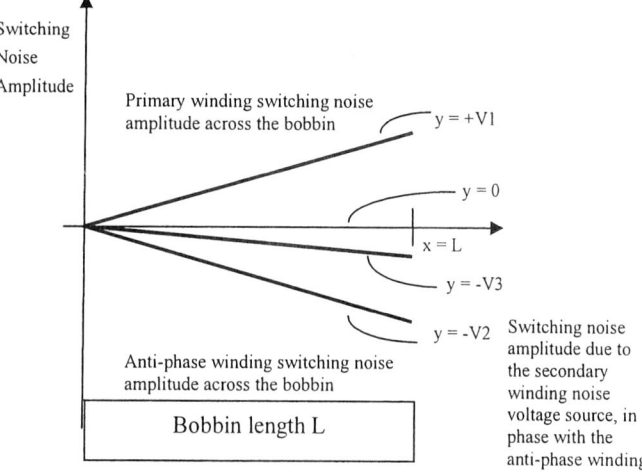

Equivalent Circuit Model
V1 = Switching noise source from the Primary Winding **P**
V2 = Switching noise source from the Secondary Winding **S**
V3 = Switching noise source from the Anti-Phase Winding **A**
C1 = C_{PS}
C2 = C_{PS} +C_{AS}
C3 = C_{AS}
Y = A bypass capacitor used for measurement purpose

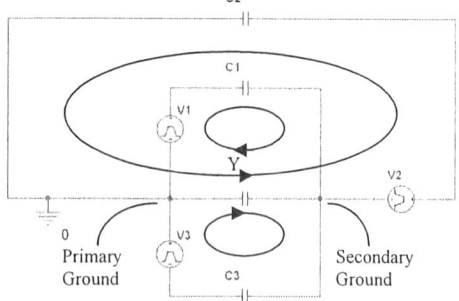

Fig.3a & 3b. A graph showing the noise amplitudes along the bobbin with the secondary noise source and an equivalent circuit model

$$V_P(C_1) - V_S(C_2) - V_A(C_3) = 0 \qquad (1)$$

Where $V_P = N_{PS}(V_S) = N_{PA}(V_A)$

and $N_{PS} = \dfrac{N_P}{N_S} \qquad N_{PA} = \dfrac{N_P}{N_A}$

$\therefore V_P(C_1) - \dfrac{1}{N_{PS}}V_P(C_2) - \dfrac{1}{N_{PA}}V_P(C_3) = 0$

$\therefore C_1 = \dfrac{1}{N_{PS}}C_2 + \dfrac{1}{N_{PA}}C_3$

Now, $C_2 = C_1 + C_3$

$\therefore C_1 = \dfrac{1}{N_{PS}}(C_1 + C_3) + \dfrac{1}{N_{PA}}C_3$

$\therefore C_{AS} = \dfrac{(N_{PS}-1)}{(\frac{N_{PS}}{N_{PA}}+1)}C_{PS} \qquad (2)$

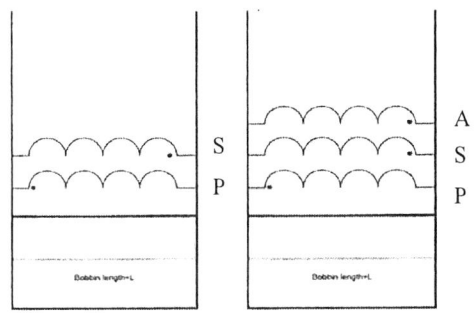

Fig 4a and 4b, winding constructions in a physical Transformer of a common flyback converter

III. EXPERIMENTS

A flyback converter is built as described in figure 1 & 2 to test the proposed method. The experiments concentrate on meeting the zero equipotential line along the bobbin. The switching frequency is 100kHz, input at 110 and 230Vac, output 25Vdc with 1.5A resistive load. Figure 4a and 4b show the transformer constructions in figure 1 & 2. The turns ratio is $N_{PS} = 3.92$. For a transformer with $C_{PS} = 80pF$ @ 100kHz. Equation (2) suggested $C_{AS} = 47.5pF$ @ 100kHz..

A conducted EMI test from 100kHz to 8MHz is performed and a RF current probe (HP 11967A) is employed to measure the noise current passing through the transformer primary – secondary coupling path. The setup is shown in figure 5. Three tests were performed for comparison.

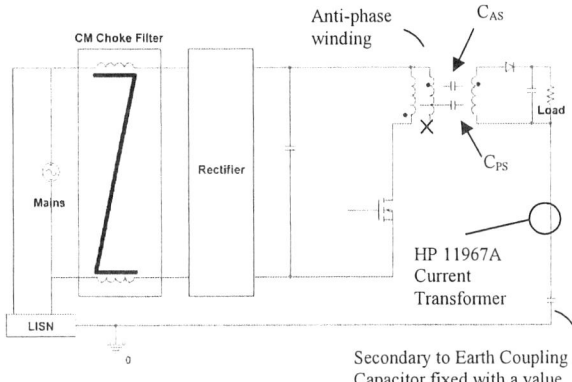

Fig.5. Conducted EMI setup for CM noise measurement with HP 11967A current transformer

In figure 6, trace 1 shows the original transformer performance as constructed in figure 4a with a 2mH CM choke filter, but without the anti-phase winding **A**. When the anti-phase winding **A** is employed as constructed in figure 4b, the Electro-Magnetic Interference (EMI) has dramatically improved by around 20dB at the low frequency end and effective up to 8MHz. The experimental result shows the theory proposed works effectively to reduce common mode noise.

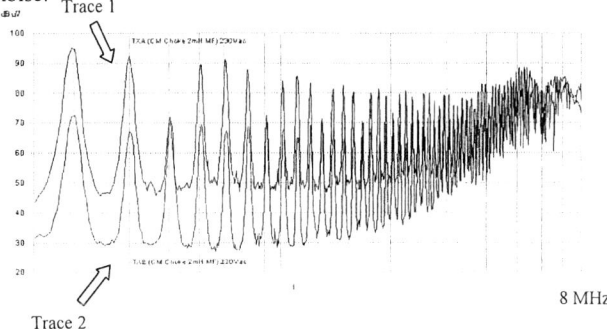

Fig.6 Conducted EMI tests showing different CM noise reduction

IV. CONCLUSION

In this paper a transformer construction technique is proposed. This technique employs the Zero Equipotential Line theory to construct an anti-phase winding. It effectively cuts down CM noise by eliminating the noise voltage across the isolated primary and secondary windings. The concept of maintaining an equipotential line along the bobbin and quiet node connections are justified by analysis. The anti-phase winding is very easy to design and it does not carry high current currents, this has definite advantage over the conventional CM noise filter. Experimental results proved the effectiveness of this method and common mode noise is reduced considerably. This method facilitates and provides a useful way to cancel the noise passing through the isolated transformer, confirmed by the conducted EMI tests. A well considered transformer design with the proposed CM noise cancellation technique can achieve high conversion efficiency as well as good noise immunization.

References

[1] F.-Y. Shih and D.Y. Chen, "A procedure for designing EMI filters for AC line applications," *IEEE Trans. Power Electron.*, vol. 11, pp. 170–181, Jan. 1996.
[2] M. Damnjanovic, G. Stojanovic, V. Desnica, L. Zivanov, R. Raghavendra, P. Bellew, N. Mcloughlin, "Analysis, design, and characterization of ferrite EMI suppressors" *IEEE Trans. Magnetics*, vol 42, issue 2, Part 2, Feb. 2006 pp. 270 – 277
[3] M. Damnjanovic, L. Zivanov, G. Stojanovic, "Common Mode Chokes for EMI Suppression in Telecommunication Systems" EUROCON, 2007. The International Conference on "Computer as a Tool" 9-12 Sept. 2007 pp. 905 – 909
[4] M. Damnjanovic, L. Zivanov, G. Stojanovic, "Analysis of effects of material and geometrical characteristics on the performance of SMD common mode choke" 26th International Conference on Microelectronics, 2008. MIEL 2008. 11-14 May 2008 pp 267 – 270
[5] A. Roc'h, H. Bergsma, D. Zhao, B. Ferreira, F. Leferink, "A new behavioural model for performance evaluation of common mode chokes" 18th International Zurich Symposium on Electromagnetic Compatibility, 2007. EMC Zurich 2007. 24-28 Sept. 2007 pp. 501 – 504
[6] A. Roc'h, H. Bergsma, D. Zhao, B. Ferreira, F. Leferink, "Comparison of evaluated and measured performances of common mode chokes" International Symposium on Electromagnetic Compatibility - EMC Europe, 2008, 8-12 Sept. 2008 pp. 1 – 5
[7] N. Mortensen, G. Venkataramanan, "An Active Common Mode EMI Filter for Switching Converters" IEEE Industry Applications Society Annual Meeting, 2008. IAS '08 5-9 Oct. 2008 pp. 1 – 7

978-1-4244-4782-4/10 $26.00 © 2010 IEEE

[8] P. Chen; H. Zhong; Z. Qian; Z. Lu, "The passive EMI cancellation effects of Y capacitor and CM model of transformers used in switching mode power supplies (SMPS)" IEEE 35th Annual Power Electronics Specialists Conference, 2004. PESC 04. 2004, vol 2, 20-25 June 2004 pp. 1076 – 1079

[9] D. Cochrane, D. Y. Chen, D. Boroyevic, "Passive cancellation of common-mode noise in power electronic circuits," *IEEE Trans. Power Electron.*, vol. 18, issue 3, pp. 756–763, May. 2003.

[10] Edward Herbert, "Transformer for switched mode power supplies and similar applications," in *U.S. Pat. No. 6,137,392*, Oct. 24, 2000.

[11] S. Wang, F. C. Lee, "Common-Mode Noise Reduction for Power Factor Correction Circuit With Parasitic Capacitance Cancellation" IEEE Transactions on Electromagnetic Compatibility, vol 49, issue 3, Aug. 2007 pp. 537 – 542

[12] Wu Xin, N. K. Poon, C. M. Lee, M. H. Pong, Zhaoming Qian, "A study of common mode noise in switching power supply from a current balancing viewpoint" in *Proc. IEEE Power Electronics and Drive System Conf. (PEDS)*, vol. 2, pp. 621–625, Jul. 1999.

[13] C. P. Liu, M. H. Pong, N. K. Poon, "Apparatus for reducing common mode noise current in power converters," in *U.S. Pat. No. 6,490,181*, Dec. 3, 2002.

[14] W. Xin, M. H. Pong, Z. Y. Lu, and Z. M. Qian, "Novel boost PFC with low common-mode EMI: Modeling and design," in *Proc. IEEE Appl. Power Electron. Conf. (APEC)*, New Orleans, LA, 2000, pp. 178–181.

[15] Shuo Wang, Pengju Kong and Fred C. Lee, "Common mode noise reduction for boost converters using general balance technique," in *Proc. of IEEE Power Electronics Specialists Conference*, 18-22, June, 2006. pp. 3142-3147.

50W Power Device (PD) Power in Power over Ethernet (PoE) System with Input Current Balance in Four-pair Architecture with Two DC/DC Converters

Haimeng Wu, Zhengshi Wang, Jiande Wu, Xiangning He, Yan Deng

College of Electrical Engineering, Zhejiang University

Hangzhou, 310027, P.R. China

Abstract—A four-pair architecture of two DC/DC converters with input current balance control strategy in Power over Ethernet (PoE) system is proposed in this paper to get more power from Power Sourcing Equipment (PSE). The four-pair architectures' steady-state models are built to analyze the imbalance problem that exists in the structure. Two main circuit modules operate in parallel and the input current balance circuit is adopted to ensure the balance of both primary current, which can improve the output level and the current sharing performance. Finally, a 50W 36-60V-input 12V-output prototype is built and tested to verify the effectiveness of the presented converter.

I. INTRODUCTION

PoE continues to gain popularity in today's networking world. The advantages of delivering power over the Ethernet cable may simplify the wiring network, and thus does not require the use of an adapter or a second power cable. Therefore, PoE can be applied to VoIP (Voice over Internet Protocol), web cameras, wireless access points, monitoring systems, sales terminals, home automation, etc. In a typical two-pair PoE system, only 12.95W power can be delivered from PSE to Power Device (PD). However, some specialized applications require more power than the existing IEEE 802.3af standard [1] can provide. To achieve higher power levels, there are some possible approaches include raising the supply voltage, raising the supply current capacity, and reducing the cable resistance. These techniques may be used alone or in combination. Unfortunately, to maintain compatibility with existing equipment, such as communication power, the 57-V upper limit should not be changed. Increasing the allowed operating current cause excessive loss over the network cable, which leads to overheating and the voltage on output port becomes too low. Taking various factors into account, doubling the value of existing current limited to 750mA (compliant with the latest IEEE 802.3at[1] standard published in 2009) over four pairs to gain four times standard power is a good way to compromise. However, in existing

four-pair architecture with one dc-dc converter, an imbalance of current in each return path will arise because of variations in diode forward voltage drops, cable resistance, and pass FET on resistance which lies in controller of interface circuit. According to the simulation results in worst-case conditions [5], the difference between two currents becomes almost 200mA (current limit in each path is 750mA), this could result in one port exceeding its current limiting and shut down the PD because of low input voltage conditions. More critical is that the imbalance is caused by the circuit topology, for which there is no solution, the reasons are detailed in the part II.

There are many articles that discuss the modules in parallel and a variety approaches are used to achieve the load current sharing, such as droop method[9], master/slave[8], average current sharing method[10] and maximum current sharing method. [6] presents several possible structures of modules in parallel that can be used in PoE system to improve the output level. However, only with the values of the input voltages are almost the same, input current can be limited and maintain a balance within a certain range. The two modules of different input voltage generate the same power output when the input current are different. Especially in PoE applications, the input current must be strictly limited. The output current balance control mode still makes the input current susceptible to imbalance, which will reduce the system stability.

A four-pair architecture with two dc-dc converters is proposed and a control strategy of input current balance is introduced in this paper to solve the problem of the imbalance discussed above. The prototype is built and tested to verify the performance of input current balancing.

This work is sponsored by the National Nature Science Foundation of China (50777055) and the Power Electronics S&E Development Program of Delta Environmental & Education Foundation (DREM2009001)

978-1-4244-4782-4/10 $26.00 © 2010 IEEE

II. MODELING AND ANALYSIS OF PROPOSED ARCHITECTURE

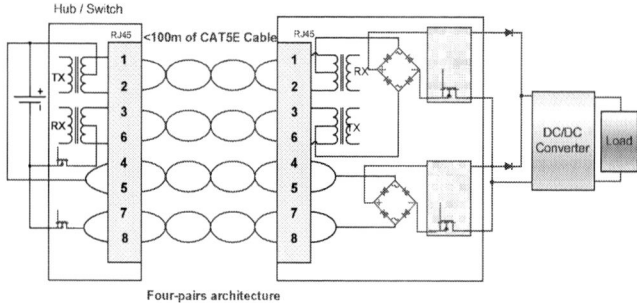

Fig.1a: Four-pair architecture with one DC/DC converter

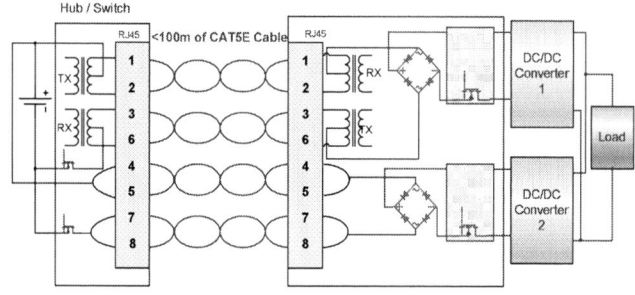

Fig.1b: Four-pair architecture with two DC/DC converters operated in parallel

Fig.1a and fig.1b show two different architectures of four pairs. Cable resistance, network transformer resistance, diode bridge resistance, pass FET on resistance and diode forward voltage drops can be represented into $R_{12}\sim R_{78}$, R_{tr} , $R_{BR1}\sim R_{BR2}$ 、 $U1\_R_{on}\sim U2\_R_{on}$ 、 $V_{BR1}\sim V_{BR2}$ 、 $V_{D1}\sim V_{D2}$ respectively as in Fig.2a and Fig.2b. In the four-pair architecture of one DC/DC converter, both current loops feed the single DC/DC converter. If the impedances of each loop were identical, current balancing would be unnecessary and each loop would provide half of the needed input current to the DC/DC converter. However, mismatches in the connectors, wires, and components will naturally cause one loop to carry more current than the other. As shown in Fig.2a, assuming that the value of total return current is I_R, and the value of return path1 and path2 are I_{P1} and I_{P2} respectively. According to the relationship between the voltage and current the following equations can be obtained:

$$\begin{cases} I_{P1} \times (R_{36} + R_{tr} + R_{BR1} + U1\_R_{on}) + V_{BR1} \\ = I_{P2} \times (R_{78} + R_{BR2} + U2\_R_{on}) + V_{BR2} \\ I_{P1} + I_{P2} = I_R \end{cases} \quad (1)$$

To simplify the expression, the total resistance of each path can be represented as R_A and R_B, as shown in equation (2).

$$\begin{cases} R_A = R_{36} + R_{tr} + R_{BR1} + U1\_R_{on} \\ R_B = R_{78} + R_{BR2} + U2\_R_{on} \end{cases} \quad (2)$$

By equation (1) and (2), solving for the value of I_{P2} and I_{P1} yield:

$$\begin{cases} I_{P1} = \dfrac{V_{BR2} - V_{BR1}}{R_A + R_B} + \dfrac{R_B}{R_A + R_B} I_R \\ I_{P2} = \dfrac{V_{BR1} - V_{BR2}}{R_A + R_B} + \dfrac{R_A}{R_A + R_B} I_R \end{cases} \quad (3)$$

From equation (3), taking into account maximum variations in diodes' forward voltage and the on resistance of network transformer's winding, 1% resistor tolerances, maximum pass FET on-resistance tolerance, and the 3% maximum cable resistance in accordance with the IEEE standard, the maximum current difference can be calculated at different load conditions. The value of current difference in worst-case condition can be close to 200mA.

To ensure reliability, a Current-booster Circuit [2] was proposed to handle the worst-case imbalance so that the current flow through controller of interface circuit below its current limitation. But it does not really solve the problem of current balance as the current flow through path1 and path2 are not identical. Also, this circuit topology results in a considerable amount of loss, affecting the overall efficiency.

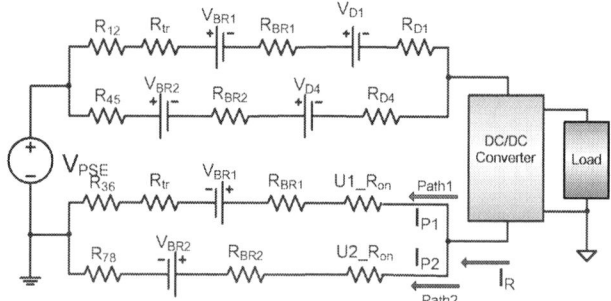

Fig.2a Model of four-pair architecture with one DC/DC converter

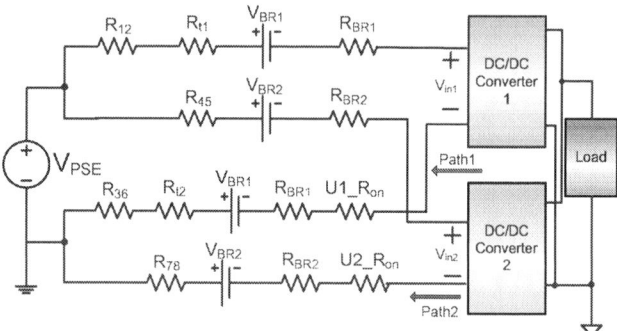

Fig.2b: Model of four-pair architecture with two DC/DC converters

Fig.2b shows that each DC/DC converter has its own loop return to PSE in the architecture of two DC/DC converters. Therefore, a control strategy of current balance can be introduced to ensure the current return in each path is identical. If we choose a output current sharing method described in the introduction and assume that the converter efficiency is μ (Due to the different input conditions, slight differences in the actual circuit) and output power P_O, for each module to share half the load at steady state that is $0.5 P_O$. The values of input current and input voltage are I_{in1} and I_{in2}, V_{in1} and V_{in2}, respectively. The resistance of each loop is represented as R_C and R_D, as shown in equation (4).

978-1-4244-4782-4/10 $26.00 © 2010 IEEE

$$\begin{cases} R_C = R_{12} + R_{36} + R_{t1} + R_{t2} + 2R_{BR1} + U1\_R_{on} \\ R_D = R_{45} + R_{78} + 2R_{BR2} + U2\_R_{on} \end{cases} \quad (4)$$

Two output power equations can be obtained as follows:

$$\begin{cases} (V_{PSE} - I_{in1}R_C - 2V_{BR1}) \times I_{in1} \times \mu = 0.5P_o \\ (V_{PSE} - I_{in2}R_D - 2V_{BR2}) \times I_{in2} \times \mu = 0.5P_o \end{cases} \quad (5)$$

If we equate to equation (3), taking into account maximum variations in actual parameters, from equation (5), the input current values can be calculated in certain load conditions and converter efficiency. It can therefore be shown that the difference of input current is less than 100mA in the worst-case. If considering the output power control precision, the result value may be higher. Therefore, the input current balance can be limited within a certain range by an output sharing control method. However, under this control, there are still some unstable factors in the system because of this imbalance, thus the input current balance control method is more suitable for PoE system.

III. MAIN CIRCUIT AND PROPOSED CONTROL STRATEGY

The system architecture is shown in Fig.3. There are two modules which operate in parallel and consist of PSE-PD interface circuit and DC-DC converter respectively. The input current balance control circuit is introduced between the PSE-PD interface circuits. The power from PSE (Power Sourcing Equipment) flows through the PSE-PD interface circuit and DC-DC converter to the load.

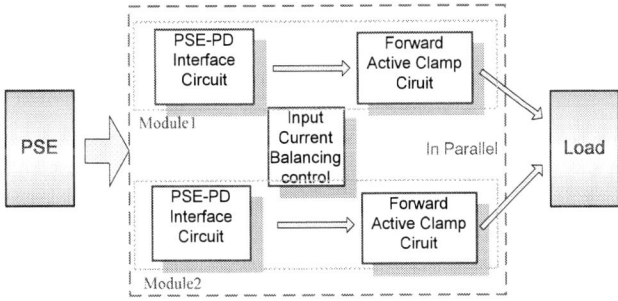

Fig.3 System architecture

As shown in Fig4, the PSE-PD interface circuit consists of a RJ45 connector with integrated network transformer, active bridge and LM5073 PD interface. The current from the PSE is injected into a transformer's center tap where it divides and passes through one of two pairs of transmission lines, at the far end, the two currents pass into a transformer in which they recombine to PD. The device's return path uses the other pair of transmission lines. The supply voltage between two transmission lines is the common-mode signal, therefore, there is no effect to send and receive differential data in pairs. In order to improve the efficiency, an active bridge [7] used for polarity protection is used to instead the diode rectifier bridge as shown in Fig.4. LM5073-based interface circuit is designed to ensure PSE to identify and supply power to PD.

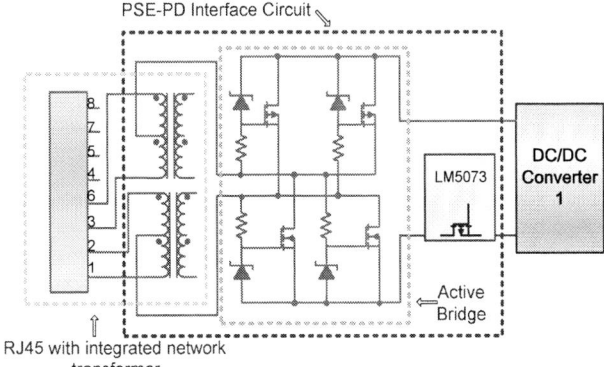

Fig.4 PSE-PD interface circuit

Active clamp forward is employed as the main circuit topology and the control strategy is shown in Fig.5. For steady state operation the net volt-second product applied to the magnetizing inductance over a complete cycle must equal zero:

$$V_{IN} \times D \times T_s = (V_{IN} - V_c) \times (1 - D) \times T_s \quad (6)$$

Solving for clamp capacitor voltage yields:

$$V_c = V_{IN} / (1 - D) \quad (7)$$

Here, D is the on-time duty cycle and T_s is the switching period, V_C is the clamp capacitor voltage. This important feature minimizes the voltage stress across the main switch for all operating conditions, thus allowing use of lower rating devices which lead to lower on resistance and lower gate charge [4]. The magnetizing and leakage energies are recycled and returned to the source by the active clamp switch, instead of the reset winding.

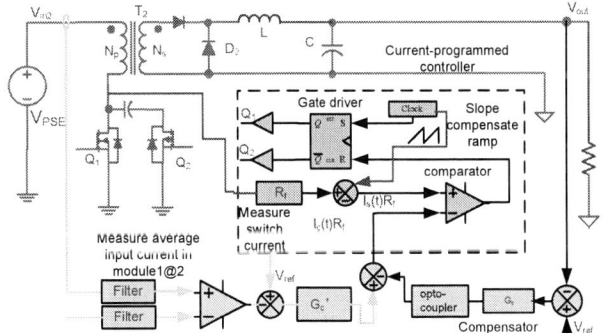

Fig.5: Main topology and control strategy

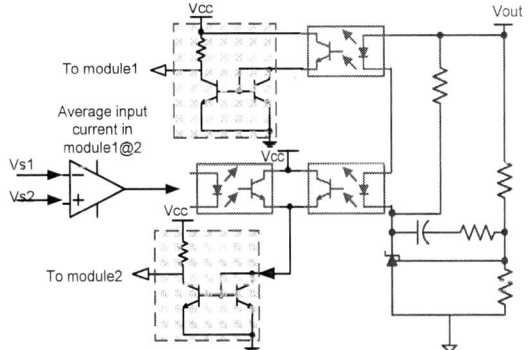

Fig.6 Feedback control circuit

Current programmed mode is used in control strategy as shown in Fig.5. The chief advantage of this mode is its simpler dynamics. Mosfet failures due to excessive switch current can then be prevented simply by limiting the maximum value of feedback control signal $i_c(t)$. This ensures that the Mosfet will turn off whenever the switch current becomes too large. The feedback control circuit is shown in Fig.6, the diodes of two opto-couplers are connected in series. In this way, with the diode-side current changing trends, the output of the transistors' emitter current is larger or smaller simultaneously. If linear opto-couplers are used, given that the opto-coupler transfer ratio is basically the same, the transistor emitter will output the same amount of current. The error current signal generated from the input current balance control circuit is added to this feedback current signal as compensation to adjust the PWM duty cycle. This is done to ensure the stable output voltage and the input current also remains constant.

The control process is as follows: input current of the two modules are sampled respectively to produce the average input current signals by the filters, then the error signal is magnified by the analogy amplify. Next, the output signal from analogy amplify is compared with the reference voltage signal to produce a compensation control signal. This compensation signal is added in feedback signal as combination control signal. Finally, this combination signal is compared with transient input current signal to adjust the duty cycle of one module.

It also should be noted that the control signal is inputted into the controller in the form of a control current which is internally mirrored by a matched pair of NPN transistors as shown in Fig.6. Therefore, greater system loop bandwidth can be realized, since the bandwidth-limiting pole associated with the opto-coupler is now at a much higher frequency. By adding an additional fixed slope voltage ramp signal (slope compensation) to the current sense signal, the sub-harmonic oscillation can be avoided when duty cycle is greater than 50 percent.

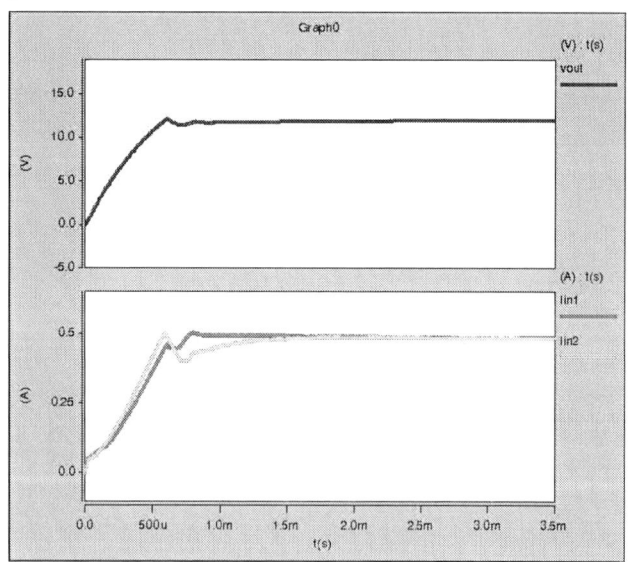

Fig.7   Input current balance control simulation results in Saber

In order to verify the current balance effect, simulation experiments are done using Saber. for the conditions Vin1=54V, Vin2=42V, Iout=4A, the simulation results of input current balance control are shown in Fig.7. Two curves for the average input current sampled value are drawn in the graph. It can be seen from the graph that input currents achieve good balance effect in the steady state and the dynamic response is also desirable.

IV.   EXPERIMENTAL VERIFICATIONS

In order to further validate the effect of the proposed control strategy of current balance, a prototype of 36-60V input and 12V output is built. The specifications are as follows: V_{in}: 36-60V; V_{out}: 12V; Pout: 48W; fs: 250 kHz; Np/Ns: 2:1; L: 60μH; C: 220μF; D1, D2: 12CWQ06FN; Q1:FDD2572; Q2:Si7119.

The experimental results of two DC/DC modules operated in parallel with current balance circuit are shown in Fig.8a and Fig.8b. These graphs demonstrate that it is ideal to maintain the current balance in two modules at different input voltages or at different load conditions. The current balance circuit performs well to ensure the difference of input current within 30mA (4% of maximum current limit).

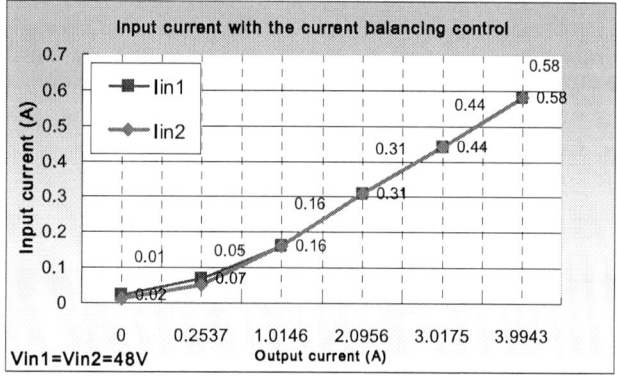

Fig.8a: Input current at different load conditions.

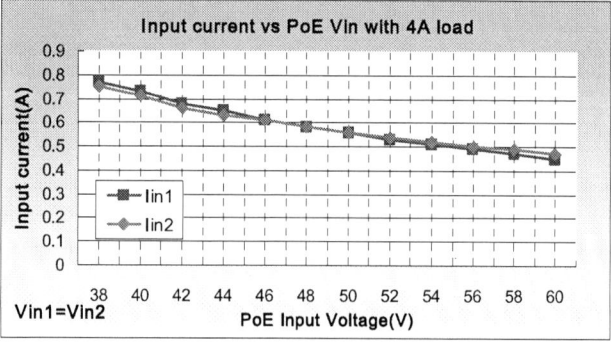

Fig.8b: Input current at different input voltage with 4A load

To verify the steady-state current balance effect under different input voltages, fix Vin1 to 42V,48V and 54V, and Vin2 is varied from 36 to 60V in steps of 2V, the value of input current between two modules are recorded and illustrated in Fig.9(a)(b)(c).The results show that the

978-1-4244-4782-4/10 $26.00 © 2010 IEEE   578

performance of current balance is good over most of the input range. Input current imbalance becomes larger with the bigger difference between the input voltages. However, even in extreme conditions, the results are acceptable because the values of input current are both under the maximum current limit. When Vin2=54V the converter efficiency can reach nearly 90% at 48W full load under different Vin1.

The picture of prototype is shown in Fig.10.

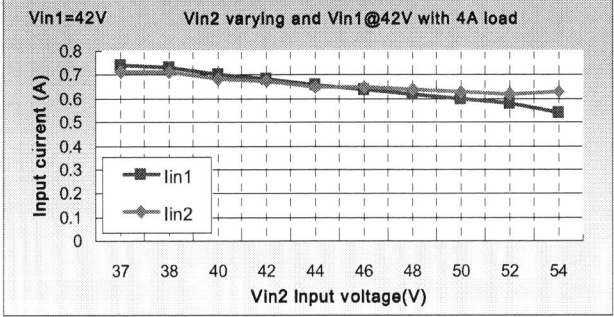

Fig.9a Input current at Vin2 varying and Vin1@42V with 4A load

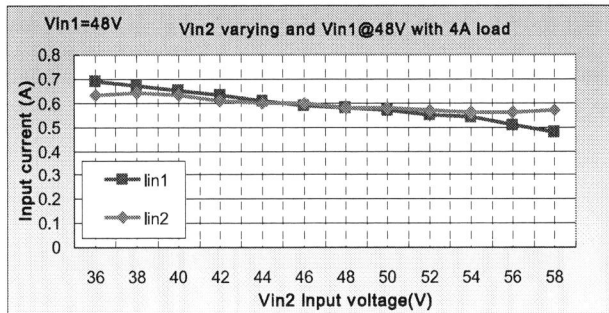

Fig.9b Input current at Vin2 varying and Vin1@48V with 4A load

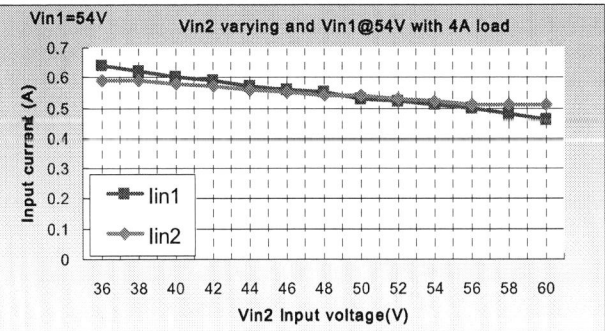

Fig.9c Input current at Vin2 varying and Vin1@54V with 4A load

Fig.10: Prototype picture

V. CONCLUSION

To get more power from PoE system, this paper introduces four-pair architecture with two DC/DC converters instead of one DC/DC topology in which imbalance can't be controlled because of the single DC/DC converter topology. A control strategy of input current balance is introduced to solve the problem of imbalance and ensure each return path shares the current identically to void one port exceed its current limiting and shut down the PD. Experimental results prove the proposed topology and control strategy perform well.

ACKNOWLEDGMENT

This work is supported by the National Semiconductor Corporation.

REFERENCES

[1] IEEE 802.3 standard, http://standards.ieee.org/getieee802/802.3.html

[2] Martin Patoka, "High-Power PoE PD Using TPS2375/77-1", Application Report, 2007.

[3] "High Power PoE Applications", AND8333, Application Report, On Semiconductor, April 2008.

[4] Bob Bell, "Operation and Benefits of Active-Clamp Forward Power Converters" Power Designer, National Semiconductor , 2005.

[5] Steven R. Tom "Current balancing in four-pair, high-power PoE applications" Power Management, Texas Instruments, 2007.

[6] Brian King, Robert Kollman "To Get more power from Ethernet" Today's electronic Feb 2008.

[7] Grant Smith "LM5073HE Evaluation Board With Active Bridge" National Semiconductor Application Note 1875 July 2, 2008

[8] Rajagopalan, J.; Xing, K.; Guo, Y.; Lee, F.C.; Manners, B.; "Modeling and dynamic analysis of paralleled DC/DC converters with master-slave current sharing control" Applied Power Electronics Conference and Exposition, 1996. APEC '96. Conference Proceedings 1996., Eleventh Annual Volume 2, 3-7 March 1996 Page(s):678 - 684 vol.2

[9] Jung-Won Kim; Hang-Seok Choi; Bo Hyung Cho;"A novel droop method for converter parallel operation" Power Electronics, IEEE Transactions on Volume 17, Issue 1, Jan. 2002 Page(s):25 – 32

[10] Supatti, U.; Boonto, S.; Prapanavarat, C.; Moneyakul, V.; "Design of an H ∞ robust controller for multi-module parallel DC-DC buck converters with average current mode control"Industrial Technology, 2002. IEEE ICIT '02. 2002 IEEE International Conference on Volume 2, 11-14 Dec. 2002 Page(s):992 - 997 vol.2 [1]

High-resolution Physically-windowed Sensors for Power Electronics Applications

Warit Wichakool, James Paris, Al-Thaddeus Avestruz, Dr. Steven B. Leeb

Abstract—This paper presents a high-resolution, physically-windowed sensor architecture that is well-suited for energy scorekeeping and diagnostic applications. The sensor can track a large-scale main signal while capturing small-scale variations. The prototype system measures a small current signal using a closed-loop Hall sensor, and extends the range by driving a compensation current through an auxiliary winding. The system combines the compensation command and the sampled output of the residual sensor to reconstruct the input signal. Results show that the prototype can measure both dc and ac currents with 10 mA resolution over a 160 A current range.

I. INTRODUCTION

In energy scorekeeping and diagnostic applications, current sensors are often deployed to collect and analyze current waveforms from a collection of loads [1]. Analysis provides load disaggregation and detection, power consumption profiling, and diagnostics based on electrical signatures [1]–[5]. Many current sensors are available according to dynamic range and sensitivity. Hall sensors, fluxgate-based sensors, and Rogowski coils have all been used for non-contact current measurement [6]–[12]. As monitoring systems grow to include more loads and to provide more detailed information about the loads, the scalability and utility of the system depends on the quality of data acquired by the current sensor [13]. When monitoring large loads or collections of loads, some relevant features may be found in harmonic or aperiodic content that is small compared to the current drawn at the fundamental line frequency. For other features, the full amplitude signal may be required.

A physically-windowed sensor architecture is introduced that allows for a more flexible tradeoff between sensitivity and dynamic range. Large-scale variations are cancelled such that the measured signal remains within a small operating window, while the residual small-scale signals are sensed conventionally with an accurate sensor. This architecture is similar to that of pipelined analog-to-digital converters [14], but utilizes a physical cancellation approach that can be applied to magnetic flux-based current sensors, strain gauges, pressure transducers, and many other physical systems. The cancellation is software-controlled by an embedded microcontroller, permitting a variety of windowing techniques and flexible processing and analysis.

In this paper, we present an initial application of this concept to power electronics by developing a physically-windowed current sensor that demonstrates high accuracy over a wide input range.

II. SYSTEM DESIGN

Our architecture uses an additional physical input to apply a cancellation signal to a sensor. This enables an accurate

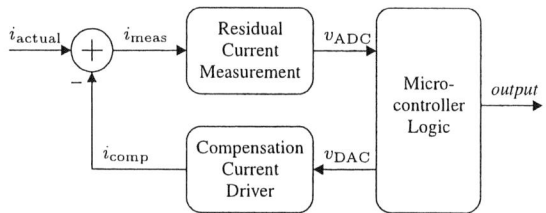

Fig. 1. System block diagram. The system consists of three primary components: the compensation current driver, the residual current measurement, and the microcontroller logic. The compensation current i_{comp} is subtracted from the primary input current i_{actual} by physical cancellation of magnetic flux.

but narrow-range sensor to measure effectively beyond its specified operating range. The initial implementation applies this approach to a current sensor and follows the overall design shown in Fig. 1. A compensation current i_{comp} is driven anti-parallel to large input currents such that the effective total current i_{meas} seen by the sensing element remains within its designated operating range. The microcontroller coordinates and controls the system, performing calibration at startup and adjusting the compensation as necessary to keep the sensor at the desired operating point.

A. Signal Reconstruction

Fig. 2 depicts the signal reconstruction used to determine the total current using the windowed measurement. The total input current i_{actual} is calculated from the instantaneous compensation current and sensor measurement as

$$i_{\text{actual}} = k_c \cdot i_{\text{comp}} + k_m \cdot i_{\text{meas}}, \qquad (1)$$

where k_c and k_m are calibration values determined by physical factors such as the number of turns on the sensing core and the amount of magnetic coupling between coils.

The compensation current i_{comp} is generated by an operational transconductance amplifier (OTA), detailed in §IV-B. This current i_{comp} is set by the microcontroller using a digital-to-analog converter (DAC) command voltage, v_{DAC}. The total compensation current is given by

$$i_{\text{comp}} = k_{\text{DAC}} \cdot v_{\text{DAC}}, \qquad (2)$$

where k_{DAC} is determined by the OTA design.

Residual current is measured using a closed-loop Hall sensor, detailed in §IV-A. This current i_{meas} is read from an analog-to-digital converter (ADC) as the voltage v_{ADC}, and is given by

$$i_{\text{meas}} = k_{\text{ADC}} \cdot v_{\text{ADC}}, \qquad (3)$$

where k_{ADC} is determined by the sensor front-end design.

978-1-4244-4782-4/10 $26.00 © 2010 IEEE

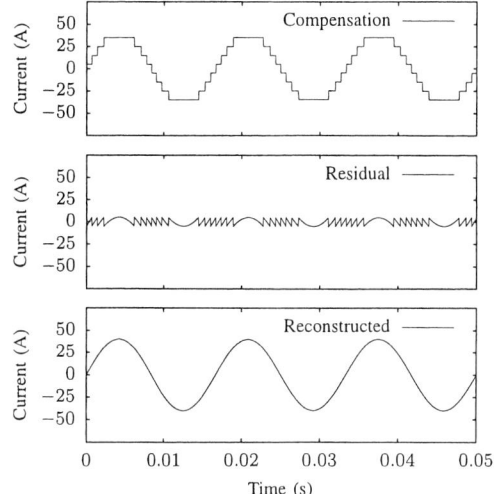

Fig. 2. Signal reconstruction. The compensation current and measured residual current are combined to determine the total current through the full sensor.

Combining these equations, the complete reconstruction is

$$i_{\text{actual}} = k_c \cdot k_{\text{DAC}} \cdot v_{\text{DAC}} + k_m \cdot k_{\text{ADC}} \cdot v_{\text{ADC}}.$$

The constants can be simplified as:

$$i_{\text{actual}} = k_s(v_{\text{DAC}} + k_r \cdot v_{\text{ADC}}), \tag{4}$$

where k_r represents the ratio between the DAC command voltage and the corresponding change in ADC input voltage, and k_s represents a scaling to convert to actual current. This simplified form is used both for discussion and by the internal calibration and windowing procedures described in §IV-C.

B. Resolution and Range

The performance of the overall physically-windowed sensor system is determined by the parameters of its components. The ranges and resolutions of the compensation current and the residual measurement overlap, as depicted in Fig. 3. In this example, the ADC is accurate to 11 bits over a range of 5 A, while the DAC command is accurate to 10 bits over a range of 160 A. The amount of overlap directly relates to the parameters k_r and k_s in (4).

A key requirement for physically windowed sensing is that the compensation output must remain stable and predictable to the full system resolution. In Fig. 3, this requirement is depicted as a dashed line on the DAC output. Here, for the lowest-order bits of the combined result to be accurate, each of the 2^{10} possible DAC commands must result in a voltage stability of one part in 2^{16}. Certainly, a 16-bit DAC would suffice. However, only stability is needed, not accuracy. If a lower-resolution DAC is, or can be made to be, similarly stable in output, it is sufficient for the sensor architecture. Using such

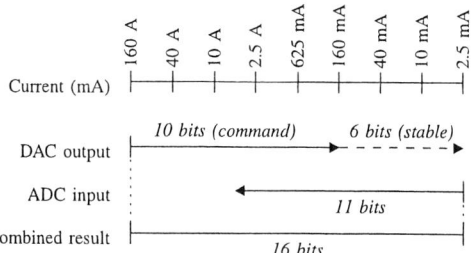

Fig. 3. Overlap of the DAC output, relating to compensation current, and the ADC input, relating to the residual measurement. The ranges and overlapped positions are related to the parameters in (4). The combined result shows high accuracy over the full range.

a DAC may provide cost or performance benefits.[1]

Given that the stability requirement is met, then the actual output voltage v_{DAC} can be related to the DAC command x as

$$v_{\text{DAC}}(x) \propto \frac{x}{2^{10}} + \frac{\text{LOOKUP}[x]}{2^{16}} \tag{5}$$

where $\text{LOOKUP}[x]$ is a 2^{10}-entry table that stores these 6 extra stable bits. This table can be populated by the microcontroller in a calibration step that uses the ADC input to determine the low-order bits of each DAC output.

C. Windowing

The front-end current measurement is "windowed" by the compensation current in the sense that the compensation sets a particular operating point, and the Hall sensor measures a small window of current around this point. The microcontroller has significant flexibility in the windowing approach, and the behavior can be adjusted based on expected workloads and system parameters.

A basic approach to windowing is to continuously recenter the window so that the ADC measurement is zeroed; that is, the residual current is driven to zero after each sample. However, this requires that the OTA change its current output nearly continuously as the input signal changes, increasing the bandwidth requirements and potentially making the data less accurate if changes in compensation current are slow to settle.

The approach demonstrated by the reconstruction in Fig. 2 is to change the DAC command when the residual current in the sensor approaches the limits of the front-end. The compensation current will remain constant for small input signal changes, and only change for larger input signals that exceed the window. For many input signals, this may allow the compensation to change relatively slowly, reducing bandwidth requirements for the compensation driver.

More advanced approaches are possible, particularly for loads with known characteristics. A predictive estimator in the microcontroller can perform an anticipatory change in the compensation current so that the residual sensor current would

---

[1]The prototype implementation in §IV simulates this stability by using a 16-bit DAC with fixed random low-order bits on a 10-bit command. The low-order bits are set by the microcontroller and can be adjusted for testing purposes.

be expected to fall within the sensor limits at the next sample interval. Such techniques can potentially increase the slew rate capability of the system.

D. Bandwidth

The bandwidth of the physically windowed sensor system depends on the input signals and their relation to the sensor window. There are two fundamental regions of operation: the first, within the windowed range of the residual current measurement, and the second, over the full range of the compensation current. For input currents that fall entirely within the window, the bandwidth performance of the system is equal to that of the residual current sensor front-end, as the compensation current is held constant. For full-scale input signals, the bandwidth is instead limited by how fast the compensation current can track the input change.

Maximum slew rate may be further affected by the windowing algorithm in use. Once the residual current exceeds the range of the sensor window, the compensation command must be adjusted. In the absence of prediction, the microcontroller will not know by how much the residual current exceeded the window, and will be limited to stepping the compensation by one "window" worth of current at a time. This, combined with the sampling rate of the residual sensor and the bandwidth of the compensation driver, will set the maximum $\frac{di}{dt}$ that can be accurately tracked. For slew rates outside this limit, the subsequent front-end sample will still exceed the window, and the microcontroller can report the potential inaccuracy as part of the output data stream.

The bandwidth and slew rate limits are a function of the resolution, range, and bandwidth of the system components. Flexible tradeoffs can be made by, for example, adjusting the system to increase k_r in (4). This would have the effect of increasing the relative size of the sensor window, increasing the region in which the recorded signal retains full bandwidth, and increasing the maximum slew rate. Conversely, increasing k_r increases the overall resolution of the reconstructed signal.

III. BENEFITS AND MOTIVATION

In many physical systems, large-scale changes occur at relatively slow speed while small-scale details can change rapidly. For example, an electric motor draws a 60 Hz fundamental current from the utility, but it may be desirable to observe a principal slot harmonic (PSH) at several hundreds or thousands of Hertz to track the motor speed [15]. These small, high-frequency details are superimposed on top of the 60 Hz current and need to be examined without saturating the sensor front-end. Conventional current sensors like the closed-loop Hall-effect sensor utilize a single compensation circuit to measure current. The physically-windowed sensing system, instead, divides the measurement into two subsystems, the compensation current driver and the residual current measurement. By dividing the problem and taking advantage of the fundamental differences between the requirements of the large-scale and small-scale measurements, the windowed system can utilize power and bandwidth trade-offs in the design of design each subsystem. This section describes these trade-offs and their design considerations.

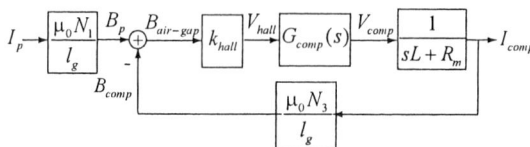

Fig. 4. Typical design for a closed-loop hall sensor, with one compensation circuit.

A. Resolution

To obtain an accurate measurement, the system needs a reference that is stable to the required resolution specification. A conventional current sensor can utilize a single high-resolution ADC to perform the measurement, or it can use a single high-resolution DAC as a reference against which to compare a measurement. In both cases, it is required that the ADC or DAC be both stable and accurate to the full resolution.

With the physically-windowed approach, it is sufficient that the DAC be stable, but not necessarily accurate. A DAC with fewer controllable bits, but stable to the full resolution, can still be used. Initial experiments, testing the output voltage of a 16-bit AD7846 DAC with a HP34401A multimeter, demonstrated an accuracy of approximately 26 μV on a ±5 V range, or approximately 18.5 bits, in a controlled environment. This example shows that, under some conditions, the output of the DAC is more stable than the controllable input.

The physically-windowed sensor design can then use a moderately accurate DAC and a moderately accurate ADC to create a compound data acquisition system that can accurately resolve more than the number of bits provided by either the DAC or ADC alone. For example, assuming proper calibration is performed, the system may be able to use one 10-bit DAC and one 10-bit ADC to create an effective 12-bit data acquisition system.

B. Bandwidth

A key element in the design process for analog circuits is the trade-off between power consumption and bandwidth. If the signal of interest is comprised of both low-frequency and high-frequency content, the current sensor may be able to take advantage of this separation by utilizing two separate compensation circuits, each optimized for one frequency region.

The basic topology of a typical current sensor based on zero-flux sensing consists of three parts: a magnetic flux sensor, a compensation circuit, and a compensation winding. The system block diagram is shown in Fig. 4. The input current creates a magnetic flux which is focused in the air gap of the gapped magnetic core. The magnetic flux sensor senses any magnetic flux in the air gap, and provides an output signal for the compensation circuit. This circuit drives a cancellation current to cancel the magnetic flux. Effectively, the system forces the magnetic flux in the air gap to zero, keeping the magnetic core around the zero-flux operating point and away from the saturation region.

In order to measure a fast dynamic signal accurately, the measuring system is required to have a wide bandwidth and a large open-loop gain. In another words, the gain-bandwidth product of the open-loop transfer function must be very large. As shown in Fig. 4, the system can be designed to meet the large gain-bandwidth product by choosing an appropriate compensation circuit $G_{\text{comp}}(s)$. A typical compensator is an integrator with a lead compensation. The integrator provides a large open-loop gain at low frequency, whereas the lead compensation provides stability for the system. A typical transfer function of the lead-compensated integrator circuit can be described as

$$G_{comp}(s) = \frac{(s\tau_2 + 1)}{(s\tau_1 + 1)}, \qquad (6)$$

where $\frac{1}{\tau_1}$ is a low frequency pole of a practical integrator circuit, and $\frac{1}{\tau_2}$ is the compensated zero. In this case, we assume that all high frequency poles of the op-amp and parasitics are negligible at the cross-over frequency. The compensation circuit would be designed to meet the required bandwidth of the input signal.

The compensation circuit must cancel the flux by driving an equivalent current through the magnetic core, typically using an output stage consisting of an operational amplifier and push-pull buffer circuit. If the compensation current coil consists of N_3 turns, the primary winding consists of N_1 turns, and the maximum input current is $i_{\text{actual\_max}}$, then the compensation circuit must be able to drive $\frac{N_1}{N_3} i_{\text{actual\_max}}$ through the coil.

As N_3 is increased, the drive current is lowered, but this also affects the inductance that the compensation circuit has to drive. Specifically, the inductance of the compensation coil is proportional to $N_3{}^2$. Large inductance will limit the maximum slew rate that the buffer circuit can provide for compensation. The slew rate $\frac{di}{dt}$ of the current in an inductor is given by

$$\frac{di}{dt} = \frac{v_L}{L} \propto \frac{v_{supply}}{N_3{}^2}, \qquad (7)$$

where v_L represents the voltage across the inductor, which is limited by the supply voltage of the system. Given the same supply voltage, a smaller inductance will allow the system to follow the input current more accurately.

The physically-window sensing system, on the other hand, divides the compensation circuit into two parts: a large, slow compensation circuit for the bulk of the flux cancellation, and a small, fast compensation circuit for measuring the residual. Block diagrams of the proposed system are shown in Fig. 5.

For the smaller residual measuring range, the compensation circuit can have the same design as the single compensation circuit case. Since the maximum current output is reduced, the number of winding turns N_3 can be reduced, effectively lowering the inductance of the coil. The lower inductance allows this compensation circuit to follow the higher $\frac{di}{dt}$ rate according to (7), given the same supply voltage.

The auxiliary compensation circuit must provide a cancellation current for a larger portion of the magnetic flux in the core. Specifically, the auxiliary compensation circuit is responsible for providing a cancellation current of $\frac{N_1}{N_2} i_{\text{actual\_max}}$ where N_2 is the number of turns in the auxiliary winding. This is analogous to the requirement of the first compensation

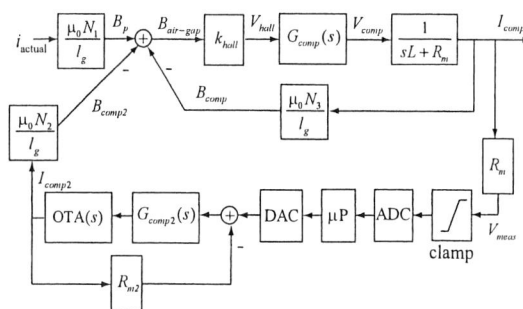

Fig. 5. Physically windowed current sensor, with two compensation circuits.

circuit, and the circuit can be made similarly, with minor adjustments. In our prototype, the auxiliary compensation circuit is implemented with a current source, which provides a high output impedance as seen by the other loop across the current transformer. As a result, when the auxiliary compensation circuit is providing a constant current output, interaction between the two compensation circuits is minimized.

By separating the compensation current into two subsystems, the design process can be divided into two problems which may be tailored to take advantage of the input signal characteristics. In this case, two compensation circuits add complexity to the system, but provide more flexible power and bandwidth trade-offs.

C. Feedback

Feedback is one of the key concepts that enables the proposed sensing system to work properly. This proposed system consists of two analog feedback loops and one digital feedback loop. The small compensation current uses analog feedback to keep the magnetic core in the zero-flux region. The second feedback loop ensures an accurate conversion between the command voltage and the compensation current. Both of these feedback loops act as minor loops within the digital feedback loop. The digital feedback enables flexible control of the entire process, and can be easily adapted for different input signal characteristics.

IV. PROTOTYPE IMPLEMENTATION

The prototype system was implemented according to the design introduced in §II. The system block diagram is shown in Fig. 6. The physical coupling of the subsystems occurs on a single toroidal core. The primary current to be measured passes through N_1 turns on the core. The cancellation current passes through N_2 turns, wound in the opposite direction. The residual current is measured by a closed-loop Hall sensor that utilizes N_3 additional turns. The N_2 and N_3 loops are co-wound to minimize leakage inductance. Typical values for our testing are $N_1 = 50$ and $N_2 = N_3 = 200$.

A. Residual Current Measurement

The residual current measurement is based on the closed-loop Hall-effect sensor shown in Fig. 4. The sensor drives an

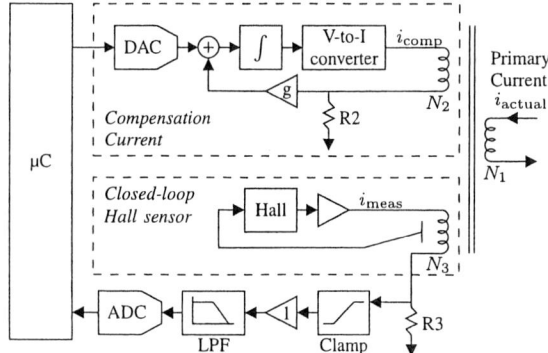

Fig. 6.   Detailed block diagram of the prototype system.

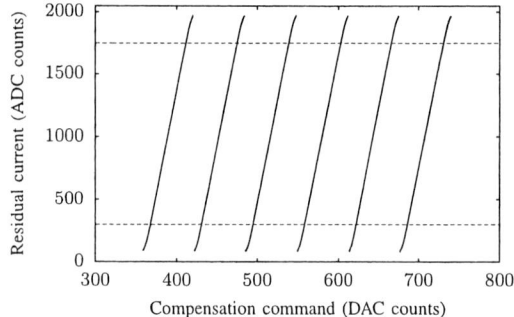

Fig. 7.   Calibration curves showing 11-bit residual current measurement versus 10-bit compensation current command, at various fixed primary currents. The curves are generally linear with slope k_r and flatten out as the clamps begin to activate outside of the dashed lines.

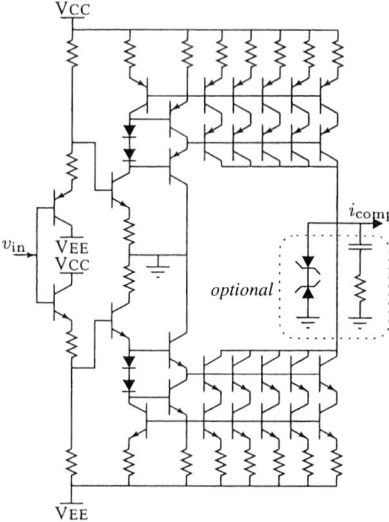

Fig. 8.   Output stage of the compensation current subsystem, showing the OTA implementation.

output current such that the flux perceived by the Hall element is near zero. At this operating point, the temperature drift and offset of the Hall sensor are minimized. The output current is read by an 11-bit ADC to produce the residual measurement.

In our system, the Hall sensor is designed to measure over a small current range of approximately ±2.5 A. If the input signal starts to exceed this range, the compensation current driver is separately commanded to cancel a portion of the flux in order to keep the residual sensor operating normally.

Fig. 7 shows measured calibration curves between the compensation current output and residual sensor input, for various operating points set by the primary current. The slope of each line corresponds to the constant k_r in (4). The dashed lines indicate the approximate measuring range of the ADC for the residual current. The digital controller attempts to maintain the window so that the residual current always falls within this range.

During a high current transient, the input current may temporarily exceed the ability of the system to compensate. To prevent the residual sensor from overloading in this condition, a clamp circuit is added at the output of the Hall sensor. This extra clamp current serves to cancel the primary current and limit maximum residual. The clamp current is not reflected in the ADC reading, causing the measured values to saturate, as shown shown in Fig. 7. After the overload condition, the clamp

deactivates and the Hall sensor returns to normal operation. The system utilizes a magnetic core with a low remnant flux to further minimize the offset error after experiencing such a transient.

B. Compensation Current

The compensation current driver uses a highly stable digital-to-analog converter to establish a voltage command reference. A closed-loop circuit, shown in Fig. 6, is designed to scale and convert the voltage command into the desired output current. This output current is co-wound on the core with the output from the closed-loop Hall sensor circuit. To minimize interaction between the two feedback loops, the output stage of the compensation circuit is high-impedance and appears as an open circuit to the residual sensor circuit.

The OTA design is shown in Fig. 8. In this implementation, the OTA includes a voltage buffer front-end to receive the voltage command from the compensation current feedback op-amp. The buffered command is used to establish a reference output current in the second stage. This reference current is replicated through a current mirror structure. The current mirror uses a cascode topology to improve the output impedance. The emitter degeneration resistors are added to scale the current and to prevent a thermal runaway condition. The OTA structure includes multiple output branches connected in parallel, in order to minimize the power dissipation per branch. Within each cascode branch, the transistor next to the rail sets up the mirrored current, and the cascode transistor acts as a current buffer. The power dissipation in each cascode branch will be concentrated at the cascode transistor. Therefore, the thermal effects on the mirrored current are reduced. Finally, the β-helper transistors are included to provide additional base current for the output stage.

The OTA output passes through N_2 turns on the core and is measured with a sense resistor. The analog feedback loop in the compensation current subsystem serves to minimize error within the output stage. The output current measurement can optionally be provided back to the microcontroller through a low-bandwidth 24-bit ADC for calibration purposes.

C. Microcontroller

Control logic for the prototype is implemented using a Microchip dsPIC33FJ256GP710 microcontroller. It controls the sampling of the residual current measurement, implements the windowing algorithm used to set the compensation current, and communicates all data to a computer for analysis. The microcontroller also performs calibration at startup and on request.

1) Sampling: The ADC is sampled at 8 kHz, a rate chosen to match that used in existing non-intrusive load monitoring systems [13]. At each sample, the microcontroller calculates the total reconstructed current from the compensation command and the residual measurement, and transmits this data to the computer. If necessary, the compensation command is then changed to adjust measurement window.

The sampling interval has a direct influence on the slew rate capability of the system. If the residual current exceeds the window range, the measurement is clamped and the recombined output will be inaccurate. For a given operating point and windowing strategy, a maximum current excursion i_w can be observed within the current window before saturation. Given the sampling interval Δt and primary input slew rate $\frac{di}{dt}$, it is necessary that

$$\Delta t \cdot \frac{di}{dt} < i_w. \tag{8}$$

Thus, decreasing Δt, by increasing the sampling rate, has a corresponding linear effect on the maximum slew rate $\frac{di}{dt}$.

In practice, the limit on Δt is set by the response time of the hardware components and the processing speed of the microcontroller. Regardless of the sampling rate, the data reporting rate to the computer can be maintained at 8 kHz for compatibility.

2) Windowing Strategy: Using (4), the microcontroller can determine by how much a given change in DAC command will affect the sensor measurement at the ADC input. Assuming that the primary current i_{actual} remains constant, a pair of DAC and ADC values are related by

$$k_s(v_{\text{DAC},1} + k_r \cdot v_{\text{ADC},1}) = k_s(v_{\text{DAC},2} + k_r \cdot v_{\text{ADC},2}) \tag{9}$$

$$v_{\text{DAC},1} - v_{\text{DAC},2} = k_r(v_{\text{ADC},1} - v_{\text{ADC},2}) \tag{10}$$

Thus, to cause a change of ΔADC at the residual measurement, the DAC command should be changed by $k_r\Delta$ADC. The prototype implementation uses this approach to "recenter" the window whenever the ADC value begins to approach the clamp limits. This is depicted in Fig. 9. For example, when a sample x from the ADC exceeds a fixed upper limit of 1536, the compensation command is increased by $k_r(x - 768)$ so that the next sample is near 768.

The chosen target ADC values intentionally overshoot the center of the ADC range because it is expected that, in most cases, an increasing current will continue to increase. This

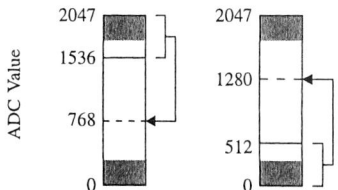

Fig. 9. Windowing approach in the prototype implementation. For input ADC values greater than 1536, the compensation current is changed such that the target ADC value is 768. For input ADC values less than 512, the target ADC value is 1280. The clamps are active in the shaded regions.

provides some extra "headroom" for the common case, which in turn increases the maximum slew rate that the prototype can handle.

3) Communication: All data, including raw DAC and ADC values and calibration constants, are continuously sent to a computer via a full-speed USB link. In some configurations, the microcontroller may read the ADC more frequently than the samples are sent to the computer, and so status flags that indicate error states are also included independently. For example, one flag denotes whether the ADC value was ever observed in the clamped region, which indicates that the returned data for that sample may not be accurate to full resolution.

4) Calibration: In order to perform windowing accurately, the microcontroller needs to know the calibration constant k_r (§II-A) and the table LOOKUP (§II-B).

The value of k_r can be determined using the relationship in (10), if the primary current is constant. In the prototype implementation, the microcontroller assumes constant current and performs calibration at startup, or when triggered by the connected computer. Using an initial estimate $k_r = 1$, the calibration algorithm adaptively adjusts the estimate as it changes v_{DAC} to seek two specific v_{ADC} values corresponding to ADC inputs 512 and 1536. Once these DAC commands are found, each v_{ADC} is oversampled to reduce noise and a final accurate estimate of k_r is computed.

Table LOOKUP is used to map each low-resolution DAC command to its corresponding high-resolution stable output voltage. In the current prototype, this extended stability is simulated using an accurate DAC, and so the lookup is hard-coded to match the randomized low-order bits written to the DAC. However, the system supports an additional low-bandwidth 24-bit ADC for measuring the output compensation current. This calibration ADC could be used to measure and fill the LOOKUP entries for each possible DAC output, in the absence of a hard-coded table.

Finally, to convert the final output of the physically-windowed current sensor to amperes, the scaling factor k_s is used. It is not directly needed by the microcontroller logic, and is currently calibrated by the computer in post-processing using known test values from a Keithley 2400 Sourcemeter.

V. PROTOTYPE RESULTS

Various aspects of the prototype physically-windowed current sensor system have been tested.

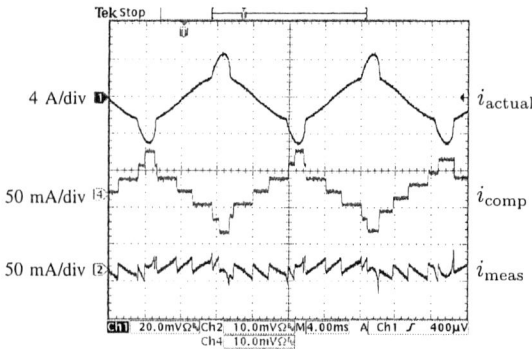

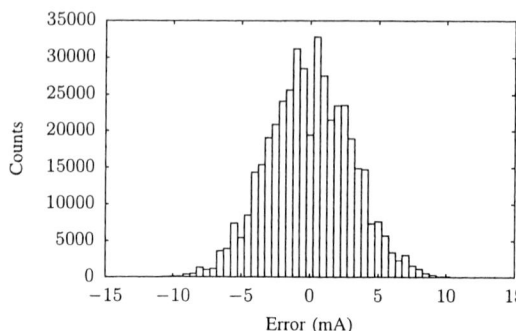

Fig. 10. Oscilloscope traces showing the measured primary, compensation, and residual measurement currents while the prototype physically-windowed current sensor is in normal operation.

Fig. 12. Histogram of the measured errors during a test of dc performance. 91% of the samples fall within ±5 mA, which gives an accuracy over the full ±80 A range of approximately one part in 2^{14}.

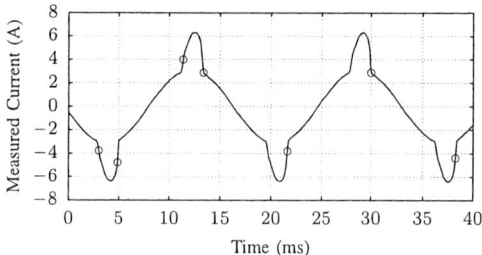

Fig. 11. Reconstructed current from the sensor. Circled points indicate individual samples that are known to be potentially inaccurate because the residual current was in the clamped region of the sensor window.

A. Full system functionality

Basic functionality was tested by constructing a test load consisting of an incandescent light bulb and a personal computer, which together draw power at both the fundamental and third harmonics of the line frequency. Fig. 10 shows the waveforms as measured by external test equipment. The input current is i_{actual}, the generated compensation current is i_{comp}, and the residual current is i_{meas}.

The reconstructed output of the sensor system for the same test load, based on data reported by the microcontroller, is shown in Fig. 11. In some cases, the high slew rate associated with the third harmonic content in the load caused the residual current measurement to exceed its window and enter the region where the clamps are active. The recombined data at these samples is known to be potentially inaccurate because of this. In the figure, these specific samples are circled. Note that they do not occur uniformly on every line cycle, because such excursions from the window depend on the varying relationship between sample time, slew rate, and the current window position.

B. Static resolution tests

The maximum resolution of the sensor system was characterized by analyzing the static performance. For this test, a Keithley 2400 Sourcemeter was used to supply various dc currents. These currents were passed through the sensor $N_1 = 50$ times to create the primary current i_{actual}. Each effective

current level was chosen at random from a ±25 A range. Once the test current stabilized, the reconstructed output from the prototype system was sampled at 8 kHz for approximately 110 ms. Approximately 500 unique test current levels were applied in total. A histogram of the resulting error between the Keithley reported output and the reconstructed sensor output is shown in Fig. 12. Typical errors for each sample are less than ±5 mA. Over the full 160 A range of the compensation current driver, this 10 mA range translates into an effective resolution of

$$\log_2(160/0.01) = 13.996 \text{ bits} \tag{11}$$

In the prototype system, the DAC command for the current compensation is 10 bits, while the ADC input from the residual measurement is 11 bits. This result of nearly 14 bits demonstrates the concept of using physical windowing to extend the sensor resolution over a larger range.

C. Dynamic resolution tests

The ability to resolve small signals while tracking a large signal was evaluated by creating a test load consisting of an incandescent lamp bulb in parallel with the Keithley 2400 Sourcemeter. For this test, the change in the envelope of the measured lamp ac waveform was examined as various small test currents were injected through the sensor using the Keithley source. The injected current was cycled between 0 mA, 20 mA, and 10 mA. The resulting reconstructed waveform is shown in Fig. 13. The small change in dc level can be seen in the detail shown in part (c). Like the static resolution test, this shows a resolution of approximately 10 mA.

D. Residual sensor transient response

In order to prevent drift in the output, the residual sensor employs a current clamp that prevents the toroidal core from saturating. The transient response of the residual sensor was tested by applying a large instantaneous current of 20 A through $N_1 = 5$ turns. This results in an effective primary current of 100 A, significantly greater than the typical operating window for the residual. This test simulates the case where a large input transient is seen before the compensation current is

978-1-4244-4782-4/10 $26.00 © 2010 IEEE

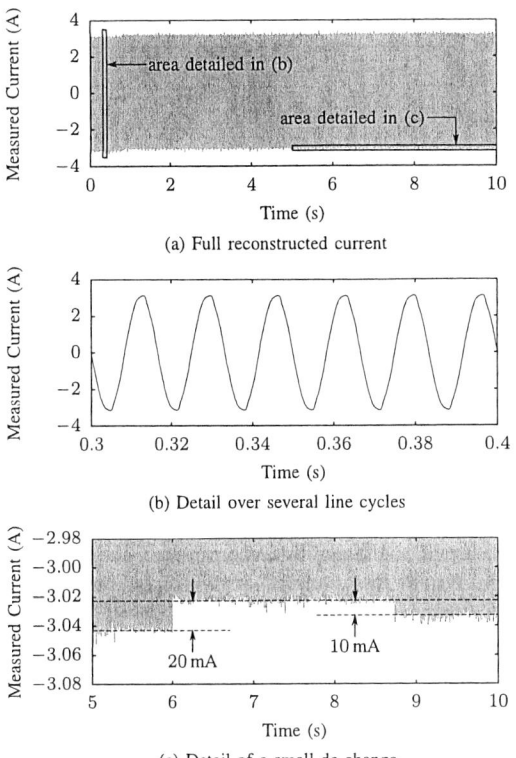

(a) Full reconstructed current

(b) Detail over several line cycles

(c) Detail of a small dc change

Fig. 13. Example of measuring a small dc current on top of a large ac current. The ac current is from a light bulb and the dc offset is added by a Keithley 2400 Sourcemeter, measured in parallel. Steps of 10 mA are resolvable in the reconstructed sensor output.

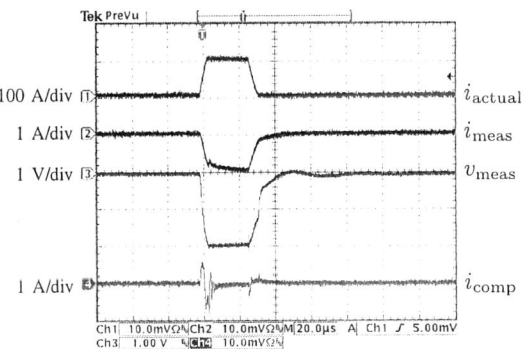

Fig. 14. System response during a high current transient. A pulse of primary current i_{actual} =100 A is applied for 25 μs. The clamp activates, limiting the residual current in the core. After deactivating, the offset measured at the core is less than 5 mV. The compensation current is also shown, demonstrating that it quickly stabilizes back to the commanded value after an induced change.

applied. The results are shown in Fig. 14. During the transient, the clamp activates, providing necessary current to keep the core near the zero-flux operating point. When the transient disappears, due to either a change in input or a change of the compensation current, the clamp deactivates and the residual current measurement resumes normal behavior. The offset drift in the residual sensor output V_{meas} is less than 5 mV after the transient, demonstrating that hysteresis loss in the core was minimized.

VI. CONCLUSIONS AND FURTHER WORK

The prototype sensor has demonstrated that the physical windowing approach is able to extend the range of an 11-bit current sensor with a 10-bit compensation command to provide approximately 14-bit resolution in the reconstructed output.

This windowing technique has many uses in smart grid and other applications where there is a need to measure fast, small signals on top of slow, large signals. The non-intrusive load monitor suggests many example uses, such as finding the high-frequency principal slot harmonics for motor diagnostics, or examining very small loads in aggregate power metering. The same technique is expected to be applicable to other physical systems, such as pressure monitors and strain gauges, with example uses including water system monitoring and wing flutter measurement.

ACKNOWLEDGEMENTS

This research was funded by The Grainger Foundation, the BP-MIT Research Alliance, the US Department of Energy ARPA-E, the MIT Sea Grant College Program, and by Dr. Manny Landsman.

REFERENCES

[1] S. B. Leeb, S. R. Shaw, and J. L. Kirtley, "Transient event detection in spectral envelope estimates for nonintrusive load monitoring," *IEEE Trans. Power Del.*, vol. 10, no. 3, pp. 1200–1210, Jul 1995.

[2] J. S. Ramsey, S. B. Leeb, T. DeNucci, J. Paris, M. Obar, R. Cox, C. Laughman, and T. J. McCoy, "Shipboard applications of non-intrusive load monitoring," in *American Society of Naval Engineers Reconfigurability and Survivability Symposium*, Atlantic Beach, Florida, February 2005.

[3] T. DeNucci, R. Cox, S. B. Leeb, J. Paris, T. J. McCoy, C. Laughman, and W. Greene, "Diagnostic indicators for shipboard systems using non-intrusive load monitoring," in *IEEE Electric Ship Technologies Symposium*, Philadelphia, Pennsylvania, July 2005.

[4] C. R. Laughman, S. R. Shaw, S. B. Leeb, L. K. Norford, R. W. Cox, K. D. Lee, and P. Armstrong, "Power signature analysis," *IEEE Power and Energy Magazine*, pp. 56–63, March 2003.

[5] R. W. Cox, P. Bennett, D. McKay, J. Paris, and S. B. Leeb, "Using the non-intrusive load monitor for shipboard supervisory control," in *IEEE Electric Ship Technologies Symposium*, Arlington, VA, May 2007.

[6] D. Son and J. D. Seivert, "A new current sensor based on the measurement of the apparent coercive field strength," *IEEE Trans. Instrum. Meas.*, vol. 38, no. 6, pp. 1080–1082, Dec 1989.

[7] S. Ogasawara, K. Murata, and H. Akagi, "A digital current sensor for pwm inverters," in *Industry Applications Society Annual Meeting, 1992, Conference Record of the 1992 IEEE*, vol. 1, Oct 1992, pp. 949–955.

[8] T. Sonoda, R. Ueda, and K. Koga, "An ac and dc current sensor of high accuracy," *IEEE Trans. Ind. Appl.*, vol. 28, no. 5, pp. 1087–1094, Sept/Oct 1992.

[9] J. Pankau, D. Leggate, D. Schlegel, R. Kerkman, and G. Shibiniski, "High frequency modeling of current sensors," *IEEE Trans. Ind. Appl.*, vol. 35, no. 6, pp. 1374–1382, Nov/Dec 1999.

[10] D. Li and G. Chen, "A wide bandwidth current probe based on rogowski coil and hall sensor," in *Power Electronics and Motion Control Conference, 2006. IPEMC 2006. CES/IEEE 5th International*, vol. 2, Aug 2006, pp. 1–5.

[11] J. Lenz and A. S. Edelstein, "Magnetic sensors and their applications," *IEEE Sensors J.*, vol. 6, no. 3, pp. 631–649, Jun 2006.

[12] M. M. Ponjavić and R. Durić, "Nonlinear modeling of the self-oscillating fluxgate current sensor," *IEEE Sensors J.*, vol. 7, no. 11, pp. 1546–1553, Nov 2007.

[13] J. Paris, Z. Remscrim, K. Douglas, S. B. Leeb, R. W. Cox, S. T. Gavin, S. G. Coe, J. R. Haag, and A. Goshorn, "Scalability of non-intrusive load monitoring for shipboard applications," in *American Society of Naval Engineers Day 2009*, National Harbor, Maryland, April 2009.

[14] C. Moreland, F. Murden, M. Elliott, J. Young, M. Hensley, and R. Stop, "A 14-bit 100-msample/s subranging adc," *IEEE J. Solid-State Circuits*, vol. 35, no. 12, pp. 1791–1798, Dec 2000.

[15] K. D. Hurst and T. G. Habetler, "Sensorless speed measurement using current harmonic spectral estimation in induction machine drives," *IEEE Trans. Power Electron.*, vol. 11, no. 1, pp. 66–73, 1996.

Edison Revisited: Impact of DC Distribution on the Cost of LED Lighting and Distributed Generation

Brinda A. Thomas
Carnegie Mellon University
Department of Engineering & Public Policy
Pittsburgh, PA, USA
brindat@cmu.edu

Abstract— **This paper models the case of a new construction 277V commercial lighting system, one of the many possible applications of a DC distribution system, using fluorescent and LED fixtures in a conventional AC circuit and in a DC lighting circuit. The levelized cost of lighting for the AC fluorescent, DC fluorescent, AC LED, and DC LED cases are calculated for projected LED and solar PV performance parameters, and a variety of DC bus voltages incorporating its impact on fluorescent ballast and LED driver efficiencies. Results show that with present fluorescent and LED efficacies, fluorescent lighting systems are the lowest annualized cost lighting option. However, if one has decided to install a DC distributed generation technology such as solar PV due to environmental policy, new-construction DC LEDs will be the lowest (annualized) cost lighting option by about 2012.**

I. INTRODUCTION

In recent years there has been a proliferation of devices such as computers, consumer electronics, and LEDs in the home and office that use direct current (DC). Similarly, plug-in electric hybrid vehicles that may be charged through commercial and residential building electrical systems use DC to store energy in batteries. On the supply side, distributed generation (DG) such as solar photovoltaics (PV), fuel cells, and some cogeneration all produce DC. Recent applications of DC electrical circuits include power plant auxiliary systems [1], telecom facilities [2], and data centers [3, 4]. The nexus of these developments suggest that the time may have come to reassess the use of DC distribution circuits in homes, offices and other commercial buildings. Other potential applications of DC distribution include microgrids, especially when supported with DC distributed generation and energy storage [5]. The cost and performance of AC-DC and DC-DC converters and the development of supporting standards will play a pivotal role in determining the feasibility of DC circuits at the building level.

II. BACKGROUND

Previous research on DC distribution has focused on network-level design, economics and feasibility in specific

applications, and integration with distributed generation. Initial techno-economic analysis of a DC distribution system to residential customers shows that medium voltage DC (± 750 V) distribution networks may have lower total costs than 400 V AC networks due to lower capital costs and outage repair costs [6]. However, medium and low-voltage DC cables have higher energy losses (due to power electronics semiconductor losses) for typical distribution network scales of a few miles [6,7]. Sannino et al. have shown that 326 V DC distribution is feasible (due to lower voltage drop) and more efficient than 48 V DC for a mid-size commercial building in Sweden [8]. The Electric Power Research Institute (EPRI) has also conducted laboratory tests that demonstrate that most household devices can readily accept DC power, except for certain applications with voltage-doubling circuits that are designed for AC input voltages [9].

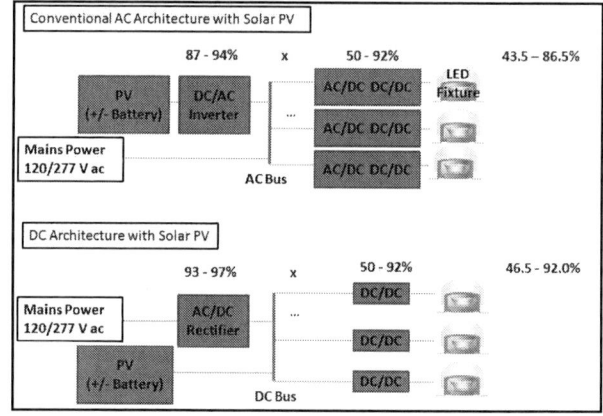

Figure 1. AC vs. DC building distribution with distributed generation

The economic case for DC circuits is strongest in applications where electricity reliability concerns lead to the use of energy storage such as batteries or uninterruptible power supplies (UPSs), as in the telecom and data center

---

Sponsored by the Gordon Moore Foundation and National Science Foundation Graduate Research Fellowship 2009-2012.

applications [2, 3, 4], or environmental considerations lead to the installation of DC distributed generation such as solar PV or fuel cells. DC circuits for applications such as lighting lead to greater end-use energy efficiency and reliability with energy storage or distributed generation, such as solar PV, by eliminating the inverter stage since it is expensive, relatively inefficient than other components, as shown in Fig. 1 [2]. EPRI reports that the use of DC distribution circuits with PV systems can yield up to 25% reduction in the capital cost of a PV system, by eliminating the inverter and thereby downsizing the PV array [10, 11]. Hammerstrom also demonstrates that a 3% energy efficiency improvement is possible with DC distribution integrated with a DG source such as fuel cells, otherwise DC distribution imposes a 2% energy efficiency penalty [12].

However, these previous studies focused on system efficiencies or capital costs with DC distribution, and have not assessed the economics of DC building electrical systems integrated with solar PV from a lifecycle perspective using realistic commercial discount rates, or the considerable uncertainty in cost savings in actual designs given the range of efficiencies and costs of components available. In addition, previous studies have not systematically analyzed the impact of DC bus voltage on the design and energy efficiency of centralized and load power converters. This research also explores the questions of whether and if an optimal DC bus voltage can be found and the economic implications of using DC distribution with solar PV.

Given a relatively stable lighting load, the use of solar PV with DC distribution should decrease system capital and lifecycle costs, and increase the system efficiency and reliability of PV-powered lighting. Without the use of solar PV, fuel cells or energy storage and holding power converter efficiencies equal, DC circuits are less efficient and possibly less reliable than conventional AC circuits. However, DC distribution systems with lower lifetime costs and greater energy efficiency can be designed if high-efficiency AC-DC and DC-DC converters are chosen for centralized rectification and load conversion, or if the power factor correction, electromagnetic interference protection, and regulation requirements for conventional AC-DC load power supplies lead to greater cost and/or an energy efficiency penalty compared to DC-DC load power supplies.

III. MODEL DESCRIPTION

We compare the economics of AC and DC distribution systems in commercial buildings by focusing on the case of a lighting system, one of the many possible applications of a DC building circuit. We model four scenarios: 1) a base case of 277 V fluorescent fixtures in a conventional AC circuit, 2) fluorescent fixtures in a DC lighting circuit, 3) 277 V LED fixtures in an AC circuit, and 4) LED fixtures in a DC lighting circuit. These scenarios were chosen to compare the costs of the DC circuit and an AC circuit for fluorescent and LED lighting systems as approaches to lower lifecycle costs and electricity consumption and associated emissions. The lighting systems are modeled as part of a hypothetical four-story, 48,000 square foot (4,400 m^2) commercial building with a lighting load of 19-31 kW for 640 occupants, where lighting

requirements are based on IESNA (Illumination Engineering Society of North America) illuminance requirements for office spaces [13]. The base case annual lighting electricity usage of 3.5 kWh/sqft is lower than Commercial Building Energy Consumption Survey (CBECS) 2003 average for office buildings of 6.7 kWh/sqft since lighting system was not modeled to include less efficient incandescent lighting [14]. Incandescent lamps were excluded from the base case for two reasons: (1) as resistive elements, their efficiencies are not impacted by the use of AC versus DC circuits, and (2) it is well known that replacing incandescent lamps with fluorescent and even LED lamps has positive net benefits, and including incandescent in the base case would artificially inflate the benefits of DC circuits.

TABLE I.   FLUORESCENT LIGHTING SYSTEM PARAMETERS

	AC FL		DC FL	
	Ambient	*Task*	*Ambient*	*Task*
Number Fixtures	360	672	360	672
Wattage (W)	100	20	95	19
Lamp/Device Efficacy (lm/W)	86	78	86	78
Fixture Efficacy (lm/W)	54	24	57	26
Fixture Efficiency (%)	73%	37%	73%	37%
Thermal Efficiency (%)	--	--	--	--
Ballast/Driver Efficiency (%)	97%	94%	102%	99%
Ballast Factor	0.90	0.90	0.90	0.90
Annual Operating Hours (h)	4100	2000	4100	2000

[15, 16] Note: Ballast Efficiency is defined as the ratio of rated lamp power over lamp and ballast power consumption. If the ballast is designed to run lamps at less than their rated power, the ballast's nominal efficiency will be greater than 100%, yet the lamp will output fewer lumens than if run at rated power. Ballast efficiency, together with ballast factor, determines the light output and power consumption of the fluorescent lamp and ballast system.

TABLE II.   LED LIGHTING SYSTEM PARAMETERS

	AC LED		DC LED	
	Ambient	*Task*	*Ambient*	*Task*
Number Fixtures	360	672	360	672
Wattage (W)	71	7	64	6
Lamp/Device Efficacy (lm/W)	122	122	122	122
Fixture Efficacy (lm/W)	77	70	85	78
Fixture Efficiency (%)	87%	80%	87%	80%
Thermal Efficiency (%)	85%	85%	85%	85%
Ballast/Driver Efficiency (%)	85%	85%	95%	94%
Ballast Factor	--	--	--	--
Annual Operating Hours (h)	4100	2000	4100	2000

[15, 16]

The lighting system consists of 360 recessed ambient lighting fixtures and about 670 desk-level undercabinet fixtures, incorporating independently measured fixture efficiencies [15]. Lighting fixture parameters are shown in Tables I-II. The LED driver and centralized AC-DC rectifier needed for a DC circuit were modeled as non-isolated buck converters with efficiencies that change according to the DC bus voltage. While the buck-boost, sepic, and Cuk converters were also modeled, the buck converter was the most efficient option for the range of operating voltages studied. Further evaluation of options for centralized ac-dc power conversion such as diode rectifiers and voltage source converters are described in Salomonsson [18]. LED driver, centralized rectifier, and fluorescent ballast parameters are shown in Table III.

TABLE III. LIGHTING SYSTEM POWER ELECTRONICS PARAMETERS

	AC-DC Central Power Supply		DC-DC Load Power Supply		Fluorescent Ballasts	
	FL	LED	LED Ambient	LED Task	AC	DC
Vin (V)	249	249	242	242	277	242
Vout (V)	242	242	24	24		
Efficiency (%)	93 ± 1.1%	93 ± 1.1%	95 ± 1.3%	94 ± 1.3%	85-109%	91-112%
Converter Type	Buck	Buck	Buck	Buck		
L (H)	8.0×10^{-7}	1.8×10^{-6}	5.3×10^{-3}	5.3×10^{-2}		
C (F)	9.8×10^{-8}	7.3×10^{-8}	2.3×10^{-7}	2.8×10^{-8}		
(R_{ON}, R_L, R_D)/Rload (%)	1%	1%	1%	1%		
Rload (Ohm)	1.3 ± .08	2.2 ± .33	7.3 ± 1.7	69 ± 16		

The levelized or annualized cost of lighting (LAC) for the AC and DC fluorescent and LED cases are calculated taking into account capital and installation (PV, LED) maintenance (M), wiring and circuit breakers (W), and direct (lighting) electricity costs (E), as defined in Equation 1. All costs were levelized over their respective lifetimes.

$$LAC = PV/CRF_P + LED/CRF_L + W/CRF_W + M + E. \quad (1)$$

$$CRF = i/[1-(1+i)^{-lifetime.}] \quad (2)$$

The commercial building was modeled for a climate such as in Pittsburgh, PA using solar radiation data from the National Renewable Energy Laboratory (NREL)'s National Solar Radiation Database [19]. The consequences of the building site should not impact the relative levelized costs of DC versus AC distribution since parameters that vary by region such as solar radiation level and electricity prices are held constant across all scenarios. The solar PV module was sized to power the ambient lighting system during the sunniest month of the year. Thus, in the remaining months of the year with less solar radiation, the remainder of the load not powered by solar PV would be supplied with grid electricity, rectified in the DC cases. An important consideration for the use of DC distribution systems is knowledge of the building's daily load curves. As EPRI notes, it is important to that the DC PV system is optimized to the load so that all available PV electricity is consumed; otherwise the cost of electricity from the PV system will increase [10].

An actual DC circuit will have a range of possible costs given the range of efficiencies and costs of DC circuit components and existing AC circuit infrastructure. Thus, to represent this uncertainty and provide error bounds on the LAC calculated for each of the scenarios, the LAC metric was calculated using a Monte Carlo simulation, which randomly samples from defined distributions of input parameters to generate a range of output values from which statistics can be generated [20].

IV. RESULTS & DISCUSSION

A. Optimal DC Voltage

We first sought to determine whether and if an optimal DC voltage can be found for a DC lighting system. The costs of the DC lighting circuit may vary according to voltage depending on the cost of circuit protection and switchgear at various operating voltages, the cost of wiring systems, wiring losses, power converter losses, and installation costs, and other design costs. In this work we consider the cost of circuit breakers, wiring systems and losses, and power converters, neglecting installation and other design cost differences that may arise between lower and higher voltage DC circuits. Table IV shows that the LACs of fluorescent or LED lighting systems without solar PV vary by 1-5% with respect to DC operating voltage. Given the considerable uncertainty component costs and efficiencies, the design engineer would gain little to no cost savings by optimizing with respect to DC operating voltage. Thus, the DC circuit is assumed to operate at full-bridge rectified 277 Vac, or approximately 250 Vdc for the remainder of the paper.

TABLE IV. LEVELIZED ANNUAL COST VS. OPERATING VOLTAGE

Operating Voltage	Levelized Annual Costs ($/yr)			
	FL	LED	FL + PV	LED + PV
277 V AC	46,400 ± 4,300*	84,200 ± 10,000*	111,000 ± 14,200	124,000 ± 16,200
250 V DC	48,200 ± 4,500	86,400 ± 10,200	89,100 ± 11,800*	116,000 ± 15,200 *
60 V DC	49,500 ± 4,500	87,000 ± 10,200	89,500 ± 11,700	117,000 ± 15,100
48 V DC	50,800 ± 4,500	87,300 ± 10,200	90,600 ± 11,600	117,000 ± 15,100

Note: Lowest LAC options are starred*.

B. 2. AC vs. DC Circuits, without and with Solar PV

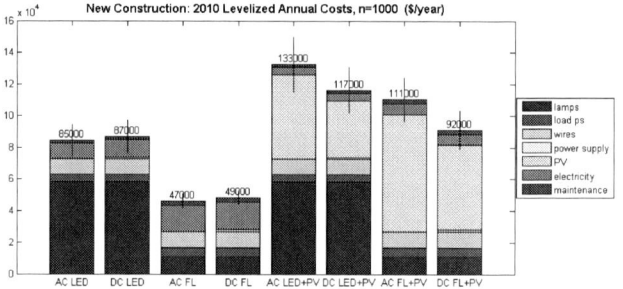

Figure 2. Levelized Annual Costs for AC and DC lighting systems, discount rate = 12%, electricity price = $0.10/kWh. Error bars are plus or minus one standard deviation from the mean.

Results in Fig. 2 show that with 2010 fluorescent and LED efficacies and costs as projected by DOE, LEDs with either 277 Vac or 250 Vdc circuits yield higher levelized annual costs (LAC) of lighting compared to fluorescents for new construction projects, at an electricity price of $0.10/kWh and a discount rate of 12%. The LACs are relatively similar within each technology whether using AC or DC circuits, with a slightly higher mean for DC LACs. Although the DC circuit LEDs and fluorescents have a higher fixture efficacies and lower fixture wattages than their AC counterparts, as seen in Table 1, the cost of the central power supply leads to a slightly higher LAC on average.

DC circuits integrated with solar PV seem to lower the LAC and capital costs of powering LEDs and fluorescent lighting systems on average, as seen in Table 4, although there is considerable overlap in AC and DC LACs and capital costs given uncertainties in component costs and efficiencies. The error bars, which represent plus and minus a standard deviation from the mean, are larger for the solar PV cases due to the considerable variation in solar PV costs from $2.5-6.5/Wp for just the module itself, with similar ranges for balance of plant (including installation and power electronics) costs [21]. The lower total capital costs for either fluorescent or LED lighting systems with DC circuits are due to eliminating the inverter, which increases system efficiencies such that a smaller solar PV module can power the ambient lighting systems. However, as in the case for the DC fluorescent lighting system shown in Table V, a smaller PV module may lead to slightly higher grid electricity consumption over the year, leading to higher CO2, SO2 and NOx emissions.

TABLE V.   COSTS, LOADS, AND EMISSIONS FROM AC VS. DC FLUORESCENT AND LED LIGHTING SYSTEMS WITH SOLAR PV

	AC FL + PV	DC FL + PV	AC LED + PV	DC LED + PV
Capital + Installation Cost (All But PV) ($)	162,000 ± 17,000	171,000 ± 17,000	485,000 ± 42,000	490,000 ± 42,000
Capital + Installation Cost (PV+Inv+BOP) ($)	583,000 ± 99,000	417,000 ± 82,000	417,000 ± 93,000	282,000 ± 67,000
Total Capital + Installation Costs ($)	745,000 ± 99,000	588,000 ± 82,000	902,000 ± 112,000	772,000 ± 88,000
Levelized Annual Cost ($/yr)	111,000 ± 14,000	92,000 ± 12,000	133,000 ± 18,000	117,000 ± 15,000
Lighting Load (kW)	50 ± 4	51 ± 4	31 ± 5	30 ± 5
PV System Size (kW)	56 ± 5	48 ± 4	40 ± 7	33 ± 6
Grid Electricity Consumption (kWh/yr)	67,200 ± 5,600	68,600 ± 5,700	48,100 ± 7,900	46,300 ± 7,600
CO2 Emissions (tons/yr)	45 ± 4	46 ± 4	32 ± 5	31 ± 5
SO2 Emissions (kg/yr)	177 ± 15	180 ± 15	126 ± 21	122 ± 20
Nox Emissions (kg/yr)	65 ± 5	66 ± 6	47 ± 8	45 ± 7

C.   AC vs. DC Circuits: 2008-2030

However, LED and solar PV are evolving technologies that should experience considerable efficiency improvements and cost reductions in the future, given the current pace of research and development. Fig. 3 shows that with a 12% discount rate, AC LEDs should achieve LAC parity with AC

or DC fluorescents by 2015, according to mean estimates. Not shown for clarity, are the considerable error bars on these projections, which may hasten or delay the breakeven point between fluorescent and LED lighting systems. DC LEDs have higher LACs that converge with AC LEDs over time as the proportion of LACs due to electricity increases, and the slight efficiency gains with DC circuits become larger.

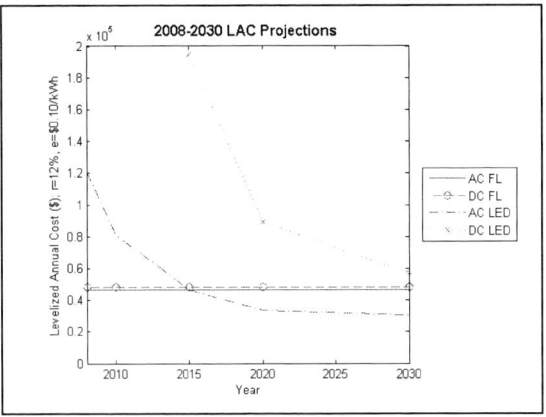

Figure 3.   LAC Projections for AC vs. DC fluorescents and LEDs. LED projections from [17].

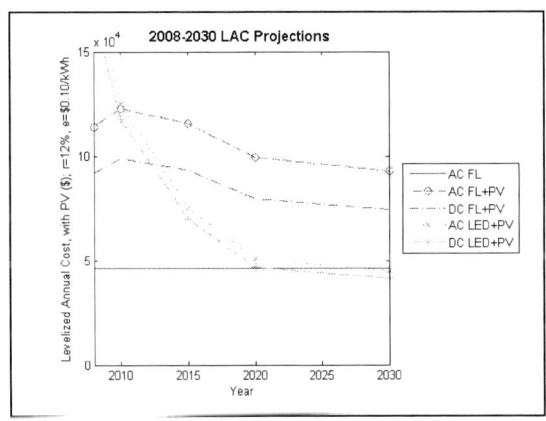

Figure 4.   LAC Projections for AC vs. DC fluorescents and LEDs integrated with solar PV. LED projections from [17] and for solar PV projections from [21]

For the solar PV integrated case in Fig. 4, DC and AC LEDs appear to have very similar LAC projections that should achieve parity with AC fluorescents integrated with PV by 2010 and parity with DC fluorescents and PV by about 2012. However, all of the PV-integrated cases are well above the levelized costs of AC fluorescents until near 2020 for AC or DC LEDs. These results have more to do with the efficiency and cost trajectories of LEDs and solar PV than with DC circuits, which appear to have a marginal effect. DC circuits do seem to reduce the levelized cost of fluorescents by about $5,000, but these lighting technologies will not achieve parity with AC fluorescents without solar PV since fluorescent lamps are a mature technology that are not expected to experience dramatic efficacy (lumen per watt) gains in the future.

V. SENSITIVITIES AND LIMITATIONS

These LAC calculations are relatively sensitive to the choice of discount rate, as shown in Fig. 5. For example, the conclusion that DC LEDs with solar PV will achieve LAC parity to AC fluorescents without PV holds as long as discount rates are under about 10%. At higher discount rates, long-term electricity and maintenance savings are less valued than initial capital costs in the decision to invest in a particular lighting system design, whether with AC or DC circuits, with or without solar PV. The LAC for each lighting system also scales according to building size and corresponding lighting load.

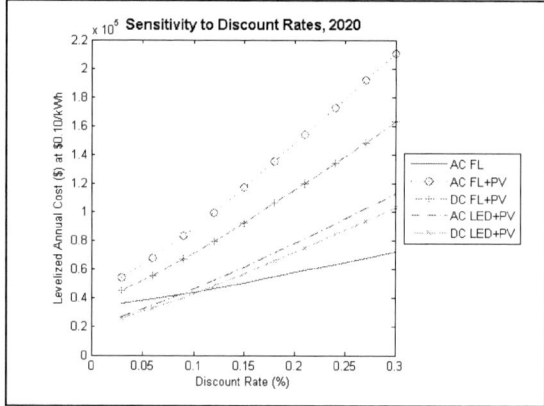

Figure 5. LAC Sensitivity to discount rate

This work is limited by its narrow scope on the lighting systems as an application for DC circuits. In addition, the model results have considerable uncertainties given the uniform distributions chosen to represent the input parameters. Realistic distributions, and allow for correlations between parameters, such as efficiency and cost, could lower the uncertainties in LAC results and given better information on the decision to invest in DC circuits.

VI. CONCLUSION

This work demonstrates that the choice of an "optimal voltage" may not be a critical decision variable in the design of a DC building electrical system because of the relatively limited sensitivity of levelized annual costs of lighting to DC bus voltage. However, the choice of optimal DC bus voltage may depend upon factors not considered in this analysis such as the difference in installation and switch costs at high (> 48 V) and low voltage (≤ 48 V), and cost-difference between AC-DC converters and DC-DC converters.

This study echoes the message of previous studies on DC circuits to conclude that they are most feasible when applied with distributed generation. However, DC distributed generation such as solar PV, fuel cells and energy storage in the form of batteries are so expensive than DC circuits are unlikely to significantly accelerate their adoption. DC circuits

could also be used with microturbines, which generate high frequency AC power which is then converted to DC and then back to 120 Hz for the electric grid. Future work on DC circuits could explore microturbines with other load applications.

ACKNOWLEDGMENTS

The author thanks M. Granger Morgan, Ines Azevedo, and Jay Apt from Carnegie Mellon University, and Ihor Lys and Kevin Dowling from Philips Lighting Company, and Clark Gellings from EPRI, and Le Tang from ABB for useful discussions and references to relevant literature.

REFERENCES

[1] Ton, M., Fortenbery, B., and Tschudi, W. "DC power for Improved Data Center Efficiency." Berkeley, CA. 2008.

[2] Pratt, A., Kumar, P., and Aldridge, T. V. "Evaluation of 400 V DC Distribution in Telco and Data Centers to Improve Energy Efficiency." *IEEE Explore.* 2007.

[3] Jancauskas, J. R. and Guthrie, R. L. "Elimination Of Direct Current Distribution Systems From New Generating Stations." IEEE. 1995.

[4] Yamashita, T., Muroyama, S., Furubo, S., and Ohtsu, S. "270 V DC System – A Highly Efficient and Reliable Power Supply System for Both Telecom and Datacom Systems." *Telecommunications Energy Conference* 1999.

[5] Kakigano, H. Miura, Y. Ise, T. Momose, T. Hayakawa, H. "Fundamental Characteristics of DC Microgrid for Residential Houses with Cogeneration System in Each House." *Power and Energy Society General Meeting - Conversion and Delivery of Electrical Energy in the 21st Century, 2008 IEEE*

[6] T. Kaipia, P. Salonen, J. Lassila, and J. Partanen, "Possibilities of the low voltage dc distribution systems," in *Proceedings NORDAC 2006 conference*, 2006. [Online]. Available: http://www.lut.fi/fi/ technology/lutenergy/electrical engineering/research/electricitymarkets/research/net workbusiness /Documents/DCdistribution Kaipia.pdf

[7] D. Nilsson and A. Sannino, "Efficiency analysis of low- and medium- voltage dc distribution systems," in *Power Engineering Society General Meeting*, 2004. IEEE, 2004, pp. 2315–2321 Vol.2. [Online]. Available: http://dx.doi.org/10.1109/PES.2004.1373299.

[8] A. Sannino, G. Postiglione, and M. H. J. Bollen, "Feasibility of a dc network for commercial facilities." *Industry Applications, IEEE Transactions on*, vol. 39, no. 5, pp. 1499-1507, 2003. [Online]. Available: http://dx.doi.org/10.1109/TIA.2003.816517

[9] George, K. "DC Power Production, Delivery and Utilization." Electric Power Research Institute White Paper. June 2006.

[10] Jimenez, A. "Photovoltaic Power Generation With Innovative Direct Current Applications: Feasibility and Example Site Evaluation." EPRI Technical Update# 1011533. February 2005.

[11] DTI. 2002. "The Use of Direct Current Output From PV Systems in Buildings." http://www.berr.gov.uk/files/file17277.pdf

[12] D. J. Hammerstrom, "Ac versus DC distribution systems: Did we get it right?" in *IEEE Power Engineering Society General Meeting, 2007.* [Online]. Available: http://ieeexplore.ieee.org/xpl/freeabs all.jsp?arnumber=4275896

[13] IESNA Lighting Standard referenced in Table 4-1 in *Lighting Market Characterization, Vol 1.*, Navigant Consulting, 2002.

[14] EIA. 1999, 2003. Commercial Buildings Energy Consumption Survey. Washington, D.C.: Department of Energy. http://www.eia.doe.gov/emeu/cbecs/

[15] DOE Commercially Available LED Product Evaluation Reporting program: CALiPER 3, 5 (SSL recessed fixture, fluorescent recessed troffer), *Under-Cabinet Factsheet*, DOE (fluorescent undercabinet fixture), LED undercabinet fixture efficiency is assumed to be 80%. In *Under-Cabinet Factsheet*, DOE states that LED fixture efficiencies cannot be determined in the absence of an LED lumen rating standard.

[16] Assumes discount rate of 12%, LED lifetime of 50,000 hours, fluorescent lifetime of 20,000 hours, and LED device efficacies and costs from DOE SSL MYPP 2009, Table 4.3.2, fixture efficiencies from CALiPER, and 2000 annual operating hours for undercabinet fixtures from *Analysis of Standard Options for Under Cabinet Fluorescent Fixtures Attached to Office Furniture*, Energy Solutions, 2004; and 3900 annual operating hours for ambient lighting fixtures from retail applications in *Lighting Market Characterization, Vol 1.*, Navigant Consulting, 2002.

[17] DOE Solid-State Lighting (SSL) Multi-Year Program Plan (MYPP) 2009

[18] D. Salomonsson and A. Sannino, Centralized ac/dc power conversion for electronic loads in a low-voltage dc power system,â€ in Power Electronics Specialists Conference, 2006. PESC 2006. 37th IEEE, 2006, pp. 17. [Online]. Available: http://dx.doi.org/10.1109/PESC.2006.1712251

[19] NREL. National Solar Radiation Database. http://rredc.nrel.gov/solar/old_data/nsrdb/

[20] M. G. Morgan and M. Henrion. *Uncertainty: A Guide to Dealing with Uncertainty in Quantitative Risk and Policy Analysis.* Cambridge University Press, 1990.

[21] Curtright, A. E., Morgan, M. G., and Keith, D. W. Forthcoming. "Expert Assessments of Future Photovoltaic Technologies." In press at Environmental Science & Technology.

A Novel Passive Off-line Light-Emitting Diode (LED) Driver with Long Lifetime

S.Y.R. Hui, *Fellow IEEE*, S.N. Li, X.H. Tao, W. Chen, *Member IEEE* and W.M. Ng, *Member IEEE*

Center for Power Electronics
City University of Hong Kong
Hong Kong
eeronhui@cityu.edu.hk

Abstract— **This paper describes a patent-pending passive off-line LED driver that has no semiconductor switches, electrolytic capacitors, auxiliary power supply and control board. It can provide a fairly smooth current from the ac mains to drive LED strings. The new circuit has the advantages of high input power factor, high energy efficiency and luminous efficacy, long lifetime, stable luminous output and high robustness against extreme weather conditions. In addition, over 90% of the driver material is recyclable, leading to reduction of electronic waste. It is particularly suitable public LED lighting systems such as road lighting systems. Experimental results based on a 50W system are included in the paper to confirm the validity of the proposal. Due to the circuit simplicity, an energy efficiency exceeding 93.6% has been achieved.**

I. INTRODUCTION *(HEADING 1)*

LED technology has emerged as a promising lighting technology to replace the energy-inefficient incandescent lamps and mercury-based fluorescent lamps [1]. While LED devices enjoy relatively long lifetime of typically 80,000 h [2], the relatively short lifetime of LED drivers, which is limited by the use of electrolytic capacitors [3], remains a limiting factor to the lifetime of the overall LED systems. Although electronic LED drivers without using electrolytic capacitors have been proposed [4-6], the use of active power electronic switches requires extra control electronics and auxiliary power supplies that will increase circuit complexity and reduce system reliability. In addition, these extra circuit boards may need electrolytic capacitors, although electrolytic capacitor is not used in the power circuits. Particularly for outdoor applications such as road lighting systems, the ballasts (or drivers for LED lighting systems) must be highly reliable. Take Hong Kong as an example. The number of lightning could be 10,000 times or higher in a stormy day in the summer [7]. With about 130,000 street lamps in Hong Kong, 1% of the system failure means problems in 1,300 street lamps. So reliability is a paramount issue in road lighting systems.

Existing street lamps primarily use high-intensity-discharge (HID) lamps and magnetic ballasts. Magnetic ballasts are highly reliable with lifetime of 20 years, recyclable and hence highly environmentally friendly [8]. Such environmental friendliness cannot be matched by electronic ballasts due to their short lifetime (typically < 5 years) and their use of toxic and/or non-biodegradable components. In the International Forum on Novel Light & Energy Sources held in Shanghai, China, in April 2009, several road lighting management institutions have expressed their needs to have LED drivers with lifetime higher than 10 years. This request arises from the experience learnt from previous trials of LED street lighting products in the last 3 years in China. Sustainable lighting technology should meet at least three criteria (i) high efficiency or energy saving, (ii) long product lifetime and (iii) recyclability.

In this paper, a novel and patent-pending [9] passive LED driver for off-line applications that meet these 3 criteria is proposed. This passive LED driver consists of passive components and diodes only, without using any power electronic switches, auxiliary power supply and control boards. The proposal features circuit simplicity, reliability and long product lifetime. A circuit analysis and practical confirmation of this passive LED driver for a 50W LED system are included in this paper.

II. PRINCIPLES OF THE PASSIVE LED DRIVER

A. Existing concept

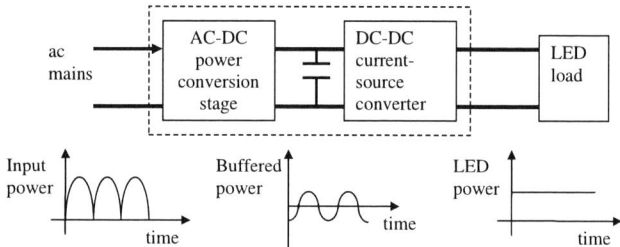

Fig.1 Schematic & power profiles of a traditional offline LED system

Fig.1 shows the schematic of a traditional AC-DC power conversion system for offline LED applications, and the typical power profiles in the input, intermediate and output stages. Both single- and two- power stage approaches have been addressed in [10,11]. For a two-stage approach, a front

978-1-4244-4782-4/10 $26.00 © 2010 IEEE

power stage converts the ac mains voltage into a stable dc voltage with the help of a large electrolytic capacitor as a energy storage and buffer. A second power stage then converts the dc voltage source into controllable dc current source for driving the LED load. Since the second stage provides a constant current source, the output power is constant, meaning that the capacitor in the intermediate stage has to be large enough to absorb the energy buffer. Therefore, electrolytic capacitor with large capacitance is usually used as the energy buffer. The single-stage approach [11-15] essentially combines part of the two stage circuits together to form a single-stage circuit. Large storage capacitor is needed as in the two-stage approach.

B. New Concept

The new concept proposed in this study [9] is illustrated in Fig.2. A single-stage passive circuit with power factor correction is proposed to replace the two stage power circuit. Instead of using a large capacitor to ensure that the output current is constant, it is proposed that a small current ripple may be allowed in the output current. In this way, the requirement for the energy buffer can be reduced and consequently non-electrolytic capacitors can be used to enable long lifetime of the overall system. This small current ripple may cause power variation in the LED load. However, such power variation will not cause noticeable luminous variation to human eyes.

With the help of the general photo-electro-thermal theory for LED systems [16], thermal designs can be made so that the luminous flux and power of a LED system can follow the profiles in Fig.3b or in Fig.3c. It can be seen that the slope of this curve is small at and around the peak flux value in Fig.3b and Fig.3c. This means that a relatively large power fluctuation will only lead to a small flux change, i.e. the sensitivity of the luminous flux with the changes in LED power is small.

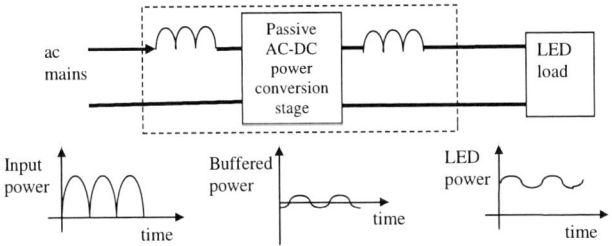

Fig.2 Schematic & power profiles of a traditional offline LED system

Fig.4a shows the schematic of a passive LED driver for offline applications. It consists of an input inductor, a diode rectifier, a valley-fill circuit, an output inductor and a LED load. The input inductor is used to limit the power output of the load and also to reduce the load power sensitivity against fluctuation of the ac mains voltage. The diode rectifier turns the ac voltage into a dc one. Unlike the previous use of the valley-fill circuit for improving the input power factor [17,18], a major function of the valley-fill circuit is to reduce the output voltage ripple [9] so as to reduce the size of the output filter inductor. This output inductor turns the dc voltage source

into a current source for driving the LED load. As an alternative, this output inductor can be replaced by a current ripple cancellation circuit, which consists of a coupled inductor and a capacitor [19-21]. This study focuses on the LED driver performance. Current balancing techniques for parallel LED strings [22] are not the scope of this study.

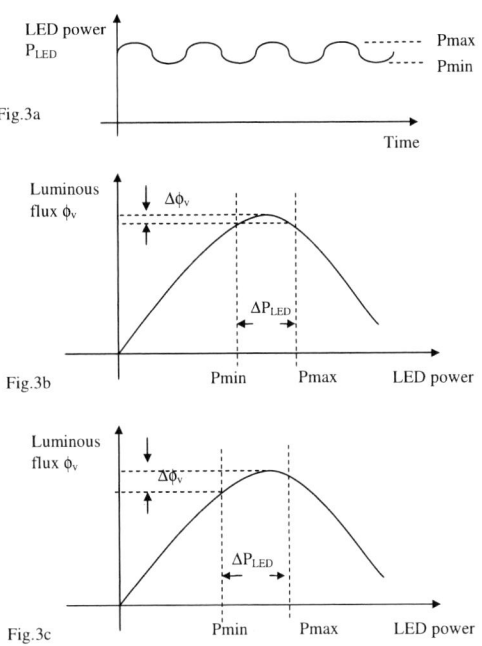

Fig.3 Variation of LED power and luminous flux in this proposal

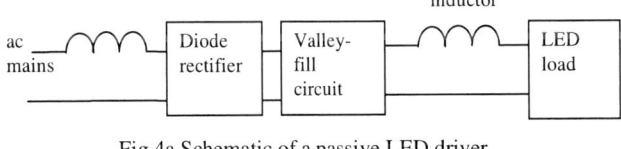

Fig.4a Schematic of a passive LED driver

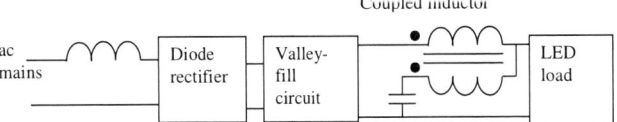

Fig.4b Schematic of a passive LED driver with a current ripple cancellation circuit

III. PASSIVE LED DRIVER AND ANALYSIS

A. Passive LED Driver & Circuit Operation

Fig.5 shows the circuit diagram of one patent-pending passive LED driver for off-line applications. It consists of only 11 components, namely 7 power diodes, two inductors and two capacitors. The input inductor L_s is used to (i) to filter the input current in order to reduce the input current harmonics and (ii) to control the power sensitivity of the LED load. The diode-bridge is to rectify the input ac voltage into dc one and

978-1-4244-4782-4/10 $26.00 © 2010 IEEE   595

the valley-fill circuit is used to reduce the voltage ripple in the output dc voltage V_3. The output inductor L is used to convert the dc voltage source V_3. into a smooth dc current source I_o to drive the LED load. The LED load can be a LED string or a parallel of LED strings. In order to increase the lifetime of this circuit, non-electrolytic capacitors can be used for the valley-fill circuit. Variants of this basic circuit can be used to further reduce the ac ripple in the output current. For example, an extra capacitor (the 12th component) can be connected across V3 so as to further reduce the output voltage ripple in V3.

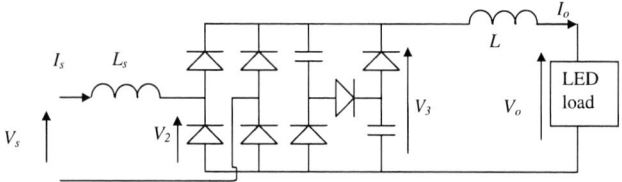

Fig.5 Basic circuit of the passive LED driver.

In this study, we use the basic circuit in Fig.4a for analysis. The idealized input voltage V_s and input current I_s waveforms are shown in Fig.6. Due to the use of the input inductor, the input current is expected to lag behind the input voltage. For further improvement of the input power factor, a standard solution is to add an input capacitor across the ac mains. However, this method is not included in the following circuit operation description.

The idealized waveforms of the input voltage to the diode rectifier with the valley-fill circuit and the input current are shown in Fig.7. Since the equivalent LED load reflected to the input side of the diode bridge is resistive, V_2 and I_s are in phase. The output voltage of the valley-fill circuit should be a rectified version of V_2. Thus, the idealized waveforms of V_3 and I_o are shown in Fig.8. Assuming that the voltage across the LED load V_o does not change significantly, the idealized output voltage, current and power waveforms are included in Fig.9.

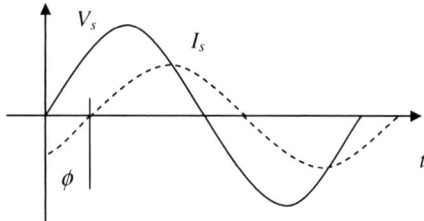

Fig.6 Idealized waveforms of input ac mains voltage and current (with a phase shift (ϕ) between Vs and Is)

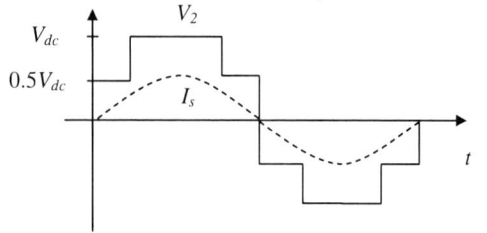

Fig.7 Idealized waveforms of input voltage V_2 and current I_s of the diode rectifier (with V_2 and I_s in phase)

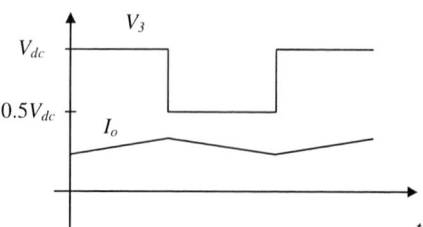

Fig.8 Idealized waveforms of output voltage V_3 and current I_o of the valley-fill circuit (with V_3 a rectified version of V_2)

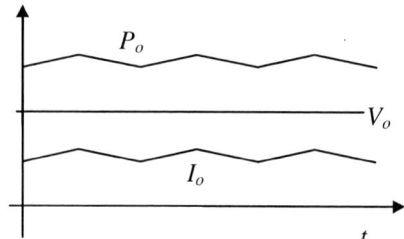

Fig.9 Idealized waveforms of voltage across LED load (V_o), output load current (I_o) and the output load power (P_o)

B. Circuit Analysis

This circuit analysis starts from the load side. If the voltage output V3 is considered as an equivalent voltage source, a simplified circuit of Fig.5 is shown in Fig.10, where R is the winding resistance in the output inductor.

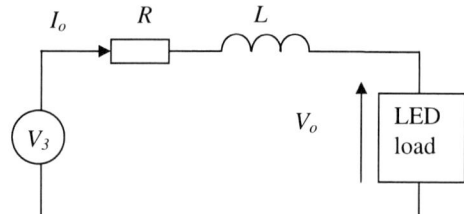

Fig.10 Simplified equivalent circuit of Fig.5 (output side)

The average output current $\bar{I}_o$ can be expressed as:

$$\bar{I}_o = \frac{\bar{V}_3 - V_o}{R} \qquad (1)$$

where $\bar{V}_3$ is the average voltage of V_3. From the waveform of V_3 in Fig.8,

$$\bar{V}_3 = \frac{3}{4} V_{dc} \qquad (2)$$

Rearranging (2) gives:

$$V_{dc} = \frac{4}{3} \bar{V}_3 = \frac{4}{3} \left(V_o + \bar{I}_o R \right) \qquad (3)$$

Note that the total voltage drop of the LED load is approximated as a constant V_o. Therefore, V_{dc} does not change significantly if $\bar{I}_o$ does not change significantly. In general, V_o is much bigger than $\bar{I}_o R$. Thus V_{dc} is close to $1.33 V_o$ and V_{dc} can be considered as a function of the V_o which is determined in the LED load.

The next issue is to find out a way to reduce the change of I_o due to fluctuation in the input mains voltage (i.e. to reduce the sensitivity of the load power with the fluctuation of the ac mains voltage. By the law of conservation of energy, input power is equal to the power entering the diode bridge, assuming that the input inductor L_s has negligible resistance. Form the waveforms in Fig.6 and note that V_{21} and I_S are in phase as shown in Fig.7.

$$V_S I_S \cos\phi = V_{21} I_S \tag{4}$$

where V_{21} is the fundamental component of V_2.

Similarly, the input power is also equal to the output power of the valley-fill circuit, assuming that the power loss in the diode rectifier and valley-fill circuit is negligible.

$$V_S I_S \cos\phi = \bar{V}_3 \bar{I}_o = \frac{3}{4} V_{dc} \bar{I}_o = \bar{I}_o^2 R + \bar{I}_o V_o \tag{5}$$

Using Fourier analysis on the waveform of V_2, the fundamental component V_{21} of V_2 can be determined as:

$$V_{21} = \frac{\left(2+\sqrt{2}\right) V_{dc}}{\pi} \sin(\omega t - \phi) = 1.087 \cdot V_{dc} \sin(\omega t - \phi) \tag{6a}$$

The root-mean-square value of V_{21} is therefore

$$V_{21\_rms} = \frac{1.087}{\sqrt{2}} \cdot V_{dc} = 0.77 \cdot V_{dc} \tag{6b}$$

Based on (4), (5) and (6) and assuming that winding resistance is negligible, one can relate I_s and $\bar{I}_o$.

$$V_{21} I_s = \bar{I}_o^2 R + \bar{I}_o V_o$$

$$0.77 V_{dc} I_s = 0.77 \times \frac{4}{3}\left(V_o + \bar{I}_o R\right) \times I_s = \bar{I}_o^2 R + \bar{I}_o V_o$$

$$\Rightarrow I_s = 0.98 \bar{I}_o \tag{7}$$

Now consider the equivalent circuit and the vectorial relationship between V_s and V_{21} as shown in Fig.11.

$$V_S^2 = V_{21}^2 + \left(\omega L_S I_s\right)^2 \tag{8}$$

and

$$\vec{I}_s = \frac{\vec{V}_S - \vec{V}_{21}}{j\omega L_s} \tag{9}$$

From (6), it can be seen that V_{21} depends on V_{dc}, which is approximately close to V_o (approximated as a constant value). With the help of (7) and (9),

$$\bar{I}_o = \frac{V_S - V_{21}}{0.98 \cdot \omega L_S} \tag{10}$$

Differentiating (10) will lead to

$$\Delta \bar{I}_o = \frac{\Delta V_S}{0.98 \cdot \omega L_S} \tag{11}$$

Equation (11) is the important equation which shows that the input inductance L_s can be used to reduce the change of average output load current $\Delta \bar{I}_o$ for a given change in the input ac mains voltage ΔV_S. That is, the power sensitivity of the LED load, which is a function of the output current Io, can be controlled by the inductance of the input inductor L_s.

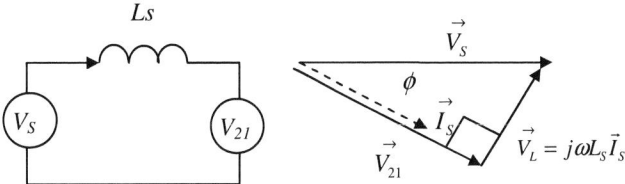

Fig.11 Simplified equivalent circuit & vector diagram of Fig.5 (input side).

In order to relate $\bar{I}_o$ with V_s, using (6), (7) and (8) gives:

$$V_s^2 = \left[(0.77)\left(\frac{4}{3}\right)(V_o)\right]^2 + \left[\omega L_s\left(0.98 \bar{I}_o\right)\right]^2 \tag{12}$$

Solving (12) gives:

$$\bar{I}_o = \frac{\sqrt{V_s^2 - \left(1.024 \cdot V_o\right)^2}}{0.98 \cdot \omega L_s} \tag{13}$$

Note that V_o can be determined from the number of LED devices in the LED strings. If L_s is chosen, then (13) provides the relationship between the average output current and the input ac mains voltage. The LED load power is therefore:

$$\bar{P}_o = V_o \cdot \frac{\sqrt{V_s^2 - \left(1.024 \cdot V_o\right)^2}}{0.98 \cdot \omega L_s} \tag{14}$$

IV. EXPERIMENTAL VERIFICATION

A passive LED driver based on the circuit in Fig.5 has been designed and built for a LED load. The load consists of a LED string using 16 Sharp LED (model number: GW5BWC15L02) in series. Its peak current rating is 400mA. The total voltage across this LED load is V_o=158V. The input inductor L_s is 1.47 H and it has a winding resistance R_{Ls}=2.7Ω. Two polypropylene capacitors of 20μF each are used in the valley-fill circuit. The output inductor L is 2.3H and it has a winding resistance R_L= 3Ω. Based on these reactive parameters (with all resistance ignored), the theoretical output current should be about 0.36A and the LED load power 57W.

A 50W LED system has been developed and tested with the proposed passive offline LED driver. Fig.11 shows the measured input voltage V_s and current I_s of the entire system. It can be seen that the input current is highly sinusoidal. The measured V_s and I_s agree with predictions in Fig.6. Fig.12 shows the measured V_2 and I_s. V_s is found to be a stepped ac voltage which is in line with the idealized stepped voltage in Fig.7, and is in phase with I_s as predicted. The measured V_3 and I_o are captured in Fig.13. V_3 is found to be a rectified version of V_2 as expected in Fig.8. The output current is fairly smooth with a small ripple only. The measured output current I_o, LED string voltage V_o and output power of one LED string P_o are shown in Fig.14. These results are highly consistent with the theoretical predictions.

The energy efficiency of the LED driver is also measured. Power measurements are made with the use of Voltech PM6000 power analyzer. [Note: readings from the Power Analyzer are slightly different from and more accurate than those displayed in a digital storage oscilloscope.] With an ac mains voltage of 230V, the total input power is 49.12W and the output power consumed by the LED load is 46W. The system loss is only 3.12W. Thus, a high efficiency of 93.6% has been achieved.

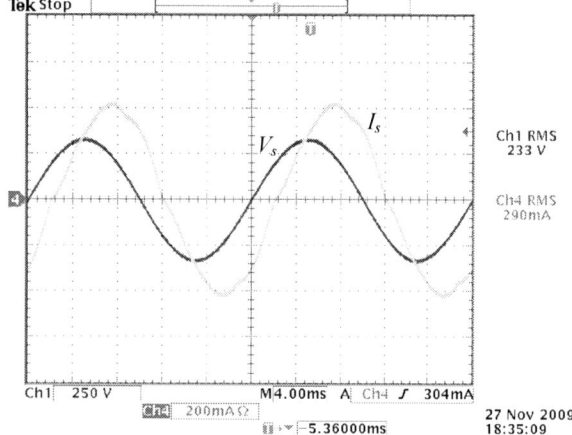

Fig. 11 Measured input voltage V_s & current I_s

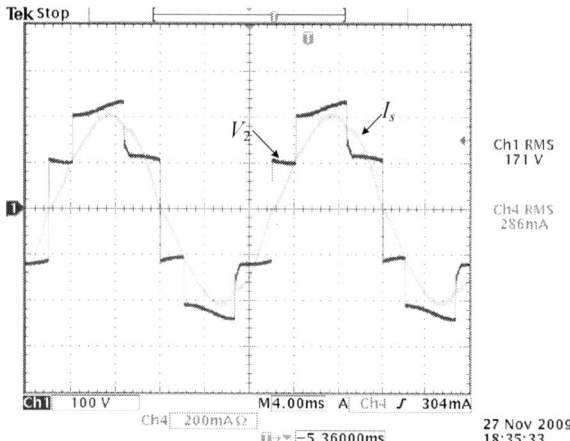

Fig. 12 Measured V_2 (stepped) and input current I_s (sinusoidal)

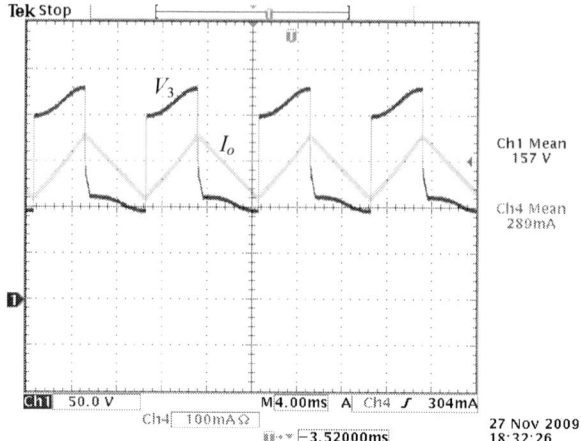

Fig. 13 Measured V_3 and output current I_o (L=2.3H)

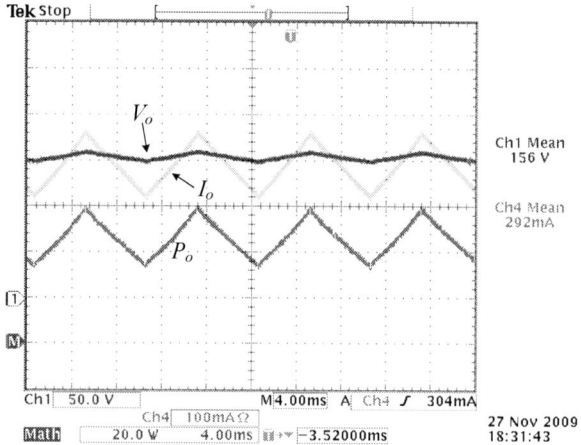

Fig. 14 Measured output current I_o, LED string voltage V_o & output power P_o (L=2.3H)

The output current ripple can be reduced further by either using a larger output inductor or a current ripple cancellation circuit. To illustrate this point, a larger output inductor L of about 5H is used. At an ac mains voltage of 230V, the total input power is 52.29W and the output power consumed by the LED load is 48.64W. A high efficiency of 93.0% can still be achieved.

The corresponding measurements are recorded in Fig.15-Fig.18. The waveforms of V_s, I_s, V_2 and V_3 in Fig.15 and Fig.16 are of the same forms as those in the previous case. However, it can be seen from the waveforms of the output current in Fig.17 and Fig.18 that the current ripple has been reduced. Consequently, the variation of the load power as shown in Fig.18 is also reduced.

It is increasing to note that the change of the output inductor does not significantly change the input and output power. The total input power is just slightly higher than the previous case with L=2.3H. The system loss has increased from 3.12W to 3.65W due to the increase in the winding resistance of the inductor. As illustrated in (14), it is the input inductor L_s that controls the load power for a given input voltage Vs and load voltage V_o.

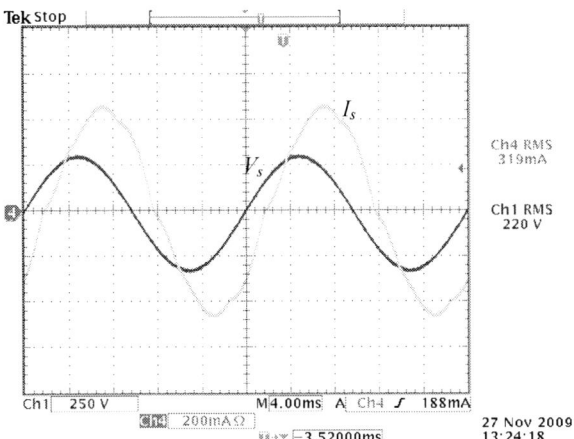

Fig. 15 Measured input voltage V_s & current I_s

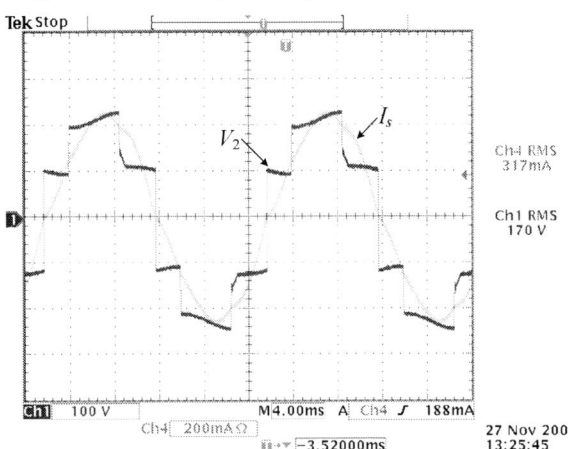

Fig. 16 Measured V_2 (stepped) and input current I_s (sinusoidal)

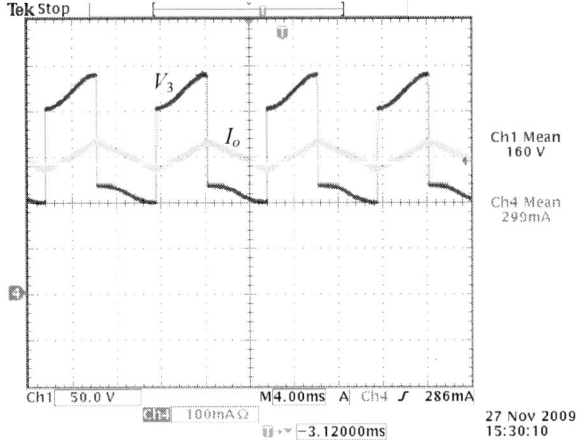

Fig. 17 Measured V_3 and output current I_o (L=5H)

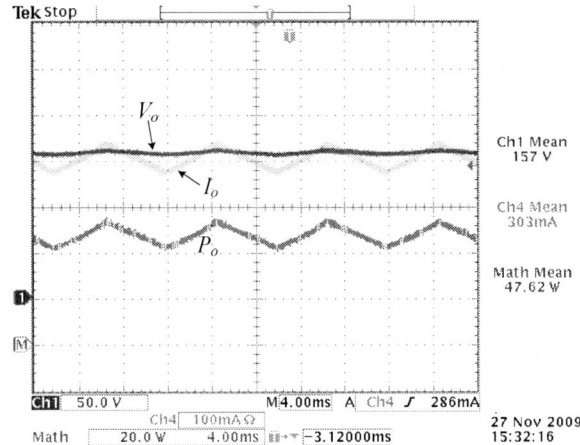

Fig. 18 Measured output current I_o, LED string voltage V_o & output power P_o (L=5H)

V. CONCLUSION

A novel single-stage passive LED driver for offline applications has been presented. This driver contains only passive and robust components without using any power switches, auxiliary power supply and control boards. Circuit analysis and experimental verification on a 50W prototype has been provided to confirm the feasibility of this proposal. As the driver consists of only a few components and there is no switching loss, a high efficiency of 93.6% has been achieved. Since only a few robust components are used in the passive LED driver, it is envisaged that this circuit provides other advantageous features such as low cost, low maintenance requirements and good robustness against extreme weather conditions such as lightning and wide temperature variation.

Since the metallic materials of the cores and windings of the two inductors, which contribute to the majority of the product material, can be recycled, this passive driver offers high efficiency, recyclability and long lifetime. These three factors are the essential criteria for sustainable lighting technology.

ACKNOWLEDGMENT

The authors would like to thank the Hong Kong Research Grant Council for its support for Project CityU 123508 and also the Centre for Power Electronics, City University of Hong Kong for the support provided for this project.

REFERENCES

[1] E.F. Schubert, "Light-emitting diodes", Cambridge, Second Edition, 2006

[2] "Datasheet of Luxeon Emitter", DS51, LUEXON POWER LEDS. http://www.lumileds.com/pdfs/DS51.pdf

[3] H. S. H Chung; N. M. Ho; W. Yan; P. W. Tam; S. Y. Hui: "Comparison of Dimmable Electromagnetic and Electronic Ballast Systems—An Assessment on Energy Efficiency and Lifetime", IEEE Transactions on Industrial Electronics, Volume 54, Issue 6, Dec. 2007 Page(s):3145 – 3154

[4] Y.X. Qin,; H.S.H. Chung,; D.Y. Lin,; S.Y.R. Hui,; "Current source ballast for high power lighting emitting diodes without electrolytic capacitor", 34th Annual Conference of IEEE Industrial Electronics, 2008. IECON 2008. 10-13 Nov. 2008 Page(s):1968 – 1973

[5] P. T. Krein, R. S. Balog, "Cost-Effective Hundred-Year Life for Single-Phase Inverters and Rectifiers in Solar and LED Lighting Applications Based on Minimum Capacitance Requirements and a Ripple Power Port", in Applied Power Electronics Conference and Exposition, 2009, APEC'09, Feb. 2009, pp. 620 – 625.

[6] L. Gu, X. Ruan, M. Xu, K. Yao, "Means of Eliminating Electrolytic Capacitor in AC/DC Power Supplies for LED Lightings", IEEE Trans. Power Electronics, vol.24, no.5, May.2009, pp.1399-1408

[7] Hong Kong Observatory official Website http://www.hko.gov.hk/ Chung, H. S.-H.; Ho, N.-M.; Yan, W.; Tam, P. W.; Hui, S. Y.; "Comparison of Dimmable Electromagnetic and Electronic Ballast Systems—An Assessment on Energy Efficiency and Lifetime", IEEE Transactions on Industrial Electronics, Volume 54, Issue 6, Dec. 2007 Page(s):3145 – 3154

[8] S.Y.R. Hui and W. Yan, "Re-examination on Energy Saving & Environmental Issues in Lighting Applications", Proceedings of the 11th International Symposium on Science 7 Technology of Light Sources, May 2007, Shanghai, China (Invited Landmark Presentation), pp.373-374

[9] S,Y,R, Hui, "Apparatus and methods of Operation of Passive LED Lighting Equipment", US patent application 12/429,792, 24 April 2009

[10] Zhongming Ye; Greenfeld, F.; Zhixiang Liang; "A topology study of single-phase offline AC/DC converters for high brightness white LED lighting with power factor pre-regulation and brightness dimmable feature", 34th Annual Conference of IEEE Industrial Electronics, 2008. IECON 2008. 10-13 Nov. 2008 Page(s):1961 – 1967

[11] Xiaohui Qu; Wong, S.C.; Tse, C.K.; Xinbo Ruan; "Isolated PFC Pre-Regulator for LED Lamps", 34th Annual Conference of IEEE Industrial Electronics, 2008. IECON 2008. 10-13 Nov. 2008 Page(s):1980 – 1987

[12] F.S.D. Reis,; J.C. Lima,; R., Jr. Tonkoski,; V.M. Canalli,; F.M. Ramos,; A. Santos,; M. Toss,; U. Sarmanho,; F. Edar,; L. Lorenzoni,;, "Single stage ballast for high pressure sodium lamps", IECON 2004. 30th Annual Conference of IEEE Industrial Electronics Society, 2004. Volume 3, 2-6 Nov. 2004 Page(s):2888 - 2893 Vol. 3

[13] Jinrong Qian; F.C. Lee,; "A high efficient single stage single switch high power factor AC/DC converter with universal input", Twelfth Annual Applied Power Electronics Conference and Exposition, 1997. APEC '97 Conference Proceedings 1997., Volume 1, 23-27 Feb. 1997 Page(s):281 - 287

[14] C. Qiao,; K.M. Smedley,; "A topology survey of single-stage power factor corrector with a boost type input-current-shaper", IEEE Transactions on Power Electronics, Volume 16, Issue 3, May 2001 Page(s):360 – 368

[15] C.K. Tse,; M.H.L. Chow,; "Single stage high power factor converter using the Sheppard-Taylor topology", 27th Annual IEEE Power Electronics Specialists Conference, 1996. PESC '96 Record., Volume 2, 23-27 June 1996 Page(s):1191 - 1197 vol.2

[16] S.Y.R Hui. and Y.X. Qin, "General photo-electro-thermal theory for light-emitting diodes (LED) systems", IEEE Transactions on Power Electronics, Volume 24, Issue 8, Aug. 2009 Page(s):1967 – 1976

[17] K. Kit Sum, "Improved Valley-Fill Passive Current Shaper", Power System World 1997, p.1-8

[18] Lam, J.; Praveen, K.; "A New Passive Valley Fill Dimming Electronic Ballast with Extended Line Current Conduction Angle", INTELEC '06. 28th Annual International Telecommunications Energy Conference, 2006. 10-14 Sept. 2006 Page(s):1 – 7

[19] Hamill, D.C.; Krein, P.T.; "A `zero' ripple technique applicable to any DC converter", 30th Annual IEEE Power Electronics Specialists Conference, 1999. PESC 99. Volume 2, 27 June-1 July 1999 Page(s):1165 - 1171

[20] Schutten, M.J.; Steigerwald, R.L.; Sabate, J.A.; "Ripple current cancellation circuit" Eighteenth Annual IEEE Applied Power Electronics Conference and Exposition, 2003. APEC '03. Volume 1, 9-13 Feb. 2003 Page(s):464 - 470

[21] Cheng, D.K.W.; Liu, X.C.; Lee, Y.S.; "A new improved boost converter with ripple free input current using coupled inductors", Seventh International Conference on Power Electronics and Variable Speed Drives, 1998. (Conf. Publ. No. 456) 21-23 Sept. 1998 Page(s):592 - 599

[22] Hwu K. and Chou S., "A simple current-balancing converter for LED lighting", IEEE Applied Power Electronics Conference, Feb. 2009, Washington DC, USA, paper: 16.7.

Improving Current Regulation for Offline LED Driver

Jianwen Shao
STMicroelectronics
Schaumburg, USA
jianwen.shao@st.com

Abstract—**A buck converter is a very common choice for non-isolated offline LED applications, using peak current regulation. The problem for peak current regulation is that the average current of the LED string will vary with different numbers of LEDs. This paper presents two simple and cost effective ways to compensate the average current variation without losing the simplicity of peak current regulation.**

I. INTRODUCTION

The simple buck converter with peak current regulation is shown in Figure 1. The peak current of the LEDs is regulated at a constant value. With the same value of peak current, the average current of the LEDs differs when the number of LEDs is different in the string [1] [2]. One of the solutions is to sense the average current of the LEDs in the high side, but it is costly to add such a current sensor. Reference [3] proposes to put a MOSFET in the high side so current sensing can be done in the low side. This increases the cost of the driver by adding high side gate driver. Reference [4] uses the integration of a sensing current to improve the accuracy, but it loses the instant current protection. In this paper, two simple solutions are presented to improve the average current regulation without changing the topology or significantly increasing the cost.

II. IMPROVE CURRENT REGULATION BY MODULATING PEAK CURRENT

For the Buck converter in Figure1, the controller can only control the peak current of the

inductor or the LED current. If the peak current value Vs is fixed, we can calculate the average current.

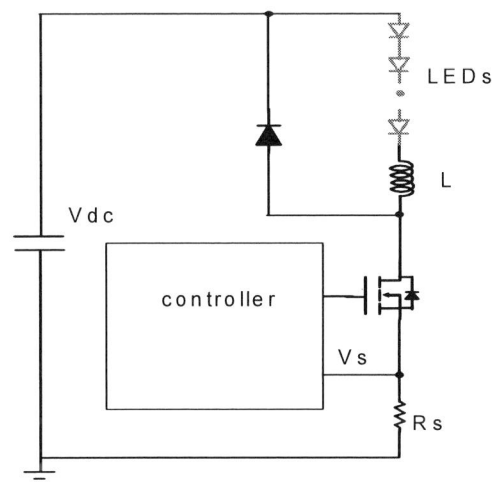

Fig. 1 typical buck converter with peak current control

$$Ipeak = Vs\,/\,Rs \qquad (1)$$

$$Iave = Ipeak - \frac{1}{2} * \frac{1-D}{L*F} * Vled \qquad (2)$$

$$Iave = Vs\,/\,Rs - \frac{1}{2} * \frac{1-D}{L*F} * Vled \qquad (3)$$

Where *Ipeak*: peak current of the LEDs; *Iave*: average current of the LEDs; *Vled*: voltage drop of the LED string; *Vs*: internal reference value for current setting; *Vdc*: dc voltage; *D*: duty cycle

$D=Vled/Vdc$; L: inductance; F: switching frequency.

We can see from equation (2) that the average current *Iave* will heavily depend on *Vled*. When the number of LEDs changes, the average current of the LEDs varies, even though the peak current of the LEDs is the same. The test result is shown in Figure 2.

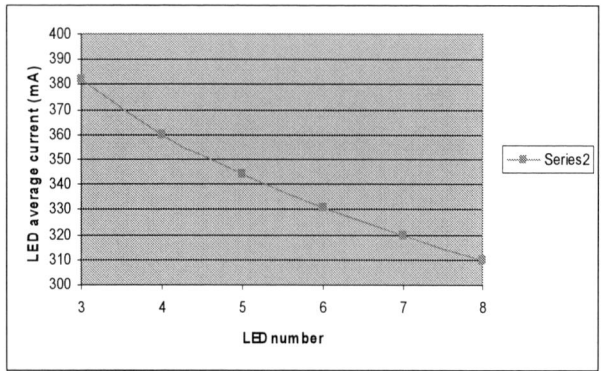

Fig.2 Average LED current with different number of LEDs

If the peak current value can be adjusted according to the LED voltage, then it is possible to keep the average LED current constant. There are numerous ways to implement the improvement. The following section gives two examples.

A. Improvement method 1:
Modulate the peak current by dc voltage and LED voltage feedback

Figure3 shows the modification of the circuit.

We can get equations as following.

$$Vs = Ipeak * Rs * \frac{R1}{R1+R2} + (Vdc - Vled) * \frac{R2}{R1+R2} \quad (4)$$

$$Ipeak = \frac{1}{Rs} * [Vs * (1+\frac{R2}{R1}) - Vdc * \frac{R2}{R1} + Vled * \frac{R2}{R1}] \quad (5)$$

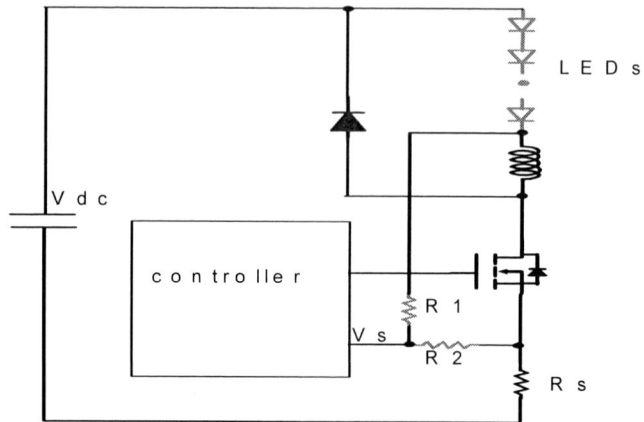

Fig. 3: modification of the circuit by adding voltage feedback

$$Iave = \frac{1}{Rs} * [Vs * (1+\frac{R2}{R1}) - Vdc * \frac{R2}{R1} + Vled * \frac{R2}{R1}] - \frac{1}{2} \frac{1-D}{L*F} * Vled$$

(6)

$$Iave = \frac{Vs}{Rs} * (1+\frac{R2}{R1}) - Vdc * \frac{R2}{R1*Rs} + (\frac{R2}{R1*Rs} - \frac{1}{2} * \frac{1-D}{L*F}) * Vled$$

(7)

If we can make the third term $\frac{R2}{R1*Rs} - \frac{1}{2} \frac{1-D}{L*F}$ of (7) as small as possible, and if Vdc is constant, then voltage drop of the LEDs will not have a significant impact on the average current. Figure 4 is the test result.

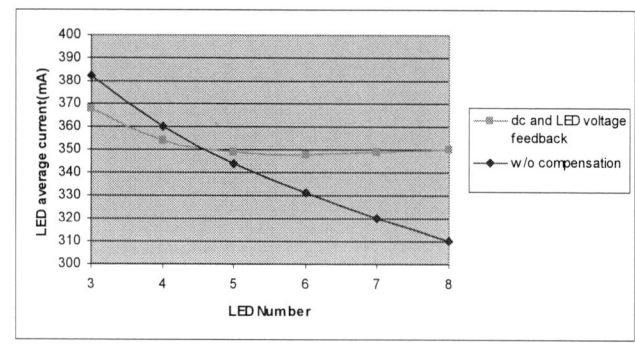

Fig.4: average current variation before and after the modification

Figure 5 shows the current waveform. The peak current is modulated according to the number of LEDs to improve the average current regulation.

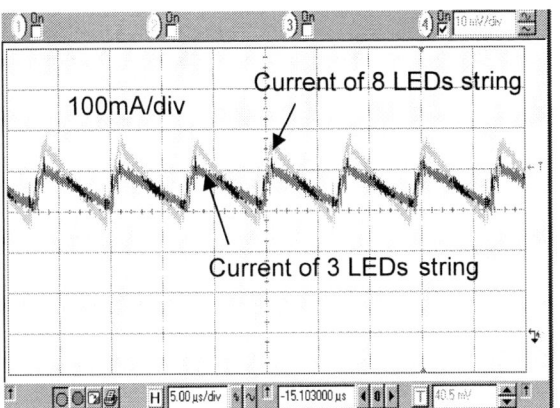

Fig.5: current waveforms for different LED numbers

B. Improvement method 2: Modulate the peak current by LED voltage feedback only

If the dc voltage is not well regulated, then the above method will not be effective, since the Vdc term will affect the average current, too. In this case, only the LED voltage is used to modulate the peak current, as shown in Figure 6. A coupled inductor is used to sense the LED voltage drop.

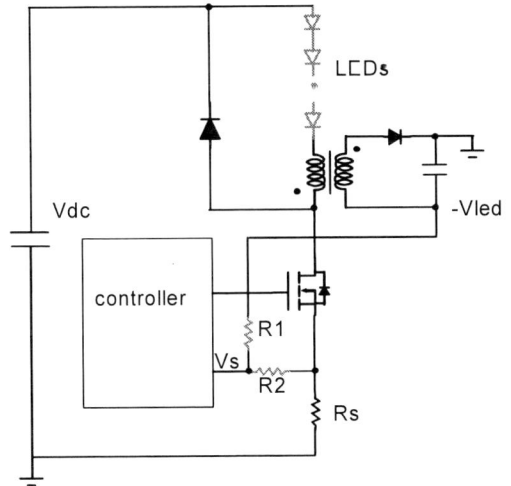

Fig.6: LED voltage feedback for peak current modulation

From Figure 6, we can have the following equation:

$$Vs = Ipeak * Rs * \frac{R1}{R1+R2} - Vled * \frac{R2}{R1+R2}$$

(8)

$$Ipeak = \frac{1}{Rs} * [Vs * (1 + \frac{R2}{R1}) + Vled * \frac{R2}{R1}]$$

(9)

$$Iave = \frac{1}{Rs} * Vs * (1 + \frac{R2}{R1}) + (\frac{R2}{R1 * Rs} - \frac{1}{2} * \frac{1-D}{L*F}) * Vled$$

(10)

Again, if we select R1 and R2 so that $\frac{R2}{R1 * Rs} - \frac{1}{2} * \frac{1-D}{L*F}$ is close to zero, we can get a constant average current. Figure 7 shows the test results.

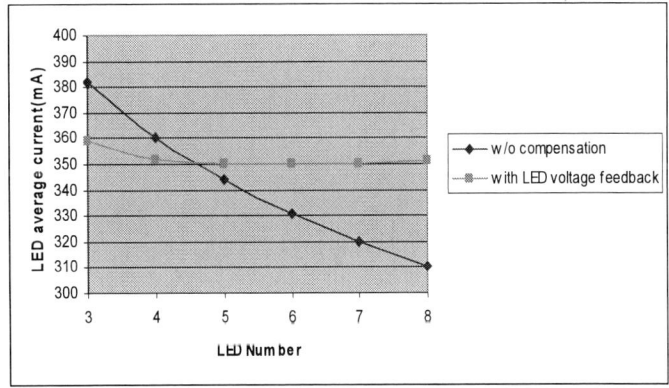

Fig. 7: average current variation before and after the LED voltage feedback

The above two methods are very simple and are easy to implement for a typical PWM controller. From equation (3), if $\frac{1}{2} * \frac{1-D}{L*F} * Vled$ can be kept constant by modulating switching frequency, it is possible to achieve the constant average current, too.

III. CONCLUSION

In offline LED applications, the simple fixed peak current regulation will cause an average LED current variation when the number of LEDs is different in the string. This paper presents two simple and low cost methods to correct the variation by modulating the peak current according to the dc voltage and the LED voltage drop. One method is good for applications where the dc voltage is well regulated; the other one is applicable when dc voltage is not well regulated. The test results prove the effectiveness of the proposed methods.

REFERENCES

[1] LED Drivers Tutorial, http://www.st.com/stonline/products/applications/blocks/lighting/lite017.shtml

[2] HV9910, datasheet from Supertex, www.suptertex.com

[3] Odile Ronat, Peter Green, Scott Ragona, "Accurate current control to drive high power LED strings," in *IEEE 2006 Applied Power Electronics Conference*, 2006, pp. 376-380.

[4] Yuan Fang1, Siu-Hong Wong2, and Lawrence Hok- Sun Ling, "A Power Converter with Pulse-Level-Modulation Control for Driving High Brightness LEDs," in *IEEE 2009 Applied Power Electronics Conference*, 2009, pp. 577-581.

Appendix: schematic of test circuit:

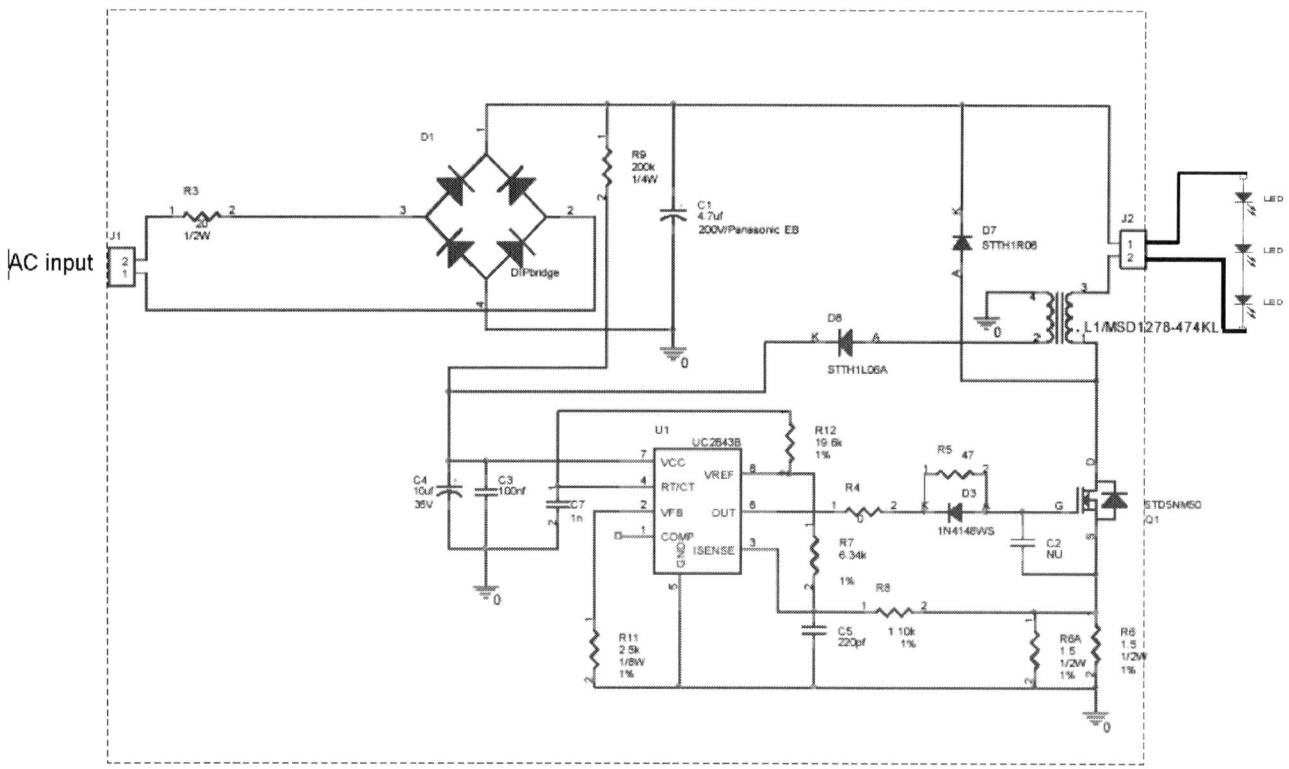

LED Driver Circuit with Inherent PFC

D. Aguilar

University of Minnesota
Department of Electrical and Computer Engineering
Minneapolis, MN 55455
aguilarda@msn.com

C. P. Henze

Analog Power Design, Inc.
16220 Hudson Ave.
Lakeville, MN 55044
chrisapdi@charter.net

Abstract— **A buck-boost topology operating in discontinuous conduction mode (DCM) is used as an off line LED driver for lighting applications. Operating from a full wave rectified ac-voltage with minimum input capacitance, with a constant switching frequency and constant on-time, the utility current has near unity power factor. This LED driver is suitable for cost sensitive applications because this circuit has a minimal parts count and a very simple control circuit. Disadvantages are that the LED drive current is modulated at twice the utility frequency and DCM operation increases component stress levels.**

I. INTRODUCTION

In today's world of lighting applications everyone is striving to make a more energy efficient and cost effective way of driving a light source. One type of light source that is growing in popularity these days is the LED because of its ability to deliver a high quantity of lumens with relatively low power consumption. This interest in LEDs has prompted many power electronic designers to come up with different ways to drive these devices. If such a lighting application is going to use the standard home utility voltage as the source then the challenge the designer faces is making sure that the driving circuit is low cost, robust, and does not adversely interact with the power grid by generating non-unity power factor and harmonic distortion currents.

LED drive circuits are typically designed to control the current running through a string of LEDs independent of the voltage drop across the LED string [1-4]. This study here presents a non-conventional method for achieving the same goal of lighting a string of LEDs by using alternating current voltage source where the circuit appears as a resistive load to the utility and the LEDs are operated with a controlled current and limited current without having to use any complex feedback control loops and a minimizing the number of components that are required. An additional benefit is that the LED light output, when driven with the proposed circuit, can be dimmed using conventional phase controlled lamp dimmers which are widely used with incandescent lamps.

II. CIRCUIT CONCEPT

The simplified circuit and how it operates is shown in Figure 1. The utility voltage is full wave rectified to create the familiar pulsating dc-waveform which for a typical 120 Vrms utility would have a peak of 167 V and a frequency of 120 Hz. This fully rectified waveform does not get filtered with a large capacitor as only a small capacitor C1 of 1.0uF is placed across the output of the rectifying bridge. The sizing of this capacitor is important. The capacitor needs to be large enough to carry the switching ripple and thereby avoid putting a high frequency disturbance current back into the utility grid. The capacitor needs to be small enough to present a relatively high impedance at the utility frequency to avoid creating a leading power factor or creating harmonic distortion in the utility current in response to any harmonics that may be present in the utility voltages.

A switching frequency of 37 kHz was selected because it is well above the utility frequency, above the audio band and well below the onset of conducted emission requirements.

For the converter stage a single inductor L1 is used – this converter is similar to the conventional non-isolated buck-boost topology. The converter operates in the discontinuous conduction mode (DCM). This is because in DCM the average of the input current (switching waveform) ends up being in-phase with the input voltage. We will see this is true when the equations that govern the converter are explained. A single series-string of LEDs is present at the output - it operates at (approximately) a constant voltage when forward biased. A high-speed silicon rectifier D1 is also included as the first diode in the series string. This diode disconnects the LED from the inductor when being charged by the power switch. In practice a resistor on the order of tens of thousands of Ohms may be required in parallel with the LED string to force the reverse voltage to appear across the high-speed silicon diode D1.

Power converters typically have relatively large capacitors at the output to minimize the output ripple, however in this application an output filter capacitor is not used. This is because it is not necessary to filter the alternating current dumped by the inductor into the LEDs. The LEDs run in a switched-mode with pulsating current and therefore pulsating light output. The switching frequency (37 KHz) is far too fast for a human to see the blinking. In addition, the ripple current at the output will be varying with the 120 Hz cycle of the full rectified wave but is also too fast for the human eye to notice

978-1-4244-4782-4/10 $26.00 © 2010 IEEE

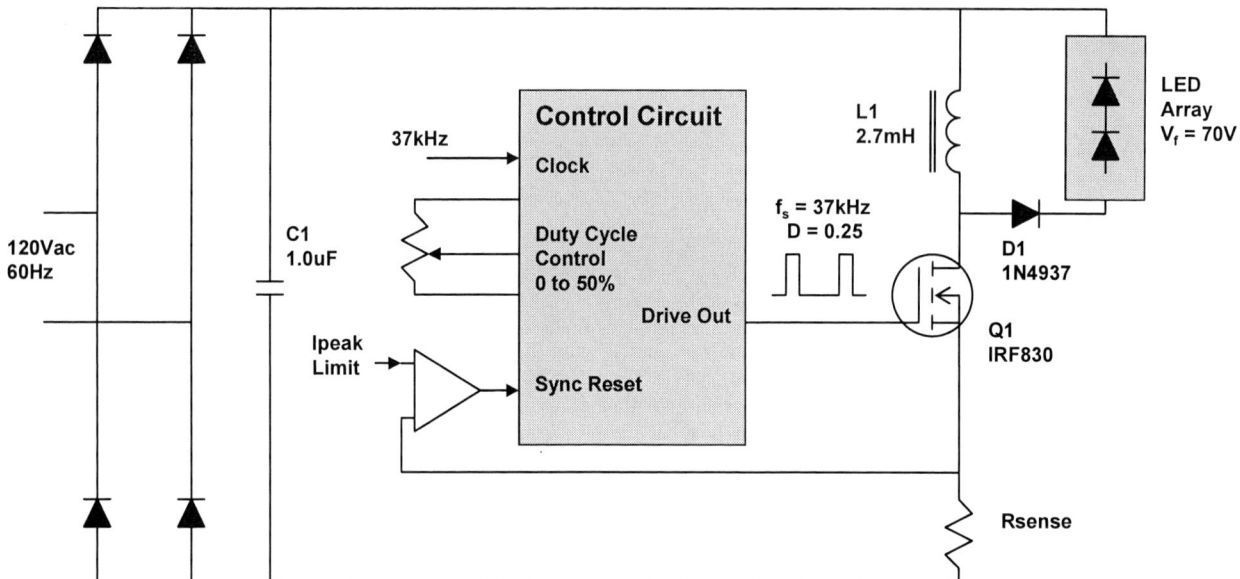

Figure 1. High level schematic of the LED driver circuit based on a buck-boost topology operating in discontinuous conduction mode.

the variation. The light output is proportional to the current delivered to the LED averaged over the utility period.

The inductor is sized so at the maximum allowed duty ratio and with the maximum input voltage and minimum LED

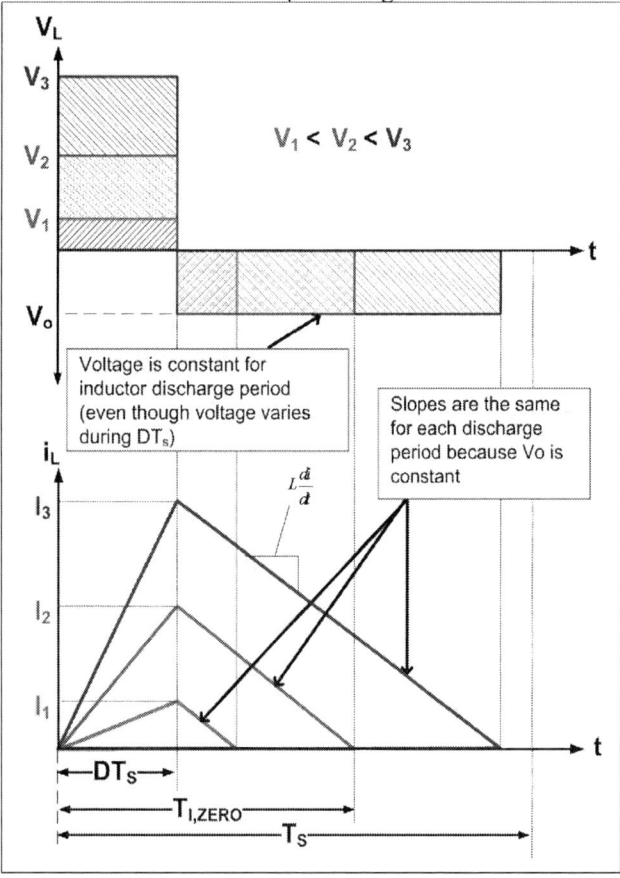

Figure 2. Switching waveforms of the inductor voltage and current.

voltage, the inductor current is discontinuous. For these conditions, during a portion of the switching cycle, the inductor current will be zero. The duty ratio is fixed (or slowly varying) compared to the utility period. The basic operation of this power stage portion of the circuit is that the inductor L1 is charged when the switch Q1 is on (just like a conventional buck-boost circuit). When the switch Q1 is turned off, the inductor will discharge back to zero through the LEDs. Because the input voltage (full rectified wave) varies with time so does the peak ripple current (see Figure 2). So over one cycle the peak ripple current in the inductor will be maximum when the input voltage is maximum and likewise for the minimum peak ripple current (when the input voltage is minimum). Because the ripple current magnitude follows the input voltage and the period is not varying over the utility cycle, the instantaneous average input current is proportional to the instantaneous input voltage (shown in Figure 3). In this fashion, power factor correction is inherent in the circuit.

Figure 2 shows the details of the inductor waveforms for several different instantaneous input voltages. In Figure 2, notice the discharge rate is constant for each voltage. This is due to the simplifying assumption that the forward voltage drop of the LED string is constant and independent of the forward current. Because the discharge rate is constant the discharge time period will vary with the input voltage.

This discharge period does need to be limited in order for the inductor to discharge completely. So the duty ratio needs to be small (about 0.25 or less) in order to operate in this mode. It is also necessary that the total conduction period of the inductor, $T_{i,zero}$, be less than the switching period, T_s (which is 27 micro-seconds in this case). Figure 4 illustrates this point – it is a graph of the time period $T_{i,zero}$ vs. V_{in}. The second portion of the switching cycle needs to be long enough so that there is enough time to discharge the inductor current completely otherwise there will be a build up of charge over the utility cycle and the waveform will not be consistent.

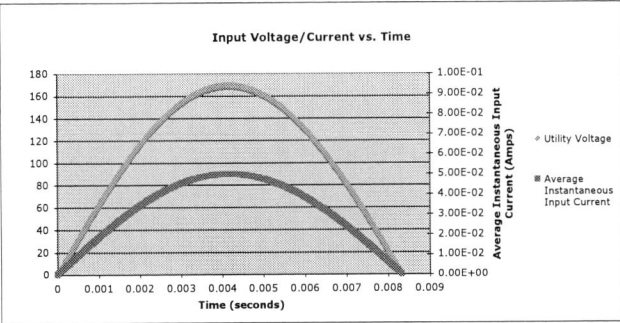

Figure 3. The instantaneous average input current is proportional to the input voltage.

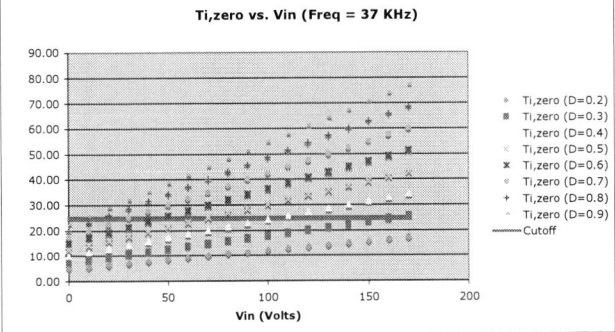

Figure 4. Inductor charge and discharge time as a function of input voltage

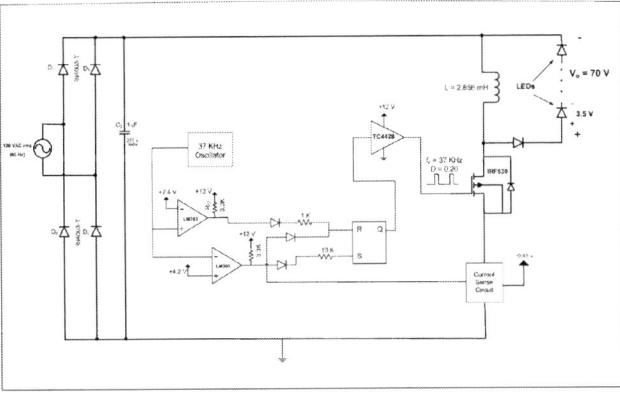

Figure 5. Control circuit block diagram

The time interval in which the inductor is active and returns to zero current, $T_{i,zero}$ is defined as a time that is equal to the sum of the charge and discharge periods of the inductor current. This time interval, $T_{i,zero}$ also varies with input voltage over the utility cycle. See the bottom of Figure 2 for an illustration of the time interval $T_{i,zero}$.

III. CIRCUIT ANALYSIS

Analyzing the circuit similar to the way we would analyze a standard non-isolated buck-boost converter we can derive a relationship between the input voltage V_{IN} and the peak inductor current $i_{L,peak}$. This is done by starting with the definition of the current and voltage relationship of an inductor. In DCM the ripple current is equal to the peak ripple

current of the inductor. So through derivations (not shown) the peak ripple current is equal to the following:

$$ i_{L,peak} = \frac{V_{IN}DT_s}{L} \qquad (1) $$

where D is the duty ratio, T_s is the switching period and L is the inductance value.

For off line applications, the input voltage will be a sinusoidal voltage with a peak value Vpk at a frequency f. In an ideal circuit, the input current will consist of a series of triangular pulses of current of a fixed width for each switching cycle determined by the duty cycle D. Averaging the input current over a number of high frequency switching cycles, the instantaneous average input current $i_{IN(avg)}$ is given by:

$$ i_{IN(avg)} = \frac{V_{PK}\sin(2\pi f t)D^2 T_s}{2L} \qquad (2) $$

The (average) input current is proportional to the input voltage; therefore this circuit presents a resistive load to the utility.

IV. CONTROL CIRCUIT

In a switch mode converter a gate driver circuit is needed to switch the power MOSFET Q1 on and off. This is accomplished with an oscillator that generates a saw tooth like waveform. The minimum and maximum values of this saw tooth can be set by using appropriate resistor values in the oscillator. The bias supply voltage used for these circuits is 12 volts and so the maximum value for this waveform is set to 8 volts and a minimum is set to 4 volts. The waveform's frequency can be set by using specific resistor and capacitor values. The converter is operating at a frequency of 37 kHz. The block diagram of the control circuit shown in Figure 5 (see Appendix) uses four low cost integrated circuits--three dual comparators (LM393) and a FET driver TC4428. The latch and overcurrent detection circuit are implemented with LM393 sections. This control circuit could be implemented as a low cost application specific integrated circuit (ASIC) which could also include the power FET Q1 and the high speed switching diode D1 if a high voltage was used.

The duty ratio can be varied (if desired) to control the brightness of the LEDs to create a dimming LED ballast. Since the duty ratio has a maximum limit, the minimum time available for the inductor to discharge into the LED string is also known. Therefore, a peak current sense circuit is also provided to prevent the inductor from being over charged (for example during an input voltage surge) to the point where the LED string could be damaged.

The way the peak current sense circuit operates is that the device Q1 shuts off when a current greater than 450 mA is drawn through the transistor and inductor. This operation is accomplished by putting a resistor at the source of Q1 and feeding that node into one input of a comparator. The comparator will also have a reference voltage fed into the other input, this input voltage will be proportional to the peak current limit. Whenever the current through the switch Q1 reaches currents of 450 mA or greater the voltage at the source

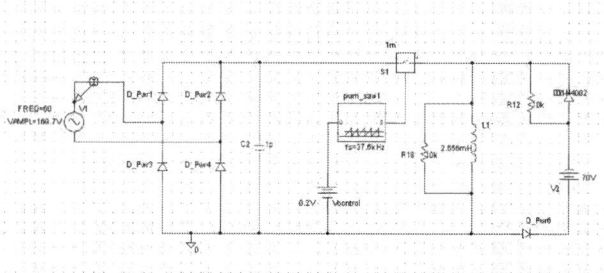

Figure 6. Pspice simulation circuit of the LED driver with an ac-input.

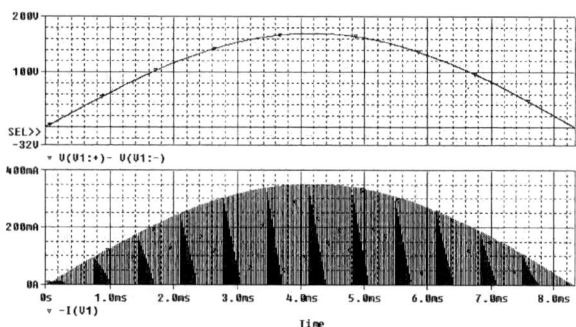

Figure 7. Simulated input voltage and current waveforms over one half of the utility cycle.

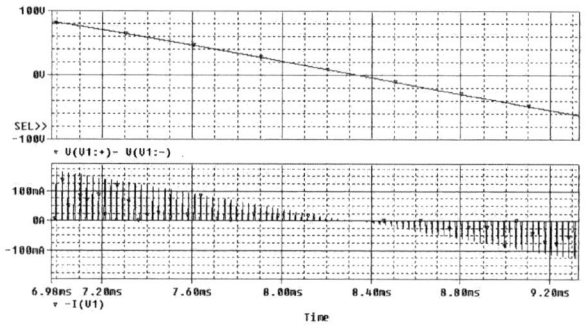

Figure 8. Simulation results for the input voltage and current near the zero crossing of the utility voltage

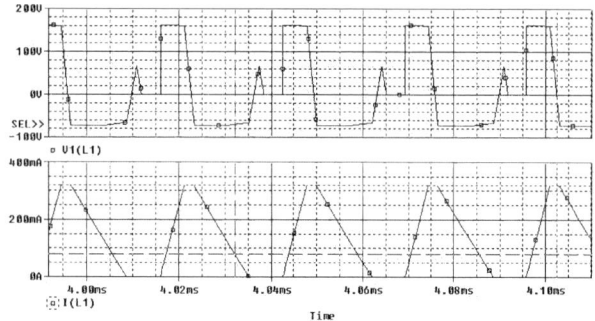

Figure 9. Simulation results for the inductor voltage and current waveforms near the peak of the utility voltage

terminal (of the MOSFET) will rise above the reference voltage (approximately 0.45 V) and the output of the comparator will go high (meaning it will reset the SR flip-flop).

V.   SIMULATION RESULTS

The circuit was simulated with an AC voltage source as shown in Figure 6. The switch Q1 is replaced with a voltage controlled switch and the overcurrent detection circuit is not included. The input voltage is the standard utility voltage of 120 Vrms at 60 Hz. The simulated input current waveform is shown along with the input voltage for one half of the utility cycle in Figure 7. It can be seen that the peak value of the inductor current is proportional to the input voltage. The triangular shape of the input current pulses is shown in Figure 8 where the simulation results near the zero crossing are displayed. The inductor current and voltage when operating near the peak of the line are shown in Figure 9.

VI.   EXPERIMENTAL RESULTS

The LED driver circuit was tested initially with a dc power supply in place of the utility at the input as shown in Figure 10. This allows measurements of the circuits operation under steady state conditions over the range of dc-input voltages that are encountered during ac-operation.   One important note that should be mentioned about the experiment results is that a series string of twenty Zener diodes (5 Watt, 3.3V, 1N5333) in reverse bias at the output instead of a LED array. This was done to avoid damaging the costly LEDs during testing. The Zener diode string is indicated in Figure 10.

The screen captures of the DC measurements show that the circuit behaves as expected. Figure 11 shows several captured waveforms when the input voltage was set at approximately 80 volts and the output is driving a 70-Volt series string of Zener diodes. The operation is at fairly light load. The oscilloscope is set for a quad display mode so each waveform can be displayed separately while using a large portion of the available vertical input. The vertical scale is for each "tick" on the center graticule.   Channel 4 (green) is the gate drive pulse (at 5V per division) at the gate of the power transistor Q1. The duty ratio is measured at 19.3%. The waveform of Channel 2 (magenta) is the inductor current at 50mA per division. The peak inductor current is 170mA and the average inductor current is measure to be 35mA. The inductor current waveform shows that the inductor is charging and discharging linearly as expected. Furthermore, the inductor current is zero for a portion of each switching cycle indicating that the circuit is operating well into the DCM mode. The waveform of Channel 1 (yellow) is the voltage across the output Zener diode string that is used to simulate the LEDs. The output voltage during the discharge is approximately constant at 70 volts, although, the voltage does sag as the current becomes lighter. When the current goes to zero, the output voltage collapses. The output power is calculated by multiplying the inductor current waveform by the output voltage waveform and is displayed as Waveform A (orange) at 5 Watts per division. The oscilloscope is also

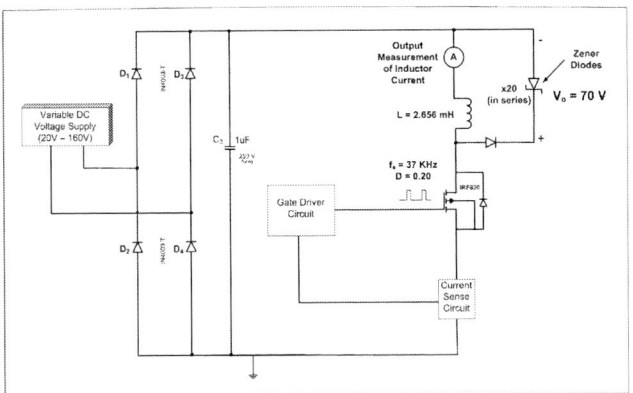

Figure 10. Test set up using Zener diodes in place of the LEDs

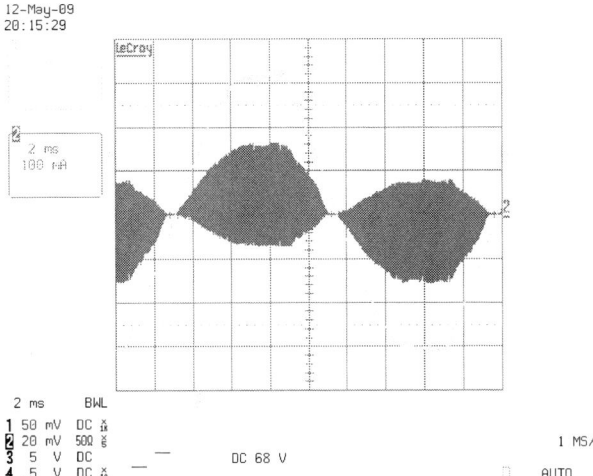

Figure 13. Measured input current and input voltage when operating on 120Vrms ac-voltage.

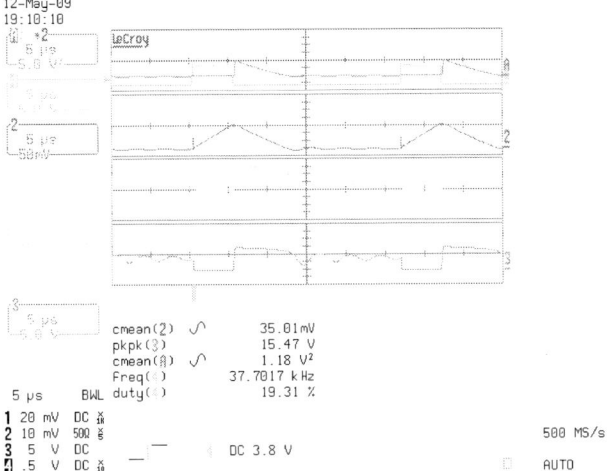

Figure 11. Experimental Waveforms when operating at light load with an 80Vdc input

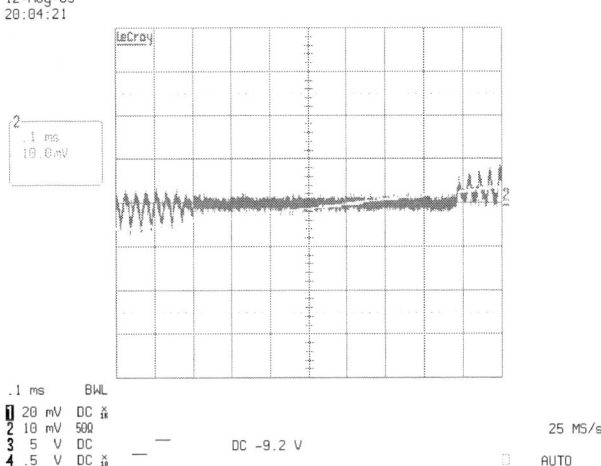

Figure 14. Measured input voltage and current near the zero crossing of the utility voltage waveform.

calculating the average output power at 1.2 Watts. The switching frequency is measured at 37.7kHz.

The input impedance of the converter is calculated from a measurement of the average (dc-) input current with a digital multi-meter as a function of the input voltage. The input impedance as a function of input voltage is plotted in Figure 12. This plot shows that the input impedance in nearly constant and independent of the input voltage. This is the condition that will result in a power factor near unity when operating on from an AC source.

In Figure 13, input current waveform (Channel 2 at 100mA per division) is shown along with the input voltage waveform (Channel 1 at 50V per division). This current is measured on the converter side of the input capacitor C1 and contains a large switching ripple. In Figure 14, the same waveforms are shown near the zero crossing.

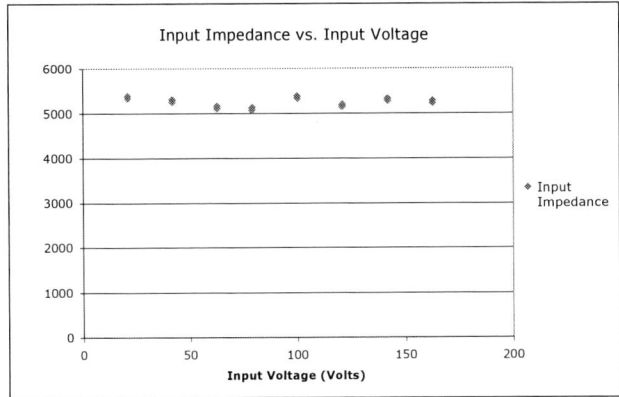

Figure 12. Measured input impedance of the LED Driver as a function of input voltage. For a unity power factor, a constant input impedance is required which is independent of input voltage.

978-1-4244-4782-4/10 $26.00 © 2010 IEEE   609

VII. CONCLUSIONS

Our conclusion is that the circuit does indeed operate successfully. We have showed that this simple LED driver circuit, operating in DCM, provides power factor correction in an open loop mode. The converter operates with pulsating current, at the output, rather than with constant current, this reduces the component count and cost of the circuit compared to a conventional PFC (which uses complex feedback control). This also causes high peak stress in the power components.

VIII. REFERENCES

[1] M. Day, "LED-Driver Considerations Texas Instruments Power Management Application Note, page 1.

[2] M.G. Craford, "LED's shallenge the incandescents", *IEEE circuits and devices Mag.*, Vol 8. pp 24-29, Sept. 1992.

[3] S. Muthu and J. Gaines, "Red, Green and Blue LED-based White Light Source: Implementation Challenges and Control Design", *Industry Application Conference IAS Record*, 2003. Vol 1, pp 515-522. 2003.

[4] F.E. Bisogono, S. Nittayarumphong, M. Radecker, A.V. Carazo and R. N. do Pardo, "A Line Power-Supply for LED Lighting using Piezoelectric Transformers in Class-E Topology", *IEEE International Power Electronics and Motion Control Conference*, Aug. 2006.

IX. APPENDIX DETAILED SCHEMATIC

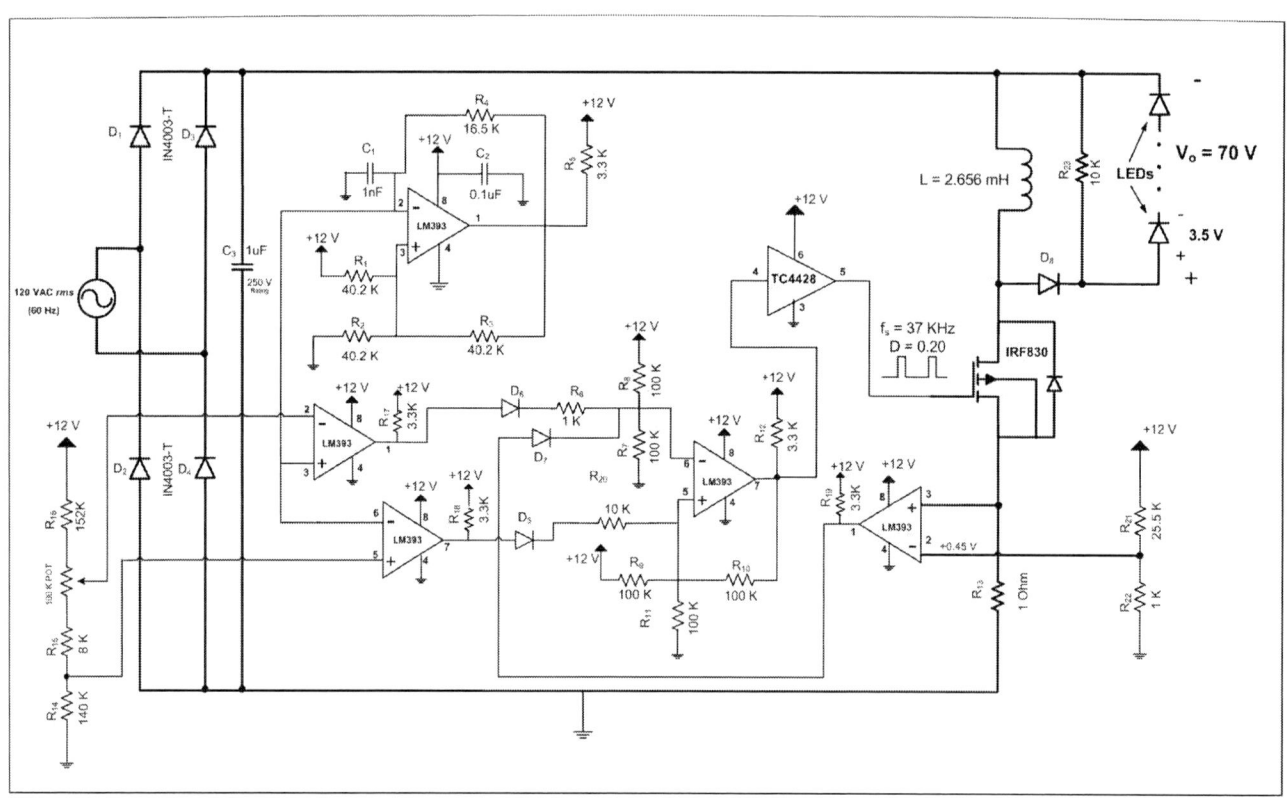

A New Circuit Design and Control to Reduce Input Harmonic Current for a Three-phase AC Machine Drive System having a very Small DC-link Capacitor

Hyunjae Yoo
Electric Power Control Part, Digital Business Division
Samsung Heavy Industries Co., Ltd.
Hwasung-City Gyeonggi-Do Korea
hyunjae.yoo@samsung.com

Seung-Ki Sul
School of Electrical Engineering & Computer Science
Seoul National University
Seoul, Korea
sulsk@plaza.snu.ac.kr

Abstract— **This paper presents a new circuit topology to meet the input harmonic current standard for an ac machine drive system which has a very small dc-link capacitor. The proposed circuit topology is based on a harmonic current injection method, and it keeps up size and cost competitiveness of an ac machine drive system having a very small dc-link capacitor. Also, this paper proposes an appropriate control algorithm and a stability analysis for the proposed circuit topology. Experimental results reveal the validity of the proposed circuit topology and its control method. Also, it is confirmed that the harmonic current standard can be satisfied with the proposed circuit and its control method.**

I. INTRODUCTION

Recently, reducing the size of a dc-link capacitor has been key issues especially for low/medium power applications [1-6]. If the size of a dc-link capacitor in an ac machine drive system is reduced, not only the size of a capacitor itself is reduced, but also a pre-charging circuit can be saved. Moreover, an electrolytic capacitor which is less reliable than a film one because of its comparably short lifetime, can be replaced by film capacitors which generally have much longer lifetime. Thus, the reliability of an ac machine drive system can be enhanced [7-8]. Furthermore, the system having a very small dc-link capacitor contains much less input harmonic current contents than the drive system having a large one. Nevertheless, this system has a few inherent drawbacks such that the dc-link voltage may easily become unstable and it is not possible to supply enough energy during even short term input voltage interruption. Moreover, though the input harmonic currents are reduced, still the harmonics cannot meet the standards such as IEC 61000 [9-10] without adding any additional hardware.

As the dc-link voltage instability can be covered by adopting one of recent active researches, the stable operation regardless of the quite small capacitance of the dc-link can be achieved. Also, the issue related to the energy storage role

during short term input voltage interruption is not always essential to every application. Applications such as a fan, a pump or a compressor, that a re-starting operation right after the input voltage interruption would not be a severe problem, may not need this energy storage role. However, the applicability of this drive system having a very small dc-link capacitor cannot be expanded if the international input harmonic current standard is not satisfied.

A lot of researches based on a harmonic current injection method have been presented in the past to reduce input harmonic current of a conventional diode rectifier-fed inverter system [11-21]. These methods have possibilities to comply with the given harmonic current standard with keeping up size and cost competitiveness of the drive system having a very small dc-link capacitor. In [19-20], new topologies using zero sequence coupling between machine and grid sides has been introduced. These topologies can greatly reduce the overall system cost as far as the neutral point of an ac machine is accessible. However, the ac machine should share the voltage for controlling the machine itself and that for controlling injection current. Therefore, the machine may not be operated at its rated speed simultaneously injecting harmonic current into the input ac side. Large size dc reactors for a harmonic current injection circuit have been replaced by smaller size ac ones in [21] to reduce the overall cost of the drive system. However, it still has a lot of passive elements in the current injection circuit as well as the dc side current smoothing purpose inductor.

This paper presents a new circuit topology to comply with the harmonic current standard with minimizing the additional hardware cost. The proposed harmonic current injection circuit consists of one switch arm, a single phase inductor and three bi-directional switches. The proposed circuit minimizes passive device which are less competitive in terms of size and cost than active one. The proposed circuit also makes it possible that the current rating of each element is reduced. In

978-1-4244-4782-4/10 $26.00 © 2010 IEEE

this paper, an appropriate control method of an injection current and a stability analysis for the proposed circuit configuration are also proposed.

II. AC MACHINE DRIVE SYSTEM HAVING A VERY SMALL DC-LINK CAPACITOR

An example of ac machine drive system having a very small dc-link capacitor is shown in Fig. 1. Though the input side rectifier that is connected to the ac source can be composed of either a diode rectifier or a PWM rectifier, the former one is only taken into account in this paper. The input ac line current waveform under constant load condition can be described theoretically as shown in Fig. 2[6]. Input harmonic contents of the drive system having a very small dc-link capacitor are much less than that of the drive system having a large dc-link capacitor [6]. Nevertheless, it still cannot comply with the international harmonic current standards. Thus, additional hardware to meet the standard is necessary. There are plenty of possibilities of hardware composition such as a PWM rectifier, a VIENNA rectifier [22,23], active or passive filters [24,25], a harmonic current injection method and etc. Note here, the target harmonic current standard [9,10] does not require strict THD specification but does limit only each order of harmonic current magnitude. Hence, in this paper, a harmonic current injection method has been adopted for the drive system having a very small dc-link capacitor since it is more competitive than others in terms of size and cost.

III. CONFIGURATION OF CURRENT INJECTION TOPOLOGY

Fig. 3 shows the basic configuration of the conventional current injection topology for a three-phase diode rectifier fed ac machine drive system. It consists of a harmonic current controller part and a current injection circuit part. Two boost converters, one series connected switch leg, a machine side zero sequence current controller and etc. can be such examples of the harmonic current controller part. A star-delta transformer with a neutral point, a zigzag auto-transformer, a resonant filter and etc. can be such examples of the injection circuit part. Fig. 4 shows a simplified block diagram of harmonic current injection topology. The desired harmonic current is synthesized by two current sources (I_h), and it is divided into 1/3 by the injection circuit part. Fig. 5 shows its harmonic current injection principle. Fig. 5(a) and 5(b) show

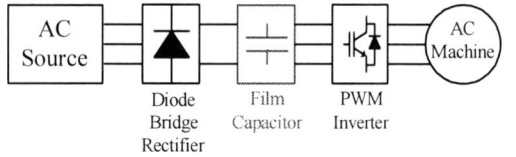

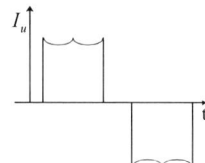

Figure 1. AC machine drive system having a very small dc-link capacitor.

Figure 2. Input ac line current waveform.

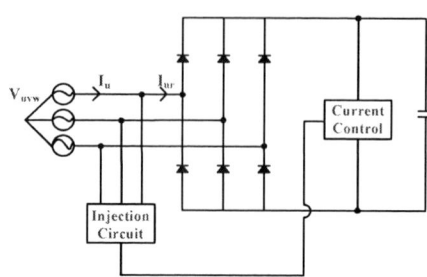

Figure 3. Basic configuration of the conventional current injection topology.

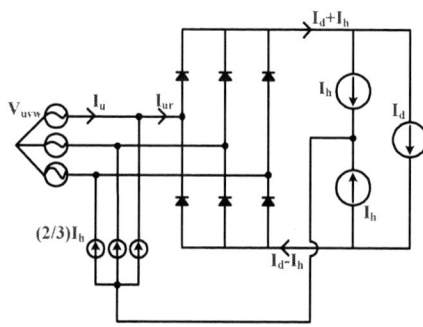

Figure 4. Simplified configuration of the current injection method.

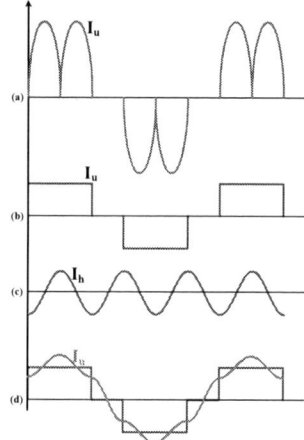

Figure 5. Current shaping principle of the conventional current injection method

input current waveforms for the drive system having large capacitance and that having very small capacitance respectively. Fig. 5(c) shows an injection current waveform based on the 3rd harmonic. Its optimal magnitude based on FFT analysis can be derived as (1) [11-13].

$$I_{h\_mag} = 0.74 \cdot I_{d\_mag} \qquad (1)$$

Lastly, Fig. 5(d) shows a resultant input current waveform which is the summation of Fig. 5(b) and Fig. 5(c) sub-period by sub-period (1/6 period). Note that, the input current should be near square wave before a harmonic current is injected as shown in Fig. 5(b), or otherwise, an injection current should contain too much harmonic content to shape an input current into sinusoidal as expected from Fig. 5. Thus, inductors on the ac side or the dc side are necessary if the dc-link capacitance

978-1-4244-4782-4/10 $26.00 © 2010 IEEE

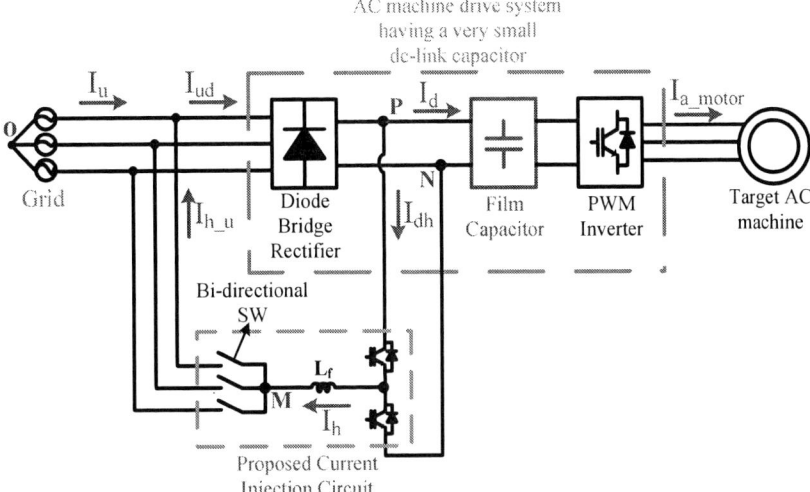

Figure 6. Configuration of the proposed harmonic current injection circuit.

is large enough to reduce a lot of low order harmonic contents preliminary. However, the input current waveform of a drive system having a very small dc-link capacitor is already near square wave as shown in Fig. 2. Therefore, an additional inductor except a current control purpose one is not necessary.

IV. PROPOSED CIRCUIT AND ITS CONTROL METHOD

In order to reduce an input harmonic current, the harmonic current injection method has been adopted to keep up size and cost competitiveness. This harmonic current injection method has a large variety of circuit composition. This paper proposes a new circuit composition which optimizes size and cost but complying with the given harmonic current standard.

A. Circuit Configuration

Fig. 6 shows the configuration of the proposed harmonic current injection circuit for an ac machine drive system having a very small dc-link capacitor. Injection of a harmonic current is achieved by one switch arm and a single phase inductor, and three bi-directional switches. The proposed circuit makes it possible that the three phase input current can be shaped simultaneously with only a single phase inductor. Also, by using three bi-directional switches instead of passive elements such as an LC tuned filter, a star delta transformer and a zig-zag auto-transformer, not only the injection of the three phase input side is possible but also the magnitude of circulating current via the current injection circuit and the diode rectifier is reduced to 1/3 of that of the circuit using passive elements. Hence, the current ratings of all elements including the single phase inductor in the proposed circuit are reduced by 1/3 of the conventional one. Thus, the circuit has competitiveness in terms of size and cost over other conventional circuit based on passive elements.

B. Current Path Generation through Three bi-directional Switches

As shown in Fig. 6, three bi-directional switches are located between the three phase grid side and the single phase inductor to constitute the injection current path. Each switch is

turned ON when the magnitude of input phase voltage is medium among three phases, and that is described in detail in [14-15,20].

C. Proposed Optimal Current Reference Generation

The main idea of the injection current reference generation method is the same as introduced in [18], but the proposed optimal current reference generation method is able to save both the complex calculation effort and the discrimination effort of magnitude and phase of the input voltages. The optimal injection current reference for the proposed current injection circuit can be simply obtained from

$$ i_{opt\_ref} = \frac{v_{mid}}{V_p} \cdot I_p \qquad (2) $$

where, i_{opt_ref} : optimal injection current reference, : input phase voltage whose magnitude is medium value among three input phase voltages, V_p : peak input voltage magnitude, I_p :peak input current magnitude. The current reference forms quasi-triangular wave and its magnitude is 1/3 of that introduced in [18] as mentioned in the previous section A.

D. Proposed Current Control Method

The injection current control can be easily achieved by a simple Proportional Integral (PI) control method. An equivalent circuit diagram of the current injection control circuit is shown in Fig 7. The controller consists of one switch arm, a single phase inductor and three bi-directional switches which directly connected to the grid side. Thus, the current controller should consider the grid side voltage change according to the switching sequence of three bi-directional switches in order to get desired control performance. In other words, the inductor current can be decided by voltage difference between the dc-link voltage and the voltage between the neutral point of three bi-directional switches and

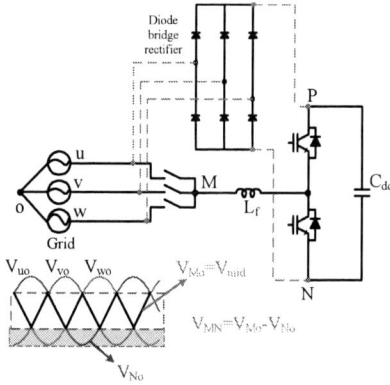

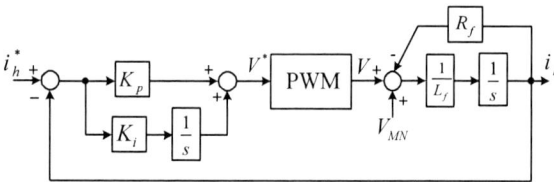

Figure 7. Voltage relation of the proposed current injection circuit.

Figure 8. An injection current controller without considering voltage V_{MN}.

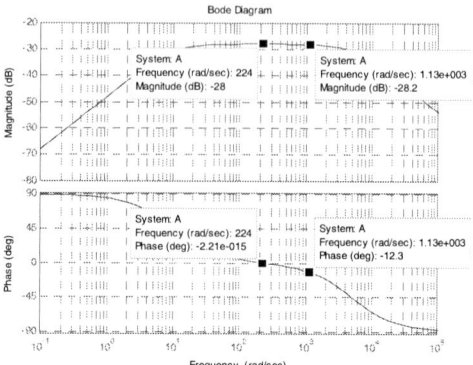

Figure 9. Bode plot of unwanted band pass filter.

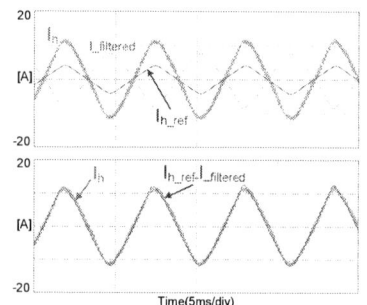

Figure 10. Simulation result of current control performance without considering voltage disturbance.

the minus point of the dc-link voltage (V_{MN}) shown in Fig. 7. A block diagram of the current control system where this voltage is not considered is shown in Fig. 8. Its mathematical representation is also given by (3). Substituting the proportional and integral gains of current controller, (4) into (3), then (5) can be derived.

$$i_h = \frac{K_p s + K_i}{s^2 L_f + (R_f + K_p)s + K_i} i_h^* - \frac{s}{s^2 L_f + (R_f + K_p)s + K_i} V_{MN}$$
(3)

$$K_p = L_f \omega_{cc}$$
$$K_i = R_f \omega_{cc}$$
(4)

$$i_h = \frac{\omega_{cc}}{s + \omega_{cc}} i_h^* - \frac{s}{s^2 L_f + (R_f + L_f \omega_{cc})s + R_f \omega_{cc}} V_{MN}$$
(5)

The first term in (5) is desired current control dynamic, and the second term in (5) is unwanted band pass filtered dynamic resulted from the disturbance voltage (V_{MN}). A Bode plot of this unwanted filtered dynamic where cut-off frequency of the current controller (ω_{cc}) is given by $2\pi \cdot 800$ [rad/s] is shown in Fig. 9. The phase difference of the unwanted filtered dynamic where the fundamental frequency of the disturbance voltage (V_{MN}) is given by 180Hz (3rd harmonic), can be obtained from Fig. 9. Also, its magnitude can be obtained from (6). A simulation result where the disturbance voltage has not been considered is shown in Fig. 10. The actual current (I_h) cannot track its reference (I_{h_ref}) in spite of applying large controller gains (800Hz), but is almost the same as the current reference minus unwanted filtered value as analyzed above.

$$-28.2 = 20 \log \frac{x}{465}$$
$$I_{filtered} = 10^{(-\frac{28.2}{20} + \log 465)} \approx 18 [A]$$
(6)

Therefore, the voltage (V_{MN}) should be considered to get rid of this unwanted filtered dynamic. Even though this voltage (V_{MN}) can be directly measured, it results in additional cost for a measurement circuit. Hence, this paper also introduces a voltage disturbance estimation method using given measured information. The disturbance voltage (V_{MN}) shown in Fig. 7 can be expressed by (7).

$$V_{MN} = V_{Mo} - V_{No}$$
(7)

The voltage between the neutral point of three bi-directional switches and that of the input three phase voltages can be defined by (8) because each bi-directional switch is turned ON when the magnitude of the corresponding input phase voltage is medium value. Also, the voltage between the dc-link minus point and the neutral point of the input phase voltage can be defined by (9) because of the commutation action of the three phase diode bridge rectifier.

$$V_{Mo} = V_{mid}$$
(8)

$$V_{No} = V_{min}$$
(9)

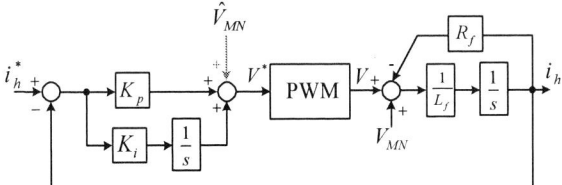

Figure 11. Block diagram of the proposed injection current control.

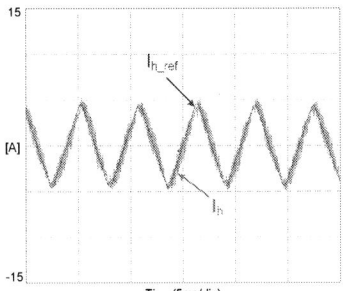

Figure 12. Proposed injection current control performance.

where V_{mid} is the input phase voltage whose magnitude is the medium value and V_{min} is the input phase voltage whose magnitude is the minimum value

Actually, these voltages cannot be defined clearly if the size of a dc-link capacitor is relatively large. In this system however, they can be defined because the size of a dc-link capacitor is very small and thus the dc-link voltage roughly follows to the maximum value among three input line-to-line voltages. As a consequence, this disturbance voltage can be estimated only by compounding input phase voltage information which is already known. A block diagram of the proposed injection current control including this input voltage disturbance rejection by feed-forward manner is shown in Fig. 11. Its simulation result is also shown in Fig. 12. The actual current tracks its reference much better than the previous case.

V. STABILITY ANALYSIS

The proposed current injection circuit and its control algorithm make it possible to reduce an input harmonic current. However, if the stability of this system cannot be guaranteed, it is very hard to apply the system into a practical system. Hence, a stability of the drive system including the proposed harmonic current injection circuit is analyzed in this section. As mentioned above, an ac machine drive system having a very small dc-link capacitor is subject to being unstable. Thus, the stability and the minimum size of the dc-link capacitance which makes the system stable have been analyzed in [26]. Fig. 13 shows an equivalent circuit diagram of a diode rectifier front ended typical ac machine drive system under a constant load condition neglecting the PWM inverter action. The load is modeled as a current source (P_{load}/v_{dc}), and L_{src}, R_{src} stand for the equivalent line impedance. Also, C_{dc} stands for the dc-link capacitance. Thus, the dynamic equation of the system is derived in (10).

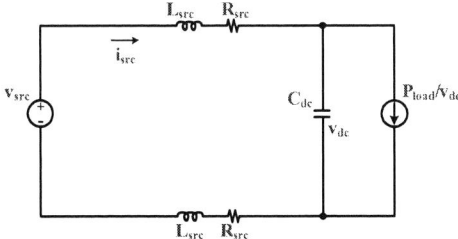

Figure 13. Equivalent circuit diagram of typical diode rectifier front ended ac machine drive system.

$$
2 \cdot L_{src} \cdot \frac{di_s}{dt} = v_{src} - 2 \cdot R_{src} \cdot i_{src} - v_{dc}
$$
$$
C_{dc} \cdot \frac{dv_{dc}}{dt} = i_{src} - \frac{P_{load}}{v_{dc}}
$$
(10)

The source current and the capacitor voltage can be expressed by (11) which contain dc and ac components respectively.

$$
i_{src} = \bar{i}_{src} + \tilde{i}_{src}
$$
$$
v_{dc} = \bar{V}_{dc} + \tilde{v}_{dc}
$$
(11)

where $\bar{i}_{src}$ and $\bar{V}_{dc}$ are mean values of the source current and the dc-link voltage respectively, $\tilde{i}_{src}$ and $\tilde{v}_{dc}$ are the source current and the dc-link voltage variations respectively. If the dc-link voltage variation term is relatively small enough, the load current (P_{load}/v_{dc}) can be linearized at the operating point as derived in (12).

$$
\begin{cases} L \cdot \dfrac{d\tilde{i}_{src}}{dt} = -R \cdot \tilde{i}_{src} - \tilde{v}_{dc} \\ C_{dc} \dfrac{d\tilde{v}_{dc}}{dt} = \tilde{i}_{src} + \dfrac{P_{load}}{\bar{V}_{dc}^{\,2}} \cdot \tilde{v}_{dc} \end{cases}
$$
(12)

Then, the resultant characteristic equation is given by (13)

$$
P(s) = s^2 + (\frac{R}{L} - \frac{P_{load}}{C_{dc}\bar{V}_{dc}^{\,2}})s + (\frac{\bar{V}_{dc}^{\,2} - RP_{load}}{LC_{dc}\bar{V}_{dc}^{\,2}}) = 0
$$
(13)

where, $R = 2 \cdot R_{src}$, $L = 2 \cdot L_{src}$.

Consequently, if the dc-link capacitance value satisfies the condition (14), the system is remained in stable operation area [26].

$$
C_{dc} > \frac{L \cdot P_{load}}{R \cdot \bar{V}_{dc}^{\,2}}
$$
(14)

The stability of the proposed drive system including harmonic current injection circuit can be analyzed with the same criterion. Its equivalent circuit diagram is shown in Fig. 14, and also its dynamic equation is given by (15).

978-1-4244-4782-4/10 $26.00 © 2010 IEEE

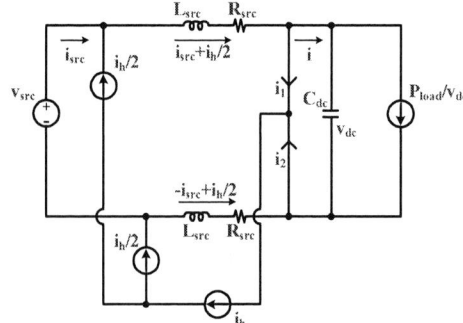

Figure 14. Equivalent circuit diagram of typical diode rectifier front ended ac machine drive system including proposed current injection circuit.

$$L_{src} \frac{d(i_{src} + i_h/2 + i_{src} - i_h/2)}{dt} = v_{src} - R_{src}(i_{src} + i_h/2 + i_{src} - i_h/2) - v_{dc}$$

$$2L_{src} \frac{di_{src}}{dt} = v_{src} - 2R_{src}i_{src} - v_{dc}$$

$$C_{dc} \cdot \frac{dv_{dc}}{dt} = i - P_{load}/v_{dc} \qquad (15)$$

where i_h is an injection current, and i is a dc-link current. If the injection current i_h satisfies (16), then the characteristic equation of the drive system including the harmonic current injection circuit becomes the same as (13).

$$i_1 + i_2 = i_h \qquad (16)$$

As a result, if the original drive system having a very small dc-link capacitor satisfies the condition (14) and the harmonic current controller also satisfies (16), the overall drive system is stable.

VI. EXPERIMENTAL RESULTS

An experimental setup for the proposed current injection circuit is shown in Fig. 15. It includes an ac machine drive system having a very small dc-link capacitor, bi-directional switches, a single phase inductor, one switch arm and its drive circuit, a target ac machine and a control board. Several core parameters of the experimental setup are listed in Table I.

Fig. 16 shows an experimental result of the drive system having a very small dc-link capacitor without applying the harmonic current injection where the target machine has been operated at its rated speed yielding 83% of its rated power. It shows the input current, the dc-link voltage, the injection current reference, the injection current, the dc side current, the switch arm current and the machine current waveforms respectively. Abbreviation of each waveform is represented in Fig. 6. The input ac current (I_u) forms quasi-square wave as expected in Fig. 2 which still contains a lot of low order harmonic components. The dc-link voltage (V_{dc}) follows to the maximum value of three-phase line-to-line voltage. The injected harmonic current (I_h) is zero in this case.

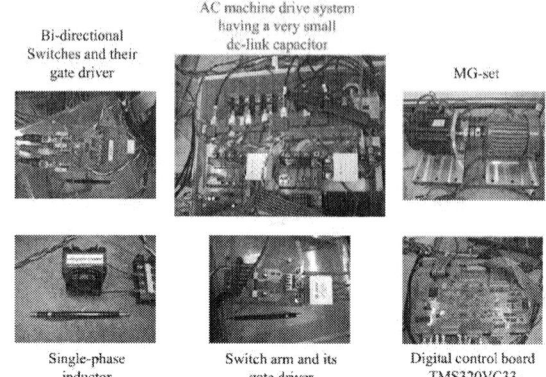

Figure 15. Experimental setup.

TABLE I
SYSTEM PARAMETERS

Input voltage	380Vrms
DC-link capacitance	$6.6\mu F$
Machine's rating	3.7kW
Bi-directional switches	50A, 1200V
One switch arm	50A, 1200V
Single phase inductor	5mH, 7Arms

The dc-link current (I_d) contains a dc-link voltage ripple component as well as mean dc value since the system is operating under the constant load condition. Also, the switch arm current (I_{dh}) is zero in this case. Lastly, the machine current forms sinusoidal at its rated operating frequency.

Fig. 17 shows the same waveforms where the proposed harmonic current injection has been applied. The injection current reference is obtained from (2) which forms quasi-triangular wave. The input ac current (I_u) forms almost sinusoidal wave due to the proposed harmonic current injection (I_h). The switch arm current (I_{dh}) contains an injected harmonic current component in this case. Lastly, the machine current is not affected by the proposed harmonic current injection method but controlled independently. Consequently, the harmonic contents of the input ac current (I_u) has been reduced considerably without affecting the performance of the machine control.

Fig. 18 shows that the frequency spectrum analysis results (FFT) of all harmonic currents of the above two cases. The results have been compared with the target harmonic current standard (IEC 61000-3-2). Fig. 18(a) shows the result for the case without harmonic current injection.

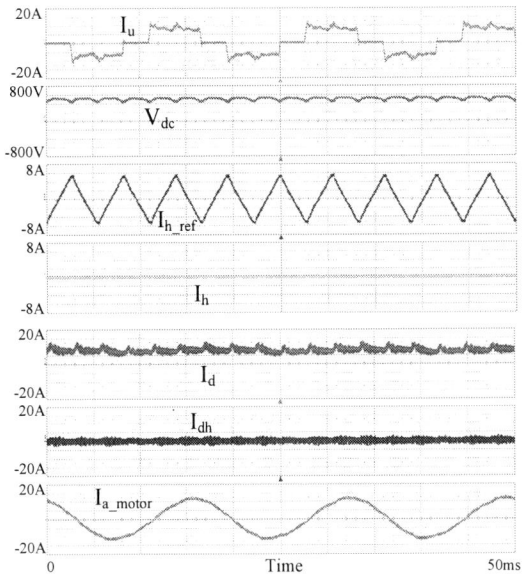

Figure 16. Experimental result for an ac machine drive system having a very small dc-link capacitor.

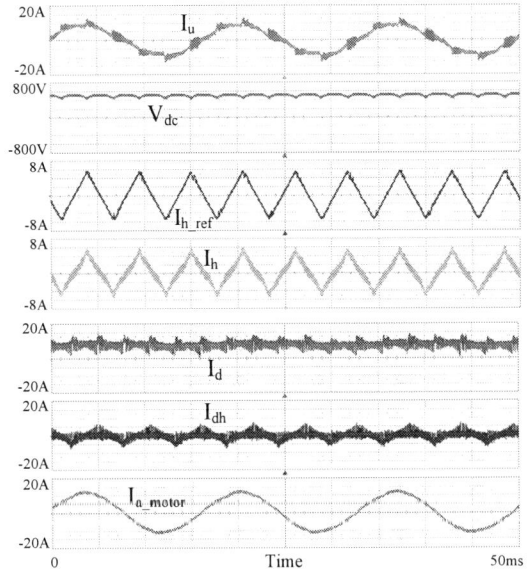

Figure 17. Experimental result for the proposed harmonic current injection circuit.

Almost all harmonic current orders are not satisfied with the standard. However, as shown in Fig. 18(b), all harmonic current orders of the input ac current have been suppressed conspicuously. Also, it is confirmed that the given harmonic current standard is satisfied by the proposed harmonic current injection method.

Fig. 19 and Fig. 20 are experimental results to verify the validation of the stability analysis. The harmonic current injection has been started at 0.2s abruptly while the drive system having a very small dc-link capacitor is operating at its 75% of the rated power as shown in Fig. 19. The dc-link voltage and the machine current are stable both before and after the injection even under abrupt harmonic current

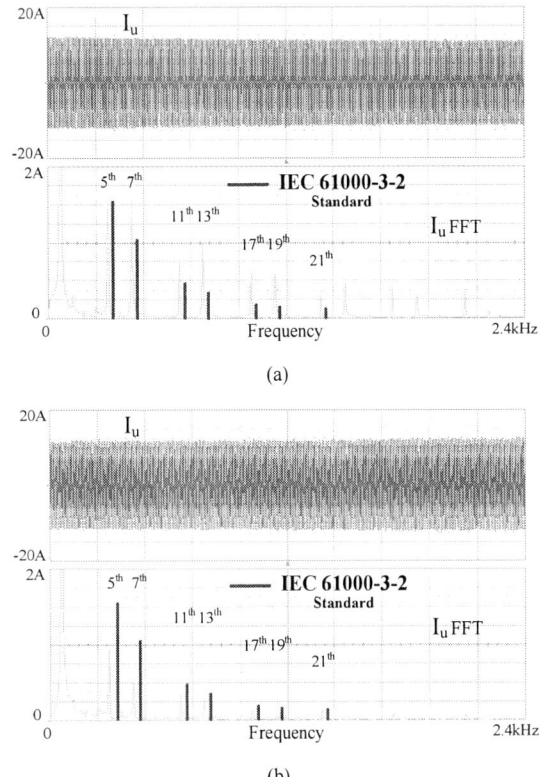

Figure 18. Frequency spectrum results for the input line current: (a) without harmonic current injection; (b) with proposed harmonic current injection.

injection because the dc-link capacitor has been designed based on (14) and the harmonic current control system satisfies (16). Fig. 20 shows another experimental result for the case that the load has been abruptly changed while the harmonic current is injecting. The system is also stable both before and after the abrupt load variation. Therefore, it is verified that the stability analysis in the previous section is valid.

VII. CONCLUSION

In this paper, a new circuit topology to reduce an input harmonic current of a three-phase ac machine drive system having a very small dc-link capacitor, has been proposed based on a harmonic current injection method. In order to minimize the additional cost and size of this harmonic current injection circuit, the proposed harmonic current injection circuit minimizes the size and the number of passive elements. Thus only a single phase inductor as a passive element has been used. Moreover, the current rating of the proposed injection circuit is 1/3 of that of the conventional current injection topologies. This also minimized the additional cost considerably.

This paper also presents an appropriate current control method for the proposed current injection circuit considering the voltage disturbance due to the switching of the bi-directional switches. Lastly, the stability of ac machine drive system including the proposed current injection circuit has been analyzed. By experimental results and their frequency analyses, it is confirmed that not only the harmonic current

978-1-4244-4782-4/10 $26.00 © 2010 IEEE   617

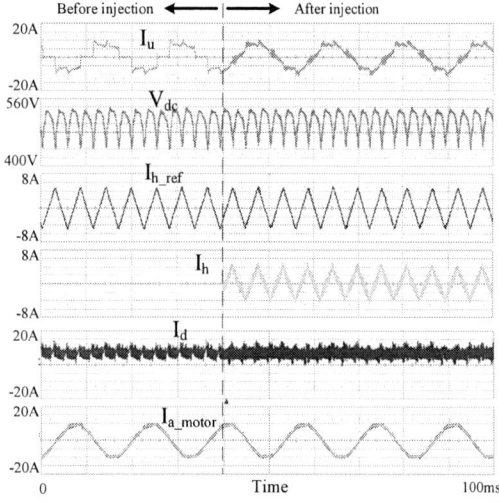

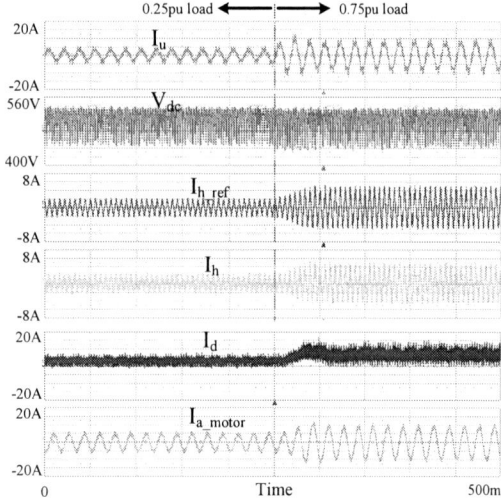

standard can be satisfied with the proposed harmonic current injection circuit and its control method, but also the stability analysis of the proposed drive system is valid.

REFERENCES

[1] B. K. Bose, D. Kastha, "Electrolytic capacitor elimination in power electronics system by high frequency active filter," in Conf. Rec. of 1991 IEEE IAS Annual Meeting, vol. 1, 1991.

[2] J. S. Kim, S. K. Sul, "New control scheme for AC-DC-AC converter without DC link electrolytic capacitor," in Proc. IEEE PESC 93, pp.300-306, 1993.

[3] M. Hinkkanen and J. Luomi, "Induction motor drives equipped with diode rectifier and small dc-link capacitance," IEEE Trans. Ind. Electron., vol. 55, no. 1, pp. 312-320, Jan. , 2008.

[4] K. Oietilainen, L. Harnefors, A. Petersson, et al., "DC-link stabilization and voltage sag ride-through of inverter drives," IEEE Trans. Ind. Electron., vol. 53, no. 4, pp. 1261-1268, Aug. 2006.

[5] Bon-Gwan Gu, Kwanghee Nam, "A DC-link capacitor minimization method through direct capacitor current control," IEEE Trans. on Industry Applications, vol.42, no. 2, pp. 573-581, March/April, 2006

[6] H. Yoo, S. Sul, H. Jang and Y. Hong, "Design of a variable speed compressor drive system for air-conditioner without electrolytic capacitor," IEEE 42nd IAS Annual meeting, pp. 305-310, Sep., 2007.

[7] Afroz M. Imam, "Condition Monitoring of Electrolytic Capacitors for Power Electronics Applications," Georgia Institute of Technology, Ph. D. Dissertation, 2007.

[8] Siyoung Kim, Seung-Ki Sul and T. A. Lipo, "AC/AC Power Conversion Based on Matrix Converter Topology with Unidirectional Switch," IEEE Trans. on Industry Applications, vol. 36, no. 1, pp. 139-145, Jan./Feb., 2000.

[9] IEC 61000-3-2 : 2005.11.

[10] IEC 61000-3-12 : 2004-11.

[11] N. Mohan, M. Rastogi, and R. Naik, "Analysis of a new power electronics interface with approximately sinusoidal 3-phase utility currents and a regulated dc output," IEEE Trans. on Power Delivery, pp. 540-546, Apr. 1993.

[12] N. Mohan, "A novel approach to minimize line-current harmonics in interfacing power electronics equipment with 3-phase utility systems," IEEE Trans. on Power Delivery, pp. 1395-1401, Jul. 1993.

[13] R. Naik, M. Rastogi, and N. Mohan, "Third-harmonic modulated power electronics interface with three-phase utility to provide a regulated DC output and to minimize line-current harmonics," IEEE Trans. on Industry Applications., vol. 31, pp. 598-602, May/June 1995.

[14] P. Pejovic, "A novel low-harmonic three-phase rectifier," IEEE Trans. on Circuits and Systems, vol. 49, pp. 955-965, July 2002.

[15] J. C. Salmon, "Operating a three-phase diode rectifier with a low-input current distortion using a series-connected dual boost converter" IEEE Trans. on Power Electronics, vol. 11, no. 4., pp. 592-603, July 1996.

[16] P. Pejovic and Z. Janda, "An Analysis of Three-Phase Low-Harmonic Rectifiers Applying the Third-Harmonic Current Injection," IEEE Trans. on Power Electronics, vol. 14, no. 3, pp. 397-407, May 1999.

[17] S. Kim, P. N. Enjeti, and P. Packebush, "A new approach to improve power-factor and reduce harmonic in a three-phase diode rectifier type utility interface," IEEE Trans. on Industry Applications, vol. 30, no. 6, pp. 1557-1564, Nov./Dec. 1995.

[18] P. Pejovic and Z. Janda, "Optimal Current Programming in Three-Phase High-Power Factor Rectifier based on two Boost Converters," IEEE Trans. on Power Electronics, vol. 13, no. 6, pp. 1152-1163, Nov. 1998.

[19] N. R. Raju, "A low-harmonic diode rectifier-fed drive using zero-sequence coupling between machine and grid," IEEE 42nd IAS Annual meeting, pp. 1582-1584, Sep., 2007.

[20] H. Yoo and S. Sul, "A novel approach to reduce line harmonic current for a three-phase diode rectifier-fed electrolytic capacitor-less inverter," IEEE APEC, 2009.

[21] J. Itoh and I. Ashida, "A novel three-phase PFC rectifier using a harmonic current injection method," IEEE Trans. on Power Electronics, vol. 23, no. 2., pp. 715-722, Mar. 2008.

[22] J. W. Kolar and F. C. Zach, "A Novel Three-Phase Utility Interface Minimizing Line current Harmonics of High-Power Telecommunications Rectifier Modules," IEEE Trans. on Industrial Electronics, vol. 44, no. 4, pp. 456-467, Aug. 1997.

[23] C. Qiao and K. M. Smedley, "Three-phase unity-power-factor star-connected switch(VIENNA) rectifier with unified constant-frequency integration control," IEEE Trans. on Power Electronics, vol. 18, no. 4, pp. 952-957, July 2003.

[24] R. P. Stratford, "Rectifier harmonics in power systems," IEEE Trans. on Industry Applications, vol. 1A-16, no. 2, pp. 271-276, Mar./Apr. 1980.

[25] F. Z. Peng, H. Akagi, and A. Nabae, "A new approach to harmonic compensation in power systems-A combined system of shunt passive and series active filters," " IEEE Trans. on Industry Applications, vol. 26. pp. 983-990, 1990.

[26] W. Lee and S. Sul, "DC-link voltage stabilization for reduced dc-link capacitor inverter," IEEE conf. ECCE 09, pp. 1740-1744, Sep. 2009.

State-Space Modeling, Analysis, and Implementation of Paralleled Inverters for Microgrid Applications

Chien Liang Chen, Jih-Sheng Lai, and Daniel Martin
Virginia Polytechnic Institute and State University
Future Energy Electronics Center
Blacksburg, VA 24060

Yuang-Shung Lee
Fu-Jen Catholic University
Dept. of Electronics Eng. and Applied Science and Eng. Inst.,
Taipei County 242, Taiwan

Abstract—The state-space model and implementation results of a power conditioning system are presented in this paper. Eigenvalues with different controller gains and load conditions for grid-tie mode and standalone mode are utilized to analyze the system stability. For standalone mode, higher controller gain and higher load resistance tend to make system more unstable. In grid-tie mode, higher controller gain is also found to make the system more unstable but load variation will not change the system stability much. Experimental and simulation results also verify the model. The state-space model is extended to a parallel-inverter system for investigation of the load variation and current-sharing controller effects to the system stability. A time-domain current ripple criterion is also suggested for light-load operation of the parallel-inverter microgrid system.

I. INTRODUCTION

Parallel inverters are widely used in uninterruptible power system (UPS) and microgrid systems because of its flexibility of system expansion using the existing power-conditioning system modules [1-5]. The current sharing control method of a paralleled inverter system is crucial to proper load distribution and system stability. In some reported large inverter-based microgrid systems, the droop control method was utilized [1,2] for load sharing without communication lines. The use of communication was found in some low-power-rated paralleled inverter systems, which utilized different balancing control schemes including centralized, master-slave, average-load sharing, and circular-chain controls to achieve better current sharing capability [3,4].

In order to extend the transmission distance, a load sharing scheme was proposed to transmit the dc quantity through a controller area network (CAN) bus among paralleled inverters [5]. The stability of a single inverter can be determined by the phase margin and gain margin of the designed compensated loop gains. This method, however, cannot determine the stability of the entire parallel-inverter system due to the possible interactions among different control loops and the current-sharing controller. In [6-8], the stability of paralleled power conditioning systems is analyzed by the impedance characteristic of input ports and output

ports. This method leads to complex measurements of the impedances for all the loads and the sources.

In [9-11], the stability of large-scale distributed generation systems was analyzed by the state-space model. This model is relatively easy to expand by combining single inverter state-space equations to build state-space equations for parallel inverters. Once the state-space model is constructed, many modern controller design techniques, such as pole assignment [12-14] and eigenvalue sensitivity [15,16] can be utilized to further optimize the system. Some studies are conducted to analyze the stability of the droop control or instantaneous active current sharing for a paralleled inverter system. However, the stability of a parallel system with average-current sharing control is rarely addressed.

In this paper, the state-space model is adopted to investigate the stability of the paralleled inverter system with the average current-sharing scheme described in [5]. The state-space model of a single inverter in both standalone and grid-tie mode will be derived. Based on the derived eigenvalues from the constructed model, the system stability is evaluated with different load and controller conditions. With the average-current sharing controller, the complete parallel-inverter model is derived to investigate the system stability. A time-domain current criteria is also suggested for proper operation of the parallel-inverter microgrid system. Simulation and experiments are conducted to verify the model and criteria.

II. SYSTEM CONFIGURATION OF PARALLEL MULTI-INVERTER SYSTEM

Figs. 1(a) and 1(b) show the hardware and control block diagram of the microgrid system running in islanding operation where all inverters in the system supply the load together when the grid is not available. In this system, one of the inverters needs to operate in dual-loop control and serve as a voltage source, while the rest inverters operate in single current loop to share the current properly. The selection of the inverter running in dual-loop mode or single-loop mode is determined by the upper level controller through the CAN bus.

978-1-4244-4782-4/10 $26.00 © 2010 IEEE

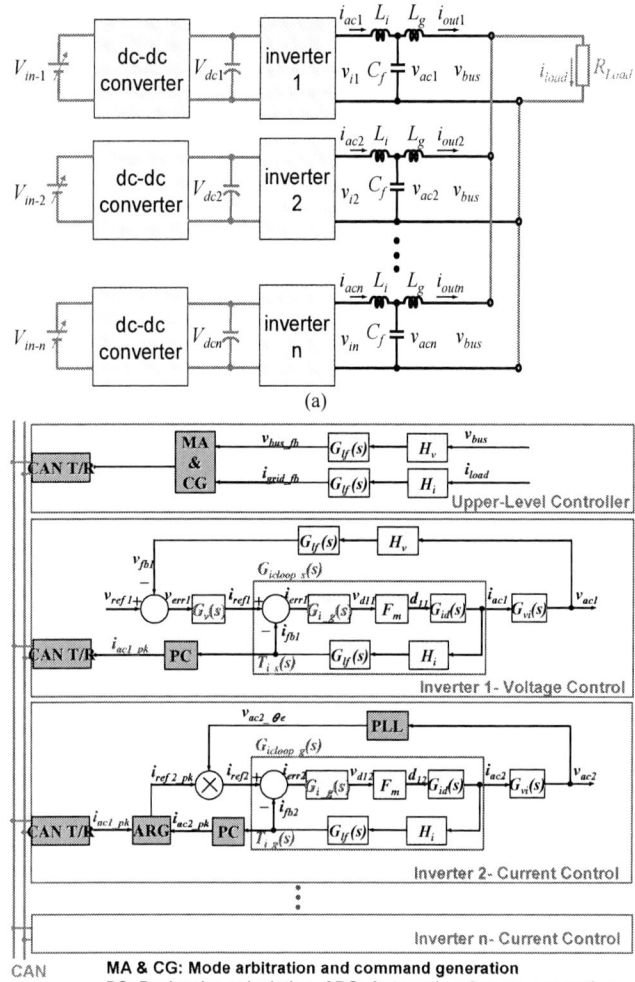

(b)

MA & CG: Mode arbitration and command generation
PC: Peak value calculation, ARG: Automatic reference generation

Fig. 1. The parallel inverter system in standalone mode (a) system configuration, and (b) control block diagram.

III. STATE-SPACE MODELING

A. Model of Single-Inverter in Grid-Tie Mode

Fig. 2 shows a single inverter running in grid-tie mode and standalone mode. As shown in Fig. 2(a), a LCL filter is commonly used for filtering the output current ripple in the grid-tie mode. The resistances of the LCL filters and the grid-impedance are also modeled here for a general case. As expressed in (1), this LCL filter makes the first 3rd order dynamic equations in the grid-tie inverter system.

$$\begin{bmatrix} L_i & 0 & R_c C_f \\ 0 & L_g + L_s & -R_c C_f \\ 0 & 0 & C_f \end{bmatrix} \begin{bmatrix} \frac{di_{ac}}{dt} \\ \frac{di_g}{dt} \\ \frac{dv_c}{dt} \end{bmatrix} = \begin{bmatrix} R_{Li} & 0 & -1 \\ 0 & -R_{Lg} - R_{Ls} & 1 \\ 1 & -1 & 0 \end{bmatrix} \begin{bmatrix} i_{ac} \\ i_g \\ v_c \end{bmatrix} + \begin{bmatrix} V_{dc} & 0 \\ 0 & -1 \\ 0 & 0 \end{bmatrix} \begin{bmatrix} d \\ v_s \end{bmatrix} \quad (1)$$

Three state variables i_{ac}, i_g and v_c are the inverter-side-inductor current, grid-side inductor current and capacitor voltage, respectively. The excitation signal d and the grid voltage v_s are the control gating signal and the grid voltage.

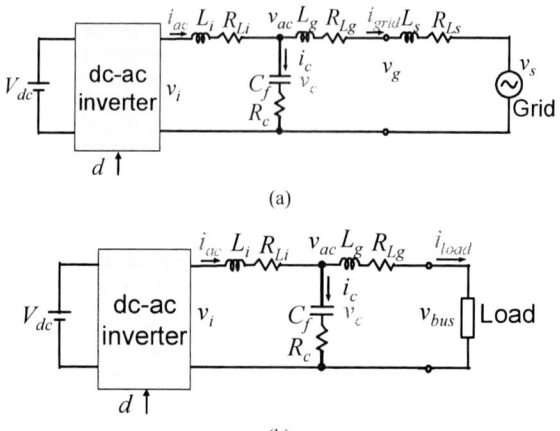

Fig. 2. Single inverter hardware configuration: (a) grid-tie mode, and (b) standalone mode.

Fig. 3(a) shows the control block diagram of the grid-tie inverter. With admittance compensation [5], the duty cycle v_d consists of the feedback duty cycle v_{d1} and admittance compensation term v_{d2} to obtain two corresponding gate signals d_1 and d_2 are expressed in (2).

$$d = F_m * v_d = F_m * (v_{d1} + v_{d2}) = d_1 + d_2 \quad (2)$$

where

$$v_{d1} = i_{err} * G_{i\_g}(s), \ v_{d2} = v_{ac} * H_v * G_c$$
$$= (v_c + R_c * (i_{ac} - i_g)) * H_v * \frac{1}{H_v V_{dc} F_m}$$

Fig. 3(b) is the simplified equivalent circuit with ideal admittance compensation. Here H_i and $G_{lf}(s)$ are the current feedback gain, and the hardware low-pass filters' transfer function which contribute to another third-order differential equation shown in (3).

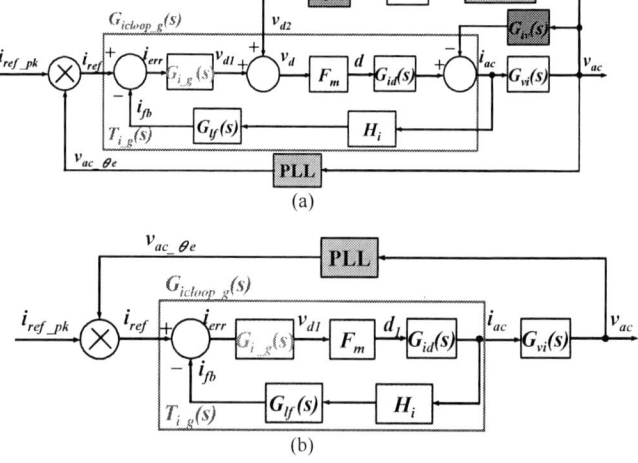

Fig. 3. Control block diagrams of a single inverter in grid-tie mode with admittance compensation (a) complete block diagram, and (b) simplified block diagram.

$$\begin{bmatrix} \dfrac{di_{fb}}{dt} \\ \dfrac{di_{HA1}}{dt} \\ \dfrac{di_{HA2}}{dt} \end{bmatrix} = \begin{bmatrix} -\omega_{HWF} & \omega_{HWF} & 0 \\ 0 & -\omega_{HWF} & \omega_{HWF} \\ 0 & 0 & -\omega_{ANF} \end{bmatrix} \begin{bmatrix} i_{fb} \\ i_{HA1} \\ i_{HA2} \end{bmatrix} + \begin{bmatrix} 0 \\ 0 \\ \omega_{ANF} H_i i_{ac} \end{bmatrix} \quad (3)$$

Here H_i is a pure gain while the $G_{lf}(s)$ contains a second-order hardware filter and a first-order anti-aliasing filter. Variables ω_{HWF} and ω_{ANF} are the cut-off angular frequencies for the hardware filter and the anti-aliasing filter, respectively. Currents i_{HA1} and i_{HA2} are the two intermediate states of the hardware filters. In Fig. 3(b), the $G_{i\_g}(s)$ and F_m are the current controller in grid-tie mode and the DSP PWM modulation gain, respectively. $G_{i\_g}(s)$ is designed to be a proportional-resonant (PR) controller to achieve a low steady-state-error output [5]. As shown in (4), the PR controller can be represented as a second-order equation.

$$\begin{bmatrix} \dfrac{di_{pr1}}{dt} \\ \dfrac{di_{pr2}}{dt} \end{bmatrix} = \begin{bmatrix} 0 & 1 \\ -\omega_1^2 & -2\omega_c \end{bmatrix} \begin{bmatrix} i_{pr1} \\ i_{pr2} \end{bmatrix} + \begin{bmatrix} 0 & 0 \\ 1 & -1 \end{bmatrix} \begin{bmatrix} i_{ref} \\ i_{fb} \end{bmatrix} \quad (4)$$

$$v_{d1} = k_{pg} * (i_{ref} - i_{fb}) + 2\omega_c k_{rg} * i_{pr2}$$

$$i_{ref} = i_{ref\_pk} * v_{ac\_\theta e} = i_{ref\_pk} * \sin\theta_e$$

Here k_{pg} and k_{rg} are the proportional and resonant gains for the grid-tie mode controller, and ω_c and ω_1 are the equivalent bandwidth, and fundamental angular frequency, respectively. State variables i_{pr1} and i_{pr2} are two intermediate variables that represent the state-space equation in canonical form. i_{ref}, i_{fb}, i_{ref_pk} and $v_{ac\_\theta e}$ are the current reference, current feedback, peak current reference and synchronized unity sine term obtained from the phase-lock loop (PLL) [13]. As shown in Fig. 4, a PLL block with the peak-value calculation (PC) is designed using an all-pass filter and D-Q transformation.

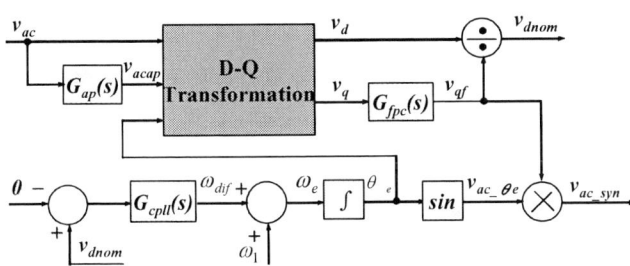

Fig. 4. Control block diagrams of the phase-lock loop with the peak-value calculation.

The d-axis and q-axis components are calculated by the estimated angle, θ_e, and filter capacitor voltage, v_{ac}, and output of all-pass filter, v_{acap} as shown in (5).

$$\begin{bmatrix} v_d \\ v_q \end{bmatrix} = \begin{bmatrix} \cos\theta_e & \sin\theta_e \\ -\sin\theta_e & \cos\theta_e \end{bmatrix} \begin{bmatrix} v_{ac} \\ v_{acap} \end{bmatrix} = \begin{bmatrix} 0 \\ -V_m \end{bmatrix} \text{if } \theta_e = \theta \quad (5)$$

where

$$v_{ac} = V_m \sin\theta, \; v_{acap} = \frac{\omega_1 - s}{\omega_1 + s} v_{ac} = -V_m \cos\theta, \; v_d = -V_m \sin(\theta_e - \theta) \cong V_m(\theta - \theta_e)$$

The output of the PLL/PC block is the synchronized term v_{ac_syn} which is obtained by multiplying the calculated magnitude v_{qf} term and synchronized unity sine term $v_{ac\_\theta e}$. Since the synchronized term v_{ac_syn} will be used for the admittance compensation, a ripple-free magnitude v_{qf} is desirable to ensure smooth operations. A low-pass filter shown in (6) can be designed to damp the double-fundamental-frequency ripple caused by (5).

$$\frac{dv_{qf}}{dt} = -\omega_{pc} v_{qf} - \omega_{pc} v_q \quad (6)$$

Here ω_{pc} is the cut-off frequency of the peak-value-calculation filter. Notice that equation (6) is a low-pass filter with a negative sign which makes the state variable v_{qf} in steady state equal to the positive voltage magnitude, V_m. In order to design a universal PLL with wide input voltage range, the normalized signal v_{dnom} is utilized as the feedback signal for the PLL. With an integrator as the PLL plant, a controller $G_{cpll}(s)$ is designed to have a 28-Hz cross-over frequency with a $36°$ phase margin. A second-order equation for the PLL controller and plant is shown in (7).

$$\begin{bmatrix} \dfrac{d\omega_{dif}}{dt} \\ \dfrac{d\theta_e}{dt} \end{bmatrix} = \begin{bmatrix} -\omega_{pll} & k_{pll}\omega_{pll} \\ 1 & 0 \end{bmatrix} \begin{bmatrix} \omega_{dif} \\ v_{dnom} \end{bmatrix} + \begin{bmatrix} 0 & 0 \\ 1 & 0 \end{bmatrix} \begin{bmatrix} \omega_1 \\ \theta \end{bmatrix}$$

$$= \begin{bmatrix} -\omega_{pll} & -k_{pll}\omega_{pll} \\ 1 & 0 \end{bmatrix} \begin{bmatrix} \omega_{dif} \\ \theta_e \end{bmatrix} + \begin{bmatrix} 0 & k_{pll}\omega_{pll} \\ 1 & 0 \end{bmatrix} \begin{bmatrix} \omega_1 \\ \theta \end{bmatrix} \quad (7)$$

where

$$v_{dnom} = \frac{v_d}{v_{qf}} = \frac{V_m \sin(\theta - \theta_e)}{V_m} = \theta - \theta_e \text{ if } \theta \cong \theta_e$$

Here ω_{pll}, k_{pll} are the cut-off frequency and gain of the PLL controller. Equations (1) to (7) describe the dynamic behavior of a single phase inverter in grid–tie mode. The above functions can be linearized and the small-signal perturbation can be applied to derive the small signal model of the grid-tie inverter shown in (8).

$$\Delta \dot{X}_g = A_g \Delta X_g + B_g \Delta U_g \quad (8)$$

where

$$\Delta X_g = \begin{bmatrix} \Delta i_{ac} \Delta i_g \Delta v_c \Delta i_{fb} \Delta i_{HA1} \Delta i_{HA2} \Delta i_{pr1} \Delta i_{pr2} \Delta\omega_{dif} \Delta\theta_e \Delta v_{qf} \end{bmatrix}^T$$

$$\Delta U_g = \begin{bmatrix} \Delta i_{ref\_pk} \Delta v_s \end{bmatrix}^T$$

Here X_g, A_g, B_g and U_g are the state variable, state matrix, input matrix and system input to the grid-tie inverter system. In order to find potential instability causes of the constructed model, the complete system eigenvalues are plotted in figure 5(a) and 5(b). With the designed proportional gain $k_{pg} = 0.78$, Fig. 5(a) shows that none of the eigenvalues are located in the right-hand plane, and the system is stable. Notice that the system eigenvalues keep at almost the same positions when the current reference magnitude i_{ref_pk} changes from 20% to full load, which suggests that the stability of grid-tie mode is not much related to the load change. By varying k_{pg}, however, it can be observed from Fig. 5(b) that the system becomes unstable when k_{pg} exceeds 16.

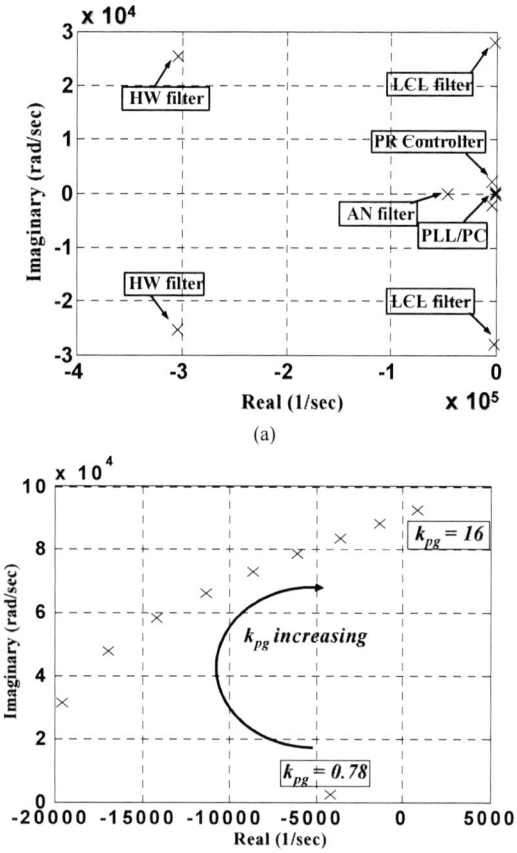

(a)

(b)

Fig. 5. Eigenvalue of grid-tie inverter (a) complete eigenvalue for $k_{pg} = 0.78$, and (b) the critical eigenvalue changes with varying k_{pg} from 0.78 to 16.

B. Model of Single-Inverter in Standalone Mode

Fig. 2.(b) shows the hardware configuration of a single inverter in standalone mode. If the load is only a pure resistor with resistance R_{load}, then the plant equation can be expressed by a third-order system, as shown in (9).

$$\begin{bmatrix} L_i & 0 & R_cC_f \\ 0 & L_g & -R_cC_f \\ 0 & 0 & C_f \end{bmatrix} \begin{bmatrix} \frac{di_{ac}}{dt} \\ \frac{di_{load}}{dt} \\ \frac{dv_c}{dt} \end{bmatrix} = \begin{bmatrix} R_{Li} & 0 & -1 \\ 0 & -R_{Lg}-R_{load} & 1 \\ 1 & -1 & 0 \end{bmatrix} \begin{bmatrix} i_{ac} \\ i_{load} \\ v_c \end{bmatrix} + \begin{bmatrix} V_{dc} \\ 0 \\ 0 \end{bmatrix} [d] \quad (9)$$

Fig. 6 shows that an outer voltage loop is used to regulate the output voltage while an inner current loop is adopted to damp the LCL resonance poles. Fig. 6 (a) shows the complete model of the inverter running in standalone mode while Fig. 6(b) shows the simplified model if the admittance compensation is applied. As shown in (10), the current controller in standalone mode $G_{i\_s}(s)$ is designed as a first-order low-pass filter to damp the LCL resonance.

$$\frac{dv_{d1}}{dt} = -\omega_{SWF} * v_{d1} + 0.5\omega_{SWF} * i_{err} \quad (10)$$

The voltage loop also contains a third-order filter $G_{lf}(s)$ with the similar equation expressed in (3) except that the

currents variables change to the voltage variables. In order to achieve high loop gain at the fundamental frequency to minimize the steady-state error, the voltage controller $G_v(s)$ is also designed to be a PR which leads to a second-order equation that has a similar form in (4) except that the current variables change to voltage ones. The current reference generation can be represented in (11).

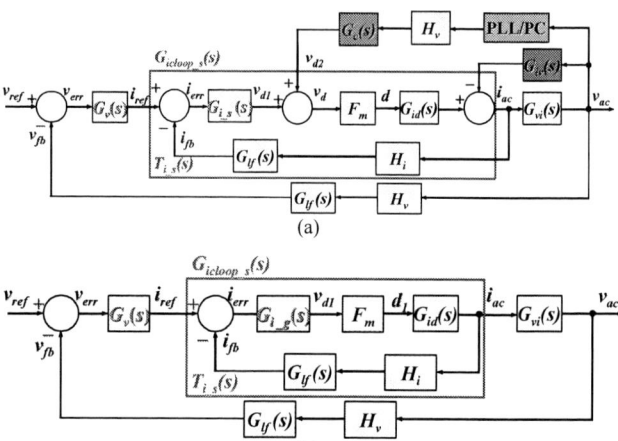

(a)

(b)

Fig. 6. Control block diagrams of a single inverter in standalone mode with admittance compensation (a) complete block diagram, and (b) simplified block diagram.

$$i_{ref} = k_{ps} * (v_{ref} - v_{fb}) + 2\omega_c k_{rs} * v_{pr2} \quad (11)$$

Here v_{ref} is the voltage reference. The functions, (10) and (11), can be linearized and the small-signal perturbations can be applied to derive the small-signal model of the standalone-mode inverter shown in (12).

$$\Delta \dot{X}_s = A_s \Delta X_s + B_s \Delta U_s \quad (12)$$

where

$$\Delta X_s = \begin{bmatrix} \Delta i_{ac} \Delta i_{load} \Delta v_c \Delta i_{fb} \Delta i_{HA1} \Delta i_{HA2} \Delta v_{d1} \Delta v_{fb} \\ \Delta v_{HA1} \Delta v_{HA2} \Delta v_{pr1} \Delta v_{pr2} \Delta \omega_{dif} \Delta \theta_e \Delta v_{qf} \end{bmatrix}^T$$

$$\Delta U_s = \begin{bmatrix} \Delta v_{ref} \end{bmatrix}^T$$

Here X_s, A_s, B_s and U_s are the state variable, input matrix and system input corresponding to the inverter in standalone mode. The complete system eigenvalues are shown in Fig. 7 to validate the state-space model. Fig. 7(a) shows that the system is stable with the designed proportional gain $k_{ps} = 0.02$. With varying k_{ps} and R_{load}, Fig. 7(b) suggests that the system become more unstable when R_{load} increases or when k_{ps} increases. With $k_{ps} = 2$, the seventh points on Fig. 7(b) indicates that the system is stable for $R_{load} = 13.5 \Omega$, but not for $R_{load} = 27 \Omega$.

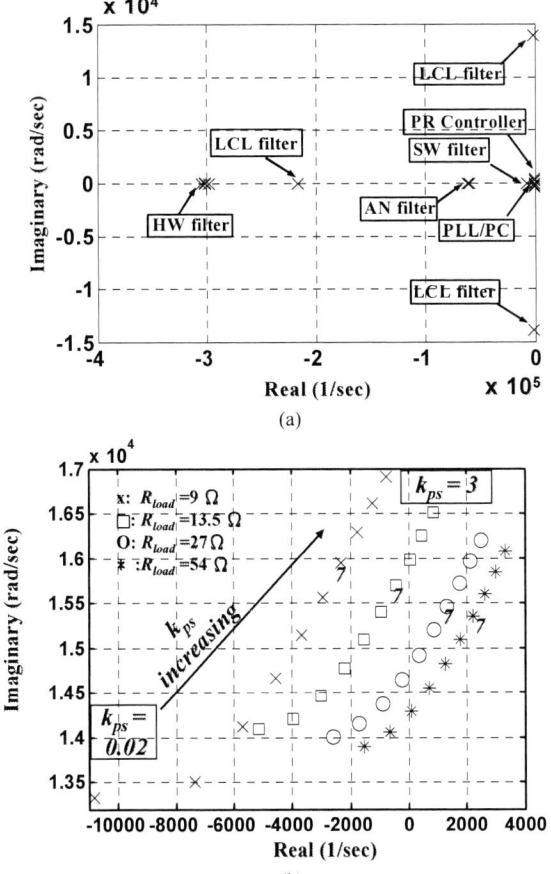

(a)

(b)

Fig. 7. Eigenvalue of standalone inverter (a) complete eigenvalue for k_{pg} = 0.02, and (b) the critical eigenvalue changes with the varying proportional gain k_{pg} and load resistance R_{load}.

C. Model of Parallel Inverters with the Current Sharing Controller

As shown in Fig. 1(a), the model of the mircogrid system in islanding mode or parallel-inverter operation mode can be obtained by combining the derived equations above sections and the current sharing controller. In order to extend the transmission distance, the transmitted signal is a frequency-decoupled current magnitude [5]. The frequency information does not need to be transmitted because it will be automatically tracked by the PLL. As shown in Fig. 1(b), a PC block is used to calculate the frequency-decoupled current magnitude information for the current sharing.

Assume that the load is only a pure resistance, then estimated phases in voltage PLL can be used to calculate the current magnitude similar to the approach shown in (6) except the input becomes currents, i_{ac1} and i_{ac2}. After the D-Q transformations, two low-pass filters with similar expression in (7) are utilized to filter the double-fundamental-frequency ripples. Once the transmitted current magnitude of the voltage-controlled inverter (i_{out1_pk}) and the calculated magnitude of the current-controlled inverter (i_{out2_pk}) are obtained, the automatic reference generation (ARG) adjusts the current reference magnitude of the current-controlled

inverter i_{ref2_pk} to share the inverter currents. The ARG control algorithm can be implemented digitally as described in [5] or a simple proportional-integral (PI) controller in (13).

$$i_{ref2\_pk} = i_{ac1\_pk} + i_{ref2\_offset},$$

$$\text{where } i_{ref2\_offset} = (i_{ac1\_pk} - i_{ac2\_pk}) * \left[\frac{sT_w}{1+sT_w}(k_{pcs} + \frac{1}{s}k_{ics}) \right] \quad (13)$$

Here i_{ref2_offset} is the offset of the current reference. T_w, k_{pcs}, k_{pis} are the washout-term time constant, proportional gain and integral gain of the PI controller. By combining state-state equations in sections A, B, and C and applying the perturbations to these equations, the complete small-signal model of the parallel-inverter system can be derived in (14).

$$\Delta \dot{X}_i = A_i \Delta X_i + B_i \Delta U_i \quad (14)$$

where

$$\Delta X_i = \Big[\Delta i_{ac1} \Delta i_{g1} \Delta v_{c1} \Delta i_{fb1} \Delta i_{HA11} \Delta i_{HA21} \Delta v_{d1} \Delta v_{fb1} \Delta v_{HA11} \Delta v_{HA21}$$

$$\Delta v_{pr11} \Delta v_{pr21} \Delta \omega_{dif1} \Delta \theta_{e1} \Delta v_{qf1} \Delta i_{ac2} \Delta i_{g2} \Delta v_{c2} \Delta i_{fb2} \Delta i_{HA12}$$

$$\Delta i_{HA22} \Delta i_{pr12} \Delta i_{pr22} \Delta \omega_{dif2} \Delta \theta_{e2} \Delta v_{qf2} \Delta i_{ac1\_pk} \Delta i_{ac2\_pk} \Delta i_{ref2\_pk} \Big]^T$$

$$\Delta U_i = \Big[\Delta v_{ref} \Big]^T$$

Here X_i, A_i, B_i and U_i are the state variable, state matrix, input matrix, and system input correspondingly of the microgrid system in islanding mode. Fig. 8(a) shows that the system is stable with the designed controller gains at about 16% load or $R_{load} = 27\ \Omega$. Fig. 8(b) suggests that the system stays stable from 8% load ($R_{load} = 54\ \Omega$) to 100% load ($R_{load} = 4.5\ \Omega$). However, for very light load cases, for example, at 0.1% load or $R_{load} = 5\ k\Omega$, the inductor current apparently runs into discontinuous conduction mode (DCM), and the frequency-domain analysis using the above average models can no longer predict the stability, but Fig. 8(b) still suggests that the system is stable. Such a wrong prediction should be corrected by additional time-domain constraint on current ripple magnitude, as indicated in (15).

$$\Delta I_{Li} \leq I_{load}$$

$$\text{where } \Delta I_{Li} = \frac{\Delta V_{Li}}{L_i} * \Delta t \quad (15)$$

Here ΔI_{Li}, ΔV_{Li} are the ripple current and voltage applied to the inverter-side inductor L_i, and Δt is the time when ΔV_{Li} applies to L_i during switching cycle. Under very light load cases, a relatively large circulating current occurs and results in system instability.

Another observation is that the eigenvalues in Fig. 8 stay almost unchanged regardless of the variations of current-sharing controller parameters k_{pcs} and k_{pis}. As shown in (13), i_{ref_offset} is close to zero because i_{ac1_pk} almost equals i_{ac2_pk} during the steady state. However, the current-sharing controller parameters, k_{pcs} and k_{pis}, will change the dynamic response when the output current magnitudes, i_{ac1_pk} and i_{ac2_pk}, are not equal.

978-1-4244-4782-4/10 $26.00 © 2010 IEEE

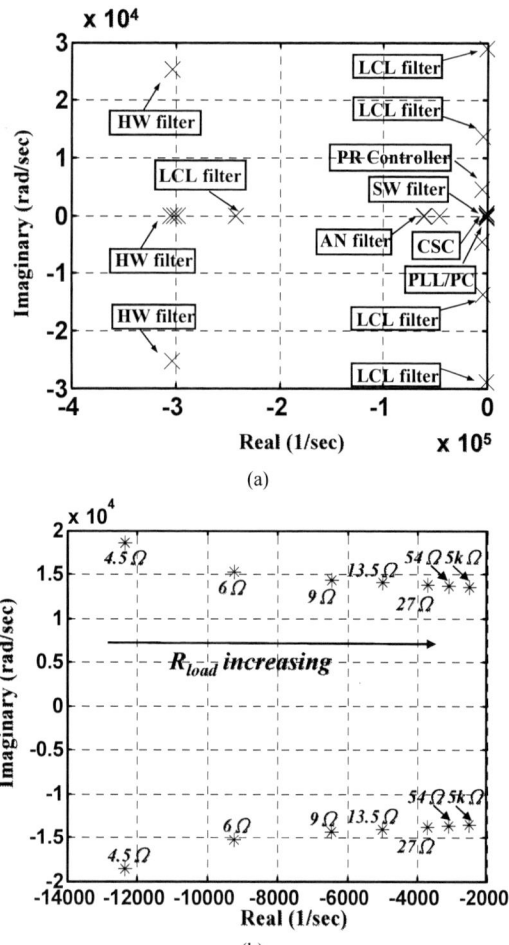

(a)

(b)

Fig. 8 Eigenvalues of the microgrid system in islanding mode (a) complete eigenvalues for $R_{load} = 27\ \Omega$, and (b) the critical mode eigenvalue changes with varying the load resistance, R_{load}.

IV. IMPLEMENTATION RESULTS

Fig. 9(a) and 9(b) shows the experimental results of two paralleled inverters running under steady-state standalone mode and grid-tie mode at 3.2kW condition. Waveforms indicate that the system is stable under both operation modes with the designed controllers.

Fig. 10 shows the simulation and experimental results of two paralleled inverters running under standalone-mode dynamic load step with load changing from $R_{load} = 27\ \Omega$ to $R_{load} = 13.5\ \Omega$ using $k_{ps} = 0.02$ and $k_{rs} = 12$. The waveforms show that the system is stable with this set of controller parameters. Both simulation and experimental results agree very well, and the system stability is well predicted by the eigenvalue analysis shown in Fig. 7.

To further verify that the system can run into unstable with an increased proportional gain, as predicted in Fig. 7, the system was simulated using $k_{ps} = 2$ and $k_{rs} = 12$. Simulation results shown in Fig. 11 indicate that initially when the load is 27 Ω, the system is unstable. With the same controller, when the load is increased to $R_{load} = 13.5\ \Omega$, the system

becomes stable. This simulation result verifies the validity of the eigenvalue analysis shown in Fig. 7.

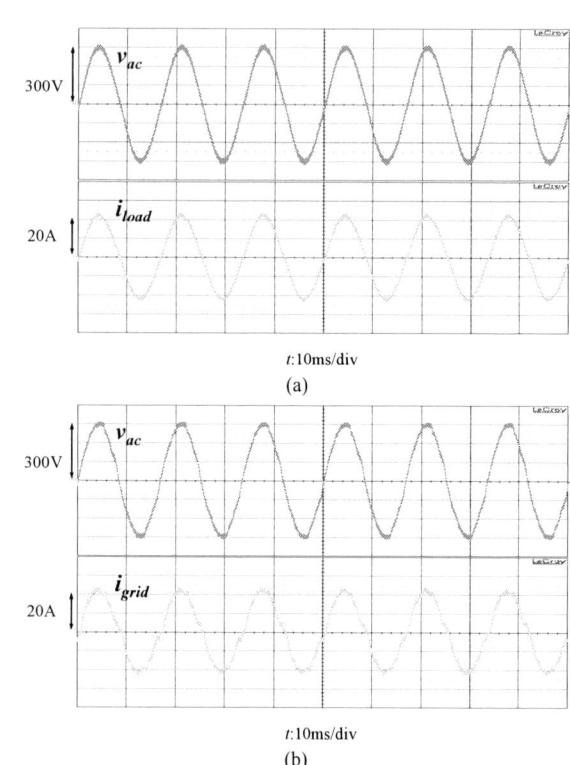

(a)

(b)

Fig. 9. Steady-state inverter test results (a) standalone test at 3.2 kW, and (b) grid-tie test at 3.2 kW.

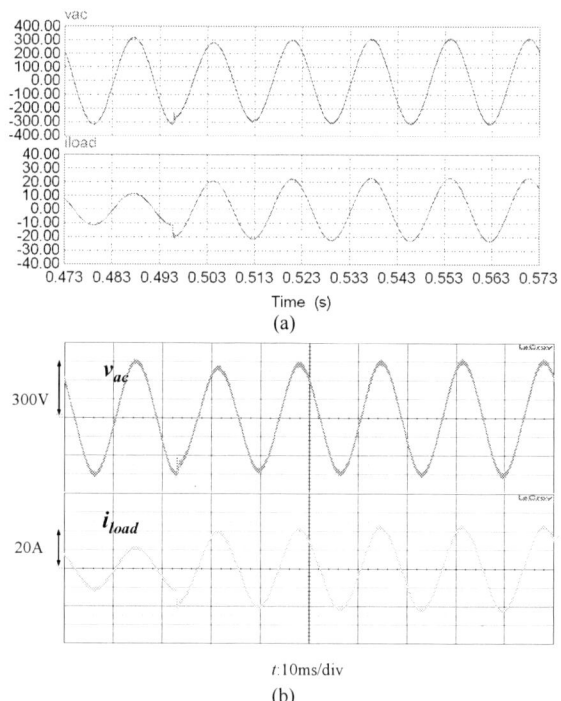

(a)

(b)

Fig.10. Standalone mode load-step test from $R_{load} = 27\ \Omega$ to $R_{load} = 13.5\ \Omega$ with $k_{ps} = 0.02$ and $k_{rs} = 12$: (a) simulation result, (b) experimental result.

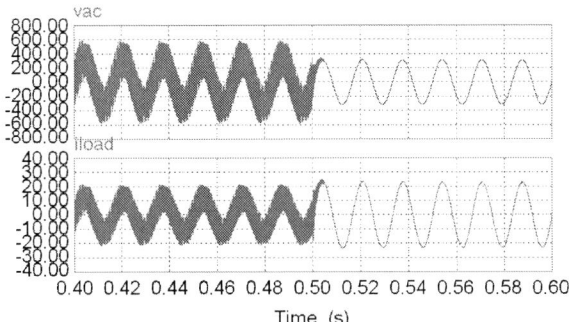

Fig. 11. Simulation result for standalone mode load-step test from $R_{load} = 27$ Ω to $R_{load} = 13.5$Ω with $k_{ps} = 2$ and $k_{rs} = 12$.

Fig. 12 shows the simulation results for grid-tie test with different proportional gain k_{pg}. As can be seen from the waveform, the system is stable with $k_{pg} = 0.78$ but becomes unstable with $k_{pg} = 16$. This time-domain simulation result again agrees with the frequency-domain prediction shown in Fig. 5.

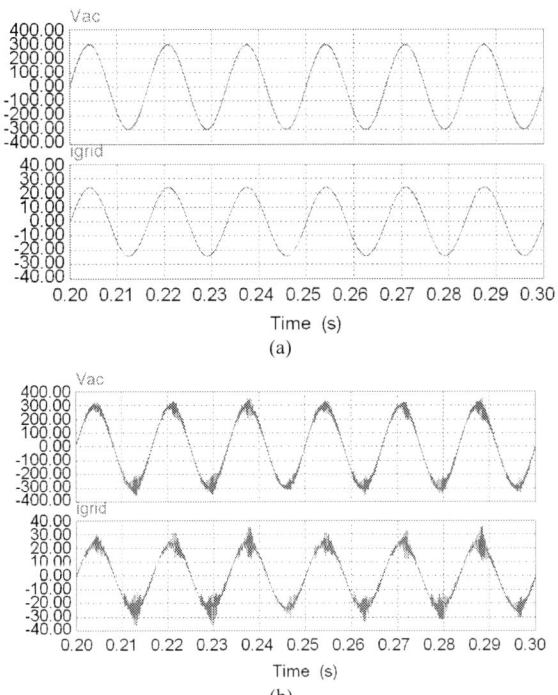

Fig. 12. Grid-tie simulation mode at 3.2kW with different controller gains: (a) $k_{pg} = 0.78$ and $k_{rg} = 97.5$, (b) $k_{pg} = 16$ and $k_{rg} = 97.5$.

Fig. 13 shows the experimental results of the mircogrid system under islanding mode at light load ($R_{load} = 54$ Ω) and heavy load ($R_{load} = 6$ Ω), respectively. Waveforms suggest that the system is stable within this load range which agrees with the eigenvalue analysis in Fig. 8. Fig. 14 shows the simulation results of the mircogrid system under islanding mode from very light load ($R_{load} = 5$ kΩ) to middle load ($R_{load} = 27$ Ω). Results show that the parallel-inverter system has a large circulating current with $R_{load} = 5$ kΩ and gradually goes back to normal after the load switches to $R_{load} = 27$ Ω. When

the load is too light and makes the inequality in (15) not satisfied, the inductor current on L_i will goes into discontinuous conducting mode which makes the capacitor voltage v_{ac} in individual inverter distorted. Due to the asynchronous PWM switching on different DSP chips, a large circulating current and instability will happen by the uneven voltage level shown in each individual inverter set. Fig. 15 shows the experimental results with lower dc-bus voltage and lower output-voltage command, which verifies the existence of this circulating issue under the very light load case.

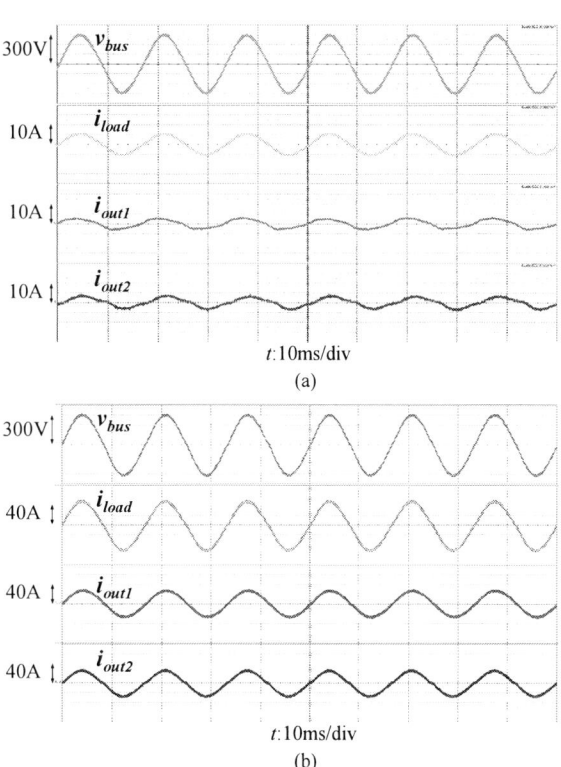

Fig. 13. Experimental results for the microgrid system in islanding mode: (a) $P_{out} = 800$ W ($R_{load} = 54$ Ω) and, (b) $P_{out} = 7.4$ kW ($R_{load} = 6$ Ω).

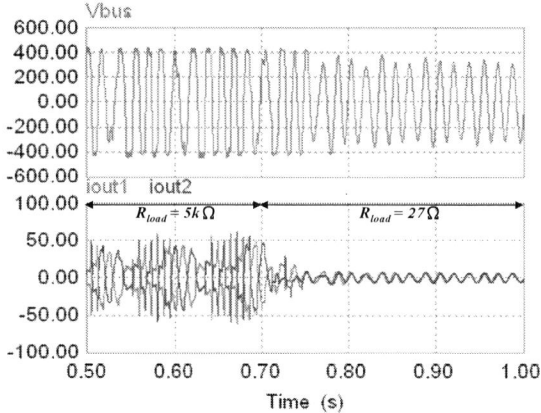

Fig. 14. Simulation results for the microgrid system in islanding mode from $R_{load} = 5$ kΩ to $R_{load} = 27$ Ω with $v_{bus\_cmd} = 208$ V_{rms} and $V_{dc} = 420$ V.

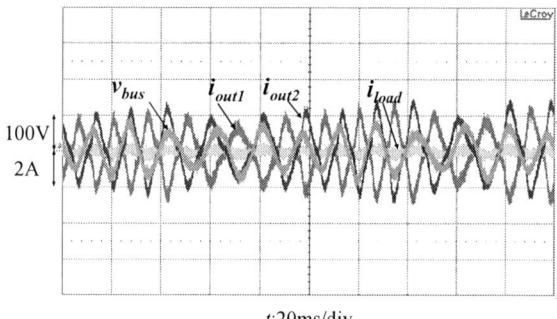

Fig. 15. Experimental results for the microgrid system in islanding mode with R_{load} = 5kΩ, v_{bus_cmd} = 35V_{rms} and V_{dc} = 210V.

V. CONCLUSION

The state-space model and implementation results of a power conditioning system operating in grid-tie mode and standalone mode were presented in this paper. Through the state-space analysis of the model, the following conclusions can be drawn:

1. Eigenvalue analysis for grid-tie mode

   The grid-tie system tends to be more unstable with increased proportional gain k_{pg} since the critical eigenvalue moves toward the right-half plane. The system eigenvalues do not change much with the increased current reference magnitude which implies the stability of grid-tie mode is not much related to the load change.

2. Eigenvalue analysis for standalone mode

   With the increased proportional gain k_{ps} and load resistance R_{load}, the standalone system becomes more unstable; the critical eigenvalue moves closer to the right-half plane.

3. Eigenvalue analysis for islanding mode

   With the increased load resistance, R_{load}, the microgrid system in islanding mode stays stable but the critical eigenvalue moves closer to the right-half plane. Changes in current-sharing controller gains have very little effects on the steady-state eigenvalue results.

Experimental and simulation results show the stable output waveforms in both grid-tie and standalone modes with the designed controller parameters. With higher controller gain in standalone mode, the simulation results show that the system is stable for heavier load condition but will become unstable in lighter load condition. This agrees with the results in the eigenvalue analysis. For grid-tie mode, the simulation results again verify the validation of the model by changing the controller gain.

In addition to the frequency-domain eigenvalue analysis, a time-domain current ripple analysis is needed to predict the stability under very light load condition where the inductor current runs into DCM. This paper has provided both simulation and experimental results sufficient to verify the predicted stability through frequency-domain analysis and time-domain current ripple calculation.

ACKNOWLEDGMENT

The authors would like to thank Dr. Y. R. Chang and his group from the Institute of Nuclear Energy Research, Atomic Energy Council, Taiwan for both financial and technical support for the project.

REFERENCES

[1] J. M. Guerrero, J. Metas, L. G. de Vicuna, M. Castilla, and J. Miret, "Decentralized control for parallel operation of distributed generation inverters using resistive output impedance," *IEEE Trans. Ind. Electron.*, vol. 54, pp. 994-1004, Apr. 2007.

[2] K. De Brabandere, B. Bolsens, J. Van den Keybus, A. Woyte, J. Driesen, and R. Belmans, "A voltage and frequency droop control method for parallel inverters," *IEEE Trans. Power Electron.*, vol. 22, pp. 1107-1115, July 2007.

[3] T. F. Wu, Y. K. Chen, and Y. H. Huang, "3C strategy for inverters in parallel operation achieving an equal current distribution," *IEEE Trans. Ind. Applicat.*, vol. 47, pp. 273-281, Apr. 2000.

[4] J. M. Guerrero, L. Hang, and J. Uceda, "Control of distributed uninterruptible power supply systems," *IEEE Trans. Ind. Electron.*, vol. 55, pp. 2845-2859, Aug. 2008.

[5] C. L. Chen, Y. B. Wang, J. S. Lai,, Y. S. Lee, and D. Martin, "Design of parallel inverters for smooth mode transfer microgrid applications ," *IEEE Trans. Power Electron.*, to be published.

[6] X. G. Feng, J. J. Liu, and F.C. Lee, "Impedance specifications for stable DC distributed power systems," *IEEE Trans. Power Electron.*, vol. 17, pp. 157-162, Mar. 2002.

[7] X. Sun, Y. S. Lee, and D. H. Xu, "Modeling, analysis, and implementation of parallel multi-inverter systems with instantaneous average-current-sharing scheme," *IEEE Trans. Power Electron.*, vol. 18, pp. 844-856, May 2003.

[8] L. Corradini, P. Mattavelli, M. Corradin, and F. Polo, "Analysis of parallel operation of uninterruptible power supplies loaded through long wiring cables," in *Proc. of IEEE APEC*, Feb. 2009, pp. 1276-1282.

[9] F. Katiraei, and M. R. Iravani, "Power management strategies for a microgrid with multiple distributed generation units," *IEEE Trans. Power Syst.*, vol. 21, pp. 1821-1831, Nov. 2006.

[10] N. Pogaku, M. Prodanovic, and T. C. Green, "Modeling, analysis and testing of autonomous operation of an inverter-based microgrid," *IEEE Trans. Power Electron.*, vol. 22, pp. 613-625, Mar. 2007.

[11] Y. Mohamed, and E. F. El-Saadany, "Adaptive decentralized droop controller to preserve power sharing stability of paralleled inverters in distributed generation microgrids," *IEEE Trans. Power Electron.*, vol. 23, pp. 2806-2816, Nov. 2008.

[12] B. S. Chen and Y. Y. Hsu, "A minimal harmonic controller for a STATCOM," *IEEE Trans. Ind. Electron.*, vol. 55, pp. 655-664, Feb. 2008.

[13] H. S. Bae, S. J. Lee, K. S. Choi, B. H. Cho, and S. S. Jang, "Current control design for a grid connected photovoltaic/fuel cell dc-ac inverter," in *Proc. of IEEE APEC*, Feb. 2009, pp. 1945-1950.

[14] N. Hirose, M. Iwasaki, M. Kawafuku, and H. Hirai, "Initial value compensation using additional input for semi-closed control systems," *IEEE Trans. Ind. Electron.*, vol. 56, pp. 635-641, Mar. 2009.

[15] J. Rommes, and N. Martins, "Computing large-scale system eigenvalues most sensitive to parameter changes, with applications to power system small-signal stability," *IEEE Trans. Power Syst.*, vol. 23, pp. 434-442, May 2008.

[16] X. Wang, W. Freitas, V. Dinavahi, and W. Xu, "Investigation of positive feedback anti-islanding control for multiple inverter-based distributed generators," *IEEE Trans. Power Syst.*, vol. 24, pp. 785-795, May 2009.

Efficiency Improvement of Grid-tied Inverters at Low Input Power Using Pulse Skipping Control Strategy

Haibing Hu, Wisam Al-Hoor, Nasser Kutkut, Issa Batarseh, Z. John Shen

School of Electrical Engineering and Computer Science
University of Central Florida, Orlando, FL 32826

Abstract- **Pulse skipping control strategy is applied to improve efficiency of grid-tied inverter at light load. To maximize the efficiency of pulse skipping operation mode, three key parameters are identified and can be optimized based on a loss model, which is developed to find the maximal efficiency points using three-dimension searching technique. To reduce the potential of pulsating on the power grid, the synchronization of the power pulse to the grid is addressed by varying the DC bus voltage window, namely varying the upper and lower DC voltage limits. A 200W prototype is setup to verify the proposed strategy. The experimental results show that the proposed strategy greatly improves the efficiency at light load and match the simulation results fairly well, thus verifying the validity of the proposed optimization method for pulse skipping.**

I. INTRODUCTION

Due to their environmental friendliness, renewable energy sources, generated from natural resources- such as sunlight, wind, tides and geothermal heat, have attracted great attention in recent years from scientific and government communities all over the world. . A fast growing number of distributed, grid-tied renewable energy generation is being deployed at residential sites, commercial buildings, and brown fields with generation power levels ranging from several hundred to several million Watts [1]. However, most renewable energy sources are intermittent in nature as their power levels are determined by sunlight, wind, and/or temperature, which can vary considerably throughout the day or the year.

In many renewable energy applications, a two-stage power conversion scheme is employed where the first DC-DC stage implements a maximum power point tracking (MPPT) algorithm while a second DC-AC stage controls the AC power and grid interface as shown in Figure 1. The energy generated from these renewable sources is directly fed to the grid. Since most renewable energy sources are intermittent, their output power, and subsequently the AC power generated, varies considerably and randomly. The fluctuation in the power generation level causes variations in the power stage efficiency as power converters do not process energy at their peak efficiency throughout the operating power range resulting in power losses (waste). Figure 2 shows the efficiency curve of a typical grid-tied inverter. As shown in Figure 2, at low input power levels, the inverter

efficiency drops dramatically resulting in greater power losses.

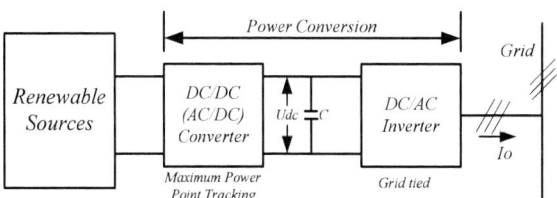

Figure 1: Two-stage power conversion architecture

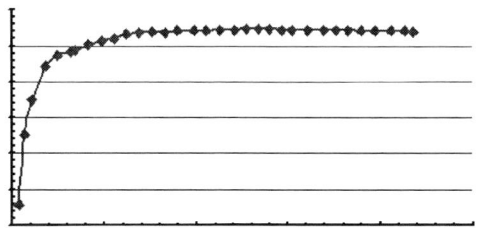

Figure 2: A typical efficiency curve of a grid-tied inverter

The overall efficiency of renewable energy systems is one of the key parameters that need to be optimized in order to reduce the overall system size and cost. Many techniques for improving the efficiency have been reported in the open literature, most of which focus on developing new topologies to achieve higher efficiencies [2-4]. The pulse skipping (burst mode) operation control method proposed in [5] aims at improving the inverter efficiency at low power levels without the need for any additional circuitry or hardware changes. The pulse skipping control strategy is aimed at operating the inverter in pulse mode where the power level of each pulse corresponds to typical operation at maximum efficiency while the average power corresponds to the low power level being generated. This can be achieved by storing energy in the intermediate DC bus capacitors of the two stage power conversion architecture. By operating the inverter at the maximum efficiency point for shorter intervals, the efficiency at low power levels can be greatly improved.

Pulse skipping control techniques are widely used in the DC-DC converters to obtain high efficiency at light loads [6-16]. Various publications address issues related to efficiency improvement at light load or stand-by mode [6-10], ripple control [11], as well as analysis and stability of pulse

978-1-4244-4782-4/10 $26.00 © 2010 IEEE

skipping control techniques [12-16]. Although pulse skipping mode of control for DC-AC inverters has been proposed in [5], no further research was reported on how to optimize the key parameters for pulse skipping operation. This paper presents a detailed analysis and optimization of pulse skipping operation mode for grid-tied DC-AC inverters.

II. PULSE-SKIPPING CONTROL STRATEGY

When the input power generated by a renewable energy source is low, the current injected into the grid by the DC-AC grid tie inverter will be low as well. In this case, the grid-tied inverter operates at very low efficiencies due to the light load condition. Since most grid tied inverters employ a two-stage power conversion architecture with an intermediate high voltage capacitor bank to smooth out the intermediate DC bus voltage, the DC bus capacitors can serve as an energy storage tank to store power from the renewable source, which can then be fed to the grid. As shown in Fig. 3, when the input power drops below a certain level, the inverter stops feeding power into the grid continuously and enters the pulse skipping operation mode. During pulse skipping mode, the DC bus is charged and the DC-AC inverter is disabled. When DC bus voltage researches an upper limit, the DC-AC inverter starts converting the energy stored in the DC bus capacitors to the grid at its maximum efficiency operating point until the DC bus voltage drops below a certain lower limit. Figure 3 shows a one-cycle pulse skipping mode of operation where the energy stored in the DC bus capacitor is enough to generate a single AC power pulse.

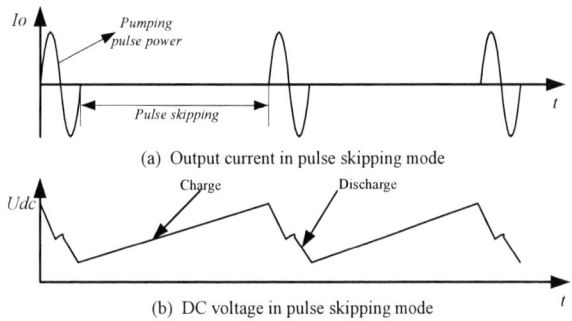

(a) Output current in pulse skipping mode

(b) DC voltage in pulse skipping mode

Figure 3: Pulse skipping operation mode

Several key parameters are closely related to the inverter efficiency during pulse skipping pulse operation mode. The first parameter is the power level below which pulse skipping operation mode is initiated to yield a boost in inverter efficiency. The second parameter is the amplitude of the power pulse that's fed into the grid, which corresponds to the maximal inverter efficiency over a given input power range. And finally, the impact of the variation of the DC bus voltage due to charging and discharging on the inverter efficiency is the third parameter.

These parameters need to be optimized in order to maximize the inverter efficiency during the pulse skipping operation mode. A loss-model based approach is proposed to address the related issues of parameter optimization.

III. LOSS-MODEL BASED OPTIMIZATION

A. Loss Model Selection

To fully explore the optimal efficiency when operating in pulse skipping mode, a loss model should be accurately established first. Losses in a grid-tied inverter usually consist of power device losses and reactive component losses. For reactive components, empirical formulas are used to calculate the losses, while for power devices many loss models have been previously proposed to achieve better accuracy and shorter simulation time [16]. Generally, three loss models can be identified. The first is a physical-based loss model which is verifiable by experimentation. However, such a model will be time-consuming to derive and verify. The second level is a circuit simulation-based loss model, which is widely used in the loss analysis due to its good trade-off between the accuracy and the simulation time. In typical simulation packages, such as Saber and Pspice, the power devices are described by some key parameters. However, such circuit simulation-based model is still not suitable for recursive optimization algorithms. The last is a mathematical-based loss model, which is based on loss expressions derived from equivalent circuits. Compared with the aforementioned models, this method is the fastest and most suitable for recursive optimization algorithms. As a result, a mathematical loss model is employed to fast calculate and compare the losses under different conditions. Note that in order to derive the optimal parameters for pulse skipping operation, the mathematical model needs to accurately model the inverter under different operating conditions, e.g. different loads, different DC bus voltages, different power levels, P_s, and different pulse power levels, P_{pulse}.

B. Mathematical Loss Model Derivation

The inverter losses primarily consist of power device losses, driver losses, and reactive component losses. Many loss mechanisms contribute to power devices' losses during inverter normal operation such as:

- Power device conduction loss
- Gate driving loss
- Diode recovery loss
- Diode conduction loss
- Switching loss
- Inductor loss

The accurate loss expressions for the aforementioned power devices and reactive components can be derived from datasheets provided by manufactures under normal operating conditions. Once these mathematical expressions are derived, a loss model can be easily built to calculate the efficiency using various software packages, e.g. Matlab or MathCad. In order to ensure the accuracy of the loss model, the model should be verified using experimental measurements under various operation conditions, e.g. different loads and different input voltage conditions. Next, the pulse skipping control strategy can be implemented and its impact on the inverter efficiency can be assessed using the developed loss model, which can be then used optimize the key parameters using recursive processing.

C. Optimization Algorithm

The proposed mathematical model, shown in Fig. 4, consists of two functional blocks, where the first block implements the pulse skipping strategy and the second block performs loss calculations. As the input power P_{input}, pulse power level P_{pulse}, and initial DC bus voltage $V_{dc\_initial}$ are varied, one can calculate the inverter efficiency at the varying operating points, from which the maximum efficiency point can be computed.

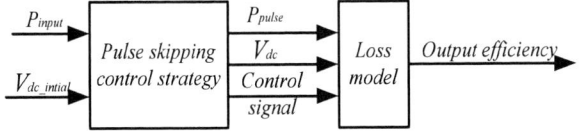

Figure 4: Model-based optimization process

Due to the complexity of the loss model in such applications, it is impractical to use an analytical approach to accurately find the optimal parameters. As such, a recursive search method is utilized to optimize the parameters. The optimization procedure is divided into four steps, namely:

1. The first step is to determine the circuit parameter operating boundaries according to power rating and specifications of the inverter.

2. The second step is to discretisize the search space since it uses numeric method to search optimal parameters.

3. The third step is to feed the input power P_{input} and initial DC voltage $V_{dc\_initial}$ into the pulse skipping control block, which in turn will generate the real-time DC bus voltage,

the pulse skipping enable signal. This signal corresponds to whether the inverter is in pulse skipping mode or not, and the power level of pulse.

4. Using the outputs of the pulse skipping control block as inputs to the loss model, the fourth step is to calculate the efficiency for one pulse skipping period base. By repeating steps three and four above to calculate all the discrete points, the maximal efficiency can be obtained.

D. Power Pulse Synchronization

In grid tied inverters, the power pulses need to be synchronized with the grid voltage zero-crossings as illustrated in Fig. 5. Due to the variation of the input power level, the start point and stop point of pulse skipping will not be fixed. Therefore, two DC bus voltage window limits, an upper window and a lower window, are used to ensure that the pulse start and stop instants are synchronized with the grid. The upper window is set for synchronization during the DC capacitor charging period. The range for the upper window can be determined as follows:

$$\Delta V_{upper\_window} > \frac{P_{in\_max}T_f}{CV_{min\_upper}} \qquad (1)$$

where P_{in_max} is the upper boundary of the power level to start pulse skipping P_{input}, T_f is the fundamental period of the grid voltage, V_{min_upper} is the low-limit voltage of the upper window, and C is the main storage capacitor

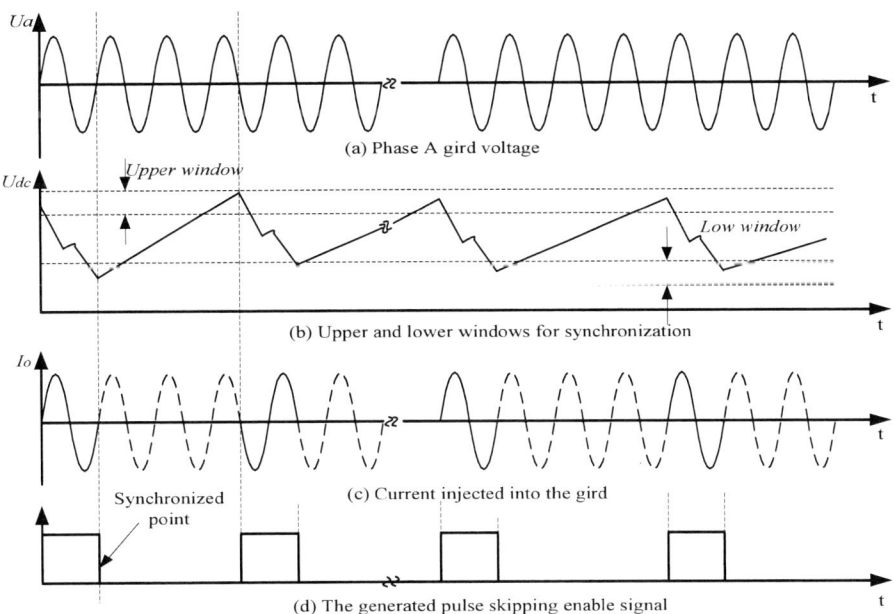

Figure 5: Pulse skipping grid synchronization

The lower window is also set for synchronization during the DC capacitor discharging period. However, the size of lower window can not be pre-determined, since the pulse power level P_{pulse}, the number of cycles, n, and the input

power level, P_{input}, are varied greatly during the optimization procedure. For a given pulse power level, input power level, the capacitor voltage level at which the inverter starts pulse skipping can be calculated as:

$$V_{syn\_point} = \sqrt{V_{dc\_initial}^2 - \frac{2(P_{pulse} - P_{input})nT_f}{C}} \qquad (2)$$

where V_{syn_point} is the DC bus voltage, at which point, grid synchronization point is reached, and the $V_{dc\_initial}$ is the initial DC bus voltage.

The minimum DC bus voltage V_{min_lower} is easy to set for the onset of charge according to the amplitude of grid voltage and the controller's maximal duty cycle. On the other hand, determining the high-side voltage of the lower window V_{max_lower} is not straightforward. The rationale for determining the V_{max_lower} follows the following rule: Ensure that the size of the lower window is as small as possible such that synchronization can be achieved regardless of the input power P_{input}, $V_{dc\_initial}$ and P_{pulse}. In this manner, the bulk of the energy stored in the DC bus can be pumped into the grid.

It can be seen from expression 2 that the synchronization instants vary with the number of cycles, n, as shown in Fig. 6. In addition, Fig. 6 shows that for a given pulse power level, P_{pulse}, input power level, P_{input}, and initial DC voltage, a number of synchronization points exist, which correspond to a synchronization voltage level U_{sys_point1} right above the minimum lower DC bus voltage window, V_{min_lower}. Different sets of initial conditions will result in different synchronization voltage levels, V_{sys_pointi}.

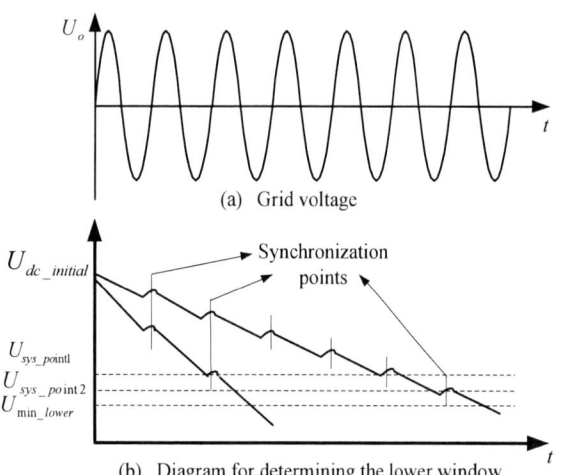

(a) Grid voltage

(b) Diagram for determining the lower window

Figure 6: Determination of the lower window

In order to make sure that the synchronization DC voltage will fall in the range of the lower window for different P_{input}, P_{pulse} and $V_{dc\_initial}$, the minimum upper voltage limit of the lower window V_{max_lower} should meet the following condition:

$$V_{max\_lower} = \max(V_{sys\_point1}, V_{sys\_point2}, ..., V_{sys\_pointn}) \qquad (3)$$

From above analysis, it is not difficult to use Matlab to compute the minimum upper voltage limit of the lower window V_{max_lower}.

Once the upper and lower windows are determined, every output AC pulse is guaranteed to be synchronized with the grid and the pulse skipping enable signal is set/reset to allow the loss model to accurately include the relevant losses in each state.

E. Optimization Procedure

Since the three parameters to be optimized are closely related, it would be quite difficult to simultaneously identify all three optimal parameters in one simulation. As a result, a recursive optimization procedure is proposed.

1. First, the initial DC bus voltage is fixed first while the optimal pulse power level, $P_{optimal\_pulse}$, is found out by varying the pulse power level, P_{pulse}, for different input power levels through a recursive search method.

2. Next, the pulse power level is set to the optimal pulse power point, $P_{optimal\_pulse}$, and DC window is varied by adjusting the DC bus voltage within the normal operating range to compute the optimal DC voltage windows.

3. By using the optimal pulse power level computed in (1) and optimal DC window computed in (2) to implement the pulse skipping operation, a new efficiency curve can be obtained. The optimal power level to start pulse skipping, $P_{optimal\_s}$, can be found by intersecting these two efficiency curves.

F. Pulse Skipping Control Cost

Based on above analysis, three variables (input power from the intermittent renewable source, intermediate DC voltage, and grid-synchronization signal) need to be measured to implement the pulse skipping control scheme. For the intermittent renewable source, a maximum power point tracking (MPPT) algorithm is usually employed to maximize the input power, which means that the input power is already measured. In the analyzed two-stage power converter, the intermediate DC voltage is usually measured to regulate the current reference for grid-connected inverters. A grid-synchronization signal is required to synchronize the phase of the inverter current with the grid voltage, which is normally derived by employing a software or hardware phase-locked loop (PLL) techniques. Therefore, all the relevant control variables are already measured in the original system and no additional circuitry is needed to implement the proposed pulse skipping control strategy. All that's required are few additional lines of code.

IV. AN OPTIMIZATION EXAMPLE

A 200W full bridge prototype DC-AC inverter was setup to verify the parameter optimization for pulse skipping control strategy as shown in Fig. 8. Using the loss based model, the conduction losses of the power devices, the switching losses, and the inductor and capacitor losses are all taken into consideration using the mathematical expressions derived from datasheets offered by their manufactures. The loss model was implemented in Matlab and was verified by experiment with different DC voltages and across all range of load. Figure 9 shows one of typical loss model verification results.

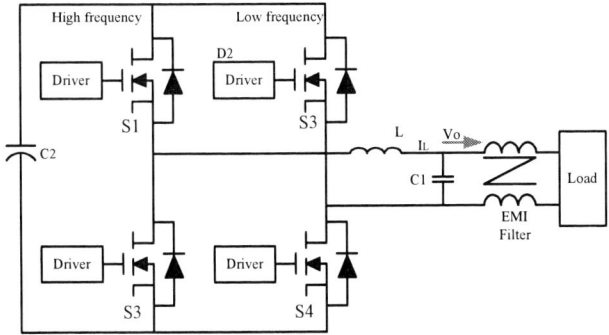

Figure 8: The prototype inverter setup

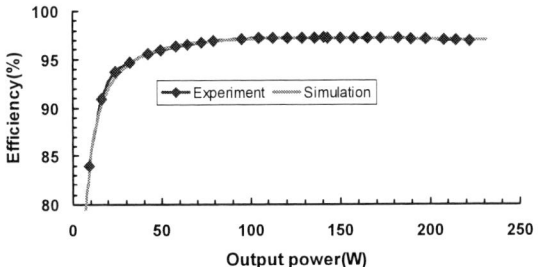

Figure 9: Loss model verification

After model verification, the loss model operating in the pulse skipping mode was developed according to Figure 4. To facilitate the optimal parameter search, the parameter ranges should be first specified. As shown in Fig. 9, the maximal efficiency point occurs around 120W. As such, we can expect a boost in efficiency if the pulse skipping control algorithm is applied below the 120W power level. As a result, the input power level, P_{in}, is varied over the range of 5W-120W. Next, the pulse power level values are varied from 30W 200W. Note that having pulse power levels below 30W will generate lower efficiencies and as such are excluded. Finally, and for the safe and normal operation of the grid-tied inverter, the DC bus voltage operating range is limited to 190-250V.

To compute the impact of DC bus voltage windows, the upper and lower DC bus voltage windows are first set to 225V-247V and 190V-215V, respectively, based on the synchronization constraint illustrated in Section III. Then, the DC bus voltage windows are varied and the difference between lower and upper windows is also varied to compute their impact on efficiency improvement. Figure 10 shows three-dimensional efficiency curves for the above specified parameter ranges. As seen from Fig. 10, the maximal efficiency points marked blue do not correspond to a fixed pulse power level, P_{pulse}. The optimal pulse power level, $P_{optimal\_pulse,}$ varies with the input power level. However the optimal pulse power levels are clustered around 140W with only slight variations across the input power range. As such, the optimal pulse power level can be set to 140W.

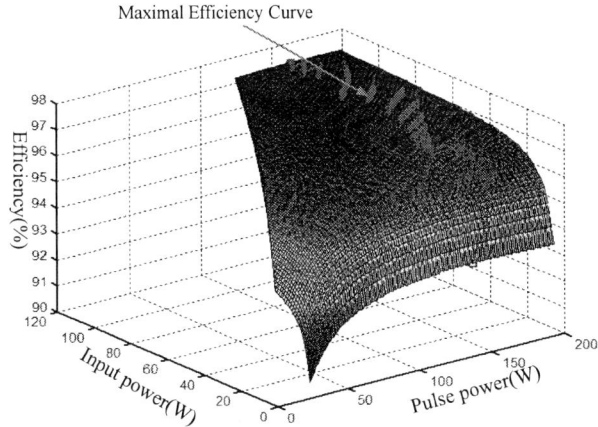

Figure 10: Pulse skipping optimization results

Figure 11 shows the efficiency improvements using pulse skipping at different pulse power levels, namely 50W, 90W and 140W. It can be seen from Fig. 11 that although the inverter efficiency is improved as the pulse power level increases, the improvement margin is not as significant increase for pulse power levels of 90W to 140W. As for the optimal input power level below which to start pulse skipping, it can be seen from Fig. 11 that this level is around 70W.

Figure 12 shows the efficiency curves for different DC bus voltage ranges utilizing the optimum pulse power level of 140W. As shown in Fig. 12, we can conclude that the DC bus voltage variation has little impact on inverter efficiency. As a result, we did not make any further optimization of the DC voltage range.

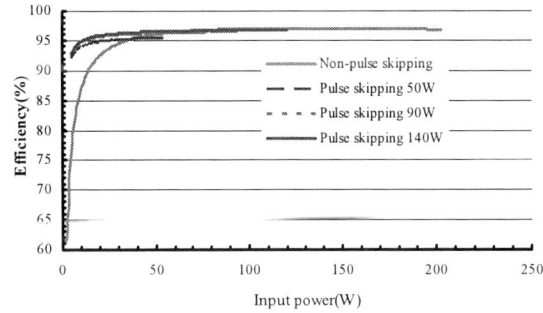

Figure 11: Simulation results of efficiency improvement

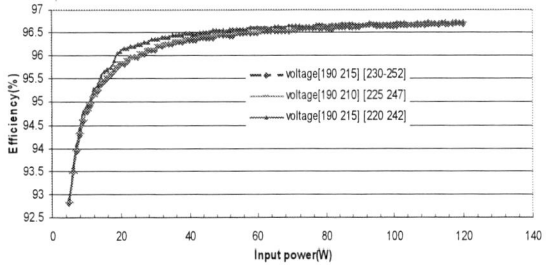

Figure 12: The comparison of efficiency curves with DC bus voltage ranges at pulse power level of 140W

V. EXPERIMENTAL RESULTS

In order to verify the findings of the previous section, a 200W, two stage DC/AC inverter consisting of a front DC-DC converter stage followed by the DC-AC prototype is used. A solar array simulator, which is set to operate in fixed mode with set voltage and current, serves as a DC power supply for the first power stage. By limiting the output current of the simulator, the input stage can be easily controlled to operate in constant power mode, which can simulate the intermittent renewable source. A power analyzer is used to measure the input and output powers of the inverter as well as the inverter efficiency.

Figure 13 shows the inverter operation in pulse skipping operation mode with input power 30W, pulse power level of 90W, a lower voltage window of 190V-215V, and an upper voltage window of 225V-245V.

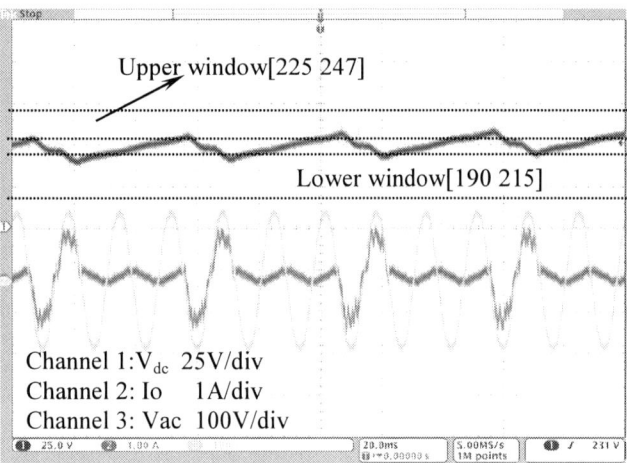

Figure 13: Pulse skipping operation mode with input power of 30W and pulse power level of 90W

Due to the output current in pulse state, the output power measurement, and thus the inverter efficiency, needs to be done carefully. the advanced functionality of the power analyzer was used to accurately measure the inverter efficiency by using a 4s time range with 25kbps sampling rate and averaging 60-time measured values. As shown in Fig. 14, the efficiency curves are greatly improved to above 90% by employing a pulse skipping strategy at light loads. The best efficiency curve corresponds to 140W pulse power level, while the lowest efficiency curve corresponds to a pulse power level of 49W, while there is slight difference in efficiency curves between the ones corresponding to pulse power levels of 140W and 90W. All the experimental results match the simulation results pretty well across the whole range of measured data except at extremely light load conditions of less than 5W where the measurement errors of the power analyzer contribute a significant part to the total losses.

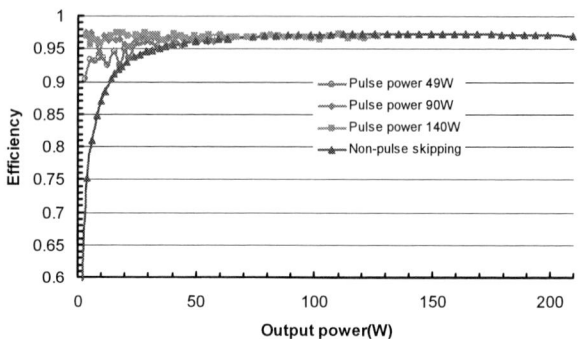

Figure 14: Efficiency improvement with pulse skipping with different pulse power levels

VI. GRID FLICKER CONSIDERATIONS

One of the concerns of employing pulse skipping schemes is that they may give rise to grid flicker problems. This may well be the case if the penetration of renewable sources becomes significant.

Although it is quite difficult to assess to what extend does pulse skipping related flicker impact the grid when a large number of inverters operating in pulse skipping mode are connected to the grid, a simple analysis reveals that this may not be a significant problem after all. This is due to the fact that the pulse skipping control strategy can be viewed as a random system when the input power is not an integer multiple of the pulse power level. In addition, tolerances in individual inverter component values and measurements variations between inverters will result in inverters synchronizing to the grid at various instants. One could implement a random process to slightly vary the lower and upper DC bus voltage windows, which will vary the synchronization instants thus randomly spreading the pulses generated by inverters in pulse skipping mode. With such a scheme, it is highly plausible that the generated power pulses from a large number of inverters operating independently in parallel will be averaged to a continuous pulse power stream with tolerable amplitude fluctuations.

Finally, and as seen from simulation and experimental results, the varying the pulse power level do not result in significant efficiency variations and thus the pulse power level can also be randomly varied resulting in a real random system.

VII. CONCLUSIONS

In this paper, a pulse skipping control strategy is proposed to improve the efficiency of the grid-tied inverter at light loads. To better implement the proposed pulse skipping control strategy, three key parameters need to be optimized. A loss-model based optimization algorithm is developed to derive the optimal parameters for pulse skipping mode of operation. A design example is given to verify the proposed optimization algorithm. Simulation and experimental results show that the optimization results yielded a significant efficiency improvement at light load, where the inverter is maintained above 90% even at very light loads.

[1] Juan manuel Carrasco, Leopoldo Garcia Franquelo, Jan T. Bialasiewicz and etc. Power-Electronic Systems for the Grid Integration of renewable Energy Sources: A Survey. IEEE Tran. On Industrial Electronics, Vol. 53, No. 4 Aug. 2006.

[2] Quan Li, Peter Wolfs. A Review of the Single Phase Photovoltaic Module Integrated Converter Topologies with Three Different DC Link Configurations. IEEE Trans. On Power Electronics, Vol. 23, No. 3, May 2008, pp. 1320-1333.

[3] Soeren Baekhoej Kjaer, John K. Pedersen, Frede Blaabjerg. A Review of Signle-Phase Grid-Connected Inverters for Photovoltaic Modules. IEEE Trans. Indudtry Applications, Vol. 41, No. 5, 2005, pp. 1292-1306.

[4] Bruno Burger, Dirk Kranzer, Extreme High Efficiency PV-Power Converters, Power Electronics and Applications, 2009, 13th European Conference on EPE Sept. 2009, pp. 1-13.

[5] M.Jantsch, C.W.G. Verhoeve. AC PV Module Inverters with Full Sine Wave Burst Operation Mode for Improved Efficiency of Grid Connected Systems at Low Irradiance. 14th EC PVSEC, Barcelona, Spain, 1997.

[6] Yu Fang, Dehong Xu,Yanjun Zhang, Fengchuan Gao, Lihong Zhu, Yi Chen. Standby Mode Control Circuit Design of LLC Resonant Converter. IEEE Power Electronics Specialists Conference, PESC2007, Orlando, USA, pp. 726-730.

[7] Bin Wang, Xiaoni Xin, Stone Wu, Hongyang Wu, Jianping Ying. Analysis and Implementation of LLC Burst Mode for Light Load Efficiency Improvement. The 24th IEEE Applied Power Electronics Conference and Exposition, APEC2009, Washington, DC, USA, pp. 58-64.

[8] Yu-Kang Lo; Shang-Chin Yen, Jin-Yuan Lin. A High-Efficiency AC-to-DC Adaptor with a Low Standby Power Consumption. The 37th IEEE Power Electronics Specialists Conference, 2006, Jeju, Korea, pp. 1-4.

[9] Jin-ho Choi, Dong-young Huh, Young-seok Kim. The Improved Burst Mode in the Stand-by Operation of Power Supply. The 19th IEEE Applied Power Electronics Conference and Exposition, 2004. Anaheim, USA, pp. 426-432.

[10] Chen, B.-Y, Lai, Y.-S. Switching Control Technique of Phase-Shift Controlled Full Bridge Converter to Improve Efficiency Under Light Load and Standby Conditions Without Additional Auxiliary Components. IEEE Trans. On Power Electronics (To be appeared)

[11] Jian Sun, Characterization and Performance Comparison of Ripple-Based Control for Voltage Regulator Modules. IEEE Trans. On Power Electronics, Vol. 21, No. 2, March 2006, pp. 346-353.

[12] S. Angkititrakul, H.Hu. Design and Analysis of Buck Converter with Pulse-skipping Modulation, Proc. IEEE Power Electronics Specialists Conference, 2008, Rhodes, Greece, pp. 1151-1156.

[13] Angel V. Peterchev, Seth R. Sanders. Digital Multimode Buck Converter Control with Loss-Minimizing Synchronous Rectifier Adaptation. IEEE Trans. On Power Electronics, Vol. 21, No. 6, Nov. 2006, pp. 1588-1599.

[14] Santanu Kapat, Soumitro Banerjee and Amit Patra. Modeling and Analysis of DC-DC Converters Under Pulse Skipping Modulation. Proc. IEEE TENCON 2008, pp. 1-6.

[15] Luo Ping, Ming Xin, Zhang Bo and Li Zhaoji. Analysis of the Stability and Ripple of PSM Converter in DCM by EB Model. International Conference on Communications, Circuits and Systems, 2007, ICCCAS 2007, pp. 1235-1239.

[16] A Saiz-Vela, P.Miribel-Catala, J.Colomer, M. Puig-Vidal, J.Samitier. Pulse Skipping Switching mode: a case study of Efficiency Improvement on a switched-capacitor DC-DC step-up converter IC. Proc. IEEE International Symposium on Industrial Electronics, 2006, , pp. 1178-1181.

[17] Yuancheng Ren, Ming Xu, Jinghai Zhou, Fred C. Lee. Analytical Loss Model of Power MOSFET. IEEE Trans. Power Electronics, Vol. 21, No. 2, March 2006, pp. 310-319.

Phase Locked Loop for Unbalanced Utility Conditions

Carlos D. Rodríguez-Valdez, Russ J. Kerkman, *Fellow, IEEE*
Standard Drives
Rockwell Automation
Milwaukee, WI 53092, USA

Abstract—A Phase-Locked-Loop (PLL) to deal with unbalanced utility conditions is introduced in this paper. The classical 3-phase PLL structure [1], [2] is studied first to gain a better understanding of the effect of the unbalance on the PLL [3]-[8]. Then, enhancements are proposed to improve the ability of the PLL to deal with unbalance. Finally, simulation and experimental results are presented.

I. INTRODUCTION

Industrial system demands are stringent and costly. Users insist on increased uptime and efficiency. A typical industrial grid, like the one shown in Fig. 1, is quite complex. This system, for instance, employs several Adjustable Speed Drives (ASDs), a switch capacitor bank, common bus inverters, isolation transformers, motors controlled by motor starters, motors started across-the-line. An ASD, consisting of a converter and inverter, might or might not be tied to the Point of Common Coupling (PCC) through a line-reactor. Each motor and associated controller has electric cables to make the necessary power connections. Power factor compensation may be achieved through switched capacitor banks. Common bus inverters share a dc-link powered by a high performance Active Front End (AFE) or a semi-controlled rectifier, for example an SCR-based rectifier.

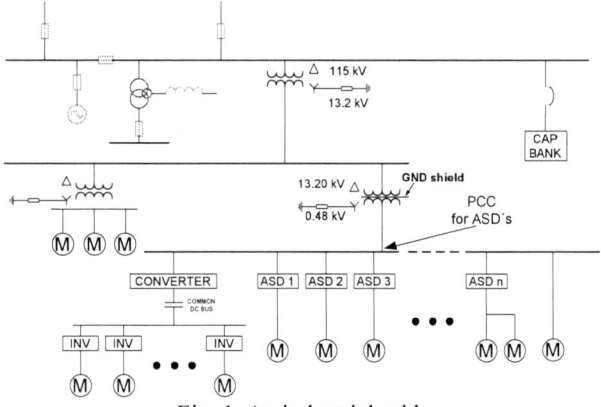

Fig. 1. An industrial grid

Complex industrial systems inevitably result in unequal loading of the utility phases. Practical systems involve various single phase loads with unequal loading of the feeders, loads being continuously connected or disconnected, unbalancing the voltage supply at the PCC, non-linear loads distorting the network. From the point of view of a customer connected to the grid through an isolation transformer (Fig. 2.a) this is often intolerable; and it is further aggravated if the PCC is shared with other customers which behave as other sources of disturbance from the point of view of the customer, as depicted in Fig. 2.b.

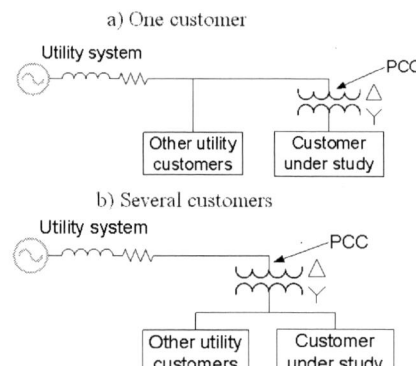

Fig. 2. Connection to the secondary of an isolation transformer (PCC)

The unbalance of the line voltage stresses the power electronics and other devices present at the customer site. It generates larger ripple in the DC-link capacitors which translates to torque ripple in the load. Voltage unbalance and disturbances present in a network are hard to mitigate and require additional hardware such as an Active Power Filter (APF) which effectiveness largely depends on the synchronizing angle to the line. Similarly, an Active-Front-End (AFE) can be part of an ASD, in processes which nature makes it economically attractive to regenerate power back to the mains. As in the case of an APF, the effectiveness of an AFE depends on the quality of the synchronization mechanism.

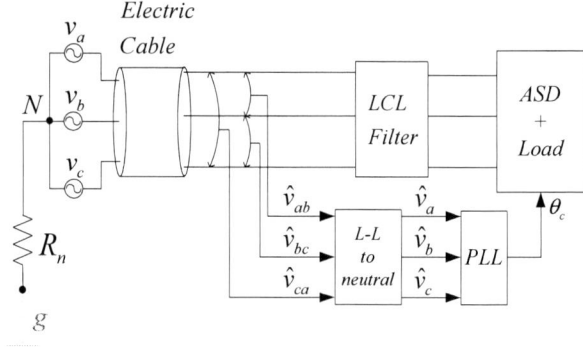

Fig. 3. PLL synchronization mechanism of an ASD

Synchronization to the mains (50 Hz or 60 Hz) is typically achieved using a zero-crossing detector or a Phase-Locked-Loop (PLL). The zero-crossing technique has the disadvantage of either being too sensitive to line notches and inaccurately detecting zero crossings or being completely blind during portions of an electric cycle. Today

978-1-4244-4782-4/10 $26.00 © 2010 IEEE

PLLs have largely displaced zero-crossing-detector mechanisms [7].

Fig. 3 depicts an ASD with AFE that requires a synchronizing mechanism (PLL). The grounding system could be one out of several possibilities [9]; and it is represented by R_n. The source impedance is not zero and it is depicted lumped with the one of the electric cable. The inputs to the PLL can be line-to-line or line-to-neutral voltages measured at some point on the converter side. Although in Fig. 3 this point is depicted right after the source impedance, such a measurement could be done at other locations like between the inductors of the LCL [10]. The disturbances at the voltage source can be of several types [11]. In this paper, only voltage sags are considered. A voltage sag is defined as a jump in magnitude and/or in phase in any of the voltage sources. A voltage sag during steady-state can be analyzed as an unbalanced occurring on the magnitude and/or in the phase of the voltage sources.

This paper will refer to the case depicted in Fig. 3; however, the findings for the synchronization mechanism can be easily extended to other power converter systems like APFs, for instance. The next section describes the limit of operation of the ASD depicted in Fig. 3, which in turn limits the angle required from the PLL.

II. BOUNDARIES OF THE SYNCHRONIZATION ANGLE

The control of an AFE requires knowledge of the angle and sequence of the source it is synchronizing to. The ability of an AFE to maintain synchronism is limited by the physical constraints of current and voltage of the converter, and the impedance between mains and converter. A single phase diagram of a simple converter mains system is depicted in Fig. 4.

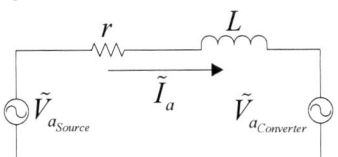

Fig. 4. Simplified view of the SZCS

To find the operating region of the converter, the converter voltage is initially set at a fixed value along the q-axis. Thus, the operating region of the converter is found through the feasible locus of the source voltage. The locus is obtained by applying the constraints imposed by maximum converter current, and voltage.

Fig. 5 displays the case when the converter voltage is the reference voltage. The locus of $\tilde{V}_S$ will be anywhere within the circle with center at $\tilde{V}_C$, such that the maximum possible radius is given by $\max\left|\tilde{V}_C - \tilde{V}_S\right|$. The angle ϕ is calculated accordingly. The value of $\max\left|\tilde{V}_C - \tilde{V}_S\right|$ is given by (1)

$$\max\left|\tilde{V}_C - \tilde{V}_S\right| = |Z|\max\left|\tilde{I}_a\right| \tag{1}$$

From (1) it is clear that the current and the impedance limits the locus of $\tilde{V}_S$. This region is called region of current

limit. The operation of the PLL for this paper is limited to the region of current limit.

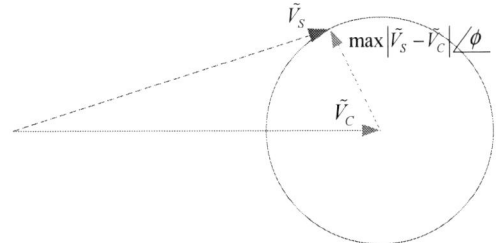

Fig. 5. Locus of $\tilde{V}_S$

III. THE 3-PHASE PLL

For the analysis of the 3-phase PLL, the abc-to-qd0 transformation, as defined in (3.3-4) in [12], is used. This is depicted in Fig. 6. Particularly, the converter voltage is regulated such that $v_{dc}=0$ (i.e., converter voltage is placed in the q-axis).

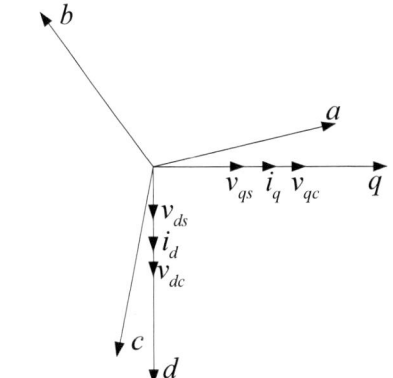

Fig. 6. abc-qd0 reference frame transformation

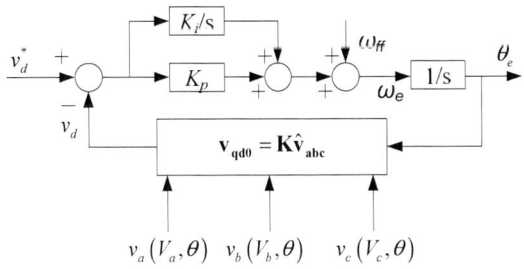

Fig. 7. Standard 3-phase PLL topology

A. PLL block diagram

Fig. 7 depicts the block diagram of a standard PLL [1], [2]. The inputs to the PLL are the line-to-line or the line-to-ground voltage measurements at the converter side of the ASD. Assuming small line impedance, such voltages are approximately v_a, v_b, v_c. The PI-controller gains (K_p, K_i) are tuned to comply with the required Band-Width (BW) and dynamic stiffness. The Feed-Forward (FF) term (ω_{ff}) is

978-1-4244-4782-4/10 $26.00 © 2010 IEEE

added to increase command tracking capability. **K** is the abc-to-qd0 transformation as defined in (3.3-4) in [12].

B. Dynamic stiffness

Distortion in the measured voltages (magnitude, phase and angle) is processed by the dynamics of the PLL, which in turn generates deviations on the estimated angular frequency, angle and phase. In [1], [2] a small signal model analysis for the PLL is presented for the case in which the mains is unbalanced. The balanced case is defined as the nominal trajectory for the unbalanced case. The unbalance is treated as disturbance. The dynamic stiffness (disturbance per unit of response) plot is presented. However, there is an error in the expression (1) for the dynamic stiffness given in [1]. The right expression for the dynamic stiffness is given in (2). Terms Q_{21} and Q_{25} are given in (3-4).

$$\frac{\Delta V_a}{\Delta \theta_e} = \frac{s^2 + Q_{25}K_p s + Q_{25}K_i}{-Q_{21}K_p s - Q_{21}K_i} \tag{2}$$

$$Q_{21} = \frac{1}{3}\sin(2\theta_{nom}) \tag{3}$$

$$Q_{25} = V_{nom} \tag{4}$$

$$V_{nom} = \bar{V}_a = \bar{V}_b = \bar{V}_c \tag{5}$$

$$\theta_{nom} = \bar{\theta} = \bar{\theta}_e \tag{6}$$

where $\bar{V}_a$, $\bar{V}_b$, $\bar{V}_c$, $\bar{\theta}$ and $\bar{\theta}_e$ are the nominal trajectories for the magnitudes of v_a, v_b, v_c and for the angles θ and θ_e, respectively.

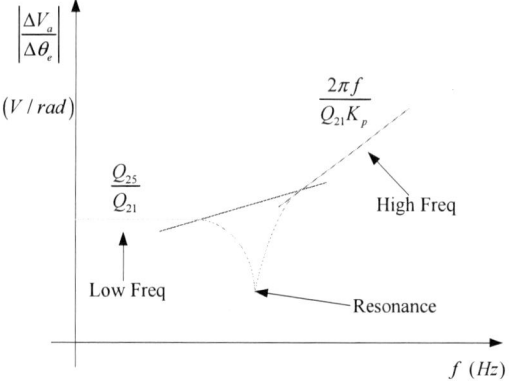

Fig. 8. Dynamic stiffness

In Fig. 8, the corrected dynamic stiffness for the case of unbalanced mains (magnitude and phase) analyzed in small signal regimen, assuming the unbalance is a disturbance from the balanced case is presented. The dynamic stiffness plot suggests that the estimation of $\hat{\theta}$ is rather prone to low-frequency disturbances. The stiffness can be increased (i.e., robustness can be provided to the system) by adding feedback loops for added integral states. The drawback, however, is that increasing robustness could translate in the ASD becoming a source of disturbance for other components in the case the ASD were placed in a more complex system.

Another possible solution to this problem is to use a rather classical adaptive scheme. This would likely include

a model and/or estimation of parameters with the added difficulty of ensuring right convergence of the estimated parameters, but with the added difficulty of modeling a changing system. Moreover, several controller gains could require tuning.

C. Nature of the oscillations on the d-axis voltage

Finding the nature of the oscillations on v_d would allow for proper compensation. An expression for v_d can be obtained using the transformation **K**:

$$v_d = \frac{2}{3}\left[\sin(\theta_e) \quad \sin(\theta_e - 120^o) \quad \sin(\theta_e + 120^o)\right]\begin{bmatrix} V_a \cos(\theta + \phi_a) \\ V_b \cos(\theta - 120^o + \phi_b) \\ V_c \cos(\theta + 120^o + \phi_c) \end{bmatrix} \tag{7}$$

Operating and regrouping, (7) can be written as (8). Phases ϕ_1 and ϕ_2 are defined in (9) and in (10)

$$v_d = \sin\left(\theta_e + \theta + \frac{\phi_1 + \phi_2}{2}\right)\cos\left(\frac{\phi_1 - \phi_2}{2}\right) + \cos\left(\theta_e - \theta - \frac{\phi_1 + \phi_2}{2}\right)\sin\left(\frac{-\phi_1 + \phi_2}{2}\right) \tag{8}$$

$$\phi_1 = \tan^{-1}\left(-\frac{2V_a \cos(\phi_a) - V_b \cos(\phi_b) - V_c \cos(\phi_c)}{2V_a \sin(\phi_a) - V_b \sin(\phi_b) - V_c \sin(\phi_c)}\right) \tag{9}$$

$$\phi_2 = \tan^{-1}\left(-\frac{V_b \cos(\phi_b) - V_c \cos(\phi_c)}{V_b \sin(\phi_b) - V_c \sin(\phi_c)}\right) \tag{10}$$

In general, the reference voltage in the d-axis, v_d^*, is 0. Then, from the dynamics of the PLL, another relationship between v_d and θ_e can be found. This is the integral-equation presented in (11).

$$\int\left(-K_p v_d + \omega_{ff}\right)dt - K_i \int\int v_d dt^2 = \theta_e \tag{11}$$

Table 1. Angular frequency estimated (ω_e in rad/s)

TABLE I. ANGULAR FREQUENCY ESTIMATED (*IN RAD/S*)

Harm	2nd	4th	6th	8th
Mag	237	41.25	6.07	1.03

TABLE II. TABLE 2. ANGLE ESTIMATED (*IN RAD*)

Harm	2nd	4th	6th	8th
Mag	0.3076	0.0274	0.005	0.0023

The solution of the system constituted by (8) and (11) is expected to be of the form of exponential modulation, and solved through Bessel functions. In Fig. 9, the angular frequency (actual and estimated) and the angle (actual and estimated) for the case in which the source voltages are 3 p.u., 1p.u. and 0 are presented. By exaggerating the unbalance a more accurate understanding of the problem's complexity was discerned. The harmonic components of the angular frequency are presented in Table I while the harmonic components of the angle are presented in Table II. Notice that the angular frequency and the angle are comprised of even harmonics that are neglected in the traditional analysis of classical PLL [1]-[7]. Thus, under extreme grid conditions, phase synchronization and converter stability could be compromised by an inadequately compensated PLL.

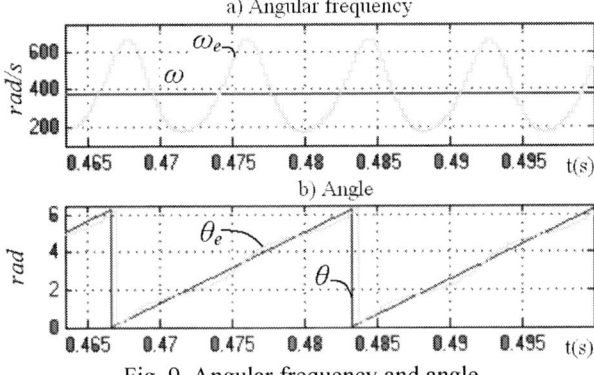

Fig. 9. Angular frequency and angle

For this paper, in (8), $\phi_1 - \phi_2 \approx 0$ is assumed. This holds for the case of small phase disturbances ($\phi_a \approx 0$, $\phi_b \approx 0$, $\phi_c \approx 0$) and approximately equal phase voltage amplitudes ($V_a \approx V_b \approx V_c$). This is similar to the case of small signal analysis. Therefore, (8) takes the form given in (12). Under the consideration of small signal and the assumption that the PLL is adequately tuned, (12) can be approximated by (13). The definition for θ_{nom} is similar to the one given in (6).

$$v_d \approx \sin\left(\hat{\theta} + \theta + \frac{\phi_1 + \phi_2}{2}\right) \quad (12)$$

$$v_d \approx \sin\left(2\theta_{nom} + \frac{\phi_1 + \phi_2}{2}\right) \quad (13)$$

A few conclusions can be drawn from (13) and the PLL structure presented in Fig. 7. For small unbalance, v_d is basically a sinusoid with frequency twice the fundamental frequency of the 3-phase system. This oscillation in the angular speed requires mitigation, not only for the potential it has to create disturbances, but also for the necessity of complying with frequency standards for power generation or grid tie power generation, which in general have stringent requirements. A few of these standards are given in Table III [13]

TABLE III.   STANDARDS FOR ANGULAR FREQUENCY

Standard	Frequency criteria
GermanyE.ON	49 Hz…50.5 Hz (normal operation) 47.5 Hz…51.5 Hz (emergency condition)
UK	47.5 Hz…52 Hz (normal operation)
US FERC	Not defined
AESO (Canada)	59.4 Hz…60.6 Hz (normal operation) 57 Hz…61.7 Hz (emergency condition)
Hydro-Quebec (Canada)	59.4 Hz…60.6 Hz (normal operation) 55.5 Hz…61.7 Hz (emergency condition)

As pointed out, an adaptive solution that requires calculating parameters or tuning controller gains is problematical in a polluted and changing environment like the one where this synchronization mechanism is required

to operate. On the other hand, a robust solution would increase the ability of the PLL to reject disturbances within a predefined range of frequencies. A robust and adaptive solution is proposed in the next section.

IV.   3-PHASE PLL ENHANCEMENTS

Paying attention to the requirements presented in the previous section and starting from (13), enhancements to the classical PLL are possible. The sinusoidal term v_d has a frequency $2\omega_{nom}$ ($\theta_{nom} = \omega t$) and a phase $(\phi_1 + \phi_2)/2$. The next two subsections will focus on attenuating the effects of these.

A.   Elimination of the sinusoidal oscillation on v_d

Eliminating the effect of the voltage unbalance and oscillating angular frequency should be done selectively to avoid affecting a broad range of frequencies that could generate distortion in other portions of the system. Furthermore, since the system frequency in distributed generation systems can vary, adaptation should be considered. . To accomplish this a tracking filter is proposed. This filter, having analogous characteristics to the single phase compensator in [14], is postioned in the PLL as illustrated in Fig. 10. Notice that this is a more general solution, where not only the 2nd harmonic is being treated as unwanted oscillation but, in fact, any frequency component could be filtered out.

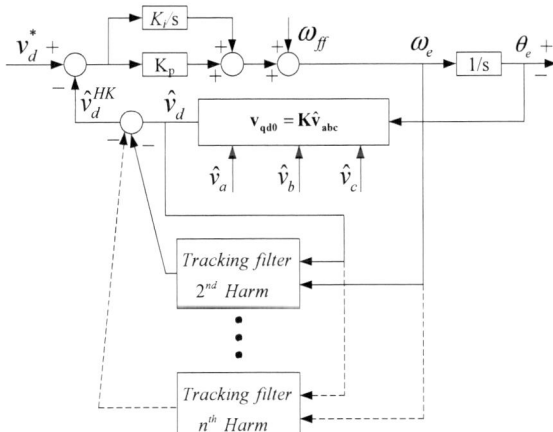

Fig. 10. Enhancement to eliminate oscillations on v_d

A state-space representation of the tracking filter is given in (14). The A-matrix is time-dependent. The case in which the A-matrix is time-independent is presented in [15], [16]. Furthermore, if $\omega(t)$ becomes an input, (14) becomes non-linear. Assuming that a is a constant, $u=u(t)$ is a signal source, $\omega = \omega(t)$ is the angular frequency, and t is the independent variable, (14) can be formally transformed into the second order differential equation (16). A power series expansion can be attempted for x_2 to solve (15), provided that ω and u admit a power series expansion around some point (typically, $t=0$ if the initial values are given at $t=0$). Finding a general closed-form solution for time-varying non-linear systems is difficult. Solutions found are usually particular. Instead, (16) can be numerically solved and characterized. The conclusions are that the tracking filter presented in (14-16) is a non-linear, time-varying system. Furthermore, it is capable of tracking instantaneous changes of frequency (adaptive capability).

$$\begin{bmatrix} \dot{x}_1 \\ \dot{x}_2 \end{bmatrix} = \begin{bmatrix} -a & -\omega(t) \\ \omega(t) & 0 \end{bmatrix} \begin{bmatrix} x_1 \\ x_2 \end{bmatrix} + \begin{bmatrix} a \\ 0 \end{bmatrix} u \qquad (14)$$

$$\begin{cases} \dot{x}_1 = -ax_1 - \omega x_2 + au \\ \dot{x}_2 = \omega x_1 \end{cases} \qquad (15)$$

$$\ddot{x}_2 + \left(a - \frac{\dot{\omega}}{\omega}\right)\dot{x}_2 + \omega^2 x_2 = a\omega u \qquad (16)$$

The inputs to the tracking filters are $\hat{v}_d$ and the estimated angular frequency, ω_e. $\hat{v}_d^{HK}$ is the difference between $\hat{v}_d$ and the output of the tracking filter (the superscript HK stands for harmonic killer). Thus, the effect of the filter is only on those frequency components contained in $\hat{v}_d$ (selective capability).

The advantage of this solution is its simplicity. No model parameters need to be calculated; and no controller gains other than the ones of the standard PI require tuning or mdification. The only settable parameter is the BW-associated parameter, a. After setting, it does not require further adjustment.

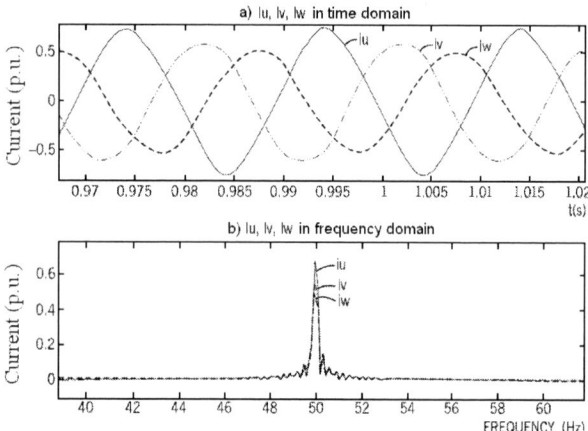

Fig. 11. Input phases to the PLL

The proposed solution has been implemented in the rti1005 dSPACE platform. Inverter currents were selected as input quatities to be tracked. This lent itself to a broad investigation with minimal hardware requirements. Unbalanced conditions and rapid frequency changes were esily implemented. Typical results in both time and frquency are depicted in Fig. 11. The pu values are: phase a 0.57 pu., phase b 0.5 pu, and phase c 0.7 pu. The operating frequency, f, has been chosen to be 50 Hz. The PI-parameters have been tuned utilizing the small signal analysis described in section 6.2. The values found are K_p=186.1892 rad/s, K_i=8666.6 rad/s^2. The BW-associated parameter a has been made 5 rad^2/s. The criteria for selecting a are: selectivity and sufficient BW to provide transient response to errors in the input frequency.

In Fig. 12, the comparison of the angular frequency before and after the proposed enhanced solution is enabled, in both time and frequency domains, is presented. Notice that the 2nd harmonic oscillation on the angular speed, generated by the PLL, is eliminated. Furthermore, after the HK is enabled, there is left remaining harmonic content. This remaining harmonic content such as the sub-harmonics components that are depicted in the magnitude frequency response of Fig. 12 have a source other than the unbalance in phase or in magnitude. They are due to DC offset, sampling, PWM modulation, distortion in the line, inaccuracy of the angular speed calculated, etc.

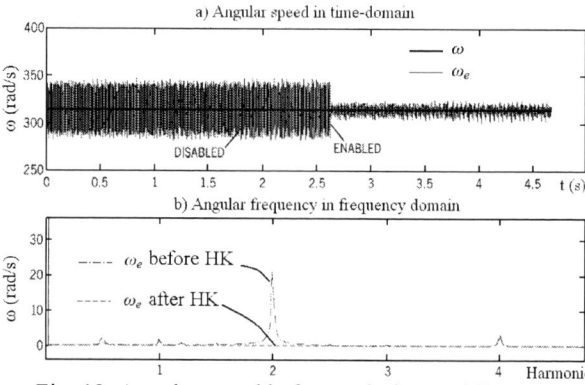

Fig. 12. Angular speed before and after enabling HK

By mitigating oscillations on v_d, the potential exponential modulation of the angular speed is mitigated. Thus, compliance with standards presented in Table 3 is possible.

B. Phase tracking: Selecting the Positive Sequence

The problem of synchronization using a 3-phase PLL is two-fold: 1) reducing angle oscillations and 2) phase selection. In the previous section the angle oscillation problem has been addressed. In this section, the focus is on the phase that the PLL should lock in. As was discussed previously in (13), v_d shows the phase that is linked to the amount of unbalance.

The requirements for the PLL in addition to the classical requirements of PLLs are: 1) dynamic and steady state tracking and lock within the region of current limit, 2) the converter should operate at a controlled displacement factor (i.e. power factor), and 3) not be a source for grid disturbance when subjected to disturbances. Furthermore, negative sequence effects should be minimized. The negative sequence is potentially destructive. For instance, a 3% negative sequence voltage could drive a 50% nominal current in an induction motor. This is because such rotating machines have a very small impedance to negative sequence [17]. Therefore, the effect of the corresponding negative sequence torque could be highly destructive for the shaft of the machine. Consequently a positive sequence synchronizer would be beneficial.

The +/-/0 sequence can be expressed as a function of the abc voltage sets, as in (17), (18). Matrices **S** and **T** are given in (19), (20)

$$\begin{pmatrix} \tilde{E}_{a1} \\ \tilde{E}_{b1} \\ \tilde{E}_{c1} \end{pmatrix} = \mathbf{S} \begin{pmatrix} \tilde{E}_a \\ \tilde{E}_b \\ \tilde{E}_c \end{pmatrix} + \mathbf{T}j \begin{pmatrix} \tilde{E}_a \\ \tilde{E}_b \\ \tilde{E}_c \end{pmatrix} \qquad (17)$$

$$\begin{pmatrix} \tilde{E}_{a2} \\ \tilde{E}_{b2} \\ \tilde{E}_{c2} \end{pmatrix} = \mathbf{S} \begin{pmatrix} \tilde{E}_a \\ \tilde{E}_b \\ \tilde{E}_c \end{pmatrix} - \mathbf{T}j \begin{pmatrix} \tilde{E}_a \\ \tilde{E}_b \\ \tilde{E}_c \end{pmatrix} \qquad (18)$$

$$\mathbf{S} = \frac{1}{3} \begin{pmatrix} 1 & -\frac{1}{2} & -\frac{1}{2} \\ -\frac{1}{2} & 1 & -\frac{1}{2} \\ -\frac{1}{2} & -\frac{1}{2} & 1 \end{pmatrix} \qquad (19)$$

$$\mathbf{T} = \frac{1}{2\sqrt{3}} \begin{pmatrix} 0 & 1 & -1 \\ -1 & 0 & 1 \\ 1 & -1 & 0 \end{pmatrix} \qquad (20)$$

In the above formulations, the expressions containing "j" correspond to a 90-degree phase shift. A properly designed filter will provide a rapid decomposition of the feedback into +/-/0 sets. The non-linear time-varying tracking filter presented in the previous section provides one implementation. The states of this filter, x_1 and x_2, are 1) the selected component and 2) the selected component phase shifted 90 degrees. Therefore, the positive component synthesizer and the negative component synthesizer can be built as illustrated in Figs. 13-14.

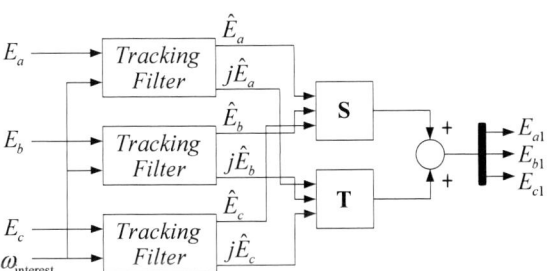

Fig. 13. Positive component synthesizer

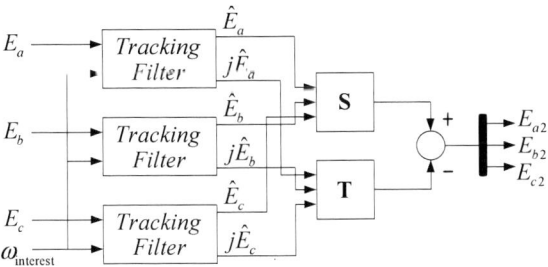

Fig. 14. Negative component synthesizer

Symmetrical components as defined by Fortescue [17] refers to a phasor representation of a 3-phase system. To differentiate, the positive and negative component synthesizers depicted in Figs. 13-14 are called quasi-positive sequence and quasi-negative sequence estimators. In the next section, the joint result from mitigating the oscillation on v_d and tracking the phase of v_d is presented.

C. The 3-phase PLL with phase selection

In this section a PLL providing independent selectivity of frequency and phase (i.e., independent selectivity of the frequency estimator and the positive sequence calculator) is introduced. In Fig. 15 the angular speed for the quasi-positive sequence filter is estimated by the HK. With such an angular speed, the quasi-positive sequence calculator provides a balanced sequence to the abc-to-qd0 transformation. The advantage of having the HK to estimate the frequency instead of PLL's frequency is independent selectivity of their respective a's within the filter of the HK and quasi-positive calculator. The classical PLL and its simple PI controller can now easily synthesize the angle to lock. This solution is simple; and tuning nothing other than the gains of the classical PLL loop is required. Moreover, the selectivity of this solution allows for avoiding introducing disturbance back to the network in a broad frequency range.

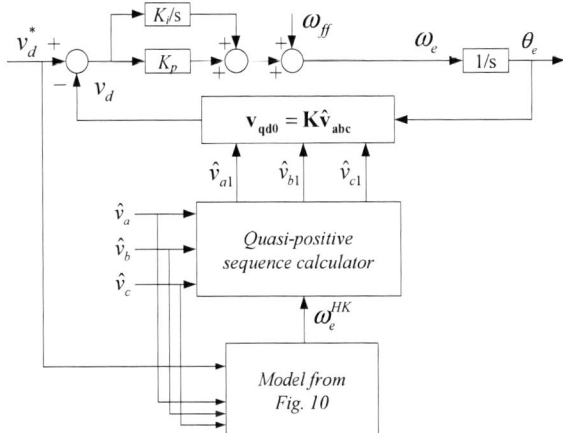

Fig. 15. Proposed solution to the problem of line-synchronization under unbalanced measured voltages condition

V. EXPERIMENTAL AND SIMULATION RESULTS

Experiments were conducted to test the tracking capability of the PLLs. To explore rates of change in fundamental frequency, a set of line currents with 3% unbalance were used as input signal instead of a set of phase voltages. In Fig. 16, a comparison is made between the calculated angular speeds using: 1) classical PLL, 2) PLL with oscillation elimination on v_d and 3) PLL with oscillation elimination on v_d and phase tracking. In Fig. 17, the same comparison is made in the frequency domain. All the tracking filters are implemented digitally with T_s=250 µs and and BW-associated parameter a=5 rad^2/s. The frequency of the unbalanced set is 50Hz. As pointed out the criteria for selecting a are: filter selectivity and sufficient BW to provide transient response to errors in the input frequency.

Figs. 16 and 17 show unaccounted for harmonic components. Note Fig. 17 shows a subharmonic, 1st, 2nd, 3rd, 4th, 6th, 8th harmonic components, are present.. The tracking filter is tuned to mitigate the 1st harmonic (i.e., DC-component in abc variables), 2nd harmonic (i.e., largest harmonic component due to phase voltage unbalance), 3rd and 4th harmonics. The remaining spectrum in the angular frequency is used as angular frequency input to the symmetrical components calculator. By combining the benefits of harmonic rejection and phase synchronization with the classical 3-phase PLL(Fig. 16 and Fig. 17) a robust PLL is achieved.

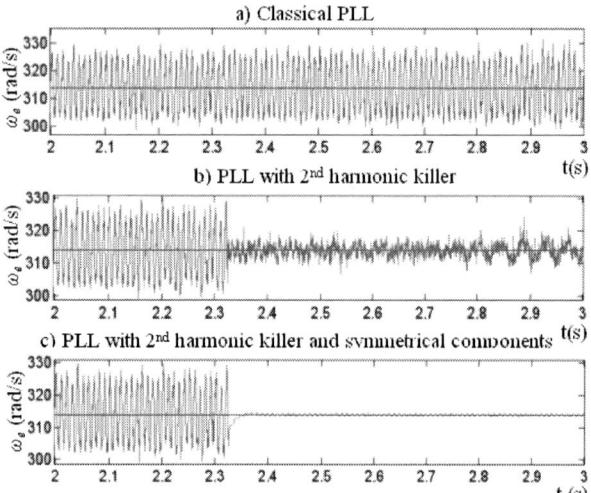

Fig. 16. Angular speed (amplitude magnitude frequency response with f_{ref} = 50 Hz)

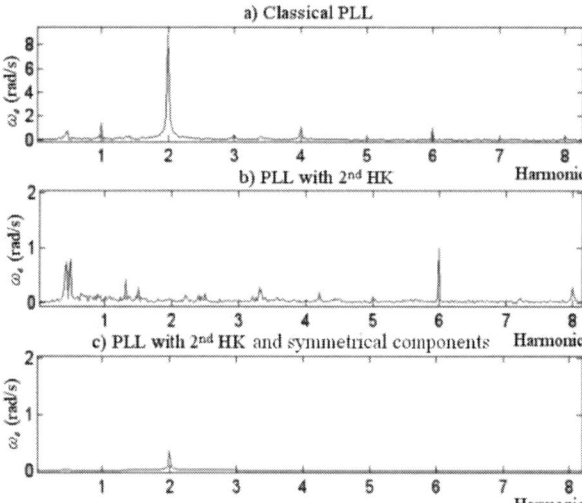

Fig. 17. Amplitude magnitude response of angular speed (amplitude magnitude frequency response with f_{ref} = 50 Hz)

Two other test scenarios were employed to test transient and steady-state capabilities of the PLL of Fig. 15. In both cases, the actual angular frequency and angle are compared to thosed synthesized by the PLL. 1) Initially, inputs are balanced. At t=1 s, amplitude of phase a, V_a=0; phase of phase b is stepped by 180 degrees; and phase of phase c is stepped by 45 degrees. 2) The frequency is ramped up and down from 50 Hz to 70 Hz every 0.6 s. At t=1 s, the same event as test 1) occurs.

The results are depicted in Figs. 18-19. The angular frequency is presented in 2 different time scales in parts a) and b), respectively.

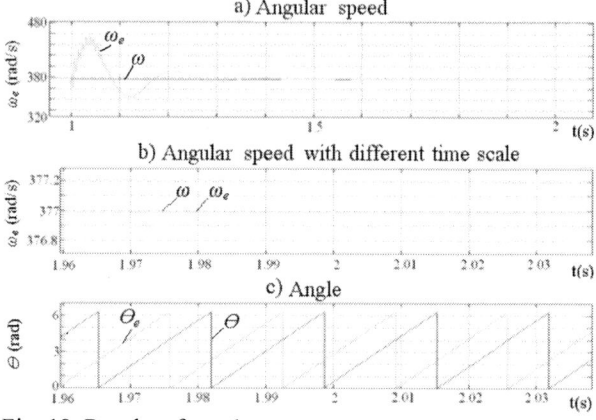

Fig. 18. Results of test 1

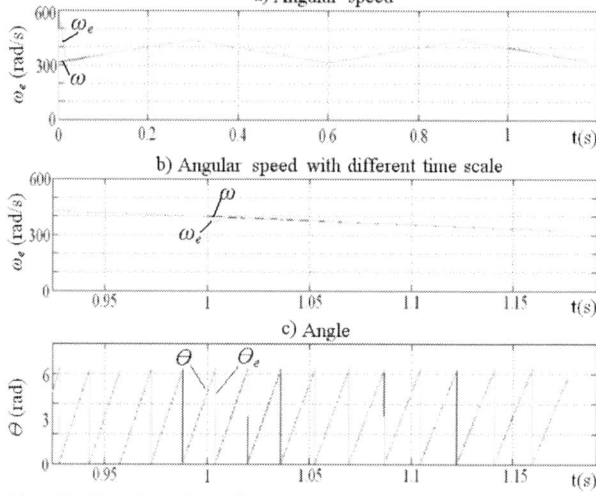

Fig. 19. Results of test 2

VI. CONCLUSION

In this paper a synchronization mechanism for an AFE under magnitude or phase unbalanced conditions at the mains is introduced. The effect of such unbalanced conditions is oscillation on the synthesized angular frequency and offset in the phase.

The solution has two parts: 1) the oscillation on the synthesized angular frequency is mitigated using a time-varying non-linear tracking filter, and 2) the offset in the phase offset is detected by using symmetrical components. The result is an easy to tune, robust and adaptive method to keep synchronism with the mains.

REFERENCES

[1] L.N. Arruda, B.J.C. Filho, S.M. Silva, S.R. Silva, A.S.A.C. Diniz. (2002, September). Wide bandwidth single and three-phase PLL structures for grid-tied PV systems. 28th IEEE Photovoltaic Specialists Conference. Conf. Rec., pp. 1660 – 1663

[2] L. Neto Arruda, S. Magalhães Silva and B.J. Cardoso Filho. (2001, Sep-Oct). PLL Structures for Utility Connected Systems. *IEEE 36th Industry Appl. Conf.,* Vol. 4, pp 2655 – 2660

[3] L.G. Barbosa Rolim, D. Rodrigues da Costa Jr., M. Aredes. (2006, December). Analysis and Software Implementation of a Robust Synchronizing PLL Circuit Based on the pq Theory. *IEEE Trans. on Ind. Electron.,* Vol. 53, No. 6, pp. 1919-1926

[4] Se-Kyo Chung. (2000, May). A phase tracking system for three phase utility interface inverters. IEEE Trans. on Pow. Electron. Vol. 15, No. 3, pp. 431 – 438

[5] S.A.O. da Silva, E. Tomizaki, R. Novochadlo, E.A. Alves Coelho. (2006, November). PLL Structures for Utility Connected Systems under Distorted Utility Conditions, *IEEE 32nd Ann. Conf. on Ind. Electr.,* pp. 2636 - 2641

[6] S. R. Naidu, A.W. Mascarenhas and D.A. Fernandes. (2004, November). A Software Phase-Locked Loop for Unbalanced and Distorted Utility Conditions. *IEEE Int. Conf. on Pow. Syst. Technol.,* Proc., pp. 1055-1060

[7] A.M. Salamah, S.J. Finney and B.W. Williams. (2007, November) Three-phase phase-lock loop for distorted utilities. *IET Electr. Power Appl.,* Vol. 1, No. 6, pp. 937-945

[8] F. Blaabjerg, R. Teodorescu, M. Liserre and A.V. Timbus. (2006, October). Overview of Control and Grid Synchronization for Distributed Power Generation Systems. *IEEE Trans. on Ind. Electron.,* Vol. 53, No. 5, pp. 1398-1409

[9] G. Skibinski. (1996, October) Design Methodology of a Cable Terminator to Reduce Reflected Voltage on AC Motors. IEEE Ind. App. Conf, Vol. 1, pp. 153-161

[10] M. Liserre, F. Blaabjerg, S. Hansen. (2005, September)Design and control of an LCL-filter-based three-phase active rectifier. IEEE Trans. on Ind. App., Vol. 41, No. 5, pp. 1281–1291

[11] M. H. J. Bollen.Understanding Power Quality Problems: Voltage Sags and Interruptions, IEEE Press Series on Power Engineering, USA, 2000

[12] P.C. Krause, O. Wasynczuk, S. D. Sudhoff. Analysis of Electric Machinery and Drive Systems. 2nd Ed., 2002, Ed. Wiley-Interscience

[13] IEC standard 61400-21

[14] R.J. Kerkman, T..M. Rowan. (1987, November). U.S. patent 4,706,012

[15] P. Rodríguez, R. Teodorescu, I. Candela, A.V. Timbus, M. Liserre, F. Blaabjerg. (June, 2006), New Positive-sequence Voltage Detector for Grid Synchronization of Power Converters under aulty Grid Conditions. *IEEE 37th Power Electronics Specialists Ann. Conf.,* pp. 1-7

[16] P. Rodríguez, A.V. Timbus, R. Teodorescu, M. Liserre, F. Blaabjerg. (2007, October). Flexible Active Power Control of Distributed Systems During Grid Faults. IEEE Trans. on Ind. Electron., Vol. 54, No. 5, pp. 2583-2592

[17] C.F. Wagner and R.D. Evans. Symmetrical Components: As Applied to the Analysis of Unbalanced Electrical Circuits. Pub. Robert E. Krieger Publishing Company, Inc., 1933, Rep. 1982, Copyright by McGraw-Hill Book Company, Inc.

Carlos D. Rodríguez Valdez (S' 2007) joined the Standard Drives Division of Rockwell Automation, Mequon, WI, USA in May 2006.

His current interests include Power Converters, Adjustable Speed Drives, Optimization and Robust Control.

Russel J. Kerkman (S'67-M'76-SM'88-F'98) received the B.S.E.E., M.S.E.E., and Ph.D degrees in electrical engineering from Purdue University, West Lafayette, IN, in 1971, 1973, and 1976, respectively. From 1976 to 1980, he was an Electrical Engineer in the Power Electronics Laboratory of Corporate Research and Development, General Electric Company, Schenectady, NY. He is currently an Engineering Consultant at Rockwell Automation / Allen Bradley Company, Mequon, WI. His career spans thirty years of industrial experience in power electronics. His current interests include: modeling and control of general purpose industrial drives, adaptive control applied to field oriented induction machines, application of observers to ac machines, design of ac motors for adjustable speed applications, and EMI from PWM inverters. He is a co-holder of forty patents, all in adjustable speed drives.

Dr. Kerkman is a Fellow of the IEEE and the recipient of ten IEEE Prize Paper Awards. He is a member of Industry Applications Society, Industrial Electronics Society, and Power Electronics Society. He was a recipient of Rockwell International's Engineer of the Year award in 1986. In 1998 he was selected as a Corporate Inventor of the year. Dr. Kerkman was a recipient of Purdue University's Outstanding Electrical Engineer award in 2000. In 2003 Kerkman received Rockwell Automation's Odo J. Struger award honoring engineers for long-term outstanding technical achievement / innovation, technical contribution and technical leadership in the field of automation. In 2005 he received the IEEE-IAS first annual Gerald Kliman Innovator Award for his contributions to power conversion.

Analysis and Design Considerations for EMI and Losses of RCD Snubber in Flyback Converter

Peipei Meng, Xinke Wu, Jianyou Yang, Henglin Chen, Zhaoming Qian (IEEE Senior Member)
College of Electrical Engineering
Zhejiang University, Hangzhou 310027, China
Email: cathleen@zju.edu.cn

Abstract—The RCD snubber is usually used in flyback converter, in order to limit the voltage spikes caused by leakage inductance of the transformer. The design considerations of RCD snubber are commonly on the power losses dissipated in the snubber and the voltage spikes of the transistor. In fact, the design of RCD snubber will also affect the EMI performance of the converter. But the RCD design consideration for EMI performance is usually in contradiction with that for losses. In this paper, a detailed model is built up to analyze the voltage spikes, the residual energy and the power losses dissipated in the RCD snubber. And a novel design procedure is derived for selecting the resistor in the RCD circuit to make an optimal tradeoff between CM noises and losses. Experimental results consist well with theoretical analysis.

I. INTRODUCTION

Flyback switch mode power supply derivatives have been widely used in the industrial applications [1]-[3] because of their simplicity and low cost. One challenge in the design of flyback converter is handling the high leakage inductance of flyback transformer that causes high voltage spike and could damage the main transistor when it is turned off. To solve this problem, a variety of turn-off snubbers were reported to limit the rate of rise voltage across the switching device. Snubbers can be either passive [4]-[9] or active networks [10]-[14]. Passive snubbers may be either dissipative [4]-[6] or non-dissipative [7]-[9]. Dissipative RCD snubber is widely used in low power DC-DC applications for its simplicity and low cost.

In [5], [6], [15], [16], dissipative RCD snubber optimization was reported for the considerations of the dissipative energy in the snubber and the voltage spike across the MOSFET. Both of these principles do not take the EMI problem into consideration. In fact, the design of RCD snubber will affect the EMI performance of the converter effectively. In flyback converter, the voltage spike is extremely serious when the main transistor is turned off due to the energy stored in the leakage inductor. The RCD snubber can clamp the voltage spike to an acceptable level to protect the main transistor, and the CM noise can also benefited at the same time. But the energy stored in the leakage inductance of the transformer can't entirely be absorbed by the RCD snubber. The residual energy left in the circuit after the

clamping diode of the RCD snubber turns off will cause high frequency oscillations between the leakage inductance and the intrinsic capacitance of the transistor. These oscillations together with the high dv/dt when the transistor turns off are both noise sources and will cause serious CM noises.

Different from the general analysis that assume the voltage of the clamping capacitor keeps invariant when the clamping diode is on [5] [6]. In this paper, in order to study into the working principles of the RCD snubber, a detailed model is built up, which can evaluate the dv/dt of the main transistor when it turns off, the residual energy and the power losses precisely. And a novel design procedure is derived for selecting the resistor in the RCD snubber to make an optimal tradeoff between CM noises and losses. Experimental results consist well with theoretical analytical results.

II. RELATIONSHIP BETWEEN RCD SNUBBER, CM NOISES AND POWER LOSSES

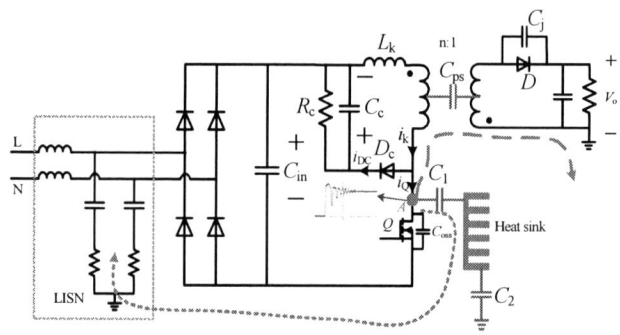

Figure 1. CM noise caused by the main transistor Q

A. CM noise caused by the main transistor Q

The flyback converter with dissipative RCD snubber is shown in Fig. 1. One of the most serious *hot-voltage* points is the drain of main transistor Q, which is shown as point A in Fig. 1. C_{ps} represents the capacitive effect between the primary winding and the secondary winding, C_1 represents

---

This work is supported by Natural Science Foundation of China under Grant 50807047 and 50907061

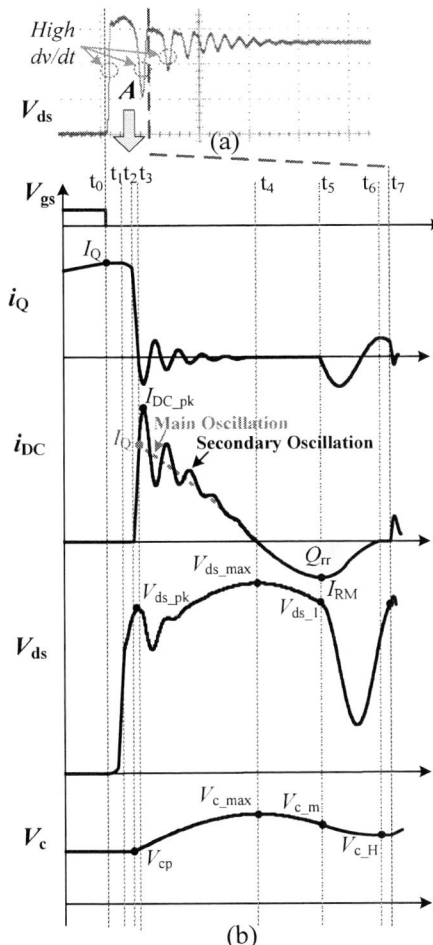

Figure 2. (a) The tested voltage waveform across Q when it turns off. (b) Operating waveforms when the RCD snubber is charged

the parasitic capacitance between the MOSFET drain and the heat sink, C_2 represents the parasitic capacitance between the heat sink and the ground. Since the MOSFET Q is operated as a switch, the drain voltage swings in every switching cycle. This voltage swing in turn causes the charging and discharging of these parasitic capacitances. The charging and discharging current will return through the ground path and show up as CM noise. The two main CM noise paths caused by the main transistor Q is illustrated in Fig. 1.

The voltage spike of the *hot-voltage* point A is extremely serious when Q is turned off due to the energy stored in the leakage inductance of the transformer. This high voltage spike may damage the MOSFET if no snubber is added in this converter. Large CM noise will also induced by the high voltage spike and the following oscillations between the parasitic parameters of the circuit.

The RCD snubber can clamp the voltage spike to an acceptable level to protect the main transistor. Take the flyback converter with RCD snubber shown in Fig. 1 as example, the tested voltage waveform across the main transistor Q when it turns off is shown in Fig. 2(a). It's easy to find that, the voltage spike is clamped by the RCD

snubber. So, less current will be induced in C_{ps} and C_1 compared with the converter without RCD snubber. Thus less CM noise will be caused. But besides this voltage spike appeared immediately after Q is turned off, there are some other voltage swings with high dv/dt in the voltage waveform of the main transistor Q, as shown in Fig. 2(a). These voltage swings are caused by the oscillations between the parasitic parameters of the circuit due to the residual energy left in the circuit that can't entirely be absorbed by the RCD snubber. These voltage swings will also cause CM noises.

But what's the relationship between these voltage swings and the RCD snubber? What's the relationship between these voltage swings and the losses? Can the RCD snubber be optimal designed for CM noise consideration? To answer these questions, the operation principles of RCD snubber should be analyzed in detail first.

B. Operation principles of RCD snubber and residual energy calculating

The interval A, as marked in Fig. 2(a), is the first time that the RCD snubber is charged, which will be studied in detail in the following. The main waveforms during interval A is drawn in Fig. 2(b). i_k, i_Q and i_{DC} are the currents in the leakage inductor L_k, transistor Q and diode D_c respectively. V_{ds} and V_c are the voltage across the transistor Q and the capacitor C_c respectively.

Q turns off at t_0, the current in the magnetizing inductance of the transformer begins to charge and discharge the intrinsic output capacitor C_{oss} and the junction capacitor C_j, respectively. At t_1 all the charges in C_j are released and the primary side of the transformer is clamped to $n \cdot V_o$ by the output voltage. After t_1 the transformer begins to transfer the energy stored during the on-time to the output. But the energy stored in the leakage inductor can't transfer to the secondary-side, so the current in the leakage inductor still charges C_{oss} until t_2 when V_{ds} equals to $V_{in} + V_{cp}$, V_{cp} is the clamping voltage of C_c. Then D_c turns on.

For $C_{oss} \cdot \dfrac{dV_{ds}}{dt} = i_Q$, so the voltage rising slope of V_{ds} during $t_0 \sim t_1$ is determined by the peak current of the transistor Q and has nothing to de with the RCD snubber. But the clamping value of V_{ds} is determined by the design of RCD snubber.

Stage 1 (t_2-t_4): The current in the transistor transfers to the RCD circuit during $t_2 \sim t_3$, which occurs in a short interval. So it can be assumed that this stage begins at t_3. There are two oscillations in this stage, as shown in Fig. 2(b). The main oscillation is formed by L_k and C_c, the secondary oscillation is formed by C_{oss} and the parasitic inductance. Fig. 3(a) shows the equivalent circuit for both the two oscillations; Fig. 3(b) shows the equivalent circuit only for the main oscillation. In which, $R_{eq}=R_{p1}+R_{trans-ac}+R_{esr}$, L_k is the leakage inductance, $R_{trans-ac}$ is the ac resistance of the transformer. R_{p1}, R_{p2} and L_{p1}, L_{p2} are the parasitic resistances and inductances in the circuit respectively. R_{esr} and L_{esl} are the ESR and ESL of the clamping resistor C_c respectively.

The secondary oscillation is mainly associated with the

978-1-4244-4782-4/10 $26.00 © 2010 IEEE

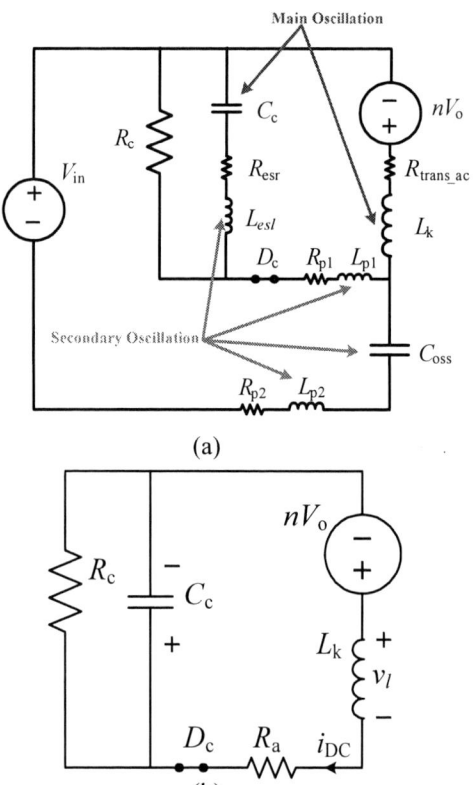

(a)

(b)

Figure 3. (a) Equivalent circuit with high frequency parameters (b) Equivalent circuit for main oscillation, of Stage 1 and Stage 2

parasitic parameters in the circuit. So, if the PCB layout is well-designed to limit the parasitic parameters, the secondary oscillation will be minimized or even disappear. Radiated emission can benefited from it because the frequency of the secondary oscillation is commonly high.

Apply Kirchoff's voltage law (KVL) around the loop in Fig. 3(b) gives

$$\frac{\partial^2 v_c}{\partial t^2} + \left(\frac{R_{eq}}{L_k} + \frac{1}{R_c C_c}\right) \cdot \frac{\partial v_c}{\partial t} + \frac{R_{eq} + R_c}{R_c L_k C_c} \cdot v_c = \frac{nV_o}{L_k C_c} \quad (1)$$

Assume that the current in the leakage inductance remains constant during t_0-t_3. So $i_{DC}(t_3) = i_Q(t_1) = I_Q$. Then the initial voltage $v_c(t_3)$ and its derivative $dv_c(t_3)/dt$ can be derived

$$v_c(t_3) = V_{cp} \quad (2)$$

$$\frac{dv_c(t_3)}{dt} = \frac{I_Q}{C_c} - \frac{V_{cp}}{R_c C_c} \quad (3)$$

The voltage $v_c(t)$ is obtained as

$$v_c(t) = \frac{R_c \cdot nV_o}{R_{eq} + R_c} + e^{-b(t-t_3)}\left[A_1 \cos(\omega_d(t-t_3)) + A_2 \sin(\omega_d(t-t_3))\right] \quad (4)$$

In which:

$$b = \frac{R_{eq} R_c C_c + L_k}{2R_c L_k C_c} \quad (5)$$

$$\omega_d = \sqrt{\frac{2R_{eq} R_c C_c L_k + 4R_c^2 L_k C_c - R_{eq}^2 R_c^2 C_c^2 - L_k^2}{4R_c^2 L_k^2 C_c^2}} \quad (6)$$

$$A_1 = \frac{V_{cp}(R_{eq} + R_c) + nV_o R_c}{R_{eq} + R_c} \quad A_2 = \frac{bA_1 R_c C_c + I_Q R_c - V_{cp}}{\omega_d R_c C_c} \quad (7)$$

The damping factor b in Equation (5) is relatively small, so the maximal value of $v_c(t)$ at t_4 can be evaluated by

$$V_{c\_max} = \frac{R_c \cdot nV_o}{R_{eq} + R_c} + \sqrt{A_1^2 + A_2^2} = f(V_{cp}) \quad (8)$$

In Equation (4) and (8), except for parameter V_{cp}, which is the clamping voltage of capacitor C_c, all the other parameters are known for a particular converter.

Stage 2 (t_4-t_5): Although i_{DC} reaches zero at t_4, the diode continues to conduct until t_5, when the recovery current reaches the maximum value I_{RM}. So this stage could also be modeled by the equivalent circuit shown in Fig. 3(a) (b).

Stage 3 (t_5-t_6): At t_5 the charge stored has been removed from the diode, and then the diode starts blocking. Reverse recovery losses occur during this stage since negative current is present while the blocking voltage across the diode increases. In this stage the circuit can be divided into two parts: the reverse recovery of the diode and the ringing between L_k and C_{oss1}, the equivalent circuits are shown in Fig. 4.

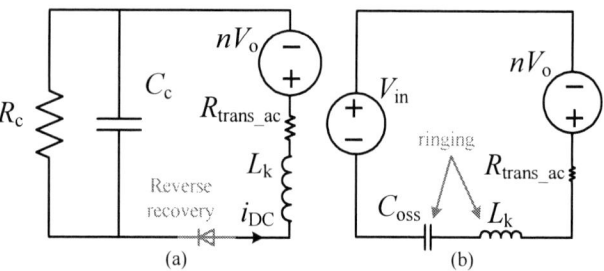

Figure 4. Equivalent circuit of Stage 3

It's shown in Fig. 2 that, the voltage swing of V_{ds} during this stage is serious, which can cause unexpected CM noises. This voltage swing is caused by the energy stored in L_k and C_{oss} at t_5, which is named residual energy in this paper. The residual energy E_r can be derived by

$$E_r = \frac{1}{2} \cdot C_{oss} \cdot \left[V_{ds\_1}^2 - (V_{in} + nV_o)^2\right] + \frac{1}{2} \cdot L_k \cdot I_{RM}^2 - E_{dio\_rr} \quad (9)$$

$$V_{ds\_1} = V_{in} + V_{c\_m} \quad (10)$$

In which, V_{ds-1} is the voltage of V_{ds} at t_5, V_{c-m} is the voltage of V_c at t_5, I_{RM} is the maximum reverse recovery current of diode D_c, E_{dio-rr} is the reverse recovery losses caused by D_c during this stage.

During the period t_4-t_6 when reverse recovery of the diode takes place, it can be derived that

$$C_c \cdot (V_{c\_max} - V_{c\_H}) = Q_{rr} \qquad (11)$$

In which, $V_{c\text{-}H}$ is the voltage of V_c at t_6, Q_{rr} is the reverse recovery charges in the diode.

Now the relationship between RCD snubber and the voltage swing then the CM noises has been revealed. Firstly, the clamping voltage of the main transistor Q is determined by the clamping voltage of RCD snubber by

$$V_{ds\_pk} \approx V_{in} + V_{cp} \qquad (12)$$

Secondly, the residual energy E_r, which will determine the large voltage swing, is given in Equation (9) and (10). So the problem now is to calculate the value of V_{cp} and $V_{c\text{-}m}$ precisely for a given RCD snubber. [17] presents an analytical method for approximating the turn-off losses and reverse recovery charges in power diodes, which isn't included in this paper.

Stage 4 (t_6-t_7): At t_6 the recovery current reaches zero, thus V_c keeps invariant during this stage. The equivalent circuit of this stage is the same with Fig. 4(b). The ringing between L_k and C_{oss} goes on until t_7 when V_{ds} equals to $V_{c\_H} + V_{in}$. Then the diode D_c conducts again, another RCD charging cycle similar with t_2-t_7 beagins. Otherwise, if the peak value of the ringing is smaller than $V_{c\text{-}H} + V_{in}$, the RCD circuit will not conduct again. Then the ringing between L_k and C_{oss} keeps on until V_{ds} was damped to $nV_o + V_{in}$ by the parasitic resistance.

Stage 5 [t_7-(t_0+T)]: Assuming that C_c is mainly charged by the first RCD charging cycle, so after t_7 the energy stored in C_c will be dissipated by R_c until Q turns on again.

The reference directions are shown in Fig. 5, so

$$i_c(t) = C_c \cdot \frac{dv_c(t)}{dt} = -\frac{v_c(t)}{R_c} \Rightarrow \frac{1}{v_c(t)} \cdot dv_c(t) = -\frac{1}{C_c R_c} \cdot dt \quad (13)$$

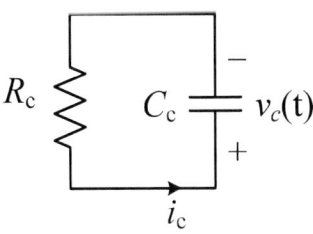

Figure 5. Equivalent circuit of Stage 5

Apply integration to (13) in one period T yields:

$$\int_{t_7}^{t_7+T} \frac{1}{v_c(t)} \cdot dv_c(t) = \int_{t_7}^{t_7+T} -\frac{1}{C_c R_c} \cdot dt$$

$$\Rightarrow \ln v_c(t_7 + T) - \ln v_c(t_7) = -\frac{T}{C_c R_c} \qquad (14)$$

According to the capacitor charging balance, there has $v_c(t_7 + T) = V_{cp}$. Substitute this equations into (14) yields

$$\ln V_{cp} - \ln V_{c\_H} = -\frac{T}{C_c R_c} \qquad (15)$$

Combine Equation (8), (11) and (15) can solve the values of $V_{c\text{-max}}$, V_{cp} and $V_{c\text{-}H}$ easily. To solve $V_{c\text{-}m}$, apply energy conservation law to the equivalent circuit shown in Fig. 3 (b) during the time interval t_3-t_5 gives

$$\frac{1}{2} \cdot L_k (I_Q^2 - I_{RM}^2) + C_c(V_{c\_m} - V_{cp}) \cdot nV_o = \frac{1}{2} \cdot C_c(V_{c\_m}^2 - V_{cp}^2) \qquad (16)$$

I_Q is the current in the leakage inductor at t_3, so $\frac{1}{2} \cdot L_k (I_Q^2 - I_{RM}^2)$ indicates the energy that provided by the leakage inductance. $C_c(V_{c\_m} - V_{cp})$ is the total charge that followed through the voltage source nV_o, so $C_c(V_{c\_m} - V_{cp}) \cdot nV_o$ indicates the energy that is provided by the voltage source nV_o. $\frac{1}{2} \cdot C_c(V_{c\_m}^2 - V_{cp}^2)$ indicates the energy that stored in the capacitor C_c.

By the detailed RCD model provided above, all the variables in the first RCD charging cycle can be evaluated. Thus, both the residual energy and the clamping voltage of the main transistor can be calculated by Equations (9), (10) and (12), respectively.

C. Calculating the losses in the RCD snubber

The losses in the RCD circuit can be divided into two parts: the losses caused during t_3-t_7 when the diode is on, and the losses caused during t_7-(t_3+T) when the energy in C_c is dissipated by R_c. The former losses W_1 can be determined from the energy balance requirement [15]:

Loss = energy from source – increase in energy stored in capacitor – increase in energy stored in inductor. So,

$$W_1 = nV_o \int_{t_3}^{t_7} i \, dt - C_c \int_{v(t_3)}^{v(t7)} v \, dv - L_k \int_{i(t_3)}^{i(t_7)} i \, di$$

$$= nV_o \cdot C_c(V_{c\_H} - V_{cp}) - \frac{1}{2} C_c(V_{c\_H}^2 - V_{cp}^2) + \frac{1}{2} L_k I_Q^2 \qquad (17)$$

The losses dissipated by R_c is

$$W_2 = \frac{1}{2} C_c(V_{c\_H}^2 - V_{cp}^2) \qquad (18)$$

Then the total power losses dissipated in the RCD circuit is

$$P_{loss} = f \cdot (W_1 + W_2) = f \cdot \left[nV_o \cdot C_c(V_{c\_H} - V_{cp}) + \frac{1}{2} L_k I_Q^2 \right] (19)$$

III. OPTIMAL DESIGN FOR RCD SNUBBER

In order to illustrate the optimal design procedure for R_c in the RCD snubber, a 25 Watts, 2.1A/12V output, variable frequency flyback converter is taken as an example. The key components and parameters are shown in table I.

TABLE I.   KEY COMPONENTS AND PARAMETERS

Transformer	n	D_c	L_k	C_c
PQ26/20/TP4	10	FR107	6.5 uH	2.2nF

For this converter, the residual energy left in the circuit E_r, the peak voltage of the main transistor V_{ds-pk} and the total power losses dissipated in the RCD circuit P_{loss} are calculated by Equations (9), (12) and (19) respectively. In order to fit them in the same coordinates, all of these three variables are normalized to a unit range and plotted in Fig. 6 as a function of R_c.

It's easy to find that, both of V_{ds-pk} and E_r increase almost linearly as R_c increases. That means both the voltage spike and the CM noises will suffer from larger value of R_c. At the beginning, P_{loss} decreases extremely as R_c increases. The decreasing slop of P_{loss} diminishes as R_c increases. At last, P_{loss} remains almost the same.

To illustrate the optimal design for R_c, the slope difference between E_r and P_{loss}, i.e. $dE_r/dR_c - dP_{loss}/dR_c$, is also plotted in Fig. 6. This curve indicates region II as the optimal designed region for both considerations of EMI and losses because it makes a tradeoff between EMI and losses. But it doesn't mean that region II should always be a good choice for a converter. If efficiency is the key problem of a converter and EMI noises is less important, then region I should be a good choice; If EMI noises is the key problem of a converter and efficiency is less important, then region III should be a good choice. But region III should never be a good choice, because in this region EMI performance will sacrifice much while the efficiency remains almost the same.

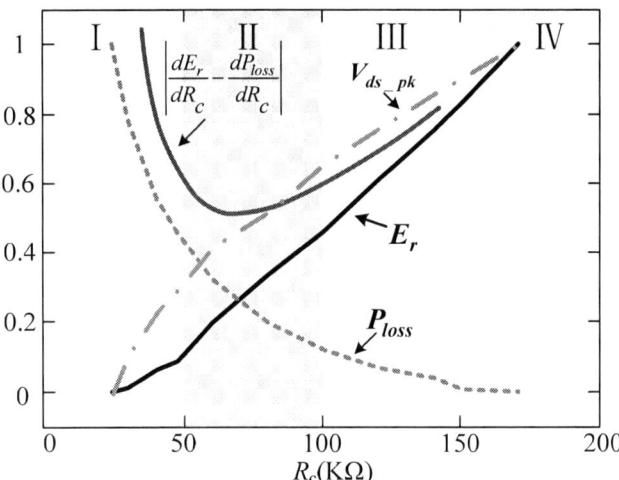

Figure 6.   Optimal design of R_c for both CM noises and losses considerations

IV. EXPERIMENTAL VERIFICATION

Fig. 7 shows the key experimental waveforms when the RCD snubber is charged for the first time with optimal designed R_c=51kΩ. It consists well with the working principle analysis in section II.

Fig. 8 shows the measured efficiency of the converter with different R_c, which verify the calculated power losses in the RCD snubber.

The CM noises of this flyback converter with different R_c are tested, as shown in Fig. 9. In Fig. 9 (a), the CM noise doesn't reduce much when R_c=24kΩ compare with the well designed R_c=51kΩ, but the efficiency of the converter will decrease more than one percent. As shown in Fig. 9 (b), the spikes in the CM noise curve when R_c=200kΩ can be suppressed effectively in the frequency rang higher than 2MHz when R_c is changed to 51kΩ. But the efficiency of the converter when R_c=51kΩ only will decrease about 0.8 percent compare with the efficiency of the converter when R_c=200kΩ. So, when R_c=51kΩ it makes an optimal tradeoff between CM noises and efficiency of the converter.

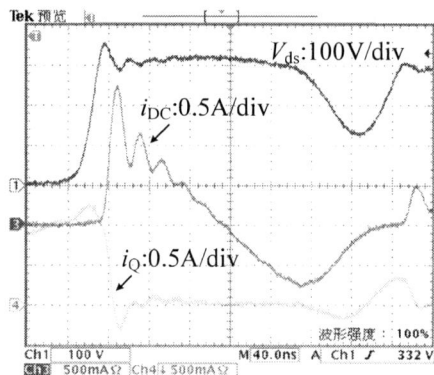

Figure 7.   Key waveforms with R_c=51kΩ

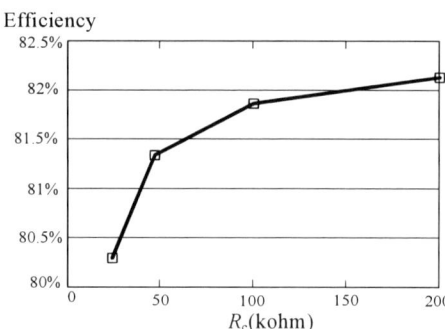

Figure 8.   Measured efficiency with different R_c

dBμV

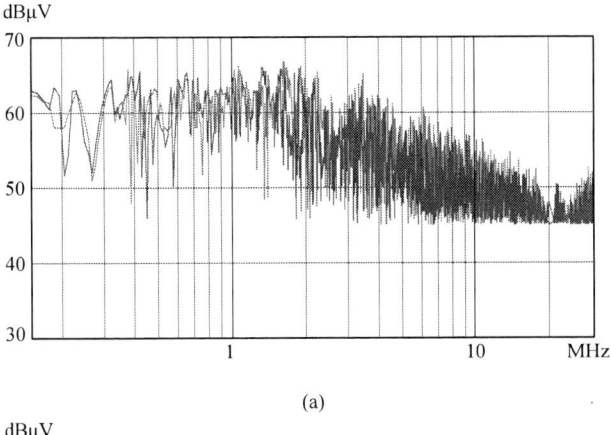

(a)

dBμV

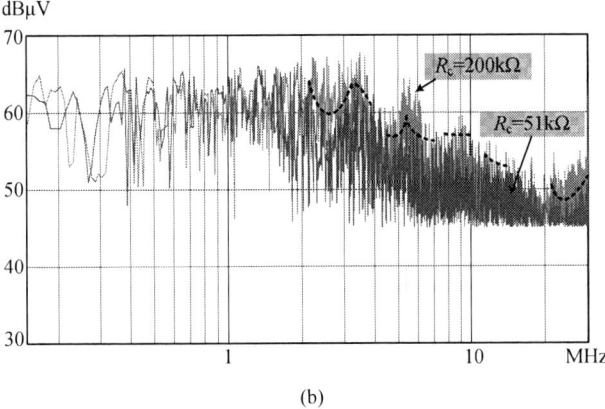

(b)

Figure 9. Comparison the effect of CM noise reduction (a) R_c=24kΩ vs R_c=51kΩ (b) R_c=51kΩ vs R_c=200kΩ

V.   Conclusion

The resistor in the RCD snubber will affect both the EMI performance and the efficiency of the converter. But the RCD snubber design consideration for EMI performance is usually in contradiction with that for efficiency. A detailed model is built up to study into the relationship between the RCD snubber and the EMI performance, and to calculate the power losses in the RCD snubber precisely. Then a novel design procedure is derived for selecting the resistor in the RCD circuit to make an optimal tradeoff between CM noises and efficiency. Experimental results consist well with theoretical analytical results.

ACKNOWLEDGMENT

This work is supported by Natural Science Foundation of China under Grant 50807047 and 50907061.

REFERENCES

[1] Y. Gu, X. Gu, L. Hang, Z. Lu, and Z. Qian, "Improved wide range dual switch flyback DC/DC converters," Applied Power Electronics Conference and Exposition, vol. 1, pp. 654-660, 2004.
[2] H. Chen, W. Dong, Y. He, and Z. Qian, "Secondary side post regulation application in multiple outputs flyback converter," Power Electronics and Drives Systems, vol. 2, pp. 1273-1277, 2005.
[3] C.C. Wen and C.L. Chen, "Magamp application and limitation for multiwinding flyback converter," Electric Power Applications, IEE, vol. 152, pp. 517-525, 2005.
[4] Alenka Hren, Joze Korelic, and Miro Milanovic "RC-RCD clamp circuit for ringing losses reduction in a flyback converter" IEEE Trans. On Circuits and systems—II:Express Briefs. Vol. 53. No. 5 May 2006.
[5] Song-Yi Lin, Chern-Lin Chen, "Analysis and design for RCD clamped snubber used in output rectifier of phase-shift full-bridge ZVS converters," IEEE Transactions on Industrial Electronics, Volume 45, Issue 2, April 1998 pp. 358 – 359.
[6] Patel, H.K., "Voltage transient spikes suppression in flyback converter using dissipative voltage snubbers," in Proc. IEEE ICIEA 2008, pp. 897-901.
[7] He, X., Finney, S.J., Williams, B.W., Green, T.C., "An improved passive lossless turn-on and turn-off snubber," in Proc. IEEE APEC 1993, pp. 385 – 392.
[8] Finney, S.J., Williams, B.W., Green, T.C., "RCD snubber revisited," IEEE Transactions on Industry Applications, Volume 32, Issue 1, Jan.-Feb. 1996 pp. 155 – 160.
[9] Chih-Sheng Liao, Smedley. K.M., "Design of high efficiency Flyback converter with energy regenerative snubber," in Proc. IEEE APEC 2008, pp. 796 – 800.
[10] R. Watson, F. C. Lee, and G. C. Hua, "Utilization of an active clamp circuit to achieve soft switching in flyback converters", IEEE Transactions on Power Electronic, vol. 11, no. 1, pp. 162-169, 1996.
[11] Y.-S. Lee and B.-T. Lin, "Adding active clamping and soft switching to boost-flyback single-stage isolated power factor-corrected power supplies", IEEE Transactions on Power Electronics, vol. 12, no. 6, pp. 1017-1027, 1997.
[12] C. T. Choi, C. K. Li and S. K. Kok, "Modeling of an active clamp discontinuous conduction mode flyback converter under variation of operating conditions", IEEE-PEDS, vol. 2, pp. 730-733, 1999.
[13] Bor-Ren Lin, Huann-Keng Chiang, Kao-Cheng Chen, Wang, D., "Analysis, design and implementation of an active clamp flyback converter," in International Conference of PEDS 2005, pp. 424 – 429.
[14] Rahnamaee, A., Milimonfared, J., Malekian, K., Abroushan, M., "Reliability consideration for a high power zero-voltage-switching flyback power supply," in Proc. IEEE EPE-PEMC 2008, pp. 365 – 371.
[15] W. McMurray, "Optimum snubbers for power semiconductors," IEEE Transactions on Industry Applications, Oct. 1972, pp 593-600.
[16] W. Mcmurray, "Selection of snubbers and clamps to optimize the design of transistor switching converters," IEEE Trans. Ind. Appl., vol. IA-16, Jul–Aug. 1980, pp. 513–523.
[17] Schonberger, J., Feix, G., "Modelling turn-off losses in power diodes," Control and Modeling for Power Electronics 2008, pp. 1 – 6.
[18] Xiangcheng Wang, Qingshui Li and Issa Batarseh, "Transient response improvement in isolated DC/DC converter with current injection circuit", IEEE Applied Power Electronics Conference 2005, pp: 706-710

978-1-4244-4782-4/10 $26.00 © 2010 IEEE

A High Output Power Density 400/400V Isolated DC/DC Converter with Hybrid Pair of SJ-MOSFET and SiC-SBD for Power Supply of Data Center

Rejeki Simanjorang, Hiroshi Yamaguchi,
Hiromichi Ohashi
Energy Semiconductor Electronics Research Laboratory
National Institute of Advanced Industrial Science and
Technology (AIST), Tsukuba, Japan
r-simanjorang@aist.go.jp

Takashi Takeda, Mikio Yamazaki, H. Murai
NTT Facilities Inc.
Research and Development Headquarters
Tokyo, Japan

Abstract— In this paper, a design of high output power density 400/400V isolated dc/dc converter for power supply in data center is discussed. The isolated DC/DC converter uses the hybrid pair of SJ-MOSFET and SiC-SBD power devices. A high accuracy of DC/DC converter analysis is carried out by considering the intrinsic and extrinsic loss parameter of power devices and circuit board. Based on this analysis, the efficiency and power density of isolated DC/DC converter are estimated for various high switching frequencies (100 kHz, 200 kHz and 300 kHz) and various junction temperatures of power switching devices (125^0C and 150^0C). The analysis estimates that the isolated DC/DC converter with efficiency around 97% and power density equal to or larger than 10 W/cm^3 is possible to be realized. To verify the estimation results, 10-W/cm^3 class power density of 5kW isolated DC/DC converter is fabricated and current results are shown in this paper.

I. INTRODUCTION

Growth of electricity consumption for data center increases rapidly as deployment of ICT equipments. It is important to adopt a high efficiency of distribution system in data center. A method to obtain the high efficiency of distribution system is to implement the high voltage (400Vdc) distribution system [1]. The high voltage dc (HVDC) distribution system impacts on reduction of ohmic power loss of distribution system. Reduction of ohmic power loss should be followed by utilization of high efficiency and high power density power converter. The high power density converter results in a minimum size of data center room. The minimum size leads to a minimum energy for lighting and cooling system (air conditioner) of room. As a result, energy conservation in data center is achieved.

The high power density isolated DC/DC converter for telecom or ICT application has been published in [2][3]. This converter uses a soft-switching converter topology, which needs a resonance capacitance and inductance. Power density and voltage rating of the converter are 10-W/cm^3 and

400V/48V, respectively. Application of this converter is aimed to be used at low voltage dc distribution system (48 Vdc).

Considering on the efficiency of 400Vdc distribution system in [1] and demand on the power converters for application in 400Vdc system, we develop a high efficiency and a high power density isolated DC/DC converter for HVDC system of Data Center. This converter has a role as galvanic isolation and voltage stabilizer of DC bus. To realize this converter, we proposed the isolated DC/DC converter (DC/DC converter) with hybrid pair Silicon Super Junction MOSFET (SJ-MOSFET) and Silicon Carbide Schottky Barrier Diode (SiC-SBD). A SJ-MOSFET offers a low on-resistance and low gate charge. Low on-resistance and low gate charge result in low conduction and low switching power loss. The SiC-SBD offers a low on-state voltage drop and its fast switching behavior, which result in low conduction and low switching power loss as well. Both of these power switching represent the new achievement in power switching device developments.

In this paper, we prefer to implement a hard switching converter topology for omitting resonance capacitance and inductance as we found in [2][3]. Combination of these power switching devices and converter topology are aimed to achieve the high efficiency and the high power density of power converter, which have enough possibility to realize 10-W/cm^3 or higher power density.

To design the proposed DC/DC converter, analysis of power loss and estimation of power density of the converter are carried out as a theoretical base. In the analysis, power loss and efficiency of the proposed DC/DC converter is calculated with regard to intrinsic and extrinsic power loss [4]. The power density of proposed DC/DC converter is estimated by considering the volume of each components of DC/DC converter which are available in the market. Moreover, the impact of junction temperature increase for efficiency and power density of proposed DC/DC converter is analyzed as well.

To verify analytical results of the proposed converter, a prototype of high output power density of the DC/DC converter is fabricated. Some of the current results are shown in this paper.

II. APPLICATION OF THE ISOLATED DC/DC CONVERTER FOR HVDC SYSTEM OF DATA CENTER

Application of 400/400V isolated DC/DC converter in HVDC system of data center is shown in Fig. 1. An AC source supplies power to IT racks of data center through a power factor corrector (PFC) and isolated DC/DC converter (DC/DC converter). The DC/DC converter is discussed in this paper.

The DC/DC converter is built from some elements such as single-phase full bridge inverter, high frequency center tap transformer, rectifier and LC filter as shown in Fig. 2. Single phase full bride inverter has two arms.

The hybrid pair of switching pair devices is represented by utilization of SJ-MOSFET and SiC-SBD. The SJ-MOSFET is a silicon power switching device and SiC-SBD is a silicon carbide power devices. In the other words, it can be said that the DC/DC converter combines two varieties of power switching devices based on their material.

III. POWER LOSS ANALYSIS AND ESTIMATION OF POWER DENSITY

A. Analysis Power Loss and Efficiency

Analysis of power loss of the proposed converter is carried out for elements of DC/DC converter [5]. In the analysis, detail parameters of power devices (800V/25A SJ-MOSFET, 180V/30A Si-SBD and 1.2kV/20A SiC-SBD) and circuit board of inverter are considered for a high accuracy of power loss estimation method [4][5]. The considered parameters of power devices are gate voltage, threshold voltage, gate resistance, conductance, capacitance between gate and source, and capacitance between drain and source, respectively [6]. Some parameters of power devices can be obtained in their data sheets [7][8]. The considered parameter of circuit board is stray impedance. These parameters are used in calculation of the intrinsic and the extrinsic power loss of inverter of the DC/DC converter. The intrinsic power loss is defined as the loss relevant to power device parameters, and the extrinsic power loss is defined as loss relevant to stray parameters of circuit board. In the power loss analysis of passive components (transformer and LC filter), the high accuracy of calculation is also adopted [5]. As the result, power loss (P_{LOSS}) of the proposed DC/DC converter can be expressed in general by (1).

$$P_{LOSS} = P_{MOS} + P_{FWD} + P_{OUTCAP} + \qquad (1)$$
$$P_{STRAY} + P_{TRANS} + P_{FILTER} + P_{REC}$$

where P_{MOS}, P_{FWD}, and P_{OUTCAP} are power loss of SJ-MOSFET, blocking diode and output capacitance of SJ-MOSFET, respectively. All of three power losses are categorized as the intrinsic power loss of inverter. P_{STRAY} is

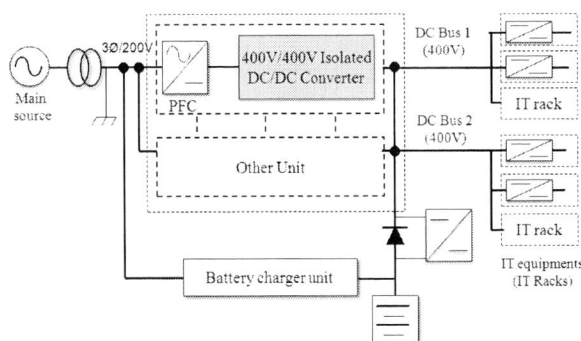

Fig. 1. Distribution system of data center with 400/400V isolated DC/DC converter

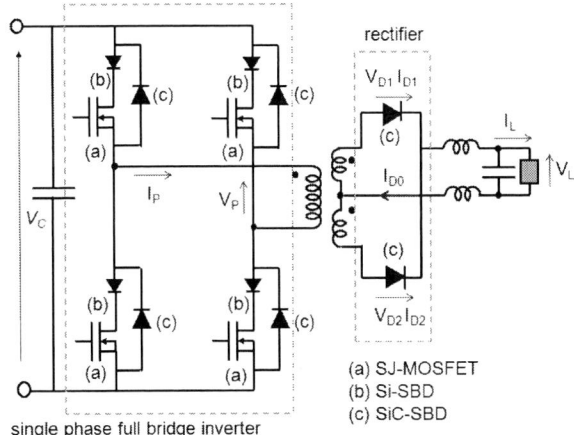

Fig. 2. Main circuit of 400/400V isolated DC/DC converter with hybrid pair of SJ-MOSFET and SiC-SBD

the extrinsic power loss. P_{TRANS}, P_{FILTER} and P_{REC} are power loss of transformer, LC filter and rectifier, respectively. The calculation results of power loss and efficiency are shown in Fig. 3. Here, the junction temperature is 125^0C.

Impact of junction temperature increase for power loss and efficiency of the proposed DC/DC converter is shown Fig. 4. The calculation of Fig. 4 based on the parameters of power switching devices on 150^0C of junction temperature. The power loss increases and the efficiency decreases for changing of junction temperature to be 150^0C. Increase of power loss is caused by increasing on-resistance of power switching devices in high junction temperature.

B. Component Volumes and Power Density

Based on Fig. 3, we define the total volume of elements (power module, transformer, LC filter, rectifier module and heat sink) for various switching frequency as shown in Fig. 5. Volume of components of element is based on commercial data sheet and a design of power module. The power module combines bare chips of SJ-MOSFET, Si-SBD and SiC-SBD to present the single-phase full bridge inverter and rectifier of main circuit in Fig. 2.

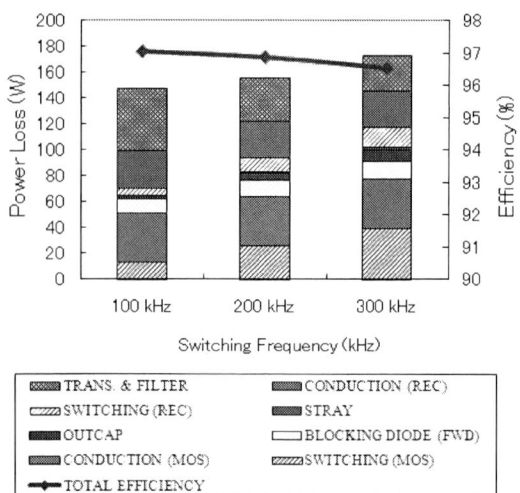

Fig. 3. Analysis of power loss and efficiency of 5 kW isolated DC/DC converter for junction temperature 125°C

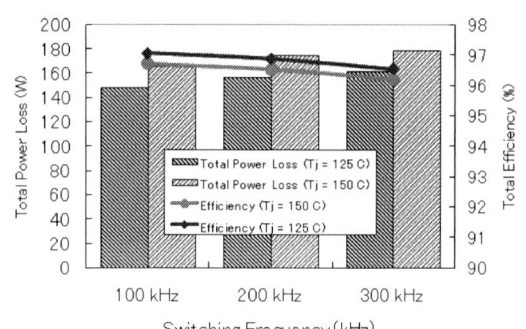

Fig. 4. Power loss and efficiency of 5kW DC/DC converter for different junction temperature (Tj)

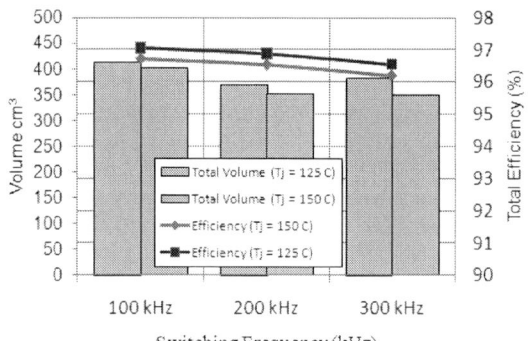

Fig. 5. Volume and efficiency of 5kW DC/DC converter for different junction temperature (Tj)

In our design, a thickness of power module base (Direct Bonded Copper circuit board) and its material layer are taken into account for calculating a required thermal resistance of heat sinks. According to these thermal resistances, the

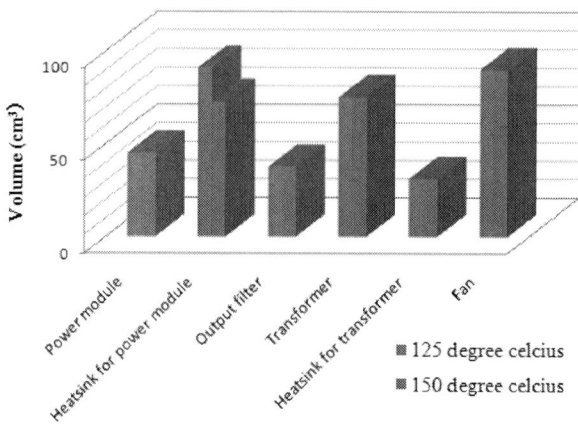

Fig. 6. Distribution of element volumes of 5kW isolated DC/DC converter for different junction temperatures (f_s 200 kHz)

suitable heat sinks are selected from commercial data sheets. To confirm the heat distribution in power model and heat sinks, thermal analysis by ANSYS is carried out.

In Fig. 5, the total volume of the DC/DC converter for two kinds of junction temperatures (125°C and 150°C) is also estimated. According to the estimation, increase of junction temperature can decrease the total volume of the DC/DC converter. This means that the power density increases by high junction temperature. In contrary, the efficiency decreases. The difference of total volume between two junction temperatures is large for the high switching frequency. However, the difference of volume in both junction temperatures is not significant as shown in Fig. 5.

In accordance with Fig. 5, the distribution of element volumes of the DC/DC converter for 200 kHz switching frequency is shown in Fig. 6. In this figure, it is assumed that size of fans is constant for two junction temperatures. As a result of this assumption, this figure shows that increase of junction temperature operation contributes to reduce the heat sink volume of power module (module of inverter and rectifier). But, for the other components, their volumes are constant values.

Volume for gate driver of SJ-MOSFET and controller is not included in Fig. 5 since its size is almost equal for each switching frequencies. According to Fig. 5, the highest possibility to realize high power density is operation of DC/DC converter by 200 kHz. At this switching frequency operation, the total volume of element of converter is estimated around 13.5W/cm³ for junction temperature 125°C and 14.2W/cm³ for junction temperature 150°C, respectively. By considering volume of gate driver, controller and dead-space, 10-W/cm³ or larger power density of DC/DC converter can be estimated for this operation. To realize this power density, the layout of DC/DC converter such components and wiring structure is indispensible and must be considered.

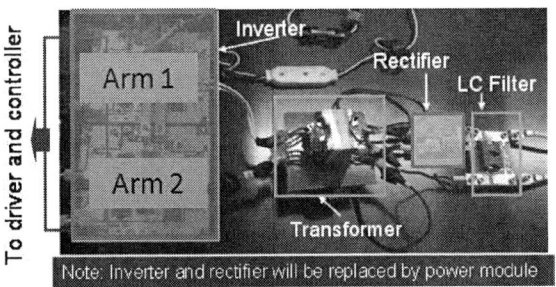

Fig. 7. 400/400V 5kW class of high output power density DC/DC converter prototype

IV. EXPERIMENTAL WORKS AND DISCUSSION

A. Experimental Works

To verify the estimation results in the previous section, we have fabricated the prototype of the DC/DC converter for 10-W/cm^3 5kW class. However, it is still in unbundle elements as shown in Fig. 7. To minimize the occupied volume by inverter and rectifier, the inverter and rectifier of Fig. 7 will be replaced by a power module in the next prototype. The prototype of Fig. 7 is used to study characteristics of the designed converter before built-up a bundled DC/DC converter. The parameters of the fabricated prototype are shown in table I.

In table I, cut-off frequency of LC filter is still high and this cut-off frequency will be reduced in the next prototype by exchanging the current inductor with a larger one.

The current experiment results of the prototype in Fig. 7 are shown in Fig. 8 with duty ratio 42.5 %. The symbols in Fig. 8 are corresponded to Fig. 2. In this experiment, we still limited the input voltage of the DC/DC converter around 200 V with resistive load 33.2 ohm due to the presence of high magnitude oscillation in the rectifier side of the DC/DC converter.

TABLE I.   PARAMETER OF EXPERIMENT

No	Inverter and Rectifier	Rating
1.1	SJ-MOSFET	800V/25A
1.2	Si-SBD	180V/30A
1.3	SiC-SBD	1200V/20A
	Center Tap Transformer	
2.1	Winding ratio	10:12
2.2	Leakage inductor (measured at 100kHz)	≅ 0.5 * 500 nH
	LC filter	
3.1	L = 4.4 uH	C=3.2 uF

Experimental results are shown in Fig. 8 under the following parameters. Switching frequency (f) = 100 kHz, Input voltage (V_C) = 200.17V, Input power = 1.66 kW, Output voltage (Vo) = 231.04V, power (Po) = 1.57 kW and

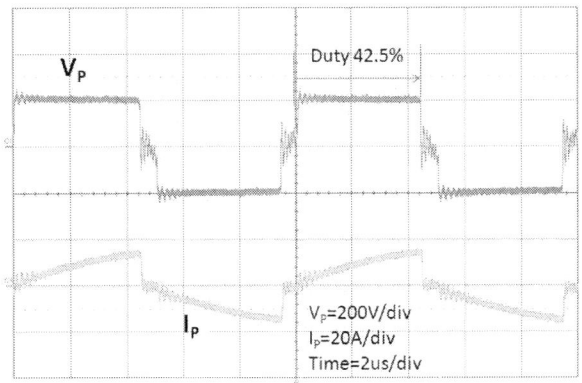

(a)   Output voltage and current waveforms of single phase full bridge inverter

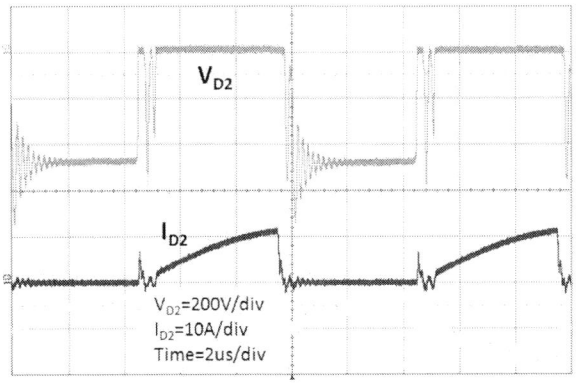

(b)   Voltage and current waveforms of diode rectifier

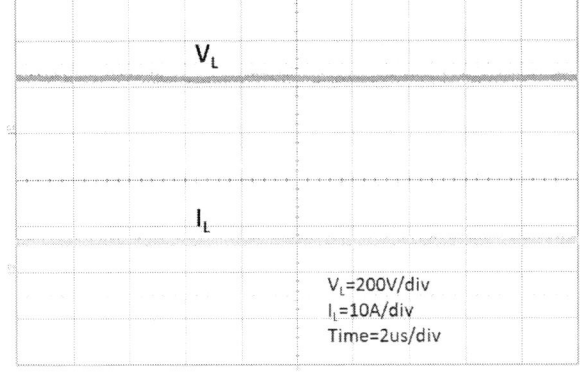

(c)   Output voltage and current at load

Fig. 8. Voltage and current waveforms of the isolated DC/DC Converter

load = 32.2 ohm (31.4% of 5 kW). The efficiency of DC/DC converter obtained under this experimental condition is 94.57 %. The calculated efficiency under the same condition is 96%. This difference seems to be caused by the high frequency transient oscillation at primary side of transformer (V_p and I_p) and

978-1-4244-4782-4/10 $26.00 © 2010 IEEE

secondary side of transformer (V_{D2} and I_{D2}).

Waveform of output voltage and output current of Fig. 8a shows that the high frequency transient oscillation during dead time. Magnitudes of these oscillations are small.

However, while SJ-MOSFET turns ON and rectifier diode turns OFF, magnitude of the oscillation of reverse recovery voltage of rectifier diode is quite high. This reverse recovery voltage can destroy the diode when it is higher than blocking voltage of diode. This oscillation should be minimized and this matter is discussed in the next section in this paper.

Waveform output voltage and current of the DC/DC converter are shown in Fig. 8c. It seems that the ripple of output voltage and current are still large due to small LC filter. In the next prototype, this LC filter will be replaced by the larger capacity, which is considered in the calculation, than capacity of LC filter in this experiment.

B. Discussion

According to Figs. 8b and 9a, the magnitude of high frequency oscillation in diode D2 is about 160 % of OFF state voltage of diode D2 (V_{D2}). Then, consider that OFF state voltage of V_{D2} is 800V for full load condition (V_L at full power is 400V) and blocking voltage of SiC-SBD is 1200V, the percentage of magnitude of high frequency oscillation while diode turn OFF must be lower than 150% of 800V.

Based on the matter above, the full power of the DC/DC converter of Fig. 7 is still not applied since a large magnitude of high frequency oscillation of voltage at diode rectifiers. This oscillation seems to be caused by the leakage inductance of transformer and junction capacitance of SiC-SBD of rectifier.

To investigate the effect of the leakage inductance and the junction capacitance, three configurations (a, b and c) of circuit at secondary side of transformers are shown in Fig. 9. Condition (a) is the same condition with Fig. 8b. Condition (b) is the condition while small inductor (1.1 uF) is added at primary side of high switching frequency of transformers and condition (c) is the condition while resistance and capacitance (RC) snubbers are added at SiC-SBD of rectifier diode.

According to Fig. 9b, it can be explained that addition of inductor can suppress the magnitude of voltage oscillation, but it results in a low frequency oscillation. The low frequency isolation is not good since it may produce another power loss.

According to Fig. 9c, addition of RC snubber can suppress oscillation better than addition inductor and magnitude of oscillation is reduced as well. However, RC snubber results in power loss and needs space. Thus, it may reduce the efficiency and power density of the DC/DC converter.

Another method to suppress the oscillation is to minimize the leakage inductance of transformer or reduce the junction capacitance of SiC-SBD which becomes the next works in this research.

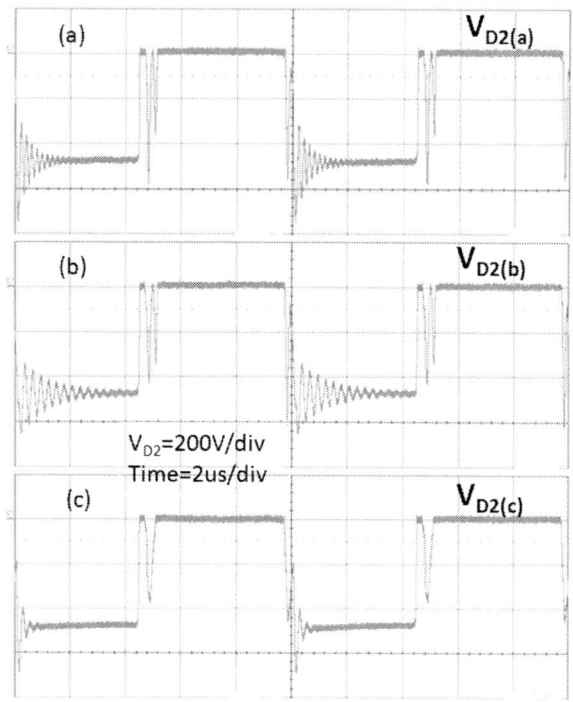

Fig. 9. Reverse blocking voltage of SiC-SBD at rectifier for three conditions ((a). without additional leakage inductance and snubber, (b). with additional leakage inductance and (c) with additional RC snuber)

V. CONCLUSION

Investigation of 400/400V class of the isolated DC/DC converter has been carried out briefly this paper. The output power density of isolated DC/DC converter has been investigated analytically, and it is confirmed that the power density larger than 10-W/cm^3 can be expected by using the high accuracy loss estimation method. This analysis is verified by experimental works.

In the analysis, we have analyzed that the high switching frequency does not always lead to the high power density. Operation at 200 kHz has the highest power density than others. The change of junction temperature from 125 to 150 degree Celsius does not contribute to the significant impact on power density.

For continuing this works, minimization magnitude of high frequency oscillation and design layout of DC/DC converter are in progress and development of the proposed DC/DC converter for a larger capacity than 5kW is studied. Moreover, field test by using the bundled prototype will be done in cooperation with data center provider in Japan.

ACKNOWLEDGMENT

This work was (partially) supported by New Energy and an Industrial Technology Development Organization (NEDO) project, Development of Next-generation Power Electronics Technology.

REFERENCES

[1] Pratt, A, Kumar, P, Adridge, TV: "Evaluation of 400V DC Distribution in Telco and Data Centers to Improve energy Efficiency, " *in Proc. IEEE Intelec* Conf. 2007, pp. 32-39

[2] Juergen Biela, Uwe Badstuener, Johann W. Kolar: "Impact of Power Density Maximation on Efficiency of DC-DC Converter System", IEEE Trans. On Power Electronics, Vol. 24, No. 1, January 2009

[3] Juergen Biela, U Badstuener, Johann W. Kolar: "Design of a 5-kW, 1-U, 10-kW/dm^3 Resonant DC-DC Converter for Telecom Application", IEEE Trans. On Power Electronics, Vol. 24, No. 7, July 2009

[4] Y Hayashi, K. Takao, K. Adachi, H. Ohashi: "Design Consideration for High Output Power Density (OPD) Converter based on Power-loss Limit Analysis Method", *in Proc. Europe Power Electronic Conf. 2005.*

[5] Rejeki Simanjorang, Hiroshi Yamaguchi, Hiromichi Ohashi, Takashi Takeda, Mikio Yamazaki, H. Murai: "Estimating Performance of High Output Power Density 400/400V Isolated DC/DC Converter with Hybrid Pair SJ-MOSFET and SiC-SBD," *in Proc. IEEE Intelec Conf. 2009*

[6] B.J.Baliga: Power Semiconductor devices, PWS publishing company, pp.388–395 (1995).

[7] http://www.ixys.com/Product_portfolio/power_devices.asp

[8] http://www.cree.com/products/power.asp

A 500 W Push-pull Dc-dc Power Converter with a 30 MHz Switching Frequency

John S. Glaser, Juan M. Rivas
ELECTRONIC POWER CONVERSION LAB
GENERAL ELECTRIC GLOBAL RESEARCH
glaser@ieee.org

Abstract—

DESIGNERS of power conversion circuits are under relentless pressure to increase power density while maintaining high efficiency. Increased switching frequency is a primary path to higher power density. Prior work has shown that the use of switching frequencies in the VHF band (30 MHz-300 MHz) are a viable path to the achievement of gains in power density. A promising topology for VHF operation is the voltage-fed Class EF_2 (Class Φ_2) inverter based topology, where the use of controlled impedance at the switching frequency and its 2nd and 3rd harmonics provides both full soft switching and substantially reduced voltage stress compared to topologies such as Class E. However, such converters contain multiple resonant elements, and the tuning of the converter can be complicated due in part to the interaction of said elements. It is proposed that a push-pull version of the Class EF_2 inverter can alleviate some of these difficulties. In particular, it is shown that odd and even frequency components can be independently tuned without interaction, and furthermore that center-tapped inductors may be used to reduce the total volume occupied by said inductors. The benefits include simplified design and increased power density. Evidence is presented in the form of a push-pull Class EF_2 (Class Φ_2) unregulated 500 W prototype dc-dc converter with a 30 MHz switching frequency, an input voltage 150 VDC, and an output voltage of 65 VDC. This converter has an efficiency of $> 81\%$ under nominal conditions, including gate drive power.

I. INTRODUCTION

Designers of dc-dc power converters are under relentless pressure to increase power density, efficiency, reliability, improve transient response, and reduce cost, preferably achieving all these goals simultaneously. In reality, certain goals are more important than others. For example, aerospace applications often have restrictions on overall system mass, and some loads require fast dynamic response. The use of switching frequencies in the VHF (very high frequency, 30 MHz-300 MHz) band is a promising approach to provide substantial gains in power density and bandwidth.

The chief operational principle of efficient power conversion is the periodic controlled storage and release of energy, whereby one regulates the average flow of power from one port to another. In principle, power processing thus accomplished is lossless, and in practice, low losses can be achieved [1]. One of the key contributors to the volume of a power processing circuit is the required energy storage, normally implemented with capacitors and inductors. For a given technology, the size of the energy storage elements is a monotonic, increasing

function of the energy to be stored. Thus, increases in power density require a reduction in energy stored, or an increase in energy storage density. The latter is influenced in part by the physical structure of the energy storage elements, but it is fundamentally dependent upon the materials available. Limitations include the finite electrical and thermal conductivity of relevant materials; dielectric breakdown voltage and permittivity for capacitors, and saturation flux density and permeability for inductors. Improvement in the material properties of conductive, magnetic and dielectric components is generally a slow, incremental process, and major breakthroughs happen infrequently. The only alternative to increased energy density is reduction of the required amount of stored energy per operating cycle. For a circuit processing a specified amount of power, this is accomplished by increasing the switching frequency. Reduction in energy storage will also improve the transient response of the converters.

Up to a point, increasing switching frequency results in increased power density. As switching frequency further increases, issues arise which detract from expected gains. These issues include increased switching losses; proximity and core losses in magnetic components; dielectric losses; and problems with parasitic components. These can be mitigated to some extent, but at a large enough switching frequency they dominate the converter performance. Further increases in switching frequency increase cost and power loss with no attendant increase, or even a decrease, in power density [1]–[3].

Thus, how does one attain the potential power density benefit resulting from high switching frequencies? Magnetic components and semiconductors dominate the design in this frequency range, with semiconductors dominating the efficiency and magnetic components dominating the converter power density. It is shown in [4] that if the frequency is high enough that air core magnetics can be used, the scaling of magnetics is favorable. In particular, it has been shown that for constant heat flux and fixed impedance, inductor volume is inversely proportional to $1/\sqrt{F_{SW}}$. However, the semiconductor switches still present a number of problems. Semiconductor switches have two dominant loss mechanisms: conduction losses and switching losses. A full discussion of semiconductor losses is beyond the scope of this paper, and is well known; see [5] for detailed discussion. The key point is that in conventional hard switching topologies, as semiconductor die size

978-1-4244-4782-4/10 $26.00 © 2010 IEEE

increases, conduction losses are reduced, but switching losses are increased, primarily due to increased device capacitance. Soft-switching topologies have reduced switching losses at the expense of higher semiconductor stresses and conduction losses.

There exist several circuit topologies with low switching loss capable of VHF frequencies. The most promising topologies to date are topologies based on the Class E inverter [6]. Furthermore, with the deliberate control of circuit impedances at the harmonics of F_{SW}, i.e. Class F operation [7]–[9], it is possible to garner additional benefits including reduction of peak stresses and higher F_{SW} for a given semiconductor switch technology. By combining Class E and F operation, the benefits of both may be obtained, and this has been demonstrated analytically and experimentally [10]–[14]. A particularly promising topology for power conversion applications is the voltage-fed Class EF_2 converter (also referred to as Class Φ_2), due to its combination of low transistor voltage stress and greater capability to absorb transistor output capacitance relative to Class E. However, this circuit suffers from some drawbacks, including the use of four inductances which must be tuned for proper operation, and whose tuning interacts. In an effort to alleviate the latter issue and to gain additional advantages, the authors propose a push-pull version of the Class EF_2 dc-dc converter. The benefits of push-pull operation in RF amplifiers are well known [7], [14], [15]. A key advantage relevant to this paper is the ability to independently tune the MOSFET's impedance at the even and odd harmonics, which simplifies the design. Additional advantages include the possibility of dc flux cancellation in magnetic cores, reduction of the total number and size of magnetic components, and a doubling of the ripple frequency. Some of these advantages have been demonstrated in the context of Class EF_x^{-1} RF amplifiers [14], [16].

Section II presents an overview of converter topologies suitable for VHF operation and highlights the advantages a push-pull configuration has in power density and transient response. Section III, briefly reviews the single-ended Class EF_2 converter design and operation, and extends the concept to a push-pull design. This section also shows how the impedance at the odd and even harmonics can be independently tuned to achieve soft-switching and low device voltage stress. Additionally, this section discusses other non-obvious benefits such as the reduction of inductor number and improvements in converter size. Details on the design of a Push-pull Class EF_2 dc-dc converter with a 30 MHz switching frequency, an input voltage 150 VDC, and an output voltage of 65 VDC are presented in Section IV. Experimental evidence on the performance of this approach is presented in Section V in the form of an unregulated 500 W converter with an efficiency of $> 81.5\%$ under nominal conditions, including gate drive power. Conclusions and future work are presented in Section VI.

II. MOTIVATION FOR A PUSH-PULL CLASS EF_2 CONVERTER

Let us first review the basic concepts of VHF dc-dc converters. Dc-dc converters can in general be modeled as an inverter which generates an ac power signal, followed by a rectifier and filter to convert the ac power signal back to a dc signal. In the VHF range, inverters and rectifiers employ soft switching for both turn-on and turn-off semiconductor transitions, keeping switching losses at acceptable levels.

This paper is focused on the inverter and most of the following discussion addresses the challenges in the design of the push-pull implementation of a VHF inverter. There is some discussion of the rectifier, but this topic is open research and a thorough discussion of the rectifier will need to be addressed by another paper.

The most common inverter topologies used in the HF or VHF band are based on Class D, E, or F topologies [2], [6], [13], [17]–[20]. Class D refers to switch mode amplifiers using two transistors, and these may be arranged as either a half-bridge or a push-pull arrangement, such that square waves are generated. According to the conventional definition, Class D does not imply soft-switching; thus, it will not be considered further. Class E generally refers to a single-ended converter with a resonant network that gives soft-switching on both switch transitions. Class F generally refers to a single-ended converter with control of impedance seen by the transistor at one or more harmonics of the switching frequency, shaping the waveforms to reduce stress and/or enhance efficiency. Both Class D and F can be combined with Class E to obtain both soft-switching and wave-shaping.

Class DE operation has desirable characteristics. When implemented with a half-bridge switch, voltage stress is limited to the bus voltage. However, driving the high-side switch with the relatively precise timing required becomes difficult as the frequency increases beyond 10 MHz-20 MHz, due in large part to common-mode currents in the high-side gate drive that flow through the parasitic impedance (usually capacitance) that connects the high side gate drive to the gate control circuit. This circuit can also be implemented as a push-pull design with a center-tapped inductor. In this case, voltage stress is twice the bus voltage, but two ground-referenced switches simplify gate drive design. A major difficulty with this is that true Class D or DE operation requires the coupling of the two halves of the center-tapped inductor to approach unity. As frequency increases, this becomes difficult to achieve, in part due the reduced permeability of practical magnetic cores at VHF frequencies.

The single-ended Class EF_2 (Φ_2) inverter has voltage stress of approximately twice the bus voltage, similar to the push-pull Class DE, but achieves this with inductance between the dc bus and the transistor drain. This suggests that a push-pull version of the Class EF_2 could achieve the relatively low voltage stress of the Class DE inverter, but with the ability to handle a center-tapped inductor with low to moderate coupling, i.e. considerable flux leakage. This allows the

978-1-4244-4782-4/10 $26.00 © 2010 IEEE

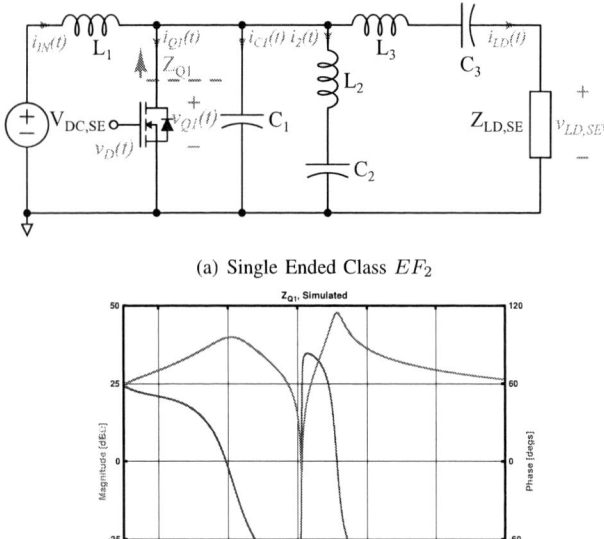

(a) Single Ended Class EF_2

(b) Drain-Source impedance

Fig. 1: Single ended Class EF_2 and typical drain source impedance.

advantageous properties of the of the Class EF_2 (Φ_2) inverter with push-pull operation to achieve the benefits of Class DE operation at VHF switching frequencies.

There are several additional benefits. One anticipated benefit is the doubling of ripple frequencies on both the inputs and outputs, thereby reducing the input and output filter sizes.

III. PUSH-PULL CLASS EF_2 CONVERTER DEVELOPMENT

This section describes the development of the push-pull Class EF_2 converter from the single-ended version. The discussion begins with inverters in order to better illustrate the key points, and the rectifier is discussed later.

A. Single-ended VHF EF_2 inverter

Figure 1(a) shows a single-ended Class EF_2 inverter. Details of operation and a design procedure are given in [12]. The key components are switch Q_1 which turns on and off at frequency F_{SW}; the resonant network formed by L_1, L_2, L_3, C_1, and C_2; and the load $Z_{LD,SE}$ (a resistor in this case). Capacitor C_1 includes any switch output capacitance that may exist. The main point is that the periodic waveforms on switch Q_1 are completely determined by the impedance Z_{Q1} seen by Q_1 at F_{SW} and its harmonics. One of the difficult aspects of the single-ended EF_2 inverter is the proper tuning of all the reactive components. For the Class EF_2 inverter, the impedances $Z_{Q1}(F_{SW})$, $Z_{Q1}(2F_{SW})$ and $Z_{Q1}(3F_{SW})$ are deliberately controlled to provide wave-shaping and ZVS at the semiconductor drain-source terminals. The impedance at higher harmonic components is dominated by the impedance of C_1.

To achieve ZVS the resonant network is tuned to present an inductive impedance at the switching frequency. The series network formed by L_2-C_2 is tuned to be series resonant at $2F_{SW}$ which provides a low impedance value of Z_{Q1} at the drain. Besides providing dc blocking, C_3 along with L_3 set the desired power to the load while at the same time interact with the other resonant elements to shape the switch voltage and lower the voltage stress. As shown in [12] a desirable operating condition occurs when the resonant components are tuned such $Z_{Q1}(F_{SW}) > Z_{Q1}(3F_{SW}) \gg Z_{Q1}(2F_{SW})$. A typical drain to source impedance of a tuned Class EF_2 inverter tuned to operate at 30 MHz is shown in Fig. 1(b). The tuning procedure given in [12] is useful because it allows for the nonlinear output capacitance of a real transistor, e.g. a MOSFET, but requires trial-and-error tuning to get to the final design. This can be difficult because adjustment of any reactive component affects the impedance at all frequencies of interest, i.e. all the resonant components interact with each other.

B. Push-Pull Class EF_2 Inverter

This section develops the push-pull Class EF_2 inverter from the single-ended version. Section III-A points out that a difficult aspect of the single-ended EF_2 inverter design is the tuning of all the reactive components. This requires some trial-and-error tuning to get to the final design, and relies heavily upon the designer's experience because adjustment of any reactive component affects the impedance at all frequencies of interest. Nevertheless, with a transistor model of sufficient accuracy it is possible to tune a design in simulation with a high probability that said design can be translated into functional hardware.

The construction of working hardware has practical difficulties. A key reason for this is that it is extremely difficult to know and model all important circuit board parasitic components prior to the design, and yet these parasitics have a strong influence on the inverter operation. While a major advantage of the Class EF_2 inverter is that these parasitics can be absorbed into the design, the lack of prior knowledge means that once a hardware prototype is constructed, the main resonant components need adjustment to account for the parasitic components. At this point, the fact that each reactive component affects the impedance at all frequencies of interest makes tuning of the circuit a long and iterative process.

The first step in developing the push-pull version of the inverter is to combine two inverters as shown in Fig. 2. The two inverters, denoted by A and B for the top and bottom inverter, respectively, are identical to each other and to Fig. 1(a). The total power processed by this system is now twice that of the single-ended system. The position of C_1 has been changed to better illustrate the separation between switch Q_1 and the rest of the circuit, but the connectivity is unchanged. The key differences are that the input dc voltage V_{DC} is shared, and more importantly, gate drive signals are related by

$$v_{DB}(t) = v_{DA}\left(t - \frac{T_{SW}}{2}\right) \qquad (1)$$

where $T_{SW} = 1/F_{SW}$.

We begin to modify the circuit of Fig. 2 towards a single push-pull converter with a single input voltage source and a single load. The main point to keep in mind is that to achieve Class EF_2 operation, the transistor voltage waveforms must be ideally identical to those for the single-ended inverter. This requires that the drain impedances for the push-pull case be identical to the drain impedance for the single-ended case at multiples of the switching frequency, i.e., $Z_{Q1A}(n\omega_{SW}) = Z_{Q1B}(n\omega_{SW}) = Z_{Q1,SE}(n\omega_{SW})$, $n \in \{0, 1, 2, \ldots\}$.

This appears straightforward until it is understood that when the circuit of Figure 2 is modified to make some components common to both halves, the effective impedance in one half of the push-pull circuit becomes influenced by the voltages and currents in the other half, the latter being driven by a time-shifted gate signal. One way to handle this is by considering the odd and even multiples of the switching frequency separately.

Equation (1) is used because it is not precise enough to say that v_{DA} and v_{DB} are 180° out of phase *in the case where waveforms have harmonic content*. Since periodic waveforms with period T_{SW} are assumed, all waveforms in inverters A and B have frequency components at nF_{SW} where $n \in \{0, 1, 2, \ldots\}$. Furthermore that all waveforms in B will be identical to those in A except for a delay of $T_{SW}/2$. These facts imply that the even components of waveforms of A and B will be in phase, and the odd components will be 180° out of phase, i.e.

$$V_B(n\omega_{SW}) = \begin{cases} V_A(n\omega_{SW}) & \text{if } n \text{ is even} \\ -V_A(n\omega_{SW}) & \text{if } n \text{ is odd} \end{cases} \quad (2)$$

$$I_B(n\omega_{SW}) = \begin{cases} I_A(n\omega_{SW}) & \text{if } n \text{ is even} \\ -I_A(n\omega_{SW}) & \text{if } n \text{ is odd} \end{cases} \quad (3)$$

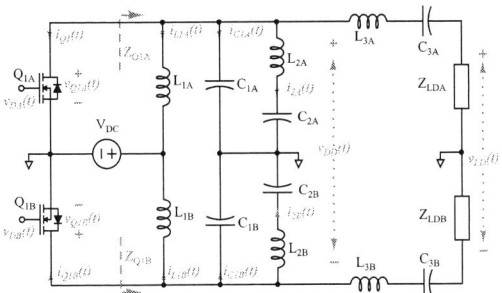

Fig. 2: Two Class EF_2 inverters operating in push-pull fashion.

We can now define odd- and even-mode quantities in the frequency domain (equivalent to common and differential mode) and relate them to the values for a single-ended converter. For voltages $V(\omega)$ and currents $I(\omega)$, the A and B subscripts designate the corresponding nodes in the push-pull circuit, and the SE subscript, the corresponding node on the single-ended

circuit.

$$V_{odd}(\omega) = V_A(\omega) - V_B(\omega) = 2V_{SE}(\omega) \quad (4)$$

$$I_{odd}(\omega) = \frac{1}{2}[I_A(\omega) - I_B(\omega)] = I_{SE}(\omega) \quad (5)$$

$$V_{even}(\omega) = \frac{1}{2}[V_A(\omega) + V_B(\omega)] = V_{SE}(\omega) \quad (6)$$

$$I_{even}(\omega) = I_A(\omega) + I_B(\omega) = 2I_{SE}(\omega) \quad (7)$$

At this point the ω will be implied unless it is necessary to specify a particular frequency. Since the component values are assumed identical in the A and B sections, we can define the even and odd mode impedances in terms of the single-ended impedances [14]:

$$Z_{odd} = \frac{V_{odd}}{I_{odd}} = Z_A + Z_B = 2Z_{SE} \quad (8)$$

$$Z_{even} = \frac{V_{even}}{I_{even}} = \frac{1}{2}(Z_A + Z_B) = \frac{1}{2}Z_{SE} \quad (9)$$

The next step is to merge the two loads $Z_{LDA} = Z_{LDB} = Z_{LD,SE}$ into a single load $Z_{LD} = 2Z_{LD,SE}$, as Fig. 3 shows. Clearly $I_{LDA} = -I_{LDB}$ in all cases, so that $I_{LD,even} \equiv 0$, i.e. the well-known cancellation of even harmonics. This means that the load now only affects the transistor drain impedance at odd harmonics. For a typical Class EF_2 inverter, this has little effect on the drain impedance due to the combined effect of the series network L_3-C_3, the second harmonic short L_2-C_2, and the dominant effect of C_1 at the third and higher harmonics.

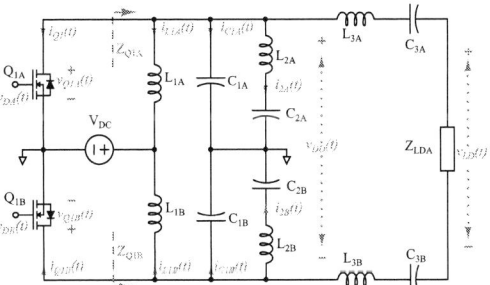

Fig. 3: Two Class EF_2 inverters operating in push-pull fashion with a single load.

The next step is to merge the two dc feed inductors L_{1A} and L_{1B} into a single center-tapped inductor. This is commonly done in push-pull RF amplifiers. In the ideal case, the two halves of the inductor would have perfect coupling, i.e. no leakage inductance. If this could be achieved, the even-mode impedance due to this component would vanish, thereby causing it to act as a short circuit to all even harmonics. This would do away with the need for a separate second harmonic short, and would allow the circuit to operate as a push-pull Class DE inverter. Unfortunately, good coupling is difficult to achieve for high power components at VHF frequencies; the second harmonic short will be addressed below. Another benefit is a result of the fact that the inductance is a super-linear function function of the number of turns. Recall that the odd-mode inductance of L_1 is simply the inductance with the center-tap open. All else being equal, the push-pull (odd-mode) value of this inductance should be twice the single-ended value. However, this will not require twice the turns, hence the push-pull inductor will not double in size compared to the single-ended inductor, even though the push-pull power is double the single-ended power. This gives the push-pull inverter a power density advantage over the single-ended inverter. Finally, for cases that employ magnetic cores, dc flux cancellation occurs. This allows a larger flux swing and thus a smaller core. The latter assumes that such a design would not be core loss dominated; there are promising developments in magnetic materials that may allow this benefit to be realized [21].

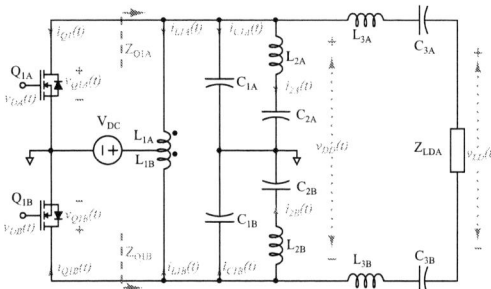

Fig. 4: Push-pull Class EF_2 inverter with a center-tapped dc feed inductor L_1.

Finally, we can address the second harmonic short. In the single-ended Class EF_2 inverter, the series resonant network formed by C_2 and L_2 forms a short circuit at the second harmonic, but it still has a substantial effect on the impedance at other components of the switching frequency, especially at the fundamental frequency. It is documented in [14] that a network can be specifically designed to have different odd- and even-mode impedances. There are two simple ways that this can be accomplished for the push-pull Class EF_2 inverter.

Figure 5 shows the first way. The second harmonic short network is accomplished with C_2, L_{2A}, and L_{2B}. Since L_{2A} and L_{2B} are identical, $I_{C_2,odd} \equiv 0$, and so the odd mode components of the drain voltage see only $Z_{L_{2A}} + Z_{L_{2B}}$. The even mode components of the drain voltage see $Z_{C_2} + \frac{1}{2}(Z_{L_{2A}} + Z_{L_{2B}})$. Thus, we can adjust the value of L_{2A} and L_{2B} to get the desired impedance at the transistor drains for F_{SW} and $3F_{SW}$ without regard to the second harmonic. Once this is done, C_2 is selected to form a series resonant short for the second harmonic $2F_{SW}$, and this can be done without affecting any odd frequency components. Higher harmonics, both even and odd, are dominated by the drain capacitances, as with the single-ended inverter. It is possible to implement L_{2A} and L_{2B} as a single center-tapped inductor. This will beneficially present a higher odd-mode impedance, and the smaller even-mode impedance would allow a larger C_2 with a lower voltage rating.

Figure 6 shows the second way, which is dual to the first. The second harmonic short network is accomplished with L_2, C_{2A}, and C_{2B}. Since C_{2A} and C_{2B} are identical, $I_{L_2,odd} \equiv 0$, and so the odd components of the drain voltage see $Z_{C_{2A}} + Z_{C_{2B}}$. The even components of the drain voltage

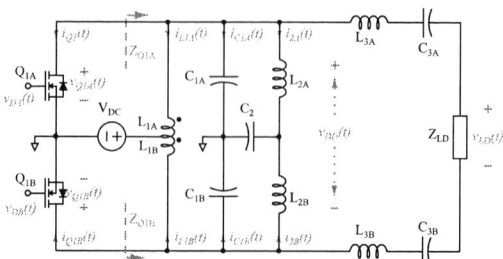

Fig. 5: Push-pull Class EF_2 inverter with center-tapped dc feed inductor and single capacitor (C_2) for second harmonic shunt.

see $Z_{L_2} + \frac{1}{2}(Z_{C_{2A}} + Z_{C_{2B}})$. Thus, we can adjust the value of C_{2A} and C_{2B} to get the desired transistor drain impedance at F_{SW} and $3F_{SW}$ without regard to $2F_{SW}$. Once done, L_2 is selected to form a series resonant short for the second harmonic $2F_{SW}$, and this can be done without affecting odd frequency components. Again, all higher harmonics are dominated by C_{1A} and C_{1B}.

In theory, both methods should work. In practice, the selection depends on the specific design. Presently it is unknown how to select the best method *a priori*. At this point, the key concepts to the push-pull inverter have been discussed. This is far from a systematic design procedure, but is sufficient to implement a design.

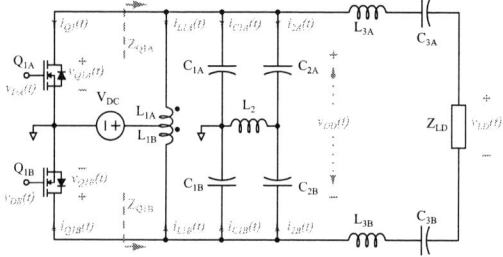

Fig. 6: Push-pull Class EF_2 inverter with center-tapped dc feed inductor and single inductor (L_2) for second harmonic shunt.

C. Push-Pull Rectifier

The development of the push-pull resonant rectifier from the single-ended version follows the same path as for the inverter [13]. The single-ended resonant rectifier is shown in Fig. 7(a), and the push-pull version in Fig. 7(b). Figure 7(b) shows two separate inductors L_{4A} and L_{4B}. It is possible to combine these into a single center-tapped inductor as with the inverter inductors L_{1A} and L_{1B}, and realize similar benefits.

IV. CONVERTER DESIGN

This section describes the design of push-pull converter with the nominal specifications in Table I. These specifications were chosen based on the desire to demonstrate the converter at power levels consistent with conventional printed-circuit board techniques. A systematic procedure for the optimal design of Class EF_2 converters does not yet exist. However, a general set of guidelines combined with some iteration can yield useful results. This section illustrates this design process.

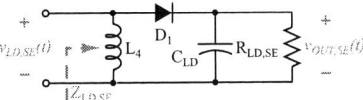

(a) Single-ended resonant rectifier.

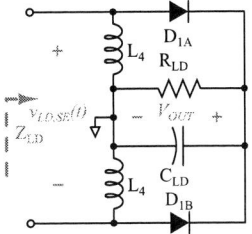

(b) Push-pull resonant rectifier.

Fig. 7: Single-ended and push-pull resonant rectifier implementations

The overall process consists of designing a simulation model of a converter, getting the simulation running and then building the converter based on the simulation model. It is likely that there will be numerous parasitic components not accounted for in the simulation. These parasitics can and should be minimized via the use of RF design techniques, but it is difficult to anticipate all the parasitics and pre-assess their values, and some cannot be eliminated due to the physical size of parts required to handle the design power. Fortunately, the Class EF_2 converter is highly tolerant of parasitics, and they can be accounted for by using the simulator to plot the expected drain impedance under a selected bias condition, and then measuring the same impedance on the real circuit board. Components are then adjusted in value until one obtains a close match between simulation and measurement of even- and odd-mode impedance values at F_{SW}, $2F_{SW}$, and $3F_{SW}$.

TABLE I: Nominal specifications for prototype push-pull Class EF_2 dc-dc converter

Parameter	Name	Value	Units
DC input voltage	V_{IN}	150	[V]
DC output voltage	V_{OUT}	65	[V]
Output power	P_{OUT}	500	[W]
Switching frequency	F_{SW}	30	[MHz]

First we choose power semiconductor devices, namely transistors Q_{1A} and Q_{1B} and rectifiers D_{1A} and D_{1B}. The transistor is a Microsemi ARF475FL [22], which is actually a pair of 500V RF MOSFETs in an RF package designed for push-pull operation. A single ARF475FL will be used for both Q_{1A} and Q_{1B}. The rectifier is a silicon carbide Schottky diode, a Cree CSD10060A rated for 10A, 600V, which is a proven performer at the frequency of interest. A Schottky diode is required to eliminate reverse recovery and associated losses. Each diode D_{1A} and D_{1B} comprises two CSD10060As in parallel, so that the complete rectifier uses four physical diodes. The use of parallel diodes reduces the rectifier losses considerably [13].

Good models of the semiconductor devices are necessary. These devices have inherently nonlinear I-V and C-V char-

acteristics. Behavioral models are sufficient, possessing the benefit of relatively simple parameter extraction and fast simulation speed. For the ARF475FL, a close comparison with the Microsemi ARF521 [23] shows that each transistor in the ARF475FL package likely uses the same die as the ARF521. This allows the use of the ARF521 Spice^{TM} model developed and used successfully in [13]. The diode model was developed using I-V and C-V measurements as input to the modeling procedure in [24]. A small stray series inductance (7nH) was incorporated into the diode model.

The next step is to design the rectifier. A simulation of the rectifier of Fig. 7(b) is built using the diode model and driven by a sinusoidal current source of frequency F_{SW}. The current amplitude and inductance value of L_{4A} and L_{4B} are adjusted until the correct power is obtained and the fundamental component of V_{LD} is in phase with the driving current. At this time, an effective equivalent load resistance can be determined.

The next choice is which second harmonic network to use. For this converter, the network in Fig. 6 yielded smaller inductor values, and had the practical benefit that a single L_2 was easier to tune in the prototype than a capacitor. Now the push-pull inverter is designed based on the load resistance from the rectifier, following the same procedure as given in [12]. Fig. 8 gives the simulation schematic. The key difference in tuning from the single-ended version is the simulated measurement of the even and odd drain impedance. Two 1 amp current AC sources are used to inject a test current at each drain, and the voltages measured. Due to the A-B symmetry, if the currents are injected in phase, a measurement of the drain voltages will yield the even-mode drain impedance

$$Z_{D,even} = \frac{1}{2(1A)}\left(V_{Q1A} + V_{Q1B}\right) \qquad (10)$$

The odd-mode impedance is determined by changing the phase of the test current sources so that they are out of phase, which will give the odd-mode impedance

$$Z_{D,odd} = \frac{1}{(1A)}\left(V_{Q1A} - V_{Q1B}\right) \qquad (11)$$

The final impedance magnitude plots thus obtained are shown in Fig. 11, where they are plotted along with the experimental results. The values are of meaning only at the corresponding components of the switching frequency (pointed out in the figure). Similar plots are used to aid the design iteration process. The final plots are used to as a reference to guide the tuning of the hardware prototype.

Figure 9 gives the key simulated waveforms of the of the full converter.

V. EXPERIMENTAL RESULTS

Figure 10 shows a photograph of the 500 W push-pull Class EF_2 dc-dc converter prototype with a 30 MHz switching frequency, an input voltage of 150 VDC and an output voltage of 65 VDC.

Figure 11 shows the measured even- and odd-mode impedances. These were measured by applying 100VDC to

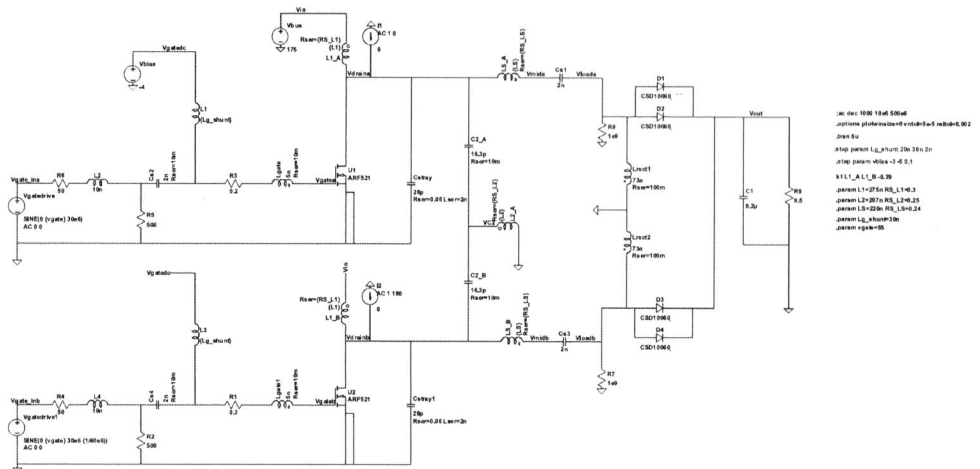

Fig. 8: Simulation schematic of push-pull Class EF_2 dc-dc converter including push-pull resonant rectifier.

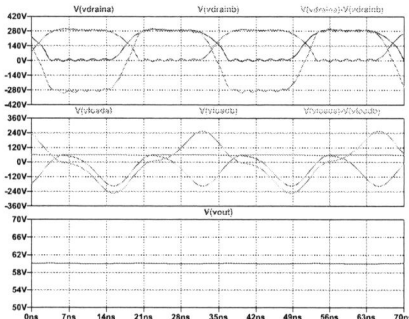

Fig. 9: Simulation results for converter of Fig. 8. Top pane shows the two drain-source voltages and the drain-drain voltage. Middle pane shows comparable waveforms for rectifier input. Bottom plane shows the output voltage. The time scale is 7 ns/div.

Fig. 10: Photograph of 500 W, 150 VDC to 65 VDC with 30 MHz switching frequency.

the drain and 50VDC to the output in order to bias the semiconductor junctions to an approximate operating point. These bias conditions are the same used for the simulation. These were measured using a network analyzer in conjunction with a balun and dc blocking capacitors to allow the odd-mode measurement to be made. Note the close match at F_{SW}, $2F_{SW}$, and $3F_{SW}$. At the other less critical frequencies the impedance plots, though similar, are not identical. This is due to parasitic component value estimation errors the range of about 30MHz to 100MHz. The additional resonances below

30MHz have been traced to changes in the gate drive circuit from the model. Nevertheless, the critical frequencies match, and the converter operates as expected.

Figure 12 shows the waveforms for the converter of Fig. 10. Note the out of phase waveforms for the two drain voltages, as expected. Note also that they are not completely identical, which indicates that the circuit is not completely balanced; nevertheless, it appears to tolerate the slight imbalance.

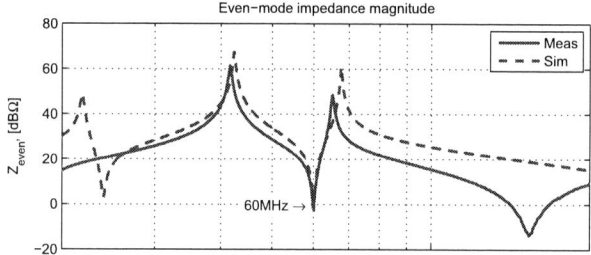

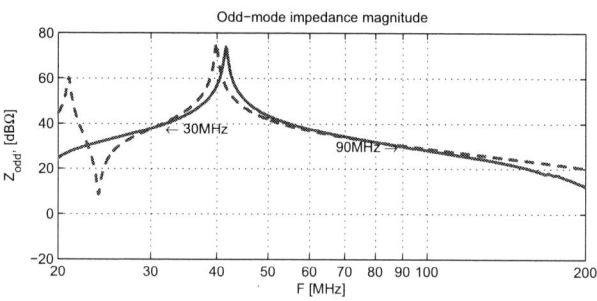

Fig. 11: Experimental and simulated push-pull Class EF_2 drain impedance plots. DC bias voltages set to 100VDC for the drain and 50VDC for the rectifier.

The electrical performance of the prototype converter of Fig. 10 was measured and results are given in Table II. Gate power given is the difference of forward and reflected power. Total efficiency η_{Total} includes the effect of gate drive power; drain efficiency η_{Drain} does not. Gate power was supplied via an RF amplifier, and was calculated from the difference of forward and reverse power.

TABLE II: Electrical performance of converter of Fig. 10 at two operating points.

V_{IN} [V]	I_{IN} [A]	V_{OUT} [V]	I_{OUT} [I]	P_{GATE} [W]	$P_{IN,DC}$ [W]	$P_{IN,Total}$ [W]	P_{OUT} [W]	η_{Drain} [%]	η_{Total} [%]
140.0	4.04	65.430	7.212	1.88	565.6	567.5	471.9	83.4	83.2
150.1	4.48	65.575	8.391	1.7	672.4	674.1	550.2	81.8	81.6

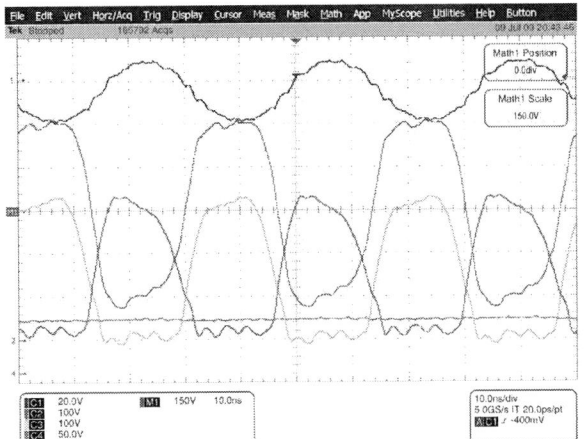

Fig. 12: Waveforms of converter of Fig. 10. Chan. 1 trace (20V/div) shows V_{gs} drive waveform for one transistor. Chan. 2 and 3 traces (100V/div) show V_{ds} for each for each transistor. Chan. 4 shows the output voltage (50 V/div). Chan. M1 (150 V/div) shows the drain-to-drain voltage V_{dd}. The time scale is 10 ns/div.

VI. CONCLUSION

A push-pull Class EF_2 (Class Φ_2) dc-dc power converter topology suitable for operation in the VHF frequency range has been proposed along with some considerations and advantages of the converter. The concept is verified via an experimental 500 W, 150 VDC to 65 VDC converter with a 30 MHz switching frequency. This converter has an efficiency $> 81.5\%$ at full power and $> 83\%$ at slightly reduced power.

The topic of efficient power processing with switching frequencies in the VHF band is an area of open research. The main goal of this paper is to extend the art of VHF power stage topologies, and to experimentally demonstrate feasibility at power levels of 0.1 to 1 kW. An additional goal is to stimulate research into areas that need to be addressed. First, the converter of this paper is a power stage only, without a practical gate drive. Second, the design of such converters is not very systematic and depends heavily upon the experience of the designer. Third, it is an unregulated power stage, and this needs to be addressed for widespread adoption. Finally, the main goal of this work is not only to increase the power density entitlement of power conversion, but to achieve this benefit in practice. Although progress is being made on all fronts, much work remains.

REFERENCES

[1] R. Erickson and D. Maksimovic, *Fundamentals of Power Electronics*. Springer, second ed., January 2001.

[2] C. Xaio, *An Investigation of Fundamental Frequency Limitations for HF/VHF Power Conversion*. PhD thesis, Virginia Polytechnic Institute, July 2006.

[3] J. Kassakian and M. Schlecht, "High-frequency high-density converters for distributed power supply systems," *Proceedings of the IEEE*, vol. 76, pp. 362–376, April 1988.

[4] D. Perreault, J. Hu, J. Rivas, Y. Han, O. Leitermann, R. Pilawa-Podgurski, A. Sagneri, and C. Sullivan, "Opportunities and challenges in very high frequency power conversion," in *Applied Power Electronics Conference and Exposition, 2009. APEC 2009. Twenty-Fourth Annual IEEE*, pp. 1–14, Feb. 2009.

[5] B. J. Baliga, *Power Semiconductor Devices*. PWS Publishing Company, 1996.

[6] N. Sokal and A. Sokal, "Class E-A new class of high-efficiency tuned single-ended switching power amplifiers," *IEEE Journal of Solid-State Circuits*, vol. 10, pp. 168–176, June 1975.

[7] S. C. Cripps, *RF Power Amplifiers for Wireless Communications*. Artech House Publishers, second ed., 2006.

[8] F. Raab, "Class-f power amplifiers with maximally flat waveforms," *Microwave Theory and Techniques, IEEE Transactions on*, vol. 45, no. 11, pp. 2007–2012, 1997.

[9] F. Raab, "Maximum efficiency and output of class-f power amplifiers," *Microwave Theory and Techniques, IEEE Transactions on*, vol. 49, no. 6, pp. 1162–1166, 2001.

[10] J. Glaser, J. Nasadoski, and R. Heinrich, "A 900w, 300v to 50v dc-dc power converter with a 30mhz switching frequency," in *Applied Power Electronics Conference and Exposition, 2009. APEC 2009. Twenty-Fourth Annual IEEE*, pp. 1121–1128, Feb. 2009.

[11] Z. Kaczmarczyk, "High-efficiency class E, EF_2, and E/F_3 inverters," *IEEE Transactions on Industrial Electronics*, vol. 53, pp. 1584–1593, Oct. 2006.

[12] J. M. Rivas, Y. Han, O. Leitermann, A. Sagneri, and D. J. Perreault, "A high-frequency resonant inverter topology with low voltage stress," *2007 IEEE Power Electronics Specialists Conference (PESC 2007)*, pp. 2705–2717, June 2007.

[13] J. Rivas, *Radio Frequency dc-dc Power Conversion*. PhD thesis, Massachusetts Institute of Technology, September 2006.

[14] S. Kee, I. Aoki, A. Hajimiri, and D. Rutledge, "The class-e/f family of zvs switching amplifiers," *Microwave Theory and Techniques, IEEE Transactions on*, vol. 51, no. 6, pp. 1677–1690, 2003.

[15] H. L. Krauss, C. W. Bostian, and F. H. Raab, *Solid State Radio Engineering*. Wiley, first ed., 1980.

[16] Z. Kaczmarczyk and W. Jurczak, "A pushpull class-e inverter with improved efficiency," *Industrial Electronics, IEEE Transactions on*, vol. 55, pp. 1871–1874, April 2008.

[17] R. Redl, B. Molnar, and N. Sokal, "Class E resonant regulated dc/dc power converters: Analysis of operations, and experimental results at 1.5MHz," *IEEE Transactions on Power Electronics*, vol. 1, pp. 111–120, April 1986.

[18] J. Rivas, R. Wahby, J. Shafran, and D. Perreault, "New architectures for radio-frequency DC/DC power conversion," in *IEEE 35th Annual Power Electronics Specialists Conference PESC 04*, vol. 5, pp. 4074–4084, June 2004.

[19] S.-A. El-Hamamsy, "Design of high-efficiency RF class-D power amplifier," *IEEE Transactions on Power Electronics*, vol. 9, pp. 297–308, May 1994.

[20] M. Kazimierczuk and J. Jozwik, "Resonant dc/dc converter with class-E inverter and class-E rectifier," *IEEE Transactions on Industrial Electronics*, vol. 36, no. 4, pp. 468–478, 1989.

[21] S. Lu, Y. Sun, M. Goldbeck, D. Zimmanck, and C. Sullivan, "30-MHz power inductor using nano-granular magnetic material," in *Power Electronics Specialists Conference, 2007. PESC 2007. IEEE*, pp. 1773–1776, 2007.

[22] Microsemi Corporation, "ARF475FL Datasheet," June 2007.

[23] Microsemi Corporation, "ARF521 Datasheet," February 2007.

[24] R. Kielkowski, *SPICE: Practical Device Modeling*. McGraw-Hill Professional Publishing, August 1995.

978-1-4244-4782-4/10 $26.00 © 2010 IEEE

Input-Series Connnected High Frequency DC-DC Converters with One Transformer

Deshang Sha, Zhiqiang Guo, and XiaoZhong Liao

School of Automation.
Beijing Institute of Technology
Beijing 100081, China
shadeshang@bit.edu.cn, Guozq@bit.edu.cn,liaoxiaozhong@bit.edu.cn

Abstract—**Input -series connected DC/DC converters based on one transformer is proposed for high voltage input and high power applications. Phase-shifted full-bridge (PS-FB) control strategy with common duty cycle control is adopted for each converter. Operation principles of the proposed converter are analyzed. Although primary currents vary for magnetic coupling and turns mismatch, combined peak current control is implemented to limit current sharp increase due to magnetic saturation. The effectiveness of the proposed topology and its control strategy are verified by simulation and experimental results of a 20kw-50kHz prototype.**

I. INTRODUCTION

Switching devices have to sustain high voltage stress with conventional full-bridge topology when the input voltage is very high. An insulated gate bipolar transistor (IGBT) should be used instead of a MOSFET because of the high input voltage [1], but switching frequency of an IGBT is limited for its turning off characteristics. High switching frequency is helpful to improve dynamic performance of a converter and a MOSFET is a good candidate. However, it can not be used in a conventional topology for the high input voltage. A multilevel converter is a solution for this purpose, but the reliability can not be guaranteed for large quantity of diodes or flying capacitors [2]. Input- series output - parallel (ISOP) converter consists of several DC-DC converter modules connected in series at the input and in parallel at the output, but separated control is needed for each module and what's more, to make each modular input voltage and output current the same, voltage balance and current sharing are necessary, which complicate the control system further [3]-[4]. Common-duty-ratio control results in stable operation for ISOP converters [5], but control of each modular converter is separated and output current sharing must be considered. Master/slave control schemes can only be used in small applications for electrical isolation problem [6]. In high voltage input and large power applications, operation safety of switches should also been guaranteed to improve the reliability of the whole converter. Switching loss can be minimized for full-bridge converters with FB-ZVS control[7].

This paper proposes a new high frequency DC-DC converter using MOSFETs as switches. The proposed converter is composed of two PS-FB DC-DC converters connected in series. Both of them not only shares one transformer but also have the common duty cycle control. To enhance the stability of the converter, combined peak current control is also implemented to enhance the reliability under large power or transient conditions.

II. CIRCUIT OF DESRIPTION AND OPERATION PRINCIPLES

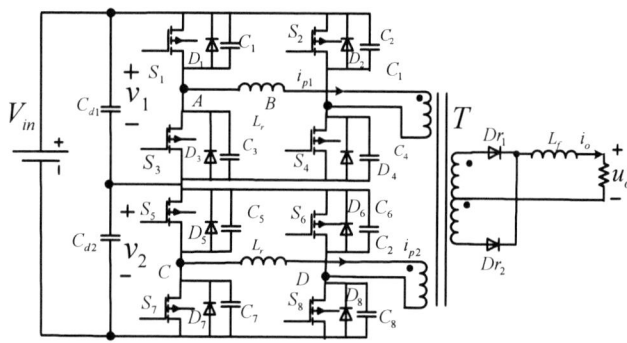

Figure 1. Input-series connected full-bridge converters with one transformer

A. Main Cirucit

The schematic of the proposed converter is shown in Fig.1. In this configuration, two PS-FB DC-DC converters are connected in series at the input and share the same output based on one transformer T with center-tapped secondary side. The input voltage V_{in} is divided by the input-series connected capacitors C_{d1} and C_{d2} .Actually, turns and leakage inductance of the two primary windings are not exactly the same for manufacturing reasons while the two secondary turns can be viewed the same for fewer turns. Turns ratio is $N_1/N_2/N_3/N_3$. L_{r1} and L_{r2} are primary leakage inductances and they are supposed to equal L_r. To understand the operation principles, some assumptions were made,

---

This work was supported by National Natural Science Foundation of China (No.50807005)

B. Moded anlaysis

- Output current I_o is assumed to be constant, thus output inductor current ripple is zero.

- All circuit parameters are ideal.

- Turns N_1 and N_2 are roughly equal but not exactly the same and the magnetizing inductance of the two primary windings is infinite.

- Input voltage V_{in} and voltage V_o are unchanged.

- All snubber(C_1-C_8) capacitors have the equal value C_r.

- Both dividing capacitors C_{d1} and C_{d2} equal C_d.

Since phase-shifted PWM control is taken for both the two converters with common duty cycle control. Thus, the operation modes of the proposed prototype can be classified as follows,

- Mode 1: The input power is transferred from the primary side to the secondary side either through MOSFETS S_1, S_4, S_5 and S_8 or through MOSFETS S_2, S_3, S_6 and S_7.

- Mode 2: Snubber capacitors of leading leg switches are charged and discharged during dead time interval.

- Mode 3: Freewheeling mode happens and no load is connected to input capacitors C_{d1} and C_{d2}.

- Mode 4: Snubber capacitors of lagging leg switches are charged and discharged during dead time interval.

- Mode 5: Duty cycle loss occurs and induced voltage of the transformer primary windings are short circuited for both diodes D_{r1} and D_{r2} conduct at the same time.

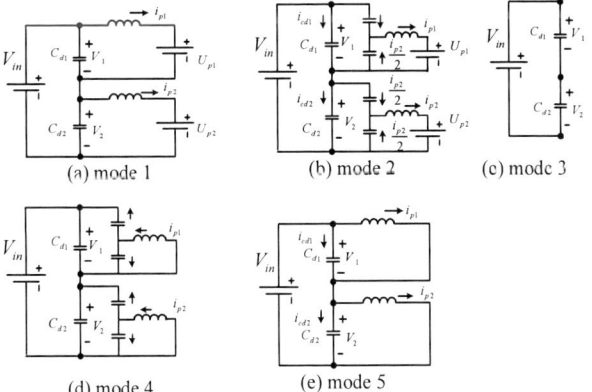

(a) mode 1   (b) mode 2   (c) mode 3

(d) mode 4   (e) mode 5

Fig.2 Equivalent circuits of different working modes

Equivalent circuits of each working mode can be seen from Fig.2. During mode 1,by Faraday's law, we have

$$v_1 = L_{r1}\frac{di_{p1}}{dt} + N_1\frac{d\phi}{dt} \qquad (1)$$

$$v_2 = L_{r2}\frac{di_{p2}}{dt} + N_2\frac{d\phi}{dt} \qquad (2)$$

Note that for the ideal transformer the total flux Φ links all the windings. The ideal transformer obeys the following relationship,

$$N_1 i_{p1} + N_2 i_{p2} = N_3 I_o \qquad (3)$$

Differentiation of (3) on both sides leads to,

$$N_1\frac{di_{p1}}{dt} + N_2\frac{di_{p2}}{dt} = N_3\frac{dI_0}{dt} = 0 \qquad (4)$$

According to (1),(2) and(4),we can have

$$\frac{d\phi}{dt} = \frac{N_1 v_1 + N_2 v_2}{N_1^2 + N_2^2} \qquad (5)$$

Eliminating voltage drops across the leakage inductors L_{r1} and L_{r2} yields

$$\frac{v_1}{v_2} = \frac{N_1}{N_2} \qquad (6)$$

Mode2 begins when mode 1 is finished and in this mode, snubber capacitors of leading leg switches are charged and discharged. It should be noted that magnetic coupling among all windings still exits, so (3) is also satisfied in this mode. Current ripple over output filtering inductor L_f can be neglected and all snubber capacitors are the same, and therefore voltage error between the two dividing voltages can be expressed as

$$v_1 - v_2 = V_{1A} - V_{2A} + \frac{1}{2C_d}\int_0^{\Delta t_1}(i_{p2} - i_{p1})dt \qquad (7)$$

where V_{1A} and V_{2A} are the steady state values of v_1 and v_2 in mode 1 respectively. Δt_1 is time duration of mode2. From the value of ip1 and ip2 at the end of mode1, we can know the variation of v_1 and v_2.Because capacitance of C_r is very small, so the time duration of mode 2 is so short that can be neglected compared with one complete switching period. During mode 3 each of primary currents flow through one switch and body diode of another switch. No loads are connected to the two input capacitors. Thus, both v_1 and v_2 do not vary at all. Mode 4 starts at the beginning of charge and discharge of subber capacitors parallel connected with lagging switches. This mode is similar to mode 2 except that transformer primary windings are short circuited in this mode.

$$v_1 - v_2 = V_{1B} - V_{2B} + \frac{1}{2C_d} \int_0^{\Delta t_2} (i_{p1} - i_{p2})dt \qquad (8)$$

where V_{1B} and V_{2B} are the values of v_1 and v_1 at the end of mode 3.Small time interval Δt_2, time duration of this period, is so short that impact on the input voltages during this period can be eliminated too. It can be concluded from above, influence of primary turns on voltage sharing can be neglected in mode 2,mode 3 and mode 4. But that is not the case in mode 5 in which duty cycle loss occurs, with all transformer windings being short circuited. Only leakage inductances of primary windings work as loads for C_{d1} and C_{d2} in this mode, assuming the a small voltage variation over C_{d1} is Δv, then we get

$$v_1 = V_{1C} + \Delta v \qquad (9)$$

$$v_2 = V_{2C} - \Delta v \qquad (10)$$

where V_{1C} and V_{2C} are the initial values at the beginning of this mode. The voltage variation can be expressed as

$$\Delta v = \frac{1}{2C_d} \int_0^{\Delta t_3} (i_{p2} - i_{p1})dt \qquad (11)$$

where Δt_3 is very small referring to the time duration of duty cycle loss sustaining. Thereby, Δv is negligible compared to V_{1C} or V_{2C}.Therefore, derivatives of primary currents can be written as follows

$$V_{1c} = L_r \frac{di_{p1}}{dt} \qquad (12)$$

$$V_{2C} = L_r \frac{di_{p2}}{dt} \qquad (13)$$

Supposing initial values of i_{p1} and i_{p2} in this mode are the same. If $V_{1c} > V_{2c}$, then $i_{p1} > i_{p2}$,causing Δv negative. Hence, v_1 decreases while v_2 increases until in the end new equilibrium that v_1 becomes V_{1c} and v_2 reaches V_{2c} is got again. On the other hand, this applies when $V_{1c} < V_{2c}$. This phenomenon in this mode can therefore mitigate the problem of voltage sharing inequality due to turns mismatch of primary windings.

C. *Stability of phase-shited PWM control based on common duty cycle.*

If a disturbance Δv is superimposed on input dividing voltages, as stated before, mode 5 in which duty cycle loss happens can overcome the disturbance by changing charging currents for input capacitors. The time duration of mode 2 and mode 4 is so short during one complete switching period

that can be ignored for stability analysis under disturbances. It is worth noting during mode 3, once equilibrium has been reached, no disturbance can be made because charging currents of the two input capacitors C_{d1} and C_{d2} are always same at any time. Before being disturbed, voltage sharing of steady state is V_{1d} and V_{2d} which can be expressed as shown in (6) .Assuming v_1 and v_2 are perturbed according to(9) and (10),if $\Delta v > 0$ then i_{p1} increases while i_{p1} decrease, which causes v_1 decreasing and v_2 increasing until eventually they reach their steady state condition before the disturbance. It can be concluded from above, voltage disturbances over dividing capacitors can be overcome for the proposed PS-FB converter with common duty cycle control.

III. PEAK CURRENT CONTROL STRATEGY

A. *Peak Current Control*

Although the waveforms of primary currents are totally different from those of single-input PS-FB DC/DC converters, primary currents still have close relationship with output current because of magnetic coupling. This can be revealed through (3) without consideration of magnetizing current when magnetic core is not saturated . But when the core becomes saturated, both the primary current i_{p1} and i_{p2} containing magnetizing currents which increase abruptly and even dominate, thus (3) is not satisfied anymore. As we know, with peak current control, cycle by cycle current limit can be achieved, which can improve the reliability of switches especially under large power or transient applications.

B. *Implementation of Triple Closed Loop Control*

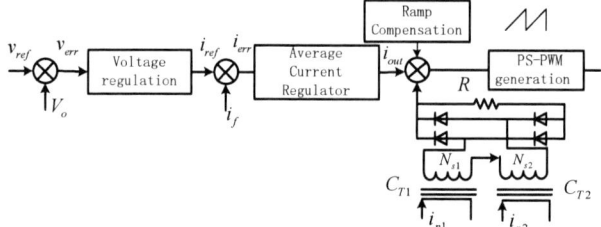

Figure 3.Block diagram of the control

Realization of peak current control for the proposed converter can be seen from Fig.3, in which current reference i_{ref} compares with feedback current proportional to the current through inductor L_f. Primary currents i_{p1} and i_{p2} are measured by current transformers CT_1 and CT_2 whose turns ratio are proportional reversely to the primary windings of transform T, then outputs of CT_1 and CT_2 connected in series are rectified by full-bridge diodes and combined peak current i_{peak} can be got across the resistor R. Combined peak current together with the ramp compensation i_{ramp} compare with sum of current regulator output i_{out} and DC offset, generating PS-FB-PWM gating signals for all the switches of the proposed converter. Apart from the inner loop made up of peak current control, there are another two closed loops, which is indicated through Fig.5.The voltage v_{ref} compares with the voltage feedback v_{of} and we can get voltage error v_{err}, which is then calculated by voltage regulator. This is the outer closed loop and the output of this loop works as the current reference for the middle close loop, whose feedback is proportional to output current through filtering inductor L_f.

IV. SIMULATION AND EXPERIMENTAL RESULTS

To confirm the validity of the proposed converter with common duty cycle control, simulation results were carried out using SABER. The parameters chosen for simulation are V_{in}=600V; V_o=40V; I_o=500A; C_{d1}= C_{d2}=10µF; S_1 to S_8: two IXFB100N50P type MOSFETs in parallel. Turns ratio $N_1/N_2/N_3$ = 41:40:10, transformer leakage inductance L_{r1}=8µH,L_{r2}=7µH; D_{r1} and D_{r2}: DAC2F150N4; C_{d1}=C_{d2}=12µF; L_f=25µH and switching frequency f_s=50kHz.

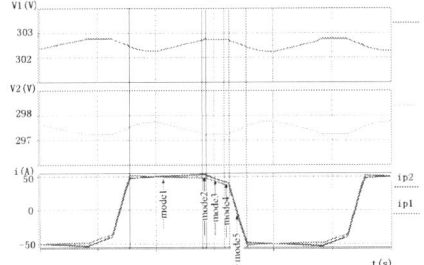

Figure 4.Simulation results of steady state

Fig.4 shows some simulation results of steady state during one complete duty cycle. The average value of dividing voltages are 303V and 297 V respectively, Their ratio is approximately proportional to the ratio of their corresponding primary turns as indicated in (6).During mode 1,small voltage ripple exist because of the filtering inductance is not infinite. Both v_1 and v_2 are kept unchanged during mode 3 in which no loads are connected to input capacitors. The impact on sharing of input voltage during mode 2 and mode 4 can be negligible for very short time transition. In mode 5, duty cycle loss occurring, voltage balance due to turns ratio inequality can be alleviated as seen from Fig.4

 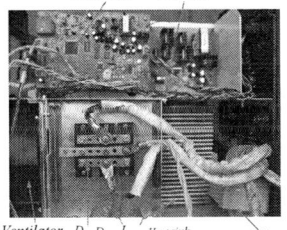

(a) one side   (b) another side

Figure 5.Prototype of the proposed 20kW DC/DC converter

Experiments of a 20kW DC/DC converter indicated in Fig.5 with the same parameters mentioned above were conducted to verify the performance and the peak current control strategy. The measured primary currents are shown in Fig.6, although each primary current changes differently, their sum according to the turns ratio is similar to that of a single input PS-FB DC-DC converter. Fig.7 illustrates gating signals generation with the peak current control, because of the 0.8V DC offset of the controller, gating signals for lagging leg switches are generated while the gating signals for leading leg switches are synchronous with the ramp waveform. Fig.8 gives the voltage across the two input

dividing capacitors at full load, each of them shares almost the half of the input voltage.

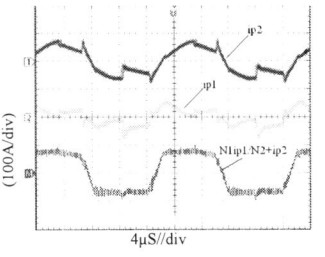 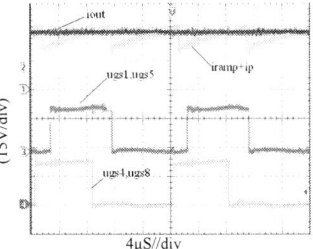

Figure 6.Transformer primary curents   Figure 7. PS-FB PWM generation

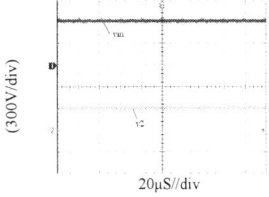

Figure 8. Voltage sharing of the inpout voltage

V. CONCLUSION

Input-series connected converters sharing one transformer enable the use of MOSFETs with low voltage ratings working at high switching frequency and thereby improving dynamic response of the system. With common duty cycle control for each modular converter, input voltage balance can be achieved well although turns mismatch of the primary windings. The voltage sharing is determined to the extent that corresponding turns ratio of primary windings. It is also stable under disturbances. Combined peak current control based on all both primary currents of the proposed converter is effective in generating PS-PWM signals for the converter, improving reliability of the converter.

REFERENCES

[1] D. S. Sha and X. Z. Liao, "Digital control of switch mode pulsed welding power," in Proc. IEEE ECCE, 2009, pp 2746-2749

[2] F.H. Khan and L. M. Tolbert, "A Multilevel Modular Capacitor-Clamped DC–DC Converter," IEEE Trans. Ind. Appl.,2007,43(6), pp 334–337

[3] J. W. Kim, J. S. You and Cho, B.H. , " Modeling , control, and design of input-series-output-parallel-connected converter for high-speed-train power system ,"IEEE Trans. Ind. Electron.,2001,48(3), pp. 536–544

[4] R. Giri, V. Choudhary, R. Ayyanar and N. Mohan, "Common duty ratio control of input-series connected modular DC-DC converters with active input voltage and load-current sharing," IEEE Trans. Ind. Appl., 2006, 42(4), pp. 1101–1111

[5] X. B. Ruan, W. Chen, L.Chen and C. K. Tse, "Control strategy for input-series-output-parallel converters," IEEE Trans. Ind. Electron., 2009, 56(4), pp. 1174–1185

[6] P. J. Grbovic, "Master/slave control of input-series-and output-parallel-connected converters: concept for low-cost high-voltage auxiliary power supplies," IEEE Trans. Power. Electron.,2009,24(2),pp.316–328

[7] G. B. Koo, G. W. Moon and M. J. Youn , "New Zero voltage-switching phase-shift full-bridge converter with low conduction losses," IEEE Trans. Ind. Electron., 2005, 52(1), pp. 228–235

978-1-4244-4782-4/10 $26.00 © 2010 IEEE

Simple Photovoltaic Solar Cell Dynamic Sliding Mode Controlled Maximum Power Point Tracker for Battery Charging Applications

Emil A. Jimenez Brea, Eduardo I. Ortiz-Rivera, *IEEE Member*, Andres Salazar-Llinas, Jesus Gonzalez-Llorente

Electrical Engineering Department,

University of Puerto Rico, Mayaguez Campus

Mayaguez, Puerto Rico

emil. jimenez@ece.uprm.edu, eduardo.ortiz@ece.uprm.edu, andres.salazar@ece.uprm.edu, jesus.gonzalez@ece.uprm.edu

Abstract - **In this paper, we present a maximum power point tracker and estimator for a PV system to estimate the point of maximum power, to track this point and force it to reach this point in finite time and to stay there for all future time in order to provide the maximum power available to the load. The load will be composed of a battery bank. This is obtained by controlling the duty cycle of a DC-DC converter using sliding mode control. The sliding mode controller is given the estimated maximum power point as a reference for it to track that point and force the PV system to operate in this point. This method has the advantage that it will guarantee the maximum output power possible by the array configuration while considering the dynamic parameters temperature and solar irradiance and delivering more power to charge the battery. The procedure of designing, simulating and results are presented in this paper.**

I. Introduction

In the actuality a lot of research work has been conducted to improve the use of the sun's energy. The generation of electricity using photovoltaic solar cells has been one of the most researched and studied. Photovoltaic is the technology that uses solar cells or an array of them to convert solar light directly into electricity. The power produced by the array depends directly form factors that are not controlled by the human being as the cell's temperature and solar irradiance. Usually the energy generated by these solar cells is used to provide electricity to a load and the remaining energy is saved into batteries.

Photovoltaic cells have a single operating point where the values of the current and voltage of the cell result in a maximum power output. These values correspond to a particular resistance which is equal to the division of the maximum voltage and maximum current. By connecting the PV cell directly to a load or a battery, the output power can be severely reduced due to load mismatching or, in case of a battery, load voltage mismatching. Since this operating point depends on factors like temperature, solar irradiance and load impedance, a device capable of tracking the maximum power point and force the PVM to operate at that point is required. A maximum power point tracker (MPPT) is a device capable of search for the point of maximum power and, using DC-DC converters, extracts the maximum power available by the cell. By controlling the duty cycle of the switching frequency of the converter we can change the equivalent voltage of the cell and by that, its

equivalent resistance into the one in which the PVM is in the maximum power operating point.

Several methods have been designed and implemented to search for this operation point. A common method is the Perturb and Observe (P&O) algorithm [1-2]. Classical P&O algorithms tend to measure the converter's output power in order to modify the input voltage by modifying the converter's duty cycle. Another common method is the hill climbing method [3-4].This method is based on a trial and error algorithm in where the voltage is increased until you reach such voltage where the PV exhibits maximum power. Other MPPT algorithms sample the open circuit voltage and operate the PV module at a fixed percent of this voltage. Incremental conductance algorithms are another method to track the MPP [5-7]. Other methods that have been used to obtain the maximum power are parameters estimations [9], neural networks [10] and linear reoriented method [12].

Some of the disadvantages with these methods are that some of them require doing a lot of iterations to calculate the optimal steady state duty ratio. Some of them use approximate values that do not guarantee near maximum power output. Some of them can be very complex, can be slow and can become instable if the MPP moves abruptly.

In this paper, we present an implementation of a maximum power point tracker, based in reaching a reference open circuit voltage, using a sliding mode controller to control the duty cycle of a DC-DC converter in order to force the PV module to operate at its maximum power point, for a given temperature and irradiance, to improve the utilization of the produced energy when connected to a load. For this case the load it's a battery and a resistance.

II. Proposed System

Figure 1 shows the proposed scheme for the MPPT.

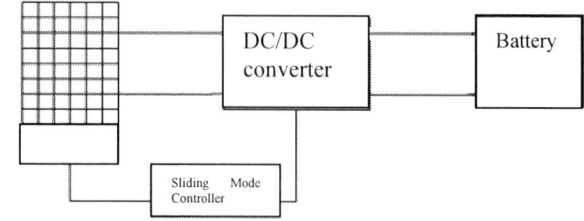

Figure 1: Proposed system scheme

This system use a PV array (*s* x *p*) composed of *s* in series cells and *p* in parallel cells. It is then connected to a

DC-DC converter in order to increase or decrease the desired voltage. It is then connected directly to the load, which is composed of a 12 V battery. The duty cycle of the converter is controlled by a sliding mode controller.

The proposed model will guarantee the extraction of the maximum power that can be produced by the PVM while regulating the load voltage to the battery's voltage. That way we can have a workable load voltage that can be connected to an inverter while matching the load resistance to the PV optimal resistance.

III. Mathematical Background
This section presents mathematical terms used in the paper.

- Re( ) Function - It extracts the real part a of a complex number written in the form $a+bi$. A complex number is a number which can be formally defined as an ordered pair of a real number
- Lambert's W Function, $lambertw(x)$ - The Lambert's W function, which was named after Johann Heinrich Lambert, is defined to be the solution $W(x)$ of the non linear equation, $W(x)\exp(W(x)) = x$.

IV. PV Model equations
In the past, there have been different types of models to estimate the non-linear equations of the photovoltaic module (PVM). Some of these models are the Anderson's, Bleasser and, the most common, the one diode model. All these models present a good approach into estimating the solar cell voltage and currents but most of them need too much computational power or need information not available in the manufacturer's sheet. A more suitable model to simulate a PV module is proposed by [13][15]. The PVM model is know as the Ortiz PVM model. In that work, a PV model was proposed where analytical equations relates the PV output current with the PV output voltage, temperature and solar irradiance over the PV module. It also shows experimental results validating the accuracy and effectiveness of the proposed model. An advantage of this model is that all the needed information can be found in the manufacturer's data sheet. Also it shows how the PV power is affected by changes in the temperature and solar irradiance. The equations are the followings:

$$I(V) = \frac{Ix}{1 - \exp\left(\frac{-1}{b}\right)}\left[1 - \exp\left(\frac{V}{b \cdot Vx} - \frac{1}{b}\right)\right] \quad (1)$$

$$P(V) = V \cdot I(V) = \frac{V \cdot Ix}{1 - \exp\left(\frac{-1}{b}\right)}\left[1 - \exp\left(\frac{V}{b \cdot Vx} - \frac{1}{b}\right)\right] \quad (2)$$

$$Vx = s \cdot \frac{E_i}{E_{in}} \cdot TCV \cdot (T - T_N) + s \cdot V\max$$
$$- s \cdot (V_{max} - V_{min}) \cdot \exp\left(\frac{E_i}{E_{in}} \cdot \ln\left(\frac{V_{max} - V_{oc}}{V_{max} - V_{min}}\right)\right) \quad (3)$$

$$Ix = p \cdot \frac{E_i}{E_{in}} \cdot \left[I_{SC} + TCi \cdot (T - T_N)\right] \quad (4)$$

Ix and Vx represent the short circuit current and open circuit voltage at a given temperature and solar irradiance. V is the PVM output voltage, T is the PVM temperature, T_N is the standard conditions temperature, E_i is the effective solar irradiance at the PVM, E_{in} is the standard condition solar irradiance, TCV is the open circuit voltage temperature coefficient and TCi is the short circuit current temperature coefficient. V_{max} is the open-circuit voltage at 25°C and more than 1200W/m². V_{min} is the open-circuit voltage at 25°C and less than 1000W/m².

Figure 1 shows the non linear relation between the current and the voltage given by the (1). Figure 2 shows the non linear relation between the power and the voltage, given by (2) under standard conditions. We can see that the maximum power produced by the PVM occurs at a certain voltage level. Since the function of power depends only of the voltage and it is differentiable for all values of voltage, then maximum power that can be extracted from the PVM will occur when the partial derivate of the power with respect to the voltage is equal to cero. The partial derivate of the power against voltage is given in the following equation:

$$\frac{\partial P(V)}{\partial V} = \frac{Ix - Ix \cdot \exp\left(\frac{V}{b \cdot Vx} - \frac{1}{b}\right)}{1 - \exp\left(\frac{-1}{b}\right)} - V \cdot \frac{-Ix \cdot \exp\left(\frac{V}{b \cdot Vx} - \frac{1}{b}\right)}{b \cdot Vx - b \cdot Vx \cdot \exp\left(-\frac{1}{b}\right)} \quad (5)$$

By equaling (5) to zero and solving by the voltage we can obtain the optimal voltage which is given by (6).

$$Vop = \text{Re}\left(b \cdot Vx\left(lambertw\left(-0.36787944 \ e^{\frac{1}{b}}\right) + 1\right)\right) \quad (6)$$

From (6) we can obtain a very approximate estimate of PV cell's output voltage at which maximum power occurs. The advantage of this equation is its dynamic property. The only variant term is Vx. Given that we can express the equation as the following:

$$Vop = C \cdot Vx \quad (7)$$

$$C = \text{Re}\left(b \cdot \left(lambertw\left(-0.36787944 \ e^{\frac{1}{b}}\right) + 1\right)\right) \quad (8)$$

Since it depends on Vx and Vx vary with respect to temperature and solar irradiance, the optimal voltage will vary with respect to the conditions of the temperature and irradiance, giving always an estimation of the required voltage necessary to extract the maximum power from the PV cell for all external conditions.

V. Sliding Mode Controller Surface
A sliding mode controller is a variable structure control where the dynamics of a non linear system is altered via the application of a high frequency switching control. In sliding mode control, the trajectories of the system are forced to reach a sliding manifold of surface, where it exhibit desirable features, in finite time and to stay on the manifold for all future time. It is achieved by suitable control strategy. To apply sliding mode control we have to know if the system can reach the sliding manifold. Once the systems reach the sliding manifold, the controller has to force the system to stay in the manifold for all future time.

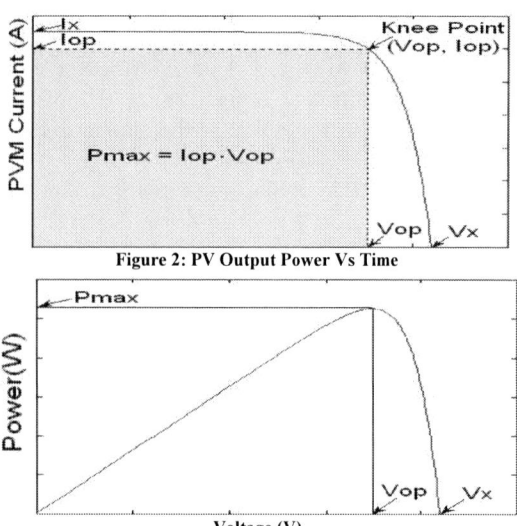

Figure 2: PV Output Power Vs Time

Figure 3: PV Output Power Vs Time

Sliding Mode Control is widely use for a lot of applications including control systems for DC/DC converters [8][14], power supply, electric grid connections[9], motors speed regulator[14], position control system, among others. To design the sliding mode controller we have to select the desired surface. We want to obtain the maximum power that can be extracted from the PV module at the given temperature and irradiance conditions. From (6) we can relate that maximum power to an optimal voltage. Since we know the output voltage we have to have in order to extract the maximum power from the PV system, we choose a surface that will force the system to reach that voltage in a finite time and stay there for infinite time. With that in mind, we chose the following sliding manifold:

$$\sigma = V - V_{op} \qquad (9)$$

V is the output voltage of the PV cell and Vop is the optimal voltage. This sliding manifold will assure us to force all the trajectories of the system to reach the optimal voltage and to keep it in the optimal voltage for all future time. Since the optimal voltage is dynamic since it change when changes occur in the temperature and irradiance this sliding surface is also changing with respet to the temperature and irradiance giving us a dynamic sliding surface. The sliding mode will be controlling the duty cycle of a switching device. So the switchin device will have two operation state:

$$\begin{cases} On & V - V_{op} > 0 \\ Off & V - V_{op} < 0 \end{cases}$$

Now, the controller will behave in the following way:

$$u = \begin{cases} 1 & V - V_{op} > 0 \\ 0 & V - V_{op} < 0 \end{cases}$$

A control law that guarantees us that our controller will behave in that way is given by the following equation:

$$u = \frac{1}{2} + \frac{1}{2} \cdot sign\left(V - V_{op}\right) \qquad (10)$$

This law of control also guarantees us that the system trajectories will reach the proposed manifold and will stay there for all future time. This can be explained in a practical way. At first, because the PVM is not connected, the PVM output voltage will be equal to its open circuit voltage. Since the open circuit voltage is greater than the optimal voltage the switching device will be on. When the switch is on, the PVM output voltage will begin to drop because of the load mismatching until it reaches below the optimal voltage. Then the switch will turn off creating an open circuit condition forcing the PVM output voltage to increase up to its open circuit voltage. When the output voltage passes the optimal voltage then the switch will turn off and the sequence will start again and will continue for all future time. This control law works for a variety of DC-DC converters like the Buck converter, SEPIC converter and Buck-Boost converter.

VI. Simulation

The system was simulated using Matlab's Simulink software with the power system toolbox. With this software we simulate and test the sliding mode controller and the proposed model. The simulink model is shown at figure 4.

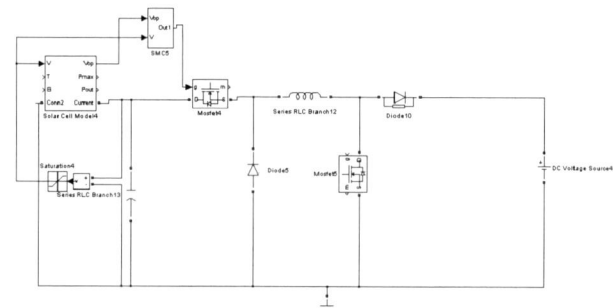

Fig 4: Simulation Scheme for the Proposed Model.

The solar Cell model was represented by a single block composed by a Matlab Embedded function containing the equations of a solar cell [13] [15]. The system was simulated under constant ambient temperature and solar irradiance and under varying ambient temperature and solar irradiance in order to validate the effectiveness of the controller. (1) and (6) were calculated and compare to the manufacturer's data sheet for several PV commercial models. The simulation results of the PV system, with constant ambient conditions, connected directly to the battery and connected to the battery by a non-inverting Buck-Boost converter, in Buck mode, are shown in figure 5 and 6. Simulations of the PV system, with varying ambient conditions, connected directly to the battery and connected to the battery by a non-inverting Buck-Boost converter, in Buck mode, were done. The converter in Buck mode only uses one Mosfet, the left one, and the second Mosfet is turn off. Finally, the simulations showed in figure 5, are for standard conditions for each PVM and figure 6, are for varying temperature and solar irradiance.

VII. Results

Table 1 and 2 shows the results of the estimated optimal voltage *Vop, Iop* and *Pmax* compared to the manufacturer's datasheet value. These tables validate (6) and (1) for the optimal voltage and optimal current estimation. The error percent for the voltage and current stays within acceptable values. Since the estimation are very near to the optimal values given by the manufacturer, by forcing the system to operate at the estimated voltage, guarantee us to be working in a near maximum power point. Figure 5 shows the temperature and irradiance over the PVM. Figures 6 and 7 validate the sliding mode controller and ensure that the PV operation point is in the knee point of the power vs. voltage graph were the PV operates at its maximum power even under standard conditions(STC) and under varying ambient temperature and solar irradiation condition while supplying a higher power to the battery. Table 3 shows the percentage of increment in the power given to the battery. It can be seen that the proposed method increase significantly the available power delivered to the battery. Figures 8 and 9 show the simulation results of the power that is given to the battery for two different connection modes, directly to the battery and through a converter.

PV Model	Vop Data-sheet	Vop Esti-mated	Error %	Iop Data-sheet	Iop Esti-mated	Error %
SiemenSP75	17.0	17.593	3.49	4.40	4.2524	3.35
Shell SQ80	17.5	18.156	3.74	4.58	4.4305	3.27
SLK60M6	30.6	30.762	0.53	6.86	6.8193	0.59
Solare SX-5	16.5	16.673	1.05	0.27	0.2669	1.17
SolarxSX-10	16.8	17.098	1.774	0.59	0.5789	1.87

Table 1: Comparison of PV voltage and current estimated values vs. Datasheet values

PV Model	Pmax Datasheet	Pmax Estimated	Error %
Siemens SP75	74.8	74.815	0.02
Shell SQ80	80.15	80.44	0.36
SLK60M6	209.92	209.777	0.068
Solarex SX-5	4.455	4.449	0.135
Solarex SX-10	9.912	9.8989	0.132

Table 2: Comparison of PV Power estimated value vs. Datasheet value

PV Model	Power at battery connected directly (W)	Power at battery connected through converter (W)	Increment in Power %
Siemens SP75	57.26	71.61	25.06
Shell SQ80	58.12	76.96	32.24
SLK60M6	90.23	191	111.68
Solarex SX-5	3.573	4.43	32.98
Solare SX-10	7.753	9.815	26.59

Table 3: Comparison of the power supplied to the battery

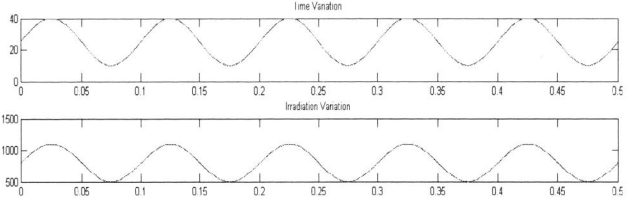

Fig 5: Temperature and irradiance variation over PVM

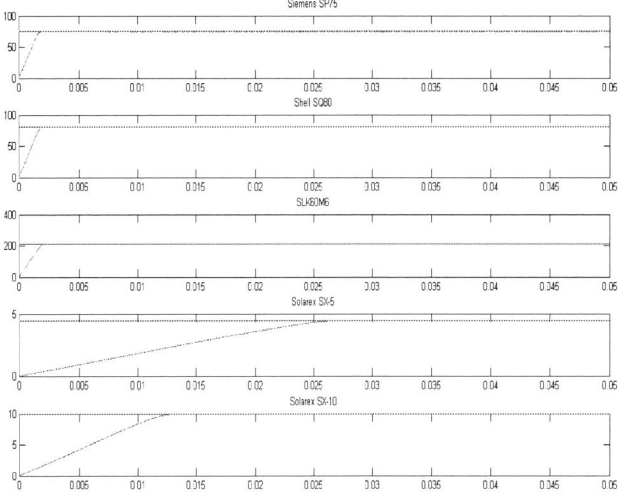

Fig 6: PV maximum possible output power (blue) vs actual PV Power when connected to proposed MPPT (green) at STC

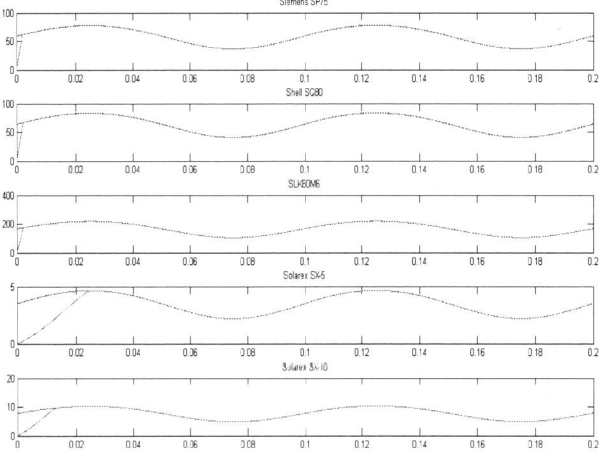

Fig 7: PV maximum possible output power (blue) vs actual PV Power when connected to proposed MPPT (green) at varying conditions

Those graphs reflects the importance of the use of a MPPT device since it can be seen that the power given to the battery is greater when connected through the MPPT converter than connected directly to the PV cell.

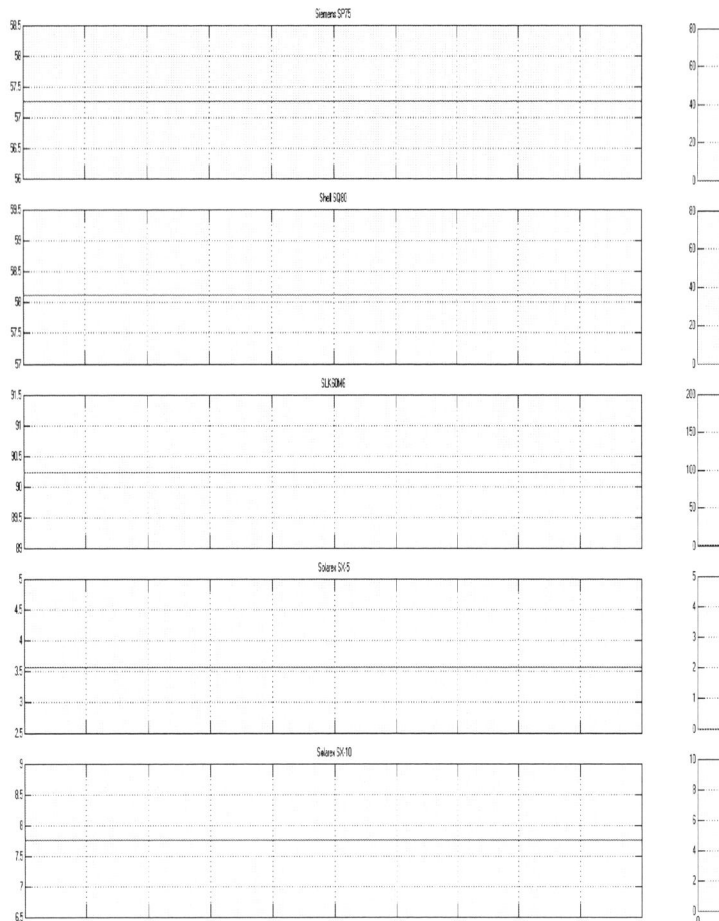

Fig 7: Power at the Battery for PV Modules connected directly to battery.

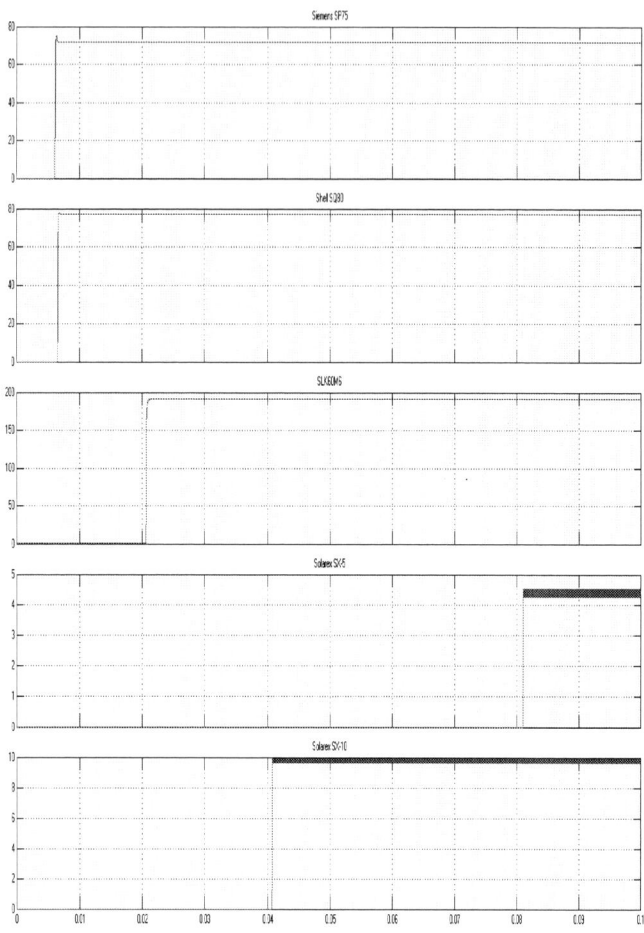

Fig 8: Power at the Battery for PV Modules connected to battery through DC/DC Converter.

VIII. Conclusion

This paper presents a simple photovoltaic solar cell dynamic sliding mode controlled maximum power point tracker for battery charging applications capable of compute the maximum power point under constant and varying ambient temperature and solar irradiation. The proposed controller is capable of changing the duty cycle of the Mosfet switch in order to move the operation point of the PV system to the optimal operation point and to maintain this operation point with time. The proposed algorithm uses a non inverting Buck-Boost converter in order to easily change the operation mode of the converter that can be necessary if the optimal voltage of the PV module is lower than the battery voltage. The proposed algorithm is capable of calculating the optimal voltage with little error. The proposed controller only requires the array output voltage and the optimal voltage which is continuously computed. From the simulation results is evident that a maximum power is tracked and achieved by the proposed sliding mode controller under constant and varying ambient temperature and solar irradiance and delivered, with the losses in the converter, to the battery increasing the current that is charging the battery which, eventually, will reduce the charging time.

References

[1]Femia, N.; Petrone, G.; Spagnuolo, G.; Vitelli, M. 'Perturb and Observe MPPT technique robustness improved', 2004 IEEE International Symposium on Industrial Electronics, Volume 2, 4-7 May 2004 Page(s):845 - 850 vol. 2

[2]Femia, N.; Petrone, G.; Spagnuolo, G.; Vitelli, M. 'Optimization of perturb and observe maximum power point tracking method',IEEE Tran actions on Power Electronics, Volume 20, Issue 4, July 2005 Page(s):963 – 973

[3]Weidong Xiao; Dunford, W.G. 'A modified adaptive hill climbing MPPT method for photovoltaic power systems' Power Electronics Specialists Conference, 2004. PESC 04. 2004 IEEE 35th Annual Volume 3, 20-25 June 2004 Page(s):1957 - 1963 Vol.3

[4]Atrash, H.; Batarseh, I.; Rustom, K. 'Statistical modeling of DSP-based Hill-climbing MPPT algorithms in noisy environments' , Applied Power Electronics Conference and Exposition, 2005. APEC 2005. Twentieth Annual IEEE Volume 3, 6-10 March 2005 Page(s):1773 - 1777 Vol. 3

[5]Yushaizad Yusof, Siti Hamizah Sayuti, Muhammad Latif, Zamri Che Wanik, 'Modeling and Simulation of Maximum Power Point Tracker for Photovoltaic System' , National Power & Energy Conference (PECon) 2004 Proceedings, Kuala Lumpur, Malysa. Pag 88-93.

[6]Jae Ho Lee; HyunSu Bae; Bo Hyung Cho, ' Advanced Incremental Conductance MPPT Algorithm with a Variable Step Size' Power Electronics and Motion Control Conference, 2006. EPE-PEMC 2006. 12th International Aug. 2006 Page(s):603 – 607

[7] Wu Libo; Zhao Zhengming; Liu Jianzheng, ' A Single-Stage Three Phase Grid-Connected Photovoltaic System With Modified MPPT Method and Reactive Power Compensation', Energy Conversion, IEEE Transaction on Volume 22, Issue 4, Dec. 2007 Page(s):881 - 886

[8] Hanifi Guldemir, ' Sliding Mode Control of DC/DC Boost Converter', Journal of Applied Sciences 5(3), ISSN 1812-5654

[9]Il-Song Kim, 'Robust Maximum power point tracker using sliding mode controller for the three-phase grid-connected photovoltaic system'. Solar Energy 81(2007) pag 415-414

[10]Il-Song Kim, Myung-Bok Kim, Myung-Joong Youn, 'New Maximum Power Point Tracker Using Sliding-Mode Observer for Estimation of Solar Array Current in the Grid-Connected Photovoltaic System', IEEE Transactions on Industrial Eletronics, Vol. 53, No.4, August 2006, pag.1027-1036

[11]Alexis de Medeiros Torres, Fernando Antunes, Fernando Soares, 'An artificial Neural Network-Based Real Time Maximum Power Tracking Controller for connecting a PV System to the grid', 1998 IEEE, 0-7803-4503-7/98 pag. 554-559

[12]Eduardo Ortiz, Fang Peng, 'A Novel Method to Estimate the Maximum Power for a Photovoltaic Inverter System',2000 35th Annual IEEE Power Electronics Specialists Conference.

[13]Eduardo Ivan Ortiz Rivera, 'Modeling and Analysis of solar distributed generation', Ph. D. dissertation, Michigan State University, 2006.

[14]A.M. Sharaf, Liang Yang, 'An efficient Photovoltaic DC Village Electricity Scheme Using a Sliding Mode Controller', IEEE Conference on Control Applications, 2005, pp 1325-1330.

[15]Ortiz-Rivera, Eduardo I; Peng, F.Z., "Analytical Model for a Photovoltaic Module using the Electrical Characteristics provided by the Manufacturer Data Sheet', Power Electronics Conference, 2005.PESC'05. IEE 36th, vol. pp 2087-2091, 11-14 Sept.2005.

An Enhanced Circuit-Based Model for Single-Cell Battery

Jiucai Zhang, Song Ci, Hamid Sharif
Department of Computer and Electronics Engineering
University of Nebraska-Lincoln
NE 68182, USA
Email: jczhang@huskers.unl.edu, {sci, hsharif}@unl.edu

Mahmoud Alahmad
Department of Architecture Engineering
University of Nebraska-Lincoln
NE 68182, USA
Email: malahmad2@unl.edu

Abstract—Battery performance prediction is crucial for battery-aware power management, battery maintenance, and multi-cell battery design. However, the existing battery models cannot capture the circuit characteristics and nonlinear battery effects, especially recovery effect. This paper aims to fill this gap by developing an enhanced circuit-based model for single-cell battery. The proposed model is validated by comparing simulation results with experimental data collected through battery testbed. The comparison shows that the proposed model can accurately characterize and predict the single-cell battery performance with considerations of various nonlinear battery effects under both constant and variable loads.

I. INTRODUCTION

Battery has been widely used in various mobile devices such as PDA, laptop, battery-powered electric vehicle, and battery energy storage system. An accurate battery model, which can capture complicated and dynamic battery circuit features and nonlinear capacity effects, is very crucial for circuit simulation, multi-cell battery analysis, battery performance prediction and optimization, and battery maintenance.

So far, many battery models have been proposed in literature [1]. In general, existing battery models can be divided into physical models, analytical models, and circuit-based models. In physical models, differential equations have been used to capture the complex electrical-chemical process in a battery [2]. Therefore, physical models are accurate and generic, which can be used to characterize battery behaviors. However, physical models require intensive computations to solve the interdependent partial differential equations. In addition, due to lack of battery model parameters such as battery structure and chemical composition, physical models are difficult to be configured and used [1], [3], [4]. To reduce the computational complexity, analytical models have been developed, where an equivalent mathematical representation is used to approximate the battery performance [5]–[8]. Analytical models are accurate and simple enough for power management, but they ignore circuit features such as voltage and internal resistance, making them infeasible for multi-cell battery design and analysis as well as circuit simulation. In circuit-based models, battery nonlinear circuit behaviors can be emulated by using capacitors, voltage and current resources, and resistors from the circuit analysis point of view. Circuit-based models can capture the complicated battery properties, which can be

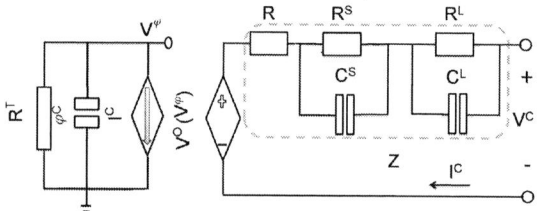

Fig. 1. Existing battery model [9]

TABLE I
SUMMARY OF NOTATIONS

V^o	Open-circuit voltage
V_i^C	Output voltage
V^F	Cutoff voltage of the single-cell battery
R^T	Self discharge resistance
R	Internal resistance
R^S	Short-transient resistance
R^L	Long-transient resistance
C^S	Short-transient capacitance
C^L	Long-transient capacitance
φ	State of charge
α^f	Full capacity of a single cell
α^A	Consumed capacity of a single-cell battery
I^C	Discharge current rate
μ	Recoverable capacity

easily implemented in electronic design automation (EDA) tools at different levels of abstraction [9]–[12]. However, current circuit-based models cannot estimate the impact of nonlinear behaviors on battery available capacity, leading to an inaccurate prediction of remaining battery capacity [8].

In this paper, we propose a new circuit-based battery model to capture the battery circuit features and nonlinear battery capacity effects, especially recovery effect. The model can accurately capture the battery performance both at constant and variable loads. We have validated the proposed battery model with experimental data collected through the ARBIN battery testing equipment.

The rest of this paper is organized as follows. Section II presents the related work. The battery model is proposed in Section III. The proposed battery model is validated in Section IV. We conclude the paper in Section V.

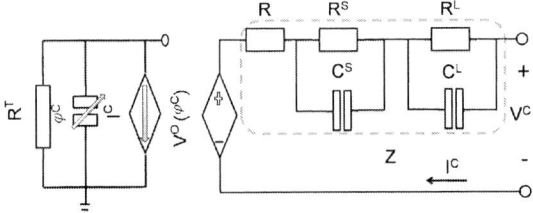

Fig. 2. The proposed battery model

II. RELATED WORK

Figure 1 illustrates the existing circuit-based battery model [9]. Here, the voltage-controlled voltage source is used to represent State of Charge (SOC) and open-circuit voltage. A current-controlled current source is used to represent battery capacity and SOC. The RC network emulates the transient voltage response. All model parameters, such as open-circuit voltage, resistors, and capacitors, can be approximated by mathematic equations listed as follows.

$$
\begin{cases}
\alpha^A(I^C, t_s, t_e) = I^C(t_e - t_s) \\
\varphi^C = 1 - \frac{\alpha^A}{c^f} \\
V^o(\varphi^C) = a_1 e^{a_2\varphi^C} + a_3\varphi^C - a_4\varphi^{C2} + a_5\varphi^{C3} + a_6 \\
R(\varphi^C) = b_1 e^{b_2\varphi^C} + b_3\varphi^C - b_4\varphi^{C2} + b_5\varphi^{C3} + b_6 \\
R^S(\varphi^C) = d_1 e^{-d_2\varphi^C} + d_3 \\
C^S(\varphi^C) = f_1 e^{f_2\varphi^C} + f_3 \\
R^L(\varphi^C) = g_1 e^{g_2\varphi^C} + g_3 \\
C^L(\varphi^C) = l_1 e^{l_2\varphi^C} + l_3
\end{cases}
\tag{1}
$$

where, α^A is the accumulated capacity during time period $[t_s, t_e]$ at rate of I^C; R, V^o, c^f, and φ^C are battery internal resistance, open-circuit voltage, the full capacity, and SOC, respectively; R^S, R^L, C^S, and C^L are resistances and capacitors to capture the transient response of battery voltage. $a_1 \sim a_6$, $b_1 \sim b_6$, $d_1 \sim d_3$, $f_1 \sim f_3$, $g_1 \sim g_3$, and $l_1 \sim l_3$ are coefficients of the model.

In general, the circuit-based model can accurately capture the dynamic circuit characteristics of a battery such as non-linear open-circuit voltage, temperature, cycle number, and self-discharge. However, the existing circuit-based model uses constant capacitor to model battery capacity, meaning that it is unable to capture and model the capacity relaxation process such as battery recovery effect. In addition, the accuracy of the existing model is very sensitive to the load variation rate, making it unable to handle dynamic battery load.

III. PROPOSED BATTERY MODEL

A. The Proposed Circuit-based Model

We propose an enhanced circuit-based model by replacing the consistent capacitor by a variable capacitor, as shown in Figure 2. The proposed model enables us to capture both battery circuit features and nonlinear battery capacity effects, making it a comprehensive and accurate model. The proposed

battery model can be denoted as:

$$
\begin{cases}
\alpha^A(I^C, \beta, L, t_s, t_e) = I^C F(L, t_s, t_e, \beta) \\
F(L, t_s, t_e, \beta) = t_s - t_e \\
\qquad + 2\sum_{m=1}^{\infty} \frac{e^{-\beta^2 m^2(L-t_s)} - e^{-\beta^2 m^2(L-t_e)}}{\beta^2 m^2} \\
\varphi^C = 1 - \frac{\alpha^A}{c^f} \\
V^o(\varphi^C) = a_1 e^{a_2\varphi^C} + a_3\varphi^C - a_4\varphi^{C2} + a_5\varphi^{C3} + a_6 \\
R(\varphi^C) = b_1 e^{b_2\varphi^C} + b_3\varphi^C - b_4\varphi^{C2} + b_5\varphi^{C3} + b_6 \\
R^S(\varphi^C) = d_1 e^{-d_2\varphi^C} + d_3 \\
C^S(\varphi^C) = f_1 e^{f_2\varphi^C} + f_3 \\
R^L(\varphi^C) = g_1 e^{g_2\varphi^C} + g_3 \\
C^L(\varphi^C) = l_1 e^{l_2\varphi^C} + l_3 \\
V^C(\varphi^C) = V_i^o(\varphi^C) - R(\varphi^C)I^C - \frac{R^S(\varphi^C)}{R^S(\varphi^C)\cdot j\omega\cdot C^S(\varphi^C)+1}I^C \\
\qquad - \frac{R^L(\varphi^C)}{R^L(\varphi^C)\cdot j\omega\cdot C^L(\varphi^C)+1}I^C
\end{cases}
\tag{2}
$$

where, $V^C(\varphi^C)$ denotes battery output voltage; ω means the current variation rate.

B. Remaining Capacity

In the proposed model, the accumulated capacity is denoted by an analytical expression [6] to capture the battery recovery effect. The consumed capacity $\alpha^C(I, \beta, L, t_s, t_e)$, which is dissipated during the load period $[t_s, t_e]$ at the discharge current I^C, can be written as [6]:

$$
\begin{cases}
\alpha^A(I^C, \beta, L, t_s, t_e) = I^C F(L, t_s, t_e, \beta) \\
F(L, t_s, t_e, \beta) = t_s - t_e \\
\qquad + 2\sum_{m=1}^{\infty} \frac{e^{-\beta^2 m^2(L-t_s)} - e^{-\beta^2 m^2(L-t_e)}}{\beta^2 m^2}
\end{cases}
\tag{3}
$$

In this equation, the first term $I^C(t_s - t_e)$ is the consumed capacity by the load I^C during the load period $[t_s, t_e]$. The second term $2I^C \sum_{m=1}^{\infty} \frac{e^{-\beta^2 m^2(L-t_s)} - e^{-\beta^2 m^2(L-t_e)}}{\beta^2 m^2}$ is the amount of discharging loss due to the current effect, which is the maximum recoverable battery capacity at t_e. It can be observed that the discharge loss will increase as the discharge current increases. β^2 is a constant related to the diffusion rate within battery. The larger the β^2, the faster the battery diffusion rate is, thus the less the discharging loss. L is the total operating time of the battery. m determines the computational complexity and accuracy of the model.

When a fully charged battery is discharged over time $\tau = \{t_0, t_1, \cdots t_N\}$, the remaining capacity can be denote as:

$$
\alpha^C = \alpha^f - \sum_{i=1}^{N} \alpha^A(I_i^C, \beta, L, t_{i-1}, t_i)
\tag{4}
$$

where, α^f is the full capacity of the battery.

As seen from Eq. 4, α^C will change with current variation, which is accurately capture the battery current effect. When the output voltage of the battery reaches cutoff voltage, the battery state of charge gets to 0.

The capacity loss could be recovered. Considering a constant load with profile defined as:

$$
i(t) = \begin{cases} I & t \in [0, T) \\ 0 & T \end{cases}
\tag{5}
$$

The maximum recoverable capacity occurs at $t = T$ can be denoted as:

$$\mu_{\max}(L, I, \beta, T, T) = 2I \sum_{i=1}^{\infty} \frac{e^{-\beta^2 i^2 L_1} - e^{\beta^2 i^2 (L_1 - T)}}{\beta^2 i^2}$$
$$= 2I \sum_{i=1}^{\infty} \frac{e^{-\beta^2 i^2 L_1}(1 - e^{\beta^2 i^2 T})}{\beta^2 i^2} \quad (6)$$

The recovery rate of a battery ε with a constant profile over time Δt is:

$$\varepsilon(T, \Delta t, \beta) = \frac{\mu(L, 0, \beta, T, T + \Delta t)}{\mu_{\max}(L, 0, \beta, T, T)}$$
$$= \frac{2I \sum_{i=1}^{\infty} \frac{e^{-\beta^2 i^2 (L_1 - T)}(1 - e^{-\beta^2 i^2 \Delta T})}{\beta^2 i^2}}{2I \sum_{i=1}^{\infty} \frac{e^{-\beta^2 i^2 L_1}(1 - e^{-\beta^2 i^2 T})}{\beta^2 i^2}} \quad (7)$$
$$= \frac{\sum_{i=1}^{\infty} \frac{e^{\beta^2 i^2 T}(1 - e^{-\beta^2 i^2 \Delta T})}{i^2}}{\sum_{i=1}^{\infty} \frac{1 - e^{-\beta^2 i^2 T}}{i^2}}$$

Therefore, the recovered capacity of the battery over time Δt is:

$$\mu(L, I, \beta, T, \Delta t) = \mu_{\max} \times \varepsilon(T, \Delta t, \beta)$$
$$= 2I \sum_{i=1}^{\infty} \frac{e^{-\beta^2 i^2 L_1}(1 - e^{-\beta^2 i^2 T})}{\beta^2 i^2} \times \frac{\sum_{i=1}^{\infty} \frac{e^{\beta^2 i^2 T}(1 - e^{-\beta^2 i^2 \Delta T})}{i^2}}{\sum_{i=1}^{\infty} \frac{1 - e^{-\beta^2 i^2 T}}{i^2}} \quad (8)$$

which not only relies on the discharging current, but only determined by the discharging time T, rest time Δt, and battery parameter β.

Self-discharge resistor R_T is used to characterize the self-discharge energy loss when battery are stored for a long time, which is a function of SOC, temperature, and cycle number. The usable capacity decreases slowly with time when no load is connected to the battery. So, in this paper we ignore the self-discharge.

IV. MODEL VALIDATION

A. Experiment Setup

We have validated the proposed battery model under both constant currents and variable currents by using HE18650 battery whose full capacity, nominal voltage, and cutoff voltage are $2600mAH$, $3.7V$ and $3V$, respectively. All parameters of the proposed battery model, as shown in Table II, can be obtained by using the standard least-square estimator [8], [12]. The simulation results of the battery are obtained by using MATLAB, and the experimental data are collected through the ARBIN battery testing instrument BT2000 as shown in Figure 3 [13]. The battery is first charged to its full capacity through Constant Current Constant Voltage (CCCV), and then it will be rested for 30 minutes [14]. Then, the battery will be discharged under different predefined profiles, respectively.

B. Simulation and Experiment Results

Figure 4 shows the battery performance at constant discharge current rate of 0.25A and 1A, respectively. When the battery output voltage goes from full capacity voltage to cutoff voltage, the state of charge of battery drops from a certain value to 0. The SOC of the full charged battery varies with discharge current rate, which reflects the current effect. Experiment results match experimental data well.

Both simulation and experiment results for a four-phase dynamic load profile at discharge current rate of $1A$, $2A$, $0.2A$,

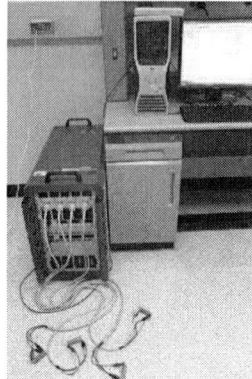

Fig. 3.   ARBIN battery testing instrument BT2000

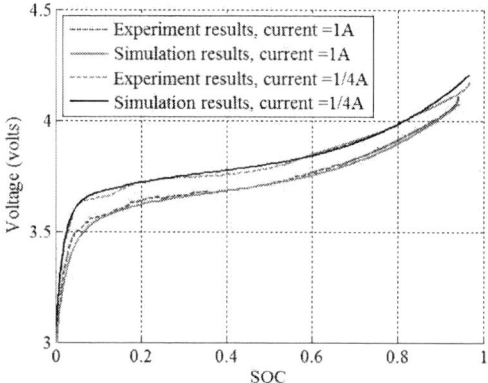

Fig. 4.   Battery model validation at constant loads

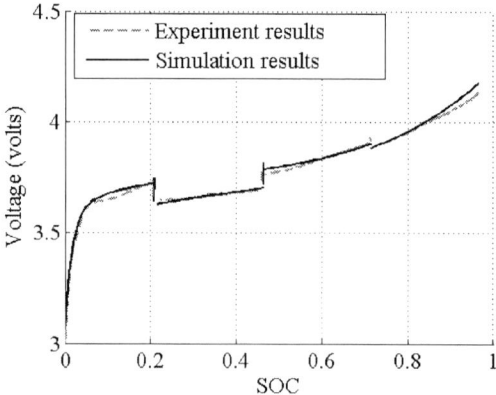

Fig. 5.   Battery model validation at variable loads

TABLE II

BATTERY MODEL PARAMETERS

a_1	-0.402	a_2	-50.58	a_3	0.8849	a_4	-1.662	a_5	1.482
a_6	3.574	b_1	-0.1726	b_2	-20.07	b_3	0.0944	b_4	-0.2301
b_5	0.1772	b_6	0.06105	f_1	-752.9	f_2	-13.51	f_3	703.6
g_1	6.603	g_2	-155.2	g_3	0.04984	l_1	-6056	l_2	-27.12
l_3	4475	d_1	0.3208	d_2	-29.14	d_3	0.04669	β	0.35

and $1A$ are shown in Figure 5. As the battery discharged over time, the battery voltage will reach the cutoff voltage, and battery SOC will be 0. This means that the proposed battery model can accurately quantify the battery characteristic.

From both figures, we can observe that the proposed model generate voltage response less than $20mV$. Therefore, we can conclude that the simulation results of the proposed battery model matches well with the experiment data. The close agreement between simulation results and experimental data indicates that the battery parameters have been accurately extracted to predict run-time battery behaviors in both steady state and transient state voltage responses.

V. CONCLUSION

In this work, an accurate and comprehensive circuit-based battery model has been proposed to capture circuit features and nonlinear battery effects such as current effect and recovery effect. Both simulation results and experiment results show that the proposed models can be used to accurately model and predict battery performance. The computational complexity of the proposed model could be controlled, which provides a way to tradeoff computational complexity and model accuracy. Therefore, the proposed model will greatly help research on circuit simulation, multi-cell battery analysis, battery performance prediction and optimization, and battery maintenance.

ACKNOWLEDGMENT

This research was supported in part by NSF ECCS Grant # 0801736.

REFERENCES

[1] R. Rao, S. Vrudhula, and D. Rakhmatov, "Battery modeling for energy aware system design," *Computer*, vol. 36, no. 12, pp. 77–87, Dec. 2003.
[2] M. Doyle, T. Fuller, and J. Newman, "Modeling of Galvanostatic Charge and Discharge of the Lithium PolymerInsertion Cell," *Journal Electrochemical Society.*, vol. 140, pp. 1526–1533, 6 1993.
[3] W. Gu and C. Wang, "Thermal-electrochemical modeling of battery systems," *Journal of Electrochemical Society*, vol. 147, no. 8, pp. 2910–2922, 2000.
[4] M. Doyle, J. Newman, A. S. Gozdz, C. N. Schmutz, and J.-M. Tarascon, *Journal Electrochemical Society*, vol. 143, pp. 18–90, 1996.
[5] P. Rong and M. Pedram, "An Analytical Model for Predicting the Remaining Battery Capacity of Lithium-Ion Batteries," *IEEE Journal Very Large Scale Integrated*, vol. 14, pp. 441–451, May 2006.
[6] D. Rakhamtov, S. Vrudhula, and D. A. WallachAn, "Analytical High-Level Battery Model for use in energy Management of Portable Electronics Systems," *IEEE Journal Very Large Scale Integrated*, vol. 11, Dec. 2003.
[7] D. N. Rakhmatov and S. B. K. Vrudhula, "An analytical high-level battery model for use in energy management of portable electronic systems," in *IEEE/ACM international conference on Computer-aided design*, Piscataway, NJ, USA, 2001, pp. 488–493.
[8] P. Rong and M. Pedram, "An analytical model for predicting the remaining battery capacity of lithium-ion batteries," *IEEE Transactions on Very Large Scale Integration Systems*, vol. 14, no. 5, pp. 441–451, May 2006.
[9] M. Chen and G. A. Rinc'on-Mora, "Accurate electrical battery model capable of predicting runtime and icv performance," *IEEE Transactions on Energy Conversion*, vol. 21, no. 2, pp. 504–511, 2006.
[10] L. Benini, G. Castelli, A. Macci, E. Macci, M. Poncino, and R. Scarsi, "Discrete-time battery models for system-level low-power design," *IEEE Transactions on Very Large Scale Integrated System*, vol. 9, no. 5, pp. 630–640, 2001.
[11] L. Gao, S. Liu, and R. A. Dougal, "Dynamic lithium-ion battery model for system simulation," *IEEE Transactions on Component Package Technology*, vol. 25, no. 3, pp. 495–505, 2002.
[12] W. X. Shen, C. C. Chan, E. W. C. Lo, and K. T. Chau, "Estimation of battery available capacity under variable discharge currents," *Journal of Power Sources*, vol. 103, no. 2, pp. 180 – 187, 2002.
[13] "Bt2000 battery testing system," Arbin Inc., Tech. Rep., 2009.
[14] M. Alahmad and H. Hess, "Evaluation and analysis of a new solid-state rechargeable micro-scale lithium battery," *IEEE Transactions on Industrial Electronics*, vol. 55, no. 9, pp. 3391 – 3401, 2008.

A High Frequency Battery Model for Current Ripple Analysis

Jin Wang* Ke Zou
Department of Electrical and Computer Engineering
The Ohio State University
Columbus, OH, USA
*Wang@ece.osu.edu

Chingchi Chen* Lihua Chen
Ford Motor Company
Dearborn, MI, USA
*Cchen4@ford.com

Abstract—**In applications where batteries work together with power electronic circuits, the current ripple generated by the power electronics will be shared by both the battery and passive components in the circuit. The amount of ripple absorbed by the battery depends on its impedance at the switching frequency of power electronics. This paper presents an impedance based high frequency battery model derived from test results of a Ni-MH battery using a novel battery impedance tester. The possible reasons for the battery impedance characteristics in high frequency region, including skin effect and proximity effect, are also discussed. This battery model can be directly used in current ripple analysis, passive components design and control strategy optimization of power electronic circuits. The effect of the passive component values on the battery current ripple is analyzed using the ac equivalent circuit of the test setup.**

I. INTRODUCTION

In hybrid electric vehicles (HEV) and other applications, batteries usually work together with power electronics circuits such as dc/dc converters and dc/ac inverters, which generate significant amount of switching frequency related current ripple. The current ripple absorbed by the battery is determined by the impedance of the battery itself as well as other passive components such as its paralleled capacitor. Since the battery impedance changes with its operation frequency and current [1], the knowledge of the battery impedance at switching related frequencies and high current ripple conditions is the key in designing the passive components and optimizing the control strategy of the power electronic circuits.

Most battery impedance testing methods available today use small signal testers, such as the electrochemical impedance spectroscopy (EIS) [2], [3]. This may not be the best approach for power electronics related applications where high amplitude dc and ac are both present. This paper employs a novel impedance tester which uses a dc/dc converter to create a dc offset and ac ripple to produce an environment that mimics the operation of real power electronics.

For the battery modeling, equivalent circuit based battery models have been extensively studied [4], [5]. However, most of them are not suitable for power electronics circuit analysis since they are derived from test frequencies lower than 5 kHz and ac current ripples whose amplitude is much less than 5 A. The battery model developed in this paper focuses on the normal switching frequency range (5 kHz ~ 20 kHz) of power electronics in HEV operations and is developed from high AC current ripple tests (larger than 10 A peak to peak).

In section IV, the model is briefly analyzed. Possible reasons for the battery impedance behavior in switching frequency regions, including skin effect and proximity effect are investigated in the analysis. The effect of stray inductance in the accuracy of the model is also discussed.

Based on this model, the battery current ripple is analyzed using the ac equivalent circuit method with different passive component values. This analysis will help the passive components design and system stability analysis of power electronics circuits.

II. TEST SETUP AND TEST METHOD

The test setup diagram is shown in Fig. 1, which consists of a boost dc/dc converter, a high-accuracy film capacitor and the battery under test. In this test setup, the boost converters are realized using an integrated power module (IPM) rated at 1200V/200A. It is controlled by a TI TMS320F2812 DSP and functions as an ac ripple generator to produce triangular ripple on the inductor.

The high accuracy film capacitor works as an impedance reference. Under the assumption that the capacitance does not change with frequency, the impedance of the battery can be calculated from the current sharing relationship between the battery current and capacitor current. The capacitor current and battery current are measured according to the direction notation in Fig.1. Then the battery impedance Z can be calculated using (1),

$$|Z| = \frac{|i_c|}{|i_b|} \times \frac{1}{2\pi f C}, \angle Z = \angle i_c - \angle i_b + 90° \tag{1}$$

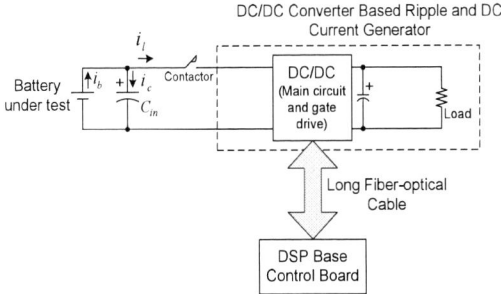

Figure 1. The general diagram of test setup.

where i_c, i_b are the switching frequency components of capacitor current and battery current, f is the switching frequency and C is the capacitance of the capacitor.

Experiments were conducted within a switching frequency range of 5 kHz to 20 kHz. A 6.5V Ni-MH battery from a current mass production hybrid vehicle was discharged to a resistive load through the boost converter. In the experiments, the battery is charged to 50% SOC and the discharging time for each test is limited to less than 3 seconds to ensure that in the whole experiment the SOC variation is less than 3%. The connection cable between the battery and the capacitor is intentionally twisted to reduce its stray inductance.

Fig.2 shows the test setup on the battery side. Fig. 3 shows the current test result under the condition of 10 kHz switching frequency and 100A dc offset. The battery current is almost in phase with the capacitor current. According to the notation of Fig. 1, this implies the battery impedance is inductive. Fig. 4 shows the battery impedance amplitudes and angles at different frequencies and 5A dc offset. The detailed test results can be found in [6].

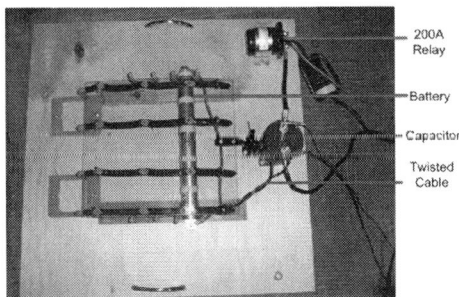

Figure 2. The test setup on the battery side.

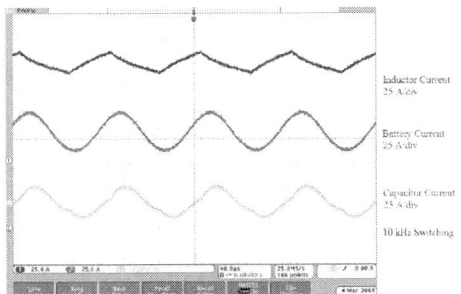

Figure 3. Current waveforms at 10 kHz, 100 A dc offset.

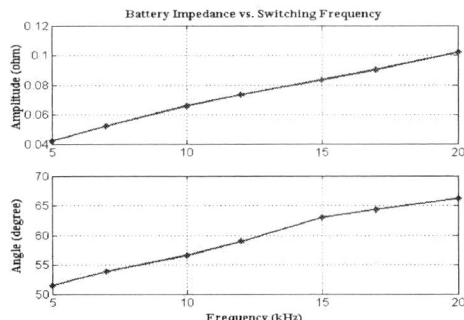

Figure 4. Impedance vs. switching frequency at 5A dc offset.

Current measurements were made with Tektronix TCP0150 current probes which have a typical dc error less than 1%. The measured current signal was recorded by a Tektronix 4054 digital oscilloscope operating in high resolution (Hi-res) mode. The digitizer of the oscilloscope is 11 bit at this mode. In the measurement the waveforms are adjusted to posses at least 4 vertical divisions (one half of the screen) so the digitization error is less than 0.1%. The recorded current data contains at least 1500 points per cycle which is processed by the computer using DFT to find out the switching frequency component. The overall measurement error could be controlled to be less than 3%. Since this system employs a current measurement method, it can test a battery cell or a string of batteries with the same level of accuracy. Other types of batteries and other energy storage devices such as fuel cells could also be tested using this tester.

III. DERIVATION OF HIGH FREQUENCY, HIGH CURRENT BATTERY MODEL

The impedance based battery model can be derived using an approximation of the frequency response curve shown in Fig. 5 and Fig. 6. This paper employs an advanced vector fitting method introduced by Gustavsen [7]-[9]. This method approximates a frequency response f(s) with a rational function, expressed in the form of a sum of partial fractions:

$$f(s) \approx \sum_{m=1}^{N} \frac{c_m}{s - u_m} + d + se \qquad (2)$$

where terms d and e are optional. c_m and a_m are the residuals and poles, respectively.

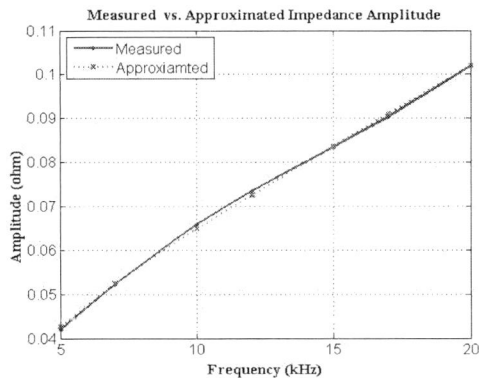

Figure 5. Measured and approximated impedance amplitude.

978-1-4244-4782-4/10 $26.00 © 2010 IEEE

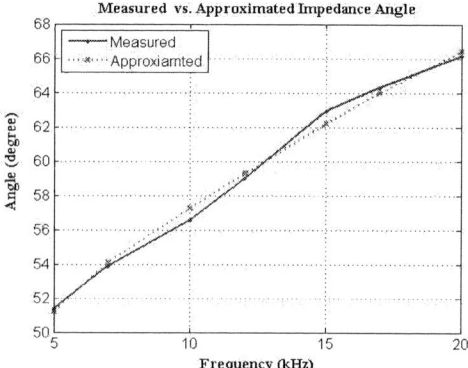
Figure 6. Measured and approximated impedance angle.

As shown in Fig. 6, in the frequency range from 5 kHz to 20 kHz, the battery impedance is inductive so d and e could not be omitted. Since the frequency response (both amplitude and angle) is quite linear, this implies one pole is enough to describe the battery impedance characteristic at this frequency range, so m=1. Using the vector fitting code provided by Gustavsen [10] and the above restrictions, parameters in (2) are found to be:

$$d = 0.0437, e = 6.8014 \times 10^{-7}, a_1 = -4.484 \times 10^4, c_1 = -1131.5.$$

Then the battery current-voltage transfer function can be written as (3).

$$Z_{batt}(s) \approx \frac{-1131.5}{s + 44842} + 0.0437 + 6.801 \times 10^{-7} s \qquad (3)$$

The approximated and measured curve of battery impedance amplitude and angle are compared in Fig. 5 and Fig. 6, respectively. Since (3) is a second order transfer function, the impedance based battery model is proposed as shown in Fig. 7, which consists of two inductors and two resistors. From Fig.7, the battery impedance can be expressed as (4):

$$Z_{batt}(s) = L_1 s + R_1 + \frac{L_2 R_2 s}{L_2 s + R_2} = \frac{L_1 L_2 s^2 + (L_1 R_2 + L_2 R_1)s + R_1 R_2}{L_2 s + R_2} \quad (4)$$

Comparing (3) and (4), each component in Fig. 7 can be calculated as shown in Table I.

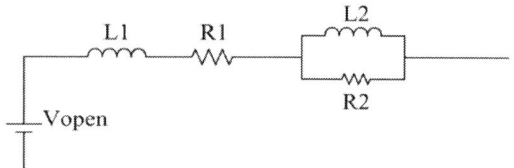
Figure 7. The proposed high frequency battery model.

TABLE I.   PASSIVE COMPONENTS PARAMETERS IN THE PROPSED BATTERY MODEL

Component	Value
L1	0.680 µH
R1	0.0184 Ω
L2	0.562 µH
R2	0.0252 Ω

IV. MODEL ANALYSIS

The model presented in Fig.7 is a simple one aiming to help with the design and control of power electronic circuits. The obvious differences of this model from normal low-frequency models are the addition of inductive components and the elimination of capacitive components. These reflect the battery's inductive behavior at normal switching frequency range. The paralleled L-R branch represents the fact that the equivalent inductance decreases as the frequency increases. It should be noted that extra capacitive components, such as the traditional R-C branches, can be added in the model to make it more accurate. However, this model makes the process easier for ripple analysis and power electronics design.

The battery's inductive behavior at high frequency is due to its intrinsic inductance which is usually been neglected in the low frequency region. Previous studies on battery's impedance at high frequency also include inductive components in the model. For example, in [1], an inductor is added to the circuit model. However, those works were not focused on the power electronics switching frequency so the frequency range is usually from 0.01 Hz to several kHz and the ac ripple is much less than 10 A. So their results either did not cover the switching frequency region, or had different impedance response, in both amplitude as well as phase, from the one described in this paper.

The stray inductance of the connection cables between the battery and the capacitor also contributes to part of the total inductance in the model. The amount of stray inductance is decided by the cable-battery loop area and the width of the cables. Using twisted cable or specially designed busbar could eliminate most of the stray inductance. For example, if the cables in Fig. 2 are not twisted together, the stray inductance could be as large as several hundreds of nanohenry. However, if they are twisted together, the stray inductance could be reduced to several tens of nanohenry which will significantly increase the accuracy of the battery impedance measurement in the test.

The test results also show that, together with the decrease or equivalent inductance, the battery resistance actually increases with the frequency. These facts provide us clues as to the possible reasons for this battery impedance characteristic in switching frequency region:

A. Skin Effect

Skin effect is the unequal distribution of ac current within a conductor. The surface of the conductor tends to have more current density than in the middle. This leads to an increase of resistance and decrease of inductance of the conductor in ac conditions compared to dc. For the battery, since its diameter is much larger than the skin depth, the resistance is approximately proportional to the square root of switching frequency.

A frequency sweeping test is performed to find out the possible relationship between the skin effect and battery impedance characteristic at switching frequency region. In this test, the battery is place far from any conductors and the test frequency ranges from 4 kHz to 20 kHz with a step of 1 kHz. The battery resistance is calculated and normalized with respect to the resistance at 4 kHz, i.e. if the resistance at 4 kHz is r_0, the normalized resistance r_n for a measured resistance r is $r_n = r / r_0$. The square root of switching frequency is also normalized with respect to the square root of 4000 Hz and plotted together with the normalized resistance in Fig.8.

It can be seen in Fig.8 that the change of resistance of the battery in the tested frequency region generally follows the square root of switching frequency. This indicates that the skin effect could be a possible source for the high frequency behavior of battery.

B. Proximity Effect

The proximity effect accounts for the fact that the current distribution within one ac-carrying conductor is affected by other conductors close to it. The proximity effect will also increase the equivalent resistance of the conductor. A frequency sweeping test is performed in which the battery is bounded together with a cable that is carrying same amount of current as the battery but in the opposite direction. Fig. 9 shows the test results. It can be seen that at lower frequencies (less than 15 kHz), the resistance follows the curve of square root of frequency. Then at the region above 15 kHz, the resistance soars, which possibly implies the influence of the proximity effect.

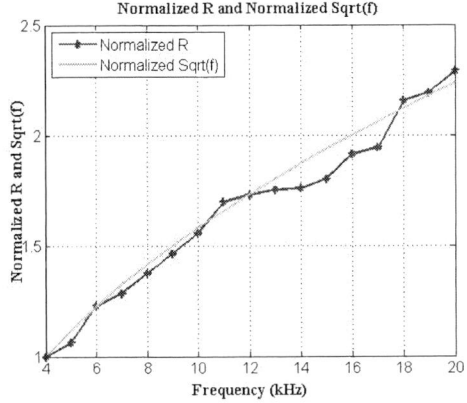

Figure 8. The normalized R and normalized square root of switching frequency without nearby conductors.

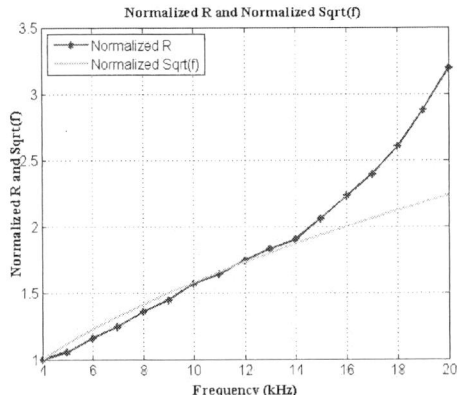

Figure 9. The normalized R and normalized square root of switching frequency with nearby conductors.

The proximity effect increases the inaccuracy of battery impedance testing. So in the tests, the battery should be placed away from other conductors to reduce their influence.

V. AC EQUIVALENT CIRCUIT ANALYSIS

The purpose of the derived battery model is to help with battery ripple analysis. In [11], an ac small signal circuit is introduced to analyze the voltage ripple on photovoltaic cells. Similarly, here the ac equivalent circuit of the test setup is used to analyze the ac ripple on the battery. For the battery discharge mode, the boost converter is replaced by a square voltage source V_{sq} operating at switching frequency to generate ac ripple, as shown in Fig. 10 (a).

The boost converter inductor L, the capacitor C and equivalent battery inductance work together as a LCL filter. So the square wave voltage source V_{sq} can be replaced by its fundamental V_1, as shown in Fig. 10(b). The amplitude of V_1 can be calculated using (5):

$$V_1 = \frac{2}{\pi} \sin((1-D)\pi) V_{out} \tag{5}$$

where V_{out} is the amplitude of boost converter output voltage.

This replacement will cause error since V_1 does not include battery current harmonics, which depends on the duty ratio D. For cases with large THD, V_{sq} could be simulated by several sinusoidal sources including V_1 and low order harmonics. In real HEV applications, a 285V battery, consisting of a string of 44 batteries, is used. So all of the parameters in Table I are multiplied by 44 in the simulation. For the component values specified here, the THD of I_{batt} is 0.8% when D=0.5 and is 12.6% when D=0.2.

The impedances of C, L and the whole circuit are

$$Z_c(s) = \frac{1}{Cs}, Z_L(s) = Ls, Z_{total}(s) = Z_L(s) + Z_c(s) // Z_{batt}(s),$$

respectively. Thus, the amplitude of battery current ripple can be expressed as (6):

$$I_{batt} = \left| \frac{V_1}{Z_{total}(s)} \cdot \frac{Z_C(s)}{Z_{batt}(s) + Z_C(s)} \right| \tag{6}$$

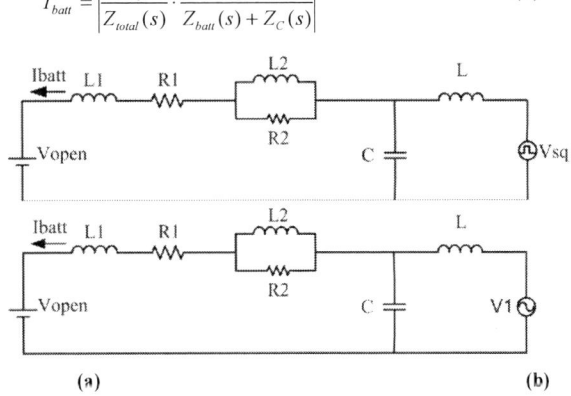

Figure 10. The ac equivalent circuit (a) square waveform source (b) sinusoidal waveform source.

Fig. 11 shows the battery current ripple magnitude as a function of L and C at a switching frequency of 10 kHz and 0.5 duty ratio. This result gives a direct indication on C and L selection. For example, if the maximum battery current ripple allowed is 0.5 A, the C could be selected as 260 μF and L is 260 μH. To verify the current ripple calculated here, PSIM was used to simulate the circuit in Fig.10 (a) under the following condition: C=260 μF and L=260 μH. The simulation result is shown in Fig. 12 and the current ripple amplitude is about 0.49A, which is consistent with the result of the previous calculation.

VI. CONCLUSIONS

This paper presents a high frequency, high current battery model based on the test results on a Ni-MH battery. This model only consists of two resistors and two inductors so it can be easily used in battery ripple analysis and the control strategy design for power electronics systems. The possible reasons for the battery impedance under high frequency, high ac current conditions, including the skin effect and the proximity effect are also investigated. Using the ac equivalent circuit of the battery impedance test apparatus, the influence of passive component values on battery current ripples is studied.

This paper is the first in a series of papers to investigate the impedance based battery model for high current and high frequency conditions. Fig. 13 shows the main circuit of the automatic tester which will be built in the next step. A grid-

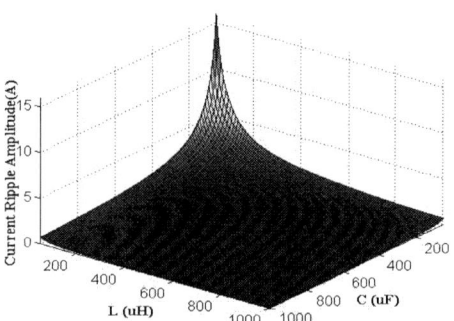

Figure 11. Battery current ripple vs. C and L.

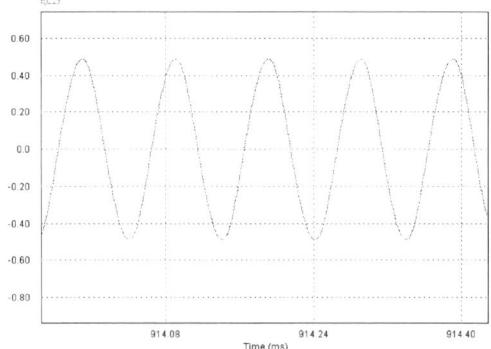

Figure 12. Simulated battery current ripple (C=260 μF, L=260 μH).

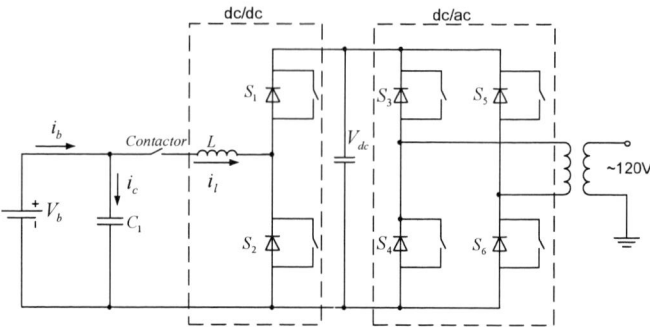

Figure 13. The main circuit for the automatic battery tester.

-tied H-bridge inverter is added into the automatic tester. In this scheme, during the battery discharge mode, the battery works as a dc source by providing power to the utility grid and during the charge mode it absorbs power from the utility grid. A dSPACE controller is employed to control the battery current and realizing SOC tracking. Effort will also be made on detailed analysis of high frequency battery characteristics, power electronics system stability analysis and control strategy optimization.

REFERENCES

[1] S. Buller, M, Doncker and E. Karden, "Impedance-based simulation models of supercapacitors and Li-Ion batteries for power electronic applications," IEEE Trans. Industry Applications, vol. 41, no. 3, MAY/JUNE 2005, pp. 742-747.

[2] P. Mauracher and E. Karden, "Dynamic modelling of lead/acid batteries using impedance spectroscopy for parameter identification," J. Power Sources, vol.67, Aug. 1997,pp. 69-84.

[3] X.Feng and Z. Sun, "A battery model including hysteresis for State-of-Charge estimation in Ni-MH battery," in IEEE 2008 Vehicle Power and Propulsion Conference,Sept. 2008, pp. 1–5.

[4] M.Chen, G.Rincon-Mora, "Accurate Electrical Battery Model Capable of Predicting Runtime and I–V Performance," IEEE Trans. Energy Conversion, vol.21, no.2, June 2006, pp.504-511.

[5] R. Kroeze and P. Krein, "Electrical Battery Model for Use in Dynamic Electric Vehicle Simulations," in IEEE 2008 Power Electronics Specialists Conference, June 2008, pp.1336-1342.

[6] Stephen Nawrocki, Renxiang Wang, Ke Zou and Jin Wang ,"High Current Battery Impedance Testing for Power Electronics Circuit Design Optimization," in IEEE 2009 Vehicle Power and Propulsion Conference, Sept. 2009.

[7] B. Gustavsen and A. Semlyen, "Rational approximation of frequency domain responses by Vector Fitting," IEEE Trans. Power Delivery, vol. 14, no. 3, pp. 1052-1061, July 1999.

[8] B.Gustavsen, "Improving the pole relocating properties of vector fitting," IEEE Trans. Power Delivery, vol. 21, no. 3, pp. 1587-1592, July 2006.

[9] D. Deschrijver, M. Mrozowski, T. Dhaene, and D. De Zutter, "Macromodeling of multiport systems using a fast implementation of the vector fitting method," IEEE Microwave and Wireless Components Letters, vol. 18, no. 6, pp. 383-385,June 2008.

[10] B. Gustavsen, User's guide for vectfit3.m. SINTEF Energy Research, N-7465 Trondheim, Norway. [Online]. Available: http://www.energy.sintef.no/Produkt/VECTFIT/index.asp : VFIT3.zip

[11] N.Benavides. P. Chapman, "Modeling the effect of voltage ripple on the power output of photovoltaic modules," IEEE Trans. Industrial Electronics, vol 55, Issue 7,July 2008 pp. 2638-2643.

A Novel Power Line Communication technique based on Power Electronics circuit Topology

Jiande Wu, Chushan Li, Xiangning He
College of Electrical Engineering
Zhejiang University
Hangzhou, China
eewjd@zju.edu.cn

Abstract—**This paper presents a novel method of designing power line communication circuits, which consolidate power and data in a single bus. This method is based on power electronics topology and can derive new application circuits. The principle of the Buck-type powered-bus circuit is introduced. Taking into account the effect of transmission line, the steady-state characteristics of the circuit as well as dynamic switching characteristics are analyzed. The effectiveness of this method is verified by experimental result.**

I. INTRODUCTION

Traditionally, power line communication (PLC) refers to the techniques that transmit data over already existing power wires, which can be divided into high-voltage power line communication and low-voltage power line communication. In recent years, low-voltage power line communication (LV PLC) has been widely investigated and apply in many areas [1-2].

PLC technique can also be applied in DC power systems [3-5]. A widely used example of DC PLC is the powered bus communication compliant with IEC61158-2 [6]. IEC61158 is an international fieldbus standard, while IEC61158-2 is the fieldbus physical layer standard which provides the physical implementation of powered bus. Similar with traditional PLC, powered bus communication is also combined a low-pass power channel with a high-pass signal channel, but the difference between powered bus communication and traditional PLC is that the former is based on base-band communication, and the latter is normally based on broadband communication. Fieldbus protocols such as Profibus, FF have adopted it as one of the physical layer standard.

Another technique similar to PLC is 1-wire bus and M-bus [7-8]. 1-Wire bus is Dallas(Maxim)'s patented technology. Its structure and circuits are very simple (Fig.1): A spare MCU port pin with a pull-up resistor is programmed as 1-wire master, which supplies power for slave devices and communications with it. The drive capacity of 1-wire bus is very weak, so it is usually applied in ultra low power system such as temperature monitor. M-bus is another technique powered from communication bus which is applied in remote

meter reading. Compared with 1-wire bus, M-bus can provide more current and transmit much longer, but the total current that source from M-bus is still small.

According to the analysis above, we conclude that the circuits of DC PLC and IEC61158-2 powered bus technique are relatively complex, but can supply higher power; the communication circuits of 1-Wire bus and M-bus are very simple, but the capacity of power supply is very weak.

This paper investigate a new method of finding circuits suitable for power line communication, which is based on Power Electronics circuit topology and can supply more power than 1-wire bus and M-bus.

II. BOOST-TYPE BUS-POWERED CIRCUIT

From Fig. 1, it can be found that the weakness of 1-wire bus' power supply capacity is due to the present of resistance R. If the resistance R is replaced by an inductor L (Fig. 2), the drive capacity of 1-wire bus should be greatly enhanced. In the modified circuit, it is found that the inductor L, switch M, diode D and capacitor C form a standard Boost circuit.

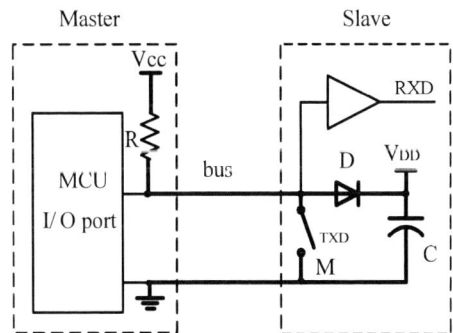

Fig. 1 1-Wire bus communication structure

Fig. 2 can be considered as a derived circuit from Boost topology: Dividing Boost circuit into two parts (master and slave), Inductor L is put in the master device, and the switch M, diode D and capacitor C is put in the slave device. The

---

This work is sponsored by the Power Electronics S&E Development Program of Delta Environmental & Education Foundation

connection wire between master and slave device is powered communication bus.

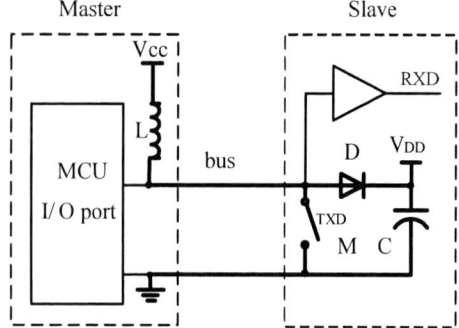

Fig. 2 Replace R with L in 1-Wire bus communication

Boost-type bus-powered circuit is very simple in principle, but in practical application, many auxiliary protection and filter circuits should be added, the detailed implementation of this circuit will not be discussed in this paper.

III. BUCK-TYPE BUS-POWERED CIRCUIT

Applying the same way to Buck circuit, a new structure of powered bus communication circuit is derived.

1) Dividing Buck circuit into two parts (Fig. 3): DC power supply E1 and the switch M1 comprise the master device, and the other components comprise the slave devices. The master and slave are connected by powered communication bus.

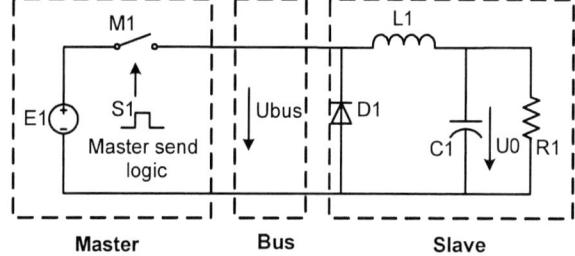

Fig. 3 Structure of Buck-type communication circuit

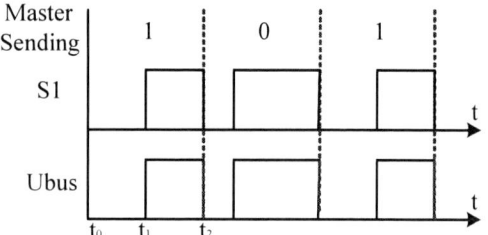

Fig. 4 Timing diagram of Master Sending data

The control logic of switch M1 should be designed carefully to transmit not only power supply but also data signals. It is assumed that in the state of bus free (no data exchange), switch M1 is always on and E1 supply power for load R1. When master sending data, switch M1 turns on and off with difference duty cycle to represent logic "0" and "1".

In Fig.4, The logic "0" is represented by duty ratio D1 and the logic "0" is represented by duty ratio D0. (D0> D1)

2) Circuit in Fig.3 can only send data from master to slave. To achieve duplex communication, Diode D2 and Switch M2 are added and the circuit is modified as Fig. 5.

In this circuit, slave device can send data to master in accordance with proper timing diagram which is shown in Fig.6. We define duty ratio D1 = 0.5 as signal "1", duty ratio D0 = 0.75 as signal "1".

In every bit cycle of slave sending data, the master must initiate a "1" signal in the bus. If the slave sends signal "1", switch M2 stays off and does nothing; if the slave sends signal "0", switch M2 turns on at the moment of t_1, so the voltage in the bus will be pulled high. Receiver circuit in the master board checks the bus voltage and demodulates the data come from the slave.

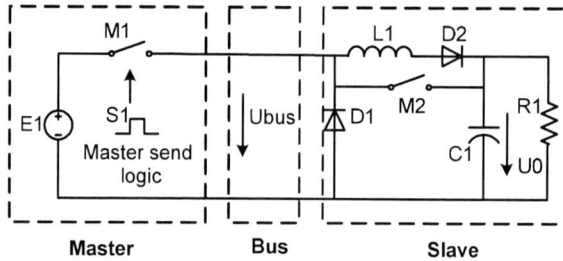

Fig. 5 Structure of Buck-type duplex communication circuit

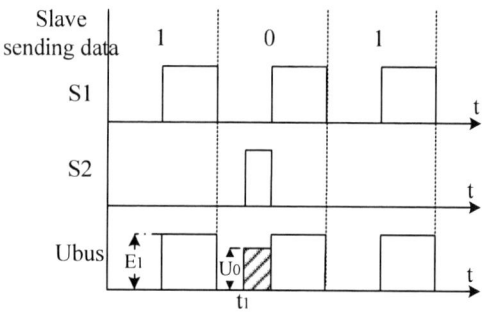

Fig. 6 Timing diagram of Slave sending data

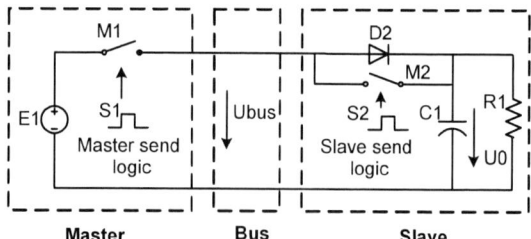

Fig. 7 Simplified Buck-type communication circuit

3) Simplified Buck-type bus-powered circuit

One of the disadvantages of Fig. 5 is that the load voltage U_0 varies with the communication data. According to the characteristics of Buck circuit, if the inductor works in discontinuous current mode, the load voltage U_0 will increase significantly and have little influence with duty ratio.

Therefore, assuming L1 = 0 and canceling diode D1, this circuit is simplified as Fig. 7. In this new circuit, the load voltage U_0 is close to E_1. This circuit is very simple and practical.

IV. ANALYSIS OF SIMPLIFIED BUCK-TYPE COMMUNICATION CIRCUIT

A. Steady-State equivalent circuit modeling

In a distributed system with powered bus communication, the effective of the DC resistor R_{bus} in transmission line must be considered. Assuming the capacitor C_1 is large enough and $R_{bus}C_1 >> T_c$, where T_c is the period of a bit, the output voltage in slave node can be derived by following: (Fig. 8)

1) stage 1[t_0,t_1]: M1 is off,
$$i_{c1}(t)= -I_0 \qquad (1)$$

2) stage 2[t_1,t_2]: M1 is on,
$$i_{d2}(t)=(E_1 - V_{D2} - U_0)/ R_{bus} \qquad (2)$$
$$i_{c1}(t)= i_{d2}(t) - I_0 \qquad (3)$$

V_{D2} is the diode voltage drop of D2.
In a bit period, the average current of C1 is
$$<i_{c1}> = \frac{1}{T}\int_{t_0}^{t_2} i_{c1}(t) = \frac{d(E_1 - V_{D2} - U_0)}{R_{bus}} - I_0 \qquad (4)$$

Assuming $<i_{c1}>=0$,so the load voltage of slave node is
$$U_0 = E_1 - V_{D2} - I_0 R_{bus} / d \qquad (5)$$

Where d is the average duty ratio in the bus. The steady-state equivalent circuit of power supply is show in Fig. 9.

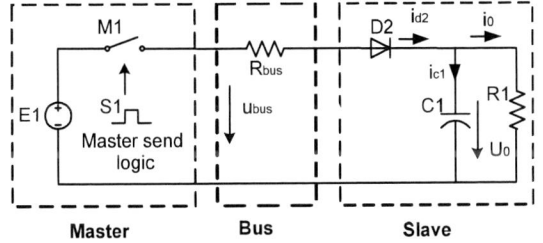

Fig. 8 Steady-State Power Supply circuit

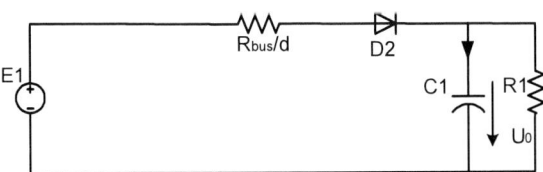

Fig. 9 Steady-State Power Supply Equivalent circuit

B. Transmission-Line effect in switching period

One of the difference between Buck-Type powered bus circuit and traditional switching circuit is that a transmission-line has been inserted. The length of the communication bus varies in different applications and the transmission line effect must be considered.

Ignoring the process of slave node, the equivalent circuit of long-distance transmission is shown as Fig. 10.

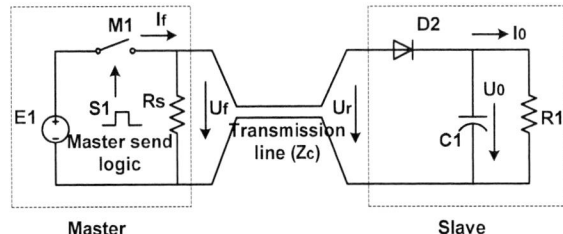

Fig. 10 Equivalent circuit with transmission line

1) Switch M1 turning on

According to the theory of transmission line, the waveform propagates through the bus and reflects at the end of the line [9]. When switch T1 turns on at t_0, a forward-traveling wave propagates down the bus. Before it reaches the edge of the bus, the bus output current is

$$i_f(t_0) = \frac{E_1}{Z_c} \qquad (6)$$

where Z_c is the characteristic impendence of the transmission line.

After a one-way transit time T_d= L/v, the forward-traveling wave reach the end of the bus and a backward-traveling wave is initiated. The value of the backward-traveling wave is determined by $u'_r(T_d) = \Gamma_L u_r(T_d)$, and the total voltage at the right edge of the bus is:

$$u_r(T_d) + u'_r(T_d) = u_r(T_d) + \Gamma_L u_r(T_d) = U_0 + V_{D2} \qquad (7)$$

Assuming the bus is a lossless transmission line, it can be derived:

$$u_r(T_d) = E_1 \qquad (8)$$

$$\Gamma_L u_r(T_d) = U_0 + V_{D2} - E_1 < 0 \qquad (9)$$

$$\Gamma_L < 0 \qquad (10)$$

The backward-traveling wave will arrive at the left edge of the bus after a one-way transit time, and a new reflection is initiated. It can be calculated that the source reflection coefficient is $\Gamma_s = -1$. So $\Gamma_s \Gamma_L > 0$, it means that the bus voltage will build up steadily and ringing phenomenon is avoided in nature.[8]

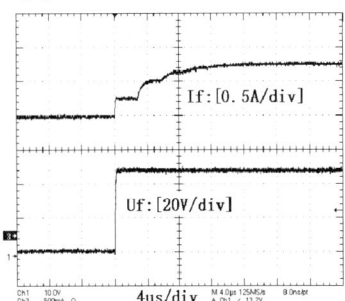

Fig. 11 Waveform of M1 Turning on

978-1-4244-4782-4/10 $26.00 © 2010 IEEE

Fig. 11 shows the current waveform and voltage waveform at the front edge of bus, where bus length L=200m, E_1=24V. It can be calculated that the characteristic impendence Z_c is about 80Ω.

2) Switch turning off

Before switch T1 turns off at t_1, the voltage at the front edge of bus is E_1 and the current is I_0/d. For transmission line, the output voltage can be considered as the sum of forward-traveling wave and forward-traveling wave from load, and the output current can be considered as the difference between forward-traveling current and reflection current. So

$$U_f(t_1^-) = V_s(t_1^-) + V_r(t_1^-) \qquad (11)$$

$$I_f(t_1^-) = I_s(t_1^-) - I_r(t_1^-) \qquad (12)$$

where $V_s(t_1^-)$ is the voltage of forward-traveling wave, $V_r(t_1^-)$ is the voltage of backward-traveling wave, $I_s(t_1^-)$ is the current of forward-traveling wave, $I_r(t_1^-)$ is the current of backward-traveling wave. The relation between current and voltage is:

$$V_s(t_1^-) = I_s(t_1^-)Z_c \qquad (13)$$

$$V_r(t_1^-) = I_r(t_1^-)Z_c \qquad (14)$$

Substituting (13)(14) into (11)(12), it can be get:

$$V_s(t_1^-) = \frac{1}{2}[U_f(t_1^-) + I_s(t_1^-)Z_c] \qquad (15)$$

$$V_r(t_1^-) = \frac{1}{2}[U_f(t_1^-) - I_s(t_1^-)Z_c] \qquad (16)$$

or

$$V_s(t_1^-) = \frac{1}{2}[E_1 + I_0 Z_c / d\ ] \qquad (17)$$

$$V_r(t_1^-) = \frac{1}{2}[E_1 - I_0 Z_c / d\ ] \qquad (18)$$

After switch T1 turns off at t_1, the forward-traveling waveform shuts down and the backward-traveling waveform remains unchanged. The equations are:

$$V_s(t_1^+) = 0 \qquad (19)$$

$$V_r(t_1^+) = V_r(t_1^-) = \frac{1}{2}[E_1 - I_0 Z_c / d\ ] \qquad (20)$$

The backward-traveling waveform initiates a new reflection waveform whose value is:

$$V_s'(t_1^+) = \Gamma_s' V_r(t_1^+) \qquad (21)$$

For open end line (the resistor Rs is removed in Fig. 10), reflection coefficient of transmission line is $\Gamma_s' = 1$. So the total voltage in front edge of the bus is:

$$U_f(t_1^+) = V_r(t_1^+) + V_s'(t_1^+) = E_1 - I_0 Z_c / d \qquad (22)$$

If the front edge of bus is terminated by a resistor Rs (Fig. 10), the reflection coefficient is:

$$\Gamma_s' = \frac{R_s - Z_c}{R_s + Z_c} \qquad (23)$$

And the total voltage in front edge of the bus is:

$$U_f(t_1^+) = V_r(t_1^+) + V_s'(t_1^+) = \frac{R_s}{R_s + Z_c}(E_1 - I_0 Z_c / d) \qquad (24)$$

From the equation (24), it can be inferred that:

1) if $E_1 = I_0 Z_c / d$, transmission line matched and output voltage steps to zero when T1 turns off (Fig. 12);

2) if $E_1 > I_0 Z_c / d$, a positive pulse is initiated when T1 turns off (Fig. 13);

3) if $E_1 < I_0 Z_c / d$, a negative pulse is initiated when T1 turns off (Fig. 14);

In Fig. 13 and Fig. 14, the width of the pulse is

$$T_p = \frac{2L}{v} \qquad (25)$$

where L is the length of bus, v is the velocity of propagation.

From the equation (24), it can be found that if the load current is much large, the absolute value of the pulse in the bus will be very high. For example, E1=24V, Z_c=80Ω, I_0=2A, d=0.5, Rs = ∞, the peak value of the pulse will be −296V, so that a terminator resistor or TVS should be added on the bus to protect the circuit. But on the other hand, the terminator changes the reflection wave and may reduce the transmission distance of the bus. The detailed process of transmission will not be discussed here.

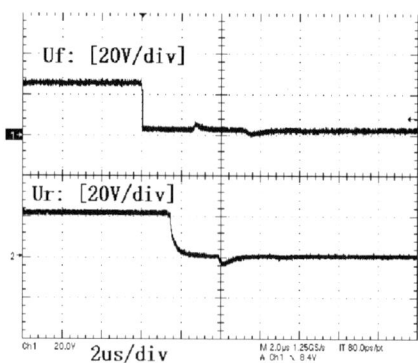

Fig. 12 Waveforms of M1 Turning off

（E1=24V，L=200m，d=0.5，R1=160Ω，Rs=110Ω）

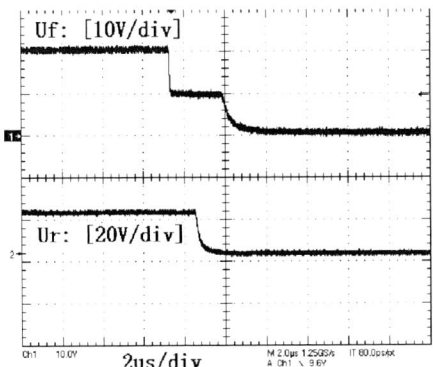

Fig. 13 Waveforms of M1 Turning off

(E1=24V，L=200m，D=0.5，R1=10kΩ，Rs=80Ω)

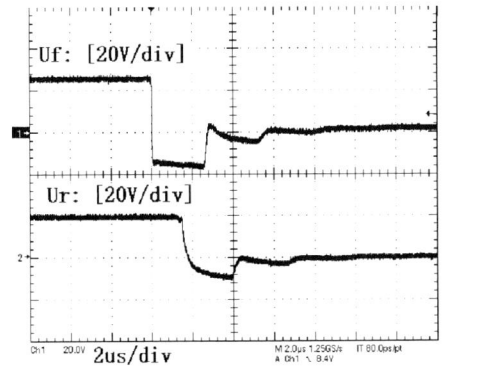

Fig. 14 Waveforms of M1 Turning off

(E1=24V，L=200m，D=0.5，R1=80Ω，Rs=110Ω)

When slave node sends data "0" back to the master node, switch M2 turns on and bus voltage increases steadily with no oscillation (Fig. 15), which can easily be detected and read by master node.

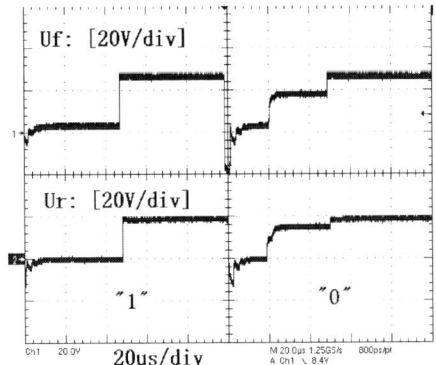

Fig. 15 Waveforms of Slave node sending data

V. CONCLUSIONS

This paper applying power electronics circuit in a new area of designing circuits for power line communication (powered bus communication). The evolution of the Buck-type communication circuit is presented in detail. In the same way, this method can be applied in other circuit such as Buck-boost, Forward and Bridge topology. Simplified Buck-type communication circuit has been applied in many areas such as alarm system and LED lighting system and the reliability of the circuit has been proved.

REFERENCES

[1] Niovi Pavlidou, A.J.Han.Vinck, et al., "Power Line Communications: State of the Art and Future trends", IEEE Communications Magazine, April 2003, pp.34-40

[2] Lotito A, Fiorelli R, Arrigo D, Cappelletti R., "A complete Narrow-Band Power Line Communication node for AMR",Power Line Communications and Its Applications, 2007. ISPLC '07. IEEE International Symposium on, 26-28 March 2007, pp.161 - 166

[3] Lienard M, Carrion M.O., Degardin V, Degauque P, "Modeling and Analysis of In-Vehicle Power Line Communication Channels", Vehicular Technology, IEEE Transactions on, Volume 57, Issue 2, March 2008 ,pp.670 - 679

[4] Eric R. Wade, Haruhiko Harry Asada, "Design of a Broadcasting Modem for a DC PLC Scheme", IEEE TRANSACTIONS ON MECHATRONICS, 2006, 11(5), pp.533-540

[5] Stefanutti W, Saggini S, Mattavelli P., Ghioni M, "Power Line Communication in Digitally Controlled DC–DC Converters Using Switching Frequency Modulation", Industrial Electronics, IEEE Transactions on Volume 55, Issue 4, April 2008, pp.1509 – 1518

[6] IEC61158-2-2002，"Fieldbus Standard For Use In Industrial Control Systems-Part 2：physical layer specification and service definition"，2002

[7] 1-Wire products mixed-signal design guide, www.maxim-ic.com

[8] The M-bus: An Overview, www.m-bus.com

[9] Clayton R. Paul, Introduction To ElectroMagnetic Compatibility,2nd ed. Wiley, 2006

978-1-4244-4782-4/10 $26.00 © 2010 IEEE

Compact Temperature Compensation of Inductive Fly-back Clamps for Integrated Power Switches Using a High-Voltage Base-Current-Compensated V_{be} Multiplier

Timothy P. Duryea[1,2] and Hoi Lee[2]

[1]Mixed Signal Automotive
Texas Instruments
Dallas, TX 75265-0311

[2]Dept. of Electrical Engineering
University of Texas at Dallas
Richardson, TX 75080-3021

Abstract— **In advanced power systems controlling inductive loads, integrating a power switch with clamping functionality that protects against inductive fly-back with other control circuitry can achieve a lower system cost. However, the conventional active clamp typically consists of many reverse-biased diodes that exhibit a positive temperature coefficient which is counterbalanced by several forward-biased diodes that consume a large chip area. A compact high-voltage V_{be} multiplier with base-current compensation circuit is presented to introduce a negative temperature coefficient within the clamp that occupies 50% less chip area than an equivalent clamp compensated with forward-biased diodes. The active clamp along with an LDMOS power transistor and driving circuitry have been fabricated in an 180nm high-voltage BiCMOS process and measurement results show that a negative temperature coefficient was introduced to achieve a 4V reduction in clamping voltage over a -40°C to 160°C range.**

I. INTRODUCTION

Evolving power electronic systems, such as automotive anti-lock brake and stability control systems, have moved the power transistors controlling high currents through solenoids, motors, and other inductive loads from discrete devices to integrated solutions that include additional system functions for cost reduction [1]–[9]. These highly integrated smart power IC's offer advantages and new features over traditional discrete devices, but also pose many significant challenges both from a process technology and circuit design perspective.

The simplest way to control an inductive load is to use a low-side switch to ground. Because the current flowing through an inductor cannot change instantaneously, some means of safely discharging the inductor current when the switch is turned off must be used to prevent damaging the transistor. Discrete power diodes can be used to clamp the inductor voltage but introduce additional system costs.

Therefore, integrated solutions that include both the power transistor and clamping circuitry are more attractive. An integrated device will generally employ active clamping techniques that use feedback to hold the power transistor in an on-state during inductor current discharge [1]–[3]. This allows the re-use of the power transistor area to dissipate the energy stored in the inductor during clamping rather than being required to allocate large chip area for both the power transistor and clamping devices.

A conventional active clamping circuit is shown in Figure 1 that breaks down from gate to drain when the power switch is turned off with a magnetized load and is held at the desired drain voltage through this feedback mechanism until all the inductor current is discharged. These reverse-biased Zener

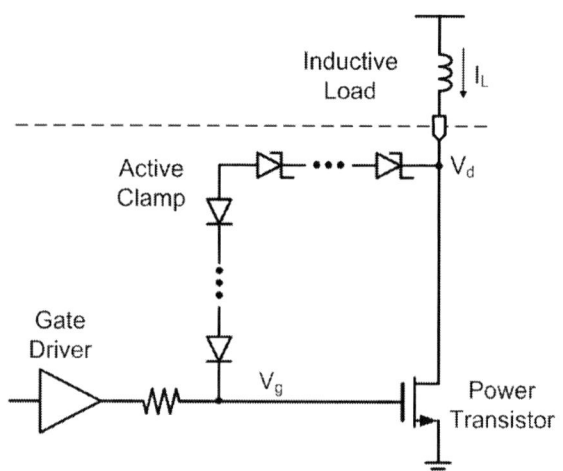

Figure 1. Standard low-side driver active clamp.

This research is supported by Texas Instruments, Inc.

diodes typically exhibit a positive temperature coefficient (PTAT voltage) that can be quite large. In many applications it is desirable to control the clamping voltage within a fixed range. The upper limit is typically restricted by the drain-to-source breakdown voltage (BV_{dss}) of the power transistor or by the peak junction temperatures reached during clamping. Conversely, the lower specification can be set by either the maximum supply voltage or the inductor current decay time requirements which, neglecting series resistance, is approximately governed by

$$\frac{dI_L}{dt} = \frac{V_{clamp} - V_{sup}}{L}, \qquad (1)$$

where V_{clamp} is the clamping voltage, V_{sup} is the supply voltage, I_L is the inductor current, and L is the inductance of the load.

To achieve better voltage control over temperature, forward-biased diodes are commonly added to the clamp to introduce a negative temperature coefficient (CTAT voltage). However, it is possible that the required number of forward-biased diodes to introduce the desired amount of CTAT voltage can become prohibitive in terms of the die area. To address this issue we introduce a high-voltage V_{be} multiplier circuit into the clamp structure to add an arbitrary amount of CTAT voltage with significantly less area than adding numerous forward-biased diodes. The proposed V_{be} multiplier-based clamping circuit will be discussed in detail in the following section.

II. PROPOSED HIGH-VOLTAGE V_{BE} MULTIPLIER CLAMP

A precise CTAT voltage is commonly generated by using the base-emitter voltage, V_{be}, of a bipolar transistor that typically exhibits a negative 1 to 2mV/°C behavior for a voltage bandgap reference [10]. Figure 2 shows the proposed concept of incorporating the V_{be} multiplier within the active clamping circuit in place of the forward-biased diode stack such that only one forward-biased diode, D_1, will still be necessary to block the current path from the gate driver to ground when the power transistor is in the "on" state. Other forward-biased diodes are replaced by the compact V_{be} multiplier circuit, where arbitrary temperature compensation can be achieved by adjusting the resistor ratio between R_2 and R_1 thereby saving significant implementation area.

Many modern BiCMOS processes are optimized towards the CMOS devices resulting in poor bipolar performance without additional processing steps [11], [12]. For example, the 180nm BiCMOS process used in this work has a NPN forward-current gain, β, that is nominally 10 with worst case corners as low as 2. Therefore, the V_{be} multiplier circuit should be designed assuming a very poor NPN performance. With β values so low, the standard V_{be} multiplier does not function properly due to high base currents that are provided through the resistor divider. The voltage equation including the finite base current error is given as

$$V_x = V_{be}\left(1 + \frac{R_2}{R_1}\right) + R_2 \frac{i_c}{\beta}, \qquad (2)$$

where the second term is the unwanted β error that is a function of collector current, i_c. This error term will cause wide variation in voltage V_x due to uncorrelated process and temperature variations of R_2 and β. Even worse, R_2 will generally have to be quite large so that the leakage current through the resistor divider does not activate the clamp prematurely.

To eliminate this error, a novel base-current compensation scheme is proposed in Figure 3. An additional NPN, Q_2, is added in series with Q_1 to obtain a replica of the base current of Q_1. Current mirrors composed of M_1, M_2, M_4, and M_5 copy and subtract the base current from the emitter of Q_2 so that the collector current of Q_1 and Q_2 are equal. The base current of Q_2 also provides a base current replica of Q_1 through

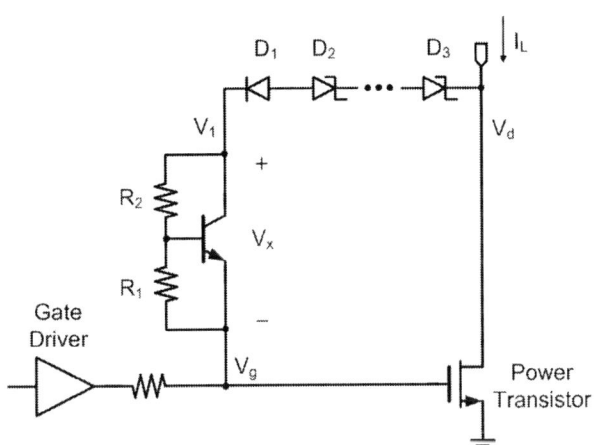

Figure 2. Concept of using V_{be} multiplier clamp circuit.

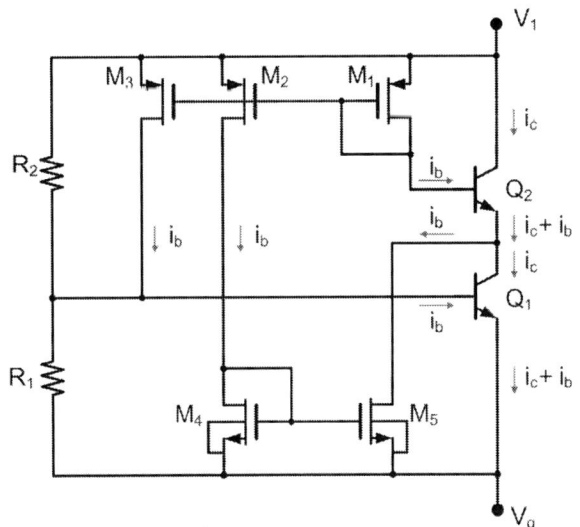

Figure 3. Base-current compensated V_{be} multiplier cell.

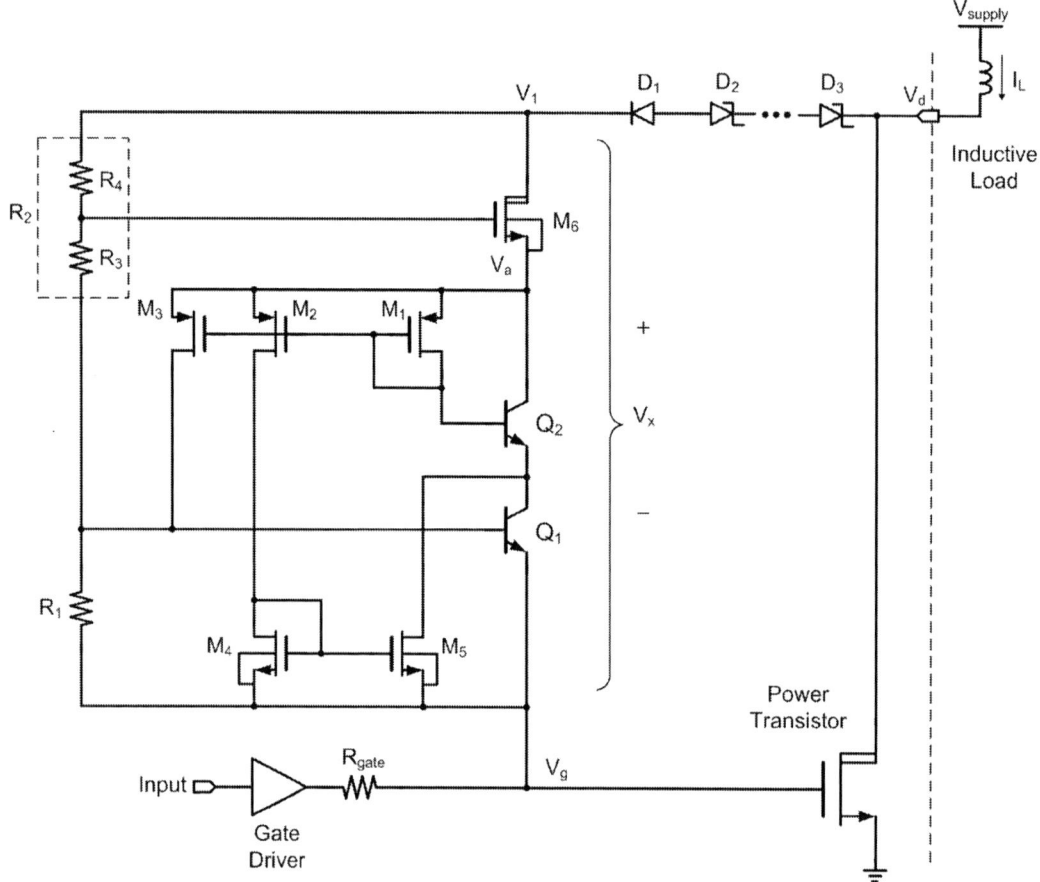

Figure 4. Proposed high voltage V_{be} multiplier active clamp with base current compensation.

transistors M_1–M_3 so that the Q_1 base current is not supplied from R_2. Assuming ideal matching, the base current error term of (1) is completely eliminated such that the circuit functions similarly to the classic V_{be} multiplier and is compatible with low β devices.

To save area, the current mirror devices M_1–M_5 are low-voltage components and must be protected against over-voltage conditions. Figure 4 shows entire active clamping circuit along with the integrated power device. The modified V_{be} multiplier cell incorporates high-voltage stand-off circuitry to allow large V_{be} multiplication factors to be achieved. A high-voltage drain-extended or laterally double-diffused MOS (LDMOS) device, M_6, is introduced as a shunt regulator to control the voltage at V_a to a level that is safe for the other low voltage CMOS devices. To set this regulation voltage, the original R_2 of Figure 3 is broken up into two separate resistors, R_3 and R_4, to define the gate voltage of M_6 during clamping. It should also be noted that this circuit is completely biased from the drain pin of the power transistor during clamping such that proper clamping behavior can still be achieved under loss of supply voltage conditions.

With the V_{be} multiplier circuit incorporated into the active clamping scheme shown in Figure 4, the total clamping voltage can be expressed as

$$V_{clamp} = M \cdot V_Z + V_{be}\left(1 + \frac{R_2}{R_1}\right) + V_D + V_{gs}, \qquad (3)$$

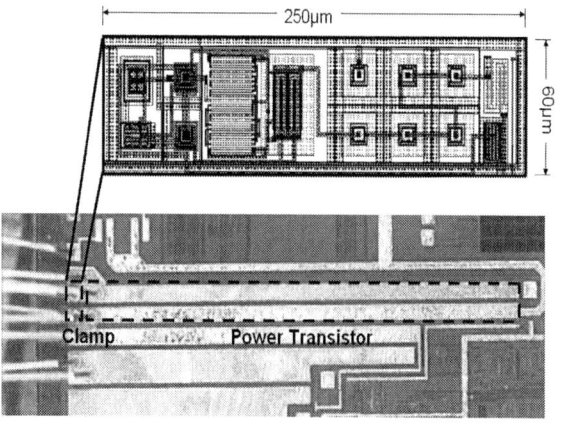

Figure 5. Die photo and layout of the proposed clamp.

where V_Z is the Zener voltage of the reverse-biased diodes, M is the number of reverse-biased diodes represented by D_2–D_3, V_D is the forward diode voltage drop of D_1, and V_{gs} is the gate-to-source voltage of the power transistor during clamping. The values of M and R_2/R_1 can be selected to achieve the desired temperature behavior.

III. RESULTS

The conventional and proposed clamping circuits were fabricated in an 180nm high-voltage BiCMOS process to achieve a clamping voltage around 42V. A 500mΩ LDMOS power transistor is also fabricated and integrated with the clamping and gate drive circuitry. Figure 5 shows the die photo with the detailed active clamping layout. The proposed circuit occupies 50% less area than a conventional clamp with forward-biased diodes adding equivalent CTAT voltage to the clamp.

The clamping voltage was measured at a constant current over an ambient temperature range of -40°C to 160°C using a thermal chamber. Figure 6 shows the temperature results where it is observed that sufficient CTAT voltage has been added to overcome the PTAT of the reversed-biased diodes to an extent where about a 4V drop is observed. This can be

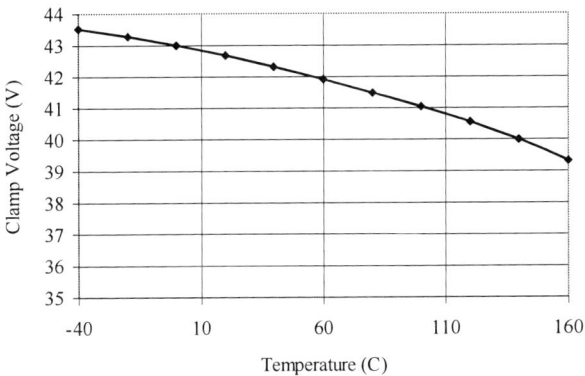

Figure 6. Measured temperature behavior of the proposed clamp.

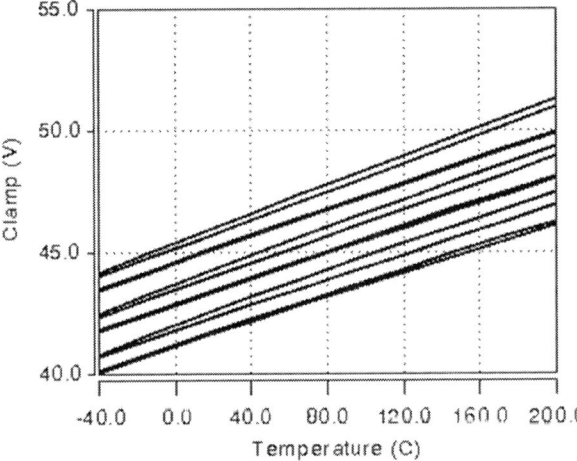

Figure 7. Simulated temperature behavior of an uncompensated clamp consisting of only reversed-biased Zener diodes over all process corners.

compared with the simulation results of a clamp of similar voltage consisting of all reversed-biased diodes shown in Figure 7.

Ideally, a flatter temperature curve was desired; however, discrepancies between the simulation modeling and actual silicon performance of the Zener voltage temperature coefficient resulted in over-compensation of the PTAT component. It should be noted that more or less CTAT voltage can be added by simply adjusting the ratio of R_1 and R_2 to achieve an arbitrary temperature coefficient of the clamping voltage.

To test the circuit in a real application, the integrated power transistor was used to switch high currents through an inductive load. Figure 8 shows the measured inductor current, I_L, and the drain voltage, V_d, of the power transistor when switching a 25mH 800mΩ inductive load at one ampere. When the power transistor is on, V_d drops to near zero volts placing the full supply voltage across the inductor allowing the current to ramp up. When the switch is turned off, V_d is clamped to the desired voltage while the inductor current is discharged through the power device that is held in the saturation region. No oscillations are observed in the clamping voltage indicating no stability issues in the clamp feedback loop.

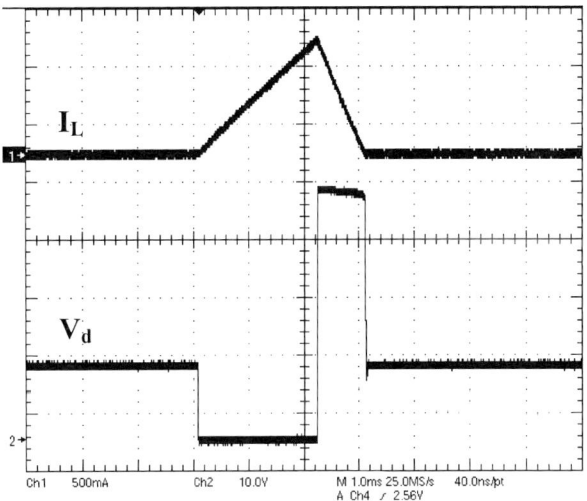

Figure 8. Measured switching waveforms of integrated power transistor with the proposed V_{be} multiplier clamp at 1A.

IV. CONCLUSION

A compact means of generating a CTAT voltage in high-voltage active clamping circuits has been introduced that utilizes a V_{be} multiplier structure. A novel base-current compensation scheme is proposed that allows the circuit to be used on processes with poor NPN performance while still achieving accurate multiplication. In addition, high-voltage stand-off circuitry is incorporated into the V_{be} multiplier to allow the use of a high multiplication factor while still maintaining low-voltage CMOS devices for the base-current compensation.

The proposed active clamp circuit reduces the required number of forward-biased diodes by 21, resulting in 50% area savings compared to a conventional clamp of similar CTAT compensation. Experimental results have verified the functionality of the proposed high-voltage base-current-compensated V_{be} multiplier circuit across wide range of temperatures. Discrepancies between the modeled temperature coefficient and actual device performance of the Zener diodes resulted in over compensation of the PTAT component. However, future circuits can be easily modified to achieve a flatter temperature behavior using the measured diode performance. In addition, the clamp was tested in high current solenoid switching applications and has shown well-behaved clamping characteristics with no oscillations.

Although the high-voltage V_{be} multiplier circuit was described in the context of active clamps for switching inductive loads, it should be noted that the proposed circuit could also be applied in many other applications requiring temperature compensation of high voltage references.

ACKNOWLEDGMENT

The authors would like to thank Bill Grose, Ben Amey, and Texas Instruments for supporting this research.

REFERENCES

[1] R. W. Adams, J. H. Carpenter, and T. Tanaka, "Low-side power output drive stage design and development concern," in *Proc. IEEE*

Bipolar/BiCMOS Circuits and Technology Meeting, Sep. 2000, pp. 74 - 81.

[2] A. Danchiv, "Protection functions in integrated low side switches," in *Proc. Int. Semiconductor Conf. CAS*, Sep. 2007, vol. 2, pp. 513 - 516.

[3] S. Havanur, "Quasi-clamped inductive switching behavior of power MOSFETs" in *Proc. IEEE Power Electronics Specialists Conf.* Jun. 2008, pp 4349 - 4354.

[4] W. Horn, and P. Singerl, "Thermally optimized demagnetization of inductive loads," in *Proc. Euro. Solid-State Circuits*, Sep. 2004, pp. 243 - 246.

[5] M. Han, "A new soft self-clamping scheme for improving the self-clamped inductive switching (SCIS) capability of automotive ignition IGBT," in *Proc. Int. Symp. Power Semiconductor Devices and ICs*, May 2007, pp. 145 - 148.

[6] C. Ionascu, "Design aspects for gate driver of power switch," in *Proc. Iint. Semiconductor Conf. CAS*, vol. 2, Sep. 2007, pp. 505 - 508.

[7] M. Wendt, L. Thoma, B. Wicht, and D. Schmitt-Landsiedel, "A configurable high-side/low-side driver with fast and equalized switching delay," *IEEE J. Solid-State Circuits*, vol. 43, no. 7, pp. 1617 - 1625, Jul. 2008.

[8] W. C. Dunn, "Driving and protection of high side NMOS power switches," *IEEE Trans. Industry Applications*, vol. 28, no.1, pp. 26 - 30, Jan. 1992.

[9] R. Gariboldi and F. Pulvirenti, "A 70m intelligent high side switch with full diagnostics," *IEEE J. Solid-State Circuits*, vol. 31, no. 7, pp. 915 - 923, Jul. 1996.

[10] P. R. Gray, P. J. Hurst, S. H. Lewis, and R. G. Meyer, *Analysis and Design of Analog Integrated Circuits*, 4th Ed., New York: Wiley, 2001.

[11] A. Marshall and R. Teggatz, "The role of bipolar structures in a power MOS process," in *Proc. IEEE Bipolar/BiCMOS Circuits and Technology Meeting*, Sep. 1996, pp. 208 - 211.

[12] T. Efland, J. Devore, A. Hastings, S. Pendharkar, and R. Teggatz, "Bipolar issues in advanced power BiCMOS technology," in *Proc. IEEE Bipolar/BiCMOS Circuits and Technology Meeting*, Sep. 2000, pp. 20 - 27.

Optimal Design for the Damping Resistor in RCD-R Snubber to Suppress Common-mode Noise

Peipei Meng, Henglin Chen, Sheng Zheng, Xinke Wu, Zhaoming Qian (IEEE Senior Member)
College of Electrical Engineering
Zhejiang University, Hangzhou 310027, China
Email: cathleen@zju.edu.cn

Abstract—The RCD snubber is usually used in flyback converter, in order to limit the voltage spikes caused by leakage inductance of the transformer. But after the clamping diode in the RCD snubber turns off, the high frequency oscillations caused by the leakage inductance and intrinsic capacitance of the transistor is still serious due to the residual energy that can't be entirely absorbed by the clamping capacitor. These oscillations will generate unexpected CM noises at some particular frequencies. In practical applications, a damping resistor in series with the RCD circuit which is called RCD-R snubber in this paper can be used to suppress the CM noises. The goal of this paper is to analyze the RCD-R snubber circuit operations and to develop design guideline to optimal choose the value of the damping resistor. Both the theoretical analysis and experimental results verify that the RCD-R snubber circuit can suppress CM noise effectively and have no cost on the efficiency of the converter.

I. INTRODUCTION

Dissipative RCD snubber is widely used in low power DC-DC applications for its simplicity and low cost. It can suppress voltage transient spikes effectively with some acceptable losses [1]-[4]. But due to the physical limitations, the energy stored in the leakage inductance of the transformer can't be entirely absorbed by the RCD snubber, which will cause high frequency oscillations between the leakage inductance and the intrinsic capacitance of the transistor. These oscillations not only can cause extra losses [5], but also will generate unexpected CM noises at some particular frequencies. Choosing smaller dissipative resistor in the RCD snubber can lower the clamping voltage of the capacitor and reduce the residual energy left in the circuit. But the efficiency of the converter will suffer from it.

The RCD-R snubber shown in Fig. 1 can decrease the residual energy left in the circuit. If the damping resistor R_d is well chosen, there won't be any residual energy left in the circuit after the clamping diode D_c turns off. So no oscillation between the leakage inductance and the intrinsic capacitance of the transistor can take place after D_c turns off and the CM noise caused by these oscillations can be suppressed. If R_d continues increasing, it will make no help to EMI performance, but the voltage spikes of the transistor Q_1 will increase.

In this paper, the working principle of the RCD-R snubber is studied in detail. A helpful method to optimal design the damping resistor is proposed. Both the theoretical analysis and experimental results verify that damping resistor can suppress CM noise effectively and have no cost on the efficiency of the converter.

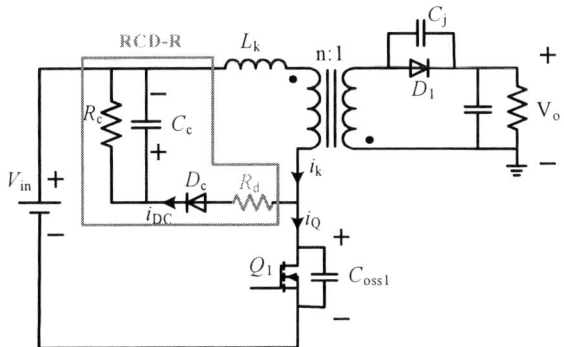

Figure 1. Flyback converter with RCD-R snubber.

II. OPERATION PRINCIPLES OF RCD-R SNUBBER

Fig. 2 shows the key operation waveforms of the flyback converter with RCD-R snubber shown in Fig. 1. i_{DC} is the current in the clamping diode D_c, V_{ds} and V_c are the voltage across the main transistor Q_1 and the capacitor C_c respectively. When Q_1 turns off, the transformer begins to transfer the energy stored during the on-time to the output. The voltage across the transformer is clamped to nV_o by the output voltage. But the energy stored in the leakage inductance can't transfer to the secondary-side, which will charge the intrinsic capacitance of the transistor C_{oss1} until t_0 when V_{ds} is charged to $V_{in} + V_{cp}$, V_{cp} is the clamping voltage of C_c. Then the diode D_c turns on. During the interval $t_0 - t_1$, the current is transferred from the transistor Q_1 to the RCD circuit.

Stage 1 (t_1-t_2): At t_1 the current in the transistor reaches the peak value I_F. The equivalent circuit of this stage is shown in Fig. 3. R_{trans_ac} is the ac resistance of the transformer. Because the RCD circuit is operating in high frequency, so R_{trans_ac} can't be ignored.

This work is supported by Natural Science Foundation of China under Grant 50807047 and 50907061

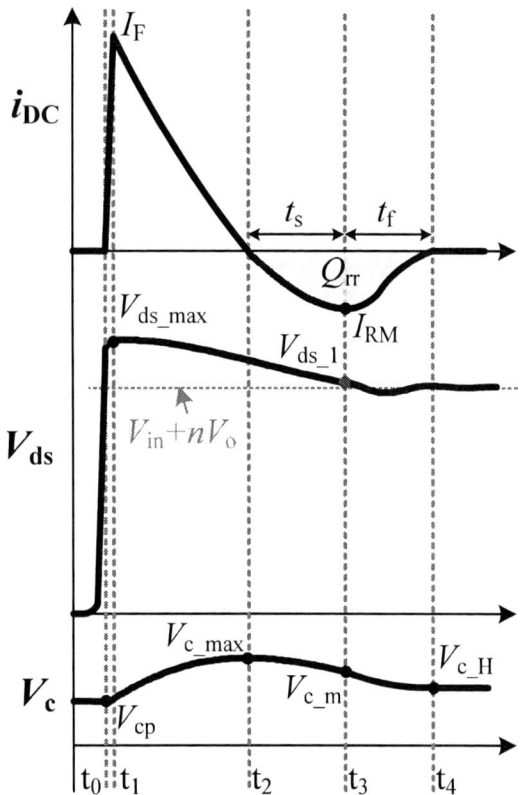

Figure 2. Operating waveforms

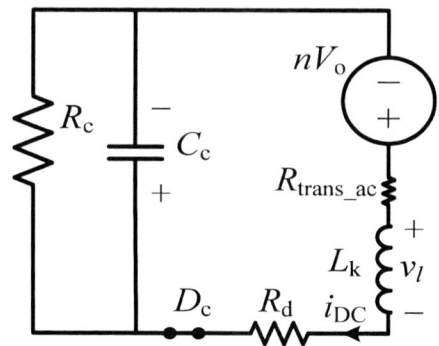

Figure 3. Equivalent circuit of Stage 1 and Stage 2

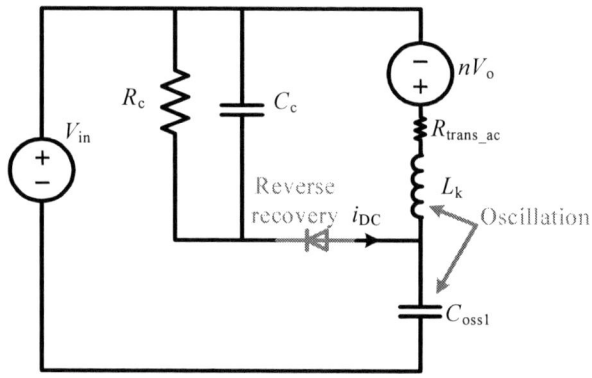

Figure 4. Equivalent circuit of Stage 3

To evaluate the voltage of the clamping capacitor $v_c(t)$, apply Kirchoff's voltage law (KVL) around the loop in Fig. 3 gives

$$\frac{\partial^2 v_c}{\partial t^2} + \left(\frac{R_{eq}}{L_k} + \frac{1}{R_c C_c}\right) \cdot \frac{\partial v_c}{\partial t} + \frac{R_{eq} + R_c}{R_c L_k C_c} \cdot v_c = \frac{nV_o}{L_k C_c} \qquad (1)$$

In which, $R_{eq} = R_d + R_{trans\_ac}$.

Assume that $v_c(t_1) \approx v_c(t_0) = V_{cp}$. V_{cp} is the clamping voltage of capacitor C_c, the initial voltage $v_c(t_1)$ and its derivative can be derived

$$v_c(t_1) = V_{cp} \qquad (2)$$

$$\frac{dv_c(t_1)}{dt} = \frac{I_F}{C_c} - \frac{V_{cp}}{R_c C_c} \qquad (3)$$

Then voltage $v_c(t)$ can be obtained as

$$v_c(t) = \frac{R_c \cdot nV_o}{R_{eq} + R_c} + e^{-b(t-t_1)}\left[A_1 \cos(\omega_d(t-t_1)) + A_2 \sin(\omega_d(t-t_1))\right] \qquad (4)$$

In which,

$$b = \frac{R_{eq} R_c C_c + L_k}{2 R_c L_k C_c} \qquad (5)$$

$$\omega_d = \sqrt{\frac{2 R_{eq} R_c C_c L_k + 4 R_c^2 L_k C_c - R_{eq}^2 R_c^2 C_c^2 - L_k^2}{4 R_c^2 L_k^2 C_c^2}} \qquad (6)$$

$$A_1 = \frac{V_{cp}(R_{eq} + R_c) + nV_o R_c}{R_{eq} + R_c} \quad A_2 = \frac{bA_1 R_c C_c + I_F R_c - V_{cp}}{\omega_d R_c C_c} \qquad (7)$$

From Equation (4), the maximal value of $v_c(t)$ at t_2 can be evaluated

$$V_{c\_max} = \frac{R_c \cdot nV_o}{R_{eq} + R_c} + e^{\frac{\pi b}{2\omega_d}}\sqrt{A_1^2 + A_2^2} = f(V_{cp}) \qquad (8)$$

The voltage spike of the main transistor Q_1 can be evaluated approximately by

$$V_{ds\_max} \approx V_{in} + V_{cp} + I_F \cdot R_d \qquad (9)$$

The current in the diode D_c can be evaluated similarly

$$i_{DC}(t) = \frac{nV_o}{R_{eq} + R_c} + e^{-b(t-t_1)}\left[B_1 \cos(\omega_d(t-t_1)) + B_2 \sin(\omega_d(t-t_1))\right] \qquad (10)$$

In which,

$$B_1 = \frac{I_F(R_{eq} + R_c) - nV_o}{R_{eq} + R_c} \quad B_2 = \frac{bB_1 L_k + nV_o - V_{cp} - I_F R_{eq}}{\omega_d L_k} \quad (11)$$

In Equation (4), (8), (9) and (10), only V_{cp} is the unknown variable that related with the RCD-R snubber, all the other parameters are constants for a particular converter.

Stage 2 (t_2-t_3): Although i_{DC} reaches zero at t_2, the diode continues to conduct until t_3 at which point all the stored charge has been removed from the diode. So this stage could also be modeled by the equivalent circuit shown in Fig. 3. Equations (4) and (10) are also valid for this stage.

Stage 3 (t_3-t_4): At t_3 the charges stored have been removed from the diode, and then the diode starts blocking. Reverse recovery losses occur during this stage since negative current is present while the blocking voltage across the diode increases. The equivalent circuit of this stage is shown in Fig. 4. The reverse recovery of the diode and the ringing between L_k and C_{oss1} takes place at the same time.

It can be found that, the reason that caused oscillation between L_k and C_{oss1} is the residual energy left in the circuit when D_c starts blocking. This residual energy is mainly stored in C_{oss1} at t_3. The voltage of V_{ds} at t_3, marked as V_{ds-1}, can be evaluated by

$$V_{ds\_1} = V_{in} + V_{c\_m} - I_{RM} \cdot R_d \quad (12)$$

In which, V_{c-m} is the voltage of V_c at t_3, I_{RM} is the peak reverse recovery current D_c at t_3.

So, if V_{ds-1} is equal to $V_{in} + nV_o$, there will be almost no residual energy left in the circuit and little oscillation can take place any more.

Stage 4 [t_4-(t_0+T)]: After t_4 the energy stored in C_c will be dissipated by R_c until Q_1 turns on again.

The reference directions are shown in Fig. 5, so

$$i_c(t) - C_c \cdot \frac{dv_c(t)}{dt} - \frac{v_c(t)}{R_c} \Rightarrow \frac{1}{v_c(t)} \cdot dv_c(t) = -\frac{1}{C_c R_c} \cdot dt \quad (13)$$

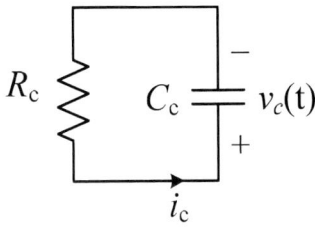

Figure 5. Equivalent circuit of Stage 4

Apply integration to (13) in one period T yields:

$$\int_{t_4}^{t_4+T} \frac{1}{v_c(t)} \cdot dv_c(t) = \int_{t_4}^{t_4+T} -\frac{1}{C_c R_c} \cdot dt$$
$$\Rightarrow \ln v_c(t_4 + T) - \ln v_c(t_4) = -\frac{T}{C_c R_c} \quad (14)$$

According to the capacitor charging balance, there has $v_c(t_4 + T) = V_{cp}$. Substitute this equation into (14) yields

$$\ln V_{cp} - \ln V_{c\_H} = -\frac{T}{C_c R_c} \quad (15)$$

In which, V_{c-H} is the voltage of V_c at t_4, T is the working period of the converter.

III. OPTIMAL DESIGN FOR RCD-R SNUBBER AND POWER LOSSES ANALYSE

A. Diode reverse recovery modeling

The reverse recovery of the diode is an important issue. Only the reverse recovery of the diode is precisely modeled, can the RCD-R snubber be well designed. [6] indicates that the reverse recovery charge is a function of the diode's forward current at the initiation of the turn-off process and the di/dt when forward current decreases. In this circuit the current in the diode oscillates to zero when the diode turns off. So the reverse recovery charge Q_{rr} can be fitted to di_{DC}/dt by using a high order polynomial which provides a good fit of the original data, so

$$Q_{rr} = f(x) = a_n x^n + a_{n-1} x^{n-1} \cdots a_1 x + a_0 \quad (16)$$

x refers to the value of di_{DC}/dt at t_2, which can be evaluated from

$$\left. \frac{di_{DC}}{dt} \right|_{t=t_2} = \frac{v_{l_k}(t_2)}{L_k} = \frac{nV_o - V_{c\_max}}{L_k} \quad (17)$$

In which V_{c-max} is the value of $v_c(t)$ at t_2. During the period t_2-t_4 when reverse recovery of the diode takes place, it can be derived that

$$C_c \cdot (V_{c\_max} - V_{c\_H}) = Q_{rr} \quad (18)$$

In which, V_{c_H} is the voltage of V_c at t_4.

Combine (16) and (18) gives

$$V_{c\_max} - V_{c\_H} = \frac{Q_{rr}}{C_c} = g(y) = b_n y^n + b_{n-1} y^{n-1} \cdots b_1 y + b_0 \quad (19)$$

In which, $y = nV_o - V_{c\_max}$, $b_i = a_i / C_c L_k$

To apply polynomial regression to (19), data points must first be selected to adequately describe the Q_{rr} surface. Solving (19) then produces a surface which provides a best fit by minimizing the squared error between the surface and the original data.

978-1-4244-4782-4/10 $26.00 © 2010 IEEE

Then the time interval t_s when the stored charge is extracted from the diode, as shown in Fig. 2, can also be modeled by a high order polynomial similarly

$$t_s = h(y) = c_n y^n + c_{n-1} y^{n-1} \cdots c_1 y + c_0 \tag{20}$$

B. Solving the variables related with the RCD-R snubber

Combine the equations (8), (15) and (19) can solve the three unknown variables V_{cp}, V_{c_max} and V_{c_H}.

Substitute equation (20) into equations (4) and (10) can solve the values of v_c and i_{DC} at t_3, respectively.

$$V_{c\_m} = v_c(t_2 + t_s) \tag{21}$$

$$I_{RM} = i_{DC}(t_2 + t_s) \tag{22}$$

Up till now, all the variables related with the RCD-R snubber have been evaluated. It's important to notice that, Different from the conventional analysis that assume the voltage of the clamping capacitor keeps invariant when the clamping diode is on, in this paper the charging procedure of the clamping capacitor is modeled precisely. Moreover, the reverse recovery of the clamping diode is also taken into consideration. So the calculating result in this paper is much closer with the reality.

C. Design procedure for R_d

It has been mentioned in section II that, if V_{ds-1} is equal to $V_{in} + nV_o$, there will be almost no residual energy left in the circuit and little oscillation can take place any more. Thus, let Equation (12) equal to $V_{in} + nV_o$ yields

$$V_{c\_m} - I_{RM} \cdot R_d = nV_o \tag{23}$$

Substitute (21) and (22) into (23) yields

$$v_c(t_2 + t_s) - R_d \cdot i_{DC}(t_2 + t_s) = nV_o \tag{24}$$

Solving Equation (24) can evaluate the optimal designed value of R_d as

$$R_d = \frac{v_c(t_2 + t_s) - nV_o}{i_{DC}(t_2 + t_s)} \tag{25}$$

D. Calculating the power losses in the RCD circuit

The losses in the RCD circuit can be divided into two parts: the losses caused during t_0-t_4 when the diode is on, and the losses caused during t_4-(t_0+T) when the energy in C_c is dissipated by R_c. The former losses W_1 can be determined from the energy balance requirement [1]:

Loss = energy from source − increase in energy stored in capacitor − increase in energy stored in inductor. So,

$$W_1 = nV_o \int_{t_0}^{t_4} idt - C_c \int_{v(t_0)}^{v(t_4)} vdv - L_k \int_{i(t_0)}^{i(t_4)} idi$$

$$= nV_o \cdot C_c(V_{c\_H} - V_{cp}) - \frac{1}{2}C_c(V_{c\_H}{}^2 - V_{cp}{}^2) + \frac{1}{2}L_k I_Q{}^2 \tag{26}$$

The losses dissipated by R_c is

$$W_2 = \frac{1}{2}C_c(V_{c\_H}{}^2 - V_{cp}{}^2) \tag{27}$$

Then the total power losses dissipated in the RCD circuit is

$$P_{loss} = f \cdot (W_1 + W_2) = f \cdot \left[nV_o \cdot C_c(V_{c\_H} - V_{cp}) + \frac{1}{2}L_k I_Q{}^2 \right] \tag{28}$$

IV. Experimental Verification

A 24 Watts, 100kHz, 170V input, 16V output flyback converter is built up to illustrate the design procedure for R_d in the RCD-R circuit and verify the analysis given above. The key components and parameters are shown in table I.

TABLE I.   KEY COMPONENTS AND PARAMETERS

Q_1	D_1	Transformer	n	D_c
SPA11N60C	STPS20S100	PQ26/20/TP4	24:4	FR107

R_c	L_k	C_c	V_o	f
47kΩ	5.7uH	2.2nF	16V	100kHz

V_{ds-1} and P_{loss} can be evaluated by Equations (12) and (28), respectively. They are plotted in Fig. 6 as a function of the damping resistor R_d. If V_{ds_1} equals to $V_{in} + nV_o$, there will be almost no residual energy left in the circuit and little oscillation can take place any more. Fig. 6 indicates the optimal design point $R_d = 62\Omega$ that can eliminate almost all the oscillations caused by the leakage inductance and intrinsic capacitance of the transistor.

The calculated power losses in the RCD circuit P_{loss} is also plotted in Fig. 6, which indicates that as R_d increases, the power losses in the RCD circuit will decrease slightly. But the decrement is too small that the measured efficiency remains almost the same with different R_d as shown in Fig. 7, which indicates that the RCD-R snubber won't cause extra power losses compare with the RCD snubber with the same parameters.

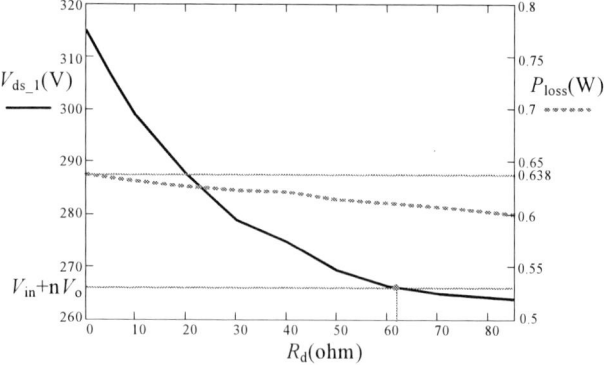

Figure 6. Optimal design for R_d and the calculated power losses in the RCD-R snubber

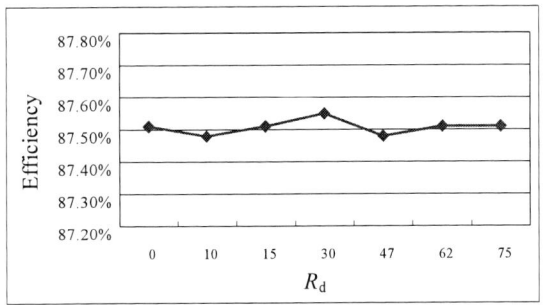

Figure 7. Measured efficiency with different R_d

Fig. 8 (a) and (b) shows the experimental waveforms of this flyback converter with optimal designed $R_d = 62\Omega$ and without R_d respectively. V_{ds} is voltage across the main transistor, i_{DC} is the current in the clamping diode. In Fig. 8 (a) no oscillation takes place in the waveform of V_{ds} any more. On the contrary, in Fig. 8 (b) there are serious high frequency oscillations appear in the waveform of V_{ds} after the clamping diode turns off. The frequency of these oscillations is determined by the leakage inductance and intrinsic capacitance of the transistor at around 8.6MHz, which is measured in Fig. 8 (b).

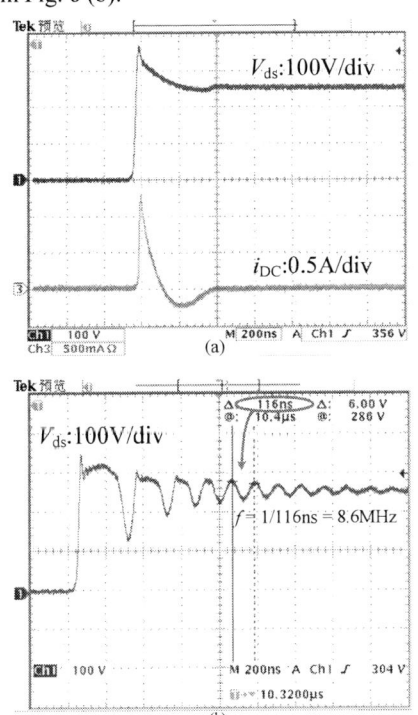

Figure 8. (a) Experimental waveforms of V_{ds} and i_{DC} when R_c=47kΩ, R_d=62Ω (b) Experimental waveform of V_{ds} when R_c=47kΩ, R_d=0

The CM noises of this topology when the converter is with and without damping resistor R_d are both tested. The curves of the tested EMI noise are shown in Fig. 9, in which curve A is tested without R_d and curve B is tested with R_d=62Ω. Please be aware that all the CM noises are tested without any EMI filter. Curve A has CM noise spikes at around 8.6 MHz, which is

consistent with the oscillation frequency between the leakage inductance and intrinsic capacitance of the transistor as shown in Fig. 8 (b). The CM noise spikes around this frequency point are suppressed in curve B, because with well designed R_d no oscillation between the leakage inductance and intrinsic capacitance of the transistor can take place any more.

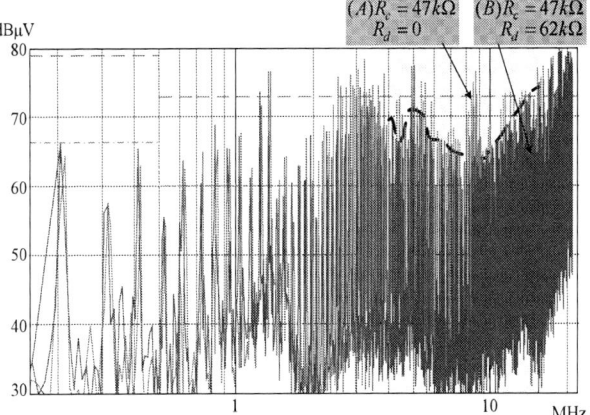

Figure 9. Comparison the effect of CM noise reduction with damping resistor R_d

V. CONCLUSION

RCD-R snubber can suppress the CM noise caused by the high frequency oscillation between the leakage inductance and intrinsic capacitance of the transistor. Different from the general concept that the improvement on EMI performance always at the cost of efficiency. The power loss in the RCD-R snubber is almost the same as in the RCD snubber with the same parameters, but the CM noise can benefits much from the RCD-R snubber. The working principle of RCD-R snubber is studied in detail. A novel design procedure is obtained to choose the value of the damping resistor.

ACKNOWLEDGMENT

This work is supported by Natural Science Foundation of China under Grant 50807047 and 50907061.

REFERENCES

[1] W. McMurray, "Optimum snubbers for power semiconductors," IEEE Transactions on Industry Applications, Oct. 1972, pp 593-600.

[2] Finney, S.J., Williams, B.W., Green, T.C., "RCD snubber revisited," IEEE Transactions on Industry Applications, Volume 32, Issue 1, Jan.-Feb. 1996 pp. 155 – 160.

[3] Patel, H.K., "Voltage transient spikes suppression in flyback converter using dissipative voltage snubbers," in Proc. IEEE ICIEA 2008, pp. 897-901.

[4] Song-Yi Lin, Chern-Lin Chen, "Analysis and design for RCD clamped snubber used in output rectifier of phase-shift full-bridge ZVS converters," IEEE Transactions on Industrial Electronics, Volume 45, Issue 2, April 1998 pp. 358 – 359.

[5] Alenka Hren, Joze Korelic, and Miro Milanovic "RC-RCD clamp circuit for ringing losses reduction in a flyback converter" IEEE Trans. On Circuits and systems—II:Express Briefs. Vol. 53. No. 5 May 2006.

[6] Schonberger, J., Feix, G., "Modelling turn-Off losses in power diodes," Control and Modeling for Power Electronics 2008, pp. 1 – 6.

A High-Efficient *LLCC* Series-Parallel Resonant Converter

Christian P. Dick, Furkan K. Titiz, Rik W. De Doncker
Institute for Power Electronics and Electrical Drives (ISEA)
RWTH Aachen University
Jaegerstr. 17-19, 52066 Aachen, Germany
E-mail: di@isea.rwth-aachen.de

Abstract—A high efficient *LLCC*-type resonant dc-dc converter is discussed in this paper for a low-power photovoltaic application. Emphasis is put on the different design mechanisms of the resonant tank. At the same time soft switching of the inverter as well as the rectifier bridge are regarded. Concerning the design rules, a new challenge is solved in designing a *LLCC*-converter with voltage-source output. Instead of the resonant elements, ratios of them, e.g. the ratio of inductances L_s/L_p is considered as design parameters first. Furthermore, the derived design rule for the transformer-inductor device fits directly into the overall *LLCC*-design. Due to the nature of transformers, i.e. the relation of the inductances L_s/L_p is only a function of geometry, this design parameter is directly considered by geometry. Experimental results demonstrate the high efficiency.

I. INTRODUCTION

A. Application Concept

A highly efficient dc-dc converter is proposed as module integrated converter for photovoltaic applications, where PV voltage v_{PV}, in the tens of volts, is boosted to a dc-distribution line voltage of $V_{dstr} = 700$ V, as indicated in Fig. 1.

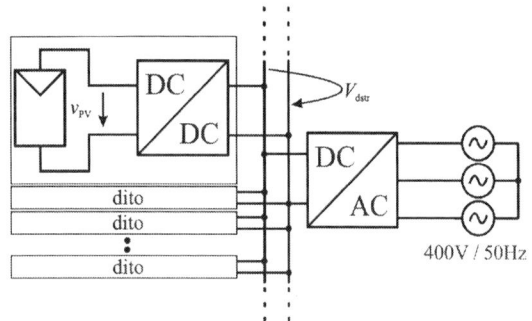

Figure 1. Parallel module-integrated converter concept

The advantages of this kind of parallel converter concept with central dc-ac converter, compared to other module-integrated solutions, are as follows:

- Lifetime: The critical component of the system exposed to harsh environment at the module is the module-integrated converter. In comparison to single-phase AC-modules, no low frequency energy buffering passives as electrolytic capacitors are applied [1]. Furthermore, the effort for the converter functionality at the module is minimized. Only maximum power point tracking (MPPT) and safety features are realized at high efficiency, also reducing costs. Thus, a potentially high lifetime is achieved.

- Costs: All grid-related functionalities like grid current control, disconnection from the grid in case of failures, metering etc. are only implemented once in a central unit, which is necessary at least for metering anyway. Furthermore, the module-integrated converter concept only shows two power stages. Most solutions show more, or higher effort [2].

- Flexibility: All kinds of modules can be connected via a specific module-integrated converter. With the high step-up ratio a high-frequency transformer will be part of the topology. Thus, also classical thin-film modules can be connected to ground to avoid deterioration coming from small leakage currents in case of a negative bias voltage. The system concept can be combined with classical string or central converter concepts. In case only parts of the PV-generator suffer from shading, these specific modules might be connected via a module-integrated converter [2].

- Safety: The proposed system concept in Fig. 1 allows grounding of the dc-distribution wires for the installation and for maintenance work on the building facade. The module-integrated converters are programmed to operate only at a certain range of V_{dstr}. Latter is a major safety improvement compared to classical string or central converter concepts using a dc-distribution, carrying the short circuit photovoltaic dc-current when being grounded [2].

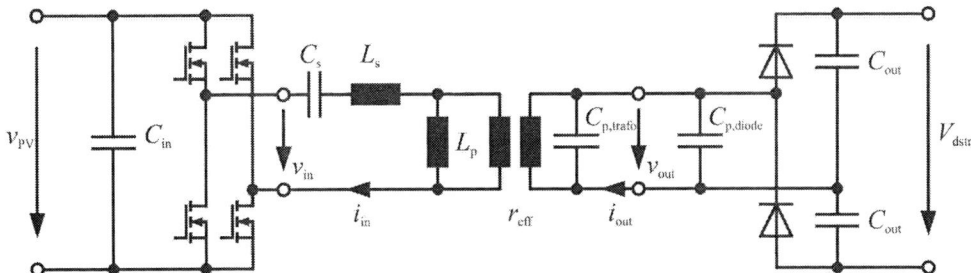

Figure 2. Single-phase *LLCC*-type Series-Parallel Resonant Converter

The DC-AC converter controls the dc-distribution voltage to a constant value of $V_{dstr} = 700$ V. Thus, the module-integrated dc-dc converter is clamped to a fixed voltage of the distribution line and performs MPPT by maximization of the output current.

B. Fundamentals on LLCC-type Converter

The critical component, exposed to the harsh environment in the application, is the module-integrated converter itself. Efficiency is maximized to maximize energy output and to reduce operation temperature enhancing lifetime. The single-phase *LLCC*-type series-parallel resonant converter as depicted in Fig. 2 is chosen, since this converter potentially shows high efficiency. It is operated at 50% duty cycle and 180° phase shift of the inverter legs. The converter is controlled by small variation of the operation frequency *f*. The topology suits the requirements for the following reasons:

- Low turn-off currents: Due to the nature of the resonance, the load-resonant current comes down before the turn-off instant. Thus, high frequencies can be achieved resulting reduced component size.

- Resonant-pole principle: The resonant tank, consisting of the four elements C_s, L_s, L_p and C_p, is designed to show an inductive behavior for the input MOSFET bridge at operation frequency. Thus, the resonant pole principle is applied resulting in zero-voltage switching [3],[4]. Additional capacitive snubbers are installed across the MOSFETs.

- Low diode stress: The parallel capacitance C_p is the sum of the parasitic capacitances of the diode, the transformer, and an external capacitor. It acts as a snubber for the rectifier diodes, since the diode's voltage slopes are limited.

- High part-load efficiency: Due to the nature of the series resonance, the rms-current in the resonant tank is reduced significantly at part-load, reducing component stress at reduced load [5]. This is a major advantage in photovoltaic applications, since part-load efficiency has major impact on "European Efficiency" η_{euro}. Latter takes the regular existence of reduced solar irradiation into account. It is defined as the weighted sum (1), with $\eta_{x\%}$ being the efficiency of the converter operating at x% of nominal load.

$$\eta_{euro} = 0.03\eta_{5\%} + 0.06\eta_{10\%} + 0.13\eta_{20\%} + 0.1\eta_{30\%} + 0.48\eta_{50\%} + 0.2\eta_{100\%} \quad (1)$$

- Controllability: Due to the nature of parallel resonance the converter can be controlled by a small operation frequency variation [6],[7],[8]. A wide input voltage range $v_{PV,max} = 2v_{PV,min}$ is designed.

Often series-parallel resonant converters are found comprising a current-source output. Due to high output voltage of $V_{dstr} = 700$ V, a voltage source dc-link is installed to minimize the stresses for the parallel resonant components L_p and C_p [5]. The rectifier is realized as voltage doubler, reducing the ratio of secondary side numbers of turns on the transformer.

This paper focuses on the most important parts of the design of the five degrees of freedom of the resonant tank, i.e. r_{eff}, C_s, L_s, L_p and C_p.

II. DESIGN OF RESONANT TANK ELEMENTS

Optimization is carried out in all steps for high efficiency at the boundary conditions of the specifications in all operational points. In this first step the choice of the components is qualified with the goal of minimum apparent power in the resonant tank, i.e. minimum rms-currents when using voltage-source inverter and rectifier as given in Fig.2.

A. Converter Design Rules based on First Harmonic Approximation (FHA)

Minimum currents in the resonant tank are the key to high efficiency. This can be directly read from the loss models of the different components as, on-state MOSFET losses, resonant capacitor losses and copper losses of the transformer-inductor device.

For the minimization of rms-currents, FHA is used as converter model to derive design rules. Here, the design method of a previous work on *LLC*-type resonant converters [9],[10] is extended to *LLCC*-type converters. Under classical ac-operation, i.e. describing the pulsed voltage waveforms only with their fundamental component in (2) [5], the FHA converter model is given by (4) and (5), with the definition of resonant frequencies in (3):

$$\text{FHA}: \quad V_{in} = \frac{2\sqrt{2}}{\pi} v_{PV} \quad \text{and} \quad V_{out} = \frac{\sqrt{2}}{\pi} V_{dstr} \quad (2)$$

978-1-4244-4782-4/10 $26.00 © 2010 IEEE

$$\omega_s = 2\pi f_s = \frac{1}{\sqrt{L_s C_s}} \quad \text{and} \quad \omega_p = 2\pi f_p = \frac{1}{\sqrt{L_p C_p}} \tag{3}$$

$$I_{in} = \frac{P}{r_{eff} V_{out}} + j r_{eff} V_{out} \frac{1}{\omega_p L_p} \left(\frac{\omega}{\omega_p} - \frac{\omega_p}{\omega} \right) \tag{4}$$

$$\left| \frac{V_{out}}{V_{in}} \right| = \frac{r_{eff}}{\sqrt{\left[1 + \frac{L_s}{L_p} \left(1 + \frac{\omega_s^2}{\omega_p^2} - \frac{\omega_s^2}{\omega^2} - \frac{\omega^2}{\omega_p^2} \right) \right]^2 + \frac{P^2}{r_{eff}^4 V_{out}^4} \frac{L_s}{C_s} \left(\frac{\omega}{\omega_s} - \frac{\omega_s}{\omega} \right)^2}} \tag{5}$$

It can be read from (5) that for an operation frequency of $\omega = \omega_s$, i.e. converter operation in the *Load Independent Point*, the voltage ratio is independent of transferred power P and equals the effective transformer ratio r_{eff}. Regrouping the five degrees of freedom to the new five parameters r_{eff}, ω_s, ω_s/ω_p, L_s/L_p and L_s/C_s allows to visualize the voltage transfer function using normalized quantities. An exemplary plot is given in Fig. 3.

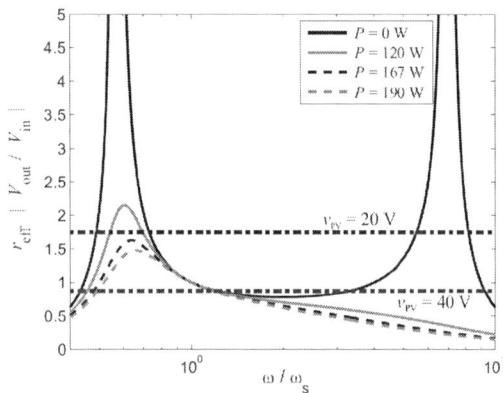

Figure 3.   Voltage conversion gain @ r_{eff}=0.1, ω_s/ω_p=0.25, L_s/L_p=0.5 and $L_s/C_s = 12\mu H/\mu F$

At an operation frequency around the series resonant frequency, the resonant tank is inductive resulting in ZVS of the MOSFET bridge. In this example it is observed, that zero power can be transferred at high input voltages, but 190 W cannot be transferred at low PV input voltages of only 20 V. With the boundary condition of being capable to operate the PV-module in all its possible operation points, now parameters can be varied to minimize rms-currents. As proposed in [9],[10] for *LLC*-type converters, it is figured out:

- I_{in} is reduced for minimum L_s/L_p, see Fig. 4
- I_{in} is reduced for maximum L_s/C_s, see Fig. 4

This coherence was evaluated numerically using FHA and later on also using a circuit simulator for a variety of operation points. Only one operation point is illustrated in Fig. 4. For the calculation of I_{in} using (4), the operation frequency ω using (5) is necessary. Since the latter calculation is of 8th order, it is carried out numerically.

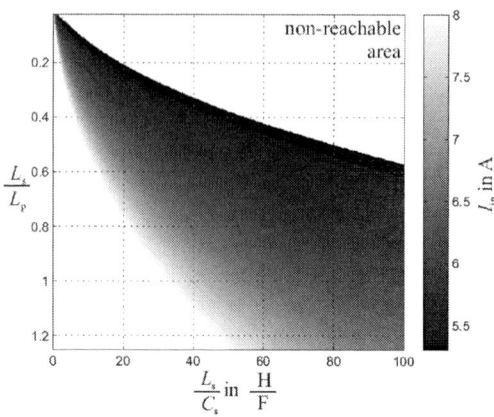

Figure 4.   Resonant current I_{in} (clamped to 8 A) @ r_{eff}=0.11, ω_s/ω_p=0.1, v_{PV}=35V, P=167W

Fig. 4 furthermore indicates a non reachable area, representing that the specific operation point cannot be driven at even more extreme values of the parameters L_s/L_p and L_s/C_s. In that case the same would happen as illustrated in Fig. 3, i.e. that 167 W cannot be transferred at $v_{PV} = 20$ V. Thus, in a good converter design the resonant tank limits the operation capability of the converter to the specified operation region. If the converter would be capable to transfer more power than necessary, rms-currents are increased in the specified operation region.

With the knowledge on how and in which direction to vary parameters, the design procedure in Fig. 5 is developed as described below.

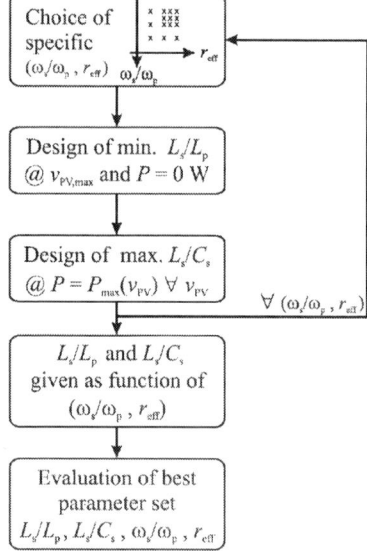

Figure 5.   Consecutive design steps of the resonant-tank parameters

At zero power, (5) indicates that there is no dependence on L_s/C_s. Thus, L_s/L_p is minimized first for the operation points at zero power. As second step L_s/C_s is maximized for the specified maximum power levels as function of v_{PV}. Since the result is still a function of r_{eff} and ω_s/ω_p, multiple combinations are iterated. The resonant current can easily be calculated using (4).

B. Visualization of MPP-tracking Capability by Frequency Variation using FHA

In the application of a photovoltaic module-integrated converter, the irradiation and temperature operation point is in interaction with the *LLCC*-converter transfer function. As indicated before, the converter should perform tracking the maximum power by variation of operation frequency. For one example at standard test conditions (STC), i.e. at 1000 W/m^2 and 25°C, the module shows its terminal behavior $P(v_{PV})$ being characterized by a maximum power point $P_{MPP}= 167$ W at the MPP voltage $v_{STC,MPP}= 35$ V. Since $P(v_{PV})$ is linked to both the PV-characteristics and the converter transfer function, $v_{PV,STC}$ can be calculated as a function of ω. For the purpose of visualization an extreme set of parameters is used for the determination of the characteristics in Fig. 6.

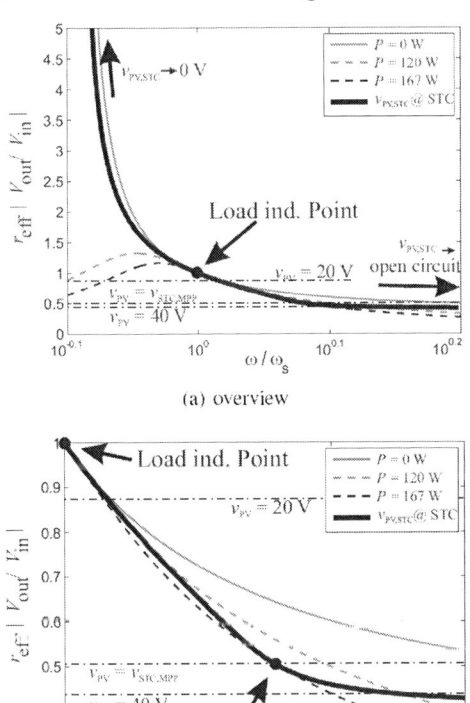

(a) overview

(b) zoomed view

Figure 6. Voltage conversion gain and resulting characteristic $v_{PV,STC}$ with PV module connected @ STC, r_{eff}=0.05, ω_s/ω_p=0.25, L_s/L_p=2 and $L_s/C_s = 25\mu H/\mu F$

Fig. 6.a indicates that for small and large frequency values, the operation point moves to a short- or open circuit of the module, both resulting in zero power and the fact that $v_{PV,STC}$ is approaching the constant power characteristic P =0 W respectively. In Fig. 6.b it is visualized, that on the one hand $v_{PV,STC}$ goes through the load independent point, but also has its maximum power point $P_{MPP}= 167$ W when $v_{PV,STC}$ tangents to the constant power characteristic P =167 W.

Varying the operation frequency ω, the MPP can be tracked for example using a simple hill climbing algorithm aiming at maximum converter output current with operation frequency as parameter.

C. Fast Numerical Design Procedure based on Derived Design Rules

Since FHA is only an approximation, simulations based on exact calculations were used for the design. With the above derived knowledge of how to design the parameters to the limits, only a few simulations have to be carried out, following the procedure in Fig. 5. Thus, a fast design is established.

Exact simulations were used since an analytical model could not be derived, even when looking at the extended First Harmonic Approximation (eFHA) [11] or the State Plane Analysis [12]. The latter ones either have an approximation included again, or are derived for *LLCC* converters with current-source output. The analytic description of the given converter with voltage-source output always shows the challenge of describing the rectifier in the discontinuous conduction mode due to snubbering with C_p. The significant deviation between exact simulation, FHA and eFHA respectively is illustrated in Fig.7.

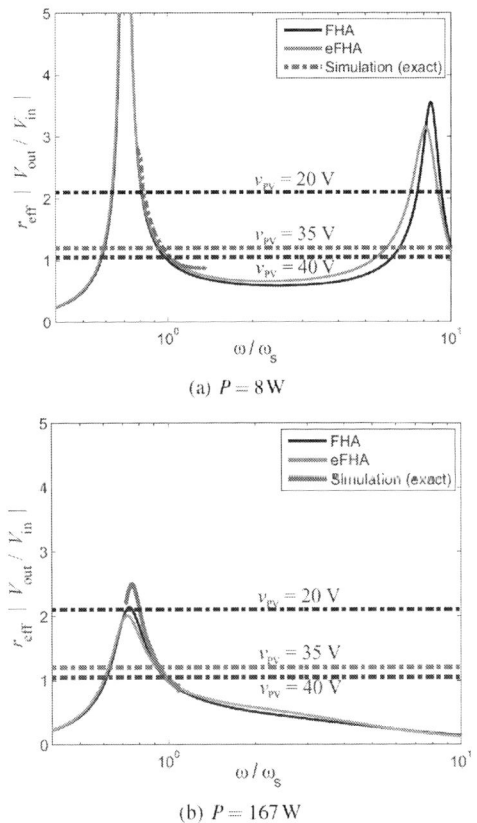

(a) $P = 8$ W

(b) $P = 167$ W

Figure 7. Comparison of FHA, eFHA and exact simulations @ f_s=200 kHz, r_{eff}=0.12, ω_s/ω_p=0.167, L_s/L_p=1 and $L_s/C_s = 36\mu H/\mu F$

The derived design method identified results in a design given in Tab. 1:

TABLE I.   TARGET QUANTITIES FOR THE IMPLEMENTATION OF THE
RESONANT TANK

L_s/L_p	L_s/C_s	ω_s/ω_p	r_{eff}
1	77 µH/µF	0.5	0.13

D. Operation Frequency

Out of the set of five degrees of freedom only four are determined yet, see Tab. 1. The remaining parameter is the series-resonant frequency ω_s. Operation frequency ω varies and covers this load independent point. By now, all quantities are related to this parameter such that ω_s is still free to choose. For gaining maximum efficiency the frequency dependent loss mechanisms quantified based on different loss models like the Improved Generalized Steinmetz Equation [13]. Results are depicted in Fig. 8.

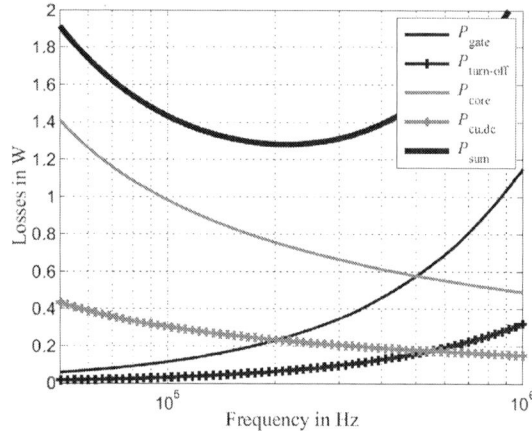

Figure 8.   Frequency dependent loss contributors and their sum

The reduction of copper and core losses of the transformer-inductor device are traced back to the reduction of material at higher frequencies. An optimum series resonant target frequency of 215 kHz is identified for the implementation.

III. CONSTRUCTION OF THE TRANSFORMER-INDUCTOR DEVICE

The most critical parameter in the design and construction of the transformer-inductor device is the parameter L_s/L_p. Furthermore, r_{eff} is integrated into the component. All other parameters can be adjusted afterwards, maybe leading to an overall increased or decreased converter operation frequency. However, not meeting the design requirement from Tab. 1 would mean additional losses coming from increased component stress due to increased rms-currents in the resonant tank. It has to be noted that leakage inductance is in the same order of magnitude as the main inductance. Thus, a leakage path must allow a high leakage flux and a reluctance in the main magnetic path has to limit the main inductance. This functionality is realized with a setup with multiple airgaps using ferrite core material as depicted in Fig. 9.

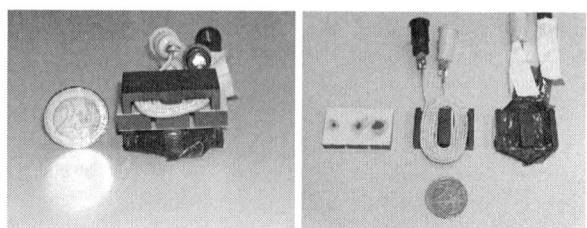

Figure 9.   Transformer-inductor device

This structure allows to independently design the leakage and the main flux path. However, there is a potential to be tapped to optimize the device. Since increased field strength leads to proximity losses in the windings close to the airgaps, the combination of outer ferrite cores with inner low- or medium-µ materials seem to gain higher efficiencies. A distributed airgap material, especially metal-powder composites, would even show better performance in theory, but the necessary thin plates of such material were not available to the authors.

In every magnetic component the inductances referred to the primary side are given by:

$$L = N_{pri}^2 A_L \qquad (6)$$

Thus, the important related parameter L_s/L_p, being both defined for the primary side, neither depend on primary side-nor on secondary side number of turns (N_{pri} and N_{sec}). Hence, the parameter important for the resonant operation L_s/L_p is only a function of material and geometry. Hence, the remaining parameter r_{eff} can be set at the end to quasi arbitrary values by choosing N_{pri} and N_{sec}. This coherence is important e.g. in the design of contactless energy transmission systems using rotating transformers [14].

IV. EXPERIMENTAL RESULTS

The proposed converter is constructed and measurements are presented here. Measurements visualizing the basic operation and the controllability by frequency variation are depicted in Fig. 10.

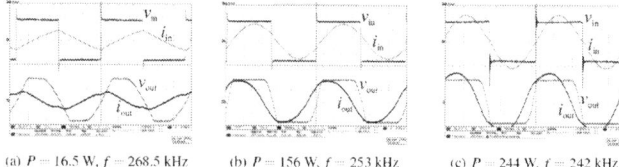

(a) $P = 16.5$ W, $f = 268.5$ kHz   (b) $P = 156$ W, $f = 253$ kHz   (c) $P = 244$ W, $f = 242$ kHz

Figure 10.   Exemplary *LLCC* characteristics at $v_{PV} = 35$ V

The inductive behavior of the resonant tank leads to the intended zero-voltage switching, i.e. that there is some small current in the resonant tank remaining at the switching instant. Furthermore, the soft commutation of the diodes can be identified by the limited dv_{out}/dt. At low power levels, see Fig. 10.a, the remaining current in the resonant tank results in a triangular shape. From the conduction time of the diodes it can be read, that mainly reactive power circulating in the resonant tank.

978-1-4244-4782-4/10 $26.00 © 2010 IEEE

Concerning the design goal of limiting the transferrable power to the specifications at low input voltages Fig. 11 shows the border to hard-switching operation.

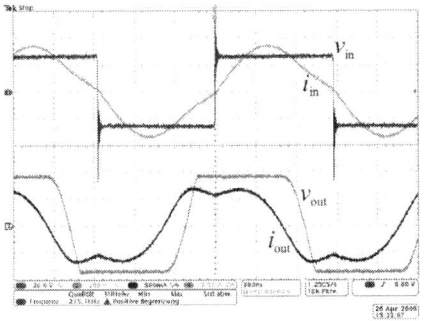

Figure 11. *LLCC* characteristics at the border to hard switching operation for minimum $v_{PV} = 27.5$ V and corresponding maximum $P = 144$ W

As seen in Fig. 10.c, considerably higher power levels can be transferred at increased input voltage.

The overall efficiency is characterized for the MPP voltage of $v_{PV} = 35$ V. Due to a design for European efficiency, see (1) with emphasis on a high part-load efficiency, Fig. 12 depicts the results including error propagation through the accuracy of the measurement equipment.

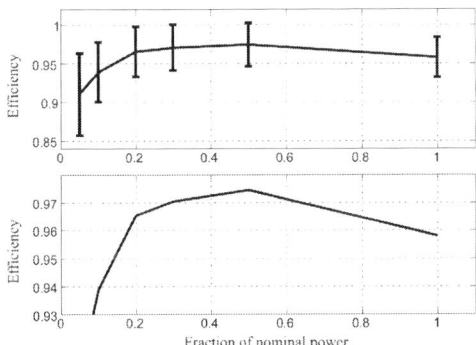

Figure 12. Measurements for European efficiency

An efficiency drop in the higher power level is tolerated and wanted, since part load efficiency was optimized in the numerical optimization using the design procedure based on FHA. A European efficiency of

$$\eta_{euro} = 96\% \qquad (7)$$

including control losses, could be demonstrated at rated power of 167 W.

V. CONLUSIONS AND FUTURE WORK

The design of a highly efficient *LLCC* series-parallel resonant converter is presented together with experimental results for a photovoltaic application. Design rules on the design of the resonant tank elements are motivated and qualified. The method leads to the high efficiency for a wide specified input region. In a future step the leakage path in the transformer-inductor device should be substituted by a distributed airgap material. However, such thin plates consisting of brittle metal-powder composites were not available to the author yet.

REFERENCES

[1] M. Meinhardt, M. Hoffmann, S.C. O'Mathuna, "Reliability of module integrated converters for photovoltaic converters," *Proceedings of the PCIM*, May 1998

[2] C. P. Dick, "Multi-resonant converters as photovoltaic module-integrated maximum power point tracker," *PhD-thesis at Institute for Power Electronics and Electrical Drives, RWTH Aachen University*, to be published in 2010

[3] D. M. Divan, G. Skibinski, "Zero-switching-loss inverters for high-power applications," *IEEE Transactions on Industry Applications*, vol. 25, no. 4, pages 634-643, July 1989.

[4] D. M. Divan, G. Venkataramanan, R. W. A. A. DeDoncker, "Design methodologies for soft switched inverters," *IEEE Transactions on Industry Applications*, vol. 29, pages 126-135, January 1993.

[5] R. L. Steigerwald, "A comparison of half-bridge resonant converter topologies," *IEEE Transactions on Power Electronics*, vol. 3, no. 2, pages 174-182, April 1988.

[6] R. P. Severns, "Topologies for three-element resonant converters," *IEEE Transactions on Power Electronics*, vol. 7, no. 1, pages 89-98, January 1992.

[7] I. Batarseh, C. Q. Lee, "Steady-state analysis of the parallel resonant converter with LLCC-type commutation network," *IEEE Transactions on Power Electronics*, vol. 6, no. 3, pages 525-538, July 1991.

[8] A. K. S. Bhat, "Analysis and design of a series-parallel resonant converter with capacitive output filter," *IEEE Transactions on Industry Applications*, vol. 27, no. 3, pages 523-530, May 1991.

[9] T. Duerbaum, G. Sauerlaender, "Analysis of the series-parallel multi-resonant LLC Converter - Comparison between first harmonic approximation and measurement," *European Conference on Power Electronics and Applications*, volume 2, pages 2174-2179, European Power Electronics Association, 1997.

[10] T. Duerbaum, "First harmonic approximation including design constraints," *Telecommunications Energy Conference 1998, INTELEC, Twentieth International*, pages 321-328, October 1998.

[11] A. Bucher, T. Duerbaum, D. Kuebrich, "Comparison of methods for the analysis of the parallel resonant converter with capacitive output filter," *European Conference on Power Electronics and Applications 2007*, pages 1-10, September 2007.

[12] Ramesh Oruganti, Fred C. Lee, "Resonant Power Processors: Part 1 - State Plane Analysis," *Industry Applications Society Annual Meeting, 1984, Conference Record of the 1984 IEEE*, pages 860-867, 1984.

[13] K. Venkatachalam, C. R. Sullivan, H A T Tacca "Accurate pediction of ferrite core loss with nonsinusoidal waveforms using only steinmetz parameters," *Workshop on Computers in Power Electronics*, 3-4 June 2002 Page(s):36 – 41, June 2002

[14] D. Hirschmann, C.P. Dick, S. Richter, R. De Doncker, "Design of a Contactless Rotary Energy Transmission for an Industrial Application," *Power Electronics Specialist Conference*, PESC 08 - IEEE 39th Annual, June 15-19 Pages 4314-4319, Rhodes, Greece

978-1-4244-4782-4/10 $26.00 © 2010 IEEE

Accurate Switching Loss Model and Optimal Design of A Current Source Driver Considering the Current Diversion Problem

Jizhen Fu(*Student Member IEEE*) *, Zhiliang Zhang(*Member IEEE*) **,

Andrew Dickson (*Student Member IEEE*) *, Yan-Fei Liu (*Senior Member IEEE*) *

and P.C. Sen (*Life Fellow IEEE*) *

* Queen' Power Group, Department of Electrical and Computer Engineering
Queen's University, Kingston, Ontario, Canada, K7L 3N6
** Aero-Power Sci-tech Center, College of Automation Engineering
Nanjing University of Aeronautics and Astronautics, Nanjing, P. R. China
jizhen.fu@queensu.ca, zlzhang@nuaa.edu.cn, andrew.dickson@queensu.ca, yanfei.liu@queensu.ca, senp@post.queensu.ca

Abstract— A new analytical switching loss model for power MOSFETs driven by Current Source Drivers (CSDs) is presented in this paper. The gate current diversion problem, which commonly exists in CSDs, is analyzed. In addition, the proposed loss model considers the Miller Plateau. The optimal design of current source driver is achieved which minimizes the total power loss for the Buck converter. The experimental result verifies the theoretical analysis. Compared with previous work, the efficiency at 1MHz with the optimal current source inductor is improved from 86.1% to 87.6% at 1MHz switching frequency, with 12V input, 1.3V/20A output, and from 82.4% to 84.0% at 1MHz switching frequency, with 12V input, 1.3V/30A output.

I. INTRODUCTION

Next generation Voltage Regulation Modules (VRMs) feature high current, low voltage and high power density [1]. In order to facilitate the complete integration of VRMs on the mother board, switched capacitors are proposed to replace the magnetic-based converter [2]. However, the large current spike, low efficiency and narrow range of the voltage regulation limit the application of the switched capacitor [3]. Another practical way to improve the dynamic performance and reduce the size of the passive components is by increasing the operating frequency of the Voltage Regulation Modules (VRMs) into MHz range [4] [5].

As the frequency increases, however, frequency dependent losses such as switching loss and gate drive loss become a penalty for switching converters driven by conventional voltage source drivers [6] [7]. In order to recover the gate driver loss that is dissipated in the charge and discharge path in the conventional voltage source driver, Resonant Gate Drive (RGD) techniques are proposed [8]-[10]. However, RGD only focuses on the gate energy loss while neglecting the potentials for minimizing switching loss, which is the dominant loss especially in high frequency applications. Recently, Current Source Drivers (CSDs) are proposed to

reduce the switching loss by charging and discharging the MOSFET with a nearly constant current [11]-[14]. For example, the CSD shown in Fig 1 can turn on and turn off the power MOSFET with a discontinuous current, minimizing the circulating current and conduction loss [12].

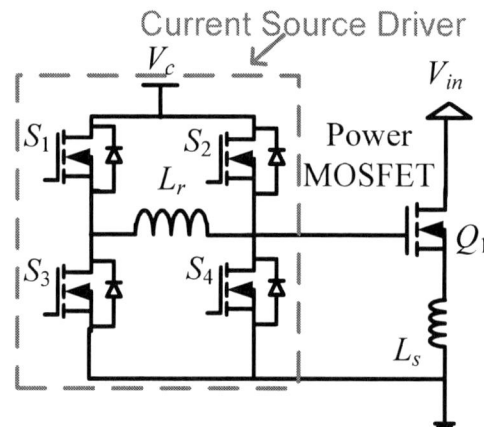

Fig 1 Topology of Current Source Driver in [12]

However, during switching transitions the current in the current source inductor is diverted, which reduces the effective current to charge or discharge the MOSFET. This is known as the Gate Current Diversion Problem and commonly exists in CSDs. Fig 2 shows the common equivalent circuits of the CSDs during turn-on and turn-off. Due to the effect of the common source inductance L_s, the gate terminal of the MOSFET is either clamped to V_c through the body diode of S_2 (D_2) during turn on or to ground through the body diode of S_4 (D_4) during turn off, causing part of driver current i_{Lr} to be diverted through D_2 or D_4 limiting the switching speed. The turn-off waveform simulated in LTspice is elucidated in Fig 3, from which it is noted that 1.8A current is diverted through D_4

in spite of a 3ampere of current in the current source driver. The CSD shown in Fig 4 presents a new concept to alleviate this problem by creating a negative voltage with D_{s1}-D_{s5} to accelerate the turn off speed [15].

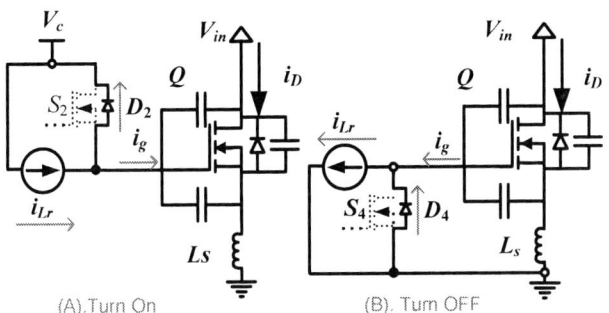

(A).Turn On   (B). Turn OFF

Fig 2   Equivalent Switching Circuit of CSDs

In order to evaluate the performance of the CSDs, an analytical loss model, which thoroughly analyzes the impact of the parasitic inductance in CSDs, is presented in [16]. More importantly, according to the proposed model, a generalized way to optimize the overall performance of the buck converter driven by a CSD is analyzed. A piecewise model that enables easy calculation and estimation of the switching loss is also proposed in [17]. However, the current diversion problem, which reduces the effective drive current and the switching speed, has not been analyzed in either of the two models. Therefore, a new analytical switching loss model considering every interval is presented in this paper in which the current diversion problem is analyzed and the effective charging and discharging current is accurately determined. Moreover, the optimal current source inductor is obtained in order to maximize the overall efficiency of the buck converter.

The proposed switching loss model that analyzes the current diversion problem is presented in Part II of this paper. Part III explains the procedures to obtain the optimal driver inductor of a CSD. The experimental results are shown in Part IV and finally, the conclusions are given in Part V.

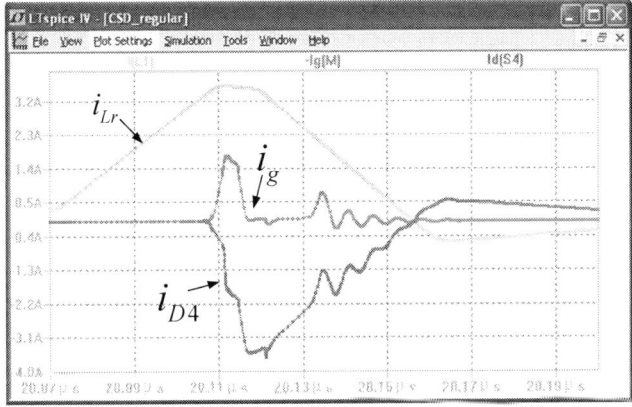

Fig 3   Simulation Waveforms of the Discharging Current of the Current Source Driver in Fig 1

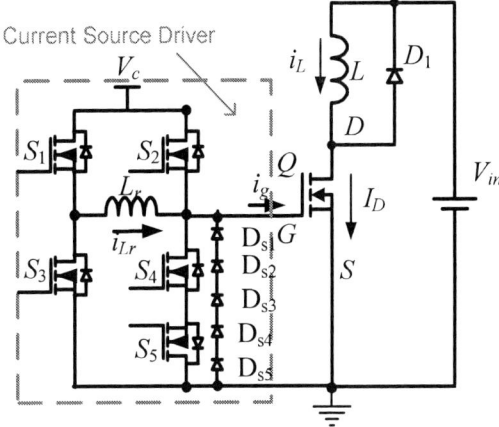

Fig 4   Inductive Clamped Load driven by CSD

II.   PROPOSED SWITCHING LOSS MODEL CONSIDERING THE CURRENT DIVERSION

The following sub-parts will present the operation principles of the CSD and a new switching loss model which considers the gate current diversion problem.

The equivalent circuit of the MOSFET driven by proposed CSD is shown in Fig 5 , where the power MOSFET Q is represented by a typical capacitance model, L_S is the parasitic inductance including the PCB track and the bonded wire inside the MOSFET package and L_D is the switching loop inductance. For the purpose of the transient analysis, the following assumptions are made [18]:

1) $i_D = g_{fs}(v_{CGS} - V_{th})$ and MOSFET is ACTIVE, provided $v_{CGS} > V_{th}$ and $v_{DS} > i_D R_{DS(on)}$

2) For $v_{CGS} < V_{th}$, $I_D = 0$, and MOSET is OFF

3) When $g_{fs}(v_{CGS} - V_{th}) > v_{DS}/R_{DS(on)}$, the MOSFET is fully ON

Where i_D is the drain current of the Q, g_{fs} is the transconductance, v_{DS} is the voltage across the drain-source capacitance of the Q, v_{CGS} is the voltage across the gate-source capacitance of the Q, V_{th} is the threshold voltage of Q , $R_{DS(on)}$ is the drain-source on state resistance. During the Active State when switching loss happens,

$$i_D = g_{fs}(v_{CGS} - V_{th}) \qquad (1)$$

According to Fig 5, i_G is the effective current to charge or discharge Q as shown below,

$$i_G = (C_{GS} + C_{GD})\frac{dv_{GS}}{dt} - C_{GD}\frac{dv_{DS}}{dt} \qquad (2)$$

v_{DS} is given as,

$$v_{DS} = V_{in} - L_D\frac{di_D}{dt} - L_S\frac{d(i_D + i_G)}{dt} \qquad (3)$$

The detailed switching waveforms are illustrated in Fig 6, where v_{gs1}-v_{gs5} are the gate drive signals for driver switches S_1-S_5 in Fig 5, i_{Lr} is the driver inductor current of L_r; v_{GS}, as shown in Equation (4), is the gate source voltage of Q including

978-1-4244-4782-4/10 $26.00 © 2010 IEEE

the effect of the common source inductance and the gate resistance, p_{SW} is the switching loss of the Q.

$$v_{GS'} = -i_G R_g + V_{CGS} - L_S \frac{d}{dt}(i_{DS} + i_G) \qquad (4)$$

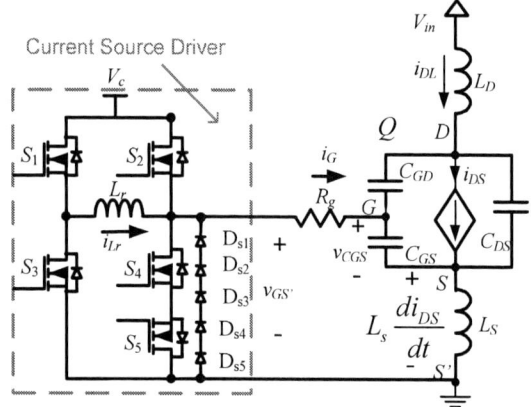

Fig 5   Equivalent circuit of MOSFET with proposed CSD

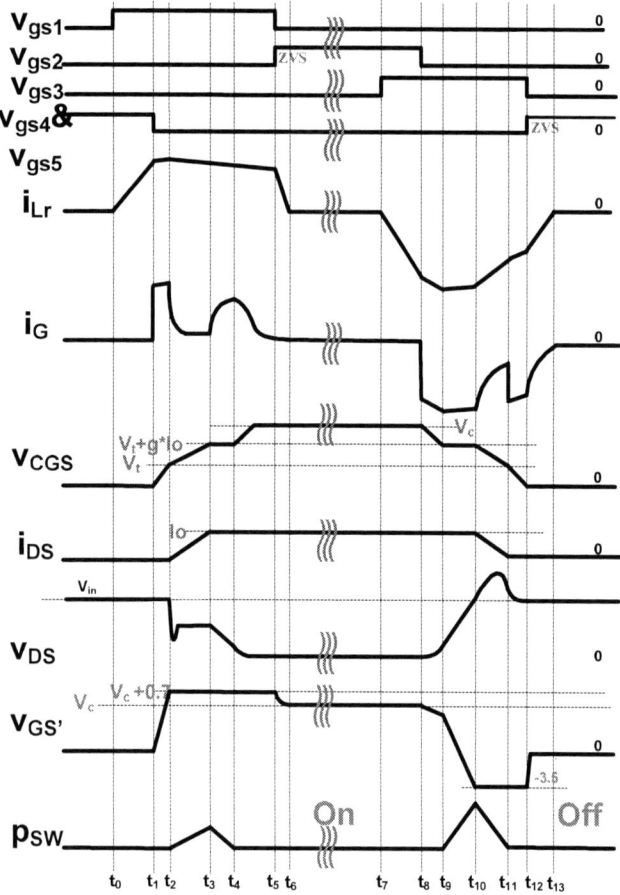

Fig 6   MOSFET Switching Transition Waveforms

The operation principle of the turn on transition is illustrated as follows. Prior to t_0, the power MOSFET is clamped in the OFF state by S_4 and S_5.

A. Turn-ON Transition:

Precharge[t_0, t_1]: At t_0, S_1 is turned on, and the inductor current i_{Lr} rises almost linearly and the interval ends at t_1 which is preset by the designer. The equivalent circuit is given in Fig7 (a).The inductor current i_{Lr} is given in Equation (5).

$$i_{Lr} \approx \frac{V_c \cdot (t - t_0)}{L_r} \qquad (5)$$

Turn-on Delay [t_1, t_2]: At t_1, S_4 &S_5 are turned off; the inductor current i_{Lr} starts to charge the gate capacitance of Q - the equivalent circuit is given in Fig7 (b).At this interval, the effective charge current i_G equals i_{Lr}. This interval ends when v_{CGS} reaches V_{th}. In the s-domain, the KVL equation for the circuit is given in Equation (6).

$$V_c = L_r \frac{d}{dt}(i_{Lr}) + i_{Lr}R_{on} + v_{CGS}, \; i_{Lr} = C_G \frac{d}{dt}(v_{CGS}) \qquad (6)$$

Mathematically, there are three possible forms for the equation of the inductor current: over damped, critically damped and under damped – for practical situations, $\omega_0 > \alpha_0$, the equations for i_G, i_{Lr} and v_{CGS} are given in Equation (7) ~ (8).

$$v_{CGS} = [A_0 \cos(\sqrt{\omega_0^2 - \alpha_0^2}t) + B_0 \sin(\sqrt{\omega_0^2 - \alpha_0^2}t)] \times e^{-\alpha_0 t} + C_0 \qquad (7)$$

$$\begin{aligned} i_G = i_{Lr} = C_G \times &[(-A_{01}\sqrt{\omega_0^2 - \alpha_0^2} + B_0 \alpha_0) \times \sin(\sqrt{\omega_0^2 - \alpha_0^2}t) \\ &+ (-A_0 \alpha - B_0 \sqrt{\omega_0^2 - \alpha_0^2}) \times \cos(\sqrt{\omega_0^2 - \alpha_0^2}t)] \times e^{-\alpha_0 t} \end{aligned} \qquad (8)$$

Where $\alpha_0 = \dfrac{R_{on}}{2(L_r + L_s)}, \omega_0 = \dfrac{1}{\sqrt{(L_r + L_s)C_g}}$

$A_0 = -[2\alpha_0 i_{Lr\_0} + (V_c - 2i_{Lr\_0}R_{on})/(L_r + L_s)]/C_G \omega_0^2$

$B_0 = (i_{Lr\_0} + C_G \alpha_0 A_0)/C_G \omega_0^2, C_0 = -A_0$

Drain Current Rising [t_2, t_3]: At t_2, $v_{CGS} = V_{th}$. During this interval, v_{CGS} keeps increasing, and i_{DS} starts to rise according to the relationship in Equation (1). Since i_{DS} flows through L_S, according to Equation (4), the large voltage induced across L_S makes $v_{GS'}$ far larger than the driver supply voltage V_c. Therefore, D_2, the body diode of the driver switch S_2, is driven on to clamp $v_{GS'}$ at $V_c + 0.7$.The equivalent circuit is shown in Fig7 (c).At this interval, i_G drops sharply because of the voltage clamping. The subtraction of i_{Lr} and i_G is diverted into D_2. The initial condition of this interval is $I_{G\_t2} = i_G(t_2 - t_1)$, and $V_{CGS\_t2} = V_{th}$. The interval ends at t_3 when i_{DS} equals the load current, I_o. The equations for i_G, i_{Lr}, v_{CGS} and v_{DS} are given in Equation (9)~(12).

$$v_{CGS} = A_1 e^{(-\alpha_1 - \sqrt{\alpha_1^2 - \omega_1^2})t} + B_1 e^{(-\alpha_1 + \sqrt{\alpha_1^2 - \omega_1^2})t} + C_1 \qquad (9)$$

$$\begin{aligned} i_G = A_1(-\alpha_1 - \sqrt{\alpha_1^2 - \omega_1^2})e^{(-\alpha_1 - \sqrt{\alpha_1^2 - \omega_1^2})t} \\ + B_1(-\alpha_1 + \sqrt{\alpha_1^2 - \omega_1^2})e^{(-\alpha_1 + \sqrt{\alpha_1^2 - \omega_1^2})t} \end{aligned} \qquad (10)$$

$$i_{Lr} = (I_{G\_t2} + 0.7/R)e^{(-R/Lr)t} - 0.7/R \quad (11)$$

$$v_{DS} = V_{in} - L_S \times d\frac{i_G}{t} \quad (12)$$

where $\alpha_1 = (RC_G + L_S g_{fs})/2L_S C_G$, $\omega_1 = 1/\sqrt{L_S C_G}$

$A_1 = [(V_{th} - V_c - 0.7)(-\alpha_1 + \sqrt{\alpha_1^2 - \omega_1^2}) - I_{G\_t2}/C_G]/2\sqrt{\alpha_1^2 - \omega_1^2}$

$B_1 = V_{th} - V_c - 0.7 - A_1$, $C_2 = V_c + 0.7$

Miller Plateau [t_3, t_4]: At t_3, $i_{DS} = I_o$. During this interval, v_{CGS} is held at the Miller Plateau voltage. i_G mainly flows through the gate-to-drain capacitance of Q, and v_{DS} decreases accordingly. It is noted that i_G starts to rapid increase since the EMF across L_s falls sharply due to the unchanged i_{DS}, however part of the inductor current is still diverted through D_2. The equivalent circuit is given in Fig7 (d). The initial values of the interval are $I_{G\_t3} = i_G(t_3 - t_2)$, $V_{CGS\_t3} = v_{CGS}(t_3 - t_2)$, and $V_{DS\_t3} = v_{DS}(t_3 - t_2)$. The interval ends when v_{DS} equals zero at t_4. The equations for i_G, v_{CGS} and v_{DS} are given in Equation (13)~(15). i_{Lr} remains the same as the previous interval.

$$v_{CGS} = V_{CGS\_t3} \quad (13)$$

$$i_G = (I_{G\_t3} - (V_c + 0.7 - V_{CGS\_t3})/R)e^{(-R/Ls)t} + (V_c + 0.7 - V_{CGS\_t3})/R \quad (14)$$

$$v_{DS} = (\frac{I_{G\_t3} - (V_c + 0.7 - V_{CGS\_t3})/R}{C_{GD}R/L_s})e^{(-R/Ls)t}$$
$$- \frac{(I_{G\_t3} - (V_c + 0.7 - V_{Cgs\_t3})/R)t}{C_{GD}}$$
$$+ (V_{DS\_t3} - \frac{I_{G\_t3} - (V_c + 0.7 - V_{CGS\_t3})/R}{C_{GD}R/L_s}) \quad (15)$$

Remaining Gate Charging [t_4, t_5]: At t_4, $v_{DS} = 0$ and v_{CGS} starts to rise again until it reaches V_c. v_{GS} remains at $V_c + 0.7$, and due to the rising of the v_{CGS}, i_G decreases gradually. The equivalent circuit is given in Fig7 (f). The initial values of this interval are: $I_{G\_t4} = i_G(t_4 - t_3)$, $V_{CGS\_t4} = V_{CGS\_t3}$, and $V_{DS\_t4} = v_{DS}(t_4 - t_3)$. This interval ends at t_5 when $v_{CGS} = V_c$. The equations for i_G, v_{CGS} and v_{DS} are given in Equation (16) ~ (18) and i_{Lr} is the same as the previous interval.

$$v_{CGS} = [A_2 \cos(\sqrt{\omega_1^2 - \alpha_1^2}\,t) + B_2 \sin(\sqrt{\omega_1^2 - \alpha_1^2}\,t)] \times e^{-\alpha_1 t} + C_2 \quad (16)$$

$$i_G = C_G \times [(-A_1\sqrt{\omega_1^2 - \alpha_1^2} + B_1\alpha_1) \times \sin(\sqrt{\omega_1^2 - \alpha_1^2}\,t) + (-A_1\alpha - B_1\sqrt{\omega_1^2 - \alpha_1^2}) \times \cos(\sqrt{\omega_1^2 - \alpha_1^2}\,t)] \times e^{-\alpha_1 t} \quad (17)$$

$$v_{DS} = V_{DS\_t3} - \frac{(v_{CGS} - V_{DS\_t3})(V_{DS\_t3} - I_o R_{on@Vc})}{V_c - V_{CGS\_t3}} \quad (18)$$

where $C_2 = V_c + 0.7$, $A_2 = V_{CGS\_t3} - C_2$, $B_2 = (I_{G\_t4}/C_G - A_2\alpha_1)/\sqrt{\alpha_1^2 - \omega_1^2}$ and $R_{on@Vc}$ means the on-resistance of the MOSFET when $V_{CGS} = V_c$

Energy Recovery [t_5, t_6]: At t_5, S_2 is turned on to recover the energy stored in the inductor to the source as well as actively clamping Q to V_c. The initial value of this interval is $i_{Lr\_t5} = i_{Lr}(t_5 - t_2)$, and this interval ends when i_{Lr} becomes zero. The equivalent circuit is illustrated in Fig7 (f). The equation for i_{Lr} is in Equation (19).

$$i_{Lr} = [I_{G\_t5} + (V_c + 0.7)/(R_{on} + R_{lr})]e^{(-R/Lr)t} - (V_c + 0.7)/(R_{on} + R_{lr}) \quad (19)$$

Prior to t_7, the power MOSFET is clamped in the ON state by S_2.

B. Turn-OFF Transition:

Predischarge[t_7, t_8]: At t_7, S_3 is turned on, and the inductor current i_{Lr} rises almost linearly and the interval ends at t_8 which is preset by the designer. The equivalent circuit is shown in Fig8 (a). The equation for i_{Lr} is given in Equation (20).

$$i_{Lr} \approx -\frac{V_c \cdot (t - t_7)}{L_r} \quad (20)$$

Turn-off Delay [t_8, t_9]: At t_8, S_2 is turned off. In this interval, v_{CGS} decreases until $V_{th} + I_o * g_{fs}$ which ends the interval. The equivalent circuit is given in Fig8 (b). The way to calculate the equations for i_G, i_{Lr} and v_{CGS} are the same as the *Turn-on Delay* interval.

Miller Plateau [t_9, t_{10}]: At t_9, $v_{CGS} = V_{th} + I_o * g_{fs}$. In this interval, v_{CGS} holds at the Miller plateau voltage, $V_{th} + I_o * g_{fs}$. i_G (equal to i_{Lr}) strictly discharges the gate-to-drain capacitance C_{gd} of Q, and v_{DS} rises until it reaches V_{in} at t_{10}. The equivalent circuit is illustrated in Fig8 (c). The equations of this interval can be obtained in the same way as the *Miller Plateau* in turn-on interval.

Drain Current Drop [t_{10}, t_{11}]: At t_{10}, $v_{DS} = V_{in}$ and v_{CGS} continues to decrease from $V_{th} + I_o * g_{fs}$ to V_{th}. i_{DS} falls from I_o to zero according to relationship in Equation (1). According to Equation (4), due to the induction EMF across L_s, the series connected diodes D_{s1}-D_{s5} are driven on to clamp v_{GS} at around -3.5V. The voltage across the current source inductor becomes -3.5V, so i_{Lr} decreases at a higher rate than in the turn on transition. The equivalent circuit of this interval is given in Fig8 (d). It is emphasized is that the CSD proposed in [12] only can clamp v_{GS} to -0.7V. This means that the turn off speed of the CSD proposed in this paper (Fig 4) is more than three times that of the CSD in [12]. It is worth mentioning that v_{DS} in this interval will keep rising due to effect of the L_s. Therefore, the derivation of the equations in this interval needs to solve the 3rd order differential equations in Equation (21).

Remaining Gate Discharging [t_{11}, t_{12}]: At t_{10}, $v_{CGS} = V_{th}$. In this interval, v_{CGS} continues to decrease until it equals zero; it is noted that v_{DS} continues to rise during this interval. The equivalent circuit is shown in Fig8 (e). The equations in this interval have the same form as the equations in *Remaining Gate Charging* Interval.

$$\begin{cases} L_S \times \dfrac{d}{dt}(i_{DS} + i_G) + v_{CGS} + i_G \times R_g + V_f = 0 \\[2mm] i_{DS} = g_{fs} \times (v_{CGS} - V_{th}) \\[2mm] i_G = C_g \times \dfrac{d}{dt}(v_{CGS}) - C_{gd}\dfrac{d}{dt}(v_{DS}) \\[2mm] L_S \times \dfrac{d}{dt}(i_G) + (L_D + L_S) \times \dfrac{d}{dt}(i_{DS}) + v_{DS} - V_{in} - 0.7 = 0 \end{cases} \quad (21)$$

Energy Recovery [t_{12}, t_{13}]: At t_{11}, S_4 & S_5 are turned on to recover the energy stored in the inductor to the source as well as actively clamping Q to ground. The equivalent circuit is given in Fig8 (f). This interval is the same as the *Gate Energy Recovery* at this turn-on transition.

The switching loss of power MOSFET P_{sw}, which consists of turn on loss P_{sw_on} and turn off loss P_{sw_off}, is derived according to in Equation (22). The driver loss P_{dr} is made up of conduction loss P_{dr_con}, gate drive loss P_{dr_gate} and output loss P_{dr_out} as given in Equation (23). The sum of the switching loss and the driver loss, P_{sum}, is given in Equation (24).

$$P_{sw} = \int_{t2}^{t4}(i_{DS} \cdot v_{DS} \cdot f_s)dt + \int_{t9}^{t11}(i_{DS} \cdot v_{DS} \cdot f_s)dt \quad (22)$$

$$P_{dr} = P_{dr\_con} + P_{dr\_gate} + P_{dr\_out} \quad (23)$$

$$P_{sum} = P_{dr} + P_{sw} \quad (24)$$

III. OPTIMAL DESIGN OF CURRENT SOURECE DRIVER

According to Equation (5), the RMS current of the current source driver, I_{Lr_RMS}, is calculated in Equation (25). The conduction loss at this interval is proportional to the precharge time T_{pre}, and it is the same for gate energy recovery interval since during charging and discharging the current in the current source inductor roughly remains constant. Therefore, T_{pre} should be set as short as possible within the practical limits of the driver to minimize the conduction loss. Taken the logic limits into consideration, T_{pre} is set to be 20ns. And it needs to be pointed out that the design procedure presented here is also applicable to other conditions.

$$I_{Lr\_RMS} \approx \frac{V_c \cdot T_{pre}}{L_r}\sqrt{\frac{T_{pre}f_s}{3}} \quad (25)$$

In order to maximize the overall efficiency of the buck converter with the proposed CSD, P_{sum} should be minimized. The optimal design of the current source driver involves a tradeoff between driver loss and switching loss, and there exists an optimal inductor current, I_{Lr_opt}, where P_{sum} reaches the minimum value. With T_{pre} fixed to 20ns and according to Equation (25), it can also be inferred that there also exists an optimal current source inductor, L_{r_opt} as given in Equation (26).

$$L_{r\_opt} = \frac{V_c \cdot T_{pre}}{I_{Lr\_opt}} \quad (26)$$

In order to validate the analysis, the following specifications are employed: V_{in}=12V, V_o=1.3V, I_o=30A, V_c=5V, f_s=1MHz, Q: SI7386DP. Typically, the parasitic

inductance value for Power PAK SO-8 package is tested by the semiconductor manufacturers in [19] [20]and range from approximately 250pH-1nH. In the models of this paper, L_s=1nH.

Fig 9 illustrates the plot of the equation P_{sum} versus the current source inductor value within practical range using MathCAD. It is noted that, in comparison with the driver loss, the switching loss is the dominant loss of the power MOSFET. It can also be observed that the optimal driver inductor is around 25nH, where P_{sum} is the minimum.

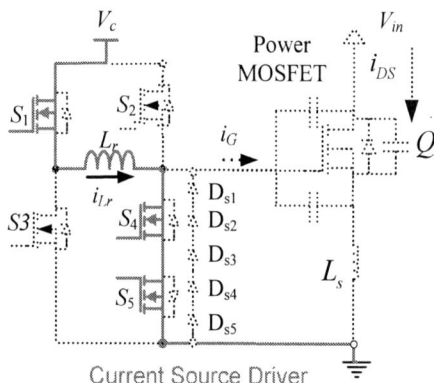

Fig 7(a): (t0, t1): Precharge

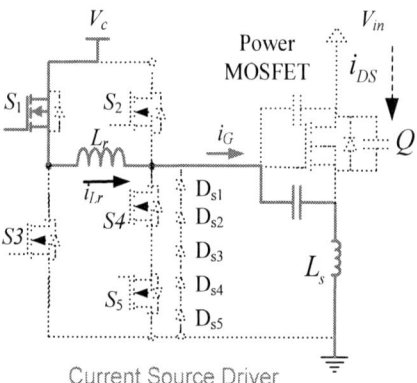

Fig 7(b): (t1 ,t2):Turn-on Delay

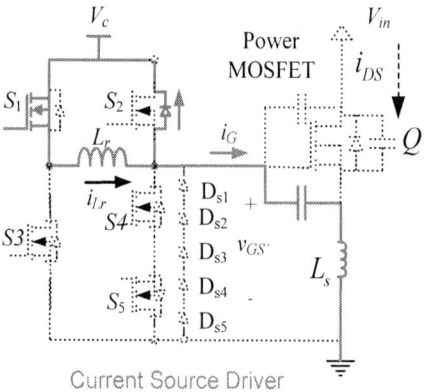

Fig 7(c): (t2, t3): Drain Current Rising

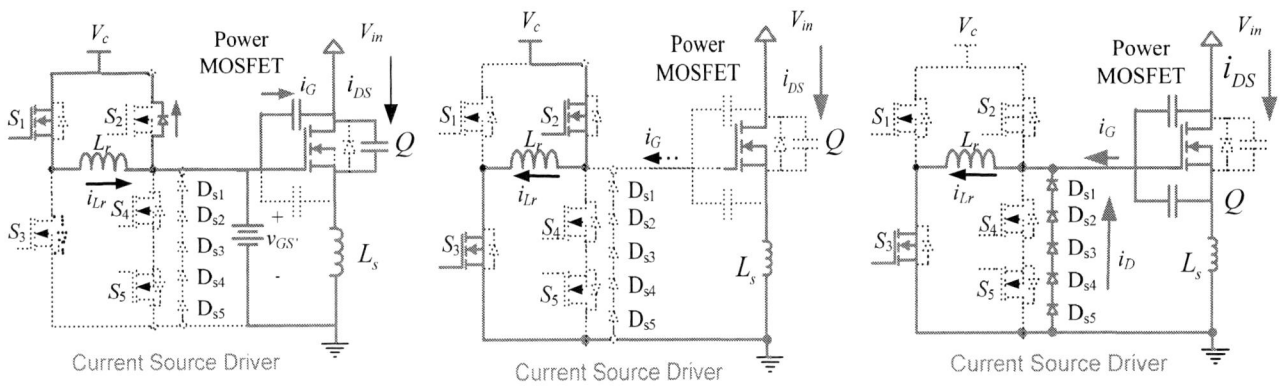

Fig 7(d): (t3, t4): Miller Plateau   Fig 8(a): (t7, t8): Predischarge   Fig 8(d): (t10, t11) Drain Current Drop

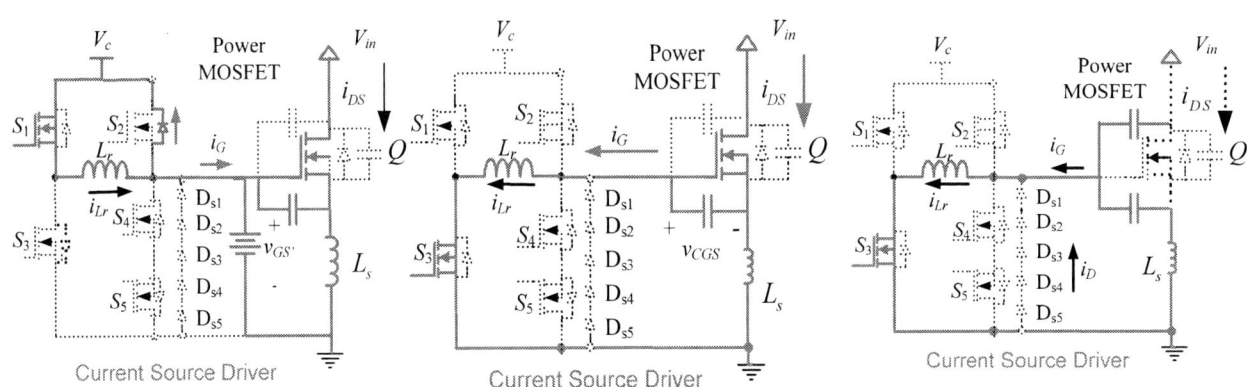

Fig 7(e): (t4, t5): Remaining Gate Charging   Fig 8(b): (t8 ,t9): Turn-off Delay   Fig 8(e): (t11, t12):Remaining Gate Discharge

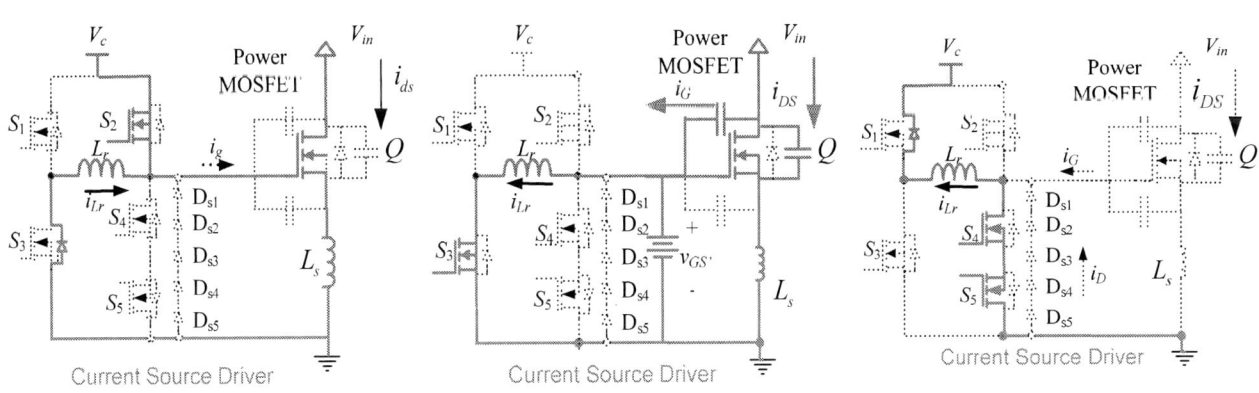

Fig 7(f): (t5, t6): Energy Recovery   Fig 8(c): (t9 ,t10): Miller Plateau   Fig 8(f): (t12,t13):Energy Recovery

978-1-4244-4782-4/10 $26.00 © 2010 IEEE

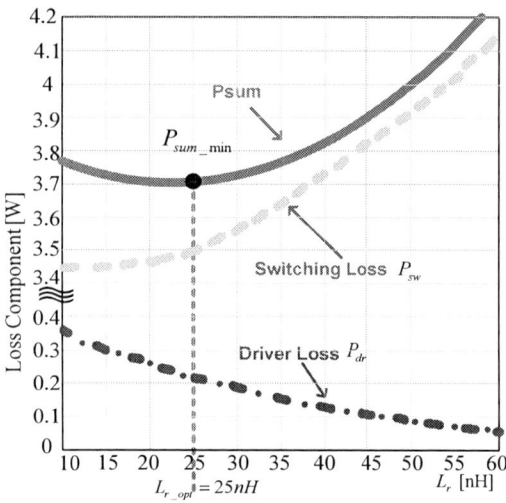

Fig 9 Total Loss Versus. Current Source Inductor

IV. EXPERIMENTAL RESULTS AND DISCUSSION

A prototype of a synchronous buck converter as shown in Fig 10 was built to verify the optimal design of the current source inductor. The control FET of the converter is driven with the proposed CSD and the SR is driven with a conventional voltage source driver for simplicity.

The PCB consists of 6 layer 4 oz copper, and the picture of the prototype is shown in Fig 11. The components used in the circuit are: Q_1: Si7386DP; Q_2: IRF6691; output filter inductance: L_f=330nH (IHLP-5050CE-01); current-source inductor: L_r=23nH (Coilcraft 2508-23N_L); drive switches S_1-S_4: FDN335; Anti-diodes $D_{s1} \sim D_{s5}$: MBR0520. For common practice, the driver voltages for the control FET and SR are both set to be 5V. The operating conditions are: input voltage V_{in}: 12V; output voltage V_o: 1.2V~1.5V; switching frequency f_s: 500kHz~1MHz.

The gate driver signals for V_{gs_Q1} and V_{gs_Q2} are shown in Fig 12. The current waveform of the current source inductor is impossible to obtain without breaking the setup of the prototype.

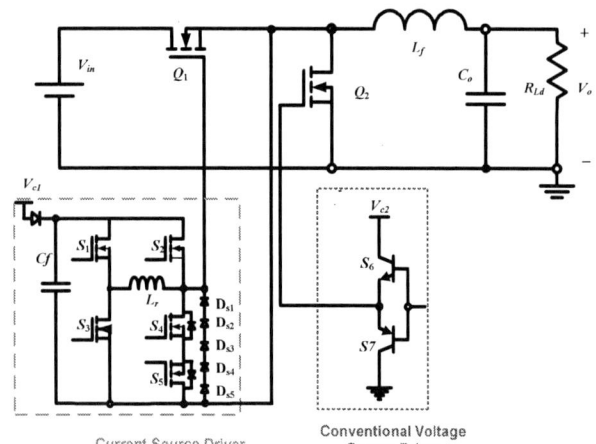

Fig 10 Buck Converter with proposed CSD

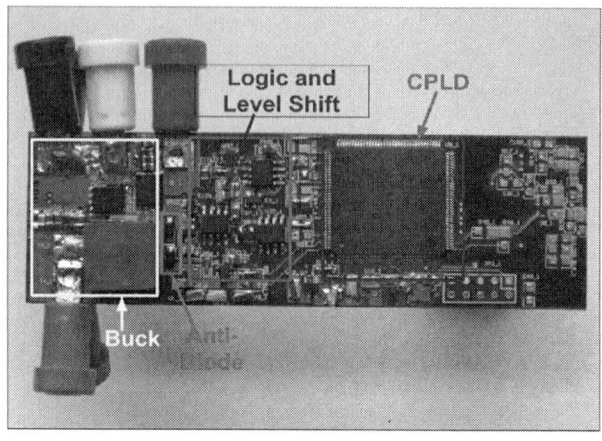

Fig 11 Photo of the buck converter with CSD in Fig 1

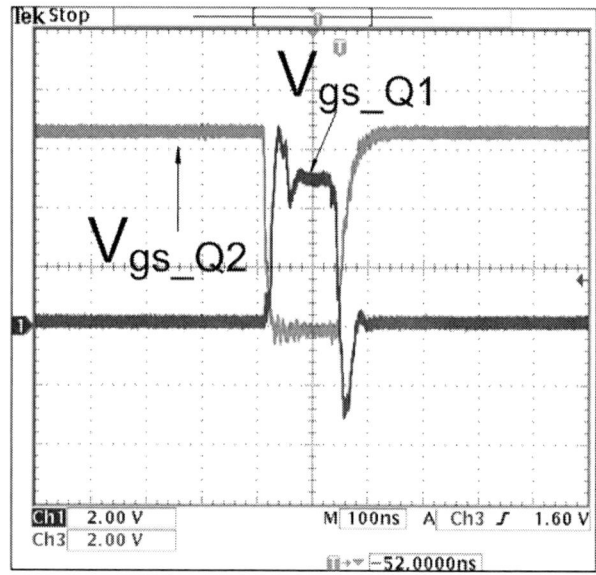

Fig 12 The waveforms of driver signals Vgs_Q1&Vgs_Q2

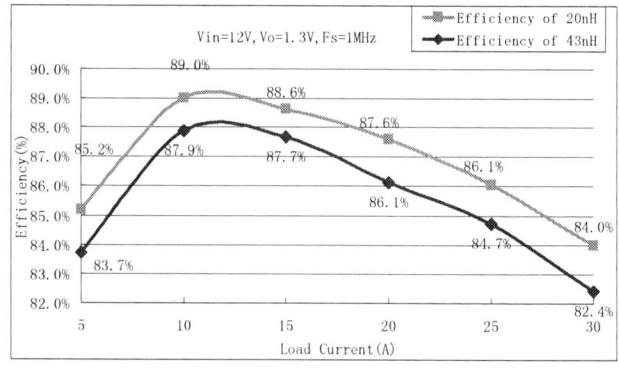

Fig 13 Efficiency comparison at 1.3V output@1MHz
(Top: CSD with 23nH, Bottom :CSD with 43nH)

978-1-4244-4782-4/10 $26.00 © 2010 IEEE

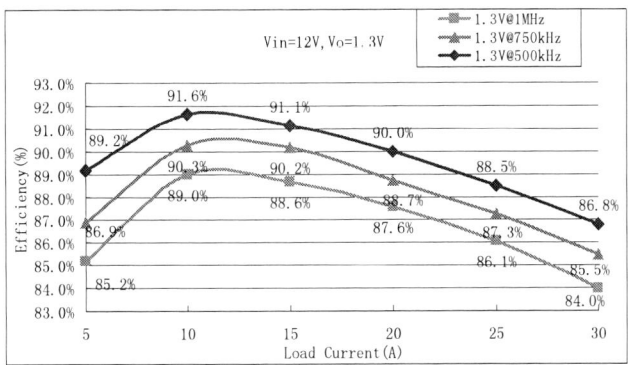

Fig 14 Efficiencies of 1.3V output @1MHz, 750kHz, 500kHz

(Top: 1MHz; Middle: 750kHz;Bottom: 500kHz)

Fig 15 Efficiencies of 1.2V, 1.3V, 1.5Voutput

(Top: 1.5V; Middle:1.3V;Bottom:1.2V)

To provide a fair comparison, a similar prototype is assembled except the current source inductor is changed to 43nH. Fig 13 illustrates the efficiency comparison at 1.3V/1MHz output. It is noted that, comparing to the CSD with 43nH, the CSD with L_r=23nH increases the efficiency from 86.1% to 87.6% at a 20A load, and from 82.4% to 84.0% at 30A load.

Fig 14 also shows the efficiencies of 1.3V output at 1MHz, 750 kHz, and 500 kHz respectively. Fig 15 summarizes the efficiencies of the CSD with the optimal inductor at 1.2V, 1.3V and 1.5V output respectively. It can be observed that the highest efficiency at 1.5V output is 89.8% for a 15A load.

V. CONCLUSIONs

In this paper, a new analytical switching loss model for power MOSFET driven by a Current Source Driver which considers the current diversion is presented, and detailed equations for each interval are derived. Based on this model, the optimal current source inductor is obtained to achieve the maximum overall efficiency of switching converter. The experimental results verify the proposed switching loss model and optimal design.

ACKNOWLEDGMENT

The authors would like to thank Power Sources Manufactures Association (PSMA) for their generous travel grant.

REFERENCES

[1] Ed Stanford, "Power Technology Roadmap for Microprocessor Voltage Regualtors", in Proc. IEEE Applied Power Electronics Conf., 2004.

[2] M. D. Seeman and S. R. Sanders, "Analysis and optimization of switched-capacitor dc-dc power converters," IEEE Trans. on Power Electronics, vol. 23, no. 2, pp. 841-851, March.2008.

[3] Ioinovici, A, "Switched-capacitor power electronics circuits", IEEE Circuits and Systems Magazine, Volume 1, Issue 3, pp.37 – 42,2001.

[4] D.J. Perreault, J. Hu, J.M. Rivas, Y. Han, O. Leitermann, R.C.N. Pilawa-Podgurski, A. Sagneri, and C.R. Sullivan, "Opportunities and Challenges in Very High Frequency Power Conversion," in Proc. 2009 IEEE Applied Power Electronics conference, Feb. 2009, pp. 1-14.

[5] L. Yao, H. Mao, and I. Batarseh, "A Rectification Topology for High Current Isolated DC-DC Converters," IEEE Transaction on Power Electronics, vol. 22, no. 4, pp. 1522-1530, July.2007.

[6] R. W. Erickson and D. Maksimovic, "Fundamentals of Power Electronics", 2nd. Edition, Kluwer Academic Publishers, 2001.

[7] T. Lopez, G. Sauerlaender, T. Duerbaum, and T. Tolle, "A detailed analysis of a resonant gate driver for PWM Applications," in Proc. IEEE Applied Power Electronics Conf., 2003, pp. 873-878.

[8] D. Maksimovic, "A MOS gate drive with resonant transitions," in Proc. IEEE Power Electronics Specialists Conf., 1991, pp. 527-532.

[9] K. Yao, and F. C. Lee, "A novel resonant gate driver for high frequency synchronous buck converters," IEEE Trans. Power Electronics, Vol. 17, No. 2, Mar. 2002, pp.180-186.

[10] Y. Chen, F. C. Lee, L. Amoroso, and H. Wu, "A resonant MOSFET gate driver with efficient energy recovery," IEEE Trans. Power Electronics, Vol. 19, No.2, Mar. 2004, pp. 470-477.

[11] Z. Yang, S. Ye and Y. F. Liu, "A new resonant gate drive circuit for synchronous buck converter," IEEE Trans. Power Electronics, Vol. 22, No.4, Jul. 2007, pp. 1311-1320.

[12] W. Eberle, Z. Zhang, Y. F. Liu and P. C. Sen, "A current source gate driver achieving switching loss savings and gate energy recovery at 1-MHz," IEEE Trans. Power Electron., Vol. 23, No. 2, pp. 678-691, Mar. 2008.

[13] Z. Zhang, W. Eberle, Z. Yang, Y.F. Liu and P.C. Sen, "A new hybrid gate drive scheme for buck voltage regulators," in Proc. IEEE Power Electronics Society Conference (PESC), June, 2008, pp.2498-2503.

[14] Z. Zhang,J. Fu, Y.F. Liu and P.C. Sen, "A New Discontinuous Current Source Driver for High Frequency Power MOSFETs",Proceeding of IEEE Energy Conversion Congress and Exposition (ECCE), Sep 2009, pp. 1655-1662

[15] J.Fu, Z.Zhang, W.Eberle,Y.F.Liu and P.C.Sen, "A high efficiency current source driver with negative gate voltage for buck voltage regulators", in Proc.IEEE Energy Conversion Congress and Exposition (ECCE), Sep.2009, pp. 1663-1670.

[16] Z. Zhang, W. Eberle, Z. Yang, Y.F. Liu and P.C. Sen, "Optimal design of current source gate driver for a buck voltage regulator based on a new analytical loss model," IEEE Trans. Power Electron., Vol. 23, No. 2, pp.653-666, Mar. 2008.

[17] W. Eberle, Z. Zhang, Y.F. Liu, P.C. Sen, " A Practical Switching Loss Model for Buck Voltage Regulators", IEEE Transactions on Power Electronics, Vol. 24, No. 3, Mar. 2009, pp. 700-713

[18] D.A.Grant and J. Gowar, "Power MOSFET Theory and Applications". New York: Wiley,1989

[19] M. Pavier, A. Woodworth, A. Green, R. Monteiro, C. Blake, J. Chiu, "Understanding the Effects of Power MOSFET Package Parasitics on VRM Circuit Efficiency at Frequencies above 1MHz," International Rectifier Application Note, www.irf.com

[20] J. Lee, "Package Parasitics Influence Efficiency," Power Electronics Technology Magazine, Nov. 2005, pp. 14-21

978-1-4244-4782-4/10 $26.00 © 2010 IEEE

Bidirectional Operation of Resonant Voltage Divider

K. I. Hwu, *Member, IEEE*, and Y. T. Yau, *Student Member, IEEE*

Department of Electrical Engineering, National Taipei University of Technology, Taiwan

Abstract–In this paper, bidirectional operation of the resonant voltage divider (RVD) is presented. As generally recognized, the RVD is used to transfer energy from the high voltage to the low voltage to feed the point of load (POL) or the voltage-regulated module (VRM). As for the bidirectional operation of the RVD, it can reversely transfer the energy to the DC bus, so as to effectively to suppress the voltage overshoot as the download transient response occurs. In the paper, the basic principles of bidirectional operation of the RVD are discussed, along with some experimental results to verify the feasibility of the control topology.

I. INTRODUCTION

Conventionally, for the converter with the high-voltage input to the low-voltage high-current output, the single-stage isolated converter has been widely used in the industry, such as the brick VRM (Voltage-Regulated Module) converter with the input voltage of 48V or 24V transferred to the output voltage between 1.2V and 3.3V. There are many literatures presented to discuss various types of structures for this converter, such as current double [1], synchronous rectification [2], resonance [3], etc. However, by doing so, the corresponding cost is expensive, because in general the overall system requires several or many different voltages. Especially, if such voltages come from individual isolated converters with the same input voltage, then the cost is too expensive and the size is too large. Consequently, the bus converter concept [4-7] is presented and hence two-stage converter is adopted. The first-stage is the bus converter whereas the second-stage is the VRM converter. The bus converter with high efficiency is presented, which is the isolated converter without the output voltage regulated, that is to say, the output voltage is varied with the input voltage. The purpose of the bus converter is used to perform isolation and voltage reduction and the purpose of the VRM converter is used to execute voltage regulation. Generally, the bus converter generally transfers 48V or 24V to 12V, then the VRM converter, generally implemented by the traditional buck converter, transfers 12V to 3.3V or lower.

However, most of today's VRM topologies are multi-phase buck converters. Due to the very low output voltage, the buck converter has the extremely small duty cycle, which impacts the performance dramatically. Both the transient response and the efficiency suffer a lot [8-10]. Consequently, there are some literatures [11-15] presented to discuss how to increase duty cycles based on tapped inductors. After this, the two-stage structure for the VRM converter [16-25], based on the bus converter without isolation, is presented, which transfers the input voltage to between 6V and 8V in the first stage without any control and then to the desired output voltage in the second stage with closed-loop control. It is found that the performance of the single-stage structure is worse than that of the second-

stage structure in efficiency and size [1][4-7][11-15]. In [21][25], the voltage divider, used in the first-stage, is implemented by the charge pump. In this case, the efficiency of such a topology can reach up to 96% at full load and 97.5% at light load. Besides, the variable switching frequency, changed from 350kHz to 100kHz, is used at light load to further improve the efficiency up to 98%. In this paper, a resonant voltage divider (RVD) is presented, where all the switches can operate in zero voltage switching (ZCS) under the fixed switching frequency. This makes the filter designed easily. By doing so, the performance in efficiency is improved, especially for light load up to about 99%. Above all, the bidirectional operation of the RVD is taken into consideration. That is to say, the RVD operates normally such that it can provide energy for the point of load (POL) or voltage-regulated module (VRM) and works bidirectionally during the download transient period so as to reversely transfer the corresponding energy to the DC bus and hence to suppress the resulting voltage overshoot as well as upgrade the download transient response. In this paper, only backward operation of the RVD is discussed herein. As for the forward operation of the RVD, please refer to [26].

II. PROPOSED CIRCUIT CONFIGURATION

Fig. 1 shows the proposed resonant voltage divider under backward operation, which is constructed by four switches Q_1, Q_2, Q_3, Q_4, one resonant inductor L_r and one resonant capacitor C_r. As this circuit operates at resonance, the output voltage is double the input voltage. Fig. 2 shows the waveforms for the circuit operating under the ideal condition, the voltage on C_r swings between zero and double the input voltage, which limits the maximum output transfer. In order to realize ZCS operation at resonance, the zero-current detector for I_{Lr} is used, thereby making the switching frequency entirely determined by the resonant frequency. It is noted that in theory the load current is large enough to let I_{Lr} force a negative value on V_{Cr}. If V_{Cr} is negative, then V_{Cr} will be over double the input voltage next cycle. By doing so, the divergent current will occur after several cycles. And hence one diode D_c is added to remove the negative voltage on C_r, but D_c does not work most of time if the RVD operates in the normal condition.

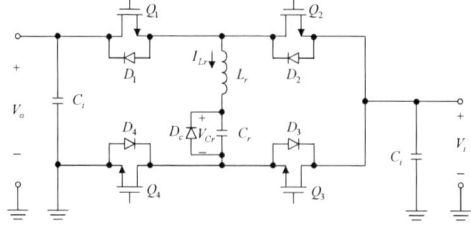

Fig. 1. Backward-operating RVD.

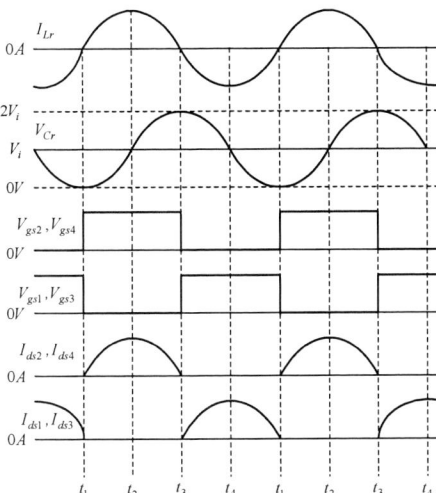

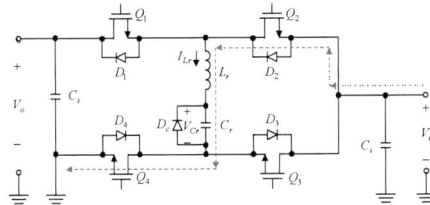

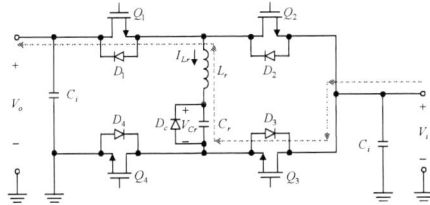

Fig. 2. Ideal waveforms for the backward-operating RVD.

Fig. 3. Backward-operating RVD in modes 1 and 2.

Fig. 4. Backward-operating RVD in modes 3 and 4.

III. BASIC OPERATING PRINCIPLES

Fig. 2 shows the ideal waveforms for the backward-operating RVD. According to Fig. 2, there are four operating modes to be described as follows.

A. Mode 1: (t₁~t₂)

As shown in Fig. 3, Q_1 and Q_3 are switched off, and Q_2 and Q_4 are switched on. In this mode, V_{Cr} is increasing from zero, whereas I_{Lr} is also increasing from zero. The current flow is from the input via Q_2 through L_r and then to C_r, Q_4 and the output. As soon as V_{Cr} reaches V_i or I_{Lr} reaches the maximum value, the operating mode proceeds to mode 2.

B. Mode 2: (t₂~t₃)

As shown in Fig. 3, Q_1 and Q_3 are still in the off-state, and Q_2 and Q_4 are still in the on-state. In this mode, V_{Cr} is still increasing from V_i, whereas I_{Lr} is decreasing from the

maximum value. The current flow is the same as that in mode 1. As soon as V_{Cr} reaches $2V_i$ or I_{Lr} is reduced to zero, the operating mode goes to mode 3. At this moment, Q_2 and Q_4 are switched off with ZCS.

C. Mode 3: (t₃~t₄)

As shown in Fig. 4, Q_1 and Q_3 are switched on, and Q_2 and Q_4 are switched off. In this mode, V_{Cr} is decreasing from $2V_i$, whereas I_{Lr} is increasing in the opposite direction. The current flow is from the ground via Q_3 through C_r and then to L_r, Q_1 and the output. As soon as V_{Cr} is reduced to V_i or I_{Lr} reaches the minimum value, the operating mode proceeds to mode 4.

D. Mode 4: (t₄~t₁)

As shown in Fig. 4, Q_1 and Q_3 are still in the on-state, and Q_2 and Q_4 are still in the off-state. In this mode, V_{Cr} is decreasing from V_i, whereas I_{Lr} is decreasing in the opposite direction. The current flow is the same as that in mode 3. As soon as V_{Cr} is reduced to zero or I_{Lr} reaches zero, the operating mode goes back to mode 1 and the next cycle is repeated. At this instant, Q_1 and Q_3 are switched off with ZCS.

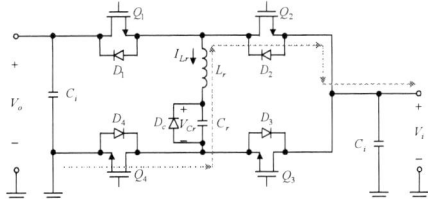

Fig. 5. Backward-operating RVD inserted in modes 2 and 3.

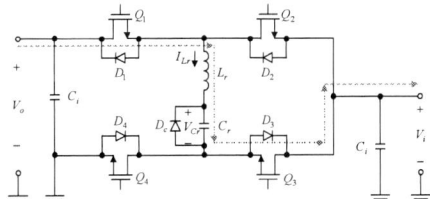

Fig. 6. Backward-operating RVD inserted in modes 4 and 1.

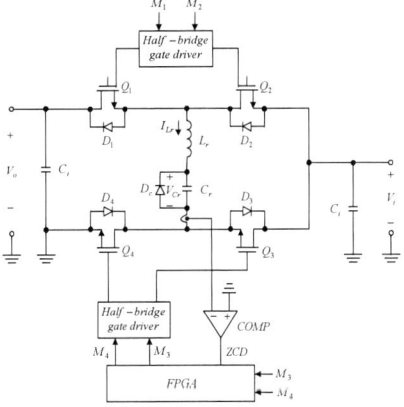

Fig. 7. Proposed overall system block diagram.

978-1-4244-4782-4/10 $26.00 © 2010 IEEE

On the other hand, in practice the delay time is indispensable in the system. This is because detecting the zero crossing of I_{Lr} will be delayed by the gate driver, FPGA calculation time, etc., thereby causing the switch not to be turned off at the zero current of I_{Lr} and hence the current flowing through L_r in the opposite direction to occur. Therefore, there are two required modes added to four ideal modes mentioned above. One mode shown in Fig. 5 is inserted between mode 2 and mode 3, and the other mode shown in Fig. 6 is inserted between mode 4 and mode 1.

IV. CONTROL METHOD APPLIED

Fig. 7 shows the proposed overall system block diagram for the backward-operating RVD. The PWM control signals M_1, M_2, M_3 and M_4 used to drive the four switches Q_1, Q_2, Q_3 and Q_4 are created from the field-programmable gate array (FPGA), respectively. And the current flowing through L_r is sensed by the current transformer (CT) and sent to one comparator COMP to get information on zero crossing of I_{Lr}, and the output result of COMP, ZCD, is sent to FPGA to get the desired PWM control signals. Besides, there are two half-bridge gate drivers used herein. One is for Q_1 and Q_2, and the other is for Q_3 and Q_4.

V. DESIGN OF KEY PARAMETERS

Prior to going into this topic, there are some specifications required as follows: (i) rated DC input voltage V_i is 12V; (ii) rated DC output voltage V_o is 24V; (iii) rated output power is 120W; (iv) maximum output power P_{max} is 144W; (v) switching frequency f_s is initially set to 120kHz; (vi) both values of the input and output capacitors are identical, with two 470μF electrolytic capacitors connected in parallel with two MLCC capacitors for each; (vii) product name of the half-bridge gate drivers is IR2011; (viii) product name of the switches is IRL3705ZS; (ix) product name of FPGA is EP1C3T100; (x) product name of COMP is LT1719; and (xi) product name of D_c is 1N5819.

Since the resonant capacitor C_r and the resonant inductor L_r are the key parameters of the proposed resonant voltage divider, how to design C_r and L_r is to be described as follows. Based on the basic principles of the backward-operating RVD, the following equations can be obtained:

$$2P_{max} = \frac{1}{2}C_r \cdot V_i^2 \cdot f_s \tag{1}$$

$$f_s = \frac{1}{2\pi \cdot \sqrt{L_r \cdot C_r}} \tag{2}$$

From (1) and (2), the values of C_r and L_r can be represented by (3) and (4), respectively:

$$C_r = \frac{4P_{max}}{V_i^2 \cdot f_s} \tag{3}$$

$$L_r = \frac{C_r \cdot V_i^2}{4\pi \cdot P_{max}^2} \tag{4}$$

Based on the given specifications, and (3) and (4), the value of C_r can be figured out to be 1.91μF and the value of L_r can be worked out to be 0.96μH. And eventually, the values of C_r and L_r are chosen to be 2μF and 1μH, respectively, and hence the corresponding switching frequency is 112kHz.

VI. EXPERIMENTAL RESULTS

Figs. 8 to 15 show the experimental waveforms. Fig. 8 displays the resonant waveforms I_{Lr} and V_{Cr}. It is obvious that the maximum value of V_{Cr} approaches to 24V and its minimum value approaches to zero. This implies that the power transfer of the backward-operating RVD is close to the maximum. Fig. 9 depicts the waveforms of I_{Lr}, ZCD, V_{gs1} and V_{gs2}. Since Q_1 and Q_3 operate simultaneously whereas Q_2 and Q_4 work simultaneously, V_{gs3} and V_{gs4} are not shown again. During the positive cycle of I_{Lr}, ZCD is high whereas during the negative cycle of I_{Lr}, ZCD is low. And hence, based on the rising or falling edge of ZCD, FPGA can determine whether the switches operate or not. That is to say, Q_1 and Q_3 are turned on and Q_2 and Q_4 are turned off as I_{Lr} is negative, whereas Q_1 and Q_3 are turned off and Q_2 and Q_4 are turned on as I_{Lr} is positive. Figs. 10 and 11 display the associated waveforms I_{Lr}, ZCD, V_{gs1} and V_{gs2} in the neighborhood of the rising and falling edges of the zero crossing of I_{Lr}, respectively. It is obvious that the instants that switches are turned off are delayed by about 200ns from the zero crossing of I_{Lr}. At rated load, Figs. 12 to 15 show the resonant currents, the voltages on and PWM control signals for Q_1, Q_2, Q_3 and Q_4. It is evident that each switch has ZCS operating characteristics. Above all, Fig. 16 shows that the highest efficiency, 99%, occurs at light load and the efficiency can reach 94.3% at rated load.

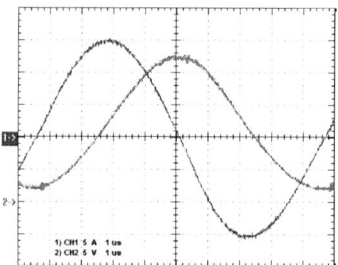

Fig. 8. Resonant waveforms at rated load: (1) I_{Lr}; (2) V_{Cr}.

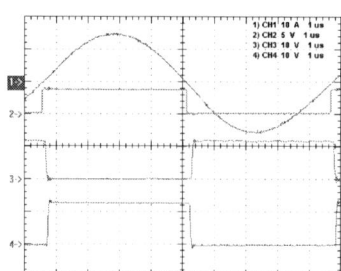

Fig. 9. Related ZCS waveforms at rated load: (1) I_{Lr}; (2) ZCD; (3) V_{gs1}; (4) V_{gs2}.

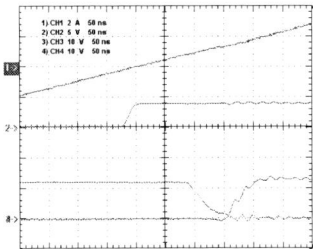

Fig. 10. Zoomed ZCS waveforms of Fig. 9 in the neighborhood of the falling edge of *ZCD*.

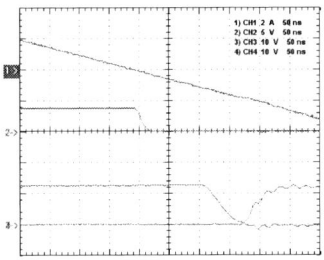

Fig. 11. Zoomed ZCS waveforms of Fig. 9 in the neighborhood of the rising edge of *ZCD*.

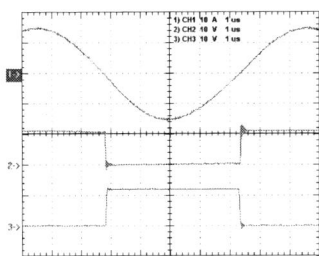

Fig. 12. Related ZCS waveforms at rated load for Q_1: (1) I_{Lr}; (2) V_{ds1}; (3) V_{gs1}.

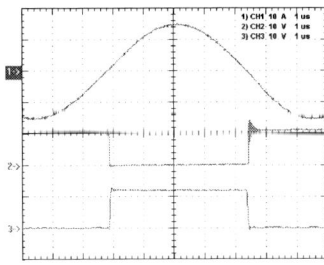

Fig. 13. Related ZCS waveforms at rated load for Q_2: (1) I_{Lr}; (2) V_{ds2}; (3) V_{gs2}.

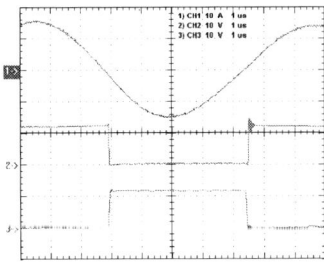

Fig. 14. Related ZCS waveforms at rated load for Q_3: (1) I_{Lr}; (2) V_{ds3}; (3) V_{gs3}.

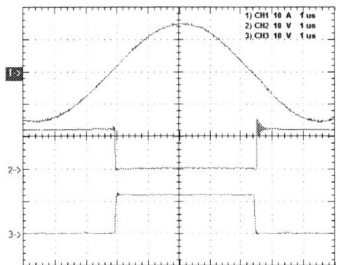

Fig. 15. Related ZCS waveforms at rated load for Q_4: (1) I_{Lr}; (2) V_{ds4}; (3) V_{gs4}.

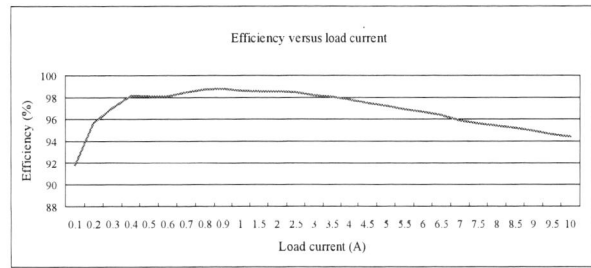

Fig. 16. Efficiency versus load current.

VII. CONCLUSION

In this paper, the bidirectional operation of the RVD is presented, which possesses high efficiency and is very suitably used in reversely transferring the energy to the DC bus, so as to effectively to suppress the voltage overshoot as the download transient response occurs in the POL or VRM. From the experimental results based on the backward-operating RVD, the efficiency at light load is up to 98.7%. Besides, the switching frequency all over the load range is fixed, thereby causing the filter design to be easy.

REFERENCES

[1] Ming Xu, Yuancheng Ren, Jinghai Zhou and F. C. Lee, "1-MHz self-driven ZVS full-bridge converter for 48-V power pod and DC/DC brick," *IEEE Trans. Power Electron.*, vol. 20, no. 5, pp. 997-1006, 2005.

[2] Dianbo Fu, Bing Lu and F. C. Lee, "1MHz high efficiency LLC resonant converters with synchronous rectifier," *IEEE PESC'07*, pp. 2404-2410, 2007.

[3] M. M. Jovanovic, M. T. Zhang and F. C. Lee, "Evaluation of synchronous-rectification efficiency improvement limits in forward converters," *IEEE Trans. Ind. Electron.* vol. 42, no. 4, pp. 387-395, 1995.

[4] L. Balogh, "A new cascaded topology optimized for efficient DC/DC conversion with large step-down ratios," *Proc. Intel Symp.*, 2000.

[5] P. Alou, J. Oliver, J. A. Cobos, O. Garcia and J. Uceda, "Buck+half bridge (d=50%) topology applied to very low voltage power converters," *IEEE APEC'01*, vol. 2, pp. 715-721, 2001.

[6] Y. Ren, M. Xu, C.-S. Leu and F. C. Lee, "A family of high power density bus converters," *IEEE PESC'04*, vol. 1, pp. 527-532, 2004.

[7] Yuancheng Ren, Ming Xu, Julu Sun and F. C. Lee, "A family of high power density unregulated bus converters," *IEEE Trans. Power Electron.* vol. 20, no. 5, pp. 1045-1054, 2005.

[8] Xunwei Zhou, Pit-Leong Wong, Peng Xu, F. C. Lee and A. Q. Huang, "Investigation of candidate VRM topologies for future microprocessors," *IEEE Trans. Power Electron.*, vol. 15, no. 6, pp. 1172-1182, 2000.

[9] Y. Panvo and M. M. Jovanovic, "Design consideration for 12-V/1.5-V, 50-A voltage regulator modules," *IEEE Trans. Power Electron* vol. 16, no. 6, pp. 776-783, 2001.

978-1-4244-4782-4/10 $26.00 © 2010 IEEE

[10] R. Miftakhutdinov, "Optimal design of interleaved synchronous buck converter at high slew-rate load current transients," *IEEE PESC'01*, vol. 3, pp. 1714-1718, vol. 3, 2001.

[11] Kaiwei Yao, Yang Qiu, Ming Xu and F. C. Lee, "A novel winding-coupled buck converter for high-frequency, high-step-down DC-DC conversion," *IEEE Trans. Power Electron.* vol. 20, no. 5, pp. 1017-1024, 2005.

[12] Jinghai Zhou, Ming Xu, Julu Sun and F. C. Lee, "A self-driven soft-switching voltage regulator for future microprocessors," *IEEE Trans. Power Electron.*, vol. 20, no. 4, pp. 806-814, 2005.

[13] Yan Dong, Julu Sun, Ming Xu, Fred. C. Lee and Milan M. Jovanovic, "The light load issue of coupled inductor laptop voltage regulators and its solutions," *IEEE APEC'07*, pp. 1581-1587, 2007.

[14] Mao Ye, Ming Xu and F. C. Lee, "Tapped-inductor buck converter for high-step-down DC-DC conversion," *IEEE Trans. Power Electron.* vol. 20, no. 4, pp. 775-780, 2005.

[15] Yungtaek Jang, Milan M. Jovanovic, and Yuri Panov, "Multi-phase buck converters with extended duty cycle," *IEEE APEC'06*, pp. 38-44, 2006.

[16] Yuancheng Ren, Ming Xu, Kaiwei Yao, Yu Meng, F. C. Lee, Jinghong Guo and Y. Ren, "Two-stage approach for 12 V VR," *IEEE APEC'04*, vol. 2, pp. 1306-1312, 2004.

[17] Yuancheng Ren, Ming Xu, Kaiwei Yao and F. C. Lee, "Two-stage 48 V power pod exploration for 64-bit microprocessor," *IEEE APEC'03*, vol. 1, pp. 426-431, 2003.

[18] Yuancheng Ren, Ming Xu, Yu Meng and F. C. Lee, "12V VR efficiency improvement based on two-stage approach and a novel gate driver," *IEEE PESC'05*, pp. 2635-2641, 2005.

[19] Kisun Lee, Jia Wei, Ming Xu and F. C. Lee, "Adaptive bus voltage positioning system for two stage laptop voltage regulators," *IEEE PESC'07*, pp. 2-8, 2007.

[20] Yuancheng Ren, Kaiwei Yao, Ming Xu and F. C. Lee, "Analysis of the power delivery path from the 12-V VR to the microprocessor," *IEEE Trans. Power Electron.*, vol. 19, no. 6, pp. 1507-1514, 2004.

[21] Julu Sun, Ming Xu, F. C. Lee and Yucheng Ying, "High power density voltage divider and its application in two-stage server VR," *IEEE PESC'07*, pp. 1872-1877, 2007.

[22] Julu Sun, Ming Xu, Yucheng Ying and F. C. Lee, "High power density, high efficiency system two-stage power architecture for laptop computers," *IEEE PESC'06*, pp. 1-7, 2006.

[23] Julu Sun, Yuancheng Ren, Ming Xu and Fred C. Lee, "Light load efficiency improvement for laptop VRs," *IEEE APEC'07*, pp. 1120-126, 2007.

[24] Yuancheng Ren, Ming Xu, Kaiwei Yao, Yu Meng and F. C. Lee, "Two-stage approach for 12-V VR," *IEEE Trans. Power Electron.* vol. 19, no. 6, pp. 1498-1506, 2004.

[25] Ming Xu, J. Sun, J. and F. C. Lee, "Voltage divider and its application in the two-stage power architecture," *IEEE APEC'06*, pp. 499-505, 2006.

[26] K. I. Hwu and Y. T. Yau, "A simple resonant voltage divider," *IEEE APEC'09*, pp. 1115-1120, 2009.

Multiple-input buck converter optimized for accurate envelope tracking in RF power amplifiers

M. Rodríguez, P.F. Miaja, A. Rodríguez, J. Sebastián

University of Oviedo, Department of Electrical and Electronic Engineering

Power Supply Systems Group

Edificio departamental 3, Campus de Viesques s/n. 33204 Gijón, Asturias. Spain

Tel.: + 34 985 182 578. Fax: +34 985 182 138. e-mail: rodriguezmiguel.uo@uniovi.es

Abstract—**Envelope Tracking techniques are used to increase the efficiency of modern radiofrequency transmitters. These techniques are based on varying the supply voltage of the radiofrequency power amplifier according to the envelope of the signal to be transmitted. This paper presents an Envelope Tracking system based on a multilevel topology that targets relatively low voltage systems (12-28 volt range) and medium to high power applications (> 50 W peak power). The proposed converter achieves low output voltage ripple, high tracking bandwidths, high efficiency and high output power capabilities. An appropriate characterization of the load behavior of the radiofrequency power amplifier allows the system to work in an open loop manner. This paper shows that the envelope tracking system proposed is capable of increasing the efficiency of linear, Class-A and Class-B commercial radiofrequency power amplifiers between 10 and 15 %.**

I. INTRODUCTION

Modern radiofrequency transmitters use linear radiofrequecy power amplifiers (RFPA) to deliver the required output power to the antenna. Linear amplifiers are preferred over switching amplifiers due to their inherent high linearity, that allows modern communication systems to use non-constant envelope transmission schemes, like Enhanced Data Rates for Global System Mobile Evolution (EDGE) or Wideband Code Division Multiplex Access (WCDMA). The major drawback of such systems is that linear RFPAs have very low efficiencies; for instance, the maximum theoretical efficiency of a class A linear RFPA is 50 %, while in conventional operation with amplitude modulated signals its average efficiency can be in the 20 % range. Fig. 1a shows a conventional linear RFPA amplifying a non-constant envelope modulated signal. Fig. 1b shows the corresponding waveforms; a considerable amount of power, represented by the shaded area, is lost in the amplifier. From these figures it can be deduced that the closer the supply voltage is to the envelope of the signal, the higher the efficiency will be.

Envelope Tracking (ET) is a technique based on varying the supply voltage of the amplifier according to the envelope of the signal being transmitted; thus, the efficiency is always very close to the theoretical maximum. Fig. 1c shows a conventional ET system, and Fig. 1d shows its operating waveforms; in the ET system the supply voltage is always very close to the envelope of the signal, which allows us to obtain the maximum theoretical efficiency.

The DC-DC converter shown in Fig. 1c should have several especial features to be suitable for ET systems:

- High efficiency
- Low output voltage ripple
- High bandwidth
- High output power capability

These requirements cause the design of an appropriate DC-DC converter to be a challenging task. Several topologies have been used for this application [1]–[7], achieving different results in terms of the output signal bandwidth that can be reproduced, the power handling capability and other figures of merit. The highest tracking bandwidth in high power applications has been reported using a SEPIC converter (nearly 1.25 MHz) in [2], but the efficiency in this case was very low (around 80 %). High tracking bandwidths compatible with high efficiencies were achieved in [3], and were in the range of 50 kHz.

This paper describes an ET system based in the Multiple Input Buck Converter (MIBuck) topology first proposed in [9]. Fig. 2 shows a block diagram of the system: an analog to digital converter acquires the envelope signal, and an FPGA generates the appropriate control signals for the multiple input topology, represented as a high frequency voltage selector plus a low-pass filter. As Fig. 2 shows, the system works in an open loop manner. This is possible because in ET techniques the varying output voltage supply that has to be generated does not need to be very accurate: there is a supply voltage that maximizes the RFPA efficiency, and the closer the supply voltage is to this value, the higher the efficiency will be (as long as the supply voltage is higher than a certain minimum value that ensures appropriate RFPA operation). Thus, a small mismatch between the output and the desired voltage simply lowers the overall transmitter efficiency. However, in order to ensure the feasibility of the system, the load behavior of the RFPA has to be carefully analyzed.

This paper is organized as follows: in section II the main features of the proposed multilevel converter are summarized. Section III presents several results concerning the behavior of Class A and Class B amplifiers as loads. Section IV shows the experimental setup built using the proposed multilevel converter and two different commercial RFPAs; a 200 W, 27 MHz, Class B amplifier and a 300 W, 27 MHz, Class A push-

978-1-4244-4782-4/10 $26.00 © 2010 IEEE

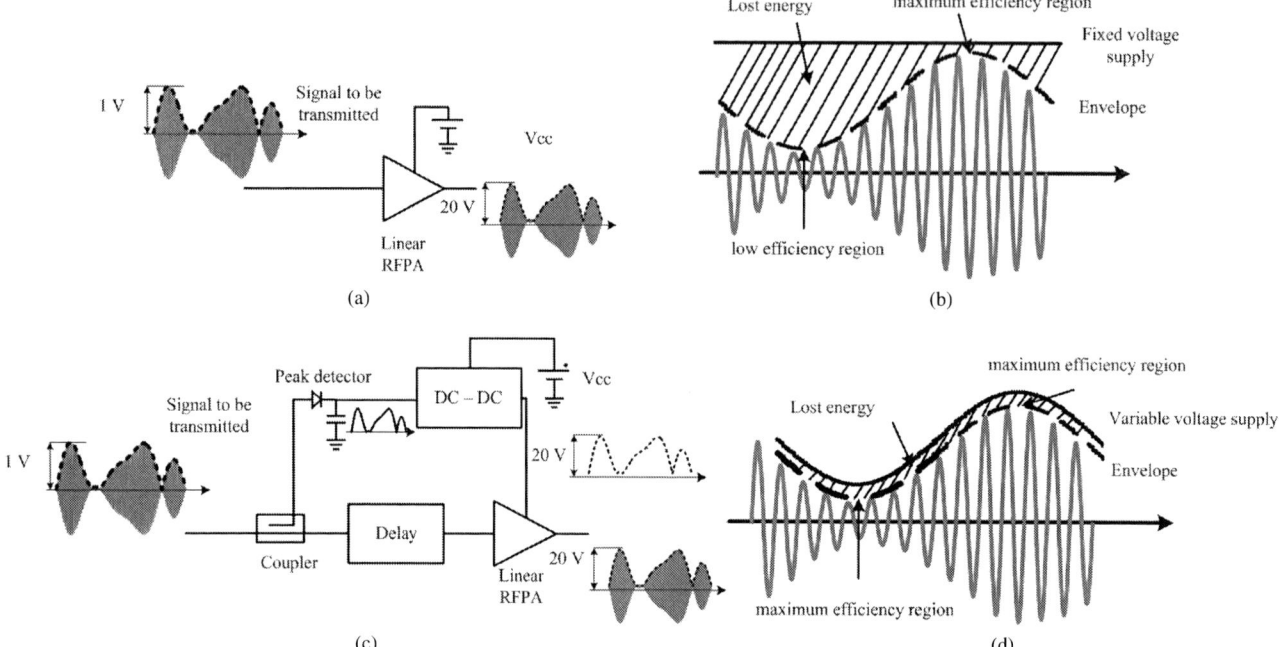

Fig. 1. (a) Diagram of a conventional transmitter without ET; (b) characteristic waveforms; (c) diagram of a transmitter that uses ET; (d) characteristic waveforms.

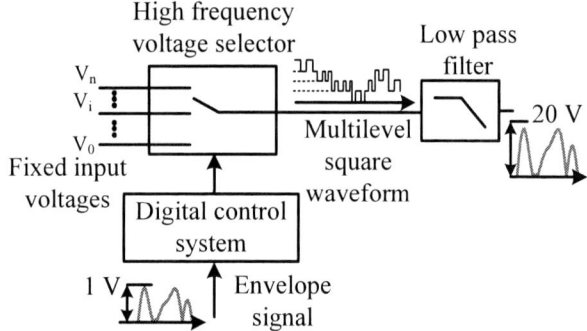

Fig. 2. General scheme of the proposed topology: a multilevel square waveform is filtered to obtain the desired envelope.

pull amplifier. Finally, the conclusions are presented in section V.

II. MULTILEVEL CONVERTER MAIN CHARACTERISTICS

Figure 3 shows the proposed MIBuck converter. Two switches are turned on and off alternatively, in order to generate a square voltage waveform at the input node of the low pass filter. Said waveform thus changes between the two input voltages that have been selected. Fig. 4 shows the equivalent circuit in the aforementioned situation and the corresponding waveforms.

The MIBuck topology has several advantages over other previously proposed topologies for ET applications: the square voltage waveform shown in Fig. 4b is easy to filter, thus allowing low output voltage ripple. As smaller filter component values can be used in comparison with a conventional

buck topology, the tracking bandwidth can also be increased. Furthermore, the voltage stresses over the semiconductors are smaller, thus diminishing switching losses, increasing the efficiency and allowing higher switching frequencies. Efficiencies above 90 % were reported in [9], when the MIBuck was handling around 35 W of average output power and more than 90 W of peak power, and the switching frequency was above 3 MHz. More details regarding the operating principles of the MIBuck topology can also be found in [9].

III. LOAD BEHAVIOR OF LINEAR RFPAS

In Fig. 3 the MIBuck converter is loaded with a constant resistor. As the actual load is in fact the RFPA, its behavior becomes a major concern; for instance, knowledge of the load value allows us to select L and C in order to fulfill certain bandwidth requirements and to ensure that the system is appropriately damped. This section qualitatively analyzes the load behavior of the linear RFPA depending on its class of operation.

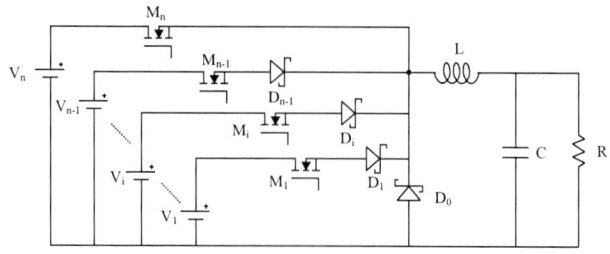

Fig. 3. Multiple input buck converter topology.

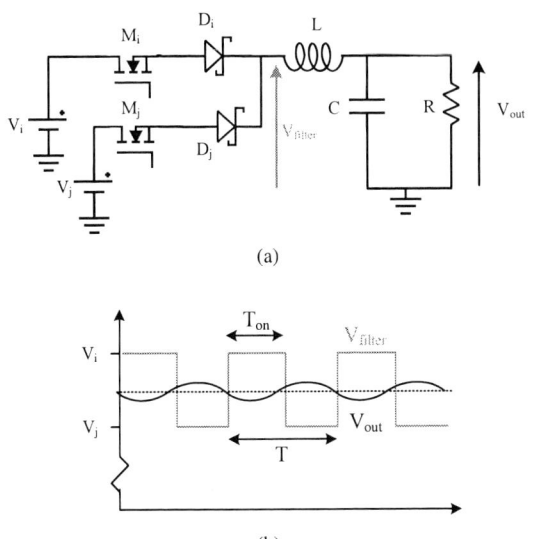

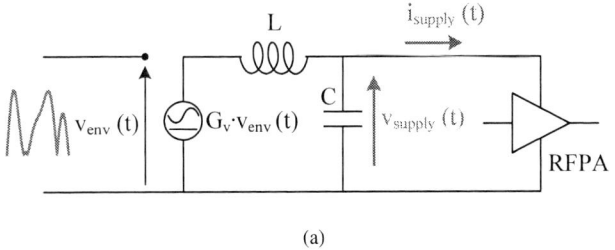

(a)

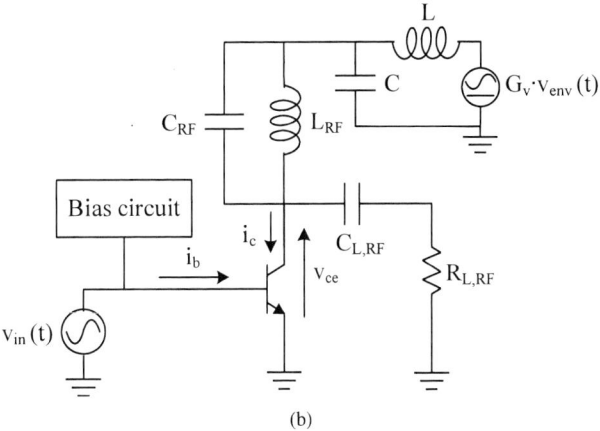

(b)

Fig. 4. (a) Equivalent circuit used to calculate the conversion ratio; (b) typical static waveforms when branches i and j are switching alternatively. The dashed line shows the average output voltage.

Fig. 5. (a) Equivalent simplified large signal model of the system: the envelope signal v_{env} is amplified and applied to the load; (b) simplified linear RFPA using a single transistor. The bias circuit defines the class of operation of the amplifier.

Fig. 5a shows a large signal simplified equivalent circuit of the system. Ideally, the reference envelope signal, $v_{env}(t)$, is amplified and then applied to the RFPA. L and C are the output filter of the converter. Fig. 5b shows a simplified linear amplifier using a single bipolar transistor and supplied by the envelope tracking system: $C_{L,RF}$ is a bypass capacitance and thus has a very low impedance at the RFPA operating frequency. C_{RF} and L_{RF} form a resonant circuit at the operating frequency, ensuring a sinusoidal output voltage.

Considering the transistor as a two-port network and taking into account that, at the beginning of an arbitrary RF cycle, $C_{L,RF}$ is charged to $G_v v_{env}(t)$ (neglecting L and C), circuit in Fig. 6 can be obtained (only the output port is shown). Note the explicit dependence of the parameter i_{scc} (short circuit colector current) with v_{in}. As C_o and y_o change within an RF cycle, due to the inherent large-signal operation of the RFPA, they should be considered non-linear impedances. However, in the calculations that follow, such non linearity will be neglected by calculating an average value for both impedances.

In the circuit of Fig. 6, L_{RF} causes the average value of the voltage v_1 to be zero over one RF cycle:

$$\langle v_1 \rangle = \int_0^{T_{RF}} v_1 dt = 0 \,, \tag{1}$$

thus allowing us to obtain the equivalent circuit in Fig. 7, $\overline{i_{scc}}$ being the averaged short circuit colector current, and $\overline{C_o}$ and $\overline{g_o}$ being averaged values of C_o and g_o, respectively. Furthermore, $\overline{C_o}$ will be typically much smaller than C, the output capacitance of the converter, and thus it can be neglected for practical purposes. At this moment, it is worth to remark that, due to the averaging procedure, the circuit in Fig. 7 is valid for the relatively low frequency variations of $v_{env}(t)$.

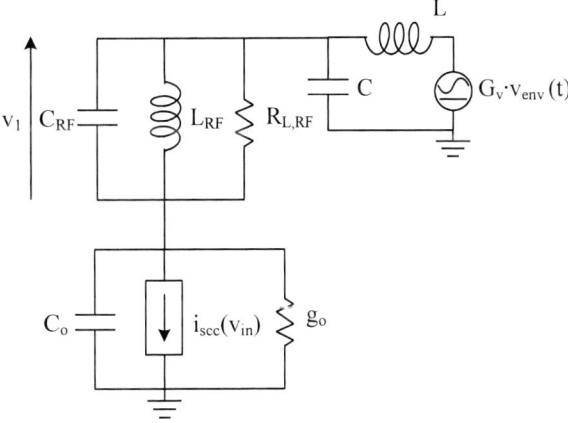

Fig. 6. Equivalent circuit based on admittance parameters.

The equivalent load seen by the converter can then be expressed as:

$$Z_{eq} = \frac{v_{supply}(t)}{i_{supply}(t)} = \frac{v_{supply}(t)}{v_{supply}(t)\,\overline{g_o} + \overline{i_{sc}}} \,. \tag{2}$$

Neglecting the dynamic effect of the low pass filter, (2) can be rewritten as:

$$Z_{eq} = \frac{G_v v_{env}(t)}{i_{supply}(t)} = \frac{G_v v_{env}(t)}{G_v v_{env}(t)\,\overline{g_o} + \overline{i_{scc}}} \,. \tag{3}$$

The following sections are dedicated to draw more specific

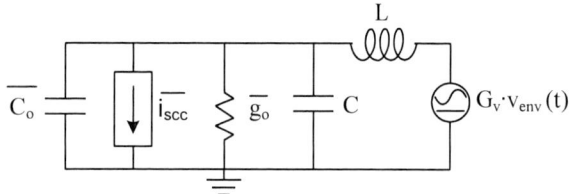

Fig. 7. Final equivalent circuit for the load behavior analysis of the RFPA with the averaged admittance parameters.

expressions from (3), depending on the class of operation of the RFPA.

A. Class-A RFPAs

In Class-A amplifiers, the bias circuit keeps the transistor in its active region even in absence of the input signal. Fig. 8 shows a set of typical transistor $i_c - v_{ce}$ curves, along with its colector current. It is assumed that $\overline{i_{scc}}$ can be approximated by the point where the actual $i_c - v_{ce}$ curve defined by the bias circuit intersects the i_c axis. Thus, it depends only on the bias current of the transistor:

$$\overline{i_{scc,A}} = k \cdot I_{bias} \, , \tag{4}$$

where the dependence on v_{in} has been eliminated. The average output conductance, $\overline{g_o}$, is now approximated by the slope of the actual $i_c - v_{ce}$ curve defined by the bias current:

$$\overline{g_{o,A}} = \frac{\overline{i_{c,t_1}} - \overline{i_{c,t_2}}}{v_{supply}(t_1) - v_{supply}(t_2)} = \tan \alpha \, . \tag{5}$$

It is apparent that the equivalent load in Class-A operation is finally given by a conductance, $\overline{g_{o,A}}$, and by a constant current source, $\overline{i_{scc,A}}$. Thereby, the main factor that determines the dynamic response and the damping factor of the LC output filter of the converter is $\overline{g_{o,A}}$.

Fig. 9 shows actual measured curves of a Class-A commercial amplifier whose main characteristics will be described in detail in the experimental results section. The figure shows several $i_c - v_{ce}$ curves. The noticeably high values found for $\overline{g_{o,A}}$ (i.e. low values for the resistance $\frac{1}{\overline{g_{o,A}}}$) show that an appropriate damping of the output filter will be possible. It is

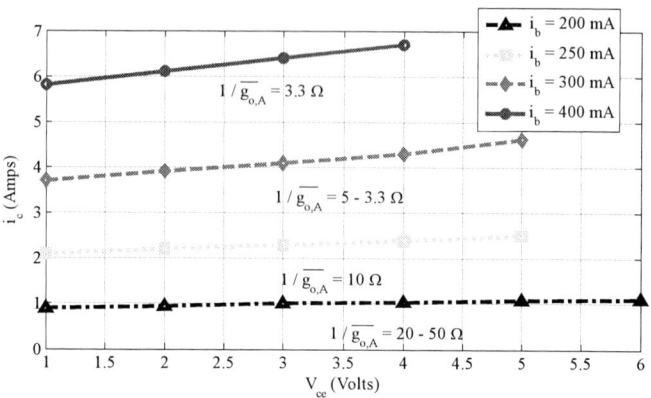

Fig. 9. Measured $i_c - v_{ce}$ curves in an actual Class-A RFPA (model KL-300). The relatively constant slope of the curves for each bias current can be clearly appreciated.

also worth to remark that, for a fixed bias current, the slope of each $i_c - v_{ce}$, i.e. $\overline{g_{o,A}}$, remains approximately constant.

B. Class-B RFPAs

In Class-B RFPAs the bias circuit keeps the transistor at the boundary between its active region and cut-off, being driven into the active region by the input signal. Fig. 10 shows a set of typical transistor $i_c - v_{ce}$ curves, along with its colector current.

It can be easily shown that the peak and the average colector currents are related as follows [10]:

$$\overline{i_{scc}} = \frac{i_{scc,peak}}{\pi} \, . \tag{6}$$

Assuming perfect linear operation of the RFPA, the colector current and the input voltage can be related as follows:

$$i_{scc}(t) = g_m \cdot v_{in}(t) \, , \tag{7}$$

g_m being the transconductance of the RFPA. Equation (7) can be rewritten in terms of the peak values:

$$i_{scc,peak} = g_m \cdot v_{in,peak}(t) \, . \tag{8}$$

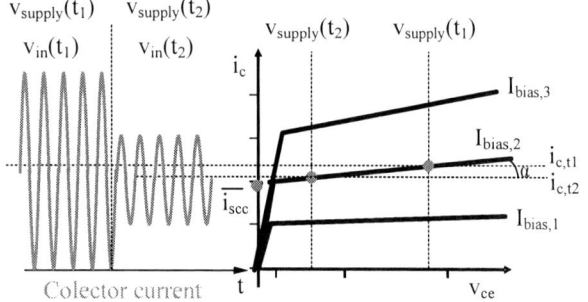

Fig. 8. Typical RF power bipolar transistor $i_c - v_{ce}$ curves and basic operating waveforms in Class-A.

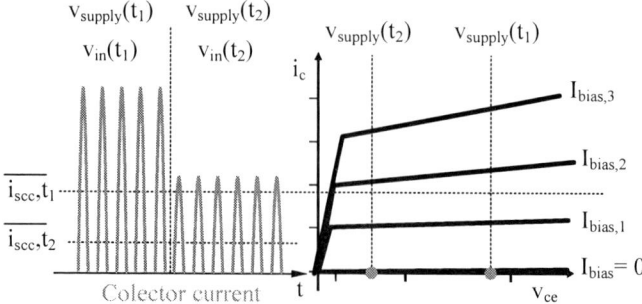

Fig. 10. Typical RF power bipolar transistor $i_c - v_{ce}$ curves and basic operating waveforms in Class-B.

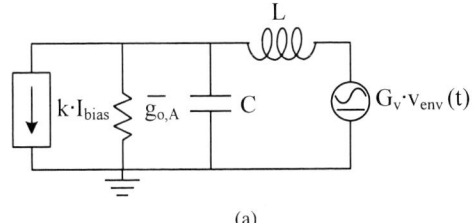

(a)

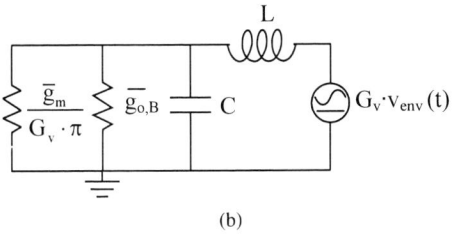

(b)

Fig. 11. (a) Final equivalent circuit of the RFPA in Class-A operation; (b) final equivalent circuit of the RFPA in Class-B operation.

Taking into account that the peak of the input voltage is by definition $v_{env}(t)$, the following relationship can be written:

$$i_{scc,peak} = g_m \cdot v_{env}(t) . \qquad (9)$$

Finally, averaging (9) yields:

$$\overline{i_{scc,peak}} = \overline{g_m} \cdot v_{in,peak}(t) = \overline{g_m} \cdot v_{env}(t) , \qquad (10)$$

$\overline{g_m}$ being the averaged value of g_m. Combining (6) and (10), the short circuit average colector current can be obtained:

$$\overline{i_{scc,B}} = \frac{\overline{g_m} \cdot v_{env}(t)}{\pi} . \qquad (11)$$

Substituting (11) into (3), the equivalent impedance in Class-B

operation is obtained:

$$
Z_{eq,B} = \frac{G_v v_{env}(t)}{G_v v_{env}(t) \overline{g_{o,B}} + \dfrac{\overline{g_m} v_{env}(t)}{\pi}} =
$$
$$
= \frac{1}{\overline{g_{o,B}} + \dfrac{\overline{g_m}}{G_v \pi}} . \qquad (12)
$$

As can be deduced from (12), the final equivalent circuit is made up of two resistances situated in parallel, $\dfrac{1}{\overline{g_{o,B}}}$ and $\dfrac{G_v \pi}{\overline{g_m}}$. Fig. 11 summarizes the previous results for each class of operation.

IV. EXPERIMENTAL RESULTS

A block diagram of the experimental setup used to test the performance of the proposed envelope tracking system is shown in Fig. 12. An Agilent 33210A function generator generated the envelope reference signal, that was sampled by a 12 bit Analog to Digital Converter (TI THS1230) and feed to the digital control system, implemented in a Virtex 4 FPGA. The Mibuck prototype was supplied using three separate voltage sources of 12 V, 8 V and 4 V, and its switching frequency was set to 1.6 MHz. The inductance of the low pass filter was 5.6 μH, and the capacitance was 500 nF, yielding a tracking bandwidth of 95 kHz.

A. ET system performance using a Class-A RFPA

A commercial Class-A amplifier (model KL-300 from RM Italy [12]) was used to test the proposed ET system. The amplifier operated at 27 MHz, and was capable of providing 300 W of peak RF power to a 50 Ω load. It used two SD1446 RF power bipolar transistors in push-pull configuration. As it was intended to operate in Class-B, an appropriate bias circuit was incorporated. Fig. 13 shows the experimental setup.

Fig. 14 shows several results when the ET system was tracking different envelope waveforms. Figs. 14a and 14b show 10 kHz and 50 kHz sinusoids, respectively. Fig. 14c shows an

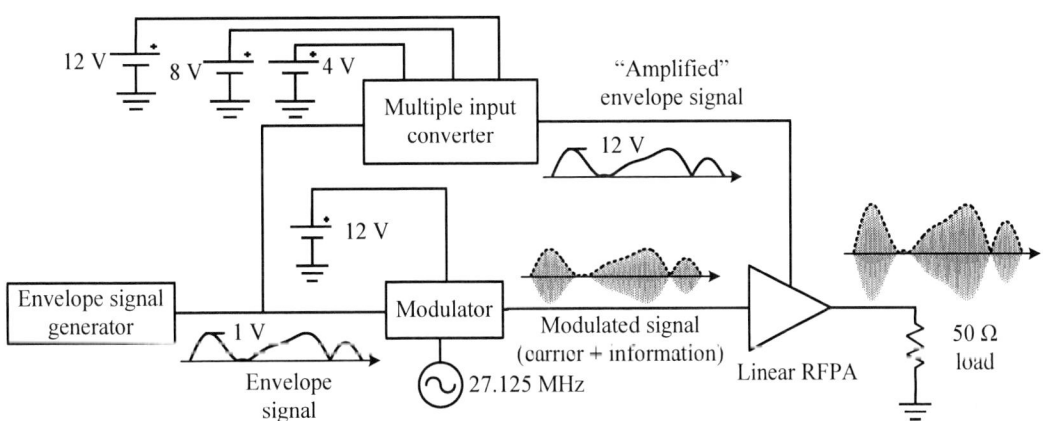

Fig. 12. Experimental setup used to test the proposed envelope tracking system.

978-1-4244-4782-4/10 $26.00 © 2010 IEEE

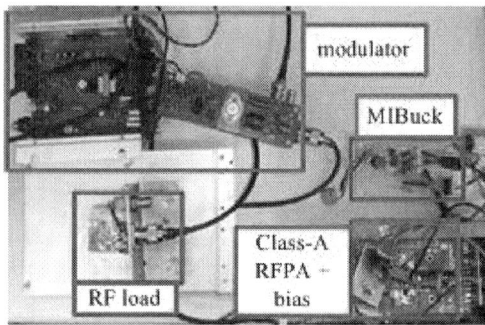

Fig. 13.   Experimental setup used for Class-A measurements.

EDGE envelope signal. It is apparent that the ET system is capable of appropriately supplying the RFPA, performing an accurate envelope tracking in open loop conditions. A slight

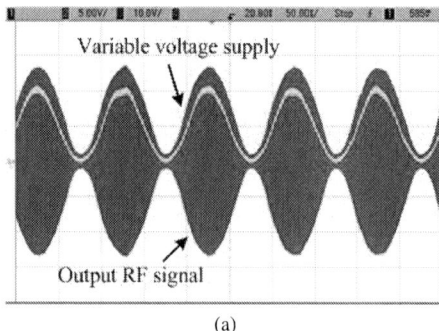

(a)

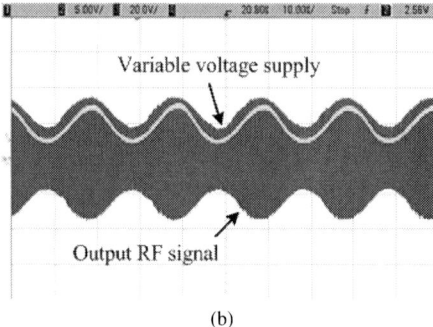

(b)

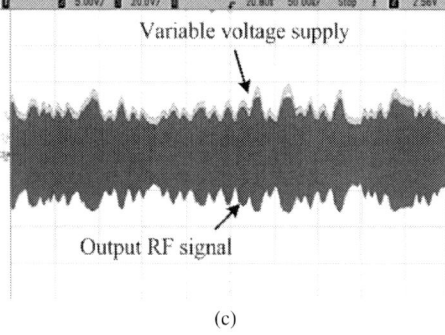

(c)

Fig. 14.   Experimental results obtained with the setup shown in Fig. 13. The output RF signal and the supply voltage are shown: (a) 10 kHz sinusoid envelope signal; (b) 50 kHz sinusoid envelope signal; (c) EDGE envelope signal.

distortion can be observed in the supply signal, probably due to the poor performance of the modulator and also to the actual non-linear behavior of the parameters derived in section III-A. However, such distortion does not noticeably affect the output RF signal. Table I shows the efficiency of the RFPA supplied with a constant voltage and the added efficiency of the converter and the RFPA while ET was being carried out. The average efficiency improvement achieved was around 10 % when the peak output RF power was around 60 W.

Fig. 15 shows the supply voltage waveform under a step reference in the envelope signal. As the output filter was designed with no overshoot, it is apparent that in Class-A operation $\overline{g_{o,A}}$ provides enough damping.

B. ET system performance using a Class-B RFPA

A commercial Class-B amplifier (model KL-200 from RM Italy [12]) was also used to test the proposed ET system. The amplifier operated at 27 MHz, and was capable of providing 200 W of peak RF power to a 50 Ω load. It used one SD1446 RF power bipolar transistor. Fig. 16 shows the experimental setup.

Fig. 17 shows the results when the ET system was supplying the amplifier with the same envelope waveforms shown in Fig. 14. Once again, the ET system is capable of appropriately supplying the RFPA, and some distortion can also be appreciated. The ET system performs as expected in open loop conditions. Table II shows efficiency measurements with and without the ET system. The average efficiency improvement achieved was around 15 % when the peak output RF power was around 75 W.

According to section III-B, the equivalent load seen by the converter in Class-B operation is made up of two terms, $\overline{i_{scc,B}}$ and $\overline{g_{o,B}}$. In Fig. 18a a step reference was set as the input to the ET system, while in Fig. 18b the step reference was

TABLE I
OVERALL EFFICIENCY WITHOUT AND WITH ET: CLASS-A RFPA.

	Modulating signal		
	10 kHz sinusoid	50 kHz sinusoid	EDGE
Without ET	13.3 %	14.2 %	12.7 %
With ET	24.2 %	25.5 %	22.1 %

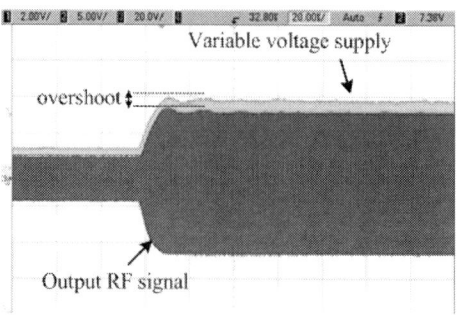

Fig. 15.   Supply voltage of the Class-A RFPA under a step change in the envelope signal.

978-1-4244-4782-4/10 $26.00 © 2010 IEEE

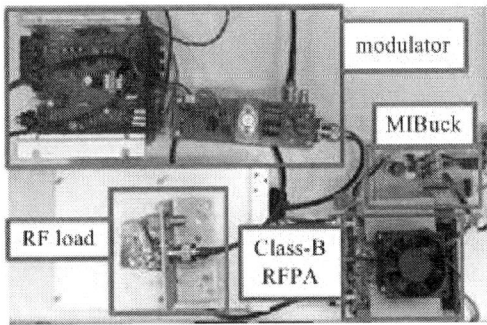

Fig. 16.   Experimental setup used for Class-B measurements.

only applied to the supply of the RFPA, but not to its input. Thereby, in the latter situation $\overline{i_{scc,B}}$ will become constant and only $\overline{g_{o,B}}$ will remain; the system will become less damped.

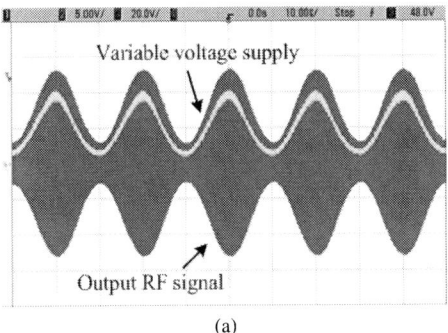

(a)

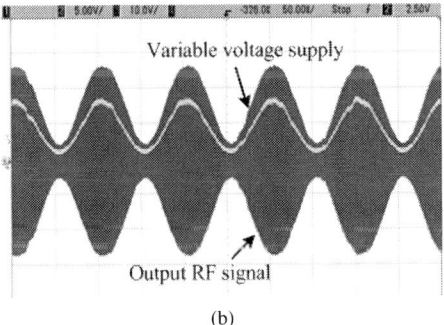

(b)

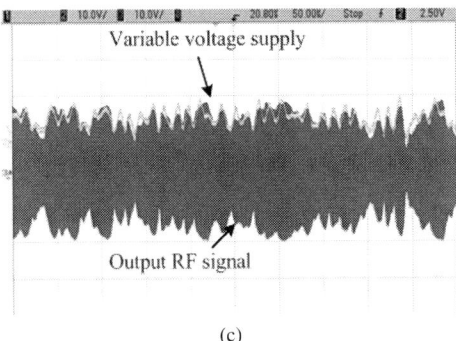

(c)

Fig. 17.   Experimental results obtained with the setup shown in Fig. 16. The output RF signal and the supply voltage are shown: (a) 10 kHz sinusoid envelope signal; (b) 50 kHz sinusoid envelope signal; (c) EDGE envelope signal.

TABLE II
OVERALL EFFICIENCY WITHOUT AND WITH ET: CLASS-B RFPA.

	Modulating signal		
	10 kHz sinusoid	50 kHz sinusoid	EDGE
Without ET	38.7 %	37.4 %	13 %
With ET	57.5 %	55.3 %	27 %

Such effect can be clearly appreciated in Fig. 18b.

V. CONCLUSION

This paper presents an Envelope Tracking system based on a Multiple Input Buck converter topology. On one hand, this topology is specially suitable for this application, as it achieves low output voltage ripple, high efficiency and high tracking bandwidth. On the other hand, it requires several input voltage sources and a more complex control system.

As the proposed system is intended to work in an open loop manner, the load behavior of the RFPA has to be taken into account. A simple, average qualitative model was derived using admittance parameters, showing that the output filter of the converter was damped, disregarding the actual class of operation of the RFPA. Experimental measurements of the amplifier $i_c - v_{ce}$ curves showed an appropriate damping due to the $\overline{g_{o,A}}$ parameter. Ideally, in Class-B the damping is guaranteed by the completely resistive behavior derived in section III-B.

Section IV showed several experimental results of the ET system in open loop conditions and using two different commercial amplifiers. As Figs. 14 and 17 show, the ET system was capable of tracking different envelope signals, increasing the efficiency of the system between 10 and 15 %.

Future work will be focused on determining the equivalent load model parameters from RFPA datasheet, as long as on accurately characterizing the distortion introduced by the ET system, in order to evaluate the limits of open loop operation.

ACKNOWLEDGMENT

This work has been funded by the Spanish Ministry of Science and Education under FPU program (refs. AP2006-04777 and AP2008-03380), and by project TEC-2007-66917.

REFERENCES

[1] F. Wang, A. Yang, D.Y.C. Lie, D. Kimball, L. Larson, P. Asbeck, "Design of Wide-Bandwidth Envelope Tracking Power Amplifiers for OFDM Applications". *IEEE Transactions on Microwave Theory and Techniques*, vol. 53, no. 4, pp. 1244–1255, 2005.

[2] D. Anderson, W. Cantrell, "High-Efficiency High-Level Modulator for Use in Dynamic Envelope Tracking CDMA RF Power Amplifiers". *IEEE MTT-S International Microwave Symposium Digest*, vol. 3, pp. 1509–1512, 2001.

[3] M. Høyerby, M. Andersen, "Self Oscillating Soft Switching Envelope Tracking Power Supply for Tetra2 Base Station". 29th *International Telecommunications Energy Conference (INTELEC)*, pp. 53–60, 2007.

[4] A. Soto, J.A. Oliver, J.A. Cobos, J. Cezón, F. Arévalo, "Power supply for a radio transmitter with modulated supply voltage". *IEEE Applied Power Electronics Conference (APEC) 2004*, pp. 392-398.

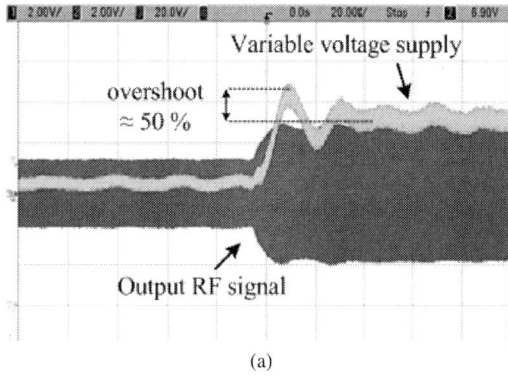

(a)

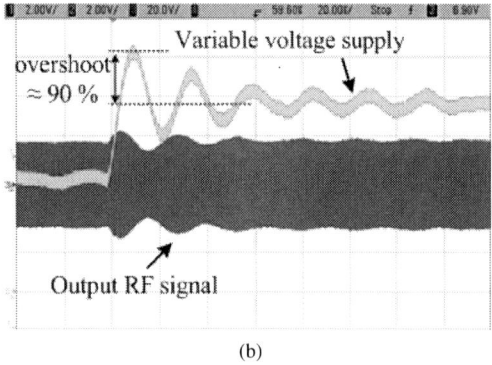

(b)

Fig. 18. (a) Supply voltage and output RF voltage when a step reference signal is given to the ET system; (b) supply voltage when the step reference is applied only to the supply of the RFPA. The increase of the overshoot suggests a decrease of the damping factor.

[5] M. Hoyerby, M. Andersen,, "High-Bandwidth, High-Efficiency Envelope Tracking Power Supply for 40W RF Power Amplifier Using Paralleled Bandpass Current Sources". *IEEE Power Electronics Specialist Conference (PESC 2005)*, pp. 2804-2809.

[6] V. Yousefzadeh, E. Alarcón, D. Maksimovic, "Three-level buck converter for Envelope Tracking Applications". *IEEE Transactions on Power Electronics*, vol. 21, no. 2, march 2006.

[7] M. Vasic, O. García, J.A. Oliver, P. Alou, D. Díaz, J.A. Cobos, "Multilevel power supply for high efficiency RF amplifiers". *IEEE Applied Power Electronics Conference (APEC 09)*, pp. 1233-1238, February 2009.

[8] J. Sebastián, P. Villegas, F. Nuno, M. Hernando, "High-Efficiency and Wide-Bandwidth Performance Obtainable from a Two-Input Buck Converter". *IEEE Transactions on Power Electronics*, vol. 13, no. 4, pp. 706–717, july 1998.

[9] M. Rodríguez, P. Fernández, A. Rodríguez, J. Sebastián, "Multilevel converter for Envelope Tracking in RF power amplifiers". *IEEE Energy Conversion Congress and Exposition 2009 (ECCE 2009)*, pp. 503-510, September 2009.

[10] Steve C. Cripps, "RF Power amplifiers for wireless communications". Artech House, Inc., 2nd edition, 2006, Chap. 3.

[11] "KL200 schematic diagram". *www.rmitaly.com/download/manuals/KL200-manual_rel_410.pdf.*

[12] "KL300 schematic diagram". *www.rmitaly.com/download/manuals/KL300-manual_rel_410.pdf.*

Switching Capacities based Envelope Amplifier for High Efficiency RF Amplifiers

M.Vasić, O. García, J.A. Oliver, P. Alou, D. Diaz, J.A. Cobos

Centro de Electrónica Industrial
Universidad Politécnica de Madrid
Madrid, Spain
miroslav.vasic@upm.es

Abstract— **Modern transmitters usually have to amplify and transmit signals with simultaneous envelope and phase modulation. Due to this property of the transmitted signal, linear power amplifiers (class A, B or AB) are usually used as a solution for the power amplifier stage. These amplifiers have high linearity, but suffer from low efficiency when the transmitted signal has high peak-to-average power ratio (PAPR). The Kahn envelope elimination and restoration (EER) technique is used to enhance efficiency of RF transmitters, by combining highly efficient, nonlinear RF amplifier (class E) with a highly efficient envelope amplifier in order to obtain linear and highly efficient RF amplifier. This paper presents a solution for the envelope amplifier based on a multilevel converter in series with a linear regulator. The multilevel converter is implemented by employing voltage dividers based on switching capacities. The implemented envelope amplifier can reproduce any signal with maximum spectral component of 2 MHz and give instantaneous maximum power of 50 W. The efficiency measurements show that when the signals with low average value are transmitted, the implemented prototypes have up to 20% higher efficiency than linear regulator that is used as a conventional solution.**

I. INTRODUCTION

In the modern radio communications it is necessary to transmit as much data as possible for the given bandwidth. The best signal modulations are those that perform simultaneous phase and amplitude modulation, and therefore the linearity of the power amplifiers that are used in these systems is essential. Due to this demand, linear power amplifiers, such as class A or class B, are employed, but, unfortunately, they suffer from extremely poor efficiency when the transmitted signal has low values. For example, ideal class A and class B amplifiers have average efficiency of only 5% and 28%, respectively, when signals with Rayleigh's envelope distribution are transmitted [1]. One of the techniques that are used to enhance the efficiency of the power amplifiers is the Kahn's technique. The Kahn's envelope estimation and restoration (EER) technique proposes use of dc-dc converter (envelope amplifier) that should modulate the voltage supply of a highly efficient but nonlinear power amplifier (class E or class D) [2], Figure 1. This idea is based

on the fact that any narrow band signal can be presented as simultaneous amplitude (envelope) and phase modulation:

$$V_{RF}(t) = I(t)\cos(2\pi ft) - Q(t)\sin(2\pi ft) =$$

$$= A(t)\cos(2\pi ft + \theta(t)) \qquad (1)$$

$$\theta(t) = arctg\left(\frac{Q(t)}{I(t)}\right) \quad A(t) = \sqrt{I(t)^2 + Q(t)^2} \qquad (2)$$

where f is the carrier frequency, $Q(t)$ and $I(t)$ are modulated signals.

In the state of the art, several solutions for the envelope amplifier can be found, such as a simple buck converter (class S modulator) in [3-5], multiphase buck converter in [6], three-level converter in [7] or linear assisted switching amplifier [8, 9]. These solutions do not exceed the bandwidth of few hundred kHz and the output power is from the range of mW up to several tens of watts. Their use for applications that require bandwidth in the MHz range is limited, for example, a buck converter with 1MHz bandwidth should have switching frequency of, at least, 5MHz.

The envelope amplifier should have high linearity, fast dynamic response, high efficiency and small interference with the spectrum of the output signal. Regarding all these restrictions, a solution that is based on a multilevel converter in series with a high slew rate linear regulator is presented in [10], Figure 2. The multilevel converter in this paper is implemented by voltage dividers based on switching capacities. Converters that employ switching capacities offer high efficiency and do not need any bulky magnetic component, thus their size is significantly smaller comparing them with classical converters and gives a possibility for integration [11]. The implemented envelope amplifier can reproduce a sine wave or any other reference of 2 MHz, and give the maximum power of 50 W. Commercial (or academic) solutions to obtain this bandwidth for the envelope amplifier are based on a linear regulator. Unfortunately, this means low efficiency for the envelope amplifier, especially when the output signal has low voltage levels.

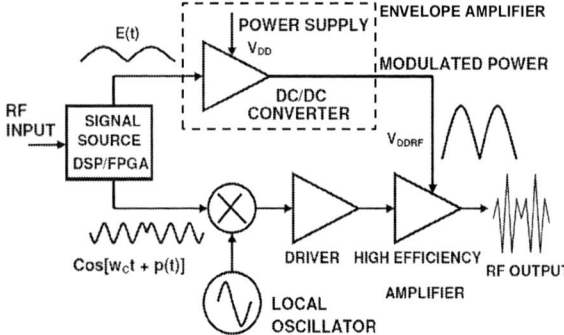

Figure 1. Block Scheme of Kahn-technique Transmitter

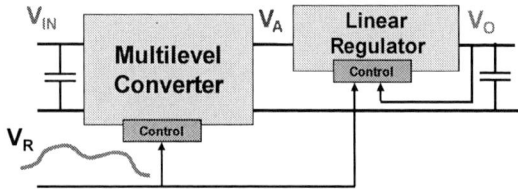

Figure 2. Simplified schematic of the proposed envelope amplifier

II. PROPOSED SOLUTION

The multilevel converter has to supply the linear regulator and it has to provide discrete voltage levels that are as close as possible to the output voltage of the envelope amplifier. If this is fulfilled, the power losses on the linear regulator will be minimal, because they are directly proportional to the difference of its input and output voltage. However, in order to guarantee correct work of the linear regulator, the output voltage of the multilevel converter always has to be higher than the output voltage of the linear regulator. Time diagrams of the multilevel converter and linear regulator voltage are shown in Figure 3.

The linear regulator can be designed to have very high bandwidth, and it should filter all the noise that could come from the multilevel converter. Therefore, the multilevel converter does not need any filter at its output and the design of the complicated filter as in the case of switched converters is avoided.

The multilevel converter presented in [10] is based on independent voltage cells that are put in series and turned on and off depending on the level of the sent reference. The multilevel solution in this paper is proposed in [10] and it is based on the independent voltage sources that supply the linear regulator through an analog multiplexer. The selection of the active voltage source will depend on the level of the envelope reference. A simplified schematic of the multilevel converter is presented in Figure 4.

The efficiency of the linear regulator depends on the number of the voltage levels that are applied and their distribution. In [10] it is shown that optimizing voltage levels leads to significant increase of efficiency comparing it with the solution with equidistant voltage levels. Due to the trade-off between the complexity of the layout, PCB parasitic and possible system efficiency, three voltage levels are selected as

an efficient and feasible solution. The optimal distribution for three voltage levels when a signal with high PAPR is transmitted is approximately: V_{MAX}, $\frac{3}{4} V_{MAX}$, $\frac{1}{2} V_{MAX}$, where V_{MAX} is the maximum level of the output signal's envelope. Due to convenient voltage distribution, these three voltage levels can be produced by two voltage dividers that are based on switching capacities. The input terminals of the first voltage divider should be connected to V_{MAX} and ground and the voltage at its output would be $\frac{1}{2} V_{MAX}$. The input terminals of the second voltage divider should be connected to V_{MAX} and $\frac{1}{2} V_{MAX}$. Its output voltage would be $\frac{1}{4} V_{MAX}$ referring it to the output of the first voltage divider, i.e. $\frac{3}{4} V_{MAX}$ referring it to the ground of the system. Both voltage dividers are implemented in the same way, using the same topology presented in [11] and shown in Figure 5.

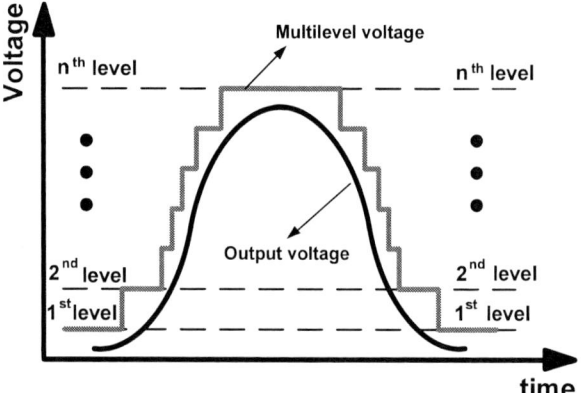

Figure 3. Time diagrams of the proposed envelope amplifier

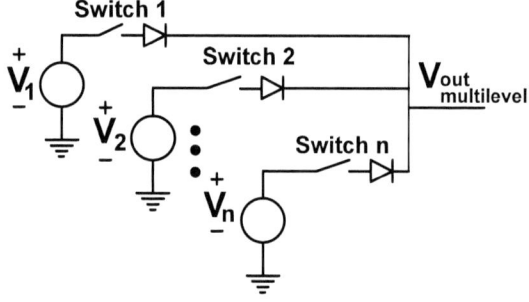

Figure 4. Multilevel converter realized with independent supplies and analog multiplexer

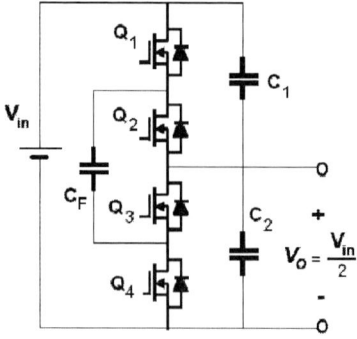

Figure 5. Voltage divider based on switching capacities

Solutions based on switching capacities offer high efficiency at very wide load profiles and low switching frequencies. Additionally, this topology does not use any inductive element and, therefore, are convenient for integration. This is significant improvement comparing proposed solution with the multilevel solution in [10], where it was necessary to implement a flyback converter with three outputs, and where the used transformer was a very bulky, low-efficient component. The possible problems in this solution are increased switching noise due to lack of filtering and complex designs that are used for closed loop solutions. In the proposed solutions the voltage dividers do not need to have precise control of its output voltages, because the linear regulator that is put in series will perform the fine regulation, therefore, the voltage dividers can work in open loop, while the linear regulator should filter all the switching noise that comes from the voltage dividers.

The only part of the system that uses high switching frequency is the analog multiplexer. If the sent reference is a 2 MHz sine wave, the MOSFET's switching frequency inside the multiplexer will be 2 MHz as well. Therefore, it can be said that even in the cases when the reference is a high frequency signal, the maximum switching frequency in the system is relatively low.

III. DESIGNED SYSTEM

In order to prove the concept a prototype of envelope amplifier has been made. The specifications of the envelope amplifier prototype are as follows:

- The input voltage is 24 V

- The output voltage can be changed from 0 V to 24 V

- The maximum power of the prototype is 50 W

- The maximum frequency of the reference signal is 2 MHz

As it is aforementioned, the voltage dividers will work in open loop, and, in order to guarantee stable voltages at their outputs, at each output there are two ceramic capacitors in parallel (each one of 22uF). The power losses in the voltage divider can be divider in two parts. The first part represents the power losses due to parasitic capacitance between the gate and source of the used MOSFETs, while the second part of the losses is due to finite resistance of the used switches and voltage difference between the flying capacitor and output capacitors. The gate losses are directly proportional to the switching frequency of the voltage divider. In order to conclude how the power losses due to the second mechanism depend on the switching frequency, it is necessary to conduct the analysis presented in [12]. It can be shown that theses losses can be modeled with output resistance of the voltage divider that is equal to:

$$R_{out} = \frac{1}{C_F f_{SW}} \qquad (3)$$

where C_F is the value of the "flying" capacitor and f_{SW} is the switching frequency of the converter. Therefore, for low power losses it is necessary to apply low switching frequency

and use flying capacitor as big as possible (in this design there are five 22uF ceramic capacitors in parallel). The switching frequency for both dividers is 50 kHz. After the voltage dividers have been implemented, the efficiency of 12 V and 18 V outputs has been measured. The result of the measurement is shown in Figure 6. It can be seen that the efficiency of both voltage dividers is higher than 95% in very wide range of output power.

As it is explained in [11], the drivers that are used in order to govern the employed MOSFETS are supplied by the voltage that is equal to the half of the voltage divider's input voltage. The input terminals of one driver are connected between the ground and the output terminal of the voltage divider, while the input terminals of the second driver are connected between the input and output voltage. Having this in mind, in the first voltage divider the used driver has to be able to support ½ V_{MAX} as its supply voltage and in the second divider the maximum voltage supply is ¼ V_{MAX}. In the first voltage divider the used drivers are IR2181 and in the second LM27222. The MOSFETs are the same in both voltage dividers and they are Si4864. The analog multiplexer is implemented using MOSFETs and diodes like in Figure 7.

The diodes are necessary in order to prevent current flow between the voltage sources due to parasitic diodes of the MOSFETs. As it can be seen, the MOSFETs from the analog multiplexer it have "floating" source and it is necessary to use additional power sources and isolation chips in order to control them, Figure 8. The isolation chips are ISO721 and the drivers are EL7156.

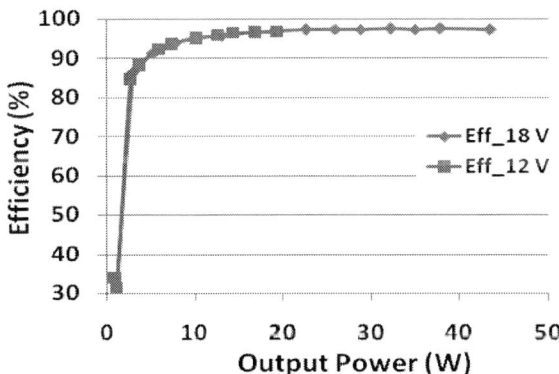

Figure 6. Measured efficiency of the implemented multilevel converter

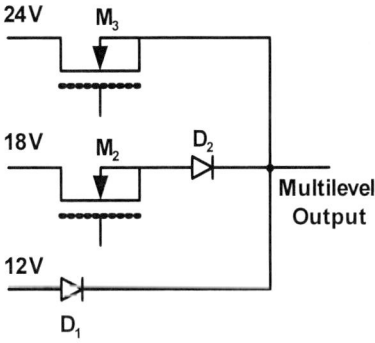

Figure 7. Implemented analog multiplexer

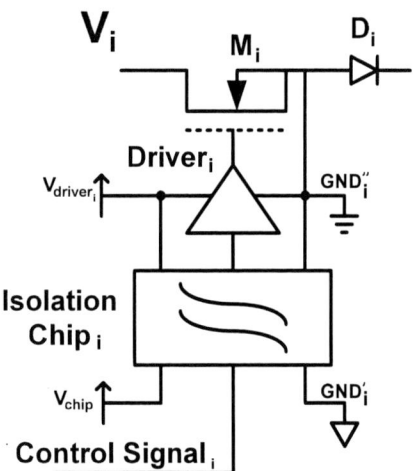

Figure 8. Simplified schematic of the circuit that is used to govern the MOSFETs in the implemented analog multiplexer

The linear regulator that is used as the last stage of the envelope amplifier should have high bandwidth and the components are selected in order to accomplish this request. The MOSFET that is used as a pass element for the linear regulator (BLF 177) is from HF/VHF power MOS family of transistors. The operational amplifier is LM6172 and it is selected because of its high bandwidth. The systems schematic is shown in Figure 9.

The triggering logic is implemented in a FPGA that is used as a source of the digital signal reference. The digitalized signal reference is sent to a D/A converter and from there to the linear regulator. The same reference signal is sent to the triggering logic and to the linear regulator, but it is of the most importance that the reference that the linear regulator receives is synchronized with the output voltage of the multilevel converter. Only when these two voltages are synchronized, the system's output voltage will be always lower than the output voltage of the multilevel converter. Therefore, a digital delay filter is implemented in the FPGA as well, in order to compensate the delays in the system and synchronize the

multilevel output voltage with linear regulator's reference. The load of the system is a 10 Ω resistance.

Figure 10 shows a photograph of the implemented envelope amplifier.

IV. EXPERIMENTAL RESULTS

In order to characterize the envelope amplifier, it has been tested with different sine waves. In table 1, the efficiency of the implemented envelope amplifier is shown depending on the frequency of the reproduced sine wave and its dc offset and amplitude. In the same table, the measured efficiency is compared with the efficiency of the ideal linear regulator and with the efficiency of the envelope amplifier that is implemented with independent voltage cells [10]. As it can be seen, the efficiency of the hybrid solution has up to 20% better efficiency than an ideal linear amplifier when a transmitted signal has low average values, and that is mostly the case when the EER technique is applied. It is important to note that the envelope amplifier based on the switching capacities has better efficiency than the envelope amplifier based on independent voltage cells [10]. Figures 11 and 12 show the multilevel and system's output voltage in the case when a reference is a sine wave of 500 kHz and 2 MHz respectively.

TABLE I. MEASURED EFFICIENCY OF THE IMPLEMENTED ENVELOPE AMPLIFIER FOR DIFFERENT SINE WAVES COMPARED WITH THE THEORETICAL EFFICIENCY OF AN IDEAL LINEAR REGULATOR SUPPLIED BY 23 V

V_{sin}(V)	Sine wave frequency (MHz)	Measure efficiency of the envelope amplifier based on switching capacities	Measure efficiency of the envelope amplifier based on the independent voltage cells	Theoretical efficiency of an ideal linear regulator supplied by 23V
0-9	0.5	47.5%	43.3%	29.3%
5-14	0.5	61.5%	58.9%	45.9%
0-22.5	0.5	75.2%	69.7%	73.0%
0-9	2	48.2%	43.7%	28.1%
5-14	2	59.0%	58.8%	45.9%
0-22.5	2	70.9%	68.3%	73.0%

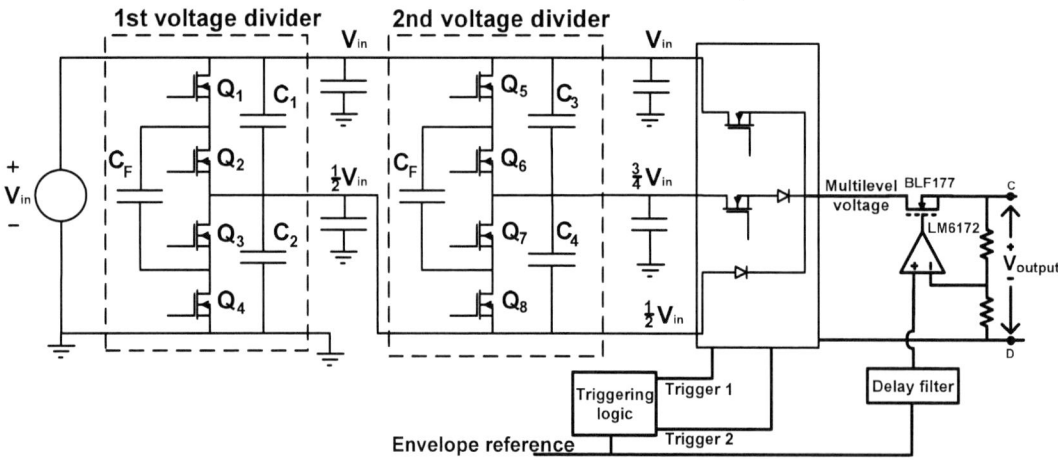

Figure 9. Simplified schematic of the implemented envelope amplifier

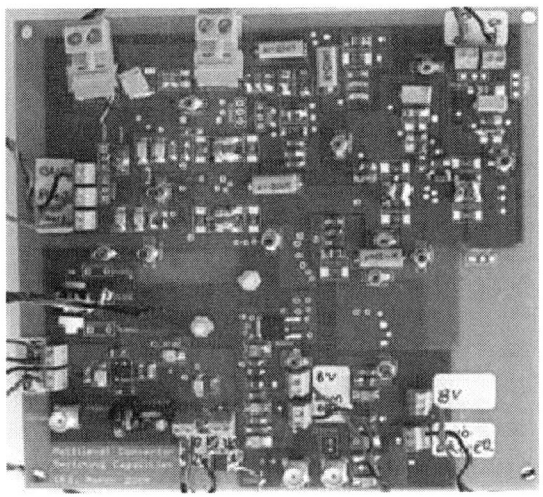

Figure 10. Photograph of the implemented envelope amplifier

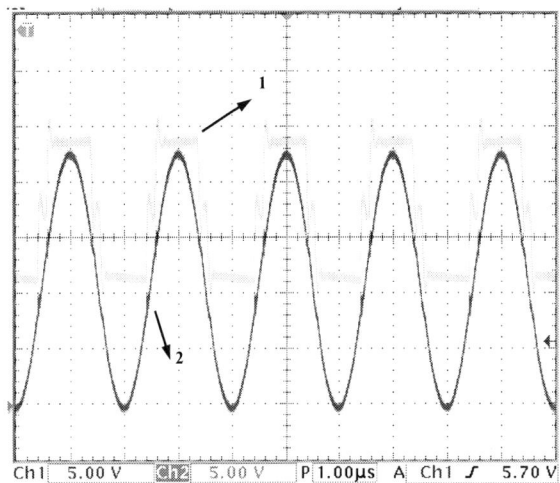

Figure 11. Multilevel output voltage (label 1) and envelope amplifier's output voltage (label 2) in the case of a 500 kHz sine wave

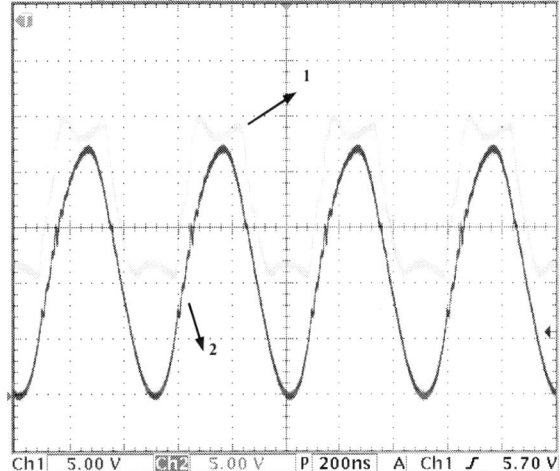

Figure 12. Multilevel output voltage (label 1) and envelope amplifier's output voltage (label 2) in the case of a 2 MHz sine wave

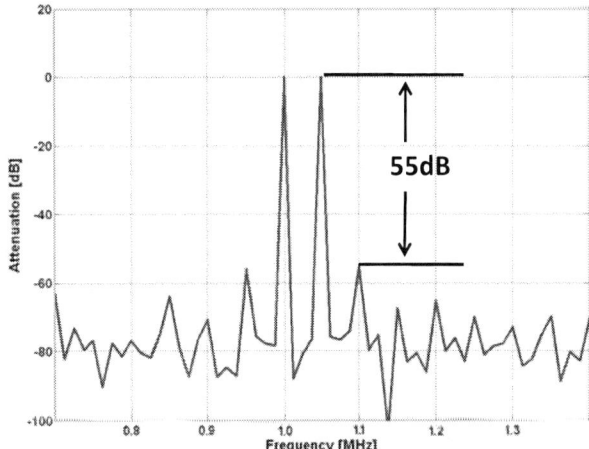

Figure 13. Attenuation of the intermodulation harmonics in the case when a reference signal is a two tone signal composed of 1 MHz and 1.05 MHz sine waves. The measurement is processed in MATLAB

The linearity and the bandwidth of the envelope amplifier are crucial in order to obtain high linearity of the power amplifier based on EER technique, therefore, these measurements have been performed as well. The linearity measurements are conducted by two tone tests, where two sine waves of the same amplitude are used as a reference signal and at the output of the envelope amplifier the ratio between the reproduced amplitudes and the intermodulation harmonics that are produced by the envelope amplifier is observed. Figure 13 shows spectral content at the output of the implemented envelope amplifier in the case when a two tone signal composed of 1 MHz and 1.05 MHz is used. It can be observed that the attenuation of the intremodulation components is higher than 50dB, what means high linearity of the envelope amplifier.

In [1] it has been explained that the bandwidth of the envelope amplifier has to be, at least, two times higher than the bandwidth of the RF signal. The reproduced envelope should not have any attenuation up to 2 MHz and it has been shown that the proposed envelope amplifier can reproduce 2 MHz sine wave of maximum amplitude. However, this does not mean that the implemented envelope amplifier cannot reproduce higher harmonics. The higher harmonics that are very important for high linearity of Kahn's transmitter usually are of much smaller amplitudes than the maximum amplitude that can be reproduced by the envelope amplifier. Based on the analysis presented in [1], a test with rectified sine wave has been conducted. If the reference signal is a rectified sine wave of frequency f, its spectrum is infinite and consists of tones that are placed on frequencies $2f, 4f, 6f...$ A rectified 500 kHz sine wave of maximum amplitude has been used as the reference and the response of the envelope amplifier has been measured, Figure 14. The spectrum of the output signal is compared with the spectrum of the reference signal, Figure 15. It can be seen that the proposed envelope amplifier admits even the harmonic higher than 2 MHz.

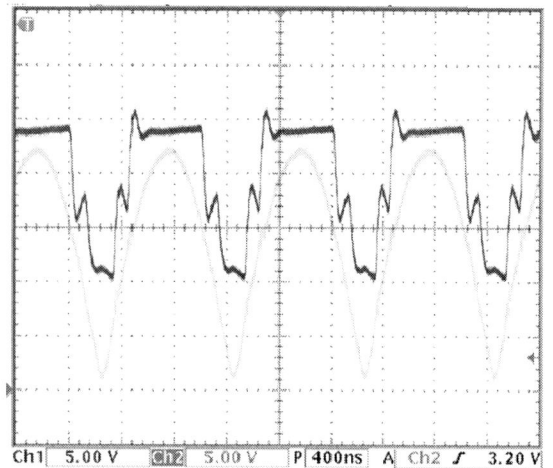

Figure 14. Waveform of multilevel's output voltage and linear regulator's output voltage when a rectified 500 kHz sine wave is used as the reference

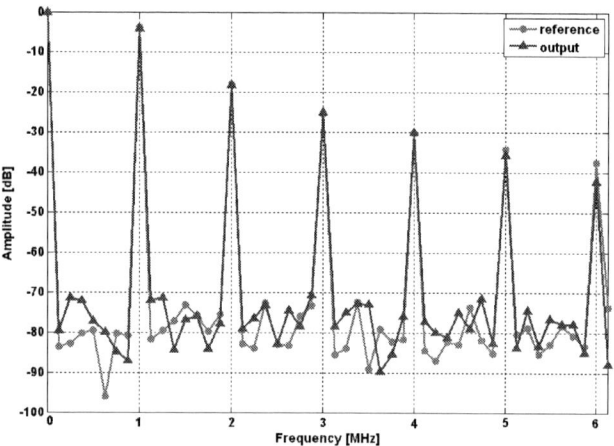

Figure 15. Spectrum of the reference and output signal when a rectified 500 kHz sine wave is used as the reference. All the values are scaled to the dc value of the signal

V. CONCLUSIONS

In this paper a solution for envelope amplifier in EER technique is presented. The solution consists of a multilevel converter in series with a linear regulator. The proposed multilevel converter is implemented by voltage dividers based on switching capacities. The designed envelope amplifier can provide up to 50 W of instantaneous power and reproduce a sine wave up to 2 MHz. The switching frequency of voltage dividers is low (50 kHz) comparing it to the bandwidth of the envelope amplifier. Therefore the proposed solution uses lower switching frequency than a conventional dc-dc converter for the same given bandwidth.

The multilevel voltage levels are selected in order to maximize the efficiency of the linear regulator for the signals with high PAPR. The efficiency of the envelope amplifier has been measured for different sine waves and its efficiency is up to 20% higher than in the case of an ideal linear regulator when sine waves have low average value (what is usually the case in the RF systems). The linearity of the prototype is measured as well and the attenuation of the intermodulation

harmonics is higher than 50dB. The implemented prototype does not have any magnetic component and it can be integrated.

VI. REFERENCES

[1] F.H. Raab, "Intermodulation Distortion in Kahn-Technique Transmitters", IEEE Transactions on Microwave Theory and Techniques, Volume 44, Issue 12, Part 1, Decemeber 1996, Pages: 2273-2278

[2] F.H. Raab, P. Asbeck, S. Cripps, P.B. Kenington, Z.B. Popovic, N. Pothecary, J.F. Sevic, N.O. Sokal, "Power amplifiers and transmitters for RF and microwave," IEEE Trans. on Microwave Theory and Techniques, Volume: 50, Issue: 3, March 2002, Pages: 814-826.

[3] P. Midya, K. Haddad, L. Connell, S. Bergstedt, B. Roeckner, "Tracking power converter for supply modulation of RF power amplifiers," IEEE Power Electronics Specialists Conference, PESC. 2001, Vol. 3, Pages:1540 –1545

[4] J. Staudinger, B. Gilsdorf, D. Newman, G. Norris, G. Sadowniczak, R. Sherman, T. Quach, "High efficiency CDMA RF power amplifier using dynamic envelope tracking technique," Microwave Symposium Digest., IEEE MTT-S International, Vol. 2, June 2000, Pages: 873-876

[5] M.C.W. Hoyerby, M.A.E. Andersen, "Envelope tracking power supply with fully controlled 4th order output filter", Power Electronics Conference, APEC '06, March. 2006

[6] A. Soto, J.A. Oliver, J.A. Cobos, J. Cezon, F. Arevalo, "Power supply for a radio transmitter with modulated supply voltage", Applied Power Electronics Conference, APEC '04, Volume: 1, Feb. 2004 Pages:392 – 398

[7] V. Yousefzadeh, E. Alarcon, D. Maksimović, "Three-level buck converter for envelope tracking in RF power amplifiers," IEEE Trans. on Power Electronics, Volume:21, Issue: 2, March 2006, Pages:549 – 552

[8] US Patent No. 6084468, Method and Apparatus for High Efficiency Wideband Power Amplification, July 2000.

[9] V. Yousefzadeh, E. Alarcon, D. Maksimović, "Efficiency optimization in linear assisted switching power converters for envelope tracking in RF power amplifiers", IEEE International Symposium on Circuits and Systems, ISCAS 2005, 23-26 May, pages:1302-1305 Vol. 2

[10] M.Vasić, O.Garcia, J.A.Oliver, P.Alou, D.Diaz, J.A.Cobos, "Multilevel Power Supply for High Efficiency RF Amplifier", Proc. of the 24th Annual IEEE Applied Power Electronics Conference, APEC '09, February 2009

[11] J.Sun, M.Xu, Y.Ying, F.C.Lee, "High Power Density, High , Efficiency System Two-stagePower Architecture for Laptop Computers", 37th IEEE Power Electronics Specialists Conference / June 18 - 22, 2006, Jeju, Korea

[12] S. Ben-Yaakov, "Switched Capacitors Converters", Professional Education Seminar at the 24th Annual IEEE Applied Power Electronics Conference, APEC '09, February 2009

978-1-4244-4782-4/10 $26.00 © 2010 IEEE

High Efficiency Power Amplifier for High Frequency Radio Transmitters

M.Vasić, O. García, J.A. Oliver, P. Alou, D. Diaz, J.A. Cobos
Centro de Electrónica Industrial (CEI)
Universidad Politécnica de Madrid
Madrid, Spain
miroslav.vasic@upm.es

A.Gimeno, J.M.Pardo, C.Benavente, F.J.Ortega
Radio Engineering Group (GIRA)
Universidad Politécnica de Madrid
Madrid, Spain

Abstract— **Modern transmitters usually have to amplify and transmit complex communication signals with simultaneous envelope and phase modulation. Due to this property of the transmitted signal, linear power amplifiers (class A, B or AB) are usually employed as a solution for the power amplifier stage. These amplifiers have high linearity, but suffer from low efficiency when the transmitted signal has high peak-to-average power ratio. The Kahn envelope elimination and restoration (EER) technique is used to enhance efficiency of RF transmitters, by combining highly efficient, nonlinear RF amplifier (class D or E) with a highly efficient envelope amplifier in order to obtain linear and highly efficient RF amplifier. This paper presents solutions for the power supply that acts as the envelope amplifier and class E amplifier that is used as a non-linear amplifier. The envelope amplifier is implemented as a multilevel converter in series with a linear regulator and can provide up to 100 W of peak power and reproduce sine wave of 2 MHz, while the implemented class E amplifier operates at 120 MHz with an efficiency near to 90%. The envelope amplifier and class E amplifier have been integrated in order to implement the Kahn's technique transmitter and series of experiments have been conducted in order to characterize the implemented transmitter.**

I. INTRODUCTION

In the modern radio communication systems high efficiency and high linearity are the essential requirements for the employed power amplifiers. High linearity is needed due to complex modulation techniques where phase and amplitude modulations are applied simultaneously in order to transmit as much data as possible for the given bandwidth, while the high efficiency improves thermal management, reliability and cost. Due to the need of high linearity, linear power amplifiers (classes A, B or AB) are usually used, but this leads to low average efficiency of the power amplifier, especially when signals with high Peak to Average Power Ratio (PAPR) are transmitted [1]. In [2] is explained that class A and class B amplifiers have efficiency of 5% and 28% respectively if a signal with a Rayleigh's distribution is transmitted.

One of the techniques that offer high efficiency and high linearity is the Kahn's technique or Envelope Elimination and Restoration (EER) technique [3]. This method proposes linearization of highly efficient, but nonlinear power amplifier (class E or D) by modulation of its supply voltage. The modulation of the supply voltage is done through an envelope amplifier according to the reference signal that is proportional to the envelope of the transmitted signal, while the phase modulation of the transmitted signal is conducted through the nonlinear amplifier. The basis for EER is the equivalence of any narrowband signal to simultaneous amplitude (envelope) and phase modulation:

$$V_{RF}(t) = I(t)\cos(2\pi ft) - Q(t)\sin(2\pi ft) =$$
$$= A(t)\cos(2\pi ft + \theta(t)) \quad (1)$$

$$\theta(t) = arctg\left(\frac{Q(t)}{I(t)}\right) \quad A(t) = \sqrt{I(t)^2 + Q(t)^2} \quad (2)$$

where f is the carrier frequency, $Q(t)$ and $I(t)$ are modulated signals. The block scheme of an EER system is shown in Figure 1.

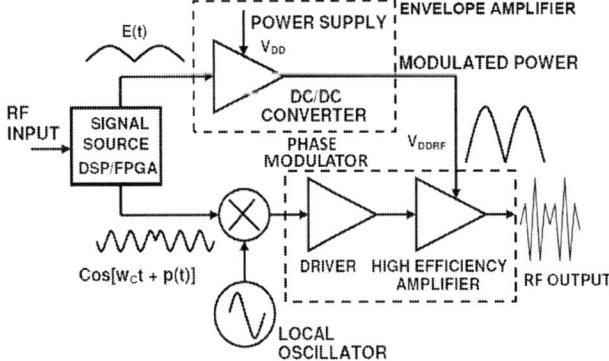

Figure 1. Block Scheme of Kahn Technique Transmitter

Thanks to this technique, the efficiency of the transmitter is almost constant for wide load range and it does not depend heavily on the level of the transmitted signal, as in the case of linear amplifiers [4]. Average efficiency three to five times those of linear amplifiers have been demonstrated from HF to L band [5,6]. In [7] a prototype of Kahn's transmitter for HF Band is presented. Its output power was about 50 W and its overall efficiency was up to 15% better than the efficiency of the conventional linear power amplifier.

In this paper solutions for the power supply that acts as the envelope amplifier and class E amplifier that is used as a non-linear amplifier are presented. The envelope amplifier is based on a multilevel converter in series with a linear regulator and its bandwidth is in range of 2 MHz, while the class E amplifier operates at 120 MHz. The instantaneous output power provided by the class E power amplifier is in range of 90 W. The proposed solutions for the envelope amplifier and class E amplifier are integrated into an EER transmitter and series of tests have been conducted in order to characterize the transmitter

II. ENVELOPER AMPLIFIER

Due to high bandwidth requirements, conventional solutions for tracking power supplies [8-10] are not energy efficient. The envelope amplifier that will be used consists of a multilevel converter and a linear regulator in series, [11]. Its block schematic is shown in Figure 2. The multilevel converter has to supply the linear regulator and it has to provide discrete voltage levels that are as close as possible to the output voltage of the envelope amplifier. If this is fulfilled, the power losses on the linear regulator will be minimal, because they are directly proportional to the difference of its input and output voltage. However, in order to guarantee correct work of the linear regulator, the output voltage of the multilevel converter has to be higher than the output voltage of the linear regulator. Time diagrams of the multilevel and linear regulator voltage are shown in Figure 3

The efficiency of the linear regulator depends on the number of the voltage levels that are used to supply it and on their distribution as well. Usually, the transmitted signal has high Peak to Average Power Ratio (PAPR), and it can be described by its probability density function [4]. Using the information about the signals probability the efficiency of the linear regulator can be presented as:

$$\eta = \int_0^{V_{max}} \frac{a}{V_{in}(a)} p(a) da \qquad (3)$$

where a is the voltage level of the generated envelope (the output voltage of the linear regulator), V_{max} is the maximum value of the signal's envelope, $V_{in}(a)$ is the linear regulator's input voltage generated by the multilevel converter that depends on the value of the linear regulator's output voltage and $p(a)$ is the probability density function of the envelope that is generated by the power supply. By optimizing the voltage levels of the multilevel converter in order to maximize average efficiency of the linear regulator, it is possible to enhance its average efficiency up to 6% comparing it with the solution that employ equidistant voltage levels [11].

In [11] several solutions for the multilevel converter are proposed. The schematic of the envelope amplifier that will be integrated with class E amplifier can be seen in Figure 4. There are three stages that can be distinguished in the proposed solution:

- single input multiple output converter (a flyback converter in our case)

- multilevel converter based on two-level independent voltage cells

- high slew rate linear regulator

The task of the flyback converter is to provide stable voltages that will supply two-level voltage cells. The bandwidth of this stage does not have to be high; therefore, the switching frequency of the multiple-outputs flyback can be very low in order to increase its efficiency In the case of the prototype in this paper, flyback's switching frequency was only 50 kHz.

The switching frequency of the MOSFETs that are inside the two-level cells will depend on the signal that is reproduced. If the signal is a sine wave of 1 MHz, the switching frequency of the multilevel converter is 1MHz as well. Therefore, with this topology high frequency signals can be reproduced by applying the switching frequency equal to the frequency of the signal, instead of several times higher, like in the case of PWM converters.

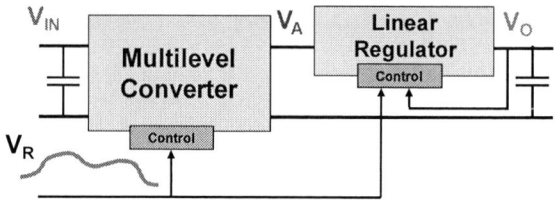

Figure 2. Simplified schematic of the proposed envelope amplifier

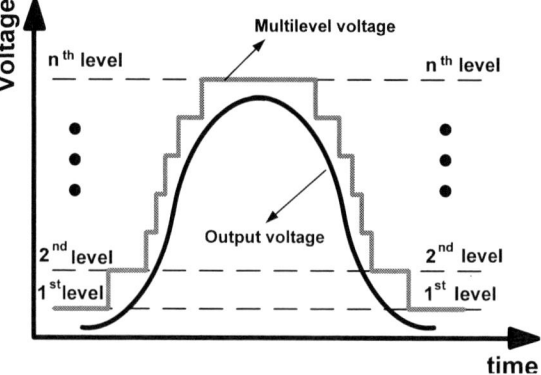

Figure 3. Time diagrams of the proposed envelope amplifier

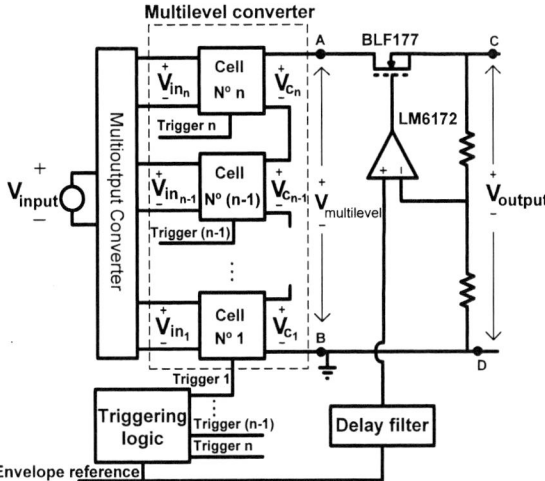

Figure 4. Simplified schematic of the implemented envelope amplifier

An envelope amplifier with the following specifications has been built:

- Variable output voltage from 0 V to 23 V

- The maximum instantaneous power is 50 W

- The maximum frequency of the reference signal is 2 MHz

- The multilevel converter is made with three optimized voltage levels (12 V, 18 V and 24 V)

The output voltages of the multilevel converter are selected by maximizing equation 3, while the MOSFET for the pass element and operational amplifier in the linear regulator have been selected in the way to obtain high bandwidth of the envelope amplifier. The bandwidth of the selected operational amplifier (LM6172) is 100 MHz in open loop, while the MOSFET (BLF177) is from HF/VHF power MOS family of transistors. The output voltages of the envelope amplifier and the multilevel converter in the case when a 2 MHz sine wave is reproduced are shown in Figure 5.

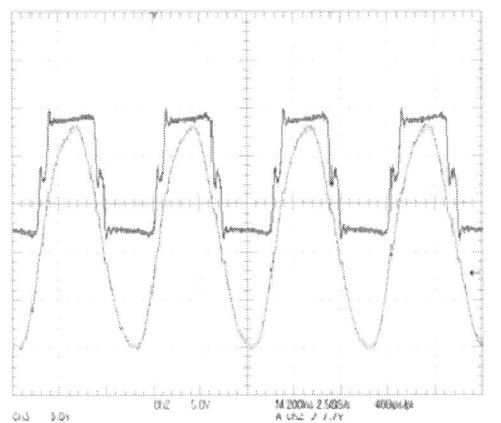

Figure 5. Output voltage of the multilevel converter and envelope amplifier when a 2MHz sine wave is reproduced

TABLE I. MEASURED EFFICIENCY OF THE IMPLEMENTED ENVELOPE AMPLIFIER FOR DIFFERENT SINE WAVES COMPARED WITH THE THEORETICAL EFFICIENCY OF AN IDEAL LINEAR REGULATOR SUPPLIED BY 23 V

Vsin(V)	Sine wave frequency (MHz)	Measured efficiency of the envelope amplifier based on the independent switching cells	Theoretical efficiency of an ideal linear regulator supplied by 23V
0-9	2	44.1%	29.3%
5-14	2	56.8%	45.9%
0-22.5	2	70.2%	73.4%
0-9	0.5	43.6%	29.3%
5-14	0.5	59.5%	45.9%
0-22.5	0.5	71.2%	73.4%

The efficiency of the prototype is measured for different sine waves and the results are summarized in Table 1. The efficiency is shown depending on the frequency of the reproduced sine wave and its DC offset and amplitude. The measured efficiency is compared with the efficiency of an ideal linear regulator and, as it can be seen, the efficiency of the hybrid solution is almost 50% higher than the efficiency of an ideal linear amplifier when a transmitted signal has low average values, and that is mostly the case when the EER technique is applied.

III. CLASS E AMPLIFIER

The class E amplifier that is used to amplify the constant envelope, phase modulated, component of the signal, operates at the VHF band and exhibit wide fractional bandwidth (from 95 MHz to 120 MHz). The drain to source voltage of the implemented class E amplifier can be seen in Figure 6 at 120 MHz showing the amplifier is operating near nominal Class-E conditions.

The design of the amplifier and its load network has been optimized to reduce power losses to a minimum. The results of efficiency measurements are shown in Figure 7. When the class E amplifier is supplied with constant voltage of 24 V, its output power exhibits a peak value of 90 W operating between 100MHz and 110MHz. The drain efficiency of the amplifier in that frequency range is around 92% (A Bird 5000EX wattmeter has been used to measure output power, the accuracy of this instrument is 5%).

The driver of the amplifier has been specially designed to reduce driving power as much as possible. It is based on using an additional Class-E amplifier at the input of the main amplifier. This driver uses the drain voltage of its RF power MOSFET to charge and discharge the gate of the main power MOSFET. Implemented in this way, the driver shows significant improvement in efficiency comparing it to the conventional driving solutions using sine waves to drive RF power MOSFET into switching conditions.

The RF power MOSFET used with this amplifier is a MRF6V2300N from Freescale Semiconductors. Input and output ports have been modeled for switching operation using the model proposed in [12].

This amplifier has been designed based upon the load impedance shynthesis design technique as shown in [13] and simulated and optimized using Advanced Design System (ADS) software form Agilent.

IV. INTEGRATION OF THE ENVELOPE AMPLIFIER WITH CLASS E AMPLIFIER

In order to obtain high linearity of an EER transmitter, it is necessary that the envelope injection produced by the envelope amplifier is synchronized with phase modulated signal component amplified by the E amplifier and that the envelope amplifier is highly linear. For the sake of simplicity, the linearity tests of the envelope amplifier are conducted by two-tone tests, where two sine waves of the same amplitude are used as a test signal and at the output of the envelope amplifier the ratio between the desired components and the intermodulation products that are produced by the envelope amplifier is observed. Figure 8 presents the output voltage of the multilevel converter and the envelope amplifier's output voltage during a two tone test. Measurements of the attenuation of the intermodulation products show that the attenuation is higher than 50 dB, which means that the system exhibits high linearity Figure 9.

In [1] it has been explained that the bandwidth of the envelope amplifier has to be, at least, two times higher than the bandwidth of the RF signal. The reproduced envelope should not have any attenuation up to 2 MHz and it has been shown that the proposed envelope amplifier can reproduce 2 MHz sine wave of maximum amplitude. However, this does not mean that the implemented envelope amplifier cannot reproduce signals of wider bandwidth. The higher harmonics that are very important for high linearity of Kahn's transmitter usually are of much smaller amplitudes than the maximum amplitude that can be reproduced by the envelope amplifier. Based on the analysis presented in [1], a test with rectified sine wave has been conducted. If the reference signal is a rectified sine wave of frequency f, its spectrum is infinite and consists of tones that are placed on frequencies *2f, 4f, 6f...* A rectified 500 kHz sine wave of maximum amplitude has been used as the reference and the response of the envelope amplifier has been measured, Figure 10. The spectrum of the output signal is compared with the spectrum of the reference signal, Figure 11. It can be seen that the proposed envelope amplifier admits even the harmonic higher than 2 MHz.

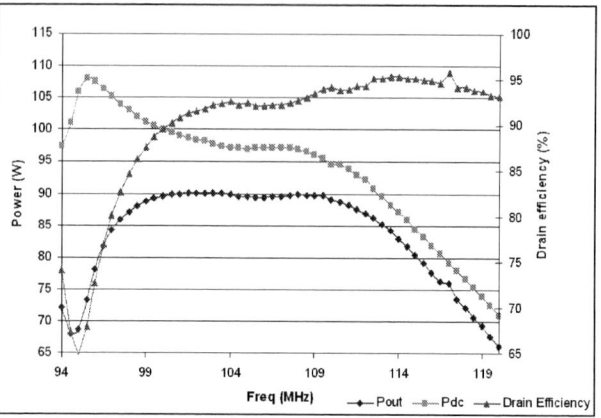

Figure 7. Measured output power and the efficiency of the implemented class E for different switching frequencies (instrument accuracy: 5%).

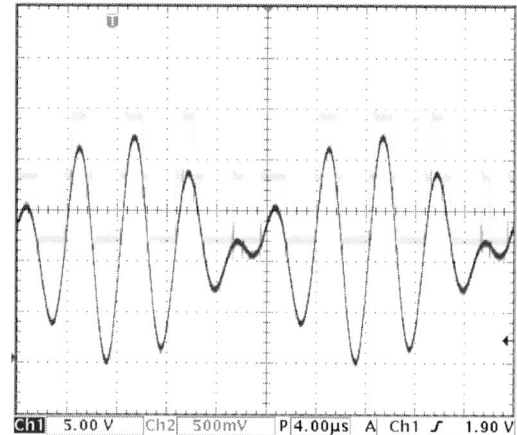

Figure 8. Output voltage of the multilevel converter and envelope amplifier when a two tone test is performed. Two tone signal consists of two sine waves of 200 kHz and 250kHz

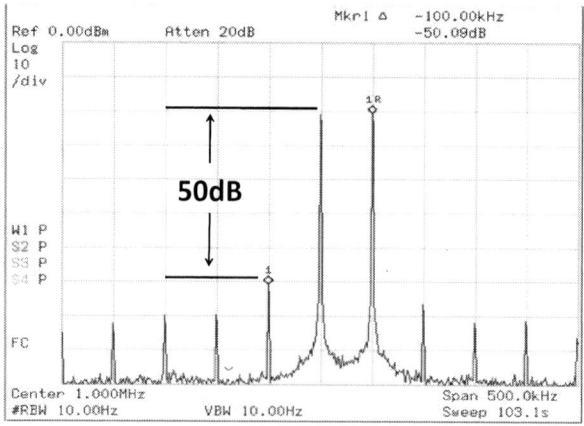

Figure 9. Attenuation of the intermodulaiton components measured by a spectral analyzer in the two tone test, when a sine waves of 1 MHz and 1.05 MHz are used.

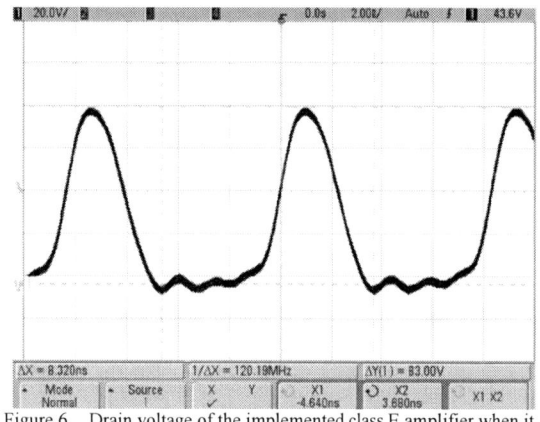

Figure 6. Drain voltage of the implemented class E amplifier when it operates at 120MHz

978-1-4244-4782-4/10 $26.00 © 2010 IEEE

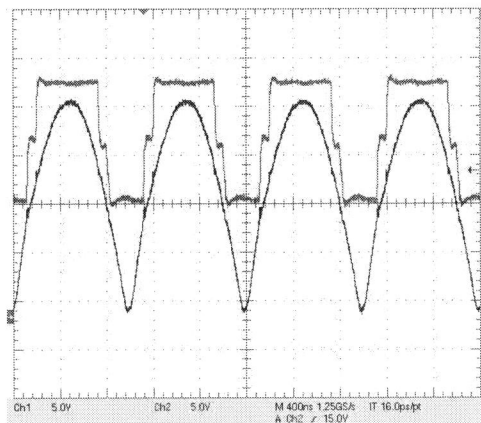

Figure 10. Waveform of multilevel's output voltage and linear regulator's output voltage when a rectified 500 kHz sine wave is used as the reference

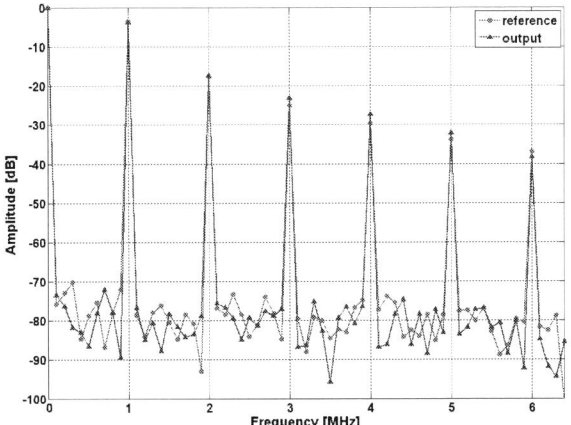

Figure 11. Spectrum of the reference and output signal when a rectified 500kHz sine wave is used as the reference. All the values are scaled to the dc value of the signal

One of the problems that can occur is that the high frequency signal amplified by the class E amplifier goes back towards the envelope amplifier. Due to this problem, the envelope amplifier supplies the class E amplifier through simple LC filter, which has been designed in the way that there is not attenuation up to 2 MHz (i.e. up to the desired bandwidth of the envelope amplifier). In the case of our design, the values of the inductance and capacitance are 200 nH and 10 nF respectively.

Besides the envelope amplifier and class E amplifier, the implemented transmitter has a part that is used to receive reference signal and than to generate needed phase and envelope references. This part consists of a FPGA Virtex 4 development board fitted with D/A and A/D converters. In this way, the transmitter can receive the RF reference signal through A/D converters, and the envelope and phase references are extracted inside a FPGA. The digitalized references are converted to analog signals using high speed D/A converters, and sent to envelope and class E amplifiers. Additionally, two delay filters and the envelope's amplifier triggering logic are implemented in the employed FPGA. The

first delay filter is used in order to synchronize the envelope reference signal sent to the linear regulator with the output voltage of the multilevel converter and its aim is to avoid the distortion of the reproduce envelope. In [1] it has been shown that the differential delay between the produced envelope and phase modulation should not be higher than one tenth of the bandwidth of the RF signal. The second delay filter is used to adjust the differential delay in order to obtain high overall linearity of the system. The triggering logic needed by the multilevel converter is a simple set of comparators that has to regulate the state of the voltage cells.

Figure 12 shows the implemented EER transmitter.

V. EXPERIMENTAL RESULTS

The first tests with the implemented EER transmitter have been conducted in open loop in order to characterize it. The simplest test is the one when only the amplitude modulation is performed. Figure 13 shows the relevant time diagrams when the envelope reference is 500 kHz sine wave. It can be seen the envelope reference is noisy due to the RF signal that is present in the class E amplifier. However, this noise does not have big influence on the envelope amplifier because its frequency is much higher than its bandwidth and, in that way, it is filtered. Another important fact is that every time when there is a change of the multilevel's output voltage, there is a small glitch in the output voltage of the envelope amplifier. These glitches are the consequence of the finite bandwidth of the implemented linear regulator. The higher the current of the envelope amplifier is, the bigger are the glitches. Fortunately, thanks to the LC filter that is placed between the envelope amplifier and the class E amplifier, these glitches are negligible after the envelope amplifier's output voltage is filtered.

Figure 14 shows the relevant waveforms when a 2 MHz sine wave is used as the envelope reference. It can be observed that the glitches in supply voltage of the class E converter are negligible. The supply voltage of the class E converter is delayed compared to the envelope amplifier's output voltage due to the employed LC filter.

Figure 12. Photogtaph of the implemented EER Transmitter (on the right) with FPGA board, A/D and D/A converters (on the left)

The transmitter's linearity when only the amplitude modulation is performed is measured as well. Figure 15 shows important oscilloscope waveforms in the case when the envelope reference is composed of 100 kHz and 150 kHz sine waves. The measured attenuation of the third order harmonic was about 30 dB.

In order to measure the linearity of the complete EER transmitter a simple double-sideband suppressed-carrier (DSB-SC) transmission is performed, like in [1]. The carrier frequency is 125 MHz, while the modulated signal is 100 kHz sine wave. This corresponds to the reference that is composed of two tones, 124.9 MHz and 125.1 MHz respectively. The envelope reference is a rectified sine wave, while the phase changes for 180 degrees every time when the envelope reaches zero. Figure 16 shows the oscilloscope waveforms for this test. High frequency noise in the envelope reference can be clearly observed, and it can be seen that it is filtered by the linear regulator due to its finite bandwidth.

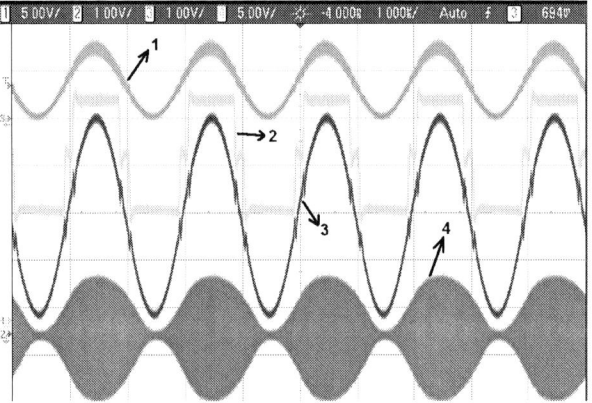

Figure 13. Oscilloscope waveforms of the envelope reference (label 1), multilevel's output voltage (label 2), envelope's amplifier output (label 3) and the EER transmitter's output (label 4), when the envelope reference is a 500 kHz sinewave

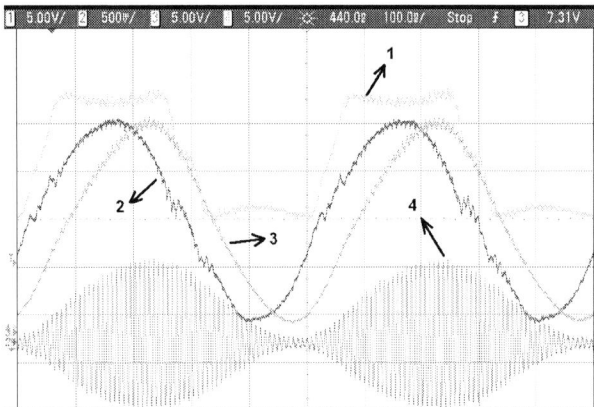

Figure 14. Oscilloscope waveforms of the multilevel's output (label 1), envelope amplifier's output (label 2), supply voltage of the class E amplifier (label 3) and the output of the implemented EER transmitter (label 4) when the envelope reference is a 2 MHz sine wave

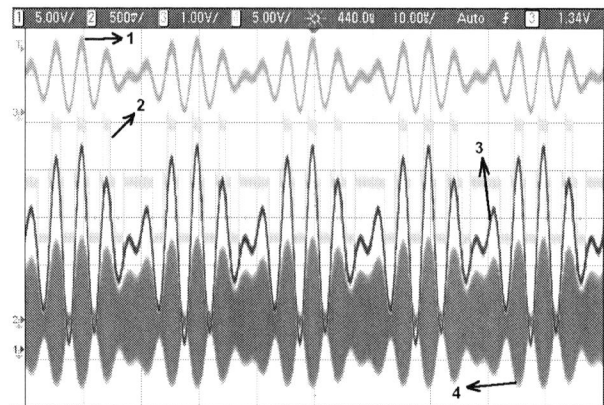

Figure 15. Oscilloscope waveforms of the envelope reference (label 1), multilevel's output voltage (label 2), envelope amplifier's output voltage (label 3) and the output of the implemented EER transmitter (label 4)when the reference signal is composed of two tones (200 kHZ and 250 kHz)

The rectified sine wave has frequency of 200 kHz and the spectrum of the output signal is shown in Figure 17. It can be seen that the attenuation of the third harmonic is just 18 dB. The reason for such a small attenuation, althought the envelope and phase modulation are synchronized is parasitic AM-PM modulation. The work in order to resolve this issue is in progress.

The last tests were conducted in order to measure the efficiency of the implemented EER transmitter. The efficiency is measured in the case when only amplitude modulation is performed, and the envelope reference is a sine wave. The sine wave reference has different amplitudes, average values and frequencies. One set of efficiency measurements have been perfromed in order to validate the proposed solution for the envelope amplifier with the efficiency of the system if the envolpe amplifier were a linear regulator with constant supply voltage. The transmitter has a 50 Ω load. All the results are summarized in Table 2.

It can be seen that the efficiency is almost constant when the propossed envelope amplifier is applied, and it goes from 36.4 to 43.8 percents. Comparing the EER transmitter when the proposed envelope amplifier and a linear regulator supplied by constant voltage are used, the advantage of the hybrid solution for the envelope amplifier is obvious, especially when the transmitted signal has low average value.

Having in mind the efficiencies that have been measured for envelope amplifier and the class E amplifier when they worked separately, the measured efficiency of the EER transmitter is lower than the expected one. There are two reasons for this. The first one is that the envelope amplifier has been designed and charaterized for the 10 Ω load. Nevertheless, the input impedance of the class E amplifier is significantly lower, just 4 Ω. Therefore, the envelope amplifier has to manage much higher currents, and the switching losses in the multilevel converter and the first stage (flyback converter) are higer. The second reason is the input filter of the class E amplifier. When the measurements of the class E amplifier were done, it was supplied by constant voltage, and at its input, there was a low-pass filter that is used in order to filter all the high harmonic from the input current. In order to supply the class E amplifier with variable voltage, this filter was changed, and this change lead to the detoriation of the efficiency.

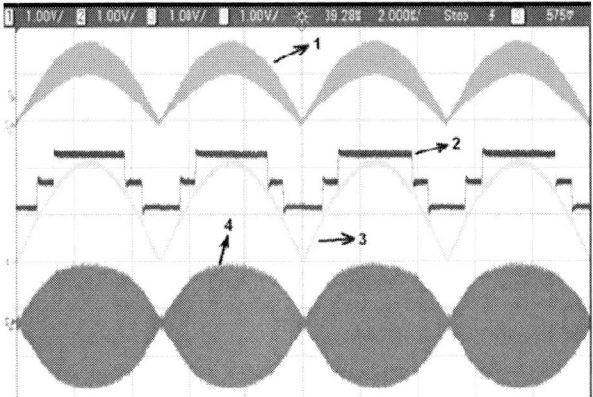

Figure 16. Oscilloscope waveforms of envelope reference (label 1), multilevel's output voltage (label 2), envelope amplifier's output (label 3) and the output of the implemented EER transmitter (label 4) when the transmitter's refrence is a simple DSB-SC signal with carrier of 125 MHz and a 100 kHz sine wave as the modualted signal

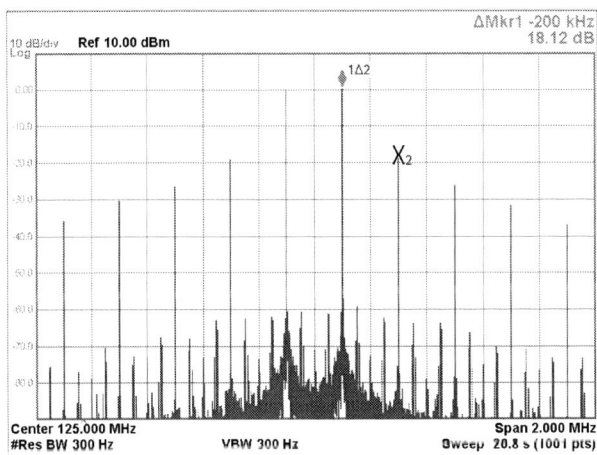

Figure 17. Spectrum of the transmitter's output signal when the system's reference is a simple DSB-SC signal with carrier of 125 MHz and a 100 kHz sine wave as the modualted signal

TABLE II.   MEASURED EFFICIENCY OF THE IMPLEMENTED EER TRANSMITTER WHEN THE AMPLITUDE MODULATION IS APPLIED AND THE LINEAR REGULATOR IS SUPPLIED BY CONSTANT AND VARIABLE VOLTAGE

Vsin(V)	Sine wave frequency (MHz)	Linear regulator's supply voltage	Output Power (dBm)	Measured efficiency
0-9	0.5	Variable	36.4	38.5%
5-14	0.5	Variable	24.3	41.5%
0-22.5	0.5	Variable	43.7	39.1%
0-9	2	Variable	36.4	38.9%
5-14	2	Variable	41.5	37.9%
0-22.5	2	Variable	43.8	35.1%
0-9	2	Constant	35.0	16.1%
5-14	2	Constant	40.3	26.9%
0-22.5	2	Constant	42.5	37.8%

VI. CONCLUSIONS

In this paper a radio transmitter based on the Kahn's technique is presented. The implemented transmitter is composed of envelope amplifier (responsible for envelope injection) and class-E amplifier (performs phase modulation). The envelope amplifier has to fulfill strict requirements regarding its bandwidth and linearity, and, therefore, it is implemented as a multilevel converter in series with a linear regulator. In this way, high frequency signals can be reproduced applying relatively low switching frequency. The levels of the multilevel converter have been optimized in order to increase the efficiency of the system when the signals with high PAPR are transmitted. It has been shown that the proposed solution for the envelope amplifier can reproduce a sine wave of 2 MHz, and provide up to 100 W of peak power. When the transmitted signal has low average value, the efficiency of the proposed envelope amplifier is almost 50% higher than the efficiency of an ideal linear regulator supplied by constant voltage. The linearity of the envelope amplifier has been measured as well and the attenuation of the intermodulation products is about 50 dB.

The class-E amplifier operates at VHF band and its efficiency is about 90%. In order to reduce the driving power, the driver for the class-E amplifier is implemented as another class-E amplifier.

Different test have been performed with the implemented EER transmitter in order to characterize it. It has been shown that the envelope can be modulated with a 2 MHz sine wave. The attenuation of the intermodulation products when only AM is applied is about 30 dB. The efficiency of the implemented transmitter has been measured for different sine wave envelopes and it goes from 35% to 41%.If the envelope amplifier were implemented as a linear regulator supplied by constant voltage, the efficiency of the EER transmitter would be 50% to 100% lower than in the case when the envelope amplifier is made as a hybrid solution. Due to parasitic AM-PM modulation, the linearity of the transmitter when AM and PM are performed is only 18 dB and the work towards enhancement of the transmitter's linearity has been in progress. By implementing this transmitter we have proved that the Kahn's technique is possible to be implemented in this frequency and power range. This approach opens the possibility to enhance the efficiency of RF amplifiers.

VII. REFERENCES

[1] F.H. Raab, "Intermodulation Distortion in Kahn-Technique Transmitters", IEEE Transactions on Microwave Theory and Techniques, Volume 44, Issue 12, Part 1, Decemeber 1996, Pages: 2273-2278

[2] F.H.Raab, P.Asbeck, S.Cripps, P.B.Kenington ,Z.B.Popovic, N.Pothecary, J.F.Sevic, N.O.Sokal ,"RF and Microwave Power Amplifier and Transmitter Technologies – Part 1", High Frequency Electronics, Vol. 2, No.3, Pages:22-36, May 2003

[3] Kahn, L.R. "Single-Sideband Transmission by Envelope Elimination and Restoration",Proceedings of the IRE, Vol. 40, No. 7, July 1952, pp. 803 - 806

[4] F.H.Raab; P. Asbeck, S. Cripps, P.B. Kenington, Z.B. Popovic,N. Pothecary, J.F. Sevic, N.O. Sokal, "Power amplifiers and transmitters for RF and microwave",IEEE Transactions on Microwave Theory and Techniques, Volume 50, Issue 3, March 2002 Page(s):814 – 826

[5] F. H. Raab and D. J. Rupp, "Highefficiency single-sideband HF/ VHF transmitter based upon envelope elimination and restoration," Proc. Sixth Int. Conf. HF Radio Systems and Techniques (HF '94) (IEE CP 392), York, UK, pp. 21-25, July 4 - 7, 1994.

[6] F. H. Raab, B. E. Sigmon, R. G. Myers, and R. M. Jackson, "L-band transmitter using Kahn EER technique," IEEE Trans. Microwave Theory Tech., pt. 2, vol. 46, no. 12, pp. 2220-2225, Dec. 1998.

[7] F.J.O. González, A.G. Martín, J.M.P. Martín, C.B. Peces, "Amplificador de Potencia de Alto Rendimiento para Transmisores EER" Seminario Anual de Automática Electrónica Industrial SAAEI 2008, Cartagena, September 2008

[8] J. Staudinger, B. Gilsdorf, D. Newman, G. Norris, G. Sadowniczak, R. Sherman, T. Quach, "High efficiency CDMA RF power amplifier using dynamic envelope tracking technique," Microwave Symposium Digest., IEEE MTT-S International, Vol. 2, June 2000, Pages: 873-876

[9] A. Soto, J.A. Oliver, J.A. Cobos, J. Cezon, F. Arevalo, "Power supply for a radio transmitter with modulated supply voltage", Applied Power Electronics Conference, APEC '04, Volume: 1, Feb. 2004 Pages:392 – 398

[10] V. Yousefzadeh, E. Alarcon, D. Maksimović, "Three-level buck converter for envelope tracking in RF power amplifiers," IEEE Trans. on Power Electronics, Volume:21, Issue: 2, March 2006, Pages:549 – 552

[11] M.Vasić, O.Garcia, J.A.Oliver, P.Alou, D.Diaz, J.A.Cobos, "Multilevel Power Supply for High Efficiency RF Amplifier", Proc. of the 24th Annual IEEE Applied Power Electronics Conference, APEC '09, February 2009

[12] N. O. Sokal, R. Redl, "Power Transistor Output Port Model", RF Design, Vol. 10, No. 6, pp. 45-48, 50, 51, 53, June 1987.

[13] Ortega-Gonzalez,F.J, "Load-Pull Wideband Class-E Amplifier", Microwave and Wireless Components Letters, Vol. 17, No. 3, pp. 235-237, March 2007

Applying One-Comparator Counter-Based Sampling to Current Sharing Control of Multi-Channel LED Strings

K. I. Hwu[1], *Member, IEEE*, and Y. T. Yau[1,2], *Student Member, IEEE*

[1]Department of Electrical Engineering, National Taipei University of Technology, Taiwan
[2]Industrial Technology Research Institute, Taiwan

Abstract—In this paper, a fully-digitalized current sharing strategy without any analog-to-digital converter (ADC) is presented and applied to light-emitting diode (LED) lighting. By doing so, the cost for digital control is reduced significantly. Above all, the proposed current sharing method has the capability of resisting variations in input voltage and component parameter. And this is verified by experiments and very suitable for LED street lighting.

I. INTRODUCTION

The Kyoto protocol to the united nations framework convention on climate charge has been held since January 2005, and this is getting more and more attractive in the world. In order to achieve low power dissipation and to reduce total carbon dioxide emissions, reduction of the power dissipation in lighting is an important policy in the advance industrialized country. Hence, the light-emitting diode (LED) street light becomes a pre-evaluated next-generation light. Such a light has many advantages, such as energy saving, environmental protection, long life, high reliability, etc., thereby causing the maintenance cost to be reduced greatly as compared to the traditional high pressure sodium lamp. The LED driver topology in [1] [2] is widely used. The main is first rectified and regulated to the low voltage, and then linear current source circuits are used to drive LED strings. Each LED string has its own linear current source. Consequently, there are two power losses in electric conversion. One is in the AC-DC converter, and the other is in the current source. And hence, one-stage power-factor-corrected (PFC) AD-DC converter with high-voltage output [3] is presented to directly drive LED strings, so as to avoid the energy loss in two-stage electric conversion.

On the other hand, the number of LED strings can be reduced by increasing the voltage across LED strings. For example, LED strings are directly driven by PFC AC-DC converters. This is practical in theory. However, the power for the commercialized LED street light locates about between 60W and 250W. As to the present technique for the LED, 1W (350mA) high-power LED has the efficiency up to 110Lm/W for mass products. That is to say, the required number of LEDs for only one LED string is about between 60 and 250. Therefore, LED strings connected in parallel are indispensable.

In order to obtain safety identification [4] and to increase reliability, the output voltage of the AC-DC converter is commonly designed to be smaller than 60V, generally between 24V and 48V. Hence, the number of LED strings is about

between 5 and 30, so the voltage across the LED string voltage is about between 20V and 40V.

For the traditional digital control of LED strings [5][6] to be considered, LED strings need individual analog-to-digital converters (ADCs) in individual current feedback loops. The more the number of LED strings is, the more the number of ADCs. For the digital integrated circuit (IC) to be considered, the digital logic gate is cheap. However, for the analog component process, it is not easy to reduce the cost via miniature, and hence the number of ADCs will take most of area and cost. In [7], digital control is presented based on micro control. As for feedback sampling, one comparator is used to compare the sensed LED current with the given current command, and the corresponding control method is the pulse number modulation (PNM). As generally recognized, the interleaved control is indispensable for the multi-channel LED with the multi-driver circuit, so as to reduce electromagnetic interference (EMI) and the output current ripple of the first-stage AC-DC converter and hence to upgrade the system reliability. However, although this circuit used in [7] is very simple, the switching frequency is not constant such that the interleaved control is not easy to realize. Besides, the capability of dealing with variations in input voltage and component parameter is not so good. In [8], the sliding control is used and has the same drawbacks as described in [7], and besides the buck converter in this case must be operated in the discontinuous conduction mode (DCM). In [9], the hysteresis control with two comparators is used. Although this case possesses robustness against variations in input voltage and component parameter, it can be used only in a single LED string and hence still has a problem in the phase shift control under multi-channel operation. To overcome this problem mentioned above, the LED driver, operating in the continuous conduction (CCM) without any ADC used, is proposed herein.

II. PROPOSED OVERALL SYSTEM CONFIGURATION

Fig. 1 shows the overall system configuration of the proposed FPGA-based high-power six-channel LED driver, which is built up by six buck converters. Buck converters have synchronous rectifiers which make these converters always operated in CCM and hence controlled easily, half-bridge driver ICs, inductance current feedback control loops, and pulse-width-modulated (PWM) generators. Each sensed output inductance current, belonging to a triangular wave, is sent to the corresponding differential amplifier with a DC gain equal

978-1-4244-4782-4/10 $26.00 © 2010 IEEE

to 7V/1A, and then the result is passed to the negative terminal of the corresponding comparator with its positive terminal connected to the current reference of 2.5V. Therefore, the information on each output inductance current can be obtained by the one-comparator counter-based sampling method.

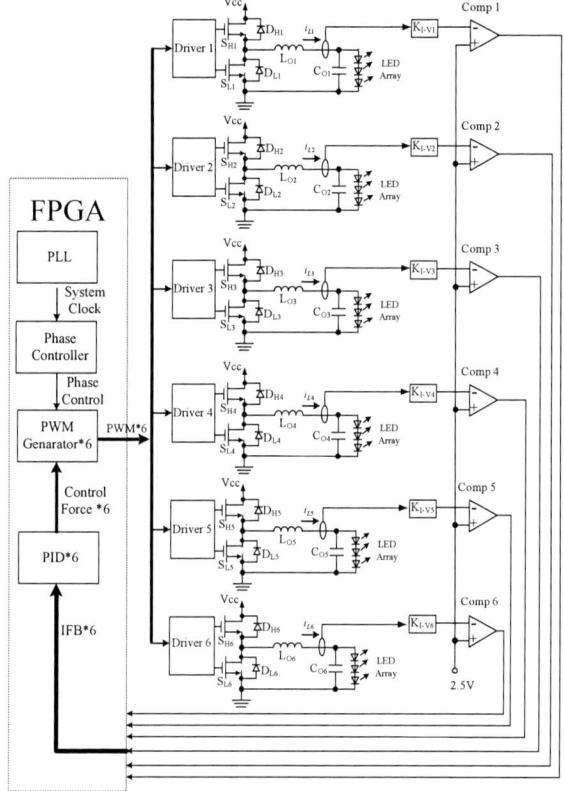

Fig. 1. Overall System Configuration.

Additionally, there are some specifications given as follows: (i) DC input voltage range is from 28V to 48V; (ii) switching frequency is 195kHz; (iii) DC current for each phase is 360mA; (iv) output capacitor for each phase is built up by one 10μF/50V X5R MLCC capacitor made by TDK Corporation; (v) product name of high and low side MOSFETs is IRF7380; (vi) product name of high and low side drivers is HIP2101; (vii) product name of current differential amplifiers is INA 169; (viii) product name of FPGA is Altera EP1C3T100-8 with the system clock of 200MHz provided by the phase lock loop and with 290 logic elements occupied.

III. BASIC PRINCIPLES OF ONE-COMPARATOR COUNTER-BASED SAMPLING AND CONTROL LOOP DESIGN

A. Basic Operating Principles

Prior to taking up this section, there are two counters created in FPGA whose system clock frequency is set to 200MHz, i.e., one clock period is 5ns. One is PWM_COUNT and the other is $COUNT$. The unit for the clock is CLK or LSB or none. And, there are three cases to describe how to get the information on

output inductance current based on 10-bit PWM control with a constant switching period T_s that corresponds to 1024CLK or 1024LSB or 1024, i.e., 5.12μs, counted by PWM_COUNT. That is to say, the corresponding switching frequency is 195kHz. Above all, for the convenience of analysis, it is assumed that the sensed output inductance current ripple, shown in Figs. 2 to 4, is triangular. The moment PWM_COUNT becomes zero, $COUNT$ is set to zero and the main switch S_H gets turned on. Besides, a comparator is utilized to determine the relationship between the sensed output inductance current i_{L2V} sent to the negative terminal of the comparator and the current reference I_{ref} sent to the positive terminal of the comparator. As S_H is turned on, i_{L2V} is increasing. As soon as i_{L2V} reaches I_{ref}, the comparator output signal IFB changes its status from the high level to the low level, i.e. $IFB = '0'$, thereby creating a negative-edged signal that is sent to FPGA. At this moment, $COUNT$ starts counting from zero. As soon as IFB changes its status from the low level to the high level, i.e., $IFB = '1'$, $COUNT$ stops counting and the resulting value of $COUNT$ is saved as REG, which is utilized to represent information on the output inductance current.

In the case of Fig. 2, the central point of the ripple of i_{L2V}, corresponding to the DC value of i_{L2V}, locates at I_{ref}. At this moment, the value of REG is set to 512CLK or 512, i.e., the corresponding elapsed time of $COUNT$ is 2.56μs, which is half of T_s. In the case of Fig. 3, the central point of the ripple of i_{L2V} is below the level of I_{ref}, thus causing the resulting value of REG to be smaller than 512CLK or 512LSB or 512, i.e., the corresponding error in the sensed output inductance current, 512 minus REG, is positive. In the case of Fig. 4, the central point of the ripple of i_{L2V} is beyond the level of i_{L2V}, thereby causing the resulting value of REG to be larger than 512CLK or 512LSB or 512, i.e. the corresponding error in the sensed output inductance current is negative. Thus, the larger the error in the sensed output inductance current is, the larger the control effort that determines the next duty cycle.

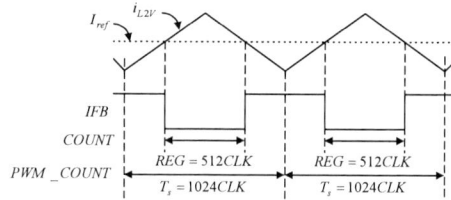

Fig. 2. Central point of the ripple of i_{L2V} equal to I_{ref}.

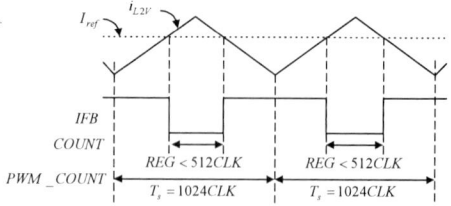

Fig. 3. Central point of the ripple of i_{L2V} lower than I_{ref}.

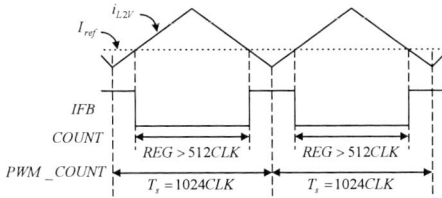

Fig. 4. Central point of the ripple of i_{L2V} higher than I_{ref}.

B. Control Loop Design

In Fig. 5, how to design the control loop for the buck converter is described. As described in Sec. III.A, the value of *REG* is obtained from the sensed output inductance current via one comparator and *COUNT*, and such a value is subtracted from 512 to get an output inductance current error $i_{o-error}$, which is sent to the proportional-integral (PI) controller to obtain control force i_f. After this, i_f is added with 512 and restricted to 940, corresponding to the maximum duty cycle of 91.8% to drive the main switch S_H.

Sequentially, how to determine controller parameters is described. All of the following experimental results to be measured have the same parameters of the PID controller, which are determined under the DC input voltage of 28V and double-checked under the DC input voltage of 48V. First, the value of the proportional gain k_p is tuned to be 0.5, so as to obtain the actual output inductance current equal to 75% of the rated DC output inductance current in the steady state. After this, the integral gain k_i is added to reduce the steady-state error as small as possible, and the value of k_i does not stop being tuned up until oscillation occurs and then reduces this resulting value to some extent to avoid oscillation, and eventually the value of k_i is obtained to be 0.125.

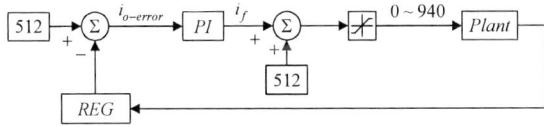

Fig. 5. Proposed control loop.

IV. EXPERIMENTAL RESULTS

Prior to taking up this section, there are some specifications for the used white LED to be described as follows. The cool white LEDs made by Seoul Semiconductor are used to do experiments. Each LED has a color temperature of 3600K, and has 100Lm at a forward conduction voltage of 3.25V and a forward conduction current of 350mA, according to the corresponding datasheet. But in actuality, each LED has about 95Lm at forward conduction of 3V and a forward conduction current of 360mA, according to measurements. Besides, each LED string has six or eight LEDs, corresponding to a series voltage of 18V or 24V. In order to highlight the proposed technique, there are six buck converters used to drive six channels of LED strings. These buck converters have individual output inductances, which are all different and with measured values of 98.3μH, 101μH, 103.9μH, 114.5μH,

124.9μH and 127.8μH from buck converter 1 to buck converter 6, respectively.

Figs. 6 and 7 show the currents flowing through six channels with six LEDs per channel at 50% of the rated load under the input voltages of 28V and 48V, respectively, whereas Figs. 8 and 9 show the currents flowing through six channels with eight LEDs per channel at 50% of the rated load under the input voltages of 28V and 48V, respectively. Aside from this, Figs. 10 and 11 show the currents flowing through six channels with six LEDs per channel at rated load under the input voltages of 28V and 48V, respectively, whereas Figs. 12 and 13 show the currents flowing through six channels with eight LEDs per channel at rated load under the input voltages of 28V and 48V, respectively. Besides, in all measured figures mentioned above, there is a synchronous signal at the bottom created by the first channel. From these results, it can be seen that under a given input voltage and DC output current the larger the output inductance is, the smaller the current ripple, but the corresponding average values of the output inductance currents are almost the same. Moreover, it can be also seen that the higher the input voltage is, the larger the current ripple under some fixed average value of the inductance current. Especially for the input voltage of 48V, the corresponding inductance currents below half load may goes to the negative values, but their corresponding average values are almost the same. Also, the larger the number of LEDs per channel is, the larger the corresponding duty cycle. Eventually, at half load, Figs. 14(a) and 14(b) show the DC currents in all channels under different input voltages with six and eight LEDs per channel, respectively, whereas at rated load, Figs. 15(a) and 15(b) show the DC currents in all channels under different input voltages with six and eight LEDs per channel. From these results, the percentages of current sharing errors are all within 5%.

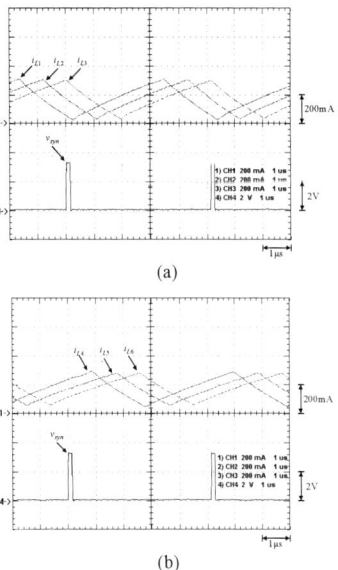

(a)

(b)

Fig. 6. Output inductance currents with six LEDs per channel at half load under the input voltage of 28V with channel 1 synchronous signal: (a) the first three; (b) the last three.

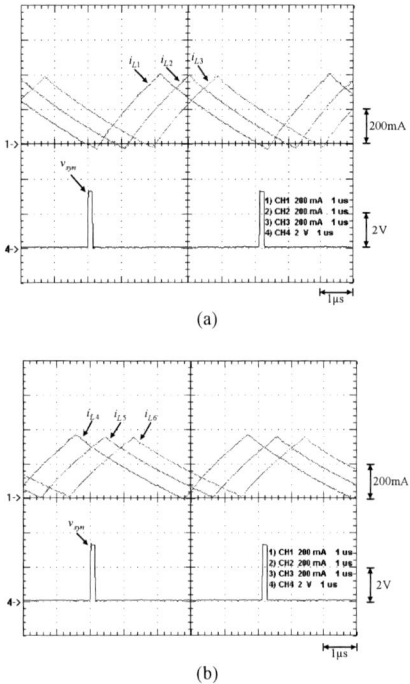

(a)

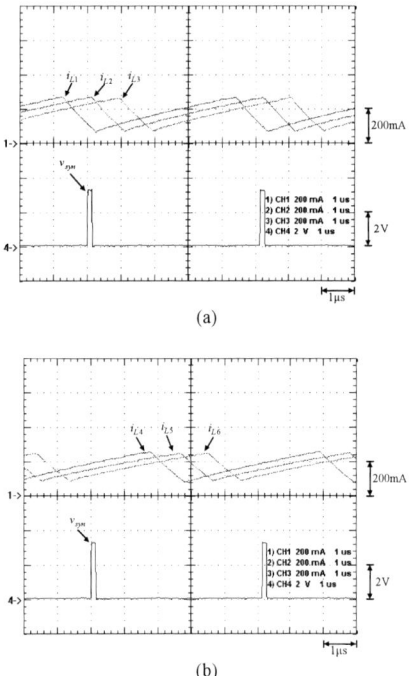

(b)

Fig. 7. Output inductance currents with six LEDs per channel at half load under the input voltage of 48V with channel 1 synchronous signal: (a) the first three; (b) the last three.

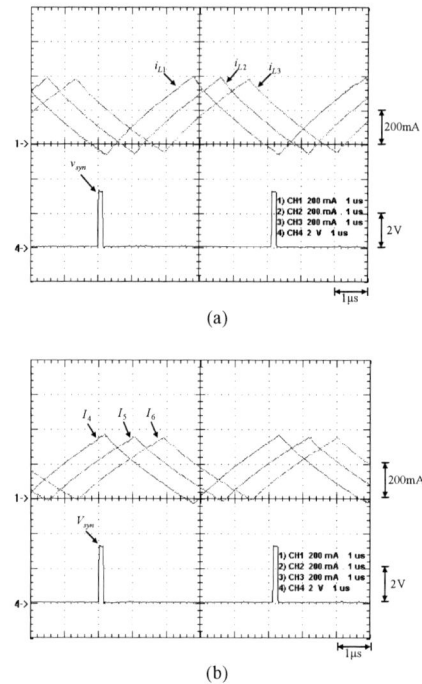

(a)

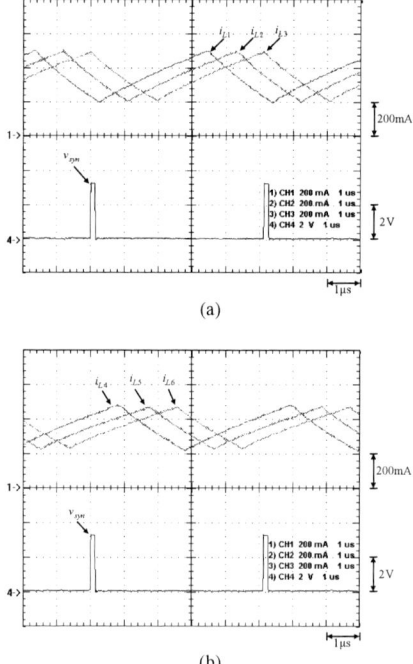

(b)

Fig. 9. Output inductance currents with eight LEDs per channel at half load under the input voltage of 48V with channel 1 synchronous signal: (a) the first three; (b) the last three.

Fig. 8. Output inductance currents with eight LEDs per channel at half load under the input voltage of 28V with channel 1 synchronous signal: (a) the first three; (b) the last three.

Fig. 10. Output inductance currents with six LEDs per channel at rated load under the input voltage of 28V with channel 1 synchronous signal: (a) the first three; (b) the last three.

978-1-4244-4782-4/10 $26.00 © 2010 IEEE

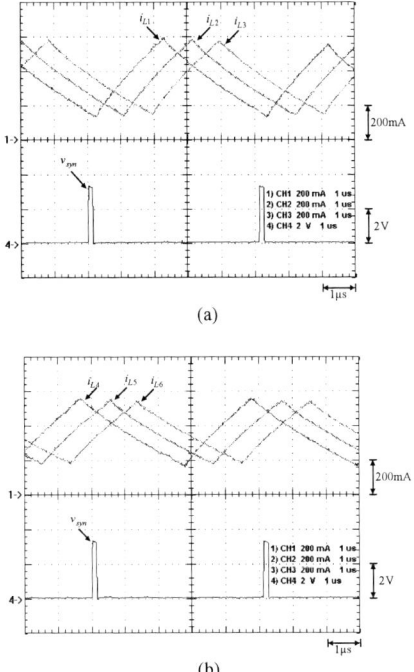

(a)

(b)

Fig. 11. Output inductance currents with six LEDs per channel at rated load under the input voltage of 48V with channel 1 synchronous signal: (a) the first three; (b) the last three.

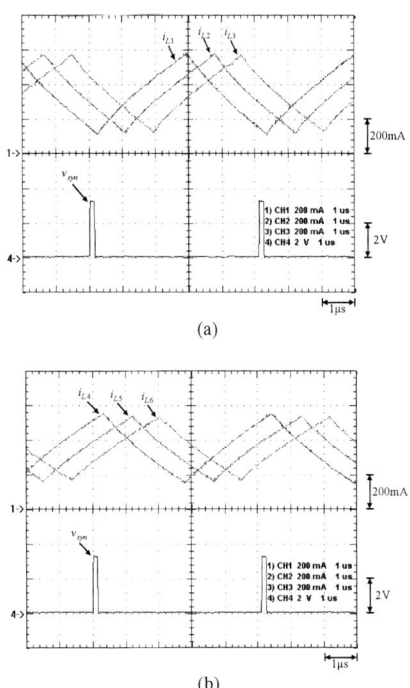

(a)

(b)

Fig. 13. Output inductance currents with eight LEDs per channel at rated load under the input voltage of 48V with channel 1 synchronous signal: (a) the first three; (b) the last three.

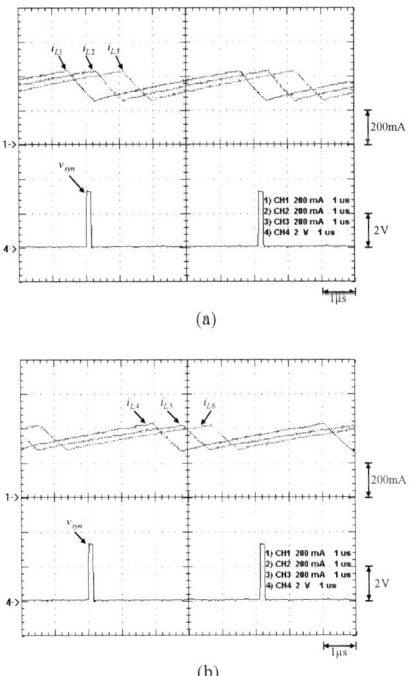

(a)

(b)

Fig. 12. Output inductance currents with eight LEDs per channel at rated load under the input voltage of 28V with channel 1 synchronous signal: (a) the first three; (b) the last three.

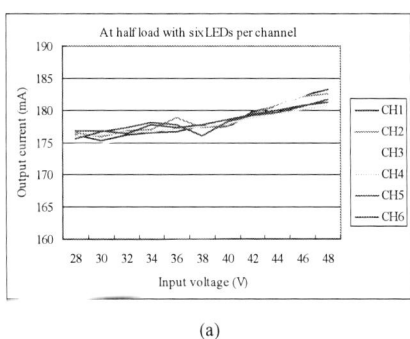

(a)

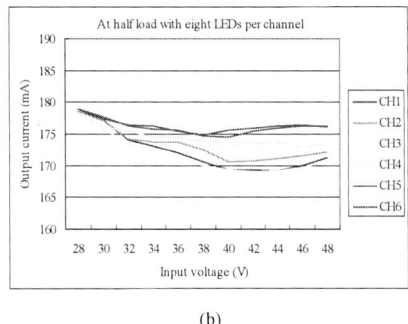

(b)

Fig. 14. Output currents versus input voltage at half load with: (a) six LEDs per channel; (b) eight LEDs per channel.

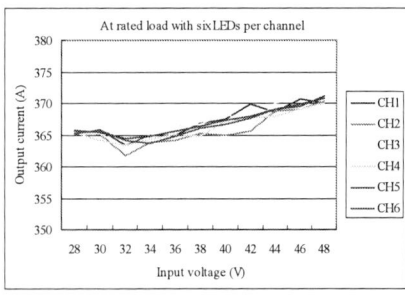

(a)

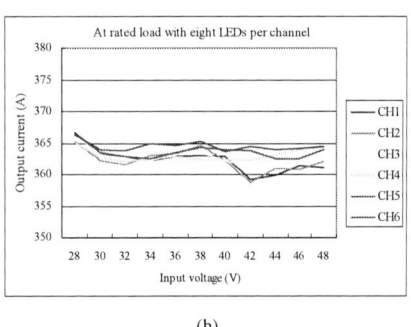

(b)

Fig. 15. Output currents versus input voltage at rated load with: (a) six LEDs per channel; (b) eight LEDs per channel.

V. CONCLUSION

The proposed fully-digitalized DC LED driver has some features as summarized in the following:

(1) No ADCs are used, thereby causing the corresponding cost to be reduced.

(2) Variations in output inductance parameter affect the DC output current imperceptibly.

(3) Variations in input voltage influence the DC output current negligibly.

(4) Such a converter is always operated in CCM, thereby causing the PWM duty cycle not to be changed too much for any load and hence rendering this converter controlled easily.

ACKNOWLEDGMENT

The authors would like to thank the National Science Council for supporting this work under Grant NSC-98-2221-E-027-109.

REFERENCES

[1] Yuequan Hu and M. M. Jovanovic, "A LED driver with self-adaptive drive voltage," *IEEE Trans. on Power Electronics*, vol. 23, no. 6, pp. 3116-3125, 2008.

[2] Xingming Long and Jing Zhou, "An intelligent driver for light emitting diode street lighting," *WAC '08*, pp. 1-5, 2008.

[3] D. R. Nuttall, R. Shuttleworth and G. Routledge, "Design of a LED street lighting system," *IEEE PEMD'08*, pp. 436-440, 2008.

[4] UL8750: *Underwriters Laboratories Inc*, 2009.

[5] A. Torres, J. Garcia, M. R. Secades, A. J. Calleja and J. Ribas, "Advancing towards digital control for low cost high power LED drivers," *IEEE ISIE '07*, pp. 3053-3056, 2007.

[6] Po-Yen Chen, Yi-Hua Liu, Yeu-Torng Yau and Hung-Chun Lee, "Development of an energy efficient street light driving system," *IEEE ICSET'08*, pp. 761-764, 2008.

[7] J. Jacobs, Jie Shen and D. Hente, "A simple digital current controller for solid-state lighting," *IEEE PESC'08*, pp. 2417-2422, 2008.

[8] A. Bhattacharya, B. Lehman, A. Shteynberg and H. Rodriguez, "Digital sliding mode pulsed current averaging IC drivers for high brightness light emitting diodes," *IEEE COMPEL '06*, pp. 136-41, 2006.

[9] In-Hwan Oh, "An analysis of current accuracies in peak and hysteretic current controlled power LED drivers," *IEEE APEC'08*, pp. 572-577, 2008.

High Frequency PWM Dimming Technique for High Power Factor Converters in LED Lighting

D. Gacio, J. M. Alonso, J. Garcia, L. Campa, M. Crespo and M. Rico-Secades

Efficient Energy Conversion, Industrial Electronics and Lighting Group
University of Oviedo
Campus Viesques, Edificio 3, ES-33204 - Gijón, Spain
gacio@ate.uniovi.es

Abstract—This paper deals with the capability of dimming operation added to the Integrated Buck-Flyback Converter (IBFC) topology developed in previous works. Firstly, the three main dimming methods that have been identified by the authors such as enable dimming, shunt dimming and series dimming will be briefly commented. Afterwards, the IBFC topology will be tested performing enable dimming at 100 Hz and 200 Hz. Secondly, besides the main dimming methods, a new proposal is introduced: high frequency series PWM dimming, which overcomes the main issues faced when developing series PWM dimming in constant-current fixed-frequency controlled converters. The dimming control loop of this technique will be presented, as well as the laboratory tests performed, obtaining good results in terms of dimming ratio, harmonic content of the input current and power factor.

I. INTRODUCTION

Since the High Brightness LEDs (HB-LEDs) industry is continually overcoming, these devices have raised as an interesting solution for most general lighting applications. However, there are still not only some issues to be solved, but also some capabilities that HB-LEDs provide which have to be explored and researched. One of these points could be the light output regulation, or dimming, which may be interesting for power saving in those conditions where the maximum light output is not required. This feature has a good application field in ambient intelligence, where light regulation is needed for sustainable lighting or lighting scenes among other purposes.

This paper follows the research developed in [1], where the High Power Factor Integrated Buck-Flyback Converter (HPF IBFC) was used for a street lighting application. The objective of this new research consists in presenting and testing a new dimming method. This topology was developed to be supplied from a universal ac source (from 90 Vrms to 265 Vrms) performing power factor correction (PFC) and achieving low enough harmonic contents for complying with European IEC 61000-3-2:2000 mandatory. The prototype developed was optimized for obtaining a fast enough response to allow the converter to perform enable dimming while keeping a low output current ripple.

PWM dimming operation consists in switching on and off the LED light, making it flicker. If this flickering is performed fast enough, the human eye can only perceive a decrease in the luminous output due to the stroboscopic effect. The human eye does not perceive the flickering if the dimming frequency is at least 50 Hz, but the recommended minimum frequency is 200 Hz so moving objects do not seem to be still. The authors have identified three main methods for PWM dimming operation [2], [3], [4].

Firstly, the so-called enable dimming is based on applying a variable reference to the controller from an external pulse signal. In this way, the obtained result is that the whole converter is switched on and off. The main drawback of this solution is the slow dynamic response that PFC converters feature, which enlarges the turn-on transient and therefore, limits the dimming frequency.

Secondly, the so-called series dimming uses a transistor in series with the load. In this case, the dimming pulsed operation is achieved by leaving the load in open-circuit. This dimming operation is usually performed at 1 kHz. Its main drawback is the high electrical stresses that appear in the load, in the dimming transistor, and in other components of the converter if the dynamic response of the converter is not fast enough. In this way, a current control might be employed in voltage mode controlled PFC converters due to their slow dynamic response.

Finally, the so-called shunt dimming employs a transistor in parallel with the load. If this transistor is switched at a medium frequency, the load is dimmed by being short-circuited. The main drawback of this solution is the high power dissipation that takes place in the dimming switch, generally a MOSFET transistor, which could even cause its destruction.

All the solutions explained above showed big issues in the simulations run using the IBFC topology designed. Therefore, a new dimming method is proposed in this paper, using the series dimming but performed at a frequency far above from the converter bandwidth. The results obtained in the

---

This work has been supported by the Spanish Government, Education and Science Office, under research grants number DPI-2007-61267 and DPI-2007-63129)

978-1-4244-4782-4/10 $26.00 © 2010 IEEE

simulations and at the lab show that the main drawback of the low frequency series dimming could be avoided in constant-current fixed-frequency controlled converters.

II. TOPOLOGY AND LOAD EMPLOYED

This work continues the one developed in [1], consisting in a High Power Factor (HPF) Integrated Buck-Flyback Converter (IBFC) [5] for off-line applications, which runs a 72 W LED load [1], and is designed for meeting the IEC 61000-3-2 European regulations. This converter, shown in Figure 1, is composed by a buck regulator integrated with a flyback converter. The former performs the PFC whereas the latter provides the output current through the LED load. The components used in the laboratory prototype for the dimming tests are summarized in Table I. In this paper, the output capacitor has been decreased from the former converter developed in [1] so the dynamic response could be faster while keeping a reasonably low output current ripple, below 15%.

On the other hand, the 60 LEDs-72 W LED load from the previous work is kept [1]. It is composed by 10 DragonTape connected in series. Each tape contains 6 Golden Dragon LEDs featuring a luminous efficiency of 21 lm/W at 350 mA. The design current was set to 350 mA, emitting 1500 lumens at a nominal power of 72 W [6].

LED loads are non-linear devices, so they had to be modeled. The model followed was the linear one shown in Figure 2, consisting in a forward voltage that represents the threshold voltage and a series resistor, acting as the dynamic resistance [7]. Thus, the output voltage of the whole string will be:

$$V_O = N \cdot (R_{Di} \cdot I_D + V_{\gamma i}) = R_D \cdot I_D + V_\gamma \qquad (1)$$

Where N is the number of LEDs in series, R_{Di} and $V_{\gamma i}$ are the dynamic resistance and the threshold voltage of each LED respectively, and I_D is the LED current. Finally, R_D and V_γ are the dynamic resistance and the threshold voltage of the whole string, respectively. In the lab tests several measurements were taken each 50 mA, starting at 100 mA. The devices were let to cool down for a few minutes between measurements so heating effects on the threshold voltage were negligible. The results obtained, and the interpolated linear model, are shown in Figure 2 and Table II.

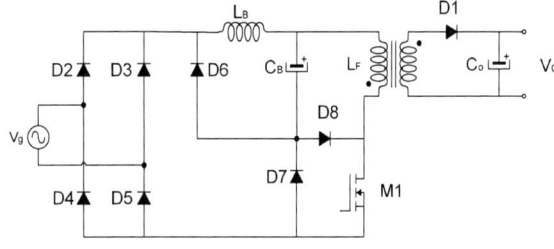

Figure 1. HPF Integrated buck-flyback ac-dc converter.

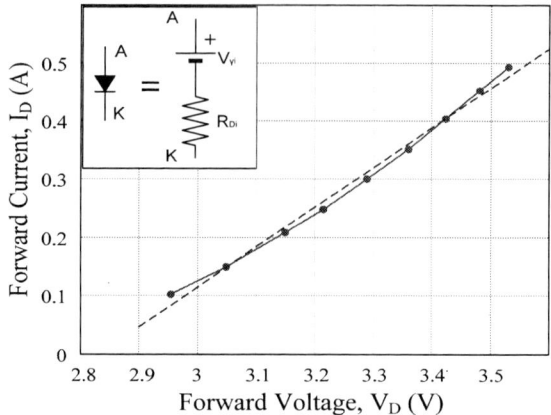

Figure 2. Forward current I_D and forward voltage V_D values obtained from the LED lamp built and tested in the laboratory (solid line) and linear model (dashed line). In the upper left box: linear model of the equivalent diode circuit.

TABLE I.   COMPONENTS OF THE LABORATORY PROTOTYPE

Component	Value
Buck	L=26.1 µH EFD25 N87 N=32
Flyback	L=19.3 µH ETD29 F44 n=4, N1=10, N2=40
M_1	SPW17N80C2
M_2	IRFP840
D_1	STTH512
D_6	MUR860
D_7	MUR860
D_8	MUR840
C_B	570 µF/250 V
C_o	1 µF/250 V MKT

III. CONTROL LOOP AND ENABLE DIMMING TEST

The IBFC presented in this paper is constant-current fixed-frequency controlled [8]-[9]. This means that the converter is controlled at a fixed frequency varying the switch duty cycle so that the output voltage is adjusted in order to keep the output current constant. The frequency chosen is 100 kHz so that the magnetic elements are as small as possible while losses are kept at a low level.

TABLE II.   TEST FORWARD CURRENT AND VOLTAGE VALUES

$I_D(A)$	$V_O(V)$	$R_D(\Omega)$	$V_\gamma(V)$
0.102	177.2		
0.150	182.9		
0.208	189.0		
0.249	192.9		
0.301	197.4	87.2	170.1
0.352	201.6		
0.403	205.4		
0.453	208.9		
0.492	211.8		

The controller was designed in order to achieve the maximum possible bandwidth, so that the control loop was adequate to perform PWM Enable dimming. The secondary of the flyback operating in DCM behaves as a current source. In this way, if the output capacitor is considered large enough, the ac component of the output current can be neglected. Thus, the equivalent circuit is the one shown in Figure 3, where only the dc component of the output current is taken into account to model the converter. Using the values gathered in Table III, the following transfer function is obtained:

$$G(s) = \frac{i_{O(s)}}{d(s)} = \frac{2 \cdot \dfrac{I_O}{D \cdot C_O \cdot R_D}}{s + \dfrac{V_O + I_O \cdot R_D}{V_O \cdot R_D \cdot C_O}} = \frac{6.75 \cdot 10^4}{s + 13219} \quad (2)$$

The controller has been designed using Sisotool toolbox from Matlab, obtaining the following PI controller with an additional pole:

$$C(s) = C_R \cdot \frac{s + s_z}{s \cdot (s + s_p)} = 8.7 \cdot 10^4 \cdot \frac{s + 9091}{s \cdot (s + 6.25 \cdot 10^5)} \quad (3)$$

It was built using the OTA included in LM3524 IC. Deeper information about the design process is provided in [1], where the design of the topology and the control loop was dealt with. The feedback loop has been implemented using a third-order low-pass Butterworth filter, which senses the average current flowing through the 1 Ω measuring resistor placed in series with the load. The cut-off frequency was set to 10 kHz in order to get rid of high frequency noise. The gain was set to 6.6 as in [1] so the LM3524 OTA common mode voltage in nominal operation is within the allowable range.

Once the feedback loop was set, the Enable dimming tests were performed at the laboratory. Those tests consisted in applying a variable reference to the converter so it could be switched on and off alternatively at such frequency that the human eye could not perceive the flickering. This variable reference was implemented by a 0 to 2.31 V 200 Hz square wave, which corresponds to a peak current of 350 mA. Finally, the dimming operation duty cycle was set from 20% to 80%

TABLE III. NOMINAL OPERATION POINT

$I_O(A)$	0.350
$V_O(V)$	200.605
D	0.119
$C_O(\mu F)$	1.0
$R_D(\Omega)$	87.151

However, these tests showed that the converter was too slow for being operated at that given frequency, so it was tested at 100 Hz as well even though the recommended minimum frequency is 200 Hz [10], although other authors recommend frequencies above 125 Hz [2]. The results obtained for both frequencies are depicted in Figure 4, for the case of dimming at 200 Hz, and Figure 5 for the case of 100 Hz. As can be seen, the converter, controlled in average current mode, is not fast enough for providing high dimming ratios even at 100 Hz. Additionally, in the tests done in the laboratory, it was stated that, as dimming ratio rose, the total harmonic distortion of the input current (THD$_I$) rose too, showing that this technique is not feasible at such a low frequency.

IV. HIGH FREQUENCY DIMMING AND BLOCK DIAGRAM

The conclusion obtained from the former enable dimming test was that the IBFC controlled in average current mode is not fast enough for performing this dimming operation. The main idea for the High Frequency Series Dimming (HFSD) operation mode comes from the tests done in the laboratory while trying Enable dimming at frequencies within the converter bandwidth. It was stated that whereas the dimming frequency was set below the converter bandwidth, i.e., below 1.4 kHz, the whole converter tended to oscillate due to the fact that the controller was trying to keep the current constant.

On the contrary, at frequencies far beyond the closed loop cut-off frequency, the controller is not capable to compensate such high frequency variations in the output current. In this way, when operated at a high enough dimming frequency (100 kHz in this case), the converter is not sensitive to this switching, perceiving only a decrease in the average value that comes from the current sensor, which is a third order-low pass filter.

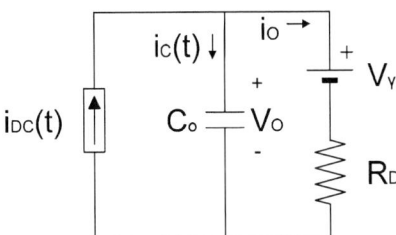

Figure 3. Dc current output equivalent circuit.

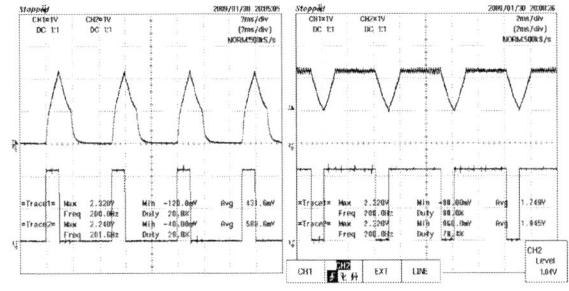

Figure 4. Enable dimming test at 200 Hz. Left: 20% dimming. Top: voltage across the 1 Ω current sensing resistor. Vertical scale: 1 Volt/div. Bottom: enable signal. Vertical scale: 1 Volt/div. Horiz. scale: 2 ms/div. Right: 80% dimming. Top: voltage across the 1 Ω current sensing resistor. Vertical scale: 1 Volt/div. Bottom: enable signal. Vertical scale: 1 Volt/div. Horiz. scale: 2 ms/div.

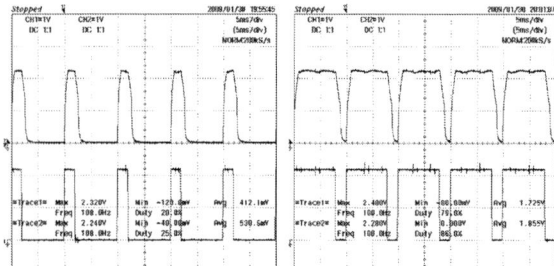

Figure 5. Enable Dimming test at 100 Hz. Left: 20% dimming. Top: voltage across the 1 Ω current sensing resistor. Vertical scale: 1 Volt/div. Bottom: enable signal. Vertical scale: 1 Volt/div. Horiz. scale: 5 ms/div. Right: 80% dimming. Top: voltage across the 1 Ω current sensing resistor. Vertical scale: 1 Volt/div. Bottom: enable signal. Vertical scale: 1 Volt/div. Horiz. scale: 5 ms/div.

Many authors consider that the higher the dimming frequency, the worse the behavior due to EMI issues, although a 25 kHz dimming frequency is considered adequate, since it is out from the audible range [11].

It was stated that as the dimming ratio is increased at constante peak current, the regulator perceives a decrease in the average output current constant peak current, since the voltage level coming from the current sensor is decreased. Therefore, the controller works raising the output average current, which, in dimming mode, means that the peak value is increased so that the average voltage level coming from the current sensor is kept constant and equal to the reference. What is more, this variation in the measured average output current was checked to be linear, so it was easy to keep the peak value constant and fixed at 350 mA. The block diagram of the dimming control circuit proposed is shown in Figure 6, and it allows the converter to adjust both the dimming ratio and the current peak value.

The operation of this block diagram is quite simple. As said before, when the converter is being operated in dimming mode, the pulsating output current is perceived as a lowered average level by the current sensor, proportional to the dimming duty cycle, or dimming ratio:

$$I_{Avg} = \frac{1}{T_{Dim}} \cdot \int_{0}^{D_{Dim} \cdot T_{Dim}} I_{Peak} \cdot dt = I_{Peak} \cdot D_{Dim} \quad (4)$$

Where I_{Peak} is the peak value of the output current, D_{Dim} stands for the dimming operation duty cycle and T_{Dim} is the dimming period, i.e. the inverse of the dimming frequency. As mentioned above, the average level of the output current is lowered in dimming operation mode and is not equal to the reference for the selected peak value. Therefore, the controller raises the output average current by increasing the peak value to let the average current match the reference in steady state. This can easily be seen in the following equation, as well as the linearity between dimming duty cycle and reference voltage variation:

$$V_{Sens} = V_{Ref} = H_O \cdot D_{Dim} \cdot I_{Peak} \quad (5)$$

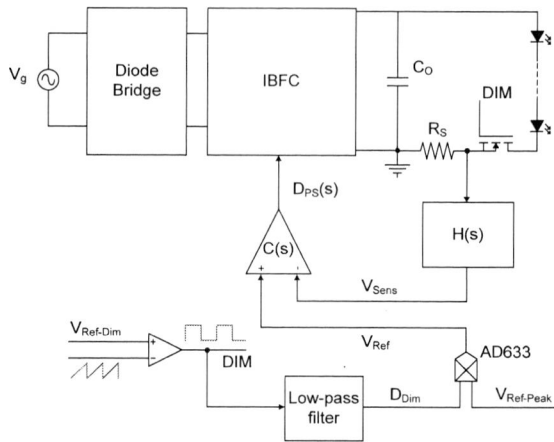

Figure 6. Block diagram of the proposed dimming method allowing both dimming level and peak current adjustment.

Where V_{Sens} is the voltage level at the current sensor and H_O is the low-pass filter gain, in terms of V/A.

There are two strategies to keep the peak value constant. Firstly, by changing the gain of the low-pass sensor as the dimming duty cycle is varied. Second, by adjusting the reference level as the dimming duty cycle is changed. The strategy followed in this work was to change the reference level, so peak current could be set as well. In this way, the implementation of this strategy is quite simple, because the dc voltage level provided by the current sensor is proportional to the peak current and the dimming duty cycle. Therefore, in order to keep constant the peak value of the output current while allowing the peak current to be adjustable, the reference voltage and dimming duty cycle have to be varied accordingly. This is achieved by multiplying the dimming duty cycle and the peak value reference using the AD633 IC, so that a dimming-duty cycle-dependant reference is generated. Afterwards, this signal obtained is compared to the current sensor output by the regulator in order to generate the error and the control signal.

There is a simplification for keeping the peak value constant avoiding the use of the multiplier, although this value will not be adjustable. In this case, there is no external reference for the peak current value, but only for the dimming ratio. Thus, the idea is to generate a reference that varies with the dimming ratio using a conditioning circuit, which provides a linear output proportional to the average value of the pulse signal applied to the dimming transistor. Afterwards, this signal obtained from the conditioning circuit is used as the reference for the peak value. A block diagram of this second proposed scheme is shown in Figure 7.

V. DEVELOPED PROTOTYPE

A prototype was built and tested in the laboratory. The dimming frequency was set to 100 kHz, the same as used for the main power switch, since it must be far above the closed loop cut-off frequency.

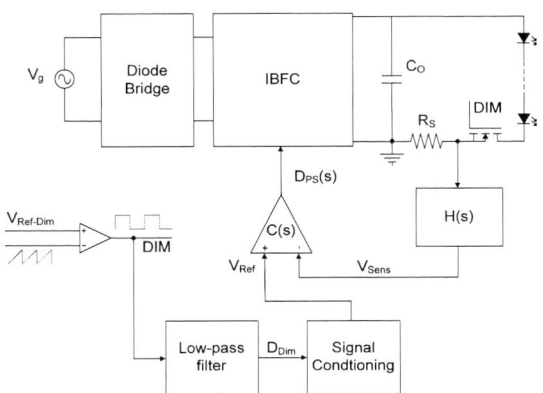

Figure 7. Block diagram of the proposed dimming method for keeping the peak current value constant.

Besides, this allows the two switches to be operated synchronously, avoiding the generation of additional EMI harmonics. The entire control circuit is shown in Figure 8.

The 100 kHz signal is generated by the LM3524 control IC and is made by a 0.6 V to 3.8 V triangle wave which is compared, using an LM393 IC, to a dimming reference to select the dimming duty cycle, this is, the dimming ratio. This block is highlighted as Dimming Generator in the figure. It creates the pulsed dimming control signal, which will be applied to the dimming switch, an IRF840 MOSFET transistor.

The average value from the dimming control signal is

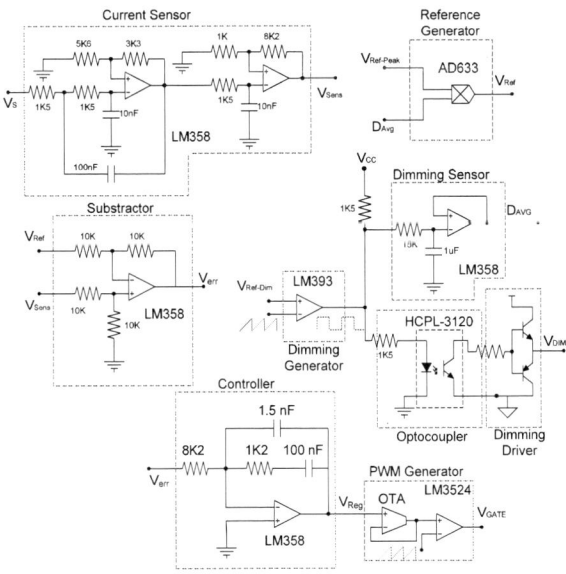

Figure 8. Control circuit for the proposed dimming mode operation, including the current sensor (up, left) and the regulator (bottom, right) The LM3524 IC is also depicted, since it is used as a voltage follower and a triangle wave generator.

multiplied by the peak current reference in the Reference Generator, using an AD633 IC analog multiplier. This reference is created after having conditioned that signal in a 0-10V range proportionally to the dimming duty cycle. That average value is extracted by filtering the pulse signal, using a 100 Hz low-pass Butterworth filter implemented with an LM358 IC, whereas the conditioning is performed by another LM358 IC, applying the needed gain to the average value obtained after filtering the pulse signal. This generates the new reference to compare to the sensor signal. It is represented as Dimming Sensor in Figure 8.

However, the peak current reference could be close to the common mode voltage limit due to the low voltage levels obtained at high dimming ratios. In this way, the gain of the feedback loop, this is, the Current Sensor, was set to 14.6. The new peak current reference is 5.11 V for an output current of 350 mA at dc current operation mode, i. e., maximum output.

The reference dc level obtained after multiplying the peak value reference and the dimming ratio is compared to the voltage level coming from the current sensor in the Substractor, generating the error signal, which will be applied to the Controller. Because of the reference value could drop below the common mode voltage of the LM3524 IC, a new external controller was developed. However, the LM3524 IC was kept as voltage follower and PWM Generator.

The new Controller was built using an LM358 IC, employing one of the operational amplifiers (OA) included for performing the comparison between the reference and the current measurement, and the other OA for building the controller. This PI controller with an additional pole was implemented as an inverting amplifier. Its Bode diagram can be seen in Figure 9. Finally, in order to switch on and off the dimming transistor, a decoupled Dimming Driver was built employing the HCPL3120 optocoupler and an isolated voltage supply.

The new open loop transfer function in dc output current mode is shown in Figure 10, whereas the closed loop transfer function is depicted in Figure 11. Finally, Figure 12 shows the topology employed.

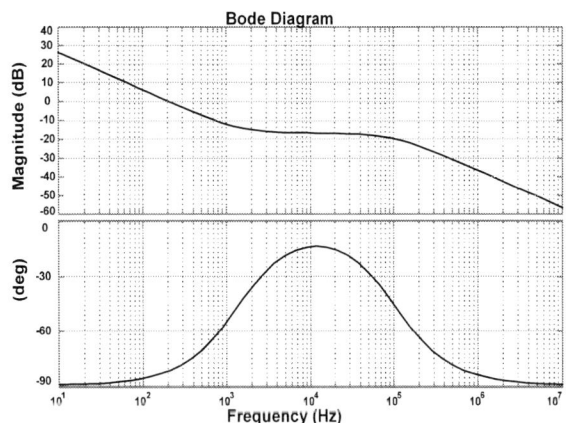

Figure 9. PI controller with additional pole Bode diagram.

978-1-4244-4782-4/10 $26.00 © 2010 IEEE

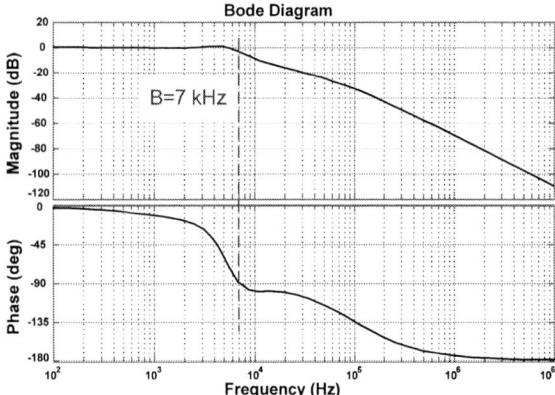

Figure 10. Closed loop transfer function for the whole converter in no dimming mode operation. The bandwidth obtained with the new controller theoretically raises up to 7 kHz.

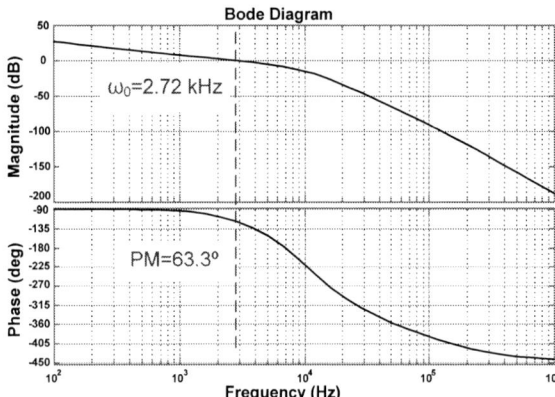

Figure 11. Open loop transfer function for the whole converter in no dimming mode operation. The phase margin achieved is 63.3 degrees and the cross-over frequency is 2.7 kHz.

VI. EXPERIMENTAL RESULTS

Simulations were carried out using PSIM 6.0, showing a highly dimmable converter. The prototype was tested both in open and closed loop at 150 Vrms in order to check the dimming capability. Only closed loop experimental results will be included in the paper, since those are the most important.

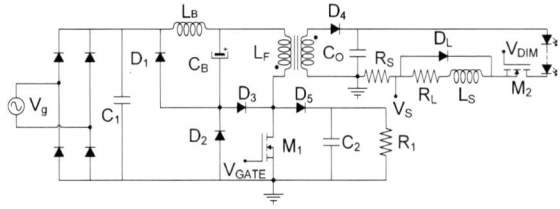

Figure 12. HPF Integrated buck-flyback ac-dc converter employed for allowing PWM dimming operation mode using a dimming transistor placed in series to the LED load.

TABLE IV.   HIGH FREQUENCY SERIES AND ENABLE DIMMING TESTS

HFS Dimming					Enable Dimming (100Hz)	
D(%)	THD_1(%)	PF	P_0(W)	η(%)	THD_1(%)	PF
100.0	31.8	0.944	69.9	75.7	-	-
90.0	31.9	0.944	61.0	74.4	-	-
80.0	32.1	0.944	52.8	72.7	34.7	0.934
70.0	32.5	0.943	45.1	71.0	37.3	0.922
60.0	32.5	0.942	37.9	69.6	41.5	0.899
50.0	32.4	0.942	31.1	67.2	47.5	0.858
40.6	33.1	0.940	25.0	64.9	54.6	0.796
30.7	32.4	0.941	18.6	62.5	62.5	0.707
20.0	33.7	0.934	12.8	58.1	70.0	0.595
10.0	33.0	0.927	8.1	55.3	-	-

The first issue faced was an instantaneous high peak current flowing through the parasitic capacitance of the LED diodes and the sensing resistor due to the instantaneous discharge of the output capacitor through the load when the diming transistor is switched on in dimming operation mode, i.e. the instant when the lamp is switched on. This high peak value could even reach double the dc output current level, being hazardous for the LEDs chosen. In this way, an overcurrent snubber was implemented employing a ferrite core inductor and a free-wheeling diode, as shown in Figure 12, where L_S is the snubber inductor, R_L is its series resistance and D_L is the free-wheeling diode. This snubber has been proven to cancel that overcurrent for a highly wide dimming range.

The first test consisted in checking the ac input current, comparing it to the values obtained in the enable dimming test. The results obtained are gathered in Table IV, compared to the enable dimming test. Figure 13 shows how the input current is hardly distorted even at high dimming ratios, whereas enable dimming technique performed at 100 Hz caused the THDI to rise dramatically. This figure shows four cases: no dimming operation, 80%, 50% and 20 % of full output power. Figure 14 shows the output current measured in the 1 Ω series resistor at four different dimming ratios. As can be seen, the converter is highly dimmable using this technique, achieving quite fast rising and falling edges.

The converter has been proven to be at least 1:10 dimmable, which is a reasonable value. Finally, Figure 15 shows the input power, output power and efficiency obtained as a function of the dimming ratio in terms of dimming duty cycle at 150 Vrms input voltage. The efficiency is low supposedly due to the universal operation range of the converter (90-265 Vrms), which makes quite difficult to optimize the topology for the entire range.

VII. CONCLUSIONS

This paper deals with the design process that started testing the Enable Dimming method in an IBFC converter controlled in voltage mode by means of the average output current. These tests showed its unfeasibility due to a not fast enough dynamic response. Then, a new technique for dimming constant-current fixed-frequency controlled converters at high frequency is introduced and the design process is dealt with.

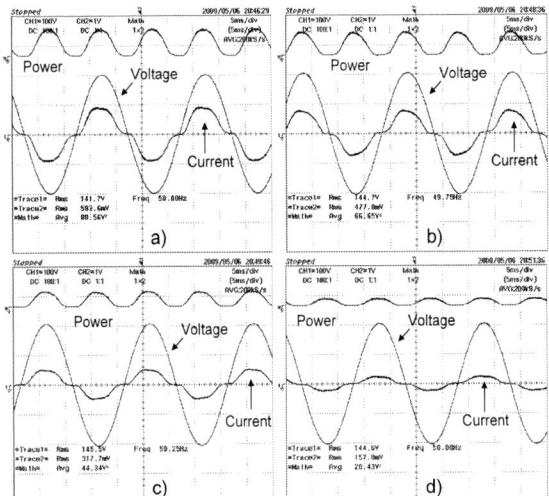

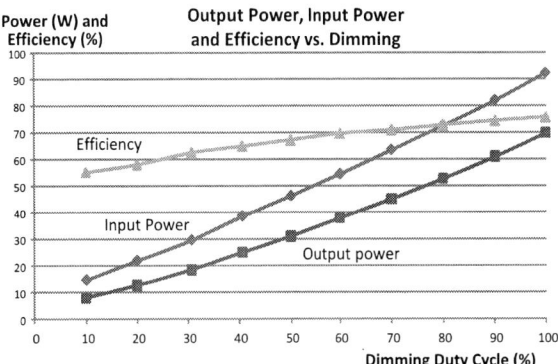

Figure 15. Input power, output power and efficiency as a function of dimming duty cycle at 150 Vrms input voltage

Figure 13. Line current, voltage and power for different dimming ratios at 150 Vrms input voltage. A) 100% of full output light (1:1 dimming ratio). B) 80% of full output light (4:5 dimming ratio). C) 50% of full output light (1:2 dimming ratio). D) 20% of full output light (1:5 dimming ratio). Vertical scale: 100 V/div for voltage line, 1 V/div for current line. Horizontal scale: 5 ms/div.

The proposed technique consists in a series dimming method performed at such a high frequency that the regulator can not handle the pulsating waveform sensed at the LED load, thus processing the average value from the output current only. Since the regulator responds raising the output current peak value in order to match the sensed value with the current reference, a control circuit is proposed in order to adapt the output current reference for keeping its peak value constant. There are two solutions proposed: the first proposed solution allows the user to set the output current peak value, whereas the second solution keeps that current peak at the design value.

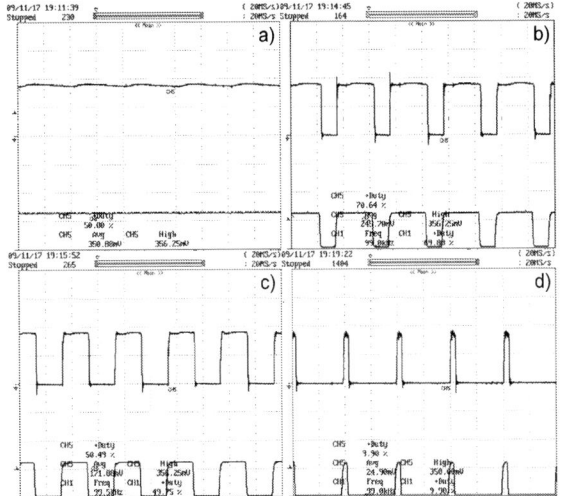

Figure 14. Output current (top) and dimming signal (bottom). A) 100% of full output light (1:1 dimming ratio). B) 70% of full output light (7:10 dimming ratio). C) 50% of full output light (1:2 dimming ratio). D) 10% of full output light (1:10 dimming ratio). Vertical scales: 5 V/div for dimming signal, 200 mV/div for output current. Horizontal scale: 5 µs/div

The converter used, the IBFC developed in previous works, is modeled and a constant-current fixed-frequency output current control is implemented using the LM 3524 IC as signal generator, voltage follower and in order to allow the converter to operate in open loop, whereas the regulator and the remaining system blocks are built using additional circuitry. The prototype built in the laboratory was tested at 150 Vrms obtaining the expected results, since the converter was proven to achieve high dimming ratios, from at least 10% to full output luminous flux, and a good reliability of the proposed technique.

REFERENCES

[1] Gacio, D.; Alonso, J.M.; Calleja, A.J.; Garcia, J.; Rico-Secades, M.; "A Universal-Input Single-Stage High-Power-Factor Power Supply for HB-LEDs Based on Integrated Buck-Flyback Converter", Applied Power Electronics Conference and Exposition 2009. APEC 2009. 15-19 Feb. 2009, pp. 570 - 576.

[2] Sameh Sarhan, "High-Frequency Wide-Range Dimming Schemes for High Power LEDs", National Semiconductors application notes, Exhibitors Seminars, APEC 2008.

[3] Prathyusha N.; Zinger, D.S.; "An Effective LED Dimming Approach", Industry Applications Conference 2004, IAS 2004. 3-7 October 2004, Vol. 3, pp. 1671 - 1676.

[4] Xiaoru X.; Xiaobo W.; "High Dimming Ratio LED Driver with Fast Transient Boost Converter", Power Electronics Specialists Conference, 2008. PESC 2008, 15-19 June 2008, pp. 4192 - 4195.

[5] Alonso, J.M.; Dalla Costa, M.A.; Ordiz, C., "Integrated Buck-Flyback Converter as a High-Power-Factor Off-Line Power Supply", IEEE Transactions on Industrial Electronics, Volume 55, Issue 3, March 2008, pp. 1090 – 1100.

[6] Golden Dragon, datasheet LW W5SG.

[7] Bhattacharya, A.; Lehman, B.; Shteynberg, A.; Rodriguez, H.; "A Probabilistic Approach of Designing Driving Circuits for Strings of High-Brightness Light Emitting Diodes", Power Electronics Specialists Conference, 2007, PESC 2007, 17-21 June 2007, pp. 1429-1435.

[8] Clique, M.; Fossard, A. J.; "A General Model for Switching Converters"; IEEE Transactions on Aerospace and Electronic Systems", Volume AES-13, Issue 4, July 1977, pp. 397 - 400.

[9] Tang, W.; Lee, F.C.; Ridley, R.B.; "Small-Signal Modeling of Average Current-Mode Control"; IEEE Transactions on Power Electronics, Volume 8, Issue 2, April 1993, pp. 112 - 119.

[10] Dimming InGaN LEDs, Application Note. Osram Opto-Semiconductors (2003-01-08).

[11] Michael Jennings, "Driver IC Design Ensures Optimum Performance of LED Backlighting", LEDs Magazine, Issue 25, pp. 51 - 53, January/February 2009.

A RGB-Driver for LED Display Panels

Jaber Hasan, Do Hung Nguyen, and Simon S. Ang
Department of Electrical Engineering
University of Arkansas
Fayetteville, AR 72701

Abstract—Red-Green-Blue (RGB) light-emitting diodes display panels are finding widespread use due to recent advances in the light-emitting diodes (LEDs) and their driver technology. This paper investigates a digital microcontroller based RGB-driver for application in display panels. The RGB-driver uses three different voltage sources from switch-mode power converters (SMPCs) as each RGB color requires different drive voltages. The proposed driver selects the minimum drain voltage of the MOSFETs of the current controllers for each color and uses this voltage to control the duty cycle of the SMPCs, and thus, maintains the minimum output voltages required to keep the MOSFETs in the current controllers in regulation. With a 5-V supply voltage, the efficiencies of the RGB LEDs are 91.5%, 95.7%, and 95.7%, respectively. Even though the driver was experimentally verified in a 3 x 3 RGB pixels, the concept can be extended to a larger display matrix.

I. INTRODUCTION

Due to the recent advancement in light-emitting diode (LED) technologies, LEDs are increasingly being used in LCD backlighting, automobiles, traffic lights, and general-purpose lighting [1] [2]. As the cost of LEDs is decreasing, LEDs are finding new applications such as in display panels and signage.

The complexity of driving RGB based LEDs in display panel is due to the large number of control nodes and RGB LEDs needed. For each RGB pixel, there are three constant current controls for dimming and three different supply voltages. Since each color of the RGB LEDs requires different drive voltages, each requires its independent driver. The most common and straightforward method is to employ independent current controllers for each RGB color [1] [2] as shown in Fig.1. The constant-current controllers can be either a linear or a switch-mode type. Linear current controller suffers from excessive power dissipation in its series-pass devices [1]. However, switch mode current controller requires a number of storage devices such as capacitors and inductors which lead to higher parts count and adding to the cost of the driver.

The LED-drivers implemented in TVs are either used for backlight or displays panels. For backlight, the LED driver operates at frequency lower than the

frequency that is required for display panels since the refresh rate in a LED display panels is about 240Hz, whereas in a backlight it is about 120Hz.

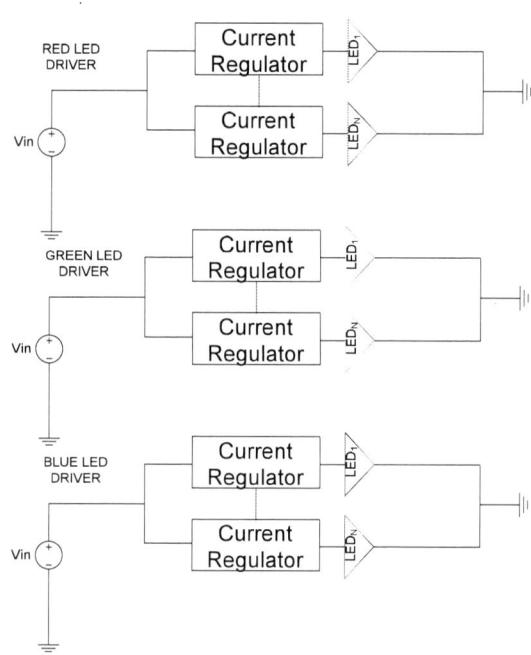

Figure 1. Independent current controlled LED-driver.

As shown in Fig. 2, linear current controllers are being used to control the current in the LED together with a SMPC supplying the required operating voltage [1] [2]. In this implementation, the output voltage of the SMPC, which is set by the feedback voltage, ensures that the linear current controllers for each RGB color LED are operating at the desired currents [1] [2]. However, this often yields the worst-case operating voltage, leading to a lower efficiency. Also, three different SMPCs are required due to different drive voltage of each color of RGB LEDs. In order to improve the efficiency of the RGB-driver using a SMPC, the lowest drain voltage of the MOSFET in the linear current controllers was detected using diodes for LED backlight applications [3]. The efficiency of this driver depends on the selection of the reference voltage and the

978-1-4244-4782-4/10 $26.00 © 2010 IEEE

temperature dependence of the sensing diodes [1] [2]. In order to maintain the linear current controllers in regulation, the reference voltage has to be selected for the worst-case condition, and hence leading to reduced efficiency. Additionally, the sensed voltage of the diode changes with operating temperature causing detrimental effects [1] [2].

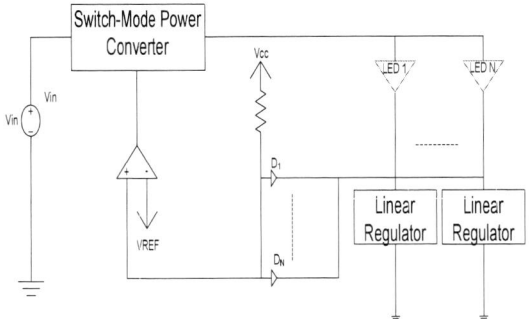

Figure 2. LED-driver implemented in [3].

In this paper, a different approach to detect the minimum drain voltage using a microcontroller with a multiplexer circuit for a RGB driver in display applications is presented. In this implementation, the efficiency of the driver is increased by eliminating the need for sensing diodes to detect the minimum drain voltage of the current controller. Additionally, this proposed circuit is implemented with dimming capabilities thus enabling each color pixel to be individually dimmed from 1% to 100%, thus enabling high contrast colors which are required for state-of-the-art display panels.

II. Proposed Driver

The proposed LED-driver is shown in Fig. 3 with N (N=1, 2, 3…) RGB LED strings. In this implementation, the proposed driver has two modes of operation – startup and operation modes. A PIC18F4431 microcontroller is used to control the duty cycles of the three switching converters using the minimum drain voltages of the MOSFET in the current controllers in each RGB color LEDs.

A digital microcontroller is being used to control the SMPC rather than an analog controller as implemented in [1] [2]. The primary reasons for implementing the driver digitally are its capability of adding extra features without additional circuitry, easy reconfiguration of control schemes by software, and accurate calculations [4]. However, digital control suffers from limitation due to analog-to-digital (ADC) as well as digital-to-analog (DAC) conversions.

During startup mode, in order to maintain the MOSFETs of the current controller in saturation – each

switching converter first outputs a high voltage. The gate-to-source (V_{gs}) voltages of the MOSFETs of all the current controllers are sensed and then the maximum V_{gs} of each color of RGB is selected by the microcontroller together with a multiplexer circuit and compare against reference voltage. If the value of the maximum V_{gs} of the MOSFETs of each color of the RGB is greater than the reference voltage, the output voltage of the SMPC is reduced until the maximum V_{gs} of the MOSFETs in the current controller of each color of the RGB is equal to the reference voltage. When the maximum V_{gs} of the MOSFET of each color of RGBs is equal to the reference voltage, the driver leaves the startup mode and enters the operation mode.

In the operation mode, the drain voltages of the MOSFETs in the linear-regulators of each color of RGBs are sensed directly using the multiplexer. The minimum drain voltage of each LED color is determined from the sensed voltages and set as the reference voltage for the SMPC. A proportional-integral-derivative (PID) control is implemented to maintain this reference voltage thus forcing the MOSFETs to operate at the boundary of the saturation region. Since the controller uses this voltage to control the duty cycle of the switching converters to output a minimum voltage necessary to maintain the MOSFETs in the current controller in regulation, an improved efficiency is thus achieved for each RGB color LED independently. Analog dimming is implemented to control the brightness of each individual RGB color by adjusting the reference voltage of the error amplifiers in the current controllers.

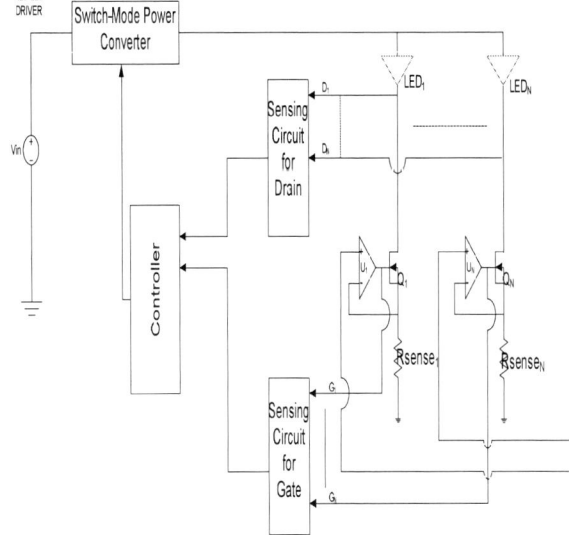

Figure 3. RGB LED-driver proposed in this paper. (Note – Similar circuits for green, and blue LED-driver.

III. Dimming Control

Dimming is being implemented to provide brightness and contrast adjustments which are required in a display panel. There are two types of dimming - analog and pulse-width-modulation (PWM). In analog dimming, LED brightness of brightness of 50% is achieved by maintaining 50% of the maximum LED current in the string. For PWM dimming on the other hand, LED brightness of 50% is achieved by providing the maximum LED current at 50% duty cycle.

The main advantage of analog dimming is that it can be easily implemented by changing the reference voltage of the error amplifiers in the current controller. However, this method suffers from shifting of LED colors as the current across the LED are changing in this method [5]. Whereas, in PWM dimming, this problem does not exist as the current across the LED are not changing since the maximum LED current is provided at different duty cycles in this method to achieve dimming. However, this method suffers from noise and EMI interferences [6].

In this implementation of a 3x3 RGB pixel for display, it requires individual dimming control of each color of each pixel to create multiple spectrums of colors. A total of 27 dimming control signals are required for a 3x3 RGB pixel and for ease of implementation, an analog dimming is implemented in this LED-driver system.

The dimming control signal is generated using one 3-wire data bus (active-low Chip Select, Clock, and DATA) by connecting the serial-data-in (SI) of one digital-potentiometer (DP) to the serial-data-out (SO) of the other DP in order to communicate with multiple DPs as shown in Fig. 4. This form of connecting multiple devices in a same data bus in series is called daisy-chain. Daisy-chaining multiple DPs allows freeing up I/O pins in the microcontroller and, thus requiring only two pins from the microcontroller i.e., clock and chip select pins, to communicate with multiple slave devices connected in series in the same data bus. It would be difficult to implement dimming in this LED-driver system since it requires 27 dimming control signals. This is because the microcontroller will require 3 individual outputs to control each slave devices.

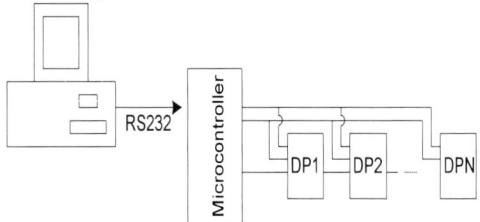

Figure 4. Daisy Chained Digital-Potentiometers.

The DP receives its dimming control signal directly from a personal computer (PC) via the serial communication ports. Therefore, a communication is established between the PC and the PIC18F4431 dimming controller. In this approach, as soon as DP1 receives its data from the dimming controller, this data is clocked into the DP1's shift register, as long as its active-low chip select pin remain low [7]. This data is then processed by DP1 and arrives at SO output pin. As the DP's slave devices are connected in series, the SO of DP1 is connected to SI of DP2, therefore the data is clocked into DP2's shift register and propagates through DP2's SO output. This process continues until the data is propagated through the entire daisy-chain until each of the DP's has received its command signal while the active-low chip select is low.

In this demonstration, a SPI interfaced daisy-chained DPs is used to generate the dimming control signals. In an integrated circuit implementation, an inter-integrated circuit (I2C) interface with each DACs having its own I2C address can be easily implemented.

IV. Experimental Results

A prototype of the proposed LED driver was designed and verified using buck converters operating at 50 kHz. The buck converter was chosen for this application because the output voltage in a buck is lower than its input voltage i.e., the nominal forward voltages of red, green, and blue LEDs are 2.2V, 3.6V, and 3.6V, respectively [7]. The output of the converters was used to drive a 3x3 pixel for LED display panels. The measured efficiency of the each driver is shown in Fig. 3 with respect to input voltages. At an input voltage of 5V, the efficiencies are 91.5%, 95.7%, and 95.7% for the RGB color LEDs. These efficiencies are higher than those reported in [8]. The measured change in output current with respect to input voltage, i.e., the line regulation, is shown in Fig. 4. As can be seen from Fig. 4, the output currents of the red, green, and blue LEDs remain constant at 100mA at input voltages from 5V to 13V. Fig. 5 shows the oscilloscope waveforms of the dimming signal as it changes from 100% to 0% and the voltage at the drain of the current-controlled MOSFET. As shown, the drain voltage changes from 0 mV to 250mV. Figure 6 shows the RGB 9 pixel display in different colors.

978-1-4244-4782-4/10 $26.00 © 2010 IEEE   752

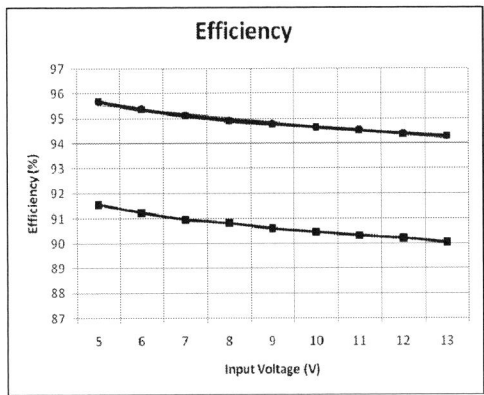

Figure 3 – Efficiency of the red, green, and blue LED drivers with respect to input voltage variations.

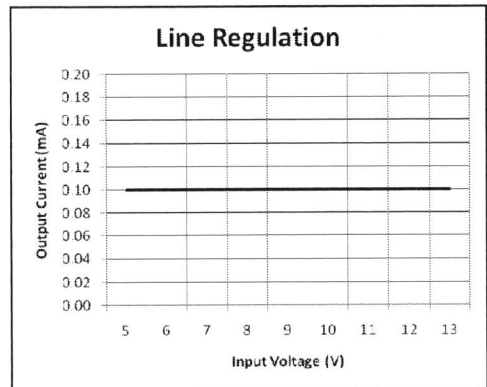

Figure 4 – Load current of red, green, and blue LED drivers with respect to input voltage variations.

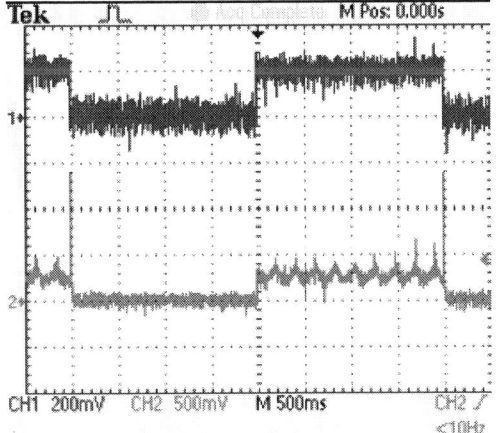

Figure 5. Analog dimming signal and drain voltage of the current controller.

Figure 6. A 3x3 RGB Pixel.

III. Conclusion

A RGB-driver for driving multiple pixels in LED display panels using a microcontroller is demonstrated in this paper. In this approach, the output voltage of the switching converters and the current in the individual LED are controlled separately, thus increasing the stability of the driver. The use of a digital microcontroller to control the output voltages of the switching converters for each color LEDs leads to reduction in size, weight, and cost of the driver. In this driver, the efficiency is maximized by selecting the minimum drain voltage of the MOSFETs in the current controllers to regulate the duty cycle, and by removing the need for external sensing of the voltage drop across the multiple current controllers. The driver was successfully implemented to show an efficiency of greater than 92% in a 3x3 RGB pixel.

VI. References

[1] Yuequan Hu, and M. M. Jovanovic, "A Novel LED Driver with Adaptive Drive Voltage," *Applied Power Electronics Conference and Exposition, APEC.* pp. 565-571. Feb 2008.

[2] Yuequan Hu, and M.M. Jovanovic, "LED Driver with Self-Adaptive Drive Voltage," *IEEE Transactions on Power Electronics*, Vol. 23, No. 6, pp. 3116 – 3125, Nov 2008.

[3] M. Doshi and R. Zane, "Digital architecture for driving large LED arrays with dynamic bus voltage regulation and phase shifted PWM," *IEEE Applied Power Electronics Conference (APEC) Proc.,* pp. 287-293, 2007.

[4] A. Torres, J. Garcia, M. Rico Secades, A. J. Calleja, and J. Ribas, "Advancing towards digital control for high power LED driver," IEEE Symposium on Industrial Electronics (ISIE), pp. 3053-3056, 2007

[5] M. Day. "LED-driver Considerations," Analog Application Journal, pp. 14-17, 2004

[6] S. S. Ang and Alejandro Oliva, *Power-Switching Converters*, Second Edition CRC Press, 2005.

[7] Y. K. Lo, K. H. Wu, K. J. Pai, and H. J. Chiu, "Design and Implementation of RGB LED Drivers for LCD Backlight Modules," *IEEE International Symposium on Industrial Electronics (ISIE)*, pp. 584-587, June 2007.

A Low Investment Single-phase to Three-phase Converter Operating with Reduced Losses.

José A. A. Dias[1], Euzeli C. dos Santos[2], Cursino B. Jacobina[2]

[1,2]Departamento de Engenharia Elétrica, Universidade Federal de Campina Grande
Caixa postal 10105, CEP 58109-970 Campina Grande - PB - Brazil.
[1]Instituto Federal de Educação, Ciência e Tecnologia da Paraíba - IFPB
Rua 1 de maio 720, Bairro Jaguaribe
CEP 58015-430, João Pessoa, PB, Brasil
e-mails: arturad@ifpb.edu.br,{euzeli,jacobina}@dee.ufcg.edu.br

Abstract: **This article proposes an active filter configuration used for single-phase to three-phase conversion. The proposed system permits to minimize IGBT dual modules losses and consequently minimizing the cost of the system. Such cost reduction is due to the operation of proposed system in shunt way, thus only part of power follows through the converter. The input and output voltages of the converter operate synchronized, and the synchronization angle between single-phase and three-phase systems is the control variable used to minimize system's losses. A transformer is used to operate as series active filter. Despite the additional cost associated with transformer, this configuration presents only four arms, the investment can be lower than the direct solution with five arms. Steady state analysis, simulated and experimental results are presented as well.**

I. INTRODUCTION

Electric single-phase distribution system presents lower investment than that of three-phase distribution, because of that, largely rural area in Brazil is covered by a system like SWER- Single Wire Earth Return or phase-neutral (two wires) [1], [2]. Moreover, it is important to emphasize the advantage of using three-phase loads instead of single-phase ones. Thus, low cost single-phase to three-phase converter has much scope application in the Brazil rural areas, which can be applied to drive irrigation pumps, mills, cooling and some rectifiers. Many of these configuration can be found in the technical literature [3], [4], [5], [6], [7], [8], [9] [10].

This article proposes an eight switches single-phase to three-phase configuration with four dual IGBT arms and a transformer, called 8C-T, shown in Fig. 1. This configuration comes from the configuration with ten switches contained in the papers [9], [8].This proposed configuration uses a series filter to set the synchronization between the single-phase and three-phase voltages, and a shunt filter to control DC link voltage and control the power factor of the grid. The proposed configuration is used to increase the efficiency of the converter, operating with low IGBT dual module losses. Steady state analysis is performed to determine the DC link voltage and

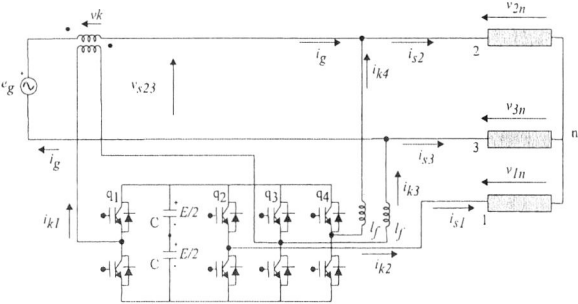

Fig. 1. Single-phase to three-phase proposed converter 8C-T.

current of the arms in order to determine the region of lower losses. The converter losses modules are estimated using loss instantaneous functions determined experimentally for a specific IDM, as proposed in [9], [11]. These functions are used in dynamic simulations to estimate the IGBT dual module losses of the converter. Dynamic simulation and experimental results of the converter are shown for minimum losses operation point. A comparison with two already published configurations, Figs. 2(a) and 2(b) , is performed.

II. PROPOSED SINGLE-PHASE TO THREE-PHASE CONVERTER

The configuration, seen in Fig. 1, has eight switches and a transformer between grid voltage e_g and line voltage v_{s23}. This configuration is studied to suply three-phase loads from single-phase sources. The arm 3 is shared by the shunt filter and series filter. This configuration has two degrees of freedom for optimizing the angle of synchronism between the single-phase and three-phase voltages (γ) and transformer turns ratio (n_e). This angle of synchronism can be seen in phasorial diagram of steady state voltages Fig. 3(a). This angle γ can be changed by control of the configuration, while transformer turns ratio is a physical freedom degree. The initial hypothesis is: it is possible to reduce losses in the converter switches from the angle of synchronism. In other words, it is possible to reduce the flow of power by the converter from the angle of synchronism.

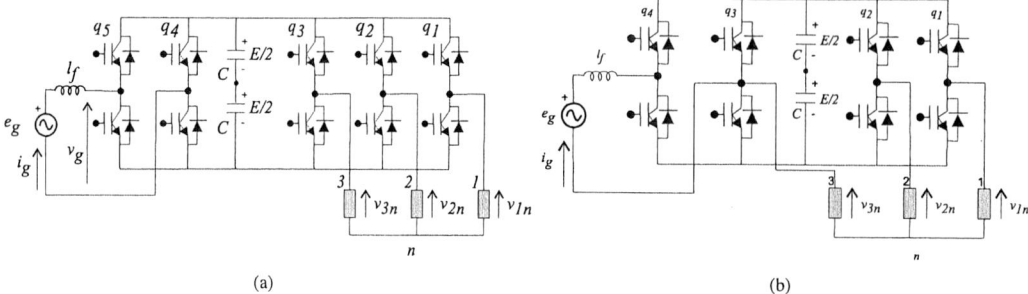

(a)   (b)

Fig. 2.   Configurations already published (a)- 10C, (b) 8C.

A. Steady state analysis

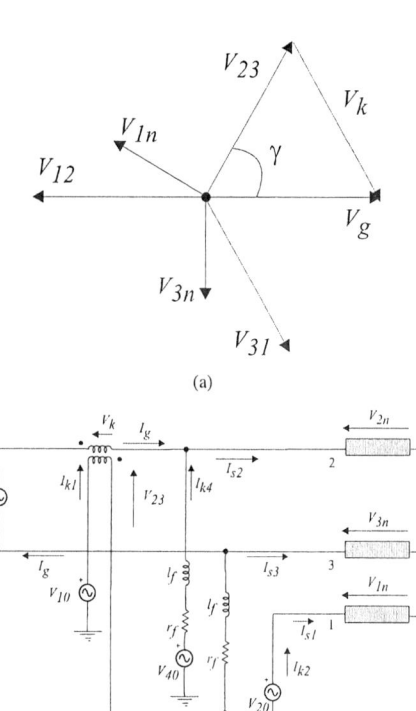

(a)

(b)

Fig. 3.   (a) Phasorial diagram of the sinlge-phase and three-phase systems, (b) Converter steady state circuit.

This model, shown in Fig. 3(b), uses only the fundamental components predominate, ignoring harmonics. The equations of the steady state model (1) to (8) are used to determine the operational characteristics. These equations are solved iteratively by imposing unity power factor at single-phase source. The steady state model is used to determine DC link voltage versus γ, shown in Fig. ??(a) and the arms current (RMS) average versus γ, shown in Fig.??(b).These

two characteristics influency IDM losses.

$$I_{k1} = I_g n_e \tag{1}$$
$$I_{k2} = I_{s1} \tag{2}$$
$$I_{k3} = I_{s3} + I_g(1 - n_e) \tag{3}$$
$$I_{k4} = I_{s2} - I_g \tag{4}$$
$$V_{10} - V_{30} = -\frac{Vg - V_{23}}{n_e} \tag{5}$$
$$V_{20} - V_{30} = -(r_f + jx_f)(I_{k3} + I_{k1}) - V_{31} \tag{6}$$
$$V_{40} - V_{30} = (r_f + jx_f)(I_{k4} - I_{k3} - I_{k1}) + V_{23} \tag{7}$$
$$V_{30} = V_x \tag{8}$$

Where V_x is generic reference voltage.

III. CONTROL OF THE CONFIGURATION

The control of the proposed converter is shown in Fig. 4. The control of the arms 1 and 3 controls the line voltage V_{s23} and the arms 2 and 3 controls line voltage V_{s31}. The arms 3 and 4 controls the DC link voltage and compensates the reactive. These line voltages and DC link control are controlled by close loop control. Controlers v_{s23}, v_{s31} and current controler are based on a unbalanced three-phase current control [12]. Four voltage sensors and one currents sensors are used for this control. The PLL used in experimental practice was based on detection of zero crossing.

A. PWM strategy

The characteristics presented were determined with sinusoidal modulation. However, it is possible to obtain gains in DC link voltage and THD using modulation strategy global SYPWM [13], [10]. The reference poles voltages are shown in (9) to (12).

$$v_{10}^* = v_{13}^* + v_h^* \tag{9}$$
$$v_{20}^* = -v_{32}^* + v_h^* \tag{10}$$
$$v_{30}^* = v_h \tag{11}$$
$$v_{40}^* = v_{43}^* + v_h^* \tag{12}$$

978-1-4244-4782-4/10 $26.00 © 2010 IEEE

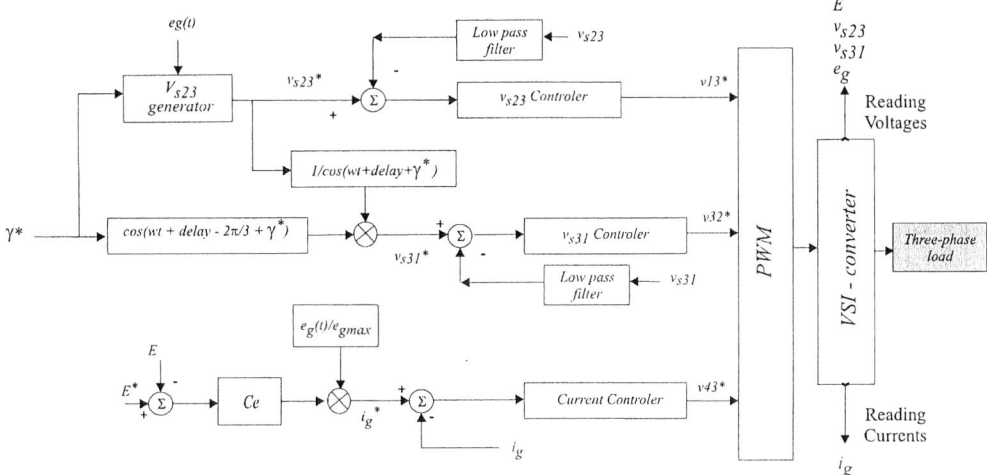

Fig. 4.   Control of the proposed converter.

The generic reference is determined by (13).

$$v_h = -\frac{(V_{\max} + V_{\min})}{2} \qquad (13)$$

Where $V_{\max} = \max\{v_{13}^*, -v_{32}^*, 0, v_{43}^*\}$ and $V_{\min} = \min\{v_{13}^*, -v_{32}^*, 0, v_{43}^*\}$

This procedure allows pole voltages more equalized, leading to lower DC link voltages. The DC link voltage and arms current (RMS) average are shown in Figs. 5(a) and 5(b). Observed in the Fig. 5(a), the point of minimum voltage. The right of this inflection point dominates the filter serie and left the inflection point dominates the shunt filter. The relation between the line voltage and the single-phase source voltage studied in this work is unit, that is to say, satisfies (15). A value of $x_f = 0.1pu$ produces a good grid current wave form and at the same time does not increase much DC link voltage. The value of synchronism angle that minimizes the all arms currents is shown in (14). The values of the pu bases are $\frac{P_{load}}{3} = 1.0pu$ and $V_{phase(RMS)} = 1.0pu$.

$$\gamma = \varphi + \frac{\pi}{6} \qquad (14)$$

Where φ is load phase angle.

$$\frac{v_{line}(RMS)}{e_g(RMS)} = 1 \qquad (15)$$

In this range that minimizes the DC link voltage and the currents of the arms is searched the operating point of lower IDM losses.

IV. ESTIMATION OF THE CONVERTER IGBT DUAL MODULE LOSSES

The converter switches losses are influencied by two variables, the angle of synchronism (γ) and transformers turns ratio (n_e). The determination of a IGBT dual module losses

instantaneous functions helps to determine the points of lower operating IDM losses. The instantaneous losses function of a IGBT dual module CM50DY-24H manufactured by *POWEREX* driven by driver SKHI-10 manufactured by *SEMIKRON* is determined, [9]. The following losses are gotten by experimental procedure: IGBT and diode conduction, IGBT turn-on energy losses, IGTB turn-off energy losses and diode turn-off energy losses (reverse recovery) [11]. A function of six parameters describes the conduction losses and a function of nine parameters describes the switching losses. Such parameters are obtained by statistical regression from the discrete points. Statistics R^2 (correlation coefficient) and S^2 (residual mean square) are used to verify the efficiency of the adjustment [14].

The operational points those minimizes the IGBT dual modules losses are estimated by dynamic simulation carried out in *Matlab-Simulink 7.6.0*, with global SYPWM modulation. Values of synchronism angle that minimize IGBT dual module losses are shown in Fig. 6, for $n_e = 0.5$, 0.75 and 1.0, inductive load with $\cos(\varphi) = 0.8$.

V. COMPARISON OF CONFIGURATIONS

The proposed configuration, 8C-T, is compared with the configurations already published, 10C and 8C. The three configurations are operated on the same conditions, minimum losses in IGBT modules, voltage ratio equal to eq. 15, with inductive load with power factor 0.8, frequency modulation of $10kHz$ and junction temperature of $25°C$.

The point of minimum losses in modules for configuration 8C-T is $n_e = 1$ and $\gamma = 60°$, with $x_f = 0.1pu$.

The point of minimum losses in modules for configuration 8C is when the angle between the phase voltage 1 is ahead of $60°$ with respect to single-phase voltage source, with $x_f = 0.2pu$.

Three criteria are used to compare the the alternatives, losses in IGBT modules, the minimum DC link voltage and arms

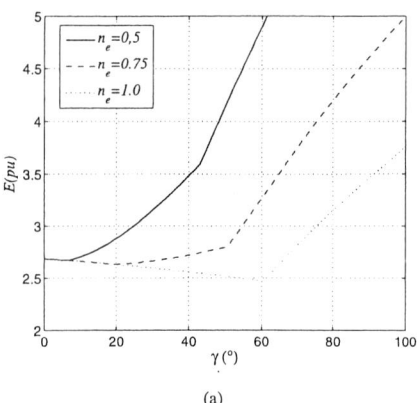

(a)

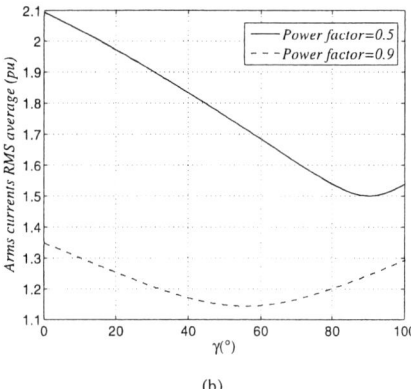

(b)

Fig. 5. (a)Minimum DC link voltage versus γ, (b) Arms current (RMS) average versus γ..

TABLE I

PERFORMANCE OF EACH CONFIGURATION OF THE CRITERIA DISCUSSED.

	$10C$	$8C$	$8C - T$
$Losses(\%)$	10.25	7.74	7.50
$E_{\min}(pu)$	2.44	2.44	2.50
Arms $i_{RMS}(pu)$	1.44	1.32	1.17

current (RMS) average. Losses influence the cost over the useful life of the converter. The minimum bus voltage and arms current (RMS) average influences the initial investment in IGBT modules. It can be seen in Tab. I the performance of each configuration of the criteria discussed.

The configurations $8C$ and $8C - T$ have similar losses but significantly less than the setting $10C$. This is due to a module unless the configuration 10C.

Configuration $8C$, operating synchronously in the operating point of minimum losses, has the same minimum DC link voltage that configuration $10C$. This value is very close to the configguração proposal $8C - T$.

Configuration 8C-T operates at arm's current 12.4% less than the setting $8C$ and 22.2% less than the setting $10C$. This effect is due to configuration $8C - T$ operate in shunt.

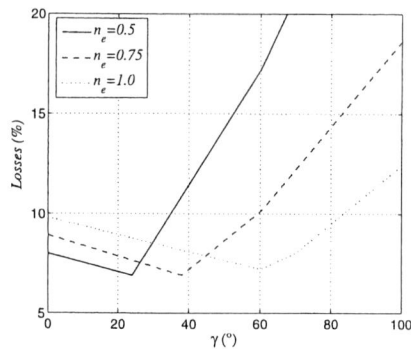

Fig. 6. Estimated losses in the modules.

VI. DINAMIC SIMULATION AND EXPERIMENTAL RESULTS

The following parameters were used in the simulation and experimental results:
* Four IGBT Dual module (F_{PWM}=10kHz).
* PC-based platform and programming language based on C.
* Load of 500 W linear (power factor = 0.6 inductive).
* Transformer $n_e = 1$.
* $V_{line} = 220V$, $e_g = 220V$, $\gamma = 60°$.
* $l_f = 20mH$ (0.08pu).
* $E^* = 450V$ (3.54pu) ($E_{\min}$ obtained by steady stante simulation 2.96pu).
* Dynamic simulation performed in the *Matlab-Simulink* 7.6.0 with Euler's method and step $1.0E - 6s$.

In Fig. 7 has shown dynamic simulation results and Fig. 8 experimental results. The DC link voltage control behaved satisfactorily. The single phase souce power factor is practically unitary and the grid current has an acceptable wave form The line voltages have acceptable balance and correct sequence.

In Fig. 9 are shown the amplitude spectrum of single phase current and line voltages v_{s23} and v_{s31} with THD (%) for the first $50th$ harmonics. These results indicate for a good waveform for both current and for the line voltages.

VII. CONCLUSIONS

The performance of the proposed configuration is very close to the configuration 8C, however has current average RMS lower arm. This is due to the proposed configuration operate in shunt. This can be an advantage in the purchase of dual IGBT modules. Another advantage is the distribution of losses in the modules. Losses in the modules in the proposed configuration is more uniform. Settings 8C and 10C have losses concentrated in modules connected to the single-phase side.

Therefore, it is an interesting option for use in rural areas in Brazil.

978-1-4244-4782-4/10 $26.00 © 2010 IEEE   758

REFERENCES

[1] M. A. Kashern, "Distribuited generation as voltage support for single wire earth return systems," *IEEE- Transaction on Power Delivery*, vol. 19, pp. 1002–1011, july 2004.

[2] P. J. Wolfs, "Capacity improvement for rural single wire earth return systems," *IEEE-IPEC 2005*, vol. -, pp. 1–8, december 2005.

[3] M. D. Bellar, B. K. Lee, and J. L. S. Neto, "Tolology selection of ac motor drive systems with soft-stating for rural application," *IEEE-Power Electronics Specialists Conference - PESC 2005*, vol. 36, pp. 2698–2704, - 2005.

[4] M. D. Bellar, B. K. Lee, B. Fahimi, and M. Ehsani, "An ac motor drive with power fator control for low cost applications," *IEEE-Applied Power Eletronics Conference - Piscataway*, vol. 1, pp. 601–607, - 2001.

[5] R. J. Cruise, C. F. Landy, and M. Culloch, "Evaluation of a reduce topology phase-converter for rural areas in southern africa," *IEEE-AFRCON 1999*, vol. 2, pp. 859–864, setember 1999.

[6] R. Madorell and J. Pou, "Modulation techniques for a low-cost single-phase to three-phase converter," *IEEE-International Symposium on Industrial Electronics*, vol. 2, pp. 1279–1284, may 2004.

[7] E. N. Tshivhilinge and M. Malengret, "A practical control of a cost reduced single-phse to three-phase converter," *IEEE-International Symposiun Industry Electronics - 99*, vol. 02, pp. 445–449, july 1998.

[8] J. A. A. Dias, E. C. dos Santos, and C. B. Jacobina, "Aplicao de filtro universal ao acionamento trifsico a partir de fontes monofsicas," *XVII Congresso Brasileiro de Automtica-2008 Juiz de Fora-MG*, vol. -, pp. , setember 2008b.

[9] J. A. A. Dias, E. C. dos Santos, C. B. Jacobina, and E. R. C. da Silva, "Application of single-phase to three-phase converter motor drive systems with switches losses reduction," *IEEE- 10th Brasilian Power Electronics Conference - COBEP 2009*, vol. -, pp. –, setember 2009.

[10] C. B. Jacobina, E. R. C. Silva, M. B. R. Corra, and A. M. Lima, "Ac motor drive systems with a reduced-switch-count converter," *IEEE-Transactions on industry applications*, vol. 39, pp. 1333–1342, setember/october 2003.

[11] S. M. Nielsen, L. N. Tutelea, and U. Jaeger, "Simulation with ideal switch models combined with measured loss data provided a goog estimate of power loss," *IEEE-Industry applications conference*, vol. 5, pp. 2915–2922, october 2000.

[12] C. B. Jacobina, E. R. C. Silva, T. M. M. B. R. Corra, and A. M. Lima, "Current control of unbalanced electrical systems," *IEEE-Transactions on industry applications*, vol. 48, pp. 517–525, june 2001.

[13] E. C. Santos, *Sistema de converso esttico com nmero reduzido de componentes - Tese de doutorado*. PB: UFCG- Universidade Federal de Campina Grande, 2006.

[14] N. R. Draper and H. Smith, *Applied regression analysis 3rd ed.* USA: John Wiley and Sons, 1998.

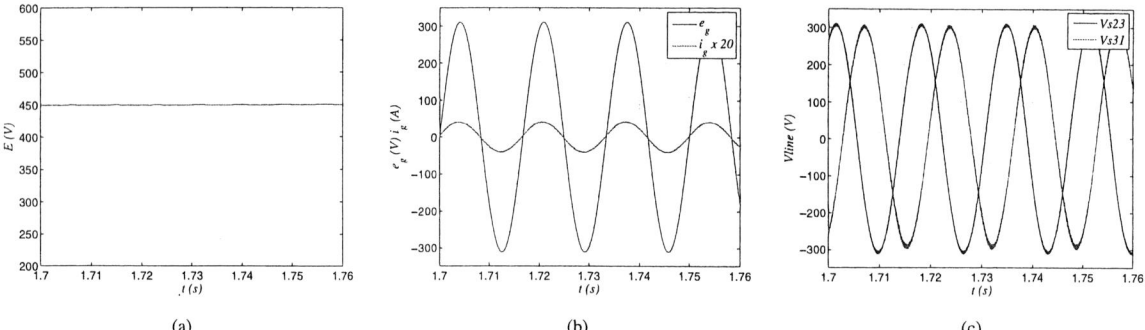

Fig. 7. Dynamic simulation results (a) DC link voltage, (b) Single-phase voltage source e_g and currents i_g , (c)- Line voltages v_{s23} and v_{s31}.

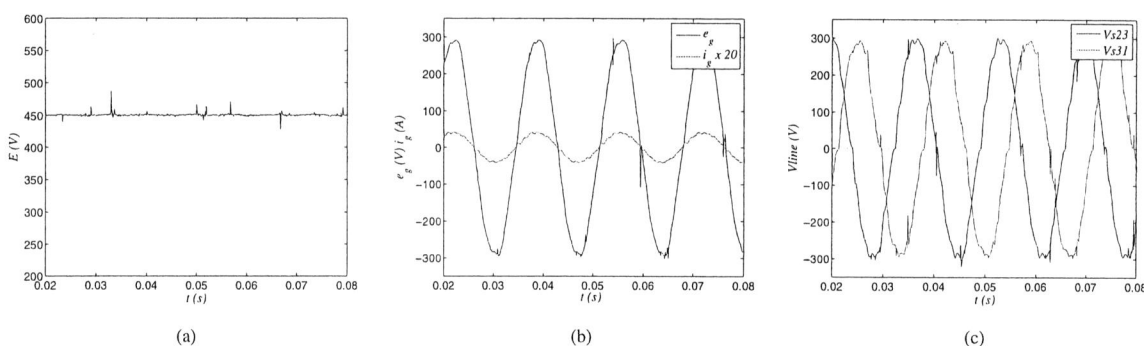

Fig. 8. Experimental results (a) DC link voltage, (b) Single-phase voltage source e_g and currents i_g , (c)- Line voltages v_{s23} and v_{s31}.

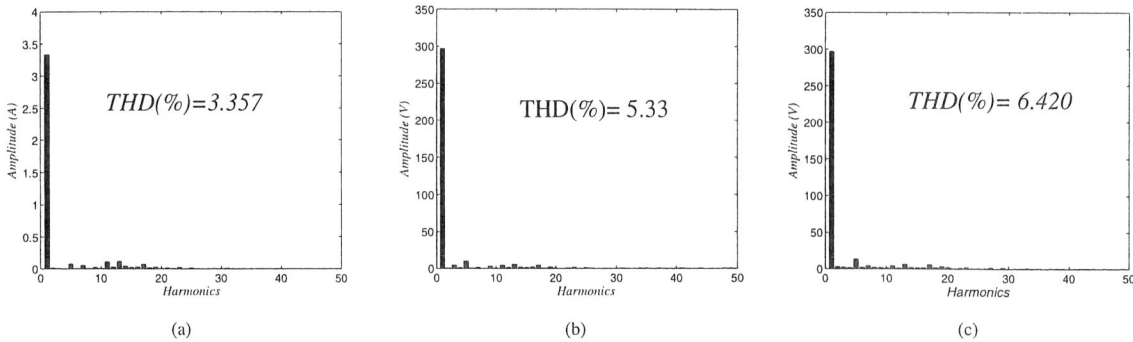

Fig. 9. Spectrum of amplitude (experimental): (a)- Single-phase currente ig, (b)- Line voltage v_{s23} (c)- Line voltage v_{s31}.

978-1-4244-4782-4/10 $26.00 © 2010 IEEE   760

Voltage and Power Balance Control for a Cascaded Multilevel Solid State Transformer

Tiefu Zhao, Gangyao Wang, Jie Zeng, Sumit Dutta, Subhashish Bhattacharya and Alex Q. Huang
Future Renewable Electric Energy Delivery and Management (FREEDM) Systems Center
North Carolina State University
Raleigh, NC 27695, USA
Email: {tzhao, gwang3, jzeng2, sdutta2, sbhatta4, aqhuang}@ncsu.edu

Abstract— In this paper, a 20kVA Solid State Transformer (SST) based on 6.5kV IGBT is proposed for interface with 7.2kV distribution system voltage. The proposed SST consists of a cascaded multilevel AC/DC rectifier stage, a Dual Active Bridge (DAB) converter stage with high frequency transformers and a DC/AC inverter stage. Based on the single phase d-q vector control, a novel control strategy is proposed to balance the rectifier capacitor voltages and the real power through the DAB parallel modules. Furthermore, the power constraints of the voltage balance control are analyzed. The SST switching model simulation demonstrates the effectiveness of the proposed voltage and power balance controller. A 3kW SST scale-down prototype is implemented. The experiment results verify the single phase d-q vector controller for the SST cascaded multilevel rectifier.

I. INTRODUCTION

The Solid State Transformer (SST) is one of the key elements in the proposed Future Renewable Electric Energy Delivery and Management (FREEDM) Systems. In the electric configuration of the FREEDM system shown in Fig.1, low voltage (120V), residential class Distributed Renewable Energy Resource (DRER), Distributed Energy Storage Device (DESD), and loads are connected to the distribution bus (12kV) through a power electronics based Intelligent Energy Management (IEM) subsystem.

The solid state transformer is within the IEM and used to enable active management of DRER, DESD and loads, rather than a 60Hz conventional transformer. The SST has the features of instantaneous voltage regulation, voltage sag compensation, fault isolation, power factor correction, harmonic isolation and DC output [1-3]. The SST will have a 400V DC port that will facilitate more efficient connection of certain classes of DRERs and DESDs. Acting very much like an energy router, each SST will have bi-directional energy flow control capability allowing it to control active and reactive power flow and to manage the fault currents on both the low voltage and high voltage sides. Its large control bandwidth provides the plug-and-play feature for distributed resources to rapidly identify and respond to changes in the system.

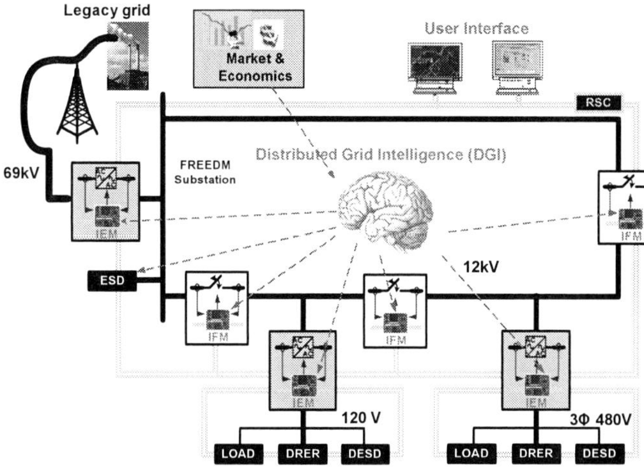

Fig.1. FREEDM Systems Diagram (IEM: Intelligent Energy Management, IFM: Intelligent Fault Management, DRER: Distributed Renewable Energy Resource, DESD: Distributed Energy Storage Device)

In order to direct interface with the 12kV distribution voltage level, series devices or multilevel converter modules are still required due to today's semiconductor voltage level (6.5kV for silicon device and 10kV for SiC MOSFET). This paper proposes a 20kVA SST based on the cascaded H-Bridge multilevel rectifier to reach the required voltage levels. One of the main disadvantages of the cascaded H-bridge rectifier is the voltage unbalance on the DC bus voltages of different H Bridges. The similar voltage unbalance issue was addressed in the cascaded H-Bridge inverter based STATCOM and drive applications [4-7]. In the STATCOM application, where the rectifier is used for reactive power compensation, the references use low frequency optimal PWM modulation technique, so the voltage balance is realized by shifting the voltage waveforms. In contrast, drive application generally transfers real power. In the references, the DC bus voltage is balanced by using different switching patterns to charge and discharge each H-Bridge capacitor, but the reactive power is not controlled. Differently from the STATCOM and drive application, the SST requires a high frequency modulation and both real and reactive power control. Therefore, the voltage

---

This work was supported by ERC Program of the National Science Foundation under Award Number EEC-08212121.

balance and power balance control is indispensible for the SST controller design.

In this paper, the modeling of the SST, including AC/DC rectifier, dual active bridge converter are developed. A single phased d-q vector controller is applied in the rectifier stage to control both the real and reactive power. Based on the single phase d-q control, a voltage balance control method is proposed to solve the voltage unbalance that could appear on the DC voltages of different H-bridges. Meanwhile, a power balance control method is proposed to regulate the real power transferring through the DAB parallel modules. The proposed voltage and power control is verified by the switching model simulation. A 3kW SST scale-down prototype is implemented by using 600V IGBT. The SST prototype experiment also verifies the single phase d-q vector control.

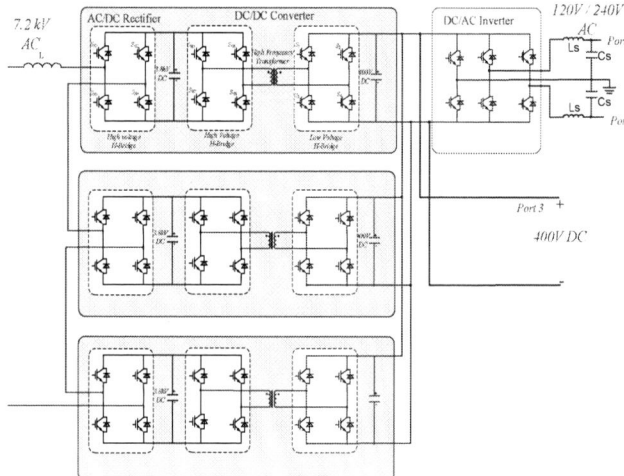

Fig. 2 Topology of Solid State Transformer

II. SST MODELING AND CONTROL

The solid-state transformer converts AC to AC for step-up or step-down function the same as a conventional transformer, but with much more advanced functionalities [3]. As shown in Fig. 2, the solid-state transformer consists of a cascaded multilevel AC/DC rectifier, Dual Active Bridge (DAB) converters with high frequency transformers and a DC/AC inverter.

The basic configuration of the proposed 20kVA SST interfaced to 12 kV distribution voltage with center-tapped 120V single-phase output is shown in Fig.2. The SST is rated as single phase input voltage 60 Hz, 7.2kV, output voltage 60 Hz, 240/120V, 1 phase/3 wires. The SST consists of a high voltage high frequency cascaded H-Bridge AC/DC rectifier that converts 60Hz, 7.2kV AC to three cascaded 3.8kV DC buses, three high voltage high frequency DC-DC converters that convert 3.8kV to 400V DC bus and a voltage source inverter (VSI) that inverts 400V DC to 60 Hz, 240/120V, 1 phase/3 wires. The switching devices in high voltage H-bridges and low voltage H-bridges are 6.5kV IGBT and 600V IGBT respectively. The switching frequency of the HV-IGBT devices is 1080Hz, because the device current is very low which results in low switching losses. The 20 kVA SST unit is

envisioned as a building block of IEM (shown in Fig.1) and also for construction of a larger rated SST.

A. Rectifier single phase d-q vector control

The AC/DC rectifier stage converts the single phase 7.8kV AC voltage to three DC output while controlling the reactive power at the input side. The rectifier consists of three cascaded H-bridges with each reference DC bus voltage 3.8kV. The average differential equations of the rectifier are:

$$\frac{di_a}{dt} = \frac{3E}{L_s}d_a - \frac{V_{pcca}}{L_s} - \frac{R_s}{L_s}i_a \tag{1}$$

$$\frac{dE}{dt} = -\frac{E}{R_L C} - \frac{d_a i_a}{C} \tag{2}$$

Where, i_a is the input side current, V_{pcca} is the input voltage, R_s is the input line resistance, L_s is the input inductor, E is the DC bus voltage, C is the rectifier DC capacitor, d_a is the rectifier PWM duty cycle.

The single phase d-q vector control is applied in the rectifier controller. An imaginary phase M which is 90 degree lagging the original phase A is hypothesized. The differential equations for the imaginary phase are:

$$\frac{di_m}{dt} = \frac{3E}{L_s}d_m - \frac{V_{pccm}}{L_s} - \frac{R_s}{L_s}i_m \tag{3}$$

$$\frac{dE_m}{dt} = -\frac{E_m}{R_L C} - \frac{d_m i_m}{C} \tag{4}$$

Where, i_m is the input current of the imaginary phase, V_{pccm} is the input voltage of the imaginary phase, E_m is the DC bus voltage of the imaginary phase, d_m is the rectifier PWM duty cycle of the imaginary phase. Based on the small ripple approximation, $E_m = E$. Then combine the equations for two phases, and rewrite the equations:

$$\frac{d\vec{i}_{am}}{dt} = \frac{3E}{L_s}\vec{d}_{am} - \frac{\vec{V}_{pccam}}{L_s} - \frac{R_s}{L_s}\vec{i}_{am} \tag{5}$$

$$\frac{dE}{dt} = -\frac{E}{R_L C} - \frac{\vec{d}_{am}^{\,T}\vec{i}_{am}}{2C} \tag{6}$$

Where,

$$\vec{i}_{am} = \begin{bmatrix} i_a \\ i_m \end{bmatrix}, \vec{d}_{am} = \begin{bmatrix} d_a \\ d_m \end{bmatrix}, \vec{V}_{pccam} = \begin{bmatrix} V_{pcca} \\ V_{pccm} \end{bmatrix}$$

The single phase d-q transformation is applied to equations (5) and (6) [8], and the differential equations in d-q coordinates are derived.

$$[x]_{dq} = [T] \cdot [x]_{am} \tag{7}$$

Where,

$$T = \begin{bmatrix} \sin(\theta) & -\cos(\theta) \\ \cos(\theta) & \sin(\theta) \end{bmatrix} , \quad \theta = 2\pi f_L , \quad f_L \text{ is line}$$

frequency.

Then the d-q axis equation of the single phase H-bridge rectifier is given in equation (8) and (9).

$$\frac{d}{dt}\begin{bmatrix} i_d \\ i_q \end{bmatrix} = \frac{3E}{L_s}\begin{bmatrix} d_d \\ d_q \end{bmatrix} - \frac{1}{L_s}\begin{bmatrix} v_{pccd} \\ v_{pccq} \end{bmatrix} - \begin{bmatrix} \dfrac{R_s}{L_s} & -\omega \\ \omega & \dfrac{R_s}{L_s} \end{bmatrix}\begin{bmatrix} i_d \\ i_q \end{bmatrix} \quad (8)$$

$$\frac{dE}{dt} = -\frac{E}{R_L C} - \frac{1}{2C}\begin{bmatrix} d_d \\ d_q \end{bmatrix}^T \begin{bmatrix} i_d \\ i_q \end{bmatrix} \quad (9)$$

With the chosen PLL, the voltage vector is aligned with the direction of the d-axis during steady state. The grid voltage component in the d-direction is equal to its peak value and the q-component of the grid voltage is equal to zero. Thus, the d-component of the current vector (in steady state parallel to the grid voltage vector) becomes the active current component (d-current) and the q-component of the current vector becomes the reactive current component (q-current) [9]. The decoupled d-q vector controller for each H-bridge is shown in Fig.3. The three SPWM carriers for the cascaded H-bridge are phase shifted so that the rectifier has seven voltage levels to reduce the voltage stress and harmonics.

The control aim of the controller is to control the reactive power (or power factor) and regulate the DC bus voltage. The two PI regulators in d-q axis loop are designed based on the bode plots.

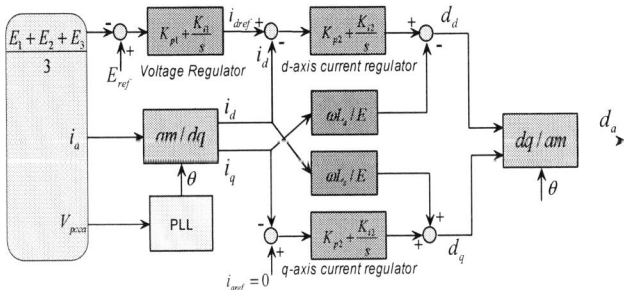

Fig. 3 Rectifier single phase d-q decoupled controller

B. Modeling and Control of DAB

The Dual Active Bridge (DAB) consists of a high voltage H-Bridge, a high frequency transformer and a low voltage H-bridge. The rectifier regulates the high voltage DC link voltage and controls the input current to be sinusoidal from the AC input. The low voltage DC link is regulated by the DAB converter.

The dual active bridge topology offers zero voltage switching for all the switches, relatively low voltage stress for the switches, low passive component ratings and complete symmetry of configuration that allows seamless control for

bidirectional power flow. Real power flows from the bridge with leading phase angle to the bridge with lagging phase angle, the amount of power transferred being controlled by the phase angle difference and the magnitudes of the DC voltages at the two ends as given by equation (10). [10].

$$P_o = \frac{V_{dc} V_{dc\_link}}{2L f_H} d_{dc}(1 - d_{dc}) \quad (10)$$

where, V_{dc} is input side high voltage DC voltage, f_H is switching frequency, L is leakage inductance, V_{dc_link} is output side low voltage DC link voltage referred to input side and d_{dc} is ratio of time delay between the two bridges to one-half of switching period.

For the DAB converter, the phase shift control is used to regulate the low voltage DC voltage to the reference 400V under different load conditions. First the difference between the low voltage DC voltage V_{dc} and the reference voltage is compared. Then the phase shift angle is adjusted by the PI controller to regulate V_{dc} according to this voltage error.

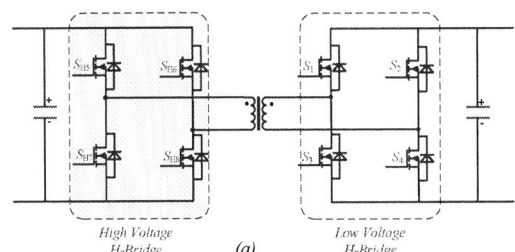

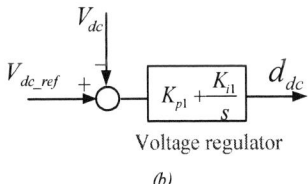

Voltage regulator

(b)

Fig. 4 (a) Dual active bridge circuit (b) DAB voltage controller

III. VOLTAGE AND POWER BALANCE CONTROL

Since the rectifier stage of the SST consists of three H-Bridges in series, the voltage unbalance could appear on the DC bus voltages (E1, E2, and E3, shown in Fig. 5) due to the device loss mismatching and H-Bridge real power difference. The unbalanced voltage may result in capacitor over-voltage.

The DAB stage consists of three DAB modules in parallel. The power unbalance (P1, P2 and P3, shown in Fig.5) can be caused by the transformer parameter (leakage inductance or turns ratio) mismatching and DC bus voltage differences. The power unbalance may cause device over-current issue.

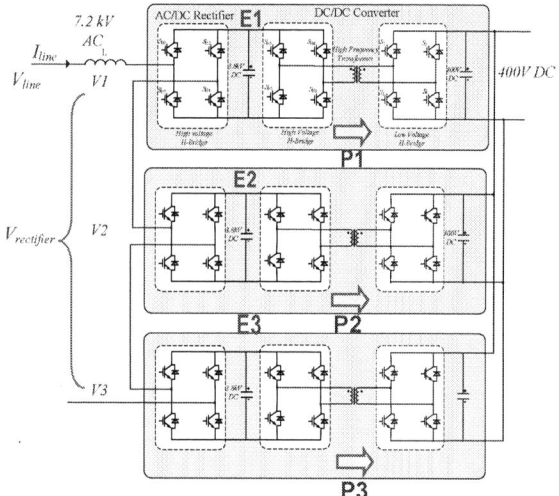

Fig. 5 Voltage and power unbalance in SST topology

A. Voltage balance control

The single phase d-q vector controller for the rectifier stage regulates the total DC bus voltage and controls the reactive power. Based on the single phase d-q control, a voltage balance control method is proposed to solve the voltage unbalance on the DC voltages of different H-bridges. In Fig.6, the d_d and d_q are calculated according to the single phase d-q vector control. Fig.7 is the voltage balance controller, the individual DC bus voltages of the first two H-Bridges, E_1 and E_2 are compared with the DC bus voltage reference E_{ref} to generate a d-axis compensation Δd_{d1} and Δd_{d2} by a PI regulator. Then Δd_{d1} and Δd_{d2} are added to the original d_d. Therefore, d_{d1} for the first H-Bridge and d_{d2} for the second H-Bridge are adjusted so that the real power of each H-Bridge can be changed. The real power of the H-Bridge with a lower (or higher) DC bus voltage is increased (or decreased) to eliminate the voltage unbalance. For the third H-Bridge, $\Delta d_{d3} = -\Delta d_{d1} - \Delta d_{d2}$, so the total DC bus voltage is still regulated.

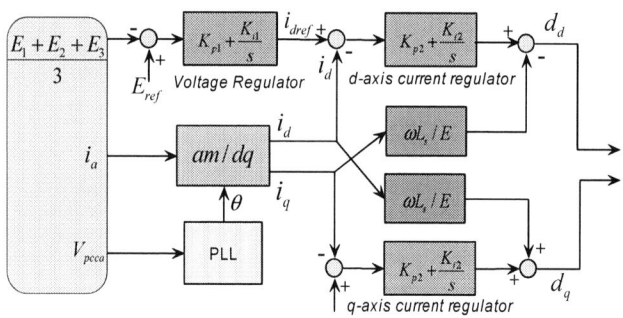

Fig.6 Rectifier single phase d-q vector controller

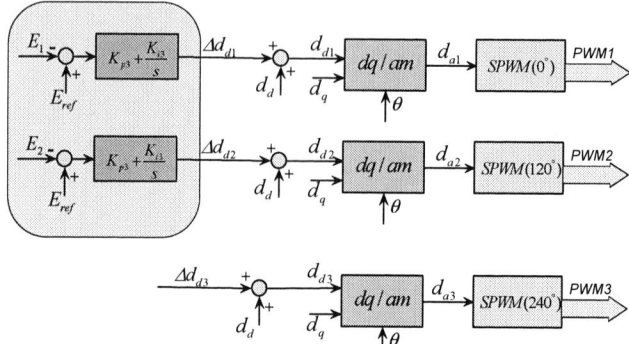

Fig. 7 Voltage balance control based on single phase d-q vector

The switching model simulation of the 20kVA SST cascaded H-Bridge rectifier is implemented in Matlab/Simulink. In the simulation, different resistor loads are connected to the DC buses, E1, E2, and E3. The resistor loads R1=1.0 pu, R2 = 0.8 pu, R3 = 1.0 pu (20% load unbalance). Fig. 8 shows the three DC bus voltages without balance control. The H-Bridge which transfers more power has the highest DC bus voltage. Fig. 9 shows the three DC bus voltages with balance control. The three DC bus voltages are equal in the steady state.

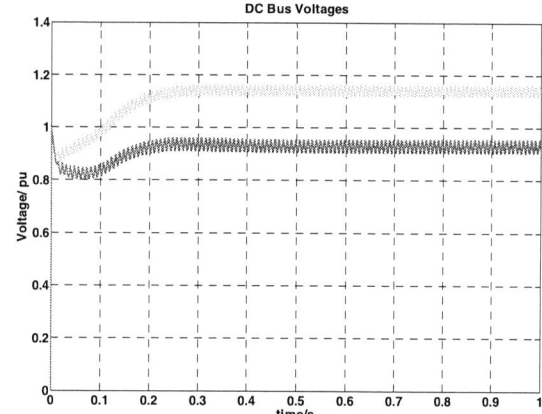

Fig. 8 DC bus voltages without voltage balance control

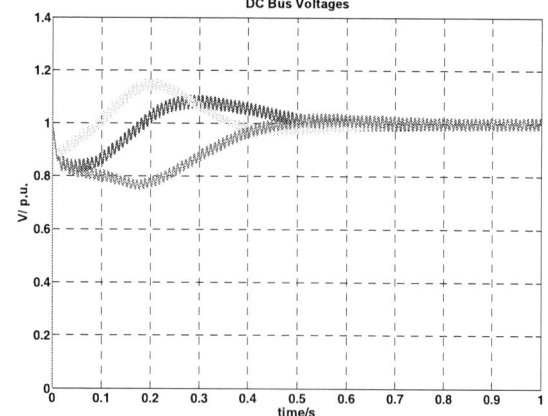

Fig. 9 DC bus voltages with voltage balance control

B. Voltage balance constraints

In the proposed voltage balance controller, the real power of each H-Bridge is adjusted to eliminate the DC bus voltage unbalance. However, the adjustable real power range of each H-Bridge is limited by the input AC voltage and the DC bus voltage reference. In the designed SST system, the input voltage is 7.2kV and each DC bus voltage reference is 3.8kV. As shown in Fig. 10, the constraints of the voltage vectors are given by equation (10)-(15).

$$V_1 + V_2 + V_3 = V_{line} + j\omega L \tag{11}$$

$$V_{d1} + V_{d2} + V_{d3} = V_{line} = 3*3.8kV \tag{12}$$

$$V_{d1} \leq |V|_1 = d_1 E_1 = d_1 *3.8kV \leq 3.8kV \tag{13}$$

$$V_{d3} \leq |V|_3 = d_3 E_3 = d_3 *3.8kV \leq 3.8kV \tag{14}$$

$$V_{d2} \leq |V|_2 = d_2 E_2 = d_2 *3.8kV \leq 3.8kV \tag{15}$$

From the above equations, the voltage vector constraints for the voltage d-axis components Vd1, Vd2 and Vd3 are given by equation (16)-(18).

$$2.6kV \leq V_{d1} \leq 3.8kV \tag{16}$$

$$2.6kV \leq V_{d2} \leq 3.8kV \tag{17}$$

$$2.6kV \leq V_{d3} \leq 3.8kV \tag{18}$$

The real power of each H-Bridge is calculated as:

$$P_1 = I_{line} V_{d1}, P_2 = I_{line} V_{d2}, P_3 = I_{line} V_{d3} \tag{19}$$

Therefore, the real power range of each H-Bridge is:

$$2.6kV * I_{line} \leq P_1 \leq 3.8kV * I_{line} \tag{20}$$

$$2.6kV * I_{line} \leq P_2 \leq 3.8kV * I_{line} \tag{21}$$

$$2.6kV * I_{line} \leq P_3 \leq 3.8kV * I_{line} \tag{22}$$

In order to balance the DC bus voltages, the real power of each H-Bridge has to be within the range in equation (20) - (21). Based on the above analysis, when Vd1=Vd2=3.2kV, Vd3=3.8kV, a maximum 15% power unbalance is allowed to maintain the balanced DC buses. Therefore, the power balance control is needed to guarantee the H-Bridge power meets the constraints.

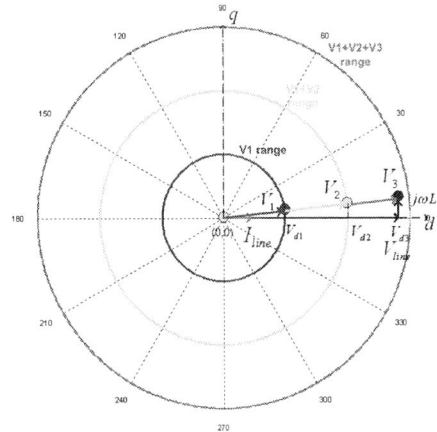

Fig. 10 Rectifier voltage vector constrains

C. Power balance control

Due to the parameter variation of the high frequency transformers, such as leakage inductance and turns ratio, the three DAB currents can be different, which results in a power unbalance of the three DAB modules. A power balance control method is proposed to regulate the real power transferring through the DAB parallel modules. As shown in Fig.11, the voltage regulator compares the low voltage DC voltage V_{dcL} with the reference V_{dcL_ref} and generates the power references P_{ref} for the three DAB modules. Then the power regulator compares the calculated average power of each DAB module with P_{ref} and generates the phase shift angles φ_1, φ_2, φ_3 for the three DAB modules. Fig. 12 shows how the average power calculator calculates the average power in each switching cycle (3 kHz). In the calculation, $P = \int_0^\pi V_{dcH} i_p dt$, the primary DC voltage V_{dcH} can be considered constant in a switching cycle. So the power calculation only involves the summation of current, which is easy to be implemented in DSP.

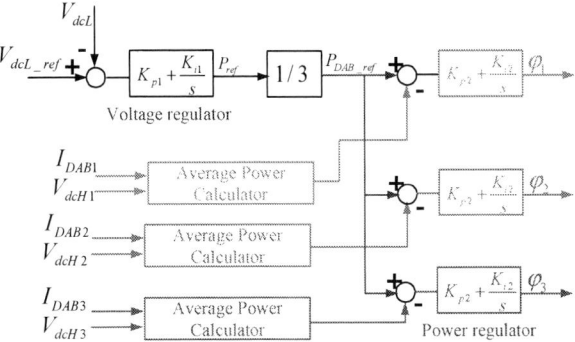

Fig. 11 Power balance controller

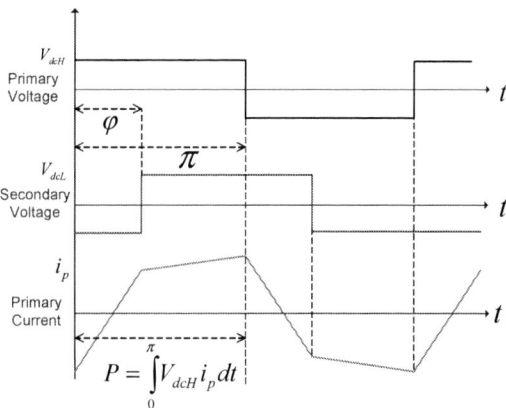

Fig. 12 DAB average power calculation

The switching model simulation is implemented to verify the proposed power balance control. In the simulation, different leakage inductance values are set for the three transformers: 165mH, 165mH, 115mH (30% variation). Fig. 13 and Fig. 15 are the DAB primary currents and power without power balance control. The DAB module with smaller leakage inductance has large current and transfers more power. Fig. 14 and Fig. 16 illustrate the DAB primary currents and power with power balance control. The power transferring through each DAB module is balanced.

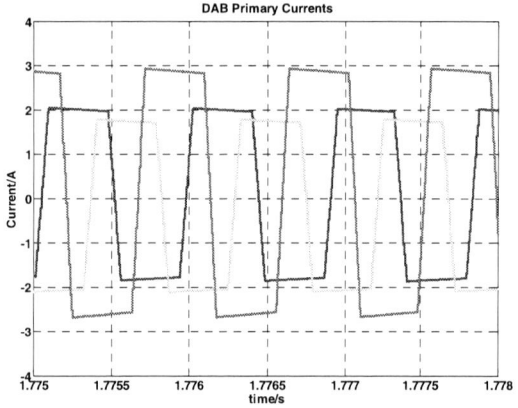

Fig. 13 DAB current without power balance control

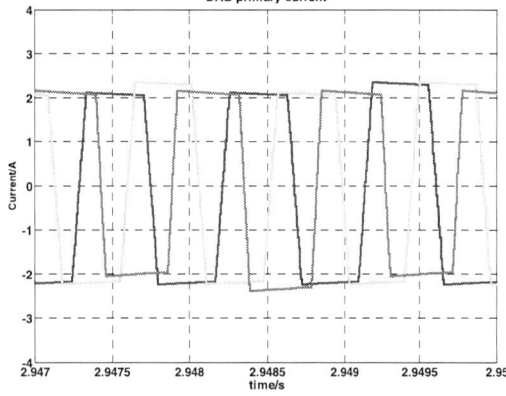

Fig. 14 DAB current with power balance control

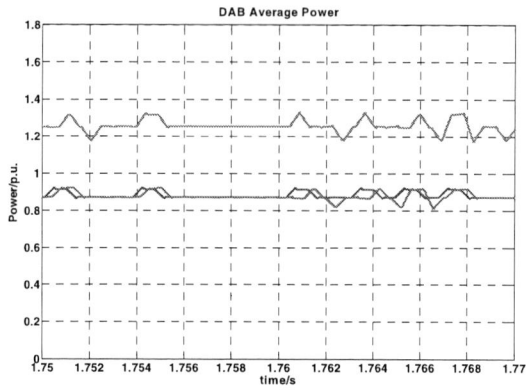

Fig. 15 DAB power without power balance control

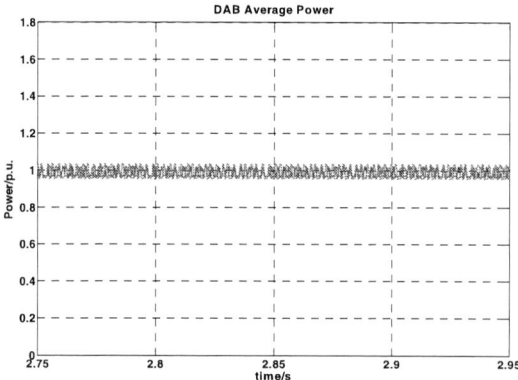

Fig. 16 DAB power with power balance control

IV. SST HARDWARE PROTOTYPE

To develop the SST controller and verify the proposed strategy, a scale-down SST prototype is implemented. Fig. 17 and Fig. 18 are the SST single module topology and prototype photo respectively.

The SST module prototype is designed as single phase input voltage 60Hz, 240V, and DC output 400V. Each module consists of an AC/DC rectifier that converts 60Hz, 240V AC to 400V DC bus, a DC/DC converter that convert 400V to 400V DC bus with 1:1 high frequency transformer. The SST module output will connect to a DC/AC inverter that converts 400V DC to 120/240V AC. The prototype is implemented by using 600V 75A Intelligent Power Modules (IPM). The SST control algorithm is programmed in DSP TMS320F28335. The experiment parameters are shown in Table. I.

The single phase d-q vector control is implemented in the DSP controller by using model based program. The phase A voltage is delayed by 1/4 cycle to synthesize the imaginary phase M. In the experiment, the DC bus voltage reference is 50V and the reactive power reference is zero, which means a unity power factor. Fig. 19 demonstrates the experiment results of the SST rectifier stage with single phase d-q vector controller. The DC bus voltage is regulated to the reference and the input current is in phase with the input voltage. The experiment results verify the single phase d-q vector control for the SST.

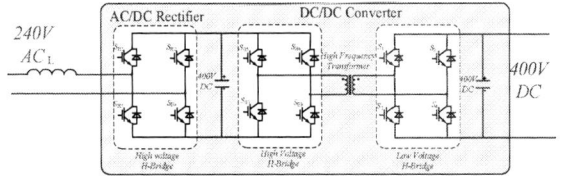

Fig. 17 Scale-down solid state transformer module topology

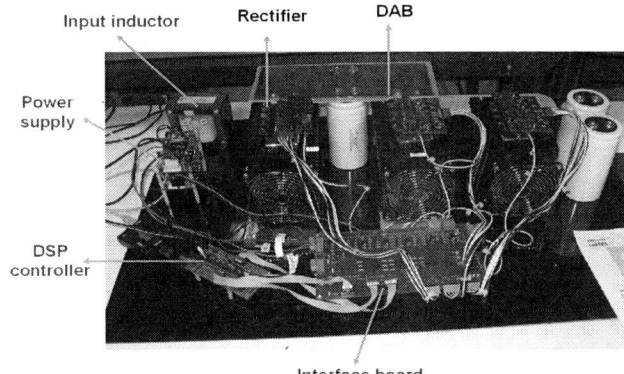

Fig. 18 Scale-down solid state transformer prototype

TABLE I.   SST PROTOTYPE PARAMETERS

Input inductance	1mH
Primary DC Capacitor	900uF
Primary DC voltage reference	400V
Secondary DC Capacitor	900uF
Secondary DC voltage reference	400V
Transformer turns ratio	1:1
Transformer magnetizing inductance	27mH
Transformer leakage inductance	3mH
Switching frequency	10kHz

V. CONCLUSIONS

In this paper, a 20kVA Solid State Transformer (SST) based on 6.5kV IGBT is proposed to interface with 7.2kV distribution system and enable active power management of DRER, DESD and loads in the Future Renewable Electric Energy Delivery and Management (FREEDM) System. The single phase d-q vector control is applied to the cascaded H-Bridge rectifier in SST. A novel voltage control strategy is proposed to balance the rectifier capacitor voltages. The constraints of the voltage balance control for the cascaded H-Bridge rectifier is analyzed in details. A power balance control is proposed to balance the real power through the DAB parallel modules. The SST switching model simulation verifies the proposed voltage and power balance controller.

The 3kW SST scale-down prototype experiment is implemented and verifies the single phase d-q vector controller.

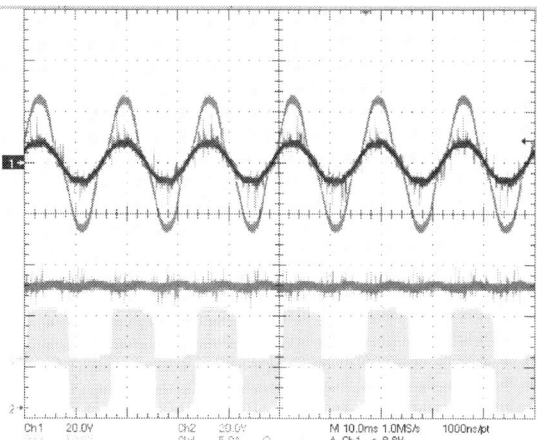

Fig. 19 Experiment results of SST rectifier stage (Ch1, input AC voltage, 20V/div; Ch2, DC bus voltage, 20V/div; Ch3, rectifier PWM voltage, 50V/div; Ch4, input current, 5A/div)

REFERENCES

[1] E. R. Ronan, S. D. Sudhoff, S. F. Glover and D. L. Galloway, "A Power Electronic-Based Distribution Transformer", IEEE Transactions on Power Delivery, vol. 17, pp. 537 - 543 , April 2002.

[2] Jih-Sheng Lai, A. Maitra, A. Mansoor and F. Goodman, "Multilevel Intelligent Universal Transformer for Medium Voltage Applications", Conference Record of Industry Applications Conference, Fourtieth IAS Annual Meeting. vol. 3, pp: 1893 – 1899, Oct, 2005.

[3] Tiefu Zhao, Jie Zeng, S. Bhattacharya, M.E. Baran, A.Q. Huang, "An average model of solid state transformer for dynamic system simulation", IEEE Power & Energy Society General Meeting, 2009. PES '09

[4] J.A. Barrena, L. Marroyo, M.A.R. Vidal, J.R.T. Apraiz, "Individual Voltage Balancing Strategy for PWM Cascaded H-Bridge Converter-Based STATCOM", IEEE Transactions on Industrial Electronics, vol. 55, no. 1, pp.21 - 29, Jan, 2008

[5] Yu Liu, A. Q. Huang, Wenchao Song, S. Bhattacharya, Guojun Tan, "Small-Signal Model-Based Control Strategy for Balancing Individual DC Capacitor Voltages in Cascade Multilevel Inverter-Based STATCOM", IEEE Transactions on Industrial Electronics, vol. 56, no. 6, pp. 2259 – 2269, June 2009

[6] Q. Song, W. Liu, Z. Yuan, W. Wei, and Y. Chen, "DC voltage balancing technique using multi-pulse optimal PWM for cascade H-bridge inverters based STATCOM," in Proc. IEEE 35th Annu. PESC, Aachen, Germany, 2004, pp. 4768–4772.

[7] C. Cecati, A. Dell'Aquila, M. Liserre and V. G. Monopoli, "Design of H-bridge multilevel active rectifier for traction systems," IEEE Trans. Ind. Appl., vol. 39, no. 5, pp. 1541–1550, Sep./Oct. 2003.

[8] R. Zhang, M. Cardinal, P. Szczesny, M. Dame, "A grid simulator with control of single-phase power converters in D-Q rotating frame", IEEE 33rd Annual Power Electronics Specialists Conference, 2002. PESC 02. vol. 3, pp. 1431- 1436.

[9] S. Sirisukprasert, A.Q. Huang, J.-S. Lai, "Modeling, analysis and control of cascaded-multilevel converter-based STATCOM", IEEE Power Engineering Society General Meeting, 13-17 July 2003, vol. 4

[10] H.K. Krishnamurthy, R. Ayyanar, "Building Block Converter Module for Universal (AC-DC, DC-AC, DC-DC) Fully Modular Power Conversion Architecture", IEEE PESC 2007, pp. 483-489

AUTHOR INDEX

A

Abdel-Rahman, Osama 2073
Abe, S. 30
Abu Qahouq, Jaber A. 19, 120, 1723, 1778, 1800
Acero, J. 92, 439, 1328
Agarwal, Pankaj 512
Agelidis, Vassilios G. 295
Aggeler, Daniel 1584
Agostinelli, Matteo 170
Agostini Junior, Eloi 1911
Aguilar, D. 605
Ahmad, Hani 1871
Ahmadi, Damoun 1038
Ahmed, S. 63, 881
Ahmed, Tarek 1825
Ahsanuzzaman, S.M. 980
Akhavan Fomani, Armin 132
Akin, Bilal 1990
Al Mamun, Mostafa 1261
Alahmad, Mahmoud 672
Alcalá, Janeth 1651
Alderman, Arnold 525
Alesi, Larry 849
Al-Hoor, Wisam 627, 1723
Alico, Jurgen 1113
Aller, J.M. 343, 1139
Almukhtar, Basil 1922
Alonso, J.M. 743
Alonso, Rafael 92, 439
Alou, P. 271, 723, 729, 781
Amin, Mahmoud M.N. 1640
Ang, Simon S. 750
Arias, Manuel 196
Arnet, Beat 474
Aroca, J. 1792
Arthur, Stephen 401, 1598
Asadi, Peyman 1578
Avestruz, Al-Thaddeus 444, 580, 2305
Ayana, Elias 1804
Ayyanar, Raja 2149
Azcona, R. 1300
Azongha, S. 216

B

Badstuebner, U.	773
Bae, Chae-Bong	112
Baek, Seunghun	1666
Baggu, Murali M.	2121
Bai, Sanzhong	1145
Baiju, M.R.	1963
Baker, Jonathan	143
Bakhshai, A.	149, 155, 768, 1209
Bakkaloglu, Bertan	1871
Balda, Juan Carlos	321
Ball, Arthur	533
Bao, Jianyu	1097
Bao, Weibing	1097
Barbaroux, Jean	248
Barbi, Ivo	550, 1911
Barlow, Fred	1108
Barrado, A.	1026, 1131, 1279
Barragán, L.A.	309, 439
Barreto, L.H.S.C.	837
Basu, Supratim	244
Batarseh, Issa	143, 627, 1723, 2073
Bates, John	474
Batschauer, Alessandro L.	909
Bazzi, Ali M.	256
Beaupre, Richard A.	1591
Beaupre, Richard A.	1603
Beechner, Troy	2174
Benavente, C.	729
Bendl, Jiri	895
Benfatto, I.	1622, 1810
Ben-Yaakov, Sam	928
Berzoy, Alberto	343
Bhattacharya, Subhashish	761, 1010, 1243, 1666
Bianco, A.	1287
Biela, J.	773, 1397, 1584, 1865
Bing, Zhonghui	336
Birdane, E.	1197
Boel, R.K.	1711
Boroyevich, D.	355, 408, 881, 1272, 1378, 1487, 1521
Brockerhoff, Philip	1970
Bucher, A.	557, 1763
Bueno, A.	1139
Bueno, Alexander	343
Bull, Chris	398
Burdío, J.M.	92, 309, 439, 1328
Burgos, R.	355, 408, 881

C

Cai, Jun	1018
Campa, L.	743

Canales, Francisco 550
Cao, Dong 1365
Cao, Lingling 920
Cao, Yue 968
Cárdenas, Víctor 1651
Carp, C. 1177
Carretero, Claudio 92
Casady, Jeff 1838
Celanovic, Ivan L. 961
Céspedes, Mauricio 2174
Ceyhan, Adil 1197
Cha, Hanju 1659
Chai, Jianyun 2104
Chan, Walker R. 961
Chan, Yick Po 571
Chang, Soon-Jyh 1043
Chang, Wei-Hsu 1727
Chang-Chien, Le-Ren 1043
Chapman, Patrick L. 2138, 2294
Cheah, Sze Kwan 1804
Chen, Baifeng 887, 1704
Chen, Chien Liang 619
Chen, Chingchi 676
Chen, Ching-Jan 1727
Chen, Dan 1727
Chen, Dong 818
Chen, Emil 1081
Chen, Fu-Zen 188
Chen, Guozhu 361, 1514
Chen, Henglin 642, 691
Chen, Jhih-Han 420
Chen, Jifeng 2091
Chen, Lihua 676, 1119, 1124
Chen, Min 935, 1204, 1340, 1674
Chen, Qianhong 920
Chen, Qing Su 948
Chen, W. 594, 1238, 1358
Chen, Yan 1534
Chen, Yang 1578
Chen, Yu 1435, 1441, 1464
Chen, Zheng 1572
Chen, Zhong 1448, 1471, 1616, 1627
Cheung, Chun 1081
Cheung, Victor Sui-pung 491
Chiang, T.-Y. 1413
Chinthavali, Madhu S. 1108
Chiu, Chen-Hua 1727
Chiu, Huang-Jen 948
Cho, B.-H. 1373
Cho, Un-Kwan 1561
Choi, Hangseok 36
Choi, Mun-Gi 1833

Choi, Seung-deog	1990
Choi, Sewan	1934
Choi, Sungjin	512
Choi, W.P.	2251
Choi, Woojin	466, 2166
Choi, Woo-Young	42, 1494
Chomat, Miroslav	895
Choo, Fook Hoong	2143
Chou, Welly	2057
Chowdhury, Badrul H.	2121
Chuang, S.-A.	1413
Chung, Bong-Gun	1698, 1833
Chung, Henry Shu-hung	491, 1214, 1904
Ci, Song	672
Clifford, Zachary	444
Cobos, J.A.	271, 723, 729, 781
Cochran, Travis	2154
Coelho, Enane Antônio Alves	2258
Cooke, Philip	183
Cooley, John J.	444, 2264, 2305
Corradini, L.	277
Corrêa, M.B.R.	239
Corzine, Keith	58, 452
Costabeber, A.	1287
Cox, Robert	1547
Crebier, Jean-Christophe	248, 2238
Crespo, M.	743
Cruz, C.M.T.	837
Cui, Xizhi	1610, 2042

D

da Câmara, Raphael A.	837
da Silva, Fábio Vincenzi Romualdo	2258
Das, Pritam	564, 1222
Davoudi, Ali	2138, 2294
Day, Jon	1243
de Britto, Jonas Reginaldo	2258
De Doncker, Rik	696
de Freitas, Luiz Carlos	2258
de M. Fernandes, Eisenhawer	1984
de Nie, Robert	1
De Novaes, Yales R.	550
Deboy, G.	1397
DeCarlo, Ray	480
Delgado, Eladio C.	1603
Delicado, Bernardo	1300
Deng, Yan	575, 1069, 1266
Deng, Zhe	2021
Deng, Zhiquan	1018
Dhople, Sairaj V.	2138, 2294
Dias, José A.A.	755

Diaz, D. 723, 729
Dick, Christian P. 696
Dickson, Andrew 702
Ding, Xiaodong 519
Ding, Yi 1555
Djabbari, Ali 1056
Domahidi, Alexander 1995
Dominguez-Garcia, Alejandro 256
Dong, Yan 79
Donlon, John F. 392
dos Santos Girio, J.A. 1306
dos Santos, Jr., E.C. 1191
dos Santos, Euzeli 755, 1183
Du, Chengrui 861
Du, Weijing 823, 1392
Du, Xiaoli 1732
Du, Yu 1145, 1666
Duerbaum, T. 557, 1763
Duryea, Timothy P. 686
Dutta, Sumit 761, 1666

E

Eckler, Kyle Roger 2202
Edrington, C.S. 216
Effler, Simon 315, 1087
Egan, Michael G. 787
Egelkraut, S. 231
Elasser, Ahmed 1598
El-Barbari, Said 1487
Elmes, John 143
Emadi, Ali 1957
Emanuel, Alexander E. 2096
Endredy, John 1172
Englehart, Amy 2305
Enjeti, Prasad 63
Erb, Dylan C. 2066
Eren, Suzan 149, 768
Ertl, H. 986

F

Fahimi, Babak 68, 1498, 2231
Falcones, Sixifo 2149
Fan, Haifeng 210
Fan, S.-Y. 1842
Fang, Xiong 1745
Farias, Valdeir José 2258
Fei, Wanmin 1034, 1732
Feng, Yupeng 915, 1093
Feng, Zhuomin 935
Ferdowsi, Mehdi 58, 452, 2111

Fernández, A. 196, 1313, 1792
Fernández, C. 1131, 1279
Fernández, Carlos 1300
Fernández, Cristina 1026
Ferrieux, J.-P. 1817
Filho, Faete 968
Firmansyah, E. 30
Fleming, F. 216
Foley, Raymond 525
Forsyth, A.J. 1306
Frey, D. 1817, 2238
Frey, L. 231
Friedli, Thomas 1527
Fu, Dianbo 940
Fu, Jizhen 702, 1482
Fu, P. 1622, 1810
Fujita, Atsushi 1825
Fukuda, Kenji 2030
Fukushima, Kentaro 289

G

Gacio, D. 743
Galperti, C. 2281
Gamboa, Gustavo 143
Gao, Feng 1555
Gao, G. 1622
Gao, G. 1810
Gao, Mingzhi 1204
Gao, Mingzhi 1674
Garces, Luis 1295
Garcia, J. 743
García, O. 271, 723, 729, 781
Gargoom, A. 162, 2132
Garrett, Jerome 1598
Gazel, Nicolas 1272
Ge, Baoming 1124
Ge, Qiongxuan 1419
Geng, Hua 2126
Gillmor, Colin 1384
Gimeno, A. 729
Glaser, John S. 401, 654
Gong, Jinwu 887, 1704
Gonzalez, M.C. 271, 781
Gonzalez-Llorente, Jesus 666, 1062, 2161, 2226
Gowda, Arun V. 1591, 1603
Graham, Jeff 2220
Green, R. 1568
Grogan, S.A.S. 873
Guépratte, K. 1817
Guerrero, Josep M. 380
Guo, Rong 1172

Guo, Suxuan 887, 1704
Guo, Zhiqiang 662
Gupta, Ranjan K. 901

H

Ha, Dong Sam 2154
Habetler, T. 343, 1139
Halton, Mark 1075, 1087, 2207
Hamilton, Christopher 143
Han, Baikhee 512
Han, Byung M. 303
Han, Jung Hee 1108
Han, Sangmin 2288
Han, Yunlong 1534
Hang, Lijun 2021
Haque, M.E. 162, 2132
Harada, Shinsuke 2030
Harada, Yosuke 289
Harbaugh, Mark 425
Harris, John H. 1048
Hartmann, L.V. 239
Hartmann, M. 986
Hartnett, Kevin J. 787
Haruni, A.M.O. 162, 2132
Hasan, Jaber 750
Hayes, John G. 787
He, Chao 361
He, Xiangning 575, 681, 801, 1069, 1266, 1454, 1534, 2080, 2300
He, Yanhui 915, 1093
He, Yingjie 1692
Hegarty, Tim 1056
Heldwein, Marcelo L. 909
Henze, C.P. 605
Herbert, Edward 1048
Herbsommer, Juan A. 398
Hewson, Christopher R. 2050
Hirose, Fumitoshi 1879
Ho, W.C. 994, 2251
Holmes, D.G. 873
Hong, Xiaoyuan 1238, 1358, 2021
Hosoda, Hiromi 1261
Hsu, G.-W. 1849
Hu, Haibing 519, 627, 1785, 2073
Hu, Qingcong 2314
Hu, Yuequan 203
Huang, Alex 761, 849, 1010, 1145, 1172, 1477, 1666, 1875, 2181
Huang, Hong 1770
Huang, Jing-Yi 1043
Huang, Xiucheng 433, 823, 1392
Huber, Laszlo 203
Huemer, Mario 170

Huh, Dong-Young 1885, 1949
Hui, Joanne 155
Hui, S.Y.R. 86, 594, 994, 1346, 2251
Hung, Chung-Wen 420
Husain, Iqbal 2007
Hutchens, Chris 1056
Hwu, K.I. 507, 710, 737, 1942
Hyeon, B.-C. 1373

I

Ide, Kozo 103
Inman, Daniel J. 2154
Ioinovici, Adrian 1214, 1904
Ishii, Kenichiro 74
Ishizuka, Yoichi 1879
Itoh, Jun-Ichi 1684
Izquierdo, D. 1300

J

Jacobina, C.B. 755, 1183, 1191, 1984
Jain, P.K. 14, 149, 155, 499, 544, 768, 1209, 1248, 1334, 2321
Jakobsen, Uffe 98
Jang, Minsoo 295
Jang, Sang-Ho 1833
Jang, Yungtaek 23
Jayakanthan, Gnanavel 794
Jeannin, P.-O. 248, 1817, 2238
Jeon, Yong-Seog 1949
Jeong, In Wha 1166
Jeong, Yu-Seok 303
Ji, Biao 1448, 1471
Ji, Feng 1448, 1471
Ji, Young-Hyok 2275
Jia, Liang 124
Jiang, Dong 408
Jiang, Wei 68
Jiménez, O. 309
Jimenez-Brea, Emil 666, 1062, 2161
Jing, Wei 1010
Jovanović, Milan M. 23, 203
Jung, Ha-Jin 1678
Jung, Sungyoon 2002
Jung, Yong-Chae 1885, 2275

K

Kanai, Takeo 1101
Kang, Sung-In 1885
Kang, Yong 1272, 1435, 1441, 1464
Kato, Koji 1684

Kazimierczuk, Marian K. 2212
Kelleher, Paul 1922
Kelley, Robin 1838
Kelly, Anthony 328, 1922, 2189
Keogh, Bernard 1384
Kerkman, Russ J. 634
Kesler, Metin 374
Khaligh, Alireza 1755, 2066, 2245
Khazarei, Mostafa 452, 58
Khoobroo, Amir 2231
Kim, Deuk-Soo 1678
Kim, Eun-Soo 1698, 1833, 1885, 1949
Kim, Gi-Taek 1678
Kim, Gyeong-Hun 2085
Kim, Hyun-Cheol 112
Kim, Jang-Mok 112
Kim, Jin-Tae 540
Kim, Joo-Hoon 1698, 1885, 1949
Kim, Jun-Gu 2275
Kim, Rae-Young 1890
Kim, Sang-Hyun 466, 2166
Kim, Sungmin 103
Kim, Tae-Hoon 466, 2166
Kim, Teahoon 512
Kim, Wook 466, 2166
Kim, Young-Gook 112
Kim, Young-Ho 2275
Kim, Young-Ju 2085
Kimball, Jonathan W. 1508, 2121, 2202
Kirlin, R. Lynn 1749
Kirtley, Jr., James L. 1547
Kisacikoglu, Mithat C. 458
Knight, Andy 2013
Koellner, Walter 1158
Kolar, J.W. 773, 986, 1397, 1527, 1584, 1865
Kong, Na 2154
Kong, Pengju 1424
Koo, Gwan-Bon 540
Krein, Philip T. 256
Krishnamurthy, Mahesh 1957
Kutkut, Nasser 627
Kwon, Bong-Hwan 1494
Kwon, Soon Kurl 1230
Kye, Moon-Ho 1698, 1833

L

Lai, Jih-Sheng 42, 387, 474, 619, 1056, 1494, 1890
Lai, Pengjie 1927
Lai, Rixin 355, 408
Lam, John 2321
Lamar, Diego G. 196

Lascu, Cristian 1749
Laughman, Christopher 1547
Lázaro, A. 1026, 1131, 1279
Leão, J.F. Araujo 239
Lee, Beomseok 2002
Lee, C.K. 86
Lee, Fred C. 79, 176, 533, 940, 1424, 1927
Lee, Hoi 686
Lee, Jae-Sam 1885, 1949
Lee, Jong-Hak 466, 2166
Lee, Jun-Young 303
Lee, Kwang-Ho 1698, 1949
Lee, Sangwon 1934
Lee, Ting-Peng 948
Lee, Tzung-Lin 380
Lee, Yuang-Shung 619
Leeb, Steven B. 444, 580, 1547, 2194, 2264, 2305
Lei, Qin 844, 854, 1002
Leslie, Scott 474
Li, Chushan 681, 2300
Li, Duo 935
Li, Hong 1740
Li, Hui 210, 223, 807
Li, Jian 176
Li, Jin 1521
Li, Jiping 2036
Li, Jun 1010
Li, Ming 1745
Li, Qiang 79, 533
Li, S.N. 594
Li, Weichen 1454, 2080
Li, Wuhua 801, 1069, 1454, 2021, 2080
Li, Xiao 861
Li, Yaohua 1419
Li, Yong 349
Li, Yongdong 1736, 2104
Liang, Xiaoguo 794
Liang, Zhigang 849, 1477
Liao, XiaoZhong 662
Lim, Michele 533
Lim, Sungkeun 1875
Lima, A.M.N. 239, 1984
Lin, Cheng-Tsung 420
Lin, D.Y. 1346
Lin, Fei 1740
Lin, Hai 818
Lin, Hung-Chih 2154
Lisi, Gianpaolo 1056
Litvinov, Alexander 474
Liu, Chih-Wen 420
Liu, Chui Pong 571
Liu, Congwei 1419

Liu, Jingbo 425
Liu, Jinjun 915, 1093, 1521, 1633, 1692, 2116
Liu, Kun 1674
Liu, Liming 223
Liu, Ting 2116
Liu, X. 86, 994, 2251
Liu, Yan-Fei 124, 702, 1482
Liu, Zeyuan 1018
Lo, Yu-Kang 948
Loh, Poh Chiang 1555, 2143
López del Cerro, F.J. 1300
Lopez, Osvaldo 398
Lorduy, Abad 1026
Losee, Peter 401, 1598
Lu, Kaiyuan 98
Lu, Ying 1785
Lu, Zhengyu 519, 1238, 1358, 1542, 2021, 2091
Lucena, Carlos 1026
Lucía, O. 92, 309, 439, 1328
Lukic, Srdjan 1145
Lukić, Zdravko 1, 315
Luo, Fang 1272
Luo, Yingpeng 1616, 1627

M

Ma, Chongguang 1340
Ma, Dongsheng 284, 813
Maciel, A.M. 1191
Maddaleno, Franco 1166
Mahdi, Abdulhussain E. 2207
Maksimović, Dragan 188, 277, 980
Mankani, A.D. 1622, 1810
Mantooth, H. Alan 321
Mao, Jingxin 1740
Mao, Xiaojing 1405
Mariéthoz, Sébastien 1995
Marsili, Stefano 170
Martin, Daniel 619
Marxgut, Christoph 1865
März, M. 231
Massoud, Ahmed 63
Matocha, Kevin S. 401
Matsuo, Hirofumi 1879
Mattavelli, P. 953, 1287, 2281
Mazumdar, Joy 1158
McGrath, B.P. 873
Melkebeek, J.A.A. 1711
Meng, Peipei 433, 642, 691
Metwally, M.K. 414
Meyer, Eric 124
Miaja, P.F. 715

Miftakhutdinov, Rais 1897
Milivojevic, Nikola 1957, 2245
Millan, Ignacio 92, 439, 1328
Miller, Greg 1081
Millner, Alan 1572
Min, Chen 1460, 1503
Ming, Zhengfeng 1919
Ming, Zheng-Feng 109
Mishima, Tomokazu 1230
Mo, Qiong 1674
Moghe, Rohit 1158
Mohammed, O.A. 1640
Mohan, Ned 901, 1804
Mohapatra, Krushna K. 901
Molina Cardozo, Diogenes D. 321
Moon, Sang-Cheol 540
Mooney, James 2207
Morari, Manfred 1995
Moschopoulos, Gerry 564, 829, 1222, 1320
Motto, Eric R. 392
Mourra, O. 1313, 1792
Mousavi, Ahmad 564, 1222
Muller, Sean 2194
Murai, H. 648
Muralidhar, Gautam 19
Mussa, Samir A. 909

N

Nakano, Shinya 74
Nakaoka, Mutsuo 1230, 1825
Nam, Kwanghee 2002
Narveson, Brian 525
Nasadoski, Jeffrey J. 401
Navarro, D. 309
Neely, Jason 480
Negnevitsky, M. 162, 2132
Neuman, Sabrina 2194
Neumeyer, C. 1810
Ng, W.M. 594, 1346
Ng, Wai Tung 132
Ngo, K.D.T. 533, 2036
Ngo, Phong 813
Nguyen, Do Hung 750
Nguyen, The-Van 2238
Ni, Guang-Zheng 109
Nilles, Gerald 2294
Ning, Puqi 1378
Ninomiya, T. 30, 289
Nishi, Mariko 1879
Nondahl, Thomas 425
Noquil, Jonathan 398

Norford, Les K. 1547
Norigoe, Isami 289
Núñez, Ciro 1651

O

Oh, J.S. 1810
Ohashi, Hiromichi 648, 1101, 2030
Oliveira, Jr., D.S. 837
Oliveira, Alexandre C. 1984
Oliver, J.A. 271, 723, 729, 781
O'Malley, Eamon 328, 1922, 2189
Ó'Mathúna, Cian 525
Omori, Hideki 1825
Onar, Omer C. 1755, 2066
Ongaro, F. 2281
Orabi, Mohamed 1778, 1800
Orietti, E. 953
Orji, Uzoma A. 1547
Ortega, F.J. 729
Ortiz-Rivera, Eduardo I. 666, 1062, 2161, 2226
Otsuki, Etsuo 74
Ozdemir, Engin 367, 374
Ozdemir, Sule 367
Ozpineci, Burak 458

P

Page, Sarah 2194
Pahlevaninezhad, Majid 149, 768
Pallo, Nathan A. 961
Pan, S. 499, 1248
Pang, H.M. 973, 1857
Parayandeh, Amir 980
Pardo, J.M. 729
Paris, James 580, 1547, 2194, 2305
Park, Jinseok 2181
Park, Jun-Ho 1885
Park, Minwon 2085
Park, Sung-Yeul 387
Parkhideh, Babak 1666
Parto, Parviz 1578
Pasquesoone, Gregory 2007
Pautsch, Adam G. 1591
Pawellek, A. 557
Pei, Xuejun 1435, 1441, 1464
Pei, Yunqing 1610, 2042, 2060
Pekarek, Steve 480
Peng, Fang Z. 818, 844, 854, 1002, 1119, 1124, 1365, 2288
Peng, Li 1435, 1441, 1464
Pepper, Michael 143
Perin, Arnaldo J. 909

Perreault, David J. 961
Pilawa-Podgurski, Robert C.N. 961
Pokryvailo, A. 1177
Pong, M.H. Bryan 973, 1857
Pong, Man Hay 571
Poon, Ngai Kit 571
Poucand, M. 1306
Praça, P.P. 837
Priewasser, Robert 170
Prodić, Aleksandar 1, 315, 980, 1113, 1256

Q

Qi, Tao 2220
Qian, Hao 474, 1056
Qian, Wei 2288
Qian, Zhaomin 1674
Qian, Zhaoming 642, 691, 818, 823, 935, 1002, 1204, 1340, 1392
Qian, Zhijun 2073
Qiu, Weihong 1081
Quesada, Isabel 1026

R

Radić, Aleksandar 1, 315
Rahimian, Mina M. 1990
Ramamurthy, Anand 1243
Rauch, M. 231
Ray, William F. 2050
Remscrim, Zachary 444, 1547, 2194
Ren, Xiaoyong 920
Ren, Zheng 1204, 1674
Restrepo, J. 343, 1139
Reutzel, Evan 1430
Rico-Secades, M. 743
Rinne, Karl 328, 1075, 1087, 1922, 2189
Ritenour, Andrew 1838
Rivas, Juan M. 654
Rocha, Nady 1183
Rodríguez, A. 715
Rodríguez, M. 715
Rodríguez-Valdez, Carlos D. 634
Rosas, Emanuel 1651
Royak, Semyon 425
Ruan, Xinbo 920, 1214, 1405
Rylko, Marek S. 787

S

Sabate, Juan 1487
Sadakata, Hideki 1825
Sagawa, Natsumi 2212

Saggini, S.	953, 1287, 2281
Saha, Bishwajit	1825
Salah Morsy, Ahmed	63
Salazar-Llinas, Andres	666, 1062, 2161, 2226
Salem, T.E.	1568
Salmon, John	2013
Samsi, Rohan	183
Sanders, Seth	1430
Sanz, M.	1279
Satoh, K.	392
Sayed, Khairy Fathy	1230
Scanlan, Tony	1075
Scapellati, C.	1177
Schantz, Christopher	1547, 2194
Scharrer, Martin	1075
Schletz, A.	231
Schmidt, Peter	425
Schreier, Ludek	895
Schulz, Martin	1970
Schutten, Michael	1598
Schweizer, Mario	1527
Sebastián, J.	196, 715
Seger, Eric	2264
Seidlitz, Steve	1804
Sekiya, Hiroo	2212
Sen, P.C.	702, 1482, 1719
Sepahvand, Hossein	58, 452
Sha, Deshang	662
Shao, Jianwen	601
Sharif, Hamid	672
Shaw, Steven R.	2194, 2264
Shen, Guoqiao	861
Shen, John	627
Shen, Weixiang	2143
Sheng, Honggang	1572
Sheng, Z.	1622, 1810
Sheridan, David	1838
Shi, Lei	1448, 1471
Shi, Wei	1785
Shih, Frank	948
Shim, Won-Sul	1678
Shinohe, Takashi	1101, 2030
Shiny, G.	1963
Shoyama, M.	30, 289
Shuai, Peng	550
Silva, C.E.A.	837
Simanjorang, Rejeki	648
Singh, Bhim	1976
Singh, Sanjeev	1976
Slepchenkov, Mikhail	1166
Slowey, John	525
Smedley, Keyue	47, 264, 1166

Smith, Chris 474
Smith, Greg 1056
Solovitz, Stephen A. 1591
Somani, Apurva 901
Somayajula, Deepak 2111
Song, Bo 2060
Song, Byeong-Mun 2085
Song, Z.Q. 1810
Spiazzi, G. 953
Stamenkovic, Igor 1957, 2245
Steimer, Peter K. 1865
Steiner, Reto 1865
Stephan, H. 1817
Steurer, M. 216
Stevanovic, Ljubisa D. 401, 1591, 1603
Straeussnigg, Dietmar 170
Stum, Zachary 1598
Su, Gui-Jia 1152
Su, Jen-Ta 420
Su, Y.-H. 1842
Su, Y.P. 86
Sugimura, Hisayuki 1230
Sul, Seung-Ki 103, 611, 1561
Sullivan, Charles R. 1048
Sumiyoshi, Shinichiro 1825
Sun, Jian 336, 2174, 2220
Sun, Jianjun 887, 1704
Sun, Julu 533, 1927
Sun, Pengwei 387, 474
Sun, Yi 176
Sung, Kyungmin 1101
Suzuki, K. 392
Szczesny, Paul 1295

T

Takahashi, T. 392
Takao, Kazuto 1101, 2030
Takeda, Takashi 648
Tamyurek, B. 1197
Tan, Kai 1419
Tan, Kuan Khoon 1555, 2143
Tanaka, Yasunori 1101
Tang, Lixin 1152
Tang, Yu 867
Tao, J. 1622, 1810
Tao, X.H. 594
Thomas, Brinda A. 588
Thompson, Chris 1243
Titiz, Furkan Kaan 696
Tjokrorahardjo, Andre 1352
Todd, R. 1306

Todeschini, Grazia 2096
Tolbert, Leon M. 458, 968, 1108
Toliyat, Hamid A. 1990
Tomioka, S. 30
Tomita, Koji 103
Tonicello, F. 1313, 1792
Torrico-Bascopé, R.P. 837
Tran, Manh Hung 248
Trowler, Derik 321
Trzynadlowski, Andrzej M. 1749
Tschirhart, Darryl J. 14, 544, 1334
Tseng, S.-Y. 1413, 1842, 1849
Tsukakoshi, Kenta 289

U
Ucar, Mehmet 367
Undeland, Tore M. 244
Urciuoli, D.P. 1568
Urriza, I. 309

V
Vafakhah, Behzad 2013
Vagnon, Eric 2238
Vaks, Nir 480
Valdivia, V. 1131, 1279
Vasić, M. 271, 723, 729
Veillette, Robert J. 2007
Vickery, Dan 444, 2305
Vieira, Jr., João Batista 2258
Viola, Julio C. 343
Visairo, H. 271, 781
Vodyakho, O. 216
Volfson, Oleg 138
Vu, Trung-Kien 1659
Vyncke, T.J. 1711

W
Wada, Keiji 1101
Waldron, Finbarr 525
Wang, Dong 124
Wang, Fred 355, 408, 881, 1272, 1378, 1487, 1572
Wang, Gangyao 761, 1666
Wang, Huai 491, 1904
Wang, Jin 676, 1038
Wang, Jun 1266
Wang, K.-C. 1413, 1849
Wang, Ke 1745
Wang, Kunrong 7
Wang, Laili 1610, 2042, 2060

Wang, Meng 794
Wang, Mingliang 2036
Wang, Peng 1555, 2143
Wang, Ruxi 1378
Wang, Shunqing 1627
Wang, Shuo 940, 1272
Wang, Siran 1097, 1238, 1358, 1542, 2021, 2091
Wang, Yen-Ching 380
Wang, Yousheng 818
Wang, Yue 915, 1093, 1745
Wang, Zhan 807
Wang, Zhaoan 1610, 1692, 2042, 2060
Wang, Zhengshi 575
Wang, Zhiqiang 361, 1514
Watson, Luke 2121, 2202
Wegner, Hagen 1384
Wei, Jukui 867
Wen, Jun 47
Wichakool, Warit 580, 1547
Wijeratne, Dunisha 829
Wilson, Jr., Thomas G. 183
Wolbank, T.M. 414
Won, Chung-Yuen 2275
Wood, R.A. 1568
Wu, Bin 1034, 1732, 2126
Wu, Chun-Hsun 1043
Wu, D. 1306
Wu, Guan-Hong 948
Wu, Haimeng 575
Wu, Jiande 575, 681, 2300
Wu, Jinlong 915, 1093
Wu, W.-C. 1842
Wu, Xinke 642, 691
Wu, Zhichao 223

X

Xiao, Xi 1736
Xie, Chuan 361, 1514
Xie, Huikai 2036
Xie, Shaojun 867
Xing, Lei 2174
Xing, Yan 1785
Xu, Biwen 1340
Xu, Chunchun 1295
Xu, Dehong 861
Xu, Dewei 2126
Xu, L. 1622
Xu, L.W. 1810
Xu, Ligang 920
Xue, Jianren 1785
Xue, Tao 519

Y

Yamada, Yusuke 1879
Yamaguchi, Hiroshi 648
Yamazaki, Mikio 648
Yan, W. 1346
Yang, Bing-Zhong 109
Yang, Binjian 1266
Yang, Bo 801
Yang, C.-M. 1849
Yang, Geng 2126
Yang, Jianyou 642
Yang, Joonhyun 512
Yang, Liyu 1172, 2181
Yang, Shuitao 844, 854, 1002
Yang, Xinyi 1340
Yang, Xu 1610, 2042, 2060
Yao, Kai 1405
Yao, Wei 1204, 1674
Yao, Wenxi 519, 2091
Yau, Y.T. 507, 710, 737, 1942
Yazdani, D. 1209
Ye, Shaoshi 1238, 1358
Ye, Zhihong 1405
Yim, Jung-Sik 1561
Yin, Zhenggang 1419
Ying, Yucheng 533
Yisheng, Yuan 1460, 1503
Yoo, Hyunjae 611
Yoon, Young-Doo 103
York, Ben 1890
Yoshihiura, Y. 392
Yoshino, Teruo 1261
You, Xiaojie 1740
Young, George 1384
Youssef, Mohamed 1778, 1800
Yu, Haidong 1498, 2025
Yu, In-Keun 2085
Yu, Wen Long 948
Yu, Wensong 387, 474, 1056
Yu, Wen-Song 42
Yuan, Wei 433, 823, 1392
Yuan, Xibo 2104

Z

Zane, Regan 2314
Zawodniok, Maciej 1508
Zeltser, Ilya 928
Zeng, Jie 761
Zha, Xiaoming 887, 1704
Zhang, Di 1487
Zhang, Guoxing 433

Zhang, Hui	1108
Zhang, Jianhui	1056
Zhang, Jing	1514
Zhang, Jiucai	672
Zhang, Jun	861
Zhang, Junming	433, 823, 1392
Zhang, Leqiang	1745
Zhang, Xin	1214
Zhang, Xuan	2116
Zhang, Yanli	1034
Zhang, Yi	284
Zhang, Yingqi	1719
Zhang, Zhe	935
Zhang, Zhiliang	702, 1482
Zhang, Zhongchao	1097
Zhao, April	132
Zhao, Guopeng	1610, 1633, 1745, 2042, 2060
Zhao, Jing	1266, 1534
Zhao, Rongxiang	1534
Zhao, Tiefu	761, 1666
Zhao, Yi	801, 1069, 1454, 2080
Zhao, Zhenyu	1256
Zhaoming, Qian	1460, 1503
Zheng, Cong	2245
Zheng, Feng	1919
Zheng, Jerry	349
Zheng, Sheng	691, 818
Zheng, Trillion Q.	1740
Zheng, Yuzhen	2080
Zhixin, Xu	1266
Zhong, W.X.	994
Zhou, Liang	47, 264
Zhou, Linyuan	2116
Zhou, Xia	2091
Zhou, Xiaohu	849, 1145, 1666
Zhou, Xin	1477
Zhu, Guangyong	7
Zhu, Haipeng	2143
Zhu, Hao	1736
Zhu, Yinyu	1616, 1627
Zou, Ke	676, 1038
Zou, Yunping	1692
Zumel, P.	1131, 1279

9781424447824